HOW TO LEARN CALCULUS OF ONE VARIABLE

HOW TO LEARN CALCULUS OF ONE VARIABLE

J. D. GHOSH
MD. ANWARUL HAQUE

PUBLISHING FOR ONE WORLD

NEW AGE INTERNATIONAL (P) LIMITED, PUBLISHERS
New Delhi • Bangalore • Chennai •Cochin • Guwahati • Hyderabad
Jalandhar • Kolkata • Lucknow • Mumbai • Ranchi
Visit us at **www.newagepublishers.com**

Published by New Age International (P) Ltd., Publishers
First Edition: 2004
Reprint : 2009

Branches:

- 36, Malikarjuna Temple Street, Opp. ICWA, Basavanagudi, **Bangalore**. ✆ (080) 26677815
- 26, Damodaran Street, T. Nagar, **Chennai**. ✆ (044) 24353401
- Hemsen Complex, Mohd. Shah Road, Paltan Bazar, Near Starline Hotel, **Guwahati**. ✆ (0361) 2543669
- No. 105, 1st Floor, Madhiray Kaveri Tower, 3-2-19, Azam Jahi Road, Nimboliadda, **Hyderabad**. ✆ (040) 24652456
- RDB Chambers (Formerly Lotus Cinema) 106A, Ist Floor, S.N. Banerjee Road, **Kolkata**. ✆ (033) 22275247
- 18, Madan Mohan Malviya Marg, **Lucknow**. ✆ (0522) 2209578
- 142C, Victor House, Ground Floor, N.M. Joshi Marg, Lower Parel, **Mumbai**. ✆ (022) 24927869
- 22, Golden House, Daryaganj, **New Delhi**. ✆ (011) 23262370, 23262368

ISBN (10) : 81-224-1524-5
ISBN (13) : 978-81-224-1524-7

Rs. 375.00

C-08-12-3261

Printed in India at A.H. Print Pack, Delhi.
Typeset at Le Studio, Delhi.

PUBLISHING FOR ONE WORLD
NEW AGE INTERNATIONAL (P) LIMITED, PUBLISHERS
4835/24, Ansari Road, Daryaganj, New Delhi-110002
Visit us at **www.newagepublishers.com**

Preface

As the title of the book 'How to learn calculus of one variable' Suggests we have tried to present the entire book in a manner that can help the students to learn the methods of calculus all by themselves we have felt that there are books written on this subject which deal with the theoretical aspects quite exhaustively but do not take up sufficient examples necessary for the proper understanding of the subject matter thoroughly. The books in which sufficient examples are solved often lack in rigorous mathematical reasonings and skip accurate arguments some times to make the presentation look apparently easier.

We have, therefore, felt the need for writing a book which is free from these deficiencies and can be used as a supplement to any standard book such as 'Analytic geometry and calculus' by G.B. Thomas and Finny which quite thoroughly deals with the proofs of the results used by us.

A student will easily understand the underlying principles of calculus while going through the worked-out examples which are fairly large in number and sufficiently rigorous in their treatment. We have not hesitated to work-out a number of examples of the similar type though these may seem to be an unnecessary repetition. This has been done simply to make the students, trying to learn the subject on their own, feel at home with the concepts they encounter for the first time. We have, therefore, started with very simple examples and gradually have taken up harder types. We have in no case deviated from the completeness of proper reasonings.

For the convenience of the beginners we have stressed upon working rules in order to make the learning all the more interesting and easy. A student thus acquainted with the basics of the subject through a wide range of solved examples can easily go for further studies in advanced calculus and real analysis.

We would like to advise the student not to make any compromise with the accurate reasonings. They should try to solve most of examples on their own and take help of the solutions provided in the book only when it is necessary.

This book mainly caters to the needs of the intermediate students whereas it can also used with advantages by students who want to appear in various competitive examinations. It has been our endeavour to incorporate all the finer points without which such students continually feel themselves on unsafe ground.

We thank all our colleagues and friends who have always inspired and encouraged us to write this book everlastingly fruitful to the students. We are specially thankful to Dr Simran Singh, Head of the Department of Lal Bahadur Shastri Memorial College, Karandih, Jamshedpur, Jharkhand, who has given valuable suggestions while preparing the manuscript of this book.

Suggestions for improvement of this book will be gratefully accepted.

DR JOY DEV GHOSH

MD ANWARUL HAQUE

Contents

1

Function

To define a function, some fundamental concepts are required.

Fundamental Concepts

Question: What is a quantity?

Answer: In fact, anything which can be measured or which can be divided into parts is called a quantity. But in the language of mathematics, its definition is put in the following manner.

Definition: Anything to which operations of mathematics (mathematical process) such as addition, subtraction, multiplication, division or measurement etc. are applicable is called a quantity.

Numbers of arithmetic, algebraic or analytic expressions, distance, area, volume, angle, time, weight, space, velocity and force etc. are all examples of quantities.

Any quantity may be either a variable or a constant.

Note: Mathematics deals with quantities which have values expressed in numbers. Number may be real or imaginary. But in real analysis, only real numbers as values such as -1, 0, 15, $\sqrt{2}$, π etc. are considered.

Question: What is a variable?

Answer:

***Definitions* 1**: (*General*): If in a mathematical discussion, a quantity can assume more than one value, then the quantity is called a variable quantity or simply a variable and is denoted by a symbol.

Example: **1.** The weight of men are different for different individuals and therefore height is a variable.

2. The position of a point moving in a circle is a variable.

Definition: **2.** (*Set theoritic*): In the language of set theory, a variable is symbol used to represent an unspecified (not fixed, i.e. arbitrary) member (element or point) of a set, i.e., by a variable, we mean an element which can be any one element of a set or which can be in turn different elements of a set or which can be a particular unknown element of a set or successively different unknown elements of a set. We may think of a variable as being a "place-holder" or a "blank" for the name of an element of a set.

Further, any element of the set is called a value of the variable and the set itself is called variable's domain or range.

If x be a symbol representing an unspecified element of a set D, then x is said to vary over the set D (i.e., x can stand for any element of the set D, i.e., x can take any value of the set D) and is called a variable on (over) the set D whereas the set D over which the variable x varies is called domain or range of x.

Example: Let D be the set of positive integers and $x \in D = \{1, 2, 3, 4, \dots\}$, then x may be 1, 2, 3, 4, … etc.

Note: A variable may be either (1) an independent variable (2) dependent variable. These two terms have been explained while defining a function.

Question: What is a constant?

Answer:

***Definition* 1**. (*General*): If in a mathematical discussion, a quantity cannot assume more than one vale, then the quantity is called a constant or a constant quantity and is denoted by a symbol.

Examples: **1.** The weights of men are different for different individuals and therefore weight is a variable. But the numbers of hands is the same for men of different weights and is therefore a constant.

2. The position of a point moving in a circle is a variable but the distance of the point from the centre of the circle is a constant.

3. The expression $x + a$ denotes the sum of two quantities. The first of which is variable while the second is a constant because it has the same value whatever values are given to the first one.

Definition: **2.** (*Set theoritic*): In the language of set theory, a constant is a symbol used to represent a member of the set which consists of only one member, i.e. if there is a variable 'c' which varies over a set consisting of only one element, then the variable 'c' is called a constant, i.e., if 'c' is a symbol used to represent precisely one element of a set namely D, then 'c' is called a constant.

Example: Let the set D has only the number 3; then $c = 3$ is a constant.

Note: Also, by a constant, we mean a fixed element of a set whose proper name is given. We often refer to the proper name of an element in a set as a constant. Moreover by a relative constant, we mean a fixed element of a set whose proper name is not given. We often refer to the "alias" of an element in a set as a relative constant.

Remark: The reader is warned to be very careful about the use of the terms namely variable and constant. These two terms apply to symbols only not to numbers or quantities in the set theory. Thus it is meaningless to speak of a variable number (or a variable quantity) in the language of set theory for the simple reason that no number is known to human beings which is a variable in any sense of the term. Hence the 'usual' text book definition of a variable as a quantity which varies or changes is completely misleading in set theory.

Kinds of Constants

There are mainly two kinds of constants namely:

1. Absolute constants (or, numerical constants).

2. Arbitrary constants (or, symbolic constants).

Each one is defined in the following way:

1. Absolute constants: Absolute constants have the same value forever, e.g.:

(i) All arithmetical numbers are absolute constants. Since $1 = 1$ always but $1 \neq 2$ which means that the value of 1 is fixed. Similarly $-1 = -1$ but $-1 \neq 1$ (Any quantity is equal to itself. this is the basic axiom of mathematics upon which foundation of equations takes rest. This is why $1 = 1, 2 = 2, 3 = 3, \ldots x = x$ and $a = a$ and so on).

(ii) π and logarithm of positive numbers (as log2, log 3, log 4, … etc) are also included in absolute constants.

2. Arbitrary constants: Any arbitrary constant is one which may be given any fixed value in a problem and retains that assigned value (fixed value) throughout the discussion of the same problem but may differ in different problems.

An arbitrary constant is also termed as a parameter.

Note: Also, the term "parameter" is used in speaking of any letter, variable or constant, other than the coordinate variables in an equation of a curve defined by $y = f(x)$ in its domain.

Examples: **(i)** In the equation of the circle $x^2 + y^2 = a^2$, x and y, the coordinates of a point moving along a circle, are variables while 'a' the radius of a circle may have any constant value and is therefore an arbitrary constant or parameter.

(ii) The general form of the equation of a straight line put in the form $y = mx + c$ contains two parameters namely m and c representing the gradient and y-intercept of any specific line.

Symbolic Representation of Quantities, Variables and Constants

In general, the quantities are denoted by the letters $a, b, c, x, y, z, \ldots$ of the English alphabet. The letters from "a to s" of the English alphabet are taken to represent constants while the letters from "t to z" of the English alphabet are taken to represent variables.

Question: What is increment?

Answer: An increment is any change (increase or growth) in (or, of) a variable (dependent or independent). It is the difference which is found by subtracting the first value (or, critical value) of the variable from the second value (changed value, increased value or final value) of the variable.

That is, increment

= final value – initial value = F.V – I. V.

Notes: **(i)** Increased value/changed value/final value/ second value means a value obtained by making addition, positive or negative, to a given value (initial value) of a variable.

(ii) The increments may be positive or negative, in both cases, the word "increment" is used so that a negative increment is an algebraic decrease.

Examples on Increment in a Variable

1. Let x_1 increase to x_2 by the amount Δx. Then we can set out the algebraic equation $x_1 + \Delta x = x_2$ which $\Rightarrow \Delta x = x_2 - x_1$.

2. Let y_1 decrease to y_2 by the amount Δy. Then we can set out the algebraic equation $y_1 + \Delta y = y_2$ which $\Rightarrow \Delta y = y_2 - y_1$.

Examples on Increment in a Function

1. Let $y = f(x) = 5x + 3$ = given value ... (i)

Now, if we give an increment Δx to x, then we also require to give an increment Δy to y simultaneously.

Hence, $y + \Delta y = f(x + \Delta x) = 5(x + \Delta x) + 3 = 5x + 5\Delta x + 3$ ... (ii)

$\therefore$ (ii) – (i) $\Rightarrow y + \Delta y - y = (5x + 5\Delta x + 3) - (5x + 3) = 5x + 5\Delta x + 3 - 5x - 3 = 5\Delta x$

i.e., $\Delta y = 5\Delta x$

2. Let $y = f(x) = x^2 + 2$ = given value,

then $y + \Delta y = f(x + \Delta x) = (x + \Delta x)^2 + 2 = x^2 + \Delta x^2 + 2x\,\Delta x + 2$

$\Rightarrow \Delta y = x^2 + \Delta x^2 + 2x\,\Delta x + 2 - x^2 - 2 = 2x\,\Delta x + \Delta x^2$

Hence, increment in $y = f(x + \Delta x) - f(x)$ where $f(x) = (x^2 + 2)$ is $\Delta y = x^2 + \Delta x^2 + 2x\,\Delta x + 2 - x^2 - 2 = 2x\,\Delta x + \Delta x^2$

3. Let $y = \dfrac{1}{x}$ = given value.

Then, $y + \Delta y = \dfrac{1}{x + \Delta x}$

Hence, increment in $y = f(x + \Delta x) - f(x)$ where $f(x) = \dfrac{1}{x}$

$$\Rightarrow \Delta y = \frac{1}{x + \Delta x} - \frac{1}{x} = \frac{x - (x + \Delta x)}{x(x + \Delta x)} = \frac{-\Delta x}{x(x + \Delta x)}$$

4. Let $y = \log x$ = given value.

Then, $y + \Delta y = \log(x + \Delta x)$

and $\Delta y = \log(x + \Delta x) - \log x = \log\left(\dfrac{x + \Delta x}{x}\right) = \log\left(1 + \dfrac{\Delta x}{x}\right)$

5. Let $y = \sin\theta$, given value

Then, $y + \Delta y = \sin(\theta + \Delta\theta)$

and $\Delta y = \sin(\theta + \Delta\theta) - \sin\theta = 2\cos\left(\dfrac{2\theta + \Delta\theta}{2}\right) \cdot \sin\left(\dfrac{\Delta\theta}{2}\right)$

Question: What is the symbol used to represent (or, denote) an increment?

Answer: The symbols we use to represent small increment or, simply increment are Greak Letters Δ and δ (both read as delta) which signify "an increment/ change/growth" in the quantity written just after it as it increases or, decreases from the initial value to another value, i.e., the notation Δx is used to denote a fixed non zero, number that is added to a given number x_0 to produce another number $x = x_0 + \Delta x$. if $y = f(x)$ then $\Delta y = f(x_0 + \Delta x) - f(x_0)$.

Notes: **If $x, y, u. v$ are variables, then increments in them are denoted by $\Delta x, \Delta y, \Delta u, \Delta v$ respectively signifying how much x, y, u, v increase or decrease, i.e., an increment in a variable (dependent or independent) tells how much that variable increases or decreases.**

Let us consider $y = x^2$

When $x = 2, y = 4$

$x = 3, y = 9$

$\therefore \Delta x = 3 - 2 = 1$ and $\Delta y = 9 - 4 = 5$

$\Rightarrow$ as x increases from 2 to 3, y increases from 4 to 9.

$\Rightarrow$ as x increases by 1, y increases by 5.

Question: What do you mean by the term "function"?
Answer: In the language of set theory, a function is defined in the following style.

A function from a set D to a set R is a rule or, law (or, rules, or, laws) according to which each element of D is associated (or, related, or, paired) with a unique (i.e., a single, or, one and only one, or, not more than one) element of R. The set D is called the domain of the function while the set R is called the range of the function. Moreover, elements of the domain (or, the set D) are called the independent variables and the elements of the range (or, range set or, simply the set R) are called the dependent variables. If x is the element of D, then a unique element in R which the rule (or, rules) symbolised as f assigns to x is termed "the value of f at x" or "the image of x under the rule f" which is generally read as "the f-function of x" or, "f of x". Further one should note that the range R is the set of all values of the function f whereas the domain D is the set of all elements (or, points) whose each element is associated with a unique elements of the range set R.

Functions are represented pictorially as in the accompanying diagram.

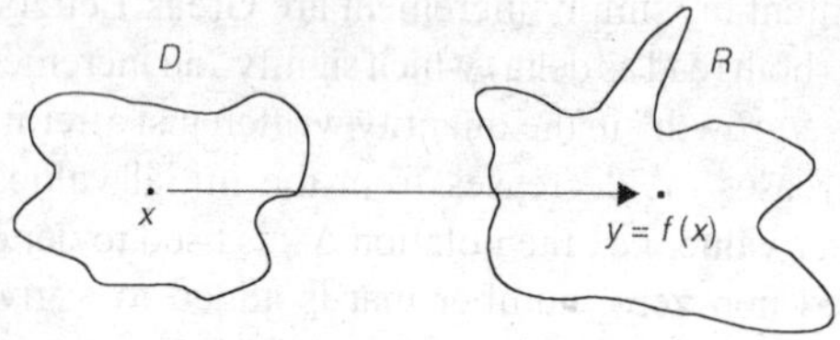

One must think of x as an arbitrary element of the domain D or, an independent variable because a value f of x can be selected arbitrarily from the domain D as well as y as the corresponding value of f at x, a dependent variable because the value of y depends upon the value of x selected. It is customary to write $y = f(x)$ which is read as "y is a function of x" or, "y is f of x" although to be very correct one should say that y is the value assigned by the function f corresponding to the value of x.

Highlight on the Term "The Rule or the Law".

1. The term "rule" means the procedure (or procedures) or, method (or, methods) or, operation (or, operations) that should be performed over the independent variable (denoted by x) to obtain the value the dependent variable (denoted by y).

Examples:

1. Let us consider quantities like

(i) $y = \log x$ **(iv)** $y = \sin x$
(ii) $y - x^3$ **(v)** $y = \sin^{-1} x$
(iii) $y = \sqrt{x}$ **(vi)** $y = e^x$, ... etc.

In these log, cube, square root, sin, $\sin^{-1}$, e, ... etc are functions since the rule or, the law, or, the function f = log, $(\ )^3$, $\sqrt{\ }$, sin, $\sin^{-1}$ or, e, ... etc has been performed separately over (or, on) the independent variable x which produces the value for the dependent variable represented by y with the assistance of the rule or the functions log, $(\ )^3$, $\sqrt{\ }$, sin, $\sin^{-1}$ or, e, ... etc. (***Note***: An arbitrary element (or point) x in a set signifies any specified member (or, element or point) of that set).

2. The precise relationship between two sets of corresponding values of dependent and independent variables is usually called a law or rule. Often the rule is a formula or an equation involving the variables but it can be other things such as a table, a list of ordered pairs or a set of instructions in the form of a statement in words. The rule of a function gives the value of the function at each point (or, element) of the domain.

Examples:

(i) The formula $f(x) = \dfrac{1}{1 + x^2}$ tells that one should square the independent variable x, add unity and then divide unity by the obtained result to get the value of the function f at the point x, i.e., to square the independent variable x, to add unity and lastly to divide unity by the whole obtained result (i.e., square of the independent variable x plus unity).

(ii) $f(x)=x^2+2$, where the rule f signifies to square the number x and to add 2 to it.
(iii) $f(x)=3x-2$, where the rule f signifies to multiply x by 3 and to subtract 2 from $3x$.
(iv) $C = 2\pi r$ an equation involving the variables C (the circumference of the circle) and r (the radius of the circle) which means that $C = 2\pi r$ = a function of r.
(v) $y=\sqrt{64s}$ an equation involving y and s which means that $y=\sqrt{64s}$ a functions of s.

3. A function or a rule may be regarded as a kind of machine (or, a mathematical symbol like $\sqrt{\ }$, log, sin, cos, tan, cot, sec, cosec, $\sin^{-1}$, $\cos^{-1}$, $\tan^{-1}$, $\cot^{-1}$, $\sec^{-1}$, cosec^{-1}, ... etc indicating what mathematical operation is to be performed over (or, on) the elements of the domain) which takes the elements of the domain D, processes them and produces the elements of the range R.

Example of a function of functions:

Integration of a continuous function defined on some closed interval $[a. b]$ is an example of a function of functions, namely the rule (or, the correspondence) that associates with each object $f(x)$ in the given set of objects, the real number $\int_a^b f(x)\,dx$.

Notes: **(i)** We shall study functions which are given by simple formulas. One should think of a formula as a rule for calculating $f(x)$ when x is known (or, given), i.e., of the rule of a function f is a formula giving y in terms of x say $y=f(x)$, to find the value of f at a number a, we substitute that number a for x wherever x occurs in the given formula and then simplify it.

(ii) For $x \in D$, $f(x) \in R$ should be unique means that f can not have two or more values at a given point (or, number) x.
(iii) $f(x)$ always signifies the effect or the result of applying the rule f to x.
(iv) Image, functional value and value of the function are synonymes.

Notations:

We write **1.** "$f: D \to R$" or "$D \xrightarrow{f} R$" for "f is a function with domain D and range R" or equivalently, "f is a function from D to R".

2. $f: x \to y$ or, $x \xrightarrow{f} y$ or, $x \to f(x)$ for "a function f from x to y" or "f maps (or, transforms) x into y or $f(x)$".
3. $f: D \to R$ defined by $y=f(x)$ or, $f: D \to R$ by $y=f(x)$ for "(a) the domain $= D$, (b) the range $= R$, (c) the rule : $y=f(x)$.
4. $D(f)$ = The domain of the function f where D signifies "domain of".
5. $R(f)$ = The range of the function f where R signifies "range of".

Remarks:
(i) When we do not specify the image of elements of the domain, we use the notation (1).
(ii) When we want to indicate only the images of elements of the domain, we use the notation (2).
(iii) When we want to indicate the range and the rule of a function together with a functional value $f(x)$, we use the notation (3).
(iv) In the language of set theory, the domain of a function is defined in the following style:

$D(f)$: $\{x: x \in D_1\}$ where, D_1 = the set of independent variables (or, arguments) = the set of all those members upon which the rule 'f' is performed to find the images (or, values or, functional values).
(v) In the language of set theory, the range of a function is defined in the following way:

$R(f)=\{f(x): x \in D, f(x) \in R\}$ = the set of all images.
(vi) The function f^n is defined by $f^n(x)=f(x)\cdot f(x)\ldots$ n. times
$=[f(x)]^n$, where n being a positive integer.
(vii) For a real valued function of a real variable both x and y are real numbers consisting of.
(a) Zero

(b) Positive or negative integers, e.g.: 4, 11, 9, 17, –3, –17, ... etc.

(c) Rational numbers, e.g.: $\frac{9}{5}, \frac{-17}{2}, \ldots$ etc.

(d) Irrational numbers e.g.: $\sqrt{7}, -\sqrt{14}, \ldots$ etc.

(viii) Generally the rule/process/method/law is not given in the form of verbal statements (like, find the square root, find the log, exponential, ... etc.) but in the form of a mathematical statement put in the form of expression containing x (i.e. in the form of a formula) which may be translated into words (or, verbal statements).

(ix) If it is known that the range R is a subset of some set C, then the following notation is used:

$f: D \to C$ signifying that

(a) f is a function

(b) The domain of f is D

(c) The range of f is contained in C.

Nomenclature: The notation $"f: D \to C"$ is read f is a function on the set D into the set C."

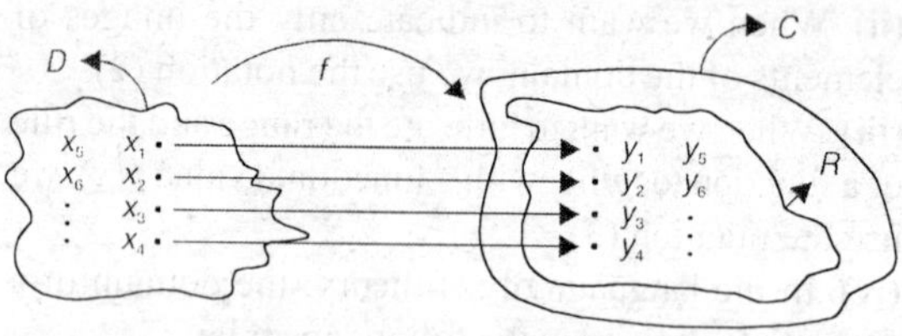

N.B: To define some types of functions like "into function and on to function", it is a must to define a function $"f: D \to C"$ where C = codomain and hence we are required to grasp the notion of co-domain. Therefore, we can define a co-domain of a function in the following way:

Definition of co-domain: A co-domain of a function is a set which contains the range or range set (i.e., set of all values of f) which means $R \subseteq C$, where R = the set of all images of f and C = a set containing images of f.

Remember:

1. If $R \subset C$ (where R = the range set, C = co-domain) i.e., if the range set is a proper subset of the co-domain, then the function is said to be an "into function".

2. If $R = C$, i.e., if the range set equals the co-domain, then the function is said to be an "onto function".

3. If one is given the domain D and the rule (or formula,) then it is possible (theoretically at least) to state explicitly a function as any ordered pair and one should note that under such conditions, the range need not be given. Further, it is notable that for each specified element $'a' \in D$, the functional value $f(a)$ is obtained under the function 'f'.

4. If $a \in D$, then the image in C is represented by $f(a)$ which is called the functional value (corresponding to a)and it is included in the range set R.

Question: Distinguish between the terms "a function and a function of x".

Answer: A function of x is a term used for "an image of x under the rule f" or "the value of the function f at (or, for) x" or "the functional value of x" symbolised as $y = f(x)$ which signifies that an operation (or, operations) denoted by f has (or, have) been performed on x to produce an other element $f(x)$ whereas the term "function" is used for "the rule (or, rules)" or "operation (or, operations)" or "law (or, laws)" to be performed upon x, x being an arbitrary element of a set known as the domain of the function.

Remarks: **1.** By an abuse of language, it has been customary to call $f(x)$ as function instead of f when a particular (or, specifies) value of x is not given only for convenience. Hence, wherever we say a "function $f(x)$ what we actually mean to say is the function f whose value at x is $f(x)$. thus we say, functions x^4, $3x^2 + 1$, etc.

2. The function 'f' also represents operator like $\sqrt[n]{\ }$, $(\ )^n$, $|\ |$, log, e, sin, cos, tan, cot, sec, cosec, $\sin^{-1}$, $\cos^{-1}$, $\tan^{-1}$, $\cot^{-1}$, $\sec^{-1}$ or cosec^{-1} etc.

3. Function, operator, mapping and transformation are synonymes.

4. If domain and range of a function are not known, it is customary to denote the function f by writing $y = f(x)$ which is read as y is a function of x.

Question: Explain the terms "dependent and independent variables".

Answer:

1. ***Independent variable***: In general, an independent variable is that variable whose value does not depend upon any other variable or variables, i.e., a variable in a mathematical expression whose value determines the value of the whole given expression is called an independent variable: in $y = f(x)$, x is the independent variable.

In set theoretic language, an independent variable is the symbol which is used to denote an unspecified member of the domain of a function.

2. *Dependent variable*: In general a dependent variable is that variable whose value depends upon any other variable or variables, i.e., a variable (or, a mathematical equation or statement) whose value is determined by the value taken by the independent variable is called a dependent variable: in $y = f(x)$, y is the dependent variable.

In set theoretic language, a dependent variable is the symbol which is used to denote an unspecified member of the range of a function.

e.g.: In $A = f(r) = \pi r^2$, r is an independent variable and A is a dependent variable.

***Question*:** Explain the term "function or function of x" in terms of dependency and independency.

***Answer*:** When the values of a variable y are determined by the values given to another variable x, y is called a function of (depending on) x or we say that y depends on (or, upon) x. Thus, any expression in x depends for its value on the value of x. This is why an expression in x is called a function of x put in the form: $y = f(x)$.

***Question*:** What are the symbols for representing the terms "a function and a function of a variable"?

***Answer*:** Symbols such as f, F, ϕ etc are used to denote a function whereas a function of a variable is denoted by the symbols $f(x)$, $\phi(x)$, $f(t)$, $F(t)$, $\phi(t)$ and can be put in the forms: $y = f(x)$; $y = \phi(x)$; $y = f(t)$; $y = F(t)$; $y = \phi(t)$, that y is a function of (depending on) the variable within the circular bracket ( ), i.e., y depends upon the variable within circular bracket.

i.e., $y = f(x)$ signifies that y depends upon x, i.e., y is a function of x.

$S = f(t)$ signifies that s depends upon t, i.e., s is a function of t.

$C = \phi(r)$ signifies that c depends upon r, i.e., c is a function of r.

***Notes*:**

1. Any other letter besides f, ϕ, F etc may be used just for indicating the dependence of one physical quantity on an other quantity.

2. The value of f/functional value of f corresponding to $x = a$ / the value of the dependent variable y for a particular value of the independent variable is symbolised as $(f(x))_{x=a} = f(a)$ or $[f(x)]_{x=a} = f(a)$ while evaluating the value of the function $f(x)$ at the point $x = a$.

3. One should always note the difference between "a function and a function of".

4. *Classification of values of a function at a point $x = a$.*

There are two kinds of the value of a function at a point $x = a$ namely

(i) The actual value of a function $y = f(x)$ at $x = a$.

(ii) The approaching or limiting value of a function $y = f(x)$ at $x = a$, which are defined as:

(i) The actual value of a function $y = f(x)$ at $x = a$: when the value of a function $y = f(x)$ at $x = a$ is obtained directly by putting in the given value of the independent variable $x = a$ wherever x occurs in a given mathematical equation representing a function, we say that the function f or $f(x)$ has the actual value $f(a)$ at $x = a$.

(ii) The approaching value of a function $y = f(x)$ at $x = a$: The limit of a function $f(x)$ as x approaches some definite quantity is termed as the approaching (or, limiting) value of the function $y = f(x)$ at $x = a$. This value may be calculated when the actual value of the function $f(x)$ becomes indeterminate at a particular value 'a' of x.

5. When the actual value of a function $y = f(x)$ is anyone of the following forms: $\frac{0}{0}$, 0^0, $0 \times \infty$, $\frac{\infty}{\infty}$, $\infty - \infty$, ∞^0, 1^∞, imaginary, $\frac{\text{any real number}}{0}$ for a particular value 'a' of x, it is said that the function $f(x)$ is not defined or is indeterminate or is meaningless at $x = a$.

6. To find the value of a function $y = f(x)$ at $x = a$ means to find the actual value of the function $y = f(x)$ at $x = a$.

Pictorial Representation of a Function, its Domain and Range.

1. ***Domain***: A domain is generally represented by any closed curve regular (i.e., circle, ellipse, rectangle, square etc) or irregular (i.e. not regular) whose members are represented by numbers or alphabets or dots.

2. ***Range***: A range is generally represented by another closed curve regular or irregular or the some closed curve regular or irregular as the domain.

3. ***Rule***: A rule is generally represented by an arrow or arc (i.e., arc of the circle) drawn from each member of the domain such that it reaches a single member or more than one member of the codomain, the codomain being a superset of the range (or, range set).

Remarks:

1. We should never draw two or more than two arrows from a single member of the domain such that it reaches more than one member of the codomain to show that the venn-diagram represents a function. Logic behind it is given as follows.

If the domain are chairs, then one student can not sit on more than one chair at the same time (i.e., one student can not sit on two or more than two chairs at the same time)

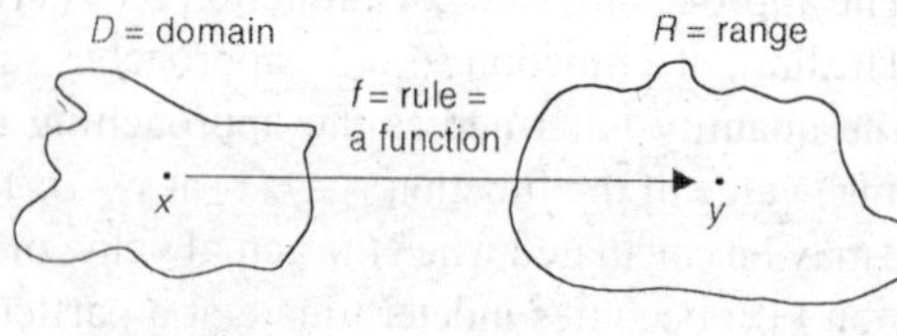

Fig. 1.1 Represents a function

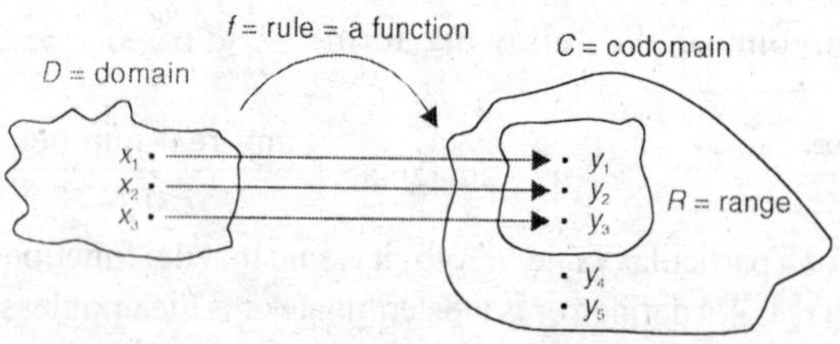

Fig. 1.2 Represents a function

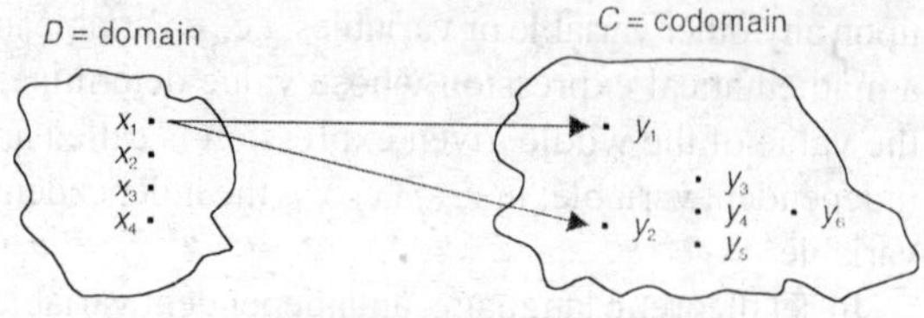

Fig. 1.3 Does not represent a function

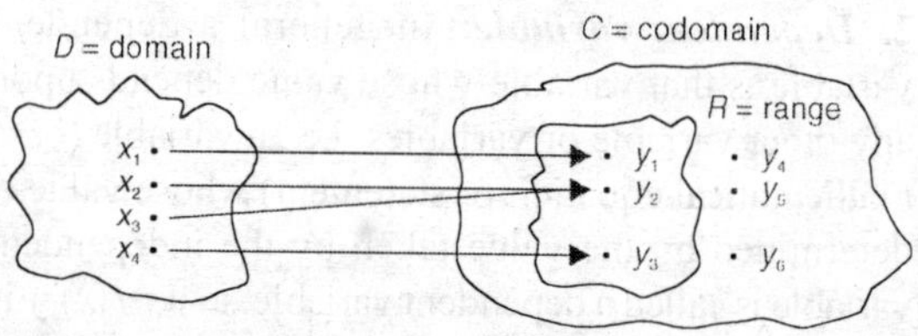

Fig. 1.4 Represents a function

2. In the pictorial representation of a function the word "rule" means.

(i) Every point/member/element in the domain D is joined by an arrow ($\rightarrow$) or arc ($\frown$) to some point in range R which means each element $x \in D$ corresponds to some element $y \in R \subseteq C$.

(ii) Two or more points in the domain D may be joined to the same point in $R \subseteq C$ (See Fig. 1.4 where the points x_2 and x_3 in D are joined to the same point y_2 in $R \subseteq C$.

(iii) A point in the domain D can not be joined to two or more than two points in C, C being a co-domain. (See Fig. 1.3)

(iv) There may be some points in C which are not joined to any element in D (See Fig. 1.4 where the points y_4, y_5 and y_6 in C are not joined to any point in D.

Precaution: It is not possible to represent any function as an equation involving variables always. At such circumstances, we define a function as a set of ordered pairs with no two first elements alike e.g., $f = \{(1, 2), (2, 4), (3, 6), (4, 8), (5, 10), (6, 12), (7, 14)\}$ whose D = domain = $\{1, 2, 3, 4, 5, 6, 7\}$, R = range = $\{2, 4, 6, 8, 10, 12, 14\}$ and the rule is: each second element is twice its corresponding first element.

But $f = \{(0, 1), (0, 2), (0, 3), (0, 4)\}$ does not define a function since its first element is repeated.

Note: When the elements of the domain and the range are represented by points or English alphabet with subscripts as $x_1, x_2, \ldots$ etc and $y_1, y_2, \ldots$ etc respectively, we generally represent a function as a set of ordered pairs with no two first elements alike, i.e., f: $\{x, f(x)$: no two first elements are same$\}$ or, $\{x, f(x)$: no two first elements are same$\}$ or, $\{(x, y)$: $x \in D$ and $y = f(x) \in R\}$ provided it is not possible to represent the function as an equation $y = f(x)$.

Question: What is meant whenever one says a function $y = f(x)$ exist at $x = a$ or $y = f(x)$ is defined at (or, f or) $x = a$?

Answer: A function $y = f(x)$ is said to exist at $x = a$ or, $y = f(x)$ is said to be defined at (or, f or) $x = a$ provided the value of the function $f(x)$ at $x = a$ (i.e. $f(a)$) is finite which means that the value of the function $f(x)$ at $x = a$ should not be anyone of the following forms $\frac{0}{0}, 0^0, 0 \times \infty, \frac{\infty}{\infty}, \infty - \infty, \infty^0, 1^\infty$, imaginary value, $\frac{\text{a real number}}{0}$.

Remarks:

(i) A symbol in mathematics is said to have been defined when a meaning has been given to it.

(ii) A symbol in mathematics is said to be undefined or non-existance when no meaning is attributed to the symbol.

e.g.: The symbols 3/2, –8/15, $\sin^{-1}(1/2)$, $\log(1/2)$ are defined or they are said to exist whereas the symbols $\sqrt{-9}$, $\cos^{-1} 5$, $5 \div 0$, $\log(-3)$, 5_2 are undefined or they are said not to exist.

(iii) Whenever we say that something exists, we mean that it has a definite finite value.

e.g.:

(i) $f(a)$ exists means $f(a)$ has a finite value.

(ii) $\lim_{x \to a} f(x)$ exists means $\lim_{x \to a} f(x)$ has a finite value.

(iii) $f'(a)$ exists means $f'(a)$ has a finite value.

(iv) $\int_a^b f(x)\,dx$ exists means that $\int_a^b f(x)\,dx$ has a finite value.

Classification of Functions

We divide the function into two classes namely:

(i) Algebraic

(ii) Transcendental which are defined as:

(i) ***Algebraic function***: A function which satisfies the equation put in the form:

$$A_0[f(x)]^m + A_1[f(x)]^{m-1} + A_2[f(x)]^{m-2} + \ldots + A_m = 0,$$

where $A_0, A_1, \ldots A_m$ are polynomials is called an algebraic function.

Notes:

1. A function $f: R \to R$ defined by $f(x) = a_0 x^n + a_1 x^{n-1} + \ldots + a_{m-1} x + a_m$ where $a_0, a_1, a_2, \ldots a_m$ are constants and n is a positive integer, is called a polynomial in x or a polynomial function or simply a polynomial. One should note that a polynomial is a particular case of algebraic function as we see on taking $m = 1$ and $A_0 = a$ constant in algebraic function.

2. The quotient of two polynomials termed as a rational function of x put in the form:

$$\frac{a_0 x^n + a_1 x^{n-1} + \ldots + a_{m-1} x + a_m}{b_0 + b_1 x + \ldots + b_m x^n}$$

is also an algebraic function. It is defined in every interval only in which denominator does not vanish. If $f_1(x)$ and $f_2(x)$ are two polynomials, then general rational functions may be denoted by $R(x) = \frac{f_1(x)}{f_2(x)}$ where R signifies "a rational function of". In case $f_2(x)$ reduces itself to unity or any other constant (i.e., a term not containing x or its power), $R(x)$ reduces itself to a polynomial.

3. Generally, there will be a certain number of values of x for which the rational function is not defined and these are values of x for which the polynomial in denominator vanishes.

e.g.: $R(x) = \frac{2x^2 - 5x + 1}{x^2 - 5x + 6}$ is not defined when $x = 2$ or $x = 3$.

4. Rational integral functions: If a polynomial in x is in a rational form only and the indices of the powers of x are positive integers, then it is termed as a rational integral function.

5. A combination of polynomials under one or more radicals termed as an irrational functions is also an algebraic function. Hence, $y = \sqrt{x} = f(x)$; $y = x^{5/3} = f(x)$; $y = \frac{x}{\sqrt{x^2+4}}$ serve as examples for irrational algebraic functions.

6. A polynomial or any algebraic function raised to any power termed as a power function is also an algebraic function. Hence, $y = x^n$, $(n \in R) = f(x)$; $y = \left(x^2+1\right)^3 = f(x)$ serve as examples for power functions which are algebraic.

Remarks:

1. All algebraic, transcendental, explicit or implicit function or their combination raised to a fractional power reduces to an irrational function. Hence, $y = x^{5/3} = f(x)$; $y = (\sin x + x)^{\frac{1}{2}} = f(x)$ serve as examples for irrational functions.

2. All algebraic, transcendental, explicit or implicit function or their combination raised to any power is always regarded as a power function. Hence, $y = \sin^2 x = f(x)$; $y = \log^2 |x| = f(x)$ serve as examples for power functions.

Transcendental function: A function which is not algebraic is called a transcendental function. Hence, all trigonometric, inverse trigonometric, exponential and logarithmic (symoblised as "TILE") functions are transcendental functions. hence, $\sin x$, $\cos x$, $\tan x$, $\cot x$, $\sec x$, $\operatorname{cosec} x$, $\sin^{-1} x$, $\cos^{-1} x$, $\tan^{-1} x$, $\cot^{-1} x$, $\sec^{-1} x$, $\operatorname{cosec}^{-1} x$, $\log |f(x)|$, $\log |x|$, $\log x^2$, $\log (a+x^2)$, a^x (for any $a > 0$), e^x, $[f(x)]^{g(x)}$ etc serve as examples for transcendental functions.

Notes: (In the extended real number system)

(A)

(i) $e^x = \infty$ when $x = \infty$

(ii) $e^x = 1$ when $x = 0$

(iii) $e^x = 0$ when $x = -\infty$.

(B) One should remember that exponential functions obeys the laws of indices, i.e.,

(i) $x^e : e^y = e^{x+y}$

(ii) $x^e / e^y = e^{x-y}$

(iii) $(e^x)^m = e^{mx}$

(iv) $e^{-x} = \frac{1}{e^x}$

(C)

(i) $\log 0 = -\infty$

(ii) $\log 1 = 0$

(iii) $\log \infty = \infty$

Further Classification of Functions

The algebraic and the transcendental function are further divided into two types namely (i) explicit function (ii) implicit function, which are defined as:

(i) *Explicit function*: An explicit function is a function put in the form $y = f(x)$ which signifies that a relation between the dependent variable y and the independent variable x put in the form of an equation can be solved for y and we say that y is an explicit function of x or simply we say that y is a function of x. hence, $y = \sin x + x = f(x)$; $y = x^2 - 7x + 12 = f(x)$ serve as examples for explicit function of x's.

Remark: If in $y = f(x)$, f signifies the operators (i.e., functions) sin, cos, tan, cot, sec, cosec, $\sin^{-1}$, $\cos^{-1}$, $\tan^{-1}$, $\cot^{-1}$, $\sec^{-1}$, $\operatorname{cosec}^{-1}$, log or e, then $y = f(x)$ is called an explicit transcendental function otherwise it is called an explicit algebraic function.

(ii) *Implicit function*: An implicit function is a function put in the form: $f(x, y) = c$, c being a constant, which signifies that a relation between the variables y and x exists such that y and x are in seperable in an equation and we say that y is an implicit function of x. Hence, $x^3 + y^2 = 4xy$ serves as an example for the implicit function of x.

Remark: If in $f(x, y) = c$, f signifies the operators (i.e., functions) sin, cos, tan, cot, sec, cosec, $\sin^{-1}$, $\cos^{-1}$, $\tan^{-1}$, $\cot^{-1}$, $\sec^{-1}$, $\operatorname{cosec}^{-1}$, log, e and the ordered pain (x, y) signifies the combination of the variables x and y, then $f(x, y) = c$ is called an implicit algebraic function of x, i.e., y is said to be an implicit algebraic function of x, if a relation of the form:

$y^m + R_1 y^{m-1} + \ldots + R_m = 0$ exists, where $R_1, R_2, \ldots R_m$ are rational function of x and m is a positive integer.

Note: Discussion on "the explicit and the implicit functions" has been given in detail in the chapter "differentiation of implicit function".

On Some Important Functions

Some types of functions have been discussed in previous sections such as algebraic, transcendental, explicit and implicit functions. In this section definition of some function used most frequently are given.

1. *The constant function*: A function $f: R \to R$ defined by $f(x) = c$ is called the "constant function".

Let $y = f(x) = c$

$\therefore$ $y = c$ which is the equation of a straight line parallel to the x-axis, i.e., a constant function represents straight lines parallel to the x-axis.

Also, domain of the constant function $= D(f) =$ {real numbers} $= R$ and range of the constant function $= R(f) = \{c\} =$ a singleton set for examples, $y = 2$; $y = 3$ are constant functions.

Remarks:

(i) A polynomial $a_0 x^n + a_1 x^{n-1} + \ldots a_{m-1} x + a_m$ (whose domain and range are sets of real numbers) reduces to a constant function when degree of polynomial is zero.

(ii) In particular, if $c = 0$, then $f(x)$ is called the "zero function" and its graph is the x-axis itself.

2. *The identity function*: A function $f: R \to R$ defined by $f(x) = x$ is called the "identity function" whose domain and range coincide with each other, i.e., $D(f) = R(f)$ in case of identity function.

Let $y = f(x) = x$

$\therefore$ $y = x$ which is the equation of a straight line passing through the origin and making an angle of 45° with the x-axis, i.e., an identity function represents straight lines passing through origin and making an angle of 45° with the x-axis.

3. *The reciprocal of identity function*: A function $f: R - \{0\} \to R$ defined by $f(x) = \dfrac{1}{x}$ is called the reciprocal function of the identity function $f(x) = x$ or simply reciprocal function.

Let $y = f(x) = \dfrac{1}{x}$

$\therefore$ $xy = 1$ which is the equation of a rectangular hyperbola, i.e., the reciprocal of an identity function represents a rectangular hyperbola.

Also, $D(f) =$ {real number except zero} $= R - \{0\}$ and $R(f) =$ {real numbers}

4. *The linear function*: A function put in the form: $f(x) = mx + c$ is called a "linear function" due to the fact that its graph is a straight line.

Also, $D(f) =$ {real numbers except $m = 0$} and $R(f) =$ {real number except $m = 0$}

Question: What do you mean by the "absolute value function"?

Answer: A function $f: R \to R$ defined by $f(x) = |x|$

$$= \begin{cases} x, & x \geq 0 \\ -x, & x < 0 \end{cases}$$ is called absolute value (or, modulus or, norm) function.

Notes: **(A)** A function put in the form $|f(x)|$ is called the "modulus of a function" or simply "modulus of a function" which signifies that:

(i) $|f(x)| = f(x)$, provided $f(x) \geq 0$, i.e., if $f(x)$ is positive or zero, then $|f(x)| = f(x)$.

(ii) $|f(x)| = -f(x)$, provided $f(x) < 0$, i.e., if $f(x)$ is negative, then $|f(x)| = -f(x)$ which means that if $f(x)$ is negative, $f(x)$ should be multiplied by -1 to make $f(x)$ positive.

(B) $|f(x)| = \text{sgn} f(x) \times f(x)$ where sgn

$$f(x) = \frac{|f(x)|}{f(x)}, \; f(x) \neq 0$$

$= 0, f(x) = 0$

i.e., $\text{sgn} f(x) = 1$ when $f(x) > 0$

$= -1$ when $f(x) < 0$

$= 0$ when $f(x) = 0$

where 'sgn' signifies "sign of" written briefly for the word "signum" from the Latin. Also, domain of absolute value function $= D(f) =$ {real numbers} and range of absolute value function $= R(f) =$ {non negative real numbers} $= R^+ \cup \{0\}$.

(C) 1. (i) $|x - a| = (x - a)$ when $(x - a) \geq 0$

$|x - a| = -(x - a)$ when $(x - a) < 0$

(ii) $|3| = 3$ since 3 is positive.

$|-3| = -(-3)$ since -3 is negative. For this reason, we have to multiply -3 by -1.

2. If the sign of a function $f(x)$ is unknown (i.e., we do not know whether $f(x)$ is positive or negative), then we generally use the following definition of the absolute value of a function.

$$|f(x)| = \sqrt{[f(x)]^2} = \sqrt{f^2(x)}$$

3. Absolute means to have a magnitude but no sign.

4. Absolute value, norm and modulus of a function are synonymes.

5. ***Notation***: The absolute value of a function is denoted by writing two vertical bars (i.e. straight lines) within which the function is placed. Thus the notation to signify "the absolute value of" is "| |".

6. $|f^2(x)| = f^2(x) = |f(x)|^2 = (-f(x))^2$

7. In a compact form, the absolute value of a function may be defined as $|f(x)| = \sqrt{f^2(x)}$

$= f(x)$, when $f(x) \geq 0$

$= -f(x)$, when $f(x) < 0$

8. $|f_1(x)| = |f_2(x)| \Leftrightarrow f_1(x) = \pm f_2(x)$

e.g.: $|x - 2| = |x + 3| \Leftrightarrow (x - 2) = \pm (x + 3)$ which is solved as under this line. $(x - 2) = (x + 3) \Rightarrow -2 = 3$ which is false which means this equation has no solution and $(x - 2) = -(x + 3) \Rightarrow x - 2 = -x - 3 \Rightarrow x + x = 2 - 3 \Rightarrow 2x = -1 \Rightarrow x = -\frac{1}{2}$.

9. $|f(x)| \leq k \Leftrightarrow -k \leq f(x) \leq k$ which signifies the intersection of $f(x) \geq -k$ and $f(x) \leq k$, $\forall k > 0$.

10. $|f(x)| \geq k \Leftrightarrow f(x) \geq k$ or $f(x) \leq -k$ which signifies the union of $f(x) \geq k$ and $f(x) \leq -k$, $\forall k > 0$

11. $|f(x)|^n = (f(x))^n$, where n is a real number.

12. $|f(x)| \geq 0$ always means that the absolute value of a functions is always non-negative (i.e., zero or positive real numbers)

13. $|f(x)| = |-f(x)|$

14. $|f(x)| \geq f(x)$

15. $|f_1(x) \cdot f_2(x)|$
$= |f_1(x)| \cdot |f_2(x)|$

16. $\left|\frac{f_1(x)}{f_2(x)}\right| = \frac{|f_1(x)|}{|f_2(x)|}, f_2(x) \neq 0$

17. $|f_1(x) + f_2(x)| \leq ||f_1(x)| + |f_2(x)||$

18. $|f_1(x) - f_2(x)| \geq ||f_1(x)| - |f_2(x)||$

19. $|0| = 0$, i.e. absolute value of zero is zero.

20. Modulus of modulus of a function (i.e. mod of $|f(x)|$) $= |f(x)|$

Remarks: When

(a) $|x| = x$, when $x \geq 0 \Leftrightarrow |x| = x,\ \forall x \in (0, \infty)$ and $|x| = -x$, when $x < 0 \Leftrightarrow |x| = -x$, $\forall x \in (-\infty, 0)$.

(b) $|x| = |-x| = x$, for all real values of x

(c) $|x| = \sqrt{x^2}$

(d) $|x| \leq a \Leftrightarrow -a \leq x \leq a$ and $|x| \geq a \Leftrightarrow x \geq a$ and $x \leq -a$.

Geometric Interpretation of Absolute Value of a Real Number *x*, Denoted by | *x* |

The absolute value of a real number x, denoted by $|x|$ is undirected distance between the origin O and the point corresponding to a (i.e. $x = a$) i.e, $|x|$ signifies the distance between the origin and the given point $x = a$ on the real line.

Explanation: Let $OP = x$

If $x > o$, P lies on the right side of origin 'O', then the distance $OP = |OP| = |x| = x$

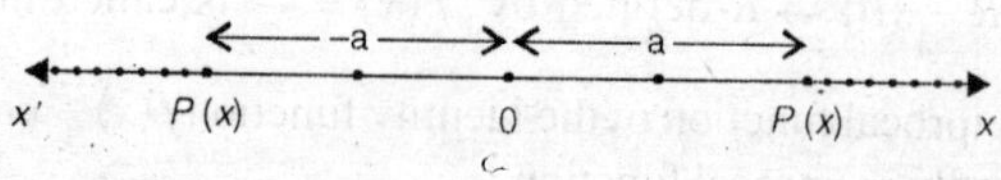

If $x = O$, P coincides with origin, the distance $OP = |x| = |o| = o$

If $x > O$, P lies on the left side of origin 'o', then the distance $OP = |OP| = |-OP| = |-x| = x$

Hence, $|x| =$

x, provided $x > o$ means that the absolute value of a positive number is the positive number itself.

o, provided $x = o$ means the absolute value of zero is taken to be equal to zero.

$-x$, provided $x < o$ means that the absolute value of a negative number is the positive value of that number.

Notes:

1. x is negative in $|x| = -x$ signifies $-x$ is positive in $|x| = -x$ e.g.: $|-7| = -(-7) = 7$.

2. The graphs of two numbers namely a and $-a$ on the number line are equidistant from the origin. We call the distance of either from zero, the absolute value of a and denote it by $|a|$.

3. $|x| = a \Leftrightarrow x = \pm a$

4. $x^2 = a^2 \Leftrightarrow \sqrt{x^2} = \sqrt{a^2} \Leftrightarrow |x| = |a| \Leftrightarrow x = \pm a \Leftrightarrow x = \pm\sqrt{a^2} \Leftrightarrow |x| = \sqrt{a^2}$.

5. $|x| = \sqrt{x^2}$ signifies that if x is any given number, then the symbol $\sqrt{x^2}$ represents the positive square root of x^2 and be denoted by $|x|$ whose graph is symmetrical about the y-axis having the shape of English alphabet 'V'. which opens (i) upwards if $y = |x|$ (ii) downwards if $y = -|x|$ (iii) on the right side if $x = |y|$ (iv) on the left side if $x = -|y|$.

An Important Remark

1. The radical sign "$\sqrt{\ }$" indicates the positive root of the quantity (a number or a function) written under it (radical sign) e.g.: $\sqrt{25} = +5$.

2. If we wish to indicate the negative square root of a quantity under the radical sign, we write the negative sign (–) before the radical sign. e.g.: $-\sqrt{4} = -2$.

3. To indicate both positive square root and negative square root of a quantity under the radical sign, we write the symbol $\pm$ (read as "plus or minus") before the radical sign.

e.g.: $\pm\sqrt{1} = \pm 1$

$\pm\sqrt{4} = \pm 2$

$\pm\sqrt{16} = \pm 4$

Remember:

1. In problems involving square root, the positive square root is the one used generally, unless there is a remark to the contrary. Hence, $\sqrt{100} = 10$; $\sqrt{169} = 13$; $\sqrt{x^2} = |x|$.

2. $x^2 + y^2 = 1 \Leftrightarrow x^2 = 1 - y^2 \Leftrightarrow \sqrt{x^2} = \sqrt{1 - y^2} \Leftrightarrow |x| = \sqrt{1 - y^2} \Leftrightarrow x = \pm\sqrt{1 - y^2}$

e.g.: $\cos^2\theta = 1 - \sin^2\theta \Leftrightarrow |\cos\theta| = \sqrt{1 - \sin^2\theta} \Leftrightarrow \cos\theta = \pm\sqrt{1 - \sin^2\theta}$

one should note that the sign of $\cos\theta$ is determined by the value of the angle 'θ' and the value of the angle 'θ' is determined by the quadrant in which it lies. Similarly for other trigonometrical functions of θ, such as, $\tan^2\theta = \sec^2\theta - 1 \Leftrightarrow \tan\theta = \pm\sqrt{\sec^2\theta - 1} \Leftrightarrow |\tan\theta| = \sqrt{\sec^2\theta - 1}$

$\cot^2\theta = \operatorname{cosec}^2\theta - 1 \Leftrightarrow \cot\theta = \pm\sqrt{\operatorname{cosec}^2\theta - 1} \Leftrightarrow |\cot\theta| = \sqrt{\operatorname{cosec}^2\theta - 1}$

$\sec^2\theta = 1 + \tan^2\theta \Leftrightarrow \sec\theta = \pm\sqrt{1 + \tan^2\theta} \Leftrightarrow |\sec\theta| = \sqrt{1 + \tan^2\theta}$, where the sign of angle 'θ' is determined by the quadrant in which it lies.

3. The word "modulus" is also written as "mod" and "modulus function" is written as "mod function" in brief.

On Greatest Integer Function

Firstly, we recall the definition of greatest integer function.

Definition: A greatest integer function is the function defined on the domain of all real numbers such that with any x in the domain, the function associates algebraically the greatest (largest or highest) integer which is less than or equal to x (i.e., not greater than x) designated by writing square brackets around x as $[x]$.

The greatest integer function has the property of being less than or equal to x, while the next integer is greater than x which means $[x] \le x < [x] + 1$.

Examples:

(i) $x = \frac{3}{2} \Rightarrow [x] = \left[\frac{3}{2}\right] = 1$ is the greatest integer in $\frac{3}{2}$.

(ii) $x = 5 \Rightarrow [x] = [5] = 5$ is the greatest integer in 5.

(iii) $x = \sqrt{50} \Rightarrow [x] = \left[\sqrt{50}\right] = 7$ is the greatest integer in $\sqrt{50}$.

(iv) $x = 2.5 \Rightarrow [x] = [-2.5] = -3$ is the greatest integer in -2.5.

(v) $x = 4.7 \Rightarrow [x] = [-4.7] = -5$ is the greatest integer in -4.7.

(vi) $x = -3 \Rightarrow [x] = [-3] = -3$ is the greatest integer in -3.

To Remember:

1. The greatest integer function is also termed as "the bracket, integral part or integer floor function".

2. The other notation for greatest integer function is $\lfloor\ \rfloor$ or [[ ]] in some books inspite of [ ].

3. The symbol [ ] denotes the process of finding the greatest integer contained in a real number but not greater than the real number put in [ ].

Thus, in general $y = [f(x)]$ means that there is a greatest integer in the value $f(x)$ but not greater than the value $f(x)$ which it assumes for any $x \in R$.

This is why in particular $y = [x]$ means that for a particular value of x, y has a greatest integer which is not greater than the value given to x.

4. The function $y = [x]$, where $[x]$ denotes integral part of the real number x, which satisfies the equality $x = [x] + q$, where $0 \le q < 1$ is discontinuous at every integer $x = 0, \pm 1, \pm 2, \ldots$ and at all other points, this function is continuous.

5. If x and y are two arbitrary real numbers satisfying the inequality $n \le x < n + 1$ and $n \le y < n + 1$, where n is an integer, then $[x] = [y] = n$.

6. $y = [x]$ is meaningless for a non-real value of x because its domain is the set of all real numbers and the range is the set of all integers, i.e. $D[x] = R$ and $R[x] = \{n: n \text{ is an integer}\}$ = The set of all integers, … $-3, -2, -1, 0, 1, 2, 3, \ldots$, i.e., negative, zero or positive integer.

7. $[f(x)] = 0 \Leftrightarrow 0 \le f(x) < 1$. Further the solution of $0 \le f(x) < 1$ provides us one of the adjacent intervals where x lies. The next of the a adjacent intervals is determined by adding 1 to the left and right end point of the solution of $0 \le f(x) < 1$. This process of adding 1 to the left and right end point is continued till we get a finite set of horizontal line segments representing the graph of the function $y = [f(x)]$

More on Properties of Greatest Integer Function.

(i) $[x + n] = n + [x]$, $n \in I$ and $x \in R$

(ii) $[-x] = -[x]$, $x \in I$

(iii) $[-x] = -[x] - 1$, $x \notin I$

(iv) $[x] \ge n \Rightarrow x \ge n$, $n \in I$

(v) $[x] \le n \Rightarrow x < n + 1$, $n \in I$

(vi) $[x] > n \Rightarrow x \ge n$

(vii) $[x] < n \Rightarrow x < n$, $n \in I$ and $x \in R$

(viii) $[x + y] \ge [x] + [y]$, $x, y \in R$

(ix) $\left[\frac{[x]}{n}\right] = \left[\frac{x}{n}\right]$, $n \in N$ and $x \in R$

(x) $x = [x] + \{x\}$ where { } denotes the fractional part of x, $\forall\, x \in R$

(xi) $x - 1 < [x] \le x,\ \forall\, x \in R$

(xii) $[x] \le x < [x] + 1$ for all real values of x.

Question: Define "logarithmic" function.

Answer: A function $f: (0, \infty) \to R$ defined by $f(x) = \log_a x$ is called logarithmic function, where $a \ne 1$, $a > 0$. Its domain and range are $(0, \infty)$ and R respectively.

Question: Define "Exponential function".

Answer: A function $f: R \to R$ defined by $f(x) = a^x$, where $a \ne 1$, $a > 0$. Its domain and range are R and $(0, \infty)$ respectively.

Question: Define the "piece wise function".

Answer: A function $y = f(x)$ is called the "piece wise function" if the interval (open or closed) in which the given function is defined can be divided into a finite number of adjacent intervals (open or closed) over each of which the given function is defined in different forms. e.g.:

1. $f(x) = 2x + 3,\ 0 \le x < 1$
   $f(x) = 7,\ x = 1$
   $f(x) = x^2,\ 1 < x \le 2$
2. $f(x) = x^2 - 1,\ 0 < x < 2$
   $f(x) = x + 2,\ x \ge 2$
3. $f(x) = 1 + x,\ -1 \le x < 0$
   $f(x) = x^2 - 1,\ 0 < x < 2$
   $f(x) = 2x,\ x \ge 2$

Notes:

1. Non-overlapping intervals: The intervals which have no points in common except one of the end points of adjacent intervals are called non overlapping intervals whose union constitutes the domain of the piece wise function. e.g.: $\left[0, \frac{1}{3}\right]$, $\left[\frac{1}{3}, \frac{2}{3}\right]$ and $\left[\frac{2}{3}, 1\right]$ serve as an example of non-overlapping intervals whose union [0, 1] is the domain of the piece wise function if it is defined as:

$$f(x) = 2x + 1,\ 0 \le x \le \frac{1}{3}$$

$$f(x) = x^2 + 2,\ \frac{1}{3} \le x \le \frac{2}{3}$$

$$f(x) = 4x^2 - 1,\ \frac{2}{3} \le x \le 1$$

2. A function $y = f(x)$ may not be necessarily defined by a single equation for all values of x but the function $y = f(x)$ may be defined in different forms in different parts of its domain.

3. Piecewise function is termed also "Piecewise defined function" because function is defined in each piece. If every function defined in adjacent intervals is linear, it is termed as "Piecewise linerar function" and if every function defined in adjacent intervals is continuous, it is called "piecewise continuous function."

Question: What do you mean by the "real variables"?

Answer: If the values assumed by the independent variable 'x' are real numbers, then the independent variable 'x' is called the "real variable".

Question: What do you mean by the "real function (or, real values of function) of a real variable"?

Answer: A function $y = f(x)$ whose domain and range are sets of a real numbers is said to be a real function (or more clearly, a real function of a real variable) which signifies that values assumed by the dependent variable are real numbers for each real value assumed by the independent variable x.

Note: The domain of a real function may not be necessarily a subset of R which means that the domain of a real function can be any set.

Examples:

1. Let $A = \{\theta, \{a, b\}, \{a\}, \{b\}\}$ and $B = \{1, 2, 3, 4, 5\}$

$\therefore f = \{(\theta, 1), (\{a, b\}, 2), (\{a\}, 4), (\{b\}, 3)\}$ is a real function since B is a subset of the set of real numbers.

2. If $f: R \to R$ such that $f(x) = 2x - 1,\ \forall\, x \in R$, then f is a real function.

Remarks:

1. In example (i) The domain of f is a class of sets and in example (ii) The domain of f is R. But in both examples, the ranges are necessarily subsets of R.

2. If the domain of a function f is any set other than (i.e. different from) a subset of real numbers and the range is necessarily a subset of the set of real numbers, the function must be called a real function (or real valued function) but not a real function of a real variable because a function of a real variable signifies that it is a function $y = f(x)$ whose domain and range are subsets of the set of real numbers.

Question: What do you mean by a "single valued function"?

Answer: When only one value of function $y = f(x)$ is achieved for a single value of the independent variable $x = a$, we say that the given function $y = f(x)$ is a single valued function, i.e., when one value of the independent variable x gives only one value of the function $y = f(x)$, then the function $y = f(x)$ is said to be single valued, e.g.:

1. $y = 3x + 2$

2. $y = x^2$

3. $y = \sin^{-1} x, \ -\frac{\pi}{2} \le y \le \frac{\pi}{2}$

serves as examples for single valued functions because for each value of x, we get a single value for y.

Question: What do you mean by a "multiple valued function"?

Answer: when two or more than two values of the function $y = f(x)$ are obtained for a single value of the independent variable $x = a$, we say that the given function $y = f(x)$ is a multiple (or, many) valued function, i.e. if a function $y = f(x)$ has more than one value for each value of the independent variable x, then the function $y = f(x)$ is said to be a multiple (or, many) valued function, e.g.:

1. $x^2 + y^2 = 9 \Rightarrow x = \pm\sqrt{9 - x^2} \Rightarrow y$ has two real values, $\forall\ x < 3$.

2. $y^2 = x \Rightarrow |y| = \sqrt{x}$ is also a multiple valued function since $x = 9 \Rightarrow y^2 = 9 \Rightarrow |y| = \sqrt{y^2} = \sqrt{9} \Rightarrow y = \pm 3$ ($\because |y| = \sqrt{y^2} = y$ for $y > 0$ and $|y| = \sqrt{y^2} = -y$ for $y > 0$).

Question: What do you mean by standard functions?

Answer: A form in which a function is usually written is termed as a standard function.

e.g.: $y = x^n$, $\sin x$, $\cos x$, $\tan x$, $\cot x$, $\sec x$, $\operatorname{cosec} x$, $\sin^{-1} x$, $\cos^{-1} x$, $\tan^{-1} x$, $\cot^{-1} x$, $\sec^{-1} x$, $\operatorname{cosec}^{-1} x$, $\log a^x$, $\log e^x$, a^x, e^x, etc. are standard functions.

Question: What do you mean by the "inverse function"?

Answer: A function, usually written as f^{-1} whose domain and range are respectively the range and domain of a given function f and under which the image $f^{-1}(y)$ of an element y is the element of which y was the image under the given function f, that is, $f^{-1}(y) = x \Leftrightarrow f(x) = y$.

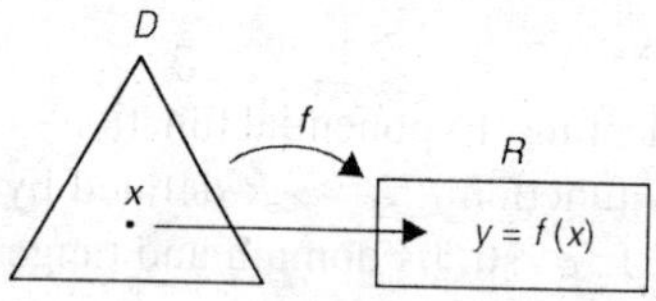

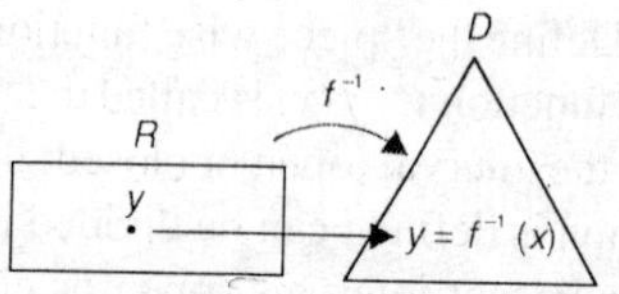

Remarks:

1. A function has its inverse $\Leftrightarrow$ it is one-one (or, one to one) when the function is defined from its domain to its range only.

2. Unless a function $y = f(x)$ is one-one, its inverse can not exist from its domain to its range.

3. If a function $y = f(x)$ is such that for each value of x, there is a unique values of y and conversely for each value of y, there is a unique value of x, we say that the given function $y = f(x)$ is one-one or we say that there exists a one to one (or, one-one) relation between x and y.

4. In the notation f^{-1}, (-1) is a superscript written at right hand side just above f. This is why we should not consider it as an exponent of the base f which means it can not be written as $f^{-1} = \frac{1}{f}$.

5. A function has its inverse $\Leftrightarrow$ it is both one-one and onto when the function is defined from its domain to its cò-domain.

Pictorial Representation of Inverse Function

To have an arrow diagram, one must follow the following steps.

1. Let $f: D \to R$ be a function such that it is one-one (i.e. distinct point in D have distinct images in R under f).

2. Inter change the sets such that original range of f is the domain of f^{-1} and original domain of f is the range of f^{-1}.

3. Change f to f^{-1}.

Therefore, $f: D \to R$ defined by $y = f(x)$ s.t it is one-one $\Leftrightarrow f^{-1}: R \to D$ defined by $f^{-1}(y) = x$ is an inverse function.

On Intervals

1. Values and range of an independent variable x: If x is a variable in (on/over) a set C, then members (elements or points) of the set C are called the values of the independent variable x and the set C is called the range of the independent variable x, whereas x itself signifies any unspecified (i.e., an arbitrary) member of the set C.

2. Interval: The subsets of a real line are called intervals. There are two types (or, kinds) of an interval namely (i) Finite and (ii) Infinite.

(i) Finite interval: The set containing all real numbers (or, points) between two real numbers (or, points) including or excluding one or both of these two real numbers known as the left and right and points is said to be a finite interval. A finite interval is classified into two kinds namely (a) closed interval and (b) open interval mainly.

(a) Closed interval: The set of all real numbers x subject to the condition $a \leq x \leq b$ is called closed interval and is denoted by $[a, b]$ where a and b are real numbers such that $a < b$.

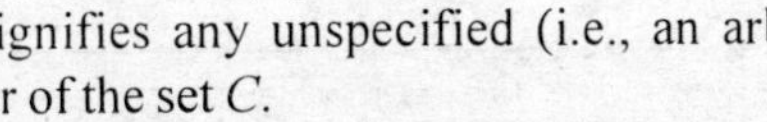

In set theoretic language, $[a, b] = \{x: a \leq x \leq b, x$ is real$\}$, denotes a closed interval.

Notes:

1. The notation $[a, b]$ signifies the set of all real numbers between a and b including the end points a and b, i.e., the set of all real from a to b.

2. The pharase "at the point $x = a$" signifies that x assumes (or, takes) the value a.

3. A neighbourhood of the point $x = a$ is a closed interval put in the form $[a - h, a + h]$ where h is a positive number, i.e.,

$[a - h, a + h] = \{x: a - h \leq x \leq a + h,\ h$ is a small positive number$\}$

4. All real numbers can be represented by points on a directed straight line (i.e., on the x-axis of cartesian coordinates) which is called the number axis. Hence, every number (i.e. real number) represents a definite point on the segment of the x-axis and conversely every point on the segment (i.e., a part) of the x-axis represents only one real number. Therefore, the numbers and points are synonymes if they represent the members of the interval concerned.(***Notes* 1.** It is a postulate that all the real numbers can be represents by the points of a straight line. **2.** Neigbourhood roughly means all points near about any specified point.)

(b) Open interval: The set of all real numbers x subject to the condition $a < x < b$ is called an open interval and is denoted by (a, b), where a and b are two real numbers such that $a < b$.

In the set theoretic language, $(a, b) = \{x: a < x < b, x$ is real$\}$

Notes:

1. The notation (a, b) signifies the set of all real numbers between a and b excluding the end points a and b.

2. The number 'a' is called the left end point of the interval (open or closed) if it is within the circular or square brackets on the left hand side and the number b is called the right end point of the interval if it is within the circular or square brackets on the right h and side.

3. Open and closed intervals are represented by the circular and square brackets (i.e., ( ) and [ ]) respectively within which end points are written separated by a comma.

(c) Half-open, half closed interval (or, semi-open, semi closed interval): The set of all real numbers x such that $a < x \leq b$ is called half open, half closed interval (or, semi-open, semi closed interval), where a and b are two real numbers such that $a < b$.

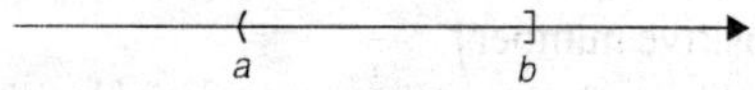

$(a, b) = \{x: a < x \leq b, x \text{ is real}\}$

Note: The notation $(a, b]$ signifies the set of all real numbers between a and b excluding the left end point a and including the right end point b.

(d) Half closed, half open (or, semi closed, semi open interval): The set of all real numbers x such that $a \leq x < b$ is called half-closed, half open interval (or, semi closed, semi open interval), where a and b be two real numbers such that $a < b$.

In set theoretic language, $[a, b) = \{x: a \leq x < b, x \text{ is real}\}$

Note: The notation $[a, b)$ signifies the set of all real numbers between a and b including the left end point a and excluding the right end point b.

2. Infinite interval

(a) The interval $(-\infty, \infty)$: The set of all real numbers x is an infinite interval and is denoted by $(-\infty, \infty)$ or R.

In set theoretic language,

$R = (-\infty, \infty) = \{x: -\infty < x < \infty, x \text{ is real}\}$

(b) The interval (a, ∞): The set of all real numbers x such that $x > a$ is an infinite interval and is denoted by (a, ∞).

In set theoretic language,

$(a, \infty) = \{x: x > a, x \text{ is real}\}$

or, $(a, \infty) = \{x: a < x < \infty, x \text{ is real}\}$

(c) The interval $[a, \infty)$: The set of all real numbers x such that $x \geq a$ is an infinite interval and is denoted by $[a, \infty)$.

In set theoretic language,

$[a, \infty) = \{x: x \geq a, x \text{ is real}\}$

or, $[a, \infty) = \{x: a \geq x > \infty, x \text{ is real}\}$

(d) The interval $(-\infty, a)$: The set of all real numbers x such that $x < a$ is an infinite interval and is denoted by $(-\infty, a)$.

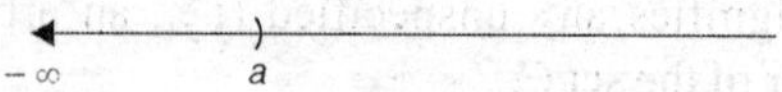

$(-\infty, a) = \{x: x < a, x \text{ is real}\}$

or, $(-\infty, a) = \{x: -\infty < x < a, x \text{ is real}\}$

(e) The interval $(-\infty, a]$: The set of all real numbers x such that $x \leq a$ is an infinite interval and is denoted by $(-\infty, a]$.

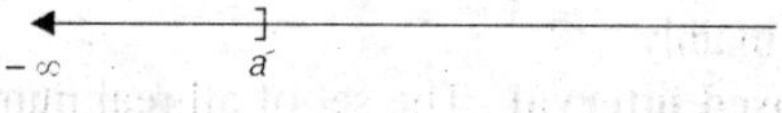

In set theoretic language,

$(-\infty, a] = \{x: x \leq a, x \text{ is real}\}$

$(-\infty, a] = \{x: -\infty < x \leq a, x \text{ is real}\}$

Remember:

1. In any finite interval, if a and/b is (or, more) replaced by ∞ and $/-\infty$, we get what is called an infinite interval.

2. $a \le x \le b$ signifies the intersection of the two sets of values given by $x \ge a$ and $x \le b$.

3. $x \ge a$ or $x \le b$ signifies the union of the two sets of values given by $x \ge a$ and $x \le b$.

4. The sign of equality with the sign of inequality (i.e., $\ge$ or $\le$) signifies the inclusion of the specified number in the indicated interval finite or infinite. The square bracket (i.e., [,)also (put before and/after any specified number) signifies the inclusion of that specified number in the indicated interval finite or infinite.

5. The sign of inequality without the sign of equality (i.e. > or <) signifies the exclusion of the specified number in the indicated interval finite or infinite. The circular bracket (i.e. (,) also (put before and/after any specified number) signifies the exclusion of that specified number indicated interval finite or infinite.

6. (i) $: x \in [a, b] \Leftrightarrow a \le x \le b$ and $x \notin [a, b] \Leftrightarrow x \in [a, b]^c \Leftrightarrow x \in (-\infty, a) \cup (b, \infty)$ where $[a, b]^c = R - [a, b]$ complement of $[a, b] = (-\infty, a) \cup (b, \infty)$.

(ii) $: x \in (a, b) \Leftrightarrow a < x < b$ and $x \notin (a, b) \Leftrightarrow x \in (-\infty, a] \cup [b, \infty)$.

7. Intervals expressed in terms of modulus: Many intervals can be easily expressed in terms of absolute values and conversely.

(i) $|x| < a \Leftrightarrow -a < x < a \Leftrightarrow x \in (-a, a)$, where '$a$' is any positive real number and $x \in R$.

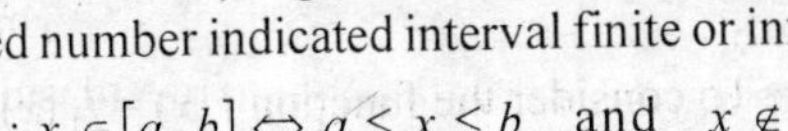

(ii) $|x| \le a \Leftrightarrow -a \le x \le a \Leftrightarrow x \in [-a, a]$ where 'a' is any positive real number and $x \in R$.

(iii) $|x| > a \Leftrightarrow x \notin [-a, a] \Leftrightarrow x \in (-\infty, -a) \cup (a, \infty) \Leftrightarrow$ either $x < -a$ or $x > a$, a being any positive real number and $x \in R$.

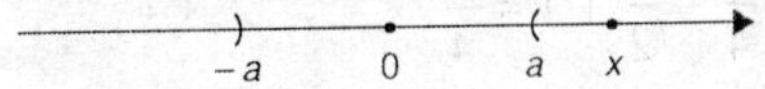

(iv) $|x| \ge a \Leftrightarrow$ either $x \le -a$ or $x \ge a \Leftrightarrow x \in (-\infty, -a] \cup [a, \infty)$.

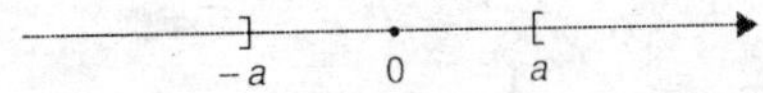

Evaluation of a Function at a Given Point

Evaluation: To determine the value of a function $y = f(x)$ at a given point $x = a$, is known as evaluation (or, more clearly evaluation of the function $y = f(x)$ at the given point $x = a$)

Notation: $[f(x)]_{x=a} = (f(x))_{x=a} = f(a)$ is a notation to signify the value of the function f at $x = a$.

Type 1: To evaluate a function $f(x)$ at a point $x = a$ when the function $f(x)$ is defined by a single expression, equation or formula.

Working rule: The method of finding the value of a function $f(x)$ at the given point $x = a$ when the given function $f(x)$ is defined by a single expression, equation or formula containing x consists of following steps.

Step 1: To substitute the given value of the independent variable (or, argument) x wherever x occurs in the given expression, equation, or formula containg x for $f(x)$

Step 2: To simplify the given expression, equation or formula containg x for $f(x)$ after substitution of the given value of the independent variable (or, argument) x.

Solved Examples

1. If $f(x) = x^2 - x + 1$, find $f(0)$, $f(1)$ and $f\left(\frac{1}{2}\right)$.

Solution: $\because f(x) = x^2 - x + 1$

$\therefore f(0) = 0^2 - 0 + 1 = 1$

$f(1) = 1^2 - 1 + 1 = 1$

and $f\left(\frac{1}{2}\right) = \left(\frac{1}{2}\right)^2 - \left(\frac{1}{2}\right) + 1$

$= \frac{1}{4} - \frac{1}{2} + 1 = \frac{3}{4}$

2. If $f(x) = \frac{1}{x}$, find $\frac{f(1+h) - f(1)}{h}$.

Solution: $\therefore\ f(x) = \frac{1}{x}$ ...(1)

$\therefore\ f(1+h) = \frac{1}{1+h}$ ...(2)

$f(1) = \frac{1}{1} = 1$ ...(3)

$\therefore\ f(1+h) - f(1) = \frac{1}{1+h} - 1 = \frac{-h}{1+h}$ ...(3)

$\therefore\ \frac{(4)}{h} = \frac{f(1+h) - f(1)}{h}$

$= \frac{-h/(1+h)}{h} = \frac{1}{1+h}$

Type 2: (To evaluate a piecewise function $f(x)$ at a point belonging to different intervals in which different expression for $f(x)$ is defined). In general, a piece wise function is put in the form

$f(x) = f_1(x)$, when $x > a$
$= f_2(x)$, when $x = a$
$f_3(x)$, when $x < a$, $\forall\ x \in R$

and one is required to find the values (i) $f(a_1)$ (ii) $f(a)$ and (iii) $f(a_0)$, where a, a_0 and a_1 are specified (or, given) values of x and belong to the interval $x > a$ which denote the domains of different function $f_1(x)$, $f_2(x)$ and $f_3(x)$ etc for $f(x)$.

Note: The domains over which different expression $f_1(x), f_2(x)$ and $f_3(x)$ etc for $f(x)$ are defined are intervals finite or infinite as $x > a$, $x < a$, $x \ge a$, $x \le a$, $a < x < b$, $a \le x < b$, $a < x \le b$ and $a \le x \le b$ etc and represent the different parts of the domain of $f(x)$.

Working rule: It consists of following steps:

Step 1: To consider the function $f(x) = f_1(x)$ to find the value $f(a_1)$, provided $x = a_1 > a$ and to and to put $x = a_1$ in $f(x) = f_1(x)$ which will provide one the value $f(a_1)$ after simplification.

Step 2: To consider the function $f(x) = f_2(x)$ to find the value $f(a)$, provided $x = a$ is the restriction against $f_2(x)$ and put $x = a$ in $f_2(x)$. If $f(x) = f_2(x)$ when the restrictions imposed against it are $x \ge a$, $x \le a$, $a \le x < b$, $a < x \le b$, $a \le x \le b$ or any other interval with the sign or equality indicating the inclusion of the value 'a' of x, we may consider $f_2(x)$ to find the value $f(a)$. But if $f(x) = f_2(x) =$ constant, when $x = a$ is given in the question, then $f_2(x) =$ given constant will be the required value of $f(x)$ i.e. $f(x) =$ given constant when $x = a$ signifies not to find the value other than $f(a)$ which is equal to the given constant.

Step 3: To consider the function $f(x) = f_3(x)$ to find the value $f(a_2)$ provided $x = a_2 < a$ and $x = a_2$ in $f(x) = f_3(x)$ which will provide one the value $f(a_2)$ after simplification.

Remember:

1. $f(x) = f_1(x)$, when (or, for, or, if) $a \le x < a_2$ signifies that one has to consider the function $f(x) = f_1(x)$ to find the functional value $f_1(x)$ for all values of x (given or specified in the question) which lie in between a_1 and a_2 including $x = a_1$.

2. $f(x) = f_2(x)$, when (or, for, or, if) $a_2 < x \le a_3$, signifies that one has to consider the function $f(x) = f_2(x)$, to find the functional value $f_2(x)$ for all values of x (given or specified in the question) which lie in between a_2 and a_3 including $x = a_3$.

3. $f(x) = f_3(x)$, when (or, for, or, if) $a_4 < x < a_5$ signifies that one has to consider the function $f(x) = f_3(x)$ to find the functional value $f_3(x)$ for all values of x (given or specified in the question) which lie in between a_4 and a_5 excluding a_4 and a_5.

Solved Examples

1. If $f : R \to R$ is defined by

$f(x) = x^2 - 3x$, when $x > 2$
$= 5$, when $x = 2$
$= 2x + 1$, when $x < 2$, $\forall\ x \in R$

find the values of (i) $f(4)$ (ii) $f(2)$ (iii) $f(0)$ (iv) $f(-3)$ (v) $f(100)$ (vi) $f(-500)$.

Solution: **1.** $\because 4 > 2, \therefore$ by definition, $f(4) = (x^2 - 3x)$ for $x = 4 = 4^2 - 3(4) = 16 - 12 = 4$

(ii) $\because 2 = 2, \therefore$ by definition, $f(2) = 5$

(iii) $0 < 2, \therefore$ by definition, $f(0) = 2(0) + 1 = 1$

(iv) $-3 < 2, \therefore$ by definition, $f(-3) = 2(-3) + 1 = -5$

(v) $100 > 2, \therefore$ by definition, $f(100) = (100)^2 - 3(100) = 10000 - 300 = 9700$

(vi) $-500 < 2, \therefore$ by definition $f(-500) = 2(-500) + 1 = -1000 + 1 = -999$

2. If $f(x) = 1 + x$, when $-1 \le x < 0$

$= x^2 - 1$, when $0 < x < 2$

$2x$, when $x \ge 2$

find $f(3), f\left(\frac{1}{2}\right), f\left(-\frac{1}{2}\right)$

Solution: $\because f(x) = 2x$ for $x \ge 2$

$\therefore f(3) = 2 \times 3 = 6 \; (\because x = 3 \ge 2)$

$\because f(x) = 1 + x$, for $-1 \le x < 0$

$$\therefore f\left(-\frac{1}{2}\right) = 1 + \left(-\frac{1}{2}\right) = \frac{1}{2}$$

$$\left(\because x = -\frac{1}{2} \in [-1, 0)\right)$$

$\because f(x) = x^2 - 1$, for $0 < x < 2$

$$\therefore f\left(\frac{1}{2}\right) = \left(\frac{1}{2}\right)^2 - 1 = -1 + \frac{1}{4} = -\frac{3}{4}$$

$$\left(\because x = \frac{1}{2} \in (0, 2)\right)$$

Refresh your memory:

1. If a function $f(x)$ is defined by various expressions $f_1(x), f_2(x), f_3(x)$ etc, then $f(a_0)$ denotes the value of the function $f(x)$ for $x = a_0$ which belongs to the domain of the function $f(x)$ represented by various restrictions $x > a, x < a, x \ge a, x \le a, a < x < b, a \le x \le b, a \le x < b$, and $a < x \le b$ etc.

2. Supposing that we are required to find the value of the function $f(x)$ for a point $x = a_0$ which does not belong to the given domain of the function $f(x)$, then $f(a_0)$ is undefined, i.e., we cannot find $f(a_0)$, i.e., $f(a_0)$ does not exist.

$f(x) = x^2 - 1$, when $0 < x < 2$

$= x + 2$, when $x \ge 2$

find $f(-1)$

Solutions: $-1 \notin$ domain of $f(x)$ represented by the union of the restrictions $0 < x < 2$ and $x \ge 2$ (i.e., $0 < x < 2$ or $x \ge 2$). For this reason $f(-1)$ is undefined (i.e., $f(x)$ is undefined at $x = -1$).

3. Sometimes we are required to find the value of a piecewise function $f(x)$ for $a_0 \pm h$ where $h > 0$, in such cases, we may put $h = 0.0001$ for easiness to guess in which domain (or, interval) the point represented by $x = a \pm h$ lies.

e.g.: If a function is defined as under

$f(x) = 1 + x$, when $-1 \le x < 0$

$= x^2 - 1$, when $0 < x < 2$

$2x$, when $x \ge 2$

find $f(2 - h)$ and $f(-1 + h)$

(***Footnotes***: 1. $f(a)$ exists or $f(a)$ is defined $\Leftrightarrow$ 'a' lies in the domain of f. 2. $f(a)$ does not exist or $f(a)$ is undefined $\Leftrightarrow$ 'a' does not lie in the domain of f.)

Solution: **1.** Putting $h = 0.001$, we get $2 - h = 2 - 0.001 = 1.999$ and $1.999 \in (0, 2) = 0 < x < 2$

$\therefore f(2 - h) = (2 - h)^2 - 1 = 2^2 + h^2 - 4h - 1 = 4 + h^2 - 4h - 1$

$= 3 + h^2 - 4h = h^2 - 4h + 3$

2. Putting $h = 0.001$, we get $-1 + h = -1 + 0.001 = 0.999$ and $0.999 \in [-1, 0) = -1 \le x < 0$

$\therefore f(-1 + h) = (1 + x)_{x = -1 + h} = 1 + h - 1 = h$

Domain of a Function

Sometimes a function of an independent variable x is described by a formula or an equation or an expression in x and the domain of a function is not explicitly stated. In such circumstances, the domain of a function is understood to be the largest possible set of real numbers such that for each real number (of the largest possible set), the rule (or, the function) gives a real number or for each of which the formula is meaningful or defined.

Definition: If $f: D \rightarrow R$ defined by $y = f(x)$ be a real valued function of a real variable, then the domain of the function f represented by $D(f)$ or dom (f) is defined as the set consisting of all real numbers representing the totality of the values of the independent variable x such that for each real value of x, the function or the equation or the expression in x has a finite value but no imaginary or indeterminate value.

Or, in set theoretic language, it is defined as:

If $f: D \rightarrow R$ be real valued function of the real variable x, then its domain is D or $D(f)$ or dom (f)

$= \{x \in R: f(x)$ has finite values $\}$

$= \{x \in R: f(x)$ has no imaginary or indeterminate value.$\}$

To remember:

1. Domain of sum or difference of two functions $f(x)$ and $g(x) = \text{dom}\,[f(x) \pm g(x)] = \text{dom}\,(f(x)) \cap \text{dom}\,(g(x))$.
2. Domain of product of two functions $f(x)$ and $g(x)$ $g(x) = \text{dom}\,[f(x) \cdot g(x)] = \text{dom}\,(f(x)) \cap \text{dom}\,(g(x))$.
3. Domain of quotient of two functions $f(x)$ and $g(x)$ $= \text{dom}\left[\dfrac{f(x)}{g(x)}\right] = \text{dom}\, f(x) \cap \text{dom}\,(g(x)) \cap$ $\{x: g(x) \neq 0\}$

$= \text{dom}\,(f(x)) \cap \text{dom}\,(g(x)) - \{x: g(x) \neq 0\}$ i.e., the domain of a rational function or the quotient function is the set of all real numbers with the exception of those real numbers for which the function in denominator becomes zero.

Notes: **1.** The domain of a function defined by a formula $y = f(x)$ consists of all the values of x but no value of y (i.e., $f(x)$).

2. (i) The statement "$f(x)$ is defined for all x" signifies that $f(x)$ is defined in the interval $(-\infty, \infty)$.

(ii) The statements "$f(x)$ is defined in an interval finite or infinite" signifies that $f(x)$ exists and is real for all real values of x belonging to the interval. Hence, the statement "$f(x)$ is defined in the closed interval $[a, b]$" means that $f(x)$ exists and is real for all real values of x from a to b, a and b being real numbers such that $a < b$. Similarly, the statement "$f(x)$ is defined in the open interval (a, b)" means that $f(x)$ exists and is real for all real values of x between a and b (excluding a and b)

3. (i) $f(x) \cdot g(x) = 0 \Leftrightarrow \left[\begin{array}{l} f(x) = 0 \\ g(x) = 0 \end{array}\right.$, or,

(ii) $f(x) \cdot g(x) \geq 0 \Leftrightarrow \left[\begin{array}{l} \begin{cases} f(x) \geq 0 \\ g(x) \geq 0 \end{cases} \\ \begin{cases} f(x) \leq 0 \\ g(x) \leq 0 \end{cases} \end{array}\right.$, or

(iii) $f(x) \cdot g(x) \leq 0 \Leftrightarrow \left[\begin{array}{l} \begin{cases} f(x) \geq 0 \\ g(x) \leq 0 \end{cases} \\ \begin{cases} f(x) \leq 0 \\ g(x) \geq 0 \end{cases} \end{array}\right.$, or

(iv) $\dfrac{f(x)}{g(x)} \geq 0 \Leftrightarrow \left[\begin{array}{l} \begin{cases} f(x) \geq 0 \\ g(x) < 0 \end{cases} \\ \begin{cases} f(x) \leq 0 \\ g(x) > 0 \end{cases} \end{array}\right.$, or

(v) $\dfrac{f(x)}{g(x)} \leq 0 \Leftrightarrow \left[\begin{array}{l} \begin{cases} f(x) \geq 0 \\ g(x) < 0 \end{cases} \\ \begin{cases} f(x) \leq 0 \\ g(x) > 0 \end{cases} \end{array}\right.$, or

4. (i) $\left(x^2 - a^2\right) < 0 \Leftrightarrow -a < x < a$

(ii) $\left(x^2 - a^2\right) \leq 0 \Leftrightarrow -a \leq x \leq a$

(iii) $\left(x^2 - a^2\right) > 0 \Leftrightarrow x < -a$ or $x > a$

(iv) $\left(x^2 - a^2\right) \geq 0 \Leftrightarrow x \leq -a$ or $x \geq a$

5. (i) $(x-a_1)(x-b_1)<0 \Leftrightarrow a_1<x<b_1 \ (a_1<b_1) \Leftrightarrow$

$x \in (a_1, b_1)$

(ii) $(x-a_2)(x-b_2)\le 0 \Leftrightarrow a_2 \le x \le b_2 \ (a_2<b_2) \Leftrightarrow$

$x \in [a_2, b_2]$

They mean the intersection of

(a) $x > a_1$ and $x < b_1$

(b) $x \ge a_2$ and $x \le b_2$

(iii) $(x-a)(x-b) > 0 \Leftrightarrow x < a$ or $x > b(a < b)$

$\Leftrightarrow x \in (-\infty, a) \cup (b, \infty)$

(iv) $(x-a)(x-b) \ge 0 \Leftrightarrow x \le a$ or $x \ge b(a < b)$

$\Leftrightarrow x \in (-\infty, a] \cup [b, \infty)$

They mean the union of

(a) $x < a$ and $x > b$

(b) $x \le a$ and $x \ge b$.

Question: How to represent the union and intersection on a number line?

Answer: Firstly, we recall the definitions of union and intersection of two sets.

Union: The union of two sets E and F is the set of elements belonging to either E or F.

Intersection: The set of all elements belonging to both sets E and F is called intersection of E and F.

Method of Representation of Union and Intersection on Real Lines

If the set of the points on the line segment AB be the set E and the set of the point on segment CD be the F, then the union of E and F is the segment $AD = AB + BD$ = sum or union and the intersection of E and F is the segment CB = common segment.

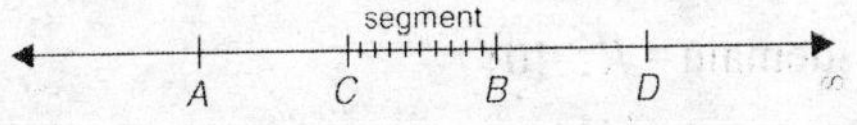

Now some rules to find the domain of real valued functions are given. They are useful to find the domain of any given real valued function.

Finding the Domain of Algebraic Functions

Type 1: Problems based on finding the domains of polynomial functions.

Working rule: One must remember that a polynomial in x has the domain R (i.e., the set of the real numbers) because any function f of x which does not become undefined or imaginary for any real value of x has the domain R. Hence, the linear $y = ax + b$; the quadratic $y = ax^2 + bx + c$; and the square functions $y = x^2$ have the domain R.

Solved Examples

Find the domain of each of the following functions:

1. $y = 11x - 7$

Solution: $y = 11x - 7$ is a linear function and we know that a linear function has the domain R.

Hence, domain of $y\ (= 11x - 7) = R = (-\infty, +\infty)$

2. $y = x^2 - 3x + 7$

Solution: $y = x^2 - 3x + 7$ is a quadratic function and we know that a quadratic function has the domain R.

Hence, domain of $y = (= x^2 - 3x + 7) = R = (-\infty, +\infty)$

3. $y = x^2$

Solution: $y = x^2$ is a square function and we know that a square function has the domain R.

Hence, domain of $y = (= x^2) = R = (-\infty, +\infty)$

Type 2: Problems based on finding the domain of a function put in the form:

(i) $y = \dfrac{f(x)}{g(x)},\ g(x) \ne 0$

or, **(ii)** $y = \dfrac{1}{g(x)},\ g(x) \ne 0$

Working rule: It consists of following steps:

1. To put the function (or, expression in x) in the denominator = 0, i.e., $g(x) = 0$

2. To find the values of x from the equation $g(x) = 0$

3. To delete the valued of x from R to get the required domain, i.e., domain of $\dfrac{f(x)}{g(x)}$ or $\dfrac{1}{g(x)} = R - \{$roots of the equation $g(x) = 0\}$, where $f(x)$ and $g(x)$ are polynomials in x.

Note: When the roots of the equation $g(x) = 0$ are imaginary then the domain of the quotient function put in the form: $\frac{f(x)}{g(x)}$ or $\frac{1}{g(x)} = R$

Solved Examples

Find the domain of each of the following functions:

1. $y = \frac{x^2 - 3x + 2}{x^2 + x - 6}$

Solution: $y = \frac{x^2 - 3x + 2}{x^2 + x - 6}$

Now, putting $x^2 + x - 6 = 0$

$\Rightarrow x^2 + 3x - 2x - 6 = 0$

$\Rightarrow x(x+3) - 2(x+3) = 0$

$\Rightarrow (x+3)(x-2) = 0$

$\Rightarrow x = 2, -3$

$\therefore$ domain $= R - \{2, 3\}$

2. $y = \frac{x^2 - 2x + 4}{x^2 + 2x + 4}$

Solution: $y = \frac{x^2 - 2x + 4}{x^2 + 2x + 4}$

Now, putting, $x^2 + 2x + 4 = 0$

$\Rightarrow x^2 + 2x + 4 = 0$

$\Rightarrow (x+1)^2 + 3 = 0$

$\Rightarrow (x+1)^2 = -3$

$\Rightarrow (x + 1) = \pm\sqrt{-3}$

$\Rightarrow x = -1 \pm \sqrt{-3}$ imaginary or complex numbers.

$\therefore$ domain $= R$

3. $y = \frac{x}{5 - x}$

Solution: $y = \frac{x}{5 - x}$

Now, putting, $5 - x = 0$

$\Rightarrow x = 5$

$\therefore$ domain $= R - \{5\}$

4. $y = \frac{2x - 4}{2x + 4}$

Solution: $y = \frac{2x - 4}{2x + 4}$

Now, putting, $2x + 4 = 0$

$\Rightarrow 2x - 4 \Rightarrow x = \frac{-4}{2} = -2$

$\therefore$ domain $R - \{2\}$

5. $f(x) = \frac{1}{(x - 1)(x - 2)}$

Solution: $f(x) = \frac{1}{(x - 1)(x - 2)}$

Now, putting $(x - 1)(x - 2) = 0$

$\Rightarrow x = 1, 2$

$\therefore$ domain $= R - \{1, 2\}$

6. $y = \frac{1}{x^2 - 1}$

Solution: $y = \frac{1}{x^2 - 1}$

Now, putting, $x^2 - 1 = 0$

$\Rightarrow x^2 = 1 \Rightarrow x = \pm 1$

$\therefore$ domain $= R - \{-1, 1\}$

7. $y = \frac{1}{x}$

Solution: $y = \frac{1}{x}$

Now, putting $x = 0 \Rightarrow x = 0$ i.e. y is undefined at $x = 0$

$\therefore$ domain $= R - \{0\}$

8. $y = \frac{x^2 - 3x + 2}{x^2 + x - 6}$

Solution: $y = \frac{x^2 - 3x + 2}{x^2 + x - 6}$

Now, putting, $x^2+x-6=0 \Rightarrow x^2+3x-2x-6=0$
$\Rightarrow x(x+3)-2(x+3)=0 \Rightarrow (x-2)(x+3)=0 \Rightarrow x=2,-3$

$\therefore$ domain $= R-\{2,-3\}$

9. $y=\dfrac{1}{2x-6}$

Solution: $y=\dfrac{1}{2x-6}$

Now, putting $2x-6=0$

$\Rightarrow x=\dfrac{6}{2}=3$

$\therefore$ domain $= R-\{3\}=(-\infty,3)\cup(3,+\infty)$

10. $y=\dfrac{1}{x^2-5x+6}$

Solution: $y=\dfrac{1}{x^2-5x+6}$

$\Rightarrow x^2-5x+6=0$
$\Rightarrow x^2-3x-2x+6=0$
$\Rightarrow x(x-3)-2(x-3)=0$
$\Rightarrow (x-3)(x-2)=0$
$\Rightarrow x=2,3$

$\therefore$ domain $= R-\{2,3\}=(-\infty,2)\cup(2,3)\cup(3,\infty)$

Type 3: Problems based on finding the domain of the square root of a function put in the forms:

(i) $\sqrt{f(x)}$

(ii) $\sqrt{\dfrac{f(x)}{g(x)}}$

(iii) $\dfrac{1}{\sqrt{g(x)}}$

(iv) $\dfrac{f(x)}{\sqrt{g(x)}}$

Now we tackle each type of problem one by one.

1. Problems based on finding the domain of a function put in the form: $\sqrt{f(x)}$.

It consists of two types when:

(i) $f(x)=ax+b=a$ linear in x.
(ii) $f(x)=ax^2+bx+c=a$ quadratic in x.
(i) Problems based on finding the domain of a function put in the form: $\sqrt{f(x)}$, when $f(x)=ax+b$.

Working rule: It consists of following steps:

Step 1: To put $ax+b\geq 0$

Step 2: To find the values of x for which $ax+b\geq 0$ to get the required domain.

Step 3: To write the domain = [root of the inequation $ax+b\geq 0,+\infty)$

Notes: **1.** The domain of a function put in the form $\sqrt{f(x)}$ consists of the values of x for which $f(x)\geq 0$.

2. $x\geq c \Leftrightarrow x\in[c,+\infty)$.

Solved Examples

Find the domain of each of the following fucntions:

1. $y=\sqrt{x}$

Solution: $y=\sqrt{x}$

Now, putting $x\geq 0 \Rightarrow x\geq 0$

$\therefore$ domain $=[0,+\infty)$

2. $y=\sqrt{2x-4}$

Solution: $y=\sqrt{2x-4}$

Now, putting $2x-4\geq 0 \Rightarrow x\geq \dfrac{4}{2}=2$

$\therefore$ domain $=[2,+\infty)$

3. $y=\sqrt{x}+\sqrt{x-1}$

Solution: $y=\sqrt{x}+\sqrt{x-1}$

Putting $y_1=\sqrt{x}$ and $y_2=\sqrt{x}$, we have $y=y_1+y_2$

$\therefore$ domain of y = dom $(y_1) \cap$ dom (y_2)

Now, domain of $y_1 \left(= \sqrt{x}\right) = D_1$ (say) $= [0, +\infty)$ [from example 1.] again, we require to find the domain of $y_2 \left(= \sqrt{x-1}\right)$.

Putting $x - 1 \geq 0 \Rightarrow x \geq 1 \Rightarrow$ domain y_2 $\left(= \sqrt{x-1}\right) = D_2$ (say) $= [1, +\infty)$ Hence, domain of $y = D$ (say) = dom $(y_1) \cap$ dom (y_2)

$= D_1 \cap D_2$

$= [0, +\infty) \cap [1, +\infty)$

$= [1, \infty)$

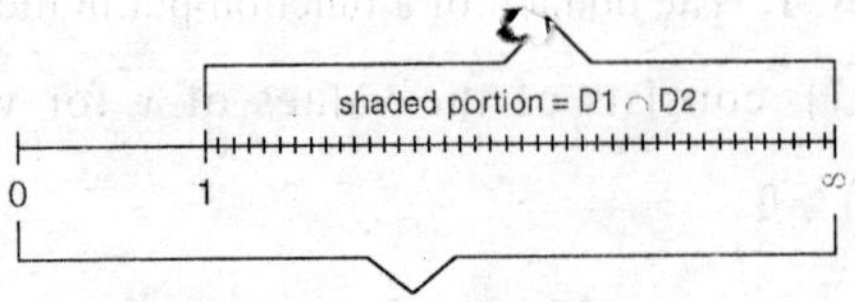

(ii) Problems based on finding the domain of a function put in the form:

$y = \sqrt{f(x)}$, when $f(x) = ax^2 + bx + c$ and α, β are the roots of $ax^2 + bx + c = 0$ $(\alpha < \beta)$

working rule: It consists of following steps:

Step 1: To put $ax^2 + bx + c \geq 0$

Step 2: To solve the in equation $ax^2 + bx + c \geq 0$ for x by factorization or by completing the square.

Step 3: To write the domain of $\sqrt{ax^2 + bx + c} =$ $\alpha \leq x \leq \beta$ only when the coefficient of $x^2 = a = -ve$ and $ax^2 + bx + c = a(x - \alpha)(x - \beta)$ and to write the domain of $\sqrt{ax^2 + bx + c} = R - (\alpha, \beta)$ only when the coefficient of $x^2 = a$ +ve and $ax^2 + bx + c = a(x - \alpha)(x - \beta)$.

Notes: **(i)** a = coefficient of $x^2 = -ve$ (and, $ax^2 + bx + c = a(x - \alpha)(x - \beta) \geq 0) \Rightarrow x$ lies between α and $\beta \Rightarrow$ domain of $\sqrt{ax^2 + bx + c} = \alpha \leq x \leq \beta$ $(\alpha < \beta)$.

(ii) a = coefficient of x^2 = + ve (and, $ax^2 + bx + c = a(x - \alpha)(x - \beta) \geq 0) \Rightarrow x$ does not lie between α and $\beta \Rightarrow$ domain of $\sqrt{ax^2 + bx + c} = R - (\alpha, \beta) =$ $(-\infty, \alpha) \cup (\beta, +\infty)$.

(iii) $(x + \alpha)(x + \beta)$ should be written as $(x - (-\alpha)), (x - (-\beta))$ while finding the domain of the square root of $ax^2 + bx + c = a(x + \alpha)(x + \beta)$.

Solved Examples

Find the domain of each of the following functions:

1. $y = \sqrt{x^2 - 3x + 4}$

Solution: $y = \sqrt{x^2 - 3x + 4}$

Now $x^2 - 3x + 4 \geq 0$

$\Rightarrow \left(x^2 - 4x + x - 4\right) \geq 0$

$\Rightarrow x(x - 4) + (x - 4) \geq 0$

$\Rightarrow (x + 1)(x - 4) \geq 0$

$\Rightarrow (x - 4)(x - (-1)) \geq 0 \Rightarrow x$ does not lie between -1 and $4 \Rightarrow x \leq -1$ or $x \geq 4$.

$\therefore$ domain $= R - (-1, 4)$

2. $y = \sqrt{(x - 2)(x - 5)}$

Solution: $y = \sqrt{(x - 2)(x - 5)}$

Now, $(x-2)(x-5) \geq 0 \Rightarrow x$ does not lie between 2 and 3.

$\Rightarrow x \leq 2$ or $x \geq 5$

$\therefore$ domain $= R - (2, 5)$

3. $y = \sqrt{x^2 - 5x + 6}$

Solution: $y = \sqrt{x^2 - 5x + 6}$

Now, $x^2 - 5x + 6 \geq 0$

$\Rightarrow (x-2)(x-3) \geq 0$

$\Rightarrow x \leq 2$ or $x \geq 3$

$\therefore$ domain $= R-(2,3)$

4. $y = \sqrt{-x^2 + 5x - 6}$

Solution: $y = \sqrt{-x^2 + 5x - 6}$

Now, $-x^2 + 5x - 6 \geq 0$

$\Rightarrow x^2 - 5x + 6 \leq 0$

$\Rightarrow (x-2)(x-3) \leq 0$

$\Rightarrow$ x lies between 2 and 3

$\Rightarrow 2 \leq x \leq 3$

$\therefore$ domain $= [2, 3]$

5. $y = \sqrt{-16x^2 - 24x}$

Solution: $y = \sqrt{-16x^2 - 24x}$

Now, $-16x^2 - 24x \geq 0$

$\Rightarrow -2x^2 - 3x \geq 0$

$\Rightarrow 2x^2 + 3x \leq 0$

$\Rightarrow x(2x+3) \leq 0$

$\Rightarrow x(2x-(-3)) \leq 0$

$\Rightarrow$ x lies between $-\frac{3}{2}$ and 0

$\Rightarrow -\frac{3}{2} \leq x \leq 0$

$\therefore$ domain $= \left[-\frac{2}{3}, 0\right]$

6. $y = \sqrt{-5 - 6x - x^2}$

Solution: $y = \sqrt{-5 - 6x - x^2}$

Now, $-5 - 6x - x^2 \geq 0$

$\Rightarrow x^2 + 6x + 5 \leq 0$

$\Rightarrow x^2 + 5x + x + 5 \leq 0$

$\Rightarrow x(x+5) + (x+5) \leq 0$

$\Rightarrow (x+5)(x+1) \leq 0$

$\Rightarrow (x-(-1))(x-(-5)) \leq 0 \Rightarrow$ x lies between -5 and $-1 \Rightarrow -5 \leq x \leq -1$.

$\therefore$ domain $= [-5, -1]$

7. $y = \sqrt{(1-x)(x+3)}$

Solution: $y = \sqrt{(1-x)(x+3)}$

Now, $(1-x)(x+3) \geq 0$

$\Rightarrow -(x-1)(x+3) \geq 0$

$\Rightarrow (x-1)(x+3) \leq 0$

$\Rightarrow (x-1)(x-(-3)) \leq 0$

$\Rightarrow -3 \leq x \leq 1$

$\therefore$ domain $= [-3, 1]$

8. $y = \sqrt{1 - x^2}$

Solution: $y = \sqrt{1 - x^2}$

Now, $1 - x^2 \geq 0$

$\Rightarrow -\left(1 - x^2\right) \leq 0$

$\Rightarrow x^2 - 1 \leq 0$

$\Rightarrow (x-1)(x+1) \leq 0 \Rightarrow (x-1)(x-(-1)) \leq 0 \Rightarrow x$ lies between -1 and $+1$

$\Rightarrow -1 \leq x \leq 1$

$\therefore$ domain $= [-1, 1]$

9. $y = -\sqrt{4 - x^2}$

Solution: $y = -\sqrt{4 - x^2}$

Now, putting $4 - x^2 \geq 0$

$\Rightarrow x^2 - 4 \leq 0$

$\Rightarrow (x - 2)(x + 2) \leq 0$

$\Rightarrow (x - 2)(x - (-2)) \leq 0$

$\Rightarrow$ x lies between –2 and 2 $\Rightarrow -2 \leq x \leq 2$

$\therefore$ domain = $[-2, 2]$

10. $y = \frac{1}{2}\sqrt{4 - x^2}$

Solution: $y = \frac{1}{2}\sqrt{4 - x^2}$

Now, putting $4 - x^2 \geq 0$

$\Rightarrow x^2 - 4 \leq 0$

$\Rightarrow (x - 2)(x + 2) \leq 0$

$\Rightarrow (x - 2)(x - (-2)) \leq 0$

$\Rightarrow$ x lies between –2 and 2 $\Rightarrow -2 \leq x \leq 2$

$\therefore$ domain = $[-2, 2]$

11. $y = -\frac{1}{2}\sqrt{4 - x^2}$

Solution: $y = -\frac{1}{2}\sqrt{4 - x^2}$

Now, putting $4 - x^2 \geq 0$

$\Rightarrow x^2 - 4 \leq 0$

$\Rightarrow (x - 2)(x + 2) \leq 0$

$\Rightarrow (x - 2)(x - (-2)) \leq 0$

$\Rightarrow -2 \leq x \leq 2$

$\therefore$ domain = $[-2, 2]$

12. $y = \sqrt{x^2 - 4x + 3}$

Solution: $y = \sqrt{x^2 - 4x + 3}$

Now, $x^2 - 4x + 3 \geq 0$

$\Rightarrow x^2 - 3x - x + 3 \geq 0$

$\Rightarrow (x - 3) - (x - 3) \geq 0$

$\Rightarrow (x - 1)(x - 3) \geq 0$

$\Rightarrow$ x does not lie between 1 and 3

$\Rightarrow x \leq 1$ or $x \geq 3$

$\therefore$ domain = $R - [1, 3] = (-\infty, 1] \cup [3, +\infty)$

13. $y = \sqrt{(x - 2)(x - 3)}$

Solution: $y = \sqrt{(x - 2)(x - 3)}$

Now, $(x - 2)(x - 3) \geq 0$

$\Rightarrow x \geq 2$ or $x \geq 3$

$\Rightarrow$ x does not lie between 2 and 3

$\therefore$ domain = $R - [2, 3] = (-\infty, 2] \cup [3, +\infty)$

14. $y = \sqrt{x^2 + 2x + 3}$

Solution: $\because y = \sqrt{x^2 + 2x + 3}$

Now, $x^2 + 2x + 3 \geq 0$

$\Rightarrow (x + 1)^2 + 2 \geq 0, \forall x$

$\Rightarrow (x + 1)^2 \geq -2$, which is true for all $x \in R$

$\therefore$ domain = $R = (-\infty, \infty)$

Type (ii): Problems based on finding the domain of a function put in the form : $y = \sqrt{\frac{f(x)}{g(x)}}$

While finding the domain of the square root of a quotient function (i.e; $y = \sqrt{\frac{f(x)}{g(x)}}$) one must remember the following facts:

1. The domain of $y \left(= \sqrt{\frac{f(x)}{g(x)}}\right)$ consists of those values of x for which $\frac{f(x)}{g(x)} \geq 0$

2. $\frac{f(x)}{g(x)} \geq 0 \Leftrightarrow f(x) \geq 0,\ g(x) > 0$, or $f(x) \geq 0$, $g(x) < 0$.

3. $\left(\frac{x-\alpha}{x-\beta}\right) \leq 0 \Leftrightarrow \alpha \leq x < \beta$ or $\beta < x \leq \alpha$ according as $\alpha < \beta$ or $\beta < \alpha$.

4. $\left(\frac{x-\alpha}{x-\beta}\right) \geq 0 \Leftrightarrow x \geq \alpha$ or $x < \beta$ if $\beta < \alpha$ and $\Leftrightarrow$ $x \leq \alpha$ or $x > \beta$ if $\alpha < \beta$.

5. The function in the denominator $\neq 0$ always.

Solved Examples

Find the domain of each of the following functions:

1. $y = \sqrt{\frac{3x-2}{2x+6}}$

Solution: y is defined for those x for which

$$\frac{3x-2}{2x+6} \geq 0$$

$$\Leftrightarrow \left.\begin{matrix}(1)\ (3x-2) \geq 0 \\ (2x+6) > 0\end{matrix}\right\},\ \text{i.e; } x \geq \frac{2}{3}$$

$$\text{or, } \left.\begin{matrix}(2)\ (3x-2) \leq 0 \\ (2x+6) < 0\end{matrix}\right\},\ \text{i.e; } x < -3$$

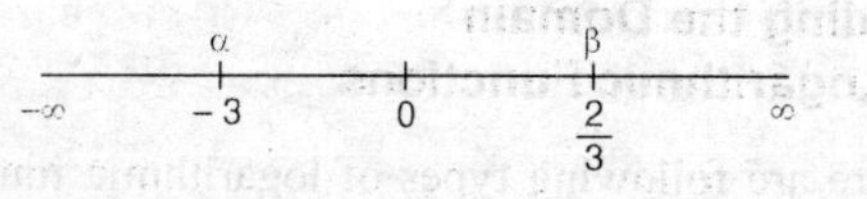

(1) and (2) $\Rightarrow x \geq \frac{2}{3}$, or $x < -3 \Leftrightarrow x$ $(-\infty, -3) \cup \left[\frac{2}{3}, +\infty\right)$

Hence, domain $\Rightarrow R - \left[-3, \frac{2}{3}\right) = (-\infty, -3) \cup \left[\frac{2}{3}, +\infty\right)$

or, alternatively:

$$\frac{3x-2}{2x+6} \geq 0 \Leftrightarrow \frac{x-\frac{2}{3}}{x+3} \geq 0 \Leftrightarrow x < -3 \text{ or}$$

$$x \geq \frac{2}{3} \Leftrightarrow x \in (-\infty, -3) \cup \left[\frac{2}{3}, +\infty\right)$$

Hence, domain $= (-\infty, -3) \cup \left[\frac{2}{3}, +\infty\right)$

2. $y = \sqrt{\frac{x-1}{x+1}}$

Solution: y is defined for all those x for which $\frac{x-1}{x+1} \geq 0 \Leftrightarrow \frac{x-1}{x-(-1)} \geq 0 \Leftrightarrow x < -1$ or $x \geq 1 \Leftrightarrow$ $x \in (-\infty, -1) \cup [1, +\infty)$

Hence, domain $= (-\infty, -1) \cup [1, +\infty)$

3. $y = \sqrt{\frac{x-2}{x+2}}$

Solution: y is defined for all those x for which $\frac{x-2}{x+2} \geq 0 \Leftrightarrow \frac{x-2}{x-(-2)} \geq 0 \Leftrightarrow x < -2$ or, $x \geq 2$ $\Leftrightarrow x \in (-\infty, -2) \cup [2, +\infty)$

Hence, domain $= (-\infty, -2) \cup [2, +\infty)$

Type (iii): Problems on finding the domain of a function put in the form: $y = \frac{1}{\sqrt{g(x)}}$

Working rule: It consists of following steps:

1. To put $g(x) > 0$
2. To find the values of x for which $g(x) > 0$

3. To form the Domain with the help of the roots of the in equation $g(x) > 0$.

Note: The domain of a function put in the form $y = \frac{1}{\sqrt{g(x)}}$ consists of all those values of x for which $g(x) > 0$.

Solved Examples

1. Find the domain of each of the following functions:

$$y = \frac{1}{\sqrt{(2-x)(x+3)}}$$

Solution: y is defined for all those values of x for which $(2-x)(x+3) > 0 \Leftrightarrow (x-2)(x+3) < 0 \Leftrightarrow x$ lies between -3 and $2 \Leftrightarrow -3 < x < 2 \Leftrightarrow x \in (-3, 2)$ hence, domain $= (-3, 2)$

2. $y = \frac{1}{\sqrt{(1-x)(x+2)}}$

Solution: y is defined for all those values of x for which $(1-x)(x+2) > 0 \Leftrightarrow (x-1)(x-(-2)) < 0 \Leftrightarrow x$ lies between -2 and $1 \Leftrightarrow -2 < x < 1 \Leftrightarrow x \in (-2, 1)$ hence, domain $= (-2, 1)$.

3. $y = \frac{1}{\sqrt{x^2 - 5x + 6}}$

Solution: y is defined for all those values of x for which $x^2 - 5x + 6 > 0 \Leftrightarrow x^2 - 3x - 2x + 6 > 0 \Leftrightarrow (x-3)x - 2(x-3) > 0 \Leftrightarrow (x-3)(x-2) > 0 \Leftrightarrow x < 2$ or $x > 3 \Leftrightarrow x \in (-\infty, 2) \cup (3, +\infty)$

Hence, domain $= R - [2, 3] = (-\infty, 2) \cup (3, +\infty)$

4. $y = \frac{1}{\sqrt{-x}}$

Solution: y is defined for all those values of x for which $-x > 0 \Leftrightarrow x < 0 \Leftrightarrow x \in (-\infty, 0)$

Hence, domain $= (-\infty, 0)$.

Type (iv): Problems on finding the domain of a function put in the form: $y = \frac{f(x)}{\sqrt{g(x)}}$.

Working rule: The rule to find the domain of a function of the form $y = \frac{f(x)}{\sqrt{g(x)}}$ is the same as for the domain of a function of the form $y = \frac{1}{\sqrt{g(x)}}$ which means.

1. To put $g(x) > 0$ and to find the values of x from the in equality $g(x) > 0$.

2. To form the domain with the help of obtained values of x.

Solved Examples

Find the domain of each of the following functions:

1. $y = \frac{x}{\sqrt{x^2 - 3x + 2}}$

Solution: y is defined when $x^2 - 3x + 2 > 0 \Leftrightarrow x^2 - 2x - x + 2 > 0 \Leftrightarrow x(x-2) - (x-2) > 0 \Leftrightarrow (x-1)(x-2) > 0 \Leftrightarrow x < 1$ or $x > 2$.

Hence, domain $= R - [1, 2] = (-\infty, 1) \cup (2, +\infty)$

2. $y = \frac{x}{\sqrt{(1-x)(x-2)}}$

Solution: y is defined when $(1-x)(x-2) > 0 \Leftrightarrow (x-1)(x-2) < 0 \Leftrightarrow x$ lies between 1 and $2 \Leftrightarrow 1 < x < 2 \Leftrightarrow x \in (1, 2)$.

Hence, domain $= (1, 2)$

Finding the Domain of Logarithmic Functions

There are following types of logarithmic functions whose domains are required to be determined.

(i) $y = \log f(x)$

(ii) $y = \log \sqrt{f(x)}$

(iii) $y = \log |f(x)|$

(iv) $y = \log\left(\frac{f(x)}{g(x)}\right)$

(v) $y = \frac{f(x)}{\log g(x)}$

(vi) $y = \log\log\log f(x)$

Now we tackle each type of problem one by one.

Type 1: Problems based on finding the domain of a function put in the form: $y = \log f(x)$.

Working rule: It consists of following steps:

Step 1: To put $f(x) > 0$ and to solve the in equality $f(x) > 0$ for x.

Step 2: To form the domain with the help of obtained values of x.

Notes: **1.** The domain of the logarithmic function $y = \log f(x)$ consists of all those values of x for which $f(x) > 0$.

2. Log $f(x)$ is defined only for positive $f(x)$.

Solved Examples

Find the Domain D of each of the following functions:

1. $y = \log(4 - x)$

Solution: y is defined when $4 - x > 0 \Leftrightarrow -x > -4 \Leftrightarrow x < 4$.

$\therefore\ D(y) = (-\infty, 4)$

2. $y = \log(8 - 2x)$

Solution: y is defined when $(8 - 2x) > 0 \Leftrightarrow -2x > -8 \Leftrightarrow x < 4$.

$\therefore\ D(y) = (-\infty, 4)$

3. $y = \log(2x + 6)$

Solution: y is defined when $(2x + 6) > 0 \Leftrightarrow 2x > -6 \Leftrightarrow x > -3 \Leftrightarrow x \in (-3, +\infty)$

Hence, $D(y) = (-3, +\infty)$

4. $y = \log\{(x + 6)(6 - x)\}$

Solution: y is defined when $(x + 6)(6 - x) > 0 \Leftrightarrow (x + 6)(x - 6) < 0 \Leftrightarrow x$ lies between -6 and $6 \Leftrightarrow -6 < x < 6 \Leftrightarrow x \in (-6, 6)$

Hence, $D(y) = (-6, 6)$

5. $y = \log(3x^2 - 4x + 5)$

Solution: Method (1)

y is defined when $(3x^2 - 4x + 5) > 0 \Leftrightarrow$

$$3\left(x^2 - \frac{4}{3}x + \frac{5}{3}\right) > 0$$

$$\Leftrightarrow 3\left[\left(x^2 - \frac{4}{3}x + \left(\frac{4}{6}\right)^2 - \left(\frac{4}{6}\right)^2 + \frac{5}{3}\right)\right] > 0$$

$$\Leftrightarrow 3\left[\left(x - \frac{2}{3}\right)^2 + \left(\frac{5}{3} - \frac{4}{9}\right)\right] > 0$$

$$\Leftrightarrow 3\left(x - \frac{2}{3}\right)^2 + (3)\left(\frac{11}{9}\right) > 0$$

$$\Leftrightarrow 3\left(x - \frac{2}{3}\right)^2 > -\frac{11}{9} \text{ which is true } \forall\ x \in R$$

$$\therefore\ D(y) = R = (-\infty, +\infty)$$

Notes: **1.** Imaginary or a complex numbers as the roots of an equation $ax^2 + bx + c = 0 \Leftrightarrow$ domain of $\log f(x) = R = (-\infty, +\infty)$ as in the above example roots are complex.

2. The method adopted in the above example is called "if method".

3. A perfect square is always positive which is greater than any negative number.

Method 2. This method consists of showing that $ax^2 + bx + c > 0, \forall x$ if $a > 0$ and discriminant $= b^2 - 4ac < 0$ here $3 > 0$, and discriminant $= 16 - 60 = -34 < 0$

$\therefore$ y is defined $\forall\ x \in R$

Therefore, $D(y) = R = (-\infty, +\infty)$

6. $y = \log(x^3 - x)$

Solution: y is defined when $(x^3 - x) > 0 \Leftrightarrow x(x^2 - 1) > 0 \Leftrightarrow x(x + 1)(x - 1) > 0 \Leftrightarrow (x - 0)(x + 1)(x - 1) > 0 \Leftrightarrow (x - (1))(x - 0)(x - 1) > 0$

Now let $f(x) = (x - (-1))(x - 0)(x - 1)$

If $x < -1$, then $f(x) < 0$ as all the three factors are < 0.

If $-1 < x < 0$, then $f(x) > 0$
If $0 < x < 1$, then $f(x) < 0$ and if $x > 1$, then $f(x) > 0$

Hence, $f(x) > 0 \Leftrightarrow x \in (-1, 0) \cup (1, \infty)$

$\therefore\ D(y) = (-1, 0) \cup (1, \infty)$

Type 2: Problems based on finding the domain of a function put in the form $y = \log \sqrt{f(x)}$

Working rule: One must remember that the function $y = \log \sqrt{f(x)}$ is defined when $y = \sqrt{f(x)} > 0$, i.e., $f(x) > 0$ which means the domain of the function $y = \log \sqrt{f(x)}$ consists of all those values of x for which $f(x) > 0$.

Solved Examples

Find the domain of each of the following functions:

1. $y = \log \sqrt{x - 4}$

Solution: $y = \log \sqrt{x - 4}$ is defined when $(x - 4) > 0 \Leftrightarrow x > 4$

$\therefore\ D(y) = (4, \infty)$

2. $y = \log \sqrt{6 - x}$

Solution: $y = \log \sqrt{6 - x}$ is defined when $(6 - x) > 0 \Leftrightarrow 6 > x \Leftrightarrow x < 6$

$\therefore\ D(y) = (-\infty, 6)$

3. $y = \log \left(\sqrt{x - 4} + \sqrt{6 - x}\right)$

Solution: $y = \log \left(\sqrt{x - 4} + \sqrt{6 - x}\right)$ is defined when $\left(\sqrt{x - 4} + \sqrt{6 - x}\right) > 0 \Leftrightarrow (x - 4)$ and $(6 - x) \geq 0 \Leftrightarrow x \geq 4$ and $x \leq 6 \Leftrightarrow 4 \leq x \leq 6$

$\therefore\ D(y) = [4, 6]$

Type 3: Problems based on finding the domain of a function put in the form: $y = \log |f(x)|$

Working rule: It consists of following steps:

1. To put $f(x) = 0$ and to find all the values of x from the equation $f(x) = 0$.

2. To form the domain of y which is the set of all real values of x excluding those values of x at which y is undefined , i.e.,

$D(y) = R - \{\text{values of } x \text{ at which } f(x) = 0\}$

Note: While finding the domain of a function of the form: $y = \log |f(x)|$, one must find all those values of x at which the function $f(x)$ (i.e., the function under the symbol of absolute value) becomes zero and then those values of x should be deleted from R (i.e., the set of all real numbers).

Solved Examples

Find the domain of each of the following functions:

1. $y = \log |x|$

Solution: $\log |x|$ is undefined only when $|x| = 0$, i.e., $x = 0$

$\therefore\ D(y) = R - \{0\}$

2. $y = \log |4 - x^2|$

Solution: $\log |4 - x^2|$ is undefined only when $|4 - x^2| = 0$; i.e., $4 - x^2 = 0 \Leftrightarrow x^2 - 4 = 0 \Leftrightarrow (x - 2)(x + 2) = 0 \Leftrightarrow (x - 2)(x - (-2)) = 0 \Leftrightarrow x = 2$ or $x = -2$

$\therefore\ D(y) = R - \{2, -2\}$

Type 4: Problems based on finding the domain of a function put in the form: $y = \log \left(\dfrac{f(x)}{g(x)}\right)$.

Working rule: It consists of following steps:

1. To put $\dfrac{f(x)}{g(x)} > 0$ and to solve the in equalities $f(x) > 0$ and $g(x) > 0$ or $f(x) < 0$ and $g(x)$,0 seperately.

2. To form the domain of y with the help of obtained values of x for which $\dfrac{f(x)}{g(x)} > 0$.

Notes: **1.** The domain of a logarithmic function of the form $y = \log \left(\dfrac{f(x)}{g(x)}\right)$ consists of all those values of x for which $\dfrac{f(x)}{g(x)} > 0$.

2. One must test whether $g(x)$ is positive or negative by the following scheme:

(i) If $a > 0$ and the discriminant (i.e.; $D = b^2 - 4ac$) of $g(x) = ax^2 + bx + c$ is < 0 (i.e., $D = -ve$), then $g(x) = ax^2 + bx + c$ is positive $\forall\ x$ and in such case, one must consider only the function in numerator $= f(x) > 0$ to find the solution set of $y = \dfrac{f(x)}{g(x)} > 0$.

Solved Examples

Find the domain of each of the following functions:

1. $y = \log\left(\dfrac{5x - x^2}{4}\right)$

Solution: y is defined when $\left(\dfrac{5x - x^2}{4}\right) > 0 \Leftrightarrow 5x - x^2 > 0 \Leftrightarrow x(5 - x) > 0 \Leftrightarrow (x - 0)(x - 5) < 0 \Leftrightarrow 0 < x < 5$ $x \in (0, 5)$

$\therefore\ D(y) = (0, 5)$

2. $y = \log\left(\dfrac{x}{10}\right)$

Solution: y is defined when $\left(\dfrac{x}{10}\right) > 0 \Leftrightarrow x > 0 \Leftrightarrow x \in (0, +\infty)$

$\therefore\ D(y) = (0, +\infty)$

N.B.: In the above two examples the functions in denominator are positive. This is why considerable function to be greater than zero is only the function in numerator.

3. $y = \log\left(\dfrac{x^2 - 5x + 6}{x^2 + 4x + 6}\right)$

Solution: y is defined when $\left(\dfrac{x^2 - 5x + 6}{x^2 + 4x + 6}\right) > 0 \Leftrightarrow$ $x^2 - 5x + 6 > 0 \Leftrightarrow x^2 - 3x - 2x + 6 > 0 \Leftrightarrow x(x - 3) - 2(x - 3) > 0 \Leftrightarrow (x - 2)(x - 3) > 0 \Leftrightarrow x$ does not lie between 2 and 3 $\Leftrightarrow x < 2$ or $x > 3 \Leftrightarrow x \in (-\infty, 2) \cup (3, +\infty)$.

$\therefore D(y) = (-\infty, 2) \cup (3, +\infty)$

N.B.: In the above example (3), the discriminant $D = 16 - 4 \times 1 \times 6 = 16 - 24 = -ve$ for the function $x^2 + 4x + 6$ in denominator which $\Rightarrow x^2 + 4x + 6 > 0$. For this reason, we considered only the function $x^2 - 5x + 6$ in numerator > 0.

4. $y = \log\left(\dfrac{x - 5}{x^2 - 10x + 24}\right)$

Solution: Method (1)

y is defined when $\left(\dfrac{x - 5}{x^2 - 10x + 24}\right) = \dfrac{(x - 5)}{(x - 4)(x - 6)} > 0 \Leftrightarrow (x - 4)(x - 5)(x - 6) > 0$

(multiplying both sides by $(x - 4)^2 (x - 6)^2$)

But $(x - 4)(x - 5)(x - 6) > 0$ when (a) all the above factors > 0 (b) one of the three factors > 0 and each of the other two factors < 0.

Hence, we have the following four cases:

Case (i): When $(x - 4) > 0$ and $(x - 5) > 0$ and $(x - 6) > 0$

$\Rightarrow x > 4$ and $x > 5$ and $x > 5 \Rightarrow x \in (6, \infty)$

Case (ii): When $(x - 4) > 0$ and $(x - 5) < 0$ and $(x - 6) < 0$

$\Rightarrow x > 4$ and $x < 5$ and $x < 6 \Rightarrow 4 < x < 5 \Rightarrow x \in (4, 5)$

Case (iii): When $(x - 4) < 0$ and $(x - 5) > 0$ and $(x - 6) < 0$

$\Rightarrow x < 4$ and $x > 5$ and $x < 6 \Rightarrow 5 < x < 6 \Rightarrow x \in \phi$

Case (iv): When $(x-4)<0$ and $(x-5)<0$ and $(x-6)>0$

$\Rightarrow x<4$ and $x>5$ and $x>6 \Rightarrow x \in \phi$

$\therefore D(y) = (4, 5) \cup (6, \infty)$

Method (2)

$$\frac{x-5}{x^2-10x+24} > 0 \Leftrightarrow \frac{(x-5)}{(x-4)(x-6)} > 0$$

If $x<4$, then $(x-4)<0, (x-5)<0, (x-6)<0$
If $4<x<5$, then $(x-4)>0, (x-5)<0, (x-6)<0$
If $5<x<6$, then $(x-4)>0, (x-5)>0, (x-6)<0$
and if $x>6$, then $(x-4)>0, (x-5)>0, (x-6)>0$

Hence, $\dfrac{x-5}{(x-4)(x-6)} > 0 \Leftrightarrow 4<x<5$ or $x>6$

$\therefore D(y) = (4, 5) \cup (6, \infty)$

Method (3)

For y to be defined

(i) $x^2 - 10x + 24 \neq 0 \Rightarrow (x-4)(x-6) \neq 0 \Rightarrow$

$x \neq 4, x \neq 6$ (A1)

(ii) $\dfrac{x-5}{x^2-10x+24} > 0 \Rightarrow$

(a) $x-5>0$ and $x^2-10x+24>0$ or
(b) $x-5$ and $x^2-10x+24<0$
from (a), $x-5>0$ and $(x-4)(x-6)>0$
$\Rightarrow x>5$ and ($x<4$ or $x>6$)
$\Rightarrow$ ($x>5$ and $x<4$) or ($x>5$ and $x>6$)
But $x>5$ and $x<4$ is not possible
$\therefore x>5$ and $x>6 \Rightarrow x>6$ (A2)
from (b), $x<5$ and $(x-4)(x-6)<0$
$\Rightarrow x<5$ and ($x>4$ and $x<6$)
$\Rightarrow x<5$ and ($4<x<6$)
$\Rightarrow 4<x<5$ (A3)
Now, combining (A2) and (A3), we get
$x>6$ or $4<x<5$ which $\Rightarrow x \in (4, 5) \cup (6, \infty)$

$\therefore D(y) = (4, 5) \cup (6, \infty)$

N.B.: On must note that in the above example discriminant of the function $x^2 - 10x + 24$ in the denominator is $10^2 - 4 \times 1 \times 24 = 100.96 = 4 = +ve$. For this reason both the functions $(x-5)$ and $(x^2-10x+24)$ simultaneously are considerable.

Type 5: Problems based on finding the domain of a function put in the form:

$$y = \frac{f(x)}{\log g(x)}$$

working rule: one must remember that a function having the form $y = \dfrac{f(x)}{\log g(x)}$ is defined when $g(x) > 0$ and $g(x) \neq 1$ which means that domain of y consists of all those values of x for which $g(x) > 0$ and $g(x) \neq 1$ i.e., $D(y) = D + (g(x)) - \{$roots of $g(x) = 1\}$, where $D + (g(x))$ signifies the solution set of $g(x) > 0$.

Note: $\log f(x) \neq \log 1 \Leftrightarrow f(x) \neq 1$

Solved Examples

Find the domain of each of the following functions:

(i) $y = \dfrac{x}{\log(1+x)}$

Solution: y is defined when $(1+x)>0$ and $\log(1+x) \neq 0$. $(1+x) > 0 \Leftrightarrow x > -1 \Leftrightarrow x \in (-1, +\infty)$.

$\log(1+x) = 0 \Leftrightarrow \log(1+x) \neq \log 1$ ($\because \log 1 = 0$)

$\Leftrightarrow 1 + x \neq 1$

$\Leftrightarrow x \neq 0$

$\therefore D(y) = (-1, +\infty) - \{0\}$

Type 6: Problems based on finding the domain of a function put in the form: $y = \log \log \log f(x)$

To remember: One must remember the following facts:

1. Inequalities of the form $\log_a x > c$, $\log_a x < c$, where $a > 0$ and $a \neq 1$ are called simplest logarithmic inequalities.

2. $\log_a x > c \Leftrightarrow \left[\begin{array}{l} \begin{cases} a > 1 \Rightarrow \\ x > a^c \end{cases} \\ \begin{cases} 0 < a < 1 \Rightarrow \\ 0 < x < a^c \end{cases} \end{array}\right.$

3. $\log_a x < c \Leftrightarrow \left[\begin{array}{l}\begin{cases} a > 1 \Rightarrow \\ 0 < x < a^c \end{cases} \\ \begin{cases} 0 < a < 1 \Rightarrow \\ x > a^c \end{cases}\end{array}\right.$

4. $\log_{f(x)} g_1(x) > \log_{f(x)} g_2(x) \Leftrightarrow \left[\begin{array}{l}\begin{cases} g_1(x) > 0 \\ g_2(x) > 0 \end{cases} \\ \left[\begin{array}{l}\begin{cases} f(x) > 1 \Rightarrow \\ g_1(x) > g_2(x) \end{cases} \\ \begin{cases} 0 < f(x) < 1 \Rightarrow \\ g_1(x) < g_2(x) \end{cases}\end{array}\right.\end{array}\right.$

Working rule: There are following working rules to find the domain of a function put in the forms:

(i) $y = \log_a \log_b f(x)$ $(a > 0, a \neq 1, b > 1)$

(ii) $y = \log_a \log_b \log_c f(x)$

$(a > 0, a \neq 1, b > 1, c > 0, c \neq 1)$

Rule 1: $\log_a \log_b f(x)$ exists

$\Leftrightarrow \log_b f(x) > 0$

$\Leftrightarrow \log_b f(x) > \log 1$

$\Leftrightarrow f(x) > 1$ and solve for x

Rule 2: $\log_a \log_b \log_c f(x)$ exists

$\Leftrightarrow \log_b \log_c f(x) > 0$

$\Leftrightarrow \log_b \log_c f(x) > \log 1$

$\Leftrightarrow \log_c f(x) > 1$

$\Leftrightarrow f(x) > c^1$ if $c > 1$; $f(x) < c$ if $c < 1$ and solve for x.

Aid to memory: To find the domain of a given function put in the above mentioned form.

1. One must remove first log operator from left hand side of the functions of the forms: $y = \log_a \log_b \log_c f(x)$ or $\log_a \log_b f(x)$ and the rest log of a function (i.e., $\log_b \log_c f(x)$ or $\log_b f(x)$) should be put > 0.
2. Use the rules:

(a) $\log_a x > c \Leftrightarrow \left[\begin{array}{l}\begin{cases} a > 1 \\ x > a^c \end{cases} \\ \begin{cases} 0 < a < 1 \\ 0 < x < a^c \end{cases}\end{array}\right.$

(b) $\log_a x < c \Leftrightarrow \left[\begin{array}{l}\begin{cases} a > 1 \\ 0 < x < a^c \end{cases} \\ \begin{cases} 0 < a < 1 \\ x > a^c \end{cases}\end{array}\right.$

Solved Examples

Find the domain of each of the following functions:

1. $y = \log_2 \log_3 (x-4)$

Solution: $y = \log_2 \log_3 (x-4)$ exists only for $\log_3 (x-4) > 0$

$\Leftrightarrow (x-4) > 3^0$

$\Leftrightarrow x - 4 > 1$

$\Leftrightarrow x > 5$

$\therefore D(y) = (5, +\infty)$

2. $y = \log_2 \log_3 \log_4 (x)$

Solution: $y = \log_2 \log_3 \log_4 (x)$ is defined when $\log_3 \log_4 (x) > 0$

$\Leftrightarrow \log_3 \log_4 (x) > \log_3 1$

$\Leftrightarrow \log_4 x > 1$

$\Leftrightarrow x > 4^1 \Leftrightarrow x > 4$

$\therefore D(y) = (4, +\infty)$

3. $y = \log_{10} [1 - \log_{10} (x^2 - 5x + 16)]$

Solution: $y = \log_{10} [1 - \log_{10} (x^2 - 5x + 16)]$ is defined when $[1 - \log_{10} (x^2 - 5x + 16)] > 0$

$\Leftrightarrow -\log_{10} (x^2 - 5x + 16) > -1$

$\Leftrightarrow \log_{10} (x^2 - 5x + 16) < 1$

$\Leftrightarrow x^2 - 5x + 16 < 10^1$ ($\because \log_a x < c \Leftrightarrow x < a^c$ when $a > 1$)

$\Leftrightarrow x^2 - 5x + 16 - 10 < 0$

$\Leftrightarrow x^2 - 5x + 6 < 0$

$\Leftrightarrow (x-2)(x-3) < 0$

$\Leftrightarrow 2 < x < 3 \Leftrightarrow x \in (2, 3)$

$\therefore D(y) = (2, 3)$

Domain of Trigonometric Functions

Question: Find the domain and range of the following functions:

1. $y = \sin x$ **2.** $y = \cos x$ **3.** $y = \tan x$
4. $y = \cot x$ **5.** $y = \operatorname{cosec} x$ **6.** $y = \sec x$

Answer: 1. $y = \sin x$

Since we know that the value of the function $y = \sin x$ is undefined and imaginary for no real value of x. Hence, the domain of $y = \sin x$ is the set of all real numbers. Again we know that $|\sin x| \le 1 \Leftrightarrow -1 \le \sin x \le 1$ for any real value of x which means that the range of $y = \sin x$ is the closed interval $[-1, 1]$.

Aid to memory:

$D(\sin x) = (-\infty, +\infty) = R =$ the set of all real numbers.

$R(\sin x) = [-1, 1]$

2. $y = \cos x$

since we know that the value of the function $y = \cos x$ is undefined and imaginary for no real value of x. Hence, the domain of $y = \cos x$ is the set of all real numbers, again we know that $|\cos x| \le 1 \Leftrightarrow -1 \le \cos x \le 1$ for any real value of x which means that the range of $y = \cos x$ is the closed interval $[-1, 1]$.

Aid to memory:

$D(\cos x) = (-\infty, +\infty) = R =$ the set of all reai numbers.

$R(\cos x) = [-1, 1]$

3. $y = \tan x$

since we know that the value of the function $y = \tan x = \dfrac{\sin x}{\cos x}$ is undefined for those values of x for which the function $\cos x$ in denominator is zero (i.e., for those values of x for which the function $\cos x$ in denominator is zero (i.e.; for $\cos x = 0 \Leftrightarrow x = (2n+1)\dfrac{\pi}{2}$ = odd multiple of $\dfrac{\pi}{2}$, n being an integer).

Hence, the domain of $y = \tan x$ is the set of all real numbers excluding $(2n+1)\dfrac{\pi}{2}$, n being an integer.

Again, we know that $\tan x$ can assume any value however large or small for real value of x which means that $-\infty < \tan x < \infty$ is the range of $y = \tan x$.

Aid to memory:

$D(\tan x) = R = \{x : x = (2n+1)\dfrac{\pi}{2}, n$ being an integer$\}$

$R(\tan x) = R = (-\infty, \infty)$ = the set of all real numbers.

4. $y = \cot x$

since, we know that the value of the function $y = \cot x = \dfrac{\cos x}{\sin x}$ is undefined for all those values of x for which the function $\sin x$ in denominator is zero (i.e.; for $\sin x = 0 \Leftrightarrow x = n\pi$ = any integral multiple of π, n being an integer). Hence, the domain of the function $y = \cot x$ is the set of all real numbers excluding $n\pi$, n being an integer. Again, we know that $\cot x$ can assume any value however large or small for real value of x which means that $-\infty < \tan < \infty$ is the range of the function $y = \cot x$.

Aid to memory:

$D(\cot x) = R - \{x : x = n\pi, n$ being an integer$\}$

$R(\cot x) = R = (-\infty, \infty)$ = the set of all real numbers.

5. $y = \text{cosec } x$

Since we know that the value of the function $y = \text{cosec } x = \dfrac{1}{\sin x}$ is undefined for those values of x for which the function $\sin x$ in denominator is zero (i.e., for $\sin x = 0 \Leftrightarrow x = n\pi$ = any integral multiple of π, n being an integer). Hence, the domain of $y = \text{cosec } x$ is the set of all real numbers excluding $n\pi$, n being an integer. Again we know that $|\text{cosec } x| \ge 1 \Leftrightarrow \text{cosec } x \ge 1$ or $\text{cosec } x \le -1$, $\forall x \ne n\pi$. Thus, $\forall x \ne n\pi$, we have $\text{cosec } x \ge 1$ or $\text{cosec } x \le -1$. Hence, the range of $y = \text{cosec } x$ is the set of all real numbers not in the open interval $(-1, 1)$.

Aid to memory:

$D(\text{cosec } x) = R - \{x : x = n\pi, n$ being an integer$\}$

$R(\text{cosec } x) = R - (-1, 1)$

6. $y = \sec x$

Since we know that the value of the function $y = \sec x = \dfrac{1}{\cos x}$ is undefined for those values of x for which the function $\cos x$ in the denominator is zero (i.e., for $\cos x = 0 \Leftrightarrow x = (2n+1)\frac{\pi}{2}$, n being an integer). Hence, the domain of $y = \sec x$ is the set of all real numbers excluding $(2n+1)\frac{\pi}{2}$, n being an integer. Again we know that $|\sec x| \ge 1 \Leftrightarrow \sec x \ge 1$ or $\sec x \le -1$, , $\forall\, x \ne (2n+1)\frac{\pi}{2}$, n being a integer.

Thus $\forall\, x \ne (2n+1)\frac{\pi}{2}$, we have $\sec x \ge 1$ or $\sec x \le -1$. Therefore, the range of $y = \sec x$ is the set of all real numbers not in the open interval (–1, 1).

Aid to memory:

$D(\sec x) = R - \{x{:}\ x = (2n+1)\frac{\pi}{2}, n \in I\}$ I = the set of integers = $\{0, \pm 1, \pm 2, \pm 3, ...\}$

$R(\sec x) = R - (-1, 1)$

Refresh your memory:

1. $|\sin x| \le 1 \Leftrightarrow \frac{1}{|\sin x|} \ge 1 \Leftrightarrow |\text{cosec } x| \ge 1$

2. $|\cos x| \le 1 \Leftrightarrow \frac{1}{|\cos x|} \ge 1 \Leftrightarrow |\sec x| \ge 1$

Now we consider different types or problems whose domains are required to be determined.

Type 1: Problems based on finding the domain of a function put in the form: $y = \sin^n x \pm \cos^m x$ (n and m being integers) = sum or difference of power of $\sin x$ and $\cos x$;

Working rule: One must remember that domain of the function $y = \sin^n x \pm \cos^m x$ is $R = (-\infty, \infty) =]-\infty, +\infty[$ since $\sin x$ and $\cos x$ are real valued functions of the real variables $x \Leftrightarrow \sin^n x$ and $\cos^m x$ are real valued functions of the real variable $x \Leftrightarrow$ the sum of $\sin^n x$ and $\cos^m x$ are real valued functions of the real variable x.

Solved Examples

Find the domain of the following:

1. $y = \sin^2 x + \cos^4 x$

Solution: Since, domain of $\sin^2 x = R = \{x{:}\ x \in R\}$

and domain of $\cos^4 x = R = \{x{:}\ x \in R\}$

$$\therefore\ D(y) = R \cap R = R$$

Type 2: Problems based on finding the domain of a function put in the form: $y = a \sin x \pm b \cos x$ or, $y = a \cos x \pm b \sin x$, where $a, b, x \in R$.

Working rule: One must remember that the domain of the function of the form: $y = a \sin x \pm b \cos x$ or $y = a \cos x \pm b \sin x$ is the set of all real numbers since $a \sin x$ or $b \cos x$ is defined for all $a, b, x \in R$ as well as $a \cos x$ or $b \cos x$ is defined for all $a, b\ x \in R$ which means that this sum and/difference is (or, are) defined for all $a, b, x \in R$.

$$\therefore\ D(y) = R = (-\infty, +\infty) =]-\infty, +\infty[$$

Solved Examples

Find the domain of each of the following functions:

1. $y = \sin x - \cos x$

Solution: $y = \sin x - \cos x$ is defined for $x \in R$ since $\sin x$ and $\cos x$ are defined for $x \in R$

$$\therefore\ D(y) = R = (-\infty, +\infty)$$

2. $y = 3\cos x + 4\sin x$

Solution: $y = 3\cos x + 4\sin x$ is defined for all $x \in R$ since $3\cos x$ and $4\sin x$ are defined for all, $x \in R$

$$\therefore\ D(y) = R = (-\infty, \infty)$$

Type 3: Problems based on finding the domain of trigonometric rational functions:

Working rule: It consists of following steps:

Step 1: To put the functional value (or, simply function) in denominator = 0 and to find the valued of the independent variable.

Step 2: To delete the values of the independent variable from R to get the required domain, i.e., domain of trigonometric rational functions = $R - \{$ real values of the argument for which functional value in denominator = 0$\}$.

Notes: **1.** If no real solution is available after putting the functional value in denominator = 0, then domain of trigonometric rational functions is $R = (-\infty, +\infty)$. We face this circumstances generally when we obtain cos mx or sin $mx = k$; $|k| > 1$, (after putting the functional value in denominator = 0) from which it is not possible to find out the values of x since maximum and/minimum value of sin mx and/cos mx = +1 and −1 respectively.

2. $\sin\theta = 0 \Leftrightarrow \theta = n\pi,\ n \in I$

$$\cos\theta = 0 \Leftrightarrow \theta = (2n+1)\frac{\pi}{2},\ n \in I$$

$$\sin\theta = \sin\alpha \Leftrightarrow \theta = n\pi + (-1)^n\,\alpha,\ n \in I$$

$$\cos\theta = \cos\alpha \Leftrightarrow \theta = 2n\pi \pm \alpha,\ n \in I$$

3. $\left.\begin{array}{l}\sin^2\theta = \sin^2\alpha\\ \cos^2\theta = \cos^2\alpha\\ \tan^2\theta = \tan^2\alpha\end{array}\right\} \Rightarrow \theta = n\pi \pm \alpha$ (n being an integer)

4. $\cos n\pi = (-1)^n$, n being an integer.

$\sin n\pi = 0$, n being an integer.

5. The domain of the function put in the form:

$y = \sqrt{f(x)},\ \sqrt{\dfrac{f(x)}{g(x)}},\ \dfrac{1}{\sqrt{g(x)}},\ \dfrac{f(x)}{\sqrt{g(x)}}$ where $f(x)$ and $g(x)$ denote trigonometric functions, is obtained by the same working rule as the case when $f(x)$ and $g(x)$ are algebraic functions.

Solved Examples

Find the domain of each of the following functions:

1. $y = \dfrac{1}{1+\cos x}$

Solution: Putting $1 + \cos x = 0$

$$\Rightarrow \cos x = -1 \Rightarrow x = (2n+1)\pi$$

$$\therefore D(y) = R - \{(2n+1)\pi : n \in I\}$$

2. $y = \dfrac{1}{2 - \cos 3x}$

Solution: Putting $2 - \cos 3x = 0$

$\Rightarrow \cos 3x = 2$ which is not true for any real x

$$\therefore D(y) = R = (-\infty, +\infty)$$

3. $y = \dfrac{1}{2 - \sin 3x}$

Solution: Putting $2 - \sin 3x$

$\Rightarrow \sin 3x = 2$ which is not true for any real x

$$\therefore D(y) = R = (-\infty, \infty)$$

Domain of Inverse Trigonometric Functions

Before studying the method of finding the domain of inverse trigonometric (or, arc) functions, we discuss the domain on which each trigonometric functions is reversible.

1. $y = \sin x \Leftrightarrow \sin^{-1} y = x,\ x \in \left[-\dfrac{\pi}{2}, \dfrac{\pi}{2}\right], y \in [-1, 1]$

which signifies that the function $y = \sin x$ defined on the interval $\left[-\dfrac{\pi}{2}, \dfrac{\pi}{2}\right]$ has an inverse function defined on the interval $[-1, 1]$.

Notes:

(i) $D\left(\sin^{-1} y\right) = [-1, 1]$, $R\left(\sin^{-1} y\right) = \left[-\dfrac{\pi}{2}, \dfrac{\pi}{2}\right]$

(ii) $\sin\left(\sin^{-1} y\right) = y,\ \forall\, y \in [-1, 1]$ and $\sin^{-1}(\sin x) = x,\ \forall\, x \in \left[-\dfrac{\pi}{2}, \dfrac{\pi}{2}\right]$.

(iii) The notation of the inverse of the sine function is $\sin^{-1}$ (or, arc sin).

(iv) $y = \sin x,\ x \in \left[-\frac{\pi}{2}, \frac{\pi}{2}\right]$, is one-one and onto functions. this is why it is possible to define its inverse on the interval [–1, 1].

2. $y = \cos x \Leftrightarrow \cos^{-1} y = x,\ x \in [0, \pi],\ y \in [-1, 1]$ which signifies that the functions $y = \cos x$ is defined on the interval $[0, \pi]$ has an inverse function $x = \cos^{-1} y$ defined on the interval [–1, 1].

Notes:

(i) $D\left(\cos^{-1} y\right) = [-1, 1],\ R\left(\cos^{-1} y\right) = [0, \pi]$.

(ii) $\cos\left(\cos^{-1} y\right) = y,\ \forall\, y \in [-1, 1]$ and $\cos^{-1}(\cos x) = x,\ \forall\, x \in [0, \pi]$.

(iii) The notation of the inverse of the cosine function is $\cos^{-1}$ (or, arc cos).

(iv) $y = \cos x,\ x \in [0, \pi],\ y \in [-1, 1]$ is a one-one and on-to function. this is why it is possible to define its inverse on the interval [–1, 1].

3. $y = \tan x \Leftrightarrow \tan^{-1} y = x,\ x \in \left(-\frac{\pi}{2}, \frac{\pi}{2}\right),\ y \in R$ which signifies that the function $y = \tan x$ defined on the interval $\left(-\frac{\pi}{2}, \frac{\pi}{2}\right)$ has an inverse function $x = \tan^{-1} y$ defined on the set of all real numbers (i.e.; R)

Notes:

(i) $D\left(\tan^{-1} y\right) = R = (-\infty, +\infty),\ R\left(\tan^{-1} y\right) = \left(-\frac{\pi}{2}, \frac{\pi}{2}\right)$.

(ii) $\tan\left(\tan^{-1} y\right) = y,\ \forall\, y \in R$ and $\tan^{-1}(\tan x) = x,\ \forall\, x \in \left(-\frac{\pi}{2}, \frac{\pi}{2}\right)$.

(iii) The notation of the inverse of the tangent function is $\tan^{-1}$ (or, arc tan).

(iv) $y = \tan x,\ x \in \left(-\frac{\pi}{2}, \frac{\pi}{2}\right)$ is a one-one and on-to function. This is why it is possible to define its inverse on the set of al real numbers (i.e.; R).

4. $y = \cot x \Leftrightarrow \cot^{-1} y = x,\ x \in (0, \pi), y \in R$ which signifies that the function $y = \cot x$ defined on the interval $(0, \pi)$ has an inverse function $x = \cot^{-1} y$ defined on the set of all real numbers.

Notes:

(i) $D\left(\cot^{-1} y\right) = R = (-\infty, +\infty),\ R\left(\cot^{-1} y\right) = (0, \pi)$

(ii) $\cot\left(\cot^{-1} y\right) = y,\ \forall\, y \in R$ and $\cot^{-1}(\cot x) = x,\ \forall\, x \in (0, \pi)$

(iii) The notation of the inverse of the contangent function is $\cot^{-1}$ (or, arc cot).

(iv) $y = \cot x,\ x \in (0, \pi),\ y \in R$ is a one-one and on-to function. this is why it is possible to define its inverse on the interval $(-\infty, \infty)$.

5 $y = \sec x \Leftrightarrow \sec^{-1} y = x,\ x \in [0, \pi] - \left\{\frac{\pi}{2}\right\}, y \in R - (-1, 1)$ which signifies that the function $y = \sec x$ is reversible on the interval $[0, \pi] - \left\{\frac{\pi}{2}\right\}$, that is, it has an inverse function $x = \sec^{-1} y$ defined on the interval $R - (-1, 1)$.

Notes:

(i) $D\left(\sec^{-1} y\right) = R - (-1, 1),\ R\left(\sec^{-1} y\right) = [0, \pi] - \left\{\frac{\pi}{2}\right\}$.

(ii) The notation of the inverse of the secant function is $\sec^{-1}$ (or, arc sec).

(iii) $\sec\left(\sec^{-1} y\right) = y, \forall y \in R - (-1, 1); \sec^{-1}$

$(\sec x) = x, \ \forall x \in [0, \pi] - \left\{\frac{\pi}{2}\right\}$.

(iv) $y = \sec x, \ x \in [0, \pi] - \left\{\frac{\pi}{2}\right\}, y \in R - (-1, 1)$ is a one-one and on-to function, this is why it is possible to define its inverse in the interval $R - (-1, 1)$.

6. $y = \text{cosec}\, x \Leftrightarrow x = \text{cosec}^{-1} y, \ x \in \left[-\frac{\pi}{2}, \frac{\pi}{2}\right] - \{0\}, \ y \in R - (-1, 1)$ which signifies that the function $y = \text{cosec}\, x$ is reversible on the interval $\left[-\frac{\pi}{2}, \frac{\pi}{2}\right] - \{0\}$ that is, it has an inverse function $x = \text{cosec}^{-1} y$ defined on the interval $R - (-1, 1)$.

Notes:

(i) $D\left(\text{cosec}^{-1} y\right) = R - (-1, 1), \ R\left(\text{cosec}^{-1} y\right) = \left[-\frac{\pi}{2}, \frac{\pi}{2}\right] - \{0\}$

(ii) $\text{cosec}\left(\text{cosec}^{-1} y\right) = y, \ \forall y \in R - (-1, 1)$ and $\text{cosec}^{-1} (\text{cosec}\, x) = x, \ \forall x \in \left[-\frac{\pi}{2}, \frac{\pi}{2}\right] - \{0\}$.

(iii) The notation of the inverse of the cosecant function is cosec^{-1} (or, arc cosec).

(iv) $y = \text{cosec}\, x, x \in \left[-\frac{\pi}{2}, \frac{\pi}{2}\right] - \{0\}, y \in R - (-1, 1)$ is a one-one and on-to function. This is why it is possible to define its inverse on the interval $R - (-1, 1)$.

Now we discuss the method of finding the domain of different types of problems.

Type1: Problems based on finding the domain of a function put in the form: $y = \sin^{-1} (f(x))$ or, $y = \cos^{-1} (f(x))$.

Working rule: It consists of following steps:

1. To put $f(x)$ in between -1 and 1, i.e., to form the inequality $-1 \le f(x) \le 1$ and to solve two inequalities $f(x) \ge -1$ and $f(x) \le 1$.

2. To find the intersection of the solution set of the inequalities $f(x) \ge -1$ and $f(x) \le 1$ to form the domain of the function of the form: $y = \sin^{-1} (f(x))$ or, $y = \cos^{-1} (f(x))$.

Notes:

(i) $\sin x \ge 0 \Leftrightarrow 2n\pi \le x \le (2n + 1)\pi$, n being an integer.

(ii) $\sin x > 0 \Leftrightarrow 2n\pi < x < 2n\pi + \pi$, n being an integer $\Leftrightarrow x$ lies in the first or second quadrant.

Solved Examples

Find the domain of each of the following functions:

1. $y = \cos^{-1}\left(\frac{2}{2 + \sin x}\right)$

Solution: y is defined when $-1 \le \frac{2}{2 + \sin x} \le 1$

$\Leftrightarrow -2 - \sin x \le 2 \le 2 + \sin x$

$\Leftrightarrow -2 - \sin x \le 2$ ...(i)

and $2 \le 2 + \sin x$ ...(ii)

(i) $\Rightarrow -2 - \sin x \le 2 \Leftrightarrow -\sin x \le 4 \Leftrightarrow \sin x \ge -4$ which is true $\forall x \in R \Leftrightarrow D_1 = -\infty < x < \infty = R$.

(ii) $\Rightarrow 2 \le 2 + \sin x \Leftrightarrow 0 \le \sin x \Leftrightarrow \sin x \ge 0 \Leftrightarrow$ $2n\pi \le x \le (2n + 1)\pi \Leftrightarrow D_2 = [2n\pi, (2n + 1)\pi]$

$\therefore D(y) = D_1 \cap D_2$

$= [2n\pi, (2n + 1)\pi]$, n being an integer.

2. $y = \cos^{-1} (2x + 3)$

Solution: y is defined when $-1 \le 2x + 3 \le 1$

$\Leftrightarrow -1 \le 2x + 3$ ...(i)

and $2x + 3 \le 1$ ...(ii)

Now, (i) $\Rightarrow 2x + 3 \geq -1 \Leftrightarrow 2x \geq -1 - 3 \Leftrightarrow x \geq \frac{-4}{2} \Leftrightarrow x \geq -2 \Leftrightarrow D_1 = [-2, \infty)$

(ii) $\Rightarrow 1 \geq 2x + 3 \Leftrightarrow 1 - 3 \geq 2x \Leftrightarrow -2 \geq 2x \Leftrightarrow \frac{-2}{2} \geq x \Leftrightarrow -1 \geq x \Leftrightarrow x \leq -1 \Leftrightarrow D_2 = (-\infty, -1]$

$\therefore D(y) = D_1 \cap D_2$

$= [-2, -1]$

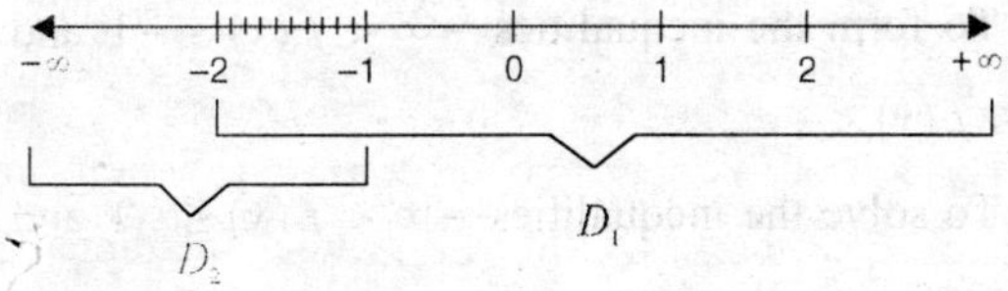

3. $y = \cos^{-1}(1 - 2x)$

Solution: y is defined when $-1 \leq 1 - 2x \leq 1$

$\Leftrightarrow -1 \leq 1 - 2x$ ...(i)

and $1 - 2x \leq 1$ ...(ii)

Now, (i) $\Rightarrow -1 \leq 1 - 2x \Leftrightarrow -1 - 1 \leq -2x \Leftrightarrow -2 \leq -2x \Leftrightarrow x \leq 1 \Rightarrow D_1 = (-\infty, 1]$

(ii) $\Rightarrow -1x \leq 1 \Leftrightarrow -2x \leq 0 \Leftrightarrow x \geq 0 \Rightarrow D_2 = [0, \infty)$

$\therefore D(y) = D_1 \cap D_2$

$= [0, 1]$

4. $y = \sin^{-1}(\log_2 x)$

Solution: y is defined when $-1 \leq \log_2 x \leq 1$

$\Leftrightarrow -1 \leq \log_2 x$ ...(ii)

and $\log_2 x \leq 1$ ...(i)

Now, (i) $\Rightarrow -1 \leq \log_2 x \Leftrightarrow 2^{-1} \leq x \Leftrightarrow \frac{1}{2} \leq x \Leftrightarrow x \geq \frac{1}{2} \Leftrightarrow D_1 = \left[\frac{1}{2}, \infty\right)$.

(ii) $\Rightarrow \log_2 x \leq 1 \Leftrightarrow 2^1 \geq x \Leftrightarrow x \leq 2 \Leftrightarrow D_2 = [-\infty, 2)$

$\therefore D(y) = D_1 \cap D_2$

$= \left[\frac{1}{2}, 2\right]$

Remember:

1. $\log_a x > c \Leftrightarrow x < a^c$, if $0 < a < 1$.

2. $\log_a x > c \Leftrightarrow x < a^c$, if $a > 1$.

Type 2: Problems based on finding the domain of a function put in the form:

$y = \sin^{-1}\sqrt{f(x)}$

or $y = \cos^{-1}\sqrt{f(x)}$

Working rule: It consists of following steps:

1. To put $f(x) \geq 0$ and to solve for x.

2. To consider $|\sin y| \leq 1 \Leftrightarrow \sin^2 y \leq 1 \Leftrightarrow \left\{\sin \sin^{-1}\sqrt{f(x)}\right\}^2 \leq 1 \Leftrightarrow \left\{\sqrt{f(x)}\right\}^2 \leq 1 \Leftrightarrow f(x) \leq 1$ and to solve for x.

3. (1) and (2) $\Leftrightarrow 0 \leq f(x) \leq 1$, domain of y is the intersection of the solution set of (1) and (2).

Note: $\sqrt{f(x)}$ is defined for $f(x) \geq 0$

$\therefore y = \sin^{-1}\sqrt{f(x)}$ or $\cos^{-1}\sqrt{f(x)}$ is defined for $0 \leq \sqrt{f(x)} \leq 1$ and $f(x) \geq 0 \Rightarrow 0 \leq f(x) \leq 1$, i.e.; domain of the function put in the form $y = \sin^{-1}\sqrt{f(x)}$ or $\cos^{-1}\sqrt{f(x)}$ consists of all those values of x for which $f(x) \geq 0$ and $0 \leq \sqrt{f(x)} \leq 1$.

Solved Examples

Find the domain of the following:

1. $y = \sin^{-1}\sqrt{x}$

Solution: ***Method (1)***

$\sqrt{x}$ is defined for $x \ge 0$

$\therefore$ y is defined for $0 \le \sqrt{x} \le 1$ and $x \ge 0$.

$\Rightarrow D(y) = [0, 1]$

$\therefore D(y) = [0, 1]$

Method (2)

(1) Putting $x \ge 0 \Leftrightarrow D_1 = [0, \infty)$

2. $|\sin y| \le 1 \Leftrightarrow \sin^2 y \le 1 \Leftrightarrow \left[\sin \sin^{-1} \sqrt{x}\right]^2 \le$

$1 \Leftrightarrow x \le 1 \Leftrightarrow D_2 = (-\infty, 1]$

$\therefore D(y) = D_1 \cap D_2$

$= [0, 1]$

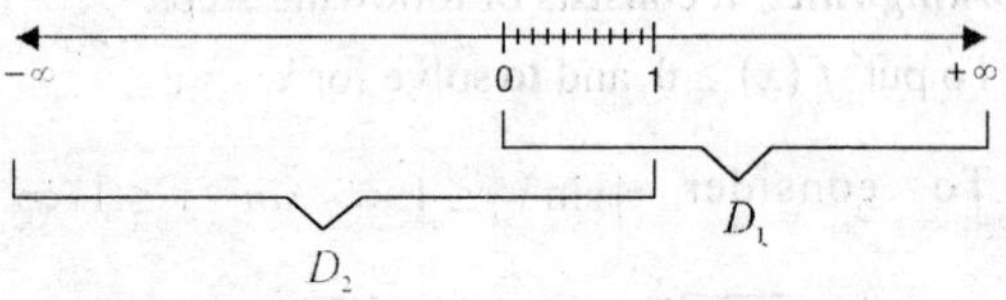

Type 3: Problems based on finding the domain of a function put in the form: $y = \tan^{-1}(f(x))$ or $y = \cot^{-1}(f(x))$.

Working rule: It consists of following steps:

1. To put $f(x)$ in between $-\infty$ and $+\infty$ i.e., to form the inequality $-\infty < f(x) < +\infty$.

2. To find the solution set of the inequality $-\infty < f(x) < +\infty$ to form the domain of the function of the form $y = \tan^{-1}(f(x))$ or $y = \cot^{-1}(f(x))$.

Note: $y = \tan^{-1}(f(x))$ or $y = \cot^{-1}(f(x))$ defined for all those real values of x for which $-\infty < f(x) < +\infty$ i.e.; the domain of the function of the form $y = \tan^{-1}(f(x))$ or $y = \cot^{-1}(f(x))$ consists of all those real values of x for which $-\infty < f(x) < +\infty$.

Solved Examples

Find the domain of the following:

1. $y = \tan^{-1}(2x + 1)$

Solution: $\because$ y is defined when $-\infty < 2x + 1 < +\infty$

$\Leftrightarrow -\infty < 2x < \infty$

$\Leftrightarrow \dfrac{-\infty}{2} < x < \dfrac{\infty}{2}$

$\Leftrightarrow -\infty < x < \infty$

$\therefore D(y) = (-\infty, +\infty) = R$

Type 4: Problems based on finding the domain of a function put in the form:

$y = \sec^{-1}(f(x))$ or $y = \text{cosec}^{-1}(f(x))$

Working rule:

1. To form the inequalities $-\infty < f(x) \le -1$ and $1 \le f(x) < +\infty$.

2. To solve the inequalities $-\infty < f(x) \le -1$ and $1 \le f(x) < +\infty$ for x to form the domain of the function of the form $y = \sec^{-1}(f(x))$ or $y = \text{cosec}^{-1}(f(x))$.

Domain of a Function Put in the Form

$y = f_1(x) \pm f_2(x)$

Working rule: It tells to find the domains of two functions, say $f_1(x)$ and $f_2(x)$ separately whose intersection is the domain of this sum or difference.

Notes: By considering the two domains (i.e., intervals) on the scale, we find their intersection (i.e., the interval of common points).

Solved Examples

Find the domain of each of the following functions:

1. $y = \sqrt{1 - x^2} + \dfrac{x - 3}{2x + 1}$

Solution: Let $f_1(x) = \sqrt{1 - x^2}$

$\because f_1(x)$ is defined when $1 - x^2 \ge 0$

$\Leftrightarrow x^2 \le 1 \Leftrightarrow |x| \le 1 \Leftrightarrow -1 \le x \le 1$

$\therefore D(f_1(x)) = [-1, 1] = D_1$ (say)

Again, let $f_2(x) = \dfrac{x-3}{2x+1}$

$\because f_2(x)$ is defined when $2x + 1 \neq 0$

$\Leftrightarrow x \neq -\dfrac{1}{2}$

$\therefore D(f_2(x)) = R - \left\{-\dfrac{1}{2}\right\} = D_2$ (say)

$= \left(-\infty, -\dfrac{1}{2}\right) \cup \left(-\dfrac{1}{2}, \infty\right)$

Thus, $D(y) = \left[-1, -\dfrac{1}{2}\right) \cup \left(-\dfrac{1}{2}, 1\right] = D_1 \cap D_2$

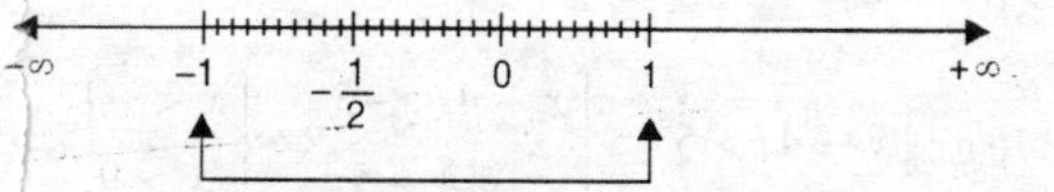

2. $y = \sqrt{4-x} + \sqrt{x-5}$

Solution: Let $f_1(x) = \sqrt{4-x}$

and $f_2(x) = \sqrt{x-5}$

$f_1(x)$ is defined when $4 - x \geq 0 \Leftrightarrow x \leq 4$

$\therefore D(f_1(x)) = (-\infty, 4] = D_1$

$f_2(x)$ is defined when $x - 5 \geq 0 \Leftrightarrow x \geq 5$

$\therefore D(f_2(x)) = [5, \infty) = D_2$ (say)

Thus, $D(y) = D_1 \cap D_2 = \varnothing$

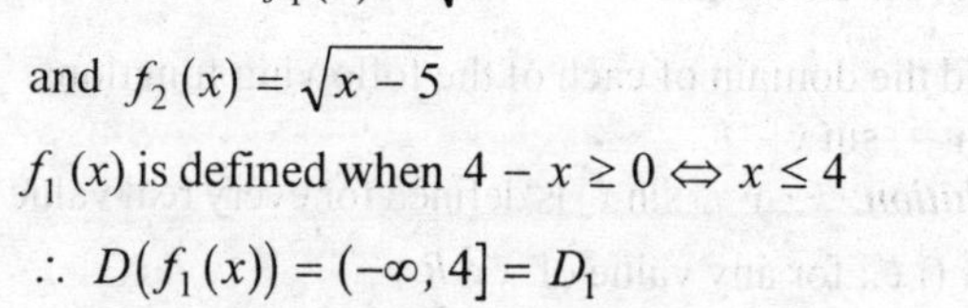

3. $y = \sqrt{\dfrac{x-2}{x+2}} + \sqrt{\dfrac{1-x}{1+x}}$

Solution: Let $f_1(x) = \sqrt{\dfrac{x-2}{x+2}}$

and $f_2(x) = \sqrt{\dfrac{1-x}{1+x}}$

$\because f_1(x)$ is defined when $\left(\dfrac{x-2}{x+2}\right) \geq 0$

$\Leftrightarrow x \neq -2$ and $(x-2)(x+2) \geq 0 \Leftrightarrow$

$x \neq -2$ and $(x-2)(x-(-2)) \geq$

$0 \Leftrightarrow x \geq 2$ or $x < -2$

$\therefore D(f_1(x)) = (-\infty, -2) \cup [2, +\infty) = D_1$ (say)

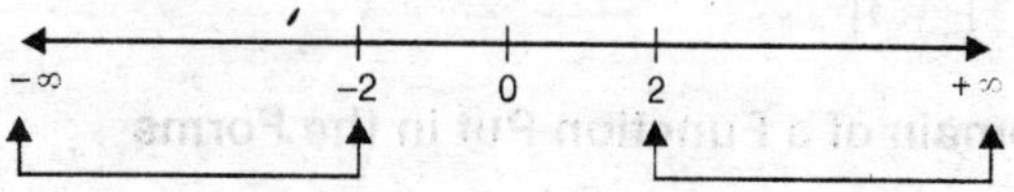

$f_2(x)$ is defined when $\left(\dfrac{1-x}{1+x}\right) \geq 0$

$\Leftrightarrow x \neq -1$ and $(1-x)(1+x) \geq 0$

$\Leftrightarrow x \neq -1$ and $(x-1)(x+1) \leq 0$

$\Leftrightarrow (x-1)(x-(-1)) \leq 0 \Leftrightarrow -1 < x \leq 1$ $(\because x \neq -1)$

$\therefore D(f_2(x)) = (-1, 1] = D_2$ (say)

Thus, $D(y) = D_1 \cap D_2 = \varnothing$

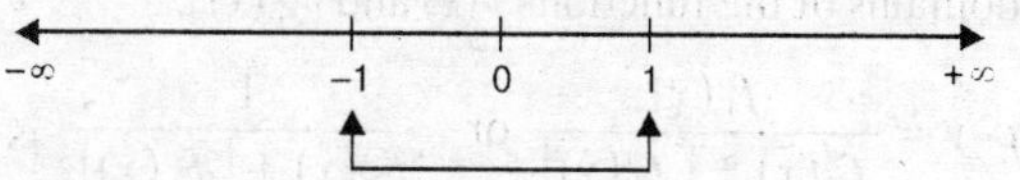

4. $y = \sqrt{3-x} + \cos^{-1}\left(\dfrac{x-2}{3}\right)$

Solution: Let $f_1(x) = \sqrt{3-x}$

and $f_2(x) = \cos^{-1}\left(\dfrac{x-2}{3}\right)$

$\because f_1(x)$ is defined when $3 - x \geq 0 \Leftrightarrow x \leq 3$

$\therefore D(f_1(x)) = (-\infty, 3] = D_1$ (say)

$f_2(x)$ is defined when $-1 \leq \dfrac{x-2}{3} \leq 1$

$\Leftrightarrow -3 \leq x - 2 \leq 3$

$\Leftrightarrow -3 + 2 \leq x - 2 + 2 \leq 3 + 2$

$\Leftrightarrow -1 \leq x \leq 5$

$\therefore\ D(f_2(x)) = [-1, 5] = D_2$ (say)

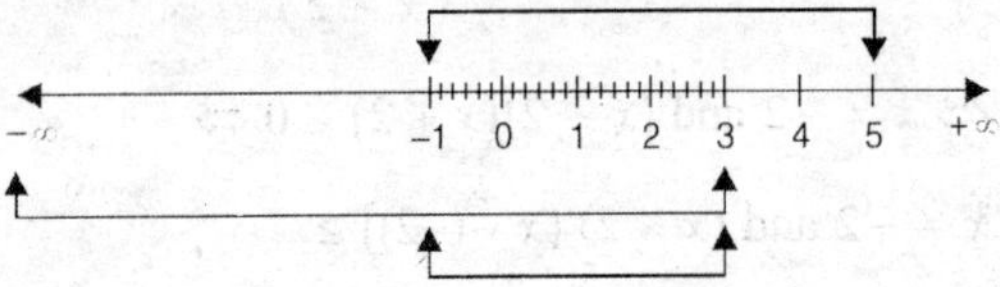

Thus, $D(y) = D_1 \cap D_2$

$= (-\infty, 3] \cap [-1, 5]$

$= [-1, 3]$

Domain of a Function Put in the Forms

(i) $y = |f(x)|$

(ii) $y = f(x) \pm |g(x)|$

(iii) $y = \dfrac{f_1(x)}{f_2(x) \pm |f_3(x)|}$ or, $\dfrac{1}{f_2(x) \pm |f_3(x)|}$

Working rule: To find the domain of a function involving absolute value function, one must remember that:

(i) $y = f(x) \pm |g(x)|$ is the sum or difference of two functions. Hence, its domain is the intersection of domains of the functions $f(x)$ and $|g(x)|$.

(ii) $y = \dfrac{f_1(x)}{f_2(x) \pm |f_3(x)|}$ or, $\dfrac{1}{f_2(x) \pm |f_3(x)|}$ is a rational function. Hence, its domain is dom $f_1(x) \cap$ dom $f_2(x) \cap$ dom $|f_3(x)| - \{x: f_2(x) \pm |f_3(x)| = 0\}$.

Notes:

1. When the sum of two non-negative numbers (or, functions) is zero, then both the numbers (or, functions) are separately zero.

e.g: $x^2 + |x| = 0 \Rightarrow x^2 = 0$ and $|x| = 0$; $|x| + |x^3| = 0 \Rightarrow |x| = 0$ and $|x^3| = 0$.

2. $(x-1)(x-2)(x-3)(x-4)$ is > 0 for $x > 4$, or $2 < x < 3$ or $x < 1$; and ≤ 0 for $3 \le x < 4$ or $1 < x < 2$

(+ always at right end and then alternatively + and –)

3. $|x| = x$ for $x \ge 0$; $|x| = -x$ for $x < 0$ (where $x = -ve \Leftrightarrow -x = +ve$, i.e; $x < 0 \Leftrightarrow -x > 0$).

4. $|f(x)| = f(x)$ for $f(x) \ge 0$; $|f(x)| = -f(x)$ for $f(x) < 0$ (where $f(x) = -ve \Leftrightarrow -f(x) = +ve$, i.e; $f(x) < 0 \Leftrightarrow 0 - f(x) > 0$).

5. $|x| < a \Leftrightarrow -a < x < a \Leftrightarrow x \in (-a, a)$

6. $|x| > a \Leftrightarrow x < -a$ or $x > a \Leftrightarrow x \in R - [-a, a]$

7. $|f(x)| > a \Leftrightarrow -a < f(x) < a$

8. $|f(x)| > a \Leftrightarrow f(x) > a$ or $f(x) < -a$

9. $x < a$ or $x > b \Leftrightarrow x \in R - [a, b] \Leftrightarrow x \in (-\infty, a) \cup (b, +\infty)$

e.g.: $|x - 4| > 5 \Leftrightarrow \left[\begin{array}{l} x - 4 < -5 \\ x - 4 > 5 \end{array}\right. \Leftrightarrow \left[\begin{array}{l} x < -1 \\ x > 9 \end{array}\right. \Leftrightarrow$

$x \in R = [-1, 9] \Leftrightarrow x \in (-\infty, -1) \cup (9, \infty)$

Solved Examples

Find the domain of each of the following functions:

1. $y = |\sin x|$

Solution: $\because y = |\sin x|$ is defined for every real value of x (i.e.; for any value of $x \in R$)

$\therefore\ D(y) = R\{x : x \in R\} = (-\infty, +\infty)$

2. $y = 1 - |x|$

Solution: $y = 1 - |x|$ is defined for every real value of x (i.e.; for any value of $x \in R$).

$\therefore\ D(y) = R\{x : x \in R\} = (-\infty, +\infty)$

3. $y = \dfrac{x^2 + 3}{x^2 + |x|}$

Solution: $y = \dfrac{x^2 + 3}{x^2 + |x|}$ a rational function.

Putting $x^2 + |x| = 0$

$\Rightarrow x^2 = 0$ and $|x| = 0$ and from each equation, we get

$x = 0$ ($\because x = 0 \Leftrightarrow x^2 = 0$ and $|x| = 0 \Leftrightarrow x = 0$)

$\therefore D(y) = R - \{0\}$

4. $y = \dfrac{1}{x - |x|}$

Solution: $y = \dfrac{1}{x - |x|}$ a rational function

Putting $x - |x| = 0$

$\Leftrightarrow |x| = x$

$\Leftrightarrow x \geq 0$ ($\because |x| = x$ provided $x \geq 0$)

$\therefore D(y) = R - \{x: x \geq 0\}$

Domain of a Function Containing Greatest Integer Function

Rule: Domains of functions involving greatest integer function are obtained by using different properties of greatest integer function.

Solved Examples

1. Find the domain of $y = \sin^{-1}[x]$

Solution: $y = \sin^{-1}[x]$ is defined when $-1 \leq [x] \leq 1$

Now $-1 \leq [x] \leq 1$

$\Leftrightarrow -1 \leq x < 2$

$\Leftrightarrow x \in [-1, 2)$

$\Leftrightarrow D(y) = [-1, 2]$

2. Find the domain of $y = \sin^{-1}[2 - 3x^2]$.

Solution: $y = \sin^{-1}[2 - 3x^2]$ is defined when

$-1 \leq \left[2 - 3x^2\right] \leq 1$

Now $-1 \leq \left[2 - 3x^2\right] \leq 1$

$\Leftrightarrow -1 \leq 2 - 3x^2 < 2$

Again, $2 - 3x^2 < 2$

$\Rightarrow -3x^2 < 0$

$\Rightarrow -x^2 < 0$

$\Rightarrow x^2 > 0 \Rightarrow x \in R$ ...(i)

Next, $2 - 3x^2 \geq -1$

$\Rightarrow -3x^2 \geq -3$

$\Rightarrow x^2 \leq 1$

$\Rightarrow |x| \leq 1 \Rightarrow x \in [-1, 1]$ ...(ii)

on finding the intersection of (i) and (ii), it is obtained $D(y) = [-1, 0) \cup (0, 1]$.

Problems on the Range of a Function

As discussed earlier, the range of a function defined by $y = f(x)$ in its domain is the set of values of $f(x)$ which it attains at points belonging to the domain. For a real function, the codomain is always a subset of R, so the range of a real function f is the set of all points y such that $y = f(x)$, where $x \in D(f)$ = domain of f.

In general, a function is described either by a single expression in x in its domain or by various expressions defined in adjacent intervals denoting different parts of the domain of the function and neither its domain nor range is mentioned. In such cases, it is required to be found out the domain and the range of the given function.

Already, how to find out the domains of different types of functions has been discussed. Now the methods of finding the range of a given function will be explained.

Firstly, domains and range sets of standard functions will be put in a tabular form.

Funtion defined by an expression	*Domain*	*Range*
1. $y = kx,\ k \neq 0$	$(-\infty, \infty)$	$(-\infty, \infty)$
2. $y = kx + l$	$(-\infty, \infty)$	$(-\infty, \infty)$
3. $y = \dfrac{k}{x},\ k \neq 0$	$R - \{0\}$	$R - \{0\}$
4. $y = x^{2n}$	$(-\infty, \infty)$	$(0, \infty)$
5. $y = x^{2n+1}$	$(-\infty, \infty)$	$(-\infty, \infty)$
6. $y = \sqrt{x}$	$[0, \infty)$	$[0, \infty)$
7. $y = ax^2 + bx + c$, $a > 0$	$(-\infty, \infty)$	$\left[-\dfrac{D}{4a}, +\infty\right)$, $D = b^2 - 4ac$
8. $y = ax^2 + bx + c$, $a < 0$	$(-\infty, \infty)$	$\left(-\infty, \dfrac{-D}{4a}\right)$, $D = b^2 - 4ac$

Now the methods to find the range of a given function are provided, when its domain is an infinite interval.

How to Find the Range of Function

Step 1: Put $y=f(x)$

Step 2: Solve the equation $y=f(x)$ for x to obtain $x=g(y)$.

Step 3: Find the values of y for which the values of x obtained from $x=g(y)$ are in the domain of f, i.e. find the domain of $g(y)$ in the same way as the domain of $f(x)$ is obtained considering $g(y)$ as inverse of $f(x)$.

Step 4: The set of all values of y obtained in step (3) is the required range, i.e. the domain of $g(y)$ is the required range of the given function $y=f(x)$.

Remarks:

1. The method mentioned above is fruitful only when the domain of a given function is infinite, i.e. the domain of a given functions is not a closed interval $[a, b]$, $a, b \in R =$ the set of reals.

2. When a function defined by a single formula $y=f(x)$ does not become imaginary or undefined for any value of independent variable x, the domain of the function $y=f(x)$ is the set of all real numbers denoted by R. To obtain its range, one should consider the domain $-\infty < x < \infty$ using the axioms of inequality in $-\infty < x < \infty$.

3. When it is possible to put a function in the form of Px^2+Qx+R, where P, Q and R are linear expressions in y, one should use the rule of discriminat, i.e., $D = b^2 - 4ac \geq 0$ for real x.

4. In case the domain of a function $y=f(x)$ is a finite set $D=\{a_1, a_2, a_3, \ldots, a_n\}$, then its range is obtained by forming the set whose members are the values of $[f(x)]\, x = a_1, a_2, a_3, \ldots, a_n$.

Type 1: Functions put in the forms: (i) $y=ax+b$ (ii) $y=ax^2+bx+c$ whose domains are not given.

Rule: When the domain of a function is not given and the question says to determine the range of a functions, one is required to find out its domain at first in the following way:

It should be checked whether the given function becomes imaginary or undefined for any value of $x \in R$ i.e. given function $y=f(x)$ does not become imaginary or undefined for any value of $x \in R \Rightarrow$ domain of the given function is the set of all real numbers denoted by R.

Lastly, one should find the range of the given functions using the axioms of inequality in $-\infty < x < \infty$.

Example worked out:

1. Find the domain and range of each of the following functions:

(i) $y=x$ **(ii)** $y=x+2$

Solution: **(i)** $y=x$ does not become imaginary or undefined for any $x \in R \Rightarrow y=x$ is defined for all $x \in R$.

$\Rightarrow D(y) = R \Rightarrow -\infty < x < \infty \Rightarrow -\infty < y < \infty$

$(\because y=x$ is given)

$\Rightarrow R(y) = R =$ the set of reals.

(ii) $y=x+2$ does not become imaginary or undefined for and $x \in R \Rightarrow y=x+2$ is defined for all $x \in R \Rightarrow D(y) = R$ Now, $D(y) = R \Rightarrow -\infty < x < \infty \Rightarrow -\infty < x+2 < \infty$ (on using the axiom of inequality) $\Rightarrow -\infty < y < \infty \Rightarrow R(y) = R =$ the set of reals.

Note:

In case one is required to find out the range of linear function $y=ax+b$ whose domain is a given subset of the set of reals namely R, the range of $y=ax+b$ is obtained with the help of given domain and the use of various axioms of inequality.

Examples: **(i)** Find the range of $f(x) = 4x - 5$ for $-6 \leq x \leq 3$.

Solution: $-6 \leq x \leq 3$

$\Rightarrow -24 \leq 4x \leq 12$

$\Rightarrow -24 - 5 \leq 4x - 5 \leq 12 - 5$

$\Rightarrow -29 \leq 4x - 5 \leq 7$

$\Rightarrow f(x) \in [-29, 7]$

$\Rightarrow R(f) = \{y: -29 \leq y \leq 7\}$ where $y=f(x)$

Similarly, the range of each of the following functions:

(ii) $f_1(x) = 2x + 3$ for $-1 \le x \le 7$ is $\{y_1 : 1 \le y_1 \le 17\}$, where $y_1 = f_1(x)$.

(iii) $g(x) = 5 - 6x$ for $-3 \le x \le 4$ is $\{y_2 : -19 \le y_2 \le 23\}$, where $y_2 = g(x)$.

(iv) $h(x) = 5x - 6$ for $-2 \le x \le 5$ is $\{y_3 : -16 \le y_3 \le 19\}$, where $y_3 = h(x)$.

2. Find the domain and range of each of the following functions:

(i) $y = x^2$ (ii) $y = x^2 - 4$.

Solution: (i) $y = x^2$ does not become imaginary or undefined for any $x \in R \Rightarrow y = x^2$ is defined for all $x \in R \Rightarrow D(y) = R$ Now, $y = x^2 \Rightarrow x^2 - y = 0 \Rightarrow$ $D = 0 - 4 \times 1 \times (-y) = 4y \ge 0 \left(\because D = b^2 - 4ac\right)$

$\Rightarrow y \ge 0 \Rightarrow y \in [0, \infty) \Rightarrow R(y) = [0, \infty)$.

(ii) $y = x^2 - 4$ does not become imaginary or undefined for any $x \in R \Rightarrow y$ is defined for all $x \in R \Rightarrow D(y) = R$.

Now, $y = x^2 - 4$

$\Rightarrow x^2 - 4 - y = 0$

$\Rightarrow x^2 - (y + 4) = 0$

$\Rightarrow D = 0 - 4 \times 1 \times \{-(y + 4)\} = 4(4 + y) \ge 0$

$\Rightarrow y + 4 \ge 0$

$\Rightarrow y \ge -4$

$\Rightarrow y \in [-4, \infty)$

$\Rightarrow R(y) = [-4, \infty)$

Type 2: Functions put in the forms:

(i) $y = \sqrt{f(x)}$ **(ii)** $y = \dfrac{1}{\sqrt{f(x)}}$

Rule: Find the domain of y and then express x in terms of y. Lastly put $g(y)$ in the domain of y and solve it to find the range of y.

Solved Examples

1. Find the range of the following functions:

(i) $y = \sqrt{x}$ **(ii)** $y = \sqrt{x - 3}$ **(iii)** $y = \sqrt{3 - 2x}$

(iv) $y = \dfrac{1}{\sqrt{x + 2}}$

Solution: (i) $y = \sqrt{x}$

$\Rightarrow x \ge 0$

Again $y = \sqrt{x}$

$\Leftrightarrow y^2 = x, (\because y \ge 0)$

but $x \ge 0$

$\Leftrightarrow y^2 \ge 0$

$\Leftrightarrow |y| \ge 0$

$\Leftrightarrow y \ge 0$ if y is non-negative which is given (since $y = \sqrt{x}$).

$\Leftrightarrow y \in (0, \infty)$

$\Rightarrow R(y) = [0, \infty)$

(ii) $y = \sqrt{x - 3}$

$\Rightarrow x - 3 \ge 0$

$\Rightarrow x \ge 3$

Again $y = \sqrt{x - 3}$

$\Leftrightarrow y^2 = x - 3, (\because y \ge 0)$

$\Leftrightarrow y^2 + 3 = x$

but $x \ge 3$

$\Leftrightarrow y^2 + 3 \ge 3 \left(\because x = y^2 + 3\right)$

$\Leftrightarrow y^2 \ge 0$

$\Leftrightarrow |y| \ge 0$

$\Leftrightarrow y \ge 0$ if y is non-negative which is given.

$\Rightarrow y \in [0, \infty) \Rightarrow R(y) = [0, \infty)$

(iii) $y = \sqrt{3 - 2x}$

$\Rightarrow 3 - 2x \geq 0$

$\Rightarrow -2x \geq -3$

$\Rightarrow 2x \leq 3$

$\Rightarrow x \leq \frac{3}{2}$

Again $y = \sqrt{3 - 2x}$

$\Leftrightarrow y^2 = 3 - 2x, (\because y \geq 0)$

$\Leftrightarrow y^2 - 3 = -2x$

$\Leftrightarrow 3 - y^2 = 2x$

$\Leftrightarrow \left(\frac{3 - y^2}{2}\right) = x$

but $x \leq \frac{3}{2}$

$\Rightarrow \left(\frac{3 - y^2}{2}\right) \leq \frac{3}{2}$

$\Rightarrow 3 - y^2 \leq 3$

$\Rightarrow -y^2 \geq 0$

$\Rightarrow |y| \geq 0$

$\Rightarrow y \geq 0$ if y is non-negative which is given.

$\Rightarrow y \in [0, \infty) \Rightarrow R(y) = [0, \infty)$

(iv) $y = \frac{1}{\sqrt{x + 2}}$

$\Rightarrow x + 2 > 0 \Rightarrow x > -2$

Again $y = \frac{1}{\sqrt{x + 2}}$

$\Leftrightarrow y^2 = \frac{1}{x + 2}, (\because y \geq 0)$

$\Leftrightarrow x + 2 = \frac{1}{y^2}$

$\Leftrightarrow x = \frac{1}{y^2} - 2$

But $x > -2$

$\Leftrightarrow \frac{1}{y^2} - 2 > -2$

$\Leftrightarrow \frac{1}{y^2} > 0$

$\Leftrightarrow y^2 > 0 \Rightarrow |y| > 0$

$\Leftrightarrow y > 0$ if y is non-negative which is given

$\Rightarrow y \in (0, \infty) \Rightarrow R(y) = (0, \infty)$

Type 3: Functions put in the forms:

(i) $y = \frac{C}{Ax + B}$ **(ii)** $y = \frac{ax + b}{Ax + B}$

Rule: Express x in terms of y by cross multiplication and simplification. Lastly use the rule:

$$R(y) = R - \left\{\text{roots of deminator of } \frac{f_1(y)}{f_2(y)} = 0\right\}$$

Solved Examples

1. Find the range of the following functions:

(i) $y = \frac{1}{x}$ **(ii)** $y = \frac{1}{x - 1}$ **(iii)** $y = \frac{x}{x + 2}$

(iv) $y = \frac{x}{5 - x}$ **(v)** $y = \frac{x - 1}{x + 3}$

Solutions:

(i) $y = \frac{1}{x}$

$\Leftrightarrow x = \frac{1}{y}, y \neq 0$

$\Rightarrow R(y) = R - \{0\}$

(ii) $y = \frac{1}{x - 1}$

$\Leftrightarrow x - 1 = \frac{1}{y}, y \neq 0$

$\Rightarrow R(y) = R - \{0\}$

(iii) $y = \dfrac{x}{x+2}, x \neq -2$

$\Leftrightarrow yx + 2y = x$

$\Leftrightarrow 2y = x - xy = x(1-y)$

$\Leftrightarrow x = \dfrac{2y}{1-y}, y \neq 1$

$\Rightarrow R(y) = R - \{1\}$

(iv) $y = \dfrac{x}{5-x}, x \neq 5$

$\Leftrightarrow 5y - xy = x$

$\Leftrightarrow 5y = x + xy = x(1+y)$

$\Leftrightarrow x = \dfrac{5y}{1+y}, y \neq -1$

$\Rightarrow R(y) = R - \{-1\}$

(v) $y = \dfrac{x-1}{x+3}, x \neq -3$

$\Leftrightarrow xy + 3y = x - 1$

$\Leftrightarrow 3y + 1 = x - xy = x(1-y)$

$\Leftrightarrow x = \dfrac{3y+1}{1-y}, y \neq 1$

$\Rightarrow R(y) = R - \{1\}$

Type 4: Functions put in the forms:

(i) $y = \dfrac{D}{Ax^2 + Bx + C}$

(ii) $y = \dfrac{ax+b}{Ax^2 + Bx + C}$

(iii) $y = \dfrac{ax^2 + bx + c}{Ax^2 + Bx + C}$

Rule: Cross multiply and obtain $Px^2 + Qx + R = 0$ where P, Q and R are functions of Y (i.e. an expression in y). Lastly use the rule $D \geq 0$, where $D = b^2 - 4ac$.

Remark: The above rule is valid in $y = \dfrac{ax^2 + bx + c}{Ax^2 + Bx + C}$ provided its numerator and denominator do not have one common factor.

Solved Examples

(1) Find the range of the following functions:

(i) $y = \dfrac{1}{x^2 - 4}$

(ii) $y = \dfrac{x}{1+x^2}$

(iii) $y = \dfrac{x^2 - 2x + 4}{x^2 + 2x + 4}$

(iv) $y = \dfrac{1}{x^2 - 3x + 2}$

Solutions:

(i) $y = \dfrac{1}{x^2 - 4}, x \neq \pm 2$

$\Leftrightarrow x^2 y - 4y = 1$

$\Leftrightarrow x = \dfrac{0 \pm \sqrt{y(1+4y)}}{y}, y \neq 0$

Now, $D \geq 0$

$\Rightarrow y(1+4y) \geq 0$

$\Rightarrow y \leq -\dfrac{1}{4}$ or $y \geq 0$

$\Rightarrow y \leq -\dfrac{1}{4}$ or $y > 0$ since $y \neq 0$

$\Rightarrow R(y) = \left(-\infty, -\dfrac{1}{4}\right] \cup (0, \infty)$

(ii) $y = \dfrac{x}{1+x^2}$

$\Leftrightarrow y + yx^2 = x$

$\Leftrightarrow yx^2 - x + y = 0$

$$\Leftrightarrow x = \frac{1 \pm \sqrt{(-1)^2 - 4 \times y \times y}}{2y}, \; y \neq 0$$

$$= \frac{1 \pm \sqrt{1 - 4y^2}}{2y}, \; y \neq 0$$

Now, $D \geq 0$

$\Rightarrow 1 - 4y^2 \geq 0$

$\Rightarrow (1 + 2y)(1 - 2y) \geq 0$

$\Rightarrow (2y + 1)(2y - 1) \leq 0$

$\Rightarrow -\frac{1}{2} \leq y \leq \frac{1}{2}, \; y \neq 0$

Also, $y = 0 \Leftrightarrow x = 0$. Hence, $R(y) = \left[-\frac{1}{2}, \frac{1}{2}\right]$

(iii) $y = \frac{x^2 - 2x + 4}{x^2 + 2x + 4}$

$\Leftrightarrow y(x^2 + 2x + 4) = x^2 - 2x + 4$

$\Leftrightarrow x^2 y - x^2 + 2xy + 2x + 4y - 4 = 0$

$\Leftrightarrow (y - 1)x^2 + 2(y + 1)x + (4y - 4) = 0$

$$\Leftrightarrow x = \frac{-2(y+1) \pm \sqrt{4(y+1)^2 - 4(y-1)(4y-4)}}{2(y-1)},$$

$y \neq 1$

Now, $D \geq 0$

$\Rightarrow 4(y + 1)^2 - 4(y - 1)(4y - 4) \geq 0$

$\Rightarrow 4(y + 1)^2 - 16(y - 1)^2 \geq 0$

$\Rightarrow (y + 1)^2 - 4(y - 1)^2 \geq 0$

$\Rightarrow \{y + 1 + 2(y - 1)\}\{y + 1 - 2(y - 1)\} \geq 0$

$\Rightarrow (y + 1 + 2y - 2)(y + 1 - 2y + 2) \geq 0$

$\Rightarrow (3y - 1)(3 - y) \geq 0$

$\Rightarrow (3y - 1)(y - 3) \geq 0$

$\Rightarrow \frac{1}{3} \leq y \leq 3$

Also $y = 1 \Leftrightarrow x = 0$

$\therefore R(y) = \left[\frac{1}{3}, 3\right]$

(iv) $y = \frac{1}{x^2 - 3x + 2}, \; x \neq 1, 2$

$\Leftrightarrow yx^2 - 3yx + 2y - 1 = 0$

$\Leftrightarrow yx^2 - 3yx + (2y - 1) = 0$

$$\Leftrightarrow x = \frac{-(-3y) \pm \sqrt{(-3y)^2 - 4y(2y-1)}}{2y}, \; y \neq 0$$

Now, $D \geq 0$

$\Rightarrow 9y^2 - 8y^2 + 4y \geq 0$

$\Rightarrow y^2 + 4y \geq 0$

$\Rightarrow |y + 2| \geq 2$

$\Rightarrow$ either $y = 2 \leq -2$ or $y + 2 \geq 2$

$\Rightarrow$ either $y \leq -4$ or $y \geq 0$ but $y \neq 0$

$\Rightarrow R(y) = (-\infty, -4] \cup (0, \infty)$

Type 5: Functions put in the forms:

(i) $y = \frac{ax + b}{Ax^2 + Bx + C}$

(ii) $y = \frac{Ax^2 + Bx + C}{ax + b}$

(iii) $y = \frac{ax^2 + bx + c}{Ax^2 + Bx + C}$

(iv) $y = \frac{x^n - a^n}{x - a}$ whose numerator and denominator contain a common factor.

Rule: Cancell the common factor present in numerator and denominator. After cancellation of common factor,

(i) If $y = mx + c$ (linear in x), then $R(y) = R - \{$ value of $y = mx + c$ at the zero of common factor$\}$, where $R =$ the set of reals.

(ii) If $y = a_1 x^2 + b_1 x + c_1$ (quadratic in x), then $R(y)$ = range of $y = a_1 x^2 + b_1 x + c - \{$value of $y = a_1 x_2 + b_1 x + c_1$ at the zero of the common factor$\}$

(iii) If $y = \dfrac{a_1 x + b_1}{a_2 x + b_2}$ (linear rational in x), then $R(y)$

$= R - \{$ zero of the denominator of $x = \dfrac{f_1(y)}{f_2(y)}$ and

the value of $y = \dfrac{a_1 x + b_1}{a_2 x + b_2}$ at the zero of common factor$\}$.

Solved Examples

1. Find the range of the following functions:

(i) $y = \dfrac{x^2 - 1}{x - 1}$

(ii) $y = \dfrac{x^3 - 8}{x - 2}$

(iii) $y = \dfrac{x^2 - 3x + 2}{x^2 + x - 6}$

Solutions: (i) $y = \dfrac{x^2 - 1}{x - 1} = \dfrac{(x-1)(x+1)}{(x-1)}, x \neq 1$

$\Rightarrow y = x + 1,\ x \neq 1$

$\Rightarrow x = y - 1$

$\Rightarrow x$ is defined for all $y \in R - \{2\}$

$\Rightarrow R(y) = R - \{2\}$

since $x + 1 = 2$ for $x = 1$ and $y = 2 \Rightarrow x = 1$

(ii) $y = \dfrac{x^3 - 8}{x - 2} = \dfrac{(x-2)(x^2 + 2x + 4)}{(x-2)}, x \neq 2$

$\Rightarrow y = x^2 + 2x + 4$

$\Rightarrow x^2 + 2x - y + 4 = 0$

$\Rightarrow x^2 + 2x - (y - 4) = 0$

$\Rightarrow x = \dfrac{-2 \pm \sqrt{4 - 4 \times 1\{-(y-4)\}}}{2} = \dfrac{-2 \pm \sqrt{4 + 4(y-4)}}{2}$

Now, $D \geq 0$

$\Rightarrow 4\{1 + (y - 4)\} \geq 0$

$\Rightarrow y - 3 \geq 0$

$\Rightarrow y \geq 3$

$\Rightarrow R(y) = [3, \infty)$ since $(x^2 + 2x + 4)$ for $x = 2 = 4 + 4 + 4 = 12$ and

$y = 12$ gives a point $x = -4 \in D(f)$

Since $12 - 4 = x^2 + 2x \Rightarrow x^2 + 2x - 8 = 0 \Rightarrow x = \dfrac{-2 \pm \sqrt{4 - 4 \times 1(-8)}}{2} = \dfrac{-2 \pm 6}{2} \Rightarrow x = 2, -4$

(iii) $y = \dfrac{x^2 - 3x + 2}{x^2 + x - 6} = \dfrac{x^2 - 2x - x + 2}{x^2 + 3x - 2x - 6}$

$= \dfrac{x(x-2) - (x-2)}{x(x+3) - 2(x+3)} = \dfrac{(x-1)(x-2)}{(x-2)(x+3)}$

$\Rightarrow yx + 3y = x - 1,\ x \neq 2$

$\Rightarrow 3y + 1 = x - xy = x(1 - y)$

$\Rightarrow x = \dfrac{3y + 1}{1 - y},\ y \neq 1$

Again, $\left(\dfrac{x-1}{x+3}\right)_{x=2} = \dfrac{1}{5}$ and $x = 2$ for $y = 1$, or $y = \dfrac{1}{5}$

Hence, $R(y) = R - \left\{1, \dfrac{1}{5}\right\}$

(iv) Find the domain and range of the function defined as

$$y = \frac{(x^2 + 3x - 4)(x^2 - 9)}{(x^2 + x - 12)(x + 3)}$$

Solution: $y = \frac{(x+4)(x-1)(x-3)(x+3)}{(x+4)(x-3)(x+3)}$

$= (x-1)$ for $x \neq -4, -3, 3$

... should note that denominator is zero for $x = -$ ... $+3$. This means that y is undefined for these three values of x. for values of $x \neq -4, -3$ or 3, one may divide numerator and denominator by common factors and obtain $y = (x-1)$ if $x \neq -4, -3$ or 3. Therefore, the domain of y is the set of all real numbers except $-4, -3$ and $+3$, i.e. $D(y) = R - \{-4, -3, +3\}$ and the range of y is the set of all real numbers except those values of $y = (x-1)$ obtained by replacing x by $-4, -3$ or 3, i.e. all real numbers except $-5, -4$ and 2, i.e. $R(y) = R - \{-5, -4, 2\}$.

Type 6: Functions put in the form: $y = \log f(x)$.

Rule: $y = \log_a f(x) \Leftrightarrow a^y = f(x) > 0$

i.e. change the given logarithmic form into the exponential form and then solve it using the in equation: $a^y > b\ (a > 0, a \neq 1) \Leftrightarrow$

(i) $y > \log_a b$ for $a > 1, b > 0$

(ii) $y < \log_a b$ for $0 < a < 1, b > 0$

(iii) $y \notin R$ for $a > 0, b < 0$.

Remark: For, $a > 1$ $a^y > a^{g(x)} \Leftrightarrow y > g(x)$ ...(A)(say)

i.e. when it is possible to change $a^y + f(x) > 0$ into the form $a^y > a^{g(x)}$, one should use (A).

Solved Examples

1. Find the range of the function $y = \log(3x^2 - 4x + 5)$.

Solution: $y = \log(3x^2 - 4x + 5)$ where $3x^2 - 4x + 5 > 0$

$$\Rightarrow e^y = 3x^2 - 4x + 5 = 3\left(x^2 - \frac{4}{3}x + \frac{5}{3}\right) =$$

$$3\left\{x^2 - \frac{4}{3}x + \left(\frac{4}{6}\right)^2 - \left(\frac{4}{6}\right)^2 + \frac{5}{3}\right\}$$

$$\Rightarrow e^y = 3\left\{\left(x - \frac{2}{3}\right)^2 + \frac{11}{9}\right\} = 3\left(x - \frac{2}{3}\right)^2 + \frac{11}{3}$$

$$\Rightarrow e^y - \frac{11}{3} = 3\left(x - \frac{2}{3}\right)^2 \geq 0$$

$$\Rightarrow e^y \geq \frac{11}{3}$$

$$\Rightarrow y \geq \log\left(\frac{11}{3}\right)$$

$$\Rightarrow y \in \left[\log\frac{11}{3}, \infty\right)$$

$$\Rightarrow R(y) = \left[\log\frac{11}{3}, \infty\right)$$

2. Find the range of the function

$$y = \log_2\left(\sqrt{x-4} + \sqrt{6-x}\right)$$

Solution: $y = \log_2\left(\sqrt{x-4} + \sqrt{6-x}\right)$

$$\Rightarrow 2^y = \left(\sqrt{x-4} + \sqrt{6-x}\right)$$

$$\Rightarrow 2^{2y} = x - 4 + 6 - x + 2\sqrt{(x-4)(6-x)}$$

...(i) (on squaring)

$$\Rightarrow \left(2^{2y} - 2\right) = 2\sqrt{(x-4)(6-x)} \geq 0 \quad \text{...(ii)}$$

$$\Rightarrow 2^{2y} - 2 \geq 0$$

$$\Rightarrow 2^{2y} \geq 2^1 \Rightarrow 2y \geq 1 \Rightarrow y \geq \frac{1}{2}$$

Again from (ii) $2^{2y-1} - 1 = \sqrt{-x^2 + 10x - 24}$

$$\Rightarrow \left(2^{2y-1} - 1\right)^2 = \left(-x^2 + 10x - 24\right)$$

$$\Rightarrow x^2 - 10x + 24 + \left(2^{2y-1} - 1\right)^2 = 0,$$ which is a quadratic in x and whose $D = b^2 - 4ac$ is

$$(10)^2 - 4\left\{24 + \left(2^{2y-1} - 1\right)^2\right\} \geq 0$$ which

$$\Rightarrow \left(2^{2y-1} - 1\right)^2 \leq 1$$

$\Rightarrow \left|2^{2y} - 1\right| \le 1$

$\Rightarrow -1 \le 2^{2y-1} - 1 \le 1$

$\Rightarrow 0 \le 2^{2y-1} - 1 \le 1$ (from (ii))

$\Rightarrow 1 \le 2^{2y-1} \le 2$

$\Rightarrow 2^0 \le 2^{2y-1} \le 2^1$

$\Rightarrow 0 \le 2y - 1 \le 2$

$\Rightarrow 1 \le 2y \le 2$

$\Rightarrow \frac{1}{2} \le y \le 1$

$\Rightarrow R(y) = \left[\frac{1}{2}, 1\right]$

Type 7: Functions put in the form: $y = a\cos x + b\sin x$.

Rule: The range of $y = a\cos x + b\sin x$ is $\left[-\sqrt{a^2 + b^2}, \sqrt{a^2 + b^2}\right]$

Solved Examples

1. Find the range of $y = \sin x - \cos x$

Solution: $y = \sin x - \cos x$

$= \sqrt{2}\left[\frac{1}{\sqrt{2}}\sin x - \frac{1}{\sqrt{2}}\cos x\right]$

$= \sqrt{2}\left[\cos\left(\frac{\pi}{4}\right)\sin x - \sin\left(\frac{\pi}{4}\right)\cos x\right]$

$= \sqrt{2}\sin\left(x - \frac{\pi}{4}\right)$

Again, it is know that

$-1 \le \sin\left(x - \frac{\pi}{4}\right) \le 1$

i.e. $-\sqrt{2} \le \sqrt{2}\sin\left(x - \frac{\pi}{4}\right) \le \sqrt{2}$

Hence, range of $y = R(y) = \left[-\sqrt{2}, \sqrt{2}\right]$

Note: Range of $a\cos x + b\sin x$

$= \left[-\sqrt{a^2 + b^2}, \sqrt{a^2 + b^2}\right]$

$a = -1, b = 1 \Rightarrow \sqrt{a^2 + b^2} = \sqrt{2}$

$\therefore R(y) = \left[-\sqrt{2}, \sqrt{2}\right]$ in the above question.

2. Find the range of $y = \cos\theta + \sqrt{3}\sin\theta$

Solution: $y = \cos\theta + \sqrt{3}\sin\theta$

$= 2\left[\frac{1}{2}\cos\theta + \frac{\sqrt{3}}{2}\sin\theta\right]$

$= 2\left[\sin\left(\frac{\pi}{6}\right)\cos\theta + \cos\left(\frac{\pi}{6}\right)\sin\theta\right]$

Again it is known that

$-1 \le \sin\left(\theta - \frac{\pi}{6}\right) \le 1$

i.e. $-2 \le 2\sin\left(\theta - \frac{\pi}{6}\right) \le 2$

Hence, $R(y) = [-2, 2]$

Note: In this question, $a = 1$, $b = \sqrt{3} \Rightarrow \sqrt{1+3} = \sqrt{4}$

hence, range of $\cos\theta + \sqrt{3}\sin\theta = [-2, 2]$ on using, range of a $a\cos\theta + b\sin\theta =$

$\left[-\sqrt{a^2 + b^2}, \sqrt{a^2 + b^2}\right]$.

Type 8: Functions put in the forms:

(i) $y = a \pm b\sin x$ **(ii)** $y = a \pm b\cos x$

(iii) $y = \frac{C}{a \pm b\sin x}$ **(iv)** $y = \frac{C}{a \pm b\cos x}$, where a, b and c are constants.

Rule: Start from $|\sin x| \le 1$ or $|\cos x| \le 1$ and form:

(i) $k \le a \pm b\sin x \le L$

(ii) $k \le a \pm b\cos x \le L$

(iii) $k \le \dfrac{C}{a \pm b \sin x} \le L$

(iv) $k \le \dfrac{C}{a \pm b \cos x} \le L$ where k and L are constants, by using the axioms of inequality.

Solved Examples

1. Find the range of $y = 2 + \sin x$.

Solution: $-1 \le \sin x \le 1$

$\Rightarrow 2 - 1 \le 2 + \sin x \le 1 + 2$

$\Rightarrow 1 \le 2 \sin x \le 3$

$\Rightarrow 1 \le y \le 3$

$\Rightarrow y \in [1, 3]$

$\Rightarrow R(y) = [1, 3]$

2. Find the range of $y = \dfrac{1}{2 - \sin 3x}$

Solution: $-1 \le \sin 3x \le 1$

$\Rightarrow -1 \le -\sin 3x \le 1$

$\Rightarrow 2 - 1 \le 2 - \sin 3x \le 1 + 2$

$\Rightarrow 1 \le 2 - \sin 3x \le 3$

$\Rightarrow 1 \ge \dfrac{1}{2 - \sin 3x} \ge \dfrac{1}{3}$

$\Rightarrow \dfrac{1}{3} \le \dfrac{1}{2 - \sin 3x} \le 1$

$\Rightarrow \dfrac{1}{3} \le y \le 1$

$\Rightarrow y \in \left[\dfrac{1}{3}, 1\right]$

$\Rightarrow R(y) = \left[\dfrac{1}{3}, 1\right]$

Type 9: Finding the range of a function containing greatest integer function.

Rule: Range of those functions containing greatest integer function is obtained by using different properties of greatest integer function.

Solved Examples

1. Find the range of $y = [\cos x]$

Solution: $-1 \le \cos x \le 1$

$$\Rightarrow y = [\cos x] = \begin{cases} -1, \text{ for } -1 \le \cos x < 0 \\ 0, \text{ for } 0 \le \cos x < 1 \\ 1, \text{ for } \cos x = 1 \end{cases}$$

$\Rightarrow R(y) = [-1, 0, 1]$

2. Find the range of $y = 1 + x - [x - 2]$

Solution: On using the property:

$[t] \le t < [t] + 1$, it is seen that

$[x - 2] \le x - 2 < [x - 2] + 1$

$\Rightarrow [x - 2] - [x - 2] \le x - 2 - [x - 2] < [x - 2] - [x - 2] + 1$

$\Rightarrow 0 \le x - 2 - [x - 2] < 1$

$\Rightarrow 3 \le x - 2 + 3 - [x - 2] < 1 + 3$ (adding 3 to each side)

$\Rightarrow 3 \le x + 1 - [x - 2] < 4$

$\Rightarrow 3 \le f(x) < 4$

$\Rightarrow R(y) = [3, 4)$

Type 10: Finding the domain and range of a piecewise defined functions.

1. $y = f_1(x)$, when $x < a$
$= f_2(x)$, when $x \ge a$

i.e. two or more functions of an independent variable namely x defined in adjacent intervals.

2. $y = c_1$, when $x < a$
$= c_2$, when $a \le x < b$
$= c_3$, when $b < x$

i.e. two or more different functions defined in adjacent intervals.

3. $y = f_1(x)$, when $x \ne a$
$= f_2(x)$, when $x = a$

Now, it will be discussed in detail how to find the domain and range of each type of piecewise function.

Rule: The domain of each type of a piecewise function is the union of each given interval whereas the range of each type of a piecewise function is the union of different range of each given function determined by considering each different given intervals and applying the axioms of inequality to obtain the different given functions in the form of inequalities.

Note: The range of a piecewise function put in the form:

$y = c_1$, when $x < a$
$= c_2$, when $a \le x < b$
$= c_3$, when $b \le x$

i.e. the range of a piecewise function defined by different constants in adjacent intervals is the set whose members are given constants (constant functions) defined in given adjacent intervals, i.e. $R(y) = \{c_1, c_2, c_3\}$, where c_1, c_2 and c_3 are different constants defined in adjacent intervals.

Solved Examples

1. Find the domain and range of the function defined by

$$y = \begin{cases} 3x - 2, \text{ if } x < 1 \\ x^2, \text{ if } 1 \le x \end{cases}$$

Solution: $x < 1 \Rightarrow x \in (-\infty, 1)$

$x \ge 1 \Rightarrow x \in [1, \infty)$

$\therefore D(y) = (-\infty, 1) \cup [1, \infty) = (-\infty, \infty)$

Again, $x < 1$
$\Rightarrow 3x < 3$
$\Rightarrow 3x - 2 < 3 - 2$
$\Rightarrow 3x - 2 < 1$ ...(i)
Also, $x \ge 1$
$\Rightarrow x^2 \ge 1$ ...(ii)
Hence, from (i) and (ii), it is concluded that $R(y) = (-\infty, 1) \cup [1, \infty) = (-\infty, \infty)$

2. Find the domain and range of the function defined by

$$y = \begin{cases} x - 1, \text{ if } x < 3 \\ 2x + 1, \text{ if } 3 \le x \end{cases}$$

Solution: $x < 3 \Rightarrow x \in (-\infty, 3)$

$x \ge 3 \Rightarrow x \in [3, \infty)$

$\therefore D(y) = (\infty, 3) \cup [3, \infty) = (-\infty, \infty)$

Again, $x < 3 \Rightarrow x - 1 < 3 - 1 \Rightarrow x - 1 < 2$ ...(i)

Also, $x \ge 3 \Rightarrow 2x \ge 6 \Rightarrow 2x + 1 \ge 7$ ...(ii)

Hence, (i) and (ii) $\Rightarrow R(y) = (-\infty, 2) \cup [7, \infty) = R - [2, 7)$ i.e. all real numbers not in $[2, 7)$.

3. Find the domain and range of the function defined as

$$y = \begin{cases} x + 3, \text{ when } x \ne 3 \\ 2, \text{ when } x = 3 \end{cases}$$

Solution: $x \ne 3$

$\Rightarrow x > 3$ or $x < 3 \Rightarrow$

$x \in (-\infty, 3) \cup (3, \infty) = R - \{3\}$

Next, $y = 2$, when $x = 3$

$\therefore D(y) = R - \{3\} \cup \{3\} = R$

Now, $x \ne 3$

$\Rightarrow x + 3 \ne 6$

$\Rightarrow R(y|x \ne 3) = R - \{6\}$

Also, $y = 2$, when $x = 3$

$\Rightarrow R(y|x = 3) = \{2\}$

$\therefore R(y) = R - \{6\} \cup \{2\} = R - \{6\}$

i.e. the range of the given function consists of all real numbers except $y = 6$.

4. Let there be a function defined as

$$y = \begin{cases} x^2, \text{ if } x \ne 2 \\ 7, \text{ if } x = 2 \end{cases}$$

find its domain and range.

Solution: $x \ne 2$

$\Rightarrow x < 2$ or $x > 2$

$\Rightarrow x \in (-\infty, 2) \cup (2, \infty) = R - \{2\}$

Next, $y = 7$, for $x = 2$

$\therefore D(y) = R - \{2\} \cup \{2\} = R$

Now, $x \neq 2$

$\Rightarrow x^2 \neq 4$

$\Rightarrow R(y|x \neq 2)$

$= R^+ - \{4\} \cup \{0\}, \because x^2 \geq 0$

Also, $y = 7$, for $x = 2$

$\Rightarrow R(y|x = 2) = \{7\}$

$\therefore R(y) = R^+ - \{4\} \cup \{7\} \cup \{0\}$

$= R^+ - \{4\} \cup \{0\}$

i.e. the range consists of all non-negative real numbers except $y = 4$.

Note: One should note that $y = c$, for $x = a$, where c and a are constants, represents a point $P(a, c)$, i.e. a point whose abscissa is 'a' and whose ordinate is 'c'.

5. If the domain of a function is $A = \{x: x \in R, -1 < x < 1\}$ and the function is defined as

$$f(x) = \begin{cases} 1, \text{ when } x > 0 \\ 0, \text{ when } x = 0 \\ -1, \text{ when } x < 0 \end{cases}$$

Find the range of $f(x)$.

Solution: The range of a piecewise function whose each function is constant defined in its domain is the union of different constants.

Therefore, $R(y) = [-1, 0, 1\}$

Type 11: A function $y = f(x)$ whose domain s a finite set.

Rule: If the domain D of $y = f(x)$ is a finite set, i.e. $D = \{a_1, a_2, a_3, \ldots, a_n\}$, then its range $R(f) = \{f(a_1), f(a_2), f(a_3), \ldots, f(a_n)\}$.

Solved Examples

1. If $A = \{1, 2, 3, 4, 5\}$, $B = \{a, b, c, d, e\}$ and $f = \{(1, b), (2, d), (3, a), (4, b), (5, c)\}$ be a mapping from A to B, find $f(A)$.

Solution: $f(A) = \{f(1), f(2), f(3), f(4), f(5)\}$
$= (b, d, a, c\}$

2. If $A = \{0, 1, -1, 2\}$ and $f: A \to R$ is defined by $f(x) = x^2 + 1$, find the range of f.

Solution: $f(x) = x^2 + 1$

$\Rightarrow f(0) = 1$
$f(1) = 2$
$f(-1) = 2$
$f(2) = 5$
$\therefore f(A) = \{f(0), f(1), f(-1), f(2)\}$
$= (1, 2, 5\}$

3. If $A = \{0, 1, 2, -3\}$ and $f(x) = 3x - 5$ is a function from A on to B, find B.

Solution: $f(x) = 3x - 5$

$\Rightarrow f(0) = -5$
$f(1) = -2$
$f(2) = 1$
$f(-3) = -14$
$\therefore f(A) = B = \{1, -2, -5, -14\}$

4. If $A = \{1, 2, 3, 4\}$ and $f(x) = x^2 + x - 1$ is a function from A on to B, find B.

Solution: $f(x) = x^2 + x - 1$

$\Rightarrow f(1) = 1$
$f(2) = 5$
$f(3) = 11$
$f(4) = 19$
$\therefore f(A) = B = \{1, 5, 11, 19\}$

Type 12: A function $y = f(x)$ defined in an open interval (a, b).

Rule 1: $a < x < b \Rightarrow f(a) < f(x) < f(b)$ if $f(x)$ is increasing in (a, b).

Rule 2: $a \leq x \leq b \Rightarrow f(a) \leq f(x) \leq f(b)$ if $f(x)$ is increasing in $[a, b]$.

Rule 3: $a < x < b \Rightarrow f(b) < f(x) < f(a)$ if $f(x)$ is decreasing in (a, b).

Rule 4: $a \leq x \leq b \Rightarrow f(b) \leq f(x) \leq f(a)$ if $f(x)$ is decreasing in $[a, b]$.

Notes:

1. When $y = f(x)$ is an increasing function in the open interval (a, b) or in the closed interval $[a, b]$, then $f(a) = L$ (say) is least and $f(b) = G$ (say) is greatest value of the given function $y = f(x)$ in (a, b) or $[a, b]$.

2. When $y = f(x)$ is a decreasing function in the open interval (a, b) or in the closed interval $[a, b]$, then $f(a) = G$ (say) is greatest and $f(b) = L$ (say) is least value of the function $y = f(x)$ in (a, b) or $[a, b]$.

3. Range of $y = f(x) = R(f) =$ (least $f(x)$, greatest $f(x)$) if the domain of $f(x)$ is an open interval (a, b) where $f(x)$ is continuous and increasing or decreasing for $x \in D(f) = (a, b)$

In the same fashion, $R(f) =$ [least $f(x)$, greatest $f(x)$] if the domain of $f(x)$ is a closed interval $[a, b]$ where $f(x)$ is continuous, and increasing or decreasing for $x \in D(f) = [a, b]$.

4. In case a function $y = f(x)$ is defined in its domain is neither increasing nor decreasing but it is continuous in its domain then its range is also determined by the rule of finding greatest and least value of the given function $f(x)$ which will be explained in the chapter namely maxima and minima of a function.

Solved Examples

1. Find the range of the function $f(x) = x^3$ whose domain is $D = \{x : x \in R, -2 < x < 2\}$.

Solution: $f(x) = x^3$ is increasing in $(-2, 2)$

$\therefore f(-2) = (-2)^3 = -8$

and $f(2) = (2)^3 = 8$

Therefore, $R(y) = (f(-2), f(2)) = (-8, 8)$

2. If $y = \tan x$ defined in $\left(-\frac{\pi}{2}, \frac{\pi}{2}\right)$, find its range.

Solution: $y = \tan x$ is increasing in $\left(-\frac{\pi}{2}, \frac{\pi}{2}\right)$ $\therefore -\frac{\pi}{2} < x < \frac{\pi}{2} \Rightarrow -\infty < \tan x < \infty$ since $\underset{x \to \frac{\pi}{2}}{L} \tan x = \infty$ and $\underset{x \to \left(-\frac{\pi}{2}\right)}{L} \tan x = -\infty$.

3. If $A = \left\{x : \frac{\pi}{6} \le x \le \frac{\pi}{3}\right\}$ and $f(x) = \cos x - x(1 + x)$, find $f(A)$.

Solution: $f(x) = \cos x - x(1 + x)$, is decreasing in $\left[0, \frac{\pi}{2}\right]$ and so in $\left[\frac{\pi}{6}, \frac{\pi}{3}\right]$

$$\therefore f\left(\frac{\pi}{3}\right) \le f(x) \le f\left(\frac{\pi}{6}\right)$$

Now, $f\left(\frac{\pi}{3}\right) = \cos\left(\frac{\pi}{3}\right) - \frac{\pi}{3}\left(1 + \frac{\pi}{3}\right) = \frac{1}{2} - \frac{\pi}{3} - \frac{\pi^2}{9}$

$$f\left(\frac{\pi}{6}\right) = \cos\left(\frac{\pi}{6}\right) - \frac{\pi}{6}\left(1 + \frac{\pi}{6}\right) = \frac{\sqrt{3}}{2} - \frac{\pi}{6} - \frac{\pi^2}{36}$$

$$\therefore f(A) = \left[\frac{1}{2} - \frac{\pi}{3} - \frac{\pi^2}{9}, \frac{\sqrt{3}}{2} - \frac{\pi}{6} - \frac{\pi^2}{36}\right]$$

Note: If $y = f(x)$ is a continuous function whose domain $D = [a, b] = [a, c) \cup [c, b]$ where $a < c < b$ and $f(x)$ increasing in (a, c) and decreasing in $[c, b)$, then to find its range $R(f)$, one is required to find out $f(a)$, $f(b)$, $f(c)$ and

$R(y) =$ (greatest $f(x)$, least $f(x)$)

In the same way, $R(y) =$ [greatest $f(x)$, least $f(x)$] if $y = f(x)$ is defined in the domain $D = [a, b] = [a, c) \cup [c, b]$ such that $f(x)$ is decreasing in $[a, c)$ and increasing in $[c, b]$.

Solved Examples

1. Find the range of the function $f(x) = x^2 + 1$ in the domain $(-5, 2)$.

Solution: $f(x) = x^2 + 1$ with its domain $(-5, 2)$ and $f(x)$ is decreasing in $(-5, 0)$ and increasing in $[0, 2)$.

$f(-5) = 26$

$f(0) = 1$

$f(2) = 5$

Therefore, it is clear that $f(x)$ lies between 1 and 26.

$\therefore R(f) = (1, 26)$

2. Find the range of the function $f(x) = x^2$ in $(-2, 2)$.

Solution: $f(x) = x^2$ with its domain $(-2, 2)$ $f(x)$ is decreasing in $(-2, 0)$ and increasing in $[0, 2)$.

$f(-2) = 4$

$f(0) = 0$

$f(2) = 4$

Therefore, it is seen that $f(x)$ between 0 and 4.

$\therefore R(f) = (0, 4)$

Composite Function

Definition: If given functions are $f: D_1 \to C_1$ and $g: D_2 \to C_2$ such that $f(D_1) \subset D_2$, then the composite function *gof* is the function from D_1 to D_2 defined by $(gof)(x) = gof(x) = g(f(x))$, $\forall\, x \in D_1$ and $f(x) \in D_2$.

N.B.: **(i)** $f(D_1) \subset D_2$ means that range of $f \subset$ domain of g.

(ii) $D_{gof} = \{x : x \in D_f \text{ and } f(x) \in D_g\}$ where *gof* represents the composite of f and g defined by $(gof)(x) = g(f(x))$ having the domain $D_{gof} = D(gof)$.

(iii) A function does not exist whenever its domain is an empty set.

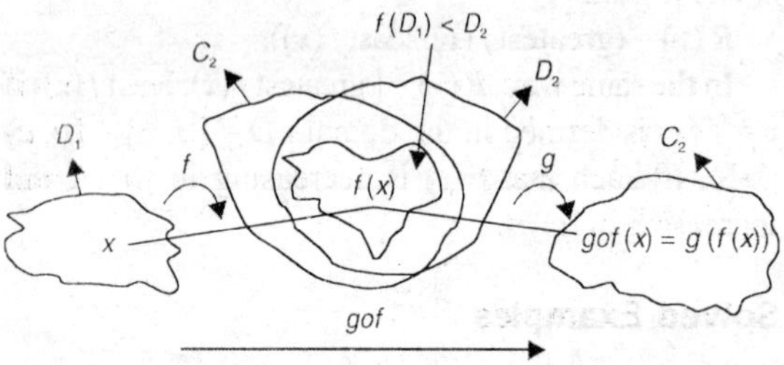

Remember: One must remember that $g(f(x)) = gof(x)$ means that g is a function of $f(x)$ which is itself a function of x. This is why $gof(x) = g(f(x))$ is called a function of a function of the independent variable x. Further, one should note that $gof(x)$ signifies the value of the function g at $f(x)$ = given analytic expression in x (or, simply given expression in x).

Type 1: Formation of composite of two functions of x whose analytic expressions in x are given.

Working rule: The rule to compute $gof(x)$ for two analytic expressions in x for $f(x)$ and $g(x)$ says.

1. Firstly to replace $f(x)$ by its given analytic expression in x.

2. Secondly, in the given analytic expression in x for $g(x)$, to replace each x by the function $f(x)$ and then to put $f(x)$ = analytic expression in x for $f(x)$.

Thus, $gof(x)$

= value of g at $f(x)$

= $[\text{analytic expression for } g(y)]_{y = f(x)}$, which signifies that the independent variable x in the analytic expression for $g(x)$ should be replaced by the analytic expression in x for $f(x)$ whereas the constant in the analytic expression in x for $g(x)$ remains unchanged.

Note: Very often the law establishing the relationship between the independent variable and dependent variable is specified by means of a formula. This method of representation of function is called analytical. Further, the expression in x is called analytic expression.

Solved Examples

1. If $f(x) = x^2$ and $g(x) = x + 1$, find $(gof)(x)$.

Solution: On applying the definition,

$(gof)(x) = gof(x)$

$= g\left(x^2\right)$

$= g(y)]_{y=x^2}$

$= (y + 1)_{y=x^2} = x^2 + 1$

2. If $f(x) = x + 3$ and $g(x) = \sqrt{x}$ find $(gof)(7)$.

Solution: $(gof)(7) = gof(7)$

$= g(3 + 7)$

$= g(10)$

$= \left(\sqrt{x}\right)_{10}$

$= \sqrt{10}$

3. If $f(x) = x^3$ and $g(x) = \sqrt[3]{x}$ find $g(f(x)$.

Solution: $g(f(x)$

$= g(x^3)$

$= \left(\sqrt[3]{x}\right)_{x^3}$

$= \sqrt[3]{x^3}$

$= x$

4. If $f(x) = x + 1$ and $g(x) = \sqrt{x}$ find $g(f(x))$.

Solution: $g(f(x)) = gof(x)$

$= g((x+1))$

$= \left(\sqrt{x}\right)_{(x+1)}$

$= \sqrt{x+1}$

5. If $f(x) = \sin x$ and $g(x) = x^2$, find $(gof)(x)$.

Solution: $(gof)(x) = g(f(x)) = g(f(x)) = g(\sin x)$

$= (\sin x)^2$

6. If $f(x) = \dfrac{x}{1-x}$ find $f(f(f(x)))$.

Solution: Given $f(x) = \dfrac{x}{1-x},\ x \neq 1$ ...(i)

On replacing x by $f(x)$ in (i), we have

$$f(f(x)) = \frac{f(x)}{1-f(x)} = \frac{\left(\frac{x}{1-x}\right)}{1-\left(\frac{x}{1-x}\right)} =$$

$$\frac{\left(\frac{x}{1-x}\right)}{\left(\frac{1-2x}{1-x}\right)} = \frac{x}{1-2x},\ x \neq \left\{1, \frac{1}{2}\right\}$$

Again, replacing x by $f(f(x))$ in (i), we get

$$f(f(f(x))) = \frac{f(f(x))}{1-f(f(x))}$$

$$= \frac{\left(\frac{x}{1-2x}\right)}{1-\left(\frac{x}{1-2x}\right)}$$

$$= \frac{x}{1-3x},\ x \neq \left\{1, \frac{1}{2}, \frac{1}{3}\right\}$$

N.B.: Whenever $f(x)$ = an analytic expression in x and we are required to find $f[f\{f(x)\}]$, we adopt the following procedure:

1. We replace x by $f(x)$ in the given expression in x which provides us $f\{f(x)\}$.

2. Again we replace x by $f\{f(x)\}$ in the given expression in x which provides us $f[f\{f(x)\}]$.

7. If $f(x) = \dfrac{3x+1}{x-3}$ and $\varnothing(x) = \dfrac{x-3}{3x+1}$ find $f(\phi(x))$ and $\phi(f(x))$.

Solution: Given

$$f(x) = \frac{3x+1}{x-3},\ x = 3,\ \phi(x) = \frac{x-3}{3x+1},\ x \neq -\frac{1}{3}$$

$$\therefore f[\phi(x)] = \frac{3\phi(x)+1}{\phi(x)-3}$$

$$= \frac{3\left(\frac{x-3}{3x+1}\right)+1}{\left(\frac{x-3}{3x+1}\right)-3} = \frac{6x-8}{-8x-6} = \frac{4-3x}{3+4x},\ \text{for}$$

$$x \neq -\frac{1}{3}, \frac{-3}{4}, 3$$

Now to find $\phi\{f(x)\}$, we consider the given function $\phi(x) = \dfrac{x-1}{3x+1}$ whose independent variable x is replaced by $f(x)$.

$$\therefore \phi[f(x)] = \frac{3x+1}{x-3} - 3 \Big/ 3\left(\frac{3x+1}{x-3}\right)+1$$

$$= \frac{3x+1-3(x-3)}{3(3x+1)+(x-3)} = \frac{10}{10x} = \frac{1}{x};\ x \neq -1, 3, 0$$

N.B.: To find $f\{\phi(x)\}$, we replace x by $\phi(x)$ in the given function for $f(x)$ and to find $\phi\{f(x)\}$, we replace x by $f(x)$ in the given expression for $\phi(x)$. Further we should note that $f\{\phi(x)\}$ means the value of f at $\phi(x)$ and $\phi\{f(x)\}$ means the value of ϕ at $f(x)$.

Refresh your memory: If given functions are $f: D_1 \to C_1$ and $g: C_1 \to C_2$, then the composite function gof is the function from D_1 to C_2 defined by $(gof)(x) = gof(x)$ for every x in D_1 and $f(x) \in C_1$ domain of g.

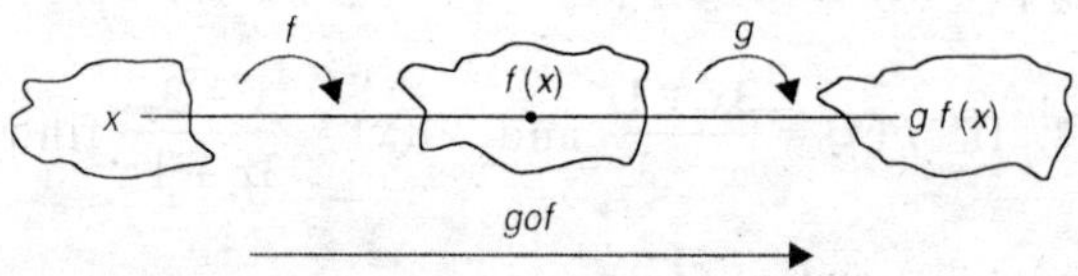

Note: The working rule to fund (gof) is the same provided the given functions are:

(i) $f: D_1 \to C_1$ and $g: D_2 \to C_2$ such that $f(D_1) \subset D_2$.

(ii) $f: D_1 \to C_1$ and $g: C_1 \to C_2$

Solved Examples

1. If the mapping $f: D_1 \to C_1$ is defined by $f(x) = \log(1 + x)$ and the mapping $g: C_1 \to C_2$ is defined by $g(x) = e^x$ find $(gof)(x)$.

Solution: $(gof)(x) = g(\log(1+x)); x > -1$ $(\because f(x) = \log(1+x))$

$= e^{\log(1+x)}$ $(\because g(x) = e^x)$

$= (1+x); x > -1$ $(\because e^{\log f(x)} = f(x))$

2. If $f: D_1 \to C_1$ is defined by $f(x) = x + 1$, $x \in R$ and $g: C_1 \to C_2$ is defined by $g(x) = x^2$, find $(gof)(x)$.

Solution: $(gof)(x) = g(f(x)) = g(1+x)$ $(\because f(x) = 1+x)$

$= (1+x)^2$ $\left(\because g(x) = x^2 \Rightarrow g(f(x)) = \left[f(x)^2\right]\right)$

3. If $f: R \to R$ is defined by $f(x) = 2x^2 - 1$ and $g: R \to R$ is defined by $g(x) = 4x - 3$, $x \in R$, compute $(gof)(x)$ and $(gof)(3)$.

Solution: $(gof)(x) = g(f(x)) = g(2x^2 - 1)$ $(\because f(x) = 2x^2 - 1)$

$= 4(f(x)) - 3$ $(\because g(x) = 4x - 3 \Rightarrow g(f(x)) = 4f(x) - 3)$

$= 4(2x^2 - 1) - 3$

$= 8x^2 - 4 - 3$

$= 8x^2 - 7$

$\therefore (gof)(2) = (8x^2 - 7)_2$

$= 8(2)^2 - 7$

$= 8 \times 4 - 7$

$= 32 - 7 = 25$

Notes: **(i)** If we are required to find $(fog)(x)$ and $(fog)(-1)$ for the just above defined functions, then $(fog)(x) = f(g(x)) = f(4x - 3)$ $(\because g(x) = 4x - 3)$

$= 2(4x - 3)^2 - 1$ $(\because f(x) = 2x^2 - 1 \Rightarrow f(g(x)) = 2(f(x))^2 - 1)$

$= 2(16x^2 - 24x + 9) - 1$

$= 32x^2 - 48x + 17$

$\therefore (fog) - 1 = 32(-1)^2 - 48(-1) + 17$

$= 32 + 48 + 17 = 97$

(ii) In general $(gof)(x) \neq (fog)(x)$

5. If the mapping $f: R \to R$ be given by $f(x) = x^2 + 2$ and the mapping $g: R \to R$ be given by $g(x) = 1 - \dfrac{1}{1-x}$ compute $(gof)(x)$ and $(fog)(x)$ and show that $(gof)(x) \neq (fog)(x)$.

Solution: $(gof)(x) = g(f(x))$

$= g(x^2 + 2)$

$$= 1 - \frac{1}{1-(x^2+2)} = 1 - \frac{1}{-x^2-1} = 1 + \frac{1}{x^2+1} \quad \text{...(i)}$$

$(fog)(x) = f(g(x))$

$$= f\left(1 - \frac{1}{1-x}\right) = f\left(\frac{1-x-1}{1-x}\right) = f\left(\frac{-x}{1-x}\right)$$

$$= \left(\frac{-x}{1-x}\right)^2 + 2 = \frac{x^2}{(1-x)^2} + 2 \quad \text{...(ii)}$$

In the light of (i) and (ii), it is clear that $(gof)(x) \neq (fog)(x)$

Note: The operation that forms a single function from two given functions by substituting the second function for the argument of the first fucntion (for the independent variable of the first function) is also termed as composition. It is only defined when the range of the first is contained in the domain of the second. Repeated composition is denoted by a superscript numeral as $f^{(n)}$; so far example $fofofof = f^{(4)}$.

6. Find $fofof$ if $f(x) = \dfrac{x}{\sqrt{1+x^2}}$

Solution: $\because f(x) = \dfrac{x}{\sqrt{1+x^2}}$

$$\therefore fof(x) = \frac{f(x)}{\sqrt{1+f^2(x)}} = \frac{\frac{x}{\sqrt{1+x^2}}}{\sqrt{1+\frac{x^2}{1+x^2}}}$$

$$= \frac{\frac{x}{\sqrt{1+x^2}}}{\sqrt{\frac{1+x^2+x^2}{1+x^2}}}$$

$$= \frac{\frac{x}{\sqrt{1+x^2}}}{\frac{\sqrt{1+2x^2}}{\sqrt{1+x^2}}} = \frac{x}{\sqrt{1+2x^2}} = G(x)\,(\text{say})$$

$$\text{and } fofof(x) = fo\left(\frac{x}{\sqrt{1+2x^2}}\right)$$

$$= foG(x) = \frac{G(x)}{\sqrt{1+G^2(x)}}$$

$$= \frac{\frac{x}{\sqrt{1+2x^2}}}{\sqrt{1+\frac{x^2}{1+2x^2}}}$$

$$= \frac{\frac{x}{\sqrt{1+2x^2}}}{\frac{\sqrt{1+2x^2+x^2}}{\sqrt{1+2x^2}}}$$

$$= \frac{x}{\sqrt{1+3x^2}}$$

Type 2: Problems based on finding the composite of two functions whenever the domains are mentioned in the form of intervals:

Question: Define the composite of two functions namely f and g whose domains are D_1 and D_2 respectively.

Answer: **1.** If f and g are two functions whose domains are D_1 and D_2 respectively, then gof is the composite of two functions namely f and g defined by $(gof)(x) = g(f(x))$.

Further, we should note that the domain of gof is the set of all those $x \in D_1$ (domain of f) for which $f(x) \in D_2$ (domain of g). But if the range of f is a subset of the domain of g, then the domain of gof is the same as the domain of f, i.e $D_{gof} = \{x : x \in D_f \text{ and } f(x) \in g\}$ and $D_{gof} = Df$ when $R_f \subset D_g$.

2. If f and g are two functions whose domains are D_1, and D_2 respectively, then fog is the composite of two functions namely f and g defined by $(fog)(x) = f(g(x))$.

Further, we should note that the domain of fog is the set of all those $x \in D_2$ (domain of g) for which $g(x) \in D_1$ (domain of f). But if the range of g is a subset of the domain of f, then the domain of fog is the same as the domain of i.e; $D_{fog} = \{x : x \in D_g \text{ and } g(x) \in D_f\}$ and $D_{fog} = D_g$ when $R_g \subset D_f$.

Notes: **1.** If $f(x)=f_1(x), x\geq a$ and $g(x)=f_2(x), x\geq b$ then f is defined for $x\geq a$ and $gof(x)$ is defined for $f(x)\geq b$ which is solved for x to find the domain of $(gof)(x)$ i.e; domain of $(gof)(x)$ is the intersection of the solution sets of the inequalities $x\geq a$ and $f(x)\geq b$. Similarly, g is defined for $x\geq b$ and $fog(x)$ is defined for $g(x)\geq a$ which is solved for x to find the domain of $(fog)(x)$, i.e; domain of $(fog)(x)$ is the intersections of the solution sets of the inequalities $x\geq b$ and $g(x)\geq a$.

2. If $f(x)=f_1(x), a\leq x\leq b$ and $g(x)=f_2(x), c\leq x\leq b$ then f is defined for $a\leq x\leq b$ and $gof(x)$ is defined for $c\leq f(x)\leq d$ which is solved for x to find the domain of $gof(x)$, i.e; domain of $(gof)(x)$ is the intersection of the solution sets of the inequalities $a\leq x\leq b$ and $c< f(x)\leq d$. Similarly, g is defined for $c\leq x\leq d$ and $fog(x)$ is defined for $a\leq g(x)\leq b$ which is solved for x to find the domain of $(fog)(x)$, i.e; domain of $(fog)(x)$ is the intersection of the solution sets of the inequalities $c\leq x\leq d$ and $a\leq g(x)\leq b$.

3. In general $fog\neq gof$ which $\Rightarrow (fog)(x)\neq (gof)(x)$.

4. It should be noted that the notations fog and fg represent two different functions namely a function f of a function g and the product of two functions f and g respectively.

Now we explain the rules of finding the composition of two functions of x's whenever their domains are mentioned as intervals.

Working rule to find (gof) (x): It consists of following steps:

Step 1: Finding $gof(x)$ (i.e; value of g at $f(x)$ as usual.

Step 2: **(i)** Considering the inequality obtained on replacing x by $f(x)$ in the domain of g which is given in the form of an interval finite or infinite.

(ii) Considering the inequality which represents the domain of f.

Step 3: Finding the intersection of the inequalities (i) and (ii) to get the domain of $(gof)(x)$.

Working rule to find (fog) (x): It consists of following steps:

Step 1: Finding $(fog)(x)$ (i.e; value of f at $g(x)$ as usual.

Step 2: **(i)** Solving the inequality obtained on replacing x by $g(x)$ in the domain of f which is given in the form of an interval.

(ii) Solving the inequality which represents the domain of g.

Steps 3: Finding the intersection of inequalities (i) and (ii) to get the domain of $(fog)(x)$.

Solved Examples

1. If $f(x) = \sqrt{x+4}, x\geq -4$ and $g(x)\ \sqrt{x-4}$, $x\geq 4$ find $(gof)(x)$.

Solutions:

(i) $(gof)(x)=gof(x)$

$=\sqrt{f(x)-4},\ f(x)\geq 4$ and $x\geq -4$

$=\sqrt{\sqrt{x+4}-4},\ \sqrt{x+4}\geq 4$

(ii) We solve the inequality $\sqrt{x+4}\geq 4$ for x:

$\therefore x+4\geq 16$

$\Leftrightarrow x\geq 16-4$

$\Leftrightarrow x\geq 12$

(iii) $D(gof)=[-4,\infty)\cap[12,\infty)=[12,\infty)$

(iv) $(gof)(x)=\sqrt{\sqrt{x+4}-4},\ \begin{cases} x\geq -4 \text{ and} \\ x\geq 12 \end{cases}$, i.e. $x\geq 12$ or, $(gof)(x)=\sqrt{\sqrt{x+4}-4},\ \forall x\in[12,\infty)$

2. If $f(x)=1+x, 0\leq x\leq 1$ and $g(x)=2-x, 1\leq x\leq 2$ find $(gof)(x)$.

Solutions:

(i) $(gof)(x)=gof(x)$

$=2-f(x), 1\leq f(x)\leq 2$ and $0\leq x\leq 1$

$=2-(1+x), 1\leq 1+x\leq 2$ and $0\leq x\leq 1$

$(\because f(x)=1+x)$

$=1-x, 1\leq 1+x\leq 2$ and $0\leq x\leq 1$

(ii) We solve the inequality $1\leq 1+x\leq 2$ for x:

$1\leq 1+x\leq 2$

$\Leftrightarrow 1-1\leq 1+x-1\leq 2-1$

$\Leftrightarrow 0\leq x\leq 1$

(iii) $D(gof)=[0,1]\cap[0,1]=[0,1]$

(iv) $(gof)(x)=1-x, 0\leq x\leq$

3. If $f(x)=x^2, 0\leq x\leq 1$ and $g(x)=1-x, 0\leq x\leq 1$ find $(gof)(x)$.

Solutions:

(i) $(gof)(x) = gof(x)$
$= 1 - f(x), 0 \le f(x) \le 1$ and $0 \le x \le 1$
$= 1 - x^2, 0 \le x^2 \le 1$ and $0 \le x \le 1$ $(\because f(x) = x^2)$

(ii) We solve the inequality $0 \le x^2 \le 1$ for x

$\therefore 0 \le x^2 \le 1$

$\Leftrightarrow -1 \le x \le 1$

(iii) $D(gof) = [-1, 1] \cap [0, 1]$

(iv) $(gof)(x) = 1 - x^2,\ x \in [-1, 1] \cap [0, 1]$, i.e; $x \in [0, 1]$

4. If $f : [0, 1] \to [0, 1]$ be defined by $f(x) = \dfrac{1-x}{1+x}$, $0 \le x \le 1$ and $g : [0, 1] \to [0, 1]$ be defined by $g(x) = 4x(1 - x), 0 \le x \le 1$, find $(gof)(x)$.

Solutions:

(i) $(gof)(x) = gof(x)$

$$= go\left(\frac{1-x}{1+x}\right), 0 \le \frac{1-x}{1+x} \le 1 \text{ and } 0 \le x \le 1$$

$$= 4\left(\frac{1-x}{1+x}\right)\left[1 - \left(\frac{1-x}{1+x}\right)\right], 0 \le \frac{1-x}{1+x} \le 1 \text{ and } 0 \le x \le 1$$

$$= 4\left(\frac{1-x}{1+x}\right)\left(\frac{2x}{1+x}\right), 0 \le \frac{1-x}{1+x} \le 1 \text{ and } 0 \le x \le 1$$

(ii) We solve the inequality $0 \le \dfrac{1-x}{1+x} \le 1$

$$0 \le \frac{1-x}{1+x} \le 1 \Leftrightarrow 0 \le \frac{1-x}{1+x} \text{ and } \frac{1-x}{1+x} \le 1$$

Now considering the inequality $\dfrac{1-x}{1+x} \le 1$,

$$\frac{1-x}{1+x} \le 1 \Leftrightarrow 1 - x \le 1 + x \Leftrightarrow 0 \le 2x \Leftrightarrow 0 \le x \quad \text{...(a)}$$

Again considering the inequality $\dfrac{1-x}{1+x} \ge 0$,

$$\frac{1-x}{1+x} \ge 0 \Leftrightarrow (1 - x) \ge 0\ (\because 0 \le x \le 1) \Leftrightarrow -x \ge -1 \Leftrightarrow x \le 1 \quad \text{...(b)}$$

Hence, (a) and (b)

$\Rightarrow x \in [0, 1]$, i.e. $0 \le x \le 1$.

(iii) $D(gof) = [0, 1] \cap [0, 1] = [0,1]$

(iv) $(gof)(x) = 4\left(\dfrac{1-x}{1+x}\right)\left(\dfrac{2x}{1+x}\right), 0 \le x \le 1$

Type 3: **Problems based on finding the composite of two piecewise functions.**

Rule: For finding $gof(x)$ defined as under:

$$f(x) = \begin{cases} f_1(x), a < x < b \\ f_2(x), b < x < c \end{cases}$$

and

$$g(x) = \begin{cases} g_1(x), \alpha < x < \beta \\ g_2(x), \beta < x < \delta \end{cases}$$

one must put $y = f(x)$ and hence to find $g(y)$ when $\alpha < y < \beta$ and $\beta < y < \delta$ for which it is required to be determined the intersection of each two intervals given below:

(i) $a < x < b$ and $\alpha < y < \beta$ where $y = f_1(x)$ defined in the given interval namely $a < x < b$.

(ii) $b < x < c$ and $\alpha < y < \beta$ where $y = f_2(x)$ defined in the given interval namely $b < x < c$.

(iii) $a < x < b$ and $\beta < y < \delta$ where $y = f_1(x)$ defined in the given interval namely $a < x < b$.

(iv) $b < x < c$ and $\beta < y < \delta$ where $y = f_2(x)$ defined in the given interval namely $b < x < c$.

If the intersection of any two intervals mentioned from (i) to (iv) is finite, then $g(y) = gof(x)$ is defined and if their intersection is ϕ, then $g(y) = gof(x)$ is not defined.

The union of finite intersection of any two intervals mentioned from (i) to (iv) is the required domain of the composite of two given piecewise functions.

Similarly for finding $fog(x)$, one must put $y = g(x)$ and so to find $f(y)$ where $a < y < b$ and $b < y < c$ for which it is required to be determined the intersection of the intervals given below:

(i) $\alpha < y < \beta$ and $a < y < b$ where $y = g_1(x)$ defined in the given interval $\alpha < y < \beta$.

(ii) $\beta < y < \delta$ and $a < y < b$ where $y = g_2(x)$ defined in the given interval $\beta < y < \delta$.

(iii) $\alpha < y < \beta$ and $b < y < c$ where $y = g_1(x)$ defined in the given interval $\alpha < y < \beta$.

(iv) $\beta < y < \delta$ and $b < y < c$ where $y = g_2(x)$ defined in the given interval $\beta < y < \delta$.

If the intersection of any two intervals mentioned from (i) to (iv) is finite, then $f(y) = fog(x)$ is defined and if their intersection is Ø, then $f(y) = fog(x)$ is not defined.

Solved Examples

1. Two functions are defined as under:

$$f(x) = \begin{cases} x + 1, & x \le 1 \\ 2x + 1, & 1 < x \le 2 \end{cases} \text{ and}$$

$$g(x) = \begin{cases} x^2, & -1 \le x < 2 \\ x + 2, & 2 \le x \le 3 \end{cases} \text{ find } fog \text{ and } gof.$$

Solution: $f(g(x)) = f(y)$ where $y = g(x)$

$f(y) = y + 1, y \le 1$ ...(a)

$= 2y + 1, 1 < y < 2$ ...(b)

$g(x) = x^2, -1 \le x < 2$ ...(c)

$= x + 2, 2 \le x \le 3$ ...(d)

Case (i): Considering the intervals in (a) and (c), $-1 \le x < 2$ and $y \le 1$

$\Rightarrow -1 \le x < 2$ and $g(x) \le 1$

$\Rightarrow -1 \le x < 2$ and $x^2 \le 1$

$(\because g(x) = x^2 \text{ in } -1 \le x < 2)$

$\Rightarrow -1 \le x < 2$ and $-1 \le x \le 1$

$\Rightarrow [-1, 2) \cap [-1, 1] = [-1, 1] = -1 \le x \le 1$

$\therefore$ For $-1 \le x \le 1$, $f(g(x)) = f(y) = y + 1$

$= g(x) + 1$

$= x^2 + 1 \left(\because g(x) = x^2 \text{ in } -1 \le x < 2\right)$

Case (ii): Considering the intervals in (a) and (d), $2 \le x \le 3$ and $y \le 1$

$\Rightarrow 2 \le x \le 3$ and $g(x) \le 1$

$\Rightarrow 2 \le x \le 3$ and $x + 2 \le 1$ ($\because g(x) = x + 2$ in $2 \le x \le 3$)

$\Rightarrow 2 \le x \le 3$ and $x \le -1$

$\Rightarrow [2, 3] \cap (-\infty, -1] = \phi$

$\Rightarrow f(y) = f(g(x))$ is not defined

Case (iii): Considering the intervals in (b) and (c), $-1 \le x < 2$ and $1 < y \le 2$

$\Rightarrow -1 \le x < 2$ and $1 < x^2 \le 2$ ($\because g(x) = x^2$ in $-1 \le x < 2$)

$\Rightarrow -1 \le x < 2$ and $\{1 < x < \sqrt{2}$ or $-\sqrt{2} \le x < -1\}$

$\Rightarrow [-1, 2) \cap \left(1, \sqrt{2}\right) = \left(1, \sqrt{2}\right] = 1 < x < \sqrt{2}$

In $\left(1 < x < \sqrt{2}\right]$, $f(g(x)) = f(y) = 2y + 1$ for $1 < y \le 2$

$= 2g(x) + 1 = 2x^2 + 1$

Case (iv): Considering the intervals in (b) and (d), $2 \le x \le 3$ and $1 < y \le 2$

$\Rightarrow 2 \le x \le 3$ and $1 < g(x) \le 2$

$\Rightarrow 2 \le x \le 3$ and $1 < x + 2 \le 2$ ($\because g(x) = x + 2$ in $2 \le x \le 3$)

$\Rightarrow 2 \le x \le 3$ and $-1 < x \le 0$

$\Rightarrow [2, 3] \cap (-1, 0] = \varnothing$

$\Rightarrow f(g(x)) = f(y)$ is not defined.

Hence $fog(x) = x^2 + 1$, for $-1 \le x \le 1$, $= 2x^2 + 1$ for $1 < x \le \sqrt{2}$.

Next, $g(f(x)) = g(y)$ where $y = f(x)$

$f(x) = x + 1, x \le 1$ ...(a)

$= 2x + 1, 1 < x \le 2$ ...(b)

$g(y) = y^2, -1 \le y < 2$ ...(c)

$= y + 2, 2 \le y \le 3$ ...(d)

Case (i): Considering the intervals in (a) and (c), $x \le 1$ and $-1 \le y < 2$

$\Rightarrow x \le 1$ and $-1 \le f(x) < 2$

$\Rightarrow x \le 1$ and $-1 \le x + 1 < 2$

$\Rightarrow x \le 1$ and $-2 \le x < 1$

$\Rightarrow (-\infty, 1] \cap [-2, 1] = [-2, 1] = -2 \le x < 1$

In $(-2 \le x < 1)$, $g(f(x)) = g(y)$ for $-1 \le y < 2$

$= y^2$

$= (x + 1)^2$ ($\because y = f(x) = x + 1$ in $x \le 1$).

Case (ii): Considering the intervals in (a) and (d), $x \le 1$ and $2 \le y \le 3$

$\Rightarrow x \le 1$ and $2 \le f(x) \le 3$

$\Rightarrow x \le 1$ and $2 \le f(x) \le 3$ ($\because f(x) = x + 1$ in $x \le 1$)

$\Rightarrow x \le 1$ and $1 \le x \le 2$

$\Rightarrow (-\infty, 1] \cap [1, 2] = \{1\}$

$\therefore$ when $x = 1$, $g(f(x) = g(y)$

$= y + 2$ for $2 \le y \le 3$

$= x = 1 + 2$ ($\therefore y = x + 1$ in $x \le 1$)

$= x + 3 = 4$ for $x = 1$

Case (iii): Considering the intervals in (b) and (c), $1 \le x \le 2$ and $-1 \le y \le 2$

$\Rightarrow 1 \le x \le 2$ and $-1 \le f(x) \le 2$

$\Rightarrow 1 \le x \le 2$ and $-1 \le 2x + 1 \le 2$ ($\because f(x) = 2x + 1$ in $1 \le x \le 2$)

$\Rightarrow 1 \le x \le 2$ and $-2 \le 2x \le 1$

$\Rightarrow 1 \le x \le 2$ and $-1 \le x \le \frac{1}{2}$

$\Rightarrow [1, 2] \cap \left[-1, \frac{1}{2}\right] = \phi$

$\Rightarrow g(f(x)) = g(y)$ is not defined.

Case (iv): Considering the intervals in (b) and (d), $1 < x \le 2$ and $2 \le y \le 3$

$\Rightarrow 1 < x \le 2$ and $2 \le f(x) \le 3$

$\Rightarrow 1 < x \le 2$ and $2 \le 2x + 1 \le 3$ ($\because f(x) = 2x + 1$ in $1 \le x \le 2$)

$\Rightarrow 1 < x \le 2$ and $\frac{1}{2} \le x \le 1$

$\Rightarrow (1, 2] \cap \left[\frac{1}{2}, 1\right] = \varnothing$

$\Rightarrow g(f(x)) = g(y)$ is not defined.

Hence $gof(x) = (x + 1)^2$ for $-2 \le x < 1$ and $= 4$ for $x = 1$.

2. If $f(x) = x^3 + 1, x < 0$

$= x^2 + 1, x \ge 0$

and $g(x) = (x - 1)^{\frac{1}{3}}, x < 1$

$= (x - 1)^{\frac{1}{2}}, x \ge 1$

compute $gof(x)$.

Solution: $g(f(x)) = g(y)$ where $y = f(x)$

$g(y) = (y - 1)^{\frac{1}{3}}, y < 1$ ...(a)

$= (y - 1)^{\frac{1}{2}}, y \ge 1$ ...(b)

and $f(x) = x^3 + 1, x < 0$ ...(c)

$= x^2 + 1, x \ge 1$ ...(d)

Case (i): Considering the intervals in (a) and (c), $x < 0$ and $y < 1$

$\Rightarrow x < 0$ and $f(x) < 1$

$\Rightarrow x < 0$ and $x^3 + 1 < 1$ ($\because f(x) = x^3 + 1$ in $x < 0$)

$\Rightarrow x < 0$ and $x^3 < 0$

$\Rightarrow$ common region $= (-\infty, 0) = x < 0$

$\therefore$ in $(x < 0)$, $g(f(x)) = g(y)$

$= (y - 1)^{\frac{1}{3}}$ for $y < 1$

$= (f(x) - 1)^{\frac{1}{3}}$

$= \left(x^3 + 1 - 1\right)^{\frac{1}{3}}$, ($\because f(x) = x^3 + 1$ in $x < 0$)

$= x$

Case (ii): Considering the intervals in (b) and (c), $x < 0$ and $y \geq 1$

$\Rightarrow x < 0$ and $f(x) \geq 1$

$\Rightarrow x < 0$ and $x^3 + 1 \geq 1$ ($\because f(x) = x^3 + 1$) in $x < 0$)

$\Rightarrow x < 0$ and $x^3 \geq 0$

$\Rightarrow x < 0$ and $x \geq 0$

$\Rightarrow (-\infty, 0) \cap [0, \infty) = \phi$

$\Rightarrow g(y) = g(f(x))$ is not defined.

Case (iii): Considering the intervals in (a) and (d), $x \geq 0$ and $y < 1$

$\Rightarrow x \geq 0$ and $f(x) < 1$

$\Rightarrow x \geq 0$ and $x^2 + 1 < 1$ ($\because f(x) = x^2 + 1$ in $x \geq 0$)

$\Rightarrow x \geq 0$ and $x^2 < 0$

$\Rightarrow x \geq 0$ and $x < 0$

$\Rightarrow [0, \infty) \cap (-\infty, 0) = \phi$

$\Rightarrow g(y) = g(f(x))$ is not defined.

Case (iv): Considering the intervals in (b) and (d), $x \geq 0$ and $y \geq 1$

$\Rightarrow x \geq 0$ and $f(x) \geq 1$

$\Rightarrow x \geq 0$ and $x^2 + 1 \geq 1$ ($\because f(x) = x^2 + 1$ in $x \geq 0$)

$\Rightarrow x \geq 0$ and $x^2 \geq 0$

$\Rightarrow x \geq 0$ and $x \geq 0$

$\Rightarrow$ common region $= [0, \infty) \Rightarrow x \geq 0$

$\therefore$ in $(x \geq 0)$, $(g(f(x)) = g(y)$

$= (y - 1)^{\frac{1}{2}}$ for $x \geq 0$

$= (f(x) - 1)^{\frac{1}{2}}$

$= \left(x^2 + 1 - 1\right)^{\frac{1}{2}}$ ($\because f(x) = x^2 + 1$ in $x \geq 0$)

$= x$, Hence $gof(x) = x$ for all x.

3. If $f(x) = x^2 - 4x + 3, x < 3$
$= x - 4, x \geq 3$
and $g(x) = x - 3, x < 4$
$= x^2 + 2x + 2, x \geq 4$
describe the function *fog*.

Solution: $fog(x) = f(y)$, where $y = g(x)$

$f(y) = y^2 - 4y + 3, y < 3$ ...(a)

$= y - 4, y \geq 3$ ...(b)

$g(x) = x - 3, x < 4$ ...(c)

$= x^2 + 2x + 2, x \geq 4$ ...(d)

Case (i): Considering the intervals in (a) and (c), $x < 4$ and $y < 3$

$\Rightarrow x < 4$ and $g(x) < 3$

$\Rightarrow x < 4$ and $x - 3 < 3$ ($\because g(x) = x - 3$ in $x < 4$)

$\Rightarrow x < 4$ and $x < 6$

$\Rightarrow (-\infty, 4) \cap (-\infty, 6) = (-\infty, 4) \Rightarrow x < 4$

$\therefore$ in $(x < 4)$, $fog(x) = f(y)$

$= y^2 - 4y + 3$ for $y < 3$

$= (x - 3)^2 - 4(x - 3) + 3$

($\because y = g(x) = x - 3$ in $x < 4$)

$= x^2 - 10x + 24$

Case (ii): Considering the intervals in (a) and (d), $x \geq 4$ and $y < 3$

$\Rightarrow x \geq 4$ and $g(x) < 3$

$\Rightarrow x \geq 4$ and $x^2 + 2x + 2 < 3$ ($\because g(x) = x^2 + 2x + 2$ in $x \geq 4$)

$\Rightarrow x \geq 4$ and $(x^2 + 2x - 1) < 0$

$\Rightarrow x \geq 4$ and $(x + 1)^2 - 2 < 0$

$\Rightarrow x \geq 4$ and $(x + 1)^2 < 2$

$\Rightarrow x \geq 4$ and $|x + 1| < \sqrt{2}$

$\Rightarrow x \geq 4$ and $-\sqrt{2} < x + 1 < \sqrt{2}$

$\Rightarrow x \geq 4$ and $-\sqrt{2} - 1 < x < \sqrt{2} - 1$

$\Rightarrow [4, \infty) \cap \left(-\sqrt{2} - 1, \sqrt{2} - 1\right) = \phi$

$\Rightarrow f(g(x)) = f(y)$ is not defined.

Case (iii): Considering the intervals in (b) and (d), $x \geq 4$ and $y \geq 3$

$\Rightarrow x \geq 4$ and $g(x) \geq 3$

$\Rightarrow x \geq 4$ and $x^2 + 2x + 2 \geq 3$

$\Rightarrow x \geq 4$ and $\left(x^2 + 2x - 1\right) \geq 0$

$\Rightarrow x \geq 4$ and $(x + 1)^2 - 2 \geq 0$

$\Rightarrow x \geq 4$ and $(x + 1)^2 \geq 2$

$\Rightarrow x \geq 4$ and $|x + 1| \geq \sqrt{2}$

$\Rightarrow x \geq 4$ and $\{x + 1 \geq \sqrt{2}$ or $x + 1 \leq -\sqrt{2}\}$

$\Rightarrow x \geq 4$ and $\{x \geq \sqrt{2} - 1$ or $x \leq -\sqrt{2} - 1\}$

$\Rightarrow x \geq 4$ and $x \geq \sqrt{2} - 1$

or $x \geq 4$ and $x \leq -\sqrt{2} - 1$

$\Rightarrow [4, \infty) \cap \left[\sqrt{2} - 1, \infty\right) = [4, \infty)$

or $[4, \infty) \cap \left(-\infty, -\sqrt{2} - 1\right] = \phi$

$\therefore$ in $(x \geq 4)$, $f(g(x)) = f(y)$

$= y - 4$ for $y \geq 3$

$= x^2 + 2x + 2 - 4$ ($\because y = x^2 + 2x + 2$ in $x \geq 4$)

$= x^2 + 2x + 2$

Case (iv): Considering the intervals in (b) and (c), $x < 4$ and $y \geq 3$

$\Rightarrow x < 4$ and $g(x) \geq 3$

$\Rightarrow x < 4$ and $x - 3 \geq 3$ ($\because g(x) = x - 3$ in $x < 4$)

$\Rightarrow x < 4$ and $x \geq 6$

$\Rightarrow (-\infty, 4) \cap [6, \infty) = \varnothing$

$\Rightarrow f(g(x)) = f(y)$ is not defined

Hence $fog(x) = x^2 - 10x + 24$ for $x < 4$ and $x^2 + 2x + 2$ for $x \geq 4$.

4. Find $f(f(x))$ if $f(x) = \begin{cases} 1 + x, & 0 \leq x \leq 2 \\ 3 - x, & 2 < x \leq 3 \end{cases}$.

Solution: Given:

$$f(x) = \begin{cases} 1 + x, & 0 \leq x \leq 2 \quad \ldots(a) \\ 3 - x, & 2 < x \leq 3 \quad \ldots(b) \end{cases}$$

To find: $f(f(x))$.

$$\text{Let } y = f(x) = \begin{cases} 1 + x, & 0 \leq x \leq 2 \\ 3 - x, & 2 < x \leq 3 \end{cases}$$

$$\therefore f(y) = \begin{cases} 1 + y, & 0 \leq y \leq 2 \quad \ldots(c) \\ 3 - y, & 2 < y \leq 3 \quad \ldots(d) \end{cases}$$

Case (i): Considering the intervals in (a) and (c), $0 \leq x \leq 2$ and $0 \leq y \leq 2$

$\Rightarrow 0 \leq x \leq 2$ and $0 \leq 1 + x \leq 2$

$\Rightarrow 0 \leq x \leq 2$ and $-1 \leq x \leq 1$

$\Rightarrow [0, 2] \cap [-1, 1] = [0, 1]$

$\therefore$ in $[0, 1]$, $f(y) = f(f(x)) = 1 + y$

$= 1 + 1 + x$

$= 2 + x$

Case (ii): Considering the intervals in (a) and (d), $0 \leq x \leq 2$ and $2 < y \leq 3$

$\Rightarrow 0 \leq x \leq 2$ and $2 < 1 + x \leq 3$

$\Rightarrow 0 \leq x \leq 2$ and $1 < x \leq 2$

$\Rightarrow [0, 2] \cap (1, 2] = (1, 2]$

$\therefore f(y) = f(f(x)) = 3 - y = 3 - 1 - x = 2 - x$ for $1 < x \le 2$

Case (iii): Considering the intervals in (b) and (c), $2 < x \le 3$ and $0 \le 3 - x \le 2$

$\Rightarrow 2 < x \le 3$ and $-3 \le -x \le -1$

$\Rightarrow 2 < x \le 3$ and $1 \le x \le 3$

$\Rightarrow (2, 3] \cap [1, 3] = (2, 3]$

$\therefore$ in $(2, 3]$, $f(y) = f(f(x)) = 1 + y = 1 + 3 - x = 4 - x$

Case (iv): Considering the intervals in (b) and (d), $2 \le x \le 3$ and $2 \le y \le 3$

$\Rightarrow 2 \le x \le 3$ and $2 \le 3 - x \le 3$

$\Rightarrow 2 \le x \le 3$ and $-1 \le -x \le 0$

$\Rightarrow 2 \le x \le 3$ and $0 \le x \le 1$

$\Rightarrow [2, 3] \cap [0, 1] = \varnothing$

$\Rightarrow f(y) = f(f(x))$ is not defined.

Hence,

$$f(f(x)) = \begin{cases} 2 + x, & 0 \le x \le 1 \\ 2 - x, & 1 < x \le 2 \\ 4 - x, & 2 < x \le 3 \end{cases}$$

***Type 4*: Determination of domain and range of composite function**

Rule: Determination of domain of composite function namely *gof* consists of following steps:

1. Determination of the interval namely S = domain of outer function namely $g(x)$ $\cap$ range of inner function namely $f(x)$.
2. Determination of all the elements present in the domain of $f(x)$ whose images form the interval 'S' is the determination of all the elements forming the set which is the required, domain of the composite function namely '*gof*' i.e. the solution set of the interval 'S' where $f(x)$ lies (i.e. the solution set of $f(x) \in S$) is the required domain of the composite function namely '*gof*'.

Next to determine the range of the composite function '*gof*' one should use the rule:

$$a \le f(x) \le b \Rightarrow g(a) \le g(f(x)) \le g(b) \text{ if } g(x)$$

is increasing in $[a, b] = D(gof)$

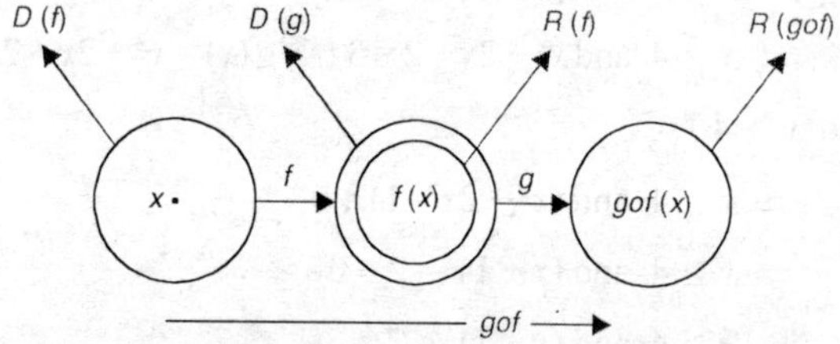

Note: One should note that the rule mentioned in type (4) holds true to determine the domain and range of the composite function if the given composite function is defined by a single formula $y = gof(x)$.

Solved Examples

Find the domain and range of each of the following functions:

(i) $y = \sin \cos x$ **(ii)** $y = \tan \cos x$ **(iii)** $y = \cos \tan x$ **(iv)** $y = \tan \sqrt{\cos x}$ **(v)** $y = \log \sin x$

Solutions: **(i)** $y = \sin \cos x$

$R(\cos x) = [-1, 1]$

$D(\sin x) = R$

$\therefore S = R \cap [-1, 1] = [-1, 1]$

and so $\cos x \in S \Leftrightarrow -1 \le \cos x \le 1$

$\Leftrightarrow x \in R$

Again, $x \in R \Rightarrow -1 \le \cos x \le 1$

$\Rightarrow \sin(-1) \le y \le \sin(1)$ since $\sin x$ is continuous and increasing in $[-1, 1]$.

(ii) $y = \tan \cos x$

$R(\cos x) = [-1, 1]$

$D(\tan x) = R$

$\therefore S = R \cap [-1, 1] = [-1, 1]$

and so $\cos x \in S \Leftrightarrow -1 \le \cos x \le 1$

$\Leftrightarrow x \in R$

Again, $x \in R \Rightarrow -1 \le \cos x \le 1$

$\Rightarrow \tan(-1) \le \tan \cos x \le \tan 1$

(iii) $y = \cos \tan x$

$R(\tan x) = (-\infty, \infty)$

$D(\cos x) = (-\infty, \infty)$

$\therefore \tan x \in S \Rightarrow -\infty < \tan x < \infty$

$\Rightarrow x \in R$

Again, $x \in R \Rightarrow -\infty < \tan x < \infty \Rightarrow \tan x \in R$

$\Rightarrow -1 \le \cos(\tan x) \le 1$

$\therefore R(\cos \tan x) = [-1, 1]$

(iv) $y = \tan \sqrt{\cos x}$

$R\left(\sqrt{\cos x}\right) = [0, 1]$

$D(\tan x) = R$

$\therefore S = R \cap [0, 1] = [0, 1]$

$\therefore \sqrt{\cos x} \in S \Rightarrow 0 \le \sqrt{\cos x} \le 1$

$\Rightarrow 0 \le \cos x \le 1$

$\Rightarrow 2n\pi - \frac{\pi}{2} \le x \le 2n\pi + \frac{\pi}{2}$

$\Rightarrow x \in \left[2n\pi - \frac{\pi}{2}, 2n\pi + \frac{\pi}{2}\right]$

Again, $2n\pi - \frac{\pi}{2} \le x \le 2n\pi + \frac{\pi}{2}$

$\Rightarrow 0 \le \cos x \le 1$

$\Rightarrow 0 \le \sqrt{\cos x} \le 1$

$\Rightarrow \tan 0 \le \tan \sqrt{\cos x} \le \tan 1$ as $\tan x$ is continuous and increasing in [0, 1].

$\therefore R\left(\tan \sqrt{\cos x}\right) = [0, \tan 1]$

(v) $y = \log \sin x$

$R(\sin x) = [-1, 1]$

$D(\log x) = (0, \infty)$

$\therefore S = (0, \infty) \cap [-1, 1] = (0, 1]$

$\therefore \sin x \in S \Rightarrow 0 < \sin x \le 1$

$\Rightarrow 2n\pi < x \le (2n + 1)\pi$

$\Rightarrow x \in (2n\pi, (2n + 1)\pi]$

Again $2n\pi < x \le (2n + 1)\pi$

$\Rightarrow 0 < \sin x \le 1$

$\Rightarrow -\infty < \log \sin x \le 0$ as $\log x$ is continuous and increasing in (0, 1] and $\lim\limits_{\in \to 0+} (\log \in) = -\infty$

$\therefore R(\log \sin x) = (-\infty, 0]$

Even and Odd Functions

Firstly, the definitions of even and odd functions are provided.

(i) Even function: $\forall\, x \in D(f): f(x) = f(-x)$ $\Rightarrow f(x)$ is even, i.e. for $x = a$ = any real number belonging to the domain of definition of $f(x)$, $f(x)$ is defined at $x = a \Rightarrow f(x)$ is also defined at $x = -a$ and $f(a) = f(-a) \Rightarrow f(x)$ is even or more simply, it is the function of an independent variable, changing neither the sign nor absolute value when the sign of the independent variable is changed.

e.g: $x^4 + 2x^2 + 1$; $\cos x$; $\frac{\sin t}{t}$ etc.

Notes:

1. Any algebraic function (or, expression) which contains only even power of x is even.

2. n is even $\Rightarrow x^n$ is even.

3. $y = \cos^n x$ is even whether n is odd or even.

4. $y = \sin^n x$ is even only when n is even.

5. $y = |x|$ is an even function.

(ii) Odd function $\forall\, x \in D(f): f(x) = -f(-x)$ $\Rightarrow f(x)$ is odd, i.e. for $x = a$ = any real number belonging to the domain of definition of $f(x)$, $f(x)$ is defined at $x = a \Rightarrow f(x)$ is also defined at $x = -a$ and $f(a) = -f(-a) \Rightarrow f(x)$ is odd, or more simply, it is the function of an independent variable changing the sign but not absolute value when the sign of the independent variable is changed.

Notes:

1. Any algebraic function (or, expression) which contains only odd power of x is an odd function. (***Note***: The constant function $f(x) = c$ is even and when $c = 0$, i.e. $y = 0$ which is also termed as zero function representing the x-axis is an even function).

2. x^{2n+1} = an odd function.
3. $y = \sin^{(2n+1)} x$ is an odd function.

(iii) ***Properties of even functions***: The sum difference, product and quotient of two even functions is again an even function, i.e. $f(x)$ and $g(x)$ are even

$\Rightarrow f(x) \pm g(x), f(x) \cdot g(x)$ and $\dfrac{f(x)}{g(x)}$ are even.

(iv) ***Properties of odd functions***: **1.** The sum and difference of two odd functions is again an odd function, i.e. $f(x)$ and $g(x)$ are odd $\Rightarrow f(x) \pm g(x)$
2. The product and quotient of two odd functions is again an even function; i.e. $f(x)$ and $g(x)$ are odd

$\Rightarrow f(x) \cdot g(x)$ and $\dfrac{f(x)}{g(x)}$ are even.

How to test whether a given function $y = f(x)$ is odd or even.

Working rule: The rule to examine (or, test) a given function $y = f(x)$ to be odd and even is (i) to replace x by $(-x)$ in the given function $f(x)$ and (ii) to inspect whether $f(x)$ changes its sign or not. It $f(x)$ changes its sign, one must declare it to be odd and if $f(x)$ does not change its sign, one must declare it to be even.

Notes:

1. $f(x) \neq \pm f(x) \Rightarrow f(x)$ is neither even nor odd.

e.g: If the oddness and eveness of the function $f(x) = e^{2x} \sin x$ is examined, it is seen that $f(-x) = e^{-2x} + \sin(-x) \neq \pm f(x)$ which means that $f(x)$ is neither odd nor even.

2. Any function $y = f(x)$ can be uniquely expressed as the sum of an even and odd function as follows:

$$f(x) = \frac{1}{2}[f(x) + f(-x)] + \frac{1}{2}[f(x) - f(-x)]$$

3. A piecewise function is even if each function defined in its domain is even and a piecewise function is odd if each function defined in its domain is odd.

4. f and g are two functions such that

(i) f is even and g is also even $\Rightarrow fog$ is an even function.

(ii) f is odd and g is also odd $\Rightarrow fog$ is an odd function.

(iii) f is even but g is odd $\Rightarrow fog$ is an even function.

(iv) f is odd but g is even $\Rightarrow fog$ is an even function.

Solved Examples

Examine the oddness and eveness of the following functions.

1. $f(x) = x^4 + 2x^2 + 7$

Solution: $f(x) = x^4 + 2x^2 + 7$

$$\Rightarrow f(-x) = (-x)^4 + (-x)^2 + 7$$
$$= x^4 + 2x^2 + 7$$
$$= f(x)$$

Therefore, $f(x)$ is an even function.

2. $g(x) = x^5 - 16x^3 + 11x - \dfrac{92}{x}$

Solution: $g(x) = x^5 - 16x^3 + 11x - \dfrac{92}{x}$

$$\Rightarrow g(-x) = (-x)^5 - 16(-x)^3 + 11(-x) - \frac{92}{(-x)}$$
$$= -\left(x^5 - 16x^3 + 11x - \frac{92}{x}\right)$$
$$= -g(x)$$

Therefore, $g(x)$ is an odd function.

3. $f(x) = x^2 - |x|$

Solution: $f(x) = x^2 - |x|$

$$\Rightarrow f(-x) = (-x)^2 - |-x|$$
$$= x^2 - |x|$$
$$= f(x)$$

Therefore, $f(x)$ is an even function.

4. $f(x) = \log\left(x + \sqrt{x^2 + 1}\right)$

Solution: $f(x) = \log\left(x + \sqrt{x^2 + 1}\right)$

$$\Rightarrow f(-x) = \log\left(-x + \sqrt{(-x)^2 + 1}\right)$$

$$= \log\left(\frac{-x + \sqrt{x^2 + 1}}{x + \sqrt{x^2 + 1}} \cdot x + \sqrt{x^2 + 1}\right)$$

$$= \log\left(\frac{-x^2 + x^2 + 1}{x + \sqrt{x^2 + 1}}\right)$$

$$= \log\left(\frac{1}{x + \sqrt{x^2 + 1}}\right)$$

$$= \log 1 - \log\left(x + \sqrt{x^2 + 1}\right)$$

$$= -\log\left(x + \sqrt{x^2 + 1}\right)$$

$= -f(x)$

So, $f(x)$ is an odd function.

5. $f(x) = x\left(\dfrac{a^x - 1}{a^x + 1}\right)$

Solution: $f(x) = x\left(\dfrac{a^x - 1}{a^x + 1}\right)$

$$\Rightarrow f(-x) = (-x)\left(\frac{a^{-x} - 1}{a^{-x} + 1}\right)$$

$$= -x\left(\frac{a^{\frac{1}{x}} - 1}{a^{\frac{1}{x}} + 1}\right) = -x\left(\frac{1 - a^x}{1 + a^x}\right)$$

$$= x\left(\frac{a^x - 1}{a^x + 1}\right) = f(x)$$

So, $f(x)$ is an even function.

6. $f(x) = \sin x + \cos x$

Solution: $f(x) = \sin x + \cos x$

$\Rightarrow f(-x) = \sin(-x) + \cos(-x)$

$= -\sin x + \cos x$ which is clearly neither equal to $f(x)$ nor equal to $-f(x)$.

Therefore, $f(x)$ is neither even not odd.

7. $f(x) = \begin{cases} x\,|x|, & x \le -1 \\ [1 + x] + [1 - x], & -1 < x < 1 \\ -x\,|x|, & x \ge 1 \end{cases}$

Solution: The given function can be rewritten as under:

$$f(x) = \begin{cases} -x^2, & x \le - \\ 2 + [x] + [-x], & -1 < x < 1 \\ -x^2, & x \ge 1 \end{cases}$$

since $[1 + x] = 1 + [x]$ and $[1 - x] = 1 + [-x]$

Also, it is known that $[x] + [-x] = 0$ when x is an integer and $[x] + [-x] = -1$ when x is not an integer.

Hence, again in the light of above facts, the given function can be rewritten as:

$$f(x) = \begin{cases} -x^2, & x \le -1 \\ 1, & -1 < x < 0 \\ 2, & x = 0 \\ 1, & 0 < x < 1 \\ -x^2, & x \ge 1 \end{cases}$$

which is clearly an even function.

Periodic Functions

Definition: When $f(x) = f(x + P) = f(x + 2P) = \ldots = f(x + nP)$, then $f(x)$ is said to be periodic function of x, for its values repeat, its period being P (where $P \ne 0$, $x \in$ domain of definition of f) which is the smallest positive number satisfying the above property $f(x) = f(x + P) = f(x + 2P) = \ldots = f(x + nP)$ where $n \in Z$, $n \ne 0$.

Notes:

1. The numbers of the form nP, $n \in Z$, $n \ne 0$ are also called period of the function. But generally the smallest positive number P is called the period of the function (or, fundamental period of the function) unless nothing is mentioned about the period of the function.

2. The smallest positive period for sine and cosine is equal to 2π and for the tangent and cotangent, it is equal to π. Since, $\sin x = \sin(x + 2\pi) = \sin(x + 4\pi) = \ldots$, is periodic, its period being 2π. Similarly, $\cos x$ is periodic, its period being 2π. Hence, the periodicity property of the trigonometric functions can be expressed by the following identities.

$$\sin x = \sin(x + 2n\pi),\ n \in Z$$

$$\cos x = \cos(x + 2n\pi),\ n \in Z$$

$$\tan x = \tan(x + n\pi),\ n \in Z$$

$$\cot x = \cot(x + n\pi),\ n \in Z$$

3. The following formulas of the trigonometric functions to find out fundamental period of the trigonometric functions are very useful.

If we have an equation of the form $y = T(Kx)$, $aT(Kx)$, $aT(Kx + b)$, where a = amplitude of trigonometric function, K = any constant, b = any other constant, T = any trigonometric function sin, cos, tan, cot, sec, cosec, then the fundamental period P of the trigonometric function having the form:

(i) $y = \sin(Kx)$, $a\sin(Kx)$ or $a\sin(Kx + b)$ is given by the formula $P = \frac{2\pi}{K}$, where K is any constant.

(ii) $y = \cos(Kx)$, $a\cos(Kx)$ or $a\cos(Kx + b)$ is given by the formula $P = \frac{2\pi}{K}$, where K is any constant.

(iii) $y = \tan(Kx)$, $a\tan(Kx)$ or $a\tan(Kx + b)$ is given by the formula $P = \frac{\pi}{K}$, where K is any constant.

(iv) $y = \cot(Kx)$, $a\cot(Kx)$ or $a\cot(Kx + b)$ is given by the formula $P = \frac{\pi}{K}$, where K is any constant.

(v) $y = \sec(Kx)$, $a\sec(Kx)$ or $a\sec(Kx + b)$ is given by the formula $P = \frac{2\pi}{K}$, where K is any constant.

(vi) $y = \operatorname{cosec}(Kx)$, $a\operatorname{cosec}(Kx)$ or $a\operatorname{cosec}(Kx + b)$ is given by the formula $P = \frac{2\pi}{K}$, where K is any constant.

***Remember*:**

1. Period of any trigonometric function, its co-function and its reciprocal is the same. Thus,

(i) The period of $\sin x$ and $\operatorname{cosec} x = 2\pi$.

(ii) The period of $\cos x$ and $\sec x = 2\pi$.

(iii) The period of $\tan x$ and $\cot x = \pi$.

2. Only those trigonometric functions are periodic whose angles are linear expressions in x (i.e. angle = $ax + b$). For examples, $\sin 3x$, $\cos 4x$, $\tan(4x + 5)$, etc.

3. Those trigonometric functions are not periodic whose angles are not linear expressions in x (i.e., angle $\neq ax + b$). For examples, $\sin\left(\frac{1}{x}\right)$, $\cos\sqrt{x}$, etc.

4. No periodic function other than a constant can be algebraic which means that algebraic functions can not be periodic excepting a constant function.

5. If the function $f_1(x)$ has the period P_1, and the function $f_2(x)$ has the period P_2, then the function $y = a f_1(x) \pm b f_2(x)$ a and b being given numbers, has the period equal to least common multiple (i.e.; l.c.m) of numbers of the set $\{P_1, P_2\}$

e.g.: $y = 2\sin x - 3\tan x$ has the period 2π since

period of $\sin x = 2\pi$

period of $\tan x = \pi$

$\therefore$ L.c.m of $\{2\pi, \pi\} = 2\pi$

6. A function of trigonometric periodic function is also periodic provided the angle = $ax + b$; i.e.; $f(T(ax + b))$ is periodic where f signifies $\sqrt[n]{\ }$, $(\ldots)^n$, $|\ldots|$, log, etc and T signifies sin, cos, tan, cot, sec and cosec.

e.g.: $\sqrt{\tan x}$, $|\cos x|$, $\sin^2(x + b)$, etc are functions of $\sin x$, $\cos x$ and $\tan x$ and are periodic.

7. The sum and difference of periodic and non periodic function is non-periodic. Moreover, the product of a periodic and non-periodic function is non-periodic. e.g.,

(i) $f(x) = \sin x + \cos\sqrt{x}$ is non-periodic since $\sin x$ is periodic and $\cos\sqrt{x}$ is non-periodic.

(ii) $f(x) = x\cos x$ is non-periodic since x being an algebraic function is non-periodic and $\cos x$ is periodic.

8. Whenever, we use the term period, we always mean the fundamental period which is the smallest positive value of P for which the relation:

$f(x) = f(x+P)$ holds true for all values of x, P being a constant, i.e.; the values of $f(x)$ at the points x and $(x+P)$ are same.(***Note***: If $f(x)$ is a periodic function with period T and $g(x)$ is any function such that domain of f is a proper subset of domain g, then gof is period with period T).

e.g.: $y = \sin(x - [x])$ is periodic with period 1, because $(x - [x])$ is periodic with period 1.

Working rule to fund the period: It consists of following steps: (Trigonometric functions)

1. To denote the desired period by P and to replace x by $(P+x)$.

2. To put $T[a(x+P)+b] = T(ax)$

3. To put aP = a constant multiple of $P = 2\pi$ or π according as the given function is sin, cos, tan, cot sec or cosec.

4. To solve $aP = 2\pi$ or π to get the required (or, desired) period of the trigonometric periodic function.

Solved Examples

Find the period of each of the following functions:

1. $y = \sin 3x$

Solution: Method (1)

On denoting the period of the function $y = \sin 3x$ by P, we get $y = \sin 3(x+P) = \sin 3x$

$$\Rightarrow \sin(3x+3P) = \sin 3x$$

$$\Rightarrow 3P = 2\pi$$

$$\Rightarrow P = \frac{2\pi}{3}$$

Method (2)

On using the formula of period of trigonometric function, we get $P = \frac{2\pi}{K}$ where K = multiple of x.

$$= \frac{2\pi}{3}$$

2. $y = \cos\left(\frac{x}{2}\right)$

Solution: Method (1)

On, denoting the period of the function $y = \cos\left(\frac{x}{2}\right)$ we get $y = \cos\left(\frac{x}{2}\right)$

$$\Rightarrow \cos\left(\frac{x+P}{2}\right) = \cos\left(\frac{x}{2}\right)$$

$$\Rightarrow \frac{P}{2} = 2\pi \Rightarrow P = 4\pi$$

Method (2)

Using the formula of period of trigonometric function, we get $P = 2\pi/K$, where $K = \frac{1}{2} =$ multiple of $x = 2\pi / \frac{1}{2} = 4\pi$.

3. $y = \sin 2x + \cos 3x$

Solution: Method (1)

We are required to find out the period of each addend of the given sum function $y = \sin 2x + \cos 3x$

Now, $\sin 2(x+P_1) = \sin 2x$

$$\Rightarrow \sin(2x+2P_1) = \sin 2x \Rightarrow 2P_1 = 2\pi \Rightarrow P_1 = \pi$$

similarly, $\cos 3(x+P_2) = \cos 3x$

$$\Rightarrow \cos(3x+3P_2) = \cos 3x \Rightarrow 3P_2 = 2\pi \Rightarrow P_2 = \frac{2\pi}{3}$$

$$\therefore \text{L.c.m of } \left\{\pi, \frac{2\pi}{3}\right\} = \frac{2\pi}{1} = 2\pi$$

Hence, period of $y = \sin 2x + \cos 3x = 2\pi$

Method (2)

On using the formula of period of trigonometric function $\sin(Kx)$, $P_1 = \frac{2\pi}{K}$, where $K = 2$ = a constant multiple of x.

$$= \frac{2\pi}{2} = \pi$$

Again using the formula of period of trigonometric function $\cos(Kx)$, $P_2 = \frac{2\pi}{K}$, where $K = 3 = a$ constant multiple of x.

$$= \frac{2\pi}{3}$$

$$\therefore \text{ L.c.m of } \left\{\pi, \frac{2\pi}{3}\right\} = 2\pi$$

Hence, the required period of the given sum function $= 2\pi$

4. $y = \sin\left(\frac{3x}{2}\right) + \sin\left(\frac{2x}{3}\right)$

Solution: Method (1)

We find the period of each addend of the given sum function.

$$\sin\left[\frac{3}{2}(x + P_1)\right] = \sin\left(\frac{3x}{2}\right)$$

$$\Rightarrow \sin\left(\frac{3}{2}x + \frac{3}{2}P_1\right) = \sin\left(\frac{3x}{2}\right) \Rightarrow \frac{3P_1}{2} = 2\pi \Rightarrow$$

$$P_1 = \frac{4\pi}{3}$$

Similarly, $\sin\left[\frac{2}{3}(x + P_2)\right] = \sin\left(\frac{2x}{3}\right)$

$$\Rightarrow \sin\left(\frac{2x}{3} + \frac{2}{3}P_2\right) = \sin\left(\frac{2x}{3}\right) \Rightarrow P_2 = 3\pi$$

$$\therefore \text{ L.c.m of } \left\{\frac{4\pi}{3}, 3\pi\right\} = \frac{12\pi}{1} = 12\pi$$

(***Note***: $f(x) = x - [x] \Rightarrow f(x+1) = x + 1 - [x+1] = x + 1 - ([x] + 1) = x - [x] = f(x) \Rightarrow f(x)$ is periodic with period 1).

Method (2)

On using the formula of period of trigonometric function $\sin(Kx)$, $P_1 = \frac{2\pi}{K}$, where $K = \frac{3}{2}$

$$= \frac{2\pi}{\frac{3}{2}} = 2\pi \times \frac{2}{3} = \frac{4}{3}\pi$$

and $P_2 = \frac{2\pi}{K}$, where $K = \frac{2}{3}$

$$= \frac{2\pi}{\frac{2}{3}} = 2\pi \times \frac{3}{2} = 3\pi$$

$$\therefore \text{ L.c.m of } \left\{\frac{4\pi}{3}, 3\pi\right\} = 12\pi$$

Remember:

L.c.m of two or more fractions

$$= \frac{\text{l.c.m of numerators}}{\text{h.c.f of denominators}}$$

Question: How to show the following function to be not periodic:

Answer: To show that a given trigonometric function $f(x)$ is not periodic, we adopt the rule consisting of following steps.

Step 1: Replacing x by $(T + x)$ in the given function and equating it to $f(x)$ which means one should write $f(T + x) = f(x)$.

Step 2: Considering the general solution of

(i) $\sin[f(x+T)] = \sin[f(x)]$, where $f(x) \neq ax + b$ and f signifies the operators $(\ )^n$, $\sqrt[n]{\ }$, etc, is

$$f(x+T) = n\pi + (-1)^n f(x) \quad (\because \sin x = \sin\alpha \Rightarrow x = n\pi + (-1)^n \alpha \ (n = \pm 1, \pm 2, \ldots))$$

(ii) $\cos[f(x+T)] = \cos[f(x)]$, where $f(x) \neq ax + b$ and f signifies the operators $(\ )^n$, $\sqrt[n]{\ }$, etc, is

$$f(x+T) = 2n\pi \pm f(x) \quad (\because \cos x = \cos\alpha \Rightarrow x = 2n\pi \pm \alpha \ (n = \pm 1, \pm 2, \ldots))$$

(iii) $\tan[f(x+T)] = \tan[f(x)]$, where $f(x) \neq ax + b$ and f signifies the operators $(\ )^n$, $\sqrt[n]{\ }$, etc, is

$$f(x+T) = n\pi + f(x) \quad (\because \tan x = \tan\alpha \Rightarrow x = n\pi + \alpha \ (n = \pm 1, \pm 2, \ldots))$$

Step 3: If $T = F(x)$ $f(x)$ is not periodic which means that if there is a least positive value of T not independent of x, then $f(x)$ will not be a periodic function with period of T.

Or, $T = F(x) \Rightarrow f(x) \neq$ periodic. (***Note***: A monotonic function can never be periodic and a periodic function can never be monotonic. That is, monotoncity and periodicity are two properties of functions which can not coexist).

Solved Examples

Show that each of the following functions is not periodic.

1. $f(x) = \sin\sqrt{x}$

Solution: $f(x) = \sin\sqrt{x}$

$$\therefore f(x+T) = f(x) \Rightarrow \sin\sqrt{x+T} = \sin\sqrt{x} \Rightarrow$$

$\sqrt{T+x} = n\pi + (-1)^n\sqrt{x}$ which does not provide us the value of T independent of x. Hence, $f(x)$ is not a periodic function.

2. $f(x) = \cos x^2$

Solution: $f(x) = \cos x^2$

$$\therefore f(x+T) = f(x) \Rightarrow \cos(x+T)^2 = \cos x^2$$

$$\Rightarrow (x+T)^2 = 2n\pi \pm x^2$$

$$\Rightarrow x + T = \sqrt{2n\pi \pm x^2}$$

$$\Rightarrow T = \sqrt{2n\pi \pm x^2} - x$$

which does not provide us the value of T independent of x. Hence, $f(x)$ is not periodic.

Question: Explain what you mean when we say "two functions are equal".

Answer: Two functions $f: D \to C$ and $g: D \to C$ are said to be equal (written as $f = g$) if $f(x) = g(x)$ for all $x \in D$ which signifies that two functions are equal provided their domains are equal as well as their functional values are equal for all values of the argument belonging to their common domain, i.e.; two functions f and g are equal $\Leftrightarrow$ the following two conditions are satisfied.

(i) f and g have the same domain D.

(ii) f and g assume the same (or, equal) value at each point of their common domain, i.e.; $f(x) = g(x)$, $\forall\, x \in D$.

Notes:

1. The above two conditions are criteria to show that given two functions namely f and g are equal.

2. $\dfrac{x}{x} = 1$, provided $x \neq 0$

3. $\dfrac{f(x)}{f(x)} = 1$, provided $f(x) \neq 0$

Working rule to show two functions to be equal: It consists of following steps:

1. To find the domain of each function f and g.

2. To inspect whether dom (f) = dom (g) as well as $f(x) = g(x)$, $\forall\, x \in D$, D being the common (or, same) domain of each function f and g, i.e.; if dom (f) = dom (g) and $f(x) = g(x)$, $\forall\, x \in D$, then we say that $f = g$ and if any one of the two conditions namely dom (f) = dom (g) and $f(x) = g(x)$, $\forall\, x \in D$ is not satisfied, we say that $f \neq g$.

Solved Examples

Examine whether the following functions are equal or not:

1. $f(x) = \sqrt{x^2}$, $g(x) = |x|$

Solution: $f(x) = \sqrt{x^2}$, $g(x) = |x|$

$\sqrt{x^2}$ is defined for all real values of x

$\Rightarrow$ dom $(f) = R$ = the set of all real numbers ...(i)

Again, $\because g(x) = |x|$

$|x|$ is defined for all real values of x

$\Rightarrow$ dom $(g) = R$ = the set of all real numbers ...(ii)

Also, we inspect that $f(x) = g(x) = |x|$ for every real values of x ...(iii) (i), (ii) and (iii) $\Rightarrow \begin{cases} \text{dom}(f) = \text{dom}(g) = R \\ f(x) = g(x), \ \forall \ x \in R, R \end{cases}$

being the common domain

$\Rightarrow f = g$

2. $f(x) = \dfrac{x^2+1}{x^2+1}$, $g(x) = 1$

Solution: $f(x) = \dfrac{x^2+1}{x^2+1}$, $g(x) = 1$

Now, we find the domain of f.

Putting $x^2 + 1 = 0 \Rightarrow x^2 = -1 \Rightarrow x = \pm i =$ imaginary

$\Rightarrow f(x)$ is defined for all real values of x

$\Rightarrow$ dom (f) = R = the set of all real numbers ...(i)

Again, $g(x) = 1$ is defined for all real values of x

$\Rightarrow$ dom (g) = R = the set of all real numbers ...(ii)

Also, we inspect that $f(x) = g(x)$, $\forall \ x \in R$ ...(iii)

(i), (ii) and (iii) $\Rightarrow \begin{cases} \text{dom}(f) = \text{dom}(g) = R \\ f(x) = g(x), \ \forall \ x \in R, R \end{cases}$

being the common domain

$\Rightarrow f = g$

Note: To find the values of the argument x for which two given functions f and g may be equal (i.e.; whenever $f(x)$ and $g(x)$ may be equal by performing any operation or rule like cancellation, extracting the root, etc on any one of the given functional value) means to find the common domain of each function f and g over which they are defined.

Working rule: We have the rule to find for what values of the argument given functions are identical. It says to find the common domain of each function f and g to examine whether their functional values are equal for all values of the argument belonging to the common domain of f and g, i.e.; whether $f(x) = g(x)$, $\forall \ x \in D$, D being the common domain of each function f and g should be examined.

Solved Examples

Find for what values of x following functions are identical.

(i) $f(x) = x$, $g(x) = \sqrt{x^2}$

(ii) $f(x) = \dfrac{x^2}{x}$, $g(x) = x$

(iii) $f(x) = \log_{10} x^2$, $g(x) = 2 \log_{10} x$

Solution: **(i)** $f(x) = x \Rightarrow$ dom (f) = R ...(a)

$g(x) = \sqrt{x^2} = x$ for $x \geq 0$ and $= -x$ for $x < 0$

$\Rightarrow g(x)$ is defined for all real values of x

$\Rightarrow$ dom (g) = R ...(b)

Hence, (a) and (b) $\Rightarrow f(x) = g(x)$, $\forall \ x \in [0, \infty)$

(ii) $f(x) = \dfrac{x^2}{x} = x$, provided $x \neq 0$

$\Rightarrow$ dom (f) = $R - \{0\}$ ...(a)

$g(x) = x$

$\Rightarrow$ dom (g) = R ...(b)

Hence, (a) and (b) $\Rightarrow f(x) = g(x)$, $\forall \ x \in R - \{0\}$

(iii) $f(x) = \log_{10} x^2$, $g(x) = 2 \log_{10} x$, provided $x > 0$ and $2 \log(-x)$ for $x < 0$

$\Rightarrow f(x)$ is defined for all real values of x

$\Rightarrow$ dom (f) = R ...(a)

$g(x) = 2 \log_{10} x$, provided $x > 0$

$\Rightarrow g(x)$ is defined for all positive values of x

$\Rightarrow$ dom(g) = $R^+ = (0, \infty)$ ...(b)

Hence, (a) and (b) $\Rightarrow f(x) = g(x)$, $\forall \ x \in (0, \infty)$

Problems on one-to-one Function

Firstly, we recall the definition of one-to-one (or, simply one-one) function.

***Definition* 1**: It the given function $f: D \rightarrow R$ defined by $y = f(x)$ is such that there is unique y in the range R for each x in the domain D and conversely there is a unique x in the domain D for each y in the range R, then it is said that given function $y = f(x)$ is one-to-one.

***Definition* 2**: (*Set theoretic*): If different elements of domain of the function have different images (or, values) in the range (or, codomain), then it is said that the function is one-to-one.

Or, in other words, a one-one function is a function whose no two different elements in the domain have the same image in the range. Symbolically, this definition is expressed as:

$f: D \to R \subseteq C$ is $1 - 1$ if $f(x_1) = f(x_2) \Rightarrow x_1 = x_2$ or equivalently, $f: D \to R \subseteq C$ is 1–1 when $x_1 \neq x_2 \Rightarrow f(x_1) \neq f(x_2)$

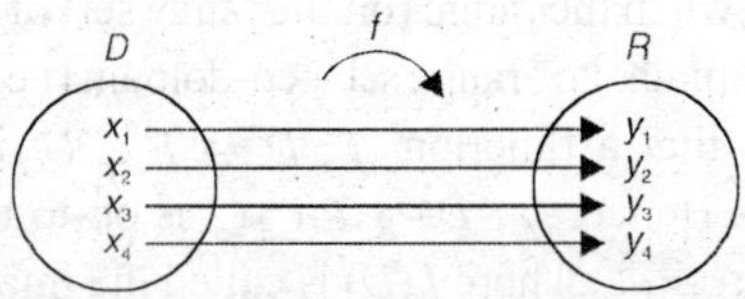

Note: While drawing an arrow diagram fore 1–1 function, all different elements of domain are paired with different elements of the range. If the number of elements of the range are more than the number of elements of domain, then some members of the range will be left.

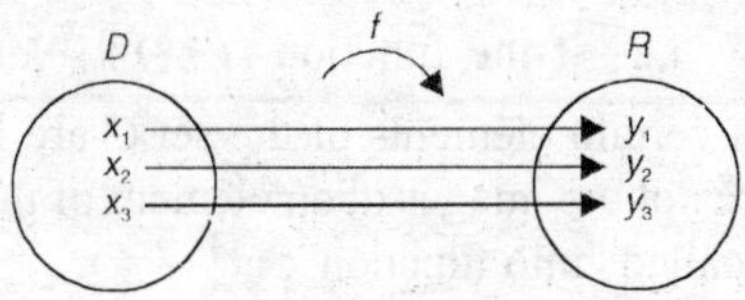

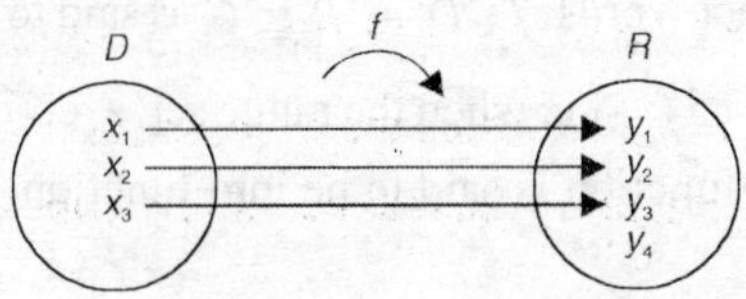

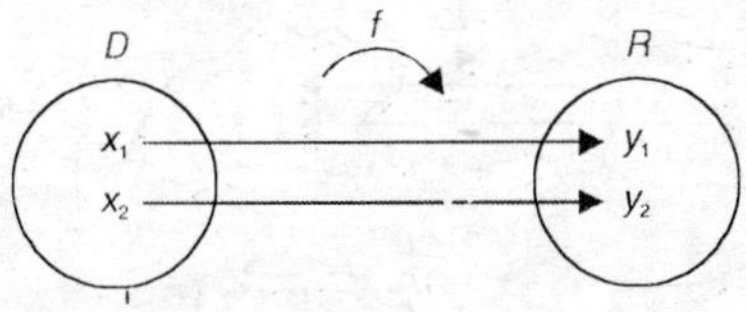

How to Show that a Given Function is one-one.

Method 1: To show that a given function is 1–1, the first rule to be adopted consists of the following steps:

Step 1: To consider the given function $y = f(x)$ for replacing its x by x_1 and x_2 to obtain $f(x_1)$ and $f(x_2)$.
Step 2: To show that $x_1 = x_2$ after solving the equation $f(x_1) = f(x_2)$.

Method 2: It says to show that there are two different images (or, values) in the rang for any two different elements in the domain, i.e. to show $x_1 \neq x_2 \Rightarrow f(x_1) \neq f(x_2),\ \forall\ x_1, x_2 \in D(f) =$ domain of f.

Solved Examples

1. Show that the function $f: R - \{3\} \to R - \{1\}$ defined by $f(x) = \dfrac{x-2}{x-3}$ is one-one.

Solution: For $x_1, x_2 \in D(f)$,

$$f(x_1) = f(x_2) \Rightarrow \frac{x_1 - 2}{x_2 - 3} = \frac{x_2 - 2}{x_2 - 3}$$

$\Rightarrow (x_1 - 2)(x_2 - 3) = (x_1 - 3)(x_2 - 2) \Rightarrow x_1 x_2 - 3x_1 - 2x_2 + 6 = x_1 x_2 - 2x_1 - 3x_2 + 6 \Rightarrow -3x_1 + 2x_1 = -3x_2 + 2x_2 \Rightarrow x_1 = x_2$ f is one-one.

(***Note***: A method to test that a function $y = f(x)$ is not 1–1 is to show that two different elements in the domain have the same image in the range, i.e. $x_1 \neq x_2 \Rightarrow f(x_1) = f(x_2),\ \forall\ x_1, x_2 \in D(f)$).

2. Find whether the following functions are one-one or not in the specified domains:

(i) $f: \left[0, \dfrac{\pi}{2}\right] \to [0, 1]$ defined by $f(x) = \sin^2 x$

(ii) $f: [-3, 3] \to [0, 9]$ defined by $f(x) = x^2$

(iii) $f: R^+ \to R^+$ defined by $f(x) = |x|$ where $R+$ denotes the set of all positive real numbers.

(iv) $f: \left(-\dfrac{\pi}{2}, \dfrac{\pi}{2}\right) \to (-\infty, \infty)$ defined by $f(x) = \tan x$.

Solutions: **(i)** For $x_1, x_2 \in \left[0, \dfrac{\pi}{2}\right]$

$$f(x_1) = f(x_2) \Rightarrow \sin^2 x_1 = \sin^2 x_2$$

$$\Rightarrow \sin x_1 = \sin x_2 \Rightarrow x_1 = x_2$$

$\therefore f$ is one-one in $\left[0, \dfrac{\pi}{2}\right]$

Remark: It the domain of $f(x) = \sin^2 x$ is extended to $[0, \pi]$ then its range is same, i.e. [0, 1], but $f(x) = \sin^2 x$ is not one-one, because for two different points namely $\frac{\pi}{6}$ and $\frac{5\pi}{6}$, $f(x) = \sin^2 x$ has the same value, as it is clear from the following:

$$f\left(\frac{\pi}{6}\right) = \sin^2\left(\frac{\pi}{6}\right) = \left(\frac{1}{2}\right)^2 = \frac{1}{4}$$

$$f\left(\frac{5\pi}{6}\right) = \sin^2\left(\frac{5\pi}{6}\right) = \sin^2\left(\pi - \frac{\pi}{6}\right)$$

$$= \sin^2\left(\frac{\pi}{6}\right) = \frac{1}{4}$$

(ii) $\because -2, 2 \in [-3, 3]$

$\therefore f(-2) = (-2)^2 = 4$

and $f(2) = (2) = 4$

$\Rightarrow f(2) = f(-2)$

Hence, $f(x) = x^2$ is not one-one because for two different values of x in $[-3, 3]$, $f(x) = x^2$ has the same value 4.

(iii) For $x_1, x_2 \in R^+$,

$f(x_1) = f(x_2)$

$\Rightarrow |x_1| = |x_2|$

$\Rightarrow x_1 = x_2$

Hence f is one-one in R^+.

(iv) For $x_1, x_2 \in \left(\frac{-\pi}{2}, \frac{\pi}{2}\right)$;,

$f(x_1) = f(x_2)$

$\Rightarrow \tan x_1 = \tan x_2$

$\Rightarrow x_1 = x_2$

Hence f is one-one in $\left(\frac{-\pi}{2}, \frac{\pi}{2}\right)$.

Problems on on-to Functions

Before doing the problems on on-to functions, we recall its definition. On-to function: If $f: D \to R \subseteq C$ is a function such that every element in C occurs as the image of at least one element of D, then $f: D \to R \subseteq C$ is called an on-to function from D to C.

In other words, $f: D \to R \subseteq C$ is called on-to function when the range (or, the range set) of f equals the co-domain, i.e., range set = co-domain (i.e., $R = C$) signifies that a function $f: D \to R \subseteq C$ is on-to function. Hence, $f: D \to R \subseteq C$ is on-to function $\Leftrightarrow f(D) = C$, where $f(D)$ is called the image of D signifying the set of images of all elements in D and is defined as:

$f(D) = \{f(x): x \in D\}$ = range set = $R(f) = R$ (simply)

Notes:

1. If a function is not on-to, it is called "into function", i.e., if the function $f: D \to R \subseteq C$ is such that certain elements of the set C are left out, which are not the images of an element of the set D, then f is called "into function".

In other words $f: D \to R \subseteq C$ is said to be into, if $f(D) \subseteq C$, i.e; when the range set ≠ co-domain, then the function is said to be into function.

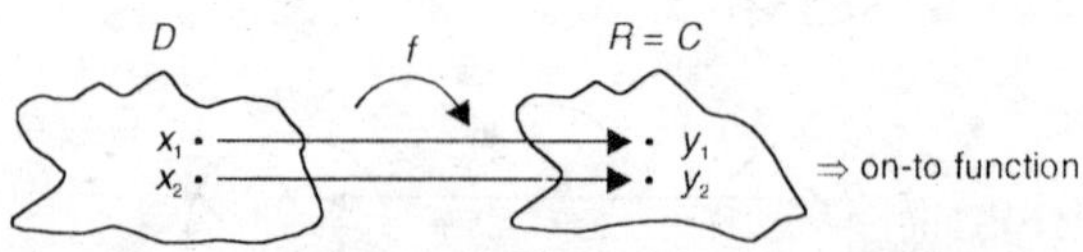

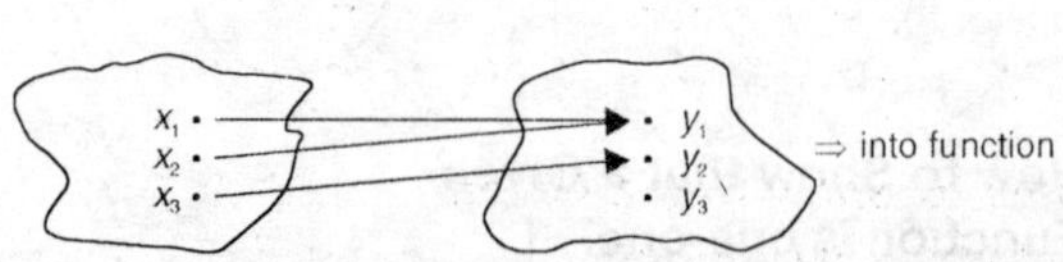

2. A one-to-one function may be into or onto. i.e., there are two types of one-one function namely (i) one-one onto (ii) one-one into, which are defined as:

(i) one-one (symbolised as 1–1) function is called onto provided there is no-element in the range set R which does not appear as an image of a certain element of the domain D or, simply, a function which is both one-one and onto is called one-one onto. Shortly, we write one-one onto ≡ one-one + onto. Further it is notable that one-one onto function satisfies the following properties.

(a) No two element of the domain have the same image.

(b) Every element of the range (or, co-domain) is the image of some element of the domain which means alternatively there is no element on the range (or, co-domain) which is not the image of any element of the domain.

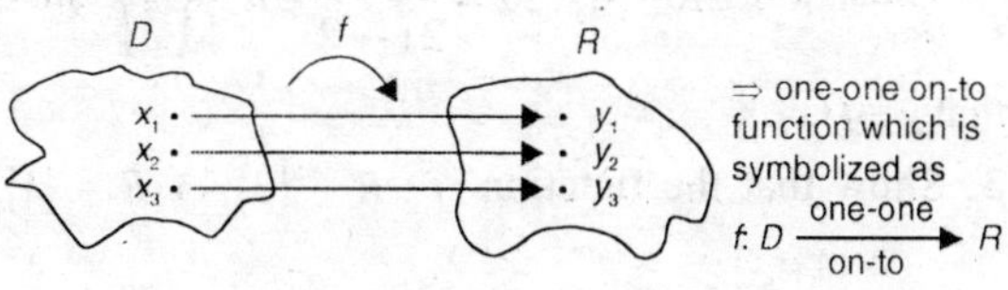

one-one on-to function which is symbolised as

$$f : D \xrightarrow[\text{on-to}]{\text{one-one}} R.$$

(ii) one-one into function: A one-one function is called into provided the range set is contained in the co-domain (i.e., $R \subset C$) such that co-domain contains at least one element which is not an image of any one element in the domain D, then the function is called one-one into, or simply a one-one function is called into provided it is not onto. Shortly we write one-one into ≡ one-one + into. It is notable that one-one into function satisfies the following properties.

(a) No two elements of domain have the same image.

(b) There is at least one element in co-domain which is not the image of any element of the domain.

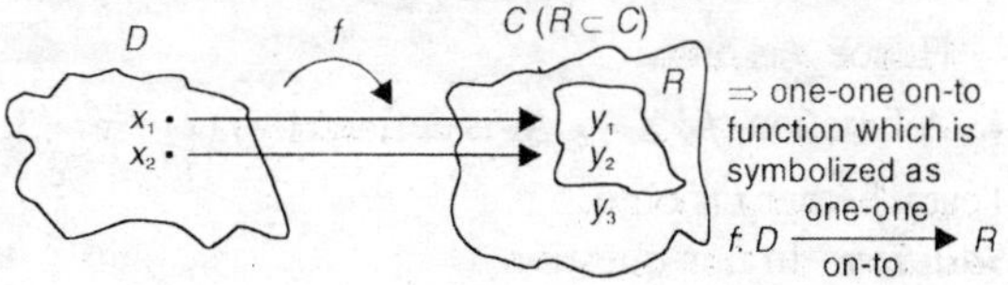

One-one in-to function which is symbolised as:

$$f : D \xrightarrow[\text{in-to}]{\text{one-one}} R \subset C.$$

Remark: When we say that a function is one-one and onto, it is assumed that number of elements of domain and range are equal such that each member of the domain has a different image in the range set whether the domain is a finite set or an infinite set.

Question: How would you show that a given function defined by a single formula $y = f(x)$ in its domain is onto.

Answer: There are mainly two methods to examine whether a given function defined by a single formula $y = f(x)$ in its domain is onto or not.

***Method* 1**: If the range of the function = codomain of the function, then given function f is onto. If the range is proper subset of codomain of the function f, then f is into.

Note: Method (1) is fruitful to examine whether a given function is onto or not only when the domain of the given function is finite and contains a very few elements.

Method 2: To show that f is onto, it is required to be shown that $\forall\ y \in B,\ \exists\ x \in A$ such that $y = f(x)$, where A = domain of f and B = codomain of f.

i.e. one should assume $y \in B$ and should show that $\exists\ x \in A$ such that $y = f(x)$.

Step 1: Choose an arbitrary elements y in B (codomain).

Step 2: Put $f(x) = y$.

Step 3: Solve the equation $y = f(x)$ for x, say $x = g(y)$.

Step 4: If $x = g(y)$ is defined for each $y \in$ codomain of f and $x = g(y) \in$ domain of f for all $y \in$ codomain of f, then f is declared to be on to.

If this requirement is not fulfilled by at least one value of y in the codomain of f, then f is decalred to be into (i.e. not onto).

i.e. to show that f is not onto (i.e. in to), one should point out a single element in the codomain of f which is not the image of any element in the domain.

Notes:

1. Method (2) is fruitful to examine whether a given function is onto or not only when the domain of the given function is infinite or finite but containing very large number of elements.

2. When either domain or codomain or both is not mentioned, it is always understood to be R while examining a given function to be onto or into.

3. An into function can be made onto by redefining the codomain as the range of the original function.

4. Any polynomial function $f: R \to R$ is onto if its degree is odd and any polynomial function $f: R \to R$ is into if its degree is even.

Solved Examples

1. A function $f: R \to R$ is defined by $f(x) = 2x + 3$ examine whether f is onto.

Solution: In this question,

domain of $f = R$

codomain of $f = R$

Let $y \in R$

Now $y = f(x) = 2x + 3 \Leftrightarrow 2x = y - 3 \Leftrightarrow x = \dfrac{y-3}{2}$

$\therefore x = \dfrac{y-3}{2} \in R$

Thus, $\forall\, y \in R$ (codomain), $\exists\, x = \dfrac{y-3}{2} \in R$ (domain) such that $y = f(x)$.

Hence, f is onto.

2. A function $f: R - \left\{\dfrac{3}{2}\right\} \to R - \{0\}$ is defined by $f(x) = \dfrac{5}{2x-3}$. Show that f is onto.

Solution: In this question,

domain of $f = R - \left\{\dfrac{3}{2}\right\}$

codomain of $f = R - \{0\}$

Let $y \in R - \{0\}$, $\therefore y \in R$ and $y \neq 0$

Now $y = f(x) = \dfrac{5}{2x-3} \Leftrightarrow 2x - 3 = \dfrac{5}{y} \Leftrightarrow$

$2x = \dfrac{5}{y} + 3 \Leftrightarrow x = \dfrac{5}{2y} + \dfrac{3}{2}$,

Again, $y \in R$ and $y \neq 0$, $\therefore \dfrac{5}{2y} + \dfrac{3}{2} \in R$

Also, $\dfrac{5}{2y} \neq 0$, $\forall\, y \Rightarrow \dfrac{5}{2y} + \dfrac{3}{2} \neq \dfrac{3}{2}$

$\therefore x = \dfrac{5}{2y} + \dfrac{3}{2} \in R - \left\{\dfrac{3}{2}\right\}$

Thus, $\forall\, y \in R - \{0\}, \exists\, x = \dfrac{5}{2y} + \dfrac{3}{2} \in R - \left\{\dfrac{3}{2}\right\}$ such that $y = f(x)$.

3. Show that the function $f: R - \{3\} \to R - \{1\}$ defined by $f(x) = \dfrac{x-2}{x-3}$ is onto.

Solution: Let $y \in R - \{1\}$, $\therefore y \in R$ and $y \neq 1$

Now $y = f(x) = \dfrac{x-2}{x-3}$

$\Rightarrow yx - 3y = x - 2 \Rightarrow x - yx = 2 - 3y$

$\Rightarrow x(1-y) = 2 - 3y \Rightarrow x = \dfrac{2-3y}{1-y} = g(y)$ (say)

Now $y \in R$ and $y \neq 1$

$\therefore x = \dfrac{2-3y}{1-y}$ is defined $\forall\, y \in R - \{1\}$

Also, $x = \dfrac{2-3y}{1-y} \in R - \{3\}, \forall\, y \in R - \{1\}$

Hence, f is onto.

4. A function $f: Z \to N$ is defined by $f(x) = x^2 + 3$. Test whether f is onto.

Solution: In this question,

Domain of f = the set of integers = Z

Codomain of f = the set of natural numbers = Z

Let $y \in N$

Now, $y = x^2 + 3$

$\Rightarrow x^2 = y - 3 \Rightarrow x = \pm\sqrt{y-3}$ which is not defined for $y < 3$, i.e. $x = \pm\sqrt{y-3} = g(y)$ is not defined $\forall\ y \in N$

Hence, $x = \pm\sqrt{y-3} \notin Z$ if $y<3$

For example $I \in N$ is not the f-image of any $x \in Z$ and hence f is not onto.

5. Let $A = \{x: -1 \le x \le 2\} = B$ and a function $f: A \to B$ is defined by $f(x) = x^2$. Examine whether f is onto.

Solution: In this question,

Domain of f = codomain of $f = -1 \le x \le 2$

Let $x \in [-1, 2] = B$

Now $y = x^2 \Rightarrow x = \pm\sqrt{y}$ which is not defined for $y < 0$.

Clearly, $x \notin [-1, 2] = A$ for all $y \in [-1, 2] = B$

For example $-\dfrac{1}{2} \in B$ is not the f-image of any $x \in A$ and hence f is not onto.

6. If $f: R \to R$ be defined $f(x) = \cos(5x + 2)$, show that $f(x) = \cos(5x+2)$ is not onto.

Solution: Since it is known that

$-1 \le \cos(5x+2) \le 1$ which $\Rightarrow$ range of $\cos(5x+2) = [-1, 1]$

$\therefore$ range of $f = [-1, 1] \ne R$ = codomain of f

$\Rightarrow f: R \to R$ defined by $f(x) = \cos(5x+2)$ is not onto.

7. Let $A = \{x: -1 \le x \le 1\} = B$ and a function $f: A \to B$ is defined by (i) $f(x) = |x|$ (ii) $f(x) = x|x|$. Examine whether it is onto or not.

Solution: **(i)** In this question,

Domain of f = codomain of $f = -1 \le x \le 1$

Let $y \in [-1, 1] = B$ and $y = f(x) = |x|$

Now $y = |x| \Rightarrow y$ is always non-negative

$\Rightarrow$ range of $f = [0, 1]$ which is $\subset B$

$\Rightarrow f$ is not onto.

(ii) $\because y = x|x|$

$\therefore y = x \cdot x$ for $x \ge 0$

$\Rightarrow y = x^2$ for $x \in [0, 1]$, according to question.

$\Rightarrow |x| = \sqrt{y}$

$\Rightarrow x = \sqrt{y}$ for $x \ge 0$

$\Rightarrow x$ is defined for $y \ge 0$

$\Rightarrow x$ is defined for $y \in [0, 1]$ according to question. ...(a)

Again, $\because y = x|x|$

$\therefore y = x(-x)$ for $x < 0$

$\Rightarrow x^2 = -y$ for $x < 0$

$\Rightarrow |x| = \sqrt{-y}$

$\Rightarrow x = -\sqrt{-y}$ for $x < 0$

$\Rightarrow x = -\sqrt{-y}$ for $x \in [-1, 0)$

$\Rightarrow x$ is defined for $y \le 0$

$\Rightarrow x$ is defined for $y \in [-1, 0]$ according to question. ...(b)

From (a) and (b), it is concluded that range of $f = [-1, 0] \cup [0, 1] = [-1, 1] = B$ = codomain of f which

$\Rightarrow f$ is onto.

8. Check whether the function $f: R \to R$ defined by $f(x) = \dfrac{x}{1+|x|}$ is onto or into. Also find $R(f)$.

Solution: Let $y \in R$ = codomain of f

and $y = f(x) = \dfrac{x}{1+|x|}$

Now, $y = \dfrac{x}{1+|x|}$

$\Rightarrow y = \dfrac{x}{1+x}$ for $x \ge 0$

$\Rightarrow y + yx = x$

$\Rightarrow y = x - yx = x(1 - y)$

$\Rightarrow x = \dfrac{y}{1 - y}, (\because y \neq -1)$

Now, $x \geq 0$

$\Rightarrow \dfrac{y}{1 - y} \geq 0$

$\Rightarrow y(1 - y) \geq 0$

$\Rightarrow y(y - 1) \leq 0$

$\Rightarrow 0 \leq y \leq 1$ but $y \neq 1$

$\Rightarrow y \in [0, 1)$ ...(a)

Again, $y = \dfrac{x}{1 + |x|}$

$\Rightarrow y = \dfrac{x}{1 - x}$ for $x < 0$

$\Rightarrow y - yx = x$

$\Rightarrow y = x + yx = x(1 + y)$

$\Rightarrow x = \dfrac{y}{1 + y}, (\because y \neq -1)$

Now, $x < 0$

$\Rightarrow \dfrac{y}{1 + y} < 0$

$\Rightarrow y(y + 1) < 0$

$\Rightarrow y(y - (-1)) < 0$

$\Rightarrow -1 < y < 0 \Rightarrow y \in (-1, 0)$ ...(b)

From (a) and (b), it is concluded that range of f = $(-1, 0) \cup [0, 1) = (-1, 1) \neq R$ = codomain of f

Hence, f is not onto

9. Prove that the function $f : (-1, 1) \to R$ defined by

$f(x) = \dfrac{x}{1 + x}, -1 < x < 0$

$= 0, x = 0$

$= \dfrac{x}{1 - x}, 0 < x < 1$ is onto.

Solution: Onto test:

Case 1: To show that for any $y < 0$, there is $x \in (-1, 0)$ such that $f(x) = y$.

Now, $y = \dfrac{x}{1 + x}$

$\Rightarrow y + xy = x$

$\Rightarrow x(1 - y) = y$

$\Rightarrow x = \dfrac{y}{1 - y}, (\because y \neq 1)$

Now if y is negative (i.e. $y = -z$), then $x = \dfrac{-z}{1 + z}$ (negative) and $|x| = \dfrac{z}{1 + z}$

$\therefore 0 < |x| < 1 \Rightarrow -1 < x < 0$

Hence, for $y < 0$, $\exists x \in (-1, 0)$ such that $f(x) = y$.

Case 2: $y = 0$ for $x = 0$.

Case 3: To show that for any $y > 0$, there is $x \in (0, 1)$ such that $f(x) = y$

Now, $y = \dfrac{x}{1 - x}$

$\Rightarrow y - yx = x$

$\Rightarrow x(1 + y) = y$

$\Rightarrow x = \dfrac{y}{1 + y}, (\because y \neq -1)$

$\therefore$ for $y > 0, 0 < x < 1$

$\Rightarrow$ for any $y > 0$, $\exists x \in (0, 1)$ such that $f(x) = y$

Hence, for any $y \in (-\infty, \infty), \exists x \in (-1, 1)$ such that $f(x) = y$.

Hence, the given function is onto.

Notes:

1. In fact this function is bijective whose one-oneness can be shown in the following way.

To test whether f is one-one, different possible cases are as follows:

Case 1: When $x_1, x_2 \in (-1, 0)$

$f(x_1) = f(x_2)$

$$\Rightarrow \frac{x_1}{1+x_1} = \frac{x_2}{1+x_2}$$

$$\Rightarrow x_1 + x_1 x_2 = x_2 + x_1 x_2$$

$$\Rightarrow x_1 = x_2$$

Case 2: When $x_1, x_2 \in (0, 1)$

$f(x_1) = f(x_2)$

$$\Rightarrow \frac{x_1}{1-x_1} = \frac{x_2}{1-x_2}$$

$$\Rightarrow x_1 - x_1 x_2 = x_2 - x_1 x_2$$

$$\Rightarrow x_1 = x_2$$

Case 3: When $x_1 = 0$ and $x_2 \neq 0$ then $f(x_1) \neq f(x_2)$ as $f(x_1) = 0$, $f(x_2) \neq 0$

Case 4: When $x_1 \in (-1, 0)$ and $x_2 \in (0, 1)$ then in this case clearly $x_1 \neq x_2 \Rightarrow f(x_1) \neq f(x_2)$, as

$$f(x_1) = \frac{x_1}{1+x_1} < 0 \text{ and } f(x_2) = \frac{x_2}{1-x_2} > 0.$$

Hence, the given function is one-one.

2. A piece-wise function is one-one $\Leftrightarrow$ each function defined in its respective sub domains is one-one.
3. A piece-wise function is on-to $\Leftrightarrow$ each function defined in its respective sub domains is on-to.

1. Finding the Value of a Given Function

Type 1: *When the given function is not a piecewise function.*

Exercise 1.1

1. If $f(x) = 3x - 2$, find $f(-1)$.
2. If $f(x) = 3x^2 - 5x + 7$, find $f(-2)$.
3. If $f(x) = x^4 - 3x^2 + 7$, find $f(-1)$ and $f(2)$.
4. If $f(x) = \sqrt{1 - x^2}$, find $f(\sin x)$ and $f\left(\frac{1-x^2}{1+x^2}\right)$.
5. If $f(x) = \sin x + \cos x$, find $f\left(\frac{\pi}{4}\right)$.
6. If $f(x) = x^3 - 2x^2 + x - 1$, find $f(0)$, $f(1)$, $f(-1)$ and $f(2)$.
7. If $f(x) = x^4 - x^3 + 2x^2 + 4$, find $f(0)$, $f(-1)$ and $f(2)$.
8. Given the function $s(t) = t^2 - 6t + 8$, find $s(0)$, $s(2)$ and $s(-1)$.
9. Given the function $f(x) = \frac{3x-5}{x+7}$, find $f(-3)$ and $f(2)$.
10. If $f(x) = 2x^2 - 4x + 1$, find $f(1)$, $f(0)$, $f(2)$, $f(-2)$, $f(a)$ and $f(x+8)$.
11. If $f(x) = (x-1)(x+5)$, compute $f(2)$, $f(1)$, $f(0)$ $f(a+1)$, $f\left(\frac{1}{a}\right)$ and $f(-5)$.
12. If $f(x) = \cos x$, find $f\left(\frac{\pi}{2}\right)$, $f(0)$, $f\left(\frac{\pi}{3}\right)$, $f\left(\frac{\pi}{6}\right)$ and $f(\pi)$.
13. If $f(x) = \frac{\sin x - \cos x}{\sin x + \cos x}$, find $f\left(\frac{\pi}{3}\right)$.
14. If $f(x) = \frac{x^2 - x + 1}{x^2 + x + 1}$, find $f(1+h)$.
15. If $f(x) = \frac{x+1}{x-1}$, find $f(x^2)$ and $(f(x))^2$.
16. When $f(x) = \frac{9}{3+x^2}$ determine $f(0)$, $f(3)$ and $f\left(\frac{1}{x}\right)$.
17. A function f is defined on R as follows

    $f(x) = c$, c being a constant, $x \in R$

    find $f(-1)$, $f(1)$ and $f\left(\frac{3}{2}\right)$.

Answers:

1. $f(-1) = -5$
2. $f(-2) = 29$

3. $f(-1)=5, f(2)=11$

4. $f(\sin x) = \cos x,\ f\left(\dfrac{1-x^2}{1+x^2}\right) = \dfrac{2x}{1+x^2}$

5. $f\left(\dfrac{\pi}{4}\right) = \sqrt{2}$

6. $f(0)=-1, f(1)=-1, f(-)=-5, f(2)=1$
7. $f(0)=4, f(-1)=8, f(2)=20$
8. $s(0)=8, s(2)=0, s(-1)=15$

9. $f(-3) = 3\dfrac{1}{2},\ f(2) = \dfrac{1}{9}$

10. $f(1) = -1, f(0) = 1, f(2) = 1, f(-2) = 17, f(a) = 2a^2 - 4a + 1, f(x+8) = 2x^2 + 28x + 97$
11. $f(2) = 7, f(1) = 0, f(0) = -5, f(a+1) = a^2 + 6a,$

$f\left(\dfrac{1}{a}\right) = \dfrac{1 + 4a - 5a^2}{a^2},\ f(-5) = 0$

12. $f\left(\dfrac{\pi}{2}\right) = 0, f(0)=1,\ f\left(\dfrac{\pi}{3}\right) = \dfrac{1}{2},\ f\left(\dfrac{\pi}{6}\right) = \dfrac{\sqrt{3}}{2},\ f(\pi) = -1$

13. $f\left(\dfrac{\pi}{3}\right) = 2 - \sqrt{3}$

14. $f(1+h) = \dfrac{h^2 + h + 1}{h^2 + 3h + 3}$

15. $f\left(x^2\right) = \dfrac{x^2 + 1}{x^2 - 1}$ and $(f(x))^2 = \left(\dfrac{x+1}{x-1}\right)^2$

16. $f(0) = 3,\ f(3) = \dfrac{3}{4},\ f\left(\dfrac{1}{x}\right) = \dfrac{9x^2}{3x^2 + 1}$

17. $f(-1) = c,\ f(1) = c,\ f\left(\dfrac{3}{2}\right) = 2$

Type 2: When the given function is a piecewise function.

Exercise 1.2

1. If $f(x) = 1 + x$, when $-1 \le x < 0$
$= x^2 - 1$, when $0 < x < 2$
$= 2x$, when $x \geq 0$

find $f(3), f(-2), f(0),\ f\left(\dfrac{1}{2}\right), f(2-h), f(-1+h),$

$f\left(f\left(\dfrac{1}{2}\right)\right)$, where $h > 0$ is sufficiently small.

2. If $f(x) = 2 + 3x$, when $-1 \leq x < 1$
$= 3 - 2x$, when $1 < x \leq 2$
find $f(0), f(1), f(1+h), f(2-h), f(2), f(f(1.5)$
3. If $f(x) = 3^x$, when $-1 < x < 0$
$= 4$, when $0 \leq x < 1$
$= 3x - 1$, when $1 \leq x \leq 3$

find $f(2), f(0), f(0.5), f(-5), f(3)$ and $f\left(f\left(-\dfrac{1}{2}\right)\right)$

4. If $f(x) = \dfrac{|x|}{x}$, when $x \neq 0$
$= 1$, when $x = 0$
find $f(0), f(2), f(-2)$
5. If $f(x) = x$, when $x < 0$
$= x^2$, when $x \leq x < 2$
$= 2x$, when $2 \leq x$

find $f(3), f(-2)$ and $f\left(\dfrac{1}{2}\right)$

6. If $f(x) = \dfrac{\sin x}{x}$, when $x \neq 0$
$= 0$, when $x = 0$

find $f\left(\dfrac{\pi}{3}\right), f(0)$ and $f\left(-\dfrac{\pi}{6}\right)$

7. If $f: R \to R$ where
$f(x) = 2x + 5$, when $x > 4$
$= x^2 - 1$, when $x \in [-9, 9]$
$= x - 4$, when $x < -9$
find $f(-15)$ and $f(f(5))$

8. A function f is defined on R is follows:
$f(x) = 0$ if x is rational
$= 1$ if x is irrational

find $f\left(-\frac{1}{3}\right)$, $f\left(\sqrt{2}\right)$ and $f(0)$

9. If $f(x) = x^2 + 2$, when $0 \le x < 2$
$= 5$, when $x = 2$
$= x - 1$, when $2 < x < 5$
$= x + 1$, when $x > 5$
find $f(-1), f(0), f(2), f(3), f(5), f(7), f(2+h), f(5+h)$ and $f(2-h)$.

10. A function f is defined as follows:
$f(x) = 2x + 6$ for $-3 \le x \le 0$
$= 6$ for $0 < x < 2$
$= 2x - 6$ for $2 \le x \le 5$

find $f(-1)$, $f\left(\frac{1}{2}\right)$ and $f(4)$

Answers:

1. $f(3) = 6, f(-2) =$ not defined, $f(0) =$ not defined
$f\left(\frac{1}{2}\right) = -\frac{3}{4}, f\left(f\left(\frac{1}{2}\right)\right) = \frac{1}{4}, f(-1+h), f(2-h) = h^2 - 4h + 3.$
2. $f(0) = 2, f(1) =$ not defined, $f(1+h) = 1 - 2h$, $f(2-h) = 2h - 1, f(2) = -1, f(f(1.5)) = 2.$
3. $f(2) = 5, f(0) = 4, f(0.5) = 4, f(-5) =$ not defined, $f(3) = 8$, $f\left(f\left(-\frac{1}{2}\right)\right) = f\left(3^{-\frac{1}{2}}\right) = f\left(\frac{1}{\sqrt{3}}\right)\left(\text{as } 0 < \frac{1}{\sqrt{3}} < 1\right)$
4. $f(0) = 1, f(2) = 1, f(-2) = -1.$
5. $f(3) = 6, f(-2) = -2, f\left(\frac{1}{2}\right) = \frac{1}{4}.$
6. $f\left(\frac{\pi}{2}\right) = \frac{2}{\pi}, f(0) = 0, f\left(-\frac{\pi}{6}\right) = \frac{3}{\pi}.$
7. $f(-15) = -19, f(f(5)) = 53.$
8. $f\left(\frac{-1}{3}\right) = 0, f\left(\sqrt{2}\right) = 1, f(0) = 0.$
9. $f(-1)$ not defined, $f(0) = 2, f(1) = 3, f(2) = 5, f(3) = 2, f(5) =$ not defined, $f(7) = 8, f(2+h) = (2+h) - 1, f(5+h) = (5+h) - 1, f(2-h) = (2-h)^2 + 2.$
10. $f(-1) = 4, f\left(\frac{1}{2}\right) = 6, f(4) = 2$

***Type 3**: Problems on showing $f(a) = f(b)$.*

Exercise 1.3

1. Given the function $f(x) = x^4 - x^2 + 1$, show that $f(1) = f(-1)$.
2. Given the function $f(x) = x^4 + x^2 + 5$, show that $f(2) = f(-2)$.
3. Given the function $f(x) = x^3 + x$, show that $f(1) = -f(-1)$.
4. Given the function $f(x) = x^5 + x^3$, show that $f(2) = -f(-2)$.
5. If $f(x) = x^4 - x^2 + 1$, show that $f(-x) = f(x)$.
6. If $f(x) = \sin x + \tan x$, show that $f(-x) = -f(x)$.
7. Given that $f(t) = a^t$, show that $f(x) \cdot f(y) = f(x+y)$.
8. If $f(x) = \frac{1+x^2}{x}$, show that $f\left(\frac{1}{x}\right) = f(x)$.
9. If $f(x) = \frac{x^2}{1-x^2}$, show that $f(\sin\theta) = \tan^2\theta$.
10. If $f(x) = \frac{1-x}{1+x}$, show that $f(\cos\theta) = \tan^2\left(\frac{\theta}{2}\right)$.
11. If $f(x) = \frac{1+x}{1-x}$, show that $f(\tan\theta) = \tan\left(\frac{\pi}{4} + \theta\right)$.
12. Given that $f(x) = \frac{x-1}{x+1}$ show that

    (i) $f\left(\frac{x-1}{x+1}\right) = -\frac{1}{x}$

    (ii) $\frac{f(a) - f(b)}{1 + f(a) \cdot f(b)} = \frac{a-b}{a+b}$
13. If $f(t) = \frac{t}{a^t - 1} + \frac{1}{2}t$, show that $f(t) \neq f(-t)$.

14. If $f(x) = \log x$, show that $f(x \cdot y) = f(x) + f(y)$ and $f(x^m) = mf(x)$.

15. If $f(x) = \dfrac{a^x - a^{-x}}{a^x + a^{-x}}$, show that $f(x + y)$

$$= \frac{f(x) + f(y)}{1 + f(x) \cdot f(y)}.$$

16. If $f(x) = \dfrac{(2x - 1)(x - 2)}{x^2 - 2x + 1}$, show that $f\left(\dfrac{1}{c}\right) =$

$f(c)$.

17. If $f(x) = \sec x + \cos x$, show that $f(-x) = f(x)$.

18. If $f(x) = x^4 + x^2 - 2\cos x$, show that $f(-x) = f(x)$.

19. If $y = f(x) = \dfrac{2x - 1}{x - 2}$, show that $f(y) = x$.

20. If $y = f(x) = \dfrac{x + 1}{2x + 3}$, show that

$$f(y) = \frac{3x + 4}{8x + 11}.$$

21. If $y = f(x) = \dfrac{ax + b}{cx - a}$, show that $f(y) = x$.

22. If $y = f(x) = \dfrac{4x - 3}{3x - 4}$, show that $f(y) = x$.

23. If $f(x) = b \cdot \left(\dfrac{x - a}{b - a}\right) + a\left(\dfrac{x - b}{a - b}\right)$, show that ,

$f(a) + f(b) = f(a + b)$.

24. If $f(x) = \log x$, $x > 0$, show that

(i) $f(x \cdot y) = f(x) + f(y)$

(ii) $f\left(\dfrac{x}{y}\right) = f(x) - f(y)$

(iii) $f(e \cdot x) = f(x) + 1$

(iv) $f(x^n) = nf(x)$

25. If $f(x) = \cos x$, $g(x) = \sin x$, show that

(i) $f(x + y) = f(x) \cdot f(y) - g(x) \cdot g(y)$

(ii) $g(x + y) = g(x) \cdot f(y) + g(y) \cdot f(x)$

2. Examining the Existance of a Function.

Exercise 1.4

1. Show that $f(x) = \dfrac{x^2 + 2x + 3}{x^2 + 4x - 5}$ is non-existent for $x = 1$.

2. Show that $f(x) = \dfrac{x^2 + 5x + 9}{x + 1}$ is not defined for $x = -1$.

3. Show that $f(x) = \dfrac{x^2 + 3x + 1}{4x^2 - 4x + 1}$ is undefined for $x = \dfrac{1}{2}$.

4. Show that $f(x) = \dfrac{x^2 - 3x + 5}{2x^2 + 5x - 3}$ is indeterminate at $x = \infty$.

5. Show that $f(x) = \dfrac{1}{x^2 - 5x + 6}$ is not defined for $x = 2$ and for $x = 3$.

6. Show that $f(x) = \sqrt{(x - 2)(x - 3)}$ is non-existent for any value of x lying between 2 and 3.

7. Prove that $f(x) = \dfrac{1}{\sqrt{(1 - x)(x - 2)(x - 3)}}$ is not defined for $1 < x < 2$.

8. Prove that $f(x) = \sqrt{(1 - x)(x - 2)(x - 3)}$ is not defined for any value of x lying between 1 and 2 but defined for any value of x lying between 2 and 3.

9. Prove that $f(x) = \dfrac{\sin x - \cos x}{1 - \tan x}$ is not defined for $x = \dfrac{\pi}{4}$.

10. Show that $f(x) = \dfrac{\sin x}{1 - \cos x}$ is not defined for $x = 0$.

11. Show that $f(x) = \dfrac{\sin x - \cos x}{1/\sqrt{2} - \cos x}$ does not exist for $x = \dfrac{\pi}{4}$.

12. Show that $f(x) = \dfrac{\sin x - \cos x}{1 - \tan x}$ is non-existent for $x = \dfrac{\pi}{4}$.

13. Show that $f(x) = \dfrac{1}{\cos x} - \tan x$ is undefined at $x = \dfrac{\pi}{2}$.

14. Show that $f(x) = \dfrac{1}{x}$ is not defined at $x = 0$.

15. Examine whether $f(x) = \dfrac{\cos x - \sin x}{1 + \sqrt{2}\cos x}$ exists at $x = \dfrac{\pi}{4}$ and $x = \dfrac{3\pi}{4}$.

16. Find x for which the following functions are not defined.

(i) $y = \dfrac{x + 5}{3x - 4}$

(ii) $y = \dfrac{2}{x}$

(iii) $y = \dfrac{x - 1}{2x - 5}$

(iv) $y = \dfrac{x^2 + 1}{x + 1}$

(v) $y = \dfrac{x^2 + 3x + 2}{x - 2}$

(vi) $y = \dfrac{x^2 + 4x + 3}{x + 3}$

(vii) $y = \dfrac{x^2 + 2}{x^2 - 3x + 2}$

(viii) $y = \tan x$

(ix) $y = \sec x$

(x) $y = \operatorname{cosec} x$

Hint: To find the points at which rational functions of x are not defined, one should put denominator of rational function of x equal to zero and solve for x.

Answers:

15. $f(x)$ exists at $x = \dfrac{\pi}{4}$ and $f(x)$ does not exist at $x = \dfrac{3\pi}{4}$.

16. (i) $x = \dfrac{4}{3}$ **(ii)** $x = 0$ **(iii)** $x = \dfrac{5}{2}$ **(iv)** $x = -1$ **(v)** $x = 2$ **(vi)** $x = -3$ **(vii)** $x = 1, 2$ **(viii)** $x = \dfrac{\pi}{2}$ or $x = \dfrac{3\pi}{2}$ and in general x = any odd multiple of $\dfrac{\pi}{2}$ **(ix)** $x = (2n + 1)\dfrac{\pi}{2}, n \in Z$ **(x)** $x = n\pi, n \in Z$

3. Finding the Domain of a Given Function

3.1. *Finding the domain of a given algebraic function*

Type 1: *When the given function is a polynomial in x.*

Exercise 1.5

Find the domain of each of the following functions:

1. $y = x^2$

2. $y = x^2 - 1$

3. $y = x^3 + 1$

Answers:

1. $(-\infty, \infty)$ 2. $(-\infty, \infty)$ 3. $(-\infty, \infty)$

Type 2: *When the given function is a rational function of x.*

Exercise 1.6

Find the domain of each of the following functions:

1. $y = \dfrac{1}{4x - 2}$

2. $y = \dfrac{x + 2}{2x - 8}$

3. $y = \dfrac{x^2 - 4}{x + 2}$

4. $y = \dfrac{1}{1 - x^2}$

5. $y = \dfrac{1}{x^2 - x - 12}$

6. $y = \dfrac{4x - 1}{3x^2 - 5x - 2}$

7. $y = \dfrac{x - 1}{x^2 - 9x + 20}$

8. $y = \dfrac{x}{x^2 - 3x + 2}$

9. $y = \dfrac{x^2 - 3x + 2}{x^2 + x - 6}$

10. $y = \dfrac{x^2 - 4x + 9}{x^2 + 4x + 9}$

11. $y = \dfrac{x + 5}{3x - 4}$

12. $y = \dfrac{2}{x}$

13. $y = \dfrac{x - 1}{2x - 5}$

14. $y = \dfrac{x^2 + 1}{x + 1}$

15. $y = \dfrac{x^2 + 3x + 2}{x - 2}$

16. $y = \dfrac{x^2 + 4x + 3}{x + 3}$

17. $y = \dfrac{x^2 + 2}{x^2 - 3x + 2}$

Answers:

1. $\left(-\infty, \frac{1}{2}\right) \cup \left(\frac{1}{2}, +\infty\right)$

2. $(-\infty, 4) \cup (4, +\infty)$

3. $(-\infty, -2) \cup (-2, +\infty)$

4. $(-\infty, -1) \cup (-1, 1) \cup (1, +\infty)$

5. $(-\infty, -3) \cup (-3, 4) \cup (4, +\infty)$

6. $\left(-\infty, -\frac{1}{3}\right) \cup \left(-\frac{1}{3}, 2\right) \cup (2, +\infty)$

7. $(-\infty, 4) \cup (4, 5) \cup (5, +\infty)$

8. $R - \{1, 2\}$

9. $(-\infty, -3) \cup (-3, 2) \cup (2, +\infty)$

10. R

11. $R - \left\{\frac{4}{3}\right\}$

12. $R - \{0\}$

13. $R - \left\{\frac{5}{2}\right\}$

14. $R - \{-1)$

15. $R - \{2\}$

16. $R - \{-3\}$

17. $R - \{1, 2\}$.

Type 3: *When the given function is put in the form:*

$y = \sqrt{f(x)}$, *where* $f(x) = ax + b$.

Exercise 1.7

Find the domain of each of the following functions:

1. $y = \sqrt{1 - x}$
2. $y = \sqrt{18 - 6x}$
3. $y = \sqrt{3x - 12}$
4. $y = \sqrt{x - 3}$
5. $y = \sqrt{3 - 2x}$
6. $y = \sqrt{2x - 3}$

Answers:

1. $(-\infty, 1]$ **2.** $(-\infty, 3]$ **3.** $[4, +\infty)$ **4.** $[3, +\infty)$

5. $\left(-\infty, \dfrac{3}{2}\right]$ **6.** $\left[\dfrac{3}{2}, +\infty\right)$.

Type 4: *When the given function is put in the form:*

$y = \sqrt{f(x)}$, *where* $f(x)\ ax^2 + bx + c$.

Exercise 1.8

Find the domain of each of the following functions:

1. $y = \sqrt{9 - 4x^2}$
2. $y = \sqrt{x^2 - 1}$
3. $y = \sqrt{x^2 - 2x - 8}$
4. $y = \sqrt{x^2 + 8x + 15}$
5. $y = \sqrt{(2 - x)(5 + x)}$
6. $y = \sqrt{x^2 - 4x + 3}$
7. $y = \sqrt{3x^2 - 4x + 5}$
8. $y = \sqrt{x^2 - 3x + 2}$
9. $y = \sqrt{2x - x^2}$

Answers:

1. $\left[-\dfrac{3}{2}, \dfrac{3}{2}\right]$

2. $|x| \geq 1$

3. $(-\infty, -2] \cup [4, +\infty)$

4. $(-\infty, -5] \cup [-3, +\infty)$

5. $[-5, 2]$

6. $(-\infty, 1] \cup [3, +\infty) = R - [1, 3]$

7. $(-\infty, \infty)$

8. $(-\infty, 1] \cup [2, +\infty)$.

Type 5: *When the function is put in the form:*

$y = \dfrac{1}{\sqrt{f(x)}}$, *where* $f(x) = ax + b$ *or* $ax^2 + bx + c$.

Exercise 1.9

Find the domain of each of the following functions:

1. $y = \dfrac{1}{\sqrt{x + 2}}$
2. $y = \dfrac{1}{\sqrt{16 - 9x^2}}$
3. $y = \dfrac{1}{\sqrt{x^2 - 3x + 2}}$
4. $y = -\dfrac{1}{\sqrt{-4x^2 + 8x - 3}}$
5. $y = \dfrac{1}{\sqrt{3 + 2x - x^2}}$

6. $y = \dfrac{1}{\sqrt{(1-x)(x-2)}}$

7. $y = \dfrac{1}{\sqrt{6-x}}$

Answers:

1. $(-2, +\infty)$
2. $\left(-\dfrac{4}{3}, \dfrac{4}{3}\right)$
3. $(-\infty, 1) \cup (2, +\infty)$
4. $\left(\dfrac{1}{2}, \dfrac{3}{2}\right)$
5. $(-1, 3)$
6. $(1, 2)$
7. $(-\infty, 6)$.

Type 6: *When the given function is put in the form*:

$$y = \sqrt{\frac{f(x)}{g(x)}}$$

Exercise 1.10

Find the domain of each of the following functions:

1. $y = \sqrt{\dfrac{x-1}{x+1}}$

2. $y = \sqrt{\dfrac{x-8}{12-x}}$

3. $y = \sqrt{\dfrac{4x-8}{3-6x}}$

4. $y = \sqrt{\dfrac{(x+1)(x-3)}{(x-2)}}$

Hint: $\dfrac{(x+1)(x-3)}{x-2} \geq 0,\ x \neq 2$

$\Leftrightarrow (x+1)(x-2)(x-3) \geq 0,\ x \neq 2$

$\Leftrightarrow D(y) = [-1, 2) \cup [3, \infty)$

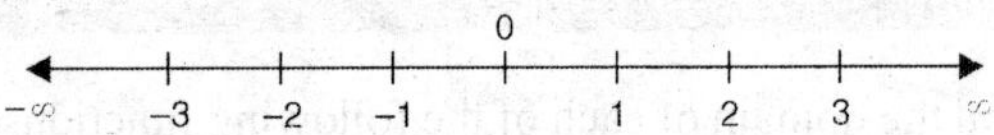

5. $y = \sqrt{\dfrac{x-1}{x-3}}$

6. $y = \sqrt{\dfrac{(x+3)}{(2-x)(x-5)}}$

Hint: $(x+3)(2-x)(x-5) \geq 0$

$\Leftrightarrow x \geq -3, 2, 5$ but $x \neq 2, 5$

$\Leftrightarrow D(y) = (-\infty, -3) \cup (2, 5)$

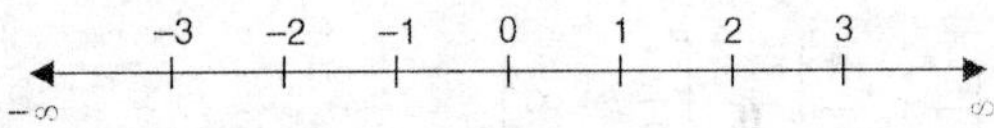

Answers:

1. $(-\infty, -1) \cup [1, \infty)$
2. $[8, 12)$
3. $\left(\dfrac{1}{2}, 2\right]$
4. $[-1, 2) \cup [3, +\infty)$
5. $(-\infty, 1] \cup (3, \infty)$
6. $(-\infty, -3] \cup (2, 5)$

Type 7: *When the given function is put in the form*:

$$y = \frac{f_1(x)}{\sqrt{f_2(x)}}$$

Exercise 1.11

Find the domain of the following:

1. $y = \dfrac{x}{\sqrt{x^2 - 3x + 2}}$

***Answer*:**

1. $R-[1,2]$

3.2. *Problems based on finding the domain of a given logarithmic function*:

***Type 1*:** *When the given function is put in the form*: $y = \log f(x)$.

Exercise 1.12

Find the domain of each of the following functions:

1. $y = \log(6-4x)$
2. $y = \log(4x-5)$
3. $y = \log(x+8) + \log(4-x)$
4. $y = \log(x-2)$
5. $y = \log(3-x)$
6. $y = \log(3x^2-4x+5)$
7. $y = \log(x^2-x-6)$
8. $y = \log(5x-x^2-6)$

***Answers*:**

1. $\left(-\infty, \dfrac{3}{2}\right)$
2. $\left(\dfrac{5}{4}, +\infty\right)$
3. $(-8, 4)$
4. $(2, \infty)$
5. $(-\infty, 3)$
6. $(-\infty, \infty)$
7. $(-\infty, -2) \cup (3, \infty)$
8. $(2, 3)$.

***Type 2*:** *When the given function is put in the form*; $y = \log|f(x)|$.

Exercise 1.13

Find the domain of each of the following functions:

1. $y = \log|x|$
2. $y = \log|x-2|$
3. $y = \log|4-x^2|$

***Answers*:**

1. $R-\{0\}$ **2.** $R-\{2\}$ **3.** $R-\{-2,2\}$

***Type 3*:** *When the given function is put in the form*: $y = \log_a \log_b \log_c f(x)$.

Exercise 1.14

1. $y = \log_2 \log_3 \log_4 x$
2. $y = \log \log \dfrac{x}{2}$

***Answers*:**

1. $(4, \infty)$ **2.** $(2, \infty)$

3.3. *Finding the domain of inverse circular functions*

***Type 1*:** *When the given function is put in the form*: $y = \sin^{-1} f(x)$ or $\cos^{-1} f(x)$.

Exercise 1.15

Find the domain of each of the following functions:

1. $y = \sin^{-1}(1-2x)$
2. $y = \sin^{-1}\left(\dfrac{2x}{3}\right)$
3. $y = \cos^{-1} 4x$
4. $y = \cos^{-1}\left(\dfrac{x}{2}\right)$
5. $y = \cos^{-1}(3x-1)$
6. $y = \cos^{-1} 2x$
7. $y = \sin^{-1}(x-2)$

***Answers*:**

1. $[0, 1]$
2. $\left[-\dfrac{3}{2}, \dfrac{3}{2}\right]$
3. $\left[-\dfrac{1}{4}, \dfrac{1}{4}\right]$
4. $[-2, 2]$
5. $\left[0, \dfrac{2}{3}\right]$ 6. Find 7. $[1, 3]$.

3.4. *Finding the domain of a sum or difference of two functions*:

Exercise 1.16

Find the domain of each of the following functions:

1. $y = \sqrt{x-3} + \sqrt{1-x}$
2. $y = \sqrt{4-x} + \sqrt{x-5}$
3. $y = \sqrt{x-1} + \sqrt{6-x}$
4. $y = \sqrt{x^2 - 1} + \dfrac{1}{\sqrt{x^2 - 3x + 2}}$
5. $y = \sqrt{2x - x^2} + \dfrac{1}{\sqrt{8x - 4x^2 - 3}}$

Hint: Domain of $\sqrt{2x - x^2} = [0, 2] = D_1$ (say) and

domain of $\dfrac{1}{\sqrt{8x - 4x^2 - 3}} = \left(\dfrac{1}{2}, \dfrac{3}{2}\right) = D_2$ (say)

$\therefore D(y) = D_1 \cap D_2 = \left(\dfrac{1}{2}, \dfrac{3}{2}\right)$

6. $y = \sqrt{x-1} + 2\sqrt{1-x} + \sqrt{x^2+1}$
7. $y = \sqrt{x^2 - 3x + 2} + \dfrac{1}{\sqrt{3 + 2x - x^2}}$
8. $y = \sqrt{\dfrac{x-2}{x+2}} + \sqrt{\dfrac{1-x}{1+x}}$
9. $y = \dfrac{3}{4 - x^2} + \log_{10}\left(x^3 - x\right)$
10. $y = \sqrt{x} + \sqrt{\dfrac{1}{x-2}} - \log_{10}(2x - 3)$

Answers:

1. ϕ **2.** ϕ **3.** $[1, 6]$ **4.** $(-\infty, -1] \cup (2, \infty)$

5. $\left(\dfrac{1}{2}, \dfrac{3}{2}\right)$ **6.** $\{1\}$ **7.** $(-1, 1] \cup [2, 3)$

8. Defined no where

9. $(-1, 0) \cup (1, 2) \cup (2, \infty)$ **10.** $(2, \infty)$.

3.5. *Finding the domain of a function put in the forms*:

1. $y = |f(x)|$
2. $y = f(x) \pm |g(x)|$
3. $y = \dfrac{f_1(x)}{f_2(x) \pm |f_3(x)|}$

Exercise 1.17

Find the domain of each of the following functions:

1. $y = |x - 2|$
2. $y = \dfrac{|x|}{x}$
3. $y = \cos^{-1}[x]$

Answers:

1. R **2.** $R - \{0\}$ **3.** $[-1, 2)$

3.6. *Finding the range of a function*:

Exercise 1.17.1

Find the range of the following functions:

1. $y = x|x|$
2. $y = 11 - 7\sin x$
3. $y = 3\sin x + 4\cos x$
4. $y = \sin\left\{\log\left(\dfrac{\sqrt{4 - x^2}}{1 - x}\right)\right\}$
5. $y = \dfrac{x}{|x|}$
6. $y = \dfrac{1}{3 - \cos 2x}$
7. $y = \dfrac{1}{2 - \cos 3x}$
8. $y = \log_3(5 + 4x - x^2)$
9. $y = x - [x]$
10. $y = [x] - x$

Answers:

1. $(-\infty, \infty)$ 2. [4. 18] 3. [–5, 5] 4. [–1, 1] 5. {–1, 1} 6. $\left[\frac{1}{4}, \frac{1}{2}\right]$ 7. $\left[\frac{1}{3}, 1\right]$ 8. $(-\infty, \log_3 9)$ 9. [0, 1) 10. (–1, 0].

Hint for (2): $-1 \le \sin x \le 1$

$$\Rightarrow -7 \le -7 \sin x \le 7$$
$$\Rightarrow 11 - 7 \le 11 - 7 \sin x \le 7 + 11$$
$$\Rightarrow 4 \le 11 - \sin x \le 18$$

Hint for (8): $y = \log_3\left(5 + 4x - x^2\right) \Leftrightarrow 3^y = 5 + 4x - x^2$

$$= 9 - (x-2)^2 > 0$$
$$\therefore 0 < 3^y \le 9$$
$$\therefore -\infty < y \le \log_3 9$$

Exercise 1.17.2

Find the domain and range of each of the following functions:

1. $y = \begin{cases} -2, \text{ if } x \le 3 \\ 2, \text{ if } x > 3 \end{cases}$

2. $y = \begin{cases} -4, \text{ if } x < -2 \\ -1, \text{ if } -2 \le x \le 2 \\ 3, \text{ if } x > 2 \end{cases}$

3. $y = \begin{cases} 2x - 1, \text{ if } x \ne 3 \\ 0, \text{ if } x = 3 \end{cases}$

4. $y = \begin{cases} \sqrt{25 - x^2}, \text{ if } x \le 5 \\ x - 5, \text{ if } x \ge 5 \end{cases}$

5. $y = \begin{cases} x^2 - 4, \text{ if } x < 3 \\ 2x - 1, \text{ if } x \ge 3 \end{cases}$

6. $y = \begin{cases} 6x + 7, \text{ if } x \le -2 \\ 4 - x, \text{ if } x > -2 \end{cases}$

Answers:

1. $D(y) = R$;
   $R(y) = \{-2, 2\}$
2. $D(y) = R$
   $R(y) = \{-4, -1, 3\}$
3. $D(y) = R$
   $R(y) = R - \{3\}$
4. $D(y) = R$
   $R(y) = [0, \infty)$
5. $D(y) = R$
   $R(y) = [-4, \infty)$
6. $D(y) = R$
   $R(y) = (-\infty, 6)$

4. Finding the composite of two function

Exercise 1.18

1. If $f(x) = \tan x$, $x \in \left(-\frac{\pi}{2}, \frac{\pi}{2}\right)$ and $g(x) = \sqrt{1 - x^2}$, find $(gof)(x)$.

2. If $f(x) = \sqrt{x}$ and $g(x) = |x|$, find $(gof)(x)$.

3. If $f(x) = e^{2x}$ and $g(x) = \log\sqrt{x}$, $x > 0$ find $(gof)(x)$.

4. If $f(x) = \frac{x+1}{x+2}$, $x \ne -2$ x being real and $g(x) = x^2$, find $(gof)(x)$.

5. If $f(x) = \frac{|x|}{x}$ and $g(x) = \frac{1}{x}$ find $(gof)(x)$.

6. If $f: R \to R$ is defined by $f(x) = \sin x$, $x \in R$ and $g: R \to R$ is defined by $g(x) = x^2$, compute $(gof)(x)$ and $(fog)(x)$.

7. If $f: R \to R$ is defined by $f(x) = 2x^2 - 1$ and $g: R \to R$ by $g(x) = 4x - 3$, $x \in R$ compute $(gof)(x)$ and $(fog)(x)$. Also find $(gof)(2)$ and $(fog)(-1)$.

8. If $f: R \to R$ is defined by $f(x) = x^2 - 3x + 2$ and $g: R \to R$ by $g(x) = 4x + 3$, $x \in R$ compute $(gof)(x)$ and $(fog)(x)$. Hence find the values of $(gof)(3)$ and $(fog)(3)$.

9. If $f: A \to B$ is defined by $f(x) = x + 1$, $x \in R$ and $g: B \to C$ by $g(x) = x^2$, find $(gof)(x)$.

10. If $f(x) = \cos x$, $g(x) = x^3$, $x \in R$, find $(gof)(x)$ and $(fog)(x)$.

11. If $f(x) = x^2 + 2$ and $g(x) = x - 1$, $x \in R$, find $(fog)(x)$ and $(gof)(x)$. Hence, find $(fog)(-2)$ and $(gof)(-2)$.

12. If $f(x) = 2x + 3$ and $g(x) = 3x^2 - 2$, $x \in R$, find $(gof)(x)$ and $(fog)(x)$. Hence, find the values of $(gof)(2)$ and $(fog)(2)$.

13. If the mapping $f: A \to B$ is define by $f(x) = \log(1 - x)$ and the mapping $g: B \to C$ is defined by $g(x) = e^{2x}$, find $(gof)(x)$.

14. If the mapping $f: R \to R$ be given by $f(x) = 1 + \frac{1}{x-1}$ and the mapping $g: R \to R$ be given by $g(x) = x^2 + 1$, show that $(gof)(x) \neq (fog)(x)$.

15. If f is defined as $f(x) = \frac{1}{1-x}$, show that $f[f\{f(x)\}] = x$, $x \neq 0, 1$.

16. If $f(x) = |x|$, find $f[f(x)]$.

Answers:

1. $\sqrt{1 - \tan^2 x}$
2. $\left|\sqrt{x}\right| = \sqrt{x}$
3. $\log \sqrt{e^{2x}} = x$
4. $\left(\frac{x+1}{x+2}\right)^2$
5. $\frac{x}{|x|}$, $x \neq 0$
6. $(gof)(x) = \sin^2 x$ and $(fog)(x) = \sin x^2$
7. $(gof)(x) = 8x^2 - 7$, $(fog)(x) = 32x^2 - 48x + 17$, $(gof)(2) = 25$ and $(fog)(-1) = 97$
8. $(gof)(x) = 4x^2 - 12x + 11$, $(fog)(x) = 16x^2 + 12x + 2$, $(gof)(3) = 11$ and $(fog)(3) = 182$
9. $(1 + x)^2$
10. $(gof)(x) = \cos^3 x$ $(fog)(x) = \cos x^3$
11. $(fog)(x) = x^2 - 2x + 3$; $(gof)(x) = x^2 + 1$; $(fog)(-2) = 11$ and $(gof)(-2) = 5$
12. $(gof)(x) = 12x^2 + 36x + 25$; $(fog)(x) = 6x^2 - 1$; $(gof)(2) = 145$ and $(fog)(2) = 23$
13. $(gof)(x) = (1 - x)^2$
16. $|x|$.

Exercise 1.19

Find the domain and range of each of the following ones:

1. $\sin(1 + \cos x)$
2. $\sqrt{\sin(\cos x)}$
3. $\cos\sqrt{\sin x}$
4. $\tan \sin^2 x$
5. $\log\sqrt{1 - x^2}$
6. $\sqrt{\log\sqrt{1 - x^2}}$
7. $\log\left(\frac{1-x}{1+x}\right)$
8. $\sin \log\left(\frac{1-x}{1+x}\right)$
9. $\tan(\sin x + \cos x)$
10. $\tan(\sin x + 2\cos x)$
11. $\cos|\sin^{-1} x|$
12. $\log_e \log_e x$
13. $\log_e \log_e \log_e x$
14. $\log_e \log_e \log_e \log_e x$
15. $\sec^{-1} \sin x$
16. $\sin \sec^{-1} \sin x$
17. $\log(1 + \sin \sec^{-1} \sin x)$,
18. $\cos \sin^{-1} x$

Answers:

1. $D = (-\infty, \infty)$; $R = [0, 1]$
2. $D = \left[2n\pi - \frac{\pi}{2}, 2n\pi + \frac{\pi}{2}\right] n \in Z$; $R = \left[0, \sqrt{\sin 1}\right]$
3. $D = [2n\pi, (2n+1)\pi]$; $R = [\cos 1, 1]$

4. $D = (-\infty, \infty)$; $R = [0, \tan 1]$
5. $D = (-1, 1)$; $R = (-\infty, 0)$
6. $D = \{0\}, R = \{0\}$
7. $D = (-1, 1)$; $R = (-\infty, \infty)$
8. $D = (-1, 1)$; $R = [-1, 1]$
9. $D = (-\infty, \infty)$; $R = \left[-\tan\sqrt{2}, \tan\sqrt{2}\right]$
10. $D = (-\infty, \infty)$; $R = (-\infty, \infty)$
11. $D = [-1, 1]$; $R = [0, 1]$
12. $D = (1, \infty)$; $R = (-\infty, \infty)$
13. $D = (e, \infty)$; $R = (-\infty, \infty)$
14. $D = \left(e^{e}, \infty\right)$; $R = (-\infty, \infty)$
15. $D = (2n+1)\frac{\pi}{2}$; $R = \{0, \pi\}$
16. $D = (2n+1)\frac{\pi}{2}$; $R = \{0\}$
17. $D = (2n+1)\frac{\pi}{2}$; $R = \{0\}$
18. $D = [-1, 1]$; $R = [0, 1]$

Exercise 1.20

1. If the functions f, g and h are defined from the set of real numbers R to R such that $f(x) = x^2 - 1$,

$$g(x) = \sqrt{x^2 + 1}$$

$$h(x) = \begin{cases} 0, \text{ when } x \geq 0 \\ x, \text{ when } x < 0 \end{cases}$$

then find composite function *hofog*.

2. (i) Find $f(f(x))$ if

$$f(x) = \begin{cases} x^2, x \geq 0 \\ x, x < 0 \end{cases}$$

(ii) Find $g(g(x))$ if

$$g(x) = \begin{cases} 2 + x, x \geq 0 \\ 2 - x, x < 0 \end{cases}$$

(iii) Find $h(h(x))$ if

$$h(x) = \begin{cases} x + 1, x \leq 1 \\ 5 - x^2, x > 1 \end{cases}$$

(iv) Find $i(i(x))$ if

$$i(x) = \begin{cases} -x, x < 0 \\ x, 0 \leq x \leq 1 \\ 2 - x, x > 1 \end{cases}$$

Answers:

1. ***Hint***: $fog(x) = f(g(x)) = f\left(\sqrt{x^2 + 1}\right)$

$$= \left(\sqrt{x^2 + 1}\right)^2 - 1 = x^2 + 1 - 1 = x^2$$

$\therefore$ $hofog(x) = h(x^2) = 0$

2. (i) $f(f(x)) = \begin{cases} x^4, x \geq 0 \\ x, x < 0 \end{cases}$

(ii) $g(g(x)) = \begin{cases} 4 + x, \text{ when } x \geq 0 \\ 4 - x, \text{ when } x < 0 \end{cases}$

(iii) $h(h(x)) = \begin{cases} x + 2, x \leq 0 \\ 5 - (x+1)^2, 0, x \leq 1 \\ 5 - \left(5 - x^2\right)^2, 1 < x < 2 \\ 6 - x^2, x \geq 2 \end{cases}$

(iv) $i(i(x)) = \begin{cases} 2 + x, x < -1 \\ -x, -1 \leq x < 0 \\ x, 0 \leq x \leq 1 \\ 2 - x, 1 < x \leq 2 \\ -(2 - x), x > 2 \end{cases}$

5. Eveness and oddness of $y = f(x)$.

Exercise 1.21

(A) By considering $f(-x)$, discover which of the following are even functions, which are odd functions and which are neither even nor odd.

1. $f(x)=x^4+7x^2+9$

2. $f(x)=x^3+x+\dfrac{1}{x^3}$

3. $f(x)=\sqrt{x^4+16}$

4. $f(x)=\dfrac{1}{4x^2+9}$

5. $f(x)=\sin x$

6. $f(x)=\cos x$

7. $f(x)=\tan x$

8. $f(x)=\sin x+\operatorname{cosec} x$

9. $f(x)=\sin x+\tan x$

10. $f(x)=\operatorname{cosec} x+\tan x$

11. $f(x)=\tan^2 x$

12. $f(x)=\tan^3 x$

13. $f(x)=\sin x+\cos x$

14. $f(x)=x+\dfrac{\cos x}{x}$

(B) Express the following functions as the sum of an even and odd functions.

1. $f(x)=e^x$

2. $f(x)=(1+x)^{100}$

3. $f(x)=\sin 2x+\cos\left(\dfrac{x}{2}\right)+\tan x$

4. $f(x)=x^2+3x+2$

5. $f(x)=1-x^3-x^4-2x^5$

(C) Let $f:[-2,2]\to R$ be a function. If for $x\in[0,2]$,

$$f(x)=\begin{cases} x\sin x, & 0\le x\le \dfrac{\pi}{2}\\ \dfrac{\pi}{2}(x), & \dfrac{\pi}{2}<x\le 2\end{cases}$$

define *f* for $x\in[-2,0]$ when

(i) *f* is an odd function.

(ii) *f* is an even function.

Answers:

(A) 1. Even **2.** Odd **3.** Even. **4.** Even **5.** Odd **6.** Even **7.** Odd **8.** Odd **9.** Odd **10.** Odd **11.** Even **12.** Odd **13.** Neither even nor odd **14.** Odd

(B) 1. $e^x=\dfrac{1}{2}\left[e^x+e^{-x}\right]+\dfrac{1}{2}\left[e^x-e^{-x}\right]$

2. $(1+x)^{100}=\dfrac{1}{2}\left[(1+x)^{100}+(1-x)^{100}\right]+\dfrac{1}{2}\left[(1+x)^{100}-(1-x)^{100}\right]$

3. $\sin 2x+\cos\left(\dfrac{x}{2}\right)+\tan x$

$=\dfrac{1}{2}[\sin 2x+\cos\left(\dfrac{x}{2}\right)+\tan x+\sin(-2x)+\cos\left(-\dfrac{x}{2}\right)+\tan(-x)]$

$+\dfrac{1}{2}[\sin 2x+\cos\left(\dfrac{x}{2}\right)+\tan x-\{\sin(-2x)+\cos\left(-\dfrac{x}{2}\right)+\tan(-x)\}]$

4. $x^2+3x+2=\dfrac{1}{2}\left[x^2+3x+2+x^2-3x+2\right]+\dfrac{1}{2}\left[x^2+3x+2-x^2+3x-2\right]$

5. $1-x^3-x^4-2x^5=\dfrac{1}{2}[1-x^3-x^4-2x^5+(1+x^3-x^4+2x^5)]+\dfrac{1}{2}[1-x^3-x^4-2x-(1+x^3-x^4+2x^5)]$

(C) For oddness,

$$f(x)=\begin{cases} -\dfrac{\pi}{2}(x), & -2\le x\le -\dfrac{\pi}{2}\\ -x\sin x; & -\dfrac{\pi}{2}\le x\le 0\end{cases}$$

for eveness,

$$f(x) = \begin{cases} \frac{\pi}{2}(x), & -2 \le x < -\frac{\pi}{2} \\ x \sin x, & -\frac{\pi}{2} \le x \le 0 \end{cases}$$

5. Periodic Functions

***Type 1**: Finding the periods of trigonometric functions.*

Exercise 1.22

(A) Find the periods of the following functions:

1. $f(x) = \cos 3x$
2. $f(x) = \cos\left(\frac{x}{4}\right)$
3. $f(x) = \tan 2x$
4. $f(x) = \tan 3x$
5. $f(x) = \cot\left(\frac{x}{5}\right)$
6. $f(x) = \cot\left(\frac{x}{2}\right)$
7. $f(x) = \sin 10x$
8. $f(x) = 10 \sin 3x$
9. $f(x) = 2\sin\left(3x + \frac{\pi}{10}\right)$
10. $f(x) = 4\sin\left(3x + \frac{\pi}{4}\right)$
11. $f(x) = \sqrt{\tan x}$
12. $f(x) = |\cos x|$
13. $f(x) = |\sin x|$
14. $f(x) = \tan^{-1}(\tan x)$
15. $f(x) = \cos^2 x$
16. $f(x) = \sin^2 x$
17. $f(x) = \sin^3 x$
18. $f(x) = \log(2 + \cos 3x)$
19. $f(x) = e^{\cot 3x}$

(B) Find the periods of the following functions:

1. $y = \sin 5x \cos 4x + 1$
2. $y = \sin\left(\frac{x}{4}\right) - 3\sin\left(\frac{x}{3}\right)$
3. $y = \tan\left(\frac{2x}{3}\right) - 4\cot\left(\frac{3x}{2}\right) - 2$
4. $y = \sin\left(\frac{3x}{4}\right) - 3\cos\left(\frac{5x}{8}\right) + \cos 5x$
5. $y = 2\sin x + 3\cos 2x$
6. $y = a\sin\lambda x + b\cos\lambda x$
7. $y = 3\cos\left(\frac{x}{2}\right) + 2\sin\left(\frac{x}{3}\right)$
8. $y = 3\sin x - 4\cos 2x$
9. $y = 1 + \tan x$
10. $y = \tan 2x - \cos 3x$

(C) The periods of $f(x) = |\sin x| + |\cos x|$ will be

(a) π **(b)** $\frac{\pi}{2}$ **(c)** 2π **(d)** none of these

***Answers**:*

(A) 1. $\frac{2\pi}{3}$ **2.** 8π **3.** $\frac{\pi}{2}$ **4.** $\frac{\pi}{3}$ **5.** 5π **6.** 2π **7.** $\frac{\pi}{5}$ **8.** $\frac{2\pi}{3}$ **9.** $\frac{2\pi}{3}$ **10.** $\frac{2\pi}{3}$ **11.** π **12.** π **13.** π **14.** π **15.** π **16.** π **17.** 2π **18.** $\frac{2\pi}{3}$ **19.** $\frac{\pi}{3}$

(B) 1. 2π **2.** 24π **3.** 6π **4.** 16π **5.** 2π **6.** $\frac{2\pi}{\lambda}$ **7.** 12π **8.** 2π **9.** π **10.** 2π

(C) (b) $\frac{\pi}{2}$

***Type 2**: Problems on showing that a given function is not periodic.*

Exercise 1.23

1. Show that the following functions are not periodic:

(i) $f(x) = \sin\sqrt{x}$

(ii) $f(x) = \sin\left(\frac{1}{x}\right), x \neq 0, f(0) = 0$

(iii) $f(x) = \sin x^2$

(iv) $f(x) = \cos\sqrt{x}$

(v) $f(x) = \cos x^2$

2. Show that the following functions are not periodic:

(i) $f(x) = x\cos x$

(ii) $f(x) = x\sin\left(\frac{1}{x}\right)$

(iii) $f(x) = \cos x^2 + \sin x^2$

(iv) $f(x) = \sin x + \cos\sqrt{x}$

(v) $f(x) = x + \sin x$

(vi) $f(x) = x + \cos x$

3. Examine which of the following functions are periodic:

(i) $f(x) = [x]$ (ii) $f(x) = 5$ (iii) $f(x) = x[x]$

Answers:

3. (i) Non-periodic (ii) Periodic but has no fundamental period (iii) Not periodic.

7. Examining a one-one and on-to function

Exercise 1.24

1. Is the function $f(x) = \frac{x^2 - 8x + 18}{x^2 + 4x + 30}$ a one-one function?

2. Verify whether the functions $f(x) = \frac{1}{x^2 + x + 1}$ is one-one.

3. Let $f: R \to R$ be defined by $f(x) = ax + b$, a, b being fixed real numbers and $a \neq 0$, show that f is one-one and on-to.

4. If $f(x) = \frac{x^2}{1 + x^2}$ is the function one-to-one?

5. If $f: [0, 2\pi] \to [-1, 1]$ be given by $f(x) = \sin x$, show that f is onto but not one-one.

6. If $f: [0, 2\pi] \to [-1, 1]$ be defined by $f(x) = \sin x$, show that f is on-to but not one-one.

7. Show that the following functions $f: R \to R$ are both one-one and on-to:

(i) $f(x) = x^3$ (ii) $f(x) = 3x + 4$

8. Let $D_1 = R - \{3\}, D_2 = R - \{1\}$ and $f: D_1 \to D_2$ be given by $f(x) = \frac{x - 2}{x - 3}$ is f bijective? Give reasons.

9. Let $D_1 = \{x : -1 \leq x \leq 1\} = D_2$ for each of the following functions from D_1 to D_2, find whether it is surjective, injective or bijective:

(i) $f(x) = \frac{x}{2}$ (ii) $h(x) = x\,|x|$ (iii) $k(x) = x^2$

(iv) $\phi(x) = \sin \pi x$ (v) $g(x) = |x|$

10. Let $f: R \to R$ be defined by $f(x) = \frac{x^2}{1 + x^2}$.

Is f one-one-onto?

11. Show that the mapping

(i) $f: R \to R$ defined by $f(x) = x$, $\forall\, x \in R$ is one-one-onto.

(ii) $f: R \to R$ defined by $f(x) = x^3$, $\forall\, x \in R$ is one-one-onto.

(iii) $f: Q \to Q$ defined by $f(x) = 2x + 3$, $\forall\, x \in Q$ is one-one-onto.

(iv) $f: R \to R$ defined by $f(x) = 4x + 3$, $\forall\, x \in R$ is one-one-onto.

12. Is the map $f: R \to R$ defined by $f(x) = \frac{x^2 + 4x + 30}{x^2 - 8x + 18}$ one-one?

13. Let $D_1 = \left[-\frac{\pi}{2}, \frac{\pi}{2}\right]$, $D_2 = [-1, 1]$, show that the map $f: D_1 \to D_2$ defined by $f(x) = \sin x$ is bijective.

Answers:

1. No 2. No 4. No

8. Yes, since $f(x_1) = f(x_2) \Rightarrow x_1 = x_2$ and $x = \frac{3y - 2}{y - 1} = f(y)$ which is true, $\forall\, y \in R - \{1\}$ which means f is onto.

9. (i) Injective but not surjective (ii) bijective (iii) Neither injective nor surjective (iv) surjective but not injective (v) Neither injective nor surjective
10. Not 12. Not

On the Graph of a Function

First of all we would like to define domain and range of a function in terms of projection of the graph of $y=f(x)$ on axes.

1. *Domain of a function*: The projection of the graph of $y=f(x)$ on the x-axis is called the domain of the function $y=f(x)$.
2. *Range of a function*: The projection of the graph of $y=f(x)$ on the y-axis is called the range of the function $y=f(x)$.

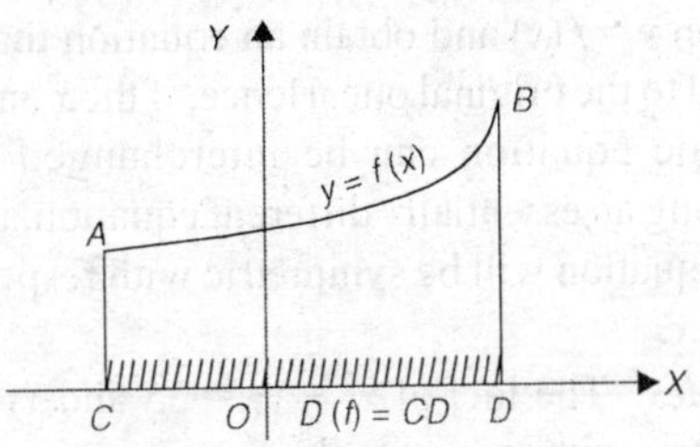

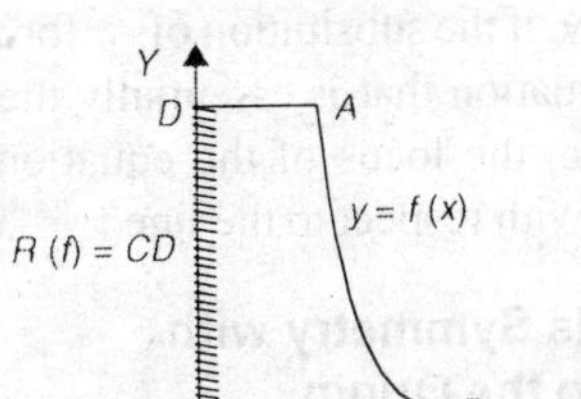

We speak of the graph of a point $P(x, y)$ meaning the point representing the ordered pair (x, y). We also speak of graphing a point $P(x, y)$, meaning to construct and locate the point P on a coordinate plane (a plane with axes).

Now we define what is the graph of a function.

Definition: The graph of a function f defined on its domain in a coordinate plane is the graph of the set $G\{(x, f(x)$: x is in the domain of $f\}$

or, equivalently, the graph of a function $f(x)$ defined on its domain is the graph of the equation $y=f(x)$.

That is, the point $P(a, b)$ is on the graph of $y=f(x)$ $\Leftrightarrow b=f(a)$.

If the domain of a function f is a finite interval, the graph of f can be explicitly plotted in a plane, but if the domain of f is an infinite set, it is not possible to plot all these points. In such case, we plot enough point to get an idea of the general shape of the graph of $y=f(x)$.

The following characteristics of a graph of a function are worth noting.

1. The graph of a function is a subset of the plane and it is uniquely determined by the function.
2. For each 'a' in the domain of f, there exists exactly one point $(a, f(a))$ on the graph of the function f, since by our definition of the function, the value of f at 'a' is uniquely determined. Geometrically, it means that $a \in D(f) \Leftrightarrow$ the vertical line $x=a$ meets the graph of f in one point only. again, $a \notin D(f) \Leftrightarrow$ the vertical line $x=a$ does not meet the graph anywhere at all, i.e. the point $(a, f(a))$ is missing (absent) on the graph of f, i.e. there is a hole in the graph of f at a, i.e. the graph of f is broken (not unbroken) at $x=a$ since there is a point on the graph of f whose abscissa is a but there is no ordinate corresponding to the abscissa $x=a$ and such points with no ordinate at an abscissa $x=a$ can not be located on the graph of a function.

On the Method of Graphing a Function

Mainly there are two methods of graphing a function which are:

1. The method of plotting a graph "point by point".
2. The method of plotting a graph "by derivative".

Hence, we use "point by point" method to draw the graph of a function defined by the formula $y=f(x)$ in its domain unless we learn how to find the derivative of a function.

How to Draw the Graph Using "Point by Point" Method

1. Find the domain of the given function $y=f(x)$
2. Find the zeros of the given function, i.e. solve the equation $f(x)=o$ whose solutions divide its domain into intervals where the function has the constant sign.
3. Examine whether the given function passes through the origin, i.e. check whether (o, o) satisfied the equation $y=f(x)$.
4. Find the intercepts on the axes, i.e. the points where the given curve cuts the x-axis and/y-axis.

5. Find whether the given function is odd or even.

6. Find whether the given function is periodic.

7. Plot a few additional points to get an idea of the general shape of the graph of $y = f(x)$.

Notes:

1. If the function is periodic, it is sufficient to investigate the behavior of the function on any closed interval whose length is equal to the period of the function and then constructing the graph on that interval, extend it to the whole of the domain of the function.

2. The position of a straight line is determined if any two points on it are known. Consequently in the case of straight lines, it is advised to find two points only instead of many to economise time, only with this care that the points are not very close to each other.

3. One should give such values to x as will enable him to get integral values of y since it is easier to plot integral units than the fractional units.

4. The graph of an even function is symmetric with respect to the axis of ordinates (i.e. y-axis) and the graph of an odd function is symmetric with respect to the origin.

On What is Symmetry of a Curve with Respect to a Line

A curve is said to be symmetric with respect to a line or symmetrical about a line) when all chords of the curve drawn perpendicular to the line are bisected by it. The lines of most importance in our "discussion" of an algebraic curve are the x-axis, the y-axis the lines bisecting the first and second quadrants which are $y = x$ and $y = -x$.

If the curve is symmetric with respect to x-axis, $y = o$, two points on the curve with the same abscissa, x, will have the ordinates y and $-y$. That means that substitution of $-y$ for $+y$ in the equation $y = f(x)$ will result in an equation that is same as or can be reduced to the original equation. Thus, in an algebraic equation, if only even powers of y are present, or every term of the equation contains an odd power of y, the locus of the equation is symmetric with respect to the x-axis.

Examples: The loci of the following equations are symmetric with respect to the x-axis:

$x^2 + y^2 = 25$ (circle)

$y^2 = x$ (parabola)

$y^2 = x^3$ (semi-cubical parabola)

$xy - y^3 = 0$

A similar discussion will bring us to the conclusion that if only even powers of x are present in the equation $y = f(x)$, or every term of the equation $y = f(x)$ contains an odd power of x, its locus is symmetric with respect to the y -axis.

Examples: The loci of the following equations are symmetric with respect to the y-axis:

$x^2 + y^2 = 25$ (circle)

$x^2 = y$ (parabola)

$y^3 - x^2 y = 0$

If the curve is symmetric with respect to the line $y = x$, a pair of symmetric points, lying on opposite sides of this line will have their coordinates reversed. This can be shown geometrically. It means that we should be able to interchange the x and y in the equation $y = f(x)$ and obtain an equation that can be reduced to the original one. Hence, if the x and y in an algebraic equation can be interchanged without producing an essentially different equation, the locus of the equation will be symmetric with respect to the line $y = x$.

Examples: The loci of $x^2 + y^2 = 25$ and $xy = 1$ are symmetric with respect to the line $y = x$.

Similarly, if the subsitution of $-y$ for x and $-x$ for y gives an equation that is essentially the same as the original one, the locus of the equation $y = f(x)$ is symmetric with respect to the line $y = -x$.

On What is Symmetry with Respect to the Origin

A curve is said to be symmetric with respect to a point (about a point) when the point bisects all chords of the curve drawn through it. Such a point of symmetry is called the centre of the curve. The most important point for our study is, perhaps, the origin. If the curve is symmetric with respect to the origin and the point (x, y) is on the curve, then the point $(-x, -y)$ will be on the curve, a fact that is made apparent by a figure. Hence, if substitution in an algebraic equation of $-x$ for x and $-y$ for y results in an equation that can be reduced to the original, the locus of the equation is symmetric with respect to the origin.

Examples: The loci of the following equations are symmetric with respect to the origin.

$x^2 + y^2 = 25$ (circle)
$xy = 5$ (hyperbola)
$x^2 - y^2 = 25$ (hyperbola)
$y = x^3$ (cubical parabola)

N.B.: A parabola is a U-shaped curve which cups (curves into the shape of a cup) up or cups down.

The Graphs of *y* = *f* (*x*)

1. $y = k$

x	y
.	k
.	k
.	k
–1	k
0	k
1	k
2	k

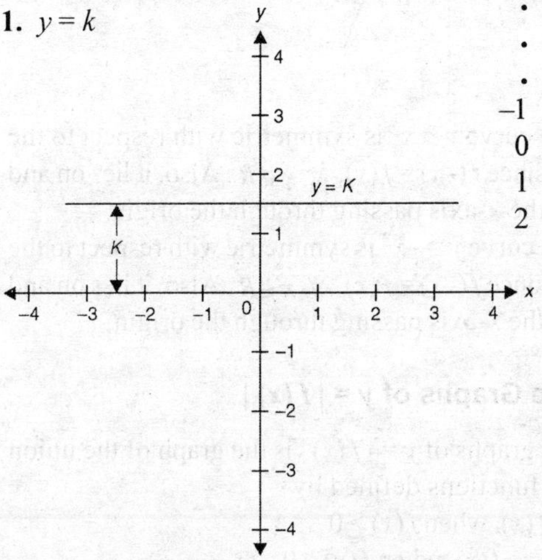

N.B.: A graph of a constant functions $y = k$ is a line.
(i) Parallel to the x-axis.
(ii) Above the x-axis at a distance k from it if $k > o$.
(iii) Below the x-axis at a distance $|k|$ from it if $k > o$.
(iv) On the x-axis if $k = o$.

2. $y = x$

x	y
.	.
.	.
.	.
–1	–1
0	0
1	1
2	2

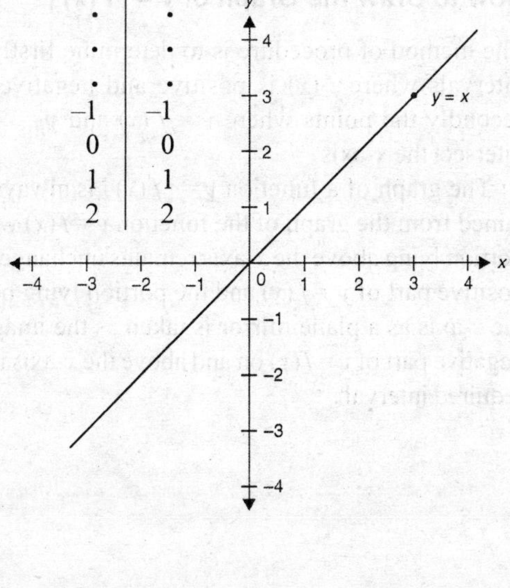

N.B.: A graph of the identity function $y = x$ is a line which
(i) Passes through the origin.
(ii) Bisects the angle between first and third quadrant or it is a line with slope 1.

3. $y = -x$

x	y
.	.
.	.
.	.
–1	1
–2	2
1	–1
2	–2

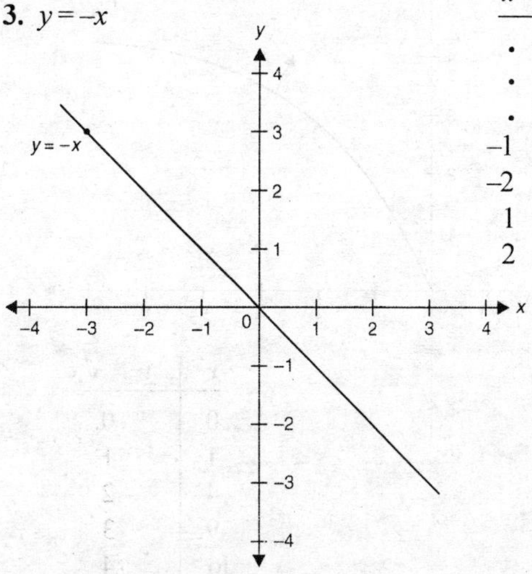

N.B.: The graph of the function $y = -x$ is a line
(i) Passing through the origin and
(ii) bisecting the angle between the second and the fourth quadrants, i.e. it is a line with slope –1.

4. $y = mx + c$.

x	y
.	.
.	.
.	.
.	.
.	.
.	.
.	.
.	.

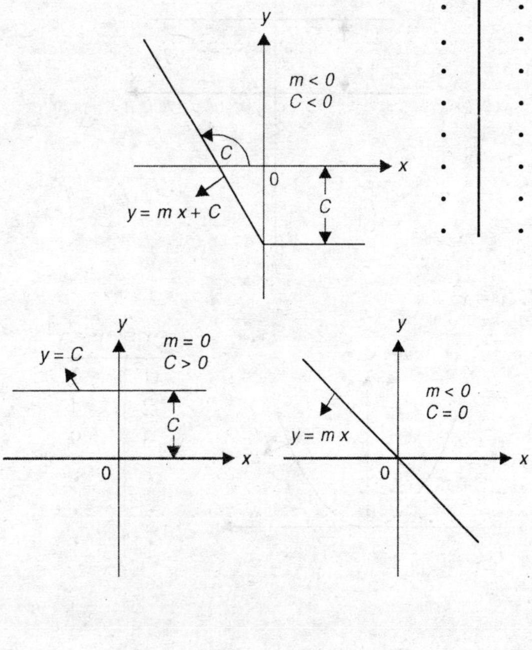

5. $y = \sqrt{x}$

$\Rightarrow y^2 = x$

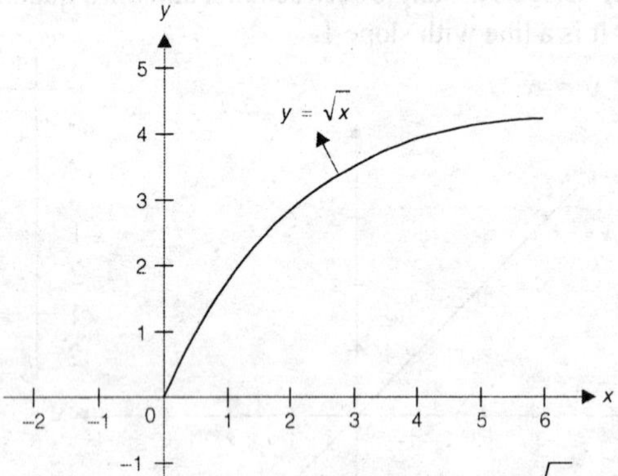

x	$y = \sqrt{x}$
0	0
1	1
4	2
9	3
16	4

6. $y = \sin^2 x + \cos^2 x$

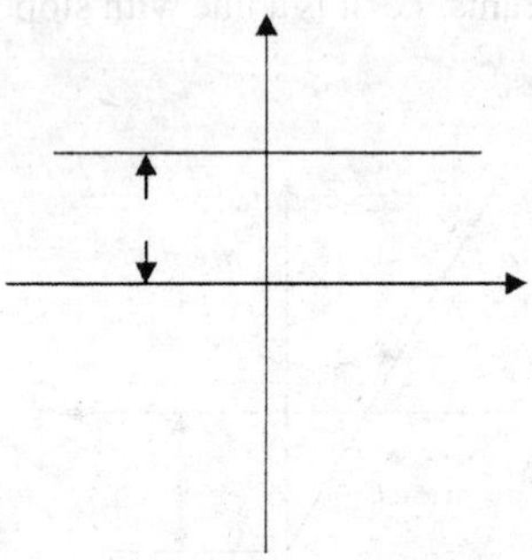

7. $y = x^2$

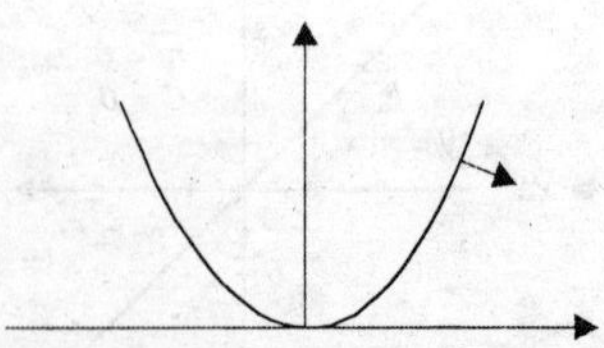

x	$y = x^2$
1	1
2	4
3	9
.	.
.	.
.	.

8. $y = -x^2$

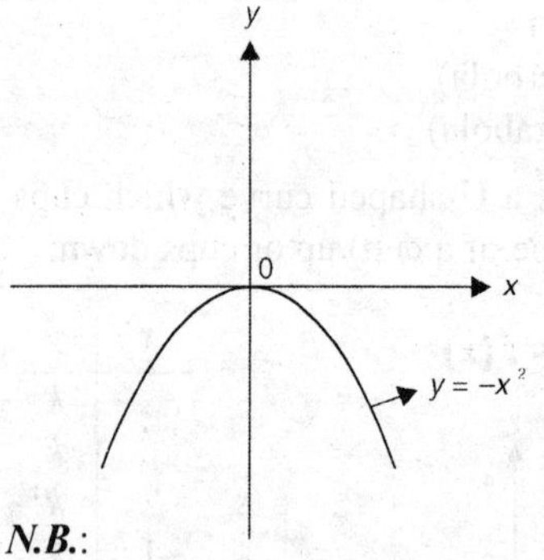

x	$y = -x^2$
1	–1
2	–4
3	–9
.	.
.	.

N.B.:

1. The curve $y = x^2$ is symmetric with respect to the y-axis since $f(-x) = f(x)$, $\forall\ x \in R$. Also, it lies on and above the x-axis passing through the origin.

2. The curve $y = -x^2$ is symmetric with respect to the y-axis since $f(-x) = f(x)$, $\forall\ x \in R$. Also, it lies on and below the x-axis passing through the origin.

On the Graphs of $y = | f(x) |$

The graphs of $y = | f(x) |$ is the graph of the union of two functions defined by

$y = f(x)$, when $f(x) \geq 0$

or $y = -f(x)$, when $f(x) < 0$

Note: The conditions $f(x) > 0$ and $f(x) < 0$ imposed on $y = f(x)$ and $y = -f(x)$ determine the intervals where $f(x)$ is positive or negative whereas the condition $f(x) = 0$ on $f(x) > 0$ determines the points where two curves $y = f(x)$ and $y = -f(x)$ intersect the x-axis.

How to Draw the Graph of $y = | f(x) |$

The method of procedure is to determine firstly the intervals where $f(x)$ is positive and negative and secondly the points where $y = f(x)$ and $y = -f(x)$ intersect the x-axis.

The graph of a function $y = | f(x) |$ is always obtained from the graph of the function $y = f(x)$ whose portion lying above the x-axis remains unchanged for positive part of $y = f(x)$ and the portion lying below the x-axis as a plane mirror is taken as the image of negative part of $y = f(x)$ on and above the x-axis in the required interval.

How to Determine the Intervals Where *f* (*x*) is Positive or Negative

1. Find the zeros of $f(x)$, i.e. the roots of $f(x) = o$.
2. Partition the real line by zeros of $f(x)$.
3. Consider the intervals:

$(-\infty, x_1)$, (x_1, x_2), (x_2, x_3) ... (x_{n-1}, x_n), (x_n, ∞), if $x_1, x_2, x_3, \ldots, x_n$ are the zeros of $f(x)$ such that x_1 = the smallest number among all the zeros of $f(x)$. x_n = the greatest number among all the zeros of $f(x)$.

4. Check the sign (i.e. positivity and negativity) of $f(x)$ in each interval determined by the zeros of $f(x)$.

How to Check the Sign of *f* (*x*) in Different Intervals

1. Take one particular point '*c*' belonging to each of the adjacent intervals.
2. Put the particular point *c* in $f(x)$.
3. Use the facts:

(i) : $f(c) > 0 \Leftrightarrow f(x)$ i.e. *y* is positive for every *x* in that interval where *c* belongs.

(ii) : $f(c) < 0 \Leftrightarrow f(x)$ i.e. *y* is negative for every *x* in that interval where *c* belongs.

For example, let us consider

$y = x^2 - 4x$ and $y = 4x - x^2$

In $(-\infty, 0)$

$f(x) = x^2 - 4x > 0$ for every $x \in (-\infty, 0)$

since $f(-1) = 5 > 0$ where $-1 \in (-\infty, 0)$

In $(4, \infty)$:

$f(x) = x^2 - 4x > 0$ for every $x \in (4, \infty)$

since $f(5) = 5 > 0$ where $5 \in (4, \infty)$

In $(0, 4)$:

$f(x) = -x^2 + 4x > 0$ for every $x \in (0, 4)$

since $f(1) = 3 > 0$ where $1 \in (0, 4)$

Notes:

1. $y = f(x)$ is positive in an interval
   $\Rightarrow$ is positive in its subinterval.
2. $y = f(x)$ is negative in an interval
   $\Rightarrow$ is negative in its subinterval.
3. The graph of $y = |f(x)|$ always lies on and above the x-axis.
4. The graph of $y = -|f(x)|$ always lies on and below the x-axis.

5. The same method of procedure is applicable to draw the graph of those functions containing absolute value function.

The Graph of *y* = | *f* (*x*) |

1. $y = |x|$

The graphs of $y = |x|$ is the graphs of the union of two functions defined by

$y = x$ and $x \geq 0$

or $y = -x$ and $x < 0$

on using the slope-intercept method, we graph $y = x$ where the domain of *y* is $\{x: x \geq 0\}$.

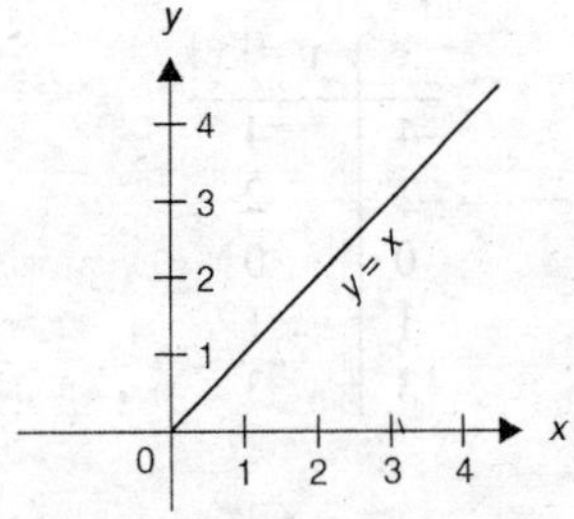

The equation $y = -x$ where $x < 0$ defines a function whose domain is $\{x: x < 0\}$. Again using the slope-intercept method, we graph $y = -x$.

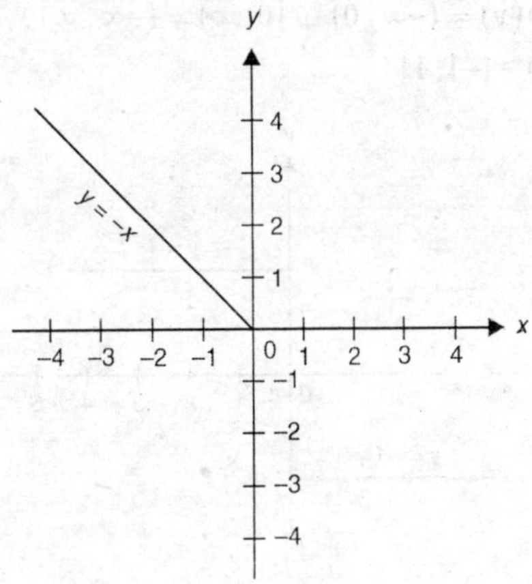

Also, $\{(x, y): y = |x|\} = [(x, y): y = x$ and $x \geq 0] \cup [(x, y): y = -x$ and $x < 0]$

Now, we combine the last two graph to obtain the graph as under:

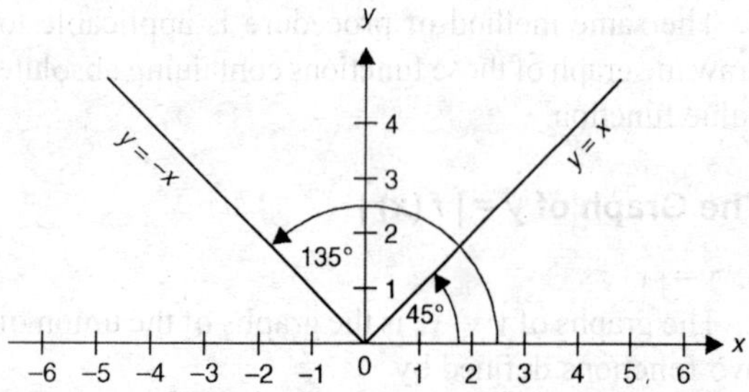

Note: A table of ordered pairs also easily can be constructed and the absolute value function $y = |x|$ can be graphed as given below:

x	$y = \|x\|$
−4	4
−2	2
0	0
1	1
3	3

2. $y = \dfrac{|x|}{x}$

$\Rightarrow y = 1$ for $x \geq 0$

or $y = -1$ for $x < 0$

$\therefore D(y) = (-\infty, 0) \cup [0, \infty) = (-\infty, \infty)$

$R(y) = \{-1, 1\}$

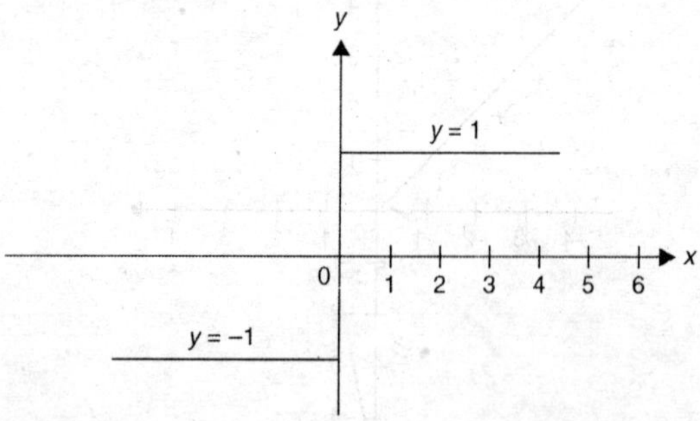

3. $y = \dfrac{|x-2|}{2}$

$\Rightarrow y = \dfrac{x-2}{2}$ for $x \geq 2$

or $y = \dfrac{2-x}{2}$ for $x < 2$

$\therefore D(y) = (-\infty, 2) \cup [2, \infty) = (-\infty, \infty)$ and

$R(y) = [0, \infty)$

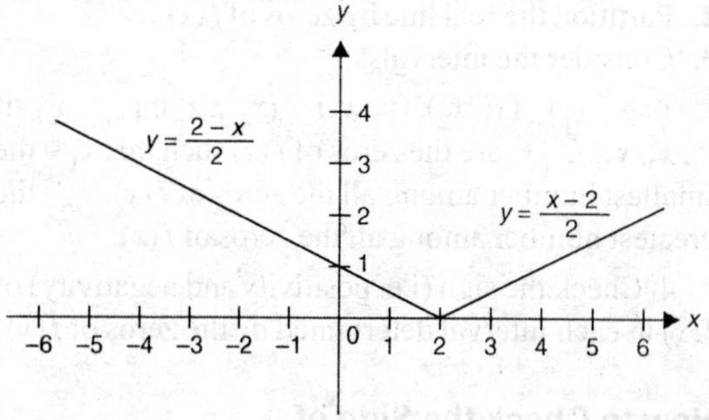

4. $y = \dfrac{2|x| - x}{3}$

$\Rightarrow y = \dfrac{x}{3}$ for $x \geq 0$

and $y = -x$ for $x < 0$

$\therefore D(y) = (-\infty, 0) \cup [0, \infty) = (-\infty, \infty)$

$R(y) = [0, \infty)$

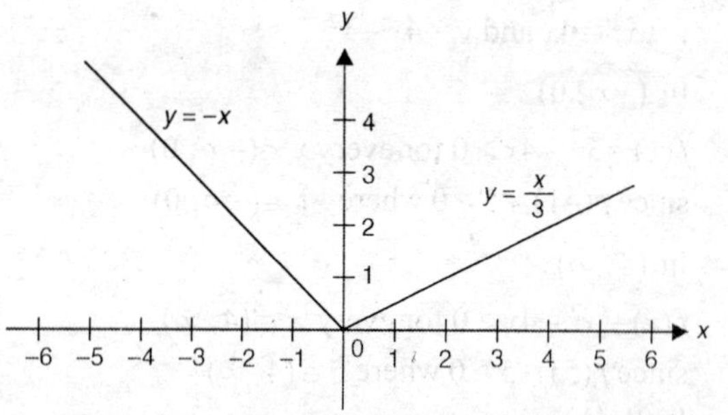

5. $y = |x^2 - 4x|$

$\Rightarrow y = x^2 - 4x$ for $x^2 - 4x \geq 0$, i.e. $x(x-4) \geq 0$

i.e. $x \leq 0$ or $x \geq 4$

and $y = -(x^2 - 4x)$ for $x^2 - 4x < 0$, i.e. $x(x-4) < 0$, i.e. $0 < x < 4$

$\therefore D(y) = (-\infty, 0] \cup (0, 4) \cup [4, \infty) = (-\infty, \infty)$

and

$R(y) = [0, \infty)$

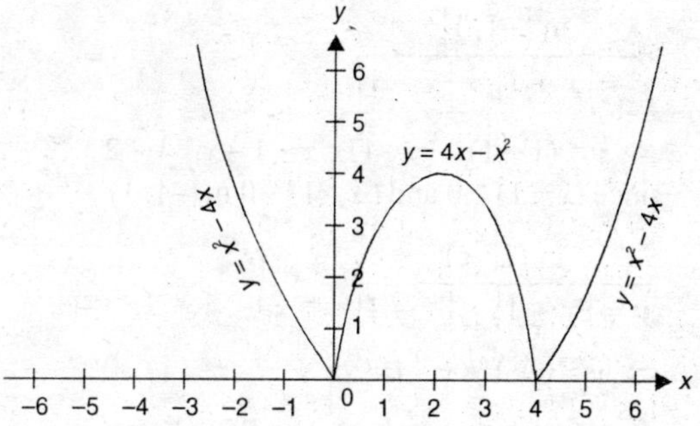

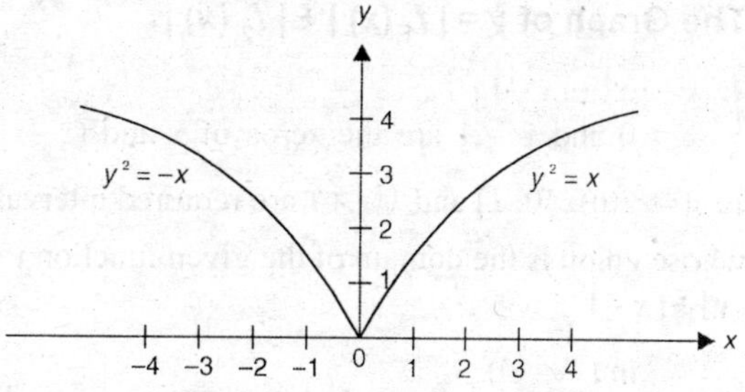

Note: To draw the graph of $y = |f(x)|$ or a function containing absolute value function, it is a must firstly to determine its domain which is the union of each interval determined by the zeros of $f(x)$.

6. $y = x|x|$

$\Rightarrow y = x \cdot x = x^2$ for $x \geq 0$

or $y = x \cdot (-x) = -x^2$ for $x < 0$

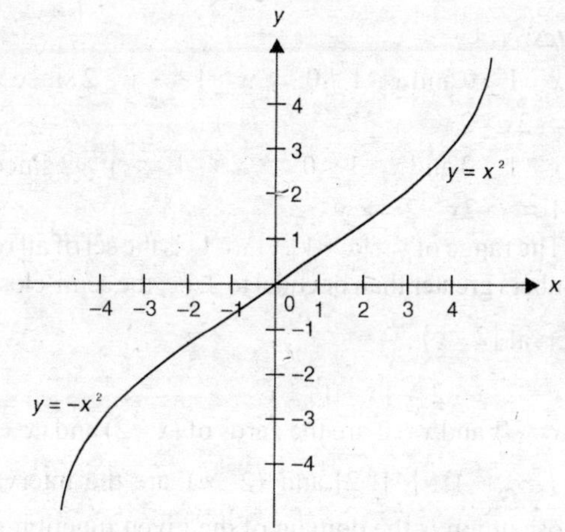

7. $y = \sqrt{|x|}$

$\Rightarrow y = \sqrt{x}$ for $x \geq 0 \Rightarrow y^2 = x$ for $x \geq 0$

or $y = \sqrt{-x}$ for $x < 0 \Rightarrow y^2 = -x$ for $x < 0$

8. $y = x^2 - 2|x|$

$\Rightarrow y = x^2 - 2x$ for $x > 0$

or $y = x^2 + 2x$ for $x < 0$

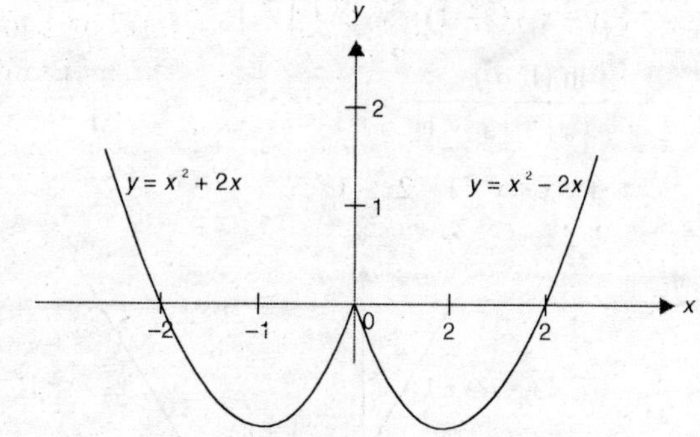

How to Draw the Graph of $y = |f_1(x)| + |f_2(x)|$

1. Partition the real line by zeros of $f_1(x)$ and $f_2(x)$.

2. Consider the intervals:

$(-\infty, x_1), (x_1, x_2), \ldots (x_n, \infty)$ if $x_1, x_2, x_3, \ldots, x_n$ are zeros of $f_1(x)$ and $f_2(x)$ such that

x_1 = the smallest number among all the zeros of $f_1(x)$ and $f_2(x)$.

x_n = the greatest number among all the zeros of $f_1(x)$ and $f_2(x)$.

3. Check the sign of $f_1(x)$ and $f_2(x)$ in each interval determined by the zeros of $f_1(x)$ and $f_2(x)$ and find a new function in each interval after simplification since the sign changes the character (the form) of a function.

The Graph of $y = |f_1(x)| + |f_2(x)|$

1. $y = |x| + |x - 1|$

$x = 0$ and $x = 1$ are the zeros of x and $(x - 1)$

$\Rightarrow (-\infty, 0)$, $[0, 1]$ and $(1, \infty)$ are required intervals whose union is the domain of the given function $y = |x| + |x - 1|$.

$$\frac{\text{in } (-\infty, 0)}{y = |x| + |x - 1|}$$

$\Rightarrow y = -x - (x - 1) = -x - x + 1 = -2x + 1$

$$\frac{\text{in } [0, 1]}{y = |x| + |x - 1|}$$

$\Rightarrow y = x - (x - 1) = x - x + 1 = 1$

$$\frac{\text{in } (1, \infty)}{y = |x| + |x - 1|}$$

$\Rightarrow y = x + x - 1 = 2x - 1$

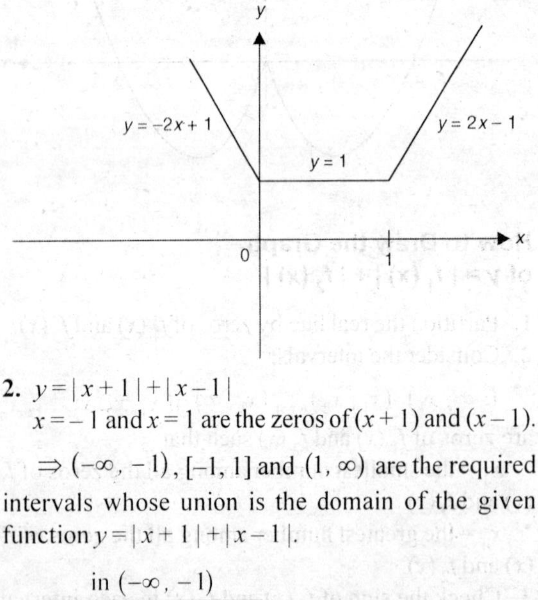

2. $y = |x + 1| + |x - 1|$

$x = -1$ and $x = 1$ are the zeros of $(x + 1)$ and $(x - 1)$.

$\Rightarrow (-\infty, -1)$, $[-1, 1]$ and $(1, \infty)$ are the required intervals whose union is the domain of the given function $y = |x + 1| + |x - 1|$.

$$\frac{\text{in } (-\infty, -1)}{y = |x + 1| + |x - 1|}$$

$\Rightarrow y = -(x + 1) - (x - 1) = -x - 1 - x + 1 = -2x$

since $(x + 1) < 0$ and $(x - 1) < 0$ in $(-\infty, -1)$

$$\frac{\text{in } [-1, 1]}{y = |x + 1| + |x - 1|}$$

$\Rightarrow y = (x + 1) - (x - 1) = x + 1 - x + 1 = 2$

since $(x + 1) > 0$ and $(x - 1) < 0$ in $(-1, 1)$

$$\frac{\text{in } (1, \infty)}{y = |x + 1| + |x - 1|}$$

$\Rightarrow y = x + 1 + x - 1 = 2x$

since $(x + 1) > 0$ and $(x - 1) > 0$ in $(1, \infty)$

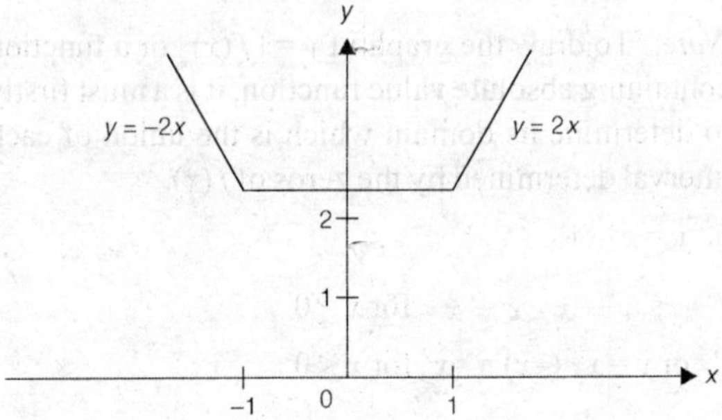

***Notes*:**

1. $x + 1 \geq 0$ and $x - 1 \geq 0 \Rightarrow x \geq 1 \Rightarrow y \geq 2$ since $x \geq 1 \Rightarrow 2x \geq 2 \Rightarrow y \geq 2$.

2. $x + 1 < 0$ and $x - 1 < 0 \Rightarrow x < -1 \Rightarrow y > 2$ since $x < -1 \Rightarrow -2x > 2 \Rightarrow y > 2$.

3. The range of $y = |x + 1| + |x - 1|$ is the set of all real numbers greater than or equal to 2, i.e. the semi-closed interval $[2, \infty)$.

3. $y = 2|x - 2| - |x + 1| + x$

$x = -1$ and $x = 2$ are the zeros of $(x - 2)$ and $(x + 1)$

$\Rightarrow (-\infty, -1)$, $[-1, 2]$ and $(2, \infty)$ are the intervals whose union is the domain of the given function $y = 2|x - 2| - |x + 1| + x$.

$$\frac{\text{in } (-\infty, -1):}{y = 2|x - 2| - |x + 1| + x}$$

$\Rightarrow y = -2(x - 2) + (x + 1) + x = 5$

$$\frac{\text{in } [-1, 2]:}{y = 2|x - 2| - |x + 1| + x}$$

$\Rightarrow y = -2(x - 2) - (x + 1) + x = -2x + 3$

in $(2, \infty)$:

$y = 2|x-2| - |x+1| + x$

$\Rightarrow y = 2(x-2) - (x+1) + x = 2x - 5$

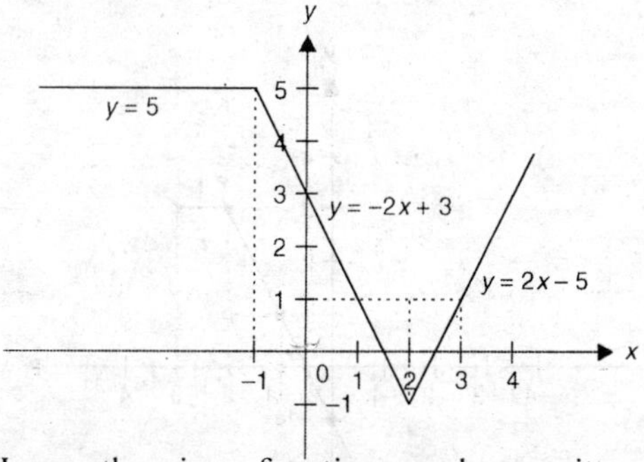

Hence, the given function can be rewritten as under:

$$y = \begin{cases} 5 & x < -1 \\ -2x + 3 & -1 \le x \le 2 \\ 2x - 5 & x > 2 \end{cases}$$

Therefore, the graph of $y = 2|x-2| - |x+1| + x$ is a polygonal line as above.

On the Graph of $y = [f(x)]$

The graph of $y = [f(x)]$ is a set of horizontal line segments (i.e. a set of line segments, each being parallel to x-axis), each of which includes the left end point but excludes its right end point. A small shaded circle is put at the left end point of a horizontal line to show the inclusion of that point and a small unshaded circle o is put at the right end point of a horizontal line to show the exclusion of that point as 

Moreover one should note that each horizontal line representing the graph of $y = [f(x)]$ always lies on and below a straight line.

The graph of a function $y = [f(x)]$ is always obtained from the graph of $y = f(x)$ where $y = [f(x)] \in I$ is marked on the y-axis of unit length such as $[-2, 1)$, $[-1, 0)$, $[0, 1)$, $[1, 2)$ etc. for which horizontal lines are drawn through integers till they intersect the graph.

Further, one should note that on y-axis for the form $[n, n+1)$, $y = n$ if y increases in its domain.

The graph of $y = [f(x)]$

1. $y = [x]$

$\therefore y = [x] = 0 \Leftrightarrow x \le -x < 1$

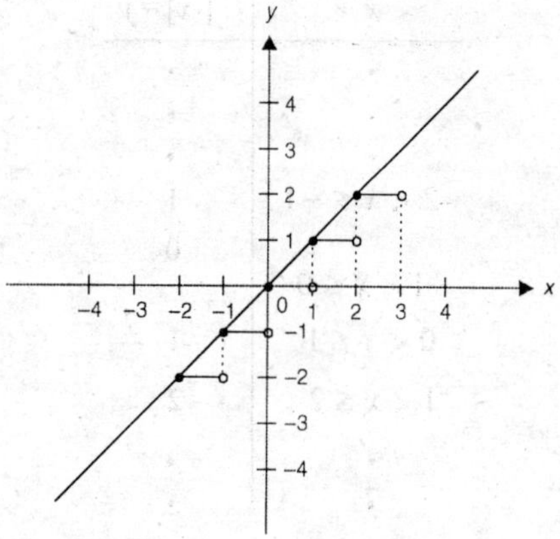

x	$[x] = y$
.	.
.	.
.	.
$0 \le x < 1$	0
$1 \le x < 2$	1
$2 \le x < 3$	2
$-1 \le x < 0$	−1
$-2 \le x < -1$	−2
.	.
.	.

N.B.: The graph of $y = [x]$ lies on and below the line $y = x$.

2. $y = [-x]$

$\therefore y = [-x] = 0 \Leftrightarrow 0 \le -x < 1 \Leftrightarrow 0 \ge x > -1$

$\Leftrightarrow -1 < x \le 0$

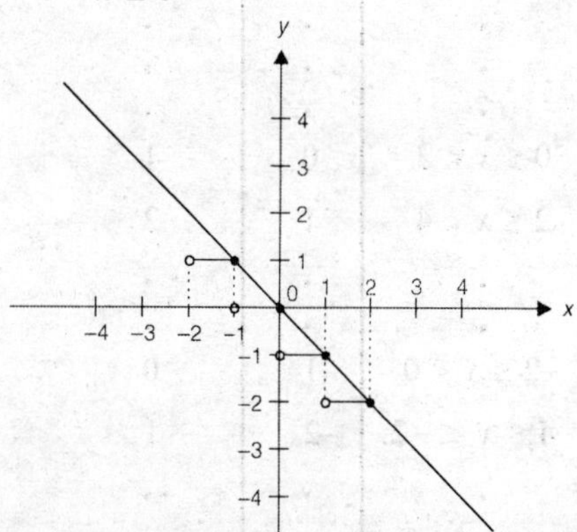

x	$[-x]=y$
.	.
.	.
.	.
$-2 < x \le -1$	1
$-1 < x \le 0$	0
$0 < x \le 1$	–1
$1 < x \le 2$	–2
.	.
.	.

N.B.: The graph of $y=[-x]$ lies on and below the line $y=-x$.

3. $y = \left[\dfrac{x}{2}\right] + 1$

$$y = \left[\frac{x}{2}\right] = 0 \Leftrightarrow 0 \le \frac{x}{2} < 1 \Leftrightarrow 0 \le x < 2$$

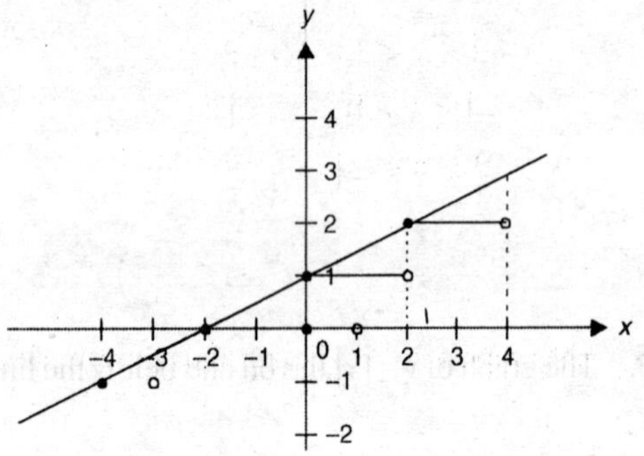

x	$\left[\frac{x}{2}\right]$	$\left[\frac{x}{2}\right]+1=y$
.	.	.
.	.	.
.	.	.
$0 \le x < 2$	0	1
$2 \le x < 4$	1	2
.	.	.
.	.	.
$-2 \le x < 0$	–1	0
$-4 \le x < -2$	–2	–1
.	.	.
.	.	.

N.B.: The graph of $y = \left[\dfrac{x}{2}\right] + 1$ lies on and below the line $y = x + 1$.

4. $y = 2[x] - 1$

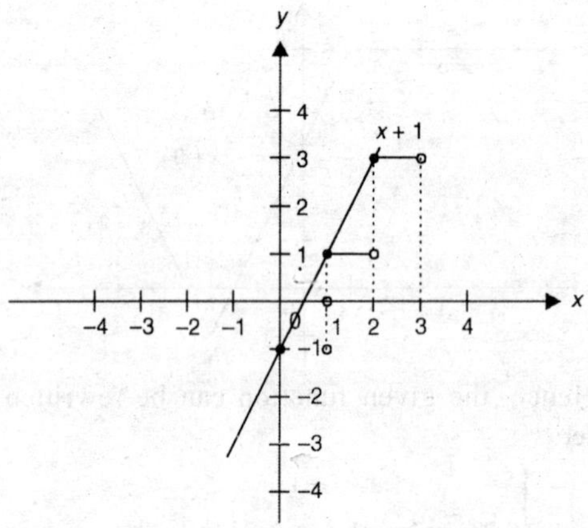

x	$2[x]$	$[2x]-1=y$
.	.	.
.	.	.
.	.	.
$0 \le x < 1$	0	–1
$1 \le x < 2$	2	1
$2 \le x < 3$	4	3
.	.	.
.	.	.

N.B.: The graph of $y=[2x]-1$ lies on and below the line $y=x-1$.

On the Graph of y = | [x] |

The graph of $y = |[x]|$ consists of all parallel line segments each being parallel to x-axis which lie on and above the x-axis such that all parallel line segment on the right side are on and below the line $y=x$ and all the parallel line segments on the left are on and above the line $y=-x$.

1. $y=|[x]|$

$$\therefore\ y = |[x]| = 0 \Leftrightarrow [x] = 0 \Leftrightarrow 0 \le x < 1$$

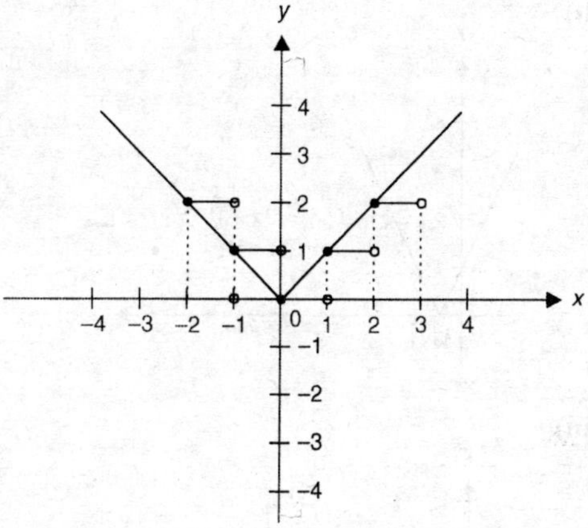

x	$\lvert[x]\rvert = y$
.	.
.	.
.	.
$-2 \le x < -1$	2
$-1 \le x < 0$	1
$0 \le x < 1$	0
$2 \le x < 3$	2

Exercise on Graphing the Functions

1. Graph the following functions:

(i) $f(x) = \dfrac{x + |x|}{2}$

(ii) $f(x) = |x-3| - 4$

(iii) $f(x) = 2x - [x]$, where $[x]$ = greatest integer function.

(iv) $f(x) = [x-2] + 2$, where $[x-2]$ = greatest integer function.

(v) $f(x) = 2x^2 - 12x + 20$

(vi) $f(x) = -x^2 + 8x - 16$

2. Construct graphs for the following functions.

(i) $y = \sqrt{x}$

(ii) $f(x) = |2x - 1|$

(iii) $f(x) = 2x|x-1|$

(iv) $f(x) = \left[\dfrac{x+4}{2}\right]$

(v) $f(x) = |x| - \dfrac{1}{[x]}$

Answers:

1. (i)

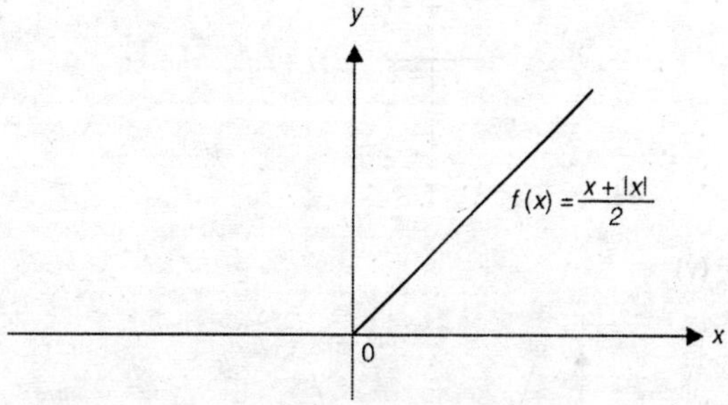

(ii)

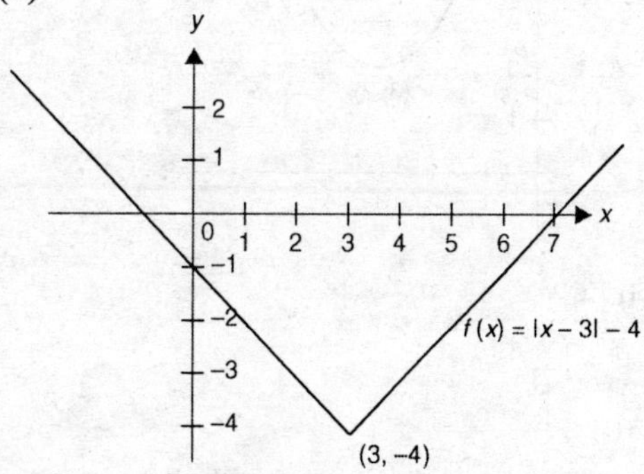

(iii)

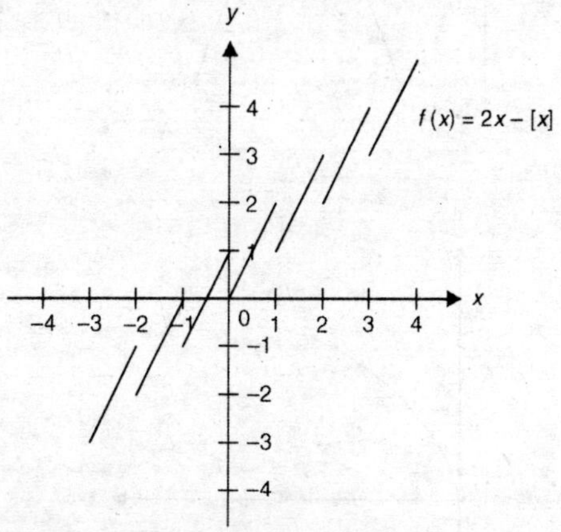

(iv)

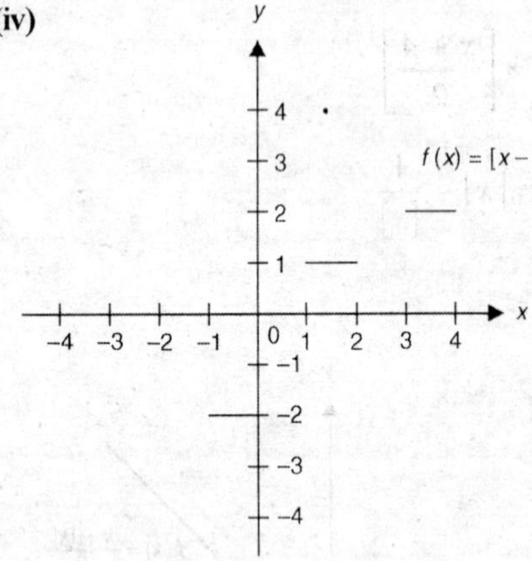

(v)

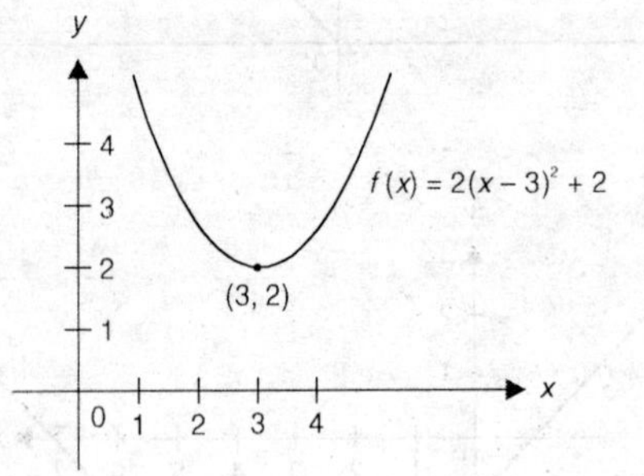

(vi)

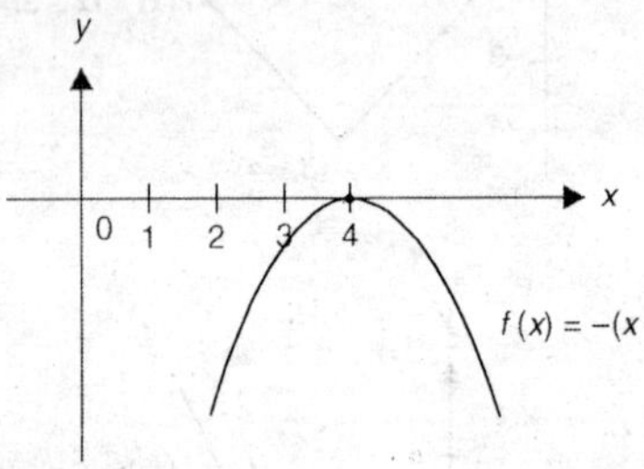

2. (i)

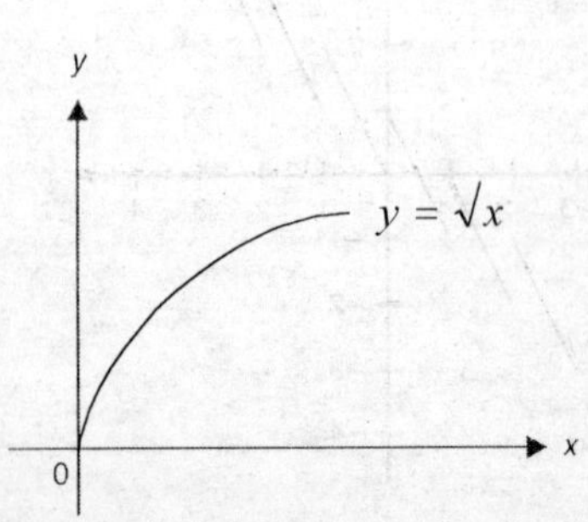

(ii)

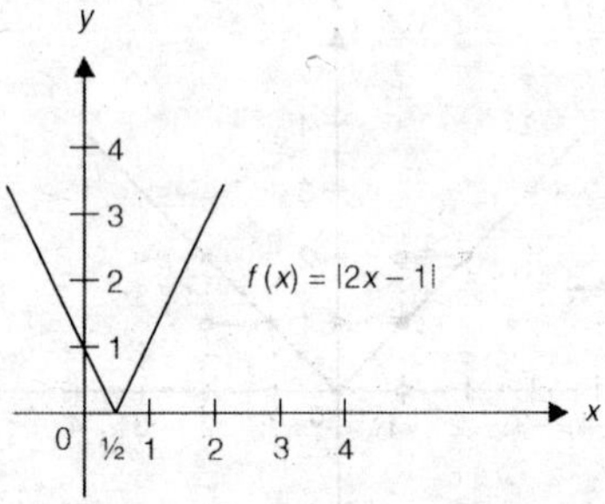

(iii)

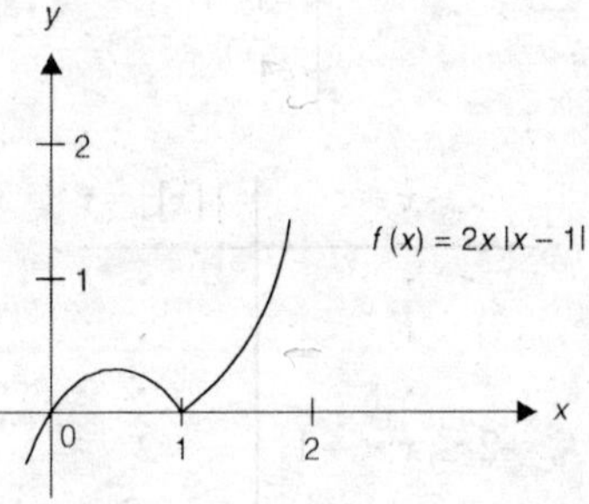

(iv)

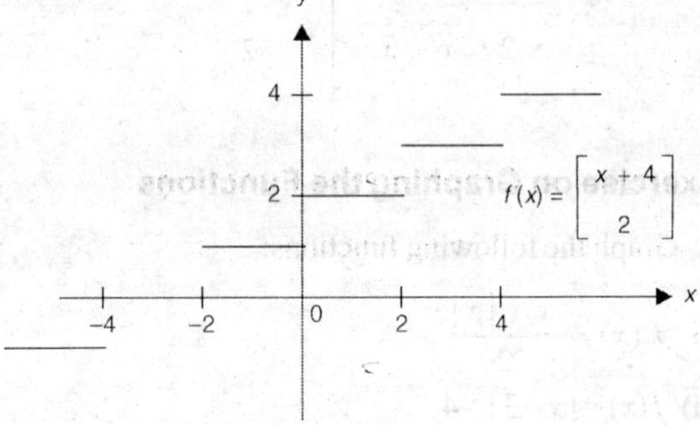

(v)

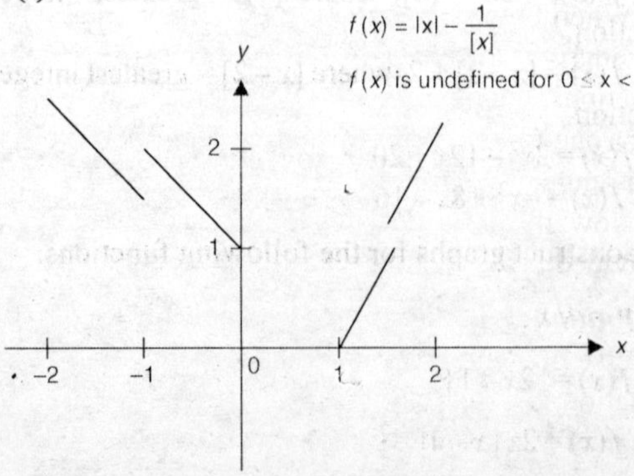

On the Inverse of a Function

Before defining the inverse of a function with respect to different aspects, one must know the following facts:

1. A function $f: A \to f(A)$ is always on-to function.

2. $f: A \to B$ is one-one function $\Rightarrow f: A \to f(A) \subset B$ is a bijection, where

A = domain of f

B = codomain of f

$f(A)$ = range of f, also denoted by $R(f)$.

3. If A and B are finite sets and $f: A \to B$ is a bijection, then $n(A) = n(B)$, i.e. number of elements in domain = number of elements in co-domain.

4. When there is only one value of the function $y = f(x)$ for every value of $x = a$ in its domain, then the function $y = f(x)$ is said to be single valued function in its domain.

The polynomial, the rational fraction, exponential and logarithmic functions are important functions which are single valued.

5. When there are two values of the function $y = f(x)$ for each value $x = a$ in its domain, the function $y = f(x)$ is said to be two valued (or, double valued) function in its domain.

Examples of double valued functions are:

(i) $y^2 = a_0 x^n + a_1 x^{n-1} + \ldots + a_{n-1} x + a_n \ (n \geq 1)$

(ii) $y^2 = \dfrac{N}{D}$ where N and D are polynomials in x.

6. When both $y = f(x)$ and $x = f^{-1}(y)$ obtained by solving $y = f(x)$ for x in terms of y, are single valued functions, then the function f establishes a bijection (or, a one-to-one correspondence) between its domain and range.

Now, the definition of the inverse of a function is provided.

Definition 1: (In terms of one-one function): The inverse of a one-one function f, denoted by f^{-1} is the function which is defined for every $y = f(x)$ in the range of f by $f^{-1}(y) = x$, i.e. if f is a function which is one-one in a part D of its domain and R is the set of values taken by f at points of D, then the function f^{-1} with domain R and range D, denoted by $f^{-1}(y) = x$ if $y = f(x)$ for every $y \in R$ is said to be the inverse of the function f on D.

Notes:

1. In some cases when the given function f is not one-one function in the entire domain, a part D of its domain is selected where the function f is one-one and the inverse of a function will exist over $f(D)$ for the new domain of f.

2. A function has an inverse $\Rightarrow$ it is a one-one function in its domain and the equation $y = f(x)$ can be solved for x in terms of y, i.e. $x = f^{-1}(y)$ which must be single valued. For an example, $y = 3x + 2$ is a 1-1 function.

$$\Rightarrow x = \frac{y-2}{3} \text{ is a single valued function.}$$

3. It may or may not be possible to find the inverse of a given function in the following cases:
If the given equation $y = f(x)$ gives $x = f^{-1}(y)$ which is double valued function or the given equation is double valued function, then in both cases, the given function (or, equation) $y = f(x)$ has no inverse, as for an example,

$$y = x^2 + 9 \Rightarrow x = \pm\sqrt{y-9} \text{ which determines}$$

two different values of x for each value of $y \neq 9$ in the range of f.

4. When it is said that the inverse $x = f^{-1}(y)$ of the function $y = f(x)$ is single valued for $y = b$ in range of the original function $y = f(x)$, it is meant that for each member $y = b$ of the range of the original function $y = f(x)$, there is exactly only one element in the domain of the original function $y = f(x)$.

Definition 2: *In terms of one-one and on-to function*: If f is a one-one and onto from A to B, then there exists a unique function $f^{-1}: B \to A$ such that for each $y \in B$ there exists exactly only one element $x \in A$ such that $f(x) = y$, then $f^{-1}(y) = x$. The function f^{-1} so defined is called the inverse of f.

Further, if f^{-1} is the inverse of f, then f is the inverse of f^{-1} and the two functions f and f^{-1} are said to be the inverse of each other.

Notes:

1. A function $f: A \to B$ has an inverse $f^{-1} \Leftrightarrow f: A \to B$ is one-one and on-to.

2. If a function f is continuous, monotonic and defined on a real interval working as a domain of the given function f, then a continuous monotonic inverse f^{-1} exists. For example, $f(x) = y = 2x + 3$ where $0 \le x < 1$, has an inverse $f^{-1}(y) = x = \frac{1}{2}(y-3)$ where $3 \le y \le 5$.

The variables x and y are often interchanged in the inverse function, so that in this example $f(x) = y = 2x + 3$ is said to have the inverse.

$$f^{-1}(x) = y = \frac{1}{2}(x-3)$$

This can be written

$$f: x \to 2x + 3 \text{ on } [0, 1]$$

$$f^{-1}: x \to \frac{1}{2}(x-3) \text{ on } [3, 5]$$

How to Find f^{-1} as a Function of x

Step 1: The equation $y = f(x)$ should be solved for x in terms of y.
Step 2: x and y should be interchanged. The resulting equation will be $y = f^{-1}(x)$.

Solved Examples

1. Find the inverse of $y = \frac{1}{2}x + 1$.

Solution:
Step 1: On solving for x in terms of y:

$$y = \frac{1}{2}x + 1$$

$$\Rightarrow 2y = x + 2$$

$$\Rightarrow x = 2y - 2$$

Step 2: On interchanging x and y:
$y = 2x - 2$

Hence, the inverse of the function $f(x) = \frac{1}{2}x + 1$ is the function $f^{-1}(x) = 2x - 2$.

2. Find the inverse of the function $y = x^2$, for $x \ge 0$.
Solution:
Step 1: On solving for x in terms of y:
$y = x^2$

$\Rightarrow \sqrt{y} = \sqrt{x^2} = |x| = x$ ($\because |x| = x$ because $x \ge 0$)

Step 2: On interchanging x and y:

$$y = \sqrt{x}$$

The inverse of a function $y = x^2$, $x \ge 0$ is the function $y = \sqrt{x}$.

One should note that, unlike the restricted function $y = x^2, x \ge 0$ the unrestricted function $y = x^2$ is not one-one and on-to and therefore has no inverse.

On the Criteria to Test Whether a Given Function f has its Inverse

There are following criteria to test whether a given function $y = f(x)$ from its domain D to its co-domain C has an inverse.

Criterion 1: One-oneness and ontoness of the function, i.e. one should show that the given function $y = f(x)$ from its domain D to its codomain C is a one-one and on-to function.

Note:
Criterion 1: Is fruitful to test the existance of an inverse of the function $y = f(x)$ whose domain D and whose codomain C are known.
Criterion 2: Single valuedness of both the function f and f^{-1}, i.e. one should show that both the given function $y = f(x)$ defined on its domain and $x = f^{-1}(y)$ defined on the range of $y = f(x)$ are single valued.
Criterion 3: One-oneness and single valuedness of a function, i.e., one should show that the function $y = f(x)$ from its domain D to its range R (in case range is known but its codomain is not known) is one-one and then show that the function $x = f^{-1}(y)$ obtained

by solving the given function $y=f(x)$ for x in terms of y is a single valued function.

Notes:

1. In case of function whose domain and range are known but whose codomain is not known, one should suppose that range is coincident with codomain and then use any one of the criteria to show whether the given function has the inverse.

2. In case of a function whose domain and range are known, one can use either the criterion (2) or the criterion (3) which is easy.

Criterion 4: One should see whether a given function $y=f(x)$ is continuous, monotonic and defined on a real interval working as a domain of the given function f.

Note: Criterion (4) is fruitful to test the existance of an inverse of the function $y=f(x)$ whose domain is known and its range can be determined by using any mathematical manipulation.

To remember:

1. The domain of the inverse of a function f^{-1} is the set of all values of y for which $x=f^{-1}(y)$, i.e. the range of the function f, i.e. the domain of f^{-1} is the range of f.

2. The range of the inverse of a function f^{-1} is the set of all values of x for which $y=f(x)$, i.e. the domain of f, i.e. the range of f^{-1} is the domain of f.

3. A function which has an inverse is said to be invertible.

4. The symbol denoted by f^{-1} is read as "eff inverse".

Solved Examples

1. Test whether the function $f: x \to y$ defined by

$f(x)=\dfrac{x-1}{x-3}$ where.

$x=R-\{3\}$ and $Y=R-\{1\}$, R being the set of reals, has its inverse.

Solution: For $x_1, x_2 \in R-\{3\}$,

$f(x_1)=f(x_2)$

$$\Rightarrow \frac{x_1-1}{x_1-3}=\frac{x_2-1}{x_3-3}$$

$$\Rightarrow (x_1-1)(x_2-3)=(x_1-3)(x_2-1)$$

$$\Rightarrow -3x_1-x_2+3=-3x_2-x_1+3$$

$$\Rightarrow 2x_2=2x_1$$

$$\Rightarrow x_1=x_2$$

$\therefore f$ is $1-1$

Also, any $y \in Y$, $y=\dfrac{x-1}{x-3}$

$$\Rightarrow y(x-3)=x-1$$

$$\Rightarrow x(y-1)=-1+3y$$

$$\Rightarrow x=\frac{3y-1}{y-1},\ y\neq 1$$

$$\therefore y \in Y \Rightarrow \exists x=\frac{3y-1}{y-1} \in x \text{ such that } y=f(x)$$

$\Rightarrow$ all the elements of y are f-images of an element in x, i.e., f is on-to.

Hence, f is one-one and on-to $\Rightarrow f$ has an inverse.

Let the inverse of f be g, i.e. $g=f^{-1}$

Then $g(y)=x=\dfrac{3y-1}{y-1} \Rightarrow x=f^{-1}(y)=\dfrac{3y-1}{y-1}$.

Thus $f^{-1}(y)=\dfrac{3y-1}{y-1} \Rightarrow f^{-1}(x)=\dfrac{3x-1}{x-1}$ is the inverse function of f.

2. Test whether the function $f:\left[-\dfrac{\pi}{2},\dfrac{\pi}{2}\right]\to[-1,1]$ defined by $f(x)=\sin x$, $x\in\left[-\dfrac{\pi}{2},\dfrac{\pi}{2}\right]$ has its inverse if so, find f^{-1}.

Solution: $x_1 \neq x_2$ and $x_1, x_2 \in \left[-\dfrac{\pi}{2},\dfrac{\pi}{2}\right]$

$$\Rightarrow \sin(x_1)\neq\sin(x_2) \Rightarrow f(x_1)\neq f(x_2)$$

$\Rightarrow f$ is one-one.

Again any $y\in[-1,1] \Rightarrow \exists x \in\left[-\dfrac{\pi}{2},\dfrac{\pi}{2}\right]$ such that $f(x)=y$, i.e. $y=\sin x \Rightarrow f$ is on-to.

Hence, f is one-one and on-to $\Rightarrow$ f has its inverse which is given by

$$f^{-1}:[-1, 1] \to \left[-\frac{\pi}{2}, \frac{\pi}{2}\right]$$

defined by

$$f^{-1}(y) = \sin^{-1}(y),\ y \in [-1, 1]$$

$\Rightarrow f^{-1}(x) = \sin^{-1}(x),\ x \in [-1, 1]$ is the required inverse of the given function.

3. Does the function $y = x^2$ have an inverse in the interval $[-1, 1]$?

Solution: $y = f(x) = x^2$ is the given function whose domain is R and $[-1, 1] \subseteq R$

$-1 \le x \le 1$

$\Rightarrow -1 \le x < 0$ and $0 \le x \le 1$

$\Rightarrow 1 \ge x^2 \ge 0$ and $0 \le x^2 \le 1$

$\Rightarrow 0 \le x^2 \le 1$

$\Rightarrow 0 \le f(x) \le 1$

$\Rightarrow$ Range of f is $[0, 1]$

Now $y = x^2 \Rightarrow x = \pm\sqrt{y}$ for $y \in [0, 1]$

$\therefore x = \pm\sqrt{y}$

$\Rightarrow$ for each value of y in $[0, 1]$, the range of f, x does not have a unique value.

$\Rightarrow x = \pm\sqrt{y}$ is not a single valued function.

Hence, f has no inverse.

4. Show that $y = |x|$ has no inverse. Restrict its domain suitably so that f^{-1} may exist and find f^{-1}.

Solution: $y = |x|$

$$\Rightarrow y = \begin{cases} x,\ x \ge 0 \\ -x,\ x < 0 \end{cases}$$

Clearly, $D(y) = R$ and $R(y) = R^{+} \cup \{0\} = [0, \infty)$

now, for every $x \in D(f)$, a unique value of $y \in R(y)$ is determined $\Rightarrow y = |x|$ is a single valued function for every $x \in D(f)$.

Now, $y = |x|$ for $y \in [0, \infty)$

$\Leftrightarrow y^2 = x^2$

$\Leftrightarrow x = \pm y$ which is a double valued function for every $y \in R^{+} \cup \{0\}$.

$\Rightarrow y = |x|$ has no inverse.

However, if the domain of f is restricted to $[0, \infty)$ or $(-\infty, 0)$, f will have the inverse.

Case 1: When the restricted domain is $[0, \infty)$ then $y = |x| = x,\ x \ge 0$

$\Rightarrow y = x,\ D(f) = [0, \infty)$ and $R(f) = [0, \infty)$

Now, for every value of $x \in D(f)$, a unique value of $y \in R(f)$ is determined.

$\Rightarrow y = x$ is a single valued function $\forall\ x \in D(f)$.

Again, $y = f(x) = x$

$\Rightarrow x = y \Rightarrow f^{-1}(y) = x = y$

$\Rightarrow D(f^{-1}) = R(f) = [0, \infty)$

$\Rightarrow$ For ever value of $y \in R(f)$, a unique value of $x \in D(f)$ is determined.

$\Rightarrow x = y$ is a single valued function for every $y \in R(f)$.

Hence, $y = f(x) = x$ is single valued function for every $x \in D(f)$ as well as $x = f^{-1}(y) = y$ is a single valued function for every $y \in R(f)$.

$\Rightarrow y = f(x)$ has its inverse in its restricted domain $[0, \infty)$.

Now, since $x = f^{-1}(y) = y$

$\Rightarrow y = f^{-1}(x) = x$ is the required inverse for $y = f(x) = x$.

Case 2: When the restricted domain is $(-\infty, 0)$, then $y = -x,\ x < 0$

$\Rightarrow y = -x,\ D(f) = (-\infty, 0)$ and $R(f) = (0, \infty)$

Now, it is observed that for every $x \in D(f)$, a unique value of $y \in R(f)$ is determined.

$\Rightarrow y = f(x) = -x$ is a single valued function for every value of $x \in D(f) = (-\infty, 0)$.

Again, $y = f(x) = -x$, $D(f) = (-\infty, 0)$ and $R(f) = (0, \infty)$ which

$$\Rightarrow -x = y \Rightarrow x = -y$$

$$\Rightarrow f^{-1}(y) = x = -y$$

$\Rightarrow f^{-1}(y) = -y$ which is a single valued function for every $y \in R(f)$.

$\Rightarrow$ Both $y = -x$ in its domain and $x = -y$ in the range of f are single valued functions.

$\Rightarrow y = -x$ has its inverse which is $x = f^{-1}(y) = -y$,

i.e., $f^{-1}(x) = -x$, $D\left(f^{-1}\right) = R(f) = (0, \infty)$.

5. Does the function $f(x) = 1 - 2^{-x}$ have an inverse?

Solution: $f(x) = 1 - 2^{-x}$ is defined for every value of $x \in R$

Further, $f(x) = 1 - 2^{-x}$

$$\Rightarrow y = 1 - 2^{-x} \text{ where } y = f(x)$$

$$\Rightarrow y - 1 = -2^{-x}$$

$$\Rightarrow 1 - y = 2^{-x}$$

$\Rightarrow x = -\log_2(1 - y)$ which is defined for $y < 1$

$R(y) = (-\infty, 1)$

Now, for every value of $x \in D(f) = R$, a unique value of $y \in R(f) = (-\infty, 1)$ is determined.

$\Rightarrow f(x) = 1 - 2^{-x}$ is single valued in its domain.

Also, for every value of $y \in R(f)$, a unique value of $x \in D(f)$ is determined from the equation $x = f^{-1}(y) = -\log_2(1 - y)$ is a single valued function in the range of the given function $y = 1 - 2^{-x}$.

Hence, $y = 1 - 2^{-x}$ exhibits a one-one correspondence between its domain and range.

$\Rightarrow y = 1 - 2^{-x}$ has an inverse.

Now, f^{-1} is found in the following way:

$x = f^{-1}(y) = -\log(1 - y)$

$\Rightarrow f^{-1}(x) = -\log(1 - x)$ which is the required inverse of the given function $f(x) = 1 - 2^{-x}$.

6. Find the inverse of the function defined as:

$$f(x) = \begin{cases} x, & x < 1 \\ x^2, & 1 \le x \le 4 \\ 8\sqrt{x}, & x > 4 \end{cases}$$

Solution: The given function is piecewise function defined as:

$$f(x) = \begin{cases} x, & x < 1 \\ x^2, & 1 \le x \le 4 \\ 8\sqrt{x}, & x > 4 \end{cases}$$

Now, $y = f(x) \Rightarrow x = f^{-1}(y)$

$y = x, x < 1$

$\Rightarrow x = y, y < 1$

$y = x^2, 1 \le x \le 4$

$\Rightarrow x = \sqrt{y}, 1 \le y \le 16$

$\left(\because 1 \le x \le 4 \Rightarrow 1 \le x^2 \le 16\right)$

$y = 8\sqrt{x}, x > 4$

$\Rightarrow x = \dfrac{y^2}{64}, y > 16$

$\left(\because x > 4 \Rightarrow \sqrt{x} > 2 \Rightarrow 8\sqrt{x} > 16\right)$

Hence,

$$x = \begin{cases} y, & y < 1 \\ \sqrt{y}, & 1 \le y \le 16 \\ \dfrac{y^2}{64}, & y > 16 \end{cases}$$

$$\Rightarrow f^{-1}(y) = \begin{cases} y; & y < 1 \\ \sqrt{y}, & 1 \le y \le 16 \\ \dfrac{y^2}{64}, & y > 16 \end{cases}$$

$$\Rightarrow f^{-1}(x) = \begin{cases} x, \ x < 1 \\ \sqrt{x}, \ 1 \le x \le 16 \\ \dfrac{x^2}{64}, \ x > 16 \end{cases}$$

is the required inverse of the given piecewise function.

Note: While finding the inverse of a piecewise function, one should find the inverse of each function defined in its respective sub domain.

Exercises on finding the inverse of a function

Exercise 1.25

Find the inverse of the given function, if there is one and determine its domain.

1. $f(x) = x^3$

2. $f(x) = x^2 + 5$

3. $f(x) = \dfrac{1}{x^2}$

4. $f(x) = (x+2)^3$

5. $f(x) = \dfrac{2x-1}{x}$

6. $f(x) = \dfrac{x+4}{2x-3}$

7. $f(x) = \dfrac{8}{x^3+1}$

8. $f(x) = \dfrac{2}{8x^2-1}$

9. $f(x) = 2|x| + x$

10. $f(x) = \dfrac{3}{1+|x|}$

Answers:

1. $f^{-1}(x) = \sqrt[3]{x}$

domain: $(-\infty, \infty)$

2. No inverse

3. No inverse

4. $f^{-1}(x) = \sqrt[3]{x} - 2$

domain: $(-\infty, \infty)$

5. $f^{-1}(x) = \dfrac{1}{(2-x)}$

domain: $R - \{2\}$

6. $f^{-1}(x) = -\left(\dfrac{3x+4}{1-2x}\right)$

domain: $R - \left\{\dfrac{1}{2}\right\}$

7. $f^{-1}(x) = \sqrt[3]{\dfrac{8-x}{x}}$

domain: $R - \{0\}$

8. $f^{-1}(x) = \sqrt{\dfrac{2+x}{8x}}$

domain: $x \le -2$ or $x > 0$

9. No inverse

10. No inverse

Exercise 1.26

Perform each of the following steps on the given function:

(a) Solve the equation for y in terms of x and express y as one or more functions of x:

(b) For each of the functions obtained in (a), determine if the function has an inverse, and if it does, determine the domain of the inverse function.

1. $x^2 + y^2 = 9$

2. $x^2 - y^2 = 16$

3. $xy = 4$

4. $y^2 - x^3 = 0$

5. $2x^2 - 3xy + 1 = 0$

6. $2x^2 + 2y + 1 = 0$

Answers:

1. (a) $f_1(x) = \sqrt{9 - x^2}$, $f_2(x) = -\sqrt{9 - x^2}$;

(b) Neither has an inverse.

2. (a) $f_1(x) = \sqrt{x^2 - 16}$, $f_2(x) = -\sqrt{x^2 - 16}$;

(b) Neither has an inverse.

3. (a) $f(x) = \dfrac{4}{x}$

(b) $f^{-1}(x) = \dfrac{4}{x}$, domain: $R-\{0\}$.

4. (a) $f_1(x) = \sqrt{x^3}$, $f_2(x) = -\sqrt[3]{x^3}$

(b) $f_1^{-1}(x) = \sqrt[3]{x^2}$; domain: R

$f_2^{-1}(x) = \sqrt[3]{x^2}$; domain: R.

5. (a) $f(x) = \dfrac{2x^2 + 1}{3x}$

(b) No inverse.

6. (a) $f(x) = -\dfrac{2x^2 + 1}{2}$

(b) No inverse.

Exercise 1.27

Determine if the given function has an inverse, and if it does, determine the domain and range of the given function.

1. $f(x) = \sqrt{x - 4}$

2. $f(x) = (x+3)^3$

3. $f(x) = x^2 - \dfrac{1}{x}$, $x > 0$

4. $f(x) = \dfrac{1}{x^2 + 4}$, $x \le 0$

5. $f(x) = x^5 + x^3$

6. $f(x) = x^3 + x$

Answer:

1. domain of f^{-1}: $[0, \infty)$ range of f^{-1}: $[4, \infty)$

2. domain of f^{-1}: R range of f^{-1}: R

3. domain of f^{-1}: R; range of f^{-1}: $(0, \infty)$

4. domain of f^{-1}: $\left(0, \dfrac{1}{4}\right]$; range of f^{-1}: $(-\infty, 0]$

5. domain of f^{-1}: R; range of f^{-1}: R

6. domain of f^{-1}: R; range of f^{-1}: R

2

Limit and Limit Points

These two concepts are defined based on the following concepts.

1. Open ε-neighbourhood of a Given Point

The set of all points (on the number line, or in a plane, or in an n-dimensional space, or in any space where the distance between any two points can be measured) whose distances from a given point are less than a given positive number 'ε' is called an open ε-neighbourhood of a given point.

In notation, we express an open ε-neighbourhood of a given point x_0 on the number line as:

$$N_\varepsilon(x_0) = \{x \in R : |x - x_0| < \varepsilon, \varepsilon > 0\}$$

Notes: **1.** $x \in N_\varepsilon(x_0) \Leftrightarrow |x - x_0| < \varepsilon$

$\Leftrightarrow x \in (x_0 - \varepsilon, x_0 + \varepsilon)'$

2. (i) on the real line, an open e-neighbourhood of a given point is a line segment (a part of the real line) without (not counting) the end points whereas the given point whose open e-neighbourhood is sought is the midpoint of the line segment, i.e., an open interval with a midpoint x_0 and without left end point x_0 – e and right end point x_0 + e represented as (x_0 – e, x_0 + e) is an e-neighbourhood of a given point x_0.

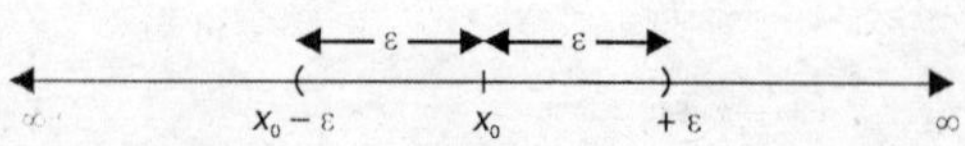

(ii) In two dimensional space (in the real plane or complex plane), an open ε-neighbourhood of a given point is the set of all those points excluding the circumference which is also termed as boundary of the circle in real analysis. In other words, a circle centred at a given point (whose open ε-neighbourhood is sought) and whose radius is the given number ∈ is an open ε-neighbourhood of the given point if we exclude all those points whose distances from the center equal the radius. In real analysis, a circle in a plane is termed as circular neighbourhood of the given point.

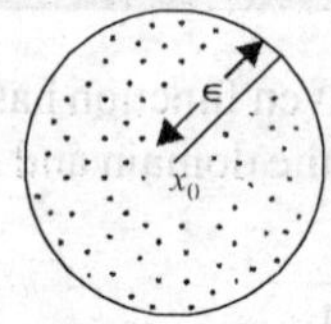

(iii) In three dimensional space, it is the set of all those points inside the sphere. A sphere is also termed as spherical neighbourhood of a given point.

3. An open ε-neighbourhood of a given point is also termed as:

(i) Open sphere (centred at the given point with a given radius ε) and it is symbolized as $S_\varepsilon(x_0)$ or $S(\varepsilon, x_0)$.

(ii) Open ball (centred at the given point with a given radius ε) and it is symbolized as $B_\varepsilon(x_0)$ or $B(\varepsilon, x_0)$.

2. A Closed ε-neighbourhood of a Given Point

The set of all points (in any space) whose distances from a given point are less than or equal to a given positive number 'ε' is called a closed ε-neighbourhood of a given point.

Notation: The notation used for a closed ε-neighbourhood of a given point x_0 is $N_\varepsilon [x_0]$. Hence, in notation, we express a closed ε-neighbourhood of a given point x_0 on the real line as:

$$N_\varepsilon [x_0] = \{x \in R : |x - x_0| \leq \varepsilon, \varepsilon > 0\}$$

Notes:

1. $x \in N_\varepsilon (x_0) \Leftrightarrow |x - x_0| \leq \varepsilon \Leftrightarrow x \in [x_0 - \varepsilon, x_0 + \varepsilon]$

2. (i) On the real line a closed ε-neighbourhood of a given point is a line segment (a part of the real line) with the end points whereas the given point whose closed ε-neighbourhood is sought is the mid point of the line segment, i.e., a closed interval with the midpoint x_0, left end point $x_0 - \varepsilon$ and right end point $x_0 + \varepsilon$ represented as $[x_0 - \varepsilon, x_0 + \varepsilon]$ is a closed ε-neighbourhood of a given point x_0.

(ii) In two dimensional space (in the real plane or a complex plane), a closed ε-neighbourhood of a given point is the set of all those points which are within the circle (circumference of the circle) together with all those points which are on the circle (circumference) of a given radius ε and whose center is the point whose closed ε-neighbourhood is sought (required).

(iii) In three dimensional space, a closed ε-neighbourhood of a given point is the set of all those points inside the sphere together with all those points which are on the sphere (the set of all points at a distance ε from a given point x_0)

3. (i) The set of all those points whose distances from a given point x_0 are less than a given positive number ε (termed as radius of the sphere) is called interior of the closed ε-neighbourhood of the given point x_0 or simply interior of the sphere (circle),

(ii) The set of all those points whose distances from a given point are equal to a given positive number (also termed as the radius of the sphere) is called boundary of the sphere, i.e., a closed sphere is the union of the interior and the boundary of the circle or sphere.

(iii) The set of all those points whose distances from a given point x_0 is greater than a given positive number ε is called exterior of the sphere (a closed ε-neighbourhood of a given point x_0)

4. A closed ε-neighbourhood of a given point x_0 is also termed as:

(i) Closed sphere or simply sphere (centred at the given point with a given radius ε) and it is symbolized as $S_\varepsilon [x_0]$ or $S[\varepsilon, x_0]$.

(ii) Closed ball (centred at the given point x_0 with a given radius ε) and it is symbolized as $B_\varepsilon [x_0]$ or $B_\varepsilon [x_0, \varepsilon]$.

Remarks: **(i)** An ε-neighbourhood of a given point can be closed as well as open. But in general, an open ε-neighbourhood is understood whenever an ε-neighbourhood of a given point is written or spoken.

(ii) An ε-neighbourhood of a given point is determined by a given point x_0 and a given small positive number ε. x_0 is called the center of the sphere and ε is called the radius of the sphere.

(iii) Roughly speaking neighbourhood of a given point x_0 means all points near about x_0.

(iv) The distance between any two points x and x_0 is denoted by:

(a) $d(x, x_0)$ in a metric space

(b) $|x - x_0|$ on a real line (one dimensional space)

(c) $\| x - x_0 \|$ in a normed linear space.

(v) We say that two sets A and B meet (intersect) $\Leftrightarrow A \cap B \neq \phi$

Note: We may divide an ε-neighbourhood of a given point x_0 into two halves namely

(a) Right ε-neighbourhood of a given point and

(b) Left ε-neighbourhood of a given point in a space R.

These two concepts are defined in the following way:

(a) Right ε-neighbourhood of a given point x_0

$N_\varepsilon (x_0) = \{x \in R : x_0 \leq x < x_0 + \varepsilon\}$ is called right ε-neighbourhood of a given point x_0, i.e., the interval $[x_0, x_0 + \varepsilon)$, where $x_0 \in R$ and $\varepsilon > 0$ is called a right ε-neighbourhood of the given point x_0.

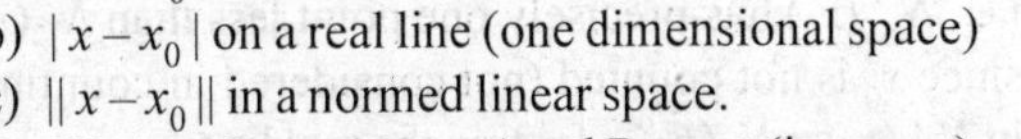

(b) Left ε-neighbourhood of a given point x_0

$N_\varepsilon (x_0) = \{x \in R : x_0 - \varepsilon < x \leq x_0\}$ is called left ε-neighbourhood of a given point x_0, i.e., the interval $(x_0 - \varepsilon, x_0]$, where $x_0 \in R$ and $\varepsilon > 0$ is called a left ε-neighbourhood of a given point x_0.

Lastly in connection with ε-neighbourhood of a given point, there is one more concept known as deleted ε-neighbourhood of a given point.

Deleted ε-neighbourhood of a given point

An ε-neighbourhood of a given point from which the given point itself is removed (from counting) usually written with a prime to denote deletion as $N_\varepsilon'(x_0)$ is called punctured ε-neighbourhood of the given point x_0, i.e., an ε-neighbourhood $N_\varepsilon(x_0)$ without (not counting) the point x_0 is called a deleted ε-neighbourhood of the given point x_0.

In notation, we express on the real line

$$N_\varepsilon'(x_0) = N_\varepsilon(x_0) - \{x_0\} = \{x \in R : 0 < |x - x_0| < \varepsilon\}$$
$$= \{x \in R : x_0 - \varepsilon < x < x_0 + \varepsilon, x \neq x_0\}$$

Notes: **1.** One should note that the deleted ε-neighbourhood of a given point has precisely one point less than the ε-neighbourhood of a given point, i.e., $N_\varepsilon'(x_0)$ has precisely one point less than $N_\varepsilon(x_0)$ since x_0 is not counted (not considered in counting) in $N_\varepsilon'(x_0) = N_\varepsilon(x_0) - \{x_0\}$ on the real line.

2. In general, if A is a neighbourhood containing the point x_0, then $A - \{x_0\}$ is called deleted neighbourhood of the point x_0, e.g.,

$1 < x < 3, x \neq 2$ is deleted neighbourhood of the point $x = 2$.

3. We may divide a deleted ε-neighbourhood of a given point x_0 say $N_\varepsilon'(x_0)$ into two halves namely (a) deleted right ε-neighbourhood of a given point and (b) deleted left ε-neighbourhood of a given point x_0 in a space R defined as:

(a) Deleted right ε-neighbourhood of a given point
$N_\varepsilon'(x_0) = \{x \in R : x_0 < x < x_0 + \varepsilon\}$, i.e., the interval $(x_0, x_0 + \varepsilon)$ where $x_0 \in R$ and $\in > 0$ is a given small number, is called a deleted right ε-neighbourhood of a given point x_0, e.g:

$2 < x < 3$ is a right neighbourhood of 2.

(b) Deleted left ε-neighbourhood of a given point
$N_\varepsilon'(x_0) = \{x \in R : x_0 - \varepsilon < x < x_0\}$, i.e., the interval $(x_0 - \varepsilon, x_0)$ where $x_0 \in R$ and $\varepsilon > 0$ is a given number, is called a deleted left ε-neighbourhood of a given point x_0, e.g:

$-1 < x < 2$ is a left deleted neighbourhood of 2.

Question 1: Explain the meaning of the statement: "Two points are close to each other in a space".

Answer: When one says that two points are close (near) to each other in a space, it means that the distance between two points in space is small which is less than some small given positive number ε, i.e., a point x is close to another point x_0 in a space R means that $|x - x_0| < \varepsilon$, i.e., $x_0 - \varepsilon < x < x_0 + \varepsilon$, or $x \in (x_0 - \varepsilon, x_0 + \varepsilon)$, e.g:

(i) $\varepsilon = 0.5 \Rightarrow x$, is close to 1 only if $|x - 1| < 0.5$, i.e., $0.5 < x < 1.5$.

(ii) $\varepsilon = 0.3 \Rightarrow x$, is close to 1 only if $|x - 1| < 0.3$, i.e., $0.7 < x < 1.3$.

One should observe that with $\varepsilon = 0.5$, 1.4 is near 1 but if $\varepsilon = 0.3$, 1.4 is not near 1.

1. Space: A set of points endowed with a structure which is usually defined by specifying a set of axioms to be satisfied by the points, e.g: Real line (one dimensional space), a plane (two dimensional space), three dimensional space, *n*-dimensional space, metric space, topological space, Banach space, Vector space, etc.

2. Closure point of a set: It is defined with respect to different aspects in the following ways:

Definition 1: (Intuitive concept): A point x_0 in a space X is called a closure point of the set A contained in the space X $(A \subseteq X) \Leftrightarrow$ there is in the set A, at least one point x which is as close to x_0 as we please.

Definition 2: (In terms of distance): A point x_0 in a space X is called a closure point of the set A contained in the space X $(A \subseteq X) \Leftrightarrow$ there is in the set A, at least one point x whose distance from the point x_0 is less than any given small positive number ε, i.e., a point x_0 in a space X $(x_0 \in X)$ is called a point of closure of the set $A \subseteq X \Leftrightarrow$ for each $\varepsilon > 0$, there is in the set A, at least one point x such that

(a) $|x - x_0| < \varepsilon$ on a number line

(b) $d(x - x_0) < \varepsilon$ in a metric space

(c) $\|x - x_0\| < \varepsilon$ in a normed space

Definition 3: (In terms of neighbourhood): A point x_0 in a space X $(x_0 \in X)$ is closure or adherent to a set $A \subseteq X \Leftrightarrow$ each neighbourhood of the point x_0 includes in itself (contains) at least one point of the set A (which may be the point x_0). Or in other words, a closure point of a set of a point in a space $\Leftrightarrow$ every neighbourhood of the point in the space intersects the set at some point which may be the point itself whose neighbourhood we consider. Hence, in notation we can express the definition of closure point of a set in different spaces in the following ways:

(a) A point x_0 on the number line R $(x_0 \in R)$ is called a closure point of the set $A \subseteq R \Leftrightarrow \forall (x_0 - \varepsilon, x_0 + \varepsilon), (x_0 - \varepsilon, x_0 + \varepsilon) \cap A \neq \phi$.

(b) A point x_0 in a metric space X $(x_0 \in X)$ is called a closure point of the set $A \subseteq X \Leftrightarrow \forall\, S_\varepsilon(x_0)$, $S_\varepsilon(x_0) \cap A \neq \phi$.

(c) A point x_0 in a topological space X $(x_0 \in X)$ is called a closure point of the set $A \subseteq X \forall \Leftrightarrow N_x, N_x \cap A \neq \phi$.

Notes: **1.** The set of all closure points of a set A is called closure of the set A and it is denoted as . Hence,

(a) $\overline{A} = \{x_0 \in R \mid \forall (x_0 - \varepsilon, x_0 + \varepsilon), (x_0 - \varepsilon, x_0 + \varepsilon) \cap A \neq \phi\}$ on the number line.

(b) $\overline{A} = \{x_0 \in X \mid \forall S_\varepsilon(x_0), S_\varepsilon(x_0) \cap A \neq \phi\}$ in a metric space.

(c) $\overline{A} = \{x_0 \in X \mid \forall N(x_0), N(x_0) \cap A \neq \phi\}$ in a topological space.

2. (a) $x_0 \in \overline{A} \Leftrightarrow S_\varepsilon(x_0) \cap A \neq \phi$ in a metric space.

(b) $x_0 \in \overline{A} \Leftrightarrow (x_0 - \varepsilon, x_0 + \varepsilon) \cap A \neq \phi$ on the number line.

(c) $x_0 \in \overline{A} \Leftrightarrow N(x_0) \cap A \neq \phi$ in a topological space.

3. A closure point of the set may or may not belong to the set but it must be in (on) the space which contains the set whose closure point is sought.

4. A closure point of a set is also termed as:

(a) a point of closure

(b) adherent point

(c) contact point.

Kinds of Closure Point of a Set

There are two types of closure point of a set namely

1. Limit point of a set.

2. Isolated point of a set.

Now we define each one in a space with respect to different aspects.

1. Limit point of a set

Definition: (Intuitive concept): A point P in a space is called the limit point of a set contained in the space $\Leftrightarrow$ There is in the set at least one point which is not P (not the same point P) arbitrarily close to $P \Leftrightarrow$ There are in the set infinitely many points, no point of which is P, arbitrarily close to P.

Explanation: a point P in a space X, not necessarily a point of the set A contained in X $(A \subseteq X)$ is a limit point of the set $A \Leftrightarrow$ There exists at least one point x in the set which does not coincide with (equal) the point P while the point x in the set is as close to P as we please $\Leftrightarrow$ There are infinitely many points of the set, no point of which coincides with (equals) the point P while these points of (in) the set are arbitrarily close to P which may or may not belong to the set.

Definition 2: (In terms of distance): A given point P in a space is called the limit point of the set contained in a space $\Leftrightarrow$ There is in the set at least one point whose distance from the point P in the space is nonzero and less than any given small positive number ε.

Hence, in notation, we can express the definition of the limit point of a set in terms of distance in different spaces in the following ways:

(a) A point P on the number line (in or on a real line R) is called the limit point of the set $A \subseteq R \Leftrightarrow \forall \in > 0, \exists$ an $x \in A$ such that $0 < |x - P| < \varepsilon$.

(b) A point P in a metric space X is called the limit point of the set $A \subseteq X \Leftrightarrow \forall \in > 0, \exists$ an $x \in A$ such that $0 < d(x, P) < \varepsilon$.

(c) A point P in a normed space X is a limit point of the set $A \subseteq X \Leftrightarrow \forall \in > 0, \exists$ an $x \in A$ such that $0 < \| x - P \| < \varepsilon$.

Definition: (In terms of neighbourhood): A point P in a space is called the limit point of the set contained in a space $\Leftrightarrow$ every neighbourhood of the point P includes in itself (contains) at least one point of the set which is not P (not the same point P, different from, distinct from or other than P), or in other words, a point P in a space is called the limit point of a set contained in a space $\Leftrightarrow$ Every neighbourhood of the point P intersects the set at a point which is not P (not the point whose neighbourhood we consider).

Hence, in notation, we can express the definition of the limit point of a set in different spaces in the following ways:

(a) A point P on the number line R ($P \in R$) is called the limit point of the set $A \subseteq R \Leftrightarrow \forall (P-\varepsilon, P+\varepsilon)$, there is at least one point $x \in (P-\varepsilon, P+\varepsilon)$, $x \in A$, $x \neq P \Leftrightarrow \forall \varepsilon > 0$, $(P-\varepsilon, P+\varepsilon) \cap A$ contains infinitely many points of the set A different from (distinct from, other than, or not the same point) P.

(b) A point P in a metric space X is called the limit point of the set $A \subseteq X \Leftrightarrow \forall S_\varepsilon(P)$, there is at least one point $x \in S_\varepsilon(P)$, $x \in A$, $x \neq P \Leftrightarrow \forall \varepsilon > 0$, $S_\varepsilon(P) \cap A$ contains infinitely many points of the set A different from P.

Remark: The reason for using the phrase "distinct from, different from, other than or which is not (not the same) point P whose neighbourhood we consider" is to emphasize that if P is in the set, at least one point of the set which is not the given point P should be in every neighbourhood of the given point P, i.e. the use of the above phrases emphasize that if P is in the set, at least one more point of the set besides the point P should be in every neighbourhood of the point P. This is why we can restate the definition of the limit point of a set in two parts in the following way to get the definition clarified.

(i) A point P belonging to the set is a limit point of the subset $A \Leftrightarrow$ Any neighbourhood of the P point contains at least one point which is not the same point P and is a member (point) of the set A, i.e., a given point $P \in$ the set A ($P \in A \subseteq X$) is a limit point of $A \Leftrightarrow$ Any neighbourhood of the point P, ($N(P)$), contains at least one point which is not the point P and is a member (point) of the set $A \Leftrightarrow \bar{N}(P) \cap (A - \{x\}) \neq \phi$.

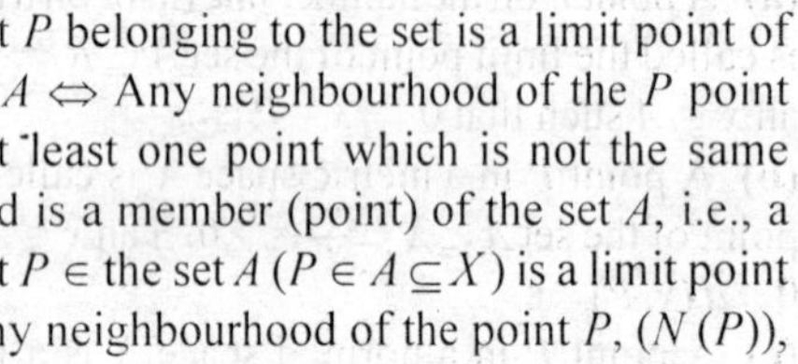

(ii) A given point not belonging to the set (not in) the set is a limit point of the set $\Leftrightarrow$ Any neighbourhood of the given point contains at least one point which is a member (point) of the set, i.e. a given point $P \notin$ the set A ($P \notin A$) is a limit point of $A \Leftrightarrow$ Any neighbourhood of the point P ($N(P)$) contains at least one point which is the member (point) of the set $A \Leftrightarrow N(P) \cap A \neq \phi$.

However, we can avoid the use of the phrase "other than, different (distinct) from, or not the same" in two ways:

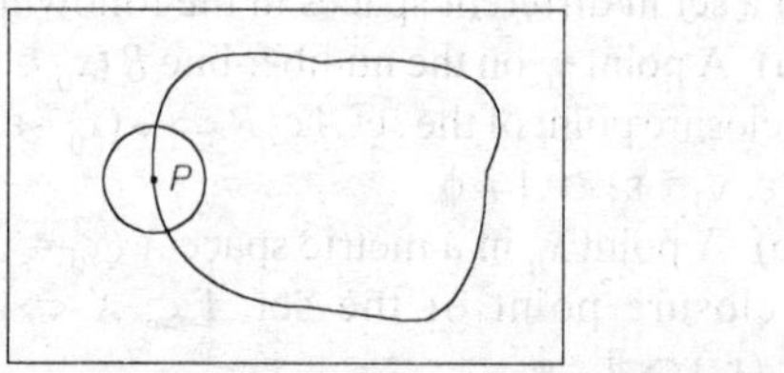

(a) using any of the words "an other point, a distinct (different) point" of the set as: a limit point of a set is a point, not necessarily in the set, in whose neighbourhood lies at least one another point of the set.

(b) using the concept of deleted neighbourhood of a given point P in a space.

Definition (iv): (in terms of deleted neighbourhood): A point P in a space X is called the limit point of a set A contained in a space X ($A \subseteq X$) $\Leftrightarrow$ Every deleted neighbourhood of the point P contains at least one point of the set $\Leftrightarrow$ Every deleted neighbourhood of the point P intersects the set A at a point

Hence, in notation, we can express the definition of the limit point in terms of deleted neighbourhood in different spaces in the following ways:

(a) A point P on the number line is called the limit point of the set $A \subseteq R \Leftrightarrow \forall \varepsilon > 0$, $(P-\varepsilon, P+\varepsilon) - \{P\} \cap A \neq \phi$.

(b) A point P in a metric space X is called the limit point of the set $A \subseteq X \Leftrightarrow \forall \varepsilon > 0$, $S_\varepsilon(P) - \{P\} \cap A \neq \phi$.

(c) A point P in a topological space X is called the limit point of the set $A \subseteq X \Leftrightarrow N(P) - \{P\} \cap A \neq \phi$, $\forall$ deleted neighbourhood of the point P, i.e., $N(P) - \{P\}$.

Definition (v): (In terms of closure of a set): A point P in a space X is a limit point of A ($A \subseteq X$) $\Leftrightarrow$ The point P belongs to the closure of $A - \{P\}$ ($P \in$ closure of $A - \{P\}$) $\Leftrightarrow N(P) \cap A - \{P\} \neq \phi$.

Remarks: **1.** P is the limit point of the set $A \Leftrightarrow P$ is the limit point of the set $A - \{P\}$. Hence, a point P in a space X ($P \in X$) is a limit point of the set A contained in the space X ($A \subseteq X$) $\Leftrightarrow$ Each neighbourhood of the point P ($N(P)$) contains at least one point of the set $A - \{P\} \Leftrightarrow N(P) \cap A - \{P\} \neq \phi$.

2. The world "every, each or all" used before ε-neighbourhood or neighbourhood emphasizes firstly that the said properly must hold even if the ε-neighbourhood or neighbourhood of the given point P is arbitrarily small and secondly that if there exists even one ε-neighbourhood or neighbourhood of the point P which does not contain at least one point of the set which is not P, P can not be said to be a limit point of the set A.

Notes: **1.** The limit points of a set may or may not belong to the set.

2. The limit point of a set is also termed as:

(i) Limiting point

(ii) Accumulating point and

(iii) Cluster point.

3. The set of all the limit points of a set A is called the derived set which is denoted by A' or $D(A)$, i.e.,

(i) $A' = D(A) = \{x_0 \in R : N_\varepsilon(x_0) - \{x_0\} \cap A \neq \phi, \forall N_\varepsilon(x_0)\} = \{x_0 \in R : x_0$ is the limit point of $A\}$ on the number line R.

(ii) $A' = D(A) = \{x_0 \in X : S_\varepsilon(x_0) - \{x_0\} \cap A \neq \phi, \forall S_\varepsilon(x_0)\}$ in a metric space X.

(iii) $A' = D(A) = \{x_0 \in X : N(x_0) - \{x_0\} \cap A \neq \phi, \forall N(x_0)\}$ in a topological space X.

4. $x_0 \in D(A) \Leftrightarrow x_0$ is a limit point of A.

$\Leftrightarrow S_\varepsilon(x_0) - \{x_0\} \cap A \neq \phi$ in a metric space

$\Leftrightarrow ((x_0 - \varepsilon, x_0) \cup (x_0, x_0 + \varepsilon)) \cap A \neq \phi$ on the real line

$\Leftrightarrow N(x_0) - \{x_0\} \cap A \neq \phi$ in a topological space

5. We may divide a limit point of a set into two halves namely.

(a) Left limit point and

(b) Right limit on the number line R.

Now we define each one on the number line R in the following way:

(a) Left limit point of a set: A point $x_0 \in R$ is a limit point of the set $A \subseteq R \Leftrightarrow \forall \in > 0$, there is at least one point $x \in A$ such that $0 < x_0 - x < \varepsilon$ $(x_0 - \varepsilon < x < x_0)$.

i.e. **(i)** A point $x_0 \in R$ is right limit point (or right hand) limit point of the set $A \Leftrightarrow x_0$ is a limit point of the subset of A lying to the right of the point x_0.

(b) Right limit point of a set: A point $x_0 \in R$ is a right limit point of a set $A \subseteq R \Leftrightarrow \forall \varepsilon > 0$, there is at least one point $x \in A$ such that $0 < x - x_0 < \varepsilon$ $(x_0 < x < x_0 + \varepsilon)$ Further, one should keep in mind that $x_0 \in R$ is a limit point of a set $A \subseteq R \Leftrightarrow x_0 \in R$ is a left limit point or a right limit point of the set $A \subseteq R$.

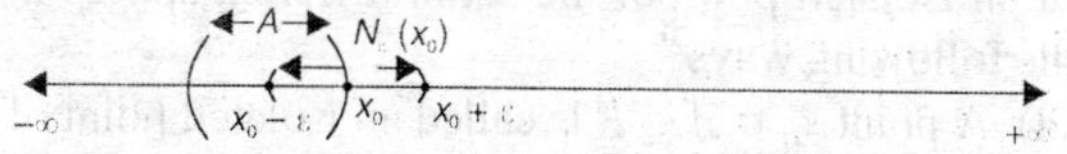

i.e. **(ii)** A point x_0 Î R is left limit point (or left hand) limit point of he set A Û x_0 is a limt point of the subset of A lying to the left of the point x_0.

6. A set A is closed $\Leftrightarrow$ each limit point of the set A is a member (point) of the set $A \Leftrightarrow$ A point x is a limit point of a set A and $x \in A \Leftrightarrow$ the set A contains all its limit points $\Leftrightarrow D(A) \subset A$.

7. $x_0 \notin D(A) \Leftrightarrow x_0 \notin$ closure of $A - \{x_0\} \Leftrightarrow x_0$ is not a limit point of the set A.

8. More on closure point and closure of a set: A closure point or closure of a set is also defined in terms of the limit point of a set.

(i) Closure point of the set (in terms of the limit point): A point x_0 in a space X is called a closure point of the set A contained in the space X $(A \subseteq X) \Leftrightarrow$ $x_0 \in A$ or x_0 is a limit point of A.

(ii) Closure of a set: Closure of a set denoted by   is the set of all points of A together with all those points (in space) which are arbitrarily close to A. That is, the closure of A, denoted by , is the union of the set A and the set of all its limits points, i.e., $= A \cup D(A)$.

Now, we give the definition of an isolated point of a set.

2. Isolated Point of a Set

It is also defined with respect to different aspects.

Definition (i): (In terms of neighbourhood): A point of a set is called an isolated point of the set $\Leftrightarrow$ there exists a neighbourhood of that point in which there is no other point of the set. That is, a point of the set whose one neighbourhood includes in itself (contains) no other point of the set, i.e., $N_{x_0} = \{x_0\}$, is called an isolated point of the set.

Definition (ii): (In terms of deleted neighbourhood): A point of a set whose one deleted neighbourhood does not intersect the given set is called an isolated point of the set. That is, a point belonging to the set which is not the limit point of the set is called an isolated point of the set.

Hence, in notation we can express the definition of an isolated point of the set in different spaces in the following ways:

(a) A point $x_0 \in A \subseteq R$ is called an isolated point of the set $A \Leftrightarrow N_\varepsilon(x_0) - \{x_0\} \cap A = \phi \Leftrightarrow N_\varepsilon(x_0) \cap A = \{x_0\}$ for some $N_\varepsilon(x_0)$.

(b) A point $x_0 \in A \subseteq X$, where X is a metric space, is called an isolated point of the set $A \Leftrightarrow S_\varepsilon(x_0) - \{x_0\} \cap A = \phi \Leftrightarrow S_\varepsilon(x_0) \cap A = \{x_0\}$ for some $S_\varepsilon(x_0)$.

(c) A point $x_0 \in A \subseteq X$, where X is a topological space, is called an isolated point of the set $A \Leftrightarrow N_{x_0} - \{x_0\} \cap A = \phi \Leftrightarrow N_{x_0} \cap A = \{x_0\}$, for some N_{x0}

Notes: **1.** An isolated point of a set is a point of the set.

2. x_0 is an isolated point of the set $A \Leftrightarrow x_0$ is not a limit point of the set $A \Leftrightarrow x_0 \notin D(A)$.

3. Roughly speaking, an isolated point of a set is a point x_0 of the A around which there is no point of the set A which is different from (distinct from, or other than) the point itself namely x_0.

Kinds of the Limit Point of a Set

There are two types of the limit point of a set namely:

(a) Interior point of a set.

(b) Boundary point of a set.

Each one is defined with respect to different aspects:

(a) Interior point:

Definition (i): (In terms of the limit point): An interior point of the set is a point of the set which is not the limit point of the complement of the set. That is, a limit point x_0 of the set A is an interior point of the set $A \Leftrightarrow$ there are only the points of the set A in some neighbourhood of the point x_0.

Definition (ii): (In terms of neighbourhood): A point x_0 in a space X is called the interior point of the set A contained in the space X $(A \subseteq X) \Leftrightarrow$ There is a neighbourhood of the point x_0 which is a subset of the set A whenever the point x_0 is in the set $A \Leftrightarrow x_0 \in A$ and $\exists$ a N_{x_0} such that $N_{x_0} \subseteq A$.

Hence, in notation, we can express the definition of the interior point of a set in different spaces in the following ways:

(i) On the number line, a point x_0 in a set $A \subseteq R$ is called an interior point of the set $A \Leftrightarrow x_0$ can be enclosed in an interval $(x_0 - \varepsilon, x_0 + \varepsilon)$, $\in > 0$, such that all the point of this interval are the points of the set A, e.g: every point of a line segment other than (not counting, or without) the end points of the line segment is the interior point of the set composed of all points of the line segment.

(ii) In a metric space X, a point x_0 in a set $A \subseteq X$ is called an interior point of the set $A \Leftrightarrow x_0$ can be enclosed in an ε-neighbourhood $N_\varepsilon(x_0)$ such that all the points of this ε-neighbourhood $N_\varepsilon(x_0)$ are the points of the set A.

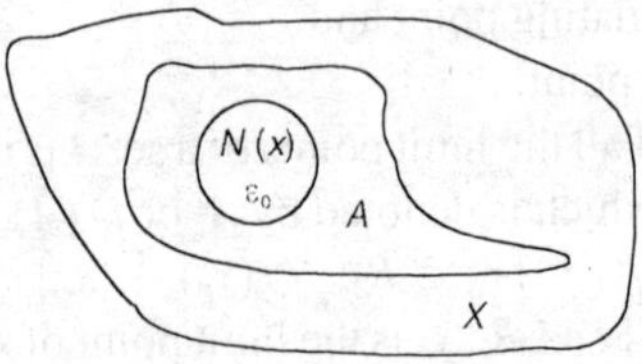

(b) Boundary point:

Definition (i): (Intuitive concept): A point x_0 in a space X is called a boundary point of the set A contained in a space X $(A \subseteq X) \Leftrightarrow x_0$ is arbitrarily close to both the set A and its complement A^C. That is, the points which are arbitrarily close to both the set and the complement of the set are called boundary points of the set.

Definition (ii): (In terms of limit point): A point x_0 in a space X is called a boundary point of the set A contained in a space $X(A \subseteq X) \Leftrightarrow x_0$ is the limit point of both the set A and its complement A^C.

Definition (iii): (In terms of neighbourhood): A point x_0 in a space X is called boundary point of the set A contained in a space X $(A \subseteq X) \Leftrightarrow$ every neighbourhood of the point x_0 intersects both the set A and the complement of the set A (A^C) at some points $\Leftrightarrow x_0$ is a point of closure of both the set A and its complement A^C.

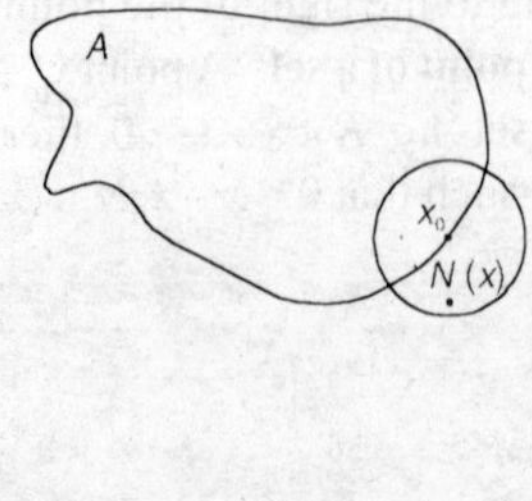

Hence, in notation we can express the definition of the boundary point of a set in different spaces in the following ways:

(i) A point x_0 on the number line R is a boundary point of the set A contained in a space R $(A \subseteq R) \Leftrightarrow \forall$ $N_\varepsilon(x_0), N_\varepsilon(x_0) \cap A \neq \phi, N_\varepsilon(x_0) \cap A^C \neq \phi$.

(ii) A point x_0 in a metric space X is a boundary point of the set A contained in the space X $(A \subseteq X) \Leftrightarrow \forall$ $S_\varepsilon(x_0), S_\varepsilon(x_0) \cap A \neq \phi, S_\varepsilon(x_0) \cap A^C \neq \phi$.

(iii) A point x_0 in a topological space X is a boundary point of the set A contained in a space X $(A \subseteq X) \Leftrightarrow \forall$ $N_x, N_x \cap A \neq \phi, N_x \cap A^C \neq \phi$.

Besides these, in connection with the interior points of the set and the boundary points of the set, there are two more sets namely:

1. **Interior of a set** and
2. **Boundary of a set**

Which are defined in the following ways.

1. Interior of a set: A set whose members are all the interior points of the set is called interior of the set, that is, the set $A = \{x : x \in A$ and some $N_x \subseteq A\}$ is called the interior of the set A.

The symbol to denote the interior of a set A is int (A), A^i or A^0.

2. Boundary of a set

Definition (i): (intuitive concept): The boundary of a set A is the set of all those points which are arbitrarily close to both the set A and its complement A^C.

Definition (ii): A set of all those points which are boundary points of a set A is called the boundary of a set A.

The boundary of a set A is symbolized as:

$b(A)$, A^b or $\partial(A)$

Hence, in terms of interior point and boundary point of a set, we can say,

A set is closed $\Leftrightarrow$ all the limit points (interior and boundary points both) of a set belong to the set.

Similarly, in terms of the interior and the boundary of a set, we can express the closure of a set A as:

$\overline{A} = \text{int}(A) \cup b(A)$

Now we define one more important concept know as "open set" in the following way:

Open set: A set A of points (on the number line, or in the plane, or in ordinary space, or in n-space or in any space) is called open if, whenever it contains a point P, it also contains all points (on the number line, or in the plane, and so forth) near P, that is, all points of some interval with midpoint P. That is, a set A contained in a space X $(A \subseteq X)$ is called open $\Leftrightarrow$ every point of the set A is an interior point of the set A (i.e., $A = \text{int}(A)$), or in other words, A set A contained in a space X $(A \subseteq X)$ is called open given any point x in A, $\exists$ a neighbourhood of the point x (Nx) such that $Nx \subseteq A$.

Notes: **1.** Roughly speaking, the 'boundary' of a region (a set of points which is either a non empty open set or such a set together with some or all of the points forming its boundary), if it exists, is the set of points in the region from those not in (Simply to avoid clumsiness of language we often say 'points are in a region' instead of 'the points are points of region'. All the regions considered by us will be what are known as 'open region'; so that in the cases in which we use it, our definition of boundary agrees with the usual definition, in which points of the region may also be the points of the boundary).

2. In 3-pace, the closure of a set A is the set A together with all its skin, whether the skin is part of the set or not. The interior of the set A is the set A minus any part of the skin which it contains. The boundary of a set A is its skin.

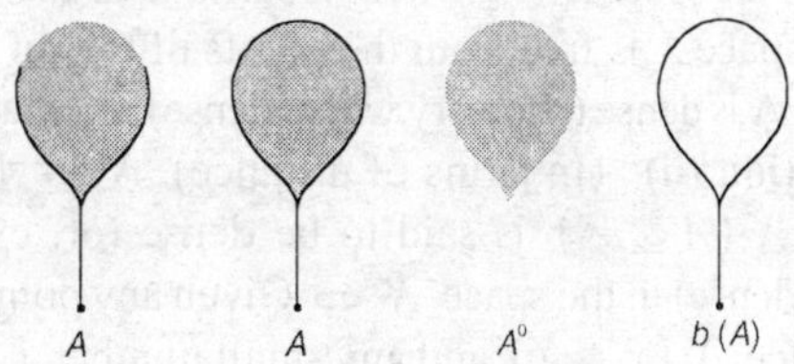

3. End points of any interval on the number line are the boundary points of the interval.

4. A boundary point may belong to either the set A or its complement A^C but the boundary point is the limit point of both of them.

5. Every point of an open set is an interior point lying in the set itself.

6. x_0 is a limit point of $A \Rightarrow x_0 \in A$ and $x_0 \notin A$.

7. $x_0 \in A \Rightarrow x_0$ is a closure point of the set A trivially.

8. Every limit point of the set A is also a point of closure of the set A but not conversely.

9. Every interior point of the set A is also a limit point.

Dense Set

We consider three kinds of sets:
1. Dense set or dense in itself set.
2. Everywhere dense set in a space.
3. Non-dense set or nowhere dense set.

Each one is defined in the following ways:

1. Dense set (dense in itself set)

Definition (i): (Intuitive concept): A set in a space is said to be dense in itself (or simply dense set) $\Leftrightarrow$ For any two distinct arbitrary points in the set, there is at least one distinct point between the two given points.

Definition (ii): (In terms of neighbourhood): A set A in a space X is said to be dense in itself $\Leftrightarrow$ Every neighbourhood of every point x of the set contains at least one point of the set which is not x.

Definition (iii): (In terms of the limit point): A set A in a space X is said to be dense in itself $\Leftrightarrow$ Every point of the set A is the limit point of the set A.

2. Everywhere dense set in a space

Definition (i): (Intuitive concept): A dense set A in a space X means that the points of the set A are distributed 'thickly' throughout the space X. In other words, the set A contains points as near as we like to each point of the space $X \Leftrightarrow$ No neighbourhood of any point in the space X is free from the points of the set $A \Leftrightarrow$ The set A is dense (or, every where dense) in a space X.

Definition (ii): (In terms of distance): A set A in a space $X\ (A \subseteq X)$ is said to be dense (or, everywhere dense) in the space $X \Leftrightarrow$ Given any point x in the space $X\ (x \in X)$ and any small number $\varepsilon > 0$, there is at least one an other number x_0 in the set $A\ (x_0 \in A)$ such that the distance between the point x in the space X and the point x_0 in the set A is less than the given small positive number ε, i.e.,

(a) $|x - x_0| < \varepsilon$ on the number line

(b) $d(x, x_0) < \varepsilon$ in a metric space

(c) $\|x - x_0\| < \varepsilon$ in a normed space

Definition (iii): (In terms of neighbourhood): A set A in a space X is dense (or, everywhere dense) in the space $X \Leftrightarrow$ Every neighbourhood of every point $P(N(P))$ in a space X contains at least one point of the set A.

Definition (iv): (in terms of closure point): A set A in a space X is dense (or, everywhere dense) in the space $X \Leftrightarrow$ Every point of the space X is a point of closure of the set $A \Leftrightarrow$ Every point of the space X is a point of the set A or a limit point of the set A (i.e. $X = \overline{A} = A \cup D(A)$).

Definition (v): (In terms of limit point): A set A in a space $X(A \subseteq X)$ is said to be dense (or everywhere dense) in the space $X \Leftrightarrow$ Every point of the space $X - A$ is a limit point of the set A).

3. Nowhere dense (or, non-dense) set in a space

Definition (i): (In terms of neighbourhood): A set A in a space $X(A \subseteq X)$ is said to be nowhere dense (non-dense) in the space $X \Leftrightarrow$ Every neighbourhood of every point in a space X contains a certain neighbourhood of a point in a space such that this certain neighbourhood is free from the points of the set $A \Leftrightarrow$ For every point x in the space $X(x \in X)$ and each neighbourhood of x ($N(x)$), there is a neighbourhood of an other point y ($N(y)$) in the space $X(y \in X)$ such that $N(y) \subset N(x)$ and $N(y) \cap A = \phi \Leftrightarrow$ interior of the closure of the set A is empty (i.e., $(\overline{A})^{\circ} = \text{int}(\overline{A}) = \phi$).

Definition (ii): (In terms of dense exterior): A set with dense exterior is said to be a non-dense set, i.e., A set A in a space $X(A \subseteq X)$ is said to be nowhere dense (or, non-dense) in the space $X \Leftrightarrow$ The complement of the closure of the set A is dense in the space X.

Definition (iii): (In terms of closure and boundary): A set whose closure is a boundary set is a non-dense set.

Notes: **1.** The statement :The interior of the closure of the set A is empty" means that the closure of the set A has no interior point.

2. The complement of the closure of a set is its exterior.

3. One should note that nowhere dense sets are closed sets with no interior points, i.e., nowhere dense sets are closed sets with only boundary points whereas more generally closed sets are sets with interior points and boundary points. A nowhere dense set is thought of as a set which does not cover very much of the space.

4. A set A is nowhere dense in R (real line) $\Leftrightarrow$ For each x and each neighbourhood $N(x) = (x - \varepsilon, x + \varepsilon)$, there is another neighbourhood $N(y) = (y - \varepsilon, y + \varepsilon), \forall y \in X$ such that $N(y) \subset N(x)$ and $N(y) \cap A = \phi$.

Perfect set: A set A in a space X is perfect $\Leftrightarrow$ The set A is dense in itself and closed $\Leftrightarrow$ Every point of the set A is a limit point of the set A and every limit point of the set A belongs to the set $A \Leftrightarrow D(A) = A$.

Sequence and Its Related Terms

A sequence is nothing but a special kind of the function whose domain is the set of all natural numbers and the range is a set contained in a space, i.e., A function $f: N \to S$ is called a sequence, where

N = the set of natural numbers,

S = a set contained in a space X, i.e.,

$S \subseteq X$, where

X = any space, namely, a real line, a metric space, a normed spaced or a topological space, etc.

$S = R_1 \subseteq R \Rightarrow$ the sequence is real sequence, i.e., A real sequence is a function from the set N of natural numbers into the set R_1 of real numbers where $R_1 \subseteq R$.

Hence, **1.** $S \subseteq X$ where X = a metric space $\Rightarrow$ the sequence is said to be in a metric space.

2. $S \subseteq X$, where X = a normed space $\Rightarrow$ the sequence is said to be in a normed space.

3. $S \subseteq X$, where X = a topological space $\Rightarrow$ the sequence is said to be in a topological space.

One should understand real sequence wherever the term sequence is used (as we will consider them only).

Notation: A sequence with general term x_n is written as: $\{x_n\}$, (x_n) or $<x_n>$

Nomenclature of Terms of the Sequence

The term written on the extreme left is called 'first term' next to it 'second term' and so on and a term whose subscript is n, i.e., x_n = nth term which is a function of n always.

Different Ways of Describing a Sequence

Generally, there are two ways to describe a sequence.

1. A sequence is described by listing its first few terms till we get a rule for writing down the other different terms of the sequence, e.g. $\left\{1, \frac{1}{2}, \frac{1}{3}, \frac{1}{4}, \cdots\right\}$ is the sequence whose nth term is $\frac{1}{n}$.

2. An other way of representing the terms of the sequence is to specify the rule for its nth term, e.g: the sequence $\left\{1, \frac{1}{2}, \frac{1}{3}, \frac{1}{4}, \cdots\right\}$ can be written as $\{x_n\}$ where $x_n = \frac{1}{n}, \forall n \in N$ gives a rule for the nth term of the sequence.

Different Types of Sequence

1. Constant sequence: The sequence $\{x_n\}$ where $x_n = c \in R, \forall n \in N$ is called a constant sequence.

In this case $\{x_n\} = \{c, c, c, c, \cdots\}$.

2. Cauchy sequence: A sequence $\{x_n\}$ is called cauchy sequence $\Leftrightarrow$ After a certain term of the sequence, the numerical difference between any two terms of the sequence is less than any given small positive number ε.

In notation, a sequence $\{x_n\}$ is a cauchy sequence $\Leftrightarrow$ given any small number $\varepsilon > 0$, there is an integer N depending on ε (i.e. N_ε) such that $|x_i - x_j| < \varepsilon, \forall i > N, j > N$, i.e. after a certain term namely the Nth term, the difference or the distance between any two terms of the sequence is less than ε.

Notes: **(i)** It is not necessary that all the terms of a sequence should be different from each other.

(ii) Care must be taken to distinguish between the range of the sequence and the sequence itself, e.g. the sequence $\{x_n\}$, where $x_n = (-1)^n, \forall n \in N$ is given by $\{x_n\} = \{1, -1, 1, -1, \cdots\}$ whose range is $\{-1, 1\}$, i.e. the range of this sequence $\{x_n\}$ is a finite set whereas the sequence is an infinite set.

(iii) A sequence, by definition, is always infinite while the range of the sequence need not be infinite, e.g: The sequence $\{x_n\}$ for which $x_n = 1, \forall n \in N$, i.e. $\{x_n\} = \{1, 1, 1, 1, \cdots\}$ is an infinite sequence whose range is $\{1\}$ which is a finite set.

(iv) One should always remember that whenever it is written "a term (or, terms) of a sequence", it always means a member (or, members) of the sequence.

Boundedness and Unboundedness

In the light of definitions of a bounded set and an unbounded set, boundedness and unboundedness of a sequence and a function are defined. This is why firstly the definition of a bounded set is presented.

Boundedness of a Set

1. Bounded below set (or a set bounded on the left)

Definition: A set D is said to be bounded below or bounded on the left $\Leftrightarrow \exists$ a number m such that no member of the set is less than the number $m \Leftrightarrow \exists$ an $m \in R : x \geq m, \forall x \in D \Leftrightarrow \exists$ a point m such that no point of the set lies to the left of m.

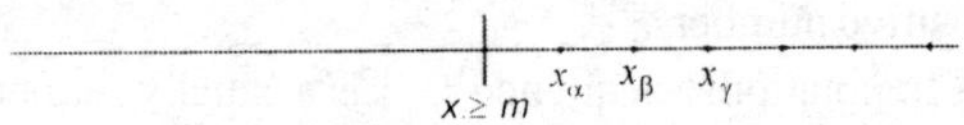

Where $x_\alpha, x_\beta, x_\gamma, \ldots$ are points of the set D.

2. Bounded above set (or a set bounded on the right)

Definition: A set D is said to be bounded above or bounded on the right $\Leftrightarrow \exists$ a number k such that all the members are less than or equal to the number $\Leftrightarrow \exists$ a $k \in R : x \leq k, \forall x \in D \Leftrightarrow \exists$ a point k such that no points of the set lie to the right of k.

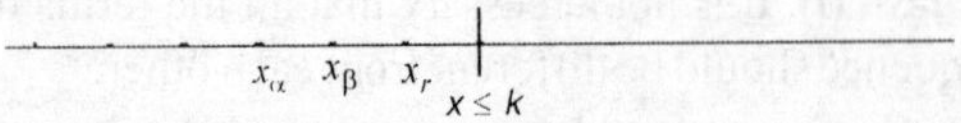

Where $x_\alpha, x_\beta, x_\gamma, \ldots$ are points of the set D.

3. Bounded set

Definition: A set D is bounded $\Leftrightarrow$ it is bounded above and below $\Leftrightarrow \exists$ two numbers k and m such that all the members of the set are contained in the closed interval $[k, m]$, i.e., $k \leq x \leq m, \forall x \in D$ and $k \leq m$; or, in other words: A set D is bounded $\Leftrightarrow \exists$ an $m > 0$ such that $|x| \leq m, \forall x \in D$.

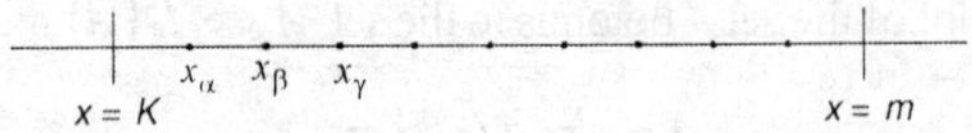

Where $x_\alpha, x_\beta, x_\gamma, \ldots$ are points of the set D.

Unboundedness of a Set

1. Unbounded above set (or, a set unbounded on the right)

Definition: A set D is said to be unbounded above or unbounded on the right $\Leftrightarrow$ whatever the number k is chosen (or taken) however large, some member of the set D is $> k$.

2. Unbounded below set (or, a set unbounded on the left)

Definition: A set D is said to be unbounded below or unbounded on the left $\Leftrightarrow$ however large a number m is chosen (or taken), there is some member of the set D which is $< -m$.

3. Unbounded Set

Definition: A set D is said to be unbounded (not bounded) if it is not a bounded set, i.e., for any $m > 0$, $\exists\, x \in D$ such that $|x| > m$.

Boundedness of a Sequence

1. Bounded above sequence (or a sequence bounded on the right)

Definition: A sequence $\{x_n\}$ is said to be bounded above or bounded on the right $\Leftrightarrow$ the range of a sequence is a set bounded above $\Leftrightarrow \exists$ a $k \in R : x_n \leq k, \forall n \in N \Leftrightarrow \exists$ a point k such that no terms of the sequence lie to the right of k.

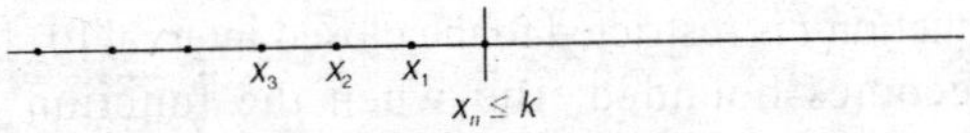

Where $x_1, x_2, x_3, \ldots$ are terms of the sequence.

2. Bounded below sequence (or a sequence bounded on the left)

Definition: A sequence $\{x_n\}$ is said to be bounded below or bounded on the left $\Leftrightarrow$ the range of a sequence is a set bounded below $\Leftrightarrow \exists$ an $m \in R : x_n \ge m, \forall m \in N \Leftrightarrow \exists$ a point m such that no terms of the sequence lie to the left of m.

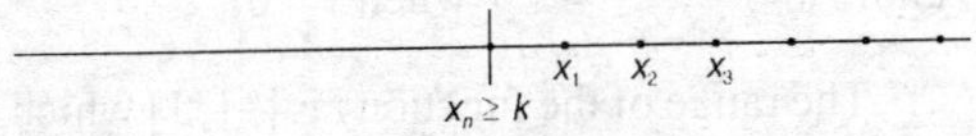

Where $x_1, x_2, x_3, \ldots$ are terms of the sequence.

3. Bounded sequence

Definition: A sequence $\{x_n\}$ is said to be bounded $\Leftrightarrow$ the range of the sequence is bounded $\Leftrightarrow$ the range of a sequence is a set bounded above and below both at the same time $\Leftrightarrow \exists$ a positive number m such that $|x_n| \le m, \forall n \in N \Leftrightarrow$ in other words, the sequence is bounded by two number $-m$ and m, i.e. $x_n \in [-m, m], \forall n \in N \Leftrightarrow$ geometrically, all the terms of the sequence lie in a certain neighbourhood (m-neighbourhood) of the point $x = 0$

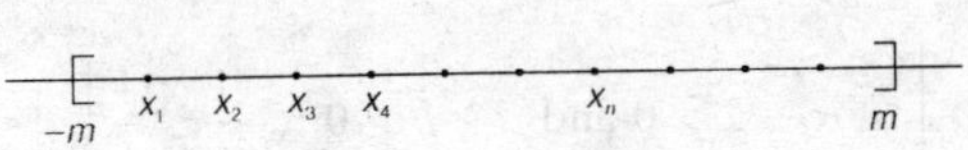

Where $x_1, x_2, x_3, \ldots x_n, \ldots$ are points of the sequence.

Unboundedness of a Sequence

1. Unbounded above sequence (or a sequence unbounded on the right)

Definition: A sequence $\{x_n\}$ is said to be unbounded above or unbounded on the right $\Leftrightarrow$ whatever the number k is chosen, however large, there is some member of the sequence $> k \Leftrightarrow$ the range of the sequence is unbounded above or unbounded on the right.

2. Unbounded below sequence (or a sequence unbounded on the left)

Definition: A sequence $\{x_n\}$ is said to be unbounded below or unbounded on the left $\Leftrightarrow$ how large a number m is taken, there is some member of the sequence $< -m \Leftrightarrow$ the range of the sequence is unbounded below or unbounded on the left.

3. Unbounded sequence

Definition: A sequence $\{x_n\}$ is unbounded $\Leftrightarrow$ the range of the sequence is unbounded $\Leftrightarrow \forall m > 0, \exists$ an $n : |x_n| > m$.

Boundedness of a Function

1. Bounded above function (or a function bounded on the right)

Definition: A function $y = f(x)$ defined on its domain D is said to be bounded above or bounded on the right $\Leftrightarrow$ the range of the function f is bounded above or bounded on the right $\Leftrightarrow \exists$ a real number m such that $f(x) \le m$ for all $x \in D$, where the number m itself is termed as an upper bound of the function f.

2. Bounded below function (or a function bounded on the left)

Definition: A function $y = f(x)$ defined on its domain D is said to be bounded below or bounded on the left $\Leftrightarrow$ the range of the function f is bounded below or bounded on the left $\Leftrightarrow \exists$ a real number k such that $f(x) \ge k$ for all $x \in D$, where the number k itself is termed as a lower bound of the function f.

3. Bounded function

Definition: A function $y = f(x)$ defined on its domain D is bounded $\Leftrightarrow$ the range of the function f is bounded $\Leftrightarrow$ the range of the function f is bounded above and bounded below $\Leftrightarrow \exists$ two real numbers k and m such that $k \le f(x) \le m$ for all $x \in D$ and $k \le m$. In other words there exists $M > 0$ such that $|f(x)| \le M, \forall x \in D$.

In the language of geometry, a function $y = f(x)$ whose domain is D is bounded $\Leftrightarrow$ the curve (the graph of the function) $y = f(x)$ defined on its domain D is situated between two horizontal lines.

Unbounded Function

1. Unbounded above function (or a function unbounded on the right)

Definition: A function $y = f(x)$ defined on its domain D is said to be unbounded above or unbounded on the right $\Leftrightarrow$ whatever the number k is chosen, however large, $f(x) > k$ for some $x \in D \Leftrightarrow$ the range of the function is unbounded above or unbounded on the right.

2. Unbounded below function (or a function unbounded on the left):

Definition: A function $y = f(x)$ defined on its domain D is said to be unbounded below or unbounded on the left $\Leftrightarrow$ however large a number m is taken, $f(x) < -m$ for some $x \in D \Leftrightarrow$ the range of the function f is unbounded below or unbounded on the left.

3. Unbounded function (The function $f(x)$ is said to be unbounded $\Leftrightarrow$ one or both of the upper and lower bounds of the function are infinite)

Definition: A function $y = f(x)$ defined on its domain D is unbounded $\Leftrightarrow$ the range of the function is unbounded $\Leftrightarrow$ the range of the function is unbounded above or unbounded below or both at the same time.

i.e. $\forall\ M > 0,\ |f(x)| > M$, for some $x \in D$.

Notes: **1.** One should note that the set, the variable, the sequence or the function is said to be bounded above, bounded below or bounded whereas the constant or the number which bounds (keeps on the left or on the right side of itself) the set, the variable, the sequence or the function is termed as lower bound, upper bound or simply bound (plural bounds) of these.

2. The fact that a sequence, a function, a variable or a set is bounded by two numbers k and m ($k \le m$) geometrically means that all the terms of the sequence, all the values of a function or a variable or all the members of the set are contained in a closed interval $[k, m]$.

3. If the domain of a bounded function is restricted, the function remains bounded.

4. The restriction of an unbounded function may or may not be bounded.

For example, $f(x) = x^2$ is unbounded on R but if the function f is restricted to the closed interval $[0, 1]$, it becomes bounded. But when the function f is restricted to the positive real numbers, it remains unbounded.

Illustrations:

1. $f(x) = \dfrac{|x|}{x}, x \ne 0$

$\Rightarrow f(x) = \dfrac{x}{x} = 1$ when $x > 0$

and $f(x) = \dfrac{-x}{x} = -1$ when $x < 0$

$\therefore$ The range of the function f is $[-1, 1]$ which is a finite set containing only two members -1 and 1.

Therefore f is a bounded function.

2. $f(x) = \dfrac{1}{x-2}, 2 < x \le 5$

$\because 2 < x \le 5$

$\therefore 0 < x - 2 \le 5$

Now, $x - 2 > 0 \Rightarrow \dfrac{1}{x-2} \to +\infty$ as $x \to 2$

Also, $x - 2 \le 3 \Rightarrow \dfrac{1}{x-2} \ge \dfrac{1}{3}$

$$\left[\because x - 2 > 0 \text{ and } a \ge b > 0 \Rightarrow \frac{1}{a} \le \frac{1}{b}\right]$$

$\therefore 2 < x \le 5 \Rightarrow \dfrac{1}{x-2} \ge \dfrac{1}{3} \Rightarrow f(x) \ge \dfrac{1}{3}$

$\therefore$ The range of the function f is bounded below and $\dfrac{1}{3}$ is its greatest lower bound and f is not bounded above.

3. $f(x) = \dfrac{1}{x}, 0 < x < \infty$ is unbounded because by choosing x sufficiently small, the function $f(x) = \dfrac{1}{x}$ can be made as large as required, i.e., infinitely great also $\frac{1}{x} > 0$ in $(0, \infty)$. Hence this function is bounded below but not bounded above.

4. $f(x) = x \sin x$ defined on domain $0 < x < \infty$ takes positive and negative values and is unbounded below and above, because by choosing sufficiently large values of x, $f(x)$ can be made sufficiently large and positive or large and negative.

Note: One must remember that a mathematical quantity or entity is called bounded $\Leftrightarrow$ its absolute value does not exceed some constant positive number M. For example, $\cos x$ is bounded for all real values of the variable x because $|\cos x| \leq 1$.

Limit of a Sequence

It is defined in various ways:

Definition (i): (In terms of ε -neighbourhood): A fixed number 'l' is the limit of a sequence $\{x_n\}$ $\Leftrightarrow$ Any ε -neighbourhood of that fixed number 'l' denoted by $N_\varepsilon(l)$ contains all the terms of the sequence $\{x_n\}$ after a certain term namely the Nth term (x_N) of the sequence $\{x_n\}$ N depending on ε.

Definition (ii): ($\varepsilon - N$ definition): A fixed number 'L' is the limit of a sequence $\{x_n\}$ $\Leftrightarrow$ Given any small positive number ε, it is possible to find out a term namely x_N such that all the terms after the Nth term (x_N) of the sequence differ from the fixed number 'l' by a number which is less than ε $\Leftrightarrow$ Given a small number $\varepsilon > 0$, $\exists$ an integer N such that $|x_r - l| < \varepsilon$ for every value of r which is greater than N (i.e. $\forall r > N$).

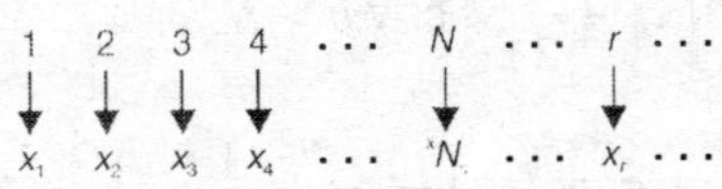

Notes: **(i)** when a variable takes on the values of a sequence which has a limit 'l', it is said that the variable has the limit 'l'. If x is a variable and 'L' is the limit of the sequence $\{x_n\}$ defined by $x = x_n$, $n = 1, 2, 3, \ldots$, one must indicate that x has the limit l by the notation $x \to l$ instead of $x_n \to l$ as $n \to \infty$, where the notation $n \to \infty$ means that the numbers of the terms of the sequence $\{x_n\}$ becomes very great. Hence, a constant 'l' is the limit of the variable x defined by $x = x_n$, $n = 1, 2, 3, \ldots$ $\Leftrightarrow$ Any given small number $\varepsilon > 0$, there exists a positive integer N_ε (N depends on ε) so that for all values of n greater than N_ε, x differs from L in absolute value by a number less than ε.

Further, the definition is symbolized as follows: Given $x = x_n$, $n = 1, 2, 3, \ldots$, then 'l' is the limit of the variable x $\Leftrightarrow$ Given any small number $\varepsilon > 0$, an integer N_ε exists such that $|x - l| < \varepsilon$ for all $n > N_\varepsilon$.

(ii) To indicate that N is a function of ε or N depends on ε, it is usual to write N_ε instead of merely N.

(iii) When a sequence has a limit, it is said to be convergent.

(iv) Already the terms limit point of a set, and limit of a sequence have been discussed. But there is a little difference among them. One should note that a set has a limit point whereas a convergent sequence has a limit which is unique. There is another term 'limit point of a sequence' which is used and discussed in real analysis. By limit point (limit points) of a sequence, one means the limit (limits) of a convergent subsequence (convergent subsequences) of a sequence. In case the sequence itself is convergent, to the limit l, any sub-sequence also converges to the same limit l.

Note: The limit of a convergent sequence may or may not by a member (or, term) of the sequence itself. For example, if there is a sequence $\{x_n\}$ where $x_n = \frac{1}{n}$, then $\lim_{n \to \infty} x_n = 0$ but it can be seen that there is no member (or, the term) of the sequence whose value is zero.

Use of $(\varepsilon - N)$ Definition

The $(\varepsilon - N)$ definition of the limit of a sequence does give us a criterion to check whether a given fixed number obtained by any mathematical manipulation or method is the limit of the sequence or not.

How to show that a given number is the limit of a sequence?

Solve the inequality $|x_n - L| < \varepsilon$, where L = a given number which is required to be shown the limit of the sequence and x_n = nth term of the sequence, and obtain the inequality $n > f(\varepsilon)$ using if method.

Examples: 1. Show that the sequence $\{x_n\}$ where $x_n = \frac{1}{n}, \forall n \in N$ converges to '0'.

Solution: In order to show that $\{x_n\}$ converges to 0, it is required to be shown that for an $\varepsilon > 0$, it is possible to obtain a positive integer N_ε such that $|x_n - 0| < \varepsilon$ for all $n > N_\in$.

Now $|x_n - 0| < \varepsilon$

If $\left|\frac{1}{n} - 0\right| < \varepsilon$

or, if $\left|\frac{1}{n}\right| < \varepsilon$

or, if $\frac{1}{n} < \varepsilon$

or, if $n > \frac{1}{\varepsilon}$

or if $n > N_\in$ where $N_\in = \left[\frac{1}{\in}\right] + 1$, i.e., whatever ε is chosen, the absolute difference between the nth term and 0 can be made as small as one likes after a certain term namely x_{N_ε}, i.e., by definition $\lim_{n \to \infty} x_n = 0$, i.e. $\{x_n\}$ converges to 0.

2. Show that the sequence $\{x_n\}$ where $x_n = \frac{n^2 + 1}{2n^2 + 5}$, $\forall n \in N$ converges to $\frac{1}{2}$.

Solution: To show that $\{x_n\}$ converges to $\frac{1}{2}$, one is required to show that for an $\varepsilon > 0$, it is possible to obtain an N_ε such that $\left|x_n - \frac{1}{2}\right| < \varepsilon \; \forall \; n > N$.

Now $x_n = \frac{n^2 + 1}{2n^2 + 5}, \forall n \in N$

and $x_n - \frac{1}{2} = \frac{n^2 + 1}{2n^2 + 5} - \frac{1}{2} = \frac{-3}{2(2n^2 + 5)}$

$$\therefore \left|x_n - \frac{1}{2}\right| = \left|\frac{-3}{2(2n^2 + 5)}\right| = \frac{3}{2(2n^2 + 5)} < \varepsilon$$

if, $\frac{3}{4n^2 + 10} < \varepsilon$

or, if, $\frac{4n^2 + 10}{3} > \frac{1}{\varepsilon}$

or, if, $4n^2 + 10 > \frac{3}{\varepsilon}$

or, if, $4n^2 > \frac{3}{\varepsilon} - 10$

or, if, $4n^2 > \frac{3 - 10\varepsilon}{\varepsilon}$

or, if, $n^2 > \frac{3 - 10\varepsilon}{4\varepsilon}$

or, if, $n > \sqrt{\frac{3 - 10\varepsilon}{4\varepsilon}} = f(\text{say})$

or if $n > N_\in$ where $N_\in = [f] + 1$

Hence, for any $\varepsilon > 0$, $\exists$ a positive integer N_ε such that $\left|x_n - \frac{1}{2}\right| < \varepsilon, \forall n > N_\in$.

$\Rightarrow$ by definition $\lim_{n \to \infty} x_n = \frac{1}{2}$

$\Rightarrow$ $\{x_n\}$ converges to $\frac{1}{2}$.

How to Find N_ε Algebraically for a Given Epsilon

Solve the inequality $|x_n - l| < \varepsilon$ for n after substituting the given expression in n for x_n using if method.

Examples: 1. Find N if $\{x_n\} = \left\{1 - \frac{1}{2^n}\right\}$ and $\varepsilon = \frac{1}{128}$.

Solution: $\lim_{n\to\infty} x_n = 1$

$\therefore |x_n - 1| < \varepsilon$ where $x_n = 1 - \frac{1}{2^n}$ and $\varepsilon = \frac{1}{128}$

if $\left|1 - \frac{1}{2^n} - 1\right| < \frac{1}{128}$

or, if $\left|-\frac{1}{2^n}\right| < \frac{1}{128}$

or, if $\frac{1}{2^n} < \frac{1}{128}$

or, if $2^n > 2^7$

or, if $n > 7 = N_\varepsilon$ (or, simply $N = 7$)

Hence, for a given $\varepsilon = \frac{1}{128}$, an integer $N = 7$ was found such that $|x_n - 1| < \varepsilon$ for $n > N(=7)$ N depends on the choice of ε and so N is a function of ε.

Theorems of the Limit of a Sequence

The following results can be proved by making use of the $(\varepsilon - \delta)$ definition of the limit of a sequence and use of these can be made in working out problems of finding the limits of the given sequences.

1. $\lim_{n\to\infty} \frac{1}{n} = 0$

2. $\lim_{n\to\infty} \frac{a}{n^p} = 0$, where a is any real number and p is positive.

3. $\lim_{n\to\infty} \left(\frac{\sin n}{n}\right) = 0$; $\lim_{n\to\infty} \left(\frac{\cos n}{n}\right) = 0$

The following theorems on limits of sequences are also stated without proofs and can be made use of them in working out examples while finding the limits of given sequences.

If $\lim_{n\to\infty} x_n = l$ and $\lim_{n\to\infty} y_n = m$, then

Theorem 1: $\lim_{n\to\infty} (x_n + y_n)$

$= \lim_{n\to\infty} x_n + \lim_{n\to\infty} y_n = l + m$

Theorem 2: $\lim_{n\to\infty} (x_n - y_n)$

$= \lim_{n\to\infty} x_n - \lim_{x\to\infty} y_n = l - m$

Theorem 3: $\lim_{n\to\infty} (x_n \cdot y_n)$

$= \left(\lim_{n\to\infty} x_n\right) \cdot \left(\lim_{n\to\infty} y_n\right) = l \cdot m$

Theorem 4: $\lim_{n\to\infty} \left(\frac{x_n}{y_n}\right)$

$= \frac{\left(\lim_{n\to\infty} x_n\right)}{\left(\lim_{n\to\infty} y_n\right)} = \frac{l}{m}$, if $m \neq 0$

Theorem 5: $\lim_{n\to\infty} (c\, x_n)$

$= c\left(\lim_{n\to\infty} x_n\right) = c \cdot l$, where c is a constant.

Note: To find the limit of a sequence whose nth term is given means that one is required to find out the limit of its nth term as $n \to \infty$ which can be determined by using the same method of evaluation of $\lim_{x\to\infty} f(x)$ replacing x by n which will be explained in methods of finding the limit of a function $y = f(x)$ at a point $x = c$.

Geometrical Meaning of the Limit of a Sequence

Geometrically $\lim_{n\to\infty} x_n = l$ means that however close the horizontal lines $y = l + \varepsilon$ and $y = l - \varepsilon$ are taken, there exists a vertical line at $x = N$ such that all the points (n, x_n) to the right of the vertical line $x = N$ lie within the horizontal lines $Y = l \pm \varepsilon$.

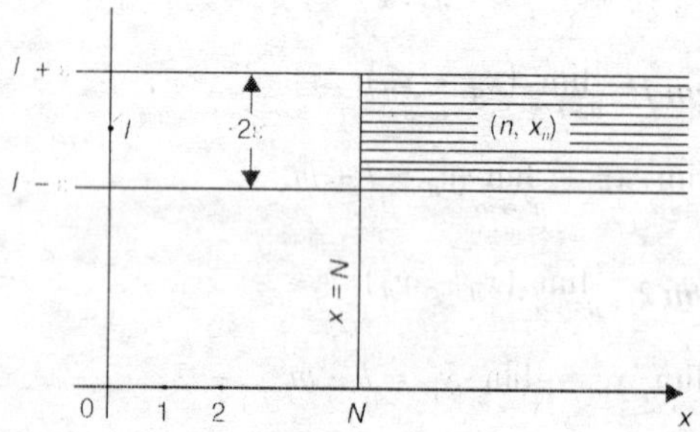

On the Relation Between the Limit of a Sequence and Limit Point of the Range Set (or, Simply Range) of the Sequence

One should note that there is no term limit point of a sequence because in fact only an infinite set which is dense has a limit point whereas a finite set has no limit point. But there is a theorem which described the relationship in between the terms "limit of a sequence and the limit point of the range of the sequence in a space".

Statement of the theorem: Let $\{x_n\}$ be a sequence in a space such that $\lim_{n\to\infty} x_n = x$. Let A be the range of the sequence $\{x_n\}$. Then

(a) If A is a finite set, then $x_n = x$ for infinitely many 'n'.

(b) If A is an infinite set, then 'x' is the limit point of A.

Remarks: **1.** the limit points of a sequences $\{x_n\}$ are either the points of the range of the sequence or the limit points of the range of the sequence.

2. If a point is a limit point of the range of a sequence, then it is also a limit point of the sequence but the converse may not always be true.

3. If a sequence has a limit 'l', then it is the limit point of the sequence but converse is not usually true.

Examples: **1.** The constant sequence $x_n = c$, $\forall n \in N$ has only one limit point namely c in the sense that $c \in (c - \varepsilon, c + \varepsilon)$, i.e. any ε-neighbourhood of C contains at least one term of the sequence which is not necessarily different from 'C'. Further one should note that the constant sequence $x_n = c$, $\forall n \in N$ has its range $\{c\}$ which is a finite set and so the range of the constant sequence has no limit point.

2. The sequence $x_n = \frac{1}{n}$, $\forall n \in N$ has '0' as a limit point which is also a limit point of its range $\left\{1, \frac{1}{2}, \frac{1}{3}, \frac{1}{4}, \ldots\right\}$.

3. The sequence $x_n = 1 + (-1)^n$, $n \in N$ has only two limit points namely '0 and 2' whereas its range $\{0, 2\}$ has no limit point.

4. The sequence $x_n = (-1)^n\left(1 + \frac{1}{n}\right)$, $\forall n \in N$ has only two limit points namely 1 and -1 which is also the limit points of its range $\left\{-2, 2, -\frac{3}{2}, \frac{3}{2}, -\frac{4}{3}, \frac{4}{3}, \ldots\right\}$.

Definition of Limit Point of a Sequence

In the above, the meaning of the term 'limit points of a sequence' has been explained. Now the definitions available in connection with 'limit points of a sequence' are provided.

The concept of the limit points of a sequence is defined in two ways.

Definition 1: (In terms of neighbourhood): A point p is the limit point of the sequence $\Leftrightarrow$ For any given small number $\varepsilon > 0$ and any given integer N, $\exists$ an other integer $n \geq N$ such that $|x_n - p| < \varepsilon$.

Or, in words, a limit point of a sequence is a point p such that for any given integer N, each neighbourhood of p contains at least one term of the sequence after the Nth term.

Or, it can be said that a limit point of a sequence is a point p such that for any given integer N, each neighbourhood of p contains infinite number of terms of the sequence after the Nth term in the sense that the terms having the same value are counted as often as they occur as terms of the sequence.

Remarks: **1.** One should note that elements of the sequence need not be distinct appearing in definite order as various distinct terms of the sequence like first term, second term, a third term and so on. As a consequence, all the infinitely many terms x_n of this definition may be the same number and so any number that occurs an infinite number of times in a given sequence or in a subsequence (subsequences) of a sequence is a limit point of the sequence according to the definition 1 since what the definition 1 requires is that there should be at least one term, i.e., an infinite number of terms of the sequence in the sense that the same element is counted as often as it occurs as a term of the sequence.

2. A constant sequence $x_n = c, \forall n \in N$ has only one limit point namely the constant 'c'.

Example: Let there be a sequence defined by $x_n = (-1)^n, \forall n \in N$, i.e.

$\{x_n\} = \{-1, 1, -1, 1, \cdots\}$. This sequence has 1 and -1 as limit points. The reason for which is that every neighbourhood of 1 contains an infinite number of terms $x_2, x_4, x_6, \ldots$ and every neighbourhood of -1 contains an infinite number of terms $x_1, x_3, x_5, \ldots$ in the sense that the same element is counted as often as it occurs as terms (first term, second term, third term and so on) of the sequence.

Moreover one should note that each of the terms x_2 = second term, x_4 = fourth term, x_6 = sixth term, ... is 1 and each of the terms x_1 = first term, x_3 = third term, x_5 = fifth term, ... is -1.

Lastly one should note that the sequence $x_n = (-1)^n, \forall n \in N$ is itself not convergent.

Definition 2: (in terms of limit of a subsequence): The limit (limits) of a convergent subsequence (convergent subsequences) of a sequence is (are) called the limit point (limit points) of the sequence.

Remarks: **1.** There is a convergent sequence $\Rightarrow$ The limit and the limit point of the sequence both are same.

2. A sequence may contain one or more convergent subsequences.

Examples: **1.** The sequence $\{x_n\}$ where $x_n = \frac{1}{n}$, $\forall n \in N$ has only one limit point namely the real number '0' since $x_n = \frac{1}{n}$ is a convergent sequence.

2. The sequence $\{x_n\}$ for which $x_n = (-1)^n$, $\forall n \in N$, i.e. $\{x_n\} = \{-1, 1, -1, 1, -1, \ldots\}$ has got two limit points namely 1 and -1 since there are two subsequences $x_{2n} = (-1)^{2n}$ and $x_{(2n+1)} = (-1)^{2n+1}$, $\forall n \in N$, whose limits are 1 and -1 respectively.

3. The sequence $\{x_n\}$ for which $x_n = 1 + \frac{1}{n}$, $\forall n \in N$, has got only one limit point namely 1 since it is a convergent sequence.

4. The sequence $\{x_n\}$ where $x_n = 1, \forall n \in N$ has only one limit point namely 1 since it is a convergent sequence.

5. The sequence $\{x_n\}$ where $x_n = n, \forall n \in N$ has no limit point since it is not a convergent sequence.

Difference Between Limit and Limit Point of a Sequence

There are some distinctions between a limit point and the limit of a sequence.

1. If all the members of the sequence $\{x_n\}$ from a certain term $x_{N(\varepsilon)}$ onwards (i.e. $x_{N+2}, x_{N+3}, \cdots$) lie within the interval $(l - \varepsilon, l + \varepsilon)$, then l is the limit of the sequence. But when l is the limit point of the sequence $\{x_n\}$, then it is sufficient that at least one term of the sequence (not necessarily different from 'l') lie within the interval $(l - \varepsilon, l + \varepsilon)$. e.g.: For any $\varepsilon > 0$, $x_n = 1 \in (1 - \varepsilon, 1 + \varepsilon)$, $\forall n \in N$. This is why 1 is the limit point of the constant sequence. Further it should be noted that 1 is also limit of the constant sequence $x_n = 1$.

2. Limit of a sequence is unique if it exists whereas a limit point of a sequence is not unique since a limit point (limit points) of a sequence is the limit (limits) of a convergent subsequence (convergent subsequences) of a sequence which means that there may be one or more than one limits according as there is one or more than one convergent subsequences of a sequence whereas the sequence whose convergent subsequence (convergent subsequences) is (are) considered, need not be convergent and in case the sequence is convergent, its limit and limit point both are same, e.g.: The sequence defined as $x_n = (-1)^n$, $\forall n \in N$ is divergent, i.e. it has no limit. But it has two subsequences $x'_n = (-1)^{2n}$ and $x''_n = (-1)^{2n+1}$ whose limits are 1 and -1 respectively which are termed as the limit points of the sequence $x_n = (-1)^n$, $\forall n \in N$ noting that the sequence ($x_n = (-1)^n$, $\forall n \in N$) itself has no limit but it has got two limit points namely 1 and -1 respectively accordingly as n is even or odd.

Limit of a Function

The concept of the limit of a function is defined in various ways:

Definition (i): (In terms of neighbourhood): we say that a fixed point p is a limit of the function f at a limit point 'a' of the domain of the function f if there is a fixed point 'p' such that if we choose any ε-neighbourhood of the point 'p' denoted by $N_\varepsilon(p)$, it is possible to find a ε-neighbourhood of the limit point 'a' of the domain of the function f denoted by $N_\delta(a)$ such that the values of the function lie in $N_\varepsilon(p)$ for every value of the independent variable x which lies in δ-deleted neighbourhood of the limit point 'a' of domain of the function denoted by $N'_\delta(a) = 0 < |x - a| < \delta$, i.e. There is a fixed point 'p' such that for every $N_\varepsilon(p)$, $\exists$ a $N_\delta(a)$ such that $f(x) \in N_\varepsilon(p)$ for all $x \in N'_\delta(a) \Leftrightarrow$ The fixed point 'p' is the limit of the function f at the limit point 'a' of the domain of the function f.

Definition (ii): (ε - δ -definition): A fixed point 'p' is the limit of a function f at a limit point 'a' of the domain of the function $f \Leftrightarrow \forall \varepsilon > 0, \exists a\ \delta > 0$ (δ depends on ε) such that for every value of x in the deleted neighbourhood $0 < |x - a| < \delta$, the value of the function $f(x)$ lies in the neighbourhood $(p - \varepsilon, p + \varepsilon)$ i.e. $\forall \varepsilon > 0$, $\exists \delta > 0$ (δ depends on ε) such that $\forall x\ 0 < |x - a| < \delta \Rightarrow |f(x) - p| < \varepsilon \Rightarrow \lim_{x \to a} f(x) = p$. Which means p is the limit of the function f at $x = a$, where a is a limit point of the domain of the function.

Note: It is common to say that $y = f(x)$ has a limit p at a point $x = a$ instead of saying that $y = f(x)$ has a limit p at the limit point $x = a$ of its domain.

Explanation of $(\varepsilon - \delta)$ definition

The following example as an explanation of $(\varepsilon - \delta)$ definition of the limit of a function is presented.

$$f(x) = \frac{5(x^2 - 4)}{(x - 2)} \text{ is not defined at } x = 2$$

Now $f(x) = \dfrac{5(x-2)(x+2)}{(x-2)} = 5(x+2)$ when $x \neq 2$

[$\because x = 2 \Rightarrow f(2) = 5(2+2)$, i.e. $f(2) \neq 20$]

In fact, $|f(x) - 20| = |5(x+2) - 20|$, for $x \neq 2$

$= |5x - 10| = 5|x - 2|$, $x \neq 2$

$$\therefore\ 0 < |x - 2| < \frac{1}{50} \Rightarrow |f(x) - 20| < \frac{1}{10}$$

Similarly, $0 < |x - 2| < \dfrac{1}{500}$

$$\Rightarrow |f(x) - 20| < \frac{1}{100}$$

$$0 < |x - 2| < \frac{1}{5000} \Rightarrow |f(x) - 20| < \frac{1}{1000}$$

Hence, we see that for every small positive number ε ,

$$0 < |x - 2| < \frac{1}{5\varepsilon} \Rightarrow |f(x) - 20| < \varepsilon$$

In other words, for any small $\varepsilon > 0, \exists$ a δ

$\left(= \frac{1}{5\varepsilon}\right.$ in this example) such that

$$|f(x) - 20| < \varepsilon \text{ whenever } 0 < |x - 2| < \delta.$$

This fact is expressed by saying that 20 is the limit of $f(x) = \frac{5(x^2 - 4)}{(x - 2)}$ as x tends to 2. it is written as

$$\lim_{x \to 2} \frac{5(x^2 - 4)}{(x - 2)} = 20$$

Note: The $(\varepsilon - \delta)$ definition does not provide a technique to calculate the limit which is a fixed number 'l'. What the $(\varepsilon - \delta)$ definition does is supply a criterion which one uses to test a number 'l' to see whether it is actually the limit of a function f as x tends to a, where 'a' is the limit point of the domain of the function.

How to find a $\delta > 0$ for a given *f*, *l*, *a* and $\varepsilon > 0$ algebraically

The process of finding a $\delta > 0$ such that for all x

$$0 < |x - a| < \delta \Rightarrow |f(x) - l| < \varepsilon$$

Consists of the following steps:

Step 1: Firstly, one should suppose that $|f(x) - l| = u$ and simplify it.

Step 2: Secondly, on supposing that 'k' is a small positive number and letting $|x - a| = k$, i.e., $x = a \pm k$, one should make the substitution $x = a + k$ and $x = a - k$ respectively in $|f(x) - l| = u$ and simplify it which gives two expressions in K namely $f_1(k)$ and $f_2(k)$ (say).

Step 3: Thirdly, out of the two expressions $f_1(k)$ and $f_2(k)$ obtained after simplification, one should choose that one which is greater.

Step 4: Fourthly, one should form the inequality $f_1(k) < \varepsilon$, where $f_1(k) > f_2(k)$ and lastly solve $f_1(k) < \varepsilon$ for k.

Note: When $f_1(k) = f_2(k)$, anyone can be chosen to form the in equality $f_1(k) < \varepsilon$.

Solved Examples

1. Show that $\lim_{x \to 2} (3x + 2) = 8$

Proof:

Method 1: Let $|3x + 2 - 8| = u$ ....(i)

and $|x - 2| = k$, i.e., $x = 2 \pm k$

$\therefore\ x = 2 + k \Rightarrow u = |3(2 + k) + 2 - 8|$, from (i).

$= |6 + 3k + 2 - 8| \quad = |3k + 2 - 2| \quad = |3k|$

$= 3k \quad (\because\ k > 0)$

and $x = 2 - k \Rightarrow u = |3(2 - k) + 2 - 8|$, from (i)

$= |6 - 3k + 2 - 8|$

$= |-3k + 2 - 2|$

$= |-3k| = 3k \quad (\because\ k > 0)$

Now, $3k < \varepsilon \Leftrightarrow k < \frac{\varepsilon}{3} = \delta$ (say)

$\therefore\ |x - 2| < \delta = \frac{\varepsilon}{3}$

$\Rightarrow |(3x + 2) - 8| < \varepsilon \Leftrightarrow \lim_{x \to 2} (3x + 2) = 8$,

Hence proved.

Method 2: $\lim_{x \to 2} (3x + 2) = 8$

if $|3x + 2 - 8| < \varepsilon$

or, if $|3x - 6| < \varepsilon$

or, if $3|x-2|<\varepsilon$

or, if $|x-2|<\frac{\varepsilon}{3}=\delta$ (say)

Thus $0<|x-2|<\delta=\frac{\varepsilon}{3}$

$$\Rightarrow |(3x+2)-8|<\varepsilon$$

$\therefore \lim_{x\to 2}(3x+2)=8$. Hence, proved.

2. Show that $\lim_{x\to 2}\left(\frac{2x-3}{3x+4}\right)=\frac{1}{10}$

Proof: Let $\left|\frac{2x-3}{3x+4}-\frac{1}{10}\right|=u$

and $|x-2|=k \Rightarrow x=2\pm k$

Now, $u=\left|\frac{2x-3}{3x+4}-\frac{1}{10}\right|$

$$=\left|\frac{20x-30-3x-4}{(3x+4)\,10}\right|=\left|\frac{17x-34}{(3x+4)\,10}\right|$$

$$=\frac{17|x-2|}{|(3x+4)|\,10} \quad \text{... (i)}$$

$\therefore x=2+k \Rightarrow u=\frac{17k}{(10+3k)\,10}$

and $x=2-k \Rightarrow u=\frac{17k}{(10-3k)\,10}$

But $\frac{17k}{(10-3k)\,10}>\frac{17k}{(10+3k)\,10}$

Now, $\frac{17k}{(10-3k)\,10}<\varepsilon$, for sufficiently small k.

$\Leftrightarrow 17k<10\,\varepsilon(10-3k)$

$\Leftrightarrow 17k<100\varepsilon-30\varepsilon k$

$\Leftrightarrow 17k+30\,\varepsilon\,k<100\,\varepsilon$

$\Leftrightarrow (17+30\varepsilon)k<100\varepsilon$

$\Leftrightarrow k<\frac{100\varepsilon}{17+30\varepsilon}=\delta$ (say)

$\therefore\ 0<|x-2|<\frac{100\varepsilon}{17+30\varepsilon}=\delta$

$$\Rightarrow \left|\frac{2x-3}{3x+4}-\frac{1}{10}\right|<\varepsilon$$

$\Leftrightarrow \lim_{x\to 2}\left(\frac{2x-3}{3x+4}\right)=\frac{1}{10}$ Hence, proved.

3. Show that $\lim_{x\to 2}\left(\frac{x^2-4}{x-2}\right)=4$

Proof:

Method 1: Let $\left|\left(\frac{x^2-4}{x-2}\right)-4\right|=u$ ... (i)

and $|x-2|=k$, i.e., $x=2\pm k$

Now $u=\left|\left(\frac{x^2-4}{x-2}\right)-4\right|$, from (i).

$$=\left|\frac{(x-2)(x+2)}{(x-2)}-4\right|=|(x+2)-4|$$

$$=|x-2|$$

$\therefore\ x=2+k \Rightarrow u=|2+k-2|=k$

and $x=2-k \Rightarrow u=|2-k-2|=k$

$\therefore\ k<\varepsilon=\delta$ (say)

Hence, $0<|x-2|<\varepsilon=\delta$

$$\Rightarrow \left|\left(\frac{x^2-4}{x-2}\right)-4\right|<\varepsilon$$

$$\Leftrightarrow \lim_{x \to 2} \left(\frac{x^2 - 4}{x - 2} \right) = 4 \text{. Hence, proved.}$$

Method 2: $\lim_{x \to 2} \left(\frac{x^2 - 4}{x - 2} \right) = 4$

If $\left| \left(\frac{x^2 - 4}{x - 2} \right) - 4 \right| < \varepsilon$

or, if $\left| \frac{(x-2)(x+2)}{(x-2)} - 4 \right| < \varepsilon$

or, if $|(x+2) - 4| < \varepsilon$

or, if $|x - 2| < \varepsilon = \delta$ (say)

$\therefore 0 < |x - 2| < \varepsilon = \delta$

$$\Rightarrow \left| \left(\frac{x^2 - 4}{x - 2} \right) - 4 \right| < \varepsilon$$

$$\Leftrightarrow \lim_{x \to 2} \left(\frac{x^2 - 4}{x - 2} \right) = 4 \text{. Hence, proved.}$$

4. Show that $\lim_{x \to 2} x^2 = 4$

Proof:

Method 1: Let $u = |x^2 - 4|$ ...(i)

and $|x - 2| = k$, i.e., $x = 2 \pm k$

$\therefore x = 2 + k \Rightarrow u = |(2+k)^2 - 4|$

$= |4 + 4k + k^2 - 4| = |k^2 + 4k|$

$= k^2 + 4k \; (\because \; k > 0)$

and $x = 2 - k \Rightarrow u = |(2-k)^2 - 4|$

$= |4 - 4k + k^2 - 4| = |k^2 - 4k|$

$= 4k - k^2$ for small k,

But $k^2 + 4k > 4k - k^2$

Now, $k^2 + 4k < \varepsilon$

$\Leftrightarrow k^2 + 4k - \varepsilon < 0$

$$\Leftrightarrow k < \frac{-4 + \sqrt{16 + 4\varepsilon}}{2} = \frac{-4 + 2\sqrt{4 + \varepsilon}}{2}$$

$= -2 + \sqrt{4 + \varepsilon} = \delta$ (say)

Thus, $0 < |x - 2| < -2 + \sqrt{4 + \varepsilon} = \delta$

$\Rightarrow |x^2 - 4| < \varepsilon$

$\Leftrightarrow \lim_{x \to 2} x^2 = 4$. Hence, proved.

Method 2: $\lim_{x \to 2} x^2 = 4$ *

(* $\because \; x^2 - 4 = (x-2)^2 + 4x - 4 - 4$

$= (x-2)^2 + 4x - 8$

$\therefore \; |x^2 - 4| = |(x-2)^2 + 4x - 8|$

$= |(x-2)|^2 + 4|x-2|$)

$|x^2 - 4| < \varepsilon$

if $|(x-2)|^2 + 4|x-2| < \varepsilon$

or, if $|x-2|^2 + 4|x-2| - \varepsilon < 0$

or, if $|x - 2| < \frac{-4 + \sqrt{16 + 4\varepsilon}}{2}$

or, if $|x - 2| < \frac{-4 + 2\sqrt{4 + \varepsilon}}{2}$

$= -2 + \sqrt{4 + \varepsilon} = \delta$ (say)

Thus, $|x^2 - 4| < \varepsilon$

if $0 < |x - 2| < \delta = -2 + \sqrt{4 + \varepsilon}$

$\Leftrightarrow \lim_{x \to 2} x^2 = 4$. Hence proved.

5. Show that $\lim_{x \to 3} \left(2x^2 - 3x + 4\right) = 13$.

Proof:

Method 1: Let $\left|2x^2 - 3x + 4 - 13\right| = u$ ... (i)

and $|x - 3| = k$, i.e., $x = 3 \pm k$

Now $u = \left|2x^2 - 3x + 4 - 13\right|$

$= \left|2x^2 - 3x - 9\right|$, from (i)

$\therefore \; x = 3 + k \Rightarrow u$

$= \left|2(3 + k)^2 - 3(3 + k) - 9\right|$

$= \left|2\left(9 + 6k + k^2\right) - 9 - 3k - 9\right|$

$= \left|18 + 12k + 2k^2 - 3k - 18\right|$

$= \left|2k^2 + 9k\right| = 2k^2 + 9k \; (\because \; k > 0)$

and $x = 3 - k$

$\Rightarrow u = \left|2(3 - k)^2 - 3(3 - k) - 9\right|$

$= \left|2\left(9 - 6k + k^2\right) - 9 + 3k - 9\right|$

$= \left|18 - 12k + 2k^2 + 3k - 18\right|$

$= \left|2k^2 - 9k\right| = 9k - 2k^2$, for small k.

But $2k^2 + 9k > 2k^2 - 9k$

Now $2k^2 + 9k < \varepsilon$

$\Leftrightarrow 2k^2 + 9k - \varepsilon < 0$

$\Leftrightarrow k < \dfrac{-9 + \sqrt{81 + 8\varepsilon}}{4} = \delta$ (say)

$\therefore \; 0 < |x - 3| < \delta = \dfrac{-9 + \sqrt{81 + 8\varepsilon}}{4}$

$\Rightarrow \left|\left(2x^2 - 3x + 4\right) - 9\right| < \varepsilon$

$\Leftrightarrow \lim_{x \to 3} \left(2x^2 - 3x + 4\right) = 13$. Hence, proved.

Method 2: Here $f(x) = 2x^2 - 3x + 4$, $l = 13$ and $a = 3$

$f(x) - l = 2x^2 - 3x + 4 - 13 = 2x^2 - 3x - 9$

$= 2(x - 3)^2 + 2(x - 3)$

$\therefore$ for any $\varepsilon > 0$, $|f(x) - l| < \varepsilon$

if $\left|2(x - 3)^2 + 9(x - 3)\right| < \varepsilon$

or, if $2|x - 3|^2 + 9|x - 3| < \varepsilon$

or, if $2|x - 3|^2 + 9|x - 3| - \varepsilon < 0$

or, if $|x - 3| < \dfrac{-9 + \sqrt{81 + 8\varepsilon}}{4} = \delta$ (say)

$\therefore \; 0 < |x - 3| < \delta = \dfrac{-9 + \sqrt{81 + 8\varepsilon}}{4}$

$\Rightarrow \left|\left(2x^2 - 3x + 4\right) - 9\right| < \varepsilon$

$\Leftrightarrow \lim_{x \to 2} \left(2x^2 - 3x + 4\right) = 13$

The use of $(\varepsilon - \delta)$ Definition to Prove Theorems

The $(\varepsilon - \delta)$ definition does not only give a criterion to check the value obtained whether it is limit or not but also it enables us to prove many theorems and many useful results.

Theorem 1: $\lim_{x \to a} f(x) = f(a) \Rightarrow \lim_{x \to a} |f(x)| = |f(x)|$

Proof: $\left||f(x)| - |f(x)|\right| \le |f(x) - f(a)|$ ... (i)

Also, $\lim_{x \to a} f(x) = f(a)$

$\Rightarrow$ for any given

$\varepsilon > 0, \exists$ a $\delta > 0$ s.t $|f(x) - f(a)| < \varepsilon, \forall x$ for which $0 < |x - a| < \delta$ ... (ii)

$\therefore$ (i) and (ii) $\Rightarrow \forall \varepsilon > 0, \exists$ a $\delta > 0$ s.t

$||f(x)| - |f(a)|| < \in, \quad \forall x$ for which $0 < |x - a| < \delta$

$\therefore \lim_{x \to a} |f(x)| = |f(a)|$

Theorem 2: Show that $\lim_{x \to a} c = c$

Proof: Here, $f(x) = c$ and $l = c$

$\therefore |f(x) - l| = |c - c| = 0$

$\therefore \forall \varepsilon > 0, |f(x) - l| < \varepsilon$ for

$0 < |x - a| < \delta$, where $\delta > 0$ is any number.

Theorem 3: $\lim_{x \to a} x = a$

Proof: $= |f(x) - l| = |x - a|$

$\therefore |f(x) - l| < \varepsilon$ when $0 < |x - a| < \delta$

Hence, the result.

Now, we state (without proof) some results on limits

Let $\lim_{x \to a} f(x) = l$ and $\lim_{x \to a} g(x) = m$

And let C be any constant. Then

(i) $\lim_{x \to a} [f(x) + g(x)]$

$= \lim_{x \to a} f(x) + \lim_{x \to a} g(x) = l + m$

(ii) $\lim_{x \to a} [C\, f(x)] = C\left[\lim_{x \to a} f(x)\right] = C \cdot l$

(iii) $\lim_{x \to a} [f(x) \cdot g(x)]$

$\left[\lim_{x \to a} f(x)\right] \cdot \left[\lim_{x \to a} g(x)\right] = l \cdot m$

(iv) $\lim_{x \to a} \left[\dfrac{f(x)}{g(x)}\right] = \dfrac{\lim_{x \to a} f(x)}{\lim_{x \to a} g(x)} = \dfrac{l}{m}$, provided $m \neq 0$

(v) For any positive integer $n = 1, 2, 3, \ldots$

$$\lim_{x \to a} [f(x)]^n = \left[\lim_{x \to a} x\right]^n = l^n.$$

[If follows that $\lim_{x \to a} x^n = \left[\lim_{x \to a} x\right]^n = a^n$]

(vi) When m and n are positive integers, then

(a) if m is even $\lim_{x \to a} x^{\frac{n}{m}} = a^{\frac{n}{m}}$ for $0 < a < \infty$

(b) if m is odd $\lim_{x \to a} x^{\frac{n}{m}} = a^{\frac{n}{m}}$ for $-\infty < a < \infty$

(vii) If $f_1(x) \leq f_2(x) \leq f_3(x)$ for all x in an open interval containing 'a' except possibly at $x = a$ and if $\lim_{x \to a} f_1(x) = \lim_{x \to a} f_3(x) = l$, then $\lim_{x \to a} f_2(x) = l$

(This theorem is called '**Sandwitch Theorem** or **Pinching Theorem**')

Note: The expression 'except possibly at $x = a$' means that $f(x)$ may or may not be defined at $x = a$.

Now, we can make use of these results in evaluating limits of polynomials, rational functions and powers of such functions.

Examples: Evaluate

1. $\lim_{x \to 2} (2x^3 - 3x^2 + 6x + 5)$

Solution: $\lim_{x \to 2} (2x^3 - 3x^2 + 6x + 5)$

$= \lim_{x \to 2} (2x^3) - \lim_{x \to 2} (3x^2) + \lim_{x \to 2} (6x) + \lim_{x \to 2} (5)$

$= 2 \lim_{x \to 2} (x^3) - 3 \lim_{x \to 2} (x^2) + 6 \lim_{x \to 2} (x) + \lim_{x \to 2} (5)$

$= 2 \cdot 2^3 - 3 \cdot 2^2 + 6 \cdot 2 + 5$

$= 16 - 12 + 12 + 5$

$= 21$

Note: We see in the above example that if $f(x) = 2x^3 - 3x^2 + 6x + 5$, then, $\lim\limits_{x \to 2} f(x) = 21$

Also, $f(2) = 21$

Hence, $\lim\limits_{x \to 2} f(x) =$ value of $f(x)$ at $x = 2 = f(2)$.

However, $\lim\limits_{x \to a} f(x) = f(a)$ is not in general true. We have already seen that if

$f(x) = \dfrac{5(x^2 - 4)}{(x-2)}$, then $\lim\limits_{x \to 2} f(x) = 20$

But $f(2)$ is undefined. We will see in the definition of the concept of continuity that if $\lim f(x) = f(a)$, then $f(x)$ is said to be continuous at $x = a$.

As seen above, $f(x) = 2x^3 - 3x^2 + 6x + 5$ is continuous at $x = 2$.

In fact, any polynomial function is continuous for each value of $x \in R$.

2. $\lim\limits_{x \to 3} \left(\dfrac{2x^2 - 3x + 4}{x^2 + x + 5} \right)$

Solution: Limit of the denominator is

$\lim\limits_{x \to 3} (x^2 + x + 5)$

$= \lim\limits_{x \to 3} (x^2) + \lim\limits_{x \to 3} (x) + \lim\limits_{x \to 3} (5)$

$= 3^2 + 3 + 5 = 17$ which is not zero.

$\therefore \lim\limits_{x \to 3} \left(\dfrac{2x^2 - 3x + 4}{x^2 + x + 5} \right)$

$= \dfrac{\lim\limits_{x \to 3} (2x^2 - 3x + 4)}{\lim\limits_{x \to 3} (x^2 + x + 5)}$

$= \dfrac{\lim\limits_{x \to 3} (2x^2) - \lim\limits_{x \to 3} (3x) + \lim\limits_{x \to 3} (4)}{17}$

$= \dfrac{(2.3^2 - 3.3 + 4)}{17} = \dfrac{13}{17}$

3. $\lim\limits_{x \to 4} (x^2 + 2x - 5)^3$

Solution: $\lim\limits_{x \to 4} (x^2 + 2x - 5)^3$

$= \left[\lim\limits_{x \to 4} (x^2 + 2x - 5) \right]^3$

$= \left[\lim\limits_{x \to 4} (x^2) + \lim\limits_{x \to 4} (2x) + \lim\limits_{x \to 4} (-5) \right]^3$

$= (4^2 + 2.4 - 5)^3$

$= 19^3$

4. $\lim\limits_{x \to 64} \left(5\sqrt{x} + 3x^{\frac{3}{2}} - 2x^{\frac{-5}{4}} \right)$

Solution: $\lim\limits_{x \to 64} \left(5\sqrt{x} + 3x^{\frac{3}{2}} - 2x^{\frac{-5}{4}} \right)$

$= 5 \lim\limits_{x \to 64} x^{\frac{1}{2}} + 3 \lim\limits_{x \to 64} x^{\frac{3}{2}} - 2 \lim\limits_{x \to 64} x^{\frac{-5}{4}}$

$= 5(64)^{\frac{1}{2}} + 3 \cdot (64)^{\frac{3}{2}} - 2 \cdot \dfrac{1}{(64)^{\frac{5}{4}}}$

$= 5 \cdot 8 + 3 \cdot 8^3 - 2 \cdot \dfrac{1}{4^5} = 40 + 1536 - \dfrac{2}{1024}$

$= 1576 - \dfrac{2}{1024} = \dfrac{1576 \times 1024 - 2}{1024}$

$= \dfrac{1613822}{1024}$

5. Prove that $\lim\limits_{x \to 0} \sin x = 0$

Proof: From the figure, for $-\dfrac{\pi}{2} < x < \dfrac{\pi}{2}$

$|\sin x| = \dfrac{\text{length of } MP}{OP}$

$= \frac{|MP|}{1}$

$= |MP|$

$\leq$ length of AP

$\leq$ length of arc AP

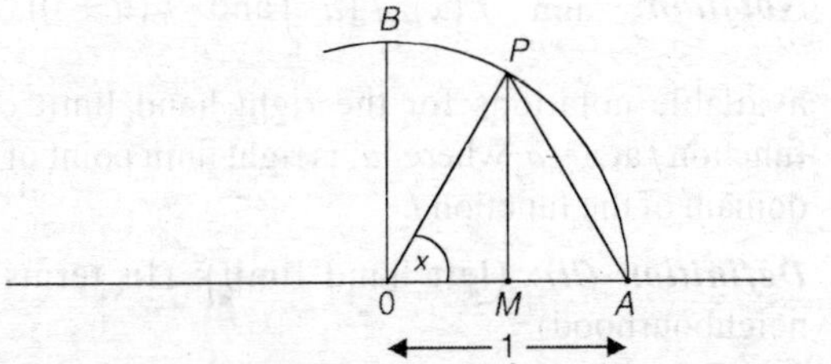

$\therefore\ |\sin x| \leq |x|$

$\therefore\ |\sin x - 0| = |\sin x| \leq |x - 0|$

$\therefore$ For any $\varepsilon > 0$, $\exists$ a $\delta > 0$ s.t $|\sin x - 0| < \varepsilon$ for $0 < |x - 0| < \delta$

Hence, $\lim_{x \to 0} \sin x = 0$

6. Prove that $\lim_{x \to 0} x \sin\left(\frac{1}{x}\right) = 0$

Proof: $\because\ 0 \leq \left| x \sin\left(\frac{1}{x}\right) \right| \leq |x| \left| \sin\left(\frac{1}{x}\right) \right| \leq |x| \cdot 1 = |x|$

$\therefore\ 0 \leq \left| x \sin\left(\frac{1}{x}\right) \right| \leq |x|$

Now, since $\lim_{x \to 0} |x| = 0$

$\therefore$ By the Pinching theorem

$$\lim_{x \to 0} \left| x \sin\left(\frac{1}{x}\right) \right| = 0 \quad \ldots \text{(i)}$$

Again,

$$-\left| x \sin\left(\frac{1}{x}\right) \right| \leq x \sin\left(\frac{1}{x}\right) \leq \left| x \sin\left(\frac{1}{x}\right) \right|$$

$\therefore$ By Pinching theorem and (i), we get

$$\lim_{x \to 0} x \sin\left(\frac{1}{x}\right) = 0$$

Note: The above result (results) can be shown in various ways which is (are) shown in this book.

7. Prove that $\lim_{x \to 0} \cos x = 1$

Proof: For $-\frac{\pi}{2} < x < \frac{\pi}{2}$,

$$\cos x = \sqrt{1 - \sin^2 x} \geq 1 - \sin^2 x$$

$(\because\ -\frac{\pi}{2} < x < \frac{\pi}{2} \Rightarrow 0 < \cos x < 1$ and $0 \leq a \leq 1 \Rightarrow a \leq \sqrt{a}\,)$

Hence, we have for $-\frac{\pi}{2} < x < \frac{\pi}{2}$,

$$1 - \sin^2 x \leq \cos x \leq 1 \quad \ldots \text{(i)}$$

Since, $\lim_{x \to 0} \left(1 - \sin^2 x\right) = 1 - \left(\lim_{x \to 0} \sin x\right)^2 = 1 - 0 = 1$, It follows from Pinching theorem and (i) that $\lim_{x \to 0} \cos x = 1$

8. $\lim_{x \to 0} \left(\frac{\sin x}{x}\right) = 1$

Note 1: This limit plays a significant role in mathematical analysis. It is often called first remarkable limit.

Note 2: A strict proof of this limit depends upon defining $\sin x$ as a power series in x and upon certain properties of power series. Therefore, its proof by geometrical argument will be presented in the chapter to find the limits of trigonometrical function using practical methods.

One Sided Limits

It is recalled that in the definition of the limit of $f(x)$ as x tends to a, it was required that the function $f(x)$ should be defined in some deleted neighbourhood of

'a' (i.e., a is an interior point of an open interval where $f(x)$ is defined and $f(x)$ may or may not be defined at $x = a$).

Now, let us consider the function $f(x) = \sqrt{x-2}$ clearly, $f(x)$ is not defined for $x < 2$. Hence, $f(x)$ is not defined in any deleted neighbourhood of 2.

Hence, $\lim\limits_{x \to 2} \sqrt{x-2}$ does not exist.

Similarly, $\lim\limits_{x \to 0} \sqrt{x}$ and $\lim\limits_{x \to 3} \sqrt{x^2 - 9}$ do not exist.

Note: If $f(x) = \dfrac{\sin\left(\frac{\pi}{x}\right)}{\sin\left(\frac{\pi}{x}\right)}$, then $\lim\limits_{x \to 0} f(x)$ does not exist. In fact for $x = \frac{1}{n}$ (where n = any nonzero integer) $\sin\left(\frac{\pi}{x}\right) = 0$ and so $f(x)$ is not defined.

Hence, in any deleted neighbourhood of $x = 0$, we have points where $f(x)$ is undefined.

Definition (i): (Right hand limit): (In terms of neighbourhood):

(a) A fixed point 'p' is a right hand limit of the function f at the right limit point 'a' of the domain of the function f if there is a fixed point p such that if we choose any ε-neighbourhood of p denoted by $N_\varepsilon(p)$, it is possible to find a δ-neighbourhood of the right limit point 'a' of the domain of the function f denoted by $N_\delta(p)$ such that the values of the function $f(x)$ lie in $N_\varepsilon(p)$ for every value of independent variable x which lies in a right hand δ-deleted neighbourhood of the right limit point 'a' of the domain of the function f denoted by $N'_\delta(a) = 0 < x - a < \delta$.

(b) (In terms of $(\varepsilon - p)$ definition): A fixed point 'p' is the right hand limit of a function f at the right limit point 'a' of the domain of the function $f \Leftrightarrow \forall\, \varepsilon > 0$, $\exists$ a $\delta > 0$ (δ depends on ε) such that for every value of x in the right hand δ-deleted neighbourhood $0 < x < \delta$, the value of the function $f(x)$ lies in the ε-neighbourhood $(p - \varepsilon,\ p + \varepsilon)$ i.e. given any $\varepsilon > 0$, it is possible to find a $\delta > 0$ such that $0 < x - a < \delta$ $\Rightarrow |f(x) - p| < \varepsilon$ that is $a < x < a + \delta$ $\Rightarrow p - \varepsilon < f(x) < p + \varepsilon$.

Notation: $\lim\limits_{x \to a^+} f(x)$, $f(a^+)$ and $f(a+0)$ are available notations for the right hand limit of a function f at $x = a$, where 'a' is right limit point of the domain of the function f.

Definition (ii): (left hand limit): (In terms of neighbourhood)

(a) A fixed point p is the left hand limit of the function f at the left limit point 'a' of the domain of the function f if there is a fixed point p such that if we choose any ε-neighbourhood of p denoted by $N_\varepsilon(p)$, it is possible to find a δ-neighbourhood of the left limit point 'a' of the domain of the function f denoted by $N_\delta(p)$ such that the values of the function $f(x)$ lie in $N_\varepsilon(p)$ for every value of the independent variable x which lies in the left hand δ-deleted neighbourhood of the left limit point 'a' of the domain of the function f denoted by $N'_\delta(a) = 0 < a - x < \delta$ i.e., $a - \delta < x < a$.

(b) (In terms of $(\varepsilon - \delta)$ definition): A fixed point 'p' is the left hand limit of a function f at the left limit point 'a' of the domain of the function $f \Leftrightarrow \forall\, \varepsilon > 0$, $\exists$ a $\delta > 0$ (δ depends on ε) such that for every value of x in the left deleted δ-neighbourhood $0 < a - x < \delta$, the value of the function $f(x)$ lies in the ε-neighbourhood $(p - \varepsilon,\ p + \varepsilon)$.

That is, given $\varepsilon > 0$, it is possible to find $\delta > 0$ such that

$$0 < a - x < \delta \Rightarrow |f(x) - p| < \varepsilon$$

or, $a - \delta < x < a \Rightarrow p - \varepsilon < f(x) < p + \varepsilon$

Notation: $\lim\limits_{x \to a^-} f(x)$, $f(a^-)$ and $f(a-0)$ are available notations for the left hand limit of a function f at $x = a$, where 'a' is the left limit point of the domain of a function.

Notes: **1.** Limit of a function f is said to exist at an interior point of its domain or at the limit point not in its domain $\Leftrightarrow$ left hand limit and right hand limit of the function are finite and equal at the interior point of the domain of the function.

2. Limit of a function f is said to exist at the right limit point of its domain $\Leftrightarrow$ right hand limit of the function f is finite at the right limit point of the domain of the function f.

3. Limit of a function f is said to exist at the left limit point of its domain $\Leftrightarrow$ left hand limit of the function f is finite at the left limit point of the domain of the function f.

4. The δ-neighbourhood of a point 'a' excluding the point 'a', is divided into two parts by the point 'a' on the real line. These are intervals $(a-\delta, a)$ and $(a, a+\delta)$ where it is required to restrict x to lie to find the left hand limit and right hand limit respectively.

Limit of a Function as $x \to \infty$

Cauchy definition: (Also, termed as "$(\varepsilon - M)$ definition): To define the limit of a function f as $x \to \infty$, first of all the domain of a function f is fixed.

Let a function f be defined out side of some interval $[c, d]$, or it can be said that the function f is defined in the neighbourhood of infinity (symbolized as ∞), i.e. the function f is defined for all x satisfying the inequality $|x| > K$, (i.e. $x > K$ or $x < -K$, where $K > 0$), i.e. the domain of the function f is not bounded.

$\therefore$ A number $L = \lim\limits_{x \to \infty} f(x) \Leftrightarrow \forall\, \varepsilon > 0, \exists$ a number $M_\varepsilon > 0\ (M > K)$ such that $|x| > M$ $\Rightarrow |f(x) - L| < \varepsilon$.

It is sometimes of interest to consider two separate cases of seeking the limits of a function f, viz., when x tends to $+\infty$ and when x tends to $-\infty$. We define each one in the following way.

Limit of a Function f as $x \to +\infty$: Let the function f be defined on the interval $(a, +\infty)$, then it is said that a number $L = \lim\limits_{x \to +\infty} f(x) \Leftrightarrow \forall\, \varepsilon > 0, \exists$ a number $M_\varepsilon > 0$ such that $\forall\, x > M_\varepsilon$ $\Rightarrow |f(x) - L| < \varepsilon$.

Limit of a function f as $x \to -\infty$: Let the function f be defineđ on the interval $(-\infty, a)$, then it is said that a number $L = \lim\limits_{x \to -\infty} f(x) \Leftrightarrow \forall\, \varepsilon > 0, \exists$ a number $M_\varepsilon > 0$ such that

$$\forall\, x < -M \Rightarrow |f(x) - L| < \varepsilon$$

Notes: **1.** One must note that if $\lim\limits_{x\to\infty} f(x) = L$, then

$$\lim_{x \to \infty} f(x) = \lim_{x \to +\infty} f(x) = \lim_{x \to -\infty} f(x) = L$$

2. (i) One must use $\lim\limits_{x\to 0^+} \left(\dfrac{K}{x}\right) = +\infty$ (K is a positive number)

(ii) $\lim\limits_{x\to 0^-} \left(\dfrac{K}{x}\right) = -\infty$ (K is a positive number)

(iii) $\lim\limits_{x\to 0} \left(\dfrac{K}{|x|}\right) = \infty$ (K is a positive number)

(iv) $\lim\limits_{x\to +\infty} \left(\dfrac{K}{x}\right) = 0$ (K is a positive number)

(v) $\lim\limits_{x\to -\infty} \left(\dfrac{K}{x}\right) = 0$ (K is a positive number)

(vi) $\lim\limits_{x\to \infty} \left(\dfrac{K}{|x|}\right) = 0$ (K is a positive number),

$\lim\limits_{x\to\infty} \dfrac{K}{x} = 0$

(vii) The notation

$$\lim_{x \to a} f(x) = \infty \Leftrightarrow \lim_{x \to a} |f(x)| = \infty$$

(viii) A function f for which $\lim\limits_{x \to a} f(x) = 0$ is called infinitesimal as $x \to a$ i.e. a function whose limit is zero at the limit point 'a' of the domain of a function f is called infinitesimal.

Solved Examples

1. Prove that $\lim_{x \to \infty} \frac{1}{x} = 0$

Proof: Let $f(x) = \frac{1}{x}$ and it is given $L = 0$

$\therefore \ |f(x) - L| < \varepsilon,$

i.e., $\Rightarrow \left|\frac{1}{x} - 0\right| < \varepsilon$

if $\left|\frac{1}{x}\right| < \varepsilon$

or, if $\frac{1}{|x|} < \varepsilon$

or, if $|x| > \frac{1}{\varepsilon} = M$ (say)

Hence, $\left|\frac{1}{x} - 0\right| < \varepsilon$ for $|x| > \frac{1}{\varepsilon}$

$\Rightarrow \lim_{x \to \infty} \frac{1}{x} = 0$

2. Prove that $\lim_{x \to \infty} \left(\frac{x+5}{x+2}\right) = 1$

Proof: Let $f(x) = \frac{x+5}{x+2}$ and it is given $L = 1$

$\therefore \ |f(x) - L| < \varepsilon,$

i.e., $\Rightarrow \left|\frac{x+5}{x+2} - 1\right| < \varepsilon$

if $\left|\frac{x+5-x-2}{x+2}\right| < \varepsilon$

or, if $\left|\frac{3}{x+2}\right| < \varepsilon$

or, if $\frac{3}{|x| - 2} < \varepsilon$

or, if $3 < \varepsilon|x| - 2\varepsilon$

or, if $\varepsilon|x| > 3 + 2\varepsilon$

or, if $|x| > \frac{3+2\varepsilon}{\varepsilon} = M$ (say)

Hence $\left|\frac{x+5}{x+2} - 1\right| < \varepsilon$ for $|x| > \frac{3+2\varepsilon}{\varepsilon}$

$\Rightarrow \lim_{x \to \infty} \left(\frac{x+5}{x+2}\right) = 1$

Remark: 1. $\lim_{x \to +\infty} f(x) \neq \lim_{x \to -\infty} f(x)$

$\Leftrightarrow \lim_{x \to \infty} f(x)$ does not exist.

Question: Prove that $\lim_{x\to\infty} \left(\frac{\sqrt{2x^2+3}}{4x+2}\right)$ does not exist.

Proof: $\lim_{x \to +\infty} \left(\frac{\sqrt{2x^2+3}}{4x+2}\right)$

$$= \lim_{x \to +\infty} \frac{x\sqrt{2 + \frac{3}{x^2}}}{x\left(4 + \frac{2}{x}\right)} = \frac{\sqrt{2}}{4} \quad \ldots\text{(i)}$$

and $\lim_{x \to -\infty} \left(\frac{\sqrt{2x^2+3}}{4x+2}\right)$

$$= \lim_{x \to -\infty} \left[\frac{-x\sqrt{2 + \frac{3}{x^2}}}{x\left(4 + \frac{2}{x}\right)}\right] = -\frac{\sqrt{2}}{4} \quad \ldots\text{(ii)}$$

Hence, (i) and (ii) $\Rightarrow \lim_{x \to \infty} \left(\frac{\sqrt{2x^2 + 3}}{4x + 2} \right)$ does not exist.

2. The notation $\lim_{x \to a} f(x) = \infty$ means that

$\lim_{x \to a} |f(x)| = +\infty$.

Examples: **(i)** $\lim_{x \to 0} \left(\frac{1}{x}\right) = \infty$ since $\lim_{x \to 0} \left|\frac{1}{x}\right| = +\infty$

(ii) $\lim_{x \to 2} \left(\frac{1}{x-2}\right) = \infty$ since $\lim_{x \to 2} \left|\frac{1}{x-2}\right| = +\infty$

3. $\lim_{x \to \infty} f(x) = \infty \Leftrightarrow$ for every positive number M, however large, there exists a positive number P such that $|f(x)| > M$ whenever $|x| > P$.

In other words $\lim_{x \to \infty} |f(x)| = +\infty$, i.e.

$\lim_{|x| \to +\infty} |f(x)| = +\infty$.

Example: $\lim_{x \to \infty} \left(\frac{2x^2}{x^2+1}\right) = \infty$

Since $\left|\frac{2x^3}{x^2+1}\right|$

$= |x| \left(\frac{2x^2}{x^2+1}\right)$

$= |x| \cdot \left(\frac{2}{1 + \frac{1}{x^2}}\right) \to +\infty$ as $|x| \to +\infty$

N.B.: $\lim_{x \to \infty} f(x) = \infty$ is possible for the function $y = f(x)$ whose domain is a subset of R and the range is unbounded.

Hein's Definition of the Limit of a Function

Let $f: X \to R$, $X \subseteq R$, and supposing that 'a' is a limit point of the set X which is the domain of the function f.

Definition: (Hein's): A number l is the limit of the function f at a limit point 'a' of the domain of the function f, i.e., $\lim_{x \to a} f(x) = L \Leftrightarrow$ For any sequence of values of x converging to the number 'a'

($x_1, x_2, x_3, \ldots x_n, \ldots$ belonging to the domain of definition of the function and being different from a, i.e., $x_n \in X$, $x_n \neq a$).

The corresponding sequence of values of y

$y_1 = f(x_1)$, $y_2 = f(x_2)$, $y_3 = f(x_3)$, ..., $y_n = f(x_n)$, ... has a limit, which is the number L.

Notes: **(i)** It is emphasized that the concept of the limit of a function at a point 'a' is possible only for a limit point 'a' of the domain of the function.

(ii) Hein's definition of the limit of a function is conveniently applied when it is required to show that a function $f(x)$ has no limit. For this it is sufficient to show that there exist two sequences $[x'_n]$ and $[x''_n]$ such that $\lim_{n \to \infty} x'_n = \lim_{n \to \infty} x''_n = a$ but the corresponding sequences $\{f(x'_n)\}$ and $\{f(x''_n)\}$ do not have identical limits, i.e. if $\lim_{x'_n \to 0} \{f(x'_n)\} \neq \lim_{x''_n \to 0} \{f(x''_n)\}$, then $\lim_{x \to 0} f(x)$ does not exist and if $\lim_{x \to 0} f(x) = L$, then $\lim_{x_n \to 0} f(x_n) = L$ for every sequence $x_n \to 0$

Example: **1.** Show that $\lim_{x \to 0} \sin\left(\frac{1}{x}\right)$ does not exist.

Solution: On choosing two sequences $x'_n = \frac{1}{2n\pi}$

and $x''_n = \frac{1}{2n\pi + \frac{\pi}{2}}$ $(n = 1, 2, 3, \ldots)$

We get

$$x'_n \to 0 \equiv n \to \infty$$

$$x''_n \to 0 \equiv n \to \infty$$

Now,

$$\lim_{x \to 0} \sin\left(\frac{1}{x}\right)$$

$$= \lim_{(x'_n \to 0)} \sin\left(\frac{1}{x'_n}\right)$$

$$= \lim_{n \to \infty} (\sin(2n\pi)) = 0$$

Also, $\lim_{x \to 0} \sin\left(\frac{1}{x}\right)$

$$= \lim_{(x''_n \to 0)} \sin\left(\frac{1}{x''_n}\right)$$

$$= \lim_{n \to \infty} \left(\sin\left(2n\pi + \frac{\pi}{2}\right)\right) = 1$$

Hence, $\lim_{x'_n \to 0} f(x'_n) = 0$ and $\lim_{x''_n \to 0} f(x''_n) = 1$

$\therefore \lim_{x \to 0} \sin\left(\frac{1}{x}\right)$ does not exist.

Limit of The Product of an Infinitesimal and a Bounded Function

Definition (i): A function f is bounded in a δ-neighbourhood of a point $x = a \Leftrightarrow \exists$ a real number 'm' such that $|f(x)| \le m, \forall\ x$ in $|x - a| < \delta$.

Definition (ii): A function f is bounded in a deleted δ-neighbourhood of a point $x = a \Leftrightarrow \exists$ a real number m such that $|f(x)| \le m$, $\forall\ x$ in $0 < |x - a| < \delta$.

Now an important theorem which provides us a relation between an infinitesimal and a bounded function.

Theorem: $\lim_{x \to a} f(x) = 0$ and $g(x)$ is bounded in a deleted neighbourhood of the point $\Leftrightarrow \lim_{x \to a} f(x) \cdot g(x) = 0$.

Solved Examples

Evaluate the following ones:

(i) $\lim_{x \to 0} x \cdot \sin\left(\frac{1}{x}\right)$

(ii) $\lim_{x \to 0} x \cdot \cos\left(\frac{1}{x}\right)$

Solution: (i) Let $f(x) = x$ and $g(x) = \sin\left(\frac{1}{x}\right)$

Now, $\lim_{x \to 0} f(x) = \lim_{x \to 0} x = 0$

and $|g(x)| = \left|\sin\left(\frac{1}{x}\right)\right| \le 1, \forall\ x, x \ne 0$

$\Rightarrow g(x)$ is bounded in a deleted neighbourhood of the point '0'.

Hence, $\lim_{x \to 0} x \sin\left(\frac{1}{x}\right) = 0$

(ii) Let $f(x) = x$ and $g(x) = \cos\left(\frac{1}{x}\right)$

Now $\lim_{x \to 0} f(x) = \lim_{x \to 0} x = 0$

and $|g(x)| = \left|\cos\left(\frac{1}{x}\right)\right| \le 1, \forall\ x, x \ne 0$.

$\Rightarrow g(x)$ is bounded in a deleted neighbourhood of the point '0'.

Hence, $\underset{x \to 0}{\text{L}}\ x \cos\left(\frac{1}{x}\right) = 0$

Geometrical Meaning of the Limit of a Function

On the Limit: In the language of geometry, the limit of a function $y = f(x)$ at a point $x = c$, where c is a limit point of the domain of the function f, can be stated as:

"If given any $\varepsilon > 0$, $\exists$ a $\delta_\varepsilon > 0$ (i.e. a positive real number δ depending upon ε): such that a part (i.e. a portion) of the graph of the function $y = f(x)$ lies in the rectangle bounded by the lines

$x = c - \delta, x = c + \delta$ (vertical lines)

$y = l - \varepsilon, y = l + \varepsilon$ (horizontal lines)

then 'L' is the limit of the function $y = f(x)$ at the point $x = c$.

That is, a portion of the graph of the function $y = f(x)$ can be obtained in a rectangle whose height can be taken as small as one pleases by choosing the width as small as one requires.

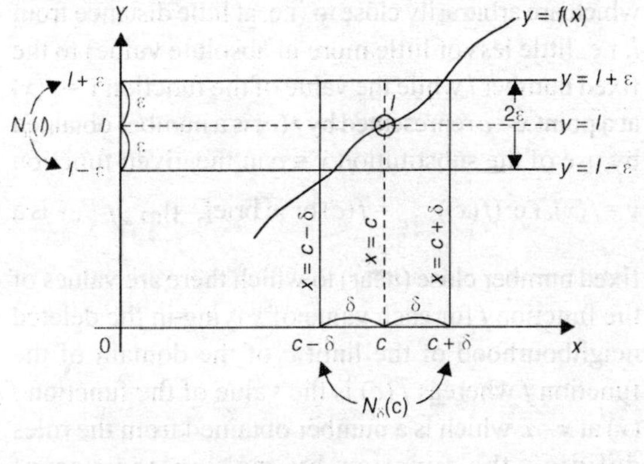

Left (or left hand) Limit

Geometrically, the left limit of the function $y = f(x)$ at a point $x = c$, where 'c' is the left limit point of the domain of the function f, can be stated as:

"Given an $\varepsilon > 0, \exists$ a $\delta_\varepsilon > 0$: a portion of the graph of the function $y = f(x)$ lies in the rectangle bounded by the lines

$x = c - \delta, x = c$ (vertical lines)

$y = l - \varepsilon, y = l + \varepsilon$ (horizontal lines)

then it is said that $f(x)$ has the left limit at the point $x = c$.

That is, a portion of the graph of the function $y = f(x)$ can be obtained in a rectangle whose height is arbitrarily small by choosing the width (on the left of the point $x = c$) as small as one requires.

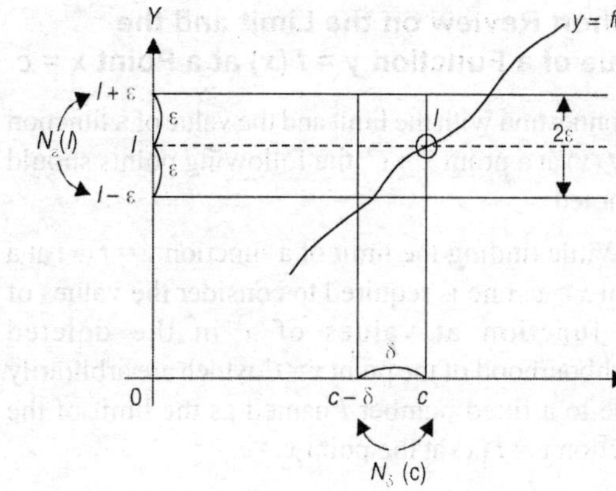

Right (or right hand) Limit

In the point of the view of geometry, the right limit of the function $y = f(x)$ at a point $x = c$, where c is the right limit point of the domain of the function f, is narrated as:

"Given any $\varepsilon > 0, \exists$ a $\delta_\varepsilon > 0$: a portion of the graph of the function $y = f(x)$ lies in the rectangle bounded by the lines

$x = c, x = c + \delta$

$y = L - \varepsilon, y = L + \varepsilon$

then $f(x)$ is said to have the right limit at $x = c$.

That is, a portion of the graph of the function $y = f(x)$ can be obtained in a rectangle whose height is arbitrarily small by choosing the width (on the right of the point $x = c$) as small as one requires.

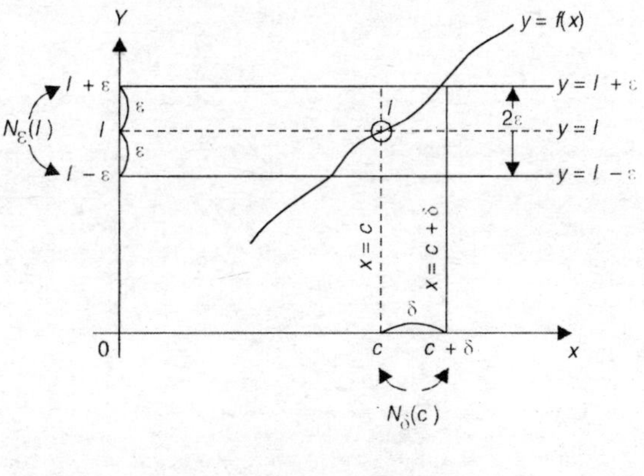

A Short Review on the Limit and the Value of a Function y = f (x) at a Point x = c

In connection with the limit and the value of a function $y = f(x)$ at a point $x = C$, the following points should be noted.

1. While finding the limit of a function $y = f(x)$ at a point $x = c$, one is required to consider the values of the function at values of x in the deleted neighbourhood of the point $x = C$ which are arbitrarily close to a fixed number l named as the limit of the function $y = f(x)$ at the point $x = c$.

2. The value of the function $y = f(x)$ at the point $x = c$ is left out of the discussion. This (i.e. $f(a)$) may or may not exist. Even if $f(a)$ exists, $f(a)$ need not be equal to or even close to the limit l of the function f at c.

For example,

$$\text{Let } f(x) = \frac{x^2 - 4}{x - 2}, x \neq 2$$

$$= 6, x = 2$$

In this example $f(2) = 6$ is given but the limit of $f(x)$ at $x = 2$ is 4.

Difference Between the Limit and the Value of a Function y = f (x) at a Point x = c

The main difference between the limit and the value of function $y = f(x)$ at a point $x = c$ is the following: The limit of a function $y = f(x)$ at a point $x = c$ is a fixed number L in whose ε -neighbourhood lie the values of the function f (at each value of the independent variable x situated in the δ -deleted neighbourhood of the limit point of the domain of the functions f) which are arbitrarily close to (i.e. at little distance from l, i.e. little less or little more in absolute value) to the fixed number l while the value of the function $y = f(x)$ at a point $x = c$ represented by $f(c)$ is a number obtained by use of the substitution $x = c$ in the given function $y = f(x)$, i.e. $(f(x))_{x=c} = f(c)$ or in brief, $\lim_{x \to c} f(x)$ is a fixed number close (near) to which there are values of the function f for each value of x lying in the deleted neighbourhood of the limit c of the domain of the function f whereas $f(c)$ is the value of the function $f(x)$ at $x = c$ which is a number obtained from the rules defining the function by making the use of substitution $x = c$ in it.

3

Continuity of a Function

In general, the Limit of a function $y = f(x)$ at a limit point of its domain namely $x = a$ need not be equal to the value of the given function $y = f(x)$ at the limit point $x = a$ which means that the limit of the given function may or may not be equal to the value of the given function, i.e. $\lim_{x \to a} f(x)$ is not necessarily equal to $f(a)$.

However, there is an important class of functions for which the limit and the value are same. Such functions are called continuous functions.

Definition: If $x = a$ is a limit point in the domain of a given function $y = f(x)$ and the limit $y = f(x)$ at $x = a$ is $f(a)$, then the function $y = f(x)$ is said to be continuous at the given limit point $x = a$ and 'a' is termed as the point of continuity of the given functions $y = f(x)$, i.e. $f(x)$ is continuous at the limit point $x = a$ of its domain $\Leftrightarrow \lim_{x \to a} f(x) = f(a)$. Hence, in words, the continuity of a given function at a given limit point in the domain of the given function $\Leftrightarrow$ limit of the same function at the given limit point = value of the given function at the given limit point.

Notes: **1.** The definition of continuity says that given function should be defined both for the limit point $x = a \in D(f)$ and for all other points near $x = a$ (near $x = a$ means in the open interval $(a - h, a + h)$, where 'h' is a small positive number) in $D(f)$.

2. In more abstract form, the definition of continuity of a function at the limit point in its domain tells us that a function f is continuous at the limit point $x = a \Leftrightarrow$

(i) $f(a)$ is defined, i.e., the limit point 'a' lies in the domain of f.

(ii) $\lim_{x \to a} f(x)$ exists

(iii) $\lim_{x \to a} f(x) = f(a)$

3. One should note that

(i) The definition of $\lim_{x \to a} f(x) = p$ requires that 'a' is the limit point of $D(f)$ where 'a' is not necessarily in $D(f)$ while the definition of continuity of a function at the limit point 'a' requires that the limit point 'a' must be in $D(f)$ which means that it is a must for 'a' to be an interior point of $D(f)$.

(ii) By definition of continuity given above it is possible for a function to be continuous at a limit point in its domain $D(f)$ but not to have a limit as $x \to a$. Situation arises in the case of a function that is continuous at an isolated point of its domain $D(f)$.

$(\varepsilon - \delta)$ Definition of Continuity of a Function at the Limit Point of its Domain

It says that a function $y = f(x)$ is continuous at the limit point $x = a \Leftrightarrow \forall$ given $\varepsilon > 0$, $\exists\, \delta(\varepsilon)$ such that $|x - a| < \delta \Rightarrow |f(x) - f(a)| < \varepsilon$.

The definition of continuity of a functions $y = f(x)$ at the limit point $x = a$ of its domain, given above, implies that, if any neighbourhood $(f(a) - \varepsilon, f(a) + \varepsilon)$ is chosen of the number $f(a)$, of arbitrary length, 2ε where ε is any positive number, however small, then for this ε-neighbourhood of $f(x)$, there is always a δ-neighbourhood $(a - \delta, a + \delta)$ of the limit point 'a' such that for every value of x in the δ-neighbourhood, the value of $f(x)$ lies in ε-neighbourhood $(f(a) - \varepsilon, f(a) + \varepsilon)$.

Remarks: In connection with $(\varepsilon - \delta)$ definition of limit and continuity of a function at the limit point of (and in) its domain, one should note that

(i) $(\varepsilon - \delta)$ definition of continuity is same as $(\varepsilon - \delta)$ definition of limit of a function in which "l = limit of $f(x)$ at the limit point $x = a$" is replaced by $f(a)$ = value of the function $f(x)$ at $x = a$.

(ii) $\varepsilon > 0$, however small, means that one may take $\varepsilon = 0.1$, 0.01, 0.001, 0.0001 and so on according to degree of accuracy which one proposes to adopt. The key point is that ε is an arbitrary (not fixed) number of our own selection, and that it may be taken as small as we please.

(iii) It is common to say that $y = f(x)$ is continuous at a point $x = a \in D(f)$ instead of saying that $y = f(x)$ is continuous at a limit point $x = a$ in the domain of the given function $y = f(x)$.

Continuity of a Function at a Limit Point in its Domain in the Language of a Sequence

In the language of sequences, the definition of continuity of a function at a point may be stated in the following way:

"A function $y = f(x)$ is continuous at the point $x = a$ if for any sequence of the values of the independent variable $x = a_1, a_2, a_3, \ldots, a_n, \ldots$ which converges to 'a', the sequences of the corresponding values of the function $f(x) = f(a_1), f(a_2), f(a_3), \ldots, f(a_n), \ldots$ converges to $f(a)$.

Theorems on continuous functions at a point: Some theorems on continuous functions at a point of its domain are direct results of definition of continuity of a function and theorems on limits at a point.

A. The sum, difference, product and quotient of two continuous functions at any point $x = a$ in their domain is continuous at the same point $x = a$, provided in the case of quotient, the divisor (the function in denominator) at the same point $x = a$ is not zero, i.e. if $f(x)$ and $g(x)$ are two functions continuous at any point $x = a$, then the functions (i) $y = f(x) + g(x)$ (ii) $y = f(x) - g(x)$ (iii) $y = f(x) \cdot g(x)$ (iv) $y = \dfrac{f(x)}{g(x)}$, $g(a) \neq 0$ are also continuous at the same point $x = a$.

B. The scalar multiple, modulus and reciprocal of a function continuous at a point in their domain are continuous at the same point, provided in case of reciprocal of a functions, the function in denominator is not zero at the point where continuity of y is required to be tested, i.e., if $f(x)$ is a function continuous at a point $x = a$ and k is any constant, then the functions

(i) $y = kf(x)$, (ii) $y = |f(x)|$, (iii) $y = \dfrac{1}{f(x)}$, $f(a) \neq 0$

are also continuous at the same point $x = a$.

C. Continuity of a composite function at a point: If the inner function of a composite function is continuous at a point a in its domain and the outer function is continuous at the point representing the value of the inner function at the point a, then the composite function is continuous at the point a of continuity of the inner function, i.e. if $y = f(x)$ is continuous at the point $x = a$ and the function $u = g(y)$ is continuous a the point $f(a) = b$ (s a y), then the composite function $u = g(f(x)) = F(x)$ (say) is continuous at the point $x = a$. i.e., $\lim_{x \to a} g(f(x)) = g\{f(a)\}$.

Remember: **1.** Every point at which the given function is continuous is called a point of continuity of the function.

2. Every point at which the condition of continuity of the given function is not satisfied is called a point of discontinuity of the function.

Question: When a given function $y = f(x)$ is continuous or discontinuous at a point $x = a$? Mention the commonest functions continuous or discontinuous at a point (points).

Answer: *1.* On continuity: A function $y = f(x)$ is continuous at $x = a \Leftrightarrow \lim_{x \to a} f(x) = f(a)$

The Commonest Functions Continuous at a Point(s)

(i) All standard functions (algebraic polynomial, rational, irrational, trigonometric, inverse trigonometric, exponential, logarithmic and constant functions (symbolised as APRL-CIT E functions) at each point of their domain Hence, $\lim_{x \to a} \sin x = \sin a$ for all real numbers in the domain of sin x which means sin x is continuous at every point which lies in its domain.

(ii) $\lim_{x \to a} \cos x = \cos a$ for all real numbers in the domain of cos x which means cos x is continuous at every point which lies in its domain.

(iii) $\lim_{x \to a} \tan x = \tan a$ where 'a' is a real number other than $(2n+1)\frac{\pi}{2}$ odd multiple of $\frac{\pi}{2}$ $(n = 0, \pm 1, \pm 2, ...)$ which means tan x is continuous for all real values of x excepting $x = (2n+1)\frac{\pi}{2}$, n being an integer.

(iv) $\lim_{x \to a} \cot x = \cot a$ where a is a real number other than (different from) $n\pi$, multiple of π $(n = 0, \pm 1, \pm 2, ...)$, which means cot x is continuous for all real values of x excepting $x = n\pi$, n being an integer.

(v) $\lim_{x \to a} \sec x = \sec a$ where a is real number other than $(2n+1)\frac{\pi}{2}$ = odd multiple of $\frac{\pi}{2}$ $(n = 0, \pm 1, \pm 2, ...)$ which means sec x is continuous for all real values of x excepting $x = (2n+1)\frac{\pi}{2}$, n being an integer.

(vi) $\lim_{x \to a} \operatorname{cosec} x = \operatorname{cosec} a$ where 'a' is a real number other than $n\pi$, multiple of π $(n = 0, \pm 1, \pm 2, ...)$, which means cosec x is continuous for all values of x excepting $x = n\pi$, n being an integer.

(vii) $\lim_{x \to a} \sin^{-1} x = \sin^{-1} a$ where $-1 \le a \le 1$ which means $\sin^{-1} x$ is continuous for every value of x from -1 to $+1$.

(viii) $\lim_{x \to a} \cos^{-1} x = \cos^{-1} a$ where $-1 \le a \le 1$ which means $\cos^{-1} x$ is continuous for every value of x from -1 to $+1$.

(ix) $\lim_{x \to a} \tan^{-1} x = \tan^{-1} a$ for all real number in the domain of $\tan^{-1} x$ which means $\tan^{-1} x$ is continuous for all real values of x.

(x) $\lim_{x \to a} \cot^{-1} x = \cot^{-1} a$ for all real numbers which means $\cot^{-1} x$ is continuous for all real values of x.

(xi) $\lim_{x \to a} \sec^{-1} x = \sec^{-1} a$ where a is a real number ≥ 1 or ≤ -1, which means $\sec^{-1} x$ is continuous for all values of x which do not belong to the open interval $(-1, 1)$.

(xii) $\lim_{x \to a} \operatorname{cosec}^{-1} x = \operatorname{cosec}^{-1} a$ where a is a real ≥ 1 or < -1 which means $\operatorname{cosec}^{-1} x$ is continuous for all real values of x which do not belong to the open interval $(-1, 1)$.

(xiii) $\lim_{x \to a} x^n = a^n$ for all real number, provided $n \ge 0$ and for all real numbers other than zero, provided $n < 0$ which means x^n is continuous for all real values of x when n (i.e., index or exponent of the base of power function x^n) is non-negative and x^n is continuous for all real values of x excepting zero when n is negative.

(xiv) $\lim_{x \to a} e^x = e^a$, for all real numbers which means e^x is continuous for all real values of x.

(xv) $\lim_{x \to c} a^x = a^c$ $(a > 0)$ for all real number which means a^x $(a > 0)$ is continuous for all real values of x.

(xvi) $\lim_{x \to a} \log x = \log a$ provided $a > 0$ which means $\log x$ $(x > 0)$ is continuous for all positive values of x.

(xvii) $\lim_{x \to a} \log |x| = \log |a|$ provided $a \ne 0$ which means $\log |x|$ $(x > 0$ or $x < 0)$ is continuous for all positive and negative values of x but not at $x = 0$.

(xviii) $\lim_{x \to a} \alpha = \alpha$, for all real numbers which means a constant function α is continuous for all real values of x.

2. *Discontinuity*: A function $y = f(x)$ is discontinuous at a point $x = a \Leftrightarrow \lim_{x \to a} f(x) \neq f(a)$, i.e., a function $y = f(x)$ is discontinuous at a point $x = a$ (or, a function $y = f(x)$ has a point of discontinuity namely 'a') if and only if limit and value of the function are not equal at the same point $x = a$.

The Commonest Functions Discontinuous at a Point(s)

1. One of the commonest cases of discontinuity which occurs in practice is when a function is defined by a single formula $y = \dfrac{f(x)}{g(x)}$ with a restriction $g(x) \neq 0$ y assumes the form a fraction with a zero denominator for a value (or, a set of values) of x provided by the equation $g(x) = 0$, e.g.:

(i) $f(x) = \dfrac{1}{x}$, $x \neq 0$ i.e., it is discontinuous at $x = 0$.

(ii) $f(x) = \dfrac{\sin x}{x}$, $x \neq 0$ i.e., it is discontinuous at $x = 0$.

(iii) $f(x) = \sin\left(\dfrac{1}{x}\right)$, $x \neq 0$ i.e., it is discontinuous at $x = 0$.

2. All standard and non-standard functions defined by a single formula $y = f(x)$ are continuous at each point of their domain excepting a finite set of points at which they are undefined which means all standard and non-standard functions are discontinuous at a point (or, a set of points) at which they are not defined, e.g.:

(i) $f(x) = \dfrac{x^2 - 1}{x - 1}$, $x \neq 1$ is discontinuous at $x = 1$.

(ii) $f(x) = \tan x$ and $\sec x$ are discontinuous at $x = (2n \pm 1)\dfrac{\pi}{2}$, and continuous at all other values of x.

(iii) $f(x) = \cot x$ and $\text{cosec}\, x$ are discontinuous at $x = n\pi$, $(n \varepsilon I)$ and continuous at all other values of x.

3. When a function is a piecewise function, then it has a chance of having both a point of continuity and a point of discontinuity at a common point (points) of adjacent intervals, e.g.:

(i) $f(x) = \begin{cases} -2x^2, & \text{for } x \leq 3 \\ 3x, & \text{for } x > 3 \end{cases}$ is discontinuous at $x = 3$.

(ii) $f(x) = \begin{cases} \frac{1}{5}\left(2x^2 + 3\right) & \text{for } -\infty < x \leq 1 \\ 6 - 5x & \text{for } 1 < x < 3 \\ x - 3 & \text{for } 3 \leq x < \infty \end{cases}$

is continuous at $x = 1$ and discontinuous at $x = 3$.

4. Whenever a function is defined by

$$f(x) = f_1(x), \text{ if } x \neq a$$
$$= \text{a constant, if } x = a \text{ (i.e. } f(a) = \text{constant)}$$

then there is a chance of having, a point of continuity or a point of discontinuity at $x = a$, e.g.:

(i) $f(x) = x + 2$, if $x \neq 2$
$= 3$, if $x = 2$
has a point of discontinuity at $x = 2$.

(ii) $f(x) = x + 2$, $x \neq 2$
$= 4$, $x = 2$
has a point of continuity at $x = 2$.

Remember: **1.** A function of undefined quantity is always undefined. For this reason $\sin\left(\dfrac{1}{x}\right)$, $\cos\left(\dfrac{1}{x}\right)$ and $\tan\left(\dfrac{1}{x}\right)$ are undefined at $x = 0$.

On Limits of a Continuous Function

1. The limit of a continuous function of variable = that function of the limit of the variable, i.e.

$\lim_{x \to a} f(x) = f\left(\lim_{x \to a} x\right)$ where $f(x)$ is a continuous functions at $x = a$.

2. The limit of a continuous function of a continuous function of an independent variable = outer continuous function of the limit of the inner continuous function of the independent variable, i.e.,

$\lim_{x \to a} f(g(x)) = f\left(\lim_{x \to a} g(x)\right)$ provided $y = g(x)$ is continuous at the point $x = a$ and the function $u = f(y)$ is continuous at the point $g(a) = b$ (say).

Remember: **1.** Finding the limit of a continuous function may be replaced by finding the value of the function of the limit of the independent variable, i.e., if $f(x)$ is continuous function, then $\lim_{x \to a} f(x) = f\left(\lim_{x \to a} x\right)$. This is sometimes expressed briefly thus: the limit sign of a continuous function can be put before the independent variable, e.g.:

(i) $\lim_{x \to a} x = a$

(ii) $\lim_{x \to a} \sin x = \sin\left(\lim_{x \to a} x\right) = \sin a$, etc.

2. The concept of continuity of a function can be used to find its limit, i.e., if the function $f(x)$ is continuous at $x = a$, then in order to find its limit $\lim_{x \to a} f(x)$, it is sufficient to calculate its value at the point $x = a$ since $\lim_{x \to a} f(x) = f(a)$.

Example: Evaluate: $\lim_{x \to \frac{\pi}{2}} \left(\frac{1 + 2\sin x}{3x + \cos x}\right)$

Solution: The functions in numerator and denominator are continuous for all positive values of x. so the quotient function $y = \left(\frac{1 + 2\sin x}{3x + \cos x}\right)$ is also continuous for all positive values of x which means it is continuous at $x = \frac{\pi}{2}$ and hence,

$$\lim_{x \to \frac{\pi}{2}} \left(\frac{1 + 2\sin x}{3x + \cos x}\right) = \frac{1 + 2\sin\left(\frac{\pi}{2}\right)}{3\left(\frac{\pi}{2}\right) + \cos\left(\frac{\pi}{2}\right)} = \frac{2}{\pi}$$

3. A function $y = f(x)$ defined in an interval is said to be piecewise continuous if the interval (in which given function $f(x)$ is defined) can be divided into a finite number of non overlaping open subintervals over each of which the functions $f(x)$ is continuous.

Geometrical Meaning of Continuity of a Function $y = f(x)$ at a given Point $x = a$ in its Domain

From the point of view of geometry, a function $y = f(x)$ is continuous at a given point $x = a$ in its domain means that the graph of the functions $y = f(x)$ is unbroken at $P(a, f(a)) \Leftrightarrow$

1. The point $P(a, f(a))$ lies on the graph of the function $y = f(x)$, i.e. $f(x)$ is defined at $x = a$.

2. If $Q(x, y)$ is a point on the graph of $y = f(x)$ and nearer to the point $P(a, f(a))$, then on which ever side of the point P, the point Q may be, it must be possible to make the distance between P and Q as small as one wants along the graph of the function by making the distance between x and a small accordingly.

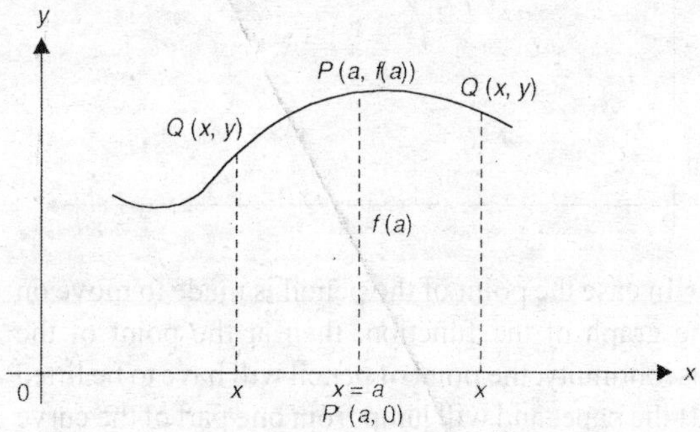

If a function $y = f(x)$ is continuous throughout an interval (a, b), the graph of the function in this interval is without any gap, break or jump, i.e. the graph of the function is unbroken in this interval, i.e. the graph of the function has no point missing corresponding to each value of the independent variable in this interval. In rough language, if the point of a pencil is placed at one end of the graph, we can move the pencil on the graph to the other end of the graph without ever having to lift the pencil off the paper. Further, if a line is drawn across the graph, it will pass through at least one point on the graph.

Note: It is better to say that f is continuous at a point $x = a \Leftrightarrow$ the graph of the function $y = f(x)$ is unbroken at and in the neighbourhood of the point $(a, f(a))$.

If a function is discontinuous at a point $x = a$, then it is a must that the graph of the function has a gap, a break, or hole at the point whose abscissa is $x = a$, i.e. the point $(a, f(a))$ will be missing on the graph of $y =$

$f(x)$. If the function $y = f(x)$ is not defined at $x = a$, then there is a gap, a break or a hole in the graph as there is no point on the graph whose abscissa is $x = a$. Moreover, if a function is defined by different formulas (different expression in x) in different intervals whose left and right end points are same known as adjacent intervals in succession, there is usually a possibility of discontinuity at the common points of the adjacent intervals. Thus, if we have one functional formula $y = f_1(x)$ for $x \geq a$ and another $y = f_2(x)$ for $x < a$ we have the possibility of a discontinuity at $x = a$.

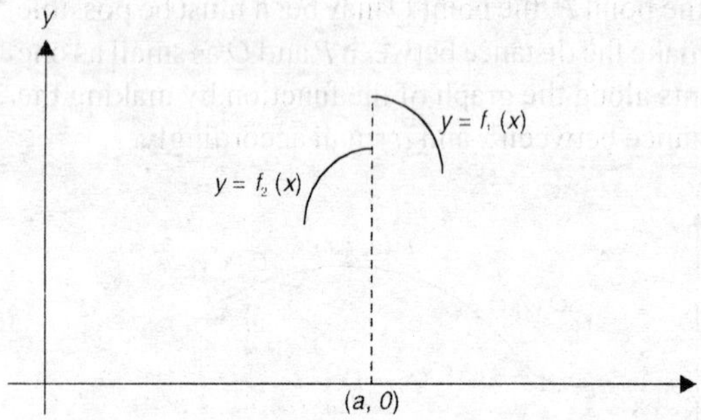

In case the point of the pencil is made to move on the graph of the function, then at the point of the discontinuity, the point of pencil will have to be lifted off the paper and will jump from one part of the curve to the other. i.e., while drawing a graph when the point of the pencil leaves contact with the paper, the function becomes discontinuous at the point where contact is left.

Illustrations on Discontinuity of a Function at a Given Point

1. $f(x) = \begin{cases} a, \text{ when } x < 0 \\ 2a, \text{ when } x = 0 \\ 3a, \text{ when } x > 0 \end{cases}$

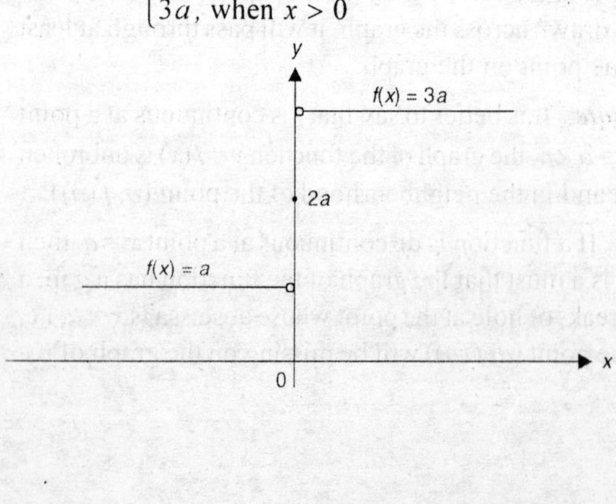

The function $f(x)$ has a discontinuity at $x = 0$.

2. $f(x) = \dfrac{x^2 - a^2}{x - a}, x \neq a$

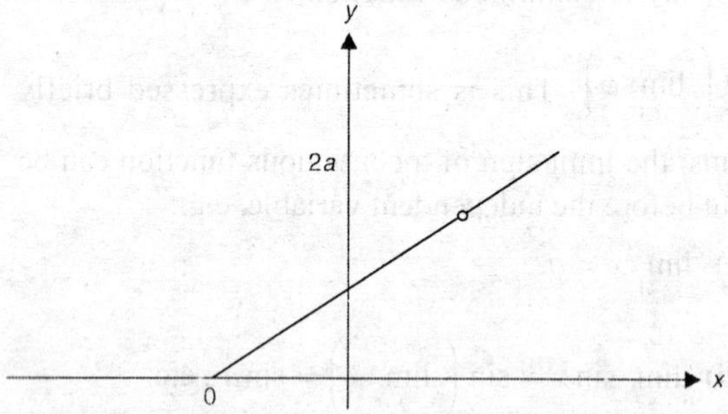

Here the function $f(x)$ is discontinuous at $x = a$.

Remember: If a function $y = f(x)$ is not defined (has no finite value) for any particular value of its independent variable, then the corresponding point on the graph of the function will be missing and the graph will have a hole (a break) at that point.

Two Sided Continuity of a Function at a Point

In general, a function $y = f(x)$ defined on its domain is right continuous (continuous from the right) at the right limit point $x = c$ in its domain $\Leftrightarrow \lim\limits_{x \to c^+} f(x) = f(c)$ and it is left continuous (continuous from the left) at the left limit point $x = c$, in its domain $\Leftrightarrow \lim\limits_{x \to c^-} f(x) = f(c)$.

Hence, a function $y = f(x)$ is continuous at the limit point $x = c$ in its domain $\Leftrightarrow$. It is both right continuous and left continuous at $x = c$.

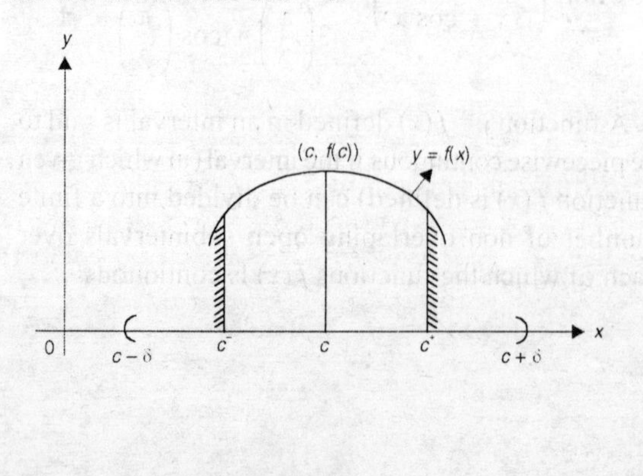

Continuity of a Function in an Open Interval

A function $y = f(x)$ is said to be continuous in an open interval $(a, b) \Leftrightarrow$ it is continuous at every point of the interval $(a, b) \Leftrightarrow$ Geometrically, the graph of the function $y = f(x)$ is unbroken between the points $(a, f(a))$ and $(b, f(b))$.

Continuity of a Function in a Closed Interval

A function $y = f(x)$ is said to be continuous in a closed interval $[a, b] \Leftrightarrow$ it is continuous in the open interval (a, b) and is continuous at the left end point $x = a$ from the right and is continuous at the right end point $x = b$ from the left.

Geometrically, a function $y = f(x)$ is continuous in a closed interval $(a, b) \Leftrightarrow$ The graph of the function $y = f(x)$ is an unbroken line (curved or straight) from the point $(a, f(a))$ to the point $(b, f(b))$.

Furthers, one should note that a function $y = f(x)$ defined over $[a, b]$ is continuous at the left end point $x = a$ from the right $\Leftrightarrow \lim_{x \to a^+} f(x) = f(a)$ and the function $y = f(x)$ defined over (a, b) is continuous at the right end point $x = b$ from the left $\Leftrightarrow \lim_{x \to b^-} f(x) = f(b)$.

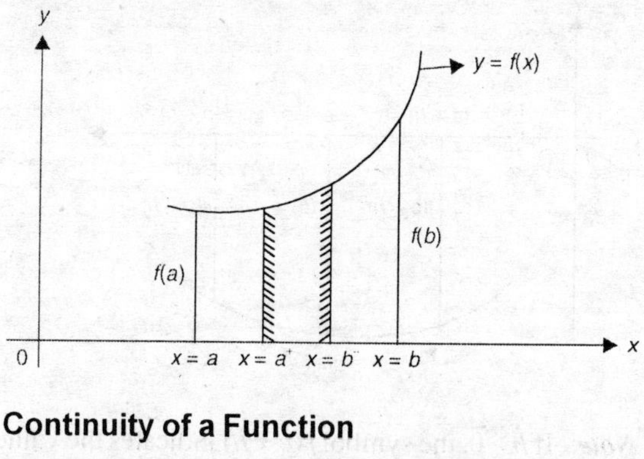

Continuity of a Function

A function $y = f(x)$ is called continuous $\Leftrightarrow$ It is continuous at every point on its domain, e.g.: e^x, sin x, cos x, any polynomial function in x are continuous functions, i.e., these are functions continuous at every point on its domain.

Classification of Points of Discontinuity of a Function

Let $x = a$ be the limit point of the domain of the function $y = f(x)$. The point 'a' is a point of discontinuity of the function $f(x)$ if at this point, $f(x)$ is not continuous.

Let $f(x)$ be defined in a deleted neighbourhood of the point 'a' then 'a' is

1. A point of removal discontinuity of the function $f(x)$ if there is a limit $\lim_{x \to a} f(x) = b$, but either $f(x)$ is not defined at the point a or $f(a) \neq b$. If we set $f(a) = b$, then the function $f(x)$ becomes continuous at the point 'a', i.e., the discontinuity will be removed.

2. A point a of discontinuity is of the first kind of the function $f(x)$ if there are $\lim_{x \to a^+} f(x)$ and $\lim_{x \to a^-} f(x)$ but $\lim_{x \to a^+} f(x) \neq \lim_{x \to a^-} f(x)$.

3. A point a of discontinuity is of the second kind of the function $f(x)$ if at least one of the one sided limits of the function $f(x)$ does not exist at the point $x = a$, i.e., either or both of $\lim_{x \to a^+} f(x)$ and $\lim_{x \to a^-} f(x)$ do not exist.

4. A point a is of mixed discontinuity if one of the one sided limits exists, i.e., if one of the limits namely $\lim_{x \to a^+} f(x)$ or $\lim_{x \to a^-} f(x)$ exists.

5. A point is of infinite discontinuity of the function $f(x)$ if either or both of the one sided limits are infinite, i.e., if either or both of $\lim_{x \to a^+} f(x)$ and $\lim_{x \to a^-} f(x)$ are infinite.

Notes: One should note that 'a' is

(i) A point of discontinuity of the first kind from the left at a if a $\lim_{x \to a^-} f(x)$ exists but $\lim_{x \to a^-} f(x) \neq f(a)$.

(ii) A point of discontinuity is of the first kind from the right at a if $\lim_{x \to a^+} f(x)$ exists but $\lim_{x \to a^+} f(x) \neq f(a)$.

(iii) A point of discontinuity is of the second kind from the left at a if $\lim_{x \to a^-} f(x)$ does not exist.

(iv) A point of discontinuity is of the second kind from the right at a if $\lim_{x \to a^+} f(x)$ does not exist.

Properties of Continuous Functions

Now we state without proof some important properties of continuous functions.

1. If $f(x)$ is continuous in a closed interval $[a, b]$, then range of $f(x)$ is bounded. In other words, if $f(x)$ is continuous in $[a, b]$, then we can find two numbers m and M such that $m < f(x) < M, \forall x \in [a, b]$

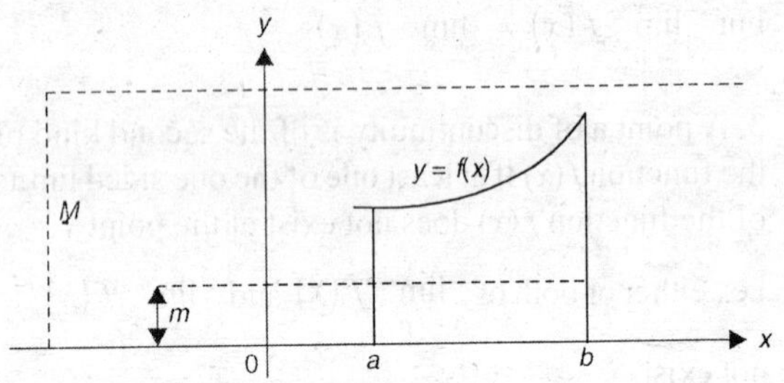

Note: This property may not be true if the domain of $f(x)$ is not a closed interval or if $f(x)$ discontinuous even at a single point in its domain. For example, if $f(x) = \frac{1}{x}$, then $f(x)$ is continuous in the open interval (0, 1), its range consists of all real numbers > 1 and evidently no number M can be found such that $\frac{1}{x} < M$ for all x in $0 < x < 1$. Again consider the function $f(x)$ defined in (–1, 1) as follows:

$f(x) = \frac{1}{x}$, when $x \neq 0$, $f(0) = 1$. Then $f(x)$ is defined in (–1, 1) and continuous at every point in this interval except at $x = 0$. Evidently no two fixed numbers can be found such that $m < \frac{1}{x} < M$ for all x in (–1, 1)

2. If $f(x)$ is continuous and positive at $x = c$ in its domain, then for all sufficiently small values of h, $f(c + h)$ and $f(c - h)$ are both > 0. In other words, if $f(x)$ is continuous and positive at $x = c$, then a h-neighbourhood of the point 'c' can be found throughout which $f(x)$ is positive.

Similarly, if $f(x)$ is continuous and negative at $x = c$ in its domain, then $f(c + h)$ and $f(c - h)$ are both, negative for all sufficiently small values of h.

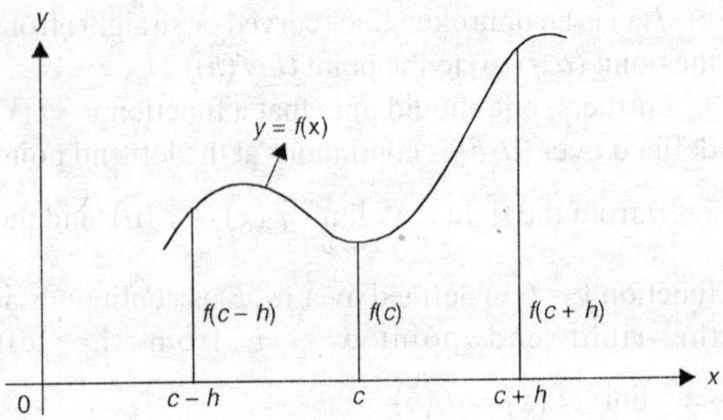

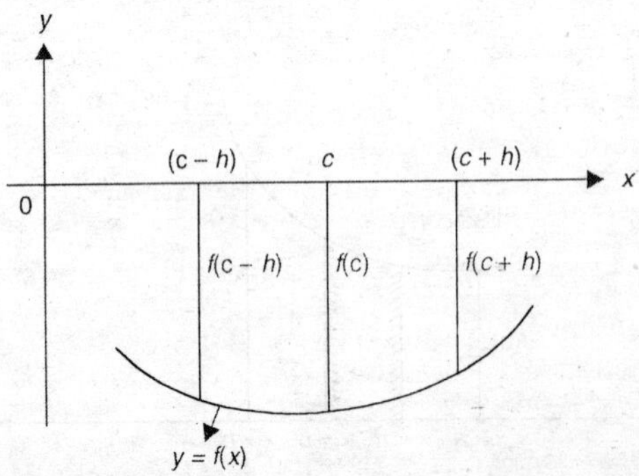

Note: If $h > 0$, the symbol $f(c + h)$ indicates the value of the function $f(x)$ for a value of x greater than c, whereas the symbol $f(c - h)$ indicates the value of the functions $f(x)$ for a value of x less than c.

4

Practical Methods of Finding the Limits

Practical Methods on Limits of a Function *f (x)* as $x \to a$, where *f (x)* is Expressed in a Closed Form

In practice, the classical definition or $(\varepsilon - \delta)$ definition of the limit of a function at $x = a$ (or, as $x \to a$) is not used in finding out the limit of a function at $x = a$ (or, as $x \to a$) whenever a single function whose limit is required to be found out is provided to us. This is why for most practical purposes, we can adopt the following method to find the limit of the function $f(x)$ at $x = a$ (or, as $x \to a$), in many examples.

1. $\lim_{x \to a} f(x) = f(a)$ if $f(a)$ is finite (where $f(a)$ = value of the given function $f(x)$ at $x = a$). But when $f(a)$ = any indeterminate form, then

2. $\lim_{x \to a} f(x) = \lim_{x \to a} g(x) = g(a)$ if $g(a)$ is finite (where $g(a)$ = value of the function $g(x)$ at $x = a$) where $f(x)$ is simplified to avoid its in determinate form and $g(x)$ is the simplified form of the given function $f(x)$.

$\Rightarrow$ The above method (1) tells us in words that whenever we are required to find out $\lim_{x \to a} f(x)$, we first put $x = a$ in the given function $f(x)$ and find $f(a)$. If $f(a)$ does not assume meaningless form, then this is the required limit.

Similarly, the above method (2) tells us that if $f(a)$ assumes any meaningless form, then various mathematical techniques are applied to simplify $f(x)$ such that when we put $x = a$ in the simplified form of the given function, it does not assume any meaningless form and so we get the required result having a finite value obtained by putting $x = a$ in the simplified form of the given function $f(x)$.

N.B.: Simplified form $g(x)$ of $f(x)$ is obtained by using any mathematical manipulation (or, technique) like factorization, rationalization, substitution, changing all trigonometric functions in terms of sine and cosine of an angle or using any formula of trigonometry or algebra, etc. whichever we need.

Problems based on the limit when the value of the function is not indeterminate:

Working rule: To evaluate $\lim_{x \to a} f(x)$, firstly, we check on putting $x = a$ in the given function whether it assumes a meaningless form or not. If $f(a)$ does not assume indeterminate form, this will be required limit, i.e., $\lim_{x \to a} f(x) = f(a)$ if $f(a)$ is finite.

Examples:

1. $\lim_{x \to 0} (3x^2 + 4x^2 - 5x + 6) = 3.0 + 4.0 - 5.0 + 6 = 6$

2. $\lim_{x \to 2} \dfrac{3 + 2x - x^2}{x^2 + 2x - 3} = \dfrac{3 + 2 \times 2 - 2^2}{2^2 + 2 \times 2 - 3} = \dfrac{3}{5}$

3. $\lim_{x\to 2} \frac{x^3-8}{x+2} = \frac{2^3-8}{2+2} = \frac{8-8}{4} = \frac{0}{4} = 0$

4. $\lim_{x\to 0} \frac{x^2+5x+6}{|x^3|+|x|} = \frac{0+5\times 0+6}{0+0} = \frac{6}{0} = \infty$

5. $\lim_{x\to \frac{\pi}{2}} \frac{1+\sin 2x}{1-\cos 4x} = \frac{1+\sin\left(2\cdot\frac{\pi}{2}\right)}{1-\cos\left(\frac{\pi}{4}\cdot 4\right)}$

$= \frac{1+\sin \pi}{1-\cos \pi} = \frac{1+0}{1-(-1)} = \frac{1}{2}$

To evaluate $\lim_{x\to a} \frac{f_1(x)}{f_2(x)}$ where both $f(x)$ and $g(x)$ are zero when $x = a$ is put in the given function:

Working rule:

1. Reduce the indeterminate form to a determinate form by using various mathematical techniques namely:

(i) Method of factorization or method of division.
(ii) Method of substitution or differential method.
(iii) Method of rationalization.
(iv) Method of expansion.
(v) Method of simplification by using any mathematical manipulation which are generally use by trigonometrical formulas, algebraic formulas or changing all trigonometrical ratios in terms of sine and cosine of an angle.

2. Put $x = a$ in the determinate form of the function $\frac{f_1(x)}{f_2(x)}$.

Remember:

1. For most practical purposes, we can obtain the limit in case of an indeterminate form as the value which is obtained by reducing the indeterminate form into a determinate form by some algebraic operations like removing common factor or using expansion such as binomial, exponential, logarithmic or trigonometric substitution, etc.

2. All the indeterminate form can be reduced to $\frac{0}{0}$ or $\frac{\infty}{\infty}$.

3. $\frac{0}{0}$ or any indeterminate form may contain a common factor which makes the given function indeterminate whose removal by various techniques provides us a determinate form.

Question: When to use which method?

Answer: **1.** Method of factorization is generally used when given function is in the quotient form whose numerator and denominator contains algebraic or trigonometric functions which can be factorized.

2. Method of rationalization is generally used when the given function is in the quotient form whose numerator and denominator contains algebraic or trigonometric expression under the square root symbol $\sqrt{\ }$

3. Method of substitution or differential method is generally used when given function can not be factorized easily or factorization of the given function is difficult or not possible. The given function may be algebraic or trigonometric expression in the quotient form.

4. Method of simplification is generally used when the given function contains trigonometric functions. This method tells us to modify the given trigonometric function by simplification in such a fashion that when we put x = given limit of the independent variable, we must get a finite value. This method is also used when given function is in the difference form like $\left[\frac{1}{f_1(x)} - \frac{1}{f_2(x)}\right]$ providing us $[\infty - \infty]$ form at $x = a$.

5. Expansion method is applied when given function contains trigonometric functions like $\sin x$, $\cos x$, $\tan x$, exponential function e^x, logarithmic functions like $\log x$, $\log (1 + x)$ or binomial expression like $(x + a)^n$, etc., whose expansion is known to us and it is quite possible to remove the common factor from numerator and denominator after expansion.

Problems based on algebraic functions

Form: $\frac{f_1(x)}{f_2(x)}$ where $f_1(x)$ and $f_2(x)$ are polynomials in x and $x \to a$.

The above form is evaluated in the following way.

Rule 1: If $g(a)$ = value of the function appearing in denominator $\neq 0$, then $\lim_{x \to a} \frac{f(x)}{g(x)} = \frac{f(a)}{g(a)}$ which means we put $x = a$ = limit of given independent variable in the numerator and denominator of the given fraction (or, rational function) provided the value of the denominator at $x = a$ is not equal to zero.

Rule 2: If $f(a) = g(a) = 0 \Rightarrow$ value of the function at $x = a$ in *Nr* = value of the function at $x = a$ in *Dr* = 0, then we adopt the following working rule.

Working rule to evaluate the limit if $\left[\frac{f_1(x)}{f_2(x)}\right]_{x=a} = \frac{0}{0}$.

1. Numerator and denominator are divided by the common factor appearing in numerator and denominator of the given fraction.

2. Remove the common factor from numerator and denominator by the rule of cancellation.

3. Put $x = a$ = given limit of the given independent variable in the simplified form of the function (i.e., the function free from common factor appearing in *Nr* and *Dr* directly or indirectly) which gives us the required limit of the function as $x \to a$.

Facts to know:

1. When $f(x)$ assumes the indeterminate form $\frac{0}{0}$ for $x = a$, it does so on account of $(x - a)$ or power of $(x - a)$ occurring as a factor in both numerator and denominator of the given function $f(x)$ under consideration. Such a common factor which produces $\frac{0}{0}$ form or any indeterminate form is called "The vanishing factor" because this factor always vanishes. In order to find the limit of $f(x)$ in such a case, our first aim is to remove the vanishing factor from the numerator and denominator of $f(x)$ with understanding that $(x - a)$ is not zero, however small it may be and then in the resulting expression, which is determinate, we put $x = a$ = limit of the independent variable. Thus the required limit is obtained.

2. If we put $x = +a$ and the given function becomes $\frac{0}{0}$, then $(x - a)$ is a common factor appearing in numerator and denominator of the given function (or, rational algebraic function) under consideration.

3. If we put $x = -a$ and the given function becomes $\frac{0}{0}$, then $(x + a)$ is a common factor appearing in both numerator and denominator of the given function under consideration.

4. The vanishing factor $(x - a) \neq 0$ because $x \to a$ $\Rightarrow x \neq a \Rightarrow (x - a) \neq 0$.

5. The phrase "at the point $x = a$ or for the value $x = a$" means "when x assumes or takes the value a"

Remember:

1. Let $f(x) = \frac{f_1(x)}{f_2(x)}$ and we require $\lim_{x \to a} \frac{f_1(x)}{f_2(x)}$.

$f_1(a) = f_2(a) = 0 \Rightarrow (x - a)$ is a common factor of the given function appearing in numerator and denominator of the given function $f(x)$.

2. Simplification of the expression $\frac{f_1(x)}{f_2(x)}$ is done whenever it assumes the form $\frac{0}{0}$ or $\frac{\infty}{\infty}$ on putting 'a' for x while finding the limit of $f(x)$ as $x \to a$.

Problems Based on Method of Factorization

Examples worked out:

Find (or, evaluate)

1. $\lim_{x \to 1} \frac{x^2 - 4x + 3}{x^2 - 5x + 4}$

Solution: $\because \left[\dfrac{x^2 - 4x + 3}{x^2 - 5x + 4}\right]_{x=1} = \dfrac{1 - 4 + 3}{1 - 5 + 4}$

$$= \frac{4 - 4}{5 - 5} = \frac{0}{0}$$

$\therefore$ $(x - 1)$ is a common factor in *Nr* and *Dr* of the given fraction. Dividing *Nr* and *Dr* by the common factor $(x - 1)$, we get

$$\frac{x^2 - 4x + 3}{x^2 - 5x + 4} = \frac{(x-1)(x-3)}{(x-1)(x-4)} = \frac{x-3}{x-4};\ (x \neq 1)$$

$$\therefore \lim_{x\to1}\left[\frac{x^2 - 4x + 3}{x^2 - 5x + 4}\right] = \lim_{x\to1}\left[\frac{x-3}{x-4}\right] = \frac{1-3}{1-4}$$

$$= \frac{-2}{-3} = \frac{2}{3}$$

2. $\displaystyle\lim_{x\to4}\frac{x^3 - 2x^2 - 9x + 4}{x^2 - 2x - 8}$

Solution: $\because \left[\dfrac{x^3 - 2x^2 - 9x + 4}{x^2 - 2x - 8}\right]_{x=4}$

$$= \frac{64 - 32 - 36 + 4}{16 - 8 - 8} = \frac{68 - 68}{16 - 16} = \frac{0}{0}$$

$\therefore$ $(x - 4)$ is a common factor in *Nr* and *Dr* of the given fraction. Dividing *Nr* and *Dr* by the common factor $(x - 4)$, we get

$$\frac{x^3 - 2x^2 - 9x + 4}{x^2 - 2x - 8} = \frac{(x-4)(x^2 + 2x - 1)}{(x-4)(x+2)}$$

$$= \frac{x^2 + 2x - 1}{x + 2};\ (x \neq 4).$$

Now taking the limits on both sides as $x \to 4$ since both sides are equal (i.e., if two functions are equal, their limits are equal), we get,

$$\lim_{x\to4}\frac{x^3 - 2x^2 - 9x + 4}{x^2 - 2x - 8} = \lim_{x\to4}\frac{x^2 + 2x - 1}{x + 2}$$

$$= \frac{16 + 8 - 1}{4 + 2} = \frac{24 - 1}{6} = \frac{23}{6}.$$

Alternative method: By direct division

$$= \frac{x^3 - 2x^2 - 9x + 4}{x^2 - 2x - 8} = x - \frac{x - 4}{x^2 - 2x - 8}$$

$$= x - \frac{(x-4)}{(x-4)(x+2)} = x - \frac{1}{(x+2)}\ (\because x \neq 4).$$

Now, taking the limits on both sides as $x \to 4$, we get

$$\lim_{x\to4}\frac{x^3 - 2x^2 - 9x + 4}{x^2 - 2x - 8}$$

$$= \lim_{x\to4}\left[x - \frac{1}{x+2}\right]$$

$$\lim_{x\to4} x - \lim_{x\to4}\frac{1}{x+2}$$

$$= 4 - \frac{1}{4+2} = 4 - \frac{1}{6} = \frac{24 - 1}{6} = \frac{23}{6}.$$

3. $\displaystyle\lim_{x\to3}\left[\frac{x^3 - 5x^2 + 7x - 3}{x^3 - x^2 - 5x - 3}\right]$

Solution: $\left[\dfrac{x^3 - 5x^2 + 7x - 3}{x^3 - x^2 - 5x - 3}\right]_{x=3} = \dfrac{0}{0}$

$\Rightarrow$ *Nr* has $(x - 3)$ as a factor and *Dr* has $(x - 3)$ as a factor.

Now, dividing *Nr* and *Dr* by the common factor $(x - 3)$, we get

$$\frac{x^3 - 5x^2 + 7x - 3}{x^3 - x^2 - 5x - 3} = \frac{(x-3)(x^2 - 2x + 1)}{(x-3)(x^2 + 2x + 1)}$$

$$= \frac{x^2 - 2x + 1}{x^2 + 2x + 1};\ (x \neq 3)$$

$$\therefore \lim_{x\to 3}\left[\frac{x^3-5x^2+7x-3}{x^3-x^2-5x-3}\right]$$

$$= \lim_{x\to 3}\left[\frac{x^2-2x+1}{x^2+2x+1}\right]$$

$$= \frac{3^2-2\times 3+1}{3^2+2\times 3+1} = \frac{4}{16} = \frac{1}{4}$$

4. $\lim_{x\to 1}\frac{x^7-2x^5+1}{x^3-3x^2+2}$

Solution: $\because \left[\frac{x^7-2x^5+1}{x^3-3x^2+2}\right]_{x=1} = \frac{0}{0}$

$\therefore$ $(x-1)$ is a common factor in *Nr* and *Dr* of the given fraction.

Now, dividing *Nr* and *Dr* by $(x-1)$ we get

$$\frac{x^7-2x^5+1}{x^3-3x^2+2}$$

$$= \frac{(x-1)\left(x^6+x^5-x^4-x^3-x^2-x-1\right)}{(x-1)\left(x^2-2x-2\right)}$$

$$= \frac{x^6+x^5-x^4-x^3-x^2-x-1}{x^2-2x-2};\ (x\neq 1)$$

Now taking the limits on both sides as $x\to 1$

$$\lim_{x\to 1}\left[\frac{x^7-2x^5+1}{x^3-3x^2+2}\right]$$

$$= \lim_{x\to 1}\left[\frac{x^6+x^5-x^4-x^3-x^2-x-1}{x^2-2x-2}\right]$$

$$= \frac{1+1-1-1-1-1-1}{1-2-2}$$

$$= \frac{2-5}{1-4} = 1$$

5. $\lim_{x\to 1}\frac{x^2-1}{x^2-3x+2}$

Solution: $\left[\frac{x^2-1}{x^2-3x+2}\right]_{x=1} = \frac{0}{0}$

$\Rightarrow$ $(x-1)$ is a common factor in *Nr* and *Dr* of the given fraction.

Now, dividing *Nr* and *Dr* by $(x-1)$, we get

$$\frac{x^2-1}{x^2-3x+2} = \frac{(x+1)(x-1)}{(x-1)(x-2)}$$

$$= \frac{x+1}{x-2};\ (x\neq 1).$$

Now taking the limits on both sides as $x\to 1$, we get

$$\lim_{x\to 1}\frac{x^2-1}{x^2-3x+2} = \lim_{x\to 1}\left[\frac{x+1}{x-2}\right]$$

$$= \frac{1+1}{1-2} = \frac{2}{-1} = -2$$

6. $\lim_{x\to 2}\left[\frac{x^2-3x+2}{x-2}\right]$

Solution: $\left[\frac{x^2-3x+2}{x-2}\right]_{x=2} = \frac{0}{0}$

$\Rightarrow$ *Nr* has $(x-2)$ as a factor and *Dr* has $(x-2)$ as a factor

Now dividing *Nr* and *Dr* by the common factor $(x-2)$, we have

$$\frac{x^2-3x+2}{x-2} = \frac{(x-1)(x-2)}{(x-2)} = (x-1);$$

$(x\neq 2)$

Now, taking the limits on both sides as $x\to 2$, we have

$$\lim_{x\to 2}\frac{x^2-3x+2}{x-2}$$

$$= \lim_{x\to 2} (x-1) = 2-1 = 1$$

7. $\lim\limits_{x\to 3} \dfrac{x^3 - 3x^2 + 5x - 15}{x^2 - x - 12}$

Solution: $\left[\dfrac{x^3 - 3x^2 + 5x - 15}{x^2 + x - 12}\right]_{x=3} = \dfrac{0}{0}$

$\Rightarrow$ *Nr* has $(x-3)$ = a factor and *Dr* has $(x-3)$ = a factor: $\Rightarrow$ $(x-3)$ is a common factor in *Nr* and *Dr*.

Now, dividing *Nr* and *Dr* by the common factor $(x-3)$, we get

$$\frac{x^3 - 3x^2 + 5x - 15}{x^2 - x - 12} = \frac{(x-3)(x^2+5)}{(x-3)(x+4)}$$

$$\frac{x^2+5}{x+4}; (\because x \neq 3)$$

$$\therefore \lim_{x\to 3}\left[\frac{x^3 - 3x^2 + 5x - 15}{x^2 - x - 12}\right]$$

$$\therefore \lim_{x\to 3}\left[\frac{x^2+5}{x+4}\right]$$

$$= \frac{\lim\limits_{x\to 3}(x^2+5)}{\lim\limits_{x\to 3}(x+4)}$$

$$= \frac{3^2+5}{3+4} = \frac{9+5}{7} = \frac{14}{7} = 2$$

8. $\lim\limits_{x\to 2} \dfrac{x^2 - 4}{x - 2}$

Solution: $\left[\dfrac{x^2-4}{x-2}\right]_{x=2} = \dfrac{0}{0}$

$\Rightarrow$ *Nr* has $(x-2)$ as a factor and *Dr* has $(x-2)$ as a factor: $\Rightarrow$ $(x-2)$ is a common factor of *Nr* and *Dr* of the given fraction.

Now dividing *Nr* and *Dr* by the common factor $(x-2)$, we get,

$$\frac{x^2-4}{x-2} = \frac{(x-2)(x+2)}{(x-2)}$$

$$= (x+2); (\because x \neq 2)$$

$$\therefore \lim_{x\to 2}\frac{x^2-4}{x-2} = \lim_{x\to 2}(x+2)$$

$$= 2+2 = 4.$$

9. $\lim\limits_{x\to -1} \dfrac{x^3+1}{x+1}$

Solution: $\left[\dfrac{x^3+1}{x+1}\right]_{x=-1} = \dfrac{0}{0}$

$\Rightarrow$ *Nr* has $(x+1)$ as a factor and *Dr* has $(x+1)$ as a factor: $(x+1)$ is a common factor in *Nr* and *Dr* of the given fraction.

Now, dividing *Nr* and *Dr* by the common factor $(x+1)$, we get

$$\frac{x^3+1}{x+1} = \frac{(x+1)(x^2-x+1)}{x+1}$$

$$= x^2 - x + 1 \text{ (for } x \neq -1)$$

$$\therefore \lim_{x\to -1}\frac{x^3+1}{x+1} = \lim_{x\to -1}(x^2-x+1)$$

$$= 1-(-1)+1 = 1+1+1 = 3.$$

Exercise 4.1

Problems set on method of factorization or division

Find the limits of the following functions:

Answers

1. $\underset{x\to 1}{\mathrm{L}} \dfrac{x^7 - 2x^5 + 1}{x^3 - 3x^2 + 2}$ (Bhag—65A) (1)

2. $\underset{x\to 1}{\mathrm{L}} \dfrac{x-1}{(x^3 - 3x^2 + 2)}$ (I.I.T.—1976) $\left(-\dfrac{1}{3}\right)$

3. $\underset{x \to 2}{\text{L}} \dfrac{2x^2 - 7x + 6}{5x^2 - 11x + 2}$ (Bombay—65) $\left(\dfrac{1}{9}\right)$

4. $\underset{x \to 4}{\text{L}} \dfrac{x^2 - 7x + 12}{x^3 - 64}$ (Bombay—69, 70) $\left(\dfrac{1}{48}\right)$

5. $\underset{x \to a}{\text{L}} \dfrac{x^4 - a^4}{x - a}$ (R.U.—65) $(4a^3)$

6. $\underset{x \to 3}{\text{L}} \dfrac{x^2 - 4x + 3}{2x^2 - 11x + 15}$ (2)

7. $\underset{x \to -3}{\text{L}} \dfrac{x^3 + 27}{x^5 + 243}$ $\left(\dfrac{1}{15}\right)$

8. $\underset{x \to -3}{\text{L}} \dfrac{x + 3}{x^3 + 27}$ $\left(\dfrac{1}{27}\right)$

9. $\underset{x \to \frac{1}{2}}{\text{L}} \dfrac{1 - 32x^5}{1 - 8x^3}$ $\left(\dfrac{5}{3}\right)$

Type 2: Limits of irrational functions as $x \to a$.

Form 1: $\sqrt{f(x)}$ or $[f(x)]^{\frac{m}{n}}$

Form 2: $\sqrt{f_1(x)} \pm \sqrt{f_2(x)}$ occurring in the *Nr* or in *Dr* or in both *Nr* and *Dr*.

Working rule for form 1:

If the given function has the form 1 mentioned above, we adopt the following working rule:

Use the following formulas:

1. $\lim\limits_{x \to a} \sqrt{f(x)} = \sqrt{\lim\limits_{x \to a} f(x)}$

2. $\lim\limits_{x \to a} [f(x)]^{\frac{m}{n}} = \left[\lim\limits_{x \to a} f(x)\right]^{\frac{m}{n}}$

Working rule for from 2:

1. If only *Nr* contains the radical of the above form, rationalize *Nr* by multiplying *Nr* and *Dr* by rationalizing factor of *Nr*.
2. If only *Dr* contains the radical of the form mentioned above, rationalize *Dr* by multiplying *Nr* and *Dr* by rationalizing factor of *Dr*.
3. If *Nr* and *Dr* both contain radicals of the above type, rationalize *Nr* and *Dr* both separately by multiplying and dividing *Nr* and *Dr* by rationalizing factor of *Nr* and *Dr*.

Facts to know about rationalizing factor:

(a) $\left(a + \sqrt{b}\right)$ and $\left(a - \sqrt{b}\right)$ are rationalizing factors of each other.

(b) $\left(\sqrt{a} + \sqrt{b}\right)$ and $\left(\sqrt{a} - \sqrt{b}\right)$ are rationalizing factors of each other.

(c) $\left(p\sqrt{a} + q\sqrt{b}\right)$ and $\left(p\sqrt{a} - q\sqrt{b}\right)$ are rationalizing factors of each other.

N.B.:

1. Any one of $\left(a + \sqrt{b}\right)$, $\left(\sqrt{a} + \sqrt{b}\right)$ or $\left(p\sqrt{a} + q\sqrt{b}\right)$ may be provided in the question whose limit is required to find out where a and b indicate a function of x.
2. Our main aim is to remove the common factor by rationalization of *Nr* or *Dr* or both *Nr* and *Dr*.
3. After removing the common factor, we put $x = a =$ limit of the independent variable x in the irrational function free from the common factor known as simplified form.
4. The above rule is valid when $\sqrt{f_1(x)} \pm \sqrt{f_2(x)}$ becomes $\dfrac{0}{0}$ at $x = a$.

Problems based on the form 1 $\sqrt{f(x)}$ **or** $[f(x)]^{\frac{m}{n}}$

Solved Examples

Evaluate:

1. $\lim\limits_{x \to 1} \sqrt{\dfrac{8x + 1}{x + 3}}$

Solution: $\lim\limits_{x \to 1} \sqrt{\dfrac{8x + 1}{x + 3}}$

$$= \sqrt{\lim_{x \to 1} \left(\frac{8x + 1}{x + 3}\right)} = \sqrt{\frac{\lim\limits_{x \to 1} (8x + 1)}{\lim\limits_{x \to 1} (x + 3)}}$$

$$= \sqrt{\frac{\lim\limits_{x\to1} 8x + \lim\limits_{x\to1} 1}{\lim\limits_{x\to1} x + \lim\limits_{x\to1} 3}} = \sqrt{\frac{8(1)+1}{1+3}}$$

$$= \sqrt{\frac{9}{4}} = \frac{3}{2}.$$

Exercise 4.2

On the form $\sqrt{f(x)}$ or $[f(x)]^{\frac{m}{n}}$

Find the limit of the following:

Answers

1. $\underset{x\to1}{\text{L}} \sqrt{2x^2+1}$ $\qquad (\sqrt{3})$
2. $\underset{x\to1}{\text{L}} \sqrt{16-x^2}$ $\qquad (\sqrt{15})$
3. $\underset{x\to-1}{\text{L}} (x-4)\sqrt{1-3x}$ $\qquad (-10)$

Problems based on the form 2 $\sqrt{f_1(x)} \pm \sqrt{f_2(x)}$

Examples worked out:

Evaluate:

1. $\lim\limits_{x\to1} \dfrac{\sqrt{x^2-1}+\sqrt{x-1}}{\sqrt{x^2-1}}$

Solution: $\left[\dfrac{\sqrt{x^2-1}+\sqrt{x-1}}{\sqrt{x^2-1}}\right]_{x=1} = \dfrac{0}{0}$

Now, $\dfrac{\sqrt{x^2-1}+\sqrt{x-1}}{\sqrt{x^2-1}}$

$$= \frac{\sqrt{(x-1)(x+1)}+\sqrt{x-1}}{\sqrt{(x-1)(x+1)}}$$

$$= \frac{\sqrt{(x-1)}\left[\sqrt{x+1}+1\right]}{\sqrt{x-1}\left(\sqrt{x+1}\right)}$$

$$= \frac{\left[\sqrt{x+1}+1\right]}{\left(\sqrt{x+1}\right)}; \text{ (for } x \neq 1)$$

Hence, $\lim\limits_{x\to1} \dfrac{\sqrt{x^2-1}+\sqrt{x-1}}{\sqrt{x^2-1}}$

$$= \lim_{x\to1} \frac{\sqrt{x+1}+1}{\sqrt{x+1}}$$

$$= \frac{\sqrt{1+1}+1}{\sqrt{1+1}} = \frac{\sqrt{2}+1}{\sqrt{2}}$$

2. $\lim\limits_{x\to1} \dfrac{\sqrt{x^3-1}-(x-1)}{\sqrt{x-1}}$

Solution: $\left[\dfrac{\sqrt{x^3-1}-(x-1)}{\sqrt{x-1}}\right]_{x=1} = \dfrac{0}{0}$

Now, $\dfrac{\sqrt{x^3-1}-(x-1)}{\sqrt{x-1}}$

$$= \frac{\sqrt{(x-1)(x^2+x+1)}-(x-1)}{\sqrt{x-1}}$$

$$= \frac{\sqrt{x-1}\left[\sqrt{x^2+x+1}-\sqrt{x-1}\right]}{\sqrt{x-1}}$$

$$= \sqrt{x^2+x+1}-\sqrt{x-1}; \text{ (for } x \neq 1)$$

Hence, $\lim\limits_{x\to1} \dfrac{\sqrt{x^3-1}-(x-1)}{\sqrt{x-1}}$

$$= \lim_{x\to1} \left[\sqrt{x^2+x+1}-\sqrt{x-1}\right]$$

$$= \sqrt{1+1+1}-\sqrt{1-1} = \sqrt{3}-0 = \sqrt{3}$$

Note: The above examples gives as a hint that we should not rationalize the given irrational function blindly but we must check whether it is possible to remove the common factor (or, not) by factorization which is present in *Nr* and *Dr* of the given irrational function, i.e., if it is possible to remove the common factor from the numerator and denominator of irrational function after factorization and then using cancellation, we must remove it.

3. $\lim_{x \to a} \frac{\sqrt{x} - \sqrt{a}}{(x - a)}$

Solution: $\left[\frac{\sqrt{x} - \sqrt{a}}{(x - a)}\right]_{x = a} = \frac{0}{0}$

Now, $\frac{\sqrt{x} - \sqrt{a}}{(x - a)} = \frac{\sqrt{x} - \sqrt{a}}{\left(\sqrt{x}\right)^2 - \left(\sqrt{a}\right)^2}$

$= \frac{\sqrt{x} - \sqrt{a}}{\left(\sqrt{x} + \sqrt{a}\right)\left(\sqrt{x} - \sqrt{a}\right)}$

$= \frac{1}{\sqrt{x} + \sqrt{a}}$; (for $x \neq a$)

Hence, $\lim_{x \to a} \frac{\sqrt{x} - \sqrt{a}}{(x - a)}$

$= \lim_{x \to a} \frac{1}{\sqrt{x} + \sqrt{a}}$

$= \frac{1}{\sqrt{a} + \sqrt{a}} = \frac{1}{2\sqrt{a}}$

4. $\lim_{x \to 5} \frac{2 - \sqrt{x - 1}}{x^2 - 25}$

Solution: $\left[\frac{2 - \sqrt{x - 1}}{x^2 - 25}\right]_{x = 5} = \frac{0}{0}$ and *Nr* contains as irrational expression which means rationalizing of *Nr* is required.

Hence, $\frac{\left(2 - \sqrt{x - 1}\right)}{\left(x^2 - 25\right)}$

$= \frac{2 - \sqrt{x - 1}}{x^2 - 25} \times \frac{2 + \sqrt{x - 1}}{2 + \sqrt{x - 1}}$

$= \frac{2^2 - (x - 1)}{(x - 5)(x + 5)} \times \frac{1}{2 \times \sqrt{x - 1}}$

$= \frac{4 - x + 1}{(x + 5)(x - 5)\left(2 + \sqrt{x - 1}\right)}$

$= \frac{-(x - 5)}{(x + 5)(x - 5)\left(2 + \sqrt{x - 1}\right)}$

$= \frac{-1}{(x + 5)\left(2 + \sqrt{x - 1}\right)}$

$\therefore \lim_{x \to 5} \frac{2 - \sqrt{x - 1}}{x^2 - 25}$

$= \lim_{x \to 5} \frac{-1}{(x + 5)\left(2 + \sqrt{x - 1}\right)}$

$= \frac{-1}{(5 + 5)\left(\sqrt{5 - 1} + 2\right)}$

$= \frac{-1}{10\left(\sqrt{4} + 2\right)} = \frac{-1}{10 \times (2 + 2)}$

$= \frac{-1}{10 \times 4} = -\frac{1}{40}$

5. $\lim_{x \to 4} \frac{3 - \sqrt{5 + x}}{1 - \sqrt{5 - x}}$

Solution: $\left[\frac{3 - \sqrt{5 + x}}{1 - \sqrt{5 - x}}\right]_{x = 4} = \frac{0}{0}$

Since *Nr* and *Dr* both contain radicals whose factorization is not possible which means we are required to rationalize *Nr* and *Dr* both separately on multiplying and dividing by the rationalizing factor.

$$Nr = \left(3 - \sqrt{5+x}\right)$$

$$= \frac{\left(3 - \sqrt{5+x}\right)\left(3 + \sqrt{5+x}\right)}{\left(3 + \sqrt{5+x}\right)}$$

$$= \frac{(3)^2 - \left(\sqrt{5+x}\right)^2}{\left(3 + \sqrt{5+x}\right)}$$

$$= \frac{9 - 5 - x}{\left(3 + \sqrt{5+x}\right)}$$

$$= \frac{4 - x}{\left(3 + \sqrt{5+x}\right)}$$

$$Dr = \left(1 - \sqrt{5-x}\right)$$

$$= \frac{\left(1 - \sqrt{5-x}\right)\left(1 + \sqrt{5-x}\right)}{\left(1 + \sqrt{5-x}\right)}$$

$$= \frac{(1)^2 - \left(\sqrt{5-x}\right)^2}{\left(1 + \sqrt{5-x}\right)}$$

$$= \frac{1 - (5 - x)}{1 + \sqrt{5-x}}$$

$$= \frac{x - 4}{1 + \sqrt{5-x}}$$

Now, given expression $= \dfrac{Nr}{Dr}$

$$= \frac{\dfrac{(4 - x)}{\left(3 + \sqrt{5+x}\right)}}{\dfrac{(x - 4)}{\left(1 + \sqrt{5-x}\right)}}$$

$$= \frac{(4 - x)\left(1 + \sqrt{5-x}\right)}{-(4 - x)\left(3 + \sqrt{5+x}\right)} \text{ for } x \neq 4$$

$$= \frac{\left(1 + \sqrt{5-x}\right)}{\left(3 + \sqrt{5+x}\right)}$$

$$\therefore \text{ Required limit} = \lim_{x \to 4} \frac{-\left(1 + \sqrt{5-x}\right)}{\left(3 + \sqrt{5+x}\right)}$$

$$= (-1)\left(\frac{1 + \sqrt{5-4}}{3 + \sqrt{5+4}}\right)$$

$$= \frac{(-1)(1 + 1)}{(3 + 3)} = \frac{-2}{6} = -\frac{1}{3}$$

6. $\displaystyle \lim_{x \to 1} \frac{x - \sqrt{2 - x^2}}{2x - \sqrt{2 + 2x^2}}$

Solution: $\displaystyle \lim_{x \to 1} \frac{x - \sqrt{2 - x^2}}{2x - \sqrt{2 + 2x^2}}$

Now on rationalizing the *Nr* and *Dr* with the help of rationalizing factor of *Nr* and *Dr*.

$$= \lim_{x \to 1} \frac{\left\{x^2 - \left(2 - x^2\right)\right\}\left\{2x + \sqrt{2 + 2x^2}\right\}}{\left(2x - \sqrt{2 + 2x^2}\right)\left(x + \sqrt{2 - x^2}\right)\left(2x + \sqrt{2 + 2x^2}\right)}$$

$$= \lim_{x \to 1} \frac{\left\{x^2 - \left(2 - x^2\right)\right\}\left\{2x + \sqrt{2 + 2x^2}\right\}}{\left\{4x^2 - \left(2 + 2x^2\right)\right\}\left\{x + \sqrt{2 - x^2}\right\}}$$

$$= \lim_{x \to 1} \frac{\left(2x^2 - 2\right)\left(2x + \sqrt{2 + 2x^2}\right)}{\left(2x^2 - 2\right)\left(x + \sqrt{2 - x^2}\right)}; \text{ (for } x \neq 1)$$

$$= \lim_{x\to 1} \frac{2x + \sqrt{2 + 2x^2}}{x + \sqrt{2 - x^2}}$$ [on removing the common factor]

$=2$

7. $\lim_{x\to 2} \frac{4 - x^2}{3 - \sqrt{x^2 + 5}}$

Solution: $\lim_{x\to 2} \frac{(4 - x^2)}{3 - \sqrt{x^2 + 5}}$

Now on rationalizing the *Dr* only since radical appears in *Dr* only

$$= \lim_{x\to 2} \frac{(4 - x^2)\left(3 + \sqrt{x^2 + 5}\right)}{\left(3 - \sqrt{x^2 + 5}\right)\left(3 + \sqrt{x^2 + 5}\right)}$$

$$= \lim_{x\to 2} \frac{(4 - x^2)\left(3 + \sqrt{x^2 + 5}\right)}{(4 - x^2)}; \text{ (for } x \neq 2)$$

$$= \lim_{x\to 2} \left(3 + \sqrt{x^2 + 5}\right)$$

$=6.$

Problems based on method of rationalization

Exercise 4.3

Find the limit of the following **_Answers_**

1. $\underset{x\to 1}{\mathrm{L}} \frac{\sqrt{2 - x} - 1}{1 - x}$ (M.U. 68) $\left(\frac{1}{2}\right)$

2. $\underset{x\to 1}{\mathrm{L}} \frac{x - 1}{\sqrt{x^2 - 1} + \sqrt{x - 1}}$ (0)

3. $\underset{x\to 4}{\mathrm{L}} \frac{2 - \sqrt{x}}{x - 4}$ $\left(\frac{1}{4}\right)$

4. $\underset{x\to 1}{\mathrm{L}} \frac{x^2 - \sqrt{x}}{\sqrt{x} - 1}$ (3)

5. $\underset{x\to 1}{\mathrm{L}} \frac{\sqrt{x^2 + 8} - \sqrt{10 - x^2}}{\sqrt{x^2 + 3} - \sqrt{5 - x^2}}$ $\left(\frac{2}{3}\right)$

6. $\underset{x\to 2}{\mathrm{L}} \frac{\sqrt[3]{x - 1} - 1}{(x - 2)}$ $\left(\frac{1}{3}\right)$

7. $\underset{x\to 1}{\mathrm{L}} \frac{(2x - 3)\left(\sqrt{x} - 1\right)}{\left(2x^2 + x - 3\right)}$ $\left(\frac{1}{10}\right)$

8. $\underset{x\to 1}{\mathrm{L}} \frac{\sqrt{x^2 + 3} - 1}{(x + 2)}$ $\left(\frac{1}{3}\right)$

9. $\underset{x\to 3}{\mathrm{L}} \frac{\sqrt{2x + 3} - 3}{(x - 6)}$ (0)

10. $\underset{x\to 1}{\mathrm{L}} \frac{\sqrt{x + 4} - \sqrt{5}}{(x - 1)}$ $\left(\frac{1}{2\sqrt{5}}\right)$

11. $\underset{x\to 4}{\mathrm{L}} \frac{3 - \sqrt{5 + x}}{1 - \sqrt{5 - x}}$ $\left(-\frac{1}{3}\right)$

Type 3:

Form: $[f_1(x) - f_2(x)] = \infty - \infty$ as $x \to a$

Working rule: To evaluate $[f_1(x) - f_2(x)] = \infty - \infty$ as $x \to a$ our main aim is to reduce the form $(\infty - \infty)$ to the form $\frac{0}{0}$ which can be explained in the notational form in the following way:

1. Write $[f_1(x) - f_2(x)] = \frac{1}{\frac{1}{f_1(x)}} - \frac{1}{\frac{1}{f_2(x)}}$

2. Write $1 = \dfrac{\dfrac{1}{f_2(x)} - \dfrac{1}{f_1(x)}}{\dfrac{1}{f_1(x)} \cdot \dfrac{1}{f_2(x)}} = \dfrac{0}{0}$ whose limit is found by the method already explained.

Note: **1.** In practice, to reduce the form $(\infty - \infty)$ to the form $\frac{0}{0}$, we generally simplify the given expression by using any mathematical operation (like taking l.c.m or changing all trigonometric functions in to sin and cosine of an angle, etc.). After obtaining the form $\frac{0}{0}$, we are able to find its limit, e.g.:

(i) $\lim_{x \to \frac{\pi}{2}} (\tan x - \sec x)$, $(= \infty - \infty$ form$)$

$$\tan x - \sec x = \frac{\sin x}{\cos x} - \frac{1}{\cos x}$$

$$= \frac{\sin x - 1}{\cos x} = \frac{0}{0} \text{ as } \to \frac{\pi}{2}$$

(ii) $\lim_{\theta \to 0} \left(\frac{1 - \cos\theta}{\theta^2} \right)$, $(= \infty - \infty$ form$)$

$$\frac{1 - \cos\theta}{\theta^2} = \frac{1 - 1 + 2\sin^2 \frac{\theta}{2}}{\theta^2}$$

$$= \frac{2\sin^2 \frac{\theta}{2}}{\theta^2} = \frac{0}{0} \text{ as } \theta \to 0.$$

(iii) $\left[\frac{2}{1 - x^2} - \frac{1}{1 - x} \right]$ $(= \infty - \infty$ form$)$ as $x \to 1$

$$= \frac{2}{(1 + x)(1 - x)} - \frac{1}{(1 - x)}$$

$$= \frac{2 - (1 + x)}{(1 - x^2)} = \frac{(1 - x)}{(1 - x)(1 + x)} = \frac{0}{0} \text{ as } x \to 1$$

Note 2: If $\lim_{x \to a} \left[\frac{1}{f_1(x)} \pm \frac{1}{f_2(x)} \right] = \infty \pm \infty$, we take first l.c.m and then we subtract. Taking l.c.m and subtracting reduces the form $(\infty - \infty)$ to the form $\frac{0}{0}$ for which we adopt the usual method of removing common factor of *Nr* and *Dr* as factorization of cancellation, etc.

Problems based on the form:

$$\left[\frac{1}{f_1(x)} \pm \frac{1}{f_2(x)} \right] (= \infty \pm \infty \text{ as } x \to a)$$

***Examples worked out*:**

Evaluate

1. $\lim_{x \to 1} \left[\frac{2}{1 - x^2} + \frac{1}{x - 1} \right]$

***Solution*:** $\left[\frac{2}{1 - x^2} + \frac{1}{x - 1} \right]$

$$= \left[\frac{2}{1 - x^2} - \frac{1}{1 - x} \right], (= \infty - \infty \text{ as } x \to 1)$$

Now, $\frac{2}{1 - x^2} - \frac{1}{1 - x}$

$$= \frac{2 - (1 + x)}{(1 - x)(1 + x)}$$

$$= \frac{\cancel{(1 - x)}}{\cancel{(1 - x)}(1 + x)} = \frac{1}{(1 + x)}$$

Thus, $\left[\frac{2}{1 - x^2} + \frac{1}{x - 1} \right] = \frac{1}{1 + x}$ ...(1)

Now taking the limit on both sides of (i) as $x \to 1$, we get

$$\lim_{x \to 1} \left[\frac{2}{1 - x^2} + \frac{1}{x - 1} \right]$$

$$= \lim_{x\to 1} \frac{1}{1+x} = \frac{1}{1+1} = \frac{1}{2}$$

2. $\lim_{x\to 1}\left[\frac{1}{x^2-1} - \frac{2}{x^4-1}\right]$

Solution: $\left[\frac{1}{x^2-1} - \frac{2}{x^4-1}\right]$; (= ∞ − ∞ as $x\to 1$)

Now, $\frac{1}{x^2-1} - \frac{2}{x^4-1} = \frac{x^2+1-2}{(x^2-1)(x^2+1)}$

$$= \frac{(x^2-1)}{(x^2-1)(x^2+1)} = \frac{1}{x^2+1}$$

Thus, $\left[\frac{1}{x^2-1} - \frac{2}{x^4-1}\right] = \frac{1}{x^2+1}$ ...(1)

Now taking the limits on both sides of (1) as $x \to 1$, we have

$$\lim_{x\to 1}\left[\frac{1}{x^2-1} - \frac{2}{x^4-1}\right] = \lim_{x\to 1}\left[\frac{1}{x^2+1}\right]$$

$$= \frac{1}{1+1} = \frac{1}{2}$$

3. $\lim_{x\to a}\left[\frac{2a}{x^2-a^2} - \frac{1}{x-a}\right]$

Solution: $\left[\frac{2a}{x^2-a^2} - \frac{1}{x-a}\right]$, (= ∞ − ∞ as $x\to a$)

Now, $\left[\frac{2a}{x^2-a^2} - \frac{1}{x-a}\right] = \frac{2a-x-a}{(x^2-a^2)}$

$$= \frac{a-x}{x^2-a^2} = -\frac{1}{x+a} \quad ...(1)$$

Now taking the limits on both sides of (1) as $x \to a$

$$\lim_{x\to a}\left[\frac{2a}{x^2-a^2} - \frac{1}{x-a}\right] = \lim_{x\to a}\left[-\frac{1}{x+a}\right]$$

$$= -\lim_{x\to a} \frac{1}{x+a} = -\frac{1}{a+a} = \frac{-1}{2a}$$

4. $\lim_{x\to 2}\left[\frac{1}{x-2} - \frac{4}{x^2-4}\right]$

Solution: $\left[\frac{1}{x-2} - \frac{4}{x^2-4}\right]$(= ∞ − ∞ as $x\to 2$)

Now, $\left[\frac{1}{x-2} - \frac{4}{x^2-4}\right] = \frac{x+2-4}{x^2-4}$

$$= \frac{x-2}{x^2-4} = \frac{(x-2)}{(x+2)(x-2)}$$

$$= \frac{1}{(x+2)}; \text{(for } x \neq 2\text{)}$$

Thus, $\left[\frac{1}{x-2} - \frac{4}{x^2-4}\right] = \frac{1}{x+2}$ ...(1)

Now taking the limits on both sides of (1) as $x \to 2$, we have

$$\lim_{x\to 2}\left[\frac{1}{x-2} - \frac{4}{x^2-4}\right] = \lim_{x\to 2} \frac{1}{x+2}$$

$$= \frac{1}{2+2} = \frac{1}{4}$$

5. $\lim_{x\to 2}\left[\frac{1}{x-2} - \frac{1}{x^2-3x+2}\right]$

Solution: $\left[\frac{1}{x-2} - \frac{1}{x^2-3x+2}\right]$(= ∞ − ∞ form) as $x\to 2$

Now, $\left[\dfrac{1}{x-2}-\dfrac{1}{x^2-3x+2}\right]$

$$=\frac{1}{x-2}-\frac{1}{(x-2)(x-1)}$$

$$=\frac{x-1}{(x-2)(x-1)}-\frac{1}{(x-2)(x-1)}$$

$$=\frac{x-2}{(x-2)(x-1)}$$

$$=\frac{1}{x-1};\ (\text{for } x \neq 2)$$

Thus, $\left[\dfrac{1}{x-2}-\dfrac{1}{x^2-3x+2}\right]=\dfrac{1}{x-1}$ ...(1)

Now, taking the limits on both sides of (1) as $x \to 2$, we have

$$\lim_{x\to 2}\left[\frac{1}{x-2}-\frac{1}{x^2-3x+2}\right]$$

$$=\lim_{x\to 2}\left[\frac{1}{x-1}\right]=\frac{1}{2-1}=1$$

Problems based on the form:

$$\left[\frac{1}{f_1(x)}-\frac{1}{f_2(x)}\right]\to\frac{1}{0}-\frac{1}{0}\ (\text{or, } \infty-\infty) \text{ as } x\to a$$

Exercise 4.4

Find the limit of the following **_Answers_**

1. $\underset{x\to 1}{\mathrm{L}}\left[\dfrac{1}{1-x}-\dfrac{3}{1-x^3}\right]$ (–1)

2. $\underset{x\to 1}{\mathrm{L}}\ \dfrac{1}{x-1}\left[\dfrac{1}{x+3}-\dfrac{2}{3x+5}\right]$ $\left(\dfrac{1}{32}\right)$

3. $\underset{x\to \frac{1}{2}}{\mathrm{L}}\left[\dfrac{8x-3}{2x-1}-\dfrac{4x^2+1}{4x^2-1}\right]$ $\left(\dfrac{7}{2}\right)$

4. $\underset{x\to 3}{\mathrm{L}}\left[\dfrac{1}{x-3}-\dfrac{3}{x^2-3x}\right]$ $\left(\dfrac{1}{3}\right)$

5. $\underset{x\to 2}{\mathrm{L}}\left[\dfrac{1}{x-2}+\dfrac{6x}{8-x^3}\right]$ (0)

6. $\underset{x\to 2}{\mathrm{L}}\left[\dfrac{1}{x-2}-\dfrac{2}{x(x-1)(x-2)}\right]$ $\left(\dfrac{3}{2}\right)$

7. $\underset{x\to 8}{\mathrm{L}}\left[\dfrac{1}{x-8}-\dfrac{8}{x^2-8x}\right]$ $\left(\dfrac{1}{8}\right)$

8. $\underset{x\to a}{\mathrm{L}}\left[\dfrac{1}{x-a}-\dfrac{a}{x^2-ax}\right]$ $\left(\dfrac{1}{a}\right)$

Problems based on the formulas:

(i) $\lim\limits_{x\to a}\dfrac{x^n-a^n}{x-a}=n\,a^{n-1}$; (where n is an integer > 1)

(ii) $\lim\limits_{x\to a}\dfrac{x^m-a^m}{x^n-a^n}=\dfrac{m}{n}\,a^{m-n}$; (where m and n are integers > 1)

(i) To show: $\lim\limits_{x\to a}\dfrac{x^n-a^n}{x-a}=n\,a^{n-1}$ provided n is an integer > 1

Proof: ∵ we know that

$$x^n-a^n$$

$$=(x-a)\left(x^{n-1}+a\,x^{x-2}+a^2\,x^{n-3}+\ldots+a^{n-1}\right) \quad \ldots(1)$$

on dividing both sides of (1) by $(x-a)$, we get

$$\frac{x^n-a^n}{x-a}$$

$$= \frac{(x-a)\left(x^{n-1} + a x^{n-2} + a^2 x^{n-3} + \ldots + a^{n-1}\right)}{(x-a)}$$

$$= \left(x^{n-1} + a x^{n-2} + a^2 x^{n-3} + \ldots + a^{n-1}\right) \ldots (2)$$

Now, on taking the limits on both sides of (2) as $x \to a$, we get

$$\lim_{x \to a} \frac{x^n - a^n}{x - a}$$

$$= \lim_{x \to a} \left(x^{n-1} + a x^{n-2} + a^2 x^{n-3} + \ldots + a^{n-1}\right)$$

$$= \lim_{x \to a} x^{n-1} + \lim_{x \to a} \left(a x^{n-2}\right) + \lim_{x \to a} a^2 x^{n-3} + \ldots + \lim_{x \to a} a^{n-1}$$

$$= a^{n-1} + a \cdot a^{n-2} + a^2 \cdot a^{n-3} + \ldots + a^{n-1}$$

$$= a^{n-1} + a^{n-1} + a^{n-1} + \ldots + a^{n-1} \text{ (up to } n \text{ terms)}$$

$$= n\, a^{n-1}$$

N.B.: This relation is true for n = any rational number whose proof is provided with the help of binomial expansion.

$x = a + h \Rightarrow x - a = h$ and as $x \to a,\ h \to 0$

$$\therefore (a + h)^n - a^n$$

$$a^n \left\{\left(1 + \frac{h}{a}\right)^n\right\} - a^n = a^n \left\{\left(1 + \frac{h}{a}\right)^n - 1\right\}$$

$$= a^n \left[\left\{1 + \frac{n}{\lfloor 1}\left(\frac{h}{a}\right) + \frac{n(n-1)}{\lfloor 2}\left(\frac{h}{a}\right)^2 + \ldots\right\} - 1\right]$$

$$= a^n \cdot n \cdot \frac{h}{a}\left[1 + \frac{n-1}{\lfloor 2} \cdot \frac{h}{a} + \text{terms containing high powers of } h\right] \ldots (1)$$

On dividing both sides of (1) by h, we get

$$\frac{x^n - a^n}{h} = \frac{a^n \cdot n \cdot \frac{h}{a}\left[1 + \frac{n-1}{\lfloor 2} \cdot \frac{h}{a} + \text{terms containing higher powers of } h\right]}{h} \ldots (2)$$

On putting $x - a = h$ on the l.h.s of (2), we get

$$\frac{x^n - a^n}{x - a} = \frac{a^n \cdot n \cdot \frac{h}{a}\left[1 + \frac{n-1}{\lfloor 2} \cdot \frac{h}{a} + \text{terms containing higher powers of } h\right]}{h} \ldots (3)$$

Lastly on taking the limits on both sides of (3) as $h \to 0$

$$\lim_{h \to 0} \frac{x^n - a^n}{x - a}$$

$$= \lim_{h \to 0} \frac{a^n \cdot n \cdot \frac{h}{a}\left[1 + \frac{n-1}{\lfloor 2} \cdot \frac{h}{a} + \text{terms containing higher powers of } h\right]}{h}$$

$$= a^n \cdot n \cdot \frac{1}{a} = n\, a^{n-1}$$

$$\therefore \lim_{x \to a} \frac{x^n - a^n}{x - a} = n\, a^{n-1} \quad (\because x \to a \Leftrightarrow h \to 0)$$

(ii) To show: $\lim_{x \to a} \frac{x^m - a^m}{x^n - a^n} = \frac{m}{n} a^{m-n}$ where m and n are integers > 1

Proof: $$\frac{x^m - a^m}{x^n - a^n} = \frac{\dfrac{\left(x^m - a^m\right)}{(x-a)}}{\dfrac{\left(x^n - a^n\right)}{(x-a)}}$$

$$= \frac{(x-a)\left(x^{m-1} + a x^{m-2} + \ldots + a^{m-1}\right)}{(x-a)\left(x^{n-1} + a x^{n-2} + \ldots + a^{n-1}\right)}$$

$$= \frac{x^{m-1} + ax^{m-2} + \ldots + a^{m-1}}{x^{n-1} + ax^{n-2} + \ldots + a^{n-1}} \qquad \ldots (1)$$

Now, on taking the limits on both sides of (1) as $x \to a$, we get

$$\lim_{x \to a} \frac{x^m - a^m}{x^n - a^n}$$

$$= \lim_{x \to a} \frac{x^{m-1} + ax^{m-2} + \ldots + a^{m-1}}{x^{n-1} + ax^{n-2} + \ldots + a^{n-1}}$$

$$= \frac{ma^{m-1}}{na^{n-1}} = \frac{m}{n} a^{m-n}$$

N.B.: This relation holds true even if m and n are rational numbers.

Aid to memory: **1.** $\lim_{x \to a} \frac{x^n - a^n}{x - a}$ = index of power 'a^n' times base 'a' raised to the power n minus 1.

2. $\lim_{x \to a} \frac{x^m - a^m}{x^n - a^n}$

$= \frac{m = \text{index of the power of the contant appearing in } Nr}{n = \text{index of the power of the contant appearing in } Dr}$ × the constant raised to the power m minus n.

Problems based on the formulas

(i) $\mathop{L}_{x \to a} \frac{x^n - a^n}{x - a} = na^{n-1}$

(ii) $\mathop{L}_{x \to a} \frac{x^m - a^n}{x^n - a^n} = \frac{m}{n} a^{m-n}$

Working rule: We should use the formulas (i) and (ii) directly provided that given function has the form either (i) $\frac{x^n - a^n}{x - a}$, or (ii) $\frac{x^m - a^m}{x^n - a^n}$ and we are required to find the limit of these functions as $x \to a$.

Examples worked out:

Evaluate:

1. $\lim_{x \to a} \left[\frac{x^{5/2} - a^{5/2}}{x^{1/2} - a^{1/2}} \right]$

Solution: $\lim_{x \to a} \frac{x^{5/2} - a^{5/2}}{x^{1/2} - a^{1/2}}$

$$= \frac{5/2\, a^{(5/2 - 1/2)}}{1/2}$$

$$= 5a^{\frac{4}{2}}$$

$$= 5a^2$$

2. $\lim_{x \to 64} \frac{x^{\frac{1}{6}} - 2}{x^{\frac{1}{3}} - 4}$

Solution: $\lim_{x \to 64} \frac{x^{\frac{1}{6}} - 2}{x^{\frac{1}{3}} - 4} = \lim_{x \to 64} \frac{x^{\frac{1}{6}} - (64)^{\frac{1}{6}}}{x^{\frac{1}{3}} - (64)^{\frac{1}{3}}}$

$$= \frac{\frac{1}{6} \cdot a^{\left(\frac{1}{6} - \frac{1}{3}\right)}}{\frac{1}{3}} \text{ (using formula)}$$

$$= \frac{1}{6} \times \frac{3}{1} \times a^{\left(\frac{1}{6} - \frac{1}{3}\right)} = \frac{1}{2} a^{\left(\frac{1-2}{6}\right)}$$

$$= \frac{1}{2} a^{-\frac{1}{6}} = \frac{1}{2} \cdot \frac{1}{a^{\frac{1}{6}}}$$

$$= \frac{1}{2} \cdot \frac{1}{\sqrt[6]{a}}$$

$$= \frac{1}{2} \cdot \frac{1}{\sqrt[6]{64}} \text{ (as } a = 64)$$

$$= \frac{1}{2} \cdot \frac{1}{(2^6)^{\frac{1}{6}}} = \frac{1}{2} \times \frac{1}{2} = \frac{1}{4}$$

3. $\lim_{x\to a} \frac{x^4 - a^4}{x - a}$

Solution: $\lim_{x\to a} \frac{x^4 - a^4}{x - a} = 4a^{4-1}$ (using the formula)

4. $\lim_{x\to a} \left[\frac{x^{\frac{3}{2}} - a^{\frac{3}{2}}}{x^2 - a^2}\right] = 4a^3$

Solution: $\lim_{x\to a}\left(\frac{x^{\frac{3}{2}} - a^{\frac{3}{2}}}{x^2 - a^2}\right) = \frac{\frac{3}{2}\cdot a^{\left(\frac{3}{2}-2\right)}}{2}$

$= \frac{3}{2} \times \frac{1}{2} \times a^{\left(\frac{3-4}{2}\right)}$

$= \frac{3}{4} a^{\left(-\frac{1}{2}\right)} = \frac{3}{4} \times \frac{1}{\sqrt{a}} = \frac{3}{4\sqrt{a}}$

5. $\lim_{x\to 1}\left[\frac{x - 1}{x^{\frac{1}{4}} - 1}\right]$

Solution: $\lim_{x\to 1}\left[\frac{x - 1}{x^{\frac{1}{4}} - 1}\right] = \lim_{x\to 1} \frac{1}{\frac{x^{\frac{1}{4}} - (1)^{\frac{1}{4}}}{x - 1}}$

$= \frac{1}{\frac{1}{4}\cdot (1)^{\left(\frac{1}{4}-1\right)}}$

$= \frac{1}{\frac{1}{4} \times (1)^{\left(-\frac{3}{4}\right)}}$

$= \frac{1}{\frac{1}{4} \times 1}$

$= 1 \times \frac{4}{1}$

$=4$

or alternatively,

$\lim_{x\to 1} \frac{x - 1}{x^{\frac{1}{4}} - 1} = \frac{1}{\frac{1}{4}}(1)^{\left(1-\frac{1}{4}\right)} = \frac{1}{\frac{1}{4}}\cdot(1)^{\left(-\frac{3}{4}\right)}$ ($\because$ Here,

$m = 1, n = \frac{1}{4}$) (according to formula)

$= 4 \cdot \frac{1}{(1)^{\frac{3}{4}}}$

$= \frac{4}{1}$

$=4$

Problems based on the form

(i) $\frac{x^m - a^m}{x^n - a^n}$ **(ii)** $\frac{x^n - a^n}{x - a}$

Exercise 4.5

Find the limit of the following: ***Answers***

1. $\underset{x\to a}{\text{L}} \frac{x^3 - a^3}{x^5 - a^5}$ $\left(\frac{3}{5}a^{-2}\right)$
2. $\underset{x\to 2}{\text{L}} \frac{x^5 - 32}{x^3 - 8}$ $\left(\frac{20}{3}\right)$
3. $\underset{x\to 2}{\text{L}} \frac{x^2 - 4}{x\sqrt{x} - 2\sqrt{2}}$ $\frac{4\sqrt{2}}{3}$
4. $\underset{x\to 2}{\text{L}} \frac{x^5 - 32}{x - 2}$ (80)
5. $\underset{x\to 2}{\text{L}} \frac{x^5 - 32}{x^4 - 16}$ $\left(\frac{5}{2}\right)$
6. $\underset{x\to 2}{\text{L}} \frac{\sqrt{x} - \sqrt{2}}{x - 2}$ $\left(\frac{1}{2\sqrt{2}}\right)$
7. $\underset{x\to 3}{\text{L}} \frac{\sqrt{x} - \sqrt{3}}{x - 3}$ $\left(\frac{1}{2\sqrt{3}}\right)$

8. $\underset{x\to 1}{\mathrm{L}} \dfrac{1-x^7}{1-x}$ (7)

9. $\underset{x\to -2}{\mathrm{L}} \dfrac{x^3+8}{x+2}$ (12)

Problems based on types of functions mentioned earlier but $x \to 0$ instead of $x \to a$

Working rule: When $x \to 0$ (zero) and types of the functions are same as mentioned earlier when $x \to a$ (any constant)

Rule: When $x \to 0$, the same rule is applied to find out the limit of a function as $x \to a$ which means rule to find out the limit of a function as $x \to a$ (constant) = rule to find out the limit of a function as $x \to 0$ (zero).

Theorem: If $f(x)$

= a polynomial in $x = a_0 x^n + a_1 x^{n-1} + \ldots + a_n$ (where $a_0, a_1, a_2, \ldots a_n$ are constants and n is a +ve integer) then $\lim_{x\to 0} f(x) = a_n$ = the constant present in the given polynomial in x which is free from the variable raised to any +ve index.

Examples worked out:

Evaluate: **1.** $\lim_{x\to 0}\left(3x^2+4x^2-5x+6\right)$

Solution: $\lim_{x\to 0}\left(3x^2+4x^2-5x+6\right)$

$$= \lim_{x\to 0} 3x^2 + \lim_{x\to 0} 4x^2 - \lim_{x\to 0} 5x + \lim_{x\to 0} 6$$

$$\Rightarrow \lim_{x\to 0}\left(3x^2+4x^2-5x+6\right)$$

$$= 3\lim_{x\to 0} x^2 + 4\lim_{x\to 0} x^2 - 5\lim_{x\to 0} x + 6$$

$$= 3.0 + 4.0 - 5.0 + 6$$

$$= 6.$$

2. $\lim_{x\to 0}\left[\dfrac{x^2+5x}{4x}\right]$

Solution: $\left[\dfrac{x^2+5x}{4x}\right] = \dfrac{0}{0}$ form as $x \to 0 \Rightarrow (x-0)$ is a factor of *Nr* and *Dr* on dividing *Nr* and *Dr* by $(x-0) = x$, we get

$$\frac{x^2+5x}{4x} = \frac{\not{x}(x+5)}{4\not{x}} = \frac{x+5}{4} \quad \ldots(1)$$

On taking the limits on both sides of (i) as $x \to 0$, we get

$$\lim_{x\to 0}\left[\frac{x^2+5x}{4x}\right] = \lim_{x\to 0}\left[\frac{0+5}{4}\right] = \frac{0+5}{4} = \frac{5}{4}$$

3. $\lim_{x\to 0} \dfrac{\sqrt{x+1}-1}{x}$

Solution: $\left[\dfrac{\sqrt{x+1}-1}{x}\right] = \dfrac{0}{0}$ form as $x \to 0$ on rationalizing the *Nr*, we get

$$\lim_{x\to 0} \frac{\sqrt{x+1}-1}{x}$$

$$= \lim_{x\to 0}\left[\frac{\sqrt{x+1}-1}{x} \times \frac{\sqrt{x+1}+1}{\sqrt{x+1}-1}\right]$$

$$= \lim_{x\to 0} \frac{(x+1)-1}{x\left(\sqrt{x+1}+1\right)} = \lim_{x\to 0} \frac{x}{x\left(\sqrt{x+1}+1\right)}$$

$$= \lim_{x\to 0} \frac{1}{\left(\sqrt{x+1}+1\right)}$$

$$= \frac{1}{\sqrt{0+1}+1} = \frac{1}{1+1} = \frac{1}{2}$$

4. $\lim_{x\to 0} \dfrac{\sqrt{x+a}-\sqrt{a}}{x}$

Solution: $\left[\dfrac{\sqrt{x+a}-\sqrt{a}}{x}\right] = \dfrac{0}{0}$ form as $x \to 0$ on, rationalizing the *Nr*, we get

$$\lim_{x\to 0}\frac{\sqrt{x+a}-\sqrt{a}}{x}$$

$$= \lim_{x\to 0}\left[\frac{\sqrt{x+a}-\sqrt{a}}{x}\times\frac{\sqrt{x+a}+\sqrt{a}}{\sqrt{x+a}+\sqrt{a}}\right]$$

$$\Rightarrow \lim_{x\to 0}\frac{\sqrt{x+a}-\sqrt{a}}{x}$$

$$= \lim_{x\to 0}\frac{(x+a)-a}{x\left(\sqrt{x+a}+\sqrt{a}\right)}$$

$$= \frac{1}{\sqrt{0+a}+\sqrt{a}} = \frac{1}{2\sqrt{a}}$$

5. $\lim_{x\to 0}\frac{x}{1-\sqrt{1-x}}$

Solution: $\left[\frac{x}{1-\sqrt{1-x}}\right] = \frac{0}{0}$ form as $x\to 0$, on rationalizing the *Dr*, we get

$$\lim_{x\to 0}\left[\frac{x}{1-\sqrt{1-x}}\right]$$

$$= \lim_{x\to 0}\left[\frac{x}{1-\sqrt{1-x}}\times\frac{1+\sqrt{1-x}}{1+\sqrt{1-x}}\right]$$

$$= \lim_{x\to 0}\frac{x\left(1+\sqrt{1-x}\right)}{x}$$

$$\Rightarrow \lim_{x\to 0}\left(\frac{x}{1-\sqrt{1-x}}\right) = \lim_{x\to 0}\left(1+\sqrt{1-x}\right)$$

$$= 1+\sqrt{1-0}$$
$$= 1+1 = 2.$$

6. $\lim_{x\to 0}\frac{\sqrt{x+1}-1}{\log(1+x)}$

Solution: $\left[\frac{\sqrt{x+1}-1}{\log(1+x)}\right] = \frac{0}{0}$ form as $x\to 0$

$(\because \log 1 = 0)$ on rationalizing the *Nr*, we get

$$\lim_{x\to 0}\left[\frac{\sqrt{x+1}-1}{\log(1+x)}\right]$$

$$= \lim_{x\to 0}\left[\frac{\sqrt{x+1}-1}{\log(1+x)}\times\frac{\sqrt{x+1}+1}{\sqrt{x+1}+1}\right]$$

$$= \lim_{x\to 0}\frac{x}{\log(1+x)\left(\sqrt{x+1}+1\right)}$$

$$\Rightarrow \lim_{x\to 0}\left[\frac{\sqrt{x+1}-1}{\log(1+x)}\right]$$

$$= \lim_{x\to 0}\frac{1}{\log(1+x)^{\frac{1}{x}}\cdot\left(\sqrt{x+1}+1\right)}$$

$$= \frac{1}{\left(\sqrt{0+1}+1\right)}\times\frac{1}{\log_e e}$$

$$\Rightarrow \lim_{x\to 0}\left[\frac{\sqrt{x+1}-1}{\log(1+x)}\right] = \frac{1}{2}\times 1$$

$$\left(\because \lim_{x\to 0}(1+x)^{\frac{1}{x}} = e\right)$$

$$\Rightarrow \lim_{x\to 0}\left[\frac{\sqrt{x+1}-1}{\log(1+x)}\right] = \frac{1}{2}$$

7. $\lim_{x\to 0}\left[\frac{5+3x}{7-2x}\right]$

Solution: $\left[\frac{5+3x}{7-2x}\right] \neq \frac{0}{0}$ which $\Rightarrow$ given function $\left(\frac{(5+3x)}{(7-2x)}\right)$ does not assume $\left(\frac{0}{0}\right)$ form at $x = 0$

$$\therefore \lim_{x\to 0}\left[\frac{5+3x}{7-2x}\right] = \frac{\lim_{x\to 0}(5+3x)}{\lim_{x\to 0}(7-2x)}$$

$$= \frac{\lim_{x\to 0} 5 + \lim_{x\to 0} 3x}{\lim_{x\to 0} 7 - \lim_{x\to 0} 2x} = \frac{5 + 3.0}{7 - 2.0} = \frac{5}{7}$$

Some Important Forms

First important form:

$$\frac{(a \pm x)^n - a^n}{x\ (= \text{independent variable})} \text{ as } x \to 0$$

Working rule:

1. Put $a \pm x = z$
2. Change the limit as $x \to 0 \Leftrightarrow z \to a$
3. The above substitution transforms the given problem in to the form:

$$\frac{z^n - a^n}{z - a} \text{ as } z \to a$$

4. Use the formula: $\lim_{z\to a} \frac{z^n - a^n}{z - a} = na^{n-1}$

Examples worked out:

Evaluate:

1. $\lim_{x\to 0} \frac{(1+x)^{\frac{1}{n}} - 1}{x}$

Solution: $\left[\frac{(1+x)^{\frac{1}{n}} - 1}{x}\right] = \frac{0}{0}$ form as $x \to 0$

We put $1 + x = z$

Now, $1 + x = z \Rightarrow z \to 1$ as $x \to 0$

$$\therefore \lim_{x\to 0} \frac{(1+x)^{\frac{1}{n}} - (1)^{\frac{1}{n}}}{(1+x) - 1} = \lim_{z\to 1} \left[\frac{z^{\frac{1}{n}} - (1)^{\frac{1}{n}}}{z - 1}\right]$$

$$= \frac{1}{n} \times (1)^{\frac{1}{n}-1} = \frac{1}{n}$$

2. $\lim_{x\to 0} \frac{(1+x)^{\frac{1}{5}} - 1}{x}$

Solution: $\left[\frac{(1+x)^{\frac{1}{5}} - 1}{x}\right] = \frac{0}{0}$ form as $x \to 0$

We put $1 + x = z$

Now, $1 + x = z \Rightarrow z \to 1$ as $x \to 0$

$$\therefore \lim_{x\to 0} \frac{(1+x)^{\frac{1}{5}} - 1}{x} = \lim_{x\to 0} \frac{(1+x)^{\frac{1}{5}} - 1}{(1+x) - 1}$$

$$= \lim_{z\to 1} \frac{z^{\frac{1}{5}} - 1^{\frac{1}{5}}}{z - 1}$$

$$= \frac{1}{5} (1)^{\frac{1}{5}-1} = \frac{1}{5}$$

3. $\lim_{x\to 0} \frac{\sqrt{1+x} - 1}{x}$

Solution: We put $1 + x = z$

Now, $1 + x = z \Rightarrow z \to 1$ as $x \to 0$

$$\therefore \lim_{x\to 0} \left[\frac{\sqrt{1+x} - 1}{x}\right] = \lim_{x\to 0} \frac{\sqrt{1+x} - \sqrt{1}}{(1+x) - 1}$$

$$= \lim_{z\to 1} \frac{z^{\frac{1}{2}} - 1^{\frac{1}{2}}}{z - 1}$$

$$= \frac{1}{2} \cdot (1)^{\frac{1}{2}-1}$$

$$= \frac{1}{2}$$

4. $\lim_{h\to 0} \frac{(x+h)^{\frac{1}{2}} - x^{\frac{1}{2}}}{h}$

Solution: We put $x + h = z$

Now, $x + h = z \Rightarrow z \to x$ as $h \to 0$

$$\therefore \lim_{h\to 0} \frac{(x+h)^{\frac{1}{2}} - x^{\frac{1}{2}}}{h} = \lim_{z\to x} \frac{z^{\frac{1}{2}} - x^{\frac{1}{2}}}{z - x}$$

$(\because h = x + h - x = z - x)$

$$= \frac{1}{2} x^{-\frac{1}{2}} = \frac{1}{2x^{\frac{1}{2}}} = \frac{1}{2\sqrt{x}}$$

5. $\lim\limits_{h\to 0} \dfrac{(x+h)^2 - x^2}{h}$

Solution: We put $x + h = z$

Now, $x + h = z \Rightarrow z \to x$ as $h \to 0$

$$\therefore \lim_{h\to 0} \frac{(x+h)^2 - x^2}{h} = \lim_{h\to 0} \frac{(x+h)^2 - x^2}{(x+h) - x}$$

$$= \lim_{z\to x} \frac{z^2 - x^2}{z - x} = 2x$$

Second important form: A rational function in x whose *Nr* and *Dr* consist of a polynomial of fractional indices and the independent variable $x \to 0$.

Working rule: Divide *Nr* and *Dr* (each term of *Nr* and *Dr*) by the lowest power of x occurring in the given function

Examples worked out:

Evaluate

1. $\lim\limits_{x\to 0} \dfrac{x^{\frac{7}{10}} + 3x^{\frac{4}{5}} + 2x}{x^{\frac{1}{3}} + 4x^{\frac{2}{3}} + 2x^{\frac{1}{5}}}$

Solution: $\left[\dfrac{x^{\frac{7}{10}} + 3x^{\frac{4}{5}} + 2x}{x^{\frac{1}{3}} + 4x^{\frac{2}{3}} + 2x^{\frac{1}{5}}}\right] = \dfrac{0}{0}$ form as $x \to 0$.

Now, on dividing *Nr* and *Dr* by the lowest power $x^{\frac{1}{5}}$, we get

$$\frac{x^{\frac{7}{10}} + 3x^{\frac{4}{5}} + 2x}{x^{\frac{1}{3}} + 4x^{\frac{2}{3}} + 2x^{\frac{1}{5}}} = \frac{x^{\frac{1}{2}} + 3x^{\frac{3}{5}} + 2x^{\frac{4}{5}}}{x^{\frac{2}{15}} + 4x^{\frac{7}{15}} + 2} \quad \ldots(1)$$

Lastly, on taking the limits on both sides of (1), as $x \to 0$, we get

$$\lim_{x\to 0} \frac{x^{\frac{7}{10}} + 3x^{\frac{4}{5}} + 2x}{x^{\frac{1}{3}} + x^{\frac{2}{3}} + 2x^{\frac{1}{5}}} = \lim_{x\to 0} \frac{x^{\frac{1}{2}} + 3x^{\frac{3}{5}} + 2x^{\frac{4}{5}}}{x^{\frac{2}{15}} + 4x^{\frac{7}{15}} + 2}$$

$$= \frac{0+0+0}{0+0+2} = \frac{0}{2} = 0$$

Problems based on limits of a function as $x \to 0$

Exercise 4.6

Find the following limits: ***Answers***

1. $\lim\limits_{x\to 0} \left(x^4 + 9x^3 - 7x + 4\right)$ (4)
2. $\lim\limits_{x\to 0} \dfrac{x}{x}$ (1)
3. $\lim\limits_{x\to 0} \dfrac{5x^3}{3x^3}$ $\left(\dfrac{5}{3}\right)$
4. $\lim\limits_{x\to 0} \dfrac{x^2}{x}$ (0)
5. $\lim\limits_{x\to 0} \dfrac{x}{\sqrt{x}}$ (0)
6. $\lim\limits_{x\to 0} \dfrac{1}{x}$ does not exist
7. $\lim\limits_{x\to 0} \dfrac{x}{x^3}$ does not exist
8. $\lim\limits_{x\to 0} \dfrac{3x+4}{5x+6}$ $\left(\dfrac{2}{3}\right)$
9. $\lim\limits_{x\to 0} \dfrac{2x^2 + x - 2}{3x^2 - x + 1}$ (–2)
10. $\lim\limits_{x\to 0} \dfrac{6x^2 - 2x + 5}{x^3 + x^2}$ (∞)
11. $\lim\limits_{x\to 0} \dfrac{(x-1)(2x+3)}{(x+5)(3x-2)}$ $\left(\dfrac{3}{10}\right)$
12. $\lim\limits_{x\to 0} \dfrac{x^2 - 3x + 2}{x^2 + x - 6}$ $\left(-\dfrac{1}{3}\right)$
13. $\lim\limits_{x\to 0} \dfrac{2x^2 + 3x}{3x^2 - 5x}$ $\left(-\dfrac{3}{5}\right)$
14. $\lim\limits_{x\to 0} \dfrac{x^2 - 4x - 5}{x^2 + x - 2}$ $\left(\dfrac{5}{2}\right)$

15. $\lim\limits_{x\to 0} \dfrac{x^2 + 5x}{x^2 + x}$ (5)

16. $\lim\limits_{x\to 0} \dfrac{\sqrt{1 + x} - \sqrt{1 - x}}{x}$ (1)

17. $\lim\limits_{x\to 0} \dfrac{\sqrt{1 + 2x} - \sqrt{1 - 3x}}{x(1 + x)}$ $\left(\dfrac{5}{2}\right)$

18. $\lim\limits_{x\to 0} \dfrac{\sqrt{1 + x} - \sqrt{1 + x^4}}{\sqrt{1 - x^4} - \sqrt{1 - x}}$ (1)

19. $\lim\limits_{x\to 0} \dfrac{x^2}{a - \sqrt{a^2 - x^2}}$ $(a+|a|)$

20. $\lim\limits_{x\to 0} \dfrac{\sqrt{1 + x} - 1}{\sqrt[3]{1 + x} + 1}$ $\left(\dfrac{3}{2}\right)$

21. $\lim\limits_{x\to 0} \dfrac{\sqrt{9 - x} - 3}{x}$ $\left(\dfrac{1}{6}\right)$

22. $\lim\limits_{x\to 0} \dfrac{x^2}{1 - \sqrt{1 - x^2}}$ (2)

23. $\lim\limits_{x\to 0} \dfrac{x^2}{x^5}$ does not exist

24. $\lim\limits_{x\to 0} \dfrac{1 - \sqrt{1 - x}}{x}$ $\left(\dfrac{1}{8}\right)$

25. $\lim\limits_{h\to 0} \dfrac{\sqrt[3]{x + h} - \sqrt[3]{x}}{h}$ $\left(\dfrac{1}{3\sqrt[3]{x^2}}\right)$

26. $\lim\limits_{h\to 0} \dfrac{\sqrt{x + h} - \sqrt{x}}{h}$ $\left(\dfrac{1}{2\sqrt{x}}\right)$

To evaluate $\lim\limits_{x\to\infty} \dfrac{f_1(x)}{f_2(x)}$, where $f_1(x)$ and $f_2(x)$ are polynomials in x.

Working rule: *One should*:

1. Divide $f_1(x)$ and $f_2(x)$ by the highest power of x occurring in the given fraction, i.e., divide each term of the numerator and denominator of the given fraction by the highest power of x occurring either in numerator or in denominator. After division by highest power of x present either in numerator or in denominator of the given fraction, each term of the numerator and denominator of the given fraction will be reduced to the forms:

$$a, \frac{b}{x}, \frac{c}{x^2}, \frac{d}{x^3}, \ldots, \text{ etc. where } a, b, c, d, \ldots, \text{ are}$$

constants.

2. Take the limits of each term of numerator and denominator both as $x \to \infty$ noting that $\dfrac{b}{c}, \dfrac{c}{x^2}, \dfrac{d}{x^3}, \ldots$, etc. (appearing in numerator and denominator) all $\to 0$ excepting a constant 'a_1' in numerator and another constant 'a_2' in denominator whose quotient $\left(\dfrac{a_1}{a_2}\right)$ will give us the required limit of the given fraction, if $a_2 \neq 0$.

Notes: **1.** Highest power of x may be present in either numerator or in denominator.

2. Highest power of x of numerator and denominator may be the same.

3. The determination of limit of a function $y = f(x)$ as $x \to +\infty$ and $x \to -\infty$ are also sometimes used to find out the range of $y = f(x)$ when $x \in (-\infty, +\infty)$ and $f(x)$ is continuous in R.

Explanation: **1.** $f(x) = x$

$$\Rightarrow \lim_{x\to-\infty} f(x) = \underset{x\to-\infty}{\mathrm{L}} (x) = -\infty$$

and $\lim\limits_{x\to+\infty} f(x) = \underset{x\to\infty}{\mathrm{L}} (x) = +\infty$

Hence, $R(f) = (-\infty, +\infty)$, since $f(x)$ is continuous.

2. $f(x) = x^2$

$$\Rightarrow \underset{x\to-\infty}{\mathrm{L}} \left(x^2\right) = +\infty \text{ and } \underset{x\to+\infty}{\mathrm{L}} \left(x^2\right) = +\infty$$

Further $x^2 \geq 0$ and $x^2 = 0$ for $x = 0$

Hence, $R(f) = [0, +\infty)$, since $f(x)$ is continuous.

3. $f(x) = x^3$

$$\Rightarrow \underset{x \to -\infty}{\mathrm{L}} \left(x^3\right) = -\infty \text{ and } \underset{x \to \infty}{\mathrm{L}} \left(x^3\right) = +\infty$$

Hence, $R(f) = (-\infty, \infty)$, since $f(x)$ is continuous.

4. $f(x) = \left(\dfrac{e^x - e^{-x}}{2}\right)$

$$\Rightarrow \underset{x \to -\infty}{\mathrm{L}} f(x) = -\infty \text{ and } \underset{x \to \infty}{\mathrm{L}} f(x) = +\infty$$

Hence, $R(f) = (-\infty, \infty)$, since $f(x)$ is continuous.

Problems based on the form:

$\lim\limits_{x \to \infty} \dfrac{f_1(x)}{f_2(x)}$, where $f_1(x)$ and $f_2(x)$ are polynomials in x.

Examples worked out:

Evaluate:

1. $\lim\limits_{x \to \infty} \dfrac{2x^2 - 3x + 5}{3x^2 + 27x - 29}$

Solution: $\lim\limits_{x \to \infty} \dfrac{2x^2 - 3x + 5}{3x^2 + 27x - 29}$

$= \lim\limits_{x \to \infty} \dfrac{2 - \dfrac{3}{x} + \dfrac{5}{x^2}}{3 + \dfrac{27}{x} - \dfrac{29}{x^2}}$ (on dividing *Nr* and *Dr* by the highest power of x, i.e., x^2)

$= \dfrac{2}{3}$

N.B.: As $x \to \infty$, $\dfrac{1}{x} \to 0$, $\dfrac{1}{x^2} \to 0$

2. $\lim\limits_{x \to \infty} \left[\dfrac{2x}{x - 1} - \dfrac{x}{x + 1}\right]$

Solution: $\lim\limits_{x \to \infty} \left[\dfrac{2x}{x - 1} - \dfrac{x}{x + 1}\right]$

$= \lim\limits_{x \to \infty} \left[\dfrac{2x}{x - 1}\right] - \lim\limits_{x \to \infty} \left[\dfrac{x}{x + 1}\right]$

$= \lim\limits_{x \to \infty} \left[\dfrac{2}{x - \dfrac{1}{x}}\right] - \lim\limits_{x \to \infty} \left[\dfrac{1}{1 + \dfrac{1}{x}}\right]$

$= 2 - 1 = 1$

3. $\lim\limits_{x \to \infty} \dfrac{3e^{2x} + 2e^{-2x}}{4e^{2x} - e^{-2x}}$

Solution: $\lim\limits_{x \to \infty} \dfrac{3e^{2x} + 2e^{-2x}}{4e^{2x} - e^{-2x}}$

$= \lim\limits_{x \to \infty} \dfrac{3 \cdot e^{2x} + \dfrac{2}{e^{2x}}}{4 \cdot e^{2x} + \dfrac{1}{e^{2x}}}$

$= \lim\limits_{x \to \infty} \dfrac{3 + \dfrac{2}{e^{4x}}}{4 + \dfrac{1}{e^{2x}}}$

$= \dfrac{3 + 0}{4 + 0}$

$= \dfrac{3}{4}$

[***N.B.***: $\because$ as $x \to \infty$, $e^{4x} \to \infty \Rightarrow \dfrac{2}{e^{4x}} \to 0$]

Problems based on the form:

$$\lim_{x \to \infty} \frac{a_0 x^n + a_1 x^{n-1} + a_2 x^{n-2} + \dots + a}{b_0 x^n + b_1 x^{n-1} + b_2 x^{n-2} + \dots + b}$$

Exercise 4.7

Find the following limits: ***Answers***

1. $\lim\limits_{x \to \infty} \dfrac{6x^3 - 5x^2 + 4}{4x^3 - 4x + 7}$ $\left(\dfrac{3}{2}\right)$

2. $\lim_{x \to \infty} \frac{ax^2 + bx + c}{lx^2 + mx + n}$ $\left(\frac{a}{l}\right)$

3. $\lim_{x \to \infty} \frac{x^4 - x^2 + 3}{x^2 + 5x + 13}$ (∞)

4. $\lim_{x \to \infty} \frac{x^3 + x^2 + x + 1}{x^5 + 1}$ (0)

5. $\lim_{x \to \infty} \frac{5x - 3}{7x + 8}$ $\left(\frac{5}{7}\right)$

6. $\lim_{x \to \infty} \frac{ax + b}{cx}$ $\left(\frac{a}{c}\right)$

7. $\lim_{x \to \infty} \frac{5x^2 + x + 1}{6x^2 - 3x - 5}$ $\left(\frac{5}{6}\right)$

8. $\lim_{x \to \infty} \frac{2x^2 - x + 100}{4x^3 + 2}$ (0)

9. $\lim_{x \to \infty} \frac{(x + 1)(2x + 3)}{(x + 2)(3x + 4)}$ $\left(\frac{2}{3}\right)$

Problems based on rationalization when $x \to \infty$

Remember:

1. If only *Nr* contains radical, rationalize *Nr* by its rationalizing factor (rationalizing factor is also known as conjugate).

2. If only *Dr* contains radical, rationalize *Dr* by its conjugate.

3. If *Nr* and *Dr* both contain radicals, rationalize *Nr* and *Dr* both by its conjugate.

4. After rationalization, we have

Nr = an expression in *x*

Dr = an expression in *x* with radical or without radical, then we divide the rationalized function by the square root of highest power of *x* (i.e. $\sqrt{\text{highest power of } x}$) seeing the power of *x* under the radical sign of the radical signs in *Nr* and *Dr* contain an expression in *x*, i.e. the process of division of rationalized function by $\sqrt{\text{highest power of } x}$ makes the coefficient of highest power free from highest power of *x* in the following examples.

5. Remember that $\lim_{x \to \infty} \frac{a}{x^n} = 0$ if $n > 0$

Examples worked out:

Evaluate:

1. $\lim_{x \to \infty} \frac{\sqrt{x^4 + 1} - \sqrt{x^4 - 1}}{x}$

$$= \lim_{x \to \infty} \frac{x^4 + 1 - (x^4 - 1)}{x\left(\sqrt{x^4 + 1} + \sqrt{x^4 - 1}\right)};$$

(on rationalizing *Nr*)

$$= \lim_{x \to \infty} \frac{x^4 + 1 - x^4 + 1}{x\left(\sqrt{x^4 + 1} + \sqrt{x^4 - 1}\right)}$$

$$= \lim_{x \to \infty} \frac{2}{x\left(\sqrt{x^4 + 1} + \sqrt{x^4 - 1}\right)}$$

$= 0$

2. $\lim_{x \to \infty} \left[\sqrt{x^2 + 1} - \sqrt{x^2 - 1}\right]$

Solution: $\left[\sqrt{x^2 + 1} - \sqrt{x^2 - 1}\right]$

$$= \frac{\sqrt{x^2 + 1} - \sqrt{x^2 - 1}}{\sqrt{x^2 + 1} + \sqrt{x^2 - 1}} \times \sqrt{x^2 + 1} + \sqrt{x^2 - 1}$$

$$= \frac{x^2 + 1 - (x^2 - 1)}{\sqrt{x^2 + 1} + \sqrt{x^2 - 1}}$$

$$= \frac{2}{\sqrt{x^2 + 1} + \sqrt{x^2 - 1}}$$

$\therefore \lim_{x \to \infty} \left(\sqrt{x^2 + 1} - \sqrt{x^2 - 1}\right)$

$$= \lim_{x \to \infty} \frac{2}{\sqrt{x^2+1} + \sqrt{x^2-1}}$$

$$= \lim_{x \to \infty} \frac{\frac{2}{x}}{\sqrt{1+\frac{1}{x^2}} + \sqrt{1-\frac{1}{x^2}}}$$

(dividing by $\sqrt{x^2} = x$)

$= 0$

3. $\lim_{x \to \infty} x\left(x - \sqrt{x^2+1}\right)$

Solution: $\lim_{x \to \infty} x\left(x - \sqrt{x^2+1}\right)$

$$= \lim_{x \to \infty} \left[\frac{x\left(x - \sqrt{x^2+1}\right) \times \left(x + \sqrt{x^2+1}\right)}{\left(x + \sqrt{x^2+1}\right)}\right]$$

$$= \lim_{x \to \infty} \frac{x\left[x^2 - \left(x^2+1\right)\right]}{\left[x + \sqrt{x^2+1}\right]}$$

$$= \lim_{x \to \infty} \frac{x(-1)}{x\left[1 + \sqrt{\frac{x^2}{x^2} + \frac{1}{x^2}}\right]}$$

(dividing by $\sqrt{x^2} = x$)

$$= \lim_{x \to \infty} \frac{(-1)}{\left[1 + \sqrt{1 + \frac{1}{x^2}}\right]}$$

$$= \frac{(-1)}{\left[1 + \sqrt{1+0}\right]} = -\frac{1}{2}$$

4. $\lim_{x \to \infty} x\left(\sqrt{x^2+1} - \sqrt{x^2-1}\right)$

Solution: $\lim_{x \to \infty} x\left(\sqrt{x^2+1} - \sqrt{x^2-1}\right)$

$$= \lim_{x \to \infty} \left[\frac{x\left(\sqrt{x^2+1} - \sqrt{x^2-1}\right) \times \left(\sqrt{x^2+1} + \sqrt{x^2-1}\right)}{\left(\sqrt{x^2+1} + \sqrt{x^2-1}\right)}\right]$$

$$= \lim_{x \to \infty} \frac{x\left(x^2 + 1 - x^2 + 1\right)}{\left(\sqrt{x^2+1} + \sqrt{x^2-1}\right)}$$

$$= \lim_{x \to \infty} \frac{2x^*}{\left(\sqrt{x^2+1} + \sqrt{x^2-1}\right)}$$

$$= \lim_{x \to \infty} \frac{\frac{2x}{x}}{\sqrt{1+\frac{1}{x^2}} + \sqrt{1-\frac{1}{x^2}}};$$

(dividing by $\sqrt{x^2} = x$)

$$= \frac{2}{\sqrt{1} + \sqrt{1}}$$

$$= \frac{2}{2}$$

$= 1$

Note: * Highest power of x in $Nr = x$

Highest power of x in $Dr = \sqrt{x^2} = x$

This is why we divide each term of Nr and Dr by x

5. $\lim_{x \to \infty} \frac{\sqrt{16x^2 - 9x + 5} - \sqrt{9x^2 + 5x - 7}}{(2x-3)}$

Solution: $\lim_{x \to \infty} \frac{\sqrt{16x^2 - 9x + 5} - \sqrt{9x^2 + 5x - 7}}{(2x-3)}$

$$\lim_{x \to \infty} \frac{\sqrt{16 - \frac{9}{x} + \frac{5}{x^2}} - \sqrt{9 + \frac{5}{x} - \frac{7}{x^2}}}{\left(2 - \frac{3}{x}\right)};$$

(dividing by $\sqrt{x^2} = x$)

$$= \frac{\sqrt{16 - 0 + 0} - \sqrt{9 + 0 - 0}}{2 - 0}$$

$$= \frac{4 - 3}{2}$$

$$= \frac{1}{2}$$

N.B.: $\left[\because \frac{1}{x}, \frac{1}{x^2} \to 0 \text{ as } x \to \infty\right]$

6. $\lim\limits_{x \to \infty} \frac{\sqrt{3x^2 + 4x + 5} - \sqrt{2x^2 + 3}}{(3x + 7)}$

Solution: $\lim\limits_{x \to \infty} \frac{\sqrt{3x^2 + 4x + 5} - \sqrt{2x^2 + 3}}{(3x + 7)}$

Note that highest power of x in $Nr = \sqrt{x^2} = x$ and highest power of x in $Dr = x$.

$$\lim_{x \to \infty} \frac{\sqrt{3x^2 + 4x + 5} - \sqrt{2x^2 + 3}}{(3x + 7)}$$

$$= \lim_{x \to \infty} \frac{\sqrt{\left(\frac{3x^2}{x^2} + \frac{4x}{x^2} + \frac{5}{x^2}\right)x^2} - \sqrt{\left(\frac{2x^2}{x^2} + \frac{3}{x^2}\right)x^2}}{x\left(\frac{3x}{x} + \frac{7}{x}\right)}$$

$$= \lim_{x \to \infty} \frac{x\sqrt{3 + \frac{4}{x} + \frac{5}{x^2}} - x\sqrt{2 + \frac{3}{x^2}}}{x\left(3 + \frac{7}{x}\right)}$$

$$= \lim_{x \to \infty} \frac{\sqrt{3 + \frac{4}{x} + \frac{5}{x^2}} - \sqrt{2 + \frac{3}{x^2}}}{\left(3 + \frac{7}{x}\right)}$$

$$= \frac{\sqrt{3 + 0 - 0} - \sqrt{2 + 0}}{3}$$

$$= \frac{\sqrt{3} - \sqrt{2}}{3}$$

Some Miscellaneous Problems

Evaluate:

1. $\lim\limits_{n \to \infty} \frac{\sqrt{(3n^2 - 1)} - \sqrt{(2n^2 + 1)}}{4n + 3}$

Solution: Let us put $n = \frac{1}{x}$

$\therefore x \to 0$ as $n \to \infty$

Making this substitution and after simplification, we get

$$\lim_{n \to \infty} \frac{\sqrt{(3n^2 - 1)} - \sqrt{(2n^2 + 1)}}{4n + 3}$$

$$\lim_{x \to 0} \frac{\sqrt{(3 - x^2)} - \sqrt{(2 + x^2)}}{4 + 3x}$$

$$= \frac{\sqrt{3} - \sqrt{2}}{4}$$

$$= \frac{1}{4}\left(\sqrt{3} - \sqrt{2}\right)$$

An important form:

Form: $\frac{\sqrt{f_1(x)}}{\sqrt[n]{f_2(x)}}$ or $\frac{f_1(x)}{\sqrt{f_1(x)}}$

Examples Worked Out:

Evaluate:

1. $\lim\limits_{x \to \infty} \frac{\sqrt{x^2 - 3}}{\sqrt[3]{(x^3 + 1)}}$

Solution: $\lim\limits_{x \to \infty} \frac{\sqrt{x^2 - 3}}{\sqrt[3]{(x^3 + 1)}}$

$$= \lim_{x \to \infty} \frac{\sqrt{x^2\left(\frac{x^2-3}{x^2}\right)}}{\sqrt[3]{x^3\left(\frac{x^3+1}{x^3}\right)}}$$

$$= \lim_{x \to \infty} \frac{x\sqrt{1-\frac{3}{x^2}}}{x\sqrt[3]{1+\frac{1}{x^3}}}$$

$$= \frac{\sqrt{\lim_{x \to \infty}\left(1-\frac{3}{x^2}\right)}}{\sqrt[3]{\lim_{x \to \infty}\left(1+\frac{1}{x^2}\right)}}$$

$$= \frac{\sqrt{1-0}}{\sqrt[3]{1+0}}$$

$$= \frac{1}{1}$$

$$= 1$$

2. $\lim_{x \to \infty} \frac{4x}{\sqrt{2x^2+1}}$

Solution: $\lim_{x \to \infty} \frac{4x}{\sqrt{2x^2+1}}$

$$= \lim_{x \to \infty} \frac{x \cdot 4}{x\sqrt{2+\frac{1}{x^2}}}$$

$$= \lim_{x \to \infty} \frac{4}{\sqrt{2+\frac{1}{x^2}}}$$

$$= \frac{\lim_{x\to\infty}(4)}{\lim_{x\to\infty}\sqrt{2+\frac{1}{x^2}}}$$

$$= \frac{4}{\sqrt{\lim_{x\to\infty}\left(2+\frac{1}{x^2}\right)}}$$

$$= \frac{4}{\sqrt{2+0}}$$

$$= \frac{4}{\sqrt{2}}$$

$$= 2\sqrt{2}$$

Problems on irrational functions when $x \to \infty$

Exercise 4.8

Find the following limits: ***Answers***

1. $\lim_{x\to\infty} \left(\sqrt{1+x} - \sqrt{1-x}\right)$ (∞)

2. $\lim_{x\to\infty} \left(\sqrt{x^2+8x-7} - \sqrt{x^2+2x+5}\right)$ (3)

3. $\lim_{x\to\infty} \frac{x}{\sqrt{1+x^2}-1}$ Find

4. $\lim_{x\to\infty} \frac{\sqrt{x^2+4}-\sqrt{x^2+1}}{\sqrt{x^2+16}-\sqrt{x^2+9}}$ $\left(\frac{3}{7}\right)$

5. $\lim_{x\to\infty} \sqrt{x}\left(\sqrt{x+c}-\sqrt{x}\right)$ (P.U. 66) $\left(\frac{c}{2}\right)$

6. $\lim_{x\to\infty} \left(\sqrt{x^2+4x}-\sqrt{x^2-4x}\right)$ (L.N. 86) (4)

7. $\lim_{x\to\infty} \sqrt{x}\left(\sqrt{x^3+4x}-\sqrt{x^3-4x}\right)$ (M.U. 86) (4)

8. $\lim_{x\to\infty} \left(x-\sqrt{x^2+x}\right)$ (I.I.T 75) $\left(-\frac{1}{2}\right)$

Problems based on summation of series $\left(\frac{\infty}{\infty}\text{ form}\right)$

Working rule:

1. Use the formulas for the sum of n-natural number, square of n-natural numbers or cube of n-natural numbers for which the following formulas are very fruitful.

(i) $1 + 2 + 3 + \ldots + n$ = sum of n-natural numbers

$= \frac{n}{2} \cdot (n + 1)$

(ii) $1^2 + 2^3 + 3^2 + \ldots + n^2$ = sum of square of n-natural numbers $= \frac{n(n+1)(2n+1)}{6}$.

(iii) $1^3 + 2^3 + 3^3 + \ldots + n^3$ = sum of cube of n-natural numbers $= \left[\frac{n(n+1)}{2}\right]^2$.

(iv) $\sum ar^{n-1} = a + ar + ar^2 + \ldots + ar^{n-1} = \frac{a(1-r^n)}{(1-r)}$

Refresh your memory: Method of finding the limit of $f(n)$ as $n \to \infty$.

We divide the numerator and denominator by the highest power of n-occurring in $f(n)$ an then use the idea that $\frac{1}{n}, \frac{1}{n^2}, \ldots$ all $\to 0$ as $n \to \infty$

Examples worked out:

Evaluate

1. $\lim\limits_{n\to\infty} \left(\frac{1^2}{n^3} + \frac{2^2}{n^3} + \frac{3^2}{n^3} + \ldots + \frac{n^2}{n^3}\right)$

Solution: $\lim\limits_{n\to\infty} \left(\frac{1^2}{n^3} + \frac{2^2}{n^3} + \frac{3^2}{n^3} + \ldots + \frac{n^2}{n^3}\right)$

$= \lim\limits_{n\to\infty} \left(\frac{1^2 + 2^2 + 3^2 + \ldots + n^2}{n^3}\right)$

$= \lim\limits_{n\to\infty} \left(\frac{n(n+1)(2n+1)}{6n^3}\right)$

$= \lim\limits_{n\to\infty} \frac{n\left(1 + \frac{1}{n}\right)\left(2 + \frac{1}{n}\right) \cdot n}{6n^2}$

$= \lim\limits_{n\to\infty} \frac{\left(1 + \frac{1}{n}\right)\left(2 + \frac{1}{n}\right)}{6} = \frac{2}{6} = \frac{1}{3}$

2. $\lim\limits_{n\to\infty} \frac{n}{1 + 2 + 3 + \ldots + n}$

Solution: $\lim\limits_{n\to\infty} \frac{n}{1 + 2 + 3 + \ldots + n}$

$= \lim\limits_{n\to\infty} \frac{n}{\frac{n}{2}(n+1)}$

$= \lim\limits_{n\to\infty} \frac{2}{(n+1)} = \lim\limits_{n\to\infty} \frac{\frac{2}{n}}{\left(1 + \frac{1}{n}\right)} = \frac{0}{1+0} = 0$

3. Show that $\lim\limits_{n\to\infty} \left[\frac{\sum\limits_{n=1}^{n} n^3}{n^4}\right] = \frac{1}{4}$

Solution: We know that $\sum\limits_{n=1}^{n} n^3 = \frac{n^2(n+1)^2}{4}$ ... (i)

$\therefore \frac{\sum\limits_{n=1}^{n} n^3}{n^4} = \frac{\frac{n^2(n+1)^2}{4}}{n^4}$

$\Rightarrow \lim\limits_{n\to\infty} \left[\frac{\sum\limits_{n=1}^{n} n^3}{n^4}\right]$

$= \frac{\lim\limits_{n\to\infty} \left\{\frac{n^2(n+1)^2}{4}\right\}}{\lim\limits_{n\to} n^4}$

$$= \lim_{n \to \infty} \left\{ \frac{n^2 (n+1)^2}{4n^4} \right\}$$

$$= \lim_{n \to \infty} \left\{ \frac{1}{4} \left(1 + \frac{1}{n} \right)^2 \right\}$$

$$= \frac{1}{4} \lim_{n \to \infty} \left(1 + \frac{1}{n} \right)^2$$

$$= \frac{1}{4}.$$

Problems based on the limits of $f(n)$ as $n \to \infty$

Exercise 4.9

Find the limits of the following functions:

Answers

1. $\lim_{n \to \infty} \frac{\Sigma n^3}{n^4}$ $\left(\frac{1}{4}\right)$

2. $\lim_{n \to \infty} \frac{1 + 2 + 3 + \dots + n}{n^2}$ $\left(\frac{1}{2}\right)$

3. $\lim_{n \to \infty} \frac{n(n-1)(n-2)(n-3)}{(n-4)(n-5)(n-6)(n-7)}$ (1)

4. $\lim_{n \to \infty} \frac{1^3 + 2^3 + 3^3 + \dots + n^3}{n^4 + 1}$ $\left(\frac{1}{4}\right)$

5. $\lim_{n \to \infty} \frac{n^4 + 5n^2 + 7n - 3}{n^2 (n^2 + 2n - 7)}$ (1)

6. $\lim_{n \to \infty} \frac{n}{1 + 2 + 3 + \dots + n}$ (0)

7. $\lim_{n \to \infty} \frac{6n^5 + n^4 - 7n^3 + 5}{n^5 + 7n^3 + 6}$ (6)

Limits of trigonometric functions as $x \to a$

Form: $\frac{0}{0}$ or $\frac{\infty}{\infty}$ at $x = a$

Type 1: To find $\lim_{x \to a} f(x) = \lim_{x \to a} \frac{t_1(x)}{t_2(x)}$

Where $f(x)$ = a trigonometric function whose Numerator = $Nr = t_1(x)$ = a trigonometric function or trigonometric expression and denominator = $Dr = t_2(x)$ = a trigonometric function or trigonometric expression.

Moreover, x = the angle of trigonometric function which tends to a finite number.

Working rule: When numerator and denominator both are trigonometric functions or trigonometric expressions which can be expressed in terms of sin θ and cos θ and cancellation of common factor from *Nr* and *Dr* is possible, we adopt the following procedure.

1. Express all trigonometrical terms into sin θ and cos θ and cancel the common factor from numerator and denominator.

2. Put $x = a$ = given limit of the independent variable in the expression free from common factor (i.e., a factor which makes $f(a)$ meaningless) which gives us the required limit of the given function $f(x)$ as $x \to a$,

where $a = \frac{\pi}{2}, \frac{\pi}{6}, \frac{\pi}{4}, \frac{\pi}{3}, \pi$, or 1, etc., for example.

Facts to know:

1. We may face the circumstances where changing given function in terms of sin x and cos x to remove the common factor does not provide us a common factor which means further modification is required which is done by using formulas of trigonometrical ratios of submultiple angle. Thus firstly changing of *t*-ratios in terms of sin x and cos x and secondly using the formulas of *t*-ratios of submultiple angles, we are able to find out the common factor which is cancelled from numerator and denominator.

2. After cancellation of common factor from numerator and denominator, we get the determinate value of the simplified function at a given value $x = a$ = limit of the independent variable.

3. Modification of the given function is not stopped unless we get a determinate value of the simplified function which provides a finite value = required limit of the given function as $x \to a$.

4. Various mathematical manipulations can be adopted to remove the common factor which is not apparent directly in the given function.

Remember: When the methods discussed above fail to give the limit of a function, a general method of evaluation known as method of substitution or *h*-method or differential method is adopted. All questions solved by various methods can be solved with the help of this method too.

Method of substitution:

1. Put $x = a \pm h$ (when $x \to a$) where $h \to 0$ through +ve values.

2. Find $\lim_{h \to 0} f(a \pm h)$ when $y = f(x)$ = algebraic or trigonometric or mixed transcendental functions or any type of function.

N.B.: 1. We always put $x = a \pm h$ so that when $x \to a$, $h \to 0$ through +ve values [but for simplicity of calculation we put $x = a + h$ (or $x = a - h$)] because the limit of $f(x)$ is said to exist at $x = a$ if right hand limit (or, right limit) $\lim_{\substack{x \to a \\ x > a}} f(x)$ and left hand limit (or, left limit) $\lim_{\substack{x \to a \\ x < a}} f(x)$ exist and are equal, e.g.:

(i) Evaluate $\lim_{x \to 0} \frac{x+2}{x+1}$

Solution: Put $x = 0 \pm h$, then for $h > 0$, we have

$$\lim_{x \to 0} \frac{x+2}{x+1} = \frac{\lim_{h \to 0} (0 \pm h + 2)}{\lim_{h \to 0} (0 \pm h + 1)} = 2$$

(ii) Evaluate $\lim_{x \to \frac{\pi}{2}} (\sec x - \tan x)$

Solution: Put $x = \frac{\pi}{2} \pm h$, then for $h > 0$, we have

$$\lim_{x \to \frac{\pi}{2}} (\sec x - \tan x) = \lim_{x \to \frac{\pi}{2}} \left(\frac{1 - \sin x}{\cos x} \right)$$

$$= \lim_{h \to 0} \frac{1 - \sin\left(\frac{\pi}{2} \pm h\right)}{\cos\left(\frac{\pi}{2} \pm h\right)} = \lim_{h \to 0} \frac{1 - \cos h}{(\pm \sin h)}$$

$$= \pm \lim_{h \to 0} \frac{2 \sin^2 \frac{h}{2}}{2 \sin \frac{h}{2} \cdot \cos \frac{h}{2}} = \pm \lim_{h \to 0} \frac{\tan\left(\frac{h}{2}\right)}{\left(\frac{h}{2}\right)} \cdot \left(\frac{h}{2}\right)$$

$$= \pm 1 \times 0 = 0.$$

2. We never put $x = a$ while finding the limit of $f(x) = \frac{f_1(x)}{f_2(x)}$ as $x \to a$ but we always put the limit of the independent variable $x = a$ in the simplified form of the function $f(x) = \frac{f_1(x)}{f_2(x)}$ whose limit is required as $x \to a$, when $f_1(a) = f_2(a) = 0$.

Problems based on type 1

Examples worked out:

Evaluate:

1. $\lim_{x \to \frac{\pi}{4}} \frac{1 + \sin 2x}{1 - \cos 4x}$

Solution: $\left(\frac{1 + \sin 2x}{1 - \cos 4x} \right)$ for $x \Rightarrow \frac{\pi}{4} \neq \frac{0}{0}$ (i.e., given function does not assume meaningless form as $x \to \frac{\pi}{4}$)

Hence, $\lim_{x \to \frac{\pi}{4}} \left(\frac{1 + \sin 2x}{1 - \cos 4x} \right) = \frac{1 + \sin\left(\frac{\pi}{4} \cdot 2\right)}{1 - \cos\left(\frac{\pi}{4} \cdot 4\right)}$

$$= \frac{1 + \sin\left(\frac{\pi}{2}\right)}{1 - \cos \pi} = \frac{1 + 1}{1 - (-1)} = \frac{2}{2} = 1$$

2. $\lim\limits_{x \to \frac{\pi}{2}} \dfrac{\tan x + \cot x}{\tan x - \cot x}$

Solution: $\left(\dfrac{\tan x + \cot x}{\tan x - \cot x}\right)_{x\to\frac{\pi}{2}} = \dfrac{\infty}{\infty} \Rightarrow$ form is meaning-less

Now, $\dfrac{\tan x + \cot x}{\tan x - \cot x} = \dfrac{\dfrac{\sin x}{\cos x} + \dfrac{\cos x}{\sin x}}{\dfrac{\sin x}{\cos x} - \dfrac{\cos x}{\sin x}}$, $x \neq \dfrac{n\pi}{4}$; $n \in z$

$$= \frac{\dfrac{\sin^2 x + \cos^2 x}{\sin x \cos x}}{\dfrac{\sin^2 x - \cos^2 x}{\sin x \cos x}}$$

$$= \frac{\left(\sin^2 x + \cos^2 x\right) \times (\sin x \cdot \cos x)}{(\sin x \cdot \cos x) \times \left(\sin^2 x - \cos^2 x\right)}$$

$$= \frac{1}{\left(\sin^2 x - \cos^2 x\right)} = -\frac{1}{\cos 2x}$$

Thus, we get, $\dfrac{\tan x + \cot x}{\tan x - \cot x} = -\dfrac{1}{\cos 2x}$ for $x \neq \dfrac{n\pi}{4}$ ... (i)

Now, on taking the limits on both sides of (1) as $x \to \dfrac{\pi}{2}$, we get

$\lim\limits_{x \to \frac{\pi}{2}}$ (given function) = $\lim\limits_{x \to \frac{\pi}{2}}$ (simplified form of the given function)

$$\Rightarrow \lim_{x \to \frac{\pi}{2}} \left(\frac{\tan x + \cot x}{\tan x - \cot x}\right)$$

$$= \lim_{x \to \frac{\pi}{2}} \left(-\frac{1}{\cos 2x}\right) = -\frac{1}{\cos \pi} = \frac{-1}{-1} = 1$$

3. $\lim\limits_{x \to \frac{\pi}{2}} \left(\dfrac{1 - \sin x}{\cos 2x}\right) \tan x$

Solution: $\left[\dfrac{(1 - \sin x)}{\cos 2x} \cdot \tan x\right]_{x \to \frac{\pi}{2}} = 0 \times \infty \Rightarrow$ form is meaningless.

Now, $\left(\dfrac{1 - \sin x}{\cos 2x}\right) \cdot \tan x = \left(\dfrac{1 - \sin x}{\cos 2x}\right) \cdot \dfrac{\sin x}{\cos x}$

$$= \frac{\left(\sin^2 \frac{x}{2} + \cos^2 \frac{x}{2} - 2\sin\frac{x}{2} \cdot \cos\frac{x}{2}\right) \sin x}{\left(\cos\frac{x}{2} + \sin\frac{x}{2}\right)\left(\cos\frac{x}{2} - \sin\frac{x}{2}\right) \cos 2x}$$

$$= \frac{\left(\cos\frac{x}{2} - \sin\frac{x}{2}\right)^2 \cdot \sin x}{\left(\cos\frac{x}{2} + \sin\frac{x}{2}\right)\left(\cos\frac{x}{2} - \sin\frac{x}{2}\right) \cdot \cos 2x}$$

$$= \frac{\left(\cos\frac{x}{2} - \sin\frac{x}{2}\right) \cdot \sin x}{\left(\cos\frac{x}{2} + \sin\frac{x}{2}\right) \cdot \cos 2x} \text{ for } x \neq \frac{n\pi}{2} + \frac{\pi}{4}$$

or $n\pi + \dfrac{\pi}{2}$; $n \in z$

Thus, we get,

$$\left(\frac{1 - \sin x}{\cos 2x}\right) \cdot \tan x = \frac{\left(\cos\frac{x}{2} - \sin\frac{x}{2}\right) \cdot \sin x}{\left(\cos\frac{x}{2} + \sin\frac{x}{2}\right) \cos 2x}$$

which does not give us meaningless form at $x = \dfrac{\pi}{2} \Rightarrow$ This is required simplified form ... (i)

Now, on taking the limits on both sides of (1) as $x \to \dfrac{\pi}{2}$, we get $\lim\limits_{x \to \frac{\pi}{2}}$ (given function) = $\lim\limits_{x \to \frac{\pi}{2}}$ (simplified form of the given function)

$$\Rightarrow \lim_{x \to \frac{\pi}{2}} \left(\frac{1 - \sin x}{\cos 2x}\right) \cdot \tan x$$

$$= \lim_{x \to \frac{\pi}{2}} \frac{\left(\cos\frac{x}{2} - \sin\frac{x}{2}\right) \cdot \sin x}{\left(\cos\frac{x}{2} - \sin\frac{x}{2}\right) \cdot \cos 2x}$$

$$= \frac{\left(\cos\frac{\pi}{2} - \sin\frac{\pi}{4}\right) \cdot \sin\frac{\pi}{2}}{\cos\pi \cdot \left(\cos\frac{\pi}{4} + \sin\frac{\pi}{4}\right)}$$

$$\Rightarrow \lim_{x \to \frac{\pi}{2}} \left(\frac{1 - \sin x}{\cos 2x}\right) \cdot \tan x$$

$$= \frac{\left(\frac{1}{\sqrt{2}} - \frac{1}{\sqrt{2}}\right) \cdot 1}{(-1)\left(\frac{1}{\sqrt{2}} + \frac{1}{\sqrt{2}}\right)} = 0$$

4. $\lim_{x \to \frac{\pi}{2}} \left(\frac{\operatorname{cosec} x - 1}{\cot^2 x}\right)$

Solution: $\left(\frac{\operatorname{cosec} x - 1}{\cot^2 x}\right)_{x \to \frac{\pi}{2}} = \frac{0}{0}$ form

Now, $\frac{\operatorname{cosec} x - 1}{\cot^2 x} = \frac{\operatorname{cosec} x - 1}{\operatorname{cosec}^2 x - 1}$

$$= \frac{(\operatorname{cosec} x - 1)}{(\operatorname{cosec} x - 1)(\operatorname{cosec} x + 1)}$$

$$= \frac{1}{(\operatorname{cosec} x + 1)} \text{ for } x \neq \frac{n\pi}{2}; n \in z$$

which does not give us a meaningless form at $\frac{\pi}{2} \Rightarrow$ this is the required simplified form ...(i)

Now, taking the limits on both sides of (i) as $x \to \frac{\pi}{2}$, we get $\lim_{x \to \frac{\pi}{2}}$ (given function) $= \lim_{x \to \frac{\pi}{2}}$ (simplified form of the given function).

$$\Rightarrow \lim_{x \to \frac{\pi}{2}} \left(\frac{\operatorname{cosec} x - 1}{\cot^2 x}\right) = \lim_{x \to \frac{\pi}{2}} \left(\frac{1}{\operatorname{cosec} x + 1}\right)$$

$$= \frac{1}{\operatorname{cosec}\frac{\pi}{2} + 1} = \frac{1}{1 + 1} = \frac{1}{2}$$

5. $\lim_{x \to \frac{\pi}{2}} \left(\frac{1 - \sin^3 x}{1 - \sin^2 x}\right)$

Solution: $\left(\frac{1 - \sin^3 x}{1 - \sin^2 x}\right)_{x \to \frac{\pi}{2}} = \frac{0}{0}$ form

Now, $\frac{1 - \sin^3 x}{1 - \sin^2 x} = \frac{(1 - \sin x)(1 + \sin x + \sin^2 x)}{(1 - \sin x)(1 + \sin x)}$

$$= \frac{1 + \sin x + \sin^2 x}{1 + \sin x} \text{ for } x \neq (2n+1)\frac{\pi}{2}, n \in z$$

Now, on taking the limits on both sides of (i), we get

$\lim_{x \to \frac{\pi}{2}}$ (given function) $= \lim_{x \to \frac{\pi}{2}}$ (simplified form of the given function)

$$\Rightarrow \lim_{x \to \frac{\pi}{2}} \left(\frac{1 - \sin^3 x}{1 - \sin^2 x}\right)$$

$$= \lim_{x \to \frac{\pi}{2}} \left(\frac{1 + \sin x + \sin^2 x}{1 + \sin x}\right)$$

$$= \left(\frac{1 + \sin\frac{\pi}{2} + \left(\sin\frac{\pi}{2}\right)^2}{1 + \sin\frac{\pi}{2}}\right)$$

$$= \frac{1+1+1}{1+1} = \frac{3}{2}$$

6. $\lim_{\theta \to \frac{\pi}{2}} \left(\frac{\sin 2\theta}{\cos\theta} \right)$

Solution: $\left(\frac{\sin 2\theta}{\cos\theta} \right)_{\theta \to \frac{\pi}{2}} = \frac{0}{0}$ form

Now, $\frac{\sin 2\theta}{\cos\theta} = \frac{2\sin\theta \cdot \cos\theta}{\cos\theta} = 2\sin\theta$ for

$\theta \neq n\pi + \frac{\pi}{2}$ ...(i)

Now, on taking the limits on both sides of (i), we get

$\lim_{\theta \to \frac{\pi}{2}}$ (given function) = $\lim_{\theta \to \frac{\pi}{2}}$ (simplified form of the given function)

$$\Rightarrow \lim_{\theta \to \frac{\pi}{2}} \left(\frac{\sin 2\theta}{\cos\theta} \right) = \lim_{\theta \to \frac{\pi}{2}} (2\sin\theta)$$

$$= 2\sin\frac{\pi}{2} = 2.1 = 2$$

7. $\lim_{x \to \pi} \left(\frac{1+\cos x}{\tan^2 x} \right)$

Solution: $\left(\frac{1+\cos x}{\tan^2 x} \right)_{x \to \pi} = \frac{0}{0}$

Now, $\frac{1+\cos x}{\tan^2 x} = \frac{1+\cos x}{\sec^2 x - 1} = \frac{(1+\cos x)}{\left(\frac{1}{\cos^2 x} - 1 \right)}$;

$x \neq \frac{n\pi}{2}$

$$= \frac{(1+\cos x)}{\frac{(1-\cos^2 x)}{\cos^2 x}} = \frac{(1+\cos x)\cos^2 x}{(1-\cos^2 x)}$$

$$= \frac{(1+\cos x)\cos^2 x}{(1-\cos x)(1+\cos x)} = \frac{\cos^2 x}{1-\cos x}$$

Thus, we get $\frac{1+\cos x}{\tan^2 x} = \frac{\cos^2 x}{1-\cos x}$ for

$x \neq \frac{n\pi}{2}, n \in z$ ...(i)

Now on taking the limits on both sides of (i) as $x \to \pi$, we get

$\lim_{x \to \pi}$ (given function) = $\lim_{x \to \pi}$ (simplified form of the given function)

i.e., $\lim_{x \to \pi} \left(\frac{1+\cos x}{\tan^2 x} \right) = \lim_{x \to \pi} \left(\frac{\cos^2 x}{1-\cos x} \right)$

$$= \frac{\cos^2 \pi}{1-\cos\pi} = \frac{(-1)^2}{1-(-1)} = \frac{1}{2}$$

8. $\lim_{x \to \frac{\pi}{4}} \left(\frac{\sin x - \cos x}{\cos 2x} \right)$

Solution: $\left(\frac{\sin x - \cos x}{\cos 2x} \right)_{x \to \frac{\pi}{4}} = \frac{0}{0}$

Now, $\frac{\sin x - \cos x}{\cos 2x} = \frac{\sin x - \cos x}{\cos^2 x - \sin^2 x}$; $x \neq \frac{n\pi}{2} + \frac{\pi}{4}$

$$= \frac{(\sin x - \cos x)}{(\sin x - \cos x)(\sin x + \cos x)} = \frac{-1}{(\sin x + \cos x)}$$

Thus, we get, $\frac{\sin x - \cos x}{\cos 2x} = \frac{-1}{(\sin x + \cos x)}$ ...(i)

Now, on taking the limits on both sides of (i) as $x \to \frac{\pi}{4}$, we get

$\lim_{x \to \frac{\pi}{4}}$ (given function) = $\lim_{x \to \frac{\pi}{4}}$ (simplified form of the given function)

i.e., $\lim_{x \to \frac{\pi}{4}} \left(\frac{\sin x - \cos x}{\cos 2x}\right) = \lim_{x \to \frac{\pi}{4}} \left(\frac{-1}{\sin x + \cos x}\right)$

$$= \frac{-1}{\sin\frac{\pi}{4} + \cos\frac{\pi}{4}} = \frac{-1}{\frac{1}{\sqrt{2}} + \frac{1}{\sqrt{2}}}$$

$$= \frac{-1}{\left(\frac{2}{\sqrt{2}}\right)} = -\frac{\sqrt{2}}{2} = -\frac{1}{\sqrt{2}}$$

9. $\lim_{x \to \frac{\pi}{4}} \left[\frac{\sin 2x - (1 + \cos 2x)}{\cos x - \sin x}\right]$

Solution: $\left[\frac{\sin 2x - (1 + \cos 2x)}{\cos x - \sin x}\right]_{x \to \frac{\pi}{4}} = \frac{0}{0}$

Now, $\frac{\sin 2x - (1 + \cos 2x)}{\cos x - \sin x}$

$$= \frac{2 \sin x \cos x - 2\cos^2 x}{\cos x - \sin x};\ x \neq \frac{n\pi}{4},\ n \in z$$

$$= \frac{-2\cos x\,(\cos x - \sin x)}{(\cos x - \sin x)} = -2\cos x$$

Thus, we get, $\frac{\sin 2x - (1 + \cos 2x)}{\cos x - \sin x} = -2\cos x$ … (i)

Now, on taking the limits on both sides of (i) as $x \to \frac{\pi}{4}$, we get

$\lim_{x \to \frac{\pi}{4}}$ (given function) $= \lim_{x \to \frac{\pi}{4}}$ (simplified form of the given function)

i.e., $\lim_{x \to \frac{\pi}{4}} \left[\frac{\sin 2x - (1 + \cos 2x)}{\cos x - \sin x}\right]$

$$= \lim_{x \to \frac{\pi}{4}} (-2\cos x) = -2\cos\frac{\pi}{4}$$

$$= -2 \times \frac{1}{\sqrt{2}} = -\sqrt{2}$$

10. $\lim_{x \to \frac{\pi}{2}} (\sec x - \tan x)$

Solution: $(\sec x - \tan x)$

$$= \left(\frac{1}{\cos x} - \frac{\sin x}{\cos x}\right) = \left(\frac{1 - \sin x}{\cos x}\right);\ x \neq n\pi + \frac{\pi}{2}$$

$$\therefore (\sec x - \tan x)_{x = \frac{\pi}{2}} = \left(\frac{1 - \sin x}{\cos x}\right)_{x = \frac{\pi}{2}} = \frac{0}{0}$$

Now, $\frac{1 - \sin x}{\cos x} = \frac{\sin^2\frac{x}{2} + \cos^2\frac{x}{2} - 2\sin\frac{x}{2}\cos\frac{x}{2}}{\cos^2\frac{x}{2} - \sin^2\frac{x}{2}}$

$$= \frac{\left(\cos\frac{x}{2} - \sin\frac{x}{2}\right)^2}{\left(\cos^2\frac{x}{2} - \sin^2\frac{x}{2}\right)}$$

$$= \frac{\left(\cos\frac{x}{2} - \sin\frac{x}{2}\right)\left(\cos\frac{x}{2} - \sin\frac{x}{2}\right)}{\left(\cos\frac{x}{2} - \sin\frac{x}{2}\right)\left(\cos\frac{x}{2} + \sin\frac{x}{2}\right)}$$

$$= \frac{\left(\cos\frac{x}{2} - \sin\frac{x}{2}\right)}{\left(\cos\frac{x}{2} + \sin\frac{x}{2}\right)}$$

Thus, we get, $(\sec x - \tan x) = \frac{1 - \sin x}{\cos x}$

$$= \frac{\cos\frac{x}{2} - \sin\frac{x}{2}}{\cos\frac{x}{2} + \sin\frac{x}{2}} \quad \ldots \text{(i)}$$

Now, on taking the limits on both sides of (i) as $x \to \frac{\pi}{2}$, we get

$\lim_{x \to \frac{\pi}{2}}$ (given function) = $\lim_{x \to \frac{\pi}{2}}$ (simplified form of the given function), i.e. $\lim_{x \to \frac{\pi}{2}} (\sec x - \tan x)$

$$= \lim_{x \to \frac{\pi}{2}} \left(\frac{\cos\frac{x}{2} - \sin\frac{x}{2}}{\cos\frac{x}{2} + \sin\frac{x}{2}} \right) = \frac{\cos\frac{\pi}{4} - \sin\frac{\pi}{4}}{\cos\frac{\pi}{4} + \sin\frac{\pi}{4}}$$

$$= \frac{\frac{1}{\sqrt{2}} - \frac{1}{\sqrt{2}}}{\frac{1}{\sqrt{2}} + \frac{1}{\sqrt{2}}} = \frac{0}{\left(\frac{2}{\sqrt{2}}\right)} = 0$$

Notes: We observe from the just above solution that

(i) Firstly sec x and tan x are changed into sin x and cos x to remove the common factor but on changing sec x and tan x in terms of sin x and cos x, no common factor is cancelled which means further modification is required.

(ii) Secondly, using formulas of *t*-ratios of submultiple angles, we are able to find out the common factor and after cancellation of common factor, we get a finite value '0' for the simplified form (of the given function) at the given value $x = \frac{\pi}{2}$. (i.e. at the limit of the independent variable x).

Problems based on type 1

Exercise 4.10

Evaluate ***Answers***

1. $\lim_{x \to \frac{\pi}{4}} \left(\frac{\sin^2 x - \cos^2 x}{\sin x - \cos x} \right)$ $(\sqrt{2})$

2. $\lim_{x \to \frac{\pi}{4}} \left(\frac{\tan^2 x - \cot^2 x}{\sec x - \operatorname{cosec} x} \right)$ $(2\sqrt{2})$

3. $\lim_{x \to \frac{\pi}{4}} \left(\frac{2 - \operatorname{cosec}^2 x}{1 - \cot x} \right)$ (2)

4. $\lim_{x \to \frac{\pi}{4}} \left(\frac{2\cos^2 x - \sin 2x}{\cos 2x} \right)$ (1)

5. $\lim_{x \to \frac{\pi}{4}} \left(\frac{\sin x - \cos x}{\tan x - \cot x} \right)$ $\left(\frac{1}{2\sqrt{2}}\right)$

6. $\lim_{\theta \to \frac{\pi}{2}} \frac{(1 - \sin\theta)}{\cos^2\theta}$ $\left(\frac{1}{2}\right)$

7. $\lim_{x \to \frac{\pi}{2}} \frac{(\cot x - \cos x)}{\cos^3 x}$ $\left(\frac{1}{2}\right)$

8. $\lim_{x \to \frac{\pi}{2}} \frac{1 - \sin^3 x}{\cos^2 x}$ $\left(\frac{3}{2}\right)$

9. $\lim_{x \to \frac{\pi}{2}} \frac{\cos^2 x}{1 - \sin^2 x}$ (1)

10. $\lim_{x \to \frac{\pi}{2}} \frac{\operatorname{cosec} x - 1}{\cot^2 x}$ $\left(\frac{1}{2}\right)$

11. $\lim_{x \to \pi} \frac{1 = \cos^3 x}{1 + \cos x}$ (3)

12. $\lim_{x \to \frac{\pi}{4}} \frac{2 - \sec^2 x}{1 - \tan x}$ (2)

13. $\lim_{x \to \pi} \frac{1 + \cos^3 x}{\sin^2 x}$ $\left(\frac{3}{2}\right)$

14. $\lim_{x \to \frac{\pi}{3}} \left(\frac{3\sin^2 x - \cos^2 x - 2}{1 - 2\cos x} \right)$ (2)

15. $\lim_{x \to 1} \frac{1 + \cos\pi x}{\tan^2 \pi x}$ $\left(\frac{1}{2}\right)$

16. $\lim_{x \to \alpha} \left(\frac{\cos x - \cos\alpha}{\cot x - \cot\alpha} \right)$ $(\sin^3\alpha)$

Type 2: To find $\lim_{x \to a} f(x)$ where $f(x)$ = a trigonometrical function (or expression) mixed with an algebraic function in any way (generally algebraic function appears as addend, subtrahend, minuend, multiplicand or divisor of trigonometric function or expression) i.e.; to find

(i) $\lim_{x \to a} f(x) = \lim_{x \to a} \frac{a_1(x) \pm t_1(x)}{a_2(x) \pm t_2(x)}$

(ii) $\lim_{x \to a} f(x) = \lim_{x \to a} \frac{a_1(x) \cdot t_1(x)}{t_2(x)}$

or $\lim_{x \to a} \frac{t_1(x)}{a_1(x) \cdot t_2(x)}$

(iii) $\lim_{x \to a} f(x) = \lim_{x \to a} \frac{t_1(x)}{a_1(x)}$

or $\lim_{x \to a} [a_1(x) \cdot t_1(x)]$

(iv) $\lim_{x \to a} f(x) = \lim_{x \to a} \frac{a_1(x)}{t_1(x)}$

Where $a_1(x)$ and $a_2(x)$ = algebraic functions (or, expression)

$t_1(x)$ and $t_2(x)$ = trigonometric functions (or, expression)

a = a finite value

$= \frac{\pi}{2}, \frac{\pi}{6}, \frac{\pi}{4}, \frac{\pi}{3}, \pi$ or 1 for instance

We adopt the following working rule:

Working rule:

1. Put $x = a + h$ (or, $a - h$) (where $h \to 0$) in the given function.

2. Modify the given function obtained after substitution $x = a + h$ or $x = a - h$ and remove the common factor if it occurs.

3. Take the limit of the modified form which is a function of h as $h \to 0$ since $\lim_{x \to a} f(x) = \lim_{h \to 0} f(a+h)$ (where $h \to 0$).

Question: When to use method of change of limit, method of substitution, h-method or differential method?

Answer: Method of change of limit is used when

1. The given trigonometric function can not be simplified easily.

2. The factorization of the given trigonometric function is not possible or difficult.

3. A trigonometric function is provided which does not contain a common factor.

Facts to Know:

1. If there exists a factor of t-ratios of angle θ as $\sin\theta$, $\cos\theta$, $\tan\theta$ or $\cot\theta$, etc, we are required to write it as

$$\sin\theta = \frac{\sin\theta}{\theta} \cdot \theta$$

$\tan\theta = \frac{\tan\theta}{\theta} \cdot \theta$ so that standard results of limits of t-ratios may be used.

2. Standard results of limits of t-ratios are following

(i) $\lim_{x \to 0} \frac{\sin mx}{x} = m$

(ii) $\lim_{x \to 0} \frac{\tan mx}{x} = m$

(iii) $\lim_{x \to 0} \cos mx = 1$

(iv) $\lim_{x \to 0} \frac{\sin x}{x} = 1$

(v) $1 = \lim_{x \to 0} \frac{\tan x}{x}$

(vi) $\lim_{x \to 0} \cos x = 1$

Remember:

1. $\theta \to 0 \Rightarrow 2\theta \to 0 \Rightarrow \frac{\theta}{2} \to 0$

In general, $\theta \to 0 \Rightarrow m\theta \to 0 \Rightarrow \frac{\theta}{m} \to 0$ where m = any integer

2. Modification of the function obtained after substitution $x = a + h$ in the given function is done by simplification using various trigonometrical formulas and mathematical manipulations so that standard formulas of limits of trigonometrical function mentioned above may be applied.

3. $f(a+h)$ = a function obtained by putting the independent variable $= x = a + h$ in the given function where $h \to 0$.

4. The method of substitution is sometimes termed as substitution and modification method since firstly we substitute $x = a \pm h$ in the given function and then we modify the function containing $(a \pm h)$.

Problems based on type 2

***Examples worked out*:**

Evaluate

1. $\lim_{x \to \frac{\pi}{2}} (\sec x - \tan x)$

***Solution*:** $\lim_{x \to \frac{\pi}{2}} (\sec x - \tan x)$

$$= \lim_{x \to \frac{\pi}{2}} \left(\frac{1}{\cos x} - \frac{\sin x}{\cos x} \right)$$

$$= \lim_{x \to \frac{\pi}{2}} \left(\frac{1 - \sin x}{\cos x} \right)$$

$$= \lim_{h \to 0} \frac{1 - \sin\left(\frac{\pi}{2} + h\right)}{\cos\left(\frac{\pi}{2} + h\right)}$$

$$= \lim_{h \to 0} \frac{1 - \cos h}{(-\sin h)}$$

$$= - \lim_{h \to 0} \frac{2\sin^2 \frac{h}{2}}{2\sin\frac{h}{2} \cdot \cos\frac{h}{2}}$$

$$= - \lim_{h \to 0} \frac{\tan\left(\frac{h}{2}\right)}{\left(\frac{h}{2}\right)} \cdot \left(\frac{h}{2}\right)$$

$$= 1 \times 0 = 0$$

2. $\lim_{x \to a} \sin\left(\frac{x - a}{2}\right) \cdot \tan\left(\frac{\pi x}{2a}\right)$

***Solution*:** Putting $x = a + h$ where $h \to 0$ as $x \to a$

$\therefore \lim_{x \to a}$ (given function)

$$= \lim_{h \to 0} \left[\sin\frac{h}{2} \cdot \tan\left(\frac{\pi}{2} + \frac{\pi h}{2a}\right) \right]$$

$$= \lim_{h \to 0} \left[\sin\frac{h}{2} \cdot \left(-\cot\frac{\pi h}{2a}\right) \right]$$

$$= \lim_{h \to 0} \left[\frac{-\left(\sin\frac{h}{2}\right)}{\dfrac{\tan\left(\frac{\pi h}{2a}\right) \cdot \left(\frac{\pi h}{2a}\right)}{\left(\frac{\pi h}{2a}\right)}} \right]$$

$$= \frac{- \lim_{h \to 0} \dfrac{\sin\left(\frac{h}{2}\right)}{\left(\frac{h}{2}\right)}}{\lim_{h \to 0} \dfrac{\tan\left(\frac{\pi h}{2a}\right)}{\left(\frac{\pi h}{2a}\right)} \cdot \left(\frac{\pi}{a}\right)}$$

$$= \frac{-1}{\left(\frac{\pi}{a}\right)} = \frac{-1}{1} \times \frac{a}{\pi} = -\frac{a}{\pi}$$

3. $\lim_{x \to \frac{\pi}{2}} \frac{\cos x}{\left(\frac{\pi}{2} - x\right)}$

***Solution*:** Putting $x = \frac{\pi}{2} + h \Rightarrow h \to 0$ as $x \to \frac{\pi}{2}$

$$\therefore \lim_{x \to \frac{\pi}{2}} \frac{\cos x}{\left(\frac{\pi}{2} - x\right)} = \lim_{h \to 0} \frac{\cos\left(\frac{\pi}{2} + h\right)}{\frac{\pi}{2} - \left(\frac{\pi}{2} + h\right)}$$

$$= \lim_{h \to 0} \left(\frac{-\sin h}{-h} \right) = \lim_{h \to 0} \frac{\sin h}{h} = 1$$

N.B.: Here we observe that Nr = a trigonometric function while Dr = an algebraic function. Hence, they can not have any factor in common. This is why we must make use of method of substitution.

4. $\lim_{x \to \frac{\pi}{2}} \left(\frac{2x - \pi}{\cos x} \right)$

Solution: Putting $x = \frac{\pi}{2} + h \Rightarrow 2x - \pi = 2h$

$$\therefore \lim_{x \to \frac{\pi}{2}} \left(\frac{2x - \pi}{\cos x} \right) = \lim_{h \to 0} \left(\frac{2h}{\cos\left(\frac{\pi}{2} + h \right)} \right)$$

$$= -2 \lim_{h \to 0} \left(\frac{h}{\sin h} \right)$$

$$= -2 \times 1 = -2.$$

N.B.: Here we observe that Dr = a trigonometric function while Nr = an algebraic function. Hence, they can not have any factor in common. This is why we must make use of method of substitution.

5. $\lim_{x \to \frac{\pi}{2}} \frac{(1 - \sin x)}{\left(\frac{\pi}{2} - x \right)^2}$

Solution: Putting $x = \frac{\pi}{2} + h \Rightarrow h \to 0$ as $x \to \frac{\pi}{2}$

$$\therefore \lim_{x \to \frac{\pi}{2}} \frac{(1 - \sin x)}{\left(\frac{\pi}{2} - x \right)^2}$$

$$= \lim_{h \to 0} \left[\frac{1 - \sin\left(\frac{\pi}{2} + h \right)}{(-h)^2} \right] = \lim_{h \to 0} \left(\frac{1 - \cos h}{h^2} \right)$$

$$= \lim_{h \to 0} \left(\frac{1 - 1 + 2\sin^2 \frac{h}{2}}{h^2} \right) = \lim_{h \to 0} \frac{2\sin^2\left(\frac{h}{2} \right)}{h^2}$$

$$= 2 \lim_{h \to 0} \frac{\left(\sin \frac{h}{2} \right)^2}{4 \cdot \left(\frac{h}{2} \right)^2}$$

$$= \frac{2}{4} \lim_{h \to 0} \left[\frac{(\sin h/2)}{h/2} \right]^2$$

$$= \frac{1}{2} \left[\lim_{h \to 0} \frac{\sin \frac{h}{2}}{\frac{h}{2}} \right]^2 = \frac{1}{2} \cdot (1)^2 = \frac{1}{2}$$

6. $\lim_{x \to \frac{\pi}{2}} \left[\frac{3\cos x + \cos 3x}{(2\pi - x)^2} \right]$

Solution: Putting $x = \frac{\pi}{2} + h \Rightarrow h \to 0$ as $x \to \frac{\pi}{2}$

$$\therefore \lim_{x \to \frac{\pi}{2}} \left[\frac{3\cos x + \cos 3x}{(2\pi - x)^2} \right]$$

$$= \lim_{x \to \frac{\pi}{2}} \left[\frac{3\cos x + 4\cos^3 x - 3\cos x}{(2\pi - x)^2} \right]$$

$$= \lim_{x \to \frac{\pi}{2}} \frac{4\cos^3 x}{(2x - \pi)^2}$$

$$= \lim_{h \to 0} \frac{4\cos^3\left(\frac{\pi}{2} + h \right)}{\left[2\left(\frac{\pi}{2} + h \right) - \pi \right]^2}$$

$$= 4 \lim_{h \to 0} \frac{-\sin^3 h}{4h^3} \cdot h$$

$$= -\left[\lim_{h\to 0}\left(\frac{\sin h}{h}\right)\right]^3 \cdot \lim_{h\to 0} h$$

$$= -1 \times 0 = 0$$

7. $\lim\limits_{x\to\frac{\pi}{4}} \dfrac{\sqrt{2}-\cos x - \sin x}{(4x-\pi)^2}$

Solution: Putting $x = \dfrac{\pi}{4} + h$ where $h \to 0$ as $x \to \dfrac{\pi}{4}$

$\therefore \lim\limits_{x\to\frac{\pi}{4}}$ (given function)

$$= \lim_{h\to 0}\frac{\sqrt{2}-\left[\cos\left(\frac{\pi}{4}+h\right)+\sin\left(\frac{\pi}{4}+h\right)\right]}{\left[4\left(\frac{\pi}{4}+h\right)-\pi\right]^2}$$

$$= \lim_{h\to 0}\frac{\sqrt{2}-\left[\cos\frac{\pi}{4}\cdot\cos h-\sin\frac{\pi}{4}\cdot\sin h+\sin\frac{\pi}{4}\cos h+\cos\frac{\pi}{4}\sin h\right]}{16h^2}$$

$$= \lim_{h\to 0}\frac{\sqrt{2}-\frac{1}{\sqrt{2}}(\cos h - \sin h + \cos h + \sin h)}{16h^2}$$

$$= \lim_{h\to 0}\frac{\sqrt{2}-\sqrt{2}\cos h}{16h^2} = \lim_{h\to 0}\frac{\sqrt{2}(1-\cos h)}{16h^2}$$

$$= \lim_{h\to 0}\frac{\sqrt{2}\times\left[2\sin^2\left(\frac{h}{2}\right)\right]}{16h^2}$$

$$= \lim_{h\to 0}\frac{\sqrt{2}}{8}\cdot\frac{\sin^2\left(\frac{h}{2}\right)}{\frac{h^2}{4}}\times\frac{1}{4}$$

$$= \frac{\sqrt{2}}{8}\times 1\times\frac{1}{4} = \frac{\sqrt{2}}{32}$$

8. $\lim\limits_{x\to\frac{\pi}{4}}\left(\dfrac{\sin x - \cos x}{x-\frac{\pi}{4}}\right)$

Solution: Putting $x = \dfrac{\pi}{4} + h$ where $h \to 0$ as $x \to \dfrac{\pi}{4}$

Now, given function $= \dfrac{\sin x - \cos x}{x - \frac{\pi}{4}}$

$$= \frac{\sin\left(\frac{\pi}{4}+h\right)-\cos\left(\frac{\pi}{4}+h\right)}{\frac{\pi}{4}+h-\frac{\pi}{4}}$$

$$= \frac{\sin\frac{\pi}{4}\cdot\cos h+\cos\frac{\pi}{4}\cdot\sin h-\left[\cos\frac{\pi}{4}\cdot\cos h-\sin\frac{\pi}{4}\sin h\right]}{h}$$

$$= \frac{\sqrt{2}\cos h+\sqrt{2}\sin h-\sqrt{2}\cos h+\sqrt{2}\sin h}{2h}$$

$= \dfrac{2\sqrt{2}\sin h}{2h}$ = simplified form of the given function.

Now, $\lim\limits_{x\to\frac{\pi}{4}}$ (given function $= \lim\limits_{h\to 0}$ (simplified form of the given function)

$$\Rightarrow \lim_{x\to\frac{\pi}{4}}\left(\frac{\sin x-\cos x}{x-\frac{\pi}{4}}\right) = \lim_{h\to 0}\frac{\sqrt{2}\sin h}{h}$$

$$= \sqrt{2}\lim_{h\to 0}\frac{\sin h}{h} = \sqrt{2}\cdot 1 = \sqrt{2}$$

9. $\lim\limits_{x\to 1}\left[(1-x)\tan\left(\dfrac{\pi x}{2}\right)\right]$

Solution: Method 1

Putting $x = 1 + h \Rightarrow h \to 0$ as $x \to 1$

$$\therefore \text{ given function } = (1 - x) \tan\left(\frac{\pi x}{2}\right)$$

$$= (-h) \tan\frac{\pi}{2}(1 + h)$$

$$(\because 1 - x = 1 - (1 + h) = 1 - 1 - h = -h)$$

$$= (-h)\left[-\cot\frac{\pi}{2}h\right] = h \cot\frac{\pi h}{2}$$

$$= \frac{h}{\tan\left(\frac{\pi h}{2}\right)}$$

$$= \frac{2}{\pi} \times \frac{\frac{\pi h}{2}}{\tan\left(\frac{\pi h}{2}\right)} \left[\because h = \frac{2}{\pi} \times \frac{\pi}{2} \times h\right]$$

$$\therefore \lim_{x \to 1}\left[(1 - x) \tan\left(\frac{\pi x}{2}\right)\right]$$

$$= \lim_{h \to 0}\left(\frac{2}{\pi}\right) \cdot \lim_{h \to 0} \frac{\left(\frac{\pi h}{2}\right)}{\tan\left(\frac{\pi h}{2}\right)} = \frac{2}{\pi} \times 1 = \frac{2}{\pi}$$

or, alternatively:

Putting $x = 1 - h$ in the given function, we have

$$(1 - x) \tan\left(\frac{\pi x}{2}\right) = [1 - (1 - h)] \cdot \left[\tan\frac{\pi}{2}(1 - h)\right]$$

$$= h \tan\left(\frac{\pi}{2} - \frac{\pi h}{2}\right)$$

$$= h \cot\left(\frac{\pi h}{2}\right) = \frac{h}{tan\left(\frac{\pi h}{2}\right)} = \left[\frac{\frac{2}{\pi} \times \frac{\pi}{2} \times h}{\tan\left(\frac{\pi h}{2}\right)}\right]$$

$$\therefore \lim_{h \to 1}\left[(1 - x) \tan\left(\frac{\pi x}{2}\right)\right]$$

$$= \lim_{h \to 0} \frac{2}{\pi} \times \lim_{h \to 0}\left(\frac{\frac{\pi}{2} \cdot h}{\tan\left(\frac{\pi h}{2}\right)}\right) = \frac{2}{\pi} \cdot 1 = \frac{2}{\pi}$$

N.B.: **1.** This example gives us light that we may put $x = a \pm h$ in the given function while adopting h-method, the result is same. But when $(a - x)$ appears in the question, we prefer to put $x = a - h$ for easiness.

2. The above function $= (1 - x) \tan\left(\frac{\pi x}{2}\right)$ whose limit is required can be done by expressing it in $\sin x$ and $\cos x$.

Method 2:

$$\lim_{x \to 1}\left[(1 - x) \tan\left(\frac{\pi x}{2}\right)\right]$$

$$= \lim_{x \to 1}\left[(1 - x) \cdot \frac{\sin\left(\frac{\pi x}{2}\right)}{\cos\left(\frac{\pi x}{2}\right)}\right]$$

We get,

$$\lim_{x \to 1}\left[(1 - x) \tan\left(\frac{\pi x}{2}\right)\right]$$

$$= \lim_{h \to 0} \frac{(-h) \sin\frac{\pi}{2}(1 + h)}{\cos\frac{\pi}{2}(1 + h)}$$

$$= \lim_{h \to 0} \frac{(-h) \cos\left(\frac{\pi h}{2}\right)}{(-1) \sin\left(\frac{\pi h}{2}\right)}$$

$$(\because \cos(90 + \theta) = -\sin\theta)$$

$$= \lim_{h \to 0} \frac{h \cos\left(\frac{\pi h}{2}\right)}{\sin\left(\frac{\pi h}{2}\right)}$$

$$= \lim_{h \to 0} \cos\left(\frac{\pi h}{2}\right) \cdot \lim_{h \to 0} \frac{h}{\sin\left(\frac{\pi h}{2}\right)}$$

$$= \lim_{h \to 0} \cos\left(\frac{\pi h}{2}\right) \cdot \lim_{h \to 0} \frac{\frac{2}{\pi} \times \left(\frac{\pi h}{2}\right)}{\sin\left(\frac{\pi h}{2}\right)}$$

$$= \lim_{h \to 0} \cos\left(\frac{\pi h}{2}\right) \cdot \frac{2}{\pi} \times \lim_{h \to 0} \frac{\left(\frac{\pi h}{2}\right)}{\sin\left(\frac{\pi h}{2}\right)}$$

$$= 1 \times \frac{2}{\pi} \times 1 = \frac{2}{\pi}$$

Problems based on type 2

Exercise 4.11

Evaluate **Answers**

1. $\lim_{x \to \pi} \frac{1 + \cos x}{1 - \sin x}$ (0)
2. $\lim_{x \to \pi} \frac{1 + \cos x}{\pi - x}$ (0)
3. $\lim_{x \to \pi} \frac{1 + \cos x}{(x - \pi)^2}$ $\left(\frac{1}{2}\right)$
4. $\lim_{x \to 1} \frac{\cos\left(\frac{\pi x}{2}\right)}{1 - \sqrt{x}}$ (π)
5. $\lim_{x \to \frac{\pi}{2}} \frac{\sec x - \tan x}{\frac{\pi}{2} - x}$ $\left(\frac{1}{2}\right)$
6. $\lim_{x \to \frac{\pi}{2}} \left(\frac{\pi}{2} - x\right) \tan x$ (1)
7. $\lim_{x \to \pi} \frac{\sin 3x - 3\sin x}{(x - \pi)^3}$ (–4)
8. $\lim_{x \to \frac{\pi}{2}} \frac{2x \sin x - \pi}{\cos x}$ (–2)
9. $\lim_{x \to \frac{\pi}{2}} \frac{1 - \sin x}{(\pi - 2x)^2}$ $\left(\frac{1}{8}\right)$
10. $\lim_{x \to \frac{\pi}{3}} \frac{1 - 2\cos x}{\sin\left(x - \frac{\pi}{3}\right)}$ $(\sqrt{3})$
11. $\lim_{x \to \frac{\pi}{6}} \frac{2 - \sqrt{3}\cos x - \sin x}{(6x - \pi)^2}$ $\left(\frac{1}{36}\right)$
12. $\lim_{x \to \frac{\pi}{2}} \frac{3\cos x + \cos 3x}{(2x - \pi)^3}$ $\left(-\frac{1}{2}\right)$
13. $\lim_{x \to \pi} \frac{\sqrt{2 + \cos x} - 1}{(\pi - x)^2}$ $\left(\frac{1}{4}\right)$
14. $\lim_{\theta \to \frac{\pi}{4}} \frac{\sin\theta - \cos\theta}{\left(\theta - \frac{\pi}{4}\right)}$ $(\sqrt{2})$
15. $\lim_{x \to \frac{\pi}{3}} \frac{\sqrt{3} - \tan x}{\pi - 3x}$ $\left(\frac{4}{3}\right)$
16. $\lim_{x \to 1} \frac{(1 - x)^2}{\sin \pi x}$ (0)
17. $\lim_{x \to 1} \frac{1 + \cos \pi x}{(1 - x)^2}$ $\left(\frac{\pi^2}{2}\right)$
18. $\lim_{x \to 1} \frac{\sin \pi x}{(x - 1)}$ $(-\pi)$

19. $\lim_{x \to 1} \dfrac{\cos \pi x + \sin\left(\dfrac{\pi x}{2}\right)}{(x-1)^2}$ $\left(\dfrac{3\pi^2}{8}\right)$

20. $\lim_{x \to 1} \dfrac{x^2 - 3x + 2}{x^2 - x + \sin(x-1)}$ $\left(-\dfrac{1}{2}\right)$

21. $\lim_{x \to 1} \dfrac{\sin(x^2 - 1)}{(x-1)}$ (2)

Special types of functions:

1. $F(x) = \dfrac{x f(a) - a f(x)}{x - a}$ whose limit as $x \to a$ is required

2. $F(x) = \dfrac{f(x) - f(y)}{x - y}$ whose limit as $x \to y$ is required or $F(x) = \dfrac{f(x) - f(y)}{x - y} = \dfrac{f(y) - f(x)}{y - x}$ whose limit as $y \to x$ is required.

***Remember*:**

***Definition*:**

1. If $\lim_{x \to a} \dfrac{f(x) - f(a)}{x - a} = L$ = a fixed value then L is called the differential coefficient of $f(x)$ and it is denoted as $f'(x)$.

2. If $\lim_{h \to 0} \dfrac{f(x+h) - f(x)}{h}$, $= L$ = a fixed value, then L is called the differential coefficient of $f(x)$ and it is denoted as $f'(x)$.

***Note*:**

1. In the above definition of $f'(a)$, we denote a particular value of the independent variable x by a while in the definition of $f'(x)$, we denote a particular value of x by itself x instead of a. Thus, we observe x has to play two roles at a time, one of which is of the independent variable and the second is of a particular value of the independent variable, i.e.; the first role is of a variable while the other role is of a constant.

2. $L = \lim_{x \to a} \dfrac{x f(a) - a f(x)}{x - a}$ ...(i)

$$\therefore L = \lim_{x \to a} \frac{x f(a) - a f(a) - a[f(x) - f(a)]}{x - a}$$

$$= \lim_{x \to a} \frac{f(a)(x-a)}{(x-a)} - a \lim_{x \to a} \frac{f(x) - f(a)}{x - a}$$

$$= f(a) - a f'(a)$$

= value of the function $f(x)$ at $x = a - a$ times $d.c$ of $f(x)$ at $x = a$, e.g.:

1. $\lim_{x \to a} \dfrac{x \sin a - a \sin x}{x - a} = \sin a - a \cos a$

***Type 1*:** $F(x) = \dfrac{x f(a) - a f(x)}{x - a}$ whose limit is required as $x \to a$.

***Working rule*:**

1. Put $x = a + h \Rightarrow (x - a) = h$ where $h \to 0$ ($h > 0$ or $h < 0$).

2. If f = sin or cos, use $C \pm D$ formulas to convert it into product form and if f = tan, cot, sec or cosec, then we are required to transform tan, cot, sec or cosec into sin and cos and then use $C \pm D$ formulas to convert it into product form.

***Examples working out*:**

Evaluate:

1. $\lim_{x \to a} \dfrac{x \sin a - a \sin x}{x - a}$

***Solution*:** Putting $x = a + h \Leftrightarrow (x - a) = h \Leftrightarrow (x - a) \to 0$ as $x \to a$.

Now, $\lim_{x \to a} \dfrac{x \sin a - a \sin x}{x - a}$

$$= \lim_{h \to 0} \frac{(a+h)\sin a - a \sin(a+h)}{a + h - a}$$

$$= \lim_{h \to 0} \frac{(a+h)\sin a - a \sin(a+h)}{h}$$

$$= \lim_{h \to 0} \left[\frac{a\{\sin a - \sin(a+h)\} + h \sin a}{h} \right]$$

$$= \lim_{h \to 0} \frac{a \cdot 2\cos\left(\frac{2a+h}{2}\right) \cdot \frac{\sin\left(-\frac{h}{2}\right)}{\left(-\frac{h}{2}\right)} \cdot \left(-\frac{h}{2}\right) + h \sin a}{h}$$

using $C \pm D$ formula

$$= \lim_{h \to 0} a \cdot 2\cos\left(\frac{2a+h}{2}\right) \cdot \frac{\sin\left(-\frac{h}{2}\right)}{\left(-\frac{h}{2}\right)} \cdot \left(-\frac{h}{2}\right) + \sin a$$

$= -a \cos a + \sin a = \sin a - a \cos a.$

Notes: **1.** When we put $x = a \pm h$, then $h \to 0$ through positive values which means $h > 0$ and $-h < 0$ and when we put $x = a + h$, then $h \to 0$ means $h > 0$ or $h < 0$ (both possibilities remain).

2. In questions, in case a function is given defined by a single formula:

(i) $y = f(x)$

(ii) $y = \frac{1}{f(x)}, f(x) \neq 0$

(iii) $y = \frac{f(x)}{g(x)}, g(x) \neq 0$ and one is required to find its limit at a given point, there is no need to calculate the right hand and left hand limit separately, i.e. it is sufficient to use the substitution either $x = a + h$ or $x = a - h$ in the given function to obtain a function h and put $h = 0$ after simplification.

3. In case a function is defined by a single formula into its domain, it is turmed as uniform function.

4. When a given function is a non uniform function or a piecewise function and the question says to examine the existence of the limit of the function, then it is a must to calculate the right hand and left hand limit separately, i.e. it is necessary to use the substitution $x = a + h$ and $x = a - h$ both in the given function to obtain a function of h and lastly put $h = 0$ in the function of h after simplification.

Type 2: $F(x) = \frac{f(x) - f(y)}{x - y}$ whose limit is required as $x \to y$ or,

$$F(x) = \frac{f(x) - f(x)}{x - y} = \frac{f(y) - f(x)}{y - x}$$

whose limit is required as $y \to x$.

Working rule:

First method:

1. If $f = \sin$ or $\cos$, we use $C \pm D$ formula to convert the sum or difference into the product form and if $f = \tan, \cot, \sec$ or cosec, we are required to transform tan, cot, sec or cosec into sin and cos and then use $C \pm D$ formula to convert it into product form.

N.B.: as $x \to a$, $\frac{x-a}{2} \to 0$

Second method:

1. Put $x = a + h \Leftrightarrow (x - a) = h$ where $h \to 0$.

2. If $f = \sin$ or $\cos$, we use $C \pm D$ formula to convert the sum or difference into the product form and if $f = \tan, \cot, \sec$ or cosec, we are required to transform tan, cot, sec or cosec into sin and cos and then use $C \pm D$ formula to convert it into product form.

Examples worked out:

Evaluate:

1. $\lim_{x \to a} \frac{\sin x - \sin a}{x - a}$

Solution: First method:

$$\sin x - \sin a = 2\cos\frac{x+a}{2} \cdot \sin\frac{x-a}{2}$$

$$\Rightarrow \frac{\sin x - \sin a}{x - a}$$

$$= \frac{2\cos\left(\frac{x+a}{2}\right) \cdot \sin\left(\frac{x-a}{2}\right)}{\left(\frac{x-a}{2}\right)} \cdot \frac{1}{2}$$

$$\Rightarrow \lim_{x \to a} \frac{\sin x - \sin a}{x - a}$$

$$= \lim_{x \to a} \frac{\cos\left(\frac{x+a}{2}\right) \cdot \sin\left(\frac{x-a}{2}\right)}{\left(\frac{x-a}{2}\right)}$$

$$= \lim_{x \to a} \cos\left(\frac{x+a}{2}\right) \cdot \lim_{x \to a} \frac{\sin\left(\frac{x-a}{2}\right)}{\left(\frac{x-a}{2}\right)}$$

$$= \cos\left(\frac{a+a}{2}\right) \cdot 1 \left(\because \text{as } x \to a, \frac{x-a}{2} \to 0\right)$$

$$= \cos\left(\frac{2a}{2}\right)$$

$= \cos a.$

Second method:

Putting $x = a + h \Rightarrow x - a = h \to 0$

$\Rightarrow x \to a \Rightarrow x - a \to 0 \Rightarrow h \to 0$

Now, $\frac{\sin x - \sin a}{x - a}$

$$= \frac{\sin(a+h) - \sin a}{h}$$

$$= \frac{2\cos\left(\frac{a+h+a}{2}\right) \cdot \sin\left(\frac{a+h-a}{2}\right)}{h}$$

$$= \frac{2\cos\left(a + \frac{h}{2}\right) \cdot \sin\left(\frac{h}{2}\right)}{h}$$

$$= \frac{2\cos\left(a + \frac{h}{2}\right) \cdot \frac{\sin\left(\frac{h}{2}\right)}{\left(\frac{h}{2}\right)} \cdot \left(\frac{h}{2}\right)}{h}$$

$$= \frac{\cos\left(a + \frac{h}{2}\right) \cdot \frac{\sin\left(\frac{h}{2}\right)}{\left(\frac{h}{2}\right)} \cdot (h)}{h}$$

$$= \cos\left(a + \frac{h}{2}\right) \cdot \frac{\sin\left(\frac{h}{2}\right)}{\left(\frac{h}{2}\right)} \qquad \ldots\text{(i)}$$

Now taking the limit on both sides of (i) as $h \to 0$, we get

$$\lim_{x \to a} \frac{\sin x - \sin a}{x - a}$$

$$= \lim_{h \to 0} \cos\left(a + \frac{h}{2}\right) \cdot \lim_{h \to 0} \frac{\sin\left(\frac{h}{2}\right)}{\left(\frac{h}{2}\right)}$$

$$= \cos a \cdot 1 = \cos a.$$

2. $\lim_{x \to a} \frac{\cos x - \cos a}{x - a}$

Solution: $\lim_{x \to a} \frac{\cos x - \cos a}{x - a}$

$$= \lim_{x \to a} \frac{-2\sin\left(\frac{x+a}{2}\right) \cdot \sin\left(\frac{x-a}{2}\right)}{(x-a)}$$

$$= \lim_{x \to a}\left(-2\sin\frac{x+a}{2}\right) \cdot \lim_{x \to a} \frac{\sin\left(\frac{x-a}{2}\right)}{\left(\frac{x-a}{2}\right)} \cdot \frac{1}{2}$$

$$= -2\sin a \times 1 \times \frac{1}{2} \left(\because \text{as } x \to a, \frac{x-a}{2} \to 0\right)$$

$$= \frac{1}{2} \times (-2) \sin a$$

$$= -\sin a$$

3. $\lim_{x \to y} \frac{\tan x - \tan y}{x - y}$

Solution: Putting $x = y + h \Rightarrow h = (x - y)$

$\therefore x \to y \text{ (given)} \Rightarrow x - y \to 0 \Rightarrow h \to 0$

Now, $\lim_{x \to y} \frac{\tan x - \tan y}{x - y}$

$$= \lim_{h \to 0} \frac{\tan(y+h) - \tan y}{y + h - y}$$

$$= \lim_{h \to 0} \frac{1}{h} \left[\frac{\sin(y+h)}{\cos(y+h)} - \frac{\sin y}{\cos y} \right]$$

$$= \lim_{h \to 0} \frac{1}{h} \left[\frac{\sin(y+h)\cos y - \cos(y+h)\sin y}{\cos(y+h) \cdot \cos y} \right]$$

$$= \lim_{h \to 0} \frac{1}{h} \left[\frac{\sin(y+h-y)}{\cos(y+h)\cos y} \right]$$

$$= \lim_{h \to 0} \left[\frac{1}{h} \cdot \sin h \cdot \frac{1}{\cos(y+h)} \cdot \frac{1}{\cos y} \right]$$

$$= \lim_{h \to 0} \left(\frac{\sin h}{h} \right) \cdot \lim_{h \to 0} \frac{1}{\cos(y+h)} \cdot \lim_{h \to 0} \frac{1}{\cos y}$$

$$= 1 \cdot \frac{1}{\cos^2 y}$$

$$= \sec^2 y.$$

Exercise 4.12

Evaluate **Answers**

1. $\lim_{x \to a} \left(\frac{\sin x - \sin a}{x - a} \right)$ $(\cos a)$

2. $\lim_{x \to a} \left(\frac{\cos x - \cos a}{x - a} \right)$ $(-\sin a)$

3. $\lim_{x \to a} \left(\frac{\tan x - \tan a}{x - a} \right)$ $(\sec^2 a)$

4. $\lim_{x \to a} \left(\frac{\sec x - \sec a}{x - a} \right)$ $(\sec a \cdot \tan a)$

5. $\lim_{x \to a} \left(\frac{\operatorname{cosec} x - \operatorname{cosec} a}{x - a} \right)$ $(-\operatorname{cosec} a \cdot \cot a)$

6. $\lim_{x \to a} \left(\frac{\cot x - \cot a}{x - a} \right)$ $(-\operatorname{cosec}^2 a)$

Method of Rationalization

Whenever a square root of a trigonometrical function appears in the given function whose limit is required as $x \to a$, we adopt the following working rule:

Working rule:

1. Rationalize the *Nr* or *Dr* or both whose square root appears.

2. Put $x = a + h$ where $h \to 0$

3. Simplify the function of $(a + h)$ and put $h = 0$ in the simplified function of h.

Examples worked out:

Evaluate:

1. $\lim_{x \to \pi} \frac{\sqrt{2 + \cos x} - 1}{(\pi - x)^2}$

Solution: Given function $= \frac{\sqrt{2 + \cos x} - 1}{(\pi - x)^2}$

Or rationalizing the *Nr*, we get

$$\frac{\sqrt{2 + \cos x} - 1}{(\pi - x)^2}$$

$$= \frac{\left(\sqrt{2 + \cos x} - 1\right)\left(\sqrt{2 + \cos x} + 1\right)}{(\pi - x)^2 \left(\sqrt{2 + \cos x} + 1\right)}$$

$$= \frac{1+\cos x}{\left(\sqrt{2+\cos x}+1\right)(\pi - x)^2}$$

Now, we put $x = \pi + h$, where $h \to 0$ as $x \to \pi$ then the above expression becomes

$$= \frac{1+\cos(\pi + h)}{\left(\sqrt{2+\cos(\pi+h)}+1\right)\cdot h^2}$$

$$= \frac{1-\cos h}{\left(\sqrt{(2-\cos h)}+1\right)\cdot h^2}$$

$\therefore$ Required limit $= \lim\limits_{x\to\pi} \dfrac{\sqrt{2+\cos x}-1}{(\pi - x)^2}$

$$= \lim_{h\to 0} \frac{(1-\cos h)}{\left(\sqrt{(2-\cos h)}+1\right)\cdot h^2}$$

$\because (\pi - x)^2 = (\pi - \pi - h)^2 = h^2$

$$= \lim_{h\to 0}\frac{(1-\cos h)}{h^2}\cdot \lim_{h\to 0}\frac{1}{\left(\sqrt{(2-\cos h)}+1\right)}$$

$$= \lim_{h\to 0}\frac{2\cdot \sin^2\left(\frac{h}{2}\right)}{h^2}\cdot \lim_{h\to 0}\frac{1}{\left(\sqrt{(2-\cos h)}+1\right)}$$

$$= 2\lim_{h\to 0}\left(\frac{\sin\left(\frac{h}{2}\right)}{\frac{h}{2}}\right)^2\cdot\frac{1}{4}\cdot\frac{1}{\sqrt{2-1}+1}$$

$$= 2\cdot 1\cdot\frac{1}{4}\cdot\frac{1}{2} = \frac{1}{4}.$$

To find the limits of trigonometric functions of an angle θ as $\theta \to 0$

We have already derived the $\lim\limits_{x\to 0}\sin x = 0$ and $\lim\limits_{x\to 0}\cos x = 1$ on pages 142 and 143 but here we are going to provide the same results with different methods and some more results on limits.

Derivation: Let us consider a circle $\odot OAP$ and $PN \perp$ drawn form P to the radius OA.

Now, letting $\angle POA = \theta > 0$

As $\theta \to 0$, then $P \to A$

$PN \to 0$ (zero)

$ON \to OA$

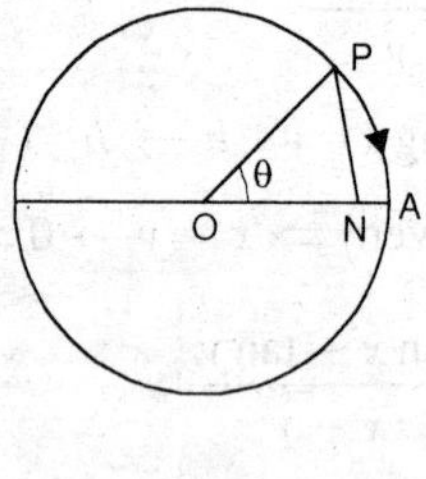

But length $OP = r$ = radius remains constant.

Now,

$$\lim_{\theta\to 0}\sin\theta = \lim_{\theta\to 0}\frac{p}{h} = \lim_{\theta\to 0}\frac{PN}{OP} = \frac{O}{OP} = 0 \text{ (zero)}$$

$$\lim_{\theta\to 0}\cos\theta = \lim_{\theta\to 0}\frac{b}{h} = \lim_{\theta\to 0}\frac{ON}{OP} = \frac{OA}{OA} = 1\ (\because OA = OP = r)$$

$$\lim_{\theta\to 0}\tan\theta = \lim_{\theta\to 0}\frac{p}{b} = \lim_{\theta\to 0}\frac{PN}{ON} = \frac{O}{OA} = 0.$$

Notes:

1. When $\theta < 0$, let $\theta = \angle POA = -\theta'$ (i.e; $\theta' = -\theta$)

$\therefore \theta \to 0 \Leftrightarrow \theta' \to 0$

As $\theta \to 0$, , then $P \to A$

$PN \to 0$ (zero)

$ON \to OA$

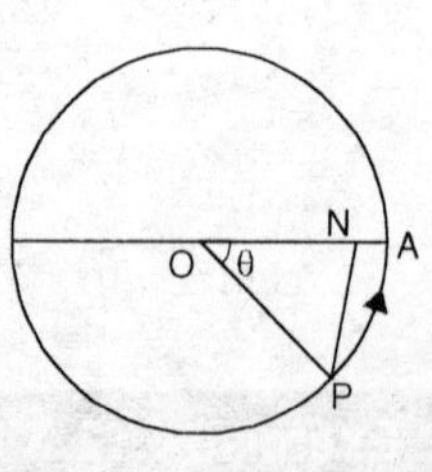

But length $OP = r$ = radius remains constant.

$$\therefore \lim_{\theta'\to 0}\sin(-\theta') = \lim_{\theta\to 0}\sin\theta = \lim_{\theta\to 0}\frac{p}{h} = \lim_{\theta\to 0}\frac{-PN}{OP}$$

$$= \frac{-O}{OP} = 0 \text{ (zero)}$$

$$\lim_{\theta'\to 0} \cos(-\theta') = \lim_{\theta\to 0} \cos\theta = \lim_{\theta\to 0} \frac{b}{h} = \lim_{\theta\to 0} \frac{ON}{OP}$$

$$= \lim_{\theta\to 0} \frac{OA}{OA} = 1$$

$$\lim_{\theta'\to 0} \tan(-\theta) = \lim_{\theta\to 0} \tan\theta = \lim_{\theta\to 0} \frac{p}{b} = \lim_{\theta\to 0} \frac{-PN}{ON}$$

$$= \frac{-O}{OA} = 0 \text{ (zero)}$$

2. The limits of trigonometric functions of an angle θ as $\theta \to 0$ can also be found by noting that (i) sin x, cos x and tan x are continuous functions at $x = 0$ (ii) $\lim_{x\to c} x = C$ (iii) the limit sign of a continuous function can be referred to the independent variable (or, argument)

i.e; $\lim_{x\to c} f(x) = f\left(\lim_{x\to c} x\right) = f(c)$ provided $f(x)$ is a continuous function of x at $x = c$.

Hence, $\lim_{\theta\to 0} \sin\theta = \sin\left(\lim_{\theta\to 0} \theta\right) = \sin 0 = 0$

$$\lim_{\theta\to 0} \cos\theta = \cos\left(\lim_{\theta\to 0} \theta\right) = \cos 0 = 1$$

$$\lim_{\theta\to 0} \tan\theta = \tan\left(\lim_{\theta\to 0} \theta\right) = \tan 0 = 0$$

3. $\lim_{\theta\to 0} \cos\theta = \lim_{\theta\to 0} \sqrt{1 - \sin^2\theta} = \sqrt{\lim_{\theta\to 0}(1 - \sin^2\theta)}$

$$= \sqrt{\lim_{\theta\to 0} 1 - \lim_{\theta\to 0} \sin^2\theta} = \sqrt{1 - 0} = 1$$

4. When θ is very small, vertical segment drawn from one end point of the radius of the circle = arc of the circle opposite to the central angle.

5. When $\theta \to 0$, vertical segment $\to 0$

Remember:

1. If $\lim_{x\to a} \frac{f_1(x)}{f_2(x)} = 1$, then $f_1(x)$ and $f_2(x)$ are called equivalent functions as $x \to a$ which means

$$\lim_{x\to a} \frac{f_1(x)}{f_2(x)} = 1 \Leftrightarrow f_1(x) \equiv f_2(x) \text{ as } x \to a$$

2. As $\theta \to 0$, we have

(i) $\sin\theta \equiv \theta$ **(ii)** $\tan\theta \equiv \theta$ **(iii)** $\log(1 + x) \equiv x$

2. To show that $\lim_{\theta\to 0} \frac{\sin\theta}{\theta}$, where θ is measured in circular measure (or, radian measure)

Proof: Let us consider a circular arc AB of radius 'r' which subtends a positive acute angle θ at the center C.

Now we draw

1. the chord AB

2. the tangent AD at A and extend it until it meets CB at D. Then AD is perpendicular to the radius $CA = r$

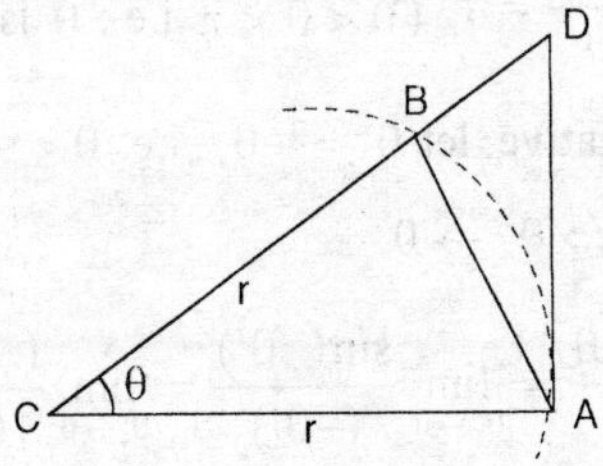

We have

$$A_1 = \text{area of } \Delta CAB = \frac{1}{2} CA \cdot CB \sin\theta = \frac{1}{2} r^2 \sin\theta \quad \ldots \text{(i)}$$

$$A_2 = \text{area of the sector } CAB = \frac{1}{2} r^2 \theta \quad \ldots \text{(ii)}$$

$$A_3 = \text{area of the } \Delta CAD = \frac{1}{2} \cdot AD \cdot CA = \frac{1}{2} CA \cdot \tan\theta \; CA = \frac{1}{2} r^2 \tan\theta \quad \ldots \text{(iii)}$$

($\because \angle ACD = \theta$, $\therefore \tan\theta = \frac{AD}{AC}$ in ΔACD which is a right angled triangle since AD being a tangent is $\perp$ to the radius $OA = r$)

Again we have, area of ΔCAB < area of the sector CAB < area of ΔCAD ... (iv)

On putting the values of (i), (ii) and (iii) in (iv), we get $A_1 < A_2 < A_3$

$$\Rightarrow \frac{1}{2}r^2 \sin\theta < \frac{1}{2}r^2\,\theta < \frac{1}{2}r^2 \tan\theta \quad \ldots \text{(v)}$$

On dividing (v) by $\frac{1}{2}r^2 \sin\theta$, we get

$$1 < \frac{\theta}{\sin\theta} < \frac{1}{\cos\theta}$$

$$\Rightarrow 1 > \frac{\sin\theta}{\theta} > \cos\theta$$

$$\Rightarrow \lim_{\theta\to 0} 1 > \lim_{\theta\to 0}\frac{\sin\theta}{\theta} > \lim_{\theta\to 0}\cos\theta$$

$$\Rightarrow 1 \ge \lim_{\theta\to 0}\frac{\sin\theta}{\theta} \ge 1$$

$$\Rightarrow \lim_{\theta\to 0}\frac{\sin\theta}{\theta} = 1,\ (0 < \theta < \pi \text{ i.e.; } \theta \text{ is positive})$$

If θ is negative, let $\theta' = -\theta$, i.e. $\theta = -\theta'$

$\therefore \theta \to 0 \Leftrightarrow \theta' \to 0$

$$\therefore \lim_{\theta\to 0}\frac{\sin\theta}{\theta} = \lim_{\theta'\to 0}\frac{\sin(-\theta')}{(-\theta')} = \lim_{\theta'\to 0}\frac{(-\sin\theta')}{(-\theta')} =$$

$$\lim_{\theta'\to 0}\frac{\sin\theta'}{\theta'} = 1 \text{ (by previous result)}$$

$$\therefore \lim_{\theta\to 0}\frac{\sin\theta}{\theta} = \lim_{\substack{\theta\to 0\\ \theta>0}}\frac{\sin\theta}{\theta} = \lim_{\substack{\theta\to 0\\ \theta<0}}\frac{\sin\theta}{\theta} = 1$$

Remember:

1. When θ is small, $\sin\theta = \theta$ (approximately)

2. $\lim_{\theta\to 0}\frac{\theta}{\sin\theta} = 1$

Proof: $\lim_{\theta\to 0}\frac{\theta}{\sin\theta} = \lim_{\theta\to 0}\frac{1}{\left(\frac{\sin\theta}{\theta}\right)} = \frac{1}{\lim_{\theta\to 0}\left(\frac{\sin\theta}{\theta}\right)} = \frac{1}{1} = 1$

3. Prove that $\lim_{\theta\to 0}\frac{\tan\theta}{\theta} = 1$

Proof: $\lim_{\theta\to 0}\frac{\tan\theta}{\theta} = \lim_{\theta\to 0}\left(\frac{\sin\theta}{\cos\theta} \times \frac{1}{\theta}\right)$

$$= \left(\lim_{\theta\to 0}\frac{\sin\theta}{\theta} \times \lim_{\theta\to 0}\frac{1}{\cos\theta}\right) = 1 \times \frac{1}{1} = 1$$

$$\left(\because \lim_{\theta\to 0}\frac{\sin\theta}{\theta} = 1 \text{ and } \lim_{\theta\to 0}\cos\theta = 1\right)$$

4. $\lim_{\theta\to 0}\frac{\theta}{\tan\theta}$

Proof: $\lim_{\theta\to 0}\frac{\theta}{\tan\theta} = \lim_{\theta\to 0}\frac{1}{\left(\frac{\tan\theta}{\theta}\right)} = \frac{1}{\lim_{\theta\to 0}\left(\frac{\tan\theta}{\theta}\right)} = \frac{1}{1} = 1$

N.B.: Limit of the reciprocal of a function = reciprocal of its limit provided that the limit of the function is not equal to zero.

Limits of trigonometric functions as $x \to 0$

Form 1: To find $\lim_{\theta\to 0} f(x) = \lim_{x\to 0}\frac{t_1(x)}{t_2(x)}$ where $f(x)$ = a trigonometric function whose numerator = $t_1(x)$ = trigonometric function or trigonometric expression.

And denominator = $t_2(x)$ = a trigonometric function or trigonometric expression.

Working rule: To find the limit of a trigonometric function (whose both *Nr* and *Dr* are trigonometric functions or trigonometric expressions) as the independent variable tends to zero, we adopt the following working rule:

1. We express all trigonometric functions in *Nr* and *Dr* in terms of $\sin\theta$ and $\cos\theta$ or in terms of product of $\sin\theta$ and $\cos\theta$ by using $C \pm D$ formulas of trigonometry or we may use any formula which is required for simplification and cancel the common factor (which makes $f(0)$ meaningless) from *Nr* and *Dr*.
2. Lastly, we use the results gives below:

(i) $\lim_{\theta\to 0}\frac{\sin\theta}{\theta} = 1$

(ii) $\lim_{\theta\to 0}\frac{\tan\theta}{\theta} = 1$

(iii) $\lim_{\theta\to 0}\cos\theta = 1$

Note:

1. Never forget to write:

(a) $\sin\theta = \frac{\sin\theta}{\theta}\cdot\theta$

(b) $\tan\theta = \frac{\tan\theta}{\theta}\cdot\theta$ so that we may use the standard formulas of limits of trigonometrical ratios mentioned above.

Problems based on the form 1

***Examples worked out*:**

Evaluate:

1. $\lim_{x\to 0}\frac{\sin 3x}{\tan 4x}$

***Solution*:** $\lim_{x\to 0}\frac{\sin 3x}{\tan 4x}$

$$= \lim_{x\to 0}\frac{3}{4}\cdot\left[\frac{\frac{\sin 3x}{3}}{\frac{\tan 4x}{4}}\right]$$

$$= \frac{3}{4}\cdot\left[\frac{\lim_{x\to 0}\left(\frac{\sin 3x}{3x}\right)}{\lim_{x\to 0}\left(\frac{\tan 4x}{4x}\right)}\right]$$

$$= \frac{3}{4}\times\frac{1}{1}$$

$$= \frac{3}{4}$$

2. $\lim_{x\to 0}\frac{\sin 2x}{2x}$

***Solution*:** $\lim_{x\to 0}\frac{\sin 2x}{2x} = 2\left(\because \lim_{x\to 0}\frac{\sin m\theta}{m\theta} = m\right)$

3. $\lim_{x\to 0}\frac{\sin ax}{x}$

***Solution*:** $\lim_{x\to 0}\left[\frac{\sin ax}{x}\right]$

$$= \lim_{x\to 0}\left[\frac{\sin ax}{ax}\cdot a\right]$$

$$= a\lim_{x\to 0}\left[\frac{\sin ax}{ax}\right] = a\cdot 1 = a$$

4. $\lim_{x\to 0}\frac{\sin 3x}{\sin 5x}$

***Solution*:** $\lim_{x\to 0}\frac{\sin 3x}{\sin 5x}$

$$= \lim_{x\to 0}\left[\frac{\sin 3x}{3x}\cdot\frac{5x}{\sin 5x}\cdot\frac{3}{5}\right]$$

$$= \left(\lim_{x\to 0}\frac{\sin 3x}{3x}\right)\cdot\left(\lim_{x\to 0}\frac{5x}{\sin 5x}\right)\cdot\lim_{x\to 0}\left(\frac{3}{5}\right)$$

$$= 1\cdot 1\cdot\frac{3}{5}$$

$$= \frac{3}{5}$$

5. $\lim_{x\to 0}\frac{\tan 5x}{x}$

***Solution*:** $\lim_{x\to 0}\frac{\tan 5x}{x}$

$$= \lim_{x\to 0}\frac{\sin 5x}{x\cdot\cos 5x}$$

$$= \lim_{x\to 0}\frac{\sin 5x}{5x}\cdot 5\cdot\lim_{x\to 0}\frac{1}{\cos 5x}$$

$$= 1\cdot 5\cdot 1$$

$$= 5$$

6. $\lim_{x\to 0}\frac{\sin^n 6x}{x^n}$

***Solution*:** $\lim_{x\to 0}\frac{\sin^n 6x}{x^n}$

$$= \lim_{x\to 0}\frac{\left(\frac{\sin 6x}{6x}\right)^n\cdot(6x)^n}{x^n}$$

$$= \lim_{x\to 0}\left(\frac{\sin 6x}{6x}\right)^n\cdot\lim_{x\to 0}\left(\frac{6^n\cdot x^n}{x^n}\right)$$

$$= \left(\lim_{x\to 0} \frac{\sin 6x}{6x}\right)^n \cdot 6^n$$

$$= 1 \cdot 6^n = 6^n$$

7. $\lim_{x\to 0} \frac{\tan\alpha x}{\tan\beta x}$

Solution: $\lim_{x\to 0} \frac{\tan\alpha x}{\tan\beta x}$

$$= \lim_{x\to 0}\left[\frac{\sin\alpha x}{\cos\alpha x}\cdot\frac{\cos\beta x}{\sin\beta x}\right]$$

$$= \lim_{x\to 0}\left[\frac{\sin\alpha x}{\alpha x}\cdot\frac{1}{\cos\alpha x}\cdot\cos\beta x\cdot\frac{\beta x}{\sin\beta x}\cdot\frac{\alpha x}{\beta x}\right]$$

$$= \lim_{x\to 0}\left[\frac{\sin\alpha x}{\alpha x}\cdot\frac{1}{\cos\alpha x}\cdot\cos\beta x\cdot\frac{\beta x}{\sin\beta x}\cdot\frac{\alpha}{\beta}\right]$$

$$= \frac{\alpha}{\beta}\left[\lim_{x\to 0}\left(\frac{\sin\alpha x}{\alpha x}\right)\cdot\lim_{x\to 0}\left(\frac{1}{\cos\alpha x}\right)\cdot\lim_{x\to 0}\left(\frac{\beta x}{\sin\beta x}\right)\cdot\lim_{x\to 0}(\cos\beta x)\right]$$

$$= \frac{\alpha}{\beta}\cdot[1\cdot 1\cdot 1\cdot 1] = \frac{\alpha}{\beta}.$$

8. $\lim_{x\to 0} \frac{\sin\alpha x}{\sin\beta x}$

Solution: $\lim_{x\to 0} \frac{\sin\alpha x}{\sin\beta x}$

$$= \lim_{x\to 0}\left[\frac{\sin\alpha x}{\alpha x}\cdot\frac{\beta x}{\sin\beta x}\cdot\frac{\alpha}{\beta}\right]$$

$$= \left[\lim_{x\to 0}\left(\frac{\sin\alpha x}{\alpha x}\right)\times\left(\lim_{x\to 0}\frac{1}{\frac{\sin\beta x}{\beta x}}\right)\cdot\frac{\alpha}{\beta}\right]$$

$$= 1\times\frac{1}{1}\times\frac{\alpha}{\beta} = \frac{\alpha}{\beta}$$

9. $\lim_{x\to 0} \frac{\tan x - \sin x}{1-\cos x}$

Solution: $\lim_{x\to 0}\left(\frac{\tan x - \sin x}{1-\cos x}\right)$

$$= \lim_{x\to 0}\left(\frac{\frac{\sin x}{\cos x} - \sin x}{1-\cos x}\right) = \lim_{x\to 0}\left(\frac{\frac{\sin x - \sin x\cos x}{\cos x}}{1-\cos x}\right)$$

$$= \lim_{x\to 0}\left[\frac{\sin x}{\cos x}\cdot\left(\frac{1-\cos x}{1-\cos x}\right)\right] = \lim_{x\to 0}\frac{\sin x}{\cos x}$$

$$\frac{\lim_{x\to 0}\frac{\sin x}{x}\cdot x}{\lim_{x\to 0}\cos x} = \frac{\lim_{x\to 0}\left(\frac{\sin x}{x}\right)\cdot\lim_{x\to 0} x}{\lim_{x\to 0}\cos x}$$

$$= \frac{1\cdot 0}{1} = 0 \quad \left(\because \lim_{x\to 0} x = 0\right)$$

10. $\lim_{x\to 0}\frac{\cos 7x - \cos 9x}{\cos 3x - \cos 5x}$

Solution: $\lim_{x\to 0}\frac{\cos 7x - \cos 9x}{\cos 3x - \cos 5x}$

$$= \lim_{x\to 0}\frac{2\sin 8x\cdot\sin x}{2\sin 4x\cdot\sin x} = \lim_{x\to 0}\frac{2\sin 8x}{2\sin 4x}$$

$$= \lim_{x\to 0}\left(\frac{\frac{\sin 8x}{8x}\cdot 8\cancel{x}}{\frac{\sin 4x}{4x}\cdot 4\cancel{x}}\right) = \frac{8}{4} = 2$$

11. $\lim_{x\to 0}\frac{\sin 7x - \sin x}{\sin 6x}$

Solution: $\lim_{x\to 0}\frac{\sin 7x - \sin x}{\sin 6x}$

$$= \lim_{x\to 0}\frac{2\cos 4x\cdot\sin 3x}{2\sin 3x\cos 3x}$$

$$= \lim_{x\to 0}\frac{\cos 4x}{\cos 3x}$$

$= \frac{1}{1}$

$= 1$

Problems based on the form 1

Exercise 4.13

Find the limits of the following functions as $x \to 0$.

Answers

1. $\dfrac{\sec 4x - \sec 2x}{\sec 3x - \sec x}$ $\left(\dfrac{3}{2}\right)$

2. $\dfrac{1 - \cos ax}{1 - \cos bx}$ $\left(\dfrac{a^2}{b^2}\right)$

3. $\dfrac{\sin 2x - \sin 4x}{\sin 4x - \sin 6x}$ Find

4. $\dfrac{\cos 2x - \cos 8x}{\sin 3x}$ Find

5. $\dfrac{\tan x - \sin x}{\sin^3 x}$ $\left(\dfrac{1}{2}\right)$

6. $\dfrac{\sin x - \tan x}{\sin^3 x}$ $\left(-\dfrac{1}{2}\right)$

7. $\sin x \cdot \cos\dfrac{1}{x}$ (0)

8. $\dfrac{1 - \cos x}{1 + \cos x}$ (0)

9. $\dfrac{1 - \cos 2x}{\cos 2x - \cos 8x}$ $\left(\dfrac{1}{15}\right)$

10. $\dfrac{\tan x - \sin x}{\sin 3x - 3\sin x}$ $\left(-\dfrac{1}{8}\right)$

11. $\dfrac{1 - \cos x}{\sin^2 (2x)}$ $\left(\dfrac{1}{8}\right)$

12. $\dfrac{\sin 7x - \sin x}{\sin 6x}$ (1)

13. $\dfrac{\sin 5x - \sin x}{\sin 4x}$ (1)

14. $\dfrac{\sin(\alpha + x) - \sin(\alpha - x)}{\cos(\alpha + x) - \cos(\alpha - x)}$ $(-\cot \alpha)$

15. $\dfrac{\tan x - \sin x}{1 - \cos x}$ (0)

16. $\dfrac{1 - \cos 2\theta}{1 - \cos 5\theta}$ $\left(\dfrac{4}{25}\right)$

Form 2: To find $\lim_{x \to 0} f(x)$ where $f(x) = a$ trigonometrical expression mixed with an algebraic function in any way (generally algebraic function appears as addend, subtrahend, minuend, multiplicand or divisor of trigonometric function or expression), i.e.' To find

(i) $\lim_{x \to 0} f(x) = \lim_{x \to 0} \dfrac{a_1(x) \pm t_1(x)}{a_2(x) \pm t_2(x)}$

(ii) $\lim_{x \to 0} f(x) = \lim_{x \to 0} \dfrac{a_1(x) \cdot t_1(x)}{t_2(x)}$ or $\lim_{x \to 0} \dfrac{t_1(x)}{a_1(x)\, t_2(x)}$

(iii) $\lim_{x \to 0} f(x) = \lim_{x \to 0} \dfrac{t_1(x)}{a_1(x)}$

(iv) $\lim_{x \to 0} f(x) = \lim_{x \to 0} \dfrac{a_1(x)}{t_1(x)}$

where $a_1(x)$ and $a_2(x)$ = algebraic functions
$t_1(x)$ and $t_2(x)$ = trigonometric function

We adopt the following working rule:

Working rule: Modify the given function by using trigonometric formulas if required so that standard results of limits of trigonometric functions may be used.

N.B.: Standard results of limits of trigonometric functions

1. $\lim_{x \to 0} \dfrac{\sin x}{x} = 1$

2. $\lim_{x\to 0} \frac{\tan x}{x} = 1$

Problems based on the form 2

Examples worked out

Evaluate:

1. $\lim_{x\to 0} x\left(\frac{\cos x + \cos 2x}{\sin x}\right)$

Solution: $\lim_{x\to 0} x\left(\frac{\cos x + \cos 2x}{\sin x}\right)$

$$= \lim_{x\to 0}\left[\left(\frac{x}{\sin x}\right)\cdot(\cos x + \cos 2x)\right]$$

$$= \lim_{x\to 0}\left[\frac{(\cos x + \cos 2x)}{\left(\frac{\sin x}{x}\right)}\right]$$

$$= \frac{\lim_{x\to 0}(\cos x + \cos 2x)}{\lim_{x\to 0}\left(\frac{\sin x}{x}\right)}$$

$$= \frac{1+1}{1} = 2$$

2. $\lim_{x\to 0} \frac{2\sin x - \sin 2x}{x^3}$

Solution: $\lim_{x\to 0} \frac{2\sin x - \sin 2x}{x^3}$

$$= \lim_{x\to 0} \frac{2\sin x - 2\sin x\cos x}{x^3}$$

$$= \lim_{x\to 0} \frac{2\sin x(1-\cos x)}{x^3}$$

$$= \lim_{x\to 0} 2\sin x \cdot \frac{2\sin^2 \frac{x}{2}}{x^3}$$

$$= 4\cdot \lim_{x\to 0}\left[\left(\frac{\sin x}{x}\right)\cdot\left(\frac{\sin \frac{x}{2}}{\frac{x}{2}}\right)^2\cdot\frac{1}{4}\right]$$

$$= \lim_{x\to 0}\left(\frac{\sin x}{x}\right)\cdot\left(\lim_{x\to 0}\frac{\sin\frac{x}{2}}{\frac{x}{2}}\right)^2$$

$$= 1 \times (1)^2$$

$$= 1$$

3. $\lim_{x\to 0} \frac{\tan x - \sin x}{x^3}$

Solution: $\lim_{x\to 0} \frac{\tan x - \sin x}{x^3}$

$$= \lim_{x\to 0}\left[\frac{\tan x\,(1-\cos x)}{x^3}\right]$$

$$[\because \tan x\,(1-\cos x) = \tan x - \sin x]$$

$$= \lim_{x\to 0}\left[\left(\frac{\tan x}{x}\right)\cdot\left(\frac{2\sin^2\frac{x}{2}}{x^2}\right)\right]$$

$$= \lim_{x\to 0}\left(\frac{\tan x}{x}\right)\cdot\lim_{x\to 0}\left(2\cdot\frac{\sin^2\frac{x}{2}}{\frac{x^2}{4}\cdot 4}\right)$$

$$= 1\cdot 2\cdot\left[\lim_{x\to 0}\left(\frac{\sin\frac{x}{2}}{\frac{x}{2}}\right)\right]\cdot\frac{1}{4}$$

$$= 1\cdot 2\cdot\frac{1}{4}$$

$$= \frac{1}{2}$$

4. $\lim_{x\to 0} \dfrac{\tan 2x - \sin 2x}{x^3}$

Solution: $\lim_{x\to 0} \dfrac{\tan 2x - \sin 2x}{x^3}$

$$= \lim_{x\to 0} \left[\frac{\tan 2x\,(1-\cos 2x)}{x^3}\right]$$

$$= \lim_{x\to 0} \frac{\tan 2x \cdot 2\sin^2 x}{x^3}$$

$$= \lim_{x\to 0} \left(\frac{\tan 2x}{2x}\right)\cdot 2 \cdot \lim_{x\to 0}\left(\frac{\sin x}{x}\right)^2 \cdot 2$$

$$= 2\cdot 1\cdot 2\,(1)^2$$
$$= 4.$$

5. $\lim_{x\to 0} \dfrac{1-\cos kx}{x^2}$

Solution: $\lim_{x\to 0} \dfrac{1-\cos kx}{x^2}$

$$= \lim_{x\to 0} \frac{2\sin^2\left(\frac{kx}{2}\right)}{x^2}$$

$$= 2\cdot \lim_{x\to 0}\left[\frac{\sin\left(\frac{kx}{2}\right)}{\left(\frac{kx}{2}\right)}\right]^2 \cdot \frac{k^2}{4}$$

$$= 2\cdot (1)^2 \cdot \left(\frac{k^2}{4}\right)$$

$$= \frac{k^2}{2}.$$

6. $\lim_{x\to 0} \dfrac{\cos 5x - 1}{x}$

Solution: $\lim_{x\to 0} \dfrac{\cos 5x - 1}{x}$

$$= \lim_{x\to 0} \frac{-2\sin^2 \frac{5x}{2}}{x}$$

$$= \lim_{x\to 0} \left[\frac{\sin^2 \frac{5x}{2}\cdot(-2)}{\left(\frac{5x}{2}\right)^2}\cdot\left(\frac{5}{2}\right)^2\cdot x\right]$$

$$= \lim_{x\to 0}\left[\frac{\sin\left(\frac{5x}{2}\right)}{\left(\frac{5x}{2}\right)}\right]^2 \cdot \lim_{x\to 0}\left[(-2)\cdot\left(\frac{5}{2}\right)^2\cdot x\right]$$

$$= 1\times 0 = 0$$

7. $\lim_{x\to 0} \dfrac{\tan x - x}{x}$

Solution: $\lim_{x\to 0} \dfrac{\tan x - x}{x}$

$$= \lim_{x\to 0}\left[\frac{\sin x}{x}\cdot\frac{1}{\cos x} - 1\right]$$

$$= \lim_{x\to 0}\left(\frac{\sin x}{x}\right)\cdot \lim_{x\to 0}\left(\frac{1}{\cos x}\right) - \lim_{x\to 0}(1)$$

$$= 1\times 1 - 1 = 0$$

8. $\lim_{x\to 0} \dfrac{(\cos x - \cos 3x)}{x\,(\sin 3x - \sin x)}$

Solution: $\lim_{x\to 0} \dfrac{(\cos x - \cos 3x)}{x\,(\sin 3x - \sin x)}$

$$= \lim_{x\to 0} \frac{2\cdot \sin\left(\frac{x+3x}{2}\right)\cdot \sin\left(\frac{3x-x}{2}\right)}{x\cdot 2\cdot \cos\left(\frac{x+3x}{2}\right)\cdot \sin\left(\frac{3x-x}{2}\right)}$$

$$= \lim_{x\to 0} \frac{2\sin 2x\cdot \sin x}{2x\cdot \cos 2x\cdot \sin x}$$

$$= \lim_{x\to 0} \frac{\sin 2x}{x\cdot \cos 2x}$$

$$= \lim_{x\to0}\left(\frac{\sin 2x}{x}\right)\cdot\lim_{x\to0}\left(\frac{1}{\cos 2x}\right)$$

$$= \lim_{x\to0}\left(\frac{\sin 2x}{2x}\right)\cdot 2\cdot\lim_{x\to0}\left(\frac{1}{\cos 2x}\right)$$

$$= 1\times2\times1 = 2$$

9. $\lim_{x\to0}\dfrac{\tan 2x - x}{3x - \sin x}$

Solution: $\dfrac{\tan 2x - x}{3x - \sin x}$

$$= \frac{\left(\frac{\tan 2x}{2x}\right)\cdot 2x - x}{3x - \left(\frac{\sin x}{x}\right)\cdot x} = \frac{x\left(\frac{\tan 2x}{2x}\cdot 2 - 1\right)}{x\left(3 - \frac{\sin x}{x}\right)}$$

$$= \frac{\left(\frac{\tan 2x}{2x}\cdot 2 - 1\right)}{\left(3 - \frac{\sin x}{x}\right)} \text{ for } x\neq0$$

Now, $\lim_{x\to0}\left[\dfrac{\tan 2x - x}{3x - \sin x}\right]$

$$= \lim_{x\to0}\left[\frac{\left(\frac{\tan 2x}{2x}\cdot 2 - 1\right)}{\left(3 - \frac{\sin x}{x}\right)}\right]$$

$$= \frac{2\left(\lim_{x\to0}\frac{\tan 2x}{2x}\right) - \lim_{x\to0} 1}{\lim_{x\to0} 3 - \lim_{x\to0}\frac{\sin x}{x}} = \frac{2-1}{3-1} = \frac{1}{2}$$

10. $\lim_{x\to0}\left(\dfrac{\operatorname{cosec} x - \cot x}{x}\right)$

Solution: $\lim_{x\to0}\left(\dfrac{\operatorname{cosec} x - \cot x}{x}\right)$

$$= \lim_{x\to0}\left[\frac{1}{x}\left(\frac{1}{\sin x} - \frac{\cos x}{\sin x}\right)\right]$$

$$= \lim_{x\to0}\left[\frac{1}{x}\left(\frac{1-\cos x}{\sin x}\right)\right]$$

$$= \lim_{x\to0}\left[\frac{x}{\sin x}\times\frac{2\sin^2\frac{x}{2}}{x^2}\right]$$

$$= \lim_{x\to0}\left[\frac{x}{\sin x}\times\frac{2}{4}\cdot\left(\frac{\sin^2\frac{x}{2}}{\frac{x}{2}}\right)^2\right]$$

$$= 1\times\frac{2}{4}\times1$$

$$= \frac{1}{2}$$

Problems based on form 2

Exercise 4.14

Find the limits of the following functions as $x\to0$

Answers

1. $\dfrac{1-\cos x}{x^2}$ $\left(\dfrac{1}{2}\right)$
2. $\dfrac{1-\cos x}{x}$ (0)
3. $\dfrac{\operatorname{cosec} x - \cot x}{x}$ $\left(\dfrac{1}{2}\right)$
4. $\dfrac{1-\cos 2x}{x}$ (0)
5. $\dfrac{\sin x - \tan x}{x}$ (0)
6. $\dfrac{\cos x - \sec x}{x^2}$ (–1)

7. $\dfrac{x\tan x}{1-\cos x}$ (2)

8. $\dfrac{1-\cos 2x}{x^2}$ (2)

9. $\dfrac{x(1+\cos mx)}{\sin mx}$ $\left(\dfrac{2}{m}\right)$

10. $\dfrac{x^2}{\cos bx-\cos ax}$ $\left(\dfrac{2}{a^2-b^2}\right)$

11. $\dfrac{\cos x-\cos 2x}{x^2}$ $\left(\dfrac{3}{2}\right)$

12. $\dfrac{\sin 2x\,(\cos 2x-\cos 3x)}{x^3}$ (5)

13. $\dfrac{8}{x^3}\cdot\left[1-\cos\dfrac{x^2}{2}-\cos\dfrac{x^2}{4}+\cos\dfrac{x^2}{2}\cdot\cos\dfrac{x^2}{4}\right]$ $\left(\dfrac{1}{32}\right)$

14. $\dfrac{x^2+3x^3+(2x-\sin x)^2}{2x^2\cos^2 x-5x^3}$ (1)

15. $\dfrac{2\cos x-2+x^2}{x^4}$ $\left(\dfrac{1}{12}\right)$

Form 3: Limits of irrational trigonometric functions as $x\to 0$.

To find the limits of irrational trigonometrical functions when it assumes an indeterminate form $\left(\dfrac{0}{0}\right)$ at $x=0$, we adopt the following working rule:

Working rule: Method of rationalization is adopted to find the limit of irrational trigonometric functions which means removal of radical sign ($\sqrt{\ }$) from numerator or denominator or both which may be done by using trigonometrical substitution or multiplying and dividing by the conjugate of irrational trigonometric expressions.

Note: When given irrational trigonometric functions doe not assume $\left(\dfrac{0}{0}\right)$ form at $x=0$, we find the limit of irrational trigonometric functions by directly putting $x=0$ in the given function, e.g.:

1. $\lim\limits_{x\to 0}\sqrt{\dfrac{1-\cos x}{1+\cos x}}=\sqrt{\lim\limits_{x\to 0}\left(\dfrac{1-\cos x}{1+\cos x}\right)}=\sqrt{\dfrac{0}{2}}=0$

2. $x\to 0\Rightarrow \dfrac{x}{m}\to 0\Rightarrow mx\to 0$

3. After rationalization, we put $x=0$ in the rationalized form of the given irrational function provided independent variable 'x' tends to zero.

4. If we have $\sqrt{f^2(x)}=|f(x)|$, we should find *l.h.l* and *r.h.l* and we should remove mod symbol by using the definition

$|f(x)|=f(x)$ if $f(x)\ge 0$

and $|f(x)|=-f(x)$ if $f(x)<0$

and lastly if *l.h.l* = *r.h.l*, we say limit of the given irrational trigonometric function exists and if *l.h.l* ≠ *r.h.l*, we say that limit of irrational trigonometric function does not exist.

Examples worked out:

Evaluate:

1. $\lim\limits_{x\to 0}\dfrac{x}{\sqrt{1-\cos x}}$

Solution: $\lim\limits_{x\to 0}\dfrac{x}{\sqrt{1-1+2\sin^2\dfrac{x}{2}}}$

$=\lim\limits_{x\to 0}\dfrac{x}{\sqrt{2\sin^2\dfrac{x}{2}}}$

$=\lim\limits_{x\to 0}\dfrac{x}{\sqrt{2}\left|\sin\dfrac{x}{2}\right|}$

$$l.h.l = \lim_{\substack{x\to 0\\ x<0}} \frac{x}{\sqrt{2}\left|\sin\frac{x}{2}\right|} = \lim_{x\to 0} \frac{x}{\sqrt{2}\left(-\sin\frac{x}{2}\right)}$$

$$= -\frac{1}{\sqrt{2}} \lim_{x\to 0}\left(\frac{x}{\sin\frac{x}{2}}\right)$$

$$= -\frac{1}{\sqrt{2}} \lim_{x\to 0}\left(\frac{\frac{x}{2}\cdot 2}{\sin\frac{x}{2}}\right)$$

$$= -\frac{2}{\sqrt{2}} \cdot \lim_{x\to 0}\left(\frac{\frac{x}{2}}{\sin\frac{x}{2}}\right) = -\frac{2}{\sqrt{2}} \times 1 = -\sqrt{2}$$

$$r.h.l = \lim_{\substack{x\to 0\\ x<0}} \frac{x}{\sqrt{2}\left|\sin\frac{x}{2}\right|}$$

$$= \lim_{x\to 0} \frac{x}{\sqrt{2}\left(\sin\frac{x}{2}\right)} = \lim_{x\to 0} \frac{\frac{x}{2}\cdot 2}{\sqrt{2}\left(\sin\frac{x}{2}\right)}$$

$$= \frac{2}{\sqrt{2}} \lim_{x\to 0}\left(\frac{\frac{x}{2}}{\sin\frac{x}{2}}\right)$$

$$= \frac{2}{\sqrt{2}} \cdot 1 = \sqrt{2}$$

Hence, $l.h.l \neq r.h.l$

Which $\Rightarrow \lim_{x\to 0} \frac{x}{\sqrt{1-\cos x}}$ does not exist.

Note:

(i) $r.h.l = \lim_{\substack{x\to 0\\ x>0}} f(x)$ means the variable x in $f(x)$ is restricted only to positive value of x.

(ii) $l.h.l = \lim_{\substack{x\to 0\\ x<0}} f(x)$ means the variable x in $f(x)$ is restricted only to negative values of x.

2. $\lim_{x\to 0} \frac{\sqrt{1-\cos x}}{\sin x}$

Solution: Let $y = \frac{\sqrt{1-\cos x}}{\sin x} = \frac{\sqrt{1-1+2\sin^2 x}}{\sin x}$

$$= \frac{\sqrt{2\sin^2 x}}{\sin x} = \frac{\sqrt{2}\,|\sin x|}{\sin x}$$

$$l.h.l = \lim_{\substack{x\to 0\\ x<0}} \frac{\sqrt{2}\,|\sin x|}{\sin x}$$

$$= \lim_{x\to 0} \frac{\sqrt{2}(-\sin x)}{\sin x}$$

$$= \left(-\sqrt{2}\right) \lim_{x\to 0}\left(\frac{\sin x}{\sin x}\right)$$

$$= \left(-\sqrt{2}\right) \times 1$$

$$= -\sqrt{2}$$

$$r.h.l = \lim_{\substack{x\to 0\\ x>0}} \frac{\sqrt{2}\,|\sin x|}{\sin x}$$

$$= \lim_{x\to 0} \frac{\sqrt{2}\,\sin x}{\sin x}$$

$$= \sqrt{2} \lim_{x\to 0} \frac{\sin x}{\sin x}$$

$$= \sqrt{2} \times 1$$

$$= \sqrt{2}$$

Hence, $l.h.l \neq r.h.l$ which $\Rightarrow \lim_{x\to 0} \frac{\sqrt{1-\cos x}}{\sin x}$ does not exist.

3. $\lim_{x \to 0} \frac{\sqrt{2} - \sqrt{1+\cos x}}{\sin^2 x}$

Solution: Let $y = \frac{\sqrt{2} - \sqrt{1+\cos x}}{\sin^2 x}$

$$\therefore y = \frac{\left(\sqrt{2} - \sqrt{1+\cos x}\right)\left(\sqrt{2} + \sqrt{1+\cos x}\right)}{\sin^2 x\left(\sqrt{2} + \sqrt{1+\cos x}\right)}$$

$$= \frac{2 - (1+\cos x)}{(1-\cos x)(1+\cos x)\left(\sqrt{2} + \sqrt{1+\cos x}\right)}$$

$$= \frac{(1-\cos x)}{(1-\cos x)(1+\cos x)\left(\sqrt{2} + \sqrt{1+\cos x}\right)}$$

$$= \frac{1}{(1+\cos x)\left(\sqrt{2} + \sqrt{1+\cos x}\right)} \text{ for } x \neq 0$$

Thus, we get $y = \frac{1}{(1+\cos x)\left(\sqrt{2} + \sqrt{1+\cos x}\right)}$

for $x \neq 0$ ... (i)

Now, on taking the limits on both sides of (i), as $x \to 0$, we get

$$\lim_{x \to 0} y = \lim_{x \to 0} \frac{1}{(1+\cos x)\left(\sqrt{2} + \sqrt{1+\cos x}\right)}$$

$$= \frac{1}{(1+1)\left(\sqrt{2} + \sqrt{1+1}\right)}$$

$$= \frac{1}{2\left(\sqrt{2} + \sqrt{2}\right)}$$

$$= \frac{1}{2 \times 2\sqrt{2}}$$

$$= \frac{1}{4\sqrt{2}}$$

Problems based on form 3

Exercise 4.15

Find the limits of the following functions as $x \to 0$

Answers

1. $\frac{|x|}{\sqrt{1-\cos x}}$ $(\sqrt{2})$

2. $\frac{1-\sqrt{1+\tan x}}{\tan x}$ $\left(-\frac{1}{2}\right)$

3. $\frac{\sin x}{\sqrt{1+\sin x} - 1}$ (2)

4. $\frac{\sqrt{\sin x}}{\sqrt{x}}$ (1)

5. $\frac{\sqrt{2+\cos z} - 1}{(\pi - z)^2}$; $z = \pi - x$ $\left(\frac{1}{4}\right)$

Limits of trigonometric functions as $x \to \infty$

To find the limits a function involving a trigonometric function as the independent variable tends to infinity, we adopt the following working rule:

Working Rule:

1. Put $x = \frac{1}{t}$ or $t = \frac{1}{x}$ where $t \to 0$ when $x \to \infty$

N.B.: **1.** The rule of putting $x = \frac{1}{t}$ or $t = \frac{1}{x}$ is known as reciprocal substitution.

2. Reciprocal substitution is useful when independent variable x appears as a factor either in numerator or denominator or when the angle of trigonometrical function (or, ratio) is the reciprocal of x.

Examples worked out:

Evaluate:

1. $\lim_{x \to \infty} x \sin\left(\frac{1}{x}\right)$

Solution: Let $y = x\sin\left(\frac{1}{x}\right)$

On putting $x = \frac{1}{t} \Rightarrow t = \frac{1}{x}$, we have $x \to \infty \Leftrightarrow t \to 0$

$$\therefore\ y = \frac{1}{t}\cdot \sin t = \frac{\sin t}{t} \qquad \ldots \text{(i)}$$

Now, on taking the limits on both sides of (i) as $x \to \infty$, we get

$$\lim_{x\to\infty} y = \lim_{t\to 0}\frac{\sin t}{t} = 1 \text{ which}$$

$$\Rightarrow \lim_{x\to\infty} x\sin\left(\frac{1}{x}\right) = 1$$

2. $\lim_{x\to\infty}\left[x\cdot\cos\left(\frac{\pi}{4x}\right)\cdot\sin\left(\frac{\pi}{4x}\right)\right]$

Solution: Let $y = \left[x\cdot\cos\left(\frac{\pi}{4x}\right)\cdot\sin\left(\frac{\pi}{4x}\right)\right]$

$$= \left[\cos\left(\frac{\pi}{4x}\right)\frac{\sin\left(\frac{\pi}{4x}\right)}{\frac{\pi}{4x}}\cdot\frac{\pi}{4}\right]$$

Now, on putting $t = \frac{\pi}{4x}$ where $t \to 0$ as $x \to \infty$, we have

$$y = \cos t\left[\frac{\sin t}{t}\right]\cdot\frac{\pi}{4} \qquad \ldots \text{(i)}$$

Now, on taking the limits on both sides of (i) as $x \to \infty$, we get

$$\lim_{x\to\infty} y = \lim_{t\to 0}\left[\cos t\cdot\left(\frac{\sin t}{t}\right)\cdot\frac{\pi}{4}\right]$$

$$= \lim_{t\to 0}\cos t\cdot\lim_{x\to 0}\left(\frac{\sin t}{t}\right)\cdot\lim_{t\to 0}\left(\frac{\pi}{4}\right)$$

$$= 1\times 1\times\frac{\pi}{4} = \frac{\pi}{4} \text{ which}$$

$$\Rightarrow \lim_{x\to\infty}\left[x\cdot\cos\left(\frac{\pi}{4x}\right)\cdot\sin\left(\frac{\pi}{4x}\right)\right] = \frac{\pi}{4}$$

An important fact to know:

Theorem: If $f(x) < g(x) < h(x)$ and $\lim_{x\to a} f(x) = L$ and $\lim_{x\to a} h(x) = L$, then $\lim_{x\to a} g(x) = L$.

N.B.: The result of this theorem remains true of either or both of the given strict inequalities are replaced by $\leq$.

Many important results of limits can be easily obtained with the help of above theorem.

Examples worked out:

1. $-1 \leq \sin x \leq 1$

$$\Rightarrow -\frac{1}{x} \leq \frac{\sin x}{x} \leq \frac{1}{x} \text{ for } x > 0$$

$$\Rightarrow \lim_{x\to\infty}\left(-\frac{1}{x}\right) \leq \lim_{x\to\infty}\left(\frac{\sin x}{x}\right) \leq \lim_{x\to\infty}\left(\frac{1}{x}\right)$$

$$\Rightarrow 0 \leq \lim_{x\to\infty}\left(\frac{\sin x}{x}\right) \leq 0\ \left(\because \lim_{x\to\infty}\left(\pm\frac{1}{x}\right) = 0\right)$$

$$\Rightarrow \lim_{x\to\infty}\left(\frac{\sin x}{x}\right) = 0$$

2. $-1 \leq \sin\left(\frac{1}{x}\right) \leq 1$

$$\Rightarrow -x \leq x\sin\left(\frac{1}{x}\right) \leq x \text{ for } x > 0,$$

$$\text{and } x \leq x\sin\left(\frac{1}{x}\right) \leq -x \text{ for } x < 0$$

$$\Rightarrow 0 \leq \lim_{x\to 0}\left(x\cdot\sin\frac{1}{x}\right) \leq 0\ \left(\because \lim_{x\to 0}(\pm x) = 0\right)$$

$$\Rightarrow \lim_{x\to 0}\left(x\sin\frac{1}{x}\right) = 0$$

3. $-1 \leq \cos x \leq 1$

$$\Rightarrow -\frac{1}{x} \leq \frac{1}{x}\cos x \leq \frac{1}{x} \text{ for } x > 0$$

$$\Rightarrow \lim_{x\to\infty}\left(-\frac{1}{x}\right) \le \lim_{x\to\infty}\left(\frac{1}{x}\cos x\right) \le \lim_{x\to\infty}\left(\frac{1}{x}\right)$$

$$\Rightarrow 0 \le \lim_{x\to\infty}\frac{\cos x}{x} \le 0$$

$$\Rightarrow \lim_{x\to\infty}\frac{\cos x}{x} = 0.$$

Examples:

Evaluate:

1. $\lim_{n\to\infty} n\cdot\sin\frac{\theta}{n}$, θ being measured in radian.

Solution: Let $y = n\sin\frac{\theta}{n}$

Now multiplying *Nr* and *Dr* by $\frac{\theta}{n}$ since we require the same angle in denominator, we get

$$y = n\cdot\frac{\theta}{n}\cdot\left(\frac{\sin\frac{\theta}{n}}{\frac{\theta}{n}}\right) = \theta\left(\frac{\sin\frac{\theta}{n}}{\frac{\theta}{n}}\right) \quad \ldots \text{(i)}$$

Now, on taking the limits on both sides of (i) as $n\to\infty$, we get

$$\lim_{n\to\infty} y = \lim_{n\to\infty}\theta\cdot\left(\frac{\sin\frac{\theta}{n}}{\frac{\theta}{n}}\right)$$

$$= \theta\cdot\lim_{\frac{\theta}{n}\to 0}\left(\frac{\sin\frac{\theta}{n}}{\frac{\theta}{n}}\right)$$

$$\left(\because n\to\infty \Rightarrow \frac{1}{n}\to 0 \Rightarrow \frac{\theta}{n}\to 0\right)$$

$$= \theta\cdot 1$$

$$= \theta \text{ which} \Rightarrow \lim_{n\to\infty} n\cdot\sin\left(\frac{\theta}{n}\right) = \theta.$$

Problems based on finding the limits of trigonometric functions as $n \to \infty$

Exercise 4.16

	Answers
1. $\lim_{x\to\infty} x\cdot\sin\left(\frac{1}{x}\right)$	(1)
2. $\lim_{x\to\infty} x\cdot\tan\left(\frac{1}{x}\right)$	(Find)
3. $\lim_{x\to\infty}\sin x$	(does not exist)
4. $\lim_{x\to\infty}\left(\frac{\sin^2 x}{x^2}\right)$	(0)
5. $\lim_{x\to\infty}\frac{\sin^3 x}{x^2}$	(0)
6. $\lim_{x\to\infty}\left(\frac{\tan x}{x}\right)$	(does not exist)
7. $\lim_{x\to\infty} 3x\cdot\sin\left(\frac{1}{x}\right)$	(3)
8. $\lim_{x\to\infty} 2\pi x\cdot\sin\left(\frac{\pi}{x}\right)$	$(2\pi^2)$
9. $\lim_{x\to\infty}\frac{\cos x}{x}$	(0)

Form 2: When the independent variable or its power appears in the given function as an addend or minuend, we adopt the following working rule.

Working rule: It consists of following steps.

Step 1: To divide *Nr* and *Dr* by the highest power of *x* appearing in *Nr* and *Dr*.

Step 2: To take the limit as $x\to\infty$ noting that $\frac{a_1}{x}, \frac{a_2}{x^2}, \frac{a_3}{x^3}, \ldots$, etc. all $\to 0$ as $x\to\infty$, provided $a_1, a_2, a_3, \ldots$, etc. all are constants.

Examples worked out

Evaluate:

1. $\lim_{x\to\infty} \sqrt[3]{\frac{x+\sin x}{x^4+\cos^2 x}}$

Solution: $y = \sqrt[3]{\frac{x+\sin x}{x^4+\cos^2 x}}$

$$= \sqrt[3]{\frac{\frac{1}{x^3}+\frac{1}{x^4}\sin x}{1+\frac{1}{x^4}\cos^2 x}} \quad \ldots \text{(i)}$$

Now, on taking the limits on both sides of (i) as $x\to\infty$, we get

$$\lim_{x\to\infty} y = \lim_{x\to\infty} \sqrt[3]{\frac{\frac{1}{x^3}+\frac{1}{x^4}\sin x}{1+\frac{1}{x^4}\cos^2 x}}$$

$$= \sqrt[3]{\lim_{x\to\infty}\left(\frac{\frac{1}{x^3}+\frac{1}{x^4}\sin x}{1+\frac{1}{x^4}\cos^2 x}\right)}$$

$$= \sqrt[3]{\frac{\lim_{x\to\infty}\left(\frac{1}{x^3}\right)+\lim_{x\to\infty}\left(\frac{\sin x}{x^4}\right)}{\lim_{x\to\infty}(1)+\lim_{x\to\infty}\left(\frac{\cos x}{x^2}\right)^2}}$$

$$= \sqrt[3]{\left(\frac{0+0}{1+0}\right)} = \sqrt[3]{0} = 0$$

Note: Now we state a theorem which has a wide use.

Theorem: If $\lim_{x\to c} f(x) = 0$ and $g(x)$ is bounded, then $\lim_{x\to c} f(x)\cdot g(x) = 0$.

Which is expressed in the following way also.

"The product of a bounded quantity and an infinitesimal is an infinitesimal".

e.g. Let $f(x) = \frac{1}{x^4}$

$$\therefore \lim_{x\to\infty}\left(\frac{1}{x^4}\right) = \left(\lim_{x\to\infty}\frac{1}{x}\right)^4 = 0$$

and let $g(x) = \sin x$ which is bounded since $-1 \le \sin x \le 1$

Hence, $\lim_{x\to\infty} f(x)\cdot g(x) = \lim_{x\to\infty}\frac{1}{x^4}\sin x = 0$

Remember: The above theorem is also true even if $g(x)$ is bounded in a deleted neighbourhood of c, e.g.:

(i) $\lim_{x\to 0} x\sin\left(\frac{1}{x}\right) = 0$

(ii) $\lim_{x\to 0} x\cos\left(\frac{1}{x}\right) = 0$

(iii) $\lim_{x\to a} (x-a)\sin\left(\frac{1}{x-a}\right) = 0$

(iv) $\lim_{x\to c} (x-c)\cos\left(\frac{1}{x-c}\right) = 0$

Problems based on the form 2

Exercise 4.17

Evaluate:

Answers

1. $\lim_{x\to\infty}\sqrt{\frac{x+\sin x}{x+\cos x}}$ (1)

2. $\lim_{x\to\infty}\sqrt{\frac{x-\sin x}{x+\cos^2 x}}$ (1)

3. $\lim_{x\to\infty}\sqrt{\frac{x-\sin x}{9x^2+\cos x}}$ (0)

On limits of a function containing are function

Type 1: When a single inverse circular function of an independent variable (i.e. $t^{-1} x$, where t^{-1} stands for $\sin^{-1}$, $\cos^{-1}$, $\tan^{-1}$, $\cot^{-1}$, $\sec^{-1}$, cosec^{-1}) appears in a given function whose limit is required as $x \to 0$ or $x \to a$, then we adopt the following working rule:

Working rule:

1. Put the inverse circular function of an independent variable which appears in the given function = θ and change the inverse circular function into circular function, i.e.; put $t^{-1}(x) = \theta$, where $t^{-1} = \sin^{-1}, \cos^{-1}, \tan^{-1}$, etc. and write $t(\theta) = x$ where t = sin, cos, tan, cot, sec, cosec.

2. Change the limit of independent variable x in terms of θ.

Note: **1.** The general method of finding the limits of a given function containing the inverse circular function of an independent variable x consists of changing the inverse circular function into the circular function (direct trigonometric function) for which it is better to substitute for an arc function (or, inverse trigonometric function) or a relation a number 'θ'.

2. Trigonometric function, trigonometrical functions and circular functions are synonymus.

3. After changing inverse circular function into circular function, we use the rule of trigonometric function as $x \to a$ if required, method of substitution is adopted.

Examples worked out:

Evaluate:

1. $\lim\limits_{x \to 0} \dfrac{\sin^{-1} x}{x}$

Solution: $y = \dfrac{\sin^{-1} x}{x}$

We put $\sin^{-1} x = \theta, \therefore x = \sin\theta$

$\therefore x \to 0 \Leftrightarrow \theta \to 0$

$$\therefore y = \frac{\theta}{\sin\theta} \qquad \ldots \text{(i)}$$

Now taking the limits on both sides of (i) as $\theta \to 0$

$$\Rightarrow \lim_{x \to 0} y = \lim_{\theta \to 0} \frac{\theta}{\sin\theta}$$

$$\Rightarrow \lim_{x \to 0} y = 1$$

2. $\lim\limits_{x \to 1} \left[\dfrac{1 - x}{\pi - 2\sin^{-1} x}\right]$

Solution: Let $y = \dfrac{1 - x}{\pi - 2\sin^{-1} x}$

We put $\sin^{-1} x = \theta \therefore x = \sin\theta, -\dfrac{\pi}{2} \le \theta \le \dfrac{\pi}{2}$

$\therefore x \to 1 \Leftrightarrow \theta \to \dfrac{\pi}{2} -$

$$\therefore y = \frac{1 - \sin\theta}{\pi - 2\theta} \qquad \ldots \text{(i)}$$

Now, taking the limits on both sides of (i) as $\theta \to \dfrac{\pi}{2}$

$$\Rightarrow \lim_{x \to 1} y = \lim_{\theta \to \frac{\pi}{2}} \left[\frac{1 - \sin\theta}{\pi - 2\theta}\right]$$

Again putting $\theta = \dfrac{\pi}{2} - z \Rightarrow z \to 0$ as $\theta \to \dfrac{\pi}{2}$

$$\therefore \lim_{z \to 0} \frac{1 - \sin\left(\frac{\pi}{2} - z\right)}{\pi - 2\left(\frac{\pi}{2} - z\right)}$$

$$= \lim_{z \to 0} \left[\frac{1 - \cos z}{-2z}\right] = \lim_{z \to 0} \left(\frac{2\sin^2 \frac{z}{2}}{+2z}\right)$$

$$= \lim_{z \to 0} \left[\frac{1}{z} \cdot \left(\frac{\sin\frac{z}{2}}{\frac{z}{2}}\right)^2 \cdot \left(\frac{z}{2}\right)^2\right]$$

$$= 1 \cdot 0 = 0$$

3. $\lim_{x\to 1} \frac{1-x}{\left(\cos^{-1} x\right)^2}$

Solution: Let $y = \frac{1-x}{\left(\cos^{-1} x\right)^2}$

We put $\cos^{-1} x = \theta \therefore x = \cos\theta,\ 0 \le \theta \le \pi$

$\therefore x \to 1 \Leftrightarrow \theta \to 0$

$\therefore$ given limit $= \lim_{\theta\to 0} \frac{1-\cos\theta}{\theta^2}$

$$= \lim_{\theta\to 0}\left(\frac{2\sin^2\frac{\theta}{2}}{\theta^2}\right)$$

$$= \lim_{\theta\to 0}\left[2\cdot\frac{\sin\frac{\theta}{2}}{\frac{\theta}{2}}\times\frac{\sin\frac{\theta}{2}}{\frac{\theta}{2}}\times\frac{1}{4}\right]$$

$$= \frac{1}{2}\times\left[\lim_{\theta\to 0}\left(\frac{\sin\frac{\theta}{2}}{\frac{\theta}{2}}\right)\cdot\lim_{\theta\to 0}\left(\frac{\sin\frac{\theta}{2}}{\frac{\theta}{2}}\right)\right] = \frac{1}{2}\times 1\times 1 = \frac{1}{2}$$

4. $\lim_{x\to 0}\left[\frac{x\left(1-\sqrt{1-x^2}\right)}{\sqrt{1-x^2}\left(\sin^{-1} x\right)^3}\right]$

Solution: Let $y = \frac{x\left(1-\sqrt{1-x^2}\right)}{\sqrt{1-x^2}\left(\sin^{-1} x\right)^3}$

We put $\sin^{-1} x = \theta \therefore x = \sin\theta,\ -\frac{\pi}{2} \le \theta \le \frac{\pi}{2}$

$\therefore x \to 0 \Leftrightarrow \theta \to 0$

$\therefore y = \frac{\sin\theta\,(1-\cos\theta)}{\theta^3\cos\theta}$ as $\sqrt{1-x^2} = |\cos\theta| = \cos\theta$ in $-\frac{\pi}{2} \le \theta \le \frac{\pi}{2}$

$$= \tan\theta\cdot 2\cdot\frac{\sin^2\frac{\theta}{2}}{\theta^3}$$

$$= \frac{\tan\theta}{\theta}\cdot 2\cdot\left(\frac{\sin\frac{\theta}{2}}{\theta}\right)^2 \quad \ldots(i)$$

Now taking the limits on both sides of (i) as $x \to 0$

$$\lim_{x\to 0} y = \lim_{\theta\to 0}\left[\frac{\tan\theta}{\theta}\times\frac{1}{2}\times\left(\frac{\sin\frac{\theta}{2}}{\frac{\theta}{2}}\right)^2\right]$$

$$= \lim_{\theta\to 0}\left[\frac{\tan\theta}{\theta}\right]\times\lim_{\theta\to 0}\left(\frac{1}{2}\right)\times\lim_{\theta\to 0}\left(\frac{\sin\frac{\theta}{2}}{\frac{\theta}{2}}\right)^2$$

$$= 1\times\frac{1}{2}\times 1$$

$$= \frac{1}{2}$$

5. $\lim_{x\to 1}\left[\frac{1-\sqrt{x}}{\left(\cos^{-1} x\right)^2}\right]$

Solution: $y = \frac{1-\sqrt{x}}{\left(\cos^{-1} x\right)^2}$

We put $\cos^{-1} x = \theta \therefore x = \cos\theta,\ 0 \le \theta \le \pi$

$\therefore x \to 1 \Leftrightarrow \theta \to 0$

$$\therefore y = \frac{1 - \sqrt{\cos\theta}}{\theta^2}$$

$$= \frac{\left(1 - \sqrt{\cos\theta}\right)\left(1 + \sqrt{\cos\theta}\right)}{\theta^2 \cdot \left(1 + \sqrt{\cos\theta}\right)}$$

$$= \frac{1 - \cos\theta}{\theta^2 \cdot \left(1 + \sqrt{\cos\theta}\right)} = \frac{2\sin^2\frac{\theta}{2}}{\theta^2\left(1 + \sqrt{\cos\theta}\right)} \quad \ldots \text{(i)}$$

Now, on taking the limits on both sides of (i) as $x \to 1$, we get

$$\lim_{x\to1} y = \lim_{\theta\to0}\left[\frac{2\sin^2\frac{\theta}{2}}{\theta^2\left(1 + \sqrt{\cos\theta}\right)}\right]$$

$$= \lim_{\theta\to0}\left[\frac{2\sin^2\frac{\theta}{2}}{4\cdot\left(\frac{\theta}{2}\right)^2}\right] \times \lim_{\theta\to0}\left[\frac{1}{1 + \sqrt{\cos\theta}}\right]$$

$$\Rightarrow \lim_{x\to1} y = \frac{2}{4} \times 1 \times \frac{1}{\left(1 + \sqrt{1}\right)} = \frac{2}{4} \times 1 \times \frac{1}{2} = \frac{1}{4}$$

6. $\displaystyle\lim_{x\to0} \frac{\tan^{-1} x}{x}$

Solution: Let $y = \dfrac{\tan^{-1} x}{x}$

We put $\tan^{-1} x = \theta \therefore x = \tan\theta, -\frac{\pi}{2} \le \theta \le \frac{\pi}{2}$

$\therefore x \to 0 \Leftrightarrow \theta \to 0$

$$\therefore y = \frac{\theta}{\tan\theta} \quad \ldots \text{(i)}$$

Now, taking the limits on both sides of (i) as $x \to 0$, we get

$$\lim_{x\to0} y = \lim_{\theta\to0} \frac{\theta}{\tan\theta} = 1$$

Or, alternatively

$$y = \frac{\theta}{\tan\theta} = \theta \cdot \cot\theta = \theta \cdot \frac{\cos\theta}{\sin\theta} = \frac{\theta}{\sin\theta} \times \cos\theta \quad \ldots \text{(i)}$$

Now on taking the limits on both sides of (i) as $x \to 0$, we get

$$\lim_{x\to0} y = \lim_{\theta\to0} \frac{\theta}{\tan\theta} = \lim_{\theta\to0}\left[\frac{\theta}{\sin\theta} \times \cos\theta\right]$$

$$= \lim_{\theta\to0}\left(\frac{\theta}{\sin\theta}\right) \times \lim_{x\to0} \cos\theta = 1 \times 1 = 1$$

7. $\displaystyle\lim_{x\to0} \frac{x}{\tan^{-1} x}$

Solution: Let $y = \dfrac{x}{\tan^{-1} x}$

We put $\tan^{-1} x = \theta \therefore x = \tan\theta, -\frac{\pi}{2} \le \theta \le \frac{\pi}{2}$

$\therefore x \to 0 \Leftrightarrow \theta \to 0$

$$\therefore y = \frac{\tan\theta}{\theta} \quad \ldots \text{(i)}$$

Now on taking the limits on both sides of (i) as $x \to 0$, we get

$$\lim_{x\to0} y = \lim_{\theta\to0} \frac{\tan\theta}{\theta} = 1$$

Problems based on type 1

Exercise 4.18

Find the limits of the following functions

Answers

1. $\displaystyle\lim_{x\to0} \frac{\sin^{-1} x}{x}$ (1)
2. $\displaystyle\lim_{x\to0} \frac{\tan^{-1} x}{x}$ (1)

3. $\lim\limits_{x\to 0} \dfrac{\sin^{-1} x}{\sin x}$ (1)

4. $\lim\limits_{x\to 0} \dfrac{\sqrt{1+x} - \sqrt{1-x}}{\sin^{-1} x}$ (1)

5. $\lim\limits_{x\to 0} \dfrac{x\left(1-\sqrt{1-x^2}\right)}{\sqrt{1-x^2}\left(\sin^{-1} x\right)^3}$ $\left(\dfrac{1}{2}\right)$

6. $\lim\limits_{x\to 0} \dfrac{\sin^{-1} x}{\tan^{-1}\left(\dfrac{\pi x}{2}\right)}$ $\left(\dfrac{2}{\pi}\right)$

7. $\lim\limits_{x\to \frac{1}{\sqrt{2}}} \dfrac{x - \cos\left(\sin^{-1} x\right)}{1-\cot\left(\sin^{-1} x\right)}$ $\left(\dfrac{1}{\sqrt{2}}\right)$

8. $\lim\limits_{x\to 0} \dfrac{\tan^{-1} 2x}{\sin 3x}$ $\left(\dfrac{2}{3}\right)$

9. $\lim\limits_{x\to 1} \dfrac{\left(\cos^{-1} x\right)^2}{1-x}$ (2)

10. $\lim\limits_{x\to 0} \dfrac{\left(\sin^{-1} x - 2x\right)}{\sin^{-1} x + 2\sin\left(\dfrac{1}{2}\sin^{-1} x\right)\left\{3-4\sin^2\left(\dfrac{1}{2}\sin^{-1} x\right)\right\}}$

$\left(-\dfrac{1}{4}\right)$

11. $\lim\limits_{x\to 1} \dfrac{1-\sqrt{x}}{\left(\cos^{-1} x\right)^2}$ $\left(\dfrac{1}{4}\right)$

12. $\lim\limits_{x\to -1} \dfrac{\sqrt{\pi} - \sqrt{\cos^{-1} x}}{\sqrt{x+1}}$ $\left(\dfrac{1}{\sqrt{2\pi}}\right)$

Hint: Put $\cos^{-1} x = \theta \therefore x = \cos\theta,\ 0 \le \theta \le \pi$

$\therefore x \to -1 \Leftrightarrow \theta \to \pi$

$\therefore y = \dfrac{\sqrt{\pi} - \sqrt{\theta}}{\sqrt{1+\cos\theta}} = \dfrac{(\pi - \theta)}{\sqrt{2}\cos\dfrac{\theta}{2}\cdot\left(\sqrt{\pi}+\sqrt{\theta}\right)}$ as

$\left|\cos\dfrac{\theta}{2}\right| = \cos\dfrac{\theta}{2}$ in $0 \le \theta \le \pi$

Again putting $\theta = \pi + z \Rightarrow z \to 0$ as $\theta \to \pi$

$\therefore \lim\limits_{z\to 0} \dfrac{-z}{\sqrt{2}\cos\dfrac{1}{2}(\pi+z)\left(\sqrt{\pi}+\sqrt{\pi+z}\right)}$

$= \lim\limits_{z\to 0} \dfrac{-z}{-\sqrt{2}\sin\left(\dfrac{z}{2}\right)\left(\sqrt{\pi}+\sqrt{\pi+z}\right)]}$

13. $\lim\limits_{x\to \frac{1}{\sqrt{2}}} \dfrac{x - \cos\left(\sin^{-1} x\right)}{1-\tan\left(\sin^{-1} x\right)}$ $\left(-\dfrac{1}{\sqrt{2}}\right)$

Type 2: If the given function contains a function of the type t^{-1} $[f(x)]$, where t^{-1} stands for $\sin^{-1}$, $\cos^{-1}$, $\tan^{-1}$, $\cot^{-1}$, $\sec^{-1}$, cosec^{-1} and $f(x)$ = an algebraic function of x.

Remember: We should remember the following formulas which gives the idea where to use which substitution:

1. (a) $\sqrt{1-\sin^2\theta} = \cos\theta$ **(b)** $\sqrt{1-\cos^2\theta} = \sin\theta$

2. (a) $\sqrt{1+\tan^2\theta} = \sec\theta$ **(b)** $\sqrt{\sec^2\theta - 1} = \tan\theta$

3. (a) $\sqrt{1+\cot^2\theta} = \text{cosec}\,\theta$ **(b)** $\sqrt{\text{cosec}^2\theta - 1} = \cot\theta$

4. (a) $1+\cos\theta = 2\cos^2\dfrac{\theta}{2}$ **(b)** $1-\cos\theta = 2\sin^2\dfrac{\theta}{2}$

5. (a) $1-2\sin^2\theta = \cos 2\theta$ **(b)** $2\cos^2\theta - 1 = \cos 2\theta$

6. $\cos 2\theta = \cos^2\theta - \sin^2\theta = 1-2\sin^2\theta = 2\cos^2\theta - 1$

7. $\sin 2\theta = 2\sin\theta\cos\theta$

8. $\sin 2\theta = \dfrac{2\tan\theta}{1+\tan^2\theta}$

9. $\cos 2\theta = \dfrac{1 - \tan^2\theta}{1 + \tan^2\theta}$

10. $\tan 2\theta = \dfrac{2\tan\theta}{1 - \tan^2\theta}$

11. $\sin 3\theta = 3\sin\theta - 4\sin^3\theta$

12. $\cos 3\theta = 4\cos^3\theta - 3\cos\theta$

13. $\tan 3\theta = \dfrac{3\tan\theta - \tan^3\theta}{1 - 3\tan^2\theta}$

14. (a) $\tan(\theta + \phi) = \dfrac{\tan\theta + \tan\phi}{1 - \tan\theta\tan\phi}$

(b) $\tan(\theta - \phi) = \dfrac{\tan\theta - \tan\phi}{1 + \tan\theta\tan\phi}$

15. (a) $\tan\left(\dfrac{\pi}{4} + \theta\right) = \dfrac{1 + \tan\theta}{1 - \tan\theta}$

(b) $\tan\left(\dfrac{\pi}{4} - \theta\right) = \dfrac{1 - \tan\theta}{1 + \tan\theta}$

2. (i) $\sin^{-1}[\sin x] = x$ for $-\dfrac{\pi}{2} \le x \le \dfrac{\pi}{2}$

(ii) $\cos^{-1}[\cos x] = x$ for $0 \le x \le \pi$

(iii) $\tan^{-1}[\tan x] = x$ for $-\dfrac{\pi}{2} < x < \dfrac{\pi}{2}$

Problems based on type 2

***Examples worked out*:**

Evaluate:

1. $\lim\limits_{x\to 0} \dfrac{1}{x}\sin^{-1}\left(\dfrac{2x}{1+x^2}\right)$

***Solution*:** We put $x = \tan\theta$, $-\dfrac{\pi}{4} \le \theta \le \dfrac{\pi}{4}$

$x \to 0 \Leftrightarrow \theta \to 0$

$$\therefore \frac{1}{x}\sin^{-1}\left(\frac{2x}{1+x^2}\right)$$

$$\Rightarrow \frac{1}{\tan\theta} \times \sin^{-1}\left(\frac{2\tan\theta}{1+\tan^2\theta}\right)$$

$$\Rightarrow \frac{1}{\tan\theta} \times \sin^{-1}(\sin 2\theta)$$

$$\Rightarrow \frac{1}{\tan\theta} \times 2\theta \text{ as } -\frac{\pi}{2} \le 2\theta \le \frac{\pi}{2}$$

$$\Rightarrow \frac{2\theta}{\tan\theta}$$

$$\therefore \lim_{x\to 0} \frac{1}{x}\sin^{-1}\left(\frac{2x}{1+x^2}\right) = \lim_{\theta\to 0} \frac{2\theta}{\tan\theta} = 2$$

2. $\lim\limits_{x\to 0} \dfrac{1}{x}\tan^{-1}\dfrac{2x}{1-x^2}$

***Solution*:** We put $x = \tan\theta$, $-\dfrac{\pi}{4} < \theta < \dfrac{\pi}{4}$

$x \to 0 \Leftrightarrow \theta \to 0$

$$\therefore \frac{1}{x}\tan^{-1}\frac{2x}{1-x^2}$$

$$\Rightarrow \frac{1}{\tan\theta} \times \tan^{-1}\left(\frac{2\tan\theta}{1-\tan^2\theta}\right)$$

$$\Rightarrow \frac{1}{\tan\theta} \times \tan^{-1}(\tan 2\theta)$$

$$\Rightarrow \frac{1}{\tan\theta} \times 2\theta \text{ as } -\frac{\pi}{2} < 2\theta < \frac{\pi}{2}$$

$$\Rightarrow \frac{2\theta}{\tan\theta}$$

$$\therefore \lim_{x\to 0} \frac{1}{x}\tan^{-1}\frac{2x}{1-x^2} = \lim_{\theta\to 0} \frac{2\theta}{\tan\theta}$$

$$= 2 \times \lim_{\theta\to 0} \frac{\theta}{\tan\theta} = 2 \times 1 = 2$$

3. $\lim_{x\to 0} \frac{1}{x} \tan^{-1}\left(\frac{3x - x^3}{1 - 3x^2}\right)$

Solution: we put $x = \tan\theta$, $-\frac{\pi}{6} < \theta < \frac{\pi}{6}$

$x \to 0 \Leftrightarrow \theta \to 0$

$$\therefore \frac{1}{x}\tan^{-1}\left(\frac{3x - x^3}{1 - 3x^2}\right)$$

$$\Rightarrow \frac{1}{\tan\theta}\tan^{-1}\left(\frac{3\theta - \tan^3\theta}{1 - 3\tan^2\theta}\right)$$

$$\Rightarrow \frac{1}{\tan\theta}\tan^{-1}(\tan 3\theta)$$

$$\Rightarrow \frac{1}{\tan\theta} \times 3\theta \text{ as } -\frac{\pi}{2} < 3\theta < \frac{\pi}{2}$$

$$\Rightarrow \frac{3\theta}{\tan\theta}$$

$$\therefore \lim_{x\to 0}\frac{1}{x}\tan^{-1}\left(\frac{3x - x^3}{1 - 3x^2}\right) = \lim_{\theta\to 0}\frac{3\theta}{\tan\theta}$$

$$= 3\cdot\lim_{\theta\to 0}\frac{\theta}{\tan\theta} = 3 \times 1 = 3.$$

Problems based on type 2

Exercise 4.19

Evaluate *Answers*

1. $\lim_{x\to 0}\frac{1}{x}\sin^{-1}\left(\frac{2x}{1 + x^2}\right)$ (2)

2. $\lim_{x\to 0}\frac{1}{x}\cos^{-1}\left(\frac{1 - x^2}{1 + x^2}\right)$ (2)

3. $\lim_{x\to 0}\frac{1}{x}\tan^{-1}\left(\frac{2x}{1 + x^2}\right)$ (2)

4. $\lim_{x\to 0}\frac{1}{x}\tan^{-1}\left(\frac{3x - x^3}{1 - 3x^2}\right)$ (3)

Type 3: If the given function contains the function $\sin^{-1}[f(x)]$ or $\tan^{-1}[f(x)]$, where $f(x)$ = an expression in x s.t $f(x) \to 0$ as $x \to a$ for 'a' being a constant.

Working rule:

1. Replace $\sin^{-1}[f(x)]$ by $f(x)$ and $\tan^{-1}[f(x)]$ by $f(x)$ in the given function provided $f(x) \to 0$ as $x \to a$.

2. Find the limit of the modified form of the given function (i.e.; the function obtained by substitution $\sin^{-1}[f(x)] = f(x)$ or $\tan^{-1}[f(x)]$ in the given function) as $x \to a$, where a = any constant = given limit of the independent variable x.

Remember:

1. If $f(x)$ is an infinitesimal as $x \to a$, then

(i) $\sin^{-1}[f(x)] \sim f(x)$ as $x \to a$

(ii) $\tan^{-1}[f(x) \sim f(x)$ as $x \to a$

2. The function $f(x)$ is called an infinitesimal as $x \to a$ if $\lim_{x\to a} f(x) = 0$ which means $\lim_{x\to a+0} f(x) = 0 = \lim_{x\to a-0} f(x)$.

i.e. *r.h.l* of $f(x)$ at $x = a$ is equal to the *l.h.l* of $f(x)$ at $x = a$.

From the definition of an infinitesimal, it follows that if $\lim_{x\to a} f(x) = b$, we may write $f(x) = b + f_1(x)$ where $f_1(x)$ is an infinitesimal (i.e. $\lim_{x\to a} f(x) = 0$)

Now, we shall state some basic results in the form of theorems.

Theorem 1: The algebraic sum or difference of two or more infinitesimals is an infinitesimal function.

Theorem 2: The product of an infinitesimal function (or simply infinitesimal) and a bounded function is an infinitesimal. This is most important theorem on infinitesimal which is widely used to find the limit of bounded function times a function tending to zero as the independent variable $x \to a$.

Theorem 3: The product of a finite number of infinitesimal function as $x \to a$ are also infinitesimals as $x \to a$.

Theorem 4: The quotient of an infinitesimal divided by a variable quantity tending to a non-zero limit is an infinitesimal.

N.B.: In particular, the product of a constant quantity by an infinitesimal is an infinitesimal.

Examples worked out

Evaluate:

1. $\lim\limits_{x\to 1} \dfrac{\tan^{-1}(1-x)^2}{x^2-2x+1}$

Solution: $\lim\limits_{x\to 1} \dfrac{\tan^{-1}(1-x)^2}{x^2-2x+1}, (x \neq 1)$

$$=\lim_{x\to 1} \frac{(1-x)^2}{x^2-2x+1} \left[\because f(x)=(1-x)^2 \to 0 \text{ as } x\to 1\right]$$

$$= \lim_{x\to 1} \frac{x^2-2x+1}{x^2-2x+1}$$

$$= \lim_{x\to 1} 1$$

$$= 1$$

2. $\lim\limits_{x\to 0} \dfrac{\left(\sqrt{1+x^2}-1\right)\sin^{-1}x}{\left(\tan^{-1}x\right)^3}$

Solution: $\lim\limits_{x\to 0} \dfrac{\left(\sqrt{1+x^2}-1\right)\sin^{-1}x}{\left(\tan^{-1}x\right)^3}$

$$= \lim_{x\to 0} \frac{\left(\sqrt{1+x^2}-1\right)x}{x^3} \quad [\because \sin^{-1}x \text{ and } \tan^{-1}x$$

are $\sim x$ as $x\to 0$]

$$= \lim_{x\to 0} \frac{\sqrt{1+x^2}-1}{x^2}$$

$$= \lim_{x\to 0} \frac{\left(\sqrt{1+x^2}-1\right)\left(\sqrt{1+x^2}+1\right)}{x^2\left(\sqrt{1+x^2}+1\right)}$$

$$= \lim_{x\to 0} \frac{1+x^2-1}{x^2\left(\sqrt{1+x^2}+1\right)}$$

$$= \lim_{x\to 0} \frac{x^2}{x^2\left(\sqrt{1+x^2}+1\right)}$$

$$= \lim_{x\to 0} \frac{1}{\left(\sqrt{1+x^2}+1\right)}$$

$$= \frac{1}{\sqrt{1}+1}$$

$$= \frac{1}{2}$$

3. $\lim\limits_{x\to 1} \dfrac{\tan^{-1}(x-1)^2}{1-\sin\dfrac{\pi x}{2}}$

Solution: $\lim\limits_{x\to 1} \dfrac{\tan^{-1}(x-1)^2}{1-\sin\dfrac{\pi x}{2}}$

$$= \lim_{x\to 1} \frac{(x-1)^2}{1-\sin\dfrac{\pi x}{2}}$$

We put $x = 1 + h$, $x\to 1 \Leftrightarrow h\to 0$

Hence, $\lim\limits_{x\to 1} \dfrac{(x-1)^2}{1-\sin\dfrac{\pi x}{2}}$

$$= \lim_{h \to 0} \frac{(1+h-1)^2}{1 - \sin\left(\frac{\pi}{2} + \frac{\pi h}{2}\right)}$$

$$= \lim_{h \to 0} \frac{h^2}{1 - \cos\frac{\pi h}{2}}$$

$$= \lim_{h \to 0} \frac{h^2}{2\sin^2\left(\frac{\pi h}{2}\right)} = \lim_{h \to 0} \frac{1}{\left[\frac{2\sin^2\left(\frac{\pi h}{2}\right)}{h^2}\right]}$$

$$= \lim_{h \to 0} \frac{1}{\left[\frac{\frac{2 \cdot \pi^2}{4} \cdot \sin^2\left(\frac{\pi h}{2}\right)}{\frac{h^2 \cdot \pi^2}{4}}\right]}$$

$$= \frac{1}{\frac{\pi^2}{2}} \cdot \lim_{x \to 1} \frac{1}{\frac{\sin^2\left(\frac{\pi h}{2}\right)}{\left(\frac{\pi h}{2}\right)^2}}$$

$$= \frac{1}{\frac{\pi^2}{2}} \times \frac{1}{\lim_{h \to 0}\left\{\frac{\sin\left(\frac{\pi h}{2}\right)}{\frac{\pi h}{2}}\right\}^2}$$

$$= \frac{1}{\frac{\pi^2}{2}} \times \frac{1}{1}$$

$$= \frac{2}{\pi^2}$$

4. $\lim_{x \to a} \frac{\sin^{-1}\sqrt{x} - \sin^{-1}\sqrt{a}}{x - a}$

Solution: $\sin^{-1}\sqrt{x} - \sin^{-1}\sqrt{a}$

$= \sin^{-1}\left(\sqrt{x}\sqrt{1-a} - \sqrt{a}\sqrt{1-x}\right)$ as

$\sqrt{x} > 0,\ \sqrt{a} > 0$

$= \left(\sqrt{x}\sqrt{1-a} - \sqrt{a}\sqrt{1-x}\right)$ as $x \to a$

$\left[\because f(x) = \left(\sqrt{x}\sqrt{1-a} - \sqrt{a}\sqrt{1-x}\right) \to 0 \text{ as } x \to a\right]$

$$\therefore \lim_{x \to a} \frac{\sin^{-1}\sqrt{x} - \sin^{-1}\sqrt{a}}{x - a}$$

$$= \lim_{x \to a} \frac{\left(\sqrt{x}\sqrt{1-a} - \sqrt{a}\sqrt{1-x}\right)}{x - a}$$

Now, $\frac{\sqrt{x}\sqrt{1-a} - \sqrt{a}\sqrt{1-x}}{x - a}$

$$= \frac{\left(\sqrt{x}\cdot\sqrt{1-a} - \sqrt{a}\cdot\sqrt{1-x}\right)\cdot\left(\sqrt{x}\sqrt{1-a} + \sqrt{a}\sqrt{1-x}\right)}{(x-a)\left(\sqrt{x}\sqrt{1-a} + \sqrt{a}\sqrt{1-x}\right)}$$

$$= \frac{\left(\sqrt{x}\sqrt{1-a}\right)^2 - \left(\sqrt{a}\sqrt{1-x}\right)^2}{(x-a)\left(\sqrt{x}\sqrt{1-a} + \sqrt{a}\sqrt{1-x}\right)}$$

$$= \frac{x(1-a) - a(1-x)}{(x-a)\left(\sqrt{x}\sqrt{1-a} + \sqrt{a}\sqrt{1-x}\right)}$$

$$= \frac{x - ax - a + ax}{(x-a)\left(\sqrt{x}\sqrt{1-a} + \sqrt{a}\sqrt{1-x}\right)}$$

$$= \frac{(x-a)}{(x-a)\left(\sqrt{x}\sqrt{1-a} + \sqrt{a}\sqrt{1-x}\right)}$$

$$= \frac{1}{\left(\sqrt{x}\sqrt{1-a} + \sqrt{a}\sqrt{1-x}\right)}$$

Hence, $\lim_{x\to a} \dfrac{\sqrt{x}\sqrt{1-a}-\sqrt{a}\sqrt{1-x}}{x-a}$

$$= \lim_{x\to a} \frac{1}{\left(\sqrt{x}\sqrt{1-a}+\sqrt{a}\sqrt{1-x}\right)}$$

$$= \frac{1}{\sqrt{a}\sqrt{1-a}+\sqrt{a}\sqrt{1-a}}$$

$$= \frac{1}{2\sqrt{a}\cdot\sqrt{1-a}}$$

5. $\lim_{x\to a} \dfrac{\tan^{-1} x - \tan^{-1} a}{x-a}$

Solution: $\tan^{-1} x - \tan^{-1} a = \tan^{-1}\left[\dfrac{x-a}{1+ax}\right]$ for $a\cdot x > -1$

$$= \frac{x-a}{1+ax}\left(\because f(x) = \frac{x-a}{1+ax} \to 0 \text{ as } x \to a\right)$$

$$\therefore \lim_{x\to a} \frac{\tan^{-1} x - \tan^{-1} a}{x-a}$$

$$= \lim_{x\to a}\left(\frac{x-a}{1+ax} \times \frac{}{x-a}\right)$$

$$= \lim_{x\to a} \frac{1}{1+ax}$$

$$= \frac{1}{1+a\cdot a} = \frac{1}{1+a^2}$$

Problems based on type 3

Exercise 4.20

Find the limits of the following functions:

Answers

1. $\lim_{x\to 0} \dfrac{2x - \sin^{-1} x}{2x + \tan^{-1} x}$ $\left(\dfrac{1}{3}\right)$

2. $\lim_{x\to 1} \dfrac{\tan^{-1}(x-1)^2}{x^2 - 2x + 1}$ (1)

3. $\lim_{x\to 1} \dfrac{\tan^{-1} x - \tan^{-1} a}{x-a}$ $\left(\dfrac{1}{1+a^2}\right)$

4. $\lim_{x\to 0} \dfrac{\left(\sqrt{1+x^2}-1\right)\sin^{-1} x}{\left(\tan^{-1} x\right)^3}$ $\left(\dfrac{1}{2}\right)$

5. $\lim_{x\to 0} \dfrac{\tan^{-1} 2x}{\sin 3x}$ $\left(\dfrac{2}{3}\right)$

6. $\lim_{x\to 0} \dfrac{\sin 2x + \left(\sin^{-1} x\right)^2 - \left(\tan^{-1} x\right)^2}{3x}$ $\left(\dfrac{2}{3}\right)$

7. $\lim_{x\to 0} \dfrac{\sin^{-1} 3x}{2x}$ $\left(\dfrac{3}{2}\right)$

8. $\lim_{x\to 0} \dfrac{\sin^{-1} 2x}{7x}$ $\left(\dfrac{2}{7}\right)$

9. $\lim_{x\to 0} \dfrac{\tan^{-1} 5x}{3x}$ $\left(\dfrac{5}{3}\right)$

10. $\lim_{x\to 0} \dfrac{\sin^{-1} 2x}{\tan^{-1} 3x}$ $\left(\dfrac{2}{3}\right)$

On limits of exponential function

Evaluation of $\lim_{x\to a} (f(x))^{g(x)}$, where a = 0/any constant/ ∞

One may adopt any one of the two methods to find the limit of the exponential function $(f(x))^{g(x)}$ as $x \to a$, a being either a constant or ∞.

Method 1: It consists of following steps.

Step 1: Put $y = (f(x))^{g(x)}$

Step 2: Use $\log y = \log (f(x))^{g(x)} = g(x) \cdot \log f(x)$

Step 3: Write $\lim_{x\to a} \log y = \lim_{x\to a} g(x) \cdot \log f(x) = b$ (say)

Step 4: Required limit of the given exponential function

$$= \lim_{x \to a} (f(x))^{g(x)} = e^{b}$$

Notes: **1.** The limit of logarithm of a function is the logarithm of the limit of the function as logarithm is a continuous function, i.e., $\lim_{x \to a} \log (f(x))$

$= \log \left\{ \lim_{x \to a} f(x) \right\}$ as logarithmic function is a continuous function.

2. $a^{x} = m \Leftrightarrow x = \log_a m$, for a positive real number "m" and a positive real number "$a \neq 1$".

3. The method of expansion for evaluating limits is applicable to the functions which can be expanded in series.

4. Method (1) is applicable commonly to find the limit of an exponential function $(f(x))^{g(x)}$ as $x \to a$ or $x \to \infty$.

5. One must note that if $f(x)$ is not throughout positive in the neighbourhood of $x = a$, then $\lim_{x \to a} (f(x))^{g(x)}$ does not exist because in this case the function is not defined in the neighbourhood of $x = a$.

Method 2: It consists of following steps.

Step 1: Put $x = a + h$ in $y = (f(x))^{g(x)}$

Step 2: Use $\log y = \log (f(a+h))^{g(a+h)} = g(a+h) \cdot \log (f(a+h))$

Step 3: Write $\lim_{h \to 0} \log y = \lim_{h \to 0} (g(a+h) \cdot \log f(a+h)) = b$ (say)

Step 4: Required limit of the given exponential function

$$= \lim_{x \to a} (f(x))^{g(x)} = e^{b}$$

Notes: **1.** One can use the following expansion if required in step (2) in any method mentioned above.

(i) $\log(1 + x) = x - \frac{x^2}{2} + \frac{x^3}{3} - \frac{x^4}{4} + \ldots \infty$ for $-1 < x \leq 1$

(ii) $\log(1 - x) = -x - \frac{x^2}{2} - \frac{x^3}{3} - \frac{x^4}{4} + \ldots \infty$ for $-1 \leq x < 1$

(iii) $e^{x} = 1 + \frac{x}{\lfloor 1} + \frac{x^2}{\lfloor 2} + \frac{x^3}{\lfloor 3} + \ldots \infty$

(iv) $e^{-x} = 1 - \frac{x}{\lfloor 1} + \frac{x^2}{\lfloor 2} - \frac{x^3}{\lfloor 3} + \ldots \infty$

(v) $a^{x} = 1 + \frac{x \log a}{\lfloor 1} + \frac{x^2 \log^2 a}{\lfloor 2} + \frac{x^3 \log^3 a}{\lfloor 3} + \ldots \infty$

2. Method (2) is applicable commonly to find the limit of the exponential function $(f(x))^{g(x)}$ as $x \to$ any constant other than (different from) zero.

Examples worked out:

Evaluate the following:

1. $\lim_{x \to 0} (1 + x)^{\frac{1}{x}}$

Solution: Let $y = (1 + x)^{\frac{1}{x}}$ ...(i)

On taking logarithm on both sides of (i), we get

$$\log y = \log(1 + x)^{\frac{1}{x}} = \frac{1}{x} \log(1 + x)$$

$$= \frac{1}{x} \left(x - \frac{x^2}{\lfloor 2} + \frac{x^3}{\lfloor 3} - \frac{x^4}{\lfloor 4} + \ldots \right)$$

$$= \left(1 - \frac{x}{\lfloor 2} + \frac{x^2}{\lfloor 3} - \frac{x^3}{\lfloor 4} + \ldots \right) \quad \text{...(ii)}$$

On taking limit on both sides of (ii) as $x \to 0$, we get

$$\lim_{x \to 0} (\log y) = \lim_{x \to 0} \left(1 - \frac{x}{\lfloor 2} + \frac{x^2}{\lfloor 3} - \frac{x^3}{\lfloor 4} + \ldots \right) = \left(1 - \frac{0}{\lfloor 2} + \frac{0^2}{\lfloor 3} + \ldots \right)$$

$$\Rightarrow \log \left(\lim_{x \to 0} y \right) = 1 = \log e \ (\because \log e = \log_e e = 1)$$

$$\Rightarrow \lim_{x \to 0} y = e$$

$$\Rightarrow \lim_{x \to 0} (1 + x)^{\frac{1}{x}} = e$$

2. $\lim\limits_{x\to\infty}\left(1+\dfrac{1}{x}\right)^x$

Solution: Letting $y=\left(1+\dfrac{1}{x}\right)^x$

$$\Rightarrow \log y = x\log\left(1+\frac{1}{x}\right)=x\left(\frac{1}{x}-\frac{1}{2x^2}+\frac{1}{2x^3}+\ldots\right)$$

$$=\left(1-\frac{1}{2x}+\frac{1}{3x^2}+\ldots\right)$$

$$\Rightarrow \lim_{x\to\infty}(\log y)=\lim_{x\to\infty}\left(1-\frac{1}{2x}+\frac{1}{3x^2}+\ldots\right)=1$$

$$\Rightarrow \log\left(\lim_{x\to\infty} y\right)=1=\log e$$

$$\Rightarrow \lim_{x\to\infty} y = e$$

$$\Rightarrow \lim_{x\to\infty}\left(1+\frac{1}{x}\right)^x=e$$

Remark: One should note that

(i) $\lim\limits_{x\to 0}(1+x)^{\frac{1}{x}}=e$

(ii) $\lim\limits_{x\to 0}(1-x)^{\frac{1}{x}}=\dfrac{1}{e}$

(iii) $\lim\limits_{x\to\infty}\left(1+\dfrac{1}{x}\right)^x=e$ and

(iv) $\lim\limits_{x\to\infty}\left(1-\dfrac{1}{x}\right)^x=\dfrac{1}{e}$; Further we must note that a function in x appearing as an index is always reciprocal of the function in x within the bracket.

3. $\lim\limits_{x\to\frac{\pi}{2}}(\sin x)^{\tan x}$

Solution: Letting $y=(\sin x)^{\tan x}$ and putting $x=\dfrac{\pi}{2}+h$

$$\Rightarrow y=\left(\sin\left(\frac{\pi}{2}+h\right)\right)^{\tan\left(\frac{\pi}{2}+h\right)}=(\cos h)^{-\cot h}$$

$$\Rightarrow \log y=\log(\cos h)^{-\cot h}=-\cot h\log(\cos h)=-\frac{1}{\tan h}\cdot\log(\cos h)$$

$$\Rightarrow \lim_{x\to\frac{\pi}{2}}(\log y)=\lim_{h\to 0}\left(-\frac{1}{\tan h}\cdot\log(\cos h)\right)$$

$$=\lim_{h\to 0}\left(\frac{-\log\left(1-\dfrac{h^2}{\underline{|2}}+\ldots\right)}{h+\dfrac{h^3}{3}+\dfrac{2}{15}h^5+\ldots}\right)$$

$$=\lim_{h\to 0}\left(\frac{\dfrac{h^2}{2}+\ldots}{h+\dfrac{h^3}{3}+\dfrac{2}{15}h^5+\ldots}\right)$$

$$\left(\because \log(1-x)=-x-\frac{x^2}{\underline{|2}}-\frac{x^3}{\underline{|3}}-\ldots\right)$$

$$=\lim_{h\to 0}\left(\frac{1}{2}h\left(1+\frac{h^2}{3}\right)^{-1}\right)$$

$$=0$$

4. $\lim\limits_{x\to 1}(x)^{\frac{1}{x-1}}$

Solution: Putting $x=1+h$ in $(x)^{\frac{1}{x-1}}$, we have

$$\lim_{x\to 1}(x)^{\frac{1}{x-1}}=\lim_{h\to 0}(1+h)^{\frac{1}{h}}=e\ (\because x\to 1\Leftrightarrow h\to 0)$$

Exercise 4.21

Evaluate the following:

1. $\lim\limits_{x\to 0}(1+x)^{\frac{1}{x}}$

2. $\lim_{x \to 0} (1 + \alpha x)^{\frac{1}{x}}$

3. $\lim_{x \to 0} \left(1 - \frac{3x}{2}\right)^{\frac{4}{x}}$

4. $\lim_{x \to 1} (1 + \sin \pi x)^{\cot \pi x}$

5. $\lim_{x \to 1} \left(1 - \cos\frac{\pi x}{2}\right)^{\tan\frac{\pi x}{x}}$

6. $\lim_{x \to \frac{\pi}{2}} (1 + \cos x)^{3\sec x}$

7. $\lim_{x \to 0} (1 - x)^{\frac{1}{x}}$

8. $\lim_{x \to 0} (1 + mx)^{\frac{m}{x}}$

Answers:

1. e **2.** e^{α} **3.** e^{-6} **4.** $\frac{1}{e}$ **5.** $\frac{1}{e}$ **6.** e^3 **7.** $\frac{1}{e}$ **8.** e^{m^2}

Problems reducible to log $(f(x))^{g(x)}$

Evaluate the following.

1. $\lim_{x \to 0} \frac{\log(1 + x)}{x}$

Solution: Let $y = \frac{\log(1 + x)}{x}$

$$\therefore y = \log(1 + x)^{\frac{1}{x}}$$

$$\Rightarrow \lim_{x \to 0} y = \lim_{x \to 0} \left(\log(1 + x)^{\frac{1}{x}}\right)$$

$$= \log\left(\lim_{x \to 0} (1 + x)^{\frac{1}{x}}\right)$$

$$= \log e \left(\because \lim_{x \to 0} (1 + x)^{\frac{1}{x}} = e\right)$$

$$= 1 \; (\because \log e = \log_e e = 1)$$

2. $\lim_{x \to 0} \left(\frac{a^x - 1}{x}\right)$ for $a > 0$

Solution: let $f(x) = \left(\frac{a^x - 1}{x}\right)$ ...(i)

And $a^x = 1 + h$, where $h \to 0$ as $x \to 0$

$\left(\because a^0 = 1\right)$ ...(ii)

On taking logarithm on both sides of (ii) with base 'e', we get

$$\log_e (a^x) = \log_e (1 + h)$$

$$\Rightarrow x \log_e a = \log_e (1 + h)$$

$$\Rightarrow x = \frac{\log(1 + h)}{\log a}$$

$$\Rightarrow f(x) = \frac{1 + h - 1}{\left(\frac{\log(1 + h)}{\log a}\right)} = \frac{h \cdot \log a}{\log(1 + h)} = \frac{\log a}{\frac{1}{h} \cdot \log(1 + h)} =$$

$$\frac{\log a}{\log(1 + h)^{\frac{1}{h}}}$$

$$\Rightarrow \lim_{x \to 0} (f(x)) = \lim_{h \to 0} \left(\frac{\log a}{\log(1 + h)^{\frac{1}{h}}}\right) = \frac{\log a}{\log e}$$

$$\left(\because \lim_{h \to 0} (1 + h)^{\frac{1}{h}} = e\right)$$

$$= \log a \; (\because \log e = \log_e e = 1)$$

3. $\lim_{x \to 0} \left(\frac{e^x - 1}{x}\right)$

Solution: Putting $a = e$ in the above solution of (2), we have

$$\lim_{x \to 0} \left(\frac{e^x - 1}{x}\right) = \log e = 1$$

Problems put in the form: $y = \left(\frac{1 \pm m_1 x}{1 \pm m_2 x}\right)^{\frac{m}{x}}$ and $x \to 0$

Working rule: To evaluate $\lim_{x \to 0} \left(\frac{1 \pm m_1 x}{1 \pm m_2 x}\right)^{\frac{m}{x}}$, one may adopt the rule consisting of following steps.

Step 1: To put the given index $\frac{m}{x} = \frac{1}{m_1 x} \cdot m_1 m$ in *Nr* and $\frac{m}{x} = \frac{1}{m_2 x} \cdot m_2 m$ in *Dr*.

Step 2: To evaluate $\left\{\lim_{x\to 0} (1 \pm m_1 x)^{\frac{1}{m_1 x}}\right\}^{m_1 \cdot m}$ … (A_1) (say)

and $\left\{\lim_{x\to 0} (1 \pm m_2 x)^{\frac{1}{m_2 x}}\right\}^{m_2 \cdot m}$ … (A_2) (say)

Step 3: To find the quotient of (A_1) and (A_2) to obtain the required limit, $\lim_{x\to 0}\left(\frac{(1 \pm m_1 x)}{(1 \pm m_2 x)}\right)$

Examples worked out:

Evaluate the following:

1. $\lim_{x\to 0}\left(\frac{1+x}{1-x}\right)^{\frac{1}{x}}$

Solution: $\because \left(\frac{1+x}{1-x}\right)^{\frac{1}{x}} = \frac{(1+x)^{\frac{1}{x}}}{(1-x)^{\frac{1}{x}}}$

$$= \frac{(1+x)^{\frac{1}{x}}}{\left\{(1-x)^{\left(-\frac{1}{x}\right)}\right\}^{(-1)}}$$

$$\therefore \lim_{x\to 0}\left(\frac{1+x}{1-x}\right)^{\frac{1}{x}} = \lim_{x\to 0}\frac{(1+x)^{\frac{1}{x}}}{\left\{(1-x)^{\left(-\frac{1}{x}\right)}\right\}^{(-1)}}$$

$$= \frac{\lim_{x\to 0}(1+x)^{\frac{1}{x}}}{\lim_{x\to 0}\left\{(1-x)^{\left(-\frac{1}{x}\right)}\right\}^{(-1)}}$$

$$= \frac{e}{\left\{\lim_{x\to 0}(1-x)^{\left(-\frac{1}{x}\right)}\right\}^{(-1)}} = \frac{e}{e^{-1}} = e^2$$

2. $\lim_{x\to 0}\left(\frac{1+2x}{1-2x}\right)^{\frac{1}{x}}$

Solution: $\because \left(\frac{1+2x}{1-2x}\right)^{\frac{1}{x}} = \frac{(1+2x)^{\frac{1}{x}}}{(1-2x)^{\frac{1}{x}}} = \frac{(1+2x)^{\left(\frac{1}{2x}\right)\cdot 2}}{(1-2x)^{\left(-\frac{1}{2x}\right)(-2)}}$

$$\therefore \lim_{x\to 0}\left(\frac{1+2x}{1-2x}\right)^{\frac{1}{x}} = \lim_{x\to 0}\frac{(1+2x)^{\left(\frac{1}{2x}\right)\cdot 2}}{(1-2x)^{\left(-\frac{1}{2x}\right)(-2)}}$$

$$= \frac{\left\{\lim_{x\to 0}(1+2x)^{\frac{1}{2x}}\right\}^2}{\left\{\lim_{x\to 0}(1-2x)^{\left(-\frac{1}{2x}\right)}\right\}^{(-2)}}$$

$$= \frac{e^2}{e^{-2}} = e^4$$

3. $\lim_{x\to 0}\left(\frac{2+x}{2-x}\right)^{\frac{1}{x}}$

Solution: $\because \left(\frac{2+x}{2-x}\right)^{\frac{1}{x}} = \left(\frac{1+\frac{x}{2}}{1-\frac{x}{2}}\right)^{\frac{1}{x}}$

$$= \frac{\left(1+\frac{x}{2}\right)^{\left(\frac{2}{x}\right)\cdot\left(\frac{1}{2}\right)}}{\left(1-\frac{x}{2}\right)^{\left(-\frac{2}{x}\right)\left(-\frac{1}{2}\right)}}$$

$$\therefore \lim_{x\to 0}\left(\frac{2+x}{2-x}\right)^{\frac{1}{x}} = \frac{\left\{\lim_{x\to 0}\left(1+\frac{x}{2}\right)^{\frac{2}{x}}\right\}^{\frac{1}{2}}}{\left\{\lim_{x\to 0}\left(1-\frac{x}{2}\right)^{\left(-\frac{2}{x}\right)}\right\}^{\left(-\frac{1}{2}\right)}}$$

$$= \frac{e^{\frac{1}{2}}}{e^{-\frac{1}{2}}} = e$$

4. $\lim_{x\to 0}\left(1+\frac{5}{7}x\right)^{\frac{2}{x}}$

Solution: $\because \left(1+\frac{5}{7}x\right)^{\frac{2}{x}} = \left(1+\frac{5x}{7}\right)^{\left(\frac{7}{5x}\right)\cdot\left(\frac{10}{7}\right)}$

$$\therefore \lim_{x\to 0}\left(1+\frac{5}{7}x\right)^{\frac{2}{x}} = \lim_{x\to 0}\left\{\left(1+\frac{5x}{7}\right)^{\left(\frac{7}{5x}\right)\cdot\left(\frac{10}{7}\right)}\right\}$$

$$= \left\{\lim_{x\to 0}\left(1+\frac{5x}{7}\right)^{\left(\frac{7}{5x}\right)}\right\}^{\left(\frac{10}{7}\right)} = e^{\frac{10}{7}}$$

5. $\lim_{x\to 0}\left(1-\frac{3x}{5}\right)^{\frac{5}{x}}$

Solution: $\because \left(1-\frac{3x}{5}\right)^{\frac{5}{x}} = \left(1-\frac{3x}{5}\right)^{\left(-\frac{5}{3x}\right)(-3)}$

$$\therefore \lim_{x\to 0}\left(1-\frac{3x}{5}\right)^{\frac{5}{x}} = \lim_{x\to 0}\left\{\left(1-\frac{3x}{5}\right)^{\left(-\frac{5}{3x}\right)(-3)}\right\}$$

$$= \left\{\lim_{x\to 0}\left(1-\frac{3x}{5}\right)^{\left(-\frac{5}{3x}\right)}\right\}^{(-3)} = e^{-3}$$

Exercise 4.22

Evaluate the following;

1. $\lim_{x\to 0}(1+2x)^{\frac{1}{x}}$

2. $\lim_{x\to 0}\left(\frac{1+3x}{1-3x}\right)^{\frac{1}{x}}$

3. $\lim_{x\to 0}(1-x)^{\frac{4}{x}}$

Answers:

1. e^2 **2.** e^6 **3.** e^{-4}

Problems put in the form:

$$y = \left(\frac{1\pm\frac{m_1}{x}}{1\pm\frac{m_2}{x}}\right)^{mx} \text{ and } x\to\infty$$

Working rule: To evaluate $\lim_{x\to\infty}\left(\frac{1\pm\frac{m_1}{x}}{1\pm\frac{m_2}{x}}\right)^{mx}$, one may adopt the rule consisting of following steps.

Step 1: To put the given index $mx = \left(\frac{x}{m_1}\right)\cdot(m m_1)$ in *Nr* and $mx = \left(\frac{x}{m_2}\right)\cdot(m\cdot m_2)$ in *Dr*.

Step 2: To evaluate $\left\{\lim_{x\to\infty}\left(1\pm\frac{m_1}{x}\right)^{\left(\frac{x}{m_1}\right)}\right\}^{(m m_1)}$ ...(A_1) (say)

and $\left\{\lim_{x\to\infty}\left(1\pm\frac{m_2}{x}\right)^{\left(\frac{x}{m_2}\right)}\right\}^{(m m_2)}$ ... (A_2) (say)

Step 3: To find the quotient of (A_1) and (A_2) to obtain the required limit, $\lim_{x\to\infty}\left(\frac{1\pm\frac{m_1}{x}}{1\pm\frac{m_2}{x}}\right)^{mx}$

Examples worked out:
Evaluate the following:

1. $\lim_{x\to\infty}\left(\frac{2x-1}{2x+1}\right)^{x}$

Solution: $\because \left(\frac{2x-1}{2x+1}\right)^{x} = \frac{\left(1-\frac{1}{2x}\right)^{2x}}{\left(1+\frac{1}{2x}\right)^{2x}}$

$$= \frac{\left(1-\frac{1}{2x}\right)^{(-2x)\left(-\frac{1}{2}\right)}}{\left(1+\frac{1}{2x}\right)^{(2x)\left(\frac{1}{2}\right)}}$$

$$\therefore \lim_{x\to\infty}\left(\frac{2x-1}{2x+1}\right)^x = \frac{\left\{\lim_{x\to\infty}\left(1-\frac{1}{2x}\right)^{(-2x)}\right\}^{\left(-\frac{1}{2}\right)}}{\left\{\lim_{x\to\infty}\left(1+\frac{1}{2x}\right)^{(2x)}\right\}^{\left(\frac{1}{2}\right)}}$$

$$= \frac{e^{-\frac{1}{2}}}{e^{\frac{1}{2}}} = e^{-\frac{1}{2}-\frac{1}{2}} = e^{-1}$$

2. $\lim_{x\to\infty}\left(\frac{2x+3}{2x+1}\right)^{x+1}$

Solution: $\left(\frac{2x+3}{2x+1}\right)^{x+1} = \frac{\left(1+\frac{3}{2x}\right)^{x+1}}{\left(1+\frac{1}{2x}\right)^{x+1}}$

$$= \frac{\left(1+\frac{3}{2x}\right)^x\cdot\left(1+\frac{3}{2x}\right)}{\left(1+\frac{1}{2x}\right)^x\cdot\left(1+\frac{1}{2x}\right)}$$

$$= \frac{\left(1+\frac{3}{2x}\right)^{\left(\frac{2x}{3}\right)\left(\frac{3}{2}\right)}\cdot\left(1+\frac{3}{2x}\right)}{\left(1+\frac{1}{2x}\right)^{(2x)\cdot\left(\frac{1}{2}\right)}\cdot\left(1+\frac{1}{2x}\right)}$$

$$\therefore \lim_{x\to\infty}\left(\frac{2x+3}{2x+1}\right)^{x+1}$$

$$= \frac{\left\{\lim_{x\to\infty}\left(1+\frac{3}{2x}\right)^{\left(\frac{2x}{3}\right)}\right\}^{\left(\frac{3}{2}\right)}\cdot\lim_{x\to\infty}\left(1+\frac{3}{2x}\right)}{\left\{\lim_{x\to\infty}\left(1+\frac{1}{2x}\right)^{2x}\right\}^{\left(\frac{1}{2}\right)}\cdot\lim_{x\to\infty}\left(1+\frac{1}{2x}\right)}$$

$$= \frac{e^{\frac{3}{2}}\times 1}{e^{\frac{1}{2}}}$$

$$= e$$

3. $\lim_{x\to\infty}\left(\frac{x+4}{x-3}\right)^x$

Solution: $\left(\frac{x+4}{x-3}\right)^x = \frac{\left(1+\frac{4}{x}\right)^x}{\left(1-\frac{3}{x}\right)^x}$

$$= \frac{\left(1+\frac{4}{x}\right)^{\left(\frac{x}{4}\right)(4)}}{\left(1-\frac{3}{x}\right)^{\left(-\frac{x}{3}\right)(-3)}}$$

$$\therefore \lim_{x\to\infty}\left(\frac{x+4}{x-3}\right)^x = \frac{\left\{\lim_{x\to\infty}\left(1+\frac{4}{x}\right)^{\left(\frac{x}{4}\right)}\right\}^{(4)}}{\left\{\lim_{x\to\infty}\left(1-\frac{3}{x}\right)^{\left(-\frac{x}{3}\right)}\right\}^{(-3)}}$$

$$= \frac{e^4}{e^{-3}} = e^7$$

4. $\lim_{x\to\infty}\left(\frac{x+3}{x+2}\right)^x$

Solution: $\because \left(\frac{x+3}{x+2}\right)^x = \frac{\left(1+\frac{3}{x}\right)^x}{\left(1+\frac{2}{x}\right)^x} = \frac{\left(1+\frac{3}{x}\right)^{\left(\frac{x}{3}\right)(3)}}{\left(1+\frac{2}{x}\right)^{\left(\frac{x}{2}\right)(2)}}$

$$\therefore \lim_{x\to\infty}\left(\frac{x+3}{x+2}\right) = \frac{\left\{\lim_{x\to\infty}\left(1+\frac{3}{x}\right)^{\left(\frac{x}{3}\right)}\right\}^{(3)}}{\left\{\lim_{x\to\infty}\left(1+\frac{2}{x}\right)^{\left(\frac{x}{2}\right)}\right\}^{(2)}}$$

$$= \frac{e^3}{e^2} = e$$

5. $\lim\limits_{x\to\infty}\left(1+\frac{a}{x}\right)^x$

Solution: $\because \left(1+\frac{a}{x}\right)^x = \left(1+\frac{a}{x}\right)^{\left(\frac{x}{a}\right)(a)}$

$$\therefore \lim_{x\to\infty}\left(1+\frac{a}{x}\right)^x = \left\{\lim_{x\to\infty}\left(1+\frac{a}{x}\right)^{\left(\frac{x}{a}\right)}\right\}^{(a)} = e^a$$

6. $\lim\limits_{x\to\infty}\left(1-\frac{5}{x}\right)^x$

Solution: $\because \left(1-\frac{5}{x}\right)^x = \left(1-\frac{5}{x}\right)^{\left(-\frac{x}{5}\right)(-5)}$

$$\therefore \lim_{x\to\infty}\left(1-\frac{5}{x}\right)^x = \left\{\lim_{x\to\infty}\left(1-\frac{5}{x}\right)^{\left(-\frac{x}{5}\right)}\right\}^{(-5)} = e^{-5}$$

7. $\lim\limits_{x\to\infty}\left(\frac{x}{1+x}\right)^x$

Solution: $\because \left(\frac{x}{1+x}\right)^x = \left(\frac{1}{1+\frac{1}{x}}\right)^x = \frac{1}{\left(1+\frac{1}{x}\right)^x}$

$$\therefore \lim_{x\to\infty}\left(\frac{x}{1+x}\right)^x = \frac{1}{\lim\limits_{x\to\infty}\left(1+\frac{1}{x}\right)^x} = \frac{1}{e} = e^{-1}$$

Exercise 4.23

Evaluate the following:

1. $\lim\limits_{x\to\infty}\left(1+\frac{1}{x}\right)^{x+\alpha}$, $(\alpha \in R)$

2. $\lim\limits_{x\to\infty}\left(1+\frac{1}{x}\right)^{\alpha x}$, $(\alpha \in R)$

3. $\lim\limits_{x\to\infty}\left(1+\frac{\alpha}{x}\right)^{x}$, $(\alpha \in R)$

4. $\lim\limits_{x\to\infty}\left(1-\frac{1}{x}\right)^{x}$

5. $\lim\limits_{x\to\infty}\left(\frac{3x+2}{3x-1}\right)^{x+2}$

6. $\lim\limits_{x\to\infty}\left(\frac{2x+3}{2x+1}\right)^{2x+5}$

7. $\lim\limits_{x\to\infty}\left(\frac{x}{1+x}\right)^{x}$

8. $\lim\limits_{x\to\infty}\left(1+\frac{1}{x}\right)^{x+5}$

Answers:

1. e **2.** e^α **3.** e^α **4.** e^{-1} **5.** e **6.** e^2 **7.** e^{-1} **8.** e

Problems reducible to the form:

$$y = \frac{a^x - 1}{x}, (a > 0) \text{ and } / \frac{e^x - 1}{x}$$

Remember:

(i) $\lim\limits_{x\to 0}\left(\frac{a^x - 1}{x}\right) = \log_e a$, $(a > 0)$ and

(ii) $\lim\limits_{x\to 0}\left(\frac{e^x - 1}{x}\right) = \log_e e = 1$

which mean $\lim\limits_{x\to 0}\frac{(\text{any positive constant})^x - 1}{x}$

$= \log_e$ (the same positive constant)

Examples worked out:

Evaluate the following ones.

1. $\lim\limits_{x\to 0}\left(\frac{e^{ax} - 1}{x}\right)$

Solution: $\because \frac{e^{ax} - 1}{x} = \frac{\left(e^a\right)^x - 1}{x}$

$$\therefore \lim_{x\to 0}\left(\frac{e^{ax}-1}{x}\right)=\lim_{x\to 0}\frac{\left(e^{a}\right)^{x}-1}{x}$$

$$=\log_e e^{a}=a\log_e e=a\cdot 1=a$$

2. $\lim_{x\to 0}\left(\frac{e^{-x}-1}{x}\right)$

Solution: $\because \frac{e^{-x}-1}{x}=\frac{\left(e^{-1}\right)^{x}-1}{x}$

$$\therefore \lim_{x\to 0}\left(\frac{e^{-x}-1}{x}\right)=\lim_{x\to 0}\frac{\left(e^{-1}\right)^{x}-1}{x}$$

$$=\log_e e^{-1}=(-1)\cdot\log_e e=-1$$

3. $\lim_{x\to 0}\left(\frac{a^{2x}-1}{x}\right)$

Solution: $\because \frac{a^{2x}-1}{x}=\frac{\left(a^{2}\right)^{x}-1}{x}$

$$\therefore \lim_{x\to 0}\left(\frac{a^{2x}-1}{x}\right)=\lim_{x\to 0}\frac{\left(a^{2}\right)^{x}-1}{x}$$

$$=\log_e a^{2}=2\log_e a=2\log a$$

4. $\lim_{x\to 0}\left(\frac{a^{mx}-1}{x}\right)$ for $a>0$

Solution: $\because \frac{a^{mx}-1}{x}=\frac{\left(a^{m}\right)^{x}-1}{x}$

$$\therefore \lim_{x\to 0}\left(\frac{a^{mx}-1}{x}\right)=\lim_{x\to 0}\frac{\left(a^{m}\right)^{x}-1}{x}$$

$$=\log_e a^{m}=m\log_e a=m\log a$$

5. $\lim_{x\to 0}\left(\frac{3^{5x}-1}{x}\right)$

Solution: $\because \frac{3^{5x}-1}{x}=\frac{\left(3^{5}\right)^{x}-1}{x}$

$$\therefore \lim_{x\to 0}\left(\frac{3^{5x}-1}{x}\right)=\lim_{x\to 0}\frac{\left(3^{5}\right)^{x}-1}{x}$$

$$=\log_e 3^{5}=5\log_e 3=5\log 3$$

6. $\lim_{x\to 0}\left(\frac{e^{ax}-e^{a}}{x}\right)$

Solution: $\because \frac{e^{ax}-e^{a}}{x}=\frac{e^{a}\cdot e^{x}-e^{a}}{x}=\frac{e^{a}\left(e^{x}-1\right)}{x}$

$$\therefore \lim_{x\to 0}\left(\frac{e^{ax}-e^{a}}{x}\right)=\lim_{x\to 0}\frac{e^{a}\left(e^{x}-1\right)}{x}$$

$$=e^{a}\lim_{x\to 0}\frac{e^{x}-1}{x}=e^{a}$$

7. $\lim_{x\to 0}\left(\frac{a^{m+x}-a^{m}}{x}\right)$ for $a>0$

Solution: $\frac{a^{m+x}-a^{m}}{x}=\frac{a^{m}\cdot a^{x}-a^{m}}{x}=\frac{a^{m}\left(a^{x}-1\right)}{x}$

$$\therefore \lim_{x\to 0}\left(\frac{a^{m+x}-a^{m}}{x}\right)=\lim_{x\to 0}\frac{a^{m}\left(a^{x}-1\right)}{x}=a^{m}\lim_{x\to 0}\left(\frac{a^{x}-1}{x}\right)$$

$$=a^{m}\log_e a=a^{m}\log a$$

Exercise 4.24

Evaluate the following:

1. $\lim_{x\to 0}\left(\frac{5^x - 1}{x}\right)$

2. $\lim_{x\to 0}\left(\frac{e^{mx} - 1}{x}\right)$

3. $\lim_{x\to 0}\left(\frac{7^x - 1}{x}\right)$

4. $\lim_{x\to 0}\left(\frac{2^{-x} - 1}{x}\right)$

5. $\lim_{x\to 0}\left(\frac{e^{3x} - 1}{x}\right)$

6. $\lim_{x\to 0}\left(\frac{2^x - 1}{x}\right)$

Answers:
1. log 5
2. m
3. log 7
4. – log 2
5. 3
6. log 2

Problems put in the form:

$$y = \left(\frac{a^x - b^x}{g(x)}\right) \text{ and } x\to 0\text{, where } a > 0,\, b > 0.$$

Working rule: The rule to evaluate $\lim_{x\to 0}\left(\frac{a^x - b^x}{g(x)}\right)$

says to write $\frac{a^x - b^x}{g(x)} = \frac{a^x - b^x}{x}\cdot\frac{x}{g(x)}$

$$= \frac{(a^x - 1) - (b^x - 1)}{x}\cdot\frac{x}{g(x)} = \left(\frac{a^x - 1}{x} - \frac{b^x - 1}{x}\right)\cdot\left(\frac{x}{g(x)}\right)$$

whose limit as $x \to 0$ is the required limit.

Notes: **(i)** $\lim_{x\to 0}\left(\frac{a^x - b^x}{x}\right) = log_e\left(\frac{a}{b}\right), (a, b > 0)$

(ii) $\lim_{x\to 0}\left(\frac{a_1^x + a_2^x + a_3^x + \ldots + a_n^x - n}{x}\right)$

$$= \log (a_1\, a_2\, a_3 \ldots a_n), (a_1, \ldots, a_n > 0)$$

Examples worked out:

1. $\lim_{x\to 0}\left(\frac{a^x - b^x - 2}{x}\right), (a, b > 0)$

Solution: $\because \frac{a^x + b^x - 2}{x} = \frac{(a^x - 1) + (b^x - 1)}{x}$

$$= \frac{a^x - 1}{x} + \frac{b^x - 1}{x}$$

$$\lim_{x\to 0}\left(\frac{a^x + b^x - 2}{x}\right) = \lim_{x\to 0}\left(\frac{a^x - 1}{x}\right) + \lim_{x\to 0}\left(\frac{b^x - 1}{x}\right)$$

$$= \log a + \log b = \log(ab).$$

2. $\lim_{x\to 0}\left(\frac{a^x + b^x + c^x - 3}{x}\right), (a, b, c > 0)$

Solution: $\because \frac{a^x + b^x + c^x - 3}{x} = \frac{(a^x - 1) + (b^x - 1) + (c^x - 1)}{x}$

$$= \frac{a^x - 1}{x} + \frac{b^x - 1}{x} + \frac{c^x - 1}{x}$$

$$\therefore \lim_{x\to 0}\left(\frac{a^x + b^x + c^x - 3}{x}\right)$$

$$= \lim_{x\to 0}\left(\frac{a^x - 1}{x}\right) + \lim_{x\to 0}\left(\frac{b^x - 1}{x}\right) + \lim_{x\to 0}\left(\frac{c^x - 1}{x}\right)$$

$$= \log a + \log b + \log c = \log(abc)$$

3. $\lim\limits_{x\to 0}\left(\dfrac{a^x - b^x}{x}\right)$

Solution: $\because \dfrac{a^x - b^x}{x} = \dfrac{(a^x - 1) - (b^x - 1)}{x}$

$$= \frac{(a^x - 1)}{x} - \frac{(b^x - 1)}{x}$$

$$\therefore \lim_{x\to 0}\left(\frac{a^x - b^x}{x}\right) = \lim_{x\to 0}\left(\frac{a^x - 1}{x}\right) - \lim_{x\to 0}\left(\frac{b^x - 1}{x}\right)$$

$$= \log a - \log b = \log\left(\frac{a}{b}\right)$$

4. $\lim\limits_{x\to 0}\left(\dfrac{a^x - b^x}{\sin x}\right)$

Solution: $\because \dfrac{a^x - b^x}{\sin x} = \dfrac{a^x - b^x}{x}\cdot\dfrac{x}{\sin x}$

$$\therefore \lim_{x\to 0}\left(\frac{a^x - b^x}{x}\right) = \lim_{x\to 0}\left(\frac{a^x - b^x}{x}\right)\cdot\lim_{x\to 0}\left(\frac{x}{\sin x}\right)$$

$$= \log\left(\frac{a}{b}\right)\cdot 1 = \log\left(\frac{a}{b}\right)$$

5. $\lim\limits_{x\to 0}\dfrac{(a^{2x} - 1)(a^x - 1)}{x\sin x}$ for $a > 0$

Solution: $\because \dfrac{(a^{2x} - 1)(a^x - 1)}{x\sin x}$

$$= \frac{a^{2x} - 1}{x}\cdot\frac{a^x - 1}{x}\cdot\frac{x}{\sin x}$$

$$\therefore \lim_{x\to 0}\left(\frac{a^{2x} - 1}{x}\right)$$

$$= \lim_{x\to 0}\left(\frac{a^{2x} - 1}{x}\right)\cdot\lim_{x\to 0}\left(\frac{a^x - 1}{x}\right)\cdot\lim_{x\to 0}\left(\frac{x}{\sin x}\right)$$

$$= \log a^2\cdot\log a\cdot 1$$

$$= 2\log a\cdot\log a = 2\log^2 a$$

6. $\lim\limits_{x\to 0}\dfrac{(e^{2x} - 1)(e^x - 1)}{x\tan x}$

Solution: $\because \dfrac{(e^{2x} - 1)(e^x - 1)}{x\tan x}$

$$= \frac{e^{2x} - 1}{x}\cdot\frac{e^x - 1}{x}\cdot\frac{x}{\tan x}$$

$$\therefore \lim_{x\to 0}\frac{(e^{2x} - 1)(e^x - 1)}{x\tan x}$$

$$= \lim_{x\to 0}\left(\frac{e^{2x} - 1}{x}\right)\cdot\lim_{x\to 0}\left(\frac{e^x - 1}{x}\right)\cdot\lim_{x\to 0}\left(\frac{x}{\tan x}\right)$$

$$= \log e^2\cdot 1\cdot 1$$

$$= 2\log e = 2$$

7. $\lim\limits_{x\to 0}\left(\dfrac{a^{\sin x} - 1}{x}\right)$ for $a > 0$

Solution: $\because \dfrac{a^{\sin x} - 1}{x} = \dfrac{a^{\sin x} - 1}{\sin x}\cdot\dfrac{\sin x}{x}$

$$\therefore \lim_{x\to 0}\left(\frac{a^{\sin x} - 1}{x}\right)$$

$$= \lim_{x\to 0}\left(\frac{a^{\sin x} - 1}{\sin x}\right)\cdot\lim_{x\to 0}\left(\frac{\sin x}{x}\right) \quad \ldots\text{(i)}$$

Now putting $\sin x = h$ so that as $x \to 0$, $h \to 0$, we have from (i)

$$\lim_{x\to 0}\left(\frac{a^{\sin x} - 1}{x}\right) = \lim_{x\to 0}\left(\frac{a^h - 1}{h}\right)\cdot 1$$

$$= \log a \cdot 1 = \log a \left(\because \lim_{x\to 0} \left(\frac{\sin x}{x} \right) = 1 \right)$$

8. $\lim_{x\to 0} \left(\frac{e^{ax} - e^{bx}}{x} \right)$

Solution: $\because \frac{e^{ax} - e^{bx}}{x} = \frac{(e^{ax} - 1) - (e^{bx} - 1)}{x}$

$$= \frac{e^{ax} - 1}{x} - \frac{e^{bx} - 1}{x}$$

$$\therefore \lim_{x\to 0} \left(\frac{e^{ax} - e^{bx}}{x} \right) = \lim_{x\to 0} \left(\frac{e^{ax} - 1}{x} \right) - \lim_{x\to 0} \left(\frac{e^{bx} - 1}{x} \right)$$

$$= \log_e e^a - \log_e e^b = a \log_e e - b \log_e e = a - b$$

$$(\because \log_e e = \log e = 1)$$

9. $\lim_{x\to 0} \left(\frac{5^x - 3^x}{4^x - 2^x} \right)$

Solution: $\because \frac{5^x - 3^x}{4^x - 2^x} = \frac{\left(\frac{5^x - 3^x}{x} \right)}{\left(\frac{4^x - 2^x}{x} \right)}$

$$\therefore \lim_{x\to 0} \left(\frac{5^x - 3^x}{4^x - 2^x} \right) = \frac{\lim_{x\to 0} \left(\frac{5^x - 3^x}{x} \right)}{\lim_{x\to 0} \left(\frac{4^x - 2^x}{x} \right)}$$

$$= \frac{\log \left(\frac{5}{3} \right)}{\log \left(\frac{4}{2} \right)} = \frac{\log \left(\frac{5}{3} \right)}{\log 2}$$

10. $\lim_{x\to 0} \left(\frac{a^x + b^x - 2^{x+1}}{x} \right)$ for $a > 0, b > 0$.

Solution: $\because \frac{a^x + b^x - 2^{x+1}}{x} = \frac{a^x + b^x - 2^x \cdot 2}{x}$

$$= \frac{(a^x - 1) + (b^x - 1) - 2(2^x - 1)}{x}$$

$$= \frac{a^x - 1}{x} + \frac{b^x - 1}{x} - \frac{2(2^x - 1)}{x}$$

$$\therefore \lim_{x\to 0} \left(\frac{a^x + b^x - 2^{x+1}}{x} \right)$$

$$= \lim_{x\to 0} \left(\frac{a^x - 1}{x} \right) + \lim_{x\to 0} \left(\frac{b^x - 1}{x} \right) - 2 \lim_{x\to 0} \left(\frac{2^x - 1}{x} \right)$$

$$= \log a + \log b - 2 \log 2 = \log a + \log b - \log 2^2$$

$$= \log a + \log b - \log 4 = \log \left(\frac{ab}{4} \right)$$

11. $\lim_{x\to 0} \frac{x \sin x}{e^x + e^{-x} - 2}$

Solution: $\because \frac{x \sin x}{e^x + e^{-x} - 2} = \frac{x \sin x}{e^x + \frac{1}{e^x} - 2}$

$$= \frac{x e^x \sin x}{e^{2x} + 1 - 2e^{ex}} = \frac{x e^x \sin x}{(e^x - 1)^2}$$

$$= e^x \cdot \left(\frac{x}{e^x - 1} \right)^2 \left(\frac{\sin x}{x} \right)$$

$$\therefore \lim_{x\to 0} \left(\frac{x \sin x}{e^x + e^{-x} - 2} \right)$$

$$= \lim_{x\to 0} e^x \cdot \lim_{x\to 0} \left(\frac{x}{e^x - 1} \right)^2 \cdot \lim_{x\to 0} \left(\frac{\sin x}{x} \right)$$

$$= e^0 \cdot (1)^2 \cdot 1 = 1$$

Exercise 4.25

Evaluate the following ones:

1. $\lim_{x\to 0} e^x$

2. $\lim_{x\to 0} \dfrac{e^x}{\log x}$

3. $\lim_{x\to 0} \dfrac{e^{tx}-1}{x}$

4. $\lim_{x\to 0} \dfrac{e^{ax}-e^{bx}}{x}$

5. $\lim_{x\to 0} \dfrac{e^{ax}-1}{\sin x}$

6. $\lim_{x\to 0} \dfrac{e^{\sin x}-1}{x}$

7. $\lim_{h\to 0} \dfrac{e^{(x+h)^2}-e^{x^2}}{h}$

8. $\lim_{x\to \frac{\pi}{2}} \dfrac{e^{\tan x}-1}{e^{\tan x}+1}$

9. $\lim_{x\to 0} \dfrac{a^{-\sin x}-1}{x}, a>0$

10. $\lim_{x\to 0} \dfrac{e^{x^2}-1}{\sin^2 x}$

11. $\lim_{x\to 0} \dfrac{e^{3x}-1}{x}$

12. $\lim_{x\to 0} \dfrac{2^x-1}{x}$

13. $\lim_{x\to 0} \dfrac{e^x-e^{-x}}{x}$

14. $\lim_{x\to 0} \dfrac{e^x-1-x}{x^2}$

15. $\lim_{x\to 0} \dfrac{a^x-b^x}{x}, a>0, b>0$

16. $\lim_{x\to 0} \dfrac{3^x-1}{2^x-1}$

17. $\lim_{x\to 0} \dfrac{7^x-3^x}{x}$

18. $\lim_{x\to 0} \dfrac{a^{\sin x}-1}{\sin x}, a>0$

19. $\lim_{x\to 0} \dfrac{e^{\sin x}-1-\sin x}{x^2}$

20. $\lim_{x\to 0} \dfrac{2^x-1}{(1+x)^{\frac{1}{2}}-1}$

21. $\lim_{x\to 0} \dfrac{x2^x-x}{1-\cos x}$

22. $\lim_{h\to 0} \dfrac{e^{x+h}-e^x}{h}$

Answers:

1. 1
2. 0
3. t
4. $(a-b)$
5. a
6. 1
7. $2x\,e^{x^2}$
8. Does not exist
9. $-\log a$
10. 1
11. 3
12. $\log 2$
13. 2

14. $\frac{1}{2}$

15. $\log\left(\frac{a}{b}\right)$

16. $\log\frac{3}{2}$

17. $\log\left(\frac{7}{3}\right)$

18. $\log a$

19. $\frac{1}{2}$

20. Hint: Divide *Nr* and *Dr* by *x*. Answer: 2 log 2

21. 2 log 2

22. e^x

Problems on $R(e^x)$

Working rule: The rule we may adopt to evaluate $\lim_{x\to 0} R\left(e^x\right)$ is (1) to change $R(e^x)$ into the combination (sum, difference, product and/quotient) of (i) x^n (ii) e^{mx} (iii) $\frac{e^{mx}-1}{x}$ by using the method of rationalization or any mathematical manipulation and (2) to find the limit of the combination of (i), (ii) and (iii) respectively.

Notes: **(a)** one should write $\frac{1}{e^{mx}}$ for e^{-mx} whenever it occurs in the given rational functions of e^x.
(b) $R(e^x)$ is the notation for rational function of e^x.

Examples worked out:

Evaluate the following:

1. $\lim_{x\to 0}\left(\frac{e^x-e^{-x}}{e^{2x}+e^{-2x}}\right)$

Solution: $\lim_{x\to 0}\frac{\left(e^x-1\right)-\left(e^{-x}-1\right)}{\left(e^{2x}-1\right)-\left(e^{-2x}-1\right)}$

$$=\lim_{x\to 0}\frac{\frac{e^x-1}{x}-\frac{e^{-x}-1}{x}}{\frac{e^{2x}-1}{x}-\frac{e^{-2x}-1}{x}}$$

$$=\frac{\log e-\log e^{-1}}{\log e^2-\log e^{-2}}$$

$$=\frac{\log e+\log e}{2\log e+2\log e}$$

$$=\frac{1}{2}$$

2. $\lim_{x\to\infty}\left(\frac{3e^{2x}+2e^{-2x}}{4e^{2x}-e^{-2x}}\right)$

Solution: $\lim_{x\to\infty}\left(\frac{3e^{2x}+2e^{-2x}}{4e^{2x}-e^{-2x}}\right)$

$$=\lim_{x\to\infty}\left(\frac{3e^{2x}+\frac{2}{e^{2x}}}{4e^{2x}-\frac{1}{e^{2x}}}\right)$$

$$=\lim_{x\to\infty}\left(\frac{3+\frac{2}{e^{4x}}}{4-\frac{1}{e^{4x}}}\right)=\frac{3+0}{4+0}=\frac{3}{4}$$

N.B.: As $x\to\infty$, $e^{mx}\to\infty$ if $m>0$.

Exercise 4.26

Evaluate the following ones:

1. $\lim_{x\to 0}\frac{a\cdot e^x+b\cdot e^{-x}}{e^x+e^{-x}}$

2. $\lim_{x\to\infty}\frac{ae^x+be^{-x}}{e^x+e^{-x}}$

3. $\lim_{x\to\infty}\frac{ae^x+be^{-x}}{e^x+e^{-x}}$

4. $\lim_{x\to\infty}\left\{\left(\frac{e^x-e^{-x}}{e^x+e^{-x}}\right)'+\tan\left(\frac{1}{x}\right)\right\}$

Answers:

1. $\dfrac{a+b}{2}$
2. a
3. b
4. 2

On Limits of function of a function

Evaluation of $\lim_{x\to a} g(f(x))$, where $a = 0$, any constant other than zero and/ ∞

Theorem: (Calculus by Burkey)

Let $\lim_{x\to a} f(x) = L$ exists and g be continuous at L, then

(i) $\lim_{x\to a} g(f(x)) = g\left(\lim_{x\to a} f(x)\right) = g(L)$

(ii) $\lim_{x\to a^+} g(f(x)) = g\left(\lim_{x\to a^+} f(x)\right) = g(L)$

(iii) $\lim_{x\to a^-} g(f(x)) = g\left(\lim_{x\to a^-} f(x)\right) = g(L)$

Working rule: One may adopt the following procedure to evaluate $\lim_{x\to a} g(f(x))$, a being zero, any constant different from zero and/infinity.

Step 1: Put the inner function $= f(x)$ and find $\lim_{x\to a}$ (inner function) $= \lim_{x\to a} f(x) = L$ (say)

Step 2: Put the outer function $= g(x)$ and find $\lim_{x\to L}$ (outer function) $= \lim_{x\to L} g(x) = g(L)$ which will be the required limit of the given composition of two functions, say $g(f(x))$ as $x \to a$.

Examples worked out:

Evaluate the following ones:

1. $\lim_{x\to 0} \cos(\sin x)$

Solution: Let $f(x) = \sin x$ and $g(x) = \cos x$

$\therefore \lim_{x\to 0} f(x) = \lim_{x\to 0} \sin x = \sin 0 = 0$ and

$\lim_{x\to 0} g(x) = \lim_{x\to 0} \cos x = \cos 0 = 1$

Hence, $\lim_{x\to 0} \cos(\sin x) = \cos \lim_{x\to 0} (\sin x)$

$= \cos 0 = 1$

2. $\lim_{x\to 0} \sin\left(\cos^3 x\right)$

Solution: Let $f(x) = \cos^3 x$ and $g(x) = \sin x$

Note: g is continuous at $L \Leftrightarrow \lim_{x\to L} g(x) = g(L)$ which means the functional value of $g(x)$ for $x = L$ and the limit of $g(x)$ as x tends to L are equal.

$\therefore \lim_{x\to 0} f(x) = \lim_{x\to 0} \cos^3 x = \left(\lim_{x\to 0} \cos x\right)^3 = (1)^3 = 1$

and $\lim_{x\to 1} g(x) = \lim_{x\to 1} \sin x = \sin 1$

Hence, $\lim_{x\to 0} g(f(x)) = \lim_{x\to 0} \sin\left(\cos x^3\right) = \sin 1$

3. $\lim_{x\to -1} \sin\left(e^x\right)$

Solution: Let $f(x) = e^x$ and $g(x) = \sin x$

$\therefore \lim_{x\to -1} f(x) = \lim_{x\to -1} e^x = e^{-1} = \dfrac{1}{e}$

and $\lim_{x\to \frac{1}{e}} g(x) = \lim_{x\to \frac{1}{e}} \sin x = \sin\left(\dfrac{1}{e}\right)$

Hence, $\lim_{x\to -1} g(f(x)) = \lim_{x\to -1} \sin\left(e^x\right) = \sin\left(\dfrac{1}{e}\right)$

4. $\lim_{x\to a} \sin\left(\dfrac{1}{x}\right)$, $a \neq 0$

Solution: Let $f(x) = \dfrac{1}{x}$ and $g(x) = \sin x$

$\therefore \lim_{x\to a} f(x) = \lim_{x\to a}\left(\dfrac{1}{x}\right) = \dfrac{1}{a}$

and $\lim_{x\to \frac{1}{a}} g(x) = \lim_{x\to \frac{1}{a}} \sin x = \sin\left(\dfrac{1}{a}\right)$

Hence, $\lim_{x\to a} g(f(x)) = \lim_{x\to a} \sin\left(\frac{1}{x}\right) = \sin\left(\frac{1}{a}\right)$

5. $\lim_{x\to 2}\left(e^{1+x+2x^2}\right)$

Solution: Let $f(x) = 1 + x + 2x^2$ and $g(x) = e^x$

$\therefore \lim_{x\to 2} f(x) = \lim_{x\to 2} (1+x+2x^2) = 1+2+2\times 4 = 11$

and $\lim_{x\to 11} g(x) = \lim_{x\to 11} e^x = e^{11}$

Hence, $\lim_{x\to 2} g(f(x)) = \lim_{x\to 2}\left(e^{1+x+2x^2}\right) = e^{11}$

Exercise 4. 27

Evaluate the following ones: **Answers**

		Answers
1.	$\lim_{x\to 0} \sin\left(\frac{x}{2}\right)$	0
2.	$\lim_{x\to c} \sin(ax+b)$	$\sin(ac+b)$
3.	$\lim_{x\to 0} e^{2x}$	1
4.	$\lim_{x\to 0} e^{\cos x}$	e
5.	$\lim_{x\to c} [x]^2,\ c \notin I$	$[c]^2$
6.	$\lim_{x\to \frac{\pi}{4}} \sin[x]$	0
7.	$\lim_{x\to \frac{\pi}{4}} \cos[x]$	1
8.	$\lim_{x\to c} \left(a^x\right),\ a>0$	a^c
9.	$\lim_{x\to c} \lvert f(x)\rvert$, when $\lim_{x\to c} f(x) = l$	$\lvert l\rvert$
10.	$\lim_{x\to 0} \lvert \cos x\rvert$	1
11.	$\lim_{x\to 0} e^{-\lvert x\rvert}$	1

Method of Expansion

Question: Where to use expansion method for evaluating limits?

Answer: The method of expansion for evaluating limits of a given function at a given point is applicable to the function (or, functions) which can be expanded in series, i.e., if the given function (whose limit is required to be found out) contains some function (or, functions) whose expansion in series is known to us, then firstly we make proper expansion for those functions which are capable of being expanded.

Working rule:

Step 1: We write the expanded form of the function (or, the functions present in the given whose limit is required) whose expansion is known to us.

Step 2: After expansion in series, we simplify and cancel the common factor (or, factors) present in the numerator and denominator of the given quotient function if any one common factor exists. If there is no common factor in the expanded form of the given quotient function (or, functions), we leave the given quotient function in the expanded form.

Step 3: Lastly, i.e., after expansion and simplification, we put the limit of the independent variable (since $x \to a$ means $\lim_{x\to a} x = a$, or $\lim x = a$) to find the value of the limit of the given function in the quotient form (or, in the product form).

Remember: Following expansions are widely used for evaluating limits by method of expansion.

1. $\sin x = x - \frac{x^3}{\lfloor 3} + \frac{x^5}{\lfloor 5} - \frac{x^7}{\lfloor 7} + \ldots\infty$

2. $\cos x = 1 - \frac{x^2}{\lfloor 2} + \frac{x^4}{\lfloor 4} - \frac{x^6}{\lfloor 6} + \ldots\infty$

3. $\tan x = x + \frac{x^2}{3} + \frac{2}{15}x^5 + \ldots\infty$

4. $\log(1+x) = x - \frac{x^2}{\lfloor 2} + \frac{x^3}{\lfloor 3} + \ldots\infty,\ (-1 < x \le 1)$

5. $\log(1-x) = -x - \frac{x^2}{\lfloor 2} - \frac{x^3}{\lfloor 3} - \ldots\infty,\ (-1 \le x < 1)$

6. $e^x = 1 + \frac{x}{\lfloor 1} + \frac{x^2}{\lfloor 2} - \frac{x^3}{\lfloor 3} + \ldots \infty$

7. $e^{-x} = 1 - \frac{x}{\lfloor 1} + \frac{x^2}{\lfloor 2} - \frac{x^3}{\lfloor 3} + \ldots \infty$

8. $(1+x)^n = 1 + nx + \frac{n(n-1)}{\lfloor 2} x^2 + \ldots \infty, (-1 < x < 1)$, where n is a negative integer, a fraction and/any real number.

9. $a^x = e^{x \log a} = 1 + x \log a + \frac{(x \log a)^2}{\lfloor 2} + \ldots \infty$

10. For positive integer, $(1+x)^n = 1 + nc_1 x + nc_2 x^2 + \ldots + nc_n x^n$

N.B.: **1.** We put $(-x)$ in $(1+x)^n$ to obtain the expansion of $(1-x)^n$.

2. The method of expansion is also widely used when

(a) the given function is the product of reciprocal of power function x^n (i.e. $\frac{1}{x}, \frac{1}{x^2}, \ldots$, etc.) and a trigonometric, logarithmic or exponential function of x which can be expanded in a series of power of x.

(b) the integrand contains $(a+x)^n$ or $(1 \pm x)^n$ and a function of x.

(c) all the expansions are valid even if x is replaced by any other variable or a function of x, e.g.

(i) $\log(1+\sin x) = \sin x - \frac{\sin^2 x}{\lfloor 2} + \frac{\sin^3 x}{\lfloor 3} + \ldots \infty$, $(-1 < \sin x \leq 1)$

(ii) $\log(1-\sin x) = -\sin x - \frac{\sin^2 x}{\lfloor 2} - \frac{\sin^3 x}{\lfloor 3} - \ldots \infty$, $(-1 < \sin x \leq 1)$

Examples worked out:

Evaluate the following ones:

1. $\lim_{x \to 0} \frac{(1+x)^5 - 1}{3x + 5x^2}$

Solution: $\lim_{x \to 0} \frac{(1+x)^5 - 1}{x(3+5x)}$

$$= \lim_{x \to 0} \frac{\left(1 + \frac{5}{\lfloor 1} x + \frac{5 \cdot 4 \cdot x^2}{\lfloor 2} + \ldots + x^5\right) - 1}{x(3+5x)}$$

$$= \lim_{x \to 0} \frac{x(5 + 10x + 10x^2 + \ldots + x^4)}{x(3+5x)} = \frac{5}{3}$$

2. $\lim_{x \to 0} \frac{(1+x)^{\frac{1}{n}} - 1}{x}$

Solution: $\lim_{x \to 0} \frac{(1+x)^{\frac{1}{n}} - 1}{x}$

$$= \lim_{x \to 0} \frac{\left(1 + \frac{1}{n} \cdot x + \text{terms having higher powers of } x\right) - 1}{x}$$

$$= \lim_{x \to 0} \left(\frac{\frac{1}{n} x + \text{terms having higher powers of } x}{x}\right)$$

$$= \lim_{x \to 0} \left(\frac{1}{n} + \text{terms having } x \text{ and its power}\right) = \frac{1}{n}$$

3. $\lim_{x \to 0} \left(\frac{\tan x}{x}\right)^{\frac{3}{x^2}}$

Solution: $\lim_{x \to 0} \left(\frac{\tan x}{x}\right)^{\frac{3}{x^2}}$

$$= \lim_{x \to 0} \left(\frac{x + \frac{x^2}{3} + \frac{2}{15} x^5 + \ldots \infty}{x}\right)^{\frac{3}{x^2}}$$

$$= \lim_{x \to 0} \left(1 + \frac{x^2}{3}\right)^{\frac{3}{x^2}} = e^{\frac{1}{3}}$$

4. $\lim\limits_{x\to 0}\left(\dfrac{1}{\sin x}-\dfrac{1}{x}\right)$

Solution: $\lim\limits_{x\to 0}\left(\dfrac{1}{\sin x}-\dfrac{1}{x}\right)=\lim\limits_{x\to 0}\left(\dfrac{x-\sin x}{x\sin x}\right)$

$$=\lim_{x\to 0}\left\{\frac{x-\left(x-\dfrac{x^3}{\lfloor 3}+\dfrac{x^5}{\lfloor 5}-\ldots\infty\right)}{x\left(x-\dfrac{x^3}{\lfloor 3}+\ldots\infty\right)}\right\}$$

$$=\lim_{x\to 0}\left\{\frac{\dfrac{x^3}{\lfloor 3}-\dfrac{x^5}{\lfloor 5}+\ldots\infty}{x^2-\dfrac{x^4}{\lfloor 3}+\ldots\infty}\right\}$$

$$=\lim_{x\to 0}\left\{\frac{x^3\left(\dfrac{1}{\lfloor 3}-\dfrac{x^2}{\lfloor 5}+\ldots\infty\right)}{x^2\left(1-\dfrac{x^2}{\lfloor 3}+\ldots\infty\right)}\right\}$$

$$=\lim_{x\to 0}\left\{\frac{x\left(\dfrac{1}{\lfloor 3}-\dfrac{x^2}{\lfloor 5}+\ldots\infty\right)}{\left(1-\dfrac{x^2}{\lfloor 3}+\ldots\infty\right)}\right\}=0$$

5. $\lim\limits_{x\to 0}\left(\dfrac{3\sin x-\sin 3x}{x-\sin x}\right)$

Solution: $\lim\limits_{x\to 0}\left(\dfrac{3\sin x-\sin 3x}{x-\sin x}\right)$

$$=\lim_{x\to 0}\frac{3\left(x-\dfrac{x^3}{\lfloor 3}+\dfrac{x^5}{\lfloor 5}-\ldots\infty\right)-\left(3x-\dfrac{(3x)^3}{\lfloor 3}+\ldots\infty\right)}{x-\left(x-\dfrac{x^3}{\lfloor 3}+\ldots\infty\right)}$$

$$=\lim_{x\to 0}\frac{\dfrac{x^3}{\lfloor 3}\left(3^2-3\right)+\ldots\infty}{\dfrac{x^3}{\lfloor 3}+\ldots\infty}=24$$

6. $\lim\limits_{x\to 0}\left(\dfrac{e^x-1+\log(1-x)}{\sin^3 x}\right)$

Solution: $\lim\limits_{x\to 0}\left(\dfrac{e^x-1+\log(1-x)}{\sin^3 x}\right)$

$$=\lim_{x\to 0}\frac{\left\{\left(1+x+\dfrac{x^2}{\lfloor 2}+\dfrac{x^3}{\lfloor 3}+\ldots\right)-1\right\}+\left(-x-\dfrac{x^2}{2}-\dfrac{x^3}{3}\ldots\right)}{\left(x-\dfrac{x^3}{\lfloor 3}+\dfrac{x^5}{\lfloor 5}-\ldots\right)^3}$$

$$=\lim_{x\to 0}\left\{\frac{x^3\left(\dfrac{1}{6}-\dfrac{1}{3}\right)+\ldots}{x^3\left(1-\dfrac{x^2}{\lfloor 3}+\dfrac{x^4}{\lfloor 4}-\ldots\right)^3}\right\}=-\frac{1}{6}$$

7. $\lim\limits_{x\to 0}\left(\dfrac{\tan x-\sin x}{\sin^3 x}\right)$

Solution: $\lim\limits_{x\to 0}\left(\dfrac{\tan x-\sin x}{\sin^3 x}\right)$

$$=\lim_{x\to 0}\frac{\left(x+\dfrac{x^3}{3}+\dfrac{2}{15}x^5+\ldots\infty\right)-\left(x-\dfrac{x^3}{\lfloor 3}+\dfrac{x^5}{\lfloor 5}-\ldots\infty\right)}{\left(x-\dfrac{x^3}{\lfloor 3}+\dfrac{x^5}{\lfloor 5}+\ldots\infty\right)^3}$$

$$=\lim_{x\to 0}\frac{x^3\left(\dfrac{1}{3}+\dfrac{1}{6}\right)+x^5\left(\dfrac{2}{15}-\dfrac{1}{\lfloor 5}+\ldots\infty\right)}{\left(x-\dfrac{x^3}{\lfloor 3}+\dfrac{x^5}{\lfloor 5}+\ldots\infty\right)^3}$$

$$= \lim_{x\to 0} \frac{x^3\left(\frac{1}{3}+\frac{1}{6}\right)+x^5\left(\frac{2}{15}-\frac{1}{\lfloor 5}+\ldots\infty\right)}{x^3\left(1-\frac{x^2}{\lfloor 3}+\frac{x^4}{\lfloor 5}-\ldots\infty\right)^3}$$

$$= \lim_{x\to 0} \frac{\left(\frac{1}{3}+\frac{1}{6}\right)+x^2\left(\frac{2}{15}-\frac{1}{\lfloor 5}\right)-\ldots\infty}{\left(1-\frac{x^2}{\lfloor 3}+\frac{x^4}{\lfloor 5}-\ldots\infty\right)^3}$$

$$= \left(\frac{1}{3}+\frac{1}{6}\right) = \frac{1}{2}$$

8. $\lim_{x\to 0} \left(\frac{e^x - e^{-x}}{x}\right)$

Solution: $\lim_{x\to 0} \left(\frac{e^x - e^{-x}}{x}\right)$

$$= \lim_{x\to 0} \frac{\left(1+x+\frac{x^2}{\lfloor 2}+\frac{x^3}{\lfloor 5}+\ldots\infty\right)-\left(1-x+\frac{x^2}{\lfloor 2}-\frac{x^3}{\lfloor 3}+\ldots\infty\right)}{x}$$

$$= \lim_{x\to 0} \frac{2\left(x+\frac{x^2}{\lfloor 3}+\frac{x^5}{\lfloor 6}+\ldots\infty\right)}{x}$$

$$= \lim_{x\to 0} 2\left(1+\frac{x^2}{\lfloor 3}+\frac{x^4}{\lfloor 5}+\ldots\infty\right) = 2$$

9. $\lim_{x\to 0} \left(\frac{e^x + e^{-x} - 2}{x^2}\right)$

Solution: $\lim_{x\to 0} \left(\frac{e^x + e^{-x} - 2}{x^2}\right)$

$$= \lim_{x\to 0} \frac{\left(1+x+\frac{x^2}{\lfloor 2}+\ldots\infty\right)+\left(1-x+\frac{x^2}{\lfloor 2}-\ldots\infty\right)-2}{x^2}$$

$$= \lim_{x\to 0} \frac{2\left(\frac{x^2}{\lfloor 2}+\frac{x^4}{\lfloor 4}+\ldots\infty\right)}{x^2}$$

$$= \lim_{x\to 0} 2\left(\frac{1}{\lfloor 2}+\frac{x^2}{\lfloor 4}+\frac{x^4}{\lfloor 6}+\ldots\infty\right)$$

$$= 2\left(\frac{1}{\lfloor 2}+0+0\right) = \frac{2}{\lfloor 2} = 1$$

10. $\lim_{x\to 0} \left(\frac{a^x - 1}{x}\right)$ for $a > 0$.

Solution: $\lim_{x\to 0} \left(\frac{a^x - 1}{x}\right)$

$$= \lim_{x\to 0} \left(\frac{e^{x\log a} - 1}{x}\right)$$

$$= \lim_{x\to 0} \frac{\left(1+x\log a+\frac{x^2}{\lfloor 2}(\log a)^2+\frac{x^3}{\lfloor 3}(\log a)^3+\ldots\infty-1\right)}{x}$$

$$= \lim_{x\to 0} \left(\log a + \frac{x}{\lfloor^2}(\log a)^2 + \ldots\infty\right)$$

$$= \log a$$

11. $\lim_{x\to\infty} x\left(e^{\frac{1}{x}} - e^{-\frac{1}{x}}\right)$

Solution: $\lim_{x\to\infty} x\left(e^{\frac{1}{x}} - e^{-\frac{1}{x}}\right)$

$$= \lim_{x\to\infty} x\left\{\left(1+\frac{\left(\frac{1}{x}\right)}{\lfloor 1}+\frac{\left(\frac{1}{x}\right)^2}{\lfloor 2}+\frac{\left(\frac{1}{x}\right)^3}{\lfloor 3}+\ldots\infty\right)-\left(1-\frac{\left(\frac{1}{x}\right)}{\lfloor 1}+\frac{\left(\frac{1}{x}\right)^2}{\lfloor 2}-\frac{\left(\frac{1}{x}\right)^3}{\lfloor 3}+\ldots\infty\right)\right\}$$

$$= \lim_{x\to\infty} x\cdot 2\left(\frac{\frac{1}{x}}{\lfloor 1}+\frac{\left(\frac{1}{x}\right)^2}{\lfloor 3}+\frac{\left(\frac{1}{x}\right)^5}{\lfloor 5}+\ldots\infty\right)$$

$$= \lim_{x\to\infty} x\cdot 2\cdot\frac{1}{x}\left(1+\frac{1}{\lfloor 3}\left(\frac{1}{x^2}\right)+\frac{1}{\lfloor 5}\left(\frac{1}{x}\right)^4+\ldots\infty\right) = 2$$

12. $\lim_{x\to 0}\left(\frac{e^x-1-x}{x^2}\right)$

Solution: $\lim_{x\to 0}\left(\frac{e^x-1-x}{x^2}\right)$

$$= \lim_{x\to 0}\frac{\left(1+\frac{x}{\lfloor 1}+\frac{x^2}{\lfloor 2}+\frac{x^3}{\lfloor 3}+\ldots\infty\right)-1-x}{x^2}$$

$$= \lim_{x\to 0}\frac{x^2\left(\frac{1}{\lfloor 2}+\frac{x}{\lfloor 3}+\ldots\infty\right)}{x^2} = \frac{1}{\lfloor 2} = \frac{1}{2}$$

13. $\lim_{x\to 0}\left(\frac{e^{\sin x}-\sin x-1}{x^2}\right)$

Solution: $\lim_{x\to 0}\left(\frac{e^{\sin x}-\sin x-1}{x^2}\right)$

$$= \lim_{x\to 0}\frac{\sin^2 x\left(\frac{1}{\lfloor 2}+\frac{\sin x}{\lfloor 3}+\ldots\infty\right)}{x^2}$$

$$= \lim_{x\to 0}\left(\frac{\sin x}{x}\right)^2\cdot\left(\frac{1}{\lfloor 2}+\frac{\sin x}{\lfloor 3}+\ldots\infty\right) = \frac{1}{\lfloor 2} = \frac{1}{2}$$

14. $\lim_{x\to 1}\frac{\log x}{x-1}$

Solution: Method 1

$$\lim_{x\to 1}\frac{\log x}{x-1} = \lim_{x\to 1}\frac{\log\{1+(x-1)\}}{(x-1)}$$

$$= \lim_{x\to 1}\left\{\frac{(x-1)-\frac{(x-1)^2}{2}+\frac{(x-1)^2}{3}-\ldots\infty}{(x-1)}\right\}$$

$$= \lim_{x\to 1}\left\{1-\frac{(x-1)}{2}+\frac{(x-1)^2}{3}-\ldots\infty\right\}$$

$$= 1-0+0 = 1$$

Method 2

Let $x = 1+h$ where $h\to 0$ as $x\to 1$

$$\therefore \lim_{x\to 1}\frac{\log x}{x-1} = \lim_{h\to 0}\frac{\log(1+h)}{1+h-1} = \lim_{h\to 0}\frac{\log(1+h)}{h}$$

$$= \lim_{h\to 0}\frac{h-\frac{h^2}{2}+\frac{h^3}{3}-\ldots\infty}{h}$$

$$= \lim_{h\to 0}\frac{h\left(1-\frac{h}{2}+\frac{h^3}{3}-\ldots\infty\right)}{h}$$

$$= \lim_{h\to 0}\left(1-\frac{h}{2}+\frac{h^2}{3}-\ldots\infty\right) = 1$$

15. $\lim_{x\to e}\left(\frac{\log x-1}{x-e}\right)$

Solution: Let $x = e+h$, where $h\to 0$ as $x\to e$

$$\therefore \lim_{x\to e}\left(\frac{\log x - 1}{x - e}\right) = \lim_{h\to 0}\frac{\log(e+h) - 1}{h}$$

$$= \lim_{h\to 0}\frac{\log(e+h) - \log e}{h} \quad (\because \log e = 1)$$

$$= \lim_{h\to 0}\frac{\log\left(\frac{e+h}{e}\right)}{h} = \lim_{h\to 0}\log\left(1+\frac{h}{e}\right)^{\frac{1}{h}}$$

$$= \lim_{h\to 0}\log\left\{\left(1+\frac{h}{e}\right)^{\frac{e}{h}}\right\}^{\frac{1}{e}}$$

$$= \lim_{h'\to 0}\log\left\{(1+h')^{\frac{1}{h'}}\right\}^{\frac{1}{e}}, \text{ where } h' = \frac{h}{e}$$

$$= \log e^{\frac{1}{e}} = \frac{1}{e}\log e = \frac{1}{e}\left(\because \lim_{x\to 0}(1+x)^{\frac{1}{x}} = e\right)$$

Exercise 4.28

Evaluate the following ones: **Answers**

1. $\lim_{t\to 0}\frac{\log(1+xt)}{t}$ (x)
2. $\lim_{x\to 0}\frac{\left(e^x - 1\right)\log(1+x)}{\sin x}$ (0)
3. $\lim_{x\to 0}\frac{e^{\alpha x} - 1}{\sin x}$ (α)
4. $\lim_{x\to 4}\frac{x^{\frac{7}{2}} - 4^{\frac{7}{2}}}{\log(x-3)}$ (112)
5. $\lim_{x\to 0}\frac{e^{x^2} - 1}{\sin^2 x}$ (1)
6. $\lim_{\theta\to 0}\frac{1}{\theta}\log(1+2\sin\theta)$ (2)
7. $\lim_{x\to\sqrt{e}}\frac{2\log x - 1}{x - \sqrt{e}}$ $\left(\frac{2}{\sqrt{e}}\right)$
8. $\lim_{x\to 1}\frac{\log x}{x - 1}$ (1)
9. $\lim_{x\to 0}\frac{\log(1+\sin x)}{x}$ (1)
10. $\lim_{x\to 0}\frac{\sin x}{\log(1+x)}$ (1)
11. $\lim_{x\to 0}\frac{1 - \cos x}{x\log(1+x)}$ $\left(\frac{1}{2}\right)$

On existence of the limit of a function at a given point

Before one knows how to examine the existence of the limit of a function $y = f(x)$ defined on its domain at an indicated point, it is required to be known some more concepts given below.

1. *Adjacent intervals*: Two intervals are said to be adjacent $\Leftrightarrow$ The left end point of one is the same as the right end point of the other.

Hence, the intervals

(i) $x < 0$ and $x > 0$
(ii) $0 < x < 1$ and $1 < x < 2$
(iii) $1 < x < 2$ and $2 \le x < 3$
(iv) $1 < x \le 3$ and $x > 3$, etc.

are examples of adjacent intervals because their left and right end points are same.

2. *Nonadjacent intervals*: Two intervals are said to be nonadjacent $\Leftrightarrow$ The left end point of one is different from the right end point of the other.

Hence, the intervals

(i) $x < 0$ and $x > 1$
(ii) $0 < x \le 1$ and $2 < x < 3$
(iii) $0 < x < 2$ and $x \ge 3$
(iv) $x < 3$ and $x > 4$, etc.

are examples of nonadjacent intervals because their left and right end points are different (not same)

How to examine the existence of the limit of a function at a given point

To examine the existence of the limit of a function f at a given point in its domain, one is required to use the following rule:

Left hand limit (*L.H.L* or *l.h.l*) at a given point = Right hand limit (*R.H.L* or *r.h.l*) at a given point $\Leftrightarrow$ the limit of a given function at a given (indicated in the question) point exists.

How to examine the nonexistence of the limit of a function at a given point

To examine the nonexistence of the limit of a function f at a given point in its domain, one is required to use the following rule:

Left hand limit at a given point $\neq$ Right hand limit at a given point $\Leftrightarrow$ The limit of a given function at a given point does not exist.

Notes: **1.** In case, a function f is defined by a single formula $y = f(x)$ in a δ-neighbourhood of a point $x = c$, there is no need to find out the left hand and right hand limits separately for the given function f at a given point 'c' where the given function f may or may not be defined while evaluating the limit of a function at a given point 'c' but while examining the existence of the limit in the case of functions denoted by a single formula $y = f(x)$ defined in a δ-neighbourhood of a given point 'c', it is a must to calculate both the limits at a given point 'c' where the given function may or may not be defined.

2. In case, a function f is defined by two or more than two formulas (different forms or expressions in x) in adjacent intervals, it is necessary to find out both the left hand limit (for the expression in x which is one form of the given function f defined in an interval indicating the left end point) and the right hand limit (for the expression in x which is an other form of the given function f defined in an other interval indicating the right end point) noting that the left and right end points of adjacent intervals are same (common) where the left and right hand limits are calculated whether the question says to evaluate the limit or examine the existence of the limit of a given function at the common point of adjacent intervals.

3. In questions, where $f(x)$ contains modulus or greatest integer functions, one is required to find out both the left and right hand limits at a given point or at the point where $|f(x)| = 0$ or greatest integer function is zero, i.e. $[f(x)] = 0$.

4. A piecewise function is always defined in adjacent intervals with different forms (expressions in x).

5. General models of a piecewise function:

A:

1. $f(x) = f_1(x)$, when $x < a$
$f(x) = f_2(x)$, when $x > a$
$f(x) = f_3(x)$, when $x = a$

2. $f(x) = f_1(x)$, when $x \leq a$
$f(x) = f_2(x)$, when $x > a$

3. $f(x) = f_1(x)$, when $x < a$
$f(x) = f_2(x)$, when $x \geq a$

B:

1. $f(x) = f_1(x)$, when $c < x < a$, $c \leq x < a$ or $c < x \leq a$
$f(x) = f_2(x)$, when $a \leq x \leq d$, $a < x < c$ or $c < x \leq a$

2. $f(x) = f_1(x)$, when $c < x < a$
$f(x) = f_2(x)$, when $a < x < d$
$f(x) = f_3(x)$, when $x = a$

3. $f(x) = f_1(x)$, when $x \neq a$
$f(x) = f_2(x)$, when $x = a$

Remarks: **1.** $f(x) = f_1(x)$, when $x \leq a$ or $c < x \leq a \Rightarrow$ The function $f_1(x)$ is to be selected to find out left hand limit as well as the value of the function $f(x)$ at $x = a$.

That is, in the case of piecewise function,

The left hand limit of a function at the right end point of an interval = $\lim\limits_{x \to a^-}$ (the function f_1 defined in an interval whose left end point is a) and the value of the function f at $x = a$ is $(f_1(x))_{x = a}$.

2. $f(x) = f_2(x)$, when $x \geq a$ or $a \leq x \leq d \Rightarrow$ The function $f_2(x)$ is to be selected to find out right hand limit as well as the value of the function $f(x)$ at $x = a$.

That is, in the case of piecewise function, the right hand limit of a function f at the left end point of an interval = $\lim\limits_{x \to a^+}$ (the function f_2 defined in an interval whose right end point is a) and the value of the function f at $x = a$ is $(f_2(x))_{x = a}$.

3. $f(x) = f_3(x)$, when $x = a \Rightarrow$ The function $f_3(x)$ is to be selected to calculate the value of the function $f(x)$ at $x = a$.

4. $f(x) = f_2(x)$ when $x \neq a \Rightarrow$ The same function $f_2(x)$ is to be selected to find out the left and right hand limits both since $\Leftrightarrow x < a$ and $x > a$.

That is, (the left hand limit at $x = a$) = (the right hand limit at $x = a$) = $\lim\limits_{x \to a}$ (the function of x opposite to which $x \neq a$ is written) = $\lim\limits_{x \to a} f_3(x)$.

5. $f(x)$ = a constant 'c' when $x = a \Rightarrow$ The function $f(x)$ has the value 'c' at $x = a$, i.e., one is given the value of a function f at $x = a$ which is 'c' and for this reason, there is no need to calculate the value of the function.

6. The value of a function $f(x)$ at a point $x = a$ is not required to be calculated while examining the existence of the limit of a function at a given point.

7. The value of a function $f(x)$ at a point $x = a$ is required to be calculated while testing or examining the continuity of a function f at a given point $x = a$.

On existence and nonexistence of the value and the limit of a function at a given point

The following possibility may arise.

1. The value of a function at a given point exists but the limit of a function at a given point does not exist.

2. The limit of a given function at a given point exists but the value of a given function at a given point does not exist.

3. The value and the limit both of a function at a given point exist but are not equal.

4. The value and the limit both of a function at a given point exist and are equal.

5. The value and the limit both of a function at a given point do not exist.

Notes: **1.** When the value and the limit both of a function at a given point exist and are equal, then the function is said to be continuous at a given point and in the rest cases, the function is said to be discontinuous at a given point.

2. A function $y = f(x)$ defined on its domain may not always have a value for each value given to the independent variable x. It is quite possible that for a particular value or a set of values of x, there is no value of the function $y = f(x)$, (i.e. there is a value of $y = f(x)$ like ∞, $\frac{0}{0}$ or imaginary). In such a case, it is said that $y = f(x)$ is undefined, undetermined, interminate, meaningless or does not exist for those values of x.

Hence, the value of a function f at a point $x = a$ exists $\Leftrightarrow$ the value of a function f at $x = a$ is a finite number.

How to find left hand limit of a piecewise function f at a point $x = a$

Step I: Replace x by $(a - h)$ in the given form of a function f and also in an interval whose right end point is a, where $h \to 0$ through positive values (i.e. h is a small positive number).

Step II: Simplify the function $f(a - h)$, i.e. a function in h and cancel out the common factor h (if any).

Step III: Substitute $h = 0$ in the simplified function and its further simplification provides us the required left hand limit of the given function f at the right end point of adjacent intervals.

How to find right hand limit of a piecewise function f at a point $x = a$

Step I: Replace x by $(a + h)$ in the given form of a function f and also in an interval whose left end point is also a, where $h \to 0$ through positive values (i.e. h is a small positive number).

Step II: Simplify the function $f(a + h)$, i.e. a function in h and cancel out the common factor h (if any).

Step III: Substitute $h = 0$ in the simplified function and its further simplification provides us the required right hand limit of the given function f at the left end point of adjacent intervals.

Notes: **1.** In the case of modulus and greatest integer functions, the above method of procedure is applicable since indeed, it is these functions which are piecewise functions.

2. In the case of functions defined by a single formula in a neighbourhood of the given point, there is a need to calculate left and right hand limit by making the substitution $x = a \pm h$ in the given single formula (expression in x) defined in the neighbourhood of a given point $x = a$.

Problems based on functions containing no modulus function

1. Show that $\lim\limits_{x \to 0} \left(\frac{\cos x}{x}\right)$ does not exist.

Solution: Let $h \to 0$ through positive values.

$$\therefore\ l.h.l = \lim_{h\to 0} \frac{\cos(0-h)}{(0-h)} = \lim_{h\to 0} \frac{\cos(-h)}{(-h)}$$

$$= \lim_{h\to 0} \frac{\cos h}{(-h)}$$

$$= \lim_{h\to 0}\left(-\frac{1}{h}\right)\cdot \lim_{l\to a} (\cos h) = -\infty \cdot 1 = -\infty$$

$$r.h.l = \lim_{h\to 0} \frac{\cos(0+h)}{(0+h)} = \lim_{h\to 0} \frac{\cos(+h)}{(+h)}$$

$$= \lim_{h\to 0} \frac{\cos h}{h}$$

$$= \lim_{l\to 0}\left(\frac{1}{h}\right)\cdot \lim_{l\to 0} (\cos h) = \infty \cdot 1 = \infty$$

Hence, neither $l.h.l$ nor $r.h.l$ exists

$\Rightarrow \lim_{x\to 0}\left(\frac{\cos x}{x}\right)$ does not exist.

2. If $f(x) = \frac{x\cdot e^{\frac{1}{x}}}{1+e^{\frac{1}{x}}}$, show that $\lim_{x\to 0} f(x)$ exists.

Solution: $l.h.l = \lim_{h\to 0} \frac{(0-h)\cdot e^{\left(\frac{1}{0-h}\right)}}{1+e^{\left(\frac{1}{0-h}\right)}}, h$

$$= \lim_{h\to 0} \frac{(-h)\cdot e^{\left(-\frac{1}{h}\right)}}{1+e^{\left(-\frac{1}{h}\right)}} = \frac{0}{1+0} = 0$$

$$r.h.l = \lim_{h\to 0} \frac{(0+h)\cdot e^{\left(\frac{1}{0+h}\right)}}{1+e^{\left(\frac{1}{0+h}\right)}},\ h>0$$

$$= \lim_{h\to 0} \frac{(+h)\cdot e^{\left(\frac{1}{h}\right)}}{1+e^{\left(\frac{1}{h}\right)}}$$

$= \lim_{h\to 0} \frac{h}{e^{-\frac{1}{h}}+1}$ (multiplying Nr and Dr by $e^{-\frac{1}{h}}$)

$$= \frac{0}{0+1} = 0 \left[e^{-\frac{1}{h}} \to 0 \text{ as } h\to 0\right]$$

Hence, $l.h.l = r.h.l = 0$

$\Rightarrow \lim_{x\to 0} f(x)$ exists.

3. If $f(x) = \frac{e^{\frac{1}{x}}}{1+e^{\frac{1}{x}}}$, show that $\lim_{x\to 0} f(x)$ does not exist.

Solution: $l.h.l = \lim_{h\to 0} \frac{e^{\left(\frac{1}{0-h}\right)}}{1+e^{\left(\frac{1}{0-h}\right)}}$

$$= \lim_{h\to 0} \frac{e^{-\frac{1}{h}}}{1+e^{-\frac{1}{h}}} = 0$$

Let $h>0$ $r.h.l = \lim_{h\to 0} \frac{e^{\left(\frac{1}{0+h}\right)}}{1+e^{\left(\frac{1}{0+h}\right)}}$

$$= \lim_{h\to 0} \frac{e^{\frac{1}{h}}}{1+e^{\frac{1}{h}}}$$

$= \lim_{h\to 0} \frac{e^{\frac{1}{h}}\cdot e^{-\frac{1}{h}}}{1\cdot e^{-\frac{1}{h}} + e^{-\frac{1}{h}}\, e^{\frac{1}{h}}}$ (multiplying Nr and Dr by $e^{-\frac{1}{h}}$)

$$= \lim_{h\to 0} \frac{1}{e^{-\frac{1}{h}}+1} = 1$$

Hence, $l.h.l \neq r.h.l$

$\Rightarrow \lim_{x\to 0} f(x)$ does not exist.

5. If $f(x) = \frac{e^{\frac{1}{x}}-1}{e^{\frac{1}{x}}+1}$, show that $\lim_{x\to 0} f(x)$ does not exist.

Solution: $l.h.l = \lim_{h\to 0} = \frac{e^{-\frac{1}{h}}-1}{e^{-\frac{1}{h}}+1} = \frac{0-1}{0+1}, h>0$

$$= -1 \left[\because \lim_{h\to 0} e^{-\frac{1}{h}} = 0 \right]$$

$$r.h.l = \lim_{h\to 0} \frac{e^{\frac{1}{h}} - 1}{1 + e^{\frac{1}{h}}}, h > 0$$

$$= \lim_{h\to 0} \frac{1 - e^{-\frac{1}{h}}}{1 + e^{-\frac{1}{h}}}$$ (multiplying *Nr* and *Dr* by $e^{-\frac{1}{h}}$).

$$= \frac{1-0}{1+0} = 1$$

$\therefore l.h.l \neq r.h.l$

$\Rightarrow \lim_{x\to 0} f(x)$ does not exist.

6. If $f(x) = e^{\frac{1}{x}}, x \neq 0$ then show that $\lim_{x\to 0} f(x)$ does not exist.

Solution: Let $h > 0$;

$$l.h.l = \lim_{h\to 0} e^{\left(\frac{1}{0-h}\right)} = \lim_{h\to 0} e^{\left(-\frac{1}{h}\right)} = 0$$

$$\left(\because e^{-\frac{1}{h}} \to 0 \text{ as } h \to 0 \right)$$

$$r.h.l = \lim_{h\to 0} e^{\left(\frac{1}{0+h}\right)} = \lim_{h\to 0} e^{\left(\frac{1}{h}\right)} = \infty$$

$$\left(\because e^{\frac{1}{h}} \to \infty \text{ when } h \to 0 \right)$$

$\therefore l.h.l \neq r.h.l$

$\Rightarrow \lim_{x\to 0} f(x)$ does not exist.

Problems based on modulus function

1. Examine the existence of $\lim_{x\to 0} |x|$.

Solution: $y = |x|$

$\Rightarrow y = x$, when $x > 0$ and

$y = -x$, when $x < 0$

That is,

$$y = \begin{cases} -x, \text{ when } x < 0 \\ x, \text{ when } x \geq 0 \end{cases}$$

Now let $h > 0$;

$\therefore x = 0 - h \Rightarrow y = -(0-h) = +h$, and $x = 0 + h \Rightarrow y = (0+h) = h$,

Further, $x = 0 - h \Rightarrow h \to 0$ when $x \to 0$ and $x = 0 + h \Rightarrow h \to 0$ when $x \to 0$

$$\therefore \lim_{x\to 0+} |x| = \lim_{x\to 0+} x = \lim_{h\to 0} h = 0$$

$$\lim_{x\to 0^-} |x| = \lim_{x\to 0^-} -x = \lim_{h\to 0} h = 0$$

Hence, $\lim_{x\to 0^+} |x| = \lim_{x\to 0^-} |x|$

$\Rightarrow \lim_{x\to 0} |x|$ does exist.

2. Show that $\lim_{x\to 0} \frac{\sqrt{x^2}}{x}$ does not exist.

Solution: $y = \frac{\sqrt{x^2}}{x} = \frac{|x|}{x}$

$\Rightarrow y = \frac{x}{x} = 1$, when $x > 0$ and

$y = \frac{-x}{x} = -1$, when $x < 0$

That is, $y = \begin{cases} -1, \text{ when } x < 0 \\ 1, \text{ when } x > 0 \end{cases}$

Now let $h > 0$;

$\therefore x = 0 - h \Rightarrow y = -1$, and $x = 0 + h \Rightarrow y = 1$

Further, $x = 0 - h \Rightarrow h \to 0$ when $x \to 0^-$ and $x = 0 + h \Rightarrow h \to 0$ when $x \to 0^+$

$$\therefore \lim_{x\to 0^+} \frac{|x|}{x} = \lim_{x\to 0^+} (1) = \lim_{h\to 0} 1 = 1$$

$$\lim_{x\to 0^-} \frac{|x|}{x} = \lim_{x\to 0^-} (-1) = \lim_{h\to 0} (-1) = -1$$

Hence, $\lim_{x \to 0^+} \frac{|x|}{x} \neq \lim_{x \to 0^-} \frac{|x|}{x}$

$\Rightarrow \lim_{x \to 0} \frac{|x|}{x}$ does not exist.

3. Examine whether $\lim_{x \to 0} \frac{|x^3|}{x}$ exists.

Solution: $y = \frac{|x^3|}{x} = \frac{|x|^3}{x} = \frac{(-x)^3}{x}$, when $x < 0$

$= -\frac{x^3}{x} = -x^2$, when $x < 0$

Also, $y = \frac{|x^3|}{x} = \frac{|x|^3}{x} = \frac{x^3}{x}$, when $x > 0$

$= x^2$, when $x > 0$

Hence $y = \frac{|x^3|}{x}$ can be rewritten as

$$y = \begin{cases} -x^2, \text{ when } x < 0 \\ x^2, \text{ when } x > 0 \end{cases}$$

Now let $h > 0$;

$\therefore x = 0 - h \Rightarrow y = -(0-h)^2, = -h^2$, and $x = 0 + h$
$\Rightarrow y = (0+h)^2, = h^2$, when $h \geq 0$

Further $x = 0 - h \Rightarrow h \to 0$ when $x \to 0^-$ and $x = 0 + h \Rightarrow h \to 0$ when $x \to 0^+$

$\therefore \lim_{x \to 0^+} \frac{|x|^3}{x} = \lim_{x \to 0^+} x^2 = \lim_{h \to 0} h^2 = 0$

$\lim_{x \to 0^-} \frac{|x|^3}{x} = \lim_{x \to 0^-} \left(-x^2\right) = \lim_{h \to 0} \left(-h^2\right) = 0$

Hence, $\lim_{x \to 0^+} \frac{|x|^3}{x} = \lim_{x \to 0^-} \frac{|x|^3}{x}$

$\Rightarrow \lim_{x \to 0} \frac{|x|^3}{x}$ does exist.

4. Examine the existence of $\lim_{x \to 2} \frac{\left(x^2 - 4\right)}{|x - 2|}$

Solution: $y = \frac{\left(x^2 - 4\right)}{|x - 2|}$

$\Rightarrow y = \frac{(x-2)(x+2)}{(x-2)}$, when $x - 2 > 0$

$= (x + 2)$, when $x > 2$

and $y = \frac{(x-2)(x+2)}{-(x-2)}$, when $x - 2 < 0$

$= -(x - 2)$, when $x < 2$

That is, $y = \begin{cases} (x+2), \text{ when } x > 2 \\ -(x+2), \text{ when } x < 2 \end{cases}$

Now, $x = 2 + h, h > 0$
$\Rightarrow y = (2 + h + 2),$
$= (4 + h),$
and $x = 2 - h, h > 0$
$\Rightarrow y = -(2 - h + 2),$
$= -(4 - h),$

Further, $x = 2 + h \Rightarrow h \to 0$, when $x \to 2^+$ and $x = 2 - h \Rightarrow h \to 0$, when $x \to 2^-$

$\therefore \lim_{x \to 2^+} \frac{\left(x^2 - 4\right)}{|x-2|} = \lim_{x \to 2^+} (x + 2) = \lim_{h \to 0} (4 + h) = 4$

and $\lim_{x \to 2^-} \frac{\left(x^2 - 4\right)}{|x-2|} = \lim_{x \to 2^-} -(x + 2) = \lim_{h \to 0} -(4 - h) = -4$

Hence, $\lim_{x \to 2^+} \frac{\left(x^2 - 4\right)}{|x-2|} \neq \lim_{x \to 2^-} \frac{\left(x^2 - 4\right)}{|x-2|}$

$\Rightarrow \lim_{x \to 2} \frac{\left(x^2 - 4\right)}{|x-2|}$ does not exist.

5. Show that $\lim_{x \to 0} \frac{|\sin x|}{x}$ does not exist.

Solution: $y = \dfrac{|\sin x|}{x}$

$\Rightarrow y = \dfrac{\sin x}{x}$, when $\sin x \geq 0$ and

$y = -\dfrac{\sin x}{x}$, when $\sin x < 0$

That is, $y = \begin{cases} \dfrac{\sin x}{x}, & \text{when } \sin x \geq 0,\ x \neq 0 \\ -\dfrac{\sin x}{x}, & \text{when } \sin x < 0 \end{cases}$

Now $x = 0 + h$, $h > 0$ and $\sin h > 0$

$\Rightarrow y = \dfrac{\sin (0+h)}{(0+h)} = \dfrac{\sin h}{h}$,

and $x = 0 - h$, $h > 0$ and $\sin h > 0$

$\Rightarrow y = \dfrac{-\sin (0-h)}{(0-h)} = \dfrac{\sin h}{h}$

Again $x = 0 + h \Rightarrow h \to 0$ when $x \to 0^+$ and $x = 0 - h \Rightarrow h \to 0$ when $x \to 0^-$

Hence, $\lim\limits_{x \to 0^+} \left(\dfrac{\sin x}{x}\right) = \lim\limits_{h \to 0} \left(\dfrac{\sin h}{h}\right) = 1$

$\lim\limits_{x \to 0^-} \left(-\dfrac{\sin x}{x}\right) = \lim\limits_{h \to 0} \left(-\dfrac{\sin h}{h}\right) = -1$

$\therefore \lim\limits_{x \to 0^+} \left(\dfrac{\sin x}{x}\right) \neq \lim\limits_{x \to 0^-} \left(\dfrac{\sin x}{x}\right) = -1$

$\Rightarrow \lim\limits_{x \to 0} \left(\dfrac{|\sin x|}{x}\right)$ does not exist.

6. Does $\lim\limits_{x \to 0} \dfrac{\sqrt{1 - \cos 2x}}{x}$ exist?

Solution: $y = \dfrac{\sqrt{1 - \cos 2x}}{x}$

$= \dfrac{\sqrt{2 \sin^2 x}}{x} = \sqrt{2}\, \dfrac{|\sin x|}{x}$

$\Rightarrow y = \sqrt{2}\left(\dfrac{\sin x}{x}\right)$ when $\sin x \geq 0$ and

$y = -\sqrt{2}\left(\dfrac{\sin x}{x}\right)$ when $\sin x < 0$

That is, $y = \begin{cases} \sqrt{2}\left(\dfrac{\sin x}{x}\right), & \text{when } \sin x \geq 0,\ x \neq 0 \\ -\sqrt{2}\left(\dfrac{\sin x}{x}\right), & \text{when } \sin x < 0 \end{cases}$

Now, $x = 0 + h$, $h > 0$ and $\sin h > 0$

$\Rightarrow y = +\sqrt{2}\left(\dfrac{\sin h}{h}\right)$, and $x = 0 - h$, $h > 0$ and $\sin h > 0$

$\Rightarrow y = -\sqrt{2}\left(\dfrac{\sin h}{h}\right)$

Further, $x = 0 + h \Rightarrow h \to 0$ when $x \to 0^+$ and $x = 0 - h \Rightarrow h \to 0$ when $x \to 0^-$

Hence, $\lim\limits_{x \to 0^+} \sqrt{2}\left(\dfrac{\sin x}{x}\right) = \lim\limits_{h \to 0} \sqrt{2}\left(\dfrac{\sin h}{h}\right) = \sqrt{2}$

and $\lim\limits_{x \to 0^-} \sqrt{2}\left(\dfrac{\sin x}{x}\right) = \lim\limits_{h \to 0} -\sqrt{2}\left(\dfrac{\sin h}{h}\right) = -\sqrt{2}$

$\therefore \lim\limits_{x \to 0^+} \sqrt{2}\left(\dfrac{\sin x}{x}\right) \neq \lim\limits_{x \to 0^-} -\sqrt{2}\left(\dfrac{\sin x}{x}\right)$

$\Rightarrow \lim\limits_{x \to 0} \dfrac{\sqrt{1 - \cos 2x}}{x}$ does not exist.

7. Examine the existence of $\lim\limits_{x \to 0} e^{-|x|}$

Solution: $y = e^{-|x|}$

$\Rightarrow y = e^{-x}$, when $x \geq 0$ and $y = e^{x}$, when $x < 0$

That is, $y = \begin{cases} e^{-x}, & \text{when } x \geq 0 \\ e^{x}, & \text{when } x < 0 \end{cases}$

Now, $x = 0 + h$, $h > 0$

$\Rightarrow y = e^{-h}$, and $x = 0 - h, h > 0$

$\Rightarrow y = e^{h}$

Further, $x = 0 + h \Rightarrow h \to 0$ when $x \to 0^+$ and $x = 0 - h \Rightarrow h \to 0$ when $x \to 0^-$

$\therefore \lim_{x \to 0^+} e^{-x} = \lim_{h \to 0} e^{-h} = e^{0} = 1$

$\lim_{x \to 0^-} e^{x} = \lim_{h \to 0} e^{h} = e^{0} = 1$

Hence, $\lim_{x \to 0^+} e^{-x} = \lim_{x \to 0^-} e^{x} = 1$

$\Rightarrow \lim_{x \to 0} e^{-|x|}$ does exist and = 1

Problems based on piecewise function

1. Examine the existence of the limit of the function $f(x)$ as $x \to 0$ if it exists where

$f(x) = x$, when $x < 0$
$= 1$, when $x = 0$
$= x^2$, when $x > 0$

Solution: Let $h > 0$;

$\therefore x = 0 + h \Rightarrow f(x)$
$= f(0 + h) = (0 + h)^2 = h^2 = h^2$, and $x = 0 - h \Rightarrow f(x)$
$= f(0 - h) = (0 - h) = -h$,

Further, $x = 0 - h \Rightarrow h \to 0$ when $x \to 0^-$ and $x = 0 + h \Rightarrow h \to 0$ when $x \to 0^+$

$\therefore \lim_{x \to 0^+} x^2 = \lim_{h \to 0} h^2 = 0$

$\lim_{x \to 0^-} x = \lim_{h \to 0} (-h) = 0$

Hence, $\lim_{x \to 0^+} x^2 = \lim_{x \to 0^-} x = 0$

$\Rightarrow \lim_{x \to 0} f(x)$ does exist.

N.B.: The value of the function $f(x) = 1$ at $x = 0$ is not required to be considered to show the existence of the limit of a given function at the point $x = 0$.

2. Examine the existence of the limit of the function $f(x)$ as $x \to 2$ if it exists where

$$f(x) = \begin{cases} x^2 + 1, \text{ when } 0 < x < 2 \\ 5, \text{ when } x \geq 0 \end{cases}$$

Solution: Let $h > 0$;

$\therefore x = 2 + h \Rightarrow f(x)$
$= f(2 + h) = 5$, and $x = 0 - h \Rightarrow f(x)$
$= f(2 - h) = (2 - h)^2 + 1$
$= 4 - 4h + h^2 + 1$
$= h^2 - 4h + 5$, when $-2 < h \leq 0$

Further $x = 2 - h \Rightarrow h \to 0$ when $x \to 2^-$ and $x = 2 + h \Rightarrow h \to 0$ when $x \to 2^+$

$\therefore \lim_{x \to 2^+} 5 = \lim_{h \to 0} 5 = 5$

$\lim_{x \to 2^-} (x^2 + 1) = \lim_{h \to 0} (h^2 - 4h + 5) = 5$

Hence, $\lim_{x \to 2^+} 5 = \lim_{x \to 2^-} (x^2 + 1)$

$\Rightarrow \lim_{x \to 2} f(x)$ does exist.

3. Examine the existence of the limit of $f(x)$ at $x = a$,

where $f(x) = \dfrac{x^2 - a^2}{x - a}$, when $0 \leq x < a$

$= 2a$, when $x \geq a$

Solution: Let $h > 0$;

$\therefore x = a - h$

$$\Rightarrow f(x) = f(a - h) = \frac{(a-h)^2 - a^2}{(a-h) - a}$$

$$= \frac{a^2 - 2ah + h^2 - a^2}{-h}$$

$$= \frac{-2ah + h^2}{-h} \text{ and } x = a + h$$

$\Rightarrow f(x) = f(a + h) = 2a$

Further, $x = a - h \Rightarrow h \to 0$ when $x \to a^-$ and $x = a + h \Rightarrow h \to 0$ when $x \to a^+$

$$\therefore \lim_{x \to a^+} \left(\frac{x^2 - a^2}{x^2 + a^2}\right) = 2a$$

$$\lim_{x \to a} \left(\frac{x^2 - a^2}{x - a}\right) = \lim_{h \to 0} \left(\frac{-2ah + h^2}{-h}\right) = \lim_{h \to 0} h\left(\frac{-2a + h}{-h}\right)$$

$= \lim_{h \to 0} -(-2a + h) = 2a$

$= \lim_{x \to a^-} \left(\frac{x^2 - a^2}{x - a} \right) = 2a$

$\Rightarrow \lim_{x \to a} f(x)$ exists.

Problems based on redefined function

1. Examine the existence of the limit of $f(x)$ at $x = 0$, if it exists where $f(x) = 1$, when $x \neq 0$

$= 2$, when $x = 0$

Solution: The given function

$f(x) = \begin{cases} 1, \text{ when } x \neq 0 \\ 2, \text{ when } x = 0 \end{cases}$ can be rewritten as under:

$$f(x) = \begin{cases} 1, \text{ when } x < 0 \\ 1, \text{ when } x > 0 \\ 2, \text{ when } x = 0 \end{cases}$$

Let $h > 0$;

$\therefore x = 0 - h$

$\Rightarrow f(x) = f(0 - h) = 1$ and $x = 0 + h$

$\Rightarrow f(x) = f(0 + h) = 1$

Further, $x = 0 - h \Rightarrow h \to 0$ when $x \to 0^-$ and $x = 0 + h \Rightarrow h \to 0$ when $x \to 0^+$

$\therefore \lim_{x \to 0^-} f(x) = \lim_{h \to 0} f(h) = 1$

and $\lim_{x \to 0^+} f(x) = \lim_{h \to 0} f(h) = 1$

$\Rightarrow \lim_{x \to 0} f(x)$ does exist.

Note: $x \neq a \Leftrightarrow x > a$ or $x < a$, where $a \in R$.

2. Examine the existence of the limit of $f(x)$ at $x = 1$, where $f(x) = x^2$, when $x \neq 1$

$= 2$, when $x = 1$

Solution: The given function

$f(x) = \begin{cases} x^2, \text{ when } x \neq 1 \\ 2, \text{ when } x = 1 \end{cases}$ can be rewritten as under:

$$f(x) = \begin{cases} x^2, \text{ when } x < 1 \\ x^2, \text{ when } x > 1 \\ 2, \text{ when } x = 1 \end{cases}$$

Let $h > 0$;

$\therefore x = 1 - h$

$\Rightarrow f(x) = f(1 - h) = (1 - h)^2$

$= (1 - 2h + h^2)$

and $x = 1 + h$

$\Rightarrow f(x) = f(1 + h) = (1 + h)^2$, when $1 + h > 1$

$= (1 + 2h + h^2)$

Further $x = 1 - h \Rightarrow h \to 0$ when $x \to 0^-$ and $x = 1 + h \Rightarrow h \to 0$ when $x \to 0^+$

Now $\lim_{x \to 1^+} x^2 = \lim_{h \to 0} \left(1 + 2h + h^2\right) = 1$

and $\lim_{x \to 1^-} x^2 = \lim_{h \to 0} \left(1 - 2h + h^2\right) = 1$

$\therefore \lim_{x \to 1^+} x^2 = \lim_{x \to 1^-} x^2 = 1$

$\Rightarrow \lim_{x \to 1} f(x)$ does exist.

Exercise 4.29.1

Problems on functions defined by a single formula with no modulus function

Do the limits of the following functions exist?

1. $\lim_{x \to 0} \sin\left(\frac{1}{x}\right)$ [***Ans***: Does not exist.]

2. $\lim_{x \to 0} \cos\left(\frac{1}{x}\right)$ [***Ans***: Does not exist.]

3. $\lim_{x \to 0} e^{\frac{1}{x}}$ [***Ans***: Does not exist.]

4. $\lim_{x \to 0} \tan^{-1}\left(\frac{1}{x}\right)$ [***Ans***: Does not exist.]

5. $\lim\limits_{x\to 0} \dfrac{\sqrt[3]{x^3}}{x}$ [*Ans*: Does exist.]

Exercise 4.29.2

Problems on functions defined by a single formula with the modulus function

Examine the existence of the limit of each given function.

1. $\lim\limits_{x\to 1} \dfrac{|x-1|}{(x-1)}$ [*Ans*: Does not exist.]

2. $\lim\limits_{x\to 0} \dfrac{x}{\sqrt{1-\cos x}}$ [*Ans*: Does not exist.]

3. $\lim\limits_{x\to 0} \dfrac{\sqrt{1+\cos 2x}}{\sqrt{2}\cos x}$ [*Ans*: Does exist.]

4. $\lim\limits_{x\to 0} \dfrac{\sqrt{1-\cos 2x}}{\sin x}$ [*Ans*: Does not exist.]

Exercise 4.29.3

Problems on piecewise functions

Examine the existence of the limit of each function defined as given below:

1. $f(x)=x^2+x+2$, when $x<1$
   $f(x)=x^4+3$, when $x>1$ [*Ans*: Exists at $x=1$]
2. $f(x)=x^3$, when $x<-1$
   $f(x)=x^5$, when $x>-1$
   $f(x)=-1$, when $x=-1$ [*Ans*: Exists at $x=-1$]
3. $f(x)=x^2-2x+3$, when $x\le 1$
   $=x+1$, when $x>1$ [*Ans*: Exists at $x=1$]
4. $f(x)=\begin{cases} x-3, \text{ when } x<4 \\ 5-x, \text{ when } x\ge 4 \end{cases}$ [*Ans.* Exists at $x=4$]

Exercise 4.29.4

Problems on redefined functions

Examine the existence of the limit of each function defined by $y=f(x)$ at the indicated point $x=a$, $a\in R$

1. $f(x)=\begin{cases} \dfrac{x^2-4x+3}{x^2-1}, \text{ when } x\ne 1 \\ 2, \text{ when } x=1 \end{cases}$ [*Ans.* Exists]

2. $f(x)=\begin{cases} \dfrac{|x|}{x}, x\ne 0 \\ 0, x=0 \end{cases}$ [*Ans.* Does not exists]

3. $f(x)=x\cdot\sin\left(\dfrac{1}{x}\right)$, when $x\ne 0$ [*Ans.* Exists]
   $=0$, when $x=0$

4. $f(x)=\dfrac{\sin x}{x}$, $x\ne 0$ [*Ans.* Exists]
   $=0, x=0$

On existence of limit of greatest integer function

Firstly one should note the following key points:

1. $[x]$ denotes the first integer less than or equal to x (given real number) lying on the number line to the left side of x (given real number) $\Leftrightarrow$ The integer on the number line which is nearest to x on the left side of the given real number. Hence, to find $[x]$ numerically at any given real number for x, we always consider the first integer (or, nearest integer) which is on the left side of the given real number lying on the number line. In the light of this explanation, we are able to provide the following useful rules:

(a_1) $[x]=x$ provided $x=$ an integer +ve, –ve or zero i.e., $0, \pm 1, \pm 2, \ldots$, e.g.:
(i) $[0]=0$
(ii) $[5]=5$
(iii) $[-3]=-3$

(a_2) $[x]$ = an integer immediately to the left of x provided x is positive or negative fraction (i.e., x is not an integer), e.g.:
(i) $[5/2]=2$
(ii) $[-5/2]=-3$
(iii) $[-0.1]=-1$
(iv) $[(-0.75)]=-1$
(v) $[-6.35]=-7$

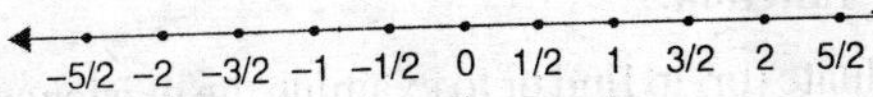

(a_3) $[x] = 0$ for all positive real values of x just less than unity ($0<x<1$), e.g.:

(i) $x = \frac{1}{2} \Rightarrow [x] = \left[\frac{1}{2}\right] = 0$

(ii) $x = \frac{3}{12} \Rightarrow [x] = \left[\frac{3}{12}\right] = 0$

(iii) $x = \frac{2}{3} \Rightarrow [x] = \left[\frac{2}{3}\right] = 0$

(a_4) $[x] = 1$ for all positive real values of x just greater than unity ($1 < x < 2$), e.g.:

(i) $x = \frac{3}{2} \Rightarrow [x] = \left[\frac{3}{2}\right] = 1$

(ii) $x = 1 \cdot 1 \Rightarrow [x] = [1 \cdot 1] = 1$

(a_5) $[x] = -1$ for all negative values of x whose absolute value is first greater than unity ($-1 < x < 0$), e.g.:

(i) $[-0.1] = -1$
(ii) $[-0.0001] = -1$
(iii) $[-0.0009] = -1$
(iv) $[-0.23.49] = -1$
(v) $[-0.9999] = -1$

(a_6) If n be any integer and m be any real number, then for $0 < h < 1$ (i.e. $[h]$ does not exceed zero) and $0 < m\,h < 1$ (i.e. $[m\,h]$ does not exceed zero) where h is sufficiently small positive number greater than zero, we have

(i) $[n+h] = n$
(ii) $[n + m\,h] = n$ as $[8 + 12\,h] = 8$
(iii) $[n-1+h] = n-1$
(iv) $[n-1+m\,h] = n-1$
(v) $[0+h] = 0$
(vi) $[n-h] = n-1$
(vii) $[n-m\,h] = n-1$ as $[8-12\,h] = 7$
(viii) $[n-1-h] = n-1-1 = n-2$
(ix) $[0-h] = 0-1 = -1$

(a_7) If n = an improper positive fraction and m be any real number, then for $0<h<1$, and $0 < m\,h < 1$, where h is sufficiently small positive real number > 0 then we have

$[n \pm h]/[n \pm m\,h]$ = integral part of the improper fraction

= the whole number before the decimal

e.g.:

(i) $[1.5-h] = 1$
(ii) $[1.5+h] = 1$
(iii) $[2.4+h] = 2$
(iv) $[2.4-h] = 2$
(v) $[2.000001-h] = 2$

(a_8) If n = *a* negative improper fraction and m be any real number, then for $0<h<1$ and $0<m\,h<1$, where h is sufficiently small positive number > 0, then we have $[n-h]/[n-m\,h]$ = integral part of the proper fraction -1

= the whole number before the decimal -1

e.g.:

(i) $[-2.3-h] = -2-1 = -3$
(ii) $[-5.0006-h] = -5-1 = -6$
(iii) $[-7.00001-12\,h] = -7-1 = -8$
(iv) $[-1.1234-2.3\,h] = -1-1 = -2$

Some important facts about the function $y = [x]$.

(i) $[x] = x \Leftrightarrow x \in I$

(ii) $[x] < x \Leftrightarrow x \notin I$

(iii) $[x] = k\,(k \in I) \Leftrightarrow k \le x < k+1$, i.e. $[x] = k \Leftrightarrow k \le x < k+1,\ k \in N$ and $[x] = -(k+I) \Leftrightarrow -(k+1) \le x < -k,\ k \in N$

2. To calculate arithmetically $[n \pm h]/[n \pm m\,h]$ where n and m are any two arbitrary numbers and h lies between 0 and 1 (i.e.; $0<h<1$), we may adopt the following working rule, in the problems on limits.

(i) Put h = any one of the numbers 0.1, 0.01, 0.001, 0.0001, 0.00001, …, etc. whose number of zeros = number of digits after decimal in the given value of n provided we calculate $[x] = [n \pm h]$ where $x = n \pm h$.

(ii) Put h = any one of the numbers 0.1, 0.01, 0.0001, 0.001, 0.00001, …, etc. whose number of zeros = the sum of number of digits in the integral part and decimal part of m provided we calculate $[x] = [n \pm mh]$ where $x = n \pm mh$.

(iii) Calculate $(n \pm mh)$ arithmetically when n and m are known particular numbers.

(iv) Use the fact: the integer on the number line which is to the left side of $(n \pm mh)/(n \pm h)$ will represent $[n \pm mh]/[n \pm h]$.

Examples

(i) $[1+h]=[1+0.1]=[1.1]=1$
(ii) $[1-h]=[1-0.1]=[0.9]=0$
(iii) $[1.5+h]=[1.5+0.01]=[1.51]=1$
(iv) $[1.5-h]=[1.5-0.01]=[1.49]=1$
(v) $[2.000001-h]=[2.000001-0.0000001]=[2.000009]=2$
(vi) $[-2.201-h]=[-2.201-0.0001]=[-2.2009]=-3$
(vii) $[-2-h]=[-2-0.1]=[-2.1]=-3$
(viii) $[8-12h]=[8-12\times 0.001]=[8-0.012]=[7.988]=7$
(ix) $[8+12h]=[8+12\times 0.001]=[8+0.012]=[8.012]=8$
(x) $[0-h]=[0-0.1]=[-0.1]=-1$
(xi) $[0+h]=[0+0.1]=[0.1]=0$
(xii) $[1+h/2]=[1+0.5h]=[1+0.5\times 0.01]=[1+0.005]=[1.005]=1$
(xiii) $[2.3-233.6h]=[2.3-233.6\times 0.00001]=[2.3-0.002336]=[2.2977]=2$

Precaution

1. While calculating $[n \pm h]$ arithmetically, 'h' should be chosen so small that it does not increase or decrease $[n]$ by unity, n is not integer.
e.g.: If we choose $h=0.1$ to calculate $[2.000001-h]$, we get $[2.000001-0.1]=[1.900001]=1$ as a greatest integer contained in $[2.000001-h]$ = which is false because $[2.000001-h]=2$.

2. While calculating $[n \pm mh]$, h should be chosen so small that '$m\,h$' must be slightly greater than zero.
e.g.: If we choose h = 0.1 to calculate $[2.3-299h]$, we get $[2.3-299h]=[2.3-299\times 0.1]=[2.3-29.1]=[-26.8]=-27$ which is false because $[2.3-299h]=2$.
Similarly, $[2.3+299h]=[2.3+299\times 0.1]=[2.3+29.1]=[31.4]=31$ which is false because $[2.3+299h]=2$.

3. The above working rule is valid to find the limit of $[n \pm mh]/[n \pm h]$ as $h \to 0$.

Limit method to examine the existence of greatest integer function

To evaluate (or, to find or to examine the existence of $\lim_{x\to n}[f(x)] = \lim_{x\to n}$ (greatest integer/bracket function), where n = any real number, we adopt the following working rules.

Working rule 1

1. Find the left hand limit using the following scheme:
(a) put $x=n-1+h$ (where $0<h<1$) in $f(x)$
(b) find the greatest integer in $f(n-1+h)$ where $0<h<1$ and remove the symbol of square bracket from $[f(n-1+h)]$
(c) use $\lim_{h\to 0}$ (greatest integer) = greatest integer

2. Find the right hand limit using the following scheme:
(a) put $x=n+h'$ (where $0<h'<1$) in $f(x)$
(b) find the greatest integer in $f(n+h')$ where $0<h'<1$ and the symbol of square bracket be removed from $[f(n+h')]$
(c) use $\lim_{h\to 0}$ (greatest integer) = greatest integer

Working rule 2

1. Find the left hand limit using the following scheme:
(a) Put $x=n-h$ in $f(x)$ where $0<h<1$
(b) Find the greatest integer in $f(n-h)$ and remove the symbol of square bracket from $[f(n-h)]$
(c) Use $\lim_{h\to 0}$ (greatest integer) = greatest integer.

2. Find the right hand limit using the following scheme:
(a) Put $x=n+h$ in $f(x)$ where $0<h<1$
(b) Find the greatest integer in $f(n+h)$ and the symbol of square bracket be removed from $[f(n+h)]$
(c) Use $\lim_{h\to 0}$ (greatest integer) = greatest integer.

Note:

1. If we are asked to examine the existence of $\lim_{h\to n}[f(x)]$, we have to examine *l.h.l* and *r.h.l*. If they are equal, it is declared that $\lim_{h\to n}[f(x)]$ exists and of they are not equal, it is declared that $\lim_{h\to n}[f(x)]$ does not exist.

2. We recall that

(i) $[n-1+h]=n-1$, provided $0<h<1$ and n is an integer.

(ii) $[n+h]=n$, provided $0<h<1$ and n is an integer

(iii) $[n-h]=(n-1)$, provided $0<h<1$ and n is an integer

Examples

(i) $[2-1+h]=2-1=1$

(ii) $[2+h]=2$

(iii) $[2-h]=2-1=1$

Note: Existence of a function in limit ⇔ Existence of a function in the sense of limiting value ⇔ Existence of limit of a function for the limit of an independent variable.

3. Method (1) is applicable only when we have greatest integer function $[f(x)]$ only whose limit is required whereas method (2) is applicable in every case and particularly when the given function is the combination (sum, difference, product and /quotient) of $f_1(x)$ and $[f_2(x)]$.

Type 1: To evaluate (or, to find or, to examine the existence of)

$\lim_{h\to n}[f(x)]$, where 'n' = an integer

Examples worked out:

Examine the existence of the following limits

1. $\lim_{x\to 1}[x]$

Solution: Method 1

$l.h.l = \lim_{x\to 1}[x]$

$= \lim_{h\to 0}[1-1+h] = \lim_{h\to 0}[h] = \lim_{h\to 0}[1-1] = \lim_{h\to 0}0 = 0$

$\because [n-1+h]=n-1$ for $0<h<1$ and n = an integer

$r.h.l = \lim_{x\to 1^-}[x] = \lim_{h'\to 0}[1+h'] = \lim_{h'\to 0}(1) = 1$

$\because [n+h']=n$ for $0<h<1$ and n = an integer

$\therefore l.h.l \neq r.h.l \Rightarrow \lim_{x\to 1}[x]$ does not exist

Method 2

$l.h.l = \lim_{x\to 1-}[x]$

$= \lim_{h\to 0}[1-h] = \lim_{h\to 0}(1-1) = \lim_{h\to 0}0 = 0$

$\because [n-h]=n-1$ provided $0<h<1$ and n = an integer

$r.h.l = \lim_{x\to 1^+}[x] = \lim_{h\to 0}[1+h] = \lim_{h\to 0}1 = 1$

$\because [n+h]=n$ provided $0<h<1$ and n = an integer

$\therefore l.h.l \neq r.h.l \Rightarrow \lim_{x\to 1}[x]$ does not exist

2. $\lim_{x\to 2}[x]$

Solution: Method 1

$l.h.l = \lim_{x\to 2^-}[x] = \lim_{h\to 0}[2-1+h] = \lim_{h\to 0}(2-1) = 1$

$\because [n-1+h]=n-1$ for any integer n

$r.h.l = \lim_{x\to 2^+}[x] = \lim_{h'\to 0}[2+h'] = \lim_{h'\to 0}2 = 2$

$\because [n+h']=n$ for any integer n

$\therefore l.h.l \neq r.h.l \Rightarrow \lim_{x\to 2}[x]$ does not exist.

Method 2

$l.h.l = \lim_{x\to 2^-}[x] = \lim_{h\to 0}[2-h] = \lim_{h\to 0}1 = 1$

$[n-h]=n-1$ for any integer n

$r.h.l = \lim_{x\to 2^+}[x] = \lim_{h\to 0}[2+h] = \lim_{h\to 0}2 = 2$

$\because [n+h]=n$ for any integer 'n'

$\therefore l.h.l \neq r.h.l \Rightarrow \lim_{x\to 2}[x]$ does not exist.

3. $\lim_{x \to n} [x]$ for any integer 'n'.

$l.h.l = \lim_{x \to n^-} [x] = \lim_{h \to 0} [n-1+h] = \lim_{h \to 0} (n-1) = n-1$

$\because [n-1+h] = n-1$ for $0 < n < 1$ and n being an integer

$r.h.l = \lim_{x \to n=} [x] = \lim_{h' \to 0} [n+h'] = \lim_{h' \to 0} n = n$

$\because [n+h'] = n$ for $0 < h' < 1$ and n being an integer

Thus, we observe that $\lim_{x \to n-} [x] \neq \lim_{x \to n+} [x] \Leftrightarrow$ $l.h.l \neq r.h.l.$

$\therefore \lim_{x \to n} [x]$ does not exist for any integer 'n'.

Method 2

$l.h.l = \lim_{x \to n-} [x]$

$= \lim_{h \to 0} [n-h] = \lim_{h \to 0} (n-1) = n-1$

$\because [n-h] = n-1$ for $0 < h < 1$ and n being an integer

$r.h.l = \lim_{x \to n+} [x] = \lim_{h \to 0} [n+h] = \lim_{h \to 0} n = n$

$[n+h] = n$ for $0 < h < 1$ and n being an integer

Thus, we observe, $\lim_{x \to n-} [x] \neq \lim_{x \to n+} [x] \Leftrightarrow l.h.l \neq r.h.l.$

$\therefore \lim_{x \to n} [x]$ does not exist.

4. $\lim_{x \to 1} [1-x]$

Solution: $l.h.l = \lim_{x \to 1-} [1-x]$

$\lim_{h \to 0} [1-(1-h)] = \lim_{h \to 0} [h] = \lim_{h \to 0} 0 = 0 \quad (\because [h] = 0)$

$r.h.l = \lim_{x \to 1+} [1-x]$

$= \lim_{h \to 0} [1-(1+h)] = \lim_{h \to 0} [-h] = \lim_{h \to 0} (-1) = -1 \quad (\because [-h] = -1)$

Thus, we observe, $l.h.l \neq r.h.l$ which means $\lim_{x \to 1} [1-x]$ doe not exist.

5. $\lim_{x \to 1} [x-1]$

Solution: $l.h.l = \lim_{x \to 1-} [x-1]$

$= \lim_{h \to 0} [1-h-1] = \lim_{h \to 0} [-h] = \lim_{h \to 0} (-1) = -1 \quad (\because [-h] = -1)$

$r.h.l = \lim_{x \to 1+} [x-1]$

$= \lim_{h \to 0} [1+h-1] = \lim_{h \to 0} [h] = \lim_{h \to 0} 0 = 0 \quad (\because [h] = 0)$

$\therefore l.h.l \neq r.h.l$ which means $\lim_{x \to 1} [x-1]$ does not exist.

6. $\lim_{x \to 2-} \left[x^3\right]$

Solution: $\lim_{x \to 2-} \left[x^3\right]$

$= \lim_{h \to 0} \left[(2-h)^3\right] = \lim_{h \to 0} \left[8 - 12h + 6h^2 - h^3\right]$

$= \lim_{h \to 0} [8 - 12h]$

$= \lim_{h \to 0} 7 \quad (\because 7 < 8 - 12h < 8)$

$= 7$

Type 2: To evaluate (or, to find or to examine the existence of) $\lim_{x \to n} [f(x)]$ when $n \neq$ an integer (i.e., n = a fraction +ve or –ve), we adopt the working rules mentioned earlier in type 1.

Examples worked out:

Evaluate if the following limit exists.

1. $\lim_{x \to 1 \cdot 5} [x]$

Solution: $l.h.l = \lim_{x \to 1 \cdot 5-} [x]$

$= \lim_{h \to 0} [1 \cdot 5 - h] = \lim_{h \to 0} 1 = 1 \quad (\because 1 < 1 \cdot 5 - h < 2)$

$r.h.l = \lim_{x \to 1.5+} [x]$

$= \lim_{h \to 0} [1.5 + h] = \lim_{h \to 0} 1 = 1 \; (\because 1 < 1.5 + h < 2)$

Thus, we observe, $l.h.l = r.h.l \Rightarrow$ limit exists and

$\lim_{x \to 1.5} [x] = 1.$

2. $\lim_{x \to 2.4} [x]$

Solution: $l.h.l = \lim_{x \to 2.4-} [x]$

$= \lim_{h \to 0} [2.4 - h] = \lim_{h \to 0} 2 = 2 \; (\because 2 < 2.4 - h < 3)$

$r.h.l = \lim_{x \to 2.4+} [x]$

$= \lim_{h \to 0} [2.4 + h] = \lim_{h \to 0} 2 = 2 \; (\because 2 < 2.4 + h < 3)$

$\therefore l.h.l = r.h.l \Rightarrow$ limit exists and $\lim_{x \to 2.4} [x] = 2.$

3. $\lim_{x \to x_0} [x]$, where $n < x_0 < n + 1$ for some integer n.

Solution: $l.h.l = \lim_{x \to x_0 - 0} [x] = \lim_{h \to 0} [x_0 - h] = n$

$r.h.l = \lim_{x \to x_0 + 0} [x] = \lim_{h \to 0} [x_0 + h] = n$

($\because$ for sufficiently small h $n < x_0 - h < n + 1$ and $n < x_0 + h < n + 1$)

$\therefore \lim_{x \to x_0} [x]$ exists and $= n$.

4. $\lim_{x \to \frac{1}{2}-} [2x]$

Solution: $\lim_{x \to \frac{1}{2}-} [2x]$

$$= \lim_{h \to 0} \left[2\left(\frac{1}{2} - h\right)\right] = \lim_{h \to 0} [1 - 2h] = \lim_{h \to 0} 0 = 0$$

$(\because 0 < 1 - 2h < 1)$

5. $\lim_{x \to \frac{1}{2}+} \left[1 + x + x^2\right]$

Solution: $\lim_{x \to \frac{1}{2}+} \left[1 + x + x^2\right]$

$$= \lim_{h \to 0} \left[1 + \left(\frac{1}{2} + h\right) + \left(\frac{1}{2} + h\right)^2\right]$$

$$= \lim_{h \to 0} \left[1 + \frac{1}{2} + h + \frac{1}{4} + \frac{1}{2} \cdot 2 \cdot h + h^2\right]$$

$$= \lim_{h \to 0} \left[\frac{3}{2} + \frac{1}{4} + 2h + h^2\right] = \lim_{h \to 0} \left[\frac{7}{4} + 2h\right] = \lim_{h \to 0} 1$$

$$\left(\because 1 < \frac{7}{4} + 2h < 2\right)$$

6. $\lim_{x \to \sqrt{2}} \left[x^2\right]$

Solution: Method 1

Let $f(x) = x^2 = z$

$\because x \to \sqrt{2} \Rightarrow x^2 \to 2 \Rightarrow z \to 2$

$\therefore \lim_{Z \to 2} [z] = \lim_{x \to \sqrt{2}} \left[x^2\right]$

Now, let $h > 0$,

$l.h.l = \lim_{Z \to 2^-} [z]$

$= \lim_{h \to 0} [2 - h] = 1 \; (\because 1 < 2 - h < 2$ for sufficiently small $h > 0)$

$r.h.l. = \lim_{Z \to 2^+} [z]$

$= \lim_{h \to 0} [2 + h] = 2 \; (\because 2 < 2 + h < 3$ for sufficiently small $h > 0)$

Thus, we observe, $l.h.l \neq r.h.l \Rightarrow \lim_{Z \to 2} [z]$

$= \lim_{Z \to \sqrt{2}} \left[x^2\right]$ does not exist.

Method 2

$l.h.l = \lim_{x \to \sqrt{2}^-} \left[x^2\right]$

$$= \lim_{h \to 0} \left[\left(\sqrt{2} - h\right)^2\right] = \lim_{h \to 0} \left[2 + h^2 - 2\sqrt{2}h\right]$$

$$= \lim_{h \to 0} \left[2 - 2\sqrt{2}h\right]$$

($\because 2 + h^2 - 2\sqrt{2}h \equiv 2 - 2\sqrt{2}h$ for sufficiently small $h > 0$)

$= \lim_{h\to 0} (1)$ ($\because$ for sufficiently small $h > 0$, $[2 - 2\sqrt{2}h] = 1$)

$= 1$

$r.h.l = \lim_{x\to\sqrt{2}^+} [x^2]$

$= \lim_{h\to 0} \left[(\sqrt{2} + h)^2\right] = \lim_{h\to 0} \left[2 + h^2 + 2\sqrt{2}h\right]$

$= \lim_{h\to 0} \left[2 + 2\sqrt{2}h\right]$

($\because 2 + h^2 + 2\sqrt{2}h \equiv 2 + 2\sqrt{2}h$ neglecting higher powers of h for h being sufficiently small > 0)

$= \lim_{h\to 0} 2$ ($\because$ for small $h > 0$, $\left[2 + 2\sqrt{2}h\right] = 2$)

$= 2$

$\therefore$ $l.h.l \neq r.h.l$ which means $\lim_{x\to\sqrt{2}} [x^2]$ does not exist.

7. $\lim_{x\to\frac{\pi}{2}} [\sin x]$

Solution; $l.h.l = \lim_{x\to\frac{\pi}{2}-} [\sin x] = \lim_{h\to 0} \left[\sin\left(\frac{\pi}{2} - h\right)\right]$

$= \lim_{h\to 0} 0$

$= 0$

$\because \left[\sin\left(\frac{\pi}{2} - h\right)\right] = 0$ as $h \to 0$ and $h > 0 \Rightarrow -h < 0$

$\Rightarrow 0 < \frac{\pi}{2} - h < \frac{\pi}{2}$

$\Rightarrow 0 < \sin\left(\frac{\pi}{2} - h\right) < 1$

$r.h.l = \lim_{x\to\frac{\pi}{2}+} [\sin x] = \lim_{h\to 0} \left[\sin\left(\frac{\pi}{2} + h\right)\right]$

$= \lim_{h\to 0} 0$

$= 0$

$\left[\sin\left(\frac{\pi}{2} + h\right)\right] = 0$, as $0 < h \Rightarrow 0 < \frac{\pi}{2} + h$

$\Rightarrow \sin 0 < \sin\left(\frac{\pi}{2} + h\right) \Rightarrow 0 < \sin\left(\frac{\pi}{2} + h\right)$

$\Rightarrow 0 < \sin\left(\frac{\pi}{2} + h\right) < 1.$

$\therefore \lim_{x\to\frac{\pi}{2}} [\sin x] = 0$

8. $\lim_{x\to\frac{\pi}{4}-} [\sin x]$

Solution: $\lim_{x\to\frac{\pi}{4}-} [\sin x]$

$= \lim_{h\to 0} \left[\sin\left(\frac{\pi}{4} - h\right)\right]$

$= \lim_{h\to 0} 0$

$= 0$

$\because \left[\sin\left(\frac{\pi}{4} - h\right)\right] = 0$, as $0 < \frac{\pi}{4} - h < \frac{\pi}{4}$

$\Rightarrow \sin 0 < \sin\left(\frac{\pi}{4} - h\right) < \sin\frac{\pi}{4}$

$\Rightarrow 0 < \sin\left(\frac{\pi}{4} - h\right) < \frac{1}{\sqrt{2}} < 1$

Note: The real number π is approximately equal to 3.14.

Type 3: To evaluate (or, to find or to examine the existence of) the function which is the combination (i.e.; sum, difference, product or quotient) or composition (i.e. a function of a function) of $f_1(x)$ and $[f_2(x)]$ (i.e.; a function and greatest integer function) as $x \to a$ a real number.

Examples worked out:

***Question* 1**: Prove that (i) $\lim_{x \to x_0} (x - [x])$ does not exist when x_0 is an integer (ii) $\lim_{x \to x_0} (x - [x])$ exists when x_0 is not an integer.

Solution: Let $f(x) = x - [x]$

(i) If $x_0 = n$ (an integer), then for $h > 0$ (h being sufficiently small) $f(x_0 - h) = f(n - h) = n - h - [n - h] = n - h - (n - 1) = 1 - h$

$\therefore \lim_{h \to 0} f(x_0 - h) = \lim_{h \to 0} (1 - h) = 1$

and $f(x_0 + h) = f(n + h) = n + h - [n + h] = n + h - n = h$

$\therefore \lim_{h \to 0} f(x_0 + h) = \lim_{h \to 0} h = 0$

$\therefore$ r.h.l. $\neq$ l.h.l.

$\therefore \lim_{x \to x_0} (x - [x])$ does not exist

(ii) If $n < x_0 < n + 1$ for some integer n, then for some sufficiently small $h > 0$, $f(x_0 - h) = x_0 - h - [x_0 - h] = x_0 - h - n \to x_0 - n$ as $h \to 0$ and $f(x_0 + h) = x_0 + h - [x_0 + h] = x_0 + h - n \to x_0 - n$ as $h \to 0$

$\therefore r.h.l = l.h.l = x_0 - [x_0]$

$\therefore$ the $\lim_{x \to x_0} (x - [x])$ exists and $= x_0 - [x_0]$

Now we will solve such problems which are all particular cases of Q. no. (1) for the beginners independently.

2. Evaluate (if it exists) the following

(i) $\lim_{x \to 0} (x - [x])$

Solution: Let $h > 0$

$$l.h.l = \lim_{x \to 0^-} (x - [x])$$

$$= \lim_{h \to 0} (0 - h - [0 - h]) = \lim_{h \to 0} (-h - [-h])$$

$$= \lim_{h \to 0} (-h - (-1)) \quad (\because [-h] = -1)$$

$$= \lim_{h \to 0} (-h) + \lim_{h \to 0} 1 = 0 + 1 = 1$$

$$r.h.l = \lim_{x \to 0^+} (x - [-x])$$

$$= \lim_{h \to 0} (0 + h - [0 + h]) = \lim_{h \to 0} (h - [h])$$

$$= \lim_{h \to 0} (h - 0) \quad (\because [h] = 0)$$

$$= \lim_{h \to 0} h = 0$$

$\therefore r.h.l \neq l.h.l \therefore \lim_{x \to 0} (x - [x])$ does not exist.

(ii) $\lim_{x \to 2} (x - [x])$

Solution: $l.h.l = \lim_{x \to 2^-} (x - [x])$

$$= \lim_{h \to 0} (2 - h - [2 - h])$$

$= \lim_{h \to 0} (2 - h - 1)$ ($\because [n - h] = n - 1$ when n is an integer)

$$= \lim_{h \to 0} (1 - h) = 1$$

$$r.h.l = \lim_{x \to 2^+} (x - [x]) = \lim_{h \to 0} (2 + h - [2 + h])$$

$= \lim_{h \to 0} (2 + h - 2)$ ($\because [n + h] = n$ when n is an integer)

$$= \lim_{h \to 0} h = 0$$

$\therefore l.h.l \neq r.h.l$ which means $\lim_{x \to 2} (x - [x])$ does not exist.

(iii) $\lim_{x \to \frac{3}{2}} (x - [x])$

Solution: $l.h.l = \lim_{h \to 0} \left(\frac{3}{2} - h - \left[\frac{3}{2} - h\right]\right)$

$$= \lim_{h \to 0} \left(\frac{3}{2} - h - 1\right)$$

($\because$ for sufficiently small $h > 0$, $\left[\frac{3}{2} - h\right] = 1$)

$$= \lim_{h \to 0} \left(\frac{1}{2} - h\right) = \frac{1}{2}$$

$$r.h.l = \lim_{x \to \frac{3}{2}+} (x - [x])$$

$$\lim_{h \to 0} \left(\frac{3}{2} + h - \left[\frac{3}{2} + h\right]\right)$$

$= \lim_{h\to 0}\left(\frac{3}{2}+h-1\right)$ ($\because$ for sufficiently small $h>0, \left[\frac{3}{2}+h\right]=1$)

$= \lim_{h\to 0}\left(\frac{1}{2}+h\right)=\frac{1}{2}$

$\therefore$ $l.h.l = r.h.l = \frac{1}{2}$

$\therefore$ $\lim_{x\to \frac{3}{2}}(x-[x])$ exists and $=\frac{1}{2}$.

3. $\lim_{x\to 2\cdot 3}(x+[x])$

Solution: $l.h.l = \lim_{x\to 2\cdot 3^-}(x+[x])$

$\lim_{h\to 0}(2\cdot 3-h+[2\cdot 3-h])$

$= \lim_{h\to 0}(2\cdot 3-h+2)$ ($\because$ for sufficiently small $h>0, [2.3-h]=2$)

$= \lim_{h\to 0}(4\cdot 3-h) = 4.3$

$r.h.l = \lim_{x\to 2\cdot 3^+}(x+[x])$

$= \lim_{h\to 0}(2\cdot 3+h+[2\cdot 3+h])$

$= \lim_{h\to 0}(2\cdot 3+h+2)$ ($\because$ for sufficiently small $h>0, [2.3+h]=2$)

$= \lim_{h\to 0}(4\cdot 3+h) = 4\cdot 3$

Thus, we observe, $l.r.l = r.h.l = 4.3 \Rightarrow \lim_{x\to 2\cdot 3}(x+[x])$ exists and $= 4.3$.

4. $\lim_{x\to 3}\frac{x}{[x]}$

Solution: $l.h.l = \lim_{x\to 3^-}\frac{x}{[x]}$

$= \lim_{h\to 0}\frac{3-h}{[3-h]} = \lim_{h\to 0}\frac{3-h}{2} = \frac{3}{2}$

($\because$ for sufficiently small $h>0, [3-h]=2$)

$r.h.l = \lim_{x\to 3+}\frac{x}{[x]}$

$= \lim_{h\to 0}\frac{3+h}{[3+h]} = \lim_{h\to 0}\frac{3+h}{3}$ ($\because$ for small $h>0$, $[3+h]=3$)

$= \frac{3}{3} = 1$

Thus, we observe, $l.h.l \neq r.h.l \Rightarrow \lim_{x\to 3}\frac{x}{[x]}$ does not exist.

5. Evaluate each of the following one sided limits

(i) $\lim_{x\to 2+0}\left[\frac{x}{2}\right]$

(ii) $\lim_{x\to 0-0}\frac{[x]}{x}$

(iii) $\lim_{x\to 1+0}\frac{[x-1]}{x-1}$

(iv) $\lim_{x\to 2-0}(x-2+[x+2])$

(v) $\lim_{x\to 1+0}\left([x]+\frac{|x-1|}{x-1}+2\right)$

(vi) $\lim_{x\to \frac{1}{3}+}x\left[\frac{1}{x}\right]$

(vii) $\lim_{x\to -\frac{1}{3}-}\frac{1}{x}\left[\frac{1}{x}\right]$

(viii) $\lim_{x\to k+\frac{3}{4}-0}\left(x+\frac{1}{4}-\left[x+\frac{1}{4}\right]\right)$, where K is an integer and $[t]$ denotes the greatest integer less than or equal to t.

Solutions: **(i)** $\lim_{x\to 2+0}\left[\frac{x}{2}\right]$

$= \lim_{h\to 0}\left[\frac{2+h}{2}\right] = \lim_{h\to 0}\left[1+\frac{h}{2}\right], h>0$

$= \lim_{h\to 0} 1$

$= 1$

$\because 1 + \frac{h}{2} < 2 \Rightarrow 1 < 1 + \frac{h}{2} < 2$ from the definition of greatest integer function

(ii) $\lim_{x\to 0-0} \frac{[x]}{x}$

$= \lim_{h\to 0} \frac{[0-h]}{0-h} = \lim_{h\to 0} \frac{[-h]}{-h}, h>0$

$= \lim_{h\to 0} \left(\frac{-1}{-h}\right)$

$= \lim_{h\to 0} \left(\frac{1}{h}\right)$

$= \infty$

$\because [-h] = -1$ as $[n-h] = n-1$ when n is an integer

(iii) $\lim_{x\to 1+0} \frac{[x-1]}{x-1}$

$= \lim_{h\to 0} \frac{[1+h-1]}{1+h-1} = \lim_{h\to 0} \frac{[h]}{h}, h>0$

$= \lim_{h\to 0} \frac{0}{h}$

$= \lim_{h\to 0} 0 = 0$

$\because [h] = 0$ as $h \to 0$ and $h > 0 \Rightarrow 0 < h < 1$

(iv) $\lim_{x\to 2-0} (x - 2 + [x+1])$

$= \lim_{h\to 0} (2 - h - 2 + [2 - h + 1]), h>0$

$= \lim_{h\to 0} (-h+[3-h]) = \lim_{h\to 0} (-h) + \lim_{h\to 0} (3-h)$

$= 0 + \lim_{h\to 0} 2$

$= 0 + 2 = 2$

$\because [3-h] = 2$ as $[n-h] = n-1$ when $n =$ an integer

(v) $\lim_{x\to 1+0} \left([x] + \frac{|x-1|}{x-1} + 2\right)$

$= \lim_{h\to 0} \left([1+h] + \frac{|1+h-1|}{1+h-1} + 2\right), h>0$

$= \lim_{h\to 0} \left(1 + \frac{|h|}{h} + 2\right)$

$\because [1+h] = 1$ as $[n+h] = n$ when $n =$ an integer

$= \lim_{h\to 0} \left(1 + \frac{h}{h} + 2\right)$ $\because |h| = h$ as $h > 0$ and $h \to 0$

$= \lim_{h\to 0} (1 + 1 + 2)$

$= \lim_{h\to 0} 4 = 4$

(vi) $\lim_{x\to \frac{1}{3}+} x \left[\frac{1}{x}\right]$

$= \lim_{h\to 0} \left(\frac{1}{3} + h\right) \cdot \left[\frac{1}{\frac{1}{3} + h}\right], h>0$

$= \lim_{h\to 0} \left(\frac{1}{3} + h\right)\left[\frac{3}{1+3h}\right]$

$= \lim_{h\to 0} \left(\frac{1}{3} + h\right) \cdot \lim_{h\to 0} \left[\frac{3}{1+3h}\right]$

$= \frac{1}{3} \times \lim_{h\to 0} (2)$

$= \frac{1}{3} \times 2$

$= \frac{2}{3}$

$$\because \left[\frac{3}{1+3h}\right] = 2 \text{ as } h\to 0 \text{ and } h>0 \Rightarrow 1+3h>1$$
$$\Rightarrow \frac{1}{1+3h}<1 \Rightarrow \frac{3}{1+3h}<3$$

(vii) $\lim_{x\to -\frac{1}{3}-} \frac{1}{x}\left[\frac{1}{x}\right]$

$$= \lim_{h\to 0}\left\{\frac{1}{-\frac{1}{3}-h}\left[\frac{1}{-\frac{1}{3}-h}\right]\right\}, h>0$$

$$= \lim_{h\to 0}\left\{\left(\frac{-3}{1+3h}\right)\left[\frac{-3}{1+3h}\right]\right\}$$

$$= \lim_{h\to 0}\left(\frac{-3}{1+3h}\right)\cdot \lim_{h\to 0}\left[\frac{-3}{1+3h}\right]$$

$$= \lim_{h\to 0}\left(\frac{-3}{1+3h}\right)\cdot \lim_{h\to 0}(-3)$$

$= (-3)\times(-3)$

$= 9$

$$\because \left[\frac{-3}{1+3h}\right] = -3, \text{ as } h\to 0 \text{ and } h>0 \Rightarrow 1+3h>1$$
$$\Rightarrow \frac{1}{1+3h}<1 \Rightarrow \frac{-3}{1+3h}>-3,$$
$$\therefore -3<\frac{-3}{1+3h}<-2$$

(viii) $\lim_{x\to k+\frac{3}{4}-0}\left\{x+\frac{1}{4}-\left[x+\frac{1}{4}\right]\right\}$

$$\lim_{h\to 0}\left\{\left(k+\frac{3}{4}-h\right)+\frac{1}{4}-\left[\left(k+\frac{3}{4}-h\right)+\frac{1}{4}\right]\right\},$$

$h>0$

$$= \lim_{h\to 0}\{k-h+1-k\}$$

$$= \lim_{h\to 0}\{-h+1\}$$

$= 1$

$\because [k+1-h]=k$ as $[n-h]=n-1$ when $n=$ an integer

6. Evaluate each of the following if it exists.

(i) $\lim_{x\to 0}\cos[x]$

(ii) $\lim_{x\to 0+} e^{[x]}$

(iii) $\lim_{x\to \frac{\pi}{4}+}\cos[x]$

Solutions: **(i)** *r.h.l* $= \lim_{x\to 0^+}\cos[x]$

$$= \lim_{h\to 0}\cos[0+h] = \lim_{h\to 0}\cos[h]$$

$$= \lim_{h\to 0}\cos 0$$

$$= \lim_{h\to 0} 1$$

$= 1$

$\because [h]=0$ as $[n+h]=n$ when $n=$ an integer

l.h.l $= \lim_{x\to 0-}\cos[x]$

$$= \lim_{h\to 0}\cos[0-h] = \lim_{h\to 0}\cos[-h]$$

$$= \lim_{h\to 0}\cos(-1)$$

$$= \lim_{h\to 0}\cos 1$$

$= \cos 1$

$\because [-h]=-1$ as $[n-h]=n-1$ when $n=$ an integer

$\therefore$ *r.h.l* $\neq$ *l.h.l* and so the limit does not exist.

(ii) $\lim_{x\to 0^+} e^{[x]}$

$$= \lim_{h\to 0} e^{[0+h]} = \lim_{h\to 0} e^{[h]}$$

$$= \lim_{h\to 0} e^{0}$$

$$= \lim_{h\to 0} 1 = 1$$

(iii) $\lim\limits_{x\to\frac{\pi}{4}+} \cos[x] = \lim\limits_{h\to 0} \cos\left[\frac{\pi}{4}+h\right]$

$= \lim\limits_{h\to 0} \cos 0$

$= \lim\limits_{h\to 0} 1$ $\quad \boxed{\because \left[\frac{\pi}{4}+h\right] = 0 \text{ as } 0 < \frac{\pi}{4}+h < 1}$

$= 1$

Note: In those problems which are the combination or composition of functions whose one function is the greatest integer function, we should consider two necessary facts while finding their limits as $h \to 0$

(i) find the greatest integer contained in [the function of h] before putting $h = 0$

(ii) cancel the highest common power of h from the numerator and denominator in the quotient of two functions of h before putting $h = 0$

How to find the limit of a piecewise functions containing greatest integer function

When the limit of a piecewise function containing at any integral point $x = a$ is sought, it must be redefined in adjacent intervals whose left and right end points are the same namely the integral point $x = a$.

How to find the adjacent intervals containing the integral point $x = a$

Step 1: Start towards left from the integral point $x = a$ and stop at the first integer you arrive at (say b), i.e., obtain $b \le x < a$ where b = the integer just on the left of the integral point $x = a$.

Step 2: Start towards right from the integral point $x = a$ and stop at the first integer you arrive at (say c), i.e. obtain $a \le x \le c$, where c = the integer just on the right of the integral point $x = a$.

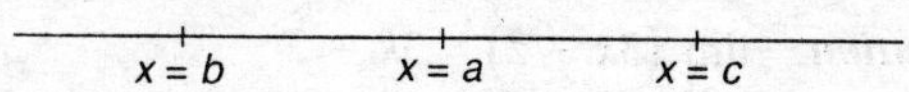

Rule: To find the limit of a piecewise function containing greatest integer function and redefined in adjacent intervals $[b, a) \cup [a, c)$ is determined by removing the symbol [ ] with the help of the definition.

$[f(x)] = n$, when $n \le f(x) < n+1$

$[f(x)] = n-1$, when $n-1 \le f(x) < n$ where $n \in I$

and lastly put $x = a$ in each different forms of the expression free from greatest integer function.

Examples worked out:

1. If $f(x) = \dfrac{\sin[x]}{[x]}$, $[x] \ne 0$

$= 0, [x] = 0$

then $\lim\limits_{x\to 0} f(x) =$

(a) 1 (b) 0 (c) –1 (d) none of them.

Solution: On redefining f, it is as under

$f(x) = \dfrac{\sin[x]}{[x]}, \; -1 \le x < 0$

$= 0, \; 0 \le x < 1$

Again on using the definition for $[x]$,

$[x] = -1$ when $-1 \le x < 0$

$[x] = 0$ when $0 \le x < 1$

the function 'f' becomes

$f(x) = \dfrac{\sin(-1)}{(-1)}$, when $-1 \le x < 0$, i.e.

$f(x) = \dfrac{-\sin 1}{-1} = \sin 1$ for $-1 \le x < 0$ and $f(x) = 0$, when $0 \le x < 1$

Hence, $\lim\limits_{x\to 0^-} f(x) = \lim\limits_{x\to 0^-} \sin 1 = \sin 1$

and $\lim\limits_{x\to 0^+} f(x) = \lim\limits_{x\to 0^+} 0 = 0$

Since $\lim\limits_{x\to 0^-} f(x) \ne \lim\limits_{x\to 0^+} f(x)$

$\Rightarrow$ the limit of $f(x)$ at $x = 0$ does not exist

Hence, (d) is true.

2. Examine the existence of limit of a function defined by $f(x) = \dfrac{[x^2] - 1}{x^2 - 1}$, for $x^2 \ne 1$ = 0, for $x^2 = 1$; at $x = 1$.

Solution: It is given

$f(x) = \dfrac{[x^2] - 1}{x^2 - 1}$, for $x^2 \ne 1$

$= 0$, for $x^2 = 1$

On redefining the given function f,

$$f(x) = \frac{[x^2] - 1}{x^2 - 1}, \text{ when } 0 < x^2 < 1$$

$$= \frac{[x^2] - 1}{x^2 - 1}, \text{ when } 1 < x^2 < 2$$

$= 0$, when $x^2 = 1$

Again on using the definition for $[x^2]$,

$[x^2] = 0$ for $0 \le x^2 < 1$

$[x^2] = 1$ for $1 \le x^2 < 2$

the function 'f' becomes

$$f(x) = \frac{-1}{x^2 - 1}, \text{ for } 0 < x^2 < 1$$

$= 0$, for $x^2 = 1$

$$= \frac{1 - 1}{x^2 - 1} = 0, \text{ for } 1 < x^2 < 2$$

Now, $\lim_{x \to 1^+} f(x) = \lim_{x \to 1^+} 0 = 0$

and $\lim_{x \to 1^-} f(x) = \lim_{x \to 1^-} \left(\frac{-1}{x^2 - 1} \right) = \infty$

i.e. $\lim_{x \to 1^-} f(x)$ does not exist

$\Rightarrow \lim_{x \to 1} f(x)$ does not exist.

alternative method

$x^2 = z \Rightarrow \sqrt{x^2} = \sqrt{z} \Rightarrow |x| = \sqrt{z} \Rightarrow x = \sqrt{z}$ for $x \ge 0$

Also, $x \to 1 \Rightarrow z \to 1$

Hence the given function becomes

$$f(\sqrt{z}) = \frac{[z] - 1}{z - 1}, \text{ for } z \ne 1$$

$= 0$ for $z = 1$

Which can be redefined as under:

$$f(\sqrt{z}) = \frac{-1}{z - 1}, \text{ for } 0 \le z < 1$$

$$= \frac{1 - 1}{z - 1} = 0, \text{ for } 1 < z < 2$$

$= 0$, for $z = 1$

Now, $\lim_{z \to 1^+} f(\sqrt{z}) = \lim_{z \to 1^+} 0 = 0$

but $\lim_{z \to 1^-} f(\sqrt{z}) = \lim_{z \to 1^-} \left(\frac{-1}{z - 1} \right) = -\infty$

$\Rightarrow \lim_{z \to 1^-} f(\sqrt{z})$ does not exist

$\Rightarrow \lim_{z \to 1} f(\sqrt{z})$ does not exist

$\Rightarrow \lim_{x \to 1} f(x)$ does not exist

Type 4: Problems based on finding the value of a constant whenever the limit of a given function is finite value/a finite number.

Examples worked out:

1. Find the value of k if

$$\lim_{x \to 2} (kx - 5) = 3$$

Solution: $\lim_{x \to 2} (kx - 5) = k \times 2 - 5 = 2k - 5$ ...(i)

and $\lim_{x \to 2} (kx - 5) = 3$ ...(ii)

Equating (i) and (ii), we have, $2k - 5 = 3$

$\Rightarrow 2k = 8 \Rightarrow k = \frac{8}{2} = 4$

2. Find the value of K if $\lim_{x \to k} (3x - 2) = 7$

Solution: $\lim_{x \to k} (3x - 2) = 3k - 2$ ...(i)

and $\lim_{x \to k} (3x - 2) = 7$ ...(ii)

(1) and (2) $\Rightarrow 3k - 2 = 7 \Rightarrow 3k = 7 + 2 = 9 \Rightarrow k = 9/3 = 3$

Problems based on existence of the limits of greatest integer function/combination of a function and the greatest integer function

Exercise 4.30.1

1. Examine the existence of the limits of the following functions as $x \to 0$

(i) $f(x) = [x] + [1-x]$, $\forall$ real x

(ii) $f(x) = x + [x]$, $\forall$ real x

(iii) $f(x) = [x] + [-x]$, when $x \neq 0$

$f(0) = 1$

(iv) $f(x) = x[x]$

2. Examine the existence of the limits of the following functions at the indicated points.

(i) $f(x) = [1-x] + [x-1]$ at $x = 1$

(ii) $f(x) = [x+2] - |2+x|$ at $x = 2$

(iii) $f(x) = \dfrac{\left[\frac{1}{2} + x^2\right] - \frac{1}{2}}{x^2}$ at $x = 1$

(iv) $f(x) = \dfrac{\left[x + \frac{1}{2}\right]}{[x]}$ at $x = -\dfrac{1}{2}$

3. Show that the function f defined by

$f(x) = [x-1] + |x-1|$ for $x \neq 1$

$f(1) = 0$ has no limit at $x = 1$.

4. Show that the function f defined by $f(x) = [x-3] + [3-x]$, where $[t]$ denotes the largest integer $\leq t$ exists at $x = 3$ and is equal to 0.

5. If $f(x) = \dfrac{\sin [x]}{[x]}$, $[x] \neq 0$; $= 0$, $[x] = 0$ then find if $\lim\limits_{x\to 0} f(x)$ exists where $[x]$ denotes the greatest integer less than or equal to x.

Answers:

1. (i) exists and $\lim\limits_{x\to 0} f(x) = 0$ (ii) does not exist since $l.h.l = -1$ and $r.h.l = 0$ (iii) exists and $\lim\limits_{x\to 0} f(x) = -1$ (iv) exists and $\lim\limits_{x\to 0} f(x) = 0$

2. (i) exists and $\lim\limits_{x\to 1} f(x) = -1$ (ii) does not exist since $l.h.l = -1$ and $r.h.l = 0$ (iii) exists and $\lim\limits_{x\to 1} f(x) = \frac{1}{2}$ (iv) does not exist since $l.h.l = 1$ and $r.h.l = 0$.

5. As $l.h.l \neq r.h.l$, so the given limit does not exist.

Hint: $l.h.l = \sin 1$ and $r.h.l = 0$

Problems based on finding the value of a constant

Exercise 4.30.2

Find the value of k if ***Answers***

1. $\lim\limits_{x\to 2} (kx - 5) = 3$ (4)

2. $\lim\limits_{x\to 2} \dfrac{x^2 - k^2}{x - k} = -4$ (–6)

3. $\lim\limits_{x\to 1} (kx^2 + 5x - 3) = 4$ (2)

4. $\lim\limits_{x\to 0} \dfrac{\sin kx}{x} = 3$ (3)

5. $\lim\limits_{x\to \infty} \dfrac{kx^2 + 4x - 8}{2x^2 - 3x + 5} = 3$ (6)

6. $\lim\limits_{x\to \frac{\pi}{2}} (k + 2\cos x) = 5$ (5)

7. $\lim\limits_{x\to k} (3x - 2) = 7$ (3)

8. If $f(x) = \dfrac{\sin 3x}{x}$, when $x < 0 = \dfrac{\tan bx}{x}$, when $x > 0$ and $\lim\limits_{x\to 0} f(x)$ exists, find the value of b.

Solution: $\lim\limits_{x\to 0} \dfrac{\sin 3x}{x}$

$$l.h.l = \lim_{x\to 0} \frac{\sin 3x}{3x} \times 3 = 3 \quad \ldots\text{(i)}$$

$$\lim_{x\to 0} \frac{\tan bx}{x} = \lim_{x\to 0} \frac{\tan bx}{bx} \times b = b = r.h.l \quad \ldots\text{(ii)}$$

$\lim\limits_{x\to 0} f(x)$ exists $\Leftrightarrow l.h.l = r.h.l \Leftrightarrow b = 3$ ***Ans.***

Objective problems

Exercise 4.30.3

(a) Choose the correct answer.

(i) The value of $\lim\limits_{x\to 0} \dfrac{2x}{5+3x}$ is ...

(A) $\dfrac{2}{3}$ **(B)** $\dfrac{2}{5}$ **(C)** 0 **(D)** 2 [*Ans.* C]

(ii) The value of $\lim\limits_{x\to \frac{\pi}{2}} \dfrac{\sin x}{1+\cos x}$ is ...

(A) 0 **(B)** 1 **(C)** $\dfrac{1}{2}$ **(D)** ∞ [*Ans.* B]

(iii) The value of $\lim\limits_{x\to 0} \dfrac{\sin 5x}{x}$ is ...

(A) 1 **(B)** 0 **(C)** $\dfrac{1}{5}$ **(D)** 5 [*Ans.* D]

(iv) The value of $\lim\limits_{x\to 0} \dfrac{\sin 5x - \sin 3x}{x}$ is ...

(A) 2 **(B)** 0 **(C)** 1 **(D)** $-\dfrac{2}{15}$ [*Ans.* A]

(v) The value of $\lim\limits_{x\to 0} \dfrac{2x}{\sin x}$ is ...

(A) 1 **(B)** 2 **(C)** $\dfrac{1}{2}$ **(D)** not possible [*Ans.* B]

(vi) The value of $\lim\limits_{x\to \infty} \dfrac{3x^2-5}{x}$ is ...

(A) 0 **(B)** ∞ **(C)** 3 **(D)** -5 [*Ans.* B]

(b) State whether the following statements are true or false

(i) $\lim\limits_{x\to 0} x^2 = 0$ [*Ans.* T]

(ii) $\lim\limits_{x\to \pi} (1+\cos x) = 0$ [*Ans.* T]

(iii) $\lim\limits_{x\to \frac{\pi}{2}} (1-\sin x) = 0$ [*Ans.* T]

(iv) $\lim\limits_{x\to 0} \dfrac{2x^2-3x}{3x} = 1$ [*Ans.* F]

(v) $\lim\limits_{x\to 0} \dfrac{x}{\tan x} = 1$ [*Ans.* T]

(vi) $\lim\limits_{x\to -2} \dfrac{x^2-4}{x+2} = 4$ [*Ans.* F]

(vii) $\lim\limits_{x\to \infty} \dfrac{x^2-x}{2x} = \dfrac{1}{2}$ [*Ans.* F]

(viii) $\lim\limits_{x\to \infty} \dfrac{2x^2+5x}{x(x+1)} = 2$ [*Ans.* T]

5

Practical Methods on Continuity Test

A little more on how to test the continuity of a function at $x = a$. To test whether $f(x)$ is continuous at $x =$ the following procedure may be adopted:

1. Find $f(a)$. If $f(a)$ is undefined, the function $f(x)$ is discontinuous at $x = a$.

2. If the value of the function represented by $f(a)$ at $x = a$ has a finite value, find the $l.h.l$ and $r.h.l$ represented by $\lim_{x \to a^-} f(x) = \lim_{h \to 0} f(a - h)$ and

$\lim_{x \to a^+} f(x) = \lim_{h \to 0} f(a + h)$ respectively where

$h \to 0$ through positive values (i.e; $h > 0$ and $-h < 0$).

3. If both $l.h.l$ and $r.h.l$ are equal to $f(a)$, $f(x)$ is continuous at $x = a$.

4. If either of the $l.h.l$ and $r.h.l$ is different from $f(a)$, $f(x)$ is discontinuous at $x = a$.

Explanation

1. $\Rightarrow$ Replace x by a $f(x)$ and in the given expression in x and find the value of the given function $f(x)$ at $x = a$. Now if the value of the function $f(x)$ at $x = a$ is undefined the given function is declared to be discontinuous.

2. $\Rightarrow$ If the value of the function $f(x)$ at $x = a$ is a finite value, we are required to find $l.h.l$ and $r.h.l$ respectively by the method already explained.

3. $\Rightarrow$ If all the three (a) value of the function at $x = a$, i.e; $f(a)$ (b) $l.h.l$ and (c) $r.h.l$ of the given function obtained are equal, then $f(x)$ is continuous at $x = a$ which means $f(a) = l.h.l = r.h.l \Rightarrow f(x)$ is continuous at $x = a$.

4. $\Rightarrow$ If $f(a) \neq l.h.l \neq r.h.l$ or $l.h.l \neq r.h.l = f(a)$ or $r.h.l \neq l.h.l = f(a)$, we declare that the given function $f(x)$ whose test of continuity is required is discontinuous at the given point $x = a$.

N.B.: The above method of testing a given function to be continuous at $x = a$ is applied when the given function is defined by different equations on imposing the conditions on the independent variables x by $<$ or $\leq$ or $>$ or $\geq$ or $=$ etc.

Aid to Memory

1. $f(a) = \lim_{x \to a}$ [the function $f_1(x)$ opposite to which $x < a / x \leq a / c < x \leq a /$]

$= \lim_{x \to a}$ [the function $f_2(x)$ opposite to which $x > a / x \geq a / c > x > a / c \geq x > a$]

$\Leftrightarrow f(x)$ is continuous at $x = a$.

2. $f(x) = f_3(x)$, when $x = a$ means we should consider the function $f_3(x)$ (i.e; an expression in x) to find out the value of the function $x = a$.

3. $f(x) = a$ constant, when $x = a$ means we are provided the value of the function which is the given constant for the independent variable $x = a$ and thus further we are not required to find out the value of the function $f(x)$ at $x = a$.

4. $f(x) = f_1(x)$, when (or, if or, provided) $x \geq a$ or $c > x \geq a$ means we should consider $f_1(x)$ for finding *r.h.l* and the value of the function at $x = a$.
5. $f(x) = f_2(x)$, when $x \leq a$ or $c < x \leq a$ means we should consider $f_2(x)$ for finding *l.h.l* and the value of the function at $x = a$.
6. $f(x) = f_1(x)$ when $x < a$ or $c < x < a \Rightarrow$ we should consider $f_1(x)$ only for finding the *l.h.l* and not for the value of the function $f(x)$ at $x = a$ as the sign of equality does not appear in the given restrictions against the given function $f(x) = f_1(x)$.
7. $f(x) = f_2(x)$ when $x > a$ or $c > x > a$ means we should consider $f_2(x)$ only for finding *r.h.l* and not for the value of the function $f(x)$ at $x = a$ as the sign of equality does not appear in the given restriction against the given function $f(x) = f_2(x)$.
8. $f(x) = f_4(x)$, when $x \neq a$ means the same function $f_4(x)$ should be considered for finding the *l.h.l* and *r.h.l*.

Note:
1. In the light of above explanation, we may declare that the sign of equality '=' with the sign of inequalities "> or <" retains the possibility to consider the same function for the *l.h.l* as well as the value of the function both or *r.h.l* as well as the value of the function both whereas only the sign of inequality > or < excludes the possibility to consider the same function for the value of the given function opposite to which > or < is written.
2. $f(x) = f_4(x)$, when $x \neq a$ means there is no need to find out *l.h.l* and *r.h.l* separately but only to find out the limit of $f_4(x)$ at $x = a$ and use the definition limit = value of the function at the given point $x = a$ to test the continuity (or, to find the *l.h.l* and *r.h.l* by putting $x = a \pm h$ as $h \to 0$ through positive values in the given function $f(x) = f_4(x)$ and then use the definition *l.h.l* = *r.h.l* = value of the function (which is given at $x = a$ by imposing the condition $x = a$ against the given function $f(x)$ as $f(x) = a$ constant, when $x = a$) $\Leftrightarrow$ continuity of the given function at $x = a$.
3. A function $f(x)$ defined in an interval is called a piecewise continuous function when the interval can be divided into a finite number of non-overlapping open sub intervals over each of which the function is continuous.
4. All the points at which the function is continuous are called points of continuities and all those points at which the function is discontinuous are called points of discontinuities (or, simply discontinuities only).
5. $f(x)$ is continuous at $x = a \Leftrightarrow x = a$ is the point of continuity of $f(x) \Leftrightarrow f(x)$ has a point of continuity namely $x = a$.
6. $f(x)$ is discontinuous at $x = a \Leftrightarrow x = a$ is the point of discontinuity of $f(x) \Leftrightarrow f(x)$ has a point of discontinuity namely $x = a$.

A Highlight on Removal Discontinuity

Question: When a function is not defined for $x = a$, is it possible to give the function such a value for $x = a$ as to satisfy the condition of continuity?
Answer: When a function is not defined for the independent variable $x = a$ as to satisfy the condition of continuity if we arbitrarily suppose that value of the function which must be the limit of the given function $\Rightarrow$ If the value of the function at a point = limit of the function at the same point is supposed, then the given function becomes continuous at that point.

Example: $y = \dfrac{x^2 - 9}{x - 3}$ is not defined at $x = 3$ but for any other value of x,

$$y = \frac{(x+3)\cancel{(x-3)}}{\cancel{(x-3)}} = (x+3)$$

and $\lim_{x \to 3} (x + 3) = 6$

$$\therefore \lim_{x \to 3} \left(\frac{x^2 - 9}{x - 3} \right) = 6$$

Now, if we suppose $f(3) = 6$, i.e; the value of the function to be 6 for $x = 3$, the function becomes continuous.

Removal Discontinuity

If $\lim_{x \to a^+} f(x) = \lim_{x \to a^-} f(x) \neq f(a)$, then the function $f(x)$ as said to have removal discontinuity at

$x = a$ because the discontinuity can be removed by making the value of the function $f(x)$ at $x = a$ equal to $\lim_{x \to a} f(x)$.

There are two types of discontinuities which can be removed by assuming the value of the function $f(x)$ at a point (or, number) $x = a$ = limit of the function $f(x)$ at a point (or, number) $x = a$.

1. If the function is not defined at $x = a$, then the function is discontinuous at $x = a$.

2. If $\lim_{x \to a} f(x)$ exists finitely but the value of the function $f(a) \neq \lim_{x \to a} f(x)$, then the function $f(x)$ is discontinuous at $x = a$. Thus, there are two types of discontinuities which can be removed by assuming value = limit for the given point $x = a$.

Remember:

(A) A function is discontinuous at $x = a$ if

1. $f(x)$ is not defined at $x = a \Leftrightarrow f(x)$ = meaningless at $x = a \Leftrightarrow f(a) = \frac{0}{0} / \frac{\infty}{\infty} / 0 \times \infty$ ... etc.

2. When $\lim_{x \to a} f(x) = \infty$

3. When $\lim_{x \to a} f(x) \neq f(a)$

4. When $l.h.l \neq r.h.l = f(a)$

5. When $l.h.l = r.h.l \neq f(a)$

6. When $r.h.l \neq l.h.l \neq f(a)$

7. Limit $\neq$ value of the function at the given point.

(B) A function $f(x)$ is said not to exist at a point $x = a$ if $[f(x)]_{x=a} = \infty$ /meaningless/imaginary.

Examples:

1. Show that $y = \frac{x^2 - 9}{x - 3}$ is discontinuous at $x = a$.

Solution: $\lim_{x \to 3} y = \lim_{x \to 3} (x + 3) = 6$ ...(1)

$\because y = f(x) = \frac{x^2 - 9}{x - 3} = \frac{0}{0}$ at $x = 3$ (undefined) ...(2)

$\therefore y = \frac{x^2 - 9}{x - 3} = \frac{\cancel{(x-3)}(x+3)}{\cancel{(x-3)}}, x \neq 3$

$= (x+3)$

Thus, we see that $f(3) \neq \lim_{x \to 3} f(x) \Rightarrow f(x)$ is discontinuous at $x = a$.

N.B.: But $y = \frac{x^2 - 9}{x - 3}$ is continuous at $x = 2$

Since $\lim_{x \to 2} \left(\frac{x^2 - 9}{x - 3} \right) = 5$ and $f(2) = 5$

$\therefore f(2) = \lim_{x \to 2} f(x) = 5.$

2. $\lim_{x \to 3} \left(\frac{5x}{6 - 2x} \right) = \frac{\lim_{x \to 3} (5x)}{\lim_{x \to 3} (6 - 2x)} = \infty$

$\Rightarrow f(x) = \frac{5x}{6 - 2x}$ is discontinuous at $x = 3$.

Now, we consider the continuity of the following functions at the point $x = a$.

1. Continuity of rational functions.
2. Continuity of absolute value functions.
3. Continuity of exponential functions.
4. Continuity of logarithmic functions.
5. Continuity of trigonometric functions.

Type 1: When no condition is imposed on the independent variable against the defined function, i.e; when given function is not piecewise.

Highlight on the Working Rule

Limit of the given function as $x \to a$ = value of the given function at $x = a$, in the examples to follow.

Solved Examples

Test the continuity of the following functions at the indicated points.

1. $y = x^2 + 3x$ at $x = 2$

Solution: $y = x^2 + 3$

$$\Rightarrow \lim_{x\to 2} y = \lim_{x\to 2}\left(x^2 + 3x\right)$$

$$= 2^2 + 3\times 2 = 4 + 6 = 10$$

$f(2) = 10$ when $f(x) = x^2 + 3x$

$\therefore f(2) = \lim_{x\to 2} f(x)$

Hence, $f(x)$ is continuous at $x = 2$.

2. $y = \dfrac{x^2 - 9}{x - 3}$ for $x = 2$

Solution: $y = \dfrac{x^2 - 9}{x - 3}$

$$\Rightarrow \lim_{x\to 2} y = \lim_{x\to 2}\frac{x^2 - 9}{x - 3} = \lim_{x\to 2}\frac{(x+3)\cancel{(x-3)}}{\cancel{(x-3)}}$$

$$= \lim_{x\to 2}(x+3) = 2 + 3 = 5$$

$$f(2) = \frac{4-9}{2-3} = \frac{-5}{-1} = 5$$

$\therefore f(2) = \lim_{x\to 2} f(x)$

Hence, $f(x)$ is continuous at $x = 2$

Type 2: In case, $f(x) = f_1(x)$, when $x < a$
$= f_2(x)$, when $x > a$
$= c$, when $x = a$

i.e; when different functions are provided with different restrictions imposed on the independent variable x as $x > a$ / $x < a$ / $x = a$ / $x \ge a$ / $x \le a$ / ... etc .against each function or in otherwords, when the given function is a piecewise function, i.e. the given function is defined adjacent intervals.

Highlight on the Working Rule

$\lim_{x\to a^+} f(x) = \lim_{x\to a} f_1(x) = l_1$, where $f_1(x)$ is a form of the given function defined in an interval whose right end point is 'a'.

$\lim_{x\to a^+} f(x) = \lim_{x\to a} f_2(x) = l_2$, where $f_2(x)$ is a form of the given function defined in an interval whose left and point is also, 'a'.

If $l_1 = l_2 = c$ then $f(x)$ is said to be continuous at $x = a$.

Solved Examples

1. If $f(x) = 5x - 4$, when $0 < x \le 1$ $= 4x^3 - 3x$, when $1 < x < 2$ Show that $f(x)$ is continuous at $x = 1$

Solution: $\lim_{x\to 1^+} f(x) = \lim_{x\to 1}\left(4x^3 - 3x\right)$

$$= 4 - 3 = 1$$

$$\lim_{x\to 1-} f(x) = \lim_{x\to 1}(5x - 4)$$

$$= 5 - 4 = 1$$

$f(1) = 5 - 4$ for $f(x) = 5x - 4$ when $0 < x \le 1$

Hence, $f(1) = \lim_{x\to 1+} f(x) = \lim_{x\to 1-} f(x) \Rightarrow f(x)$ is continuous at $x = 1$.

2. If $f(x) = \dfrac{x^2}{2}$, when $0 \le x \le 1$ $= 2x^2 - 3x + \dfrac{3}{2}$, when $1 \le x \le 2$ test the continuity of $f(x)$ at the point $x = 1$.

Solution: $\lim_{x\to 1-0} f(x) = \lim_{x\to 1}\dfrac{x^2}{2} = \dfrac{1}{2}$

$$\lim_{x\to 1+0} f(x) = \lim_{x\to 1}\left(2x^2 - 3x + \frac{3}{2}\right)$$

$$= 2 - 3 + \frac{3}{2} = -1 + \frac{3}{2} = \frac{1}{2}$$

$f(x) = \dfrac{x^2}{2}$ for $0 \le x \le 1 \Rightarrow f(1) = 2$

Hence, $f(1) = \lim_{x\to 1+0} f(x) = \lim_{x\to 1-0} f(x) \Rightarrow f(x)$ is continuous at $x = 1$.

Note: If $f(x) = \dfrac{x^2}{2}$, when $0 \le x \le 1$

$= 2x^2 - 3x + \dfrac{3}{2}$, when $1 \le x \le 2$

then to find the value of the function $f(x)$ at $x = 1$, we may consider any one of the two pieces. Hence, in the above example if we consider

$f(x) = 2x^2 - 3x + \frac{3}{2}$ for $0 \le x \le 1$, then $f(1) =$

$2 - 3 + \frac{3}{2} = -1 + \frac{3}{2} = \frac{-2+3}{2} = \frac{1}{2}$ so, in general, when the condition (or, restriction) imposed on an independent variable x contains $x \ge a / x \le a / c > x \ge a / c < x \le a$, then we may consider any one of both functions to find the value of the function at $x = a$ because both functions provided us the same value.

3. Test the continuity of the function $f(x)$ at $x = 2$

$f(x) = 2x + 1$ for $x \le 2$

$= x^2 - 1$ for $x > 2$

Solution: $l.h.l = \lim_{x \to 2-0} (2x + 1) = 5$

$$r.h.l = \lim_{x \to 2+0} \left(x^2 - 1\right) = 4 - 1 = 3$$

$\therefore l.h.l \ne l.h.l \Rightarrow \lim_{x \to 2} f(x)$ does not exist $\Rightarrow$

$f(x)$ is discontinuous at $x = 2$.

4. Show that the function defined as

$$f(x) = \frac{x^2}{a} - a, 0 < x < a$$

$$= 0, x = a$$

$= a - \frac{a^3}{x^2}$, $x > a$ is continuous at $x = a$.

Solution: $f(a) = 0$ [$\because f(x) = 0$ when $x = a$ is given in the problem] ...(1)

$$\lim_{x \to a^+} f(x) = \lim_{x \to a} \left(a - \frac{a^3}{x^2}\right)$$

$$= a - \frac{a^3}{a^2} = a - a = 0 \quad ...(2)$$

$$\lim_{x \to a-} f(x) = \lim_{x \to a} \left(\frac{x^2}{a} - a\right)$$

$$= \frac{a^2}{a} - a = a - a = 0 \quad ...(3)$$

(1), (2) and (3) $\Rightarrow \lim_{x \to a+} f(x) = \lim_{x \to a-} f(x) =$

$f(a) \Rightarrow f(x)$ is continuous at $x = a$

5. Show that the function for $f(x)$ defined by

$$f(x) = x + \frac{1}{2}, \text{ when } 0 < x < \frac{1}{2}$$

$$= \frac{1}{2}, \text{ when } x = \frac{1}{2}$$

$= \left(x - \frac{1}{2}\right)$, when $\frac{1}{2} < x < 1$ is discontinuous

at $x = \frac{1}{2}$

Solution: $f\left(\frac{1}{2}\right) = \frac{1}{2}$ ...(1)

$$\lim_{x \to \frac{1}{2}+} f(x) = \lim_{x \to \frac{1}{2}} \left(x - \frac{1}{2}\right)$$

$$= \left(\frac{1}{2} - \frac{1}{2}\right) = 0 \quad ...(2)$$

$$\lim_{x \to \frac{1}{2}-} f(x) = \lim_{x \to \frac{1}{2}} \left(x + \frac{1}{2}\right)$$

$$= \frac{1}{2} + \frac{1}{2} = 1 \quad ...(3)$$

(1), (2) and (3) $\Rightarrow \lim_{x \to \frac{1}{2}+} f(x) = \lim_{x \to \frac{1}{2}-} f(x) \ne$

$f\left(\frac{1}{2}\right) \Rightarrow f(x)$ is discontinuous at $x = \frac{1}{2}$.

N.B.: In fact, the in-equality of any two of (1), (2) and (3) ensures discontinuity of the given function at $x = \frac{1}{2}$.

6. Test the continuity of the function $f(x)$ at $x = 2$

$$f(x) = \frac{x^2 - 4}{x - 2}, \text{ when } 0 < x < 2$$

$f(x) = x + 2$, when $2 \le x \le 5$

Solution: $l.h.l = \lim_{x\to 2^-} f(x) = \lim_{x\to 2} \frac{x^2-4}{x-2}$

$= \lim_{x\to 2} (x+2) = 2+2 = 4$

$r.h.l = \lim_{x\to 2^+} f(x) = \lim_{x\to 2} (x+2) = 2+2 = 4$

$f(2) = (x+2)_{x=2} = 2+2 = 4$

Hence, $l.h.l = r.h.l = f(2) \Rightarrow f(x)$ is continuous at $x = 2$.

7. Test the continuity of the given function at $x = 1$

$f(x) = 2x + 3$, when $x \le 1$ ...(1)

$= 8 - 3x$, when $1 < x < 2$ ...(2)

Solution: $f(1) = 2\times 1 + 3 = 2 + 3 = 5$ ...(1)

$\lim_{x\to 1+} f(x) = \lim_{x\to 1} (8-3x) = 8 - 3 = 5$ ...(2)

$\lim_{x\to 1-} f(x) = \lim_{x\to 1} (2x+3) = 2 + 3 = 5$ ...(3)

(1), (2) and (3) $\Rightarrow f(x)$ is continuous at $x = 1$.

8. Test the continuity of the given function $f(x)$ at $x = 2$

$f(x) = \frac{x^2-4}{x-2}$, when $0 < x < 2$

$f(x) = x + 1$, when $2 \le x \le 5$

Solution: $l.h.l = \lim_{x\to 2-} f(x) = \lim_{x\to 2} \frac{x^2-4}{x-2}$

$= \lim_{x\to 2} (x+2) = 2+2 = 4$ ...(1)

$r.h.l = \lim_{x\to 2+} f(x) = \lim_{x\to 2} (x+1)$

$= 2 + 1 = 3$ ...(2)

(1) and (2) $\Rightarrow l.h.l \ne r.h.l \Rightarrow \lim_{x\to 2} f(x)$ does not exist $\Rightarrow f(x)$ is discontinuous at $x = 2$.

9. Test the continuity of the given function $f(x)$ at $x = 1$.

$f(x) = \frac{9x}{x+2}$, when $0 < x \le 1$

$= \frac{x+2}{x}$, when $1 < x \le 2$

Solution: $l.h.l = \lim_{x\to 1^-} f(x) = \lim_{x\to 1} \frac{9x}{x+2}$

$= \frac{9}{1+2} = \frac{9}{3} = 3$ ...(1)

$r.h.l = \lim_{x\to 1^+} f(x) = \lim_{x\to 1} \frac{x+2}{x} = \frac{1+2}{1} = 3$ ...(2)

$f(1) = \left(\frac{9x}{x+2}\right)_{x=1} = \frac{9\times 1}{1+2} = \frac{9}{3} = 3$ ...(3)

(1), (2) and (3) $\Rightarrow l.h.l = r.h.l = f(1) \Rightarrow f(x)$ is continuous at $x = 1$.

10. Test the continuity of the given function $f(x)$ at $x = 1$ and $x = 3$ $f(x) = x + 2$, when $x < 1$

$= 4x - 1$, when $1 \le x \le 3$

$= x^2 + 5$, when $x > 3$

Solution: $f(x) = [f(x)]_{x=1} = (4x-1)_{x=1}$

$= 4\times 1 - 1 = 3$ ...(1)

$\lim_{x\to 1+} f(x) = \lim_{x\to 1} (4x-1)$

$= 4\times 1 - 1 = 3$ ...(2)

$\lim_{x\to 1-} f(x) = \lim_{x\to 1} (x+2)$

$= 1 + 2 = 3$ ...(3)

(1), (2) and (3) $\Rightarrow f(x)$ is continuous at $x = 1$

Now, we test the continuity at $x = 3$

$f(3) = [f(x)]_{x=3} = (4x-1)_{x=3}$

$= 4\times 3 - 1 = 12 - 1 = 11$ ...(1)

$l.h.l = \lim_{x\to 3-} (4x-1)$

$= 4\times 3 - 1 = 12 - 1 = 11$ ...(2)

$r.h.l = \lim_{x\to 3+} \left(x^2+5\right)$

$= 3^2 + 5 = 9 + 5 = 14$ ...(3)

(2) and (3) $\Rightarrow r.h.l \ne l.h.l \Rightarrow f(x)$ is discontinuous at $x = 3$.

11. Test the continuity of the given function $f(x)$ at $x = 0, 1$ defined as $f(x) = 2$, when $x \le 0$.

$= 3x+2$, when $0 < x \le 1$

$= \dfrac{x}{x-1}$, when $x > 1$

Solution: **(a)** Continuity test at $x = 0$

$$r.h.l = \lim_{x\to 0+} f(x) = \lim_{x\to 0}(3x+2) = 2 \quad ...(1)$$

$$l.h.l = \lim_{x\to 0-} f(x) = \lim_{x\to 0}(2) = 2 \quad ...(2)$$

$$f(0) = [f(x)]_{x=0} = 2 \quad ...(3)$$

(1), (2) and (3)

$$\Rightarrow f(0) = \lim_{x\to 0+} f(x) = \lim_{x\to 0-} f(x)$$

$\therefore f(x)$ is continuous at $x = 1$

(b) Test of continuity at $x = 1$

$$\lim_{x\to 1+0} f(x) = \lim_{h\to 0} f(1+h), h > 0$$

$$= \lim_{h\to 0}\frac{1+h}{1+h-1} = \lim_{h\to 0}\frac{1+h}{h} = +\infty$$

$$\therefore f(1+0) = +\infty$$

$\therefore \lim_{x\to 1} f(x)$ does not exist

Hence, $f(x)$ is not continuous at $x = 1$

$\therefore f(x)$ has a discontinuity of second kind at $x = 1$.

Type 3: When a function is defined as

$f(x) = f_1(x)$, when $x \ne a$

$=$ a constant 'c' (say), when $x \ne a$

Remember:

1. A point of removable discontinuity/removable discontinuity: If the limits of a function from the right and left exists and are equal but is not equal to the value of the function at a point, then the function is said to have (or, contain) a point of removable discontinuity (or, simply a removable discontinuity) at the considered point.

Or, more explicitly,

A point of discontinuity (or, simply a discontinuity) namely $x = a$ is called removable discontinuity if the limit of the function exists but the function either is not defined at $x = a$ or has a value different from the limit $x = a$ (i.e; $\lim_{x\to a} f(x) \ne f(a)$)

Or, in the notational form,

If at $x = a$, $\lim_{x\to a+} f(x) = \lim_{x\to a-} f(x) \ne f(a)$, then $f(x)$ is said to have (or, to contain) a point of removable discontinuity (or, simply a removable discontinuity) namely $x = a$ (or, at $x = a$), e.g:

(i) $f(x) = \dfrac{x^2 + x}{x}$ for $x \ne 0$

$=$ undefined, for $x = 0$

(ii) $f(x) = |x|$

$f(0) = 3$ are the functions having removable discontinuity at $x = 0$.

2. A function having removable discontinuity at a point $x = a$ can be made continuous by giving the function a new value 'c' equal to the limit of the function at the point $x = a$.

3. A discontinuity is called a removable discontinuity because it can be removed from the function and the function becomes continuous whereas a non-removable discontinuity is any discontinuity which is not removable.

4. A function $f(x)$ having (or, containing) removable discontinuity is also said to be redefined at $x = a$ if the function $f(x)$ is made continuous by assuming (or, setting) the value of the function $f(x)$ at $x = a$ to be equal to the limit of the function at $x = a$.

5. *Jump discontinuity*: A function $f(x)$ is said to have a jump discontinuity at $x = a$ if the left hand limit and right hand limit of the function $f(x)$ at $x = a$ exist and are finite but are not equal.

Or, in the notational form,

If $\lim_{x\to a-} f(x) = L_1$ and $\lim_{x\to a+} f(x) = L_2$ but $L_1 \ne L_2$, then we say that the function $f(x)$ has a jump discontinuity namely $x = a$ or we say that $f(x)$ has a jump discontinuity at $x = a$ or a discontinuity of the first kind (or, a point of jump discontinuity or, a point of discontinuity of the first kind) at $x = a$. e.g:,

(i) $f(x) = 1$, when $x > 0$

$= -1$, when $x < 0$

is a function having a jump discontinuity (or, discontinuity of the first kind) at $x = 0$ or we can say $x = 0$ is a point of jump discontinuity or a point of discontinuity of the first kind of $f(x)$.

6. A function $f(x)$ is said to have a discontinuity of the second kind at a point $x = a$ if at least one of one sided limit fails to exist at the considered point $x = a$.

Or, in the notational form,

If at least one of the limits $\lim_{x \to a-} f(x) / \lim_{x \to a+} f(x)$ does not exist, we say that the function $f(x)$ has a discontinuity of the second kind at $x = a$. e.g.,

(i) $f(x) = \sin\left(\frac{1}{x}\right), x \neq 0$

$f(0) = 0$

is the function having a discontinuity of the second kind at $x = 0$ or we can say $x = 0$ is a point of discontinuity of the second king of $f(x)$.

Highlight on the Working Rule

($l.h.l = r.h.l$ at $x = a$) = $\lim_{x \to a}$ [the function opposite which $x \neq a$ is written] and see whether $l.h.l = r.h.l = \lim_{x \to a}$ [the function written before $x \neq a$]

$= \lim_{x \to a} f_1(x) = f(a) =$ given value $= c$

N.B.:

1. $f(a+0)$ and $f(a-0)$ are the notations used for $l.h.l$ and $r.h.l$ at $x = a$.

2. $x \neq a \Leftrightarrow$ either $x > a$ or $x < a \Leftrightarrow x \to a$

Solved Examples

1. Show that the function $f(x)$ is discontinuous at $x = 1$

$f(x) = x^2$, when $x \neq 1$

$= 2$, when $x = 1$

Solution: ($l.h.l$ at $x = 1$)

$$= \lim_{x \to 1-} f(x) = \lim_{x \to 1} x^2 = 1 \quad \text{...(i)}$$

($r.h.l$ at $x = 1$)

$$= \lim_{x \to 1+} f(x) = \lim_{x \to 1} x^2 = 1 \quad \text{...(ii)}$$

$$f(1) = 2 \text{ (given)} \quad \text{...(iii)}$$

(i), (ii) and (iii) $\Rightarrow l.h.l = r.h.l \neq f(1) \Rightarrow f(x)$ is discontinuous at $x = 1$.

2. Test the continuity of the function $f(x)$ at $x = 1$ defined by

$$f(x) = \frac{x^2 - 4x + 3}{x^2 - 1}, x \neq 1$$

$$= 2, x = 1$$

Solution: ($l.h.l = r.h.l$ at $x = 1$)

$$= \lim_{x \to 1} \left(\frac{x^2 - 4x + 3}{x^2 - 1}\right)$$

$$= \lim_{x \to 1} \frac{(x^2 - 3x - x + 3)}{(x+1)(x-1)}$$

$$= \lim_{x \to 1} \frac{x(x-3) - (x-3)}{(x+1)(x-1)}$$

$$= \lim_{x \to 1} \frac{(x-3)(x-1)}{(x+1)(x-1)}$$

$$= \lim_{x \to 1} \frac{(x-3)}{(x+1)}$$

$$= \frac{1-3}{1+1} = \frac{-2}{2} = -1$$

$f(1) = 2$ (given)

Hence, ($l.h.l = r.h.l$ at $x = 1$) $\neq f(1) \Rightarrow f(x)$ is discontinuous at $x = 1$.

3. If $f(x) = \frac{x^2 - 3x + 2}{x^2 - 4x + 3}, x \neq 1$

$= 2, x = 1$ show that given function is discontinuous at $x = 1$.

Solution: ($l.h.l = r.h.l$ at $x = 1$)

$$= \lim_{x \to 1} \left(\frac{x^2 - 3x + 2}{x^2 - 4x + 3}\right)$$

$$= \lim_{x \to 1} \left(\frac{x^2 - x - 2x + 2}{x^2 - x - 3x + 3}\right)$$

$$= \lim_{x\to 1} \frac{(x-1)(x-2)}{(x-1)(x-3)}$$

$$= \lim_{x\to 1} \frac{(x-2)}{(x-3)}$$

$$= \frac{1-2}{1-3} = \frac{-1}{-2} = \frac{1}{2}$$

$f(1) = 2$ (given)

Hence, ($l.h.l = r.h.l$ at $x = 1$) $\neq f(1) \Rightarrow f(x)$ is discontinuous at $x = 1$.

4. If $f(x) = \dfrac{\sqrt{4+x} - \sqrt{4-x}}{x}$, $x \neq 0$ and $f(0) = \dfrac{1}{2}$, test the continuity at $x = 0$.

Solution: $l.h.l = r.h.l$ at $x = 0$

$$= \lim_{x\to 0} \frac{\sqrt{4+x} - \sqrt{4-x}}{x}$$

$$= \lim_{x\to 0} \left(\frac{\sqrt{4+x} - \sqrt{4-x}}{x} \times \frac{\sqrt{4+x} + \sqrt{4-x}}{\sqrt{4+x} + \sqrt{4-x}} \right)$$

$$= \lim_{x\to 0} \frac{(4+x-4+x)}{x\left(\sqrt{4+x} + \sqrt{4-x}\right)}$$

$$= \lim_{x\to 0} \frac{2x}{x\left(\sqrt{4+x} + \sqrt{4-x}\right)}$$

$$= \lim_{x\to 0} \frac{2}{\left(\sqrt{4+x} + \sqrt{4-x}\right)}$$

$$= \frac{2}{\sqrt{4+0} + \sqrt{4-0}} = \frac{2}{2+2} = \frac{1}{2}$$

$f(0) = \dfrac{1}{2}$ (given)

Thus, $l.h.l = r.h.l = f(0) = \dfrac{1}{2} \Rightarrow$ continuity of the function $f(x)$ at $x = 0$.

5. Test the continuity of the given function at $x = 1$

$$f(x) = \frac{\sqrt{x+3} - 2}{x^3 - 1}, \ x \neq 1$$

$$= 2, x = 1$$

Solution: $f(1-0) = f(1+0)$

$$= \lim_{x\to 1} \frac{\sqrt{x+3} - 2}{x^3 - 1}$$

$$= \lim_{x\to 1} \left(\frac{\sqrt{x+3} - 2}{x^3 - 1} \times \frac{\sqrt{x+3} + 2}{\sqrt{x+3} + 2} \right)$$

$$= \lim_{x\to 1} \frac{x+3-4}{(x-1)\left(x^2+x+1\right)\left(\sqrt{x+3}+2\right)}$$

$$= \lim_{x\to 1} \frac{x-1}{(x-1)\left(x^2+x+1\right)\left(\sqrt{x+3}+2\right)}$$

$$= \lim_{x\to 1} \frac{1}{\left(x^2+x+1\right)\left(\sqrt{x+3}+2\right)}$$

$$= \frac{1}{\left(1^2+1+1\right)\left(\sqrt{1+3}+2\right)}$$

$$= \frac{1}{3(2+2)} = \frac{1}{12}$$

$f(1) = 2$ (given)

Thus, $l.h.l = r.h.l \neq f(1) \Rightarrow$ discontinuity of the function at $x = 1$.

6. Test the continuity of the function $f(x)$ at $x = 1$

$$f(x) = \frac{\sqrt{x^2+1} - \sqrt{2}}{x - 1}, \ x \neq 1$$

$$= \frac{1}{\sqrt{2}}, \ x = 1$$

Solution: $f(1-0) = f(1+0)$

$$= \lim_{x\to 1} \frac{\sqrt{x^2+1}-\sqrt{2}}{(x-1)}$$

$$= \lim_{x\to 1} \left(\frac{\sqrt{x^2+1}-\sqrt{2}}{x-1} \times \frac{\sqrt{x^2+1}+\sqrt{2}}{\sqrt{x^2+1}+\sqrt{2}}\right)$$

$$= \lim_{x\to 1} \frac{x^2+1-2}{(x-1)\left(\sqrt{x^2+1}+\sqrt{2}\right)}$$

$$= \lim_{x\to 1} \frac{x^2-1}{(x-1)\left(\sqrt{x^2+1}+\sqrt{2}\right)}$$

$$= \lim_{x\to 1} \frac{(x+1)}{\left(\sqrt{x^2+1}+\sqrt{2}\right)}$$

$$= \frac{1+1}{\sqrt{2}+\sqrt{2}}$$

$$= \frac{2}{2\sqrt{2}} = \frac{1}{\sqrt{2}}$$

$$f(1) = \frac{1}{\sqrt{2}} \text{ (given)}$$

Hence, $f(1-0) = f(1+0) = f(1) \Rightarrow f(x)$ is continuous at $x = 1$

7. Show that the function f defined as

$$f(x) = \frac{\sqrt{x+3}-\sqrt{3}}{x} \text{ is discontinuous at } x = 0, \text{ and}$$

then determine if the discontinuity is removable.

Solution: **(a) (i)** f or $f(x)$ is not defined at $x = 0$ as

$$\left[f(x)\right]_{x=0} = \frac{0}{0} \text{ (undefined)}$$

$\therefore f(x)$ has discontinuity at $x = 0$.

(ii) $$\lim_{x\to 0} f(x) = \lim_{x\to 0}\left(\frac{\sqrt{x+3}-\sqrt{3}}{x}\right)$$

$$= \lim_{x\to 0}\left(\frac{\sqrt{x+3}-\sqrt{3}}{x} \times \frac{\sqrt{x+3}+\sqrt{3}}{\sqrt{x+3}+\sqrt{3}}\right)$$

$$= \lim_{x\to 0} \frac{x+3-3}{x\left(\sqrt{x+3}+\sqrt{3}\right)}$$

$$= \lim_{x\to 0} \frac{x}{x\left(\sqrt{x+3}+\sqrt{3}\right)}$$

$$= \lim_{x\to 0} \frac{1}{\left(\sqrt{x+3}+\sqrt{3}\right)}$$

$$= \frac{1}{\sqrt{3}+\sqrt{3}} = \frac{1}{2\sqrt{3}}$$

(b) The discontinuity is removable if we define the function as follows

$$f(x) = \frac{\sqrt{x+3}-\sqrt{3}}{x},\ x \neq 0$$

$$= \frac{1}{2\sqrt{3}},\ x = 0$$

8. A function is defined as under

$$f(x) = \frac{x^2-x-6}{x^2-2x-3}, \text{ at } x \neq 3$$

$$= \frac{5}{3}, \text{ at } x = 3$$

show that $f(x)$ is discontinuous at $x = 3$

Solution: $\lim_{x\to 3+} f(x) = \lim_{x\to 3-0} f(x)$

$$= \lim_{x\to 3}\left(\frac{x^2-x-6}{x^2-2x-3}\right)$$

$$= \lim_{x\to 3} \frac{\cancel{(x-3)}(x+2)}{\cancel{(x-3)}(x+1)}$$

$$= \lim_{x\to 3} \frac{x+2}{x+1} = \frac{3+2}{3+1} = \frac{5}{4}$$

$f(3) = \frac{5}{3}$ (given)

$\therefore \lim_{x \to 3} f(x) \neq f(3) \Rightarrow f(x)$ has discontinuity at $x = 3$.

Type 4: Problems based on continuity of an absolute value function (or, modulus function)

Firstly, we recall the definition of absolute value function.

Definition: An absolute value function is a function defined on the domain of all real numbers such that with any number x in the domain, the function associates algebraically the non-negative number $\sqrt{x^2}$, which is designated by writing two vertical lines around x as $|x|$. Therefore the value of the function at x is

$$|x| = \sqrt{x^2} = x, \text{ when } x \geq 0$$

$$|x| = \sqrt{x^2} = -x, \text{ when } x < 0$$

which means an absolute value function (or, modulus function or, simply mod function) maps every positive real number onto its positive number, zero onto zero and every negative real number onto its negative, which is the corresponding positive number. Thus $|x|$ is never negative. Hence, an absolute value function represented by the symbol '$||$' designated by writing two vertical lines is like an electric current rectifier that converts either positive or negative current into positive. Further we should note that the domain of an absolute value function is the set of all real numbers and the range is the set of all positive real number including zero which can be written in the following notational form.

Function	*Image*	*Domain*	*Range*
$\|\|$	$\|x\|$	R	$R^+ \cup \{0\}$

Remember:

1. $|f(x)| = f(x)$, when $f(x) \geq 0$

$= -f(x)$, when $f(x) < 0$

2. $|x - a| = (x - a)$, when $(x - a) \geq 0 \Rightarrow x \geq a$;

$= a - x$, when $(x - a) < 0 \Rightarrow x < a$

3. $|x| = \sqrt{x^2} = x$, when $x \geq 0$

$= -x$, when $x < 0$

4. $|x| = \sqrt{x^2}$

5. $|x - a| = \sqrt{(x - a)^2}$

6. $|x| = |-x| = x$, when x is positive

7. $\lim_{x \to c} |x| = |c|$, for any real c $\lim_{x \to c} |f(x)| = \left|\lim_{x \to c} f(x)\right|$

8. $|f(x)| = \sqrt{(f(x))^2}$

9. $x \neq 0 \Rightarrow x > 0$ and $x < 0$

10. $|(|f(x)|)|$ = mod of a mod function $f(x) = |f(x)|$ = mod of $f(x)$ since $\sqrt{|x|^2} = \sqrt{x^2} = |x|$.

Method to test the continuity of an absolute value function

To test the continuity of an absolute value function, we may adopt the following working rule.

(***Note***: **(i)** $|x|$ is read as modulus of x / mod of x/simply mod. x. **(ii)** $R + \cup \{0\}$ means a set of all positive real numbers including zero).

Working rule:

1. Convert the given mod problem in $f(x)$ and $-f(x)$ on imposing the condition (or, restriction) $\gtreqless$ which becomes the problem for finding the *l.h.l* and *r.h.l* by using the mod function definition $\Rightarrow$ when we are required to find *l.h.l*, we should put $|f(x)| = -f(x)$ which involves no mod symbol as well as we should put $|f(x)| = +f(x)$ which involves no mod symbol $||$.

2. Find the *l.h.l* and *r.h.l*.

3. See whether *l.h.l* and *r.h.l* are equal or not.

4. If *l.h.l* = *r.h.l*, then we declare the existance of limits for the given mod function at the given point $x = a$.

5. Find the value of the function at the given point $x = a$.

6. If *l.h.l* = *r.h.l* = value of the given function at a given point $x = a$ then we declare that continuity of the given mod function $|f(x)|$ holds at the given point

$x = a$ and if $l.h.l \neq r.h.l$, we declare that discontinuity of the given mod function $|f(x)|$ holds at the given point $x = a$.

Solved Examples

1. Discuss the continuity of the function $f(x)$ at $x = 8$

$$f(x) = \frac{|x-8|}{x-8}.$$

Solution: $l.h.l = \lim_{x \to 8} \frac{-(x-8)}{(x-8)} = \lim_{x \to 8} (-1) = -1$

$$r.h.l = \lim_{x \to 8} \frac{(x-8)}{x-8} = \lim_{x \to 8} 1 = 1$$

Thus, $l.h.l \neq r.h.l \Rightarrow$ given function is discontinuity at $x = 8$ or, alternatively, the function $f(x) = \frac{|x-8|}{x-8}$ is not defined at $x = 8$ and so discontinuity at $x = 8$.

2. Discuss the continuity of the function $f(x)$ at $x = 0$

$f(x) = \frac{x(x+1)}{|x|}$ for $x \neq 0$ and $f(0) = -1$.

Solution: $l.h.l = \lim_{x \to 0} \frac{x(x+1)}{-x}$ ($\because |x| = -x$ when $x < 0$)

$$= \lim_{x \to 0} \{-(x+1)\} = -(0+1) = -1$$

$r.h.l = \lim_{x \to 0} \frac{x(x+1)}{x}$ ($\because |x| = x$ when $x > 0$)

$$= \lim_{x \to 0} (x+1) = 1$$

$\therefore l.h.l \neq r.h.l \Rightarrow$ given function is discontinuous at $x = 0$.

3. Discuss the continuity of the function $f(x)$ at $x = 0$

$$f(x) = e^{-|x|}.$$

Solution: $l.h.l = \lim_{x \to 0} e^{-(-x)}$

$$= \lim_{x \to 0} e^{x} = e^{0} = 1$$

$$r.h.l = \lim_{x \to 0} e^{-x}$$

$$= e^{-0} = e^{0} = 1$$

$$f(0) = e^{-0} = 1$$

$\therefore l.h.l = r.h.l = f(0) \Rightarrow$ given function has a continuity at the origin ($x = 0$).

4. Discuss the continuity of the function $f(x)$ at $x = 0$.

(i) $f(x) = \frac{x}{|x|}$

(ii) $f(x) = \frac{|x|}{x}$

Solution: (i) The function $f(x) = \frac{x}{|x|}$ is not defined at $x = 0$ and so it is discontinuous at $x = 0$.

(ii) The function $f(x) = \frac{|x|}{x}$ is not defined at $x = 0$ and so it is discontinuous at $x = 0$.

5. Discuss the continuity of the function $f(x)$ at $x = -2$ $f(x) = x + \frac{x+2}{|x+2|}$.

Solution: The function $f(x) = x + \frac{x+2}{|x+2|}$ is not defined at $x = -2$ and so discontinuous at $x = -2$.

6. Show that $f(x) = |x|$ is continuous at the origin.

Solution: $f(x) = |x|$

$$l.h.l = \lim_{x \to 0} (-x) = 0$$

$$r.h.l = \lim_{x \to 0} (x) = 0$$

$$f(0) = |0| = 0$$

$\therefore l.h.l = r.h.l = f(0) \Rightarrow$ given function $f(x)$ is continuous at $x = 0$.

7. Discuss the continuity of the function $f(x)$ at $x = 8$

$$f(x) = \frac{|x-8|^3}{(x-8)^3},\ x \neq 8\ ;\ \text{and}\ f(8) = -1.$$

Solution: $f(x) = \dfrac{|x-8|^3}{(x-8)^3},\ x \neq 8$

$$\therefore\ l.h.l = \lim_{x \to 8} \frac{-(x-8)^3}{(x-8)^3}$$

$$= \lim_{x \to 8} (-1) = -1$$

$$r.h.l = \lim_{x \to 8} \left(\frac{x-8}{x-8}\right)^3$$

$$= \lim_{x \to 8} 1 = 1$$

Hence, $l.h.l \neq r.h.l \Rightarrow$ given function is discontinuous at $x = 8$.

Note

1. Any polynomial function, e^x, $\sin x$, $\cos x$, $|x|$ are functions which are continuous everywhere.

2. $\log x$ is continuous everywhere except $x \leq 0$ where it is not defined.

3. Ratio of two polynomials $\dfrac{p(x)}{q(x)}$ is continuous everywhere except at those points (or, those values of x) which make $q(x) = 0$.

4. $|f(x)|$ is continuous everywhere provided $f(x)$ is continuous everywhere.

Problems on $|f(x)|$

1. Discuss the continuity of $|x| + |x-1|$ at $x = 0, 1$.

Solution: Since x is continuous everywhere $\Rightarrow |x|$ is continuous everywhere similarly, $(x-1)$ being a polynomial is continuous everywhere $\Rightarrow |x-1|$ is continuous everywhere.

Hence, $|x| + |x-1|$ is continuous everywhere

$\Rightarrow |x| + |x-1|$ is continuous at $x = 0, 1$

2. A function is defined as

$$f(x) = |x| + |x-1| + |x-2|$$

Solution: Since we know that $|x|, |x-1|$ and $|x-2|$ are continuous everywhere $\Rightarrow |x| + |x-1| + |x-2|$ is continuous everywhere

$\Rightarrow |x| + |x-1| + |x-2|$ is continuous at $x = 0, 1$

Examples on redefined functions which is the combination of $f(x)$ and $|f(x)|$

1. Discuss the continuity of the function $f(x)$ at $x = 2$

$$f(x) = \frac{\left|x^3\right|}{x},\ \text{when } x \neq 0 = 0,\ \text{when } x = 0.$$

Solution: $l.h.l = \lim_{x \to 0} \left(\dfrac{-x^3}{x}\right) = \lim_{x \to 0} \left(-x^2\right) = 0$

$$r.h.l = \lim_{x \to 0} \left(\frac{x^3}{x}\right) = \lim_{x \to 0} \left(x^2\right) = 0$$

$f(0) = 0$

$\therefore\ l.h.l = r.h.l = f(0) = 0 \Rightarrow f(x)$ is continuous at $x = 0$.

2. Discuss the continuity of the function $f(x)$ at $x = 0$

$$f(x) = \frac{|x|}{x},\ \text{when } x \neq 0 = 0,\ \text{when } x = 0.$$

Solution: $l.h.l = \lim_{x \to 0} \left(\dfrac{-x}{x}\right) = -1$

$$r.h.l = \lim_{x \to 0} \left(\frac{x}{x}\right) = 1$$

$\therefore\ l.h.l \neq r.h.l \Rightarrow f(x)$ is discontinuous at $x = 0$.

3. If $f(x) = |x| + 1,\ x \neq 0 = 1$, at $x = 0$ test the continuity of $f(x)$ at $x = 0$.

Solution: $f(x) = |x| + 1$

$$l.h.l = \lim_{x \to 0} (-x+1) = (-0+1) = 1$$

$$r.h.l = \lim_{x \to 0} (x+1) = (0+1) = 1$$

$f(0) = |0| + 1 = 0 + 1 = 1$

$\therefore l.h.l = r.h.l = f(0) \Rightarrow$ given function $f(x)$ is continuous at $x = 0$.

4. Discuss the continuity of $f(x)$ at $x = 0$

$$f(x) = \frac{x - |x|}{x}, \text{ when } x \neq 0 = 2, \text{ when } x = 0.$$

Solution: $l.h.l = \lim_{x \to 0}\left(\frac{x - (-x)}{x}\right) =$

$$= \lim_{x \to 0} \frac{2x}{x} = 2$$

$$r.h.l = \lim_{x \to 0}\left(\frac{x - x}{x}\right) = 0$$

$\therefore l.h.l \neq r.h.l \Rightarrow f(x)$ is discontinuous at $x = 0$.

Examples on trigonometrical and inverse trigonometrical functions redefined

Remember:

1. $\lim_{x \to 0} \frac{\sin^{-1} x}{x} = 1$

2. $\lim_{x \to 0} \frac{\tan^{-1} x}{x} = 1$

3. $\lim_{x \to 0} \frac{\sin x}{x} = 1$

4. $\lim_{x \to 0} \frac{\tan x}{x} = 1$

5. $\lim_{x \to 0} \cos x = 1$

Now we come to the problems.

1. Test the continuity of the function $f(x)$ at $x = 0$ where

$$f(x) = \frac{\sin 3x}{2x}, x \neq 0$$

$$= \frac{2}{3}, x = 0$$

Solution: $l.h.l = r.h.l = \lim_{x \to 0}\left(\frac{\sin 3x}{2x}\right)$

$$= \lim_{x \to 0}\left(\frac{\sin 3x}{3x} \times 3x \times \frac{1}{2x}\right)$$

$$= \lim_{x \to 0}\left(\frac{\sin 3x}{3x} \times \frac{3}{2}\right)$$

$$= \frac{3}{2} \times \lim_{x \to 0}\left(\frac{\sin 3x}{3x}\right) = \frac{3}{2} \times 1$$

$$= \frac{3}{2}$$

$$f(0) = \frac{2}{3} \text{ (given)}$$

$\therefore l.h.l = r.h.l \neq f(0) \Rightarrow$ the given function $f(x)$ is discontinuous at $x = 0$

2. Test the continuity of $f(x)$ at $x = 0$, if

$$f(x) = \frac{\sin |x|}{|x|}, x \neq 0$$

$$= 1, x = 0$$

Solution: $l.h.l = \lim_{x \to 0} \frac{\sin(-x)}{(-x)}$

$$= \lim_{x \to 0}\left(\frac{-\sin x}{-x}\right) = 1$$

$$r.h.l = \lim_{x \to 0}\left(\frac{\sin x}{x}\right) = 1$$

$$f(0) = 1$$

Hence, $l.h.l = r.h.l = f(0) \Rightarrow f(x)$ is continuous at $x = 0$.

3. Test the continuity of the function $f(x)$ at $x = 0$ defined as under

$$f(x) = \begin{cases} \sin^{-1}|x| \cos\frac{1}{x}, & \text{when } x \neq 0 \\ 0, & \text{when } x = 0 \end{cases}$$

Solution: $f(x) = \begin{cases} \sin^{-1} x \cdot \cos\frac{1}{x}, & \text{when } x > 0 \\ -\sin^{-1} x \cdot \cos\frac{1}{x}, & \text{when } x < 0 \end{cases}$

$(\because \sin^{-1}(-x) = -\sin^{-1} x)$

$$l.h.l = \lim_{x\to 0}\left(-\sin^{-1} x \cdot \cos\frac{1}{x}\right)$$

$$= 0\left[\because \lim_{x\to 0} \sin^{-1} x = 0 \text{ and } \left|\cos\frac{1}{x}\right| \le 1 \text{ for } x \ne 0\right]$$

$$r.h.l = \lim_{x\to 0}\left(\sin^{-1} x \cdot \cos\frac{1}{x}\right) = 0$$

$f(0) = 0$ (given)

Thus, $l.h.l = r.h.l = f(0) \Rightarrow$ given function is continuous at $x = 0$.

4. Examine whether the following function is continuous at $x = 0$

$$f(x) = \frac{x^4 + x^3 + 2x^2}{\tan^{-1} x},\ x \ne 0 \text{ and } f(0) = 0.$$

Solution: $l.h.l = r.h.l = \lim_{x\to 0} \frac{x^4 + x^3 + 2x^2}{\tan^{-1} x}$

$$= \lim_{\theta\to 0} \frac{\tan^4 \theta + \tan^3 \theta + 2\tan^2 \theta}{\theta}$$

[on putting $x = \tan\theta$ $x \to 0, \theta \to 0$]

$$= \lim_{\theta\to 0}\left(\frac{\tan^4 \theta}{\theta}\right) + \lim_{\theta\to 0}\left(\frac{\tan^3 \theta}{\theta}\right) + \lim_{\theta\to 0}\left(2\tan^2\theta\right)$$

$$= \lim_{\theta\to 0}\frac{\tan\theta}{\theta} \times \lim_{\theta\to 0}\tan^3\theta + \lim_{\theta\to 0}\frac{\tan\theta}{\theta} \times \lim_{\theta\to 0}\tan^2\theta$$

$$+ \lim_{\theta\to 0}\frac{\tan\theta}{\theta} \times \lim_{\theta\to 0}(2\tan\theta)$$

$= 1 \times 0 + 1 \times 0 + 1 \times 0 = 0$

$f(0) = 0$ (given)

Hence, $\lim_{x\to 0} f(x) = f(0) \Rightarrow$ the given function $f(x)$ is continuous at $x = 0$. (***Note***: **(i)** If $\lim_{x\to a} f(x) = 0$ and $|g(x)| \le m$, then $\lim_{x\to a}[f(x)\cdot g(x)] = 0$.

(ii) If $\lim_{x\to a} f(x) = 0$, then $f(x)$ is said to be an infinitesimal (or, infinitely small)).

Examples on redefined exponential function of x or defined function of x

1. Test the continuity at $x = 0$ of the following function

$f(x) = e^{-\frac{1}{x}}$, $x \ne 0$ $f(0) = 0$.

Solution: For $h > 0$,

$$f(0+0) = \lim_{h\to 0} f(0+h) = \lim_{h\to 0} f(h) = \lim_{h\to 0} e^{-\frac{1}{h}} = 0$$

$$f(0-0) = \lim_{h\to 0} f(0-h) = \lim_{h\to 0} f(-h) = \lim_{h\to 0} e^{\frac{1}{h}} = +\infty$$

$\therefore f(x)$ has infinite discontinuity at $x = 0$.

2. Test the continuity of $y = e^{-|x|}$ at the origin (i.e. $x = 0$).

Solution: $l.h.l = \lim_{x\to 0} e^{-(-x)} = \lim_{x\to 0} e^{x} = e^{0} = 1$

$$r.h.l = \lim_{x\to 0} e^{-x} = \lim_{x\to 0}\frac{1}{e^x} = \frac{1}{e^0} = \frac{1}{1} = 1$$

$$f(0) = \frac{1}{e^0} = \frac{1}{1} = 1$$

Thus, $l.h.l = r.h.l = f(0) \Rightarrow$ given function $f(x)$ is continuous at $x = 0$ (or, origin).

Problems based on greatest integer function/ combination of two greatest integer function/ combination of a function and a greatest integer function defined or redefined

1. Discuss the continuity of the following functions at $x = 0$.

(i) $f(x) = [x] + [1 - x]$ for all real x

(ii) $f(x) = x + [x]$ for all real x

(iii) $f(x) = [x] + [-x]$ if $x \ne 0$, $f(0) = -1$

(iv) $f(x) = x[x]$

where $[t]$ denotes the largest integer less than or equal to t.

Solution: **(i)** $f(x) = [x] + [1 - x]$ for all real x

$$l.h.l = \lim_{x\to 0-0} f(x) = \lim_{x\to 0-0}([x]+[1+x])$$

$$= \lim_{h\to 0}([0-h]+[1-(0-h)])$$

$$= \lim_{h\to 0}([-h]+[1+h])$$

$$= \lim_{h\to 0}(-1+1) = 0$$

$\because 1 > h > 0 \Rightarrow$
$1 < 1+h < 2$
and $-1 < -h < 0$
$\therefore [-h] = -1$
$[1+h] = 1$ and $[1-h] = 0$

$$r.h.l = \lim_{x\to 0+0} f(x) = \lim_{x\to 0+0}([x]+[1-x])$$

$$= \lim_{h\to 0}([0+h]+[1-0-h])$$

$$= \lim_{h\to 0}([h]+[1-h])$$

$$= \lim_{h\to 0}(0+0) = 0$$

$\therefore \lim_{x\to 0} f(x) = 0$ and $f(0) = [0]+[1-0] = 0+1 = 1$

Hence, $\lim_{x\to 0} f(x) = 0 \neq f(0) = 1$ $f(0)=1$ which means $f(x)$ is discontinuous at $x = 0$.

(ii) $f(x) = x + [x]$ for all real x

$$l.h.l = \lim_{x\to 0-0} f(x)$$

$$= \lim_{x\to 0-0}(x+[x])$$

$$= \lim_{h\to 0}(0-h+[0-h])$$

$$= \lim_{h\to 0}(0-h-1) = 0-1 = -1$$

$$r.h.l = \lim_{x\to 0+0} f(x)$$

$$= \lim_{x\to 0+0}(x+[x])$$

$$= \lim_{h\to 0}(0+h+[0+h])$$

$$= \lim_{h\to 0}(h+[h])$$

$$= \lim_{h\to 0}(h+0) = 0$$

$f(0) = (0+[0]) = 0$

$\therefore l.h.l \neq r.h.l$ which means $f(x)$ is discontinuous at $x = 0$

(iii) $f(x) = [x]+[-x]$ if $x = 0$
$f(0) = -1$

$$l.h.l = \lim_{x\to 0-0}([x]+[-x])$$

$$= \lim_{h\to 0}([0-h]+[-(0-h)]),\ h>0$$

$$= \lim_{h\to 0}([0-h]+[-0+h])$$

$$= \lim_{h\to 0}([-h]+[h])$$

$$= \lim_{h\to 0}(-1+0) = -1$$

$\because [-h] = -1$ and $[h] = 0$

$$r.h.l = \lim_{x\to 0+0}([x]+[-x])$$

$$= \lim_{h\to 0}([0+h]+[-(0+h)])$$

$$= \lim_{h\to 0}([h]+[-h])$$

$$= \lim_{h\to 0}(0-1) = -1$$

$f(0) = -1$ (given)

$\therefore l.h.l = r.h.l = f(0) = -1 \Rightarrow f(x)$ is continuous at $x = 0$

(iv) $f(x) = x[x]$

$$r.h.l = \lim_{x\to 0+0}(x[x])$$

$$= \lim_{h\to 0}[(h)[h]],\ h>0$$

$$= \lim_{h\to 0}\{(h)\cdot 0\}\ (\because [h]=0)$$

$$= \lim_{h\to 0} 0 = 0$$

$$l.h.l = \lim_{x\to 0^-}(x[x])$$

$$= \lim_{h\to 0}\{(0-h)[0-h]\},\ h>0$$

$$= \lim_{h\to 0} \{(-h)[-h]\}$$

$$= \lim_{h\to 0} \{(-h)(-1)\} \quad (\because [-h] = -1)$$

$$= \lim_{h\to 0} h = 0$$

$f(0) = 0\,[0] = 0$

$\therefore$ $l.h.l = r.h.l = f(0) = 0$ which means the function $f(x)$ is continuous at $x = 0$.

2. Discuss the continuity of the following functions at the indicated points.

(i) $f(x) = [1-x] + [x-1]$ at $x = 1$

(ii) $f(x) = [x+2] + [2-x]$ at $x = 2$

(iii) $f(x) = \dfrac{\left[\frac{1}{2} + x^2\right] - \frac{1}{2}}{x^2}$ at $x = 1$

(iv) $f(x) = \dfrac{\left[x + \frac{1}{2}\right]}{[x]}$ at $x = -\dfrac{1}{2}$

Solution: $f(x) = [1-x] + [x-1]$

$$l.h.l = \lim_{x\to 1-0} f(x)$$

$$= \lim_{x\to 1-0} ([1-x] + [x-1])$$

$$= \lim_{h\to 0} ([1-1+h] + [1-h-1]), h>0$$

$$= \lim_{h\to 0} ([h] + [-h])$$

$$= \lim_{h\to 0} (0 + (-1))$$

$$= \lim_{h\to 0} (-1)$$

$$= -1$$

> $0 < h < 1$ and $-1 < -h < 0 \Rightarrow [h] = 0$ and $[-h] = -1$

$$r.h.l = \lim_{x\to 1+0} ([1-x] + [x-1])$$

$$= \lim_{h\to 0} ([1-1-h] + [1+h-1]), h>0$$

$$= \lim_{h\to 0} ([-h] + [h])$$

$$= \lim_{h\to 0} (0 + (-1))$$

$$= \lim_{h\to 0} (-1)$$

$$= -1$$

$f(1) = [1-1] + [1-1] = [0] + [0] = 0$

$\therefore$ $l.h.l = r.h.l \neq f(1) \Rightarrow f(x)$ is discontinuous at $x = 1$

(ii) $f(x) = [x+2] + [2-x]$

$$l.h.l = \lim_{x\to 2-0} ([x+2] + [2-x])$$

$$= \lim_{h\to 0} \{[2-h+2] + [2-(2-h)]\}, h>0$$

$$= \lim_{h\to 0} \{[4-h] + [h]\}$$

$$= 3 + 0 = 3$$

$$r.h.l = \lim_{x\to 2+0} ([x+2] + [2-x]), h>0$$

$$= \lim_{h\to 0} \{[2+h+2] + [2-(2+h)]\}$$

$$= \lim_{h\to 0} \{[4+h] + [-h]\}$$

$$= 4 + (-1) = 3$$

$f(2) = [2+2] + [2-2] = [4] + [0] = 4 + 0 = 4$

Hence, $l.h.l = r.h.l \neq f(2) \Rightarrow f(x)$ is discontinuous at $x = 2$

(iii) $f(x) = \dfrac{\left[\frac{1}{2} + x^2\right] - \frac{1}{2}}{x^2}$

$$r.h.l = \lim_{x\to 1+0} \frac{\left[\frac{1}{2} + x^2\right] - \frac{1}{2}}{x^2}$$

$$= \lim_{h\to 0} \frac{\left[\frac{1}{2} + (1+h)^2\right] - \frac{1}{2}}{(1+h)^2}, h>0$$

$$= \lim_{h\to 0} \frac{\left[\frac{1}{2} + 1 + h^2 + 2h\right] - \frac{1}{2}}{1 + h^2 + 2h}$$

$$= \lim_{h\to 0} \frac{\left[\frac{1}{2}+1+2h\right]-\frac{1}{2}}{1+2h}$$

$$= \lim_{h\to 0} \frac{\left[\frac{3}{2}+2h\right]-\frac{1}{2}}{1+2h}$$

$$= \lim_{h\to 0} \frac{1-\frac{1}{2}}{1+2h}$$

$$= \lim_{h\to 0} \frac{\frac{1}{2}}{1+2h} = \frac{1}{2}$$

$\therefore \left[\frac{3}{2}+2h\right] = 1$ as $1 < \frac{3}{2}+2h < 2$ for h being sufficinetly small>0

$$l.h.l = \lim_{x\to 1-0} \frac{\left[\frac{1}{2}+x^2\right]-\frac{1}{2}}{x^2}$$

$$= \lim_{h\to 0} \frac{\left[\frac{1}{2}+(1-h)^2\right]-\frac{1}{2}}{(1-h)^2}, h>0$$

$$= \lim_{h\to 0} \frac{\left[\frac{3}{2}-2h\right]-\frac{1}{2}}{(1-2h)}$$ [on neglecting higher power of h].

$$= \lim_{h\to 0} \frac{1-\frac{1}{2}}{1-2h}$$

$$= \lim_{h\to 0} \frac{\frac{1}{2}}{1-2h} = \frac{1}{2}$$

$\because \left[\frac{3}{2}-2h\right] = 1$ as $1 < \frac{3}{2}-2h < 2$ for h being suffuciently small>0

$$f(1) = \frac{\left[\frac{1}{2}+1\right]-\frac{1}{2}}{1} = 1-\frac{1}{2} = \frac{1}{2}$$

$\therefore l.h.l = r.h.l = f(1) \Rightarrow f(x)$ is continuous at $x=1$.

(iv) $f(x) = \dfrac{\left[x+\frac{1}{2}\right]}{[x]}$

$$l.h.l = \lim_{x\to -\frac{1}{2}-0} \frac{\left[x+\frac{1}{2}\right]}{[x]}$$

$$= \lim_{h\to 0} \frac{\left[-\frac{1}{2}-h+\frac{1}{2}\right]}{\left[-\frac{1}{2}-h\right]}, h>0$$

$$= \lim_{h\to 0} \frac{[-h]}{\left[-\frac{1}{2}-h\right]}$$

$$= \lim_{h\to 0} \frac{-1}{-1}$$

$$= \lim_{h\to 0} 1$$

$$=1$$

$\because \left[-\frac{1}{2}-h\right] = -1$ and $[-h]=-1$ as $0<h<1 \Rightarrow -1<-h<0 \Rightarrow -1 < -\frac{1}{2}-h < 0$

$$r.h.l = \lim_{h\to 0} \frac{\left[-\frac{1}{2}+h+\frac{1}{2}\right]}{\left[-\frac{1}{2}+h\right]}, h>0$$

$$= \lim_{h\to 0} \frac{[h]}{\left[-\frac{1}{2}+h\right]}$$

$$= \lim_{h\to 0} \frac{0}{-1}$$

$$= \lim_{x\to} 0 = 0$$

$\because \left[-\frac{1}{2}+h\right] = -1$ and $[h] = 0$ as $0 < h < 1 \Rightarrow$

$-1 < -h < 0 \Rightarrow -1 < -\frac{1}{2} + h < 0$

$$f\left(-\frac{1}{2}\right) = \frac{\left[-\frac{1}{2}+\frac{1}{2}\right]}{\left[-\frac{1}{2}\right]} = \frac{[0]}{\left[-\frac{1}{2}\right]}$$

$$= \frac{0}{-1} = 0$$

$\left[-\frac{1}{2}\right] = -1$ as $-1 < -\frac{1}{2} < 0$

$\therefore l.h.l \neq r.h.l = f(0) \Rightarrow f(x)$ is discontinuous at $x = -\frac{1}{2}$.

3. Show that the function f defined by $f(x) = [x-3] + [3-x]$, where $[t]$ denotes the largest integer $\leq t$, is discontinuous at $x = 3$. Modify the definition of f so as to make it continuous there.

Solution: $f(x) = [x-3] + [3-x]$

$$l.h.l = \lim_{x\to 3-0} f(x)$$

$$= \lim_{x\to 3-0} ([x-3] + [3-x])$$

$$= \lim_{h\to 0} ([3-h-3] + [3-3+h]), h>0$$

$$= \lim_{h\to 0} ([-h] + [h])$$

$$= \lim_{h\to 0} (-1+0) = -1$$

$\because [h] = 0$ and $[-h] = -1$ as $0 < h < 1 \Rightarrow$
$-1 < -h < 0$

$$r.h.l = \lim_{x\to 3+0} f(x)$$

$$= \lim_{x\to 3+0} ([x-3] + [3-x])$$

$$= \lim_{h\to 0} ([3+h-3] + [3-3-h])$$

$$= \lim_{h\to 0} = ([h] + [-h])$$

$$= \lim_{h\to 0} (-1+0) = -1$$

$\because [h] = 0$ and $[-h] = -1$ as $0 < h < 1 \Rightarrow$
$-1 < -h < 0$

$$f(3) = [3-3] + [3-3] = [0] + [0]\, 0 + 0 = 0$$

$\therefore l.h.l = r.h.l \neq f(3) \Rightarrow f(x)$ is discontinuous at $x = 3$

Further, $l.h.l = r.h.l \neq f(3) \Rightarrow f(x)$ has a removal discontinuity at $x = 3$. This is why to remove the discontinuity at $x = 3$, are must modify the definition of f as follows.

$f(x) = [x-3] + [3-x]$, when $x \neq 3$
$= -1$, when $x = 3$

4. Show that the function f defined by $f(x) = [x-1] + |x-1|$ for $x \neq 1$, and $f(1) = 0$ is discontinuous at $x = 1$. Can the definition of f at $x = 1$ be modified so as to make it continuous there?

Solution: $f(x) = [x-1] + |x-1|$ for $x \neq 1$

$f(1) = 0$ (given)

$$l.h.l = \lim_{x\to 1-0} f(x)$$

$$= \lim_{x\to 1-0} \{[x-1] + |x-1|\}$$

$$= \lim_{h\to 0} \{[1-h-1] + |1-h-1|\}, h>0$$

$$= \lim_{h\to 0} \{[-h] + |-h|\}$$

$$= \lim_{h\to 0} (-1+h)$$

$$= -1 + 0 = -1$$

$\because |-h| = |h| = h$ as $h > 0$ and $0 < h < 1$
$\Rightarrow -1 < -h < 0 \Rightarrow [-h] = -1$

$$r.h.l = \lim_{x \to 1+0} f(x)$$

$$= \lim_{x \to 1+0} \{[x-1] + |x-1|\}$$

$$= \lim_{h \to 0} \{[1+h-1] + |1+h-1|\}, h > 0$$

$$= \lim_{h \to 0} \{[h] + |h|\}$$

$$= \lim_{h \to 0} (0 + h)$$

$$= 0 + 0 = 0 \qquad \boxed{\because [h] = 0 \text{ as } 0 < h < 1}$$

$\therefore l.h.l \neq r.h.l = f(1) = 0 \Rightarrow f(x)$ is discontinuous at $x = 1$. Further, $l.h.l \neq r.h.l \Rightarrow \lim_{x \to 1} f(x)$ does not exist which $\Rightarrow$ we can not modify the definition of f at $x = 1$ in any way to make it continuous at $x = 1$.

Type 5: Problems based on finding the value of a constant 'k' if the given function it continuous at a given point $x = a$.

Working rule: To find the value of a constant 'k' if the given function is continuous at a given point $x = a$, we adopt the following procedure.

1. Find the $l.h.l$ and $r.h.l$
2. Equate $l.h.l = r.h.l = f(a)$ = a constant given in the question and solve the equation for k.

Solved Examples

1. A function f is defined as

$$f(x) = \frac{x^2 - 25}{x + 5}, \; x \neq -5 \text{ and } f(-5) = k, x = -5 \text{ if } f(x)$$

is continuous at $x = -5$, find K.

Solution: $l.h.l = r.h.l = \lim_{x \to -5} \left(\frac{x^2 - 25}{x + 5}\right)$

$$= \lim_{x \to -5} \frac{\cancel{(x+5)}(x+5)}{\cancel{(x+5)}}$$

$$= \lim_{x \to -5} (x - 5) = (-5 - 5) = -10$$

Now, since, $f(x)$ is continuous at $x = -5$ (given in the problem)

$\therefore l.h.l = r.h.l = f(-5)$ ...(i)

But $f(-5) = k$ ...(ii)

(i) and (ii) $\Rightarrow -10 = k \Rightarrow k = -10$ ***Ans.***

2. If $f(x) = \dfrac{x^3 + x^2 - 16x + 20}{(x-2)^2}, x \neq 2$

$= k$, when $x = 2$ and $f(x)$ is continuous at $x = 2$, find the value of K.

Solution: $l.h.l = r.h.l = \lim_{x \to 2} \dfrac{x^3 + x^2 - 16x + 20}{(x-2)^2}$,

$$= \lim_{x \to 2} \frac{\cancel{(x-2)}(x^2 + 3x - 10)}{\cancel{(x-2)}(x-2)}$$

$$= \lim_{x \to 2} \frac{\cancel{(x-2)}(x+5)}{\cancel{(x-2)}}$$

$$= \lim_{x \to 2} (x + 5) = 2 + 5 = 7$$

Now, since $f(x)$ is continuous at $x = 2$ (given in the problem)

$\therefore l.h.l = r.h.l = f(2)$ ...(i)

But $f(2) = k$ (given in the problem) ...(ii)

Hence, (i) and (ii) $\Rightarrow 7 = k$, i.e; $k = 7$ ***Ans.***

3. Find the value of k if the following function is continuous at $x = 0$

$$f(x) = \frac{1 - \cos kx}{x \sin x}, \; x \neq 0 \text{ and } f(0) = 2.$$

Solution: $l.h.l = r.h.l = \lim_{x \to 0} \dfrac{1 - \cos kx}{x \sin x}$

$$= \lim_{x \to 0} \frac{2\sin^2\left(\frac{kx}{2}\right)}{x \sin x}$$

$$= \lim_{x \to 0} \left[\frac{\sin^2\left(\frac{kx}{2}\right)}{\left(\frac{k^2 x^2}{4}\right)} \cdot \frac{k^2}{4} \cdot \frac{x}{\sin x} \cdot 2\right]$$

$$= 2 \times (1)^2 \times \frac{k^2}{4}$$

$$= \frac{k^2}{2}$$

Now, that given function $f(x)$ is continuous at $x = 0 \Rightarrow l.h.l = r.h.l = f(0)$ ...(i)

But $f(0) = 2$ (given) ...(ii)

(i) and (ii) $\Rightarrow l.h.l = r.h.l = f(0)$

$$\Rightarrow \frac{k^2}{2} = 2$$

$$\Rightarrow k^2 = 4$$

$$\Rightarrow k = \pm 2$$

4. Find the value of k if the following function $f(x)$ is continuous at $x = 0$

$$f(x) = \frac{\sin kx}{x},\ x \neq 0$$

$f(x) = 4 + x, x = 0$

Solution: $l.h.l = r.h.l = \lim_{x \to 0} \left(\frac{\sin kx}{x} \right)$

$$= \lim_{x \to 0} \left(\frac{\sin kx}{kx} \cdot k \right)$$

$$= \lim_{x \to 0} \frac{\sin kx}{kx} \cdot k$$

$= k \times 1 = k$

Also, $f(0) = 4 + 0 = 4$

Now, $f(x)$ is continuous at $x = 0 \Rightarrow l.h.l = r.h.l = f(0)$ ...(i)

But $f(0) = 4$ ...(ii)

(i) and (ii) $\Rightarrow k = 4$ ***Ans.***

5. A function $f(x)$ is defined as follows

$$f(x) = \frac{\sin(a+1)x + \sin x}{x}, \text{ for } x < 0$$

$= c$, for $x = 0$

$$= \frac{\sqrt{x + bx^2} - \sqrt{x}}{bx\sqrt{x}}, \text{ for } x > 0$$

Find the values of a, b, c if $f(x)$ is continuous at $x = 0$

Solution:

$$l.h.l = \lim_{x \to 0} f(x) = \lim_{x \to 0} \left[\frac{\sin(a+1)x + \sin x}{x} \right]$$

$$= \lim_{x \to 0} \left[\frac{\sin(a+1)x}{x} + \frac{\sin x}{x} \right]$$

$$= \lim_{x \to 0} \left[(a+1) \frac{\sin(a+1)x}{x(a+1)} + \frac{\sin x}{x} \right]$$

$$= \lim_{x \to 0} (a+1) \frac{\sin(a+1)x}{(a+1)x} + \lim_{x \to 0} \frac{\sin x}{x}$$

$$= (a+1) \cdot 1 + 1 = a + 1 + 1 = a + 2 \quad ...(1)$$

$$r.h.l = \lim_{x \to 0} f(x) = \lim_{x \to 0} \frac{\sqrt{x + bx^2} - \sqrt{x}}{bx\sqrt{x}}, x > 0$$

$$= \lim_{x \to 0} \left[\frac{\sqrt{x + bx^2} - \sqrt{x}}{bx\sqrt{x}} \times \frac{\sqrt{x + bx^2} + \sqrt{x}}{\sqrt{x + bx^2} + \sqrt{x}} \right]$$

$$= \lim_{x \to 0} \left[\frac{x + bx^2 - x}{bx\sqrt{x}\left(\sqrt{x + bx^2} + \sqrt{x}\right)} \right]$$

$$= \lim_{x \to 0} \left[\frac{bx^2}{bx\sqrt{x}\left(\sqrt{x + bx^2} + \sqrt{x}\right)} \right]$$

$$= \lim_{x \to 0} \left[\frac{x}{\sqrt{x}\left(\sqrt{x + bx^2} + \sqrt{x}\right)} \right]$$

$$= \lim_{x \to 0} \left[\frac{\sqrt{x} \cdot \sqrt{x}}{\sqrt{x}\left(\sqrt{x + bx^2} + \sqrt{x}\right)} \right]$$

$$= \lim_{x\to 0}\left[\frac{\sqrt{x}}{\left(\sqrt{x+bx^2}+\sqrt{x}\right)}\right]$$

$$= \lim_{x\to 0}\left[\frac{1}{\sqrt{1+bx}+1}\right] = \frac{1}{2} \qquad \text{...(2)}$$

$f(0) = c$ (given)

$f(x)$ is continuous at $x = 0 \Leftrightarrow l.h.l = r.h.l = f(0)$

$$\Leftrightarrow a + 2 = \frac{1}{2} = c$$

$$\Leftrightarrow a = \frac{1}{2} - 2 = -\frac{3}{4} \text{ and } c = \frac{1}{2}$$

Thus, we get $a = -\frac{3}{2}$

$$c = \frac{1}{2}$$

Again, no restriction is imposed on $b \Rightarrow$ it can have any non-zero finite value ($\because$ $b = 0$ makes $f(x)$ undefined)

Thus, we conclude

$$\begin{cases} a = -\frac{3}{2} \\ c = \frac{1}{2} \\ b = \text{any non-zero finite value} \end{cases}$$

6. The function

$$f(x) = \frac{\log(1+ax) - \log(1-bx)}{x}$$ is not do find at $x = 0$.

Find the value which should be assigned to f at $x = 0$ so that it is continuous at $x = 0$ is

(a) $a - b$ **(b)** $a + b$ **(c)** $\log a + \log b$ **(d)** none of these.

Solution:

$$\lim_{x\to 0} f(x) = \lim_{x\to 0}\left[\frac{1}{x}\log(1+ax) - \frac{1}{x}\log(1-bx)\right]$$

$$= \lim_{x\to 0}\left[a\cdot\frac{1}{ax}\cdot\log(1+ax) + b\cdot\frac{1}{-bx}\cdot\log(1-bx)\right]$$

$$= \lim_{x\to 0}\left[a\cdot\log(1+ax)^{\frac{1}{ax}} + b\log(1-bx)^{-\frac{1}{bx}}\right]$$

$$= a\lim_{x\to 0}\log(1+ax)^{\frac{1}{ax}} + b\lim_{x\to 0}\log(1-bx)^{-\frac{1}{bx}}$$

$$= a(1) + b(1) = a + b$$

Now, for $f(x)$ to be continuous at $x = 0$,

$$\lim_{x\to 0} f(x) = f(0)$$

$$\Rightarrow f(0) = a + b$$

Thus, the correct answer is (b).

Note: Problem (6) is a problem of a point of removable discontinuity of the function $f(x)$ which tells $x = a$ is a point of removable discontinuity of the function $f(x)$ if there is a limit $\lim_{x\to a} f(x) = b$ but either $f(x)$ is not defined at the point $x = a$ or ($f(x)$ at $x = a$) $\neq b$ and if we set $f(a) = b$, then the function $f(x)$ becomes continuous at the point $x = a$, i.e; the discontinuity is removed.

7. If $f(x) = (1+x)^{\frac{1}{x}}$, when $x \neq 0 = k$, when $x = 0$ find the value of k if $f(x)$ is continuous at $x = 0$.

Solution: $y = (1+x)^{\frac{1}{x}}$ (where $y = f(x)$)

$$\Rightarrow \log y = \log(1+x)^{\frac{1}{x}}$$

$$\Rightarrow \log y = \frac{1}{x}\log(1+x)$$

$$\Rightarrow \lim_{x\to 0}\log y = \lim_{x\to 0}\frac{1}{x}\cdot\log(1+x) = 1$$

$$\Rightarrow \lim_{x\to 0} y = e^1 = e$$

Now, $l.h.l = r.h.l = \lim_{x\to 0} y = e \qquad \text{...(i)}$

($\because x \neq 0 \Rightarrow x > 0$ and $x < 0$)

Now, $f(x)$ is given to be continuous at $x=0$

$\Rightarrow l.h.l = r.h.l = f(0)$

$\Rightarrow e = k$, i.e; $k = e$. ***Ans.***

Continuity in an Interval

1. A function f or $f(x)$ defined in the open interval (a, b) is said to be continuous in the open interval $(a, b) \Leftrightarrow f(x)$ is continuous at any arbitrary point $x = c$ where $a < c < b$. Hence, to test the continuity in an open interval (or, for all x), we simply consider an arbitrary point $x = c$ *s.t* $a < c < b = (a, b)$ and we show that $\lim_{h\to 0} f(x) = f(c)$

or, $\lim_{h\to 0} f(c+h) = \lim_{h\to 0} f(c-h) = f(c)$, where $h\to 0$ through $+ve$ values

or, $\lim_{\substack{x\to c\\ x<c}} f(x) = \lim_{\substack{x\to c\\ x>c}} f(x) = f(c)$

Note: We must remember that it is not possible to test the continuity of the function at every point of an interval, however small it may be. This is why to test the continuity of a function $f(x)$ in an interval (a, b), we always consider an arbitrary point $x = c$ *s.t* $a < c < b$ and we show that $\lim_{x\to c} f(x) = f(c)$.

This rule is applicable when the given function is not a piecewise function or when the given function is not redefined.

Explanation

1. Test the continuity of the function

$f(x) = \dfrac{x^3 - 7x + 3x - 1}{x^2 - 3x}$ in the interval

$0 < x < 3$, i.e; $x \in (0, 3)$

Solution: Let $x = a$ be any arbitrary point in $(0, 3)$ *s.t* $0 < a < 3$

Now,

$$\lim_{\substack{x\to a\\ x>a}} f(x) = \lim_{h\to 0} \frac{(a+h)^3 - 7(a+h) + 3(a+h) - 1}{(a+h)^2 - 3(a+h)}$$

(putting $x = a + h$)

$$= \frac{a^3 - 7a^2 + 3a - 1}{a^2 - 3a} \quad \ldots\text{(i)}$$

Similarly, $\lim_{\substack{x\to a\\ x<a}} f(x) = \dfrac{a^3 - 7a^2 + 3a - 1}{a^2 - 3a}$ ...(ii)

and $f(a) = \dfrac{a^3 - 7a^2 + 3a - 1}{a^2 - 3a}$ ... (iii)

Hence, (i), (ii) and (iii)

$\Rightarrow \lim_{\substack{x\to a\\ x>a}} f(x) = \lim_{\substack{x\to a\\ x<a}} f(x) = f(a) \Rightarrow$ the given function $f(x)$ is continuous at $x = a$ but 'a' is any arbitrary point in (0,3), so the given function $f(x)$ is continuous in $0 < x < 3$.

Or, alternatively,

$f(a) = \dfrac{a^3 - 7a^2 - 3a - 1}{a^2 - 3a}$

$\lim_{x\to a} f(x) = \dfrac{a^3 - 7a^2 - 3a - 1}{a^2 - 3a}$

Hence, $f(a) = \lim_{x\to a} f(x) \Rightarrow$ continuity of $f(x)$ at any arbitrary point $x = a \Rightarrow$ continuity of $f(x)$ in the given interval (0, 3).

2. Some-times taking some points in the given interval (open or closed), we test the continuity of the function at each such point separately belonging to the given interval (open or closed). This is the case when the given function is a piecewise function, i.e; if a function is defined by different formulas for different ranges of values of x, then there is a possibility of discontinuity at the values where the two ranges of x meet. Thus if we have one expression $f_1(x)$ for $x \geq a$ and another expression $f_2(x)$ for $x < a$, then the probable point of discontinuity of the function given to us is $x = a$. Similarly, if we have one expression $f_3(x)$ for $x \leq a'$ and another expression $f_4(x)$ for $x > a'$, then the probable point of the discontinuity of the function $f(x)$ given to us is $x = a'$.

Explanation

1. Consider the function $f(x)$ defined as under

$f(x) = 0$ for all values of $x > 1$

$f(x) = 1$ for all values of $x < 1$

$f(x) = \frac{1}{2}$ for $x = 1$

where, we can see that only probable point of the function $f(x)$ at which it may be discontinuous is $x = a$ since $f(x)$ (or, the value of the function 'f') changes its expression in the neighbourhood of $x = 1$.

2. Consider the function $f(x)$ defined as under

$f(x) = (2x - 1)$, when $x < 0$

$f(x) = 2x$, when $x \geq 0$

where by inspecting the behaviour of the given function $f(x)$, we can see that only probable point of $f(x)$ at which it may be discontinuous is $x = 0$ since the function $f(x)$ (or, the value of the function 'f') changes its expression in the neighbourhood of $x = 0$.

3. The functions which we face in elementary applications of the calculus are usually either continuous for values of x or have discontinuities only for a number of values which makes the function undefined/imaginary/infinite.

Explanation

(i) $y = \frac{1}{x} \Rightarrow$ this function is discontinuous at $x = 0$.

(ii) $y = \tan x \Rightarrow$ this function is discontinuous at $x = \frac{\pi}{2}$.

(iii) $y = \frac{x}{(x^2 - 4)} \Rightarrow$ this function is discontinuous at $x = \pm 2$.

4. If the continuity fails to exist for some value of x between a and b in case the interval is open, we say that the function is discontinuous in (a, b).

5. If the continuity fails to exist for some value of x between a and b (including a and b) in case the interval is closed, we say that the function is discontinuous in $[a, b]$.

Solved Examples

1. At what points of the interval, shown against each function are the following functions discontinuous?

(i) $f(x) = x$ if $x \neq 0$

$f(0) = 1$, in the interval $[-1, 1]$

(ii) $f(x) = 4x + 7$ for $x \neq 2$

$f(2) = 3$, in the interval $[-4, 4]$

(iii) $f(x) = \frac{9x^2 - 16}{27x^3 - 64}$, when $x \neq \frac{4}{3}$

$f\left(\frac{4}{3}\right) = \frac{2}{3}$, in the interval $[-1, 3]$

Solutions: **(i)** By inspecting the behaviour of the given function, we can see that the only probable point of the interval $[-1, 1]$ at which the given function may be discontinuous is $x = 0$ because the function $f(x)$ changes its expression in the neighbourhood of

$x = 0$ $\left[\begin{array}{l}\text{i.e; when } x \neq 0, \text{ the value of} \\ \text{the function } f(x) = x \text{ and} \\ f(0) = 1\end{array}\right]$

Thus, $l.h.l = r.h.l = \lim_{x \to 0} f(x) = \lim_{x \to 0} x = 0$

$f(0) = 1$ (given)

(i) and (ii) $\Rightarrow l.h.l = r.h.l \neq f(0) \Rightarrow f$(or, $f(x)$) is discontinuous at $x = 0$.

(ii) we can see that only probable point of the interval $[-4, 4]$ at which the function may be discontinuous is $x = 2$ since the value of the function f(or, $f(x)$) changes its expression in the neighbourhood of $x = 2$ $[\because f(x)_{x \neq 2} = 4x + 3$ and $(f(x))_{x=2} = 3]$

Thus, $l.h.l = r.h.l = \lim_{x \to 2} (4x + 7) = 8 + 7 = 15$

$[f(x)]_{x=2} = f(2) = 3$

$\therefore l.h.l = r.h.l \neq f(2) \Rightarrow f$ (or, $f(x)$) is discontinuous at $x = 2$.

(iii) The only probable point of discontinuity of the function is $x = \frac{4}{3}$ since the function f or $f(x)$ changes its expression in the neighbourhood of $x = \frac{4}{3}$, i.e.

$$\left[f(x)\right]_{x \neq \frac{4}{3}} = \frac{9x^2 - 16}{27x^3 - 64}$$

$$\left[f(x)\right]_{x = \frac{4}{3}} = \frac{2}{3}$$

Thus $l.h.l = r.h.l = \lim_{x \to \frac{4}{3}} \frac{9x^2 - 16}{27x^3 - 64}$

$$= \lim_{x \to \frac{4}{3}} \frac{(3x-4)(3x+4)}{(3x-4)\left(9x^2 + 12x + 16\right)}$$

$$= \lim_{x \to \frac{4}{3}} \frac{3x+4}{9x^2 + 12x + 16}$$

$$= \frac{3\left(\frac{4}{3}\right) + 4}{9\left(\frac{4}{3}\right)^2 + 12\left(\frac{4}{3}\right) + 16} = \frac{8}{48} = \frac{1}{6}$$

But $f\left(\frac{4}{3}\right) = \frac{2}{3}$

Hence, $l.h.l = r.h.l \neq f\left(\frac{4}{3}\right)$

$\Rightarrow \lim_{x \to \frac{4}{3}} f(x) \neq f\left(\frac{4}{3}\right)$

$\Rightarrow f$ (or, $f(x)$) is not continuous at $x = \frac{4}{3}$

2. Discuss the continuity of f in [0, 2] if

$f(x) = x + 1,\ 0 \le x < 1$

$= 2x + 1,\ 1 \le x \le 2$

Solution: $l.h.l = \lim_{x \to 1-} f(x) = \lim_{x \to 1} (x+1) = 1 + 1 = 2$

$r.h.l = \lim_{x \to 1+} f(x) = \lim_{x \to 1} (2x+1) = 2 \times 1 + 1$

$= 2 + 1 = 3$

$f(1) = 2 + 1 = 3$

$\therefore f(x)$ is discontinuous at $x = 1$.

Therefore, $f(x)$ is discontinuous in the given closed interval [0, 2].

3. Show that the function

$$f(x) = \begin{cases} 2x + 3, \text{ when } -3 < x < -2 \\ x - 1, \text{ when } -2 \le x < 0 \\ x + 2, \text{ when } 0 < x < 1 \end{cases}$$

is discontinuous in the open interval (–3, 1).

Solution: $l.h.l = \lim_{\substack{x \to -2 \\ x < -2}} (2x+3) = 2 \times (-2) + 3$

$= -4 + 3 = -1$

$r.h.l = \lim_{\substack{x \to -2 \\ x > -2}} (x-1) = -2 - 1 = -3$

Thus, ($l.h.l \neq r.h.l$ at $x = -2$) $\Rightarrow f(x)$ is discontinuous is [–3, 1]

Note: There is no need to test the continuity at $x = 0$ where

$l.h.l = \lim_{\substack{x \to 0 \\ x < 0}} f(x) = \lim_{x \to 0} (x-1) = -1$

$r.h.l = \lim_{\substack{x \to 0 \\ x > 0}} f(x) = \lim_{x \to 0} (x+2) = 0 + 2 = 2$

Hence, $l.h.l \neq r.h.l$ at $x = 0$ which means $f(x)$ is discontinuous at $x = 0$ and so discontinuous in the given open interval (–3, 1).

Continuity at the End Points of a Closed Interval

Definition: Set a function $f(x)$ be defined on (or, over or, in) the closed interval $a \le x \le b = [a, b]$ where a = left end point and b = right end point.

1. The function $f(x)$ is said to be continuous at the left end point 'a' of a closed interval $a \le x \le b$ iff

$r.h.l$ of $f(x)$ at $x = a$ is $= f(a)$, i.e., $\lim_{\substack{x \to a \\ x > a}} f(x) = f(a)$

/ $\lim_{x \to a^+} f(x) = f(a)$, where $f(a)$ = value of the function at $x = a$.

2. The function $f(x)$ is said to be continuous at the right end point 'b' of a closed interval $a \le x \le b$ iff $l.h.l$ of $f(x)$ at $x = b$ is $f(b)$, i.e; $\lim_{\substack{x \to a \\ x < a}} f(x) = f(b)$ / $\lim_{x \to b^-} f(x) = f(b)$, where $f(b)$ = value of the function at $x = b$.

Note:

1. Continuity at the left end point of closed interval is required only when a given function is not defined for $x < a$, where a = left end point $\Rightarrow$ If we are not provided $f(x) = \phi(x)$ for $x < a$ and we are provided $f(x) = \phi_1(x)$ for $x \ge a \Rightarrow$ we are required to test the continuity at $x = a = l.e.p$ for which only $r.h.l$ is required to find out.

2. Continuity at the right end point of a closed interval is required only when a given function is not defined for $x > b$, where b = right end point $\Rightarrow$ if we are not provided $f(x) = \phi(x)$ for $x > b$ and we are provided $f(x) = \phi_1(x)$ for $x \le b \Rightarrow$ we are required to test the continuity at $x = b = r.e.p$ for which only $l.h.l$ is required to find out. This is why we do not need to find out the right hand limit of the given function.

Aid to memory
$r.e.p \xrightarrow{\text{needs}} l.h.l$
$l.e.p \longrightarrow r.h.l$

Now, we come to the model of the questions which are provided to us. Generally the model of the question is the following:

1. $f(x) = \begin{cases} f_1(x), \text{ when } a \le x \le b \\ f_2(x), \text{ when } b \le x \le c \end{cases}$

and the interval is $[a, c]$

2. $f(x) = \begin{cases} f_1(x), \text{ when } a < x < b \\ f_2(x), \text{ when } b \le x \le c \end{cases}$

and the interval is $(a, c]$

3. $f(x) = \begin{cases} f_1(x), \text{ when } a \le x < b \\ f_2(x), \text{ when } b \le x \le c \end{cases}$

and the interval is $[a, c]$

4. $f(x) = \begin{cases} f_1(x), \text{ when } a \le x < b \\ f_2(x), \text{ when } b \le x < c \end{cases}$

and the interval is $[a, c)$

In all above types of problems 'b' may be regarded as the common point where the two ranges of values of the independent variable x meet when a given function $f(x)$ is defined by various formulas for different ranges of x as well as a and c may be regarded as the end point of a closed interval or semi (or, half) open and semi closed or semi closed and semi open interval as $(a, c]$ or $[a, c)$.

Now, a question arises how to test the continuity at $x = a, b, c$ when the given function is a piecewise function.

Question: How to test the continuity at $x = b$ = common point where the two ranges of x meet?

Answer: Continuity at the common point where the two ranges of x meet requires to find out $l.h.l$, $r.h.l$ and the value of the function at $x = b$ for which we should consider both given functions $f_1(x)$ and $f_2(x)$ against which the restriction.

$x > b$ and $x \le b$

or, $x \ge b$ and $\le b$

or, $x > b$, $x < b$ and $x = b$

or, $x \ge b$ and $x < b$ is imposed.

Question 2: How to test the continuity at $x = l.e.p$ of $[a, c] = a \le x \le c$?

Answer: Continuity at $x = l.e.p$ of $[a, c]$ or, $[a, c)$ requires to find the $r.h.l$ of the function $f(x) = f_1(x)$ against which $x \ge a$ is written and the value of the function $f(x)$ at $x = a$ and the other function $f(x) = f_2(x)$ against which the restriction $x < a$ is imposed is not defined (or, given). This is why $l.h.l$ of $f(x) = f_2(x)$ as $x \to a$ is not required which means that $\lim_{x \to a^-} f_2(x)$ is not required to find out. While testing the continuity at $l.e.p$ of the closed interval $[a, c]$ or semi closed and semi open interval $[a, c)$.

Question 3: How to test the continuity at $x = r.e.p$ of $[a, c] = a \le x \le c$?

Answer: Continuity at $x = r.e.p$ of $[a, c]$ or, $(a, c]$ requires to find out $l.h.l$ of the function $f(x) = f_2(x)$ against which $x \le c$ is written and the value of the function $f(x) = f_2(x)$ at $x = c$ and the other function $f(x) = f_1(x)$ against which $x > c$ is imposed is not defined (or, given). This is why $r.h.l$ of $f(x) = f_1(x)$ as $x \to c$ is not required which means $\lim_{x \to c^+} f_1(x)$ is not required to find out while testing the continuity at $r.e.p$ of the closed interval $[a, c]$ or semi open and semi closed $(a, c]$ interval.

Solved Examples

1. A function $f(x)$ is defined as follows

$f(x) = 0$, when $x = 0$

$= 3x - 1$, when $0 < x < 1$

$= 2x$, when $x = 1$

Test the continuity at $x = 0$ and $x = 1$.

Solution: At $x = 0$

$r.h.l = \lim_{x \to 0} (3x - 1) = 3 \cdot 0 - 1 = -1$ ...(i)

$[f(x)]_{x=0} = f(0) = 0$ ...(ii)

(i) and (ii) $\Rightarrow r.h.l \neq$ value of the function at $x = 0$ $\Rightarrow$ discontinuity of the given function at the left end point of the closed interval $[0, 1]$ at $x = 1$

$l.h.l = \lim_{x \to 1} (3x - 1) = 2$ ...(i)

$[f(x)]_{x=1} = f(1) = [2x]_{x=1} = 2 \times 1 = 2$ ...(ii)

(i) and (ii) $\Rightarrow l.h.l =$ value of the function $\Rightarrow$ continuity of the given function $f(x)$ at the right end point of the closed interval $[0, 1] = 0 \leq x \leq 1$.

2. Test the continuity of $f(x)$ at $x = 2$, when

$f(x) = x^2 + x + 1,\ 0 \leq x \leq 1$

$= x^2 + 2,\ 1 \leq x \leq 2$

Solution: Here $f(x)$ is defined on the closed interval $[0, 2] = 0 \leq x \leq 2$ Hence, to test the continuity of $f(x)$ at $x = 2$ only $l.h.l$ is required since $x = 2$ is a right end point of the closed interval [0. 2].

$l.h.l = \lim_{x \to 2} \left(x^3 + 2\right) = 8 + 2 = 10$ ...(i)

$[f(x)]_{x=2} = (x^3 + 2) = 8 + 2 = 10$ ...(ii)

(i) and (ii) $\Rightarrow l.h.l = f(2) \Rightarrow$ continuity of the given function at the right end point of the closed interval [0, 2].

Note: $f(x)$ is continuous at $x = 0$ as $r.h.l$ at $x = 0$ is $= f(0)$.

Also $\lim_{x \to 1-0} f(x) = \lim_{x \to 1+0} f(x) = f(1)$

$\therefore f(x)$ is continuous at $x = 0$.

Hence $f(x)$ is continuous in [0, 2]

3. A function $f(x)$ is defined by

$f(x) = (x + 2),\ 0 \leq x \leq 2$

$= \sqrt{8x},\ 2 \leq x \leq 4$

test the continuity at $x = 0, 2, 4$

Solution: **(1)** At $x = 2$

$l.h.l = \lim_{x \to 2} (x + 2) = 4$ ...(i)

$[f(x)]_{x=2} = [(x + 2)]_{x=2} = \left[\sqrt{8x}\right]_{x=2} = 4$ ...(ii)

$r.h.l = \lim_{x \to 2} \sqrt{8x} = \sqrt{8 \times 2} = 4$ ...(iii)

(i), (ii) and (iii) $\Rightarrow l.h.l = r.h.l =$ value of the function at $x = 2$.

(2) At $x = 0 =$ left end point of the closed interval [0, 4]

$r.h.l = \lim_{x \to 0} (x + 2) = 2$ ...(i)

$[f(x)]_{x=0} = [x + 2]_{x=0} = 2$ ...(ii)

(i) and (ii) $\Rightarrow$ the function $f(x)$ is continuous at $x = 0 =$ left end point of the closed interval [0, 4].

(3) At $x = 4 =$ right end point of the closed interval [0, 4]

$l.h.l = \lim_{x \to 4} \left(\sqrt{8x}\right) = \sqrt{8 \times 4} = 4\sqrt{2}$

$[f(x)]_{x=4} = \left[\sqrt{8x}\right]_{x=4} = \sqrt{8 \times 4} = \sqrt{32} = 4\sqrt{2}$

(1) and (2) $\Rightarrow$ continuity at $x = 4$ since $l.h.l =$ value of the function at $x = 4$

4. A function f is defined in the following way:

$f(x) = 0$, when $x = 0$

$= 2x - 1$, when $0 < x < 1$

$= 2 - x$, when $x \geq 1$

is $f(x)$ continuous at $x = 0$ and $x = 1$?

Solution: At $x = 0$

We inspect that $f(x)$ is not defined for $x < 0$ since $f(x) = \phi(x) =$ an expression in x when $x < 0$ is not provided.

$r.h.l = f(0+0) = \lim_{x \to 0} f(x) = \lim_{x \to 0} (2x - 1) = -1$

$[f(x)]_{x=0} = 0 \text{ (given)} = f(0)$

(i) and (ii) $\Rightarrow f(x)$ is not continuous at $x = 0$ = left end point (*l.e.p*) of the closed interval [0, 1] at $x = 1$

$l.h.l = f(1-0) = \lim_{x \to 1} (2x - 1) = 2 - 1 = 1$

$r.h.l = f(1+0) = \lim_{x \to 1} (2 - x) = 2 - 1 = 1$

$[f(x)]_{x=1} = [(2-x)]_{x=1} = 2 - 1 = 1$

Hence, $f(1-0) = f(1+0) = f(1) \Rightarrow$ continuity at $x = 1$.

Facts to Know

We end this chapter by mentioning some facts about continuity which the students must remember.

1. *Criterion of continuity at a real number*: The function $y = f(x)$ is continuous at the point $x = a$ provided (i) it is defined at this point (ii) there is a limit namely $\lim_{x \to a} f(x) = L$ (iii) this limit is equal to the value of the function at the point (or, real number) $x = a$.

Or, alternatively,

A function $y = f(x)$ is continuous at the real number $x = a$ / $x = a$ is a point of continuity of $f(x) \Leftrightarrow$ (i) $f(a)$ exists (ii) $\lim_{x \to a} f(x)$ exists (iii) $\lim_{x \to a} f(x) = f(a)$

If one or more of these three conditions fail to hold good at $x = a$, we say that $f(x)$ is discontinuous at $x = a$ or $x = a$ is a point of discontinuity of the function $y = f(x)$.

2. *Criterion of continuity in an interval*: The function which has continuity at every point of an interval is continuous throughout that interval.

3. *Criterion of continuity in an open interval*: The function $y = f(x)$ is continuous in an open interval (a, b) provided $f(x)$ is continuous at each point of (a, b) and the function $y = f(x)$ is discontinuous in (a, b) if $f(x)$ is discontinuous atleast at one point of (a, b).

Or, alternatively,

The function $y = f(x)$ is continuous in an open interval (a, b) provided $\lim_{x \to c} f(x) = f(c)$ where $x = c$ is an arbitrary point (or, number) *s.t.*

4. *Criterion of continuity in a closed interval*: The function $y = f(x)$ is continuous in a closed interval $[a, b] \Leftrightarrow$ (i) $f(x)$ is continuous at each point of (a, b) (ii) $\lim_{x \to a^+} f(x) = f(a)$ (iii) $\lim_{x \to b} f(x) = b$ hold good .

Or, alternatively,

If the domain of a real valued function (or, real function) f or $f(x)$ is a closed interval $[a, b]$, then f or $f(x)$ is continuous in (or, over/on/) closed interval $\Leftrightarrow$ (i) $\lim_{x \to c} f(x) = f(c)$ where $x = c$ is any arbitrary point *s.t* $a < c < b$ (ii) $\lim_{\substack{x \to a \\ x > a}} f(x) = r.h.l$ at the *l.e.p* = $f(a)$ = value of the function at the *l.e.p* (i.e; left end point) (iii) $\lim_{\substack{x \to a \\ x < b}} f(x) = l.h.l$ at the *r.e.p* = $f(b)$ = value of the function $y = f(x)$ at *r.e.p* (i.e; right end point)

5. *Criterion at the left end point of a closed interval* $[a, b]$: $f_1(x)$ is continuous at $x = a$ = left end point of the closed interval

$$[a, b] \Leftrightarrow \lim_{x \to a^+} f_1(x) = \lim_{\substack{x \to a \\ x > a}} f_1(x) = f(a) \Leftrightarrow$$

($r.h.l$ at $x = a$) = value of the function at $x = a$.

6. *Criterion at the right end point of a closed interval* $[a, b]$: $f_1(x)$ is continuous at $x = b$ = right end point of the closed interval

$$[a, b] \Leftrightarrow \lim_{x \to b^-} f_1(x) = \lim_{\substack{x \to b \\ x < b}} f_1(x) = f(b) \Leftrightarrow$$

($l.h.l$ at $x = b$) = value of the function at $x = b$.

N.B.: **(i)** To test the continuity at right end point of a closed interval, $l.h.l = \lim_{\substack{x \to a \\ x < a}} f_2(x)$ is not required since the function $f_2(x)$ is not given (or, defined) for $x < a$ where $y = f(x)$ is a piecewise function defined in a closed interval which is divided into a finite number of non-overlapping sub intervals over each of which different functions are defined.

(ii) To test the continuity at the left end point of a closed interval, $r.h.l = \lim\limits_{\substack{x \to b \\ x > b}} f_2(x)$ is not required to find out since the function $f_2(x)$ is not given (or, defined) for $x > b$ where $y = f(x)$ is a piecewise function defined in a closed interval which is divided into a finite number of non-overlapping sub intervals over each of which different functions are defined.

(iii) If the interval is closed one, the limit at the left end will mean right hand limit and the limit at the right end will mean the left hand limit.

6. If functions $f_1(x)$ and $f_2(x)$ are continuous in an interval (a, b) or $[a, b]$, then the function

(i) $c_1 f_1(x) + c_2 f_2(x)$

(ii) $f_1(x) \times f_2(x)$

(iii) $\dfrac{f_1(x)}{f_2(x)}$, provided $f_2(x) \neq 0$ for any value of x belonging to the interval (open or closed), are also continuous in the open interval (a, b) or in the closed interval $[a, b]$.

7. The continuity of the function can be used to calculate its limits which means if the function $y = f(x)$ is continuous at the point $x = a$, then, in order to find out its limit $\lim\limits_{x \to a} f(x)$, it is sufficient to calculate the value of the function at $x = a$ because $\lim\limits_{x \to a} f(x) = f(a)$ when $y = f(x)$ is continuous at $x = a$.

8. While testing the continuity of a redefined function we use $(l.h.l = r.h.l \text{ at } x = a) = \lim\limits_{x \to a} f_1(x)$, where $f(x) = f_1(x)$ for $x \neq a$ provided $f_1(x)$ is not a mod function (or a combination or composition of mod function) or a greatest integer function.

9. When a given function is a piecewise function, then to find the $l.h.l$ at a point $x = a$, we consider a function defined in an interval (or, sub interval) containg all those values of x which are less than (or, less than or equal to) a and to find the $r.h.l$ at a point $x = a$, we consider another function defined in an interval (or sub interval) containing all those values of x which are greater than (or, greater than or equal to) a.

Problems

Type1: When the function is defined.

Exercise 5.1

Examine each function for the continuity defined by

1. $f(x) = \dfrac{x^2 - 9}{x - 3}$, when $x \neq 3$

$f(x) = 6$, when $x = 3$ [***Ans***: cont. at $x = 3$]

2. $f(x) = \dfrac{x^2 + 2x - 9}{x - 1}$, when $x \neq 1$

$f(x) = 4$, when $x = 1$ [***Ans***: discont. at $x = 0$]

3. $f(x) = 4x + 3$, when $x \neq 4$

$f(x) = 3x + 7$, when $x = 4$ [***Ans***: cont. at $x = 4$]

4. $f(x) = (1 + x)^{\frac{1}{x}}$, when $x \neq 0$

$f(x) = e$, when $x = 0$ [***Ans***: cont. at $x = 0$]

5. $f(x) = \dfrac{\cos ax - \cos bx}{x^2}$, when $x \neq 0$

$f(x) = \dfrac{b^2 - a^2}{2}$ when $x = 0$ [***Ans***: cont. at $x = 0$]

6. $f(x) = \dfrac{|x^3|}{x}$, when $x \neq 0$

$f(x) = 0$, when $x = 0$ [***Ans***: cont. at $x = 0$]

7. $f(x) = x^2 \sin\left(\dfrac{1}{x}\right)$, when $x \neq 0$

$f(0) = 0$ [***Ans***: cont. at $x = 0$]

8. $f(x) = \dfrac{1 - \cos x}{x^2}$, when $x \neq 0$

$f(x) = \dfrac{1}{2}$ when $x = 0$ [***Ans***: cont. at $x = 0$]

9. $f(x) = \begin{cases} \dfrac{1-\cos mx}{1-\cos nx}, & \text{when } x \neq 0 \\ \dfrac{m^2}{n^2}, & \text{when } x = 0 \end{cases}$

[*Ans*: cont. at $x = 0$]

10. $f(x) = \begin{cases} \dfrac{\sin^{-1} x}{x}, & \text{when } x \neq 0 \\ 1, & \text{when } x = 0 \end{cases}$

[*Ans*: cont. at $x = 0$]

11. $f(x) = \begin{cases} x \sin\left(\dfrac{1}{x}\right), & \text{when } x \neq 0 \\ 0, & \text{when } x = 0 \end{cases}$

[*Ans*: cont. at $x = 0$]

12. $f(x) = \begin{cases} \dfrac{\tan x - \sin x}{x^3}, & \text{when } x \neq 0 \\ \dfrac{1}{2}, & \text{when } x = 0 \end{cases}$

[*Ans*: cont. at $x = 0$]

13. $f(x) = \dfrac{\sin 3x - 3\sin x}{(x-\pi)^3}$, when $x \neq \pi$

$f(x) = 4$, when $x = \pi$ [*Ans*: cont. at $x = \pi$]

14. $f(x) = \dfrac{x \tan x}{1 - \cos x}$, when $x \neq 0$

$f(x) = 2$, when $x = 0$ [*Ans*: cont. at $x = 0$]

15. $f(x) = \dfrac{\sqrt{1+x^2} - \sqrt{1-x^2}}{x^2}$ when $x \neq 0$

$f(x) = 2$, when $x = 0$ [*Ans*: discont. at $x = 0$]

16. Show that $f(x) = |x - 5|$ is continuous at $x = 5$.

17. Show that

$$f(x) = \begin{cases} \dfrac{|x-a|}{x-a}, & x \neq a \\ 1, & x = a \end{cases}$$

is discontinuous at $x = 1$.

18. Examine the continuity of $f(x) = |x - b|$ at $x = b$.

[*Ans*: cont. at $x = b$]

19. Examine the continuity of the function $f(x)$ at $t = \dfrac{\pi}{2}$

$f(t) = \dfrac{\cos t}{\dfrac{\pi}{2} - t}$, $t \neq \dfrac{\pi}{2}$

$f(t) = 1$, $t = \dfrac{\pi}{2}$ [*Ans*: continuous at $= \dfrac{\pi}{2}$]

20. Discuss the continuity of the function

$$\left.\begin{aligned} f(x) &= \frac{\tan 7x}{\sin 4x}, \text{ for } x \neq 0 \\ &= \frac{7}{4}, \text{ for } x = 0 \end{aligned}\right\} \text{ at } x = 0$$

[*Ans*: cont. at $x = 0$]

Type 2: Problems on piecewise function:

Exercise 5.2.1

Examine the continuity of the function $f(x)$ defined by

1. $f(x) = 2x - 1$, when $x < 2$
   $f(x) = x^2 - 4x + 5$, when $x \geq 2$
   [*Ans*: discont. at $x = 2$]
2. $f(x) = x - 3$. when $x < 4$
   $f(x) = 5 - x$, when $x \geq 4$
   [*Ans*: cont. at $x = 4$]
3. $f(x) = 3 - 2x$, when $x \leq 2$
   $f(x) = x - 1$, when $x > 2$
   [*Ans*: discont. at $x = 2$]
4. $f(x) = \dfrac{x^3}{16}$, when $x < 4$
   $f(x) = 1 - 2x - x^2$, when $x \geq 4$
   [*Ans*: discont. at $x = 4$]
5. $f(x) = x^2 - 2x + 3$, when $x \leq 1$
   $f(x) = x + 1$, when $x > 1$ [*Ans*: cont. at $x = 1$]
6. $f(x) = x^2$, when $0 < x < 1$
   $f(x) = x$, when $1 \leq x < 2$

$f(x) = -6 + x^3$, when $2 \le x < 3$

[*Ans*: cont. at $x = 1, 2$]

7. $f(x) = 1$, when $x < 0$
   $f(x) = 0$, when $x = 0$
   $f(x) = -1$, when $x > 0$ [*Ans*: discont. at $x = 0$]

8. $f(x) = \dfrac{x^2}{2}$, when $0 \le x \le 1$

   $f(x) = 2x^2 - 3x + \dfrac{3}{2}$, when $1 \le x \le 2$

   [*Ans*: cont. at $x = 1$]

9. $f(x) = 5x - 4$, when $0 \le x \le 1$

   $f(x) = 4x^3 - 3x$, when $1 < x \le 2$

   [*Ans*: cont. at $x = 1$]

10. $f(x) = x$, when $0 \le x < \dfrac{1}{2}$

    $f(x) = 1 - x$, when $\dfrac{1}{2} \le x < 1$

    [*Ans*: cont. at $x = \dfrac{1}{2}$]

11. $f(x) = x^3 + 1$, when $0 \le x < 1$

    $f(x) = 3x^2 - 1$, when $1 \le x \le 2$

    [*Ans*: cont. at $x = 1$]

Exercise 5.2.2

1. If $f(x) = 0$, when $x = 0$

   $f(x) = 1 - x$, when $0 < x < \dfrac{1}{2}$

   $f(x) = \dfrac{1}{2}$, when $x = \dfrac{1}{2}$

   then test the continuity of $f(x)$ at $x = 0$ and $x = \dfrac{1}{2}$.

   [*Ans.* cont. at $x = 0$ and $x = \dfrac{1}{2}$]

2. If $f(x) = \dfrac{x^2 - 4x + 3}{x - 4}$, when $0 \le x \le 5$

   $f(x) = \dfrac{x^2 + 1}{x - 1}$, when $5 < x \le 7$

   then test the continuity of $f(x)$ at $x = 5$.

   [*Ans*: discont. at $x = 5$]

3. If $f(x) = x^3 + 1$, when $0 \le x < 1$

   $f(x) = 3x^2 - 1$, when $1 \le x \le 2$

   then test the continuity of $f(x)$ at $x = 0$, 1 and 2.

   [*Ans*: cont. at $x = 0, 1, 2$]

4. The function $f(x)$ is defined as under

   $$\text{if } f(x) = \begin{cases} -x, \text{ for } x \le 0 \\ x, \text{ for } 0 < x < 1 \\ 2 - x, \text{ for } x \ge 1 \end{cases}$$

   is the function $f(x)$ continuous at $x = 0$ and $x = 1$

   [*Ans*: cont. at $x = 0$ and $x = 1$]

5. Examine the continuity of $f(x)$ at $x = 0$ and $x = 1$ if

   $$f(x) = \begin{cases} 1, \text{ when } x = 0 \\ x - 1, \text{ when } 0 < x < 1 \\ 0, \text{ when } x = 1 \end{cases}$$

   [*Ans*: discont. at $x = 0$ and cont. at $x = 1$]

6. Examine the continuity of $f(x)$ at $x = 1$ and $x = 2$ if

   $$f(x) = \begin{cases} x^2, \text{ when } 0 \le x < 1 \\ 2x - 1, \text{ when } 1 \le x < 2 \\ x + 1, \text{ when } 2 \le x \end{cases}$$

   [*Ans*: cont. at $x = 1$ and $x = 2$]

7. The function f is defined in the following way

   $$f(x) = \begin{cases} 3 + 2x, \text{ when } -\dfrac{3}{2} \le x < 0 \\ 3 - 2x, \text{ when } 0 \le x < \dfrac{3}{2} \\ -3 - 2x, \text{ when } x \ge \dfrac{3}{2} \end{cases}$$

   Prove that $f(x)$ is continuous at $x = 0$ and discontinuous at $x = \dfrac{3}{2}$.

Problems on greatest integer function

Exercise 5.2.3

1. Test the continuity of the function $f(x)$ at $x = \frac{2}{3}$ if $f(x) = [x]$, where $[x]$ is the greatest integer function.
2. If $f(x) = (x - [x])$, where $[x]$ is the greatest integer function, test the continuity of $f(x)$ at $x = \frac{3}{2}$.
3. Test the continuity of the function $f(x)$ at $x = 4$ $f(x) = \frac{[x]}{x-2}$.
4. Show that $f(x) = x - [x]$ is discontinuous at $x = -1$.
5. Test for continuity of $f(x) = [x]$ at $x = 2.5$.

Answers:

1. Cont. at $x = \frac{2}{3}$
2. Limit $= \frac{1}{2}$, find $f\left(\frac{3}{2}\right)$ and examine whether limit $=$ value $\left(\text{cont. at } \frac{3}{2}\right)$.
3. $l.h.l = 2$ and $r.h.l = \frac{3}{2}$; (discont. at $x = 4$)
5. ($l.h.l = r.h.l$ at $x = 2.5$) $= 2$; find $f(2.5)$ and examine whether limit $=$ value.

Type 3: Problems based on finding the value of a function at a point where the function is continuous.

Exercise 5.3

1. $f(x) = \frac{x-a}{\sqrt[3]{x} - \sqrt[3]{a}}$, $x \neq a$ is continuous at $x = a$. Find $f(a)$. [***Ans***: $3a^{\frac{2}{3}}$]
2. $f(x) = \frac{x \tan x}{1 - \cos x}$, $x \neq 0$ is continuous at $x = 0$. Find $f(0)$. [***Ans***: 2]
3. $f(x) = \frac{1 - \sin x}{(\pi - 2x)^2}$, $x \neq \frac{\pi}{2}$ is continuous at $x = \frac{\pi}{2}$. Find $f\left(\frac{\pi}{2}\right)$. [***Ans***: $\frac{1}{8}$]
4. $f(x) = \frac{\sin x - 1}{\cos^2 x}$, $x \neq \frac{\pi}{2}$ is continuous at $x = \frac{\pi}{2}$. Find $f\left(\frac{\pi}{2}\right)$. [***Ans***: $-\frac{1}{2}$]
5. $f(x) = \frac{x^3 - a^3}{x^2 - a^2}$, $x \neq a$ is continuous at $x = a$. Find $f(a)$. [***Ans***: $\frac{3a}{2}$]
6. A function $f(x)$ is defined by

   $f(x) = \frac{1 - \cos x}{x^2}$, when $x \neq 0$

   $f(x) = A$, when $x = 0$

   Find A so that $f(x)$ is continuous at $x = 0$. [***Ans***: $A = \frac{1}{2}$]
7. If $f(x)$ is continuous at $x = 3$ and is defined as

   $f(x) = \frac{x^2 - 9}{x^2 - 4x + 3} + a$, when $x < 3$

   $= 2$, when $x = 3$

   $= \frac{x^2 - 3x}{x^2 - 2x - 3} + b$, when $x > 3$

   Find a and b [***Ans***: $a = 1$ and $b = \frac{5}{4}$]
8. $f(x)$ is continuous at $x = \frac{\pi}{4}$ and is defined as

   $f(x) = \frac{\cos x - \sin x}{x - \frac{\pi}{4}} + a$, when $0 < x < \frac{\pi}{4}$

   $= 2$, when $x = \frac{\pi}{4}$

   $= x^2 + b$, when $x > \frac{\pi}{4}$

Find a and b.

[***Ans***: $a = 2 + \sqrt{2}$ and $b = 2 - \frac{\pi^2}{16}$]

9. $f(x)$ is continuous at $x = 0$, where

$f(x) = \frac{\sin kx}{x}$, for $x \neq 0$

$= 4 + x$, for $x = 0$

Find the value of k. [***Ans***: $k = 4$]

Type 4: Problems based on finding the value of a constant when the given function is continuous at a point.

Exercise 5.4

1. Find the value of k for which the functions defined below is continuous at $x = 1$.

(a) $f(x) = \frac{x^2 - 1}{x - 1}$, for $x \neq 1$

$f(1) = k$ [***Ans***: $k = 2$]

(b) $f(x) = 5x - 3k$, if $x \leq 1$

$= 3x^2 - kx$, if $x > 1$ [***Ans***: $k = 1$]

(c) $f(x) = 3x - 4k$, when $x \geq 1$

$= 2kx^2 - 3$, when $x < 1$ [***Ans***: $k = 1$]

2. If $f(x) = \frac{x^2 - 3x + 2}{x - 2}$, when $x \neq 2$

$= p$, when $x = 2$

and $f(x)$ is continuous at $x = 2$, find the value of p. [***Ans***: $p = 1$]

3. If $f(x) = \frac{a \sin 2x}{x}$, when $x > 0$

$= 2$, when $x = 0$

$= \frac{2b\left(\sqrt{1+x} - 1\right)}{x}$, when $x < 0$

and $f(x)$ is continuous at $x = 0$, find the values of a and b. [***Ans***: $a = 1$ and $b = 2$]

4. $f(x) = \frac{3x^2 - x - 2}{x - 1}$, for $x \neq 1$

$f(1) = k$

If $f(x)$ is continuous at $x = 1$, find k.

[***Ans***: $k = 5$]

5. A function $f(x)$ is defined as under

$f(x) = x^2 + A$, when $x \geq 0$

$f(x) = -x^2 - A$, when $x < 0$

what should be A so that $f(x)$ is continuous at $x = 0$. [***Ans***: $A = 0$]

Type 5: Problems based on test for continuity at one/both end points (*l.e.p* / *r.e.p* / *l.e.p* and *r.e.p* both) of a closed interval $[a, b]$ or semi open-semi closed $(a, b]$ or semi closed and semi open $[a, b)$.

Exercise 5.5

1. If $f(x) = 0$, when $x = 0$

$= \frac{1}{2} - x$, when $0 < x < \frac{1}{2}$

$= \frac{1}{2}$, when $x = \frac{1}{2}$

then test the continuity of $f(x)$ at $x = 0$ and $x = \frac{1}{2}$.

[***Ans***: discont. at $x = 0$ and $x = \frac{1}{2}$]

2. If $f(x) = x^2 + 1$, when $0 \leq x < 1$

$= 3x^2 - 1$, when $1 \leq x \leq 2$

then test the continuity of $f(x)$ at $x = 0$, 1 and 2

[***Ans***: cont. at $x = 0$, 1 and 2]

3. A function $f(x)$ is defined as follows

$f(x) = 0$, when $x = 0$

$= 3x - 1$, when $0 < x < 1$

$= 2x$, when $x = 1$

Is $f(x)$ continuous at $x = 0$ and $x = 1$?

[***Ans***: discont. at $x = 0$ and cont. at $x = 1$]

4. Test the continuity of $f(x)$ at $x = 0, 1, 2$ where

(i) $f(x) = 1 + x$, when $0 \leq x < 1$

$= 2 - x$, when $1 \leq x \leq 2$

[***Ans***: cont. at $x = 0$ and 2; discont. at $x = 1$]

(ii) $f(x) = x^2$, when $0 \le x < 1$

$= 2x - 1$, when $1 \le x < 2$

$= x + 3$, when $x \ge 2$

[***Ans***: cont. at $x = 0$ and $x = 1$ and discont. at $x = 2$]

5. A function $f(x)$ is defined as follows:

$f(x) = 0$, when $x = 0$

$= \frac{1}{2} - x$, when $0 < x < \frac{1}{2}$

$= \frac{1}{2}$, when $x = \frac{1}{2}$

$= \frac{3}{2} - x$, when $\frac{1}{2} < x < 1$

$= 1$, when $x = 1$, discuss the continuity of the function for $x = 0, \frac{1}{2}$ and 1.

[***Ans***: discont. at all points $x = 0, \frac{1}{2}, 1$]

6

Derivative of a Function

Question: What is differential Calculus?

Answer: Differential Calculus provides us rules and methods for computing the limit:

$$\lim_{\Delta x \to 0} \frac{\Delta y}{\Delta x} = \lim_{\Delta x \to 0} \frac{f(x + \Delta x) - f(x)}{\Delta x}$$, for a large class of functions where, $\Delta y = f(x + \Delta x) - f(x)$ is the increment in the value of the function $y = f(x)$ (or in dependent variable y), $\Delta x = (x + \Delta x) - x$ is the increment in independent (or in the valueof independent) variable and

$$\frac{\Delta y}{\Delta x} = \text{increment (or incremental) ratio}$$

$$= \frac{\text{increment in functional value (dependent variable)}}{\text{increment in independent variable}}$$

Question: What is the derivative of a function?

Answer: The derivative of a function *f* is that function, denoted by f', whose function (functional) value at any limit point *x* of the domain of the function *f* which is in the domain of the function *f*, denoted by $f'(x)$, is given by the limit of the incremental ratio as the increment in the independent variable tends to zero. That is in notation,

$$f'(x) = \lim_{\Delta x \to 0} \frac{f(x+\Delta x) - f(x)}{\Delta x} = \lim_{\Delta x \to 0} \frac{\Delta y}{\Delta x},$$

provided that this limit exists at each limit point *x* of the domain of the function *f* which is in the domain of the function *f* defined by the equation $y = f(x)$.

Notes:

1. The derivative f' of a function *f* is also termed as: (i) slope function (ii) derived function (iii) differential coefficient of the function *f*.
2. Instead of saying that there is a derivative f' of the function *f* at a point $x = a$ in the domain of the function *f* which is also a limit point in the domain of the function f' it is common to say that there is a derivative f' of the function *f* at a point $x = a$ in the domain of the function *f* defined by the equation $y = f(x)$.

The Domain of a Derivative

The domain of a derivative is defined with respect to different aspects:

***Definition* 1**. (In terms of limit points):
$f : D \to R$ is a function defined by $y = f(x)$, where D is a subset of reals $\Rightarrow$ domain of the derivative f' of the function $f = D \cap D' \subseteq D$, where D' is the set of all limit points of the domain D of the function *f*.

***Definition* 2**. (In terms of existance of the limit of incremental ratio):
$D(f') = \{x \in D(f) : f'(x) \text{ exists}\}$, i.e. the domain of a derived function $y = f'(x)$ is a subset of the domain of the function *f* because it (domain of f') contains all points *x* in the domain of the function *f* where $\lim_{\Delta x \to 0} \frac{\Delta y}{\Delta x}$ exists but does not contain those exceptional points where $\lim_{\Delta x \to 0} \frac{\Delta y}{\Delta x}$ does not exist.

The Range of a Derivative

The range of a derivative is the set of all values of the derivative f' (of a function f) at values of x in the domain of the function.

Thus, range of $f' = f'(D(f)) = \{f'(x) : x \in D(f)\}$.

Notes:

1. Derivatives at isolated points are not defined.
2. A function is always continuous at an isolated point of the domain of the function.
3. In fact, if the domain of a function has an isolated point, then the function would be continuous there without being differentiable.
4. If the limit point of the domain D of the function f lying in D at which derivative is sought is not mentioned, it is understood that it is required at any limit point x of the domain of the function f which is in the domain of f.
5. Instead of saying to determine the functional value $f'(x)$ of the derivative f' of a function f at $x \in D(f)$, it is common to say to determine the derivative $y' = f'(x)$ of the function $y = f(x)$

Question: What is differentiation of a function?
Answer: It is the process of finding the derivative

$$f'(x) = \lim_{\Delta x \to 0} \frac{\Delta y}{\Delta x}.$$

Question: How is the derivative of a function generally determined?
Answer: The derivative of a function of the independent variable x, say $f(x)$, is determined by the general process indicated in the limit of increment ratio as the increment in the independent variable tends to zero, i.e. indicated in:

$$\frac{dy}{dx} = \lim_{\Delta x \to 0} \frac{\Delta y}{\Delta x} \text{ or } \frac{d}{dx} f(x) = \lim_{\Delta x \to 0} \frac{f(x + \Delta x) - f(x)}{\Delta x}.$$

Question: What are the symbols to express the derivative of a function with respect to an independent variable?
Answer: D is the most common notation for indicating the operation of obtaining the derivative of a function with respect to an independent variable with the attentions: (i). Functional letter or functional value of the function is written on the right side of the symbol D. (ii) the independent variable with respect to which differentiation is to be carried out is written at the bottom and on the right side of the symbol D. Thus $D_x y$, $D_x f$, $D_x f(x)$ or $D_x (f(x))$ if $f(x)$ is the expression containing more than one term or $f(x)$ is the sum of a finite number of function.

Note: The phrase "with respect to x" is shortly written as "*w.r.t x*".

An other notation $\frac{d}{dx}$ is also in frequent use when either it is mentioned or obvious in the problem that x is the independent variable of the function $y = f(x)$.

When the symbol $\frac{d}{dx}$ is used to indicate the derivative of a function having x as its independent variable, the function or the functional value is written on the right side of the symbol $\frac{d}{dx}$. Thus: $\frac{df}{dx}, \frac{d}{dx} f(x), \frac{dy}{dx}$ or $\frac{d}{dx}(f(x))$ if $f(x)$ is the expression containing more than on term or $f(x)$ is the sum of a finite number of function.

Hence in the light of above explanation, D_x and $\frac{d}{dx}$ are the most common notations for indicating the operation of obtaining the derivative of a function with respect to x. These are notations which are prefixed to the functional letter or functional value as:

(i) $D_x y$, $D_x (y)$, $\frac{dy}{dx}$ and $\frac{d}{dx}(y)$

(ii) $D_x f$, $D_x f(x)$, $\frac{df}{dx}$ and $\frac{d}{dx} f(x)$ Lastly, the third notation employed for indicating the operation of obtaining the derivative is to put a dash or prime at the top and on the right of the functional letter as $f'(x), f'$ or y', but it is noteable that the notation f' or y' has the disadvantage of not indicating the variable with respect to which the differentiation is to be carried out. This is why $f'(x)$ or $y'(x)$ is also written if the independent variable is x.

Remark: $\frac{d}{dx}(\ )$, $D_x(\ )$ or $D(\ )$ is termed as operator which tells what operation should be carried on with the function put in the bracket to get an other function named as derivative but $D(\ )$ alone is ambiguous when the expression in x representing the functional value contains constant as well as an independent variable both put after the operator "D" as $D(x^2+x+K)$ which has a possibility or ambiguity of being which one of the letters namely x and K is constant. This is why $D(f(x))$ is also sometimes used instead of $D_x(f(x))$ when there is no ambiguity of the letter used as an independent variable.

Note: $\frac{d}{dx}$, D_x or simply D is also called differential operatir.

Therefore, there are following identical symbols frequently used:

(i) $y' = \frac{dy}{dx} = \frac{d}{dx}(y) = \frac{d}{dx}[f(x)] = \frac{d}{dx}(f(x)) = D_x f(x) = f'(x)$, when $y = f(x)$ or

(ii) $f' = \frac{df}{dx}$ to denote the derivative as an other function f' of a function f.

The Nomenclature of the Symbols

Notation	Read as
$\lim\limits_{x \to a}(\ )$	limit as *eKs* tends to a, of ( )
	or limit of ( ) as *eKs* tends to a.
$\lim\limits_{\Delta x \to 0}(\ )$	limit as delta-*eKs* tends to zero, of ( ) or limit of ( ) as delta-*eKs* tends to zero.
$D_x(\ )$	x-derivative of ( ) or derivative
or $\frac{d}{dx}(\ )$	of ( ) with respect to x.
f'	f prime or f dash

(Contd.)

Notation	Read as
y'	y prime or y dash
$f'(x)$	f prime of x or f dash of *eKs*.
$D_x f$	*dee - eKs* of f
$D_x f(x)$	*dee - eKs* of f of *eKs*
$\frac{df}{dx}$	*dee - dee - eKs* of f or df over dx
$\frac{d}{dx} f(x)$	*dee–dee–eKs* of f of x or *dee–f* of *eKs* over *dee - eKs*
$\frac{dy}{dx}$	*dee - y* over *dee - eKs*

Note: The notation $[f'(x)]\ x = a$ or $(f(x))\ x = a$ or $f'(a)$ signifies the value of the derivative f' of the function f at $x = a \in D(f)$.

Question: What do you mean by *ab-initio* differentiation?

Answer: Differentiation *ab-initio* (or *ab-initio* differentiation), differentiation from the first principle or differentiation from the delta method means that the theorems on differentiation or the results on differentiation of standard forms of the function are not to be applied for obtaining the derivatives of given functions.

Or, alternatively, finding the derivative of a function of an independent variable using the definition in terms of limit, theorems on limits and standard results on limits is called differentiation from the definition. This process has no practical utility but a knowledge of it is required.

Now there is illustration of *ab-initio* differentiation consisting of five steps before working out problems on different typed of functions.

Question: Explain the general method of finding differential coefficient of a function.

Answer: The general method of finding the differential coefficient of a function is indicated in

$$\frac{d}{dx}[f(x)] = \lim_{\Delta x \to 0} \frac{f(x+\Delta x)-f(x)}{\Delta x}, \quad \forall x \in D(f)$$

or, $\frac{d}{dx}(y) = \lim_{\Delta x \to 0} \frac{\Delta y}{\Delta x}$ which means

Step 1: To put $y = f(x)$ = given function

Step 2: To add Δy to y and Δx to x wherever it is present in the given function, i.e., to obtain $y + \Delta y = f(x + \Delta x)$.

Step 3: To find Δy by subtracting the first value (y) from the second value $(y + \Delta y)$, i.e., to obtain

$$(y + \Delta y) - y = f(x + \Delta x) - f(x)$$

$$\Rightarrow \Delta y = f(x + \Delta x) - f(x)$$

Step 4: To divide $\Delta y = f(x + \Delta x) - f(x)$ by Δx,

i.e., to obtain $\frac{\Delta y}{\Delta x} = \frac{f(x + \Delta x) - f(x)}{\Delta x}$

Step 5: To take the limit as $\Delta x \to 0$ on both sides of

$\frac{\Delta y}{\Delta x} = \frac{f(x + \Delta x) - f(x)}{\Delta x}$ to find $\frac{d}{dx}(y)$, i.e., to

obtain $\lim_{\Delta x \to 0} \frac{\Delta y}{\Delta x} = \lim_{\Delta x \to 0} \frac{f(x+\Delta x) - f(x)}{\Delta x} = \frac{d}{dx}(y)$

respectively.

Remember: **1.** The above method (or, process) to find the derivative of a function is what the delta–method says to one to find the derivative of a function.

2. $\lim_{\Delta x \to 0} \frac{\Delta y}{\Delta x}$ means that the ratio $\frac{\Delta y}{\Delta x}$ tends to a definite value which is unique as $\Delta x \to 0$ and hence it is to be carefully noticed that on the above definition, we speak of limiting value of a certain ratio $\frac{\Delta y}{\Delta x}$ and not of the ratio of the limiting values of Δy and Δx which is indeterminate put in the form $\left(\frac{0}{0}\right)$ (i.e. $\frac{\lim \Delta y}{\lim \Delta x} = \frac{0}{0}$ since $\Delta y \to 0$ as $\Delta x \to 0$ means $\lim \Delta y \to 0$ when $\lim \Delta x = 0$)

(***Note***: The differential coefficient is shortly written as d.c.)

3. x being regarded as fixed, the ratio $\frac{\Delta y}{\Delta x}$ will be a function of Δx.

Question: How to simplify the last step in delta method?

Answer: **(a)** In case of algebraic power, logarithmic or exponential function, we go on simplifying till Δx, some power of Δx and/a term containing Δx comes as a common factor if we use expansion method.

(b) On using binomial expansion for any index for a power function $(1+\Delta x)^n = 1 + n\Delta + \frac{n(n+1)(\Delta x)^2}{\underline{|2}} + \ldots$, we consider only two terms of the series viz "$1 + n\,\Delta x$" that will serve our purpose while finding limit.

N.B.: **(i)** The common factor being Δx, some power of Δx and/a term containing Δx must be separated out of the bracket provided we use expansion method to find the limit.

(ii) If we do not use expansion method to find the limit of power (algebraic) logarithmic and/exponential function, there is no need of taking Δx, some power of Δx and/the terms containing Δx as a common factor.

(iii) $(f(x))^n = (f(x + \Delta x))^n$ when $\Delta x \to 0$

(***Note***: If y be a function of x (i.e. $y = f(x)$) and the value of x changes, then the value of y will also change.)

(c) In case of trigonometric functions, we go on simplifying till we get $2 \sin \frac{\Delta x}{2}$ except for $\tan x$ and $\cot x$ where we get $\sin \Delta x$ and then we divide both sides of the equation defining Δy by Δx and lastly we take the limit as $\Delta x \to 0$ on both sides of the equation defining $\frac{\Delta y}{\Delta x}$.

(d) In case of inverse circular functions, we find Δx instead of Δy, i.e. in case of inverse circular functions, first of all we remove inverse operators (or, notations) and we proceed as in case of trigonometrical functions and lastly we take limit as $\Delta y \to 0$ (instead of limit as $\Delta x \to 0$ in case of direct functions or trigonometrical functions) on both sides of the equation defining $\frac{\Delta y}{\Delta x}$ since $\Delta x \to 0 \Leftrightarrow \Delta y \to 0$.

N.B.: **(i)** we use

(a) $\lim_{\theta \to 0} \frac{\sin r\theta}{\theta} = r$ and

(b) $\lim_{\theta \to 0} \frac{\tan r\theta}{\theta} = r$ which mean limit of sin (or, tan) of any constant multiple of an angle over the same angle when that angel tends to zero, is the same constant which is the multiple of the angle or, we use

(a_1) $\lim_{\theta \to 0} \frac{\sin \theta}{\theta} = 1$ (b_1) $\lim_{\theta \to 0} \frac{\tan \theta}{\theta} = 1$ which mean limit of sin (or, tan) of any angle over the same angle when that angle tends to zero, is unity. e.g.,

(a′) $$\lim_{\Delta x \to 0} \frac{2\sin\left(\frac{\Delta x}{2}\right)}{\Delta x} = \lim_{\Delta x \to 0} \frac{\sin\left(\frac{\Delta x}{2}\right)}{\frac{\Delta x}{2}} = 1$$

(b′) $$\lim_{\Delta x \to 0} \frac{\sin\left(\frac{\Delta x}{2}\right)}{\Delta x} = \frac{1}{2} \text{ or } \lim_{\Delta x \to 0} \frac{\sin\left(\frac{\Delta x}{2}\right)}{\Delta x}$$

$$= \lim_{\Delta x \to 0} \frac{\sin\left(\frac{\Delta x}{2}\right)}{2\left(\frac{\Delta x}{2}\right)} = \frac{1}{2} \lim_{\Delta x \to 0} \frac{\sin\left(\frac{\Delta x}{2}\right)}{\left(\frac{\Delta x}{2}\right)} = \frac{1}{2} \cdot 1 = \frac{1}{2}$$

(c′) $$\lim_{\Delta x \to 0} \frac{\sin(3\Delta x)}{\Delta x} = 3 \text{ or, } \lim_{\Delta x \to 0} \frac{\sin(3\Delta x)}{\Delta x}$$

$$= \lim_{\Delta x \to 0} \frac{3\sin(3\Delta x)}{(3\Delta x)} = 3 \lim_{\Delta x \to 0} \frac{\sin(3\Delta x)}{(3\Delta x)} = 3 \cdot 1 = 3$$

(d′) $$\lim_{\Delta x \to 0} \frac{\tan(n\Delta x)}{\Delta x} = n,\ n \in R, \text{ or}$$

$$\lim_{\Delta x \to 0} \frac{\tan(n\Delta x)}{\Delta x} = \lim_{\Delta x \to 0} \frac{n\tan(n\Delta x)}{(n\Delta x)}$$

$$= n \lim_{\Delta x \to 0} \frac{\tan(n\Delta x)}{(n\Delta x)} = n$$

(e′) Δx and Δy are increments in x and y and they are regarded as single quantity like x, y, z, ... etc. This is why they may be added, subtracted, multiplied and divided like the numbers. e.g.,

(i) $\frac{\Delta x}{\Delta y} = \Delta x \div \Delta y = \Delta x \frac{1}{\Delta y}$

(ii) $\frac{\Delta x}{\Delta y} \cdot \frac{\Delta y}{\Delta x} = 1$

(iii) $\frac{\Delta x}{\Delta y} = \frac{1}{\left(\frac{\Delta y}{\Delta x}\right)}$

Derivatives of Elementary Functions

Algebraic Functions

1. Find the derivative of a constant function $y = c$ by Δ – method. (or by first principles)

Solution: Let $y = f(x) = c$ ...(i)

$\therefore y + \Delta y = f(x + \Delta x) = c$ ...(ii)

Hence, (i) – (ii) $\Rightarrow \Delta y = c - c = 0$

$\Rightarrow \frac{\Delta y}{\Delta x} = 0$

$\Rightarrow \frac{d}{dx}(y) = \lim_{\Delta x \to 0} \frac{\Delta y}{\Delta x} = \lim_{\Delta x \to 0} 0 = 0 \left(\because \lim_{x \to a} c = c\right)$

$\Rightarrow$ derivative of a constant is zero.

$\Rightarrow \frac{d}{dx}(y) = \frac{d}{dx}(c) = 0$, provided y = c = any constant, $D(f') = R$.

2. Find the derivative of an independent variable w.r.t itself by delta method (or *ab*-initio).

Solution: Let $y = f(x) = x$ ...(i)

$\therefore y + \Delta y = f(x + \Delta x) = x + \Delta x$ ...(ii)

Hence, (i) – (ii) $\Rightarrow \Delta y = (x + \Delta x) - x = \Delta x$

$$\Rightarrow \frac{\Delta y}{\Delta x} = 1$$

$$\Rightarrow \frac{d}{dx}(y) = \lim_{\Delta x \to 0} \frac{\Delta y}{\Delta x} = \lim_{\Delta x \to 0} 1 = 1 \left(\because \lim_{x \to a} c = c \right)$$

$\Rightarrow$ derivative of an independent variable w.r.t itself is unity $\Rightarrow \frac{d}{dx}(x) = 1$, provided $y = x$ = identity function, $D(f') = R$.

3. Find the derivative of a constant multiple of an independent variable w.r.t the same independent variable.

Solution: Let $y = f(x) = ax$ ...(i)

$\therefore y + \Delta y = f(x + \Delta x) = a(x + \Delta x)$ ...(ii)

Hence, (i) – (ii)

$$\Rightarrow \Delta y = a(x + \Delta x) - ax = ax + a\Delta x - ax = a\Delta x$$

$$\Rightarrow \frac{\Delta y}{\Delta x} = a$$

$$\Rightarrow \frac{d}{dx}(y) = \lim_{\Delta x \to 0} \frac{\Delta y}{\Delta x} = \lim_{\Delta x \to 0} a = a \left(\because \lim_{x \to a} c = c \right)$$

$\Rightarrow$ derivative of a constant mulitple of the independent variable w.r.t the same independent variable is constant times d.c. of the independent variable w.r.t itself $\Rightarrow \frac{d}{dx}(y) = \frac{d}{dx}(ax)$

$= a\frac{d}{dx}(x) = a \cdot 1 = a$, provided $y = ax$, where a is any constant and x is the independent variable, $D(f') = R$.

4. Find the derivative of a power function w.r.t its base.

Solution: Method 1

Let $y = x^n$, $n \in Q$ ...(i)

$\therefore y + \Delta y = (x + \Delta x)^n$ ...(ii)

Hence, (i) – (ii) $\Rightarrow \Delta y = (x + \Delta x)^n - x^n$

$$\Rightarrow \frac{\Delta y}{\Delta x} = \frac{(x + \Delta x)^n - x^n}{(x + \Delta x) - x} \quad (\because \Delta x = x + \Delta x - x)$$

$$\Rightarrow \frac{d}{dx}(y) = \lim_{\Delta x \to 0} \frac{\Delta y}{\Delta x} = \lim_{\Delta x \to 0} \frac{(x + \Delta x)^n - x^n}{(x + \Delta x) - x}$$

Now substituting $z = x + \Delta x$, we have $z \to x$ as $\Delta x \to 0$ and $\frac{d}{dx}(y) = \lim_{\Delta x \to 0} \frac{(x + \Delta x)^n - x^n}{(x + \Delta x) - x}$

$$= \lim_{z \to x} \frac{z^n - x^n}{z - x} = n x^{n-1} \text{ (By limit theorem)}$$

$\Rightarrow$ derivative of a power function w.r.t its base is the power function whose index is decreased by unity times the original (given) index of the given base $\Rightarrow \frac{d}{dx}(y) = \frac{d}{dx}\left(x^n\right) = n x^{n-1}$, provided $y = x^n$, $n \in Q$.

Method 2

Let $y = x^n$, (n is any rational number) ...(i)

$\therefore y + \Delta y = (x + \Delta x)^n$ ...(ii)

Hence, (i) – (ii)

$$\Rightarrow \Delta y = (x + \Delta x)^n - x^n = x^n \left\{ \left(1 + \frac{\Delta x}{x}\right)^n - 1 \right\} \quad \text{...(iii)}$$

Now on using binomial theorem because $\left|\frac{\Delta x}{x}\right| < 1$ and n is a $+ve$ or $-ve$ integer or fractions, we have

$$\text{(iii)} = \Delta y = x^n \left[\left(1 + n \cdot \frac{\Delta x}{x} + \frac{n(n-1)}{\lfloor 2} \cdot \left(\frac{\Delta x}{x}\right)^2 + \ldots \right) - 1 \right]$$

which $\Rightarrow \frac{\Delta y}{\Delta x} = x^n \left[\frac{n}{x} + \frac{n(n-1)}{\lfloor 2} \cdot \left(\frac{\Delta x}{x^2}\right) + \text{terms having higher powers of } \Delta x \right]$

$$\Rightarrow \frac{d}{dx}(y) = \lim_{\Delta x \to 0} \frac{\Delta y}{\Delta x}$$

$$= \lim_{\Delta x \to 0} x^n \left[\frac{n}{x} + \frac{n(n-1)}{\lfloor 2} \left(\frac{\Delta x}{x^2}\right) + \ldots\right]$$

$$= \lim_{\Delta x \to 0} n x^{n-1} \left[1 + \frac{(n-1)}{\lfloor 2} \cdot \left(\frac{\Delta x}{x}\right) + \ldots\right]$$

$$= n x^{n-1},\ n \in Q$$

Remark: $\frac{d}{dx}\left(x^n\right) = nx^{n-1}$ for $x > 0$, $n \in R$ also holds true noting that for $x < 0$, x^n may not be real as $(-3)^{\frac{1}{2}} = \sqrt{-3} \notin R$.

Logarithmic and Exponential Functions

1. Find the derivative of logarithm of an independent variable w.r.t the same independent variable, the independent variable being positive (d.c. of logarithmic function) by delta method.

Solution: Method 1.

Let $y = \log x, x > 0$ ...(i)

$\therefore\ y + \Delta y = \log(x + \Delta x)$ ...(ii)

Hence, (i) – (ii) $\Rightarrow \Delta y = \log(x + \Delta x) - \log x$

$$= \log\left(\frac{x + \Delta x}{x}\right) = \log\left(\frac{x}{x} + \frac{\Delta x}{x}\right) = \log\left(1 + \frac{\Delta x}{x}\right)$$

$$\Rightarrow \frac{\Delta y}{\Delta x} = \frac{1}{\Delta x} \log\left(1 + \frac{\Delta x}{x}\right) = \frac{1}{x} \cdot x \cdot \frac{1}{\Delta x} \log\left(1 + \frac{\Delta x}{x}\right)$$

$$= \frac{1}{x} \log\left(1 + \frac{\Delta x}{x}\right)^{\frac{x}{\Delta x}}$$

$$\Rightarrow \frac{d}{dx}(y) = \lim_{\Delta x \to 0}\left[\frac{1}{x} \log\left(1 + \frac{\Delta x}{x}\right)^{\frac{x}{\Delta x}}\right]$$

$$= \lim_{\Delta x \to 0}\left(\frac{1}{x}\right) \cdot \lim_{\Delta x \to 0} \log\left(1 + \frac{\Delta x}{x}\right)^{\frac{x}{\Delta x}}$$

$$= \lim_{\Delta x \to 0}\left(\frac{1}{x}\right) \log \lim_{\Delta x \to 0}\left(1 + \frac{\Delta x}{x}\right)^{\frac{x}{\Delta x}}$$ (on using composit function rule on limits)

$$= \frac{1}{x} \log e = \frac{1}{x} \cdot \log_e e = \frac{1}{x} \cdot 1$$

$$\left(\because \lim_{x \to 0} (1 + x)^{\frac{1}{x}} = e \text{ and } \log e = \log_e e = 1\right)$$

$= \frac{1}{x}$ which $\Rightarrow$ derivative of logarithm of an independent variable w.r.t the same independent variable, the independent variable being positive, is reciprocal of the independent variable, the independent variable being positive $\Rightarrow \frac{d}{dx}(y) =$

$\frac{d}{x}(\log x) = \frac{1}{x}$, $x > 0$ provided $y = \log x$, $x > 0$, $D(f') = R^+$.

Remember:

(i) $n \log m = \log m^n$

(ii) $\Delta x \to 0 \Rightarrow \frac{\Delta x}{x} \to 0$

(iii) $\lim_{x \to 0} \log f(x) = \log \lim_{x \to 0} f(x)$

(iv) $\lim_{\frac{\Delta x}{x} \to 0} \left(1 + \frac{\Delta x}{x}\right)^{\frac{x}{\Delta x}} = e$

(v) $\log e = \log_e e = 1$, where e lies between 2 and 3.

These are facts which have been used to find the d.c. of $y = \log x, x > 0$, w.r.t x.

Method 2

Let $y = \log x, x > 0$ ...(i)

$\therefore\ y + \Delta y = \log(x + \Delta x)$ ...(ii)

Hence, (i) – (ii) $\Rightarrow \Delta y = \log(x + \Delta x) - \log x$

$$= \log\left(\frac{x + \Delta x}{x}\right) = \log\left(1 + \frac{\Delta x}{x}\right)$$

$$\Rightarrow \Delta y = \frac{\Delta x}{x} - \frac{1}{2}\left(\frac{\Delta x}{x}\right)^2 + \frac{1}{3}\left(\frac{\Delta x}{x}\right)^3 - \ldots$$

$\Rightarrow \frac{\Delta y}{\Delta x} = \frac{1}{x} - \frac{1}{2}\left(\frac{\Delta x}{x^2}\right)$ + terms having higher powers of Δx.

$$\Rightarrow \frac{d}{dx}(y) = \lim_{\Delta x \to 0} \frac{\Delta y}{\Delta x} = \lim_{\Delta x \to 0}\left[\frac{1}{x} - \frac{1}{2}\left(\frac{\Delta x}{x^2}\right) + \ldots\right]$$

$$\Rightarrow \frac{d}{dx}(y) = \lim_{\Delta x \to 0} \frac{\Delta y}{\Delta x} = \frac{1}{x}$$

$$\Rightarrow \frac{d}{dx}(y) = \frac{d}{dx}(\log x) = \frac{1}{x},\quad x > 0$$

Cor: Derivative of $\log_a x\,(a>0, \neq 1), x>0$

Using $\log_a x = \dfrac{\log x}{\log a}$, we have $\dfrac{d}{dx}(\log_a x)$

$$= \frac{d}{dx}\left(\frac{\log x}{\log a}\right) = \frac{1}{\log a}\,\frac{d}{dx}(\log x) = \frac{1}{\log a}\cdot\frac{1}{x}$$

$$= \frac{1}{x \log a},\ (x>0)\ (\because \log a \text{ is a constant})$$

Notes: **1.** $\log_a f(x) = \dfrac{\log f(x)}{\log a}$ where as

$\log_{10} f(x) = \log_{10} e \cdot \log_e f(x)$ and $\log_{10} e = 0.4344$

2. If the base of the logarithm of a function is e, then the logarithm of the function is called natural logarithmic function and is denoted as 'lnx' without the base 'e'. Hence, $\log_e x = lnx$.

3. Meaning of e : e^x means the infinite series (or, infinite power series) $1 + \frac{x}{\lfloor 1} + \frac{x^2}{\lfloor 2} + \frac{x^3}{\lfloor 3} + \ldots \infty$ where

$e = 1 + \frac{1}{\lfloor 1} + \frac{1}{\lfloor 2} + \frac{1}{\lfloor 3} + \ldots \infty$ and x is a real number;

e is also defined as $\lim\limits_{n \to \infty}\left(1 + \frac{1}{n}\right)^n$.

Remark: e is a transcendental irrational number. $2 < e < 3$ and $e = 2.718$ nearly.

2. Find the derivative of the function $y = \log|x|$, $(x \neq 0)$ w.r.t x *ab–initio*.

Solution: $y = \log|x| = \log\sqrt{x^2} = \frac{1}{2}\log x^2$

$$\Rightarrow y + \Delta y = \frac{1}{2}\log(x + \Delta x)^2$$

$$\Rightarrow \Delta y = \frac{1}{2}\left[\log(x + \Delta x)^2 - \log x^2\right]$$

$$= \frac{1}{2}\log\left(\frac{x + \Delta x}{x}\right)^2$$

$$\left(\because \log\frac{f(x)}{g(x)} = \log f(x) - \log g(x)\right)$$

$$\therefore \frac{\Delta y}{\Delta x} = \frac{1}{2}\cdot\frac{1}{\Delta x}\cdot\log\left(\frac{x + \Delta x}{x}\right)^2$$

$$= \frac{1}{2}\cdot\frac{x}{\Delta x}\cdot\frac{1}{x}\log\left(\frac{x + \Delta x}{x}\right)^2$$

$$= \frac{1}{2}\cdot\frac{1}{x}\cdot\log\left\{\left(\frac{x + \Delta x}{x}\right)^{\frac{x}{\Delta x}}\right\}^2 \text{ and } \lim_{\Delta x \to 0}\frac{\Delta y}{\Delta x}$$

$$= \lim_{\Delta x \to 0}\left[\frac{1}{2}\cdot\frac{1}{x}\cdot\log\left\{\left(\frac{x + \Delta x}{x}\right)^{\frac{x}{\Delta x}}\right\}^2\right]$$

$$= \frac{1}{2}\cdot\lim_{\Delta x \to 0}\left(\frac{1}{x}\right)\cdot\lim_{\Delta x \to 0}\log\left\{\left(\frac{x + \Delta x}{x}\right)^{\frac{x}{\Delta x}}\right\}^2$$

$$= \frac{1}{2}\cdot\frac{1}{x}\cdot\log\lim_{\Delta x \to 0}\left\{\left(\frac{x + \Delta x}{x}\right)^{\frac{x}{\Delta x}}\right\}^2$$

$$= \frac{1}{2}\cdot\frac{1}{x}\cdot\log\left\{\lim_{\Delta x \to 0}\left(\frac{x + \Delta x}{x}\right)^{\frac{x}{\Delta x}}\right\}^2$$

$$= \frac{1}{2}\cdot\frac{1}{x}\cdot\log\left\{\lim_{\Delta x \to 0}\left(1 + \frac{\Delta x}{x}\right)^{\frac{x}{\Delta x}}\right\}^2$$

$$= \frac{1}{2}\cdot\frac{1}{x}\cdot\log(e)^2\left(\because \lim_{x \to 0}(1 + x)^{\frac{1}{x}} = e\right)$$

$$= \frac{1}{2} \cdot \frac{1}{x} \log e^2 = \frac{1}{2} \cdot \frac{1}{x} 2 \log e = \frac{1}{2} \cdot \frac{1}{x} \cdot 2 \cdot 1$$

$$= \frac{1}{x},\ x \neq 0,\ D(f') = R - \{0\}$$

***Notes*: A.**

(i) $\log|f(x)| = \log\sqrt{f^2(x)} = \frac{1}{2}\log f^2(x)$

(ii) $\log|x| = \log\sqrt{x^2} = \frac{1}{2}\log x^2$

B.: **(i)** The logarithm of the product of positive numbers is equal to the sum of the logarithms of the factors:

$$\log_a(mn) = \log_a m + \log_a n$$

(ii) The logarithm of the quotient of two positive numbers is equal to the logarithm of the dividend minus the logarithm of the divisor:

$$\log_a\left(\frac{m}{n}\right) = \log_a m - \log_a n$$

(iii) The logarithm of the power of a positive number is equal to the exponent times the logarithm of the base:

$$\log_a m^n = n \log_a m,\ n \in R$$

(iv) The logarithm of a root of a positive number is equal to the logarithm of the number divided by the index of the root:

$$\log_a \sqrt[n]{m} = \frac{1}{n}\log_a m$$

Remark: $\log x^2 \neq 2\log x$ for all $x \neq 0$, because $x \neq 0$ $\Rightarrow$ either $x > 0$ or $x < 0$ whereas $\log x^2 = 2\log x$ only when $x > 0$ as (B) (iii) Says. This is why $\log e^2 = 2\log e$ since $e > 0$.

Cor: Derivative of $\log_a|x|$, $(a > 0, \neq 1)$, $x \neq 0$ on using $\log_a|x| = \frac{\log|x|}{\log a}$, we have $\frac{d}{dx}(\log_a|x|)$

$$= \frac{d}{dx}\left(\frac{\log|x|}{\log a}\right)$$

$$= \frac{1}{\log a} \cdot \frac{d}{dx}(\log|x|) = \frac{1}{\log a} \cdot \frac{1}{x}$$

$$= \frac{1}{x\log a},\ (x \neq 0)\ (\because \log a \text{ is a constant})$$

2. Find the derivative of an exponential function w.r.t the index, the index being an independent variable (d.c. of exponential function) by delta method.

Solution: **(a)** Method 1

Let $y = e^x$ ...(i)

$\therefore\ y + \Delta y = e^{x + \Delta x}$ ...(ii)

Hence, (i) – (ii)

$$\Rightarrow \Delta y = e^{x+\Delta x} - e^x = e^x \cdot e^{\Delta x} - e^x = e^x\left(e^{\Delta x} - 1\right)$$

$$\Rightarrow \frac{\Delta y}{\Delta x} = e^x\left(\frac{e^{\Delta x} - 1}{\Delta x}\right)$$

$$\Rightarrow \frac{d}{dx}(y) = \lim_{\Delta x \to 0}\frac{\Delta y}{\Delta x} = \lim_{\Delta x \to 0} e^x\left(\frac{e^{\Delta x} - 1}{\Delta x}\right)$$

$$\Rightarrow \frac{d}{dx}(y) = \lim_{\Delta x \to 0} e^x \cdot \lim_{\Delta x \to 0}\left(\frac{e^{\Delta x} - 1}{\Delta x}\right)$$

$$= e^x \cdot 1 \quad \left(\because \lim_{x \to 0}\left(\frac{e^x - 1}{x}\right) = \log e = 1\right)$$

$= e^x$ which $\Rightarrow$ derivative of an exponential function with base e w.r.t the index, the index being an independent variable, is itself the exponential function without any change in the given form $\Rightarrow \frac{d}{dx}(y) = \frac{d}{dx} = \left(e^x\right) = e^x$, provided $y = e^x$.

Notes to Remember: One must remember that terms having no Δx must be regarded as constants when $\Delta x \to 0$. This is why we write

(i) $\lim_{\Delta x \to 0} e^x = e^x$

(ii) $\lim_{\Delta x \to 0} x^{n-1} = x^{n-1}$, etc

Method 2

Let $y = e^x$ ...(i)

$\therefore y + \Delta y = e^{x + \Delta x}$ ...(ii)

Hence, (i) – (ii)

$\Rightarrow \Delta y = e^{x+\Delta x} - e^{x} = e^{x} \cdot e^{\Delta x} - e^{x} = e^{x}\left(e^{\Delta x} - 1\right)$

$$\Rightarrow \frac{\Delta y}{\Delta x} = \frac{e^{x}}{\Delta x}\left(1 + \Delta x + \frac{(\Delta x)^2}{\underline{|2}} + \frac{(\Delta x)^3}{\underline{|3}} + \ldots - 1\right)$$

$$\Rightarrow \frac{\Delta y}{\Delta x} = \frac{e^{x}}{\Delta x}\left(\Delta x + \frac{(\Delta x)^2}{\underline{|2}} + \frac{(\Delta x)^3}{\underline{|3}} + \ldots\right)$$

$$\Rightarrow \frac{\Delta y}{\Delta x}$$

$$= e^{x}\left(1 + \frac{\Delta x}{\underline{|2}} + \text{terms having higher powers of } \Delta x.\right)$$

$$\Rightarrow \frac{d}{dx}(y) = \lim_{\Delta x \to 0} \frac{\Delta y}{\Delta x} = \lim_{\Delta x \to 0} e^{x}\left(1 + \frac{\Delta x}{\underline{|2}} + \ldots\right) = e^{x}$$

(b) Let $y = a^x$, $x \in R, a > 0$ ...(i)

$\therefore y + \Delta y = a^{x + \Delta x}$ ...(ii)

Hence, (i) – (ii)

$\Rightarrow \Delta y = a^{x+\Delta x} - a^{x} = a^{x} \cdot a^{\Delta x} - a^{x} = a^{x}\left(a^{\Delta x} - 1\right)$

$$\Rightarrow \frac{\Delta y}{\Delta x} = a^{x}\left(\frac{a^{\Delta x} - 1}{\Delta x}\right)$$

$$\Rightarrow \frac{d}{dx}(y) = \lim_{\Delta x \to 0} \frac{\Delta y}{\Delta x} = \lim_{\Delta x \to 0} a^{x}\left(\frac{a^{\Delta x} - 1}{\Delta x}\right)$$

$$= \lim_{\Delta x \to 0} a^{x} \cdot \lim_{\Delta x \to 0}\left(\frac{a^{\Delta x} - 1}{\Delta x}\right)$$

$$= a^{x} \log_e a \left(\because \lim_{x \to 0} \frac{a^{x} - 1}{x} = \log_e a\right)$$

$$= a^{x}\, l\,n\,a \left(\because \log_e a = l\,n\,a\right)$$

Which $\Rightarrow$ derivative of an exponential function with base $a > 0$ w.r.t the index, the index being an independent variable, is the exponential function with the same base $a > 0$ times "ell – en a" $\Rightarrow \frac{d}{dx}(y) = \frac{d}{dx}(a^x) = a^x l_n a$, provided $y = a^x$, $x \in R$, $a > 0$ and

Remember: **(i)** The function $f(x) = a^x$, where a is any positive real number and x is any real number (i.e. $a > 0$, $x \in R$) is called the general exponential function with base a.

(ii) The function $f(x) = e^x$ for all $x \in R$ is called natural exponential function or simply exponential function.

(iii) $f(x) = \ln x$ is called ell – en (, y – , u) function or natural logarithmic function which is alternation to $\log x$.

(iv) We should note that d.c. of

(a constant)$^{\text{independent variable}}$

= (the constant)$^{\text{the independent variable}}$

times log (the constant), where 'constant' stands for any positive real number (i.e.; e or a)

Hence, $\frac{d}{dx}(e^x) = e^x \log_e e = e^x \log e = e^x . 1 = e^x$ and

$\frac{d}{dx}(a^x) = a^x \log_e a = a^x \log a$

(***Note***: In calculus, when no base of logarithmic function is mentioned, it is always understood to be "e". Hence, $\log_e x = \log x = lnx$)

Trigonometrical Functions

1. Find the derivatives of all elementary trigonometrical functions w.r.t their independent variables by delta method.

Solution: **1.** Differential coefficient of $\sin x$ w.r.t x

Let $y = \sin x$ ...(i)

$\therefore y + \Delta y = sin(x + \Delta x)$ ...(ii)

Hence, (i)–(ii) $\Rightarrow \Delta y = \sin(x + \Delta x) - \sin x$

$$= 2\cos\left(\frac{x + \Delta x + x}{2}\right) \cdot \sin\left(\frac{x + \Delta x + x}{2}\right)$$

$$\Rightarrow \Delta y = 2\cos\left(x + \frac{\Delta x}{2}\right) \cdot \sin\left(\frac{\Delta x}{2}\right)$$

$$\Rightarrow \frac{\Delta y}{\Delta x} = 2\cos\left(x + \frac{\Delta x}{2}\right) \cdot \sin\frac{\Delta x}{2} \cdot \frac{1}{\Delta x}$$

$$= \cos\left(x + \frac{\Delta x}{2}\right) \cdot \frac{\sin\left(\frac{\Delta x}{2}\right)}{\frac{\Delta x}{2}}$$

$$\Rightarrow \frac{d}{dx}(y)$$

$$= \lim_{\Delta x \to 0} \frac{\Delta y}{\Delta x} = \lim_{\Delta x \to 0} \cos\left(x + \frac{\Delta x}{2}\right) \cdot \frac{\sin\left(\frac{\Delta x}{2}\right)}{\frac{\Delta x}{2}}$$

$$= \lim_{\Delta x \to 0} \cos\left(x + \frac{\Delta x}{2}\right) \cdot \lim_{\Delta x \to 0} \frac{\sin\left(\frac{\Delta x}{2}\right)}{\frac{\Delta x}{2}}$$

$= \cos x \,.\, 1 = \cos x$ which $\Rightarrow$ derivative of sine of an angle w.r.t the same angle is cosine of the same angle $\Rightarrow \frac{d}{dx}(y) = \frac{d}{dx}(\sin x) = \cos x$.

***Notes to Remember*:** **1.** While finding the d.c. of trigonometrical functions using delta method, one must remember that trigonometrical function which becomes zero on putting $\Delta x = 0$ is modified by writing

(i) $\sin\theta = \frac{\sin\theta}{\theta} \cdot \theta$ **(ii)** $\tan\theta = \frac{\tan\theta}{\theta} \cdot \theta$ so that standard formulas of limits of trigonometrical functions.

(a) $\lim_{\theta \to 0} = \frac{\sin\theta}{\theta} = 1$ **(b)** $\lim_{\theta \to 0} = \frac{\tan\theta}{\theta} = 1$ may be used

2. $\sin C - \sin D = 2\cos\left(\frac{C + D}{2}\right) \cdot \cos\left(\frac{C - D}{2}\right)$ is used to find Δy in simplified form while finding d.c. of trigonometrical functions (particularly sin of an angle and co sine of an angle).

(ii) Differential coefficient of $\cos x$ w.r.t x

Let $y = \cos x$ ...(i)

$\therefore\ y + \Delta y = \cos(x + \Delta x)$ ...(ii)

Hence, (i)–(ii), $\Rightarrow \Delta y = \cos(x + \Delta x) - \cos x$

$$= 2\sin\left(\frac{x + \Delta x + x}{2}\right)\sin\left(\frac{x - x - \Delta x}{2}\right)$$

$$\Rightarrow \Delta y = 2\sin\left(x + \frac{\Delta x}{2}\right) \cdot \sin\left(-\frac{\Delta x}{2}\right)$$

$$= -2\sin\left(x + \frac{\Delta x}{2}\right)\sin\left(\frac{\Delta x}{2}\right)$$

$$\Rightarrow \frac{\Delta y}{\Delta x} = -2\sin\left(x + \frac{\Delta x}{2}\right) \cdot \frac{\sin\left(\frac{\Delta x}{2}\right)}{\Delta x}$$

$$= -\sin\left(x + \frac{\Delta x}{2}\right)\frac{\sin\left(\frac{\Delta x}{2}\right)}{\left(\frac{\Delta x}{2}\right)}$$

$$\Rightarrow \frac{d}{dx}(y) = \lim_{\Delta x \to 0} \frac{\Delta y}{\Delta x}$$

$$\Rightarrow -\lim_{\Delta x \to 0} \sin\left(x + \frac{\Delta x}{2}\right) \cdot \lim_{\Delta x \to 0} \frac{\sin\left(\frac{\Delta x}{2}\right)}{\frac{\Delta x}{2}}$$

$$= -\sin x \times 1 = -\sin x \quad \left(\because \lim_{\Delta x \to 0} \frac{\sin\left(\frac{\Delta x}{2}\right)}{\left(\frac{\Delta x}{2}\right)} = 1\right)$$

which $\Rightarrow$ derivative of cosine of an angle w.r.t the same angle is negative of sine of the same angle

$$\Rightarrow \frac{d}{dx}(y) = \frac{d}{dx}(\cos x) = -\sin x.$$

***Notes*:** One must remember that

(i) $\cos C - \cos D = -2\sin\left(\frac{C - D}{2}\right) \cdot \sin\left(\frac{C + D}{2}\right)$

(ii) $\sin(-\theta) = -\sin\theta$ are used to find Δy in simplified form while finding d.c. of $\cos x$ w.r.t x.

(iii) Differential coefficient of $\tan x$ w.r.t x

Let $y = \tan x$, $x \neq n\pi + \dfrac{\pi}{2}$ ...(i)

$\therefore\ \Delta y = \tan(x + \Delta x)$ ...(ii)

Hence, (i) – (ii)

$$\Rightarrow \Delta y = \tan(x + \Delta x) - \tan x$$

$$= \frac{\sin(x + \Delta x)}{\cos(x + \Delta x)} - \frac{\sin x}{\cos x}$$

$$\Rightarrow \Delta y = \frac{\cos x \sin(x+\Delta x) - \sin x \cos(x+\Delta x)}{\cos(x + \Delta x)\cos x}$$

$$= \frac{\sin(x + \Delta x - x)}{\cos(x + \Delta x)\cos x}$$

$$\Rightarrow \Delta y = \frac{\sin \Delta x}{\cos(x + \Delta x)\cos x}$$

$$\Rightarrow \frac{\Delta y}{\Delta x} = \frac{\sin \Delta x}{\Delta x \cos(x + \Delta x)\cos x}$$

$$\Rightarrow \frac{d}{dx}(y) = \lim_{\Delta x \to 0} \frac{\Delta y}{\Delta x}$$

$$= \lim_{\Delta x \to 0}\left[\frac{\sin \Delta x}{\Delta x} \cdot \frac{1}{\cos x} \cdot \frac{1}{\cos(x + \Delta x)}\right]$$

$$= \lim_{\Delta x \to 0}\left(\frac{\sin \Delta x}{\Delta x}\right) \cdot \lim_{\Delta x \to 0}\left(\frac{1}{\cos x}\right) \cdot \lim_{\Delta x \to 0}\left(\frac{1}{\cos(x+\Delta x)}\right)$$

$$= 1 \cdot \frac{1}{\cos x} \cdot \frac{1}{\cos x}\ \left(\because \lim_{\theta \to 0} \frac{\sin\theta}{\theta} = 1\right)$$

$= \dfrac{1}{\cos^2 x} = \sec^2 x$ which $\Rightarrow$ derivative of tangent of an angle w.r.t the same angle is square of secant of the same angle $\Rightarrow \dfrac{d}{dx}(y) = \dfrac{d}{dx}(\tan x) =$, $\sec^2 x$ provided $y = \tan x$

Notes to Remember: One must remember that (i) $\sin(A - B) = \sin A \cdot \cos B - \cos A \sin B$ is used to find Δy in simplified form while finding d.c. of $\tan x$ w.r.t. x

(iv) Differential coefficient of $\cot x$ w.r.t. x

Let $y = \cot x$, $x \neq n\pi$ ...(i)

$\therefore\ y + \Delta y = \cot(x + \Delta x)$ ...(ii)

Hence, (i) – (ii)

$$\Rightarrow \Delta y = \cot(x + \Delta x) - \cot x$$

$$= \frac{\cos(x + \Delta x)}{\sin(x + \Delta x)} - \frac{\cos x}{\sin x}$$

$$\Rightarrow \Delta y = \frac{\cos(x+\Delta x)\sin x - \sin(x+\Delta x)\cos x}{\sin(x+\Delta x)\sin x}$$

$$= \frac{\sin(x - x - \Delta x)}{\sin(x + \Delta x)\sin x}$$

$$\Rightarrow \Delta y = \frac{\sin(-\Delta x)}{\sin(x+\Delta x)\sin x} = -\frac{\sin \Delta x}{\sin(x+\Delta x)\sin x}$$

$$\Rightarrow \frac{\Delta y}{\Delta x} = -\frac{\sin \Delta x}{\Delta x \sin(x + \Delta x)\sin x}$$

$$\Rightarrow \frac{d}{dx}(y) = \lim_{\Delta x \to 0} \frac{\Delta y}{\Delta x}$$

$$= (-1) \lim_{\Delta x \to 0}\left[\frac{\sin \Delta x}{\Delta x} \cdot \frac{1}{\sin x} \cdot \frac{1}{\sin(x + \Delta x)}\right]$$

$$= (-1) \lim_{\Delta x \to 0}\left(\frac{\sin \Delta x}{\Delta x}\right) \cdot \lim_{\Delta x \to 0}\left(\frac{1}{\sin x}\right) \cdot \lim_{\Delta x \to 0}\left(\frac{1}{\sin(x+\Delta x)}\right)$$

$$= (-1) \cdot (-1) \cdot \left(\frac{1}{\sin x}\right) \cdot \left(\frac{1}{\sin x}\right)$$

$$= (-1)\left(\frac{1}{\sin^2 x}\right) = -\operatorname{cosec}^2 x$$

which $\Rightarrow$ derivative of cotangent of an angle w.r.t the same angle is minus one times square of cosecant of the same angle $\Rightarrow \frac{d}{dx}(y) = \frac{d}{dx}(\cot x)$ $= -\text{cosec}^2 x$, provided $y = \cot x$.

Remember: **(i)** $\sin(A - B) = \sin A \cos B - \cos A \sin B$ is used to find Δy in simplified form while finding d.c. of $\cot x$ w.r.t. x.

(v) Differential coefficient of $\sec x$ w.r.t. x

Let $y = \sec x,\ x \neq n\pi + \frac{\pi}{2}$ ...(i)

$\therefore\ y + \Delta y = \sec(x + \Delta x)$ ...(ii)

Hence, (i) – (ii)

$$\Rightarrow \Delta y = \sec(x + \Delta x) - \sec x$$

$$= \frac{1}{\cos(x + \Delta x)} - \frac{1}{\cos x}$$

$$\Rightarrow \Delta y = \frac{\cos x - \cos(x + \Delta x)}{\cos(x + \Delta x)\cos x}$$

$$= \frac{2\sin\left(x + \frac{\Delta x}{2}\right) \cdot \sin\left(\frac{x + \Delta x - x}{2}\right)}{\cos(x + \Delta x)\cos x}$$

$$\Rightarrow \frac{\Delta y}{\Delta x} = \frac{2\sin\left(x + \frac{\Delta x}{2}\right) \cdot \sin\left(\frac{\Delta x}{2}\right)}{\Delta x \cdot \cos(x + \Delta x)\cos x}$$

$$= \frac{2\sin\left(x + \frac{\Delta x}{2}\right)\sin\left(\frac{\Delta x}{2}\right)}{\cos(x + \Delta x)\cos x \cdot \left(\frac{\Delta x}{2}\right) \cdot 2}$$

$$\Rightarrow \frac{\Delta y}{\Delta x} = \left(\frac{\sin\left(\frac{\Delta x}{2}\right)}{\left(\frac{\Delta x}{2}\right)}\right)\left(\frac{\sin\left(x + \frac{\Delta x}{2}\right)}{\cos(x + \Delta x)\cos x}\right)$$

$$\Rightarrow \frac{d}{dx}(y) = \lim_{\Delta x \to 0} \frac{\Delta y}{\Delta x}$$

$$= \lim_{\Delta x \to 0}\left(\frac{\sin\frac{\Delta x}{2}}{\frac{\Delta x}{2}}\right) \cdot \lim_{\Delta x \to 0}\left(\frac{1}{\cos x}\right) \cdot \lim_{\Delta x \to 0}\left(\frac{\sin\left(x + \frac{\Delta x}{2}\right)}{\cos(x + \Delta x)}\right)$$

$$= (1) \cdot \left(\frac{1}{\cos x}\right) \cdot \left(\frac{\sin x}{\cos x}\right) = \sec x \cdot \tan x$$ which

$\Rightarrow$ derivative of secant of an angle w.r.t the same angle is secant of the same angle times tangent of the same angle $\Rightarrow \frac{d}{dx}(y) = \frac{d}{dx}(\sec x)\ \sec x \tan x$, provided $y = \sec x$.

Remember: **1.** $\cos C - \cos D = 2\sin\left(\frac{D - C}{2}\right) \cdot \sin\left(\frac{C + D}{2}\right)$ is used to find Δy in simplified form while finding d.c. of $\sec x$ w.r.t x.

(vi) Differential coefficient of $\text{cosec}\, x$ w.r.t. x

Let $y = \text{cosec}\, x,\ x \neq n\pi$ ...(i)

$\therefore\ y + \Delta y = \text{cosec}(x + \Delta x)$ ...(ii)

Hence, (i)–(ii)

$$\Rightarrow \Delta y = \text{cosec}(x + \Delta x) - \text{cosec}\, x$$

$$= \frac{1}{\sin(x + \Delta x)} - \frac{1}{\sin x}$$

$$\Rightarrow \Delta y = \frac{\sin x - \sin(x + \Delta x)}{\sin(x + \Delta x)\sin x}$$

$$\Rightarrow \Delta y = \frac{2\cos\left(x + \frac{\Delta x}{2}\right) \cdot \sin\left(\frac{x - x - \Delta x}{2}\right)}{\sin(x + \Delta x)\sin x}$$

$$\Rightarrow \Delta y = \frac{2\cos\left(x + \frac{\Delta x}{2}\right) \cdot \left(-\sin\frac{\Delta x}{2}\right)}{\sin(x + \Delta x)\sin x}$$

$$\Rightarrow \frac{\Delta y}{\Delta x} = \frac{-2\cos\left(x + \frac{\Delta x}{2}\right)\sin\left(\frac{\Delta x}{2}\right)}{\sin(x + \Delta x)\sin x\left(\frac{\Delta x}{2}\right) \cdot 2}$$

$$\Rightarrow \frac{\Delta y}{\Delta x} = \frac{\cos\left(x + \frac{\Delta x}{2}\right) \cdot \sin\left(\frac{\Delta x}{2}\right)}{\sin(x + \Delta x)\sin x \left(\frac{\Delta x}{2}\right)}$$

$$\Rightarrow \frac{d}{dx}(y)$$

$$= -\lim_{\Delta x \to 0}\left(\frac{\sin\frac{\Delta x}{2}}{\frac{\Delta x}{2}}\right) \cdot \lim_{\Delta x \to 0}\left(\frac{1}{\sin x}\right) \cdot \lim_{\Delta x \to 0}\left(\frac{\cos\left(x + \frac{\Delta x}{2}\right)}{\sin(x + \Delta x)}\right)$$

$$= (-1)\cdot(-1)\cdot\left(\frac{1}{\sin x}\right)\cdot\left(\frac{\cos x}{\sin x}\right)$$

$= -\operatorname{cosec} x \cdot \cot x$ which $\Rightarrow$ derivative of cosecant of an angle w.r.t the same angle is minus one times cosecant of the same angle times cotangent of the same angle $\Rightarrow \frac{d}{dx}(y) = \frac{d}{dx}(\operatorname{cosec} x) = -\operatorname{cosec} x \cdot \cot x$, provided $y = \operatorname{cosec} x$.

Recapitulation: The derivative of every co-function (i.e., cos, cot, cosec) can be obtained from the derivative of the corresponding function (i.e., sin, tan, cos) by (i) introducing a minus sing, and (ii) Replacing each function by its co-function. Hence applying this rule to $= \frac{d}{dx}(\sec x) = \sec x \tan x$, we get $= \frac{d}{dx}(\operatorname{cosec} x) = -\operatorname{cosec} x \cdot \cot x$, and so on.

Inverse Circular Functions

Question: Find the derivatives of inverse circular functions with respect to their independent variables.

Answer: The derivatives of inverse circular functions with respect to their independent variables have been derived and discussed in the chapter "Derivatives of Inverse Circular Functions" in detail and the formulas for finding their derivatives have been recapitulated in the chapter "Chain rule for the derivatives". This is why it is advised to consult those chapters.

Remarks: **1.** In particular, if $x = a$ is a point in the domain of a function defined by a single formula $y = f(x)$, then $\lim_{h \to 0} \frac{f(a+h) - f(a)}{h}$, h is a small positive number, is called the differential coefficient of $f(x)$ at $x = a$ provided this limit exists and is denoted by $f'(a)$. Hence, if $f(x) = \frac{1}{2x^3}$, then $f'(2)$

$$= \lim_{h\to 0}\frac{f(2+h) - f(2)}{h} = \lim_{h\to 0}\left(\frac{\frac{1}{2(2+h)^3} - \frac{1}{16}}{h}\right)$$

$$= \lim_{h\to 0}\frac{8 - (2+h)^3}{16(2+h)^3 \cdot h}$$

$$= \lim_{h\to 0}\frac{8 - \left[8 + 12h + 6h^2 + h^3\right]}{16 \cdot h \cdot (2+h)^3}$$

$$= \lim_{h\to 0}\frac{-h\left(12 + 6h + h^2\right)}{16h(2+h)^3}$$

$$= \lim_{h\to 0}\frac{-\left(12 + 6h + h^3\right)}{16(2+h)^3} = \frac{-12}{16 \times 8} = \frac{-3}{32}$$

2. In the case of functions defined by a single formula (rule or expression in x) in a neighbourhood of a point belonging to their domains, there is no need to calculate left hand derivative (symbolised as $L\,f'(a)$ or $f_-'(a)$) and right hand derivative (symbolised as $R\,f'(a)$ or $f_+'(a)$) separately about which, there is discussion in the chapter "differentiability at a point in the domain of the function". Further, in the case of piecewise functions defined in adjacent intervals (Intervals whose left end point and right end point are same), it is necessary to calculate both the left hand and the right hand derivatives at the common point of the adjacent intervals.

3. The domain of $f'(x)$ is a subset of points in the domain of $f(x)$ where the $= \lim_{h \to 0} \frac{f(x+h) - f(x)}{h}$, $(h > 0)$ exists excluding all those points where this limit fails to exist, i.e. excluding all those points where $f'(x)$ is undefined.

Exercise 6.1

Problems on algebraic functions

Differentiate by Δ – method (read as delta method) the following functions w.r.t their independent variables.

1. **(i)** $2x$ **(ii)** x^2 **(iii)** $5x^3$ **(iv)** $\sqrt{x}$ **(v)** $\frac{3}{x}$

(vi) $\frac{1}{\sqrt{x}}$ **(vii)** $x^{\frac{-3}{2}}$

2. $7x^3 - 5x^2 + 4x + 13$

3. $(2x+3)(3x-7)$

4. $\frac{x^2+1}{x-1}$

5. $\frac{1}{x+1}$

6. $\frac{1}{\sqrt{2x-1}}$

7. $\frac{x+5}{x-3}$

8. $\sqrt{2x-3}$

9. $\sqrt{ax+b}$

10. $\sqrt{x^2+a^2}$

11. $\frac{1}{\sqrt{x+a}}$

12. $\frac{1}{x\sqrt{x}}$

***Answers*:** **1.** **(i)** 2, **(ii)** $2x$, **(iii)** $15x^2$,

(iv) $\frac{1}{2\sqrt{x}}, (x>0)$ **(v)** $-3x^{-2}, (x>0)$

(vi) $-\frac{1}{2}x^{-\frac{3}{2}}, (x>0)$ **(vii)** $-\frac{3}{2}x^{-5/2}, (x>0)$

2. $21x^2 - 10x + 4$ **3.** $12x - 5$

4. $\frac{x^2-2x-1}{(x-1)^2}, (x \neq 1)$ **5.** $\frac{-1}{(x+1)^2}, (x \neq -1)$

6. $\frac{-1}{(2x-1)^{3/2}}, \left(x > \frac{1}{2}\right)$ **7.** $\frac{-8}{(x-3)^2}, (x \neq 3)$

8. $\frac{1}{\sqrt{2x-3}}, \left(x > \frac{3}{2}\right)$ **9.** $\frac{a}{2\sqrt{ax+b}}, \left(x > \frac{-b}{a}\right)$

10. $\frac{x}{\sqrt{x^2+a^2}}$ **11.** $-\frac{1}{2}(a+x)^{-\frac{3}{2}}$ for $x + a > 0$

12. $-\frac{3}{2x^{\frac{5}{2}}} (x > 0)$

Exercise 6.2

Problems on trigonometric functions

Find the derivatives of the following functions w.r.t x *ab - initio*.

1. $\sin x$ **2.** $\cos x$ **3.** $\sin 2x$ **4.** $\sin\left(\frac{x}{2}\right)$

5. $\sqrt{\sin x}$ **6.** $\sqrt{\cos x}$ **7.** $\cos^2 x$ **8.** $\sin^2 x$

9. $\sin x^2$ **10.** $\cos\left(\frac{x}{2}\right)$ **11.** $\tan x$ **12.** $\tan 2x$

13. $\cot\left(\frac{x}{2}\right)$ **14.** $\sqrt{\sec x}$ **15.** $\text{cosec}^2 x$

16. $\tan 4x$ **17.** $\text{cosec}\, 3x$ **18.** $\cot 2x$ **19.** $\tan ax$

20. sec $(2x+5)$ **21.** sin (x^2+1) **22.** cos (ax^2+bx+c)
23. $x^2 \cos x$ **24.** $\tan^2 ax$

Answers:

1. $\cos x$ **2.** $-\sin x$ **3.** $2\cos 2x$

4. $\frac{1}{2}\cos\left(\frac{x}{2}\right)$

5. $\frac{\cos x}{2\sqrt{\sin x}}(2n\pi < x < (2n+1)\pi)$

6. $\frac{-\sin x}{2\sqrt{\cos x}}\left(2n\pi - \frac{\pi}{2} < x < 2n\pi + \frac{\pi}{2}\right)$

7. $-\sin 2x$

8. $\sin 2x$

9. $2x \cos x^2$

10. $\frac{-1}{2}\sin\left(\frac{x}{2}\right)$

11. $\sec^2 x \left(x \neq n\pi + \frac{\pi}{2}\right)$

12. $\sec^2 2x \left(x \neq \frac{n\pi}{2} + \frac{\pi}{4}\right)$

13. $-\frac{1}{2}\operatorname{cosec}^2\left(\frac{x}{2}\right)$ for $x \neq 2n\pi$

14. $\frac{1}{2}\cdot\sqrt{\sec x}\cdot \tan x \left(2n\pi - \frac{\pi}{2} < x < 2n\pi + \frac{\pi}{2}\right)$

15. $-2\operatorname{cosec}^2 x . \cot x,\ (x \neq n\pi)$

16. $\sec^2 4x; \left(x \neq \frac{n\pi}{4} + \frac{\pi}{8}\right)$

17. $-3\operatorname{cosec} 3x . \cot 3x; \left(x \neq \frac{n\pi}{3}\right)$

18. $-2\operatorname{cosec}^2 2x; \left(x \neq \frac{n\pi}{2}\right)$

19. $a\sec^2 ax; \left(ax \neq n\pi + \frac{\pi}{2}\right)$

20. $a\sec(2x+5)\tan(2x+5); \left(2x+5 \neq n\pi + \frac{\pi}{2}\right)$

21. $2x\cos(x^2+1)$

22. $-(2ax+b)\cdot\sin(ax^2+bx+c)$

23. $2x\cos x - x^2\sin x$

24. $2a\tan ax \sec^2 ax; \left(ax \neq n\pi + \frac{\pi}{2}\right)$

Exercise 6.3

Problems on logarithmic functions

Find from the first principle the derivatives w.r.t x of:

1. $\log(3x+2)$ **2.** $\log_a x$ $(a > 0, a \neq 1)$
3. $\log(ax+b)$ **4.** $\log_a(5x+3)$

Answers:

1. $\frac{3}{3x+2}\left(x > \frac{-2}{3}\right)$

2. $\frac{1}{x\log a}(x>0)$

3. $\frac{a}{ax+b}\left(x > \frac{-b}{a}\right)$

4. $\frac{1}{(5x+3)\log a}\left(x > \frac{-3}{5}\right)$

Exercise 6.4

Problems on exponential functions

Find from the definition the derivatives of the following:

1. e^{2x} **2.** e^{mx} **3.** e^{-x} **4.** $e^{\sqrt{x}}$ **5.** e^{4x} **6.** $a^{5x}(a>0)$
7. $a^{\sin x}$ **8.** $a^{\tan x}(a>0)$ **9.** $e^{\sin 2x}$

Answers: **1.** $2e^{2x}$ **2.** me^{mx} **3.** $-e^{-x}$

4. $\frac{e^{\sqrt{x}}}{2\sqrt{x}}, (x>0)$ **5.** $4e^{4x}$ **6.** $5a^{5x}\log a$

7. $a^{\sin x}\cos x \ . \ \log a$ **8.** $a^{\tan x}\sec^2 x \log a$

9. $2e^{\sin 2x}\cdot\cos 2x$.

7

Differentiability at a Point

To clear the concepts of differentiability of a function at a limit point in the domain of a function. The definitions of some other concepts connected with it are required to be provided.

1. Difference function $f : D \xrightarrow{\text{on to}} R$ defined by $y = f(x)$ is a function and there is an attention on a δ – neighourhood of a limit point $x = a \in D$ such that for each value of x in δ – neighbourhood of the limit point $x = a \in D \Rightarrow$ there is a number $f(x) - f(a)$ which is called the value of the difference function for the given function *f* and a given limit point namely *a*, in the domain of the function *f*. The difference $f(x) - f(a)$ is the algebraic increment in $f(a)$ for the increment $(x - a)$ in the value of x at the limit point $x = a$. It is customary to denote the difference function by the symbol $\Delta f = (f + \Delta f) - f$.

Further, domain of the difference function Δf is the same as the domain of the function *f*, i.e. $D(f) = D(\Delta f)$.

2. Difference quotient (or increment – ratio) function: $f : D \xrightarrow{\text{on to}} R$ defined by $y = f(x)$ and there is an attention on a δ – neighbourhood of a limit point $x = a \in D$ such that for each x in δ – deleted neighbourhood of the limit point $x = a \in D \Rightarrow$ there is a number $\dfrac{f(x) - f(a)}{x - a}$ which is called the value of a difference – quotient function or the value of an increment ratio function for a given function *f*. Moreover, the domain of increment ratio function is of course the domain D of the function *f* with the exception of the limit point $x = a$ at which the increment – ratio function is not defined. Now, it can be readily guessed that if $f(x)$ is continuous at the limit point $x = a$, then limit of the difference function at the limit point $x = a$ is the number zero whereas the increment ratio function is not defined at the limit point $x = a$. We are to find out whether the left-hand and the right-hand limits of the increment-ratio function at the limit point $x = a$ exist or not.

The concept of the derivative of a function at a limit point $x = a$ in the domain of a function is defined in various ways:

Definition 1: (In terms of neighbourhood): we say that a fixed number $f'(a)$ is the derivative of the function $y = f(x)$ at a limit point $x = a$ in the domain of the function $f \Leftrightarrow$ There is a fixed number $f'(a)$ such that if we choose any $\in$- neighbourhood of the fixed number $f'(a)$ denoted by $N_\varepsilon(f'(a))$, it is possible to find out a δ-neighbourhood of the limit point '*a*' in the domain of the function *f* denoted by $N_\delta(a)$ such that the values of the increment-ratio function $g(x) = \dfrac{f(x) - f(a)}{x - a}$ lies in $N_\delta(f'(a))$ for every value of x which lies in the δ-deleted neighbourhood of the limit point *a* in the domain of *f* denoted by $N'_\delta(a) = 0 < |x - a| < \delta$.

That is a number denoted by $f'(a)$ is called the derivative of a function *f* at a limit point $x = a$ in the domain of the function $f \Leftrightarrow$ There is a number $f'(a)$ such that for every ε - neighbourhood N of a number $f'(a)$ denoted by $N_\varepsilon(f'(a))$, $\exists$ a δ - neighbourhood of the limit point *a* in the domain of the function f denoted by $N_\delta(a)$ such that $g(x)$ $= \dfrac{f(x) - f(a)}{x - a} \in N_\delta(f'(a))$ for every $x \in N'_\delta(a) = 0 < |x - a| < \delta = (a - \delta, a) \cup (a, a + \delta)$ = δ-deleted neighbour-N of the limit point a in the domain of the function f.

Notes: **1.** The definition of the derivative of a function can be redefined in short in terms of δ-deleted neighbourhood N of a limit point 'a' in the domain of the function f.

Definition 2: (In terms of deleted neighbourhood): A number denoted by $f'(a)$ is called the derivative of function f at a limit point $x = a \in D(f) \Leftrightarrow \forall\ \varepsilon$-neighbourhood N of a number $f'(a)$ denoted by $N_\varepsilon(f'(a))$, $\exists$ a δ_- deleted neighbourhood N' of the limit point $x = a \in D(f)$ denoted by $N'_\delta(a)$ $= (a - \delta, a) \cup (a, a + \delta)$ such that $g(x)$ $= \dfrac{f(x) - f(a)}{x - a} \in N_\varepsilon(f'(a))$, $\forall\ x \in N'_\delta(a)$.

2. One should keep in mind that the definitions of the limit of a function $y = f(x)$ at a limit point $x = a \in D(f)$ and the derivative of function $y = f(x)$ at a limit point $x = a \in D(f)$ are similar in the following sense.

(a) The derivative of a function f at a limit point $x = a \in D(f)$ is the limit of the increment ratio function

$g(x) = \dfrac{f(x) - f(a)}{x - a}$ at a limit point $x = a \in D(f)$.

(b) In the definition of the limit of a function f at a limit point $x = a \in D(f)$, the functional values lies in $N_\varepsilon(L)$, $\forall\ x \in N'_\delta(a)$ whereas in the definition of the derivative of a function f at a limit point $x = a \in D(f)$, the values of an increment ratio

function $g(x) = \dfrac{f(x) - f(a)}{x - a} \in N_\varepsilon(f'(a))$, $\forall\ x \in N'_\delta(a)$

Definition 3: ($\varepsilon - \delta$ definition): A number $f'(a)$ is the derivative of the function f at a limit point $x = a \in D(f) \Leftrightarrow \forall\ \varepsilon > 0$, $\exists\ \delta > 0$ (δ depends on ε) such that for all points x,

$$0 < |x - a| < \delta \Rightarrow \left|\frac{f(x) - f(a)}{x - a}\ f'(a)\right| < \varepsilon$$

On one Sided Derivatives

There are two types of derivatives namely (i) right hand derivative and (ii) left hand derivative.

1. *Right hand derivative*:

***Definition* (a)**: In terms of neighbourhood: A function f is said to have the right hand derivative at a point $x = a \in D(f)$, which is also the right hand limit point of $D(f)$, denoted by $f'_+(a) \Leftrightarrow \forall\ \varepsilon$-neighbourhood N of $f'_+(a)$ denoted by $N_\varepsilon(f'_+(a))$, $\exists a$ δ-neighbourhood N of the right hand limit point 'a' of the domain of the function which is in the domain of the function f denoted by $N_\delta(a)$ such that the functional values $g(x) = \dfrac{f(x) - f(a)}{x - a}$ $\in N_\varepsilon(f'_+(a))$, $\forall\ x$ which lies in a right deleted neighbourhood N' of the right hand limit point 'a' of the domain of the function f denoted by $N'_\delta(a) = 0 < x - a < \delta = a < x < a + \delta$

***Definition* (b)**: (In terms of $(\varepsilon - \delta)$ definition): $f'_+(a)$ is called right hand derivative of function f at the right hand limit point $x = a$ of the domain of the function f which is in the domain of the function $f \Leftrightarrow \forall\ \varepsilon > 0$, $\exists\ \delta > 0$ (δ depends on ε) such that

$$0 < x - a < \delta \Rightarrow \left|\frac{f(x) - f(a)}{x - a} - f'_+(a)\right| < \varepsilon$$

i.e. $a < x < a + \delta \Rightarrow \left|\dfrac{f(x) - f(a)}{x - a} - f'_+(a)\right| < \varepsilon$

Note: One should keep in mind that $f'_+(a)$ is a notation for a finite number.

(ii) *Left hand derivative*:

***Definition* (a)**: In terms of neighbourhood: A function f is said to have the left hand derivative at a point $x = a \in D(f)$ which is also the left hand limit point $D(f)$ denoted by $f'(a) \Leftrightarrow \forall\ \varepsilon$-neighbourhood N of $f'_-(a)$ denoted by $N_\varepsilon(f'_-(a))$, $\exists$ a δ-neighbourhood N of the left hand limit point 'a' of the domain of function f which is in the domain of the function f denoted by $N_\delta(a)$, such that the functional values $g(x) = \dfrac{f(x) - f(a)}{x - a}$ $N_\varepsilon(f'_-(a))$, $\forall x$ which lies in a left deleted neighbourhood N' of the left hand limit point 'a' of the domain of the function f denoted by $N'_\partial(a) = 0 < a - x < \partial = a - \partial < x < a$.

***Definition* (b)**: In terms of $\varepsilon - \partial$ definition: A fixed number $f'_-(a)$ is called left hand derivative of a function f at the left hand limit point $x = a$ of the domain of the function f which is in the domain of the function $f \Leftrightarrow \forall\, \varepsilon > 0, \exists\, a\, \partial > 0$ (∂ depends on ε) such that

$$0 < x - a < \partial \Rightarrow \left|\frac{f(x) - f(a)}{x - a} - f'_-(a)\right| < \varepsilon$$

i.e. $$a - \partial < x < a \Rightarrow \left|\frac{f(x) - f(a)}{x - a} - f'_-(a)\right| < \varepsilon$$

On other forms: To find the derivative of a function f at a limit point $x = a$ in its domain, one should avoid making use of neighbourhood definition or ($\varepsilon - \partial$) definition for the reason that these definitions have no practical utility. This is why other forms are used in terms of which the derivative of a function f at a limit point $x = a$ in its domain is defined in the following ways.

1. Right hand derivative: If the function f is defined at a right hand limit point $x = a$, then the right hand derivative of the function f at $x = a$, denoted by $f'_+(a)$, is defined by $f'_+(a) = \lim\limits_{x \to a^+} \dfrac{f(x) - f(a)}{x - a}$

or equivalently if $x = a + h$, then

$$f'_+(a) = \lim_{h \to 0} \frac{f(a+h) - f(a)}{h}, (h > 0)$$

2. Left hand derivative: If the function f is defined at a left hand limit point $x = a$, then the left hand derivative of the function f at $x = a$, denoted by $f'_-(a)$, is defined by $f'_-(a) = \lim\limits_{x \to a^-} \dfrac{f(x) - f(a)}{x - a}$

or equivalently if $x = a - h$, then

$$f'_-(a) = \lim_{h \to 0} \frac{f(a-h) - f(a)}{-h}, (h > 0).$$

3. Derivative of a function at a limit point in its domain: If a function f is defined at a limit point $x = a$, then the derivative of f at $x = a$, denoted by $f'(a)$, is defined by $f'(a) = \lim\limits_{x \to a} \dfrac{f(x) - f(a)}{x - a}$ or equivalently if $x = a + h$, then $f'(a) = \lim\limits_{h \to 0} \dfrac{f(a+h) - f(a)}{h}$. Hence, to find the derivative of f at $x = a$, one should first of all find the difference quotient $\dfrac{f(a+h) - f(a)}{h}$ and $\dfrac{f(a-h) - f(a)}{-h}$, ($h > 0$) and then proceed to find its limit as $h \to 0$. In finding out this limit, one can make use of all the theorems on limits.

Note: The symbols $L\,f'(a)$ for $f'_-(a)$ and $R\,f'(a)$ for $f'_+(a)$ are also in use.

Question: When is a function $y = f(x)$ is said to be differentiable at a limit point $x = a \in D(f)$?

Answer: The function $y = f(x)$ is differentiable at a limit point $x = a \in D(f) \Leftrightarrow$ 'a' lies in the domain of the derived function $f'(x)$, i.e. $f'(a)$ exists, i.e.

$$f'_+(a) = \lim_{h \to 0} \frac{f(a+h) - f(a)}{h}, (h > 0) = f'(a); \text{ and}$$

$$f'_-(a) = \lim_{h \to 0} \frac{f(a-h) - f(a)}{h}, (h > 0) = f'(a)$$

= a finite number.

Question: When is a function differentiable in an interval?

Answer: A function $y = f(x)$ is said to be differentiable in an open interval (finite or infinite) $\Leftrightarrow$ The function $y = f(x)$ has a derivative at each limit point in between the left and right end points of the open interval.

Also, a function $y = f(x)$ is said to be differentiable in a closed interval $\Leftrightarrow$ The function $y = f(x)$ is differentiable at and in between the left and right end limit points of the closed interval, i.e. a function $f(x)$ is differentiable on a closed interval $[a, b] \Leftrightarrow$ it is differentiable on the interior (a,b) and if the limits $f'_+(a) = \lim_{h \to 0} \frac{f(a+h) - f(a)}{h}, (h > 0)$ = right hand derivative at the left end point of the closed interval, $f'_-(b) = \lim_{h \to 0} \frac{f(b-h) - f(b)}{-h}, (h > 0)$ = left hand derivative at the right end point of the closed interval, exist.

Lastly, a function is said to be differentiable if it is differentiable at each limit point of its domain.*

Notes: **1.** If a function $y = f(x)$ is defined on a closed interval $[a, b]$, then the value of its left hand derivative on the right end point and the value of its right hand derivative on the left end point are taken, respectively, as the values of its derivatives at the end points of the closed interval.

2. A function which has a continuous derivative is called continuously differentiable.

How to test differentiability of a function at a point?

A simple rule to test the differentiability of a function f at a point $x = a$ in its domain is to find its derived function $f'(x)$ of the function $f(x)$ using the definition of the derivative f' of a function f at any limit point x in the domain of the function, i.e.

$$f'(x) = \lim_{\Delta x \to 0} \frac{f(x + \Delta x) - f(x)}{\Delta x} \quad \text{and then to}$$

check whether $f'(x)$ is defined (continuous) or undefined (discontinuous) at the given limit point $x = a$, i.e. (i) $f'(x)$ is defined at limit $x = a \Rightarrow f(x)$ is differentiable at limit point $x = a$ (ii) $f'(x)$ is undefined at limit point $x = a \Rightarrow f(x)$ may or may not be differentiable at $x = a$ for which one must make use the definition of the left hand and right hand derivative at the given point $x = a$ whose equal finite value confirms the differentiability of $f(x)$ at the limit point $x = a$ and whose unequal finite (or, infinite) value confirms the non-differentiability of $f(x)$ at the limit point $x = a \in D(f)$.

Note: If one is asked to test the differentiability of a function f at a limit point $x = a$ in its domain with the use of the definition of the derivative of a function f at a limit point $a \in D(f)$ the above method should not be adopted.

Non-differentiability of a function at a point

Question: When a function $f(x)$ is said to be non-differentiable at a limit point $x = a \in D(f)$?

Answer: A function $f(x)$ is said to be non-differentiable at a limit point $x = a \in D(f) \Leftrightarrow f'(a)$ does not exist. For instance,

1. $f'_+(a) \neq f'_-(a)$ or

2. $f'_+(a) = f'_-(a) = \infty$ or

3. $f'_+(a) = f'_-(a) = -\infty$ or

4. $f'_+(a) = +\infty$ and $f'_-(a) = -\infty$ or

5. $f'_+(a) = \infty$ or

6. $f'_-(a) = -\infty$ or

7. $f(a)$ is undefined or imaginary.

That is, in words, a function $f(x)$ is non-differentiable (not differentiable) at a limit point $x = a \in D(f) \Leftrightarrow$ 'a' does not lie in the domain of $f'(x)$, i.e. $f'(a)$ does not exist, i.e.,

1. Either left hand derivative or right hand derivative or both left hand derivative and right hand derivative do not exist at the limit point $x = a \in D(f)$.

2. right hand derivative and left hand derivative exist but are unequal at the limit point $x = a \in D(f)$.

3. $f(x)$ is undefined or imaginary at the limit point $x \notin D(f)$.

4. In most cases a modulus function $y = |f(x)|$ or a radical function $y = \sqrt[n]{f^m(x)}$ is non-differentiable at a point $x = a \in D(f) \Leftrightarrow x = a$ makes $f(x)$ zero, i.e. $(f(x))_{x=a} = 0 \Leftrightarrow |f(x)|$ or $\sqrt[n]{f^m(x)}$ is non-differentiable at $x = a \in D(f)$, $(m, n \in R)$.

Notes: **1.** Points of discontinuity of a function $f(x)$ are also points of non-differentiability of the function.

2. Points at which $f(x)$ is non-differentiable are called points of non-differentiability of $f(x)$.

* All standard functions are differentiable in their domains.

3. The function $y = f(x)$ defined for real x, by

$f(x) = x^p \sin\left(\frac{1}{x}\right),\ x \neq 0$

$f(0) = 0$

has the following properties:

(i) $f'(0)$ does not exist if $p = 1$

(ii) $f'(0)$ exists but $f'(x)$ is not continuous at $x = 0$, if $p = 2$

(iii) $f'(x)$ is continuous at $x = 0$ if $p > 2$.

4. The set of all those points where $f(x)$ is differentiable is called domain of differentiability.

5. Derivatives at isolated points are not defined whereas a function is always continuous at an isolated point.

Question: Explain the cases where to use the concepts of left and right hand derivative to test the differentiability of a function at a point on its domain?

Answer: There are mainly four cases where the concepts of one sided derivative (left hand derivative and right hand derivative) are used.

1. When a function is defined as under:

$$f(x) = \begin{cases} f_1(x), & x \neq a \\ c, & x = a \end{cases}$$

2. When a function $f(x)$ is a piecewise function, i.e. when a function $f(x)$ is defined by more than two formulas (different expression in x) in adjacent intervals.

3. When a function contains modulus, radical or greatest integer function.

4. When the question says to examine the differentiability of a function $f(x)$ at a point $x = a$ in its domain or to examine whether $f'(a)$ exists.

On methods of finding one sided derivatives

Method 1:

How to find left hand (or left side) derivative of piecewise function $f(x)$ at a common point $x = a$ in the adjacent intervals.

Step 1. Find $f(a)$ from a form (expression in x) of the function $f(x)$ defined in a semiclosed or closed interval whose left or right end point is 'a' or from a form of the function $f(x)$ with a restriction $x = a$.

Step 2. Replace x by $(a - h)$ in a given form of the function $f(x)$ and also in an interval whose right end point is 'a'. This is $f(a - h)$, for $-h < 0$.

Step 3. Simplify the function $\left[\frac{f(a-h) - f(a)}{-h}\right]$

and cancel out the common factor h (if any).

Step 4. Find $\lim_{h \to 0}\left[\frac{f(a-h) - f(a)}{-h}\right]$ which is the required left side derivative at the right end point of the adjacent intervals where a given function $f(x)$ is defined.

How to find right hand (or right side) derivative of a piecewise function $f(x)$at a common point $x = a$ in the adjacent intervals.

Step 1. Find $f(a)$ from a form (expression in x) of the function $f(x)$ in a semiclosed or closed interval whose left or right end point is 'a' or from a form of the function $f(x)$ with a restriction $x = a$.

Step 2. Replace x by $(a + h)$ in the given form of a function $f(x)$ and also in an interval whose left end point is 'a'. This is $f(a + h)$, for $h > 0$.

Step 3. Simplify the function $\left[\frac{f(a+h) - f(a)}{h}\right]$

and cancel out the common factor h (if any).

Step 4. Find $\lim_{h \to 0}\left[\frac{f(a+h) - f(a)}{h}\right]$ which is the required right side derivative at the left end point of the adjacent intervals where a given function $f(x)$ is defined.

Notes: **1.** In the case of functions containing modulus or greatest integer function, the above method of procedure is applicable since indeed, it is these functions which are piecewise functions.

2. When a function $f(x)$ is redefined, one should put $x = a \pm h$ in the same expression and in a restriction $x \neq a$ to find $f(a + h)$ and $f(a - h)$ and then it remains to find $f(a)$ from the expression with a restriction $x = a$ and lastly $f'_+(a)$ and $f'_-(a)$ are determined, i.e. $f(x) = \begin{cases} f_1(x), & \text{when } x \neq a \\ f_2(x), & \text{when } x = a \end{cases}$

should be changed into the form:

$$f(a \pm h) = \begin{cases} f_1(h), & \text{when } h \neq 0 \\ f_2(h), & \text{when } h = 0 \end{cases}$$ and finally it

requires the use of the definitions:

$$f'_+(a) = \lim_{h \to 0} \frac{f(a+h) - f(a)}{h}, (h > 0)$$

$$f'_-(a) = \lim_{h \to 0} \frac{f(a-h) - f(a)}{-h}, (h > 0).$$

Method 2:

1. Find $f(a)$ by putting $x = a$ in one of the given function $f_1(x), f_2(x), f_3(x)$ or $f_4(x)$ against which the sign of equality with the sign of inequality in the given restrictions in the form of different intervals whose union provides us the domain of the given function $f(x)$. i.e;

Find $f(a)$ by putting $x = a$ in that function (one of $f_1(x), f_2(x), f_3(x)$ or $f_4(x)$ ... etc) against which any one of the restrictions $x \geq a, x \leq a, x = a, a \leq x < c, c < x \leq a, ...$ etc is imposed.

2. Find $f(a+h)$ by putting $x = a + h$ in the different given functions $f_1(x), f_2(x), f_3(x)$, ... etc. and in the different intervals (i.e.; $x \geq a, x \leq a, x = a, a \leq x < c, c < x \leq a, ...$ etc.) as the restrictions (or, the conditions) imposed against each different functions $f_1(x), f_2(x), f_3(x)$ ... etc i.e;

Put $x = a + h$ in all the different functions $f_1(x), f_2(x), f_3(x)$ ... etc. and in all different given intervals which are imposed as a restriction against each different function $f_1(x), f_2(x), f_3(x)$ ... etc as well as in $f(x)$ to find $f(a+h)$.

3. Solve the restrictions only to have a function $[f_1(a+h), f_2(a+h), f_3(a+h), ...$ etc] for $h > 0$ and a function $[f_1(a+h), f_2(a+h), f_3(a+h), ...$ etc for $h < 0$.

4. Use the definition:

$$\text{L } f'(a) = f'_-(a) = \lim_{h \to 0^-} \frac{f(a+h) - f(a)}{h},$$

for $h < 0$

$$\text{R } f'(a) = f'_+(a) = \lim_{h \to 0^+} \frac{f(a+h) - f(a)}{h},$$

for $h > 0$ which $\Rightarrow$

(i) Put $f_1(a+h), f_2(a+h)$, or $f_3(a+h)$, which is an expression in h for $h < 0$ (already determined) by replacing $f(a+h)$ in the definition of L $f'(a)$ i.e.; put $f(a+h)$ = proper one of $f_1(a+h), f_2(a+h)$, or $f_3(a+h)$ which is an expression in h for $h < 0$ in L $f'(0)$.

(ii) Put $f_1(a+h), f_2(a+h)$, or $f_3(a+h)$, which is an expression in h for $h > 0$ (already determined) by replacing $f(a+h)$ in the definition or R $f'(a)$ i.e.; put $f(a+h)$ = proper one of $f_1(a+h), f_2(a+h), f_3(a+h)$ or $f_4(a+h)$ etc. already determined which is an expression for $h > 0$ on R $f'(0)$.

(iii) Put $f(a)$ already determined in the definition of L $f'(0)$ and R $f'(0)$.

(iv) Simplify $\frac{f(a+h) - f(a)}{h}$ for $h < 0$ and $h > 0$

(v) Find the limit of $\frac{f(a+h) - f(a)}{h}$ for $h < 0$ and $h > 0$ as $h \to 0$ after simplifying $\frac{f(a+h) - f(a)}{h}$ for $h < 0$ and $h > 0$

***Note*:** **1.** Method 2 (or, the second method) is applicable when we can obtain various functions $f_1(h), f_2(h), f_3(h)$ ... etc. (i.e. functions or expressions in h) against which $h > 0$ and $h < 0$ are imposed as the restrictions if we put $x = a + h$ in various functions $f_1(x), f_2(x), f_3(x)$, ... etc. and in the intervals $x \geq a, x \leq a, x > a, x < a, a \leq x < b$, ... etc. which $\Rightarrow$ if $f(a+h) = f_1(h), h \geq 0$
$= f_2(h), h < 0$

are obtained by putting $x = a + h$ in the given functions and in the given intervals.

2. Inequalities $x > a \equiv (a, \infty)$

$x < a \equiv (-\infty, a)$

$x \geq a \equiv [a, \infty]$

$x \leq a \equiv [-\infty, a]$

***N.B.*:** **(i)** An inequality in x only means an interval

(ii) We put $x = (a + h)$ every where in the given function i.e., in

(a) $f(x)$

(b) $f_1(x), f_2(x), f_3(x)$, ... etc.

(c) The restrictions $x > a, x \geq a, x < a, x \leq a, c < x < a, c \leq x < a, c < x \leq a, c \leq x \leq a$, ... etc. Whenever we follow the method 2.

A Theorem

Theorem: If a function possesses a finite derivative at a point, then it is continuous at that point.

Or, if a function $f(x)$ is derivable (or, differentiable) at $x = a$, then $f(x)$ is continuous at $x = a$.

Or, the function $f(x)$ is continuous at $x = a$ if its derivative $f'(x)$ at $x = a$ (i.e.; $f'(a)$) is finite.

Or, $f'(a)$ is finite $\Rightarrow f(x)$ is continuous at $x = a$.

Proof: Hypothesis is $\lim\limits_{x \to a} \dfrac{f(x) - f(a)}{x - a} = f'(a)$

which is finite [from the definition] ...(i) Now,

$$[f(x) - f(a)] = \frac{[f(x) - f(a)]}{x - a} \times (x - a)$$

Now, taking the limit as $x \to a$, we get

$$\lim_{x \to a} [f(x) - f(a)] = \lim_{x \to a} \left\{ \frac{[f(x) - f(a)]}{(x - a)} \times (x - a) \right\}$$

$$\Rightarrow \lim_{x \to a} [f(x) - f(a)] = \left\{ \lim_{x \to a} \frac{[f(x) - f(a)]}{(x - a)} \right\} \times \left\{ \lim_{x \to a} (x - a) \right\}$$

$$\Rightarrow \lim_{x \to a} [f(x) - f(a)] = f'(a) \times 0 \left[\because \lim_{x \to a} (x - a) = 0 \text{ and from (1), } f'(a) = \lim_{x \to a} \frac{f(x) - f(a)}{x - a} \right]$$

$$\Rightarrow \lim_{x \to a} [f(x) - f(a)] = 0$$

$$\Rightarrow \lim_{x \to a} f(x) - \lim_{x \to a} f(a) = 0$$

$$\Rightarrow \lim_{x \to a} f(x) = \lim_{x \to a} f(a) = f(a)$$

$$\Rightarrow f(x) \text{ is continuous at } x = a$$

Hence, the theorem is proved.

Remark: But the converse of the theorem is not true.

We cite the following example.

Let $f(x) = |x|$

We claim that $f(x)$ is continuous at the origin (i.e.; $x = 0$) but not differentiable at the origin.

Now, because,

$f(0) = |0| = 0$,

$$f(0 - 0) = \lim_{h \to 0} f(0 - h) = \lim_{h \to 0} f(-h)$$

$$= \lim_{h \to 0} |-h| = \lim_{h \to 0} (h) = 0, \quad \text{...(i)}$$

$$f(0 + 0) = \lim_{h \to 0} f(0 + h) = \lim_{h \to 0} f(h)$$

$$= \lim_{h \to 0} |h| = \lim_{h \to 0} (h) = 0 \quad \text{...(ii)}$$

Thus, from (i) and (ii), we get

$\lim\limits_{h \to 0} f(0 - h) = \lim\limits_{h \to 0} f(0 + h) = 0$ which $\Rightarrow$

$f(x)$ is continuous at $x = 0$

Now, test of differentiability:

$$\text{L } f'(0) = \lim_{h \to 0} \frac{f(0 - h) - f(0)}{-h}$$

$$= \lim_{h \to 0} \frac{f(-h) - f(0)}{(-h)}$$

$$\Rightarrow \text{L } f'(0) = \lim_{h \to 0} \frac{|(-h)| - 0}{-h} = \lim_{h \to 0} \left(\frac{h}{-h} \right) = -1 \quad \text{...(iii)}$$

$$\text{R } f'(0) = \lim_{h \to 0} \frac{f(h) - f(0)}{h}$$

$$\Rightarrow \text{R } f'(0) = \lim_{h \to 0} \frac{|(h)| - 0}{h} = \lim_{h \to 0} \frac{|h|}{h} = \lim_{h \to 0} \frac{h}{h}$$

$$= 1 \quad \text{...(iv)}$$

Thus, from (iii) and (iv), we get, L $f'(0) = -1 \neq 1 =$ R $f'(0)$ which $\Rightarrow f'(0)$ does not exist.

Conclusion:

1. Differentiability at a point $\Rightarrow$ continuity at the same point.

2. Continuity at a point $\not\Rightarrow$ differentiability at the same point.

Note: **1.** We say that $f(x)$ is differentiable at the left end point 'a' of an interval $[a, b]$ or $[a, b)$, we mean that $f'_+(a)$ exists.

Similarly, we say that $f(x)$ is differentiable at the right end point 'b' of an interval $(a, b]$ or $[a, b]$ we mean that $f'_-(b)$ exists.

2. By a closed interval $[a,b]$ (or $a \le x \le b$), we mean that the end points a and b of the interval are also to be considered while by an open interval (a,b) (or, $]a, b[$ or, $a < x < b$) we mean that end points a and b are not to be considered in any problem.

Important deductions based on the definition of derivative of a function $f(x)$ at a point $x = a$

(i) If $y = \dfrac{x f(a) - a f(x)}{x - a}$,

then $\lim\limits_{x \to a} \dfrac{x f(a) - a f(x)}{x - a} = f(a) - a f'(a)$

= value of $f(x)$ at $x = a - a$ times d.c of $f(x)$ at $x = a$.

Proof: $\lim\limits_{x \to a} \dfrac{x f(a) - a f(x)}{x - a}$

$$= \lim_{x \to a} \frac{[x f(a) - a f(a)] - a[f(x) - f(a)]}{x - a}$$

[adding and subtracting $a f(a)$ in Nr]

$$= \lim_{x \to a} \frac{f(a)(x - a)}{(x - a)} - a \lim_{x \to a} \frac{f(x) - f(a)}{x - a}$$

$= f(a) - a f'(a)$

= value of $f(x)$ at $x = a - a$ times d.c. of $f(x)$ at $x = a$

$$= [f(x)]_{x=a} - a\left[\frac{dy}{dx}\right]_{x=a}$$

e.g.:(a) $\lim\limits_{x \to \alpha} \dfrac{x \sin \alpha - \alpha \sin x}{x - \alpha} = \sin \alpha - \alpha \cos \alpha$

2. If $y = \dfrac{f(x) - f(y)}{x - y}$ $\left(\text{or}, = \dfrac{f(y) - f(x)}{y - x}\right)$,

then $= \lim\limits_{y \to x} \dfrac{f(x) - f(y)}{x - y} = f'(x)$ e.g.,

(a) $\lim\limits_{y \to x} \dfrac{\sin x - \sin y}{x - y} = \cos x = \dfrac{d \sin x}{dx}$

(b) $\lim\limits_{y \to x} \dfrac{\tan x - \tan y}{x - y} = \sec^2 x = \dfrac{d \tan x}{dx}$

3. If $y = \dfrac{f(x) - f(y)}{x - y}$,

then $\lim\limits_{x \to y} \dfrac{f(x) - f(y)}{x - y} = f'(y)$

e.g.: **(a)** $\lim\limits_{x \to y} \dfrac{\tan x - \tan y}{x - y} = \sec^2 y = \dfrac{d \tan y}{dy}$

(b) $\lim\limits_{x \to y} \dfrac{\sin x - \sin y}{x - y} = \cos y = \dfrac{d \sin y}{dy}$

N.B.: **(i)** Remember that when $x \to y$ we get $f'(y)$ and when $y \to x$, we get $f'(x)$ as in (2) and (3) i.e.; after f', converging point is written.

(ii) On the left side of the symbol "$\to$" independent variable of he function is mentioned and the right side of the symbol "$\to$" the constant quantity (i.e.; the limit of independent variable) is mentioned. i.e.;

$\lim\limits_{a \to b} f(a)$, a = independent variable, $b = \lim a$

Types of the problem

Type A

1. To find l.h.d and r.h.d or $f'(a)$ by using the definition of d.c

2. Existence and non-existence of $f'(x)$ at a point $x = a$.

3. To show whether a given function is differentiable or not at a given point.

4. Examination of differentiability at a given point or, discussion of differentiability at a given point.

All above types of problems have the following working rule.

1. Find L.H.D or R.H.D

2. Observe whether L.H.D and R.H.D are equal or not.

3. Conclude $f'(x)$ at a point to be differentiable or not according to the observation.

Type B

1. To find the value of constants.

Type 1: Problems based on piecewise function.

Examples worked out:

1. If $f(x)=3x-4,\ x\le 2$

$f(x)=2(2x-3),\ x>2$ examine $f'(2)$.

Solution: *Method 1*

$f(x)=(3x-4)$, for $x<2$

$\Rightarrow f(2-h)=3(2-h)-4$, for $h>0$ [$\because 2-h<2$]

$=6-3h-4=2-3h$ ...(1)

$\therefore f(2)=3\times 2-4=6-4=2$ ...(2)

$\therefore$ (1) – (2) $= f(2-h)-f(2)=2-3h-2=-3h$ ...(3)

Again, $f(x)=4x-6$ for $x>2$

$\Rightarrow f(2+h)=4(2+h)-6=8+4h-6=2+4h$ ...(4)

$[2+h>2]$

$\therefore$ (4) – (2) $\Rightarrow f(2+h)-f(2)=2+4h-2=4h$ ...(5)

Now, $\text{R.H.D} = \lim_{h\to 0}\dfrac{f(2+h)-f(2)}{h},\ h>0$

$= \lim_{h\to 0}\dfrac{4h}{h}=4$ ...(6)

$\text{L.H.D} = \lim_{h\to 0}\dfrac{f(2-h)-f(2)}{-h},\ h>0$

$= \lim_{h\to 0}\dfrac{-3h}{-h}=3$ ...(7)

From (6) and (7), we see that $f'_-(2)\ne f'_+(2)$

$\Rightarrow f'(2)$ does not exist.

Method 2: Given is

$f(x)=3x-4,\ x\le 2$

$f(x)=2(2x-4),\ x>2$

By definition of function $f(x)$,

$f(2+h)=[3(2+h)-4],\ 2+h\le 2$

$=2[2(2+h)-3],\ 2+h>2$

$\Rightarrow f(2+h)=[6+3h-4],\ h\le 0$

$=2[4+2h-3],\ h>0$

$\Rightarrow f(2+h)=2+4h$ when $h>0$ ...(1)

$=2+3h$ when $h\le 0$ ...(2)

and $f(2)=2$ ...(3)

Now,

$\text{L.H.D} = \lim_{h\to 0}\dfrac{f(2+h)-f(2)}{h}$ for $h<0$

$= \lim_{h\to 0}\dfrac{2+3h-2}{h}=\lim_{h\to 0}\dfrac{3h}{h}=3$ ...(4)

$\text{R.H.D} = \lim_{h\to 0}\dfrac{f(2+h)-f(2)}{h}$, for $h>0$

$= \lim_{h\to 0}\dfrac{2+4h-2}{h}=\lim_{h\to 0}\dfrac{4h}{h}=4$ ...(5)

Hence, from (4) and (5), we see that l.h.d. $\ne$ r.h.d which $\Rightarrow f'(2)$ does not exist.

2. Test the differentiability of the function

$$f(x)=\begin{cases} x, & 0\le x\le 1 \\ 2x-1, & 1<x \end{cases}\quad \text{at } x=1.$$

Solution: By using method (2), from the definition of the function $f(x)$,

$$f(1+h)=\begin{cases} 1+h, & \text{when } h<0 \\ 2(1+h)-1, & \text{when } h>0 \end{cases}$$

$f(1)=1$ (considering $f(x)=x$ for the value of the function $f(x)$ at $x=1$)

$\text{L}f'(1)=\lim_{h\to 0}\dfrac{f(1+h)-f(1)}{h},\ h<0$

$=\lim_{h\to 0}\dfrac{1+h-1}{h}=1$ and

$\text{R}f'(1)=\lim_{h\to 0}\dfrac{f(1+h)-f(1)}{h},\ h>0$

$=\lim_{h\to 0}\dfrac{2h+1-1}{h}=2$

thus, $\text{L}f'(1)\ne \text{R}f'(1)\Rightarrow f'(1)$ does not exist $\Rightarrow$ $f(x)$ is not differentiable at $x=1$

3. A function f (or, $f(x)$) is defined by $f(x)=x+2$, when $0\le x<2$

$=\sqrt{8x}$, when $2\le x\le 4$,

find the left hand derivative and right hand derivative of f at $x=2$.

Solution:

$$\text{l.h.d} = f'_{-}(2) = \lim_{h\to 0} \frac{f(2-h)-f(2)}{2-h-2}, h>0$$

$$= \lim_{h\to 0} \frac{f(2-h)-f(2)}{-h} \quad \ldots(A)$$

Now, $f(x) = x+2$ for $x<2$

$\Rightarrow f(2-h) = 2-h+2 = 4-h$ ...(1)

$[\because 2-h<2]$

$f(2) = \left[\sqrt{8x}\right]_{x=2} = \sqrt{16} = \sqrt{4\times 4} = 4$ ...(2)

Putting (1) and (2) in (A), we get

$$f'_{-}(2) = \lim_{h\to 0} \frac{4-h-4}{-h} = \lim_{h\to 0} \frac{-h}{-h} = 1 \quad \ldots(3)$$

Again,

$$\text{r.h.d} = f'_{+}(2) = \lim_{h\to 0} \frac{f(2+h)-f(2)}{2+h-2}, h>0$$

$$= \lim_{h\to 0} \frac{f(2+h)-f(2)}{h} \quad \ldots(B)$$

and $f(x) = \sqrt{8x}$ for $x>2$

$\Rightarrow f(2+h) = \sqrt{8(2+h)} = \sqrt{16+8h}$ ... (4)

$[\because 2+h>2]$

Putting (2) and (4) in (B),

$$f'_{+}(2) = \lim_{h\to 0} \frac{\sqrt{16+8h}-4}{h}$$

$$= \lim_{h\to 0} \frac{\left(\sqrt{16+8h}-4\right)\times\left(\sqrt{16+8h}+4\right)}{h\left(\sqrt{16+8h}+4\right)}$$

$$= \lim_{h\to 0} \frac{\cancel{16}+8h-\cancel{16}}{h\left(\sqrt{16+8h}+4\right)}$$

$$= \lim_{h\to 0} \frac{8\cancel{h}}{\cancel{h}\left(\sqrt{16+8h}+4\right)}$$

$$= \lim_{h\to 0} \frac{8}{\left(\sqrt{16+8h}+4\right)}$$

$$= \frac{8}{\left(\sqrt{16+0}+4\right)} = \frac{8}{4+4} = \frac{8}{8} = 1 \quad \ldots(4)$$

thus, (3) and (4) $\Rightarrow f'_{-}(2) = f'_{+}(2) = 1 \Rightarrow f'(2)$ exists and $= 1$.

4. If $f(x) = \begin{cases} -x, & x<0 \\ x^2, & 0\le x\le 1 \\ x^3-x+1, & x>1 \end{cases}$

test the differentiability at $x=0$ and $x=1$

Solution: At $x=0$

$f(x) = x^2$ ...(1)

$\Rightarrow f(0) = 0$ when $x>0$,

$f(x) = x^2$ when $x<0$, ...(2)

$f(x) = -x$ ...(3)

$\therefore f(0+h) = (0+h)^2 = h^2, h>0$

$f(0-h) = -(0-h) = h$

$$\text{Now, } f'_{-}(0) = \lim_{h\to 0} \frac{f(0-h)-f(0)}{-h}, h>0$$

$$= \lim_{h\to 0} \frac{f(-h)-f(0)}{-h} = \lim_{h\to 0} \frac{h-0}{-h} = -1 \ldots(4)$$

$$f'_{+}(0) = \lim_{h\to 0} \frac{f(0+h)-f(0)}{h}, h>0$$

$$= \lim_{h\to 0} \frac{f(h)-f(0)}{h}$$

$$= \lim_{h\to 0} \frac{h^2-0}{h} = \lim_{h\to 0} \frac{h^2}{h} = \lim_{h\to 0} h = 0$$

Thus, $f'_{-}(0) \ne f'_{+}(0) \Rightarrow f'(0)$ does not exist

$\Rightarrow f(x)$ is not differentiable at $x=0$

Similarly, we can test the differentiability at $x=1$

5. If $f(x) = \begin{cases} 2x+7, & \text{when } x\le 3 \\ 16-x, & \text{when } x\ge 3 \end{cases}$ test the differentiability at $x=3$.

Solution: Let $h>0$.

$$\text{R.H.D} = f'_{+}(3) = \lim_{h\to 0} \frac{f(3+h)-f(3)}{h}, h>0$$

$$= \lim_{h \to 0} \frac{16 - (3+h) - 13}{h} = \lim_{h \to 0} \left[\frac{-h}{h}\right]$$

$$= \lim_{h \to 0} [-1] = -1 \qquad \ldots(1)$$

$$\text{L.H.D} = f'_-(3) = \lim_{h \to 0} \frac{f(3-h) - f(3)}{-h}, h > 0$$

$$= \lim_{h \to 0} \frac{7 + 2(3-h) - 13}{-h} = \lim_{h \to 0} \left[\frac{-2h}{-h}\right]$$

$$= \lim_{h \to 0} [2] = 2 \qquad \ldots(2)$$

Thus, (1) and (2) $\Rightarrow$ L.H.D $\neq$ R.H.D $\Rightarrow f(x)$ is not differentiable at $x = 3$

Second method:

$f(3) = [2x + 7]_{x=3} = 2 \times 3 + 7 = 13$ ...(1)

Now, using the definition of the function,

$f(3 + h) = [2(3 + h) + 7],\ 3 + h \le 3$

$\Rightarrow f(3 + h) = 6 + 2h + 7,\ h \le 0$ ...(2)

$f(3 + h) = [16 - (3 + h)],\ 3 + h \ge 3$

$\Rightarrow f(3 + h) = 13 - h,\ h \ge 0$ ... (3)

Now, using the definition,

$$\text{L } f'(3) = \underset{h \to 0}{\text{L}} \frac{f(a+h) - f(a)}{h},\ h < 0$$

$$= \underset{h \to 0}{\text{L}} \frac{13 + 2h - 13}{h} = \underset{h \to 0}{\text{L}} \frac{2h}{h}$$

$$= \underset{h \to 0}{\text{L}}\ 2 = 2 \qquad \ldots(4)$$

$$\text{R } f'(3) = \underset{h \to 0}{\text{L}} \frac{f(a+h) - f(a)}{h},\ h > 0$$

$$= \underset{h \to 0}{\text{L}} \frac{13 - h - 13}{h} = \underset{h \to 0}{\text{L}} \frac{-h}{h}$$

$$= \underset{h \to 0}{\text{L}} (-1) = -1 \qquad \ldots(5)$$

Thus, (4) and (5) $\Rightarrow f'_-(3) \neq f'_+(3) \Rightarrow f'(3)$ does not exist $\Rightarrow f(x)$ is not differentiable at $x = 3$

Remember: If $f(x) = f_1(x)$, when $x \le a$

$f(x) = f_2(x)$, when $x \ge a$,

then we may consider any one of $f_1(x)$ and $f_2(x)$ for considering the value of $f(x)$ at $x = a$ because both give the same value.

6. If $f(x) = \dfrac{\sin x}{x}$ when $x > 0$

$= 1 - x\cos x$, when $x \le 0$,

show that the function $f(x)$ is not differentiable at $x = 0$.

Solution: $f'_+(0) = \lim\limits_{h \to 0} \dfrac{f(0+h) - f(0)}{h}, h > 0$

$$= \lim_{h \to 0} \frac{\frac{\sin h}{h} - 1}{h} = \lim_{h \to 0} \frac{\sin h - h}{h^2} \left[\text{form is } = \frac{0}{0}\right]$$

$$= \lim_{h \to 0} \frac{\left[h - \frac{h^3}{\lfloor 3} + \frac{h^5}{\lfloor 5} \ldots\right] - h}{h^2}$$

$$= \lim_{h \to 0} \left[-\frac{h}{\lfloor 3} + \frac{h^3}{\lfloor 5} + \ldots\right] = 0 \qquad \ldots(1)$$

$$\text{Again, } f'_-(0) = \lim_{h \to 0} \frac{[(1 + h\cos h) - 1]}{-h}$$

$$= \lim_{h \to 0} [-\cos h] = -1 \qquad \ldots(2)$$

Thus, we see that $f'_+(0) \neq f'_-(0) \Rightarrow f'(0)$ does not exist $\Rightarrow f(x)$ is not differentiable at $x = 0$.

Type 2: Problems based on the function which are redefined.

Examples worked out:

1. If $f(x) = \dfrac{x - 1}{2x^2 - 7x + 5}$, when $x \neq 1$ and $= -\dfrac{1}{3}$, when $x = 1$ find $f'(1)$.

Solution: $f(x) = \dfrac{x - 1}{2x^2 - 7x + 5}$

$$\therefore f'(1) = \lim_{h \to 0} \frac{f(1+h) - f(1)}{h}$$

$$= \lim_{h \to 0} \frac{\dfrac{(1+h)-1}{2(1+h)^2 - 7(1+h) + 5} - \left(-\dfrac{1}{3}\right)}{h}$$

$$= \lim_{h \to 0} \frac{\dfrac{h}{2h^2 - 3h} + \dfrac{1}{3}}{h}$$

$$= \lim_{h \to 0} \frac{\dfrac{1}{2h-3} + \dfrac{1}{3}}{h}$$

$$= \lim_{h \to 0} \frac{3 + 2h - 3}{3h(2h-3)}$$

$$= \lim_{h \to 0} \frac{2}{3(2h-3)} = -\frac{2}{9}$$

Note: Problems where $f(x) = f_1(x)$, when $x \neq a$ = constant, when $x = a$ are provided and we are required to find $f'(a)$, we use the definition, $f'(a) = \lim_{h \to 0} \frac{f(a+h) - f(a)}{h}$ which $\Rightarrow$ we are not required to find l.h.d. and r.h.d. separately but when we have to examine $f'(a)$ or to test the differentiability of the given function $f(x)$ at $x = a$ which is redefined, it is a must to find l.h.d and r.h.d. for the same function $f(x) = f_1(x)$, when $x \neq$ by putting $(a + h)$ and $(a - h)$ separately since $x \neq a$ means $x > a$ and $x < a$ which $\Rightarrow$ we have to consider the same function $f(x)$ whose independent variable is replaced by $(a - h)$ while finding l.h.d. and the independent variable x in $f(x)$ is replaced by $(a + h)$ while finding r.h.d.

2. If $f(x) = \dfrac{x-2}{x^2 - 3x + 2}$, $x \neq 2$ and $= 1$, $x = 2$ examine the differentiability at $x = 2$.

Solution: $f'_+(2) = \lim_{h \to 0} \dfrac{f(2+h) - f(2)}{h}$

$$= \lim_{h \to 0} \left[\frac{\dfrac{(2+h)-2}{(2+h)^2 - 3(2+h) + 2} - 1}{h} \right]$$

$$= \lim_{h \to 0} \frac{h - (2+h)^2 + 3(2+h) - 2}{h\{(2+-h)^2 - 3(2+h) + 2\}}$$

$$= \lim_{h \to 0} \frac{h - \left(4 + h^2 + 4h\right) + 6 + 3h - 2}{h\left(4 + h^2 + 4h - 6 - 3h + 2\right)}$$

$$= \lim_{h \to 0} \frac{h - \left(4 + h^2 + 4h\right) + 6 + 3h - 2}{h\left(4 + h^2 + 4h - 6 - 3h + 2\right)}$$

$$= \lim_{h \to 0} \frac{h - \left(4 + h^2 + 4h\right) + 6 + 3h - 2}{h\left(6 + h^2 + h - 6\right)}$$

$$= \lim_{h \to 0} \frac{h - 4 - h^2 - 4h + 6 + 3h - 2}{h\left(h^2 + h\right)}$$

$$= \lim_{h \to 0} \frac{h - 6 - h^2 - h + 6}{h\left(h^2 + h\right)}$$

$$= \lim_{h \to 0} \frac{-h^2}{h^2(h+1)}$$

$$= \lim_{h \to 0} \frac{-1}{(h+1)} = -1$$

$$f'_-(2) = \lim_{h \to 0} \frac{f(2-h) - f(2)}{-h}$$

$$= \lim_{h \to 0} \frac{\dfrac{(2-h)-2}{(2-h)^2 - 3(2-h) + 2} - 1}{-h}$$

$$= \lim_{h\to 0} \frac{(2-h)-2-(2-h)^2+3(2-h)-2}{-h\{(2-h)^2-3(2-h)+2\}}$$

$$= \lim_{h\to 0} \frac{\cancel{2}-h-\cancel{2}-(4+h^2-4h)+6-3h-2}{-h(6+h^2-h-6)}$$

$$= \lim_{h\to 0} \frac{-h-4-h^2+4h+6-3h-2}{-h(6+h^2-h-6)}$$

$$= \lim_{h\to 0} \frac{-4h-6-h^2+4h+6}{-h(h^2-h)}$$

$$= \lim_{h\to 0} \frac{-\cancel{h^2}}{-\cancel{h^2}(h-1)} = \lim_{h\to 0} \frac{1}{(h-1)} = \frac{1}{0-1} = -1$$

$\therefore\ f'_{+1}(2) = f'_{-}(2) = -1$ which $\Rightarrow$ gives function $f(x)$ is differentiable at $x=2$.

3. If $f(x) = \begin{cases} \dfrac{x}{1+e^{1/x}}, & x \neq 0 \\ 0, & x = 0 \end{cases}$ test the differentiability at $x=0$.

Solution: Given that $f(x) = \dfrac{x}{1+e^{1/x}}$, when $x>0$

$$f(x) = \frac{x}{1+e^{1/x}}, \text{ when } x<0$$

$f(x)=0$, when $x=0$

Now, $\because f(x) = 0$ when $x=0$

$\Rightarrow f(0)=0$ ...(1)

$$f(x) = \frac{x}{1+e^{1/x}}, \text{ when } x>0$$

$\therefore$ For $h>0$,

$$f(0+h) = \frac{0+h}{1+e^{\frac{1}{0+h}}} = \frac{h}{1+e^{\frac{1}{h}}} \quad ...(2)$$

$\therefore$ (2)−(1)

$$\Rightarrow f(0+h)-f(0) = \frac{h}{1+e^{\frac{1}{h}}} - 0 = \frac{h}{1+e^{\frac{1}{h}}} \quad ...(3)$$

Again, $\because f(x)$ (when $x<0$) $= \dfrac{x}{1+e^{\frac{1}{x}}}$

$$\therefore f(0-h) = \frac{0-h}{1+e^{\frac{1}{0-h}}} = \frac{-h}{1+e^{-\frac{1}{h}}} \quad ...(4)$$

Now, using the definition,

$$f'_{+}(0) = \lim_{h\to 0} \frac{f(0+h)-f(0)}{h};\ h>0$$

$$= \lim_{h\to 0} \frac{\dfrac{h}{1+e^{\frac{1}{h}}}}{h} = \lim_{h\to 0} \frac{1}{1+e^{\frac{1}{h}}} = \frac{1}{\infty} = 0 \quad ...(5)$$

$$f'_{-}(0) = \lim_{h\to 0} \frac{f(0-h)-f(0)}{-h}, h>0$$

$$= \lim_{h\to 0} \frac{\dfrac{(-h)}{1+e^{-\frac{1}{h}}}}{(-h)}$$

$$= \lim_{h\to 0} \frac{1}{1+e^{-\frac{1}{h}}}$$

$$= \frac{1}{1+0} = 1 \quad ...(6)$$

Thus, we see that $f'_{+}(0) \neq f'_{-}(0) \Rightarrow f'(0)$ does not exist $\Rightarrow f'(x)$ is not differentiable at $x=0$

> ***Remember***: For $h>0$ **(i)** $\lim_{h\to 0} e^{-\frac{1}{h}} = e^{-\infty} = 0$
>
> **(ii)** $\lim_{h\to 0} e^{\frac{1}{h}} = e^{\infty} = \infty$

4. Show that $f(x) = \begin{cases} x^2 \sin\dfrac{1}{x}, & \text{when } x \neq 0 \\ 0, & \text{when } x = 0 \end{cases}$

is differentiable at $x = 0$.

Solution: Given that $f(x) = x^2 \sin\dfrac{1}{x}$, when $x > 0$

$(\because x \neq 0 \Rightarrow x > 0 \text{ and } x < 0)$

$f(x) = x^2 \sin\dfrac{1}{x}$, when $x < 0$

$f(x) = 0$, when $x = 0$

Now, $f(x) = 0$, when $x = 0$

$\Rightarrow f(0) = 0$ ...(1)

$f(x)$ (when $x > 0$) $= x^2 \sin\dfrac{1}{x}$

$\therefore$ For $h > 0$, $f(0 + h) = (0 + h)^2 \cdot \sin\left(\dfrac{1}{0 + h}\right)$

$= h^2 \sin\left(\dfrac{1}{h}\right)$ ...(2)

$\therefore (2) - (1) \Rightarrow f(0 + h) - f(0) = h^2 \sin\left(\dfrac{1}{h}\right) - 0$

$= h^2 \sin\left(\dfrac{1}{h}\right)$ ...(3)

Again, $f(x)$ (when $x < 0$) $= x^2 \sin\dfrac{1}{x}$

$\therefore$ For $h > 0$

$f(0 - h) = (0 - h)^2 \cdot \sin\left(\dfrac{1}{0 - h}\right)$

$= h^2 \sin\left(-\dfrac{1}{h}\right)$ ...(4)

Now, using the definition,

$$f'_+(0) = \lim_{h \to 0} \frac{f(0 + h) - f(0)}{h}, h > 0$$

$$= \lim_{h \to 0} \frac{h^2 \sin\left(\frac{1}{h}\right)}{h}$$

$$= \lim_{h \to 0} h \sin\left(\frac{1}{h}\right) = 0 \quad \text{...(5)}$$

$$f'_-(0) = \lim_{h \to 0} \frac{f(0 - h) - f(0)}{-h}, h > 0$$

$$= \lim_{h \to 0} \frac{h^2 \sin\left(-\frac{1}{h}\right) - 0}{-h}$$

$$= \lim_{h \to 0} \frac{-h^2 \sin\left(\frac{1}{h}\right)}{-h}$$

$$= \lim_{h \to 0} h \sin\left(\frac{1}{h}\right) = 0 \quad \text{...(6)}$$

$\therefore$ (5) and (6) $\Rightarrow f'_+(0) = f'_-(0) = 0$ which means that $f'(0)$ exists. i.e.; $f'(x)$ exists at $x = 0 \Rightarrow f(x)$ is differentiable at $x = 0$.

5. If $f(x) = x \sin\left(\dfrac{1}{x}\right)$, when $x \neq 0 = 0$, when $x = 0$.

Show that the function $f(x)$ does not have the derivative at the point $x = 0$.

Solution: $\because f(x) = x \sin\left(\dfrac{1}{x}\right)$, when $x \neq 0$

$f(0) = 0$

$$\therefore f'(0) = \lim_{h \to 0} \frac{f(0 + h) - f(0)}{h}$$

$$= \lim_{h \to 0} \frac{(0 + h) \sin\left(\frac{1}{0 + h}\right) - 0}{h}$$

$$= \lim_{h \to 0} \frac{h \sin\left(\frac{1}{h}\right)}{h}$$

$= \lim\limits_{h \to 0} \sin\left(\dfrac{1}{h}\right)$ which has no finite value

which means that $f'(x)$ does not exist at $x = 0$.

6. If $f(x) = x^2$, when $x \geq 1$, and $f(x) = -x$, when $x < 1$, show that the function $f(x)$ does not have derivative $x = 1$.

Solution: $f(x) = x^2$, when $x > 1$

$\Rightarrow f(1+h) = (1+h)^2$, when $1+h > 1$, i.e. $h > 0$

$= (1+h)^2$, when $h > 0$ and $f(x) = -x$, when $x < 1$

$\Rightarrow f(1+h) = -(1+h)$, when $1+h < 1$, i.e.; $h < 0$

$= -(1+h)$, when $h < 0, f(1) = 1$

$$\therefore f'_+(1) = \lim_{h \to 0} \frac{f(1+h) - f(1)}{h}, h > 0$$

$$= \lim_{h \to 0} \frac{(1+h)^2 - 1}{h} = \lim_{h \to 0} \frac{1 + 2h + h^2 - 1}{h}$$

$$= \lim_{h \to 0} \frac{2h + h^2}{h} = \lim_{h \to 0} \frac{h(h+2)}{h}$$

$$= \lim_{h \to 0} (h+2) = \lim_{h \to 0} h + \lim_{h \to 0} 2$$

$$= 0 + 2 = 2 \quad \ldots(1)$$

$$f'_-(1) = \lim_{h \to 0} \frac{f(1+h) - f(1)}{h}, h < 0$$

$$= \lim_{h \to 0} \frac{-(1+h) - 1}{h} = \lim_{h \to 0} \frac{-1 - h - 1}{h}$$

$$= \lim_{h \to 0} \frac{-2 - h}{h} = \lim_{h \to 0} \left(-\frac{2}{h}\right) - \lim_{h \to 0} \left(\frac{h}{h}\right)$$

$$= -\infty - 1 = -\infty \quad \ldots(2)$$

$\therefore$ (1) and (2) $\Rightarrow f'_+(1) \neq f'_-(1)$ which $\Rightarrow f'(1)$ does not exist which means that $f(x)$ has no finite derivative at $x = 1$.

Type 3: Problems based on a mod function $|f(x)|$.

Whenever we want to test the differentiability of a mod function $|f(x)|$ at a point $x = a$ at which $f(x) = 0$ or $f(x) \neq 0$, we are required to find the l.h.d. and r.h.d at a given point $x = a$ because a mod function $|f(x)|$ can be expressed as a piecewise function in the following way.

$|f(x)| = f(x)$, when $f(x) \geq 0$

$= -f(x)$, when $f(x) < 0$

This is why a mod function $|f(x)|$ also is called the piecewise function.

Note: **1.** Generally, we are asked to test the differentiability of a mod function $|f(x)|$ at a point $x = a$ at which $f(x) = 0$.

2. In most cases of $|f(x)|$, point $x = a$ at which $f(x) = 0$, we have l.h.d $\neq$ r.h.d which $\Rightarrow$ in most cases, the mod function $|f(x)|$ is not differentiable at a point $x = a$ at which $f(x) = 0$.

3. To find the derivative or existence of a derivative of a mod function at a point $x = a$ where $x = a$ is a point at which $f(x) = 0$, we are required to find the l.h.d and r.h.d separately,

4. If $x = a$ be a point at which $f(x) \neq 0$ in $|f(x)|$, then we use the following formula to find the value of the derivative of $|f(x)|$ at $x = a$.

$$\left[\frac{d}{dx}|f(x)|\right]_{x=a} = \left[\frac{|f(x)|}{f(x)} \times f'(x)\right]_{x=a}$$

5. $|h| = |-h| = h$ and $|h^2| = h^2 = |h|^2$

Examples worked out:

1. If $f(x) = |x|$, show that the function $f(x)$ does not have derivative at $x = 0$

Solution: $\because f(x) = |x|$

$$\text{R } f'(0) = \lim_{h \to 0} \frac{f(0+h) - f(0)}{h}, h > 0$$

$$= \lim_{h \to 0} \frac{|0+h| - |0|}{h} = \lim_{h \to 0} \frac{|h|}{h}$$

$$= \lim_{h \to 0} \frac{h}{h} = \lim_{h \to 0} 1 = 1 \quad \ldots(\text{i})$$

$$\text{L } f'(0) = \lim_{h \to 0} \frac{f(0-h) - f(0)}{-h}$$

$$= \lim_{h \to 0} \frac{|0-h| - |0|}{-h} = \lim_{h \to 0} \frac{|-h|}{-h}$$

$$= \lim_{h \to 0} \frac{h}{-h} = \lim_{h \to 0} (-1) = -1 \quad \ldots(\text{ii})$$

(i) and (ii) $\Rightarrow$ R $f'(0) \neq$ L $f'(0)$ which $\Rightarrow f'(0)$ does not exist i.e.; $f(x)$ has no finite derivative at $x = 0$.

2. Examine the differentiability of $f(x)$ at the indicated points.

(i) $f(x)=|\cos x|$ at $x=\dfrac{\pi}{2}$

(ii) $f(x)=|x^3|$ at $x=0$

Solution: (i) $f(x)=|\cos x|$

$$\therefore \text{R } f'\left(\frac{\pi}{2}\right)=\lim_{h\to 0}\frac{\left|\cos\left(\frac{\pi}{2}+h\right)\right|-\left|\cos\frac{\pi}{2}\right|}{h}, h>0$$

$$=\lim_{h\to 0}\frac{|-\sin h|-|0|}{h}=\lim_{h\to 0}\frac{|-1||\sin h|-0}{h}$$

$$=\lim_{h\to 0}\frac{\sin h}{h}=1 \qquad \ldots\text{(i)}$$

$$\text{L } f'\left(\frac{\pi}{2}\right)=\lim_{h\to 0}\frac{\left|\cos\left(\frac{\pi}{2}-h\right)\right|-\left|\cos\frac{\pi}{2}\right|}{-h}, h>0$$

$$=\lim_{h\to 0}\frac{\left|\cos\left(\frac{\pi}{2}-h\right)\right|-|0|}{h}$$

$$=\lim_{h\to 0}\frac{|\sin h|-0}{-h}=\lim_{h\to 0}\frac{\sin h}{-h}$$

$$=(-1)\lim_{h\to 0}\frac{\sin h}{h}$$

$$=(-1)\times(1)=-1 \qquad \ldots\text{(ii)}$$

From (i) and (ii), we see that $\text{R } f'\left(\frac{\pi}{2}\right)\neq \text{L } f'\left(\frac{\pi}{2}\right)$

$\therefore f'\left(\frac{\pi}{2}\right)$ does not exist which $\Rightarrow f(x)$ is not differentiable at $x=\dfrac{\pi}{2}$.

(ii) $f(x)=|x^3|$

$$\therefore \text{R } f'(0)=\lim_{h\to 0}\frac{|(0+h)^3|-|0|}{h}, h>0$$

$$=\lim_{h\to 0}\frac{|h^3|-0}{h}$$

$$=\lim_{h\to 0}\frac{|h^3|}{h}=\lim_{h\to 0}\frac{h^3}{h}=\lim_{h\to 0}\frac{h^3}{h}$$

$$=\lim_{h\to 0}h^2=0 \qquad \ldots\text{(i)}$$

$$\text{L } f'(0)=\lim_{h\to 0}\frac{|(0-h)^3|-|0|}{-h}, h>0$$

$$=\lim_{h\to 0}\frac{|-h^3|-0}{-h}=\lim_{h\to 0}\frac{|-1||h^3|}{-h}$$

$$=\lim_{h\to 0}\frac{h^3}{-h}=(-1)\lim_{h\to 0}\frac{h^3}{h}$$

$$=(-1)\cdot\lim_{h\to 0}h^2=(-1)\times 0=0 \qquad \ldots\text{(ii)}$$

Thus from (i) and (ii), we see that $\text{R } f'(0)=\text{L } f'(0)$ which $\Rightarrow f'(0)$ exists.

$\therefore f(x)$ is differentiable at $x=0$.

3. If $f(x)=|\log x|$, find $f'_+(1)$ and $f'_-(1)$.

Solution: $\because f(x)=|\log x|$

$$\therefore \text{R } f'(1)=f'_+(1)=\lim_{h\to 0}\frac{f(1+h)-f(1)}{h}, h>0$$

$$=\lim_{h\to 0}\frac{|\log(1+h)|-|\log 1|}{h}$$

$$=\lim_{h\to 0}\frac{|\log(1+h)|-|0|}{h}$$

$$=\lim_{h\to 0}\frac{|\log(1+h)|-0}{h}$$

$$=\lim_{h\to 0}\frac{\log(1+h)}{h}=1$$

$$\text{L } f'(1)=f'_-(1)=\lim_{h\to 0}\frac{f(1-h)-f(1)}{-h}, h>0$$

$$=\lim_{h\to 0}\frac{|\log(1-h)|-|\log 1|}{-h}$$

$$= \lim_{h \to 0} \frac{|\log(1-h)| - |0|}{h}$$

$$= \lim_{h \to 0} \frac{|\log(1-h)| - 0}{-h}$$

$$= \lim_{h \to 0} \frac{|\log(1-h)|}{-h} = \lim_{h \to 0} \frac{\log(1-h)}{-h}$$

$$= (-1) \lim_{h \to 0} \frac{\log(1-h)}{h} = (-1) \times (1) = -1$$

4. If $f(x) = 1 + \sqrt[3]{(x-2)^2}$, find $f'_+(2)$ and $f'_-(2)$.

Solution: $\because \; f(x) = 1 + \sqrt[3]{(x-2)^2}$

$$\therefore \text{R} f'(2) = f'_+(2) = \lim_{h \to 0} \frac{f(2+h) - f(2)}{h}, h>0$$

$$= \lim_{h \to 0} \frac{1 + \sqrt[3]{(2+h-2)^2} - 1}{h}$$

$$= \lim_{h \to 0} \frac{\sqrt[3]{h^2}}{h} = \lim_{h \to 0} \frac{h^{2/3}}{h} = \lim_{h \to 0} h^{\left(\frac{2}{3}-1\right)}$$

$$= \lim_{h \to 0} h^{\left(\frac{2-3}{3}\right)} = \lim_{h \to 0} h^{\left(-\frac{1}{3}\right)} = \lim_{h \to 0} \frac{1}{\sqrt[3]{h}} = \infty$$

$$\text{L} f'(2) = f'_-(2) = \lim_{h \to 0} \frac{f(2-h) - f(2)}{-h}, h>0$$

$$= \lim_{h \to 0} \frac{1 + \sqrt[3]{(2-h-2)^2} - 1}{-h}$$

$$= \lim_{h \to 0} \frac{\sqrt[3]{(-h)^2}}{-h} = \lim_{h \to 0} \frac{h^{2/3}}{-h}$$

$$= (-1) \lim_{h \to 0} h^{\left(\frac{2-3}{3}\right)} = (-1) \lim_{h \to 0} h^{-\frac{1}{3}}$$

$$= (-1) \times \lim_{h \to 0} \frac{1}{\sqrt[3]{h}} = (-1) \cdot \infty = -\infty$$

5. $f(x) = 1 + |\sin x|$. Examine differentiability at $x = 0$.

$$\text{R } f'(0) = \lim_{h \to 0} \frac{1 + |\sin(0+h)| - \{1 + |\sin 0|\}}{h},$$

$h > 0$

$$= \lim_{h \to 0} \frac{1 + |\sin h| - \{1 + |0|\}}{h}$$

$$= \lim_{h \to 0} \frac{1 + |\sin h| - 1}{h} = \lim_{h \to 0} \frac{|\sin h|}{h}$$

$$= \lim_{h \to 0} \frac{\sin h}{h} = 1 \qquad \ldots\text{(i)}$$

$$\text{L } f'(0) = \lim_{h \to 0} \frac{1 + |\sin(0-h)| - \{1 + |\sin 0|\}}{-h},$$

$h > 0$

$$= \lim_{h \to 0} \frac{1 + |\sin(-h)| - \{1 + |0|\}}{-h}$$

$$= \lim_{h \to 0} \frac{1 + |-1| |\sin h| - 1}{-h} = \lim_{h \to 0} \frac{|\sin h|}{-h}$$

$$= \lim_{h \to 0} \frac{\sin h}{-h} = (-1) \cdot \lim_{h \to 0} \frac{\sin h}{h}$$

$$= (-1) \times (1) = -1 \qquad \ldots\text{(ii)}$$

Thus, from (i) and (ii), we see that R $f'(0) \neq$ L $f'(0)$ which $\Rightarrow f'(0)$ does not exist.

$\therefore f(x)$ is not differentiable at $x = 0$.

6. $f(x) = |x - 2|$. Examine differentiability at $x = 2$.

$$\text{R} f'(0) = \lim_{h \to 0} \frac{|2 + \text{h} - 2| - |0|}{h}, h>0$$

$$= \lim_{h \to 0} \frac{|\text{h}|}{h} = \lim_{h \to 0} \frac{h}{h} = \lim_{h \to 0} 1 = 1 \qquad \ldots\text{(i)}$$

$$\text{L } f'(0) = \lim_{h \to 0} \frac{|2 - \text{h} - 2| - |0|}{-h}, h>0$$

$$= \lim_{h \to 0} \frac{|-\text{h}| - 0}{-h} = \lim_{h \to 0} \frac{|-1| |h|}{-h}$$

$$= \lim_{h \to 0} \frac{h}{-h} = \lim_{h \to 0} (-1) = -1 \quad \ldots \text{(ii)}$$

Thus, from (i) and (ii) we see that $f(x)$ is not differentiable at $x = 2$ caused by R $f'(2) \neq$ L $f'(2)$.

Type 4: To find the values of the constants so that a given function $f(x)$ becomes differentiable at a given point $x = a$.

To find the values of the constants so that a given function $f(x)$ becomes differentiable at a given point $x = a$, the following two conditions must be satisfied.

1. $f(x)$should be continuous at a given point $x = a$ for which we must have L.H.L = R.H.L at athe same point $x = a$ = value of the function at $x = a = f(a)$.

L.H.D at $x = a$ = R.H.D at $x = a$

Hence, we adopt the following working rule to find the values of the constants so that a given function $f(x)$ becomes differentiable at a given point $x = a$

Working rule:

1. Find $\underset{x \to a^+}{\text{L}} f(x) = \underset{x \to a^-}{\text{L}} f(x) = f(a)$ from the definition of continuity of a function at a point $x = a$

2. Find left hand derivative and right hand derivative at the same point $x = a$ by definition and then put L.H.D = R.H.D since differentiability at a point means $f'_+(a) = f'_-(a)$.

Remember: Differentiability at a point $x = a \Rightarrow$ continuity at the same point $x = a$ which is used to find the values of the constants so that a given function $f(x)$ becomes differentiable at a given point $x = a$.

Examples worked out:

1. If $f(x) = 3x^2 + 5x$, when $x \leq 0$ and $f(x) = ax + b$, when $x > 0$ find a and b so that $f(x)$ becomes differentiable at $x = 0$.

Solution: For continuity of $f(x)$ at $x = 0$

$$\underset{x \to 0^+}{\text{Lim}} f(x) = \underset{x \to 0^-}{\text{Lim}} f(x) = f(0)$$

$$\therefore \underset{x \to 0}{\text{Lim}} \left(3x^2 + 5x\right) = \underset{x \to 0}{\text{Lim}} (ax + b)$$

$$\Rightarrow 0 = b \Rightarrow b = 0 \quad \ldots \text{(A)}$$

$$\text{L.H.D } f'_-(0) = \lim_{h \to 0} \frac{f(0-h) - f(0)}{-h}, (h > 0)$$

$$= \lim_{h \to 0} \frac{3(0-h)^2 + 5(0-h) - 0}{-h}$$

$$= \lim_{h \to 0} \frac{3h^2 - 5h}{-h} = \lim_{h \to 0} (-3h + 5) = 5 \quad \ldots \text{(B)}$$

$$\text{R.H.D } f'_+(0) = \lim_{h \to 0} \frac{f(0+h) - f(0)}{h}, (h > 0)$$

$$= \lim_{h \to 0} \frac{a(0+h) + b - 0}{h} = \lim_{h \to 0} \frac{ah + b - 0}{h}$$

$$= \lim_{h \to 0} \frac{ah}{h} \quad (\because b = 0 \text{ from (A)})$$

$$= \lim_{h \to 0} a = a \quad \ldots \text{(C)}$$

Now equating (B) and (C), we get $a = 5$ ($\because$ differentiability at $x = a \Leftrightarrow$ L.H.D = R.H.D) and $b = 0$ (from (A))

Thus, $\left.\begin{array}{l} a = 5 \\ b = 0 \end{array}\right\}$ is the required answer.

2. For what values of a and b is the function $f(x) = \begin{cases} x^2 & \text{when } x \leq 1 \\ 2ax + b & \text{when } x > 1 \end{cases}$ differentiable at $x = 1$.

Solution: For continuity of $f(x)$ at $x = 1$

$f(1) = 1^2 = 1$ ($\because f(x) = x^2$ when $x = 1$)

$$\text{L.H.L} = \lim_{x \to 1^-} f(x) = \lim_{x \to 1} x^2 = 1$$

$$\text{R.H.L} = \lim_{x \to 1^+} f(x) = \lim_{x \to 1} (2ax + b) = 2a + b$$

Now, since $f(x)$ should be continuous at $x = 1$

$\therefore$ L.H.L = R.H.L = $f(1)$

$\therefore$ $2a + b = 1$ which $\Rightarrow$ $2a + b = 1$ ...(A)

$$\text{L.H.D} = \lim_{h \to 0} \frac{f(1-h) - f(1)}{-h}, (h > 0)$$

$$= \lim_{h \to 0} \frac{(1-h)^2 - 1}{-h} = \lim_{h \to 0} \frac{h^2 - 2h}{-h}$$

$$= \lim_{h \to 0} (2 - h) = 2 \quad \text{...(B)}$$

$$\text{R.H.D} = \lim_{h \to 0} \frac{f(1+h) - f(1)}{h}, \ (h > 0)$$

$$= \lim_{h \to 0} \frac{2a(1+h) + b - 1}{h}$$

$$= \lim_{h \to 0} \frac{2ah + (2a + b) - 1}{h}$$

$$= \lim_{h \to 0} \frac{2ah + 1 - 1}{h} \quad (\because 2a + b = 1 \text{ from (A)})$$

$$= \lim_{h \to 0} \frac{2ah}{h} = \lim_{h \to 0} 2a = 2a \quad \text{...(C)}$$

Now, equating (B) and (C), we get

$2a = 2$ ($\because$ differentiability at $x = 1 \Leftrightarrow$ L.H.D = R.H.D at the same point $x = a$)

$\Rightarrow a = 1$

Now, putting this value of $a = 1$ in (A), we have

$b = 1 - 2 \cdot 1 = -1 \quad (\because 2a + b = 1)$

Hence, $\left.\begin{array}{l} a = 1 \\ b = -1 \end{array}\right\} \Rightarrow f(x)$ would be differentiable at $x = 1$ only if $a = 1$ and $b = -1$.

On continuity and differentiability of a function at a point $x = a$.

Question: Explain the cases where l.h.l at a point $x = a$ = r.h.l at the same point $x = a$ is used to test the continuity of a given function $f(x)$ at a given point $x = a$.

Answer: There are following cases where the concept of l.h.l at a point $x = a$ = r.h.l at the same point $x = a$ is used to test the continuity of a given function $f(x)$ at a given point $x = a$.

(i) When a function $f(x)$ is redefined.

(ii) When a function $f(x)$ is a piecewise function.

(iii) When a function $f(x)$ is a mod function.

Note: **1.** Whenever we want to find the limit of a mod function $|f(x)|$ at a point $x = a$, we are required to find the l.h.l and r.h.l at the same given point $x = a$ because a mod function is also a piecewise function.

2. To examine the continuity and differentiability in an interval, we are required to check the given function at the following points.

(a) The point on both sides of which two different functions $f_1(x)$ and $f_2(x)$ are defined in two different subinterval (or, part of the domain of the given function). These points are termed as turning points of the definition or interior points of the interval if the given function $f(x)$ is a piecewise function.

(b) Points (including the end points of the closed interval) at which the given function becomes infinite, imaginary, or indeterminate.

(c) The end points of the closed interval where it has one sided limiting values or derivatives. i.e.;

(i) $\lim_{x \to a^+} f(x) = f(a)$ and $\lim_{x \to b^-} f(x) = f(b)$

where a and b are the end points of the closed interval $[a, b]$

(ii) $\lim_{x \to a^+} \frac{f(x) - f(a)}{x - a}$ = a finite value and

$\lim_{x \to b^-} \frac{f(x) - f(b)}{x - b}$ = a finite value. Where a and b are the end points of the closed interval $[a, b]$.

working rule to test the continuity and differentiability of the given function at the same point $x = a$.

To test the continuity and differentiability of the given function at the same point $x = a$, we should test the continuity at the point $x = a$ and differentiability at the same point $x = a$ separately. i.e.;

(i) To test the continuity at $x = a$, we are required to show that l.h.l of the given function at the point $x = a$ = r.h.l of the given function at the point $x = a$ = value of the function at $x = a$.

(ii) To test the differentiability at $x = a$, we are required to show that l.h.d of the given function at a point $x = a$ = r.h.d of the given function at a point $x = a$.

Remember: Whenever a function is redefined in an interval (open or closed) i.e.;

$$\left.\begin{aligned} f(x) &= f_1(x), \; x \neq c \\ &= d, \; x = c \end{aligned}\right\}$$ in the interval (a,b) or $[a, b]$ where a, b, c are constants.

Then only probable point of the interval (a, b) or $[a, b]$, at which the given redefined function may be discontinuous or non-differentiable is $x = c$. For this reason, we test the continuity and differentiability at $x = c$, e.g.,

(i) $f(x) = x$ if $x \neq 0$

$f(0) = 1$ in the interval $[-1, 1]$

The only probable point of the interval $[-1, 1]$, at which the function may be discontinuous or non-differentiable is $x = 0$.

(ii) $f(x) = 4x + 7$ if $x \neq 2$

$f(2) = 3$ in the interval at $[-4, 4]$

The only probable point of the interval $[-4, 4]$, at which the function may be discontinuous or non-differentiable is $x = 2$.

(iii) $f(x) = \dfrac{9x^2 - 16}{27x^3 - 64}$ for $x \neq \dfrac{4}{3}$

$f\left(\dfrac{4}{3}\right) = \dfrac{2}{3}$ in the interval $[-1, 3]$

The only probable point of the interval $[-1, 3]$, at which the redefined function may be discontinuous or non-differentiable is $x = \dfrac{4}{3}$.

(iv) $f(x) = \dfrac{1 - \sin x}{\left(\dfrac{\pi}{2} - x\right)^2}$ for $x \neq \dfrac{\pi}{2}$

$f\left(\dfrac{\pi}{2}\right) = \dfrac{1}{2}$ in the interval $\left[0, \dfrac{3\pi}{2}\right]$.

The only probable point of the interval $\left[0, \dfrac{3\pi}{2}\right]$, at which the redefined function may be discontinuous or non-differentiable is $x = \dfrac{\pi}{2}$.

Worked out examples on continuity and differentiability at a given point $x = a$.

Type 1: Problems based on piecewise functions:

Questions: **1.** A function f is defined as follows:

$f(x) = -x$ for $x \leq 0$

$f(x) = x$ for $x \geq 0$.

Test the continuity and differentiability of the given function at $x = 0$.

Solution: **(i)** Continuity test at $x = 0$.

l.h.l at $x = 0 = \lim_{x \to 0^-} f(x)$

$= \lim_{x \to 0} (-x) = (-1) \times \lim_{x \to 0} x = 0 \times -1$

$= 0$ ...(a_1)

r.h.l at $x = 0 = \lim_{x \to 0^+} f(x) = \lim_{x \to 0} (x)$

$= 0$ ...(a_2)

$f(0) = 0$ ...(a_3)

(a_1), (a_2) and (a_3) $\Rightarrow$ l.h.l at $x = 0$ = r.h.l at $x = 0$ = value of the function at $x = 0 = 0$.

$\therefore$ $f(x)$ is continuous at $x = 0$.

(ii) Differentiability test at $x = 0$.

l.h.d at $x = 0 = \text{L} f'(0) = \lim_{h \to 0} \dfrac{-(0 - h) - 0}{-h}$

$= \lim_{h \to 0} \left(\dfrac{-h}{-h}\right) = \lim_{h \to 0} 1 = -1$ ...(b_1)

r.h.d at $x = 0 = \text{R} f'(0) = \lim_{h \to 0} \dfrac{(0 + h) - 0}{h}$

$= \lim_{h \to 0} \dfrac{h}{h} = 1$ ...(b_2)

(b_1) and (b_2) $\Rightarrow \text{L} f'(0) \neq \text{R} f'(0)$

$\therefore$ $f(x)$ is not differentiable at $x = 1$.

2. Examine the following function for continuity and differentiability:

$f(x) = x^2$ for $x \leq 0$

$f(x) = 1$ for $0 < x \leq 1$

$f(x) = \dfrac{1}{x}$ for $x > 1$

Solution: Considerable points at which we are required to test the continuity and differentiability are $x = 0$ and $x = 1$.

(i) Continuity and differentiability test at $x = 0$

$f(0) = 0$ ...(a_1)

$$\text{l.h.l at } x = 0 = \lim_{x\to 0^-} f(x)$$

$$= \lim_{x\to 0} x^2 = 0 \quad \ldots(a_2)$$

$$\text{r.h.l at } x = 0 = \lim_{x\to 0^+} f(x)$$

$$= \lim_{x\to 0} (1) = 1 \quad \ldots(a_3)$$

(a_2) and (a_3) $\Rightarrow$ l.h.l at $x = 0 \neq$ r.h.l at $x = 0$ which $\Rightarrow$ given piecewise function is discontinuous at $x = 0$. Caused by this, given piecewise function is non-differentiable at $x = 0$.

(ii) Continuity and differentiability test at $x = 1$

Now, $f(1) = 1$ ...(b_1)

$$\text{l.h.l at } x = 1 = \lim_{x\to 1^-} f(x)$$

$$= \lim_{x\to 1} (1) = 1 \quad \ldots(b_2)$$

$$\text{r.h.l at } x = 1 = \lim_{x\to 1^+} f(x)$$

$$= \lim_{x\to 1} \left(\frac{1}{x}\right) = 1 \quad \ldots(b_3)$$

(b_1), (b_2) and (b_3) $\Rightarrow$ l.h.l at $x = 1$ = r.h.l at $x = 1$ = value of the function at $x = 1$ which $\Rightarrow f(x)$ is continuous at $x = 1$

Again,

$$\text{l.h.d at } x = 1 = \text{L } f'(1) = \lim_{h\to 0} \frac{f(1-h) - f(1)}{-h},$$

$h > 0$

$$= \lim_{h\to 0} \frac{1-1}{-h} = 0 \quad \ldots(b_1)$$

$$\text{r.h.d at } x = 1 = \text{R } f'(1) = \lim_{h\to 0} \frac{f(1+h) - f(1)}{h},$$

$h > 0$

$$= \lim_{h\to 0} \frac{\frac{1}{1+h} - 1}{h} = \lim_{h\to 0} \frac{1 - 1 - h}{h(1+h)}$$

$$= \lim_{h\to 0} \frac{-h}{h(1+h)} = (-1) \cdot \lim_{h\to 0} \left(\frac{1}{1+h}\right)$$

$$= (-1) \times \left(\frac{1}{1+0}\right) = -1 \quad \ldots(b_2)$$

(b_1) and (b_2) $\Rightarrow$ L $f'(1) \neq$ R $f'(1)$ which $\Rightarrow$ given piecewise function is non-differentiable at $x = 1$.

3. Discuss the continuity and differentiability of the following function:

$f(x) = x^2$ for $x < -2$

$f(x) = 4$ for $-2 \leq x \leq 2$

$f(x) = x^2$ for $x > 2$

Solution: Considerable points at which we are required to test the continuity and differentiability of the given piecewise function $f(x)$ are $x = -2$ and 2.

(i) Continuity and differentiability test at $x = -2$

$f(-2) = 4$ ...(a_1)

$$\text{l.h.l at } x = -2 = \lim_{x\to -2^-} f(x)$$

$$= \lim_{x\to -2} (x^2) = 4 \quad \ldots(a_2)$$

$$\text{r.h.l at } x = -2 = \lim_{x\to -2^+} f(x)$$

$$= \lim_{x\to -2} 4 = 4 \quad \ldots(a_3)$$

(a_1), (a_2) and (a_3) $\Rightarrow$ l.h.l at $x = -2$ = r.h.l at $x = -2$ = value of the function at $x = -2$ which $\Rightarrow f(x)$ is continuous at $x = -2$

Again,

$$\text{L } f'(-2) = \lim_{h\to 0} \frac{f(-2-h) - f(-2)}{-h}, (h > 0)$$

$$= \lim_{h\to 0} \frac{(-2-h)^2 - 4}{-h} = \lim_{h\to 0} \frac{4 + 4h + h^2 - 4}{-h}$$

$$= \lim_{h\to 0} (-4 - h) = -4 \quad \ldots(b_1)$$

$$R\ f'(-2) = \lim_{h \to 0} \frac{4-4}{h} = \lim_{h \to 0} \frac{0}{h} = 0 \quad \ldots(b_2)$$

(b_1) and $(b_2) \Rightarrow L\ f'(-2) \neq R\ f'(-2)$.

Caused by this, the given function is non-differentiable at $x=-2$.

(ii) Continuity and differentiability test at $x = 2$

$$\text{l.h.l at } x = 2 = \lim_{x \to 2^-} f(x)$$

$$= \lim_{x \to 2^-} 4 = \lim_{x \to 2} 4 = 4 \quad \ldots(c_1)$$

$$\text{r.h.l at } x = 2 = \lim_{x \to 2^+} f(x)$$

$$= \lim_{x \to 2} \left(x^2\right) = 4 \quad \ldots(c_2)$$

$$f(2) = 4 \quad \ldots(c_3)$$

(c_1), (c_2) and $(c_3) \Rightarrow$ l.h.l at $x=2$ = r.h.l at $x=2$ = $f(2)$ which $\Rightarrow$ given piecewise function is continuous at $x=2$.

Again,

$$L\ f'(2) = \lim_{h \to 0} \frac{f(2-h) - f(2)}{-h}, (h>0)$$

$$= \lim_{h \to 0} \frac{4-4}{-h} = \lim_{h \to 0} \left(\frac{0}{-h}\right) = 0 \quad \ldots(d_1)$$

$$R\ f'(2) = \lim_{h \to 0} \frac{f(2+h) - f(2)}{h}, (h>0)$$

$$= \lim_{h \to 0} \frac{(2+h)^2 - 4}{h} = \lim_{h \to 0} \frac{4 + 4h + h^2 - 4}{h}$$

$$= \lim_{h \to 0} \frac{h(4+h)}{h} = \lim_{h \to 0} (4+h)$$

$$= \lim_{h \to 0} 4 + \lim_{h \to 0} (h) = 4 + 0 = 4 \quad \ldots(d_2)$$

(d_1) and $(d_2) \Rightarrow R\ f'(2) \neq L\ f'(2)$ which $\Rightarrow$ $f(x)$ is not differentiable at $x=2$.

4. Examine the continuity and differentiability of the following function in the interval $-\infty < x < \infty$.

$f(x) = 1$ in $-\infty < x < 0$

$f(x) = 1 + \sin x$ in $0 \leq x < \frac{\pi}{2}$

$f(x) = 2 + \left(x - \frac{\pi}{2}\right)^2$ in $\frac{\pi}{2} \leq x < \infty$

Note: In this question we are required to examine the continuity and differentiability in an interval $-\infty < x < \infty$.

Hence, we adopt the rule to examine the continuity and differentiability in an interval.

Solution: Considerable points at which it is required to test the continuity and differentiability of the given piecewise function $f(x)$ are $x = 0, \frac{\pi}{2}$.

(i) Continuity and differentiability at $x=0$.

$$\text{l.h.l at } x = 0 = \lim_{x \to 0^-} f(x) = \lim_{x \to 0} 1 = 1 \quad \ldots(a_1)$$

$$\text{r.h.l at } x = 0 = \lim_{x \to 0^+} f(x)$$

$$= \lim_{h \to 0} \{1 + \sin(0+h)\} = 1 \quad \ldots(a_2)$$

$$f(0) = 1 + \sin 0 = 1 \quad \ldots(a_3)$$

(a_1), (a_2) and $(a_3) \Rightarrow \lim_{x \to 0^-} f(x) = \lim_{x \to 0^+} f(x)$ $= f(0)$ which $\Rightarrow$ given piecewise function $f(x)$ is continuous at $x = 0$ ...(A)

$$\text{Again, } L\ f'(0) = \lim_{h \to 0} \frac{f(0-h) - f(0)}{-h}, (h>0)$$

$$= \lim_{h \to 0} \frac{1-1}{-h} = \lim_{h \to 0} \frac{0}{-h} = 0 \quad \ldots(b_1)$$

$$R\ f'(0) = \lim_{h \to 0} \frac{f(0+h) - f(0)}{h}, h>0$$

$$= \lim_{h \to 0} \frac{1 + \sin(0+h) - 1}{h}$$

$$= \lim_{h \to 0} \frac{\sin h}{h} = 1 \quad \ldots(b_2)$$

(b_1) and $(b_2) \Rightarrow L\ f'(0) \neq R\ f'(0)$ which $\Rightarrow$ given piecewise function $f(x)$ is not differentiable at $x = 0$. ...(B)

(ii) Continuity and differentiability test at $x = \frac{\pi}{2}$

$$f\left(\frac{\pi}{2}\right) = 2 + \left(\frac{\pi}{2} - \frac{\pi}{2}\right)^2 = 2 \qquad ...(c_1)$$

$$\text{l.h.l at } x = \frac{\pi}{2} = \lim_{x \to \frac{\pi}{2}^-} f(x) = \lim_{x \to \frac{\pi}{2}} (1 + \sin x)$$

$$= 1 + \sin\frac{\pi}{2} = 1 + 1 = 2 \qquad ...(c_2)$$

$$\text{r.h.l at } x = \frac{\pi}{2} = \lim_{x \to \frac{\pi}{2}^+} f(x)$$

$$= \lim_{x \to \frac{\pi}{2}} \left\{2 + \left(x - \frac{\pi}{2}\right)^2\right\} = 2 \qquad ...(c_3)$$

(c_1), (c_2) and (c_3)

$$\Rightarrow \lim_{x \to \frac{\pi}{2}^-} f(x) = \lim_{x \to \frac{\pi}{2}^+} f(x) = f\left(\frac{\pi}{2}\right)$$

which $\Rightarrow f(x)$ is continuous at $x = \frac{\pi}{2}$ ...(C)

Again,

$$\text{L } f'\left(\frac{\pi}{2}\right) = \lim_{h \to 0} \frac{f\left(\frac{\pi}{2} - h\right) - f\left(\frac{\pi}{2}\right)}{-h}, (h > 0)$$

$$= \lim_{h \to 0} \frac{1 + \sin\left(\frac{\pi}{2} - h\right) - 2}{-h} = \lim_{h \to 0} \frac{1 - \cos h}{h}$$

$$= \lim_{h \to 0} \frac{2\sin^2\left(\frac{h}{2}\right)}{h} = \lim_{h \to 0} \frac{\sin\left(\frac{h}{2}\right)}{\left(\frac{h}{2}\right)} \cdot \sin\left(\frac{h}{2}\right)$$

$$= 1 \times 0 = 0 \qquad ...(d_1)$$

$$\text{R } f'\left(\frac{\pi}{2}\right) = \lim_{h \to 0} \frac{f\left(\frac{\pi}{2} + h\right) - f\left(\frac{\pi}{2}\right)}{h}, (h > 0)$$

$$= \lim_{h \to 0} \frac{2 + \left(\frac{\pi}{2} + h - \frac{\pi}{2}\right)^2 - 2}{h}$$

$$= \lim_{h \to 0} \frac{h^2}{h} = \lim_{h \to 0} h = 0 \qquad ...(d_2)$$

(d_1) and (d_2) $\Rightarrow \text{L } f'\left(\frac{\pi}{2}\right) = \text{R } f'\left(\frac{\pi}{2}\right)$ which

$\Rightarrow$ the given piecewise function $f(x)$ is differentiable at $x = \frac{\pi}{2}$ ...(D)

In the light of the results (A), (B), (C) and (D), we observe and declare that $f(x)$ is continuous in the interval $(-\infty, \infty)$ but it is not differentiable in $(-\infty, \infty)$.

5. If $f(x) = \frac{x^2}{2}$, $0 \le x < 1$,

$= 2x^2 - 3x + \frac{3}{2}$, $1 \le x \le 2$ discuss the continuity of f, f', and f'' on [0, 2].

Solution: (i) Given is

$$f(x) = \frac{x^2}{2}, \quad 0 \le x < 1 \qquad ...(a_1)$$

$$= 2x - 3x + \frac{3}{2}, \quad 1 \le x \le 2 \qquad ...(a_2)$$

$$\therefore \; f'(x) = x, \quad 0 \le x < 1 \qquad ...(a_3)$$

$$= 4x - 3, \quad 1 \le x \le 2 \qquad ...(a_4)$$

$$f''(x) = 1, \quad 0 \le x < 1 \qquad ...(a_5)$$

$$= 4, \quad 1 < x \le 2 \qquad ...(a_6)$$

(ii) Continuity test at $x = 1$ for $f(x)$,

$$f(1) = \left[2x^2 - 3x + \frac{3}{2}\right]_{x=1}$$

$$= 2 \times 1 - 3 \times 1 + \frac{3}{2} = -1 + \frac{3}{2} = \frac{1}{2} \qquad ...(b_1)$$

$$\text{l.h.l at } x = 1 = \lim_{x \to 1^-} f(x)$$

$$= \lim_{x \to 1} \left(\frac{x^2}{2}\right) = \frac{1}{2} \quad ...(b_2)$$

r.h.l at $x = 1 = \lim_{x \to 1^+} f(x)$

$$= \lim_{x \to 1} \left(2x^2 - 3x + \frac{3}{2}\right) = 2 - 3 + \frac{3}{2} = \frac{1}{2} \quad ...(b_3)$$

(b_1), (b_2) and (b_3) $\Rightarrow$ l.h.l at $x = 1$ = r.h.l at $x = 1$ = value of the function $f(x)$ at $x = 1$ which $\Rightarrow$ $f(x)$ is continuous at $x = 1$ for $f'(x)$,

$$f'(1) = [4x - 3]_{x=1}$$

$$= 4 \times 1 - 3 = 4 - 3 = 1 \quad ...(c_1)$$

l.h.l at $x = 1 = \lim_{x \to 1^-} f'(x) = \lim_{x \to 1} (x) = 1 \quad ...(c_2)$

r.h.l at $x = 1 = \lim_{x \to 1^+} f'(x) = \lim_{x \to 1} (4x - 3)$

$$= (4 - 3) = 1 \quad ...(c_3)$$

(c_1), (c_2) and (c_3) $\Rightarrow$ $f'(x)$ is continuous at $x = 1$ for $f''(x)$

l.h.l at $x = 1 = \lim_{x \to 1^-} f''(x) = \lim_{x \to 1} 1 = 1 \quad ...(d_1)$

r.h.l at $x = 1 = \lim_{x \to 1^+} f''(x) = \lim_{x \to 1} 4 = 4 \quad ...(d_2)$

$\therefore$ (d_1) and (d_2) $\Rightarrow$ $f''(x)$ is discontinuous at $x = 1$.

Hence, in the light of above observation and the results, we declare that the function f and f'' are continuous in $[0, 2]$ but f'' is discontinuous in $[0, 2]$ since f'' is discontinuous at $x = 1 \in [0, 2]$.

Note: $f''(x)$ does not exist at $x = 1$.

6. A function f is defined as follows:

$f(x) = x$ for $0 \le x \le 1$ and $f(x) = 2 - x$ for $x \ge 1$.

Test the character of the function at $x = 1$ as regards its continuity and differentiability.

Solution: **(i)** Continuity test at $x = 1$

$f(1) = 1 \quad ...(a_1)$

l.h.l at $x = 1 = \lim_{x \to 1^-} f(x)$

$$= \lim_{x \to 1} x = 1 \quad ...(a_2)$$

r.h.l at $x = 1 = \lim_{x \to 1^+} f(x)$

$$= \lim_{x \to 1} (2 - x) = 1 \quad ...(a_3)$$

(a_1), (a_2) and (a_3)

$\Rightarrow \lim_{x \to 1^-} f(x) = \lim_{x \to 1^+} f(x) = f(1)$ which $\Rightarrow$ given piecewise function is continuous at $x = 1$

(ii) Differentiability test at $x = 1$

$$L f'(1) = \lim_{h \to 0} \frac{f(1 - h) - f(1)}{-h}, (h > 0)$$

$$= \lim_{h \to 0} \frac{(1 - h) - 1}{-h} = \lim_{h \to 0} \frac{-h}{-h}$$

$$= \lim_{h \to 0} 1 = 1 \quad ...(b_1)$$

$$R f'(1) = \lim_{h \to 0} \frac{f(1 + h) - f(1)}{h}, (h > 0)$$

$$= \lim_{h \to 0} \frac{2 - (1 + h) - 1}{h} = \lim_{h \to 0} \frac{-h}{h}$$

$$= \lim_{h \to 0} (-1) = -1 \quad ...(b_2)$$

(b_1) and (b_2) $\Rightarrow$ $L f'(1) \ne R f'(1)$ which $\Rightarrow$ given piecewise function is not differentiable at $x = 1$

Type 2: Problems based on redefined functions:

Questions: **(i)** If $f(x) = \dfrac{x}{1 + e^{\frac{1}{x}}}$, $x \ne 0$ $f(x) = 0$, $x = 0$ show that f is continuous at $x = 0$ but $f'(0)$ does not exist.

Solution: **(i)** Continuity test at $x = 0$

r.h.l at $x = 0 = \lim_{x \to 0^+} f(x)$

$$= \lim_{h \to 0} \frac{(0 + h)}{1 + e^{\left(\frac{1}{0+h}\right)}}, h > 0 = \lim_{h \to 0} \frac{h}{1 + e^{\frac{1}{h}}} = 0 \quad ...(a_1)$$

l.h.l at $x = 0 = \lim_{x \to 0^-} f(x)$

$$= \lim_{h\to 0} \frac{(0-h)}{1+e^{\left(\frac{1}{0-h}\right)}},\ h>0 = \lim_{h\to 0} \frac{-h}{1+e^{\left(-\frac{1}{h}\right)}} = 0 \quad \ldots(a_2)$$

$f(0) = 0 \quad \ldots(a_3)$

(a_1), (a_2) and (a_3) $\Rightarrow$ l.h.l at $x = 0$ = r.h.l at $x = 0$ = value of the function at $x = 0$ which $\Rightarrow$ given redefined function $f(x)$ is continuous at $x = 0$.

(ii) Differentiability test at $x = 0$

$$R\ f'(0) = \lim_{h\to 0} \frac{f(0+h) - f(0)}{h},\ h>0$$

$$= \lim_{h\to 0} \frac{\dfrac{h}{1+e^{\frac{1}{h}}} - 0}{h} = \lim_{h\to 0} \frac{h}{h\left(1+e^{\frac{1}{h}}\right)}$$

$$= \lim_{h\to 0} \frac{1}{1+e^{\frac{1}{h}}}$$

$$= 0 \quad \ldots(b_1)$$

$$L\ f'(0) = \lim_{h\to 0} \frac{f(0-h) - f(0)}{-h},\ h>0$$

$$= \lim_{h\to 0} \frac{\dfrac{-h}{1+e^{-\frac{1}{h}}} - 0}{-h} = \lim_{h\to 0} \frac{-h}{-h\left(1+e^{-\frac{1}{h}}\right)}$$

$$= \lim_{h\to 0} \frac{1}{1+e^{-\frac{1}{h}}} = \frac{1}{1+0} = 1 \quad \ldots(b_2)$$

$\therefore$ (b_1) and (b_2) $\Rightarrow R\ f'(0) \neq L\ f'(0)$ which

$\Rightarrow f'(0)$ does not exist.

2. If $f(x) = \dfrac{1}{x}\sin x^2$, $x \neq 0$ $f(x) = 0$, $x = 0$ discuss the continuity and differentiability of $f(x)$ at $x = 0$.

***Solution*:** **1.** Continuity test at $x = 0$

$f(0) = 0 \quad \ldots(a_1)$

$$\text{r.h.l at } x = 0 = \lim_{x\to 0^+} f(x) = \lim_{h\to 0} \frac{\sin(0+h)^2}{(0+h)},\ h>0$$

$$= \lim_{h\to 0} \frac{\sin h^2}{h^2}\cdot h = 1 \times 0 = 0 \quad \ldots(a_2)$$

$$\text{l.h.l at } x = 0 = \lim_{x\to 0^-} f(x) = \lim_{x\to 0} \frac{\sin(0-h)^2}{(0-h)},\ h>0$$

$$= \lim_{h\to 0}\left\{\frac{\sin h^2}{h^2}\cdot(-h)\right\}$$

$$= \lim_{h\to 0}\left(\frac{\sin h^2}{h^2}\right) \times \lim_{h\to 0}(-h) = 1 \times 0 = 0 \quad \ldots(a_3)$$

(a_1), (a_2) and (a_3) $\Rightarrow f(x)$ is continuous at $x = 0$

(ii) Differentiability test at $x = 0$

$$R\ f'(0) = \lim_{h\to 0}\left\{\frac{\left(\dfrac{\sin(0+h)}{(0+h)}\right) - 0}{(0+h)}\right\},\ h>0$$

$$= \lim_{h\to 0} \frac{\sin h^2}{h^2} = 1 \quad \ldots(b_1)$$

$$L\ f'(0) = \lim_{h\to 0}\left\{\frac{\left(\dfrac{\sin(0-h)^2}{(0-h)}\right) - 0}{(-h)}\right\},\ h>0$$

$$= \lim_{h\to 0} \frac{\sin h^2}{h^2} = 1 \quad \ldots(b_2)$$

(b_1) and (b_2) $\Rightarrow R\ f'(0) = L\ f'(0)$ which $\Rightarrow$ redefined function $f(x)$ is differentiable at $x = 0$ and $f'(0) = 1$.

3. If $f(x) = x^2 \sin\left(\frac{1}{x}\right)$ for $x \neq 0$ $f(x) = 0$ for $x = 0$

show that

(i) $f(x)$ is continuous at $x = 0$

(ii) $f'(0) = 0$

(iii) $f'(x)$ is discontinuous at $x = 0$

***Solution*:** **(i)** Continuity test at $x = 0$

l.h.l at $x = 0 = \lim\limits_{x \to 0^-} f(x)$

$$= \lim_{h \to 0} (0-h)^2 \cdot \sin\left(\frac{1}{0-h}\right), h > 0$$

$$= \lim_{h \to 0} (-1) \cdot (h^2) \cdot \sin\left(\frac{1}{h}\right) = 0 \quad \ldots(a_1)$$

r.h.l at $x = 0 = \lim\limits_{x \to 0^+} f(x)$

$$= \lim_{h \to 0} (0+h)^2 \cdot \sin\left(\frac{1}{0+h}\right), h > 0$$

$$= \lim_{h \to 0} (h^2) \sin\left(\frac{1}{h}\right) = 0 \quad \ldots(a_2)$$

$f(0) = 0 \quad \ldots(a_3)$

(a_1), (a_2) and (a_3) $\Rightarrow \lim\limits_{x \to 0^-} f(x) = \lim\limits_{x \to 0^+} f(x)$

$= f(0)$ which $\Rightarrow$ redefined function $f(x)$ is continuous at $x = 0$

(ii) To find $f'(0)$:

$$\text{R } f'(0) = \lim_{h \to 0} \frac{(0+h)^2 \cdot \sin\left(\frac{1}{0+h}\right) - 0}{h}, h > 0$$

$$= \lim_{h \to 0} h \cdot \sin\left(\frac{1}{h}\right) = 0 \quad \ldots(b_1)$$

$$\text{L } f'(0) = \lim_{h \to 0} \frac{(0-h)^2 \cdot \sin\left(\frac{1}{0-h}\right) - 0}{-h}, h > 0$$

$$= \lim_{h \to 0} \frac{h^2 \sin\left(-\frac{1}{h}\right)}{-h}$$

$$= \lim_{h \to 0} \left\{(-1)(h)(-1)\sin\left(\frac{1}{h}\right)\right\}$$

$$= \lim_{h \to 0} h \sin\frac{1}{h} = 0 \quad \ldots(b_2)$$

(b_1) and (b_2) $\Rightarrow$ L $f'(0) =$ R $f'(0) = 0$ which

$\Rightarrow f'(0) = 0$

(iii) To show that $f'(x)$ is discontinuous at $x = 0$.

$$\because f'(x) = 2x \sin\left(\frac{1}{x}\right) - \cos\left(\frac{1}{x}\right) \text{ at } x \neq 0$$

$f'(0) = 0$ (already determined) $\quad \ldots(c_1)$

Now, r.h.l at $x = 0 = \lim\limits_{x \to 0^+} f'(x)$

$$= \lim_{h \to 0} \left\{2(0+h) \sin\left(\frac{1}{0+h}\right) - \cos\left(\frac{1}{0+h}\right)\right\}, h > 0$$

$$= \lim_{h \to 0} \left\{2h \sin\left(\frac{1}{h}\right) - \cos\left(\frac{1}{h}\right)\right\}$$

$$= 2 \lim_{h \to 0} h \sin\left(\frac{1}{h}\right) - \lim_{h \to 0} \cos\left(\frac{1}{h}\right)$$

$$= 0 - \lim_{h \to 0} \cos\left(\frac{1}{h}\right)$$

$= -\lim\limits_{h \to 0} \cos\left(\frac{1}{h}\right)$ which does not exist $\Rightarrow$ r.h.l at $x = 0$ of the function $f'(x)$ does not exist. $\quad \ldots(c_2)$

Again, l.h.l at $x = 0 = \lim\limits_{x \to 0^-} f'(x)$

$$= \lim_{h \to 0} \left\{2(0-h) \sin\left(\frac{1}{0-h}\right) - \cos\left(\frac{1}{0-h}\right)\right\}, h > 0$$

$$= \lim_{h \to 0} \left\{(-2)(h) \sin\left(\frac{1}{h}\right) - \cos\left(-\frac{1}{h}\right)\right\}$$

$$= \lim_{h\to 0} (-2)(h)\sin\left(\frac{1}{h}\right) - \lim_{h\to 0}\cos\left(\frac{1}{h}\right)$$

$(\because \cos(-\theta) = \cos\theta)$

$$= (-2)\lim_{h\to 0}(h)\sin\left(\frac{1}{h}\right) - \lim_{h\to 0}\cos\left(\frac{1}{h}\right)$$

$$= (-2)\times 0 - \lim_{h\to 0}\cos\left(\frac{1}{h}\right)$$

$= -\lim_{h\to 0}\cos\left(\frac{1}{h}\right)$ which does not exist $\Rightarrow$ l.h.l at $x = 0$ of the derived function $f'(x)$ does not exist ...(c_3)

Hence, (c_2) or (c_3) $\Rightarrow$ the derived function $f'(x)$ is discontinuous at $x = 0$, (of second kind).

4. If $f(x) = x\left\{1 + \frac{1}{3}\sin\left(\log x^2\right)\right\},\ x \neq 0$

$f(x) = 0,\ x = 0$. Test the continuity and differentiability at $x = 0$

Solution: **(i)** Continuity test at $x = 0$

$f(0) = 0$ ...(a_1)

l.h.l at $x = 0 = \lim_{x\to 0^-} f(x)$

$$= \lim_{h\to 0}(0-h)\left\{1 + \frac{1}{3}\sin\log(0-h)^2\right\},\ h>0$$

$$= \lim_{h\to 0}\left[-h - \frac{1}{3}\cdot h\sin\log h^2\right]$$

$$= \lim_{h\to 0}(-h) - \frac{1}{3}\lim_{h\to 0} h\cdot\sin\log h^2$$

$$= 0 - \frac{1}{3}\lim_{h\to 0} h\cdot\sin\log h^2 = 0 - \frac{1}{3}\cdot 0$$

($\because$ Product of an infinitesimal and *a* bounded function = 0, in the limit)

$= 0 - 0 = 0$ ...(a_2)

r.h.l at $x = 0 = \lim_{x\to 0^+} f(x)$

$$= \lim_{h\to 0}(0+h)\left\{1 + \frac{1}{3}\sin\log(0+h)^2\right\},\ h>0$$

$$= \lim_{h\to 0} h\left\{1 + \frac{1}{3}\sin\log h^2\right\}$$

$$= \lim_{h\to 0}\left\{h + \frac{1}{3}h\sin\log h^2\right\}$$

$$= \lim_{h\to 0} h + \lim_{h\to 0}\frac{1}{3}h\sin\log h^2 = 0 + 0$$

$= 0$ ...(a_3)

(a_1), (a_2) and (a_3) $\Rightarrow$ l.h.l at $x = 0$ = r.h.l. at $x = 0$ = value of the function at $x = 0$ which $\Rightarrow f(x)$ is continuous at $x = 0$.

(ii) Differentiability test at $x = 0$

$$\text{L } f'(0) = \lim_{h\to 0}\frac{(0-h)\left\{1 + \frac{1}{3}\sin\log(0-h)^2\right\} - 0}{-h},$$

$h > 0$ $(\because f(0) = 0)$

$$= \lim_{h\to 0}\left\{1 + \frac{1}{3}\sin\log h^2\right\}$$

$$= \lim_{h\to 0} 1 + \frac{1}{3}\lim_{h\to 0}\sin\log h^2$$

$$= 1 + \frac{1}{3}\lim_{h\to 0}\sin\log h^2$$

but $\lim_{h\to 0}\sin\log h^2$ oscillates between +1 and –1.

$\therefore$ L $f'(0)$ does not exist. ...(b_1)

$$\text{R } f'(0) = \lim_{h\to 0}\frac{(0+h)\left\{1 + \frac{1}{3}\sin\log(0+h)^2\right\} - 0}{h}$$

$h > 0$

$$= \lim_{h\to 0}\left\{1 + \frac{1}{3}\sin\log h^2\right\}$$

$= \lim_{h\to 0} 1 + \frac{1}{3}\lim_{h\to 0}\sin\log h^2$ but $\lim_{h\to 0}\sin\log h^2$ oscillates between +1 and – 1.

$\therefore$ R $f'(0)$ does not exist. ...(b_2)

(b_1) or (b_2) $\Rightarrow f(x)$ is not differentiable at $x = 0$.

On Derivative of $y=[x]$

Let $y=f(x)=[x]$, where $D(f)=R$, i.e.; $x \in R$

$\therefore$ there are two cases:

Case 1. When x is an integer.

considering $f(x)=[x]$ at $x=N$, it is seen that for $h>0$

$$\lim_{h \to 0} [N+h] = \lim_{h \to 0} [N] + \lim_{h \to 0} [h]$$

$$= N + 0 = N \quad (\because [x+n] = n + [x], n \in I)$$

$$\lim_{h \to 0} [N-h] = \lim_{h \to 0} [N] + \lim_{h \to 0} [-h]$$

$$= \lim_{h \to 0} [N] + (-1) \lim_{h \to 0} [h] - 1$$

$$(\because [-x] = -[x] - 1 \text{ if } x \notin I)$$

$$= N + (-1) \cdot 0 - 1 = N - 1$$

$f(N)=N$

Hence, the function $y=[x]$ is discontinuous at $x=N$; through it is right continuous at this point $\Rightarrow$ $y=[x]$ can not have derivative at any integral value of x.

Case 2: When x is non-integer (or, non-integral). Considering $f(x)=[x]$ at $x=N+h$, where $0<h<1$, it is seen that for $\Delta x > 0$

$$= \lim_{\Delta x \to 0} [N+h+\Delta x]$$

$$= \lim_{\Delta x \to 0} [N] + \lim_{\Delta x \to 0} [h+\Delta x]$$

$$= N + 0 = N \,(\because [x+n] = n + [x],\ n \in I\,) \quad \ldots\text{(i)}$$

$$\lim_{\Delta x \to 0} [N+h-\Delta x]$$

$$= \lim_{\Delta x \to 0} [N] + \lim_{\Delta x \to 0} [h-\Delta x]$$

$$= N + 0 = N \text{ as } h - \Delta x < 1 \quad \ldots\text{(ii)}$$

$$f(N+h) = [N+h] = N \quad \ldots\text{(iii)}$$

Hence, (i), (ii) and (iii) $\Rightarrow$ the function $y=[x]$ is continuous at a non-integer point $x=N+h$.

Again, we know that continuity of a function $y=f(x)$ at any point $x=a \not\Rightarrow$ differentiability of the function $y=f(x)$ at the same point $x=a$.

Hence, the derivative test of $y=[x]$ at a non-integer point namely $x=N+h$, $(0<h<1)$, is required.

$$f'_+(N+h) = \lim_{\Delta x \to 0} \frac{f(N+h+\Delta x) - f(N+h)}{\Delta x}$$

$$= \lim_{\Delta x \to 0} \frac{[N+h+\Delta x] - [N+h]}{\Delta x}$$

$$= \lim_{\Delta x \to 0} \frac{[N] + [h+\Delta x] - [N] - [h]}{\Delta x}$$

$$= \lim_{\Delta x \to 0} \frac{[N] + 0 - [N] - 0}{\Delta x}$$

$$= \lim_{\Delta x \to 0} \left(\frac{N-N}{\Delta x} \right) = 0$$

$$f'_-(N-h) = \lim_{\Delta x \to 0} \frac{f(N+h-\Delta x) - f(N+h)}{-\Delta x}$$

$$= \lim_{\Delta x \to 0} \frac{[N+h-\Delta x] - [N+h]}{-\Delta x}$$

$$= \lim_{\Delta x \to 0} \frac{[N] + [h-\Delta x] - [N] - [h]}{-\Delta x}$$

$$= \lim_{\Delta x \to 0} \frac{[N] + 0 - [N] - 0}{-\Delta x}$$

$$= \lim_{\Delta x \to 0} \left(\frac{N-N}{-\Delta x} \right) = 0$$

$$\therefore f'_+(N+h) = f'_-(N-h) = 0,\ \forall\ x \notin I$$

$\Rightarrow$ $y=[x]$ has the derivative zero at any non-integral point namely $x=N+h$, where $N=$ an integer and h is small positive number such that $0<h<1$.

On Differentiability

Type 1. Problems based on piecewise functions.

Exercise 7.1

1. Test the differentiability of the following functions at the indicated points.

(i) $f(x)=\begin{cases}1, & \text{when } x<0\\ 1+\sin x, & \text{when } 0\le x<\dfrac{\pi}{2}\\ 2+\left(x-\dfrac{\pi}{2}\right)^2, & \text{when } \dfrac{\pi}{2}\le x\end{cases}$

at $x=\dfrac{\pi}{2}$ and $x=0$.

Answer: Differentiable at $x=\dfrac{\pi}{2}$ and non-differentiable at $x=0$.

(ii) $f(x)=\begin{cases}1, & \text{when } x<0\\ 1+\sin x, & \text{when } 0\le x<\dfrac{\pi}{2}\\ 2-\left(x-\dfrac{\pi}{2}\right), & \text{when } x\ge\dfrac{\pi}{2}\end{cases}$

at $x=0$ and $x=\dfrac{\pi}{2}$.

Answer: Non-differentiable at both points $x=\dfrac{\pi}{2}$ and $x=0$.

(iii) $f(x)=\begin{cases}x, & \text{when } 0<x<1\\ 2-x, & \text{when } 1\le x\le 2\\ x-\dfrac{1}{2}x^2, & \text{when } x>2\end{cases}$

at $x=1$ and $x=2$.

Answer: Non-differentiable at $x=1$ and differentiable at $x=2$

(iv) $f(x)=\begin{cases}-x, & \text{when } x<0\\ x^2, & \text{when } 0\le x\le 1\\ x^2-x+1, & \text{when } x>1\end{cases}$

at $x=0$ and $x=1$.

Answer: Non-differentiable at both point $x=0$ and $x=1$

(v) $f(x)=\begin{cases}x, & \text{when } 0\le x\le 1\\ 2x-1, & \text{when } x>1\end{cases}$

at $x=1$.

Answer: Non-differentiable at $x=1$

(vi) $f(x)=\begin{cases}x\,[x], & \text{when } 0\le x<2\\ (x-1)\,[x], & \text{when } 2\le x<3\end{cases}$

at $x=1$ and $x=2$. Where $[x]$ is the greatest integer not greater than x.

Answer: Not differentiable at both points $x=1$ and $x=2$.

2. If the function $f(x)$ is defined by

$$f(x)=\begin{cases}3+2x, & \text{when } -\dfrac{3}{2}<x\le 0\\ 3-2x, & \text{when } 0<x<\dfrac{3}{2}\end{cases}$$

show that $Df(0)$ does not exist.

3. If the function $f(x)$ is defined by

$$f(x)=\begin{cases}x^2, & \text{when } x\ge 0\\ x, & \text{when } x<0\end{cases}$$

then find left hand derivative and right hand derivative at $x=0$. Is $f(x)$ differentiable at $x=0$?

Answer: Non-differentiable at $x=0$

4. Does the differential coefficient of the following function exist at $x=0$ and $x=1$

$$f(x)=\begin{cases}1-x, & \text{when } x<0\\ 1+x^2, & \text{when } 0\le x\le 1\\ x^2-x+2, & \text{when } x>1\end{cases}$$

Answer: Non-differentiable at both points $x=0$ and $x=1$.

Type 2: Problems based on redefined functions:

Exercise 7.2

1. If $f(x)=\begin{cases}x\sin\left(\dfrac{1}{x}\right), & \text{when } x\ne 0\\ 0, & \text{when } x=0\end{cases}$ test the differentiability at $x=0$.

Answer: Non-differentiable at $x=0$

2. Show that $f(x) = \begin{cases} x^2 \sin\left(\frac{1}{x}\right), & \text{for } x \neq 0 \\ 0, & \text{for } x = 0 \end{cases}$

is differentiable at $x = 0$.

3. If $f(x) = \begin{cases} (x-2)^2 \sin\left(\frac{1}{x-2}\right), & \text{when } x \neq 2 \\ 0, & \text{when } x = 2 \end{cases}$

test the differentiability at $x = 2$.

Answer: Differentiable at $x = 2$.

4. If $f(x) = \begin{cases} x^{4/3} \cos\left(\frac{1}{x}\right), & \text{when } x \neq 0 \\ 0, & \text{when } x = 0 \end{cases}$

test the differentiability at $x = 0$

Answer: Differentiable at $x = 0$.

5. If $f(x) = \begin{cases} \dfrac{x-1}{2x^2 - 7x + 5}, & \text{when } x \neq 1 \\ -\dfrac{1}{3}, & \text{when } x = 1 \end{cases}$

test the differentiability at $x = 1$

Answer: Differentiable at $x = 1$.

6. If $f(x) = \begin{cases} \sin\left(\frac{1}{x}\right), & x \neq 0 \\ 0, & x = 0 \end{cases}$

test the differentiable at $x = 0$

Answer: Non-differentiable at $x = 0$

7. If $f(x) = \tan^{-1}\left(\frac{1}{x-1}\right)$, when $x \neq 1$; $f(1) = \frac{\pi}{2}$

and if $g(x) = (x-1)^2 \cdot f(x)$ for all x, test the differentiability of $g(x)$ at $x = 1$.

Answer: $g(x)$ is differentiable at $x = 1$

8. If $f(x) = \frac{1}{x} \sin x^2$, when $x \neq 0 = 0$, when $x = 0$

test the differentiability at $x = 0$.

Answer: Differentiable at $x = 0$.

9. Discuss the differentiability of the function

$f(x) = x\cos\left(\frac{1}{x}\right)$, for $x \neq 0$ $f(0) = 0$ at the point $x = 0$.

Answer: Not differentiable at $x = 0$.

10. Test the differentiability of the function

$$f(x) = \left(1 - e^{-\frac{1}{x}}\right)^{-1} \quad \text{for } x \neq 0$$

$f(0) = 0$ at the point $x = 0$.

Answer: Not differentiable at $x = 0$.

11. Examine the differentiability of the function

$$f(x) = \frac{1}{(x-a)} \operatorname{cosec}\left(\frac{1}{x-a}\right), \quad x \neq a$$

$f(a) = 0$ at the point $x = a$.

Answer: Find.

12. Test the function $f(x) = \frac{1}{x} \sin ax$ for $x \neq 0$

$f(0) = 1$ for differentiability at $x = 0$ where 'a' is a positive constant such that $a \neq 1$.

Answer: Find.

13. Discuss the differentiability of the function

$f(x) = e^{-\frac{1}{x^2}} \cdot \sin\left(\frac{1}{x}\right)$, when $x \neq 0$ $f(0) = 0$ at the point $x = 0$.

Answer: Differentiable at $x = 0$

14. The function f is defined as follows

$f(x) = e^{-\frac{1}{x}}$, $x \neq 0$ $f(0) = 0$ test the differentiability at $x = 0$

Answer: Differentiable at $x = 0$.

Type 3: Problems based on mod functions:

Exercise 7.3

1. Is $|x|$ differentiable at $x = 0$?

Answer: Not differentiable at $x = 0$.

2. Test the differentiability of the function $f(x) = |x - 1|$ at $x = 0$ and $x = 1$.

Answer: Differentiable at $x = 0$ and non-differentiable at $x = 1$.

3. Show that $f(x) = |x - 1| + |x + 1|$ is differentiable at all points except $x = 1$ and $x = -1$.

4. Test the differentiability of the function $f(x)=\dfrac{|x|}{x}$ at $x=1$.

Answer: Differentiable at $x=1$.

5. Is $|x-1|$ differentiable at $x=1$?

Answer: Non-differentiable at $x=1$.

6. Is $f(x)=e^{-|x|}$ differentiable at $x=0$?

Answer: Non- differentiable at $x=0$.

Type 4: To find the values of the constants so that a given function becomes differentiable at a given point $x=a$.

Exercise 7.4

1. For what choices of a, b, c if any does the function

$$f(x)=\begin{cases} x^2, & x\le 0\\ ax+b, & 0<x\le 1\\ c, & x>1\end{cases}$$

becomes differentiable at $x=0$ and $x=1$.

Answer: $a=b=c=0$.

2. For what choice of a, b, c if any does the function

$$f(x)=\begin{cases} ax^2-bx+c, & \text{when } 0\le x\le 1\\ bx-c, & \text{when } 1<x\le 2\\ c, & \text{when } x>2\end{cases}$$

becomes differentiable at $x=1$ and $x=2$.

Answer: $a=b=c=0$.

3. If $f(x)=x^2+2x$, when $x\le 0$, $=ax+b$, when $x>0$ find a and b so as to make the function $f(x)$ continuous and differentiable at $x=0$.

Answer: $a=2$ and $b=0$.

4. Find the constants a and b so as to make the following function $f(x)$ continuous and differentiable for all x where

(i) $f(x)=x^2$, when $x\le K$

$=ax+b$, when $x>K$

Answer: $a=2K$ and $b=-K^2$

(ii) $f(x)=\dfrac{1}{|x|}$, when $|x|\ge 1$

$=ax^2+b$, when $|x|<1$

Answer: $a=-\dfrac{1}{2}$ and $b=\dfrac{3}{2}$.

(iii) $f(x)=2^{\frac{1}{(x-1)}}$, when $x<1$

$=ax^2+bx$, when $x\ge 1$.

Answer: $a=b=0$.

On Continuity and Differentiability

Type 1: Problems based on piecewise functions:

Exercise 7.5

1. Test the differentiability of the function

$$f(x)=\begin{cases} x, & \text{when } 0\le x\le 1\\ 2x-1, & \text{when } 1<x\end{cases}$$

at $x=1$. Is the function continuous at $x=1$.

Answer: Non-differentiable at $x=1$ but continuous at $x=1$.

2. Prove that the function $f(x)$ defined by

$$f(x)=\begin{cases} 3+2x, & \text{when } -\dfrac{3}{2}<x\le 0\\ 3-2x, & \text{when } 0<x<\dfrac{3}{2}\end{cases}$$

is continuous at $x=0$ but $Df(0)$ does not exist.

Answer: Continuous but non-differentiable at $x=0$.

3. If $f(x)=1+x$, when $x\le 2$

$=5-x$, when $x>2$

test the continuity and differentiability of the function $f(x)$ at $x=2$.

Answer: $f(x)$ is continuous every where and differentiable every where except $x=2$.

4. Discuss the continuity and differentiability of the function where $f(x)=x^2$, when $x<-2$

$=4$, when $-2\le x\le 2$

$=x^2$, when $x>2$.

Answer: $f(x)$ is continuous every where and differentiable everywhere except $x = 2$ and $x = -2$.

5. A function $f(x)$ is defined as follows

$$f(x) = 1 \text{ for } -\infty < x < 0$$
$$= 1 + \sin x \text{ for } 0 \le x < \frac{\pi}{2}$$
$$= 2 + \left(x - \frac{\pi}{2}\right)^2 \text{ for } \frac{\pi}{2} \le x < \infty.$$

Discuss the continuity and differentiability of $f(x)$ at $x = 0$ and $x = \frac{\pi}{2}$.

Answer: Non-differentiable at $x = 0$ and differentiable at $x = \frac{\pi}{2}$ but it is continuous at both points $x = 0$ and $x = \frac{\pi}{2}$.

6. Prove that the function f defined by $f(x) = 3 - 2x$, when $x < 2 = 3x - 7$, when $x \ge 2$ is continuous but not differentiable at $x = 2$.

7. A function f is defined by

$$f(x) = \begin{cases} 5x - 4, & \text{when } x \le 1 \\ 3x^2 - 2, & \text{when } x > 1 \end{cases}$$

is the function $f(x)$ continuous and differentiable at $x = 1$?

8. If $f(x) = \begin{cases} 0, & \text{when } x < 0 \\ \sin x, & \text{when } 0 \le x \le \frac{\pi}{4} \\ \cos x, & \text{when } x > \frac{\pi}{4} \end{cases}$

show that $f(x)$ is continuous but not differentiable at $x = 0$.

9. A function $f(x)$ is defined as follows:

$f(x) = 1 + x$, when $x \le 2$, $= 5 - x$, when $x > 2$. Test the character of the function at $x = 2$ as regards its continuity and differentiability.

Answer: Continuous but non-differentiable at $x = 2$.

10. Examine whether $f(x)$ is continuous and has a derivative at the origin when $f(x) = 2 + x$, when $x \ge 0$ $f(x) = 2x$, when $x < 0$.

Answer: Continuous but not differentiable at $x = 0$.

11. A function $f(x)$ is defined as follows

$f(x) = x$ for $0 \le x \le 1$, $f(x) = 2 - x$ for $x \ge 1$. Test the character of the function at $x = 1$ regarding its continuity and differentiability.

Answer: Continuous but non-differentiable at $x = 1$.

12. Show that the function $f(x)$ defined by $f(x) = x^2 - 1$, when $x \ge 1 = 1 - x$, when $x < 1$ is continuous but not differentiable at $x = 1$.

Type 2: Problems based on redefined functions:

Exercise 7.6

1. If $f(x) = \frac{1}{x} \sin x^2$, when $x \ne 0$

$= 0$, when $x = 0$

discuss the continuity and differentiability of $f(x)$ at $x = 0$.

Answer: Continuity and differentiable at $x = 0$.

2. If $f(x) = x \tan^{-1}\left(\frac{1}{x}\right)$, when $x \ne 0$

$= 0$, when $x = 0$

examine the continuity and differentiability at $x = 0$.

Answers: Continuous but bot differentiable at $x = 0$.

3. If $f(x) = e^{-\frac{1}{x^2}} \cdot \sin\left(\frac{1}{x}\right)$, when $x \ne 0$

$= 0$, when $x = 0$

show that $f(x)$ is differentiable at $x = 0$.

4. If $f(x) = \frac{x}{1 + e^{\frac{1}{x}}}$, when $x \ne 0$

$f(0) = 0$

show that $f(x)$ is continuous at $x = 0$ but $f'(0)$ does not exist.

5. Show that $f(x) = x^2 \cdot \sin\left(\frac{1}{x}\right)$, for $x \ne 0$

$f(0) = 0$ is differentiable at $x = 0$.

6. If $f(x) = \dfrac{x\left(e^{\frac{1}{x}} - e^{-\frac{1}{x}}\right)}{\left(e^{\frac{1}{x}} + e^{-\frac{1}{x}}\right)}, \quad x \neq 0$

$= 0, \; x = 0$

show that $f(x)$ is continuous but not differentiable at $x = 0$.

Type 3: Problems based on mod functions:

Exercise 7.7

1. Let $f(x) = |x - 2|$. Is the function continuous and differentiable at $x = 2$?

2. Examine the continuity and differentiability of the following functions at the indicated points:

(i) $f(x) = |\cos x|$ at $x = \dfrac{\pi}{2}$

Answer: Continuous at $x = \dfrac{\pi}{2}$ but non-differentiable at $x = \dfrac{\pi}{2}$.

(ii) $f(x) = |x^3|$ at $x = 0$

Answer: Continuous at $x = 0$ and also differentiable at $x = 0$.

(iii) $f(x) = 1 + |\sin x|$ at $x = 0$

Answer: Continuous at $x = 0$ but non-differentiable at $x = 0$.

(iv) $f(x) = |x - 2|$ at $x = 2$

Answer: Continuous at $x = 2$ but non-differentiable at $x = 2$.

(v) $f(x) = \dfrac{|x|}{x}$ at $x = 0$

Answer: Discontinuous and non-differentiable at $x = 0$.

8

Rules of Differentiation

Before we gather the rules of differentiation, let us recapitulate some phrases and concepts frequently used in calculus.

1. Differentiation at any limit point (or, simply differentiable): A function f is said to be differentiable at any limit point x (or, simply differentiable), where x is assumed to have any finite value in $D(f)$

$$\Leftrightarrow \lim_{h\to 0} \frac{f(x+h)-f(x)}{h} \text{ is finite when}$$

$x \in D(f)$ and $(x+h) \in D(f)$ is the same whether $h \to 0$ through positive values or through negative values. The value of this limit for every finite value of x is called derivative of the function f at x and is denoted as $f'(x)$, i.e.,

$$f'(x) = \lim_{h\to 0} \frac{f(x+h)-f(x)}{h},$$ where x is supposed to have any finite value in $D(f)$, whether $h \to 0$ through positive values or through negative values.

2. Differentiability at a point: A function f is said to be differentiable at a point $x = a$, a being a particular finite value $\in D(f) \Leftrightarrow \lim_{h\to 0} \frac{f(a+h)-f(a)}{h}$ is finite for the finite given value of $x \in D(f)$ and is the same whether $h \to 0$ through positive values or through negative values. The value of this limit is called derivative of the function f at $x = a$ and is denoted as $f'(a)$, i.e.,

$$f'(a) = \lim_{h\to 0} \frac{f(a+h)-f(a)}{h},$$ where a is a particular finite value given for $x \in D(f)$ and $h \to 0$ through positive values or through negative values.

***Remark*:**

1. The result obtained after the evaluation of the limit, $\lim_{h\to 0} \frac{f(x+h)-f(x)}{h}$, for the function f at the limit point x is called derivative of f at any limit point x (or, simply at x or at the point x) or derived function of f at any limit point x because it is derived from the function f at the limit point x and is symbolised as $f'(x)$ for every finite value of x in its domain (whereas the derivative or derived function is symbolised as f' for the function f) while the function f at the limit point x is said to be derivable or differentiable at any point x (or, simply, at x) provided the limit, $\lim_{h\to 0} \frac{f(x+h)-f(x)}{h}$ is finite for every finite value of x and is the same whether $h \to 0$ through positive values or through negative values.

2. Instead of saying that the function f is differentiable at any limit point x (or, at a point, or, at a particular point $x = a$), we also say that the function f has a derivative at any limit point x (or, at a point, or, at a particular point $x = a$) in $D(f)$.

3. If the function f is written as $y = f(x)$, then its derived function is also symbolised as

$$y' = \frac{d}{dx}(y) = \frac{d}{dx}(f(x)) = \frac{d}{dx}[f(x)].$$

4. By an abuse of language, it has been customary to call $f(x)$ as function instead of f.

5. It is also said that a function $y = f(x)$ is differentiable at some point x (or, at a particular point $x = a$, or, at any point x) instead of the function f is differentiable at some point x (or, at a particular point $x = a$, or, at any point x) or $f(x)$ is a differentiable function of the independent variable x.

Now we explain the general rules of differentiation which are actually the fundamental theorems on differentiation.

1. Show that the function which is a constant times a differentiable function is differentiable and the derivative of a constant times a differentiable function is the constant times the derivative of the function.

Solution: Let $y = a f(x)$, where a is a constant and $f(x)$ is a differentiable function of the independent variable x. It is required to show that $y = a f(x)$ is differentiable and $\frac{d}{dx}(y) = a f'(x)$.

$$\because y = a f(x)$$

$$\Rightarrow y + \Delta y = a f(x + \Delta x)$$

$$\Rightarrow \Delta y = a f(x + \Delta x) - a f(x)$$

$$\Rightarrow \frac{\Delta y}{\Delta x} = \frac{a f(x + \Delta x) - a f(x)}{\Delta x}$$

$$\Rightarrow \frac{d}{dx}(y) = \lim_{\Delta x \to 0} \frac{\Delta y}{\Delta x}$$

$$= \lim_{\Delta x \to 0} \frac{a f(x + \Delta x) - a f(x)}{\Delta x}$$

$$= a \lim_{\Delta x \to 0} \frac{f(x + \Delta x) - f(x)}{\Delta x}$$

$$\Rightarrow \frac{d}{dx}(y) = a f'(x)$$

Hence, $y = a f(x)$ is a differentiable function and $\frac{d}{dx}(y) = \frac{d}{dx}(a f(x)) = a f'(x)$ which means while differentiating a constant times a differentiable function, the constant (or, numerical factor) may be taken out of the differentiation symbol $\frac{d}{dx}(\ )$.

2. Show that the sum of two separate functions differentiable on a common domain is differentiable on the same common domain and the derivative of the sum of two differentiable functions is the sum of the derivatives of the separate functions.

Solution: let $y = f(x) + g(x)$ be the sum of two functions, say $f(x)$ and $g(x)$ differentiable on a common domain (or, interval) I. It is required to show that the sum function $y = f(x) + g(x)$ is also differentiable on the same common domain I and $\frac{d}{dx}(y) = f'(x) + g'(x)$

$$\because y = f(x) + g(x)$$

$$\Rightarrow y + \Delta y = f(x + \Delta x) + g(x + \Delta x)$$

$$\Rightarrow \Delta y = f(x+\Delta x) + g(x+\Delta x) - (f(x) + g(x))$$

$$\Rightarrow \frac{\Delta y}{\Delta x} = \frac{f(x+\Delta x) + g(x+\Delta x) - (f(x) + g(x))}{\Delta x}$$

$$\Rightarrow \frac{d}{dx}(y) = \lim_{\Delta x \to 0} \frac{\Delta y}{\Delta x}$$

$$= \lim_{\Delta x \to 0} \frac{f(x+\Delta x) + g(x+\Delta x) - (f(x) + g(x))}{\Delta x}$$

$$= \lim_{\Delta x \to 0} \frac{f(x + \Delta x)}{\Delta x} + \lim_{\Delta x \to 0} \frac{g(x + \Delta x)}{\Delta x}$$

$$- \lim_{\Delta x \to 0} \frac{f(x)}{\Delta x} - \lim_{\Delta x \to 0} \frac{g(x)}{\Delta x}$$

$$= \lim_{\Delta x \to 0} \frac{f(x + \Delta x) - f(x)}{\Delta x} +$$

$$\lim_{\Delta x \to 0} \frac{g(x + \Delta x) - g(x)}{\Delta x} = f'(x) + g'(x)$$

Hence, $y = f(x) + g(x)$ is differentiable on the interval I and

$$\frac{d}{dx}(y) = \frac{d}{dx}(f(x) + g(x)) = f'(x) + g'(x)$$

Cor: The sum of a finite number of functions differentiable on a common domain is differentiable on the same domain and the derivative of the sum of a finite number of differentiable functions is the sum of their derivatives, i.e.

If $f_1, f_2, f_3, .. f_n$ be differentiable functions at the same point x, then $y = f_1(x) + f_2(x) + \ldots + f_n(x)$ is also differentiable at the same point x and $\frac{d}{dx}(y) = f_1'(x) + f_2'(x) + \ldots + f_n'(x)$ which is known as extension rule (or, formula) for the derivative of the sum of a finite number of a differentiable functions.

3. Show that the product of two functions differentiable on a common domain is differentiable on the same common domain and the derivative of the product of two differentiable functions is the first function times the derivative of the second function plus the second function times the derivative of the first function.

Solution: Let $y = f(x) \cdot g(x)$, where $f(x)$ and $g(x)$ are functions differentiable on a common interval I. It is required to show that the product function $y = f(x) \cdot g(x)$ is differentiable on the same common interval I and

$$\frac{d}{dx}(f(x) \cdot g(x)) = f(x) \cdot g'(x) + g(x) \cdot f'(x)$$

$$\because y = f(x) \cdot g(x)$$

$$\Rightarrow y + \Delta y = f(x + \Delta x) \cdot g(x + \Delta x)$$

$$\Rightarrow \Delta y = f(x + \Delta x) \cdot g(x + \Delta x) - f(x) \cdot g(x)$$

$$\Rightarrow \frac{\Delta y}{\Delta x} = \frac{f(x + \Delta x) \cdot g(x + \Delta x) - f(x)\, g(x)}{\Delta x}$$

Now adding and subtracting $f(x) \cdot g(x + \Delta x)$ in numerator, we have

$$\frac{\Delta y}{\Delta x} = \frac{f(x+\Delta x) g(x+\Delta x) - f(x) g(x+\Delta x) + f(x) g(x+\Delta x) - f(x) \cdot g(x)}{\Delta x}$$

$$\Rightarrow \frac{d}{dx}(y) = \lim_{\Delta x \to 0} \left[\frac{f(x+\Delta x) - f(x)}{\Delta x} \cdot g(x+\Delta x)\right] + \lim_{\Delta x \to 0} \left[\frac{g(x+\Delta x) - g(x)}{\Delta x} \cdot f(x)\right]$$

$$= \lim_{\Delta x \to 0} \frac{f(x+\Delta x) - f(x)}{\Delta x} \cdot \lim_{\Delta x \to 0} g(x+\Delta x) + \lim_{\Delta x \to 0} \frac{g(x+\Delta x) - g(x)}{\Delta x} \cdot \lim_{\Delta x \to 0} f(x)$$

$$= f'(x) \cdot g(x) + g'(x) \cdot f(x) = f(x) \cdot g'(x) + g(x) \cdot f'(x)$$

Hence, $y = f(x) \cdot g(x)$ is differentiable on the common interval I and

$$\frac{d}{dx}(y) = \frac{d}{dx}(f(x) \cdot g(x)) = f(x) \cdot g'(x) + g(x) \cdot f'(x)$$

Note: It is immaterial to consider any function of the two given functions as the first function and the second function, i.e., any one of the two given functions may be considered as the first function and the other one as the second function. This is why we are at liberty to consider any one of the two given functions as the first function and the other one as the second function.

Cor: The product of a finite number of functions differentiable on a common domain is differentiable on the same domain and the derivative of the product of a finite number of differentiable functions is the sum of the n terms obtained by multiplying the derivative of each one of the factors by the other $(n-1)$ factors undifferentiated, i.e. if $f_1, f_2, \ldots, f_n$ be separate functions differentiable on a common interval I (or, at any point x), then $y = f_1 \cdot f_2 \cdot f_3 \ldots f_n$ is also differentiable on the same common interval I (or, at the same point x) and

$$\frac{d}{dx}(y) = (f_1 \cdot f_2 \cdot f_3 \cdots f_{n-1})\frac{d}{dx}(f_n) +$$

$$(f_1 \cdot f_2 \cdots f_{n-1} f_n)\frac{d}{dx}(x_{n-1}) + \ldots + (f_2 \cdot f_3 \ldots f_n)\frac{d}{dx}$$

(f_1), where each product within the bracket contains $(n-1)$ factors undifferentiated. This is known as the extension rule for the derivative of the product of a finite number of differentiable functions.

Question: Find $\frac{d}{dx}(x^n)$ using the extension rule for the derivative of the product of a finite number of differentiable functions, n being a positive integer.

Solution: Let $y = x$ be a differentiable function

$$\therefore \frac{d}{dx}(y) = \frac{d}{dx}(x) = \lim_{\Delta x \to 0} \frac{\Delta x}{\Delta x} = 1$$

Now, $x^n = x \cdot x \cdot x \ldots x$ (upto 'n' factors) is differentiable and using the extension rule for the derivative of product of a finite number of differentiable functions, we have

$$\frac{d}{dx}(x^n) = (x \cdot x \ldots x)\frac{d}{dx}(x) + (x \cdot x \ldots x)\frac{d}{dx}(x) +$$

$\ldots + (x \cdot x \ldots x)\frac{d}{dx}(x)$ [each product within the bracket contains $(n-1)$ factors undifferentiated.]

$$= x^{n-1}\frac{d}{dx}(x) + \ldots + x^{n-1}\frac{d}{dx}(x) \text{ (upto } n\text{-times)}$$

$$= x^{n-1}(1 + 1 + \ldots \text{ upto } n\text{-times}) = nx^{n-1}$$

$$\left(\because \frac{d}{dx}(x) = 1\right)$$

4. Show that the quotient of two functions differentiable on a common domain is differentiable on the same common domain excepting the points (or common domain) where the function in denominator is zero and the derivative of the quotient of two differentiable functions is the function in denominator times the derivative of the function in denominator minus the function in numerator times the derivative of the function in denominator, all divided by the square of the function in denominator.

Solution: Let $y = \frac{f(x)}{g(x)}$, $g(x) \neq 0$, where $f(x)$ and $g(x)$ are functions differentiable on a common domain

I. It is required to show that $y = \frac{f(x)}{g(x)}$, $g(x) \neq 0$ is differentiable on the same common domain I and

$$\frac{d}{dx}\left(\frac{f(x)}{g(x)}\right) = \frac{g(x)\, f'(x) - f(x)\, g'(x)}{(g(x))^2}$$

$$\because y = \frac{f(x)}{g(x)},\ g(x) \neq 0$$

$$\Rightarrow y + \Delta y = \frac{f(x + \Delta x)}{g(x + \Delta x)}$$

$$\Rightarrow \Delta y = \frac{f(x + \Delta x)}{g(x + \Delta x)} - \frac{f(x)}{g(x)}$$

$$= \frac{f(x + \Delta x)\, g(x) - g(x + \Delta x)\, f(x)}{g(x + \Delta x)\, g(x)}$$

$$\Rightarrow \frac{\Delta y}{\Delta x} = \frac{f(x + \Delta x)\, g(x) - g(x + \Delta x)\, f(x)}{\Delta x\, g(x + \Delta x)\, g(x)}$$

Now adding and subtracting $f(x) \cdot g(x)$ in numerator, we have

$$\frac{\Delta y}{\Delta x} = \frac{f(x + \Delta x)\, g(x) - f(x)\, g(x) + f(x)\, g(x) - f(x)\, g(x + \Delta x)}{\Delta x\, g(x + \Delta x)\, g(x)}$$

$$\Rightarrow \frac{d}{dx}(y) = \lim_{\Delta x \to 0} \frac{\Delta y}{\Delta x} = \lim_{\Delta x \to 0} \frac{1}{g(x + \Delta x)\, g(x)}$$

$$\left[\frac{f(x + \Delta x) - f(x)}{\Delta x} \cdot g(x) - \frac{g(x + \Delta x) - g(x)}{\Delta x} \cdot f(x)\right]$$

$$= \frac{1}{g(x) \cdot g(x)} \cdot (f'(x)\, g(x) - g'(x)\, f(x))$$

$$= \frac{g(x)\, f'(x) - f(x)\, g'(x)}{(g(x))^2}$$

Hence, the quotient function $y = \frac{f(x)}{g(x)}$ is differentiable at every point of the common domain I excepting only those points (of common domain) at which the function in denominator is zero and

$$\frac{d}{dx}(y) = \frac{d}{dx}(f(x) \cdot g(x))$$

$$= \frac{g(x)\, f'(x) - f(x)\, g'(x)}{(g(x))^2},\ g(x) \neq 0$$

Notes:

1. We should note that in the formula of the derivative of quotient of the two functions differentiable on a common domain (or, at any point x), the derivative of the function in denominator occuring in numerator is always with negative sign.

2. $\frac{d}{dx}\left(\frac{1}{f(x)}\right) = -\frac{f'(x)}{(f(x))^2}$

3. The quotient $y = \frac{f(x)}{g(x)}$ of two functions differentiable at the point $x = a$, is a function which is differentiable at the same point $x = a$, provided $g(a) \neq 0$.

4. The quotient $y = \frac{f(x)}{g(x)}$ of two functions differentiable at every value of x is a function which is differentiable at every value of x, provided $g(x) \neq 0$ at those values of x at which $g(x)$ is differentiable.

5. One can differentiate a function at only those points at which it is defined. e.g.

(i) $f(x) = \frac{1}{g(x)} \Rightarrow \frac{d}{dx}(f(x)) = \frac{d}{dx}\left(\frac{1}{g(x)}\right)$

$= -\frac{g'(x)}{(g(x))^2}$, but this is not valid when $g(x) = 0$ for any value of x since $\frac{1}{g(x)}$ is undefined when $g(x) = 0$ for any value of x. So it is better to write $\frac{d}{dx}(f(x)) = \frac{d}{dx}\left(\frac{1}{g(x)}\right) = -\frac{g'(x)}{(g(x))^2}, g(x) \neq$ for any x.

(ii) $f(x) = \log g(x) \Rightarrow \frac{d}{dx}(f(x)) = \frac{d}{dx}(\log g(x)) = \frac{g'(x)}{g(x)}$, but this is not valid when $g(x) \leq 0$ for any value of x because $\log g(x)$ is not defined when $g(x) \leq 0$. So, it is proper to write $\frac{d}{dx}(f(x)) = \frac{d}{dx}(\log g(x)) = \frac{g'(x)}{g(x)}, g(x) > 0$.

(iii) $f(x) = \tan x \Rightarrow \frac{d}{dx}(f(x)) = \frac{d}{dx}(\tan x) = \sec^2 x$,

but this is not valid when $x = (2n+1)\frac{\pi}{2}$ because $\tan x$ is not defined for $x = (2n+1)\frac{\pi}{2}$. Hence, it is proper to write $\frac{d}{dx}(\tan x) = \sec^2 x, x \neq (2n+1)\frac{\pi}{2}$.

6. Even if a function is defined at a point, its derivative need not exist at that point. e.g.

(i) $f(x) = \sqrt{x}$ exists (i.e., it is defined) for $x \geq 0$ but its derivative $\frac{d}{dx}\left(\sqrt{x}\right) = \frac{1}{2\sqrt{x}}$ does not exists at $x = 0$. So, it is proper to write $\frac{d}{dx}\left(\sqrt{x}\right) = \frac{1}{2\sqrt{x}}, x > 0$.

(ii) $f(x) = \sin^{-1} x$ exists (i.e., it is defined) for $|x| \leq 1$ but its derivative $\frac{d}{dx}\left(\sin^{-1} x\right) = \frac{1}{\sqrt{1-x^2}}$ and does not exist for $x = \pm 1$. So, it is proper to write $\frac{d}{dx}\left(\sin^{-1} x\right) = \frac{1}{\sqrt{1-x^2}}, |x| < 1$.

7. One function can not have two different derivatives at a point while two different functions can have the same derivative. e.g.,

(i) $\frac{d}{dx}\left(\sin^{-1} x\right) = \frac{1}{\sqrt{1-x^2}}$ and $\frac{d}{dx}\left(-\cos^{-1} x\right)$

$$= \frac{1}{\sqrt{1 - x^2}}, \text{ for } |x| < 1.$$

8. A function or its derived function (or, derivative) may not exist at all points of an interval. e.g.,

(i) $f(x) = \frac{1}{x}$ does not exist at $x = 0$ in $[-1, 1]$.

(ii) $f(x) = \sqrt{x}$ has no derivative (or, derived function) at $x = 0$ because $\lim_{h \to 0} \frac{\sqrt{0 + h} - \sqrt{0}}{h}$ does not exist at $x = 0$ in $[-1, 2]$.

9. When a function or its derived function (or, derivative) does not exist atleast at one point belonging to the interval being open or closed, we say that the function or its derived function does not exist over (or, in, or on) that interval open or closed. e.g.,

(i) $f(x) = \frac{1}{x^2 - 1}$ does not exist in $[-1, 1]$ because $f(x)$ does not exist at $x = \pm 1 \in [-1, 1]$.

(ii) $\frac{d}{dx}\left(\sqrt{x}\right)$ does not exist in $[-1, 1]$ because

$$\lim_{h \to 0} \frac{\sqrt{0 + h} - \sqrt{0}}{h} = \lim_{h \to 0} \frac{\sqrt{h}}{h} = \lim_{h \to 0} \frac{\sqrt{h}}{|h|} \; (\because |h| = h \text{ for } h > 0)$$

$$= \lim_{h \to 0} \frac{\sqrt{h}}{\sqrt{h^2}} = \lim_{h \to 0} \frac{\sqrt{h}}{\sqrt{h} \cdot \sqrt{h}} = \lim_{h \to 0} \left(\frac{1}{\sqrt{h}}\right) = \infty$$

which means $f(x) = \sqrt{x}$ has no derivative at the point $x = 0 \in (-1, 1)$ or $[-1, 1]$.

10. The domain of the derived function $f'(x)$ is a subset (of the domain of f) which contains all elements x (in the domain of $f(x)$) at which the $\lim_{h \to 0} \frac{f(x + h) - f(x)}{h}$ exists, whether $h \to 0$ through positive values or negative values, but does not contain those exceptional points x where the derivative (or, derived function) fails to exist or is undefined. e.g.,

(i) $f(x) = \cos\sqrt{x} \Rightarrow f'(x) = -\frac{\sin\sqrt{x}}{2\sqrt{x}}, \; 0 < x \le 1$

which means domain of the derived function $f'(x)$ in the set of all $0 < x \le 1$.

11. $f(x) = \cos\sqrt{ax^2 + bx + c} \Rightarrow f'(x)$

$$= \frac{-(2ax + b)\sin\sqrt{ax^2 + bx + c}}{2\sqrt{ax^2 + bx + c}},$$

which means the domain of the derived function $f'(x)$ is the set of all real numbers, for which $0 < ax^2 + bx + c \le 1$.

Important Facts to Know

In connection with the function, differentiation of the function or integration of the function at any finite value of the independent variable, the following key points must be kept in one's mind.

1. One must consider the restriction that the function in denominator $\neq 0$ for any finite value of the independent variable against the fractional form of the function, the derivative or integral even if this restriction is not stated (or, written) explicitly because one must never divide by zero. e.g.,

(i) $f(x) = \frac{x}{(x - 2)}$ means $f(x) = \frac{x}{(x - 2)}, \; x \neq 2$

(ii) $f(x) = |x| \Rightarrow f'(x) = \frac{|x|}{x}$ means $f'(x) = \frac{|x|}{x}, x \neq 0$

(iii) $\int \sec x \tan x \, dx = \sec x$ means $\int \sec x \tan x \, dx$

$$= \sec x, \; x \neq (2n + 1)\frac{\pi}{2}$$

2. We say that a function, its derivative or its integral does not exist (or, we say that the function, its derivative or its integral is discontinuous) at any finite value of the independent variable when the function, its derivative or its integral assumes the form of a "fractional with a zero denominators" at any considered finite value of the independent variable. e.g.,

(i) $f(x) = \frac{1}{x}, x \neq 0 \Rightarrow \frac{1}{x}$ is discontinuous at $x = 0$.

(ii) $f(x) = \sqrt{x}, x \geq 0 \Rightarrow f'(x) = \frac{1}{2\sqrt{x}}, x \neq 0 \Rightarrow \frac{1}{2\sqrt{x}}$ is discontinuous at $x = 0$.

(iii) $\int \sec^2 x \, dx = \tan x, x \neq (2n+1)\frac{\pi}{2} \Rightarrow$ the integral is discontinuous at $x = (2n+1)\frac{\pi}{2}$.

3. By using the relations

(i) $f_+'(a) = \lim_{h \to 0} \frac{f(a+h) - f(a)}{h}, (h > 0)$

(ii) $f_-'(a) = \lim_{h \to 0} \frac{f(a-h) - f(a)}{h}, (h > 0)$,

One can show the non-differentiability of the function $y = f(x)$ at a finite value of the independent variable $x = a$ when its derivative assumes the form of a "fraction with a zero denominator" at the same considered finite value 'a' of the independent variables 'x'. e.g.,

(i) Show that the function $f(x) = \sqrt{x}$, for $x \geq 0$, $f(x) = \sqrt{-x}$, for $x < 0$, is non-differentiable at $x = 0$.

Solution: For differentiability, we have

$$f_+'(0) = \lim_{h \to 0} \frac{f(0+h) - f(0)}{h}, (h > 0)$$

$$= \lim_{h \to 0} \frac{\sqrt{h} - 0}{h} = \lim_{h \to 0} \frac{\sqrt{h}}{|h|}$$

$$= \lim_{h \to 0} \frac{\sqrt{h}}{\sqrt{h^2}} \quad (\because |h| = h \text{ for } h > 0)$$

$$= \lim_{h \to 0} \frac{\sqrt{h}}{\sqrt{h} \cdot \sqrt{h}} = \lim_{h \to 0} \frac{1}{\sqrt{h}} = \infty \quad \text{...(i)}$$

$$f_-'(0) = \lim_{h \to 0} \frac{f(0-h) - f(0)}{-h}, (h > 0)$$

$$= \lim_{h \to 0} \frac{\sqrt{h} - 0}{-h} = \lim_{h \to 0} -\frac{\sqrt{h}}{|h|}$$

$$= \lim_{h \to 0} \frac{\sqrt{h}}{-\sqrt{h^2}} \quad (\because |h| = h \text{ for } h > 0)$$

$$= \lim_{h \to 0} \frac{\sqrt{h}}{-\sqrt{h} \cdot \sqrt{h}} = \lim_{h \to 0} \left(-\frac{1}{\sqrt{h}}\right) = -\infty \quad \text{...(ii)}$$

Hence, (i) and (ii) $\Rightarrow$ non-differentiability of the function defined by $f(x) = \sqrt{x}$, for $x \geq 0$, $f(x) = \sqrt{(-x)}$, for $x < 0$, at the point $x = 0$.

4. When the denominator of a function, derived function or the integral of a function does not vanish for any real value of the independent variable, we state (or, write) that the function, derived function (or, simply derivative) or integral of the function exists for all real values of the independent variable. e.g.,

(i) $f(x) = \frac{1}{4x^2 + 7x + 9}, \forall x$

(ii) $\left(\tan^{-1} x\right)' = \frac{1}{1+x^2}, \forall x$

(iii) $\int \left(x^2 + 1\right)^{-2} dx^2 = \frac{1}{x^2 + 1}, \forall x$

Recapitulation

(A) One must remember that power, exponential, logarithmic, trigonometric and inverse trigonometric functions are differentiable on any interval on which they are defined and their derivatives can be found from the formulas.

(i) $\frac{d}{dx}(c) = 0$, c being any constant.

(ii) $\frac{d}{dx}\left(x^n\right) = nx^{n-1}, n \in Q$

(iii) $\frac{d}{dx}\left(a^x\right) = a^x \log a, a > 0$

(iv) $\frac{d}{dx}\left(e^x\right) = e^x$

(v) $\frac{d}{dx}(\log x) = \frac{1}{x}, x > 0$

(vi) $\frac{d}{dx}(\sin x) = \cos x$

(vii) $\frac{d}{dx}(\cos x) = -\sin x$

(viii) $\frac{d}{dx}(\tan x) = \sec^2 x$, $x \neq n\pi + \frac{\pi}{2}$

(ix) $\frac{d}{dx}(\cot x) = -\text{cosec}^2 x$, $x \neq n\pi$

(x) $\frac{d}{dx}(\sec x) = \sec x \cdot \tan x$, $x \neq n\pi + \frac{\pi}{2}$

(xi) $\frac{d}{dx}(\text{cosec}\, x) = \text{cosec}\, x \cdot \cot x$, $x \neq n\pi$

(B) The following rules of finding the derivatives are valid for differentiable functions.

(i) $\frac{d}{dx}(a f(x)) = a\frac{d}{dx}(f(x))$, a being any constant.

(ii) $\frac{d}{dx}(f(x) \pm g(x)) = \frac{d}{dx}(f(x)) + \frac{d}{dx}(g(x))$

(iii) $\frac{d}{dx}(f(x)\cdot g(x)) = g(x)\cdot\frac{d}{dx}(f(x)) \pm f(x)\frac{d}{dx}(g(x))$

(iv) $\frac{d}{dx}\left(\frac{f(x)}{g(x)}\right) = \frac{g(x)\frac{d}{dx}(f(x)) - f(x)\cdot\frac{d}{dx}(g(x))}{g^2(x)}$,

$(g(x) \neq 0)$

Notes:

1. If the question does not say to find the limits, derivatives and integrals by using their definitions, one can use their formulas derived by using their definitions.

2. If a constant appears as a constant multiple of the differentiable function, the constant can be taken out of the symbol $\frac{d}{dx}(\ )$.

3. If a constant appears as an additive quantity in a differentiable function, it vanishes while differentiating that function since *d.c.* of a constant is zero.

Solved Examples

Form 1: y = any constant

Differentiate the following w.r.t. x.

1. 1995

2. $\frac{3}{7}$

3. a^2, a being a constant.

Solutions: (i) $\frac{d}{dx}(1995) = 0$

(ii) $\frac{d}{dx}\left(\frac{3}{7}\right) = 0$

(iii) $\frac{d}{dx}\left(a^2\right) = 0$

Form 2: $y = a f(x)$, where $f(x)$ = an algebraic expression in x and the value of the function sin, cos, tan, cot, sec, cosec, $\sin^{-1}$, $\cos^{-1}$, $\tan^{-1}$, $\cot^{-1}$, $\sec^{-1}$, cosec^{-1}, log, e, | |, etc. at any point x.

Differentiate the following w.r.t. x.

1. $28x^{\frac{3}{4}}$

Solution: $\frac{d}{dx}\left(28x^{\frac{3}{4}}\right) = 28\frac{d}{dx}\left(x^{\frac{3}{4}}\right)$

$= 28 \times \frac{3}{4} \times \left(x^{\frac{3}{4}-1}\right) = 21x^{-\frac{1}{4}} = \frac{21}{x^{\frac{1}{4}}}; x > 0$

2. $3x^2$

Solution: $\frac{d}{dx}\left(3x^2\right) = 3\frac{d}{dx}\left(x^2\right) = 3 \times 2x = 6x$

3. $6 \sin x$

Solution: $\frac{d}{dx}(6\sin x) = 6\frac{d}{dx}(\sin x) = 6\cos x$

4. $a \sin^{-1} x$

Solution: $\frac{d}{dx}\left(a\sin^{-1} x\right) = a\frac{d}{dx}\left(\sin^{-1} x\right)$

$= a \cdot \frac{1}{\sqrt{1-x^2}} = \frac{a}{\sqrt{1-x^2}}; |x| < 1$

5. $m \log x$

Solution: $\frac{d}{dx}(m\log x) = m\frac{d}{dx}(\log x) = \frac{m}{x}; x > 0.$

6. $\frac{x^2}{m}$

Solution: $\frac{d}{dx}\left(\frac{x^2}{m}\right) = \frac{1}{m} \cdot \frac{d}{dx}(x^2)$

$$= \frac{1}{m} \cdot 2x = \frac{2x}{m}$$

Form 3: $y = a\, f_1(x) \pm b\, f_2(x)$, where a and b are constants and $f_1(x)$ and $f_2(x)$ are algebraic expressions in x or the values of the functions sin, cos, tan, cot, sec, cosec, $\sin^{-1}$, $\cos^{-1}$, $\tan^{-1}$, $\cot^{-1}$, $\sec^{-1}$, cosec^{-1}, log, e, | |, etc. at the same point x.

Differentiate the following w.r.t. x.

1. $(lx^2 + mx + c)$

Solution: $\frac{d}{dx}(lx^2 + mx + c)$

$$= \frac{d}{dx}(lx^2) + \frac{d}{dx}(mx) + \frac{d}{dx}(c)$$

$$= l\frac{d}{dx}(x^2) + m\frac{d}{dx}(x) + 0$$

$$= l \cdot 2x + m \cdot 1 + 0 = 2lx + m$$

2. $\left(5x^3 - 2x^{\frac{3}{2}}\right)$

Solution: $\frac{d}{dx}\left(5x^3 - 2x^{\frac{3}{2}}\right)$

$$= \frac{d}{dx}(5x^3) - \frac{d}{dx}\left(2x^{\frac{3}{2}}\right)$$

$$= 5\frac{d}{dx}(x^3) - 2\frac{d}{dx}\left(x^{\frac{3}{2}}\right)$$

$$= 5 \times 3x^2 - 2 \times \frac{3}{2} \times x^{\left(\frac{3}{2}-1\right)}$$

$$= 15x^2 - 3x^{\frac{1}{2}}$$

3. $y = \sqrt{x} + \frac{1}{\sqrt{x}}$

Solution: $y = \sqrt{x} + \frac{1}{\sqrt{x}}$

$$\Rightarrow \frac{d}{dx}(y) = \frac{d}{dx}\left(\sqrt{x} + \frac{1}{\sqrt{x}}\right)$$

$$= \frac{d}{dx}\left(\sqrt{x}\right) + \frac{d}{dx}\left(\frac{1}{\sqrt{x}}\right)$$

$$= \frac{1}{2}x^{-\frac{1}{2}} + \left(-\frac{1}{2}\right)\left(x^{-\frac{3}{2}}\right)$$

$$\left(\because \frac{1}{\sqrt{x}} = \left(\sqrt{x}\right)^{-1} = \left(x^{\frac{1}{2}}\right)^{-1} = x^{-\frac{1}{2}}\right)$$

$$= \frac{1}{2}x^{-\frac{1}{2}} - \frac{1}{2}x^{-\frac{3}{2}};\ \text{for } x > 0.$$

4. $y = \sec x + \tan x$

Solution: $y = \sec x + \tan x$

$$\Rightarrow \frac{d}{dx}(y) = \frac{d}{dx}(\sec x + \tan x)$$

$$= \frac{d}{dx}(\sec x) + \frac{d}{dx}(\tan x)$$

$$= \sec x \cdot \tan x + \sec^2 x$$

$$= \sec x\,(\sec x + \tan x);\ x \neq n\pi + \frac{\pi}{2}$$

5. $y = x^2 + 5 \sin x + \tan x$

Solution: $y = x^2 + 5 \sin x + \tan x$

$$\Rightarrow \frac{d}{dx}(y) = \frac{d}{dx}\left(x^2 + 5\sin x + \tan x\right)$$

$$= \frac{d}{dx}\left(x^2\right) + \frac{d}{dx}(5\sin x) + \frac{d}{dx}(\tan x)$$

$$= 2x + 5\cos x + \sec^2 x;\ x \neq n\pi + \frac{\pi}{2}$$

6. $y = \frac{x}{a} + \frac{b}{x}$

Solution: $y = \frac{x}{a} + \frac{b}{x}$

$$\Rightarrow \frac{d}{dx}(y) = \frac{d}{dx}\left(\frac{x}{a} + \frac{b}{x}\right) = \frac{d}{dx}\left(\frac{x}{a}\right) + \frac{d}{dx}\left(\frac{b}{x}\right)$$

$$= \frac{1}{a}\frac{d}{dx}(x) + b\frac{d}{dx}\left(x^{-1}\right)$$

$$= \frac{1}{a} + (-b)\, x^{-2} = \frac{1}{a} - b\,x^{2}; \text{for } x \neq 0$$

Form 4: $y = f_1(x) \times f_2(x)$, where $f_1(x)$ and $f_2(x)$ are algebraic expressions in x or the values of the functions sin, cos, tan, cot, sec, cosec, $\sin^{-1}$, $\cos^{-1}$, $\tan^{-1}$, $\cot^{-1}$, $\sec^{-1}$, cosec^{-1}, log, e, | | etc. at the same point x.

Differentiate the following w.r.t. x:

1. $y = (2x+1)(x-1)^2$

Solution: $y = (2x+1)(x-1)^2 = (2x+1)(x^2-2x+1)$

$$\Rightarrow \frac{d}{dx}(y) = \frac{d}{dx}\left[(2x+1)\left(x^2 - 2x + 1\right)\right]$$

$$= (2x+1)\frac{d}{dx}\left(x^2-2x+1\right) + \left(x^2-2x+1\right)\frac{d}{dx}(2x+1)$$

$$= (2x+1)(2x-2) + (x^2-2x+1)\cdot 2$$

$$= 4x^2 - 2x - 2 + 2x^2 - 4x + 2 = 6x^2 - 6x = 6x(x-1)$$

Note: On multiplying term by term, the product of two or more than two algebraic polynomial functions of x's can be always put in the form of the sum of a finite number of terms, each term having the form x^n and / ax^n and then it can be differentiated using the rule of the derivative of the sum of a finite number of differentiable functions of x's. Hence, applying this rule to the above given function, one can have its differential coefficient given below:

$y = (2x+1)(x-1)^2 = (2x+1)(x^2-2x+1) = 2x^3 - 4x^2 + 2x + x^2 - 2x + 1 = 2x^3 - 3x^2 + 1$

$$\Rightarrow \frac{d}{dx}(y) = \frac{d}{dx}\left(2x^3 - 3x^2 + 1\right)$$

$$= \frac{d}{dx}\left(2x^3\right) - \frac{d}{dx}\left(3x^2\right) + \frac{d}{dx}(1) = 2\frac{d}{dx}\left(x^3\right) - 3\frac{d}{dx}\left(x^2\right) + 0$$

$$= 2 \times 3x^2 - 3 \times 2x = 6x^2 - 6x = 6x(x-1)$$

2. $y = (x^2+1)(x^3+2)$

Solution: $y = (x^2+1)(x^3+2)$

$$\Rightarrow \frac{d}{dx}(y) = \frac{d}{dx}\left[\left(x^2+1\right)\left(x^3+2\right)\right]$$

$$= \left(x^2+1\right)\frac{d}{dx}\left(x^3+2\right) + \left(x^3+2\right)\frac{d}{dx}\left(x^2+1\right)$$

$$= (x^2+1)(3x^2+0) + (x^3+2)(2x+0)$$

$$= 3x^2(x^2+1) + (x^3+2)\cdot 2x$$

$$= 3x^4 + 3x^2 + 2x^4 + 4x$$

$$= 5x^4 + 3x^2 + 4x$$

3. $y = (1-x)^2(1-3x^2+5x^3)$

Solution: $y = (1-x)^2(1-3x^2+5x^3) = (x^2-2x+1)(5x^3-3x^2+1)$

$$\Rightarrow \frac{d}{dx}(y) = \left(x^2 - 2x + 1\right)\frac{d}{dx}\left(5x^3 - 3x^2 + 1\right) + \left(5x^3 - 3x^2 + 1\right)\frac{d}{dx}\left(x^2 - 2x + 1\right)$$

$$= (x^2-2x+1)(15x^2-6x) + (5x^3-3x^2+1)(2x-2+0)$$

$$= (x^2-2x+1)(15x^2-6x) + 2(x-1)(5x^3-3x^2+1)$$

$$= 3x(1-x)^2(5x-2) - 2(1-x)(5x^3-3x^2+1)$$

$$= (1-x)\{3x(1-x)(5x-2) - 2(5x^3-3x^2+1)\}$$

4. $y = x^4 \log x$

Solution: $y = x^4 \log x$

$$\Rightarrow \frac{d}{dx}(y) = \frac{d}{dx}\left(x^4 \log x\right)$$

$$= x^4\frac{d}{dx}(\log x) + \log x\frac{d}{dx}\left(x^4\right)$$

$$= x^4\cdot\left(\frac{1}{x}\right) + \log x\cdot\left(4x^3\right)$$

$$= x^3 + 4x^3\log x$$

$$= x^3(1 + 4\log x); \text{ for } x > 0.$$

5. $y = \sin x \cdot \log x$

Solution: $y = \sin x \cdot \log x$

$$\Rightarrow \frac{d}{dx}(y) = \frac{d}{dx}(\sin x \log x)$$

$$= \sin x\frac{d}{dx}(\log x) + \log x\frac{d}{dx}(\sin x)$$

$$= \sin x\cdot\frac{1}{x} + \log x\cdot\cos x$$

$$= \frac{\sin x}{x} + \cos x\log x; \text{ for } x > 0$$

6. $y = \cot x \cdot \cos^{-1} x$

Solution: $y = \cot x \cdot \cos^{-1} x$

$$\Rightarrow \frac{d}{dx}(y) = \frac{d}{dx}\left(\cot x\cdot\cos^{-1} x\right)$$

$$= \cos^{-1} x \frac{d}{dx}(\cot x) + \cot x \frac{d}{dx}\left(\cos^{-1} x\right)$$

$$= \cos^{-1} x \cdot \left(-\text{cosec}^2 x\right) + \cot x \cdot \left(\frac{-1}{\sqrt{1-x^2}}\right)$$

$$= -\text{cosec}^2 x \cdot \cos^{-1} x - \frac{\cot x}{\sqrt{1-x^2}}; \text{ for } 0 < |x| < 1$$

7. $y = x^3 \tan x$

Solution: $y = x^3 \tan x$

$$\Rightarrow \frac{d}{dx}(y) = \frac{d}{dx}\left(x^3 \tan x\right) = x^3 \frac{d}{dx}(\tan x) + \tan x \frac{d}{dx}\left(x^3\right)$$

$$= x^3 \sec^2 x + \tan x \cdot (3x^2) = x^3 \sec^2 x + 3x^2 \tan x$$

$$= x^2 (3 \tan x + x \sec^2 x);\ x \neq n\pi + \frac{\pi}{2}$$

8. $y = e^x \sin x$

Solution: $y = e^x \sin x$

$$\Rightarrow \frac{d}{dx}(y) = \frac{d}{dx}\left(e^x \sin x\right)$$

$$= e^x \frac{d}{dx}(\sin x) + \sin x \frac{d}{dx}\left(e^x\right)$$

$$= e^x \cdot \cos x + \sin x \cdot e^x$$

$$= e^x (\sin x + \cos x)$$

9. $y = e^x \log |x|$

Solution: $y = e^x \log |x|$

$$\Rightarrow \frac{d}{dx}(y) = \frac{d}{dx}\left(e^x \log|x|\right)$$

$$= \log|x| \frac{d}{dx}\left(e^x\right) + e^x \frac{d}{dx}(\log|x|)$$

$$= \log|x| \cdot e^x + e^x \cdot \frac{1}{x}$$

$$= e^x \left(\log|x| + \frac{1}{x}\right);\ x \neq 0$$

10. $y = \sin x \log |x|$

Solution: $y = \sin x \log |x|$

$$\Rightarrow \frac{d}{dx}(y) = \sin x \frac{d}{dx}(\log|x|) + \log|x| \frac{d}{dx}(\sin x)$$

$$= \sin x \cdot \frac{1}{x} + \log|x| \cdot \cos x$$

$$= \frac{\sin x}{x} + \cos x \cdot \log|x|;\ x \neq 0$$

11. $y = \cos x \cdot \cot x$

Solution: $y = \cos x \cdot \cot x$

$$\Rightarrow \frac{d}{dx}(y) = \frac{d}{dx}(\cos x \cdot \cot x)$$

$$= \cos x \cdot \frac{d}{dx}(\cot x) + \cot x \cdot \frac{d}{dx}(\cos x)$$

$$= \cos x \cdot (-\text{cosec}^2 x) + \cot x\,(-\sin x)$$

$$= -\text{cosec}^2 x \cdot \cos x - \cot x \sin x;\ x \neq n\pi$$

12. $y = \sin x \cdot \cos x$

Solution: $y = \sin x \cos x$

$$\Rightarrow \frac{d}{dx}(y) = \frac{d}{dx}(\sin x \cdot \cos x)$$

$$= \sin x \frac{d}{dx}(\cos x) + \cos x \cdot \frac{d}{dx}(\sin x)$$

$$= \sin x \cdot (-\sin x) + \cos x \cdot \cos x$$

$$= -\sin^2 x + \cos^2 x$$

$$= \cos^2 x - \sin^2 x = \cos 2x$$

13. $y = \sin^2 x$

Solution: $y = \sin^2 x = \sin x \cdot \sin x$

$$\Rightarrow \frac{d}{dx}(y) = \sin x \frac{d}{dx}(\sin x) + \sin x \cdot \frac{d}{dx}(\sin x)$$

$$= \sin x \cdot (\cos x) + \sin x \cdot (\cos x)$$

$$= \sin x \cos x + \sin x \cos x = 2 \sin x \cos x$$

14. $y = \sec x \cdot \tan x$

Solution: $y = \sec x \cdot \tan x$

$$\Rightarrow \frac{d}{dx}(y) = \frac{d}{dx}(\sec x \cdot \tan x)$$

$$= \sec x \cdot \frac{d}{dx}(\tan x) + \tan x \frac{d}{dx}(\sec x)$$

$$= \sec x \cdot (\sec^2 x) + \tan x \cdot (\sec x \cdot \tan x)$$

$$= \sec^3 x + \tan^2 x \cdot \sec x$$

$$= \sec x\,(\sec^2 x + \tan^2 x)$$

$$= \sec x\,(1 + \tan^2 x + \tan^2 x)$$

$= \sec x\,(1 + 2\tan^2 x);\ x \neq n\pi + \frac{\pi}{2}$

15. $y = x^n \cot x$

Solution: $y = x^n \cot x$

$$\Rightarrow \frac{d}{dx}(y) = \frac{d}{dx}\left(x^n \cot x\right)$$

$$= x^n \frac{d}{dx}(\cot x) + \cot x \frac{d}{dx}\left(x^n\right)$$

$$= x^n(-\text{cosec}^2 x) + \cot x\,(n x^{n-1})$$

$$= n x^{n-1} \cot x - x^n \text{cosec}^2 x = x^{n-1}(n \cot x - x\,\text{cosec}^2 x);\ x \neq n\pi$$

Form 5: $y = f_1(x) \times f_2(x) \times f_3(x)$, where $f_1(x), f_2(x)$ and $f_3(x)$ are algebraic expressions in x or the values of the functions sin, cos, tan, cot, sec, cosec, $\sin^{-1}$, $\cos^{-1}$, $\tan^{-1}$, $\cot^{-1}$, $\sec^{-1}$, cosec^{-1}, log, e, | | etc. at the same point x.

Refresh your memory: $\frac{d}{dx}(1 \times 2 \times 3) = (2 \times 3)\frac{d}{dx}(1) + (1 \times 3)\frac{d}{dx}(2) + (1 \times 2)\frac{d}{dx}(3)$ which means one should differentiate separately each function of x marked as 1, 2, 3 (standing for the first, second and third function considered at our liberty) and multiply each differentiated function of x by two remaining functions of x's undifferentiated and lastly add each product to get the differential coefficient of the product of three differentiable functions of x's.

Find the differential coefficient if

1. $y = (2x^2 - 5)\cot x \cdot \log|x|$

Solution: $y = (2x^2 - 5)\cot x \cdot \log|x|$

$$\Rightarrow \frac{d}{dx}(y) = \left(2x^2 - 5\right)\cot x \cdot \frac{d}{dx}(\log|x|) + \cot x \cdot \log|x| \frac{d}{dx}\left(2x^2 - 5\right) + \left(2x^2 - 5\right)\log|x| \cdot \frac{d}{dx}(\cot x)$$

$$= \left(2x^2 - 5\right)\cot x \cdot \left(\frac{1}{x}\right) + \cot x \cdot \log|x| \cdot (4x) + \left(2x^2 - 5\right)\log|x|\left(-\text{cosec}^2 x\right)$$

$$= \frac{\left(2x^2 - 5\right)\cot x}{x} + 4x \cdot \cot x \log|x| - \left(2x^2 - 5\right)\log|x|\,\text{cosec}^2 x;\ \text{for } x \neq n\pi$$

2. $y = e^x \log|x| \sec x$

Solution: $y = e^x \log|x| \sec x$

$$\Rightarrow \frac{d}{dx}(y) = \log|x| \sec x \frac{d}{dx}\left(e^x\right) + \sec x\, e^x \frac{d}{dx}(\log|x|) + \log|x|\, e^x \frac{d}{dx}(\sec x)$$

$$= \log|x| \sec x \cdot e^x + \sec x \cdot e^x \cdot \frac{1}{x} + \log|x| \cdot e^x \cdot \sec x \cdot \tan x$$

$$= e^x \sec x\left(\log|x| + \frac{1}{x} + \log|x| \tan x\right);\quad \text{for } x \neq 0,\ n\pi + \frac{\pi}{2}$$

3. $y = 2x^3 \sin x \log x$

Solution: $y = 2x^3 \sin x \log x$

$$\Rightarrow \frac{d}{dx}(y) = \sin x \log x \frac{d}{dx}\left(2x^3\right) + 2x^3 \cdot \log x \frac{d}{dx}(\sin x) + 2x^3 \sin x \frac{d}{dx}(\log x)$$

$$= \sin x \log x\left(6x^2\right) + 2x^3 \log x \cos x + 2x^3 \sin x \cdot \frac{1}{x}$$

$$= 6x^2 \sin x \log x + 2x^3 \log x \cos x + 2x^2 \sin x;\ \text{for } x > 0$$

4. $y = \sqrt{x} \cdot e^x \cdot \tan x$

Solution: $y = \sqrt{x} \cdot e^x \cdot \tan x$

$$\Rightarrow \frac{d}{dx}(y) = \sqrt{x} \cdot e^x \frac{d}{dx}(\tan x) + e^x \tan x \frac{d}{dx}\left(\sqrt{x}\right) + \sqrt{x} \tan x \frac{d}{dx}\left(e^x\right)$$

$$= \sqrt{x} \cdot e^x \sec^2 x + e^x \tan x \cdot \frac{1}{2\sqrt{x}} + \sqrt{x} \cdot \tan x \cdot e^x$$

$$= e^x \left(\sqrt{x} \sec^2 x + \frac{\tan x}{2\sqrt{x}} + \sqrt{x} \cdot \tan x \right);$$ for

$x > 0$ and $\neq n\pi + \dfrac{\pi}{2}$

Form 6: $y = \dfrac{f_1(x)}{f_2(x)}$, where $f_1(x)$ and $f_2(x)$ are algebraic expressions in x or the values of the functions sin, cos, tan, cot, sec, cosec, $\sin^{-1}$, $\cos^{-1}$, $\tan^{-1}$, $\cot^{-1}$, $\sec^{-1}$, cosec^{-1}, log, e, | | etc. at the same point x.

Refresh your memory: $\dfrac{d}{dx}\left(\dfrac{(1)}{(2)}\right) = \dfrac{d}{dx}\left((1) \times (2)^{-1}\right)$

$= \dfrac{(2) \times (1)' - (1) \times (2)'}{(2)^2}$ which means one should express the function in denominator in the form of a power function with negative index before differentiating the quotient of two differentiable functions x's by using the rule of the derivative of the product of two differentiable functions of x's noting that (1) and (2) represent functions of x's in *Nr* and *Dr* respectively.

Note: (i) $\dfrac{d}{dx}\left((2)^{-1}\right) = (-1) \times (2)^{-2} \times \dfrac{d}{dx}((2))$ which has been explained in the chapter "chain rule for the derivative".

(ii) $\dfrac{df(g(x))}{dx} = \dfrac{df(g(x))}{dg(x)} \times \dfrac{dg(x)}{dx}$ (proved later)

Differentiate the following w.r.t. x:

1. $y = \dfrac{bx^5 + c}{x^2 + a} = \left(bx^5 + c\right)\left(x^2 + a\right)^{-1}$

$$\Rightarrow \frac{d}{dx}(y) = \left(bx^5 + c\right)\frac{d}{dx}\left(\left(x^2 + a\right)^{-1}\right) + \left(x^2 + a\right)^{-1}\frac{d}{dx}\left(bx^5 + c\right)$$

$$= \left(bx^5 + c\right) \times (-1) \times \left(x^2 + a\right)^{-2} \times \frac{d}{dx}\left(x^2 + a\right) + \left(x^2 + a\right)^{-1} \times \left(5bx^4\right)$$

$$= -\left(bx^5 + c\right) \times \left(x^2 + a\right)^{-2} \times (2x) + \left(x^2 + a\right)^{-1} \times \left(5bx^4\right)$$

$$= 5bx^4\left(x^2 + a\right)^{-1} - 2x\left(bx^5 + c\right)\left(x^2 + a\right)^{-2}$$

$$= \frac{5bx^4}{\left(x^2 + a\right)} - \frac{2x\left(bx^5 + c\right)}{\left(x^2 + a\right)^2}$$

$$= \frac{5bx^4\left(x^2 + a\right) - 2x\left(bx^5 + c\right)\left(x^2 + a\right)}{\left(x^2 + a\right)^2}$$

$$= \frac{5bx^6 + 5abx^4 - 2bx^6 - 2cx}{\left(x^2 + a\right)^2}$$

$$= \frac{3bx^6 + 5abx^4 - 2cx}{\left(x^2 + a\right)^2}$$

Or, alternatively,

Directly using the rule of the derivative of the quotient of two differentiable functions of x's, the d.c. of the given function of x w.r.t. x is

$$\frac{d}{dx}(y) = \frac{\left(x^2 + a\right)\frac{d}{dx}\left(bx^5 + c\right) - \left(bx^5 + c\right)\frac{d}{dx}\left(x^2 + a\right)}{\left(x^2 + a\right)^2}$$

$$= \frac{\left(x^2 + a\right)\left(5bx^4 + 0\right) - \left(bx^5 + c\right) \cdot 2x}{\left(x^2 + a\right)^2}$$

$$= \frac{5bx^6 + 5abx^4 - 2bx^6 - 2cx}{\left(x^2 + a\right)^2}$$

$$= \frac{3bx^6 + 5abx^4 - 2cx}{\left(x^2 + a\right)^2}$$

2. $y = \dfrac{x^n}{\log x}$

Solution: $y = \dfrac{x^n}{\log x}$

$$\Rightarrow \frac{d}{dx}(y) = \frac{\log x \dfrac{d}{dx}\left(x^n\right) - x^n \dfrac{d}{dx}(\log x)}{(\log x)^2}$$

$$= \frac{nx^{n-1} \log x - x^n \cdot \dfrac{1}{x}}{(\log x)^2}$$

$$= x^{n-1} \cdot \frac{n \log x - 1}{\log^2 x}; \text{ for } x > 0 \text{ and } \neq 1$$

3. $y = \dfrac{\sin x}{x^3}$

Solution: $y = \dfrac{\sin x}{x^3}$

$$\Rightarrow \frac{d}{dx}(y) = \frac{x^3 \dfrac{d}{dx}(\sin x) - \sin x \dfrac{d}{dx}\left(x^3\right)}{\left(x^3\right)^2}$$

$$= \frac{x^3 (\cos x) - \sin x \left(3x^2\right)}{x^6}$$

$$= \frac{x \cos x - 3 \sin x}{x^4}; \text{ for } x \neq 0.$$

4. $y = \dfrac{\sin x}{\log|x|}$

Solution: $y = \dfrac{\sin x}{\log|x|}$

$$\Rightarrow \frac{d}{dx}(y) = \frac{\log|x| \dfrac{d}{dx}(\sin x) - \sin x \dfrac{d}{dx}(\log|x|)}{(\log|x|)^2}$$

$$= \frac{\log|x| \cos x - \sin x \cdot \dfrac{1}{x}}{\log^2|x|}$$

$$= \frac{x \cos x \log|x| - \sin x}{x \log^2|x|}; \text{ for } x \neq 0, \text{ and} -1.$$

5. $y = \dfrac{\sin^{-1} x}{\sin x}$

Solution: $y = \dfrac{\sin^{-1} x}{\sin x}$

$$\Rightarrow \frac{d}{dx}(y) = \frac{\sin x \cdot \dfrac{d}{dx}\left(\sin^{-1} x\right) - \sin^{-1} x \dfrac{d}{dx}(\sin x)}{(\sin x)^2}$$

$$= \frac{\sin x \cdot \dfrac{1}{\sqrt{1 - x^2}} - \sin^{-1} x \cdot \cos x}{\sin^2 x}$$

$$= \frac{\sin x - \sin^{-1} x \cos x \cdot \sqrt{1 - x^2}}{\sqrt{1 - x^2} \cdot \sin^2 x}; \text{ for } |x| < 1$$

and $x \neq 0$

6. $y = \dfrac{e^x}{1 + x^2}$

Solution: $y = \dfrac{e^x}{1 + x^2}$

$$\Rightarrow \frac{d}{dx}(y) = \frac{\left(1+x^2\right)\frac{d}{dx}\left(e^x\right) - e^x \cdot \frac{d}{dx}\left(1+x^2\right)}{\left(1+x^2\right)^2}$$

$$= \frac{\left(1+x^2\right)e^x - e^x \cdot (2x)}{\left(1+x^2\right)^2}$$

$$= \frac{e^x\left(1+x^2-2x\right)}{\left(1+x^2\right)^2}$$

$$= \frac{e^x (1-x)^2}{\left(1+x^2\right)^2}$$

Form 7: $y = \frac{z}{z_1}$, where

$z = f_1(x)$ = any single algebraic or transcendental function of x

or, $z = f_1(x) \pm f_1(x)$ = sum of two algebraic or transcendental functions of x's.

or, $z = f_1(x) \cdot f_2(x)$ = product of two algebraic or transcendental functions of x's.

similarly, $z_1 = g_1(x)$ = any single algebraic or transcendental function of x.

or, $z_1 = g_1(x) \pm g_2(x)$ = sum of two algebraic or transcendental function of x's.

or, $z_1 = g_1(x) \cdot g_2(x)$ = product of two algebraic or transcendental functions of x's.

Working rule: One can find the *d.c.* of the function of x put in the form (7) using the rule which consists of following steps.

Step 1: Put $z = f_1(x) \pm f_1(x) / f_1(x) \cdot f_2(x)$ and

$$z_1 = g_1(x) \pm g_2(x) / g_1(x) \cdot g_2(x)$$

Step 2: Find $\frac{dz}{dx}$ and $\frac{dz_1}{dx}$ using the rules for the derivative of the sum, difference or product of two differentiable functions of x's.

Step 3: Use the formula: $\frac{d}{dx}\left(\frac{z}{z_1}\right) = \frac{z_1\frac{dz}{dx} - z\frac{dz_1}{dx}}{(z_1)^2}$

Step 4: Put $\frac{dz}{dx}$ and $\frac{dz_1}{dx}$ in (4).

Note: One should do the problems directly without making substitutions z and z_1 for the functions of x's (in *Nr* and *Dr*) put in the form of the sum, difference or product function after practising the above working rule.

Differentiate the following w.r.t. x:

1. $y = \frac{e^x + \log x}{\sin x - 5x^3}$

Solution: $y = \frac{e^x + \log x}{\sin x - 5x^3}$

Putting $z = e^x + \log x$, we have

$$\frac{d}{dx}(z) = \frac{d}{dx}\left(e^x + \log x\right) = \frac{d}{dx}\left(e^x\right) + \frac{d}{dx}(\log x)$$

$$= e^x + \frac{1}{x};\ \text{for } x > 0 \qquad \ldots(1)$$

Again, putting $z_1 = \sin x - 5x^3$, we have

$$\frac{d}{dx}(z_1) = \frac{d}{dx}\left(\sin x - 5x^3\right) = \frac{d}{dx}(\sin x) - 5\frac{d}{dx}\left(x^3\right)$$

$$= \cos x - 15x^2 \qquad \ldots(2)$$

Now using the formula $\frac{d}{dx}\left(\frac{z}{z_1}\right) = \frac{z_1\frac{dz}{dx} - z\frac{dz_1}{dx}}{(z_1)^2}$,

we have

$$\frac{d}{dx}(y) = \frac{d}{dx}\left(\frac{z}{z_1}\right) =$$

$$\frac{\left(\sin x - 5x^3\right)\left(ex + \frac{1}{x}\right) - \left(\cos x - 15x^2\right)\left(e^x + \log x\right)}{\left(\sin x - 5x^3\right)^2},\ \text{for}$$

$x > 0$ and $5x^3 \neq \sin x$

$$\left(\because y = \frac{z}{z_1}\right)$$

2. $y = \dfrac{x \log x}{e^x + \sec x}$

Solution: $y = \dfrac{x \log x}{e^x + \sec x}$

Putting $z = x \log x$, we have

$$\frac{d}{dx}(z) = \frac{d}{dx}(x \log x)$$

$$= x\frac{d}{dx}(\log x) + \log x \frac{d}{dx}(x) = x \cdot \frac{1}{x} + \log x \cdot 1$$

$$= 1 + \log x \qquad \text{...(1)}$$

Again, putting $z_1 = e^x + \sec x$, we have

$$\frac{d}{dx}(z_1) = \frac{d}{dx}\left(e^x + \sec x\right) = \frac{d}{dx}\left(e^x\right) + \frac{d}{dx}(\sec x)$$

$$= e^x + \sec x \cdot \tan x \qquad \text{...(2)}$$

Now, using the formula $\dfrac{d}{dx}\left(\dfrac{z}{z_1}\right) = \dfrac{z_1 \dfrac{dz}{dx} - z\dfrac{d}{dx}(z_1)}{(z_1)^2}$,

we have

$$\frac{d}{dx}(y) = \frac{d}{dx}\left(\frac{z}{z_1}\right)$$

$$= \frac{(1+\log x)\left(e^x + \sec x\right) - \left(e^x + \sec x \tan x\right) x \log x}{\left(e^x + \sec x\right)^2},$$

for $x > 0$ and $\neq n\pi + \dfrac{\pi}{2}$

3. $y = \dfrac{x e^x}{1 + \log x}$

Solution: $y = \dfrac{x e^x}{1 + \log x}$

$$\Rightarrow \frac{d}{dx}(y) = \frac{d}{dx}\left(\frac{x e^x}{1 + \log x}\right)$$

$$= \frac{(1 + \log x)\dfrac{d}{dx}\left(x e^x\right) - x e^x \dfrac{d}{dx}(1 + \log x)}{(1 + \log x)^2}$$

$$= \frac{(1 + \log x)(x + 1)e^x - x \cdot e^x \cdot \dfrac{1}{x}}{(1 + \log x)^2}$$

$$= \frac{e^x\left[(1 + \log x)(1 + x) - 1\right]}{(1 + \log x)^2}, \text{ for } x > 0.$$

4. $y = \dfrac{x + e^x}{1 + \log x}$

Solution: $y = \dfrac{x + e^x}{1 + \log x}$

$$\Rightarrow \frac{d}{dx}(y) = \frac{d}{dx}\left(\frac{x + e^x}{1 + \log x}\right)$$

$$= \frac{(1+\log x)\dfrac{d}{dx}\left(x+e^x\right) - \left(x+e^x\right)\dfrac{d}{dx}(1+\log x)}{(1+\log x)^2}$$

$$= \frac{(1+\log x)\left(1+e^x\right) - \left(x+e^x\right)\cdot\dfrac{1}{x}}{(1+\log x)^2}$$

$$= \frac{(1+\log x)\left(1+e^x\right) - \left(1+\dfrac{e^x}{x}\right)}{(1+\log x)^2}$$

$$= \frac{1 + \log x + e^x + e^x \log x - 1 - \dfrac{e^x}{x}}{(1 + \log x)^2}$$

$$= \frac{e^x (x + x \log x - 1) + x \log x}{x (1 + \log x)^2}; \text{ for } x > 0.$$

5. $y = \dfrac{1 - \cos x}{1 + \cos x}$

Solution: $y = \dfrac{1 - \cos x}{1 + \cos x}$

$$\Rightarrow \frac{d}{dx}(y) = \frac{d}{dx}\left(\frac{1 - \cos x}{1 + \cos x}\right)$$

$$= \frac{(1+\cos x)\dfrac{d}{dx}(1-\cos x)-(1-\cos x)\dfrac{d}{dx}(1+\cos x)}{(1+\cos x)^2}$$

$$= \frac{(1 + \cos x) \cdot \sin x + \sin x (1 - \cos x)}{(1 - \cos x)^2}$$

$$= \frac{\sin x (1 + \cos x + 1 - \cos x)}{(1 + \cos x)}$$

$$= \frac{2 \sin x}{(1 + \cos x)^2}; x \neq (2n + 1)\pi.$$

6. $y = \dfrac{x^2}{1 + \cos x}$

Solution: $y = \dfrac{x^2}{1 + \cos x}$

$$\Rightarrow \frac{d}{dx}(y) = \frac{d}{dx}\left(\frac{x^2}{1 + \cos x}\right)$$

$$= \frac{(1 + \cos x)\dfrac{d}{dx}\left(x^2\right) - x^2 \dfrac{d}{dx}(1 + \cos x)}{(1 + \cos x)^2}$$

$$= \frac{x (2 + 2 \cos x + x \sin x)}{(1 + \cos x)^2}; x \neq (2n + 1)\pi.$$

7. $y = \dfrac{1 - \sin x}{1 + \cos x}$

Solution: $y = \dfrac{1 - \sin x}{1 + \cos x}$

$$\Rightarrow \frac{d}{dx}(y) = \frac{d}{dx}\left(\frac{1 - \sin x}{1 + \cos x}\right)$$

$$= \frac{(1+\cos x)\dfrac{d}{dx}(1-\sin x)-(1-\sin x)\dfrac{d}{dx}(1+\cos x)}{(1+\cos x)^2}$$

$$= \frac{(1 + \cos x)(-\cos x)-(1 - \sin x)(-\sin x)}{(1 + \cos x)^2}$$

$$= \frac{-\cos x - \cos^2 x + \sin x - \sin^2 x}{(1 + \cos x)^2}$$

$$= \frac{(\sin x - \cos x) - \left(\cos^2 x + \sin^2 x\right)}{(1 + \cos x)^2}$$

$$= \frac{\sin x - \cos x - 1}{(1 + \cos x)^2}; x \neq (2n + 1)\pi$$

8. $y = \dfrac{x \cos x}{1 + x^2}$

Solution: $y = \dfrac{x \cos x}{1 + x^2}$

$$\Rightarrow \frac{d}{dx}(y) = \frac{d}{dx}\left(\frac{x \cos x}{1 + x^2}\right)$$

$$= \frac{\left(1 + x^2\right)\dfrac{d}{dx}(x \cos x) - (x \cos x)\dfrac{d}{dx}\left(1 + x^2\right)}{\left(1 + x^2\right)^2}$$

$$= \frac{\left(1+x^2\right)\left(x\dfrac{d}{dx}(\cos x)+\cos x\dfrac{d}{dx}(x)\right)-(x\cos x)\cdot(2x)}{\left(1+x^2\right)^2}$$

$$= \frac{(1+x^2)(-x\sin x+\cos x)-2x^2\cos x}{(1+x^2)^2}$$

$$= \frac{-x\sin x+\cos x-x^3\sin x+x^2\cos x-2x^2\cos x}{(1+x^2)^2}$$

$$= \frac{(1-x^2)\cos x - x(1+x^2)\sin x}{(1+x^2)^2}$$

Form 8: $y=\frac{1}{z}$, $(z\neq 0)$ where $z=a$ single function of x, sum, difference or product of two or more than two differentiable functions of $x\neq 0$ for any finite value of the independent variable x.

Working rule: To find the *d.c.* of $y=\frac{1}{z}$, $(z\neq 0)$ one can adopt the rule which consists of following steps, provided $dz\neq f_1(x)=$ a single function of x.

Step 1: Put $z=f_1(x)\pm f_2(x)$, or $f_1(x)\cdot f_2(x)$ whichever is given.

Step 2: Find $\frac{d}{dx}(z)$

Step 3: Use $\frac{d}{dx}(y)=\frac{d}{dx}\left(\frac{1}{z}\right)=-\frac{\frac{d}{dx}(z)}{z^2}$

Notes:

1. When $z=f_1(x)=a$ single function of x, one should use directly the formula:

$\frac{d}{dx}\left(\frac{1}{z}\right)=\frac{d}{dx}\left(\frac{1}{f_1(x)}\right)=-\frac{f_1'(x)}{f_1^2(x)}$ which means the *d.c.* of a reciprocal of a function of x is negative of the d.c. of the function of x in numerator divided by the square of the function in denominator.

2. One should do the problems without making a substitution z for the given sum, difference and product functions of x in denominator after practising the above given working rule.

Differentiate the following w.r.t. x.

1. $y=\frac{1}{x^4\sec x}$

Solution: $y=\frac{1}{x^4\sec x}$

Putting $z=x^4\sec x$, we have

$$\frac{d}{dx}(z)=\frac{d}{dx}\left(x^4\sec x\right)$$

$$= x^4\cdot\frac{d}{dx}(\sec x)+\sec x\,\frac{d}{dx}\left(x^4\right)$$

$$= x^4\sec x\tan x+4x^3\sec x$$

$$\therefore \frac{d}{dx}(y)=\frac{d}{dx}\left(\frac{1}{z}\right)=-\frac{\frac{d}{dx}(z)}{z^2}$$

$$= \frac{-\left(x^4\sec x\tan x+4x^3\sec x\right)}{\left(x^4\sec x\right)^2}$$

$$= \frac{-x^3(x\sec x\tan x+4\sec x)}{x^8\sec^2 x}$$

$$= \frac{-\sec x\,(x\tan x+4)}{x^5\sec^2 x}$$

$$= \frac{-(x\tan x+4)}{x^5\sec x}$$

$$= \frac{-(x\sin x+4\cos x)}{x^5},\ (x\neq 0)\ \text{and}$$

$x\neq n\pi+\frac{\pi}{2}$

or, alternatively, it can be done in the following way:

$$y=\frac{1}{x^4\sec x}=\frac{\cos x}{x^4}$$

$$\Rightarrow \frac{d}{dx}(y) = \frac{d}{dx}\left(\frac{\cos x}{x^4}\right)$$

$$= \frac{x^4 \frac{d}{dx}(\cos x) - \cos x \frac{d}{dx}\left(x^4\right)}{\left(x^4\right)^2}$$

$$= \frac{-x^4 \sin x - 4x^3 \cos x}{x^8}$$

$$= \frac{-x^3 (x \sin x + 4 \cos x)}{x^8}$$

$$= \frac{-(x \sin x + 4 \cos x)}{x^5}$$

2. $y = \dfrac{1}{\sin x \cos x}$

Solution: $y = \dfrac{1}{\sin x \cos x}$

$$\Rightarrow \frac{d}{dx}(y) = \frac{d}{dx}\left(\frac{1}{\sin x \cos x}\right)$$

$$= -\frac{\frac{d}{dx}(\sin x \cos x)}{(\sin x \cos x)^2}$$

$$= -\frac{\left(\sin x \frac{d}{dx}(\cos x) + \cos x \frac{d}{dx}(\sin x)\right)}{(\sin x \cos x)^2}$$

$$= \frac{-\left(-\sin^2 x + \cos^2 x\right)}{\sin^2 x \cos^2 x} = \frac{-\left(\cos^2 x - \sin^2\right)}{\sin^2 x \cos^2 x}$$

$$= -\left(\frac{\cos^2 x}{\sin^2 \cos^2 x} - \frac{\sin^2 x}{\sin^2 x \cos^2 x}\right)$$

$$= -\left(\frac{1}{\sin^2 x} - \frac{1}{\cos^2 x}\right)$$

$$= -(\text{cosec}^2 x - \sec^2 x) = \sec^2 x - \text{cosec}^2 x;$$

$x \neq \dfrac{\pi}{2}$.

3. $y = \dfrac{1}{(x-a)(x-b)(x-c)}$

Solution: $y = \dfrac{1}{(x-a)(x-b)(x-c)}$

Putting $z = (x-a)(x-b)(x-c)$, we have

$$\frac{d}{dx}(z) = (x-b)(x-c)\frac{d}{dx}(x-a) + (x-a)(x-c)\frac{d}{dx}(x-b) + (x-b)(x-a)\frac{d}{dx}(x-c)$$

$$= (x-b)(x-c) + (x-a)(x-c) + (x-b))(x-a)$$

$$\therefore \frac{d}{dx}(y) = \frac{d}{dx}\left(\frac{1}{z}\right) = -\frac{\frac{d}{dx}(z)}{z^2}$$

$$= -\frac{(x-b)(x-c)+(x-a)(x-c)+(x-b)(x-a)}{[(x-a)(x-b)(x-c)]^2};$$

$x \neq a, b, c$.

4. $y = \dfrac{1}{\cos x}$

Solution: $y = \dfrac{1}{\cos x}$

$$\Rightarrow \frac{d}{dx}(y) = \frac{d}{dx}\left(\frac{1}{\cos x}\right)$$

$$= -\frac{\frac{d}{dx}(\cos x)}{(\cos x)^2}$$

$$= -\frac{(-\sin x)}{\cos^2 x}$$

$$= \frac{\sin x}{\cos^2 x} = \frac{\sin x}{\cos x} \cdot \frac{1}{\cos x}$$

$$= \tan x \cdot \sec x = \sec x \cdot \tan x,\ x \neq n\pi + \frac{\pi}{2}.$$

Form 9: y = sum of a finite number of terms such that each term is the product and/quotient function, i.e.

$$y = f_1(x) \times f_2(x) \pm g_1(x) \times g_2(x) / f_1(x) \times f_2(x) \pm \frac{g_1(x)}{g_2(x)} \quad \text{or} \quad y = f_1(x) \pm g_1(x) \times g_2(x) / f_1(x) \pm \frac{g_1(x)}{g_2(x)}$$

Working rule: To find the *d.c.* of the sum of a finite number of terms such that each term is the product and/quotient function, one may adopt the rule consisting of following steps:

Step 1: Use the rule for the derivative of the sum of a finite number of differentiable functions of x's regarding each product and/quotient function as a single function of x.

Step 2: Find the *d.c.* of the product and quotient function.

Differentiate the following w.r.t. x:

1. $y = x \sin x + \sin x \cos 3x$

Solution: $y = x \sin x + \sin x \cos 3x$

$$\Rightarrow \frac{d}{dx}(y) = \frac{d}{dx}(x \sin x + \sin x \cos 3x)$$

$$= \frac{d}{dx}(x \sin x) + \frac{d}{dx}(\sin x \cos 3x)$$

$$= x\frac{d}{dx}(\sin x) + \sin x \frac{d}{dx}(x) + \sin x \frac{d}{dx}(\cos 3x) + \cos 3x \frac{d}{dx}(\sin x)$$

$$= x \cos x + \sin x - 3 \sin x \cdot \sin 3x + \cos 3x \cdot \cos x.$$

2. $y = x^2 \sin x + \frac{\cos x}{x}$

Solution: $y = x^2 \sin x + \frac{\cos x}{x}$

$$\Rightarrow \frac{d}{dx}(y) = \frac{d}{dx}\left(x^2 \sin x + \frac{\cos x}{x}\right)$$

$$= \sin x \frac{d}{dx}\left(x^2\right) + x^2 \cdot \frac{d}{dx}(\sin x) + \frac{d}{dx}\left(\frac{\cos x}{x}\right)$$

$$= 2x \sin x + x^2 \cos x + \frac{x(-\sin x) - \cos x}{x^2}$$

$$= 2x \sin x + x^2 \cos x - \frac{\sin x}{x} - \frac{\cos x}{x^2};\ x \neq 0.$$

3. $y = \sqrt{x} + \frac{\cot x}{x}$

Solution: $y = \sqrt{x} + \frac{\cot x}{x}$

$$\Rightarrow \frac{d}{dx}(y) = \frac{d}{dx}\left(\sqrt{x} + \frac{\cot x}{x}\right)$$

$$= \frac{d}{dx}\left(\sqrt{x}\right) + \frac{d}{dx}\left(\frac{\cot x}{x}\right)$$

$$= \frac{1}{2\sqrt{x}} + \frac{x\left(-\operatorname{cosec}^2 x\right) - \cot x}{x^2}$$

$$= \frac{1}{2\sqrt{x}} - \frac{\left(x \operatorname{cosec}^2 x + \cot x\right)}{x^2};\ x \neq n\pi.$$

4. $y = \sin x + x \cos x$

Solution: $y = \sin x + x \cos x$

$$\Rightarrow \frac{d}{dx}(y) = \frac{d}{dx}(\sin x + x \cos x)$$

$$= \frac{d}{dx}(\sin x) + \frac{d}{dx}(x \cos x)$$

$$= \cos x + x\frac{d}{dx}(\cos x) + \cos x \frac{d}{dx}(x)$$

$$= \cos x + x(-\sin x) + \cos x$$

$$= \cos x - x \sin x + \cos x$$

5. $y = 2x \sec x - 2 \tan x \cdot \sec x$

Solution: $y = 2x \sec x - 2 \tan x \cdot \sec x$

$$\Rightarrow \frac{d}{dx}(y) = \frac{d}{dx}(2x \sec x - 2\tan x \sec x)$$

$$= 2\sec x \frac{d}{dx}(x) + 2x \frac{d}{dx}(\sec x) - 2\sec x \frac{d}{dx}(\tan x) - \tan x \frac{d}{dx}(\sec x)$$

$= 2\sec x + 2x \sec x \tan x - 2\sec x \cdot \sec^2 x - \tan x \cdot \sec x \tan x$

$= \sec x\,(2 + 2x \tan x - 2\sec^2 x - \tan^2 x)$

Form 1: Problems on constant functions, i.e., $y = c$, c being a constant.

Exercise 8.1

Differentiate w.r.t. x if

1. $y = e$
2. $y = \pi$
3. $y = \log e$
4. $y = 1993$

Note: Since Δ -method is not mentioned by which one has to do the problems which means one is free to do them by any other method also (i.e. one can use the method of applying directly the formulas obtained from Δ -method.

Answers:

1. 0, **2.** 0, **3.** 0, **4.** 0.

Form 2: Problems on power function and/a constant multiple of power function, i.e., $y = x^n$ and $y = cx^n$, c being a constant.

Exercise 8.2

Differentiate the following w.r.t. x

1. $y = x^9$
2. $y = \dfrac{8}{x^9}$
3. $y = 15x^{\frac{1}{2}}$
4. $y = 7x^{-\frac{2}{3}}$
5. $y = \sqrt{x^3}$
6. $y = \sqrt{x^{-3}}$
7. $y = 6\sqrt[7]{x^{-2}}$
8. $y = \dfrac{8a^2b^2}{7\sqrt[3]{27x}}$
9. $y = \left(\dfrac{7x^{-3}}{5x^7}\right)^{\frac{1}{2}}$
10. $y = \sqrt[3]{\dfrac{x^4}{7}}$

Answers:

1. $9x^8$

2. $\dfrac{-72}{x^{10}}$; $x \neq 0$.

3. $\dfrac{15}{2\sqrt{x}}$; $x > 0$.

4. $-\dfrac{14}{3} x^{-\frac{5}{3}}$; $x \neq 0$.

5. $\dfrac{3}{2}\sqrt{x}$; $x \geq 0$.

6. $-\dfrac{3}{2} x^{-\frac{5}{2}}$; $x > 0$.

7. $\dfrac{-12}{7} \cdot x^{\frac{-9}{7}}$; $x \neq 0$.

8. $\dfrac{-8a^2b^2}{63} \cdot x^{\frac{-4}{3}}$; $x \neq 0$.

9. $-\sqrt{35}\,|x| \cdot x^{-7}$, $x \neq 0$.

10. $\dfrac{4}{3}\sqrt[3]{\dfrac{x}{7}}$.

Form 3: Problems on a constant multiple of transcendental functions of x's, i.e., $y = c \times$ any one of the functions of x's say $\sin x$, $\cos x$, $\tan x$, $\cot x$, $\sec x$, cosec x, $\sin^{-1} x$, $\cos^{-1} x$, $\tan^{-1} x$, $\cot^{-1} x$, $\sec^{-1} x$, $\text{cosec}^{-1} x$, $\log x$ and e^x, c being a constant.

Exercise 8.3

Find the differential coefficients of the following functions w.r.t. x:

1. $7e^x$

2. $\dfrac{8}{(\log x)^{-1}}$

3. $\sqrt{2}\,\sin x$

4. $\dfrac{\sqrt{3}}{2}\sec x$

5. $\dfrac{\tan^{-1} x}{4}$

***Answers*:**

1. $7e^x$

2. $\dfrac{8}{x}$; $x > 0, \neq 1$

3. $\sqrt{2}\,\cos x$

4. $\dfrac{\sqrt{3}}{2}\sec x\tan x,\ x \neq n\pi + \dfrac{\pi}{2},\ n \in Z$

5. $\dfrac{1}{4}\left(\dfrac{1}{1+x^2}\right)$

***Form 4*:** Problems on the sum of a finite number of differentiable functions of x's, i.e., $y = f_1(x) \pm f_2(x) + \ldots \pm f_n(x)$ and/ $a_1 f_1(x) \pm a_2 f_2(x) \pm \ldots \pm a_n f_n(x)$, $a_1, a_2, \ldots, a_n$ being constants.

Exercise 8.4

Differentiate the following functions w.r.t. x:

1. $ax^2 + bx + c$

2. $\sqrt{x} + \dfrac{1}{\sqrt{x}}$

3. $\dfrac{17x^2 - 42x^7 + 14x^4 + 11}{5x^3}$

4. $15x^{3.1} - 16x^{4.8} - \dfrac{13}{2x^3} + 87$

5. $4x^3 + 3\sin x + 5\cos x - 2e^x$

6. $2\sin x - 5\cos x$

7. $\sqrt[4]{x} - \dfrac{5}{\sec x} + 3$

8. $\dfrac{x^3 - a^3}{x - a}$

9. $\left(\dfrac{1}{\sqrt{x}} + \sqrt{x}\right)^2$

10. $6\log x - \sqrt{x} - 7$

***Answers*:**

1. $2ax + b$

2. $\dfrac{1}{2\sqrt{x}} - \dfrac{1}{2x^{\frac{3}{2}}}$; $x > 0$

3. $\dfrac{-17x^2 - 168x^7 - 33 + 14x^2}{5x^4}$; $x \neq 0$

4. $45.5x^{2.1} - 76.8x^{3.8} - \dfrac{39}{2x^4}$; $x \neq 0$

5. $12x^2 + 3\cos x - 5\sin x - 2e^x$

6. $2\cos x - 5\sin x$

7. $\dfrac{1}{4}x^{-\frac{3}{4}} + 5\sin x$; $x \neq n\pi + \dfrac{\pi}{2}$ and $\neq 0$

8. $2x + a$; $x \neq a$

9. $\dfrac{x^2 - 1}{x^2}$; $x > 0$

10. $\dfrac{6}{x} - \dfrac{1}{2\sqrt{x}}$; $x > 0$

Form 5: Problems on the product of two differentiable functions of x's, i.e., $y = f_1(x) \times f_2(x)$.

Exercise 8.5

Differentiate the following w.r.t. x:

1. $(x+4)(5x-6)$
2. $\left(x+\dfrac{1}{x}\right)\left(\sqrt{x}+\dfrac{1}{\sqrt{x}}\right)$
3. $\dfrac{e^x \log x}{7}$
4. $e^x \log x$
5. $9\sqrt{x}\,\log x$
6. $4x^5 \cdot e^x$
7. $11 \sin x \cdot \dfrac{e^x}{9}$
8. $\sin x \cdot \log x$
9. $2x \cos x$
10. $(3x+1)\,e^x$
11. $\cos x \log x$
12. $x^3 \tan x$
13. $\sin x \tan x$
14. $e^x \cdot \tan^{-1} x$
15. $e^x \sin^{-1} x$

Answers:

1. $10x+14$
2. $\dfrac{(x-1)\left(3x^2+4x+3\right)}{2x^2\sqrt{x}}$; $x>0$
3. $\dfrac{e^x}{7}\cdot\left(\log x+\dfrac{1}{x}\right)$; $x>0$
4. $\dfrac{e^x}{x}+(\log x)\,e^x\,;x>0$
5. $\dfrac{9}{\sqrt{x}}+\dfrac{9}{2\sqrt{x}}\log x\,;x>0$
6. $(4x^5+20x^4)\,e^x$
7. $\dfrac{11}{9}(\sin x+\cos x)\,e^x$
8. $\dfrac{\sin x}{x}+\cos x\,\log x\,;x>0$
9. $2\cos x-2x\sin x$
10. $3e^x+(3x+1)\,e^x$
11. $\dfrac{\cos x}{x}-\sin x\,\log x\,;x>0$
12. $3x^2\tan x+x^3\sec^2 x\,;x\neq n\pi+\dfrac{\pi}{2},\ n\in Z$
13. $\sin x\cdot\sec^2 x+\tan x\cdot\cos x\,;x\neq n\pi+\dfrac{\pi}{2},\ n\in Z$
14. Find
15. Find

Form 6, 7 and 8: Problems on quotient function such that the function in the numerator and denominator is either the sum or the product function, i.e., $y=\dfrac{z}{z_1}$, where $z=f_1(x)/\,f_1(x)\pm f_2(x)/\,f_1(x)\times f_2(x)/$ unity and $z_1=g_1(x)/\,g_1(x)\pm g_2(x)/\,g_1(x)\times g_2(x)$

Exercise 8.6

Find $\dfrac{d}{dx}(y)$ if

1. $y=\dfrac{ax+b}{cx+d}$
2. $y=\dfrac{x^3+3}{x^3-5}$
3. $y=\dfrac{x^2+9x+10}{x^2-7x+12}$
4. $y=\dfrac{\sqrt{a}+\sqrt{x}}{\sqrt{a}-\sqrt{x}}\,;a>0.$
5. $y=\dfrac{\log x}{\sin x}$

6. $y = \dfrac{\cos x}{\log x}$

7. $y = \dfrac{x^n}{\log x}$

8. $y = \dfrac{\cos x}{x}$

9. $y = \dfrac{x^2}{\tan x}$

10. $y = \dfrac{3x^2 + 4}{\sin x + \cos x}$

11. $y = \dfrac{5 + \tan x}{8x + 9}$

12. $y = \dfrac{\tan x + \cot x}{\log x}$

13. $y = \dfrac{e^x}{1 + x}$

14. $y = \dfrac{\tan x}{x + e^x}$

15. $y = \dfrac{e^x + \tan x}{\cot x - x^n}$

16. $y = \dfrac{1}{\sin x}$

17. $y = \dfrac{1}{\log x}$

18. $y = \dfrac{x^2}{\sin x}$

19. $y = \dfrac{\cot x}{x}$

20. $y = \dfrac{1}{e^x \cdot \tan x}$

21. $y = \dfrac{\sec x + \tan x}{\sec x - \tan x}$

22. $y = \dfrac{1 - \cos x}{1 + \cos x}$

23. $y = \dfrac{1 - \tan x}{1 + \tan x}$

24. $y = \dfrac{x^2 + \sec x}{1 + \tan x}$

25. $y = \dfrac{x \cos x}{x^2 + 4}$

26. $y = \dfrac{\sin x + \cos x}{\sin x - \cos x}$

27. $y = \dfrac{\log x}{e^x}$

28. $y = \dfrac{\log x}{\cos x}$

29. $y = \dfrac{\log x}{x}$

Answers:

1. $\dfrac{ad - bc}{(cx + d)^2};\ x \neq \dfrac{-d}{c}$

2. $\dfrac{-24x^2}{\left(x^3 - 5\right)^2};\ x^3 \neq 5$

3. $\dfrac{-16x^2 + 4x + 178}{\left(x^2 - 7x + 12\right)^2};\ x \neq 3, 4$

4. $\dfrac{\sqrt{a}}{\sqrt{x}\left(\sqrt{a} - \sqrt{x}\right)^2};\ x > 0, \neq a$

5. $\dfrac{\frac{1}{x}\sin x - \cos x \log x}{\sin^2 x};\ x \neq n\pi,\ x > 0$

6. $\dfrac{-\sin x \log x - \frac{1}{x}\cos x}{\log^2 x};\ x > 0$

7. $\dfrac{n x^{n-1} \log x - x^{n-1}}{\log^2 x};\ x > 0$

8. $\dfrac{-x\sin x+\cos x}{x^2}; x\neq 0$

9. $\dfrac{2x\tan x-x^2\sec^2 x}{\tan^2 x}; x\neq \dfrac{n\pi}{2}, n\in Z$

10. $\dfrac{6x(\sin x+\cos x)-(\cos x-\sin x)\left(3x^2+4\right)}{(\sin x+\cos x)^2}$,

$x\neq n\pi-\dfrac{\pi}{4}$.

11. $\dfrac{(8x+9)\sec^2 x-8\tan x-40}{(8x+9)^2}; x\neq \dfrac{-9}{8}$

12. $\dfrac{\left(\sec^2 x-\operatorname{cosec}^2 x\right)\log x-\dfrac{1}{x}(\tan x+\cot x)}{\log^2 x}$;

$x>0, x\neq \dfrac{n\pi}{2}$

13. $\dfrac{x e^x}{(1+x)^2}; x\neq -1$

14. $\dfrac{\sec^2 x\left(x+e^x\right)-\left(1+e^x\right)\tan x}{\left(x+e^x\right)^2}; x\neq n\pi+\dfrac{\pi}{2}, -e^x$

15. $\dfrac{\left(e^x+\sec^2 x\right)\left(\cot x-x^n\right)+\left(\operatorname{cosec}^2 x+nx^{n-1}\right)\left(e^x+\tan x\right)}{\left(\cot x-x^n\right)^2}$;

$x\neq \dfrac{n\pi}{2}$ and $\cot x\neq x^n$

16. $\dfrac{-\cos x}{\sin^2 x}; x\neq n\pi$

17. $\dfrac{-1}{x\log^2 x}; x>0$

18. $\dfrac{2x\sin x-x^2\cos x}{\sin^2 x}; x\neq n\pi$

19. $\dfrac{-x\operatorname{cosec}^2 x-\cot x}{x^2}; x\neq n\pi$

20. $\dfrac{-e^x\tan x-\sec^2 x\cdot e^x}{\left(e^x\tan x\right)^2}; x\neq \dfrac{n\pi}{2}$

21. Find
22. Find
23. Find
24. Find
25. Find
26. Find

27. $\dfrac{1-x\log x}{x e^x}; x>0$

28. $\dfrac{\cos x+x\sin x\log x}{x\cos^2 x}; x>0, \neq n\pi+\dfrac{\pi}{2}$

29. $\dfrac{1-\log x}{x^2}; x>0$

Form 9: Problems on the sum of a finite number of terms such that term is the product and/quotient function, i.e.,

$$y=f_1(x)\times f_2(x)\pm g_1(x)\times g_2(x)/f_1(x)\times f_2(x)\pm\frac{g_1(x)}{g_2(x)}\text{ or, } y=f_1(x)\pm g_1(x)\times g_2(x)/f_1(x)\pm\frac{g_1(x)}{g_2(x)}\Big/\frac{f_1(x)}{f_2(x)}\pm\frac{g_1(x)}{g_2(x)}$$

Exercise 8.6

Differentiate the following functions of x's w.r.t. x:

1. $x\sin x+\sin x\cos x$

2. $x^2\sin x+\dfrac{\cos x}{x}$

3. $\dfrac{c+\sqrt[3]{x}}{c-\sqrt[3]{x}}$

4. $\left(x^2+x+1\right)+\dfrac{x^3-1}{x-1}+\dfrac{2x+1}{2x-1}$

Hint: $\dfrac{x^3-1}{x-1} = x^2 + x + 1$

$\dfrac{2x+1}{2x-1} = 1 + \dfrac{2}{2x-1}$

5. $\dfrac{\sin x}{1+\cos x}$

6. $\dfrac{x\,(x-2)}{(x+1)\,(x+4)}$

7. $\dfrac{x}{2+\cos x}$

8. $\dfrac{\sin x}{1+x^2} - \sqrt{x}\cdot e^x + 5$

9. $\dfrac{e^x \sec x - \tan x}{1+\tan x}$

10. $\dfrac{x\sin x}{\cos x - \sin x}$

11. $\dfrac{x\tan x}{\sec x + \tan x}$

12. $\dfrac{x\sin x}{\sin x + \cos x}$

13. $\dfrac{e^x}{1+x^2}$

14. $\dfrac{1}{1+x^2} + e^x \sec x$

15. $x^2 \sec x + \dfrac{x^2}{1+\sin x}$

16. $\dfrac{x\log x}{e^x \tan x}$

17. $\dfrac{e^x + \sin x}{1+\log x}$

18. $\dfrac{e^x + \log x}{e^x - \log x}$

19. $\dfrac{\sec x + \tan x}{\sec x - \tan x}$

20. $\dfrac{1-\cos x}{1+\cos x}$

21. $\dfrac{1-\tan x}{1+\tan x}$

22. $\dfrac{x^2 + \sec x}{1+\tan x}$

23. $\dfrac{x\cos x}{x^2+4}$

24. $\dfrac{\sin x + \cos x}{\sin x - \cos x}$

25. $\dfrac{\tan x}{x + \sin x}$

26. $\dfrac{x^2 + \operatorname{cosec} x}{1+\cot x}$

27. $\dfrac{x^2 \sin x}{1+\tan x}$

28. $\dfrac{\cot x}{x^2 + \sin x}$

29. $\dfrac{\sin x + \cos x}{\sqrt{x}}$

30. $\dfrac{1+\sin x}{1-\sin x}$

31. $\dfrac{1+\sin x}{1+\cos x}$

32. $\dfrac{\log x - e^x}{x^2}$

33. $\dfrac{x\log x}{e^x \tan x}$

34. $x^3 \tan x + \dfrac{e^x}{1+\cos x}$

Answers:

1. $\sin x + x\cos x + \cos^2 x - \sin^2 x$

2. $2x \sin x + x^2 \cos x - \dfrac{\sin x}{x} - \dfrac{\cos x}{x^2}; x \neq 0$

3. Find

4. $(2x+1) + (2x+1) - \dfrac{4}{(2x-1)^2}; x \neq 1, \dfrac{1}{2}$

5. $\dfrac{1}{1+\cos x}; x \neq (2n+1)\pi$

6. $\dfrac{7x^2 + 8x - 8}{(x+1)^2 (x+4)^2}; x \neq -1, -4$

7. $\dfrac{2 + \cos x + x \sin x}{(2+\cos x)^2}$

8. $\dfrac{(1+x^2)\cos x - 2x \sin x}{(1+x^2)^2} - \dfrac{e^x}{2\sqrt{x}}(2x+1)$

9. $\dfrac{2e^x \sec x \tan x - \sec^2 x}{(1+\tan x)^2}, x \neq n\pi - \dfrac{\pi}{4}$

10. $\dfrac{x + \sin x(\cos x - \sin x)}{(\cos x - \sin x)^2}, x \neq n\pi + \dfrac{\pi}{2}$

11. $\dfrac{x\cos x + \sin x(1+\sin x)}{(1+\sin x)^2}, x \neq n\pi + \dfrac{\pi}{2}$

12. $\dfrac{x + \sin x(\sin x + \cos x)}{(\sin x + \cos x)^2}; x \neq n\pi - \dfrac{\pi}{4}$

13. $\dfrac{e^x(1-x)^2}{(1+x^2)^2}$

14. $\dfrac{-2x}{(1+x^2)^2} + e^x \sec x(1+\tan x), x \neq n\pi + \dfrac{\pi}{2}$

15. $x^2 \sec x \tan x + 2x \sec x + \dfrac{2x(1+\sin x) - x^2 \cos x}{(1+\sin x)^2};$

$x \neq n\pi + \dfrac{\pi}{2}$

16. $\dfrac{e^x\left[\tan x(1+\log x) - x\log x(\tan x + \sec^2 x)\right]}{e^{2x}\tan^2 x},$

$x \neq n\pi + \dfrac{\pi}{2}$ and $x > 0$.

17. $\dfrac{(1+\log x)x(e^x + \cos x) - e^x - \sin x}{x(1+\log x)^2}; x > 0$

18. $\dfrac{2e^x\left(\dfrac{1}{x} - \log x\right)}{(e^x - \log x)^2}; x > 0$

19. $\dfrac{2\sec x(\sec x + \tan x)}{(\sec x - \tan x)}; x \neq n\pi + \dfrac{\pi}{2}$

20. $\dfrac{2\sin x}{(1+\cos x)^2}; x \neq (2n+1)\pi$

21. $\dfrac{-2\sec^2 x}{(1+\tan x)^2}, x \neq n\pi - \dfrac{\pi}{4}$

22. $\dfrac{(1+\tan x)(2x + \sec x \tan x) - (x^2 + \sec x)\sec^2 x}{(1+\tan x)^2}$

$x \neq n\pi - \dfrac{\pi}{4}$ and $x \neq n\pi + \dfrac{\pi}{2}$

23. $\dfrac{(4-x^2)\cos x - x(x^2+4)\sin x}{(x^2+4)^2}$

24. $\dfrac{-2}{1-\sin 2x}; x \neq n\pi + \dfrac{\pi}{4}$

25. $\dfrac{x\sec^2 x + \tan x(\sec x - 1) - \sin x}{(x+\sin x)^2}; x \neq 0$

26. $\dfrac{(2x - \operatorname{cosec} x \cot x)(1+\cot x) - \operatorname{cosec}^2 x(x^2 + \operatorname{cosec} x)}{(1+\cot x)^2};$

$x \neq n\pi$ and $\neq n\pi - \dfrac{\pi}{4}$

27. $\dfrac{2x\sin x}{1+\tan x}+\dfrac{x^2\left(\cos x-\sin x\tan^2 x\right)}{(1+\tan x)^2}$;

$x\neq n\pi+\dfrac{\pi}{2}$ and $\neq n\pi-\dfrac{\pi}{4}$

28. $\dfrac{-\operatorname{cosec} x\left(x^2\operatorname{cosec}+1\right)-\cot x(2x+\cos x)}{\left(x^2+\sin x\right)^2}$;

$x\neq n\pi+\dfrac{\pi}{2}$ and $x^2\neq-\sin x$

29. $\dfrac{(2x-1)\cos x-(2x+1)\sin x}{2x\sqrt{x}}, x>0$

30. $\dfrac{2\cos x}{(1-\sin x)^2}$; $x\neq 2n\pi+\dfrac{\pi}{2}$

31. $\dfrac{1+\sin x+\cos x}{(1+\cos x)^2}$; $x\neq(2n+1)\pi$

32. $\dfrac{e^x(x-2)+1-2\log x}{x^3}$; $x>0$

33. $\dfrac{e^x\tan x(1+\log x)-e^x x\log x\left(\sec^2 x+\tan x\right)}{\left(e^x\tan x\right)^2}$;

$x>0,\ x\neq n\pi$ and $x\neq n\pi+\dfrac{\pi}{2}$

34. $x^2\left(x\sec^2 x+3\tan x\right)+\dfrac{e^x(1+\sin x+\cos x)}{(1+\cos x)^2}$;

$x\neq n\pi+\dfrac{\pi}{2}$ and $x\neq(2n+1)\pi$

9

Chain Rule for the Derivative

Firstly, we recall the basic ideas of composition of differentiable functions.

1. Composition of two differentiable functions:

If $y = f_1(z_1)$ is a differentiable function of z_1 and $z_1 = f_2(x)$ is a differentiable function of x, then $y = f_1 f_2(x)$ is a differentiable function of a differentiable function of x or composition (or, composite) of two differentiable functions f_1 and f_2 whose composition in the arrow diagram can be expressed as

$$x \xrightarrow{f_2} f_2(x) \xrightarrow{f_1} f_1 f_2(x)$$

In practice, a differentiable function of a differentiable function of x is obtained by replacing the independent variable x in the differentiable function by another differentiable function of x.

Explanation

(i) $y = e^x$ is a differentiable function of x replacing x by another differentiable function $\log x$ $(x > 0)$, we have

(ii) $y = e^{\log x}$ which is a composition of two differentiable functions e and log.

2. Composition of a finite number of differentiable functions:

If $y = f_1(z_1)$ is a differentiable function of z_1;

$z_1 = f_2(z_2)$ is a differentiable function of z_2;

$z_2 = f_3(z_3)$ is a differentiable function of z_3;

$z_3 = f_4(z_4)$ is a differentiable function of z_4;

...

...

$z_{n-2} = f_{n-1}(z_{n-1})$ is a differentiable function of z_{n-1};

$z_{n-1} = f_n(z_n)$ is a differentiable function of z_n;

$z_n = f(x)$ is a differentiable function of x;

then $y = f_1 f_2 f_3 \ldots$ for $f(x)$ is a differentiable function of a differentiable function of a differentiable function ... of a differentiable function of x (or, composite/composition of $(n + 1)$ number of differentiable functions $f_1, f_2, \ldots f_n$ and f) whose composition in the arrow diagram is expressed as

$$x \xrightarrow{f} f(x) \xrightarrow{f_n} f_n f(x) \xrightarrow{f_{n-1}} f_{n-1} f(x) \ldots$$

$\xrightarrow{f_1} f_1 f_2 \cdots f_n f(x)$. But if $z_n = x =$ identity function being differentiable, we get the composition of n-number of differentiable functions $f_1, f_2, f_3, \ldots, f_n$ represented by the notation:

$y = f_1 f_2 f_3 \ldots f_n(x)$ whose arrow diagram of the composition is $x \xrightarrow{f_n} f_n(x) \xrightarrow{f_{n-1}} f_{n-1}$

$f_n(x) \longrightarrow \ldots \xrightarrow{f_1} f_1 f_2 \cdots f_n(x)$.

In practice, if the process of replacing the independent variable x in the preceeding differentiable function by another differentiable function of x is continued upto $(n + 1)$ or n- number of times, we get what is called the composition of $(n + 1)$or n-number of differentiable functions or composite of $(n + 1)$ or n-number of differentiable functions of a differentiable function ... of a differentiable function of x.

Explanation

1. $y = \sin x$ is a differentiable function of x, replacing x by a differentiable function of x, say e^x, we get

2. $y = \sin(e^x)$ which is a differentiable function of x being the composition of two differentiable functions namely sin and e. Replacing x by a differentiable function $\sqrt{x}$ $(x > 0)$, , we get

3. $y = \sin e^{\sqrt{x}}$ which is a differentiable function of x being the composition of three differentiable functions namely sin, e and $\sqrt{\ }$. Replacing x by a differentiable function of x, say $\sec^{-1} x$, we get

4. $y = \left(\sin e^{\sqrt{\sec^{-1} x}}\right)$ which is a differentiable function of x being the * or, alternatively, $y = f_1(z_1)$ is a differentiable function of z_1; $z_1 = f_2(z_2)$ is a differentiable function of z_2 and so on; $z_{n-1} = f_n(x)$ is a differentiable function of $x \Rightarrow y = f_1 f_2 f_3 \ldots f_n(x)$ is a differentiable function of x composition of four differentiable functions namely sin, e, $\sqrt{\ }$ and $\sec^{-1}$ and this composition of four differentiable functions can be expressed in the following way.

$$x \xrightarrow{\sec^{-1}} \sec^{-1} x \xrightarrow{\sqrt{\ }} \sqrt{\sec^{-1} x} \xrightarrow{e} e^{\sqrt{\sec^{-1} x}}$$

$$\xrightarrow{\sin} \sin e^{\sqrt{\sec^{-1} x}}$$

Remember: Composition of two or more than two differentiable functions is a differentiable function.

Question: What is the chain rule for the derivatives?

Answer: It is a rule (or, a theorem) which is used to find the derivatives of composite (or, composition) of two or more than two differentiable functions.

1. Chain rule for the derivatives of composite of two differentiable functions:

The chain rule for the derivatives of composite of two differentiable functions states that if y is a differentiable function of z and z is a differentiable function of x, then the derivative of a differentiable function y is the product of the derivative of y with respect to z and the derivative of z with respect to x.

Or, more explicitly,

If $y = f(z)$ is a differentiable function of z and $z = f(x)$ is a differentiable function of x, then $y = F(f(x)) = G(x)$ (say) is a differentiable function of x and $\frac{dy}{dx} = \frac{dy}{dz} \cdot \frac{dz}{dx}$.

This is known as the chain rule of differentiation since the derivative with respect to x of $y = F(f(x))$ involves the following chain of steps. Firstly differentiation with respect to z of the whole differentiable function $y = F(f(x))$. Secondly differentiation with respect to x of inner differentiable function $z = f(x)$ lastly, the product of these gives $\frac{dy}{dx}$.

2. The chain rule for the derivative of a finite number of differentiable functions:

The chain rule for the derivative of a finite number of differentiable functions is the generalised form of chain rule for the derivatives of the composite of two differentiable functions. It states that

If $y = f_1(z_1)$ is a differentiable function of z_1;
$z_1 = f_2(z_2)$ is a differentiable function of z_2;
$z_2 = f_3(z_3)$ is a differentiable function of z_3;
...
...
$z_{n-1} = f_n(z_n)$ is a differentiable function of z_n and $z_n = f(x)$ is a differentiable function of x, then, $y = f_1 f_2 f_3 \ldots f_n f(x)$ is a differentiable function of x, and

$$\frac{dy}{dx} = \frac{dy}{dz_1} \cdot \frac{dz_1}{dz_2} \cdot \frac{dz_2}{dz_3} \cdots \frac{dz_{n-1}}{dz_n} \cdot \frac{dz_n}{dx}$$

and if $y = f_1(z_1)$ is a differentiable function of z_1; $z_1 = f_2(z_2)$ is a differentiable function of z_2, and so on; $z_{n-1} = f_n(x)$ is a differentiable function of x; then $y = f_1 f_2 f_3 \ldots f_n(x)$ is a differentiable function of x and $\frac{dy}{dx} = \frac{dy}{dz_1} \cdot$

$$\frac{dz_1}{dz_2} \cdot \frac{dz_2}{dz_3} \frac{dz_3}{dz_4} \cdots \frac{dz_{n-1}}{dx}.$$

Theorem: Show that $y = F(z)$ and $z = f(x) \Rightarrow y' = F'(z) \cdot f'(x)$ provided $F(z)$ and $f(x)$ are differentiable functions.

Proof: $y = F(z)$

$$\Rightarrow y + \Delta y = F(z + \Delta z)$$

$$\Rightarrow \Delta y = F(z + \Delta z) - F(z)$$

$$\Rightarrow \frac{\Delta y}{\Delta x} = \frac{F(z + \Delta z) - F(z)}{\Delta x}$$

$$\Rightarrow \frac{\Delta y}{\Delta x} = \frac{F(z + \Delta z) - F(z)}{\Delta x} \cdot \frac{\Delta z}{\Delta z} \quad \because \frac{\Delta z}{\Delta z} = 1$$

$$\Rightarrow \frac{\Delta y}{\Delta x} = \frac{F(z + \Delta z) - F(z)}{\Delta z} \cdot \frac{\Delta z}{\Delta x}$$

shifting the place of Δz and Δx in the denominator

$$\Rightarrow \lim_{\Delta x \to 0} \frac{\Delta y}{\Delta x}$$

$$= \lim_{\Delta x \to 0} \left(\frac{F(z + \Delta z) - F(z)}{\Delta z} \cdot \frac{\Delta z}{\Delta x} \right)$$

$$= \lim_{\Delta x \to 0} \left(\frac{F(z + \Delta z) - F(z)}{\Delta z} \right) \cdot \lim_{\Delta x \to 0} \left(\frac{\Delta z}{\Delta x} \right)$$

using the product rule of limits

$$= \frac{d\,F(z)}{dz} \cdot \frac{dz}{dx} \; (\because y = F(z))$$

$$= F'(z) \cdot f'(x)$$

Remember:

(A): The following rules of finding derivatives are valid for the differentiable functions.

1. $(f_1(x) + f_2(x))' = f_1'(x) + f_2'(x)$
2. $(k\,f_1(x))' = k\,f_1'(x)$, k being a constant
3. $(f_1(x)) \cdot f_2(x)' = f_1'(x) \cdot f_2(x) + f_2'(x) \cdot f_1(x)$
4. $\left(\frac{f_1(x)}{f_2(x)}\right)' = \frac{f_1'(x) \cdot f_2(x) - f_2'(x) \cdot f_1(x)}{f_2^2(x)}$; $(f_2(x) \neq 0)$
5. $(f_1 f_2(x))' = f_1'(z) \cdot f_2'(x)$, where $z = f_2(x)$

(B): Power, exponential, logarithmic, trigonometric and inverse trigonometric functions are differentiable on any interval on which they are defined and their derivatives with respect to the inner differentiable function $f(x)$ are found from the following formulas provided their composition is represented as $y = F(f(x))$, where F and f represent $(\;)^n$, e, log, sin, cos, tan, cot, sec, cosec, $\sin^{-1}$, $\cos^{-1}$, $\tan^{-1}$, $\cot^{-1}$, $\sec^{-1}$, cosec^{-1}, etc.

1. $(c)' = 0$, where c is any constant
2. $(f(x))^n{}' = n \cdot (f(x))^{n-1} \cdot f'(x)$
3. $(\sin f(x))' = \cos f(x) \cdot f'(x)$
4. $(\cos f(x))' = -\sin f(x) \cdot f'(x)$
5. $(\tan f(x))' = \sec^2 f(x) \cdot f'(x)$
6. $(\cot f(x))' = -\text{cosec}^2 f(x) \cdot f'(x)$
7. $(\sec f(x))' = \sec f(x) \cdot \text{cosec}\, f(x)$
8. $(\text{cosec}\, f(x))' = -\text{cosec}\, f(x) \cdot \cot f(x)$
9. $\left(\sin^{-1} f(x)\right)' = \frac{1}{\sqrt{1 - f^2(x)}} \cdot f'(x)$
10. $\left(\cos^{-1} f(x)\right)' = \frac{-1}{\sqrt{1 - f^2(x)}} \cdot f'(x)$
11. $\left(\tan^{-1} f(x)\right)' = \frac{1}{1 + f^2(x)} \cdot f'(x)$
12. $\left(\cot^{-1} f(x)\right)' = \frac{-1}{1 + f^2(x)} \cdot f'(x)$
13. $\left(\sec^{-1} f(x)\right)' = \frac{1}{|f(x)|\sqrt{f^2(x) - 1}} \cdot f'(x)$
14. $\left(\text{cosec}^{-1} f(x)\right)' = \frac{-1}{|f(x)|\sqrt{f^2(x) - 1}} \cdot f'(x)$
15. $(\log f(x))' = \frac{f'(x)}{f(x)}$; $f(x) > 0$
16. $\left(e^{f(x)}\right)' = e^{f(x)} \cdot f'(x)$

(C): If y is a function of x then

1. $\frac{dy^n}{dx} = n \cdot y^{n-1} \cdot \frac{dy}{dx}$, $n \in Q$
2. $\frac{d}{dx} e^y = e^y \cdot \frac{dy}{dx}$

3. $\dfrac{d}{dx}\log y = \dfrac{1}{y}\cdot\dfrac{dy}{dx},\ y > 0$

4. $\dfrac{d\sin y}{dx} = \cos y \cdot \dfrac{dy}{dx}$

5. $\dfrac{d\cos y}{dx} = -\sin y \cdot \dfrac{dy}{dx}$

6. $\dfrac{d\tan y}{dx} = \sec y \cdot \tan y \cdot \dfrac{dy}{dx}$

7. $\dfrac{d\cot y}{dx} = -\operatorname{cosec}^2 y \cdot \dfrac{dy}{dx}$

8. $\dfrac{d\sec y}{dx} = \sec y \cdot \tan y \cdot \dfrac{dy}{dx}$

9. $\dfrac{d\operatorname{cosec} y}{dx} = -\operatorname{cosec} y \cdot \cot y \cdot \dfrac{dy}{dx}$

10. $\dfrac{d\sin^{-1} y}{dx} = \dfrac{1}{\sqrt{1-y^2}} \cdot \dfrac{dy}{dx}$

11. $\dfrac{d\cos^{-1} y}{dx} = \dfrac{-1}{\sqrt{1-y^2}} \cdot \dfrac{dy}{dx}$

12. $\dfrac{d\tan^{-1} y}{dx} = \dfrac{1}{1+y^2} \cdot \dfrac{dy}{dx}$

13. $\dfrac{d\cot^{-1} y}{dx} = \dfrac{-1}{1+y^2} \cdot \dfrac{dy}{dx}$

14. $\dfrac{d\sec^{-1} y}{dx} = \dfrac{1}{|y|\sqrt{y^2-1}} \cdot \dfrac{dy}{dx}$

15. $\dfrac{d\operatorname{cosec}^{-1} y}{dx} = \dfrac{-1}{|y|\sqrt{y^2-1}} \cdot \dfrac{dy}{dx}$

Note: All the formulas for the derivatives of the differentiable power, exponential, logarithmic, trigonometric and inverse trigonometric of a differentiable function $f(x)$ can be expressed in words in the following way.

1. $\dfrac{d}{dx}$ (any function of $x)^n$, $n \in Q$

= index · (that function of x used as a base)$^{n-1 = \text{given index minus one}}$ · $\dfrac{d}{dx}$ (base), where the base must be a differentiable function of x.

2. $\dfrac{d}{dx}$ (t-function having another function of x at the place of the angle)

$$= \frac{d\begin{pmatrix} t\text{ - function having another function} \\ \text{of } x \text{ at the place of the angle}\end{pmatrix}}{d\begin{pmatrix}\text{the function of the independent variable} \\ x \text{ after removing } t\text{ - operator}\end{pmatrix}}$$

· $\dfrac{d}{dx}$ (the function of the independent variable x after removing t-operator) where t or t-operator, means all the sex trigonometric functions namely sin, cos, tan, cot, sec, and cosec.

3. $\dfrac{d}{dx}$ (t^{-1}-function having another function as an inner function of x)

$$= \frac{d\left(t^{-1}\text{ - inner function of } x\right)}{d\begin{pmatrix}\text{inner function of } x \text{ after} \\ \text{removing } t^{-1}\text{ - operator}\end{pmatrix}} \cdot \frac{d}{dx}$$ (inner function of the independent variable x after removing t^{-1}-opertor)

4. $\dfrac{d}{dx}$ ($e^{\text{a function of } x \,=\, \text{index of the base `}e\text{'}}$)

$= e^{\text{index without any change}}$. $\dfrac{d}{dx}$ (index given as a function of x)

5. $\dfrac{d}{dx}$ log (any function of x)

$= \dfrac{1}{\text{given function of } x} \cdot \dfrac{d}{dx}$ (given function of x after removing log-operator)

(***Note***: (any function of $x)^n$ is read as 'n' power of any function of x/any function of x to the power 'n'/any function (or, any function of x) raised to the power 'n'.)

On method of finding the derivative of composite (or, composition) of two differentiable functions

A differentiable function f_1 of a differentiable function f_2 of the independent variable x is symbolically represented as $y = f_1 f_2(x)$, where

f_1 = outer differentiable function (or, outside differentiable function) which means sin, cos, tan, cot, sec, cosec, $\sin^{-1}$, $\cos^{-1}$, $\tan^{-1}$, $\cot^{-1}$, $\sec^{-1}$, cosec^{-1}, $(\ )^n$, log, $\sqrt{\ }$, or, e, etc.

f_2 = inner differentiable function (or, inside differentiable function) which means also sin, cos, tan, cot, sec, cosec, $\sin^{-1}$, $\cos^{-1}$, $\tan^{-1}$, $\cot^{-1}$, $\sec^{-1}$, cosec^{-1}, $(\ )^n$, log, $\sqrt{\ }$, or, e, etc

$f_1 f_2(x)$ = given function/whole function/dependent variable $f_2(x)$ = inner differentiable function of x/insider differentiable function of x.

There are two methods of finding the derivative of the composite of two differentiable functions.

(A): Method of substitution

(B): Method of making no substitution

(A): *On method of substitution*: This method consists of following steps:

Step 1: Put a new variable z for the function of x whose differential coefficient can be found from the formulas.

1. $\frac{d}{dx}(c) = 0$

2. $\frac{d}{dx}\left(x^n\right) = nx^{n-1}$

3. $\frac{d}{dx}(\sin x) = \cos x$

4. $\frac{d}{dx}(\cos x) = -\sin x$

5. $\frac{d}{dx}(\tan x) = \sec^2 x$

6. $\frac{d}{dx}(\cot x) = \text{cosec}^2 x$

7. $\frac{d}{dx}(\sec x) = \sec x \cdot \tan x$

8. $\frac{d}{dx}(\text{cosec}\, x) = -\text{cosec}\, x \cdot \cot x$

9. $\frac{d}{dx}\left(\sin^{-1} x\right) = \frac{1}{\sqrt{1 - x^2}}$

10. $\frac{d}{dx}\left(\cos^{-1} x\right) = \frac{1}{\sqrt{1 - x^2}}$

11. $\frac{d}{dx}\left(\tan^{-1} x\right) = \frac{1}{1 + x^2}$

12. $\frac{d}{dx}\left(\cot^{-1} x\right) = \frac{-1}{1 + x^2}$

13. $\frac{d}{dx}\left(\sec^{-1} x\right) = \frac{1}{|x|\sqrt{x^2 - 1}}$

14. $\frac{d}{dx}\left(\text{cosec}^{-1} x\right) = \frac{-1}{|x|\sqrt{x^2 - 1}}$

15. $\frac{d}{dx}(\log x) = \frac{1}{x},\ x > 0$

16. $\frac{d}{dx}\left(e^x\right) = e^x$

N.B: The formulas for the derivatives of inverse trigonometric functions have been derived in the chapter containing inverse circular functions but we have mentioned here their derivatives because their application is required here.

Note: All these results remains valid even if x is replaced by any other variable z, u, v, w or t etc.

Explanation

1. $\frac{d}{dz}\left(z^n\right) = nz^{n-1}$, $\frac{d}{du}\left(u^n\right) = nu^{n-1}$,

$\frac{d}{dt}\left(t^n\right) = nt^{n-1}$

2. $\frac{d}{dz}(\sin z) = \cos z$, $\frac{d}{du}(\sin u) = \cos u$,

$\frac{d}{dt}(\sin t) = \cos t$

3. $\frac{d}{dz}(\cos z) = -\sin z$, $\frac{d}{du}(\cos u) = -\sin u$,

$\frac{d}{dt}(\cos t) = -\sin t$

4. $\frac{d}{dz}(\tan z) = \sec^2 z,\ \frac{d}{du}(\tan u) = \sec^2 u,$

$\frac{d}{dt}(\tan t) = \sec^2 t$

5. $\frac{d}{dz}(\cot z) = -\operatorname{cosec}^2 z,\ \frac{d}{du}(\cot u) = -\operatorname{cosec}^2 u,$

$\frac{d}{dt}(\cot t) = -\operatorname{cosec}^2 t$

6. $\frac{d}{dz}(\sec z) = \sec z \cdot \tan z,\ \frac{d}{du}(\sec u) = \sec u \cdot \tan u,$

$\frac{d}{dt}(\sec t) = -\sec t \cdot \tan t$

7. $\frac{d}{dz}(\operatorname{cosec} z) = -\operatorname{cosec} z \cdot \cot z,\ \frac{d}{du}(\operatorname{cosec} u)$

$= -\operatorname{cosec} u \cdot \tan u,\ \frac{d}{dt}(\operatorname{cosec} t) = -\operatorname{cosec} t \cdot \cot t$

8. $\frac{d}{dz}\left(\sin^{-1} z\right) = \frac{1}{\sqrt{1 - z^2}},\ \frac{d}{du}\left(\sin^{-1} u\right) = \frac{1}{\sqrt{1 - u^2}},$

$\frac{d}{dt}\left(\sin^{-1} t\right) = \frac{1}{\sqrt{1 - t^2}}$

9. $\frac{d}{dz}\left(\cos^{-1} z\right) = \frac{-1}{\sqrt{1 - z^2}},\ \frac{d}{du}\left(\cos^{-1} u\right) = \frac{-1}{\sqrt{1 - u^2}},$

$\frac{d}{dt}\left(\cos^{-1} t\right) = \frac{-1}{\sqrt{1 - t^2}}$

10. $\frac{d}{dz}\left(\tan^{-1} z\right) = \frac{1}{\sqrt{1-z^2}},\ \frac{d}{du}\left(\tan^{-1} u\right) = \frac{1}{\sqrt{1-u^2}},$

$\frac{d}{dt}\left(\tan^{-1} t\right) = \frac{1}{\sqrt{1-t^2}}$

11. $\frac{d}{dz}\left(\cot^{-1} z\right) = \frac{-1}{\sqrt{1+z^2}},\ \frac{d}{du}\left(\cot^{-1} u\right) = \frac{-1}{\sqrt{1+u^2}},$

$\frac{d}{dt}\left(\cot^{-1} t\right) = \frac{1}{\sqrt{1+t^2}}$

12. $\frac{d}{dz}\left(\sec^{-1} z\right) = \frac{1}{|z|\sqrt{z^2 - 1}},\ \frac{d}{du}\left(\sec^{-1} u\right)$

$= \frac{1}{|u|\sqrt{u^2 - 1}},\ \frac{d}{dt}\left(\sec^{-1} t\right) = \frac{1}{|t|\sqrt{t^2 - 1}}$

13. $\frac{d}{dz}\left(\operatorname{cosec}^{-1} z\right) = \frac{-1}{|z|\sqrt{z^2 - 1}},\ \frac{d}{du}\left(\operatorname{cosec}^{-1} u\right)$

$= \frac{-1}{|u|\sqrt{u^2 - 1}},\ \frac{d}{dt}\left(\operatorname{cosec}^{-1} t\right) = \frac{-1}{|t|\sqrt{t^2 - 1}}$

14. $\frac{d}{dz}(\log z) = \frac{1}{z},\ (z > 0),\ \frac{d}{du}\log u = \frac{1}{u},\ (u > 0),$

$\frac{d}{dt}(\log t) = \frac{1}{t},\ (t > 0)$

15. $\frac{d}{dz}\left(e^z\right) = e^z,\ \frac{d}{du}\left(e^u\right) = e^u,\ \frac{d}{dt}\left(e^t\right) = e^t$

Step 2: Put $y = f_1$ (a new variable z) $= f_1(z)$

Step 3: Use the formula:

$$\frac{dy}{dx} = \frac{d\left(f_1(\text{new variable } z)\right)}{d(\text{new variable } z)} \cdot \frac{d(\text{new variable } z)}{dx}$$

$$= \frac{d\, f_1(z)}{dz} \cdot \frac{dz}{dx}$$

Step 4: Express the result in terms of x by using the relation $z = f_2(x)$ established in step (1).

Problems on Composition of Two Differentiable Functions

Solved Examples

Method of substitution

Find the derivative of the following

1. $y = \tan 3x^2$

Solution: $y = \tan 3x^2$

$$x^2 \xrightarrow{\text{multiply by } 3} \underset{\substack{\downarrow \\ = z_1}}{3x^2} \xrightarrow{\tan} \underset{\substack{\downarrow \\ = y}}{\tan 3x^2}$$

$$\therefore \frac{dy}{dx} = \frac{dy}{dz_1} \cdot \frac{dz_1}{dx}, \text{ where } z_1 = 3x^2$$

$$= \frac{d\left(\tan 3x^2\right)}{d\left(3x^2\right)} \cdot \frac{d\left(3x^2\right)}{dx} = 6x \sec^2 3x$$

2. $y = e^{\left(\frac{1+x}{1-x}\right)}$

Solution: $y = e^{\left(\frac{1+x}{1-x}\right)}$

$$(1+x) \xrightarrow{\text{divide by}(1-x)} \left(\frac{1+x}{1-x}\right) \xrightarrow{e} e^{\left(\frac{1+x}{1-x}\right)}$$

$$\downarrow \qquad \downarrow$$

$$= z_1 \qquad = y$$

$$\therefore \frac{dy}{dx} = \frac{dy}{dz_1} \cdot \frac{dz_1}{dx}, \text{ where } z_1 = \left(\frac{1+x}{1-x}\right)$$

$$= \frac{d\, e^{\left(\frac{1+x}{1-x}\right)}}{d\left(\frac{1+x}{1-x}\right)} \cdot \frac{d}{dx}\left(\frac{1+x}{1-x}\right)$$

$$= e^{\left(\frac{1+x}{1-x}\right)} \cdot \frac{(1-x)\frac{d}{dx}(1+x) - (1+x)\cdot\frac{d}{dx}(1-x)}{(1-x)^2}$$

$$= e^{\left(\frac{1+x}{1-x}\right)} \cdot \frac{(1-x)\cdot 1 - (1+x)(-1)}{(1-x)^2}$$

$$= e^{\left(\frac{1+x}{1-x}\right)} \cdot \left(\frac{1-x+1+x}{(1-x)^2}\right)$$

$$= \frac{2}{(1-x)^2} \cdot e^{\left(\frac{1+x}{1-x}\right)}$$

3. $y = \sec^{-1} x^2$

Solution: $y = \sec^{-1} x^2$

$$x \xrightarrow{\text{square}} x^2 \xrightarrow{\sec^{-1}} \sec^{-1} x^2$$

$$\downarrow \qquad \downarrow$$

$$= z_1 \qquad = y$$

$$\therefore \frac{dy}{dx} = \frac{dy}{dz_1} \cdot \frac{dz_1}{dx}, \text{ where } z_1 = x^2$$

$$= \frac{d\left(\sec^{-1} x^2\right)}{d\left(x^2\right)} \cdot \frac{d\left(x^2\right)}{dx}$$

$$= \frac{1}{\left|x^2\right|\sqrt{\left(x^2\right)^2 - 1}} \cdot (2x)$$

$$= \frac{2x}{x^2\sqrt{x^4 - 1}} \left(\because \left|x^2\right| = |x|^2 = x^2\right)$$

$$= \frac{2}{x\sqrt{x^4 - 1}}$$

(B) On method of making no substitution while finding the derivative of composite (or, composition) of two differentiable functions

A composite (or, composition) of two differentiable functions f_1 and f_2 is symbolically represented as $y = f_1 f_2(x)$ and its derivative making no substitution is found using the following working rule.

Working rule: The differential coefficient of the differentiable function f_1 of the differentiable function f_2 of the independent variable x is equal to the product of the differential coefficient of the whole function $y = f_1 f_2(x)$ with respect to the inner differentiable function $f_2(x)$ and the differential coefficient of the inner differentiable function $f_2(x)$ with respect to the independent variable x; i.e. if $y = f_1 f_2(x)$, then

$$\frac{dy}{dx} = \frac{d\, f_1 f_2(x)}{d\, f_2(x)} \cdot \frac{d\, f_2(x)}{dx} = f_1'\, f_2(x) \cdot f_2'(x)$$

which tells us to apply the formulas as discussed in (B) after the theorem for the chain rule.

Note:

1. $y = f_1 f_2(x) \Rightarrow$ (any differentiable function of $x)^n$, provided $f_2(x)$ = any differentiable function of x means algebraic function and/transcendental function of x and f_1 means to raise the base (being any differentiable function of x) to the nth power.

2. $y = f_1 f_2(x) \Rightarrow (f_1 f_2(x))' = a f_1'(ax+b)$, provided $f_2(x)$ = a linear algebraic function of x. e.g.,

(i) $(\sin(ax+b))' = a\cos(ax+b)$

(ii) $(\cos(ax+b))' = -a\sin(ax+b)$

(iii) $\left(\tan(ax+b)\right)' = a\sec^2(ax+b)$

(iv) $\left(\cot(ax+b)\right)' = -a\,\text{cosec}^2(ax+b)$

(v) $\left(\sec(ax+b)\right)' = a\sec(ax+b)\cdot\tan(ax+b)$

(vi) $\left(\text{cosec}(ax+b)\right)' = -a\,\text{cosec}(ax+b)\cdot\cot(ax+b)$

(vii) $\left((ax+b)^n\right)' = na\,(ax+b)^{n-1}$ and so on.

Problems on composition of two differentiable functions

Solved Examples

Making no substitution

Find the differential coefficient of the following.

1. $y = \sin x^3$

Solution: $y = \sin x^3$

$$\Rightarrow \frac{dy}{dx} = \frac{d}{dx}\left(\sin x^3\right) = \frac{d\left(\sin x^3\right)}{dx^3}\cdot\frac{dx^3}{dx}$$

$$= \cos x^3 \cdot 3x^2 = 3x^2\cdot\cos x^3$$

2. $y = \sqrt{\tan x}$

Solution: $y = \sqrt{\tan x}$

$$\Rightarrow \frac{dy}{dx} = \frac{d\left(\sqrt{\tan x}\right)}{dx} = \frac{d\sqrt{\tan x}}{d(\tan x)}\cdot\frac{d\tan x}{dx} = \frac{sec^2 x}{2\sqrt{\tan x}}$$

3. $y = \cos e^x$

Solution: $y = \cos e^x$

$$\Rightarrow \frac{dy}{dx} = \frac{d\left(\cos e^x\right)}{dx} = \frac{d\left(\cos e^x\right)}{d\left(e^x\right)}\cdot\frac{de^x}{dx}$$

$$= \left(-\sin e^x\right)\cdot e^x = -e^x\cdot\sin x$$

4. $y = \tan 4x$

Solution: $y = \tan 4x$

$$\Rightarrow \frac{dy}{dx} = \frac{d(\tan 4x)}{dx} = \frac{d(\tan 4x)}{d(4x)}\cdot\frac{d(4x)}{dx}$$

$$= \sec^2 x\cdot 4 = 4\sec^2 x$$

5. $y = \sec(\log x)$

Solution: $y = \sec(\log x)$

$$\Rightarrow \frac{dy}{dx} = \frac{d\sec(\log x)}{dx} = \frac{d\sec(\log x)}{d(\log x)}\cdot\frac{d\log x}{dx}$$

$$= \frac{\sec(\log x)\cdot\tan(\log x)}{x}$$

6. $y = \tan(\sin^{-1} x)$

Solution: $y = \tan(\sin^{-1} x)$

$$\Rightarrow \frac{dy}{dx} = \frac{d}{dx}\left(\tan\cdot\left(\sin^{-1} x\right)\right)$$

$$= \frac{d\left(\tan\left(\sin^{-1} x\right)\right)}{d\left(\sin^{-1} x\right)}\cdot\frac{d\sin^{-1} x}{dx}$$

$$= \frac{\sec^2\left(\sin^{-1} x\right)}{\sqrt{1-x^2}}$$

7. $y = \sec\left(\tan^{-1} x\right)$

Solution: $y = \sec\left(\tan^{-1} x\right)$

$$\Rightarrow \frac{dy}{dx} = \frac{d\left(\sec\left(\tan^{-1} x\right)\right)}{dx}$$

$$= \frac{d\left(\sec\left(\tan^{-1} x\right)\right)}{d\left(\tan^{-1} x\right)}\cdot\frac{d\left(\tan^{-1} x\right)}{dx}$$

$$= \sec\left(\tan^{-1} x\right)\cdot\left(\tan\left(\tan^{-1} x\right)\right)\cdot\left(\frac{1}{1+x^2}\right)$$

$$= \frac{x\sec\left(\tan^{-1} x\right)}{1+x^2}\quad\left(\because \tan\left(\tan^{-1} x\right) = x\right)$$

8. $y = \tan^{-1}(\sin x)$

Solution: $y = \tan^{-1}(\sin x)$

$$\Rightarrow \frac{dy}{dx} = \frac{d\left(\tan^{-1}(\sin x)\right)}{dx}$$

$$= \frac{d\tan^{-1}(\sin x)}{d(\sin x)} \cdot \frac{d(\sin x)}{dx} = \frac{\cos x}{1+\sin^2 x}$$

9. $y = \sin(\cos^{-1} x)$

Solution: $y = \sin(\cos^{-1} x)$

$$\Rightarrow \frac{dy}{dx} = \frac{d\left(\sin\left(\cos^{-1} x\right)\right)}{dx}$$

$$= \frac{d\left(\sin\left(\cos^{-1} x\right)\right)}{d\left(\cos^{-1} x\right)} \cdot \frac{d\left(\cos^{-1} x\right)}{dx}$$

$$= \frac{-\cos\left(\cos^{-1} x\right)}{\sqrt{1-x^2}}$$

$$= \frac{-x}{\sqrt{1-x^2}}; x \neq x \pm 1 \left(\because \cos\left(\cos^{-1} x\right) = x\right)$$

10. $y = \tan^{-1}\sqrt{x}$

Solution: $y = \tan^{-1}\sqrt{x}$

$$\Rightarrow \frac{dy}{dx} = \frac{d\left(\tan^{-1}\sqrt{x}\right)}{dx}$$

$$= \frac{d\left(\tan^{-1}\sqrt{x}\right)}{d\sqrt{x}} \cdot \frac{d\sqrt{x}}{dx}$$

$$= \frac{1}{1+\left(\sqrt{x}\right)^2} \cdot \frac{1}{2\sqrt{x}} = \frac{1}{2\sqrt{x}\left(1+x^2\right)}$$

11. $y = \sin\sin^{-1} x$

Solution: $y = \sin\sin^{-1} x$

$$\Rightarrow \frac{dy}{dx} = \frac{d\left(\sin\sin^{-1} x\right)}{dx}$$

$$= \frac{d\left(\sin\sin^{-1} x\right)}{d\left(\sin^{-1} x\right)} \cdot \frac{d\left(\sin^{-1} x\right)}{dx} = \frac{\cos\left(\sin^{-1} x\right)}{\sqrt{1-x^2}}$$

Note: $\frac{\cos\left(\sin^{-1} x\right)}{\sqrt{1-x^2}}$ may also simplified in the following way.

$$\because \sin^{-1} x = \theta \Leftrightarrow x = \sin\theta, \left(-\frac{\pi}{2} \leq \theta \leq \frac{\pi}{2}\right)$$

$$\therefore \cos\theta = |\cos\theta| = \sqrt{1-\sin^2\theta} = \sqrt{1-x^2}$$

$$\text{Hence, } \frac{dy}{dx} = \frac{\cos\left(\sin^{-1} x\right)}{\sqrt{1-x^2}} = \frac{\sqrt{1-x^2}}{\sqrt{1-x^2}} = 1.$$

12. $y = \sin^{-1}(\sin x)$

Solution: $y = \sin^{-1}(\sin x)$

$$\Rightarrow \frac{dy}{dx} = \frac{d\left(\sin^{-1}(\sin x)\right)}{dx}$$

$$= \frac{d\left(\sin^{-1}(\sin x)\right)}{d(\sin x)} \cdot \frac{d(\sin x)}{dx} = \frac{\cos x}{\sqrt{\cos^2 x}}$$

$$= \frac{\cos x}{|\cos x|}; \cos x \neq 0$$

13. $y = \text{cosec}(4-3x)$

Solution: $y = \text{cosec}(4-3x)$

$$\Rightarrow \frac{dy}{dx} = \frac{d(\text{cosec}(4-3x))}{dx}$$

$$= \frac{d(\text{cosec}(4-3x))}{d(4-3x)} \cdot \frac{d(4-3x)}{dx}$$

$= -\text{cosec}(4-3x) \cdot \cot(4-3x) \cdot (-3) = 3\text{cosec}(4-3x)\cot(4-3x)$

On method of substitution for the differential coefficient of composition (or, composite) of a finite number of differentiable functions

A differentiable function f_1 of a differentiable function f_2 of … of a differentiable function f_n on a differentiable function f of the independent variable x (or, composite/composition of a finite number of differentiable functions namely $f_1, f_2, \ldots f_n$ and f) is symbolically represented as $y = f_1 f_2 f_3 \cdots f_n f(x)$, where $f_1 f_2 f_3 \cdots f_n f(x)$ = given function/whole function/ dependent variable.

$f_2 f_3 \cdots f_n f(x)$ = inner differentiable function/ inside differentiable function/inner function/inside function/ independent variable.

$f_1, f_2, f_3, \ldots, f_n$ or f, each = constituent differentiable function/constituent function.

$f(x)$ = inner most differentiable function/inner most function.

There are two methods of substitution for finding the derivatives of composition of a finite number of differentiable functions.

Method I: The first method of substitution for finding the derivatives of composition of a finite number of differentiable functions consists of the following steps.

Step 1: Start from the left hand side introducing a new variable whenever a new function occurs untill we get a simple function of x (or, simply an identity function x) whose derivatives can be found by using the rules for the differential coefficient of power, trigonometric, inverse trigonometric, logarithmic, exponential, sum, difference, product or quotient of two (or, more than two) differentiable functions of x; i.e.;

Separate the first function and put the rest function $= z_1$

Separate the second function and put the rest function $= z_2$

...

...

We continue the process of separating the function and putting the rest = a new variable unless we get a simple function of x having the standard form which can be differentiated by using the rules for the derivatives of power, trigonometric, inverse trigonometric, logarithmic, exponential, sum, difference, product or quotient of two (or, more than two) differentiable functions of x.

Step 2: Differentiate separately each function supposed as $z_1, z, z_3, \ldots, z_n$ with respect to the independent variable x of the given function and multiply all these derivatives thus obtained to have

$$\frac{dy}{dx} = \frac{dy}{dz_1} \cdot \frac{dz_2}{dz_3} \cdots \frac{dz_{n-1}}{dz_n} \cdot \frac{dz_n}{dx}$$

Step 3: Express the result in terms of x using the relations

$f_2 f_3 f_4 \ldots f_n f(x) = z_1$,

$f_3 f_4 \ldots f_n f(x) = z_2$,

$f_4 \ldots f_n f(x) = z_3$

...

$f_n f(x) = z_{n-1}$

$f(x) = z_n$ which are stablished in step (1)

Note: Generally in practice, we put z_1 for the inner most function which is simply an identity function x or a simple function of x, say $f(x)$ having the standard form x^n, sin x, cos x, tan x, cot x, sec x, cosec x, $\sin^{-1} x$, $\cos^{-1} x$, $\tan^{-1} x$, $\cot^{-1} x$, $\sec^{-1} x$, $\text{cosec}^{-1} x$, log x, e^x etc. starting from the right hand side which means we may start introducing a new variable from the right hand side whenever a new function occurs linked in a chain unless we get the first function in the left hand side.

Explanation

(A) If the given function $y = f_1 f_2 f_3 \ldots f_n f(x)$, then

$z_1 = f(x)$

$z_2 = f_n(z_1)$

$z_3 = f_{n-1}(z_2)$

...

$z_n = f_2(z_{n-1})$

$y = f_1(z_n)$ = given function

$$\frac{dy}{dx} = \frac{dy}{dz_n} \cdot \frac{dz_n}{dz_{n-1}} \cdot \frac{dz_{n-1}}{dz_{n-2}} \cdots \frac{dz_2}{dz_1} \cdot \frac{dz_1}{dx}, \text{ provided}$$

we start making substitutions from right hand side.

(B): If the given function $y = f_1 f_2 f_3 \cdots f_n f(x)$, then

$z_1 = f_2(z_2) = f_2 f_3 f_4 \cdots f_n f(x)$

$z_2 = f_3(z_3) = f_3 f_4 \cdots f_n f(x)$

$z_3 = f_4(z_4) = f_4 \cdots f_n f(x)$

...

...

$z_{n-2} = f_{n-1}(z_{n-1})$

$z_{n-1} = f_n(z_n)$

$z_n = f(x)$

$$\frac{dy}{dx} = \frac{dy}{dz_1} \cdot \frac{dz_1}{dz_2} \cdots \frac{dz_{n-1}}{dz_n} \cdot \frac{dz_n}{dx}, \text{ provided we start}$$

making substitutions from left hand side.

Remember:

1. Total number of function in $y = f_1 f_2 f_3 \ldots f_n f(x) = (n + 1)\ n \in N$, including f whereas total number of substitutions made $= n$.

2. Total number of the derivatives of all the constituent differentiable functions $f_1, f_2, f_3, \ldots, f_n$ and $f = (n + 1)\ n \in n$.

3. We make no substitution for the given function $y = f_1 f_2 f_3 \ldots f_n f(x)$, since it is already given equal to y.

4. We must make a substitution for the expression (or, the function) in x under the radical whenever we have $\sqrt{\text{an expression in } x}$.

5. The reader should mind that the method of substitution becomes lengthy when the chain of the intermediate variable is more than one. This is why the reader is advised to derive the differential coefficient of the composition of a finite number of differentiable functions without making substitutions since method of substitution is easy to handle only in elementary cases.

Method 2: To make the substitution $z_1, z_2, \ldots, z_n$ etc while finding the derivative of composite of a finite number of differentiable functions, we may adopt the following procedure also.

Step 1: Start with a function of x (or, simply x) which can be differentiated by using the rules for the differential coefficient of power, trigonometric, inverse trigonometric, logarithmic, exponential, sum, difference, product or quotient of two (or, more than two) differentiable functions.

Step 2: Seek the way how to reach $y = f_1 f_2 \ldots f_n f(x)$ = the whole given function from x (or, the power, trigonometric, inverse trigonometric, logarithmic, exponential, sum, difference, prouct or quotient of two (or, more than two) two differentiable function (or, functions) of x.

Step 3: Name each of these steps as the variable $z_1, z_2, z_3 \ldots, z_n$ etc. and differentiate these all with respect to x.

Explanation:

$$\begin{array}{ccccccccc} x & \xrightarrow{f} & f(x) & \xrightarrow{f_n} & f_n f(x) & \xrightarrow{f_{n-1}} & f_{n-1} f_n(x) & \rightarrow \ldots \\ \downarrow & & \downarrow & & \downarrow & & \downarrow & \\ = (x) & & = (z_1) & & = (z_2) & & = (z_3) & \end{array}$$

$$\begin{array}{cc} \xrightarrow{f_1} & f_1 f_2 \ldots f_n f(x) \\ & \downarrow \\ & = (y) \end{array}$$

Solved Examples

1. Let $y = [\sin(\log x)]^2$

Here, starting from x and reaching y in successive steps can be explained in the following way.

$$\begin{array}{ccccccc} x & \xrightarrow{\text{take log}} & \log x & \xrightarrow{\text{take sin}} & \sin \log x & \xrightarrow{\text{square}} & [\sin(\log)]^2 \\ \downarrow & & \downarrow & & \downarrow & & \downarrow \\ = (x) & & = (z_1) & & = (z_2) & & = (y) \end{array}$$

Remark: We put the variables for the function of x as last one = given composite differentiable function $= y$ before $y = z_2$

before $\sin \log x = z_1$

before $\log x = x$

and in this way we reach x. Further we should note that total number of functions = 3, (log, sin, $(\ldots)^2$) whereas the substitution made = 2 in number.

2. Let $y = e^{\sin(\tan^{-1} 2x)}$

Here, again starting from x and reaching y in successive steps can be explained as below

$$\begin{array}{ccccccc} x & \xrightarrow{\text{doulble}} & 2x & \xrightarrow{\text{take } \tan^{-1}} & \tan^{-1}(2x) & \xrightarrow{\text{take sin}} & \\ \downarrow & & \downarrow & & \downarrow & & \\ = (x) & & = (z_1) & & = (z_2) & & \end{array}$$

$$\begin{array}{ccc} \sin(\tan^{-1} 2x) & \xrightarrow{\text{take exponential}} & e^{\sin \tan^{-1}(2x)} \\ \downarrow & & \downarrow \\ = (z_3) & & = (y) \end{array}$$

Remember: In each successive step while finding the derivative of a differentiable function of a differentiable function ... of a differentiable function of x, we differentiate each constituent differentiable function regarding the rest as an inner differentiable function using the rule of composite of two differentiable functions; i.e.,

$$\frac{d}{dx}(f_m f_n(x)) = \frac{d f_m f_n(x)}{d f_n(x)} \cdot \frac{d f_n(x)}{dx} \cdot \frac{dx}{dx}$$

On method for the derivative of composition of a finite number of differentiable functions without making any substitution

To find the derivative of the composition of a finite number of differentiable functions namely $f_1, f_2, f_3, \ldots, f_n$ represented as $y = f_1 f_2 f_3 \ldots f_n f(x)$, we have a rule known as the chain rule for the derivative of composite (or, composition) of $(n+1)$ number of differentiable functions.

Rule: The derivative of a differentiable function f_1 of a differentiable function f_2 ... of a differentiable function f_n of a differentiable function f of the independent variable x

$$= \prod_{i=1}^{n} f_i' \left[\begin{array}{l}\text{operand} = \text{all the remaining successive} \\ \text{differentiable functions excepting} \\ \text{the preceeding differentiable} \\ \text{function (or, functions) which} \\ \text{has (have) been differentiated.}\end{array}\right]$$

$$= f_1'(f_2 f_3 \ldots f_n f(x)) \cdot f_2'(f_3 f_4 \ldots f_n f(x)) \cdot$$
$$f_3'(f_4 f_5 \ldots f_n f(x)) \cdot f_4'(f_5 f_6 \ldots f_n f(x)) \ldots$$
$$f_n'(f(x)) \cdot f'(x)$$

The above working rule may be expressed in the following way.

1. Start from left to find the derivative of each successive (following in order/coming one after another) functions coming near and near to x with respect to the remaining all other function (or, functions) (excepting the differentiated function or functions) till we arrive at x or an expression in x being power, trigonometric, inverse trigonometric, logarithmic, exponential, sum, difference, product or quotient of two (or, more than two) differentiable function (or, functions).

2. Find the product by multiplying all the results being the derivative of each constituent differentiable function. We may express in words the facts in (1) and (2) in the following way.

$$f_1' \left(\begin{array}{l}\text{operand} = \text{the rest} \\ \text{function upto } x \\ \text{without any} \\ \text{change excepting} \\ f_1\end{array}\right) \cdot f_2' \left(\begin{array}{l}\text{operand} = \text{the rest} \\ \text{function up to } x \\ \text{without any} \\ \text{change excepting} \\ f_1 \text{ and } f_2\end{array}\right) \cdot$$

$$f_3' \left(\begin{array}{l}\text{operand} = \text{the rest} \\ \text{function up to } x \\ \text{without any change} \\ \text{excepting } f_1,\ f_2 \text{ and} \\ f_3\end{array}\right) \ldots f_n' \text{(last operand)} \cdot$$

$$\text{(last operand)}' .$$

Where f_n' (last operand) means the derivative of the last function f_n whose last operand is simply an identity function x or a simple function of x being differentiable whose derivative can be found by using the rule for the derivatives of sum, difference, product or quotient of two or more than two differentiable functions.

Notes: **1.** If the last operand = identity function = x, then $\frac{dx}{dx} = (x)' = 1$ which is generally ignored while finding the derivative.

2. $\frac{d}{dx}$ may be thought as an operator which changes the form of only that function (or, operator like sin, cos, tan, cot, sec, cosec, $\sin^{-1}$, $\cos^{-1}$, $\tan^{-1}$, $\cot^{-1}$, $\sec^{-1}$, cosec^{-1}, log, e, etc) before which it is put excepting the exponential operator 'e' which does not change its form before and after the differentiation. e.g.,

1. $\frac{d}{dx}\left(\sqrt{\cdot\cdot}\right) = \frac{1}{2\sqrt{\cdot\cdot}} \cdot (\cdot\cdot)'$

2. $\frac{d}{dx}(\cdot\cdot)^n = n(\cdot\cdot)^{n-1} \cdot (\cdot\cdot)'$

3. $\frac{d}{dx}(\sin(\cdot\cdot)) = \cos(\cdot\cdot) \cdot (\cdot\cdot)'$

4. $\frac{d}{dx}(\cos(\cdot\cdot)) = -\sin(\cdot\cdot) \cdot (\cdot\cdot)'$

5. $\frac{d}{dx}(\tan(\cdot\cdot)) = \sec^2(\cdot\cdot) \cdot (\cdot\cdot)'$

6. $\frac{d}{dx}(\cot(\cdot\cdot)) = -\text{cosec}^2(\cdot\cdot) \cdot (\cdot\cdot)'$

7. $\frac{d}{dx}(\sec(\cdot\cdot)) = \sec(\cdot\cdot) \cdot \tan(\cdot\cdot) \cdot (\cdot\cdot)'$

8. $\frac{d}{dx}(\text{cosec}(\cdot\cdot)) = -\text{cosec}(\cdot\cdot)\cdot\cot(\cdot\cdot)'\cdot(\cdot\cdot)'$

9. $\frac{d}{dx}\left(\sin^{-1}(\cdot\cdot)\right) = \frac{1}{\sqrt{1-(\cdot\cdot)^2}}\cdot(\cdot\cdot)'$

10. $\frac{d}{dx}\left(\cos^{-1}(\cdot\cdot)\right) = \frac{-1}{\sqrt{1-(\cdot\cdot)^2}}\cdot(\cdot\cdot)'$

11. $\frac{d}{dx}\left(\tan^{-1}(\cdot\cdot)\right) = \frac{1}{\sqrt{1+(\cdot\cdot)^2}}\cdot(\cdot\cdot)'$

12. $\frac{d}{dx}\left(\cot^{-1}(\cdot\cdot)\right) = \frac{-1}{1+(\cdot\cdot)^2}\cdot(\cdot\cdot)'$

13. $\frac{d}{dx}\left(\sec^{-1}(\cdot\cdot)\right) = \frac{1}{|\cdot\cdot|\sqrt{(\cdot\cdot)^2-1}}\cdot(\cdot\cdot)'$

14. $\frac{d}{dx}\left(\text{cosec}^{-1}(\cdot\cdot)\right) = \frac{-1}{|\cdot\cdot|\sqrt{(\cdot\cdot)^2-1}}\cdot(\cdot\cdot)'$

15. $\frac{d}{dx}(\log(\cdot\cdot\cdot)) = \frac{1}{(\cdot\cdot)}\cdot(\cdot\cdot)'$

16. $\frac{d}{dx}\left(e^{(\cdot\cdot)}\right) = e^{(\cdot\cdot)}\cdot(\cdot\cdot)'$

where dots within circular brackets/absolute value sign denote the operand being the same in the *l.h.s* and *r.h.s*. Further, we should note that it is only the operator 'e' which remains unaltered before and after the differentiation.

Remember: The above explanation throws light upon the fact that only function (or, operator) is differentiated successively till we get x when the given function is $y = f_1\, f_2 \ldots f_n f(x)$, where each constituent function $f_1, f_2, \ldots, f_n$ or f being differentiable represents sin, cos, tan, cot, sec, cosec, $\sin^{-1}$, $\cos^{-1}$, $\tan^{-1}$, $\cot^{-1}$, $\sec^{-1}$, cosec^{-1}, log, e, $(\ )^n$, $\sqrt[n]{\ }$, $|\,|$, etc.

Problems on the composition of a finite number of differentiable functions

Different types of problems in the composition of a finite number of differentiable functions may be guided only by the problems appearing in different forms which are

(i) $y = f_1 f_2 f_3 \cdots f_n f(x)$ or, $y = (f_1 f_2 f_3 \cdots f_n f(x))^n$

(ii) $y = f_1\sqrt{f_2\sqrt{f_3\sqrt{\ldots\sqrt{f(x)}}}}$

(iii) $y = \sqrt{f_1 + \sqrt{f_2 + \sqrt{f_3 + \ldots\sqrt{f_n + (x)}}}}$

(iv) $y = \sqrt{\sqrt{\ldots\sqrt{f(x)}}}$

(v) (a) $y = f_1(x)\cdot f_2(x) \pm g_1(x)\cdot g_2(x)$

(b) $y = f_1(x)\cdot f_2(x) \pm \frac{g_1(x)}{g_2(x)}$

(c) $y = f_1(x) \pm g_1(x)\cdot g_2(x)$

(d) $y = f_1(x) \pm \frac{g_1(x)}{g_2(x)}$

(e) $y = f_1(x) \pm f_2(x)$

Now we consider each one by one

Form: **(i)** $\rightarrow y = f_1 f_2 f_3 \cdots f_n f(x)$

(ii) $\rightarrow y = (f_1 f_2 f_3 \ldots f_n f(x))^n$

whenever we are given the problems in the above forms, we find their derivatives using the following working rule.

Working rule: To find the derivative of differentiable function having the form (1), we use directly the rule for the derivative of the composition of a finite number of differentiable functions and the derivative of a differentiable function having the form (2) is found using the following rule:

(a) Use the formula: $\left[(F(x))^n\right]' = n\cdot(F(x))^{n-1}\cdot F'(x)$ where $F(x) = f_1 f_2 \ldots f_n f(x)$.

(b) Find the derivative $F'(x)$ using the rule for the derivative of the composition of a finite number of a differentiable functions.

Do not forget

1. $f_1(f(x))^n \Rightarrow x \xrightarrow{f} f(x) \xrightarrow{(\ )^n} (f(x))^n \xrightarrow{f_1} f_1(f(x))^n$

2. $(f_1(f(x)))^n \Rightarrow x \xrightarrow{f} f(x) \xrightarrow{f_1} f_1(f(x)) \xrightarrow{(\ )^n} (f_1(f(x)))^n$

where $f(x) = f_1 f_2 f_3 \ldots f_n f(x)$/sum/difference/product or quotient of two or more than two differentiable functions.

Solved Examples

Without substitution

Find the differential coefficients of the following.

1. $y = \sqrt{\tan 2x}$

Solution: $y = \sqrt{\tan 2x}$, defined for $\tan 2x > 0$

$$\Rightarrow \frac{dy}{dx} = \frac{d\sqrt{\tan 2x}}{dx}$$

$$= \frac{d\sqrt{\tan 2x}}{d(\tan 2x)} \cdot \frac{d(\tan 2x)}{d(2x)} \cdot \frac{d(2x)}{dx}$$

$$= \frac{\sec^2 2x}{\sqrt{\tan 2x}}$$

> ***Remember***: $\sqrt{\ }$ changes into $\frac{1}{2\sqrt{\ }}$, tan 2*x* changes into $\sec^2 2x$, 2*x* changes into 2 while differentiating.

2. $y = \sqrt{\sin x^2}$

Solution: $y = \sqrt{\sin x^2}$

$$\Rightarrow \frac{dy}{dx} = \frac{d\sqrt{\sin x^2}}{dx}$$

$$= \frac{d\sqrt{\sin x^2}}{d\left(\sin x^2\right)} \cdot \frac{d\left(\sin x^2\right)}{d\left(x^2\right)} \cdot \frac{d\left(x^2\right)}{dx}$$

$$= \frac{x \cos x^2}{\sqrt{\sin x^2}};\ \sin x^2 > 0$$

> ***Remember***: sin changes into cos, while differentiating.

3. $y = \sqrt{\sin \sqrt{x}}$

Solution: $y = \sqrt{\sin \sqrt{x}}$

$$\Rightarrow \frac{dy}{dx} = \frac{d}{dx}\sqrt{\sin\sqrt{x}}$$

$$= \frac{d\sqrt{\sin\sqrt{x}}}{d\left(\sin\sqrt{x}\right)} \cdot \frac{d\left(\sin\sqrt{x}\right)}{d\left(\sqrt{x}\right)} \cdot \frac{d\sqrt{x}}{dx}$$

$$= \frac{\cos\sqrt{x}}{4\sqrt{x}\,\sin\sqrt{x}};\ \sin\sqrt{x} > 0$$

4. $y = \tan\sqrt{1 + x + x^2}$

Solution: $y = \tan\sqrt{1 + x + x^2}$

$$\Rightarrow \frac{dy}{dx} = \frac{d}{dx}\left(\tan\sqrt{1+x+x^2}\right) \cdot \frac{d\sqrt{1+x+x^2}}{d\left(1+x+x^2\right)} \cdot \frac{d\left(1+x+x^2\right)}{dx}$$

$$= \frac{(1+2x)\sec^2\sqrt{1+x+x^2}}{2\sqrt{1+x+x^2}}$$

> ***Note***: That $f_1 = \tan$, $f_2 = \sqrt{\ }$
>
> $f_3(x) = 1 + x + x^2$
>
> $f_1 f_2 f_3(x) = \tan\sqrt{1+x+x^2}$

> ***Remember***: tan changes into $\sec^2$
>
> $\sqrt{\ }$ changes into $\frac{1}{2\sqrt{\ }}$, $(1+x+x^2)$ changes into $(1+2x)$

5. $y = \sqrt{\tan(\tan x)}$

Solution: $y = \sqrt{\tan(\tan x)}$

$$\Rightarrow \frac{dy}{dx} = \frac{d}{dx}\left(\sqrt{\tan(\tan x)}\right)$$

$$= \frac{d\sqrt{\tan(\tan x)}}{d\tan(\tan x)} \cdot \frac{d\tan(\tan x)}{d\tan x} \cdot \frac{d\tan x}{dx}$$

$$= \frac{\sec^2 x \cdot \sec^2(\tan x)}{2\sqrt{\tan(\tan x)}};\ \tan(\tan x) > 0$$

> ***Note***: That $f_1 = \sqrt{\ } = (\ )^{\frac{1}{2}}$, $f_2 = \tan$, $f_3 = \tan$, $f_4 = x$, $f_1 f_2 f_3 f_4 = \sqrt{\tan(\tan x)}$

6. $y = \cos^{-1}(\tan x^2)$

Solution: $y = \cos^{-1}(\tan x^2)$

$$\Rightarrow \frac{dy}{dx} = \frac{d\left(\cos^{-1}\left(\tan x^2\right)\right)}{dx}$$

$$= \frac{d\left(\cos^{-1}\left(\tan x^2\right)\right)}{d\left(\tan x^2\right)} \cdot \frac{d\left(\tan x^2\right)}{d\left(x^2\right)} \cdot \frac{d\left(x^2\right)}{dx}$$

$$= \frac{-1}{\sqrt{1-\tan^2 x^2}} \cdot \sec^2 x^2 \cdot 2x$$

$$= \frac{-2x \sec^2 x^2}{\sqrt{1-\tan^2 x^2}},$$

$\left|\tan x^2\right| < 1.$

> ***Note***: $f_1 = \cos^{-1}$
> $f_2 = \tan$
> $f_3 = (\ )^2$
> $f_3(x) = (x)^2$
> $f_1 f_2 f_3(x) = \cos^{-1} \tan x^2$

7. $y = \tan^{-1}(a \cdot e^x \cdot x^2)$

Solution: $y = \tan^{-1}(a \cdot e^x \cdot x^2)$

$$\Rightarrow \frac{dy}{dx} = \frac{d\left(\tan^{-1}\left(a \cdot e^x \cdot x^2\right)\right)}{dx}$$

$$= \frac{d\left(\tan^{-1}\left(a \cdot e^x \cdot x^2\right)\right)}{d\left(a \cdot e^x \cdot x^2\right)} \cdot \frac{d\left(a \cdot e^x \cdot x^2\right)}{dx}$$

$$= \frac{a\left(e^x \cdot x^2 + 2x \cdot e^x\right)}{1+\left(a \cdot e^x \cdot x^2\right)^2}$$

$$= \frac{a e^x \cdot \left(2x + x^2\right)}{1+\left(a \cdot e^x \cdot x^2\right)^2}$$

> ***Note***: Here, $f_1 = \tan^{-1}$, $f_2(x) = a \cdot e^x \cdot x^2$

8. $y = \sqrt{\sec^{-1} x^2}$

Solution: $y = \sqrt{\sec^{-1} x^2}$

$$\Rightarrow \frac{dy}{dx} = \frac{d}{dx}\left(\sqrt{\sec^{-1} x^2}\right)$$

$$= \frac{d\left(\sqrt{\sec^{-1} x^2}\right)}{d\left(\sec^{-1} x^2\right)} \cdot \frac{d\left(\sec^{-1} x^2\right)}{dx^2} \cdot \frac{dx^2}{dx},$$

$$= \frac{1}{2\left(\sqrt{\sec^{-1} x^2}\right)} \cdot \frac{1}{\left|x^2\right|\left(\sqrt{\left(x^2\right)^2 - 1}\right)} \cdot 2x$$

$$= \frac{x}{x^2\left(\sqrt{\sec^{-1} x^2}\right)\left(\sqrt{x^4 - 1}\right)}$$

$$\left(\because \left|x^2\right| = |x|^2 = x^2\right)$$

$$= \frac{1}{x\left(\sqrt{\sec^{-1} x^2}\right)\left(\sqrt{x^2 - 1}\right)}, |x| > 1$$

Form 2:

$$f_1 \sqrt{f_2 \sqrt{f_3 \sqrt{\ldots \sqrt{f(x)}}}}$$

i.e. the form of the composite function being differentiable obtained by the operation of applying a differentiable function (sin, cos, tan, cot, sec, cosec, $\sin^{-1}$, $\cos^{-1}$, $\tan^{-1}$, $\cot^{-1}$, $\sec^{-1}$, cosec^{-1}, log, e, | |, etc.) and the operation of taking the square root is successively performed more than once upon a differentiable function of x.

Working rules: **1.** Method of substitution:

(a) Put $z = \sqrt{f(x)}$ which may be regarded as the inner most function

(b) Use $\left(\sqrt{z}\right)' = \frac{1}{2\sqrt{z}} \cdot (z)' \cdot \frac{dz}{dx}$

(c) Lastly express z in terms of x.

2. Method of making no substitution:
The above type of problems may be differentiated by using directly the formula for the derivative of a finite number of differentiable functions regarding firstly $\sqrt{\ }$ as f_1 and secondly f_2 as any function being differentiable like trigonometric, inverse trigonometric, power, logarithm i.e., mod or exponential in every successive step i.e. we have to use the formula:

$$[f_1 f_2 f_3 \cdots f_n f(x)]'$$

$$= f_1'(f_2 f_3 \cdots f_n f(x)) \cdot f_2'(f_3 f_4 \cdots f_n f(x)) \cdots$$

$f_n'(f(x)) \cdot f'(x)$, regarding each successive differentiable function as $\sqrt{\ }$ and trigonometric, inverse trigonometric, logarithmic, mod or exponential etc. in every step.

Solved Examples

Find the differential coefficient of the following.

1. $y = \sqrt{\sin\sqrt{1+x^2}}$

Method 1:

Solution: $y = \sqrt{\sin\sqrt{1+x^2}}$

Putting $z = \sqrt{1+x^2}$,

$$\Rightarrow \frac{dy}{dx} = \frac{d}{dx}\sqrt{\sin z}$$

$$\Rightarrow \frac{dy}{dx} = \frac{1}{2\sqrt{\sin z}} \cdot \frac{d \sin z}{dx} \cdot \frac{dz}{dx}$$

$$= \frac{1}{2\sqrt{\sin\sqrt{1+x^2}}} \cdot \cos z \cdot \frac{2x}{2\sqrt{1+x^2}}$$

$$= \frac{x}{2\sqrt{\sin\sqrt{1+x^2}}} \cdot \frac{\cos\left(\sqrt{1+x^2}\right)}{\sqrt{1+x^2}};\ \sin\left(1+x^2\right)>0$$

Method 2:

$$\frac{d}{dx}(y) = \frac{d}{dy}\left(\sqrt{\sin\sqrt{1+x^2}}\right)$$

$$= \frac{d\left(\sqrt{\sin\sqrt{1+x^2}}\right)}{d\left(\sin\sqrt{1+x^2}\right)} \cdot \frac{d\left(\sin\sqrt{1+x^2}\right)}{d\left(\sqrt{1+x^2}\right)} \cdot \frac{d\left(\sqrt{1+x^2}\right)}{dx} \cdot \frac{d\left(1+x^2\right)}{dx}$$

$$= \frac{1}{2\sqrt{\sin\sqrt{1+x^2}}} \cdot \cos\left(\sqrt{1+x^2}\right) \cdot \frac{1}{2\sqrt{1+x^2}} \cdot (2x)$$

$$= \frac{x\cos\left(\sqrt{1+x^2}\right)}{\left(2\sqrt{\sin\sqrt{1+x^2}}\right)\left(\sqrt{1+x^2}\right)}$$

2. $y = \cos\sqrt{\sin\sqrt{x}}$

Solution: $y = \cos\sqrt{\sin\sqrt{x}}$

$$\frac{dy}{dx} = \frac{d\left(\cos\sqrt{\sin\sqrt{x}}\right)}{dx}.$$

$$= \frac{d\left(\cos\sqrt{\sin\sqrt{x}}\right)}{d\left(\sqrt{\sin\sqrt{x}}\right)} \cdot \frac{d\left(\sqrt{\sin\sqrt{x}}\right)}{d\left(\sin\sqrt{x}\right)} \cdot \frac{d\left(\sin\sqrt{x}\right)}{d\left(\sqrt{x}\right)} \cdot \frac{d\left(\sqrt{x}\right)}{dx}$$

$$= -\sin\sqrt{\sin\sqrt{x}} \cdot \frac{1}{2\sqrt{\sin\sqrt{x}}} \cdot \cos\sqrt{x} \cdot \frac{1}{2\sqrt{x}}$$

$$= \frac{\left(-\sin\sqrt{\sin\sqrt{x}}\right) \cdot \left(\cos\sqrt{x}\right)}{4\left(\sqrt{\sin\sqrt{x}}\right) \cdot \left(\sqrt{x}\right)},\ \sin\sqrt{x} > 0$$

3. $y = \sqrt{\sin\sqrt{x}}$

Solution: $y = \sqrt{\sin\sqrt{x}}$

$$\Rightarrow \frac{dy}{dx} = \frac{d\left(\sqrt{\sin\sqrt{x}}\right)}{dx}$$

$$= \frac{d\left(\sqrt{\sin\sqrt{x}}\right)}{d\left(\sin\sqrt{x}\right)} \cdot \frac{d\left(\sin\sqrt{x}\right)}{d\left(\sqrt{x}\right)} \cdot \frac{d\left(\sqrt{x}\right)}{dx}$$

$$= \frac{1}{2\left(\sqrt{\sin\sqrt{x}}\right)} \cdot \cos\left(\sqrt{x}\right) \cdot \frac{1}{2\sqrt{x}}$$

$$= \frac{\cos\left(\sqrt{x}\right)}{4\left(\sqrt{\sin\sqrt{x}}\right) \cdot \left(\sqrt{x}\right)},\ \sin\sqrt{x} > 0$$

4. $y = \sin\sqrt{\cos x}$

Solution: $y = \sin\sqrt{\cos x}$

$$\Rightarrow \frac{dy}{dx} = \frac{d\left(\sin\sqrt{\cos x}\right)}{dx}$$

$$= \frac{d\left(\sin\sqrt{\cos x}\right)}{d\left(\sqrt{\cos x}\right)} \cdot \frac{d\left(\sqrt{\cos x}\right)}{d(\cos x)} \cdot \frac{d(\cos x)}{dx}$$

$$= \cos\sqrt{\cos x} \cdot \frac{1}{2\sqrt{\cos x}} \cdot (-\sin x)$$

$$= \frac{-\sin x \cdot \cos\sqrt{\cos x}}{2\sqrt{\cos x}},\ \text{for } \cos x > 0$$

5. $y = \tan\sqrt{1 + x + x^2}$

Solution: $y = \tan\sqrt{1 + x + x^2}$

$$\Rightarrow \frac{dy}{dx} = \frac{d}{dx}\left(\tan\sqrt{1 + x + x^2}\right)$$

$$= \frac{d\left(\tan\sqrt{1+x+x^2}\right)}{d\left(\sqrt{1+x+x^2}\right)} \cdot \frac{d\left(\sqrt{1+x+x^2}\right)}{d\left(1+x+x^2\right)} \cdot \frac{d\left(1+x+x^2\right)}{dx}$$

$$= \frac{\sec^2\sqrt{1 + x + x^2} \cdot (2x + 1)}{2\left(\sqrt{1 + x + x^2}\right)}$$

$$= \frac{\sec^2\left(\sqrt{1 + x + x^2}\right)}{\sqrt{1 + x + x^2}} \cdot \frac{2x + 1}{2}$$

Form 3:

$$\sqrt{f_1(x) + \sqrt{f_2(x) + \sqrt{\ldots\sqrt{f_n(x) + \sqrt{f(x)}}}}}$$

where each function of x is either a constant or a differentiable function of x excepting the inner most differentiable function of x which can never be a constant; the form of the composite function obtained by performing the operation of taking square root and addition of a differentiable function of x or a constant successively more than once upon a differentiable function of the independent variable x which can be expressed in the arrow diagram as

$$f(x) \xrightarrow{\sqrt{\ }} \sqrt{f(x)} \xrightarrow{+f_n(x)} f_n(x) + \sqrt{f(x)} \xrightarrow{\sqrt{\ }}$$

$$\sqrt{f_n(x) + \sqrt{f(x)}} \xrightarrow{+f_{n-1}(x)} \ldots$$

The method of finding the derivative of the composite differentiable function of the above form is explained by the examples done below using the method of substitution.

Solved Examples

Find the differential coefficient of the following.

1. $y = \sqrt{a + \sqrt{a + \sqrt{a + x^2}}}$

Solution: $y = \sqrt{a + \sqrt{a + \sqrt{a + x^2}}}$

Putting $\sqrt{a + x^2} = u$, we have $y = \sqrt{a + \sqrt{a + u}}$

and

$$\frac{du}{dx} = \frac{d}{dx}\left(\sqrt{a + x^2}\right)$$

$$= \frac{d\left(\sqrt{a+x^2}\right)}{d\left(a+x^2\right)} \cdot \frac{d\left(a+x^2\right)}{dx}$$

$$= \frac{x}{\sqrt{a+x^2}} \quad ...(1)$$

$\because y = \sqrt{a+\sqrt{a+u}}$

Again putting $\sqrt{a+u} = v$, we have $y = \sqrt{a+v}$ and

$$\frac{dv}{du} = \frac{d\left(\sqrt{a+u}\right)}{du}$$

$$= \frac{d\left(\sqrt{a+u}\right)}{d(a+u)} \cdot \frac{d(a+u)}{du}$$

$$= \frac{1}{2\sqrt{a+u}} \quad ...(2)$$

$\because y = \sqrt{a+v}$

Lastly, putting $\sqrt{a+v} = w$, we have

$$\frac{dw}{dv} = \frac{1}{2\sqrt{a+v}} \cdot \frac{d(a+v)}{dv}$$

$$= \frac{1}{2\sqrt{a+v}} \quad ...(3)$$

$$\therefore \frac{dy}{dx} = (1) \times (2) \times (3) = \frac{dw}{dv} \cdot \frac{dv}{du} \cdot \frac{du}{dx}$$

$$= \frac{1}{2\sqrt{a+v}} \cdot \frac{1}{2\sqrt{a+u}} \cdot \frac{x}{\sqrt{a+x^2}}$$

$$= \frac{1}{\left(2\sqrt{a+\sqrt{a+u}}\right)} \cdot \frac{1}{\left(2\sqrt{a+\sqrt{a+x^2}}\right)} \cdot \frac{x}{\left(\sqrt{a+x^2}\right)}$$

$\left(\because v = \sqrt{a+u}\right)$

$$= \frac{x}{4\left(\sqrt{a+\sqrt{a+u}}\right) \cdot \left(\sqrt{a+\sqrt{a+x^2}}\right) \cdot \left(\sqrt{a+x^2}\right)}$$

$$= \frac{x}{4\left(\sqrt{a+\sqrt{a+\sqrt{a+x^2}}}\right) \cdot \left(\sqrt{a+\sqrt{a+x^2}}\right) \cdot \left(\sqrt{a+x^2}\right)}$$

$\left(\because u = \sqrt{a+x^2}\right)$

Note: The above procedure of finding the derivative of the composite differentiable function having the form mentioned above is fruitful only when $f_1(x), f_2(x), \ldots, f_n(x)$ are all constants excepting the inner most differentiable function 'f' of x but when $f_1(x), f_2(x), \ldots, f_n(x)$ are all constants besides the inner most differentiable function 'f' of x, we directly use the chain rule for the derivative without making any substitution.

Form 4:

$$\sqrt{\sqrt{\ldots\sqrt{f(x)}}}$$

i.e. the form of the composition of a finite number of differentiable functions obtained by performing the operation of taking the square root more than once upon a differentiable function of the independent variable x which can be expressed in the arrow diagram as $x \xrightarrow{f} f(x) \xrightarrow{\sqrt{\ }} \sqrt{f(x)} \xrightarrow{\sqrt{\ }} \sqrt{\sqrt{f(x)}} \xrightarrow{\sqrt{\ }} \ldots \xrightarrow{\sqrt{\ }} \sqrt{\sqrt{\ldots\sqrt{f(x)}}}$. The method of finding the derivative of the composite differentiable function having the above form is explained by the examples done below using the method of substitution.

Solved Examples

Find the differential coefficient of the following.

1. $y = \sqrt{\sqrt{x+1}}$

Solution: $y = \sqrt{\sqrt{x+1}}$

Putting $\sqrt{x+1} = z$, we have $y = \sqrt{z}$ and

$$\frac{dz}{dx} = \frac{d\left(\sqrt{x+1}\right)}{dx}$$

$$= \frac{d\left(\sqrt{x+1}\right)}{d(x+1)} \cdot \frac{d(x+1)}{dx}$$

$$\frac{dy}{dz} = \frac{1}{2\sqrt{z}} \cdot \frac{dz}{dx}$$

$$= \frac{1}{2\sqrt{z}} \cdot \frac{1}{2\sqrt{x+1}}$$

$$= \frac{1}{2\left(\sqrt{\sqrt{x+1}}\right)} \cdot \frac{1}{2\left(\sqrt{x+1}\right)}$$

$$= \frac{1}{4\left(\sqrt{\sqrt{x+1}}\right) \cdot \left(\sqrt{x+1}\right)}, x > -1$$

Form 5: **(a)** $y = f_1(x) \cdot f_2(x) \pm g_1(x) \cdot g_2(x)$

(b) $y = f_1(x) \cdot f_2(x) \pm \frac{g_1(x)}{g_2(x)}$

(c) $y = f_1(x) \pm g_1(x) \cdot g_2(x)$

(d) $y = f_1(x) \pm \frac{g_1(x)}{g_2(x)}$

(e) $y = f_1(x) \pm f_2(x)$

where anyone or all of $f_1(x), f_2(x), g_1(x)$ and $/g_2(x)$ may be a differentiable function of a differentiable function of the independent variable x.

Working rule: The working rule consists of following steps.

Step 1: Put each addend, subtrahend and minuend equal to u, v and w respectively and then y becomes equal to $u \pm v \pm w$; i.e. $y = u \pm v \pm w$.

Step 2: Take the differential operator $\left(\frac{d}{dx}\right)$ on both sides of the equation defining y as a function of the independent variable x. i.e.,

$$\frac{dy}{dx} = \frac{d}{dx}(u \pm v \pm w) = \frac{du}{dx} \pm \frac{dv}{dx} \pm \frac{dw}{dx}$$

Step 3: Use the rules for the derivative of the product, quotient and/composite of two or more than two differentiable functions of x's for finding $\frac{du}{dx}$, $\frac{dv}{dx}$ and $\frac{dw}{dx}$.

Note: The given problem may be the combination (sum, difference, product and/quotient) of composite differentiable functions of x's or it may be the combination of a differentiable x and a differentiable function of a differentiable function of x.

Solved Examples

Find the differential coefficient of the following

1. $y = \left(\frac{\cot x}{x} + \sqrt{1 - x^2}\right)$

Solution: $y = \left(\frac{\cot x}{x} + \sqrt{1 - x^2}\right)$

Putting $u = \frac{\cot x}{x}$ and $v = \sqrt{1 - x^2}$, we have $y = u + v$ and $\frac{dy}{dx} = \frac{du}{dx} + \frac{dv}{dx}$ which means we have to find $\frac{du}{dx}$ and $\frac{dv}{dx}$ separately and then their sum is to be found out.

$$\text{Now, } \frac{du}{dx} = \frac{-\left(x \operatorname{cosec}^2 x + \cot x\right)}{x^2} \quad \ldots\text{(i)}$$

$$\text{and } \frac{dv}{dx} = \frac{-2x}{2\sqrt{1 - x^2}} \quad \ldots\text{(ii)}$$

$$\therefore \frac{dy}{dx} = \frac{du}{dx} + \frac{dv}{dx} = (1) + (2)$$

$$= \frac{-\left(x \operatorname{cosec}^2 x + \cot x\right)}{x^2} - \frac{x}{\sqrt{1 - x^2}}, 0 < |x| < 1$$

2. $y = \frac{\tan x}{x} + x\sqrt{1 - x^2}$

Solution: $y = \frac{\tan x}{x} + x\sqrt{1 - x^2}$

Putting $u = \frac{\tan x}{x}$ and $v = x\sqrt{1-x^2}$, we have

$y = u + v$ and $\frac{dy}{dx} = \frac{du}{dx} + \frac{dv}{dx}$

Now, $\frac{du}{dx} = \frac{x\sec^2 x - \tan x}{x^2}$ ...(i)

and $\frac{dv}{dx} = \left[\sqrt{1-x^2} + \frac{x}{2}\left(1-x^2\right)^{-\frac{1}{2}} \cdot (-2x)\right]$

$= \left[\sqrt{1-x^2} - \frac{x^2}{\sqrt{1-x^2}}\right]$

$= \left(\frac{1-2x^2}{\sqrt{1-x^2}}\right)$ ...(ii)

$\therefore \frac{dy}{dx} = \frac{du}{dx} + \frac{dv}{dx} = (1) + (2)$

$= \frac{x\sec^2 x - \tan x}{x^2} - \left(\frac{1-2x^2}{\sqrt{1-x^2}}\right), 0 < |x| < 1$

3. $y = \sin\sqrt{1-x^2} + x^2\cos 4x$

Solution: $y = \sin\sqrt{1-x^2} + x^2\cos 4x$

Putting $\sin\sqrt{1-x^2} = u$ and $x^2\cos 4x = v$, we have $y = u + v$ and $\frac{dy}{dx} = \frac{du}{dx} + \frac{dv}{dx}$

Now, $\frac{du}{dx} = \left(\cos\sqrt{1-x^2}\right) \cdot \frac{1}{2} \cdot \left(1-x^2\right)^{-\frac{1}{2}} \cdot (-2x)$

$= \frac{-x\cos\sqrt{1-x^2}}{\sqrt{1-x^2}}$ ...(i)

and $\frac{dv}{dx} = 2x\cos 4x + x^2(-\sin 4x)\cdot 4$

$= 2x\cos 4x - 4x^2\sin 4x$ ...(ii)

$\because \frac{dy}{dx} = \frac{du}{dx} + \frac{dv}{dx} = (1) + (2)$

$= \frac{-x\cos\sqrt{1-x^2}}{\sqrt{1-x^2}} + 2x\cos 4x - 4x^2\sin 4x, |x| < 1$

4. $y = \cos\left(ax^2 + bx + c\right) + \sin^3\left(\sqrt{ax^2 + bx + c}\right)$

Solution: $y = \cos\left(ax^2 + bx + c\right) + \sin^3\left(\sqrt{ax^2 + bx + c}\right)$

Putting $\cos(ax^2 + bx + c) = u$ and $\left(\sin\sqrt{ax^2 + bx + c}\right)^3 = v$, we have $y = u + v$ and $\frac{dy}{dx} = \frac{du}{dx} + \frac{dv}{dx}$

Now, $\frac{du}{dx} = \frac{d\cos\left(ax^2 + bx + c\right)}{d\left(ax^2 + bx + c\right)} \cdot \frac{d\left(ax^2 + bx + c\right)}{dx}$

$= -\sin(ax^2 + bx + c)\cdot(2ax + b)$

$= -(2ax + b)\sin(ax^2 + bx + c)$ ...(i)

and $\frac{dv}{dx} =$

$\frac{d\left(\sin\sqrt{ax^2 + bx + c}\right)^3}{d\left(\sin\sqrt{ax^2 + bx + c}\right)} \cdot \frac{d\left(\sin\sqrt{ax^2 + bx + c}\right)}{d\left(\sqrt{ax^2 + bx + c}\right)} \cdot$

$\frac{d\left(\sqrt{ax^2 + bx + c}\right)}{dx}$

$= 3\sin^2\sqrt{ax^2 + bx + c}\cdot\cos\sqrt{ax^2 + bx + c}\cdot$

$\frac{1}{2\sqrt{ax^2 + bx + c}}\cdot(2ax + b)$

$$= \frac{3(2ax+b)\cdot\left(\sin^2\sqrt{ax^2+bx+c}\right)\cdot\left(\cos\sqrt{ax^2+bx+c}\right)}{2\sqrt{ax^2+bx+c}} \quad \text{...(ii)}$$

$$\therefore \frac{dy}{dx} = \text{(i)} + \text{(ii)}$$

$$= -(2ax+b)\sin(ax^2+bx+c) + \frac{3(2ax+b)\left(\sin^2\sqrt{ax^2+bx+c}\right)\cdot\left(\cos\sqrt{ax^2+bx+c}\right)}{2\sqrt{ax^2+bx+c}},$$

for $ax^2 + bx + c > 0$.

Exponential Functions

We recall that exponential functions are differentiable on any interval on which they are defined and their derivatives are obtained by using the formula

$$\frac{d}{dx} e^{f(x)} = e^{f(x)} \cdot f'(x)$$

where the derivative $f'(x)$ is obtained by using the rule for differential coefficient or power, exponential, logarithmic, inverse trigonometric, sum, difference, product, quotient or composite of two or more than two differentiable functions.

Note: If $f(x) = x =$ an identity function,

$$\frac{d}{dx} e^x = e^x \cdot$$

Solved Examples

Find the differential coefficient of the following.

1. $y = e^{\sqrt{x^2+1}}$

Solution: $y = e^{\sqrt{x^2+1}}$

$$\Rightarrow \frac{dy}{dx} = \frac{d}{dx} e^{\sqrt{x^2+1}}$$

$$\Rightarrow \frac{dy}{dx} = \frac{d\, e^{\sqrt{x^2+1}}}{d\sqrt{x^2+1}} \cdot \frac{d\left(\sqrt{x^2+1}\right)}{d\left(x^2+1\right)} \cdot \frac{d}{dx}\left(x^2+1\right)$$

$$= e^{\sqrt{x^2+1}} \cdot \frac{1}{2\sqrt{x^2+1}} \cdot 2x$$

$$= \frac{x \cdot e^{\sqrt{x^2+1}}}{\sqrt{x^2+1}}$$

2. $y = e^{\left(\sqrt{x}-x^2\right)}$

Solution: $y = e^{\left(\sqrt{x}-x^2\right)}$

$$\Rightarrow \frac{dy}{dx} = \frac{d}{dx} e^{\left(\sqrt{x}-x^2\right)}$$

$$= \frac{d\, e^{\left(\sqrt{x}-x^2\right)}}{d\left(\sqrt{x}-x^2\right)} \cdot \frac{d}{dx}\left(\sqrt{x}-x^2\right)$$

$$= e^{\left(\sqrt{x}-x^2\right)} \cdot \left(\frac{d\sqrt{x}}{dx} - \frac{dx^2}{dx}\right)$$

$$= e^{\left(\sqrt{x}-x^2\right)} \cdot \left(\frac{1}{2\sqrt{x}} - 2x\right) \text{ for } x > 0.$$

3. $y = e^{(\cot x)^2}$

Solution: $y = e^{(\cot x)^2}$

$$\Rightarrow \frac{dy}{dx} = \frac{d}{dx} e^{(\cot x)^2}$$

$$\Rightarrow \frac{dy}{dx} = \frac{d\, e^{(\cot x)^2}}{d(\cot x)^2} \cdot \frac{d(\cot x)^2}{d\cot x} \cdot \frac{d\cot x}{dx}$$

$$= e^{(\cot x)^2} \cdot 2\cot x \cdot \left(-\operatorname{cosec}^2 x\right)$$

$$= -2\cot x \cdot e^{(\cot x)^2} \cdot \operatorname{cosec}^2 x,\ x \neq n\pi$$

4. $y = e^{(\sin^{-1} x)^2}$

Solution: $y = e^{(\sin^{-1} x)^2}$

$$\Rightarrow \frac{dy}{dx} = \frac{d}{dx} e^{(\sin^{-1} x)^2}$$

$$\Rightarrow \frac{dy}{dx} = \frac{d\, e^{(\sin^{-1} x)^2}}{d(\sin^{-1} x)^2} \cdot \frac{d(\sin^{-1} x)^2}{d\sin^{-1} x} \cdot \frac{d\sin^{-1} x}{dx}$$

$$= e^{(\sin^{-1} x)^2} \cdot 2(\sin^{-1} x) \cdot \frac{1}{\sqrt{1-x^2}}$$

$$= \frac{2\sin^{-1} x \cdot e^{(\sin^{-1} x)^2}}{\sqrt{1-x^2}},\ |x| < 1$$

Logarithmic Functions

We recall that logarithmic functions are differentiable on any interval on which they are defined and their derivatives are

1. $\frac{d}{dx}(\log x) = \frac{1}{x},\ x > 0$

2. $\frac{d}{dx}(\log|x|) = \frac{1}{x},\ x \neq 0$

3. $\frac{d}{dx}(\log f(x)) = \frac{f'(x)}{f(x)},\ f(x) > 0$

4. $\frac{d}{dx}(\log|f(x)|) = \frac{f'(x)}{f(x)},\ f(x) \neq 0$

i.e. of some positive differentiable function of x, say $f(x)$ is under the sign of logarithm, its derivative found using the chain rule for the derivative of the composite differentiable function is

$$\frac{d}{dx}(\log f(x)) = \frac{f'(x)}{f(x)},\ f(x) > 0$$

and if some differentiable function $|f(x)|$, provided $f(x) \neq 0$ at any point belonging to any interval on which $f(x)$ is defined is under the sign of logarithm, its derivative found using the chain rule for the derivative of the composite differentiable function is

$$\frac{d}{dx}(\log|f(x)|) = \frac{f'(x)}{f(x)},\ f(x) \neq 0.$$

Remember: **1.** The derivative $\frac{d}{dx}(\log f(x))$ for $f(x) > 0$ and/ $\frac{d}{dx}(\log|f(x)|)$ for $f(x) \neq 0$ is called logarithmic derivative of the function $f(x)$. Further we should note that either the derivative $(\log f(x))'$ for $f(x) > 0$ or the derivative $(\log f(x))'$ for $f(x) \neq 0$ is equal to the ratio of the derivative $f'(x)$ to the value of the function $f(x)$; i.e. if $y = f(x)$ is a positive differentiable function of x and / $y = f(x)$ is a differentiable function of x such that $f(x)$ may be positive and negative both, then the ratio $\frac{f'(x)}{f(x)}$ is obtained on finding the first derivative $\frac{dy}{dx}$ and then dividing it by the given value of the differentiable function represented at $f(x)$.

2. If some differentiable function of x is under the sign of logarithm, it is pre-assumed (or, understood) that $f(x)$ is positive (i.e. $f(x) > 0$ is pre-assumed or understood) or we have to mention that $f(x) > 0$ while finding the logarithmic derivative, i.e. to differentiable the function of x that can be put in the form:

$\log f(x)$ = a logarithm of the function of the independent variable x, it is pre-assumed (or, understood) that $f(x)$ is a positive differentiable function of x whenever no restriction (or, condition) which makes $f(x)$ positive is imposed on the independent variable x or we have to mention that $f(x) > 0$.

3. Logarithm of a function of x (or, logarithmic function of x) put in the form $\log f(x)$ is not defined for $f(x) \le 0$ at any point x belonging to any interval on which it is defined. This is why we take the modulus of the value of the function $f(x)$ if the value of the function concerned $f(x)$ is negative at any point x while finding the logarithmic derivative of the differentiable function of x.

4. Sometimes we are given a differentiable function of x under the sign of logarithm along with the interval (or, the quadrant) in (or, on, or, over) which the value of the differentiable function $f(x) > 0$ or the condition imposed on the independent variable x (like $x > a$, $x < a$, $x \ge a$ etc.) which makes $f(x)$ positive is given.

Notes: **1.** If the value of the function $f(x)$ is both positive and negative in the interval (or, the quadrant) in which it is defined, we take the modulus (or, absolute value) of the value of the function $f(x)$ only when the function f concerned at x is to be positive as in case of logarithmic differentiation.

2. By using the rules for differentiating the power, exponential, logarithmic, trigonometric, inverse trigonometric, sum, difference, product, quotient or composite of two or more than two differentiable functions, we are able to find the differential coefficient of the value of the differentiable function $f(x)$ written under the sign of logarithm.

3. $f(x) \ne 0 \Rightarrow |f(x)| > 0$

On Types of Problems

In general there are two types of logarithmic functions whose derivative is required to find out.

1. The value of a differentiable function $f(x)$ written under the sign of logarithm with or without an interval (or, quadrant) in which $f(x)$ is positive.

2. A differentiable function $|f(x)|$ written under the sign of logarithm.

Further we should note that type (1) has the following forms:

(i) $y = \log f(x)$ and / $y = \log \log \ldots \log f(x)$

(ii) Power of logarithmic function, i.e. $y = [\log f(x)]^n$ and / $y = [\log \log \log \ldots \log f(x)]^n$

(iii) A logarithmic function with base other than 'e', i.e. $y = \log_{\phi(x)} f(x)$.

On Language

1. It is common to say "the function $f(x)$ and / the value of the function $f(x)$ for the function f at (or, of) x, the value of the function f at (or, of) x and / the function of x namely (or, say) $f(x)$".

Hence, we say $\dfrac{f'(x)}{f(x)}$ is the ratio of the derived function f' at (or, of) x to the function f at x instead of saying $\dfrac{f'(x)}{f(x)}$ is the ratio of the derived function (or, derivative) $f'(x)$ to the function (or, the value of the function) $f(x)$.

2. Plural of "a function of x" is functions of x's. This is why whenever we want to mention more than one function of x, we write (or, say) functions of (or, at) x's. however, if $f_1(x)$ is a differentiable function of x and $f_2(x)$ is also a differentiable function of x, we write $f_1(x)$ and $f_2(x)$ are differentiable functions of x's.

Problems based on first type

Form 1: Problems on the form

$y = \log f(x)$

or, $y = \log \log \log \ldots \log f(x); f(x) > 0$

Solved Examples

Find the differential coefficient of the following.

1. $y = \log\left(\dfrac{1+\tan x}{1-\tan x}\right),\ 0 < x < \dfrac{\pi}{4}$

Solution: $y = \log\left(\dfrac{1+\tan x}{1-\tan x}\right)$

$$= \log(1+\tan x) - \log(1-\tan x)$$

$$\Rightarrow \frac{dy}{dx} = \frac{d}{dx}[\log(1+\tan x)] - \frac{d}{dx}[\log(1-\tan x)]$$

$$= \frac{\sec^2 x}{1+\tan x} + \frac{\sec^2 x}{1-\tan x}$$

$$= \sec^2\left[\frac{1-\tan x+1+\tan x}{(1+\tan x)(1-\tan x)}\right]$$

$$= \frac{2\sec^2 x}{1-\tan^2 x} = \frac{\left(\dfrac{2}{\cos^2 x}\right)}{\left(\dfrac{\cos^2 x - \sin^2 x}{\cos^2 x}\right)}$$

$$= \frac{2}{\cos^2 x - \sin^2 x} = \frac{2}{\cos 2x}$$

$$= 2\sec 2x,\ 0 < x < \frac{\pi}{4}$$

2. $y = \log\left(\dfrac{2x-6}{5-x}\right),\ 3 < x < 5$

Solution: $y = \log\left(\dfrac{2x-6}{5-x}\right) = \log(2x-6) - \log(5-x)$

$$\Rightarrow \frac{dy}{dx} = \frac{d}{dx}[\log(2x-6)] - \frac{d}{dx}[\log(5-x)]$$

$$= \frac{2}{2x-6} - \frac{-1}{5-x} = \frac{1}{x-3} - \frac{1}{x-5}$$

$$= \frac{-2}{(x-3)(x-5)}$$

3. $y = \log \cos x,\ 0 < x < \dfrac{\pi}{2}$

Solution: $y = \log \cos x$

$$\Rightarrow \frac{dy}{dx} = \frac{d}{dx}\log \cos x$$

$$= \frac{1}{\cos x}\cdot(-\sin x)$$

$$= -\tan x$$

4. $y = \log\left(x + \sqrt{x^2 - 1}\right),\ x > 1$

Solution: $y = \log\left(x + \sqrt{x^2 - 1}\right)$

$$\Rightarrow \frac{dy}{dx} = \frac{d}{dx}\log\left(x + \sqrt{x^2-1}\right)$$

$$= \frac{1}{x+\sqrt{x^2-1}}\cdot\left[1 + \frac{1}{2}\left(x^2-1\right)^{-\frac{1}{2}}\cdot 2x\right]$$

$$= \frac{1}{x+\sqrt{x^2-1}}\left(1 + \frac{x}{\sqrt{x^2-1}}\right)$$

$$= \frac{1}{x+\sqrt{x^2-1}}\left(\frac{\sqrt{x^2-1}+x}{\sqrt{x^2-1}}\right).$$

5. $y = \log\left(x + \sqrt{x^2 + a^2}\right)$

Solution: $y = \log\left(x + \sqrt{x^2 + a^2}\right)$

$$\Rightarrow \frac{dy}{dx} = \frac{1}{x+\sqrt{x^2+a^2}}\cdot\left[1 + \frac{1}{2}\left(x^2+a^2\right)^{-\frac{1}{2}}\cdot 2x\right]$$

$$= \frac{1}{x+\sqrt{x^2+a^2}}\cdot\left[1 + \frac{2x}{2\sqrt{x^2+a^2}}\right]$$

$$= \frac{1}{x+\sqrt{x^2+a^2}}\cdot\left[\frac{\sqrt{x^2+a^2}+x}{\sqrt{x^2+a^2}}\right]$$

$$= \frac{1}{\sqrt{x^2+a^2}}$$

6. $y = \log[\log(\cot x)]$

Solution: $y = \log[\log(\cot x)]$

$$\Rightarrow \frac{dy}{dx} = \frac{1}{\log \cot x}\cdot\frac{1}{\cot x}\cdot\left(-\text{cosec}^2 x\right)$$

$$= \frac{-\text{cosec}^2 x}{\cot x \cdot \log \cot x};\ (\cot x > 1)$$

$\because \cot x > 1 \Rightarrow \log \cot x > 0$
$\Rightarrow \log \log \cot x$ is defined

7. $y = \log[\log(\log x)]$

Solution: $y = \log[\log(\log x)]$

$$\Rightarrow \frac{dy}{dx}$$

$$= \frac{d\log[\log(\log x)]}{d[\log(\log x)]} \cdot \frac{d\log\log x}{d\log x} \cdot \frac{d\log x}{dx}$$

$$= \frac{1}{\log(\log x)} \cdot \frac{1}{\log x} \cdot \frac{1}{x}$$

$$= \frac{1}{x} \cdot \frac{1}{\log x} \cdot \frac{1}{\log(\log x)};$$ for $x > 0$ i.e. when y is defined.

8. $y = \log\log\log\log\log x^5$

Solution: $y = \log\log\log\log\log x^5$

$$\Rightarrow \frac{dy}{dx} = \frac{d\log\log\log\log\log x^5}{dx}$$

$$= \frac{d\log\log\log\log\log x^5}{d\log\log\log\log x^5} \times$$

$$\frac{d\log\log\log\log x^5}{d\log\log\log x^5} \times \frac{d\log\log\log x^5}{d\log\log x^5} \times$$

$$\frac{d\log\log x^5}{d\log x^5} \times \frac{d\log x^5}{dx^5} \times \frac{dx^5}{dx}$$

$$= \frac{5x^4}{x^5(\log\log\log\log x^5)\cdot(\log\log\log x^5)\cdot(\log\log x^5)\cdot\log x}$$

$$= \frac{5}{x(\log\log\log\log x^5)\cdot(\log\log\log x^5)\cdot(\log\log x^5)\cdot\log x}$$

for all those values of x at which y is defined.

Form 2: Problems on the form

$y = [\log f(x)]^n$

or, $y = [\log\log\log \ldots \log f(x)]^n$

Solved Examples

Find the differential coefficient of the following.

1. $y = (\log x)^3$; $x > 0$

Solution: $y = (\log x)^3$

$$\Rightarrow \frac{dy}{dx} = \frac{d}{dx}(\log x)^3 = \frac{d(\log x)^3}{d\log x} \cdot \frac{d\log x}{dx}$$

$$\Rightarrow \frac{dy}{dx} = 3(\log x)^2 \cdot \frac{1}{x} = \frac{3(\log x)^2}{x}$$

$$= \frac{3\log^2 x}{x}.$$

2. $y = [\log(\cos x)]^4$

Solution: $y = [\log(\cos x)]^4$

$$\Rightarrow \frac{dy}{dx} = \frac{d}{dx}[\log(\cos x)]^4$$

$$= \frac{d[\log(\cos x)]^4}{d\log(\cos x)} \cdot \frac{d\log(\cos x)}{d\cos x} \cdot \frac{d\cos x}{dx}$$

$$= 4[\log(\cos x)]^3 \cdot \frac{1}{\cos x} \cdot (-\sin x)$$

$$= -4[\log(\cos x)]^3 \cdot \frac{\sin x}{\cos x}$$

$$= -4\log^3(\cos x) \cdot \tan x; \text{ where } \cos x > 0.$$

3. $y = \left[\log\left(\cos^3 x\right)\right]^2, -\frac{\pi}{2} < x < \frac{\pi}{2}$

Solution: $y = [\log(\cos^3 x)]^2$

$$\Rightarrow \frac{dy}{dx} = \frac{d}{dx}\left[\log\left(\cos^3 x\right)\right]^2$$

$$= \frac{d\left[\log\left(\cos^3 x\right)\right]^2}{d\log\left(\cos^3 x\right)} \cdot \frac{d\log\left(\cos^3 x\right)}{d\cos^3 x} \cdot \frac{d\cos^3 x}{\cos x} \cdot \frac{d\cos x}{dx}$$

$$= 2\left[\log\left(\cos^3 x\right)\right] \cdot \frac{3\cos^2 x^2}{\cos^3 x} \cdot (-\sin x)$$

$$= -6\log\left(\cos^3 x\right) \cdot \frac{\sin x}{\cos x}$$

$$= -6\log\left(\cos^3 x\right)\cdot \tan x;\left(-\frac{\pi}{2}<x<\frac{\pi}{2}\right)$$

Remember:

$f(x)\cdot f(x)\cdot f(x)\ldots$ up to n times $=[f(x)]^n = f^n(x)$.

Problems on Second Type

Form 1: Problems on the form

$y=\log|f(x)|$ and $y=\log\log\log\ldots\log|f(x)|$; $f(x)\neq 0$

Solved Examples

Find the differential coefficient of the following

1. $y=\log\left|\dfrac{a+b\tan x}{a-b\tan x}\right|;\ |\tan x|\neq\dfrac{a}{b}$

Solution: $y=\log\left|\dfrac{a+b\tan x}{a-b\tan x}\right|$

$$\Rightarrow \frac{dy}{dx}=\frac{d}{dx}\left(\log\left|\frac{a+b\tan x}{a-b\tan x}\right|\right)$$

$$=\frac{(a-b\tan x)}{(a+b\tan x)}\cdot\frac{(a-b\tan x)\left(b\sec^2 x\right)-(a+b\tan x)\left(-b\sec^2 x\right)}{(a-b\tan x)^2}$$

$$=\frac{(a-b\tan x)}{(a+b\tan x)}\cdot\frac{(a-b\tan x)\left(b\sec^2 x\right)+(a+b\tan x)\left(b\sec^2 x\right)}{(a-b\tan x)^2}$$

$$=\frac{(a-b\tan x)}{(a+b\tan x)}\cdot\frac{b\sec^2 x(a-b\tan x+a+b\tan x)}{(a-b\tan x)^2}$$

$$=\frac{2ab\sec^2 x}{(a+b\tan x)(a-b\tan x)}$$

$$=\frac{2ab\sec^2 x}{a^2-b^2\tan^2 x}$$

2. $y=\log\log|x|, |x|>1$

Solution: $y=\log\log|x|$

$$\Rightarrow \frac{dy}{dx}=\frac{d}{dx}(\log\log|x|)$$

$$=\frac{1}{\log|x|}\cdot\frac{1}{x}=\frac{1}{x\log|x|}$$

3. $y=\log|(x^3+1)|$

Solution: $y=\log|(x^3+1)|$

$$\Rightarrow \frac{dy}{dx}=\frac{d}{dx}\left(\log\left|x^3+1\right|\right)$$

$$=\frac{1}{\left(x^3+1\right)}\cdot 3x^2=\frac{3x^2}{\left(x^3+1\right)}\text{ for } x\neq -1$$

4. $y=\log|\tan(1-x^2)|, |x|<1$

Solution: $y=\log|\tan(1-x^2)|$

$$\Rightarrow \frac{dy}{dx}=\frac{d}{dx}\left(\log\left|\tan\left(1-x^2\right)\right|\right)$$

$$=\frac{1}{\tan\left(1-x^2\right)}\cdot\sec^2\left(1-x^2\right)\cdot(-2x)$$

$$=\frac{-2x\sec^2\left(1-x^2\right)}{\tan\left(1-x^2\right)}$$

$$=2x\sec\left(x^2-1\right)\operatorname{cosec}\left(x^2-1\right)$$

5. $y=\log|\sin x|$, $x\neq a$ multiple of π

Solution: $y=\log|\sin x|,\ x\neq n\pi,\ n\in Z$

$$\Rightarrow \frac{dy}{dx}=\frac{d}{dx}(\log|\sin x|)$$

$$=\frac{1}{\sin x}\cdot\cos x$$

$$=\frac{\cos x}{\sin x}=\cot x,\ x\neq n\pi,\ n\in Z$$

Form 3: Problems on the form:

$$y=\log_{\phi(x)} f(x)$$

Working rule: To differentiate a logarithm of function of x whose base is another function of x, we adopt the following working rule.

Step 1: Write $y = \frac{\log_e f(x)}{\log_e \phi(x)} = \frac{\log f(x)}{\log \phi(x)}$

Step 2: Differentiate $y = \frac{\log f(x)}{\log \phi(x)}$ w.r.t. x by using the rule for differentiating the quotient of two differentiable functions.

i.e. $y' = \frac{DN' - ND'}{D^2}$

Remember: $\log_b a = \frac{\log_e a}{\log_e b} = \frac{\log a}{\log b}$

Solved Examples

Find the differential coefficient of the following.

1. $y = \log_{\sin x}(1 + \tan x);\ 0 < x < \frac{\pi}{2}$

Solution: $y = \log_{\sin x}(1 + \tan x)$

$$\Rightarrow y = \frac{\log(1 + \tan x)}{\log \sin x}$$

$$\Rightarrow \frac{dy}{dx}$$

$$= \frac{\log \sin x \cdot \frac{d}{dx}\log(1+\tan x) - \log(1+\tan x)\cdot\frac{d}{dx}\log\sin x}{(\log \sin x)^2}$$

$$= \frac{\log \sin x\left(\frac{1}{1+\tan x}\right)\cdot\sec^2 x - \log(1+\tan x)\cdot\frac{1}{\sin x}\cdot(-\cos x)}{\log^2 \sin x}$$

$$= \frac{\frac{\sec^2 x \cdot \log \sin x}{(1+\tan x)} + \frac{\cos x \cdot \log(1+\tan x)}{\sin x}}{\log^2 \sin x}$$

2. $y = \log_x(1 + x),\ x > 0, \neq 1$

Solution: $y = \log_x(1 + x)$

$$\Rightarrow y = \frac{\log(1 + x)}{\log x}$$

$$\Rightarrow \frac{dy}{dx} = \frac{\frac{\log x}{x+1} - \frac{\log(x+1)}{x}}{(\log x)^2}$$

$$= \frac{1}{(\log x)^2}\cdot\left[\frac{\log x}{x+1} - \frac{\log(x+1)}{x}\right]$$

Exercise Set on

Composition of two differentiable functions

Form: $y =$ (any differentiable function of $x)^n$

or, $y = f_1 f_2(x)$

Exercise 9.1.1

Find $\frac{dy}{dx}$ of each of the following functions.

1. $y = (3x + 1)^4$
2. $y = (7x^2 + x)^2$
3. $y = \sqrt{1 - x^2}$
4. $y = \sqrt{x^2 + ax + 1};\ a^2 < 4$
5. $y = (x^3 + 1)^5$
6. $y = \left(7x^2 + 11x + 39\right)^{\frac{3}{2}}$
7. $y = \sqrt{a - bx}$
8. $y = \frac{1}{(p - qx)^{\frac{3}{4}}}$
9. $y = \frac{1}{(l - mx)^n},\ n \in Z$
10. $y = (3x - 7)^{12}$
11. $y = (5x^5 - 3x)^{24}$
12. $y = (4x + 3)^{-5}$
13. $y = (3x^2 + 2x + 1)^8$

14. $y = \left(x + \frac{1}{x}\right)^2$

15. $y = \left(3x^2 + 4\right)^{\frac{5}{2}}$

16. $y = \sqrt{3x^2 + 5}$

17. $y = (1-x)^6$

18. $y = \frac{1}{3\sqrt{4x^3 + 5x^2 - 7x + 6}}; x > 0$

19. $y = \sqrt{(x+1)(x+2)(x+3)}; x > -1$

Answers

1. $12(3x+1)^3$
2. $2(7x^2+x)\,14+1$
3. $\frac{-2x}{2\sqrt{1-x^2}}; |x| < 1$
4. $\frac{(2x+a)}{2\sqrt{x^2 + ax + 1}}$
5. $5(x^3+1)^4 \cdot 3x^2$
6. $\frac{3}{2}\left(7x^2 + 11x + 39\right)^{\frac{1}{2}} \cdot (14x + 11)$
7. $\frac{-b}{2\sqrt{a - bx}}; x < \frac{a}{b}$
8. $\frac{3}{4} \cdot q \cdot (p - qx)^{-\frac{7}{4}}; x < \frac{p}{q}$
9. $\frac{m \cdot n}{(l - mx)^{n+1}}; x \neq \frac{l}{m}$ if $n > 0$
10. $36(3x-7)^{11}$
11. $24(10x-3)(5x^2-3x)^{23}$
12. $-20(4x+3)^{-6}, x \neq \frac{-4}{3}$
13. $16(3x+1)(3x^2+2x+1)^7$
14. $(2x-2x^{-3})\; x \neq 0$
15. $15x\left(3x^2 + 4\right)^{\frac{3}{2}}$
16. $\frac{3x}{\sqrt{3x^2 + 5}}$
17. $-6(1-x)^5$
18. $\frac{-\left(4x^2 + \frac{10}{3}x - \frac{7}{3}\right)}{2\left(\sqrt{4x^3 + 5x^2 - 7x + 6}\right)^3}$
19. $\frac{3x^2 + 12x + 11}{2\sqrt{(x+1)(x+2)(x+3)}}$

Exercise 9.1.2

Find $\frac{dy}{dx}$ of each of the following.

1. $y = \sin x^3$
2. $y = \tan x^2$
3. $y = \sin(\cot x)$
4. $y = \sin(\sec x)$
5. $y = \sin 5x$
6. $y = \cos 2x$
7. $y = \sin x^2$
8. $y = \sin\sqrt{x}$
9. $y = \sin\left(\frac{2}{x}\right)$
10. $y = \cot 2x$
11. $y = \cot(1-2x^2)$
12. $y = \cos(1-x^2)$
13. $y = \sqrt{\sin^{-1} x}$
14. $y = \sqrt{\cos^{-1} x}$
15. $y = \sqrt{\tan^{-1} x}$
16. $y = \operatorname{cosec}\left(\frac{1-x}{1+x}\right), 0 \leq x < 1$

17. $y = \sin\left(\dfrac{2x+7}{1-2x}\right)$

18. $y = \sin\left(\dfrac{1}{\sqrt{x}}\right)$

Answers

1. $3x^2 \cos x^2$
2. $2x \sec^2 x^2,\ x^2 \neq n\pi + \dfrac{\pi}{2},\ n \in Z$
3. $-\cos \cot x \cdot \operatorname{cosec}^2 x,\ x \neq n\pi$
4. $\cos(\sec x)(\sec x \cdot \tan x);\ x \neq n\pi + \dfrac{\pi}{2}$
5. $5 \cos 5x$
6. $-2 \sin 2x$
7. $2x \cos x^2$
8. $\dfrac{\cos\sqrt{x}}{2\sqrt{x}}, x > 0$
9. $-\left(\dfrac{2\cos\frac{2}{x}}{x^2}\right),\ x \neq 0$
10. $-2 \operatorname{cosec}^2 2x,\ x \neq \dfrac{n\pi}{2}$
11. $-4x \operatorname{cosec}^2(1-2x^2)$
12. $2x \sin(1-x^2)$
13. $\dfrac{1}{2\sqrt{\sin^{-1} x}} \cdot \dfrac{1}{\sqrt{1-x^2}},\ 0 < x < 1$
14. $\dfrac{1}{2\sqrt{\cos^{-1} x}} \cdot \left(\dfrac{-1}{\sqrt{1-x^2}}\right);\ |x| < 1$
15. $\left(\dfrac{1}{2\sqrt{\tan^{-1} x}}\right)\left(\dfrac{1}{1+x^2}\right);\ x > 0$
16. $\dfrac{2}{(x+1)^2} \cdot \operatorname{cosec}\left(\dfrac{1-x}{1+x}\right) \cdot \cot\left(\dfrac{1-x}{1+x}\right)$
17. $\dfrac{16}{(1-2x)^2} \cdot \cos\left(\dfrac{2x+7}{1-2x}\right),\ x \neq \dfrac{1}{2}$
18. $-\dfrac{1}{2x\sqrt{x}} \cdot \cos\left(\dfrac{1}{\sqrt{x}}\right),\ x > 0$

Exercise Set on

Composition of more than two differentiable functions

Form 1: $y = f_1 f_2 f_3 \ldots f_n f(x)$

or, $y = (f_1 f_2 f_3 \ldots f_n f(x))^n$

Exercise 9.2.1

1. $y = \operatorname{cosec}\left(\sqrt{3x+4}\right)$
2. $y = \cot\left(\sin\sqrt{x}\right)$
3. $y = \cot(\tan x)$
4. $y = \sin^2 x^2$
5. $y = \cos^2 x^2$
6. $y = \tan^3 x^2$
7. $y = (\log x^2)^2$
8. $y = \sqrt{\sec^{-1} x^2}$
9. $y = \sqrt{\operatorname{cosec}^{-1} x^2}$
10. $y = \sqrt{\sin^{-1} x^3}$
11. $y = (\tan^{-1} x^3)^2$
12. $y = (\sec^{-1} x^m)^n$
13. $y = (\cos^{-1} x^p)^q$
14. $y = \left(\cot^{-1} x^3\right)^{\frac{1}{3}}$
15. $y = \sqrt{\cot^{-1} e^x}$
16. $y = \sec^n(ax^2 + bx + c)$
17. $y = \sqrt{\cos(3x^2 + 4)}$
18. $y = \sec^n(mx)$
19. $y = \tan^{-1}(4e^x + 3)^2$
20. $y = \sec^3(m \sin^{-1} x)$

21. $y = \cos^5 (\log \tan^2 x^3)^4$

22. $y = \sqrt{\sin\left(\sin\sqrt{x}\right)}$

23. $y = \sqrt{\tan(\tan x)}$

24. $y = (\sin(\log x))^2$

Answers (under suitable restrictions on x)

1. $\left(-\frac{3}{2}\operatorname{cosec}\sqrt{3x+4}\cdot\cot\sqrt{3x+4}\right)\cdot\frac{1}{\sqrt{3x+4}}$

2. $-\frac{1}{2\sqrt{x}}\cdot\operatorname{cosec}^2\left(\sin\sqrt{x}\right)\cos\sqrt{x}$

3. $-\operatorname{cosec}^2 (\tan x) \sec^2 x$

4. $4x \sin x^2 \cos x^2$

5. $-4x \cos x^2 \sin x^2$

6. $6x \tan^2 x^2 \sec^2 x^2$

7. $\frac{4}{x}\log x^2$

8. $\frac{1}{2\sqrt{\sec^{-1} x^2}}\cdot\frac{1}{x^2\sqrt{x^4-1}}\cdot 2x$

9. $\frac{-1}{x\sqrt{\operatorname{cosec}^{-1} x^2}\cdot\sqrt{x^4-1}}$

10. $\frac{3x^2}{2\sqrt{\sin^{-1} x^3}}\cdot\frac{1}{\sqrt{1-x^6}}$

11. $\frac{6x^2\tan^{-1} x^3}{1+x^6}$

12. $\frac{m\cdot n\cdot x^{m-1}\cdot\left(\sec^{-1}x^m\right)^{n-1}}{\left|x^m\right|\sqrt{x^{2m}-1}}$

13. $\frac{-pq\,x^{p-1}\left(\cos^{-1}x^p\right)^{q-1}}{\sqrt{1-x^{2p}}}$

14. $\frac{-x^2\left(\cot^{-1}x^3\right)^{-\frac{2}{3}}}{1+x^6}$

15. $\frac{-e^x}{2\left(\sqrt{\cot^{-1}e^x}\right)\left(1+e^{2x}\right)}$

16. $n(2ax+b)\sec^n(ax^2+bx+c)\tan(ax^2+bx+c)$

17. $\frac{-3x\sin\left(3x^2+4\right)}{\sqrt{\cos\left(3x^2+4\right)}}$

18. $m\cdot n\sec^n(mx)\tan mx$

19. $\frac{8e^x\left(4e^x+3\right)}{1+\left(4e^x+3\right)^4}$

20. $\frac{3m\sec^3\left(m\sin^{-1}x\right)\tan\left(m\sin^{-1}x\right)}{\sqrt{1-x^2}}$

21. $-120x^2\cos^4\left(\log\tan^2 x^3\right)^4\sin\left(\log\tan^2 x^3\right)$
$\left(\log\left(\tan^2 x^3\right)\right)^3\left(\frac{\sec^2 x^3}{\tan x^3}\right)$

22. $\frac{\cos\sqrt{x}\cdot\cos\left(\sin\sqrt{x}\right)}{4\sqrt{x}\cdot\sqrt{\sin\left(\sin\sqrt{x}\right)}}$

23. $\frac{\sec^2 x\sec^2(\tan x)}{2\sqrt{\tan(\tan x)}}$

24. $(2\sin(\log x))^2\cdot\cos(\log x)\cdot\frac{1}{x}$

Exercise 9.2.2

Find $\frac{dy}{dx}$ of each of the following functions.

1. $y = \sin\cos\tan\sqrt{x}$

2. $y = \sin\cos\tan\cot x$

3. $y = \cos \tan \sqrt{x+1}$
4. $y = \cos \sin \log x$
5. $y = e^{(\sin \tan^{-1} 2x)}$
6. $y = \cot \sin \cos x$
7. $y = \log \log \log x \ (x > 0)$
8. $y = \log \sin x^2$
9. $y = e^{\sqrt{\tan x}}$
10. $y = \log (\tan^{-1} x^2)$
11. $y = e^{(\tan^{-1} x^2)}$
12. $y = \sin \cos \tan \sqrt{mx}$, $m > 0$
13. $y = \sin \cos \tan \sec x$

Answers (under proper restrictions on x)

1. $\left(-\cos \cos \tan \sqrt{x}\right)\left(\sin \tan \sqrt{x}\right) \cdot \sec^2 \sqrt{x} \cdot \dfrac{1}{2\sqrt{x}}$
2. $(-\cos \cos \tan \cot x) \cdot (\sin \tan \cot x)(\sec^2 \cot x) \cdot \operatorname{cosec}^2 x$
3. $-\left(\sin \tan \sqrt{x+1}\right) \cdot \sec^2 \left(\sqrt{x+1}\right) \cdot \dfrac{1}{2\sqrt{x+1}}$
4. $-(\sin \sin \log x) \cdot (\cos \log x) \cdot \dfrac{1}{x}$
5. $e^{\sin\left(\tan^{-1} 2x\right)} \cdot \cos\left(\tan^{-1} 2x\right) \cdot \dfrac{2}{1+4x^2}$
6. $(\operatorname{cosec}^2 \sin \cos x) \cdot (\cos \cos x) \cdot \sin x$
7. $\dfrac{1}{\log \log x} \cdot \dfrac{1}{\log x} \cdot \dfrac{1}{x}$
8. $\dfrac{2x \cos x^2}{\sin x^2}$
9. $\dfrac{1}{2} \sec^2 x \sqrt{\cot x} \cdot e^{\sqrt{\tan x}}$
10. $\dfrac{2x}{\left(1+x^4\right) \tan^{-1} x^2}$
11. $\dfrac{2x\, e^{\tan^{-1} x^2}}{1+x^4}$
12. $\dfrac{-\sqrt{m}}{2\sqrt{x}}\left(\sec^2 \sqrt{mx}\right)\cdot\left(\sin \tan \sqrt{mx}\right)\cdot\left(\cos \cos \tan \sqrt{mx}\right)$
13. $-\sec x \cdot \tan x \cdot (\sec^2 \sec x) \cdot (\sin \tan \sec x) \cdot (\cos \cos \tan \sec x)$

Form 2:

$$\sqrt{f_1 \sqrt{f_2 \sqrt{f_3 \sqrt{\ldots \sqrt{f(x)}}}}}$$

Exercise 9.3

Find $\dfrac{dy}{dx}$ of the following functions.

1. $y = \sin \sqrt{\cos \sqrt{ax}}$
2. $y = \sin \sqrt{\cos \sqrt{\tan mx}}$
3. $y = \sqrt{\sin \sqrt{x}}$
4. $y = \sqrt{\cos \sqrt{x}}$
5. $y = \sqrt{\tan \sqrt{x}}$
6. $y = \sqrt{\log \sqrt{x}}$

Answers (under suitable restrictions on x)

1. $\dfrac{-a\left(\cos \sqrt{\cos \sqrt{ax}}\right) \cdot \left(\sin \sqrt{ax}\right)}{4\left(\sqrt{\cos \sqrt{ax}}\right) \cdot \left(\sqrt{ax}\right)}$
2. $\dfrac{-m\left(\cos \sqrt{\cos \sqrt{\tan mx}}\right) \cdot \left(\sin \sqrt{\tan mx}\right) \sec^2 mx}{4\left(\sqrt{\cos \sqrt{\tan mx}}\right)\left(\sqrt{\tan mx}\right)}$
3. $\dfrac{\cos \sqrt{x}}{4\sqrt{x}\, \sqrt{\cos \sqrt{x}}}$

4. $\dfrac{-\sin\sqrt{x}}{4\sqrt{x}\sqrt{\cos\sqrt{x}}}$

5. $\dfrac{\sec^2\sqrt{x}}{4\sqrt{x}\sqrt{\tan\sqrt{x}}}$

6. $\dfrac{1}{4x\sqrt{\log\sqrt{x}}}$

Form 3:

$$y=\sqrt{f_1(x)+\sqrt{f_2(x)+\sqrt{\ldots+\sqrt{f(x)}}}}$$

Exercise 9.4

Find $\dfrac{dy}{dx}$ of each of the following functions

1. $y=\sqrt{\sin x+\sqrt{\sin x+\sqrt{\sin x}}}$

2. $y=\sqrt{\cos x+\sqrt{\cos x+\sqrt{\cos x}}}$

3. $y=\sqrt{x+\sqrt{x+\sqrt{x}}}$

4. $y=\sqrt{2+\sqrt{2x}}$

5. $y=\sqrt{a+\sqrt{ax}}$

Answers

Hint: In the above problems, we may use directly the chain rule without making any substitution as

1. $$\frac{d\left(\sqrt{\sin x+\sqrt{\sin x+\sqrt{\sin x}}}\right)}{d\left(\sin x+\sqrt{\sin x+\sqrt{\sin x}}\right)}\cdot\left\{\frac{d}{dx}(\sin x)+\left(\frac{d\sqrt{\sin x+\sqrt{\sin x}}}{d\left(\sin x+\sqrt{\sin x}\right)}\right)\cdot\left(\frac{d\sin x}{dx}+\frac{d\sqrt{\sin x}}{d\sin x}\cdot\frac{d\sin x}{dx}\right)\right\}$$

Required Answer

1. $$\frac{1}{2}\cos x\left(\frac{1}{\sqrt{\sin x+\sqrt{\sin x+\sqrt{\sin x}}}}\right)\cdot\left(1+\frac{1}{2\sqrt{\sin x+\sqrt{\sin x}}}+\frac{1}{4\left(\sqrt{\sin x+\sqrt{\sin x}}\right)\sqrt{\sin x}}\right)$$

2. Find
3. Find

4. $\dfrac{1}{2\sqrt{2x}\left(2+\sqrt{2x}\right)}$

5. $\dfrac{a}{4\sqrt{ax}\left(a+\sqrt{ax}\right)}$

Form 4:

$\sqrt{\sqrt{\ldots\sqrt{f(x)}}}$, $f(x)$ being a differentiable function of x.

Exercise 9.5

Find $\dfrac{dy}{dx}$ of each of the following functions.

1. $y=\sqrt{\sqrt{\sin x}}$

2. $y=\sqrt{\sqrt{x}}$

3. $y=\sqrt{\sqrt{x^2}}$

Answers (under suitable restrictions on x)

1. $\dfrac{\cos x}{4\left(\sqrt{\sqrt{\sin x}}\right)\left(\sqrt{\sin x}\right)}$

2. $\dfrac{1}{4\left(\sqrt{\sqrt{x}}\right)\cdot\left(\sqrt{x}\right)}$

3. $\dfrac{x}{2\left(\sqrt{\sqrt{x^2}}\right)\left(\sqrt{x^2}\right)}$ or $\dfrac{|x|}{2x\left(\sqrt{|x|}\right)}$

$$\left(\because \sqrt{x^2} = |x| \text{ and } |x|' = \frac{|x|}{x},\ x \neq 0\right)$$

Form 5:

(a) $y = f_1(x) \cdot f_2(x) \pm g_1(x) \cdot g_2(x)$

(b) $y = f_1(x) \cdot f_2(x) \pm \dfrac{g_1(x)}{g_2(x)}$

(c) $y = f_1(x) \cdot \pm \dfrac{g_1(x)}{g_2(x)}$

(d) $y = f_1(x) \pm g_1(x) \cdot g_2(x)$

Exercise 9.6

Find $\dfrac{dy}{dx}$ of each of the following functions.

1. $y = x \sin \log x - x \cos \log x,\ x > 0$

2. $y = x\left(\sin^{-1} x\right)^2 + 2\sqrt{1 - x^2}\sin^{-1} x - 2x,\ |x| < 1$

3. $y = x \sec^{-1} x - \log\left(x + \sqrt{x^2 - 1}\right),\ x > 1$

4. $y = \log\left(x^2 + 4\right) - x\tan^{-1}\left(\dfrac{x}{2}\right),\ \forall x$

5. $y = x\sqrt{x^2 + 1} + \log\left(x + \sqrt{x^2 + 1}\right),\ \forall x$

6. $y = \log\left(\dfrac{\sqrt{x+1} - 1}{\sqrt{x+1} + 1}\right) + \dfrac{\sqrt{x}}{\sqrt{1+x}}$

7. $y = \dfrac{5x}{\sqrt[3]{1 - x^2}} + \cos^2(2x + 1)$

8. $y = \dfrac{x}{2}\sqrt{a^2 + x^2} + \dfrac{a^2}{2}\log\left(x + \sqrt{x^2 + a^2}\right)$

9. $y = \dfrac{x}{2}\sqrt{x^2 - a^2} - \dfrac{a^2}{2}\log\left(x + \sqrt{x^2 - a^2}\right)$

10. $y = -\dfrac{\cos^3 x}{3\sin^3 x} + \cot x$

11. $y = -\dfrac{1}{2\sin^2 x} + \log \tan x$

12. $y = \dfrac{1}{2}\log \tan \dfrac{x}{2} - \dfrac{1}{2}\dfrac{\cos x}{\sin^2 x}$

Answers

1. $2 \sin \log x,\ x > 0$

2. $\left(\sin^{-1} x\right)^2,\ |x| < 1$

3. $\sec^{-1} x,\ x > 1$

4. $-\tan^{-1}\left(\dfrac{x}{2}\right),\ \forall x$

5. $2\sqrt{x^2 + 1},\ \forall x$

6. $\dfrac{1}{\sqrt{x+1}} \times \left[\dfrac{1}{x} + \dfrac{1}{2(x+1)\sqrt{x}}\right]$

7. $\dfrac{5\left(3 - x^2\right)}{3\left(1 - x^2\right)^{\frac{4}{3}}} - 2\sin(4x + 2)$

8. $\sqrt{a^2 + x^2}$

9. $\sqrt{x^2 - a^2}$

10. $\dfrac{\cos 2x}{\sin^4 x}$

11. $\dfrac{1}{\sin^3 x \cdot \cos x}$

12. $\dfrac{1}{\sin^3 x}$

Form 5: continued
Problems based on combination composite differentiable functions of x's.

Exercise 9.6.1

Find $\frac{dy}{dx}$ of each of the following functions.

1. $y = \sin \log x - \log \sin x$
2. $y = \sin\sqrt{x} + \cos^2\sqrt{x}$
3. $y = \cos\left(ax^2 + bx + c\right) + \sin^3\left(\sqrt{ax^2 + bx + c}\right)$
4. $y = \log \log x - e^{5x}$
5. $y = \cot^3 x - \tan\sqrt{x}$

Answers

1. $\frac{1}{x}\cos\log x - \cot x$
2. $\frac{1}{2\sqrt{x}}\cos\sqrt{x}\left(1 - 2\sin\sqrt{x}\right)$
3. $-(2ax+b)\sin\left(ax^2+bx+c\right)+\frac{3}{2}\sin^2\sqrt{ax^2+bx+c}\cdot$ $\cos\sqrt{ax^2+bx+c}\cdot\frac{(2ax+b)}{\sqrt{ax^2+bx+c}}$
4. $\frac{1}{x\log x} - 5e^{5x}$
5. $-3\cot^2 x\cdot\operatorname{cosec}^2 x - \frac{1}{2\sqrt{x}}\cdot\sec^2\sqrt{x}$

Exercise 9.6.2

Find $\frac{dy}{dx}$ of each of the following functions

1. $y = \sin^2 3x\cos^3 2x$
2. $y = x^2\cot 2x$
3. $y = x^2\sec^2 x$
4. $y = x^3\operatorname{cosec}^3 x$
5. $y = e^{\sqrt{x}}\log\cos x$
6. $y = \left(\cos\sqrt{x}\right)\log\sin x$
7. $y = \sin^m \alpha x\cdot\cos^n \beta x$
8. $y = \sin(2x+3)\cdot\cos^2(3x+4)$
9. $y = e^{\sqrt{x}}\log\left(\cos\sqrt{x}\right)$
10. $y = e^{\sqrt{x}}\sin\sqrt{x}$

Note: This type of problems can be done using the rules of logarithmic differentiation.

Answers (under proper restrictions on x)

1. $6\sin 3x\cos^2 2x\cos 5x$
2. $2x(\cot 2x - x\operatorname{cosec}^2 2x)$
3. $2x\sec^2 x(1 + x\tan x)$
4. $3x^2\operatorname{cosec}^3 x(1 - x\cot x)$
5. $\frac{1}{2\sqrt{x}}e^{\sqrt{x}}\log\cos x - e^{\sqrt{x}}\tan x$
6. $\cot x\cos\sqrt{x} - \frac{1}{2\sqrt{x}}\sin\sqrt{x}\log\sin x$
7. $\sin^{m-1}\alpha x\cos^{n-1}\beta x\,(m\alpha\cos\alpha x\cos\beta x -$ $n\beta\sin\alpha x\sin\beta x)$
8. $2\cos(3x+4)\cos(2x+3) - 3\sin(2x+3)\sin(3x+4)$
9. $\frac{e^{\sqrt{x}}}{2\sqrt{x}}\left(\log\cos\sqrt{x} - \tan\sqrt{x}\right)$
10. $\frac{1}{2\sqrt{x}}e^{\sqrt{x}}\left(\cos\sqrt{x} + \sin\sqrt{x}\right)$

Exercise 9.6.3

Find $\frac{dy}{dx}$ of each of the following functions

1. $y = \frac{2(x - \sin x)^{\frac{3}{2}}}{\sqrt{x}}$

2. $y = \dfrac{e^{\tan^{-1} x}}{1 + x^2}$

3. $y = \dfrac{(x+1)^2 \sqrt{x-1}}{(x+4)^2 e^x}$

4. $y = \dfrac{\sin 6x}{1 + \cos 6x}$

5. $y = \dfrac{e^{\sin^{-1} x}}{\sqrt{1 - x^2}}$

6. $y = \dfrac{\sin mx}{\cos nx}$

7. $y = \dfrac{\sin^4 x}{\cos^5 x}$

8. $y = \dfrac{\cos (3 - 2x)}{\sin (3 + 2x)}$

9. $y = \dfrac{\cos^m (ax)}{\sin^n (bx)}$

10. $y = \dfrac{\tan 2x}{\sec (x + 1)}$

Note: This type of problem can also be done using the rules of logarithmic differentiation.

Answers

1. $\dfrac{1}{x\sqrt{x}} (2x - 3x \cos x + \sin x) \sqrt{x - \sin x}$

2. $\dfrac{e^{\tan^{-1} x} (1 - 2x)}{\left(1 + x^2\right)^2}$

3. $\dfrac{(x+1)^2 \sqrt{x-1}}{(x+4)^2 \cdot e^x} \left[\dfrac{5x - 3}{2\left(x^2 - 1\right)} - \dfrac{x + 6}{x + 4} \right]$

4. $6 \operatorname{cosec} 6x (\operatorname{cosec} x - \cot 6x)$

5. $\dfrac{e^{\sin^{-1} x}\left(x + \sqrt{1 - x^2}\right)}{\left(1 - x^2\right)^{\frac{3}{2}}}$

6. $m \cos mx \sec nx + n \sin mx \sec nx \tan nx$

7. $(4 \sin^3 x \sec^4 x + 5 \tan^5 x \sec x)$

8. $[2 \sin (3 - 2x) - 2\cos (3 - 2x) \cot (2x + 3)] \operatorname{cosec} (2x + 3)$

9. $-(ma \sin ax \sin bx + bn \cos bx \cos ax) \cdot \left(\dfrac{\cos^{m-1} ax}{\sin^{n+1} bx} \right)$

10. $2 \sec^2 2x \cos (x + 1) - \tan 2x \sin (x + 1)$

Exercise 9.6.4

Find $\dfrac{dy}{dx}$ of each of the following functions.

1. $y = (x - 1)^3 \cdot (x + 1)^2$

2. $y = x\sqrt{1 + 2x}$

3. $y = x^5 \left(x^2 + 3x + 2\right)^3$

4. $y = 2x^2 \sqrt{2 - x}$

5. $y = x\sqrt{3 - 2x^2}$

6. $y = (x - 1) \sqrt{x^2 - 2x + 2}$

7. $y = x\sqrt{x^2 + c^2}$

Answers

1. $(x + 1)(5x + 1)(x - 1)^2$

2. $\dfrac{1 + 3x}{\sqrt{1 + 2x}}$

3. $x^4 \left(x^2 + 3x + 2\right)^2 \left(11x^2 + 24x + 10\right)$

4. $\dfrac{x(8-5x)}{\sqrt{2-x}}$

5. $\dfrac{3-4x^2}{\sqrt{3-2x^2}}$

6. $\dfrac{2x^2-4x+3}{\sqrt{x^2-2x+2}}$

7. $\dfrac{x^2}{\sqrt{x^2+e^2}}+\sqrt{x^2+c^2}$

8. $y=\dfrac{2x^2-4x+3}{\sqrt{x^2-2x+3}}$

9. $y=\sqrt{\dfrac{x-1}{x+1}}$

10. $y=\dfrac{x}{\sqrt{1-x^2}}$

11. $y=\dfrac{x}{\sqrt{1-4x^2}}$

12. $y=\dfrac{3x}{(1+3x)^{\frac{1}{2}}}$

13. $y=\sqrt{\dfrac{1+x}{2+x}}$

14. $y=\dfrac{\sqrt{1-x^2}}{1-x}$

15. $y=\dfrac{\sqrt{(x-1)(x-2)}}{\sqrt{(x-3)(x-4)}}$

Exercise 9.6.5

Find $\dfrac{dy}{dx}$ of each of the following functions.

1. $y=\dfrac{\sqrt{x}}{\sqrt{x+1}}$

2. $y=\dfrac{\sqrt{x-1}-1}{\sqrt{x+1}+1}$

3. $y=\dfrac{x}{\left(a^2-x^2\right)^{\frac{3}{2}}}$

4. $y=\dfrac{\sqrt{1-x}}{\sqrt{1+x}}$

5. $y=\dfrac{1+3x}{\sqrt{1+2x}}$

6. $y=\dfrac{x(8-5x)}{\sqrt{2-x}}$

7. $y=\dfrac{3-4x^2}{\sqrt{3-2x^2}}$

***Answer* (under proper restrictions on x)**

1. Find
2. Find
3. Find
4. $\dfrac{-\dfrac{1}{2}\left[\dfrac{\sqrt{1+x}}{\sqrt{1-x}}+\dfrac{\sqrt{1-x}}{\sqrt{1+x}}\right]}{(1+x)}$
5. Find
6. Find
7. Find
8. Find
9. $\dfrac{1}{(x+1)\sqrt{x^2-1}}$

10. $\dfrac{1}{\left(1-x^2\right)^{\frac{3}{2}}}$

11. $\dfrac{1}{\left(1-4x^2\right)^{\frac{3}{2}}}$

12. $\dfrac{3(2-3x)}{2(1+3x)^{\frac{3}{2}}}$

13. $\dfrac{1}{2(2+x)\sqrt{x^2+3x+2}}$

14. $\dfrac{1}{(1-x)\sqrt{1-x^2}}$

15. $\dfrac{-\left(2x^2-10x+11\right)}{[(x-3)(x-4)]^{\frac{3}{2}}\cdot[(x-1)(x-2)]^{\frac{1}{2}}}$

Form 6: Problems based on exponential functions of x's having the form $y = e^{f(x)}$, where $f(x)$ stands for any one of the elementary functions of x's (like $\sin x$, $\cos x$, $\tan x$, $\cot x$, $\sec x$, $\operatorname{cosec} x$, $\sin^{-1} x$, $\cos^{-1} x$, $\tan^{-1} x$, $\cot^{-1} x$, $\sec^{-1} x$, $\operatorname{cosec}^{-1} x$, $x^n \log x$, e^x or their combination) or composite of a differentiable function of a differentiable function … of a differentiable function of x.

Exercise 9.7

Find $\dfrac{dy}{dx}$ of each of the following functions.

1. $y = e^{-\frac{x}{2}}$

2. $y = e^{x^3}$

3. $y = e^{\sin x}$

4. $y = e^{-3x^2}$

5. $y = e^{\sqrt{1+x^2}}$

6. $y = e^{\left(3x^2-6x+2\right)}$

7. $y = e^{\tan x}$

8. $y = e^{\cos^{-1} x}$

9. $y = e^{\cot^{-1} x}$

10. $y = e^{\sin^{-1} x}$

11. $y = e^{\sec^{-1} x}$

12. $y = e^{\left(\sin^{-1} x\right)^2}$

13. $y = e^{\sqrt{\cot x}}$

14. $y = e^{\operatorname{cosec}^2 \sqrt{x}}$

15. $y = e^{\sqrt{x}}$

16. $y = e^{-\sqrt{x}}$

17. $y = e^{\sqrt{x^2}}$

18. $y = x \cdot e^{2x}$

Answers (Under proper restrictions on x)

1. $\dfrac{-e^{-\frac{x}{2}}}{2}$

2. $3x^2 \cdot e^{x^3}$

3. $\cos x \cdot e^{\sin x}$

4. $-6x \cdot e^{-3x^2}$

5. $\dfrac{x}{\sqrt{1+x^2}} \cdot e^{\sqrt{1+x^2}}$

6. $6(x-1)\,e^{(3x^2-6x+2)}$

7. $\sec^2 x \cdot e^{\tan x}$

8. $\dfrac{-e^{\cos^{-1} x}}{\sqrt{1-x^2}}$

9. $\dfrac{-e^{\cot^{-1} x}}{1+x^2}$

10. $\dfrac{e^{\sin^{-1} x}}{\sqrt{1-x^2}}$

11. $\dfrac{e^{\sec^{-1} x}}{|x|\sqrt{x^2-1}}$

12. $\dfrac{2\sin^{-1} x}{\sqrt{1-x^2}}\left(e^{\sin^{-1} x}\right)^2$

13. $\dfrac{-e^{\sqrt{\cot x}} \cdot \operatorname{cosec}^2 x}{2\sqrt{\cot x}}$

14. $\dfrac{-\operatorname{cosec}^2\sqrt{x} \cdot \cot\sqrt{x} \cdot e^{\operatorname{cosec}^2\sqrt{x}}}{\sqrt{x}}$

15. $\dfrac{e^{\sqrt{x}}}{2\sqrt{x}}$

16. $\dfrac{-e^{\sqrt{x}}}{2\sqrt{x}}$

17. $\dfrac{2x\,e^{\sqrt{x^2}}}{2\sqrt{x^2}}$ or $\dfrac{x\,e^{|x|}}{|x|}$, $x \neq 0$

18. $e^{2x}(2x+1)$

Form 7: Problems based on logarithmic functions of x's having the form:

$y = \log f(x)$, where $f(x)$ stands for any one of the elementary functions of x's (like $\sin x$, $\cos x$, $\tan x$, $\cot x$, $\sec x$, $\operatorname{cosec} x$, $\sin^{-1} x$, $\cos^{-1} x$, $\tan^{-1} x$, $\cot^{-1} x$, $\sec^{-1} x$, $\operatorname{cosec}^{-1} x$, x^n, $\log x$, e^x or their combination) or composition of a differentiable function of a differentiable function of… of a differentiable function of x.

Exercise 9.7.1

Find $\dfrac{dy}{dx}$ if

1. $y = \log\left(x + \sqrt{1+x^2}\right)$
2. $y = \log(x+3)^2$
3. $y = \log \sin^2 x$
4. $y = \log \sin 3x$
5. $y = \log \tan\left(\dfrac{x}{2}\right)$
6. $y = \log 4x$
7. $y = \log x^2$
8. $y = \log(5x-14)$
9. $y = \log(x^n + a)$
10. $y = \log(e^x + 1)$
11. $y = \log(e^x + e^{-x})$
12. $y = \log(\cos x + 3)$
13. $y = \log \sin x$
14. $y = \log \tan x$
15. $y = \log \tan^{-1} x$
16. $y = \log \log \tan^{-1} x$
17. $y = \log \tan\left(\dfrac{\pi}{4} + \dfrac{x}{2}\right)$
18. $y = \log\left(\dfrac{\sqrt{x+4}-9}{\sqrt{x+4}+9}\right)$
19. $y = \log\sqrt{\dfrac{1+x+x^2}{1-x+x^2}}$

20. $y = \log \sqrt{\dfrac{a \cos x - b \sin x}{a \cos x + b \sin x}}$

21. $y = \log \sqrt{\dfrac{1-4x}{1+4x}}$

22. $y = \log \left(x - \sqrt{1+x^2} \right)$

23. $y = \log \left(x - \sqrt{x^2 - 1} \right)$

24. $y = \log \sqrt{\dfrac{1-\cos x}{1+\cos x}}$

25. $y = \log \left(\dfrac{x + \sqrt{x^2-1}}{x - \sqrt{x^2-1}} \right)$

26. $y = \log \sqrt{\dfrac{1+ax}{1-ax}}$

27. $y = \log \left(\dfrac{\sqrt{1+x} - \sqrt{1-x}}{\sqrt{1+x} + \sqrt{1-x}} \right)$

28. $y = \log \sqrt{x} \cdot \log x^2$

Answers (under proper restrictions on x)

1. $\dfrac{1}{\sqrt{1+x^2}}$

2. $\dfrac{2}{x+3}$

3. $2 \cot x$

4. $3 \cot 3x$

5. $\dfrac{1}{2} \sec \dfrac{x}{2}, \operatorname{cosec} \dfrac{x}{2} = \operatorname{cosec} x$

6. $\dfrac{1}{x}$

7. $\dfrac{2}{x}$

8. $\dfrac{5}{5x - 14}$

9. $\dfrac{n x^{n-1}}{x^n + a}$

10. $\dfrac{e^x}{e^x + 1}$

11. $\dfrac{e^x - e^{-x}}{e^x + e^{-x}}$

12. $\dfrac{-\sin x}{\cos x + 3}$

13. $\cot x$

14. $\dfrac{1}{\sin x \cos x}$

15. $\dfrac{1}{\left(1+x^2\right)\tan^{-1} x}$

16. $\dfrac{1}{\left(1+x^2\right)\tan^{-1} x \log \tan^{-1} x}$

17. $\sec x$

18. $\dfrac{9}{(x-77)\sqrt{x+4}}$

19. $\dfrac{1-x^2}{x^4 + x^2 + 1}$

20. $\dfrac{-ab}{\left(a^2 \cos^2 x - b^2 \sin^2 x\right)}$

21. $\dfrac{4}{16x^2 - 1}$

22. $\dfrac{-1}{\sqrt{1-x^2}}$

23. $\dfrac{-1}{\sqrt{x^2-1}}$

24. cosec x

25. $\dfrac{2}{\sqrt{x^2-1}}$

26. $\dfrac{a}{1-a^2x^2}$

27. $\dfrac{1}{x\sqrt{1-x^2}}$

28. $\dfrac{2\log x}{x}$

Exercise 9.7.2

Find $\dfrac{dy}{dx}$ if

1. $y = \log(3x-2)$
2. $y = \log \sec x$
3. $y = \log \operatorname{cosec} 2x$
4. $y = \log\sqrt{x^2+1}$
5. $y = \log\log x$
6. $y = \log(\sec x + \tan x)$
7. $y = \log \sin x$
8. $y = \log(\operatorname{cosec} x - \cot x)$
9. $y = \log \sin^{-1} x$
10. $y = \log \cos x^2$
11. $y = \log \cos^{-1} x^4$
12. $y = \log \sin 2x$
13. $y = \log\log\log x^3$
14. $y = \log \sin e^{x^2}$
15. $y = \log \sin(x^2+1)$
16. $y = \log\left(x+\sqrt{x^2+1}\right)$
17. $y = \log\left(x+\sqrt{x^2-1}\right)$
18. $y = \log\left(\dfrac{1+\sin x}{1-\sin x}\right)$
19. $y = \log\left(\dfrac{a+b\tan x}{a-b\tan x}\right)$
20. $y = \log\sqrt{\dfrac{1+x}{1-x}}$
21. $y = \log\left[\dfrac{(4x+1)^{\frac{1}{4}}\cdot(3x+2)^{\frac{1}{3}}}{(2x+3)^{\frac{1}{2}}\cdot(6x-4)^{\frac{1}{6}}}\right]$
22. $y = \log\left[\dfrac{(x-1)^2\cdot(2x-3)^3}{(2-x)^5}\right]$
23. $y = \log\left(\dfrac{e^x-1}{e^x+1}\right)$
24. $y = \log\left(\dfrac{x^2\cdot\sqrt{x^2+1}}{\sqrt[3]{x^3+2}}\right)$
25. $y = \log\left(\dfrac{\sqrt{1+x^2}-x}{\sqrt{1+x^2}+x}\right)$

Answers (under proper restrictions on x)

1. $\dfrac{3}{3x-2}$

2. $\tan x$

3. $-2\cot 2x$

4. $\dfrac{x}{x^2+1}$

5. $\dfrac{1}{x\log x}$

6. $\sec x$

7. $\cot x$

8. cosec x

9. $\dfrac{1}{\sin^{-1} x \sqrt{1 - x^2}}$

10. $-2x \tan x^2$

11. $\dfrac{-4x^3}{\cos^{-1} x^4 \sqrt{1 - x^8}}$

12. $2 \cot 2x$

13. $\dfrac{3}{x \log x^3 \log \log x^3}$

14. $\dfrac{2x\, e^{x^2} \cos e^{x^2}}{\sin e^{x^2}}$

15. $\dfrac{2x \cos(x^2 + 1)}{\sin(x^2 + 1)}$

16. $\dfrac{1}{\sqrt{1 + x^2}}$

17. $\dfrac{1}{\sqrt{x^2 - 1}}$

18. $2 \sec x$

19. $\dfrac{2ab}{a^2 \cos^2 x - b^2 \sin^2 x}$

20. $\dfrac{1}{1 - x^2}$

21. $\dfrac{1}{4x + 1} + \dfrac{1}{3x + 2} - \dfrac{1}{2x + 3} - \dfrac{1}{6x - 4}$

22. $\dfrac{2}{x - 1} + \dfrac{6}{2x - 3} + \dfrac{5}{2 - x}$

23. $\dfrac{2e^x}{e^{2x} - 1}$

24. $\dfrac{2}{x} + \dfrac{x}{x^2 + 1} - \dfrac{x^2}{x^3 + 2}$

25. $\dfrac{-2}{\sqrt{1 + x^2}}$

Exercise 9.7.3

Find $\dfrac{dy}{dx}$ if

1. $y = \log |\cos (ax + b)|$
2. $y = \log |x^2 - 6|$
3. $y = \log |\tan x|$
4. $y = \log |x|$

Answers

1. $-a \tan(ax + b),\ ax + b \neq n\pi + \dfrac{\pi}{2}$

2. $\dfrac{2x}{x^2 - 6},\ x \neq \pm\sqrt{6}$

3. $\dfrac{\sec^2 x}{\tan x},\ x \neq n\dfrac{\pi}{2}$

4. $\dfrac{1}{x},\ x \neq 0$

Exercise 9.7.4

Differentiate the following w.r.t. x

1. $y = \log \cos x,\ \left(0 < x < \dfrac{\pi}{2}\right)$
2. $y = \tan \log x,\ \left(0 < x < e^{\frac{\pi}{2}}\right)$
3. $y = \sin \log x,\ (x > 0)$
4. $y = \dfrac{\log x}{1 + x \log x},\ (x > 0)$
5. $y = \dfrac{x \sin x}{1 + \log x},\ (x > 0)$

6. $y = [\log(\cos x)]^2, \left(-\dfrac{\pi}{2} < x < \dfrac{\pi}{2}\right)$

7. $y = \log\left(x + \sqrt{x^2 + a^2}\right)$

Answers

1. $-\tan x$

2. $\dfrac{1}{x}\sec^2(\log x)$

3. $\dfrac{1}{x}\cos(\log x)$

4. $\dfrac{\frac{1}{x} - (\log x)^2}{(1 + x\log x)^2}$

5. $\dfrac{x\cos x(1 + \log x) + \sin x \log x}{(1 + \log x)^2}$

6. $-2\tan x \log(\cos x)$

7. $\dfrac{1}{\sqrt{x^2 + a^2}}$

10

Differentiation of Inverse Trigonometric Functions

Differentiation of Inverse Circular Functions

Before explaining the techniques of finding the differential coefficient of inverse circular functions, we recall the definition of an inverse of a function which tells us that the mapping must be one-one and onto in order that an inverse of a function should exist. Hence, to make the trigonometric functions one-one and onto, we restrict the domain of each trigonometric function by the principal values of the angle (i.e., the smallest positive value of the angle or the smallest numerical value of the angle) because each trigonometric function is a many valued function and a many valued function has no inverse.

Definitions:

1. We define $\sin^{-1} x$ as an angle 'θ' measured from $-\frac{\pi}{2}$ to $\frac{\pi}{2}$ $\left(\text{i.e., } -\frac{\pi}{2} \le \theta \le \frac{\pi}{2}\right)$ whose sin is x. The angle 'θ' satisfying the inequality $-\frac{\pi}{2} \le \theta \le \frac{\pi}{2}$ is called the principal value of $\sin^{-1} x$. Hence, θ = $\sin^{-1} x$ is an angle representing an inverse circular function whose domain is $-1 \le x \le 1$ and whose range is $-\frac{\pi}{2} \le \theta \le \frac{\pi}{2}$. In the notational form, the inverse of sin function is defined as $\theta = f^{-1}(x)$ $= \sin^{-1} x \Leftrightarrow x = \sin\theta$ and $\theta \in \left[\frac{-\pi}{2}, \frac{\pi}{2}\right], x \in [-1, 1]$

Note:

(i) $\sin^{-1}(\sin\theta) = \theta$ for $\theta \in \left[\frac{-\pi}{2}, \frac{\pi}{2}\right]$

(ii) $\sin\left(\sin^{-1} x\right) = x$ for $x \in [-1, 1]$

2. We define $\cos^{-1} x$ as an angle 'θ' measured from 0 to π $(\text{i.e. } 0 \le \theta \le \pi)$ whose cosine is x. The angle 'θ' satisfying the inequality $0 \le \theta \le \pi$ is called the principal value of $\cos^{-1} x$. Hence, $\theta = \cos^{-1} x$ is an angle representing an inverse circular function whose domain is [– 1, 1] and whose range is $[0, \pi]$. In the notational form, the inverse of a cosine function is defined as $\theta = f^{-1}(x) = \cos^{-1} x \Leftrightarrow x = \cos\theta$ and $\theta \in [0, \pi]$, $x \in [-1, 1]$.

Note:

(i) $\cos^{-1}(\cos\theta) = \theta$ for $\theta \in [0, \pi]$

(ii) $\cos\left(\cos^{-1} x\right) = x$ for $x \quad [-1, 1]$

3. We define $\tan^{-1} x$ as an angle 'θ' lying between $-\frac{\pi}{2}$ and $\frac{\pi}{2}$ $\left(\text{i.e; } \frac{-\pi}{2} < \theta < \frac{\pi}{2}\right)$ whose tan gent is x. The angle θ satisfying the inequality $-\frac{\pi}{2} < \theta < \frac{\pi}{2}$ is called the principal value of $\tan^{-1} x$. Hence θ

$= \tan^{-1} x$ is an angle representing an inverse circular function whose domain is entire number line (i.e. $-\infty < x < \infty$) and whose range is $\left(\frac{-\pi}{2}, \frac{\pi}{2}\right)$ $\left(\text{i.e; } -\frac{\pi}{2} < \theta < \frac{\pi}{2}\right)$.

In the notational form, the inverse of trigonometrical tangent is defined as $\theta = f^{-1}(x) = \tan^{-1} x \Leftrightarrow x = \tan\theta$ and $\theta \in \left(-\frac{\pi}{2}, \frac{\pi}{2}\right), x \in R$.

Note:

(i) $\tan^{-1}(\tan\theta) = \theta$ for $\theta \in \left(-\frac{\pi}{2}, \frac{\pi}{2}\right)$

(ii) $\tan\left(\tan^{-1} x\right) = x$ for $x \in R$

4. Some writers define $\cot^{-1} x$ as an angle 'θ' lying between $-\frac{\pi}{2}$ and $\frac{\pi}{2}$ excluding $\theta = 0$ (i.e., $-\frac{\pi}{2} < \theta < \frac{\pi}{2}$, $\theta \neq 0$) whose cotangent is x. The angle 'θ' satisfying the inequality $-\frac{\pi}{2} < \theta \frac{\pi}{2}$, $\theta \neq 0$ is called the principal value of $\cot^{-1} x$. Hence, $\theta = \cot^{-1} x$ is an angle representing an inverse circular function whose domain is entire number line (i.e; $-\infty < x < \infty$) and whose range is $\left(-\frac{\pi}{2}, \frac{\pi}{2}\right) - \{0\} = \left(-\frac{\pi}{2}, 0\right) \cup \left(0, \frac{\pi}{2}\right)$.

In the notational form, the inverse of cotangent function is defined as $\theta = f^{-1}(x) = \cot^{-1}(x) \Leftrightarrow x = \cot\theta$ and $\theta \in \left(-\frac{\pi}{2}, \frac{\pi}{2}\right) - \{0\}, x \in R$.

An important remark: It is customary to take the open interval (as a domain) $(0, \pi)$ in (or, on, or over) which the function $x = \cot\theta$ has an inverse $\theta = \cot^{-1} x$ because in this interval $0 < \theta < \pi$, $\theta = \cot^{-1} x$ is defined at all points. For this reason, it is usual to take the interval $0 < \theta < \pi$ as the range for $\theta = \cot^{-1} x$ for practical purpose instead of considering the range $\left(-\frac{\pi}{2}, 0\right) \cup \left(0, \frac{\pi}{2}\right)$ in which $\theta = \cot^{-1} x$ is undefined at $\theta = 0$.

Hence, an alternative definition of $\cot^{-1} x$ is also available which is expressed in the notational form as follows $\theta = f^{-1}(x) = \cot^{-1} x \Leftrightarrow x = \cot\theta$ and $\theta \in (0, \pi)$, $x \in R$. We shall follow this definition.

Note:

(i) $\cot^{-1}(\cot x) = \theta$ for $\theta \in (0, \pi)$

(ii) $\cot\left(\cot^{-1} x\right) = x$ for $x \in R$

5. We define $\sec^{-1} x$ as angle 'θ' measured from 0 to π excluding $\theta = \frac{\pi}{2}$ $\left(\text{i.e; } [0, \pi] - \left\{\frac{\pi}{2}\right\}\right)$ whose secant is x. The angle θ belonging to the interval $[0, \pi] - \left\{\frac{\pi}{2}\right\}$ is called the principal value of $\sec^{-1} x$.

Hence $\theta = \sec^{-1} x$ is an angle representing an inverse circular function whose domain is $(-\infty, -1) \cup (1, \infty)$ and the range is $[0, \pi] - \left\{\frac{\pi}{2}\right\}$ $\left(\text{i.e; } \left[0, \frac{\pi}{2}\right) \cup \left(\frac{\pi}{2}, \pi\right]\right)$.

In the notational form, the inverse of trigonometrical secant function is defined as $\theta = f^{-1}(x) = \sec^{-1} x \Leftrightarrow x = \sec\theta$ and $\theta \in [0, \pi] - \left\{\frac{\pi}{2}\right\}, x \in (-\infty, -1) \cup (1, \infty)$.

6. (i) $f\left[f^{-1}(x)\right] = x$ for all x in the domain of f^{-1}.

(ii) $f^{-1}[f(x)] = x$ for all x in the domain of f.

Therefore these relations (i) and (ii) hold for the restricted trigonometric functions and their inverse.

Note:

(i) An angle denoted by $\sec^{-1} x$ given by the relation

$\theta = \sec^{-1} x = \cos^{-1}\left(\frac{1}{x}\right)$, $|x| \geq 1$ is also accepted as a definition of inverse trigonometrical secant function of an angle θ.

(ii) $\sec^{-1} x$ is not defined for $x \in (-1, 1)$

(iii) $\sec^{-1}(\sec\theta) = \theta$ for $\theta \in \left[0, \frac{\pi}{2}\right) \cup \left(\frac{\pi}{2}, \pi\right]$

(iv) $\sec\left(\sec^{-1} x\right) = x$ for $|x| \geq 1$

6. We define $\text{cosec}^{-1} x$ as an angle 'θ' measured from $-\frac{\pi}{2}$ to $\frac{\pi}{2}$ excluding $\theta = 0$ whose cosecant is x. The angle θ belonging to the interval $\left(-\frac{\pi}{2}, \frac{\pi}{2}\right) - \{0\}$, is called the principal value of $\text{cosec}^{-1} x$. Hence, $\theta = \text{cosec}^{-1} x$ is an angle representing an inverse circular function whose domain is $(-\infty, -1) \cup (1, \infty)$ and whose range is $\left[-\frac{\pi}{2}, \frac{\pi}{2}\right] - \{0\}$ $\left(\text{i.e; } \left[-\frac{\pi}{2}, 0\right) \cup \left(0, \frac{\pi}{2}\right]\right)$.

In the notational form, the inverse of the trigonometrical secant function is defined as $\theta = f^{-1}(x) = \text{cosec}^{-1} x \Leftrightarrow x = \text{cosec}\, x$ and $\theta \in \left[-\frac{\pi}{2}, 0\right) \cup \left(0, \frac{\pi}{2}\right]$, $x \in (-\infty, -1) \cup (1, \infty)$

Note:

(i) An angle denoted by $\text{cosec}^{-1} x$ given by the relation $\theta = \text{cosec}^{-1} x = \sin\left(\frac{1}{x}\right)$, $|x| \geq 1$ is also accepted as a definition of inverse of cosecant function of θ

(ii) $\text{cosec}^{-1}(\text{cosec}\,\theta) = \theta$ for $\theta \in \left[-\frac{\pi}{2}, 0\right) \cup \left(0, \frac{\pi}{2}\right]$

(iii) $\text{cosec}\left(\text{cosec}^{-1} x\right) = x$ for $|x| \geq 1$

Remember:

1. The restricted domains and the ranges over which each respective inverse circular function (defined earlier) can be summarised in the chart.

Functions	*Domain*	*Range(θ may be a real number or an angle)*
$\sin^{-1} x$	$\{x \mid -1 \leq x \leq 1\}$	$\left\{\theta \mid -\frac{\pi}{2} \leq \theta \leq \frac{\pi}{2}\right\}$
$\cos^{-1} x$	$\{x \mid -1 \leq x \leq 1\}$	$\{\theta \mid 0 \leq \theta \leq \pi\}$
$\tan^{-1} x$	$\{x \mid -\infty < x < \infty\}$	$\left\{\theta \mid -\frac{\pi}{2} < \theta < \frac{\pi}{2}\right\}$
$\cot^{-1} x$	$\{x \mid -\infty < x < \infty\}$	$\left\{\theta \mid -\frac{\pi}{2} < \theta < \frac{\pi}{2}, \theta \neq 0\right\}$ (some writers) or $\{\theta \mid 0 < \theta < \pi\}$ (we adopt)
$\sec^{-1} x$	$\{x \mid -\infty < x \leq -1;$ $1 \leq x < \infty\}$	$\left\{\theta \mid 0 \leq \theta \leq \pi, \theta \neq \frac{\pi}{2}\right\}$
$\text{cosec}^{-1} x$	$\{x \mid -\infty < x \leq -1;$ $1 \leq x \leq \infty\}$	$\left\{\theta \mid -\frac{\pi}{2} \leq \theta \leq \frac{\pi}{2}, \theta \neq 0\right\}$

2. The principal value of $\sin^{-1} x$, $\text{cosec}^{-1} x$, $\tan^{-1} x$ and $\cot^{-1} x$ is the smallest numerical value of the angle.

3. The principal value of $\cos^{-1} x$ and $\sec^{-1} x$ is the smallest positive value of the angle.

4. A circular function of an angle θ = a number.

5. An inverse circular function of a number = an angle.

6. General value of $\sin^{-1} x = n\pi + (-1)^n \theta$ whose θ stands for the principal value of $\sin^{-1} x$.

7. General value of $\cos^{-1} x = 2n\pi \pm \theta$ whose θ stands for the principal value of $\cos^{-1} x$.

8. General value of $\tan^{-1} x = n\pi + \theta$ where θ stands for the principal value of $\tan^{-1} x$.

N.B.:

1. The principal value of each circular function is obtained by putting $n = o$ in its general value.

Important Notes:

1. Unless otherwise stated by the value of each inverse circular function, we mean its principal value.

2. If $y = f(x)$, then the derivative of x with respect to y is given by the formula:

$$\frac{dx}{dy} = \frac{1}{\left(\frac{dy}{dx}\right)},$$ where 'f' is assumed to be one-one and onto as well as $\frac{dy}{dx}$ is assumed to be finite and non-zero $\left(\text{i.e; } \frac{dy}{dx} = f'(x) \neq 0\right)$ or, alternatively,

let $y = f(x)$ be a single – valued monotonic and continuous, and $x = f'(y) = g(y)$ be the inverse function, then if $f'(x)$ be finite and non – zero,

$$g'(y) = \frac{1}{f'(x)}$$ [E.G. Philips $(p-q4)$]

Proof: $y = f(x)$ ...(1)

$x = f^{-1}(y)$ ...(2)

1. $\Rightarrow y + \Delta y = f(x + \Delta x)$ ...(3)

2. $\Rightarrow x + \Delta x = f^{-1}(y + \Delta y)$ ...(4)

Also, as $\Delta x \to 0$, $\Delta y \to 0$ and as $\Delta y \to 0$, $\Delta x \to 0$ ($x = f^{-1}(y)$ is continuous)

Now, using the definition of derivative of a function, we have,

$$\left[f^{-1}(y)\right]' = g'(y)$$

$$= \lim_{\Delta y \to 0} \frac{f^{-1}(y + \Delta y) - f^{-1}(y)}{\Delta y}$$

$$= \lim_{\Delta y \to 0} \frac{f^{-1}(y + \Delta y) - f^{-1}(y)}{(y + \Delta y) - y}$$

[adding and subtracting y in denominator]

$$= \lim_{\Delta y \to 0} \frac{(x + \Delta x) - x}{f(x + \Delta x) - f(x)}$$

[$y + \Delta y = f(x + \Delta x)$ from (3) and $x + \Delta x = f^{-1}(y + \Delta y)$ from (4)]

$$= \lim_{\Delta x \to 0} \frac{\Delta x}{f(x + \Delta x) - f(x)}$$ [since $\Delta x \to 0$ when $\Delta y \to 0$]

$$= \frac{1}{\lim_{\Delta x \to 0}\left(\frac{f(x+\Delta x) - f(x)}{\Delta x}\right)}$$ [Taking reciprocal]

$$= \frac{1}{f'(x)}$$

N.B.: If $y = f(x)$ and $x = f^{-1}(y) = g(y)$

Then $f'(x) = \frac{dy}{dx}$ and $g'(y) = \frac{dx}{dy}$

$$\therefore g'(y) = \frac{1}{f'(x)}$$

$$\Rightarrow \frac{dx}{dy} = \frac{1}{\left(\frac{dy}{dx}\right)}$$ which provides us a formula to be very useful in the differentiation of inverse functions.

Differential coefficients of inverse circular functions

If $y = f^{-1}(x) \Leftrightarrow x = f(y)$, then y is called an inverse function of x and x is called the direct function of y.

Question: What is the rule for the derivative of the inverse function ?

Answer: The rule for the derivative of the inverse function is the reciprocal of the derivative of the original function (i.e. direct function) expressed as

$$\frac{d}{dx}(y)=\frac{1}{\frac{d}{dy}(x)}\Leftrightarrow\frac{d}{dx}(y)\cdot\frac{d}{dy}(x)=1$$

$$\Leftrightarrow\frac{d}{dx}(y)=1\div\frac{d}{dy}(x)$$

Recapitulation of working rule to find d.c. of inverse functions by Δ - method

The rule to find d.c of inverse functions by Δ - method consists of following steps.

Step 1. Change inverse functions into direct functions which means $y=f^{-1}(x)$ should be expressed as $x=f(y)$, where f^{-1} stands for $\sin^{-1}$, $\cos^{-1}$, $\tan^{-1}$, $\cot^{-1}$, $\sec^{-1}$, $\operatorname{cosec}^{-1}$ or any other inverse function and f stands for sin, cos, tan, cot, sec, cosec or any other direct function.

Step 2. Add Δx to x and Δy to y whereever these are present in the direct function which means forming the equation $x+\Delta x=f(y+\Delta y)$.

Step 3. Find Δx by subtracting the first value $x=f(y)$ from the second value $x+\Delta x=f(y+\Delta y)$ which means forming the equation $\Delta x=f(y+\Delta y)-f(y)$.

Step 4. Divide $\Delta x=f(y+\Delta y)-f(y)$ by Δy which means forming the equation

$$\frac{\Delta x}{\Delta y}=\frac{f(y+\Delta y)-f(y)}{\Delta y}.$$

Step 5. Take the limit as $\Delta y\to 0$ on both sides of the equation $\frac{\Delta x}{\Delta y}=\frac{f(y+\Delta y)-f(y)}{\Delta y}$ which means forming the equation

$$\frac{d}{dx}(y)=\lim_{\Delta y\to 0}\frac{f(y+\Delta y)-f(y)}{\Delta y}.$$

Step 6. Express the direct function (like sin, cos, tan, cot, sec, cosec or any other direct function) of y in terms of x using the relation $x=f(y)$.

Step 7. Lastly find $\frac{d}{dx}(y)$ using the rule $\frac{d}{dx}(y)\cdot\frac{d}{dx}(x)=1$ which means taking the reciprocal of $\frac{d}{dy}(x)$ to find $\frac{d}{dx}(y)$.

Notes: **1.** In the process of finding the d.c of inverse trigonometrical functions, we generally get $\frac{2\sin\left(\frac{\Delta y}{2}\right)}{\Delta y}$ except for $\tan^{-1}x$ and $\cot^{-1}x$. In order to find its limit as $\Delta y\to 0$, we can use any one of the following methods.

(i) $\frac{2\sin\left(\frac{\Delta y}{2}\right)}{\Delta y}=\frac{2\cdot\left(\frac{\Delta y}{2}\right)}{\Delta y}=\frac{1}{2}\cdot\frac{\Delta y\cdot 2}{\Delta y}=1$

(when θ is small, sin θ = θ)

(ii) $\lim_{\theta\to 0}\left(\frac{\sin r\,\theta}{\theta}\right)=\lim_{\theta\to 0}\left(\frac{r\sin r\,\theta}{r\,\theta}\right)$

$=r\lim_{\theta\to 0}\left(\frac{\sin r\,\theta}{r\,\theta}\right)=r\cdot 1=r$, r being a constant.

(iii) $\lim_{\Delta y\to 0}\frac{2\sin\left(\frac{\Delta y}{2}\right)}{\Delta y}=\lim_{\Delta y\to 0}\frac{2\sin\left(\frac{\Delta y}{2}\right)}{2\cdot\left(\frac{\Delta y}{2}\right)}$

$=\lim_{\Delta y\to 0}\frac{\sin\left(\frac{\Delta y}{2}\right)}{\left(\frac{\Delta y}{2}\right)}=1$

2. $\Delta x\to 0\Leftrightarrow\Delta y\to 0$

3. Finding d.c. by Δ - method means finding d.c. of the given function by making no use of rules of differentiation and d.c. of standard functions but one can use only fundamental theorems on limits and standard results on limits.

Practical methods of finding d.c. of inverse circular functions

In practice, of the question says simply to find d.c. of an inverse circular function, one can use the following methods.

Method (A): It consists of following steps.

(i) Change the given inverse circular function in to direct circular function (called commonly circular function), i.e. of $y = f^{-1}(x)$ is given, it should be changed into $x = f(y)$.

(ii) While differentiating, we must choose the variable y as an independent variable (instead of usual variable x as independent variable) w.r.t. which the direct circular function should be differentiated, i.e.,

$$x = f(y) \Rightarrow \frac{d}{dy}(x) = \frac{d}{dy}(f(y)) = f'(y)$$

(iii) Now required derivative is $\frac{d}{dx}(y)$ which is obtained by simply taking the reciprocal of $\frac{d}{dy}(x)$,

$$\text{i.e; } \frac{d}{dx}(y) = \frac{1}{\frac{d}{dy}(x)} \Leftrightarrow \frac{d}{dy}(x)\,\frac{1}{\frac{d}{dx}(y)}$$

Method B: It may be explained in the following way.

Let $y = f^{-1}(x) \Rightarrow x = f(y)$

$$\Rightarrow x = f\,f^{-1}(y) \quad [\because y = f^{-1}(x)] \quad \ldots(1)$$

or, $\quad x = f(y) \Rightarrow y = f^{-1}(x) \Rightarrow y = f^{-1}\,f(y)$ ...(2)

Hence, whenever, we have the form $y = f^{-1}\,f(y)$, it can be differentiated by using the chain rule, i.e,

$$\frac{d}{dy}(y) = \frac{d}{dy}\left(f^{-1}\,f(y)\right)$$

$$= \frac{d}{d\,f(y)}\left(f^{-1}\,f(y)\right) \cdot \frac{d}{dy}(f(y))$$

(using chain rule)

$$\Rightarrow 1 = \frac{d}{dx}\left(f^{-1}(x)\right) \cdot \frac{d}{dy}(x)$$

$$\left(\because \frac{d}{dy}(y) = 1,\ f(y) = x\right)$$

$$\Rightarrow \frac{d}{dx}\left(f^{-1}(x)\right) = \frac{1}{\left(\frac{d}{dy}(x)\right)}$$

$$\Rightarrow \frac{d}{dx}(y) = \frac{1}{\left(\frac{d}{dy}(x)\right)} \quad \left(\because f^{-1}(x) = y\right)$$

$$= \frac{1}{f'(y)} = \frac{1}{f'\left(f^{-1}(x)\right)}.$$

Standard formulas or d.c of inverse circular functions

1. If $y = \sin^{-1} x$, $y \in \left(-\frac{\pi}{2}, \frac{\pi}{2}\right)$, show that

$$\frac{dy}{dx} = \frac{1}{\sqrt{1-x^2}}$$

Proof : *First method*: (Derivation of d.c. Using the definition)

Step 1. Let $y = \sin^{-1} x$, $(|x| \le 1)$

$$\Rightarrow x = \sin y, \quad \left(-\frac{\pi}{2} \le y \le \frac{\pi}{2}\right) \quad \ldots(1)$$

Step 2. $x + \Delta x = \sin(y + \Delta y)$ ...(2)

Step 3. $\Delta x = \sin(y + \Delta y) - \sin y$ (subtracting (1) from (2))

$$= 2\cos\left(y + \frac{\Delta y}{2}\right) \cdot \sin\left(\frac{\Delta y}{2}\right)$$

$$\Rightarrow \frac{1}{\Delta x} = \frac{1}{2\cos\left(y + \frac{\Delta y}{2}\right) \cdot \sin\left(\frac{\Delta y}{2}\right)} \quad \ldots(3)$$

Step 4: $\frac{\Delta y}{\Delta x} = \frac{\Delta y}{2\cos\left(y + \frac{\Delta y}{2}\right) \cdot \sin\left(\frac{\Delta y}{2}\right)}$

(Multiplying both sides of (3) by Δy)

$$= \frac{1}{\cos\left(y + \frac{\Delta y}{2}\right) \cdot \sin\left(\frac{\Delta y}{2}\right) \Big/ \frac{\Delta y}{2}} \quad \ldots(4)$$

Step 5: $\lim_{\Delta x \to 0} \frac{\Delta y}{\Delta x} = \frac{1}{\cos y \cdot 1} = \frac{1}{\cos y}$ (Taking the limits on both sides of (4) as $\Delta x \to 0$)

Step 6: $\because |\cos y| = \sqrt{\cos^2 y}$ $\left(\because |f(x)| = \sqrt{f^2(x)}\right)$

$$= \sqrt{1 - \sin^2 y} = \sqrt{1 - x^2} \quad (\because \sin y = x)$$

Now $\cos y$ is positive for $y \in \left(-\frac{\pi}{2}, \frac{\pi}{2}\right)$

$$\Rightarrow |\cos y| = \cos y \text{ for } y \in \left(-\frac{\pi}{2}, \frac{\pi}{2}\right)$$

$$\left(\because |f(x)| = f(x) \text{ for } f(x) \ge 0\right)$$

$$\therefore \frac{dy}{dx} = \frac{1}{\sqrt{1-x^2}}$$

$$\left(\because \lim_{\Delta x \to 0} \frac{\Delta y}{\Delta x} = \frac{dy}{dx}\right), |x| < 1.$$

Second method using the formula:

$$\frac{dy}{dx} = \frac{1}{\left(\frac{dx}{dy}\right)}$$

1. $y = \sin^{-1} x$

$$\Rightarrow x = \sin y \text{ for } -\frac{\pi}{2} \le y \le \frac{\pi}{2}$$

(from the definition of $\sin^{-1} x$)

$$\Rightarrow \frac{dx}{dy} = \cos y \text{ (differentiating both sides w.r.t. } y)$$

$$\Rightarrow \frac{dy}{dx} = \frac{1}{\left(\frac{dx}{dy}\right)} = \frac{1}{\cos y} \quad ...(1)$$

Now, $|\cos y| = \sqrt{\cos^2 y}$

$$= \sqrt{1 - \sin^2 y} = \sqrt{1 - x^2} \quad ...(2)$$

and $y \in \left(-\frac{\pi}{2}, \frac{\pi}{2}\right) \Rightarrow \cos y > 0 \Rightarrow |\cos y| = \cos y$

$$(\because |f(x)| = f(x) \text{ for } f(x) > 0) \quad ...(3)$$

Equating (2) and (3), we have

$$|\cos y| = \cos y = \sqrt{1 - x^2} \quad ...(4)$$

Putting (4) in (1), we have

$$\frac{dy}{dx} = \frac{1}{\sqrt{1-x^2}}, \text{ for } |x| < 1$$

$$\Rightarrow \frac{d \sin^{-1} x}{dx} = \frac{1}{\sqrt{1-x^2}} \text{ for } |x| < 1$$

Cor 1.: On replacing x by $f(x)$ in the L.H.S. and R.H.S. of the above formula, we get

$$\frac{d \sin^{-1} f(x)}{dx} = \frac{1}{\sqrt{1 - f^2(x)}} \cdot f'(x)$$

Cor 2.: $\frac{d \sin^{-1} |f(x)|}{dx} = \frac{1}{\sqrt{1 - f^2(x)}} \times \frac{d|f(x)|}{dx}$

$$\left[\because |f(x)|^2 = f^2(x)\right]$$

2. If $y = \cos^{-1} x,\ y \in (0, \pi)$, show that

$$\frac{dy}{dx} = -\frac{1}{\sqrt{1-x^2}}.$$

Proof: *First method* (Derivation of D.C using the definition)

Step 1.: Let $y = \cos^{-1} x,\ (|x| \le 1)$

$$\Rightarrow x = \cos y,\ (0 \le y \le \pi) \quad ...(1)$$

Step 2.: $x + \Delta x = \cos(y + \Delta y)$ ...(2)

Step 3.: $\Delta x = \cos(y + \Delta y) - \cos y$ (subtracting (1) from (2))

$$= -2 \sin\left(y + \frac{\Delta y}{2}\right) \cdot \sin\left(\frac{\Delta y}{2}\right)$$

$$\Rightarrow \frac{1}{\Delta x} = -\frac{1}{2\sin\left(y + \frac{\Delta y}{2}\right) \cdot \sin\left(\frac{\Delta y}{2}\right)} \quad ...(3)$$

Step 4.: $\dfrac{\Delta y}{\Delta x} = -\dfrac{\Delta y}{2\sin\left(y+\frac{\Delta y}{2}\right) . \sin\left(\frac{\Delta y}{2}\right)}$

(multiplying both sides of (3) by Δy)

$$= -\frac{1}{\sin\left(y + \frac{\Delta y}{2}\right) . \sin\frac{\left(\frac{\Delta y}{2}\right)}{\left(\frac{\Delta y}{2}\right)}} \quad ... (4)$$

Step 5.: $\lim\limits_{\Delta x \to 0} \dfrac{\Delta y}{\Delta x} = -\dfrac{1}{\sin(y+0).1}$ (Taking the limit of both sides of (4) as $\Delta x \to 0$)

$$= \frac{1}{\sin y}$$

Step 6.: $|\sin y| = \sqrt{\sin^2 y} \left(\because \ |f(x)| = \sqrt{f^2(x)}\right)$

$$= \sqrt{1-\cos^2 y}$$

$$= \sqrt{1-x^2} \quad (\because \cos y = x) \quad ...(a)$$

Now $\sin y > 0$ in $(0, \pi)$

$\Rightarrow |\sin y| = \sin y$ for $y \in (0, \pi)$

$(\because |f(x)| = f(x)$ for $f(x) > 0)$

Equating (a) and (b), we have ...(b)

$$\frac{dy}{dx} = -\frac{1}{\sqrt{1-x^2}} \left(\because \lim_{\Delta x \to 0} \frac{\Delta y}{\Delta x} = \frac{dy}{dx}\right), |x| < 1$$

Remark:

1. The derivatives of $\sin^{-1} x$ and $\cos^{-1} x$ become undefined at $x = \pm 1$. For this reason we leave these points out of our consideration.

2. The derivatives of the inverse trigonometric functions are not transcendental but algebraic.

3. The domains of the derived functions of $\sin^{-1} x$ and $\cos^{-1} x$ are the open interval $(-1, 1)$.

Second method

Using the formula: $\dfrac{dy}{dx} = \dfrac{1}{\left(\frac{dy}{dx}\right)}$

2. $y = \cos^{-1} x$

$\Rightarrow x = \cos y$ for $0 \le y \le \pi$ (from the definition of $\cos^{-1} x$)

$\Rightarrow \dfrac{dx}{dy} = -\sin y$ (differentiating both sides w.r.t. y)

$$\Rightarrow \frac{dy}{dx} = \frac{1}{\left(\frac{dx}{dy}\right)} = \frac{1}{-\sin y} \quad ...(1)$$

$$\text{Now, } |\sin y| = \sqrt{\sin^2 y} = \sqrt{1-\cos^{2} y} = \sqrt{1-x^2} \quad ...(2)$$

and $y \in (0, \pi) \Rightarrow \sin y > 0$ for y being measured from 0 to $\pi \Rightarrow |\sin y| = \sin y$

$(\because |f(x)| = f(x)$ for $f(x) > 0)$ ...(3)

Equating (2) and (3), we have

$$|\sin y| = \sin y = \sqrt{1-x^2} \quad ... (4)$$

Putting (4) in (1), we have

$$\frac{dy}{dx} = \frac{1}{-\sqrt{1-x^2}}, \text{ for } |x| < 1$$

$$\Rightarrow \frac{d\cos^{-1} x}{dx} = \frac{1}{-\sqrt{1-x^2}}, \text{ for } |x| < 1$$

Note: **(i)** $\sin^{-1} x + \cos^{-1} x = \dfrac{\pi}{2}, \forall |x| \le 1$

$$\Rightarrow \frac{d}{dx}\left(\sin^{-1} x\right) + \frac{d}{dx}\left(\cos^{-1} x\right) = 0$$

$$\Rightarrow \frac{d}{dx}\left(\sin^{-1} x\right) = -\frac{d}{dx}\left(\cos^{-1} x\right)$$

$$= \frac{1}{\sqrt{1-x^2}}; |x| < 1.$$

(ii) $\frac{d}{dx}\sin^{-1} f(x)$

$$= -\frac{d\cos^{-1} f(x)}{dx} = \frac{-1}{\sqrt{1-f^2(x)}} \cdot \frac{d f(x)}{dx}$$

(iii) $\frac{d}{dx}\cos^{-1}|f(x)| = \frac{-1}{\sqrt{1-f^2(x)}} \cdot \frac{d|f(x)|}{dx}$

3. If $y = \tan^{-1} x$, $y \in \left(-\frac{\pi}{2}, \frac{\pi}{2}\right)$, show that $\frac{dy}{dx}$

$$= \frac{1}{1+x^2}.$$

Proof: *First method*: (Derivation of d.c using the definition)

Step 1.: $y = \tan^{-1} x \quad (\forall x \in R)$

$$\Rightarrow x = \tan y, \left(\frac{-\pi}{2} < y < \frac{\pi}{2}\right) \quad ...(1)$$

Step 2.: $x + \Delta x = \tan(y + \Delta y)$ ...(2)

Step 3.: $\Delta x = \tan(y+\Delta y) - \tan y$ (subtracting (1) from (2))

$$= \frac{\sin(y+\Delta y)}{\cos(y+\Delta y)} - \frac{\sin y}{\cos y}$$

$$= \frac{\sin(y+\Delta y)\cdot\cos y - \sin y \cdot \cos(y+\Delta y)}{\cos(y+\Delta y)\cdot \cos y}$$

$$= \frac{\sin(y+\Delta y - y)}{\cos y \cdot \cos(y+\Delta y)} = \frac{\sin \Delta y}{\cos y \cdot \cos(y+\Delta y)}$$

$$\Rightarrow \frac{1}{\Delta x} = \frac{\cos y \cdot (y+\Delta y)}{\sin \Delta y} \quad ...(3)$$

Step 4.: $\frac{\Delta y}{\Delta x} = \cos y \cos(y+\Delta y) \cdot \frac{\Delta y}{\sin \Delta y}$ (Multiplying both sides of (3) by Δy) ...(4)

Step 5.: $\lim_{\Delta x \to 0} \frac{\Delta y}{\Delta x}$

$$= \lim_{\Delta x \to 0} \left\{\cos y \cdot \cos(y+\Delta y) \cdot \frac{\Delta y}{\sin \Delta y}\right\}$$

(on taking the limit of both sides of (4)) as $\Delta x \to 0$

$$= \cos^2 y = \frac{1}{\sec^2 y} = \frac{1}{1+\tan^2 y}$$

$$= \frac{1}{1+x^2} \quad (\because \tan y = x) \quad ...(a)$$

$$\therefore \frac{dy}{dx} = \frac{1}{1+x^2}.$$

Second method (*using the formula*):

$$\frac{dy}{dx} = \frac{1}{\left(\frac{dx}{dy}\right)}$$

3. $y = \tan^{-1} x$

$$\Rightarrow x = \tan y \text{ for } -\frac{\pi}{2} < y < \frac{\pi}{2}$$

(from the definition of $\tan^{-1} x$).

$$\Rightarrow \frac{dy}{dx} = \sec^2 y$$

(differentiating both sides of (1) w.r.t. y)

$$\Rightarrow \frac{dy}{dx} = \frac{1}{\left(\frac{dx}{dy}\right)} = \frac{1}{\sec^2 y} = \frac{1}{1+\tan^2 y} = \frac{1}{1+x^2}$$

$(\because x = \tan y)$

Cor 1.: On replacing x by $f(x)$ in the L.H.S. and R.H.S of the above formula, we get

$$\frac{d\tan^{-1}[f(x)]}{dx} = \frac{1}{1+f^2(x)} \cdot f'(x)$$

Cor 2.: $\frac{d\tan^{-1}|f(x)|}{dx} = \frac{1}{1+f^2(x)} \cdot \frac{d|f(x)|}{dx}$

4. If $y = \cot^{-1} x$, $y \in (0, \pi)$, show that

$$\frac{dy}{dx} = -\frac{1}{1+x^2}.$$

Proof: *First method* (Derivation of d.c using definition)

Step 1.: Let $y = \cot^{-1} x$, $\forall x \in R$

$\Rightarrow x = \cot y$, $y \in (0, \pi)$ ...(1)

Step 2.: $x + \Delta x = \cot(y + \Delta y)$ ...(2)

Step 3.: $\Delta x = \cot(y+\Delta y) - \cot y$ (subtracting (1) from (2))

$$= \frac{\cos(y+\Delta y)}{\sin(y+\Delta y)} - \frac{\cos y}{\sin y}$$

$$= \frac{\sin y \cos(y+\Delta y) - \cos y \cdot \sin(y+\Delta y)}{\sin(y+\Delta y)\sin y}$$

$$= \frac{\sin(y-y-\Delta y)}{\sin y \cdot \sin(y+\Delta y)} = \frac{\sin \Delta y}{\sin y \cdot \sin(y+\Delta y)}$$

$$\Rightarrow \frac{1}{\Delta x} = -\frac{\sin y \cdot \sin(y+\Delta y)}{\sin \Delta y} \quad ...(3)$$

Step 4.: $\dfrac{\Delta y}{\Delta x} = -\sin y \cdot \sin(y+\Delta y) \cdot \dfrac{\Delta y}{\sin \Delta y}$

(Multiplying both sides of (3) by Δy) ...(4)

Step 5.: $\lim\limits_{\Delta x \to 0} \dfrac{\Delta y}{\Delta x}$

$$= \lim_{\Delta x \to 0} \left\{ -\sin y \cdot \sin(y+\Delta y) \cdot \frac{\Delta y}{\sin \Delta y} \right\}$$

(On taking the limit of both sides of (4) as $\Delta x \to 0$)

$$= -\sin y \cdot \sin y = -\sin^2 y$$

$$= -\frac{1}{\operatorname{cosec}^2 y} = -\frac{1}{1+\cot^2 y} = -\frac{1}{1+x^2} \quad ...(a)$$

($\because \cot y = x$)

$$\therefore \frac{dy}{dx} = -\frac{1}{1+x^2}$$

Note: The domains of the derived functions of $\tan^{-1} x$ and $\cot^{-1} x$ are the open interval $(-\infty, \infty)$ because $\dfrac{dy}{dx} = \pm \dfrac{1}{1+x^2}$ is defined for all real values of x.

Second method using the formula:

$$\frac{dy}{dx} = \frac{1}{\left(\dfrac{dx}{dy}\right)}$$

5. $y = \cot^{-1} x$

$\Rightarrow x = \cot y$ for $0 < y < \pi$ (from the definition of $\cot^{-1} x$) ...(1)

$\Rightarrow \dfrac{dx}{dy} = -\operatorname{cosec}^2 y$ (differentiating both sides of (1) w.r.t. y)

$$\Rightarrow \frac{dy}{dx} = \frac{1}{\left(\dfrac{dx}{dy}\right)}$$

$$= -\frac{1}{\operatorname{cosec}^2 y} = \frac{-1}{1+\cot^2 y} = \frac{-1}{1+x^2} \quad (\because x = \cot y)$$

Note: **(i)** $\tan^{-1} x + \cot^{-1} x = \dfrac{\pi}{2}$, $\forall x$

$$\Rightarrow \frac{d}{dx}\tan^{-1} x + \frac{d}{dx}\cot^{-1} x = 0$$

$$\Rightarrow \frac{d}{dx}\tan^{-1} x = -\frac{d}{dx}\cot^{-1} x = \frac{1}{1+x^2}$$

(ii) $\dfrac{d}{dx}\tan^{-1} f(x)$

$$= -\frac{d}{dx}\cot^{-1} f(x) = \frac{1}{1+f^2(x)} \cdot \frac{d f(x)}{dx}$$

(iii) $\dfrac{d\cot^{-1}|f(x)|}{dx} = \dfrac{-1}{1+f^2(x)} \cdot \dfrac{d|f(x)|}{dx}$

6. If $y = \sec^{-1} x$, $y \in (0, \pi) - \left\{\dfrac{\pi}{2}\right\}$, show that

$$\frac{dy}{dx} = \frac{1}{|x|\sqrt{1-x^2}}.$$

Proof: *First method* (derivation of the d.c. using the definition)

Step 1.: Let $y = \sec^{-1} x$, $(|x| > 1)$ ...(1)

$$\Rightarrow x = \sec y, \left(0 < y < \pi, y \neq \frac{\pi}{2}\right)$$

Step 2.: $x + \Delta x = \sec(y + \Delta y)$ ...(2)

Step 3.: $\Delta x = \sec(y + \Delta y) - \sec y$ (subtracting (1) from (2))

$$= \frac{1}{\cos(y+\Delta y)} - \frac{1}{\cos y} = \frac{\cos y - \cos(y+\Delta y)}{\cos y \cdot \cos(y+\Delta y)}$$

$$\frac{2\sin\left(y+\frac{\Delta y}{2}\right) \cdot \sin\left(\frac{\Delta y}{2}\right)}{\cos y \cdot \cos(y+\Delta y)}$$

$$\Rightarrow \frac{1}{\Delta x} = \frac{\cos y \cdot \cos(y+\Delta y)}{2\sin\left(y+\frac{\Delta y}{2}\right) \cdot \sin\left(\frac{\Delta y}{2}\right)} \quad \text{... (3)}$$

Step 4.: $\dfrac{\Delta y}{\Delta x} = \dfrac{\cos y \cdot \cos(y+\Delta y)}{2\sin\left(y+\frac{\Delta y}{2}\right)} \cdot \dfrac{\Delta y}{\sin\left(\frac{\Delta y}{2}\right)}$

(Multiplying both sies of (3) by Δy)

$$= \frac{\cos y \cdot \cos(y + \Delta y)}{\sin\left(y + \frac{\Delta y}{2}\right)} \cdot \frac{1}{\dfrac{\sin\left(\frac{\Delta y}{2}\right)}{\left(\frac{\Delta y}{2}\right)}} \quad \text{...(4)}$$

Step 5.: $\lim\limits_{\Delta x \to 0} \dfrac{\Delta y}{\Delta x} = \dfrac{\cos y \cdot \cos y}{\sin y \,.\, 1}$ (Taking the limits of both sides of (4) as $\Delta x \to 0$)

$$= \frac{1}{\sec y \cdot \tan y}$$

Step 6.: $|\sec y \cdot \tan y| = |\sec y| . |\tan y|$

$$= |\sec y| \left(\sqrt{\tan^2 y}\right) \left(\because |f(x)| = \sqrt{f^2(x)}\right)$$

$$= |\sec y| \left(\sqrt{\sec^2 y - 1}\right)$$

$$= |x| \cdot \sqrt{x^2 - 1}$$

$(\because \sec y = x)$ ...(a)

Now, $y \in (0, \pi) - \left\{\dfrac{\pi}{2}\right\}$

$$\Rightarrow y \in \left(0, \frac{\pi}{2}\right) \cup \left(\frac{\pi}{2}, \pi\right)$$

$$\Rightarrow y \in \left(0, \frac{\pi}{2}\right) \text{ or } y \in \left(\frac{\pi}{2}, \pi\right)$$

Again,

$y \in \left(0, \dfrac{\pi}{2}\right) \Rightarrow$ sec y and tan y both are positive

$\Rightarrow$ The product sec y . tan y is positive ...(b)

and $y \in \left(\dfrac{\pi}{2}, \pi\right) \Rightarrow$ sec y and tan y both are negative

$\Rightarrow$ the product sec $y \cdot$ tan y is positive ...(c)

Hence, from (b) and (c), we observe that the product of sec y and tan y is positive in the first quadrant and in the second quadrant for $y \in (0, \pi) - \left\{\dfrac{\pi}{2}\right\} \Rightarrow$ the product sec $y \cdot$ tan y is positive for $y \in (0, \pi) - \left\{\dfrac{\pi}{2}\right\}$. Which means that $|\sec y \cdot \tan y| = \sec y \cdot \tan y$ for $y \in (0, \pi) - \left\{\dfrac{\pi}{2}\right\}$

$(\because |f(x)| = f(x) \text{ for } f(x) > 0)$ ...(d)

Equating (a) and (d), we have

$$|\sec y \cdot \tan y| = \sec y \cdot \tan y = |x| . \sqrt{x^2 - 1}$$

$$\frac{dy}{dx} = \frac{1}{|x|\sqrt{x^2-1}} \quad \left(\because \lim_{\Delta x \to 0} \frac{\Delta y}{\Delta x} = \frac{dy}{dx}\right),$$

for $|x| > 1$.

Second method using the formula:

$$\frac{dy}{dx} = \frac{1}{\left(\frac{dx}{dy}\right)}$$

7. $y = \sec^{-1} x$

$\Rightarrow x = \sec y \quad \text{for } 0 \le y \le \pi,\ y \neq \frac{\pi}{2}$ (from the definition of $\sec^{-1} x$) ...(1)

$\Rightarrow \frac{dx}{dy} = \sec y \cdot \tan y$ (differentiating both sides of (1) w.r.t. y) ...(2)

Now,

$$y \in \left[0, \frac{\pi}{2}\right) \cup \left(\frac{\pi}{2}, \pi\right]$$

$\Rightarrow$ tan y and sec y both are positive for $0 < y < \frac{\pi}{2}$ and tan y and sec y both are negative for $\frac{\pi}{2} < y < \pi$

$\Rightarrow$ the product $\sec y . \tan y$ is always positive for, $0 < y < \pi,\ y \neq \frac{\pi}{2}$

Hence, $|\sec y \cdot \tan y| = \sec y \cdot \tan y$

$(\because |f(x)| = f(x))$

(for $f(x) > 0$) ...(3)

and $|\sec x \cdot \tan y| = |\sec x| \cdot |\tan y|$

$$= |x| \sqrt{\tan^2 y} \quad \left(\because |\tan y| = \sqrt{\tan^2 y}\right)$$

$$= |x| \sqrt{\sec^2 x - 1}$$

$$= |x| \sqrt{x^2 - 1} \quad \text{...(4)}$$

Equating (3) and (4), we have

$$|\sec y \cdot \tan y| = \sec y \cdot \tan y = |x| \cdot \sqrt{x^2 - 1}$$

$$\therefore \frac{dy}{dx} = \frac{1}{\left(\frac{dx}{dy}\right)} = \frac{1}{|x|\sqrt{x^2-1}},\ |x| > 1.$$

Third method

$$y = \sec^{-1} x = \cos^{-1}\left(\frac{1}{x}\right),\ |x| \ge 1$$

$$\Rightarrow \frac{dy}{dx} = \frac{d}{dx}\sec^{-1} x = \frac{d}{dx}\cos^{-1}\left(\frac{1}{x}\right),\ |x| \ge 1$$

$$= -\frac{1}{\sqrt{1-\frac{1}{x^2}}} \cdot \frac{d}{dx}\left(\frac{1}{x}\right),\ |x| > 1$$

$$= -\frac{1}{\sqrt{(x^2-1/x^2)}} \cdot \left(-\frac{1}{x^2}\right),\ |x| > 1$$

$$= \frac{1}{\left(\frac{\sqrt{x^2-1}}{\sqrt{x^2}}\right)} \cdot \left(\frac{1}{x^2}\right),\ |x| > 1$$

$$= \frac{|x|}{\sqrt{x^2-1}} \cdot \frac{1}{(|x|)^2} \quad (\because |x| = \sqrt{x^2} \text{ and } |x|^2 = x^2)$$

$$= \frac{1}{|x|\sqrt{x^2-1}},\ |x| > 1$$

Cor: On replacing x by $f(x)$ in the L.H.S and R.H.S of the above formula, we get

$$\frac{d \sec^{-1}[f(x)]}{dx} = \frac{1}{|f(x)|\sqrt{f^2(x)-1}} \cdot f'(x)$$

8. If $y = \text{cosec}^{-1} x$, $y \in \left(-\frac{\pi}{2}, \frac{\pi}{2}\right) - \{0\}$, show that $\frac{dy}{dx} = -\frac{1}{|x|\sqrt{x^2-1}}, \left(|x| > 1\right)$.

Proof: *First method* (derivation of d.c. using the definition)

Step 1.: Let $y = \text{cosec}^{-1} x$, $\left(|x| > 1\right)$ ...(1)

$\Rightarrow x = \text{cosec}\, y$, $\frac{-\pi}{2} < y < \frac{\pi}{2}$, $y \neq 0$

Step 2.: $x + \Delta x = \text{cosec}\,(y + \Delta y)$ ...(2)

Step 3.: $\Delta x = \text{cosec}\,(y+\Delta y) - \text{cosec}\, y$ (subtracting (1) from (2))

$$= \frac{1}{\sin(y+\Delta y)} - \frac{1}{\sin y}$$

$$= \frac{\sin y - \sin(y+\Delta y)}{\sin(y+\Delta y)\cdot \sin y}$$

$$= -\frac{2\cos(y+\Delta y)\cdot \sin\left(\frac{\Delta y}{2}\right)}{\sin(y+\Delta y)\cdot \sin y}$$

$$\Rightarrow \frac{1}{\Delta y} = -\frac{\sin(y+\Delta y)\cdot \sin y}{2\cos(y+\Delta y)\cdot \sin\left(\frac{\Delta y}{2}\right)}\cdot \Delta y \quad ...(3)$$

$$= -\frac{\sin y \cdot \sin(y+\Delta y)}{\cos\left(y+\frac{\Delta y}{2}\right)}\cdot \frac{1}{\dfrac{\sin\left(\frac{\Delta y}{2}\right)}{\left(\frac{\Delta y}{2}\right)}} \quad ...(4)$$

Step 4.: $\lim\limits_{\Delta x \to 0} \frac{\Delta y}{\Delta x} = -\frac{\sin y \cdot \sin y}{\cos y \cdot 1}$ (Taking the limits of both sides of (4) as $\Delta x \to 0$)

$$= -\frac{1}{\text{cosec}\ y \cdot \cot\ y} \quad ...(5)$$

Step 5.: $|\text{cosec}\, y \cdot \cot y| = |\text{cosec}\, y| \cdot |\cot y|$

$= |\text{cosec}\, y| \sqrt{\cot^2 y}$ $\left(\because |f(x)| = \sqrt{f^2(x)}\right)$

$= |\text{cosec}\, y| \cdot \sqrt{\text{cosec}^2 y - 1}$

$= |x| \cdot \sqrt{x^2 - 1}$ ($\because$ cosec $y = x$) ...(a)

Now,

$y \in \left(-\frac{\pi}{2}, \frac{\pi}{2}\right) - \{0\}$

$\Rightarrow y \in \left(-\frac{\pi}{2}, 0\right) \cup \left(0, \frac{\pi}{2}\right)$

$\Rightarrow y \in \left(-\frac{\pi}{2}, 0\right)$ or $y \in \left(0, \frac{\pi}{2}\right)$

Again, $y \in \left(-\frac{\pi}{2}, 0\right)$

$\Rightarrow$ cosec y and cot y both are negative

$\Rightarrow$ The product cosec y . cot y is positive

and $y \in \left(0, \frac{\pi}{2}\right)$ ...(b)

$\Rightarrow$ cosec y and cot y both are positive

$\Rightarrow$ the product cosec $y \cdot$ cot y is positive

From (b) and (c), we observe that the product of cosec y and cot y is positive for $y \in \left(-\frac{\pi}{2}, \frac{\pi}{2}\right) - \{0\}$

$\Rightarrow$ The product cosec $y \cdot$ cot y is positive for $y \in \left(-\frac{\pi}{2}, \frac{\pi}{2}\right) - \{0\}$ which means that

$|\text{cosec}\, y \cdot \cot y| = \text{cosec}\, y \cdot \cot y$

($|f(x)| = f(x)$ for $f(x) > 0$ ...(d)

Equating (a) and (d), we have

$|\text{cosec}\, y \cdot \cot y| = \text{cosec}\, y \cdot \cot y$

$= |x| \cdot \sqrt{x^2 - 1}$

and since, $\lim\limits_{\Delta x \to 0} \frac{\Delta y}{\Delta x} = \frac{dy}{dx}$, so (5) becomes equal to $\frac{dy}{dx} = -\frac{1}{|x|\sqrt{x^2-1}}$, $|x| > 1$

Second method using the formula:

$$\frac{dy}{dx} = \frac{1}{\frac{dx}{dy}}$$

9. $y = \operatorname{cosec}^{-1} x$

$\Rightarrow x = \operatorname{cosec} y$, for $-\frac{\pi}{2} \le y \le \frac{\pi}{2}$, $y \ne 0$ ... (1)

$\Rightarrow \frac{dx}{dy} = -\operatorname{cosec} y \cdot \cot y$

(differentiating (1) w.r.t. y) ...(2)

Now, $y \in \left[-\frac{\pi}{2}, \frac{\pi}{2}\right] - \{0\}$

$\Rightarrow y \in \left[-\frac{\pi}{2}, 0\right) \cup \left(0, \frac{\pi}{2}\right]$

$\Rightarrow$ cosec y and cot y both are negative for $-\frac{\pi}{2} < y < 0$ and cosec y and cot y both are positive for $0 < y < \frac{\pi}{2}$

$\Rightarrow$ the product cosec $y \cdot$ cot y is always positive for $\frac{-\pi}{2} < y < \frac{\pi}{2}$, $y \ne 0$

Hence, $|\operatorname{cosec} y \cdot \cot y| = \operatorname{cosec} y \cdot \cot y$

($|f(x)| = f(x)$ for $f(x) > 0$) ...(3)

and $|\operatorname{cosec} y \cdot \cot y| = |\operatorname{cosec} y| \cdot |\cot y|$

$= |x| \cdot \sqrt{\cot^2 y}$ $\left(\because |\tan y| = \sqrt{\tan^2 y}\right)$

$= |x| \cdot \sqrt{\operatorname{cosec}^2 y - 1}$

$= |x| \cdot \sqrt{x^2 - 1}$ ...(4)

Equating (3) and (4), we have

$|\operatorname{cosec} y \cdot \cot y| = \operatorname{cosec} y \cdot \cot y$

$= |x| \sqrt{x^2 - 1}$

(2) becomes equal to

$$\frac{dy}{dx} = \frac{1}{\frac{dx}{dy}} = \frac{-1}{|x| \cdot \sqrt{x^2-1}}, |x| > 1.$$

Third method

10. $y = \operatorname{cosec}^{-1} x = \sin^{-1}\left(\frac{1}{x}\right), |x| \ge 1$

$\Rightarrow \frac{dy}{dx} = \frac{d}{dx} \operatorname{cosec}^{-1} x = \frac{d}{dx}\left(\sin^{-1} \frac{1}{x}\right), |x| \ge 1$

$\Rightarrow \frac{dy}{dx} = \frac{1}{\sqrt{1 - \frac{1}{x^2}}} \cdot \frac{d}{dx}\left(\frac{1}{x}\right), |x| > 1$

$= \frac{-1}{\sqrt{x^2 - 1/x^2}} \cdot \left(-\frac{1}{x^2}\right), \quad |x| > 1$

$= \frac{-1}{\left(\frac{\sqrt{x^2-1}}{\sqrt{x^2}}\right)} \cdot \frac{1}{|x|^2} \quad (\because |x|^2 = x^2)$

$= \frac{-|x|}{\sqrt{x^2-1}} \cdot \frac{1}{|x|^2} \quad \left(\because \sqrt{x^2} = |x|\right)$

$= -\frac{1}{|x|\sqrt{x^2-1}}, |x| > 1$

Note: (i) $\sec^{-1} x + \operatorname{cosec}^{-1} x = \frac{\pi}{2}, \forall\, x \le -1$ or $x \ge 1$

$\Rightarrow \frac{d}{dx} \sec^{-1} x + \frac{d}{dx} \operatorname{cosec}^{-1} x = 0$

$\Rightarrow \frac{d \sec^{-1} x}{dx} = -\frac{d \operatorname{cosec}^{-1} x}{dx} = \frac{1}{|x|\sqrt{x^2-1}}$

(ii) $\frac{d}{dx} \sec^{-1} f(x) = -\frac{d}{dx} \operatorname{cosec}^{-1} f(x)$

$= \frac{1}{|f(x)|\sqrt{f^2(x) - 1}} \cdot \frac{d}{dx} f(x)$

An important remark:

The domain of the derived function of $\sec^{-1} x$ and $\text{cosec}^{-1} x$ are defined for $|x| > 1$. According to the convention we have adopted

(i) $\dfrac{d \sec^{-1} x}{dx} = \dfrac{1}{x\sqrt{x^2 - 1}}$

(ii) $\dfrac{d\, \text{cosec}^{-1} x}{dx} = \dfrac{-1}{x\sqrt{x^2 - 1}}$ are inaccurate unless we consider the restriction '$x > 1$' against the derived functions of inverse circular functions namely $\sec^{-1} x$ and / $\text{coses}^{-1} x$ because

$$\frac{d \sec^{-1} x}{dx} = \frac{1}{|x|\sqrt{x^2 - 1}},\ (|x| > 1)$$

$$\Leftrightarrow \frac{d \sec^{-1} x}{dx} = \frac{1}{x\sqrt{x^2 - 1}},\ \text{for } x > 1$$

$$= \frac{-1}{x\sqrt{x^2 - 1}},\ \text{for } x < -1$$

and $$\frac{d\, \text{cosec}^{-1} x}{dx} = -\frac{1}{|x|\sqrt{x^2 - 1}},\ (|x| > 1)$$

$$\Leftrightarrow \frac{d\, \text{cosec}^{-1} x}{dx} = \frac{-1}{x\sqrt{x^2 - 1}},\ \text{for } x > 1$$

$$= \frac{1}{x\sqrt{x^2 - 1}},\ \text{for } x < -1$$

Remember:

(i) $\dfrac{d}{dx} \sec^{-1} |x| = \dfrac{d}{dx} \cos^{-1}\left(\dfrac{1}{|x|}\right)$

$$= -\frac{1}{\sqrt{1 - \frac{1}{x^2}}} \cdot \frac{d}{dx}\left(\frac{1}{|x|}\right)\ (\because |x|^2 = x^2)$$

$$= -\frac{\sqrt{x^2}}{\sqrt{x^2 - 1}} \cdot (-1) \cdot \left(|x|^{-2}\right) \cdot \frac{d}{dx}|x|$$

$$= \frac{+|x|}{\sqrt{x^2 - 1}} \cdot \frac{1}{|x|^2} \cdot \frac{|x|}{x}$$

$$= \frac{|x|^2}{|x|^2 \sqrt{x^2 - 1}} \cdot \frac{1}{x}$$

$$= \frac{1}{x\sqrt{x^2 - 1}}$$

Cor. $$\frac{d}{dx} \sec^{-1} |f(x)| = \frac{1}{f(x)\sqrt{f^2(x) - 1}} \cdot \frac{d}{dx} f(x)$$

$$= \frac{1}{|f(x)|\sqrt{f^2(x) - 1}} \cdot \frac{d\,|f(x)|}{dx}$$

(ii) $\dfrac{d}{dx}\, \text{cosec}^{-1} |x| = \dfrac{d}{dx} \sin^{-1}\left(\dfrac{1}{|x|}\right)$

$$= \frac{1}{\sqrt{1 - \frac{1}{x^2}}} \cdot \frac{d}{dx} \frac{1}{|x|}\ \left(\because |x|^2 = x^2\right)$$

$$= \frac{\sqrt{x^2}}{\sqrt{x^2 - 1}} \cdot \frac{d}{dx} |x|^{-1} \cdot \frac{d\,|x|}{dx}$$

$$= \frac{|x|}{x^2 - 1} \cdot (-1)\left(|x|^{-2}\right) \cdot \frac{|x|}{x}$$

$$= \frac{-|x|}{\sqrt{x^2 - 1}} \cdot \frac{|x|}{|x|^2} \cdot \frac{1}{x}$$

$$= \frac{-|x|^2}{|x|^2 \cdot \sqrt{x^2 - 1} \cdot x}$$

$$= \frac{-1}{x\sqrt{x^2 - 1}}$$

Cor: $\dfrac{d\,\text{cosec}^{-1}}{dx}|f(x)| = \dfrac{-1}{f(x)\sqrt{f^2(x)-1}} \cdot \dfrac{d}{dx} f(x)$

$$= \frac{-1}{|f(x)|\sqrt{f^2(x)-1}} \cdot \frac{d|f(x)|}{dx}$$

3. The differential coefficients of those trigonometric and inverse trigonometric function which begin with '*c*' are negative.

4. Method of remembering the d.c. of standard inverse circular functions

Let $y = f^{-1}(x) \Rightarrow x = f(y) \Rightarrow \dfrac{dx}{dy} = f'(y)$

$\Rightarrow \dfrac{dy}{dx} = \dfrac{1}{\left(\dfrac{dx}{dy}\right)} = \dfrac{1}{f'(y)} = \dfrac{1}{f'\left[f^{-1}(x)\right]}$, where

(i) $f^{-1}(x) = \theta$ is required to put and simplify as well as $f(\theta) = x$ is put in the last stage to get an algebraic expression in x.

(ii) f' = d.c. of f = prime of f

(iii) f = sin / cos / tan / cot / sec / cosec

(iv) f' = d.c. of sin / cos / tan / cot / sec / cosec

$= \cos / -\sin / \sec^2 / -\text{cosec}^2 / \sec\cdot\tan / -\text{cosec}$

cot whose operand is $f^{-1}(x)$ = given inverse trigonometric function. Thus the d.c. of inverse trigonometric function = reciprocal of the derivative of the direct trigonometric function (of the given t^{-1} – function) whose operand is the given function which is put equal to θ to get an algebric expression in x.

Thus:

1. $\dfrac{d\sin^{-1}x}{dx} = \dfrac{1}{\cos(\sin^{-1}x)} = \dfrac{1}{\cos\theta} = \dfrac{1}{|\cos\theta|}$

$\left[\because \dfrac{-\pi}{2} \le \theta \le \dfrac{\pi}{2}\right]$

$$= \frac{1}{\sqrt{1-\sin^2\theta}} = \frac{1}{\sqrt{1-x^2}}\,;\ |x| < 1$$

2. $\dfrac{d\cos^{-1}x}{dx} = \dfrac{1}{-\sin(\cos^{-1}x)} = -\dfrac{1}{\sin\theta} = -\dfrac{1}{|\sin\theta|}$

$[\because 0 \le \theta \le \pi]$

$$= \frac{-1}{\sqrt{1-\cos^2\theta}} = -\frac{1}{\sqrt{1-x^2}}\,;\ |x| < 1$$

3. $\dfrac{d\tan^{-1}x}{dx} = \dfrac{1}{\sec^2(\tan^{-1}x)}$

$$= \frac{1}{\sec^2\theta} = \frac{1}{1+\tan^2\theta} = \frac{1}{1+x^2}$$

4. $\dfrac{d\cot^{-1}x}{dx} = \dfrac{-1}{\text{cosec}^2 x\,(\cot^{-1}x)}$

$$= \frac{-1}{1+\cot^2\theta} = \frac{-1}{1+x^2}$$

5. $\dfrac{d\sec^{-1}x}{dx} = \dfrac{1}{\sec(\sec^{-1}x)\cdot\tan(\sec^{-1}x)}$

$$= \frac{1}{\sec\theta\cdot\tan\theta} = \frac{1}{|\sec\theta\cdot\tan\theta|}$$

$[\because 0 < \theta < \pi]$

$$= \frac{1}{|\sec\theta|\cdot|\tan\theta|} = \frac{1}{|x|\sqrt{\sec^2\theta - 1}}$$

$$= \frac{1}{|x|\sqrt{x^2-1}}\,;\ |x| > 1$$

6. $\dfrac{d\,\text{cosec}^{-1}x}{dx}$

$$= \frac{-1}{\text{cosec}(\text{cosec}^{-1}x)\cot(\text{cosec}^{-1}x)} = \frac{-1}{\text{cosec}\,\theta\cdot\cot\theta}$$

$$= \frac{-1}{|\operatorname{cosec}\theta \cdot \cot\theta|} \quad \left[\because -\frac{\pi}{2} \le \theta \le \frac{\pi}{2}\right]$$

$$= \frac{-1}{|\operatorname{cosec}\theta||\cot\theta|} = \frac{-1}{|x|\sqrt{\operatorname{cosec}^2\theta - 1}}$$

$$= \frac{-1}{|x|\sqrt{x^2 - 1}} \; ; |x| > 1$$

Note: **1.** Insertion and removal of mod operator is performed by considering the principal values of given inverse circular function.

2. $|x| < 1$ means domains of derivatives of inverse sine and cosine function are the same namely $|x| < 1$ whereas $|x| > 1$ means domains if derivatives of inverse secant and cosecant function are the same namely $|x| > 1$

3. On replacing x by $f(x)$ in the L.H.S and R.H.S of the above furmulas, we have

(i) $$\frac{d\sin^{-1}[f(x)]}{dx} = \frac{1}{\sqrt{1 - f^2(x)}} \cdot f'(x)$$

(ii) $$\frac{d\cos^{-1}[f(x)]}{dx} = \frac{-1}{\sqrt{1 - f^2(x)}} \cdot f'(x)$$

(iii) $$\frac{d\tan^{-1} f(x)}{dx} = \frac{1}{1 + f^2(x)} \cdot f'(x)$$

(iv) $$\frac{d\cot^{-1} f(x)}{dx} = \frac{-1}{1 + f^2(x)} \cdot f'(x)$$

(v) $$\frac{d\sec^{-1} f(x)}{dx} = \frac{1}{|f(x)|\sqrt{f^2(x) - 1}} \cdot f'(x)$$

(vi) $$\frac{d\operatorname{cosec}^{-1}(x)}{dx} = \frac{-1}{|f(x)|\sqrt{f^2(x) - 1}} \cdot f'(x)$$

Which are known as derivatives of inverse trigonometric function of a function of x formulas.

4. On replacing $f(x)$ by $|f(x)|$ in the L.H.S and R.H.S and retaining $|f(x)|$ where it is in the above formulas, we get

(i) $$\frac{d\sin^{-1}|f(x)|}{dx} = \frac{1}{\sqrt{1 - f^2(x)}} \cdot \frac{d|f(x)|}{dx}$$

$$\left(\because |f(x)|^2 = f^2(x)\right)$$

(ii) $$\frac{d\cos^{-1}|f(x)|}{dx} = \frac{-1}{\sqrt{1 - f^2(x)}} \cdot \frac{d|f(x)|}{dx}$$

(iii) $$\frac{d\tan^{-1}|f(x)|}{dx} = \frac{1}{\sqrt{1 + f^2(x)}} \cdot \frac{d|f(x)|}{dx}$$

(iv) $$\frac{d\cot^{-1}|f(x)|}{dx} = \frac{-1}{\sqrt{1 + f^2(x)}} \cdot \frac{d|f(x)|}{dx}$$

(v) $$\frac{d\sec^{-1}|f(x)|}{dx} = \frac{1}{|f(x)|\sqrt{f^2(x) - 1}} \cdot \frac{d|f(x)|}{dx}$$

(vi) $$\frac{d\operatorname{cosec}^{-1}|f(x)|}{dx} = \frac{-1}{|f(x)|\sqrt{f^2(x) - 1}} \cdot \frac{d|f(x)|}{dx}$$

5. " $\sqrt{a^2 T^2(\theta)}$ " can be written equal to "$a\,T(\theta)$", if it is pre-assumed that 'a' is positive and the angle 'θ' of any trigonometric function sin, cos, tan, cot, sec or cosec is an acute angle.

N.B.: " T " in " $T(\theta)$ " means the operator sin, cos, tan, cot, sec or cosec indicating trigonometric functions.

Type 1.: $y = t^{-1}[f(x)]$

Where $t^{-1} = \sin^{-1}/\cos^{-1}/\tan^{-1}/\cot^{-1}/\sec^{-1}/\operatorname{cosec}^{-1}$

$$f(x) = \frac{f_1(x) + f_2(x)}{f_3(x) + f_4(x)} \Bigg/ \sqrt{\frac{f_1(x) + f_2(x)}{f_3(x) + f_4(x)}} \Bigg/$$

$$f_1(x) \times f_2(x) \Big/ \sqrt{f_1(x) \times f_2(x)}$$

Where $f_1(x)$, $f_2(x)$, $f_3(x)$ and $f_4(x)$ are trigonometric functions of x / algebraic expressions in x / one or two of the four functions $f_1(x)$, $f_2(x)$, $f_3(x)$ and $f_4(x)$ may be constant.

Working rule:

1. Change the inverse trigonometric function into direct function

$$\Rightarrow y = t^{-1}[f(x)] \Leftrightarrow t(y) = f(x)$$

2. Differentiate both sides w.r.t. x, i.e

$$\frac{d\,t(y)}{dx} = \frac{d\,f(x)}{dx}$$

$$\Rightarrow \frac{d\,t(y)}{dy} \cdot \frac{dy}{dx} = f'(x)$$

$$\Rightarrow \frac{dy}{dx} = f'(x) \Big/ \left(\frac{d\,t(y)}{dy}\right)$$

3. Express the d.c. of $t(y)$ in terms of $f(x)$ and simplify.

Note: The above method is known as general method to differentiate an inverse trigonometric function of the type : $y = t^{-1}[f(x)]$ which can be used to differentiate any type of inverse circular function provided given function can be put in the form $y = t^{-1}[f(x)] \Leftrightarrow t(y) = f(x)$.

Examples worked out on type (1) find the d.c. of the following

1. $y = \cos^{-1}\left(\dfrac{3+5\cos x}{5+3\cos x}\right)$

Solution: $y = \cos^{-1}\left(\dfrac{3+5\cos x}{5+3\cos x}\right)$ ...(i)

$$\Rightarrow \cos y = \left(\frac{3+5\cos x}{5+3\cos x}\right) \quad \text{...(ii)}$$

Now differentiating both sides w.r.t. x

$$\Rightarrow -\sin y \frac{dy}{dx}$$

$$= \frac{(5+3\cos x)(-5\sin x)-(3+5\cos x)(-3\sin x)}{(5+3\cos x)^2}$$

$$\Rightarrow -\sin y \frac{dy}{dx}$$

$$= \frac{(-\sin x)(25+15\cos x - 9 - 15\cos x)}{(5+3\cos x)^2}$$

$$\Rightarrow \frac{dy}{dx} = \frac{16}{|(5+3\cos x)|^2} \cdot \frac{\sin x}{\sin y},\ \sin y \ne 0.$$

$$\left(\because |f^2(x)| = f^2(x)\right) \quad \text{...(iii)}$$

Now, from (ii),

$\sin y = |\sin y|$

$$= \sqrt{1 - \left(\frac{3+5\cos x}{5+3\cos x}\right)^2} = \frac{4|\sin x|}{|(5+3\cos x)|}$$

$$(\because 0 \le y \le \pi \cdot \therefore \sin y \ge 0)$$

Which gives for $x \ne n\pi$

$$\frac{\sin x}{\sin y} = \frac{|(5+3\cos x)|\sin x}{4|\sin x|} \quad \text{...(iv)}$$

From (iii) and (iv), we get for $x \ne n\pi$

$$\frac{dy}{dx} = \frac{4\sin x}{|(5+3\cos x)||\sin x|}$$

$$= \frac{4\sin x}{(5+3\cos x)|\sin x|}$$

Note: A direct method has been mentioned later.

2. $y = \sin^{-1}\left(\dfrac{2x}{1+x^2}\right)$

Solution: $y = \sin^{-1}\left(\dfrac{2x}{1+x^2}\right)$

$$\Rightarrow \sin y = \frac{2x}{1+x^2} \quad \text{...(1)}$$

$$\Rightarrow \frac{d\sin y}{dx} = \frac{(1+x^2)\cdot 2 - 2x(2x)}{(1+x^2)^2}$$

$$= \frac{2\left(1+x^2\right)-4x^2}{\left(1+x^2\right)^2}$$

$$\Rightarrow \cos y \frac{dy}{dx} = \frac{2-2x^2}{\left(1+x^2\right)^2} = \frac{2\left(1-x^2\right)}{\left(1+x^2\right)^2}$$

$$\left[\because \frac{d \sin y}{dx} = \frac{d \sin y}{dy} \cdot \frac{dy}{dx}\right]$$

$$\Rightarrow \frac{dy}{dx} = \frac{2\left(1-x^2\right)}{\left(1+x^2\right)^2} \cdot \frac{1}{\cos y}; \cos y \neq 0 \quad \text{...(ii)}$$

Now, $\cos y = |\cos y| = \sqrt{1-\sin^2 y}$

$$\left(\because -\frac{\pi}{2} \le y \le \frac{\pi}{2} \Rightarrow \cos y \ge 0\right)$$

$$= \sqrt{1 - \frac{4x^2}{\left(1+x^2\right)^2}} \quad \text{[from (1)]}$$

$$= \sqrt{\frac{\left(1+x^2\right)^2 - 4x^2}{\left(1+x^2\right)^2}}$$

$$= \frac{\sqrt{1+x^4+2x^2-4x^2}}{\left(1+x^2\right)}$$

$\because \left(1+x^2\right)$ is always $+ve \Rightarrow \left|\left(1+x^2\right)\right| = 1+x^2$

$$= \frac{\sqrt{1+x^4-2x^2}}{\left(1+x^2\right)}$$

$$= \frac{\sqrt{\left(1-x^2\right)^2}}{\left(1+x^2\right)}$$

$$= \frac{\left|\left(1-x^2\right)\right|}{\left(1+x^2\right)} \left(\because \sqrt{f^2(x)} = |f(x)|\right)$$

Which gives

$$\frac{dy}{dx} = \frac{2\left(1-x^2\right)}{\left(1+x^2\right)^{\not 2}} \cdot \frac{\cancel{\left(1+x^2\right)}}{\left|\left(1-x^2\right)\right|}$$

$$= \frac{2\left(1-x^2\right)}{\left|\left(1-x^2\right)\right|\left(1+x^2\right)}, x \neq \pm 1.$$

3. $y = \tan^{-1}\left(\frac{1+x}{1-x}\right)$

Solution: $y = \tan^{-1}\left(\frac{1+x}{1-x}\right), x \neq 1$

$$\Rightarrow \tan y = \frac{1+x}{1-x} \quad \text{...(i)}$$

$$\Rightarrow \frac{d \tan y}{dx} = \frac{(1-x)\cdot 1 - (1+x)(-1)}{(1-x)^2}$$

$$= \frac{(1-x)+(1+x)}{(1-x)^2}$$

$$\Rightarrow \sec^2 y \cdot \frac{dy}{dx} = \frac{2}{(1-x)^2}$$

$$\Rightarrow \frac{dy}{dx} = \frac{2}{(1-x)^2} \cdot \frac{1}{\sec^2 y} \quad \text{...(ii)}$$

Now, $\sec^2 y = 1 + \tan^2 y = 1 + \left(\frac{1+x}{1-x}\right)^2$

$$= \frac{(1-x)^2 + (1+x)^2}{(1-x)^2} \text{ [from (1)]}$$

$$= \frac{1+x^2 - \cancel{2x} + 1 + x^2 + \cancel{2x}}{(1-x)^2}$$

$$= \frac{2+2x^2}{(1-x)^2}$$

$$= \frac{2\left(1+x^2\right)}{(1-x)^2} \quad \text{...(iii)}$$

Which gives

$$\frac{dy}{dx} = \frac{2}{(1-x)^2} \cdot \frac{(1-x)^2}{2(1+x^2)} = \frac{1}{(1+x^2)}; \; x \neq 1$$

4. $y = \sin^{-1}(\cos x)$

Solution: $y = \sin^{-1}(\cos x)$

$$\Rightarrow \sin y = \cos x$$

$$\Rightarrow \frac{d \sin y}{dx} = \frac{d \cos x}{dx}$$

$$\cos y \cdot \frac{dy}{dx} = -\sin x$$

$$\Rightarrow \frac{dy}{dx} = \frac{-\sin x}{\cos y} \text{ for } y \neq \pm\frac{\pi}{2}$$

$$= -\frac{\sin x}{|\cos y|}$$

$$\because -\frac{\pi}{2} \leq y \leq \frac{\pi}{2} \Rightarrow \cos y \geq 0$$

$$\therefore |\cos y| = \cos y$$

$$\therefore \frac{dy}{dx} = -\frac{\sin x}{\sqrt{\cos^2 y}}, \; y \neq \pm\frac{\pi}{2}$$

$$= \frac{-\sin x}{\sqrt{1-\sin^2 y}}$$

$$= \frac{-\sin x}{\sqrt{1-\cos^2 x}}; \; x \neq n\pi$$

$$= \frac{-\sin x}{\sqrt{\sin^2 x}} = \frac{-\sin x}{|\sin x|}, \; x \neq n\pi, \; n \in z$$

5. $y = \sin^{-1}(\sin x)$

Solution: $y = \sin^{-1}(\sin x)$

$$\Rightarrow \sin y = \sin x$$

$$\Rightarrow \frac{d \sin y}{dx} = \frac{d \sin x}{dx}$$

$$\Rightarrow \cos y \frac{dy}{dx} = \cos x$$

$$\Rightarrow \frac{dy}{dx} = \frac{\cos x}{\cos y} \quad \left(y \neq \pm\frac{\pi}{2}\right)$$

$$= \frac{\cos x}{|\cos y|}$$

$$\left(\therefore -\frac{\pi}{2} \leq y \leq \frac{\pi}{2} \Rightarrow \cos y \geq 0 \; \therefore |\cos| = \cos y\right)$$

$$= \frac{\cos x}{\sqrt{1-\sin^2 x}} \quad [\because \sin y = \sin x]$$

$$= \frac{\cos x}{\sqrt{\cos^2 x}} = \frac{\cos x}{|\cos x|}; \; x \neq n\pi + \frac{\pi}{2}$$

6. $y = \tan^{-1}\left(\dfrac{\cos x - \sin x}{\cos x + \sin x}\right)$

First method:

Solution: $y = \tan^{-1}\left(\dfrac{\cos x - \sin x}{\cos x + \sin x}\right)$

$$\Rightarrow \tan y = \frac{\cos x - \sin x}{\cos x + \sin x} \quad \text{...(i)}$$

$$\Rightarrow \frac{d \tan y}{dx}$$

$$= \frac{(\cos x + \sin x)(-\sin x - \cos x) - (\cos x - \sin x)\cdot(-\sin x + \cos x)}{(\cos x + \sin x)^2}$$

$$Nr = -\left[\left(\cos x + \sin x^2\right) + (\cos x - \sin x)^2\right]$$

$$= -2\left[\cos^2 x + \sin^2 x\right] = -2$$

$$\Rightarrow \sec^2 y \frac{dy}{dx} = -\frac{2}{(\sin x + \cos x)^2}$$

$$\Rightarrow \frac{dy}{dx} = -\frac{2}{(\sin x + \cos x)^2} \cdot \frac{1}{\sec^2 y} \quad \text{...(ii)}$$

Now, from (i),

$$\sec^2 y = 1 + \tan^2 y = 1 + \left\{\frac{\cos x - \sin x}{\cos x + \sin x}\right\}^2$$

$$= \frac{(\cos x + \sin x)^2 + (\cos x - \sin x)^2}{(\cos x + \sin x)^2}$$

$$= \frac{\cos^2 x + \sin^2 x + \cos^2 x + \sin^2 x}{(\cos x + \sin x)^2}$$

$$= \frac{2}{(\cos x + \sin x)^2} \quad \text{...(iii)}$$

Which gives

$$\frac{dy}{dx} = \frac{-2}{(\sin x + \cos x)^2} \times \frac{(\cos x + \sin x)^2}{2} = -1 \textit{ Ans.}$$

Second method:

$$\frac{\cos x - \sin x}{\cos x + \sin x} = \frac{1 - \tan x}{1 + \tan x} = \tan\left(\frac{\pi}{4} - x\right)$$

$$\therefore\ y = \tan^{-1}\left[\tan\left(\frac{\pi}{4} - x\right)\right] = n\pi + \frac{\pi}{4} - x$$

where 'n' is such that $-\frac{\pi}{2} \le n\pi + \frac{\pi}{4} - x \le \frac{\pi}{2}$

$$\therefore\ \frac{dy}{dx} = -1$$

7. $y = \tan^{-1}\left(\frac{\cos x}{1 + \sin x}\right)$

Solution: **Method (1)**

$$y = \tan^{-1}\left(\frac{\cos x}{1 + \sin x}\right)$$

$$\Rightarrow \tan y = \frac{\cos x}{1 + \sin x} \quad \text{...(i)}$$

$$\Rightarrow \frac{d \tan y}{dx}$$

$$= \frac{(1 + \sin x)\frac{d}{dx}(\cos x) - \cos x\, \frac{d}{dx}(1 + \sin x)}{(1 + \sin x)^2}$$

$$\Rightarrow \sec^2 y\, \frac{dy}{dx} = \frac{-\sin x - \sin^2 y - \cos^2 y}{(1 + \sin x)^2}$$

$$\Rightarrow \sec^2 y\, \frac{dy}{dx} = \frac{-(1 + \sin x)}{(1 + \sin x)^2} = \frac{-1}{(1 + \sin x)}$$

$$\Rightarrow \frac{dy}{dx} = -\frac{1}{(1 + \sin x)} \cdot \frac{1}{\sec^2 y} \quad \text{...(ii)}$$

$$\text{Now, } \sec^2 y = 1 + \tan^2 y = 1 + \left(\frac{\cos x}{1 + \sin x}\right)^2$$

$$= \frac{(1 + \sin x)^2 + \cos^2 x}{(1 + \sin x)^2}$$

$$= \frac{1 + \sin x^2 + 2 \sin + \cos^2 x}{(1 + \sin x)^2}$$

$$= \frac{2 + 2 \sin x}{(1 + \sin x)^2} = \frac{2\,\cancel{(1 + \sin x)}}{(1 + \sin x)^{\cancel{2}}}$$

$$= \frac{2}{(1 + \sin x)} \quad \text{...(iii)}$$

Which gives

$$\frac{dy}{dx} = \frac{1}{(1 + \sin x)} \times \frac{(1 + \sin x)}{2} = -\frac{1}{2} \quad \textit{Ans.}$$

Method (2)

$$\frac{\cos x}{1 + \sin x} = \frac{\cos^2 \frac{x}{2} - \sin^2 \frac{x}{2}}{\left(\cos\frac{x}{2} + \sin\frac{x}{2}\right)^2}$$

$$= \frac{\cos\frac{x}{2} - \sin\frac{x}{2}}{\left(\cos\frac{x}{2} + \sin\frac{x}{2}\right)} = \frac{1 - \tan\frac{x}{2}}{1 + \tan\frac{x}{2}} = \tan\left(\frac{\pi}{4} - \frac{x}{2}\right)$$

$$y = n\pi + \frac{\pi}{4} - \frac{x}{2}$$

$$\frac{dy}{dx} = -\frac{1}{2}$$

8. $y = \tan^{-1}\frac{1}{x}$

Solution: $y = \tan^{-1}\frac{1}{x}$

$$\Rightarrow \tan y = \frac{1}{x}$$

$$\Rightarrow \frac{d\tan y}{dx} = \frac{d}{dx}\left(\frac{1}{x}\right) = -\frac{1}{x^2}$$

$$\Rightarrow \sec^2 y \cdot \frac{dy}{dx} = -\frac{1}{x^2}$$

$$\Rightarrow \frac{dy}{dx} = -\frac{1}{x^2} \cdot \frac{1}{\sec^2 y}$$

$$= -\frac{1}{x^2} \cdot \left(\frac{1}{1+\tan^2 y}\right)$$

$$= -\frac{1}{x^2} \cdot \left(\frac{1}{1+\frac{1}{x^2}}\right) \quad \left[\because \tan y = \frac{1}{x}\right]$$

$$= -\frac{1}{x^2}\left\{\frac{1}{\left(\frac{x^2+1}{x^2}\right)}\right\} = -\frac{1}{1+x^2}$$

9. $y = \cos^{-1}\left(\frac{a + b\sin x}{b + a\sin x}\right)$

Solution: $y = \cos^{-1}\left(\frac{a + b\sin x}{b + a\sin x}\right)$

$$\Rightarrow \cos y = \frac{a + b\sin x}{b + a\sin x}$$

$$\Rightarrow \frac{d\cos y}{dx}$$

$$= \frac{(b+a\sin x)\frac{d}{dx}(a+b\sin x) - (a+b\sin x)\frac{d}{dx}(b+a\sin x)}{(b+a\sin x)^2}$$

$$\Rightarrow -\sin y\,\frac{dy}{dx}$$

$$= \frac{(b + a\sin x)\,b\cos x - (a + b\sin x)\,a\cos x}{(b + a\sin x)^2}$$

$$\Rightarrow \frac{dy}{dx} = \frac{b^2\cos x - a^2\cos x}{(b + a\sin x)^2} \cdot \frac{1}{-\sin y}$$

$$= \frac{(b^2 - a^2)\cos x}{(b + a\sin x)^2} \cdot \frac{-1}{\sqrt{1-\cos^2 y}}$$

$$(\because \sin y = |\sin y| \text{ as } 0 \le y \le \pi)$$

$$= \frac{(b^2 - a^2)\cos x}{(b + a\sin x)^2} \times \frac{-1}{\sqrt{1-\left(\frac{a + b\sin x}{b + a\sin x}\right)^2}}$$

$$= \frac{(b^2 - a^2)\cos x}{|(b + a\sin x)|^2} \times \frac{-|(b+a\sin x)|}{\sqrt{b^2 + a^2\sin^2 x + 2ab\sin x - (a^2 + b^2\sin^2 x + 2ab\sin x)}}$$

$$(\because |f(x)|^2 = f^2(x))$$

$$= \frac{(b^2 - a^2)(\cos x)}{\sqrt{\left[b^2 - a^2 + (a^2 - b^2)\sin^2 x\right]}} \times \frac{-1}{|(b+a\sin x)|}$$

$$= \frac{-(\cos x)(b^2 - a^2)}{\sqrt{(b^2 - a^2) - (b^2 - a^2)\sin^2 x}} \times \frac{1}{|(b+a\sin x)|}$$

$$= \frac{-(\cos x)\left(b^2 - a^2\right)}{\left(\sqrt{b^2 - a^2}\right)\left(\sqrt{1 - \sin^2 x}\right)} \times \frac{1}{|(b + a \sin x)|}$$

$$= \frac{-\left(\sqrt{b^2 - a^2}\right)(\cos x)}{|b + a \sin x|\,|\cos x|}$$

Note: $\frac{dy}{dx}$ exists only when $b^2 > a^2$.

10. $y = \sin^{-1}\left(\frac{a + b \cos x}{b + a \cos x}\right)$

Solution: $y = \sin^{-1}\left(\frac{a + b \cos x}{b + a \cos x}\right)$

$$\Rightarrow \sin y = \frac{a + b \cos x}{b + a \cos x}$$

$$\Rightarrow \frac{d \sin y}{dx}$$

$$= \frac{(b + a\cos x)\frac{d}{dx}(a + b\cos x) - (a + b\cos x)\frac{d}{dx}(b + a\cos x)}{(b + a\cos x)^2}$$

$$\Rightarrow \cos y \frac{dy}{dx}$$

$$= \frac{(b + a\cos x)(-b\sin x) - (a + b\cos x)\cdot(-a\sin x)}{(b + a\cos x)^2}$$

$$\Rightarrow \frac{dy}{dx} = \frac{-b^2 \sin x + a^2 \sin x}{(b + a \cos x)^2} \cdot \frac{1}{\cos y}$$

$$= \frac{-b^2 \sin x + a^2 \sin x}{(b + a \cos x)^2} \cdot \frac{1}{|\cos y|}$$

$$\left(\because \cos y = |\cos y| \text{ as } -\frac{\pi}{2} \le y \le \frac{\pi}{2}\right)$$

$$= \frac{\left(a^2 - b^2\right)\sin x}{(b + a\cos x)^2} \cdot \frac{1}{\sqrt{1 - \sin^2 y}}$$

$$= \frac{\left(a^2 - b^2\right)\sin x}{(b + a\cos x)^2} \cdot \frac{1}{\sqrt{1 - \left(\frac{a + b\cos x}{b + a\cos x}\right)^2}}$$

$$= \frac{\left(a^2 - b^2\right)\sin x}{|b + a\cos x|^2} \cdot \frac{|b + a\cos x|}{\sqrt{\left(b + a\cos x^2\right) - (a + b\cos x)^2}}$$

$$\left[\because \sqrt{f^2(x)} = |f(x)| \text{ and } \left|f^2(x)\right| = f^2(x)\right]$$

$$= \frac{\left(a^2 - b^2\right)\sin x}{|(b + a\cos x)|} \times$$

$$\frac{1}{\sqrt{b^2 + a^2\cos^2 x + 2ab\cos x - \left(a^2 + b^2\cos^2 x + 2ab\cos x\right)}}$$

$$= \frac{\left(a^2 - b^2\right)\sin x}{|b + a\cos x|} \cdot \frac{1}{\sqrt{\left(b^2 - a^2\right) - \left(b^2 - a^2\right)\cos^2 x}}$$

$$= \frac{\left(a^2 - b^2\right)\sin x}{|b + a\cos x|} \cdot \frac{1}{\sqrt{\left(b^2 - a^2\right)\left(1 - \cos^2 x\right)}}$$

$$= \frac{-\left(b^2 - a^2\right)\sin x}{|b + a\cos x|} \cdot \frac{1}{\sqrt{b^2 - a^2}} \cdot \frac{1}{\sqrt{1 - \cos^2 x}}$$

$$= \frac{-\sqrt{b^2 - a^2}}{|b + a\cos x|} \cdot \frac{\sin x}{|\sin x|}$$

Type 2: Method of substitution and use of chain rule.

Form A: $y =$ a constant $\times\, t^{-1}(f(x))$

or, $y =$ a constant $\times\, t^{-1}$ [a constant times $f(x)$]

Where, $f(x) = a$ function of x / $\sin x$ / $\cos x$/.... / an expression in x etc.

Working rule:

1. Put $t^{-1}(f(x))$ or t^{-1} [constant $f(x)$ = u and differentiate w.r.t. x].

2. Put $y = \dfrac{u}{\text{constant}}$ or, (a constant $\times u$) and differentiate it by using chain rule, i.e. $\dfrac{dy}{dx} = \dfrac{dy}{du} \cdot \dfrac{du}{dx}$, where

$$\frac{dy}{du} = \frac{1}{\text{constant}} \text{ or, constant.}$$

Solved Examples:

Find the d.c. of the following.

1. $y = \dfrac{1}{ab} \cdot \tan^{-1}\left(\dfrac{b}{a}\tan x\right)$

Solution: Let $u = \tan^{-1}\left(\dfrac{b}{a}\tan x\right)$ ...(i)

$$\therefore \tan u = \frac{b}{a}\tan x$$

$$\Rightarrow \sec^2 u \frac{du}{dx} = \frac{b}{a}\sec^2 x$$

$$\Rightarrow \frac{du}{dx} = \frac{b}{a}\sec^2 x \cdot \frac{1}{\sec^2 u} = \frac{b\sec^2 x}{a} \cdot \frac{1}{1+\dfrac{b^2}{a^2}\tan^2 x}$$

$$\Rightarrow \frac{du}{dx} = \frac{b\sec^2 x}{a} \cdot \frac{a^2}{a^2 + b^2\tan^2 x}$$

$$\Rightarrow \frac{du}{dx} = \frac{ab}{a^2\cos^2 x + b^2\sin^2 x} \quad \text{...(ii)}$$

Now again, let $y = \dfrac{u}{ab}$ * ...(iii)

$$\therefore \frac{dy}{dx} = \frac{1}{ab} \cdot \frac{du}{dx} = \frac{1}{ab} \cdot \frac{ab}{a^2\cos^2 x + b^2\sin^2 x}$$

[from (ii)]

$$= \frac{1}{a^2\cos^2 x + b^2\sin^2 x}$$

* or alternatively,

$$\frac{dy}{dx} = \frac{dy}{du} \cdot \frac{du}{dx} = \frac{1}{ab} \cdot \frac{ab}{a^2\cos^2 x + b^2\sin^2 x}$$

$$\Rightarrow \frac{dy}{dx} = \frac{1}{a^2\cos^2 x + b^2\sin^2 x}$$

Form B:

$$y = a \text{ constant} \times t^{-1}\left(\sqrt{f(x)}\right)$$

where, $f(x) = \dfrac{f_1(x) \pm f_2(x)}{f_1(x) \mp f_2(x)}$ a quotient of two functions/an expression in x.

Working rule:

1. Put $\left.\begin{array}{l} u = \sqrt{f(x)} \\ v = f(x) \end{array}\right\} \Rightarrow \begin{array}{l} y = a \text{ constant} \times t^{-1}(u) \\ u = \sqrt{v} \Rightarrow \dfrac{du}{dv} = \dfrac{1}{2\sqrt{v}} \end{array}$.

2. Find $\dfrac{du}{dv}$, $\dfrac{dv}{dx}$ and $\dfrac{dy}{du}$.

3. Use the chain rule:

$$\frac{dy}{dx} = \frac{dy}{du} \cdot \frac{du}{dv} \cdot \frac{dv}{dx}$$

Note: **1.** The constant may be unity. Similarly any one of $f_1(x)$ and $f_2(x)$ may be unity.

2. The above method may be termed as $u - v$ method just for easiness.

Solved Examples on form (B)

Find the d.c. of the following

1. $y = \tan^{1}\sqrt{\left(\dfrac{1+\sin x}{1-\sin x}\right)}$

Solution: $y = \tan^{1}\sqrt{\left(\dfrac{1+\sin x}{1-\sin x}\right)}$

$$\text{Let } u = \sqrt{\left(\frac{1+\sin x}{1-\sin x}\right)} \quad \text{...(i)}$$

$$v = \frac{1+\sin x}{1-\sin x} \quad \text{...(ii)}$$

$$\therefore y = \tan^{-1}\sqrt{v} \Leftrightarrow \tan^{-1} u \quad \text{...(iii)}$$

Now, from (1),

$$u = \sqrt{v}$$

$$\Rightarrow \frac{du}{dv} = \frac{1}{2\sqrt{v}} = \frac{1}{2\sqrt{\left(\frac{1+\sin x}{1-\sin x}\right)}}$$

and from (ii),

$$y = \tan^{-1} u$$

$$\Rightarrow \frac{dy}{du} = \frac{1}{1+u^2}$$

$$= \frac{1}{1+\left(\frac{1+\sin x}{1-\sin x}\right)} = \frac{1}{\left(\frac{1-\sin x+1+\sin x}{1-\sin x}\right)}$$

$$= \frac{1-\sin x}{2} \quad \text{...(iv)}$$

Hence, $\frac{dy}{dx} = \frac{dy}{du}\cdot\frac{du}{dv}\cdot\frac{dv}{dx}$

$$= \frac{1-\sin x}{2}\cdot\frac{1}{2}\cdot\sqrt{\left(\frac{1-\sin x}{1+\sin x}\right)}\cdot\frac{2\cos x}{(1-\sin x)^2}$$

$$= \frac{1}{2}\cdot\frac{\cos x}{\sqrt{(1+\sin x)(1-\sin x)}} = \frac{1}{2}\cdot\frac{\cos x}{|\cos x|},$$

$$x \neq n\pi + \frac{\pi}{2}$$

2. $y = \tan^{-1}\sqrt{\left(\frac{1-\cos x}{1+\cos x}\right)}$

Solution: $y = \tan^{-1}\sqrt{\left(\frac{1-\cos x}{1+\cos x}\right)}$

Let $u = \sqrt{\left(\frac{1-\cos x}{1+\cos x}\right)} = \sqrt{v}$ ...(i)

where $v = \frac{1-\cos x}{1+\cos x}$ ...(ii)

$\therefore y = \tan^{-1} u$ ...(iii)

From (i),

$$u = \sqrt{v} \Rightarrow \frac{du}{dv} = \frac{1}{2\sqrt{v}} = \frac{1}{2\sqrt{\left(\frac{1-\cos x}{1+\cos x}\right)}}$$

and from (ii),

$$v = \frac{1-\cos x}{1+\cos x}$$

$$\Rightarrow \frac{dv}{dx} = \frac{\sin x\,(1+\cos x) + \sin x\,(1-\cos x)}{(1+\cos x)^2}$$

$$\Rightarrow \frac{dv}{dx} = \frac{\sin x + \sin x\cos x + \sin x - \sin x\cos x}{(1+\cos x)^2}$$

Now, $y = \tan^{-1} u$

$$\Rightarrow \frac{dy}{du} = \frac{1}{1+u^2} = \frac{1}{1+\left(\frac{1-\cos x}{1+\cos x}\right)}$$

$$= \frac{1}{\left(\frac{1+\cos x+1-\cos x}{1+\cos x}\right)} = \frac{1}{\left(\frac{2}{1+\cos x}\right)}$$

$$= \frac{1+\cos x}{2} \quad \text{...(iv)}$$

Now,

$$\frac{dy}{dx} = \frac{dy}{du}\cdot\frac{du}{dv}\cdot\frac{dv}{dx}$$

$$= \frac{(1+\cos x)}{2}\cdot\frac{1}{2}\cdot\frac{\sqrt{1+\cos x}}{\sqrt{1-\cos x}}\cdot\frac{2\sin}{(1+\cos x)^2}$$

$$= \frac{1}{2}\cdot\frac{\sin x}{\sqrt{1-\cos^2 x}} = \frac{1}{2}\cdot\frac{\sin x}{|\sin x|},\ x \neq n\pi.$$

On Method of Trigonometric Substitution

Type 1: $y = t^{-1}[f(x)]$

where $t^{-1} = \sin^{-1} / \cos^{-1} / \tan^{-1} / \cot^{-1} / \sec^{-1} / \text{cosec}^{-1}$

$f(x)$ = a trigonometrical function of x / algebraic function of x / an algebric expression in x.

Note: **1.** In such types of problems mentioned above, our main aim is to remove inverse trigonometric operator $\sin^{-1}/\cos^{-1}/\tan^{-1}/\cot^{-1}/\sec^{-1}/\text{cosec}^{-1}$ by a trigonometric substitution $x = \sin\theta / \cos\theta / \tan\theta / \cot\theta / \sec\theta / \text{cosec}\,\theta$ or by any other means.

2. When $y = t^{-1}[f(x)]$ is provided, where $f(x)$ = an expression in x, we substitute x = a trigonometrical function of θ and t given expression in x becomes the direct trigonometrical function (or, trigonometrical function / circular function) of multiple angle of $\theta \Rightarrow f^{-1} f(m\theta) = m\theta$ which is differentiated w.r.t. x and lastly θ is expressed in terms of x, where $m = 1, 2, 3, \ldots$ etc.

$\Leftrightarrow$ **(a)** In $\sin^{-1}[f(x)]$, we put $\theta = \sin^{-1} x$, $-\frac{\pi}{2} \le \theta \le \frac{\pi}{2}$, then $x = \sin\theta$.

(b) In $\cos^{-1}[f(x)]$, we put $\theta = \cos^{-1} x$, $0 \le \theta \le \pi$, then $x = \cos\theta$

(c) In $\tan^{-1}[f(x)]$, we put $\theta = \tan^{-1} x$, $-\frac{\pi}{2} < \theta < \frac{\pi}{2}$, then $x = \tan\theta$ etc.

3. Remember the following formulas which give us idea of trigonometric substitution

(i) (a₁) $\sqrt{1 - \sin^2\theta} = |\cos\theta|$

(a₂) $\sqrt{1 - \cos^2\theta} = |\sin\theta|$

(b₁) $\sqrt{1 + \tan^2\theta} = |\sec\theta|$

(b₂) $\sqrt{\sec^2\theta - 1} = |\tan\theta|$

(c₁) $\sqrt{1 + \cot^2\theta} = |\text{cosec}\,\theta|$

(c₂) $\sqrt{\text{cosec}^2\theta - 1} = |\cot\theta|$, ... etc.

Where mod operator '$|(\ldots)|$' can be removed from the trigonometric function by considering the prinicipal values of $\theta = \sin^{-1} x, \cos^{-1} x, \tan^{-1} x, \cot^{-1} x, \sec^{-1} x, \text{cosec}^{-1} x$.

(ii) (a₁) $1 + \cos\theta = 2\cos^2\frac{\theta}{2}$

(a₂) $1 - \cos\theta = 2\sin^2\frac{\theta}{2}$

(b₁) $1 - 2\sin^2\theta = \cos^2\theta$

(b₂) $2\cos^2\theta - 1 = \cos^2\theta$

(iii) $\sin 2\theta = 2\sin\theta \cdot \cos\theta$

(iv) $\cos 2\theta = \cos^2\theta - \sin^2\theta = 1 - 2\sin^2\theta$

$= 2\cos^2\theta - 1$

(v) $\sin 2\theta = \dfrac{2\tan\theta}{1 + \tan^2\theta}$

(vi) $\cos 2\theta = \dfrac{1 - \tan^2\theta}{1 + \tan^2\theta}$

(vii) $\tan 2\theta = \dfrac{2\tan\theta}{1 - \tan^2\theta}$

(viii) $\sin 3\theta = 3\sin\theta - 4\sin^3\theta$

(ix) $\cos 3\theta = 4\cos^3\theta - 3\cos\theta$

(x) $\tan 3\theta = \dfrac{3\tan\theta - \tan^3\theta}{1 - 3\tan^2\theta}$

(xi) (a₁) $\tan(\theta + \phi) = \dfrac{\tan\theta + \tan\phi}{1 - \tan\theta \cdot \tan\phi}$

(a₂) $\tan(\theta - \phi) = \dfrac{\tan\theta - \tan\phi}{1 + \tan\theta \cdot \tan\phi}$

(xii) (a₁) $\tan^{-1}\dfrac{A + B}{1 - AB} = \tan^{-1} A + \tan^{-1} B$

(b₂) $\tan^{-1}\dfrac{A-B}{1+A\cdot B}=\tan^{-1}A-\tan^{-1}B$

(xiii) $\tan\left(\dfrac{\pi}{4}+\theta\right)=\dfrac{1+\tan\theta}{1-\tan\theta}$

(xiv) $\tan\left(\dfrac{\pi}{4}-\theta\right)=\dfrac{1-\tan\theta}{1+\tan\theta}$

N.B.: **1.** In the given question, we are given an expression in x which are obtained by replacing $x=\sin\theta/\cos\theta/\tan\theta/\cot\theta/\text{cosec}\,\theta$ and $\sec\theta$ in the above formulas in r.h.s. This is why the form of the expression in x gives us the idea of proper substitution $x=\sin\theta\,/\cos\theta\,/\tan\theta\,/\cot\theta\,/\sec\theta\,/\text{cosec}\,\theta$.

2. ***Remember***:

(i) $\sin^{-1}(\sin x)=x$, when $-\dfrac{\pi}{2}\le x\le\dfrac{\pi}{2}$,

$\sin^{-1}\sin\left(150^\circ\right)\ne 150^\circ\ (=30^\circ)$

(ii) $\cos^{-1}(\cos x)=x$, when $0\le x\le\pi$,

$\cos^{-1}\left(\cos\dfrac{2\pi}{3}\right)=\dfrac{2\pi}{3}$

(iii) $\tan^{-1}(\tan x)=x$, when $-\dfrac{\pi}{2}<x<\dfrac{\pi}{2}$,

$\tan^{-1}\left(\tan\dfrac{\pi}{6}\right)=\dfrac{\pi}{6}$, ... etc.

where we should note that $\sin^{-1}x$, $\tan^{-1}x$, and $\text{cosec}^{-1}x$ are angles which lie between $-\dfrac{\pi}{2}$ and $+\dfrac{\pi}{2}$ denoting their principal values and $\cos^{-1}x$, $\cot^{-1}x$ and $\sec^{-1}x$ are angles lying between 0 and π denoting their principal values.

3. (A) If $y=\sin^{-1}x$, $\text{cosec}^{-1}x$, or $\tan^{-1}x$, then x is negative means y is between $-\dfrac{\pi}{2}$ and 0. In this case $y=\sin^{-1}(-x)=-\sin^{-1}x$, $y=\text{cosec}^{-1}(-x)=-\text{cosec}^{-1}x$, $y=\tan^{-1}(-x)=-\tan^{-1}x$

Examples:

(i) $\sin^{-1}(-1)=-\sin^{-1}(1)=-\dfrac{\pi}{2}$

(ii) $\tan^{-1}(-1)=-\tan^{-1}(1)=-\dfrac{\pi}{4}$ etc.

(B) If $y=\cos^{-1}x$, $\sec^{-1}x$, or $\cot^{-1}x$, then x is negative means y is between $\dfrac{\pi}{2}$ and π. In this case

$y=\pi-\cos^{-1}x=\cos^{-1}(-x)$

$y=\pi-\sec^{-1}x=\sec^{-1}(-x)$

$y=\pi-\cot^{-1}x=\cot^{-1}(-x)$

Examples:

(i) $\cos^{-1}\left(-\dfrac{1}{2}\right)=\pi-\cos^{-1}\left(\dfrac{1}{2}\right)=\pi-\dfrac{\pi}{3}=\dfrac{2\pi}{3}$

(ii) $\sec^{-1}\left|\left(\dfrac{-2}{3}\right)\right|=\pi-\sec^{-1}(2)=\pi-\dfrac{\pi}{3}=\dfrac{2\pi}{3}$ etc.

(4) (i) $\sin\left(\sin^{-1}x\right)=x$ for $-1\le x\le 1$

(ii) $\cos\left(\cos^{-1}x\right)=x$ for $-1\le x\le 1$

(iii) $\tan\left(\tan^{-1}x\right)=x$ for $x\in R$

(iv) $\cot\left(\cot^{-1}x\right)=x$ for $x\in R$

(v) $\sec\left(\sec^{-1}x\right)=x$ for $|x|\ge 1$

(vi) $\text{cosec}\left(\text{cosec}^{-1}x\right)=x$ for $|x|\ge 1$

5. If $|x|\ge 1$, then $\text{cosec}^{-1}x=\sin^{-1}\left(\dfrac{1}{x}\right)$ or $\sin^{-1}x=\text{cosec}^{-1}\left(\dfrac{1}{x}\right)$.

(ii) If $x>0$, then $\tan^{-1}x=\cot^{-1}\left(\dfrac{1}{x}\right)$ or $\cot^{-1}x=\tan^{-1}\left(\dfrac{1}{x}\right)$.

(iii) If $|x|\ge 1$, then $\sec^{-1}x=\cos^{-1}\left(\dfrac{1}{x}\right)$ or $\cos^{-1}x=\sec^{-1}\left(\dfrac{1}{x}\right)$.

Solved Examples put in the form: $y = t^{-1}[f(x)]$

Find the d.c. of the follwoing:

1. $y = \cos^{-1}(1 - 2x^2)$

Solution: on putting $x = \sin\theta \Leftrightarrow \theta = \sin^{-1}x$, where $-\frac{\pi}{2} \le \theta \le \frac{\pi}{2}$, $1 - 2\sin^2\theta = \cos 2\theta$ suggests to put $x = \sin\theta$ in $(1 - 2x^2)$ to get

$$y = \cos^{-1}(1 - 2\sin^2\theta) = \cos^{-1}(\cos 2\theta)$$

$$= 2\theta = 2\sin^{-1}x \text{ if } 0 \le \theta \le \frac{\pi}{2}\text{; i.e; } 0 \le x \le 1$$

$$= -2\theta = -2\sin^{-1}x \quad \text{if} \quad -\frac{\pi}{2} \le \theta \le 0; \quad \text{i.e; } -1 \le x \le 0$$

$$\frac{dy}{dx} = \frac{2}{\sqrt{1 - x^2}} \text{ if } 0 < x < 1$$

$$= \frac{-2}{\sqrt{1 - x^2}} \quad \text{if } -1 < x < 0$$

Note: **(i)** y is defined for $|x| \le 1$

(ii) $\frac{dy}{dx}$ does not exist for $x = 0, 1, -1$

(iii) (Chain rule): $\frac{dy}{dx} = \frac{-1}{\sqrt{1 - (1 - 2x^2)^2}}(-4x)$

$$= \frac{2x}{|x|\sqrt{1 - x^2}}, \; x \ne 0, \pm 1$$

2. $y = \sin^{-1}(3x - 4x^3)$

Solution: on putting $x = \sin\theta \Leftrightarrow \theta = \sin^{-1}x$, where $-\frac{\pi}{2} \le \theta \le \frac{\pi}{2}$, $3\sin\theta - 4\sin^3\theta = \sin 3\theta$ suggests to put $x = \sin\theta$ in $(3x - 4x^3)$ to get,

$$y = \sin^{-1}(3\sin\theta - 4\sin^3\theta)$$

$$= \sin^{-1}(\sin 3\theta)$$

$$= 3\theta = 3\sin^{-1}x \text{ if } -\frac{\pi}{2} \le 3\theta \le \frac{\pi}{2}\text{, i.e;}$$

$$-\frac{\pi}{6} \le \theta \le \frac{\pi}{6} \text{ i.e; } -\frac{1}{2} \le x \le \frac{1}{2}$$

$$\therefore \frac{dy}{dx} = \frac{3}{\sqrt{1 - x^2}} \text{ for } |x| < \frac{1}{2}$$

Now, $y = -\pi - 3\theta$ for $-\frac{\pi}{2} \le \theta \le -\frac{\pi}{6}$ and

$$y = \pi - 3\theta \text{ for } \frac{\pi}{6} \le \theta \le \frac{\pi}{2}$$

$$\therefore \frac{dy}{dx} = \frac{-3}{\sqrt{1 - x^2}} \quad \text{for } \frac{1}{2} < |x| < 1$$

Note: By chain rule

$$\frac{dy}{dx} = \frac{3(1 - 4x^2)}{\sqrt{1 - x^2}\,|1 - 4x^2|}, \; x \ne \pm\frac{1}{2}, \pm 1.$$

3. $y = \cos^{-1}(4x^3 - 3x)$

Solution: on putting $x = \cos\theta \Leftrightarrow \theta = \cos^{-1}x$, where $0 \le \theta \le \pi$, $4\cos^3\theta - 3\cos\theta = \cos 3\theta$ suggests to put $x = \cos\theta$ to get,

$$y = \cos^{-1}(4\cos^3\theta - 3\cos\theta)$$

$$= \cos^{-1}(\cos 3\theta) = 3\theta = 3\cos^{-1}x \text{ for } 0 \le \theta \le \frac{\pi}{3}$$

$$= 3\theta - 2\pi \text{ for } -1 < x < -\frac{1}{2}$$

$$\frac{dy}{dx} = \frac{-3}{\sqrt{1 - x^2}} \text{ for } \frac{1}{2} < |x| < 1$$

Also $y = 2\pi - 3\theta$ for $|x| < \frac{1}{2}$

$$\therefore \frac{dy}{dx} = \frac{3}{\sqrt{1 - x^2}} \quad \text{for } |x| < \frac{1}{2}$$

4. $y = \tan^{-1}\dfrac{2x}{1-x^2}$

Solution: on putting $x = \tan\theta \Leftrightarrow \theta = \tan^{-1}x$ for

$-\dfrac{\pi}{2} < \theta < \dfrac{\pi}{2}$, $\dfrac{2\tan\theta}{1-\tan^2\theta} = \tan 2\theta$ suggests to put

$x = \tan\theta$.

$$\therefore\ y = \tan^{-1}\frac{2\tan\theta}{1-\tan^2\theta} = \tan^{-1}(\tan 2\theta)$$

$$= n\pi + 2\theta = n\pi + 2\tan^{-1}x$$

where $\dfrac{-\pi}{2} < y < \dfrac{\pi}{2}$.

$$\therefore\ \frac{dy}{dx} = \frac{2}{1+x^2};\ x \neq \pm 1.$$

5. $y = \sin^{-1}\dfrac{2x}{1+x^2}$

Solution: On putting $x = \tan\theta \Leftrightarrow \theta = \tan^{-1}x$;

$-\dfrac{\pi}{2} < \theta < \dfrac{\pi}{2}$

$\dfrac{2\tan\theta}{1+\tan^2\theta} = \sin 2\theta$ suggests to put

$x = \tan\theta$ in $\dfrac{2x}{1+x^2}$

$$\therefore\ y = \sin^{-1}\frac{2\tan\theta}{1+\tan^2\theta}$$

$$= \sin^{-1}\left(\frac{2\sin\theta\cos\theta}{\cos^2\theta + \sin^2\theta}\right) = \sin^{-1}(2\sin\theta\cos\theta)$$

$$= \sin^{-1}(\sin 2\theta)$$

$$= 2\theta = 2\tan^{-1}x,\ \text{if } -\frac{\pi}{4} \le \theta \le \frac{\pi}{4}\ \text{i.e; } |x| \le 1$$

$$= \pi - 2\theta\ \text{for } \frac{\pi}{4} < \theta < \frac{\pi}{2},\ \text{i.e. } x > 1$$

$$= -\pi - 2\theta\ \text{for } -\frac{\pi}{2} < \theta < -\frac{\pi}{4}\ \text{i.e; } x < -1$$

$$\frac{dy}{dx} = \frac{2}{1+x^2}\ \text{for } |x| < 1$$

$$= \frac{-2}{1+x^2}\ \text{for } |x| > 1$$

Note: **(i)** $\dfrac{dy}{dx}$ does not exist at $x = \pm 1$

(ii) To avoid different cases, we may differentiate directly using chain rule as shown below in example (6) and others

6. $y = \cos^{-1}\left(\dfrac{1-x^2}{1+x^2}\right)$

Solution: $y = \cos^{-1}\left(\dfrac{1-x^2}{1+x^2}\right)$

$$\Rightarrow \frac{dy}{dx} = \frac{-1}{\sqrt{1-\left(\frac{1-x^2}{1+x^2}\right)^2}} \times \frac{d}{dx}\left(\frac{1-x^2}{1+x^2}\right)$$

$$= \frac{-(1+x^2)}{\sqrt{4x^2}} \times \frac{-4x}{(1+x^2)^2}$$

$$= \frac{2x}{|x|(1+x^2)}\ x \neq 0. \quad \text{(very easy method)}$$

7. $y = \sin^{-1}\left(\dfrac{1-x^2}{1+x^2}\right)$

Solution: $y = \sin^{-1}\left(\dfrac{1-x^2}{1+x^2}\right)$

$$\Rightarrow \frac{dy}{dx} = \frac{1}{\sqrt{1-\left(\frac{1-x^2}{1+x^2}\right)^2}} \times \frac{d}{dx}\left(\frac{1-x^2}{1+x^2}\right)$$

$$= \frac{(1+x^2)}{\sqrt{4x^2}} \times \frac{-4x}{(1+x^2)^2}$$

$$= \frac{-2x}{|x|(1+x^2)}, \; x \neq 0$$

8. $y = \cos^{-1}\left(\dfrac{2x}{1+x^2}\right)$

Solution: $y = \cos^{-1}\left(\dfrac{2x}{1+x^2}\right), \; 0 \le y \le \pi$

$$\Rightarrow \cos y = \frac{2x}{1+x^2}$$

$$\Rightarrow \frac{d\cos y}{dx} = \frac{(1+x^2)\times 2 - 2x\,(2x)}{(1+x^2)^2}$$

$$= \frac{2(1+x^2) - 4x^2}{(1+x^2)^2}$$

$$\Rightarrow -\sin y\,\frac{dy}{dx} = \frac{2-2x^2}{(1+x^2)^2} = \frac{2(1-x^2)}{(1+x^2)^2}$$

$$\Rightarrow \frac{dy}{dx} = \frac{2(1-x^2)}{(1+x^2)^2}\cdot\frac{1}{\sin y}, \; \sin y \neq 0.$$

Now, $\sin y = \sqrt{1-\cos^2 y}$

$$= \sqrt{1 - \frac{4x^2}{(1+x^2)^2}}$$

$$= \sqrt{\frac{(1+x^2)^2 - 4x^2}{(1+x^2)^2}} = \sqrt{\frac{1+x^4+2x^2-4x^2}{(1+x^2)^2}}$$

$$= \sqrt{\frac{1+x^4-2x^2}{(1+x^2)^2}} = \sqrt{\frac{(1-x^2)^2}{(1+x^2)^2}}$$

$$= \frac{|(1-x^2)|}{|(1+x^2)|} = \frac{|(1-x^2)|}{(1+x^2)}$$

[$\because |(1+x^2)| = 1 + x^2$ because $(1+x^2)$ is always positive]

Hence, $\dfrac{dy}{dx} = \dfrac{2(1-x^2)}{(1+x^2)^{2}} \times \dfrac{(1+x^2)}{|(1-x^2)|}$

$$= \frac{2(1-x^2)}{|(1-x^2)|(1+x^2)}, \; x \neq \pm 1.$$

Note: If we do the above problem by the method of trigonometric substitution, we get complete result provided different cases are properly considered otherwise (i.e; if different cases are not properly considered) , we get an incomplete result $\dfrac{dy}{dx} = \dfrac{2}{1+x^2}$ which is only possible when $|x| < 1$

9. $y = \cos^{-1}\left(\dfrac{1-x^2}{1+x^2}\right)$

Solution: $y = \cos^{-1}\left(\dfrac{1-x^2}{1+x^2}\right)$

$$\Rightarrow \cos y = \frac{1-x^2}{1+x^2}$$

$$\Rightarrow \frac{d\cos y}{dx} = \frac{(1+x^2)\times(-2x) - (1-x^2)\times(2x)}{(1+x^2)^2}$$

$$= \frac{-2x - 2x^3 - 2x + 2x^3}{(1+x^2)^2}$$

$$\Rightarrow -\sin y \frac{dy}{dx} = \frac{-4x}{\left(1+x^2\right)^2}$$

$$\Rightarrow \frac{dy}{dx} = \frac{-4x}{\left(1+x^2\right)^2} \times \frac{1}{-\sin y} = \frac{4x}{\left(1+x^2\right)^2} \cdot \frac{1}{\sin y}$$

Now, $\sin y = \sqrt{1-\cos^2 y}$

$(\because \sin y = |\sin y| \text{ as } 0 \le y \le \pi)$

$$= \sqrt{1-\left(\frac{1-x^2}{1+x^2}\right)^2}$$

$$= \sqrt{\frac{\left(1+x^2\right)^2-\left(1-x^2\right)^2}{\left(1+x^2\right)^2}}$$

$$= \sqrt{\frac{1+x^4+2x^2-\left(1+x^4-2x^2\right)}{\left(1+x^2\right)^2}}$$

$$= \sqrt{\frac{\cancel{1}+\cancel{x^4}+2x^2-\cancel{1}-\cancel{x^4}+2x^2}{\left(1+x^2\right)^2}}$$

$$= \sqrt{\frac{4x^2}{\left(1+x^2\right)^2}}$$

$$= \frac{2|x|}{\left|\left(1+x^2\right)\right|}$$

Hence, $\dfrac{dy}{dx} = \dfrac{4x}{\left(1+x^2\right)^2} \cdot \dfrac{\left|\left(1+x^2\right)\right|}{2|x|}$

$$= \frac{4x}{\left(1+x^2\right)} \cdot \frac{\left|\left(1+x^2\right)\right|}{2|x|}$$

$\left(\because \left(1+x^2\right) \text{ is always positive}\right)$

$$= \frac{2x}{\left(1+x^2\right)\cdot|x|};\ x \ne 0$$

10. $y = \sin^{-1}\left(\dfrac{1-x^2}{1+x^2}\right)$

Solution: $y = \sin^{-1}\left(\dfrac{1-x^2}{1+x^2}\right)$

$$\Rightarrow \sin y = \left(\frac{1-x^2}{1+x^2}\right)$$

$$\Rightarrow \frac{d\sin y}{dx} = \frac{\left(1+x^2\right)\times(-2x)-\left(1-x^2\right)\times(2x)}{\left(1+x^2\right)^2}$$

$$\Rightarrow \cos y\frac{dy}{dx} = \frac{-2x-2x^3-2x+2x^3}{\left(1+x^2\right)^2} = \frac{-4x}{\left(1+x^2\right)^2}$$

$$\Rightarrow \frac{dy}{dx} = \frac{-4x}{\left(1+x^2\right)^2} \cdot \frac{1}{\cos y}$$

Now, $\cos y = \sqrt{1-\sin^2 y}$

$\left(\because \cos y = |\cos y| \text{ as } -\dfrac{\pi}{2} \le y \le \dfrac{\pi}{2}\right)$

$$= \sqrt{1-\left(\frac{1-x^2}{1+x^2}\right)^2}$$

$$= \sqrt{\frac{\left(1+x^2\right)^2-\left(1-x^2\right)^2}{\left(1+x^2\right)^2}}$$

$$= \sqrt{\frac{1+x^4+2x^2-\left(1+x^4-2x^2\right)}{\left(1+x^2\right)^2}}$$

$$= \sqrt{\frac{4x}{\left(1+x^2\right)^2}}$$

$$= \frac{2|x|}{\left|\left(1+x^2\right)\right|} = \frac{2|x|}{1+x^2}$$

($\because \left|1+x^2\right| = 1+x^2$ for $\left(1+x^2\right)$ being always positive)

$$\therefore \frac{dy}{dx} = -\frac{4x}{\left(1+x^2\right)} \times \frac{\left(1+x^2\right)}{2|x|} = \frac{-2x}{|x|\left(1+x^2\right)}$$

11. $y = \sin^{-1}\sqrt{1-x^2}$

Solution: $y = \sin^{-1}\sqrt{1-x^2}$

$$\Rightarrow \sin y = \sqrt{1-x^2}$$

$$\Rightarrow \frac{d\sin y}{dx} = \frac{1}{\sqrt{1-x^2}} \times (-2x) = \frac{-x}{\sqrt{1-x^2}}$$

$$\Rightarrow \cos y\,\frac{dy}{dx} = \frac{-x}{\sqrt{1-x^2}}$$

$$\Rightarrow \frac{dy}{dx} = \frac{-x}{\sqrt{1-x^2}} \cdot \frac{1}{\cos y}$$

$$\cos y = \sqrt{1-\sin^2 y}$$

$$\left(\because \cos y = |\cos y| \text{ as } -\frac{\pi}{2} \le y \le \frac{\pi}{2}\right)$$

$$= \sqrt{1-\left(\sqrt{1-x^2}\right)^2} = \sqrt{1-\left(1-x^2\right)} = \sqrt{1-1+x^2}$$

$$= \sqrt{x^2} = |x|$$

$$\therefore \frac{dy}{dx} = -\frac{x}{\sqrt{1-x^2}} \times \frac{1}{|x|}$$

$$= -\frac{x}{|x|\sqrt{1-x^2}},\ x \ne 0.$$

12. $y = \tan^{-1}\left(\dfrac{3x-x^3}{1-3x^2}\right)$

Solution: $y = \tan^{-1}\left(\dfrac{3x-x^3}{1-3x^2}\right)$

on putting $x = \tan\theta \Leftrightarrow \theta = \tan^{-1} x$ for $-\dfrac{\pi}{2} < \theta < \dfrac{\pi}{2}$, $\dfrac{3\tan\theta - \tan^3\theta}{1-3\tan^2\theta} = \tan 3\theta$ suggests to put $x = \tan\theta$ in $\left(\dfrac{3x-x^3}{1-3x^2}\right)$; $x \ne \pm\dfrac{1}{\sqrt{3}}$

$$\therefore y = \tan^{-1}\left(\frac{3\theta - \tan^3\theta}{1-3\tan^2\theta}\right) = \tan^{-1}(\tan 3\theta)$$

$$= 3\theta + n\pi = 3\tan^{-1} x + n\pi$$

$$\Rightarrow \frac{dy}{dx} = \frac{3}{1+x^2};\ x \ne \pm\frac{1}{\sqrt{3}}.$$

13. $y = \tan^{-1}\left(\dfrac{4x}{4-x^2}\right)$

Solution: $y = \tan^{-1}\left(\dfrac{4x}{4-x^2}\right)$; $x \ne \pm 2$

$$= \tan^{-1}\left(\frac{x}{1-\left(\frac{x}{2}\right)^2}\right) = \tan^{-1}\left(\frac{\frac{2x}{2}}{1-\left(\frac{x}{2}\right)^2}\right)$$

Now on putting, $\dfrac{x}{2} = \tan\theta \Leftrightarrow \theta = \tan^{-1}\left(\dfrac{x}{2}\right)$ for $-\dfrac{\pi}{2} < \theta < \dfrac{\pi}{2}$, we have

$$y = \tan^{-1}\left(\frac{2\tan\theta}{1-\tan^2\theta}\right) = \tan^{-1}(\tan 2\theta) = 2\theta + n\pi$$

where $-\frac{\pi}{2} < y < \frac{\pi}{2}$

$$\Rightarrow y = 2\tan^{-1}\frac{x}{2} + n\pi$$

$$\Rightarrow \frac{dy}{dx} = 2\cdot\frac{1}{1+\left(\frac{x}{2}\right)^2}\cdot\frac{1}{2}$$

$$= \frac{4}{4+x^2},\ x \neq \pm 2$$

14. $y = \sin^{-1}\left(2x\sqrt{1-x^2}\right)$

Solution: $y = \sin^{-1}\left(2x\sqrt{1-x^2}\right)$

on putting, $x = \sin\theta \Leftrightarrow \theta = \sin^{-1} x$ for

$-\frac{\pi}{2} \le \theta \le \frac{\pi}{2}$, $\sqrt{1-\sin^2\theta} = |\cos\theta|$ suggests to put

$x = \sin\theta$ in $\sqrt{1-x^2}$

$$\therefore y = \sin^{-1}\left(2\sin\theta\sqrt{1-\sin^2\theta}\right)$$

$$= \sin^{-1}(2\sin\theta\cos\theta)$$

$$\left(\because |\cos\theta| = \cos\theta \text{ as } -\frac{\pi}{2} \le \theta \le \frac{\pi}{2}\right)$$

$$\Rightarrow y = \sin^{-1}(\sin 2\theta) = 2\theta \text{ for } -\frac{\pi}{4} \le \theta \le \frac{\pi}{4}$$

i.e; $-\frac{1}{\sqrt{2}} \le x \le \frac{1}{\sqrt{2}}$ or $2x^2 \le 1$ ...(i)

$y = \sin^{-1}\sin(2\theta) = -\pi - 2\theta$,

when $-\frac{\pi}{2} \le \theta < -\frac{\pi}{4}$ i.e; $-1 \le x < -\frac{1}{\sqrt{2}}$ ...(ii)

and $y = \sin^{-1}\sin(2\theta) = \pi - 2\theta$, when $\frac{\pi}{4} < \theta \le \frac{\pi}{2}$

i.e; $\frac{1}{\sqrt{2}} < x \le 1$ ...(iii)

Hence, from (i),

$$\frac{dy}{dx} = \frac{2\,d\theta}{dx} = \frac{2\,d\left(\sin^{-1}x\right)}{dx} = \frac{2}{\sqrt{1-x^2}} \text{ for } 2x^2 < 1 \quad \text{...(iv)}$$

from (ii) and (iii),

$$\frac{dy}{dx} = -\frac{2\,d\theta}{dx} = \frac{-2}{\sqrt{1-x^2}} \text{ for } 2 > 2x^2 > 1 \quad \text{...(v)}$$

$\frac{dy}{dx}$ does not exist for $x^2 = 1, \frac{1}{2}$

or, alternatively,

$$y = \sin^{-1}\left(2x\sqrt{1-x^2}\right)$$

$$\Rightarrow \sin y = 2x\sqrt{1-x^2}$$

$$\Rightarrow \frac{d\sin y}{dx} = 2\sqrt{1-x^2} + \frac{2x}{2\sqrt{1-x^2}} \times (-2x)$$

$$\Rightarrow \cos y\,\frac{dy}{dx} = 2\sqrt{1-x^2} - \frac{2x^2}{\sqrt{1-x^2}}$$

$$= \frac{2\left(1-x^2\right) - 2x^2}{\sqrt{1-x^2}}$$

$$= 2\left\{\frac{1-x^2-x^2}{\sqrt{1-x^2}}\right\} = \frac{2\left(1-2x^2\right)}{\sqrt{1-x^2}}$$

$$\Rightarrow \frac{dy}{dx} = \frac{2\left(1-2x^2\right)}{\sqrt{1-x^2}} \times \frac{1}{\cos y}$$

Now, $\cos y = \sqrt{1-\sin^2 y} = \sqrt{1-4x^2\left(1-x^2\right)}$

$$\left[\because |\cos y| = \cos y \text{ as } -\frac{\pi}{2} \le y \le \frac{\pi}{2}\right]$$

$$= \sqrt{1-4x^2+4x^4} = \sqrt{(1-2x^2)^2} = \left|(1-2x^2)\right|$$

$$\left[\because \sqrt{f^2(x)} = |f(x)|\right]$$

$$\frac{dy}{dx} = \frac{2(1-2x^2)}{\sqrt{1-x^2}} \times \frac{1}{\left|(1-2x^2)\right|}$$

$$= \frac{2(1-2x^2)}{\left|(1-2x^2)\right|\sqrt{1-x^2}}\,;\, 2x^2 \ne 1,\, x^2 \ne 1$$

More about substitution

1. (a$_1$) If $\sqrt{a^2 - x^2}$ occurs, we put $x = a\sin\theta$ or $\cos\theta$, i.e; $\theta = \sin^{-1}\left(\frac{x}{a}\right)$, when $-\frac{\pi}{2} \le \theta \le \frac{\pi}{2}$ and 'a' is positive

(a$_2$) If $\sqrt{x^2 - a^2}$ occurs, we put $x = a\sec\theta$ or $\operatorname{cosec}\theta$, i.e; $\theta = \sec^{-1}\left(\frac{x}{a}\right)$, when $0 \le \theta \le \pi$ $\left(\theta \ne \frac{\pi}{2}\right)$ and 'a' is positive.

(a$_3$) If $\sqrt{a^2 + x^2}$ $(\text{or}, a^2 + x^2)$ occurs, we put $x = a\tan\theta$ or $a\cot\theta$, i.e; $\theta = \tan^{-1}\left(\frac{x}{a}\right)$, when $-\frac{\pi}{2} < \theta < \frac{\pi}{2}$ and 'a' is positive.

(a$_4$) If $\sqrt{a-x}$ or $\sqrt{a+x}$ or $\sqrt{\frac{a-x}{a+x}}$ or $\sqrt{\frac{a+x}{a-x}}$ occurs, we put $x = a\cos 2\theta$.

2. The preceeding square roots are $a\cos\theta$, $a\sec\theta$ and a $\tan\theta$ when these quantities are not negative and this will be true in particular when a is a positive constant and θ represents the principal values of $\sin^{-1}x$ $\left(\text{or}, \sin^{-1}\frac{x}{a}\right)$, $\cos^{-1}x$ $\left(\text{or}, \cos^{-1}\frac{x}{a}\right)$, $\tan^{-1}x$ $\left(\text{or}, \tan^{-1}\frac{x}{a}\right)$, $\cot^{-1}x$ $\left(\text{or}, \cot^{-1}\frac{x}{a}\right)$

(i) If $x = a\sin\theta$, then $\sqrt{a^2 - x^2} = \sqrt{a^2 - a^2\sin^2\theta}$ $= |a\cos\theta| = a\cos\theta$.

(ii) If $x = a\tan\theta$, then $\sqrt{a^2 + x^2} = \sqrt{a^2 + a^2\tan^2\theta}$ $= \sqrt{a^2\sec^2\theta} = |a\sec\theta| = a\sec\theta$.

(iii) If $x = a\sec\theta$, then $\sqrt{x^2 - a^2} = \sqrt{a^2\sec^2\theta - a^2}$ $= \sqrt{a^2\tan^2\theta} = |a\tan\theta|$.

3. If we are not given the expression which can be transformed into the trigonometric function of multiple angle of θ, we should avoid this substitution rule since it becomes complicated and gives no fruitful result to find d.c.. For this reason we should use chain rule to find d.c. when substitution method fails, i.e; we should use the formula.

$$\frac{dy}{dx} = \frac{1}{\left(\frac{dx}{dy}\right)}$$ or chain rule (i.e; function of a function rule).

Remember: Generally, we are provided the following form $t^{-1}\left[f\left(\sqrt{1 \pm x^2} \,/\, \sqrt{a^2 \pm x^2} \,/\, \sqrt{1 \pm x^2} \pm \text{a constant} \,/\, \sqrt{a^2 + x^2} \pm \text{a constant}\right)\right]$ Whose differential coefficient is required to find out, where $t^{-1} = \sin^{-1}, \cos^{-1}, \tan^{-1}, \cot^{-1}, \sec^{-1}$, and $\operatorname{cosec}^{-1}$.

Solved Examples:

Find the d.c. of the following inverse trigonometric functions.

1. $y = \sin^{-1}\left[\dfrac{x}{\sqrt{1+x^2}}\right]$

Solution: $y = \sin^{-1}\left[\dfrac{x}{\sqrt{1+x^2}}\right]$

Firstly, on simplification after putting

$$x = \tan\theta \Leftrightarrow \theta = \tan^{-1}x, \text{ where } -\frac{\pi}{2} \le \theta \le \frac{\pi}{2}$$

$$\therefore y = \sin^{-1}\left[\frac{\tan\theta}{\sqrt{1+\tan\theta}}\right]$$

$$\Rightarrow y = \sin^{-1}\left[\frac{\tan\theta}{\sqrt{\sec^2\theta}}\right]$$

$$\therefore y = \sin^{-1}\left[\frac{\tan\theta}{\sec\theta}\right]$$

$$\left(\because |\sec\theta| = \sec\theta \text{ for } -\frac{\pi}{2} \le \theta \le \frac{\pi}{2}\right)$$

$$\Rightarrow y = \sin^{-1}\left[\frac{\sin\theta}{\cos\theta} \times \cos\theta\right]$$

$$\Rightarrow y = \sin^{-1}[\sin\theta]$$

$$\Rightarrow y = \theta = \tan^{-1}x$$

$$\Rightarrow \frac{dy}{dx} = \frac{1}{1+x^2}$$

2. $y = \tan^{-1}\dfrac{x}{\sqrt{1-x^2}}$

Solution: Firstly, on simplification after putting

$$x = \sin\theta \Leftrightarrow \theta = \sin^{-1}x \text{ for } -\frac{\pi}{2} \le \theta \le \frac{\pi}{2}$$

$$\therefore y = \tan^{-1}\frac{\sin\theta}{\sqrt{1-\sin^2\theta}} = \tan^{-1}\left(\frac{\sin\theta}{\cos\theta}\right)$$

$$\left(\because \text{ for } -\frac{\pi}{2} \le \theta \le \frac{\pi}{2} \Rightarrow |\cos\theta| = \cos\theta\right)$$

$$\Rightarrow y = \tan^{-1}(\tan\theta) = \theta = \sin^{-1}x$$

$$\Rightarrow \frac{dy}{dx} = \frac{d\sin^{-1}x}{dx} = \frac{1}{\sqrt{1-x^2}}, x \ne \pm 1.$$

3. $y = \tan^{-1}\left[\dfrac{x}{\sqrt{a^2-x^2}}\right]$

Solution: Firstly, on simplification by putting

$$x = a\sin\theta \Leftrightarrow \theta = \sin^{-1}\left(\frac{x}{a}\right) \text{ for } -\frac{\pi}{2} \le \theta \le \frac{\pi}{2}$$

and $a > 0$ (i.e; $|a| = a$)

$$\therefore y = \tan^{-1}\left[\frac{a\sin\theta}{\sqrt{a^2 - a^2\sin^2\theta}}\right] = \tan^{-1}\left[\frac{a\sin\theta}{a\cos\theta}\right]$$

$$\left[\because |\cos\theta| = \cos\theta \text{ for } -\frac{\pi}{2} \le \theta \le \frac{\pi}{2}\right]$$

$$\Rightarrow y = \tan^{-1}[\tan\theta] = \theta = \sin^{-1}\left(\frac{x}{a}\right).$$

$$\Rightarrow \frac{dy}{dx} = \frac{1}{\sqrt{1-\frac{x^2}{a^2}}} \cdot \frac{1}{a} = \frac{1}{\left(\frac{\sqrt{a^2-x^2}}{|a|}\right)} \cdot \frac{1}{a}$$

$$= \frac{|a|}{\sqrt{a^2-x^2}} \cdot \frac{1}{a}$$

$$\Rightarrow \frac{dy}{dx} = \frac{1}{\sqrt{a^2-x^2}}, x \ne \pm a.$$

4. $y = \tan^{-1}\left[\dfrac{\sqrt{1+x^2}-1}{x}\right]$

Solution: y is defined for all $x \neq 0$. Firstly on simplification by putting

$$x = \tan\theta \Leftrightarrow \theta = \tan^{-1} x \text{ for } -\frac{\pi}{2} < \theta < \frac{\pi}{2}$$

$$\therefore y = \tan^{-1}\left[\frac{\sqrt{1+x^2}-1}{x}\right] = \tan^{-1}\left[\frac{\sqrt{1+\tan^2\theta}-1}{\tan\theta}\right]$$

$$= \tan^{-1}\left[\frac{\sec\theta - 1}{\tan\theta}\right]; \theta \neq 0, \text{ i.e; } x \neq 0$$

$$\left(\because |\sec\theta| = \sec\theta \text{ for } -\frac{\pi}{2} < \theta < \frac{\pi}{2}\right)$$

$$= \tan^{-1}\left[\frac{1-\cos\theta}{\sin\theta}\right] = \tan^{-1}\left[\frac{\frac{2\sin^2\theta}{2}}{\frac{2\sin\theta}{2}\frac{\cos\theta}{2}}\right] = \tan^{-1}\left(\tan\frac{\theta}{2}\right)$$

$$= \frac{\theta}{2} = \frac{1}{2}\tan^{-1} x$$

$$\Rightarrow \frac{dy}{dx} = \frac{1}{2}\cdot\frac{1}{1+x^2} = \frac{1}{2\left(1+x^2\right)}, (x \neq 0)$$

5. $y = \cot^{-1}\left[\dfrac{\sqrt{1+x^2}-1}{x}\right]$

Solution: $y = \cot^{-1}\left[\dfrac{\sqrt{1+x^2}-1}{x}\right]$

The function is defined for all $x \neq 0$. Firstly, on simplification by putting

$$x = \tan\theta \Leftrightarrow \theta = \tan^{-1} x \text{ for } -\frac{\pi}{2} < \theta < \frac{\pi}{2}, \theta \neq 0$$

$$y = \cot^{-1}\left[\frac{\sqrt{1+\tan^2\theta}+1}{\tan\theta}\right]$$

$$= \cot^{-1}\left[\frac{\sec\theta + 1}{\tan\theta}\right]$$

$$\left(\because |\sec\theta| = \sec\theta \text{ for } -\frac{\pi}{2} < \theta < \frac{\pi}{2}\right)$$

$$= \cot^{-1}\left(\frac{1+\cos\theta}{\sin\theta}\right) = \cot^{-1}\left[\frac{1+2\cos^2\frac{\theta}{2}-1}{2\sin\frac{\theta}{2}\cdot\cos\frac{\theta}{2}}\right]$$

$$= \cot^{-1}\left(\cot\frac{\theta}{2}\right)$$

$$= \frac{\theta}{2} = \frac{1}{2}\tan^{-1} x \text{ if } 0 < \theta < \frac{\pi}{2}$$

$$\text{and } y = \pi + \frac{\theta}{2}, \text{ if } -\frac{\pi}{2} < \theta < 0$$

$$= \pi + \frac{1}{2}\tan^{-1} x, \text{ if } -\frac{\pi}{2} < \theta < 0$$

$$\text{Hence, } \frac{dy}{dx} = \frac{1}{2\left(1+x^2\right)}; (x \neq 0).$$

Type 2:

Form: A function having the form $t\left(t^{-1}x\right)$ or $\left[t\left(t^{-1}x\right)\right]^n$ where t = sin / cos / tan / cot / sec / cosec and t^{-1} = $\sin^{-1}$ / $\cos^{-1}$ / $\tan^{-1}$ / $\cot^{-1}$ / $\sec^{-1}$ / cosec^{-1} is differentiated using the following working rule.

Working rule:

(1) Put $x = \sin\theta$ in $\sin^{-1} x$

$x = \cos\theta$ in $\cos^{-1} x$

$x = \tan\theta$ in $\tan^{-1} x$

$x = \cot\theta$ in $\cot^{-1} x$

$x = \sec\theta$ in $\sec^{-1} x$

$x = \text{cosec}\,\theta$ in $\text{cosec}^{-1} x$

So that we may obtain $t^{-1}\,t(\theta) = \theta$ where θ represents the principal values of $\sin^{-1} x$, $\cos^{-1} x$, $\tan^{-1} x$, $\cot^{-1} x$, $\sec^{-1} x$ and $\text{cosec}^{-1} x$.

2. Differentiate $\frac{dt(\theta)}{dx} = \frac{dt(\theta)}{d\theta}\cdot\frac{d\theta}{dx} = \frac{\frac{dt(\theta)}{d\theta}}{\frac{dx}{d\theta}}$

3. Express the required result in terms of x (or, in terms of x and y both if required).

Note: The above types of problems also can be done by using the chain rule $\Rightarrow \frac{dy}{dx} = \frac{dt(t^{-1}x)}{d(t^{-1}x)}\cdot\frac{d(t^{-1}x)}{dx}$

or, $\frac{dy}{dx} = \frac{d\left[t(t^{-1}x)\right]^n}{d\left[t(t^{-1}x)\right]}$

$= n\left[t(t^{-1}x)\right]^{n-1}\cdot\frac{d\left[t(t^{-1}x)\right]}{d(t^{-1}x)}\cdot\frac{d(t^{-1}x)}{dx}$

Solved Examples on the form: $y = t(t^{-1}x)$

Find the d.c. of the following.

1. $y = \sec(\tan^{-1}x)$

Solution: Put $x = \tan\theta \Leftrightarrow \tan^{-1}x = \theta$ for $-\frac{\pi}{2} < \theta < \frac{\pi}{2}$

$\therefore y = \sec(\tan^{-1}\tan\theta) = \sec\theta$

$\Rightarrow \frac{dy}{dx} = \frac{d\sec\theta}{dx} = \frac{d\sec\theta}{d\theta}\cdot\frac{d\theta}{dx} = \frac{\frac{d\sec\theta}{d\theta}}{\frac{dx}{d\theta}}$

$\Rightarrow \frac{dy}{dx} = \frac{\sec\theta\cdot\tan\theta}{\frac{d\tan\theta}{d\theta}}\quad [\because x = \tan\theta]$

$= \frac{\sec\theta\cdot\tan\theta}{\sec^2\theta} = \frac{\sin\theta}{\cos\theta}\cdot\cos\theta = \sin\theta$

$= \frac{\tan\theta}{\sec\theta} = \frac{x}{\sqrt{1+\tan^2\theta}}$

$\left[\because \sec\theta = |\sec\theta| \text{ for } -\frac{\pi}{2} < \theta < \frac{\pi}{2}\right]$

$= \frac{x}{\sqrt{1+x^2}}$

2. $y = \tan(\sin^{-1}x)$

Solution: Put $x = \sin\theta \Leftrightarrow \sin^{-1}x = \theta$ for $\left(-\frac{\pi}{2} \le \theta \le \frac{\pi}{2}\right)$

$\therefore y = \tan(\sin^{-1}x) = \tan(\sin^{-1}\sin\theta) = \tan\theta$

$\Rightarrow \frac{dy}{dx} = \sec^2\theta\cdot\frac{d\theta}{dx} = \frac{\sec^2\theta}{\left(\frac{dx}{d\theta}\right)} = \frac{\sec^2\theta}{\frac{d\sin\theta}{d\theta}} = \frac{\sec^2\theta}{\cos\theta}$

$= \sec^3\theta = \frac{1}{\cos^3\theta}\quad\left(\theta \ne \pm\frac{\pi}{2}\right)$

$= \frac{1}{\left(\sqrt{1-\sin^2\theta}\right)^3}$ as $\cos\theta > 0$ for $-\frac{\pi}{2} < \theta < \frac{\pi}{2}$

$= \frac{1}{(1-x^2)^{3/2}}$ for $|x| < 1$.

Solved Examples on the form: $y = \left[t(t^{-1}x)\right]^n$

Find the d.c. of the following

1. $y = \sec^2(\cos^{-1}x)$

Solution: $y = \sec^2(\cos^{-1}x),\ x \ne 0$

$\Rightarrow \frac{dy}{dx} = \frac{d\sec^2(\cos^{-1}x)}{dx} = \frac{d\left[\sec(\cos^{-1}x)\right]^2}{dx}$

Now, using chain rule, we have

$\frac{dy}{dy} = \frac{d\left[\sec(\cos^{-1}x)\right]^2}{d\left[\sec(\cos^{-1}x)\right]}\cdot\frac{d\sec(\cos^{-1}x)}{d\cos^{-1}x}\cdot\frac{d\cos^{-1}x}{dx}$

$$=2\sec\left(\cos^{-1}x\right)\cdot\sec\left(\cos^{-1}x\right)\cdot\tan\left(\cos^{-1}x\right)\cdot\frac{(-1)}{\sqrt{1-x^2}},$$

$|x| < 1.$

$$=\frac{-2\left[\sec\left(\cos^{-1}x\right)\right]^2\cdot\tan\left(\cos^{-1}x\right)}{\sqrt{1-x^2}}$$

$$=-\frac{2}{x^3}\ \text{for } |x|<1,\ x\neq 0.$$

or, alternatively,

$$y=\sec^2\left(\cos^{-1}x\right)$$

Put $x=\cos\theta \Leftrightarrow \cos^{-1}x=\theta$, where $0\le\theta\le\pi$

$$\Rightarrow y=\sec^2\theta$$

$$\Rightarrow \frac{dy}{dx}=2\sec\theta\frac{d\sec\theta}{dx}=2\sec\theta\cdot\sec\theta\cdot\tan\theta\cdot\frac{d\theta}{dx}$$

$$=2\sec^2\theta\cdot\tan\theta\cdot\frac{d\cos^{-1}x}{dx}$$

$$=\frac{-2\sin\theta}{\cos^3\theta\sqrt{1-x^2}},\ |x|<1,\ x\neq 0.$$

$$=\frac{-2\sqrt{1-\cos^2\theta}}{x^3\sqrt{1-x^2}}$$

$$[\because |\sin\theta|=\sin\theta \ \text{for } 0\le\theta\le\pi]$$

$$=\frac{-2\sqrt{1-x^2}}{x^3\sqrt{1-x^2}}=-\frac{2}{x^3}\ \text{for } x\neq 0,\pm 1.$$

Note:

$$y=\sec^2\left(\cos^{-1}x\right)=\left[\sec\left(\cos^{-1}x\right)\right]^2$$

$$=\left[\sec\left(\sec^{-1}\frac{1}{x}\right)\right]^2$$

$$=\frac{1}{x^2}$$

2. $y=\tan^2\left(\cos^{-1}x\right)$

Solution: $y=\tan^2\left(\cos^{-1}x\right)=\left[\tan\left(\cos^{-1}x\right)\right]^2$

Put $x=\cos\theta \Leftrightarrow \cos^{-1}x=\theta$, where $0\le\theta\le\pi$

$$\therefore y=\tan^2\theta,\ \theta\neq\frac{\pi}{2},\ \text{i.e. } x\neq 0.$$

$$\Rightarrow \frac{dy}{d\theta}=\frac{d\tan^2\theta}{dx}$$

$$=2\tan\theta\cdot\sec^2\theta\cdot\frac{d\theta}{dx}$$

$$=\frac{2\tan\theta\cdot\sec^2\theta}{\dfrac{dx}{d\theta}}$$

$$=\frac{2\tan\theta\cdot\sec^2\theta}{\dfrac{d\cos\theta}{d\theta}}$$

$$=\frac{2\tan\theta\cdot\sec^2\theta}{-\sin\theta},\ |x|<1,\ x\neq 0.$$

$$=\frac{-2}{\cos^3\theta}=\frac{-2}{x^3}\ (\because x=\cos\theta)\ \text{for}$$

$$|x|<1,\ x\neq 0$$

3. $y=\cot\left(\tan^{-1}x\right)$

Solution: Put $x=\tan\theta\Leftrightarrow\theta=\tan^{-1}x$, $\dfrac{-\pi}{2}<\theta<\dfrac{\pi}{2}$

$$\therefore y=\cot\left(\tan^{-1}\tan\theta\right)=\cot\theta,\ \theta\neq 0.$$

$$\Rightarrow \frac{dy}{dx}=\frac{d\cot\theta}{dx}=\frac{d\cot\theta}{d\theta}\cdot\frac{d\theta}{dx}=\frac{\dfrac{d\cot\theta}{d\theta}}{\dfrac{dx}{d\theta}}$$

$$= \frac{-\text{cosec}^2\theta}{\frac{d\tan\theta}{d\theta}} = -\frac{\text{cosec}^2\theta}{\sec^2\theta} = -\frac{\cos^2\theta}{\sin^2\theta} = -\cot^2\theta$$

$$= -\frac{1}{\tan^2\theta} = -\frac{1}{x^2}, \quad (x \neq 0)$$

4. $y = \sin\left(m\sin^{-1}x\right)$

Solution: Put $x = \sin\theta \Leftrightarrow \theta = \sin^{-1}x$ for $-\frac{\pi}{2} \le \theta \le \frac{\pi}{2}$

$$\therefore y = \sin\left(m\sin^{-1}x\right) = \sin\left(m\sin^{-1}\sin\theta\right) = \sin m\theta$$

$$\Rightarrow \frac{dy}{dx} = \frac{d\sin m\theta}{dx} = \frac{d\sin m\theta}{d\,m\theta}\cdot\frac{d\,m\theta}{dx} = \cos m\theta\cdot m\cdot\frac{d\theta}{dx}$$

$$\Rightarrow \frac{dy}{dx} = \frac{m\cos m\theta}{\frac{dx}{d\theta}} = \frac{m\sqrt{1-\sin^2 m\theta}}{1-\sin^2\theta}$$

if $\cos m\theta \ge 0$ and $\theta \neq \pm\frac{\pi}{2}$

$$\Rightarrow \frac{dy}{dx} = \frac{+m\sqrt{1-y^2}}{\sqrt{1-x^2}} \quad [\because \sin m\theta = y], \text{ for}$$

$|x| < 1.$

if $2n\pi - \frac{\pi}{2} \le m\sin^{-1}x \le 2n\pi + \frac{\pi}{2}$

and $\Rightarrow \frac{dy}{dx} = \frac{-m\sqrt{1-\sin^2 m\theta}}{1-\sin^2\theta}$ if $\cos m\theta < 0$

$$= -\frac{m\sqrt{1-y^2}}{\sqrt{1-x^2}} \text{ for } |x| < 1.$$

$[\because \sin\theta = x$ and $\sin m\theta = y]$,

if $2n\pi + \frac{\pi}{2} < m\sin^{-1}x < 2n\pi + \frac{3\pi}{2}$

or, alternatively,

$y = \sin\left(m\sin^{-1}x\right)$

Now

1. $\sin^{-1}y = 2n\pi + m\sin^{-1}x$

if $-\frac{\pi}{2} \le 2n\pi + m\sin^{-1}x \le \frac{\pi}{2}$

2. $\sin^{-1}y = (2n+1)\pi - m\sin^{-1}x$

if $\frac{-\pi}{2} \le (2n\pi+1)\pi - m\sin^{-1}x \le \frac{\pi}{2}$

Hence,

$$\frac{dy}{dx} = +\frac{m\sqrt{1-y^2}}{\sqrt{1-x^2}} \quad \text{in case (1), } |x| < 1.$$

$$= -\frac{m\sqrt{1-y^2}}{\sqrt{1-x^2}} \quad \text{in case (2), } |x| < 1.$$

5. $y = \cos\left(2\sin^{-1}x\right)$

Solution: Put $x = \sin\theta \Leftrightarrow \sin^{-1}x = \theta$, $\left(-\frac{\pi}{2} \le \theta \le \frac{\pi}{2}\right)$

$$\therefore y = \cos\left(2\sin^{-1}\sin\theta\right) = \cos 2\theta$$

$$\Rightarrow \frac{dy}{dx} = \frac{d\cos 2\theta}{dx} = \frac{d\cos 2\theta}{d\,2\theta}\cdot\frac{d\,2\theta}{dx} = -\sin 2\theta\cdot 2\cdot\frac{d\theta}{dx}$$

$$\Rightarrow \frac{dy}{dx} = \frac{-2\sin 2\theta}{\frac{dx}{d\theta}} = \frac{-2\sin 2\theta}{\frac{d\sin\theta}{d\theta}}$$

$$= \frac{-\sin 2\theta}{\cos\theta}, \theta \neq \pm\frac{\pi}{2}$$

$$\Rightarrow \frac{dy}{dx} = -\frac{2\times 2\sin\theta\cdot\cos\theta}{\cos\theta} = -4\sin\theta = -4x$$

for $|x| < 1.$

$[\because x = \sin\theta]$

6. $y = \cos\left(2\cos^{-1} x\right)$

Solution: using chain rule, we have

$$\frac{dy}{dx} = \frac{d\cos\left(2\cos^{-1} x\right)}{d\left(2\cos^{-1} x\right)} \cdot \frac{d\left(2\cos^{-1} x\right)}{d\cos^{-1} x} \cdot \frac{d\cos^{-1} x}{dx}$$

$$= -\sin\left(2\cos^{-1} x\right) \cdot 2\frac{d\cos^{-1} x}{d\cos^{-1} x} \cdot \frac{-1}{\sqrt{1-x^2}},$$

$|x| < 1$

$$= -\sin\left(2\cos^{-1} x\right) \cdot \frac{-2}{\sqrt{1-x^2}}$$

$$= \frac{2\sin\left(2\cos^{-1} x\right)}{\sqrt{1-x^2}}$$

$\left[\text{Let } \cos^{-1} x = \theta,\ 0 \le \theta \le \pi\right]$

$$= \frac{2\sin 2\theta}{\sqrt{1-x^2}} = \frac{2 \cdot 2\sin\theta \cdot \cos\theta}{\sqrt{1-x^2}} = \frac{4x \not{\sin\theta}}{\not{\sin\theta}}$$

$= 4x \ \left[\because |\sin\theta| = \sin\theta \text{ for } 0 \le \theta \le \pi\right]$ for $|x| < 1$.

Type 3:

Form 1: $y = t^{-1}[f(x, y)]$

Where, $t^{-1} = \sin^{-1} / \cos^{-1} / \tan^{-1} / \cot^{-1} / \sec^{-1} / \operatorname{cosec}^{-1}$ $f(x, y)$ = an algebraic implicit function of x and y or $f(x, y)$ = an algebraic expression in x and y mixed together.

Working rule:

1. Change the given inverse trigonometric function into direct trigonometric function.

2. Differentiate both sides implicitly w.r.t. x remembering that y is a function of x.

3. Finally solve for $\dfrac{dy}{dx}$.

Form 2: An algebraic implicit function $\Leftrightarrow f_1(x, y) = t^{-1}[f_2(x, y)] = t^{-1}$ (an algebraic implicit function)

Working rule:

1. Change the given inverse trigonometric function into direct trigonometric function.

2. Differentiate both sides implicitly w.r.t. x remembering that y is a function of x.

3. Finally solve for $\dfrac{dy}{dx}$.

Solved Examples on the form: $y = t^{-1}[f(x, y)]$

Find the d.c. of the following.

1. $y = \tan^{-1}(x + y)$

Solution: $y = \tan^{-1}(x + y)$

$\Rightarrow \tan y = x + y$

$$\Rightarrow \frac{d\tan y}{dx} = \frac{d(x+y)}{dx}$$

$$\Rightarrow \frac{d\tan y}{dy} \cdot \frac{dy}{dx} = \frac{dx}{dx} + \frac{dy}{dx}$$

$$\Rightarrow \sec^2 y \cdot \frac{dy}{dx} = 1 + \frac{dy}{dx}$$

$$\Rightarrow \sec^2 y \frac{dy}{dx} - \frac{dy}{dx} = 1$$

$$\Rightarrow \left(\sec^2 y - 1\right) \frac{dy}{dx} = 1$$

$$\Rightarrow \frac{dy}{dx} = \frac{1}{\left(\sec^2 y - 1\right)} = \frac{1}{\tan^2 y} = \frac{1}{(x+y)^2}$$

2. $y = \sin^{-1}(x + y)$

Solution: $y = \sin^{-1}(x + y)$

$\Rightarrow \sin y = x + y$

$$\Rightarrow \frac{d\sin y}{dx} = \frac{d(x+y)}{dx}$$

$$\Rightarrow \cos y \cdot \frac{dy}{dx} = \frac{dx}{dx} + \frac{dy}{dx}$$

$$\Rightarrow \cos y \cdot \frac{dy}{dx} - \frac{dy}{dx} = 1$$

$$\Rightarrow (\cos y - 1)\frac{dy}{dx} = 1$$

$$\Rightarrow \frac{dy}{dx} = \frac{1}{(\cos y - 1)}.$$

Solved Examples out on the form:

$f_1(x, y) = t^{-1}[f_2(x, y)]$

Find the d.c. from the following.

1. $x + y = \sin^{-1}(x + y)$

Solution: $x + y = \sin^{-1}(x + y)$

$$\Rightarrow \sin(x + y) = x + y$$

$$\Rightarrow \frac{d\sin(x + y)}{dx} = \frac{d(x + y)}{dx}$$

$$\Rightarrow \cos(x + y) \cdot \frac{d(x + y)}{dx} = 1 + \frac{dy}{dx}$$

$$\Rightarrow \cos(x + y)\left[1 + \frac{dy}{dx}\right] = 1 + \frac{dy}{dx}$$

$$\Rightarrow \cos(x + y) + \cos(x + y)\frac{dy}{dx} - \frac{dy}{dx} = 1$$

$$\Rightarrow [\cos(x + y) - 1]\frac{dy}{dx} = 1 - \cos(x + y)$$

$$\Rightarrow \frac{dy}{dx} = \frac{-[\cos(x + y) - 1]}{[\cos(x + y) - 1]} = -1$$

2. $x + y = \sec^{-1}(x - y)$

Solution: $x + y = \sec^{-1}(x - y)$

$$\Rightarrow \frac{d\sec(x + y)}{dx} = \frac{d(x - y)}{dx}$$

$$\Rightarrow \sec(x + y) \cdot \tan(x + y) \cdot \frac{d(x + y)}{dx} = 1 - \frac{dy}{dx}$$

$$\Rightarrow \sec(x + y) \cdot \tan(x + y)\left[1 + \frac{dy}{dx}\right] = 1 - \frac{dy}{dx}$$

$$\Rightarrow \sec(x + y)\tan(x + y) + \sec(x + y) \cdot \tan(x + y)\frac{dy}{dx} = 1 - \frac{dy}{dx}$$

$$\Rightarrow \sec(x + y)\tan(x + y)\frac{dy}{dx} + \frac{dy}{dx} = 1 - \sec(x + y) \cdot \tan(x + y)$$

$$\Rightarrow \frac{dy}{dx} = \frac{1 - \sec(x + y)\tan(x + y)}{1 + \sec(x + y)\tan(x + y)}$$

3. $xy = \sin^{-1}(x + y)$

Solution: $xy = \sin^{-1}(x + y)$

$$\Rightarrow \sin(xy) = x + y$$

$$\Rightarrow \frac{d\sin(xy)}{dx} = \frac{d(x + y)}{dx}$$

$$\Rightarrow \cos(xy) \cdot \frac{d(xy)}{dx} = 1 + \frac{dy}{dx}$$

$$\Rightarrow \cos(xy)\left[x\frac{dy}{dx} + y\frac{dx}{dx}\right] = 1 + \frac{dy}{dx}$$

$$\left[\because \frac{d\sin(xy)}{dx} = \frac{d\sin(xy)}{d(xy)} \cdot \frac{d(xy)}{dx}\right]$$

$$\Rightarrow x\cos(xy)\frac{dy}{dx} + y\cos(xy) \cdot 1 = 1 + \frac{dy}{dx}$$

$$\Rightarrow x\cos(xy)\frac{dy}{dx} - \frac{dy}{dx} = 1 - y\cos(xy)$$

$$\Rightarrow [x\cos(xy) - 1]\frac{dy}{dx} = 1 - y\cos(xy)$$

$$\Rightarrow \frac{dy}{dx} = \frac{1 - y\cos(xy)}{x\cos(xy) - 1} = -\frac{1 - y\cos(xy)}{1 - x\cos(xy)}$$

Type 4: Differentiation of inverse trigonometric function of a function of x w.r.t. an other inverse trigonometric function of a function of x / an other function of x $\Rightarrow$ differentiate $t^{-1}[f(x)]$ w.r.t. $t^{-1}[g(x)]$ or, differentiate $t^{-1}[f(x)]$ w.r.t. $g(x)$ where $f(x)$ = a function of x

$g(x)$ = an other function of x / $\sin x$ / $\cos x$ / $\tan x$ / $\cot x$ / $\sec x$ / $\operatorname{cosec} x$

$t^{-1} = \sin^{-1}/\cos^{-1}/\tan^{-1}/\cot^{-1}/\sec^{-1}/\operatorname{cosec}^{-1}$

t_1^{-1} = another inverse trigonometric operator excepting t^{-1}.

Working rule:

1. Put $t^{-1}[f(x)] = y$ [The function put before "w.r.t."] ...(1)

and $t^{-1}[g(x)] = z$ [the function put after w.r.t.] ...(2)

2. Find the d.c. of (1) w.r.t. x using the rule of finding d.c. of inverse trigonometric function. Similarly, find x the d.c. of (2) w.r.t. using the rule of finding the inverse trigonometric function.

3. Divide the d.c of y by the d.c. of z regarding z as a function of $x \Rightarrow \dfrac{\frac{dy}{dx}}{\frac{dz}{dx}}$ which is the required d.c.

Remember:

1. $\dfrac{d\,f(y)}{dx} = f'(y)\cdot\dfrac{dy}{dx}$

2. $\dfrac{d\,f(z)}{dx} = f'(z)\cdot\dfrac{dz}{dx}$

Solved Examples:

Find the d.c. of the following.

1. $\sin^{-1} x$ w.r.t. $\cos^{-1} x$

Solution: Let $\sin^{-1}x = y$

$$\Rightarrow \sin y = x,\ -\frac{\pi}{2} \le y \le \frac{\pi}{2} \quad \text{...(1)}$$

and $\cos^{-1}x = z \Rightarrow \cos z = x,\ 0 \le z \le \pi$ . ...(2)

$$(1) \Rightarrow \frac{d\sin y}{dx} = \frac{dx}{dx}$$

$$\Rightarrow \cos y\,\frac{dy}{dx} = 1$$

$$\Rightarrow \frac{dy}{dx} = \frac{1}{\cos y}$$

$$\Rightarrow \frac{dy}{dx} = \frac{1}{\sqrt{1-\sin^2 y}} \text{ for } |x| < 1.$$

$$\left[\because |\cos y| = \cos y \text{ as } -\frac{\pi}{2} \le y \le \frac{\pi}{2}\right]$$

$$\Rightarrow \frac{dy}{dx} = \frac{1}{\sqrt{1-x^2}} \quad \text{...(3)}$$

$$(2) \Rightarrow \frac{d\cos z}{dx} = \frac{dx}{dx}$$

$$\Rightarrow -\sin z\,\frac{dz}{dx} = 1$$

$$\Rightarrow \frac{dz}{dx} = -\frac{1}{\sin z}$$

$$= -\frac{1}{\sqrt{1-\cos^2 z}} = \frac{-1}{\sqrt{1-x^2}}$$

$$[\because |\sin z| = \sin z \text{ as } 0 \le z \le \pi] \quad \text{...(4)}$$

$$\text{Now, } \frac{(3)}{(4)} \Rightarrow \frac{dy}{dz} = \left[\frac{\frac{1}{\sqrt{1-x^2}}}{\frac{1}{\sqrt{1-x^2}}}\right] = -1$$

for $|x| < 1$.

Or, alternatively,

$$\sin^{-1} x = \frac{\pi}{2} - \cos^{-1} x$$

$$\Rightarrow \frac{d\left(\sin^{-1} x\right)}{d\left(\cos^{-1} x\right)} = -1$$

2. $\tan^{-1}\left(\dfrac{x}{\sqrt{1-x^2}}\right)$ w.r.t. $\operatorname{cosec}^{-1}\left(\dfrac{1}{\sqrt{1-x^2}}\right)$.

Solution: Let $\tan^{-1}\left(\dfrac{x}{\sqrt{1-x^2}}\right) = y$

$$\Rightarrow \tan y = \frac{x}{\sqrt{1-x^2}} \quad ...(1)$$

and $\operatorname{cosec}^{-1}\left(\dfrac{1}{\sqrt{1-x^2}}\right) = z$

$$\Rightarrow \operatorname{cosec} z = \frac{1}{\sqrt{1-x^2}} \quad ...(2)$$

$\left(0 < z \le \dfrac{\pi}{2}\right)$

Now,

$$(1) \Rightarrow \frac{d \tan y}{dx} = \frac{d}{dx}\left[\frac{x}{\sqrt{1-x^2}}\right]$$

$$\Rightarrow \sec^2 y \cdot \frac{dy}{dx} = \frac{\sqrt{1-x^2} \cdot 1 - \dfrac{x \cdot 1}{2\sqrt{1-x^2}} \cdot (-2x)}{\left(\sqrt{1-x^2}\right)^2}$$

$$\Rightarrow \frac{dy}{dx} = \frac{\sqrt{1-x^2} + \dfrac{x^2}{\sqrt{1-x^2}}}{\left(1-x^2\right)\left(1+\tan^2 y\right)}$$

$$= \frac{\left(\sqrt{1-x^2}\right) + \dfrac{x^2}{\sqrt{1-x^2}}}{\left(1-x^2\right)} \cdot \frac{1}{1+\left(\dfrac{x}{\sqrt{1-x^2}}\right)^2}$$

$$= \frac{\left(1-x^2\right)+x^2}{\left(\sqrt{1-x^2}\right)\cdot\left(1-x^2\right)} \cdot \frac{1}{\left[\dfrac{\left(1-x^2\right)+x^2}{\left(\sqrt{1-x^2}\right)^2}\right]}$$

$$= \frac{\left(\cancel{x^2} - \cancel{x^2} + 1\right)}{\left(\sqrt{1-x^2}\right)\cancel{\left(1-x^2\right)}} \cdot \frac{\cancel{\left(1-x^2\right)}}{1}$$

$$= \frac{1}{\sqrt{1-x^2}} \quad ...(3)$$

Again,

$$(2) \Rightarrow \operatorname{cosec} z = \frac{1}{\sqrt{1-x^2}}$$

$$\Rightarrow \frac{d \operatorname{cosec} z}{dx} = \frac{d}{dx}\left(\frac{1}{\sqrt{1-x^2}}\right)$$

$$= -1 \times \frac{1}{2\sqrt{1-x^2}} \cdot \frac{(-2x)}{1} \cdot \frac{1}{\left(1-x^2\right)}$$

$$\Rightarrow -\operatorname{cosec} z \cdot \cot z \, \frac{dz}{dx} = \frac{x}{\left(\sqrt{1-x^2}\right)\left(1-x^2\right)}$$

$$\Rightarrow \frac{dz}{dx} = \frac{-1}{\cancel{\sqrt{1-x^2}}} \cdot \frac{x}{\left(1-x^2\right)} \cdot \frac{1 \cdot \cancel{\sqrt{1-x^2}}}{\cot z}$$

$$= \frac{-x}{\sqrt{\operatorname{cosec}^2 z - 1}} \cdot \frac{1}{\left(1-x^2\right)} \quad (\because \cot z > 0)$$

$$= \frac{-x}{\left(1-x^2\right)} \cdot \frac{1}{\sqrt{\left(\dfrac{1}{\sqrt{1-x^2}}\right)^2 - 1}}$$

$$= \frac{-x}{(1-x^2)} \cdot \frac{1}{\sqrt{\frac{1}{(1-x^2)} - 1}}$$

$$= \frac{-x}{(1-x^2)} \cdot \frac{1}{\sqrt{\frac{1-(1-x^2)}{(1-x^2)}}}$$

$$= \frac{-x\sqrt{1-x^2}}{(1-x^2)\sqrt{x^2}}$$

$$= \frac{-\sqrt{1-x^2}}{\left(\sqrt{1-x^2}\right)^2} \cdot \frac{x}{|x|} = \frac{-x}{|x|} \cdot \frac{1}{\sqrt{1-x^2}}$$

Hence,

$$\frac{(3)}{(4)} \Rightarrow \frac{\frac{dy}{dx}}{\frac{dz}{dx}} = -\frac{1}{\sqrt{1-x^2}} \cdot \sqrt{1-x^2} \cdot \frac{|x|}{x}$$

$$\therefore \frac{dy}{dz} = \frac{-|x|}{x}, \ |x| < 1.$$

On Method of Transformation

Type 1:

Form: **1.** $t^{-1}\sqrt{\frac{1 \pm \sin x}{1 \mp \sin x}}$ or $t^{-1}\sqrt{\frac{1 \pm \cos x}{1 \mp \cos x}}$

2. $t^{-1}\sqrt{\frac{1 \pm \sin x}{1 \pm \cos x}}$ or $t^{-1}\sqrt{\frac{1 \pm \cos x}{1 \pm \sin x}}$

Working rule: Whenever $1 \pm \sin x$ and / $1 \pm \cos x$ appear (or, appears) under the radical sign $\sqrt{\ }$, we express the function within the radical as a square of some function.

Solved Examples:

Find d.c. of the following

1. $y = \tan^{-1}\sqrt{\frac{1-\cos x}{1+\cos x}}$

Solution: $\frac{1-\cos x}{1+\cos x} = \frac{2\sin^2\frac{x}{2}}{2\cos^2\frac{x}{2}} = \frac{\sin^2\frac{x}{2}}{\cos^2\frac{x}{2}}$

$$\therefore y = \tan^{-1}\sqrt{\frac{\sin^2\frac{x}{2}}{\cos^2\frac{x}{2}}} = \tan^{-1}\left\{\left|\frac{\sin^2\frac{x}{2}}{\cos^2\frac{x}{2}}\right|\right\}$$

$$= \tan^{-1}\left\{\left|\tan\frac{x}{2}\right|\right\}$$

$$\Rightarrow \tan y = \left|\tan\frac{x}{2}\right|$$

$$\Rightarrow \sec^2 y \frac{dy}{dx} = \frac{\left|\tan\frac{x}{2}\right|}{\tan\frac{x}{2}} \cdot \frac{\sec^2\frac{x}{2}}{2}$$

$$\Rightarrow \frac{dy}{dx} = \frac{1}{2}\frac{\left|\tan\frac{x}{2}\right| \cdot \sec^2\frac{x}{2}}{\tan\frac{x}{2} \cdot \sec^2 y}$$

$$= \frac{1}{2} \cdot \frac{\left|\tan\frac{x}{2}\right| \cdot \sec^2\frac{x}{2}}{\tan\frac{x}{2}\left(1 + \left|\tan\frac{x}{2}\right|^2\right)}$$

$$= \frac{1}{2}\frac{\left|\tan\frac{x}{2}\right| \cdot \sec^2\frac{x}{2}}{\tan\frac{x}{2} \cdot \left(1 + \tan^2\frac{x}{2}\right)} = \frac{1}{2} \cdot \frac{\left|\tan\frac{x}{2}\right| \cdot \sec^2\frac{x}{2}}{\tan\frac{x}{2} \cdot \sec^2\frac{x}{2}}$$

$$= \frac{1}{2} \cdot \frac{\left|\tan\frac{x}{2}\right|}{\tan\frac{x}{2}}, \ x \neq n\pi$$

Or, alternatively,

$\tan y = \sqrt{\frac{1-\cos x}{1+\cos x}}$, Now using logarithmic differentiation,

$$\Rightarrow \log \tan y = \log \sqrt{\frac{1-\cos x}{1+\cos x}} = \frac{1}{2}\log\left(\frac{1-\cos x}{1+\cos x}\right)$$

$$= \frac{1}{2}\{\log(1-\cos x) - \log(1+\cos x)\}$$

$$\Rightarrow \frac{1}{\tan y}\cdot\frac{d\tan y}{dx}$$

$$= \frac{1}{2}\left\{\frac{1}{1-\cos x}\cdot\sin x - \frac{1}{1+\cos x}\cdot(-\sin x)\right\}$$

$$\Rightarrow \frac{1}{\tan y}\cdot\sec^2 y\cdot\frac{dy}{dx} = \frac{1}{2}\left\{\frac{\sin x}{1-\cos x} + \frac{\sin x}{1+\cos x}\right\}$$

$$\Rightarrow \frac{1+\tan^2 y}{\tan y}\cdot\frac{dy}{dx}$$

$$= \frac{1}{2}\sin x\left\{\frac{1+\cos x+1-\cos x}{1-\cos^2 x}\right\} = \frac{1}{2}\cdot\frac{\sin x\cdot 2}{\sin^2 x}$$

$$\Rightarrow \frac{1+\frac{1-\cos x}{1+\cos x}}{\sqrt{\frac{1-\cos x}{1+\cos x}}}\cdot\frac{dy}{dx} = \frac{1}{\sin x}$$

$$\Rightarrow \frac{\frac{2}{1+\cos x}}{\sqrt{\frac{1-\cos x}{1+\cos x}}}\cdot\frac{dy}{dx} = \frac{1}{\sin x}$$

$$\Rightarrow \frac{dy}{dx} = \frac{\sqrt{\frac{1-\cos x}{1+\cos x}}}{\sin x\cdot\left(\frac{2}{1+\cos x}\right)} = \frac{\sqrt{\frac{1-\cos x}{1+\cos x}}}{\frac{2\sin x}{1+\cos x}}$$

$$= \sqrt{\frac{1-\cos x}{1+\cos x}}\cdot\frac{(1+\cos x)}{2\sin x} = \frac{(1+\cos x)}{2\sin x}\cdot\sqrt{\frac{1-\cos x}{1+\cos x}}$$

2. $y = \tan^{-1}\sqrt{\dfrac{1+\sin x}{1-\sin x}}$

***Solution*:** $\dfrac{1+\sin x}{1-\sin x} = \dfrac{\left(\cos\frac{x}{2}+\sin\frac{x}{2}\right)^2}{\left(\cos\frac{x}{2}-\sin\frac{x}{2}\right)^2}$

$$\Rightarrow \sqrt{\frac{1+\sin x}{1-\sin x}} = \sqrt{\frac{\left(\cos\frac{x}{2}+\sin\frac{x}{2}\right)^2}{\left(\cos\frac{x}{2}-\sin\frac{x}{2}\right)^2}} = \frac{\left|\cos\frac{x}{2}+\sin\frac{x}{2}\right|}{\left|\cos\frac{x}{2}-\sin\frac{x}{2}\right|}$$

$$= \sqrt{\frac{1+\sin x}{1-\sin x}} = \frac{\left|\left(1+\tan\frac{x}{2}\right)\right|}{\left|\left(1-\tan\frac{x}{2}\right)\right|}$$

$$\because \tan(A+B) = \frac{\tan A+\tan B}{1-\tan A\cdot\tan B}$$

$$\Rightarrow \tan^{-1}\sqrt{\frac{1+\sin x}{1-\sin x}} = \tan^{-1}\left|\tan\left(\frac{\pi}{4}+\frac{x}{2}\right)\right|$$

$$\Rightarrow y = \tan^{-1}\left|\tan\left(\frac{\pi}{4}+\frac{x}{2}\right)\right|$$

$$\Rightarrow \frac{dy}{dx} = \frac{d}{dx}\tan^{-1}\left|\tan\left(\frac{\pi}{4}+\frac{x}{2}\right)\right|$$

$$= \frac{1}{1+\left\{\left|+\tan\left(\frac{\pi}{4}+\frac{\pi}{2}\right)\right|\right\}^2}\cdot\frac{\left|\tan\left(\frac{\pi}{4}+\frac{x}{2}\right)\right|}{\tan\left(\frac{\pi}{4}+\frac{x}{2}\right)}\cdot\frac{\sec^2\left(\frac{\pi}{4}+\frac{x}{2}\right)}{2}$$

$$= \frac{1}{1+\tan^2\left(\frac{\pi}{4}+\frac{x}{2}\right)}\cdot\frac{\left|\tan\left(\frac{\pi}{4}+\frac{x}{2}\right)\right|}{\tan\left(\frac{\pi}{4}+\frac{x}{2}\right)}\cdot\frac{\sec^2\left(\frac{\pi}{4}+\frac{x}{2}\right)}{2}$$

$$= \frac{1}{\sec^2\left(\frac{\pi}{4}+\frac{x}{2}\right)}\cdot\frac{\left|\tan\left(\frac{\pi}{4}+\frac{x}{2}\right)\right|}{\tan\left(\frac{\pi}{4}+\frac{x}{2}\right)}\cdot\frac{\sec^2\left(\frac{\pi}{4}+\frac{x}{2}\right)}{2}$$

$$= \frac{1}{2} \cdot \frac{\left|\tan\left(\frac{\pi}{4} + \frac{x}{2}\right)\right|}{\tan\left(\frac{\pi}{4} + \frac{x}{2}\right)}$$

Note: $\frac{d}{dx} \tan^{-1} |f(x)| = \frac{1}{\{1 + f^2(x)\}} \cdot \frac{d|f(x)|}{dx}$

Type 2:

Form: t^{-1} (an expression in direct $t-$ function involving sin x and cos x only)

Working rule:

1. Transform given trigonometrical expression in sin x and cos x in such a way that it becomes equal to $t^{-1} t\left(\frac{\pi}{4} \pm \frac{x}{2}\right) = \frac{\pi}{4} \pm \frac{x}{2}$ for which are required to

(i) Use $\sin x = \cos\left(\frac{\pi}{2} \pm \frac{x}{2}\right)$

(ii) Use multiple and / submultiple angle formulas of trigonometric functions.

Or alternatively,

Put y = given function and change it into direct function and then find $\frac{dy}{dx}$ regarding y as a function of x, i.e; find $\frac{dy}{dx}$ implicitly.

Example: **1.** $y = \tan^{-1}\left(\frac{\cos x}{1 + \sin x}\right)$

$\Rightarrow \tan y = \frac{\cos x}{1 + \sin x}$, Now find $\frac{dy}{dx}$ implicitly

2. $y = \tan^{-1}\left(\frac{1 + \sin x}{\cos x}\right)$

$\Rightarrow \tan y = \frac{1 + \sin x}{\cos x}$. Now find $\frac{dy}{dx}$ implicit function rule:

Solved Examples:

Find the d.c. of the following.

1. $y = \tan^{-1}\left(\frac{\cos x}{1 + \sin x}\right)$

Solution: $y = \tan^{-1}\left(\frac{\cos x}{1 + \sin x}\right)$

$$\Rightarrow y = \tan^{-1}\left[\frac{\sin\left(\frac{\pi}{2} - x\right)}{1 + \cos\left(\frac{\pi}{2} - x\right)}\right]$$

$$\Rightarrow y = \tan^{-1}\left[\frac{2\sin\left(\frac{\pi}{4} - \frac{x}{2}\right) \cdot \cos\left(\frac{\pi}{4} - \frac{x}{2}\right)}{1 + 2\cos^2\left(\frac{\pi}{4} - \frac{x}{2}\right) - 1}\right]$$

$$\Rightarrow y = \tan^{-1}\left[\tan\left(\frac{\pi}{4} - \frac{x}{2}\right)\right]$$

$$\Rightarrow y = n\pi + \frac{\pi}{4} - \frac{x}{2} \text{ such that } -\frac{\pi}{2} < y < \frac{\pi}{2}$$

Now, differentiating both sides w.r.t. x

$$\Rightarrow \frac{dy}{dx} = 0 - \frac{1}{2} = -\frac{1}{2}$$

2. $y = \tan^{-1}\left(\frac{1 + \sin x}{\cos x}\right)$

Solution: $y = \tan^{-1}\left(\frac{1 + \sin x}{\cos x}\right)$

$$\Rightarrow y = \tan^{-1}\left[\frac{\cos^2\frac{x}{2} + \sin^2\frac{x}{2} + 2\sin\frac{x}{2} + \cos\frac{x}{2}}{\cos^2\frac{x}{2} - \sin^2\frac{x}{2}}\right]$$

$$\Rightarrow y = \tan^{-1}\left[\frac{\left(\cos\frac{x}{2} + \sin\frac{x}{2}\right)^2}{\left(\cos\frac{x}{2} + \sin\frac{x}{2}\right)\left(\cos\frac{x}{2} - \sin\frac{x}{2}\right)}\right]$$

$$\Rightarrow y = \tan^{-1}\left[\frac{\cos\frac{x}{2} + \sin\frac{x}{2}}{\cos\frac{x}{2} + \sin\frac{x}{2}}\right]$$

$$\Rightarrow y = \tan^{-1}\left[\frac{1+\tan\frac{x}{2}}{1-\tan\frac{x}{2}}\right]$$

$$\Rightarrow y = \tan^{-1}\left[\frac{\tan\frac{\pi}{4}+\tan\frac{x}{2}}{1-\tan\frac{\pi}{4}\cdot\tan\frac{x}{2}}\right]$$

$$\Rightarrow y = \tan^{-1}\left[\tan\left(\frac{\pi}{4}+\frac{x}{2}\right)\right] = n\pi + \frac{\pi}{4} + \frac{x}{2}$$

such that $-\frac{\pi}{2} < y < \frac{\pi}{2}$

$$\Rightarrow \frac{dy}{dx} = 0 + \frac{1}{2} = \frac{1}{2}$$

Type 3:

Form: $y = \tan^{-1}\left[\frac{a \pm b}{a \mp ab}\right]$

The following points should be noted about the bracketed expressions in the above.

(i) The signs connecting the two terms in the numerator and denominator are opposite.

(ii) One of the two term in the denominator is 1 and the other terms is the product of two terms in the numerator.

(iii) Somtimes at the place of 1, another constant is provided, then that constant should be changed into 1 by using the mathematical manipulation of dividing numerator and denominator by that constant.

Example: $y = \tan^{-1}\left(\frac{2+3\tan x}{3-2\tan x}\right)$

$$= \tan^{-1}\left(\frac{\frac{2}{3}+\tan x}{1-\frac{2}{3}\tan x}\right)$$

$$\Rightarrow y = \tan^{-1}\left(\frac{\tan\alpha+\tan x}{1-\tan\alpha\ \tan}\right) \text{ Where } \tan\alpha = \frac{2}{3}$$

$$\Rightarrow y = \tan^{-1}\left[\tan(\alpha+x)\right] = n\pi + (\alpha+x) \text{ where}$$

n is such that $-\frac{\pi}{2} < y < \frac{\pi}{2}$

$$\Rightarrow \frac{dy}{dx} = 1 \text{ (} \alpha \text{ being a constant)}$$

Remember:

1. If $y = \tan^{-1}\left(\frac{K-x}{1+Kx}\right)$, find $\frac{dy}{dx}$

Solution: Let $y = n\pi + \tan^{-1} K - \tan^{-1} x$

$$\Rightarrow \frac{dy}{dx} = \frac{1}{1+x^2}$$

Similarly, $\frac{dy}{dx}$ can be obtained if

$$y = \tan^{-1}\left(\frac{K+x}{1-Kx}\right)$$

Solved Examples:

Find the d.c. of the following.

1. $y = \tan^{-1}\left(\frac{x-a}{x+a}\right)$

Solution: $y = \tan^{-1}\left(\frac{x-a}{x+a}\right)$

$$\Rightarrow y = \tan^{-1}\left(\frac{x}{a}\right) - \tan^{-1}(1) + n\pi$$

$$\Rightarrow \frac{dy}{dx} = \frac{\frac{1}{a}}{1+\frac{x^2}{a^2}} = \frac{a}{a^2+x^2}$$

2. $y = \tan^{-1}\left(\frac{a+bx}{b-ax}\right)$

Solution: $y = \tan^{-1}\left(\frac{a+bx}{b-ax}\right) = \tan^{-1}\left[\frac{\frac{a}{b}+x}{1-\left(\frac{a}{b}\right)x}\right]$

Let $\frac{a}{b} = \tan\alpha$ and $x = \tan B$

$$\therefore y = \tan^{-1}\left[\frac{\tan\alpha + \tan B}{1 - \tan\alpha \tan B}\right]$$

$$= \tan^{-1}[\tan(\alpha + B)] = n\pi + \alpha + B$$

$$\Rightarrow y = n\pi + \tan^{-1}\frac{a}{b} + \tan^{-1}x$$

$$\Rightarrow \frac{dy}{dx} = 0 + 0 + \frac{1}{1+x^2} = \frac{1}{1+x^2}$$

3. $y = \tan^{-1}\left(\frac{5+2x}{2-5x}\right)$

Solution: $y = \tan^{-1}\left(\frac{5+2x}{2-5x}\right)$

$$\Rightarrow y = \tan^{-1}\left(\frac{\frac{5}{2} + x}{1 - \left(\frac{5}{2}\right)x}\right)$$

Let $\frac{5}{2} = \tan\alpha$ and $x = \tan B$

$$\therefore y = \tan^{-1}\left[\frac{\tan\alpha + \tan B}{1 - \tan\alpha \tan B}\right] = \tan^{-1}[\tan(\alpha + B)]$$

$$\Rightarrow y = n\pi + \alpha + B$$

$$\Rightarrow y = n\pi + \tan^{-1}\frac{5}{2} + \tan^{-1}x \text{ Where } -\frac{\pi}{2} < y < \frac{\pi}{2}$$

$$\Rightarrow \frac{dy}{dx} = \frac{1}{1+x^2}$$

4. $y = \tan^{-1}\left[\frac{\sqrt{x} + \sqrt{a}}{1 - \sqrt{x}\cdot\sqrt{a}}\right]$

Solution: $y = \tan^{-1}\left[\frac{\sqrt{x} + \sqrt{a}}{1 - \sqrt{x}\cdot\sqrt{a}}\right]$

Now, $\frac{\sqrt{x} + \sqrt{a}}{1 - \sqrt{x}\cdot\sqrt{a}}$ gives us an idea of the formula

$$\tan(A + B) = \frac{\tan A + \tan B}{1 - \tan A \cdot \tan B}$$

Hence, on putting, $\sqrt{x} = \tan A$

$\sqrt{a} = \tan B$, we get

$$y = \tan^{-1}\left[\frac{\tan A + \tan B}{1 - \tan A \cdot \tan B}\right] = \tan^{-1}[\tan(A+B)]$$

$$\Rightarrow y = n\pi + A + B = n\pi + \tan^{-1}\sqrt{x} + \tan^{-1}\sqrt{a}$$

$$\Rightarrow \frac{dy}{dx} = \frac{1}{1 + \left(x^{\frac{1}{2}}\right)^2}\cdot\frac{d}{dx}\left(x^{\frac{1}{2}}\right) + 0$$

$$\Rightarrow \frac{dy}{dx} = \frac{1}{1+x}\cdot\frac{1}{2}\cdot x^{\left(-\frac{1}{2}\right)} = \frac{1}{2\sqrt{x}\,(1+x)}$$

Note: $\tan^{-1}x = \cot^{-1}\frac{1}{x}$ only for $x > 0$

$$\therefore \tan^{-1}\left(\frac{5+2x}{2-5x}\right) \neq \cot^{-1}\left(\frac{2-5x}{5+2x}\right)$$

for $x < \frac{-5}{2}$ or $x > \frac{2}{5}$.

Problems on inverse circular functions

Exercise 10.1

1. Prove the following results by Δ_ method.

(i) $\frac{d}{dx}\left(\cos^{-1}x\right) = -\frac{1}{\sqrt{1-x^2}}$, $(|x| < 1)$

(ii) $\frac{d}{dx}\left(\cot^{-1}x\right) = -\frac{1}{1+x^2}$, $\forall x$

(iii) $\frac{d}{dx}\left(\operatorname{cosec}^{-1}x\right) = -\frac{1}{|x|\sqrt{x^2-1}}$, $(|x| > 1)$

2. Find the differential coefficients of the following functions w.r.t. their independent variables using Δ_ method.

(i) $\sin^{-1} x$ **(ii)** $\tan^{-1} x$ **(iii)** $\sec^{-1} x$

(iv) $\text{Log} \sin^{-1} x$

Problems based on type 1:

Form: $t^{-1}[f(x)] \,/\, \left[t^{-1}\{f(x)\}\right]^n \,/\, f\left[t^{-1}\{f(x)\}\right]$

where $t^{-1} = \sin^{-1}/\cos^{-1}/\tan^{-1}/\cot^{-1}/\sec^{-1}/\text{cosec}^{-1}$

f = sin / cos / tan / cot / sec / cosec / log / e / .....

etc. in $f\left[t^{-1}\{f(x)\}\right]$

Exercise 10.2

Find $\dfrac{dy}{dx}$ of the following functions.

1. $y = \left(\sin^{-1} x\right)^3$

2. $y = \tan^{-1} \sqrt{3x}$

3. $y = \left(\sin^{-1} \sqrt{x}\right)^2$

4. $y = \cos^{-1} \phi(x)$

5. $y = \tan^{-1} (\sec x + \tan x)$

6. $y = \cos^{1} \left(\tan x^2\right)$

7. $y = \left(\tan^{-1} 2x\right)^3$

8. $y = \sec^{-1} \sqrt{x}$

9. $y = \tan^{-1} \sqrt{1 + x^2}$

10. $y = \sin\left(\cos^{-1} x\right)$

11. $y = \cos\left(\sin^{-1} x\right)$

12. $y = \tan^{-1} (\sin x)$

13. $y = \tan\left(\sin^{-1} x\right)$

14. $y = \tan^{-1} \sqrt{x}$

15. $y = \sin^{-1}\left(a \sin^2 x\right)$

16. $y = \tan^{-1}\left(x^2 \cdot e^{-x}\right)$

17. $y = \sin^{-1}\left(e^{\tan^{-1} x}\right)$

18. $y = \tan^{-1}\left(\dfrac{x}{\sqrt{1 - x^2}}\right)$

19. $y = \left(\tan^{-1} 5x\right)^2 + \sin^{-1}\left(\dfrac{1}{\sqrt{1 - x^2}}\right)$

20. $y = e^{10x} \tan^{-1}\left(1 + x^2\right)$

21. $y = \sin^{-1} \sqrt{x} + \cos^{-1} \sqrt{x}$

22. $y = \cos^{-1}\left(1 - \dfrac{1}{x} - \dfrac{1}{x^2}\right)^{\frac{1}{2}}$

23. $y = \tan^{-1}\left(1 + e^x + 2 \log x\right)$

24. $y = \tan^{-1}\left(\dfrac{x}{a - \sqrt{b^2 - x^2}}\right)$

25. $y = \sec^{-1} |x|$

26. $y = \left(\sin^{-1} x^2\right)^2$

27. $y = \left(\tan^{-1} 5x\right)^2$

28. $y = \left[\cos^{-1}\left(x^2 - 2\right)\right]^2$

29. $y = \left[\cot^{-1} x^2\right]^{\frac{1}{3}}$

30. $y = \left[\tan^{-1} \sqrt{x}\right]^2$

31. $y = a \cot^{-1}\left[m \tan^{-1}(bx)\right]$

32. $y = \cos\left[a \sin^{-1}\left(\frac{1}{x}\right)\right]$

***Type 2*:**

Problems based on substitution and change of form

***Form*:** $t^{-1}[f(x)]$

Where $t^{-1} = \sin^{-1}/\cos^{-1}/\tan^{-1}/\cot^{-1}/\sec^{-1}/\text{cosec}^{-1}$ and $f(x)$ = algebraic function of x

Exercise 10.3

Find $\frac{dy}{dx}$ of the following functions.

1. $y = \sin^{-1}\sqrt{1 - x^2}$

2. $y = \sin^{-1}\left(\frac{\sqrt{a^2 - x^2}}{a}\right)$

3. $y = \sin^{-1}(3x - 4x^3)$

4. $y = \sin^{-1}\left(\frac{2x\sqrt{a^2 - x^2}}{a}\right)$

5. $y = \sin^{-1}\left(\frac{\sqrt{a + x} - \sqrt{a - x}}{2\sqrt{a}}\right)$

6. $y = \sin^{-1}\left(x\sqrt{1 - x} - \sqrt{x}\sqrt{1 - x^2}\right)$

7. $y = \sin^{-1}\left(\frac{2x}{1 + x^2}\right)$

8. $y = \sin^{-1}\left(\frac{1 - x^2}{1 + x^2}\right)$

9. $y = \sin^{-1}\left[2\tan^{-1}\sqrt{\frac{1 - x}{1 + x}}\right]$.

10. $y = \cos^{-1}\sqrt{1 - x^2}$

11. $y = \cos^{-1}\left(\frac{\sqrt{a^2 - x^2}}{a}\right)$

12. $y = \cos^{-1}(2x^2 - 1)$

13. $y = \cos^{-1}\sqrt{\frac{1 + x^2}{2}}$

14. $y = \cos^{-1}\left(\sqrt{\frac{1 - x^2}{1 + x^2}}\right)$

15. $y = \cos^{-1}\left(\sqrt{\frac{x - x^{-1}}{x + x^{-1}}}\right)$

16. $y = \tan^{-1}\left(\frac{2x}{1 - x^2}\right)$

17. $y = \tan^{-1}\left(\frac{x}{\sqrt{a^2 - x^2}}\right)$

18. $y = \tan^{-1}\left[\frac{\sqrt{1 + x^2} - 1}{x}\right]$

19. $y = \tan^{-1}\left[\frac{\sqrt{a^2 + x^2} + x}{\sqrt{a^2 + x^2} - x}\right]$

20. $y = \tan^{-1}\left(\frac{a - x}{1 + ax}\right)$

21. $y = \tan^{-1}\left(\dfrac{\sqrt{x} - x}{1 + \frac{3}{2}}\right)$

22. $y = \tan^{-1}\left(\dfrac{1 + x}{1 - x}\right)$

23. $y = \tan^{-1}\left(\dfrac{3x - x^3}{1 + 3x^2}\right)$

24. $y = \tan^{-1}\left(\dfrac{a + bx}{b - ax}\right)$

25. $y = \tan^{-1}\sqrt{\dfrac{x - a}{b - x}}$

26. $y = \tan^{-1}\sqrt{\dfrac{\sqrt{x} + x}{1 - x^{\frac{3}{2}}}}$

27. $y = \tan^{-1}\sqrt{\dfrac{\sqrt{x} - x}{1 + x^{\frac{3}{2}}}}$

28. $y = \tan^{-1}\left[\dfrac{3ax}{a^2 - 2x^2}\right]$

Hint: dividing *Nr* and *Dr* by a^2, we have

$$y = \tan^{-1}\left[\frac{3 \cdot \frac{x}{a}}{1 - 2 \cdot \frac{x^2}{a^2}}\right] = \tan^{-1}\left[\frac{3 \cdot \frac{x}{a}}{1 - 2 \cdot \frac{x^2}{a^2}}\right]$$

$$y = \tan^{-1}\left[\frac{\frac{2x}{a} + \frac{x}{a}}{1 - \frac{2x}{a} \cdot \frac{x}{a}}\right] = \tan^{-1}\frac{2x}{a} + \tan^{-1}\frac{x}{a} = \ldots\text{etc.}$$

29. $y = \tan^{-1}\left(\dfrac{5ax}{a^2 - x^2}\right)$

30. $y = \tan^{-1}\left[\dfrac{3a^2 x - x^3}{a\left(a^2 - 3x^2\right)}\right]$

31. $y = \tan^{-1}\left[\dfrac{1}{\sqrt{x^2 - 1}}\right]$

32. $y = \tan^{-1}\left(\dfrac{x}{\sqrt{1 - x^2}}\right)$

33. $y = \tan^{-1}\left(\dfrac{4x}{4 - x^2}\right)$

34. $y = \tan^{-1}\left[\dfrac{\sqrt{1 + x} - \sqrt{1 - x}}{\sqrt{1 + x} + \sqrt{1 - x}}\right]$

35. $y = \cot^{-1}\left(\dfrac{2 - 5x}{5 + 2x}\right)$

36. $y = \cot^{-1}\left(\dfrac{1 + x}{1 - x}\right)$

37. $y = \sec^{-1}\left(\dfrac{x^2 + 1}{x^2 - 1}\right)$

38. $y = \sec^{-1}\left(\dfrac{1}{2x^2 - 1}\right)$

39. $y = \text{cosec}^{-1}\left(\dfrac{1 + x^2}{2x}\right)$

***Answers** (under suitable restrictions on x)*:

1. $-\dfrac{1}{\sqrt{1 - x^2}}$ **2.** $-\dfrac{1}{\sqrt{a^2 - x^2}}$ **3.** $\dfrac{3}{\sqrt{1 - x^2}}$

4. $\dfrac{2}{\sqrt{a^2 - x^2}}$ **5.** $\dfrac{1}{2\sqrt{1 - x^2}}$ **6.** $\dfrac{1}{\sqrt{1 - x^2}}$

7. $\frac{2}{1+x^2}$ **8.** $\frac{-2}{1+x^2}$ **9.** $\frac{-x}{\sqrt{1-x^2}}$

10. $\frac{1}{\sqrt{1-x^2}}$ **11.** $\frac{1}{\sqrt{a^2-x^2}}$ **12.** $\frac{-2}{\sqrt{1-x^2}}$

13. $\frac{-1}{2(1-x^2)^{\frac{1}{2}}}$ **14.** $\frac{2}{1+x^2}$ **15.** $\frac{-2}{1+x^2}$

16. $\frac{2}{1+x^2}$ **17.** $\frac{1}{\sqrt{a^2-x^2}}$ **18.** $\frac{2}{1+x^2}$

19. $\frac{1}{\sqrt{a^2-x^2}}$ **20.** $\frac{-1}{1+x^2}$

21. $\frac{1}{1+x}\cdot\frac{1}{2\sqrt{x}}-\frac{1}{1+x^2}$ **22.** $\frac{1}{1+x^2}$

23. $\frac{3}{1+x^2}$ **24.** $\frac{1}{1+x^2}$ **25.** $\frac{1}{2\sqrt{(x-a)(x-b)}}$

26. $\frac{1}{2\sqrt{x}}\cdot\frac{1}{1+x}+\frac{1}{1+x^2}$

27. $\frac{1}{1+x}\cdot\frac{1}{2\sqrt{x}}-\frac{1}{1+x^2}$ **28.** -1

29. $\frac{a^2(2+13x^2)}{(a^2+9x^2)(a^2+4x^2)}$ **30.** $\frac{3a}{a^2+x^2}$

31. $\frac{-1}{x\sqrt{x^2-1}}$ **32.** $\frac{1}{\sqrt{1-x^2}}$ **33.** $\frac{4}{4+x^2}$

34. $\frac{1}{2}\frac{1}{\sqrt{1-x^2}}$ **35.** $\frac{1}{1+x^2}$ **36.** $\frac{-1}{1+x^2}$

37. $\frac{-2}{1+x^2}$ **38.** $\frac{-2}{\sqrt{1-x^2}}$ **39.** $\frac{2}{\sqrt{1+x^2}}$

Type 2 continued

Problems based on substitution and change of form

Form: $t^{-1}[t(x)]$

Where $t^{-1} = \sin^{-1}/\cos^{-1}/\tan^{-1}/\cot^{-1}/\sec^{-1}/\text{cosec}^{-1}$

$t(x) =$ a trigonometrical function of x / a combination of trigonometrical functions of x.

Exercise 10.4

Find $\frac{dy}{dx}$ of the following functions.

1. $y = \cos^{-1}\left(\frac{a+b\cos x}{b+a\cos x}\right)$

2. $y = \tan^{-1}\left(\frac{\sin x}{1-\cos x}\right)$

3. $y = \tan^{-1}\sqrt{\frac{1-\cos x}{1+\cos x}}$

4. $y = \tan^{-1}(\sec x + \tan x)$

5. $y = \cot^{-1}(\text{cosec}\, x - \cot x)$

6. $y = \tan^{-1}\left[\frac{a\cos x - b\sin x}{b\cos x + a\sin x}\right]$

Hint: Put $a = r\sin\alpha$ and $b = r\cos\alpha \Rightarrow \frac{a}{b} = \tan\alpha$

$$\therefore y = \tan^{-1}\left[\frac{r(\sin\alpha\cos x - \cos\alpha\sin x)}{r(\cos\alpha\cos x + \sin\alpha\sin x)}\right]$$

$$= \tan^{-1}\left[\frac{\sin(\alpha - x)}{\cos(\alpha - x)}\right] = \tan^{-1}\tan(\alpha - x)$$

$$= \alpha - x = \tan^{-1}\frac{a}{b} - x$$

7. $y = \tan^{-1}\left[\frac{\cos x - \sin x}{\cos x + \sin x}\right]$

8. $y = \tan^{-1}\left[\frac{a - b\tan x}{a + b\tan x}\right]$

9. $y = \tan^{-1}[\tan x + \sec x]$

10. $y = \tan^{-1}\sqrt{\frac{1-\sin x}{1+\sin x}}$

11. $y = \tan^{-1}\left[\frac{\cos x}{1+\sin x}\right]$

12. $y = \tan^{-1}\left[\frac{b}{a}\tan x\right]$

13. $y = \cot^{-1}[\operatorname{cosec} x + \cot x]$

14. $y = \tan^{-1}(\sec x)$

15. $y = \cot^{-1}(\operatorname{cosec} x)$

16. $y = \sec^{-1}(\tan x)$

17. $y = \sec^{-1}\left[\frac{a + b\cos x}{b + a\cos x}\right]$

18. $y = \tan^{-1}\sqrt{\frac{1 + \cos x}{1 - \cos x}}$

Answers (under suitable restrictions on x):

1. $\frac{-\sqrt{b^2 - a^2}}{b + a\cos x}$ 2. $-\frac{1}{2}$ 3. $\frac{1}{2}$ 4. $\frac{1}{2}$ 5. $-\frac{1}{2}$

6. -1 7. -1 8. -1 9. $\frac{1}{2}$ 10. $-\frac{1}{2}$ 11. $-\frac{1}{2}$

12. $\frac{ab\sec^2 x}{a^2 + b^2\tan^2 x}$ 13. $\frac{1}{2}$ 14. $\frac{\sin x}{1 + \cos^2 x}$

15. $\frac{\cos x}{1 + \sin^2 x}$ 16. $\frac{1}{|\sin x|\sqrt{\sin^2 x - \cos^2 x}}$

17. $\frac{-\sqrt{b^2 - a^2}}{(a + b\cos x)}$ 18. $-\frac{1}{2}$

Type 3:

Differentiation of a function w.r.t. an other function

Form: $t^{-1}[f_1\{f_2(x)\}]$ w.r.t. $t[t^{-1}\{g_1(g_2(x))\}]$ or,

$t[t^{-1}\{f(x)\}]$ w.r.t. $t[t^{-1}(x)]$

where t = sin / cos / tan / cot / sec / cosec

$t^{-1} = \sin^{-1}/\cos^{-1}/\tan^{-1}/\cot^{-1}/\sec^{-1}/\operatorname{cosec}^{-1}$

$f(x)$ = a function of x.

Exercise 10.5

Differentiate:

1. $\sec^2(\tan^{-1} x)$ w.r.t $(1 - x^2)$

2. $\tan^{-1}\left(\sqrt{1 + x^2} - x\right)$ w.r.t $\tan^{-1} x$

3. $\cot^{-1}\left[\frac{\sqrt{x^2 + 1} - 1}{x}\right]$ w.r.t $\cot^{-1} x$

4. $2\sin^{-1}\left(\frac{2x}{1 + x^2}\right)$ w.r.t $\sin^{-1}\left(\frac{1 + x^2}{1 - x^2}\right)$

5. $\operatorname{cosec}(\cot^{-1} x)$ w.r.t $\sec(\tan^{-1} x)$

6. $\tan^{-1}\sqrt{\frac{1 - x^2}{1 + x^2}}$ w.r.t $\cos^{-1} x^2$

7. $\tan^{-1}\left(\frac{2x}{1 - x^2}\right)$ w.r.t $\sin^{-1}\left(\frac{2x}{1 + x^2}\right)$

8. $\tan^{-1}\left(\frac{\sqrt{1 + x^2}}{x}\right)$ w.r.t $\cos^{-1}\sqrt{\frac{\sqrt{1 + x^2} + 1}{2\sqrt{1 + x^2}}}$

9. $\sec^{-1}\left(\frac{1}{2x^2 - 1}\right)$ w.r.t $\sqrt{1 - x^2}$

10. $\tan^{-1}\left(\frac{2x}{1 - x^2}\right)$ w.r.t $\cos^{-1}\left(\frac{1 - x^2}{1 + x^2}\right)$

11. $\tan^{-1}\left(\frac{2\sqrt{x}}{1 - x}\right)$ w.r.t $\sin^{-1}\left(\frac{2\sqrt{x}}{1 + x}\right)$

12. $\sin^{-1}\left(\frac{1 - x}{1 + x}\right)$ w.r.t $\sqrt{x}$

13. $\tan^{-1}\left(\frac{\sqrt{1+x^2}-1}{x}\right)$ w.r.t $\tan^{-1}x$

14. $\sin^{-1}\left(\frac{2x}{1+x^2}\right)$ w.r.t $\cos^{-1}\left(\frac{1-x^2}{1+x^2}\right)$

15. $\tan^{-1}\left(\frac{\sqrt{1-x^2}}{x}\right)$ w.r.t $\cos^{-1}x$

16. $\tan^{-1}\left(\frac{\sqrt{1+x^2}-1}{x}\right)$ w.r.t $\sin^{-1}\left(\frac{x}{\sqrt{1+x^2}}\right)$

Answers: **1.** -1 **2.** $-\frac{1}{2}$ **3.** $\frac{1}{2}$ **4.** 2 **5.** 1 **6.** $\frac{1}{2}$ **7.** 1 **8.** 1 **9.** 4 **10.** 1 **11.** 1 **12.** $\frac{2}{1+x}$ **13.** $\frac{1}{2}$ **14.** $\frac{1}{2}$ **15.** 1 **16.** $\frac{1}{2}$.

11

Differential Coefficient of Mod Functions

Differentiation of Mod Functions

Definition: $|f(x)| = \sqrt{(f(x))^2}$

Formulas:

1. $\dfrac{d}{dx}|f(x)| = \dfrac{|f(x)|}{f(x)} \cdot f'(x)$; for $f(x) \neq 0$

2. $\dfrac{d}{dx}|f(x)|^n = \dfrac{n|f(x)|^n}{f(x)} \cdot f'(x)$; $f(x) \neq 0,\ n \in Q$

3. $\dfrac{d|x|}{dx} = \dfrac{|x|}{x}$; $x \neq 0$

4. $\dfrac{d|x|^n}{dx} = \dfrac{n|x|^n}{x}$; $x \neq 0,\ n \in Q$

Proof:

1. $\dfrac{d}{dx}|f(x)| = \dfrac{d}{dx}\sqrt{[f(x)]^2}$

$$= \frac{d}{dx}\sqrt{f^2(x)} \left[\because |f(x)| = \sqrt{[f(x)]^2}\right]$$

$$= \frac{1}{2\sqrt{[f(x)]^2}} \cdot \frac{d[f(x)]^2}{dx}; \text{ when } f(x) \neq 0$$

$$= \frac{2f(x)}{2\sqrt{[f(x)]^2}} \cdot \frac{df(x)}{dx} = \frac{f(x)}{|f(x)|} \cdot f'(x)$$

$$= \frac{|f(x)|}{f(x)} \times f'(x)$$ which is the general formula of (or, for) d.c. of mod function $f(x)$ (at the point x where $f(x) \neq 0$).

Note: At the points where $f(x) = 0$, the value of $\dfrac{d|f(x)|}{dx}$ is to be found out by first principles. It may or may not exist at such a point because generally "mod functions" are not differentiable at their roots i.e. at the roots of $|f(x)| = 0$.

2. $\dfrac{d|f(x)|^n}{dx}$, $n \in Q$

$$|f(x)| = z \Rightarrow |f(x)|^n = z^n$$

Now $\dfrac{dz^n}{dx} = nz^{n-1} \times \dfrac{dz}{dx}$

$$\Rightarrow \frac{d|f(x)|^n}{dx} = n|f(x)|^{n-1} \times \frac{d|f(x)|}{dx}$$

$$= n|f(x)|^{n-1} \times \frac{|f(x)|}{f(x)} \times f'(x)$$

$$= \frac{n|f(x)|^n}{f(x)} \times f'(x) \text{ when } f(x) \neq 0$$

3. $\dfrac{d|x|}{dx} = \dfrac{d\sqrt{x^2}}{dx}$

$= \dfrac{1}{2\sqrt{x^2}} \cdot \dfrac{dx^2}{dx}$ for $x \neq 0$

$= \dfrac{2x}{2\sqrt{x^2}} \times \dfrac{\sqrt{x^2}}{\sqrt{x^2}}$

$= \dfrac{2x|x|}{2x^2}$

$= \dfrac{|x|}{x}\ (x \neq 0)$

Further, $f'_+(0) = 1$, $f'_-(0) = -1$ where $f(x) = |x|$

$\therefore f'(0)$ does not exist.

4. $\dfrac{d|x|^n}{dx}$, $n \in Q$

$|x| = z \Rightarrow |x|^n = z^n$ which means on differentiating

$\dfrac{dz^n}{dx} = nz^{n-1}\dfrac{dz}{dx}$

$\Rightarrow \dfrac{d|x|^n}{dx} = n|x|^{n-1} \cdot \dfrac{d|x|}{dx}$

$= n|x|^{n-1} \cdot \dfrac{|x|}{x}$

$= \dfrac{n|x|^n}{x}\ (x \neq 0)$

Further, $f'(0) = 0$ for $n > 1$ where $f(x) = |x|^n$.

Remember:

$\dfrac{d|f(x)|}{dx} = \dfrac{|f(x)|}{f(x)} \times f'(x)$ for $f(x) \neq 0$

or, $\dfrac{d|f(x)|}{dx} = \dfrac{f(x)}{|f(x)|} \times f'(x)$ for $f(x) \neq 0$

Problems based on algebraic functions

Solved Examples

Find the d.c. of

1. $y = |x^3|$

Solution: $y = |x^3|$

$\therefore \dfrac{dy}{dx} = \dfrac{d|x^3|}{dx}$

$= \dfrac{d|x^3|}{dx^3} \cdot \dfrac{dx^3}{dx}$

$= \dfrac{|x^3|}{x^3} \cdot 3x^2$

$= \dfrac{3x^2|x^3|}{x^3}$; $x \neq 0 = 3x|x|$, $x \neq 0$

Further,

$f'_+(0) = f'_-(0) = 0$

$\therefore \dfrac{dy}{dx} = 0$ for $x = 0$

2. $y = |x|^2 - 4|x| + 2$

Solution: $y = |x|^2 - 4|x| + 2$

$\therefore \dfrac{dy}{dx} = \dfrac{d}{dy}\left[|x|^2 - 4|x| + 2\right]$

$= \dfrac{d}{dx}|x|^2 - 4\dfrac{d|x|}{dx} + \dfrac{d(2)}{dx}$

$= 2|x| \cdot \dfrac{d|x|}{dx} - \dfrac{4|x|}{x}$, $x \neq 0$

$= 2|x| \cdot \dfrac{|x|}{x} - 4\dfrac{|x|}{x}$

$= \dfrac{2x^2}{x} - \dfrac{4|x|}{x}$

$= 2x - \dfrac{4|x|}{x}$; $(x \neq 0)$

Further $y'(0)$ does not exist.

3. $y = x|x|$

Solution: $\because y = x|x|$

$$\therefore \frac{dy}{dx} = \frac{d}{dx}[x|x|]$$

$$= x\frac{d|x|}{dx} + |x|\frac{dx}{dx}$$

$$= x \cdot \frac{|x|}{x} + |x|, (x \neq 0)$$

$$= |x| + |x|$$

$$= 2|x|, x \neq 0$$

Also, $f'_+(0) = f'_-(0) = 0$

$$\therefore \frac{dy}{dx} = 2|x| \text{ for all } x.$$

Problems based on transcendental functions

Solved Examples

Find d.c. of

1. $y = \log|f(x)|$

Solution: $y = \log|f(x)|$ which is defined for $f(x) \neq 0$

For $f(x) \neq 0$, $\frac{dy}{dx} = \frac{d}{dx}\left[\log|f(x)|\right]$

$$= \frac{d\log|f(x)|}{d|f(x)|} \times \frac{d|f(x)|}{dx}$$

$$= \frac{1}{|f(x)|} \times \frac{|f(x)|}{f(x)} \times f'(x)$$

$$= \frac{f'(x)}{f(x)}; f(x) \neq 0$$

2. $y = \log|x-1|$

Solution: $y = \log|x-1|$; which is defined for $x \neq 1$

$\therefore$ For $x \neq 1$, $\frac{dy}{dx} = \frac{d\log|x-1|}{d|x-1|} \cdot \frac{d|x-1|}{dx}$

$$= \frac{1}{|x-1|} \cdot \frac{|x-1|}{(x-1)} \cdot \frac{d}{dx}(x-1)$$

$$= \frac{1}{(x-1)}$$

3. $y = \log|x|$

Solution: $y = \log|x|$ which is defined for $x \neq 0$

$\therefore$ For $x \neq 0$; $\frac{dy}{dx} = \frac{d\log|x|}{dx}$

$$= \frac{d\log|x|}{d|x|} \cdot \frac{d|x|}{dx}$$

$$= \frac{1}{|x|} \cdot \frac{|x|}{x}$$

$$= \frac{1}{x}; x \neq 0x$$

4. $y = |\log x|$

Solution: $y = |\log x|$, which is defined for $x > 0$

$\therefore$ For $x > 0$, $\frac{dy}{dx} = \frac{d|\log x|}{dx}$

$$= \frac{d|\log x|}{d\log x} \cdot \frac{d\log x}{dx}$$

$$= \frac{|\log x|}{\log x} \cdot \frac{1}{x}; (\text{for } x \neq 1)$$

Also $f'_+(1) = \lim_{h \to 0} \frac{\log(1+h) - \log 1}{h}; h > 0$

$= 1$

$$f'_-(1) = \lim_{h \to 0} \frac{\log(1-h) - \log 1}{h}; h > 0$$

$$= \lim_{h \to 0} \frac{\log(1-h)}{h}$$

$= -1$

$\therefore \frac{dy}{dx}$ does not exist at $x = 1$

5. $y = |\sin x|$

Solution: $y = |\sin x|$

$$\Rightarrow \frac{dy}{dx} = \frac{d}{dx}|\sin x|$$

$$= \frac{|\sin x|}{\sin x} \cdot \frac{d \sin x}{dx}; (\sin x \neq 0, \text{ i.e.; } x \neq n\pi)$$

$$= \frac{|\sin x|}{\sin x} \cdot \cos x$$

$$= |\sin x| \cdot \cot x$$

Also, $f'_+(n\pi) = \lim_{h \to 0} \frac{|\sin(n\pi + h)| - |\sin n\pi|}{h};$

$h > 0$

$$= \lim_{h \to 0} \frac{|\sin h|}{h}$$

$$= \lim_{h \to 0} \frac{\sin h}{h}$$

$$= 1$$

$$f'_-(n\pi) = \lim_{h \to 0} \left(\frac{\sin h}{-h}\right), h > 0$$

$$= -1$$

$\therefore \frac{dy}{dx}$ does not exist for $x = n\pi$

6. $y = |\cos x|$

Solution: $y = |\cos x|$

$$\Rightarrow \frac{dy}{dx} = \frac{d|\cos x|}{dx}$$

$$= \frac{|\cos x|}{\cos x} \cdot \frac{d \cos x}{dx}; (\cos x \neq 0)$$

$$= \frac{|\cos x|}{\cos x} \cdot (-\sin x)$$

$$= |\cos x| \cdot (-\sin x); \left(x \neq (2n+1)\frac{\pi}{2}\right)$$

Also, $f'_+\left[(2n+1)\frac{\pi}{2}\right]$

$$= \lim_{h \to 0} \frac{\left|\cos\left(n\pi + \frac{\pi}{2} + h\right)\right| - \left|\cos\left(n\pi + \frac{\pi}{2}\right)\right|}{h};$$

$(h > 0)$

$$= \lim_{h \to 0} \frac{|\sin h|}{h}$$

$$= 1$$

$$f'_-\left\{(2n+1)\frac{\pi}{2}\right\} = \lim_{h \to 0} \frac{\sin h}{-h} = -1$$

$\therefore \frac{dy}{dx}$ does not exist for $x = (2n+1)\frac{\pi}{2}$

7. $y = \sin|x|$

Solution: $y = \sin|x|$

$$\Rightarrow \frac{dy}{dx} = \frac{d \sin|x|}{dx}$$

$$= \frac{d \sin|x|}{d|x|} \cdot \frac{d|x|}{dx}; (x \neq 0)$$

$$= \frac{\cos|x| \cdot |x|}{x} \text{ for } x \neq 0$$

Also, $f'_+(0) = \lim_{h \to 0} \frac{\sin(h) - \sin(0)}{h}; (h > 0)$

$$= 1$$

$$f'_-(0) = \lim_{h \to 0} \frac{\sin h}{-h}, (h > 0)$$

$$= -1$$

$\therefore \frac{dy}{dx}$ does not exist for $x = 0$

8. $y = \left|\cos \frac{x}{2}\right|$

Solution: $y = \left|\cos \frac{x}{2}\right|$

$$\Rightarrow \frac{dy}{dx} = \frac{d\left|\cos \frac{x}{2}\right|}{dx}$$

$$= \frac{\left|\cos \frac{x}{2}\right|}{\cos \frac{x}{2}} \cdot \frac{d}{dx}\left(\cos \frac{x}{2}\right)$$

$$= \frac{\left|\cos\frac{x}{2}\right|}{\cos\frac{x}{2}} \cdot \left(-\sin\frac{x}{2}\right) \cdot \frac{d\left(\frac{x}{2}\right)}{dx}; \; (x \neq (2n+1)\pi)$$

$$= \frac{-\sin\frac{x}{2}\left|\cos\frac{x}{2}\right|}{\cos\frac{x}{2}} \cdot \frac{1}{2}$$

$$= \frac{-\sin\frac{x}{2}\left|\cos\frac{x}{2}\right|}{2\cos\frac{x}{2}}; \; (x \neq (2n+1)\pi)$$

Also, $\frac{dy}{dx}$ does not exist for $x = (2n+1)\pi$.

9. $y = \sin|x-2|$

Solution: $y = \sin|x-2|$

$$\Rightarrow \frac{dy}{dx} = \frac{d\sin|x-2|}{dx}$$

$$= \cos|x-2| \frac{d}{dx}[|x-2|]$$

$$= \cos|x-2| \cdot \frac{|x-2|}{(x-2)} \cdot \frac{d(x-2)}{dx}; \; (x \neq 2)$$

$$= \cos|x-2| \cdot \frac{|x-2|}{(x-2)} \cdot 1; \; (x \neq 2)$$

$$= \frac{|x-2|\cos|x-2|}{(x-2)}; \; (x \neq 2)$$

$\frac{dy}{dx}$ does not exist at $x = 2$

10. $y = \tan|x|$

Solution: $y = \tan|x|$

$$\Rightarrow \frac{dy}{dx} = \frac{d\tan|x|}{dx}$$

$$= \sec^2|x| \cdot \frac{d|x|}{dx}$$

$$= \sec^2|x| \cdot \frac{|x|}{x}; \; (x \neq 0)$$

$$= \frac{|x| \cdot \sec^2|x|}{x}; \; (x \neq 0)$$

Also, $f'_+(0) = \lim_{h \to 0} \frac{\tan h}{h}; \; (h > 0)$

$= 1$

$f'_-(0) = \lim_{h \to 0} \frac{\tan h}{-h}, \; (h > 0)$

$= -1$

$\therefore \frac{dy}{dx}$ does not exist at $x = 0$

11. $y = \sin^{-1}|x|$

Solution: $y = \sin^{-1}|x|$ which is defined for $|x| \leq 1$

$$\Rightarrow \frac{dy}{dx} = \frac{d}{dx}\sin^{-1}|x|$$

$$= \frac{d\sin^{-1}|x|}{d|x|} \cdot \frac{d|x|}{dx}$$

$$= \frac{1}{\sqrt{1-(|x|)^2}} \cdot \frac{|x|}{x}; \; x \neq 0, \; x \neq \pm 1$$

$$= \frac{|x|}{x\sqrt{1-x^2}} \left(\because |x|^2 = |x^2| = x\right); \; x \neq 0, \; |x| < 1$$

$f'_+(0) = \lim_{h \to 0} \frac{\sin^{-1} h}{h}; \; h > 0$

$= 1$

and $f'_-(0) = \lim_{h \to 0} \frac{\sin^{-1} h}{-h}, \; h > 0$

$= -1$

$\therefore \frac{dy}{dx}$ does not exist at $x = 0$

The derivative is $+\infty$ at $x = 1$; $-\infty$ at $x = -1$

12. $y = |\sin^{-1} x|$

Solution: $y = |\sin^{-1} x|$

$$\Rightarrow \frac{dy}{dx} = \frac{d\left|\sin^{-1} x\right|}{dx}$$

$$= \frac{|\sin^{-1} x|}{\sin^{-1} x} \times \frac{d}{dx}\left(\sin^{-1} x\right); \ (x \neq 0)$$

$$= \frac{|\sin^{-1} x|}{\sin^{-1} x} \times \frac{1}{\sqrt{1-x^2}}; \ (x \neq 0, |x| < 1)$$

$\frac{dy}{dx}$ does not exist at $x = 0$ and the derivative is $+\infty$ and $-\infty$ at $x = 1, -1$ respectively.

13. $y = \tan^{-1}|x|$

Solution: $y = \tan^{-1}|x|$

$$\Rightarrow \frac{dy}{dx} = \frac{d}{dx}\left[\tan^{-1}|x|\right]$$

$$= \frac{1}{1+(|x|)^2} \cdot \frac{d}{dx}(|x|)$$

$$= \frac{1}{1+x^2} \cdot \frac{|x|}{x}; \ x \neq 0 \left(\because \left|x^2\right| = |x|^2 = x^2\right)$$

$$= \frac{|x|}{x\left(1+x^2\right)}$$

$\frac{dy}{dx}$ does not exist for $x \neq 0$

14. $y = \log\left(\left|\tan\frac{x}{2}\right|\right)$

Solution: $y = \log\left(\left|\tan\frac{x}{2}\right|\right)$ which is defined for $x \neq n\pi$

$\therefore$ For $x \neq n\pi$, $\frac{dy}{dx}$

$$= \frac{1}{\tan\frac{x}{2}} \cdot \frac{d}{dx}\left(\tan\frac{x}{2}\right)\left[\because \frac{d\log|f(x)|}{dx} = \frac{f'(x)}{f(x)}\right]$$

$$= \frac{1}{\tan\frac{x}{2}} \cdot \sec^2\left(\frac{x}{2}\right) \cdot \frac{d}{dx}\left(\frac{x}{2}\right)$$

$$= \frac{\cos\frac{x}{2}}{\sin\frac{x}{2}} \cdot \frac{1}{\cos^2\frac{x}{2}} \cdot \frac{1}{2}$$

$$= \frac{1}{2\sin\frac{x}{2} \cdot \cos\frac{x}{2}}$$

$$= \frac{1}{\sin x}$$

$= \operatorname{cosec} x$

15. $y = \log(\log|x|)$

Solution: $y = \log(\log|x|)$ is defined for $|x| > 1$

$$\therefore \text{For } |x| > 1, \ \frac{dy}{dx} = \frac{d\log(\log|x|)}{dx}$$

$$= \frac{1}{\log|x|} \cdot \frac{d}{dx}(\log|x|)$$

$$= \frac{1}{\log|x|} \cdot \frac{1}{x}$$

$$\left[\because \frac{d\log|f(x)|}{dx} = \frac{f'(x)}{f(x)} \text{ and } \frac{d\log|x|}{dx} = \frac{1}{x}\right]$$

$$= \frac{1}{x\log|x|}$$

16. $y = \log\left|\frac{a+b\tan x}{a-b\tan x}\right|$

Solution: $y = \log\left|\frac{a+b\tan x}{a-b\tan x}\right|$ which is defined for $a \neq \pm b\tan x$ i.e. $a^2 \neq b^2\tan^2 x$

$\therefore$ For $a^2 \neq b^2\tan^2 x$,

$$\frac{dy}{dx} = \frac{d}{dx}\log\left|\frac{a+b\tan x}{a-b\tan x}\right|$$

$$= \frac{(a - b\tan x)}{(a + b\tan x)} \times \left[\frac{b\sec^2 x(a - b\tan x) - \left(-b\sec^2 x\right)(a + b\tan x)}{(a - b\tan x)^2}\right]$$

$$= \frac{ab\sec^2 x - b^2\sec^2 x\tan x + ab\sec^2 x + b^2\sec^2 x\tan x}{a^2 - b^2\tan^2 x}$$

$$= \frac{2ab\sec^2 x}{a^2 - b^2\tan^2 x}$$

17. $y = e^{-|x|}$

Solution: $y = e^{-|x|}$

$$\Rightarrow \frac{dy}{dx} = \frac{d}{dx}e^{-|x|}$$

$$= \frac{d}{d|x|}e^{-|x|} \cdot \frac{d(d|x|)}{dx}$$

$$= e^{-|x|} \times (-1) \cdot \frac{d|x|}{dx}$$

$$= e^{-|x|} \times (-1) \times \frac{|x|}{x}$$

$$= \frac{-e^{-|x|} \cdot |x|}{x} = \frac{-|x|\, e^{-|x|}}{x}; \ (x \neq 0)$$

Remember: Rule to differentiate mod function $|a\cos\theta + b\sin\theta|$. To differentiate mod of a function $a\cos\theta + b\sin\theta$, we may adopt the following working rule:

1. Express $a\cos\theta + b\sin\theta$ (or, $a\sin\theta + b\cos\theta$) as a single cosine (or, single sine).

2. Use $\frac{d|f(x)|}{dx} = \frac{|f(x)|}{f(x)} \times f'(x); \ (f(x) \neq 0)$

Question: How to express $a\cos\theta + b\sin\theta$ or $a\sin\theta + b\cos\theta$ as a single cosine (or, single cosine)

Answer: **(i)** Multiply and divide the expression $(a\cos\theta + b\sin\theta)$ or $(a\sin\theta + b\cos\theta)$ by $\sqrt{a^2 + b^2}$ which means to multiply and to divide the given expression $(a\cos\theta + b\sin\theta)$ or $(a\sin\theta + b\cos\theta)$ by

$$\sqrt{(\text{coefficient of } \cos\theta)^2 + (\text{coefficient of } \sin\theta)^2}$$

(ii) Use the "$A + B$ or $A - B$" formula as the case may require.

Solved Examples

Find the d.c. of

1. $y = |\cos x - \sin x|$

Solution: $y = |\cos x - \sin x|$

$$= \left|\frac{\sqrt{2}(\cos x - \sin x)}{\sqrt{2}}\right|$$

$$= \left|\sqrt{2}\right|\left|\frac{1}{\sqrt{2}}\cos x - \frac{1}{\sqrt{2}}\sin x\right|$$

$$= \left|\sqrt{2}\right|\left|\cos 45^\circ \cos x - \cos 45^\circ \sin x\right|$$

$$= \left|\sqrt{2}\right|\left|\cos\left(x + 45^\circ\right)\right| = \sqrt{2}\left|\cos\left(x + \frac{\pi}{4}\right)\right|$$

Now differentiating both sides w.r.t. x, we get

$$\frac{dy}{dx} = \frac{d}{dx}\left\{\sqrt{2}\left|\cos\left(x + \frac{\pi}{4}\right)\right|\right\}$$

$$= \sqrt{2}\,\frac{d}{dx}\left|\cos\left(x + \frac{\pi}{4}\right)\right|$$

$$= \sqrt{2}\,\frac{d\left|\cos\left(x+\frac{\pi}{4}\right)\right|}{d\cos\left(x+\frac{\pi}{4}\right)} \cdot \frac{d\cos\left(x+\frac{\pi}{4}\right)}{d\left(x+\frac{\pi}{4}\right)} \cdot \frac{d\left(x+\frac{\pi}{4}\right)}{dx}$$

$$= \sqrt{2}\left|\cos\left(x+\frac{\pi}{4}\right)\right| \cdot \left\{-\sin\left(x+\frac{\pi}{4}\right)\right\} \cdot 1 \cdot \frac{1}{\cos\left(x+\frac{\pi}{4}\right)}$$

$$= -\sqrt{2}\,\frac{\left|\cos\left(x+\frac{\pi}{4}\right)\right|\cdot\sin\left(x+\frac{\pi}{4}\right)}{\cos\left(x+\frac{\pi}{4}\right)}$$

for $\cos\left(x+\frac{\pi}{4}\right) \neq 0$

Further, y is not differentiable at x when

$\cos\left(x+\frac{\pi}{4}\right) = 0$

Or, alternatively,

by using general method

$$\frac{dy}{dx} = \frac{d|\cos x - \sin x|}{dx}$$

Now, putting $(\cos x - \sin x) = f(x)$

$$\Rightarrow \frac{dy}{dx} = \frac{d|f(x)|}{dx}$$

$$= \frac{|f(x)|}{f(x)} \times f'(x),\ \text{for } f(x) \neq 0$$

$$= \frac{|\cos x - \sin x|}{(\cos x - \sin x)} \times \frac{d(\cos x - \sin x)}{dx}$$

when $\cos x \neq \sin x$

$$= \frac{|\cos x - \sin x|}{(\cos x - \sin x)} \times (-\sin x - \cos x)$$

$$= \frac{|\cos x - \sin x|}{(\cos x - \sin x)} \times (-1) \times (\sin x + \cos x)$$

$$= \frac{-|\cos x - \sin x|\cdot(\sin x + \cos x)}{(\cos x - \sin x)}\ \text{for } \tan x \neq 1$$

i.e. $x \neq n\pi = \frac{\pi}{4}$

Further y is not differentiable at $x = n\pi + \frac{\pi}{4}$

2. $y = |\sec x - \tan x|$

Solution: $\because y = |\sec x - \tan x|$ which is defined for

$x \neq n\pi + \frac{\pi}{2}$

For $x \neq n\pi = \frac{\pi}{2}$,

$$\frac{dy}{dx} = \frac{d|f(x)|}{dx} = \frac{d|f(x)|}{d\,f(x)} \times \frac{d\,f(x)}{dx}$$

$$= \frac{|f(x)|}{f(x)} \times f'(x)$$

$$= \frac{|\sec x - \tan x|}{(\sec x - \tan x)} \cdot \frac{d}{dx}(\sec x - \tan x)$$

$$= \frac{|\sec x - \tan x|}{(\sec x - \tan x)} \cdot \left(\sec x \cdot \tan x - \sec^2 x\right)$$

$$= \frac{|\sec x - \tan x|\cdot(-\sec x)\cdot(\sec x - \tan x)}{(\sec x - \tan x)}$$

$$= (-\sec x)\,|\sec x - \tan x|$$

Problems based on the substitution $|f(x)| = \sqrt{f^2(x)}$

Refresh your memory: In calculus while differentiating a given function under the square root symbol, when we simplify a given function under the square root symbol and after simplification, we get $\sqrt{[f(x)]^2} = \sqrt{f^2(x)}$, this should be replaced by $|f(x)|$ and then we should find $\frac{d|f(x)|}{dx}$.

Solved Examples

Find the d.c. of

1. $y = \sqrt{1 - \cos^2 x}$

Solution: $y = \sqrt{1 - \cos^2 x} = \sqrt{\sin^2 x}$

$$\Rightarrow y = |\sin x|$$

$$\Rightarrow \frac{dy}{dx} = \frac{d|\sin x|}{dx}$$

$$= \frac{|\sin x|}{\sin x} \cdot \frac{d \sin x}{dx}$$

$$= \frac{|\sin x|}{\sin x} \cdot \cos x,\ x \neq n\pi,\ n \in Z$$

$y'(x)$ does not exist for $x = n\pi$.

Precaution: It is a common mistake to write down

$$y = \sqrt{1 - \cos^2 x} = \sqrt{\sin^2 x} = \sin x$$

$\Rightarrow \frac{dy}{dx} = \cos x$ which is completely wrong.

2. $y = \sqrt{1 - \sin^2 x}$

Solution: $y = \sqrt{1 - \sin^2 x} = \sqrt{\cos^2 x} = |\cos x|$

$$\Rightarrow \frac{dy}{dx} = \frac{d|\cos x|}{dx}$$

$$= \frac{|\cos x|}{\cos x} \cdot \frac{d \cos x}{dx}$$

$$= -\sin x \cdot \frac{|\cos x|}{\cos x}$$

$$= -\tan x\, |\cos x|,\ x \neq n\pi + \frac{\pi}{2},\ n \in Z$$

3. $y = \sqrt{\frac{1 - \sin x}{1 + \sin x}}$

Solution: $y = \sqrt{\frac{1 - \sin x}{1 + \sin x}}$, defined for

$$x \neq (2n + 1)\,\pi + \frac{\pi}{2}$$

Rationalizing the denominator, we get

$$= \sqrt{\frac{(1 - \sin x) \times (1 - \sin x)}{(1 + \sin x) \times (1 - \sin x)}} = \sqrt{\frac{(1 - \sin x)^2}{1 - \sin^2 x}}$$

$$= \sqrt{\frac{(1 - \sin x)^2}{\cos^2 x}}$$

$$= \sqrt{\left(\frac{1 - \sin x}{\cos x}\right)^2} = \sqrt{(\sec x - \tan x)^2}$$

$$= |\sec x - \tan x|$$

$$\therefore \frac{dy}{dx} = \frac{d}{dx} |\sec x - \tan x|$$

$$= \frac{|\sec x - \tan x|}{(\sec x - \tan x)} \cdot \frac{d}{dx} (\sec x - \tan x)$$

$$= \frac{|\sec x - \tan x|}{(\sec x - \tan x)} \times \left(\sec x \tan x - \sec^2 x\right)$$

$$= \frac{|\sec x - \tan x|}{(\sec x - \tan x)} \cdot (-\sec x) \cdot (\sec x - \tan x)$$

$$= -\sec x\, |\sec x - \tan x|,\ x \neq n\pi + \frac{\pi}{2}.$$

4. $y = \sqrt{\left(\frac{1 + \cos x}{2}\right)}$

Solution: $y = \sqrt{\left(\frac{1 + \cos x}{2}\right)}$

$$= \sqrt{\frac{2 \cos^2 \frac{x}{2}}{2}} = \sqrt{\cos^2 \frac{x}{2}} = \left|\cos \frac{x}{2}\right|$$

$$\Rightarrow \frac{dy}{dx} = \frac{d}{dx} \left|\cos \frac{x}{2}\right|$$

$$= \frac{\left|\cos \frac{x}{2}\right|}{\cos \frac{x}{2}} \cdot \frac{d}{dx}\left(\cos \frac{x}{2}\right)$$

$$= \frac{\left|\cos \frac{x}{2}\right|}{\cos \frac{x}{2}} \left(-\sin \frac{x}{2}\right) \cdot \frac{d}{dx}\left(\frac{x}{2}\right)$$

$$= \frac{-\sin\frac{x}{2}\left|\cos\frac{x}{2}\right|}{\cos\frac{x}{2}} \cdot \frac{1}{2}$$

$$= \frac{-\sin\frac{x}{2}\left|\cos\frac{x}{2}\right|}{2\cos\frac{x}{2}}, \; x \neq (2n+1)\pi$$

5. $y = \sqrt{1+\sin 2x} - \sqrt{1-\sin 2x}$

Solution: $y = \sqrt{1+\sin 2x} - \sqrt{1-\sin 2x}$

$$\Rightarrow y = \sqrt{(\cos x + \sin x)^2} - \sqrt{(\cos x - \sin x)^2}$$

$$\Rightarrow y = |\cos x + \sin x| - |\cos x - \sin x|$$

$$\Rightarrow \frac{dy}{dx} = \frac{d|\cos x + \sin x|}{dx} - \frac{d|\cos x - \sin x|}{dx}$$

$$= \frac{|\cos x + \sin x|}{(\cos x + \sin x)} \times (\cos x - \sin x) + \frac{|\cos x - \sin x|}{(\cos x - \sin x)} \times (\sin x + \cos x)$$

$$= \frac{(\cos x - \sin x)|\cos x + \sin x|}{(\cos x + \sin x)} + \frac{(\sin x + \cos x)|\cos x - \sin x|}{(\cos x - \sin x)},$$

for $x \neq n\pi \pm \frac{\pi}{4}, \; n \in Z$

Problems based on differentiation of mod of a function

Exercise 11.1

Find the differential coefficients of the following functions.

1. $|5x+3|$
2. $|x^2 - a^2|$
3. $|x+1|$
4. $|x-1|$
5. $x+|x|$
6. $|x|+\cos x$
7. $\sin x - |x|$
8. $\dfrac{\sin x}{|x|}$
9. $\dfrac{|x|}{\sin x}$
10. $\dfrac{|x-1|}{x-1}$
11. $\dfrac{|x|}{x}$
12. $\dfrac{(x-2)|x|}{|x-2|}$
13. $\dfrac{x(x-1)}{|x-1|}$
14. $\dfrac{1}{|x|}$
15. $x|x|$
16. $\log|x|$
17. $|\log x|$
18. $\dfrac{\sqrt{1+\cos 2x}}{\sqrt{2}\cos x}$
19. $\dfrac{x}{\sqrt{1-\cos x}}$
20. $\sqrt{1+\sin x} - \sqrt{1-\sin x}$
21. $\sqrt{\dfrac{1+\cos 2x}{1-\cos 2x}}$
22. $\log|\sin x|$
23. $\log|\cos(ax+b)|$
24. $|\sin x|$
25. $|\cos x|$
26. $|\tan x|$
27. $|\cot x|$
28. $|\sec x|$
29. $|\text{cosec}\, x|$
30. $\sin|x|$

31. $\cos|x|$
32. $\tan|x|$
33. $\cot|x|$
34. $\sec|x|$
35. $\text{cosec}\,|x|$

Answers (with suitable restrictions on x).

1. $\dfrac{5|5x+3|}{(5x+3)}$

2. $\dfrac{2x\left|x^2-a^2\right|}{\left(x^2-a^2\right)}$

3. $\dfrac{|x+1|}{(x+1)}$

4. $\dfrac{|x-1|}{(x-1)}$

5. $1+\dfrac{|x|}{x}$

6. $\dfrac{|x|}{x}-\sin x$

7. $\cos x-\dfrac{|x|}{x}$

8. $\dfrac{x|x|\cos x-\sin x|x|}{x|x|^2}$

9. $\dfrac{\sin x|x|-x|x|\cos x}{x(\sin x)^2}$

10. 0

11. 0

12. $\dfrac{\{(x-2)|x|+x|x|\}|x-2|-x|x||x-2|}{|x-2|^2}$

13. $\dfrac{|x-1|(2x-1)-x|x-1|}{|x-1|^2}$

14. $\dfrac{-|x|}{x^3}$

15. $2|x|$

23. $-a\tan(ax+b)$

24. $\dfrac{|\sin x|}{\sin x}\cdot\cos x$

25. $\dfrac{|\cos x|}{\cos x}\cdot(-\sin x)$

26. $\dfrac{|\tan x|}{\tan x}\cdot\sec^2 x$

27. $\dfrac{|\cot x|}{\cot x}\cdot\left(-\text{cosec}^2 x\right)$

28. $\dfrac{|\sec x|}{\sec x}\cdot(\sec x\cdot\tan x)$

29. $\dfrac{|\text{cosec}\,x|}{\text{cosec}\,x}\cdot(-\text{cosec}\,x\cdot\cot x)$

30. $\cos|x|\cdot\dfrac{|x|}{x}$

31. $-\sin|x|\cdot\dfrac{|x|}{x}$

32. $\sec^2|x|\cdot\dfrac{|x|}{x}$

33. $-\text{cosec}^2|x|\cdot\dfrac{|x|}{x}$

34. $\sec|x|\cdot\tan|x|\cdot\dfrac{|x|}{x}$

35. $\text{cosec}|x|\cdot\cot|x|\left(-\dfrac{|x|}{x}\right)$

Change of Form before Differentiation

In certain cases the given function can be reduced into a simple form before it is differentiated so that process of differentiation becomes simple and easier. Notable cases are those of fractions whose denominator is a surdic quantity, which are simplified by rationalizing the denominator.

Type 1

Form:

1. $\dfrac{\sqrt{f(x)} \pm \sqrt{g(x)}}{\sqrt{f(x)} \mp \sqrt{g(x)}}$

2. $\dfrac{1}{\sqrt{f(x)} \pm \sqrt{g(x)}}$

3. $\dfrac{f(x) \pm \sqrt{g(x)}}{f(x) \mp \sqrt{g(x)}}$

4. $\dfrac{\sqrt{f(x)}}{\sqrt{f(x)+a} \pm \sqrt{f(x)-a}}$

5. $\dfrac{f(x)}{f_1(x) \pm \sqrt{f_2(x)}}$

⇔ If numerator and denominator are irrational functions of x, rationalization helps to get the d.c. in easy way.

N.B.: Irrational function is always rationalized by substitution or rationalization while finding limit/d.c./ integral/value of the function at a particular point/ ... etc.

Working rule: To find the d.c. of the forms mentioned above, we rationalize the denominator.

Type 2: $t^{-1}[f(x)]$

Where $t^{-1} = \sin^{-1}/\cos^{-1}/\tan^{-1}/\cot^{-1}/\sec^{-1}/\text{cosec}^{-1}$

$f(x)$ = a quotient of trigonometrical functions of x/ algebraic function of x which can be reduced into a simple and easy form by simplification or substitution before differentiation.

N.B.: The second type has been explained in the chapter of d.c. of inverse circular functions.

Solved Examples on rationalization

Find the d.c. of the following

1. $y = \dfrac{\sqrt{x+a} + \sqrt{x-a}}{\sqrt{x+a} - \sqrt{x-a}}$

Solution: $y = \dfrac{\sqrt{x+a} + \sqrt{x-a}}{\sqrt{x+a} - \sqrt{x-a}}$, defined for $x > |a| > 0$

$$= \frac{\sqrt{x+a}+\sqrt{x-a}}{\sqrt{x+a}-\sqrt{x-a}} \times \frac{\sqrt{x+a}+\sqrt{x-a}}{\sqrt{x+a}+\sqrt{x-a}}$$

[Rationalizing the denominator]

$$= \frac{\left(\sqrt{x+a} + \sqrt{x-a}\right)^2}{(x+a)-(x-a)}$$

$$= \frac{x+a+x-a+2\sqrt{x^2-a^2}}{2a}$$

$$= \frac{\left(x+\sqrt{x^2-a^2}\right)}{a}$$

$$\therefore \frac{dy}{dx} = \frac{1}{a}\left\{1 + \frac{d}{dx}\left(x^2-a^2\right)^{\frac{1}{2}}\right\}$$

$$\Rightarrow \frac{dy}{dx} = \frac{1}{a}\left\{1 + \frac{1}{2\sqrt{x^2-a^2}} \times 2x\right\},$$

$$\Rightarrow \frac{dy}{dx} = \frac{1}{a}\left\{1 + \frac{x}{\sqrt{x^2-a^2}}\right\} \text{ for } x > |a| > 0.$$

2. $y = \dfrac{\sqrt{x^2+a^2} + \sqrt{x^2-a^2}}{\sqrt{x^2+a^2} - \sqrt{x^2-a^2}}$

Solution: $y = \dfrac{\sqrt{x^2+a^2} + \sqrt{x^2-a^2}}{\sqrt{x^2+a^2} - \sqrt{x^2-a^2}}$, defined for $x^2 > a^2 > 0$

$$= \frac{\sqrt{x^2+a^2}+\sqrt{x^2-a^2}}{\sqrt{x^2+a^2}-\sqrt{x^2-a^2}} \times \frac{\sqrt{x^2+a^2}+\sqrt{x^2-a^2}}{\sqrt{x^2+a^2}+\sqrt{x^2-a^2}}$$

$$= \frac{(x^2+a^2)+(x^2-a^2)+2\sqrt{x^2+a^2}\cdot\sqrt{x^2-a^2}}{(x^2+a^2)-(x^2-a^2)}$$

$$= \frac{2x^2+2\sqrt{x^4-a^4}}{x^2+a^2-x^2+a^2}$$

$$= \frac{2x^2+2\sqrt{x^4-a^4}}{2a^2}$$

$$= \frac{x^2+\sqrt{x^4-a^4}}{a^2}$$

$$= \frac{x^2}{a^2}+\frac{\sqrt{x^4-a^4}}{a^2}$$

Now, differentiating both sides w.r.t. x.

$$\Rightarrow \frac{dy}{dx}$$

$$= \frac{2x}{a^2}+\frac{1}{2a^2}(x^4-a^4)^{\left(\frac{1}{2}-1\right)}\frac{d}{dx}(x^4-a^4)$$

$$\Rightarrow \frac{dy}{dx} = \frac{2x}{a^2}+\frac{4x}{2\left(\sqrt{x^4-a^4}\right)a^2}$$

$$\Rightarrow \frac{dy}{dx} = \frac{2x}{a^2}+\frac{2x^3}{a^2\left(\sqrt{x^4-a^4}\right)} \text{ for } x^2 > a^2.$$

3. $y = \dfrac{x+\sqrt{x^2+a^2}}{x-\sqrt{x^2+a^2}}$

Solution: $y = \dfrac{x+\sqrt{x^2+a^2}}{x-\sqrt{x^2+a^2}}$

$$= \frac{\left(x+\sqrt{x^2+a^2}\right)\times\left(x+\sqrt{x^2+a^2}\right)}{\left(x-\sqrt{x^2+a^2}\right)\times\left(x+\sqrt{x^2+a^2}\right)}$$

$$= \frac{\left(x+\sqrt{x^2+a^2}\right)^2}{x^2-(x^2-a^2)}$$

$$= \frac{x^2+x^2+a^2+2x\sqrt{x^2+a^2}}{x^2-x^2-a^2}$$

$$= \frac{2x^2+a^2+2x\sqrt{x^2+a^2}}{-a^2}$$

Now, differentiating both sides w.r.t. x.

$$\Rightarrow \frac{dy}{dx} = \frac{-2}{a^2}\cdot 2x + 0 + \left\{\frac{d}{dx}\left(\frac{-2x}{a^2}\sqrt{x^2+a^2}\right)\right\}$$

$$\Rightarrow \frac{dy}{dx} = \frac{-4x}{a^2}-\frac{2}{a^2}\sqrt{x^2+a^2}-\frac{1\times 2x}{2\sqrt{x^2+a^2}}\times\frac{2x}{a^2}$$

$$= -\left(\frac{4x}{a^2}+\frac{2}{a^2}\sqrt{x^2+a^2}+\frac{2x}{\left(\sqrt{x^2+a^2}\right)a^2}\right)$$

4. $y = \dfrac{1}{\sqrt{x^2+a^2}+\sqrt{x^2+b^2}}$

Solution: $y = \dfrac{1}{\sqrt{x^2+a^2}+\sqrt{x^2+b^2}}$

$= \dfrac{\sqrt{x^2+a^2}-\sqrt{x^2+b^2}}{a^2-b^2}$ [Rationalizing the denominator]

Now, on differentiating both sides w.r.t. x.

$$\Rightarrow \frac{dy}{dx} = \frac{1}{(a^2 - b^2)} \cdot \left[\frac{1}{2}(x^2 + a^2)^{(\frac{1}{2}-1)} \frac{d}{dx}(x^2 + a^2) - \frac{1}{2}(x^2 + b^2)^{(\frac{1}{2}-1)} \frac{d}{dx}(x^2 + b^2)\right]$$

$$\Rightarrow \frac{dy}{dx} = \frac{1}{(a^2 - b^2)} \cdot \left[\frac{2x}{2\sqrt{x^2 + a^2}} - \frac{2x}{\sqrt{x^2 + b^2}}\right]$$

$$= \frac{x}{(a^2 - b^2)} \cdot \left[\frac{1}{\sqrt{x^2 + a^2}} - \frac{1}{\sqrt{x^2 + b^2}}\right]$$

5. $y = \dfrac{\sqrt{a+x} - \sqrt{a-x}}{\sqrt{a+x} + \sqrt{a-x}}, a > 0.$

Solution: $y = \dfrac{\sqrt{a+x} - \sqrt{a-x}}{\sqrt{a+x} + \sqrt{a-x}}$, defined for $|x| < a.$

$$= \frac{(\sqrt{a+x} - \sqrt{a-x}) \cdot (\sqrt{a+x} - \sqrt{a-x})}{(\sqrt{a+x})^2 - (\sqrt{a-x})^2}, x \neq 0$$

$$= \frac{a + x + (a-x) - 2\sqrt{a+x} \cdot \sqrt{a-x}}{a + x - (a-x)}$$

$$= \frac{2a - 2\sqrt{a^2 - x^2}}{a + x - a + x} = \frac{2a - 2\sqrt{a^2 - x^2}}{2x}$$

$$= \frac{2\left(a - \sqrt{a^2 - x^2}\right)}{2x}$$

$$= \frac{a}{x} - \frac{\sqrt{a^2 - x^2}}{x}$$

$$\Rightarrow \frac{dy}{dx} = -\frac{a}{x^2} - \left[\frac{1}{2x} \cdot \frac{(-2x)}{\sqrt{a^2 - x^2}} - \frac{1}{x^2} \cdot \sqrt{a^2 - x^2}\right]$$

$$= -\frac{a}{x^2} + \left[\frac{1}{\sqrt{a^2 - x^2}} + \frac{\sqrt{a^2 - x^2}}{x^2}\right], x \neq 0.$$

6. $y = \dfrac{\sqrt{x+1} + \sqrt{x-1}}{\sqrt{x+1} - \sqrt{x-1}}$

Solution: $y = \dfrac{\sqrt{x+1} + \sqrt{x-1}}{\sqrt{x+1} - \sqrt{x-1}}$, defined for $x > 1$

$$= \frac{(\sqrt{x+1} + \sqrt{x-1})^2}{(\sqrt{x+1})^2 - (\sqrt{x-1})^2}$$ [Rationalizing the denominator]

$$= \frac{x + 1 + x - 1 + 2\sqrt{x^2 - 1}}{x + 1 - (x - 1)}$$

$$= \frac{2x + 2\sqrt{x^2 - 1}}{x + 1 - x + 1}$$

$$= \frac{2\left(x + \sqrt{x^2 - 1}\right)}{2}$$

$$= x + \sqrt{x^2 - 1}$$

$$\Rightarrow \frac{dy}{dx} = 1 + \frac{1}{2\sqrt{x^2 - 1}} \times \frac{d}{dx}(x^2 - 1)$$

$$= 1 + \frac{2x}{2\sqrt{x^2 - 1}} = 1 + \frac{x}{\sqrt{x^2 - 1}}, x > 1.$$

7. $y = \dfrac{\sqrt{x}}{\sqrt{x+2} + \sqrt{x-2}}$

Solution: $y = \dfrac{\sqrt{x}}{\sqrt{x+2} + \sqrt{x-2}}$, defined for $x > 2$

$$= \frac{\sqrt{x} \cdot (\sqrt{x+2} - \sqrt{x-2})}{(\sqrt{x+2})^2 - (\sqrt{x-2})^2}$$

$$= \frac{\sqrt{x}\cdot\left(\sqrt{x+2}-\sqrt{x-2}\right)}{(x+2)-(x-2)}$$

$$= \frac{\sqrt{x}\left(\sqrt{x+2}-\sqrt{x-2}\right)}{4} \quad ...(i)^*$$

$$\Rightarrow \frac{dy}{dx}$$

$$= \frac{1}{4}\left[\sqrt{x}\frac{d}{dx}\left(\sqrt{x+2}-\sqrt{x-2}\right)+\left(\sqrt{x+2}-\sqrt{x-2}\right)\cdot \frac{d}{dx}\sqrt{x}\right]$$

$$= \frac{1}{4}\left[\sqrt{x}\left(\frac{1}{2\sqrt{x+2}}-\frac{1}{\sqrt{x-2}}\right)+\frac{\left(\sqrt{x+2}-\sqrt{x-2}\right)}{2\sqrt{x}}\right],$$

for $x > 2$

Note:

* ***Or, alternatively*** (1) $\Rightarrow y = \frac{1}{4}\left(\sqrt{x^2+2x}-\sqrt{x^2-2x}\right)$

$$\Rightarrow \frac{dy}{dx} = \frac{1}{4}\left[\frac{1\cdot(2x+2)}{2\sqrt{x^2+2x}}-\frac{1\cdot(2x-2)}{2\sqrt{x^2-2x}}\right]$$

$$\Rightarrow \frac{dy}{dx} = \frac{1}{4}\times\frac{1}{2}\left[\frac{(2x+2)}{\sqrt{x^2+2x}}-\frac{(2x-2)}{\sqrt{x^2-2x}}\right]$$

$$= \frac{1}{8}\left[\frac{(2x+2)}{\sqrt{x^2+2x}}-\frac{(2x-2)}{\sqrt{x^2-2x}}\right]$$

$$= \frac{1}{4}\left[\frac{(x+1)}{\sqrt{x^2+2x}}-\frac{(x-1)}{\sqrt{x^2-2x}}\right] \text{ for } x > 2.$$

8. $y = \dfrac{x}{a-\sqrt{a^2-x^2}}$, $a \neq 0$

Solution: $y = \dfrac{x}{a-\sqrt{a^2-x^2}}$, defined for $x^2 < a^2$

Now rationalizing the denominator, we have

$$y = \frac{x\cdot\left(a+\sqrt{a^2-x^2}\right)}{\left(a-\sqrt{a^2-x^2}\right)\left(a+\sqrt{a^2-x^2}\right)},\ x \neq 0$$

$$= \frac{x\left(a+\sqrt{a^2-x^2}\right)}{a^2-a^2+x^2}$$

$$= \frac{a+\sqrt{a^2-x^2}}{x}$$

$$\Rightarrow \frac{dy}{dx}$$

$$= \frac{x\cdot\frac{d}{dx}\left(a+\sqrt{a^2-x^2}\right)-\left(a+\sqrt{a^2-x^2}\right)\frac{dx}{dx}}{x^2}$$

$$= \frac{\frac{1}{2}\left(a^2-x^2\right)^{-\frac{1}{2}}\times(-2x)-\left(a+\sqrt{a^2-x^2}\right)}{x^2}$$

$$= \frac{\left[\frac{-x^2}{\sqrt{a^2-x^2}}-a-\sqrt{a^2-x^2}\right]}{x^2}$$

$$= \frac{\left[-x^2-a\sqrt{a^2-x^2}-a^2+x^2\right]}{x^2\sqrt{a^2-x^2}}$$

$$= \frac{-a\left(a+\sqrt{a^2-x^2}\right)}{x^2\sqrt{a^2-x^2}};\ x \neq 0,\ x^2 < a^2$$

9. $y = \dfrac{\sqrt{x^2-a^2}-x}{\sqrt{x^2+a^2}+x}$

Solution: $y = \dfrac{\sqrt{x^2 - a^2} - x}{\sqrt{x^2 + a^2} + x}$, defined for $x^2 > a^2$

$$= \frac{\left(\sqrt{x^2 + a^2} - x\right)^2}{\left(\sqrt{x^2 + a^2} + x\right)\left(\sqrt{x^2 + a^2} - x\right)}$$

$$= \frac{2x^2 + a^2 - 2x\sqrt{x^2 + a^2}}{a^2}$$

$$\Rightarrow \frac{dy}{dx}$$

$$= \frac{1}{a^2}\left[4x - 2\sqrt{x^2 + a^2} - 2x \cdot \frac{1}{\not{2}\sqrt{x^2 + a^2}} \cdot \not{2}x\right]$$

$$= \frac{1}{a^2}\left[4x - \frac{2\left(x^2 + a^2\right) + 2x^2}{\sqrt{x^2 + a^2}}\right]$$

$$= \frac{2}{a^2}\left[2x - \frac{2x^2 + a^2}{\sqrt{x^2 + a^2}}\right] \text{ for } x^2 > a^2.$$

Problems based on Trigonometrical Transformation

Type 1: Whenever $1 \pm \cos x$ appears under the radical sign $\sqrt{\ }$, we always express the function within the radical as a square of some function.

Type 2: Wherever $1 \pm \sin x$ appears under the radical sign $\sqrt{\ }$, we always express the function within the radical as a square of some function.

N.B.: The above method may be remembered as "expressing the function within the radical as a square of some function".

Solved Examples

Find the d.c. of the following.

1. $y = \sqrt{\dfrac{1 - \cos x}{1 + \cos x}}$

Solution: $y = \sqrt{\dfrac{1-\cos x}{1+\cos x}}$ defined for $x \ne (2n+1)\pi$

$$= \frac{\sqrt{2}\sqrt{\sin^2 \frac{x}{2}}}{\sqrt{2}\sqrt{\cos^2 \frac{x}{2}}}$$

$$= \frac{\left|\sin \frac{x}{2}\right|}{\left|\cos \frac{x}{2}\right|} = \left|\tan \frac{x}{2}\right|$$

$$\Rightarrow \frac{dy}{dx} = \frac{d}{dx}\left|\tan \frac{x}{2}\right|$$

$$= \frac{\left|\tan \frac{x}{2}\right|}{\tan \frac{x}{2}} \cdot sec^2 \frac{x}{2} \cdot \frac{1}{2}$$

$$= \frac{1}{2}\sec^2 \frac{x}{2} \cdot \frac{\left|\tan \frac{x}{2}\right|}{\tan \frac{x}{2}};\ x \ne n\pi,\ n \in Z$$

Note: **(i)** This problem can be done by substitution method also but that method becomes lengthy.

$$y = \sqrt{\frac{1 - \cos x}{1 + \cos x}} = \sqrt{u} \text{ where '}u\text{' } = \frac{1 - \cos x}{1 + \cos x}$$

$\Rightarrow$ We are required to find

$$\frac{du}{dx} = \frac{du^{\frac{1}{2}}}{dx} = \frac{1}{2}u^{\left(\frac{1}{2}-1\right)} \cdot \frac{du}{dx} = \frac{1}{2\sqrt{u}} \cdot \frac{du}{dx}$$

(ii) This problem also can be done by logarithmic differentiation.

$$\log y = \log\left(\frac{1 - \cos x}{1 + \cos x}\right)^{\frac{1}{2}} = \frac{1}{2}\log\left(\frac{1 - \cos x}{1 + \cos x}\right)$$

$$\Rightarrow 2\log y = \log\left(\frac{1 - \cos x}{1 + \cos x}\right)$$

$$= \log(1 - \cos x) - \log(1 + \cos x)$$

$$\Rightarrow 2\frac{d\log y}{dx}=\frac{d}{dx}\log(1-\cos x)-\frac{d}{dx}\log(1+\cos x)$$

$$\Rightarrow 2\cdot\frac{1}{y}\frac{dy}{dx}=\frac{\sin x}{1-\cos x}+\frac{\sin x}{1+\cos x}$$

$$=\sin x\left(\frac{1+\cos x+1-\cos x}{1-\cos^2 x}\right)=\sin x\cdot\frac{2}{\sin^2 x}$$

$$\Rightarrow \frac{dy}{dx}=\frac{2}{\sin x}\times\frac{y}{2}=\frac{y}{\sin x}$$

$$=\frac{1}{\sin x}\times\sqrt{\frac{1-\cos x}{1+\cos x}};\ x\neq n\pi$$

2. $y=\sqrt{\dfrac{1-\sin x}{1+\cos x}}$

Solution: $y=\sqrt{\dfrac{1-\sin x}{1+\cos x}}$, defined for $x\neq(2n+1)\pi$

$$=\frac{\sqrt{\left(\sin\frac{x}{2}-\cos\frac{x}{2}\right)^2}}{\sqrt{2\cos^2\frac{x}{2}}}$$

$$=\frac{\left|\sin\frac{x}{2}-\cos\frac{x}{2}\right|}{\sqrt{2}\left|\cos\frac{x}{2}\right|}$$

$$\Rightarrow y=\frac{1}{\sqrt{2}}\left|\frac{\sin\frac{x}{2}-\cos\frac{x}{2}}{\cos\frac{x}{2}}\right|$$

$$=\frac{1}{\sqrt{2}}\left|\tan\frac{x}{2}-1\right|$$

$$\Rightarrow \frac{dy}{dx}=\frac{1}{\sqrt{2}}\frac{d}{dx}\left|\tan\frac{x}{2}-1\right|$$

$$=\frac{1}{\sqrt{2}}\cdot\frac{\left|\tan\frac{x}{2}-1\right|}{\left(\tan\frac{x}{2}-1\right)}\cdot\frac{d}{dx}\left(\tan\frac{x}{2}-1\right),$$

$$x\neq 2n\pi+\frac{\pi}{2}$$

$$=\frac{1}{\sqrt{2}}\frac{\left|\tan\frac{x}{2}-1\right|}{\left(\tan\frac{x}{2}-1\right)}\cdot\sec^2\frac{x}{2}\cdot\frac{1}{2}$$

$$=\frac{1}{2\sqrt{2}}\sec^2\frac{x}{2}\cdot\frac{\left|\tan\frac{x}{2}-1\right|}{\left(\tan\frac{x}{2}-1\right)};$$

$$\left(x\neq(2n+1)\pi \text{ and } x\neq\left(2n\pi+\frac{\pi}{2}\right)\right)$$

3. $y=\sqrt{1-\sin x}$

Solution: $y=\sqrt{1-\sin x}$

$$=\sqrt{\left(\sin\frac{x}{2}-\cos\frac{x}{2}\right)^2}$$

$$=\left|\sin\frac{x}{2}-\cos\frac{x}{2}\right|$$

$$\Rightarrow \frac{dy}{dx}=\frac{d\left|\sin\frac{x}{2}-\cos\frac{x}{2}\right|}{dx}$$

$$=\frac{\left|\sin\frac{x}{2}-\cos\frac{x}{2}\right|}{\left(\sin\frac{x}{2}-\cos\frac{x}{2}\right)}\cdot\frac{1}{2}\cdot\left(\sin\frac{x}{2}+\cos\frac{x}{2}\right)$$

$$=\frac{1}{2}\cdot\frac{\left(\sin\frac{x}{2}+\cos\frac{x}{2}\right)\left|\sin\frac{x}{2}-\cos\frac{x}{2}\right|}{\left(\sin\frac{x}{2}-\cos\frac{x}{2}\right)};\quad\text{for}$$

$$x\neq 2n\pi+\frac{\pi}{2}.$$

4. $y = \sqrt{1 + \cos x}$

Solution: $y = \sqrt{1 + \cos x}$

$$= \sqrt{2\cos^2 \frac{x}{2}} = \sqrt{2}\left|\cos\frac{x}{2}\right|$$

$$\Rightarrow \frac{dy}{dx} = \sqrt{2}\,\frac{d\left|\cos\frac{x}{2}\right|}{dx}$$

$$= \sqrt{2}\cdot\frac{\left|\cos\frac{x}{2}\right|}{\cos\frac{x}{2}}\cdot\left(-\sin\frac{x}{2}\right)\cdot\frac{1}{2}$$

$$= -\frac{1}{2}\cdot\sqrt{2}\cdot\sin\frac{x}{2}\cdot\frac{\left|\cos\frac{x}{2}\right|}{\cos\frac{x}{2}}$$

$$= -\frac{1}{\sqrt{2}}\sin\frac{x}{2}\cdot\frac{\left|\cos\frac{x}{2}\right|}{\cos\frac{x}{2}},\ x \neq (2n+1)\pi$$

N.B.: The above problem also can be done by using the chain rule for $\sqrt{u}$, where u = given function $x = f(x)$ or by using logarithmic differentiation.

5. $y = \tan^{-1}\sqrt{\dfrac{1 - \cos x}{1 + \cos x}}$

Solution: $y = \tan^{-1}\sqrt{\dfrac{1 - \cos x}{1 + \cos x}}$

$$= \tan^{-1}\sqrt{\frac{2\sin^2\frac{x}{2}}{2\cos^2\frac{x}{2}}};\ x \neq (2n+1)\pi$$

$$\Rightarrow y = \tan^{-1}\left\{\frac{\left|1\sin\frac{x}{2}\right|}{\left|\cos\frac{x}{2}\right|}\right\} = \tan^{-1}\left\{\left|\tan\frac{x}{2}\right|\right\}$$

$$\Rightarrow \tan y = \left|\tan\frac{x}{2}\right|$$

$$\Rightarrow \frac{d\tan y}{dx} = \frac{d\left|\tan\frac{x}{2}\right|}{dx}$$

$$\Rightarrow \sec^2 y\cdot\frac{dy}{dx} = \frac{\left|\tan\frac{x}{2}\right|}{\tan\frac{x}{2}}\cdot\frac{\sec^2\frac{x}{2}}{2},\ x \neq n\pi$$

$$\Rightarrow \frac{dy}{dx} = \frac{\left|\tan\frac{x}{2}\right|}{\tan\frac{x}{2}}\cdot\frac{\sec^2\frac{x}{2}}{2\sec^2 y}$$

$$= \frac{\left|\tan\frac{x}{2}\right|}{\tan\frac{x}{2}}\cdot\frac{\sec^2\frac{x}{2}}{2\left(1 + \tan^2\frac{x}{2}\right)}\left\{\because \left|\tan\frac{x}{2}\right|^2 = \tan^2\frac{x}{2}\right\}$$

$$\Rightarrow \frac{dy}{dx} = \frac{\left|\tan\frac{x}{2}\right|}{\tan\frac{x}{2}}\cdot\frac{\sec^2\frac{x}{2}}{2\sec^2\frac{x}{2}}$$

$$= \frac{1}{2}\,\frac{\left|\tan\frac{x}{2}\right|}{\tan\frac{x}{2}},\ x \neq n\pi$$

or, alternatively,

$$\tan y = \sqrt{\frac{1 - \cos x}{1 + \cos x}}$$

$$\Rightarrow \log\tan y = \log\sqrt{\frac{1 - \cos x}{1 + \cos x}},\ x \neq n\pi$$

$$= \frac{1}{2}\log\left(\frac{1 - \cos x}{1 + \cos x}\right)$$

$$= \frac{1}{2}[\log(1 - \cos x) - \log(1 + \cos x)]$$

$$\Rightarrow \frac{1}{\tan y} \cdot \frac{d \tan y}{dx} = \frac{1}{2}\left[\frac{\sin x}{1-\cos x} + \frac{\sin x}{1+\cos x}\right]$$

$$= \frac{\sin x}{2}\left[\frac{1 + \cos nx + 1 - \cos ex}{1 - \cos^2 x}\right]$$

$$\Rightarrow \frac{1}{\tan y} \cdot \sec^2 y \cdot \frac{dy}{dx} = \frac{\sin x}{2} \cdot \frac{2}{\sin^2 x}$$

$$\Rightarrow \frac{1 + \tan^2 y}{\tan y} \cdot \frac{dy}{dx} = \frac{1}{\sin x}$$

$$\Rightarrow \frac{1 + \left(\frac{1 - \cos x}{1 + \cos x}\right)}{\sqrt{\frac{1 - \cos x}{1 + \cos x}}} \cdot \frac{dy}{dx} = \frac{1}{\sin x}$$

$$\Rightarrow \frac{\frac{2}{1 + \cos x}}{\sqrt{\frac{1 - \cos x}{1 + \cos x}}} \cdot \frac{dy}{dx} = \frac{1}{\sin x}$$

$$\Rightarrow \frac{dy}{dx} = \frac{\sqrt{\frac{1 - \cos x}{1 + \cos x}}}{\sin x \cdot \left(\frac{2}{1 + \cos x}\right)}$$

$$= \frac{\sqrt{\frac{1 - \cos x}{1 + \cos x}}}{\left(\frac{2 \sin x}{1 + \cos x}\right)} = \sqrt{\frac{1 - \cos x}{1 + \cos x}} \cdot \frac{(1 + \cos x)}{2 \sin x}$$

$$\Rightarrow \frac{dy}{dx} = \frac{(1 + \cos x)}{2 \sin x} \cdot \sqrt{\frac{1-\cos x}{1 + \cos x}}, \ x \neq n\pi$$

Note: The first method is more simple than the second method. But a general method to find differential coefficient of

1. $y = f\sqrt{\frac{g_1(x)}{g_2(x)}}$

2. $y = f^{-1}\sqrt{\frac{g_1(x)}{g_2(x)}}$, where $f^{-1} = \sin^{-1}, \cos^{-1}, \tan^{-1}$, $\cot^{-1}$, $\sec^{-1}$, and cosec^{-1}.

3. $y = f^{-1}\sqrt{\frac{1 \pm \sin x}{1 \mp \sin x}}$, $f^{-1}\sqrt{\frac{1 \pm \cos x}{1 \mp \cos x}}$,

$f^{-1}\sqrt{\frac{1 \pm \sin x}{1 \pm \cos x}}$, $f^{-1}\sqrt{\frac{1 \pm \cos x}{1 \pm \sin x}}$

4. $y = \sqrt{\frac{1 - \sin x}{1 + \sin x}}$, $\sqrt{\frac{1 + \sin x}{1 - \sin x}}$, $\sqrt{\frac{1 + \cos x}{1 - \cos x}}$,

$\sqrt{\frac{1 \pm \sin x}{1 \pm \cos x}}$, $\sqrt{\frac{1 \pm \cos x}{1 \pm \sin x}}$,

5. $y = \sqrt{1 \pm \sin x}$, $\sqrt{1 \pm \cos x}$ is logarithmic differentiation.

6. $y = \log\sqrt{\frac{1 + \sin x}{1 - \sin x}}$,

Solution: First method:

$$y = \log\sqrt{\frac{1 + \sin x}{1 - \sin x}}, \text{ defined for } x \neq n\pi + \frac{\pi}{2}$$

$$= \frac{1}{2}\{\log(1 + \sin x) - \log(1 - \sin x)\}$$

$$\Rightarrow \frac{dy}{dx} = \frac{1}{2}\left\{\frac{\cos x}{1 + \sin x} - \frac{(-\cos x)}{1 - \sin x}\right\}$$

$$= \frac{1}{2}\left\{\frac{\cos x}{1 + \sin x} + \frac{\cos x}{1 - \sin x}\right\}$$

$$= \frac{1}{2} \cdot \cos x \cdot \left\{\frac{1 - \sin x + 1 + \sin x}{1 - \sin^2 x}\right\}$$

$$= \frac{1}{2} \cdot \cos x \cdot \frac{2}{\cos^2 x} = \frac{1}{\cos x}, \ x \neq n\pi + \frac{\pi}{2}$$

Second method

$$\sqrt{\frac{1 + \sin x}{1 - \sin x}} = \frac{\left|\sin\frac{x}{2} + \cos\frac{x}{2}\right|}{\left|\sin\frac{x}{2} - \cos\frac{x}{2}\right|}$$

$$\therefore\ y = \log\sqrt{\frac{1+\sin x}{1-\sin x}} = \log\frac{\left|\sin\frac{x}{2}+\cos\frac{x}{2}\right|}{\left|\sin\frac{x}{2}-\cos\frac{x}{2}\right|}$$

$$= \log\left|\sin\frac{x}{2}+\cos\frac{x}{2}\right| - \log\left|\sin\frac{x}{2}-\cos\frac{x}{2}\right|$$

$$\Rightarrow \frac{dy}{dx} = \frac{1}{\left(\sin\frac{x}{2}+\cos\frac{x}{2}\right)}\cdot\left(\frac{1}{2}\cdot\cos\frac{x}{2}-\frac{1}{2}\cdot\sin\frac{x}{2}\right) - \frac{1}{\left(\sin\frac{x}{2}-\cos\frac{x}{2}\right)}\cdot\left(\frac{1}{2}\cos\frac{x}{2}+\frac{1}{2}\sin\frac{x}{2}\right)$$

$$= \frac{1}{2}\left\{\frac{\left(\cos\frac{x}{2}-\sin\frac{x}{2}\right)}{\left(\sin\frac{x}{2}+\cos\frac{x}{2}\right)} - \frac{\left(\cos\frac{x}{2}+\sin\frac{x}{2}\right)}{\left(\sin\frac{x}{2}-\cos\frac{x}{2}\right)}\right\}$$

$$= \frac{1}{2}\left\{\frac{\left(\cos\frac{x}{2}-\sin\frac{x}{2}\right)}{\left(\sin\frac{x}{2}+\cos\frac{x}{2}\right)} + \frac{\left(\cos\frac{x}{2}+\sin\frac{x}{2}\right)}{\left(\cos\frac{x}{2}-\sin\frac{x}{2}\right)}\right\}$$

$$= \frac{1}{2}\left\{\frac{\left(\cos\frac{x}{2}-\sin\frac{x}{2}\right)^2+\left(\cos\frac{x}{2}+\sin\frac{x}{2}\right)^2}{-\left(\sin^2\frac{x}{2}-\cos^2\frac{x}{2}\right)}\right\}$$

$$= \frac{1}{2}\left\{\frac{1-2\sin\frac{x}{2}\cos\frac{x}{2}+1+2\sin\frac{x}{2}\cos\frac{x}{2}}{\cos x}\right\}$$

$$= \frac{1}{2}\cdot\frac{2}{\cos x} = \frac{1}{\cos x},\ x \neq n\pi + \frac{\pi}{2}$$

Remark: The above problem has not been solved by logarithmic differentiation.

Problems based on Change of Form before Differentiation

Exercise 11.2

Find the differential coefficients of

1. $\frac{1}{\sqrt{x+a}+\sqrt{x+b}}$
2. $\frac{\sqrt{(1+x)}+\sqrt{(1-x)}}{\sqrt{(1+x)}-\sqrt{(1-x)}}$
3. $\frac{\sqrt{(a^2+x^2)}+\sqrt{(a^2-x^2)}}{\sqrt{(a^2+x^2)}-\sqrt{(a^2-x^2)}}$
4. $\frac{x}{1-\sqrt{a^2-x^2}}$
5. $\frac{\sqrt{x}}{\sqrt{x+2}+\sqrt{x-2}}$
6. $\frac{x}{\sqrt{x+2}-1}$
7. $\frac{\sqrt{(a+bx)}+\sqrt{(a-bx)}}{\sqrt{(a+bx)}-\sqrt{(a-bx)}}$
8. $\frac{\sqrt{2+x^2}+x}{\sqrt{2+x^2}-x}$

$$\left[\textit{Hint}: \frac{\left(\sqrt{2+x^2}+x\right)^2}{2+x^2-x^2} = \frac{2+x^2+2x\sqrt{2+x^2}+x^2}{2}\right]$$

Answers (with proper restrictions on x)

1. $\frac{1}{2(a-b)}\cdot\left[\frac{1}{\sqrt{x+a}}-\frac{1}{\sqrt{x+b}}\right]$

2. $-\frac{1}{x^2\sqrt{1-x^2}} - \frac{1}{x^2}$

3. $-\frac{2a^2}{x^3}\left\{1+\frac{a^2}{\sqrt{a^4-x^4}}\right\}$

4. Find

5. Find

6. $-\frac{1}{x^2}\left\{1+\frac{1}{\sqrt{(1+x^2)}}\right\}$

7. $-\frac{a}{bx^2}\left\{1+\frac{a}{\sqrt{a^2-b^2x^2}}\right\}$

Exercise 11.3

Differentiate the following functions.

1. $y=\sqrt{1+\sin 2x}-\sqrt{1-\sin 2x}$

2. $y=\tan^{-1}\sqrt{\frac{1-\cos x}{1+\cos x}}$

3. $y=\sqrt{\frac{1+\sin x}{1-\sin x}}$

Answers (with proper restrictions on x)

1. $\frac{|(\sin x+\cos x)|}{(\sin x+\cos x)}\cdot(\cos x-\sin x)-\frac{|(\sin x-\cos x)|}{(\sin x-\cos x)}\cdot(\cos x+\sin x)$

2. $y=\frac{x}{2}$ if $\tan\frac{x}{2}>0$, then $\left|\tan\frac{x}{2}\right| = \tan\frac{x}{2}\ \therefore\ \frac{dy}{dx}=\frac{1}{2}$

or, $\frac{dy}{dx}=\frac{1}{2}\frac{\left|\tan\frac{x}{2}\right|}{\tan\frac{x}{2}}$, $(x\neq n\pi)$

3. $\frac{1}{2\sqrt{2}}\sec^2\frac{x}{2}\cdot\frac{\left|\tan\frac{x}{2}+1\right|}{\left(\tan\frac{x}{2}+1\right)}$

12

Implicit Differentiation

Firstly, we recall the basic definitions in connection with implicit differentiation.

1. *Explicit function*: Whenever it is possible to equate the dependent variable y directly to a function of the independent variable x as $y = f(x)$, we say that the dependent variable y is an explicit function of the independent variable x. Here f (in the equation $y = f(x)$) stands for all elementary functions (sin, cos, tan, cot, sec, cosec, $\sin^{-1}$, $\cos^{-1}$, $\tan^{-1}$, $\cot^{-1}$, $\sec^{-1}$, cosec^{-1}, log, e, $(\,)^n$, $\sqrt[n]{\ }$, $|\,|$, etc.). examples,

(i) $y = x^2 - 1$

(ii) $y = \sin x^3$, etc.

N.B.: All algebraic, trigonometric, inverse trigonometric, logarithmic and exponential functions of x's are explicit functions of x's.

2. *Implicit function*: Whenever it is not possible to equate the dependent variable y directly to a function of the independent variable x as $y = f(x)$, we say that the dependent variable y is an implicit function of the independent variable x.

Examples:

(i) $x^3 y^4 = (x + y)^7$

(ii) $x y = c$

(iii) $xy + x^2 y^2 = c$

(iv) $x^m y^m = (x + y)^{m+n}$

(v) $y = \cos(x - y)$

(vi) $y = \tan(x + y)$

(vii) $\cot xy + xy = 3$

(viii) $\sin(x + y) + \sin(x - y) = 1$

Notation: The notation for the implicit function is either

1. $F(x, y) = c$, c being a constant, or, (***Note***: Implicit function is also defined in the following way: If the dependence of the dependent variable y on the independent variable x is expressed by the equation: $F(x, y) = c$ or $F(x, y) = 0$ not solvable for y, then y is called an implicit function of x.)

2. $F(x, y) = 0$ where $F(x, y)$ denotes

(i) A rational integral function of x and y, i.e. it is the sum of a finite series of the terms $C_{m,n} x^m, y^n$, where m, n may have the values 0, 1, 2, 3, ….

(ii) A function (sin, cos, tan, cot, sec, cosec, $\sin^{-1}$, $\cos^{-1}$, $\tan^{-1}$, $\cot^{-1}$, $\sec^{-1}$, cosec^{-1}, log, e, $(\,)^n$, $\sqrt[n]{\ }$, $|\,|$, etc.) of a rational integral function of x and y.

(iii) A combination of (i) and (ii) and the equation $F(x, y) = c$ or $F(x, y) = 0$ denotes the dependence of the dependent variable y on the independent variable x not solvable for y.

Remember:

1. The equation $F(x, y) = c$ or, $F(x, y) = 0$ determines one or more values of y to be associated with the given value of the independent variable x provided a definite number from some domain is substituted for x.

2. If the set $\{(x, y) \mid F(x, y) = c\}$ or, $\{(x, y) \mid F(x, y) = 0\}$ is the graph of a function (or, union of graphs or more than one function), we say that the equation $F(x, y) = c$ or $F(x, y) = 0$ defined implicitly y as a function x. Further we should note that if $y = f(x)$ is a function of x, then $F(x, y) = F(x, f(x)) = c$ or $F(x, y) = F(x, f(x)) = 0$ is an identity, i.e. $F(x, f(x))$ is a constant function.

Kinds of Implicit Function

1. Implicit algebraic function: y is said to be an implicit algebraic function of x if a relation of the form:

$Y^m + R_1 Y^{m-1} + \ldots + R_m = 0$ exists, where $R_1, R_2, \ldots, R_m$ are rational functions of x's and m is a positive integer. e.g.,

(i) $y^2 - 2xy + x = 0$

(ii) $y^2 - \dfrac{x}{x-1}y + 1 = 0$

2. Implicit transcendental function: y is said to be an implicit transcendental function of x if a relation of the form: $T(x, y) = 0$ exists, where T denotes trigonometric, inverse trigonometric, logarithmic and exponential functions and the ordered pair (x, y) denotes a rational integral function of x and y. e.g.,

(i) $\tan(x, y) = 0$
(ii) $\cos(x + 2y) = 0$
(iii) $\sin(a + y) = 0$
(iv) $e^{y-x} = 0$
(v) $\tan^{-1}(x + y) = 0$
(vi) $\log(xy) = 0$

Question: What is implicit differentiation?

Answer: Differentiation of an implicit function is called implicit differentiation, or more explicitly it is defined as "Finding the derivative $\dfrac{dy}{dx}$ of an implicit function put in the form $F(x, y) = 0$ or $F(x, y) = c$ without explicitly determining the function $y = f(x)$ is called the implicit differentiation.

N.B.: Whenever we differentiate an implicit function put in the form $F(x, y) = c$ or, $F(x, y) = 0$, we assume that the given implicit function is differentiable and y is a differentiable function of x.

Facts to Know:

1. Whenever we differentiate the algebraic implicit function of x (or, the rational integral function of x and y), we simply need do is to differentiate each term of it with respect to x remembering that y is itself a differentiate function of x, say $g(x)$ and the derivative of any function of y, say $\varnothing(y)$ with respect to x is equal to its derivative with respect to y multiplied by $\dfrac{dy}{dx}$, i.e. $\dfrac{d\varnothing(y)}{dx} = \dfrac{d}{dx}\varnothing(y) \cdot \dfrac{dy}{dx}$

Hence, $\dfrac{dy^n}{dx} = \dfrac{dy^n}{dx} \cdot \dfrac{dy}{dx}$, provided y is a differentiable function of x.

2. The derivative of a function (sin, cos, tan, cot, sec, cosec, $\sin^{-1}$, $\cos^{-1}$, $\tan^{-1}$, $\cot^{-1}$, $\sec^{-1}$, cosec^{-1}, log, e, $(\ )^n$, $|\ |$, etc.) of a rational integral function of x and y with respect to the independent variable x is equal to the derivative of the whole given implicit function with respect to the rational integral function of x and y times the derivative of the rational integral function of x and y with respect to the independent variable x only, i.e. $\dfrac{d}{dx}T(x, y) = \dfrac{dT(x, y)}{d(x, y)} \cdot \dfrac{d(x, y)}{dx}$ provided denotes sin, cos, tan, cot, sec, cosec, $\sin^{-1}$, $\cos^{-1}$ $\tan^{-1}$, $\cot^{-1}$, $\sec^{-1}$, cosec^{-1}, log, e, $(\ )^n$, $|\ |$, etc and the ordered pair (x, y) denotes the rational integral function of x and y. e.g.,

(i) $\dfrac{d}{dx}\sin(x + y) + \dfrac{d\sin(x + y)}{d(x + y)} \cdot \dfrac{d(x + y)}{dx}$

$= \cos(x + y) \cdot \left(1 + \dfrac{dy}{dx}\right)$

(ii) $\dfrac{d}{dx}\log(xy) = \dfrac{d\log(xy)}{d(xy)} \cdot \dfrac{d(xy)}{dx}$

$= \dfrac{1}{xy}\left(x\dfrac{dy}{dx} + y\right)$, $xy > 0$

3. ***Question***: When would you differentiate as an implicit function?

Answer: (i) When it is neither convenient nor possible to find y (dependent variable) in terms of x only, i.e. if it is convenient or impossible to write $y =$ an expression in x only or $y = f(x)$, where 'f' stands for sin, cos, tan, cot, sec, cosec, $\sin^{-1}$, $\cos^{-1}$, $\tan^{-1}$, $\cot^{-1}$, $\sec^{-1}$, cosec^{-1}, log, e, etc. then we differentiate the given function as an implicit function of x. e.g.,

$x^3 + y^3 + 3axy = 0$

$\sin(x^2 + y^2) = y$

$\sin^{-1}\left(\dfrac{x}{y}\right) = y$

$\log(x + y) = y$

$e^{x^2+\frac{x}{y}} = y$

$x \cdot \log y = y$, etc.

(ii) Whenever we would like to find out the derivatives of an implicit function of x without solving its given equation for y, we differentiate the given relationship between the variables x and y as an implicit function of x. e.g.,

$x^2 + y^2 = a^2$

$\tan(x+y) = 0$

$\log\left(\frac{x}{y}\right) = 0$

$e^{\frac{x}{y}} = 0$, etc.

4. In the implicit function, y is said to be defined implicitly as the function of x or x is said to be defined implicitly as a function of y.

5. An implicit function expresses always an unsolved relationship between the variables. e.g.,

$$\left.\begin{array}{l}\frac{x}{y}=\tan 60^\circ \\ x^2+y^2=a^2 \\ Z+y^2\tan x=a\end{array}\right]$$ are functions expressing unsolved relationship between variables.

6. An explicit function expresses always a solved relationship between the variables. One variable is solved in terms of the other. e.g.,

$y = \tan 60^\circ \Rightarrow y$ is an explicit function of x.

$x = \frac{y}{\tan 60^\circ} \Rightarrow x$ is an explicit function of y.

$x = \pm\sqrt{a^2 - y^2} \Rightarrow x$ is an explicit function of y.

$Z = a - y^2 \tan x \Rightarrow Z$ is an explicit function of y and x.

The dependent variable is therefore the value of the explicit function.

Problems on Implicit Algebraic Functions

To find the derivative of an implicit algebraic function of x without solving its given equation for y, we adopt the rule which consists of following steps.

Step 1: Take $\frac{d}{dx}(\ )$ on both sides of the given equation.

Step 2: Differentiation each term of the given equation with respect to the independent variable x using the rules for the derivatives of sum, difference, product, quotient, composite of differentiable functions and a constant multiple of the differentiable function remembering that

$$\frac{dF(y)}{dx} = \frac{dF(y)}{dy} \cdot \frac{dy}{dx}$$

which means wherever with respect to the independent variable x, we differentiate a differentiable algebraic, trigonometric, inverse trigonometric, logarithmic and exponential function of y being preassumed (or, understood) to be a differentiable function of x, we should multiply the differential coefficient of the differentiable function of y with respect to y by $\frac{dy}{dx}$.

Step 3: Collect the terms involving $\frac{dy}{dx}$ on the left hand side and the terms without $\frac{dy}{dx}$ on the right hand side.

Step 4: Finally, solve the equation for $\frac{dy}{dx}$.

Notes: **1.** The final result for $\frac{dy}{dx}$ is an expression in terms of both x and y.

2. While finding the derivatives of an implicit algebraic function of x, often required derivatives obtained from $\frac{dF(y)}{dx} = \frac{dF(y)}{dy} \cdot \frac{dy}{dx}$ are the following.

(i) $\frac{d}{dx} y^n - ny^{n-1}\frac{dy}{dx}$

(ii) $\frac{d}{dx}\left(x^n \pm y^n\right) = \frac{dx^n}{dx} + \frac{dy^n}{dx}$

$= nx^{n-1} + ny^{n-1}\frac{dy}{dx}$

(iii) $\frac{d}{dx}\left(x^n y^n\right) = x^n \frac{dy^n}{dx} + y^n \frac{dx^n}{dx}$

$= n x^n y^{n-1} \frac{dy}{dx} + n x^{n-1} y^n$

(iv) $\frac{d}{dx}\left(\frac{x^n}{y^n}\right) = \frac{y^n \frac{dx^n}{dx} - x^n \frac{dy^n}{dx}}{\left(y^n\right)^2}$

$= \frac{n y^{n-1} x^{n-1} - n x^{n-1} y^{n-1} \frac{dy}{dx}}{(yn)^2}$

Solved Examples

Find $\frac{dy}{dx}$ if

1. $y^8 - 5x^2 y^6 + x^8 = 11$

Solution: $y^8 - 5x^2 y^6 + x^8 = 11$

$\Rightarrow \frac{d}{dx}\left(x^8 - 5x^2 y^6 + y^8\right) = \frac{d}{dx}(11)$

$\Rightarrow \frac{d}{dx}\left(x^8\right) - \frac{d}{dx}\left(5x^2 y^6\right) + \frac{d}{dx}\left(y^8\right) = 0$

$\Rightarrow 8x^7 - 10x y^6 - 30x^2 y^5 \frac{dy}{dx} + 8y^7 \frac{dy}{dx} = 0$

$\Rightarrow \left(8y^7 - 30x^2 y^5\right) \frac{dy}{dx} = 10x y^6 - 8x^7$

$\Rightarrow \frac{dy}{dx} = \frac{x\left(5y^6 - 4x^6\right)}{y^5\left(4y^2 - 15x^2\right)}$

2. $x^2 + y^2 = a^2$

Solution: $x^2 + y^2 = a^2$

$\Rightarrow \frac{d}{dx}\left(x^2 + y^2\right) = \frac{d}{dx}\left(a^2\right)$

$\Rightarrow \frac{dx^2}{dx} + \frac{dy^2}{dx} = 0$

$\Rightarrow 2x + \frac{dy^2}{dy} \cdot \frac{dy}{dx} = 0$

$\Rightarrow 2x + 2y \cdot \frac{dy}{dx} = 0$

$\Rightarrow 2y \frac{dy}{dx} = -2x$

$\Rightarrow \frac{dy}{dx} = \frac{-2x}{2y} = \frac{-x}{y}$

3. $xy + x^2 y^2 = c$

Solution: $xy + x^2 y^2 = c$

$\Rightarrow \frac{d}{dx}\left(xy + x^2 y^2\right) = \frac{d}{dx}(c)$

$\Rightarrow \frac{d}{dx}(xy) + \frac{d}{dx}\left(x^2 y^2\right) = 0$

$\Rightarrow \left(x\frac{dy}{dx} + y\frac{dx}{dx}\right) + \left(x^2 \frac{dy^2}{dx} + y^2 \frac{dx^2}{dx}\right) = 0$

$\Rightarrow x\frac{dy}{dx} + y + 2x^2 y\frac{dy}{dx} + y^2 \cdot 2x = 0$

$\Rightarrow \left(x\frac{dy}{dx} + 2x^2 y\frac{dy}{dx}\right) = -\left(y + 2x y^2\right)$

$\Rightarrow \left(x + 2x^2 y\right)\frac{dy}{dx} = -\left(y + 2x y^2\right)$

$\Rightarrow \frac{dy}{dx} = \frac{-\left(y+2xy^2\right)}{\left(x+2x^2 y\right)} = \frac{-y(1+2x y)}{x(1+2x y)} = -\frac{y}{x}$

4. $x^5 + x^4 y^2 - y = 4$

Solution: $x^5 + x^4 y^2 - y = 4$

$\Rightarrow \frac{d}{dx}\left(x^5 + x^4 y^2 - y\right) = \frac{d}{dx}(4)$

$\Rightarrow \frac{d}{dx}\left(x^5\right) + \frac{d}{dx}\left(x^4 y^2\right) - \frac{dy}{dx} = 0$

$\Rightarrow 5x^4 + \left(x^4 2y \frac{dy}{dx} + y^2\, 4x^3\right) - \frac{dy}{dx} = 0$

$$\Rightarrow \frac{dy}{dx}\left(2x^4 y - 1\right) = -\left(5x^4 + 4x^3 y^2\right)$$

$$\Rightarrow \frac{dy}{dx} = \frac{-5x^4 + 4x^3 y^2}{2x^4 y - 1}$$

5. $ax^2 + by^2 + 2hxy + 2gx + 2fy + c = 0$

Solution: $ax^2 + by^2 + 2hxy + 2gx + 2fy + c = 0$

$$\Rightarrow \frac{d}{dx}\left(ax^2 + by^2 + 2hxy + 2gx + 2fy + c\right) = 0$$

$$\Rightarrow \frac{d}{dx}\left(ax^2\right) + \frac{d}{dx}\left(by^2\right) + \frac{d}{dx}(2hxy) + \frac{d}{dx}(2gx) + \frac{d}{dx}(2fy) + \frac{d}{dx}(c) = 0$$

$$\Rightarrow 2ax + 2by\frac{dy}{dx} + 2hx\frac{dy}{dx} + 2hy + 2g + 2f\frac{dy}{dx} = 0$$

$$\Rightarrow ax + by\frac{dy}{dx} + hx\frac{dy}{dx} + hy + g + f\frac{dy}{dx} = 0$$

$$\Rightarrow (by + hx + f)\frac{dy}{dx} = -(ax + hy + g)$$

$$\Rightarrow \frac{dy}{dx} = \frac{-(ax + hy + g)}{hx + by + f}$$

6. $x\sqrt{1 + y} + y\sqrt{1 + x} = 0$

Solution: $x\sqrt{1 + y} + y\sqrt{1 + x} = 0$

$$\Rightarrow \frac{d}{dx}\left(x\sqrt{1 + y} + y\sqrt{1 + x}\right) = 0$$

$$\Rightarrow x\frac{d}{dx}\left(\sqrt{1 + y}\right) + \sqrt{1 + y}\,\frac{dx}{dx} + y\frac{d}{dx}\left(\sqrt{1 + x}\right) + \sqrt{1 + x}\,\frac{dy}{dx} = 0$$

$$\Rightarrow x \cdot \frac{1}{2\sqrt{1+y}} \cdot \left(\frac{dy}{dx}\right) + \left(\sqrt{1+y}\right) \cdot 1 + \frac{y}{2\sqrt{1+x}} + \sqrt{1+x} \cdot \frac{dy}{dx} = 0$$

$$\Rightarrow \frac{x}{2\sqrt{1+y}}\frac{dy}{dx} + \sqrt{1+y} + \frac{y}{2\sqrt{1+x}} + \sqrt{1+x}\,\frac{dy}{dx} = 0$$

$$\Rightarrow \left(\frac{x}{2\sqrt{1+y}}\frac{dy}{dx} + \sqrt{1+x}\,\frac{dy}{dx}\right) + \sqrt{1+y} + \frac{y}{2\sqrt{1+x}} = 0$$

$$\Rightarrow \left(\frac{x}{2\sqrt{1+y}} + \sqrt{1+x}\right)\frac{dy}{dx}$$

$$\Rightarrow -\left(\sqrt{1 + y} + \frac{y}{2\sqrt{1 + x}}\right)$$

$$\Rightarrow \frac{dy}{dx} = \frac{-\left(\sqrt{1 + y} + \frac{y}{2\sqrt{1 + x}}\right)}{\left(\frac{x}{2\sqrt{1 + y}} + \sqrt{1 + x}\right)}$$

7. $\sqrt{x} + \sqrt{y} = 1$

Solution: $\sqrt{x} + \sqrt{y} = 1$

$$\Rightarrow \frac{d}{dx}\left(\sqrt{x} + \sqrt{y}\right) = \frac{d}{dx}(1)$$

$$\Rightarrow \frac{d}{dx}\sqrt{x} + \frac{d}{dx}\sqrt{y} = 0$$

$$\Rightarrow \frac{1}{2\sqrt{x}} + \frac{1}{2\sqrt{y}}\frac{dy}{dx} = 0$$

$$\Rightarrow \frac{1}{2\sqrt{y}}\frac{dy}{dx} = -\frac{1}{2\sqrt{x}}$$

$$\Rightarrow \frac{dy}{dx} = \frac{-2\sqrt{y}}{2\sqrt{x}} = \frac{-\sqrt{y}}{\sqrt{x}}$$

Problems based on the combination of implicit algebraic and transcendental functions

To find the derivatives of problems being the combination of implicit algebraic and transcendental functions of x's, we adopt the rule consisting of following steps.

Step 1: Take $\frac{d}{dx}(\ )$ on both sides of the given equation.

Step 2: Differentiate the transcendental functions of rational integral functions of x's and y's and implicit algebraic functions of x's with respect to x's using the rules for the derivatives of sum, difference, product, quotient, composite of differentiable functions and a constant multiple of the differentiable function remembering that

(i) $\frac{d\,F(x)}{dx} = \frac{d\,F(x)}{dy} \cdot \frac{dy}{dx}$

(ii) $\frac{d}{dx} F\,(a \cdot r \cdot i \cdot f \cdot o \cdot x \text{ and } y)$

$$= \frac{d\,F(a \cdot r \cdot o \cdot f \cdot o \cdot x \text{ and } y)}{d(a \cdot r \cdot i \cdot f \cdot o \cdot x \text{ and } y)} \cdot \frac{d(a \cdot r \cdot i \cdot f \cdot o \cdot x \text{ and } y)}{dx}$$

Where "F" stands for "power, trigonometric, inverse trigonometric, logarithmic, exponential, etc, function".

"$a \cdot r \cdot e \cdot f \cdot o \cdot x$ and y" stands for "a rational integral function of x and y; and y is pre-assumed (or, understood) to be a differentiable function of x, say $g(x)$.

Step 3: Collect the terms involving $\frac{dy}{dx}$ on the left hand side and the terms without $\frac{dy}{dx}$ on the right hand side.

Step 4: Solve the equation for $\frac{dy}{dx}$ on the right hand side.

Notes:

1. The final result for $\frac{dy}{dx}$ is an expression either (i) in terms of both x and y only or (ii) interms of both x and y with one (or, more than one) transcendental function (or, functions) of the rational integral function (or, functions) of x and y (or, x's and y's).

2. While finding the derivatives of an implicit function being the combination of implicit algebraic and transcendental functions of x's, often required derivatives from:

(i) $\frac{d\,F(y)}{dx} = \frac{d\,F(y)}{dy} \cdot \frac{dy}{dx}$

(ii) $\frac{d}{dx} F\,(a \cdot r \cdot i \cdot f \cdot o \cdot x \text{ and } y)$

$$= \frac{d\,F(a \cdot r \cdot o \cdot f \cdot o \cdot x \text{ and } y)}{d(a \cdot r \cdot i \cdot f \cdot o \cdot x \text{ and } y)} \cdot \frac{d(a \cdot r \cdot i \cdot f \cdot o \cdot x \text{ and } y)}{dx}$$

are the following ones.

(a) $\frac{d\,y^n}{dx} = n y^{n-1} \frac{dy}{dx}$

(b) $\frac{d}{dx}(x + y)^n = n(x + y)^{n-1}\left(1 + \frac{dy}{dx}\right)$

(c) $\frac{d}{dx} F\left(y^n\right) = \frac{d\,F\left(y^n\right)}{d\left(y^n\right)} \cdot \frac{dy^n}{dx}$

$$= F'\left(y^n\right) \cdot n \cdot y^{n-1} \cdot \frac{dy}{dx}$$

(d) $\frac{d}{dx} F\left(x^n \pm y^n\right) = \frac{d\,F\left(x^n \pm y^n\right)}{d\left(x^n \pm y^n\right)} \cdot \frac{d\left(x^n \pm y^n\right)}{dx}$

(e) $\frac{d}{dx} F\left(x^n\, y^n\right) = \frac{d\,F\left(x^n \cdot y^n\right)}{d\left(x^n\, y^n\right)} \cdot \frac{d\left(x^n\, y^n\right)}{dx}$

Solved Examples

Find $\frac{dy}{dx}$ if

1. $y = \sin(x + y)$

Solution: $y = \sin(x + y)$

$$\Rightarrow \frac{dy}{dx} = \frac{d\sin(x+y)}{dx} = \frac{d\sin(x+y)}{d(x+y)} \cdot \frac{d(x+y)}{dx}$$

$$\Rightarrow \frac{dy}{dx} = \cos(x+y)\left(1 + \frac{dy}{dx}\right)$$

$$= \cos(x+y) + \cos(x+y)\frac{dy}{dx}$$

$$\Rightarrow \frac{dy}{dx} - \cos(x+y) \cdot \frac{dy}{dx} = \cos(x+y)$$

$$\Rightarrow (1 - \cos(x+y))\frac{dy}{dx} = \cos(x+y)$$

$$\Rightarrow \frac{dy}{dx} = \frac{\cos(x+y)}{1 - \cos(x+y)}$$

2. $xy = \sin(x+y)$

Solution: $xy = \sin(x+y)$

$$\Rightarrow \frac{dy}{dx}(xy) = \frac{d}{dx}\sin(x+y) = \frac{d\sin(x+y)}{d(x+y)} \cdot \frac{d(x+y)}{dx}$$

$$\Rightarrow x\frac{dy}{dx} + y \cdot \frac{dx}{dx} = \cos(x+y) \cdot \left(\frac{dx}{dx} + \frac{dy}{dx}\right)$$

$$\Rightarrow x\frac{dy}{dx} + y = \cos(x+y)\left(1 + \frac{dy}{dx}\right)$$

$$\Rightarrow x\frac{dy}{dx} + y = \cos(x+y) + \cos(x+y)\frac{dy}{dx}$$

$$\Rightarrow x\frac{dy}{dx} - \cos(x+y)\frac{dy}{dx} = \cos(x+y) - y$$

$$\Rightarrow (x - \cos(x+y))\frac{dy}{dx} = \cos(x+y) - y$$

$$\Rightarrow \frac{dy}{dx} = \frac{\cos(x+y) - y}{x - \cos(x+y)}$$

3. $x^3 y^3 = \cos(xy)$

Solution: $x^3 y^3 = \cos(xy)$

$$\Rightarrow \frac{d}{dx}\left(x^3 y^3\right) = \frac{d}{dx}\cos(xy)$$

$$\Rightarrow x^3\frac{dy^3}{dx} + y^3\frac{dx^3}{dx} = \frac{d\cos(xy)}{d(xy)} \cdot \frac{d(xy)}{dx}$$

$$\Rightarrow 3x^3y^2\frac{dy}{dx} + 3x^2y^3 = -\sin xy\left(x\frac{dy}{dx} + y\right)$$

$$\Rightarrow 3x^3y^2\frac{dy}{dx} + 3x^2y^3 = -x\sin xy\frac{dy}{dx} - y\sin xy$$

$$\Rightarrow \left(3x^3y^2 + x\sin xy\right)\frac{dy}{dx} = -3x^2y^3 - y\sin xy$$

$$= -y\left(3x^2y^2 + \sin xy\right)$$

$$\Rightarrow \frac{dy}{dx} = -\frac{y\left(3x^2y^2 + \sin xy\right)}{3x^3y^2 + x\sin xy}$$

$$= -\frac{y\left(3x^2y^2 + \sin xy\right)}{x\left(3x^2y^2 + \sin xy\right)} = -\frac{y}{x}$$

4. $\frac{x}{y} = \operatorname{cosec} xy$

Solution: $\frac{x}{y} = \operatorname{cosec} xy$

$$\Rightarrow x = y \operatorname{cosec} xy$$

$$\Rightarrow \frac{dy}{dx} = \frac{d}{dx}(y \operatorname{cosec} xy)$$

$$= y\frac{d}{dx}\operatorname{cosec}(xy) + \operatorname{cosec}(xy) \cdot \frac{dy}{dx}$$

$$\Rightarrow 1 = y\frac{d\operatorname{cosec} xy}{d(xy)} \cdot \frac{d(xy)}{dx} + \operatorname{cosec} xy \cdot \frac{dy}{dx}$$

$$\Rightarrow 1 = y(-\operatorname{cosec} xy \cot xy)\left(x\frac{dy}{dx} + y\frac{dx}{dx}\right) +$$

$$\operatorname{cosec} xy \cdot \frac{dy}{dx}$$

$$\Rightarrow 1 = (-yx \operatorname{cosec} xy \cot xy)\frac{dy}{dx} + \operatorname{cosec} xy\frac{dy}{dx} -$$

$$y^2 \operatorname{cosec} xy \cot xy$$

$$\Rightarrow (-yx\,\text{cosec}\,xy\cot xy + \text{cosec}\,xy)\frac{dy}{dx}$$

$$= 1 + y^2\,\text{cosec}\,xy\cot xy$$

$$\Rightarrow \frac{dy}{dx} = \frac{1 + y^2\,\text{cosec}\,xy\cot xy}{(\text{cosec}\,xy - yx\,\text{cosec}\,xy\cot xy)}$$

5. $y = \tan^{-1}(x+y)$

Solution: $y = \tan^{-1}(x+y)$

$$\Rightarrow \frac{dy}{dx} = \frac{d\tan^{-1}(x+y)}{dx} = \frac{d\tan^{-1}(x+y)}{d(x+y)}\cdot\frac{d(x+y)}{dx}$$

$$\Rightarrow \frac{dy}{dx} = \frac{1}{1+(x+y)^2}\cdot\left(1+\frac{dy}{dx}\right)$$

$$= \frac{1}{1+(x+y)^2} + \frac{1}{1+(x+y)^2}\cdot\frac{dy}{dx}$$

$$\Rightarrow \frac{dy}{dx} - \frac{1}{1+(x+y)^2}\left(\frac{dy}{dx}\right) = \frac{1}{1+(x+y)^2}$$

$$\Rightarrow \left(1 - \frac{1}{1+(x+y)^2}\right)\frac{dy}{dx} = \frac{1}{1+(x+y)^2}$$

$$\Rightarrow \frac{1+(x+y)^2-1}{1+(x+y)^2}\cdot\frac{dy}{dx} = \frac{1}{1+(x+y)^2}$$

$$\Rightarrow \frac{(x+y)^2}{1+(x+y)^2}\,\frac{dy}{dx} = \frac{1}{1+(x+y)^2}$$

$$\Rightarrow \frac{dy}{dx} = \frac{1}{1+(x+y)^2}\cdot\frac{1+(x+y)^2}{(x+y)^2} = \frac{1}{(x+y)^2}$$

6. $y = \log(x+y)$

Solution: $y = \log(x+y)$

$$\Rightarrow \frac{dy}{dx} = \frac{d}{dx}\log(x+y) = \frac{d\log(x+y)}{d(x+y)}\cdot\frac{d(x+y)}{dx}$$

$$\Rightarrow \frac{dy}{dx} = \frac{1}{(x+y)}\cdot\left(1+\frac{dy}{dx}\right) = \frac{1}{(x+y)} + \frac{1}{(x+y)}\cdot\frac{dy}{dx}$$

$$\Rightarrow \frac{dy}{dx} - \frac{1}{(x+y)}\,\frac{dy}{dx} = \frac{1}{(x+y)}$$

$$\Rightarrow \left(1 - \frac{1}{(x+y)}\right)\frac{dy}{dx} = \frac{1}{(x+y)}$$

$$\Rightarrow \frac{x+y-1}{(x+y)}\cdot\frac{dy}{dx} = \frac{1}{(x+y)}$$

$$\Rightarrow \frac{dy}{dx} = \frac{1}{(x+y)}\cdot\frac{(x+y)}{(x+y-1)} = \frac{1}{(x+y-1)}$$

7. $y = e^{xy}$

Solution: $y = e^{xy}$

$$\Rightarrow \frac{dy}{dx} = \frac{d}{dx}\left(e^{xy}\right) = \frac{d\left(e^{xy}\right)}{d(xy)}\cdot\frac{d(xy)}{dx}$$

$$\Rightarrow \frac{dy}{dx} = e^{xy}\cdot\left(x\frac{dy}{dx} + y\frac{dx}{dx}\right) = e^{xy}\cdot\left(x\frac{dy}{dx} + y\cdot 1\right)$$

$$\Rightarrow \frac{dy}{dx} = xe^{xy}\cdot\frac{dy}{dx} + y\cdot e^{xy}$$

$$\Rightarrow \frac{dy}{dx} - xe^{xy}\cdot\frac{dy}{dx} = y\cdot e^{xy}$$

$$\Rightarrow \left(1 - xe^{xy}\right)\frac{dy}{dx} = y\cdot e^{xy}$$

$$\Rightarrow \frac{dy}{dx} = \frac{ye^{xy}}{1 - x\cdot e^{xy}}$$

8. $x^2 + y^2 = \log(xy)$

Solution: $x^2 + y^2 = \log(xy)$

$$\Rightarrow \frac{d}{dx}\left(x^2 + y^2\right) = \frac{d}{dx}(\log(xy))$$

$$= \frac{d(\log(xy))}{d(xy)}\cdot\frac{d(xy)}{dx}$$

$$\Rightarrow \frac{dx^2}{dx} + \frac{dy^2}{dx} = \frac{1}{xy}\cdot\left(x\frac{dy}{dx} + y\cdot 1\right)$$

$$\Rightarrow 2x + 2y\cdot\frac{dy}{dx} = \frac{1}{y}\cdot\frac{dy}{dx} + \frac{1}{x}$$

$$\Rightarrow 2y\frac{dy}{dx} - \frac{1}{y}\frac{dy}{dx} = \frac{1}{x} - 2x$$

$$\Rightarrow \left(2y - \frac{1}{y}\right)\frac{dy}{dx} = \frac{\left(1-2x^2\right)}{x}$$

$$\Rightarrow \frac{\left(2y^2-1\right)}{y}\cdot\frac{dy}{dx} = \frac{\left(1-2x^2\right)}{x}$$

$$\Rightarrow \frac{dy}{dx} = \frac{\left(1-2x^2\right)}{x}\cdot\frac{y}{\left(2y^2-1\right)} = \left(\frac{-y}{x}\right)\cdot\left(\frac{1-2x^2}{1-2y^2}\right)$$

9. $e^{xy} = \log(xy)$

Solution: $e^{xy} = \log(xy)$

$$\Rightarrow \frac{d}{dx}\left(e^{xy}\right) = \frac{d}{dx}(\log(xy))$$

$$\Rightarrow \frac{d\left(e^{xy}\right)}{d(xy)}\cdot\frac{d(xy)}{dx} = \frac{d(\log(xy))}{d(xy)}\cdot\frac{d(xy)}{dx}$$

$$\Rightarrow \left(x\frac{dy}{dx} + y\cdot 1\right)\cdot e^{xy} = \frac{1}{xy}\cdot\frac{dy}{dx} + \frac{y}{xy}$$

$$\Rightarrow xe^{xy}\cdot\frac{dy}{dx} - \frac{1}{y}\cdot\frac{dy}{dx} = \frac{1}{x} - ye^{xy}$$

$$\Rightarrow \left(xe^{xy} - \frac{1}{y}\right)\cdot\frac{dy}{dx} = \frac{1-xye^{xy}}{x}$$

$$\Rightarrow \frac{dy}{dx} = \frac{\left(1-xye^{xy}\right)}{x}\cdot\frac{y}{\left(xye^{xy}-1\right)}$$

$$= \left(\frac{-y}{x}\right)\cdot\left(\frac{1-xye^{xy}}{1-xye^{xy}}\right) = -\left(\frac{y}{x}\right)$$

10. $y = x\log y$

Solution: $y = x\log y$

$$\Rightarrow \frac{dy}{dx} = \frac{d}{dx}(x\log y) = x\frac{d\log y}{dx} + \log y\cdot\frac{dx}{dx}$$

$$\Rightarrow \frac{dy}{dx} = x\frac{d\log y}{dy}\cdot\frac{dy}{dx} + \log y\cdot 1$$

$$= x\cdot\frac{1}{y}\cdot\frac{dy}{dx} + \log y$$

$$\Rightarrow \frac{dy}{dx} - \frac{x}{y}\frac{dy}{dx} = \log y$$

$$\Rightarrow \left(1 - \frac{x}{y}\right)\frac{dy}{dx} = \log y$$

$$\Rightarrow \left(\frac{y-x}{y}\right)\frac{dy}{dx} = \log y$$

$$\Rightarrow \frac{dy}{dx} = \left(\frac{y\log y}{y-x}\right)$$

11. $e^{xy} = \cos(x^2+y^2)$

Solution: $e^{xy} = \cos(x^2+y^2)$

$$\Rightarrow \frac{d}{dx}\left(e^{xy}\right) = \frac{d}{dx}\left(\cos\left(x^2+y^2\right)\right)$$

$$\Rightarrow \frac{d\left(e^{xy}\right)}{d(xy)}\cdot\frac{d(xy)}{dx} = \frac{d\left(\cos\left(x^2+y^2\right)\right)}{d\left(x^2+y^2\right)}\cdot\frac{d\left(x^2+y^2\right)}{dx}$$

$$\Rightarrow e^{xy}\cdot\left(x\frac{dy}{dx} + y\right)$$

$$= -\sin\left(x^2+y^2\right)\cdot\left(2x + 2y\frac{dy}{dx}\right)$$

$$\Rightarrow \left(xe^{xy}\cdot\frac{dy}{dx} + y\cdot e^{xy}\right)$$

$$= -2x\sin\left(x^2+y^2\right) - 2y\sin\left(x^2+y^2\right)\cdot\frac{dy}{dx}$$

$$\Rightarrow xe^{xy}\cdot\frac{dy}{dx} + 2y\cdot\sin\left(x^2+y^2\right)\cdot\frac{dy}{dx}$$

$$= -2x\sin\left(x^2+y^2\right) - y\cdot e^{xy}$$

$$\Rightarrow \left(xe^{xy} + 2y\sin\left(x^2+y^2\right)\right)\frac{dy}{dx}$$

$$= -2x\sin\left(x^2 + y^2\right) - ye^{xy}$$

$$\Rightarrow \frac{dy}{dx} = -\left(\frac{2x\sin\left(x^2 + y^2\right) + ye^{xy}}{2y\sin\left(x^2 + y^2\right) + xe^{xy}}\right)$$

Problems based on implicit algebraic functions

Exercise 12.1

Find $\frac{dy}{dx}$ if

1. $y^2 = 5x^2 + 1$
2. $y^2 = 4ax$
3. $x^2 + y^2 = 9$
4. $2x^2 + 3y^3 = a^2$
5. $\frac{x^2}{a^2} + \frac{y^2}{b^2} = 1$
6. $\frac{x^2}{a^2} - \frac{y^2}{b^2} = 1$
7. $x^n + y^n = a^n$
8. $ax^2 + by^2 = (x + y)$
9. $xy = a$
10. $3x^2 y = 16'$
11. $x^2 + y^2 - xy = a$
12. $ax^2 + 2hxy + by^2 = 1$
13. $x^2 + y^2 + 2hxy + 2gx + 2fy + c = 0$
14. $x^3 + y^3 = 3axy$
15. $\sqrt{x} + \sqrt{y} = \sqrt{a}$
16. $x^{\frac{3}{2}} + y^{\frac{3}{2}} = a^{\frac{3}{2}}$
17. $x\sqrt{y} + y\sqrt{x} = 1$
18. $x^3 + y^3 = xy$
19. $x + y = xy^2$
20. $xy + x^2 y^2 = c$
21. $x^2 y + xy^2 = c^3$
22. $x^2 y + xy^2 = 3x^3 + 4y^3$
23. $y = (x + y)^2$
24. $x^2 y = (x + 2y)^3$
25. $x^3 y = (2x + 3y)^2$
26. $(x^3 + y^3)\, xy = x^5 - y^5$
27. $x^5 + y^5 - 5x^2 y^3 - 5x^3 y^2 = 0$
28. $x^2 y = (2x + 3y)^2$
29. $x^2 y^3 = (2x + y)^5$
30. $x^3 y^4 = (x + y)^7$
31. $x^5 y^4 = (x + y)^9$
32. $x^4 y^5 = (x - y)^9$
33. $x^a y^b = (x - y)^{a+b}$

Answers

1. $\frac{5x}{y}$
2. $\frac{2a}{y}$
3. $-\frac{x}{y}$
4. $-\frac{2x}{3y}$
5. $-\frac{b^2 x}{a^2 y}$
6. $\frac{b^2 x}{a^2 y}$
7. $-\left(\frac{x}{y}\right)^{n-1}$
8. $\frac{1 - 2ax}{2by - 1}$
9. $-\frac{y}{x}$
10. $-\frac{2y}{x}$
11. $\frac{y - 2x}{2y - x}$
12. $-\left(\frac{ax + hy}{hx + by}\right)$

13. $-\left(\dfrac{g+x+hy}{f+hx+y}\right)$

14. $\dfrac{ay-x^2}{y^2-ax}$

15. $-\sqrt{\dfrac{y}{x}}$

16. $-\sqrt{\dfrac{x}{y}}$

17. $\dfrac{y}{x}\left(\dfrac{2\sqrt{x}+\sqrt{y}}{2\sqrt{y}+\sqrt{x}}\right)$

18. $\dfrac{y-3x^2}{3y^2-x}$

19. $\dfrac{y^2-1}{1-2xy}$

20. $-\dfrac{y}{x}$

21. $-\dfrac{y}{x}\left(\dfrac{y+2x}{x+2y}\right)$

22. $\dfrac{9x^2-2xy-y^2}{x^2+2xy-12y^2}$

23. $\dfrac{2(x+y)}{1-2(x+y)}$

24. $-\left(\dfrac{3x^2+10xy+12y^2}{5x^2+24xy+6y^2}\right)$

25. $\dfrac{8x+12y-3x^2y}{x^3-12x-18y}$

26. $\dfrac{5x^4-4xy^3-y^4}{x^4+4xy^3-5y^4}$

27. $-\left(\dfrac{x^4-3x^2y^2-2xy^3}{y^4-3x^2y^2-2x^3y}\right)$

28. $\dfrac{-28x^2+54y^2+70xy}{35x^2+81y^3+108xy}$

29. $\dfrac{10(2x+y)^4-2xy^3}{3x^2y^2-5(2x+y)^4}$

30. $\dfrac{y}{x}$

31. $\dfrac{y}{x}$

32. $\dfrac{y}{x}$

33. $\dfrac{y}{x}$

Problems based on the combination of implicit algebraic and transcendental functions

Exercise 12.2

Find $\dfrac{dy}{dx}$ if

1. $y=\sin(x+y)$
2. $y=\tan(x+y)$
3. $y=\cot(x+y)$
4. $y=\sec(x+y)$
5. $y=\cos(x-y)$
6. $\sin(x+y)+\sin(x-y)=1$
7. $x=2\cos y+3\sin y$
8. $x=\sqrt{\sin y+\cos y}$
9. $x=2\sqrt{3\sin y-4\cos y}$
10. $\cos(x+y)=y\sin x$
11. $\tan(x+y)+\tan(x-y)=1$
12. $x=(2\cos^{-1}y)^2$

Answers

1. $\dfrac{\cos(x+y)}{1-\cos(x+y)}$

2. $-\dfrac{\sec^2(x+y)}{1-\sec^2(x+y)}$

3. $-\dfrac{\operatorname{cosec}^2(x+y)}{1+\operatorname{cosec}^2(x+y)}$

4. $\dfrac{\sec(x+y)\tan(x+y)}{1-\sec(x+y)\tan(x+y)}$

5. $\dfrac{\sin(x-y)}{\sin(x-y)-1}$

6. $2\cot x\cdot\cot y$

7. $\dfrac{1}{3\cos y-2\sin y}$

8. $\dfrac{2\sqrt{\sin x+\cos y}}{\cos y-\sin y}$

9. $\dfrac{\sqrt{3\sin y-4\cos y}}{3\cos y+4\sin y}$

10. $-\dfrac{y\cos x+\sin(x+y)}{\sin(x+y)+\sin x}$

11. $\dfrac{\sec^2(x+y)+\sec^2(x-y)}{\sec^2(x-y)-\sec^2(x+y)}$

12. $\dfrac{-\sqrt{1-y^2}}{4\cos^{-1}y}$

Exercise 12.3

Find $\dfrac{dy}{dx}$ if

1. $x+y=\sin(xy)$
2. $x+y=\sin(x+y)$
3. $x+y=\tan(xy)$
4. $xy=\tan(xy)$
5. $xy=\sin(x+y)$
6. $xy=\cos(x+y)$
7. $xy=\sin(2x+3y)$
8. $x-y=\sec(x+y)$
9. $xy=\sec(x+y)$
10. $x^2y=\sin y$
11. $x^2y^2=\sin(xy)$
12. $x^3y^3=\cos(xy)$
13. $x^3+y^3=\sin(x+y)$
14. $xy=\sin^2(x+y)$
15. $\sqrt{xy}=\tan(2y-x)$
16. $y^2=\tan(2y+x)$
17. $y=x+y^2\sin^3\left(\dfrac{x}{2}\right)$
18. $x\cos y+y\sin x=0$
19. $x\sin y+y\cos x=0$
20. $x\cos y+y\cos x=\tan(x+y)$
21. $y^3=(x+\sin x)(x-\cos x)$
22. $\log(xy)=x^2+y^2$
23. $e^{xy}=\log(xy)$
24. $\log y=e^{xy}$
25. $e^{xy}+xy=0$
26. $x=y\log(xy)$
27. $e^{xy}+\log(xy)+xy=0$
28. $\log|xy|=x^2+y^2$
29. $y\log x=x-y$
30. $x+y=\tan^{-1}(xy)$
31. $e^{xy}=\cos(x^2+y^2)$
32. $x=y\log|xy|$
33. $y=x\log\left(\dfrac{y}{a+bx}\right)$
34. $y^2=\dfrac{\log(x+y)+\sin(e^x)}{x^3}$
35. $xy=\log(x^2+y^2)$
36. $x^2+y^2=\log(x+y)$

Answers

1. $\dfrac{y\cos(xy)-1}{1-x\cos(xy)}$

2. -1

3. $\dfrac{1 - y\sec^2(xy)}{x \cdot \sec^2(xy) - 1}$

4. $-\dfrac{y}{x}$

5. $-\dfrac{y - \cos(x+y)}{x - \cos(x+y)}$

6. $-\dfrac{y + \sin(x+y)}{x + \sin(x+y)}$

7. $\dfrac{2\cos(2x+3y) - y}{x - 3\cos(2x+3y)}$

8. $\dfrac{1 - \sec(x+y)\tan(x+y)}{1 + \sec(x+y)\tan(x+y)}$

9. $-\dfrac{y - \sec(x+y)\cdot\tan(x+y)}{x - \sec(x+y)\cdot\tan(x+y)}$

10. $\dfrac{2xy}{\cos y - x^2}$

11. $\dfrac{y\cos(xy) - 2xy^2}{2x^2y - x\cos(xy)}$

12. $\dfrac{3x^2y^3 - y\sin(xy)}{3x^3y^2 + x\sin(xy)}$

13. $\dfrac{\cos(x+y) - 3x^2}{3y^2 - \cos(x+y)}$

14. $\dfrac{\sin 2(x+y) - y}{x - \sin 2(x+y)}$

15. $-\left(\dfrac{y + 2\tan(2y-x)\sec^2(2y-x)}{x - 4\tan(2y-x)\sec^2(2y-x)}\right)$

16. $\dfrac{\sec^2(2y+x)}{2y - 2\sec^2(2y+x)}$

17. $\dfrac{1 + \frac{3}{2}\sin^2\frac{x}{2}\left(\cos\frac{x}{2}\right)y^2}{1 - 2y\sin^3\frac{x}{2}}$

18. $\dfrac{y\cos x + \cos y}{x\sin y - \sin x}$

19. $\dfrac{y\sin x - \sin y}{x\cos y + \cos x}$

20. $\dfrac{\sec^2(x+y) - \cos y + y\sin x}{\cos x - x\sin y - \sec^2(x+y)}$

21. $\dfrac{x + (x + \sin x - \cos x)(1 + \sin x + \cos x)}{3y^2}$

22. $\dfrac{y}{x}\cdot\left(\dfrac{2x^2 - 1}{1 - 2y^2}\right)$

23. $\dfrac{y}{x}\cdot\left(\dfrac{1 - xye^{xy}}{xye^{xy} - 1}\right)$

24. $\dfrac{y^2e^{xy}}{1 - xye^{xy}}$

25. $-\dfrac{y}{x}$

26. $\dfrac{(x-y)y}{x(x+y)}$ or $\dfrac{x-y}{x(1+\log xy)}$

27. $-\dfrac{y}{x}$

28. $\dfrac{y(2x^2-1)}{x(1-2y^2)}$

29. $\dfrac{\log x}{(1+\log x)^2}$

30. $\dfrac{y - \sec^2(x + y)}{\sec^2(x + y) - x}$

31. $-\left(\dfrac{ye^{xy} + 2x\sin\left(x^2 + y^2\right)}{xe^{xy} + 2y\sin\left(x^2 + y^2\right)}\right)$

32. $\dfrac{y(x - y)}{x(x + y)}$

33. $\dfrac{-y\left(ay + bxy - bx^2\right)}{x(x - y)(a + bx)}$

34. $\dfrac{1 + \left\{e^x \cos\left(e^x\right) - 3x^2y^2\right\}(x + y)}{2x^3y(x + y) - 1}$

35. $\dfrac{2x - yx^2 - y^3}{x^3 + xy^2 - 2y}$

36. $\dfrac{1 - 2x^2 - 2xy}{2xy + 2y^2 - 1}$

Conditional identities based on a given explicit function of x

Whenever we have to form a differential equation with the help of a given explicit function of x, we adopt the following working rule provided the required differential equation contains only first derivative.

Working rule: Find the first derivative (i.e. $\dfrac{dy}{dx}$) of the given explicit function of x and use mathematical manipulations to put the first derivative into the required differential equation.

Notes:

1. In successive differentiation, we have discussed in detail the methods of procedure of forming differential equation whenever a function of x is given.

2. A given function x or an equation defining y as a function of x can put into different forms according to our need.

3. When a function of x under the radical sign appears defining y as a function of x, we should try to remove the radical sign by squaring or raising the same power both sides of the equation defining y as a function of x under the radical sign.

4. Whenever we are given a function of x put in the forms:

(i) $y = f_1(x)^{f_2(x)}$

(ii) $y = \dfrac{f_1(x)^{g_1(x)} \cdot f_2(x)^{g_2(x)} \cdot f_3(x)^{g_3(x)} \ldots}{f'(x)^{g'(x)} \cdot f''(x)^{g''(x)} \cdot f'''(x)^{g'''(x)} \ldots}$

the bases $f^1(x), f^2(x), f^3(x) \ldots, f'(x), f''(x), f'''(x), \ldots$ are assumed to be positive before taking the logarithm of both sides of the equation defining y as a function of x provided it is not mentioned in the problems that the bases are positive (i.e. > 0).

5. If we are given a function of x put in the form:

$$y = \frac{(f_1(x))^{m_1} \cdot (f_2(x))^{m_2} \cdot (f_3(x))^{m_3} \ldots}{(f'(x))^{m'} \cdot (f''(x))^{m''} \cdot (f'''(x))^{m'''} \ldots}$$

where the bases $f^1(x), f^2(x), f^3(x), \ldots; f'(x), f''(x), f'''(x), \ldots$ are functions of x's where the indices $m_1, m_2, m_3, \ldots; m', m'', m''', \ldots$ are constants, it is differentiated taking the logarithm of both sides of the given equation defining y as a function of x.

Solved Examples

1. $y = \sqrt{x} + \dfrac{1}{\sqrt{x}}$, show that $2x\dfrac{dy}{dx} + y = 2\sqrt{x}$.

Solution: $y = \sqrt{x} + \dfrac{1}{\sqrt{x}}, = x^{\frac{1}{2}} + x^{-\frac{1}{2}}$ ...(i)

$$\Rightarrow \frac{dy}{dx} = \frac{1}{2}x^{-\frac{1}{2}} - \frac{1}{2}x^{-\frac{3}{2}}$$

$$\Rightarrow 2x\frac{dy}{dx} = 2x\left(\frac{1}{2}x^{-\frac{1}{2}} - \frac{1}{2}x^{-\frac{3}{2}}\right) = x^{\frac{1}{2}} - x^{-\frac{1}{2}} \quad \text{...(ii)}$$

Adding (i) and (ii), we have

$$y + 2x\frac{dy}{dx} = 2x^{\frac{1}{2}} = 2\sqrt{x}$$

2. If $y = x + \frac{1}{x}$, show that $x\frac{dy}{dx} + y = 2x$.

Solution: $y = x + \frac{1}{x}$ ...(i)

$$\Rightarrow y = \frac{x^2 + 1}{x}$$

$$\Rightarrow \frac{dy}{dx} = \frac{x\frac{d}{dx}(x^2+1) - (x^2+1)\frac{d}{dx}(x)}{x^2}$$

$$= \frac{x \times 2x - (x^2+1)}{x^2}$$

$$\Rightarrow \frac{dy}{dx} = \frac{2x^2 - (x^2+1)}{x^2} = 2 - \left(\frac{x^2+1}{x^2}\right)$$

$$\Rightarrow \frac{dy}{dx} + \left(\frac{x^2+1}{x^2}\right) = 2$$

$$\Rightarrow x\frac{dy}{dx} + x \times \left(\frac{x^2+1}{x^2}\right) = 2 \times x$$

$$\Rightarrow x\frac{dy}{dx} + \left(\frac{x^2+1}{x}\right) = 2x$$

$$\Rightarrow x\frac{dy}{dx} + \left(\frac{x^2}{x} + \frac{1}{x}\right) = 2x$$

$$\Rightarrow x\frac{dy}{dx} + \left(x + \frac{1}{x}\right) = 2x$$

$$\Rightarrow x\frac{dy}{dx} + y = 2x \text{ (From (1))}$$

3. If $y = \sqrt{1 + x^6}$, show that $y\frac{dy}{dx} = 3x^5$.

Solution: $y = \sqrt{1 + x^6}$

$$\Rightarrow y^2 = 1 + x^6$$

$$\Rightarrow 2y\frac{dy}{dx} = 6x^5$$

$$\Rightarrow y\frac{dy}{dx} = 3x^5$$

4. $y = \frac{1}{x}$, show that $\frac{dy}{\sqrt{1+x^4}} + \frac{dx}{\sqrt{1+y^4}} = 0$.

Solution: $y = \frac{1}{x}$

$$\Rightarrow \frac{dy}{dx} = -\frac{1}{x^2}$$

Now, $\frac{\sqrt{1+y^4}}{\sqrt{1+x^4}} = \frac{\sqrt{1+\frac{1}{x^4}}}{\sqrt{1+x^4}} = \frac{\frac{\sqrt{x^4+1}}{x^2}}{\sqrt{x^4+1}} = \frac{1}{x^2} = \frac{dy}{dx}$

$$\Rightarrow \frac{\sqrt{1+y^4}}{\sqrt{1+x^4}} = -\frac{dy}{dx}$$

$$\Rightarrow \frac{dx}{\sqrt{1+x^4}} = \frac{dy}{\sqrt{1+y^4}}$$

$$\Rightarrow \frac{dx}{\sqrt{1+x^4}} + \frac{dy}{\sqrt{1+y^4}} = 0$$

Remark: Since the derivative is a limit of the quotient as $\Delta x \to 0$ (i.e. $\lim_{\Delta x \to 0} \frac{\Delta y}{\Delta x}$) which is symbolised as $\frac{dy}{dx}$ which does not indicate a quotient of *dy* and *dx* but *dx* and *dy* are so defined as to consider $\frac{dy}{dx}$ as a quotient of *dy* and *dx* which may be seperated as $dy \div dx$ or $dy \times \frac{1}{dx}$. The above example (4) provides us a fruitful example for regarding $\frac{dy}{dx}$ as a quotient of dy and dx which are known as differential of y and differential of x respectively.

Conditional identities based on a given implicit function of x

Now, we will learn how to form a differential equation with the help of a given implicit function of x. The rule to form a differential equation with the help of a given implicit function of x consists of following steps provided the required differential equation contains only first derivative.

Step 1: Find the first derivative (i.e. $\frac{dy}{dx}$) of the given implicit function of x.

Step 2: Use mathematical manipulations to put the first derivative into the required differential equation.

Refresh your memory:

1. Whenever we have functions put in the forms (i) $f_1(x)^{f_2(y)}$ (ii) $f_1(y)^{f_2(x)}$ (iii) $f_1(x)^{f_2(x)} = f_1(y)^{f_2(y)}$ (iv) $f_1(x)^{f_2(y)} = f_1(y)^{f_2(x)}$, we should assume the bases to be positive to use logarithmic differentiation.

2. Whenever we are given product or quotient of functions put in the form $f_1(x)$ and $f_2(y)$, we should take firstly modulus to use logarithmic differentiation.

3. Logarithmic differentiation is only possible when the given function is positive. This is why we take the modulus of the given function or equation if it has the possibility of being negative.

4. $\log f(x)$ always means $f(x)$ is preassumed to be positive similarly $\log f(y)$ means $f(y)$ is preassumed to be positive.

Solved Examples

1. $x\sqrt{1+y} + y\sqrt{1+x} = 0$, show that $\frac{dy}{dx} = -(1+x)^{-2}$.

Solution: $x\sqrt{1+y} + y\sqrt{1+x} = 0$

$$\Rightarrow x\sqrt{1+y} = -y\sqrt{1+x} \quad \ldots\text{(i)}$$

$$\Rightarrow x^2(1+y) = y^2(1+x)$$

$$\Rightarrow x^2 - y^2 + x^2y - y^2x = 0$$

$$\Rightarrow (x+y)(x-y) + xy(x-y) = 0$$

$$\Rightarrow (x-y)(x+y+xy) = 0$$

$\Rightarrow x + y + xy = 0$ ($\because x - y \neq 0$ since x and y have opposite signs from (i))

Now differentiating both sides w.r.t. x, we have

$$1 + \frac{dy}{dx} + x\frac{dy}{dx} + y = 0$$

$$\Rightarrow (1+x)\frac{dy}{dx} = -(1+y)$$

$$\Rightarrow \frac{dy}{dx} = -\frac{(1+y)}{(1+x)} \quad \ldots\text{(ii)}$$

Again, $\because x + y + xy = 0 \Rightarrow y(1+x)$

$$= -x \Rightarrow y = -\frac{x}{1+x} \Rightarrow 1 + y = 1 - \frac{x}{1+x}$$

$$\Rightarrow 1 + y = \frac{1}{1+x} \quad \ldots\text{(iii)}$$

Putting (iii) in (ii), we have

$$\frac{dy}{dx} = -\frac{1}{(1+x)^2} = -(1+x)^{-2}$$

2. If $y\sqrt{x^2+1} = \log\left(x + \sqrt{x^2+1}\right)$, show that $\left(x^2+1\right)\frac{dy}{dx} + xy - 1 = 0$.

Solution: $y\sqrt{x^2+1} = \log\left(x + \sqrt{x^2+1}\right)$

Now differentiating both sides w.r.t. x, we have

$$y \cdot \frac{d}{dx}\left(\sqrt{x^2+1}\right) + \sqrt{x^2+1} \cdot \frac{dy}{dx}$$

$$= \frac{1}{x + \sqrt{x^2+1}} \cdot \frac{d}{dx}\left(x + \sqrt{x^2+1}\right)$$

$$\Rightarrow y \cdot \frac{1}{2\sqrt{x^2+1}} \cdot 2x + \sqrt{x^2+1} \cdot \frac{dy}{dx}$$

$$= \frac{1}{x+\sqrt{x^2+1}} \cdot \left(1 + \frac{1}{2\sqrt{x^2+1}} \cdot 2x\right)$$

$$\Rightarrow \frac{xy}{\sqrt{x^2+1}} + \sqrt{x^2+1} \cdot \frac{dy}{dx}$$

$$= \frac{1}{x+\sqrt{x^2+1}} \cdot \frac{\sqrt{x^2+1}+x}{\sqrt{x^2+1}} = \frac{1}{\sqrt{x^2+1}}$$

$\Rightarrow xy + \left(x^2+1\right)\frac{dy}{dx} = 1$ (multiplying both sides by $\sqrt{x^2+1}$)

$$\Rightarrow \left(x^2+1\right)\frac{dy}{dx} + xy - 1 = 0$$

3. $x^y = y^x$, show that $\dfrac{dy}{dx} = \dfrac{xy\,\log y - y^2}{xy\,\log x - x^2}$.

Solution: Firstly supposing that the bases x and y both are positive and then taking the log of both sides of the given equation.

$x^y = y^x$ we have

$y \log x = x \log y$

$$\Rightarrow y \cdot \frac{d}{dx}(\log x) + \log x \cdot \frac{dy}{dx}$$

$$= x \cdot \frac{d}{dx}(\log y) + \log y$$

$$\Rightarrow \frac{y}{x} + \log x \cdot \frac{dy}{dx} = \left(\frac{x}{y}\right)\frac{dy}{dx} + \log y$$

$$\Rightarrow y^2 + xy\,\log x\frac{dy}{dx} - x^2\frac{dy}{dx} = xy\,\log y$$

(multiplying both sides by xy)

$$\Rightarrow \left(xy\,\log x - x^2\right)\frac{dy}{dx} = xy\,\log x - y^2$$

$$\Rightarrow \frac{dy}{dx} = \frac{xy\,\log y - y^2}{xy\,\log x - x^2}$$

4. If $\sin y = x \sin(a+y)$, show that

$$\frac{dy}{dx} = \frac{\sin^2(a+y)}{\sin a}$$

Solution: $\sin y = x \sin(a+y)$

$$\Rightarrow |\sin y| = |x \sin(a+y)|$$

$$\Rightarrow \log|\sin y| = \log|x| + \log|\sin(a+y)|$$

$$\Rightarrow \frac{1}{\sin y} \cdot \cos y \cdot \frac{dy}{dx} = \frac{1}{x} + \frac{\cos(a+y)}{\sin(a+y)} \cdot \frac{dy}{dx}$$

$$\Rightarrow \left(\frac{\cos y}{\sin y} - \frac{\cos(a+y)}{\sin(a+y)}\right)\frac{dy}{dx} = \frac{1}{x}$$

$$\Rightarrow \left(\frac{\cos y\,\sin(a+y) - \sin y\cos(a+y)}{\sin y \cdot \sin(a+y)}\right)\frac{dy}{dx} = \frac{1}{x}$$

$$\Rightarrow \frac{\sin(a+y-y)}{\sin y\,\sin(a+y)} \cdot \frac{dy}{dx} = \frac{1}{x}$$

$$\Rightarrow \frac{\sin a}{\sin y\,\sin(a+y)} \cdot \frac{dy}{dx} = \frac{1}{x}$$

$$\Rightarrow \frac{dy}{dx} = \frac{1}{x} \cdot \frac{\sin y \cdot \sin(a+y)}{\sin a}$$

$= \dfrac{1}{x} \cdot \dfrac{x\sin(a+y) \cdot \sin(a+y)}{\sin a}$ ($\because \sin y = x \sin(a+y)$ is given)

$$= \frac{\sin^2(a+y)}{\sin a}$$

5. If $\sqrt{1-x^2} + \sqrt{1-y^2} = a(x-y)$, show that

$$\frac{dy}{dx} = \frac{\sqrt{1-y^2}}{\sqrt{1-x^2}}$$

Solution: Given is $\sqrt{1-x^2}+\sqrt{1-y^2}=a(x-y)$

Putting $x=\sin C, \left(-\frac{\pi}{2}\le C\le\frac{\pi}{2}\right)$

$y=\sin D, \left(\frac{-\pi}{2}\le D\le\frac{\pi}{2}\right)$ in the hypothesis, we have

$$\sqrt{1-\sin^2 C}+\sqrt{1-\sin^2 D}=a(\sin C-\sin D)$$

$$\Rightarrow |\cos C|+|\cos D|=a(\sin C-\sin D)$$

$$\Rightarrow \cos C+\cos D=a(\sin C-\sin D)$$

$$\Rightarrow 2\cos\left(\frac{C+D}{2}\right)\cdot\cos\left(\frac{C-D}{2}\right)$$

$$=a\left(2\cos\left(\frac{C+D}{2}\right)\cdot\sin\left(\frac{C-D}{2}\right)\right)$$

$$\Rightarrow \cos\left(\frac{C-D}{2}\right)=a\sin\left(\frac{C-D}{2}\right)$$

$$\Rightarrow \frac{\cos\left(\frac{C-D}{2}\right)}{\sin\left(\frac{C-D}{2}\right)}=a\ (\because\ x\ne y\Rightarrow C\ne D \text{ or } \pi-D)$$

$$\Rightarrow \cot\left(\frac{C-D}{2}\right)=a$$

$$\Rightarrow \frac{C-D}{2}=\cot^{-1}a$$

$$\Rightarrow C-D=2\cot^{-1}a$$

$$\Rightarrow \sin^{-1}x-\sin^{-1}y=2\cot^{-1}a$$

$$\Rightarrow \frac{d}{dx}\left(\sin^{-1}x-\sin^{-1}y\right)=\frac{d}{dx}\left(2\cot^{-1}a\right)$$

$\Rightarrow \frac{d}{dx}\left(\sin^{-1}x\right)-\frac{d}{dx}\left(\sin^{-1}y\right)=0$ ($\because 2\cot^{-1}a=$ constant)

$$\Rightarrow \frac{1}{\sqrt{1-x^2}}-\frac{1}{\sqrt{1-y^2}}\frac{dy}{dx}=0$$

$$\Rightarrow -\frac{1}{\sqrt{1-y^2}}\frac{dy}{dx}=-\frac{1}{\sqrt{1-x^2}}$$

6. Show that $x^2+y^2=3xy\Rightarrow\frac{dy}{dx}\cdot\frac{dx}{dy}=1$.

Solution: $x^2+y^2=3xy$

$$\Rightarrow \frac{d}{dx}\left(x^2+y^2\right)=\frac{d}{dx}(3xy)$$

$$\Rightarrow 2x+2y\frac{dy}{dx}=3\left(x\frac{dy}{dx}+y\cdot 1\right)$$

$$\Rightarrow 2y\frac{dy}{dx}-3x\frac{dy}{dx}=3y-2x$$

$$\Rightarrow (2y-3x)\frac{dy}{dx}=3y-2x$$

$$\Rightarrow \frac{dy}{dx}=\frac{3y-2x}{2y-3x} \quad \text{...(i)}$$

Again $x^2+y^2=3xy$

$$\Rightarrow \frac{d}{dy}\left(x^2+y^2\right)=\frac{d}{dy}(3xy)$$

$$\Rightarrow 2x\frac{dx}{dy}+2y=3\left(x+y\frac{dx}{dy}\right)$$

$$\Rightarrow 2x\frac{dx}{dy}-3y\frac{dx}{dy}=3x-2y$$

$$\Rightarrow (2x-3y)\frac{dx}{dy}=3x-2y$$

$$\Rightarrow \frac{dx}{dy}=\frac{3x-2y}{2x-3y} \quad \text{...(ii)}$$

Hence, (i) × (ii) $\Rightarrow \frac{dy}{dx}\cdot\frac{dx}{dy}$

$$=\left(\frac{3y-2x}{2y-3x}\right)\times\left(\frac{3x-2y}{2x-3y}\right)=1$$

7. Show that $y = x + \frac{1}{y} \Rightarrow \left(x^2 - y^2 + 3\right)\frac{dy}{dx} = 1.$

Solution: $y = x + \frac{1}{y}$

$$\Rightarrow y - x = \frac{1}{y}$$

$$\Rightarrow y^2 + x^2 - 2xy = \frac{1}{y^2} \qquad \text{...(i)}$$

Again $y = x + \frac{1}{y}$

$$\Rightarrow y = \frac{xy + 1}{y}$$

$$\Rightarrow y^2 - 1 = xy$$

$$\Rightarrow 2y^2 - 2 = 2xy$$

$$\Rightarrow 2 - 2y^2 = -2xy \qquad \text{...(ii)}$$

Now, $y = x + \frac{1}{y}$

$$\Rightarrow \frac{dy}{dx} = 1 - \frac{1}{y^2}\cdot\frac{dy}{dx}$$

$$\Rightarrow \frac{dy}{dx} + \frac{1}{y^2}\frac{dy}{dx} = 1$$

$$\Rightarrow \left(1 + \frac{1}{y^2}\right)\frac{dy}{dx} = 1$$

$$\Rightarrow \left(1 + y^2 + x^2 - 2xy\right)\frac{dy}{dx} = 1 \text{ (using (i))}$$

$$\Rightarrow \left(1 + y^2 + x^2 + 2 - 2y^2\right)\frac{dy}{dx} = 1 \text{ (using (ii))}$$

$$\Rightarrow \left(x^2 - y^2 + 3\right)\frac{dy}{dx} = 1$$

8. Show that $e^x + e^y = e^{x+y} \Rightarrow \frac{dx}{dy} = -e^{y-x}$

Solution: $e^x + e^y = e^{x+y}$

Dividing both sides of the given equation by e^{x+y}, we have

$$e^{-y} + e^{-x} = 1$$

$$\Rightarrow \frac{d}{dx}\left(e^{-y} + e^{-x}\right) = \frac{d}{dx}(1)$$

$$\Rightarrow -e^{-y}\frac{dy}{dx} - e^{-x} = 0$$

$$\Rightarrow -e^{-y}\frac{dy}{dx} = e^{-x}$$

$$\Rightarrow \frac{dy}{dx} = -\frac{e^{-x}}{e^{-y}} = -e^{-x}e^{y}$$

Conditional identities based on a given implicit function of *x*

Exercise 12.4

1. If $y = x \sin y$, show that $x\frac{dy}{dx} = \frac{y}{1 - \cos y}$.

2. If $\cos y = x \cos (a + y)$, show that $\frac{dy}{dx} = \frac{\cos^2(a + y)}{\sin a}$.

3. If $\sin y = x \sin (a + y)$, show that $\frac{dy}{dx} = \frac{\sin^2(a + y)}{\sin a}$.

4. If $y = x + \frac{1}{x}$, show that $x\frac{dy}{dx} + y = 2x$.

5. If $y = \sqrt{x} + \frac{1}{x}$, show that $2x\frac{dy}{dx} + y = 2\sqrt{x}$.

6. If $y = \frac{1}{x}$, show that $\frac{dy}{\sqrt{1 + y^4}} + \frac{dx}{\sqrt{1 + x^4}} = 0$.

7. If $\sqrt{1+x^2} + \sqrt{1+y^2} = a(x - y)$, show that $\frac{dy}{dx} = \frac{\sqrt{1 + y^2}}{\sqrt{1 + x^2}}$.

8. If $y = \sqrt{\dfrac{1-x}{1+x}}$, show that $(1-x^2)\dfrac{dy}{dx} + y = 0$.

9. If $y = \sqrt{(1-x)(1+x)}$, show that $(1-x^2)\dfrac{dy}{dx} + xy = 0$.

10. If $y = x^{\frac{1}{x}}$, show that $\dfrac{dy}{dx}$ vanishes when $x = e$.

11. If $y = x + \dfrac{1}{y}$, show that $(x^2 - y^2 + 3)\dfrac{dy}{dx} = 1$.

12. $\sqrt{1-x^2} + \sqrt{1-y^2} = a(x-y)$, show that $\dfrac{dy}{dx} = \sqrt{\dfrac{1-y^2}{1-x^2}}$.

13. If $x^m y^n = (x+y)^{m+n}$, show that $\dfrac{dy}{dx} = \dfrac{y}{x}$.

14. If $y = e^{x-y}$, show that $\dfrac{dy}{dx} = \dfrac{y}{1+y}$.

15. If $e^{xy} - 4xy = 2$, show that $\dfrac{dy}{dx} = -\dfrac{y}{x}$.

16. If $x = y\log(xy)$, show that $\dfrac{dy}{dx} = \dfrac{y}{x}\left(\dfrac{x-y}{x+y}\right)$.

17. If $y = x\log(xy)$, show that $\dfrac{dy}{dx} = \dfrac{y}{x}\left(\dfrac{y+x}{y-x}\right)$.

18. If $xy - \log(xy) = \log 2$, show that $\dfrac{dy}{dx} = -\dfrac{y}{x}$.

19. If $x^y = e^{x-y}$, show that $\dfrac{dy}{dx} = \dfrac{\log x}{(1+\log x)^2}$.

20. If $y = \tan^{-1}\left(\dfrac{\sin x}{1+\cos x}\right)$, show that $\dfrac{dy}{dx} = \dfrac{1}{2}$.

21. If $x = b\cos^{-1}\left(\sqrt{\dfrac{y}{b}}\right) + (by - y^2)^{\frac{1}{2}}$, show that $\dfrac{dy}{dx} = \dfrac{\sqrt{by-y^2}}{y}$.

22. If $x\sqrt{1+y} + y\sqrt{1+x} = 0$, show that $\dfrac{dy}{dx} = -(1+x)^{-2}$.

23. If $y = \tan^{-1}\sqrt{\dfrac{1-x^2}{1+x^2}}$, and $t = \cos^{-1}$, x^2, show that $\dfrac{dy}{dx} = \dfrac{1}{2}$.

24. If $x^y = e^{-x+y}$, show that $\dfrac{dy}{dx} = \dfrac{2-\log x}{(1-\log x)^2}$.

25. If $y\sqrt{x^2+1} = \log\left(\sqrt{x^2+1} - x\right)$, show that $(x^2+1)\dfrac{dy}{dx} + xy + 1 = 0$.

26. If $\sin y = x\sin(x+m)$, show that $\dfrac{dy}{dx} = \dfrac{\sin^2(m+y)}{\sin m}$.

27. If $u = \sin^{-1}(x-y)$, $x = 3^t$, $y = 4t^3$, show that $\dfrac{du}{dt} = 3(1-t^2)^{-\frac{1}{2}}$.

28. If $y = \sin(2\sin^{-1}x)$, show that $\dfrac{dy}{dx} = 2\sqrt{\dfrac{1-y^2}{1-x^2}}$.

29. If $p = \cos^{-1}\left(\dfrac{3+5\cos x}{5+3\cos x}\right)$, show that $(5+3\cos x)\dfrac{dp}{dx} = 4$.

30. If $y = \tan^{-1}\left(\dfrac{x\sin a}{1 - x\cos a}\right)$, show that $(1 - 2x\cos a + x^2)\dfrac{dy}{dx} = \sin a$.

31. If $\left(\dfrac{x}{a}\right)^m + \left(\dfrac{y}{b}\right)^n = 1$, show that $\dfrac{dy}{dx} = -\dfrac{b}{a}$ at (a, b).

32. If $x = a\sin 2t\,(1 + \cos 2t), y = a\cos 2t\,(1 - \cos 2t)$, show that $\dfrac{dy}{dx} = 1$, when $t = \dfrac{\pi}{4}$.

33. If $y = \log\left(x - 3 + \sqrt{x^2 - 6x + 1}\right)$, show that $\dfrac{dy}{dx} = \left(x^2 - 6x + 1\right)^{-\frac{1}{2}}$.

Parametric Differentiation

Firstly, we recall the basic definitions in connection with parametric differentiation.

1. Parametric differentiation: The equations put in any one of the following forms:

(i) $x = f_1(t), y = f_2(t); t \in [T_1, T_2]$

(ii) $x = f_1(t), y = f_2(t); t \in [T_1, T_2]$

(iii) $x = f_1(t), y = f_2(t)$

which tells x and y are seperately differentiable functions of (depending upon) the same variable 't' (called the parameter) are said to be parametric equations of the curve or simply parametric equations. e.g.,

(i) $x = t, y = t^2$; t being a parameter.

(ii) $x = ct^2$, $y = ct$; where c is a constant and t is a parameter.

(iii) $x = r\cos\theta,\ y = r\sin\theta$; r being a constant and θ being a parameter.

(iv) $x = a\sec\theta,\ y = b\tan\theta,\ \theta \in \left(0, \dfrac{\pi}{2}\right)$, $a>b>0$

(v) $x = e^t(\sin t + \cos t),\ y = e^t(\sin t - \cos t),\ t \in \left(-\dfrac{\pi}{4}, \dfrac{\pi}{4}\right)$

(vi) $x = a(t - \sin t),\ y = a(1 - \cos t),\ t \in [0, 2\pi]$

Notes:

1. Actually parametric equations $x = f_1(t), y = f_2(t)$; represents the x and y co-ordinates of a variable point p on a given curve.

2. When the parameter 't' is eliminated by any process from the parametric equations $x = f_1(t), y = f_2(t)$; the cartesian equation of the curve put in the form $F(x, y) = c$ or $F(x, y) = 0$ is obtained. e.g.,

1. If we have the equations $x = r\cos\theta, y = r\sin\theta$; where r is a constant and θ is a parameter, these equations can be expressed in cartesian equation by the eliminating 'θ' using the mathematical manipulation of squaring and adding the separate equations for x and y, i.e. (***Note***: $x = f_1(t)$ and $y = f_2(t)$ must be differentiable on a common domain.)

$$x = r\cos\theta \Rightarrow x^2 = r^2\cos^2\theta \quad \ldots(1)$$

$$y = r\sin\theta \Rightarrow y^2 = r^2\sin^2\theta \quad \ldots(2)$$

$$\therefore (1) + (2) \Rightarrow x^2 + y^2 = r^2\left(\sin^2\theta + \cos^2\theta\right) = r^2$$

which is the cartesian equation of the circle.

2. The point of the curve $x = f_1(t), y = f_2(t)$; found by giving a special value, say t_1, to the parameter t is called shortly "the point t_1" whereas it is not unusual for the letter 't' itself to be used instead of t_1 for a special point.

3. Any other letter θ, s, u, etc can be used to represent the parameter instead of t in parametric equations.

Question: What is parametric differentiation?

Answer: Finding the derivative of the parametric equations is called parametric differentiation.

Or, more explicitly,

"Finding the derivative $\dfrac{dy}{dx}$ of the given parametric equations put in the form (i) $x = f_1(t), y = f_2(t)$; (ii) $x = f_2(t),\ y = f_2(t),\ t \in (T_1, T_2)$ (iii) $x = f_1(t), y = f_2(t)$; $t \in [T_1, T_2]$ without eliminating the parameter 't' (or, any other parameter θ, s, y, etc given the parametric equations) by any process is called parametric differentiation".

Question: Using the definition, find the differential coefficient of parametric equations

$x = f(t)$

$y = g(t)$;

where x and y are differentiable functions (depending upon) a single variable 't' (called parameter)

Solution: $x = f(t) \Rightarrow x + \Delta x = f(t + \Delta t)$ ...(1)

$y = g(t) \Rightarrow y = \Delta y = g(t + \Delta t)$ ...(2)

(1) $\Delta x = f(t + \Delta t) - f(t)$ ...(3)

(2) $\Rightarrow \Delta y = g(t + \Delta t) - g(t)$ ...(4)

Now dividing (3) by Δt, we obtain

$$\frac{\Delta x}{\Delta t} = \frac{f(t + \Delta t) - f(t)}{\Delta t} \quad ...(5)$$

and dividing (4) by Δt, we obtain

$$\frac{\Delta y}{\Delta t} = \frac{g(t + \Delta t) - g(t)}{\Delta t} \quad ...(6)$$

Again dividing (6) by (5), we obtain

$$\frac{\Delta y}{\Delta x} = \frac{\frac{\Delta y}{\Delta t}}{\frac{\Delta x}{\Delta t}} = \frac{\frac{g(t + \Delta t) - g(t)}{\Delta t}}{\frac{f(t + \Delta t) - f(t)}{\Delta t}} \quad ...(7)$$

lastly, taking the limit as $\Delta x \to 0, \Delta t \to 0, \Delta y \to 0$, we have form (7),

$$\lim_{\Delta x \to 0} \frac{\Delta y}{\Delta x} = \frac{\lim\limits_{\Delta t} \frac{\Delta y}{\Delta t}}{\lim\limits_{\Delta t \to 0} \frac{\Delta x}{\Delta t}}$$

$$= \frac{\lim\limits_{\Delta t \to 0} \frac{g(t + \Delta t) - g(t)}{\Delta t}}{\lim\limits_{\Delta t \to 0} \frac{f(t + \Delta t) - f(t)}{\Delta t}}$$

$$\Rightarrow \frac{dy}{dx} = \frac{\frac{dy}{dt}}{\frac{dx}{dt}} = \frac{\frac{d}{dt}(f(t))}{\frac{d}{dt}(f(t))}$$

$$= \frac{g'(t)}{f'(t)}, \; f'(t) \neq 0$$

Hence, the general rule for the differential coefficient of $g(t)$ with respect to $f(t)$ is $\frac{dy}{dx} = \frac{g'(t)}{f'(t)}$ being valid for all values of t such that $f'(t) \neq 0$ which can be stated in words in the following way:

Derivative of y with respect x is equal to the quotient of the derivative of y with to t and the derivative of x with respect to t.

Working rule of find the differential coefficient of parametric equations

To find the differential coefficient of the given parametric equations put in any one of the forms:

(i) $x = f_1(t), y = f_2(t); t \in (T_1, T_2)$

(ii) $x = f_1(t), y = f_2(t): t \in (T_1, T_2)$

(iii) $x = f_1(t), y = f_2(t)$ we have the rule which consists of the following steps.

Step 1: Find the derivative of 'y' with respect to 't'.

Step 2: Find the derivative of 'x' with respect to the same parameter 't'.

Step 3: Divide the derivative of 'y' with respect to 't' by the derivative of x with respect to 't'.

Step 4: The quotient obtained in step (3) is the required derivative of y with respect to x.

Solved Examples

Find $\frac{dy}{dx}$ from the following equations.

1. $x = 3\cos\theta$

$y = 4\sin\theta$

Solution: $y = 4\sin\theta$

$$\Rightarrow \frac{dy}{d\theta} = 4\cos\theta \quad ...(i)$$

$$\Rightarrow \frac{dx}{d\theta} = -3\sin\theta \quad ...(ii)$$

> ***Note***: Here $x = f_1(\theta)$ and $y = f_2(\theta)$ where $f_1(\theta) = 3\cos\theta$ and $f_2(\theta) = 4\sin\theta$.

$$\therefore \frac{(1)}{(2)} \Rightarrow \frac{dy}{dx} = -\frac{4\cos\theta}{3\sin\theta} = -\frac{4}{3}\cot\theta$$

2. $x = 2 - 3\sin\theta$

$y = 3 + 2\cos\theta$

Solution: $y = 3 + 2\cos\theta \Rightarrow \frac{dy}{d\theta}$

$$= 0 + (-2\sin\theta) = -2\sin\theta \qquad \text{...(i)}$$

$$x = 2 - 3\sin \Rightarrow \frac{dy}{d\theta}$$

$$= 0 - 3\cos\theta = -3\cos\theta \qquad \text{...(ii)}$$

$$\therefore \frac{(1)}{(2)} \Rightarrow \frac{dy}{dx} = \frac{-2\sin\theta}{-3\cos\theta} = \frac{2}{3}\tan\theta$$

3. $x = a(t + \sin t)$
$y = a(1 - \cos t)$

Solution: $y = a(1 - \cos t) \Rightarrow \frac{dy}{dt}$

$$= a(0 - (-\sin t)) = a\sin t \qquad \text{...(i)}$$

$$x = a(t + \sin t) \Rightarrow \frac{dx}{dt} = a(1 + \cos t) \qquad \text{...(ii)}$$

$$\therefore \frac{(1)}{(2)} \Rightarrow \frac{dy}{dx} = \frac{a\sin t}{a(1+\cos t)} = \frac{a\sin t}{a\left(1 + 2\cos^2\frac{t}{2} - 1\right)}$$

$$= \frac{a \cdot 2\sin\frac{t}{2}\cdot\cos\frac{t}{2}}{a\cdot 2\cos\frac{t}{2}\cdot\cos\frac{t}{2}} = \tan\frac{t}{2}$$

4. $x = \log t + \sin t$
$y = e^t + \cos t$

Solution: $y = e^t + \cos t \Rightarrow \frac{dy}{dx} = e^t - \sin t \qquad \text{...(i)}$

$$x = \log t + \sin t \Rightarrow \frac{dx}{dt} = \frac{1}{t} + \cos t \qquad \text{...(ii)}$$

$$\therefore \frac{(1)}{(2)} \Rightarrow \frac{dy}{dx} = \frac{e^t - \sin t}{\frac{1}{t} + \cos t}$$

$$= \frac{e^t - \sin t}{\left(\frac{1 + t\cos t}{t}\right)} = \left(e^t - \sin t\right)\cdot\frac{t}{(1 + t\cos t)}$$

$$= \frac{t\left(e^t - \sin t\right)}{(1 + t\cos t)}$$

5. $x = \frac{2t}{1+t^2},\ y = \frac{1-t^2}{1+t^2}$

Solution: $y = \frac{1-t^2}{1+t^2}$

$$\Rightarrow \frac{d}{dt}(y) = \frac{\left(1+t^2\right)\left[\frac{d}{dt}\left(1-t^2\right)\right] - \left(1-t^2\right)\left[\frac{d}{dt}\left(1+t^2\right)\right]}{\left(1+t^2\right)^2}$$

$$= \frac{\left(1+t^2\right)(-2t) - \left(1-t^2\right)\cdot 2t}{\left(1+t^2\right)^2}$$

$$= \frac{-2t - 2t^3 - 2t + 2t^3}{\left(1+t^2\right)^2} = \frac{-4t}{\left(1+t^2\right)^2} \qquad \text{...(i)}$$

$$\frac{d}{dt}(x) = \frac{\left(1+t^2\right)\cdot\frac{d}{dt}(2t) - 2t\cdot\frac{d}{dt}\left(1+t^2\right)}{\left(1+t^2\right)^2}$$

$$= \frac{\left(1+t^2\right)\cdot 2 - 2t\cdot(2t)}{\left(1+t^2\right)^2} = \frac{2 + 2t^2 - 4t^2}{\left(1+t^2\right)^2}$$

$$= \frac{2 - 2t^2}{\left(1+t^2\right)^2} = \frac{2\left(1-t^2\right)}{\left(1+t^2\right)^2} \qquad \text{...(ii)}$$

$$\therefore \frac{(1)}{(2)} \Rightarrow \frac{dy}{dx} = \frac{-4t}{\left(1+t^2\right)^2} \times \frac{\left(1+t^2\right)^2}{2\left(1-t^2\right)}$$

$$= -\frac{4t}{2\left(1-t^2\right)} = -\frac{2t}{\left(1-t^2\right)}$$

6. $x = \sin^{-1} \dfrac{2t}{1+t^2},\ y = \cos^{-1} \dfrac{1-t^2}{1+t^2}$

Solution: $y = \cos^{-1}\left(\dfrac{1-t^2}{1+t^2}\right)$

$$\Rightarrow \cos y = \frac{1-t^2}{1+t^2}$$

$$\Rightarrow \frac{d}{dt}(\cos y) = \frac{\left(1+t^2\right)\frac{d}{dt}\left(1-t^2\right) - \left(1-t^2\right)\frac{d}{dt}\left(1+t^2\right)}{\left(1+t^2\right)^2}$$

$$\Rightarrow -\sin y \cdot \frac{dy}{dt} = \frac{\left(1+t^2\right)(-2t) - \left(1-t^2\right)\cdot 2t}{\left(1+t^2\right)^2}$$

$$\Rightarrow -\sin y \cdot \frac{dy}{dt} = \frac{-2t - 2t^3 - 2t + 2t^3}{\left(1+t^2\right)^2} = -\frac{4t}{\left(1+t^2\right)^2}$$

$$\Rightarrow \frac{dy}{dt} = \frac{4t}{\left(1+t^2\right)^2} \cdot \frac{1}{\sin y} = \frac{4t}{\left(1+t^2\right)^2} \cdot \frac{1}{\sqrt{1-\cos^2 y}}$$

$$= \frac{4t}{\left(1+t^2\right)^2} \cdot \frac{1}{\sqrt{1-\left(\frac{1-t^2}{1+t^2}\right)^2}}$$

$$= \frac{4t}{\left(1+t^2\right)^2} \cdot \frac{1}{\sqrt{\frac{\left(1+t^2\right)^2 - \left(1-t^2\right)^2}{\left(1+t^2\right)^2}}}$$

$$= \frac{4t}{\left(1+t^2\right)} \cdot \frac{\left(1+t^2\right)}{\sqrt{4t^2}}$$

$$= \frac{4t}{\left(1+t^2\right)^2} \cdot \frac{\left(1+t^2\right)}{2|t|} = \frac{2t}{\left(1+t^2\right)|t|},\ t \neq 0 \quad \text{...(i)}$$

Again $x = \sin^{-1} \dfrac{2t}{1+t^2}$

$$\Rightarrow \sin x = \frac{2t}{1+t^2}$$

$$\Rightarrow \frac{d}{dt}(\sin x) = \frac{\left(1+t^2\right)\frac{d}{dt}(2t) - 2t\frac{d}{dt}\left(1+t^2\right)}{\left(1+t^2\right)^2}$$

$$= \frac{2\left(1+t^2\right) - 2t \cdot 2t}{\left(1+t^2\right)^2}$$

$$\Rightarrow \cos x \frac{dx}{dt} = \frac{2+2t^2-4t^2}{\left(1+t^2\right)^2} = \frac{2-2t^2}{\left(1+t^2\right)^2} = \frac{2\left(1-t^2\right)}{\left(1+t^2\right)^2}$$

$$\Rightarrow \frac{dx}{dt} = \frac{2\left(1-t^2\right)}{\left(1+t^2\right)^2} \cdot \frac{1}{\cos x} = \frac{2\left(1-t^2\right)}{\left(1+t^2\right)^2} \cdot \frac{1}{\sqrt{1-\sin^2 x}}$$

$$= \frac{2\left(1-t^2\right)}{\left(1+t^2\right)^2} \cdot \frac{1}{\sqrt{1-\left(\frac{2t}{1+t^2}\right)^2}}$$

$$=\frac{2(1-t^2)}{(1+t^2)^2}\cdot\frac{1}{\sqrt{\frac{1+t^4+2t^2-4t^2}{(1+t^2)^2}}}$$

$$=\frac{2(1-t^2)}{(1+t^2)^2}\cdot\frac{(1+t^2)}{\sqrt{(1-t^2)^2}},\ t^2\neq 1$$

$$=\frac{2(1-t^2)}{(1+t^2)^2}\cdot\frac{(1+t^2)}{|(1-t^2)|}=\frac{2(1-t^2)}{(1+t^2)|(1-t^2)|}\quad ...(ii)$$

$$\therefore \frac{(1)}{(2)}\Rightarrow\frac{dy}{dx}=\frac{2t}{(1+t^2)|t|}\cdot\frac{(1+t^2)|(1-t^2)|}{2(1-t^2)}$$

$$=\frac{t|(1-t^2)|}{(1-t^2)|t|},\ t\neq 0,\pm 1$$

7. $x = e^t(\sin t + \cos t)$, $y = e^t(\sin t - \cos t)$,

$t\in\left(-\frac{\pi}{4},\frac{\pi}{4}\right)$

Solution: $y = e^t(\sin t - \cos t)$

$$\Rightarrow \frac{dy}{dx} = \frac{d}{dt}\left(e^t(\sin t - \cos t)\right) = e^t(\cos t + \sin t) +$$

$(\sin t - \cos t)\, e^t$

$= e^t(\cot t + \sin t + \sin t - \cos t) = 2\sin t \cdot e^t \quad ...(1)$

and $x = e^t(\sin t + \cos t)$

$$\Rightarrow \frac{dx}{dt} = \frac{d}{dt}\left(e^t(\sin t + \cos t)\right)$$

$= e^t(\cos t - \sin t) + (\sin t + \cos t)\, e^t$

$= e^t(\cos t - \sin t + \sin t + \cos t)$

$= 2\cos t \cdot e^t \quad ...(2)$

$$\therefore \frac{(1)}{(2)}\Rightarrow\frac{dy}{dx}=\frac{2\sin t\cdot e^t}{2\cos t\cdot e^t}=\tan t,\ t\in\left(-\frac{\pi}{4},\frac{\pi}{4}\right)$$

8. $\left.\begin{aligned} x &= a\cos t + \frac{a}{2}\log\left(\tan^2\frac{t}{2}\right)\\ y &= a\sin t\end{aligned}\right|\ t\in\left(0,\frac{\pi}{2}\right)$

Solution: $y = a\sin t \Rightarrow \frac{dy}{dt}$

$$=\frac{d}{dt}(a\sin t) = a\cos t \quad ...(1)$$

$$x = a\cos t + \frac{a}{2}\log\left(\tan^2\frac{t}{2}\right)$$

$$\Rightarrow\frac{dy}{dt}=-a\sin t+\frac{a}{2}\cdot\frac{1}{\tan^2\frac{t}{2}}\cdot 2\tan\frac{t}{2}\left(\sec^2\frac{t}{2}\right)\frac{1}{2}$$

$$=-a\sin t+\frac{a\sec^2\frac{t}{2}}{2\tan\frac{t}{2}}=-a\sin t+\frac{a}{2\sin\frac{t}{2}\cos\frac{t}{2}}$$

$$=-a\sin t+\frac{a}{\sin t}$$

$$=a\left(\frac{1}{\sin t}-\sin t\right)=a\frac{(1-\sin^2 t)}{\sin t}=\frac{a\cos^2 t}{\sin t}\quad ...(2)$$

$$\therefore\frac{(1)}{(2)}\Rightarrow\frac{dy}{dx}=\frac{a\cos t\cdot\sin t}{a\cos^2 t}=\frac{\sin t}{\cos t}=\tan t,\ t\in\left(0,\frac{\pi}{2}\right)$$

9. $\left.\begin{aligned} x &= a(t-\sin t)\\ y &= a(1-\cos t)\end{aligned}\right|\ t\in[0,2\pi]$

Solution: $y = a(1-\cos t)\Rightarrow\frac{dy}{dt} = a\sin t \quad ...(1)$

$$x = a(t-\sin t)\Rightarrow\frac{dx}{dt}=a(1-\cos t)\quad ...(2)$$

$$\therefore\frac{(1)}{(2)}\Rightarrow\frac{dy}{dx}=\frac{a\sin t}{a(1-\cos t)}$$

$$= \frac{2\sin\left(\frac{t}{2}\right)\cos\left(\frac{t}{2}\right)}{\left(1-1+2\sin^2\left(\frac{t}{2}\right)\right)} = \frac{\cos\left(\frac{t}{2}\right)}{\sin\left(\frac{t}{2}\right)} = \cot\left(\frac{t}{2}\right),$$

for $t \neq 0, 2\pi$

Problems based on parametric equations

Exercise 12.5

Find $\frac{dy}{dx}$ from the following equations.

1. $x = a\left(\cos\theta + \log\tan\frac{\theta}{2}\right)$
   $y = a\sin\theta$
2. $x = 2\cos\theta - \cos 2\theta$
   $y = 2\sin\theta - \sin 2\theta$
3. $x = a\cos\varnothing$
   $y = b\sin\varnothing$
4. $x = a\sec\varnothing$
   $y = b\tan\varnothing$
5. $x = a\cos^3\theta$
   $y = b\sin^3\theta$
6. $x = a\sec^2\theta$
   $y = a\tan^3\theta$
7. $x = a(\cos t + t\sin t)$
   $y = a(\sin t - t\cos t)$
8. $x = a(\theta - \sin\theta)$
   $y = a(1 + \cos\theta)$
9. $x = a(\theta - \sin\theta)$
   $y = a(1 - \cos\theta)$
10. $x = 2\cos^2\theta$
    $y = 3\sin^2\theta$
11. $x = a\log t$
    $y = b\sin t$
12. $x = \tan^{-1} t$
    $y = t\sin 2t$
13. $x = at^2$
    $y = 2at$
14. $x = a\left(\frac{1-t^2}{1+t^2}\right)$
    $y = b\left(\frac{2t}{1+t^2}\right)$
15. $x = \frac{3at}{1+t^3}$
    $y = \frac{3at^2}{1+t^3}$
16. $x = \frac{2t}{1+t^2}$
    $y = \frac{1-t^2}{1+t^2}$
17. $x = \sqrt{1-t}$
    $y = \sqrt{1+t}$
18. $x = u^2$
    $y = \frac{u^3 - 3u + 5}{1+u^2}$
19. $x = \frac{a}{m^2}$
    $y = \frac{2a}{m}$
20. $x = a\sqrt{\frac{1-t}{1+t}}$
    $y = a\cdot t\cdot\sqrt{\frac{1-t}{1+t}}$
21. $x = 2t - |t|$
    $y = t^2 + t|t|$
22. $x = a(\sin\theta + \cos\theta)$
    $y = a(\sin\theta - \cos\theta)$

23. $x = a\sqrt{\dfrac{t^2 - 1}{t^2 + 1}}$

$y = a \cdot t \cdot \sqrt{\dfrac{t^2 - 1}{t^2 + 1}}$

Answers (with proper restrictions on the parameters)

1. $\tan\theta$

2. $\tan\dfrac{3\theta}{2}$

3. $-\dfrac{b}{a}\cot\phi$

4. $\dfrac{b}{a}\operatorname{cosec}\phi$

5. $-\dfrac{b}{a}\tan\theta$

6. $\dfrac{3}{2}\tan\theta$

7. $\tan t'$

8. $-\cot\dfrac{\theta}{2}$

9. $\cot\dfrac{\theta}{2}$

10. $-\dfrac{3}{2}$

11. $\dfrac{b}{a}t\cos t$

12. $(2t\cos 2t + \sin 2t)(1 + t^2)$

13. $\dfrac{1}{t}$

14. $-\dfrac{b}{a}\cdot\left(\dfrac{1 - t^2}{2t}\right) = \left(-\dfrac{b^2}{a^2}\cdot\dfrac{x}{y}\right)$

15. $\dfrac{t(2 - t^3)}{(1 - 2t^3)}$

16. $\dfrac{-2t}{1 - t^2}$

17. $-\sqrt{\dfrac{1 - t}{1 + t}}$

18. $\dfrac{u^4 + 6u^2 - 10u - 3}{2u(u^2 + 1)^2}$

19. m

20. $t^2 + t - 1$

21. $\dfrac{2t + 2|t|}{(2t - |t|)}\cdot t$

22. $\dfrac{\cos\theta + \sin\theta}{\cos\theta - \sin\theta}$

23. $\dfrac{t^4 + 2t^2 - 1}{2t}$

Exercise 12.6

Find $\dfrac{dy}{dx}$ from the following equations.

1. $\left.\begin{aligned} x &= a(t - \sin t) \\ y &= a(1 - \cos t)\end{aligned}\right| t \in [0, 2\pi],\ a \neq 0$

2. $\left.\begin{aligned} x &= a\cos t \\ y &= a\sin t\end{aligned}\right| t \in (0, \pi),\ a \neq 0$

3. $\left.\begin{aligned} x &= a\sec\theta \\ y &= b\tan\theta\end{aligned}\right| \theta \in \left(0, \dfrac{\pi}{2}\right),\ a > b > 0$

4. $\left.\begin{aligned} x &= e^t(\sin t + \cos t) \\ y &= e^t(\sin t - \cos t)\end{aligned}\right| t \in \left[-\dfrac{\pi}{4}, \dfrac{\pi}{4}\right]$

5. $\left.\begin{aligned} x &= \theta + \sin\theta \\ y &= 1 - \cos\theta\end{aligned}\right| \theta \in (-\pi, \pi)$

6. $\left.\begin{array}{l} x = a\cos^3 t \\ y = a\sin^3 t \end{array}\right| t \in \left[0, \frac{\pi}{2}\right), a \neq 0$

7. $\left.\begin{array}{l} x = at^2 \\ y = 2at \end{array}\right| t > 0, a \neq 0$

8. $\left.\begin{array}{l} x = \dfrac{1 - t^2}{1 + t^2} \\ y = \dfrac{2t}{1 + t^2} \end{array}\right| t > 0$

9. $\left.\begin{array}{l} x = a(\cos\theta + \theta\sin\theta) \\ y = a(\sin\theta - \theta\cos\theta) \end{array}\right| \theta \in \left(\frac{\pi}{2}, \pi\right), a \neq 0$

Answers

1. $\cot\left(\frac{t}{2}\right), t \in (0, 2\pi)$

2. $-\cot t, t \in (0, \pi)$

3. $\frac{b}{a}\operatorname{cosec}\theta, \theta \in \left(0, \frac{\pi}{2}\right)$

4. $\tan t, t \in \left[-\frac{\pi}{4}, \frac{\pi}{4}\right]$

5. $\tan\left(\frac{\theta}{2}\right), \theta \in (-\pi, \pi)$

6. $-\tan t, t \in \left(0, \frac{\pi}{2}\right)$

7. $\frac{1}{t}, t > 0$

8. $-\frac{1 - t^2}{2t}, t > 0$

9. $\tan\theta, \theta \in \left(\frac{\pi}{2}, \pi\right]$

To find the differential coefficient of a function with respect to an other function

Question: Using the definition, find the differential coefficient of a function $f(x)$ with respect to another function $g(x)$, where $f(x)$ and $g(x)$ both are differentiable functions having the same independent variable x.

Solution: Let $y = f(x)$ be a differentiable function of x ...(1)

and $z = g(x)$ be another differentiable function of x ...(2)

Now, $y + \Delta y = f(x + \Delta x)$ ...(3)

and $z + \Delta z = g(x + \Delta x)$ ...(4)

From (3), we have $\Delta y = f(x + \Delta x) - f(x)$ ...(5)

and from (4), we have $\Delta z = g(x + \Delta x) - g(x)$ ...(6)

Dividing (5) by Δx, we obtain

$$\frac{\Delta y}{\Delta x} = \frac{f(x + \Delta x) - f(x)}{\Delta x} \quad ...(7)$$

Dividing (6) by Δx, we obtain

$$\frac{\Delta z}{\Delta x} = \frac{g(x + \Delta x) - g(x)}{\Delta x} \quad ...(8)$$

Dividing (7) by (8), we obtain

$$\frac{\Delta y}{\Delta z} = \frac{\dfrac{\Delta y}{\Delta x}}{\dfrac{\Delta z}{\Delta x}} = \frac{\dfrac{f(x + \Delta x) - f(x)}{\Delta x}}{\dfrac{g(x + \Delta x) - g(x)}{\Delta x}}$$

lastly, taking the limit as $\Delta x \to 0, \Delta y \to 0, \Delta z \to 0$, we obtain from (9),

$$\lim_{\Delta x \to 0} \frac{\Delta y}{\Delta z} = \frac{\lim\limits_{\Delta x \to 0} \dfrac{\Delta y}{\Delta x}}{\lim\limits_{\Delta x \to 0} \dfrac{\Delta z}{\Delta x}}$$

$$= \frac{\lim\limits_{\Delta x \to 0} \dfrac{f(x + \Delta x) - f(x)}{\Delta x}}{\lim\limits_{\Delta x \to 0} \dfrac{g(x + \Delta x) - g(x)}{\Delta x}}$$

$$\Rightarrow \frac{dy}{dz} = \frac{\frac{dy}{dx}}{\frac{dz}{dx}} = \frac{\frac{d\,f(x)}{dx}}{\frac{d\,g(x)}{dx}} = \frac{f'(x)}{g'(x)} = \frac{dy}{dx}\cdot\frac{dx}{dz}$$

Hence, the general rule for the differentiable coefficient of a differentiable function $f(x)$ with respect to another differentiable function $g(x)$ is

$$\frac{d\,f(x)}{d\,g(x)} = \frac{\frac{d\,f(x)}{dx}}{\frac{d\,g(x)}{dx}}$$

which can be stated in words in the following way.

$$\frac{d\,(\text{the function given to be differentiated})}{d\,(\text{the function w.r.t which we have to find the d.c})}$$

$$= \frac{\text{differential coefficient of the given function w.r.t } x}{\text{differential coefficient w.r.t } x \text{ of the function w.r.t which}}$$

we have to find the d.c. of the given function

Remark: We must note that the function to be differentiated and the other function with respect to which we have to differentiate, both have the same independent variable.

Working rule to differentiate *f*(*x*) w.r.t. *g*(*x*)

To find the differential coefficient of a function $f(x)$ w.r.t. another function $g(x)$, where $f(x)$ and $g(x)$ both are differentiable functions having the same independent variable x, we adopt the rule consisting of following steps.

Step 1: Find the differential coefficient of the given function with respect to the independent variable x which the function to be differentiated has.

Step 2: Find the differential coefficient with respect to the same independent variable x of the other function w.r.t. which d.c. of the given function is required.

Step 3: Divide the differential coefficient obtained in step (1) by the differential coefficient obtained in step (2). And obtain the required differential coefficient of $f(x)$ w.r.t. $g(x)$.

Remember:

Although $\frac{dy}{dx}$ is not a quotient we may interpret $\frac{dy}{dx}$ as a quotient which means we are able to write.

(i) $\frac{dy}{dx} = dy \div dx$

(ii) $\frac{dy}{dx} = dy \times \frac{1}{dx} \Leftrightarrow dy = \frac{dy}{dx} \times dx$

(iii) $\frac{dy}{dx} = \frac{dy}{dx} \div \frac{dz}{dx} = \frac{dy}{dx} \times \frac{dx}{dz}$

Solved Examples

1. $\cos^2 x$ w.r.t. $(\log x)^3$

Solution: Putting $y = \cos^2 x$ ...(1)

and $z = (\log x), x > 0$ ...(2)

We have

$$\frac{dy}{dx} = \frac{d}{dx}\left(\cos^2 x\right) = \frac{d(\cos x)^2}{d\cos x}\cdot\frac{d\cos x}{dx}$$

$$= \cos\cdot(-\sin x) = -2\sin x\cos x = -\sin 2x \quad ...(3)$$

and $$\frac{dz}{dx} = \frac{d(\log x)^3}{d\log x}\cdot\frac{d\log x}{dx}$$

$$= 3(\log x)^2\cdot\frac{1}{x} = \frac{3(\log x)^2}{x} \quad ...(4)$$

Hence, $$\frac{(3)}{(4)} \Rightarrow \frac{dy}{dx}\Big/\frac{dz}{dx} = \frac{dy}{dz} = \frac{-x\sin 2x}{3(\log x)^2},\ x>0$$

Or, alternatively,

$$\frac{d}{dx}\left(\cos^2 x\right) = \frac{d(\cos x)^2}{d\cos x}\cdot\frac{d\cos x}{dx} = -\sin 2x \quad ...(i)$$

$$\frac{d(\log x)^3}{dx} = \frac{d(\log x)^3}{d\log x}\cdot\frac{d\log x}{dx} = \frac{3(\log x)^2}{x} \quad ...(ii)$$

Hence, $$\frac{d\left(\cos^2 x\right)}{d(\log x)^3} = \frac{(1)}{(2)} = \frac{-x\sin 2x}{3(\log x)^2},\ x>0$$

2. $\log x$ w.r.t. x^3

Solution: Putting $y = \log x$ ...(1)

and $z = x^3$ ...(2)

We have

$$\frac{dy}{dx} = \frac{d(\log x)}{dx} = \frac{1}{x}, x > 0 \quad ...(3)$$

and $$\frac{dz}{dx} = \frac{d(x^3)}{dx} = 3x^2 \quad ...(4)$$

Hence, $$\frac{(3)}{(4)} \Rightarrow \frac{dy}{dx} \bigg/ \frac{dz}{dx} = \frac{dy}{dz} = \frac{1}{x} \cdot \frac{1}{3x^2} = \frac{1}{3x^3},$$

$x > 0$.

3. e^x w.r.t. $\sqrt{x}$

Solution: Putting $y = e^x$ ...(1)

and $z = \sqrt{x}$ ...(2)

We have

$$\frac{dy}{dx} = \frac{d}{dx}(e^x) = e^x \quad ...(3)$$

and $$\frac{dz}{dx} = \frac{d}{dx}(\sqrt{x}) = \frac{1}{2\sqrt{x}}, x > 0 \quad ...(4)$$

Thus, $$\frac{(3)}{(4)} \Rightarrow \frac{dy}{dx} \bigg/ \frac{dz}{dx} = \frac{2e^x\sqrt{x}}{1} = 2e^x\sqrt{x}, x > 0$$

4. $\sqrt{\sin(1+x^2)^2}$ w.r.t $(1+x^2)$

Solution: Putting $y = \sqrt{\sin(1+x^2)^2}$ ...(1)

and $z = (1 + x^2)$ ...(2)

We have

$$\frac{dy}{dx} = \frac{d\left(\sqrt{\sin(1+x^2)^2}\right)}{dx}$$

$$= \frac{d\left(\sqrt{\sin(1+x^2)^2}\right)}{d\left(\sin(1+x^2)^2\right)} \cdot \frac{d\left(\sin(1+x^2)^2\right)}{d(1+x^2)^2} \cdot \frac{d(1+x^2)^2}{d(1+x^2)} \cdot \frac{d(1+x^2)}{dx}$$

$$= \frac{1}{2\sqrt{\sin(1+x^2)^2}} \cdot \cos(1+x^2)^2 \cdot 2(1+x^2) \cdot 2x$$

$$= \frac{2x(1+x^2)\cos(1+x^2)^2}{\sqrt{\sin(1+x^2)^2}} \quad ...(3)$$

and $$\frac{dz}{dx} = \frac{d(1+x^2)}{dx} = 2x \quad ...(4)$$

$$\therefore \frac{(3)}{(4)} \Rightarrow \frac{2x \cdot (1+x^2) \cdot \cos(1+x^2)}{\sqrt{\sin(1+x^2)^2}} \Big/ 2x, x \neq 0$$

$$\Rightarrow \frac{dy}{dz} = \frac{\dfrac{2x(1+x^2)\cos(1+x^2)^2}{\sqrt{\sin(1+x^2)^2}}}{2x}$$

$$= \frac{(1+x^2)\cos(1+x^2)^2}{\sqrt{\sin(1+x^2)^2}}, x \neq 0$$

5. $e^{\sin^{-1}x}$ w.r.t. $\sin^{-1}x$

Solution: Putting $y = e^{\sin^{-1}x}$ ...(1)

and $z = \sin^{-1}x$ ...(2)

We have

$$\frac{dy}{dx} = \frac{d\left(e^{\sin^{-1}x}\right)}{dx}$$

$$= \frac{d\left(e^{\sin^{-1}x}\right)}{d(\sin^{-1}x)} \cdot \frac{d\sin^{-1}x}{dx}$$

$$= \frac{e^{\sin^{-1}x}}{\sqrt{1-x^2}},\ x \neq \pm 1 \qquad ...(3)$$

and $\dfrac{dz}{dx} = \dfrac{d\sin^{-1}x}{dx} = \dfrac{1}{\sqrt{1-x^2}};\ x \neq \pm 1 \qquad ...(4)$

$$\therefore \frac{(3)}{(4)} \Rightarrow \frac{dy}{dx}\cdot\frac{dx}{dz} = \frac{e^{\sin^{-1}x}}{\sqrt{1-x^2}}\cdot\left(\sqrt{1-x^2}\right)$$

$$\Rightarrow \frac{dy}{dz} = e^{\sin^{-1}x};\ x \neq \pm 1$$

Problems on differentiation of a function with respect to another function

Exercise 12.7

Differentiable:

1. x^5 w.r.t. x^3
2. x^2+x+3 w.r.t. x^3
3. $\sin x$ w.r.t. $\cos x$
4. $\tan^{-1}x$ w.r.t. x^2
5. $\sin^{-1}x$ w.r.t. $\cos^{-1}x$
6. $x^{\sin^{-1}x}$ w.r.t. $\tan^{-1}x$
7. e^x w.r.t. $\sqrt{x}$
8. $e^{\sin^{-1}x}$ w.r.t. $\sin^{-1}x$
9. $\sin x$ w.r.t. x^3
10. $\sin x$ w.r.t. $\tan x$
11. $\tan x$ w.r.t. $\sin x$
12. $\operatorname{cosec} x$ w.r.t. $\cot x$
13. $\cot x$ w.r.t. $\operatorname{cosec} x$
14. $\tan x$ w.r.t. $\sec x$
15. $\tan x$ w.r.t. $\cot x$
16. $\operatorname{cosec} x$ w.r.t. $\sec x$
17. $\sin x$ w.r.t. $\cos^2 x$
18. $\cos^2 x$ w.r.t. $\sin x$
19. $\dfrac{x}{\sqrt{1+x^2}}$ w.r.t. $\tan x$
20. $\sqrt{\dfrac{1+\cos 2x}{1-\cos 2x}}$ w.r.t. x
21. $\sqrt{\dfrac{1-\tan^2 x}{1+\tan^2 x}}$ w.r.t. x
22. $\dfrac{x}{\sin x}$ w.r.t. $\sin x$
23. e^x w.r.t. $\log x$
24. $\log x^2$ w.r.t. e^x
25. $e^{\sin^{-1}x}$ w.r.t. $\log x$
26. $\sqrt{\sin\left(1+x^2\right)^2}$ w.r.t. $(1+x^2)$
27. $e^{\sin^{-1}x}$ w.r.t. $\cos^{-1}x$
28. $\tan^{-1}x$ w.r.t. $\sin^{-1}x$
29. $\log\sin x$ w.r.t. $\sin^{-1}x$
30. $\dfrac{1-\cos x}{1+\cos x}$ w.r.t. $\dfrac{1+\sin x}{1-\sin x}$

Answers

1. $\dfrac{5}{3}x^2$

2. $\dfrac{2x+1}{3x^2},\ x \neq 0$

3. $-\cot x,\ x \neq n\pi$

4. $\dfrac{1}{2x\left(1+x^2\right)},\ x \neq 0$

5. -1 for $|x| \leq 1$

6. $\left(1+x^2\right)x^{\sin^{-1}x}\left(\dfrac{\log x}{\sqrt{1-x^2}} + \dfrac{\sin^{-1}x}{x}\right),\ 0<x<1$

7. $2\sqrt{x}\,e^x,\ x \neq 0$

8. $e^{\sin^{-1}x},\ |x| \leq 1$

9. $\dfrac{\cos x}{3x^2},\ x \neq 0$

10. $\cos^3 x,\ x \neq n\pi + \dfrac{\pi}{2}$

11. $\sec^3 x,\ x \neq n\pi + \dfrac{\pi}{2}$

12. $\cos x,\ x \neq n\pi$

13. $\sec x,\ x \neq n\pi$

14. $\operatorname{cosec} x,\ x \neq \dfrac{n\pi}{2}$

15. $-\tan^2 x,\ x \neq n\pi + \dfrac{\pi}{2}$

16. $-\cot^3 x,\ x \neq \dfrac{n\pi}{2}$

17. $-\dfrac{1}{2}\operatorname{cosec} x,\ x \neq n\pi$

18. $-2 \sin x$ for all x

19. $\dfrac{\cos^2 x}{\left(1+x^2\right)^{\frac{3}{2}}},\ x \neq n\pi + \dfrac{\pi}{2}$

20. $\dfrac{2}{\cos 2x - 1} \cdot \dfrac{|\cot x|}{\cot x},\ x \neq \dfrac{n\pi}{2}$

21. $\dfrac{-2\tan x}{\sqrt{1-\tan^4 x}},\ x \neq n\pi \pm \dfrac{\pi}{4}$

22. $\dfrac{\sin x - x\cos x}{\sin^2 x \cdot \cos x},\ x \neq \dfrac{n\pi}{2}$

23. $x \cdot e^x, x > 0$

24. $\dfrac{2}{x e^x},\ x > 0$

25. $\dfrac{x \cdot e^{\sin^{-1} x}}{\sqrt{1-x^2}},\ 0 < x < 1$

26. $\dfrac{\left(1+x^2\right)\cos\left(1+x^2\right)^2}{\sqrt{\sin\left(1+x^2\right)^2}}$

27. $-e^{\sin^{-1} x},\ |x| \leq 1$

28. $\dfrac{\sqrt{1-x^2}}{1+x^2},\ |x| \leq 1$

29. $(\cot x)\sqrt{1-x^2},\ 0 < x \leq 1$

30. $\tan x\left(\dfrac{1-\sin x}{1+\cos x}\right)^2$

Differentiation of infinite series (or, continued fraction which $\to \infty$)

Before we find the derivatives of the problems on in finite series or continued fraction which tends to infinity, we must know the definitions of the following terms.

1. *Infinite series*: A series having the number of terms infinite (i.e. not finite) is called an infinite series. An infinite series is written in any one of the following forms.

(i) $x_1 + x_2 + x_3 + \ldots + x_n + \ldots$, where dots denote the existance of similar terms obeying the same rule as its previous terms.

(ii) $\sum x_n$

(iii) $\sum_{i=0}^{\infty} x_i$ or, more briefly $\sum_i x_i$, where x_n and x_i are the general term or nth term of the infinite series.

2. *Continued fraction*: A continued fraction is a fraction expressed as a number plus a fraction whose denominator is a number plus a fraction, i.e.

a continued fraction

$\Rightarrow$ a number + a fraction whose denominator = a number + a fraction

$$\Rightarrow a_1 \cfrac{b_2}{a_2 + \cfrac{b_3}{a_3 + \cfrac{b_4}{a_4 + \cfrac{b_5}{a_5 + \ldots\infty}}}}$$

Notes:

1. A continued fraction may have either a finite or in finite number or terms. A continued fraction having finite number of terms is called terminating continued fraction. A continued fraction having infinite number of terms is called non-terminating continued fraction. If a certain sequence of the a's and b's occurs periodically, the continued fraction is recurring or periodic. The terminating continued fraction

$$a_1,\ a_1+\cfrac{b_2}{a_2+\cfrac{b_3}{a_3}},$$ etc are convergents of the continued fraction whereas the quotients $\frac{b_2}{a_2}, \frac{b_3}{a_3}$ etc are partial quotients. The continued fraction is a simple continued fraction if $b_i = 1, i = 2, 3, \ldots$.

2. For the sake of saving space, it is usual to write a simple continued fraction in the more compact forms as:

$$a+\frac{1}{b+}\frac{1}{c+}\ldots \text{ or, } \frac{1}{a+}\frac{1}{b+}\frac{1}{c+}\ldots$$ instead of writing

$$a+\cfrac{1}{b+\cfrac{1}{c+\ldots}} \text{ or, } \cfrac{1}{a+\cfrac{1}{a+\cfrac{1}{b+\cfrac{1}{c+\ldots}}}},$$ where

it is to be noticed that the letters $a, b, c, \ldots$ all denote integral numbers, the signs are all positive and each of the numerator is unity.

We are now prepared to find the ways of differentiating the non-terminating continued fraction of the form

$$f(x)+\frac{1}{f(x)}+\frac{1}{f(x)}+\frac{1}{f(x)}+\ldots$$

1. If $\quad y = f(x)+\frac{1}{f(x)}+\frac{1}{f(x)}+\ldots$ then

$$y = f(x)+\frac{1}{y}.$$

2. Find $\frac{dy}{dx}$ using the method of differentiating implicit functions.

Remember:

1. In the given series being infinite, we firstly inspect what expression in x terminates or converges and then we retain only one terminating (or, converging) expression in x as it is and the rest same terminating expression is put equal to y (= $l.h.s$). e.g.,

$$y=\sqrt{\underset{\longleftrightarrow}{x+2}\sqrt{\underset{\longleftarrow\cdots\cdots\longrightarrow}{x+2\sqrt{x+\ldots\infty}}}}$$
$$\downarrow \qquad\qquad \downarrow$$

> ***Note***: We inspect that $\sqrt{x+2}$ is the expression which terminates.

retain the first and put the first $= y$

$$\therefore y = \sqrt{x+2y}$$

2. If we are given y = a non-terminating continued fraction put in the form:

$$y = f(x)+\cfrac{1}{f(x)+\cfrac{1}{f(x)+\cfrac{1}{f(x)+\cfrac{1}{f(x)+\ldots\infty}}}},$$

we retain $f(x)$ as it is and the rest is put equal to $\frac{1}{y}$, i.e. reciprocal of $l.h.s.$

$$\therefore y = f(x)+\frac{1}{y}$$

3. Differentiation of a non terminating continued fraction is also termed as differentiation of explicit functions in an "ad infinitum" form and "$\ldots\infty$" is replaced by "ad in f".

e.g.: $$y=\sqrt{x}+\cfrac{1}{\sqrt{x}+\cfrac{1}{\sqrt{x}+\cfrac{1}{\sqrt{x}+\text{ad in } f}}}$$

4. Differentiation of explicit functions in any "ad infinitum" form becomes differentiable after transformation to finite implicit form which is possible as under.

(i) $y = x^{x^{x^{\dots\infty}}} \Rightarrow y = x^y$

(ii) $y = e^{e^{e^{x\dots\infty}}} \Rightarrow y = \left(e^x\right)^y$

(iii) $y = \sqrt{x}^{\sqrt{x}^{\sqrt{x}\dots\infty}} \Rightarrow y = \left(\sqrt{x}\right)^y$ etc.

Problems based on "ad infinitum" form.

Solved Examples

Find $\dfrac{dy}{dx}$ if

1. $y = e^{x+e^{x+e^{x+e^{x\dots\infty}}}}$

Solution: $y = e^{x+e^{x+e^{x+e^{x\dots\infty}}}}$

$$\Rightarrow y = e^{x+y}$$

$$\Rightarrow \log y = \log\left(e^{x+y}\right) = (x+y)\cdot\log_e e = (x+y)$$

$(\because \log e = \log_e e = 1)$

$$\Rightarrow \frac{d}{dx}(\log y) = \frac{d}{dx}(x+y)$$

$$\Rightarrow \frac{1}{y}\frac{dy}{dx} = 1 + \frac{dy}{dx}$$

$$\Rightarrow \frac{1}{y}\frac{dy}{dx} - \frac{dy}{dx} = 1$$

$$\Rightarrow \left(\frac{1}{y} - 1\right)\frac{dy}{dx} = 1$$

$$\Rightarrow \left(\frac{1-y}{y}\right)\frac{dy}{dx} = 1$$

$$\Rightarrow \frac{dy}{dx} = \left(\frac{y}{1-y}\right) \text{ for } y \neq 1, \text{ i.e. } x \neq -1$$

2. $y = \sqrt{\cos x + \sqrt{\cos x + \sqrt{\cos x + \dots\infty}}}$

Solution: $y = \sqrt{\cos x + \sqrt{\cos x + \sqrt{\cos x + \dots\infty}}}$

$$\Rightarrow y = \sqrt{\cos x + y}$$

$$\Rightarrow y^2 = \cos x + y$$

$$\Rightarrow \frac{d}{dx}\left(y^2\right) = \frac{d}{dx}(\cos x + y)$$

$$\Rightarrow 2y\frac{dy}{dx} = -\sin x + \frac{dy}{dx}$$

$$\Rightarrow 2y\frac{dy}{dx} - \frac{dy}{dx} = \sin x$$

$$\Rightarrow (2y-1)\frac{dy}{dx} = -\sin x$$

$$\Rightarrow \frac{dy}{dx} = \frac{-\sin x}{(2y-1)} = \frac{\sin x}{1-2y};\ y \neq \frac{1}{2}, \text{ i.e. } \cos x \neq \frac{-1}{4}$$

3. $y = \sqrt{\sin x + \sqrt{\sin x + \sqrt{\sin x + \dots\infty}}}$

Solution: $y = \sqrt{\sin x + \sqrt{\sin x + \sqrt{\sin x + \dots\infty}}}$

$$\Rightarrow y = \sqrt{\sin x + y}$$

$$\Rightarrow y^2 = \sin x + y$$

$$\Rightarrow \frac{d}{dx}\left(y^2\right) = \frac{d}{dx}(\sin x + y)$$

$$\Rightarrow 2y\frac{dy}{dx} = \cos x + \frac{dy}{dx}$$

$$\Rightarrow 2y\frac{dy}{dx} - \frac{dy}{dx} = \cos x$$

$$\Rightarrow (2y-1)\frac{dy}{dx} = \cos x$$

$$\Rightarrow \frac{dy}{dx} = \frac{\cos x}{(2y-1)};\ y \neq \frac{1}{2}, \text{ i.e. } \sin x \neq -\frac{1}{4}$$

4. $y = \sqrt{x + 2\sqrt{x + 2\sqrt{x + \dots\infty}}}$

Solution: $y = \sqrt{x + 2\sqrt{x + 2\sqrt{x + \ldots \infty}}}$

$$\Rightarrow y = \sqrt{x + 2y}$$

$$\Rightarrow y^2 = x + 2y$$

$$\Rightarrow \frac{d}{dx}\left(y^2\right) = \frac{d}{dx}(x + 2y)$$

$$\Rightarrow 2y\frac{dy}{dx} = 1 + 2\frac{dy}{dx}$$

$$\Rightarrow 2y\frac{dy}{dx} - 2\frac{dy}{dx} = 1$$

$$\Rightarrow (2y - 1)\frac{dy}{dx} = 1$$

$$\Rightarrow \frac{dy}{dx} = \frac{1}{(2y - 2)} = \frac{1}{2(y - 1)};\ y \neq 1, \text{ i.e. } x \neq -1$$

5. $y = x^{x^{x^{x^{\cdots\infty}}}}$

Solution: $y = x^{x^{x^{x^{\cdots\infty}}}}$, $x > 0$

$$\Rightarrow y = x^y$$

$$\Rightarrow \log y = y \log x$$

$$\Rightarrow \frac{d}{dx}(\log y) = \frac{d}{dx}(y \log x)$$

$$\Rightarrow \frac{1}{y}\frac{dy}{dx} = \frac{y}{x} + \log x\frac{dy}{dx}$$

$$\Rightarrow \left(\frac{1}{y} - \log x\right)\frac{dy}{dx} = \frac{y}{x}$$

$$\Rightarrow \frac{1 - y \log x}{y}\frac{dy}{dx} - \frac{y}{x}$$

$$\Rightarrow \frac{dy}{dx} = \frac{y^2}{x\,(1 - y \log x)}$$

6. $y = e^{x^{e^{x^{e^{x^{\cdots\infty}}}}}}$

Solution: $y = e^{x^{e^{x^{e^{x^{\cdots\infty}}}}}}$, $x>0$

$$\Rightarrow y = e^{x^y} = e^{\left(e^{\cdot y}\right)}$$

$$\Rightarrow \log y = x^y \log e = x^y$$

Note: $y = e^{x^y}$ $\not\Rightarrow \log y = y \log e^x$

$$\Rightarrow \log \log y = y \log x$$

$$\Rightarrow \frac{d}{dx}(\log \log y) = \frac{d}{dx}(y \log x)$$

$$\Rightarrow \frac{1}{\log y}\cdot\frac{1}{y}\cdot\frac{dy}{dx} = y\cdot\frac{d}{dx}(\log x) \neq \log x\frac{dy}{dx}$$

$$= \frac{y}{x} + \log x\frac{dy}{dx}$$

$$\Rightarrow \frac{1}{y \log y}\cdot\frac{dy}{dx} - \log x\frac{dy}{dx} = \frac{y}{x}$$

$$\Rightarrow \left(\frac{1 - y \log x \log y}{y \log y}\right)\frac{dy}{dx} = \frac{y}{x}$$

$$\Rightarrow \frac{dy}{dx} = \frac{y}{x}\cdot\left(\frac{y \log y}{1 - y \log x \log y}\right) = \frac{y^2 \log y}{x(1 - y \log x \log y)}$$

Remark:

3^{3^3} means $3^{\left(3^3\right)}$

$y = e^{x^y}$ means $y = e^{\left(x^y\right)} \neq \left(e^x\right)^y = e^{x\cdot y}$

7. $x = y^{y^{y^{y^{\cdots\infty}}}}$

Solution: $x = y^{y^{y^{y^{\cdots\infty}}}}$, $y>0$

$$\Rightarrow x = y^x$$

$$\Rightarrow \log x = \log x^x - x \log y$$

$$\Rightarrow \frac{d}{dx}(\log x) = \frac{d}{dx}(x \log y)$$

$$\Rightarrow \frac{1}{x} = x \cdot \frac{d}{dx}(\log y) + \log y \cdot \frac{dx}{dx}$$

$$= x \cdot \frac{1}{y} \cdot \frac{dy}{dx} + \log y$$

$$\Rightarrow \frac{1}{x} = \frac{x}{y} \cdot \frac{dy}{dx} + \log y$$

$$\Rightarrow \frac{1}{x} - \log y = \frac{x}{y} \cdot \frac{dy}{dx}$$

$$\Rightarrow \frac{1 - x \log y}{x} = \frac{x}{y} \cdot \frac{dy}{dx}$$

$$\Rightarrow \frac{dy}{dx} = \left(\frac{1 - x \log y}{x}\right) \cdot \frac{y}{x} = \frac{y(1 - x \log y)}{x^2}$$

8. $y = \sqrt{x + \sqrt{x + \sqrt{x + \ldots \infty}}}$

Solution: $y = \sqrt{x + \sqrt{x + \sqrt{x + \ldots \infty}}}$

$$\Rightarrow y = \sqrt{x + y}$$

$$\Rightarrow y^2 = x + y$$

$$\Rightarrow \frac{d}{dx}\left(y^2\right) = \frac{d}{dx}(x + y)$$

$$\Rightarrow 2y \cdot \frac{dy}{dx} = 1 + \frac{dy}{dx}$$

$$\Rightarrow 2y \cdot \frac{dy}{dx} - \frac{dy}{dx} = 1$$

$$\Rightarrow (2y - 1)\frac{dy}{dx} = 1$$

$$\Rightarrow \frac{dy}{dx} = \left(\frac{1}{2y - 1}\right)$$

9. $y = x^{x^{x + \ldots \infty}}$

Solution: $y = x^{x^{x + \ldots \infty}}, x > 0$

$$\Rightarrow y = x^{x + y}$$

$$\Rightarrow \log y = \log x^{(x + y)} = (x + y) \log x$$

$$\Rightarrow \frac{d}{dx}(\log y) = \frac{d}{dx}((x + y) \log x)$$

$$= (x + y)\frac{d}{dx}(\log x) + \log x \frac{d}{dx}(x + y)$$

$$\Rightarrow \frac{1}{y} \cdot \frac{dy}{dx} = (x + y) \cdot \frac{1}{x} + \log x\left(1 + \frac{dy}{dx}\right)$$

$$= \frac{(x + y)}{x} + \log x + \log x \frac{dy}{dx}$$

$$\Rightarrow \frac{1}{y} \cdot \frac{dy}{dx} - \log x \frac{dy}{dx} = \frac{(x + y)}{x} + \log x$$

$$= \frac{(x + y) + x \log x}{x}$$

$$\Rightarrow \left(\frac{1 - y \log x}{y}\right)\frac{dy}{dx} = \frac{(x + y) + x \log x}{x}$$

$$\Rightarrow \frac{dy}{dx} = \left(\frac{(x + y) + x \log x}{x}\right) \cdot \left(\frac{y}{1 - y \log x}\right)$$

$$= \frac{y}{x} \cdot \frac{x + y + x \log x}{1 - y \log x}$$

10. $y = \sqrt{x}^{\sqrt{x}^{\sqrt{x} \ldots \infty}}$

Solution: $y = \sqrt{x}^{\sqrt{x}^{\sqrt{x} \ldots \infty}}, x > 0$

$$\Rightarrow y = \left(\sqrt{x}\right)^y = x^{\left(\frac{y}{2}\right)}$$

$$\Rightarrow \log y = \log x^{\left(\frac{y}{2}\right)} = \left(\frac{y}{2}\right) \log x$$

$$\Rightarrow \frac{d}{dx}(\log y) = \frac{d}{dx}\left(\frac{y}{2} \cdot \log x\right)$$

$$= \frac{y}{2} \cdot \frac{d}{dx}(\log x) + \log x \frac{d}{dx}\left(\frac{y}{2}\right)$$

$$\Rightarrow \frac{1}{y}\cdot\frac{dy}{dx} = \frac{y}{2}\cdot\frac{1}{x} + \frac{1}{2}\cdot\log x\,\frac{dy}{dx}$$

$$= \frac{y}{2x} + \frac{1}{2}\log x\,\frac{dy}{dx}$$

$$\Rightarrow \frac{1}{y}\cdot\frac{dy}{dx} - \frac{1}{2}\log x\cdot\frac{dy}{dx} = \frac{y}{2x}$$

$$\Rightarrow \left(\frac{1}{y} - \frac{1}{2}\log x\right)\frac{dy}{dx} = \frac{y}{2x}$$

$$\Rightarrow \left(\frac{2 - y\log x}{2y}\right)\cdot\frac{dy}{dx} = \frac{y}{2x}$$

$$\Rightarrow \frac{dy}{dx} = \left(\frac{y}{2x}\right)\cdot\left(\frac{2y}{2 - y\log x}\right)$$

$$= \frac{y^2}{x(2 - y\log x)}$$

11. $y = \sqrt{\cos x - \sqrt{\cos x - \sqrt{\cos x - \ldots\infty}}}$

Solution: $y = \sqrt{\cos x - \sqrt{\cos x - \sqrt{\cos x - \ldots\infty}}}$

$$\Rightarrow y = \sqrt{\cos x - y}$$

$$\Rightarrow \frac{dy}{dx} = \frac{d}{dx}\left(\sqrt{\cos x - y}\right) = \frac{d\left(\sqrt{\cos x - y}\right)}{d(\cos x - y)}\cdot\frac{d}{dx}(\cos x - y)$$

$$\Rightarrow \frac{dy}{dx} = \frac{1}{2\sqrt{\cos x - y}}\cdot\left(-\sin x - \frac{dy}{dx}\right)$$

$$= -\frac{\sin x}{2\sqrt{\cos x - y}} - \frac{1}{2\sqrt{\cos x - y}}\,\frac{dy}{dx}$$

$$\Rightarrow \frac{dy}{dx} + \frac{1}{2\sqrt{\cos x - y}}\cdot\frac{dy}{dx} = -\frac{\sin x}{2\sqrt{\cos x - y}}$$

$$\Rightarrow \left(\frac{2\sqrt{\cos x - y} + 1}{2\sqrt{\cos x - y}}\right)\frac{dy}{dx} = -\frac{\sin x}{2\sqrt{\cos x - y}}$$

$$\Rightarrow \frac{dy}{dx} = \left(-\frac{\sin x}{2\sqrt{\cos x - y}}\right)\cdot\left(\frac{2\sqrt{\cos x - y}}{2\sqrt{\cos x - y} + 1}\right)$$

$$= \frac{-\sin x}{2\sqrt{\cos x - y} + 1} = \frac{-\sin x}{2y + 1}\left(\because y = \sqrt{\cos x + y}\right)$$

$$= \frac{-\sin x}{1 + 2y}$$

12. $y = (\cos x)^{(\cos x)^{(\cos x)^{\cdots\infty}}}$

Solution: $y = (\cos x)^{(\cos x)^{(\cos x)^{\cdots\infty}}}$; $\cos x > 0$

$$\Rightarrow \log y = y\log\cos x$$

$$\Rightarrow \frac{1}{y}\cdot\frac{dy}{dx} = y\cdot\frac{1}{\cos x}\cdot(-\sin x) + \log\cos x\cdot\frac{dy}{dx}$$

$$= -y\tan x + \log\cos x\,\frac{dy}{dx}$$

$$\Rightarrow \left(\frac{1}{y} - \log\cos x\right)\frac{dy}{dx} = -y\tan x$$

$$\Rightarrow \frac{dy}{dx} = -\frac{y^2\tan x}{(1 - y\log\cos x)}$$

13. $y = \sqrt{x + \sqrt{2 + \sqrt{x + \sqrt{2 + \sqrt{x + \ldots\infty}}}}}$

Solution: $y = \sqrt{x + \sqrt{2 + \sqrt{x + \sqrt{2 + \sqrt{x + \ldots\infty}}}}}$

$$\Rightarrow y = \sqrt{x + \sqrt{2 + y}}$$

$$\Rightarrow y^2 = x + \sqrt{2 + y}$$

$$\Rightarrow y^2 - x = \sqrt{2 + y}$$

$$\Rightarrow \left(y^2 - x\right)^2 = (2 + y)$$

$$\Rightarrow \frac{d}{dx}\left(\left(y^2 - x\right)^2\right) = \frac{d}{dx}(2 + y)$$

$$\Rightarrow 2\left(y^2 - x\right)\left(2y\frac{dy}{dx} - 1\right) = \frac{dy}{dx}$$

$$\Rightarrow 4y\left(y^2 - x\right)\frac{dy}{dx} - 2\left(y^2 - x\right) = \frac{dy}{dx}$$

$$\Rightarrow \left[4y\left(y^2 - x\right) - 1\right]\frac{dy}{dx} = 2\left(y^2 - x\right)$$

$$\Rightarrow \frac{dy}{dx} = \frac{2\left(y^2 - x\right)}{4y\left(y^2 - x\right) - 1} = \frac{2\sqrt{2 + y}}{4y\sqrt{2 + y} - 1}$$

14. $y = \sqrt{x - 2\sqrt{x - 2\sqrt{x - 2\sqrt{x - 2\sqrt{\ldots\infty}}}}}$

Solution: $y = \sqrt{x - 2\sqrt{x - 2\sqrt{x - 2\sqrt{x - 2\sqrt{\ldots\infty}}}}}$

$$\Rightarrow y = \sqrt{x - 2y}$$

$$\Rightarrow y^2 = x - 2y$$

$$\Rightarrow y^2 + 2y = x$$

$$\Rightarrow 2y\frac{dy}{dx} + 2\frac{dy}{dx} = 1$$

$$\Rightarrow 2(y + 1)\frac{dy}{dx} = 1$$

$$\Rightarrow \frac{dy}{dx} = \frac{1}{2(y + 1)}$$

Problems based on non-terminating continued fraction

Solved Examples

Find $\frac{dy}{dx}$ if

1. $y = x + \cfrac{1}{x + \cfrac{1}{x + \cfrac{1}{x + \ldots\infty}}}$

Solution: $y = x + \cfrac{1}{x + \cfrac{1}{x + \cfrac{1}{x + \ldots\infty}}}$

$$\Rightarrow y = x + \frac{1}{y}$$

$$\Rightarrow y^2 = xy + 1$$

$$\Rightarrow 2y\frac{dy}{dx} = x\frac{dy}{dx} + y$$

$$\Rightarrow 2y\frac{dy}{dx} - x\frac{dy}{dx} = y$$

$$\Rightarrow (2y - x)\frac{dy}{dx} = y$$

$$\Rightarrow \frac{dy}{dx} = \frac{y}{(2y - x)}$$

2. $x = y + \cfrac{1}{y + \cfrac{1}{y + \cfrac{1}{y + \ldots\infty}}}$

Solution: $x = y + \cfrac{1}{y + \cfrac{1}{y + \cfrac{1}{y + \ldots\infty}}}$

$$\Rightarrow x = y + \frac{1}{x}$$

$$\Rightarrow x^2 = xy + 1$$

$$\Rightarrow 2x = x\frac{dy}{dx} + y$$

$$\Rightarrow 2x - y = x\frac{dy}{dx}$$

$$\Rightarrow \frac{dy}{dx} = \left(\frac{2x - y}{x}\right)$$

3. $y = \cfrac{x}{a + \cfrac{x}{b + \cfrac{x}{a + \cfrac{x}{b + \ldots\infty}}}}$

or, $y = \frac{x}{a+}\frac{x}{b+}\frac{x}{a+}\frac{x}{b+}\frac{x}{a+}\frac{x}{b+}\ldots\infty$

Solution: $y = \cfrac{x}{a + \cfrac{x}{b + \cfrac{x}{a + \cfrac{x}{b + \ldots\infty}}}}$

$$\Rightarrow y = \cfrac{x}{a + \cfrac{x}{b + y}}$$

$$\Rightarrow y = \cfrac{x}{\cfrac{a(b+y)+x}{(b+y)}}$$

$$\Rightarrow y = \frac{(b+y)x}{a(b+y)+x}$$

$$\Rightarrow aby + y^2 + xy = bx + yx$$

$$\Rightarrow ab\frac{dy}{dx} + 2y\frac{dy}{dx} = b$$

$$\Rightarrow \frac{dy}{dx} = \frac{b}{(ab+2y)}$$

4. $y = \cfrac{\sin x}{1 + \cfrac{\cos x}{1 + \cfrac{\sin x}{1 + \cfrac{\cos x}{1 + \ldots\infty}}}}$

Solution: $y = \cfrac{\sin x}{1 + \cfrac{\cos x}{1 + \cfrac{\sin x}{1 + \cfrac{\cos x}{1 + \ldots\infty}}}}$

$$\Rightarrow y = \cfrac{\sin x}{1 + \cfrac{\cos x}{1 + y}}$$

$$\Rightarrow y = \frac{(1+y)\sin x}{1 + y + \cos x}$$

$$\Rightarrow y^2 + y + y\cos x = (1+y)\sin x$$

$$\Rightarrow \frac{d}{dx}\left(y^2 + y + y\cos x\right) = \frac{d}{dx}((1+y)\sin x).$$

$$\Rightarrow 2y\frac{dy}{dx} + \frac{dy}{dx} + \cos x\frac{dy}{dx} - y\sin x$$

$$= \frac{dy}{dx}\sin x + (1+y)\frac{d}{dx}\sin x$$

$$\Rightarrow (2y + 1 + \cos x - \sin x)\frac{dy}{dx} = y\sin x + (1+y)\cos x$$

$$\Rightarrow \frac{dy}{dx} = \frac{y\sin x + (1+y)\cos x}{(1 + 2y + \cos x - \sin x)}$$

5. $y = \cfrac{\tan x}{1 + \cfrac{\cot x}{1 + \cfrac{\tan x}{1 + \cfrac{\cot x}{1 + \cfrac{\tan x}{1 + \cfrac{\cot x}{1 + \ldots\infty}}}}}}$

or, $y = \frac{\tan}{1+}\frac{\cot x}{1+}\frac{\tan}{1+}\frac{\cot x}{1+}\frac{\tan}{1+}\ldots\infty$

Solution: $y = \cfrac{\tan x}{1 + \cfrac{\cot x}{1 + \cfrac{\tan x}{1 + \cfrac{\cot x}{1 + \cfrac{\tan x}{1 + \cfrac{\cot x}{1 + \ldots\infty}}}}}}$

$$\Rightarrow y = \frac{\tan x}{1 + \frac{\cot x}{1 + y}} = \frac{\tan x}{\left(\frac{1 + y + \cot x}{1 + y}\right)}$$

$$\Rightarrow y = \frac{(1 + y)\tan x}{(1 + y + \cot x)}$$

$$\Rightarrow y(1 + y + \cot x) = (1 + y)\tan x$$

$$\Rightarrow y + y^2 + y\cot x = (1 + y)\tan x$$

$$\Rightarrow \frac{d}{dx}\left(y^2 + y + y\cot x\right) = \frac{d}{dx}((1 + y)\tan x)$$

$$\Rightarrow 2y\frac{dy}{dx} + \frac{dy}{dx} + \cot x\frac{dy}{dx} - y\,\text{cosec}^2 x$$

$$= \left(0 + \frac{dy}{dx}\right)\tan x + (1 + y)\frac{d}{dx}(\tan x)$$

$$\Rightarrow 2y\frac{dy}{dx} + \frac{dy}{dx} + \cot x\frac{dy}{dx} - \tan x\frac{dy}{dx}$$

$$= y\,\text{cosec}^2 x + (1 + y)\sec^2 x$$

$$\Rightarrow (1 + 2y + \cot x - \tan x)\frac{dy}{dx}$$

$$= y\,\text{cosec}^2 x + (1 + y)\sec^2 x$$

$$\Rightarrow \frac{dy}{dx} = \frac{y\,\text{cosec}^2 x + (1 + y)\sec^2 x}{(1 + 2y + \cot x - \tan x)}$$

6. $y = \sqrt{x} + \cfrac{1}{\sqrt{x} + \cfrac{1}{\sqrt{x} + \cfrac{1}{\sqrt{x} + \ldots\infty}}}$

Solution: $y = \sqrt{x} + \cfrac{1}{\sqrt{x} + \cfrac{1}{\sqrt{x} + \cfrac{1}{\sqrt{x} + \ldots\infty}}}$

$$\Rightarrow y = \sqrt{x} + \frac{1}{y}$$

$$\Rightarrow \frac{dy}{dx} = \frac{1}{2\sqrt{x}} - \frac{1}{y^2}\frac{dy}{dx}$$

$$\Rightarrow \frac{dy}{dx} + \frac{1}{y^2}\frac{dy}{dx} = \frac{1}{2\sqrt{x}}$$

$$\Rightarrow \left(1 + \frac{1}{y^2}\right)\frac{dy}{dx} = \frac{1}{2\sqrt{x}}$$

$$\Rightarrow \frac{dy}{dx} = \frac{y^2}{2\left(1 + y^2\right)\sqrt{x}}$$

Conditional identities on non-terminating continued fraction or 'ad infinitum' form.

To form a differential equation with the help of a given equation, we adopt the rule which consists of following steps.

Step 1: Take the given and express it as an implicit function of x.

Step 2: Find the derivative of the implicit function of x.

Step 3: Use mathematical manipulations to put the first derivatives (or, the derivative involved in the required differential equation) in to the required form of the differential equations.

Solved Examples

1. If $y = a^{x^{a^{x^{a^{x \ldots\infty}}}}}$ show that

$$\frac{dy}{dx} = \frac{y^2 \log y}{x(1 - y\log x \log y)}$$

Solution: $y = a^{x^{a^{x^{a^{x \ldots\infty}}}}}$, $a>0, x>0$

$$\Rightarrow y = a^{x^y}$$

$$\Rightarrow \log y = x^y \log a$$

$$\Rightarrow \log\log y = y\log x + \log\log a$$

$$\Rightarrow \frac{1}{\log y}\cdot\frac{1}{y}\cdot\frac{dy}{dx} = \frac{dy}{dx}\cdot\log x + \frac{y}{x}$$

$$\Rightarrow \left(\frac{1}{y\log y} - \log x\right)\frac{dy}{dx} = \frac{y}{x}$$

$$\Rightarrow \frac{dy}{dx} = \frac{y^2\log y}{x(1 - y\log x\log y)}$$

2. If $y = x^{x^{x^{\cdots\infty}}}$, show that $x\dfrac{dy}{dx} = \dfrac{y^2}{1 - y\log x}$

Solution: $y = x^{x^{x^{\cdots\infty}}}, x>0$

$$\Rightarrow y = x^y \qquad ...(1)$$

$$\Rightarrow \log y = y\log x$$

$$\Rightarrow \frac{1}{y}\cdot\frac{dy}{dx} = \log x\frac{dy}{dx} + \frac{y}{x}$$

$$\Rightarrow \left(\frac{1}{y} - \log x\right)\frac{dy}{dx} = \frac{y}{x}$$

$$\Rightarrow \frac{dy}{dx} = \frac{y^2}{x(1 - y\log x)} \qquad ...(2)$$

$$\therefore x\frac{dy}{dx} = \frac{y^2}{(1 - y\log x)}$$

3. If $y = \sqrt{\sin x + \sqrt{\sin x + \sqrt{\sin x + \sqrt{\text{etc}\ldots\text{to}\,\infty}}}}$,

show that $\dfrac{dy}{dx} = \dfrac{\cos x}{2y-1}$.

Solution: $y = \sqrt{\sin x + \sqrt{\sin x + \sqrt{\sin x + \sqrt{\ldots\infty}}}}$

$$\Rightarrow y = \sqrt{\sin x + y}$$

$$\Rightarrow y^2 = \sin x + y$$

$$\Rightarrow y^2 - y = \sin x$$

$$\Rightarrow 2y\frac{dy}{dx} - \frac{dy}{dx} = \cos x$$

$$\Rightarrow (2y-1)\frac{dy}{dx} = \cos x$$

$$\Rightarrow \frac{dy}{dx} = \frac{\cos x}{(2y-1)}$$

4. If $y = \sqrt{\log x + \sqrt{\log x + \sqrt{\log x + \ldots\infty}}}$, show that $(2y-1)\dfrac{dy}{dx} = \dfrac{1}{x}$.

Solution: $y = \sqrt{\log x + \sqrt{\log x + \sqrt{\log x + \ldots\infty}}}, x>0$

$$\Rightarrow y = \sqrt{\log x + y}$$

$$\Rightarrow y^2 = \log x + y$$

$$\Rightarrow 2y\frac{dy}{dx} = \frac{1}{x} + \frac{dy}{dx}$$

$$\Rightarrow (2y-1)\frac{dy}{dx} = \frac{1}{x}$$

5. If $y = \dfrac{x}{1 + \dfrac{x}{1 + \dfrac{x}{1 + \dfrac{x}{1 + \ldots\infty}}}}$, show that

$$\frac{dy}{dx} = \frac{1}{2y+1}$$

Solution: $y = \dfrac{x}{1 + \dfrac{x}{1 + \dfrac{x}{1 + \dfrac{x}{1 + \ldots\infty}}}}$

$$\Rightarrow y = \frac{x}{1+y}$$

$$\Rightarrow y^2 + y = x$$

$$\Rightarrow 2y\frac{dy}{dx}+\frac{dy}{dx}=1$$

$$\Rightarrow (2y+1)\frac{dy}{dx}=1$$

$$\Rightarrow \frac{dy}{dx}=\frac{1}{2y+1}$$

6. If $x = y + \cfrac{1}{y+\cfrac{1}{y+\cfrac{1}{y+\cfrac{1}{\ldots\infty}}}}$, show that

$$\frac{dy}{dx} = 2x^2 + y^2 - 3xy.$$

Solution: $x = y + \cfrac{1}{y+\cfrac{1}{y+\cfrac{1}{y+\cfrac{1}{\ldots\infty}}}}$

$$\Rightarrow x = y + \frac{1}{x} \qquad \ldots(1)$$

$$\Rightarrow 1 = \frac{dy}{dx} - \frac{1}{x^2}$$

$$\Rightarrow \frac{dy}{dx} = 1 + \frac{1}{x^2}$$

$= 1 + (x-y)^2$ (from (1))

$= 1 + x^2 + y^2 - 2xy$

$= x^2 - xy + x^2 + y^2 - 2xy$ ($\because x - y = \frac{1}{x} \Rightarrow x^2 - xy = 1$ from (1))

$= 2x^2 + y^2 - 3xy$

$$\therefore \frac{dy}{dx} = 2x^2 + y^2 - 3xy$$

Remark: Special care must be taken when implicit function or a function defined by an infinite process are differentiated for the function $y = F(x)$ or $F(x, y) = 0$ at which it is undefined. e.g.,

$$y = \sqrt{\sin x + \sqrt{\sin x + \sqrt{\sin x + \ldots}}}$$

$$\Rightarrow y = \sqrt{\sin x + y}$$

$$\Rightarrow y^2 = \sin x + y$$

$$\Rightarrow (2y-1)\frac{dy}{dx} = \cos x$$

$\Rightarrow \dfrac{dy}{dx} = \dfrac{\cos x}{(2y-1)}$ which means that we can find $\dfrac{dy}{dx}$ when ever $y \neq \dfrac{1}{2}$.

Conditional identities based on explicit function of x

1. If $y = \tan^{-1}\left(\dfrac{2-3x}{1+6x}\right)$, show $\dfrac{dy}{dx} + \dfrac{3}{1+9x^2} = 0$.

2. If and $f(x) = \sin^{-1}\dfrac{x}{\sqrt{1+x^2}}$ and $g(x) = \tan^{-1}\left(\dfrac{1+x}{1-x}\right)$, $|x| < 1$, show that $f'(x) = g'(x)$.

3. If $f(x) = \sec x + \tan x$, show that $\dfrac{f'(x)}{f(x)} = \sec x$.

4. If $y = \dfrac{x}{x+5}$, show that $x\dfrac{dy}{dx} = y(1-y)$.

5. If $y = x^{-3}$, show that $x\dfrac{dy}{dx} + 3y = 0$.

6. If $y = x^3$, show that $x\dfrac{dy}{dx} = 3y, \forall x$.

7. If $y = x + \dfrac{1}{x}$, show that $x\dfrac{dy}{dx} + y = 2x$.

8. If $y = \sin^{-1} x + \sin^{-1}\sqrt{1-x^2}$, show that $\dfrac{dy}{dx} = 0$.

(*Hint*: Put $x = \sin\theta$, $-\dfrac{\pi}{2} \leq \theta \leq \dfrac{\pi}{2}$)

9. If $x = \sin^{-1} \frac{2u}{1+u^2}$ and $y = \tan^{-1} \frac{2u}{1-u^2}$, show that $\frac{dy}{dx} = 1$.

10. If $y = \frac{\cos x + \sin x}{\cos x - \sin x}$, show that $\frac{dy}{dx} = \sec^2\left(x + \frac{\pi}{4}\right)$.

11. If $y = \cos^{-1}\sqrt{\frac{\cos 3x}{\cos^3 x}}$, show that $\frac{dy}{dx} = \sqrt{\frac{3}{\cos x \cos 3x}}$.

12. If $y = \frac{\sin^{-1} x}{\sqrt{1-x^2}}$, ,m show that $(1-x^2)\frac{dy}{dx} - xy = 1$

13. If $y = \cos^{-1}\left(\frac{3 + 5\cos x}{5 + 3\cos x}\right)$, show that $\cos x = \frac{4 - 5y_1}{3y_1}$, where $y_1 = \frac{dy}{dx}$.

14. If $y = \log\sqrt{\frac{1 + \tan x}{1 - \tan x}}$, show that $\frac{dy}{dx} = \sec 2x$.

15. If $y = e^{-x}$, show that $\frac{dy}{dx} + y = 0$

16. If $y = e^{ax} \sin(bx + c)$, show that

$$\frac{dy}{dx} = \sqrt{a^2 + b^2}\, e^{ax} \sin\left[(bx + c) + \tan^{-1}\frac{b}{a}\right]$$

Problems based on infinite series

Exercise 12.8

Find $\frac{dy}{dx}$ if

1. $y = e^{x + e^{x + e^{x + \dots \infty}}}$

2. $y = x^{x^{x^{\dots \infty}}}$

3. $y = \sqrt{\sin x + \sqrt{\sin x + \sqrt{\sin x + \dots \infty}}}$

4. $y = \sqrt{\cos x + \sqrt{\cos x + \sqrt{\cos x + \dots \infty}}}$

5. $y = \sqrt{x + \sqrt{x + \sqrt{x + \dots \infty}}}$

6. $y = \sqrt{x}^{\sqrt{x}^{\sqrt{x}^{\dots \infty}}}$

7. $x = y + \cfrac{1}{y + \cfrac{1}{y + \dots \infty}}$

8. $y = \cfrac{1}{x + \cfrac{1}{x + \cfrac{1}{x + \dots \infty}}}$

9. $x = y^{y^{y^{\dots \infty}}}$

Answers

1. $\frac{y}{1-y}$

2. $\frac{1}{x}\left(\frac{y^2}{1 - y \log x}\right)$

3. $\frac{\cos x}{2y - 1}$

4. $\frac{\sin x}{1 - 2y}$

5. $\frac{1}{2y - 1}$

6. $\frac{y^2}{x(2 - y \log x)}$

7. $1 + \frac{1}{x^2}$

8. $-\frac{y^2}{1 + y^2}$

9. $\dfrac{y(1-x\log y)}{x^2}$

Conditional problems

Exercise 12.9

1. If $y = x^{x^{x^{\cdots\infty}}}$, show that $y^2 = x(1 - y\log x)\dfrac{dy}{dx}$.

(*Hint*: Here $y = x^y$)

2. If $y = \sqrt{\sin x + \sqrt{\sin x + \sqrt{\sin x + \ldots\infty}}}$, show that

$(2y-1)\dfrac{dy}{dx} = \cos x$.

(*Hint*: $y = \sqrt{\sin x + y} \Rightarrow y^2 = \sin x + y$)

3. Show that $\dfrac{dy}{dx} = \dfrac{\sin x}{1-2y}$, when

$\sqrt{\cos x + \sqrt{\cos x + \sqrt{\cos x + \ldots\infty}}}$.

13
Logarithmic Differentiation

Question: What is logarithmic differentiation?
Answer: Taking logarithm of both sides of an identity before differentiation is known as logarithmic differentiation.

Question: Where to use logarithmic differentiation?
Answer: Logarithmic differentiation is used when

1. The given function defining y as a function of x is the product of two differentiation functions of x's, one (or both) of which may be quotient, power (or, under radical), implicit function or composite of two (or, more) functions; i.e. when $y = f_1(x) \cdot f_2(x)$ where $f_1(x)$ and $f_2(x)$ are differentiable functions of x's.

2. The given function defining y as a function of x is the product of a finite number of differentiable functions of x's, some of which may be quotients, powers (or, under radicals), implicit functions or composite of two (or, more) functions; i.e. when $y = f_1(x) \cdot f_2(x) \cdot f_3(x) \ldots f_n(x)$, where $f_1(x), f_2(x) \ldots f_n(x)$ are differentiable functions of x's.

3. The given function defining y as a function of x is the quotient of two differentiable functions of x's whose numerator and denominator may be a power function (or, a function under radical), implicit function or composite of two (or, more) functions; i.e. when

$$y = \frac{f_1(x)}{f_2(x)},$$ where $f_1(x)$ and $f_2(x)$ are differentiable functions of x's.

4. The given function defining y as a function of x is the quotient whose numerator and denominator contain a finite number of differentiable functions of x's some of which may be powers or, under radicals), implicit function or composite of two (or, more) functions; i.e. when $y = \dfrac{f_1(x) \cdot f_2(x) \cdot f_3(x) \ldots f_n(x)}{g_1(x) \cdot g_2(x) \cdot g_3(x) \ldots g_n(x)}$, where $f_1(x), f_2(x), \ldots, f_n(x)$ and $g_1(x), g_2(x), \ldots, g_n(x)$ are differentiable functions of x's.

5. The given function defining y as a function of x is a power of the function of x whose base is an implicit or explicit function of x whereas the index is a real number; i.e. when $y = [f_1(x)]^n$, where $f_1(x)$ is a differentiable functions of x whereas the index is a real number.

6. The given function defining y as a function of x is composite exponential function which is a function whose both the base and the exponent are differentiable functions of x's; i.e. when $y = [f_1(x)]^{f_2(x)}$, where $f_1(x)$ and $f_2(x)$ are differentiable functions of x's. (***Note***: Logarithmic differentiation may be defined as "differentiation after taking logarithm of both sides of an identity".

Remember:

1. Logarithmic differentiation is practically useful when the given function defining y as a function of x is a complicated one consisting of products, quotients and powers (or, radicals) of differentiable functions whose derivatives can not be found easily by using the rule of differentiating the products, quotient or power of the functions being differentiable.

2. We take logs throughout to avoide repeated applications of the rules for the differentiation of products and quotients of functions of x's.

3. We can use logarithmic differentiation only when the function concerned is positive in its domain.

Example

$$y = \sin^n x \Rightarrow \log|y| = |\sin x|^n$$

$$\Rightarrow \log|y| = n\log|\sin x| \Rightarrow \frac{1}{y}\frac{dy}{dx} = \frac{n\cos x}{\sin x}$$

$$\Rightarrow \frac{dy}{dx} = n\sin^{n-1} x\cos x$$

This result is valid when n is an integer ≥ 1, whether $\sin x$ is positive or negative.

Question: What are the rules of logarithmic differentiation?

Answer: There are two rules of logarithmic differentiation.

Rule 1: When $y = f(x)$, $f(x)$ being differentiable positive function (i.e. $f(x) > 0$),

$$\frac{d}{dx}(\log y) = \frac{d}{dx}(\log f(x))$$

$$\Rightarrow \frac{1}{y}\cdot\frac{dy}{dx} = \frac{1}{f(x)}\cdot\frac{d}{dx}f(x),\ f(x) > 0$$

Rule 2: When $y = f(x)$, $f(x)$ being differentiable and $f(x) \neq 0$

$$\frac{d}{dx}\log|y| = \frac{d}{dx}\log|f(x)|$$

$$\Rightarrow \frac{1}{y}\cdot\frac{dy}{dx} = \frac{1}{f(x)}\cdot\frac{d}{dx}f(x),\ f(x) \neq 0$$

Notes: **(A)**

(i) $y = \sqrt{f(x)}$

(ii) $y = e^{f(x)}$

(iii) $y = |f(x)|$

(iv) $y = \log f(x)$

(v) $\log_{\varnothing(x)} f(x) = y$

(vi) $y = \log\log\log \ldots \log f(x)$ are differentiable functions where $f(x) > 0$ is pre assumed but in $y = |f(x)|$, $f(x)$ may be greater than zero or less than zero remaining the possibility of being zero also whereas $|f(x)|$ means always ≥ 0. For this reason wherever $f(x) < 0$, we multiply $f(x)$ by minus one (i.e. -1) to make $f(x) > 0$.

(B) $\frac{d}{dx}[\log|f(x)|] = \frac{f'(x)}{f(x)}$ which means the derivative of the function $\log|f(x)|$ is a logarithmic derivative of the function $f(x)$. To simplify the notation in logarithmic differentiation, the sign of absolute value of the function $f(x)$ is usually omitted only when the function $f(x)$ concerned is positive. Hence, this is why we write $\frac{d}{dx}[\log f(x)] = \frac{f'(x)}{f(x)}$.

(C) Logarithmic differentiation simplifies finding the derivative of the given function.

On working rule of problems belonging to first type

When we are given the interval or the quadrant in which $f(x)$ = given function of x is positive or the condition imposed on the independent variable x makes $f(x) > 0$, we adopt the following rule to find the differential coefficient of the given function by using logarithmic differentiation.

Step 1: Take logarithm on both sides of the equation defining y as a function of x, i.e. $\log y = \log f(x)$.

Step 2: Differentiate the equation $\log y = \log f(x)$ using the rule:

(i) $\frac{d}{dx}\log y = \frac{1}{y}\cdot\frac{dy}{dx}, y > 0$

(ii) $\frac{d}{dx}\log f(x) = \frac{1}{f(x)}\cdot\frac{d\,f(x)}{dx}, f(x) > 0$

(iii) $\frac{dy}{dx} = y\cdot\frac{d}{dx}\log f(x)$

Remember: **(a)** After taking logarithm, we are required to use the following formulas.

(i) $\log u\cdot v = \log u + \log v,\ u > 0, v > 0.$

(ii) $\log\frac{u}{v} = \log u - \log v,\ u > 0, v > 0.$

(iii) $\log u^v = v\log u,\ u > 0, v \in 0.$

(iv) $\log u^n = n\log u,\ n \in R, u > 0.$

where u and v are functions of x.

(b) The above working rule may be remembered as "GLAD", the letters being in order which means

G = given function
L = take the logarithm
A = apply the logarithmic formulas
D = differentiate.

(c) We take logarithm on both sides of the equation defining y as a function of x only when the given function is positive because logarithmic differentiation is applicable only when the function concerned is positive.

(d) The above working rule is applicable to the composite exponential function $[f_1(x)]^{f_2(x)}$ provided the function in the base is positive and for this reason the above working rule is applicable to $e^{f(x)}$ since $e = 2.718\ldots > 0$.

Remarks: To take the logarithm of any quantity, we have to be sure that it is positive.

Problems based on first type

Form 1: Problems based on irrational functions.

Solved Examples

Find the differential coefficient of the following.

1. $y = \sqrt{x}$

Solution: $y = \sqrt{x} = (x)^{\frac{1}{2}}$, defined for $x > 0$

$$\Rightarrow \log y = \frac{1}{2}\log x$$

$$\Rightarrow \frac{1}{y}\cdot\frac{dy}{dx} = \frac{1}{2}\cdot\frac{1}{x}$$

$$\Rightarrow \frac{dy}{dx} = \frac{y}{2x}$$

$$\Rightarrow \frac{dy}{dx} = \frac{\sqrt{x}}{2x},\ x > 0 = \frac{1}{2\sqrt{x}},\ x > 0.$$

2. $y = \dfrac{x\cdot\sqrt{x^2+1}}{(x+1)^{\frac{2}{3}}}$

Solution: $|y| = \dfrac{|x|\cdot\sqrt{x^2+1}}{\left|(x+1)^{\frac{2}{3}}\right|}$

$$\Rightarrow \log|y| = \log|x| + \frac{1}{2}\log\left(x^2+1\right) - \frac{2}{3}\log|(x+1)|$$

$$\Rightarrow \frac{1}{y}\cdot\frac{dy}{dx} = \frac{1}{x} + \frac{1}{2}\cdot\frac{1}{x^2+1}\cdot 2x - \frac{2}{3}\cdot\frac{1}{x+1}\cdot 1$$

$$\Rightarrow \frac{dy}{dx} = y\cdot\left(\frac{1}{x} + \frac{x}{x^2+1} - \frac{2}{3(x+1)}\right)$$

$$= \frac{x\cdot\sqrt{x^2+1}}{(x+1)^{\frac{2}{3}}}\cdot\left(\frac{1}{x} + \frac{x}{x^2+1} - \frac{2}{3(x+1)}\right)$$

3. $y = \sqrt{\cos x},\ \left(-\frac{\pi}{2} < x < \frac{\pi}{2}\right)$

Solution: $y = (\cos x)^{\frac{1}{2}}$

$$\Rightarrow \log y = \frac{1}{2}\log\cos x$$

$$\Rightarrow \frac{1}{y}\cdot\frac{dy}{dx} = \frac{1}{2}\cdot\frac{1}{\cos x}\cdot(-\sin x)$$

$$\Rightarrow \frac{dy}{dx} = -\frac{1}{2}\cdot y\cdot\tan x$$

$$= -\frac{1}{2}\sqrt{\cos x}\cdot\tan x$$

4. $y = \sqrt[3]{\log x}$

Solution: $y = (\log x)^{\frac{1}{3}}$, defined for $x > 0$

$$\Rightarrow \log|y| = \frac{1}{3}\log|\log x|$$

$$\Rightarrow \frac{1}{y}\cdot\frac{dy}{dx} = \frac{1}{3}\cdot\frac{1}{\log x}\cdot\frac{1}{x} = \frac{1}{3x\log x}$$

$$\Rightarrow \frac{dy}{dx} = \frac{y}{3x\log x},\ x > 0$$

Form 2: Problems based on product of two or more than two differentiable functions.

Solved Examples

Find the differential coefficient of the following.

1. $y = x^2 \cdot e^{2x} \cdot \sin 3x$

Solution: $y = x^2 \cdot e^{2x} \cdot \sin 3x$

$$\Rightarrow |y| = x^2 \cdot e^{2x} |\sin 3x|$$

$$\Rightarrow \log|y| = \log x^2 + \log e^{2x} + \log|\sin 3x|$$

$$\Rightarrow \log|y| = 2\log|x| + 2x + \log|\sin 3x| \ (\because \log e = 1)$$

$$\Rightarrow \frac{1}{y} \cdot \frac{dy}{dx} = \frac{2}{x} + 2 + \frac{3\cos 3x}{\sin 3x}$$

$$\Rightarrow \frac{dy}{dx} = y \cdot \left(2 + 3\cot 3x + \frac{2}{x}\right)$$

$$= x^2 \cdot e^{2x} \cdot \sin 3x \left(2 + 3\cot 3x + \frac{2}{x}\right)$$

2. $y = e^x \cdot \log x$

Solution: $y = e^x \cdot \log x$, defined for $x > 0$

$$\Rightarrow \log|y| = \log e^x + \log|\log x| = x + \log|\log x|$$

$(\because \log e = 1)$

$$\Rightarrow \frac{1}{y} \cdot \frac{dy}{dx} = 1 + \frac{1}{x\log x}$$

$$\Rightarrow \frac{dy}{dx} = y\left(1 + \frac{1}{x\log x}\right)$$

$$= e^x \cdot \log x\left(1 + \frac{1}{x\log x}\right)$$

3. $y = x \cdot \sqrt[3]{\log x}$

Solution: $y = x \cdot \sqrt[3]{\log x} = x(\log x)^{\frac{1}{3}}, x > 0$

$$\Rightarrow \log|y| = \log x + \frac{1}{3}\log|\log x|$$

$$\Rightarrow \frac{1}{y} \cdot \frac{dy}{dx} = \frac{1}{x} + \frac{1}{3} \cdot \frac{1}{\log x} \cdot \frac{1}{x} = \frac{1}{x} + \frac{1}{3x\log x}$$

$$\Rightarrow \frac{dy}{dx} = y\left(\frac{1}{x} + \frac{1}{3x\log x}\right)$$

$$= x \cdot \sqrt[3]{\log x} \cdot \left(\frac{1}{x} + \frac{1}{3x\log x}\right), x > 0$$

Form 3: Problems based on quotient of two or more than two differentiable functions

Solved Examples

Find the differential coefficient of the following.

1. $y = \dfrac{4}{\log x}$

Solution: $y = \dfrac{4}{\log x}, x > 0$

$$\Rightarrow \log|y| = \log 4 - \log|\log x|$$

$$\Rightarrow \frac{1}{y} \cdot \frac{dy}{dx} = -\frac{1}{x\log x}$$

$$\Rightarrow \frac{dy}{dx} = -\frac{y}{x\log x} = -\frac{4}{x\log x\log x} = -\frac{4}{x(\log x)^2}$$

$$= -\frac{4}{x \cdot \log^2 x}, x > 0$$

2. $y = \dfrac{\log x}{x}, x > 0$

Solution: $y = \dfrac{\log x}{x}$

$$\Rightarrow \log y = \log\log x - \log x$$

$$\Rightarrow \frac{1}{y} \cdot \frac{dy}{dx} = \frac{1}{x\log x} - \frac{1}{x} = y \cdot \left(\frac{1}{x\log x} - \frac{1}{x}\right)$$

$$= \frac{\log x}{x}\left(\frac{1 - \log x}{x\log x}\right) = \frac{1 - \log x}{x^2}$$

3. $y = \dfrac{e^x - e^{-x}}{e^x + e^{-x}}$

Solution: $y = \dfrac{e^x - e^{-x}}{e^x + e^{-x}}$

$$\Rightarrow \log|y| = \log\left|\left(e^x - e^{-x}\right)\right| - \log\left(e^x + e^{-x}\right)$$

$$\Rightarrow \frac{1}{y}\cdot\frac{dy}{dx} = \frac{e^x + e^{-x}}{e^x - e^{-x}} - \frac{e^x - e^{-x}}{e^x + e^{-x}}$$

$$= \frac{\left(e^x + e^{-x}\right)^2 - \left(e^x - e^{-x}\right)^2}{\left(e^x + e^{-x}\right)\left(e^x - e^{-x}\right)}$$

$$\therefore \frac{dy}{dx} = y\left(\frac{e^{2x} + e^{-2x} + 2 - e^{2x} - e^{-2x} + 2}{e^{2x} - e^{-2x}}\right)$$

$$= \frac{e^x - e^{-x}}{e^x + e^{-x}} \cdot \frac{4}{\left(e^x - e^{-x}\right)\left(e^x + e^{-x}\right)}$$

$$= \frac{4}{\left(e^x + e^{-x}\right)^2}$$

Form 4: **Problems based on exponential composite functions. $y = f(x)^{g(x)}$ which is defined only when the base $f(x) > 0$.**

Note: **Whether the questions says or does not say about the base $f(x)$ of the exponential composite function $y = f(x)^{g(x)}$ to be positive, it is understood always that the base $f(x) > 0$ since $y = f(x)^{g(x)}$ is defined only when $f(x) > 0$.**

Solved Examples

Find the differential coefficient of the following.

1. $y = (\sin x)^{\sin x}$, $\sin x > 0$

Solution: $y = (\sin x)^{\sin x}$

$$\Rightarrow \log y = \log(\sin x)^{\sin x} = \sin x \log \sin x$$

$$\Rightarrow \frac{1}{y}\cdot\frac{dy}{dx} = \sin x \cdot \frac{1}{\sin x}\cdot \cos x + \cos x \cdot \log \sin x$$

$$\Rightarrow \frac{dy}{dx} = y(\cos x + \cos x \cdot \log \sin x)$$

2. $y = (1 + x)^{2x}$

Solution: $y = (1 + x)^{2x}$, defined for $x > -1$

$$\Rightarrow \log y = 2x \log(1 + x)$$

$$\Rightarrow \frac{1}{y}\cdot\frac{dy}{dx} = 2\cdot\left[\frac{x}{x+1} + \log(1 + x)\right]$$

$$\Rightarrow \frac{dy}{dx} = 2\cdot y\cdot\left[\frac{x}{x+1} + \log(1 + x)\right]$$

$$= 2\cdot(1 + x)^{2x}\cdot\left[\frac{x}{x+1} + \log(1 + x)\right], x > -1$$

3. $y = x^{\frac{1}{x}}$

Solution: $y = x^{\frac{1}{x}}$, defined for $x > 0$

$$\Rightarrow \log y = \frac{1}{x}\log x$$

$$\Rightarrow \frac{1}{y}\cdot\frac{dy}{dx} = -\frac{1}{x^2}\log x + \frac{1}{x}\cdot\frac{1}{x} = \frac{1 - \log x}{x^2}$$

$$\Rightarrow \frac{dy}{dx} = y\left[\frac{1 - \log x}{x^2}\right]$$

4. $y = (1 + x)^{\log x}$

Solution: $y = (1 + x)^{\log x}$, defined for $x > -1$

$$\Rightarrow \log y = \log x \cdot \log(1 + x)$$

$$\Rightarrow \frac{1}{y}\cdot\frac{dy}{dx} = \frac{\log x}{1 + x} + \frac{\log(1 + x)}{x}$$

$$\Rightarrow \frac{dy}{dx} = y\cdot\left[\frac{\log x}{1 + x} + \frac{\log(1 + x)}{x}\right]$$

$$= (1 + x)^{\log x}\cdot\left[\frac{\log x}{1 + x} + \frac{\log(1 + x)}{x}\right]$$

5. $y = x^{(x^x)}$, $x > 0$

Solution: $y = x^{(x^x)}$ ...(i)

$$\Rightarrow \log y = \left(x^x\right)\cdot \log x$$

$$\Rightarrow |\log y| = x^x |\log x| \quad ...(ii)$$

Again taking logarithm on both sides of the equation (ii), we have $\log|\log y| = \log[(x^x)|\log x|] = x\log x + \log|\log x|$

$$\Rightarrow \frac{1}{\log y}\cdot\frac{1}{y}\cdot\frac{dy}{dx} = x\cdot\frac{1}{x} + \log x + \frac{1}{\log x}\cdot\frac{1}{x}$$

$$= 1 + \log x + \frac{1}{x\log x}$$

$$\Rightarrow \frac{dy}{dx} = y\cdot\log y\left[1 + \log x + \frac{1}{x\log x}\right] \quad ...(iii)$$

Now, putting (i) and (ii) in (iii), we have

$$\frac{dy}{dx} = x^{(x^x)}\cdot(x^x)\cdot\log x\left[1 + \log x + \frac{1}{x\log x}\right]$$

Remember: The operation of taking logarithm on both sides of the equation defining y as a function of x (i.e. $y = f(x)$ may be performed more than once if we go on having an exponential composite function just before the differentiation and we differentiate the logarithmic function only when the occurrence of exponential composite function in any intermediate step is over as a factor of logarithmic function or, alternatively we make the substitution for the exponential composite function occuring in any intermediate step after differentiation which is explained in the following way.

$$y = x^{(x^x)}$$

$$\Rightarrow \log y = \log(x)^{(x^x)} = (x^x)\log x \quad ...(i)$$

$$\Rightarrow \frac{1}{y}\cdot\frac{dy}{dx} = x^x\cdot\frac{1}{x} + \log x\,\frac{d}{dx}(x^x)$$

$$\Rightarrow \frac{dy}{dx} = y\left(x^x\cdot\frac{1}{x} + \log x\cdot\frac{d}{dx}(x^x)\right) \quad ...(ii)$$

Now on supposing that $z = x^x$, $x > 0$, we have

$$\log z = \log(x^x) + x\log x$$

$$\Rightarrow \frac{1}{z}\cdot\frac{dz}{dx} = x\cdot\frac{1}{x} + \log x\cdot 1$$

$$\Rightarrow \frac{dz}{dx} = z(1 + \log x)$$

$$= (x^x)\cdot(1 + \log x) \quad ...(iii)$$

Now on putting (iii) in (ii), we have

$$\frac{dy}{dx} = y\cdot(x^x)\cdot\left(\frac{1}{x} + \log x + \log^2 x\right)$$

$$= x^{x^x}\cdot x^x\cdot\left(\frac{1}{x} + \log x + \log^2 x\right)$$

Form 5: Problems based on the sum of two more than two exponential composite (or, other) functions. One should note that logarithmic differentiation of the sum of two or more than two exponential composite functions are performed by using the theorem of differential coefficient of the sum of two or more than two differentiable functions and the differential coefficient of each addend being the function of x is obtained seperately by using the rule of logarithmic differentiation and lastly adding the differential coefficient of each addend, we get the differential coefficient of the whole function which is given as the sum of two or more than two exponential composite functions.

Note: We should remember that while finding the differential coefficient of each addend being the function of x, we must make the substitution $u, v, w, \ldots$ for each addend and then operation of taking logarithm should be performed.

Solved Examples

Find the differential coefficient of the following.

1. $y = x^{\tan x} + (\tan x)^{\cot x}, \left(0 < x < \frac{\pi}{2}\right)$

Solution: $y = x^{\tan x} + (\tan x)^{\cot x}$, defined for $x > 0$, $\tan x > 0$

$$\Rightarrow \frac{dy}{dx} = \frac{d}{dx}(x^{\tan x}) + \frac{d}{dx}((\tan)^{\cot x}) \quad ...(i)$$

Now, on letting $u = x^{\tan x}$, we have

$\log u = \log(x^{\tan x}) = \tan x\cdot\log x$

$$\Rightarrow \frac{1}{u}\cdot\frac{du}{dx} = \tan x\cdot\frac{1}{x} + \log x\cdot\sec^2 x$$

$$\Rightarrow \frac{du}{dx} = u\left(\frac{\tan x}{x} + \sec^2 x\cdot\log x\right)$$

$$= x^{\tan x}\left(\frac{\tan x}{x} + \sec^2 x\cdot\log x\right) \quad \text{...(ii)}$$

Again, on letting $v = (\tan x)^{\cot x}$, we have

$\log v = \log(\tan x)^{\cot x} = \cot x\cdot\log\tan x$

$$\Rightarrow \frac{1}{v}\cdot\frac{dv}{dx} = \frac{\cot x}{\tan x}\cdot\sec^2 x - \operatorname{cosec}^2 x\cdot\log\tan x$$

$$\Rightarrow \frac{dv}{dx} = v\cdot\left(\frac{\cot x}{\tan x}\cdot\sec^2 x - \operatorname{cosec}^2 x\cdot\log\tan x\right)$$

$$= (\tan x)^{\cot x}\cdot\left(\frac{\cot x}{\tan x}\cdot\sec^2 x - \operatorname{cosec}^2 x\cdot\log\tan x\right) \quad \text{...(iii)}$$

Putting (ii) and (iii) in (i), we have

$$\frac{dy}{dx} = x^{\tan x}\left(\frac{\tan x}{x} + \sec^2 x\cdot\log x\right) + (\tan x)^{\cot x}\cdot\left(\frac{\cot x}{\tan x}\cdot\sec^2 x - \operatorname{cosec}^2 x\cdot\log\tan x\right)$$

2. $y = (\sin x)^{\tan x} + (\tan x)^{\sin x}$, $\left(0 < x < \frac{\pi}{2}\right)$

Solution: $y = (\sin x)^{\tan x} + (\tan x)^{\sin x}$

$$\Rightarrow \frac{dy}{dx} = \frac{d}{dx}(\sin x)^{\tan x} + \frac{d}{dx}(\tan x)^{\sin x} \quad \text{...(i)}$$

Now, on letting $u = (\sin x)^{\tan x}$, we have

$\log u = \tan x\cdot\log\sin x$

$$\Rightarrow \frac{1}{u}\cdot\frac{du}{dx} = \tan x\cdot\frac{1}{\sin x}\cdot\cos x + (\log\sin x)\sec^2 x$$

$$= \tan x\cdot\cot x + \sec^2 x\cdot\log\sin x$$

$$\Rightarrow \frac{du}{dx} = u\left(1 + \sec^2 x\cdot\log\sin x\right)$$

$$= (\sin x)^{\tan x}\cdot\left(1 + \sec^2 x\cdot\log\sin x\right) \quad \text{...(ii)}$$

Again, on letting $v = (\tan x)^{\sin x}$, we have

$\log v = \sin x\log\tan x$

$$\Rightarrow \frac{1}{v}\cdot\frac{dv}{dx} = \sin x\cdot\frac{1}{\tan x}\cdot\sec^2 x + (\log\tan x)\cos x$$

$$= \sec x + \cos x\cdot\log\tan x$$

$$\Rightarrow \frac{dv}{dx} = v\cdot(\sec x + \cos x\cdot\log\tan x)$$

$$= (\tan x)^{\sin x}\cdot(\sec x + \cos x\cdot\log\tan x) \quad \text{...(iii)}$$

Putting (ii) and (iii) in (i), we have

$$\frac{dy}{dx} = (\sin x)^{\tan x}\cdot\left(1 + \sec^2 x\cdot\log\sin x\right) + (\tan x)^{\sin x}\cdot(\sec x + \cos x\cdot\log\tan x)$$

3. $y = x^{\sin x} + (\sin x)^{\frac{3}{2}}$, $0 < x < \frac{\pi}{2}$

Solution: $y = x^{\sin x} + (\sin x)^{\frac{3}{2}}$

$$\Rightarrow \frac{dy}{dx} = \frac{d}{dx}\left(x^{\sin x}\right) + \frac{d}{dx}(\sin x)^{\frac{3}{2}} \quad \text{...(i)}$$

Now on letting $u = x^{\sin x}$, we have

$\log u = \sin x\cdot\log x$

$$\Rightarrow \frac{1}{u}\cdot\frac{du}{dx} = \sin x\cdot\frac{1}{x} + (\log x)\cos x$$

$$\Rightarrow \frac{du}{dx} = u\left(\frac{\sin x}{x} + \cos x\cdot\log x\right)$$

$$= x^{\sin x}\cdot\left(\frac{\sin x}{x} + \cos x\cdot\log x\right) \quad \text{...(ii)}$$

Again, on letting $v = (\sin x)^{\frac{3}{2}}$

$$\Rightarrow \frac{dv}{dx} = \frac{3}{2}(\sin x)^{\frac{1}{2}}\cdot\cos x$$

$$\Rightarrow \frac{dv}{dx} = \frac{3}{2}\cos x\sqrt{\sin x} \quad \text{...(iii)}$$

Putting (ii) and (iii) in (i), we have

$$\frac{dy}{dx} = x^{\sin x}\left(\frac{\sin x}{x} + \cos x\cdot\log x\right) + \frac{3}{2}\cos x\cdot\sqrt{\sin x}$$

4. $y = (\sin x)^x + x\sin^{-1} x$, $0 < x < 1$

Solution: $y = (\sin x)^x + x\sin^{-1} x$

$$\Rightarrow \frac{dy}{dx} = \frac{d}{dx}(\sin x)^x + \frac{d}{dx}\left(x \sin^{-1} x\right) \quad \text{...(i)}$$

Now on letting $u = (\sin x)^x$, we have

$\log u = x \log \sin x$

$$\Rightarrow \frac{1}{u} \cdot \frac{du}{dx} = x \cdot \frac{1}{\sin x} \cdot \cos x + \log \sin x \cdot 1$$

$$\Rightarrow \frac{du}{dx} = u(x \cot x + \log \sin x)$$

$$= (\sin x)^x (x \cot x + \log \sin x) \quad \text{...(ii)}$$

Again on letting $v = x \sin^{-1} x$, we have

$\log v = \log x + \log \sin^{-1} x$

$$\Rightarrow \frac{1}{v} \cdot \frac{dv}{dx} = \frac{1}{x} + \frac{1}{\sin^{-1} x} \cdot \frac{1}{\sqrt{1 - x^2}}$$

$$\Rightarrow \frac{dv}{dx} = v \cdot \left(\frac{1}{x} + \frac{1}{\sin^{-1} x \sqrt{1 - x^2}}\right)$$

$$= x \sin^{-1} x \left(\frac{1}{x} + \frac{1}{\sin^{-1} x \sqrt{1 - x^2}}\right)$$

$$= \sin^{-1} x + \frac{x}{\sqrt{1 - x^2}} \quad \text{...(iii)}$$

Putting (ii) and (iii) in (i), we have

$$\frac{dy}{dx} = (\sin x)^x \cdot (x \cot x + \log \sin x) + \sin^{-1} x + \frac{x}{\sqrt{1 - x^2}}$$

5. $y = \left(\frac{1}{x}\right)^x + x^{\frac{1}{x}}, x > 0$

Solution: $y = \left(\frac{1}{x}\right)^x + x^{\frac{1}{x}}$

$$\Rightarrow \frac{dy}{dx} = \frac{d}{dx}\left(\frac{1}{x}\right)^x + \frac{d}{dx}\left(x^{\frac{1}{x}}\right) \quad \text{...(i)}$$

Now on letting $u = \left(\frac{1}{x}\right)^x$, we have

$$\log u = x \cdot \log\left(\frac{1}{x}\right) = x \log\left(x^{-1}\right) = -x \log x$$

$$\Rightarrow \frac{1}{u} \cdot \frac{du}{dx} = -x \cdot \frac{1}{x} + \log x(-1) = -(1 + \log x)$$

$$\Rightarrow \frac{du}{dx} = -u(\log x + 1)$$

$$\Rightarrow \frac{du}{dx} = -\left(\frac{1}{x}\right)^x (\log x + 1) \quad \text{...(ii)}$$

Again on letting $v = x^{\frac{1}{x}}$, we have

$$\log v = \frac{1}{x} \log x$$

$$\Rightarrow \frac{1}{v} \cdot \frac{dv}{dx} = \frac{1}{x} \cdot \frac{1}{x} - \frac{1}{x^2} \log x = \frac{1}{x^2} - \frac{1}{x^2} \log x$$

$$= \frac{(1 - \log x)}{x^2}$$

$$\Rightarrow \frac{dv}{dx} = \frac{v(1 - \log x)}{x^2}$$

$$= \frac{x^{\frac{1}{x}} \cdot (1 - \log x)}{x^2}$$

Putting (ii) and (iii) in (i), we have

$$\frac{dy}{dx} = -\left(\frac{1}{x}\right)^x (1 + \log x) + \frac{x^{\frac{1}{x}} (1 - \log x)}{x^2}$$

6. $y = x^{2x} + (2x)^x, x > 0$

Solution: $y = x^{2x} + (2x)^x$

$$\Rightarrow \frac{dy}{dx} = \frac{d}{dx}\left(x^{2x}\right) + \frac{d}{dx}(2x)^x \quad \text{...(i)}$$

Now on putting $u = x^{2x}$, we have

$\log u = \log (x^{2x}) = 2x \log x$

$$\Rightarrow \frac{1}{u} \cdot \frac{du}{dx} = 2 \cdot 1 \cdot \log x + \frac{2x}{x} = 2 \log x + 2$$

$$\Rightarrow \frac{du}{dx} = u(2 \log x + 2) = 2 \cdot x^{2x} \cdot (1 + \log x) \quad \text{...(ii)}$$

Again, on putting $v = (2x)^x$, we have

$\log v = \log (2x)^x = x \log 2x$

$$\Rightarrow \frac{1}{v} \cdot \frac{dv}{dx} = x \cdot \frac{1}{2x} \cdot 2 + \log 2x = (1 + \log 2x)$$

$$\Rightarrow \frac{dv}{dx} = v(1+\log 2x) = (2x)^x (1+\log 2x) \quad ...(iii)$$

Putting (ii) and (iii) in (i), we have

$$\frac{dy}{dx} = 2 \cdot x^{2x}(1+\log x) + (2x)^x \cdot (1+\log 2x)$$

7. $y = x^{\log x} + (\log x)^x, x > 1$

Solution: $y = x^{\log x} + (\log x)^x$

$$\Rightarrow \frac{dy}{dx} = \frac{d}{dx}\left(x^{\log x}\right) + \frac{d}{dx}(\log x)^x \quad ...(i)$$

Now, on putting $u = x^{\log x}$, we have

$\log u = \log x \cdot \log x = (\log x)^2$

$$\Rightarrow \frac{1}{u}, \frac{du}{dx} = 2\log x \cdot \frac{1}{x}$$

$$\Rightarrow \frac{du}{dx} = 2u \cdot \log x \cdot \frac{1}{x} = \frac{2x^{\log x} \cdot \log x}{x} \quad ...(ii)$$

Again on putting $v = (\log x)^x$, we have

$\log v = x \log \log x$

$$\Rightarrow \frac{1}{v}\cdot\frac{dv}{dx} = \frac{x}{\log x}\cdot\frac{1}{x} + \log\log x = \frac{1}{\log x} + \log\log x$$

$$\Rightarrow \frac{dv}{dx} = v\left(\frac{1}{\log x} + \log\log x\right)$$

$$= (\log x)^x\left(\frac{1}{\log x} + \log\log x\right) \quad ...(iii)$$

Putting (ii) and (iii) in (i), we have

$$\frac{dy}{dx} = \frac{2\cdot x^{\log x}\cdot\log x}{x} + (\log x)^x\cdot\left(\frac{1}{\log x} + \log\log x\right)$$

Form 6: Miscellaneous problems

Solved Examples

1. If $x^y = e^{x-y}, x > 0, x \neq \frac{1}{e}$, show that $\frac{dy}{dx} = \frac{\log x}{(1+\log x)^2}$.

Solution: $x^y = e^{x-y}$

$$\Rightarrow \log\left(x^y\right) = \log\left(e^{x-y}\right)$$

$$\Rightarrow y\log x = (x-y)\log e = (x-y) \; (\because \log e = 1)$$

$$\Rightarrow y\log x + y = x$$

$$\Rightarrow y(\log x + 1) = x$$

$$\Rightarrow y = \frac{x}{1+\log x}$$

$$\Rightarrow \frac{dy}{dx} = \frac{d}{dx}\left(\frac{x}{1+\log x}\right) = \frac{(1+\log x)\cdot 1 - x\left(0+\frac{1}{x}\right)}{(1+\log x)^2}$$

$$= \frac{1+\log x - 1}{(1+\log x)^2}$$

$$\Rightarrow \frac{dy}{dx} = \frac{\log x}{(1+\log x)^2}$$

2. If $x^m \cdot y^n = (x+y)^{m+n}, x > 0, y > 0$, show that $\frac{dy}{dx} = \frac{y}{x}, nx \neq my$.

Solution: $x^m y^n = (x+y)^{m+n}$

$$\Rightarrow \log\left(x^m y^n\right) = \log(x+y)^{m+n}$$

$$\Rightarrow m\log x + n\log y = (m+n)\log(x+y)$$

$$\Rightarrow \frac{m}{x} + \frac{n}{y}\cdot\frac{dy}{dx} = \left(\frac{m+n}{x+y}\right)\cdot\left(1+\frac{dy}{dx}\right)$$

$$\Rightarrow \left(\frac{m+n}{x+y} - \frac{n}{y}\right)\cdot\frac{dy}{dx} = \frac{m}{x} - \frac{m+n}{x+y} = \frac{mx+my-mx-nx}{x(x+y)}$$

$$\Rightarrow \frac{my+ny-nx-ny}{y(x+y)}\cdot\frac{dy}{dx} = \frac{my-nx}{x(x+y)}$$

$$\Rightarrow \frac{my-nx}{y(x+y)}\cdot\frac{dy}{dx} = \frac{my-nx}{x(x+y)}$$

$$\Rightarrow \frac{dy}{dx} = \frac{y}{x}, \text{ since, } nx \neq my$$

Note: The result holds for $x \neq 0, y \neq 0$ (shown later)

3. If $y = \log\left(\frac{x}{1+x}\right)$, $x \notin [-1, 0]$, find its derivative.

Solution: $y = \log\left(\frac{x}{1+x}\right) \neq \log x - \log(1+x)$ for $x \notin [-1, 0]$

When $x>0,\ x+1>1>0 \Rightarrow \frac{x}{1+x}>0$ which means y is defined for $x>0$.

Again when $x<-1,\ x+1<0 \Rightarrow \frac{x}{1+x}>0$ which means y is defined for $x<-1$.

$\therefore$ y is defined for $x \in R - [-1, 0] = (-\infty, -1) \cup (0, \infty)$

i.e. y is defined for $x \notin [-1, 0]$

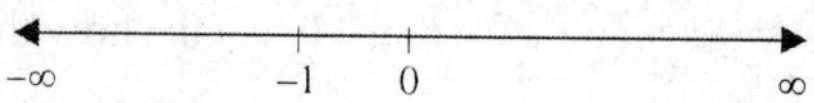

Hence, $\frac{dy}{dx} = \frac{d}{dx}\left(\log\left(\frac{x}{1+x}\right)\right)$

$$= \frac{x+1}{x} \cdot \left[\frac{(x+1)\cdot 1 - x \cdot 1}{(1+x^2)^2}\right] = \frac{(x+1)}{x} \cdot \left[\frac{x+1-x}{(1+x)^2}\right]$$

$$= \frac{1}{x(x+1)},\ x \notin [-1, 0]$$

Remark:

1. In the above problem $\log\left(\frac{x}{1+x}\right) \neq \log x - \log(1+x)$ since $\log x$ is undefined for $x \le 0$ and also $\log(x+1)$ is undefined for $x \le -1$.
2. $x \notin [-1, 0] \Leftrightarrow x \in R - [-1, 0]$

On working rule of problems belonging to second type

When the interval or the quadrant in which $f(x)$ = given function of x is positive is not given or the condition imposed on the independent variable x which makes $f(x) \neq 0$ is given (or, not given), we adopt the following working rule to find the differential coefficient of the given problems by using logarithmic differentiation.

Step 1: Take the modulus on both sides of the equation defining y as a function of x, i.e. $y = f(x) \Leftrightarrow |y| = |f(x)|$

Step 2: Take the logarithm on both sides of the equation defining modulus of y as modulus of $f(x)$, i.e. $\log|y| = \log|f(x)|$

Step 3: Differentiate the equation $\log|y| = \log|f(x)|$ using the rule:

(i) $\frac{d}{dx}\log|y| = \frac{1}{y}\cdot\frac{dy}{dx};\ y \neq 0.$

(ii) $\frac{d}{dx}\log|f(x)| = \frac{1}{f(x)}\cdot\frac{d\,f(x)}{dx};\ f(x) \neq 0.$

(iii) $\frac{dy}{dx} = y\cdot\frac{d}{dx}\log|f(x)|;\ f(x) \neq 0.$

Remember: **(a)** After taking logarithm, we are required to use the following formulas.

1. $\log|u \cdot v| = \log|u| + \log|v|\ ;\ u \neq 0,\ v \neq 0.$
2. $\log\left|\frac{u}{v}\right| = \log|u| - \log|v|;\ u \neq 0,\ v \neq 0.$
3. $\log|u^v| = \log|u|^v = v\log|u|\ ;\ u > 0,\ v \in R.$
4. $\log|u^n| = \log|u|^n = n\log|u|,\ n \in R,\ u > 0.$

where u and v are functions of x.

(b) The above working rule may be remembered as "MLAD" the letters being in order which means

M = take mod
L = take log
A = apply logarithmic formulas
D = differentiate.

(c) We take the modulus of given function only when the function may be negative because we can use logarithmic differentiation only when the function concerned is positive in its domain; i.e. the sign of absolute value is written only when the expression standing under the sign of logarithm may have a negative value in its domain.

(d) It should be noted that the derivative of any function of y, say $F(y)$, w.r.t. x is $\frac{d}{dy}[F(y)]\cdot\frac{dy}{dx}$ i.e. $F'(y)\cdot\frac{dy}{dx}$ obtained by the rule of differentiating a function of a function.

Problems based on second type

Form 1: Problems based on irrational functions.

Solved Examples

Find the differential coefficient of the following.

1. $y=\sqrt[3]{1-x^2}$

Solution: $y=\sqrt[3]{1-x^2}=\left(1-x^2\right)^{\frac{1}{3}}$

$$\Rightarrow |y|=\left|\left(1-x^2\right)^{\frac{1}{3}}\right|=\left|\left(1-x^2\right)\right|^{\frac{1}{3}}$$

$$\Rightarrow \log|y|=\frac{1}{3}\log\left|\left(1-x^2\right)\right|$$

$$\Rightarrow \frac{1}{y}\cdot\frac{dy}{dx}=\frac{-2x}{3\left(1-x^2\right)}$$

$$\Rightarrow \frac{dy}{dx}=\frac{-2y\cdot x}{3\left(1-x^2\right)}$$

$$\Rightarrow \frac{dy}{dx}=\frac{2x\sqrt[3]{1-x^2}}{3\left(1-x^2\right)}=-\frac{2x}{3}\cdot\frac{1}{\left(1-x^2\right)^{\frac{2}{3}}},\ x^2\neq 1$$

2. $y=\sqrt{\left(\frac{x+1}{x-1}\right)}$

Solution: $y=\left(\frac{x+1}{x-1}\right)^{\frac{1}{2}}$

$$\Rightarrow |y|=\left|\frac{x+1}{x-1}\right|^{\frac{1}{2}}$$

$$\Rightarrow \log|y|=\frac{1}{2}\log\left|\frac{x+1}{x-1}\right|=\frac{1}{2}\log|(x+1)|-\frac{1}{2}\log|(x-1)|$$

$$\Rightarrow \frac{1}{y}\cdot\frac{dy}{dx}=\frac{-2}{2\left(x^2-1\right)}=-\frac{1}{\left(x^2-1\right)}$$

$$\Rightarrow \frac{dy}{dx}=-\frac{y}{\left(x^2-1\right)}$$

$$=\sqrt{\left(\frac{x+1}{x-1}\right)}\cdot\frac{-1}{\left(x^2-1\right)},\ |x|>1$$

3. $y=\sqrt[3]{\cos x}$

Solution: $y=\sqrt[3]{\cos x}=(\cos x)^{\frac{1}{3}}$

$$\Rightarrow |y|=|\cos x|^{\frac{1}{3}}$$

$$\Rightarrow \log|y|=\frac{1}{3}\log|\cos x|$$

$$\Rightarrow \frac{1}{y}\cdot\frac{dy}{dx}=\frac{1}{3}\cdot\frac{(-\sin x)}{\cos x}=-\frac{1}{3}\tan x$$

$$\Rightarrow \frac{dy}{dx}=-\frac{1}{3}\cdot y\cdot\tan x=-\frac{1}{3}\cdot\sqrt[3]{\cos x}\cdot\tan x$$

4. $y=\sqrt{\frac{1-\cos x}{1+\cos x}}$

Solution: $y=\left(\frac{1-\cos x}{1-\cos x}\right)^{\frac{1}{2}}$

$$\Rightarrow |y|=\left|\frac{1-\cos x}{1+\cos x}\right|^{\frac{1}{2}}=\left|\tan^2\frac{x}{2}\right|^{\frac{1}{2}}=\left|\tan\frac{x}{2}\right|^{2\cdot\frac{1}{2}}=\left|\tan\frac{x}{2}\right|$$

$$\Rightarrow \log|y|=\log\left|\tan\frac{x}{2}\right|$$

$$\Rightarrow \frac{1}{y}\cdot\frac{dy}{dx}=\frac{1}{\tan\frac{x}{2}}\cdot\left(\sec^2\frac{x}{2}\right)\cdot\frac{1}{2}$$

$$= \frac{1}{2} \cdot \frac{\cos\frac{x}{2}}{\sin\frac{x}{2}} \cdot \frac{1}{\cos^2\frac{x}{2}} = \frac{1}{2\sin\frac{x}{2}\cos\frac{x}{2}}$$

$$\Rightarrow \frac{dy}{dx} = y \cdot \frac{1}{\sin x} = \sqrt{\frac{1-\cos x}{1+\cos x}} \cdot \operatorname{cosec} x,\ x \neq n\pi$$

5. $y = \sqrt{\dfrac{\cos x - \sin x}{\cos x + \sin x}}$

Solution: $y = \left(\dfrac{\cos x - \sin x}{\cos x + \sin x}\right)^{\frac{1}{2}}$

$$\Rightarrow |y| = \left|\frac{\cos x - \sin x}{\cos x + \sin x}\right|^{\frac{1}{2}}$$

$$\Rightarrow \log|y| = \frac{1}{2}\log\left|\frac{\cos x - \sin x}{\cos x + \sin x}\right|$$

$$= \frac{1}{2}\log|\cos x - \sin x| - \frac{1}{2}\log|\cos x + \sin x|$$

$$\Rightarrow \frac{1}{y} \cdot \frac{dy}{dx} = \frac{-(\cos x + \sin x)}{2(\cos x - \sin x)} - \frac{(\cos x - \sin x)}{2(\cos x + \sin x)}$$

$$= \frac{-(\sin x + \cos x)^2 - (\cos x - \sin x)^2}{2(\cos^2 x - \sin^2 x)}$$

$$= \frac{-(1 + 2\sin x \cos x) - (1 - 2\sin x \cos x)}{2\cos 2x} = \frac{-2}{2\cos 2x}$$

$$= -\sec 2x$$

$$\Rightarrow \frac{dy}{dx} = -y \cdot \sec 2x$$

$$= -\sqrt{\frac{\cos x - \sin x}{\cos x + \sin x}} \cdot \sec 2x$$

6. $y = |\cos x - \sin x|$

Solution: $y = |\cos x - \sin x|$

$$\Rightarrow \log y = \log|\cos x - \sin x|$$

$$\Rightarrow \frac{1}{y} \cdot \frac{dy}{dx} = \frac{(-\sin x - \cos x)}{(\cos x - \sin x)} = \frac{-(\sin x + \cos x)}{(\cos x - \sin x)}$$

$$\Rightarrow \frac{dy}{dx} = -y \cdot \frac{(\sin x + \cos x)}{(\cos x - \sin x)}$$

$$= \frac{-|\cos x - \sin x| \cdot (\sin x + \cos x)}{(\cos x - \sin x)},\ (\cos x - \sin x) \neq 0$$

Note: Differentiation of mod of a function can be performed easily by logarithmic differentiation.

Form 2: Problems based on product of two or more than two differentiable functions.

Solved Examples

Find d.c. of y from the following.

1. $x^m y^n = (x+y)^{m+n}$, $nx \neq my$

Solution: $x^m y^n = (x+y)^{m+n}$

$$\Rightarrow \left|x^m \cdot y^n\right| = \left|(x+y)^{m+n}\right|$$

$$\Rightarrow |x|^m \cdot |y|^n = |x+y|^{m+n}$$

$$\Rightarrow m\log|x| + n\log|y| = (m+n)\log|x+y|$$

$$\Rightarrow \frac{m}{x} + \frac{n}{y} \cdot \frac{dy}{dx} = \frac{m+n}{x+y} \cdot \left(1 + \frac{dy}{dx}\right)$$

$$\Rightarrow \left(\frac{n}{y} - \frac{m+n}{x+y}\right) \cdot \frac{dy}{dx} = \frac{m+n}{x+y} - \frac{m}{x}$$

$$\Rightarrow \frac{(nx+ny) - (my+ny)}{y \cdot (x+y)} \cdot \frac{dy}{dx}$$

$$= \frac{(mx+ny) - (mx+my)}{x(x+y)}$$

$$\Rightarrow \frac{nx - my}{y(x+y)} \cdot \frac{dy}{dx} = \frac{nx - my}{x(x+y)}$$

$$\Rightarrow \frac{dy}{dx} = \frac{y}{x};\ x \neq 0,\ y \neq 0$$

2. $y = x(x^2+1)$

Solution: $y = x(x^2+1)$

$$\Rightarrow |y| = \left|x \cdot (x^2+1)\right| = |x| \cdot \left|(x^2+1)\right|$$

$$\Rightarrow |y| = |x| \cdot (x^2+1)$$

$$\Rightarrow \log|y| = \log|x| + \log\left(x^2+1\right)$$

$$\Rightarrow \frac{1}{y}\cdot\frac{dy}{dx} = \frac{1}{x} + \frac{2x}{x^2+1} = \frac{1+3x^2}{x\left(x^2+1\right)}$$

$$\Rightarrow \frac{dy}{dx} = y\cdot\left[\frac{1+3x^2}{x\left(x^2+1\right)}\right] = x\cdot\left(x^2+1\right)\cdot\left[\frac{1+3x^2}{x\cdot\left(x^2+1\right)}\right]$$

$$\Rightarrow \frac{dy}{dx} = \left(1+3x^2\right)$$

Note: The above problem (2) can be differentiated in a much simpler way by using directly the product formulas but above method has been used only to show the procedure of logarithmic differentiation is also applicable.

3. $y = x\log x$

Solution: $y = x\log x, x > 0$

$$\Rightarrow |y| = |x\log x| = |x|\cdot|\log x|$$

$$\Rightarrow \log|y| = \log|x| + \log|\log x|$$

$$\Rightarrow \frac{1}{y}\cdot\frac{dy}{dx} = \frac{1}{x} + \frac{1}{x\log x} = \frac{\log x+1}{x\cdot\log x}$$

$$\Rightarrow \frac{dy}{dx} = y\cdot\left[\frac{\log x+1}{x\log x}\right] = x\log x\cdot\left[\frac{\log x+1}{x\log x}\right] = \log x+1$$

4. $y = x\log y$

Solution: $y = x\log y$

$$\Rightarrow |y| = |x|\cdot|\log y|$$

$$\Rightarrow \log|y| = \log|x| + \log|\log y|$$

$$\Rightarrow \frac{1}{y}\cdot\frac{dy}{dx} = \frac{1}{x} + \frac{1}{y\log y}\cdot\frac{dy}{dx}$$

$$\Rightarrow \left(\frac{1}{y} - \frac{1}{y\log y}\right)\frac{dy}{dx} = \frac{1}{x}$$

$$\Rightarrow \frac{dy}{dx} = \frac{1}{x}\cdot\left(\frac{x\log y\cdot\log y}{\log y-1}\right) = \frac{(\log y)^2}{\log y-1} = \frac{\log^2 y}{\log y-1}$$

Note: The problems (1) and (4) belong to problems of implicit function which can be done by using the rule of implicit differentiation but above method has been used only to show that the procedure of logarithmic differentiation is also applicable in that case.

5. $y = \sin x\cdot\sin 2x\cdot\sin 3x\cdot\sin 4x$

Solution: $y = \sin x\cdot\sin 2x\cdot\sin 3x\cdot\sin 4x$

$$\Rightarrow |y| = |\sin x|\cdot|\sin 2x||\sin 3x|\cdot|\sin 4x|$$

$$\Rightarrow \log|y| = \log|\sin x| + \log|\sin 2x| + \log|\sin 3x| + \log|\sin 4x|$$

$$\Rightarrow \frac{1}{y}\cdot\frac{dy}{dx} = \frac{\cos x}{\sin x} + \frac{2\cos 2x}{\sin 2x} + \frac{3\cos 3x}{\sin 3x} + \frac{4\cos 4x}{\sin 4x}$$

$$\Rightarrow \frac{dy}{dx} = y\cdot(\cot x + 2\cot 2x + 3\cot 3x + 4\cot 4x)$$

$$\Rightarrow \frac{dy}{dx} = (\sin x\cdot\sin 2x\cdot\sin 3x\sin 4x)\cdot(\cot x + 2\cot 2x + 3\cot 3x + 4\cot 4x)$$

6. $y = x\cdot|x|$

Solution: $y = x\cdot|x|$

$$\Rightarrow |y| = |x|\cdot|x|$$

$$\Rightarrow \log|y| = \log|x| + \log|x|$$

$$\Rightarrow \frac{1}{y}\cdot\frac{dy}{dx} = \frac{2}{x}$$

$$\Rightarrow \frac{dy}{dx} = 2\cdot\frac{y}{x} = \frac{2\cdot x|x|}{x} = 2|x|$$

Precaution:

1. Whenever we have to find the differential coefficient of mod of a function by logarithmic differentiation, the operation of taking the modulus on both sides of the equation defining y as a function of x is not performed but directly the operation of taking logarithm on both sides of the equation defining y as a function of x is performed just before the differentiation.

2. Whenever we have to find the differential coefficient of a function using the mod of a function, the operation of taking the modulus and logarithm on both sides of the equation just defining y as a function of x must be performed respectively before the differentiation.

3. Whenever we have to find the differential coefficient of a logarithmic function (i.e. $y = \log f(x)$ or $\log \log \log \ldots f(x)$), we have not to perform both the operation of taking the modulus and logarithmic function of x but simply, we have to use the formula:

$$\frac{d}{dx} \log f(x) = \frac{f'(x)}{f(x)},\ f(x) > 0.$$

Examples

Find the differential coefficient of the following.

1. $y = |x|$

Solution: $y = |x|$

$$\Rightarrow \log y = \log|x|$$

$$\Rightarrow \frac{1}{y} \cdot \frac{dy}{dx} = \frac{1}{x}$$

$$\Rightarrow \frac{dy}{dx} = \frac{y}{x} = \frac{|x|}{x},\ x \neq 0$$

2. $y = \dfrac{|x|}{x}$

Solution: $y = \dfrac{|x|}{x},\ x \neq 0$

$\Rightarrow |y| = \dfrac{|x|}{|x|} = 1$ ($\because$ mod of a mod of a function = mod of a function)

$$\Rightarrow \log|y| = \log 1 = 0\ (\because \log 1 = 0)$$

$$\Rightarrow \frac{1}{y} \cdot \frac{dy}{dx} = 0$$

$$\Rightarrow \frac{dy}{dx} = 0$$

3. $y = \log(x^2 + 1)$

Solution: $y = \log(x^2 + 1)$

$$\Rightarrow \frac{dy}{dx} = \left(\frac{1}{x^2 + 1}\right) \cdot 2x = \frac{2x}{1 + x^2}$$

Form 3: Problems based on quotient of two or more than two differentiable functions.

Solved Examples

Find the differential coefficient of the following.

1. $y = \dfrac{\sin^3 x \cdot \sqrt{x^2 + 1}}{e^{4x} \cdot (x^3 + 5)^{\frac{1}{3}}}$

Solution: $y = \dfrac{\sin^3 x \cdot \sqrt{x^2 + 1}}{e^{4x} \cdot (x^3 + 5)^{\frac{1}{3}}}$

$$\Rightarrow \log|y| = 3\log|\sin x| + \frac{1}{2}\log(x^2 + 1) - 4x - \frac{1}{3}\log|x^3 + 5|$$

$$\Rightarrow \frac{1}{y} \cdot \frac{dy}{dx} = 3\frac{\cos x}{\sin x} + \frac{2x}{2(x^2 + 1)} - 4 - \frac{3x^2}{3(x^3 + 5)}$$

$$\Rightarrow \frac{dy}{dx} = y \cdot \left(3\cot x + \frac{x}{x^2 + 1} - 4 - \frac{x^2}{x^3 + 5}\right)$$

$$= \frac{\sin^3 x \cdot \sqrt{x^2 + 1}}{e^{4x} \cdot (x^3 + 5)^{\frac{1}{3}}} \cdot \left(3\cot x + \frac{x}{x^2 + 1} - 4 - \frac{x^2}{x^3 + 5}\right)$$

2. $y = \dfrac{\sqrt{2x - 1} \cdot \sin^{-1} x}{(x^2 + 2)^{\frac{3}{2}} \cdot \sqrt[3]{\tan^{-1} 2x}}$

Solution: $y = \dfrac{\sqrt{2x - 1} \cdot \sin^{-1} x}{(x^2 + 2)^{\frac{3}{2}} \cdot \sqrt[3]{\tan^{-1} 2x}},\ x > \dfrac{1}{2}$

$$\Rightarrow \log|y| = \frac{1}{2}\log|(2x - 1)| + \log\left|\sin^{-1} x\right| - \frac{3}{2}\log(x^2 + 2) - \frac{1}{3}\log\left|\tan^{-1} 2x\right|$$

$$\Rightarrow \frac{1}{y}\cdot\frac{dy}{dx} = \frac{2}{2(2x-1)} + \frac{1}{\left(\sin^{-1}x\right)\sqrt{1-x^2}} - \frac{3\cdot 2x}{2\cdot\left(x^2+2\right)} - \frac{2}{3\tan^{-1}2x\left(1+4x^2\right)}$$

$$\Rightarrow \frac{dy}{dx} = y\cdot\left(\frac{1}{2x-1} + \frac{1}{\left(\sin^{-1}x\right)\sqrt{1-x^2}} - \frac{3x}{x^2+2} - \frac{2}{3\tan^{-1}2x\left(1+4x^2\right)}\right)$$

$$= \frac{\sqrt{2x-1}\cdot\sin^{-1}x}{\left(x^2+2\right)^{\frac{3}{2}}\cdot\sqrt[3]{\tan^{-1}2x}}\cdot k$$

where $k = \left(\frac{1}{2x-1} + \frac{1}{\left(\sin^{-1}x\right)\cdot\sqrt{1-x^2}} - \frac{3x}{x^2+2} - \frac{2}{3\tan^{-1}2x\left(1+4x^2\right)}\right)$

3. $y = \dfrac{x^2\cdot(2x+1)^2}{\sqrt{x+1}}$

Solution: $y = \dfrac{x^2\cdot(2x+1)^2}{\sqrt{x+1}}$

$$\Rightarrow \log|y| = 2\log|x| + 2\log|(2x+1)| - \frac{1}{2}\log|(x+1)|$$

$$\Rightarrow \frac{dy}{dx} = y\cdot\left(\frac{2}{x} + \frac{4}{2x+1} - \frac{1}{2(x+1)}\right)$$

$$\Rightarrow \frac{dy}{dx} = \frac{x^2\cdot(2x+1)^2}{\sqrt{x+1}}\cdot\left(\frac{2}{x} + \frac{4}{2x+1} - \frac{1}{2(x+1)}\right)$$

4. $y = \dfrac{(x+1)^{\frac{3}{4}}\cdot\left(x^2+3\right)^{\frac{5}{3}}\cdot\left(x^3+7\right)^7}{\sqrt{x}\cdot\left(\sqrt{x}+2\right)^{\frac{4}{3}}}$

Solution: $y = \dfrac{(x+1)^{\frac{3}{4}}\cdot\left(x^2+3\right)^{\frac{5}{3}}\cdot\left(x^3+7\right)^7}{\sqrt{x}\cdot\left(\sqrt{x}+2\right)^{\frac{4}{3}}}, x>0$

$$\Rightarrow \log|y| = \frac{3}{4}\log|(x+1)| + \frac{5}{3}\log\left(x^2+3\right) + 7\log\left|\left(x^3+7\right)\right| - \frac{1}{2}\log|x| - \frac{4}{3}\log\left|\sqrt{x}+2\right|$$

$$\Rightarrow \frac{1}{y}\cdot\frac{dy}{dx} = \frac{3}{4(x+1)} + \frac{5\cdot 2\cdot x}{3\left(x^2+3\right)} + \frac{7\cdot 3x^2}{\left(x^3+7\right)} - \frac{1}{2x} - \frac{4}{3\left(\sqrt{x}+2\right)\cdot 2\sqrt{x}}$$

$$\Rightarrow \frac{dy}{dx} = y\cdot\left(\frac{3}{4(x+1)} + \frac{10x}{\left(3x^2+9\right)} + \frac{21x^2}{\left(x^3+7\right)} - \frac{1}{2x} - \frac{2}{3\sqrt{x}\cdot\left(\sqrt{x}+2\right)}\right)$$

$$\Rightarrow \frac{dy}{dx} = \frac{(x+1)^{\frac{3}{4}}\cdot\left(x^2+3\right)^{\frac{5}{3}}\cdot\left(x^3+7\right)^7}{\sqrt{x}\cdot\left(\sqrt{x}+2\right)^{\frac{4}{3}}}\cdot\left(\frac{3}{4(x+1)} + \frac{10x}{\left(3x^2+9\right)} + \frac{21x^2}{\left(x^3+7\right)} - \frac{1}{2x} - \frac{2}{3\sqrt{x}\left(\sqrt{x}+2\right)}\right)$$

5. $y = \dfrac{(x+1)^{\frac{3}{4}}\cdot e^{mx}\cdot\sin^{-1}(ax)}{(x-1)^{\frac{7}{3}}\cdot(x+2)^{\frac{5}{7}}}$

Solution: $y = \dfrac{(x+1)^{\frac{3}{4}} \cdot e^{mx} \cdot \sin^{-1}(ax)}{(x-1)^{\frac{7}{3}} \cdot (x+2)^{\frac{5}{7}}}$

$$\Rightarrow \log|y| = \frac{3}{4}\log|(x+1)| + mx +$$

$$\log\left|\sin^{-1}(ax)\right| - \frac{7}{3}\log|(x-1)| - \frac{5}{7}|x+2|$$

$$\Rightarrow \frac{1}{y}\cdot\frac{dy}{dx} = \frac{3}{4(x+1)} + m + \frac{1\cdot a}{\left(\sin^{-1}(ax)\right)\sqrt{1-a^2x^2}} -$$

$$\frac{7}{3(x-1)} - \frac{5}{7(x+2)}$$

$$\Rightarrow \frac{dy}{dx} = y\cdot\left(\frac{3}{4(x+1)} + m + \frac{a}{\left(\sin^{-1}(ax)\right)\sqrt{1-a^2x^2}} - \right.$$

$$\left.\frac{7}{3(x+1)} - \frac{5}{7(x+2)}\right)$$

$$\Rightarrow \frac{dy}{dx} = \frac{(x+1)^{\frac{3}{4}} \cdot e^{mx} \cdot \sin^{-1}(ax)}{(x-1)^{\frac{7}{3}} \cdot (x+2)^{\frac{5}{7}}}\cdot$$

$$\left(\frac{3}{4(x+1)} + m + \frac{a}{\sin^{-1}(ax)\sqrt{1-a^2x^2}} - \frac{7}{3(x-1)} - \frac{5}{7(x+2)}\right)$$

Form 4: Miscellaneous problems.

Solved Examples

1. If $\log|xy| = x^2 + y^2$, show that $\dfrac{dy}{dx} = \dfrac{y(2x^2-1)}{x(1-2y^2)}$.

Solution: $\log|xy| = x^2 + y^2$

$$\Rightarrow \log(|x|\cdot|y|) = x^2 + y^2$$

$$\Rightarrow \log|x| + \log|y| = x^2 + y^2$$

$$\Rightarrow \frac{1}{x} + \frac{1}{y}\cdot\frac{dy}{dx} = 2x + 2y\cdot\frac{dy}{dx}$$

$$\Rightarrow \left(\frac{1}{y} - 2y\right)\frac{dy}{dx} = 2x - \frac{1}{x}$$

$$\Rightarrow \left(\frac{1-2y^2}{y}\right)\cdot\frac{dy}{dx} = \frac{2x^2-1}{x}$$

$$\Rightarrow \frac{dy}{dx} = \frac{y(2x^2-1)}{x(1-2y^2)}$$

2. If $y = x\sin y$, show that $\dfrac{dy}{dx} = \dfrac{y}{x(1-x\cos y)}$.

Solution: $y = x\sin y$

$$\Rightarrow |y| = |x\sin y| = |x|\cdot|\sin y|$$

$$\Rightarrow \log|y| = \log|x| + \log|\sin y|$$

$$\Rightarrow \frac{1}{y}\cdot\frac{dy}{dx} = \frac{1}{x} + \frac{1}{\sin y}\cdot\cos y\cdot\frac{dy}{dx}$$

$$\Rightarrow \frac{1}{y}\cdot\frac{dy}{dx} - \frac{\cos y}{\sin y}\frac{dy}{dx} = \frac{1}{x}$$

$$\Rightarrow \left(\frac{1}{y} - \frac{\cos y}{\sin y}\right)\frac{dy}{dx} = \frac{1}{x}$$

$$\Rightarrow \left(\frac{\sin y - y\cos y}{y\sin y}\right)\frac{dy}{dx} = \frac{1}{x}$$

$$\Rightarrow \frac{dy}{dx} = \frac{y\sin y}{x(\sin y - y\cos y)} = \frac{y\sin y}{x(\sin y - x\sin y\cos y)}$$

$$= \frac{y\sin y}{x\sin y\ 1 - x\cos y}$$

$$= \frac{y}{x\ 1 - x\cos y}$$

Problems belonging to type (1)

(i): *Problems based on irrational functions*

Exercise 13.1.1

Find the differential coefficient of each of the following.

1. $y = \sqrt{\dfrac{2x-1}{2x+1}}, |x| > \dfrac{1}{2}$

2. $y = (x+1)^{\frac{3}{2}}, x > -1$

3. $y = \sqrt[3]{\dfrac{x+1}{x-1}}, |x| \neq 1$

4. $y = \sqrt[3]{\dfrac{x(x+1)(x-2)}{(x^2+1)(2x+3)}}, x > 2$

Answers

1. $\dfrac{2y}{(4x^2-1)}$

2. $\dfrac{3}{2}\cdot\dfrac{y}{(x+1)}$

3. $\dfrac{-2y}{3(x^2-1)}$

4. $\dfrac{y}{3}\left[\dfrac{1}{x}+\dfrac{1}{x+1}+\dfrac{1}{x-2}-\dfrac{2x}{x^2+1}-\dfrac{2}{2x+3}\right]$

(ii) *Problems based on product and quotient of functions*

Exercise 13.1.2

Find the differential coefficient of each of the following.

1. $y^2 = x(x+1), x > 0$

2. $y^{\frac{2}{3}} = \dfrac{(x^2+1)\cdot(3x+4)^{\frac{1}{2}}}{\sqrt[5]{(2x-3)(x^2-4)}}, x > 2$

3. $y = \dfrac{x\cdot\sqrt{x^2+1}}{(x+1)^{\frac{2}{3}}}, x > 0$

4. $y = x^x\left(\dfrac{2x-1}{2x+1}\right), x > \dfrac{1}{2}$

Answers

1. $\dfrac{y(2x+1)}{2x(x+1)}$

2. $\dfrac{3y}{2}\left[\dfrac{2x}{x^2+1}+\dfrac{\frac{3}{2}}{3x+4}-\dfrac{\frac{2}{5}}{2x-3}-\dfrac{\frac{2x}{5}}{x^2-4}\right]$

3. $y\left[\dfrac{1}{x}+\dfrac{x}{x^2+1}-\dfrac{2}{3(x+1)}\right]$

4. $y\left[1+\log x+\dfrac{4}{4x^2-1}\right]$

(iii) *Problems based on exponential composite functions*

Exercise 13.1.3

Find the differential coefficient of each of the following.

1. $y = x^{\sin x}, x > 0$
2. $y = (\sin x)^{\tan x}, \sin x > 0$
3. $y = x^{\log x}, x > 0$
4. $y = x^x, x > 0$
5. $y = (ax)^{bx}, x > 0$
6. $x^y = e^{y-x}, x > 0$
7. $y = e^{\tan x}$
8. $y = e^{x^{x^x}}, x > 0$
9. $y = e^{e^{e^x}}$
10. $y = e^{\log\sin x}, \sin x > 0$
11. $x^y = e^{x-y}, x > 0$
12. $y = e^{x^{\frac{1}{x}}}, x > 0$

Answers

1. $y\left[\dfrac{\sin x}{x}+\cos x\cdot\log x\right]$

2. $y \cdot \left[1 + \sec^2 x \cdot \log \sin x\right]$

3. $\dfrac{2y \log x}{x}$

4. $\dfrac{2}{x} \cdot x^{\log x} \cdot (\log x)$

5. $b(ax)^{bx} \cdot [1 + \log ax], a > 0$

6. $\dfrac{2 - \log x}{(1 - \log x)^2}$

7. Find

8. $e^{x^x} \cdot x^x \cdot (1 + \log x)$

9. $e^{e^x} \cdot e^x$

10. $\cos x$

11. $\dfrac{y \log x}{x(1 + \log x)}$

12. $\left(e^{x^{\frac{1}{x}}}\right) \cdot \left(x^{\frac{1-2x}{x}}\right) \cdot (1 - \log x)$

(iv) *Problems based on the sum of two differentiable functions*

Exercise 13.1.4

Find the differential coefficient of each of the following.

1. $y = \left(\dfrac{1}{x}\right)^x + x^{\frac{1}{x}}, x > 0$

2. $y = x^{\tan x} + (\tan x)^{\cot x}, 0 < x < \dfrac{\pi}{4}$

3. $y = (\sin x)^{\tan x} + (\tan x)^{\sin x}, 0 < x < \dfrac{\pi}{2}$

4. $y = x^{2x} + (2x)^x, x > 0$

5. $y = x^{\log x} + (\log x)^x, x > 1$

6. $y = x^{\sin x} + (\sin x)^{\frac{3}{2}}, 0 < x < \dfrac{\pi}{2}$

7. $y = (\sin x)^x + x \sin^{-1} x$

Answers

1. $-\left(\dfrac{1}{x}\right)^x \cdot (1 + \log x) + x^{\frac{1}{x}} \cdot \left(\dfrac{1 - \log x}{x^2}\right)$

2. $x^{\tan x} \cdot \left(\dfrac{\tan x}{x} + \sec^2 x \cdot \log x\right) + (\tan x)^{\cot x}$

$\operatorname{cosec}^2 x \cdot (1 - \log \tan x)$

3. $(\sin x)^{\tan x} \cdot \left(1 + \sec^2 x \cdot \log \sin x\right) + (\tan x)^{\sin x} \cdot$

$(\sec x + \cos x \cdot \log \tan x)$

4. $2x^{2x} \cdot (1 + \log x) + (2x)^x \cdot (1 + \log 2x)$

5. $2x^{\log x} \cdot \left(\dfrac{\log x}{x}\right) + (\log x)^x \cdot \left(\dfrac{1}{\log x} + \log \log x\right)$

6. $x^{\sin x} \cdot \left(\dfrac{\sin x}{x} + \cos x \cdot \log x\right) + \dfrac{3}{2} \cos x \sqrt{\sin x}$

7. $(\sin x)^x \cdot (x \cot x + \log \sin x) + \dfrac{x}{\sqrt{1 - x^2}} + \sin^{-1} x,$

$0 < x < 1.$

(v) Miscellaneous problems.

Exercise 13.1.5

1. If $x = e^{\frac{x}{y}}$, show that $\dfrac{dy}{dx} = \dfrac{x - y}{x \log x}$.

2. If $x^p y^q = (x + y)^{p+q}, x > 0, y > 0$, show that $\dfrac{dy}{dx} = \dfrac{x}{y}, qx \neq py$.

3. If $e^{x+y} = e^x + e^y$, show that $\dfrac{dy}{dx} = e^{y-x} \dfrac{1 - e^y}{e^x - 1}$.

4. If $x^y = e^{x-y}$, show that $\dfrac{dy}{dx} = \dfrac{\log x}{(1 + \log x)^2}, x > 0$.

Problems belonging to type (2)

(i) *Problems based on square root of functions*

Exercise 13.2.1

Find the differential coefficient of each of the following.

1. $y = \sqrt{(x+1)(x+2)}$
2. $y = \sqrt{\dfrac{x+a}{x-a}}$
3. $y = \sqrt{\dfrac{a^2 - x^2}{a^2 + x^2}}$
4. $y = \sqrt[3]{2ax^2}$
5. $y = \left(5 - x^2\right)^{\frac{1}{2}}$
6. $y = \sqrt{5x^3}$
7. $y = \sqrt{x^2 - 7x + 4}$
8. $y = \sqrt{x^2 - 9}$
9. $y = \sqrt{x}$
10. $y = \sqrt{|x|}$
11. $y = \sqrt{\sin x}$
12. $y = \sqrt{\sin 3x}$
13. $y = \sqrt{\tan 3x}$

Answers (with proper restrictions on x)

1. $\dfrac{1}{2} \cdot \dfrac{2x+3}{\sqrt{(x+1)(x+2)}}$
2. $\left(\dfrac{a}{a^2 - x^2}\right) \cdot \left(\sqrt{\dfrac{x+a}{x-a}}\right)$
3. $\dfrac{2a^2 x}{x^4 - a^4}\left(\sqrt{\dfrac{a^2 - x^2}{a^2 + x^2}}\right)$
4. Find
5. Find
6. Find
7. Find
8. Find
9. Find
10. Find
11. Find
12. Find
13. Find

(ii) *Problems based on the product of two or more than two differentiable functions*

Exercise 13.2.2

Find $\dfrac{dy}{dx}$ of each the following.

1. $y = \cos x \cdot \cos 2x \cdot \cos 3x \cdot \cos 4x$
2. $y = (x-3)^2 \cdot \sqrt{x+2} \cdot \left(\sqrt{x^2 - 1}\right)^{-1}$
3. $y = \tan x \cdot x^x \cdot e^x \cdot \log\left(x\sqrt{x}\right)$
4. $y = (\sin^{-1} x) \cdot (\cos x)^x$
5. $y = (\sin x)^x \cdot \left(\sqrt{1 + x^2}\right) \cdot \sin x$
6. $y = x^x \cdot \log x \cdot \sin x$
7. $y = \sin x \cdot \sin 2x \cdot \sin 3x \cdot \sin 4x$

Answers

1. $-\cos x \cdot \cos 2x \cdot \cos 3x \cdot \cos 4x \cdot (\tan x + 2\tan 2x + 3\tan 3x + 4\tan 4x)$
2. $(x-3)^2 \cdot \sqrt{x+2} \cdot \dfrac{1}{2} \cdot (x^2+1)^{-\frac{3}{2}} \cdot 2x + (x-3)^2 \cdot \left(\sqrt{x^2+1}\right)^{-1} \cdot \dfrac{1}{2} \cdot (x+2)^{-\frac{1}{2}} + \dfrac{\sqrt{x+2}}{\sqrt{x^2+1}} \cdot 2(x-3)$ and simplify.
3. $y \cdot \left[2 + \log x + \dfrac{2}{\sin x} + \dfrac{3}{2x \log x}\right]$
4. $y \cdot \left[\dfrac{1}{\left(\sqrt{1-x^2}\right)\sin^{-1} x} + \log\cos x - x\tan x\right]$

5. $y \cdot \left[\log \sin x + x \cot x + \frac{x}{1+x^2} + \cot x\right]$

6. $\left[\left(1 + \log x + \cot x + \frac{1}{x \log x}\right) x^x \cdot \log x \cdot \sin x\right]$

7. $\sin x \cdot \sin 2x \cdot \sin 3x \cdot \sin 4x (\cot x + 2 \cot 2x + 2 \cot 3x + 4 \cot 4x)$

(iii) *Problems based on quotient of two or more than two differentiable functions*

Exercise 13.2.3

Find $\frac{dy}{dx}$ of each of the following.

1. $y = \frac{2(x - \sin x)^{\frac{3}{2}}}{\sqrt{x}}$

2. $y = \frac{(x+1)^2 \cdot \sqrt{x-1}}{(x+4)^2 \cdot e^x}$

3. $y = \frac{(2x+3)^{\frac{1}{3}} \cdot \left(x^2+5\right)^{\frac{1}{2}}}{\left(3x^2+1\right)^2}$

4. $y = \frac{\sqrt{3x+4} \cdot (2-3x)^{\frac{2}{3}} \cdot \left(x^2+4\right)^{\frac{1}{3}}}{(2-3x)^{\frac{1}{3}} \cdot (4x+1)^{\frac{2}{3}}}$

5. $y = \frac{(5-2x)^{\frac{3}{2}}}{(5+3x)^{\frac{5}{2}} \cdot (5-4x)^2}$

6. $y = \frac{10^x \cdot \cot x^2 \cdot x^{\frac{1}{3}}}{\sin 2x}$

Answers (with proper restrictions on x)

1. $\frac{(2x - 3x \cos x + \sin x) \cdot \sqrt{(x - \sin x)}}{x\sqrt{x}}$

2. $\frac{(x+1)^2 \cdot \sqrt{x-1}}{(x+4)^2 \cdot e^x} \cdot \left[\frac{5x-3}{2\left(x^2-1\right)} - \frac{x+6}{x+4}\right]$

3. $y \cdot \left[\frac{2}{3(2x+3)} + \frac{x}{x^2+5} - \frac{12x}{3x^2+1}\right]$

4. $y \cdot \left[\frac{3}{2(3x+4)} - \frac{3}{2(2-3x)} + \frac{2x}{3\left(x^2+4\right)} + \frac{1}{(2-3x)} - \frac{8}{3(4x+1)}\right]$

5. $\frac{(5-2x)^{\frac{3}{2}}}{(5+3x)^{\frac{5}{2}} \cdot (5-4x)^2}\left[-\frac{3}{5-2x} - \frac{15}{2(5+3x)} + \frac{8}{5-4x}\right]$

6. $\frac{10^x \cdot \cot x^2 \cdot x^{\frac{1}{3}}}{\sin 2x} \cdot \left(\log_e 10 - 4x \cdot \operatorname{cosec} 2x^2\right) + \frac{1}{3x} - 2 \cot 2x$

(iv) *Problems based on exponential composite functions*

Exercise 13.2.4

Find $\frac{dy}{dx}$ of each of the following.

1. $y = (\sin x)^{\sin x}$
2. $y = (\log x)^x$
3. $y = (\tan^{-1} x)^{\sin x}$
4. $y = x^{x^2}$
5. $y = (\log x)^{\log x}$
6. $x^y = y^{\sin x}$
7. $y = x^x$
8. $y = x^{\sin^{-1} x}$
9. $y = (1+x)^{\log x}$
10. $y = (\sin x)^{\tan 3x}$
11. $y = x^{x^x}$
12. $y = (\tan x)^{\sin x}$
13. $y = x^y$
14. $y = (\cos x)^{\log x}$
15. $y^{\sin x} = x^{\sin y}$
16. $y = \left(\cot^{-1} x\right)^{\frac{1}{x}}$
17. $y^{\tan x} = x^{\tan y}$
18. $(\sec x)^y = (\tan y)^x$

19. $y = (x \log x)^{\log \log x}$

20. $y = x^{\tan x}$

21. $y = x^{y^x}$

Answers

1. $(\sin x)^{\sin x} \cdot [\cos x \cdot (1 + \log \sin x)]$

2. $(\log x)^x \cdot \left[\frac{1}{\log x} + \log \log x\right]$

3. $\left(\tan^{-1} x\right)^{\sin x} \cdot \left[\frac{\sin x}{\tan^{-1} x (1+x)^2} + \cos x \cdot \log \tan^{-1} x\right]$

4. $x^{x^2} \cdot x \cdot (1 + 2 \log x)$

5. $(\log x)^{\log x} \cdot \frac{1}{x} \cdot (1 + \log \log x)$

6. $\frac{y}{x} \cdot \left[\frac{y - \cos x \cdot \log y}{\sin x - y \log x}\right]$

7. $x^x \cdot (1 + \log x)$

8. $x^{\sin^{-1} x} \left[\frac{\log x}{\sqrt{1 - x^2}} + \frac{\sin^{-1} x}{x}\right]$

9. $(1 + x)^{\log x} \cdot \left[\frac{\log(1 + x)}{x} + \frac{\log x}{1 + x}\right]$

10. $(\sin x)^{\tan 3x} \cdot [3 \sec^2 3x \log \sin x + \tan 3x \cdot \cot x]$

11. $x^{x^x} \cdot \left[x^x \cdot (1 + \log x)\right]$

12. Find

13. $\frac{y^2}{x(1 - y \log x)}$

14. $(\cos x)^{\log x} \cdot \left[\frac{\log \cos x}{x} - \log x \cdot \tan x\right]$

15. $\left[\frac{x \log y \cos x - \sin y}{y \log x \cos y - \sin x}\right] \cdot \frac{y}{x}$

16. $\left(\cot^{-1} x\right)^{\frac{1}{x}} \cdot \left[-\frac{\log \cot^{-1} x}{x^2} - \frac{1}{x\left(1 + x^2\right) \cot^{-1} x}\right]$

17. $\dfrac{\sec^2 x \cdot \log y - \frac{1}{x} \cdot \tan y}{\sec^2 y \cdot \log x - \frac{1}{y} \tan x}$

18. $\dfrac{\log \tan y - y \tan x}{\log \sec x - x \operatorname{cosec} y \cdot \sec y}$

19. $(x \log x)^{\log \log x} \cdot \left[\frac{\log(x \log x)}{x \log x} + \frac{(1 + \log x) \cdot \log \log x}{x \log x}\right]$

20. $x^{\tan x} \cdot \left[\sec^2 x \cdot \log x + \frac{\tan x}{x}\right]$

21. $\left(\frac{y}{x}\right) \cdot \left(\frac{\log y}{\log x}\right) \cdot \left(\frac{1 + x \log x \log y}{1 - x \log y}\right)$

(v) ***Problems based on the sum of two or more than two exponential composite (or, other) functions***

Exercise 13.2.5

Find $\frac{dy}{dx}$ of each of the following.

1. $y = x^{1+x} + \log x$

2. $y = x^{\cos x} + \sin \log x$

3. $y = x^x + e^{\tan x}$

4. $y = \log \log x + 2^{\sin x}$

5. $y = x^x + x^{\sin x}$

6. $y = x^x + x^{\frac{1}{x}}$

7. $y = x^{\sin x} + (\sin x)^x$

8. $y = x^{\tan x} + (\sin x)^{\cos x}$

9. $y = x^x + (\log x)^x$

10. $y = x^{\sin x} = 10^7$

11. $y = e^{x \sin x^3}$

12. $y = x^{\sin x} + 5x^2$

13. $y = (\sin x)^{\cos x} + (\cos x)^{\sin x}$

14. $y = (\log x)^{\tan x} + (\tan x)^{\log x}$

15. $y = x^n \cdot \log x + x(\log x)^n$

16. $y = x^3 + (\log x)^x$

Answers

1. $x^{1+x}\left[\log x + \frac{1+x}{x}\right] + \frac{1}{x}$

2. $x^{\cos x}\left[-\sin x \cdot \log x + \frac{\cos x}{x}\right] + \frac{\cos \log x}{x}$

3. $x^x(1 + \log x) + e^{\tan x} \cdot \sec^2 x$

4. $\frac{1}{x \log x} + \log 2 \cdot \cos x \cdot 2^{\sin x}$

5. $x^x(1 + \log x) + x^{\sin x}\left[\frac{\sin x}{x} + \cos x \cdot \log x\right]$

6. $x^x(1 + \log x) + x^{\frac{1}{x}-2} \cdot (1 - \log x)$

7. $x^{\sin x}\left[\cos x \cdot \log x + \frac{\sin x}{x}\right] + (\sin x)^x \cdot (\log \sin x + x \cot x)$

8. $x^{\tan x}\left[\sec^2 x \cdot \log x + \frac{\tan x}{x}\right] + (\sin x)^{\cos x} \cdot$

$(\cos x \cdot \cot x - \sin x \cdot \log \sin x)$

9. $x^x(1 + \log x) + (\log x)^x \cdot \left(\log \log x + \frac{1}{\log x}\right)$

10. $x^{\sin x}\left[\frac{\sin x}{x} + \cos x \cdot \log x\right]$

11. $e^{x \sin x^3}\left(\sin x^3 + 3x^3 \cos x^3\right) + (\tan x)^x$ [log

$\tan x + x \cot x \cdot \sec^2 x]$

12. $x^{\sin x} \cdot \left[\cos x \log x + \frac{\sin x}{x}\right] + 10x$

13. $(\sin x)^{\cos x}\left(-\sin x \log \sin x + \frac{\cos^2 x}{\sin x}\right) + (\cos x)^{\sin x} \cdot$

$\left(\cos x \cdot \log \cos x - \frac{\sin^2 x}{\cos x}\right)$

14. $(\log x)^{\tan x}\left[\frac{\tan x}{x \log x} + (\log \log x) \cdot \sec^2 x\right] + (\tan x)^{\log x} \cdot$

$\left[\sec x \cdot \log x \cdot \operatorname{cosec} x + \frac{\log \tan x}{x}\right]$

15. $x^{n-1} \cdot (n \log x + 1) + (\log x)^{n-1} \cdot (\log x + n)$

16. $3x^2 + (\log x)^x\left[\frac{1}{\log x} + \log(\log x)\right]$

(vi) *Miscellaneous problems*

Exercise 13.2.6

1. If $\sin y = x \sin(a + y)$, show that

$$\frac{dy}{dx} = \frac{\sin^2(a + y)}{\sin a}.$$

2. If $y e^y = x$, show that $\frac{dy}{dx} = \frac{y}{x(1 + y)}$.

3. If $y = x e^y$, show that $\frac{dy}{dx} = \frac{y}{x(1 - y)}$.

4. If $y = \sqrt{(1 - x)(1 + x)}$, show that

$$\left(1 - x^2\right)\frac{dy}{dx} + xy = 0$$

5. If $y = \sqrt{\frac{1 - x}{1 + x}}$, show that

$$\left(1 - x^2\right)\frac{dy}{dx} + y = 0$$

6. If $y = x^{\frac{1}{x}}$, show that $\frac{dy}{dx}$ vanishes when $x = e$.

7. If $x = \cos(xy)$, show that $\frac{dy}{dx} = \frac{-\left(1 + y\sqrt{1 - x^2}\right)}{x\sqrt{1 - x^2}}$;

for $0 < xy < \pi$.

8. If $y = xe^{xy}$, show that $\frac{dy}{dx} = \frac{y(1+xy)}{x(1-xy)}$.

9. If $e^x = x^y$, show that $\frac{dy}{dx} = \frac{\log x - 1}{(\log x)^2}$.

Problems on Binomial Coefficients

One can find many equalities with the help of a given binomial expansion in terms of binomial coefficients.

Working rule: It consists of following steps:

Step 1: Differentiate both sides of the given condition $(1+x)^n = c_0 + c_1 x + c_2 x^2 + \ldots + c_n x^n$ w.r.t. x

Step 2: Either put $x = 1, 2, 3, \ldots$ etc if each term is positive in the required result or put $x = -1, -2, -3, \ldots$ etc. if term are for each term being alternatively positive and negative in the required result.

Solved Examples

1. $(1+x)^n = c_0 + c_1 x + c_2 x^2 + \ldots + c_n x^n$, show that $c_1 + 2c_2 + 3c_3 + \ldots + n\,c_n = n\,2^{n-1}$

Solution: Given condition is $(1+x)^n = c_0 + c_1 x + c_2 x^2 + \ldots + c_n x^n$ differentiating both sides of the given condition w.r.t. x, we have $n(1+x)^{n-1}(1+0) = 0 + c_1 \cdot 1 + c_2 2x + \ldots + n x^{n-1} c_n (\because c_0 = 1)$ ...(i)

Now, putting $x = 1$ in (i), we have

$$n(1+1)^{n-1} = c_1 + 2c_2 + 3c_3 + \ldots + n \cdot c_n$$

$\Rightarrow n \cdot 2^{n-1} = c_1 + 2c_2 + 3c_3 + \ldots + nc_n$ which is the required result.

2. If $(1+x)^n = c_0 + c_1 x + c_2 x^2 + \ldots + c_n x^n$, show that

$$c_1 - 2c_2 + 3c_3 + \ldots + (-1)^{n-1} \cdot n \cdot c_n = 0$$

Solution: Given condition is $(1+x)^n = c_0 + c_1 x + c_2 x^2 + \ldots + c_n x^n$ differentiating both sides of the given condition w.r.t. x, we have

$$n(1+x)^{n-1} = 0 + c_1 \cdot 1 + c_2 \cdot 2x + \ldots + nx^{n-1} c_n \ldots\text{(i)}$$

Now putting $x = -1$ in (i) ($\because$ terms in the required result are alternatively +ve and –ve), we get

$$n(1-1)^{n-1} = c_1 - 2c_2 + 3c_3 + \ldots + c_n \cdot n(-1)^{n-1}$$

$\Rightarrow 0 = c_1 - 2c_2 + 3c_3 - \ldots + c_n \cdot n(-1)^{n-1}$ which is the required result. (***Note***: Positive and negative are shortly written +ve and –ve respectively.

Notes: **(A)**: When the last term of the given equality to be proved contains $kn + r$, where k, r and n are positive integers, then one should note that (1) x is to replaced by x^k on both sides of the given condition, k being the coefficient of n. (2) the expression obtained after x being replaced by x^k should be multiplied by x^r, where r is the addend in $kn + r$ (3) the expression obtained in step (2) should be differentiated w.r.t. x and lastly either the substitution $x = 1, 2, 3, \ldots$ etc or $x = -1, -2, -3, \ldots$ etc is made accordingly as whether terms in the required equality to be proved are positive or alternatively positive and negative.

(B): When the last term in the equality to be proved has n^2, one should multiply both sides of the equality (obtained after differentiating the given condition) by x.

3. If $(1+x)^n = c_0 + c_1 x + c_2 x^2 + \ldots + c_n x^n$, show that $c_1 + 2^2 c_2 + 3^2 c_3 + \ldots + n^2 c_n = n(n+1)\,2^{n-2}$

Solution: Given condition is $(1+x)^n = c_0 + c_1 x + c_2 x^2 + \ldots + c_n x^n$ whose last term is $n_2 c_n$.

Now, differentiating both sides of the given condition w.r.t. x, we get

$n(1+x)^{n-1} = c_1 x + 2x c_2 + 3x^2 c_3 + \ldots + nx^{n-1} c_n \ldots$(i)

Multiplying both sides of (i) by x, we get

$n(1+x)^{n-1} \cdot x = c_1 x + 2x^2 c_2 + 3x^3 c_3 + \ldots + nx^n c_n \ldots$(ii)

Again differentiating both sides of (ii), w.r.t. x, we get

$n[1 \cdot (1+x)^{n-1} + n(n-1)(1+x)^{n-2} \cdot 1] = c_1 + 2^2 x c_2 + 3^2 x^2 c_3 + \ldots + n^2 x^{n-1} c_n$ ...(iii)

Lastly, putting $x = 1$ (iii), we get

$$c_1 + 2^2 c_2 + 3^2 c_3 + \ldots + n^2 \cdot c_n$$

$$= n\left[2^{n-1} + (n-1)2^{n-2}\right] = n \cdot 2^{n-2}(2 + n - 1)$$

$= n(n+1)2^{n-2}$ which is the require result.

Remark: On putting $x = -1$ in (iii), we get

$$c_1 - 2^2 c_2 + 3^2 c_3 - 4^2 c_4 + \ldots + (-1)^{n-1} n^2 c_n = n(0+0) = 0$$

4. If $(1+x)^n = c_0 + c_1 x + c_2 x^2 + c_3 x^3 + \ldots + c_n x^n$, show that $c_0 + 2c_1 + 3c_2 + \ldots + (n+1)c_n = 2^{n-1}(2+n)$.

Solution: Given condition is $(1 + x)^n = c_0 + c_1 x + c_2 x^2 + \ldots + c_n x^n$ whose last term has $(nk + r) = n + 1$ for k and $r = 1$ multiplying both sides of the given condition by x, we get

$$x(1+x)^n = c_0 x + c_1 x^2 + c_2 x^3 + \ldots + c_n x^{n+1} \quad \ldots\text{(i)}$$

Now, differentiating both sides of (i) w.r.t. x, we get

$$1 \cdot (1+x)^n x + n \cdot (1+x)^{n-1} = c_0 + 2x\, c_1 + 3x^2 c_2 + \ldots + c_n (n+1) x^n \quad \ldots\text{(ii)}$$

Putting $x = 1$ in (ii), we get the required result

$$2^n + 1 \cdot n + 2^{n-1} = c_0 + 2c_1 + 3c_2 + \ldots + (n+1)\, c_n = 2^{n-1}(2+n)$$

5. If $(1+x)^n = c_0 + c_1 x + c_2 x^2 + \ldots + c_n x^n$, show that $c_0 + 3c_1 + 5c_2 + \ldots + (2n+1)\, c_n = 2^n (n+1)$.

Solution: Given condition is $(1 + x)^n = c_0 + c_1 x + c_2 x^2 + \ldots + c_n x^n$ whose last term has $(nk + r) = 2n + 1$ for $k = 2$ and $r = 1$ on replacing x in each term of the given condition by x^2, we get

$$(1+x^2)^n = c_0 + c_1 x^2 + c_2 x^4 + \ldots + c_n x^{2n} \quad \ldots\text{(i)}$$

on multiplying both sides of (i) by x (since additive constant in $2n + 1$ is 1), we get

$$x(1+x^2)^n = c_0 x + c_1 x^3 + c_2 x^5 + \ldots + c_n x^{2n+1} \quad \ldots\text{(ii)}$$

Now, differentiating both sides of (ii) w.r.t. x, we get

$$1 \cdot (1+x^2)^n + x \cdot n \cdot (1+x^2)^{n-1} \cdot 2x = c_0 + c_1 \cdot 3 \cdot x^2 + c_2 \cdot 5 \cdot x^4 + \ldots + c_n (2n+1)\, x^{2n} \quad \ldots\text{(iii)}$$

Lastly, on putting $x = 1$ in (iii), we get the required result $c_0 + 3c_1 + 5 \cdot c_2 + \ldots + (2n+1)\, c_n$

$$= 2^n + n \cdot 2^{n-1} \cdot 2 = 2^n (1+n)$$

Exercise 13.3

1. If $(1+x)^n = 1 + c_1 x + c_2 x^2 + c_3 x^3 + \ldots + c_n x^n$, show that

(i) $c_0 + 2c_1 + 2c_2 + 4c_3 + \ldots + (n+1)\, c_n = (n+2) \cdot 2^{n-1}$

(ii) $c_0 - 2c_1 + 3c_2 - 4c_3 + \ldots + (-1)^n (n+1)\, c_n = 0$

(iii) $c_1 - 2c_2 + 3c_3 - \ldots + (-1)^{n-1}\, n\, c_n = 0$

(iv) $c_1 - 2^2 c_2 + 3^2 c_3 - 4^2 c_4 + \ldots + (-1)^{n-1} \cdot n^2\, c_n = 0$

2. If $(1+x)^n = 1 + c_1 x + c_2 x^2 + c_3 x^3 + \ldots + c_n x^n$, find the values of

(i) $c_0 + 2c_1 x + 3c_2 x^2 + \ldots + (n+1)\, c_n x^n$

(ii) $c_1 + 2^2 c_2 + 3^2 c_3 + 4^2 c_4 \ldots + n^2 c_n$

14

Successive Differentiation

Question: What do you mean by successive differentiation?

Answer: The process of finding derivatives of the derivatives in succession is called successive differentiation.

Explanation: A function may be differentiated more than once in the following way.

We know that differential coefficient of a function $f(x)$ in general is itself a function of x known as derived function of x or first derivative of the function $f(x)$ symbolised as $f'(x)$ indicating $f(x)$ has been differentiated the first time, the first derivative $f'(x)$ can be differentiated to obtain the second derivative of the function $f(x)$ symbolised as $f''(x)$ indicating $f(x)$ has been differentiated two times. After the derived function $f'(x)$ having been differentiated second time, the second derivative $f''(x)$ can be differentiated third time providing us again a function of x knwon as third derivative symbolised as $f'''(x)$ indicating $f(x)$ has been differentiated three times. This process of getting a derived function may go on indefinitely and after the derived function $f^{(n-2)}(x)$ is differentiated $(n-1)$th derivative $f^{(n-1)}x$ is obtained which again can be differentiated providing us again a function of x known as nth derivative symbolised as $f^{n}(x)$ indicating $f(x)$ has been differentiated n-times. This process of differentiation of a function $f(x)$ repeated more han one successively or the process of finding derivatives one after the other from a given function $f(x)$ is known as successive differentiation.

Remember:

1. If y be a function of x, the derived function $f'(x)$ will be in general itself a differentiable function of x except perhaps at those points at which derived function $f'(x)$ becomes undefined.

2. Points at which $f(x)$ or $f'(x)$ are undefined are knwon as points of discontinuities of $f(x)$ or $f'(x)$ respectively.

3. Points of discontinuities of the function $f(x)$ are also the points of discontinuities of the given function $f'(x)$.

Question: How would you differentiated successively $y = x^6$?

Answer: $y = x^6$

First derivative $= y_1 = \dfrac{dy}{dx} = 6x^{(6-1)} = 6x^5$

Second derivative

$$= y_2 = \frac{d}{dx}\left(\frac{dy}{dx}\right) = \frac{d^2y}{dx^2} = 6 \times 5 \times x^{(5-1)} = 30x^4$$

Third derivative

$$= y_3 = \frac{d}{dx}\left(\frac{d^2y}{dx^2}\right) = \frac{d^3y}{dx^3}$$

$$= 30 \times 4 \times (x)^{(4-1)} = 120 \cdot x^3$$

Fourth derivative

$$= y_4 = \frac{d}{dx}\left(\frac{d^3y}{dx^3}\right) = \frac{d^4y}{dx^4}$$

$= 120 \times 3 \times (x)^{(3-1)} = 360 \cdot x^2$

Fifth derivative

$$= y_5 = \frac{d}{dx}\left(\frac{d^4 y}{dx^4}\right) = \frac{d^5 y}{dx^5}$$

$= 360 \times 2 \times (x)^{(2-1)} = 720 x^1 = 720 x$

Sixth derivative

$$= y_6 = \frac{d}{dx}\left(\frac{d^5 y}{dx^5}\right) = \frac{d^6 y}{dx^6}$$

$= 720 \times 1 \times (x)^{(1-1)} = 720 x^0 = 720 \cdot 1 = 720$

Seventh derivative

$$= y_7 = \frac{d}{dx}\left(\frac{d^6 y}{dx^6}\right) = \frac{d^7 y}{dx^7} = 0$$

Eighth, 9th, ... nth derivative = 0

Question: How would you differentiate successively $y = \sin x$?

Answer: $\because y = \sin x$

$\therefore y_1 = \cos x \quad = \sin\left(\frac{\pi}{2} + x\right)$

$y_2 = -\sin x \quad y_2 = \cos\left(\frac{\pi}{2} + x\right) = \sin\left(2 \cdot \frac{\pi}{2} + x\right)$

$y_3 = -\cos x \quad y_3 = \cos\left(2 \cdot \frac{\pi}{2} + x\right) = \sin\left(3 \cdot \frac{\pi}{2} + x\right)$

$y_4 = \sin x \quad \vdots$

$y_5 = \cos x \quad \vdots$

$y_6 = -\sin x \quad \vdots$

$y_7 = -\cos x \quad \vdots$

$y_8 = \sin x \quad y_n = \sin\left(n \cdot \frac{\pi}{2} + x\right)$

N.B.: We inspect (1) $y_1, y_3 \ldots, y_{2n+1} = \pm \cos x$, alternatively +ve and –ve (2) $y_2, y_4, \ldots,$ $y_{2n} = \pm \sin x$, alternatively + ve and – ve provided $y = \sin x$.

Notation: If $y = f(x)$ = a function of independent variable x, then the differential coefficient (or, derivatives or derived function) of first, second, third, fourth, fifth, and higher orders can be denoted by any one of the following notations.

1. $\frac{dy}{dx}, \frac{d^2 y}{dx^2}, \frac{d^3 y}{dx^3}, \frac{d^4 y}{dx^4}, \ldots, \frac{d^n y}{dx^n}$ denoting that a function $y = f(x)$ has been differentiated one time, two times, three times, four times, ..., n times as well as first, 2nd, 3rd, 4th derivative, ..., n th derivative of some function $y = f(x)$. Hence to differentiate a function $y = f(x)$ n-times means to find n th derivative of the function $y = f(x)$

2. $Dy, D^2 y, D^3 y, D^4 y, \ldots, D^n y$, ($n$ being any + ve integer) which is known as capital D - notion..

3. $f'(x), f''(x), f'''(x) \ldots\ldots, f^n(x)$ (n being any + ve integer)

4. $y', y'', y''', \ldots\ldots, y^n$ (n being + ve integer)

5. $y_1, y_2, y_3, \ldots\ldots., y_n$ (n being + ve integer)

6. $f^{(1)}(x), f^{(2)}(x), f^{(3)}(x), \ldots\ldots f^{(n)}(x)$, ($n$ being + ve integer)

7. $D_x y, D_x^2 y, D_x^3 y, \ldots\ldots D_x^n y$ (n being any + ve integer)

Nomenclature: Notation	*Read as (Nomenclature)*
$\frac{dy}{dx}$	dee wy over dee eks or, dee wy by dee eks
$\frac{d^2 y}{dx^2}$	dee two wy over dee eks two or, dee two wy by dee eks squared
$\frac{d^3 y}{dx^3}$,	dee three wy over dee eks three or, dee three wy by dee eks cubed
$\vdots$	
$\frac{d^n y}{dx^n}$	dee en wy over dee eks en or, dee en wy by dee eks en

Note: **1.** The first notation $\frac{dy}{dx}, \frac{d^2 y}{dx^2}, \frac{d^3 y}{dx^3}, \ldots, \frac{d^n y}{dx^n}$ is in common use.

2. The capital D-notation or, dash notation $\left(\text{i.e; dy, } d^2y, d^3y, \ldots, d^ny \text{ or } y', y'', y''', \ldots, y^n\right)$ is used when the independent variable w.r.t which we differentiate is understood. The capital D or dash notation has the disadvantage of not indicating the variable with respect to which the differentiation is carried out. This is why it is convenient to use the symbol $\frac{dy}{dx}$ to mean the operation of finding the derivative with respect to x.

3. $\frac{d^2y}{dx^2} = \frac{d}{dx}\left(\frac{dy}{dx}\right) = \left(\frac{d}{dx}\right)^2 \cdot (y) = \frac{d^2y}{dx^2}$

. . .

. . .

$\left(\frac{d}{dx}\right)^n = \frac{d^n}{dx^n} \Leftrightarrow \frac{d}{dx}\left(\frac{d}{dx}\right)$ (... upto n times)

which requires the operand y = a function of an independent variable $x = f(x)$ put in the last of the operator.

4. The derivative of a function of an independent variable = First derivative of the given function of an independent variable or simply first derivative.

Derivative of first derivative = second derivative of the original function or simply second derivative.

Derivative of second derivative = third derivative of the original (or, given) function or simply third derivative.

Derivative of third derivative = Fourthond derivative of the original function or simply fourth derivative.

. . .

. . .

Derivative of $(n - 1)$ derivative = nth derivative of the original function or simply nth derivative.

5. Second and higher order derivatives of a function are called higher derivatives and the process of finding them is called successive differentiation.

6. Care must be taken to distinguish between $\frac{d^2y}{dx^n}$ and $\left(\frac{dy}{dx}\right)^n \cdot \frac{d^ny}{dx^n}$ means the nth differential coefficient of function of x whereas $\left(\frac{dy}{dx}\right)^n$ means the nth power of the first differential coefficient of function of x.

Problems based on finding higher derivatives

***Examples worked out*:**

1. Differentiate $y = x^3 + 5x^2 - 7x + 2$ four times.

***Solution*:** $y = x^3 + 5x^2 - 7x + 2$

$$y_1 = 3x^2 + 10x - 7$$

$$y_2 = 6x + 10$$

$$y_3 = 6$$

$$y_4 = 0$$

2. Differentiate $y = e^x + \log x$ three times.

***Solution*:** $\because\ y = e^x + \log x, x > 0$

$$y_1 = e^x + \frac{1}{x}$$

$$y_2 = e^x - \frac{1}{x^2}$$

$$y_3 = e^x + \frac{2}{x^3},\ x > 0$$

3. If $y = \log x$ find y_4.

***Solution*:** $\because\ y = \log x, x > 0$

$$y_1 = \frac{1}{x}$$

$$y_2 = -\frac{1}{x^2}$$

$$y_3 = \frac{2}{x^3}$$

$$y_4 = -\frac{2.3}{x^4} = -\frac{6}{x^4}, x > 0$$

4. If $y = e^{mx}$, find y_2.

***Solution*:** $y = e^{mx}$

$$y_1 = m e^{mx}$$

$$y_2 = m \cdot m e^{mx} = m^2 e^{mx}$$

Working rule to find second derivative from the equations of a curve given in parametric form:

$x = x(t),\ y = y(t)$

1. Find $\dfrac{dy}{dx} = \dfrac{\frac{dy}{dt}}{\frac{dx}{dt}}$

2. $\dfrac{d^2y}{dx^2} = \dfrac{d}{dt}\left[\dfrac{dy}{dx}\right] \cdot \dfrac{dt}{dx}$ should be used

3. Write $\dfrac{dt}{dx} = \dfrac{1}{\frac{dx}{dt}}$ and simplify.

Examples worked out on finding second derivative of parametric equations of a curve

Question: 1. Find $\dfrac{d^2y}{dx^2}$ provided

$x = a\cos\theta$

$y = a\sin\theta$

Solution:

$$\frac{d^2x}{d\theta^2} = \frac{d^2(a\cos\theta)}{d\theta^2} = \frac{d}{d\theta}\left[\frac{d}{d\theta}(a\cos\theta)\right]$$

$$= \frac{d}{d\theta}[-a\sin\theta] = -a\cos\theta \quad ...(1)$$

$$\frac{d^2y}{d\theta^2} = \frac{d^2(a\sin\theta)}{d\theta^2} = \frac{d}{d\theta}\left[\frac{d}{d\theta}(a\sin\theta)\right]$$

$$= \frac{d}{d\theta}[-a\cos\theta] = -a\sin\theta \quad ...(2)$$

$$\frac{dx}{d\theta} = \frac{d(a\cos\theta)}{d\theta} = -a\sin\theta \quad ...(3)$$

Now, using the formula,

$$\frac{d^2y}{dx^2} = \frac{\frac{d^2y}{d\theta^2}\cdot\frac{dx}{d\theta} - \frac{d^2x}{d\theta^2}\cdot\frac{dy}{d\theta}}{\left[\frac{dx}{d\theta}\right]^3}$$

$$= \frac{-a\sin\theta\cdot a(-\sin\theta) - (-a\cos\theta)(a\cos\theta)}{(-a\sin\theta)^3},\ \theta \neq n\pi$$

$$= \frac{a^2\sin^2\theta + a^2\cos^2\theta}{-a^3\sin^3\theta} = \frac{a^2}{-a^3\sin^3\theta} = \frac{-1}{a\sin^3\theta}$$

Or, alternatively,

$$\frac{dy}{dx} = \frac{\frac{dy}{d\theta}}{\frac{dx}{d\theta}} = \frac{a\cos\theta}{-a\sin\theta} = -\cot\theta \quad ...(1)$$

$$\frac{d}{dx}\left[\frac{dy}{dx}\right] = \frac{d}{d\theta}[-\cot\theta]\cdot\frac{d\theta}{dx}$$

$$= \operatorname{cosec}^2\theta\cdot\frac{1}{\frac{dx}{d\theta}}$$

$$= \operatorname{cosec}^2\theta\cdot\frac{1}{a(-\sin\theta)}$$

$$= \frac{-1}{a\sin^3\theta},\ \theta \neq n\pi$$

2. If $x = 2\cos\theta - \cos 2\theta,\ y = 2\sin\theta - \sin 2\theta$, find $\dfrac{d^2y}{dx^2}$.

Solution; $x = 2\cos\theta - \cos 2\theta$

$y = 2\sin\theta - \sin 2\theta$

$$\frac{dx}{d\theta} = -2\sin\theta - \sin 2\theta \times 2 = 2(\sin 2\theta - \sin\theta) \quad ...(1)$$

$$\frac{dy}{d\theta} = 2\cos\theta - \cos 2\theta\,(2) = 2(\cos\theta - \cos 2\theta) \quad ...(2)$$

$$\frac{dy}{dx} = \frac{\frac{dy}{d\theta}}{\frac{dx}{d\theta}} = \frac{2(\cos\theta - \cos 2\theta)}{2(\sin 2\theta - \sin\theta)}\ [\text{using (1) and (2)}]$$

$$= \frac{\cos\theta - \cos 2\theta}{\sin 2\theta - \sin\theta}$$

$$= \frac{2\sin\frac{3\theta}{2}\cdot\sin\frac{\theta}{2}}{2\cos\frac{3\theta}{2}\cdot\sin\frac{\theta}{2}} = \tan\left(\frac{3\theta}{2}\right) \qquad ...(3)$$

$$\therefore \frac{d^2y}{dx^2} = \frac{d}{dx}\left(\frac{dy}{dx}\right) = \frac{d}{d\theta}\left(\frac{dy}{dx}\right)\cdot\frac{d\theta}{dx}$$

$$= \frac{d}{d\theta}\left(\tan\frac{3\theta}{2}\right)\cdot\frac{d\theta}{dx}$$

$$= \frac{3}{2}\sec^2\left(\frac{3\theta}{2}\right)\frac{1}{\left(\frac{dx}{d\theta}\right)}$$

$$= \frac{3}{2}\cdot\sec^2\left(\frac{3\theta}{2}\right)\cdot\frac{1}{2(\sin 2\theta - \sin\theta)} \quad \text{[using (1)]}$$

$$= \frac{3}{2}\cdot\frac{\sec^2\left(\frac{3\theta}{2}\right)}{2(\sin 2\theta - \sin\theta)}$$

$$= \frac{3}{4}\cdot\frac{\sec^2\left(\frac{3\theta}{2}\right)}{2\cos\frac{3\theta}{2}\cdot\sin\frac{\theta}{2}} = \frac{3}{8}\cdot\frac{1}{\sin\frac{\theta}{2}\cdot\cos^2\frac{3\theta}{2}},$$

for $\theta \neq 2n\pi, (2n+1)\frac{\pi}{3}$

3. If $x = a\sec^3\theta$ $y = a\tan^3\theta$ find $\frac{d^2y}{dx^2}$.

Solution: $x = a\sec^3\theta$

$y = a\tan^3\theta$

$$\frac{dx}{d\theta} = a\cdot 3\cdot\sec^2\theta\,(\sec\theta\cdot\tan\theta) \qquad ...(1)$$

$$\frac{dy}{d\theta} = a\cdot 3\cdot\tan^2\theta\left(\sec^2\theta\right) \qquad ...(2)$$

$$\frac{dy}{dx} = \frac{\frac{dy}{d\theta}}{\frac{dx}{d\theta}} = \frac{3a\tan^2\theta\cdot\sec^2\theta}{3a\sec^2\theta\,(\sec\theta\cdot\tan\theta)} = \frac{\tan\theta}{\sec\theta} = \sin\theta \qquad ...(3)$$

$$\frac{d^2y}{dx^2} = \frac{d\sin\theta}{d\theta}\cdot\frac{d\theta}{dx} = \cos\theta\cdot\frac{1}{\frac{dx}{d\theta}}$$

$$= \frac{\cos\theta}{a\cdot 3\cdot\sec^2\theta\cdot\sec\theta\cdot\tan\theta} = \frac{\cos^4\theta\cdot\cot\theta}{3a}, \theta \neq \frac{n\pi}{n}$$

4. If $x = a(\theta - \sin\theta)$ $y = a(1 - \cos\theta)$, Find $\frac{d^2y}{dx^2}$

Solution: $x = a(\theta - \sin\theta)$

$y = a(1 - \cos\theta)$

$$\frac{dx}{d\theta} = a(1 - \cos\theta) \Rightarrow \frac{d\theta}{dx} = \frac{1}{a(1-\cos\theta)},$$

$\theta \neq 2n\pi$ ...(1)

$$\frac{dy}{d\theta} = a\sin\theta \qquad ...(2)$$

(1) and (2)

$$\Rightarrow \frac{dy}{dx} = \frac{dy}{d\theta}\cdot\frac{d\theta}{dx} = \frac{a\sin\theta}{a(1-\cos\theta)} = \frac{\sin\theta}{(1-\cos\theta)}$$

$$\therefore \frac{d^2y}{dx^2} = \frac{d}{d\theta}\left[\frac{\sin\theta}{1-\cos\theta}\right]\cdot\frac{d\theta}{dx}$$

$$= \frac{(1-\cos\theta)\cos\theta - \sin\theta\cdot\sin\theta}{(1-\cos\theta)^2}\cdot\frac{1}{a(1-\cos\theta)}$$

$$= \frac{\cos\theta - 1}{a(1-\cos\theta)^3} = -\frac{1}{a(1-\cos\theta)^2}, \theta \neq 2n\pi$$

5. If $x = a(\cos\theta + \theta\sin\theta)$ $y = a(\sin\theta - \theta\cos\theta)$, where $0 < \theta < \frac{\pi}{2}$ prove that $\frac{d^2y}{dx^2} = \frac{\sec^3\theta}{a\theta}$.

Solution: If $x = a(\cos\theta + \theta\sin\theta)$

$y = a(\sin\theta - \theta\cos\theta)$

$$\frac{dx}{d\theta} = a\left[-\sin\theta + (\theta\cos\theta + \sin\theta)\right] = a\,\theta\cos\theta \quad ...(1)$$

$$\frac{dy}{d\theta} = a\left[\cos\theta - (-\theta\sin\theta + \cos\theta)\right] = a\,\theta\sin\theta \quad ...(2)$$

$$(1) \text{ and } (2) \Rightarrow \frac{dy}{dx} = \frac{\frac{dy}{d\theta}}{\frac{dx}{d\theta}} = \frac{a\,\theta\sin\theta}{a\,\theta\cos\theta} = \tan\theta$$

$$\therefore \frac{d^2y}{dx^2} = \frac{d}{d\theta}\left(\frac{dy}{dx}\right)\cdot\frac{d\theta}{dx} = \frac{d}{d\theta}(\tan\theta)\cdot\frac{d\theta}{dx}$$

$$= \sec^2\theta\cdot\frac{1}{\frac{dx}{d\theta}}$$

$$= \sec^2\theta\cdot\frac{1}{a\,\theta\cos\theta}$$

$$= \frac{\sec^3\theta}{a\,\theta}$$

6. If $x = at^2$, $y = 2at$ find $\frac{d^2y}{dx^2}$.

Solution: $x = at^2$

$y = 2at$

$$\frac{dx}{dt} = 2at \quad ...(1)$$

$$\frac{dy}{dt} = 2a \quad ...(2)$$

$$(1) \text{ and } (2) \Rightarrow \frac{dy}{dx} = \frac{\frac{dy}{dt}}{\frac{dx}{dt}} = \frac{2a}{2at} = \frac{1}{t},\ t \neq 0 \quad ...(3)$$

$$\text{Now, } \frac{d^2y}{dx^2} = \frac{d}{dt}\left[\frac{dy}{dx}\right]\cdot\frac{dt}{dx} = \frac{d}{dt}\left[\frac{1}{t}\right]\cdot\frac{dt}{dx}$$

$$= -\frac{1}{t^2}\times\frac{1}{2at}$$

$$= \frac{-1}{2at^3},\ t \neq 0$$

Problems based on showing that a given function $y = f(x)$ satisfies a differential equation

Question: What is a differential equation?

Answer: An equation containing one (or, more derived functions) is called a differential equation. or, in more explicit form, an equation containing the independent variable x, the function y and its derivatives or differentials is called a differential equation.

Notation: A differential equation is symbolised as follows.

1. $F(x, y, y') = 0$ or $F\left(x, y, \frac{dy}{dx}\right) = 0$

2. $F(x, y, y'') = 0$ or $F\left(x, y, \frac{d^2y}{dx^2}\right) = 0$

3. $F(x, y, y', y'') = 0$ or $F\left(x, y, \frac{dy}{dx}, \frac{d^2y}{dx^2}\right) = 0$

4. $F(x, y, y', y'', y''', ...y^n) = 0$ ......etc

Examples: **1.** $\frac{dy}{dx} = 5$

2. $4\frac{d^2y}{dx^2} + 7\frac{dy}{dx} - 5 = 0$

3. $y\left(\frac{dy}{dx}\right) + 2x\frac{dy}{dx} - y = 0$

Note: **1.** The general type of differential equation of the first order is $\frac{dy}{dx} + py = Q$ where P and Q are given functions of x.

2. The general type of differential equation of the second order is $\frac{d^2y}{dx^2} + a\frac{dy}{dx} + by = f(x)$

Now we come to our main problem.

If $y = f(x)$, then we are required to show that

1. $F(x, y, y') = 0$
2. $F(x, y, y'') = 0$
3. $F(x, y, y''') = 0$
4. $F(x, y, y', y'') = 0$
5. $F(x, y, y', y'', y''') = 0$

We adopt the following working rule:

Working rule: We proceed from the given function $y = f(x)$ in general finding those all derivatives appearing in the given differential equation by using successive differentiation and using various mathematical manipulations, we show that

1. $F(x, y, y') = 0$
2. $F(x, y, y'') = 0$
3. $F(x, y, y''') = 0$
4. $F(x, y, y', y'') = 0$
5. $F(x, y, y', y'', y''') = 0$

or we show L.H.S = R..H.S

About Mathematical Manipulation

1. If the differential equations are

(i) $F(x, y, y') = 0$

(ii) $F(x, y, y'') = 0$

(iii) $F(x, y, y''') = 0$

(iv) $F(x, y, y'''') = 0$ i.e; only one derivative, then we find that derivative only by successive differentiation and putting its value in the given differential equation, we show that

(i) $F(x, y, y') = 0$

(ii) $F(x, y, y'') = 0$

(iii) $F(x, y, y''') = 0$

(iv) $F(x, y, y'''') = 0$

2. If the differentail equation is $F(x, y, y', y'') = 0$, then we find y' and y'' by successive differentiation and then putting the expressions obtained for y' and y'' in the left hand side of differential equation, we show that $F(x, y, y', y'') = 0$

3. If the differential equation is $F(x, y, y', y'') = f(x)$, then we find y' and y'' by successive differentiation and then using various techniques, we show that L.H.S. = R.H.S.

4. If $y = f(x) = a$ rational function, inverse circular function, a function containing inverse circular function or logarithm of a function (i.e; a function whose first derivative is a fractional expression in x rational or irrational) and we are required to show

(i) $F(x, y, y') = 0$

(ii) $F(x, y, y'') = 0$

(iii) $F(x, y, y', y'') = 0$

(iv) $F(x, y, y', y'', y''') = 0$

Then firstly we find first derivative and then in most cases using the rule of cross multiplication, we change the quotient into product form so that the rule of d.c. of product of two functions

$$\left(\text{i.e; } \frac{d}{dx}\{f_1(x)\cdot f_2(x)\} = f_1(x)\cdot f_2'(x) + f_2(x)\cdot f_1'(x)\right)$$

should be applied. But this method is not always fruitful.

5. If the given differential equation $F(x, y, y', y'') = 0$ or, $F(x, y, y', y'') = f(x)$ does not contain radical sign and the first derivative of the given function contains the radical sign, then both sides of the equation containing the first derivative may be raised to the same*.

*Power in any stage to remove the radical symbol or sometimes rationalization is also fruitful device provided the given function is irrational or the first derivative of the given function is irrational.

Remember: **1.** $\dfrac{d}{dx}\left\{\left(\dfrac{dy}{dx}\right)^2\right\} = 2\left(\dfrac{dy}{dx}\right)\cdot\dfrac{d}{dx}\left(\dfrac{dy}{dx}\right)$

$= 2\dfrac{dy}{dx}\cdot\dfrac{d^2y}{dx^2}$ in which $\dfrac{dy}{dx}$ is regarded as a symbol z which is differentiated w.r.t. x which

$\Rightarrow \dfrac{d}{dx}(y_1)^2 = y_1\, y_2$ which is generally used when

$f(x)$ = a rational function of x, inverse circular function of x, a function involving inverse circular function of x, logarithm of a function of x.

2. Whenever we have $y \times \sqrt{f(x)}$ or $y' \times \sqrt{f(x)}$ we are required to square both sides provided given differential equation to be proved does not contain the square root symbol.

3. Sometimes a modification in the form of a given function is also fruitful device before finding the first derivative to get the required differential equation.

4. $y_1 = y' = \frac{dy}{dx}$

$y_2 = y'' = \frac{d^2 y}{dx^2}$

...

$y_n = y^n = \frac{d^n y}{dx^n}$

5. $\left[\frac{d^n y}{dx^n}\right]_{x=a}$, $[y_n]_{x=a}$, $[y_n]_a$, $f^n(a)$,

$\left(\frac{d^n y}{dx^n}\right)_{x=a}$, $(y_n)_{x=a}$ or $(y_n)_a$ etc. denotes the value of the nth derivative (or, nth differential coefficient) of the given function $y = f(x)$ at (or, for) $x = a$.

6. In general $\frac{d^2 y}{dx^2} = f\left(x, y, \frac{dy}{dx}\right)$

$\frac{d^3 y}{dx^3} = f\left(x, y, \frac{dy}{dx}, \frac{d^2 y}{dx^2}\right)$

...

...

$\frac{d^n y}{dx^n} = f\left(x, y, \frac{dy}{dx}, \frac{d^2 y}{dx^2}, \ldots \frac{d^{n-1} y}{dx^{n-1}}\right)$

are second, third, fourth . . . and nth differential equation.

7. A given function $y = f(x)$ is said to satisfy a differential equation if we can find (or, derive) that differential equation by differentiating the given function $y = f(x)$ and using various mathematical manipulations.

Type 1: To show that a given explicit function satisfies a given differential equation.

Examples worked out:

1. If $y = \sin x$, show that $\frac{d^4 y}{dx^4} = y$

Proof: $y = \sin x$

$\Rightarrow \frac{dy}{dx} = \cos x$

$\frac{d^2 y}{dx^2} = -\sin x$

$\frac{d^3 y}{dx^3} = -\cos x$

$\frac{d^4 y}{dx^4} = \sin x$

$\therefore \frac{d^4 y}{dx^4} = y$

2. If $y = a \cos nx + b \sin nx$, show that $\frac{d^2 y}{dx^2} + n^2 y = 0$

Proof: $y = a \cos nx + b \sin nx$

$\Rightarrow \frac{dy}{dx} = -na \sin nx + nb \cos nx$

$\frac{d^2 y}{dx^2} = -n^2 a \cos nx - n^2 b \sin nx$

$= -n^2 (a \cos nx + b \sin nx)$

$= -n^2 y$

$\frac{d^2 y}{dx^2} + n^2 y = -n^2 y + n^2 y = 0$

$\frac{d^2 y}{dx^2} + n^2 y = 0$

3. If $y = a e^{mx} + b e^{-mx}$, show that $\frac{d^2 y}{dx^2} - m^2 y = 0$

Proof: $y = ae^{mx} + be^{-mx}$

$$\Rightarrow \frac{dy}{dx} = ae^{mx} \cdot m + be^{-mx} \cdot (-m) = m\left[ae^{mx} - be^{-mx}\right]$$

$$\frac{d^2y}{dx^2} = m\left[ae^{mx} \cdot m - be^{-mx} \cdot (-m)\right]$$

$$= m^2\left[ae^{mx} + be^{-mx}\right] = m^2 y$$

$$\therefore \frac{d^2y}{dx^2} - m^2 y = m^2 y - m^2 y = 0$$

$$\therefore \frac{d^2y}{dx^2} - m^2 y = 0$$

4. If $y = \log\left(x + \sqrt{1 + x^2}\right)$, show that

$$\left(1 + x^2\right)\frac{d^2y}{dx^2} + x\frac{dy}{dx} = 0.$$

Proof: $y = \log\left(x + \sqrt{1 + x^2}\right)$

$$\Rightarrow \frac{dy}{dx} = \frac{1}{x + \sqrt{1 + x^2}} \cdot \frac{d}{dx}\left[x + \sqrt{1 + x^2}\right]$$

$$= \frac{1}{x + \sqrt{1 + x^2}} \cdot \left[1 + \frac{1}{2\sqrt{1 + x^2}} \times 2x\right]$$

$$= \frac{1}{x + \sqrt{1 + x^2}} \times \frac{\left[\sqrt{1 + x^2} + x\right]}{\sqrt{1 + x^2}}$$

$$= \frac{1}{\sqrt{1 + x^2}}$$

$$\frac{d^2y}{dx^2} = \frac{-\frac{1}{2} \times \frac{2x}{\sqrt{1 + x^2}}}{\left(\sqrt{1 + x^2}\right)^2}$$

$$\left[\because \frac{d}{dx}\left(\frac{1}{f(x)}\right) = -\frac{f'(x)}{\{f(x)\}^2}\right]$$

$$= \frac{\frac{-x}{\sqrt{1 + x^2}}}{\left(x^2 + 1\right)}$$

Putting the values of $\frac{dy}{dx}$ and $\frac{d^2y}{dx^2}$ in the L.H.S of the differential equation,

$$\left(x^2 + 1\right)\frac{d^2y}{dx^2} + x\frac{dy}{dx} = \frac{\left(x^2 + 1\right)}{\left(x^2 + 1\right)} \cdot \frac{-x}{\sqrt{x^2 + 1}} + \frac{x}{\sqrt{1 + x^2}}$$

$$= \frac{-x}{\sqrt{1 + x^2}} + \frac{x}{\sqrt{1 + x^2}} = 0$$

5. If $y = e^{ax}\sin(bx + c)$ show that

$$\frac{d^2y}{dx^2} - 2a\frac{dy}{dx} + \left(a^2 + b^2\right)y = 0$$

Proof: $y = e^{ax}\sin(bx + c)$

$$\Rightarrow \frac{dy}{dx} = ae^{ax}\sin(bx + c) + e^{ax}\, b\cos(bx + c)$$

$$= ay + be^{ax}\cos(bx + c)$$

$$\frac{d^2y}{dx^2} = a\frac{dy}{dx} + be^{ax}\cos(bx + c) - b^2 e^{ax}\sin(bx + c)$$

$$= a\frac{dy}{dx} + a\left(\frac{dy}{dx} - ay\right) - b^2 y$$

$$\left[\because y = e^{ax}\sin(bx + c)\right] = \text{given function}$$

$$= 2a\frac{dy}{dx} - a^2 y - b^2 y$$

$$= 2a\frac{dy}{dx} - \left(a^2 + b^2\right)y$$

Putting the value of $\frac{d^2y}{dx^2}$ in L.H.S of the differential equation,

$$\frac{d^2y}{dx^2} - 2a\frac{dy}{dx} + (a^2 + b^2)\,y$$

$$= 2a\frac{dy}{dx} - (a^2 + b^2)\,y + (a^2 + b^2)\,y - 2a\frac{dy}{dx}$$

$$= 0$$

6. If $(a+bx)\,e^{\frac{y}{x}} = x$, show that $x^3\frac{d^2y}{dx^2} = \left(x\frac{dy}{dx} - y\right)^2$

Proof: We have $\left|(a+bx)\cdot e^{\frac{y}{x}}\right| = |x|$ ...(1)

Now taking log of both sides of (1),

$$\log\left|(a+bx)\right|\cdot e^{\frac{y}{x}} = \log|x|$$

$$\Rightarrow \log\left|(a+bx)\right| + \frac{y}{x}\log e = \log|x|$$

$$\Rightarrow \frac{x\log\left|(a+bx)\right| + y\cdot 1}{x} = \log|x|$$

$$\Rightarrow x\cdot\log\left|(a+bx)\right| + y = x\log|x|$$

$$\Rightarrow y = x\log|x| - x\log\left|(a+bx)\right|$$

$$= x\left[\log|x| - \log\left|(a+bx)\right|\right] \quad ...(2)$$

Now, differentiating both sides of (2) w.r.t. x,

$$\frac{dy}{dx} = 1\cdot\left[\log|x| - \log\left|(a+bx)\right|\right] + x\left[\frac{1}{x} - \frac{b}{a+bx}\right]$$

$$\Rightarrow \frac{dy}{dx} = \log|x| - \log\left|(a+bx)\right| + x\left[\frac{a+bx-bx}{x\,(a+bx)}\right]$$

$$\Rightarrow \frac{dy}{dx} = \log|x| - \log\left|(a+bx)\right| + \frac{a}{a+bx} \quad ...(3)$$

Again differentiating both sides of (3) w.r.t. x,

$$\frac{d^2y}{dx^2} = \frac{1}{x} - \frac{b}{a+bx} - \frac{ab}{(a+bx)^2}$$

$$= \frac{(a+bx)^2 - xb\,(a+bx) - x\,ab}{x\,(a+bx)^2} = \frac{a^2}{x(a+bx)^2}$$

Now L.H.S. $= x^3\dfrac{d^2y}{dx^2}$

$$= x^3\cdot\frac{a^2}{x\,(a+bx)^2} = \frac{x^2\cdot a^2}{(a+bx)^2} = \left(\frac{ax}{a+bx}\right)^2$$

R.H.S. $= \left[x\cdot\dfrac{dy}{dx} - y\right]^2$,

$$= \left[x\cdot\left\{\log\left(\left|\frac{x}{a+bx}\right|\right) + \frac{a}{a+bx}\right\} - x\log\left(\left|\frac{x}{a+bx}\right|\right)\right]^2$$

$$= \left(\frac{ax}{a+bx}\right)^2$$

Hence, L.H.S. = R.H.S

7. If $y = (a+bx)\,e^{-nx}$, show that

$$\frac{d^2y}{dx^2} + 2n\frac{dy}{dx} + n^2y = 0$$

Proof: we are given $y = (a+bx)\cdot e^{-nx}$

$$\frac{dy}{dx} = (a+bx)\cdot\left(-n\cdot e^{-nx}\right) + b\,e^{-nx}$$

$$= -n\,(a+bx)\,e^{-nx} + b\,e^{-nx}$$

$$= -n\,y + b\,e^{-nx} \quad ...(1)$$

$$\left[\because y = (a+bx)\,e^{-nx}\right]$$

Now, differentiating (1) again w.r.t. x,

$$\frac{d^2y}{dx^2} + 2n\frac{dy}{dx} + n^2y = -n\frac{dy}{dx} - bne^{-nx} + 2n\frac{dy}{dx} + n^2y$$

$$= n\frac{dy}{dx} - bn\,e^{-nx} + n^2y$$

$$= n\left(-ny + b\,e^{-nx}\right) - bn\,e^{-nx} + n^2y$$

$$= 0$$

8. If $y = \left[x + \sqrt{1+x^2}\right]^n$, show that

$$(1+x^2)^2\cdot\left(\frac{dy}{dx}\right)^2 = n^2y^2.$$

Proof: $y = \left[x + \sqrt{1+x^2}\right]^n$

$$\Rightarrow \frac{dy}{dx}=n\left[x+\sqrt{1+x^2}\right]^{(n-1)}\cdot\frac{d}{dx}\left[x+\sqrt{1+x^2}\right]$$

$$= n\left[x+\sqrt{1+x^2}\right]^{(n-1)}\cdot\left[1+\frac{2x}{2\sqrt{1+x^2}}\right]$$

$$= n\left[x+\sqrt{1+x^2}\right]^{(n-1)}\cdot\left[x+\sqrt{1+x^2}\right]^{\sqrt{1+x^2}}$$

$$= \frac{ny}{\sqrt{1+x^2}}\left[\because y=\left(x+\sqrt{1+x^2}\right)^n\right]$$

$$\Rightarrow \sqrt{1+x^2}\cdot\frac{dy}{dx}=ny$$

Now, squaring both sides,

$$\left(1+x^2\right)\cdot\left(\frac{dy}{dx}\right)^2=n^2y^2$$

9. If $y=\left[x+\sqrt{1+x^2}\right]^m$, show that

$$\left(1+x^2\right)\frac{d^2y}{dx^2}+x\frac{dy}{dx}-m^2y=0.$$

Proof: $$y=\left[x+\sqrt{1+x^2}\right]^m \quad ...(1)$$

Now, differentiating the given function (1) w.r.t x

$$\frac{dy}{dx}=m\left[x+\sqrt{1+x^2}\right]^{(m-1)}\cdot\left[1+\frac{x}{\sqrt{1+x^2}}\right]$$

$$=\frac{m\left[x+\sqrt{1+x^2}\right]^m}{\sqrt{1+x^2}}$$

$$=\frac{my}{\sqrt{1+x^2}}$$

$$\Rightarrow \sqrt{1+x^2}\cdot\frac{dy}{dx}=my \quad ...(2)$$

Now, squaring both sides of (2), we get

$$\left(1+x^2\right)\cdot\left(\frac{dy}{dx}\right)^2=m^2y^2 \quad ...(3)$$

Now, differentiating both sides of (3) again w.r.t x

$$\left(1+x^2\right)\cdot 2\frac{dy}{dx}\cdot\frac{d^2y}{dx^2}+\left(\frac{dy}{dx}\right)^2\cdot 2x=2m^2y\cdot\frac{dy}{dx} \quad ...(4)$$

Now, dividing the equation (4) by $2\frac{dy}{dx}$ we get

$$\left(1+x^2\right)\cdot\frac{d^2y}{dx^2}+x\cdot\frac{dy}{dx}=m^2y$$

$$\Rightarrow \left(1+x^2\right)\cdot\frac{d^2y}{dx^2}+x\cdot\frac{dy}{dx}-m^2y=0$$

10. If $y=A\left[x+\sqrt{x^2-1}\right]^n+B\left[x-\sqrt{x^2-1}\right]^n$

show that $\left(x^2-1\right)y_2+xy_1-n^2y=0$.

Proof: $$y=A\left[x+\sqrt{x^2-1}\right]^n+B\left[x-\sqrt{x^2-1}\right]^n \quad ...(1)$$

Now, differentiating both sides of (1), we get

$$\frac{dy}{dx}=nA\left[x+\sqrt{x^2-1}\right]^{(n-1)}\cdot\left[1+\frac{x}{\sqrt{x^2-1}}\right]$$

$$+nB\left[x-\sqrt{x^2-1}\right]\left[1-\frac{x}{\sqrt{x^2-1}}\right]$$

$$=\frac{nA\left[x+\sqrt{x^2-1}\right]^n}{\sqrt{x^2-1}}-\frac{nB\left[x-\sqrt{x^2-1}\right]^n}{\sqrt{x^2-1}}$$

$$=\frac{nA\left[x+\sqrt{x^2-1}\right]^n-nB\left[x-\sqrt{x^2-1}\right]^n}{\sqrt{x^2-1}}$$

$$\Rightarrow \sqrt{x^2-1}\,\frac{dy}{dx}$$

$$=nA\left[x+\sqrt{x^2-1}\right]^n-nB\left[x-\sqrt{x^2-1}\right]^n \quad ...(2)$$

Again differentiating (2) w.r.t x to get

$$\sqrt{x^2-1}\cdot\frac{d}{dx}\left(\frac{dy}{dx}\right)+\frac{x}{\sqrt{x^2-1}}\cdot\frac{dy}{dx}$$

$$= \frac{n^2 A\left[x+\sqrt{x^2-1}\right]^n}{\sqrt{x^2-1}} + \frac{n^2 B\left[x-\sqrt{x^2-1}\right]^n}{\sqrt{x^2-1}}$$

$$\Rightarrow \sqrt{x^2-1}\cdot\frac{d^2y}{dx^2} + \frac{x}{\sqrt{x^2-1}}\cdot\frac{dy}{dx}$$

$$= \frac{n^2 A\left[x+\sqrt{x^2-1}\right]^n + n^2 B\left[x-\sqrt{x^2-1}\right]^n}{\sqrt{x^2-1}}$$

$$\Rightarrow \left(x^2-1\right)\frac{d^2y}{dx^2} + x\cdot\frac{dy}{dx}$$

$$= n^2\left[A\left(x+\sqrt{x^2-1}\right)^n + B\left(x+\sqrt{x^2-1}\right)^n\right] = n^2 y$$

$$\Rightarrow \left(x^2-1\right) y_2 + x\, y_1 = n^2 y$$

$$\Rightarrow \left(x^2-1\right) y_2 + x\, y_1 - n^2 y = 0$$

11. If $y^{\frac{1}{m}} + y^{\frac{-1}{m}} = 2x$ show that $\left(x^2-1\right) y_2 + xy_1 - m^2 y = 0$

Proof: Let us suppose that $y^{\frac{1}{m}} = Z$ ...(1)

$$y^{\frac{-1}{m}} = \frac{1}{z} \quad ...(2)$$

Now, adding (1) and (2)

$$\Rightarrow Z + \frac{1}{Z} = 2x$$

$$\Rightarrow Z^2 + 1 - 2x\,Z = 0$$

$$\Rightarrow Z = \frac{2x \pm \sqrt{4x^2-4}}{2}$$

$$\Rightarrow Z = x \pm \sqrt{x^2-1} \quad ...(3)$$

$$\Rightarrow y = \left(x \pm \sqrt{x^2-1}\right)^m \quad ...(4)$$

Now, differentiating both sides of (4) w.r.t x

$$\Rightarrow \frac{dy}{dx} = m\left[x \pm \sqrt{x^2-1}\right]^{(m-1)}\cdot\left[1 \pm \frac{x}{\sqrt{x^2-1}}\right]$$

$$= \pm m\left[x \pm \sqrt{x^2-1}\right]^{(m-1)}\cdot\left[\frac{x \pm \sqrt{x^2-1}}{\sqrt{x^2-1}}\right]$$

$$= \pm\frac{m\left[x \pm \sqrt{x^2-1}\right]^m}{\sqrt{x^2-1}}$$

$$= \pm\frac{my}{\sqrt{x^2-1}} \quad ...(5)$$

Now, squaring both sides of (5).

$$\left(x^2-1\right)\left(\frac{dy}{dx}\right)^2 = m^2 y^2 \quad ...(6)$$

on differentiating (6) w.r.t. x

$$\Rightarrow \left(x^2-1\right)\cdot 2\frac{dy}{dx}\cdot\frac{d^2y}{dx^2} + \left(\frac{dy}{dx}\right)^2\cdot 2x$$

$$= 2m^2 y\cdot\frac{dy}{dx} \quad ...(7)$$

Now, Dividing (6) by $2\cdot\frac{dy}{dx}$,

$$\Rightarrow \left(x^2-1\right)\cdot\frac{d^2y}{dx^2} + x\frac{dy}{dx} - m^2 y = 0$$

$$\Rightarrow \left(x^2-1\right)\cdot y_2 + x\, y_1 - m^2 y = 0$$

Note: In some cases the form of the function suggests simplification. For these functions, we modify the form of the function using mathematical manipulation before differentiation to save labour and time.

Now, we shall do harder and tricky problems on the explicit functions having rational, trigonometric, inverse trigonometric and logarithmic functions which satisfy the given differential equation.

Examples worked out:

1. If $y = \frac{ax+b}{cx+d}$, show that $2\,y_1\,y_3 = 3\,{y_2}^2$

Solution: $y = \dfrac{ax + b}{cx + d}, x \neq -\dfrac{d}{c}$

$$\Rightarrow y_1 = \frac{a(cx+d) - (ax+b)c}{(cx+d)^2} = \frac{ad - bc}{(cx+d)^2}$$

$$\Rightarrow y_2 = \frac{-2c(ad - bc)}{(cx+d)^3}$$

$$\Rightarrow y_3 = \frac{6c^2(ad - bc)}{(cx+d)^4}$$

Now, L.H.S $= 2 y_1 y_3$

$$= 2\frac{(ad - bc)}{(cx+d)^2} \cdot \frac{6c^2(ad - bc)}{(cx+d)^4}$$

$$= \frac{12c^2(ad - bc)^2}{(cx+d)^6}$$

R.H.S. $= 3 y_2^2$

$$= 3 \cdot \frac{4c^2(ad - bc)^2}{(cx+d)^6}$$

$$= \frac{12c^2(ad - bc)^2}{(cx+d)^6}$$

Hence, $2 y_1 y_3 = 3 y_2^2$

Second method:

$$y = \frac{ax + b}{cx + d} \Rightarrow (cx + d)\, y = (ax + b)$$

$$\Rightarrow \frac{d}{dx}[(cx + d)\, y] = \frac{d}{dx}(ax + b)$$

$$\Rightarrow (cx + d)\, y_1 + yc = a$$

$$\Rightarrow \frac{d}{dx}[(cx + d)\, y_1 + yc] = \frac{d}{dx}(a)$$

$$\Rightarrow (cx + d)\, y_2 + y_1 c + y_1 c = 0$$

$$\Rightarrow (cx + d)\, y_2 + 2 y_1 c = 0 \qquad \text{...(1)}$$

$$\Rightarrow \frac{d}{dx}[(cx + d)\, y_2 + 2y_1 c] = 0$$

$$\Rightarrow (cx + d)\, y_3 + y_2 (c) + 2 y_2 c = 0$$

$$\Rightarrow (cx + d)\, y_3 + 3 y_2 c = 0 \qquad \text{...(2)}$$

Now, (1) $\Rightarrow (cx + d)\, y_2 = -2 y_1 c$ ...(3)

(2) $\Rightarrow (cx + d)\, y_3 = -3 y_2 c$ ...(4)

Dividing (3) by (4),

$$\frac{(cx + d)\, y_2}{(cx + d)\, y_3} = \frac{-2 y_1 c}{-3 y_2 c} \rightarrow \frac{2 y_1}{3 y_2}$$

$$\Rightarrow \frac{y_1}{y_2} = \frac{2}{3}\,\frac{y_1}{y_2}$$

$$\Rightarrow 3 y_2^2 = 2 y_1 y_3$$

2. If $y = \left(\dfrac{1}{x}\right)^x$, show that y_2 at $x = 1$ is equal to zero.

Solution: $y = \left(\dfrac{1}{x}\right)^x, x > 0$

$$\Rightarrow \log y = \log\left(\frac{1}{x}\right)^x = x \log\left(\frac{1}{x}\right) = x[\log 1 - \log x]$$

$$\Rightarrow \log y = x[0 - \log x] = -x \log x \qquad \text{...(1)}$$

Now differentiating (1) w.r.t. x, we get

$$\frac{1}{y} \cdot \frac{dy}{dx} = -\log x - x \cdot \frac{1}{x} = -\log x - 1$$

$$\Rightarrow \frac{dy}{dx} = -y(\log x + 1)$$

$$\Rightarrow y_1 = -y(\log x + 1)$$

$$\Rightarrow y_2 = -y_1(\log x + 1) - y \cdot \frac{1}{x}$$

$$\therefore [y_2]_{x=1} = \left[-y_1(1 + \log x) - y \cdot \frac{1}{x}\right]_{x=1} \qquad \text{...(2)}$$

But $\left[-y_1(1 + \log x) - y \cdot \dfrac{1}{x}\right]_{x=1}$

$$= [-y_1]_{x=1} \cdot [(1 + \log x)]_{x=1} \cdot \left[\frac{y}{x}\right]_{x=1}$$

$$[-y_1]_{x=1} = -[-y(\log x + 1)]_{x=1}$$

$$= (y)_{x=1} \cdot (\log x + 1)_{x=1}$$

$$= 1(0+1) = 1 \quad \text{...(a)}$$

$$[(1+\log x)]_{x=1} = (1+\log 1) = (1+0) = 1 \quad \text{...(b)}$$

$$\left[\frac{y}{x}\right]_{x=1} = (y)_{x=1} = 1 \quad \text{...(c)}$$

Putting the values of (a), (b) and (c) in (2), we have

$$(y_2)_{x=1} = 1 \cdot 1 - 1 = 0$$

Note: If $\frac{d^2y}{dx^2} = f\left(x, y, \frac{dy}{dx}\right)$ and we are required to find out the value of $\frac{d^2y}{dx^2}$ at a point $x = a$, we are required to find the value of (i) $\frac{dy}{dx}$ at $x = 1$ (ii) The value of y at $x = 1$ (iii) The value of function of x at $x = 1$ i.e; the value of each term of $\frac{d^2y}{dx^2}$ at $x = a$

3. If $y = \left[\log\left(x + \sqrt{1+x^2}\right)\right]^2$ show that $(1+x^2)\,y_2 + x\,y_1 = 2$

Proof: $y = \left[\log\left(x + \sqrt{1+x^2}\right)\right]^2$

$$\Rightarrow \frac{dy}{dx} = y_1$$

$$= 2\left[\log\left(x+\sqrt{1+x^2}\right)\right] \cdot \frac{d}{dx}\left[\log\left(x+\sqrt{1+x^2}\right)\right]$$

$$= 2\left[\log\left(x+\sqrt{1+x^2}\right)\right] \times \frac{1}{\left(x+\sqrt{x^2+1}\right)} \times \left[1 + \frac{1}{2\sqrt{x^2+1}} \times 2x\right]$$

$$= 2\left[\log\left(x+\sqrt{1+x^2}\right)\right] \times \left[\frac{1}{\left(x+\sqrt{x^2+1}\right)} \times \frac{\left(\sqrt{x^2+1}+x\right)}{\sqrt{1+x^2}}\right]$$

$$= 2\left[\log\left(x+\sqrt{1+x^2}\right)\right] \cdot \frac{1}{\sqrt{x^2+1}}$$

Now we square both sides to get

$$y_1^2 = 4\left[\log\left(x+\sqrt{x^2+1}\right)\right]^2 \cdot \frac{1}{(x^2+1)}$$

$$= \frac{4y}{1+x^2}$$

$$\Rightarrow (1+x^2)\,y_1^2 = 4y \quad \text{...(1)}$$

Now differentiating again to get the second derivative from (1), we have

$$\frac{d}{dx}\left[(1+x^2)\,y_1^2\right] = \frac{d}{dx}(4y)$$

$$\Rightarrow (1+x^2)\frac{dy_1^2}{dx} + y_1^2 \cdot \frac{d}{dx}(1+x^2) = 4 \cdot \frac{dy}{dx}$$

$$\Rightarrow (1+x^2) \cdot 2y_1 \cdot \frac{dy_1}{dx} + y_1^2 \cdot 2x = 4 \cdot \frac{dy}{dx}$$

$$\Rightarrow (1+x^2) \cdot 2y_1 \cdot y_2 + y_1^2 \cdot 2x = 4y_1$$

Now, dividing both sides of (2) by $2y_1$ we have

$$(1+x^2)\,y_2 + x\,y_1 = 2$$

4. If $y\sqrt{1+x^2} = \log\left(x+\sqrt{1+x^2}\right)$ show that $(1+x^2)\,y_1 + xy = 1$

Proof: $y\sqrt{1+x^2} = \log\left(x+\sqrt{1+x^2}\right)$ ...(1)

Now differentiating both sides of (1) w.r.t. x, we have

$$\frac{dy}{dx} \times \left(\sqrt{1+x^2}\right) + \frac{d\sqrt{1+x^2}}{dx} \times y$$

$$= \frac{1}{\left(x+\sqrt{1+x^2}\right)} \cdot \left[1 + \frac{2x}{2\sqrt{1+x^2}}\right]$$

$$\Rightarrow \frac{dy}{dx} \cdot \left(\sqrt{1+x^2}\right) + \frac{xy}{\sqrt{1+x^2}} = \frac{1}{\sqrt{1+x^2}} \qquad ...(2)$$

Now multiplying both sides of (2) by $\sqrt{1+x^2}$ we have

$$\left(1+x^2\right) \cdot \frac{dy}{dx} + xy = 1$$

$$\Rightarrow \left(1+x^2\right) \cdot y_1 + xy = 1$$

5. If $y = a \sin \log x$, show that $x^2 y_2 + x y_1 + y = 0$

Proof: $y = a \sin(\log x)$

$$\Rightarrow \frac{dy}{dx} = a \cos(\log x) \cdot \frac{1}{x}$$

$$\Rightarrow x \cdot \frac{dy}{dx} = a \cos(\log x) \qquad ...(1)$$

Now, differentiating (1) w.r.t. x, we get

$$\Rightarrow x \cdot \frac{d}{dx}\left(\frac{dy}{dx}\right) + \frac{dy}{dx} \cdot \frac{dx}{dx} = a(-\sin \log x) \cdot \frac{1}{x}$$

$$\Rightarrow x \cdot \frac{d^2y}{dx^2} + \frac{dy}{dx} = \frac{-a \sin \log x}{x}$$

$$\Rightarrow x^2 \cdot \frac{d^2y}{dx^2} + x \cdot \frac{dy}{dx} = -y$$

$(\because a \sin \log x = y)$

$$\Rightarrow x^2 y_2 + x y_1 + y = 0$$

6. If $x = \cos(\log y)$, show that $\left(1-x^2\right) y_2 - x y_1 = y$

Proof: $x = \cos(\log y)$

$$\Rightarrow \log y = \cos^{-1} x \qquad ...(1)$$

Now, differentiating both sides of (1) w.r.t x, we have

$$\frac{1}{y} \cdot \frac{dy}{dx} = \frac{-1}{\sqrt{1-x^2}}$$

$$\Rightarrow \sqrt{1-x^2}\,\frac{dy}{dx} = -y \qquad ...(2)$$

Now squaring both sides of (2), we have

$$\left(1-x^2\right) \cdot \left(\frac{dy}{dx}\right)^2 = y^2 \qquad ...(3)$$

Again differentiating (3) w.r.t. x (to get second derivative)

$$\frac{d}{dx}\left[\frac{dy}{dx}\right]^2 \cdot \left(1-x^2\right) + \left(\frac{dy}{dx}\right)^2 (-2x) = 2y \cdot \frac{dy}{dx}$$

$$\Rightarrow 2 \cdot \frac{dy}{dx} \cdot \frac{d}{dx}\left(\frac{dy}{dx}\right) \cdot \left(1-x^2\right) - 2x \cdot \left(\frac{dy}{dx}\right)^2$$

$$= 2y \cdot \frac{dy}{dx}$$

$$\Rightarrow 2y_1 \cdot y_2\left(1-x^2\right) - 2x \cdot y_1^2 = 2y\,y_1 \qquad ...(4)$$

Now, dividing both sides of (4) by y_1, we have

$$2y_2\left(1-x^2\right) - 2x\,y_1 = 2y$$

$$\Rightarrow \left(1-x^2\right) y_2 - x y_1 = y$$

7. If $y = \sin(\sin x)$, show that

$$\frac{d^2y}{dx^2} + \tan x \cdot \frac{dy}{dx} + y \cos^2 x = 0.$$

Proof: $y = \sin(\sin x) \qquad ...(1)$

$$\Rightarrow \frac{dy}{dx} = \cos(\sin x) \cdot \cos x = \cos x \cdot \cos(\sin x) \qquad ...(2)$$

$$\Rightarrow \frac{d^2y}{dx^2} = \cos x\left[-\sin(\sin x) \cdot \cos x\right] + \cos(\sin x) \cdot (-\sin x)$$

$$= -\cos^2 x \sin(\sin x) - \sin x \cos(\sin x) \qquad ...(3)$$

$$\therefore \text{L.H.S} = \frac{d^2y}{dx^2} + \tan x \cdot \frac{dy}{dx} + y \cos^2 x$$

$= -\cos^2 x \sin(\sin x) - \sin x \cos(\sin x) +$

$\frac{\sin x}{\cos x} \cos x \cdot \cos(\sin n) + y \cos^2 x$

$= -\cos^2 x \sin(\sin x) - \sin x \cdot \cos(\sin x) +$

$\sin x \cdot \cos(\sin x) + y \cos^2 x$

$= -\cos^2 x \sin(\sin x) + y \cos^2 x$

$(\because y = \sin(\sin x))$

$= 0 =$ R.H.S

8. If $y = a\cos(\log x) + b\sin(\log x)$, show that $x^2 y_2 + x y_1 + y = 0$

Proof: $\because y = a\cos(\log x) + b\sin(\log x)$ ...(1)

Now, differentiating (1) w.r.t x, we have

$$\frac{dy}{dx} = -a\sin(\log x) \cdot \frac{1}{x} + b\cos(\log x) \cdot \frac{1}{x} \quad ...(2)$$

$$\Rightarrow x\frac{dy}{dx} = -a\sin(\log x) + b\cos(\log x) \quad ...(3)$$

Again differentiating the equation (3) (to get second derivative), we have,

$$x\frac{d}{dx}\left(\frac{dy}{dx}\right) + \frac{dy}{dx} \cdot \frac{dx}{dx}$$

$$= -a\cos(\log x) \cdot \frac{1}{x} + b(-\sin(\log x)) \cdot \frac{1}{x}$$

$$\Rightarrow x^2 \frac{d^2 y}{dx^2} + x \cdot \frac{dy}{dx} = -a\cos(\log x) - b\sin(\log x)$$

$$\Rightarrow x^2 y_2 + x y_1 = -[a\cos(\log x) + b\sin(\log x)]$$

$$\Rightarrow x^2 y_2 + x y_1 = -y$$

$$\Rightarrow x^2 y_2 + x y_1 + y = 0$$

9. If $y = \sin\left(m\sin^{-1} x\right)$ prove that

$\left(1 - x^2\right) y_2 - x y_1 + m^2 y = 0$.

Proof: $y = \sin\left(m\sin^{-1} x\right)$

$$\Rightarrow \frac{dy}{dx} = \frac{d}{dx}\left[\sin\left(\sin^{-1} x\right)\right]$$

$$= \cos\left(m\sin^{-1} x\right) \cdot m \cdot \frac{1}{\sqrt{1 - x^2}}$$

$$\Rightarrow \sqrt{1 - x^2} \cdot \frac{dy}{dx} = m\cos\left(m\sin^{-1} x\right)$$

$$\Rightarrow \left[\sqrt{1 - x^2} \cdot \frac{dy}{dx}\right]^2 = m^2 \cos^2\left(m\sin^{-1} x\right)$$

$$\Rightarrow \left(1 - x^2\right) \cdot \left(\frac{dy}{dx}\right)^2 = m^2\left[1 - \sin^2\left(m\sin^{-1} x\right)\right]$$

$$\Rightarrow \left(1 - x^2\right) \cdot \left(\frac{dy}{dx}\right)^2 = m^2\left(1 - y^2\right) \quad ...(1)$$

Now again differentiating both sides of equation (1) to get second derivative,

$$\left(1 - x^2\right) \cdot 2\frac{dy}{dx} \cdot \frac{d}{dx} \cdot \left(\frac{dy}{dx}\right) + \left(\frac{dy}{dx}\right)^2 (-2x)$$

$$= -m^2\left(2y\frac{dy}{dx}\right)$$

$$\Rightarrow \left(1 - x^2\right) 2 y_1 \cdot y_2 - 2x y_1^2 = -m^2\, 2 y y_1$$

$$\Rightarrow 2\left(1 - x^2\right) y_2 - 2x y_1 = -m^2\, 2y$$

$$\Rightarrow \left(1 - x^2\right) y_2 - x y_1 = -m^2 y$$

10. If $y = \tan^{-1} x$ prove that $\left(1 + x^2\right) y_2 + 2x y_1 = 0$.

Proof: $y = \tan^{-1} x$ ...(1)

Now differentiating both sides of (1) w.r.t. x, we have,

$$\frac{dy}{dx} = \frac{1}{1 + x^2}$$

$$\Rightarrow \left(1 + x^2\right) \frac{dy}{dx} = 1$$

Again differentiating (2) w.r.t.x to get the second derivative,

$$(1+x^2)\frac{d}{dx}\left(\frac{dy}{dx}\right)+\frac{dy}{dx}\cdot\frac{d}{dx}(1+x^2)=0$$

$$\Rightarrow (1+x^2)\cdot\frac{d^2y}{dx^2}+\frac{dy}{dx}\cdot 2x=0$$

$$\Rightarrow (1+x^2)\cdot y_2+2x\,y_1=0$$

11. If $y=\left(\tan^{-1}x\right)^2$ show that

$$(1+x^2)^2\frac{d^2y}{dx^2}+2x(1+x^2)\frac{dy}{dx}=2$$

Proof: $y=\left(\tan^{-1}x\right)^2$

$$\Rightarrow \frac{dy}{dx}=\frac{d}{dx}\left(\tan^{-1}x\right)^2$$

$$=2\tan^{-1}x\cdot\frac{d\tan^{-1}x}{dx}=\frac{2\tan^{-1}x}{1+x^2}$$

$$\Rightarrow (1+x^2)\frac{dy}{dx}=2\tan^{-1}x \qquad ...(1)$$

$$\Rightarrow (1+x^2)^2\cdot\left(\frac{dy}{dx}\right)^2=4\left(\tan^{-1}x\right)^2$$

$$\Rightarrow (1+x^2)^2\cdot\left(\frac{dy}{dx}\right)^2=4y \qquad ...(2)$$

Now again differentiating both sides of (2) w.r.t. x, we have

$$(1+x^2)^2\left[\frac{d}{dx}\left(\frac{dy}{dx}\right)^2\right]+\left(\frac{dy}{dx}\right)^2\left[\frac{d}{dx}(1+x^2)^2\right]$$

$$=\frac{d(4y)}{dx}$$

$$\Rightarrow (1+x^2)\cdot\left[2\cdot\frac{dy}{dx}\cdot\frac{d}{dx}\left(\frac{dy}{dx}\right)\right]+$$

$$\left(\frac{dy}{dx}\right)^2\left[2(1+x^2)\cdot 2x\right]=\frac{4dy}{dx}$$

$$\Rightarrow (1+x^2)^2\left[2\cdot\frac{dy}{dx}\cdot\frac{d^2y}{dx^2}\right]+$$

$$\left(\frac{dy}{dx}\right)^2\left[4x(1+x^2)\right]=4\cdot\frac{dy}{dx} \qquad ...(3)$$

Now dividing equation (3) by $2\frac{dy}{dx}$ we have

$$(1+x^2)^2\frac{d^2y}{dx^2}+2x(1+x^2)\frac{dy}{dx}=2$$

$$\Rightarrow (1+x^2)^2\frac{d^2y}{dx^2}+2x(1+x^2)\frac{dy}{dx}=2$$

12. If $y=e^{m.\sin^{-1}x}$, prove that

$$(1-x^2)y_2-x\,y_1=m^2y$$

Proof: $y=e^{m\sin^{-1}x}$

$$\Rightarrow \frac{dy}{dx}=\frac{d\,e^{m\sin^{-1}x}}{dx}$$

$$=e^{m\sin^{-1}x}\cdot\frac{m\times 1}{\sqrt{1-x^2}}=\frac{my}{\sqrt{1-x^2}}$$

$$\Rightarrow \sqrt{1-x^2}\cdot\frac{dy}{dx}=my \qquad ...(1)$$

(We square both sides of (1) so that square root symbol may be removed)

$$(1-x^2)\cdot\left(\frac{dy}{dx}\right)^2=m^2y^2 \qquad ...(2)$$

Again differentiating both sides of (2) to get second derivative,

$$(1-x^2)\cdot\frac{d}{dx}\left[\left(\frac{dy}{dx}\right)^2\right]+\left(\frac{dy}{dx}\right)^2\cdot\frac{d}{dx}(1-x^2)$$

$$=\frac{d}{dx}(m^2y^2)$$

$$\Rightarrow (1-x^2)\cdot 2\frac{dy}{dx}\cdot\frac{d}{dx}\left(\frac{dy}{dx}\right)+\left(\frac{dy}{dx}\right)^2\times(-2x)$$

$$= m^2 \frac{dy^2}{dx}$$

$$\Rightarrow \left(1 - x^2\right) \cdot 2 \frac{dy}{dx} \cdot \frac{d^2y}{dx} - 2x\left(\frac{dy}{dx}\right)^2$$

$$= m^2 \cdot 2y \frac{dy}{dx}$$

$$\Rightarrow 2\left(1 - x^2\right) \cdot y_1 \, y_2 - 2x \, y_1^2 = 2m^2 y$$

$$\Rightarrow \left(1 + x^2\right) y_2 - x y_1 = m^2 y$$

13. If $y = x \sin(\log x) + x \log x$, Show that

$$x^2 \frac{d^2y}{dx^2} - x\frac{dy}{dx} + 2y = x \log x$$

Proof: We are given $y = x \sin(\log x) + x \log x$ ...(1)

Now differentiating each terms of both sides w.r.t. x of (1), we have

$$\frac{dy}{dx} = \left[x \cos(\log x) \cdot \frac{1}{x} + \sin(\log x)\right] +$$

$$\left[x \cdot \frac{1}{x} + \log x\right]$$

$$\Rightarrow y_1 = \cos(\log x) + \sin(\log x) + 1 + \log x \quad ...(2)$$

Now, multiplying both sides of (2) by x, we have

$$x y_1 = x \cos(\log x) + x \sin(\log x) + x + x \log x \quad ...(3)$$

$\Rightarrow x y_1 = x \cos(\log x) + y + x$, (by using (1) in (3)) ...(4)

Now, differentiating again both sides of (4) w.r.t x to get y_2,

$$x y_2 + y_1 = -x \sin \log(x) \cdot \frac{1}{x} + \cos(\log x) + y_1 + 1$$

$$\Rightarrow x y_2 = -\sin \log x + \cos(\log x) + 1 \quad ...(5)$$

Now multiplying both sides of (5) by x, are get

$$x^2 y_2 = -x \sin(\log x) + x \cos(\log x) + x \quad ...(6)$$

$$\Rightarrow x^2 y_2 = -x \sin(\log x) + [xy_1 - y - x] + x$$

$\because x y_1 = x \cos(\log x) + y + x$ from (4)

$\Rightarrow x y_1 - y - x = x \cos(\log x)$

$$\Rightarrow x^2 y_2 = -(y - x \log x) + x y_1 - y$$

$\because y = x \log x + x \sin(\log x)$ from (1)

$\Rightarrow y - x \log x = x \sin(\log x)$

$$\Rightarrow x^2 y_2 = x \log x - y + x y_1 - y$$

$$\Rightarrow x^2 y_2 = x \log x + x y_1 - 2y$$

$$\Rightarrow x^2 y_2 - x y_1 + 2y = x \log x$$

14. If $y = \log\left(\frac{x}{a+bx}\right)^x$, prove that $x^3 y_2 = (y - x y_1)^2$.

Proof: $y = \log\left(\frac{x}{a+bx}\right)^x$, defined for $\frac{x}{a+bx} > 0$

$$\Rightarrow y = x \log\left(\left|\frac{x}{a+bx}\right|\right)$$

$$\Rightarrow y_1 = \left[\log|x| - \log|(a+bx)|\right] + x\left[\frac{1}{x} - \frac{b}{a+bx}\right]$$

$$\Rightarrow y_1 = \log\frac{x}{a+bx} + x\left[\frac{a+bx-bx}{x(a+bx)}\right]$$

$$= \log\frac{x}{a+bx} + \frac{a}{(a+bx)}$$

$$= \frac{y}{x} + \frac{a}{(a+bx)}$$

$$\Rightarrow x y_1 = y + \frac{ax}{a+bx}$$

$$\Rightarrow x y_1 - y = \frac{ax}{a+bx} \quad ...(1)$$

Again differentiating (1) to get y_2

$$x y_2 + y_1 - y_1 = \frac{(a + bx)\, a - ax \cdot b}{(a + bx)^2}$$

$$\Rightarrow x y_2 = \frac{a^2}{(a + bx)^2} \qquad ...(2)$$

Now squareing (1), we get

$$(x y_1 - y)^2 = \frac{x^2 a^2}{(a + bx)^2} = x^2\, x y_2 \text{ [from (2)]}$$

$\Rightarrow x^3 y_2 = (x y_1 - y)^2$ Which was required to be proved.

15. If $y = e^{\tan^{-1} x}$ show that $(1 + x^2) y_2 + (2x - 1) y_1 = 0$

Solution: $y = e^{\tan^{-1} x}$

$$\Rightarrow \frac{dy}{dx} = e^{\tan^{-1} x} \cdot \frac{d}{dx}\left(\tan^{-1} x\right) = \frac{e^{\tan^{-1} x}}{1 + x^2}$$

$$\Rightarrow (1 + x^2) \frac{dy}{dx} = e^{\tan-1} x = y \qquad ...(1)$$

Differentiating both sides of (1) w.r.t x, we have

$$(1 + x^2) \frac{d^2 y}{dx^2} + \frac{dy}{dx} \cdot 2x = \frac{dy}{dx}$$

$$\Rightarrow (1 + x^2) \frac{d^2 y}{dx^2} + (2x - 1) \frac{dy}{dx} = 0$$

$$\Rightarrow (1 + x^2) y_2 + (2x - 1) y_1 = 0$$

Second method after having

$$\frac{dy}{dx} = \frac{e^{\tan^{-1} x}}{1 + x^2}$$

We may straight way (or, directly), find

$$\frac{d^2 y}{dx^2} = \frac{(1 + x^2) \cdot e^{\tan^{-1} x}\left(\frac{1}{1 + x^2}\right) - e^{\tan^{-1} x} \cdot 2x}{(1 + x^2)^2}$$

$$= \frac{e^{\tan^{-1} x}(1 - 2x)}{(1 + x^2)^2}$$

$$\therefore \text{L.H.S.} = (1 + x^2)\, y_2 + (2x - 1)\, y_1$$

$$= \frac{(1 + x^2) \cdot e^{\tan^{-1} x}(1 - 2x)}{(1 + x^2)^2} + (2x - 1) \cdot \frac{e^{\tan^{-1} x}}{1 + x^2}$$

$$= \frac{e^{\tan^{-1} x}(1 - 2x)}{1 + x^2} + (2x - 1) \cdot \frac{e^{\tan^{-1} x}}{1 + x^2}$$

$$= \frac{e^{\tan^{-1} x}(1 - 2x)}{1 + x^2} - \frac{(1 - 2x) \cdot e^{\tan^{-1} x}}{1 + x^2}$$

$= 0$

$=$ R.H.S

N.B.: The first method is more convenient in comparision to the second method.

Type 3: Problem based on finding a differential equation when the given function is in the implicit form $f(x, y) = c$ (where c = any constant and $c = 0$ in particular) or givin a function having the form $x = f_1\, f_2\, (y)$

Working rule: Proceed as usual i.e finding the derivatives involved in the given differential equation to be satisfeid by the given function .

Example worked out:

1. If $x = \sin(\log y)$ show that

$$(1 - x^2) y_2 - x y_1 = y$$

Proof: $\because x = \sin(\log y)$

$$\Rightarrow \log y = \sin^{-1} x \qquad ...(1)$$

$\Rightarrow \dfrac{1}{y} \cdot \dfrac{dy}{dx} = \dfrac{1}{\sqrt{1 - x^2}}$ (on differentiating (1) w.r.t. x)

$$\Rightarrow \frac{y_1}{y} = \frac{1}{\sqrt{1-x^2}}$$

$$\Rightarrow y_1 \sqrt{1-x^2} = y \qquad ...(2)$$

$\Rightarrow (y_1)^2 (1-x^2) = y^2$ (on squaring both sides of (2))

$$\Rightarrow \frac{d}{dx}\left[(y_1)^2 (1-x^2)\right] = \frac{dy^2}{dx}$$

$$\Rightarrow y_1^2(-2x) + 2y_1 y_2 (1-x^2) = 2y\, y_1$$

$$\Rightarrow -x y_1 + (1-x^2) y_2 = y$$

$$\Rightarrow (1-x^2) y_2 - x y_1 = y$$

2. If $xy + 4y = 3x$ find $\frac{d^2y}{dx^2}$

Proof: $xy + 4y = 3x \qquad ...(1)$

Now, differentiating (1) w.r.t x

$$x \cdot \frac{dy}{dx} + y \cdot 1 + 4 \cdot \frac{dy}{dx} = 3$$

$$\Rightarrow (x+4) \cdot \frac{dy}{dx} = 3 - y \qquad ...(2)$$

$$\Rightarrow y_1 = \frac{dy}{dx} = \frac{(3-y)}{(x+4)} \qquad ...(3)$$

Now considering (2),

$$(x+4)\frac{d^2y}{dx^2} + \frac{dy}{dx} \cdot (1) = -\frac{dy}{dx}$$

$$\Rightarrow (x+4) \cdot y_2 + y_1 + y_1 = 0$$

$$\Rightarrow (x+4) \cdot y_2 + 2y_1 = 0$$

$\Rightarrow y_2 = \frac{-2y_1}{(x+4)} = \frac{-2}{(x+4)} \times \frac{(3-y)}{(x+4)}$ (by using (3))

$$= \frac{2y-6}{(x+4)^2} = \frac{1}{(x+4)^2} \cdot \left[2\left(\frac{3x}{x+4}\right) - 6\right]$$

$$= \frac{1}{(x+4)^3} \cdot [6x - 6x - 24]$$

$$= \frac{1}{(x+4)^3} \times (-24) = \frac{-24}{(x+4)^3}$$

Note: $y = 3 - \frac{12}{x+4} \Rightarrow y_2 = \frac{-24}{(x+4)^3}$

3. If $\log y = \log(\sin x) - x^2$, show that

$$\frac{d^2y}{dx^2} + 4y\frac{dy}{dx} + (4x^2+3)\, y = 0$$

Proof: $\log y = \log(\sin x) - x^2$

$$\Rightarrow \log y = \log(\sin x) - \log_e e^{x^2}$$

$$\left[\because \log_e e^{x^2} = x^2 \log_e e = x^2 \cdot 1 = x^2\right]$$

$$\Rightarrow \log y = \log\left[\frac{\sin x}{e^{x^2}}\right] \qquad ...(1)$$

Now, taking anti log of both sides of (1) $y = \frac{\sin x}{e^{x^2}}$

$$\Rightarrow y \cdot e^{x^2} = \sin x \qquad ...(2)$$

[***N.B.***: Product rule of derivative is easier than the quotient rule. This is why we cross multiply]

Now differentiating (2) w.r.t x to find y_1,

$$e^{x^2} \cdot y_1 + 2x\, e^{x^2} \cdot y = \cos x \qquad ...(3)$$

Again differentiating (3) w.r.t x to find y_2

$$e^{x^2} y_2 + 2x e^{x^2} y_1 + 2\left[x \cdot e^{x^2} y_1 + 2x^2 e^{x^2} y + e^{x^2} y\right]$$

$$= -\sin x = -e^{x^2} y \text{ (from (2))} \qquad ...(4)$$

Now dividing the equation (4) by e^{x^2} and then transposing,

$$y_2 + 2x y_1 + 2x y_1 + 4x^2 y + 2y + y = 0$$

$$\Rightarrow y_2 + 4x y_1 + (4x^2+3)\, y = 0$$

4. If $\cos^{-1}\left(\frac{y}{b}\right) = \log\left(\frac{x}{n}\right)^n$, show that

$x^2 y_2 + x y_1 + n^2 y = 0$.

Proof: $\cos^{-1}\left(\frac{y}{b}\right) = \log\left(\left|\frac{x}{n}\right|\right)^n$ since $\frac{x}{n} > 0$

$$\Rightarrow \cos^{-1}\left(\frac{y}{b}\right) = n\log\left(\left|\frac{x}{n}\right|\right) = n\left(\log|x| - \log|n|\right) \quad ...(1)$$

Now differentiating the equation (1) w.r.t. x

$$-\frac{1}{\sqrt{1-\left(\frac{y}{b}\right)^2}}\cdot\frac{y_1}{b} = \frac{n}{x}$$

$$\Rightarrow -\frac{y_1|b|}{b\cdot\sqrt{b^2-y^2}} = \frac{n}{x} \quad ...(2)$$

Now on cross multiplying the equation (2) and then squaring, we get

$$x^2\, y_1^2 = n^2\left(b^2 - y^2\right) \quad ...(3)$$

Now, again differentiating the equation (3) w.r.t x

$$\frac{d}{dx}\left(x^2 y_1^2\right) = \frac{d}{dx}\left[n^2\left(b^2 - y^2\right)\right]$$

$$\Rightarrow x^2\cdot 2\cdot y_1\cdot y_2 + 2x\, y_1^2 = -\,2n^2\, y\, y_1 \quad ...(4)$$

Now dividing (4) by $2y_1$, we get

$$x^2 y_2 + x\, y_1 + n^2 y = 0$$

Type 4: Problems based on finding a differential equation when the given function is in parametric form $\begin{cases} x = f_1(t) \\ y = f_2(t) \end{cases}$

Working rule: We proceed directly for finding the first derivative using the formula $\frac{dy}{dx} = \frac{\frac{dy}{dt}}{\frac{dx}{dt}}$ and then we find other derivatives (2nd, 3rd, ... etc) involved in the given differential equation using various mathematical manipulation performed on $\frac{dy}{dx}$ or directly from $\frac{dy}{dx}$ according to the need of the problem of differential equation. Lastly by simplification, cancellation, transposition and using axioms of an equation, we arrive at our target (i.e., we find the required differential equation)

Examples worked out:

1. If $y = \sin pt$, $x = \sin t$, show that

$$\left(1 - x^2\right)\frac{d^2y}{dx^2} - x\,\frac{dy}{dx} + p^2 y = 0.$$

Proof: $y = \sin pt$

$x = \sin t$

$$\Rightarrow \frac{dy}{dx} = \frac{\frac{dy}{dt}}{\frac{dx}{dt}} = \frac{p\cos pt}{\cos t}$$

$$\Rightarrow \left(\frac{dy}{dx}\right)^2 = \frac{p^2\cos^2 pt}{\cos^2 t} = \frac{p^2\left(1 - \sin^2 pt\right)}{\left(1 - \sin^2 t\right)}$$

$$= \frac{p^2\left(1 - y^2\right)}{1 - x^2} \text{ for } |x| \neq 1$$

$$\Rightarrow \left(1 - x^2\right)\left(\frac{dy}{dx}\right)^2 = p^2\left(1 - y^2\right)$$

$$\Rightarrow \frac{d}{dx}\left[\left(1 - x^2\right)\cdot\left(\frac{dy}{dx}\right)^2\right] = \frac{d}{dx}\left[p^2\left(1 - y^2\right)\right]$$

$$\Rightarrow \left(1 - x^2\right)2\cdot\frac{dy}{dx}\cdot\frac{d^2y}{dx^2} + (-\,2x)\left(\frac{dy}{dx}\right)^2$$

$$= p^2\cdot\left(-\,2y\,\frac{dy}{dx}\right)$$

Dividing by $2\,\frac{dy}{dx}$, we have for $|x| \neq 1, |y| \neq 1$;

$\left(1 - x^2\right)\frac{d^2y}{dx} - x\,\frac{dy}{dx} + p^2 y = 0$ proved.

2. If $x = a(\cos\theta + \theta\sin\theta)$ and $y = a(\sin\theta - \theta\cos\theta)$

where $0 < \theta < \frac{\pi}{2}$ prove that $\tan^{-1}(y_1) = \frac{(1 + y_1^2)^{\frac{3}{2}}}{a y_2}$

Proof: $x = a(\cos\theta + \theta\sin\theta) \Rightarrow \frac{dx}{d\theta}$

$= a(-\sin\theta + 1 \cdot \sin\theta + \theta \cdot \cos\theta) = a\theta\cos\theta$ ...(1)

$y = a(\sin\theta - \theta\cos\theta)$

$\Rightarrow \frac{dy}{d\theta} = a(\cos\theta - 1 \cdot \cos\theta - \theta(-\sin\theta)) = a\theta\sin\theta$

...(2)

Now, $\frac{dy}{dx} = \frac{\frac{dy}{d\theta}}{\frac{dx}{d\theta}} = \frac{a\theta\sin\theta}{a\theta\cos\theta} = \tan\theta$ ...(3)

Differentiating both sides of (3) w.r.t. x, we get

$$\frac{d}{dx}\left(\frac{dy}{dx}\right) = \frac{d}{dx}(\tan\theta)$$

$$\Rightarrow \frac{d^2y}{dx^2} = \sec^2\theta \cdot \frac{d\theta}{dx} = \sec^2\theta \cdot \frac{1}{a\theta \cdot \cos\theta}$$

$$\left(\because \frac{dx}{d\theta} = a\,\theta\cos\theta\right)$$

$$\Rightarrow \frac{d^2y}{dx^2} = \frac{\sec^3\theta}{a\theta}$$

$= \frac{(1 + \tan^2\theta)^{\frac{3}{2}}}{a\theta} = \frac{(1 + y_1^2)^{\frac{3}{2}}}{a\tan^{-1}(y_1)}$ hence the result.

3. If $x = \cos t$, $y = \log t$, show that

$$\frac{d^2y}{dx^2} + \left(\frac{dy}{dx}\right)^2 = 0 \quad \text{at } t = \frac{\pi}{2}.$$

Solution: $x = \cos t \Rightarrow \frac{dx}{dt} = -\sin t$

$$y = \log t \Rightarrow \frac{dy}{dt} = \frac{1}{t}$$

$\therefore \frac{dy}{dx} = \frac{\frac{dy}{dt}}{\frac{dx}{dt}} = \frac{\frac{1}{t}}{-\sin t} = -\frac{1}{t\sin t}$ ...(i)

$$\Rightarrow \frac{d}{dx}\left(\frac{dy}{dx}\right) = \frac{d}{dx}\left(-\frac{1}{t\sin t}\right)$$

$$\Rightarrow \frac{d^2y}{dx^2} = \frac{1}{(t\sin t)^2} \cdot (\sin t + t\cos t) \cdot \frac{dt}{dx}$$

$$= \frac{\sin t + t\cos t}{t^2\sin^2 t} \cdot \left(\frac{1}{-\sin t}\right)$$

$= \frac{-(\sin t + t\cos t)}{t^2\sin^3 t}$ ...(ii)

Now, squaring both sides of (1), we get

$\left(\frac{dy}{dx}\right)^2 = +\frac{1}{t^2\sin^2 t}$ ...(iii)

Adding (ii) and (iii), we get

$$\frac{d^2y}{dx} + \left(\frac{dy}{dx}\right)^2 = -\frac{(\sin t + t\cos t)}{t^2\sin^3 t} + \frac{1}{t^2\sin^2 t}$$

$$\therefore \left[\frac{d^2y}{dx^2} + \left(\frac{dy}{dx}\right)^2\right]_{t=\frac{\pi}{2}}$$

$$= \left[\frac{-(\sin t + t\cos t)}{t^2\sin^3 t} + \frac{1}{t^2\sin^2 t}\right]_{t=\frac{\pi}{2}}$$

$$= \left[\frac{-1}{\frac{\pi 2}{4}} + \frac{1}{\frac{\pi 2}{4}}\right]$$

$= 0$

Type 1: Problems based on finding second derivative

(A) Algebraic functions

Exercise 14.1

Find $\frac{d^2y}{dx^2}$ of the following functions.

1. $y = \sqrt{2x - 3}$
2. $y = (x^2 - 2x + 3)^{\frac{2}{3}}$
3. $y = \frac{1}{(3x - 4)}$
4. $y = \frac{1}{(a - x)}$
5. $y = (3 - 2x)^{-\frac{1}{2}}$
6. $y = \sqrt{x}$
7. $y = \frac{1}{\sqrt{x}}$
8. $y = \frac{a - x}{a + x}$
9. $y = \frac{1}{x^2 + 4}$
10. $y = \sqrt{x^2 - 9}$
11. $y = (2x + 3)^{\frac{2}{3}}$
12. $y = x^3 (7x + 1)^2$
13. $y = \frac{\sqrt{3x + 4}}{x}$

Answers:

1. $-\frac{1}{(2x - 3)^{\frac{3}{2}}}$ **2.** $\frac{4(x^2 - 2x + 7)}{9(x^2 - 2x + 3)^{\frac{4}{3}}}$

3. $\frac{18}{(3x - 4)^3}$ **4.** $\frac{2}{(a - x)^3}$ **5.** $3(3 - 2x)^{\frac{-5}{2}}$

6. $\frac{-1}{4x\sqrt{x}}$ **7.** $\frac{3}{4x^2\sqrt{x}}$ **8.** $\frac{4a}{(a + x)^3}$

9. $\frac{6x^3 - 8}{(x^2 + 4)^3}$ **10.** $-9(x^2 - 9)^{-\frac{3}{2}}$

11. $-\frac{8}{9}(2x + 3)^{-\frac{4}{3}}$

12. $6x(7x + 1)^2 + 84x^2(7x + 1) + 98x^3$

13. Find

Type 1: (continued)

(B) Trigonometric function:

Exercise 14.2

Find $\frac{d^2y}{dx^2}$ of the following functions.

1. $y = \sqrt{\sec 2x}$
2. $y = \text{cosec}\,(x^2 - 3)$
3. $y = \sin 3x \cdot \cos x$
4. $y = \sin^4 x \cdot \cos^2 x$
5. $y = \frac{\sin x + 2}{2\cos x + 3}$

Answers:

1. $y \cdot (2\sec^2 2x + \tan^2 2x)$
2. $2y\left[2x^2 \cot^2 (x^2 - 3) + 2x^2 y^2 - \cot (x^2 - 3)\right]$
3. $-(8\sin 4x + 2\sin 2x)$
4. $\frac{1}{8}(\cos 2x + 8\cos 4x - 9\cos 6x)$
5. $\frac{8\cos^2 x + 16\sin^2 x + 12\sin x\cos x + 12\cos x - 10\sin x}{(2\cos x + 3)^3}$

Type1: (continued)

(C) Inverse Trigonometric functions:

Exercise 14.3

Find $\frac{d^2y}{dx^2}$ of the following functions

1. $y = \sin^{-1} x$

2. $y = \tan^{-1} 2x$

3. $y = \sec^{-1} x$

4. $y = \sqrt{1 - x^2} \cdot \sin^{-1} x$

Answers:

1. $\frac{x}{(1 - x^2)^{\frac{3}{2}}}$

2. $\frac{-16x}{(1 + 4x^2)^2}$

3. $\frac{-(2x^2 - 1)|x|}{x^3(x^2 - 1)^{\frac{3}{2}}}$

4. $\frac{-x\sqrt{1 - x^2} - \sin^{-1} x}{(1 - x^2)^{\frac{3}{2}}}$

Type 1: (continued)

(D) Exponential functions:

Exercise 14.4

Find $\frac{d^2y}{dx^2}$ of the following functions.

1. $y = e^{2x + 3}$

2. $y = e^{\sin x}$

3. $y = e^{m \sin^{-1} x}$

Answers:

1. $4y$

2. $y(\cos^2 x - \sin x)$

3. $y_2 = \frac{m^2 y + x y_1}{(1 - x^2)}$

Type 1: (continued)

(E) Logarithmic functions:

Exercise 14.5

Find $\frac{d^2y}{dx^2}$ of the following functions

1. $y = \log(2x + 3)$

2. $y = \log \sin x$

3. $y = \log \cos x$

4. $y = \log\{(2x + 3)(3 - 5x)\}$

5. $y = \log\left(\frac{a + x}{a - x}\right)$

6. $y = \log\sqrt{\left(\frac{ax + b}{cx + d}\right)}$

Answers:

1. $-\frac{4}{(2x + 3)^2}$

2. $-\text{cosec}^2 x$

3. $-\sec^2 x$

Type 1: (continued)

(F) Implicit functions:

Exercise 14.6

Find $\frac{d^2y}{dx^2}$ of the following functions.

1. $x^2 + xy + y^2 = a^2$

2. $x^2 + y^2 + 3xy - 7 = 0$
3. $\sqrt{x} + \sqrt{y} = \sqrt{a}$
4. $\cos x \cdot \cos y = c$
5. $y = \tan(x + y)$
6. $y^2 + 3x = 25$
7. $\frac{x^2}{a^2} + \frac{y^2}{b^2} = 1$

***Answers*:**

1. $\frac{-6a^2}{(x + 2y)^3}$
2. $\frac{10\left(x^2 + 3xy + y^2\right)}{(2y + 3x)^3}$
3. $\frac{1}{2}\sqrt{\frac{a}{x^3}}$
4. $-\frac{\tan^2 x + \tan^2 y + 2\tan^2 x \cdot \tan^2 y}{\tan^3 y}$
5. $-2 \cdot \frac{\cot^3(x + y)}{\sin^2(x + y)}$
6. $-\frac{9}{4} \cdot (25 - 3x)^{-\frac{3}{2}}$
7. $\frac{-b^4}{a^2 y^3}$

***Type 1*: (continued)**

***(G) Parametric functions*:**

Exercise 14.7

Find $\frac{d^2y}{dx^2}$ of the following functions.

1. $x = a\cos^3\theta$

   $y = b\sin^3\theta$
2. $x = \frac{3t}{1 + t^3}$

   $y = \frac{3t^2}{1 + t^3}$
3. $x = at^2$

   $y = 2at$
4. $x = a\cos\theta$

   $y = b\sin\theta$
5. $x = a(\theta + \sin\theta)$

   $y = a(1 - \cos\theta)$
6. $x = a\sec^2\theta$

   $y = a\tan^3\theta$
7. $x = 3\cos\theta - \cos^3\theta$

   $y = 3\sin\theta - \sin^3\theta$
8. $x = t^2$

   $y = (t - 1)^2$
9. $x = \frac{1}{t}$

   $y = \frac{1}{t + 1}$
10. $x = \frac{a\left(1 - t^2\right)}{1 + t^2}$

    $y = \frac{2at}{1 + t^2}$
11. $x = a\cos t$

    $y = a\sin t$
12. $x = a\sin(2t + 3)$

    $y = a\cos(2t + 3)$
13. $x = 2(1 - \sin t)$

    $y = 4\cos t$

14. $x = a\cos^2\theta$

$y = a\sin^2\theta$

15. $x = a\cos^3\theta$

$y = a\sin^3\theta$

16. $x = a\tan\theta$

$y = \frac{1}{2}a\sin^2\theta$

17. $x = a\,\text{cosec}\,\theta$

$y = a\cot\theta$

Answers:

1. $\frac{b}{3a^2}\sec^4\theta\cdot\text{cosec}\,\theta$ 2. $\frac{2(1+t^3)^4}{3(1-2t^3)^3}$

3. $\frac{-1}{2at^3}$ 4. $-\frac{b}{a}\text{cosec}^3\theta$ 5. $\frac{1}{4a}\sec^4\left(\frac{\theta}{2}\right)$

6. $\frac{3}{4a}\cot\theta$ 7. $-\cot^2\theta\cdot\text{cosec}^5\theta$

8. $\frac{1}{2t^3}$ 9. $\frac{-2t^3}{(1+t)^3}$ 10. $-\frac{(1+t^2)^3}{8at^3}$

11. $-\frac{\text{cosec}^3 t}{a}$ 12. $-\frac{\sec^3(2t+3)}{a}$

13. $-\sec^3 t$ 14. 0 15. $\frac{\sec^4\theta\cdot\text{cosec}\,\theta}{3a}$

16. $\frac{\cos^4\theta\left(4\cos^2\theta-3\right)}{a}$ 17. $-\frac{\tan^3\theta}{a}$

***Type 2*: Problems based on finding the value of at the indicated points**

Exercise 14.8

Find $\frac{d^2y}{dx^2}$ at the indicated points for the following functions.

1. $x = a(\theta + \sin\theta)$

$y = a(1 - \cos\theta)$ at $\theta = \frac{\pi}{2}$

2. $x = a(1 - \cos 2\theta)$

$y = a(2\theta - \sin 2\theta)$ at $\theta = \frac{\pi}{4}$

3. $x = 2\cos t - \cos 2t$

$y = 2\sin t - \sin 2t$ at $\theta = \frac{\pi}{2}$

4. If $x = t^2$

$y = t^3$ at $t = 1$

5. If $x = \frac{1}{t^3}$

$y = t^2$ at $t = \frac{1}{2}$

Answers:

1. $\frac{1}{a}$ 2. $\frac{1}{a}$ 3. $-\frac{3}{2}$ 4. $\frac{3}{4}$ 5. $\frac{5}{1152}$

***Type 3*: To show that a given function satisfies a differential equation**

(A) Problems based on showing that an algebraic function satisfies a differential equation

Exercise 14.9

1. If $y^2 = 2x^2 + 3x + 5$, show that

$$\frac{d^2y}{dx^2} = \frac{31}{4\left(2x^2+3x+5\right)^{\frac{3}{2}}}$$

2. If $x^3 + y^3 = 3x^2$, show that $\frac{d^2y}{dx^2} + \frac{2x^2}{y^5} = 0$

3. If $x^2 - xy + 2y^2 - 5 = 0$, show that

$$\frac{d^2y}{dx^2} = \frac{14\left(x^2 - xy + 2y^2\right)}{(x-4y)^3}$$

4. If $y = x^2 - 3x + 5$, show that

$$\frac{d^2y}{dx^2} = \frac{2}{(2x-3)^3} \cdot \left(\frac{dy}{dx}\right)^3$$

5. If $y = \left(x - \sqrt{1+x^2}\right)^3$, show that

$$\left(1+x^2\right)\frac{d^2y}{dx^2} + x\frac{dy}{dx} - 9y = 0$$

6. If $2y = \sqrt{x+1} + \sqrt{x-1}$, Show that

$$4\left(x^2-1\right)\frac{d^2y}{dx^2} + 4x\frac{dy}{dx} - y = 0$$

7. If $y^m + y^{-m} = 2x$, show that

$$\left(x^2-1\right)\frac{d^2y}{dx^2} + x\frac{dy}{dx} - \frac{y}{m^2} = 0$$

8. If $y = \sqrt{2x+1} + \sqrt{2x-1}$, show that

$$\left(4x^2-1\right)\frac{d^2y}{dx^2} + 4x\frac{dy}{dx} - y = 0$$

9. If $x^3 + y^3 = 3axy$, show that

$$y_2 = -2 \cdot \frac{a^3 xy}{\left(y^2 - ax\right)^3}$$

10. If $y = \left(x + \sqrt{1+x^2}\right)^n$, show that

$$\left(1+x^2\right)y + xy_1 - n^2y = 0$$

11. If $y = \left(x + \sqrt{x^2-1}\right)^m$ show that

$$\left(x^2-1\right)y_2 + xy_1 - m^2y = 0$$

12. If $y^{\frac{1}{5}} + y^{-\frac{1}{5}} = 2x$, show that

$$\left(x^2-1\right)y_2 + xy_1 - 25y = 0$$

13. If $Z = \dfrac{1}{x}$, show that $x^4\dfrac{d^2y}{dx^2} + 2x^3\dfrac{dy}{dx} = \dfrac{d^2y}{dZ^2}$

14. If $\sqrt{x+y} + \sqrt{y-x} = c$, show that $y_2 = \dfrac{2}{c^2}$

15. If $ax^2 + 2h\,xy + by^2 = 1$, show that

$$\frac{d^2y}{dx^2} = \frac{h^2 - ab}{(hx+by)^3}$$

16. If $y = \left(1-x^2\right)^{\frac{3}{2}}$, show that

$$\left(1-x^2\right)y_2 + xy_1 + 3y = 0$$

17. If $y = A\left(x+\sqrt{x^2-1}\right)^m + B\left(x-\sqrt{x^2-1}\right)^n$,

show that $\left(x^2-1\right)y_2 + xy_1 - n^2y = 0$

18. If $y = ax + \dfrac{b}{x}$, show that $x^2y_2 + xy_1 - y = 0$

19. If $y = \sqrt{ax^2+b}$, show that $y_2\,y^3 = ab$

20. If $y = \dfrac{ax+b}{cx+d}$, show that $2y_1\,y_3 = 3{y_2}^2$

21. If $y = \dfrac{ax-b}{a-bx}$, show that $2y_1\,y_3 = 3{y_2}^2$

22. If $y = \dfrac{ax^2+bx+c}{1-x}$, prove that

$(1-x)\,y_3 = 3y_2$

23. If $y = \dfrac{ax+b}{a+bx}$, show that $2y_1\,y_3 = 3{y_2}^2$

24. If $y = 2ax + \dfrac{3b}{x}$, show that $x^2y_2 + xy_1 - y = 0$

25. If $y = \sqrt{ax^2+2b}$, show that $y_2\,y^3 = 2ab$

26. If $y = \dfrac{ax-b}{cx-d}$, show that $2y_1\,y_3 = 3{y_2}^2$

27. If $y = \dfrac{ax-b}{2a-bx}$, show that $2y_1\,y_3 = 3{y_2}^2$

28. If $y = \dfrac{-ax^2+bx+c}{1-x}$, prove that

$(1-x)\,y_3 = 3y_2$

29. If $y = \dfrac{4ax + b}{2a + bx}$, show that $2y_1 y_3 = 3{y_2}^2$

30. If $x = \sin(\log y)$, show that

$(1 - x^2) y_2 - xy_1 = y$

Type 3: (continued)

(B) Problems based on showing that a trigonometric function satisfies a differentiatial equation

Exercise 14.10

1. If $y = \sin(ax + b)$ show that

$$\frac{d^2 y}{dx^2} = \frac{\sec^2(ax + b)\tan(ax + b)}{a}$$

2. If $y = 2\sin 3x - 5\cos 3x$, show that $\dfrac{d^2 y}{dx^2} + 9y = 0$

3. If $y = \sin(m\cos^{-1} x)$, show that

$$(1 + x^2)\frac{d^2 y}{dx^2} - x\frac{dy}{dx} + m^2 y = 0$$

4. If $y = a\sin(\log x)$, show that

$x^2 y_1 + xy_1 + y = 0$

5. If $y = a\cos(\log x)$, show that

$x^2 y_1 + xy_1 + y = 0$

6. If $y = \sin(\sin x)$, show that

$y_2 + y_1 \tan x + y\cos^2 x = 0$

7. If $y = \cos(m\sin^{-1} x)$, show that

$(1 - x^2) y_2 - xy_1 + m^2 y = 0$

8. If $x = \sin(\log y)$, show that

$(1 + x^2) y_2 - xy_1 = y$

9. If $x = \cos(\log y)$, show that

$(1 - x^2) y_2 - xy_1 = y$

10. If $y = 2\cos(\log x) + 3\sin(\log x)$, show that

$$x^2 \cdot \frac{d^2y}{dx^2} + x\frac{dy}{dx} + y = 0$$

11. If $y = \tan^{-1} x$, show that $(1 + x^2) y_2 + 2xy_1 = 0$

12. If $y^2 = 4\cos^2 x + 9\sin^2 x$, show that

$$y + \frac{d^2y}{dx^2} = \frac{36}{y^3}$$

13. If $y = A\cos mx + B\sin mx$, show that

$y_2 + m^2 y = 0$

14. If $y = \sin^{-1} x$, show that $(1 - x^2) y_2 - xy_1 = 0$

15. If $y = (\sin^{-1} x)^2$, show that

$(1 - x^2) y_2 - x y_1 = 2$

16. If $y = (\tan^{-1} x)^2$, show that

$(x^2 + 1)^2 y_2 + 2x(x^2 + 1) y_1 = 2$

17. If $y = (\cos^{-1} x)^2$, show that

$(1 - x^2) y_2 = xy_1 + 2$

18. If $p^2 = a^2\cos^2\theta + b^2\sin^2\theta$, show that

$$p + \frac{d^2p}{d\theta^2} = \frac{a^2 b^2}{p^3}$$

19. If $y = P\sin 2x - Q\cos 2x$, where P and Q are constant, show that $\dfrac{d^2y}{dx^2} + 4y = 0$

20. If $y = a\sin x + b\cos x$ show that $\dfrac{d^2y}{dx^2} + y = 0$

21. If $y = a\cos(\log x) + b\sin(\log x)$ show that

$x^2 y_2 + xy_1 + y = 0$

22. If $y = \dfrac{\sin^{-1} x}{\sqrt{1-x^2}}$ show that

$$\left(1 - x^2\right) y_2 - 3\, xy_1 + m^2 y = 0$$

23. If $y = \sin\left(m \sin^{-1} x\right)^2$ show that

$$\left(1 - x^2\right) y_2 - xy_1 + m^2 y = 0$$

Type 3: (continued)

(C) Problems based on showing that an exponential or a logarithmic function satisfies a differential equation

Exercise 14.11

1. If $y = Ae^{nx} + Be^{-nx}$, show that $y_2 = n^2 y$

2. If $y = 2e^{-3x} + 3e^{-2x}$, Show that

$$\frac{d^2y}{dx^2} + 5 \cdot \frac{dy}{dx} + 6y = 0$$

3. If $y = e^{2\sin^{-1} x}$ show that

$$\left(1 - x^2\right) \frac{d^2y}{dx^2} - x \frac{dy}{dx} - 4y = 0$$

4. If $x + y = e^{x-y}$, show that $\dfrac{d^2y}{dx^2} = \dfrac{4\,(x+y)}{(x+y+1)^2}$

5. If $y = e^{m \tan^{-1} x}$ show that

$$\left(1 + x^2\right) \frac{d^2y}{dx^2} + (2x - m) \frac{dy}{dx} = 0$$

6. If $y = e^{\tan^{-1} x}$ show that

$$\left(x^2 + 1\right) y_2 + (2x - 1)\, y_1 = 0$$

7. If $y = e^{\tan x}$, show that

$$\left(\cos^2 x\right) y - (1 + \sin 2x)\, y_1 = 0$$

8. If $y = \log\left(x + \sqrt{x^2 + 1}\right)$, show that

$$\left(1 + x^2\right) \frac{d^2y}{dx^2} + x \frac{dy}{dx} = 0$$

9. If $y = \left[\log\left(x + \sqrt{x^2 + 1}\right)\right]^2$ show that

$$\left(1 + x^2\right) \frac{d^2y}{dx^2} + x \frac{dy}{dx} = 2$$

10. If $y = e^{a \sin^{-1} x}$, show that

$$\left(1 - x^2\right) \frac{d^2y}{dx^2} - x \frac{dy}{dx} = a^2 y$$

11. If $y = \log(\sin x)$,

show that $y_2 + (y_1)^2 + 1 = 0$

12. If $y = \log\left(x + \sqrt{a^2 + x^2}\right)$, Show that

$$\left(a^2 + x^2\right) y_2 + xy_1 = 0$$

Type 3: (continued)

(D) Problems based on showing that parametric functions satisfy differential equations

Exercise 14.12

1. If $x = a\,(\theta + \sin\theta)$ $y = a\,(1 + \cos\theta)$, show that

$$\frac{d^2y}{dx^2} = -\frac{a}{y^2}.$$

2. If $x = 2\cos\theta + 3\sin\theta$ $y = 2\sin\theta - 3\sin\theta$ show that $\dfrac{d^2y}{dx^2} = -\dfrac{13}{y^3}$.

3. If $x = a(\cos\theta + \theta\sin\theta)$ $y = a(\sin\theta - \theta\sin\theta)$, $\left(0 < \theta < \dfrac{\pi}{2}\right)$ show that $\dfrac{d^2y}{dx^2} = \dfrac{\sec^3\theta}{a\theta}$

4. If $x = \cos^{-1}\left(\dfrac{1-t^2}{1+t^2}\right)$ $y = \sin^{-1}\left(\dfrac{2t}{1+t^2}\right)$ show that $\dfrac{d^2y}{dx^2}$ is independent of t.

Hint: If $x = \cos^{-1}\left(\dfrac{1-t^2}{1+t^2}\right) = 2\tan^{-1} t$

$y = \sin^{-1}\left(\dfrac{2t}{1+t^2}\right) = 2\tan^{-1} t$

5. If $x = \cos t$ $y = \log t$ show that

$$\frac{d^2y}{dx^2} + \left(\frac{dy}{dx}\right)^2 = 0 \text{ at } t = \frac{\pi}{2}$$

Type 3: (continued)

(E) Problems based on showing that product of any two functions satisfy a differential equation

Exercise 14.13

1. If $y = x \cdot \sin x$ Show that

$$x^2 y_2 - 2xy_1 + (x^2 + 2)\, y = 0$$

2. If $y = x \cdot \tan^{-1} x$ show that $\dfrac{d^2y}{dx^2} = \dfrac{2}{(1+x^2)^2}$

3. If $y = e^{-\frac{x}{2}} \cdot \left(a\cos\dfrac{\sqrt{3}}{2}x + b\sin\dfrac{\sqrt{3}}{2}x\right)$, show that $\dfrac{d^2y}{dx^2} + \dfrac{dy}{dx} + y = 0$.

4. If $x = (a + bx) \cdot e^{\frac{y}{x}}$ show that

$$x^3 \cdot \frac{d^2y}{dx^2} = \left(x\frac{dy}{dx} - y\right)^2$$

5. If $y = e^{ax} \cdot \cos bx$ show that

$$y_2 - 2a\, y_1 + (a^2 + b^2)\, y = 0$$

6. If $y = Ae^{-ax} \cdot \cos(bx + c)$ show that

$$y_2 + 2a\, y_1 + (a^2 + b^2)\, y = 0$$

7. If $y = x^{n-1} \cdot \log x$ show that

$$x^2 y_2 + (3 - 2n)\, x\, y_1 + (n-1)^2\, y = 0$$

15

L'Hospital's Rule

Question: What is L'Hospital's rule?

Answers: L'Hospital's rule is a rule permitting the evaluation of an indeterminate quotient of functions as the quotient of the limits of their derivatives.

Example: **1.** $\lim_{x \to 0} \frac{\sin x}{x}$ is an indeterminate of the form $\frac{0}{0}$ but it can be evaluated as $\lim_{x \to 0} \frac{\cos x}{1} = 1$

Indeterminate Forms

There are eight indeterminate forms involving difference, product, quotient or power of 0 and ∞ (in connection with finding limit) which we face in practice which are (1) $\frac{0}{0}$ (2) $\frac{\infty}{\infty}$ (3) $0 \times \infty$ (4) $\infty - \infty$ (5) 0^{∞} (6) ∞^{0} (7) 0^0 (8) 1^{∞} which we face in practice are the expressions known as indeterminate forms of the given functions for the limit.

Remember: $1^0 = 1$, $\frac{0}{\infty} = 0$ or $\frac{\infty}{0} = \infty$ is the determinate form having a meaning for the given expression for the limit.

Statement of L'Hospital's Rule

If $\lim_{x \to a} f(x) = \lim_{x \to a} F(x) = 0$ and $\frac{f'(x)}{F'(x)}$, where f' and F' are the derivatives of f and F, approaches a limit as x approaches a, then $\frac{f(x)}{F(x)}$ approaches the same limit.

Proof: $\lim_{x \to a} \frac{f(x)}{F(x)}$

$$= \lim_{x \to a} \frac{\dfrac{f(x) - f(a)}{x - a}}{\dfrac{F(x) - F(a)}{x - a}}$$

[Since $f(a) = F(a) = 0$]

$$= \frac{f'(a)}{F'(a)}, \; F'(a) \neq 0$$

e.g. If $f(x) = \left(x^2 - 1\right)$

$F(x) = (x - 1)$

$a = 1$

then $\frac{f(a)}{F(a)} = \frac{0}{0}$ and $\lim_{x \to 1} \frac{f'(x)}{F'(x)} = \lim_{x \to 1} \frac{2x}{1} = 2$

which is $\lim_{x \to 1} \frac{\left(x^2 - 1\right)}{(x - 1)}$.

Method of Evaluating Limits in Indeterminate Forms

1. Evaluation of $\frac{0}{0}$ and $\frac{\infty}{\infty}$ forms:

If $\lim_{x \to a} \frac{f(x)}{g(x)} = \frac{0}{0}$ or, $\lim_{x \to a} \frac{f(x)}{g(x)} = \frac{\infty}{\infty}$

then we adopt the following working rule:

Working rule: **1.** Go on differentiating numerator and denominator separately till we get a definite value at $x = a$. This working rule may be expressed in the symbolic form in the following way

$$\lim_{x \to a} \frac{f(x)}{g(x)} = \lim_{x \to a} \frac{f'(x)}{g'(x)} = \lim_{x \to a} \frac{f''(x)}{g''(x)}$$

$$= \lim_{x \to a} \frac{f'''(x)}{g'''(x)} = \dots = \text{a definite value}$$

N.B.: Any one of the $\left[\frac{f'(x)}{g'(x)}\right]_{x=a}$, $\left[\frac{f''(x)}{g''(x)}\right]_{x=a}$ $\left[\frac{f'''(x)}{g'''(x)}\right]_{x=a}$, ...etc which first provide us a definite value (i.e.; not meaningless form) will give us the answer.

2. Working rule to evaluate $0 \cdot \infty$ at $x = a$ $\left(\lim_{x \to a} f(x) \cdot g(x) = 0 \cdot \infty\right)$ For this indeterminate form, we are required to change $f \cdot g = \frac{f}{1/g}$ which

$\Rightarrow \lim (f \cdot g)$ should be changed into $\lim \frac{f}{1/g}$ or to $\lim \frac{g}{1/f}$, then apply L'Hospital's rule.

3. Working rule to evaluate $\lim_{x \to a} f(x)^{g(x)}$ which provides us 1^{∞}, 0^{0}, ∞^{0} at $x = a$.

For these indeterminate forms, we are required to change f^g into $e^{\left(\frac{\log f}{1/g}\right)}$ or $e^{\left(\frac{g}{1/\log f}\right)}$, then we apply L'Hospital's rule.

4. $\lim_{x \to a} (f(x) - g(x)) = \infty - \infty$ form:

Working rule to evaluate $(\infty - \infty)$.

For this indeterminate form, we write

$$f - g = \frac{\left(\frac{1}{g} - \frac{1}{f}\right)}{\left(\frac{1}{f} \cdot \frac{1}{g}\right)}$$ and then we use L'Hospital's rule which provide us the required answer.

The above types of indeterminate forms and the working rule for these indeterminate forms may be put in the chart form in the following way.

Types of indeterminate froms at x = a	*working rules*
1. $\lim \frac{f}{g} = \frac{0}{0}$ **2.** $\lim \frac{f}{g} = \frac{\infty}{\infty}$	$\lim \frac{f}{g} = \lim \frac{f'}{g'} = \lim \frac{f''}{g''}$ $= \lim \frac{f'''}{g'''} = \dots$ then any one of these which is not meaningless at $x = a$ will give the required answer.
3. $\lim f \cdot g = \infty \cdot 0$	$\lim \frac{f}{\frac{1}{g}}$ or $\frac{g}{\frac{1}{f}}$ and then use L'Hospital's rule
4. $\lim f^g = 1^{\infty}$ **5.** $\lim f^g = 0^0$ **6.** $\lim f^g = \infty^0$	$\lim \frac{\log f}{\frac{1}{g}}$ or $\lim \frac{g}{\frac{1}{\log f}}$ then use L'Hospital's rule. If L'Hospital's rule. If L'Hospital's rule applied to one of these give 'b' = a finite value for the limit, then answer to the original problem in exponential form is e^b.
7. $\lim (f - g) = \infty - \infty$	$\lim \frac{\frac{1}{g} - \frac{1}{f}}{\frac{1}{g} \cdot \frac{1}{f}}$, then use L'Hospital's rule which will directly give us the answer.

Remember: **1.** Some times L'Hospital's rule may fail to lead to any fruitful result. In such situation, we use the usual method of finding limit of the given function.

Example: Let $\frac{f(x)}{g(x)} = \frac{x}{\sqrt{1 + x^2}}$ as $x \to \infty$

Here, $\lim_{x \to \infty} \frac{f(x)}{g(x)} = \lim_{x \to \infty} \frac{x}{\sqrt{1 + \frac{1}{x^2}}} = 1$

But L'Hospital's rule gives us

$$\frac{f'(x)}{g'(x)} = \frac{1}{\frac{x}{\sqrt{1+x^2}}} = \frac{\sqrt{1+x^2}}{x}$$

$$\frac{f''(x)}{g''(x)} = \frac{x/\sqrt{1+x^2}}{1} = \frac{x}{\sqrt{1+x^2}} = \frac{f(x)}{g(x)}$$

= original function which $\Rightarrow$ limiting value can not be obtained by L'Hospital's rule.

2. Generally, functions involving surds should be avoided for application of L'Hospital's rule.

For example, to evaluate $\lim_{x \to 0} \frac{\sqrt{x-a} + \sqrt{x^2 - a^2}}{\sqrt{3x+a} - 2\sqrt{x+a}}$,

we first go for rationalization instead of L'Hospital's rule though it has the form $\frac{0}{0}$.

3. L-Hospital rule is applicable only when the indeterminate expression has the form $\frac{0}{0}$ or $\frac{\infty}{\infty}$ in the limit.

4. If the indeterminate expressions of the given functions as $x \to a$ do not assume $\left(\frac{0}{0}\right)$ or $\left(\frac{\infty}{\infty}\right)$, we are required first to change that form into $\left(\frac{0}{0}\right)$ or $\left(\frac{\infty}{\infty}\right)$ by simplification or using any mathematical manipulation.

5. The two indeterminate forms $(0 \times \infty)$ and $(\infty - \infty)$ can be easily evaluated by reducing them to the form $\left(\frac{0}{0}\right)$ or $\left(\frac{\infty}{\infty}\right)$.

6. The three indeterminate exponential forms 0^0, 1^∞, ∞^0 can easily be evaluated by first taking logarithm. By taking logarithm, all these three forms can be reduced to the indeterminate form $(0 \times \infty)$ which can be further reduced to $\left(\frac{0}{0}\right)$ or $\left(\frac{\infty}{\infty}\right)$ form.

7. The students must not differentiate $\frac{f(x)}{g(x)}$ as a fraction. The numerator and the denominator have to be differentiated separately.

8. Differentiation of numerator and denominator is performed w.r.t. the variable which converges

9. L'Hospital's rule involves differentiating the functions. This is why it is useful only when the functions are easily differentiable.

10. L'Hospital's rule may be applied repeatedly till we get a definite value value at $x = a$ or when $x \to a$ (i.e.; $\lim x = a$).

***N.B.*: 1.** To evaluate limit of a function whose value at $x \to a$ is 0^0, ∞^0 or 1^∞ (in exponential form), we are required to take the logarithm first and then proceed.

e.g. $\lim_{x \to 0} x^x = \left(0^0 \text{ form}\right)$

$$\because y = x^x \Rightarrow \log y = x \log x = \frac{\log x}{\frac{1}{x}}$$

$$\Rightarrow \lim_{x \to 0} \log y = \lim_{x \to 0} \frac{\log x}{\frac{1}{x}} = \lim_{x \to 0} \frac{\frac{1}{x}}{-\frac{1}{x^2}}$$

$$= -\lim_{x \to 0} x = 0 \text{ which} \Rightarrow \lim_{x \to 0} x^x = e^0 = 1$$

2. $\lim_{x \to 0^+} \log x = -\infty$, $\lim_{x \to +\infty} e^x = +\infty$

$\log 1 = 0$, $\lim_{x \to -\infty} e^x = 0$

$\lim_{x \to +\infty} \log x = \infty$

3. $\log_a m = \log_b m \cdot \log_a b$

$$= \log_b m \cdot \frac{1}{\log_b a}$$

$$= \frac{\log_b m}{\log_b a} \quad (\because \log_b a \cdot \log_a b = 1)$$

$$= \frac{\log m}{\log a}$$ ($\because$ when the base is same in numerator and in denominator, the base may be omitted for easiness of computation)

This formula may be remembered by the equality:

$$\frac{m}{a} = \frac{\frac{m}{b}}{\frac{a}{b}}$$

(This formula is known as changing into the same base formula or base changing formula)

Type 1: Problems based on indeterminate form $\left(\frac{0}{0}\right)$.

Examples worked out:

1. Evaluate $\lim_{x \to 0} \frac{\sin x}{x}$

Solution: $\because \lim_{x \to 0} \frac{\sin x}{x} = \left(\frac{0}{0} \text{ form}\right)$

$$\therefore \lim_{x \to 0} \frac{\sin x}{x} = \lim_{x \to 0} \frac{\frac{d}{dx}[\sin x]}{\frac{d}{dx}(x)} = \lim_{x \to 0} \frac{\cos x}{1}$$

$$= [\cos x]_{x=0} = \cos 0 = 1$$

$$\therefore \lim_{x \to 0} \frac{\sin x}{x} = \lim_{x \to 0} \cos x = 1$$

2. Evaluate $\lim_{x \to a} \frac{x \sin a - a \sin x}{x - a}$.

Solution: $\because \lim_{x \to a} \frac{x \sin a - a \sin x}{x - a} = \left(\frac{0}{0} \text{ form}\right)$

$$\therefore \lim_{x \to a} \frac{x \sin a - a \sin x}{x - a}$$

$$= \lim_{x \to a} \frac{\frac{d}{dx}[x \sin a - a \sin x]}{\frac{d}{dx}(x - a)}$$

$$= \lim_{x \to a} \frac{\sin a - a \cos x}{1}$$

$$= [\sin a - a \cos x]_{x=a} = \sin a - a \cos \alpha$$

$$\therefore \lim_{x \to a} \frac{x \sin a - a \sin x}{x - a}$$

$$= \sin a - a \cos \alpha$$

3. Evaluate $\lim_{x \to y} \frac{\tan x - \tan y}{x - y}$

Solution: $\because \lim_{x \to y} \frac{\tan x - \tan y}{x - y} = \left(\frac{0}{0} \text{ form}\right)$

$$\therefore \lim_{x \to y} \frac{\tan x - \tan y}{x - y} = \lim_{x \to y} \frac{\frac{d}{dx}[\tan x - \tan y]}{\frac{d}{dx}(x - y)}$$

$$= \lim_{x \to y} \frac{\sec^2 x - 0}{1} = \lim_{x \to y} \sec^2 x$$

$$[\sec^2 x]_{x=y} = \sec^2 y$$

$$\therefore \lim_{x \to y} \frac{\tan x - \tan y}{x - y} = \sec^2 y.$$

4. Evaluate $\lim_{x \to 0} \frac{\sin^{-1} x - \tan^{-1} x}{x^3}$

Solution: $\because \lim_{x \to 0} \frac{\sin^{-1} x - \tan^{-1} x}{x^3} = \left(\frac{0}{0} \text{ form}\right)$

$$\therefore \lim_{x \to 0} \frac{\sin^{-1} x - \tan^{-1} x}{x^3}$$

$$= \lim_{x \to 0} \frac{\dfrac{d}{dx}\left[\sin^{-1} x - \tan^{-1} x\right]}{\dfrac{d}{dx}\left(x^3\right)}$$

$$= \lim_{x \to 0} \frac{\dfrac{1}{\sqrt{1-x^2}} - \dfrac{1}{1+x^2}}{3x^2} = \left(\frac{0}{0} \text{ form}\right)$$

$$= \lim_{x \to 0} \frac{\dfrac{d}{dx}\left[\dfrac{1}{\sqrt{1-x^2}} - \dfrac{1}{1+x^2}\right]}{\dfrac{d}{dx}\left(3x^2\right)}$$

$$= \lim_{x \to 0} \frac{\dfrac{2x}{2\left(1-x^2\right)^{3/2}} + \dfrac{2x}{\left(1+x^2\right)^2}}{6x}$$

$$= \lim_{x \to 0} \frac{1}{6}\left[\frac{1}{\left(1-x^2\right)^{3/2}} + \frac{2}{\left(1+x^2\right)^2}\right]$$

$$= \frac{1}{6}\left[\frac{1}{\left(1-x^2\right)^{3/2}} + \frac{2}{\left(1+x^2\right)^2}\right]_{x=0}$$

$$= \frac{1}{6} \times [1+2] = \frac{3}{6} = \frac{1}{2}$$

which $\Rightarrow \lim_{x \to 0} \dfrac{\sin^{-1} x - \tan^{-1} x}{x^3} = \dfrac{1}{2}$

5. Evaluate $\lim_{x \to 0}\left[\dfrac{\sin x - x + \dfrac{x^3}{6}}{x^5}\right]$

Solution: $\because \lim_{x \to 0}\left[\dfrac{\sin x - x + \dfrac{x^3}{6}}{x^5}\right] = \left(\dfrac{0}{0} \text{ form}\right)$

$$\therefore \lim_{x \to 0}\left(\frac{\sin x - x + \dfrac{x^3}{6}}{x^5}\right)$$

$$= \lim_{x \to 0} \frac{\dfrac{d}{dx}\left(\sin x - x + \dfrac{x^3}{6}\right)}{\dfrac{d}{dx}\left(x^5\right)}$$

$$= \lim_{x \to 0} \frac{\cos x - 1 + \dfrac{1}{2}x^2}{5x^4} = \left(\frac{0}{0} \text{ form}\right)$$

$$= \lim_{x \to 0} \frac{\dfrac{d}{dx}\left(\cos x - 1 + \dfrac{1}{2}x^2\right)}{\dfrac{d}{dx}\left(5x^4\right)}$$

$$= \lim_{x \to 0}\left(\frac{-\sin x + x}{20x^3}\right) = \left(\frac{0}{0} \text{ form}\right)$$

$$= \lim_{x \to 0} \frac{\dfrac{d}{dx}(-\sin x + x)}{\dfrac{d}{dx}\left(20x^3\right)}$$

$$= \lim_{x \to 0} \frac{-\cos x + 1}{60x^2} = \left(\frac{0}{0} \text{ form}\right)$$

$$= \lim_{x \to 0} \frac{\dfrac{d}{dx}(-\cos x + x)}{\dfrac{d}{dx}\left(60x^2\right)}$$

$$= \lim_{x \to 0} \frac{\sin x}{120x} = \left(\frac{0}{0} \text{ form}\right)$$

$$= \lim_{x \to 0} \frac{\frac{d}{dx} \sin x}{\frac{d}{dx}(120\, x)} = \lim_{x \to 0} \frac{\cos x}{120} = \left[\frac{\cos x}{120}\right]_{x=0}$$

$$= \frac{\cos 0}{120} = \frac{1}{120}$$

$$\text{which} \Rightarrow \lim_{x \to 0} \left(\frac{\sin x - x + \frac{x^3}{6}}{x^5}\right) = \frac{1}{120}$$

6. Evaluate $\lim_{x \to 0} \left(\frac{x \cos x - \sin x}{x^2 \sin x}\right)$

Solution: $\because \lim_{x \to 0} \left(\frac{x \cos x - \sin x}{x^2 \sin x}\right) = \left(\frac{0}{0} \text{ form}\right)$

$$\therefore \lim_{x \to 0} \left(\frac{x \cos x - \sin x}{x^2 \sin x}\right)$$

$$= \lim_{x \to 0} \frac{\frac{d}{dx}(x \cos x - \sin x)}{\frac{d}{dx}\left(x^2 \sin x\right)}$$

$$= \lim_{x \to 0} \frac{1 \cdot \cancel{\cos x} - x \sin x - \cancel{\cos x}}{2\, x \sin x + x^2 \cos x} = \left(\frac{0}{0} \text{ form}\right)$$

$$= \lim_{x \to 0} \frac{(-\cancel{x} \sin x)}{\cancel{x}(2 \sin x + x \cos x)}$$

$$= \lim_{x \to 0} \frac{-\sin x}{(2 \sin x + x \cos x)} = \left(\frac{0}{0} \text{ form}\right)$$

$$= \lim_{x \to 0} \frac{\frac{d}{dx}(-\sin x)}{\frac{d}{dx}(2 \sin x + (\cos x) \cdot (x))}$$

$$= \lim_{x \to 0} \frac{-\cos x}{2 \cos x + 1 \cdot \cos x - x \sin x}$$

$$= \left[\frac{-\cos x}{3 \cos x - x \sin x}\right]_{x=0} = -\frac{1}{3}$$

$$\text{which} \Rightarrow \lim_{x \to 0} \left[\frac{x \cos x - \sin x}{x^2 \sin x}\right] = -\frac{1}{3}$$

N.B.: In practice, differentiation of numerator and denominator is done separately mentally as below just for saving time.

$$\lim_{x \to 0} \left(\frac{x \cos x - \sin x}{x^2 \sin x}\right)$$

$$= \lim_{x \to 0} \frac{1 \cdot \cos x - x \sin x - \cos x}{2\, x \sin x + x^2 \cos x}$$

$$= \lim_{x \to 0} \frac{-\sin x}{2 \sin x + x \cos x}$$

$$= \lim_{x \to 0} \frac{-\cos x}{2 \cos x + \cos x - x \sin x}$$

$$= \lim_{x \to 0} \frac{-\cos x}{3 \cos x - x \sin x} = -\frac{1}{3}$$

Type 2: Problems based on indeterminate form $\left(\frac{\infty}{\infty}\right)$.

Examples worked out:

1. Find $\lim_{x \to 0} \frac{\log x}{\cot x}$

Solution: $y = \frac{\log x}{\cot x} = \left(\frac{\infty}{\infty} \text{ form}\right)$

$$\Rightarrow \lim_{x \to 0} y = \lim_{x \to 0} \frac{\log x}{\cot x} = \lim_{x \to 0} \frac{\frac{1}{x}}{-\operatorname{cosec}^2 x}$$

$$= \lim_{x \to 0} \frac{\left(-\sin^2 x\right)}{x} = \left(\frac{0}{0} \text{ form}\right)$$

$$= \lim_{x \to 0} \frac{-2 \sin x \cos x}{1} = \frac{0}{1} = 0$$

2. Find $\lim_{x \to 0} \frac{\log x^2}{\cot x^2}$

Solution: $y = \frac{\log x^2}{\cot x^2} = \left(\frac{\infty}{\infty} \text{ form}\right)$

$$\Rightarrow \lim_{x \to 0} y = \lim_{x \to 0} \frac{\log x^2}{\cot x^2}$$

$$= \lim_{x \to 0} \frac{\left(\frac{1}{x^2}\right) \cdot (2x)}{\left(-\operatorname{cosec}^2 x^2\right) \cdot (2x)}$$

$$= \lim_{x \to 0} \frac{\left(-\sin^2 x^2\right)}{x^2} = \left(\frac{0}{0} \text{ form}\right)$$

$$= \lim_{x \to 0} \frac{\left(-2 \sin x^2 \cdot \cos x^2 \cdot 2x\right)}{2x}$$

$$= \lim_{x \to 0} \left(-2 \sin x^2 \cdot \cos x^2\right) = (-2) \cdot (0) \cdot (1) = 0$$

3. Find $\lim_{x \to a} \frac{\log (x - a)}{\log \left(e^x - e^a\right)}$

Solution: $y = \frac{\log (x - a)}{\log \left(e^x - e^a\right)} = \left(\frac{\infty}{\infty} \text{ form}\right)$

$$\Rightarrow \lim_{x \to a} y = \lim_{x \to a} \frac{\log (x - a)}{\log \left(e^x - e^a\right)}$$

$$= \lim_{x \to a} \frac{\left(\frac{1}{x - a}\right)}{\left(\frac{1}{e^x - e^a}\right) \cdot e^x} = \left(\frac{\infty}{\infty} \text{ form}\right)$$

$$= \lim_{x \to a} \frac{e^x - e^a}{e^x (x - a)} = \left(\frac{0}{0} \text{ form}\right)$$

$$= \lim_{x \to a} \frac{e^x}{e^x \cdot 1 + (x - a) e^x} = \frac{e^a}{e^a + 0} = 1$$

4. Find $\lim_{x \to \frac{\pi}{2}} \frac{\log \left(x - \frac{\pi}{2}\right)}{\tan x}$

Solution: $\lim_{x \to \frac{\pi}{2}} \frac{\log \left(x - \frac{\pi}{2}\right)}{\tan x} = \left(\frac{\infty}{\infty} \text{ form}\right)$

$$\Rightarrow \lim_{x \to \frac{\pi}{2}} \frac{\log \left(x - \frac{\pi}{2}\right)}{\tan x}$$

$$= \lim_{x \to \frac{\pi}{2}} \frac{\frac{1}{\left(x - \frac{\pi}{2}\right)}}{\sec^2 x} = \left(\frac{\infty}{\infty} \text{ form}\right)$$

$$= \lim_{x \to \frac{\pi}{2}} \frac{\cos^2 x}{\left(x - \frac{\pi}{2}\right)} = \left(\frac{0}{0} \text{ form}\right)$$

$$= \lim_{x \to \frac{\pi}{2}} \frac{2 \cos x \cdot (-\sin x)}{1} = \lim_{x \to \frac{\pi}{2}} \frac{(-\sin 2x)}{1} = 0$$

5. Find $\lim_{x \to 0} \sin x \cdot \log x$

Solution: $y = \sin x \cdot \log x = \frac{\log x}{\operatorname{cosec} x}$

$$\Rightarrow \lim_{x \to 0} y = \lim_{x \to 0} \frac{\log x}{\operatorname{cosec} x} = \left(\frac{\infty}{\infty} \text{ form}\right)$$

$$= \lim_{x \to 0} \frac{\left(\frac{1}{x}\right)}{(-\operatorname{cosec} x \cdot \cot x)} \quad \text{(by L-Hospital rule)}$$

$$= \lim_{x\to 0}\left[-\frac{\sin^2 x}{x\cdot\cos x}\right]$$

$$= \lim_{x\to 0}(-1)\cdot\left(\frac{\sin x}{x}\right)\cdot(\tan x)$$

$$= (-1)\cdot(1)\cdot(0) = 0$$

6. Find $\lim_{x\to 0}\log_{\tan x}(\tan 2x)$

Solution: $y = \log_{\tan x}(\tan 2x) = \frac{\log\tan 2x}{\log\tan x}$

$\left[\text{by base changing formula } \log_a m = \frac{\log_b m}{\log_b a}\right]$

$$\Rightarrow \lim_{x\to 0} y = \lim_{x\to 0}\frac{\log\tan 2x}{\log\tan x} = \left(\frac{\infty}{\infty}\text{ form}\right)$$

$$= \lim_{x\to 0}\frac{\left(\frac{1}{\tan 2x}\cdot\sec^2 2x\cdot 2\right)}{\left(\frac{1}{\tan x}\cdot\sec^2 x\right)} \quad \text{[by L-Hospital rule]}$$

$$= \lim_{x\to 0}\frac{\left(\frac{2}{\sin 2x\cdot\cos 2x}\right)}{\left(\frac{1}{\sin x\cos x}\right)} = \lim_{x\to 0}\frac{2\sin x\cos x}{\sin 2x\cdot\cos 2x}$$

$$= \lim_{x\to 0}\frac{2\sin 2x}{\sin 4x} = \left(\frac{0}{0}\text{ form}\right)$$

$$= \lim_{x\to 0}\frac{2\cos 2x\cdot 2}{\cos 4x\cdot 4} = \frac{2\cdot 1\cdot 2}{1\cdot 4} = 1$$

Type 3: Problems based on the indeterminate form $(0\times\infty)$.

Examples worked out:

1. Determine $\lim_{x\to 0} x\log x$

Solution: $y = x\log x$

$$\Rightarrow \lim_{x\to 0} y = \lim_{x\to 0}(x\log x) = (0\times\infty\text{ form})$$

$$= \lim_{x\to 0}\frac{\log x}{\left(\frac{1}{x}\right)} = \left(\frac{\infty}{\infty}\text{ form}\right)$$

$$= \lim_{x\to 0}\frac{\frac{1}{x}}{\left(-\frac{1}{x^2}\right)} = \lim_{x\to 0}(-x) = 0$$

2. Determine $\lim_{x\to\infty} x\cdot\tan\left(\frac{1}{x}\right)$

Solution: $y = x\cdot\tan\left(\frac{1}{x}\right)$

$$\Rightarrow \lim_{x\to\infty} y = \lim_{x\to\infty} x\cdot\tan\left(\frac{1}{x}\right) = (0\times\infty\text{ form})$$

$$= \lim_{x\to\infty}\frac{\tan\left(\frac{1}{x}\right)}{\frac{1}{x}} = \left(\frac{0}{0}\text{ form}\right)$$

$$= \lim_{x\to\infty}\frac{\sec^2\left(\frac{1}{x}\right)\cdot\left(-\frac{1}{x^2}\right)}{\left(-\frac{1}{x^2}\right)}$$

$$= \lim_{x\to\infty}\left(\sec^2\frac{1}{x}\right) = 1$$

3. Determine $\lim_{x\to 1}(1-x)\tan\left(\frac{\pi x}{2}\right)$

Solution: $y = (1-x)\tan\left(\frac{\pi x}{2}\right)$

$$\Rightarrow \lim_{x\to 1} y = \lim_{x\to 1}(1-x)\tan\left(\frac{\pi x}{2}\right) = (0\times\infty\text{ form})$$

$$= \lim_{x\to 1}\frac{\tan\left(\frac{\pi x}{2}\right)}{\frac{1}{(1-x)}} = \lim_{x\to 1}\frac{\left(\sec^2\frac{\pi x}{2}\right)\cdot\left(\frac{\pi}{2}\right)}{\frac{1}{(1-x)^2}}$$

$$= \lim_{x \to 1} \frac{(1-x)^2 \cdot \left(\frac{\pi}{2}\right)}{\cos^2\left(\frac{\pi x}{2}\right)} = \left(\frac{0}{0} \text{ form}\right)$$

$$= \lim_{x \to 1} \frac{2(1-x)\cdot(-1)\cdot\left(\frac{\pi}{2}\right)}{\left(2\cos\frac{\pi x}{2}\right)\cdot\left(-\sin\frac{\pi x}{2}\right)\cdot\left(\frac{\pi}{2}\right)}$$

$$= \lim_{x \to 1} \frac{2(1-x)}{\sin \pi x} = \lim_{x \to 1} \frac{(-2)}{(\pi)(\cos \pi x)}$$

$$= \frac{-2}{(\pi)(-1)} = \frac{2}{\pi}$$

Type 4: Problem based on the indeterminate form $(\infty - \infty)$.

Examples worked out:

1. Evaluate $\lim_{x \to 0} (\operatorname{cosec} x - \cot x)$

Solution: $y = (\operatorname{cosec} x - \cot x)$

$$\Rightarrow \lim_{x \to 0} y = \lim_{x \to 0} (\operatorname{cosec} x - \cot x) = (\infty - \infty \text{ form})$$

$$= \lim_{x \to 0} \frac{(1-\cos x)}{\sin x} = \left(\frac{0}{0} \text{ form}\right)$$

$$= \lim_{x \to 0} \frac{\sin x}{\cos x} = \frac{0}{1} = 0$$

2. Evaluate $\lim_{x \to 0} \left(\frac{1}{x} - \cot x\right)$

Solution: $y = \left(\frac{1}{x} - \cot x\right)$

$$\Rightarrow \lim_{x \to 0} y = \lim_{x \to 0} \left(\frac{1}{x} - \cot x\right) = (\infty - \infty \text{ form})$$

$$= \lim_{x \to 0} \left(\frac{\sin x - x\cos x}{x \sin x}\right) = \left(\frac{0}{0} \text{ form}\right)$$

$$= \lim_{x \to 0} \frac{x \sin x}{x \cos x + \sin x} = \left(\frac{0}{0} \text{ form}\right)$$

$$= \lim_{x \to 0} \frac{(\sin x + x\cos x)}{(-x \sin x + 2\cos x)} = \frac{0+0}{0+2} = \frac{0}{2} = 0$$

3. Evaluate $\lim_{x \to 0} \left(\frac{\cot x - \frac{1}{x}}{x}\right)$

Solution: $y = \dfrac{\left(\cot x - \frac{1}{x}\right)}{x}$

$$\Rightarrow \lim_{x \to 0} y = \lim_{x \to 0} \frac{\left(\cot x - \frac{1}{x}\right)}{x}$$

$$= \lim_{x \to 0} \left(\frac{x\cos x - \sin x}{x^3}\right)\cdot\left(\frac{x}{\sin x}\right)$$

$$= \lim_{x \to 0} \left(\frac{x\cos x - \sin x}{x^3}\right)\cdot 1 = \left(\frac{0}{0} \text{ form}\right)$$

$$= \lim_{x \to 0} \frac{(-x \sin x)}{3x^2} = \lim_{x \to 0} \frac{(-\sin x)}{3x} = \left(\frac{0}{0} \text{ form}\right)$$

$$= \lim_{x \to 0} \frac{-\cos x}{3} = -\frac{1}{3}$$

Type 5: Problems based on indeterminate form ∞^0, 1^∞, 0^0 etc.

Examples worked out:

1. Find $\lim_{x \to \frac{\pi}{2}} (\sec x)^{\cot x}$

Solution: Let $y = (\sec x)^{\cot x} = \left(\infty^0 \text{ form as } x \to \frac{\pi}{2}\right)$

$\therefore \log y = \cot x \cdot \log \sec x$

$$\Rightarrow \lim_{x \to \frac{\pi}{2}} \log y = \lim_{x \to \frac{\pi}{2}} \cot x \log \sec x$$

$$= \lim_{x \to \frac{\pi}{2}} \frac{\log \sec x}{\tan x} = \lim_{x \to \frac{\pi}{2}} \frac{\sec x \cdot \tan x / \sec x}{\sec^2 x}$$

(by L'Hospital's rule)

$$= \lim_{x \to \frac{\pi}{2}} \frac{\tan x}{\sec^2 x} = \lim_{x \to \frac{\pi}{2}} \frac{\sin x}{\cos x} \cdot \cos^2 x$$

$$= \lim_{x \to \frac{\pi}{2}} \sin x \cos x = 0$$

$$\Rightarrow \log\left(\lim_{x \to \frac{\pi}{2}} y\right) = 0 = \log 1$$

as $\log x$ a is continuous function for $x > 0$.

$$\Rightarrow \lim_{x \to \frac{\pi}{2}} y = 1$$

2. Find $\lim_{x \to 0} (1 + \sin x)^{\cot x}$

Solution: Let $y = (1 + \sin x)^{\cot x}$

$\therefore \log y = \log (1 + \sin x)^{\cot x} = \cot x \log (1 + \sin x)$

$$\frac{\log (1 + \sin x)}{\tan x}$$

$$\Rightarrow \lim_{x \to 0} \log y = \lim_{x \to 0} \frac{\frac{\cos x}{(1 + \sin x)}}{\sec^2 x} = 1$$

$$\Rightarrow \lim_{x \to 0} \log y = \log e$$

$$\Rightarrow \log\left(\lim_{x \to 0} y\right) = \log e \Rightarrow \lim_{x \to 0} y = e$$

3. Find $\lim_{x \to 0} \left(\frac{\sin x}{x}\right)^{\frac{1}{x^2}}$

Solution: $y = \left(\frac{\sin x}{x}\right)^{\frac{1}{x^2}} = \left(1^\infty \text{ form as } x \to 0\right)$

$$\Rightarrow \log y = \log\left(\frac{\sin x}{x}\right)^{\frac{1}{x^2}} = \frac{1}{x^2} \log \left|\frac{\sin x}{x}\right|$$

$$\Rightarrow \lim_{x \to 0} \log y = \lim_{x \to 0} \frac{\log|\sin x| - \log|x|}{x^2}$$

$$\Rightarrow \lim_{x \to 0} \log y = \lim_{x \to 0} \frac{\cot x - \frac{1}{x}}{2x}$$

$$\Rightarrow \lim_{x \to 0} \log y = \lim_{x \to 0} \frac{\frac{1}{\tan x} - \frac{1}{x}}{2x}$$

$$= \lim_{x \to 0} \frac{x - \tan x}{2x^2 \tan x} = \lim_{x \to 0} \frac{x - \tan x}{2x^3} \cdot \frac{x}{\tan x}$$

$$= \lim_{x \to 0} \frac{x - \tan x}{2x^3} \cdot 1 = \lim_{x \to 0} \frac{1 - \sec^2 x}{6x^2}$$

$$= \lim_{x \to 0} \frac{-\tan^2 x}{6x^2} = \lim_{x \to 0} \left(-\frac{1}{6}\right)\left(\frac{\tan x}{x}\right)^2 = -\frac{1}{6}$$

which $\Rightarrow \lim_{x \to 0} \log y = -\frac{1}{6}$

$$\Rightarrow \log\left(\lim_{x \to 0} y\right) = -\frac{1}{6} \Rightarrow \lim_{x \to 0} y = e^{-\frac{1}{6}}$$

$$\Rightarrow \lim_{x \to 0} \left(\frac{\sin x}{x}\right)^{\frac{1}{x^2}} = e^{-\frac{1}{6}}$$

4. Find $\lim_{\theta \to \frac{\pi}{2}} (\cos \theta)^{\cos \theta}$

Solution: $y = (\cos \theta)^{\cos \theta} = \left(0^0 \text{ from as } \theta \to \frac{\pi}{2}\right)$

$$\Rightarrow \log y = \log (\cos \theta)^{\cos \theta} = \cos \theta \log \cos \theta$$

$$\Rightarrow \lim_{\theta \to \frac{\pi}{2}} \log y = \lim_{\theta \to \frac{\pi}{2}} (\cos \theta \log \cos \theta)$$

$$\Rightarrow \lim_{\theta \to \frac{\pi}{2}} \log y = \lim_{\theta \to \frac{\pi}{2}} \cos \theta \log \cos \theta$$

$$= \lim_{\theta \to \frac{\pi}{2}} \frac{\log \cos \theta}{\sec \theta} = \left(\frac{\infty}{\infty} \text{ form}\right)$$

$$= \lim_{\theta \to \frac{\pi}{2}} \frac{-\tan \theta}{\sec \theta \cdot \tan \theta} = \lim_{\theta \to \frac{\pi}{2}} \left[-\frac{1}{\sec \theta}\right]$$

$$= \lim_{\theta \to \frac{\pi}{2}} [-\cos \theta] = 0$$

$$\Rightarrow \log \lim_{\theta \to \frac{\pi}{2}} y = 0 \Rightarrow \lim_{\theta \to \frac{\pi}{2}} y = e^0 = 1$$

5. Find $\lim_{x \to 0} (\cos x)^{\frac{1}{x^2}}$

Solution: $y = (\cos x)^{\frac{1}{x^2}} = \left(1^\infty \text{ form as } x \to 0\right)$

$$\Rightarrow \log y = \log (\cos x)^{\frac{1}{x^2}}$$

$$= \frac{1}{x^2} \log \cos x$$

$$\Rightarrow \lim_{x \to 0} \log y = \lim_{x \to 0} \left(\frac{1}{x^2} \log \cos x\right) = (0 \times \infty \text{ form})$$

$$= \lim_{x \to 0} \frac{\log \cos x}{x^2} = \left(\frac{0}{0} \text{ form}\right)$$

$$= \lim_{x \to 0} \left(\frac{-\tan x}{2x}\right) = \left(\frac{0}{0} \text{ form}\right)$$

$$= \lim_{x \to 0} \frac{-\sec^2 x}{2} = -\frac{1}{2}$$

$$\Rightarrow \log\left(\lim_{x \to 0} y\right) = -\frac{1}{2}$$

$$\Rightarrow \lim_{x \to 0} y = e^{-\frac{1}{2}}$$

6. Find $\lim_{x \to 1} x^{\frac{1}{1-x}}$

Solution: $y = x^{\frac{1}{(1-x)}} = \left(1^\infty \text{ form, as } x \to 1\right)$

$$\Rightarrow \log y = \log \left\{x^{\frac{1}{(1-x)}}\right\} = \frac{1}{(1-x)} \cdot \log x$$

$$\Rightarrow \lim_{x \to 1} \log y = \lim_{x \to 1} \frac{\log x}{(1-x)} = \left(\frac{0}{0} \text{ form}\right)$$

$$\Rightarrow \lim_{x \to 1} \frac{\frac{1}{x}}{(-1)} = \lim_{x \to 1} \left(-\frac{1}{x}\right) = -1$$

$$\Rightarrow \log\left(\lim_{x \to 1} y\right) = (-1) \text{ which} \Rightarrow \lim_{x \to 1} y = e^{-1} = \frac{1}{e}$$

7. Find $\lim_{x \to 0} x^x$

Solution: $y = x^x = (0^0$ form in the limit)

$$\Rightarrow \log y = \log x^x = x \log x = ((0 \times \infty \text{ form}))$$

$$\Rightarrow \lim_{x \to 0} \log y = \lim_{x \to 0} (x \log x)$$

$$= \lim_{x \to 0} \frac{\log x}{\left(\frac{1}{x}\right)} = \left(\frac{\infty}{\infty} \text{ form}\right)$$

$$= \lim_{x \to 0} \frac{\frac{1}{x}}{\left(-\frac{1}{x^2}\right)} \text{ (by L-Hospital rule)}$$

$$= \lim_{x \to 0} (-x) = 0$$

8. Find $\lim_{x \to \infty} \left(1 + \frac{a}{x}\right)^x$

Solution: $y = \left(1 + \frac{a}{x}\right)^x = \left(1^\infty \text{ form in the limit}\right)$

$$\Rightarrow \log y = \log\left(1 + \frac{a}{x}\right)^x = x \log\left(1 + \frac{a}{x}\right) = (0 \times \infty \text{ form})$$

$$\Rightarrow \lim_{x \to \infty} \log y = \lim_{x \to \infty} \left[x \log\left(1 + \frac{a}{x}\right)\right]$$

$$= \lim_{x\to\infty} \frac{\log\left(1+\frac{a}{x}\right)}{\left(\frac{1}{x}\right)} = \left(\frac{0}{0}\text{ form}\right)$$

$$= \lim_{x\to\infty} \frac{\frac{1}{\left(1+\frac{a}{x}\right)}\cdot\left(\frac{-a}{x^2}\right)}{\left(-\frac{1}{x^2}\right)}$$

$$= \lim_{x\to\infty} \frac{a}{\left(1+\frac{a}{x}\right)}$$

$$= a$$

$$\Rightarrow \log\left(\lim_{x\to\infty} y\right) = a$$

$$\Rightarrow \lim_{x\to\infty} y = e^a$$

Type 6: Problems based on finding the values of constants occuring in $f(x)$ from the known limiting value.

Working rule: To find the values of the constants from the given finite limit, we adopt the following working rule:

1. find the limit using L'Hospital's rule for the given function in case of indeterminate form.

2. Equate the limit of the given function determined by using L'Hospital's rule to the given value of the limit of the given function and form an equation involving constants only.

3. Observe whether limit should assume indeterminate forms

$\frac{0}{0}, \frac{\infty}{\infty}, 0\cdot\infty, (\infty-\infty), 0^{\infty}, \infty^0, 0^0, 1^0$ etc or not.

4. If the limit is $\frac{0}{0}, \frac{\infty}{\infty}, \ldots$ etc., we have a relation between the constants we again use L'Hospital's rule and find the limit which is equatied to given limit of the given function and form another equation involving constants.

Solve the simultaneous equations involving constants.

5. That the given limit is finite may give a relation between the constants.

Examples worked out:

1. Find the values of a and b such that

$$\lim_{x\to 0} \frac{x(1+a\cos x) - b\sin x}{x^2} = 1$$

Solution: Limit $= \lim_{x\to 0} \frac{1+a\cos x + x(-a\sin x) - b\cos x}{3x^2}$

[by L-Hospital rule]

This limit $= \infty$ if $1 + a - b \neq 0$

But limit $= 1$ (given)

$\therefore 1 + a \cdot b = 0$ ...(i)

$$\therefore \text{limit} = \lim_{x\to 0} \frac{1+a\cos x - b\cos x - ax\sin x}{3x^2} = \left(\frac{0}{0}\text{ form}\right)$$

$$= \lim_{x\to 0} \frac{1+(a-b)\cos x - ax\sin x}{3x^2}$$

$$= \lim_{x\to 0} \frac{0-(a-b)\sin x - a(\sin x + x\cos x)}{6x}$$ [by using L-Hospital rule]

$$= \lim_{x\to 0} \left[\frac{(b-a)}{6}\cdot\frac{\sin x}{x} - \frac{a}{6}\cdot\frac{\sin x}{x} - \frac{a}{6}\cos x\right]$$

$$= \frac{(b-a)}{6} - \frac{a}{6} - \frac{a}{6} = \frac{(b-3a)}{6}$$

which $\Rightarrow$ limit $= \frac{(b-3a)}{6}$ ...(ii)

But this limit is given to be equal to 1 ...(iii)

Equation (ii) and (iii), we get

$\frac{(b-3a)}{6} = 1 \Rightarrow (b-3a) = 6$ ...(iv)

solving (i) and (iv) $\Rightarrow 1 + a - \not{b} = 0$

$\not{b} - 3a = 6$

$1 - 2a = 6$

$\Rightarrow -2a = 6 - 1$

$$\Rightarrow a = -\frac{5}{2} \quad \text{...(v)}$$

Now, putting $a = -\frac{5}{2}$ in (1),

$$\Rightarrow 1 - \frac{5}{2} - b = 0$$

$$\Rightarrow b = 1 - \frac{5}{2} = -\frac{3}{2}$$

Thus, $a = -\frac{5}{2}, b = -\frac{3}{2}$

Note: The following argument to have $1 + a - b = 0$ is wrong.

$$\text{Limit} = \lim_{x \to 0} \frac{1 + a\cos x + x(-a\sin x - b\cos x)}{3x^2}$$

[using L' Hospital rule]

$$= \frac{1 + a - b}{0} \quad \text{...(i)}$$

but this limit is given to be equal to 1 ...(ii)
equating (i) and (ii), we get

$$\frac{1 + a - b}{0} = 1$$

$$\Rightarrow 1 + a - b = 0 \times 1$$

$$\Rightarrow 1 + a - b = 0 \text{ (inaccurate reasoning)}$$

Type 1: Form: $\frac{0}{0}$ and $\frac{\infty}{\infty}$

Problems based on algebraic functions

Exercise 16.1

Evaluate (using L'Hospital's rule)

1. $\lim_{x \to \infty} \frac{3x^3 + 4x^2 + 5}{7x^3 - 5x + 7}$

2. $\lim_{n \to \infty} \frac{3n^2 + 4n + 5}{4n^2 + 6n - 7}$

3. $\lim_{x \to 2} \frac{(x^2 - 5x + 6)(x^2 - 3x + 2)}{x^3 - 3x^2 + 4}$

4. $\lim_{x \to 3} \frac{x^2 - 9}{x - 3}$

5. $\lim_{x \to 0} \frac{(1 + x)^5 - 1}{3x + 5x^2}$

6. $\lim_{x \to 2} \frac{x^2 - x - 2}{x^2 + x - 6}$

7. $\lim_{x \to a} \frac{\sqrt{a + 2x} - \sqrt{3x}}{\sqrt{3a + x} - 2\sqrt{x}}; (a > 0)$

Answers:

1. $\frac{3}{7}$ **2.** $\frac{3}{4}$ **3.** $-\frac{1}{3}$ **4.** 6

5. $\frac{5}{3}$ **6.** $\frac{3}{5}$ **7.** $\frac{2}{3\sqrt{3}}$

Type: continued

Problems based on a combination of algebraic and trigonometric functions

Exercise 16.2

Evaluate (using L-Hospital rule)

1. $\lim_{x \to 0} \frac{1 - \cos x}{x^2}$

2. $\lim_{x \to \frac{\pi}{4}} \frac{\sin x - \cos x}{x - \frac{\pi}{4}}$

3. $\lim_{x \to 0} \frac{\tan 2x - x}{3x - \sin x}$

4. $\lim_{x \to 0} \frac{\cos x - 1}{x^2}$

5. $\lim\limits_{x\to 0} \dfrac{\tan x - x}{x^3}$

6. $\lim\limits_{x\to 0} \dfrac{x - \sin x}{x^3}$

7. $\lim\limits_{x\to 0} \dfrac{\tan x - \sin x}{\sin^3 x}$

8. $\lim\limits_{x\to 0} \dfrac{x^2}{1 - \cos x}$

9. $\lim\limits_{x\to 0} \dfrac{x \sin \alpha - \alpha \sin x}{x - \alpha}$

10. $\lim\limits_{x\to 0} \dfrac{\tan x - \sin x}{x^3}$

11. $\lim\limits_{x\to 0} \dfrac{2x - \sin^{-1} x}{2x + \tan^{-1} x}$

12. $\lim\limits_{x\to 0} \dfrac{\sin^{-1} x}{x}$

13. $\lim\limits_{x\to 0} \dfrac{\tan^{-1} x}{x}$

14. $\lim\limits_{x\to \frac{\pi}{4}} \dfrac{\sec^2 x - 2\tan x}{1 + \cos 4x}$

15. $\lim\limits_{x\to 0} \dfrac{x - \sin x}{x^3}$

16. $\lim\limits_{x\to 0} \dfrac{\tan x - x}{x^2 \tan x}$

17. $\lim\limits_{\theta\to \frac{\pi}{4}} \dfrac{\sqrt{2} - \cos\theta - \sin\theta}{(4\theta - \pi)^2}$

18. $\lim\limits_{x\to 0} \dfrac{x(1 + a\cos x) - b\sin x}{x^3}, \ (a > b - 1)$

19. $\lim\limits_{h\to 0} \dfrac{(a+h)^2 \sin(a+h) - a^2 \sin a}{h}$

20. $\lim\limits_{x\to 0} \dfrac{1 - \cos 2x}{x^2}$

21. $\lim\limits_{x\to \frac{\pi}{2}} \dfrac{\cos x}{\pi - 2x}$

22. $\lim\limits_{y\to 1} \dfrac{\cos^{-1} y}{y - 1}$

23. $\lim\limits_{x\to \frac{\pi}{4}} \dfrac{\sec^2 x - 2\tan x}{1 + \cos 4x} \cdot x$

24. $\lim\limits_{x\to 0} \dfrac{x + \sin \pi x}{x - \sin \pi x}$

25. $\lim\limits_{x\to 0} \dfrac{\sin x}{\sqrt{x}}$

26. $\lim\limits_{x\to \pi} \dfrac{\sin(\pi - x)}{\sqrt{\pi - x}}$

Answers:

1. $\dfrac{1}{2}$ **2.** $\sqrt{2}$ **3.** $\dfrac{1}{2}$ **4.** $-\dfrac{1}{2}$ **5.** $\dfrac{1}{3}$

6. $\dfrac{1}{6}$ **7.** $\dfrac{1}{2}$ **8.** 2 **9.** $\sin\alpha - \alpha\cos\alpha$

10. $\dfrac{1}{2}$ **11.** $\dfrac{1}{3}$ **12.** 1 **13.** 1 **14.** $\dfrac{1}{2}$

15. $\dfrac{1}{6}$ **16.** $\dfrac{1}{3}$ **17.** $\dfrac{1}{4\sqrt{2}}$ **18.** ∞

19. $2a\sin a + a^2 \cos a$ **20.** 2

21. $\dfrac{1}{2}$ **22.** $-\infty$ **23.** $\dfrac{\pi}{8}$

24. $\dfrac{1+\pi}{1-\pi}$ **25.** 0 **26.** 0

Type 1: Continued.

Problems based on a combination of algebraic, exponential, logarithmic, trigonometric or inverse trigonometric functions

Exercise 16.3

Evaluate (using L'Hospital's rule)

1. $\lim_{x\to 0} \frac{xe^x - \log(1+x)}{x^2}$

2. $\lim_{x\to 0} \frac{x\cos x - \log(1+x)}{x^2}$

3. $\lim_{x\to 0} \frac{e^x - 2\cos x + e^{-x}}{x \sin x}$

Hint: $\lim_{x\to 0} \frac{e^x - 2\cos x + e^{-x}}{x \sin x}$

$\lim_{x\to 0} \frac{e^x - 2\cos x + e^{-x}}{x^2} \cdot \left(\frac{x}{\sin x}\right)$

4. $\lim_{h\to 0} \frac{\log(x+h) - \log x}{h}$

5. $\lim_{h\to 1} \frac{\sqrt{x+4} - \sqrt{5}}{x-1}$

6. $\lim_{x\to 1} \frac{\log x}{(x-1)}$

7. $\lim_{x\to 0} \frac{a^x - 1}{x}$

8. $\lim_{x\to 0} \frac{a^x - b^x}{x}$

9. $\lim_{x\to 0} \frac{\log(1+kx^2)}{1-\cos x}$

10. $\lim_{x\to 0} \frac{e^x - e^{-x} - 2}{x^2}$

11. $\lim_{x\to 0} \frac{x\log(1+x)}{1-\cos x}$

12. $\lim_{x\to 0} \frac{xe^x - \log(1+x)}{x^2}$

13. $\lim_{x\to 0} \frac{x\cos x - \log(1+x)}{x^2}$

14. $\lim_{x\to 0} \frac{\sin x - x + \frac{1}{6}x^3}{x^5}$

15. $\lim_{x\to 0} \frac{e^x + e^{-x} - 2\cos x}{x\sin x}$

16. $\lim_{x\to 0} \frac{e^x - e^{-x}}{\sin x}$

17. $\lim_{x\to 0} \frac{x\left(1 - e^{\alpha x}\right)}{\cos\alpha x - 1}$

Answers:

1. $\frac{3}{2}$ **2.** $\frac{1}{2}$ **3.** 2 **4.** $\frac{1}{x}$ **5.** $\frac{1}{2\sqrt{5}}$ **6.** 1

7. $\log a$ **8.** $\log\left(\frac{a}{b}\right)$ **9.** $2k$ **10.** 1 **11.** 2

12. 2 **13.** $\frac{1}{2}$ **14.** $\frac{1}{120}$ **15.** 2 **16.** 2 **17.** $\frac{2}{\alpha}$

Type 1: Continued

Problems based on only on the form: $\frac{\infty}{\infty}$.

Exercise 16.4

Evaluate

1. $\lim_{x\to a} \frac{\log(x-a)}{\log\left(e^x - e^a\right)}$

2. $\lim\limits_{x \to \frac{\pi}{2}} \dfrac{\log\left(x - \dfrac{\pi}{2}\right)}{\tan x}$.

3. $\lim\limits_{x \to 0} \dfrac{\log x^2}{\cot x^2}$

4. $\lim\limits_{x \to \infty} \dfrac{x^2 - 2}{3x^2 + 1}$

5. $\lim\limits_{x \to \infty} \dfrac{e^x}{x^8}$

Answers:

1. 1 **2.** 0 **3.** 0 **4.** $\dfrac{1}{3}$ **5.** ∞

Type 2: Form: $0 \cdot \infty$ and $(\infty - \infty)$

Problems based on the form: $0 \cdot \infty$

Exercise 16.5

Evaluate

1. $\lim\limits_{x \to 1} (1 - x) \cdot \tan\left(\dfrac{\pi x}{2}\right)$

2. $\lim\limits_{x \to 0} x \log x$

3. $\lim\limits_{x \to \infty} x \cdot \tan\left(\dfrac{1}{x}\right)$

4. $\lim\limits_{x \to 1} (1 + \cos \pi x) \cdot \cot^2 \pi x$

5. $\lim\limits_{x \to \infty} x^2 e^{-2x}$

6. $\lim\limits_{x \to \infty} x \cdot e^{-x}$

Answers:

1. $\dfrac{2}{\pi}$ **2.** 0 **3.** 1 **4.** $\dfrac{1}{2}$ **5.** 0 **6.** 0

Type 2: Continued

Problems based on the form: $(\infty - \infty)$

Exercise 16.6

Evaluate

1. $\lim\limits_{x \to 0} \left[\dfrac{1}{x^2} - \dfrac{1}{\sin^2 x}\right]$

2. $\lim\limits_{x \to 0} \left[\dfrac{1}{\sin^2 x} - \dfrac{1}{x^2}\right]$

3. $\lim\limits_{x \to 0} \left[\dfrac{1}{x} - \cot x\right]$

4. $\lim\limits_{x \to 1} \left[\dfrac{x}{x - 1} - \dfrac{1}{\log x}\right]$

5. $\lim\limits_{\theta \to \frac{\pi}{2}} [\sec\theta - \tan\theta]$

6. $\lim\limits_{x \to 0} [\operatorname{cosec}\theta - \cot x]$

7. $\lim\limits_{x \to 0} \left[\dfrac{a}{x} - \cot\dfrac{x}{a}\right]$

8. $\lim\limits_{x \to 1} \left[\dfrac{1}{\log x} - \dfrac{x}{x - 1}\right] \cdot \left(\dfrac{x - 2}{x}\right)$

9. $\lim\limits_{x \to \frac{1}{2}} \left[\dfrac{1}{\log 2x} - \dfrac{2x}{2x - 1}\right]$

10. $\lim\limits_{x \to 1} \left[\dfrac{2}{x^2 - 1} - \dfrac{1}{x - 1}\right]$

11. $\lim\limits_{x \to \infty} \left[x - \sqrt{x^2 - 1}\right]$

12. $\lim\limits_{x \to \infty} \left[x - x^2 \log\left(1 + \dfrac{1}{x}\right)\right]$

Answers:

1. $-\dfrac{1}{3}$ **2.** $\dfrac{1}{3}$ **3.** 0 **4.** $\dfrac{1}{2}$ **5.** 0 **6.** 0

7. 0 **8.** $\frac{1}{2}$ **9.** $-\frac{1}{2}$ **10.** $-\frac{1}{2}$ **11.** 0 **12.** $\frac{1}{2}$

Type 3: Problems based on the form: $0^{\infty}, \infty^{0}, 0^{0}, 1^{\infty}$

Problems based on the form : $0^{\infty}, \infty^{0}, 0^{0}$ (exponential form of 0 and ∞)

Exercise 16.7

Evaluate

1. $\lim_{x \to 0} \left(\frac{1}{x}\right)^{\tan x}$
2. $\lim_{x \to 0} (\operatorname{cosec} x)^{\frac{1}{\log x}}$
3. $\lim_{x \to \infty} (1 + x)^{\frac{1}{x}}$
4. $\lim_{x \to 1} \left(1 - x^2\right)^{\frac{1}{\log(1-x)}}$
5. $\lim_{x \to \infty} \left(a^{\frac{1}{x}} - 1\right)^{x}$

Answers:

1. 1 **2.** $\frac{1}{e}$ **3.** 1 **4.** e **5.** $\log a$

Type 3: Continued

Problems based on the form: $1^{\infty}, \infty^{0}$

Exercise 16.8

Evaluate

1. $\lim_{x \to \frac{\pi}{2}} (\sin x)^{\tan x}$
2. $\lim_{x \to 0} (\cot x)^{\frac{1}{\log x}}$
3. $\lim_{x \to 0} \left(\frac{2x + 1}{x + 1}\right)^{\frac{1}{x}}$
4. $\lim_{x \to \infty} \left(\cos \frac{1}{x}\right)^{x}$
5. $\lim_{x \to 0} \left(\frac{\sin x}{x}\right)^{\frac{1}{x}}$
6. $\lim_{x \to 0} (\cos x)^{\cot^2 x}$
7. $\lim_{x \to 0} \left(\frac{\tan x}{x}\right)^{\frac{1}{x}}$
8. $\lim_{x \to \frac{\pi}{2}} (\sec x)^{\cot x}$
9. $\lim_{x \to \frac{\pi}{2}} (\tan x)^{\cos x}$
10. $\lim_{x \to 1} (x)^{\frac{1}{x-1}}$
11. $\lim_{x \to 0} (1 + \sin x)^{\frac{1}{x}}$
12. $\lim_{x \to 0} (\operatorname{cosec} x)^{\sin x}$
13. $\lim_{x \to 0} (\cos x + a \sin b x)^{\frac{1}{x}}$
14. $\lim_{m \to \infty} \left(\cos \frac{x}{m}\right)^{m}$
15. $\lim_{x \to 0} \left(\frac{\tan x}{x}\right)^{\frac{1}{x^2}}$
16. $\lim_{x \to \infty} \left[\frac{3x - 4}{3x + 2}\right]^{\frac{(x+1)}{3}}$
17. $\lim_{x \to \infty} \left[\frac{x^2 - 2x + 1}{x^2 - 4x + 2}\right]^{x}$
18. $\lim_{x \to \frac{\pi}{2}} (\cos x)^{\cot x}$
19. $\lim_{x \to 1} (\log x)^{\sin \pi x}$
20. $\lim_{x \to 0} (\cot x)^{\sin x}$
21. $\lim_{x \to 0} (\cos x)^{\cot x}$
22. $\lim_{x \to 0} \left(\frac{\sin x}{x}\right)^{\frac{1}{x^2}}$

23. $\lim_{x\to 0} (1+\sin x)^{\frac{1}{x}}$

24. $\lim_{x\to 0} \left(\frac{1}{x^2}\right)^{\tan x}$

25. $\lim_{x\to 0} (\log x)^{\frac{1}{1-\log x}}$

26. $\lim_{x\to 0} \left(a^x + x\right)^{\frac{1}{x}}$

Answers:

1. 1 **2.** $\frac{1}{e}$ **3.** e **4.** 1 **5.** 1 **6.** $e^{-\frac{1}{2}}$ **7.** 1 **8.** 1 **9.** 1
10. e **11.** e **12.** 1 **13.** e^{ab} **14.** 1 **15.** $e^{\frac{1}{3}}$
16. $e^{-\frac{2}{3}}$ **17.** e^2 **18.** 1 **19.** 1 **20.** 1 **21.** 1
22. $e^{-\frac{1}{6}}$ **23.** e **24.** 1 **25.** 1 **26.** $a \cdot e$

Type 4: Problems based on finding the values of the constants from the given limit of the function of independent variable.

Exercise 16.9

Evaluate

1. If $\lim_{x\to 0} \frac{\tan x - a\sin x}{x^3}$ be finite, find a.

2. If $\lim_{x\to 0} \frac{ae^x + 3e^{2x} - b}{x} = 8$, find a and b.

3. If $\lim_{x\to 0} \frac{x(1+a\cos x) + b\sin x}{2x} = 1$, find a relation between a and b.

Hint: Limit $= \lim_{x\to 0} \frac{x(1+a\cos x)+b\sin x}{2x} = \left(\frac{0}{0} \text{ form}\right)$

$$= \lim_{x\to 0} \frac{1 + a\cos x - ax\sin x + b\cos x}{2}$$

$$= \frac{1}{2}(1+a+b) \quad \text{...(i)}$$

But this limit = 1 (given) ...(ii)

4. If $\lim_{x\to 0} \frac{\sin 2x + a\sin x}{x^3}$ be finite, find the value of 'a' and the limit.

5. Find the values of a, b and c so that

$$\lim_{x\to 0} \frac{ae^x - b\cos x + ce^{-x}}{x \sin x} = 2$$

Answers:

1. $a = 1$ **2.** $a = 2$ and $b = 5$ **3.** $a + b = 1$
4. $a = -2$ and limit $= -1$ **5.** $a = 1, b = 2$ and $c = 1$

16

Evaluation of Derivatives for Particular Arguments

Evaluation means to find the value of or to fix the value of a quantity, e.g.

1. to evaluate 8 + 3 – 4 means to reduce it to 7.

2. to evaluate $x^2 + 2x + 2$ for $x = 3$ means to replace x by 3 and collect the result which is 17.

3. to evaluate $\lim_{x \to 2} x^2$ means to find the limit of x^2 as $x \to 2$ which is 4.

4. to evaluate an integral $\int f(x)\, dx$ means to carry out integration.

5. to evaluate a definite integral $\int_a^b f(x)\, dx$ means to carry out integration and then substitute the limits of integration a and b. Now we shall learn the process of finding the value of a derived function at a given number or point $x = a$

Defination of a derived function of an independent variable x: The derivative of a function $f(x)$ is a function of x represented as $f'(x)$ which is derived from the original function $f(x)$ through a limiting process or various techniques. In this sense, the derivative $f'(x)$ is called the derived function and the graph representing it is called the derived graph.

Notation for the value of the derived function $f'(x)$ at $x = a$.

Just as the value of $f(x)$ for (or, at) $x = a$ is symbolised as $[f(x)]_{x=a} = f(a)$, $a \in D(f)$

The value of a derivative (or, derived function) $f'(x)$ for (or, at) $x = a$ is symbolised as

$$[f'(x)]_{x=a} = \left[\frac{dy}{dx}\right]_{x=a} = \left(\frac{dy}{dx}\right)_{x=a} = f'(a),\ a \in D(f')$$

Question: How to find $[f'(x)]_{x=a}$?

Answer: To find the derivative values (or, the values of the derivative) at $x = a$ (i.e. $f'(a)$), we must first differentiate the given function $f(x)$ by using the formulas to get the derived function (or, derivative) $f'(x)$ and then put $x = a$ in $f'(x)$ if 'a' belongs to the domain of $f'(x)$ obtained by using general formula or, in other words, to find $f'(a)$, we adopt the following working rule.

1. Find the derived function $f'(x)$ by using the formulas

2. Put $x = a$ in $f'(x)$ provided $f'(a)$ is not undefined $\left(\text{i.e; } \frac{a}{0}, \frac{0}{0}, \text{etc.}\right)$ and lastly simplify to get the required answer.

3. If $f'(a)$ cannot be obtained thus then we should find $f'(a)$ directly from the defination [provided $f(a)$ is defined i.e. $a \in D(f)$]

$$f'(a) = \lim_{h \to 0} \frac{f(a+h) - f(a)}{h} \text{ (if exists).}$$

Note: The process of finding $f'(a)$ by differenting a function $y = f(x)$ with the help of known formulas and then putting $x = a$ in $f'(x)$ is fruitful and gives right answer when $f'(x)$ is continuous at $x = a$.

Let $f(x) = \begin{cases} x^2 \sin\left(\frac{1}{x}\right), & x \neq 0 \\ 0 & , x = 0 \end{cases}$

Now $f'(x) = x^2 \cos\left(\frac{1}{x}\right)\left(-\frac{1}{x^2}\right) + \sin\left(\frac{1}{x}\right)(2x)$

$= -\cos\left(\frac{1}{x}\right) + 2x\sin\left(\frac{1}{x}\right)$ and $\lim_{x\to 0} f'(x)$

$= \lim_{x\to 0}\left(-\cos\left(\frac{1}{x}\right) + 2x\sin\left(\frac{1}{x}\right)\right)$ does not exist

(because $\lim_{x\to 0} \cos\left(\frac{1}{x}\right)$ does not exist), so $f'(x)$ is not continuous at $x = 0$. It is not possible to find $f'(0)$ by putting $x = 0$ in $f'(x)$

However

$$f'(0) \lim_{h\to 0} = \frac{f(h) - f'(0)}{h} = \lim_{h\to 0} \frac{h^2 \sin\left(\frac{1}{h}\right) - 0}{h}$$

$$\lim_{h\to 0} h \sin\left(\frac{1}{h}\right) = 0$$

Remember:

1. To substitute specific values (or, particular values) for $x = 1, 2, 3, \ldots a$ etc. into the original function $f(x)$ before differentiation is not permissible.

2. Original (or, given function) is also termed as primitive function denoted as $f(x)$ without dash.

3. $\left[\frac{d}{dx}(f_1(x) \pm f_2(x))\right]_{x=a} = \left[\frac{d f_1(x)}{dx}\right]_{x=a}$

$\pm \left[\frac{d f_2(x)}{dx}\right] = f'(a) \pm f_2'(a)$

4. $\left[\frac{d}{dx}(f_1(x) \times f_2(x))\right]_{x=a}$

$= f_1'(a) \cdot f_2'(a) + f_2'(a) \cdot f_1'(a)$

5. $\left[\frac{d}{dx}\frac{f_1(x)}{f_2(x)}\right]_{x=a} = \left[\frac{d}{dx}\frac{N(x)}{D(x)}\right]_{x=a}$

$$= \frac{D(a) \cdot N'(a) - N(a) \cdot D'(a)}{[D(a)]^2}$$

Where $N(x)$ means a function of x in Nr and $D(x)$ means a function of x in Dr

6. $\left[\frac{d f(x)}{dx}\right]_{x=x}$ = The value of the derivative at a general point $x \in D(f')$

7. $\left[\frac{d f(x)}{dx}\right]_{x=a}$ = The value of the derivative, not at a general point x, but at some definite point namely $a \in D(f')$

8. Letting $F(x) = \frac{f(x)}{g(x)}$ where $F(x)$ and $g(x)$ are any two differentiable functions, the derivative of such a function is another function of x as $\frac{g(x) f'(x) - f(x) \cdot g'(x)}{\{g(x)\}^2}$ and this derivative exists at $x = 0$ provided that denominator is not zero at $x = 0$ i.e., $g(0) \neq 0$

9. In general $f(x) = |g(x)|$ has no derivative at the point where $g(x) = 0$, e.g.

(i) $f(x) = |x - 1| \Rightarrow f'(1)$ does not exist

(ii) $f(x) = 1 + |\sin x| \Rightarrow f'(\pi)$ does not exist

10. If $F(x) = f(x) \cdot g(x)$ then $F'(a)$ may exist even if $f'(a)$ or $g'(a)$ does not exist.

e.g.: $f(x) = \left|x - \frac{x}{2}\right| \cos x$

$$\Rightarrow f'\left(\frac{\pi}{2}+\right) = \lim_{h\to 0} + \frac{\left|\left(\frac{\pi}{2}+h\right) - \frac{\pi}{2}\right| \cos\left(h + \frac{\pi}{2}\right)}{h}$$

$$= \lim_{h\to 0} \cos\left(h + \frac{\pi}{2}\right) = 0$$

and $f'\left(\frac{\pi}{2}-\right) = \lim_{h\to 0} - \frac{|h| \cos\left(h + \frac{\pi}{2}\right)}{h} = 0$

Hence, $f'\left(\frac{\pi}{2}+\right)=f'\left(\frac{\pi}{2}-\right)=0 \Rightarrow f'\left(\frac{\pi}{2}\right)=0$

11. $\lim_{x\to a} f'(x)=\pm\infty \Rightarrow f(x)$ is not differentiable at $x=a$, where 'a' is the root of $f(x)=0 \Rightarrow f'(a)$ does not exist.

e.g.: $f(x)=(2x-1)^{\frac{1}{2}}$

$$\Rightarrow \lim_{x\to\left(\frac{1}{2}\right)^+} f'(x)=\lim_{x\to\left(\frac{1}{2}\right)^+}\frac{1}{\sqrt{2x-1}}=+\infty$$

$\Rightarrow f(x)$ is not differentiable at $x=\frac{1}{2}$

$\Rightarrow f'\left(\frac{1}{2}\right)$ does not exist.

Type 1: To find the value of the derived function (obtained by using the power, sum, difference, product or quatient rule of d.c.) at the indicated point.

Examples worked out:

1. Find $\frac{dy}{dx}$ at $x=\frac{1}{2}$ if (or, when or, for) $y=3x^2+2x$

Solution: $y=3x^2+2x$

$$\frac{dy}{dx}=6x+2$$

$$\Rightarrow \left[\frac{dy}{dx}\right]_{x=\frac{1}{2}}=[6x+2]_{x=\frac{1}{2}}$$

$$=6\times\frac{1}{2}+2=3+2=5$$

2. Find $\frac{dy}{dx}$ at $x=2$ when $y=x^3+2x+3$

Solution: $y=x^3+2x+3$

$$\frac{dy}{dx}=3x^2+2$$

$$\Rightarrow \left[\frac{dy}{dx}\right]_{x=2}=\left[3x^2+2\right]_{x=2}=3\times 4+2=14$$

3. Find $\frac{dy}{dx}$ at $x=\frac{\pi}{2}$ when $y=5\sin x$

Solution: $\because y=5\sin x$

$$\therefore \frac{dy}{dx}=5\cos x$$

$$\Rightarrow \left[\frac{dy}{dx}\right]_{x=\frac{\pi}{2}}=5\cdot\cos\frac{\pi}{2}=5\times 1=5$$

4. Find $\frac{dy}{dx}$ at $x=t$ when $y=2+\tan x$

Solution: $\because y=2+\tan x$

$$\therefore \frac{dy}{dx}=0+\sec^2 x$$

$$\Rightarrow \left[\frac{dy}{dx}\right]_{x=t}=\sec^2 t$$

5. Find $\frac{dy}{dx}$ at $x=\frac{\pi}{2}$ when $y=x^2+\cos x+\frac{1}{2}\log x$

Solution: $\because y=x^2+\cos x+\frac{1}{2}\log x$

$$\therefore \frac{dy}{dx}=2x-\sin x+\frac{1}{2}\cdot\frac{1}{x};\ (x>0)$$

$$\Rightarrow \left[\frac{dy}{dx}\right]_{x=\frac{\pi}{2}}=2\cdot\frac{\pi}{2}-\sin\frac{\pi}{2}+\frac{1}{2}\times\frac{2}{\pi}$$

$$\Rightarrow \left[\frac{dy}{dx}\right]_{x=\frac{\pi}{2}}=\pi-1+\frac{1}{\pi}$$

6. If $f(x)=5\sin^{-1}x-3\cos^{-1}x$, find $f'\left(\frac{\sqrt{3}}{2}\right)$

Solution: $\because f(x)=5\sin^{-1}x-3\cos^{-1}x$

$$\therefore \frac{d\,f(x)}{dx}=\frac{5}{\sqrt{1-x^2}}+\frac{3}{\sqrt{1-x^2}}=\frac{8}{\sqrt{1-x^2}};x^2<1.$$

$$\Rightarrow \left[\frac{d\,f(x)}{dx}\right]_{x=\frac{\sqrt{3}}{2}}=\frac{8}{\sqrt{1-\frac{3}{4}}}$$

$$=\frac{8}{\sqrt{\frac{4-3}{4}}}=\frac{8}{\sqrt{\frac{1}{4}}}=\frac{8}{\frac{1}{2}}=\frac{8}{1}\times\frac{2}{1}=16$$

7. Find the derivative of $f(x)\left[\text{i.e. } \frac{d\,f(x)}{dx}\right]=\frac{1}{x-2}$ at $x=1$ and $x=3$.

Solution: $f(x) = \dfrac{1}{x-2}$

$$\Rightarrow f'(x) = -1\,(x-2)^{-1-1} \times \frac{d\,(x-2)}{dx}$$

$$= -1\,(x-2)^{-2} \times 1$$

$$= \frac{-1}{(x-2)^2},\ x \neq 2.$$

$$\text{Now, } f'(1) = \frac{-1}{(+1-2)^2} = \frac{-1}{(-1)^2} = -\frac{1}{1} = -1$$

$$f'(3) = \frac{-1}{(3-2)^2} = \frac{-1}{1} = -1$$

Note: While applying the result $\dfrac{d}{dx}[f(x)]^n =$ derivative of a power function rule to a radical or to the reciprocal of a function, it is useful to put the radical or reciprocal function into a power function whose index is negative or fraction as we require. e.g.

1. $\dfrac{1}{\sqrt{x^5}}$ should be put into the form $x^{-\frac{5}{2}}$

2. $\dfrac{1}{x+a}$ should be put into the form $(x+a)^{-1}$ and then apply the power function rule

$$\Rightarrow \left[\frac{d\,(f(x))^n}{dx}\right] = n\,[f(x)]^{n-1} \cdot \frac{d\,f(x)}{dx}$$

8. Find the derivative of $f(x) = \sqrt{x^2 + a^2}$ at the point $x = a$

Solution: $f(x) = \sqrt{x^2 + a^2} = (x^2 + a^2)^{\frac{1}{2}}$

$$\Rightarrow f'(x) = \frac{1}{2}(x^2 + a^2)^{\frac{1}{2}-1} \times \frac{d(x^2 + a^2)}{dx}$$

$$\Rightarrow f'(x) = \frac{1}{2\sqrt{x^2 + a^2}} \times 2x$$

$$= \frac{x}{\sqrt{x^2 + a^2}}$$

$$\text{Now, } f'(a) = \frac{a}{\sqrt{a^2 + a^2}} = \frac{a}{|a|\sqrt{2}} = \frac{1}{\sqrt{2}}\ (\text{if } a > 0)$$

9. Find the derivative of $f(x) = \sqrt{2x-3}$ at $x = 2$

Solution: $f(x) = \sqrt{2x-3} = (2x-3)^{\frac{1}{2}}$

$$\Rightarrow f'(x) = \frac{1}{2\sqrt{2x-3}} \times 2 = \frac{1}{\sqrt{2x-3}}\left(x \neq \frac{3}{2}\right)$$

$$\Rightarrow f'(2) = \frac{1}{\sqrt{2 \times 2 - 3}} = \frac{1}{\sqrt{4-3}} = \frac{1}{1} = 1$$

10. Find $\dfrac{dy}{dx}$ at $x = 0$ when $y = \dfrac{x + \cos x}{1 + \cos x}$

Solution: $\because\ N = x + \cos x$

$\therefore\ N' = 1 - \sin x$

and $D = 1 + \cos x$

$\therefore\ D' = 0 - \sin x = -\sin x$

$$\therefore \frac{dy}{dx} = \frac{DN' - ND'}{D^2}$$

$$= \frac{(1+\cos x)(1-\sin x) - (x+\cos x)(0-\sin x)}{(1+\cos x)^2}$$

$$\Rightarrow \frac{dy}{dx} = \frac{1 + \cos x - \sin x + x \sin x}{(1+\cos x)^2}$$

$$\Rightarrow \left[\frac{dy}{dx}\right]_{x=0} = \frac{1+1}{(1+1)^2} = \frac{1}{2}$$

11. Find $\dfrac{dy}{dx}$ at $x = 1$ when $y = \dfrac{\log(xe^x)}{x + e^x}$

Solution: $y = \dfrac{\log(xe^x)}{x + e^x}$

Now, $N = \log(xe^x) = \log x + \log e^x$

$= \log x + x \log_e{}^e = \log x + x$

$$\Rightarrow N' = \frac{1}{x} + 1$$

Again, $D = x + e^x$

$$\Rightarrow D' = 1 + e^x$$

$$\frac{dy}{dx} = \frac{DN' - ND'}{D^2}$$

$$\Rightarrow \frac{dy}{dx} = \frac{\left(x+e^x\right)\left(\frac{1}{x}+1\right) - \log\left(xe^x\right)\left(1+e^x\right)}{\left(1+e^x\right)^2}$$

$$\Rightarrow \frac{dy}{dx} = \frac{1+\frac{e^x}{x} + x + e^x - [\log x + x]\left(1+e^x\right)}{\left(1+e^x\right)^2}$$

$$\Rightarrow \frac{dy}{dx} = \frac{1+\frac{e^x}{x}+x+e^x - \log x - x - e^x \log x - e^x \cdot x}{\left(1+e^x\right)^2}$$

Hence, $\left[\frac{dy}{dx}\right]_{x=1} = \frac{1+e+1+e-0-1-e\cdot 0-e}{(1+e)^2}$

$$= \frac{(1+e)}{(1+e)^2} = \frac{1}{(1+e)}$$

12. Find $\frac{dy}{dx}$ at $x = 1$ when $y = \frac{2+\sqrt{x}}{2-\sqrt{x}} \cdot e^x$

Solution: $\because\ y = \frac{2+\sqrt{x}}{2-\sqrt{x}} \cdot e^x$

Taking log ,

$$\log y = \log\left(2+\sqrt{x}\right) + \log e^x - \log\left(2-\sqrt{x}\right)$$

$$\Rightarrow \frac{1}{y}\cdot\frac{dy}{dx} = \frac{1}{2+\sqrt{x}} \times \frac{1}{2\sqrt{x}} + \frac{1}{e^x} \times e^x - \frac{1}{2-\sqrt{x}} \times \left(-\frac{1}{2\sqrt{x}}\right)$$

$$\Rightarrow \frac{1}{y}\cdot\frac{dy}{dx} = \frac{1}{2\sqrt{x}\left(2+\sqrt{x}\right)} + 1 + \frac{1}{2\sqrt{x}\left(2-\sqrt{2}\right)}$$

$$\Rightarrow \frac{dy}{dx} = y\left[\frac{1\times\left(2-\sqrt{x}\right)+2\sqrt{x}\,(4-x)+\left(2+\sqrt{x}\right)}{2\sqrt{x}\left(2+\sqrt{x}\right)\left(2-\sqrt{x}\right)}\right]$$

$$\Rightarrow \frac{dy}{dx} = \left[\frac{2+\sqrt{x}}{2-\sqrt{x}}\right]\cdot e^x \cdot \left[\frac{4+2\sqrt{x}\,(4-x)}{2vx\,(4-x)}\right]$$

Hence, $\left[\frac{dy}{dx}\right]_{x=1} = \frac{2+1}{2-1}\cdot e \times \left[\frac{10}{2\cdot 1\cdot(4-1)}\right]$

$$= \frac{3}{1}\cdot e \cdot \left[\frac{10}{2\times 3}\right] = 5e$$

13. If $f(x) = x^{\frac{1}{3}}$, find $f'(0)$

Solution: $f(x) = x^{\frac{1}{3}}$

$$\therefore\ f'(x) = \frac{1}{3x^{\frac{2}{3}}},\ x \neq 0.$$

Since 0 does not belong to the domain of $\frac{1}{3x^{\frac{2}{3}}}$, we find $f'(0)$ using definition of derivative at $x = 0$.

$$\therefore\ f'(0) = \lim_{h\to 0} \frac{f(0+h) - f(0)}{h}$$

$$= \lim_{h\to 0} \frac{(0+h)^{\frac{1}{3}} - 0}{h}$$

$$\lim_{h\to 0} \frac{h^{\frac{1}{3}}}{h} = \lim_{h\to 0} h^{-\frac{2}{3}} = \lim_{h\to 0} \frac{1}{\sqrt[3]{h^2}} = \infty$$

Remember: The use and meaning of the expression $\frac{a}{0}$ must become clear in the mind of ours.

1. If a function $f(x)$ is an infinitesimal as $x \to a$ $\left(\text{i.e; if } \lim_{x \to a} f(x) = 0\right)$ then the function $\frac{1}{f(x)}$ is infinitely large which is symbolically represented as ∞ and we say that $\frac{1}{f(x)}$ has no limit or infinite limit as $x \to a$ i.e., $\lim_{x \to a} \frac{1}{f(x)} = \infty$ (no limit or infinite limit) if $\lim_{x \to a} f(x) = 0$.

2. The expression $\frac{a}{0}$ is undefined since division by zero is not allowed in mathematics. The expression $\frac{a}{0}$ always should be written as $\frac{a}{0}$ = undefined for the value of the expression $\frac{a}{x}$ for $x = 0$ which provides us the following sense.

If a function $f(x)$ is zero for any value of x, then the function $\frac{1}{f(x)}$ is undefined or meaningless for that value x at which $f(x) = 0$

Thus $\frac{a}{0}$ = undefined should be used in the sense of the actual value (or, simply value) of the function $f(x)$ at a point at which $f(x) = 0$ and $\lim_{x \to a} \frac{a}{f(x)} = \frac{a}{0} = \infty$ if $\lim_{x \to a} f(x) = 0$ in the sense of limiting value of $\frac{1}{f(x)}$ as $x \to a$, e.g.,

1. The symbols, $\lim_{x \to \frac{\pi -}{2}} \tan x = +\infty$

$\lim_{x \to \frac{\pi +}{2}} \tan x = -\infty$

$[\tan x]_{x = \frac{\pi}{2}} = \tan \frac{\pi}{2}$ = undefined.

Type 2: To find the value of the derivative of a composite of two or more than two functions at the indicated point.

Working rule:

1. Find the differential coefficient (i.e. d.c.) of a composite of two or more than two functions by using the chain rule.
2. Put $x = a$ in the derived function and see whether $f'(a)$ is defined or undefined.
3. If $f'(a)$ is defined, then simplify to get the required answer and if $f'(a)$ is undefined, then we should find $f'(a)$ directly from the defination provided $f(a)$ is defined.

Examples worked out:

1. If $f(x) = \log_x(\log x)$, find $f'(x)$ at $x = e$

Solution: $f(x) = \log_x(\log x) = \dfrac{\log_e \log x}{\log_e x}$; $(x > 1)$

$\Rightarrow f(x) = \dfrac{\log \log x}{\log x}$ ($\because f(x)$ is defined only for $x > 1$)

Now, $\dfrac{dy}{dx} = \dfrac{d\,f(x)}{dx} = f'(x)$

$$= \frac{\frac{1}{\log x} \times \frac{1}{x} \times \log x - \frac{1}{x} \cdot \log(\log x)}{(\log x)^2}$$

$$= \frac{\frac{1}{x} - \frac{1}{x} \log \log x}{(\log x)^2}$$

$$= \frac{1 - \log \log x}{x\,(\log x)^2}$$

Hence, $\left[\dfrac{dy}{dx}\right]_{x=e} = \left[\dfrac{1 - \log \log x}{x(\log x)^2}\right]_{x=e}$

$$= \frac{1 - \log \log e}{e(\log e)^2}$$

$$= \frac{1 - 0}{e\,(1)^2} = \frac{1}{e}$$

Note: Derivative of a function of the form $\log_{f(x)} g(x)$ is alwasy obtained by first changing $\log_{f(x)} g(x)$ into $\dfrac{\log_e g(x)}{\log_e f(x)}$ and then we use the quotient rule to find the d.c.

2. If $y = a^x + \sqrt{\dfrac{1+x}{1-x}}$, find d.c. at $x = 0$

Solution: $y = a^x + \sqrt{\dfrac{1+x}{1-x}}$

Now, differentiating both sides w.r.t. x, we obtain

$$\frac{dy}{dx} = \frac{d\,a^x}{dx} + \frac{d}{dx}\left(\sqrt{\frac{1+x}{1-x}}\right)$$

Now, $\dfrac{d\,a^x}{dx} = a^x \log a$ and $\dfrac{d}{dx}\left[\sqrt{\dfrac{1+x}{1-x}}\right]$

$$= \frac{\sqrt{1-x}\left[\dfrac{1}{2\sqrt{1+x}}\right] - \sqrt{1+x}\left[\dfrac{1}{2\sqrt{1-x}}(-1)\right]}{\left(\sqrt{1-x}\right)^2}$$

$$= \frac{\sqrt{1-0}\left[\dfrac{1}{2\sqrt{1+0}}\right] - \sqrt{1+0}\left[\dfrac{1}{2\sqrt{1-0}}(-1)\right]}{\left(\sqrt{1-0}\right)^2};$$

for $x = 0$

$$= \frac{\frac{1}{2} + \frac{1}{2}}{(1)^2} = \frac{1}{1} = 1$$

Hence, $\left[\dfrac{dy}{dx}\right]_{x=0} = \left[a^x \log a\right]_{x=0} + 1$

$$\Rightarrow \left[\frac{dy}{dx}\right]_{x=0} = a^0 \log a + 1$$

$$\Rightarrow \left[\frac{dy}{dx}\right]_{x=0} = \log a + 1$$

Type 3: To find the value of d.c of mod of a function at $x = a$ working rule:

1. Find d.c of $|f(x)|$

2. Put $x = a$ in the d.c of $|f(x)|$ and see whether $f'(a)$ is defined or undefined.

3. If $f'(a)$ is defined, then put $x = a$ in $f'(x)$ to get the required answer and if $f'(a)$ is not obtained thus, then we should find $f'(a)$ directly from the defination. Provided $f(a)$ is defined.

Remember. $\dfrac{d}{dx}|f(x)| = \dfrac{|f(x)|}{f(x)} \times f'(x);\ (f(x) \neq 0)$

Examples worked out:

1. If $f(x) = |\cos x|$, find $f'\left(\dfrac{3\pi}{4}\right)$

Solution: $f(x) = |\cos x|$

$$\therefore f'(x) = \frac{d}{dx}|\cos x| = \frac{|\cos x|}{\cos x} \times (-\sin x),\ \cos x \neq 0$$

$$\therefore f'\left(\frac{3\pi}{4}\right) = \frac{\left|\cos\left(\frac{3\pi}{4}\right)\right|}{\cos\frac{3\pi}{4}} \times \left(-\sin\frac{3\pi}{4}\right)$$

$$= \frac{\frac{1}{v2}}{-\frac{1}{v2}} \cdot \left(-\frac{1}{v2}\right)$$

$$= \frac{1}{v2}$$

2. If $f(x) = |\cos x - \sin x|$, find $f'\left(\dfrac{\pi}{2}\right)$

Solution: $\because\ f(x) = |\cos x - \sin x|$

$$\therefore f'(x) = \frac{|\cos x - \sin x|}{(\cos x - \sin x)} \times \frac{d}{dx}(\cos x - \sin x)$$

$$\Rightarrow f'(x) = \left|\frac{\cos x - \sin x}{(\cos x - \sin)}\right|(-\sin x - \cos x)$$

for $x \neq n\pi + \dfrac{\pi}{4}$

$$\therefore f'\left(\frac{\pi}{2}\right)=\frac{\left|\cos\frac{\pi}{2}-\sin\frac{\pi}{2}\right|}{\left(\cos\frac{\pi}{2}-\sin\frac{\pi}{2}\right)}\cdot\left[-\sin\frac{\pi}{2}-\cos\frac{\pi}{2}\right]$$

$$=\frac{|0-1|}{(0-1)}\times(-1-0)$$

$$=\frac{|-1|}{(-1)}\times(-1)$$

$$=1$$

3. If $y=|x|^2-4|x|+2$, find $\frac{dy}{dx}$ at $x=3$

Solution: $y=|x|^2-4|x|+2$

$$\frac{dy}{dx}=\frac{d}{dx}\left[|x|^2-4|x|+2\right]$$

$$=\frac{d|x|^2}{dx}-4\frac{d|x|}{dx}+\frac{d(2)}{dx}$$

$$=\frac{dx^2}{dx}-4\frac{|x|}{x}+0\left[\because |x|^2=x^2\right]$$

$$=2x-4\frac{|x|}{x},\ x\neq 0$$

$$\left[\frac{dy}{dx}\right]_{x=3}=2\cdot 3-\frac{4|3|}{3}$$

$$=6-4$$

$$=2$$

4. $f(x)=\left|\sin^3 x\right|$, find $f'(x)$ at $x=0$

Solution: $f'(x)=\frac{\left|\sin^3 x\right|}{\sin x}\cdot\cos x$ for $x\neq n\pi$

$\therefore$ This formula can not give us the value of $f'(x)$ at $x=n\pi$ and so at $x=0$.

Hence we find this value by definition:

$$f'(0)=\lim_{h\to 0}\frac{f(h)-f(0)}{h}$$

$$=\lim_{h\to 0}\frac{\sin^3 h}{h}$$

$$=\lim_{h\to 0}\left(\sin^2 h\right)\cdot\frac{\sin h}{h}$$

$$=0.1$$

$$=0$$

Type 4: To determine the value of d.c of implicit function defined by $f(x,y)=c$ at $x=a$; $y=b$

Working rule:

1. Find $\frac{dy}{dx}$ from $f(x,y)=c$

2. Put $\begin{smallmatrix}x=a\\ y=b\end{smallmatrix}$ in $\frac{dy}{dx}$ to find $\left[\frac{dy}{dx}\right]_{\substack{x=a\\ y=b}}$

Remember: In the derived function of y in $f(x,y)=c$, x - coordinate and y - coordinate both may be provided at which we require the value of derived function obtained from the given implicit function $f(x,y)=c$

Examples worked out:

1. Find d.c. of y from $xy+4=0$, at $(x,y)=(2,-2)$

Solution: $xy+4=0$

$$\Rightarrow y\frac{dx}{dx}+x\frac{dy}{dx}=0$$

$$\Rightarrow y+x\frac{dy}{dx}=0$$

$$\Rightarrow \frac{dy}{dx}=-\frac{y}{x}$$

$$\Rightarrow \left[\frac{dy}{dx}\right]_{(x,y)=(2,-2)}=\left[\frac{-y}{x}\right]_{(2,-2)}=\frac{-(-2)}{2}=1$$

2. Find $\frac{dy}{dx}$ for $\sqrt{x}=y+\sqrt{a}$ at $(a,0)$

Solution: $y+\sqrt{a}=\sqrt{x}$

$$\Rightarrow y=\sqrt{x}-\sqrt{a}$$

$$\Rightarrow \frac{dy}{dx}=\frac{d\sqrt{x}}{dx}-\frac{d\sqrt{a}}{x}$$

$$\Rightarrow \frac{dy}{dx} = \frac{1}{2\sqrt{x}}, \; x > 0.$$

$$\Rightarrow \left[\frac{dy}{dx}\right]_{(x,\,y)=(a,\,0)} = \left[\frac{1}{2\sqrt{x}}\right]_{(a,\,0)} = \frac{1}{2\sqrt{a}}$$

N.B.: Since y does not appear in the derived function of $f(x, y) = c$, there is no question of putting $y = 0$ in the derived function.

3. Find $\frac{dy}{dx}$ for $x^2 = y$ at the point (1, 1)

Solution: $y = x^2$

$$\Rightarrow \frac{dy}{dx} = 2x$$

$$\Rightarrow \left[\frac{dy}{dx}\right]_{(1,\,1)} = [2x]_{(1,\,1)} = 2 \times 1 = 2$$

4. Find $\frac{dy}{dx}$ when $x^2 - xy + y^2 = 3$ and find its value at $(1, -1)$

Solution: $x^2 - xy + y^2 = 3$

Now, differentiating both sides of the equation w.r.t x, we get,

$$2x - y - x\frac{dy}{dx} + 2y\frac{dy}{dx} = 0$$

$$\Rightarrow \frac{dy}{dx}(2y - x) = y - 2x$$

$$\Rightarrow \frac{dy}{dx} = \frac{y - 2x}{2y - x}$$

Now, the value of $\frac{dy}{dx}$ at

$$(1, 1) = \left[\frac{dy}{dx}\right]_{(x,\,y)=(1,\,-1)} = \frac{-1 \times -2 \times 1}{-2 \times 1 \times -1}$$

$$\Rightarrow \left[\frac{dy}{dx}\right]_{(1,\,-1)} = \frac{-3}{-3} = 1$$

5. Find $\frac{dy}{dx}$ when $x^2 + xy - y^2 =$ at (2, 3) and at (1,2)

Solution: Given function is $x^2 + xy - y^2 = 1$

Now, differentiating both sides of the equation w.r.t x, we get,

$$2x + y + x\frac{dy}{dx} - 2y \cdot \frac{dy}{dx} = 0$$

$$\Rightarrow \frac{dy}{dx}(x - 2y) = -2x - y$$

$$\Rightarrow \frac{dy}{dx} = \frac{-2x - y}{x - 2y} = \frac{2x + y}{(2y - x)}$$

$$\left[\frac{dy}{dx}\right]_{(2,\,3)} = \frac{2 \times 2 + 3}{2 \times 3 - 2} = \frac{7}{4}$$

(1, 2) does not satisfy the given equation and so the question of finding $\frac{dy}{dx}$ at (1, 2) does not arise.

Type 5: To find the value of differential coefficient of a given function w.r.t. another given function at $x = a$

Working rule: To find the value of differential coefficient of a given function w.r.t. another given function at a point $x = a$, we adopt the following working rule:

1. $\frac{d f_1(x)}{d f_2(x)} = \frac{\frac{d}{dx} f_1(x)}{\frac{d}{dx} f_2(x)} = \frac{f_1'(x)}{f_2'(x)}$ where $f_1(x)$ = the function whose differential coefficient is required and $f_2(x)$ = the function with respect to which differential coefficient is required.

2. Put $x = a$ in $\frac{f_1'(x)}{f_2'(x)}$

Examples worked out:

1. Find d.c. of $\sec^{-1}\frac{1}{2x^2 - 1}$ w.r.t $\sqrt{1 - x^2}$ at $x = \frac{1}{2}$

Solution: Let $f_1(x) = \sec^{-1}\left(\frac{1}{2x^2 - 1}\right)$

and $f_2(x) = \sqrt{1 - x^2}$

$$\therefore \frac{d f_1(x)}{dx} = \frac{d}{dx}\sec^{-1}\left(\frac{1}{2x^2-1}\right)$$

$$= \frac{1}{\left(\left|\frac{1}{2x^2-1}\right|\right)\cdot\sqrt{\left(\frac{1}{2x^2-1}\right)^2-1}} \times \frac{d}{dx}\left(\frac{1}{2x^2-1}\right)$$

$$= \frac{(2x^2-1)^2}{\sqrt{4x^2-4x^4}} \times \frac{-1}{(2x^2-1)^2} \times 4x$$

$$= \frac{-4x}{\sqrt{4x^2-4x^4}}$$

$$\frac{d f_2(x)}{dx} = \frac{1 \times (-2x)}{2\sqrt{1-x^2}}$$

$$= \frac{-x}{\sqrt{1-x^2}}$$

Hence, $\dfrac{d f_1(x)}{d f_2(x)} = \dfrac{\frac{d}{dx} f_1(x)}{\frac{d}{dx} f_2(x)} = \dfrac{4\sqrt{1-x^2}}{\sqrt{4x^2-4x^4}}$

$$= \frac{2\sqrt{1-x^2}}{|x|\sqrt{1-x^2}}$$

$$= \frac{2}{|x|}$$

$$\left[\frac{d f_1(x)}{d f_2(x)}\right]_{x=\frac{1}{2}} = \left[\frac{2}{|x|}\right]_{x=\frac{1}{2}} = \frac{2}{\frac{1}{2}} = 4$$

Type 6: To find the value of differential coefficient of parametric equations at a particular value of the given parameter $t\,(\text{or}, \theta) = a$.

Working rule:

1. Let $x = x(t)$ and $y = y(t)$

2. Find $\dfrac{dx}{dt}$ and $\dfrac{dy}{dt}$

3. Divide $\dfrac{dy}{dt}$ by $\dfrac{dx}{dt}$ to have $\dfrac{dy}{dx} = \dfrac{\frac{dy}{dt}}{\frac{dx}{dt}}$

4. Find $\left[\dfrac{dy}{dx}\right]_{x=a}$

Examples worked out:

1. If $x = a(\theta - \sin\theta)$ $y = a(1-\cos\theta)$, Find $\dfrac{dy}{dx}$ when $\theta = \dfrac{\pi}{2}$.

Solution: $y = a(1-\cos\theta)$ ...(1)

$x = a(\theta - \sin\theta)$ ...(2)

$(1) \Rightarrow \dfrac{dy}{d\theta} = a\,[0-(-\sin\theta)] = a\sin\theta$ ...(3)

$(2) \Rightarrow \dfrac{dx}{d\theta} = a\,[1-\cos\theta]$ ...(4)

$$\therefore \frac{dy}{dx} = \frac{\frac{dy}{d\theta}}{\frac{dx}{d\theta}} = \frac{a\sin\theta}{a(1-\cos\theta)} = \frac{\sin\theta}{(1-\cos\theta)}$$

Hence, $\left[\dfrac{dy}{dx}\right]_{\theta=\frac{\pi}{2}} = \dfrac{\sin\frac{\pi}{2}}{1-\cos\frac{\pi}{2}} = \dfrac{1}{1} = 1$

2. If $x = a(1-\cos t)$ $y = a(t + \sin t)$ Find $\dfrac{dy}{dx}$ at $t = \dfrac{\pi}{2}$

Solution: $y = a(t + \sin t)$ ...(1)

$x = a(1-\cos t)$ ...(2)

$(1) \Rightarrow \dfrac{dy}{dt} = a(1+\cos t)$ ...(3)

$(2) \Rightarrow \dfrac{dx}{dt} = a(0+\sin t)$

$= a\sin t$ ...(4)

$\therefore \frac{dy}{dx} = \frac{a(1+\cos t)}{a\sin t}$

Hence, $\left[\frac{dy}{dx}\right]_{t=\frac{\pi}{2}} = \frac{a\left(1+\cos\frac{\pi}{2}\right)}{a\sin\frac{\pi}{2}}, t \neq n\pi.$

$= \frac{a(1+0)}{a\cdot 1} = \frac{a}{a} = 1$

3. If $x = a\cos^3\theta$ $y = a\sin^3\theta$ Find $\frac{dy}{dx}$ when $\theta = \frac{\pi}{4}$

Solution: $y = a\sin^3\theta$ ...(1)

$x = a\cos^3\theta$ ...(2)

$(1) \Rightarrow \frac{dy}{d\theta} = 3a\sin^2\theta\cos\theta$ ...(3)

$(2) \Rightarrow \frac{dx}{d\theta} = 3a\cos^2\theta(-\sin\theta)$

$= -3a\sin\theta\cdot\cos^2\theta$ ...(4)

$\therefore \frac{dy}{dx} = \frac{3a\sin^2\theta\cos\theta}{-3a\sin\theta\cos^2\theta} = -\tan\theta, \theta \neq \frac{n\pi}{2}.$

Hence, $\left[\frac{dy}{dx}\right]_{x=\frac{\pi}{4}} = \left[-\tan\frac{\pi}{4}\right] = -1$

4. If $x = 3\cos\theta - 2\cos^3\theta$, $y = 3\sin\theta - 2\sin^3\theta$

Find $\frac{dy}{dx}$ when $\theta = \frac{\pi}{4}$.

Solution: $y = 3\sin\theta - 2\sin^3\theta$ ...(1)

$x = 3\cos\theta - 2\cos^3\theta$ ...(2)

$(1) \Rightarrow \frac{dy}{d\theta} = 3\cos\theta - 6\sin^2\theta\cos\theta$

$= 3\cos\theta\left(1 - 2\sin^2\theta\right)$

$\Rightarrow \frac{dy}{d\theta} = 3\cos\theta\cdot\cos 2\theta$ ...(3)

$(2) \Rightarrow \frac{dx}{d\theta} = -3\sin\theta - 6\cos^2\theta(-\sin\theta)$

$\Rightarrow \frac{dx}{d\theta} = -3\sin\theta\left(1 - 2\cos^2\theta\right)$

$= 3\sin\theta\left(2\cos^2\theta - 1\right)$

$\Rightarrow \frac{dx}{d\theta} = 3\sin\theta\cos 2\theta$ ...(4)

$\therefore \frac{dy}{dx} = \frac{3\cos\theta\cos 2\theta}{3\sin\theta\cos 2\theta} = \cot\theta$

when $\sin\theta \neq 0, \cos 2\theta \neq 0$.

Hence we cannot put $\theta = \frac{\pi}{4}$ to have $\left[\frac{dy}{dx}\right]_{\theta=\frac{\pi}{4}} = \cot\frac{\pi}{4} = 1$ (an erroneous process since for $\theta = \frac{\pi}{4}$, $\cos 2\theta \neq 0$)

However applying L'Hospital's rule we have

$= \left(\frac{dy}{dx}\right)_{\theta=\frac{\pi}{4}} = \lim_{\theta\to\frac{\pi}{4}} \frac{\Delta y}{\Delta x}$

$= \lim_{\theta\to\frac{\pi}{4}} \frac{y(\theta) - y\left(\frac{\pi}{4}\right)}{x(\theta) - x\left(\frac{\pi}{4}\right)}$

$= \lim_{\theta\to\frac{\pi}{4}} \frac{y'(\theta)}{x'(\theta)} = \lim_{\theta\to\frac{\pi}{4}} \cot\theta = 1$

Type 7: To determine the value of d.c of a function found by taking first logarithm at a point $x = a$

Working rule:

1. APPLY the rule "GLAD" to find derived function.
2. Put $x = a$ in the derived function .

Remember: Derivative of a function raised to the power another function as $y = [f(x)]^{g(x)}$ is found only with the help of logarithmic differentiation.

Examples worked out:

1. If $y = \left[2^{\log_2 x}\right]^{2x} + \left[\tan\frac{\pi x}{4}\right]^{\frac{4}{\pi x}}$ Find $\frac{dy}{dx}$ at $x = 1$

Solution: Let $u = \left[2^{\log_2 (x)}\right]^{2x}$ ...(1)

$$\Rightarrow u = x^{2x} \left[\because a^{\log_a N} = N\right]$$

Now taking the log of both sides, we get

$$\log u = \log x^{2x}$$

$$\Rightarrow \log u = 2x \log x$$

Now, differentiating w.r.t. x, we get

$$\frac{1}{u}\cdot\frac{du}{dx} = 2\left[x\cdot\frac{1}{x} + 1\cdot\log x\right] = 2(1 + \log x)$$

$$\Rightarrow \frac{du}{dx} = 2x^{2x}\,[1 + \log x]\left(\because u = x^{2x}\right)$$

$$\therefore \left[\frac{du}{dx}\right]_{x=1} = 2\cdot 1[1 + \log 1] = 2\cdot 1(1 + 0) = 2$$

Again, let $v = \left[\tan\frac{\pi x}{4}\right]^{\frac{4}{\pi x}}$

Now, taking log of both sides, we get

$$\log v = \frac{4}{\pi x}\log\left[\tan\frac{\pi x}{4}\right]$$

Now, differentiating both sides w.r.t. x, we get

$$\frac{1}{v}\cdot\frac{dv}{dx}$$

$$= \frac{4}{\pi x}\cdot\frac{1}{\tan\frac{\pi x}{4}}\left(\sec^2\frac{\pi x}{4}\right)\frac{\pi}{4} + \frac{4}{\pi}\log\left(\tan\frac{\pi x}{4}\right)\left(-\frac{1}{x^2}\right)$$

$$\Rightarrow \frac{dv}{dx} = v\left[\frac{1}{x}\cdot\cot\frac{\pi x}{4}\cdot\sec^2\frac{\pi x}{4} - \frac{4}{\pi x^2}\log\tan\left(\frac{\pi x}{4}\right)\right]$$

$$\therefore \left[\frac{dv}{dx}\right]_{x=1} = \left[\tan\frac{\pi}{4}\right]^{\frac{4}{\pi}}\cdot\left[\frac{1}{1}\cot\frac{\pi}{4}\cdot\sec^2\frac{\pi}{4} - \frac{4}{\pi}\log\tan\frac{\pi}{4}\right]$$

$$y = u + v$$

$$\frac{dy}{dx} = \frac{du}{dx} + \frac{dv}{dx}$$

$$\text{Hence, } \left[\frac{dy}{dx}\right]_{x=1} = \left[\frac{du}{dx}\right]_{x=1} + \left[\frac{dv}{dx}\right]_{x=1}$$

$$= 2 + 2$$

$$= 4$$

Type 8: To find the value of x for which d.c of a function is a constant:

Working rule:

1. Find d.c of the given function $f(x)$
2. Write $f'(x)$, i.e. d.c. of the given function = The given constant
3. Solve $f'(x)$ = the given constant. This provides us the required value or values of x.

Examples worked out:

1. If $f(x) = x^2 - 4x + 3$, Find the value of x for which the derivative is 2.

Solution: $f(x) = x^2 - 4x + 3$

$$\Rightarrow f'(x) = 2x - 4$$

$$f'(x) = 2 \Rightarrow 2x - 4 = 2$$

$$\Rightarrow 2x = 2 + 4$$

$$\Rightarrow x = \frac{6}{2} = 3$$

2. If $f(x) = x^2 - 3x + 2$ Find the value of x for which the derivative is zero.

Solution: $f(x) = x^2 - 3x + 4$

$$f'(x) = 2x - 3$$

$$f'(x) = 0$$

$$\Rightarrow 2x - 3 = 0$$

$$\Rightarrow 2x = 3$$

$\Rightarrow x = \frac{3}{2}$

Type 9: Problems based on existance of a derived function at a point $x = a$

Whenever we say that something exists, we mean that it (something) has a finite value in the world of mathematics which implies that whenever we say that $f(x)$ or $f'(x)$ at $x = a$ exists, we mean that $f(a)$ or $f'(a)$ has a finite value and whenever $[f(x)]_{x=a}$ or $\left[\frac{dy}{dx}\right]_{x=a}$ has an infinite value or meaningless or undetermined value $\left(\text{i.e; } \frac{a}{0}, \frac{0}{0}, \text{etc}\right)$ we say that $f(x)$ at $x = a$ or $f'(x)$ at $x = a$ does not exist.

Question: How to guess the existance of a derived function at a point $x = a$?

Answer: A simple rule to guess the existance of $f'(x)$ at $x = a$ is the following :

1. Differentiate the given function $f(x)$ w.r.t the independent variable x by using the formulas of d.c. of a power function, sum, difference, product, quotient, function of a function etc.

2. Find $[f'(x)]_{x=a}$

3. If $f'(a)$ is finite, we expect that $f'(x)$ exists at $x = a$ and if $f'(a)$ is undefined, we expect that $f'(x)$ does not exist at $x = a$

Question: How to test that existance of a derived function $f'(x)$ at a point $x = a$?

Answer: Whenever we are required to test (or, examine) whether $f'(x)$ exists at $x = a$ or $f'(x)$ does not exist at $x = a$, we adopt the following working rule.

1. Find the left hand derivative $= f'_{-}(a) = L_1 = a$ where 'a' is a fininte number and the right hand derivative $= f'_{+}(a) = L_2 = a$ where 'a' is a finite number.

2. If $L_1 = L_2$, then $f'(a)$ is said to exist (or, we say that $f'(x)$ exists at $x = a$) and if $L_1 \neq L_2$, then $f'(a)$ is said not to exist (or, $f'(x)$ is said not to exist at $x = a$ or we say that $f'(x)$ does not exists at $x = a$)

Remember:

1. $f'(a)$ exists $\Leftrightarrow f'_{-}(a) = f'_{+}(a) = a$ a finite number.

2. $f'(a)$ does not exists $\Leftrightarrow f'_{-}(a) \neq f'_{+}(a) \neq a$ a finite number.

or $f'(a)$ does not exist

$\Leftrightarrow f'_{-}(a) = f'_{+}(a) = \infty, -\infty;$

or $f'(a)$ does not exists

$\Leftrightarrow f'_{-}(a) = +\infty$ and $f'_{+}(a) = -\infty$

Examples worked out:

1. If $y = \frac{2x-3}{3x-4}$, examine the existance of $\frac{dy}{dx}$ at $x = -\frac{4}{3}$

Solution: Let $f(x) = \frac{2x-3}{3x-4}$

$$\therefore f\left(-\frac{4}{3}\right) = \frac{2 \times \left(-\frac{4}{3}\right) - 3}{3 \times \left(-\frac{4}{3}\right) - 4}$$

$$= \frac{\frac{-8}{3} - 3}{-4-4} = \frac{-8-9}{-8 \times 3} = \frac{17}{24}$$

$$f\left(-\frac{4}{3}+h\right) = \frac{2\left(\frac{-4}{3}+h\right)-3}{3\left(\frac{-4}{3}+h\right)-4}$$

$$= \frac{\frac{-8}{3}+2h-3}{3\left(\frac{-4}{3}+h\right)-4} = \frac{\frac{-17}{3}+2h}{-4+3h-h}$$

$$= \frac{-\frac{17}{3}+2h}{-8+3h}$$

$$f\left(-\frac{4}{3}+h\right) - f\left(-\frac{4}{3}\right) = \frac{-\frac{17}{3}+2h}{-8+3h} - \frac{17}{24}$$

$$= \frac{-17+6h}{3(-8+3h)} - \frac{17}{24} = \frac{-17+6h}{-24+9h} - \frac{17}{24}$$

$$= \frac{24(-17+6h) - 17 \times (-24+9h)}{(-24+9h) \times 24}$$

$$= \frac{-408+144h+408-153h}{(-24 \times 24 + 9 \times 24h)}$$

$$= \frac{144h - 153h}{-24 \times 24 + 9 \times 24h} = \frac{-9h}{-24 \times 24 + 9 \times 24h}$$

$$\therefore f'_+\left(\frac{-4}{3}\right) = \lim_{h \to 0} \frac{f\left(\frac{-4}{3}+h\right) - f\left(\frac{-4}{3}\right)}{h}; (h>0)$$

$$= \lim_{h \to 0} \frac{\frac{-9h}{-24 \times 24 + 9 \times 24h}}{h}$$

$$= \lim_{h \to 0} \frac{-9h}{h(-24 \times 24 + 9 \times 24h)}$$

$$= \frac{-9}{-24 \times 24} = \frac{1}{64}$$

Similarly,

$$\therefore f'_-\left(-\frac{4}{3}\right) = \lim_{h \to 0} \frac{f\left(-\frac{4}{3}-h\right) - f\left(\frac{-4}{3}\right)}{-h}$$

$$= \frac{1}{64}$$

$$f_+\left(-\frac{4}{3}\right) = f'_-\left(\frac{-4}{3}\right) = \frac{1}{64}$$

Hence, $f'(x)$, i.e; $\frac{dy}{dx}$ exists at $x = -\frac{4}{3}$

2. If $y = |x|$ examine the existance of $\frac{dy}{dx}$ at $x = 0$

Solution: $f(x) = |x|$

$$f'_+(0) = \lim_{h \to 0} \frac{f(0+h) - f(0)}{h}; h > 0$$

$$= \lim_{h \to 0} \frac{|h| - 0}{h}$$

$$= \lim_{h \to 0} \frac{h}{h}$$

$$= 1$$

$$f'_-(0) = \lim_{h \to 0} \frac{f(0-h) - f(0)}{-h}; (h > 0)$$

$$= \lim_{h \to 0} \frac{|-h| - 0}{h}$$

$$= \lim_{h \to 0} \frac{h}{-h}$$

$$= -1$$

$$f'_+(0) = f'_-(0)$$

Hence, $f(x) = |x|$ is not differentiable at $x = 0$ i.e; $\frac{dy}{dx}$ does not exist at $x = 0$

To find $f'(a)$ or some special types of functions:

Type 1: When a function is defined in the following way $y = f_1(x)$, $x \neq a$; $= c$, $x = a$ then for finding $f'(a)$ we have to calculate the derivative at $x = a$ directly from its defination.

Examples worked out:

1. If $y = \begin{cases} x^2 \sin\left(\frac{1}{x}\right) & \text{for } x \neq 0 \\ 0 \text{ for } x = 0 \end{cases}$

Solution: For $x \neq 0$ the derivative may be calculated by the formulas and the rules of differentiation (as y is the product of two differentiable functions)

$$\therefore f'(x) = \frac{d}{dx}\left(x^2 \sin\frac{1}{x}\right)$$

$$= \frac{dx^2}{dx} \cdot \sin\frac{1}{x} + x^2 \frac{d}{dx}\left(\sin\frac{1}{x}\right)$$

$$= 2x \cdot \sin\left(\frac{1}{x}\right) - \cos\left(\frac{1}{x}\right) \text{ for } x \neq 0$$

[we can not use the expression for $x = 0$. At the point $x = 0$, we can calculate the derivative using the defination of the derivative]

$$\therefore\ f_+'(0) = \lim_{h\to 0} \frac{f(0+h) - f(0)}{h},\ (h > 0)$$

$$= \lim_{h\to 0} h \cdot \sin\frac{1}{h} = 0$$

[$\because$ The product of an infinitesimal function and a bounded function is an infinitesimal]

$$f_-'(0) = \lim_{h\to 0} \frac{f(0-h) - f(0)}{-h},\ (h > 0)$$

$$= \lim_{h\to 0} h \sin\frac{1}{h} = 0$$

$$f_-'(0) = 0$$

Hence $f'(0) = 0$.

2. If $y = (x-1)^2 \cdot \sin\left(\frac{1}{x-1}\right)$, $x \neq 1$, $= 0$, when $x = 1$ Find $f'(1)$.

Solution: $f_+'(1) = \lim_{h\to 0} \frac{f(1+h) - f(1)}{h};\ (h > 0)$

$$= \lim_{h\to 0} \frac{(1+h-1)^2 \cdot \sin\left(\frac{1}{1+h-1}\right) - 0}{h}$$

$$= \lim_{h\to 0} \frac{h^2 \sin\left(\frac{1}{h}\right)}{h} = \lim_{h\to 0} h \sin\frac{1}{h} = 0$$

[$\because$ The product of an infinitesimal function and a bounded function is an infinitesimal]

Similarly, $f_-'(1) = 0$

$$\therefore\ f'(1) = 0$$

Type 2: When a function is defined on both sides of $x = a$ by different formulas in various intervals of its domains of defination and we are required to find $f'(x)$ and $f'(a)$, e.g.,

1. $f(x) = f_1(x)$, when $x < a$

$f(x) = f_2(x)$, when $x \geq a$ and we are required to find $f'(x)$ and $f'(a)$

Working rule:

1. Find $f_1'(x)$, $f_2'(x)$, ... etc. by using the formulas of d.c. of a power function, sum, difference, product, quotient etc and retain the same intervals (or, restrictions or, conditions) against $f_1'(x)$, $f_2'(x)$ ... etc. This provides us $f_1'(x)$, $f_2'(x)$, ... etc. in various given intervals.

2. Find left hand derivative $= f_-'(a) = L_1$ and right hand derivative $= f_+'(a) = L_2$ by using the defination.

3. If $L_1 = L_2$, then $f'(a) = L_1$ (or, L_2) [$\because L_1 = L_2$]

Examples worked out:

1. If $f(x) = 2x^2 + 3x$ when $x < 0$, $= 3x - x^2$, when $0 \leq x \leq 1$ $= x + 1$, when $x > 1$ Find $f'(x)$, $f'(0)$ and $f'(1)$.

Solution: (1) To find $f'(x)$

When $x < 0$,

$$f'(x) = \frac{d}{dx}\left[2x^2 + 3x\right] = 2\cdot 2x + 3\cdot 1 = 4x + 3 \quad ...(i)$$

When, $0 < x < 1$

$$f'(x) = \frac{d}{dx}\left[3x - x^2\right] = 3\cdot 1 - 2x = 3 - 2x \quad ...(ii)$$

When, $x > 1$,

$$f'(x) = \frac{d}{dx}(x+1) = 1 \quad ...(iii)$$

(2) It remains to find $f'(0)$ and $f'(1)$

$$f_-'(0) = \lim_{h\to 0} \frac{f(-h) - f(0)}{-h} = \lim_{h\to 0} \frac{(2h^2 - 3h) - 0}{-h}$$

$$= \lim_{h\to 0} (-2h + 3) = 3$$

$$f_+'(0) = \lim_{h\to 0^+} \frac{f(h) - f(0)}{h} = \lim_{h\to 0} \frac{(3h - h^2) - 0}{h}$$

$= \lim_{h \to 0} (3 - h) = 3$

$f'(0) = 3$ ...(iv)

$$f'_-(1) = \lim_{h \to 0} \frac{\left[3(1-h) - (1-h)^2\right] - (3-1)}{-h}$$

$$= \lim_{h \to 0} \frac{\left[3 - 3h - 1 - h^2 + 2h\right] - 2}{-h}$$

$$= \lim_{h \to 0} \frac{\left[3 - h - 1 - h^2\right] - 2}{-h}$$

$$= \lim_{h \to 0} \frac{\left[2 - h - h^2\right] - 2}{-h}$$

$$= \lim_{h \to 0} \frac{-h(1+h)}{-h} = \lim_{h \to 0} (1+h) = 1$$

Again $f'_+(1) = \lim_{h \to 0^+} \dfrac{f(1+h) - f(1)}{h}$

$$= \lim_{h \to 0} \frac{[(1+h)+1] - [3-1]}{h}$$

$$= \lim_{h \to 0} \frac{h}{h}$$

$$= \lim_{h \to 0} 1$$

$= 1$

$f'_-(1) + f'_+(1) = 1$

$f'(1) = 1$ ...(v)

Hence $f'(0) = 3$ and $f'(1) = 1$

Type 3: To examine the existance of a derived function $f'(x)$ at a point at which the given function $f(x)$ is undefined.

Working rule: Regarding derivative $f'(x)$ at a point $x = a$ at which the given function $f(x)$ is undefined $\left(\text{i.e; } \frac{a}{0}, \frac{0}{0}, \frac{\infty}{\infty}, \text{ etc.}\right)$ we use the following facts:

A function $f(x)$ is undefined at a point $x = a$

$\Rightarrow$ The function $f(x)$ is discontinuous at $x = a$

$\Rightarrow$ The function $f(x)$ is not differentiable at the same point $x = a$

$\Rightarrow$ $f'(x)$ does not exist at the same point $x = a$

$\Rightarrow$ $f'(x)$ can not be obtained at $x = a$ finitely.

Note:

1. We should remember that evaluating indeterminate form means finding its limits or showing that its limit does not exist.

2. The derivative is not defined at the points of discontinuities of the function.

3. When the function $y = f(x)$ is not defined for (or, at or when) $x = a$, this function $f(x)$ can not be said to take any value at $x = a$. Therefore its derivative $f'(x)$ also can not be said to take any value at $x = a$

4. The function $\sin\left(\frac{1}{x}\right)$, $\cos\left(\frac{1}{x}\right)$ and $\tan\left(\frac{1}{x}\right)$ are undefined at $x = 0$

Examples worked out:

1. If $f(x) = \dfrac{1 - \cos x}{1 - \sin x}$, does $f'\left(\frac{\pi}{2}\right)$ exist ?

Solution: $\because f(x) = \dfrac{1 - \cos x}{1 - \sin x}$

$$\therefore f\left(\frac{\pi}{2}\right) = \frac{1 - \cos\left(\frac{\pi}{2}\right)}{1 - \sin\left(\frac{\pi}{2}\right)} = \frac{1-0}{1-1} = \frac{1}{0} =$$

undefined

$f(x)$ is undefined at $x = \dfrac{\pi}{2}$

$\Rightarrow$ $f(x)$ is discontinuous at $x = \dfrac{\pi}{2}$

$\Rightarrow$ $f'(x)$ does not exist at $x = \dfrac{\pi}{2}$

2. If $y = \dfrac{x-4}{2\sqrt{x}}$, does $\dfrac{dy}{dx}$ at $x = 0$ exist ?

Solution: Let $f(x) = \dfrac{x - 4}{2\sqrt{x}}$

$$f(0) = \frac{0-4}{2\sqrt{0}} = \frac{-4}{0} = \text{undefined.}$$

$\therefore f(x)$ is undefined at $x = 0$

$\Rightarrow$ $f(x)$ is discontinuous at $x = 0$

$\Rightarrow$ $f'(x)$ does not exist at $x = 0$

Type (1), (2), (3) and (4)

Problems based on evalution of $\dfrac{dy}{dx}$ at $x = a$

Exercise 15.1

Find $\dfrac{dy}{dx}$ at the indicated points if

1. $y = \dfrac{1}{x+5}$ at $x = 2$
2. $y = \dfrac{1}{x^2+3}$ at $x = 1$
3. $y = \left(x^3 + x + 1\right)^5$ at $x = -1$
4. $y = \cos^3 x$ at $x = c$
5. $y = x^3 - 2x^2 y + 4xy^2 - 8xy + 6x - 3$ at $y = 2$
6. $y = x^3 y + xy^3 - x$ at $x = 1$
7. $y = \cos\left(2x + \dfrac{\pi}{2}\right)$ at $x = \dfrac{\pi}{3}$
8. $y = \cos\left(\sin x^2\right)$ at $x = \sqrt{\dfrac{\pi}{2}}$
9. $y = \sqrt{\sin^4 x^2 + \cos^4 x^2}$ at $x = 0, \sqrt{\dfrac{\pi}{2}}$
10. $y = \left(\operatorname{cosec}^2 x + \tan x + \cot x\right)^7$ at $x = \dfrac{\pi}{4}$
11. $y = \sec^4 x - \tan^4 x$ at $x = \dfrac{\pi}{3}$
12. $y = 5\sin x$ at $x = \dfrac{\pi}{2}$
13. $y = 2 + \tan x$ at $x = t$
14. $y = \cos^{-1} x - \cot^{-1} x$ at $x = \dfrac{\sqrt{3}}{2}$
15. $y = \sqrt{2x^3} + \sqrt{\dfrac{2}{x^3}} - 4$ at $x = 4$
16. $y = \dfrac{x + \cos x}{1 + \cos x}$ at $x = 0$
17. $y = \dfrac{\log\left(xe^x\right)}{x + e^x}$ at $x = 1$
18. $y = \dfrac{2 + \sqrt{x}}{2 - \sqrt{x}} \cdot e^x$ at $x = 1$
19. $y = (\tan x)^{\sin x}$ at $x = \dfrac{\pi}{4}$
20. $y = \dfrac{(4x+1)^{\frac{3}{4}}}{\sqrt{2+5x} \cdot (2x-3)^{\frac{5}{3}}}$ at $x = 12$
21. $y = \sqrt{2x} - \sqrt{\dfrac{2}{x}} + \dfrac{x+4}{4-x}$ at $x = 2$
22. $y = \sqrt{4x} - \sqrt{\dfrac{4}{x}} + \dfrac{4+\sqrt{x}}{4-\sqrt{x}}$ at $x = 4$
23. $y = \tan^{-1} \dfrac{\cos x}{1 + \sin x}$ at $x = 0$
24. $y = \sin^{-1}\left\{x \cdot \sqrt{1 - x^2}\right\}$ at $x = \dfrac{1}{2}$
25. $y = \tan^{-1} \dfrac{3x - x^3}{1 - 3x^2}$ at $x = 1$
26. $\sqrt{x} + \sqrt{y} = 3$ at $x = 1, y = 4$
27. $x^2 + y^2 = 10$ at $(1, 3)$

28. $x^3 - 3axy + y^3 = 0$ at $x = \frac{3a}{2}, y = \frac{3a}{2}$

29. Find $\frac{dy}{da}$ if $y = \frac{1}{\cos x + x\cos a}$ at $a = \frac{\pi}{2}$

30. If $y = |\cos x|$, find $\frac{dy}{dx}$ at $x = \pi$

31. If $y = |\sin x|$, find $\frac{dy}{dx}$ at $x = 0$

32. If $f(x) = 3\tan^{-1} x - 2\cot^{-1} x$, find $f'(2)$

33. $f(x) = 2\sin^{-1} x + \cos^{-1} x$, find $f'\left(\frac{\sqrt{2}}{2}\right)$

34. $f(x) = 4\sin^{-1} x + \cos^{-1} x$, find $f'\left(\frac{1}{2}\right)$

35. If $f(x) = \tan^{-1} x$, find $f'(\sqrt{3})$

Answers:

1. $-\frac{1}{49}$ 2. $-\frac{1}{8}$ 3. 20 4. $-3\cos^2 c \sin c$
5. Find 6. -1 7. 1 8. 0 9. 0,0 10. -114688
11. $16\sqrt{3}$ 12. 0 13. $\sec^2 t$ 14. $-\frac{10}{7}$
15. $\frac{189}{64}\cdot\sqrt{2}$ 16. $\frac{1}{2}$ 17. $\frac{1}{1+e}$ 18. $5e$
19. $\sqrt{2}$ 20. $-\frac{77}{16}$ 21. $\frac{11}{4}$ 22. $\frac{9}{8}$ 23. $-\frac{1}{2}$
24. $\frac{4}{\sqrt{39}}$ 25. $\frac{3}{2}$ 26. -2 27. $-\frac{1}{3}$ 28. $\frac{4}{5}$
29. $x\sec^2 x$ 30. 0 31. does not exist 32. 1
33. $\sqrt{2}$ 34. $2\sqrt{3}$ 35. $\frac{1}{4}$

Type 5: To find the value of d.c of a given function w.r.t another given function at $x = a$

Exercise 15.2

1. Find differentiation of

$y = \sqrt{x^2 + 16}$ w.r.t $\frac{x}{x-1}$ at $x = 3$

2. Find differentiation of

$y = \sec^{-1}\left(\frac{1}{2x^2 - 1}\right)$ w.r.t $\sqrt{1 - x^2}$ at $x = \frac{1}{2}$.

Answers:

1. $-\frac{12}{5}$ 2. 4

Type 6: Evalution of d.c. of parametric equations (or, functions) at a point (or, parameter t, θ or θ etc)

Exercise 15.3

Find $\frac{dy}{dx}$ for the parametric equations at the indicated points.

1. $\left.\begin{array}{l} x = a\left(\cos\theta + \log\tan\frac{\theta}{2}\right) \\ y = a\sin\theta \end{array}\right\}$ at $\theta = \frac{\pi}{3}$

2. $\left.\begin{array}{l} x = 2\cos\theta - \cos 2\theta \\ y = 2\sin\theta - \sin 2\theta \end{array}\right\}$ at $\theta = \frac{\pi}{2}$

3. $\left.\begin{array}{l} x = a(\cos\theta + \theta\sin\theta) \\ y = a(\sin\theta - \theta\cos\theta) \end{array}\right\}$ at $\theta = \frac{\pi}{4}$

4. $\left.\begin{array}{l} x = \frac{3at}{1+t^3} \\ y = \frac{3at^2}{1+t^3} \end{array}\right\}$ at $= \frac{1}{2}$

5. $\left.\begin{array}{l} x = 2\cos\theta - 2\cos^3\theta \\ y = 2\sin\theta - 2\sin^3\theta \end{array}\right\}$ at $\theta = \frac{\pi}{4}$

6. $\left.\begin{array}{l} x = a\cos^3\theta \\ y = a\sin^3\theta \end{array}\right\}$ at $\theta = \frac{\pi}{4}$

7. $\left.\begin{array}{l} x = a(\cos t + t\sin t) \\ y = a(\sin t - t\cos t) \end{array}\right\}$ at $t = \dfrac{\pi}{3}$

8. $\left.\begin{array}{l} x = a(t + \sin t) \\ y = a(1 - \cos t) \end{array}\right\}$ at $t = \dfrac{\pi}{2}$

9. $\left.\begin{array}{l} x = a(\theta - \sin\theta) \\ y = a(1 - \cos\theta) \end{array}\right\}$ at $\theta = \dfrac{\pi}{2}$

10. $\left.\begin{array}{l} x = a(1 - \cos t) \\ y = a(t + \sin t) \end{array}\right\}$ at $t = \dfrac{\pi}{2}$

11. $\left.\begin{array}{l} x = at^2 \\ y = 2at \end{array}\right\}$ at $t = 2$

Answers:

1. $\sqrt{3}$ 2. -1 3. 1 4. $\dfrac{5}{4}$ 5. Find 6. -1

7. $\sqrt{3}$ 8. 1 9. 1 10. 1 11. $\dfrac{1}{2}$

Miscellaneous problems on evaluation:

Exercise 15.4

1. If $y = \dfrac{x+1}{x-1}$, evaluate $f'(1)$ if possible.

2. If $f(x) = \dfrac{x^3+1}{x}$, evaluate $f'(1)$

3. If $f(x) = \dfrac{x^3+1}{x^2+1}$, evaluate $f'(0)$

4. If $f(x) = \dfrac{1}{x^2+3x+2}$, evaluate $f'(0)$

5. If $s = \sqrt{t} + \sqrt[3]{t}$, evaluate $f'(4)$

6. If $y = \dfrac{2}{\sqrt{x}} - \dfrac{3}{\sqrt[3]{x}}$, evaluate $f'(4)$

7. If $f(x) = \dfrac{\sqrt{x}+1}{\sqrt{x}}$, evaluate $f'(4)$

8. If $f(x) = \dfrac{4+\sqrt{x}}{4-\sqrt{x}}$, evaluate $f'(1)$

9. If $f(t) = \left(t^m + t^n\right)^3$, evaluate $f'(1)$

10. If $y = \left[\dfrac{x-1}{x+1}\right]^2$, evaluate $f'\left(\sqrt{3}\right)$

11. If $y = \dfrac{3x^2}{(3x-1)^3}$, evaluate $f'\left(\sqrt{3}\right)$

12. If $f(x) = \left(x^2-1\right)\cdot\sqrt{x^2+1}$, evaluate $f'\left(\sqrt{3}\right)$

13. If $f(u) = \sqrt{2+\sqrt{2u}}$, compute $f'(2)$

14. If $f(x) = \sqrt{5x^2+2x+1}$, compute $f'(-1)$

15. If $y = \sqrt{\dfrac{1+ax}{1-ax}}$, compute $f'\left(\sqrt{5}\right)$

16. If $f(z) = \dfrac{\sqrt{4+z^2}}{z}$, compute $f'\left(\sqrt{5}\right)$

17. If $f(x) = \dfrac{x^3}{\sqrt{8+x^3}}$, compute $f'(1)$

18. If $f(x) = \sqrt{\dfrac{\sqrt{x}-1}{\sqrt{x}-1}}$, evaluate $f'(4)$

19. If $f(x) = \dfrac{x}{x+\sqrt{1+x^2}}$, evaluate $f'\left(\sqrt{3}\right)$

20. If $f(x) = \dfrac{2}{x} - \dfrac{8}{\sqrt{x}} + \dfrac{6}{\sqrt[3]{x^2}} + 2x + 6x^2\sqrt{x}$,

find $f'(1)$

21. If $f(x) = \left(x^2-2\right)\sqrt{x^2+1}$, find $f'\left(\sqrt{3}\right)$

22. If $f(Z) = \dfrac{9Z}{\sqrt{Z^2+1}}$, find $f'\left(2\sqrt{2}\right)$

23. If $f(x) = \dfrac{1}{x^2} + \dfrac{3}{2\sqrt[3]{x^2}} - \dfrac{4}{\sqrt{x}} + 3x - 2x^2\sqrt{x}$,

find $f'(1)$

24. If $f(u) = (u^2+3)\cdot\sqrt{u^2-1}$, find $f'(\sqrt{2})$

25. If $f(x) = \dfrac{x}{1-\sqrt{x^2+1}}$, find $f'(\sqrt{3})$

26. If $f(x) = e^{2x}\log x^2$, find $f'(1)$

27. If $f(x) = \sqrt{e^x}\cdot\log x^2$, find $f'(1)$

Answer:

1. Does not exist **2.** 1 **3.** 0 **4.** $-\dfrac{3}{4}$

5. $\left[\left(\dfrac{1}{2}\right)t^{-\frac{1}{2}} + \left(\dfrac{1}{3}\right)t^{-\frac{2}{3}}\right]_{t=4}$

6. $\left[x^{-\frac{4}{3}} - x^{-\frac{3}{2}}\right]_{x=4}$

7. $-\dfrac{1}{16}$ **8.** $\dfrac{4}{9}$ **9.** $12(m+n)$

10. $\left[\dfrac{4(x-1)}{(x+1)^3}\right]_{x=\sqrt{3}}$ **11.** $\left[\dfrac{-9x^2}{(3x-1)^3}\right]_{x=\sqrt{3}}$

12. $18\sqrt{3}$ **13.** $\dfrac{1}{8}$ **14.** -2

15. $\left[\dfrac{a(1-ax)^{-\frac{3}{2}}}{(1+ax)^{-\frac{1}{2}}}\right]_{x=\sqrt{5}}$ **16.** $-\dfrac{4}{5}$

17. $\dfrac{17}{18}$ **18.** $\dfrac{\sqrt{3}}{36}$ **19.** $\dfrac{7-4\sqrt{3}}{2}$ **20.** 15

21. $\dfrac{9\sqrt{3}}{2}$ **22.** $\dfrac{1}{3}$ **23.** -3 **24.** $7\sqrt{2}$

25. $\dfrac{1}{2}$ **26.** $2e^2$ **27.** $2\sqrt{e}$

Special types of functions:

Type 1: Problems based on finding the derivative of function having the form:

$y = f_1(x)$, when $x \neq a$
$= c$, when $x = a$

Exercise 15.5

1. If $f(x) = e^{-\frac{1}{x^2}}\cdot\sin\left(\dfrac{1}{x}\right)$, when $x \neq 0$, $f(0) = 0$; find $f'(x)$ at $x = 0$ and $x \neq 0$.

2. If $f(x) = \sin\left(\dfrac{1}{x}\right)$, $x \neq 0$, $f(0) = 0$ find $f'(x)$ at $x = 0$ and $x \neq 0$

3. If $f(x) = (x-1)^2\sin\left(\dfrac{1}{x-1}\right)$, when $x \neq 1$, $= 0$ when $x = 1$; find $f'(x)$ at $x = 1$ and at $x \neq 1$

4. If $f(x) = \dfrac{1}{x}\cdot\sin 2x$ for $x \neq 0$, $= 1$ for $x = 0$ find $f'(x)$ at $x = 0$ and at $x \neq 0$

5. If $f(x) = x\sin\left(\dfrac{1}{x}\right)$, $x \neq 0$, $f(0) = 0$; find $f'(x)$ at $x = 0$ if derivative of $f(x)$ exists and $f'(x)$ at $x \neq 0$

6. If $f(x) = \dfrac{1}{1-e^{-\frac{1}{x}}}$, $x \neq 0$, $f(0) = 0$; find $f'(x)$ at $x = 0$ and $f'(x)$ at $x \neq 0$

7. If $f(x) = e^{-\frac{1}{x}}$, when $x \neq 0$, $f(0) = 0$; find $f'(x)$ at $x = 0$ and $f'(x)$ at $x \neq 0$

8. If $f(x) = x\cos\left(\dfrac{1}{x}\right)$, for $x \neq 0$, $f(0) = 0$; find $f'(x)$ at $x = 0$ and $f'(x)$ when $x \neq 0$

Type 2: Problems based on the function defined on both sides of $x = a$ by various formulas in different intervals of its domains of definition and we are required to find $f'(x)$ as well as $f'(a)$. Exercise set:

Exercise 15.6

1. If $f(x) = 3x - 4$, when $x \leq 2$
$f(x) = 2(2x-3)$, when $x > 2$, Find $f'(2)$

2. If $f(x) = 2x - 1,\ 0 \le x < 1$

$f(x) = 2 - x,\ 1 \le x \le 3$ Find $f'(1)$

3. If $f(x) = \dfrac{x}{1+x}$, when $x \ge 0$

$f(x) = \dfrac{x}{1-x}$, when $x < 0$ Find $f'(0)$

4. If $f(x) = 2x^2 + 7x$, when $x < 0$

$f(x) = x^3 - 4$, when $x > 0$ Find $f'(-1)$.

Answers:

1. Does not exist
2. Does not exist
3. 1 4. 4

Type 3: Problems based on finding the value of d.c of the function which is undefined at a point $x = a$

1. If $y = \dfrac{x-4}{2\sqrt{x}}$, can the value of $\dfrac{dy}{dx}$ be obtained at $x = 0$?

2. If $y = \dfrac{1-x}{1+x}$, can the value of $\dfrac{dy}{dx}$ be obtained at $x = -1$?

3. If $y = \dfrac{2x-3}{3x+4}$ can the value of $\dfrac{dy}{dx}$ be obtained at $-\dfrac{4}{3}$?

4. If $f(x) = \dfrac{|x|}{x}$, can the value of $\dfrac{dy}{dx}$ be obtained at $x = 0$?

5. Prove that the function $f(x) = \dfrac{\sin x}{1 - \cos x}$ has no derivative at $x = 0$

6. Prove that the function $f(x) = \dfrac{1}{\cos x} - \tan x$ has no derivative at $x = \dfrac{\pi}{2}$.

7. Prove that the function $f(x) = \dfrac{x^2 - 3x + 5}{2x^2 + 5x - 3}$ has no derivative at $x = -3$.

8. Prove that the following functions have no derivative at the indicated points.

(i) $f(x) = \sin\dfrac{1}{x}$ at $x = 0$

(ii) $f(x) = \cos\dfrac{1}{x}$ at $x = 0$

(iii) $f(x) = \tan\dfrac{1}{x}$ at $x = 0$

Answers:

1. No, as the function is not defined at the indicated point.
2. No, as the function is not defined at the indicated point.
3. No, as the function is not defined at the indicated point.
4. No, as the function is not defined at the indicated point.
5. No, as the function is not defined at the indicated point.
6. No, as the function is not defined at the indicated point.

17

Derivative as Rate Measurer

General definition: If a variable z is a function of another variable y, then the rate of change of z with respect y is $\frac{dz}{dy}$.

That is, $z = f(y) \Rightarrow \frac{dz}{dy} = \frac{d\,f(y)}{dy}$ which is known as rate of change of z with respect to y or rate at which z changes with y.

A case of great practical importance occurs when the independent variable represents time.

That is, we may have $y = f(t)$ and we wish to find the rate of change of y with respect to time 't'. It is sufficient to calculate $\frac{dy}{dt} = f'(t)$ directly.

Again, we may have $y = f(x)$ and $x = g(t)$. This is often the case in problems of physics. If we have been given $\frac{dx}{dt}$, th time rate of change of x, we can calculate $\frac{dy}{dx}$ from the formula.

$$\frac{dy}{dt} = \frac{dy}{dx} \cdot \frac{dx}{dt} = f'(x) \cdot \frac{dx}{dt}$$

Moreover, we may have $y = g(t), x = f(t)$.

Then $\frac{dy}{dx} = \frac{\frac{dy}{dt}}{\frac{dx}{dt}} = \frac{g'(t)}{f'(t)}, f'(t) \neq 0$

Thus, the rate of change of one variable can be calculated if the rate of change of the other variable is known.

Notes:

1. If $\frac{dy}{dx}$ is positive, then the rate of change of y with respect to x is positive. This means that if x increase, then y also increases and if x decreases, then y also decreases.

2. ·If $\frac{dy}{dx}$ is negative, then the rate of change of y with respect x is negative. This means that if x increases, then y decreases and if x decreases, then y increases.

Remember:

1. The pharase "rate of change or rate of variation or rate of increase" of a variable quantity is often used in reference to 'time' and the words "with respect to 't'" are omitted. This is why when no special mention is made of the variable with respect to which the rate is calculated, it is assumed that the rate is taken w.r.t time 't'.

2. By the rate of change or rate of variation or rate of increase of a variable quantity is meant the change in the value of a quantity per unit of time.

3. Whatever be the quantity 'Q' its derivative $\frac{dQ}{dt}$ gives how fast Q is changing with 't'.

4. If the quantity Q increases with time 't', its derivative $\frac{dQ}{dt}$ is positive.

5. One important point to be remembered is to use the same units of measurements for the same variable.

If in a problem, some distances are given in feet and others in inches, we may convert them into inches. If the rates are given as feet per minute as well as feet per second, then we may convert them all into feet per second.

Examples

Units of rate $\frac{dQ}{dt}$ = units of 'Q' / units of 't'.

Where Q = volume, area, distance, … etc.
T = time

Units of $\frac{dr}{dt}$ (= rate increase of radius w.r.t time 't' where r = radius) is cm/sec, (r is in cm and t is in secs).

Units of $\frac{dS}{dt}$ (=rate of increase of S w.r.t time 't' where S = surface) is cm^2/sec, (if distance is in cm and time is in secs) and Units of $\frac{dv}{dt}$ (= rate of increase of volume w.r.t time 't' where v = volume) is cm^3/sec.

6. When one quantity changes or grows or varies, we like to know the rate of change or rate of growth or rate of variation of another related quantity. This is why it is some times called related rates. We are concerned with the rate of changes of one quantity relative to other with which it is connected by some given relation.

7. If y is a function of x and x is varying with time 't', then $\frac{dy}{dt} = \frac{dy}{dx} \cdot \frac{dx}{dt}$ which $\Rightarrow$ rate of change of y = rate of change of x times d.c of y w.r.t x. which in turn can be written as $\frac{dy}{dx} = \frac{\text{rate of chage of } y}{\text{rate of change of } x} = \frac{dy/dt}{dy/dt}$

which means that $\frac{dy}{dx}$ compares the rate of change of both y and x.

8. If the rate of change of a quantity is not w.r.t time, it must be mentioned in the problem (question). e.g.,

(i) If $y^2 = 4x$, find the rate at which y is changing with respect to x when $x = 4$.

Solution: $y^2 = 4x$

$$\Rightarrow 2y\frac{dy}{dx} = 4$$

$$\Rightarrow \frac{dy}{dx} = \frac{4}{2y} = \frac{2}{y}$$

Now, from $y^2 = 4x$, when $x = 4$, $y^2 = 16$ or $y = \pm 4$

which $\Rightarrow$ the rate = value of $\frac{dy}{dx}$ (at the point $x = 4$),

i.e., (4, 4); (4, –4) $= \frac{1}{2}; -\frac{1}{2}$ (The negative sign shown y is decreasing with increase in x at (4, – 4).

(ii) The radius of a spherical soap bubble is uniformly increasing. Find the rate at which the volume of the bubble is increasing with radius, when its radius 'r' is 8 cm.

Solution: Volume 'v' of a sphere of a radius 'r' is given by $v = \frac{4}{3}\pi r^3$.

Rate of increase with respect to 'r' is given by

$$\frac{dv}{dr} = 4\pi r^2.$$

$$\left[\frac{dv}{dr}\right]_{r=8} = 4\pi\, 8^2 = 256\pi \text{ cm}^3/\text{cm}$$

Note:

1. Increases with, varies with or changes with means varies increase or changes w.r.t.

2. A quantity 'Q' increase with uniform rate

$$\Rightarrow \frac{dq}{dt} = k = \text{constant.}$$

3. If x and y are two variables, then the rate of change of y with respect to x at (or, when) $x = a$ means the value of $\frac{dy}{dx}$ at $x = a = \left[\frac{dy}{dx}\right]_{x=a} = [f'(x)] = f'(a)$.

4. Change in a quantity, a variable or a variable quantity = final value of the quantity (or, variable) – initial value of the quantity (or, variable).

5. If $\frac{dy}{dt} = \frac{dy}{dx} \cdot \frac{dx}{dt}$, then we can find $\frac{dy}{dx}$ provided $\frac{dy}{dt}$ and $\frac{dx}{dt}$ are known or we can find $\frac{dx}{dt}$ provided $\frac{dy}{dt}$ and $\frac{dy}{dx}$ are known.

On Language of Mathematics

1. Language of rate problems:

$\frac{dr}{dt} = a$ cm/sec means rate of increase of radius is a cm/sec where r = radius.

$\frac{dv}{dt} = a$ cm^3/sec means rate of increase (or change or variation or growth ... etc) of volume is a cm^3/sec where v = volume.

$\frac{dS}{dt} = a$ cm^2/sec means rate of increase (or change or variation or growth ... etc) of surface area or simply area is a cm^2/sec where S = surface.

... and so on

2. Language of variation:

(i) *Direct variation*: When the ratio of two variables equals a constant, the variables are said to vary directly.

If we let y and x represent any two variables, then we may state that they vary directly by writing $\frac{y}{x} = k$ where k represents a constant.

Thus, $\frac{y}{x} = k$ is also symbolised as

$$y \propto x \Leftrightarrow y = kx.$$

(ii) *Inverse variation*: When the product of two variables equals a constant, then the variables are said to vary inversely.

If we let x and y represent any two variables, we may state that they vary inversely by writing $x\,y = k$, where k represents a constant.

The relationship may also be expressed as $y = \frac{k}{x}$ or $x = \frac{k}{y}$ or alternatively we may say

y is inversely proportional to x i.e;

$y \propto \frac{1}{x} \Leftrightarrow y = \frac{k}{x}$ or x is inversely proportional to y i.e: $x \propto \frac{1}{y} \Leftrightarrow x = \frac{k}{y}$.

N.B.: We can always replace the sign '$\propto$' by the sign of equality '=' provided we introduce a multiplying constant on one side of the equation. This constant is often termed as constant or proportionality. The symbol '$\propto$' is termed as sign of variation, moreover, instead of 'is proportional to' or 'varies as' the symbol '$\propto$' if often used.

3. Joint variation: When a variable varies directly as the product of two or more than two variables, it is said to vary jointly as these variables.

If x varies jointly as y and z, then $x \propto y \cdot z$ or $x = k \cdot y \cdot z$ where k is any constant.

4. $\frac{dQ}{dt}$ is constant = Q increases or decreases with a constant rate respectively where Q = any quantity and t = time.

5. $\frac{dQ}{dt}$ tells us at what rate a physical quantity Q changes or increases or decreases if Q be a certain quantity varying with time.

Or, alternatively, if Q be a certain quantity varying with time $\frac{dQ}{dt}$ represents the rate at which that quantity Q is changing.

Type 1: Problems based on finding the rate of physical quantities like volume, area, perimeter etc.

Working rule: In such problems where time rate of change (rate of change w.r.t. time) of certain variable (variables) is (are) given and time rate of change of some other variable (variables) is (are) to be found out, we use the rule described in following four steps:

1. Find the relation by mensuration formulas (between the two variables) between that quantity whose rate of change is required and whose rate of change w.r.t time is given. The following relations are very helpful.

A = area of a square = x^2, its perimeter = $4x$

A = area of rectangle = $x\,y$, its perimeter = $2\,(x + y)$

Area of trapezium $= \frac{1}{2}$ (sum of parallel sides). $\times$ distance between them.

A = area of a circle = πr^2, its perimeter = $2\pi r$.

V = volume of right cone = $\frac{1}{3}\pi r^2 h$, its total surface = $\pi r(r + l)$ whereas its curved surface is $\pi r l$ only.

V = volume of a cylinder = $\pi r^2 h$, its total surface = $2\pi r(r + h)$. Whereas its curved surface is = $2\pi r h$.

V = volume of a box = $x\,y\,z$ and its surface or surface area = $2\,(xy + yz + zx)$

V = volume of a cube = x^3.

2. Differentiate the equation (between the quantities whose rate of change is required and whose rate of change w.r.t. time is given which exists at any instant or time during which the condition of the problem holds) w.r.t time 't'. Generally we differentiate any one of the formula of area, volume or perimeter etc of a substance having a geometrical shape like square, rectangle, circle cylinder or cone etc with respect to time 't' provided that time rate of change of a certain variable is given in the problem.

3. Substitute the known quantities in the differentiated result.

4. Solve for the required unknown.

N.B.: **1.** We may draw the figure for convenience.

2. Formulas for volumes, areas, perimeters of a substances having geometrical regular shape like square, rectangle, trapezium, circle, cone, cylinder or cube etc must be remembered.

3. While working out problems of rate measurer, we must note the two variables, one whose rate of change is given and the other one whose rate of change is to be found. Then we express either variable in terms of the other by means of an equation. (or known formula) and lastly we differentiate through out with respect to time t.

Worked out

Problems based on type (1)

1. At what rate is the area increasing when the side of an equilateral triangle is 10 ft, if the side of an equilateral triangle increases uniformly at the rate of 3 ft/sec.

Solution: Let A = area of an equilateral triangle.

$$= \frac{\sqrt{3}\,x^2}{4}, \ (x = \text{length of a side}) \qquad ...(1)$$

Now, differentiating both sides of (1) w.r.t 't', we get

$$\frac{dA}{dt} = \frac{\sqrt{3}}{4} \cdot 2x \cdot \frac{dx}{dt}$$

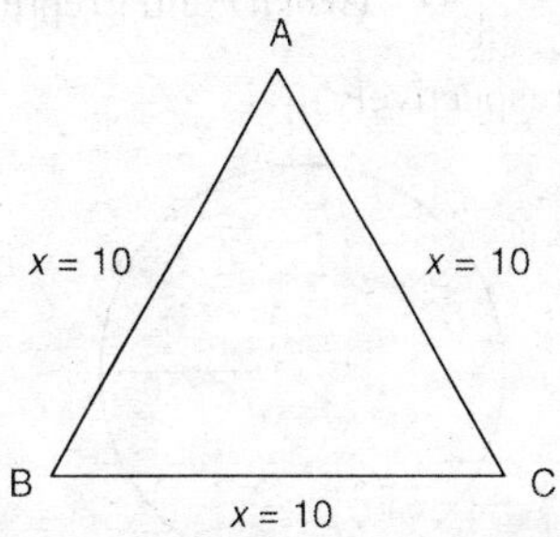

$$= \frac{\sqrt{3}}{4} \cdot 2x \cdot 3 \left(\because \frac{dx}{dt} = 3 \text{ is given}\right)$$

$$\Rightarrow \left[\frac{dA}{dt}\right]_{x=10} = \frac{\sqrt{3}}{4} \cdot 2 \times 10 \times 3$$

$$= 15\sqrt{3} \text{ sq. ft/sec.}$$

2. A spherical balloon is inflated and the radius is increasing at $\frac{1}{3}$ inches/minutes. At what rate would the volume be increasing at the instant when its radius is 2 inches.

Solution: let V = volume of a spherical balloon

$$= \frac{4}{3}\pi r^3, (\text{radius} = r). \qquad ...(1)$$

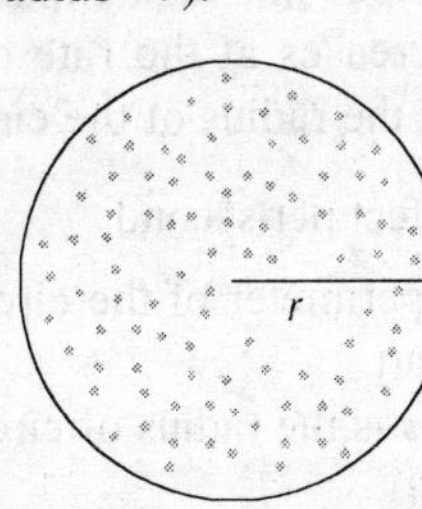

Now, differentiating (1) w.r.t 't' we get

$$\frac{dv}{dt} = \frac{4}{3}\pi \cdot 3r^2 \cdot \frac{dr}{dt}$$

$$= \frac{4\pi}{3} \cdot 3 \cdot r^2 \cdot \frac{1}{3} \left(\because \frac{dr}{dt} = \frac{1}{3} \text{ is given}\right)$$

$$\Rightarrow \left[\frac{dv}{dt}\right]_{r=2} = \left[\frac{4\pi}{3} \cdot r^2\right]_{r=2} = \frac{16\pi}{3} \text{ inch}^3\text{/minute.}$$

3. A spherical balloon is pumped at the rate of 10 cubic inches per minute. Find the rate of increase of its radius when its radius is 15 inches.

Solution: $v = \frac{4}{3}\pi r^3$ (when r and v represent radius and volume respectively)

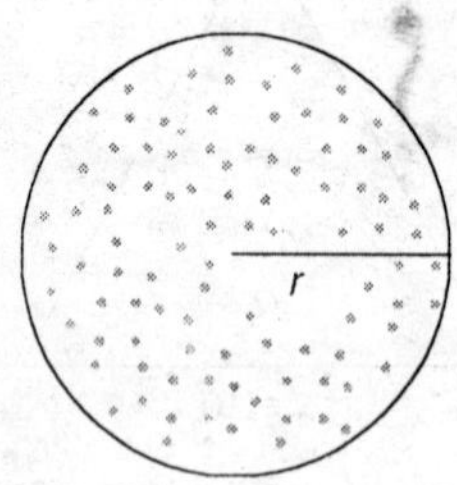

$$\Rightarrow \frac{dv}{dt} = \frac{d}{dt}\left(\frac{4}{3}\pi r^3\right) = \frac{4}{3} \times \pi \times 3 \times r^2 \times \frac{dr}{dt}$$

$$\Rightarrow 10 = 4\pi r^2 \cdot \frac{dr}{dt} \left(\because \frac{dr}{dt} = 10\right)$$

$$\Rightarrow \frac{dr}{dt} = \frac{10}{4 \times \pi \times r^2}$$

$$\Rightarrow \left[\frac{dr}{dt}\right]_{r=15} \frac{10}{4 \times \pi \times (15)^2} = \frac{10}{900\pi} = \frac{1}{90\pi} \text{ inch/}$$

minute.

4. The circular waves in a tank expand so that the circumference increases at the rate of a feet per second. Show that the radius of the circle increases at the rate of $\frac{a}{2\pi}$ feet per second.

Solution: let p = perimeter of the circular wave at time 't' from the start.

$= 2\pi r$ (where r is the radius of circular wave at time t from the start)

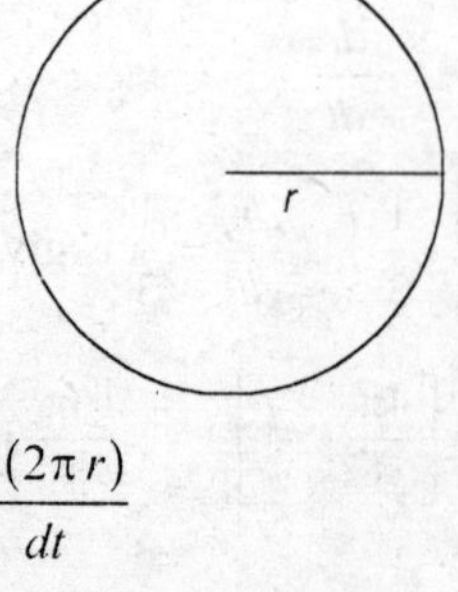

$$\Rightarrow \frac{dP}{dt} = \frac{d(2\pi r)}{dt}$$

$$\Rightarrow a = 2\pi \frac{dr}{dt} \left(\because \frac{dP}{dt} = a\right)$$

$$\Rightarrow \frac{dr}{dt} = \frac{a}{2\pi} \text{ ft/sec.}$$

5. A cylindrical vessel is held with its axis vertical. Water is poured into it at the rate of one unit per second. Given that one unit is equal to 34.66 cu. Inch. Find the rate at which the surface of water is rising in the vessel when the depth is x inches.

Solution: let A = are of cross section of cylindrical vessel.

x = height of cylindrical vessel.

V = volume = $A \cdot x$

$$\Rightarrow \frac{dv}{dt} = \frac{d}{dt}(A \cdot x)$$

$$\Rightarrow \frac{dv}{dt} = A \cdot \frac{dx}{dt}$$

$$\Rightarrow \frac{dx}{dt} = \frac{34.66}{A}$$

$$\left(\because \text{Given } \frac{dv}{dt} = 34.66 \text{ cu. inch/sec}\right)$$

$$\Rightarrow \frac{dx}{dt} = \frac{34.66}{A} \text{ inch/sec.}$$

$\left(\frac{dx}{dt} = \text{rate of change of } x = \text{rate of rising of the surface of water in the vessel}\right).$

6. A balloon which always remains spherical has a variable radius. Find the rate at which its volume is increasing with radius, when the latter is 7 cm.

Solution: Let the radius of the balloon = r and the volume of the balloon = v

$$\therefore v = \frac{4}{3}\pi r^3$$

rate of increase in volume with radius r

$$= \frac{dv}{dr} = \frac{4}{3} \times \pi \times 3r^2$$

$$\therefore \left[\frac{dv}{dr}\right]_{r=7} = 4 \times \pi \times (7)^2 = 196\pi \text{ cm}^3\text{/cm.}$$

7. The volume v and the pressure p of a gas under constant temperature are connected by $pv = c$, where c is a constant, show that $\frac{dp}{dv} = \frac{-c}{v^2}$.

Solution: $\because pv = e$

$$\therefore p \cdot 1 + v \cdot \frac{dp}{dv} = 0$$

$$\Rightarrow \frac{dp}{dv} = -\frac{p}{v} = -\frac{\left(\frac{c}{v}\right)}{v} = -\frac{c}{v^2} \text{ (proved)}$$

8. A sphere of metal is expanding under the action of heat. Compare the rate of increase of its volume with that of its radius. At what rate is the volume increasing when the radius is 2 inch and increasing at the rate of $\frac{1}{3}$ inches per minute.

Solution: let v = volume of the metalic sphere at time 't'

r = radius at time 't'

$$\therefore v = \frac{4}{3}\pi r^3$$

$$\Rightarrow \frac{dv}{dt} = 4\pi r^2 \frac{dr}{dt} \qquad ...(1)$$

given that $\frac{dr}{dt} = \frac{1}{3}$ inches/minute where $r = 2$ inches

$$\therefore \left(\frac{dv}{dt}\right)_{r=2} = \left(\frac{4\pi \times 2^2}{3}\right) \text{ cube inch/minute}$$

$$= \left(\frac{16\pi}{3}\right) \text{ cube inch/minute}$$

Comparison: $\frac{\frac{dv}{dt}}{\frac{dr}{dt}} = 4\pi r^2$

9. The volume of spherical soap bubble is denoted by v, its surface by s, the radius being r, show that

(1) $\frac{dv}{dt} = 4\pi r^2 \frac{dr}{dt}$ **(2)** $\frac{ds}{dt} = \frac{2}{3}\frac{dv}{dt}$

Solution: Let v, x and r denote the volume, surface and radius of the spherical soap bubble respectively at time 't' from the start.

$$\because v = \frac{4}{3}\pi r^3 \qquad ...(1)$$

$$s = 4\pi r^2 \qquad ...(2)$$

Differentiating (1) w.r.t 't', we get

$$\frac{dv}{dt} = 4\pi r^2 \frac{dr}{dt} \qquad ...(3)$$

Again differentiating (2) w.r.t 't', we get

$$\frac{ds}{dt} = 8\pi r \cdot \frac{dr}{dt}$$

$$= \frac{2}{r}\left(4\pi r^2 \cdot \frac{dr}{dt}\right)$$

$$= \frac{2}{r} \cdot \frac{dv}{dt} \text{ [from (3)]}$$

Problems based on cylinder, cone, cube, ... etc.

1. A cone is 10 inches in diameter and 10 inches deep water is poured into it at 4 cubic inches per minute. At what rate is water level rising at the instant when the depth is 6 inches.

Solution: Let ABC = a cone

AD = depth of the cone = 10″

BC = diameter = 10″

$\alpha = \angle CAD$

$$\tan\alpha = \frac{CD}{AD} = \frac{5}{10} = \frac{1}{2}$$

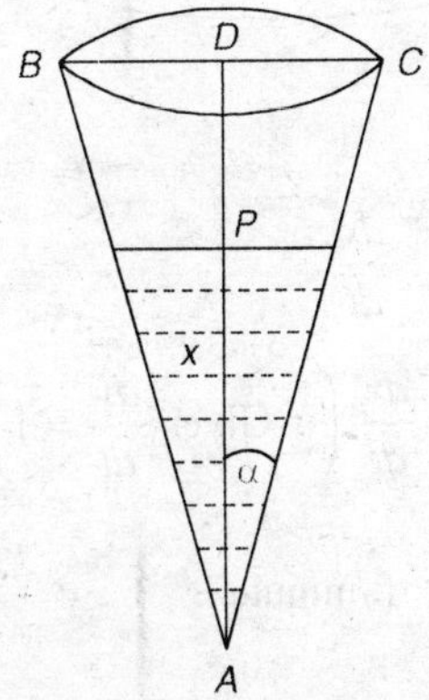

Let AP = depth of water = x and v = volume of water at time 't'

$$\therefore v = \frac{1}{3}\pi (x \tan\alpha)^2 \cdot x$$

$$\Rightarrow v = \frac{1}{3}\pi x^3 \tan^2 \alpha \quad ...(1)$$

Now, differentiating both sides of (1) w.r.t 't',

$$\frac{dv}{dt} = \frac{1}{3}\pi \cdot 3 \cdot x^2 \cdot \tan^2 \alpha \cdot \frac{dx}{dt}$$

$$\Rightarrow 4 = \frac{1}{3} \cdot \pi \cdot 3 \cdot 6^2 \cdot \left(\frac{1}{2}\right)^2 \cdot \left(\frac{dx}{dt}\right)_{x=6}$$

$$\Rightarrow \left(\frac{dx}{dt}\right)_{x=6} = \frac{4 \times 4}{36 \times \pi} = \frac{4}{9\pi} \text{ inch/min.}$$

2. Water is being poured at the rate of one cubic foot per minute into a cylindrical tube. If the tube has a circular base of radius a ft, find the rate at which water is rising in tube.

Solution: Let at any time t,
the height of water level = x ft and volume = v

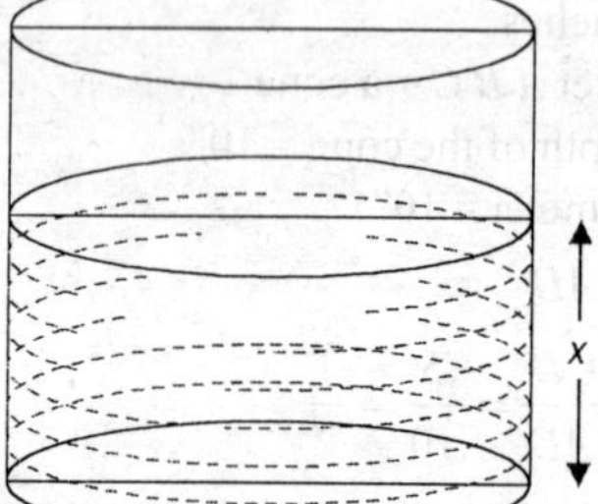

$$\therefore v = \pi a^2 x$$

$$\therefore \frac{dv}{dt} = \pi a^2 \cdot \frac{dx}{dt}$$

$$\Rightarrow 1 = \pi a^2 \cdot \frac{dx}{dt} \left(\because \text{Given } \frac{dv}{dt} = 1\right)$$

$$\Rightarrow \frac{dx}{dt} = \frac{1}{\pi a^2} \text{ ft/minute.}$$

3. Water runs into a circular conical tank at the constant rate of 2 cubic ft per minute. How fast is the water level rising when the water is 6 ft deep? It being given that the radius of the circular base of the cone = 5 ft and height of the cone = 10 ft.

Solution: let at any time 't'

$OP = h$ = height of water level

$PR = r$ = radius of water surface

v = volume of water at any time 't'.

Given that $ON = 10$ ft

$NM = 5$ ft

$$\frac{dv}{dt} = 2$$

then to find $\frac{dh}{dt}$ when $h = 6$

$$v = \frac{1}{3}\pi r^2 h$$

$$\Rightarrow v = \frac{1}{3}\pi \left(\frac{h}{2}\right)^2 \cdot h$$

$$\Rightarrow v = \frac{1}{3} \cdot \pi \cdot \frac{h^3}{4} = \frac{1}{12}\pi h^3$$

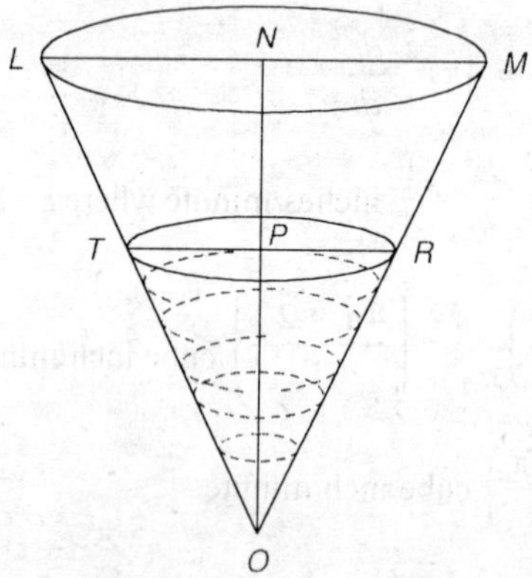

$$\because \frac{PR}{NM} = \frac{OP}{ON} \Rightarrow \frac{r}{5} = \frac{h}{10} \Rightarrow r = \frac{5h}{10} = \frac{h}{2}$$

From is Δ^s

$$\Rightarrow \frac{dv}{dt} = \frac{3\pi}{12} \cdot h^2 \cdot \frac{dh}{dt}$$

$$\Rightarrow 2 = \frac{\pi}{4} \cdot 6^2 \cdot \left(\frac{dh}{dt}\right)_{h=6}$$

$$\Rightarrow \left(\frac{dh}{dt}\right)_{h=6} = \frac{8}{36\pi} = \frac{2}{9\pi} = 0.071 \text{ ft/min.}$$

4. A hallow cone whose semi vertical angle is 30° is held with its vertex downwards and axis vertical and water is poured into it at the steady rate of 3 c. ft/min. Find the rate at which the depth (measured along the axis) of the water is increasing when the depth of the water is 3 ft.

Solution: Let v = volume of water at any time 't'

$\alpha = 30°$ = semi vertical angel (given)

Now, $\tan\alpha = \frac{r}{x}$ (from $rt \angle d\ \Delta\ OPR$)

$$\Rightarrow \tan 30 = \frac{r}{x}$$

$$\Rightarrow \frac{1}{\sqrt{3}} = \frac{r}{x}$$

$$\Rightarrow r = \frac{x}{v3} \quad ...(1)$$

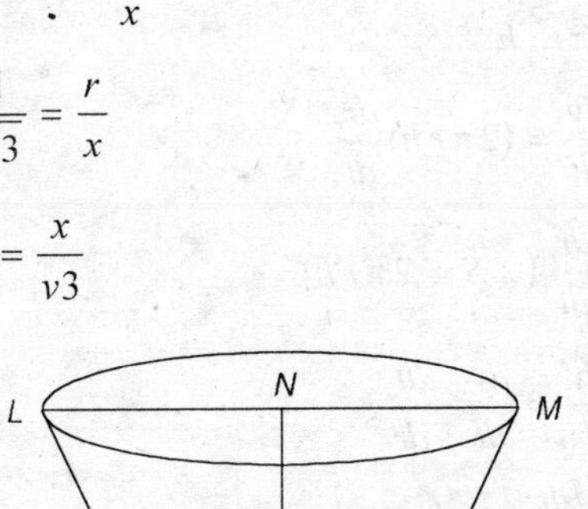

Again $\because v = \frac{1}{3}\pi r^2 x$ = volume of water at any time 't'

$\therefore v = \frac{1}{3}\pi \cdot \frac{x^2}{3} \cdot x = \frac{1}{9}\pi x^3$ (putting from (1) $r = \frac{x}{\sqrt{3}}$)

which $\Rightarrow \frac{dv}{dt} = \frac{1}{9}\pi \cdot 3x^2 \cdot \frac{dx}{dt} = \frac{1}{3}\pi x^2 \frac{dx}{dt}$

$$\Rightarrow 3 = \frac{1}{3}\pi x^2 \cdot \frac{dx}{dt}$$

$$\Rightarrow \frac{9}{\pi x^2} = \frac{dx}{dt}$$

$$\Rightarrow \left[\frac{dx}{dt}\right]_{x=3} = \left[\frac{9}{\pi x^2}\right]_{x=3}$$

$= \frac{9}{\pi \times 9} = \frac{1}{\pi}$ ft/min. = the rate at which depth of water is increasing when the depth of water = 3 ft.

5. An inverted cone has a depth of 10 cm and a base of radius 5 cm. Water is poured into it at the rate of 2.5 c.c per minute. Find the rate at which the level of the water in the cone is rising when the depth of the water is 4 cm.

Solution: Let the depth of the water at time 't' minutes from the start in x cm. If the radius of the surface of water at this time is r, then from the similar triangles COQ and COB, we have

$$\frac{r}{5} = \frac{x}{10}$$

$$\Rightarrow r = \frac{x}{2} \quad ...(1)$$

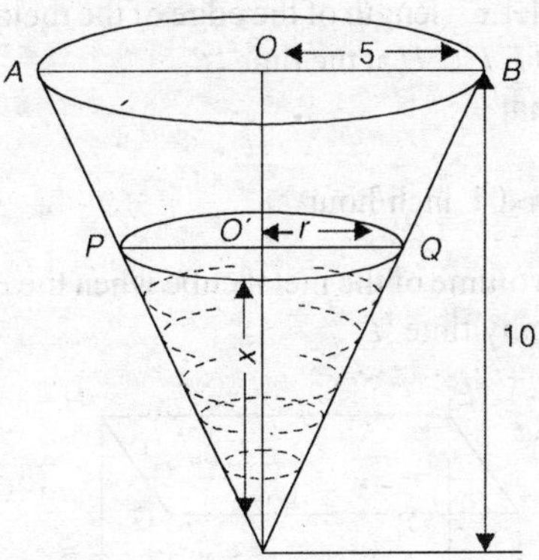

Again $\because v = \frac{1}{3}\pi r^2$ = volume of water at any time 't'

$$\Rightarrow v = \frac{1}{3}\left(\frac{x}{2}\right)^2 \cdot \pi \cdot x = \frac{1}{12}\pi x^3$$

$$(\because r = \frac{x}{2} \text{ from (1)})$$

$$\Rightarrow \frac{dv}{dt} = \frac{1}{12}\pi \cdot 3x^2 \frac{dx}{dt}$$

$$\Rightarrow \frac{dv}{dt} = \frac{1}{4}\pi x^2 \frac{dx}{dt} \quad \text{...(2)}$$

but we are given $\frac{dv}{dt} = \frac{5}{2}$ c.c. per minute.

Putting this value in (2), we get

$$\frac{dx}{dt} = \frac{4 \times 5}{\pi x^2 \times 2}$$

$$\Rightarrow \left[\frac{dx}{dt}\right]_{x=4} = \left[\frac{4 \times 5}{\pi \cdot 4^2} \cdot \frac{1}{2}\right]$$

$$= \left[\frac{5}{8\pi}\right] \text{cm/minute}$$

= rate of rising water when the depth of water is 4 cm.

6. The temperature of metal cube is being raised steadily so that each edge expands at the rate of .01 inch per hour. At what rate is the volume increasing when the edge is 2 inches.

Solution: let x = length of the edge of the metal cube $ABCDEFGH$ at the time 't'

Given that

$$\frac{dx}{dt} = .01 \text{ inch/hour}$$

Let v = volume of the metal cube when the edge is x inches at any time 't'

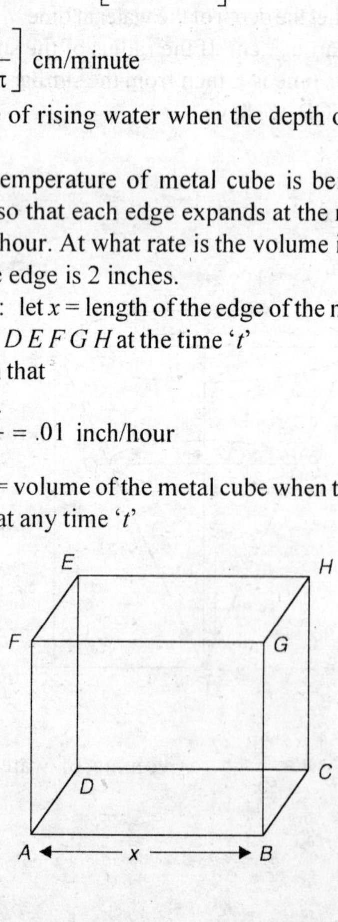

$$\therefore v = x^3$$

$$\text{which} \Rightarrow \frac{dv}{dt} = 3x^2 \cdot \frac{dx}{dt}$$

$$= 3x^2 \cdot (.01) \text{ cube inch/hour}$$

$$= .03x^2 \text{ cubic inch/hour}$$

$$\therefore \left[\frac{dv}{dt}\right]_{x=2} = .03 \times 4 = .12 \text{ cubic inch per hour}$$

7. A right circular cylinder has a constant height h but the radius r of its base varies. If v be the volume and S the curved surface of the cylinder, prove that $\frac{dv}{dr} = S$.

Solution: let r = radius of right circular cylinder and v is its volume at time 't' from the start, then

$$v = \pi r^2 h$$

$$\Rightarrow \frac{dv}{dt} = (2\pi r h)\frac{dr}{dt}$$

$$= S\frac{dr}{dt} \quad (\because S = 2\pi r h) \quad \text{...(1)}$$

$$\Rightarrow \frac{dv}{dr} = \frac{dv}{dt} \cdot \frac{dt}{dr}$$

$$= \frac{dv/dt}{dr/dt}$$

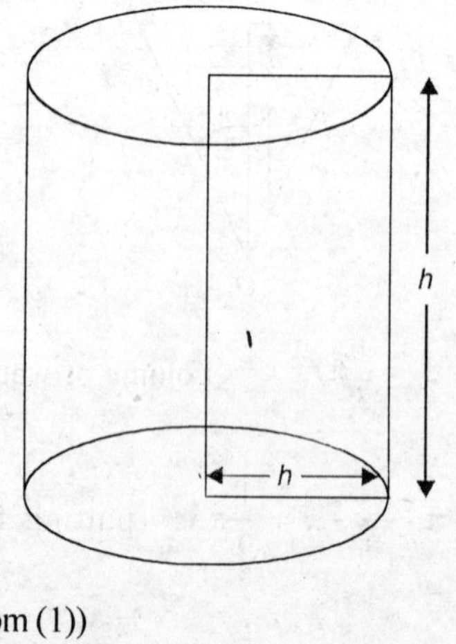

$= S$ (from (1))

(proved)

Problems based on proportionality

1. If the area of the circle increases at a uniform rate, then the rate of increase of the perimeter varies inversely as the radius.

Solution: let the area of the circle $A = \pi r^2$ ...(1)

Now, differentiating both sides of (1) w.r.t 't', we get $\frac{dA}{dt} = 2\pi r \frac{dr}{dt}$

$\Rightarrow k = 2\pi r \frac{dr}{dt}$

($\because \frac{dA}{dt}$ = uniform rate = k (say) = given)

$\Rightarrow \frac{2\pi\, dr}{dt} = \frac{k}{r}$ ...(2)

Again perimeter of the circle $P = 2\pi r$ ...(3)

Differentiating both sides of (3) w.r.t 't', we get

$\frac{dP}{dt} = 2\pi \frac{dr}{dt}$ ...(4)

Putting (2) in (4), we get

$\frac{dP}{dt} = \frac{k}{r}$

$\Rightarrow \frac{dP}{dt} \propto \frac{1}{r}$ (proved)

2. Prove that if a particle moves so that the space described is proportional to the square of the time of description, the velocity will be proportional to the time and rate of increase of the velocity will be constant.

Solution: letting S = the space described by the particle in time 't'.

$S \propto t^2$ (given)

$\Rightarrow S = k t^2$, where $k = a$ constant

$\Rightarrow \frac{dS}{dt} = 2kt$

$\Rightarrow v = 2kt$ ($\because \frac{dS}{dt} = v$ = velocity of the particle) ...(1)

$\Rightarrow v \propto t$

$\Rightarrow v$ is proportional to time 't'

Again $\frac{dv}{dt} = 2k$

$\Rightarrow \frac{dv}{dt}$ = a constant

$\Rightarrow$ rate of increase in velocity = constant (proved).

3. A spherical ball of salt is dissolving in water in such a manner that the rate of decrease in volume is proportional to the surface. Prove that the radius is decreasing at a constant rate.

Solution: let v and S denote the volume and surface of the spherical ball of salt respectively when the radius is r at time 't'.

We know that $v = \frac{4}{3}\pi r^3$ ...(1)

$S = 4\pi r^2$ ...(2)

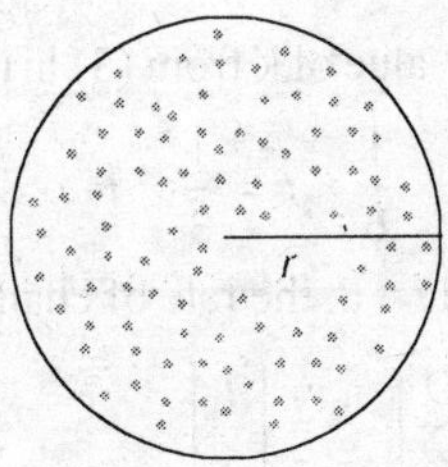

$\frac{dv}{dt} \propto S$ (given) ...(3)

$\Rightarrow \frac{dv}{dt} = -uS$ [$\because$ v decreases with increase in 't'], $u > 0$ (constant) ...(4)

Now, $\frac{dv}{dt} = 4\pi r^2 \cdot \frac{dr}{dt} = S \cdot \frac{dr}{dt}$

$\Rightarrow -uS = S\frac{dr}{dt}$ (using (4))

$\Rightarrow \frac{dr}{dt} = -u$ (constant)

$\Rightarrow$ radius is decreasing at a constant rate (–ve sign signifies the decreasing of r)

4. The diameter of an expanding smoke ring at time t is proportional to t^2. If the diameter is 6 cm after 6 sec, find at what rate it is then changing.

Solution: let D denote the diameter of an expanding smoking then according to question, $D \propto t^2$

$$\therefore D = k t^2, \text{ (where } k \text{ is a constant)} \quad \ldots(1)$$

$$\therefore \frac{dD}{dt} = 2kt \quad \ldots(2)$$

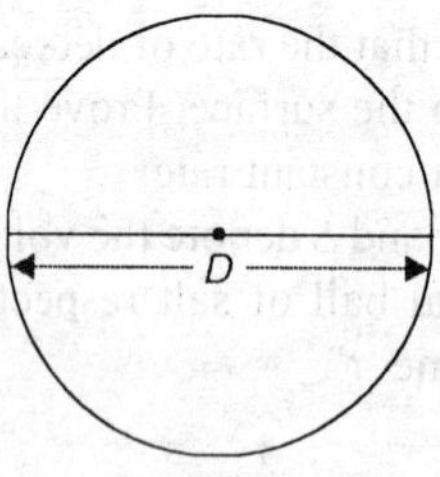

Now, since, it is given $D = 6$, when $t = 6$

$$\therefore \text{ from (1), } 6 = k \cdot 36 \Rightarrow k = \frac{6}{36} = \frac{1}{6} \quad \ldots(3)$$

Putting the value of k from (3) in (2), we have

$$\frac{dD}{dt} = 2 \times \frac{1}{6} \times t = \frac{1}{3} \times t$$

lastly, when $t = 6$, the rate of change of

$$D = \left[\frac{dD}{dt}\right]_{t=6} = \left[\frac{t}{3}\right]_{t=6}$$

$$= \frac{6}{3} = 2 \text{ cm/sec.}$$

Type 2: *Problems based on right angled triangle*

In such types of problems, we adopt the following working rule.

1. An question given in x and y stated in words should be translated into symbolic equation in x and y or we form an equation in x and y by using the properties of Δ^s.

2. Then differentiate both sides of the equation determined w.r.t time 't' which provides us a relation (or equation) between $\frac{dy}{dt}$ and $\frac{dx}{dt}$.

3. Use the given value for $\frac{dy}{dt}$ or $\frac{dx}{dt}$ which ever is given.

4. Find the other rate by solving the equation involving $\frac{dx}{dt}$ and $\frac{dy}{dt}$.

Notes:

1. Figures must be drawn for convenience, for doing rate problems based on right angled triangle as well as formulas for volume, area, perimeter of regular geometrical figures (as triangle, rectangle, square, cylinder, cone, sphere or cube etc) must be remembered.

2. We substitute the given quantities and rates in the differentiated results to get the required rate.

3. $\left[\frac{dQ}{dt}\right]_{x=a}$ = the rate of Q when (or, at) $x = a$ where Q = any quantity.

4. Rate at which two bodies are nearing (or being separated) is the rate of distance 'r' between them denoted $\frac{dr}{dt}$.

Worked out

Problems based on type (2)

1. A man 6 ft tall walks away along a straight line from the foot of a light post 24 ft high at the rate of 3. m. p. h. How fast does the end of his shadow move? Find the rate at which the length of his shadow increases.

Solution: LP = height of the lamp post = 24 ft

MN = height of the man = 6 ft

$PN = x$ = the distance between the foot of light post and the position N obtained after t second, when the man is x feet away from the foot P of light post.

$PR = y$

$NR = PR - PN = y - x$

Now from similar $\Delta\, LPR$ and $\Delta\, MNR$, we have

$$\frac{PL}{MN} = \frac{PR}{NR}$$

$$\text{which} \Rightarrow \frac{24}{6} = \frac{y}{y-x}$$

$$\Rightarrow 4 = \frac{y}{y-x}$$

$$\Rightarrow 4(y-x) = y$$

$$\Rightarrow 3y = 4x \quad \ldots(1)$$

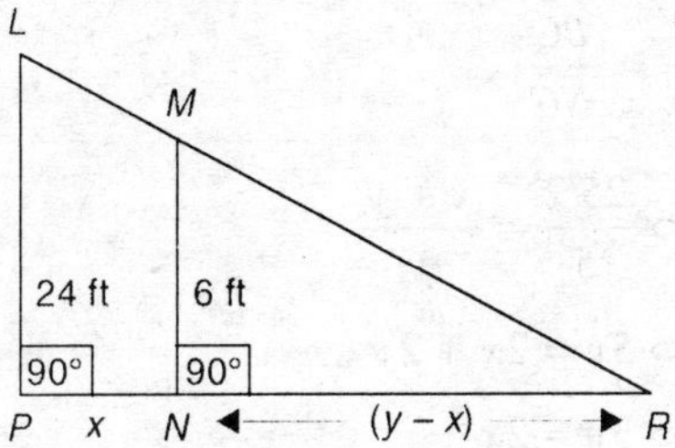

Now, differentiating both sides of (1) w.r.t 't', we have

$$3\frac{dy}{dt} = 4 \cdot \frac{dx}{dt}$$

$$\Rightarrow \frac{dy}{dt} = \frac{4}{3} \cdot \frac{dx}{dt}$$

$$\Rightarrow \frac{dy}{dt} = \frac{4}{3} \cdot 3 \left[\because \frac{dx}{dt} = 3 \text{ given}\right]$$

$$\therefore \frac{dy}{dt} = 4 \text{ miles/hour}$$

$\therefore$ The end of the shadow moves at the rate 3 miles/hour.

The rate at which the shadow lengthens

$$= \frac{d(NR)}{dt} = \frac{d(y-x)}{dt}$$

$$= \frac{dy}{dt} - \frac{dx}{dt} = (4 - 3) \text{ m.p.h.} = 1 \text{ m.p.h.}$$

2. A point source of light is hung a meter directly above a straight horizontal path on which a bot b meter in height is walking. How fast is the boy's shadow lengthening when he is walking away from the light at the rate of c meter per minute.

Solution: let y = length of the shadow in meter and x = the distance of the boy after time 't'

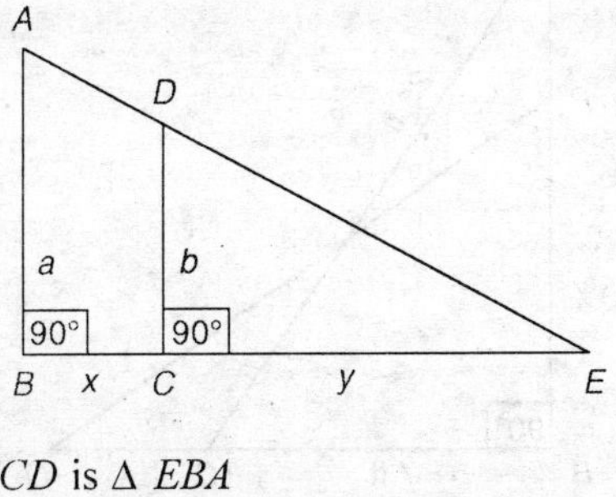

Δ ECD is Δ EBA

$$\Rightarrow \frac{y}{x+y} = \frac{b}{a}$$

$$\Rightarrow ay = bx + by$$

$$\Rightarrow (a-b)y = bx$$

$$\Rightarrow y = \frac{bx}{a-b}$$

$$\Rightarrow \frac{dy}{dt} = \frac{b}{a-b} \cdot \frac{dx}{dt} \qquad \text{...(1)}$$

But we are given $\frac{dx}{dt} = c$ = velocity of the boy

$\therefore$ rate of lengthening in the boy's shadow

$$= \frac{dy}{dt} = \frac{dy}{dx} \cdot \frac{dx}{dt}$$

$$= \left(\frac{b}{a-b}\right) \cdot c \left[\because \frac{dy}{dx} = \frac{b}{a-b} \text{ and } \frac{dx}{dt} = c\right]$$

$$= \left(\frac{bc}{a-b}\right) \text{ meter/minute.}$$

3. A lamp is at a height 376 cm from the ground. A man 188 cm tall is walking on the ground steadily at the rate of 10 cm/sec, in a straight line passing through the lamp post. Find the rate at which the end of his shadow is moving.

Solution: let A be the lamp

BC, the man

D, the point where AC produced meets OB produced,

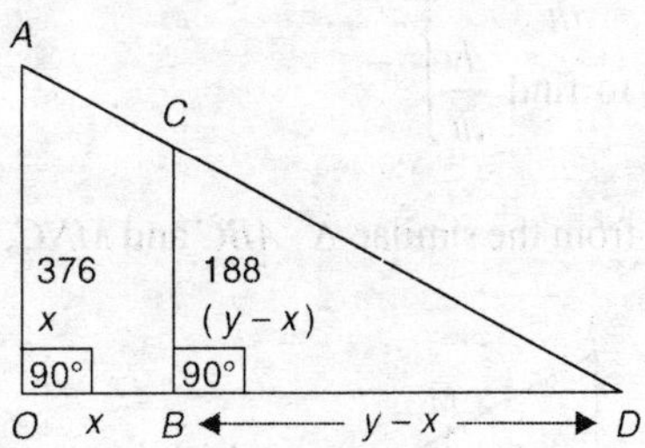

then we are required to find out the rate at which D is moving.

Again let us suppose that

$OB = x$

$OD = y$

then $\frac{dx}{dt} = 10$ (given)

and we have to find $\frac{dy}{dt}$.

From similar triangles AOD and CBD, we have

$$\frac{BD}{OD} = \frac{BC}{OA}$$

which $\Rightarrow \frac{y-x}{y} = \frac{188}{376} = \frac{1}{2}$

$\Rightarrow 2y - 2x = y$

$\Rightarrow y = 2x$

$\Rightarrow \frac{dy}{dt} = 2\frac{dx}{dt} = 2 \times 10 = 20$ cm/sec.

$(\because \frac{dx}{dt} = 10$ is given in the question)

4. A man 5 ft tall walks away from a lamp post $12\frac{1}{2}$ ft high at the rate of 3 miles per hour. Find how fast is his shadow lengthening.

Solution: let AB = lamp post = $12\frac{1}{2}$ ft

MN = height of the man = 5 ft

$BN = x$ ft = distance of a man at any time 't' from the lamp post.

$NC = y$ = length of the shadow

given $\frac{dx}{dt} = 3$
then to find $\frac{dy}{dt}$

Now from the similar Δ^s ABC and MNC, we have

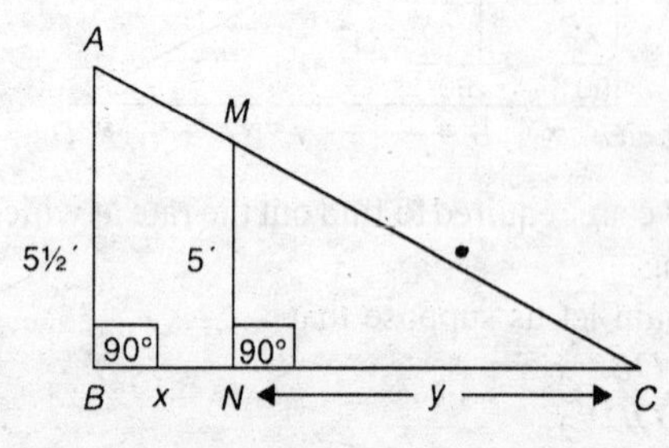

$$\frac{AB}{MN} = \frac{BC}{NC}$$

$\Rightarrow \frac{25/2}{5} = \frac{x+y}{y}$

$\Rightarrow 5y = 2x + 2y$

$\Rightarrow 3y = 2x$

$\Rightarrow \frac{dy}{dt} = \frac{2}{3}\frac{dx}{dt}$

$= \frac{2}{3} \cdot 3$

= 2 miles/hour.

5. A ladder 25 ft long reclines against a wall. A man begins to pull the lower extremity which is 7 ft distant from the bottom of the wall along the ground outwards at the rate of 3 ft/sec. At what rate does the other end begin to descend along the wall?

Solution: let AB = wall

AC = ladder

BC = 7' = 7 feet (given)

Let after t second, the foot 'c' be at 'Q' and let $CQ = x$

$BP = y$

AC = 25 feet

From Δ PBQ

$PB^2 = PQ^2 - BQ^2$

$\Rightarrow y^2 = (25)^2 - (7+x)^2$ ...(1)

Now differentiating both sides of (1) w.r.t 't', we have,

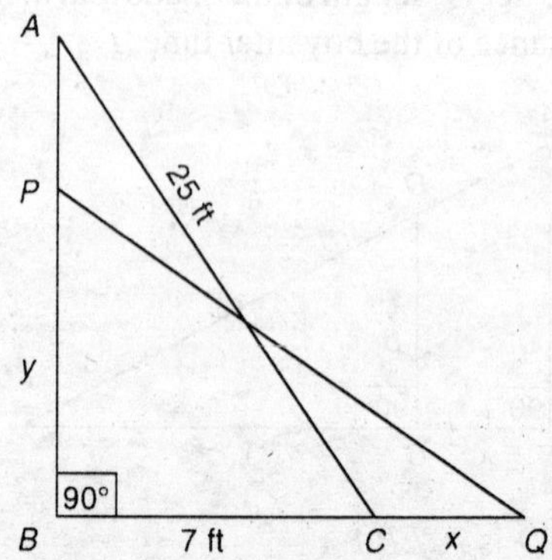

$$2y \cdot \frac{dy}{dt} = -2(7+x)\frac{dx}{dt}$$

$$\Rightarrow \frac{dy}{dt} = \frac{-(7+x)}{y} \times \frac{dx}{dt}$$

$$= -\frac{(7+x)}{y} \times 3 \quad \left(\because \frac{dx}{dt} = 3 \text{ is given}\right) \qquad ...(2)$$

Now, we know that when $x = 0, y = 24$ feet (when $t = 0$)

Again from (2), we have

$$\left(\frac{dy}{dt}\right)_{t=0} = -\frac{21}{24} = -\frac{7}{8} \text{ feet/sec.}$$

(–ve sign shows that y decreases with t i.e. the other end descends and the rate at which it descends $= \frac{7}{8}$ ft/sec)

6. The top of ladder, 20 feet long, is resting against a vertical wall and its foot on a level pavement, when the ladder begins to slide outwards. At the moment when the foot of the ladder is 12 feet from the wall, it is sliding away from the wall at the rate of 2 feet per second. How fast is the top sliding downwards at this instant? How far is the foot from the wall when it and the top are moving at the same rate.

Solution: let at any time t

AB be the position of the ladder having the length = 20 feet where

$OA = x$

$OB = y$ where O represents the foot of the wall.

Now, from the right angled triangle, we have,

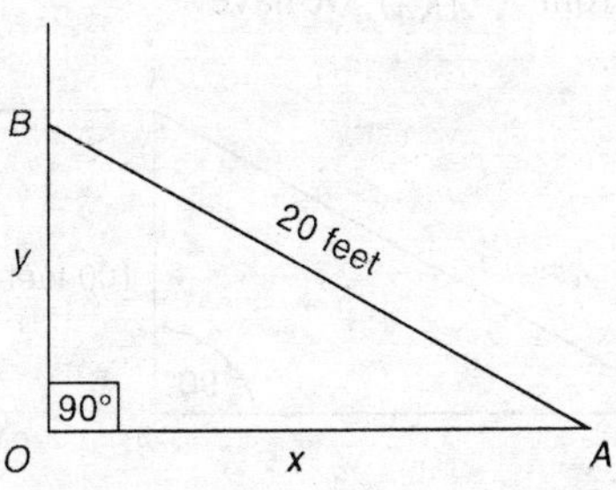

$$OA^2 + OB^2 = AB^2$$

$$\Rightarrow x^2 + y^2 = (20)^2 \quad (\because AB = 20 \text{ feet})$$

$$\Rightarrow \frac{d}{dt}\left(x^2 + y^2\right) = \frac{d}{dt}(400)$$

$$\Rightarrow 2x\frac{dx}{dt} + 2 \cdot y \cdot \frac{dy}{dt} = 0$$

$$\Rightarrow x\frac{dx}{dt} + y \cdot \frac{dy}{dt} = 0 \qquad ...(1)$$

Now, again from the relation $x^2 + y^2 = 400$, when $x = 12$, we have

$$(12)^2 + y^2 = 400$$

$$\Rightarrow 144 + y^2 = 400$$

$$\Rightarrow y^2 = 400 - 144 = 256$$

$$\Rightarrow y = \sqrt{256}$$

$$\Rightarrow y = 16$$

Also we are given $x = 12$, as well as $\frac{dx}{dt} = 2$ ft/sec.

Substituting these values in (1), we have

$$12 \cdot 2 + 16 \cdot \frac{dy}{dx} = 0$$

$$\Rightarrow \frac{dy}{dx} = \frac{24}{16} = \frac{3}{2} \text{ ft/sec.}$$

Hence B is sliding downwards (as the negative sign shows) as the rate of $\frac{3}{2}$ ft/sec at the instant under consideration.

If, at a particular instant, A and B are sliding at the same rate, then

$\frac{dx}{dt} = -\frac{dy}{dt}$ and then (2) provides us $x = y$

and for the reason, $x^2 + y^2 = 400$

$$\Rightarrow 2x^2 = 400$$

$$\Rightarrow x^2 = \frac{400}{2} = 200$$

$$\Rightarrow x = 10\sqrt{2} \text{ feet.}$$

7. A ladder is inclined to a wall making an angle of 30° with it. A man is ascending the ladder at the rate of 3 ft/sec. How fast is he approaching the wall.

Solution: AB is the position of the ladder inclined to a wall OB having the length 'L' feet.

x = distance moved along the ladder and y = distance from the wall at time t

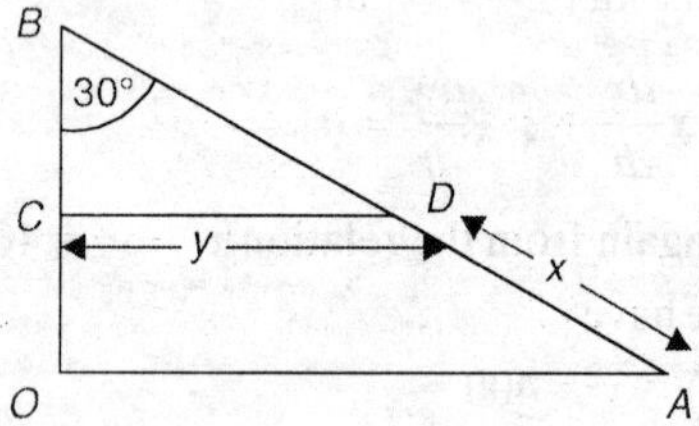

Now, $\sin 30° = \frac{CD}{BD} = \frac{y}{(L-x)}$

$\Rightarrow y = (L-x)\sin 30°$

$\Rightarrow y = (L-x)\cdot\frac{1}{2}$

$\Rightarrow \frac{dy}{dt} = -\frac{1}{2}\frac{dx}{dt}$ ($\because$ L is a constant)

$\Rightarrow \frac{dy}{dt} = -\frac{1}{2}\cdot 3$ ($\because \frac{dx}{dt} = 3$ feet/sec is given)

$\Rightarrow \frac{dy}{dt} = -\frac{3}{2}$ ft/sec

Hence, y decreases (as the –ve sign shows) at the rate of $\frac{3}{2}$ ft/sec.

8. A ladder 26 ft long leans against a vertical wall. The foot of the ladder is drawn away from the wall at the rate of 4 ft per second. How fast is the top of the ladder sliding down the wall when the foot of the ladder is 10 ft away from the wall.

Solution: let AB = ladder = 26 ft

$OA = x$ and

$OB = y$ at time t

Now from the right angled triangle, we have

$x^2 + y^2 = (26)^2$ ...(1)

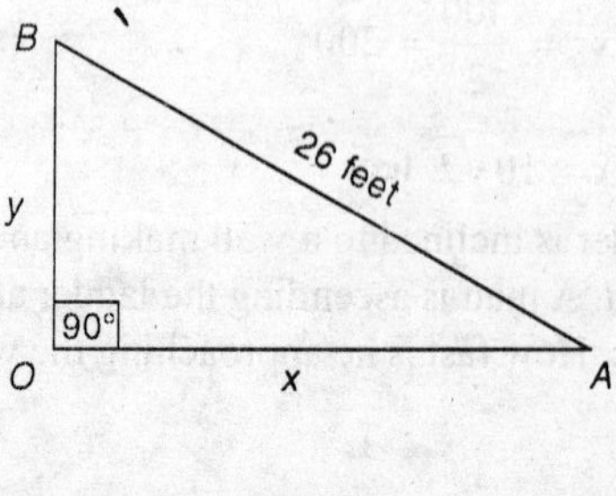

Given that $\frac{dx}{dt} = 4$ ft/sec

$x = 10$ ft $\Rightarrow y = \sqrt{26^2 - 10^2} = 24$ ft,

to find: $\frac{dy}{dt}$ when $x = 10$ ft.

From (1), $2x\frac{dx}{dt} + 2y\frac{dy}{dt} = 0$

$\Rightarrow \frac{dy}{dt} = -\frac{x}{y}\cdot\frac{dx}{dt}$

$\Rightarrow \left(\frac{dx}{dt}\right)_{x=10} = -\frac{10}{24}\times 4 = -\frac{5}{3}$ ft/sec.

which $\Rightarrow$ the top of the ladder is sliding down at the rate of $\frac{5}{3}$ ft/sec.

9. A kite is 100 ft high and a length of 260 ft of the string is out. If the kite is moving horizontally at the rate of $6\frac{1}{2}$ m.p.h directly away from the person who is flying it, how fast the string (or, cord) is being paid out.

Solution: let k be kite which is flying in the horizontal direction in the height of 100 ft.

M is the point where the man is standing and is flying the kite. Let at time t,

MK = the length of the string (or, cord) which is out = y (say)

MA = the horizontal distance of the kite from M at time 't' = x (say)

Now, from Δ AKM we have

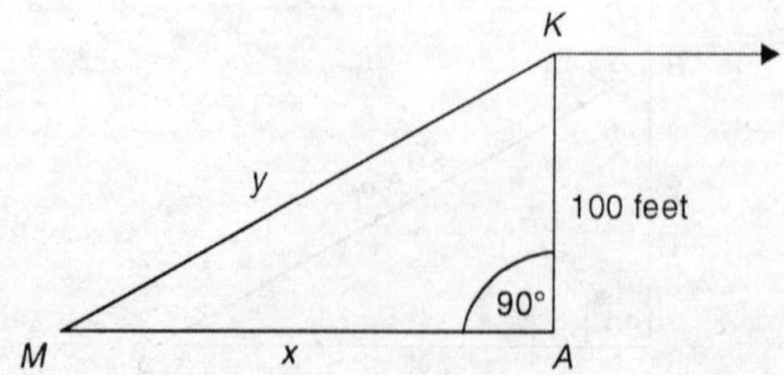

$y^2 - 100^2 = x^2$ ...(1)

since, the height of the kite is same always = 100 ft

Now, differentiating (1), we get

$$2y \cdot \frac{dy}{dt} = 2x \cdot \frac{dx}{dt}$$

$$\Rightarrow y \cdot \frac{dy}{dt} = x \frac{dx}{dt} \quad ...(2)$$

$$y = 260 \Rightarrow x = \sqrt{260^2 - 100^2} = 240$$

Substituting these values of x and y in (2), we have

$$260\frac{dy}{dt} = 240 \times \frac{13}{2}$$

$$\Rightarrow \frac{dy}{dt} = 6 \text{ m.p.h} = \text{the required rate.}$$

Second method:

$$x = MA = \sqrt{MK^2 - AK^2}$$

$$= \sqrt{(260)^2 - (100)^2} = 240$$

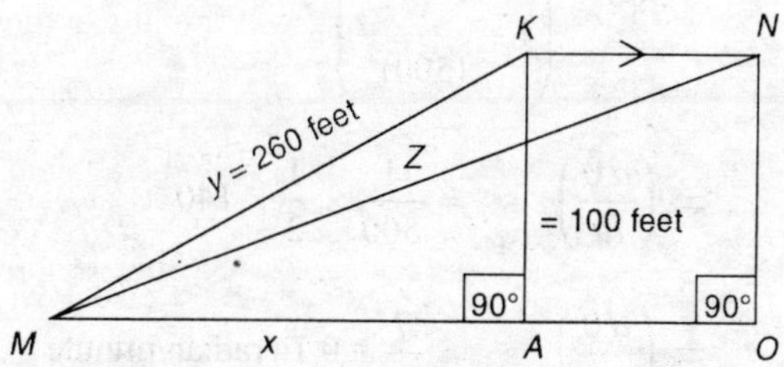

$MN = Z$

$$KN = 6\frac{1}{2} \cdot t \ (\because \text{velocity} = \frac{\text{distance}}{\text{time}} \text{ i.e, } v = \frac{S}{t})$$

$$= \frac{13}{2}t = AO$$

$$MO = MA + AO = 240 + \frac{13}{2}t$$

and $Z^2 = MN^2 = MO^2 + ON^2 = \left(240 + \frac{13}{2}t\right)^2 + (100)^2$

$$\Rightarrow 2Z\frac{dZ}{dt} = 2\left(240 + \frac{13}{2}t\right) \cdot \frac{13}{2} + 0$$

$$\Rightarrow \frac{dZ}{dt} = \frac{13\left(240 + \frac{13}{2}t\right)}{2Z}$$

Initially, when $t = 0, Z = OA = 260$

$$\text{Hence, } \frac{dZ}{dt} = \frac{13 \times 240}{2 \times 260}$$

$= 6$ m. p. h

$=$ the require rate

10. A kite is moving horizontally at a height of 151.5 meters. If the kite is moving horizontally directly away from the boy who is flying it at the rate of 10 meter/sec, how fast is the string being let out when the kite is 250 meters from the boy who is flying the kite, the height of the boy being 1.5 meters.

Solution: Let $AB = x$ and

$AC = y$ at time t.

$AE = 1.5$

$DC = 151.5$ m

$BC = 150$ m

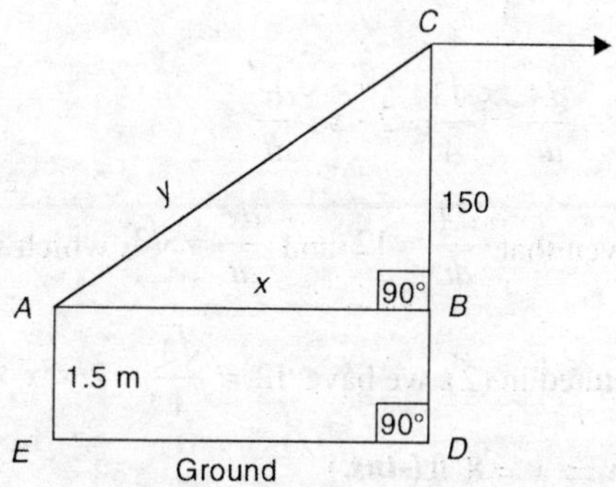

Also $\frac{dx}{dt} = 10$ m/sec (given)

Then we find $\frac{dy}{dt}$, when $y = 250$ ft.

Clearly, $y^2 = x^2 + 150^2$ ...(i)

$$\Rightarrow 2y\frac{dy}{dt} = 2x \cdot \frac{dx}{dt}$$

$$\Rightarrow \frac{dy}{dt} = \frac{x}{y} \cdot \frac{dx}{dt}$$

When $y = 250$ ft, $x = 200$ ft, from (i)

$$\therefore \left(\frac{dy}{dt}\right)_{y=250} = \frac{200}{250} \times 10 \text{ ft/sec}$$

$= 8$ ft/sec.

11. If the side of an equilateral triangle increases at the rate of $\sqrt{3}$ ft/sec and its area at the rate of 12 sq. ft/sec find the side of the triangle.

Solution: let x = each side of the equilateral Δ at time t and A = area of $\Delta\ PQR$, then

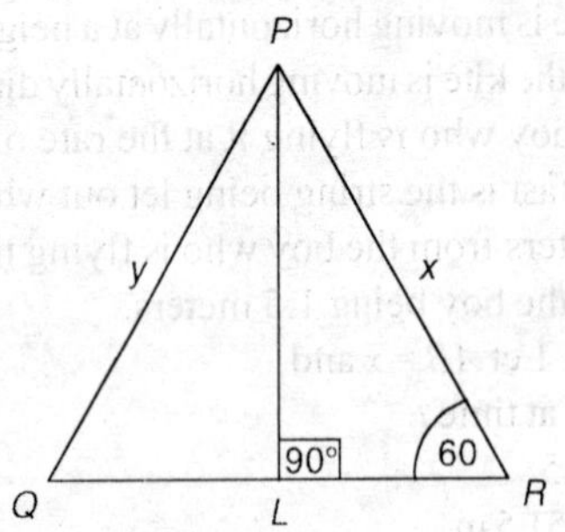

$$A = \frac{\sqrt{3}}{4} \cdot x^2 \qquad \text{...(1)}$$

$$\therefore \frac{dA}{dt} = \frac{\sqrt{3}}{4} \cdot 2 \cdot x \cdot \frac{dx}{dt} \qquad \text{...(2)}$$

Given that $\frac{dA}{dt} = 12$ and $\frac{dx}{dt} = \sqrt{3}$ which when substituted in (2), we have $12 = \frac{\sqrt{3}}{4} \times 2 \times x \times \sqrt{3}$

which $\Rightarrow x = 8$ ft (***Ans.***)

12. A balloon rising from the ground at 140 ft/minute is tracked by range finder at a point A located 500 ft from the point of lift off. Find the rate at which the angle at A and the range 'r' are changing when the balloon is 500 ft above the ground.

Solution: let the range finder be at A.

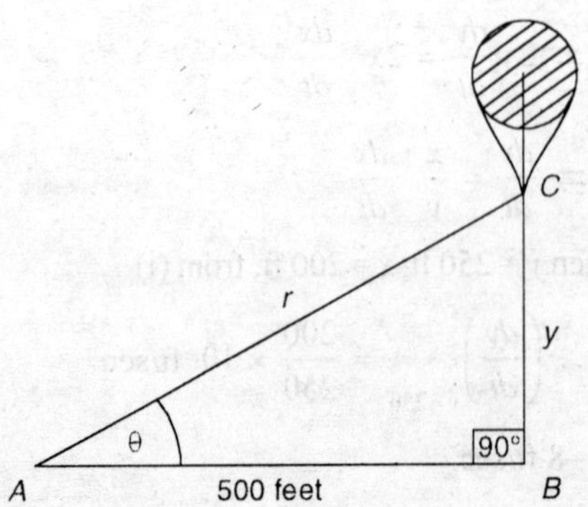

$AC = r$, $BC = y$, $\angle CAB = \theta$ at time t.

Given that $\frac{dy}{dt} = 140$ ft/min

$AB = 500$ ft

Then we have to find (i) $\frac{d\theta}{dt}$ (ii) $\frac{dr}{dt}$ when $y = 500$ ft

Now, $\tan\theta = \frac{y}{500}$

$$\Rightarrow \sec^2\theta \frac{d\theta}{dt} = \frac{1}{500} \cdot \frac{dy}{dt}$$

$$\Rightarrow \frac{d\theta}{dt} = \frac{1}{500} \times \frac{1}{\left(1 + \frac{y^2}{(500)^2}\right)} \times \frac{dy}{dt}$$

$$= \frac{1}{500} \times \frac{1}{\left(1 + \frac{y^2}{(500)^2}\right)} \times 140$$

$$\Rightarrow \left(\frac{d\theta}{dt}\right)_{y=500} = \frac{1}{500} \times \frac{1}{2} \times \overset{7}{140}$$

$$\Rightarrow \left(\frac{d\theta}{dt}\right)_{y=500} = \frac{.7}{5} = 0.14 \text{ radian/minute}$$

Also, $r^2 = y^2 + (500)^2$

$$\Rightarrow 2r \cdot \frac{dr}{dt} = 2y \cdot \frac{dy}{dt}$$

$$\Rightarrow \frac{dr}{dt} = \frac{y}{r} \cdot \frac{dy}{dt} \Rightarrow \left(\frac{dr}{dt}\right)_{y=500}$$

$$= \frac{500}{500\sqrt{2}} \times 140 = 70.\sqrt{2} \text{ ft/min.}$$

12. Two bodies start from O, one traveling along OX at the rate of 3 miles per hour and the other along OY (which is perpendicular to OX) at the rate of 4 mile per hour. Find the rate at which the distance between them is increasing at time 't'.

Solution: let r be the distance between two bodies after t hours.

$OA = 3t \left(\because \text{velocity} = \dfrac{\text{distance}}{\text{time}} \; i.e.\; v = \dfrac{S}{t} = \dfrac{OA}{t}\right)$

$OB = 4t \left(\because \text{velocity} = \dfrac{\text{distance}}{\text{time}} \; i.e.\; v = \dfrac{S}{t} = \dfrac{OB}{t}\right)$

Now, from the right angled triangle, we have

$$r^2 = (3t)^2 + (4t)^2$$

$$\Rightarrow r^2 = 9t^2 + 16t^2$$

$$\Rightarrow r = 5t$$

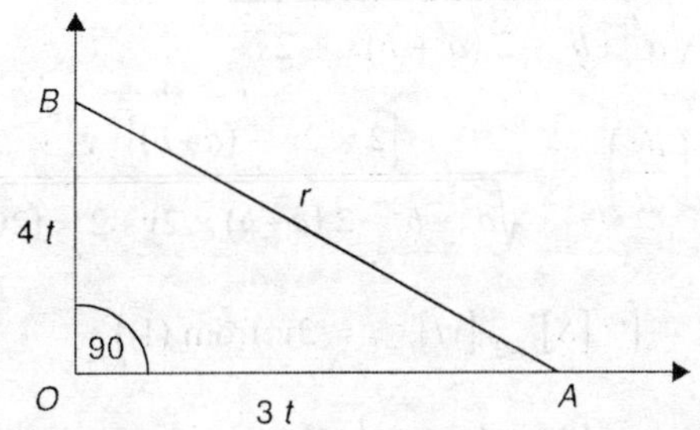

$\Rightarrow \dfrac{dr}{dt} = 5$ miles/hour which is the required rate at which the distance AB increasing.

13. Two cars started from a place, one going due north and the other due west with equal uniform speed v, find the rate at which they were being separated from each other.

Solution: let P and Q denote the position (place) of the cars after time 't'

$OP = OQ = x$ ($\because$ speed is same)

$PQ = y$

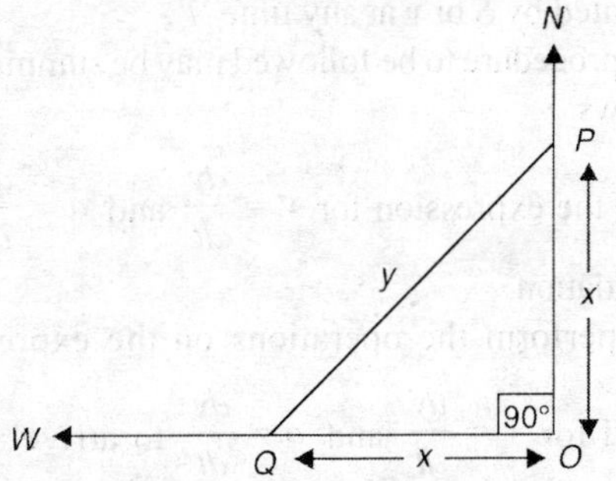

Now from the right angled triangle, we have

$$y^2 = x^2 + x^2$$

$$\Rightarrow y^2 = 2x^2 \qquad \Rightarrow y = x\sqrt{2}$$

$$\Rightarrow \frac{dy}{dt} = \sqrt{2}\frac{dx}{dt} = \sqrt{2}v \left(\because \frac{dx}{dt} = v\right)$$

$$\Rightarrow \frac{dy}{dt} = \sqrt{2}\, v$$

Second method:

$$OP = vt \left(\because v = \frac{S}{t} = \frac{OP}{t}\right)$$

$$PQ = vt \left(\because v = \frac{S}{t} = \frac{PQ}{t}\right)$$

$$y^2 = v^2 t^2 + v^2 t^2$$

$$\Rightarrow y^2 = 2v^2 t^2$$

$$\Rightarrow y = \sqrt{2}\,(v \cdot t)$$

$$\therefore \frac{dy}{dt} = \sqrt{2}\, v$$

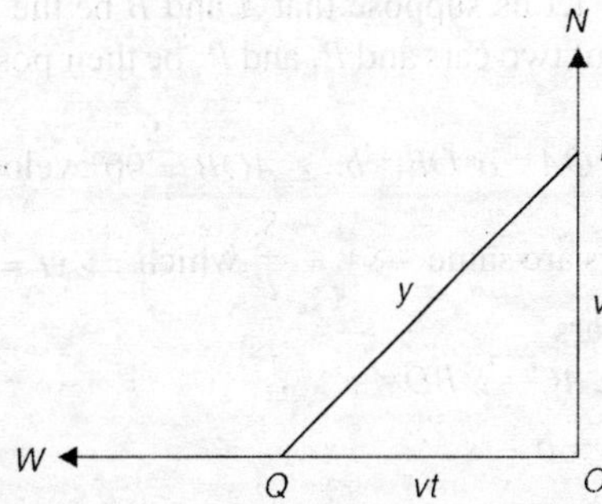

14. Obtain the rate at which the distance between two cyclists is widening out after an hour given that they start simultaneously from the junction of two roads inclined at 60°, one in each road and that they cycle with the same speed v miles per hour.

Solution: Let the distance travelled after t-second from the junction $= x$ now, according to the question,

$$x = vt \left[\because v = \frac{x}{t}\right] \qquad \ldots(1)$$

It is clear from the data that joining the junction and the position of two cyclist, we get an equilateral triangle $\Rightarrow$ the distance between two cyclist $= x$

Now, differentiating (1) w.r.t, we have

$\dfrac{dx}{dt} = v$ which is the rate at which distance between two cyclists is winding.

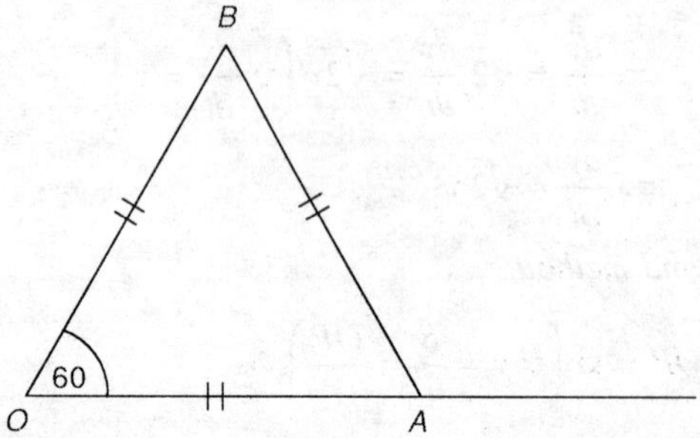

15. Two cars start with the same velocity v miles/hour from distances a miles and b miles from the junction of two roads inclined at 90° and travel towards the junction. Prove that after 2 hours, they are nearing each other at the rate of

$$\frac{4v^2 - (a+b)v}{\sqrt{a^2 + b^2 - 4v(a+b) + 8v^2}} \text{ miles per hour.}$$

Solution: let us suppose that A and B be the initial positions of two cars and P_1 and P_2 be their positions at time t.

Given: $OA = a$, $OB = b$, $\angle AOB = 90°$ velocities of two cars are same $\Rightarrow v = \frac{S}{t}$ which $\Rightarrow vt = S$ for both cars ...(1)

$CD = r, AC = s, BD = s$

$\therefore OP_1 = a - s$

$OP_2 = b - s$

To prove: $\left(\frac{dr}{dt}\right)_{t=2} = \frac{4v^2 - (a+b)v}{\sqrt{a^2+b^2-4v(a+b)+8v^2}}$

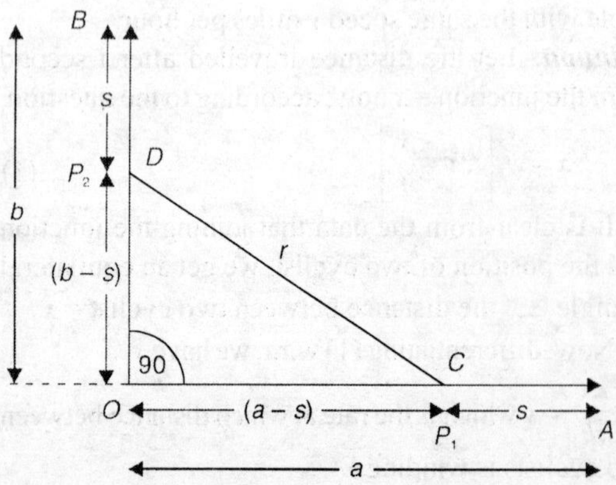

Proof: In the right angled $\Delta\ P_1 0 P_2$,

$$r = \sqrt{(a-s)^2 + (b-s)^2}$$

$$\Rightarrow \frac{dr}{dt} = \frac{1}{2\sqrt{(a-s)^2 + (b-s)^2}} \times [2(a-s) + 2(b-s)] \times \left(-\frac{ds}{dt}\right)$$

$$= \frac{[2s - (a+b)]v}{\sqrt{a^2 + b^2 - 2(a+b)s + 2s^2}}$$

$$\Rightarrow \left(\frac{dr}{dt}\right)_{t=2} = \frac{[2 \times 2v - (a+b)] \cdot v}{\sqrt{a^2+b^2-2(a+b) \times 2v + 2 \times (2v)^2}}$$

$$\left(\because [S]_{t=2}\ [vt]_{t=2} = 2v \text{ from (1)}\right)$$

$$= \frac{[4v - (a+b)]v}{\sqrt{a^2+b^2-4(a+b)v+8v^2}} \text{ miles/hour}$$

Note: They are nearing each other at the rate of ... means the rate of distance between them is....

Problems based on velocity and acceleration as a rate measure

To investigate generally the motion (velocity, acceleration or the position of a particle at any time 't')of a particle (or, body) moving in a straight line according to a law of motion give by $y = f(t)$

Where y = distance or velocity of the moving body represented by S or v at any time 't'.

The procedure to be followed may be summarised as follows.

1. Find the expression for $v = \frac{dy}{dt}$ and $a = \frac{dv}{dt}$ by differentiation.

2. We perform the operations on the expression obtained for $v = \frac{dy}{dt}$ and $a = \frac{dv}{dt}$ to arrive at out target accordingly as the question says.

Remember:

1. If we are required to find out the time when the velocity vanishes (or becomes, zero) or we are required to find out the acceleration at a point at which the velocity of the body becomes zero (or vanish) and $y = f(t)$, a given law of motion, then we put $v = 0$ in the expression for $\frac{dy}{dt}$.

2. Initial given conditions determine the values of the constants appearing in the law of motion given by $y = f(t)$.

3. If we are given a law of motion $y = f(t)$ as well as initial values of velocity k_1 and acceleration k_2 and we have to find out the velocity and acceleration at time t, then we put $t = 0$, $\frac{dv}{dt} = k_2$ $\frac{dS}{dt} = 0$ to determine the constants appearing the give law of motion $y = f(t)$.

4. Initial values of velocity and acceleration means the values of velocity and acceleration when $t = 0$ or alternatively, the initial position, initial velocity and initial acceleration correspond to time $t = 0$.

5. Uniform velocity (acceleration) means that the velocity (acceleration) is constant.

6. $\frac{ds}{dt} \times \frac{dt}{ds} = 1$ which $\Rightarrow \frac{ds}{dt} = \frac{1}{dt/ds}$

7. $\frac{dv}{ds} \cdot \frac{ds}{dt} = a =$ acceleration.

8. $a = \frac{d^2 s}{dt^2} = \frac{d}{dt}\left(\frac{ds}{dt}\right) = \frac{dv}{ds} \cdot \frac{ds}{dt} = v \cdot \frac{dv}{ds}$

9. $\left(\frac{ds}{dt}\right)_{t=c}$ means the value of velocity at time $t = c$ or after time $t = c$ or at the end of time $t = c$ or when time is $t = c$.

10. $\left(\frac{d^2 s}{dt}\right)_{t=c}$ means the value of acceleration when the time is $t = c$ or after time $t = c$ or at the end of time $t = c$ or at time $t = c$.

11. In the problems considered if the path s is expressed in meters (m), time t in seconds (s), velocity is v in metres per second (m/s) and acceleration a in metres per second per second (m/s^2) which is read "meter per second squared".

12. If a point move in a straight line, the velocity v at a given instant of time $t = t_0$ (called the instantaneous velocity) is defined as the derivative $\frac{ds}{dt}$ of the path s with respect to time t evaluated for $t = t_0$.

13. The acceleration a at a given instant of time $t = t_0$ is the derivative $\frac{dv}{dt}$ of the velocity v with respect to time t calculated for $t = t_0$.

14. The particle or body comes to rest at a point where $v = 0$.

15. It should be noted that $\frac{ds}{dt}$ is positive when s is increasing and $\frac{ds}{dt}$ is negative when s is decreasing.

16. It should be noted that $\frac{dv}{dt}$ is positive when the velocity is increasing and negative the velocity is decreasing.

Worked out problems

Based on velocity and acceleration as a rate measurer

1. A stone projected vertically upwards with initial velocity 112 ft/sec moves according to the law $s = 112t - 16t^2$, where s is the distance from the starting point. Find the velocity y and acceleration when $t = 3$ and when $t = 4$.

Solution: $s = (112t - 16t^2)$ ft

$$\Rightarrow \frac{ds}{dt} = (112 - 32t) \text{ ft/sec} \quad \ldots(1)$$

$$\Rightarrow \frac{d^2 s}{dt^2} = -32 \text{ ft/sec}^2 \quad \ldots(2)$$

Now, $\left[\frac{ds}{dt}\right]_{t=3} = 112 - 96 = 16$

and $\left[\frac{d^2 s}{dt^2}\right]_{t=3} = [-32]_{t=3} = -32$

Again $\left[\frac{ds}{dt}\right]_{t=4} = [112 - 32 \times 4] = 112 - 128 = -16$

and $\left[\frac{d^2s}{dt^2}\right]_{t=4} = [-32]_{t=4} = -32$

2. The position of a particle in motion is given by $s = 180\,t - 10\,t^2$. What is its velocity at the end of t second? At what instant would its velocity be zero.

Solution: $s = 180\,t - 10\,t^2$

Velocity at the end of t second, $\frac{ds}{dt} = 180 - 20t$

$\therefore \frac{ds}{dt} = 0 \Rightarrow 180 - 2t = 0$

$\Rightarrow t = \frac{180}{20} = 9$ sec

$\therefore$ The velocity is zero when $t = 9$ sec.

3. A body moves in a straight line according to the law of motion $s = t^3 - 4t^2 - 3t$. Find its acceleration at the instant (time) when the velocity is zero.

Solution: $s = t^3 - 4t^2 - 3t$

$$\Rightarrow \frac{ds}{dt} = 3t^2 - 8t - 3 \quad \ldots(1)$$

$$\Rightarrow \frac{d^2s}{dt^2} = 6t - 8 \quad \ldots(2)$$

Now, we are required to find out the value of t.

when $\frac{ds}{dt} = v = 0$. Now $v = 0$

$\Rightarrow 3t^2 - 8t - 3 = 0$

$\Rightarrow 3t^2 - 9t + t - 3 = 0$

$\Rightarrow (3t + 1)(t - 3) = 0$

$\Rightarrow t = 3$ secs

$\therefore$ The acceleration when the velocity is zero

$= \left(\frac{d^2s}{dt^2}\right)_{t=3}$

$= 6.3 - 8$

$= 10$ units.

Note: To find the acceleration when velocity is zero $\Leftrightarrow$.

(i) First find t, by putting $\frac{ds}{dt} = 0$ and solve the equation for t.

(ii) Put the value of t in the expression in t for $\frac{d^2s}{dt^2}$

4. A point moves in a plane according to the law $x = t^2 + 2t$ and $y = 2t^3 - 6t$. Find $\frac{dy}{dx}$ when $t = 0$, $t = 2$ and $t = 5$.

Solution: $\frac{dx}{dt} = 2t + 2 \quad \ldots(1)$

$$\frac{dy}{dt} = 6t^2 - 6 \quad \ldots(2)$$

Now, $\frac{dy}{dx} = \frac{dy/dt}{dx/dt} = \frac{6t^2 - 6}{2t + 2}$

$= \frac{6(t^2 - 1)}{2(t + 1)} = \frac{3(t^2 - 1)}{(t + 1)} = 3(t - 1)$

$\therefore \left(\frac{dy}{dx}\right)_{t=0} = 3(0 - 1)$

$= -3$ units

$\left(\frac{dy}{dx}\right)_{t=2} = 3(2 - 1)$

$= 3$ units

$\left(\frac{dy}{dx}\right)_{t=5} = 3(5 - 1)$

$= 12$ units.

5. Find the velocity and acceleration of a moving point after 10 sec if its position is given by $s = 5t^2 + 5t - 3$ if s is measured in centimeters.

Solution: $s = 5t^2 + 5t - 3$ cm

$$\Rightarrow \frac{ds}{dt} = (10t + 5) \text{ cm/sec} \quad \ldots(1)$$

$$\Rightarrow \frac{d^2s}{dt^2} = 10 \text{ cm/sec}^2 \quad \ldots(2)$$

Now, $\left[\frac{ds}{dt}\right]_{t=10} = [10t + 5]_{t=10} = 105$ cm/sec

$$\left[\frac{d^2 s}{dt^2}\right]_{t=10} = [10]_{t=10} = 10 \text{ cm/sec}^2.$$

Remember: To find the velocity and acceleration at time $t = c$ or after time $t = c$ or at the end of time $t = c$. We find

$$\left[\frac{ds}{dt}\right]_{t=c} \text{ and } \left[\frac{d^2 s}{dt^2}\right]_{t=c}$$

6. If s represents the distance which a body moves in time 't', determine its acceleration if $s = 250 - 40t - 16t^2$. Determine the acceleration and time when the velocity vanishes and the value of s then.

Solution: $s = 250 - 40t - 16t^2$

$$\Rightarrow \frac{ds}{dt} = -40 - 32t \text{ units} \quad \ldots(1)$$

$$\Rightarrow \frac{d^2 s}{dt^2} = -32 \text{ units} \quad \ldots(2)$$

Now, we are required to find out the time and acceleration when velocity $= 0$

$$\left[\frac{ds}{dt}\right] = 0$$

$$\Rightarrow -40 - 32t = 0$$

$$\Rightarrow t = -\frac{5}{4} \text{ sec} \quad \ldots(3)$$

$$\text{Now, } \left[\frac{d^2 s}{dt^2}\right]_{t=-\frac{5}{4}} = [-32]_{t=-\frac{5}{4}} = -32 \text{ units}$$

Lastly, the value of S at $t = -\frac{5}{4}$ sec

$$= [s]_{t=-\frac{5}{4}} = 250 - 40 \times \left(-\frac{5}{4}\right) - 16 \times \frac{25}{16}$$

$$= 250 + 50 - 25 = 275 \text{ units}$$

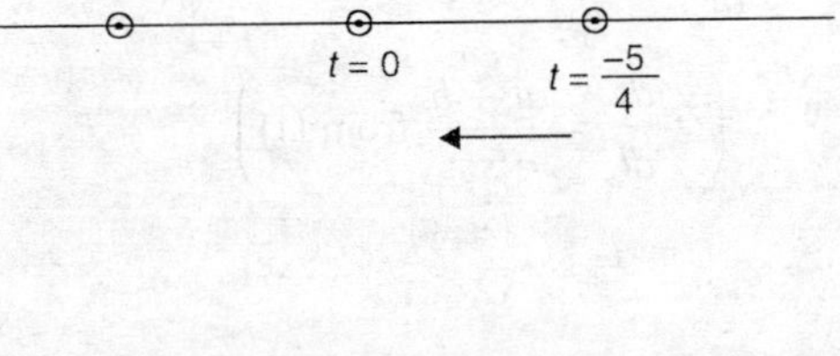

Note: Here negative time $-\frac{5}{4}$ secs points that measurement of time is started from the position where $s = 250$ and the motion started $-\frac{5}{4}$ secs earlier.

7. An aeroplane moves a distance of $(3t^2 + 2t)$ ft in t-seconds. Find its velocity when it has flown for 5 minutes.

Solution: let the distance moved by the aeroplane in t-seconds be s.

$$\therefore s = 3t^2 + 2t$$

$$\Rightarrow \frac{ds}{dt} = 3 \times 2 \times t + 2 = 6t + 2$$

$\therefore$ velocity of the aeroplane when it has flown for 5 minutes i.e. 300 seconds = the value of $\frac{ds}{dt}$ when $t = 300$.

$$= \left[\frac{ds}{dt}\right]_{t=300} = [6t + 2]_{t=300} = 1802 \text{ ft/sec.}$$

8. A point moves in accordance with the law $v = a + bt + ct^2$ and the initial values of the velocity and acceleration are 3 ft/sec and 2 ft/sec^2 respectively and at the end of the first second the acceleration is 12 ft/sec^2.

Find: **(i)** the velocity at the end of 3 seconds:

(ii) the acceleration at the end of 4 seconds.

Solution: we are given $v = a + bt + ct^2$ ...(1)

Differentiating both sides of (1) w.r.t t we have the

$$\text{acceleration} = \frac{dv}{dt} = b + 2ct \quad \ldots(2)$$

Now we will find the values of constants a, b and c from the given initial conditions of the problem.

Initial velocity = 3 ft/sec^2 which $\Rightarrow [v]_{t=0} = 3$ ...(3)

Initial acceleration = 2 ft/sec^2 which $\Rightarrow \left[\frac{dv}{dt}\right]_{t=0} = 2$

...(4)

$$(3) \Rightarrow [v]_{t=0} = \left[a + bt + ct^2\right]_{t=0} = 3$$

$$\Rightarrow a = 3$$

$(4) \Rightarrow \left[\frac{dv}{dt}\right]_{t=0} = [b + 2ct]_{t=0}$

$= b = 2$

Again, acceleration = 12 ft/sec^2 at the end of first second

$\Rightarrow$ acceleration = 12 ft/sec^2 when $t = 1$ which

$$\Rightarrow \left[\frac{dv}{dt}\right]_{t=1} = 12 \quad \ldots(5)$$

$$(5) \Rightarrow \left[\frac{dv}{dt}\right]_{t=1} = [b+2ct]_{t=1} = b+2c = 12 \quad \ldots(6)$$

Now putting the value of b in (6), we have

$2 + 2c = 12$ which $\Rightarrow 2c = 12 - 2 = 10 \Rightarrow c = 5$

$\therefore$ Required velocity when $t = 3 = [v]_{t=3}$

$= \left[a + bt + ct^2\right]_{\substack{a=3\\b=2\\c=5\\t=3}}$

$= [3 + 2t + 5t^2]_{t=3}$

$= [3 + 2 \times 3 + 5 \times 9]$

$= [3 + 6 + 45] = 54$ ft/sec.

Required acceleration when $t = 4 = \left[\frac{dv}{dt}\right]_{t=4}$

$= [b + 2ct]_{\substack{b=2\\c=5\\t=4}}$

$= 2 + 2 \times 5 \times 4$

$= 42$ ft/sec^2.

Conditional Problems

When a law of motion of a particle (or, body) is given by a formula $y = f(x)$ (or the law of motion stated in words be written in the symbolic form) and we have to show that distance, velocity or acceleration of the particle (or, body) obeys a differential equation of motion we adopt the following working rule.

1. Find $\frac{dy}{dx}$ and $\frac{d^2y}{dx^2}$ by differentiating the given formula given in words translated into symbolic from or the given law $y = f(x)$ with respect to the given independent variable.

2. After obtaining $\frac{dy}{dx}$ and $\frac{d^2y}{dx^2}$, we arrive at our target (required law or formula) using various mathematical manipulations (like simplification, cancellation or substitution etc) performed upon the expression obtained for $\frac{dy}{dx}$ and $\frac{d^2y}{dx^2}$.

Examples worked out

1. If a body moves according to the law $s = a + bt + ct^2$, show that its acceleration is constant.

2. If $s^2 = at^2 + 2bt + c$, show that acceleration varies as $\frac{1}{s^3}$.

Solution: **1.** $s = a + bt + ct^2$

$$\Rightarrow \frac{ds}{dt} = b + 2ct$$

$$\Rightarrow \frac{d}{dt}\left(\frac{ds}{dt}\right) = 2c$$

$$\Rightarrow \frac{d^2s}{dt^2} = \text{acceleration} = 2c = k \text{ (say)} = \text{constant}$$

(proved).

2. Given that $s^2 = at^2 + 2bt + c$

$$\Rightarrow 2s\frac{ds}{dt} = 2at + b$$

$$\Rightarrow s\frac{ds}{dt} = at + b \quad \ldots(1)$$

$$\Rightarrow \frac{ds}{dt} \cdot \frac{ds}{dt} + s \cdot \frac{d^2s}{dt^2} = a$$

(differentiating (1) w.r.t 't')

$$\Rightarrow \left(\frac{ds}{dt}\right)^2 + s\frac{d^2s}{dt^2} = a$$

$$\Rightarrow s\frac{d^2s}{dt^2} = a - \left(\frac{ds}{dt}\right)^2 = a - \frac{(at+b)^2}{s^2}$$

$$\left(\because \frac{ds}{dt} = \frac{at+b}{s} \text{ from (1)}\right)$$

$$\Rightarrow s\frac{d^2s}{dt^2} = \frac{as^2 - (at+b)^2}{s^2}$$

$$= \frac{a\left(at^2 + 2bt + c\right) - \left(a^2t^2 + 2abt + b^2\right)}{s^2}$$

$$\left(\because s^2 = at^2 + 2bt + c\right)$$

$$= \frac{ac - b^2}{s^2}$$

$$\Rightarrow \frac{d^2s}{dt^2} = \frac{ac - b^2}{s^3}$$

$$\Rightarrow \frac{d^2s}{dt^2} = \frac{k}{x^3} \text{ (where } k = ac - b^2 \text{ (say))}$$

$$\Rightarrow \frac{d^2s}{dt^2} \propto \frac{1}{x^2} \text{ (proved).}$$

3. If the law of motion of a point moving in a straight line be $ks = \log\frac{1}{v}$, prove that acceleration f is given by $f = -kv^2$; s and v represent distance and velocity respectively.

Solution: $ks = \log\frac{1}{v} = \log 1 - \log v = -\log v$ ($\because$ log 1 = 0)

$$\Rightarrow v = e^{-ks} \qquad \ldots(1)$$

$$\Rightarrow \frac{dv}{ds} = -ke^{-ks} = -kv \text{ (from (1)} \qquad \ldots(2)$$

$$\Rightarrow v\frac{dv}{ds} = -kv^2 \text{ (multiplying both sides of (2) by } v)$$

$$\Rightarrow f = -kv^2 \left(\because f = \frac{v\,dv}{ds}\right)$$

4. If the velocity of a point moving in a straight line is given by $v^2 = se^s$, prove that the acceleration is $\frac{1}{2}\left(1 + \frac{1}{s}\right)\cdot v^2$, where x is a constant.

Solution: $v^2 = se^s \qquad \ldots(1)$

$$\Rightarrow \frac{d\left(v^2\right)}{dt} = \frac{d}{ds}\left(s\cdot e^s\right) = s\cdot\frac{d}{ds}e^s + e^s\frac{ds}{ds} = se^s + e^s$$

$$\Rightarrow 2v\frac{dv}{ds} = (s+1)\cdot\frac{v^2}{s}\left(\because \text{ from (1) } e^s = \frac{v^2}{s}\right)$$

$$\Rightarrow v\frac{dv}{ds} = \frac{1}{2}\left(1 + \frac{1}{s}\right)\cdot v^2$$

5. If the law of motion is $t = as^2 + 2bs + c$, show that the acceleration varies as v^3 and has a sign opposite to that of a.

Solution: $t = as^2 + 2bs \qquad \ldots(1)$

$$\Rightarrow \frac{dt}{ds} = 2as + 2ab$$

$$\Rightarrow v = \frac{ds}{dt} = \frac{1}{2as + 2b} = \frac{1}{2}(as + b)^{-1} \qquad \ldots(2)$$

$$\Rightarrow \frac{dv}{dt} = \frac{d}{dt}\left(\frac{ds}{dt}\right) = \frac{1}{2}(-1)\cdot(as+b)^{-2}\cdot\left(a\frac{ds}{dt}\right)$$

$$\Rightarrow \frac{dv}{dt} = \frac{d^2s}{dt^2} = -\frac{1}{2}(as+b)^{-2}\cdot a\cdot\frac{1}{2}(as+b)^{-1}$$

$$\left(\because \frac{ds}{dt} = \frac{1}{2}(as+b)^{-1} \text{ from (2)}\right)$$

$$= -\frac{1}{4}a(as+b)^{(-3)} = -\frac{1}{4}\cdot a\cdot\left\{(as+b)^{(-1)\times(3)}\right\}$$

$$= -\frac{1}{4}a\cdot(2v)^{(3)}$$

$$= -\frac{1}{4}a\,(2v)^3 \text{ (from (2))}$$

$$= -\frac{1}{4}a\cdot 8\cdot v^3$$

$$= -2av^3$$

$$= -kv^3 \text{ (where } k = 2a)$$

which $\Rightarrow \frac{dv}{dt}$ varies as v^3

Also, it is clear that if a is +ve, the acc. is –ve and if a is –ve, the ecc. is +ve, i.e., the acc. is opposite to the sign of a. (proved).

6. If the equation of a rectilinear motion be $s = \sqrt{t+1}$ where s is the displacement and t the time, show that the acceleration is negative and proportional to the cube of velocity.

Solution: $s = \sqrt{t+1}$

$$\Rightarrow v = \frac{ds}{dt} = \frac{1}{2}(t+1)^{-\frac{1}{2}} \quad ...(1)$$

$$\Rightarrow a = \frac{dv}{dt} = \frac{d}{dt}\left(\frac{ds}{dt}\right) = \left(\frac{1}{2}\right)\left(-\frac{1}{2}\right)\cdot(t+1)^{-\frac{3}{2}}$$

$$= -\frac{1}{4}(2v)^3 \text{ [from (1)}$$

$= -2v^3$ which is negative

$= -kv^3$ (where $k = 2$)

which $\Rightarrow$ acceleration = a varies as cube of velocity and is negative.

7. Prove that if a particle moves so that the space described is proportional to the square of the time of description, the velocity will be proportional to the time and the rate of increase of the velocity will be constant.

Solution: let s = the space described by the particle in time 't'.

$s \propto t^2$

$\Rightarrow s = kt^2$ where $k = a$ constant

$$\Rightarrow \frac{ds}{dt} = 2kt$$

$$\Rightarrow v = 2kt \ \left(\frac{ds}{dt} = v = \text{velocity of the particle}\right) \quad ...(1)$$

$\Rightarrow v \propto t$

Again, $\frac{dv}{dt} = 2k$ (differentiating (1) w.r.t 't')

$\Rightarrow \frac{dv}{dt}$ a constant

$\Rightarrow$ rate of increase in velocity = constant.

8. Prove that if a particle moves so that the space described is proportional to the cube of the time of description, the velocity will be proportional to the square of the time and acceleration will be proportional to the time.

Solution: According the given law

$s \propto t^3$

$\Rightarrow s = kt^3$ (where k = constant) ...(1)

$$\Rightarrow \frac{ds}{dt} = 3kt^2 \quad ...(2)$$

$$\Rightarrow \frac{ds}{dt} \propto t^2$$

and $\frac{d^2s}{dt^2} = 6kt \Rightarrow \frac{d^2s}{dt^2} \propto t$.

9. A moving body describes a distance x which is proportional to sin at in time 't'. prove that the acceleration will be proportional to the distance travelled by the body.

Solution: let x = distance described in time 't'

Given that $x \propto \sin at$

$\Rightarrow x = k \sin at$ ...(1)

$$\Rightarrow \frac{dx}{dt} = ak \cos at$$

$$\Rightarrow \frac{d^2x}{dt^2} = -a^2 k \sin at = -a^2 x \text{ (from (1))}$$

$$\Rightarrow \frac{d^2x}{dt} \propto x \text{ (proved).}$$

N.B.: The motion is simple harmonic.

Type I: *Problems based on area, perimeter and volume*

Exercises 17.1

(A) Problems based on triangle

1. If the side of an equilateral triangle increases uniformly at the rate of 3 ft/sec, at what rate is the area increasing when the side is 10 ft?

(***Ans.*** $15\sqrt{3}$ sq. ft/sec)

2. If the side of an equilateral triangle increases at the rate of $\sqrt{3}$ ft per second and its area at the rate of 12 sq. ft per second, find the side of the triangle.

(***Ans.*** x = side of the $\Delta = 8$ ft)

3. The area of an equilateral triangle is expanding. How many time, as fast as each of its sides is the area increasing at any instant? What is the rate of increase of the area when each of the equal sides is 15 inches long and increasing at the rate of 10 inches a second.

(***Ans.*** $\frac{\sqrt{3}}{2}$ times as fast as each of its sides; $75\sqrt{3}$ square inches per sec)

4. A triangle whose sides are varying with time is always equilateral. The rates at which the area and the height increase simultaneously at an instant at an instant are 6 m^2/sec and 3 m/sec respectively. Find the rate of increase of the side at that instant.

(***Ans.*** 2 m/sec)

(B) Problems based on square and rectangle

1. The side of a square sheet of metal is increasing at 3 cm/minute. At what rate is the area increasing when the side is 10 cm long.

(***Ans.*** 60 cm per minute)

2. The sides of a square plate of metal are expanding uniformly at the rate of 0.3 cm per second. Find the rate at which its area is (i) 30 cm (ii) 50 cm.

(***Ans.*** **(i)** 18 sq cm/sec **(ii)** 30 sq cm/sec)

3. A square plate of metal is expanding and each of its sides is increasing at a uniform rate of 2 inches per minute. At what rate is the area of the plate increasing when the side is 20 inches long?

(***Ans.*** 60 sq. cm per second)

4. A rectangle is of given perimeter *p*. Find the rate of change of the area at the instant when the length equals the breadth.

(***Ans.*** 0 = zero)

5. The breadth of a rectangle is increasing at the rate of 2 cm per second and its length is always 3 times its breadth. When the breadth is 5 cm, at what rate is the area of the rectangle increasing.

(***Ans.*** 60 sq. cm per second)

(C) Problems based on circle

1. A balloon which always remains spherical has a variable diameter $\frac{3}{2}(2x+3)$. Determine the rate of change of its volume with respect to *x*.

2. The radius of a circle is increasing uniform by at the rate of 3 cm/sec. At what rate is the area increasing when the radius is 10 cm.

(***Ans.*** 60π cm/sec^2)

3. If the radius of a circle increase at a uniform rate of 6 cm per second, find the rate of increase of its area when the radius is 50 cm.

(***Ans.*** 600π cm^2/sec)

4. If the radius of a circle is increasing at the constant rate of 2 ft per second, find the rate of increase of its area when the radius is 20 ft.

(***Ans.*** 80π ft^2/sec)

5. If the circular waves in a tank expand so that the circumference increases at the rate of a ft/sec, show that the radius of the circle is increasing at the rate of $\frac{a}{2\pi}$ ft/sec.

6. The area of a circle is increasing at the uniform rate of 5 sq. cm per minute. Find the rate in cm per minute at which the radius is increasing when the circumference of the circle is 40 cm.

(***Ans.*** $\frac{1}{8}$ cm/minute)

7. A spherical balloon is inflated and the radius is increasing at $\frac{1}{3}$ inches/minute. At what rate would the volume be increasing at the instant when its radius is 2 inches.

(***Ans.*** $\frac{16\pi}{3}$ inch^3/minute)

8. A spherical balloon is pumped at the rate of 10 cubic inches per minute. Find the rate of increase of its radius when its radius is 15 inches.

(***Ans.*** $\frac{1}{90\pi}$ inch/minute)

9. If the area of a circle increase at a uniform rate, show that rate of increase of the perimeter varies inversely as the radius.

10. A spherical ball of salt is dissolving in water in such a manner that the rate of decrease in volume is proportional to the surface. Prove that the radius is decreasing at a constant rate.

11. The volume of a spherical soap bubble is denoted by v, its surface by S, the radius being r. Prove that

(i) $\frac{dv}{dt} = 4\pi r^2 \cdot \frac{dr}{dt}$ (ii) $\frac{ds}{dt} = \frac{2}{r} \cdot \frac{dv}{dt}$.

12. A sphere of metal is expanding under the action of heat. Compare the rate of increase of its volume with that of its radius. At what rate is the volume increasing when the radius is 2 inches and increasing at the rate of $\frac{1}{3}$ inches per minute.

(*Ans.* 8 π cubic inch/sec)

(D) Problems based on cube

1. An edge of a variable cube is increasing at the rat of 3 cm/sec. how fast is the volume of the cube increasing when the edge is 10 cm long?

(*Ans.* 900 cm^3/sec)

2. When a cubical block of metal is heated, each edge increases .1 percent per degree of rise in temperature. Show that the surface increases .2 percent and the volume .3 percent per degree.

3. A metal cube is heated so that its edge increases at the rate of 2 cm/minute. At what rate the volume of the cube increases when the edge is 10 cm long?

(*Ans.* 2400 cubic cm/minute)

4. The volume of a cube increases at a constant rate. Prove that the increase in its surface varies inversely as the length of the side.

5. The temperature of a metal cube is being raised stead by so that each edge expands at the rate of .01 inch per hour. At what rate is the volume increasing when the edge is 2 inches.

(*Ans.* 12 cu. in per hour)

(E) Problems based on cylinder

1. A cylindrical vessel is held with its axis vertical. Water is poured into it at the rate of one point per second. Given that one point is equal to 34.66 cubic inches, find the rate at which the surface of water is rising in the vessel when the depth is x inches.

(*Ans.* $\frac{34.66}{A}$ inch per second where A is the area of cross-section)

2. Water is running out of cistern in the form of an inverted right circular cone of semi vertical angle 45° with its axis vertical. Find the rate at which the water is flowing out at the instant when the depth of water is 2 ft; given that at that instant, the level of the water is diminishing at the rate of 3 inches per minute.

(*Ans.* π ft^3/minute)

3. The volume of a right circular cone is constant. If the height decreases at a constant rate of 8 cm/minure, how fast is the radius of the base changing at the instance when the height is 8 cm and the radius of the base is 4 cm.

(*Ans.* 2 cm/min)

***Type 2*: Problems based on right angled triangle**

Exercises 17.2

(A) Problems based on the height of a man

1. A man of height 6 ft walks directly away from a lamp post of height 15 ft at the rate of 3 miles per hour. At what rate does his shadow lengthen?

(*Ans.* 2 miles/hour)

2. A man 1.6 m high walks at the rate of 50 meters per minute away from a lamp which is 4m above the ground. How fast is the man's shadow lenghtning?

3. A man of height $5\frac{1}{2}$ ft approaches directly towards a lamp-post along a horizontal road. If the light is 8 ft above the level of the road, show that the length of this shadow decreases at a rate of $\frac{11}{25}$ times the rate at which be approaches the lamp-post.

4. A man 6ft tall walking away along a straight line from the foot of the light post, 30 ft high at the rate of 4 ft per second; find how fast be is approaching the wall.

(*Ans.* $\frac{5}{4}$ miles/hour)

5. A man of height $5\frac{1}{2}$ ft walks directly away from a lamp-post of height 120' at the rate of $4\frac{1}{2}$ miles per hour; find how fast is his shadow lenthening.

(*Ans.* $\frac{96}{58}$ miles/hour)

6. A man 160 cm tall walks from a lamp-post 4m high at the rate of 3 km/hr. Find the rate at which the shadow of his head moving on the pavement and the rate at the which his shadow is lengthening.

(*Ans.* (i) 5 km/hr (ii) 2 km/hr)

7. A man 6 ft tall walking away along a straight line from the foot of light-post, 30 ft high at the rate of 5 miles per hour. How fast does the end of this shadow move?

(*Ans.* $\frac{5}{4}$ miles/hour)

8. A man 6 ft tall walks at the rate of 5 ft/sec towards street light 16 ft above the ground. At what rate is the tip of his shadow moving? At what rate is the length of his shadow changing when ne is 10 ft from the foot of the light?

(*Ans.* 8 ft/sec and decreasing at 3 ft/sec)

(B) Ladder problems

1. A ladder 13 ft long slides down from a vertical wall remaining in a vertical plane all the time. What is the velocity of the upper end when the lower end is at a distance of 5 ft from the wall and has a velocity of 2 ft per second?

(*Ans.* 5 ft/sec)

2. A ladder is inclined to a wall making an angle of 30° with it. A man is ascending the ladder at the rate of 3 ft/sec. how fast is he approaching the wall.

(*Ans.* $\frac{2}{3}$ ft/sec)

3. A ladder is inclined to a wall making an angle of 45° with it. If a man is ascending the ladder at the rate of 4 ft/sec; find how fast he is approaching the wall.

(*Ans.* $2\sqrt{2}$ ft/sec)

4. A ladder 26 ft in length is resting on a horizontal plane inclined against a vertical wall. It slips away from the wall at the rate of 5 ft/sec. Find the velocity of the top of the ladder down the wall when it is at a height of 24 ft.

(*Ans.* $\frac{25}{12}$ ft/sec)

5. A ladder 5 meter long standing on a horizontal floor leans against a vertical wall. If the top of the ladder slides downwards at the rate of 10 cm/sec. Find the rate at which the angle between the floor and the ladder is decreasing when the lower end of the ladder is 2 meter from the wall.

(*Ans.* $\frac{1}{20}$ radius/sec)

6. A ladder 25 ft long slides down from a vertical wall remaining in a vertical plane all the time. What is the velocity of the upper end when the lower end is at a distance of 15 ft from the wall and has a velocity of 3 ft/sec.

(*Ans.* $\frac{9}{4}$ ft/sec)

(C) Kite problems

1. A kite is 45 ft high and there is 117 ft cord out. If the kite is moving horizontally at the rate of 13 m.p.h directly away form the position who is flying it, find how fast the cord is being paid out.

(*Ans.* 19 m.p.h)

2. A kite is 100 ft high and there is 260 ft of cord out. if the kite is moving horizontally at the rate of $3\frac{1}{4}$ miles per hour directly away from the person who is flying it; how fast is the cord being paid out.

(*Ans.* 3 miles/hour)

3. A kite is moving horizontally at a height of 151.5 meters. If the speed of the kite is 10 meters/sec, how fast is the string being paid out when the kite is 250 meters from the boy who is flying the kite, the height of the boy being 1.5 meters.

(*Ans.* 8 m/sec)

4. A girl flies a kite at a height of 300 ft; the wind carrying the kite horizontally away from her at a rate of 25 ft/sec. How fast must she let out the string when the kite is 500 ft away from her.

(*Ans.* 20 ft/sec)

(D) Rod Problems

1. A rod 13 ft long moves with its ends *A*, *B* on two perpendicular lines *OX* and *OY* respectively. If the end *A* is 12 ft from *O* and is slipping away at $2\frac{1}{2}$ ft/sec; find how fast end *B* is moving?

(*Ans.* –6 ft/sec)

2. Two rods AB and AC are inclined to each other at an angle of 120°. A car starts from B towards A with a velocity of 40 miles an hour and an other can starts from C towards the same place with an equal velocity at the same. If $AB = AC$, find the rate at which each is approaching the other.

(*Ans.* $40\sqrt{3}$ m/h)

3. Two straight rods OA and OB cross each other at O at right angles; If $OA = 10$ miles and $OB = 8$ miles, a man starts at the rate of 3 miles per hour from O to B. Find the rate at which his distance from A is altering.

(*Ans.* $\frac{3x}{\sqrt{10x + x^2}}$)

4. A rod AB of length 10 ft slides with ends A and B on two perpendicular OX and OY respectively. If the end A on OX moves with a constant velocity 2 ft/minute, find the velocity of its med-point at the time the rod makes an angle of 30° with OX.

(*Ans.* 2 ft/minute)

Problems based on velocity and acceleration

Exercise 17.2.1

1. A point moves in a straight line according to the law:

$S = 2t^3 + t^2 - 4$

Find its velocity and acceleration at the instant of time $t = 4$.

(*Ans.* **(i)** $v(4)$ 104 m/s **(ii)** $a(4) = 50$ m/s^2)

2. A point moves in a straight line as given by the equation:

$S = 6t - t^2$

At what instant of time will the velocity of the point be equal to zero?

(*Ans.* $t = 3$ second)

3. Find the velocity and acceleration at the indicated instants of time for a point moving in a straight line if its motion is described by the following equations.

(i) $S = t^3 + 5t^2 + 4, t = 2$

(ii) $S = \sqrt{t}, t = 1$

(iii) $S = t^2 + 11t + 30, t = 3$

(*Ans.* **(i)** 32 m/s; 22 m/s^2 **(ii)** 0.5 m/s; –0.25 m/s^2 **(iii)** 17 m/s; 2 m/s^2)

4. At time t, the distance x of a particle moving in a straight line is given by $x = 4t^2 + 2t$. Find the velocity and acceleration when $t = \frac{1}{2}$.

(*Ans.* 6, 8)

5. The distance S, at the time t, of a particle moving in a straight line is given by the equation $S = t^4 - 18t^2$. Find its speed at $t = 10$ seconds.

(*Ans.* 340 units/sec)

6. A particle is moving in a straight line in such a way that its distance in cm from a fixed point on the line after t seconds is given by $4t^3 + 2t + 5$. Find the distance, velocity and acceleration at the end of 3 seconds.

(*Ans.* **(i)** 119 cm **(ii)** 100 cm/sec **(iii)** 72 cm/sec^2)

7. The distance S meters moved by a particle travelling in a straight line in t seconds is given by $S = 45t + 11t^2 - t^3$. Find the time when the particle comes to rest.

[*Hint*: solve $\frac{ds}{dt} = 0$]

(*Ans.* 9 seconds)

8. A particle is moving on a line where its position S in meters is a function of time t in seconds given by $S = t^3 + at^2 + bt + c$, where a, b, c are constants. It is known that $t = 1$ second, the position of the particle is given by $S = 7$ meters, velocity is 7 m/sec and the acceleration is 12 m/sec^2. Find the values of a, b and c.

(*Ans.* $a = 3, b = -2, c = 5$)

9. Find the acceleration of a moving point at the indicated instants of time if the velocity of the point moving in a straight line is given by the following equation:

(i) $v = t^2 + t - 1, t = 3$

(ii) $v = t^2 + 5t + 1, t = 3$

(*Ans.* **(i)** 7 m/s^2 **(ii)** 11 m/s^2)

10. A point moves in a straight line according to the law $S = t^2 - 8t + 4$. At what instant of time will the velocity of the point turn out to be equal to zero?

(*Ans.* $t = 4s$)

11. A point moves in a straight line according to the law:

$S = \sin^2 t$

Find the instant of time at which its acceleration is equal to 1.

(**Ans.** $t = \pm\frac{\pi}{6} + \pi k,\ k \in z$)

12. A point moves in a straight line according to the law:

$S = \sin^2 t$

Find the instant t at which its acceleration is equal to zero.

(**Ans.** $t = \frac{\pi}{4} + \frac{\pi k}{2},\ k \in z$)

13. If $S = \frac{1}{3}t^3 - 16t$, find the acceleration at the time when the velocity vanishes.

(**Ans.** 8 units/sec^2)

Conditional problems

Exercise 17.2.2

1. If a particle vibrates according to the law:

$y = a \sin (Pt - e)$, show that the velocity and acceleration at any instant are $aP \cos (Pt - e)$ and $-P^2 y$ respectively.

2. The motion of a particle moving in a straight line is given by $x = 3 \cos 2t$ with usual symbols. Show that its acceleration is proportional to the distance travelled by the particle and determine the distance x when the speed is zero.

3. If a particle moves so that the space described varies as the square of the time of description, prove that the velocity varies as the time and the acceleration is constant.

4. If the law of motion is $t = s^2 + s - 1$, show that acceleration varies as v^3.

5. If the law of motion is $t = as^2 + bs + c$, show that the rate of change of velocity is proportional to the cube of the velocity and has a sign opposite to that of a.

6. If $t = 2s^2 + 3s + 1$, show that acceleration is proportional to the cube of the velocity.

More problems on physical application of derivatives

Exercise 17.2.3

1. The law of change of temperature T of a body with time is given by $T = 0.2t^3$. At what rate does this body get warm at the instant of time $t = 10$?

2. A body of mass 10 kg moves in a straight line according to the law:

$S = 3t^2 + t + 4$

Find the kinetic energy of the body $\left(\frac{mv^2}{2}\right)$ four seconds after it started.

[**Hint**: (i) $v(4) = \left[\frac{ds}{dt}\right]_{t=4} = (6t+1)_{t=4} = 25$ m/sec

(ii) Determine the kinetic energy of the body at $t = 4$ which is $\frac{mv^2}{2} = 10 \times \frac{25^2}{2} = 3125(J)$]

3. The strength of current I changes with time 't' according to the law:

$I = 0.4t^2$, (I is expressed in amperes, t in seconds)

Find the rate of change in strength of current after expiry of the 8th second.

[**Hint**: $\left(\frac{dI}{dt}\right)_{t=8} = (0.8t)_{t=8} = 0.8.8 = 6.4$ A/m]

4. The temperature T of a body changes with time t obeying the law:

$T = 0.5t^2 - 2t$

At what rate does this body get warm at the instant of time $t = 5$?

(**Ans.** 3 deg/s)

5. A body of mass 100 kg moves in a straight line according to the law:

$S = 5t^2 - 2$

Find the kinetic energy of the body in two seconds after the beginning of motion.

(**Ans.** 20,000 J)

6. The change in the strength of current I with time t is given by the equation $I = 2t^2 - 5t$ (I is amperes, t in seconds). Find the rate of change in the strength of current after the expiry of 10th second.

(**Ans.** 25 A/S)

18

Approximations

Mathematics is a language.

J. Willard Gibbs

Let $y = f(x)$

(A) We should recall that $\frac{dy}{dx} = \lim_{\Delta x \to 0} \frac{\Delta y}{\Delta x}$ ...(1)

when Δx is very small, we write,

$$\frac{dy}{dx} = \frac{\Delta y}{\Delta x} \text{ (nearly)} \quad ...(2)$$

$$\therefore \Delta y = \frac{dy}{dx} \cdot \Delta x \text{ (when } \Delta x \text{ is small)} \quad ...(3)$$

(B) Again we should recall $\frac{dy}{dx} = \lim_{\Delta x \to 0} \frac{\Delta y}{\Delta x}$ ...(1)

when Δx is not sufficiently small, then $\frac{\Delta y}{\Delta x}$ will in general differ from $f'(x)$. If $\in$ is the difference between $\frac{\Delta y}{\Delta x}$ and $\frac{dy}{dx}$, we have

$\frac{\Delta y}{\Delta x} - \frac{dy}{dx} = \in$ which implies

$$\frac{\Delta y}{\Delta x} = \frac{dy}{dx} + \in$$

$$\Rightarrow \frac{\Delta y}{\Delta x} = f'(x) + \in$$

(where $\in \to 0$ as $\Delta x \to 0$) ...(2)

$$\Rightarrow \Delta y = f'(x) \cdot \Delta x + \in \Delta x \quad ...(3)$$

where the second term on the r.h.s of the equation (3) of (B) is very small and can be neglected (since $\in \to 0$ and $\Delta x \to 0 \Rightarrow \in \cdot \Delta x \to 0$ on using the product theorems on limits) and the first term $f'(x)\,\Delta x$ is the larger part of Δy (in equation (3) of (B)) which is known as the principal part. The principal part Δy is called the differential of y and is denoted by dy i.e.

But if we let $dy = f'(x)\,\Delta x$ ...(4)

$\therefore \Delta x = dx$, using this result in (4) for $y = x$, we have, therefore, $dy = f'(x)\,dx$ which tells the differential of y is obtained by differentiating the given function $f(x)$ w.r.t its independent variable x and then multiplying d.c of that function $f'(x)$ by the differential of x (i.e. dx).

(C) We know that $\frac{dy}{dx} = \lim_{\Delta x \to 0} \frac{\Delta y}{\Delta x}$ ...(1)

and when Δx is very small,

$$\frac{dy}{dx} = \frac{\Delta y}{\Delta x} \text{ (approximately)}$$

$$\Rightarrow \Delta y = \frac{dy}{dx} \cdot \Delta x \text{ (nearly, approximately, approx)}$$

$$\Rightarrow f(x + \Delta x) - f(x) = \frac{dy}{dx} \cdot \Delta x$$

$\Rightarrow f(x + \Delta x) = f(x) + \frac{dy}{dx} \cdot \Delta x$ which provides us an approximate formula and can be written in following way:

$$f(a + h) = f(a) + f'(a) \cdot h$$

where $x = a$ and $\Delta x = h$ are given numerical values.

Remember:

1. If the increament of the variable on which y depends (i.e. Δx) is small enough then $\Delta y \simeq dy$.

2. The fact that differential dy is the approximation of Δy when dx is small may be used to approximate errors.

3. $\frac{\Delta y}{\Delta x} = \frac{dy}{dx}$ (nearly) if the increament of the variable on which y depends is very small (or small enough).

4. $\Delta y = f(x + \Delta x) - f(x) \neq dy$ in general.

5. Relative error in y; let $y = f(x)$, then relative error in y at $x = a$ is

$$= \frac{f'(a)}{f(a)} \cdot h$$

$$= \frac{\text{derivative of the function } f(x) \text{ at } x = a \text{ times the increament}}{\text{value of the function at } x = a}$$

Now we shall explain different types of problems on approximations and errors besides their techniques to solve them.

Type A

To find the approximate value of a function of an independent variable x when the independent variable x is replaced by a number, we adopt the following working rule:

First working rule:

1. For finding $f(c)$ we choose a and h such that $f(c) = f(a + h)$ where $f(a)$ is easily obtainable and h is small

2. Find $f'(x)$ by differentiating $f(x)$

3. Find $f(a)$ and $f'(a)$

4. Use the formula $f(a \pm h) = f(a) \pm h f'(a)$ when $f(a)$, $f'(a)$ and h are known. h is a small number which is positive or negative and $(a \pm h)$ = given number for the independent variable. $f(a \pm h) = \sqrt[n]{a \pm h}$, $(a \pm h)^n$, $\log(a \pm h)$, $\sin(a \pm h)$, $\cos(a \pm h)$, $\tan(a \pm h)$, $\cot(a \pm h)$, $\sec(a \pm h)$, $\operatorname{cosec}(a \pm h)$, $\sin^{-1}(a \pm h)$, $\cos^{-1}(a \pm h)$, $\tan^{-1}(a \pm h)$, $\cot^{-1}(a \pm h)$, $\sec^{-1}(a \pm h)$, $\operatorname{cosec}^{-1}(a \pm h)$, $\log(a \pm h)$, $e^{(a \pm h)}$, ...etc. some examples.

Which provides us the required approximate value of the function for the given value of the independent variable.

Second working rule:

1. In finding the value of $f(c)$ approximately we express the given number c as $(a \pm n)$ to choose $x = a$ s.t $f(a)$ can be determined as a rule as a whole number b where a = any number whose nth root, power n, t-ratio, inverse t-ratio, log, e, etc are known to us and $n = \Delta x = h$ = any number +ve or –ve. After this step, we

2. Use the formula $\Delta y = \left[\frac{dy}{dx}\right]_{x=a} \cdot \Delta x = f'(a) \cdot h$.

3. Lastly are find the required approximate value using the formula:

$$y + \Delta y = f(a) + f'(a) \cdot h$$

where $\quad y + \Delta y = f(a + h)$

Note:

1. There are two types f notations for the approximate value of a function of an independent variable.

(i) $y + \Delta y$ **(ii)** $f(x + \Delta x)$ or $f(a + h)$ where $\Delta x = h$ is written for easiness and $\Delta y = \left[\frac{dy}{dx}\right]_{x=a} \cdot \Delta x = f'(a) \cdot h$.

2. Generally $h = a$ decimal fraction +ve or –ve or an integer +ve or –ve like ± 0.001, ± 0.002, ± 0.003, ± 0.009 incase of given number c is a decimal fraction and $\pm 1, \pm 2, \ldots$ etc in case of given whole number c.

3. If h = –ve integer or –ve decimal fraction, we use the formula:

$f(a-h) = f(a) - h f'(a)$

and if h = +ve integer or +ve decimal fraction, we use the formula:

$f(a+h) = f(a) - h f'(a)$

4. Usually incase of decimal fraction c the value of a is taken as the nearest integer. $a = 2$ is to be considered so that $a + h = 1.999$ gives us $h = -0.0001$ ($\because 2 - 0.001 = 1.999$), if $c = 1.999$.

5. Usually incase of whole number a to be considered, we express the given number c as $a \pm n$ where $x = a$ is s.t. $f(a)$ is known to us already where $f \Rightarrow \sqrt[n]{\ }$, $(\ )^n$, sin, cos, $\sin^{-1}$, $\cos^{-1}$, log, e or any other operator.

6. When $y = f(x)$, then $\Delta x = dx$ and $\Delta y = dy$ provided that Δx is small enough (or, provided that Δx is nearly equal to zero which means $\Delta x \to 0$).

7. $\dfrac{\Delta y}{y} = \dfrac{dy}{y}$ nearly Δx is small enough.

8. Notation $\doteqdot$, $\simeq$, $\doteq$ or $\approx$ means approximately equal to.

Some useful hints to find the values of *a* and *h* in some problems

1. If a given number $c = b + \dfrac{1}{10^n}$ or $b + \dfrac{k}{10^n}$ where k = digits from 1 to 9, b = any whole number, n = +ve integer, then $b = a = x$.

$$\Delta x = \frac{1}{10^n} \text{ or } \frac{k}{10^n} \text{ and given number} = x + \Delta x$$

where Δx = +ve

e.g.: $3.003 = 3 + \dfrac{3}{10^n} = x + \Delta x$ or $a + h$ if $x = a = 3$

and $\Delta x = h = 0.003$

Given number = 3.003 = 3 + 0.003

2. If a given number $c = b + \cdot 9...$ where dots after recurring decimal $\cdot 9$ denote any one or more than one digit from 1 to 9, then $x = a = b + 1$ and $\Delta x = h$ = given decimal fraction $-(b+1)$

$$= \left(b + \cdot 9...\right) - (b+1)$$

given number $= (b+1) + \Delta x$ where $\Delta x = -$ve

e.g.: $31.98 + 31 + .98 \Rightarrow x = a = b + 1 = 31 + 1 = 32$ (where $b = 31$) and $\Delta x = 31.98 - 32 = -0.002 = -$ve decimal fraction given number $32 + \Delta x = 32 - 0.002$

3. If a given number $= \dfrac{\cdot 9klm...}{10^n}$ where $k, l, m, \ldots$ = digits from 1 to 9 then $x = 1$

$$\Delta x = h = \frac{\cdot 9klm...}{10^n} - 1$$

given number = $1 + \Delta x$ where $\Delta x = -$ve

e.g.: $\Rightarrow x = a = 1$ and $\Delta x = 0.998 - 1 = -0.002$

given number = $1 + \Delta x = 1 - 0.002$

N.B.: Type (A) has two types of problems which are explained below

Type 1: To find the approximate value of a function when the independent variable x is replaced by a whole number, decimal fraction or a whole number + decimal fraction.

Type 2: Conditional problems.

Problems based on finding the approximate value of logarithmic, trigonometric, inverse trigonometric function of an independent variable replaced by a number.

Examples worked out:

1. Given $\log_e 2 = 0.6931$, find the approximate value of $\log_e 2.01$.

Solution: $2.01 = 2 + 0.001$

$\therefore x = 2 (= a)$ ...(1)

$\Delta x = 2.01 - 2 = 0.01$ ...(2)

Now on letting $y = \log_e x$, we have

$$\frac{dy}{dx} = \frac{d \log_e x}{dx} = \frac{1}{x}$$

$$f'(a) = \left[\frac{dy}{dx}\right]_{x=a} = \left[\frac{1}{x}\right]_{x=2} = \frac{1}{2}$$

Now, $\Delta y = \left[\dfrac{dy}{dx}\right]_{x=2} \cdot \Delta x = f'(a) \cdot \Delta x$

$$= \frac{1}{2} \times 0.01 \text{ (from (2))}$$

hence, required approximate value $= y + \Delta y$

$$= \log 2 + \frac{1}{2} \times 0.01$$

$$= 0.6931 + 0.005$$

$$= 0.6981$$

2. Given $1^0 = 0.0175^C$, find the approximate value of $\cos 61^0$.

Solution: letting, $f(x) = \cos x$ where given number $= 60 + 1 = x + 1$ which $\Rightarrow \Delta x = 1^0 = 0.0175$,

$$x = \frac{\pi}{3} (= a)$$

Further, $f'(x) = -\sin x$

$$f(a) = \cos 60°$$

$$f'(a) = -\sin 60°$$

now, $\Delta y = f'(a) \cdot \Delta x$

$$= -\sin 60 \times 0.0175$$

$$= \frac{-\sqrt{3}}{2} \times 0.0175$$

Hence, required approximate value

$$= y + \Delta y = [\cos x]_{x=a} + \left(-\frac{\sqrt{3}}{2}\right) \times 0.0175$$

$$= \frac{1}{2} - \frac{\sqrt{3}}{2} \times 0.0175$$

$$= \frac{1}{2}\left(1 - \sqrt{3} \times 0.0175\right)$$

3. Find the approximate value of $\tan^{-1}(0.99)$.

Solution: $y = f(x) = \tan^{-1} x$

$$\Rightarrow f'(x) = \frac{1}{1 + x^2}$$

now expressing 0.99 as $1 - 0.01$, we have

$a = 1$, and $h = \Delta x = -0.01$

$$\therefore f(a) = f(1) = \tan^{-1}(1) = \frac{\pi}{4}$$

$$f'(a) = \frac{1}{1 + a^2} \Rightarrow f'(1) = \frac{1}{1+1} = \frac{1}{1+1} = \frac{1}{2}$$

Hence, $f(a+h) = \tan^{-1}(0.99) = f(a) + f'(a) \times h$

$$= \frac{\pi}{4} + \frac{1}{2} \times (-0.01)$$

$$= \frac{\pi}{4} - 0.005.$$

Derivation of a formula for approximate calculation of powers

To compute the approximate value of the function

$f(a+h) = (a+h)^n$

On applying the formula $f(a+h) = f(a) + h\, f'(a)$, we have

$$f(a+h) = (a+h)^n;\ f(x) = x^n,\ f'(a) = na^{n-1}$$

Hence, $(a+h)^n = a^n + na^{n-1} h$ (approx)

Where $x = a = a$ whole number which is nearly equal to (or, approximately equal to) given number which means a may be slightly (or, alittle) greater than or less than the given number.

and $h =$ given number $- a$.

Hence, in the light of above explanation, we provide the following working rule for approximate calculation of powers.

Working rule: For finding $f(c)$

1. Express the given number c as '$a + h$'. i.e., given number $c = a + h$ (proper a)

2. Put $f(x) = x^n$ and find $f(a)$. Moreover differentiate the function $f(x) = x^n$ and find $f'(a)$.

3. Find h from the formula:

$h =$ given number $- a$

4. Lastly, we use the formula

$$f(a + h) = f(a) + h\, f'(a)$$

i.e.; $(a + h)^n = a^n + na^{n-1}$ to get the required approximate value of the power of the given number used as base of the power.

Problems based on approximate calculations of powers

Examples worked out:

1. Find the approximate value of $(0.998)^8$

Solution: $0.998 = 1 - 0.002 = a + h$

Let $f(x)=x^8$

$\therefore f'(x) = 8x^7$

taking $x=a=1$ and $h=-0.002$

$f(1)=(1)^8$ and $f'(1) = 8\times(1)^7 = 8\times 1 = 8$

$(0.998)^8=(1-0.002)^8=f(1-0.002)$

on using the formula

$f(a-h) = f(a) - h\,f'(a)$ i.e. $(a+h)^n = a^n + n\,a^{n-1}$, we have

$f(.998) = f(1) - .002 \times f'(1)$ (approx)

$= 1 - \dfrac{2}{1000}\times 8$

$= 1 - 0.016$

$= 0.9840$ (approx).

2. Find the approx value of $(0.9999)^6$

Solution: $\because 0.9999 = 1-0.0001$

$\therefore (0.9999)^6 = (1-0.0001)^6$

Let $f(x)=x^6$

$\therefore f'(x) = 6x^5$

taking $x=a=1$ and $h=-0.0001$

$f(a)=f(1)=1^6=1$

and $f'(a) = f'(1) = 6\times(1)^5 = 5$

now, using the formula

$f(a-h) = f(a) - h\,f'(a)$ i.e. $(a+h)^n = a^n + n\,a^{n-1}$, we have

$f(1-0.0001)=(1-0.0001)^6$

$=(1)^6 - 0.0001\times 6$

$= 1\text{-}0.0006$

$= 0.9994$ (approx).

Note: 1 is the integer nearest to .999.

3. Find the approximate value of $(1.999)^6$.

Solution: *First method:*

$\because 1.999 = 2-0.001$

$\therefore (1.999)^6 = (2-0.001)^6$

let $f(x)=x^6 \therefore f'(x) = 6x^5$

Taking $a=2$ and $h=-0.001$

$f(a)=f(2)=2^6=64$

$f'(a) = f'(2) = 6\times 2^5 = 6\times 32 = 192$

now, using the formula

$f(a-h)=f(a)-h\,f'(a)$ i.e. $(a-h)^n = a^n - n\,a^{n-1}\,h$, we have

$(2-0.0001)^6 = (2)^6 - 0.001\times 192$

$= 64 - 0.192$

$= 63.808$

Second approach

Let $y=x^6$

$\therefore \dfrac{dy}{dx} = 6x^5$

$\therefore f'(a) = f'(2) = 6\times 2^5 = 192$

again $a + \Delta x = 1.999$

$\therefore 2 + \Delta x = 1.999$

$\Delta x = 1.999 - 2 = -0.001$

now, on using the formula

$\Delta y = f'(a)\cdot \Delta x$

$= 192\times(-0.001)$

$= -0.192$

$\therefore$ Required approximate value $= y + \Delta y$

$= 2^6 - 0.192$

$= 192 - 0.192$

$= 63.808$

4. Find the approximate value of $(4.012)^2$

Solution: $\because 4.012 = 4+0.012$

$\therefore (4.012)^2 = (4+0.012)^2$

let $f(x)=x^2 \therefore f'(x) = 2x$

taking $x=a=4$ and $h=0.012$, $f(4)=4^2=16$ and $f'(4) = 2\times 4 = 8$. Hence approximate value of $f(a+h)=f(a) + h\,f'(a)$ i.e.; $(a+h)^2 = a^2 + 2ah$

$\therefore (4.012)^2 = (4+0.012)^2 = 4^2 + 2\times 4\times 0.012$

$= 16.096$ (approx).

Deriving formula for approximate solution of equation

Let a root of the equation $f(x) = 0$ be approximately equal to 'a'.

We are able to obtain a better value of the root by using the concept of derivative and differential.

Let '$a + \Delta a$' be the exact root,

Then $f(a + \Delta a) = 0$ ...(1)

Again, $\Delta f(x) = f'(x) \cdot \Delta x$ (or $\Delta y = f'(x)\,\Delta x$)

$\Rightarrow f(a + \Delta a) - f(a) = f'(a)\,\Delta a$ ...(2)

[since the increament in the function is $f(a + \Delta a) - f(a)$ as x changes from a to $a + \Delta a$]

now putting $f(a + \Delta a) = 0$, from (1), into (2), we have

$$0 - f(a) = f'(a) \cdot \Delta a$$

$$\Rightarrow -f(a) = f'(a) \cdot \Delta a$$

$$\Rightarrow \Delta a = \frac{-f(a)}{f'(a)} \quad \text{...(3)}$$

on adding 'a' to both sides of (3), we have the required formula $a + \Delta a = a - \dfrac{f(a)}{f'(a)}$ which $\Rightarrow$ a better approximation of the root '$a + \Delta a$' is

$$a - \frac{f(a)}{f'(a)} \quad \text{...(4)}$$

This formula $a + \Delta a = a - \dfrac{f(a)}{f'(a)}$ is fruitful for approximate computation of roots.

Hence, in the light of above explanation, we can provide a rule to find the nth approximate root of a given number.

Working rule:

1. Let $x = (\text{given number})^{\frac{1}{n}}$ and $x = a$, a whole number whose nth power is approximately equal to the given number.

2. Solve the equation x^n = given number

3. Put $f(x) = x^n$ and differentiating it (i.e. $f(x)$), find $f'(x)$ and $f'(a)$. Moreover we find $f(a)$ from $f(x)$.

4. Last we use the formula

$a - \dfrac{f(a)}{f'(a)}$ which provides us the required approximate value of the root of the given number.

N.B.: The above method is practically more easier than any other method to find the nth (approximate value of the) root of a given number.

Second method:

By the use of the formula:

$f(a + h) = f(a) + h\, f'(a)$ also, we can find approximate nth root of the given number where $f \Rightarrow \sqrt[n]{\ }$ and hence $f(a+h) = f(a) + h\, f'(a)$ can be written as

$$(\text{given number})^{\frac{1}{n}} = (a + \Delta a)^{\frac{1}{n}} = \sqrt[n]{a} + \frac{h}{n\sqrt[n]{x^{n-1}}}$$

where h = given number $- a$ = increament and a = a whole number approximately equal to the given number which means a whole number which is a little (or, slightly) greater than or less than the given number and which can be expressed as nth power (or, perfect square incase of square root)

Remember: In the formula $f(a+h) = f(a) + h\, f'(a)$

1. $h\, f'(a)$ denotes the differential of a function at the value 'a' of the independent variable x. Thus to obtain the value of the differential of a function, it is necessary to know two numbers: the value of the independent variable x and its increament h.

e.g.: Calculate the differential of the function $y = x^2$ for a change in x from 3 to 3.1.

Solution: $dy = f'(x) \cdot \Delta x = h \cdot f'(a)$

$= [2x]_{x=a} \cdot h$

$= [2x]_{x=3}$ (the independent variable $= a = 3$ and h = final value – initial value $= 3.1 - 3 = 0.1$)

$\therefore\ dy = 2 \times 3 \times 0.1 = 6 \times 0.1 = 0.6$

2. 'a' always denotes the approximate value of an independent variable 'x' obtained as a result of measurement and $(a + h)$ denotes its true or given value. then 'a' determines the approximate value of the function $f(x)$ and '$(a + h)$' gives the value of the function $f(x + h)$.

3. Given value indicates the changed value or final value of the independent variable so it must be expressed as $(a + h)$ so that we may have the initial value as well as the increament of independent variable.

4. Increament may be positive or negative. If the final value is greater than initial value, increament is positive. If the final value is smaller than initial value, increament is negative. Increament is always determined by the formula h = final value – initial value = given number – a.

Problems based on approximate calculation of roots

***Examples worked out*:**

1. Find the approximate value of $(31.98)^{\frac{1}{5}}$ with the help of calculus.

Solution: *First method:*

Let $x = (31.98)^{\frac{1}{5}} \Rightarrow x = 2$ (approximately) = a (say)

Now we have to solve the equation

$x^5 = 31.98$

$\Rightarrow x^5 - 31.98 = 0$

again let $f(x) = x^5 - 31.98$

$\therefore f'(x) = 5x^4$

$f(2) = 32 - 31.98 = 0.02$

$f'(2) = 5 \times 2^4 = 5 \times 16 = 80$

$\therefore$ Required approximate value of the root

$$= a - \frac{f(a)}{f'(a)}$$

$$= 2 - \frac{0.02}{80}$$

$$= \frac{160 - 0.02}{80}$$

$$= \frac{159.98}{80}$$

$$= \frac{15.998}{8}$$

$= 1.99975$

Second method:

$31.98 = (32 - 0.02)$

$\Rightarrow (31.98)^{\frac{1}{5}} = (32 - 0.02)^{\frac{1}{5}}$

Now on letting $f(x) = x^{\frac{1}{5}}$

$$f'(x) = \frac{1}{5} x^{-\frac{4}{5}} = \frac{1}{5 \cdot x^{\frac{4}{5}}}$$

Now, taking $x = a = 32$ and $h = -0.02$

$f(a) = f(32) = 2$

$$f'(a) = \frac{1}{5 \times 16} = \frac{1}{80}$$

$\therefore$ Required approximate value of the root

$= f(a - h) = f(a) - h\,f'(a)$

$$= 2 - 0.02 \cdot \frac{1}{5 \times 16}$$

$$= 2 - \frac{2}{8000}$$

$$= 2 - \frac{1}{4000}$$

$= 2 - 0.00025$

$= 1.99975$

2. Find the approximate value of $(80.999)^{\frac{1}{4}}$

Solution: *First method:*

Let $x = (80.999)^{\frac{1}{4}} \Rightarrow x = 3$ (approximately) = a (say)

Now we have to solve the equation

$x^4 = 80.999$

$\Rightarrow x^4 - 80.999 = 0$

again let $f(x) = = x^4 - 80.999$

$\therefore f'(x) = 4x^3$

$f(3) = 3^4 - 80.999 = 81 - 80.999 = .001$

$f'(3) = 4 \times 3^3 = 4 \times 27 = 108$

$\therefore$ Required approximate value of the root

$$= a - \frac{f(a)}{f'(a)}$$

$$= 3 - \frac{0.001}{108}$$

$$= \frac{324 - 0.001}{108}$$

$$= \frac{323.999}{108} = 2.9999$$

Second method:

$80.999 = (81 - 0.001)$

Let $f'(x) = x^{\frac{1}{4}}$

$$f'(x) = \frac{1}{4} \cdot x^{-\frac{3}{4}} = \frac{1}{4 \cdot x^{\frac{3}{4}}}$$

Now putting $a = 81, h = -.001$ in

$f(a-h) = f(a) - h\ f'(a)$, we have

$$\sqrt[4]{80.999} = f(81) - 0.001 \cdot f'(81)$$

$$= 3 - \frac{1}{1000} \times \frac{1}{108}$$

$= 3 - 0.00009259$

$= 2.99990741$

3. Find the approximate value of $\sqrt{145}$.

Solution: *First method:*

Let $x = (145)^{\frac{1}{2}} \Rightarrow x = 12$ (approximately) = a (say)

Now we have to solve the equation

$x^2 = 145$

$$\Rightarrow x^2 - 145 = 0$$

again let $f(x) = x^2 - 145$

$\therefore f'(x) = 2x$

$f(12) = 144 - 145 = -1$

$f'(12) = 2 \times 12 = 24$

Required approximate value of the root

$$= a - \frac{f(a)}{f'(a)}$$

$$= 12 - \frac{(-1)}{24}$$

$$= 12 + \frac{1}{24}$$

$$= \frac{288 + 1}{24}$$

$$= \frac{289}{24}$$

$= 12.04$

Second method:

$\sqrt{145} = \sqrt{144 + 1} = f(a+h)$ where $f(x) = \sqrt{x}$, $a = 144, h = 1$

$$\therefore f'(x) = \frac{1}{2\sqrt{x}}$$

Now using the formula,

$f(a+h) = f(a) + h\ f'(a)$

$$\Rightarrow f(144 + 1) = 12 + 1 \times \frac{1}{2\sqrt{144}}$$

$$\Rightarrow f(145) = 12 + \frac{1}{2 \times 12}$$

$$= 12 + \frac{1}{24}$$

$= 12 + 0.04$

$= 12.04.$

4. Find the approximate value of $(33)^{\frac{1}{5}}$.

Solution: *First method:*

Let $x = (33)^{\frac{1}{5}} \Rightarrow x = 2$ (approximately) = a (say)

Now we have to solve the equation:

$x^5 = 33$

$$\Rightarrow x^5 - 33 = 0$$

Again let $f(x) = x^5 - 33$

$\therefore f'(a) = 5x^4$

$f(2) = 32 - 33 = -1$

$f'(2) = 5 \times 2 \times 4 = 80$

$\therefore$ Required approximate value of the root

$$= a - \frac{f(a)}{f'(a)}$$

$$= 2 - \frac{(-1)}{f'(2)}$$

$$= 2 - \left(\frac{-1}{80}\right)$$

$$= 2 + \frac{1}{80}$$

$$= \frac{160+1}{80}$$

$$= \frac{161}{80}$$

$$= 2.0125$$

Second method:

$$(33)^{\frac{1}{5}} = (32+1)^{\frac{1}{5}} \text{ and } (32)^{\frac{1}{5}} = 2$$

Hence, if we write $f(x) = x^{\frac{1}{5}}$, then $33^{\frac{1}{5}} = f(a+h)$ for $a = 32, h = 1$

$$\therefore f(2) = 2^{\frac{1}{5}} = 2$$

$$f'(x) = \frac{1}{5}x^{-\frac{4}{5}} = \frac{1}{5x^{\frac{4}{5}}}$$

$$f'(2) = \frac{1}{5 \times (32)^{\frac{4}{5}}}$$

Now, using the formula

$$f(a+h) = f(a) + h\, f'(a)$$

$$\Rightarrow f(32+1) = 2 + \frac{1}{5} \times \frac{1}{(32)^{\frac{4}{5}}}$$

$$= 2 + \frac{1}{5 \times 2^4}$$

$$= 2 + \frac{1}{5 \times 16}$$

$$= 2 + \frac{1}{80}$$

$$= \frac{160+1}{80}$$

$$= \frac{161}{80}$$

$$= \frac{16.1}{8}$$

$$= 2.0125.$$

5. Find the approximate value of $(215)^{\frac{1}{3}}$

Solution: *First method:*

Let $x = (215)^{\frac{1}{3}} \Rightarrow x = 6$ (approximately) $= a$ (say)

Now we have to solve the equation

$$x^3 = 215$$

$$\Rightarrow x^3 - 215 = 0$$

Again we let $f(x) = x^3 - 215$

$$\therefore f'(x) = 3x^2$$

$$f(6) = 2^3 - 215 = 216 - 215 = 1$$

$$f'(6) = 3 \times 36 = 108$$

$\therefore$ Required approximately value of the root

$$= a - \frac{f(a)}{f'(a)}$$

$$= 6 - \frac{f(6)}{f'(6)}$$

$$= 6 - \frac{1}{108}$$

$$= \frac{648-1}{108}$$

$$= \frac{647}{108}$$

$$= 5.9907$$

Second method:

$$f(x) = x^{\frac{1}{3}}$$

$$f'(x) = \frac{1}{3} \times x^{-\frac{2}{3}} = \frac{1}{3x^{\frac{2}{3}}}$$

Now, we write

$$f(a+h) = (215)^{\frac{1}{3}}$$

where $f(x) = \sqrt[3]{x}$, $a = 216$, $h = -1$

$$f(a) = f(216) = f(6^3) = \left(6^3\right)^{\frac{1}{3}} = 6$$

$$f'(a) = \frac{1}{3 \times \left(6^3\right)^{\frac{2}{3}}} = \frac{1}{3 \times 6^2}$$

$$= \frac{1}{3 \times 36} = \frac{1}{108}$$

Now, using the formula,

$$f(a+h) = f(a) + h\, f'(a)$$

$$\therefore f(216-1) = 6 - \frac{1}{1 \times 108} = \frac{648-1}{108} = \frac{647}{108}$$

$= 5.9907$ (approx)

6. Find the approximate value of $(127)^{\frac{1}{3}}$.

Solution: *First method:*

Let $x = (127)^{\frac{1}{3}}$

$\Rightarrow x^3 = 127$

$\Rightarrow x = 5$ (approximately) = a (say)

Now we have to solve the equation

$x^3 = 127$

$\Rightarrow x^3 - 127 = 0$

Again we let $f(x) = x^3 - 127$

$\therefore f'(x) = 3x^2$

$f(5) = 125 - 127 = -2$

$f'(5) = 3 \times 25 = 75$

$\therefore$ Required approximate value of the root

$$= a - \frac{f(a)}{f'(a)}$$

$$= 5 - \frac{f(5)}{f'(5)}$$

$$= 5 - \frac{(-2)}{75}$$

$$= 5 + \frac{2}{75}$$

$$= \frac{375+2}{75}$$

$$= \frac{377}{75}$$

$= 5.0266$

Second method:

Expressing given number $= 127 = 125 + 2 = 5^3 + 2$

On letting $x = 5^3 = 125 = a$ (say)

$$f(x) = x^{\frac{1}{3}}$$

$$\Rightarrow f'(x) = \frac{1}{3} \times x^{\left(\frac{1}{3}-1\right)} = \frac{1}{3} \cdot x^{-\frac{2}{3}}$$

$$\therefore f(a) = (125)^{\frac{1}{3}} = \left(5^3\right)^{\frac{1}{3}} = 5$$

$$f'(a) = f'(125) = \frac{1}{\left(5^3\right)^{\frac{2}{3}}} \times \frac{1}{3} = \frac{1}{3} \times \frac{1}{5^2} = \frac{1}{75}$$

Now, using the formula,

$f(a+h) = f(a) + h\, f'(a)$, we have

$$f(125+2) = 5 + 2 \times \frac{1}{75} = 5 + \frac{2}{75} = 5.0266$$

Third method:

$$\text{Let } y = f(x) = x^{\frac{1}{3}} = \left(5^3\right)^{\frac{1}{3}} = 5$$

$$\Delta y = \frac{dy}{dx} \cdot \Delta x = f'(a) \cdot \Delta x = \frac{1}{75} \times 2$$

$\therefore$ Required approximate value

$= y + \Delta y = 5 + .0266 = 5.0266$

7. Find approximately $\sqrt[4]{627}$.

Solution: *First method:*

Let $x = (627)^{\frac{1}{4}} \Rightarrow x = 5$ (approximately) $= a$ (say)

Now we have to solve the equation

$x^4 = 627$

$\Rightarrow x^4 - 627 = 0$

Again on setting $f(x) = x^4 - 627$

$\Rightarrow f'(x) = 4x^3$

$\therefore f(a) = f(5) = 625 - 627 = -2$

$f'(a) = f'(5) = 4 \times 5^3 = 4 \times 125 = 500$

$\therefore$ Required approximate value

$$= a - \frac{f(a)}{f'(a)}$$

$$= 5 - \frac{f(5)}{f'(5)}$$

$$= 5 + \frac{2}{500} = 5 + \frac{1}{250}$$

$= 5.004$

Second method:

We observe that 627 is close to 625 of which fourth root is 5.

$\because 627 = 625 + 2$

on setting $x = a = 625, h = 2$

$$f(x) = (x)^{\frac{1}{4}}$$

we have, $f'(x) = \frac{1}{4}x^{\left(-\frac{3}{4}\right)}$

$$\therefore f(a) = f(625) = (625)^{\frac{1}{4}} = \left(5^4\right)^{\frac{1}{4}} = 5$$

$$f'(a) = f'(625) = \frac{1}{4}(625)^{\left(-\frac{3}{4}\right)}$$

$$= \frac{1}{4} \times \left(5^4\right)^{\left(-\frac{3}{4}\right)} = \frac{1}{4} \times (5)^{-3}$$

$$= \frac{1}{4 \times 5^3} = \frac{1}{4 \times 125} = \frac{1}{500} = 0.002$$

Now, using the formula: $f(a+h) = f(a) + h\ f'(a)$

We have: $f(625+2) = f(625) + 2 \cdot f'(625)$

$= 5 + 2 \times 0.002 = 5 + 0.004 = 5.004$

8. Find the approximate value of $\sqrt{1.006}$.

Solution: *First method:*

Let $x = (1.006)^{\frac{1}{2}} \Rightarrow x = 1$ (approximately) $= a$ (say)

Now, we have to solve the equation

$x^2 = 1.006$

$\Rightarrow x^2 - 1.006$

Again, let $f(x) = x^2 - 1.006$

$\therefore f'(x) = 2x$

$f(1) = 1 - 1.006 = -.006$

$f'(1) = 2 \times 1 = 2$

> If α is small, then we can use the formula:
> $\sqrt{1+\alpha} \cong 1 + \frac{\alpha}{2}$.

$\therefore$ Required approximate value of the root

$$= a - \frac{f(a)}{f'(a)}$$

$$= 1 - \frac{-(0.006)}{2}$$

$$= 1 + \frac{0.006}{2}$$

$$= \frac{2.006}{2} = 1.003$$

Second method:

$(1.006) = 1 + 0.006 = f(a+h)$ where $f(x) = \sqrt{x}$, $a = 1$, $h = 0.006$

$$\therefore f'(x) = \frac{1}{2\sqrt{x}}$$

$f(a) = f(1) = 1$

$$f'(a) = f'(1) = \frac{1}{2}$$

Now, using the formula,

$f(a+h) = f(a) + h\ f'(a)$

$$= 1 + 0.006 \times \frac{1}{2}$$

$$= \frac{2 + 0.006}{2} = 1.003$$

Remember: The formula:

$$a + \Delta a = a - \frac{f(a)}{f'(a)}$$

is also applicable to find the approximate root of the equation $f(x) = 0$ which is nearly equal to a given root $x = a$.

Examples based on approximate solution of the equation

1. Find the root of the equation $x^4 - 12x + 7 = 0$ which is near to 2.

Solution: $\because f(x) = x^4 - 12x + 7$ and $a = 2$

$\therefore f'(x) = 4x^3 - 12$

$f(2) = 2^4 - 12 \times 2 + 7 = -1$

$f'(2) = 4 \times 2^3 - 12 = 20$

$\therefore$ Required root $= a - \frac{f(a)}{f'(a)}$ (approx)

$$= 2 - \frac{f(2)}{f'(2)}$$

$$= 2 - \frac{(-1)}{20}$$

$$= 2 + \frac{1}{20} = \frac{40+1}{20} = \frac{41}{20} = 2.05$$

2. Find the root of the equation $x^4 - 12x^2 - 12x - 3 = 0$ which is approximately 4.

Solution: Let $f(x) = x^4 - 12x^2 - 12x - 3$ and $a = 4$

$$\therefore f'(x) = 4x^3 - 24x - 12$$

$$f'(4) = 4 \times 4^3 - 24 \times 4 - 12$$

$$= 256 - 108$$

$$= 148$$

$$f(4) = 4^4 - 12 \times 4^2 - 12 \times 4 - 3$$

$$= 256 - 192 - 48 - 3$$

$$= 256 - 243 = 13$$

$\therefore$ Required approximate root

$$= a - \frac{f(a)}{f'(a)}$$

$$= 4 - \frac{f(4)}{f'(4)}$$

$$= 4 - \frac{13}{148}$$

$$= \frac{592 - 13}{148}$$

$$= \frac{579}{148}$$

$$= 3.91$$

Derivation of a formula for approximate calculation of reciprocal quantities

Let us consider the function:

$$f(x) = \frac{1}{x} \qquad \ldots(1)$$

We let the argument x receive a small increament Δx, then

$$f(x + \Delta x) = \frac{1}{x + \Delta x} \qquad \ldots(2)$$

$$f'(x) = -\frac{1}{x^2}$$

$$\Rightarrow f'(x)\,\Delta x = -\frac{1 \times \Delta x}{x^2} = \frac{\Delta x}{x^2} \qquad \ldots(3)$$

Now using the formula:

$f(x + \Delta x) = f(x) + f'(x)\,\Delta x$, we have

$$\frac{1}{x + \Delta x} \approx \frac{1}{x} - \frac{\Delta x}{x^2}$$

[i.e.; $f(a + h) \approx f(a) + h\, f'(a)$]

Problems based on approximate computation of reciprocal quantities

Examples worked out:

1. Find the approximate value of $\frac{1}{1.004}$.

Solution: $1.004 = 1 + 0.004$

On setting $x = a = 1$, $h = \Delta x = 0.004$

$$f(x) = \frac{1}{x}$$

$$f'(x) = -\frac{1}{x^2}$$

$$f(a) = f(1) = 1$$

$$f'(a) = f'(1) = -1$$

Now, using the formula,

$f(a+h) = f(a) = h\, f'(a)$, we have

$$\frac{1}{a+h} = 1 - 0.004 \times 1 = 0.996$$

2. Find the approximate value of $\frac{1}{x^4}$ when $x = 2.04$.

Solution: Let $f(x) = \frac{1}{x^4}$

$$x = 2.04 = 2 + 0.04$$

on setting, $x = a = 2$, $h = \Delta x = 0.004$

$$\because f(x) = \frac{1}{x^4}$$

$$f'(x) = -\frac{4}{x^5}$$

$$f(2) = \frac{1}{16}$$

$$f'(2) = -\frac{4}{2^5} = -\frac{1}{8}$$

Now, using the formula:

$$f(a+h) = f(a) + h\, f'(a)$$

$$= f(2) + h\, f'(2)$$

$$= \frac{1}{16} + 0.04 \times \left(-\frac{1}{8}\right)$$

$$= \frac{1}{16} - 0.005$$

$$= \frac{1 - 0.005 \times 16}{16}$$

$$= \frac{1 - 0.080}{16}$$

$$= \frac{0.920}{16}$$

$$= 0.0575$$

Conditional Problems

When $x = a + \frac{k}{10^n}$ is provided where a = an integer $= x_1$ (say) and $h = \Delta x = \frac{k}{10^n}$ = a decimal fraction, and the expression in x is to be approximated, we adopt the following working rule.

Working rule:

1. Let y = given expression in x
2. Suppose $a = x_1$ = a given integer before decimal and $h = \Delta x_1 = \frac{k}{10^n}$ the decimal fraction (the number after the decimal).
3. Use the formula:

$$f(x_1 + \Delta x_1) = f(x_1) + f'(x_1)\,\Delta x_1$$

or, $f(a+h) = f(a) + f'(a) \cdot h$

N.B.: Sometimes $\Delta x = h$ is provided in the given problems to be approximated which means there is no need of finding Δx.

Remember:

1. Approximate change in

$$y = \left[\frac{dy}{dx}\right]_{x = \text{approx value of the given value for } x} \cdot \Delta x$$

$$= f'(a) \cdot h$$

2. Approximate value of $y = [y]_{x=a} + dy = f(a) + dy$, when x = given value
3. Given number (or value of x) may be whole number or simply pure decimal fraction like 26, 65, 0.9993 etc, then to find a and h we should consult the hints given earlier in the topic on finding the approximate value of a function of an independent variable replaced by a number (i.e. in type (A)).
4. Given value of x always requires a result of applying increament or changed value or final value of the argument x in the problems of approximation.

Worked out example on conditional problems

1. Find the approximate value of $(1.001)^5 - 2(1.001)^{\frac{4}{3}} + 3$ by considering $y = x^5 - 2x^{\frac{4}{3}} + 3$.

Solution: Since, $x = 1.001$ which can be expressed as $1 + 0.001$ on setting $x_1 = a = 1$

$$h = \Delta x_1 = 0.001$$

$$f(x) = x^5 - 2x^{\frac{4}{3}} + 3$$

We have, $f(1)$

$$f(a) = 1 - 2 + 3 = 2$$

$$f'(x) = 5x^4 - \frac{8}{3}x^{\frac{1}{3}}$$

$$f'(1) = f'(a) = 5 - \frac{8}{3} = \frac{7}{3}$$

Hence, $f(1 + 0.001) = f(a+h) = f(a) + f'(a) \cdot h$

$$= 2 + \frac{7}{3} \times (0.001)$$

$$= 2.0023$$

2. Find the approximate value of $y + \Delta y$.

When $y = 2x^2 - 3x + 5$, $x = 3$ and $\Delta x = 0.1$

Solution: $y = 2x^2 - 3x + 5$, $a = 3$

$$\Rightarrow \frac{dy}{dx} = 4x - 3$$

$$\therefore \left(\frac{dy}{dx}\right)_{x=a} = f'(a) = (4x-3)_{x=3}$$

$$= 4 \times 3 - 3 = 12 - 3 = 9$$

and $\Delta y = \left[\frac{dy}{dx}\right]_{x=3} \cdot \Delta x$

$$= 9 \times 0.1 = 0.9$$

Hence, the required approximate value

$$= f(a) + \Delta y = (y)_{x=3} + \Delta y$$

$$= 14 + .9$$

$$= 14.9 \text{ (approx)}$$

3. Evaluate $3x^2 - 7x + 5$ when $x = 3.02$

Solution: $f(x) = 3x^2 - 7x + 5$

$$\therefore f'(x) = 6x - 7$$

on setting $x_1 = a = 3$

$$h = \Delta x = 0.02$$

we have $f(a+h) = f(a) + h\, f'(a)$ which $\Rightarrow$

$$f(3.02) = f(3 + 0.02) = f(3) + 0.02 \times f'(3)$$

$f(3) = 11$ and $f'(3) = 11$

$$\therefore f(3.02) = 11 + 0.02 \times 11$$

$$= 11.22 \text{ (approx)}$$

4. Find the approximate value of a function by using differential when $f(x) = 5x^3 - 2x + 3$ and $x = 2.01$.

Solution: $f(x) = 5x^3 - 2x + 3$

$$f'(x) = 15x^2 - 2$$

on setting $x_1 = a = 2$ and $\Delta x_1 = h = 0.01$, we have

$$f(a) = f(2) = 5 \times 2^3 - 2 \times 2 + 3 = 39$$

$$f'(a) = f'(2) = 15 \times 2^2 - 2 = 60 - 2 = 58$$

Now using the formula:

$$f(a+h) = f(a) + f'(a) \cdot h$$

we find

$$f(2.01) = f(2) + f'(2) \times 0.01$$

$$= 39 + 58 \times 0.01$$

$$= 39 + 0.58$$

$$= 39.58$$

N.B.: Exact value of the function for $x = 2.01$

$$= 5 \times (2.01)^3 - 2 \times 2.01 + 3$$

$$= 39.583005$$

5. Find the approximate value of $y = x^3 - 3x^2 + 2x - 1$ when $x = 1.998$.

Solution: $f(x) = x^3 - 3x^2 + 2x - 1$

$$f'(x) = 3x^2 - 6x + 2$$

Now on setting $1.998 = 2 - 0.002$

Where $x_1 = a = 2$ and $\Delta x_1 = h = -0.002$

($\because \Delta x$ = given value of $x - a$)

$$\therefore f(2) = f(a) = 2^3 - 3 \times 2^2 + 2 \times 2 - 1$$

$$= 8 - 12 + 4 - 1$$

$$= -1$$

$$f'(2) = f'(a) = 3 \times 2^2 - 6 \times 2 + 2$$

$$= 12 - 12 + 2$$

$$= 2$$

Now, using the formula:

$$f(a+h) = f(a) + h \cdot f'(a)$$

we find

$$f(1.998) = f(2 - 0.002) = f(2) + (-0.002) \times f'(2)$$

$$= -1 - 0.002 \times 2$$

$$= -1 - 0.004$$

$$= -1.004$$

N.B.: To check the closeness of this method, we substitute 1.998 in the given (original) function to get the exact value for this point and hence we get

$$y + \Delta y = f(a+h) = -1.003988008$$

Which shows that the method of approximation by differentials holds very closely (or nearly).

6. Find the approx value of $x^3 + 5x^2 - 3x + 2$ when $x = 3.003$.

Solution: $f(x) = x^3 + 5x^2 - 3x + 2$

$$\Rightarrow f'(x) = 3x^2 + 10x - 3$$

on setting $x = a = 3$ and $h = 0.003$, we have

$$f(3) = 3^3 + 5(3)^2 - 3(3) + 2 = 65$$

$$f'(x) = 3 \times (3)^2 + 10(3) - 3 = 54$$

Now, using the formula of approximation

$$f(a+h) = f(a) + h\, f'(a)$$

we find

$f(3.003) = f(3) + 0.003 \times f'(3)$

$= 65 + \frac{3}{1000} \times 54$

$= 65 + \frac{162}{1000}$

$= 65.162$

7. Find the approximate value of the increament in the function $y = 2x^3 + 5$ for $x = 2$ and $\Delta x = 0.001$.

Solution: We have $\Delta y \doteqdot dy = 6x^2\,dx = 6x^2\,\Delta x$

$= 6 \times 2^2 \times 0.001$

$= .024$

$\therefore$ The exact value of the increament:

$\Delta y = 2(x + \Delta x)^3 + 5 - 2x^3 - 5$

$= 6x^2\,\Delta x + 6x\,(\Delta x)^2 + 2(\Delta x)^3$

$= 6 \times 4 \times 0.001 + 6 \times 2 \times 0.000001 + 2 \times 0.000000001$

$= 0.024012002$

$(\because \Delta y = f(x + \Delta x) - f(x) \doteqdot f'(x)\,dx = dy)$

8. Given that $4^5 = 1024$, find the approximate value of fifth root of 1028.

Solution: let $y = x^{\frac{1}{5}} = f(x)$

Since $4^5 = 1024$, we set $x = a = 1024$ and $\Delta x =$ final value $- a$

$= 1028 - 1024$

$= 4 = h$

$\because f(x) = x^{\frac{1}{5}}$

$\therefore f'(x) = \frac{1}{5} x^{-\frac{4}{5}}$

$f(1024) = f(a) = (1024)^{\frac{1}{5}}$

$f'(1024) = f'(a) = \frac{1}{5} \times (1024)^{-\frac{4}{5}}$

Now, using the formula:

$f(a+h) = f(a) + f'(a)\,h$

we find

$f(1024 + 4) = f(1024) + h\,f'(1024)$

$= (1024)^{\frac{1}{5}} + 4 \times \frac{1}{5}(1024)^{-\frac{4}{5}}$

$= 4 + \frac{1}{5} \times (4)^{-3}$

$$\left(\because 1024 = 4^5 \Rightarrow (1024)^{\frac{1}{5}} = \left\{(4)^5\right\}^{\frac{1}{5}} = 4\right)$$

$= 4 + \frac{1}{5 \times 4^3}$

$= 4 + 0.0031$

$= 4.0031$

Verbal problems on approximation

The formula $\Delta Q = f'(q)\,\Delta q$ is practically fruitful for calculating approximate change or simply change in the function Q (i.e. dependent variable Q) due to small change in the independent variable q, i.e. if $Q = f(q)$ be a functional relation between q and Q and Δq is the small change in q, then the consequent change in Q is given by the formula:

$\Delta Q = f'(q)\,\Delta q$,

Where, Q = any dependent quantity, dependent variable or dependent physical quantity like volume, area, perimeter, ... etc.

q = independent quantity, independent variable or independent physical quantity like radius, length, height, thickness, ... etc.

$\Delta q = q_f - q_i$

q_f = final value of independent variable

q_i = initial value of independent variable

$$f'(q_i) = \left[\frac{dQ}{dq}\right]_{q=q_2}$$

Working rule:

1. Find $\frac{dQ}{dq}$ and $\Delta q = q_f - q_i$

2. Compute $\left[\frac{dQ}{dq}\right]_{q=q_i}$

3. Lastly use the formula:

$$= \Delta Q = \left[\frac{dQ}{dq}\right]_{q=q_i} \times \Delta q$$

which gives us the required change or approximate change in the dependent physical quantity Q.

Note:

1. Approximate change in a quantity Q or change in a quantity Q due to a small change in q.

$$= \Delta Q = \left[\frac{dQ}{dq}\right]_{q=q_i} \times \Delta q$$

Thus, (i) approximate change in volume V, change in volume V due to small change in r

$$= \Delta V = \left[\frac{dV}{dr}\right]_{r=ri=\text{initial value of } r} \times \Delta r$$

(ii) Approximate change in area A, change in area A

$$= \Delta A = \left[\frac{dA}{dr}\right]_{r=ri=\text{initial value of radius } r} \times \Delta r$$

2. In example 3 thickness suggests change (or, increament) in the argument (or, independent variable) q. Hence, Δq = thickness provided q is the independent variable.

3. If we have

volume of a circular cylinder $V = \pi r^2 h$

surface area of a right circular cone

$$A = \pi r \sqrt{r^2 + h^2}$$

i.e. a function of two variables, then any one of the two variables whose increament is given or can be determined should be regarded as a variable w.r.t which differentiation is performed. Moreover in the function of two variables, the variable which is a constant is always mentioned by stating that it remains fixed or by giving its numerical value.

e.g. (1) What is the approximate volume of a thin circular cylinder with fixed height h?

Explanation: The problems says h remains fixed which means h is a constant.

Remark:

1. The approximate value of a dependent physical quantity

$$= f(a+h) = f(a) + f'(a) \cdot h$$

whereas approximate change in a dependent physical quantity

$$= \Delta Q = f'(q) \cdot \Delta q$$

2. Whenever, we have a formula of a physical quantity like volume, area, perimeter etc, differentiation is performed w.r.t the variable whose increament in the problems of approximate is given to us.

Verbal problems on approximate change of a quantity

Examples worked out:

1. If the radius of the sphere changes from 3 cm to 3.01 cm, find the change in the volume.

Solution: Volume of the sphere $= V = \frac{4}{3}\pi r^3$ ...(1)

Change in the radius = increament in the radius

$= 3.01 - 3 = 0.01$ cm $= \Delta r$

Now, differentiating (1) w.r.t the variable involved in it (i.e. r)

$$\frac{dv}{dr} = \frac{4}{3} \times \pi \times 3 \times r^2$$

$$\Rightarrow \frac{dv}{dr} = 4 \times \pi \times r^2 \quad \text{...(2)}$$

$$\left[\frac{dv}{dr}\right]_{r=3} = 4 \times \pi \times 3^2 = 4 \times 9 \times \pi = 36\pi$$

N.B.: $r_1 = r$ $r_2 = r + \Delta r$

lastly, using the formula:

$$\Delta V = \text{change in } V = \left[\frac{dV}{dr}\right]_{r=3} \cdot \Delta r$$

we have

$$\Delta V = 36\pi \times 0.01$$

$$\Rightarrow \Delta V = 0.36\pi \text{ cm}^3$$

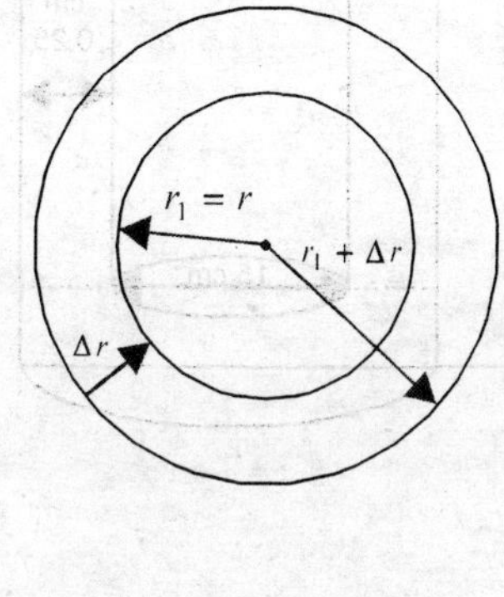

2. Find the approximate increase in the area of circular ring of inner radius 4 cm and outer radius 4.04 cm.

Solution: $\because A = \pi r^2$

$$\therefore \frac{dA}{dr} = 2\pi r$$

and $\left[\frac{dA}{dr}\right]_{r=4} = (2\pi r)_{r=4} = 2\pi \times 4 = 8\pi$

Now change in radius = change in $r = \Delta r$

$= 4.04 - 4 = 0.04$

lastly approximate increase in the area

$$= \Delta A = \left[\frac{dA}{dr}\right]_{r=4} \cdot \Delta r$$

$$\Rightarrow \Delta A = 8 \times \pi \times 0.04 = 0.32\pi \text{ cm}^2$$

3. Find approximately the volume of a metal in a hallow cylindrical pipe 60 cm in length, 7.5 cm inside radius and 0.25 cm thick.

Solution: Volume of the hallow sphere $= V = \pi r^2 h$ ...(1)

Where h = height = 60 cm (given)

$$\Rightarrow V = 60\pi r^2 \quad ...(2)$$

$$\Rightarrow \frac{dV}{dr} = 60\pi \, 2r = 120\pi r$$

$$\Rightarrow \left(\frac{dV}{dr}\right)_{r=7.5} = 120 \times \pi \times 7.5 = 12 \times \pi \times 75 = 900\pi$$

and Δr = thickness = 0.25 cm

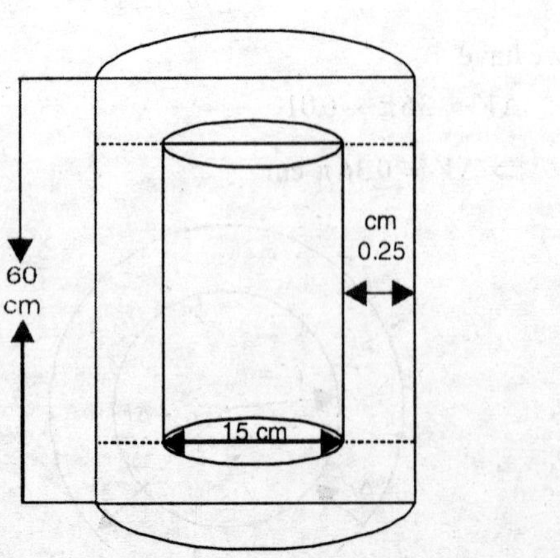

$\therefore$ Required approximate volume

$$= \Delta V = \left(\frac{dV}{dr}\right)_{r=7.5} \Delta r$$

$$= 0.25 \times 900 \times \pi$$

$$= 225\pi \text{ cm}^3$$

On small errors

Question: What do you mean by error?

Answer: Errors are increaments or changes in the values of x and y and are taken as Δx and Δy respectively,

Where y = dependent quantity, dependent variable or dependent physical quantity.

And x = independent quantity, independent variable or independent physical quantity.

Derivation of formula for calculation of small errors

The result:

$$\Delta y = f'(x) \cdot \Delta x + \in \Delta x$$

obtained in the beginning of this chapter shows that

$$\Delta y = f'(x) \cdot \Delta x \text{ approximately}$$

this fact is symbolically expressed by writing

$$\Delta y = f'(x) \, dx$$

which is useful for finding small errors in dependent variable.

Use of the formula:

$$\Delta Q = f'(q_i) \, \Delta q$$

The formula:

$$\Delta Q = f'(q_i) \, \Delta q$$

is practically fruitful for calculating small errors in the dependent physical quantity Q due to small errors in the independent quantity q, i.e. if $Q = f(q)$ be a functional relation between q and Q and Δq is the small errors in q, then the consequent small error in Q in given by the formula:

$$\Delta Q = f'(q_i) \, \Delta q$$

where, Q = any dependent quantity like volume, area, perimeter, temperature, ... etc.

q = any independent quantity like radius, length, height, thickness, ... etc.

$\Delta q = q_f - q_i$

q_f = final value of independent quantity

q_i = initial value of independent quantity

$$f'(q_i) = \left[\frac{dQ}{dq}\right]_{q=q_i}$$

ΔQ = approximate error in Q or simply error in Q.

Remember:

1. In the problems of small errors, the approximate error (or, value) of a dependent physical quantity like volume, area, ... etc is subjected to an error in the independent physical quantity like radius, length, thickness, height, ... etc.

The error in independent quantity like radius and length = Δr and ΔL respectively are generally given in the problem and we are required to find dV, dA, ... etc.

2. The approximate error (or, change) of a dependent physical quantity Q subjected to an error independent physical quantity q

$$\Delta Q = \left(\frac{dQ}{dq}\right)_{q=q_i} \Delta q$$

3. Error in $Q = \Delta Q$

4. Approx error in $Q = dQ = \Delta q$

5. The error to which a variable q is subject (or, subjected) means the error in q which is symbolised as $\Delta q = dq$.

6. Max error, possible error or greatest error in dependent quantity is to be determined means we are required to find the error in dependent physical quantity, i.e. ΔQ.

7. Δq is also called absolute error or total error in independent variable q.

8. q is not measured correctly to the extent or q is measured with uncertainty, ... etc means the increament in q, where q = independent quantity or which a physical quantity Q depends.

Verbal problems on errors

Examples worked out:

1. A box in the form of a cube has an edge of length = 4cm with a possible error of 0.05. what is the possible error in volume V of the box.

Solution: Volume of the cube = (a side)^3 = s^3 ...(1)

$$\Rightarrow V = s^3$$

$$\Rightarrow \frac{dV}{ds} = 3s^2$$

$$\Rightarrow \left(\frac{dV}{ds}\right)_{x=4} \quad 3 \times 4^2 = 3 \times 16 = 48\,\text{cm}^3/\text{cm}$$

...(2)

and we are given $\Delta s = 0.05$ cm ...(3)

Putting (2) and (3) in the formula:

$$dV = \left(\frac{dV}{ds}\right)_{x=4} \times \Delta s$$

We have $\Delta V = 48 \times 0.05 = 2.40\ \text{cm}^3$ = possible error in the volume.

2. The volume of a cone is found by measuring its height and the diameter of a base as 7" and 5" respectively. It is found that diameter is not correctly measured to the extent 0.03. Find the consequent error in the volume approximately.

Solution: The volume of a cone = $V = \frac{1}{3}\pi r^2 h$

(where h is a constant)

$$= \frac{1}{3}\pi\left(\frac{R}{2}\right)^2 \times h \text{ (where } R = \text{diameter} = 2r)$$

$$= \frac{\pi}{12} h R^2 \quad ...(1)$$

Now, differentiating (1) w.r.t R, we have

$$\frac{dV}{dR} = \frac{1}{6} \times \pi \times R \times h$$

$$\Rightarrow \left(\frac{dV}{dR}\right)_{R=5} = \frac{1}{6} \times 5 \times h \times \pi$$

and ΔR = 0.03" as well as h = 7"

Putting these values of ΔR, h and $\left(\frac{dV}{dR}\right)_{R=5}$ in the formula:

$$\Delta V = \left(\frac{dV}{dR}\right)_{R=5} \times \Delta R$$

we have

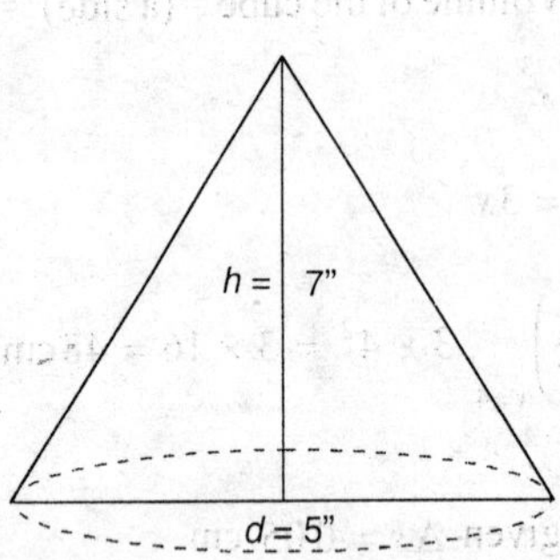

$$\Delta V = \frac{\pi}{6} \times 5 \times 7 \times 0.03 \text{ inch}^3$$

$$= \frac{22}{7} \times \frac{1}{6} \times 5 \times 7 \times \frac{3}{100}$$

$$= \frac{22}{6} \times \frac{5 \times 3}{100}$$

$$= \frac{22}{2} \times \frac{5}{100}$$

$$= \frac{110}{2 \times 100} = \frac{55}{100} = .55 \text{ (inch)}^3$$

= error in the volume.

3. The altitude of a right circular cone is 6 cm. The measurement of the radius of the base is 2 cm with an uncertainly of 0.02 cm. Find approximately the greatest possible error in the computed lateral surface area.

Solution: The lateral surface area of a right circular cone with base radius r and height

$$h = A = \pi \cdot r \cdot \sqrt{r^2 + h^2}$$

Here $h = 6$ cm (a constant)

$$\therefore\ A = \pi r \sqrt{r^2 + 36}$$

$$\text{which} \Rightarrow \frac{dA}{dr} = \pi \cdot r \cdot \frac{1 \times 2r}{2\sqrt{r^2 + 36}} + \pi \sqrt{r^2 + 36} \qquad \ldots(1)$$

$$\Rightarrow \frac{dA}{dr} = \frac{\pi r^2}{\sqrt{r^2 + 36}} + \pi \sqrt{r^2 + 36}$$

$$\therefore \left(\frac{dA}{dr}\right)_{r=2} = \left[\frac{\pi \times 4}{\sqrt{4 + 36}} + \pi \sqrt{4 + 36}\right]$$

$$= \left[\pi \sqrt{40} + \frac{4\pi}{\sqrt{40}}\right] \qquad \ldots(2)$$

and we are given $\Delta r = 0.02$ cm ...(3)

Putting the values of (2) and (3) in the formula:

$$\Delta A = \left[\frac{dA}{dr}\right]_{r=2} \cdot \Delta r$$

We have:

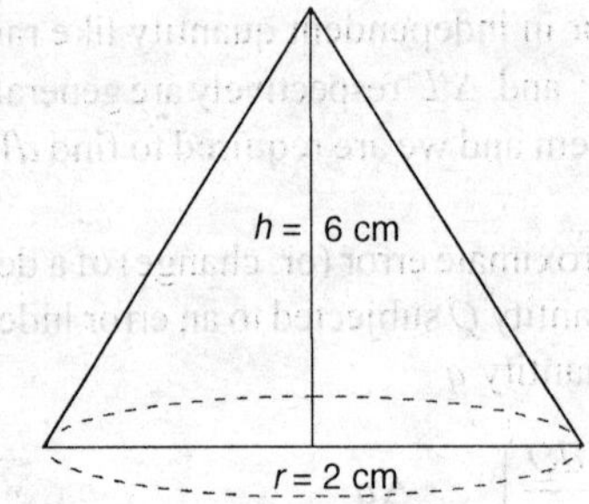

$$\Delta A = \left[\pi \sqrt{40} + \frac{4\pi}{\sqrt{40}}\right] \times (0.02)$$

$$\Rightarrow \Delta A = \left[\frac{\pi \times 40 + 4\pi}{\sqrt{40}}\right] \times (0.02)$$

$$= \frac{44\pi}{\sqrt{40}} \times (0.02) \text{ cm}^2$$

Verbal problems on relative errors

If $Q = f(q)$ be a functional relation between two quantities Q and q and Δq is a small error in q, then $\frac{dQ}{Q}$ is called the relative error in Q and $\frac{\Delta q}{q}$ is called the relative error in q. But since we use dy for an approximate value of Δy, for this reason $\frac{\Delta Q}{Q}$ suggests the convenience of finding first log Q and then calculating $d(\log Q) = \frac{dQ}{Q}$.

Hence, in the light of above explanation, we can provide the following working rule to find the relative error in Q.

First working rule:

Let $Q = f(q)$ be a functional relation between two quantities Q and q

1. Take first log of both sides of $Q = f(q)$.
2. Differentiate both sides of $\log Q = \log f(q)$.
3. Multiply both sides of the simplified form of the equation: $\frac{d}{dq}\log Q = \frac{d}{dq}\log f(q)$ by dq.

Which provides us the required relative rate in Q.

4. Put the given values of q and $\Delta q = dq$ in the expression obtained for the relative error in Q.

Second working rule:

To find the relative error in $Q = f(q)$ at $q = q_i$ = initial value of a quantity q, we proceed in the following way:

1. Find $f'(q_i)$ and then multiply it by

$$h = \Delta q = q_f - q_i$$

2. Divide the product $f'(q_i) \cdot \Delta q$ by the value of the function at $q = q_i$ (i.e. by $f(q_i)$) which provides us the required relative rate, i.e.

$$\frac{dQ}{Q} = \frac{f'(q_i) \cdot \Delta q}{f(q_i)}$$

Remember:

1. Relative error in y is defined by $\frac{\Delta y}{y}$ if Δy is the error in y and $\frac{\Delta y}{y}$ may by approximated by $\frac{dy}{y}$ if the increament of the variable (the quantity being measured) on which y depends is small enough.
2. Method (1) (or first method) is convenient. Whenever we have a formula for volume, area or perimeter etc in the form of power of an independent variable or in the form of product, whereas method (2) or, second method is applicable in all cases.

Verbal problems on relative errors

Examples worked out:

1. The radius of a sphere is found to be 10 cm with a possible error 0.02 cm, what is the relative error in the computed volume?

Solution: *First method:*

We have $V = \frac{4}{3}\pi r^3$ ...(1)

Taking log of both sides of (1), we have

$$\log V = \log\frac{4}{3}\pi + \log r^3$$

$$= \log\frac{4}{3}\pi + 3\log r \quad \text{...(2)}$$

Now, differentiating both sides of (2) w.r.t r, we get

$$\frac{1}{V}\cdot\frac{dV}{dr} = 3\cdot\frac{1}{r} \quad \text{...(3)}$$

Multiply both sides of (3) by dr, we obtain

$$\frac{dV}{V} = \frac{3\,dr}{r} \quad \text{...(4)}$$

Putting the given values $r = 10$ cm and $\Delta r = dr = 0.02$ in (4), we have the required relative rate in V.

$$\frac{dV}{V} = 3 \times \frac{0.02}{10} = \frac{0.06}{10} = 0.006$$

Second method:

$$V = \frac{4}{3}\pi r^3 \Rightarrow (V)_{r=10} = \left(\frac{4}{3}\times\pi\times r^3\right)_{r=10}$$

$$= \frac{4}{3}\times\pi\times 1000 \quad \text{...(1)}$$

$$\frac{dV}{dr} = \frac{4}{3}\times\pi\times 3\times r^2 = \frac{4}{3}\times 3\times\pi\times r^2 = 4\pi r^2 \quad \text{...(2)}$$

$$\left(\frac{dV}{dr}\right)_{r=10} = \left(4\pi r^2\right)_{r=10} = 4\times\pi\times 100 \quad \text{...(3)}$$

and we are given $dr = 0.02$ ...(4)

Putting the values of (1), (3) and (4) in the formula: $\frac{dQ}{Q} = \frac{f'(q_i)}{f(q_i)}\times\Delta q$ where $Q = V$ and $q = r\}$ in this problem i.e. $\dfrac{dV}{V} = \dfrac{f'(r_i)}{f(r_i)} \cdot \Delta r = \dfrac{4 \times \pi \times 100 \times .02 \times 3}{4 \times \pi \times 1000}$

$= \dfrac{0.06}{10}$

$= 0.006$

2. Show that the relative error in x^n is n times the relative error in x.

Solution: *First method:*

$y = x^n$

$\Rightarrow \log y = \log x^n = n \log x$

$\Rightarrow \dfrac{d}{dx} \log y = n \dfrac{d}{dx} \log x$

$\Rightarrow \dfrac{1}{y} \dfrac{dy}{dx} = n \cdot \dfrac{1}{x}$

$\Rightarrow \dfrac{dy}{y} = \dfrac{n\,dx}{x}$

Second method:

Let $y = x^n$

$\Rightarrow \dfrac{dy}{dx} = n x^{n-1}$ ...(1)

Now, we are required to show

$\dfrac{\Delta y}{y} = n \cdot \dfrac{\Delta x}{x}$ ...(2)

Now, using the definition

$\Delta y = f'(x) \cdot \Delta x = n x^{n-1} \Delta x$ ...(3)

(applying for small Δx and using (1))

$\therefore \dfrac{\Delta y}{y} = \dfrac{n x^{n-1} \Delta x}{x^{n-1} \cdot x} = n \dfrac{\Delta x}{x}$

N.B.: If y is a function of x, then $y = f(x)$ and $\dfrac{dy}{dx} = \dfrac{d f(x)}{dx}$ or $f'(x)$.

Moreover, $\Delta y = f'(x) \cdot \Delta x$ (approximately)

And $dy = f'(x) \cdot dx$ (accurately)

3. What is the relative error in the area of a circle if the diameter is found by measurement to be 10 inches, with a maximum error of 0.01 inch?

Solution: $A = \pi r^2 \Rightarrow A = \dfrac{1}{4} \pi d^2$

$\left(\because d = 2r = r \Rightarrow \dfrac{d}{2} = r \right)$

$= \dfrac{1}{4} \pi x^2$ (where x = diameter) ...(1)

Taking log of both sides of (1), we have

$\log A = \log \left(\dfrac{\pi}{4} x^2 \right)$

$= \log \dfrac{\pi}{4} + \log x^2$

$= \log \dfrac{\pi}{4} + 2 \log x$

$\Rightarrow \dfrac{d}{dx} \log A = \dfrac{d}{dx} (2 \log x)$

$\Rightarrow \dfrac{1}{A} \cdot \dfrac{dA}{dx} = 2 \cdot \dfrac{1}{x} = \dfrac{2}{x}$

$\Rightarrow \dfrac{dA}{A} = 2 \dfrac{dx}{x}$ ...(2)

Now, putting the given values: $x = 10$ and $\Delta x = dx = 0.01$ inch in (2), we have,

$\dfrac{dA}{A} = 2 \times \dfrac{0.01}{10} = \dfrac{0.02}{10} = 0.002$

Verbal problems on percentage errors

Definition: Percentage error in a quantity means 100 times relative error in that quantity, i.e.

1. $100 \dfrac{dx}{x} =$ percentage error in x

$= k_1$ (say)

= a number written before the percentage symbol %

which $\Rightarrow dx$ = error in $x = \dfrac{x \times k_1}{100}$

$= k_1$ % of x (or k_1 % in the value of x)

where x = an independent variable

2. $100 \dfrac{dy}{y}$ = percentage error in y

$= k_2$ (say)
= a number to be determined

which $\Rightarrow$ dy = error in $y = \frac{y \times k_2}{100}$

$= k_2$ % of y (or k_2 % in the value of y)

where y = dependent variable (i.e. a quantity being measured).

Note:

1. If $y = f(x)$ be a functional relation between x and y, then percentage error in dependent variable y is caused by the percentage error in independent variable x or in other words;

Let $y = f(x)$ be a functional relation between two quantities x and y.

If Δx is k_1 per cent (or, k_1 %) error in x (or, in the value of x), then Δy will be k_2 per cent (or, k_2 %) error in y (or, in the value of x).

Now in the light of above explanation, we can provide the following working rule to find the percentage error or percentage increase, ... etc in a function (i.e. dependent quantity) = y.

To find the percentage error in a dependent quantity Q implies we are required to use the formula

which is $= \frac{dQ}{Q} \times 100$ where $Q = f(q) = y$ = a dependent quantity, a dependent physical quantity like volume, area, etc.

And q = independent quantity on which Q depends like length, radius, height or thickness etc.

First working rule:

1. Find $\frac{dQ}{dq} = \frac{dy}{dx}$ by differentiating the given function, the formula for volume, formula for area, ... etc obtained by mensuration formula, ... etc w.r.t the independent variable involved in the formula or given function.

2. Use the formula $\Delta Q = \frac{dQ}{dq} \cdot \Delta q$

which $\Rightarrow dQ = \frac{dQ}{dq} \cdot \Delta q$ $(\because \Delta Q = dQ)$

3. Apply the percentage error formula

$$\text{which} \Rightarrow \frac{dQ}{Q} \times 100 = \frac{\frac{dQ}{dq} \cdot \Delta q}{Q} \times 100$$

Second working rule:

1. Find $\frac{dQ}{dq}$ by differentiating w.r.t the independent variable q.

2. Use $\frac{dQ}{dq} = \frac{\Delta Q}{\Delta q}$

3. Divide $\left[\Delta Q = \frac{dQ}{dq} \cdot \Delta q\right]$ by Q (both sides be divided by Q)

$$\text{which} \Rightarrow \frac{\Delta Q}{Q} = \frac{\frac{dQ}{dq} \cdot \Delta q}{Q}$$

where Δq = error in $q = k$% of q (or, k% in the value of q)

Remember:

1. Error in $Q = \Delta Q$ and error in $q = \Delta q$

2. Relative error in $Q = \frac{\Delta Q}{Q}$ and relative error in $q = \frac{dq}{q}$.

3. Percentage error in $Q = \frac{\Delta Q}{Q} \times 100$ and percentage error in q is $= \frac{\Delta q}{q} \times 100$.

4. $\Delta Q = dQ$ and $\Delta q = dq$

Verbal problems on percentage

Examples worked out:

1. If the radius of a spherical balloon increases by 0.1 % find approximately the percentage increase in the volume.

Solution: The volume of a sphere of radius r is

$$V = \frac{4}{3}\pi r^3 \qquad \text{...(1)}$$

$$\Rightarrow \frac{dV}{dr} = \frac{4}{3} \times \pi \times r^2 \times 3$$

But $\frac{dV}{dr} \simeq \frac{\Delta V}{\Delta r} = 4\pi r^2$

$$\Rightarrow \Delta V \simeq 4\pi r^2 \times \Delta r \quad \ldots(2)$$

There is a percentage increase in r of

$$0.1\% = \frac{r \times 0.1}{100} = \Delta r$$

$$\Rightarrow \frac{100 \times \Delta V}{V} = \frac{100 \times 4\pi \times r^2}{\frac{4}{3} \times \pi \times r^3} \times \Delta r$$

$$= \frac{400 \times 3}{4r} \times \frac{r}{1000} = 0.3$$

2. The time period for one complete oscillation of a simple pendulum of length L is given by $T = 2\pi \times \sqrt{L/g}$. Find the approximate error in T corresponding to an error of .5 % in the value of L, g is a constant.

Solution: The time for one comple oscillation is T (Which means) $\frac{T}{2}$ units of time i.e. $\frac{T}{2}$ second for "tick" and $\frac{T}{2}$ second for "tock")

And $T = 2\pi \times \sqrt{L/g} \quad \ldots(1)$

$$\frac{dL}{L} \times 100 = .5 = k \text{ (say)} \quad \ldots(2)$$

(given in the problem)

since (1) consists of product, so taking logarithm of both sides

$$\log T = \log 2\pi + \frac{1}{2} \log L - \frac{1}{2} \log g \quad \ldots(3)$$

Now differentiating both sides of (3) w.r.t 'L' noting that g is constant

$$\Rightarrow \frac{1}{T}\frac{dT}{dL} = 0 + \frac{1}{2} \cdot \frac{1}{L} - 0$$

$$\Rightarrow \frac{dT}{T} = \frac{dL}{2L} \Rightarrow \frac{dT}{T} \times 100$$

$$= \frac{dL}{2L} \times 100$$

$$= \frac{5}{2} = 0.25 \text{ (from (2))}$$

= percentage error in T.

3. If p is a small percentage error in measuring the radius of a sphere, find the percentage error in the calculated value of volume and surface.

Solution: $V = \frac{4}{3}\pi r^2 \quad \ldots(1)$

$$\frac{dr}{r} \times 100 = p \text{ (given)} \quad \ldots(2)$$

Now differentiating both sides of (1) w.r.t r after taking the logarithm of both sides, i.e.

$$\log V = \log 4 - \log 3 + \log \pi + 3 \log r$$

$$\Rightarrow \frac{1}{V}\frac{dV}{dr} = 3\frac{1}{r}$$

$$\Rightarrow \frac{dV}{V} = 3\frac{dr}{r}$$

$$\Rightarrow \frac{dV}{V} \times 100 = 3\frac{dr}{r} \times 100 \quad \ldots(3)$$

= $3p$ (on putting (2) in (3))

2nd part:

surface area $= S = 4\pi r^2 \quad \ldots(1)$

$$\frac{dr}{r} \times 100 = p \quad \ldots(2)$$

Taking logarithm of both sides of (1),

$$\log S = \log 4\pi + 2 \log r \quad \ldots(3)$$

Differentiating both sides of (3) w.r.t 'r' we get

$$\frac{1}{S}\frac{dS}{dr} = 0 + 2 \cdot \frac{1}{r}$$

$$\Rightarrow \frac{dS}{S} = 2\frac{dr}{r}$$

$$\Rightarrow \frac{dS}{S} \times 100 = 2\frac{dr}{r} \times 100 \quad \ldots(4)$$

$= 2p$ (on putting (2) in (4))

= percentage error in calculated value of required surface area.

4. Find the percentage error in x^3 corresponding to a small error of r % in the value of x.

Solution: Let $y = x^3$ ...(1)

error of r % in the value of $x \Rightarrow \frac{dx}{x} \times 100 = r$ ...(2)

$$y = x^3$$

$$\Rightarrow \log y = 3 \log x$$

$$\Rightarrow \frac{1}{y}\frac{dy}{dx} = \frac{3}{x}$$

$$\Rightarrow \frac{dy}{y} = 3\frac{dx}{x}$$

$$\Rightarrow \frac{dy}{y} \times 100 = 3\frac{dx}{x} \times 100$$

$= 3r$ (from (2), $\frac{dx}{x} \times 100 = r$)

Verbal problems on percentage errors continued:

5. If b and c are measured correctly. $A = 36°$ with a possible error of 3', find the possible % error in Δ.

Solution: $\Delta = \frac{1}{2}\ bc \sin A$

Δ = area of triangle has been given as a product, we use logarithmic differentiation.

$$\log \Delta = \log \frac{1}{2} + \log b + \log c + \log \sin A$$

Now differentiating w.r.t A, we have

$$\frac{d \log \Delta}{d\Delta} \cdot \frac{d\Delta}{dA} = 0+0+0+\frac{d \log \sin A}{d \sin A} \cdot \frac{d \sin A}{dA}$$

$$\Rightarrow \frac{1}{\Delta} \cdot \frac{d\Delta}{dA} = \frac{1}{\sin A} \cdot \cos A$$

$\Rightarrow \frac{d\Delta}{\Delta} = \frac{\cos A}{\sin A} \cdot dA = \cot A \cdot dA$ = relative error in A.

$$\Rightarrow \frac{d\Delta}{\Delta} \times 100 = \cot A \times dA \times 100$$

$$= \cot 36° \times \frac{3}{60} \times \frac{\pi}{180} \times 100$$

$= 0.12$ %

error $A = dA = 3' = \frac{3}{60} \times \frac{\pi}{180}$

Type 1: Problems based on finding the approximate values of numbers:

(A) Problems based on finding the approx values of numbers.

Exercise 18.1

1. Find approximately

(i) $\sqrt[4]{627}$

(ii) $\sqrt[3]{66}$

(iii) $\sqrt[3]{122}$

(iv) $\sqrt[4]{252}$

(v) $\sqrt[4]{82}$

(vi) $(28)^{\frac{1}{3}}$

(vii) $(63)^{\frac{1}{3}}$

(viii) $(126)^{\frac{1}{3}}$

(ix) $\sqrt{401}$

(x) $(15)^{\frac{1}{4}}$

(xi) $(235)^{\frac{1}{4}}$

2. Find the approximate values of the following number

(i) $\sqrt{0.0037}$

(ii) $\sqrt[3]{0.009}$

(iii) $(0.998)^8$

(iv) $(31.98)^{\frac{1}{2}}$

(v) $\sqrt[4]{80.999}$

(vi) $\frac{1}{\sqrt{100.5}}$

(vii) $\sqrt{4.08}$

(viii) $\sqrt[3]{65}$

(ix) $(1000.1)^{\frac{1}{3}}$

(x) $(3.998)^{\frac{3}{2}}$

(xi) $(02.97)^3$

(xii) $\sqrt{.82}$

(xiii) $\sqrt{6.23}$

(xiv) $\sqrt[3]{8.01}$

(xv) $\dfrac{1}{(2.001)^2}$

3. By use of differentials, calculate approximately the values of the following:

(i) 79^2

(ii) 103^2

(iii) $(9.06)^2$

(iv) $(1.012)^3$

(v) $(9.95)^3$

(vi) $(1.005)^{10}$

(vii) $(0.975)^4$

4. Calculate approximately the reciprocal of 997 and 102.

5. Find the approximate values of the following quantities:

(i) $\dfrac{1}{0.99}$

(ii) $\dfrac{1}{9.93}$

(iii) $\dfrac{1}{(1.004)^2}$

Answers

1. (i) 5.004

(iii) 4.96

(v) 3.009

(vi) $\dfrac{82}{27}$

(vii) $\dfrac{191}{48}$

(viii) $\dfrac{376}{75}$

(ix) 20.025

(x) 1.96875

(xi) 3.9961

2. (i) 0.060833

(ii) 0.208

(iii) 0.9840

(iv) 1.99975

(v) 2.99990741

(vi) 0.09975

(vii) 2.02

(viii) 4.02083

(ix) 10.0003

(x) 7.994

(xi) 26.19

(xii) 0.956

(xiii) 2.496

(xiv) 2.00083

(xv) 0.24975

3. (i) 6240

(iii) 82.08

(iv) 1.036

(v) 985

(vi) 1.05

(vii) 0.9

4. Approximate value of reciprocal of 102 = 0.0098.

5. (i) 1.01

(ii) 0.1007

(iii) 0.992

(B) Conditional problems

Exercise 18.2

1. Find the approximate values of the following quantities:

(i) sin 31° when 1° = 0.175 radians

(ii) cos 29° when 1° = 0.175 radians

(iii) sin 60° 2' given sin 60° = 0.86603 and 1' = 0.00029 radians

(iv) tan 44° given 1° = 0.0175 radians

(v) tan 45° 30' when 1° = 0.0175 radians

(vi) cos 30° 1' when 1° = 0.01745 radians

(vii) tan° when $1° = 0.01745$ radians
(viii) $\log_e 10.01$ when $\log_e 10 = 2.3026$
(ix) $\log_e (3.0001)$ when $\log_e{}^3 = 1.0986$
(x) $\log_e (3.001)$ when $\log_e{}^3 = 1.0986$

2. Find the approximate value of the function $f(x) = x^3 - 2x^2 + 1$ when $x = 1.0001$

3. Find the approximate value of the function $f(x) = x^3 + 5x^2 - 3x + 1$ when $x = 3.003$

4. Find the approximate value of the function $f(x) = \sin x + 2 \cos x$ when $x = 46°$

5. Find the approximate value of the function $f(x) = 3x^2 - 8x + 11$ when $x = 8.007$

6. Find the approximate value of the function $f(x) = x^4 + 2x^2 + 5$ when $x = 1.998$

Answers:

1. (i) 0.5152

(ii) $\frac{\sqrt{3}}{2}$

(iii) 0.86632
(iv) 0.965
(v) 1.0175
(vi) Find
(vii) Find
(viii) 2.3036
(ix) 1.098633
(x) 1.0989

2. –0.0001
3. 65.162
4. Find
5. 139.28
6. 28.92

Verbal problems on approximations and errors

Exercise 18.3

1. The radius of a circle is 10 inches and there is an error of 0.1 inch in measuring it. Find the consequent error in the area.

2. What is approximate error in the volume and surface of a cube of edge 8 inches, if an error of 0.03 inch is made in measuring the edge?

3. Find the approximate error in the curved surface of a cylinder of diameter one foot and height 4 feet, if there is a possible error of 1 inch in the height.

4. The radius of a sphere is found by measurement to be 10 inches with a possible error of $\frac{1}{10}$ of an inch. Find the consequent errors possible in (i) the surface area and (ii) the volume as calculated from the measurement.

5. The time of oscillation 'T' of a simple pendulum of length 'L' is given by $T = 2\pi \sqrt{L/g}$. If L is creased by 1 % (one per cent), show that the percentage error in T is 0.5.

6. The angle of elevation of the top of a tower at a point 200 ft away from its base is 45°. Find the approximate error in the height of the power due to an error of 2 % in the angle of elevation.

7. Find the percentage error in calculating the area of a triangle when one of its angles has been measured as 45° with an error of 1'.

8. A metal cube of side 3 cms is heated. Find the approximate increase in its volume if its side becomes 3.0001 cms.

9. Find the percentage error in x^3 corresponding to small error of r % in the value of x.

Answers

1. 2π sq inch
2. 5.76 cu in and 2.88 sq in
3. π in
4. (i) 8π sq inch **(ii)** 40π cu. in.
5. 2π ft
6. 0.029
8. 0.002700
9. $3r$

19

Tangent and Normal to a Curve

Give me a place to stand on and I will move the earth.

Archemedes (287–212 BC)

The concepts of tangent and normal are based on the following concepts.

1. *Inclination*: The inclination of a line is the angle between the line (or its extension) and the positive direction of the x-axis (the direction from a point on the left (the point of intersection of the line and the x-axis) to an other point on the right situated on the x-axis) measured by convention in the anticlockwise direction from the x-axis to the part of the line (intersecting the x-axis) above the x-axis.

Notation: An angle of inclination is generally denoted by the symbol 'i' or 'θ'.

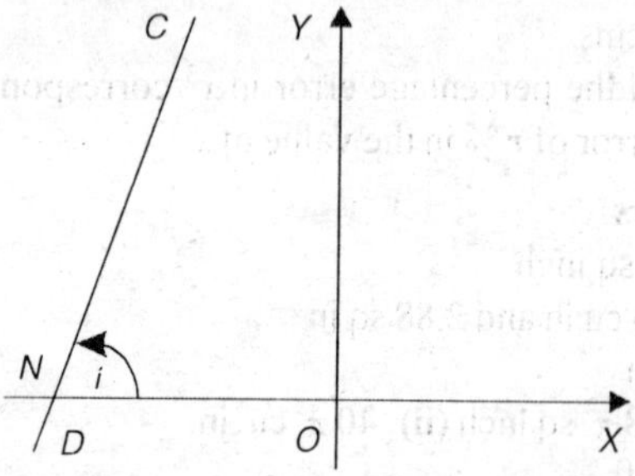

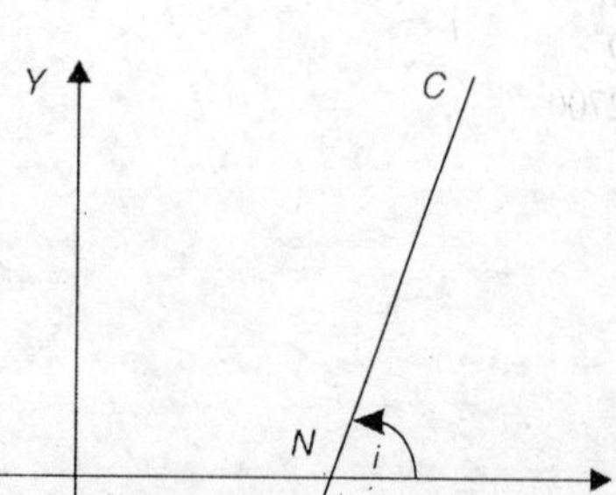

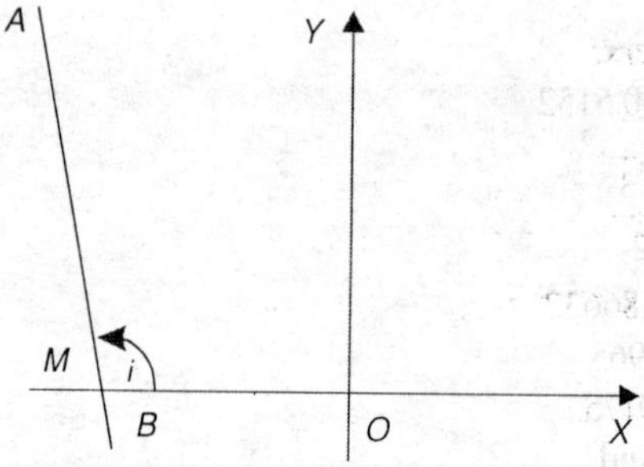

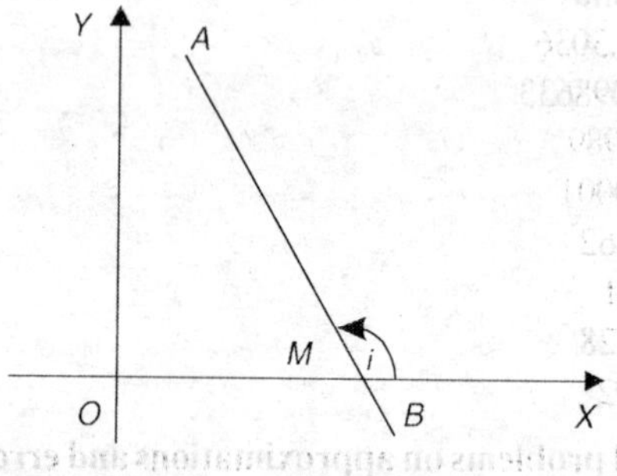

Explanation:

(i) The inclination of the line CD is the angle XNC.

(ii) The inclination of the line AB is the angle XMA.

2. Slope of a line. The slope of a line is the tangent (or trigonometric tangent) of its inclination.

Notation: The slope of a line is represented by $m = \dfrac{\Delta y}{\Delta x} = \tan i$, where $\Delta_y = y_2 - y_1$ difference of y-coordinates of two points (x_1, y_1) and (x_2, y_2) on the line whose slope is sought.

$\Delta x = x_2 - x_1$ difference of x-coordinates of two points (x_1, y_1) and (x_2, y_2) on the line whose slope is sought.

Nomenclature: Δy = delta y

Δx = delta x

Explanation: The slope of the line $FG = \tan i$

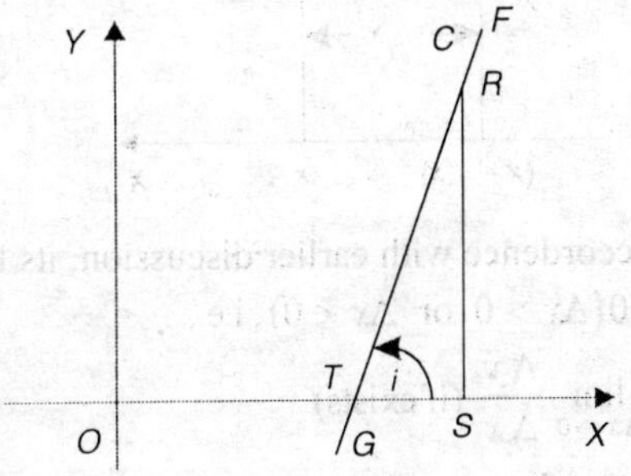

$= \dfrac{SR}{TS}$

$= m$

To remember:

(a) When the inclination of a line is an acute angle, the slope of a line is positive.

(b) When the inclination of a line is an obtuse angle, the slope of a line is negative.

(c) The inclination is acute $(0^\circ < \theta < 90^\circ)$ or obtuse $(90^\circ < \theta < 180^\circ)$ accordingly as the line leans to the right or to the left of the point of intersection of the x-axis and the line. Further, the inclination of a line parallel to the axis is 0° and the inclination of a line parallel to the y-axis is 90°.

Explanation:

(a) The slope of the line CD is positive.

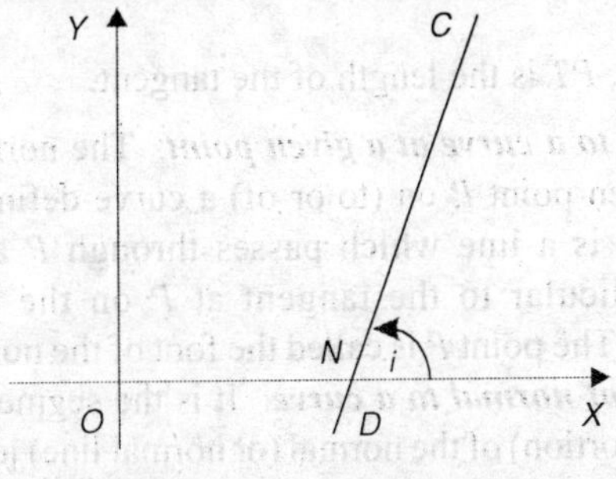

(b) The slope of the line AB is negative.

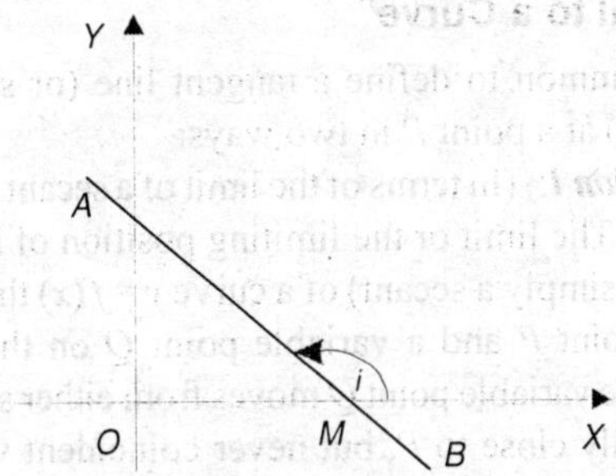

Now, let us procede to have a clear idea of the concepts of tangent and normal to a point on the curve. Ordinarily, it is known that a line l cuts a circle in two points, but a tangent line cuts it in one point only.

From this special case, we are led to describe a tangent as a line which cuts the curve at only one point.

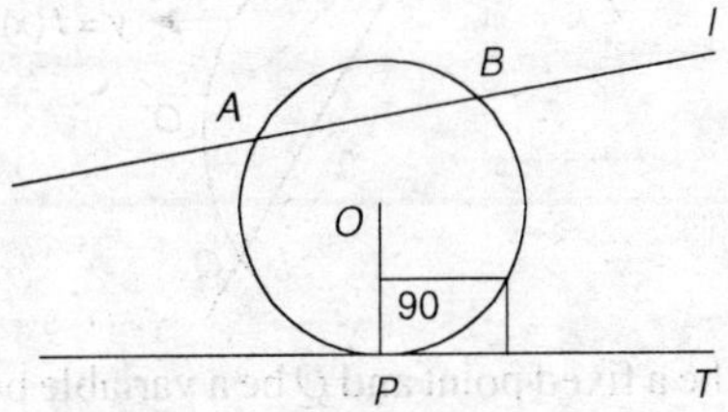

Now this above description is all right in case of a circle, but it is not applicable to a tangent to a curve in general.

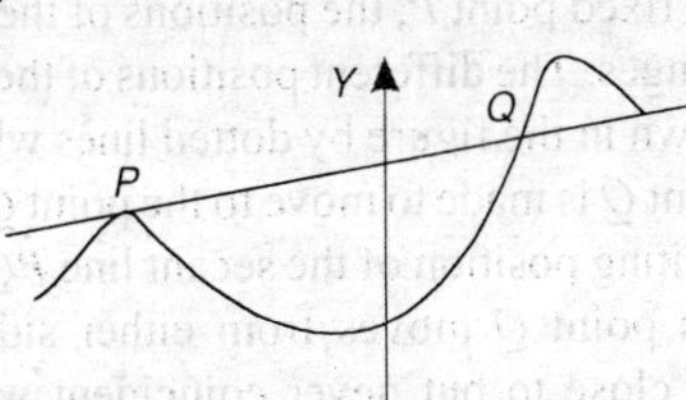

For example, in the figure given below, the tangent line to the curve at a point namely P intersects the curve at another point namely Q. This is why in Calculus "a tangent to a curve is a straight line which cuts the curve at only one point" is not applicable as a definition. But it is endeavored to arrive a suitable definition of the tangent line at a point on the graph of the function f defined by $y = f(x)$.

Definitions of "Tangent and Normal to a Curve"

It is common to define a tangent line (or simply a tangent) at a point P in two ways:

Definition I: (In terms of the limit of a secant line of a curve): The limit or the limiting position of a secant line (or simply a secant) of a curve $y = f(x)$ through a fixed point P and a variable point Q on the curve when the variable point Q moves from either side of P arbitrarily close to P, but never coincident with the fixed point P (i.e., the distance between the fixed point P and the variable point Q is non-zero and less than any given small positive number $\in$) is called the tangent to a curve $y = f(x)$ at a fixed point P on the curve.

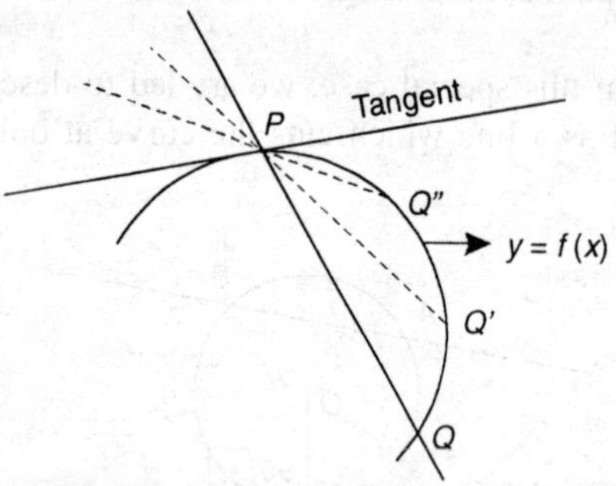

Let P be a fixed point and Q be a variable point on the same curve $y = f(x)$. Then the line PQ is called a secant line of the curve $y = f(x)$. Now, when the variable point Q is made to move along the curve $y = f(x)$, towards the fixed point P, the positions of the secant line PQ changes. The different positions of the secant line are shown in the figure by dotted lines when the variable point Q is made to move to the point Q', Q'', etc. The limiting position of the secant line PQ when the variable point Q moves from either side of P indefinitely close to but never coincident with the fixed point P is called the tangent at the fixed point P on the curve $y = f(x)$.

Definition II: (In terms of slope of a curve): A tangent line to a curve $y = f(x)$ at a point $P(x, y)$ on the curve is a line through $P(x, y)$ with a slope $\frac{dy}{dx}$ at $P(x, y)$. i.e., the tangent to a curve $y = f(x)$ at a point $P(x, y)$ on the curve is the line which passes through $P(x, y)$ and which has the same slope $\frac{dy}{dx}$ as the curve at $P(x, y)$.

Slope of $P\theta$

$$= \frac{\Delta y}{\Delta x} = \frac{f(x) - f(x - \Delta x)}{\Delta x} \quad \ldots(1)$$

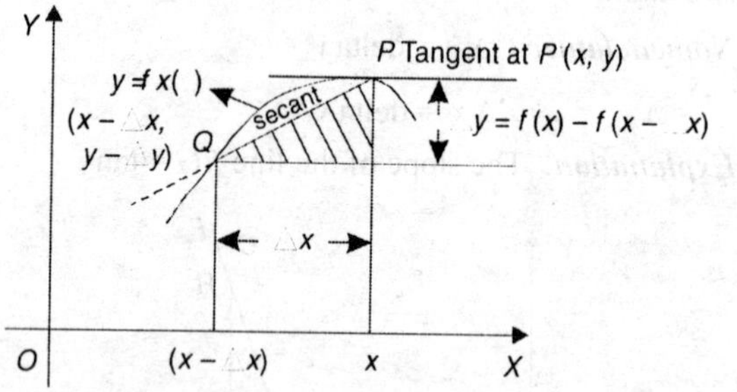

In accordance with earlier discussion, its limit as $\Delta x \to 0\,(\Delta x > 0 \text{ or } \Delta x < 0)$, i.e.,

$$\lim_{\Delta x \to 0} \frac{\Delta y}{\Delta x} \text{ (if exists)}$$

$$= \frac{dy}{dx}, \quad \ldots(2)$$

which is said to be the slope of the tangent line at (x, y) and consequently the tangent is defined to be the line through $P(x, y)$ with the slope given by (2).

Length of the tangent to a curve: It is the segment (or part or portion) of the tangent joining the point of tangency (the point where the tangent touches the curve) and the point of intersection of the tangent with the x-axis.

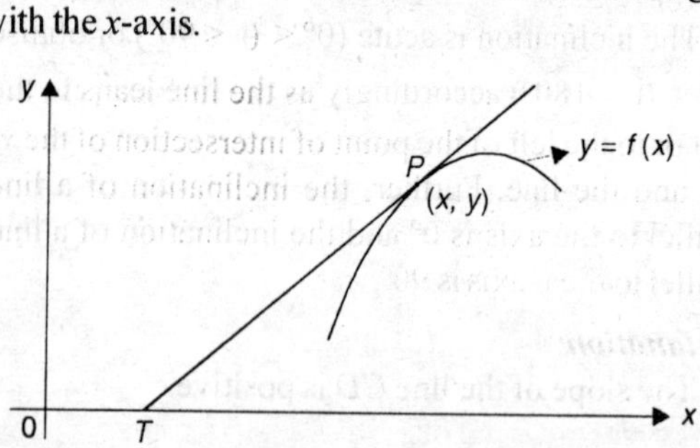

Thus, PT is the length of the tangent.

Normal to a curve at a given point: The normal at any given point P on (to or of) a curve defined by $y = f(x)$ is a line which passes through P and is perpendicular to the tangent at P on the curve $y = f(x)$. The point P is called the foot of the normal.

Length of normal to a curve: It is the segment (or part or portion) of the normal (or normal line) joining the point of tangency on the curve and the point of intersection of the normal with the x-axis. Thus PN is the length of the normal. Thus PN is the length of the normal.

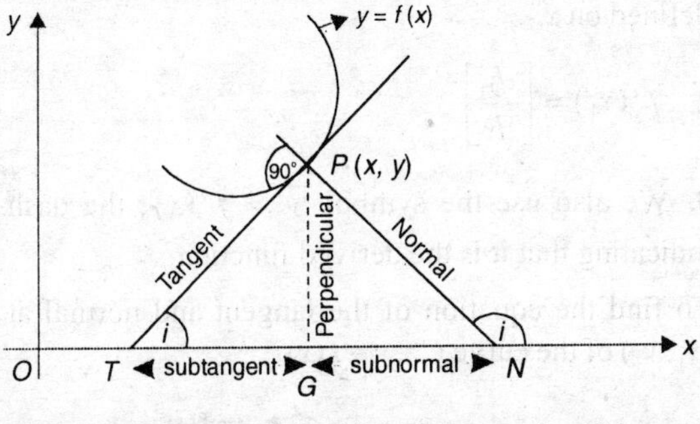

Subtangent: The projection on the x-axis, of the part (or portion or the segment) of the tangent joining the point of tangency on the curve and the point of intersection of the tangent with the x-axis is called the subtangent or length of the subtangent, *i.e.* the projection of the length of the tangent is called the length of subtangent. Thus, TG is the subtangent.

Subnormal: The projection on the x-axis, of the segment (or part or portion of the normal joining the point of tangency on the curve and the point of intersection of the normal with the x-axis is called the subnormal or length of subnormal, *i.e.* the projection of the length of the normal is called the length of the subnormal or simply the subnormal. Thus, GN is the subnormal or the length of subnormal.

Remark: The slope of the tangent line to a curve at a point is equal to the slope of the curve, *i.e.* $m = \dfrac{dy}{dx}$ at (x, y) = slope of the tangent at (x, y) on the curve = slope of the curve defined by $y = f(x)$ at a given point (x, y) where $x = a$ point on the domain of f.

Geometrical meaning of $\dfrac{dy}{dx}$: Supposing that $P(x, y)$ is a point on the continuous curve C whose equation is $y = f(x)$. Again supposing that Q is a point on the given curve $y = f(x)$, whose co-ordinates are $(x + \Delta x, y + \Delta x)$.

Construction: We draw perpendiculars PL, QM upon OX and $PN \perp QM$. We join PQ and produce it to meet OX at R.

Let PQ make (or, QR) make an angle θ with OX.

Now, $NQ = QM - MN$

$= QM - LP = \Delta y$

$= QM - LP$

$= y_1 + \Delta y - y_1$

$= \Delta y$

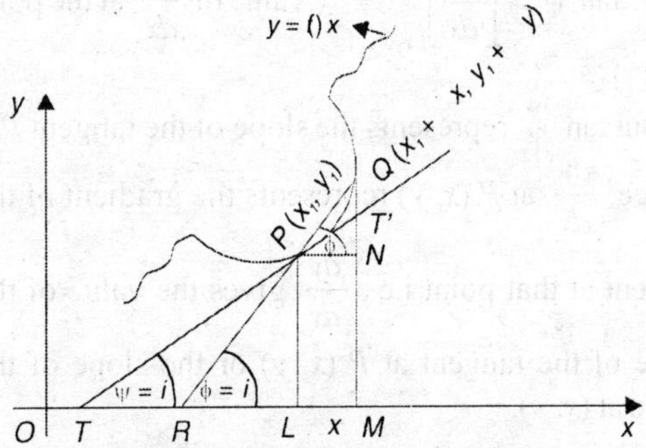

Again, $PN = LM = OM - OL = \Delta x$

$[\because OM = x + \Delta x,\ OL = x]$

$= x_1 + \Delta x - x_1$

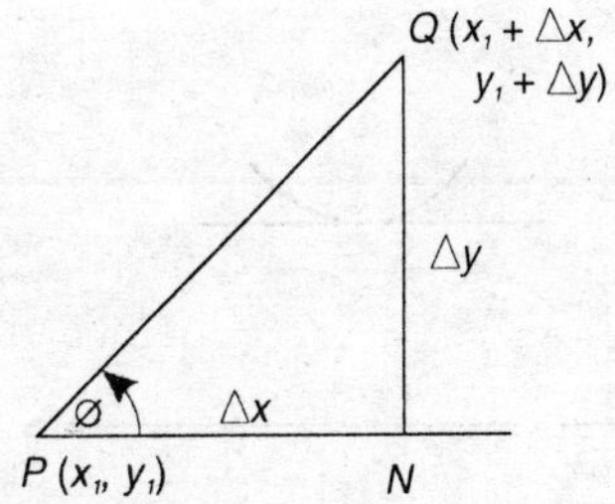

Since, PN is parallel to $OX \Rightarrow \angle NPQ = \phi$

$$\therefore \tan\phi = \tan NPQ = \frac{NQ}{PN}$$

$$= \frac{QM - LP}{LM - OL}$$

$$= \frac{\Delta y}{\Delta x} \qquad \text{...(1)}$$

Let the tangent $T'\ PT$ at P intersect the x-axis at T and makes an angle ψ with the positive direction of the x-axis.

$\therefore \angle XTP = \psi$

We observe that as Q moves along the curve C towards P, then the chord QP tends to the tangent $PT \Rightarrow$ If we suppose that $\Delta x \to 0$, then $M \to L$ (*i.e.*, $Q \to P$ along the curve C) $\Rightarrow$ The chord QP tends to the tangent at $P \Rightarrow PQ \to PT'$. Hence $\Delta x \to 0$, $\Delta y \to 0 \Rightarrow \phi \to \psi \Rightarrow \tan\phi = \tan\psi$

Hence, $\lim\limits_{\phi \to \psi} [\tan \phi] = \lim\limits_{\Delta x \to 0} \dfrac{\Delta y}{\Delta x} = \left[\dfrac{dy}{dx}\right]_{P(x_1,\, y_1)}$

$\therefore \tan \psi = \left[\dfrac{dy}{dx}\right]_{(x_1,\, y_1)}$ = value of $\dfrac{dy}{dx}$ at the point (x_1, y_1).

But tan ψ represents the slope of the tangent PT. Hence, $\dfrac{dy}{dx}$ at $P(x, y)$ represents the gradient of the tangent at that point i.e., $\dfrac{dy}{dx}$ gives the value of the slope of the tangent at $P(x, y)$ or the slope of the curve at (x, y).

Facts to know:

1. $\dfrac{dy}{dx} = 0$ at a point $\Rightarrow \tan \psi = 0 \Leftrightarrow$ The tangent at that point is parallel to the x-axis.

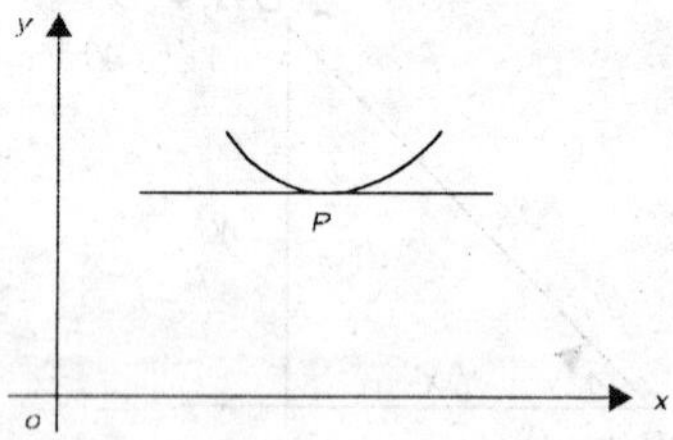

2. If $\dfrac{dx}{dy} = 0$ then $\psi = 90 \Leftrightarrow$ the tangent is parallel to the y-axis.

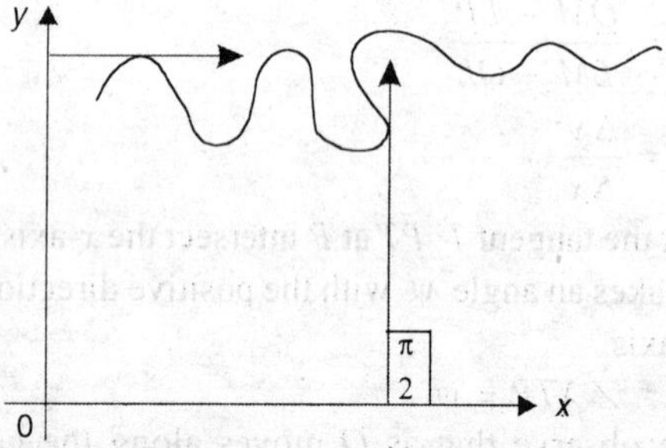

3. ***Notation***: The slope of the curve at a point $P(x_1, y_1)$ is denoted by

$$\left[\frac{dy}{dx}\right]_p \text{ or } \left(\frac{dy}{dx}\right)_p \text{ or } \left[\frac{dy}{dx}\right]_{(x_1,\, y_1)} \text{ or } \left(\frac{dy}{dx}\right)_{(x_1,\, y_1)}$$

or for the purpose of simplicity, we write $f'(x_1)$.

Note: **1.** In general, if the function $f(x)$ has a finite derivative at every point $x \in X$, then we can write the derivative $f'(x)$ as a function of x which is also defined on x.

2. $f'(x_1) = \left[\dfrac{dy}{dx}\right]_{x = x_1}$

3. We also use the symbol $\overline{y'} = f'(x)$, the dash indicating that it is the derived function.

To find the equation of the tangent and normal at (x_1, y_1) of the curve $C = y = f(x)$.

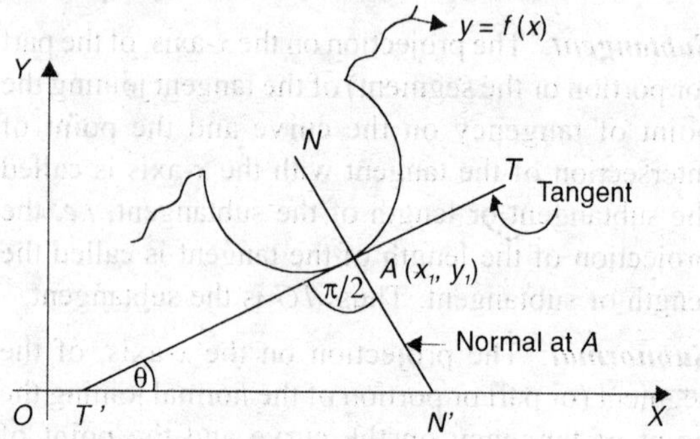

Refresh your memory:

Definition of normal to a curve at a point $P(x_1, y_1)$ on the curve:

The normal to a curve at the point $A(x_1, y_1)$ on it is the line through the point A perpendicular to the tangent at the point.

Remember:

1. If the slope of the tangent is m, then the slope of the normal is $-\dfrac{1}{m}$.

2. Equation of a line passing through (x_1, y_1) and having the slope m is $\dfrac{y - y_1}{x - x_1} = m \Leftrightarrow y - y_1 = m(x - x_1)$ which is the point slope form of the equation of a straight line.

3. The normal is a line passing through a point (x_1, y_1) lying on the curve and having the slope $= -\dfrac{1}{m} \Leftrightarrow \dfrac{y - y_1}{x - x_1} = -\dfrac{1}{m}$ (where $m = \tan \theta = \left[\dfrac{dy}{dx}\right]_p$ = derivative of a function at a general point (x, y)) is the equation of the normal (a line) passing through (x, y).

4. Two lines are perpendicular to each other if the product of their slopes is

$-1 \Rightarrow$ if $m_1 m_2 = -1$, then lines having the slopes m_1 and m_2 respectively are perpendicular to each other

$\Leftrightarrow$ Slope of one line is equal to the negative reciprocal of the slope of the other line means two lines are perpendicular to each other.

N.B.: Normal and tangent to a curve are perpendiculars to each other. Derivation of the equation of the tangent and normal at (x_1, y_1) lying on the curve $C = y = f(x)$.

TAT' and *NAN'* are respectively tangent and normal at $A\,(x_1, y_1)$ to the curve $C, y = f(x)$.

$\therefore$ By the geometrical meaning of $\frac{dy}{dx}$,

$\left[\frac{dy}{dx}\right]_{(x_1, y_1)}$ = The slope m of the tangent *TAT'*

$\therefore$ Now we know that the equation of the straight line passing through (x_1, y_1) and having a slope m is

$y - y_1 = m\,(x - x_1)$ ...(1)

A tangent is a line passing through (x_1, y_1) and having a slope $m = \left[\frac{dy}{dx}\right]_{(x_1, y_1)}$ ...(2)

$\therefore$ Putting (2) in (1), we get the required equation of the tangent, $y - y_1 = \left[\frac{dy}{dx}\right]_{(x_1, y_1)} (x_1 - x_1)$.

Again since the normal is perpendicular to the tangent, slope 'm' of the normal

$$= -\frac{1}{\left(\frac{dy}{dx}\right)_{(x_1, y_1)}} \quad ...(4)$$

Therefore, the normal has the equation

$$y - y_1 = -\frac{1}{\left[\frac{dy}{dx}\right]_{(x_1, y_1)}} \times (x - x_1)$$

$$= -\frac{1}{\left[f'(x)\right]_{\substack{x = x_1 \\ y = y_1}}} \times (x - x_1)$$

$$= -\frac{1}{\left[f'(x)\right]_{x = x_1}} \times (x - x_1)$$

Alternative proof based on definition of tangent to a curve at a point.

Let us suppose that the equation of the curve is $y = f(x)$ ...(1)

Let $P\,(x_1, y_1)$ be a point lying on the curve C given by (1) and $Q\,(x_1 + \Delta x,\ y_1 + \Delta y)$ be any point adjacent to P on C.

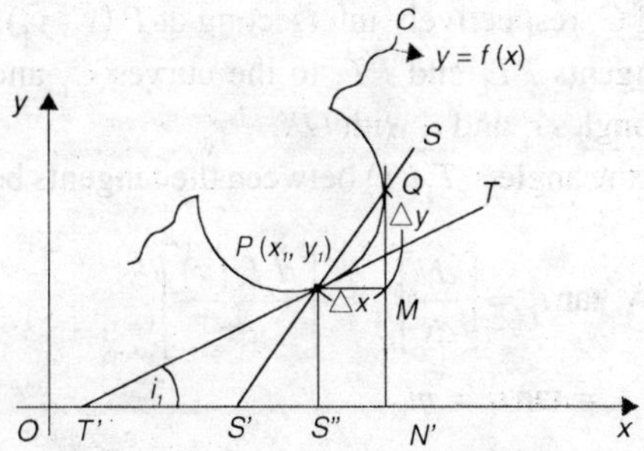

Then by the co-ordinates geometry, the equation of the secant line PQ is $(y - y_1) = \frac{\Delta y}{\Delta x}(x - x_1)$ ...(2)

Now, the slope of the tangent is the limit of the slope of the secant $\Rightarrow$ the equation of the tangent is given by $(y - y_1) = \lim\limits_{\Delta x \to 0} \frac{\Delta y}{\Delta x}(x - x_1)$

$$\Rightarrow (y - y_1) = \left(\frac{dy}{dx}\right)_P (x - x_1)$$

which is the required equation of tangent where

$\left[\frac{dy}{dx}\right]_{p(x_1, y_1)}$ = value of $\frac{dy}{dx}$ at (x_1, y_1).

Angle between two curves: The angel between two intersecting curves is the angle between the tangents at their common point of intersection.

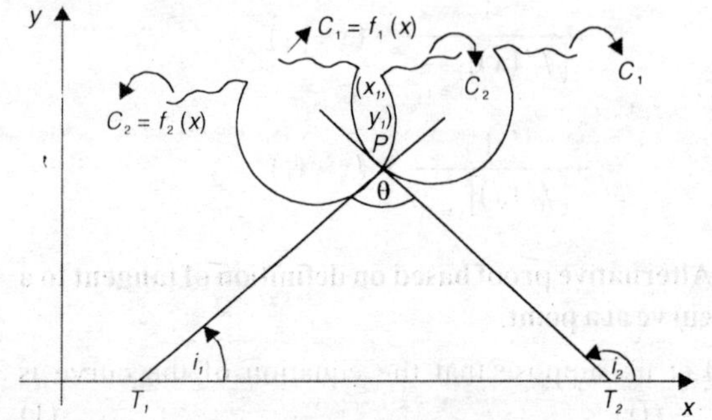

Derivation of the formula: Supposing that $y = f_1(x)$

$y = f_2(x)$ are the two equations of the two curves C_1 and C_2 respectively intersecting at $P(x_1, y_1)$. Let the tangents PT_1 and PT_2 to the curves C_1 and C_2 make angles i_1 and i_2 with OX.

Let the angle $< T_1PT_2$ between the tangents be θ.

$$\text{Now, } \tan i_1 = \left[\frac{dy}{dx}\right]_{C_1} = \left[\frac{d\, f_1(x)}{dx}\right]_{\substack{x = x_1 \\ y = y_1}}$$

$$= \tan i_1 = m_1$$

$$\text{and } \tan i_2 = \left[\frac{dy}{dx}\right]_{C_2} = \left[\frac{d\, f_2(x)}{dx}\right]_{\substack{x = x_1 \\ y = y_1}}$$

$$= \tan i_2 = m_2$$

Now, since $\theta = i_2 - i_1$

$$\tan\theta = \tan(i_2 - i_1) = \frac{\tan i_2 - \tan i_1}{1 + \tan i_1 \cdot \tan i_2}$$

$$= \frac{m_2 - m_1}{1 + m_1 m_2}$$

$$\therefore \theta = \tan^{-1}\left[\frac{m_2 - m_1}{1 + m_1 m_2}\right],$$

$$\text{where } m_1 = \left[\frac{d\, f_1(x)}{dx}\right]_{\substack{x = x_1 \\ y = y_1}}$$

= Value of d.c of $f_2(x)$ at the point of intersection

$$P(x_1, y_1) \text{ and } m_2 = \left[\frac{df_2(x)}{dx}\right]_{\substack{x = x_1 \\ y = y_1}}$$

= Value of d.c. of $f_2(x)$ at the point of intersection $P(x_1, y_1)$.

Orthogonal Curves: Two curves are said to be orthogonal if they intersect each other at right angles.

Thus, for the curves $C_1 = y = f_1(x)$ and $C_2 = y = f_2(x)$ to be orthogonal, they must make at an angle $\frac{\pi}{2}$ at their point of intersection

$$\therefore i_2 = \frac{\pi}{2} + i_1$$

$$\Rightarrow \tan i_2 = -\cot i_1$$

$$\Rightarrow \tan i_1 \cdot \tan i_2 = -1$$

$$\Rightarrow m_1 \cdot m_2 = -1$$

$\Rightarrow$ derivative of $f_1(x)$ at (x_1, y_1) × derivative of $f_2(x)$ at $(x_1, y_1) = -1$.

Remember: **1.** Angle between two curves or angle between two intersecting curves at a point of intersection or angle of intersection of two curves are synonyms by which we mean the angle between their tangents at the point of intersection of two curves.

2. If the angle between the tangents to the two curves at the point of intersection of two curves is $\frac{\pi}{2} = \theta$, then the two curves are said to be orthogonal or to cut orthogonally.

3. 'Ortho' means *right* and gonal means *angular*.

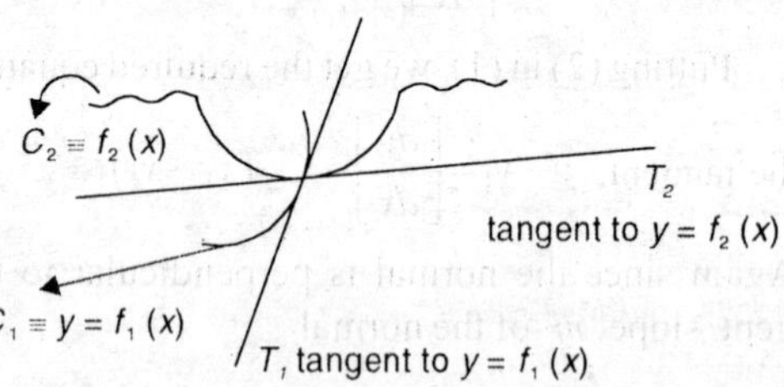

4. Tangent of an angle between two curves

$$= \frac{\text{Difference of slopes at their common point of intersection}}{1 + \text{product of slopes at their common points of intersection}}$$

5. When the two curves touch at (x_1, y_1), $\theta = 0$ $\Rightarrow \tan\theta = 0 \Rightarrow m_1 = m_2 \Rightarrow$ when the slopes of the tangents (or curves) at their common point of intersection are same, the curves touch each other.

6. The equations of the curves may be explicit or implicit. When the given equation is explicit function of x only, the derived function contains only x and when the given equation is implicit function of x and y, the derived function contains both x and y. This is why we put only x-coordinate of point of intersection in the derived function of explicit function while finding the slopes at the point of intersection of two given curves while we put x co-ordinate and y co-ordinate both of point of intersection in the derived function of implicit function while finding the slopes at the point of intersection of two given curves.

Types of the problem:

1. Finding the inclination or slopes when x_1 and $/y_1$ is given.

2. Finding the equation of the tangent and normal when x_1 and $/y_1$ is given.

3. Finding intercepts and proving the result based on equation of tangent and normal.

4. Finding (x_1, y_1) = co-ordinates of the point where the tangent ... and finding the length of perpendicular.

5. Finding the angle of intersection or proving orthogonal or touch ... etc.

6. Proving the equation of tangent and normal to the curve $y = f(x)$ at any point (x_1, y_1) to a straight line $ax + bx + c = 0$.

7. Finding the angle between two tangents to a curve at two given points.

8. Finding the area of a triangle.

9. Finding the length of subtangent and subnormal.

Refresh your memory: In every type of problem, the equation of the curves in planes may be given in three forms:

(i) Explicit form: $y = f(x)$.

(ii) Implicit form: $f(x, y) = 0$ or constant

(iii) Parametric form: $x = f_1(t)$

$y = f_2(t)$

Problems based on finding inclination:

Working rule:

1. Find y_1 i.e. find d.c of the given equation of the line.

2. Find $[f'(x)]_{\left(\substack{x=x_1 \\ y=y_1}\right)}$ which is the slope of the tangent line.

$= \tan i$

3. Find the angle of inclination using $\tan i = \tan \alpha$ which $\Rightarrow i = \alpha$.

Note: **1.** Actually (x_1, y_1) is taken to be the point of contact of the tangent and the curve which is given in the problem.

2. $[f'(x)]_{\left(\substack{x=x_1 \\ y=y_1}\right)} = \tan i = \tan \alpha$

3. If $\dfrac{dy}{dx}$ = an expression in x only, we put only the value of x coordinate of the given point and if $\dfrac{dy}{dx}$ = $f(x, y)$, we put x coordinate and y coordinate both of the given point in $f(x, y)$.

4. $[f'(x)]_{\left(\substack{x=x_1 \\ y=y_1}\right)}$ = value of $f'(x)$ at $x = x_1$ and $y = y_1$ which are the coordinates of the point of contact of the curve and the tangent.

Examples worked out:

Question 1: Find the inclination of the tangent at the point where $x = 1$ to the curve $y = \dfrac{x^3}{3} - x^2 + 2$.

Solution: Given equation of the curve is

$$y = \frac{x^3}{3} - x^2 + 2 \qquad \text{...(1)}$$

Now, differentiating both sides of (1) w.r.t x, we get

$$\frac{dy}{dx} = 3\frac{x^2}{3} - 2x + 0 = x^2 - 2x \qquad \text{...(2)}$$

$$\text{Now, } [f'(x)]_{x=1} = \left[x^2 - 2x\right]_{x=1} = 1^2 - 2 = 1 - 2 = -1$$

$$\Rightarrow \tan i = -1 = \tan 135 \Rightarrow i = 135°$$

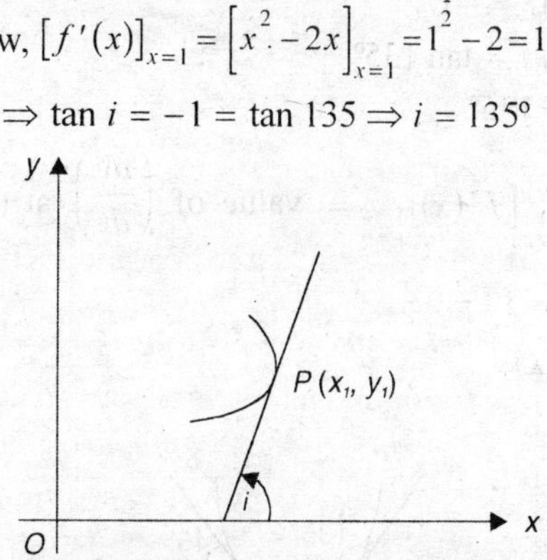

Note: If $[f'(x)]_{\substack{x=x_1 \\ y=y_1}}$ = – ve $\Rightarrow$ Angle of inclination is an obtuse angle.

Question 2: Find the inclination of the tangent at (1, 1) lying on the curve $y = x^2 - x + 1$.

Solution: Given equation is $y = x^2 - x + 1$ ...(1)

Now, differentiating both sides of (1) w.r.t x

$$\Rightarrow \frac{dy}{dx} = 2x - 1 + 0 = 2x - 1$$

Now, $[f'(x)]_{x=1} = [2x-1]_{x=1} = 2\cdot 1 - 1 = 2 - 1 = 1$

$\Rightarrow \tan i = 1 \quad \Rightarrow \tan i = \tan 45°$

$\Rightarrow i = 45°$

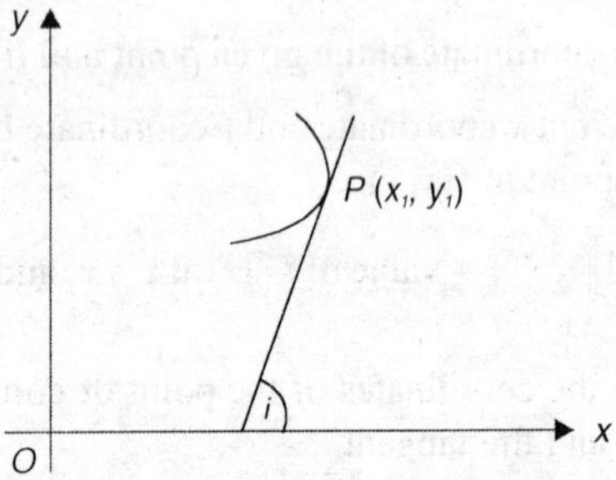

Question 3: Find the inclination of the tangents at the points (1, 0) and (2, 0) to the curve $y = (x-1)(x-2)$.

Solution: Given equation of the curve is $y = (x-1)(x-2)$ ...(1)

Now, differentiating both sides of (1) w.r.t x, we get

$$\frac{dy}{dx} = (2x - 3)$$

Now, $[f'(x)]_{\substack{x=1\\y=0}}$ = The value of $\left(\frac{dy}{dx}\right)$ at (1, 0)

$= [2x-3]_{x=1} = 2 - 3 = -1$

$\Rightarrow \tan i = -1$

$\Rightarrow \tan i = \tan 135°$

$\Rightarrow i = 135°$ ...(2)

Again, $[f'(x)]_{\substack{x=2\\y=0}}$ = value of $\left(\frac{dy}{dx}\right)$ at (2, 0)

$= [2x-3]_{x=2} = 2 \times 2 - 3 = 1$

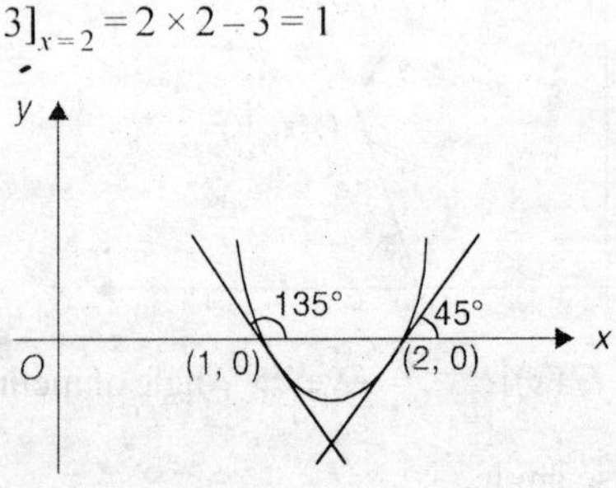

$\Rightarrow \tan i = \tan 45°$

$\Rightarrow i = 45°$ ...(3)

Hence, the inclinations of the two tangents to the curve at the points (1, 0) and (2, 0) are 135° and 45° respectively.

Problems based on finding the slope of a curve at a given point

Working rule:

1. Find $\frac{dy}{dx}$ by differentiating both sides of the given equation w.r.t x.

2. Find $[f'(x)]_{\left(\substack{x=x_1\\y=y_1}\right)}$ = Value of $\frac{dy}{dx}$ at the point (x_1, y_1) which is given in the problem.

The slope of the curve = The slope of the tangent.

Remember:

1. Slope of the given curve = the slope of the tangent at the same point.

2. If $x = a \Rightarrow y = \pm b$ and tangents are required to find out at $x = a$ then $(x_1, y_1) = (a, b)$ and $(x_2, y_2) = (a, -b)$ are the two points on the curve where the tangents are to be found out.

3. If $y = c \Rightarrow x = a, b$, then $(x_1, y_1) = (a, c), (x_2, y_2) = (b, c)$ are two points on the curve where tangents are required to find out.

Hence, we see that common x-coordinate or y co-ordinate is included in both points.

Note:

1. If $[f'(x)]$ = a function containing only x terms and no y terms, then we put only the x coordinate of the given point in derived function $\Rightarrow \left[\frac{dy}{dx}\right]_{\left(\substack{x=a\\y=b}\right)}$

$$= \left[\frac{dy}{dx}\right]_{x=a}.$$

2. If $[f'(x)]$ = a function containing only y terms and no x terms, then we put only the y coordinate of the given point in the derived function

$$\Rightarrow \left[\frac{dy}{dx}\right]_{\left(\substack{x=a\\y=b}\right)} = \left[\frac{dy}{dx}\right]_{y=b}.$$

3. If $[f'(x)]$ = derived function = a function containing both x terms and y terms, then we put x-coordinate and y-coordinate of the given point in the derived function.

$$\therefore \left[\frac{dy}{dx}\right]_{\left(\substack{x=a\\y=b}\right)} = [f'(x)]_{\left(\substack{x=a\\x=b}\right)}$$

4. If $[f'(x)]$ = derived function = a function containing only constant but no x and y terms, then we do not put x-coordinate and y-coordinates of the given point in the derived functions.

Question: How to find the x-coordinate or y-coordinate of the point lying on the curve whose equation is given and only one coordinate of the x-coordinate and y-coordinate is given?

Solution: Since the point lies on the curve, equation of the curve will satisfy the given equation, i.e. we put x-coordinate or y-coordinate in the given equation of the curve and solve for x or solve for y which will provide us the value of x or value of y representing x-coordinate or y-coordinate of the points lying on the curve.

Example: In the curve $y^2 = 4x$, obtain the slope of the curve at the point where $y = 2$.

Solution: Before finding the slope of the curve at a point whose y-coordinate is provided in the question as $y = b$ (Use $y = 2$) x-coordinate should be determined first.

Hence, $y^2 = 4x \Rightarrow y^2 = [4x] \Rightarrow 2^2$ or $4x \Rightarrow 4$
$\Rightarrow x = 1$

Therefore, required point of the curve $y^2 = 4x$ is $(x, y) = (1, 2)$

Notation: $\left[\frac{dy}{dx}\right]_{\left(\substack{x=x_1\\y=y_1}\right)}$ = value of derived function at $x = x_1$ and $y = y_1$.

Examples worked out:

Type 1: Finding the slope of a curve at a given point whose x and /y-coordinate is produced.

N.B.: In such type of problems, we must find the missing coordinate of the point by putting its value in the given equation of the curve provided that derived function does not contain the variable whose coordinate is given.

Question: In the curve $y^2 = 4x$, obtain the slope of the curve at the point where $y = 2$.

Solution: Given equation is $y^2 = 4x$ ...(1)

Now, differentiating both sides of (1) w.r.t x, we get

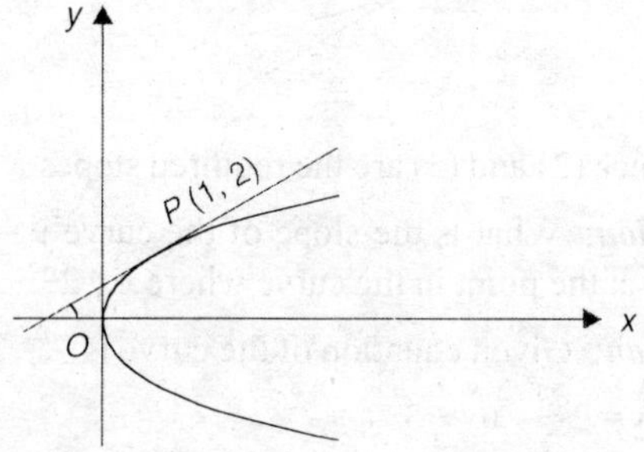

$$2y \cdot \left[\frac{dy}{dx}\right] = 4$$

$$\Rightarrow \frac{dy}{dx} = \frac{4}{2y} = \frac{2}{y}; \ (y \neq 0)$$

$$\Rightarrow \left[\frac{dy}{dx}\right]_{y=2} = \left[\frac{2}{y}\right]_{y=2} = \frac{2}{2} = 1$$

which is the required slope.

Question: Find the slope of the curve $y^2 = 4x$ at the point where $x = 1$.

Solution: Given equation is $y^2 = 4x$ ...(1)

$$x = 1 \Rightarrow y^2 = 4 \cdot 1 \Rightarrow y = \pm\sqrt{4} = \pm 2$$

Now, differentiating both sides of (1) w.r.t x, we get

$$2y \cdot \frac{dy}{dx} = 4$$

$$\Rightarrow \frac{dy}{dx} = \frac{4}{2y} = \left[\frac{2}{y}\right] \Rightarrow \left[\frac{dy}{dx}\right]_{\left(\substack{x=1\\y=2}\right)}$$

$$= \left[\frac{2}{y}\right]_{\left(\substack{x=1\\y=2}\right)} = \frac{2}{2} = 1 \qquad ...(2)$$

and $\Rightarrow \left[\dfrac{dy}{dx}\right]_{\left(\substack{x=1 \\ y=-2}\right)} = \left[\dfrac{2}{y}\right]_{\left(\substack{x=1 \\ y=-2}\right)} = \dfrac{2}{-2} = -1$ ...(3)

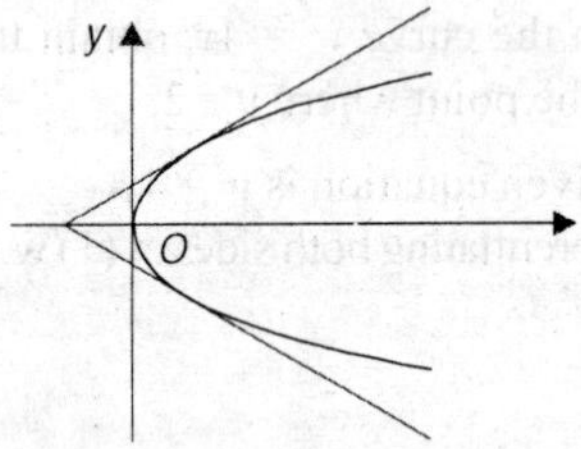

Hence (2) and (3) are the required slopes.

Question: What is the slope of the curve $y = 2x^2 - 6x + 3$ at the point in the curve where $x = 2$?

Solution: Given equation of the curve is

$$y = 2x^2 - 6x + 3 \quad \text{...(1)}$$

Now, differentiating both sides of (1) w.r.t x, we get

$$\frac{dy}{dx} = 4x - 6$$

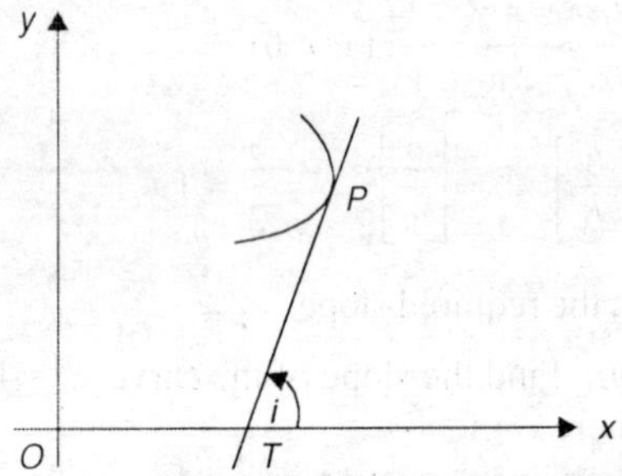

Now, the slope of the curve at the point $x = 2$

$$\therefore \left[\frac{dy}{dx}\right]_{x=2} = [4x-6]_{x=2} = 4 \times 2 - 6 = 8 - 6 = 2$$

Question: Find the slope of the ellipse $\dfrac{x^2}{a^2} + \dfrac{y^2}{b^2} = 1$ at the point where (i) $x = a$ and (ii) $y = b$.

Solution: $\because$ Given equation of the ellipse is

$$\frac{x^2}{a^2} + \frac{y^2}{b^2} = 1 \quad \text{...(1)}$$

When $x = a$, $y = 0$ [from the given equation]

When $y = b$, $x = 0$ [from the given equation]

$\because$ Given coordinates are $(a, 0)$ and $(0, b)$

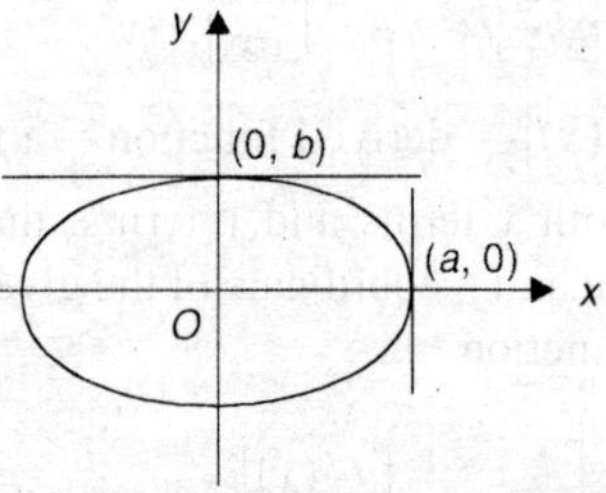

Now, differentiating both sides of (1) w.r.t x, we get

$$\frac{2x}{a^2} + \frac{2y}{b^2}\frac{dy}{dx} = 0$$

or $\left(\dfrac{x}{a^2} + \dfrac{y}{b^2}\dfrac{dy}{dx}\right) = 0$

$$\Rightarrow \frac{dy}{dx} = -\left(\frac{b^2 x}{a^2 y}\right) \Rightarrow \left[\frac{dx}{dy}\right]_{\left(\substack{x=a \\ y=0}\right)} = 0$$

$\Rightarrow \cot i = 0$

$\Rightarrow i = \dfrac{\pi}{2}$.

$\Rightarrow$ tangent at $(a, 0)$ is perpendicular to x-axis.

Similarly, $\left[\dfrac{dy}{dx}\right]_{\left(\substack{x=0 \\ y=b}\right)} = 0 \Rightarrow i = 0$ which

$\Rightarrow$ Tangent at $(0, b)$ is parallel to x-axis.

Question: Find the slope of the tangent to the curve $y = 2x^2 + 3 \sin x$ at $x = 0$.

What is the slope of the normal at $x = 0$?

Solution: Given equation is $y = 2x^2 + 3 \sin x$ ...(1)

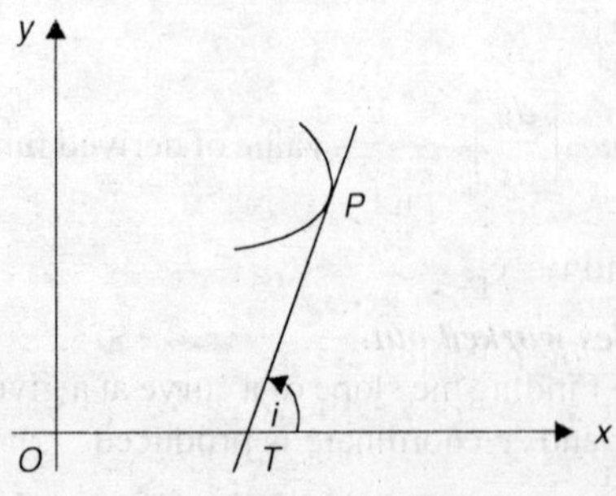

$$\Rightarrow \frac{dy}{dx} = 4x + 3\cos x$$

$$\Rightarrow \left[\frac{dy}{dx}\right]_{x=0} = [4x + 3\cos x]_{x=0}$$

$$= 4\cdot 0 + 3\cos 0 = 0 + 3\cdot 1 = 3$$

$$\Rightarrow \left[\frac{dy}{dx}\right]_{x=0} = 3 = \text{ Slope of the tangent to the}$$

curve at $x = 0$.

Now again, we know that slope of the normal is negative reciprocal of the slope of tangent at the same point which means slope of the normal

$$= -\frac{1}{\left[\frac{dy}{dx}\right]_{x=0}} = -\frac{1}{3}.$$

Question: Find the slope of the curve $y^2 = x$ at the point $x = 1$.

Solution: (1) Given equation is $y^2 = x$

When $x = 1$, $y = \pm 1 \Rightarrow$ Points are $(1, 1)$ and $(1, -1)$.

(2) Differentiating the given equation w.r.t x, we get

$$2y\frac{dy}{dx} = 1 \Rightarrow \frac{dy}{dx} = \frac{1}{2y}$$

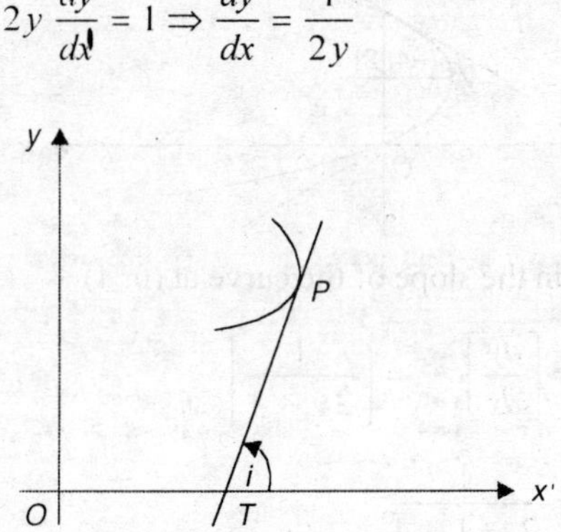

(3) Slope at $(1, 1) = \left[\frac{dy}{dx}\right]_{\substack{x=1\\y=1}} = \frac{1}{2\cdot 1} = \frac{1}{2}$

Slope at $(1, -1) = \left[\frac{dy}{dx}\right]_{\substack{x=1\\y=-1}} = \frac{1}{2\cdot(-1)} = -\frac{1}{2}$

Question: Find the slope and inclination of the curve $y = x^3 + x^2 - 4x$ at the point $(1, -2)$ lying on the curve.

Solution: Given equation of the curve is

$$y = x^3 + x^2 - 4x \qquad \ldots(1)$$

Now, differentiating both sides of (1) w.r.t x, we get

$$\frac{dy}{dx} = 3x^2 + 2x - 4 = f'(x)$$

$$\Rightarrow [f'(x)]_{\substack{x=1\\y=-2}} = 3 + 2 - 4 = 1$$

$$\Rightarrow \tan\psi = 1$$

$$\Rightarrow \tan\psi = \tan 45^\circ$$

$$\Rightarrow \psi = 45^\circ \Rightarrow \text{inclination} = 45^\circ.$$

Question: Find the slope of the curve $y^2 = 4x$ at the point $(1, 2)$.

Solution: Given equation of the curve is

$$y^2 = 4x \qquad \ldots(1)$$

Now, differentiating both sides of (1) w.r.t x

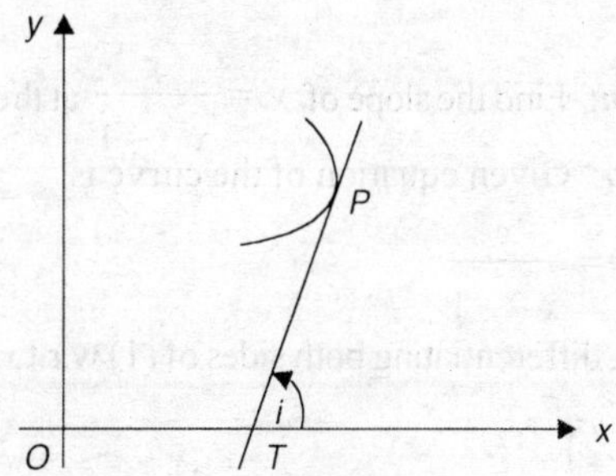

$$\Rightarrow 2y\cdot\frac{dy}{dx} = 4\cdot 1$$

$$\Rightarrow \frac{dy}{dx} = \frac{4}{2y} = \frac{2}{y}$$

$$\Rightarrow \left[\frac{dy}{dx}\right]_{\substack{x=1\\y=2}} = \left[\frac{2}{y}\right]_{\substack{x=1\\y=2}} = \left[\frac{2}{y}\right]_{y=2} = \frac{2}{2} = 1$$

Question: Given the curve $y = 6x - x^2$, find the slope of the curve at (x_1, y_1) and $(0, 0)$.

Solution: Given equation of the curve is $y = 6x - x^2$

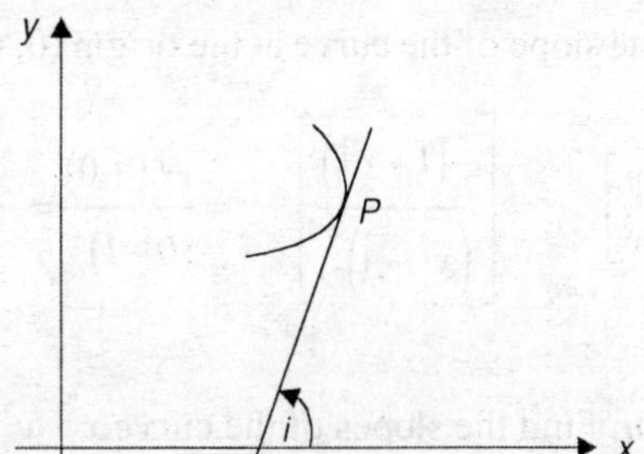

$$\Rightarrow \frac{dy}{dx} = 6 - 2x$$

$$\Rightarrow \left[\frac{dy}{dx}\right]_{\substack{x=x_1\\y=y_1}} = [6-2x]_{\substack{x=x_1\\y=y_1}} = 6 - 2x_1$$

⇒ The slope of the tangent at $(x_1, y_1) = 6 - 2x_1$

⇒ The slope of the given curve at (x_1, y_1) $= 6 - 2x_1$

Now, the slope of the tangent at (0, 0)

$$= \left[\frac{dy}{dx}\right]_{\substack{x=0\\y=0}} = [6 - 2x]_{\substack{x=0\\y=0}} = 6$$

(Slope of the curve at P is the slope of the tangent at P).

Question: Find the slope of $y = \dfrac{x}{x^2 - 1}$ at the origin.

Solution: Given equation of the curve is

$$y = \frac{x}{x^2 - 1} \quad \text{...(1)}$$

Now, differentiating both sides of (1) w.r.t x, we get

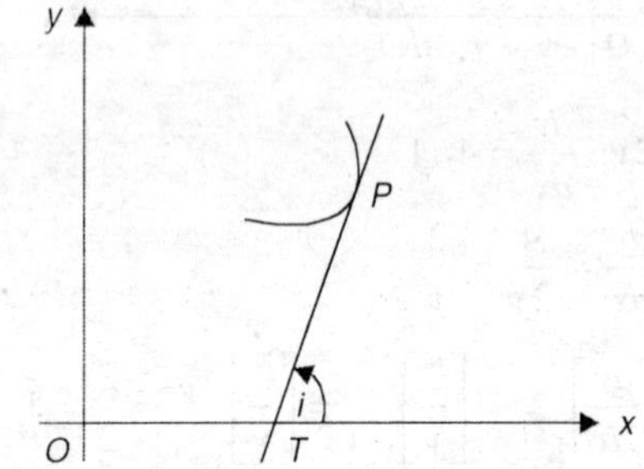

$$\frac{dy}{dx} = \frac{x^2 - 1 - 2x^2}{\left(x^2-1\right)^2} = \frac{-\left(1+x^2\right)}{\left(x^2-1\right)^2}$$

∴ The slope of the curve at the origin (0, 0)

$$= \left[\frac{dy}{dx}\right]_{\substack{x=0\\y=0}} = \left[\frac{-\left(1+x^2\right)}{\left(x^2-1\right)^2}\right]_{\substack{x=0\\y=0}} = \frac{-(1+0)}{(0-1)^2} = \frac{-1}{1} = -1$$

Question: Find the slopes of the curve $x = y^2 - 4y$ at the point where it crosses the y-axis.

Solution: Given equation of the curve is $x = y^2 - 4y$ ...(1)

Now, differentiating both sides of (1) w.r.t x, we get

$$1 = 2y\frac{dy}{dx} - 4 \cdot \frac{dy}{dx}$$

$$\Rightarrow \frac{dy}{dx} = \frac{1}{2y - 4} \quad \text{...(2)}$$

Since the curve $x = y^2 - 4y$ cuts the y-axis where $x = 0$

$$\therefore x = 0 \Rightarrow y^2 - 4y = 0 \Rightarrow y(y-4) = 0 \Rightarrow y = 0, 4$$

⇒ The two points are (0, 0)and (0, 4) where the curve crosses the y-axis.

Now, the slope of the curve at (0, 0)

$$= \left[\frac{dy}{dx}\right]_{\substack{x=0\\y=0}} = \left[\frac{1}{2y-4}\right] = \frac{1}{2 \cdot 0 - 4}$$

$$= \frac{-1}{4} \quad \text{...(3)}$$

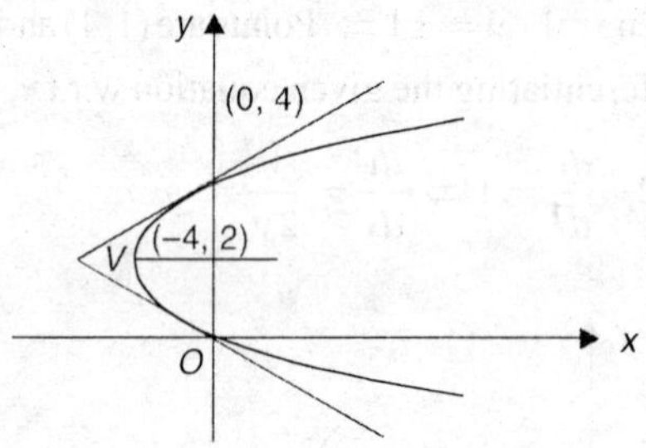

Again the slope of the curve at (0, 4)

$$= \left[\frac{dy}{dx}\right]_{\substack{x=0\\y=4}} = \left[\frac{1}{2y - 4}\right]_{\substack{x=0\\y=4}}$$

$$= \frac{1}{2 \times 4 - 4}$$

$$= \frac{1}{8 - 4} = \frac{1}{4} \quad \text{...(4)}$$

Important results to be committed to memory:

1. The geometrical meaning of differential coefficient $\dfrac{dy}{dx}$ at the point (x, y) of a curve is the tangent of the angle which the tangent line to the curve at (x, y) makes with the positive direction of the x-axis.

2. (i) The equation of the tangent to the curve $y = f(x)$ at the point (x_1, y_1) is $y - y_1 = \left[\frac{dy}{dx}\right]_{p(x_1, y_1)} (x - x_1)$,

where $\left[\frac{dy}{dx}\right]_{p(x_1, y_1)}$ stands for the value of $\frac{dy}{dx}$ at the point P whose x-coordinate $= x_1$ and y-coordinate $= y_1$.

(ii) If the tangent line to the curve $y = f(x)$ at the point (x_1, y_1) is parallel to the x-axis, then $\left[\frac{dy}{dx}\right]_{p(x_1, y_1)} = 0$.

(iii) If the tangent line to the curve $y = f(x)$ at the point (x_1, y_1) is perpendicular to the x-axis, then $\left[\frac{dx}{dy}\right]_{p(x_1, y_1)} = 0$.

3. The angle of intersection of two curves is defined to be the angle between the tangent lines to the curves at their points of intersection.

4. We know from coordinates geometry that:

(i) If two lines are parallel, their slopes are equal.

(ii) If two lines are perpendicular, the product of their slopes $= -1 \Rightarrow m_1 m_2 = -1$ where m_1 and m_2 are slopes of two lines perpendicular to each other if $m_1 =$ slope of the tangent line at (x_1, y_1) to C_1

$m_2 =$ Slope of the tangent line at (x_2, y_2) to C_2.

(iii) Given equation $y = f(x) =$ an expression containing only x terms $\Rightarrow$ Explicit function.

(iv) Given equation $f(x, y) = 0$ or c (c = any constant) = an expression containing both x and y terms $\Rightarrow$ Implicit function.

N.B.: Generally tangent means trigonometrical tangent which is written in short tan. Whereas tangent line means geometrical tangent which represents a line to a curve at a point. Moreover, whenever we say the tangent to a curve $y = f(x)$ at the point $P(x, y)$, we understand geometrical tangent or the tangent line to a curve at a point.

Type 2: Working rule to find the equation of the tangent to any curve $y = f(x)$ at the point (x_1, y_1) of the curve whose x-coordinate $x_1 = a$ is given and y-coordinate is absent.

Step 1: Find y_1 if it is not given in the following way:

Put the given x-coordinate of the point $P(x_1, y_1)$ on the curve $y = f(x)$ in the given equation of the curve $y = f(x)$ i.e., $y_1 = f(x_1)$.

Step 2: Find $\frac{dy}{dx}$ by differentiating the equation of the given curve w.r.t x using the rule of explicit and implicit function provided the given equation is explicit or implicit.

Step 3: Find the value of $\frac{dy}{dx}$ at the given point $P(x_1, y_1)$. This gives us the slope 'm' of the tangent line at the given point.

Step 4: Now put the value 'm' in the equation of the tangent line $y - y_1 = m(x - x_1)$ where x_1, y_1 are the given coordinates of the point where tangent line is drawn to the curve.

Note: **1.** The equation of the tangent line must be simplified as much as possible.

2. If x-coordinates and y-coordinate both of a point P are given which is generally represented as $P(x_1, y_1)$ or (x_1, y_1), then we directly find $\frac{dy}{dx}$ by differentiating the equation of the given curve w.r.t x and we follow the steps (3) to (4).

3. How to find the slope or the gradient of a given curve $y = f(x)$ at a point $P(x_1, y_1)$ of the curve where x-coordinate $= x_1 =$ a is given.

Method: **(i)** Find y_1 if it is not given.

(ii) Find $\frac{dy}{dx}$

(iii) Find $\left[\frac{dy}{dx}\right]_{\substack{x = x_1 = a \\ y = y_1 = f(a)}} = m$

4. ***Remember***:

(i) If $\left[\frac{dy}{dx}\right]_p = 0$, then $\tan \psi = 0 \Rightarrow \psi = 0$ at P. This means that the tangent line at P is parallel to the x-axis.

(ii) If $\left[\frac{dx}{dy}\right]_p = 0$ then the tangent line at P is perpendicular to the x-axis and parallel to y-axis.

Notation: **1.** Slope of the tangent line at P of the curve $y = f(x)$ is $\left[\frac{dy}{dx}\right]_P$.

2. The slope of the normal at P of the curve $y = f(x)$ is $-\frac{1}{\left[\frac{dy}{dx}\right]_P}$.

3. Equation of the tangent to the curve $y = f(x)$ at the point (x, y) is $Y - y = \frac{dy}{dx}(X - x)$ since (x, y) is given point, so we use X and Y for current coordinates.

N.B.: **1.** In question generally (x_1, y_1) notation are used instead of (x, y) so we have

$$y - y_1 = \left[\frac{dy}{dx}\right]_{\substack{x=x_1\\y=y_1}}(x - x_1)$$ for the equation of the tangent. This notation in tangent line equation is common in use.

2. The value of $\frac{dy}{dx}$ at $P(x, y)$ is denoted by $\frac{dy}{dx}$ itself, $\left[\frac{dy}{dx}\right]_{\substack{x=x\\y=y}} = \frac{dy}{dx}$.

3. The value of $\frac{dy}{dx}$ at $P(x_1, y_1)$ is given by

$$\left[\frac{dy}{dx}\right]_{\substack{x=x_1\\y=y_1}} = m$$

About the figure

Draw the figure provided it is easily possible to do so even though drawing a rough sketch of the figure is not necessary.

Working rule to find the equation of the normal at the point $P(x_1, y_1)$ whose x-coordinate $x_1 = a$ is given (P lies on the curves $y = f(x)$).

1. Find y_1 by putting x_1 in the given equation of the curve $y = f(x)$ i.e., $y_1 = f(x_1)$.

2. Find $\frac{dy}{dx}$ by differentiating the equation of the given curve w.r.t x using the rule of explicit and implicit function provided the given equation is explicit or implicit function.

3. Find the value of $\frac{dy}{dx}$ at the given point $P(x_1, y_1)$. This gives the slope 'm' of the tangent line at the given point.

4. Now put the value 'm' in the equation of the normal $y - y_1 = -\frac{1}{m}(x - x_1)$.

Note: **1.** If x-coordinate and y-coordinate both of a point P are given, then we directly find $\frac{dy}{dx}$ by differentiating the equation of the curve w.r.t x and we follow the steps (3) and (4).

2. The derivative of the function $y = f(x)$ at the point x_0 is equal to the slope of the tangent line to the graph of the function $y = f(x)$ at the point $M(x_0, f(x_0))$.

Question: What is the slope of the normal to the curve $y = 2x^2 - 6x - 3$ at the point in the curve where $x = 2$.

Solution: Given equation of the curve is

$$y = 2x^2 - 6x - 3 \quad \ldots(i)$$

Now, differentiating both sides of (i) w.r.t x, we get

$$\frac{dy}{dx} = 4x - 6$$

Now, $\left[\frac{dy}{dx}\right]_{x=2}$ = slope of the tangent line to the graph of the function $y = f(x)$ at the point $(2, -7) = (x_0, f(x_0))$.

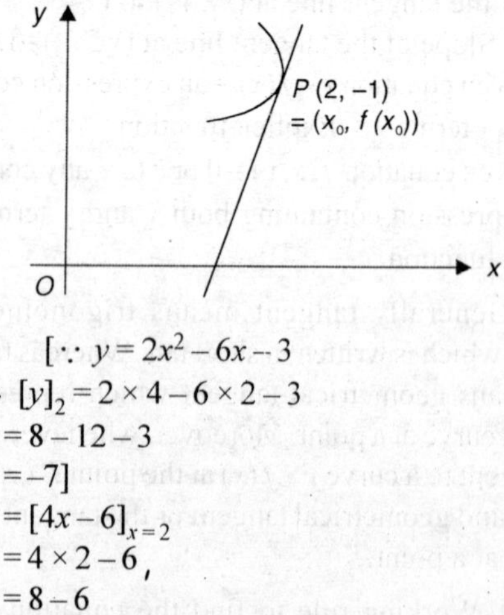

$[\because y = 2x^2 - 6x - 3$

$\Rightarrow [y]_2 = 2 \times 4 - 6 \times 2 - 3$

$= 8 - 12 - 3$

$= -7]$

$= [4x - 6]_{x=2}$

$= 4 \times 2 - 6,$

$= 8 - 6$

$= 2$

$\therefore$ slope of the normal at $(2, -7)$ is $-\frac{1}{2}$.

Examples worked out:

Question: Find the equation of the tangent to the curve $y = \sqrt{10 - x^2}$ at the point whose x-coordinate is 1.

Solution: **(1)** Given equation of the curve is

$$y = \sqrt{10 - x^2}$$

$$\therefore x = 1 \Rightarrow y = \sqrt{10 - 1} = \sqrt{9} = 3$$

$$\therefore (x_1, y_1) = (1, 3)$$

(2) Now, differentiating both sides of the given equation $y = \sqrt{10 - x^2}$ w.r.t x

$$\Rightarrow \frac{dy}{dx} = \frac{1}{2\sqrt{10 - x^2}} \times (-2x)$$

(3) $m = \left[\frac{dy}{dx}\right]_{x=1} = \frac{-1}{\sqrt{10 - 1}} = -\frac{1}{3}$

(4) Equation of the tangent is $(y - y_1) = m(x - x_1)$

i.e., $(y - 3) = -\frac{1}{3}(x - 1)$

Question: Find the equation of the normal to the curve $y = e^x$ at the point where $x = 0$.

Solution: **(1)** Given equation of the curve is $y = e^x$

$$\because x = 0 \Rightarrow y = e^0 = 1$$

$$\therefore x_1 = 0 \Rightarrow y_1 = 1$$

(2) $y = e^x \Rightarrow \frac{dy}{dx} = e^x$

(3) $m = \left[\frac{dy}{dx}\right]_{\substack{x=0\\y=1}} = \left[e^x\right]_{\substack{x=0\\y=1}} = e^0 = 1$

$$\therefore -\frac{1}{m} = \frac{1}{\left[\frac{dy}{dx}\right]_{\substack{x=0\\y=1}}} = -1$$

(4) Equation of the normal at (0, 1) is $(y - y_1) = \frac{-1}{m}(x - x_1)$

$$\Rightarrow (y - 1) = (-1)(x - 0)$$

$$\Rightarrow y - 1 = -1(x)$$

$$\Rightarrow y - 1 = -x$$

$$\Rightarrow y + x - 1 = 0$$

Question: Find the equation of the tangent to the curve $y = \sin x$ at $x = \frac{\pi}{6} = 30^\circ$.

Solution: Given equation of the curve is $y = \sin x$ ...(1)

$$\because x = \frac{\pi}{6} \Rightarrow y = \sin\frac{\pi}{6} = \frac{1}{2}$$

$$\therefore x_1 = \frac{\pi}{6} \Rightarrow y_1 = \frac{1}{2}$$

$$\therefore (x_1, y_1) = \left(\frac{\pi}{6}, \frac{1}{2}\right)$$

Now $\frac{dy}{dx} = \cos x$

$$\Rightarrow \left[\frac{dy}{dx}\right]_{\substack{x=\pi/6\\y=1/2}} = \cos\frac{\pi}{6} = \frac{\sqrt{3}}{2}$$

$\therefore$ The equation of the tangent at (x_1, y_1) is

$$y - y_1 = \left[\frac{dy}{dx}\right]_{\substack{x=x_1\\y=y_1}}(x - x_1)$$

$$\Rightarrow \left(y - \frac{1}{2}\right) = \frac{\sqrt{3}}{2}\left(x - \frac{\pi}{6}\right)$$

$$\Rightarrow y - \frac{\sqrt{3}}{2}x = \frac{1}{2} - \frac{\sqrt{3}}{2} \times \frac{\pi}{6} \Rightarrow y - \frac{\sqrt{3}}{2}x = \frac{1}{2} - \frac{\sqrt{3}\pi}{12}$$

Question: Find the equation of the normal to the curve $y = 2x + 3x^3$ at the point where $x = 3$.

Solution: Given equation is $y = 2x + 3x^3$ ...(1)

$$\because x = 3 \Rightarrow y = 2 \times 3 + 3 \times 3^3 = 87$$

$$\therefore x_1 = 3, y_1 = 87$$

$$(x_1, y_1) = (3, 87)$$

Now, we are required to find out the equation of the normal at (3, 87).

Differentiating (1) w.r.t x, we get

$$\frac{dy}{dx} = 2 + 3 \times 3x^2 = 2 + 9x^2$$

Now, $m = \left[\frac{dy}{dx}\right]_{\substack{x=x_1 \\ y=y_1}} = \left[2 + 9x^2\right]_{\substack{x=3 \\ y=87}}$

$= [2 + 9 \times 3^2] = [2 + 9 \times 9] = 83$

$\therefore$ Slope of the normal $= -\frac{1}{m} = -\frac{1}{83}$

$\therefore$ Equation of the normal is

$$(y - y_1) = -\frac{1}{m}(x - x_1)$$

$$\Rightarrow (y - 87) = -\frac{1}{83}(x - 3)$$

$$\Rightarrow 83y - 7221 = -x + 3$$

$$\Rightarrow x + 83y = 7224$$

Question: Find the equation of the normal to the curve $2x^2 - 3x + 2y^2 - xy = 0$ at the point where $x = 1$.

Solution: Given equation of the curve is

$$2x^2 - 3x + 2y^2 - xy = 0 \qquad \ldots(1)$$

Now, differentiating (1) w.r.t x, we get

$$2 \cdot 2x - 3 + 2 \cdot 2y \cdot \frac{dy}{dx} - x \cdot \frac{dy}{dx} - y \times 1 = 0$$

$$\Rightarrow \frac{dy}{dx}(4y - x) = -4x + y + 3$$

$$\Rightarrow \frac{dy}{dx} = \frac{-4x + y + 3}{(4y - x)}$$

Now putting $x = 1$ in the equation of the curve, we get

$$\left[2x^2 - 3x + 2y^2 - xy\right]_{x=1} = 0$$

$$\therefore 2 \cdot 1 - 3 \times 1 + 2 \times y^2 - 1 \cdot y = 0$$

$$\Rightarrow 2y^2 - y - 1 = 0$$

$$\Rightarrow (2y + 1)(y - 1) = 0$$

$$\Rightarrow y = 1 \text{ or } y = -\frac{1}{2}$$

$\therefore$ $x = 1$ corresponds to two points whose coordinates are $(1, 1)$ and $\left(1, -\frac{1}{2}\right)$.

Now, slope of the tangent to the curve at the point $(1, 1)$.

$$= \left[\frac{dy}{dx}\right]_{\binom{x=1}{y=1}} = \left[\frac{3 - 4x + y}{4y - x}\right]_{\substack{x=1 \\ y=1}}$$

$$= \frac{3 - 4 + 1}{4 - 1} = \frac{0}{3} = 0 \qquad \ldots(2)$$

Slope of the tangent to the curve at the point $\left(1, -\frac{1}{2}\right)$.

$$= \left[\frac{dy}{dx}\right]_{\binom{x=1}{y=-1/2}} = \left[\frac{3 - 4x + y}{4y - x}\right]_{\substack{x=1 \\ y=-1/2}}$$

$$= \frac{3 - 4 - \frac{1}{2}}{-2 - 1} = \frac{1}{2} \qquad \ldots(3)$$

(2) $\Rightarrow$ The tangent at $(1, 1)$ is parallel to x-axis

since $\left[\frac{dy}{dx}\right]_{\substack{x=x_1 \\ y=y_1}} = 0 \Rightarrow$ tangent is parallel to x-axis through $(1, 1)$.

$\Rightarrow$ the normal is parallel to y-axis through $(1, 1)$

$\Rightarrow$ The equation of this normal is $x = 1$.

(3) $\Rightarrow$ The slope of the tangent at $\left(1, -\frac{1}{2}\right) = \frac{1}{2}$

$\Rightarrow$ The slope of the normal is negative reciprocal of $\frac{1}{2} = -2$.

$\Rightarrow$ Equation of the normal at $\left(1, -\frac{1}{2}\right)$ is

$$y - y_1 = -\frac{1}{\left[\frac{dy}{dx}\right]_{\substack{x=1 \\ y=-1/2}}} \cdot (x - x_1)$$

$$\therefore y - \left(-\frac{1}{2}\right) = -2(x - 1)$$

$$\therefore y + \frac{1}{2} = -2(x - 1)$$

$\therefore y + \frac{1}{2} = -2x + 2$

$\therefore y + \frac{1}{2} + 2(x-1) = 0$

$\therefore y + 2x = \frac{3}{2}.$

Question: Find the equation of the tangent and normal to the curve $y^2(2a-x) = x^3$ at points on it where $x = a$.

Solution: Given equation of the curve is

$y^2(2a-x) = x^3$ ...(1)

Put $x = a$ in (1), we get $y^2(2a-a) = a^3$

$\Rightarrow y^2 \cdot a = a^3 \Rightarrow y^2 = a^2 \Rightarrow y = \pm a$

$\Rightarrow$ Two points $(a, +a)$ and $(a, -a)$ correspond to $x = a$

Now, differentiating the equation of the curve (1) w.r.t x, we get $2y \cdot \frac{dy}{dx}(2a-x) - y^2 = 3x^2$

$\Rightarrow 2y\frac{dy}{dx} = \frac{3x^2 + y^2}{2a-x}$

$\Rightarrow \frac{dy}{dx} = \frac{1}{2y} \cdot \frac{3x^2+y^2}{2a-x}$

$\Rightarrow \left[\frac{dy}{dx}\right]_{\substack{x=a \\ y=a}} = -\frac{1}{2a} \cdot \frac{4a^2}{a} = 2$

$\Rightarrow$ The slope of the tangent = 2 at (a, a)

$\Rightarrow$ The slope of the normal $= -\frac{1}{2}$ at (a, a)

Similarly,

The slope of the tangent $= -2$ at $(a, -a)$

$\Rightarrow$ The slope of the normal $= -\left(-\frac{1}{2}\right) = \frac{1}{2}$ at $(a, -a)$.

Now, the equation of the tangent at (a, a) is

$(y-a) = +2(x-a) \Rightarrow y - 2x + a = 0$

Equation of the normal at (a, a) is

$(y-a) = -\frac{1}{2}(x-a)$

$\Rightarrow 2y + x - 3a = 0$

Similarly, the equation of the tangent at $(a, -a)$ is

$(y+a) = -2(x-a) \Rightarrow y + 2x - a = 0$

and the equation of the normal at $(a, -a)$ is

$(y+a) = \frac{1}{2}(x-a) \Rightarrow 2y - x + 3a = 0$

Question: Find the equation of the tangent to the curve $y = x^3 - 2x^2 + 4$ at $(2, 4)$.

Solution: $y = x^3 - 2x^2 + 4$ is the equation of the curve. ...(1)

Differentiating both sides of (1) w.r.t x, we get

$\frac{dy}{dx} = 3x^2 - 4x$

$\therefore$ The slope of the tangent at $(2, 4)$

$= \left[\frac{dy}{dx}\right]_{\substack{x=2 \\ y=4}} = \left[3x^2 - 4x\right]_{x=2} = 3(2)^2 - 4(2) = 4$

Now, the equation of the tangent to the curve at $(2, 4)$ is

$y - 4 = 4(x-2) \Rightarrow y - 4x + 4 = 0$

Question: Find the equation of the tangent to the curve $xy = 1$ at the point $\left(3, \frac{1}{3}\right)$.

Solution: $xy = 1$ is the given equation of the curve. ...(1)

Differentiating both sides of this equation (1) w.r.t x, we get $y + x\frac{dy}{dx} = 0$

$\Rightarrow \frac{dy}{dx} = -\frac{y}{x}$

$\therefore$ The slope of the tangent at $\left(3, \frac{1}{3}\right)$ is

$\left[\frac{dy}{dx}\right]_{\substack{x=3 \\ y=1/3}} = \left[\frac{-y}{x}\right]_{\substack{x=3 \\ y=1/3}} = \frac{-1/3}{3} = -\frac{1}{9}$

$\therefore$ The equation of the tangent at $\left(3, \frac{1}{3}\right)$ is

$\left(y - \frac{1}{3}\right) = -\frac{1}{9}(x-3)$

$\Rightarrow x + 9y - 6 = 0$

Question: Find the equations of the tangent and normal at (1, 4) to the curve $y = 2x^2 - 3x + 5$.

Solution: $y = 2x^2 - 3x + 5$ is the given equation of the curve. ...(1)

Now, differentiating both sides of the equation (1) w.r.t x, we get $\frac{dy}{dx} = 4x - 3$

Now, the slope of the tangent at (1, 4)

$$= m = \left[\frac{dy}{dx}\right]_{\substack{x=1\\y=4}} = [4x-3]_{x=1} = 4 - 3 = 1$$

$\therefore$ Negative reciprocal of the slope of the tangent

$$= -\frac{1}{m} = -1$$

$\therefore$ Equation of the tangent to the curve at (1, 4) is

$$(y - y_1) = m(x - x_1) \Rightarrow y - 4 = 1(x - 1)$$

$$\Rightarrow y - x - 3 = 0$$

Equation of the normal to the curve at (1, 4) is

$$y - y_1 = -\frac{1}{m}(x - x_1) \Rightarrow (y - 4) = -\frac{1}{1}(x - 1)$$

$$\Rightarrow y - 4 = -(x - 1)$$

$$\Rightarrow y - 4 = -x + 1 \Rightarrow x + y - 5 = 0$$

Question: Obtain the equation of the tangent to the curve $3y = x^3 - 3x^2 + 6x$ at the point (3, 6) on it.

Solution: Given equation is $3y = x^3 - 3x^2 + 6x$ ...(1)

Now, differentiating the equation (1) w.r.t x we get

$$3\frac{dy}{dx} = 3x^2 - 6x + 6 \Rightarrow \frac{dy}{dx} = x^2 - 2x + 2$$

$\therefore$ The slope of the tangent at (3, 6) on the curve

$$= \left[\frac{dy}{dx}\right]_{\substack{x=3\\y=6}}$$

$$= [x^2 - 2x + 2]_{x=3} = 3^2 - 2 \times 3 + 2 = 9 - 6 + 2 = 5$$

$\therefore$ Equation of the tangent is $y - 6 = 5(x - 3)$

$$\Rightarrow y - 6 = 5x - 15 \Rightarrow y - 5x = -15 + 6$$

$$\Rightarrow y - 5x = -9 \Rightarrow y - 5x + 9 = 0 \Rightarrow 5x - y - 9 = 0$$

Question: Find the equation of tangents to the curve $y = (x^3 - 1)(x - 2)$ at the points where the curve cuts the x-axis.

Solution: Given equation is $y = (x^3 - 1)(x - 2)$ ...(1)

The curve cuts the x-axis where $y = 0$ and $y = 0$

$$\Rightarrow (x^3 - 1)(x - 2) = 0$$

$$\Rightarrow \text{either } (x - 2) = 0 \text{ or } (x^3 - 1) = 0$$

$$\Rightarrow x = 2 \text{ or } x^3 = 1$$

$$\Rightarrow x = 2 \text{ or } x = 1 \Rightarrow x = 1, 2$$

Thus, the points are (1, 0) and (2, 0).

Now, differentiating both sides of (1) w.r.t x, we get

$$\frac{dy}{dx} = 3x^2(x - 2) + (x^3 - 1) \cdot 1 = 4x^3 - 6x^2 - 1$$

$$\Rightarrow \left[\frac{dy}{dx}\right]_{\substack{x=1\\y=0}} = \left[4x^3 - 6x^2 - 1\right]_{x=1}$$

$$= 4 - 7 = -3$$

and $\left[\frac{dy}{dx}\right]_{\substack{x=2\\y=0}} = \left[4x^3 - 6x^2 - 1\right]_{x=2}$

$$= 4 \times 8 - 6 \times 4 - 1 = 7$$

Now, the equation of the tangents at (1, 0) is

$$(y - 0) = \left[\frac{dy}{dx}\right]_{\substack{x=1\\y=0}} (x - 1)$$

$$\Rightarrow (y - 0) = -3(x - 1)$$

$$\Rightarrow y = -3x + 3$$

$$\Rightarrow y + 3x - 3 = 0$$

The equation of the tangents at (2, 0) is

$$(y - 0) = \left[\frac{dy}{dx}\right]_{\substack{x=2\\y=0}} (x - 2)$$

$$\Rightarrow (y - 0) = 7(x - 2)$$

$$\Rightarrow y = 7x - 14$$

$$\Rightarrow y - 7x + 14 = 0$$

Question: Find the equation of the tangent to the curve $x^2 + 4y^2 - 4x = 0$ at the point (4, 0).

Solution: Given equation is $x^2 + 4y^2 - 4x = 0$ ...(1)

Now, differentiating both sides of (1) w.r.t x, we get

$$\frac{dy}{dx} = \frac{2 - x}{4y}$$

$$\therefore \left(\frac{dx}{dy}\right)_{\substack{x=4\\y=0}} = 0$$

$\Rightarrow$ The tangent is perpendicular to x-axis or parallel to y-axis.

$\Rightarrow$ In this case the tangent line is the line through the point (4, 0) parallel to y-axis.

$\Rightarrow$ Its equation is $x = 4$.

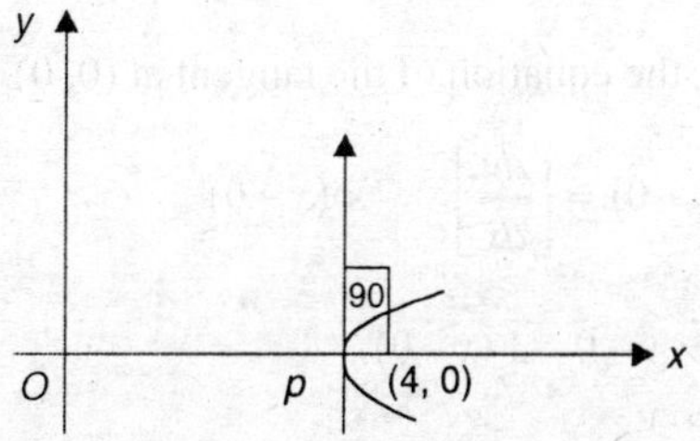

Question: Find the equation of the tangent at the point $(1, -1)$ on the curve $x^3 - xy^2 - 4x^2 - xy + 5x + 3y + 1 = 0$.

Solution: Given equation of the curve is $x^3 - xy^2 - 4x^2 - xy + 5x + 3y + 1 = 0$ ...(1)

Now, differentiating both sides of (1) w.r.t x, we get

$$3x^2 - 2xy\frac{dy}{dx} - y^2 - 8x - x\frac{dy}{dx} - y + 5 + 3\frac{dy}{dx} = 0$$

$$\Rightarrow \frac{dy}{dx}(3 - x - 2xy) + 3x^2 - y^2 - 8x - y + 5 = 0$$

$$\Rightarrow \frac{dy}{dx} = \frac{y^2 - 3x^2 + 8x + y - 5}{3 - x - 2xy}$$

$\therefore$ The slope of the tangent to the curve at $(1, -1)$

$$= \left[\frac{dy}{dx}\right]_{\substack{x=1\\y=-1}} = \frac{1 - 3 + 8 - 1 - 5}{3 - 1 + 2} = \frac{0}{4} = 0$$

Hence, the required equation of the tangent to the curve at $(1, -1)$ is $(y - (-1)) = \left[\frac{dy}{dx}\right]_{\substack{x=1\\y=-1}} \times (x - 1)$

$$\Rightarrow y + 1 = 0 \cdot (x - 1)$$

$$\Rightarrow y + 1 = 0$$

Question: Find the equations of the tangent and normal to the curve $y^2 = x$ at the point (1, 1).

Solution: Given equation of the curve is $y^2 = x$ ...(1)

and $(x_1, y_1) = (1, 1)$

Differentiating both sides of (1) w.r.t x, we get

$$2y\frac{dy}{dx} = 1 \Rightarrow \frac{dy}{dx} = \frac{1}{2y}$$

Now, the slope of the tangent at (1, 1)

$$= \left[\frac{dy}{dx}\right]_{\substack{x=1\\y=1}} = \left[\frac{1}{2y}\right]_{\substack{x=1\\y=1}} = \frac{1}{2}$$

The slope of the normal at (1, 1) = negative reciprocal of slope of tangent = -2

$\therefore$ Equation of the tangent is

$$(y - y_1) = \left[\frac{dy}{dx}\right]_{\substack{x=1\\y=1}} \times (x - x_1)$$

$$\Rightarrow (y - 1) = \frac{1}{2}(x - 1) \Rightarrow x - 2y + 1 = 0$$

Again equation, of the normal to the curve at (1, 1) is

$$(y - y_1) = -\frac{1}{\text{Slope of the tangent}} \times (x - x_1)$$

$$\Rightarrow (y - 1) = -\frac{1}{\frac{1}{2}} \times (x - 1) \Rightarrow (y - 1) = -2(x - 1)$$

$$\Rightarrow (y - 1) = -2(x - 1)$$

$$\Rightarrow (y - 1) = -2x + 2$$

$$\Rightarrow 2x + y - 3 = 0$$

Question: Find the equations of tangents and normals to the curve $yx^2 + x^2 - 5x + 6 = 0$ where it cuts the axis of x.

Solution: The curve $yx^2 + x^2 - 5x + 6 = 0$ ...(1)

Cuts the axis of x at the points where $y = 0$

$$y = 0 \Rightarrow 0 \cdot x^2 + x^2 - 5x + 6 = 0 \Rightarrow x^2 - 5x + 6 = 0$$

$$\Rightarrow (x - 2)(x - 3) = 0 \Rightarrow x = 2, 3$$

Thus, the given curve cuts the x-axis at the points (2, 0) and (3, 0).

Now, differentiating both sides of (1) w.r.t x, we get

$$y \cdot 2x + x^2\frac{dy}{dx} + 2x - 5 = 0$$

$$\Rightarrow \frac{dy}{dx} = \frac{5-2x-2xy}{x^2}$$

Now, the slope of the tangent at (2, 0)

$$= \left[\frac{dy}{dx}\right]_{\substack{x=2\\y=0}} = \left[\frac{5-2x-2xy}{x^2}\right]_{\substack{x=2\\y=0}}$$

$$= \frac{5-2\times2-2\times2\times0}{2^2} = \frac{5-4}{4} = \frac{1}{4}$$

again, the slope of the tangent at (3, 0)

$$= \frac{5-2\times3-2\times3\times0}{3^2} = -\frac{1}{9}$$

$\therefore$ The equation of the tangent to the curve at (2, 0) is

$$(y-0) = \left[\frac{dy}{dx}\right]_{\substack{x=2\\y=0}} (x-2)$$

$$\Rightarrow y-0 = \frac{1}{4}(x-2)$$

$$\Rightarrow 4y = x-2$$

Again the tangent to the curve at (3, 0) is

$$y-0 = -\frac{1}{9}(x-3)$$

$$\Rightarrow 9y = -x+3$$

$$\Rightarrow x+9y = 3$$

Now, the slopes of the normal at (2, 0) and (3, 0) are -4 and 9 respectively.

Normal at (2, 0) is $y-0 = -4(x-2) \Rightarrow y+4x = 8$

And normal at (3, 0) is $y-0 = 9(x-3)$

$\Rightarrow y = 9x-27$

Question: Find the equation of the tangent to the curve $y = \dfrac{x}{x^2+1}$ at (0, 0).

Solution: Given equation of the curve is

$$y = \frac{x}{x^2+1} \qquad \ldots(1)$$

Now differentiating both sides of the equation (1) w.r.t x, we get

$$\frac{dy}{dx} = \frac{x^2+1-2x^2}{(x^2+1)^2} = \frac{1-x^2}{(1+x^2)^2}$$

Now, the slope of the tangent to the curve at (0, 0)

$$= \left[\frac{dy}{dx}\right]_{\substack{x=0\\y=0}} = \left[\frac{1-x^2}{(1+x^2)^2}\right]_{x=0} = \left[\frac{1-0}{(0+1)^2}\right] = 1$$

Now, the equation of the tangent at (0, 0) is

$$(y-0) = \left[\frac{dy}{dx}\right]_{\substack{x=0\\y=0}} \times (x-0)$$

$$\Rightarrow y-0 = 1(x-0)$$

$$\Rightarrow y = x$$

Question: Find the equation of the normal to the curve $y = e^x$ at the point (0, 1).

Solution: Given equation is $y = e^x$ ...(1)

Now, differentiating both sides of (1) w.r.t x, we get

$$\frac{dy}{dx} = e^x$$

Now, $m = \left[\frac{dy}{dx}\right]_{\substack{x=0\\y=1}}$ = the value of $\frac{dy}{dx}$ at (0, 1)

$$= \left[e^x\right]_{\substack{x=0\\y=1}} = e^o = 1$$

$\Rightarrow$ Slope of the tangent passing through (0, 1) = 1

Hence, the equation of the normal to the curve at (0, 1) is

$$(y-y_1) = -\frac{1}{m}(x-x_1)$$

$$\Rightarrow (y-1) = -\frac{1}{m}(x-0)$$

$$\Rightarrow (y-1) = -1(x-0)$$

$$\Rightarrow y+x-1 = 0$$

Question: Find the equation of the normal to the curve $9x^2 - 4y^2 = 108$ at the point (4, 3).

Solution: Given equation of the curve is $9x^2 - 4y^2 = 108$ ...(1)

Now, differentiating both sides of (1) w.r.t x, we get

$$9 \cdot 2x - 4 \cdot 2y \frac{dy}{dx} = 0$$

$$\Rightarrow \frac{dy}{dx} = \frac{9x}{4y}$$

Now, $m = \left[\frac{dy}{dx}\right]_{\substack{x=4\\y=3}}$ = The value of $\frac{dy}{dx}$ at (4, 3)

$$= \left[\frac{9x}{4y}\right]_{\substack{x=4\\y=3}} = \frac{9 \times 4}{4 \times 3} = 3$$

$\Rightarrow$ The slope of the normal at (4, 3) $= -\frac{1}{m} = -\frac{1}{3}$

Hence, the required equation of the normal is

$$(y-3) = -\frac{1}{3}(x-4)$$

$$\Rightarrow 3y - 9 = -x + 4$$

$$\Rightarrow x + 3y = 13$$

Question: Find the equations of the tangent and normal to the circle $x^2 + y^2 = a^2$ at the point (x_1, y_1).

Solution: The point (x_1, y_1) lies on the curve

$$x^2 + y^2 = a^2 \quad \ldots(1)$$

$$\Rightarrow x_1^2 + y_1^2 = a^2 \quad \ldots(2)$$

Now, differentiating both sides of the given equation of the circle (1) w.r.t x, we get

$$\frac{d}{dx}\left[x^2 + y^2\right] = \frac{d}{dx}\left[a^2\right]$$

$$\Rightarrow 2x + 2y \cdot \frac{dy}{dx} = 0 \Rightarrow \frac{dy}{dx} = -\frac{x}{y}$$

Now $m = \left[\frac{dy}{dx}\right]_{\substack{x=x_1\\y=y_1}}$ = The value of $\frac{dy}{dx}$ at the point (x_1, y_1)

$$= \left[-\frac{x}{y}\right]_{\substack{x=x_1\\y=y_1}} = -\frac{x_1}{y_1}$$

$\therefore$ Required equation of the tangent to the curve at the point (x_1, y_1) is $(y - y_1) = -\frac{x_1}{y_1}(x - x_1)$

$\Rightarrow y y_1 - y_1^2 = -x x_1 + x_1^2 \Rightarrow x x_1 + y y_1 = x_1^2 + y_1^2$
$= a^2$ from (2)

Again the slope of the normal at (x_1, y_1) negative reciprocal of the slope of tangent

$\therefore$ Required equation of the normal to the curve at (x_1, y_1) is $(y - y_1) = \frac{y_1}{x_1}(x - x_1) \Rightarrow x_1 y - x_1 y_1$
$= x y_1 - x_1 y_1 \Rightarrow x y_1 - x_1 y = 0$

Question: Find the equation of the tangent and normal to the ellipse $\frac{x^2}{a^2} + \frac{y^2}{b^2} = 1$ at the point (x_1, y_1).

Solution: Equation of the curve is

$$b^2 x^2 + a^2 y^2 = a^2 b^2 \quad \ldots(1)$$

Differentiating both sides of (1) w.r.t x, we get

$$2b^2 x + 2a^2 y \frac{dy}{dx} = 0$$

$$\Rightarrow \frac{dy}{dx} = -\frac{b^2 x}{a^2 y}$$

$\therefore$ The value of $\frac{dy}{dx}$ at the point (x_1, y_1)

$$= \left[\frac{dy}{dx}\right]_{\substack{x=x_1\\y=y_1}} = -\frac{b^2 x_1}{a^2 y_1} = m$$

$\therefore$ Required equation of the tangent is

$$(y - y_1) = -\frac{b^2 x_1}{a^2 y_1}(x - x_1)$$

$$\Rightarrow a^2 y y_1 - a^2 y_1^2 = -b^2 x x_1 + b^2 x_1^2$$

$$\Rightarrow b^2 x x_1 + a^2 y y_1 = b^2 x_1^2 + a^2 y_1^2$$

Now, dividing both sides by $a^2 b^2$, we get

$$\frac{x x_1}{a^2} + \frac{y y_1}{b^2} = \frac{x_1^2}{a^2} + \frac{y_1^2}{b^2} = 1$$

[$\because (x_1, y_1)$ lies on the ellipse]

Now, the slope of the normal = Negative reciprocal of the slope of the tangent

$$= \frac{a^2 y_1^2}{b^2 x_1^2}$$

$\therefore$ Required equation of the normal to the curve at (x_1, y_1) is

$$(y - y_1) = \frac{a^2 y_1}{b^2 x_1}(x - x_1)$$

$$\Rightarrow \frac{a^2(x - x_1)}{x_1} = \frac{b^2(y - y_1)}{y_1}$$

$$\Rightarrow \frac{a^2 x}{x_1} - a^2 = \frac{b^2 y}{y_1} - b^2$$

$$\Rightarrow \frac{a^2 x}{x_1} - \frac{b^2 y}{y_1} = a^2 - b^2$$

Question: Find the equation of the normal to the curve $\frac{x^2}{a^2} - \frac{y^2}{b^2} = 1$ at the point (x_1, y_1).

Solution: Given equation of the curve is

$$\frac{x^2}{a^2} - \frac{y^2}{b^2} = 1 \quad \ldots(1)$$

Now, differentiating both sides of the equation of the curve (1) w.r.t x, we get,

(i) $\frac{2x}{a^2} - \frac{2y}{b^2}\frac{dy}{dx} = 0$

$$\Rightarrow \frac{dy}{dx} = \frac{b^2 x}{a^2 y}$$

$\therefore$ The value of $\frac{dy}{dx}$ at the point (x_1, y_1)

$$= \left[\frac{dy}{dx}\right]_{\substack{x = x_1 \\ y = y_1}} = \frac{b^2 x_1}{a^2 y_1}$$

$\therefore$ The slope of the normal at (x_1, y_1)

$$= -\frac{a^2 y_1}{b^2 x_1}$$

Hence, the equation of the normal at (x_1, y_1) is

$$(y - y_1) = -\frac{a^2 y_1}{b^2 x_1}(x - x_1)$$

$$\Rightarrow \frac{b^2(y - y_1)}{y_1} = \frac{-a^2(x - x_1)}{x_1}$$

$$\Rightarrow \frac{x - x_1}{\left(\frac{x_1}{a^2}\right)} = \frac{y - y_1}{\left(\frac{y_1}{-b^2}\right)}$$

Question: Find the equation of the tangent and normal to the curve $y(2a - x) = x^2$ at the point (a, a).

Solution: The equation of the curve is

$$y(2a - x) = x^2 \quad \ldots(1)$$

Now, differentiating both sides of the equation (1) w.r.t x, we get,

$$y(0 - 1) + (2a - x)\frac{dy}{dx} = 2x$$

$$\Rightarrow \frac{dy}{dx} = \frac{2x + y}{2a - x}$$

$$\therefore \left[\frac{dy}{dx}\right]_{\substack{x = a \\ y = a}} = \frac{2a + a}{2a - a} = 3$$

$\therefore$ The slope of the normal at $(a, a) = -\frac{1}{3}$

Hence, the equation of the tangent at (a, a) is

$y - a = 3(x - a)$

$\Rightarrow y = 3x - 2a$ and the equation of the normal is

$y - a = -\frac{1}{3}(x - a)$

$\Rightarrow 3y - 3a = -x + a$

$\Rightarrow x + 3y = 4a$

Question: Prove that the normal to the curve $9x^2 - 4y^2 = 128$ at the point $(4, 2)$ lying on it intersects the x-axis at a distance 13 from the origin.

Solution: Given equation of the curve is

$$9x^2 - 4y^2 = 128 \quad \ldots(1)$$

Differentiating both sides of the equation (1) w.r.t x, we get,

$$9 \times 2x - 4 \times 2y\frac{dy}{dx} = 0$$

$$\Rightarrow \frac{dy}{dx} = \frac{9x}{4y}$$

Now, the slope of the tangent at (4, 2)

$$= \left[\frac{dy}{dx}\right]_{\substack{x=4\\y=3}} = \left[\frac{9x}{4y}\right]_{\substack{x=4\\y=2}}$$

$$= \frac{9 \times 4}{4 \times 2} = \frac{9}{2}$$

and the slope of the normal at (4, 2)

$$= -\frac{1}{\text{slope of tangent}} = -\frac{2}{9}$$

$\therefore$ Required equation of the normal at (4, 2) to the curve $9x^2 - 4y = 128$ is

$$(y - y_1) = -\frac{2}{9}(x - x_1)$$

$$\Rightarrow (y-2) = -\frac{2}{9}(x-4)$$

$$\Rightarrow 9y - 18 = -2x + 8$$

$$\Rightarrow 2x + 9y = 26$$

$$\Rightarrow x = 13 \text{ when } y = 0$$

Problems based on finding the constants present in the equation of the curve.

Working rule:

1. Obtain the equation of the tangent to the curve.
2. Write the given equation of the tangent to the curve.
3. Equate the coefficients of x in both the equations of the curve which will given us the value of one constant.
4. Put this value of one constant in the equation of the curve which contains another constant passing through the given point (x_1, y_1).

N.B.: Given conditions provide us the value of required constants.

Examples worked out:

Question: The equation of the tangent at the point (2, 3) on the curve $y^2 = ax^2 + b$ is $y = 4x - 5$, find the values of a and b.

Solution: Given equation of the curve is

$$y^2 = ax^2 + b \quad \ldots(1)$$

On differentiating both sides of (1) w.r.t x, we get

$$2y\frac{dy}{dx} = 2ax$$

$$\Rightarrow \frac{dy}{dx} = \frac{ax}{y} = a \cdot \frac{x}{y}$$

$$\Rightarrow \left[\frac{dy}{dx}\right]_{\substack{x=2\\y=3}} = \left[a \cdot \frac{x}{y}\right]_{\substack{x=2\\y=3}}$$

$$= \frac{2}{3}a = \text{slope of the tangent at } (2, 3)$$

Now, the equation of the tangent is $y = 4x - 5$ which tells us the slope of the tangent = 4 ...(3)

$\therefore$ Equating equations, (2) and (3), we get

$$\frac{2}{3}a = 4$$

$$\Rightarrow a = 4 \times \frac{3}{2} = 6$$

Again equation of the curve is $y^2 = ax^2 + b$ which passes through (2, 3).

$$\therefore 3^2 = 6 \times 2^2 + b$$

$$\Rightarrow 9 = 6 \times 4 + b$$

$$\Rightarrow 9 - 24 = b$$

$$\Rightarrow b = -15$$

Hence, $a = 6, b = -15$

Note:

$$(y - y_1) = \left[\frac{dy}{dx}\right]_{\substack{x=x_1\\y=y_1}}(x - x_1)$$

$$\Rightarrow (y-3) = \frac{2}{3}a(x-2)$$

$$\Rightarrow (y-3) = \frac{2}{3}ax - \frac{4a}{3}$$

$$\Rightarrow y = \frac{2}{3}ax - \frac{4a}{3} + 3$$

$$\Rightarrow y = \frac{2}{3}ax - \left(\frac{4a}{3} - 3\right) \qquad \text{...(1)}$$

Comparing this equation (1) with the given equation of the tangent $y = 4x - 5$ and equating the coefficients of x in both equations, we get

$$\frac{2}{3}a = 4 \Rightarrow a = \frac{4 \times 3}{2} = 6.$$

We see that $\frac{4a}{3} - 3 = \frac{4.6}{3} - 3 = 5$, as it should be.

Question: If the equation of the normal to the curve $y = x^3 + x - 2$ at the point (1, 0) on it is $y = ax + b$, then find the value of a and b.

Solution: Given equation of the curve is

$$y = x^3 + x - 2 \qquad \text{...(1)}$$

Differentiating (1) w.r.t. x, we get

$$\frac{dy}{dx} = 3x^2 + 1$$

$\therefore$ Slope of the tangent at (1, 0)

$$= \left[\frac{dy}{dx}\right]_{\substack{x=1\\y=0}} = \left[3x^2 + 1\right]_{\substack{x=1\\y=0}} = 3 + 1 = 4$$

$\therefore$ Slope of the normal at (1, 0) $= -\frac{1}{4}$

The equation of the normal to the curve at (1, 0) is

$$(y - y_1) = \frac{1}{\text{Slope of tangent}} \times (x - x_1)$$

$$\Rightarrow (y - 0) = -\frac{1}{4}(x - 1)$$

$$\Rightarrow y = -\frac{1}{4}x + \frac{1}{4} \qquad \text{...(2)}$$

Also the equation of the normal is $y = ax + b$

Comparing eqn (2) with the given equation of the normal $y = ax + b$ and equating the coefficients of x,

$$a = -\frac{1}{4} \text{ and } b = \frac{1}{4} \qquad \text{...(3)}$$

Note:

Putting this value of $a = -\frac{1}{4}$ in $y = ax + b$, we get,

$$y = -\frac{1}{4} + b$$

and this line passes through (1, 0) $\Rightarrow 0 = -\frac{1}{4} + b$

$$\Rightarrow 0 + \frac{1}{4} = b$$

$$\Rightarrow b = \frac{1}{4}$$

Question: If there are two values of a such that the tangents at $x = 1$ and $x = 3$ to the curve

$$y = ax^2 - 2x - 1$$

are perpendicular, find the two values of a.

Solution: Given equation of the curve is

$$y = ax^2 - 2x - 1 \qquad \text{...(1)}$$

Differentiating (1) w.r.t. x, we get

$$\frac{dy}{dx} = 2ax - 2$$

$$\Rightarrow \left[\frac{dy}{dx}\right]_{x=1} = 2a - 2 = m_1 \text{ (say)} \qquad \text{...(2)}$$

and $\left[\frac{dy}{dx}\right]_{x=3} = 2 \times 3a - 2 = 6a - 2 = m_2$ (say) ...(3)

The tangents at $x = 1$ and $x = 3$ are perpendicular.

$$\Rightarrow m_1 m_2 = -1$$

$$\Rightarrow (2a - 2) \times (6a - 2) = -1$$

$$\Rightarrow 12a^2 - 16a + 5 = 0$$

$$\Rightarrow (2a - 1)(6a - 5) = 0$$

$$\Rightarrow a = \frac{1}{2} \text{ or } a = \frac{5}{6}$$

Problems based on perpendicularity and touching.

Working rule to show that two curves intersect at right angles:

1. Find the point of intersection of two curves by solving the given equations simultaneously provided the point of intersection is not given.

2. Find the differential coefficient $\frac{dy}{dx}$ from each given equation.

3. Find the value of $\frac{d}{dx}f(x)$ and $\frac{d}{dx}g(x)$ at the point of intersection which will represent m_1 and m_2 respectively.

4. If $m_1 m_2 = -1$, then two curves cut orthogonally or they intersect at right angles at the point of intersection.

Note: **1.** To show that two curves are perpendicular to each other at a given point (x_1, y_1) means (x_1, y_1) is the point of intersection of two given curves whose equations are given. This is why there is no need to find the point of intersection of two curves.

2. To show that two tangents to a curve $y = f(x)$ at a given point x_0 are perpendicular to each other means we have to show $m_1 m_2 = -1$, where

$$m_1 = \left[\frac{dy}{dx}\right]_{\substack{x=x_0\\ y=f(x_0)}}$$

$$m_2 = \left[\frac{dy}{dx}\right]_{\substack{x=x_0\\ y=f(x_0)}}$$

x_0 = given x-coordinate of the point where the tangents are perpendicular and $f(x_0) = [f(x)]_{x=x_0}$ is required to find out.

Working rule to show that two curves touch each other

1. Find the point of intersection of two curves by solving simultaneously by the given equations of the curves (or, the curve and a line) provided the point of intersection is not given.

2. Find the differential coefficient $\frac{dy}{dx}$ from each given equation.

3. Find the values of $\frac{d}{dx}f(x)$ and $\frac{d}{dx}g(x)$ at the point of intersection which will represent $\tan\psi_1$ and $\tan\psi_2$ respectively.

4. If $\tan\psi_1 = \tan\psi_2$ i.e. $\psi_1 = \psi_2$ then two curves touch each other.

Note: **1.** To show that two curves touch each other at a given point (x_1, y_1) means (x_1, y_1) is the point of intersection. This is why there is no need to find the point of intersection of two curves.

Question: What is the criteria to show that two curves touch each other?

Solution: The two curves touch each other provided

$$\left[\frac{d}{dx}f(x)\right]_P = \left[\frac{d}{dx}g(x)\right]_P$$

$$[\because \tan\psi_1 = \tan\psi_2 \Rightarrow \psi_1 = \psi_2]$$

Where P is the point of intersection determined by solving the given equation if it is not given.

Examples worked out:

Question: Show that the curves $2y = 3x + x^2$ and $y^2 = 2x + 3y$ intersect at right angles at the origin (0, 0).

N.B.: Here the point of intersection $(0, 0) \equiv$ origin is given.

$\therefore$ There is no need to determine the point of intersection.

Solution: The curves are $2y = 3x + x^2$ ...(1)

and $y^2 = 2x + 3y$ ...(2)

Now, differentiating (1) w.r.t x, we get

$$2\frac{dy}{dx} = 3 + 2x$$

$$\Rightarrow \frac{dy}{dx} = \frac{3}{2} + x \quad \text{...(3)}$$

Again differentiating (2) w.r.t x, we get

$$2 + 3\frac{dy}{dx} = 2y\frac{dy}{dx}$$

$$\Rightarrow \frac{dy}{dx}(2y - 3) = 2$$

$$\Rightarrow \frac{dy}{dx} = \frac{2}{2y-3} \quad \text{...(4)}$$

Now, from (3), $m_1 = \left[\frac{dy}{dx}\right]_{\substack{x=0\\ y=0}} = \left[\frac{3}{2} + x\right]_{\substack{x=0\\ y=0}} = \frac{3}{2}$

Again, from (4),

$$m_2 = \left[\frac{dy}{dx}\right]_{\substack{x=0\\y=0}} = \left[\frac{2}{2y-3}\right]_{\substack{x=0\\y=0}} = -\frac{2}{3}$$

$$\text{Now, } m_1 \times m_2 = -\frac{2}{3} \times \frac{3}{2} = -1$$

$\Rightarrow$ Two curves are perpendicular to each other at the origin.

Question: Show that the curves $y = x^3$ and $6y = 7 - x^2$ intersect orthogonally.

Solution: Given equations of the curve are

$$y = x^3 \quad \ldots(1)$$

$$\text{and } 6y = 7 - x^2 \quad \ldots(2)$$

Now, from (1) and (2), we get

$$6x^3 = 7 - x^2$$

$$\Rightarrow 6x^3 + x^2 - 7 = 0$$

$$\Rightarrow (x-1)\left(6x^2 + 7x + 7\right) = 0$$

$$\Rightarrow (x-1) = 0 \text{ or } \left(6x^2 + 7x + 7\right) = 0$$

$$\Rightarrow x = 1$$

[$\because$ The quadratic equation $6x^2 + 7x + 7 = 0$ has imaginary roots, therefore the only root under consideration is $x = 1$]

Again, differentiating both sides of (1) w.r.t x, we get

$$\frac{dy}{dx} = 3x^2 \quad \ldots(3)$$

and differentiating both sides of (2) w.r.t x, we get

$$6\frac{dy}{dx} = -2x \Rightarrow \frac{dy}{dx} = -\frac{2x}{6} \Rightarrow \frac{dy}{dx} = -\frac{x}{3} \quad \ldots(4)$$

$$\text{Now, } m_1 = \left[\frac{dy}{dx}\right]_{x=1} = \left[3x^2\right]_{x=1} = 3 \text{ from (3).}$$

$$m_2 = \left[\frac{dy}{dx}\right]_{x=1} = \left[-\frac{x}{3}\right]_{x=1} = -\frac{1}{3} \text{ from (4).}$$

$$\text{Hence, } m_1 \, m_2 = 3 \times -\frac{1}{3} = -1$$

$\Rightarrow$ The curves intersect orthogonally.

Question: Show that the tangents to the curves $y^2 = 2px$ at the points where $x = \frac{1}{2}p$ are at right angles.

Solution: Given equation of the curve is

$$y^2 = 2px \quad \ldots(1)$$

Putting $x = \frac{p}{2}$ in (1), we get

$$y^2 = 2p \cdot \frac{1}{2}p = p^2 \Rightarrow y = \pm p \quad \ldots(2)$$

$$\Rightarrow \text{Points are } \left(\frac{1}{2}p, p\right) \text{ and } \left(\frac{1}{2}p, -p\right)$$

corresponding to $x = \frac{1}{2}p$

Now, differentiating both sides of (1) w.r.t x, we get

$$2y\frac{dy}{dx} = 2 \cdot p \cdot 1$$

$$\Rightarrow \frac{dy}{dx} = \frac{p}{y}$$

$$\text{Now, } m_1 = \left[\frac{dy}{dx}\right]_{\substack{x=\frac{1}{2}p\\y=p}} = \frac{p}{p} = 1$$

$$m_2 = \left[\frac{dy}{dx}\right]_{\substack{x=\frac{1}{2}p\\y=-p}} = -\frac{p}{p} = -1$$

$$\text{Hence, } m_1 \, m_2 = 1 \times (-1) = 1$$

$\Rightarrow$ Two tangents are perpendicular to the curve $y^2 = 2px$ at the points $\left(\frac{1}{2}p, p\right)$ and $\left(\frac{1}{2}p, -p\right)$.

Question: Prove that the normals to the curve $y^2 = 4ax$ at the point where $x = a$ are perpendicular to each other.

Solution: Given equation of the curve is

$$y^2 = 4ax \quad \ldots(1)$$

Putting $x = 1$ in (1), we get $y^2 = 4a^2 \Rightarrow y = \pm 2a$

$\therefore$ Points are $(a, 2a)$ and $(a, -2a)$.

Now, we have to show that tangents at $(a, 2a)$ and $(a, -2a)$ are perpendicular.

Now, differentiating given equation (1) w.r.t x, we get $2y\dfrac{dy}{dx} = 4a \Rightarrow \dfrac{dy}{dx} = \dfrac{4a}{2y} = \dfrac{2a}{y}$

$$\Rightarrow \left[\frac{dy}{dx}\right]_{\substack{x=a\\y=2a}} = \frac{2a}{2a} = 1 = m_1$$

and $\left[\dfrac{dy}{dx}\right]_{\substack{x=a\\y=-2a}} = \left[\dfrac{2a}{y}\right]_{y=-2a} = \dfrac{2a}{-2a} = -1 = m_2$

Hence, $m_1 m_2 = -1$

$\Rightarrow$ Tangents are perpendicular to each other.

$\Rightarrow$ Normals are perpendicular to each other.

Question: Show that the normals to the curve $y^2 = 3x$ at $x = \dfrac{3}{4}$ are at right angles to each other.

Solution: Given equation of the curve is $y^2 = 3x$ ...(1)

$\therefore$ From (1), when $x = \dfrac{3}{4} \Rightarrow y^2 = 3 \cdot \dfrac{3}{4} = \dfrac{9}{4}$

$\Rightarrow y = \pm\dfrac{3}{2}$

$\therefore$ Given points are $\left(\dfrac{3}{4}, \dfrac{3}{2}\right)$ and $\left(\dfrac{3}{4}, -\dfrac{3}{2}\right)$

Now, differentiating both sides of eqn (1) w.r.t x, we get

$$2y\frac{dy}{dx} = 3 \cdot 1 \Rightarrow \frac{dy}{dx} = \frac{3}{2y}$$

Now, $\left[\dfrac{dy}{dx}\right]_{y=\frac{3}{2}} = \dfrac{3}{2 \cdot \dfrac{3}{2}} = 1 = m_1$ ...(2)

and $\left[\dfrac{dy}{dx}\right]_{y=-\frac{3}{2}} = \dfrac{3}{2 \cdot \left(-\dfrac{3}{2}\right)} = -1 = m_2$ ...(3)

Hence $\left(-\dfrac{1}{m_1}\right)\left(-\dfrac{1}{m_2}\right) = -1.$

$\therefore$ The normals are at right angles.

Alternative Method:

$$\left[\frac{dy}{dx}\right]_{\substack{x=3/4\\y=3/2}} = \tan\psi_1 = 1 = \tan 45^\circ \Rightarrow \tan\psi_1$$

$$= \tan 45 \Rightarrow \psi_1 = 45^\circ$$

Again, $\left[\dfrac{dy}{dx}\right]_{\substack{x=3/4\\y=-3/2}} = \tan\psi_2 = 1 = \tan 135^\circ$

$$\Rightarrow \tan\psi_2 = \tan 135^\circ \Rightarrow \psi_2 = 135^\circ$$

Now $\theta = \psi_2 - \psi_1$

$= 135^\circ - 45^\circ = 90^\circ \Rightarrow$ the two tangents are perpendicular to each other.

Hence the normals are also perpendicular to each other.

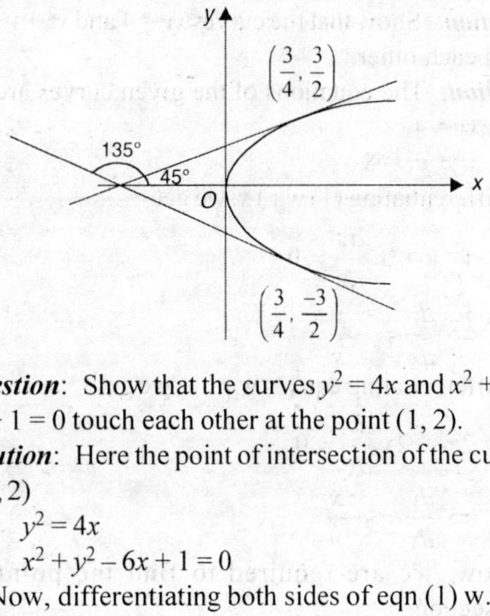

Question: Show that the curves $y^2 = 4x$ and $x^2 + y^2 - 6x + 1 = 0$ touch each other at the point (1, 2).

Solution: Here the point of intersection of the curves = (1, 2)

$y^2 = 4x$ ...(1)

and $x^2 + y^2 - 6x + 1 = 0$ ...(2)

Now, differentiating both sides of eqn (1) w.r.t x, we get

$$2y\frac{dy}{dx} = 4$$

$$\Rightarrow \frac{dy}{dx} = \frac{4}{2y} = \frac{2}{y} \quad ...(3)$$

Again, differentiating both sides of eqn (2) w.r.t x, we get

$$2x + 2y\frac{dy}{dx} - 6 = 0$$

$$\Rightarrow \frac{dy}{dx} = \frac{6-2x}{2y} = \frac{3-x}{y} \qquad \ldots(4)$$

Now the slope of the tangent at (1, 2) from eqn (3) to the curve (1),

$$m_1 = \left[\frac{2}{y}\right]_{\substack{x=1\\y=2}} = \frac{2}{2} = 1 \qquad \ldots(5)$$

And the slope of the tangent at (1, 2) from (4) to the curve (2),

$$m_2 = \left[\frac{3-x}{y}\right]_{\substack{x=1\\y=2}} = \frac{3-1}{2} = \frac{2}{2} = 1 \qquad \ldots(6)$$

$\therefore m_1 = m_2$

$\Rightarrow$ The two curves touch each other at (1, 2).

Question: Show that the curves $xy = 4$ and $x^2 + y^2 = 8$ touch each other.

Solution: The equations of the given curves are

$$xy = 4 \qquad \ldots(1)$$

$$x^2 + y^2 = 8 \qquad \ldots(2)$$

Differentiating (1) w.r.t x, we get

$$1 \cdot y + x\frac{dy}{dx} = 0$$

$$\Rightarrow \frac{dy}{dx} = -\frac{y}{x} \qquad \ldots(3)$$

Differentiating eqn (2) w.r.t. x, we get

$$2x + 2y\frac{dy}{dx} = 0$$

$$\Rightarrow \frac{dy}{dx} = -\frac{x}{y} \qquad \ldots(3)$$

Now, we are required to find the point of intersection.

$$\because xy = 4 \Rightarrow y = \frac{4}{x}$$

Putting $y = \frac{4}{x}$ in $x^2 + y^2 = 8$, we get

$$x^2 + \frac{16}{x^2} = 8 \Rightarrow x^4 + 16 = 8x^2$$

$$\Rightarrow x^4 - 8x^2 + 16 = 0 \Rightarrow \left(x^2 - 4\right)^2 = 0$$

$$\Rightarrow x^2 - 4 = 0 \Rightarrow x^2 = 4 \Rightarrow x = \pm 2$$

From (1), when $x = 2,\ y = \frac{4}{2} = 2$

When $x = -2,\ y = -\frac{4}{2} = -2$

Hence, the point of intersection of the two curves are (2, 2) and (–2, 2).

$\therefore$ The slope of the tangent to the curve $xy = 4$ at (2, 2)

$$= m_1 = \left[\frac{dy}{dx}\right]_{\substack{x=2\\y=2}} = \left[-\frac{y}{x}\right]_{\substack{x=2\\y=2}} = -\frac{2}{2} = -1 \qquad \ldots(5)$$

The slope of the tangent to the curve $x^2 + y^2 = 8$ at (2, 2)

$$= m_2 = \left[\frac{dy}{dx}\right]_{\substack{x=2\\y=2}} = \left[-\frac{x}{y}\right]_{\substack{x=2\\y=2}} = -\frac{2}{2} = -1 \qquad \ldots(6)$$

Thus, we get $m_1 = m_2 = -1 \Rightarrow$ The two curves touch each other at (2, 2). Similarly, we can show that the two curves touch each other at (–2, 2).

Hence, the two curves $xy = 4$ and $x^2 + y^2 = 8$ touch each other.

Question: Show that the curve $\left(\frac{x}{a}\right)^n + \left(\frac{y}{b}\right)^n = 2$ touches the straight line $\frac{x}{a} + \frac{y}{b} = 2$ at the point (a, b) whatever be the value of n.

Solution: Given equations of the curves are

$$\left(\frac{x}{a}\right)^n + \left(\frac{y}{b}\right)^n = 2 \qquad \ldots(1)$$

$$\frac{x}{a} + \frac{y}{b} = 2 \qquad \ldots(2)$$

Now, differentiating the equation (1) w.r.t x, we get

$$n \cdot \frac{x^{n-1}}{a^n} + n \cdot \frac{y^{n-1}}{b^n} \cdot \frac{dy}{dx} = 0$$

$$\Rightarrow \frac{dy}{dx} = -\frac{b^n}{a^n} \cdot \frac{x^{n-1}}{y^{n-1}}$$

$$\therefore m_1 = \left[\frac{dy}{dx}\right]_{\substack{x=a\\y=b}}$$

$$= -\frac{b^n}{a^n} \cdot \frac{a^{n-1}}{b^{n-1}} = -\frac{b}{a} \quad ...(3)$$

Again, differentiating the equation (2), w.r.t x, we get

$$\frac{1}{a} + \frac{1}{b}\frac{dy}{dx} = 0$$

$$\Rightarrow \frac{dy}{dx} = -\frac{b}{a} \text{ (constant)} \quad ...(4)$$

Thus, we see from (3) and (4), $m_1 = m_2$

$\Rightarrow$ The curve touches the line at (a, b) for any n.

Question: Show that the curve $y^2 = 2x$ touches the straight line $2y - x = 2$. Also find the point of intersection.

Solution: Given equation of the curves are

$$y^2 = 2x \quad ...(1)$$

$$2y - x = 2 \quad ...(2)$$

Now, differentiating eqn (1) w.r.t x, we get

$$2y\frac{dy}{dx} = 2 \Rightarrow \frac{dy}{dx} = \frac{2}{2y} = \frac{1}{y} \quad ...(3)$$

Again, differentiating eqn (2) w.r.t x, we get

$$2\frac{dy}{dx} - 1 = 0 \Rightarrow \frac{dy}{dx} = \frac{1}{2} \quad ...(4)$$

Now, we are required to find the point of intersection where, they intersect by solving the equations of the given curves.

$$2y = x + 2 \text{ [from eqn (2)]} \Rightarrow y = \frac{x+2}{2} \quad ...(5)$$

Putting (5) in (1), we get

$$\left[\frac{x+2}{2}\right]^2 = 2x \Rightarrow (x+2)^2 = 8x$$

$$\Rightarrow x^2 - 4x + 4 = 0$$

$$\Rightarrow (x-2)^2 = 0$$

$$\Rightarrow x = 2$$

$$\therefore y = \frac{x+2}{2} = \frac{2+2}{2} = 2$$

Thus, we get $P(x, y) = (2, 2)$

Now, for the curve (1),

$$m_1 = \left[\frac{dy}{dx}\right]_{\substack{x=2\\y=2}} = \left[\frac{1}{y}\right]_{\substack{x=2\\y=2}} = \frac{1}{2} \quad ...(6)$$

And for the curve (2), $m_2 = \left[\frac{dy}{dx}\right]_{\substack{x=2\\y=2}} = \frac{1}{2} \quad ...(7)$

From (6) and (7), we see that $m_1 = m_2 = \frac{1}{2}$

$\Rightarrow$ The given curves touch each other at (2, 2).

Question: Show that the curve $y = be^{-x/a}$ touches the straight line $\frac{x}{a} + \frac{y}{b} = 1$ at the point $(0, b)$.

Solution: Differentiating $y = be^{-x/a}$ w.r.t x, we get

$$\frac{dy}{dx} = be^{-x/a}\left(-\frac{1}{a}\right) = -\frac{b}{a}e^{-x/a} \quad ...(1)$$

Again differentiating $\frac{x}{a} + \frac{y}{b} = 1$ w.r.t x, we get

$$\frac{1}{a} + \frac{1}{b}\frac{dy}{dx} = 0 \Rightarrow \frac{dy}{dx} = -\frac{1}{a} \times b = -\frac{b}{a} \quad ...(2)$$

Now, from (1), $m_1 = \left[\frac{dy}{dx}\right]_{\substack{x=0\\y=b}}$

$$= -\frac{b}{a}e^{0/a} = -\frac{b}{a} \cdot 1 = -\frac{b}{a} \quad ...(3)$$

And from eqn (2), $m_2 = \left[\frac{dy}{dx}\right]_{\substack{x=0\\y=b}}$

$$= \left[-\frac{b}{a}\right]_{\substack{x=0\\y=b}} = -\frac{b}{a} \quad ...(4)$$

Eqns (3) and (4) $\Rightarrow m_1 = m_2$

$\Rightarrow$ Both the curves touch each other at $(0, b)$.

Question: Show that the curve

$$\frac{2x}{a} + \frac{2x^2}{b^2} + \frac{y^2}{b^2} = 1$$

touches the curve $y = be^{-x/a}$ at the point where the curve crosses the axis of y. Find the equation of the common tangent.

Solution: The point where the curve cuts (or crosses) the y-axis $\Rightarrow x = 0$

Now, we are required to find out y.

$$\Rightarrow y = b \cdot e^{-x/a} = b \cdot e^{-0/a} = b \cdot e^{0} = b \cdot 1 = b$$

Hence, required where the curve crosses the axis of $y = (0, b)$

Again given equations of the curves are

$$\frac{2x}{a} + \frac{2x^2}{b^2} + \frac{y^2}{b^2} = 1 \quad \ldots(1)$$

$$y = be^{-x/a} \quad \ldots(2)$$

Differentiating the equation (2) w.r.t x, we get

$$\left[\frac{dy}{dx}\right]_{\substack{x=0\\y=b}} = \left[b\left(-\frac{1}{a}e^{-x/a}\right)\right]_{\substack{x=0\\y=b}}$$

$$\Rightarrow m_1 = \left[b\left(-\frac{1}{a}\right)e^{-0/a}\right] = b\left[-\frac{1}{a}e^{0}\right] = -\frac{b}{a} \quad \ldots(3)$$

Differentiating the equation (1) w.r.t x, we get

$$\frac{2}{a} + \frac{4x}{b^2} + \frac{2y}{b^2}\frac{dy}{dx}$$

$$= 0$$

$$\therefore 2b^2 + 4ax + 2ay\frac{dy}{dx}$$

$$= 0$$

$$\therefore \frac{dy}{dx} = -\frac{(2b^2 + 4ax)}{2ay}$$

clearly $(0, b)$ is a point on (1) and from (1),

$$\left[\frac{dy}{dx}\right]_{\substack{x=0\\y=b}}$$

$$= -\frac{b}{a} = m_2$$

$$\therefore m_1 = m_2$$

Hence the curves touch at $(0, b)$. The equation of the common tangent is

$$(y - b) = -\frac{b}{a}(x - 0)$$

$$\therefore \frac{x}{a} + \frac{y}{b} = 1.$$

To find the equation of the tangent and normal to the curve whose equation is given in parametric form $x = f_1(t)$, $y = f_2(t)$ at a point 't'.

Working Rule:

1. Find $\frac{dy}{dx}$ using the formula.

$$\frac{dy}{dx} = \frac{dy}{dt} \cdot \frac{dt}{dx} = \frac{dy/dt}{dx/dt} = \frac{f_2'(t)}{f_1'(t)}$$

2. The equation of the tangent at the given point 't' is

$$[y - f_2(t)] = \frac{f_2'(t)}{f_1'(t)} \cdot [x - f_1(t)]$$

The equation of the normal at the given point t is

$$[y - f_2(t)] = -\frac{f_1'(t)}{f_2'(t)} \cdot [x - f_1(t)]$$

where the slope of the normal = negative reciprocal of the slope of the tangent

$$= -\frac{1}{\text{Slope of the tangent}} = -\frac{f_2'(t)}{f_1'(t)}$$

Examples worked out:

Question: Find the equation of the tangent and normal to the ellipse $x = a\cos\theta$, $y = b\sin\theta$ at the given point θ.

Solution: Here $\frac{dx}{d\theta} = -a\sin\theta$ and $\frac{dy}{d\theta} = b\cos\theta$

$$\therefore \frac{dy}{dx} = \frac{dy/d\theta}{dx/d\theta} = -\frac{b\cos\theta}{a\sin\theta} \quad \ldots(1)$$

Now, the equation of the tangent at θ is

$$(y - b\sin\theta) = -\frac{b\cos\theta}{a\sin\theta} \cdot (x - a\cos\theta)$$

$$\therefore ay\sin\theta - ab\sin^2\theta = -bx\cos\theta + ab\cos^2\theta$$

$$\Rightarrow bx\cos\theta + ay\sin\theta = ab(\sin^2\theta + \cos^2\theta) = ab$$

$$\Rightarrow \frac{x}{a}\cos\theta + \frac{y}{b}\sin\theta = 1$$

Again, the slope of the normal at the given point θ is $\dfrac{a\sin\theta}{b\cos\theta}$.

The equation of the normal θ is

$$(y-b\sin\theta)=\frac{a\sin\theta}{b\cos\theta}\cdot(x-a\cos\theta)$$

$$\Rightarrow by\cos\theta-b^2\sin\theta\cos\theta=ax\sin\theta-a^2\sin\theta\cos\theta$$

$$\Rightarrow ax\sin\theta-by\cos\theta=\left(a^2-b^2\right)\sin\theta\cos\theta$$

$$\Rightarrow \frac{ax}{\cos\theta}-\frac{by}{\sin\theta}=a^2-b^2$$

Question: Find the equation of the tangent and normal at $\theta=\dfrac{\pi}{2}$ to the cycloid $x=a(\theta-\sin\theta)$, $y=a(1-\cos\theta)$.

Solution: Given equations $\left.\begin{aligned}x&=a(\theta-\sin\theta)\\ y&=a(1-\cos\theta)\end{aligned}\right\}$ ...(1)

$$\therefore \frac{dx}{d\theta}=a(1-\cos\theta)=2a\sin^2\frac{\theta}{2} \quad ...(2)$$

$$\frac{dy}{d\theta}=a\sin\theta=2a\sin\frac{\theta}{2}\cdot\cos\frac{\theta}{2} \quad ...(3)$$

$$\text{Eqns (2) and (3)} \Rightarrow \frac{dy}{dx}=\frac{dy/d\theta}{dx/d\theta}=\cot\frac{\theta}{2}$$

$$\therefore \frac{dy}{dx}\text{ at }\theta=\frac{\pi}{2}\Rightarrow\left[\frac{dy}{dx}\right]_{\theta=\frac{\pi}{2}}=\cot\frac{\pi}{4}=1$$

Also for $\theta=\dfrac{\pi}{2}$, $x=a\left(\dfrac{\pi}{2}-1\right)$ and $y=1$.

Equation of the tangent at the point $\left[a\left(\dfrac{\pi}{2}-1\right),a\right]$ is $(y-a)=1\left[x-a\left(\dfrac{\pi}{2}-1\right)\right]\Rightarrow x-y=\dfrac{\pi a}{2}-2a$

and the equation of the normal is

$$(y-a)=-1\left[x-a\left(\frac{\pi}{2}-1\right)\right]$$

which implies $x+y=\dfrac{1}{2}a\pi$ [Slope of the normal $=\dfrac{1}{\text{Slope of the tangent}}=-1$]

Question: Find the equation of the tangent and normal to the parallel $x=at^2, y=2at$ at the point 't'.

Solution: $\left.\begin{aligned}\frac{dx}{dt}&=2at\ ...(1)\\ \frac{dy}{dt}&=2a\ ...(2)\end{aligned}\right]\Rightarrow \dfrac{dy}{dx}=\dfrac{dy/dt}{dx/dt}=\dfrac{2a}{2at}=\dfrac{1}{t}$

$\therefore$ Equation of the tangent at any point 't' is

$$y-2at=\frac{1}{t}\left(x-at^2\right)$$

$$\Rightarrow yt-2at^2=x-at^2\Rightarrow ty=x+at^2$$

Slope of the normal at the given point 't'

$$=\frac{1}{\text{Slope of the tangent}}$$

Hence, the required equation of the normal is

$$y-2at=-t\left(x-at^2\right)$$

$$\Rightarrow y-2at=-tx+at^3$$

$$\Rightarrow y+tx=2at+at^3$$

Question: Find the equation of the tangent to the curve $x=\sqrt{t}$, $y=t-\dfrac{1}{\sqrt{t}}$ at the point $t=4$.

Solution:

$\left.\begin{aligned}\frac{dx}{dt}&=\frac{1}{2\sqrt{t}}\ ...(1)\\ \frac{dy}{dx}&=1+\frac{1}{2(t)^{\frac{3}{2}}}\ ...(2)\end{aligned}\right\}\Rightarrow \dfrac{dy}{dx}=\dfrac{dy/dt}{dx/dt}$ which implies

$$\frac{dy}{dx}=\frac{1+\dfrac{1}{2(t)^{\frac{3}{2}}}}{\dfrac{1}{2\sqrt{t}}}=\frac{\left[2(t)^{\frac{3}{2}}+1\right]\cdot 2\sqrt{t}}{2(t)^{\frac{3}{2}}}=\frac{2(t)^{\frac{3}{2}}+1}{t}$$

$\therefore$ Slope of the tangent at $t = 4$ is $\dfrac{2(4)^{\frac{3}{2}}+1}{4}=\dfrac{17}{4}$

$\Rightarrow$ The equation of the tangent at $t = 4$ is

$$\left[y-\frac{7}{2}\right]=\frac{17}{4}(x-2)$$

$\Rightarrow 17x - 4y - 20 = 0$

Problems based on proving the equation of the tangent to a curve at any point (x_1, y_1) to be a linear equation $y = ax + b$ or finding the condition for a given line to touch a given curve.

Working rule to show the equation of the tangent line to a curve at any point (x_1, y_1) to be a linear equation $y = ax + b$.

1. First of all find $\dfrac{dy}{dx}$. This gives the slope of the curve or the slope of the tangent at the general point (x, y).

2. Find $\left[\dfrac{dy}{dx}\right]_{\substack{x=x_1\\y=y_1}}$

3. Afterwards apply the slope form equation

$$(y-y_1)=\left[\frac{dy}{dx}\right]_{\substack{x=x_1\\y=y_1}}(x-x_1)$$

which on simplification gives the required equation of the tangent or the straight line which touches the curve.

Question: How would you find the condition for a given line to touch a given curve?

Solution: **1.** Let the line be tangent to the given curve at (x, y).

2. Write the equation of the tangent at (x, y) as

$(Y-y)=\left[\dfrac{dy}{dx}\right](X-x)$ where $\dfrac{dy}{dx}=$ d.c of the equation of the curve.

3. Compare $(Y-y)=\left[\dfrac{dy}{dx}\right](X-x)$ with the given line $ax+by+c=0$ and then eliminate x and y.

On intercepts:

The equation of the tangent of the curve $y = f(x)$ at the point $P = (x_1, y_1)$ is

$$(y-y_1)=\left[\frac{dy}{dx}\right]_p(x-x_1) \qquad \ldots(1)$$

The intercept OA made by the tangent on the axis of x is obtained by putting $y = 0$ in (1) and solving for x.

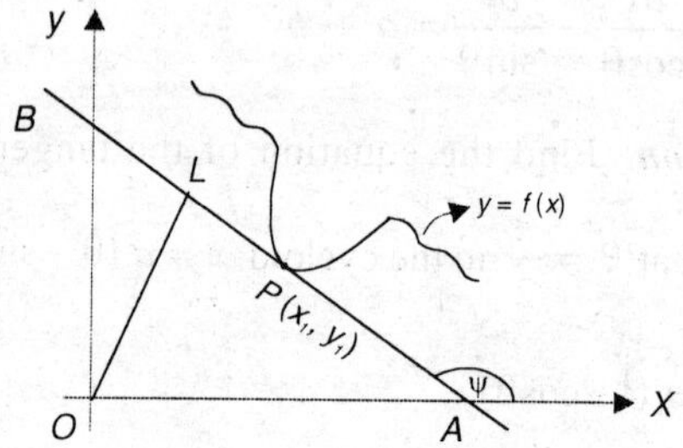

$$\therefore -y_1=\left[\frac{dy}{dx}\right]_p(x-x_1)$$

$$(x-x_1)=-\frac{y_1}{\left[\dfrac{dy}{dx}\right]_p}$$

$$\Rightarrow OA = x = x_1-\frac{y_1}{\left[\dfrac{dy}{dx}\right]_p}$$

Similarly, to obtain the intercept on y-axis, we put $x = 0$ in eqn (1) to have $y-y_1=\left[\dfrac{dy}{dx}\right]_p(-x_1)$

$$\Rightarrow y = y_1 - x_1\left[\frac{dy}{dx}\right]_p = OB$$

The portion of the tangent intercepted between the axis

$$= AB = \sqrt{OA^2+OB^2}$$

$$=\sqrt{(\text{Intercept on } x\text{-axis})^2+(\text{Intercept on } y\text{-axis})^2}$$

Substituting the values of OA and OB, we can find the length AB.

Facts to know:

1. Length of the perpendicular from the origin upon the tangent to the curve $y=f(x)$ at any point (x_1, y_1) is

$$OL = [y\text{-intercept of the tangent}] \times |\cos\psi|$$

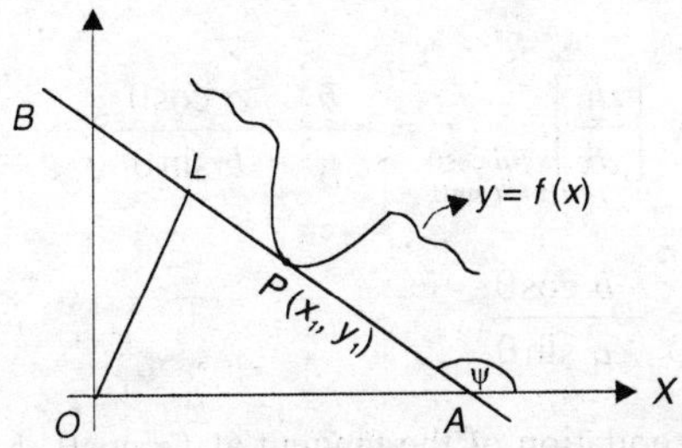

$$= \left[y_1 - x_1\left(\frac{dy}{dx}\right)_{\substack{x=x_1\\y=y_1}}\right] \Bigg/ \sqrt{1+\left[\left(\frac{dy}{dx}\right)_{\substack{x=x_1\\y=y_1}}\right]^2}$$

$$= \frac{y_1 - x_1\left(\frac{dy}{dx}\right)_{\substack{x=x_1\\y=y_1}}}{\sqrt{1+\left[\left(\frac{dy}{dx}\right)_{\substack{x=x_1\\y=y_1}}\right]^2}}$$

2. ***Length of tangent***:

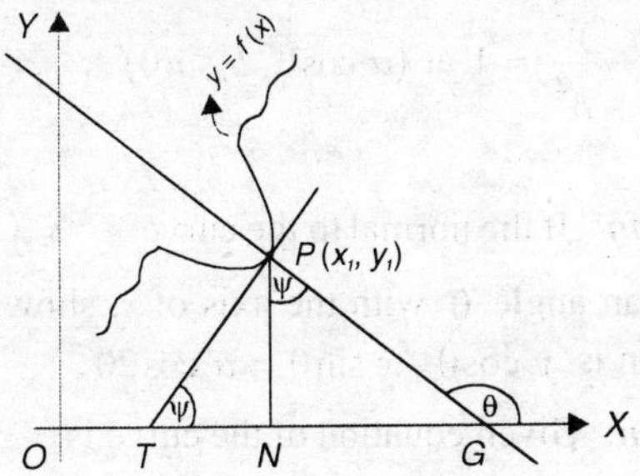

The portion of the tangent intercepted between the point of contact and the axis of x is called the length of the tangent.

In Δ PTN,

Length of tangent

$$= PT = PN \cdot \operatorname{cosec}\psi$$

$$= PN\sqrt{1+\cot^2\psi}$$

$$= PN\sqrt{1+\frac{1}{\tan^2\psi}}$$

$$= y \cdot \frac{\sqrt{\left[\left(\frac{dy}{dx}\right)_{\substack{x=x_1\\y=y_1}}\right]^2+1}}{\left(\frac{dy}{dx}\right)_{\substack{x=x_1\\y=y_1}}}$$

$$= y \cdot \frac{\sqrt{\left[\left(\frac{dy}{dx}\right)_p\right]^2+1}}{\left(\frac{dy}{dx}\right)_p}$$

3. ***Length of the normal***: The portion of the normal at any point on the curve intercepted between the curve and the axis of x is called the length of normal.

$\therefore$ In Δ NGP

$\therefore$ Length of the normal $= PG$

$$= PN \cdot \sec\psi = y\sqrt{1+\tan^2\psi}$$

$$= y \cdot \sqrt{1+\left[\left(\frac{dy}{dx}\right)_p\right]^2}$$

$$\left[\because y = PN \text{ and } \left(\frac{dy}{dx}\right)_p = \tan\psi_1 = m_1\right]$$

4. ***Angle of intersection between two curves***: The angle of intersection of two curves is the angle between the tangents drawn to the two curves at the common point of intersection of two curves.

Examples worked out:

Question: Show that the equation of the tangent to the curve $y=2x^3+2x^2-8x+7$ at the point $(1, 3)$ is $2x-y+1=0$.

Solution: Equation of the curve is

$$y = 2x^3+2x^2-8x+7 \qquad \ldots(1)$$

Now differentiating (1) w.r.t x, we get

$$\frac{dy}{dx} = 2 \cdot 3x^2 + 2 \cdot 2x - 8 \cdot 1 + 0 = 6x^2 + 4x - 8$$

$$\Rightarrow \left[\frac{dy}{dx}\right]_{\substack{x=1\\y=3}} = \left[6x^2 + 4x - 8\right]_{x=1} = 6 \cdot 1^2 + 4 \cdot 1 - 8 = 2$$

Hence, the equation of the tangent to the curve at (1, 3) is

$$(y - y_1) = \left[\frac{dy}{dx}\right]_{\substack{x=1\\y=3}} \cdot (x - x_1)$$

$$\Rightarrow y - 3 = 2(x - 1)$$

$$\Rightarrow y - 3 = 2x - 2$$

$$\Rightarrow 2x - y + 1 = 0.$$

Question: Show that the equation of the tangent to the curve $3ay^2 = x^2(x + a)$ at $(2a, 2a)$ is $3y = 4x - 2a$.

Solution: We have $3ay^2 = x^2(x + a)$ ...(1)

Now, differentiating (1) w.r.t x, we get

$$6ay \frac{dy}{dx} = 3x^2 + 2ax$$

$$\Rightarrow \frac{dy}{dx} = \frac{3x^2 + 2ax}{6ay}$$

$$\Rightarrow \left[\frac{dy}{dx}\right]_{\substack{x=2a\\y=2a}} = \text{slope of the tangent at } (2a, 2a)$$

$$= \left[\frac{3x^2 + 2ax}{6ay}\right]_{\substack{x=2a\\y=2a}} = \frac{3 \cdot 4a^2 + 2a \cdot 2a}{6a \cdot 2a} = \frac{16a^2}{12a^2} = \frac{4}{3}$$

Now, the equation of the tangent at $(2a, 2a)$ is

$$(y - y_1) = \left[\frac{dy}{dx}\right]_{\substack{x=2a\\y=2a}} \cdot (x - x_1)$$

$$\therefore (y - 2a) = \frac{4}{3}(x - 2a)$$

$$\Rightarrow 4x = 3y + 2a$$

$$\Rightarrow 4x - 2a = 3y.$$

Question: Show that the equation of the tangent at a point $(a\cos\theta, b\sin\theta)$ on the curve $\frac{x^2}{a^2} + \frac{y^2}{b^2} = 1$ is $\frac{x}{a}\cos\theta + \frac{y}{b}\sin\theta = 1$.

Solution: Differentiating the given equation of the curve $\frac{x^2}{a^2} + \frac{y^2}{b^2} = 1$ w.r.t x, we get

$$\frac{2x}{a^2} + \frac{2y}{b^2}\frac{dy}{dx} = 0 \Rightarrow \frac{dy}{dx} = -\frac{b^2 x}{a^2 y} \quad ...(1)$$

$$\text{Now, } \left[\frac{dy}{dx}\right]_{\substack{x=a\cos\theta\\y=b\sin\theta}} = -\frac{b^2}{a^2} \cdot \frac{a\cos\theta}{b\sin\theta}$$

$$= -\frac{b\cos\theta}{a\sin\theta} \quad ...(2)$$

The equation of the tangent at $(a\cos\theta, b\sin\theta)$

$$\text{is } (y - b\sin\theta) = -\frac{b\cos\theta}{a\sin\theta}(x - a\cos\theta)$$

$$\Rightarrow \frac{y}{b}\sin\theta - \sin^2\theta = -\frac{x\cos\theta}{a} + \cos^2\theta$$

$$\Rightarrow \frac{x}{a}\cos\theta + \frac{b}{y}\sin\theta = \cos^2\theta + \sin^2\theta = 1$$

$$\Rightarrow \frac{x}{a}\cos\theta + \frac{y}{b}\sin\theta = 1$$

which is the required equation of the tangent to the curve

$$\frac{x^2}{a^2} + \frac{y^2}{b^2} = 1 \text{ at } (a\cos\theta, b\sin\theta).$$

Question: If the normal to the curve $x^{\frac{2}{3}} + y^{\frac{2}{3}} = a^{\frac{2}{3}}$ makes an angle θ with the axis of x, show that its equation is $y\cos\theta - x\sin\theta = a\cos 2\theta$.

Solution: Given equation of the curve is

$$x^{\frac{2}{3}} + y^{\frac{2}{3}} = a^{\frac{2}{3}} \quad ...(1)$$

Now, differentiating the equation (1) w.r.t x, we have

$$\frac{2}{3}x^{-\frac{1}{3}} + \frac{2}{3}y^{-\frac{1}{3}}\frac{dy}{dx} = 0 \Rightarrow \frac{dy}{dx} = -\frac{y^{\frac{1}{3}}}{x^{\frac{1}{3}}}$$

Now, slope of the normal at (x_1, y_1)

$$= -\frac{1}{\text{Slope of the tangent}} = \left[\frac{x_1}{y_1}\right]^{\frac{1}{3}} \quad \ldots(2)$$

Again, as the normal makes an angle θ with x-axis,

$$\therefore \tan\theta = \text{Slope of the normal} \quad \ldots(3)$$

Equating (2) and (3) $\tan\theta = \dfrac{x_1^{\frac{1}{3}}}{y_1^{\frac{1}{3}}}$

$$\Rightarrow \frac{\sin\theta}{\cos\theta} = \frac{x_1^{\frac{1}{3}}}{y_1^{\frac{1}{3}}}$$

$$\Rightarrow \frac{x_1^{\frac{1}{3}}}{\sin\theta} = \frac{y_1^{\frac{1}{3}}}{\cos\theta} = \frac{\sqrt{\left(x^{\frac{1}{3}}\right)^2 + \left(y^{\frac{1}{3}}\right)^2}}{\sqrt{\sin^2\theta + \cos^2\theta}}$$

$$= \sqrt{a^{\frac{2}{3}}} = a^{\frac{2}{3}\times\frac{1}{2}} = a^{\frac{1}{3}}$$

$$\Rightarrow x_1^{\frac{1}{3}} = a^{\frac{1}{3}} \sin\theta$$

$$\Rightarrow x_1 = a\sin^3\theta$$

and $y_1^{\frac{1}{3}} = a_1^{\frac{1}{3}} \cos\theta \Rightarrow y = a\cos^3\theta$

Now, the equation of the normal at

$\left(a\sin^3\theta, a\cos^3\theta\right)$ is $(y - y_1) = \left(\dfrac{x}{y}\right)^{\frac{1}{3}}(x - x_1)$

$$\Rightarrow y - a\cos^3\theta = (\tan\theta)\left(x - a\sin^3\theta\right)$$

$$\Rightarrow y - a\cos^3\theta = \frac{\sin\theta}{\cos\theta}\left(x - a\sin^3\theta\right)$$

$$\Rightarrow y\cos\theta - a\cos^4\theta = x\sin\theta - a\sin^4\theta$$

$$\Rightarrow y\cos\theta - x\sin\theta = a\left(\cos^4\theta - \sin^4\theta\right)$$

$$= a\left(\cos^2\theta - \sin^2\theta\right)\left(\cos^2\theta + \sin^2\theta\right)$$

$$\Rightarrow y\cos\theta - x\sin\theta = a\cos 2\theta$$

Question: If a tangent to the curve $\sqrt{\dfrac{x}{a}} + \sqrt{\dfrac{y}{b}} = 1$ at any point on it makes the intercepts p and q along the axes, then show that $\dfrac{p}{a} + \dfrac{q}{b} = 1$.

Solution: The point (x_1, y_1) is on the curve

$$\sqrt{\frac{x}{a}} + \sqrt{\frac{y}{b}} = 1 \quad \ldots(1)$$

$\Rightarrow$ (x_1, y_1) must satisfy the equation (1)

$$\Rightarrow \frac{x_1^{\frac{1}{2}}}{a^{\frac{1}{2}}} + \frac{y_1^{\frac{1}{2}}}{b^{\frac{1}{2}}} = 1 \quad \ldots(2)$$

Now, differentiating (1) w.r.t x, we have

$$\frac{1}{2a^{\frac{1}{2}}x^{\frac{1}{2}}} + \frac{1}{2b^{\frac{1}{2}}y^{\frac{1}{2}}}\frac{dy}{dx} = 0$$

$$\Rightarrow \frac{dy}{dx} = -\frac{b^{\frac{1}{2}}\cdot y^{\frac{1}{2}}}{a^{\frac{1}{2}}\cdot x^{\frac{1}{2}}}$$

$\therefore$ The value of $\dfrac{dy}{dx}$ at $(x_1, y_1) = \left[\dfrac{dy}{dx}\right]_{\substack{x=x_1\\y=y_1}} = -\dfrac{b^{\frac{1}{2}}\cdot y_1^{\frac{1}{2}}}{a^{\frac{1}{2}}\cdot x_1^{\frac{1}{2}}}$

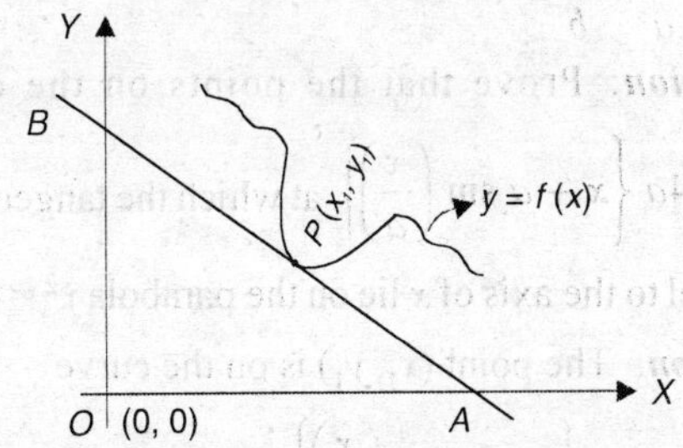

Now, the tangent at (x_1, y_1) is

$$(y - y_1) = \left[\frac{dy}{dx}\right]_{\substack{x=x_1\\y=y_1}} \cdot (x - x_1)$$

$$\Rightarrow (y - v_1) = \frac{-b^{\frac{1}{2}}\cdot y_1^{\frac{1}{2}}}{a^{\frac{1}{2}}\cdot x_1^{\frac{1}{2}}}\cdot(x - x_1)$$

$$\Rightarrow \frac{y}{b^{\frac{1}{2}} \cdot y_1^{\frac{1}{2}}} - \frac{y_1^{\frac{1}{2}}}{b^{\frac{1}{2}}\, y_1^{\frac{1}{2}}} = -\frac{x}{a^{\frac{1}{2}}\, x_1^{\frac{1}{2}}} + \frac{x_1}{a^{\frac{1}{2}} \cdot x_1^{\frac{1}{2}}}$$

$$\Rightarrow \frac{x}{a^{\frac{1}{2}} \cdot x_1^{\frac{1}{2}}} + \frac{y}{b^{\frac{1}{2}}\, y_1^{\frac{1}{2}}} = \frac{x_1^{\frac{1}{2}}}{a^{\frac{1}{2}}} + \frac{y_1^{\frac{1}{2}}}{b^{\frac{1}{2}}}$$

$$\Rightarrow \frac{x}{a^{\frac{1}{2}} \cdot x_1^{\frac{1}{2}}} + \frac{y}{b^{\frac{1}{2}}\, y_1^{\frac{1}{2}}} = 1 \text{ [from (2)]}$$

$\Rightarrow$ The intercepts of the above tangent along the axis are $a^{\frac{1}{2}}\, x_1^{\frac{1}{2}}$ and $b^{\frac{1}{2}}\, y_1^{\frac{1}{2}}$.

$\Rightarrow$ According to question, $a^{\frac{1}{2}} \cdot x_1^{\frac{1}{2}} = p \Rightarrow x_1^{\frac{1}{2}} = \dfrac{p}{a^{\frac{1}{2}}}$

and $b^{\frac{1}{2}}\, y_1^{\frac{1}{2}} = q \Rightarrow y_1^{\frac{1}{2}} = \dfrac{q}{b^{\frac{1}{2}}}$

But from (2),

$$\frac{x_1^{\frac{1}{2}}}{x^{\frac{1}{2}}} + \frac{y_1^{\frac{1}{2}}}{b^{\frac{1}{2}}} = 1$$

$$\Rightarrow \frac{p}{a^{\frac{1}{2}} \cdot a^{\frac{1}{2}}} + \frac{q}{b^{\frac{1}{2}} \cdot b^{\frac{1}{2}}} = 1$$

$$\Rightarrow \frac{p}{a} + \frac{q}{b} = 1$$

Question: Prove that the points on the curve $y^2 = 4a\left\{x + a \sin\left(\dfrac{x}{a}\right)\right\}$ at which the tangents are parallel to the axis of x lie on the parabola $y^2 = 4ax$.

Solution: The point (x_1, y_1) is on the curve

$$y^2 = 4a\left\{x + a \sin\left(\frac{x}{a}\right)\right\} \quad \ldots(1)$$

$\Rightarrow$ (x_1, y_1) must satisfy the equation (1)

$$\Rightarrow y_1^2 = 4a\left\{x_1 + a \sin\left(\frac{x_1}{a}\right)\right\} \quad \ldots(2)$$

Now, differentiating (1) w.r.t x, we get

$$\Rightarrow 2y \cdot \frac{dy}{dx} = 4a\left\{1 + a\left(\cos\frac{x}{a}\right)\frac{1}{a}\right\}$$

$$\Rightarrow \frac{dy}{dx} = \frac{2a}{y}\left\{1 + \cos\frac{x}{a}\right\}$$

$\therefore$ The value of $\dfrac{dy}{dx}$ at (x_1, y_1)

$$= \frac{2a}{y_1}\left(1 + \cos\frac{x_1}{a}\right)$$

If the tangent at (x_1, y_1) is parallel to x-axis, then

$$\left[\frac{dy}{dx}\right]_{\substack{x = x_1 \\ y = y_1}} = 0$$

$$\therefore \frac{2a}{y_1}\left(1 + \cos\frac{x_1}{a}\right) = 0 \Rightarrow \cos\frac{x_1}{a} = -1$$

$$\Rightarrow \cos^2\frac{x_1}{a} = 1 \Rightarrow 1 - \sin^2\frac{x_1}{a} = 0$$

$$\Rightarrow \sin^2\frac{x_1}{a} = 0 \Rightarrow \sin\frac{x_1}{a} = 0$$

Now, substituting the value of $\sin\dfrac{x_1}{a} = 0$ in (2), we get

$$y_1^2 = 4a\,\{x_1 + a \times 0\}$$

$$\Rightarrow y_1^2 = 4ax_1$$

$\therefore (x_1, y_1)$ lies on $y^2 = 4ax$.

Question: Tangents are drawn from the origin to the curve $y = \sin x$ show that their points of contact lie on $x^2 y^2 = x^2 - y^2$.

Solution: There is a point (x_1, y_1) on the curve

$$y = \sin x \quad \ldots(1)$$

$\Rightarrow$ (x_1, y_1) satisfies the equation $y = \sin x$

$$\Rightarrow y_1 = \sin x_1 \quad \ldots(2)$$

Now, differentiating the equation (1) w.r.t x, we get

$$\frac{dy}{dx} = \cos x$$

Again the value of $\dfrac{dy}{dx}$ at

$$(x_1, y_1) = \left[\frac{dy}{dx}\right]_{\substack{x = x_1 \\ y = y_1}} = [\cos x]_{x = x_1} = \cos x_1$$

$\therefore$ Equation of the tangent at any point (x_1, y_1) on the curve is given by

$$(y-y_1)=\left[\frac{dy}{dx}\right]_{\substack{x=x_1\\y=y_1}}\cdot(x-x_1)$$

$$\Rightarrow (y-y_1)=(\cos x_1)(x-x_1)$$

If this line passes through (0, 0), then

$$(0-y_1)=(\cos x_1)(0-x_1)$$

$$\Rightarrow -y_1=(\cos x_1)(-x_1)$$

$$\Rightarrow y_1 = x_1\cos x_1$$

$$\Rightarrow y_1^2=(x_1\cos x_1)^2$$

$$\Rightarrow y_1^2=x_1^2\left(1-\sin^2 x_1\right)$$

$$\Rightarrow y_1^2=x_1^2\left(1-y_1^2\right)\quad(\because y_1=\sin x_1)$$

$$\Rightarrow y_1^2=x_1^2-x_1^2y_1^2$$

$$\Rightarrow y_1^2-x_1^2=-x_1^2y_1^2$$

$$\Rightarrow x_1^2-y_1^2=x_1^2y_1^2$$

$\therefore (x_1, y_1)$ lies on the curve $x^2-y^2=x^2y^2$.

Question: Show that the sum of the intercepts of the tangent to the parabola $x^{\frac{1}{2}}+y^{\frac{1}{2}}=a^{\frac{1}{2}}$ on the co-ordinate axis is constant and equal to a.

Solution: Given equation of the curve is

$$x^{\frac{1}{2}}+y^{\frac{1}{2}}=a^{\frac{1}{2}} \quad \ldots(1)$$

Now, differentiating (1) w.r.t x, we get

$$\frac{1}{2}x^{-\frac{1}{2}}+\frac{1}{2}y^{-\frac{1}{2}}\frac{dy}{dx}=0\Rightarrow\frac{dy}{dx}=-\frac{x^{-\frac{1}{2}}}{y^{-\frac{1}{2}}}=-\frac{y^{\frac{1}{2}}}{x^{\frac{1}{2}}}$$

$$\left[\frac{dy}{dx}\right]_{P(x_1,y_1)}=-\frac{y_1^{\frac{1}{2}}}{x_1^{\frac{1}{2}}} \quad \ldots(2)$$

Again since (x_1, y_1) is a point where the tangent touches the curve.

$\therefore (x_1, y_1)$ satisfies the equation of the curve since it lies on the curve.

$$\therefore x_1^{\frac{1}{2}}+y_1^{\frac{1}{2}}=a^{\frac{1}{2}} \quad \ldots(3)\text{ [from (1)]}$$

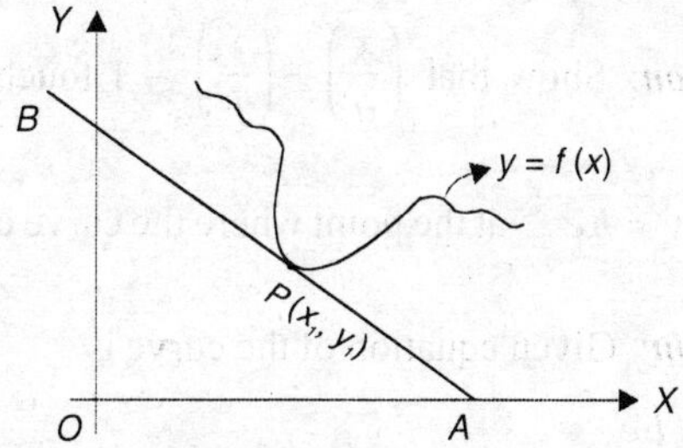

Now, the equation of the tangent at $P(x_1, y_1)$ is

$$(y-y_1)=-\frac{y_1^{\frac{1}{2}}}{x_1^{\frac{1}{2}}}(x-x_1)$$

$$\Rightarrow yx_1^{\frac{1}{2}}-y_1x_1^{\frac{1}{2}}=-y_1^{\frac{1}{2}}x+x_1y_1^{\frac{1}{2}}$$

$$\Rightarrow xy_1^{\frac{1}{2}}+yx_1^{\frac{1}{2}}=x_1y_1^{\frac{1}{2}}+x_1^{\frac{1}{2}}y_1=x_1^{\frac{1}{2}}x_1^{\frac{1}{2}}y_1^{\frac{1}{2}}+x_1^{\frac{1}{2}}y_1^{\frac{1}{2}}y_1^{\frac{1}{2}}$$

$$\Rightarrow x\sqrt{y_1}+y\sqrt{x_1}=x_1^{\frac{1}{2}}y_1^{\frac{1}{2}}\left(x_1^{\frac{1}{2}}+y_1^{\frac{1}{2}}\right)=x_1^{\frac{1}{2}}\cdot y_1^{\frac{1}{2}}a^{\frac{1}{2}}$$

$$\left[\because x_1^{\frac{1}{2}}+y_1^{\frac{1}{2}}=a^{\frac{1}{2}}\right]$$

$$\Rightarrow x\sqrt{y_1}+y\sqrt{x_1}=\sqrt{x_1y_1a} \quad \ldots(4)$$

Now, to find the intercepts on the axis, we put $x = 0$ and $y = 0$ respectively in the equation of the tangent (4)

For intercepts on x-axis, we put $y = 0$.

$$y=0\Rightarrow xy_1^{\frac{1}{2}}=\left[x_1^{\frac{1}{2}}y_1^{\frac{1}{2}}a^{\frac{1}{2}}\right]$$

$$\Rightarrow x=x_1^{\frac{1}{2}}a^{\frac{1}{2}}$$

$$\Rightarrow x=\sqrt{x_1a} \quad \ldots(5)$$

Again for the intercept on y-axis, we put $x = 0$.

$$x=0\Rightarrow x_1^{\frac{1}{2}}y=y_1^{\frac{1}{2}}x_1^{\frac{1}{2}}a^{\frac{1}{2}}$$

$$\Rightarrow y=y_1^{\frac{1}{2}}a^{\frac{1}{2}}$$

$$\Rightarrow y=\sqrt{y_1a} \quad \ldots(6)$$

Now, adding (5) and (6) ⇒ the sum of the intercepts on the axes

$$x_1^{\frac{1}{2}} a^{\frac{1}{2}} + y_1^{\frac{1}{2}} a^{\frac{1}{2}} = a^{\frac{1}{2}}\left(x_1^{\frac{1}{2}} + y_1^{\frac{1}{2}}\right) = a^{\frac{1}{2}} \cdot a^{\frac{1}{2}} = a$$

Question: Show that $\left(\frac{x}{a}\right) + \left(\frac{y}{b}\right) = 1$ touches the curve $y = be^{-\frac{x}{a}}$ at the point where the curve crosses y-axis.

Solution: Given equation of the curve is

$$y = be^{-\frac{x}{a}} \quad \ldots(1)$$

And the curve $y = be^{-\frac{x}{a}}$ meets y-axis at a point where $x = 0$ (Since $x = 0$ is the equation of y-axis).

Now, to find the point of intersection, we put $x = 0$ in the equation of the curve (1), we get

$$y = be^{-\frac{0}{a}} = b \quad \ldots(2)$$

Thus, $(0, b)$ = The point of intersection of the curve with y-axis.

Now, differentiating the given equation of the curve

$$y = be^{-\frac{x}{a}}$$

$$\Rightarrow \frac{dy}{dx} = be^{-\frac{x}{a}}\left(-\frac{1}{x}\right) = -\frac{b}{a}e^{-\frac{x}{a}}$$

Now, the value of $\frac{dy}{dx}$ at $(0, b)$

$$= \left[\frac{dy}{dx}\right]_{\substack{x=0\\y=b}} = -\frac{b}{a}e^{-\frac{0}{a}} = -\frac{b}{a}e^{0} = -\frac{b}{a}$$

Hence, the tangent at $(0, b)$

$$\Rightarrow [y - y_1] = \left[\frac{dy}{dx}\right]_{\substack{x=0\\y=b}} \cdot (x - x_1)$$

$$\Rightarrow (y - b) = -\frac{b}{a}(x - a) \Rightarrow \frac{y}{b} - 1 = -\frac{x}{a}$$

$$\Rightarrow \frac{x}{a} + \frac{y}{b} = 1$$

Question: Show that the condition that the line $x\cos\theta + y\sin\theta = p$ touches the curves $\left[\frac{x}{a}\right]^m + \left[\frac{y}{b}\right]^m = 1$ is

$$(a\cos\theta)^{\frac{m}{m-1}} + (b\sin\theta)^{\frac{m}{m-1}} = p^{\frac{m}{m-1}}.$$

Or, find the condition that the line $x\cos\theta + y\sin\theta = p$ should touch the curve $\frac{x^m}{a^m} + \frac{y^m}{b^m} = 1$.

Solution: Given equation of the curve is

$$\frac{x^m}{a^m} + \frac{y^m}{b^m} = 1 \quad \ldots(1)$$

Now, differentiating both sides of (1) w.r.t x, we get

$$\frac{mx^{m-1}}{a^m} + \frac{my^{m-1}}{b^m}\frac{dy}{dx} = 0$$

$$\Rightarrow \frac{dy}{dx} = -\frac{b^m x^{m-1}}{a^m y^{m-1}}$$

∴ Equation of the tangent at (x, y) is

$$(Y - y) = -\frac{b^m x^{m-1}}{a^m y^{m-1}}(X - x)$$

$$\Rightarrow \frac{X x^{m-1}}{a^m} + \frac{Y y^{m-1}}{b^m} = \frac{x^m}{a^m} + \frac{y^m}{b^m} = 1 \quad \ldots(2)$$

Now, since the equation of a straight line can be written as

$$X\cos\theta + Y\sin\theta = p \quad \ldots(3)$$

Thus, the equation (2) and (3) represents the same straight line

$$\Rightarrow \frac{x^{m-1}}{a^m\cos\theta} = \frac{y^{m-1}}{b^m\sin\theta} = \frac{1}{p}$$

$$\Rightarrow x = \left[\frac{a^m\cos\theta}{p}\right]^{\frac{1}{m-1}},\ y = \left[\frac{b^m\sin\theta}{p}\right]^{\frac{1}{m-1}}$$

Since the point (x, y) lies on the curve

$$\Rightarrow \frac{1}{a^m}\left[\frac{a^m\cos\theta}{p}\right]^{\frac{m}{m-1}} + \frac{1}{b^m}\left[\frac{b^m\sin\theta}{p}\right]^{\frac{m}{m-1}} = 1$$

$$\Rightarrow (a\cos\theta)^{\frac{m}{m-1}} + (b\sin\theta)^{\frac{m}{m-1}} = p^{\frac{m}{m-1}}$$

Hence proved.

Question: If $X\cos\alpha + Y\sin\alpha = p$ touches the curve $\left[\frac{x}{a}\right]^{\frac{n}{n-1}} + \left[\frac{y}{b}\right]^{\frac{n}{n-1}} = 1$, then prove that $(a\cos\alpha)^n + (b\sin\alpha)^n = p^n$.

Solution: Given equation of the curve is

$$\left[\frac{x}{a}\right]^{\frac{n}{n-1}} + \left[\frac{y}{b}\right]^{\frac{n}{n-1}} = 1 \quad \ldots(1)$$

Now, differentiating the equation of the curve, we get

$$\left(\frac{n}{n-1}\right)\left(\frac{x}{a}\right)^{\frac{1}{n-1}} \cdot \frac{1}{a} + \frac{n}{n-1}\left(\frac{y}{b}\right)^{\frac{1}{n-1}} \cdot \frac{1}{b} \cdot \frac{dy}{dx} = 0$$

$$\Rightarrow \frac{dy}{dx} = -\left(\frac{x}{y}\right)^{\frac{1}{n-1}} \cdot \left(\frac{b}{a}\right)^{\frac{n}{n-1}}$$

Hence, the equation of the tangent at any point (x, y) is

$$(Y - y) = -\left(\frac{x}{y}\right)^{\frac{1}{n-1}} \cdot \left(\frac{b}{a}\right)^{\frac{n}{n-1}} (X - x)$$

$$\Rightarrow Yy^{\frac{1}{n-1}} a^{\frac{n}{n-1}} - y^{\frac{n}{n-1}} a^{\frac{n}{n-1}}$$

$$= -\left[x^{\frac{1}{n-1}} b^{\frac{n}{n-1}} X - x^{\frac{n}{n-1}} \cdot b^{\frac{n}{n-1}}\right] \quad \ldots(2)$$

Now, dividing both sides of (2) by $a^{\frac{n}{n-1}} \cdot b^{\frac{n}{n-1}}$, we get

$$\frac{x^{\frac{1}{n-1}} X}{a^{\frac{n}{n-1}}} + \frac{y^{\frac{1}{n-1}} Y}{b^{\frac{n}{n-1}}} = \left(\frac{x}{a}\right)^{\frac{n}{n-1}} + \left(\frac{y}{b}\right)^{\frac{n}{n-1}} = 1 \quad \ldots(3)$$

Also, we are given $\cos\alpha \cdot X + \sin\alpha \cdot Y = p$ is the equation of the tangent. ...(4)

Thus, eqns (3) and (4) represent the same straight line. Comparing eqns (3) and (4), we have the coefficients of x and y in (3) and (4).

$$\Rightarrow \frac{\cos\alpha}{x^{\frac{1}{n-1}} \Big/ a^{\frac{n}{n-1}}} = \frac{\sin\alpha}{y^{\frac{1}{n-1}} \Big/ b^{\frac{n}{n-1}}} = \frac{p}{1} = p.$$

Coefficient of $X = \dfrac{x^{\frac{1}{n-1}}}{a^{\frac{n}{n-1}}}$ from (3)
Coefficient of $X = \cos\alpha$ from (4)

$$\Rightarrow \frac{a\cos\alpha}{\left(\frac{x}{a}\right)^{\frac{1}{n-1}}} = \frac{b\sin\alpha}{\left(\frac{y}{b}\right)^{\frac{1}{n-1}}} = p$$

$$\Rightarrow \frac{(a\cos\alpha)^n}{\left(\frac{x}{a}\right)^{\frac{n}{n-1}}} = \frac{(b\sin\alpha)^n}{\left(\frac{y}{b}\right)^{\frac{n}{n-1}}} = p^n$$

[Raising both sides to the nth power]

$$\Rightarrow \frac{(a\cos\alpha)^n + (b\sin\alpha)^n}{\left(\frac{x}{a}\right)^{\frac{n}{n-1}} + \left(\frac{y}{b}\right)^{\frac{n}{n-1}}} = p^n$$

$$\Rightarrow (a\cos\alpha)^n + (b\sin\alpha)^n = p^n$$

$$\left[\because \left(\frac{x}{a}\right)^{\frac{n}{n-1}} + \left(\frac{y}{b}\right)^{\frac{n}{n-1}} = 1\right]$$

which is the required condition.

Note: We should note while solving the problem involving x and y that touching point has been supposed to be (x, y) instead of (x_1, y_1) here. This is why the necessity of substituting the coordinates of the point in $\frac{dy}{dx}$ does not arise.

Question: Show that the normal to the curve

$x = a(\cos\theta + \theta\sin\theta)$

$y = a(\sin\theta - \theta\cos\theta)$ at any point θ is at a constant distance from the origin.

Solution: $\dfrac{dy}{dx} = \dfrac{dy/d\theta}{dx/d\theta}$

$$\Rightarrow \frac{dy}{dx} = \frac{a[\cos\theta - \{\theta(-\sin\theta) + 1\cdot\cos\theta\}]}{a[-\sin\theta + 1\cdot\sin\theta + \theta\cos\theta]} = \tan\theta$$

$\Rightarrow$ The equation of the normal at any point θ on the curve is given by

$$y - a(\sin\theta - \theta\cos\theta) = -\frac{\cos\theta}{\sin\theta}[x - a(\cos\theta + \theta\sin\theta)]$$

$$\Rightarrow y\sin\theta - a(\sin\theta - \theta\cos\theta)\sin\theta$$

$$= x\cos\theta + a\cos\theta(\cos\theta + \theta\sin\theta)$$

$$\Rightarrow x\cos\theta + y\sin\theta = a\left(\sin^2\theta + \cos^2\theta\right) = a$$

$$\Rightarrow x\cos\theta + y\sin\theta - a = 0 \qquad \ldots(1)$$

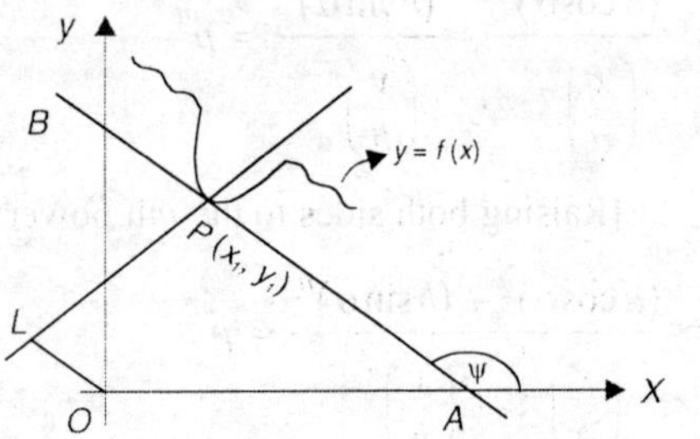

Now, the length of the perpendicular from the origin to the normal (1)

$$= \left|\frac{0\cdot\cos\theta + 0\cdot\sin\theta - a}{\sqrt{\cos^2\theta + \sin^2\theta}}\right| = a = \text{constant.}$$

Note: Length of the perpendicular from the origin upon the tangent to the curve at (x_1, y_1).

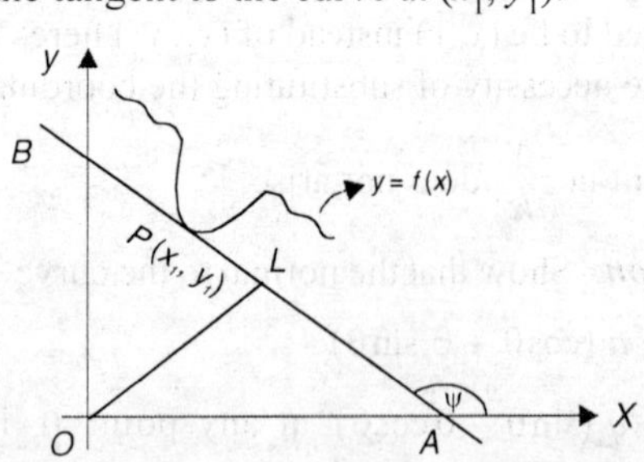

Since the equation of the tangent at (x_1, y_1) is

$$(y - y_1) = \left[\frac{dy}{dx}\right]_{\substack{x=x_1\\y=y_1}} \cdot (x - x_1)$$

$$\Rightarrow x\left[\frac{dy}{dx}\right]_{\substack{x=x_1\\y=y_1}} - y + \left(y_1 - x_1\left[\frac{dy}{dx}\right]_{\substack{x=x_1\\y-y_1}}\right) = 0$$

$\therefore$ The length of the perpendicular upon this line from the origin whose coordinates are (0, 0) is

$$= \frac{\left|y_1 - x_1\left[\frac{dy}{dx}\right]_{\substack{x=x_1\\y=y_1}}\right|}{\sqrt{1 + \left[\left(\frac{dy}{dx}\right)_{\substack{x=x_1\\y=y_1}}\right]^2}}$$

> $\because$ The length of perpendicular from (h, k) to any line $ax + by + c = 0$ is $\dfrac{|ah + bk + c|}{\sqrt{a^2 + b^2}}$

Question: Find the equation of the tangent and normal at the point 't' on the curve $x = a\cos^3 t$, $y = a\sin^3 t$. Show that portion of the tangent intercepted between the axes is of constant length a.

Solution: $\dfrac{dx}{dt} = -3a\cos^2 t\sin t \qquad \ldots(1)$

$$\frac{dy}{dx} = 3a\sin^2 t\cos t \qquad \ldots(2)$$

Now, $\dfrac{dy}{dx} = \dfrac{dy/dt}{dx/dt} = -\dfrac{3a\sin^2 t\cos t}{3a\cos^2 t\sin t} = -\dfrac{\sin t}{\cos t}$

$\Rightarrow$ Slope of the tangent at any point 't' $= -\tan t$

$\Rightarrow$ Slope of the normal at any point 't' $= \cot\theta$

$\therefore$ Equation of the tangent at 't' is

$$y - a\sin^3 t = -\frac{\sin t}{\cos t}\left(x - a\cos^3 t\right)$$

$$\Rightarrow y\cos t - a\sin^3 t\cos t = -x\sin t + a\cos^3 t\sin t$$

$$\Rightarrow x\sin t + y\cos t - a\cos^3 t\cdot\sin t - a\sin^3 t\cos t = 0$$

$\Rightarrow x\sin t + y\cos t - a\cos t\sin t\left(\cos^2 t + \sin^2 t\right) = 0$

$\Rightarrow x\sin t + y\cos t = a\sin t\cos t$ ...(1)

Again, since the slope of the normal at the given point 't' $= \dfrac{\cos t}{\sin t}$

$\therefore$ The equation of the normal at 't' is

$$y - a\sin^3 t = \frac{\cos t}{\sin t}\left(x - a\cos^3 t\right)$$

$$\Rightarrow \left(y - a\sin^3 t\right)\sin t = \left(x - a\cos^3 t\right)\cos t$$

$$\Rightarrow y\sin t - a\sin^4 t = x\cos t - a\cos^4 t$$

$$\Rightarrow x\cos t - y\sin t = a\left(\cos^4 t - \sin^4 t\right)$$

$$= a\left(\cos^2 t + \sin^2 t\right)\left(\cos^2 t - \sin^2 t\right)$$

$$\Rightarrow x\cos t - y\sin t = a\cos 2t$$

Now, since the tangent is $x\sin t + y\cos t = a\sin t\cos t$ (From eqn (1)) ...(2)

To get the intercept on x-axis, we put $y = 0$ in the equation (2) to have

$$x\sin t = a\sin t\cos t \Rightarrow x = a\cos t$$

To get the intercept on y-axis, we put $x = 0$ in the equation (2), then $y = a\sin t$

$\therefore$ The portion of the tangent intercepted between the axes.

$$= \sqrt{(\text{Intercept on x-axis})^2 + (\text{Intercept on y-axis})^2}$$

$$= \sqrt{a^2\cos^2 t + a^2\sin^2 t}$$

$= \sqrt{a^2} = a =$ Constant which implies the required result.

Note: This question may be asked as follows.

Question: Find the equation of the tangent to the curve $x^{\frac{2}{3}} + y^{\frac{2}{3}} = a^{\frac{2}{3}}$ and show that the portion of the tangent intercepted between the coordinate axes is constant and is equal to a.

Solution: Equation of the given curve is

$$x^{\frac{2}{3}} + y^{\frac{2}{3}} = a^{\frac{2}{3}} \quad \text{...(1)}$$

Now, differentiating both sides of the equation (1) w.r.t x, we get

$$\frac{2}{3}x^{-\frac{1}{3}} + \frac{2}{3}y^{-\frac{1}{3}}\frac{dy}{dx} = 0 \Rightarrow \frac{dy}{dx} = -\frac{y^{\frac{1}{3}}}{x^{\frac{1}{3}}}$$

Now, let the tangent touch the given curve at a point $P(x_1, y_1)$.

$\therefore \left[\dfrac{dy}{dx}\right]_{\substack{x=x_1\\y=y_1}} =$ The slope of the tangent at

$$P(x_1, y_1) = -\frac{y_1^{\frac{1}{3}}}{x_1^{\frac{1}{3}}}$$

$\therefore$ Equation of the tangent at (x_1, y_1) on the given curve is

$$(y - y_1) = -\frac{y_1^{\frac{1}{3}}}{x_1^{\frac{1}{3}}}(x - x_1)$$

$$\Rightarrow yx_1^{\frac{1}{3}} - y_1x_1^{\frac{1}{3}} = -y_1^{\frac{1}{3}}x + x_1y_1^{\frac{1}{3}}$$

$$\Rightarrow yx_1^{\frac{1}{3}} + y_1^{\frac{1}{3}}x = x_1y_1^{\frac{1}{3}} + y_1x_1^{\frac{1}{3}}$$

$$= x_1^{\frac{1}{3}}y_1^{\frac{1}{3}}\left(x_1^{\frac{2}{3}} + y_1^{\frac{2}{3}}\right)$$

$$\Rightarrow yx_1^{\frac{1}{3}} + xy_1^{\frac{1}{3}} = x_1^{\frac{1}{3}}y_1^{\frac{1}{3}}\left(a^{\frac{2}{3}}\right) \quad \text{...(2)}$$

[$\because x_1^{\frac{2}{3}} + y_1^{\frac{2}{3}} = a^{\frac{2}{3}}$ from the given equation since (x_1, y_1) lies on the curve $\Rightarrow$ It must satisfy given equation of the curve]

$\Rightarrow$ To find the intercept on the x-axis, we put $y = 0$, to have $0 + xy_1^{\frac{1}{3}} = x_1^{\frac{1}{3}}y_1^{\frac{1}{3}}a^{\frac{2}{3}}$

$$\Rightarrow x = x_1^{\frac{1}{3}}a^{\frac{2}{3}}$$

and to find the intercept on y-axis, we put $x = 0$ in (2) to have

$y = a^{\frac{2}{3}}y_1^{\frac{1}{3}}$ [On putting $x = 0$ in (2)]

Hence, the length of intercept between the axes

$$= \sqrt{x^2 + y^2}$$

$$= \sqrt{(\text{Intercept on x-axis})^2 + (\text{Intercept on y-axis})^2}$$

$$= \sqrt{a^{\frac{4}{3}}\left(x^{\frac{2}{3}} + y^{\frac{2}{3}}\right)} = \sqrt{a^2} = a$$

Question: Find the coordinates of the point on the curve $xy = 16$, the normal at which intersects at the origin of the coordinates.

Prove also that the portion of any tangent to the curve intercepted between the coordinate axes is bisected at the point of contact.

Solution: If (x_1, y_1) be any point on the curve $xy = 16$ ...(1)

Then (x_1, y_1) must satisfy the equation (1).

$\therefore x_1 y_1 = 16$ ...(2)

Now differentiating both sides of the given equation (1) w.r.t x, we get

$x \cdot \frac{dy}{dx} + y \cdot 1 = 0 \Rightarrow \frac{dy}{dx} = -\frac{y}{x} \Rightarrow$ the value of

$\frac{dy}{dx}$ at $(x_1, y_1) = -\frac{y_1}{x_1}$

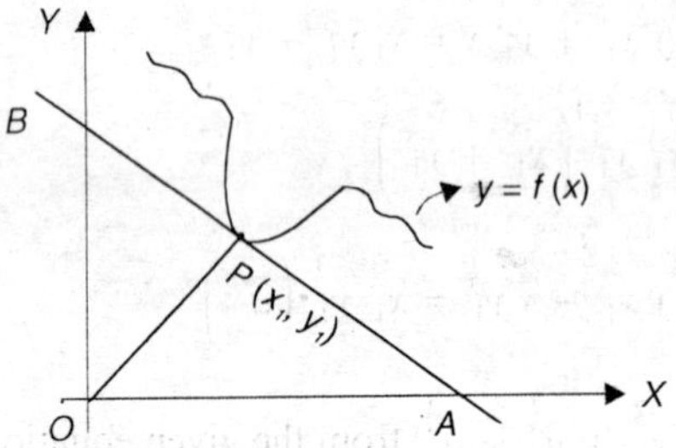

$\Rightarrow$ The slope of the tangent at $(x_1, y_1) = -\frac{y_1}{x_1}$

$\Rightarrow$ The slope of the normal at $(x_1, y_1) = \frac{x_1}{y_1}$

Now, the equation of the normal at (x_1, y_1) is

$$(y - y_1) = \frac{x_1}{y_1}(x - x_1)$$

The normal at (x_1, y_1) passes through $(0, 0)$

$$\Rightarrow (0 - y_1) = \frac{x_1}{y_1}(0 - x_1)$$

$$\Rightarrow y_1^2 = x_1^2 \quad ...(3)$$

From eqns (2) and (3) $\Rightarrow y_1^2 = \frac{16 \times 16}{y_1^2}$

$$\left[\because x_1 y_1 = 16 \Rightarrow x_1 = \frac{16}{y_1}\right]$$

$$\Rightarrow y_1^4 = 16 \times 16 \Rightarrow y_1 = \pm 4$$

Again from eqn (2), $x_1 y_1 = 16$

$$\Rightarrow x_1 = \frac{16}{y_1} = \frac{16}{y_1} = \frac{16}{\pm 4} = \pm 4$$

Hence, the required points are $(4, 4)$ and $(-4, -4)$.

Now the equation of the tangent at any point

(x_1, y_1) is $(y - y_1) = -\frac{y_1}{x_1}(x - x_1)$ ...(4)

Since the tangent meets the axis of x at the point where $y = 0$, so to find x-intercept, we put $y = 0$ in the equation (4) to have

$$(0 - y_1) = -\frac{y_1}{x_1}(x - x_1)$$

$$\Rightarrow -y_1 x_1 = -y_1 x + y_1 x_1$$

$$\Rightarrow -2y_1 x_1 + y_1 x = 0$$

$$\Rightarrow y_1(-2x_1 + x) = 0 \Rightarrow (-2x_1 + x) = 0$$

$$(\because y_1 \neq 0) \Rightarrow x = 2x_1 \Rightarrow OA = 2x_1$$

$$[\because x = OA]$$

Thus, the tangent at (x_1, y_1) meets the x-axis at the point $(2x_1, 0)$

Again, the tangent at (x_1, y_1) meets y-axis at the point where $x = 0 \Rightarrow$ To find y-intercept we put $x = 0$

in eqn (4) to have $(y - y_1) = -\frac{y_1}{x_1}(0 - x_1)$

$$\Rightarrow y - y_1 = y_1 \Rightarrow y = 2y_1 = OB \; (\because y = OB)$$

$\Rightarrow$ The tangent meets y-axis at the point $(0, 2y_1)$

Thus, we see that the tangent at (x_1, y_1) meets x-axis at $(2x_1, 0)$ and the tangent at (x_1, y_1) meets y-axis at $(0, 2y_1)$.

⇒ The mid-point of the portion of the tangent to the curve $xy = 16$ intercepted between the coordinate axes is $\left(\frac{2x_1+0}{2}, \frac{0+2y_1}{2}\right)$

⇒ Which is the point of contact.

Question: A normal is drawn to the curve $y = x^2$ at the point (1, 1) on it. Find the length of the part of the normal intercepted between the coordinate axes. Also find subtangent and subnormal.

Solution: We suppose that the tangent and the normal at the point (1, 1) on the curve $y = x^2$ (i) meet the x-axis in T and G.

Now, differentiating the given equation $y = x^2$ w.r.t x, we get $\frac{dy}{dx} = 2x$

⇒ The slope of the tangent at (1, 1) $= \left[\frac{dy}{dx}\right]_{\substack{x=1\\y=1}}$

$= [2x]_{x=1} = 2 \times 1 = 2$

Similarly, the slope of the normal at (1, 1)

$= -\frac{1}{\text{Slope of the tangent}} = -\frac{1}{2}$

∴ The equation of the normal at (1, 1) is

$(y-1) = -\frac{1}{2}(x-1)$

$\Rightarrow -2y + 2 = x - 1$

$\Rightarrow x + 2y = 3$ ...(2)

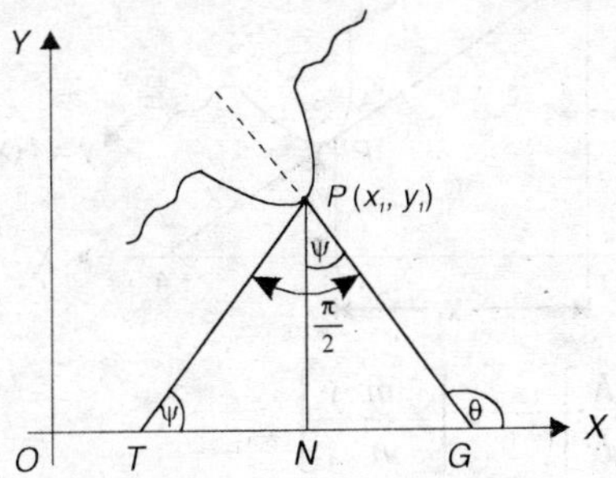

Now, the intercept x on x-axis is found by putting $y = 0$ in eqn (2) i.e., $x + 0 = 3 \Rightarrow x = 3$

Again the intercept y on y-axis is found by putting $x = 0$ in eqn (2) i.e., $0 + 2y = 3 \Rightarrow y = \frac{3}{2}$

Thus, the lengths of the intercepts made by the normal on the axes are 3 and $\frac{3}{2}$ unit.

∴ Length of the normal intercepted between the coordinate axes is

$$\sqrt{(x\text{-intercept})^2 + (y\text{-intercept})^2}$$

$$= \sqrt{3^2 + \left(\frac{3}{2}\right)^2} = \sqrt{9 + \frac{9}{4}}$$

$$= \sqrt{\frac{36+9}{4}} = \sqrt{\frac{45}{4}} = \frac{3\sqrt{5}}{2}$$

TN = Subtangent = PN $\tan\psi$ = 1.2 = 2

NG = Subnormal = PN $\cot\psi$ = 1.½ = $\frac{1}{2}$.

Question: In the curve $x = a\left(\cos t + \log\tan\frac{t}{2}\right)$,

$y = a\sin t$

Show that the portion of the tangent between the point of contact and the x-axis is of constant length.

Solution: $\frac{dx}{dt} = a\left\{-\sin t + \frac{1}{\tan\frac{t}{2}}\cdot\frac{1}{2}\sec^2\frac{t}{2}\right\}$

$$\therefore \frac{dx}{dt} = a\left\{-\sin t + \frac{1}{2}\cdot\frac{\sec^2\frac{t}{2}}{\tan\frac{t}{2}}\right\} \quad ...(1)$$

$$\frac{dy}{dt} = a\cos t \quad ...(2)$$

Now on simplifying the equation (1),

$$\frac{dx}{dt} = a\left\{-\sin t + \frac{1}{2}\cdot\frac{\sec^2\frac{t}{2}}{\tan\frac{t}{2}}\right\}$$

$$= a\left\{-\sin t + \frac{1}{2}\sec^2\frac{t}{2}\cot\frac{t}{2}\right\}$$

$$= a\left\{-\sin t + \frac{1}{2}\cdot\frac{1}{\cos^2\frac{t}{2}}\cdot\frac{\cos\frac{t}{2}}{\sin\frac{t}{2}}\right\}$$

$$= a\left\{-\sin t + \frac{1}{2\sin\frac{t}{2}\cdot\cos\frac{t}{2}}\right\}$$

$$= a\left\{-\sin t + \frac{1}{\sin t}\right\} = a\left\{\frac{-\sin^2 t + 1}{\sin t}\right\}$$

$$= \frac{a\cos^2 t}{\sin t} \qquad \ldots(3)$$

Now the slope of the tangent at (x, y) is

$$\frac{dy}{dx} = \frac{dy}{dt} \div \frac{dx}{dt}$$

$$\therefore \tan\psi = a\cos t \div \frac{a\cos^2 t}{\sin t}$$

$$= a\cos t \times \frac{\sin t}{a\cos^2 t} = \frac{\sin t}{\cos t} = \tan t$$

$$\therefore \psi = t$$

Now from the figure

$$\frac{PT}{NP} = \operatorname{cosec}\psi = \operatorname{cosec} t$$

$$\Rightarrow PT = NP\ \operatorname{cosec} t = y\operatorname{cosec} t$$

$\Rightarrow$ PT $= a\sin t\operatorname{cosec} t$

[$\because$ $y = a\sin t$ is given in the problem]

$$\Rightarrow PT = a\sin t \times \frac{1}{\sin t} = a \text{ which is a constant.}$$

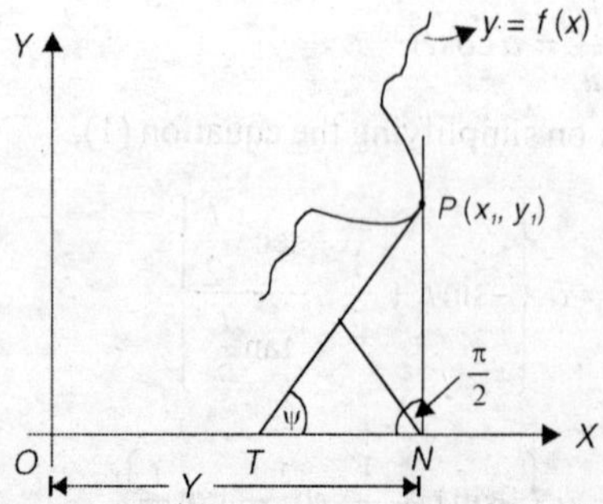

($y = a\sin t$ is given in the problem)

Question: **(a)** Show that the condition that the line $x\cos\alpha + y\sin\alpha = p$ touches the curve $x^m y^m = a^{m+n}$ is $p^{m+n}\cdot m^m\cdot n^n = (m+n)^{m+n}\cdot a^{m+n}\cdot n\cos^m\alpha\cdot\sin^n\alpha$.

Or, find the condition that the line $x\cos\alpha + y\sin\alpha = p$ may be the tangent to the curve $x^m y^n = a^{m+n}$.

(b) In the curve $x^m y^n = a^{m+n}$, prove that the portion of the tangent intercepted between the axes is divided at its points of contact into segments which are in a constant ratio.

Solution: Given equation is $x^m y^n = a^{m+n}$ …(1)

Now, differentiating both sides of eqn (1) w.r.t x, we get

$$x^m\cdot ny^{n-1}\frac{dy}{dx} + y^n\cdot m\cdot x^{m-1} = 0$$

$$\Rightarrow \frac{dy}{dx} = \frac{-mx^{m-1}\cdot y^n}{nx^m\cdot y^{n-1}} \qquad \ldots(2)$$

$$\Rightarrow \frac{dy}{dx} = -\frac{m}{n}x^{m-1-m}y^{n-n+1} = -\frac{m}{n}x^{-1}y$$

$$\Rightarrow \frac{dy}{dx} = -\frac{m}{n}\cdot\frac{y}{x} = \text{Slope of the tangent at } (x, y)$$

Now the value of $\frac{dy}{dx}$ at (x_1, y_1)

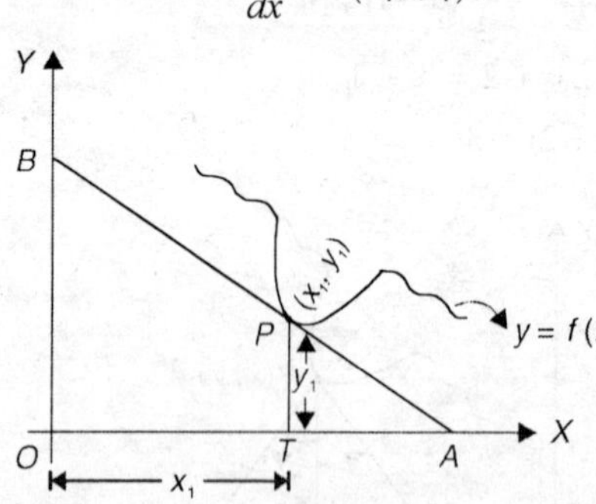

$$= \left[\frac{dy}{dx}\right]_{\substack{x=x_1\\y=y_1}} = \left[-\frac{m}{n}\frac{y}{x}\right]_{\substack{x=x_1\\y=y_1}}$$

$\Rightarrow$ The value of $\frac{dy}{dx}$ at (x_1, y_1) = Slope of the tangent at $(x_1, y_1) = -\frac{my_1}{nx_1}$

$\therefore$ Equation of the tangent at (x_1, y_1) is

$$(y-y_1) = \left[\frac{dy}{dx}\right]_{\substack{x=x_1\\y=y_1}} (x-x_1)$$

$$\Rightarrow (y-y_1) = -\frac{m}{n}\frac{y_1}{x_1}(x-x_1) \quad \text{....(3)}$$

$$\Rightarrow (y-y_1) = -\frac{m x_1^{m-1} y_1^n}{n x_1^m y_1^{n-1}}(x-x_1)$$

$$\left[\because \frac{y_1}{x_1} = \frac{x_1^{m-1}\cdot y_1^n}{x_1^m \cdot y_1^{n-1}}\right]$$

$$\Rightarrow y\cdot n\cdot x_1^m y_1^{n-1} - y_1\, n x_1^m y_1^{n-1}$$
$$= -m x_1^{m-1} y_1^n x + m x_1^{m-1} y_1^n x_1$$
$$\Rightarrow y\cdot n x_1^m y_1^{n-1} + m x_1^{m-1} y_1^n x$$
$$= m x_1^{m-1} y_1^n x_1 + y n x_1^m y_1^{n-1}$$
$$\Rightarrow y\cdot n\cdot x_1^m y_1^{n-1} + m x_1^{m-1} y_1^n x$$
$$= m x_1^m y_1^n + n x_1^m y_1^n = (m+n) x_1^m y_1^n$$
$$\Rightarrow m x x_1^{m-1} y^n + n y x_1^m y_1^{n-1}$$
$$= (m+n) a^{m+n}$$

$\because (x_1, y_1)$ lies on the curve $x^m y^n = a^{m+n}$ $\Rightarrow (x_1, y_1)$ satisfies the equation $x^m y^n = a^{m+n}$ $\Rightarrow x_1^m y_1^n = a^{m+n}$

$\therefore$ Equation of the tangent is $m x x_1^{m-1} y^n + n y x_1^m y_1^{n-1} = (m+n) a^{m+n}$ ...(4)

and the line $x \cos\alpha + y \sin\alpha = p$ ...(5)

Thus, eqns (4) and (5) represent the same straight line

$\Rightarrow$ Coefficients of x, y and constant terms are in proportion

$$\Rightarrow \frac{\cos\alpha}{m x_1^{n-1} y_1^n} = \frac{\sin\alpha}{n x_1^m y_1^{n-1}} = \frac{p}{(m+n)a^{m+n}}$$

$$\Rightarrow \frac{x_1\cos\alpha}{m x_1^m y_1^n} = \frac{y_1 \sin\alpha}{n x_1^m y_1^n} = \frac{p}{(m+n)\left(a^{m+n}\right)}$$

$$\Rightarrow \frac{x_1\cos\alpha}{m a^{m+n}} = \frac{y_1\sin\alpha}{n a^{m+n}} = \frac{p}{(m+n)\left(a^{m+n}\right)} \quad \text{...(6)}$$

$$\left[\because x_1^m y_1^n = a^{m+n}\right]$$

$$\Rightarrow x_1 \cos\alpha = \frac{p\cdot m}{(m+n)} \quad \text{...(7)}$$

$$\text{and } y_1 \cos\alpha = \frac{pn}{m+n} \quad \text{...(8)}$$

$$\text{Eqn (7)} \Rightarrow x_1^m \cos^m \alpha = \frac{p^m m^m}{(m+n)^m} \quad \text{...(9)}$$

$$\text{Eqn (8)} \Rightarrow y_1^n \sin^n \alpha = \frac{p^n \cdot n^n}{(m+n)^n} \quad \text{...(10)}$$

Now, multiplying these two results of eqns (9) and (10) together, we get

$$x_1^m y_1^n \cos^m\alpha \sin^n\alpha = \frac{p^{m+n}\cdot m^m \cdot n^n}{(m+n)^{m+n}}$$

$$\Rightarrow a^{m+n} (m+n)^{m+n} \cos^m\alpha \sin^n\alpha$$

$= p^{m+n}\cdot m^m \cdot n^n$ which is the required condition.

Now, again letting that the tangent at (x_1, y_1) cuts the axis in A and B.

To find the intercept x on x-axis of the tangent, we put $y = 0$ in the equation of tangent eqn (4).

$\therefore$ Equation of tangent is

$$m x x_1^{m-1} y_1^n + n y x_1^m y_1^{n-1} = (m+n) a^{m+n}$$

$$\Rightarrow m x x_1^{m-1} y_1^n = (m+n) a^{m+n} \quad (\text{when } y = 0)$$

$$\Rightarrow x = \frac{(m+n) a^{m+n}}{m x_1^{m-1} y_1^n} = \frac{(m+n) a^{m+n} x_1}{m x_1^m y_1^n}$$

$$= \frac{(m+n) a^{m+n} x_1}{m\cdot a^{m+n}} = \frac{(m+n) x_1}{m} = OA$$

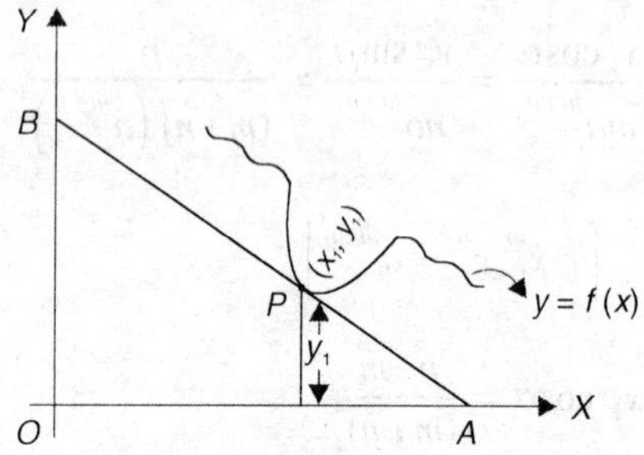

Now, $OT = x_1$ and $OT + TA = OA$

$\Rightarrow TA = OA - OT$

$$\Rightarrow TA = \frac{(m+n)\,x_1}{m} - x_1$$

$$\Rightarrow TA = \frac{(m+n)\,x_1 - m x_1}{m}$$

$$\Rightarrow TA = \frac{n x_1}{m}$$

In $\Delta\,OAB$,

$$PT \parallel OB \Rightarrow \frac{AP}{PB} = \frac{AT}{OT} = \frac{\frac{n x_1}{m}}{x_1} = \frac{n}{m}$$ which proves the required.

Problems based on finding the coordinates of a point $P(x_1, y_1)$ where the tangent touches the given curve $y = f(x)\,/\,f(x, y) = 0$ or constant.

Working rule: **1.** We suppose that there is a required point (x_1, y_1) on the given curve where the tangent touches the curve $y = f(x)$.

2. Find $\frac{dy}{dx}$ by differentiating the given equation w.r.t. x.

3. Find $\left[\frac{dy}{dx}\right]_{\substack{x = x_1\\ y = y_1}}$

4. Find $\left[\frac{dy}{dx}\right]_{\substack{x = x_1\\ y = y_1}}$ from the given condition imposed on tangential line (i.e., tangent to the given curve).

5. Solve the equation satisfied by (x_1, y_1) and the equation in x_1, and y_1 obtained after imposing the given condition on $\left[\frac{dy}{dx}\right]_{\substack{x = x_1\\ y = y_1}}$.

N.B.: **1.** Imposed condition on tangent to the curve may be: tangential line is parallel to x-axis or parallel to a line / tangential line passes through origin / Tangential line passes through the point (α, β), etc.

2. $\left[\frac{dy}{dx}\right]_{\substack{x = x_1\\ y = y_1}}$ found from the given condition on the tangential line is generally constant or zero $\Rightarrow \left[\frac{dy}{dx}\right]_{\substack{x = x_1\\ y = y_1}}$, an expression in x_1 and y_1 is equated to zero or constant according to the given condition imposed on the tangential line (i.e tangent to the curve or slope of the curve).

3. $\left[\frac{dy}{dx}\right]_{\substack{x = x_1\\ y = y_1}} = 0$ (Imposed condition) provided the condition imposed on tangent tells that tangent is parallel to x-axis.

4. We shall consider the problems on tangent at one point of the curve/two or more tangents of the curve making an angle with x-axis/y-axis/ parallel to x-axis/ y-axis.

Examples worked out:

Question: Find the points on the graph of the function defined by $y = x^2 - 6x + 9$ at which the tangents are parallel to x-axis.

Solution: Let the required point on the curve $y = x^2 - 6x + 9$ be (x_1, y_1).

$\therefore\ (x_1, y_1)$ must satisfy $y = x^2 - 6x + 9$ ...(1)

$\therefore\ y_1 = x_1^{\,2} - 6x_1 + 9$ ...(2)

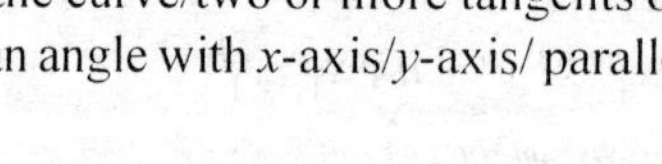

Now, differentiating the equation (1) w.r.t x, we get

$$\frac{dy}{dx} = 2x - 6$$

Now, the value of $\frac{dy}{dx}$ at (x_1, y_1) = The slope of the tangent at (x_1, y_1)

$$= \left[\frac{dy}{dx}\right]_{\substack{x=x_1\\y=y_1}} = [2x-6]_{\substack{x=x_1\\y=y_1}} = 2x_1 - 6$$

Now, since the tangent at (x_1, y_1) is parallel to x-axis

$$\therefore \left[\frac{dy}{dx}\right]_{\substack{x=x_1\\y=y_1}} = 0 \Rightarrow 2x_1 - 6 = 0 \Rightarrow x_1 = \frac{6}{2} = 3$$

Substituting this value of x_1 in eqn (2), we get $y_1 = 9 - 18 + 9 = 0$

Hence, the required point = $(x_1, y_1) = (3, 0)$.

Question: At what points on the curve $x^2 + y^2 - 2x - 4y + 1 = 0$, the tangent is parallel to (a) x-axis (b) y-axis.

Solutions: (a) Equation of the given curve is:

$$x^2 + y^2 - 2x - 4y + 1 = 0 \qquad \ldots(1)$$

Letting the required point to be (x_1, y_1) on the curve

$\Rightarrow$ (x_1, y_1) must satisfy the given equation of the curve.

$$\Rightarrow x_1^2 + y_1^2 - 2x_1 - 4y_1 + 1 = 0 \qquad \ldots(2)$$

Now, on differentiating the equation (1) w.r.t x, we get

$$2x + 2y\frac{dy}{dx} - 2 - 4\frac{dy}{dx} = 0$$

$$\Rightarrow 2y\frac{dy}{dx} - 4\frac{dy}{dx} = 2 - 2x$$

$$\Rightarrow \frac{dy}{dx}(2y - 4) = 2 - 2x$$

$$\Rightarrow \frac{dy}{dx} = \frac{2(1-x)}{2(y-2)} = \frac{1-x}{y-2}$$

Now, the value of $\frac{dy}{dx}$ at (x_1, y_1)

$$= \left[\frac{dy}{dx}\right]_{\substack{x=x_1\\y=y_1}} = \left[\frac{1-x}{y-2}\right]_{\substack{x=x_1\\y=y_1}} = \frac{1-x_1}{y_1-2}$$

= Slope of the tangent at (x_1, y_1)

Now, since the tangent is parallel to x-axis

$\therefore$ Slope of the tangent at (x_1, y_1) is zero.

$$\therefore \left[\frac{dy}{dx}\right]_{\substack{x=x_1\\y=y_1}} = 0 \Rightarrow \frac{1-x_1}{y_1-2} = 0$$

$$\Rightarrow 1 - x_1 = 0 \Rightarrow x_1 = 1 \qquad \ldots(3)$$

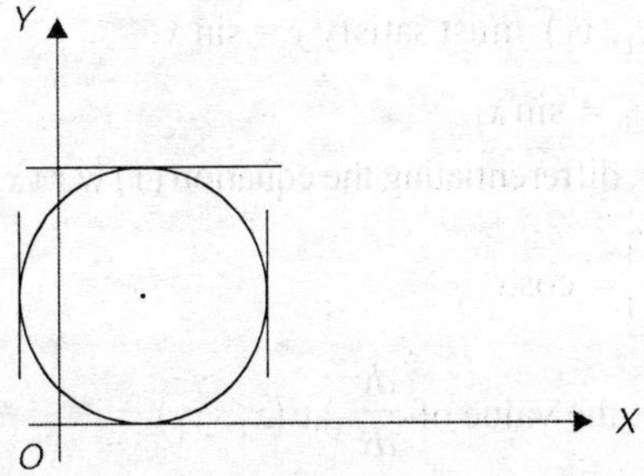

On putting $x_1 = 1$ from eqn (3) in eqn (2), we get

$$1 + y_1^2 - 2\cdot 1 - 4\cdot y_1 + 1 = 0$$

$$\Rightarrow y_1^2 - 4y_1 = 0$$

$$\Rightarrow y_1(y_1 - 4) = 0$$

$$\Rightarrow y_1 = 0 \text{ or } 4 \qquad \ldots(4)$$

Equations (3) and (4) $\Rightarrow$ $(x_1, y_1) = (1, 0)$ and $(1, 4)$ where the tangent is parallel to x-axis.

(b) Since the tangent is parallel to y-axis

$$\left[\frac{dx}{dy}\right]_{\substack{x=x_1\\y=y_1}} = 0$$

$$\Rightarrow \frac{y_1-2}{1-x_1} = 0 \Rightarrow y_1 - 2 = 0 \Rightarrow y_1 = 2 \qquad \ldots(5)$$

Putting $y_1 = 2$ from eqn (5) in eqn (2), we get,

$$x_1^2 + 4 - 2x_1 - 4\times 2 + 1 = 0$$

$$\Rightarrow x_1^2 + 4 - 2x_1 - 8 + 1 = 0$$

$$\Rightarrow x_1^2 - 2x_1 - 3 = 0$$

$$\Rightarrow x_1^2 - 3x_1 + x_1 - 3 = 0$$

$$\Rightarrow x_1(x_1 - 3) + (x_1 - 3) = 0$$

$$\Rightarrow (x_1 - 3)(x_1 + 1) = 0$$

Hence, required points where the tangent is parallel are $(x_1, y_1) = (-1, 2)$ and $(3, 2)$.

Question: At what points on the following curve are the tangents parallel to x-axis.

(a) $y = \sin x$ (b) $y^2 = 4a\left\{x + a\sin\frac{x}{a}\right\}$

Solution: (a) Let the required points on the curve $y = \sin x$ be (x_1, y_1).

$\therefore (x_1, y_1)$ must satisfy $y = \sin x$ ...(1)

$\Rightarrow y_1 = \sin x_1$

Now, differentiating the equation (1) w.r.t x, we get

$$\left[\frac{dy}{dx}\right] = \cos x$$

Now, the value of $\frac{dy}{dx}$ at (x_1, y_1)

$$= \left[\frac{dy}{dx}\right]_{\substack{x = x_1 \\ y = y_1}} = \cos x_1 = \text{Slope of the tangent at}$$

(x_1, y_1).

Since the tangent at (x_1, y_1) is parallel to x-axis.

$$\therefore \left[\frac{dy}{dx}\right]_{\substack{x = x_1 \\ y = y_1}} = 0 \Rightarrow \cos x_1 = 0$$

$$\Rightarrow x_1 = (2n+1)\frac{\pi}{2} \quad \text{...(3)}$$

where n is zero, positive or negative integer.

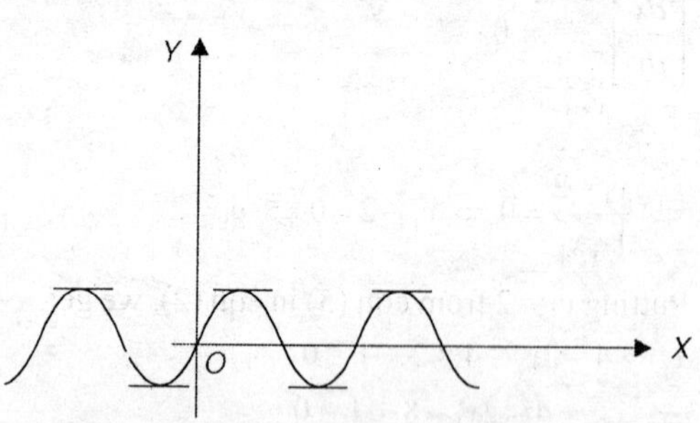

Substituting this value of x_1 in eqn (1), we get

$$y_1 = \sin(2n+1)\frac{\pi}{2}$$

$= 1$ when n is even.

$= -1$ when is odd.

Hence, the required points $(x_1, y_1) = \left[(2n+1)\frac{\pi}{2}, 1\right]$

When $n = 0, 2, 4$

and $\left[(2n+1)\frac{\pi}{2}, -1\right]$ when $n = 1, 3, 5$

(b) Let (x_1, y_1) be the required point on the curve.

$$y^2 = 4a\left\{x + a\sin\frac{x}{a}\right\} \quad \text{...(1)}$$

$$\therefore y_1 = 4a\left\{x_1 + a\sin\frac{x_1}{a}\right\} \quad \text{...(2)}$$

Now, differentiating eqn (1) w.r.t x, we get

$$2y\frac{dy}{dx} = 4a\left\{1 + a\left(\cos\frac{x}{a}\right)\cdot\frac{1}{a}\right\} = 4a\left(1 + \cos\frac{x}{a}\right)$$

$$\Rightarrow \frac{dy}{dx} = \frac{2a}{y}\left(1 + \cos\frac{x}{a}\right)$$

$\Rightarrow$ The value of $\frac{dy}{dx}$ at (x_1, y_1)

$$= \frac{2a}{y_1}\left(1 + \cos\frac{x_1}{a}\right) = \text{Slope of the tangent at}$$

(x_1, y_1)

Now, since the tangent at (x_1, y_1) is parallel to x-axis,

$$\Rightarrow \left[\frac{dy}{dx}\right]_{\substack{x = x_1 \\ y = y_1}} = \frac{2a}{y_1}\left(1 + \cos\frac{x_1}{a}\right) = 0$$

$$\Rightarrow \cos\left(\frac{x_1}{a}\right) = -1 = \cos\pi \Rightarrow \frac{x_1}{a} = 2n\pi \pm \pi$$

$\Rightarrow \frac{x_1}{a} = (2n \pm 1)\pi$ where n is zero, positive or negative integer.

$$\Rightarrow x_1 = (2n+1)a\pi$$

Now, substituting this value of x_1 in eqn (2), we get

$$y_1^2 = 4a\{(2n+1)a\pi + a\sin(2n+1)\pi\}$$

$$= 4a^2[(2n+1)\pi + 0] = 4a^2(2n+1)\pi$$

$$\Rightarrow y_1 = \pm 2a\sqrt{(2n+1)\pi}$$

Hence, the required points

$$= (x_1, y_1) = \left\{(2n+1)a\pi, \pm 2a\sqrt{(2n+1)\pi}\right\}$$

where n is zero, positive or negative integer.

Question: Find the coordinates of the point of contact of tangent on the curve $xy + 4 = 0$ at which the tangent makes an angle of 45° with x-axis. (Or, find the equations of the tangents on the curve $xy + 4 = 0$ which are inclined at an angle of 45° with the axis of x). Also find the coordinates of the points of contact.

Solution: Let (x_1, y_1) be the required point on the curve $xy + 4 = 0$ ...(1)

Now, since (x_1, y_1) lies on the curve $xy + 4 = 0$

$\therefore$ (x_1, y_1) satisfies the equation (1)

$\Rightarrow x_1 y_1 + 4 = 0$ ...(2)

Now, differentiating (1) w.r.t x, we have

$$x \cdot \frac{dy}{dx} + y = 0 \Rightarrow \frac{dy}{dx} = -\frac{y}{x} \quad ...(3)$$

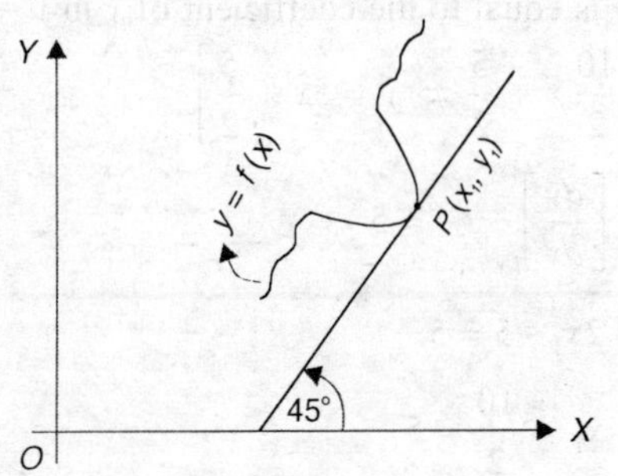

Now, the value of $\frac{dy}{dx}$ at (x_1, y_1)

$$= \left[\frac{dy}{dx}\right]_{\substack{x=x_1\\y=y_1}} = \left[-\frac{y}{x}\right]_{\substack{x=x_1\\y=y_1}} = -\frac{y_1}{x_1}$$

Now, the tangents at (x_1, y_1) are inclined at an angle of 45° with the axis of x.

$$\Rightarrow -\frac{y_1}{x_1} = \tan 45^o$$

$$\Rightarrow -\frac{x_1}{y_1} = 1 \Rightarrow y_1 = -x_1 \quad ...(4)$$

Now substituting $y_1 = -x_1$ from eqn (4) in eqn (2), we get

$$-x_1^2 + 4 = 0 \Rightarrow x_1 = \pm 2 \quad ...(5)$$

Again from (4), $y_1 = \mp 2$ ...(6)

Hence, the required points are (2, –2) and (–2, 2).

$\therefore$ The tangent at (2, –2) is $y + 2 = \tan 45^o (x - 2) \Rightarrow y + 2 = 1 \cdot (x - 2) \Rightarrow x - y = 4$ and the tangent at (–2, 2) is $y - 2 = \tan 45^o (x + 2) \Rightarrow y - 2 = x + 2 \Rightarrow x - y + 4 = 0$.

Question: At what points on the curve

$y = \frac{2}{3}x^3 + \frac{1}{2}x^2$ are the tangents equally inclined to the axis?

Solution: Let (x_1, y_1) be the required points on the curve $y = \frac{2}{3}x^3 + \frac{1}{2}x^2$ ...(1)

$$\therefore y_1 = \frac{2}{3}x_1^3 + \frac{1}{2}x_1^2 \quad ...(2)$$

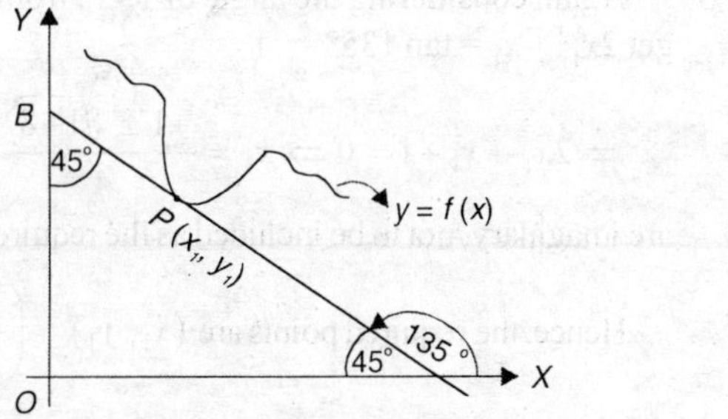

Now, differentiating (1) w.r.t x, we get

$$\frac{dy}{dx} = \frac{2}{3} \cdot 3x^2 + \frac{1}{2} \cdot 2x = 2x^2 + x$$

$\Rightarrow$ The value of $\frac{dy}{dx}$ at (x_1, y_1)

$$= 2x_1^2 + x_1 \quad ...(3)$$

The tangents at (x_1, y_1) are equally inclined to x-axis and y-axis (i.e., axis)

$\Rightarrow$ The tangent at (x_1, y_1) makes an angle of 45° or 135° with x-axis.

Now, considering angle 45°

$\tan 45^o = 2x_1^2 + x_1 \Rightarrow 2x_1^2 + x_1 = 1$

$\Rightarrow 2x_1^2 + x_1 - 1 = 0 \Rightarrow 2x_1^2 + 2x_1 - x_1 - 1 = 0$

$\Rightarrow 2x_1 (x_1 + 1) - (x_1 + 1) = 0$

$$\Rightarrow (x_1 + 1)(2x_1 - 1) = 0 \Rightarrow x_1 = -1, \frac{1}{2} \quad ...(4)$$

On putting the values of $x_1 = -1, \frac{1}{2}$ from eqn (4) in eqn (2), we get

$$y_1 = \frac{2}{3}(-1)^3 + \frac{1}{2}(-1)^2 = -\frac{2}{3} + \frac{1}{2} = \frac{-4+3}{6} = -\frac{1}{6}$$

when $x_1 = -1$ and $y_1 = \frac{2}{3}\left(\frac{1}{2}\right)^3 + \frac{1}{2}\left(\frac{1}{2}\right)^2 = \frac{2}{3} \times \frac{1}{8} + \frac{1}{4} \times \frac{1}{2} = \frac{2}{24} + \frac{1}{8} = \frac{2+3}{24} = \frac{5}{24}$ when $x_1 = \frac{1}{2}$

Thus, the required points are $(x_1, y_1) = \left(\frac{1}{2}, \frac{5}{24}\right)$ and $\left(-1, -\frac{1}{6}\right)$

Again considering the angle of 135°, from (3), we get $2x_1^2 + x_1 = \tan 135° = -1$

$$\Rightarrow 2x_1^2 + x_1 + 1 = 0 \Rightarrow x_1 = \frac{-1 \pm \sqrt{1-8}}{4} \text{ which}$$

are imaginary, not to be included as the required point.

Hence, the required points are $(x_1, y_1) = \left(\frac{1}{2}, \frac{5}{24}\right)$ and $\left(-1, -\frac{1}{6}\right)$

Question: At what points on the curve $y = (x-2)(x-3)$ is the tangent parallel to the line $2y = 10x + 5$?

Solution: Let (x_1, y_1) be the required point on the curve whose equation is

$y = (x-2)(x-3)$ ...(1)

$\Rightarrow y_1 = (x_1 - 2)(x_1 - 3)$ ...(2)

[∵ Since (x_1, y_1) satisfies the equation (1) because of lying on the curve $y = (x-2)(x-3)$]

Now, differentiating eqn (1) w.r.t x, we get

$$\frac{dy}{dx} = (x-2)\cdot 1 + (x-3)\cdot 1 = 2x - 5$$

∴ The value of $\frac{dy}{dx}$ at (x_1, y_1)

$$= \left[\frac{dy}{dx}\right]_{\substack{x=x_1\\y=y_1}} = [2x-5]_{x=x_1} = 2x_1 - 5$$

= Slope of the tangent at (x_1, y_1) ...(3)

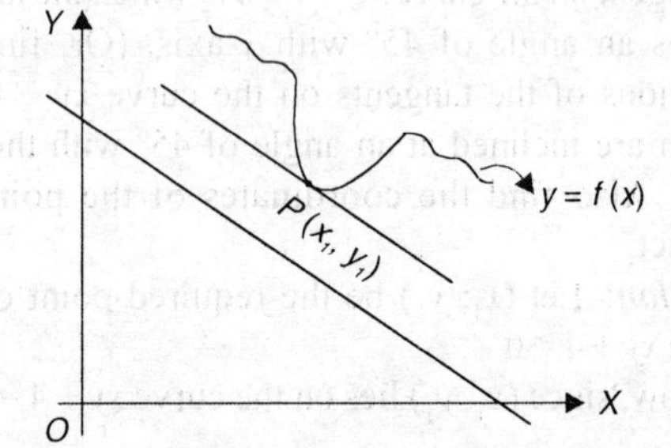

Now, since the tangent at (x_1, y_1) is parallel to the line $2y = 10x + 5$

∴ $\left[\frac{dy}{dx}\right]_{\substack{x=x_1\\y=y_1}}$ must be equal to the slope of the line which is equal to the coefficient of x in $y = mx + c$

$$\left[y = \frac{10}{2}x + \frac{5}{2} \Rightarrow y = 5x + \frac{5}{2}\right]$$

$$\Rightarrow \left[\frac{dy}{dx}\right]_{(x_1, y_1)} = 5$$

$$\Rightarrow 2x_1 - 5 = 5$$

$$\Rightarrow x_1 = \frac{10}{2} = 5 \qquad ...(4)$$

Now, putting $x_1 = 5$ from eqn (4) in eqn (2), we get

$y_1 = (x_1 - 2)(x_1 - 3) = (5-2)(5-3) = 6$

∴ Required point $(x_1, y_1) = (5, 6)$

Question: Find the points on the curves $y = x^3$ at which the tangent makes an angle of 60° with x-axis.

Solution: Let (x_1, y_1) be the coordinates of the required point on the curve $y = x^3$ ...(1)

Then (x_1, y_1) satisfies the equation (1)

$$\Rightarrow y_1 = x_1^3 \qquad ...(2)$$

Now, differentiating (1) w.r.t x, we get

$$\frac{dy}{dx} = 3x^2$$

$\Rightarrow$ The value of $\frac{dy}{dx}$ at (x_1, y_1)

$$= \left[\frac{dy}{dx}\right]_{\substack{x=x_1\\y=y_1}}$$

$= 3x_1^2$ ...(3)

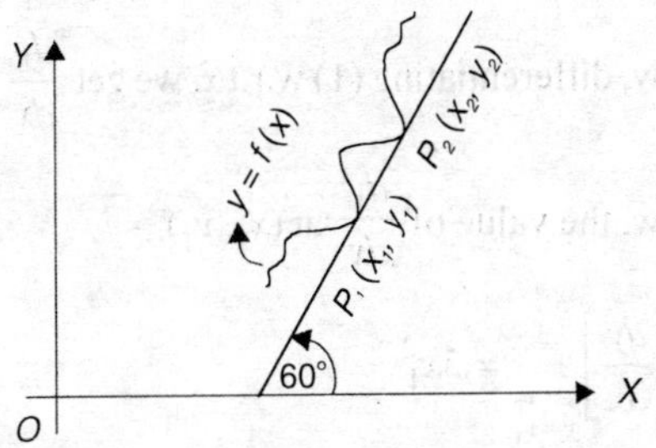

Now since, the tangent at (x_1, y_1) makes an angle of 60° with x-axis.

$$\left[\frac{dy}{dx}\right]_{\substack{x=x_1\\y=y_1}} = \tan 60^o \Rightarrow 3x_1^2 = \sqrt{3}$$

$$\Rightarrow x_1^2 = \frac{\sqrt{3}}{3} = \frac{1}{\sqrt{3}} \Rightarrow x_1^2 = (3)^{-\frac{1}{2}}$$

$$\Rightarrow x_1 = \pm 3^{-\frac{1}{4}}$$

Now, on substituting $x_1 = \pm\, 3^{-1/4}$ from (4) in eqn (2), we get $y_1 = \pm(3)^{-\frac{3}{4}}$

Hence, the required points are $(x_1, y_1) = \left(3^{-\frac{1}{4}}, 3^{-\frac{3}{4}}\right)$ and $\left(3^{-\frac{1}{4}}, -3^{-\frac{3}{4}}\right)$.

Question: Determine the coordinates of the points on the ellipse $4x^2 + 9y^2 = 40$ at which the slope of the curve is $-\frac{2}{9}$.

Solution: Let the required coordinates of the points be (x_1, y_1) on the curve

$$4x^2 + 9y^2 = 40 \quad \ldots(1)$$

$$\therefore 4x_1^2 + 9y_1^2 = 40 \quad \ldots(2)$$

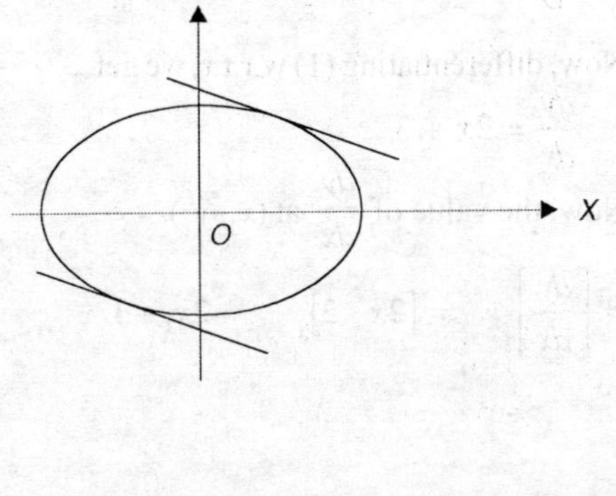

Now, differentiating both sides of (1) w.r.t. x, we get

$$8x + 18y\frac{dy}{dx} = 0 \Rightarrow \frac{dy}{dx} = -\frac{4x}{9y} \quad \ldots(3)$$

Now, the value of $\frac{dy}{dx}$ at (x_1, y_1)

$$= \left[\frac{dy}{dx}\right]_{\substack{x=x_1\\y=y_1}}$$

$\Rightarrow$ The slope of the tangent at (x_1, y_1)

$$= \left[-\frac{4x}{9y}\right]_{\substack{x=x_1\\y=y_1}} = -\frac{4x_1}{9y_1} \quad \ldots(4)$$

Again, the slope of the tangent $= -\frac{2}{9}$ (Given in the problem) ...(5)

$$\text{Eqns (4) and (5)} \Rightarrow \left[\frac{dy}{dx}\right]_{\substack{x=x_1\\y=y_1}} = -\frac{4x_1}{9y_1} = -\frac{2}{9}$$

$$\Rightarrow -\frac{4x_1}{9y_1} = -\frac{2}{9}$$

$$\Rightarrow y_1 = 2x_1 \quad \ldots(6)$$

Now, substituting (6) in (2), we get

$$4x_1^2 + 9\cdot(2x_1)^2 = 40$$

$$\Rightarrow 4x_1^2 + 36x_1^2 = 40$$

$$\Rightarrow 40x_1^2 = 40$$

$$\Rightarrow x_1 = \pm 1$$

$$\therefore y_1 = 2x_1 = \pm 2$$

$\therefore$ Required points are $(x_1, y_1) = (\pm 1, \pm 2)$

Question: Find the point on the curve $x^2 - y^2 = 2$ at which the slope of the curve is 2.

Solution: Let the required point be (x_1, y_1).

Given equation of the curve is $x^2 - y^2 = 2$ ...(1)

$$\therefore x_1^2 - y_1^2 = 2 \quad \ldots(2)$$

[$\because$ Since (x_1, y_1) lies on the curve].

Now, differentiating both sides of the equation (1) w.r.t x, we get

$$2x - 2y\frac{dy}{dx} = 0$$

$$\Rightarrow \frac{dy}{dx} = \frac{x}{y} \quad \ldots(3)$$

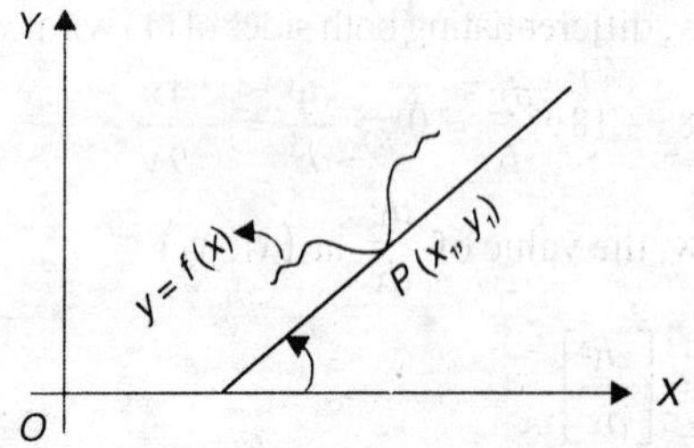

Now, the value of $\frac{dy}{dx}$ at $(x_1, y_1) = \frac{x_1}{y_1}$ Slope of the tangent ...(4)

Now, according to question, the slope of the curve at $(x_1, y_1) = 2$ ...(5)

$\because$ The slope of the curve = The slope of the tangent at (x_1, y_1).

$$\Rightarrow \frac{x_1}{y_1} = 2 \Rightarrow x_1 = 2y_1 \quad ...(6)$$

Now, putting (6) in (2), we get

$$4y_1^2 - y_1^2 = 2 \Rightarrow 3y_1^2 = 2 \quad y_1 = \pm\sqrt{\frac{2}{3}} \quad ...(7)$$

Now, from eqn (6), $x_1 = 2y_1 \Rightarrow x_1 = \pm 2\sqrt{\frac{2}{3}}$

$\therefore$ Required points are $(x_1, y_1) = \left(\pm 2\sqrt{\frac{2}{3}}, \pm\sqrt{\frac{2}{3}}\right)$

$$= \left(2\sqrt{\frac{2}{3}}, \sqrt{\frac{2}{3}}\right) \text{ and } \left(-2\sqrt{\frac{2}{3}}, -\sqrt{\frac{2}{3}}\right)$$

Question: Find the coordinates of the point on $y = x^3$, where the tangents through the point (0, 54) meet the curve.

Solution: Let (x_1, y_1) be the coordinates of the required point on the curve $y = x^3$ ...(1)

$$\therefore y_1 = x_1^3 \quad ...(2)$$

Now, differentiating (1) w.r.t x, we get $\frac{dy}{dx} = 3x^2$

Now, the value of $\frac{dy}{dx}$ at (x_1, y_1)

$$= \left[\frac{dy}{dx}\right]_{\substack{x = x_1 \\ y = y_1}} = 3x_1^2 \quad ...(3)$$

$\therefore$ The equation of the tangent at (x_1, y_1) is

$(y - y_1) = 3x_1^2 (x - x_1)$

Since this line passes through (0, 54)

$\therefore$ (0, 54) will satisfy the equation of tangent

$\therefore 54 - y_1 = 3x_1^2 (0 - x_1)$

$\Rightarrow y_1 - 54 = 3x_1^2 \cdot x_1$

$\Rightarrow y_1 = 3x_1^3 + 54$ ...(4)

Putting the value of y_1 from eqns (4) in (2), we get

$3x_1^3 + 54 = x_1^3 \Rightarrow 2x_1^3 = -54$

$\Rightarrow x_1^3 = -\frac{54}{2} = -27 \Rightarrow x_1 = -3$

$\therefore$ Required point is $(x_1, y_1) = (-3, -27)$

Question: Find the coordinates of the point where the tangent to the curve $y = x^2 + 3x + 4$ passes through the origin.

Solution: Let (x_1, y_1) be the coordinates of the required point on the curve $y = x^2 + 3x + 4$ ...(1)

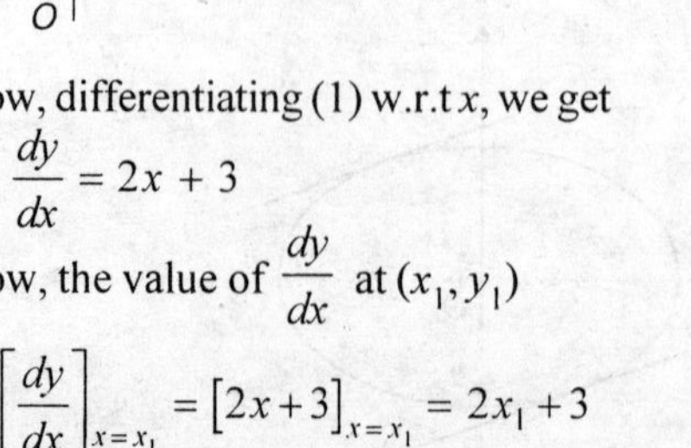

Now, differentiating (1) w.r.t x, we get

$$\frac{dy}{dx} = 2x + 3$$

Now, the value of $\frac{dy}{dx}$ at (x_1, y_1)

$$= \left[\frac{dy}{dx}\right]_{\substack{x = x_1 \\ y = y_1}} = [2x + 3]_{x = x_1} = 2x_1 + 3$$

= Slope of the tangent

Again the equation of the tangent at (x_1, y_1) is

$$(y - y_1) = \left[\frac{dy}{dx}\right]_{\substack{x = x_1 \\ y = y_1}} \cdot (x - x_1)$$

$$\Rightarrow (y - y_1) = (2x_1 + 3)\,(x - x_1) \qquad \ldots(2)$$

Now, the tangent (2) passes through the origin (0, 0)

$\Rightarrow (0 - y_1) = (2x_1 + 3)\,(0 - x_1)$

$\Rightarrow -y_1 = (2x_1 + 3)\,(-x_1)$

$\Rightarrow y_1 = (2x_1 + 3) \cdot x_1$

$\Rightarrow y_1 = (2x_1 + 3)\,x_1 = 2x_1^2 + 3x_1 \qquad \ldots(3)$

Again (x_1, y_1) lies on the curve $y = x^2 + 3x + 4$

$\therefore y_1 = x_1^2 + 3x_1 + 4 \qquad \ldots(4)$

Eqns (3) and (4) $\Rightarrow 2x_1^2 + 3x_1 = x_1^2 + 3x_1 + 4$

$\Rightarrow 2x_1^2 - x_1^2 = 3x_1 - 3x_1 + 4$

$\Rightarrow x_1^2 = 4 \Rightarrow x_1 = \pm 2$

Again, substituting the values of x_1 in (4), we get

$y_1 = 14$ for $x = 2$ and 2 for $x = -2$

Hence, the required points are (2, 14) and (–2, 2).

Question: Find the coordinates of the points at which the tangents to the curve $y = x^2 + 2x$ pass through origin.

Solution: Letting that there is a point (x_1, y_1) on the curve $y = x^2 + 2x \qquad \ldots(1)$

We have, $y_1 = x_1^2 + 2x_1 \qquad \ldots(2)$

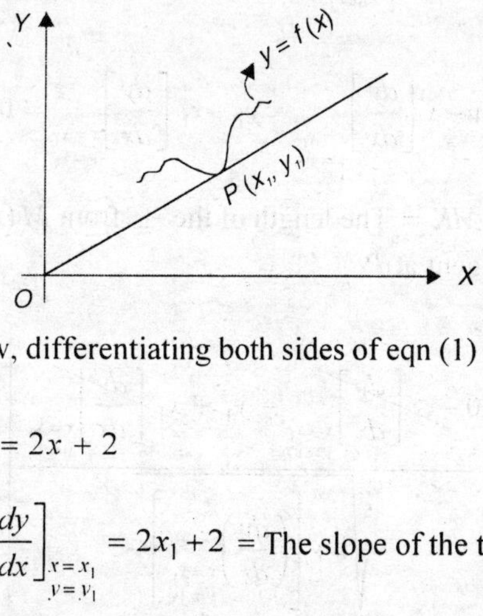

Now, differentiating both sides of eqn (1) w.r.t x, we get

$$\frac{dy}{dx} = 2x + 2$$

$$\therefore \left[\frac{dy}{dx}\right]_{\substack{x = x_1 \\ y = y_1}} = 2x_1 + 2 = \text{The slope of the tangent}$$

at $(x_1, y_1) \qquad \ldots(3)$

Now, the equation of the tangent at (x_1, y_1) is

$$(y - y_1) = \left[\frac{dy}{dx}\right]_{\substack{x = x_1 \\ y = y_1}} \cdot (x - x_1)$$

$$\Rightarrow (y - y_1) = (2x_1 + 2) \cdot (x - x_1) \qquad \ldots(4)$$

Now, the tangent passes through (0, 0)

$\Rightarrow (0 - y_1) = (2x_1 + 2)\,(0 - x_1)$

$\Rightarrow y_1 = 2x_1^2 + 2x_1$

Now eqns (2) and (5) $\Rightarrow 2x_1^2 + 2x_1 = x_1^2 + 2x_1$

$\Rightarrow 2x_1^2 - x_1^2 = 0 \Rightarrow x_1^2 = 0 \Rightarrow x_1 = 0$

Now putting $x_1 = 0$ in (2), we get,

$y_1 = 2x_1^2 + 2x_1 = 0$

$\therefore$ Required point $= (x_1, y_1) = (x_1, y_1) = (0, 0)$

Question: Find the coordinates of the points on the curve $y = 5 \log (3 + x^2)$ at which the slope is 2.

Solution: We are given $y = 5 \log (3 + x^2) \qquad \ldots(1)$

(1) $\Rightarrow y_1 = 5 \log (3 + x_1^2)$ provided (x_1, y_1) lies on the curve.

Now, differentiating (1) w.r.t x, we get

$$\frac{dy}{dx} = 5 \cdot \frac{1}{3 + x^2} \cdot (0 + 2x) = \frac{10x}{3 + x^2} \qquad \ldots(2)$$

$$\text{Again since } \frac{dy}{dx} = \tan \psi = 2 \text{ (given)} \qquad \ldots(3)$$

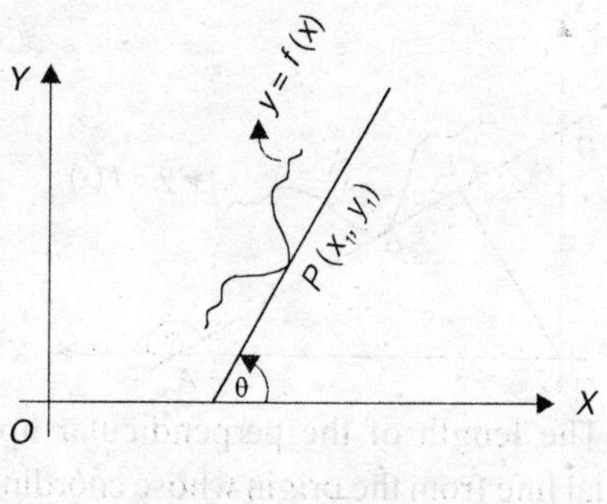

Now let the required point be (x_1, y_1)

$$\therefore \left[\frac{dy}{dx}\right]_{(x_1,\, y_1)} = 2$$

$$\text{and } \left[\frac{10x}{3 + x^2}\right]_{x = x_1} = \frac{10x_1}{3 + x_1^2} \qquad \ldots(4)$$

$$(3) \text{ and } (4) \Rightarrow \frac{10x_1}{3 + x_1^2} = 2 \Rightarrow 10x_1 = 6 + 2x_1^2$$

$\Rightarrow 5x_1 = 3 + x_1^2$

$\Rightarrow x_1^2 - 5x_1 + 3 = 0$

$\Rightarrow x_1 = \frac{5 \pm \sqrt{25-12}}{2} = \frac{5 \pm \sqrt{13}}{2}$

$\therefore$ Required points are (x_1, y_1)

$$= \left\{\frac{5+\sqrt{13}}{2}, \log\left(\frac{5+\sqrt{13}}{2}\right)^2\right\}$$

and $\left\{\frac{5-\sqrt{13}}{2}, 5\log\left\{3+\left(\frac{5-\sqrt{13}}{2}\right)^2\right\}\right\}$

Problems based on length of the perpendicular from the origin upon the tangent at (x_1, y_1).

$\because$ The equation of the tangent is

$$(y - y_1) = \left[\frac{dy}{dx}\right]_{\substack{x=x_1\\y=y_1}} (x - x_1)$$

$$\therefore x\left[\frac{dy}{dx}\right]_{\substack{x=x_1\\y=y_1}} - y + \left(y_1 - x_1\left[\frac{dy}{dx}\right]_{\substack{x=x_1\\y=y_1}}\right) = 0$$

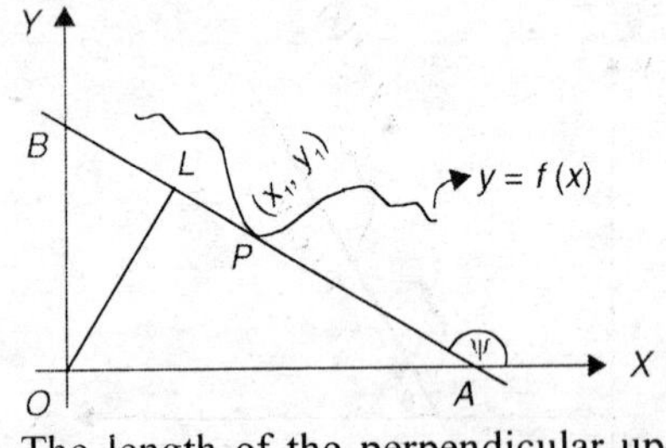

$\Rightarrow$ The length of the perpendicular upon this tangential line from the origin whose coordinates are (0, 0) is

$$= \left|\frac{y_1 - x_1\left[\frac{dy}{dx}\right]_{\substack{x=x_1\\y=y_1}}}{\sqrt{1+\left[\left(\frac{dy}{dx}\right)_{\substack{x=x_1\\y=y_1}}\right]^2}}\right|$$

> ***N.B.***: Since the length of perpendicular from (x_1, y_1) to any line $ax + by + c = 0$ is $\left|\frac{ax_1 + by_1 + c}{\sqrt{a^2 + b^2}}\right|$ since length is always positive.

***Examples worked out*:**

Question: Find the length of perpendicular from the foot of the ordinate upon the tangent to the curve $y = f(x)$.

Solution: Let us draw $PM \perp$ to x-axis $MK \perp$ to the tangent PB.

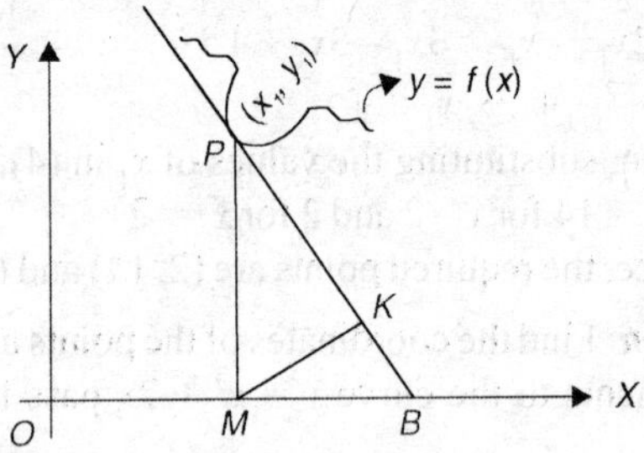

The equation of the tangent at $P(x_1, y_1)$ is

$$(y - y_1) = \left[\frac{dy}{dx}\right]_{\substack{x=x_1\\y=y_1}} \cdot (x - x_1)$$

$$\Rightarrow y - x\left[\frac{dy}{dx}\right]_{\substack{x=x_1\\y=y_1}} - y_1 + x_1\left[\frac{dy}{dx}\right]_{\substack{x=x_1\\y=y_1}} = 0$$

$\Rightarrow MK =$ The length of the $\perp$ from $M(x, 0)$ on the tangent at P

$$= \left|\frac{0 - \cancel{x_1\left[\frac{dy}{dx}\right]_{\substack{x=x_1\\y=y_1}}} - y_1 + \cancel{x_1\left[\frac{dy}{dx}\right]_{\substack{x=x_1\\y=y_1}}}}{\sqrt{1+\left[\left(\frac{dy}{dx}\right)_{\substack{x=x_1\\y=y_1}}\right]^2}}\right|$$

$$= \left| \frac{-y_1}{\sqrt{1 + \left[\left(\frac{dy}{dx}\right)_{\substack{x=x_1\\y=y_1}}\right]^2}} \right| = \frac{|y_1|}{\sqrt{1 + \left[\left(\frac{dy}{dx}\right)_{\substack{x=x_1\\y=y_1}}\right]^2}}$$

Question: Show that the normal to the curve

$x = a\,(\cos\theta + \theta \sin\theta)$

$y = a(\sin\theta - \theta\cos\theta)$

at any point θ is at a constant distance from the origin.

Solution: Let (x_1, y_1) be any point on the given curve where the parametric equation is

$$\left.\begin{aligned} x &= a(\cos\theta + \theta\sin\theta) \\ y &= a(\sin\theta - \theta\sin\theta) \end{aligned}\right\} \quad ...(1)$$

$$\left.\begin{aligned} \therefore\ x_1 &= a\left(\cos\theta_1 - \theta_1 \sin\theta_1\right) \\ y_1 &= a\left(\sin\theta_1 - \theta_1 \cos\theta_1\right) \end{aligned}\right\} \quad ...(2)$$

for some value of $\theta = \theta_1$

$$\frac{dy}{d\theta} = \frac{d\left[a(\sin\theta - \theta\cos\theta)\right]}{d\theta} = a\frac{d\sin\theta}{d\theta} - a\frac{d(\theta\cos\theta)}{d\theta}$$

$= a\cos\theta - a\left[1\cdot\cos\theta + (-\theta\sin\theta)\right]$

$= a\cos\theta - a\cos\theta + a\theta\sin\theta = a\theta\sin\theta \quad ...(3)$

Again $\frac{dx}{d\theta} = \frac{d}{d\theta}\left[a(\cos\theta + \theta\sin\theta)\right]$

$= a\left\{\frac{d\cos\theta}{d\theta}\right\} + a\left\{\frac{d(\theta\sin\theta)}{d\theta}\right\}$

$= -a\sin\theta + a\{\sin\theta + \theta\cos\theta\}$

$= -a\sin\theta + a\sin\theta + a\theta\cos\theta$

$= a\theta\cos\theta \quad ...(4)$

$\therefore \frac{dy}{dx} = \frac{dy}{d\theta} \div \frac{dx}{d\theta} = \frac{a\theta\sin\theta}{a\theta\cos\theta} = \tan\theta \quad ...(5)$

$$\left[\frac{dy}{dx}\right]_{\substack{x=x_1\\y=y_1}} = [\tan\theta]_{\theta=\theta_1} = \tan\theta_1 \quad ...(6)$$

Now, the equation of the normal at any point θ on the given curve is given by

$$(y - y_1) = -\left[\frac{1}{\text{Slope of the tangent}}\right]\cdot(x - x_1)$$

$\Rightarrow y - a(\sin\theta_1 - \theta_1\cos\theta_1)$

$= -\frac{\cos\theta_1}{\sin\theta_1}\left[x - a(\cos\theta_1 + \theta_1\sin\theta_1)\right]$

$\Rightarrow \sin\theta_1\left[y - a(\sin\theta_1 - \theta_1\cos\theta_1)\right]$

$= -\cos\theta_1\left[x - a(\cos\theta_1 + \theta_1\sin\theta_1)\right]$

$\Rightarrow y\sin\theta_1 - a\sin^2\theta_1 + a\theta_1\sin\theta_1\cos\theta_1$

$= -x\cos\theta_1 + a\cos^2\theta_1 + a\theta_1\sin\theta_1\cos\theta_1$

$\Rightarrow x\cos\theta_1 + y\sin\theta_1 = a\left(\cos^2\theta_1 + \sin^2\theta_1\right) = a \quad ...(7)$

Now, the length of $\perp$ from (0, 0) to the normal (7)

is $= \left|\frac{0\cdot\cos\theta_1 + 0\cdot\sin\theta_1 - a}{\sqrt{\cos^2\theta_1 + \sin^2\theta_1}}\right|$

$= |a|$ which is independent of the value θ_1 of θ and hence a constant for all values of θ.

Working rule to find the coordinates of a point on the curve whose parametric equations are given and the tangential line passes through that point such that some condition is imposed on that tangent.

Working rule:

1. Find $\frac{dy}{dx}$ using the formula $\frac{dy}{dx} = \frac{dy}{dx} \div \frac{dt}{dx}$

$$= \frac{dy/dt}{dx/dt} = \frac{f_1'(t)}{f_2'(t)}$$

2. Let (x_1, y_1) be a point on the curve corresponding to the parameter θ [or t, u, v etc. whichever is given in the problem $x = f_1(\theta),\ y = f_2(\theta)] = \theta_1$ [or, t_1, u_1, v_1, etc.].

3. Find $\left[\dfrac{dy}{dx}\right]_{\theta=\theta_1}$ = The slope of the tangent at θ_1 [or, u_1, t_1, v_1 etc.] from the given condition and solve for θ_1 [or, u_1, t_1, v_1 etc.].

4. Find the values of x_1 and y_1 from the equations
$x_1 = f_1(\theta_1), y_1 = f_2(\theta_2)$
by putting the values of parameter θ_1 [or, u_1, t_1, v_1 etc.].

Problems based on finding the coordinates of a point on the curve whose parametric equations are given.

***Examples worked out*:**

***Question*:** Find the coordinates of the point on the curve

$$x = a(\theta + \sin\theta),\ y = a(1 - \cos\theta)$$

where the tangent is inclined at an angle of $\dfrac{\pi}{4}$ to the x-axis.

***Solution*:** Equation of the curve is

$$\left.\begin{aligned} x &= a(\theta + \sin\theta) \\ y &= a(1 - \cos\theta) \end{aligned}\right\} \quad \ldots(1)$$

Let (x_1, y_1) be a point on the curve corresponding to $\theta = \theta_1$

$$\therefore \left.\begin{aligned} x_1 &= a(\theta_1 + \sin\theta_1) \\ y_1 &= a(1 - \cos\theta_1) \end{aligned}\right\} \quad \ldots(2)$$

Now, differentiating (1) w.r.t θ, we get

$$\left.\begin{aligned} \frac{dx}{d\theta} &= a(1 + \cos\theta) \\ \frac{dy}{d\theta} &= a\sin\theta \end{aligned}\right\}$$

$$\Rightarrow \frac{dy}{dx} = \frac{a\sin\theta}{a(1+\cos\theta)} = \frac{a\sin\theta}{a\left(1 + 2\cos^2\dfrac{\theta}{2} - 1\right)}$$

$$= \frac{a \cdot 2\sin\dfrac{\theta}{2} \cdot \cos\dfrac{\theta}{2}}{a \cdot 2\cos\dfrac{\theta}{2} \cdot \cos\dfrac{\theta}{2}} = \tan\frac{\theta}{2} \quad \ldots(3)$$

Now, the value of $\dfrac{dy}{dx}$ at $P(x_1, y_1) = \left[\tan\dfrac{\theta}{2}\right]_{\theta=\theta_1}$

$$= \tan\frac{\theta_1}{2} \text{ where } \theta_1 \text{ is the parameter at } P. \quad \ldots(4)$$

$\therefore$ The slope of the tangent at $(x_1, y_1) = \tan\dfrac{\theta_1}{2}$

$$\text{Let } \left[\frac{dy}{dx}\right]_P = \tan\frac{\pi}{4} = 1 \quad \ldots(5)$$

[At (x_1, y_1), the tangential line makes $\dfrac{\pi}{4}$ angle with x-axis]

$$\text{Eqns (4) and (5)} \Rightarrow \tan\frac{\theta_1}{2} = \tan\frac{\pi}{4}$$

$$\Rightarrow \frac{\theta_1}{2} = \frac{\pi}{4} \Rightarrow \theta_1 = \frac{\pi}{2} \quad \ldots(6)$$

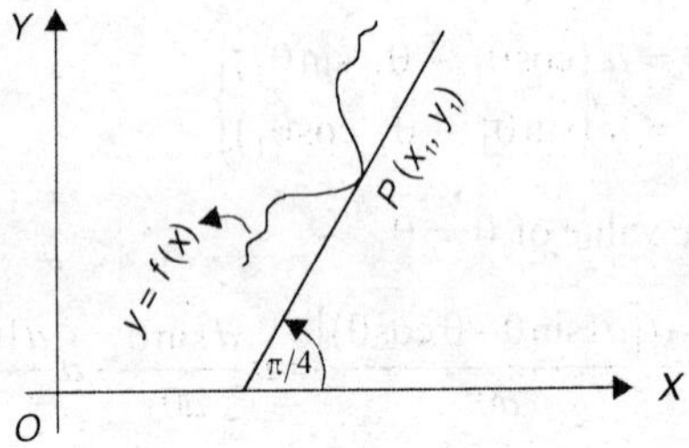

From (2), for $\theta_1 = \dfrac{\pi}{2}$

$$x_1 = a\left(\frac{\pi}{2} + \sin\frac{\pi}{2}\right) = a\left(\frac{\pi}{2} + 1\right)$$

$$y_1 = a\left(1 - \cos\frac{\pi}{2}\right) = a$$

$\therefore$ Required point is $(x_1, y_1) = \left[a\left(\dfrac{\pi}{2} + 1\right), a\right]$

***Question*:** Find the co-ordinates of the points where the tangents to the curve given in parametric form

$$x = 2u^4 + u$$

$y = u^4 - 2u^2 + 1$ are parallel to x-axis.

Solution: Let (x_1, y_1) be a point on the curve corresponding to $u = u_1$

$\therefore y = u^4 - 2u^2 + 1, x = 2u^4 + u$ ...(1)

$\Rightarrow y_1 = u_1^{\ 4} - 2u_1^{\ 2} + 1, x_1 = 2u_1^{\ 4} + u_1$ ...(2)

Now, differentiating the equation (1) w.r.t u, we get

$$\frac{dy}{du} = 4u^3 - 4u, \frac{dx}{du} = 8u^3 + 1 \quad ...(3)$$

$$\therefore \frac{dy}{dx} = \frac{dy}{du} \div \frac{dx}{du} = \frac{4(u^3 - u)}{8u^3 + 1}$$

Now, $\left[\frac{dy}{dx}\right]_{\substack{x = x_1 \\ y = y_1}}$

$$= \left[\frac{4(u^3 - u)}{8u^3 + 1}\right]_{u = u_1} = \frac{4(u_1^3 - u_1)}{8u_1^3 + 1} \quad ...(4)$$

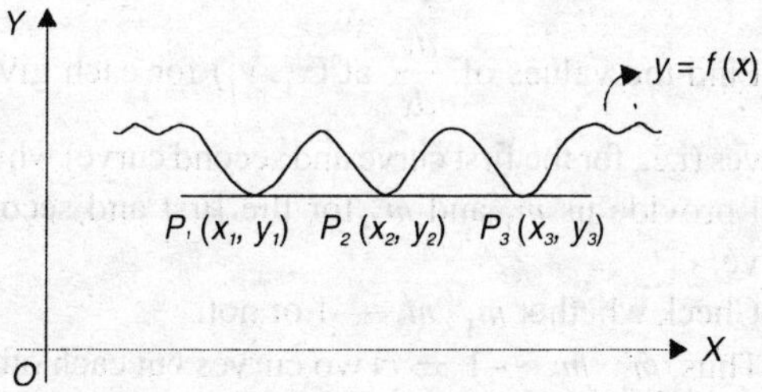

Again since the tangents are parallel to x-axis

$$\therefore \left[\frac{dy}{dx}\right]_{\substack{x = x_1 \\ y = y_1}} = 0 \quad ...(5)$$

Equating (4) and (5)

$$\frac{4(u_1^3 - u_1)}{8u_1^3 + 1} = 0 \Rightarrow u_1 = 0 \text{ or } u_1 = 1 \text{ or } u_1 = -1$$

Now $u_1 = 0 \Rightarrow x_1 = 2u_1^{\ 4} + u_1 = 0$,

$y_1 = u_1^{\ 4} - 2u_1^{\ 2} + 1 = 1 \ \therefore (x_1, y_1) = (0, 1)$

$u_1 = 1 \Rightarrow x_1 = 3, y_1 = 0 \ \therefore (x_1, y_1) = (3, 0)$

$u_1 = -1 \Rightarrow x_1 = 1, y_1 = 0 \ \therefore (x_1, y_1) = (1, 0)$

Thus, the required points are (0, 1), (3, 0) and (1, 0).

Question: Show that in the ellipse $\frac{x^2}{a^2} + \frac{y^2}{b^2} = 1$, the length of the normal varies inversely as the perpendicular from the origin.

Solution: Let (x_1, y_1) be a point of the curve whose equation is $\frac{x^2}{a^2} + \frac{y^2}{b^2} = 1$ ...(1)

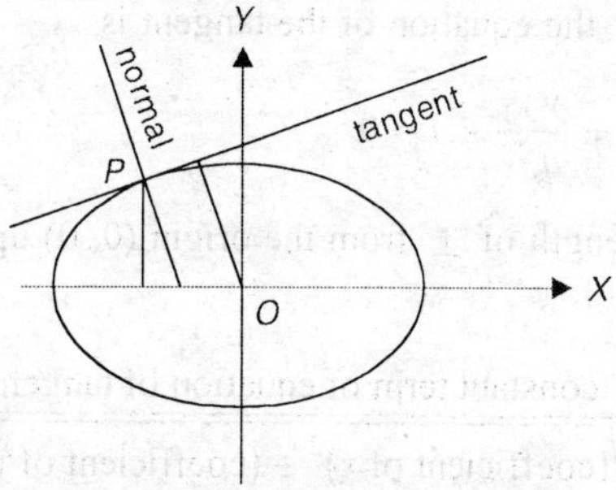

Now, differentiating (1) w.r.t x, we get

$$\frac{2x}{a^2} + \frac{2y}{b^2}\frac{dy}{dx} = 0 \Rightarrow \frac{dy}{dx} = -\frac{b^2 x}{a^2 y}$$

$\Rightarrow$ The slope at (x_1, y_1)

$$= \left[\frac{dy}{dx}\right]_{\substack{x = x_1 \\ y = y_1}} = -\frac{b^2 x_1}{a^2 y_1} \quad ...(3)$$

$\Rightarrow$ The slope of the normal at (x_1, y_1)

$$= \frac{a^2 y_1}{b^2 x_1} \quad ...(4)$$

$\therefore$ The equation of the normal is

$$(y - y_1) = \frac{a^2 y_1}{b^2 x_1}(x - x_1)$$

$\Rightarrow b^2 x_1 y - b^2 x_1 y_1 = a^2 y_1 x - a^2 x_1 y_1$

$\Rightarrow b^2 x_1 y + a^2 x_1 y_1 = a^2 y_1 x + b^2 x_1 y_1$

$\Rightarrow x_1 (b^2 y + a^2 y_1) = y_1 (a^2 x + b^2 x_1)$ ...(5)

Now, the equation of the tangent is

$$(y - y_1) = -\frac{b^2 x_1}{a^2 y_1} \cdot (x - x_1)$$

$\Rightarrow a^2 y y_1 - a^2 y_1^2 = -b^2 x x_1 + b^2 x_1^2$

$\Rightarrow b^2 x x_1 + a^2 y y_1 = b^2 x_1^2 + a^2 y_1^2$

[Dividing both sides by $a^2 b^2$]

$$\Rightarrow \frac{xx_1}{a^2} + \frac{yy_1}{b^2} = \frac{x_1^2}{b^2} + \frac{y_1^2}{a^2} = 1 \quad ...(6)$$

[from (2)]

Thus, the equation of the tangent is

$$\frac{xx_1}{a^2} + \frac{yy_1}{b^2} - 1 = 0$$

$\therefore$ Length of $\perp$ from the origin (0, 0) upon the tangent

$$= \left| \frac{\text{constant term of equation of tangent}}{\sqrt{(\text{coefficient of } x)^2 + (\text{coefficient of } y)^2}} \right|$$

$$\Rightarrow p = \left| \frac{-1}{\sqrt{\left(\frac{x_1}{a^2}\right)^2 + \left(\frac{y_1}{b^2}\right)^2}} \right| = \frac{1}{\sqrt{\frac{x_1^2}{a^4} + \frac{y_1^2}{b^4}}}$$

$$= \frac{1}{\sqrt{\frac{x_1^2 b^4 + y_1^2 a^4}{a^4 b^4}}} = \frac{1}{\frac{\sqrt{x_1^2 b^4 + y_1^2 a^4}}{a^2 b^2}}$$

$$= \frac{a^2 b^2}{\sqrt{x_1^2 b^4 + a^4 y_1^2}} \quad ...(7)$$

Now, length of the normal

$$= y_1 \sqrt{1 + \left[\left(\frac{dy}{dx}\right)_{\substack{x = x_1 \\ y = y_1}}\right]^2} = y_1 \sqrt{1 + \frac{b^4 x_1^2}{a^4 y_1^2}}$$

$$\Rightarrow l = y_1 \sqrt{\frac{a^4 y_1^2 + b^4 x_1^2}{a^4 y_1^2}} = \frac{y_1}{a^2 y_1} \sqrt{a^4 y_1 + b^4 x_1^2}$$

$$\Rightarrow l = \frac{1}{a^2} \cdot \sqrt{a^4 y_1^2 + b^4 x_1^2}$$

Hence, $l \times p = \dfrac{a^2 b^2}{\sqrt{x_1^2 b^4 + a^4 y_1}} \times \dfrac{1}{a^2} \cdot \sqrt{a^4 y_1 + b^4 x_1^2}$

which is a constant.

$$\therefore l = \frac{b^2}{p}$$

$$\therefore l \propto \frac{1}{p}$$

Problems based on showing that two curves cut each other orthogonally whose implicit equations are $f_1(x, y) = 0$ or constant and $f_2(x, y) = 0$ or constant

Working rule:

1. Mark the given equations $f_1(x, y) = 0$ or constant as first curve and $f_2(x, y) = 0$ or constant, second curve.

2. Find $\dfrac{dy}{dx}$ for the first curve and the second curve.

3. Let (x_1, y_1) be the point of intersection of two given curves.

4. Find the values of $\dfrac{dy}{dx}$ at (x_1, y_1) for each given curves (i.e., for the first curve and second curve) which will provide us m_1 and m_2 for the first and second curve.

5. Check whether $m_1 \cdot m_2 = -1$ or not.

Thus, $m_1 \cdot m_2 = -1 \Rightarrow$ Two curves cut each other orthogonally and $m_1 \cdot m_2 \neq -1 \Rightarrow$ Two curves do not cut each other orthogonally.

***N.B.*: 1.** When the given two equations can be solved simultaneously by using simultaneous equations methods (i.e., by elimination/comparison of coefficients of x and y/equating two equations/ additions and subtraction method/...etc), we should first of all find the point of intersection of two given curves at which are find the values of $\dfrac{dy}{dx}$ for the first curve and second curve and then we should check whether $m_1 \cdot m_2 = -1$ or not.

2. Whenever it is not possible to find the point of intersection by using simultaneous equations method, then we find the values of $\dfrac{dy}{dx}$ for each given equation at (x_1, y_1) which is supposed to be the point of intersection of given curves (i.e., the first curve and second curve).

Examples worked out:

Question: Show that the curves $x^3 - 3x y^2 = a$ and $3x^2 y - y^3 = b$ cut each other orthogonally where a and b are constants.

Solution: Let (x_1, y_1) be the point of intersection of two curves whose equations are

$$x^3 - 3x y^2 = a \quad \ldots(1)$$

$$3x^2 y - y^3 = b \quad \ldots(2)$$

Now, differentiating both sides of equation (1) w.r.t x, we get

$$3x^2 - 3\left(1 \cdot y^2 + x \cdot 2y\frac{dy}{dx}\right) = 0$$

$$\Rightarrow x^2 - y^2 - 2xy\frac{dy}{dx} = 0$$

$$\Rightarrow \frac{dy}{dx} = \frac{x^2 - y^2}{2xy}$$

$$\Rightarrow \left[\frac{dy}{dx}\right]_{\substack{x=x_1 \\ y=y_1}} = \frac{x_1^2 - y_1^2}{2x_1 y_1} = m_1 \quad \ldots(3)$$

Again differentiating both sides of equation (2) w.r.t x, we have

$$3\left(2xy + x^2\frac{dy}{dx}\right) - 3y^2\frac{dy}{dx} = 0$$

$$\Rightarrow \left(x^2 - y^2\right)\frac{dy}{dx} + 2xy = 0$$

$$\Rightarrow \frac{dy}{dx} = -\frac{2xy}{x^2 - y^2}$$

$$\Rightarrow \left[\frac{dy}{dx}\right]_{(x_1, y_1)} = -\frac{2x_1 y_1}{x_1^2 - y_1^2} = m_2 \quad \ldots(4)$$

Now $m_1 \cdot m_2 = \dfrac{-2x_1 y_1}{x_1^2 - y_1^2} \times \dfrac{x_1^2 - y_1^2}{2x_1 y_1} = -1$ which means given curves cut each other orthgonally.

Question: Show that the curves $y = x^3$ and $6y = 7 - x^2$ intersect orthogonally.

Solution: Given equations of the curves are

$$y = x^3 \quad \ldots(1)$$

$$6y = 7 - x^2 \quad \ldots(2)$$

Solving the equations (1) and (2) simultaneously by eliminating y, we get $6x^3 = 7 - x^2$

$\Rightarrow 6x^3 + x^2 - 7 = 0$

$\Rightarrow (x - 1)(6x^2 + 7x + 7) = 0$

$\Rightarrow$ Either $(x - 1) = 0$ or $(6x^2 + 7x + 7) = 0$

$(x - 1) = 0 \Rightarrow x = 1$

$(6x^2 + 7x + 7) = 0$

$$\Rightarrow x = \frac{-b \pm \sqrt{b^2 - 4ac}}{2a} = \frac{-7 \pm \sqrt{49 - 4 \times 6 \times 7}}{2 \times 6}$$

$$= \frac{-7 \pm \sqrt{49 - 168}}{12} \text{ which are imaginary.}$$

The above explanation implies only root under consideration is $x = 1$ since imaginary root is not considered $\Rightarrow y = 1$ ($\because y = x^3 \Rightarrow y = 1^3 = 1$)

$\therefore$ Required point of intersection $= P(x_1, y_1) = P(1, 1)$

Now, $m_1 = \left[\dfrac{dy}{dx}\right]_{P(1,1)}$ from the first curve

$$= \left[\frac{dx^3}{dx}\right]_{P(1,1)} = \left[3x^2\right]_{x=1} = 3$$

$m_2 = \left[\dfrac{dy}{dx}\right]_{P(1,1)}$ from the second curve

$$= \left[\frac{d\left(\frac{7}{6} - \frac{x^2}{6}\right)}{dx}\right]_{P(1,1)} = \left[-\frac{2x}{6}\right]_{\substack{x=1 \\ y=1}} = -\frac{1}{3}$$

$\therefore m_1 \cdot m_2 = 3 \times \left(-\dfrac{1}{3}\right) = -1$ which means two given curves cut orthogonally.

Question: Show that that curves $x^2 = 2y$ and $6y = 5 - 2x^3$ intersect orthogonally.

Solution: Letting (x_1, y_1) to be the point of intersection of the given curves

$x^2 = 2y$ ...(1)

$6y = 5 - 2x^3$ ...(2)

and solving the equations of the given curves by eliminating y from (1), (2) we find $6 \times \frac{x^2}{2} = 5 - 2x^3$

$\Rightarrow 3x^2 + 2x^3 - 5 = 0 \Rightarrow 2x^3 + 3x^2 - 5 = 0$

$2x^3 + 3x^2 - 5 = 0$

$\Rightarrow 2x^3 - 2x^2 + 5x^2 - 5x + 5x - 5 = 0$

$\Rightarrow (x - 1)(2x^2 + 5x + 5) = 0$

$\therefore x = 1$ or $x = \frac{-5 \pm \sqrt{24 - 40}}{4}$

which is imaginary.

Hence $x = 1$ and $y = \left[\frac{x^2}{2}\right]_{x=1} = \frac{1}{2}$

$\therefore$ Required point $= (x_1, y_1) = \left(1, \frac{1}{2}\right)$

Now, $2y = x^2 \Rightarrow y = \frac{x^2}{2} \Rightarrow \frac{dy}{dx} = \frac{2x}{2} = x$

$\Rightarrow m_1 = [x]_{\substack{x=1 \\ y=\frac{1}{2}}} = 1$

Again $y = \frac{5 - 2x^3}{6} = \frac{5}{6} - \frac{1}{3}x^3$

$\Rightarrow \frac{dy}{dx} = 0 - \frac{1}{3} \cdot 3x^2 = -x^2$

$\therefore m_2 = \left[-x^2\right]_{\substack{x=1 \\ y=\frac{1}{2}}} = -1$

$\therefore m_1 \cdot m_2 = -1 \times 1 = -1$ which means the given curves cut each other orthogonally.

Question: Find the condition that the curves $ax^2 + by^2 = 1$ and $a_1 x^2 + b_1 y^2 = 1$ may intersect orthogonally.

Solution: Letting (x_1, y_1) to be the point of intersection of the curves

$ax^2 + by^2 = 1$ ...(1)

$a_1 x^2 + b_1 y^2 = 1$ ...(2)

we have, $ax_1^2 + by_1^2 = 1 = a_1 x_1^2 + b_1 y_1^2$

[$\because$ lies on the curve]

$\Rightarrow ax_1^2 - a_1 x_1^2 = b_1 y_1^2 - by_1^2$

$\Rightarrow x_1^2 (a - a_1) = y_1^2 (b_1 - b)$

$\Rightarrow \frac{x_1^2}{y_1^2} = \frac{b_1 - b}{a - a_1}$

Alternatively,

(x_1, y_1) lies on the curves

$\Rightarrow (x_1, y_1)$ satisfies the equations (1) and (2)

(1) $\Rightarrow a x_1^2 + b y_1^2 - 1 = 0$ ...(3)

(2) $\Rightarrow a_1 x_1^2 + b_1 y_1^2 - 1 = 0$ ...(4)

Now, solving eqns (3) and (4) simultaneously by cross-multiplication rule:

$\Rightarrow \frac{x_1^2}{b_1 - b} = \frac{y_1^2}{a - a_1} = \frac{1}{ab_1 - a_1 b}$

$$\left.\begin{aligned} &\Rightarrow x_1^2 = \frac{b_1 - b}{ab_1 - a_1 b} \\ &\text{and } y_1^2 = \frac{a - a_1}{ab_1 - a_1 b} \end{aligned}\right\} \Rightarrow \frac{x_1^2}{y_1^2} = \frac{b_1 - b}{a - a_1} \quad \text{...(5)}$$

Now, differentiating the equation (1) w.r.t x, we get

$2ax + 2by\frac{dy}{dx} = 0$

$\Rightarrow \frac{dy}{dx} = \frac{-ax}{by} \Rightarrow m_1 = \left[\frac{-ax}{by}\right]_{\substack{x=x_1 \\ y=y_1}} = \frac{-ax_1}{by_1}$

Similarly, differentiating the equation (2) w.r.t x, we get

$2a_1 x + 2b_1 y \frac{dy}{dx} = 0$

$\Rightarrow \frac{dy}{dx} = -\frac{a_1 x}{b_1 x}$

$\Rightarrow m_2 = \left[\frac{-a_1 x}{b_1 y}\right]_{\substack{x=x_1 \\ y=y_1}} = \frac{-a_1 x_1}{b_1 y_1}$

The two curves intersect orthogonally at (x_1, y_1)

$\Rightarrow m_1 \cdot m_2 = -1 \Rightarrow \left(\frac{-a_1 x_1}{b_1 y_1}\right) \times \left(-\frac{ax_1}{by_1}\right) = -1$

$\Rightarrow a a_1 x_1^2 = -b b_1 y_1^2 \Rightarrow \frac{x_1^2}{y_1^2} = -\frac{b b_1}{a a_1}$ ...(6)

Eqns (5) and (6)

$$\Rightarrow \frac{b_1-b}{a-a_1} = -\frac{b\,b_1}{a\,a_1} \Rightarrow \frac{a-a_1}{b_1-b} = -\frac{a\,a_1}{b\,b_1}$$

$$\Rightarrow \frac{a-a_1}{a\,a_1} = \frac{b_1-b}{-b\,b_1}$$

$$\Rightarrow \frac{1}{a_1} - \frac{1}{a} = \frac{1}{b_1} - \frac{1}{b}$$

which is the required condition.

Remember: **1.** The algebraic form of a condition is always a relation (or, an equation) among the given constant. Thus when we are asked to find the condition, we mean to find the algebraic form of a condition.

2. The student should know that by the conditions under which a straight line touches a given curve is meant the relation that must exist among the constants occurring in the equations of the straight line and the curve in order that they may touch each other.

Question: Find the angle between the curves $y = f_1(x)$ and $y = f_2(x)$ at (x_1, y_1), the point of intersection of the curves.

Or, find the angle between the curves $f_1(x, y) = 0$ constant and $f_2(x, y) = 0$ at (x_1, y_1), the point of intersection of the curves.

Solution: Supposing that $y = f_1(x)$ or $f_1(x, y) = 0$ and $y = f_2(x)$ or $f_2(x, y) = 0$ are the two equations of the curves C_1 and C_2 respectively intersection at $P(x_1, y_1)$. Let the tangents PT_1 and PT_2 to the to the curves C_1 and C_2 make angles of inclination i_1 and i_2 respectively with x-axis.

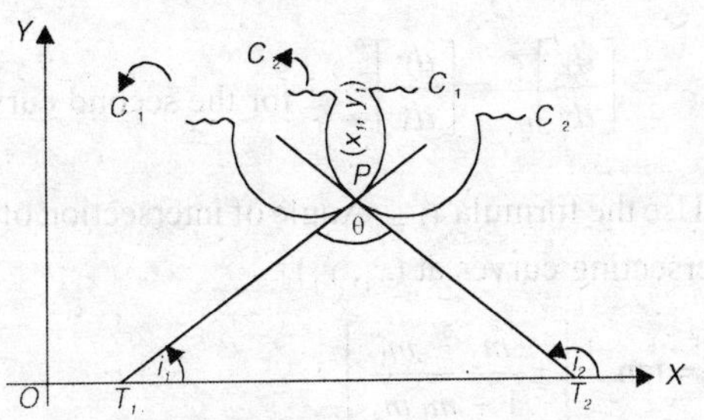

Again let the angle between the two tangents = θ

Now, the slope of the tangent to the curve $y = f_1(x)$ or $f_1(x, y) = 0$ at (x_1, y_1)

$$= \left[\frac{dy}{dx}\right]_{\substack{x=x_1\\y=y_1}} = \left[\frac{dy}{dx}\right]_{C_1} = \tan i_1 = m_1$$

Again the slope of the tangent to the curve $y = f_2(x)$ or $f_2(x, y) = 0$ at (x_1, y_1)

$$= \left[\frac{dy}{dx}\right]_{\substack{x=x_1\\y=y_1}} = \left[\frac{dy}{dx}\right]_{C_2} = \tan i_2 = m_2$$

Now, angle of intersection of the curves

$$= \theta = i_2 \sim i_1 \qquad [\because i_2 = \theta + i_1]$$

$$\Rightarrow \tan\theta = \tan(i_2 \sim i_1) = \frac{m_2 \sim m_1}{1 + m_1 m_2}$$

$$\Rightarrow \theta = \tan^{-1}\left[\frac{m_2 \sim m_1}{1 + m_1 m_2}\right] \text{ when } m_1 m_2 \neq -1$$

$$\text{Also } \tan(\pi - \theta) = -\tan\theta = \left[-\frac{m_1 \sim m_2}{1 + m_1 m_2}\right]$$

Hence, the angles between the curves

$$= \theta = \tan^{-1}\left[\pm\frac{m_1 \sim m_2}{1 + m_1 m_2}\right]$$

Remember:

1. When the curves touch, the angle of intersection $= 0 \Rightarrow$ when the slopes of the tangents are same, the curves touch each other.

2. When the angle between the tangents is a right angle, the curves are said to cut each other orthogonally $\Rightarrow$ If $m_1 m_2 = -1$, then the curves are said to cut each other orthogonally.

3. If either m_1 or m_2 (or both of them) is infinity, then it is advisable to express m_1 and m_2 in the form of $\tan\theta$. In other words,

If $m_1 = \tan i_1$ and $m_2 = \tan i_2$, then the angle between the two tangents to the two intersecting curves at (x_1, y_1) is

$\theta = |i_1 - i_2|$ = absolute value of difference of i_1 and i_2 provided m_1 or m_2 is infinity or both of them are infinity.

4. In general $\left[\frac{dy}{dx}\right]_{C_1} = \left[\frac{dy}{dx}\right]_{(x_1, y_1)} = 0 \Rightarrow \tan\psi_1 = 0$

$\Rightarrow \psi_1 = 0$ for the curve C_1. ...(i)

and $\left[\frac{dx}{dy}\right]_{C_2} = \left[\frac{dx}{dy}\right]_{(x_1, y_1)} = 0$

$\Rightarrow \psi_2 = 90°$ for the curve C_2. ...(ii)

Hence from (i) and (ii), we get the angle of intersection between the two curves is

$$\theta = \psi_2 - \psi_1 = 90°.$$

In particular (a) If the slope of the tangent line drawn to any curve $y = f_1(x)$ at the point (0, 0) is zero, then the tangent to the curve $y = f_1(x)$ at the point (0, 0) is x-axis (i.e., $m_1 = 0 \Rightarrow$ The tangent to the curve is x-axis).

(b) If the slope of the tangent line drawn to the curve $y = f_2(x)$ at the origin (0, 0) is infinity (i.e., $\frac{dx}{dy} = 0$ at origin) then the tangent to the curve $y = f_2(x)$ at the point (0, 0) is y-axis.

(a) and (b) $\Rightarrow$ The curves $y = f_1(x)$ and $y = f_2(x)$ at (0, 0) intersect orthogonally. Since $m_1 = 0 \Rightarrow \tan\psi_1 = 0 \Rightarrow \psi_1 = 0$ and $\left(\frac{dx}{dy}\right)_{(0,0)} = 0$

$\Rightarrow \psi_2 = 90°$

$\therefore \theta = \psi_2 - \psi_1 = 90°$.

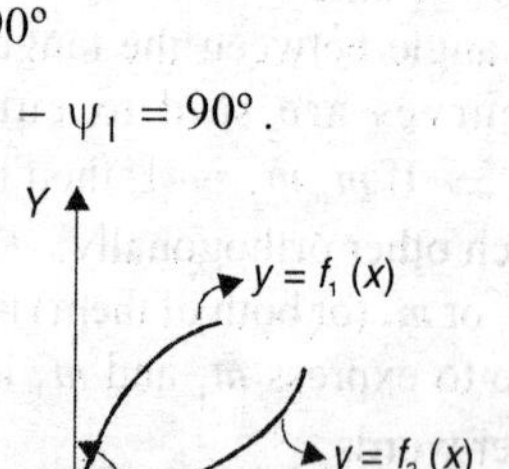

3. Angle between two curves/angle of intersection of two curves implies the angle between their tangent lines drawn to the curves at a common point known as point of intersection of two curves.

4. When two curves intersect each other, at the point of intersection, each curve has a tangent separate and distinct (unless they touch there). The tangents are always inclined to each other at some angle called the angle of intersection of two curves. When the angle between tangents is a right angle, the intersection is called orthogonal.

Question: How to find the angle of intersection between two curves $y = f_1(x)$ or $f_1(x, y) = 0$ and $y = f_2(x)$ or $f_2(x, y) = 0$

Working Rule:

1. Let (x_1, y_1) be the point or intersection of two given curves $y = f_1(x)$ or $f_1(x, y) = 0$ and $y = f_2(x)$ or $f_2(x, y) = 0$

2. Find the coordinates of the point of intersection of two given curves by putting (x_1, y_1) in the given two equations of the curves and solving these two equations to find (x_1, y_1).

3. Find $\frac{dy}{dx}$ of both curves (differentiate both equations of the curves w.r.t x).

4. Put the numerical values of the coordinates of the point of intersection (x_1, y_1) of the curves in $\frac{dy}{dx}$ for both equations of the curve and call these values as m_1 and m_2.

Where m_1 = The value of $\frac{dy}{dx}$ at (x_1, y_1)

$$= \left[\frac{dy}{dx}\right]_{C_1} = \left[\frac{dy}{dx}\right]_{\substack{x = x_1 \\ y = y_1}} \text{ for the first curve}$$

m_2 = The value of $\frac{dy}{dx}$ at (x_1, y_1)

$$= \left[\frac{dy}{dx}\right]_{C_2} = \left[\frac{dy}{dx}\right]_{\substack{x = x_1 \\ y = y_1}} \text{ for the second curve}$$

5. Use the formula θ = Angle of intersection of two intersecting curves at (x_1, y_1)

$$= \tan^{-1}\left[\pm\frac{m_2 \sim m_1}{1 + m_1 m_2}\right]$$

Note: **1.** $\left[\frac{dy}{dx}\right]_P = \left[\frac{dy}{dx}\right]_{(x_1, y_1)}$, provided (x_1, y_1) is supposed to be the point of intersection of the curves.

2. If (x, y) is supposed to be the point of intersection of the two intersecting curves, then (x, y) is found directly by solving simultaneously the two given equations of the curves $y = f_1(x)$ and $y = f_2(x)$.
3. The slopes m_1 and m_2 at the point of intersection is always found by the derivatives calculated from the two given equations at the known coordinates of the point of intersection of the two given equations for the curves.
4. Remember that the slopes m_1 and m_2 are always calculated at a given point whose coordinates are known represented as (x_1, y_1).
5. m_1 and m_2 are slopes of two tangents drawn to the two curves at the same common point of intersection of two curves.

***Examples worked out*:**

***Question*:** Find the angle of intersection of the curves $x^2 + y^2 = 8$ and $y^2 = 2x$.

***Solution*:** (a) Let (x_1, y_1) be the point of intersection of the given curves

$$\therefore\ x_1^2 + y_1^2 = 8 \quad \ldots(1)$$

and $$y_1^2 = 2x_1 \quad \ldots(2)$$

Now, solving eqns (1) and (2) by eliminating y_1 from (1), we have

$$x_1^2 + 2x_1 = 8 \Rightarrow (x_1 + 4)(x_1 - 2) = 0 \Rightarrow x_1 = 2, -4$$

when $x_1 = 2,\ y_1 = \pm 2$

when $x_1 = -4$, y_1 = imaginary for $y_1^2 = -8$

$\therefore$ Required points of intersection are (x_1, y_1) = (2, 2) and (2, –2) ...(3)

(b) The equations of the curves are

$$x^2 + y^2 = 8 \quad \ldots(4)$$

$$y^2 = 2x \quad \ldots(5)$$

Differentiating eqn (4) w.r.t x, we get

$\dfrac{dy}{dx} = -\dfrac{x}{y}$ = Slope of the tangent to the curve $x^2 + y^2 = 8$ at any point (x, y).

Again differentiating eqn (5) w.r.t x, we get

$\dfrac{dy}{dx} = \dfrac{1}{y}$ = Slope of the tangent to the curve $y^2 + 2x$ at any point (x, y).

(c) Now, the slope of the tangent to the curve $x^2 + y^2 = 8$ at any particular point (x_1, y_1)

$$= \left[\frac{dy}{dx}\right]_{C_1} = \left[\frac{dy}{dx}\right]_{\substack{x = x_1 \\ y = y_1}} = \left[-\frac{x_1}{y_1}\right]_{\substack{x_1 = 2 \\ y_1 = 2}} = -\frac{2}{2} = -1$$

$$= m_1 = \tan i_1 \quad \ldots(6)$$

Similarly, the slope of the tangent to the curve $y^2 = 2x$ at any particular point (x_1, y_1)

$$= \left[\frac{dy}{dx}\right]_{C_2} = \left[\frac{dy}{dx}\right]_{\substack{x = x_1 \\ y = y_1}} = \left[\frac{1}{y_1}\right]_{\substack{x_1 = 2 \\ y_1 = 2}} = \frac{1}{2}$$

$$= m_2 = \tan i_2 \quad \ldots(7)$$

Thus, at (2, 2) we get the slopes which are $-1 = \tan i_1$ and $\dfrac{1}{2} = \tan i_2$

Again, we find the slopes of the tangents at (2, –2) is

$$\left[\frac{dy}{dx}\right]_{C_1} = \left[\frac{dy}{dx}\right]_{\substack{x = x_1 \\ y = y_1}} = \left[-\frac{x_1}{y_1}\right]_{\substack{x_1 = 2 \\ y_1 = -2}} = \frac{-2}{-2} = 1 \quad \ldots(8)$$

and $$\left[\frac{dy}{dx}\right]_{C_2} = \left[\frac{dy}{dx}\right]_{\substack{x = x_1 \\ y = y_1}} = \left[\frac{1}{y_1}\right]_{\substack{x_1 = 2 \\ y_1 = -2}} = -\frac{1}{2} \quad \ldots(9)$$

Thus, at (2, –2), we get the slopes which are

$1 = \tan i_1$ and $-\dfrac{1}{2} = \tan i_2$

(d) Hence, the angle of intersection at (2, –2) is

$$\theta = \tan^{-1}\left[\pm\frac{m_1 \sim m_2}{1 + m_1 m_2}\right]$$

$$= \tan^{-1}\left[\pm\frac{\text{difference of slopes}}{1 + \text{product of slopes}}\right]$$

$$= \tan^{-1}\left[\pm\frac{\frac{1}{2} - (-1)}{1 + (-1)\frac{1}{2}}\right]$$

$$= \tan^{-1}\left[\pm\frac{\frac{1}{2} + 1}{1 - \frac{1}{2}}\right] = \tan^{-1}\left[\pm\frac{\frac{3}{2}}{\frac{1}{2}}\right]$$

$$= \tan^{-1}[\pm 3] \quad \ldots(10)$$

Again, the angle of intersection at (2, –2)

$$= \tan^{-1}\left[\pm\frac{1-\left(-\frac{1}{2}\right)}{1+1\left(-\frac{1}{2}\right)}\right] = \tan^{-1}\left[\pm\frac{1+\frac{1}{2}}{1-\frac{1}{2}}\right]$$

$$= \tan^{-1}\left[\pm\frac{\frac{3}{2}}{\frac{1}{2}}\right] = \tan^{-1}[\pm 3] \quad \ldots(11)$$

Thus, at (2, 2) and (2, –2), we get the angle of intersection $\theta = \tan^{-1}[\pm 3]$

Question: Find the angle of intersection between the parabolas $y^2 = ax$ and $x^2 = ay$.

Solution: Letting (x_1, y_1) to be the point of intersection of both given curves

$$y^2 = ax \ \ldots(1) \Rightarrow y_1^{\,2} = ax_1 \quad \ldots(3)$$

$$y = \frac{x^2}{a} \ \ldots(2) \Rightarrow y_1 = \frac{x_1^2}{a} \quad \ldots(4)$$

Now, solving eqns (3) and (4) by eliminating y_1 first, we find that

$$\left[\left(\frac{x_1^2}{a}\right)\right]^2 = a x_1 \Rightarrow \frac{x_1^4}{a^2} = ax_1 \Rightarrow x_1^4 = a^3 x_1$$

$$\Rightarrow x_1^{\,4} - a^3 x_1 = 0 \Rightarrow x_1 (x_1^{\,3} - a^3) = 0 \Rightarrow x_1 = 0$$

or a

$$\text{when } x_1 = 0,\ y_1 = \frac{x_1^2}{a} = \frac{0}{a} = 0$$

$$\text{and when } x_1 = a,\ y_1 = \frac{x_1^2}{a} = \frac{a^2}{a} = a$$

∴ Required points are (0, 0) and (a, a).

(b) Now considering eqns (1) and (2) differentiating (1) first, we have

$$\frac{d}{dx}\left[y^2\right] = \frac{d}{dx}[ax]$$

$$\Rightarrow 2y\frac{dy}{dx} = a \Rightarrow \frac{dy}{dx} = \frac{a}{2y}$$

= Slope of the tangent at point (x, y) ...(3)

for $y \neq 0$

Again differentiating (2) w.r.t x, we get

$\frac{dy}{dx} = \frac{2x}{a}$ = Slope of the tangent at any point (x, y) ...(4)

(c) Now the slope of the tangent to the curve $y^2 = ax$ at any particular point (x_1, y_1) is

$$\left[\frac{dy}{dx}\right]_{C_1} = \left[\frac{dy}{dx}\right]_{\substack{x=x_1\\y=y_1}}$$

$$\because y^2 = ax \Rightarrow y = \sqrt{ax}$$

$$\therefore f'(0) = \left[\frac{dy}{dx}\right]_{(0,0)} = \lim_{h\to 0}\frac{f(0+h)-f(0)}{h}$$

$$= \lim_{h\to 0}\frac{\sqrt{a(0+h)}}{h} = \infty \quad \therefore i_1 = 90° \quad \ldots(5)$$

Again, the slope of the tangent to the curve $y = \frac{x^2}{a}$ at any particular point (x_1, y_1) is

$$\left[\frac{dy}{dx}\right]_{C_2} = \left[\frac{dy}{dx}\right]_{\substack{x=x_1\\y=y_1}}$$

$$= \left[\frac{2x}{a}\right]_{\substack{x=x_1\\y=y_1}} = \left[\frac{2x_1}{a}\right]_{\substack{x_1=0\\y_1=0}} = \frac{2\cdot 0}{a}$$

$$= 0 \ \therefore i_2 = 0° \quad \ldots(6)$$

Similarly, we find the slopes of the tangents at (a, a)

$$\left[\frac{dy}{dx}\right]_{C_1} = \left[\frac{dy}{dx}\right]_{\substack{x=x_1\\y=y_1}} = \left[\frac{a}{2y_1}\right]_{\substack{x_1=a\\y_1=a}} = \frac{a}{2\cdot a}$$

$$= \frac{1}{2} = m_1 \quad \ldots(7)$$

$$\left[\frac{dy}{dx}\right]_{C_2} = \left[\frac{dy}{dx}\right]_{\substack{x=x_1\\y=y_1}} = \left[\frac{2x_1}{a}\right]_{\substack{x_1=a\\y_1=a}}$$

$$= \frac{2a}{a} = 2 = m_2 \quad \ldots(8)$$

(d) Hence, the angle of intersection at (0, 0)

$= \theta = i_1 - i_2 = 90°$

Again, the angle of intersection at (a, a)

$$= \theta = \tan^{-1}\left[\pm\frac{m_1 \sim m_2}{1 + m_1 m_2}\right]$$

$$= \tan^{-1}\left[\pm\frac{2 - \frac{1}{2}}{1 + 2 \cdot \frac{1}{2}}\right] = \tan^{-1}\left[\pm\frac{3}{4}\right]$$

Question: Find the angle of intersection of the curves $x^2 + y^2 = 8$ and $xy = 4$.

Solution: (a) Letting (x_1, y_1) to be the point of intersection of the given curves

$$x^2 + y^2 = 8 \quad ...(1) \Rightarrow x_1^2 + y_1^2 = 8 \quad ...(3)$$

and $xy = 4 \quad ...(2) \Rightarrow x_1 y_1 = 4 \quad ...(4)$

Now, solving (3) and (4) simultaneous by eliminating y_1 first from (4), we get

$$x_1^2 + \frac{16}{x_1^2} = 8 \Rightarrow x_1^4 + 16 = 8x_1^2$$

$$\Rightarrow (x_1^2 - 4)^2 = 0 \Rightarrow x_1 = \pm 2 \Rightarrow y_1 = 2(x_1 = 2)$$

$$= -2\ (x_1 = -2)$$

[from $x_1 y_1 = 4$ ...(4)]

Required points of intersection are $(x_1, y_1) = (2, 2)$ and $(-2, -2)$...(5)

(b) The equations of the curves are

$$x^2 + y^2 = 8 \quad ...(1)$$

$$xy = 4 \quad ...(2)$$

Now, differentiating (1) w.r.t x, we get

$$2x + 2y\frac{dy}{dx} = 0 \Rightarrow \frac{dy}{dx} = -\frac{2x}{2y} = -\frac{x}{y} \quad ...(6)$$

Again, differentiating (2) w.r.t x, we get

$$x \cdot \frac{dy}{dx} + y \cdot 1 = 0 \Rightarrow \frac{dy}{dx} = -\frac{y}{x} = -\frac{4}{x^2}$$

$$\left[\because xy = 4 \Rightarrow y = \frac{4}{x}\right] \quad ...(7)$$

(c) Now, the slope of the tangent to the curve $x^2 + y^2 = 8$ at any particular point $(x_1, y_1) = (2, 2)$ is

$$\left[\frac{dy}{dx}\right]_{C_1} = \left[\frac{dy}{dx}\right]_{\substack{x = x_1 \\ y = y_1}}$$

$$= \left[-\frac{x}{y}\right]_{\substack{x_1 = 2 \\ y_1 = 2}} = \left[-\frac{2}{2}\right] = -1 = m_1 = \tan i_1 \quad ...(8)$$

Similarly, the slope of the tangent to the curve $xy = 4$ at any particular point (x_1, y_1)

$$= \left[\frac{dy}{dx}\right]_{C_2} = \left[\frac{dy}{dx}\right]_{\substack{x = x_1 \\ y = y_1}}$$

$$= \left[-\frac{4}{x^2}\right]_{\substack{x_1 = 2 \\ y_1 = 2}} = \left[-\frac{4}{4}\right] = -1 = m_2 = \tan i_2 \quad ...(9)$$

Thus, at (2, 2), we get the slopes which are

$$\tan i_1 = -1 \text{ and } \tan i_2 = -1 \ \therefore i_1 = i_2$$

Again, the slope of the tangent to the curve $x^2 + y^2 = 8$ at any particular point (x_1, y_1)

$$= (-2, -2) \text{ is } \left[\frac{dy}{dx}\right]_{C_1} = \left[\frac{dy}{dx}\right]_{\substack{x_1 = -2 \\ y_1 = -2}}$$

$$= \left[-\frac{x_1}{y_1}\right]_{\substack{x_1 = -2 \\ y_1 = -2}} = \left[-\frac{2}{2}\right] = -1 = m_1 \quad ...(10)$$

and the slope of the tangent to the curve $xy = 4$ at any particular point (x_1, y_1)

$$= (-2, -2) = \left[\frac{dy}{dx}\right]_{C_2} = \left[\frac{dy}{dx}\right]_{\substack{x_1 = -2 \\ y_1 = -2}}$$

$$= \left[-\frac{4}{x_1^2}\right]_{x_1 = -2} = \left[-\frac{4}{4}\right] = -1 = m_2 \quad ...(11)$$

$$\therefore i_1 = i_2$$

(d) Hence, the angle of intersection at (2, 2) and at (–2, –2) is $i_1 - i_2 = 0$, as $i_1 = i_2$ at both the points.

Question: Find the angle at which the curves $x^2 - y^2 = a^2$ and $x^2 + y^2 = 2a^2$ intersect.

Solution: (a) Letting (x_1, y_1) to be the point of intersection of the given cuves

$x^2 - y^2 = a^2 \quad \ldots(1) \Rightarrow x_1{}^2 - y_1{}^2 = a^2 \quad \ldots(3)$

and $x^2 + y^2 = 2a^2 \quad \ldots(2) \Rightarrow x_1{}^2 + y_1{}^2 = 2a^2 \quad \ldots(4)$

Now, solving the above two equations (3) and (4)

$$x_1{}^2 - y_1{}^2 = a^2$$
$$x_1{}^2 + y_1{}^2 = 2a^2$$

$$2x_1{}^2 = 3a^2 \Rightarrow x_1 = \pm\sqrt{\frac{3a^2}{2}} = \pm\frac{\sqrt{3}}{\sqrt{2}}a$$

$$\because x_1^2 - y_1^2 = a^2$$

$$\Rightarrow \frac{3}{2}a^2 - y_1^2 = a^2$$

$$\Rightarrow y_1^2 = \frac{3a^2}{2} - a^2 = \frac{3a^2 - 2a^2}{2} = \frac{a^2}{2}$$

$$\Rightarrow y_1 = \pm\frac{\sqrt{a^2}}{\sqrt{2}} = \pm\frac{a}{\sqrt{2}}$$

$\therefore$ Required points of intersection are (x_1, y_1)

$$= \left(\pm\frac{\sqrt{3}}{\sqrt{2}}a, \pm\frac{a}{\sqrt{2}}\right), \text{(four points).}$$

(b) The given equations of the curves are

$x^2 - y^2 = a^2 \quad \ldots(1)$

$x^2 + y^2 = 2a^2 \quad \ldots(2)$

Now, differentiating (1) w.r.t x, we get

$$2x - 2y\frac{dy}{dx} = 0 \Rightarrow \frac{dy}{dx} = \frac{x}{y} \quad \ldots(5)$$

Again, differentiating (2) w.r.t x, we get

$$2x - 2y\frac{dy}{dx} = 0 \Rightarrow \frac{dy}{dx} = -\frac{x}{y} \quad \ldots(6)$$

(c) Now, the slope of the tangent to the curve $x^2 - y^2 = a^2$ at any particular point (x_1, y_1)

$$= \left(\frac{\sqrt{3}\,a}{\sqrt{2}}, \frac{a}{\sqrt{2}}\right)$$

is $\left[\frac{dy}{dx}\right]_{C_1} = \left[\frac{dy}{dx}\right]_{(x_1, y_1)} = \left[\frac{x_1}{y_1}\right]_{\substack{x_1 = \frac{\sqrt{3}a}{\sqrt{2}} \\ y_1 = \frac{a}{\sqrt{2}}}}$

$$= \frac{\sqrt{3}\,a}{\sqrt{2}} \times \frac{\sqrt{2}}{a} = \sqrt{3} = m_1 \quad \ldots(7)$$

Similarly, the slope of the tangent to the curve $x^2 + y^2 = 2a^2$ at any particular point (x_1, y_1)

$$= \left[\frac{dy}{dx}\right]_{C_2} = \left[\frac{dy}{dx}\right]_{(x_1, y_1)} = \left[-\frac{x_1}{y_1}\right]_{\substack{x_1 = \frac{\sqrt{3}a}{\sqrt{2}} \\ y_1 = \frac{a}{\sqrt{2}}}}$$

$$= \left[-1 \times \frac{\sqrt{3}}{\sqrt{2}}a \times \frac{\sqrt{2}}{a}\right] = -\sqrt{3} = m_2 \quad \ldots(8)$$

Again, the slope of the tangent to the curve $x^2 - y^2 = a^2$ at any particular point (x_1, y_1)

$$= \left(\frac{-\sqrt{3}\,a}{\sqrt{2}}, \frac{-a}{\sqrt{2}}\right)$$

is $\left[\frac{dy}{dx}\right]_{C_1} = \left[\frac{dy}{dx}\right]_{(x_1, y_1)} = \left[\frac{x_1}{y_1}\right]_{\substack{x_1 = -\frac{\sqrt{3}a}{\sqrt{2}} \\ y_1 = -\frac{a}{\sqrt{2}}}}$

$$= -\frac{\sqrt{3}a}{\sqrt{2}} \times \frac{\sqrt{2}}{-a} = \sqrt{3} = m_1 \quad \ldots(9)$$

Similarly, the slope of the tangent to the curve $x^2 + y^2 = 2a^2$ at any particular point (x_1, y_1)

$$= \left[\frac{dy}{dx}\right]_{C_2} = \left[\frac{dy}{dx}\right]_{(x_1, y_1)} = \left[-\frac{x_1}{y_1}\right]_{\substack{x_1 = \frac{\sqrt{3}a}{\sqrt{2}} \\ y_1 = -\frac{a}{\sqrt{2}}}}$$

$$= \left[-\left(-\frac{\sqrt{3}a}{\sqrt{2}} \times -\frac{\sqrt{2}}{a}\right)\right] = -\sqrt{3} = m_2 \quad \ldots(10)$$

(d) Hence, the angle of intersection at

$\left(\frac{\sqrt{3}}{\sqrt{2}}a, \frac{a}{\sqrt{2}}\right)$ is $\theta = \tan^{-1}\left[\pm\frac{m_1 \sim m_2}{1 + m_1 m_2}\right]$

$$= \tan^{-1}\left[\pm\frac{\sqrt{3} - (-\sqrt{3})}{1 + \sqrt{3}(-\sqrt{3})}\right] = \tan^{-1}\left[\pm\frac{\sqrt{3} + \sqrt{3}}{1 - 3}\right]$$

$$= \tan^{-1}\left[\pm\frac{2\sqrt{3}}{-2}\right] = \tan^{-1}\left[\mp\sqrt{3}\right]$$

$= 120°$ and $60°$

Again, the angle of intersection at

$\left(-\frac{\sqrt{3}}{\sqrt{2}}a, -\frac{a}{\sqrt{2}}\right)$ is $\theta = \tan^{-1}\left[\pm\frac{m_1 \sim m_2}{1+m_1 m_2}\right]$

$$= \tan^{-1}\left[\pm\frac{\sqrt{3}-(-\sqrt{3})}{1+\sqrt{3}(-\sqrt{3})}\right] = \tan^{-1}\left[\pm\frac{2\sqrt{3}}{1-3}\right]$$

$$= \tan^{-1}\left[\pm\frac{2\sqrt{3}}{-2}\right] = \tan^{-1}\left[\pm\sqrt{3}\right]$$

$= \tan^{-1}\left[\mp\sqrt{3}\right] = 120°$ and $60°$. Similarly the angles of intersection at the points $\left(\frac{\sqrt{3}a}{2}, \frac{-a}{\sqrt{2}}\right)$ and $\left(\frac{-\sqrt{3}a}{2}, \frac{a}{\sqrt{2}}\right)$ are the same.

Question: Find the angle between the curves $y = 4x^2$ and $y = 8x^3$.

Solution: (a) If (x, y) be the point of intersection of the given curves

$y = 4x^2$ ...(1)

$y = 8x^3$, ...(2)

then let us find the point of intersection of the given curves by solving the given equations (1) and (2),

$$\left.\begin{array}{l} y = 4x^2 \\ y = 8x^3 \end{array}\right\} \Rightarrow 8x^3 = 4x^2 \Rightarrow 8x^3 - 4x^2 = 0$$

$$\Rightarrow 4x^2(2x-1) = 0 \Rightarrow x = 0 \text{ or } x = \frac{1}{2}$$

$\therefore$ Required points of intersection are $(x, y) = (0, 0)$ and $\left(\frac{1}{2}, 1\right)$.

(b) The given equations of the curves are

$y = 4x^2$ ...(1)

and $y = 8x^3$ ...(2)

Now, differentiating (1) w.r.t x, we get

$\frac{dy}{dx} = 8x$ ...(3)

Again, differentiating (2) w.r.t x, we get

$\frac{dy}{dx} = 24x^2$ ...(4)

(c) The slope of the tangent to the curve $y = 4x^2$ at the point of intersection (x, y)

$= (0, 0)$ is $\left[\frac{dy}{dx}\right]_{C_1} = [8x]_{(0,0)} = 0 = m_1$ ...(5)

Similarly, the slope of the tangent to the curve $y = 8x^3$ at the point of intersection (x, y)

$= (0, 0)$ is $\left[\frac{dy}{dx}\right]_{C_2} = \left[24x^2\right]_{(0,0)} = 0 = m_2$ ...(6)

Again, the slope of the tangent to the curve $y^2 = 4x$ at the point of intersection (x, y)

$= \left(\frac{1}{2}, 1\right)$ is $\left[\frac{dy}{dx}\right]_{C_1} = [8x]_{\left(\frac{1}{2},1\right)} = 8 \times \frac{1}{2}$

$= 4 = m_1$ ...(7)

Similarly, the slope of the tangent to the curve $y = 8x^3$ at the point of intersection (x, y)

$$= \left(\frac{1}{2}, 1\right) = \left[\frac{dy}{dx}\right]_{C_2} = \left[\frac{dy}{dx}\right]_{\left(\frac{1}{2},1\right)}$$

$= \left[24x^2\right]_{\left(\frac{1}{2},1\right)} = 24 \times \frac{1}{4} = 6 = m_2$ ...(8)

(d) Hence, the angle of intersection of the curves at the point of intersection (x, y)

$$= (0, 0) \text{ is } \theta = \tan^{-1}\left[\pm\frac{m_1 \sim m_2}{1+m_1 m_2}\right]$$

$$= \tan^{-1}\left[\pm\frac{0-0}{1+0\cdot 0}\right] = \tan^{-1} 0 = 0$$

Again, the angle of intersection of the curves at the point of intersection (x, y)

$$= \left(\frac{1}{2}, 1\right) \text{ is } \theta = \tan^{-1}\left[\pm\frac{m_1 \sim m_2}{1+m_1 m_2}\right]$$

$$= \tan^{-1}\left[\pm\frac{6-4}{1+24}\right] = \tan^{-1}\left[\pm\frac{2}{25}\right]$$

Question: Find the angle of intersection of the curves $2y^2 = x^3$ and $y^2 = 32x$.

Solution: (a) If (x, y) be the point of intersection of the given curves

$2y^2 = x^3$ and ...(1)

$y^2 = 32x$, ...(2)

then let us find the point of intersection of the given curves by solving the above equations (1) and (2), $x^3 = 64x$ [Eliminating y form (1)]

$\Rightarrow x^3 - 64x = 0 \Rightarrow x(x^2 - 64) = 0$

$\Rightarrow x(x-8)(x+8) = 0 \Rightarrow x = 0, 8, -8$

Now, from (2), we have $y^2 = 32 \times 0, 32 \times 8, 32 \times (-8)$

$\Rightarrow y = 0, \pm 16, \pm 16\sqrt{-1}$

$\therefore$ Required points of intersection of the given curves are $(x, y) = (0, 0), (8, 16)$ and $(8, -16)$.

(b) The given equations of the curves are

$2y^2 = x^3$ ...(1)

and $y^2 = 32x$ ...(2)

Now, differentiating (1) w.r.t x, we get

$$2 \cdot 2y\frac{dy}{dx} = 3x^2 \Rightarrow 4 \cdot y \cdot \frac{dy}{dx} = 3x^2$$

$$\Rightarrow \frac{dy}{dx} = \frac{3x^2}{4y} \quad ...(3)$$

for $y \neq 0$

Again, differentiating (2) w.r.t x, we get

$$2y \cdot \frac{dy}{dx} = 32 \cdot 1 \Rightarrow \frac{dy}{dx} = \frac{16}{y} \quad ...(4)$$

(c) The slope of the tangent to the curve $2y^2 = x^3$ at the point of intersection (0, 0)

$$= \left[\frac{dy}{dx}\right]_{C_1} = \left[\frac{dy}{dx}\right]_{\substack{x_1=0\\y_1=0}}$$

$$\because 2y^2 = x^3 \Rightarrow y = \sqrt{\frac{x^3}{2}}$$

$$\Rightarrow f'(0) = \left[\frac{dy}{dx}\right]_{(0,0)} = \lim_{h\to 0} \frac{f(0+h) - f(0)}{h}$$

$$f'(0) = \lim_{h\to 0} \frac{\sqrt{(0+h)^3}}{h} = 0 = m_1$$

Similarly, the slope of the tangent to the curve $y^2 = 32x$ at the point of intersection (0, 0)

$$= \left[\frac{dy}{dx}\right]_{C_2} = \left[\frac{dy}{dx}\right]_{(0,0)} = \infty = m_2$$

Again the slope of the tangent to the curve $2y^2 = x^3$ at the point of intersection (8, 16)

$$= \left[\frac{dy}{dx}\right]_{C_1} = \left[\frac{dy}{dx}\right]_{(8,16)} = \left[\frac{3x^2}{4y}\right]_{(8,16)}$$

$$= \frac{3 \times 8 \times 8}{4 \times 16} = 3 = m_1$$

Similarly, the slope of the tangent to the curve $y^2 = 32x$ at the point of intersection (8, 16)

$$= \left[\frac{dy}{dx}\right]_{C_2} = \left[\frac{dy}{dx}\right]_{(8,16)} = \left[\frac{16}{y}\right]_{(8,16)}$$

$$= \frac{16}{16} = 1 = m_2$$

Lastly the slope of the tangent to the curve $2y^2 = x^3$ at the point of intersection (8, –16)

$$= \left[\frac{dy}{dx}\right]_{C_1} = \left[\frac{dy}{dx}\right]_{(8,-16)}$$

$$= \frac{3 \times 64}{4 \times (-16)} = -3 = m_1$$

and the slope of the tangent to the curve $y^2 = 32x$ at the point of intersection (8, –16)

$$= \left[\frac{dy}{dx}\right]_{C_2} = \left[\frac{16}{y}\right]_{(-8,-16)}$$

$$= \left[-\frac{16}{16}\right] = -1 = m_2$$

(d) Now, we consider the angle of intersection at (0, 0)

$$m_1 = 0 \Rightarrow i_1 = 0$$

$$m_2 = \infty \Rightarrow i_2 = \frac{\pi}{2}$$

$$\therefore \theta = i_1 - i_2 = \frac{\pi}{2}.$$

Similarly, the angle between the curves at (8, 16)

$$= \theta = \tan^{-1}\left[\pm\frac{m_1 \sim m_2}{1 + m_1 m_2}\right]$$

$$= \tan^{-1}\left[\pm\frac{(3-1)}{(1+3\cdot 1)}\right]$$

$$= \tan^{-1}\left[\pm\frac{2}{4}\right] = \tan^{-1}\left[\pm\frac{1}{2}\right]$$

Lastly, the angle between the curves at (8, –16)

$$= \theta = \tan^{-1}\left[\pm\frac{m_1 \sim m_2}{1 + m_1 m_2}\right]$$

$$= \tan^{-1}\left[\pm\frac{-3-(-1)}{1+(-3)(-1)}\right] = \tan^{-1}\left[\pm\frac{-3+1}{1+3}\right]$$

$$= \tan^{-1}\left[\pm\frac{-2}{4}\right] = \tan^{-1}\left[\pm\left(-\frac{1}{2}\right)\right]$$

$$= \tan^{-1}\left[\pm\frac{1}{2}\right]$$

Question: Find the angle of intersection of the curves $x^2 + y^2 - 4x - 1 = 0$ and $x^2 + y^2 - 2y - 9 = 0$.

Solution: (a) Letting (x, y) to be the point of intersection of the given curves

$$x^2 + y^2 - 4x - 1 = 0 \qquad \ldots(1)$$

$$x^2 + y^2 - 2y - 9 = 0 \qquad \ldots(2)$$

and solving these equations (1) and (2) simultaneously, we get

$$-4x + 2y + 8 = 0 \Rightarrow y = 2x - 4 \qquad \ldots(3)$$

Now, putting the value of y from (3) into (1), we get

$$x^2 + (2x-4)^2 - 4x - 1 = 0$$

$$\Rightarrow 5x^2 - 20x + 15 = 0$$

$$\Rightarrow x^2 - 4x + 3 = 0$$

$$\Rightarrow (x-1)(x-3) = 0$$

$$\Rightarrow x = 1, 3$$

Form (3), putting $x = 1, 3$ in (1), we get $y = -2, 2$

Hence, the required points of intersection are $(x, y) = (1, -2), (3, 2)$

(b) The given equations of the curves are

$$x^2 + y^2 - 4x - 1 = 0 \qquad \ldots(1)$$

$$x^2 + y^2 - 2y - 9 = 0 \qquad \ldots(2)$$

Now, differentiating (1) w.r.t x, we get

$$2x + 2y\frac{dy}{dx} - 4 = 0 \Rightarrow \frac{dy}{dx} = \frac{2-x}{y},\ y \neq 0$$

Again differentiating (2) w.r.t x, we get

$$2x + 2y\frac{dy}{dx} - 2\frac{dy}{dx} = 0$$

$$\Rightarrow x + (y-1)\frac{dy}{dx} = 0 \Rightarrow \frac{dy}{dx} = \frac{x}{1-y},\ y \neq 1.$$

(c) The slope of the tangent to the curve $x^2 + y^2 - 4x - 1 = 0$ at the point of intersection (x, y)

$$= (1, -2) \text{ is } \left[\frac{dy}{dx}\right]_{C_1} = \left[\frac{dy}{dx}\right]_{(1,-2)}$$

$$= \left[\frac{2-x}{y}\right]_{(1,-2)} = \frac{2-1}{-2} = -\frac{1}{2} = m_1$$

Similarly, the slope of the tangent to the curve $x^2 + y^2 - 2y - 9 = 0$ at the point of intersection (x, y)

$$= (1, -2) \text{ is } \left[\frac{dy}{dx}\right]_{C_2} = \left[\frac{dy}{dx}\right]_{(1,-2)} = \left[\frac{x}{1-y}\right]_{(1,-2)}$$

$$= \frac{1}{1+2} = \frac{1}{3} = m_2$$

Again the slope of the tangent to the curve $x^2 + y^2 - 4x - 1 = 0$ at the point of intersection (x, y)

$$= (3, 2) \text{ is } \left[\frac{dy}{dx}\right]_{C_1} = \left[\frac{dy}{dx}\right]_{(3,2)}$$

$$= \left[\frac{2-x}{y}\right]_{(3,2)} = \frac{2-3}{2} = -\frac{1}{2} = m_1$$

Similarly, the slope of the tangent to the curve $x^2 + y^2 - 2y - 9 = 0$ at the point of intersection (x, y)

$$= (3, 2) \text{ is } \left[\frac{dy}{dx}\right]_{C_2} = \left[\frac{dy}{dx}\right]_{(3,2)}$$

$$= \left[\frac{x}{1-y}\right]_{(3,2)} = \frac{3}{1-2} = \frac{3}{-1} = -3 = m_2$$

(d) Hence, the angle of intersection of the curves at the point of intersection (x, y)

$$= (1, -2) \text{ is } \theta = \tan^{-1}\left[\pm\frac{m_1 \sim m_2}{1 + m_1 m_2}\right]$$

$$= \tan^{-1}\left[\pm\frac{\frac{1}{3} - \left(-\frac{1}{2}\right)}{1 + \frac{1}{3}\left(-\frac{1}{2}\right)}\right] = \tan^{-1}[\pm 1]$$

$\Rightarrow \theta = 45^\circ$ or 135°

Again, the angle of intersection of the curves at the point of intersection (x, y)

$$= (3, 2) \text{ is } \theta = \tan^{-1}\left[\pm\frac{m_1 \sim m_2}{1 + m_1 m_2}\right]$$

$$= \tan^{-1}\left[\pm\frac{-\frac{1}{2} + 3}{1 + \frac{3}{2}}\right] = \tan^{-1}(\pm 1)$$

$= 45^\circ, 135^\circ$

Type: To find the angle between two tangents to a curve at two given points:

Supposing that there are two points P and Q lying on the curve c_1 represented by the equation

$f_1(x, y) = 0$ ...(1)

PT_1 = tangent at P

QT_2 = tangent at Q

R = The point of intersection of the two tangents PT_1 and QT_2.

Now, we are required to find out the angle between these two tangents.

Now, we suppose that θ = Angle between two tangents to the single curve at two given points P and Q.

$P = (x_1, y_1)$

$Q = (x_2, y_2)$

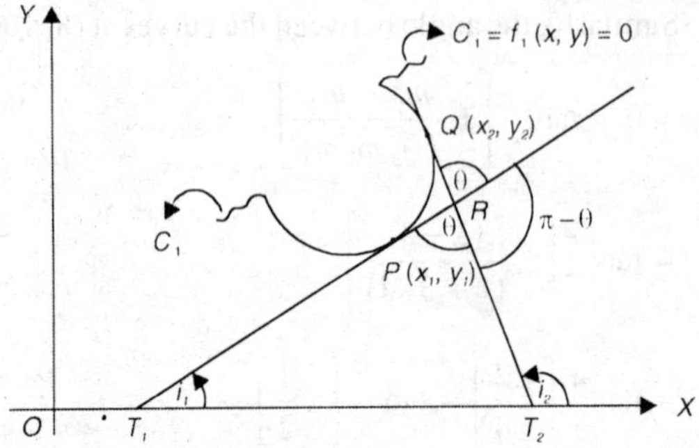

Now, differentiating the equation $f_1(x, y) = 0$ ...(1)

w.r.t x, we get $\left[\frac{dy}{dx}\right]_P = \tan i_1 = m_1$ (say) ...(2)

Similarly, $\left[\frac{dy}{dx}\right]_Q = \tan i_2 = m_2$ (say) ...(3)

Now, required angle between the two tangents to a curve at two given points $= \theta = i_2 \sim i_1$

$\Rightarrow \tan\theta = \tan(i_2 \sim i_1)$

$$= \frac{\tan i_2 \sim \tan i_1}{1 + \tan i_2 \cdot \tan i_1} = \frac{m_2 \sim m_1}{1 + m_1 m_2}$$

$$\Rightarrow \theta = \tan^{-1}\left[\frac{m_2 \sim m_1}{1 + m_1 m_2}\right]$$

Also, $\tan(\pi - \theta) = -\tan\theta = -\dfrac{m_2 \sim m_1}{1 + m_1 m_2}$

$\therefore \theta$ = Angle between the tangents

$$= \tan^{-1}\left[\pm\frac{m_2 \sim m_1}{1 + m_1 m_2}\right]$$

Working Rule:

1. Find $\frac{dy}{dx}$ by differentiating both sides of the given equation w.r.t x.

2. Find $\left[\frac{dy}{dx}\right]_{(x_1, y_1)}$ = The value of $\frac{dy}{dx}$ at $(x_1, y_1) = m_1$

3. Find $\left[\frac{dy}{dx}\right]_{(x_2, y_2)}$ = The value of $\frac{dy}{dx}$ at $(x_2, y_2) = m_2$

4. Use $\theta = \tan^{-1}\left[\pm\dfrac{m_2 \sim m_1}{1 + m_1 m_2}\right]$ or $\theta = i_2 - i_1$.

Note: **1.** $\theta = \tan^{-1}\left|\dfrac{m_2 \sim m_1}{1 + m_1 m_2}\right|$ is used to find the acute angle between the tangents to a single curve at two given points.

2. $\theta = \tan^{-1}\left|\dfrac{m_2 \sim m_1}{1 + m_1 m_2}\right|$ is used to find the obtuse angle between the tangents to a single curve at two given points.

3. If $m_1 = \tan i_1$ and $m_2 = \tan i_2$, then the angle between the two tangents is $\theta = |i_1 - i_2|$ = Absolute value of difference of i_1 and i_2. (If i_1 or $i_2 = \dfrac{\pi}{2}$)

Examples worked out:

Question: Find the angle between the tangents to the curve $x^2 = 8y + 6$ at the points $\left(0, -\dfrac{3}{4}\right)$ and $\left(4, \dfrac{5}{4}\right)$.

Solution: (a) Given equation of the curve is

$$x^2 = 8y + 6 \Rightarrow 8y = x^2 - 6 \qquad \text{...(1)}$$

Now, differentiating the equation (1) w.r.t x, we get

$$8\frac{dy}{dx} = 2x - 0 \Rightarrow 8\frac{dy}{dx} = 2x$$

$$\frac{dy}{dx} = \frac{2x}{8} = \frac{x}{4} \qquad \text{...(2)}$$

(b) The value of $\dfrac{dy}{dx}$ at $\left(0, \dfrac{3}{4}\right)$

$$= \left[\frac{dy}{dx}\right]_{\left(0, \frac{3}{4}\right)} = \left[\frac{x}{4}\right]_{\left(0, \frac{3}{4}\right)} = \frac{0}{4}$$

$$= 0 = m_1 \qquad \text{...(3)}$$

and the value of $\dfrac{dy}{dx}$ at $\left(4, \dfrac{5}{4}\right)$

$$= \left[\frac{dy}{dx}\right]_{\left(4, \frac{5}{4}\right)} = \left[\frac{x}{4}\right]_{\left(4, \frac{5}{4}\right)} = \frac{4}{4}$$

$$= 1 = m_2 \qquad \text{...(4)}$$

$$\therefore \tan\theta = \pm\frac{0 - 1}{1 + 0} = \pm 1$$

$$\therefore \theta = \frac{\pi}{4} \text{ (Acute angle between the tangents)}$$

Alternatively,

$$\because m_1 = 0 \Rightarrow \tan i_1 = \tan 0 \Rightarrow i_1 = 0 \qquad \text{...(5)}$$

$$\text{and } m_2 = 1 \Rightarrow \tan i_2 = \tan\frac{\pi}{4} \Rightarrow i_2 = \frac{\pi}{4} = 45^\circ \qquad \text{...(6)}$$

$\therefore$ Angle between the tangents $= \theta = i_2 - i_1$

$$= (45 - 0)^\circ = 45^\circ = \frac{\pi}{4}$$

Question: Find the angle between the tangents to the curve $y = x^2 + 1$ at the points $\left(\dfrac{1}{2}, \dfrac{5}{4}\right)$ and $\left(-\dfrac{\sqrt{3}}{2}, \dfrac{7}{4}\right)$.

Solution: (a) Given equation of the curve

$$y = x^2 + 1 \qquad \text{...(1)}$$

Now, differentiating equation (1) w.r.t x, we get

$$\frac{dy}{dx} = 2x + 0 = 2x \qquad \text{...(2)}$$

(b) $\left[\dfrac{dy}{dx}\right]_{\substack{x=\frac{1}{2} \\ y=\frac{5}{4}}} = [2x]_{\substack{x=\frac{1}{2} \\ y=\frac{5}{4}}}$

$$= 2 \times \frac{1}{2} = 1 = m_1 \qquad \text{...(3)}$$

and $\left[\dfrac{dy}{dx}\right]_{\substack{x=-\frac{\sqrt{3}}{2} \\ y=\frac{7}{4}}} = [2x]_{\substack{x=-\frac{\sqrt{3}}{2} \\ y=\frac{7}{4}}}$

$$= 2 \times \left(-\frac{\sqrt{3}}{2}\right) = -\sqrt{3} = m_2 \qquad \text{...(4)}$$

$$\text{Now, } m_1 = \tan i_1 = 1 = \tan\frac{\pi}{4} \Rightarrow i_1 = \frac{\pi}{4} = 45^\circ$$

$$m_2 = \tan i_2 = -\sqrt{3} = \tan 120^\circ \Rightarrow i_2 = 120^\circ$$

$\therefore$ The acute angle between the tangents to the curve is given by $\theta = (i_2 - i_1) = [120 - 45]^\circ = 75^\circ$

Alternatively, θ = Acute angle between the tangents

$$= \tan^{-1}\left[\frac{\sqrt{3}-1}{\sqrt{3}+1}\right]$$

Question: Find the angle between the tangents to the curve $\frac{x^2}{a^2}+\frac{y^2}{b^2}=1$ at the points $(a, 0)$ and $(0, b)$.

Solution: (a) Given equation of the curve is

$$\frac{x^2}{a^2}+\frac{y^2}{b^2}=1 \quad \text{...(1)}$$

Now, differentiating equation (1) w.r.t x, we get

$$\frac{dy}{dx} = -\frac{b^2}{a^2}\cdot\frac{x}{y} \quad \text{...(2)}$$

for $y \neq 0$

(b) $y = f(x) = b\sqrt{1-\frac{x^2}{a^2}}$, $x^2 \le a^2$.

$$\therefore f'(a) = \left[\frac{dy}{dx}\right]_{(a,0)} = \lim_{h\to 0}\frac{f(a+h)-f(a)}{h},$$

$h < 0$

$$= \lim_{h\to 0}\frac{b\sqrt{-h(h+2a)}}{ah} = -\infty \quad \therefore i_1 = 90° \quad \text{...(3)}$$

and $\left[\frac{dy}{dx}\right]_{(0,b)} = -\frac{b^2}{a^2}\left[\frac{x}{y}\right]_{\substack{x=0\\y=b}} = -\frac{b^2}{a^2}\cdot 0$

$$= 0 = \tan 0° \quad \therefore i_2 = 0° \quad \text{...(4)}$$

$$\therefore \theta = |i_1 - i_2| = |90° - 0| = 90°$$

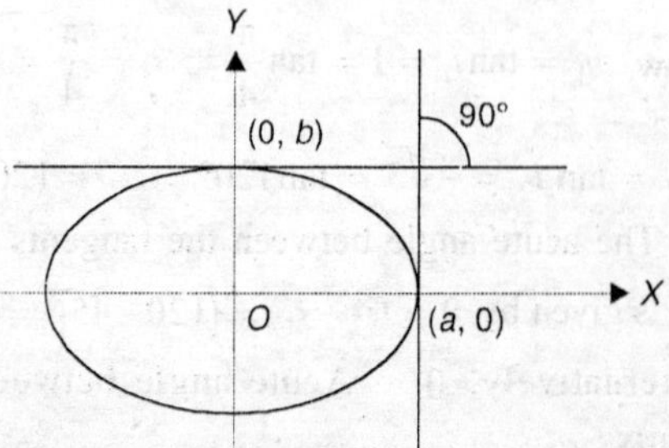

Problems based on finding the area of a triangle formed by the portion included between the axis and the tangent to a curve.

Working Rule:

1. Let (x_1, y_1) be the point of contact of the tangent to the given curve.
2. Put (x_1, y_1) in the given equation of the curve.
3. Find $\left[\frac{dy}{dx}\right]_{\substack{x=x_1\\y=y_1}}$
4. Find the equation of the tangent, using the formula at (x_1, y_1)

$$(y-y_1) = \left[\frac{dy}{dx}\right]_{(x_1, y_1)}\cdot(x-x_1)$$

5. Find x-intercept of the tangent by putting $y = 0$ in the equation of the tangent.
6. Find the y-intercept of the tangent by putting $x = 0$ in the equation of the tangent.
7. Use the formula of area of a triangle.

$$= \left|\frac{1}{2}\left[x_1(y_2-y_3)+x_2(y_3-y_2)+x_3(y_1-y_2)\right]\right|$$

Examples worked out:

Question: Find the area of the triangle formed by the portion included between the axis and the tangent to the curve $x^{\frac{2}{3}} + y^{\frac{2}{3}} = a^{\frac{2}{3}}$.

Solution: (a) Given equation of the curve is

$$x^{\frac{2}{3}} + y^{\frac{2}{3}} = a^{\frac{2}{3}} \quad \text{...(1)}$$

Letting (x_1, y_1) be the point of contact of tangent to the given curve.

$$\therefore x_1^{\frac{2}{3}} + y_1^{\frac{2}{3}} = a^{\frac{2}{3}} \quad \text{...(2)}$$

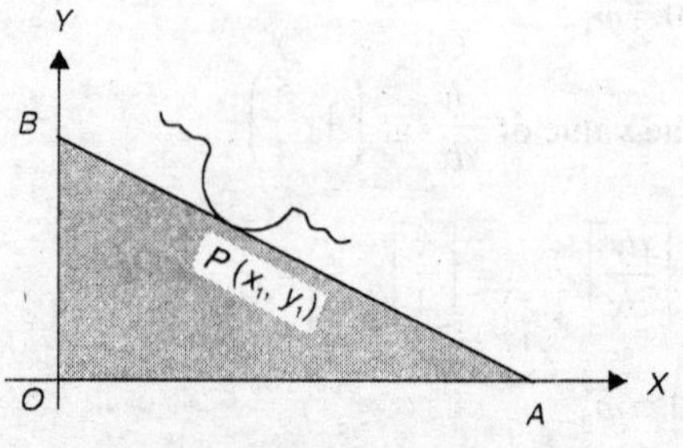

Now, differentiating (1) w.r.t x, we get

$$\frac{2}{3}x^{\frac{2}{3}-1}+\frac{2}{3}y^{\frac{2}{3}-1}\frac{dy}{dx}=0$$

$$\Rightarrow x^{-\frac{1}{3}}+y^{-\frac{1}{3}}\frac{dy}{dx}=0$$

$$\Rightarrow y^{-\frac{1}{3}}\frac{dy}{dx}=-x^{-\frac{1}{3}}$$

$$\Rightarrow \frac{dy}{dx}=-\frac{y^{\frac{1}{3}}}{x^{\frac{1}{3}}} \quad \ldots(3)$$

Now, the value of $\frac{dy}{dx}$ at (x_1, y_1)

$$=\left[\frac{dy}{dx}\right]_{\substack{x=x_1\\y=y_1}}=-\frac{y_1^{\frac{1}{3}}}{x_1^{\frac{1}{3}}} \quad \ldots(4)$$

(b) Now, the equation of the tangent at (x_1, y_1) is

$$(y-y_1)=-\frac{y_1^{\frac{1}{3}}}{x_1^{\frac{1}{3}}}\cdot(x-x_1)$$

$$\Rightarrow x_1^{\frac{1}{3}}y-x_1^{\frac{1}{3}}y_1=-y_1^{\frac{1}{3}}x+x_1y_1^{\frac{1}{3}}$$

$$\Rightarrow x_1^{\frac{1}{3}}y+y_1^{\frac{1}{3}}x=x_1^{\frac{1}{3}}y_1+x_1y_1^{\frac{1}{3}}$$

$$\Rightarrow x_1^{\frac{1}{3}}y+y_1^{\frac{1}{3}}x=x_1^{\frac{1}{3}}y_1^{\frac{1}{3}}\left(x_1^{\frac{2}{3}}+y_1^{\frac{2}{3}}\right)$$

$$=x_1^{\frac{1}{3}}y_1^{\frac{1}{3}}\left[a_1^{\frac{2}{3}}\right]$$

$$\Rightarrow x_1^{\frac{1}{3}}y+y_1^{\frac{1}{3}}x=\sqrt[3]{x_1y_1}\sqrt[3]{a^2} \quad \text{[from (2)]} \quad \ldots(5)$$

(c) x-intercept on x-axis is determined by putting $y=0$ in (5)

$\therefore\ y_1^{\frac{1}{3}}x=x_1^{\frac{1}{3}}y_1^{\frac{1}{3}}a_a^{\frac{2}{3}} \Rightarrow x=x_1^{\frac{1}{3}}a_a^{\frac{2}{3}}$ = x-coordinate of point A. And y-intercept on y-axis is determined by putting $x=0$ in (5)

$\therefore\ x_1^{\frac{1}{3}}y=x_1^{\frac{1}{3}}y_1^{\frac{1}{3}}a_a^{\frac{2}{3}} \Rightarrow y=y_1^{\frac{1}{3}}a_a^{\frac{2}{3}}$ = y-coordinate of B.

Thus, the coordinates of $A=\left(x_1^{\frac{1}{3}}a_a^{\frac{2}{3}}, 0\right)$

$=(x_1, y_1)$

Coordinates of $B=\left(0, y_1^{\frac{1}{3}}a^{\frac{2}{3}}\right)=(x_2, y_2)$

Coordinates of origin 'O' $=(0, 0)=(x_3, y_3)$

(d) Area of $\Delta\, OAB$ having the vertices (x_1, y_1), (x_2, y_2) and (x_3, y_3)

$$=\frac{1}{2}\left[x_1(y_2-y_3)+x_2(y_3-y_1)+x_3(y_1-y_2)\right]$$

$$=\frac{1}{2}\left[x_1^{\frac{1}{3}}a_1^{\frac{2}{3}}\left(y_1^{\frac{1}{3}}a^{\frac{2}{3}}-0\right)+0(0-0)+0\left(0-y_1^{\frac{1}{3}}a^{\frac{2}{3}}\right)\right]$$

$$=\frac{1}{2}\left[x_1^{\frac{1}{3}}a^{\frac{2}{3}}y_1^{\frac{1}{3}}a^{\frac{2}{3}}\right]=\frac{1}{2}x_1^{\frac{1}{3}}y_1^{\frac{1}{3}}a^{\frac{4}{3}}$$

Note: Area $=\frac{1}{2}\cdot OA\cdot OB=\frac{1}{2}x_1^{\frac{1}{3}}\cdot y_1^{\frac{1}{3}}\cdot a^{\frac{2}{3}}$.

Question: Find the area of the triangle formed by the x-axis the tangent and normal to the curve $y(2a-x)=x^2$ at the point (a, a).

Solution: (a) Given equation of the curve is

$$y(2a-x)=x^2 \quad \ldots(1)$$

Now, differentiating equation (1) w.r.t x, we get

$$y(0-1)+[2a-x]\frac{dy}{dx}=2x \Rightarrow \frac{dy}{dx}=\frac{2x+y}{2a-x}$$

Now, the value of $\frac{dy}{dx}$ at $(x_1, y_1)=(a, a)$ is

$$\left[\frac{dy}{dx}\right]_{(x_1,\, y_1)}=\left[\frac{2x+y}{2a-x}\right]_{(x_1,\, y_1)}=\frac{2a+a}{2a-a}=\frac{3a}{a}=3$$

$\therefore$ Slope of the tangent at $(a, a)=3$ and the slope of the normal at $(a, a)=-\frac{1}{3}$

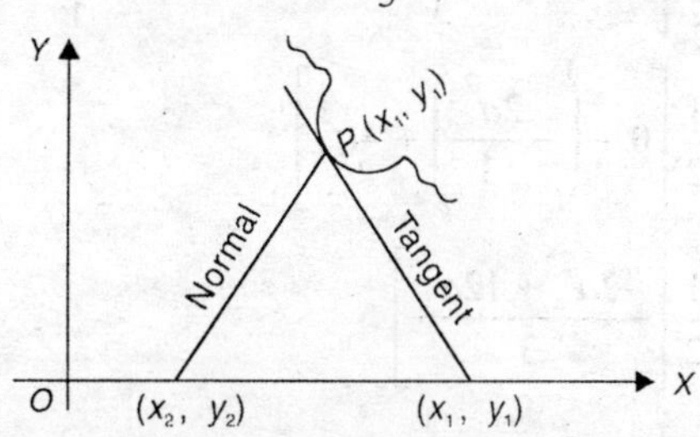

Now, the equation of the tangent at (a, a) is $(y-y_1)$ = slope of the tangent $\cdot\,(x-x_1)$

$\Rightarrow (y-a) = 3(x-a)$

$\Rightarrow (y-a) = 3x - 3a$

$\Rightarrow y - 3x = -3a + a$

$\Rightarrow y = 3x - 2a$ ...(2)

Now, the equation of the normal at (a, a) is

$$(y - y_1) = -\frac{1}{3}(x - x_1)$$

$$\Rightarrow (y-a) = -\frac{1}{3}(x-a)$$

$\Rightarrow 3y - 3a = -x + a$

$\Rightarrow x + 3y = 4a$ ...(3)

(b) Now, for the point of intersection of the tangent and x-axis.

$$y = 0 \text{ in } (2) \Rightarrow 3x - 2a = 0 \Rightarrow 3x = 2a \Rightarrow x = \frac{2a}{3}$$

$\therefore \left(\frac{2a}{3}, 0\right) = (x_1, y_1)$ = point of intersection of tangent and x-axis ...(4)

Again, for the point of intersection of normal and x-axis.

$y = 0$ in (3) $\Rightarrow x = 4a$

$\therefore (4a, 0) = (x_2, y_2)$ = point of intersection of normal and x-axis ...(5)

Also we are given the point of intersection of the tangent and normal $= (a, a) = (x_3, y_3)$ ...(6)

(c) Now, we are required to find out the area of the triangle having the vertices (x_1, y_1), (x_2, y_2) and (x_3, y_3).

$\therefore$ Area of the $\Delta = \frac{1}{2}\,[x_1(y_2 - y_3) + x_2(y_3 - y_1) + x_3(y_1 - y_3)]$

$$= \frac{1}{2}\left[a(0-0) + \frac{2a}{3}(0-a) + 4a(a-0)\right]$$

$$= \frac{1}{2}\left[0 + \left(-\frac{2a^2}{3}\right) + 4a^2\right]$$

$$= \frac{1}{2}\left[\frac{-2a^2 + 12a^2}{3}\right]$$

$$= \frac{1}{2} \cdot \frac{10a^2}{3} = \frac{5a^2}{3}$$

Refresh Your Memory:

Definitions: **1.** ***Subtangent***: The projection of the tangent on x-axis is called the subtangent $\Rightarrow$ the subtangent to a curve at the point of tangency of a tangent to a curve is the portion of the x-axis intercepted between the tangent at the point and the ordinate through the point.

In the figure, the tangent at P to the curve show intersects x-axis at T and F is the foot of the perpendicular from P to the x-axis. FT is then subtangent to the curve at P.

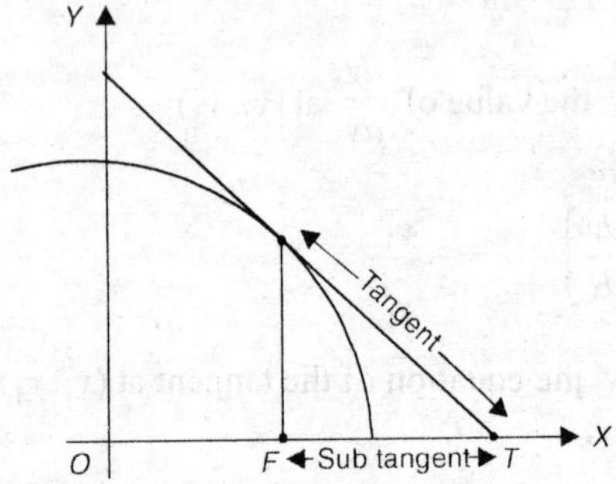

2. ***Subnormal***: The projection of the normal on x-axis is known as the subnormal $\Rightarrow$ Subnormal to the curve at any point is the portion of the x-axis intercepted between the normal and the ordinate through that point.

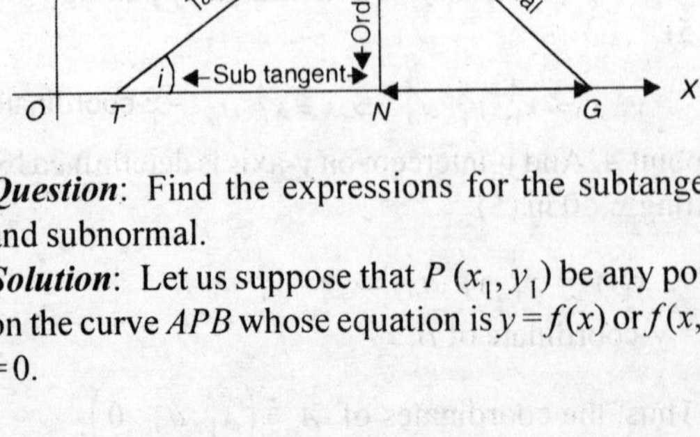

Question: Find the expressions for the subtangent and subnormal.

Solution: Let us suppose that $P(x_1, y_1)$ be any point on the curve APB whose equation is $y = f(x)$ or $f(x, y) = 0$.

The tangent and the normal at P meet the x-axis in T and G respectively. Draw the ordinate from P meeting the x-axis in N.

Let i be the angle which the tangent at P makes with the x-axis.

Then $\angle NPG = \angle NTP = i$

and $\tan i = \left[\frac{dy}{dx}\right]_{\substack{x=x_1 \\ y=y_1}}$ = Value of $\frac{dy}{dx}$ at (x_1, y_1)

Now, from the right angled ΔTPN subtangent $NT = NP \cot i$

$\left[\because NP = y_1 \text{ and } \tan i = \left(\frac{dy}{dx}\right)_{(x_1, y_1)}\right]$

$$= \frac{NT}{\tan i} = \frac{y_1}{\left[\frac{dy}{dx}\right]_{(x_1, y_1)}}$$

Again, from the rt right angled ΔNPG

Subnormal $= NG = NP \tan i$

$= y_1 \tan i$

$$= y_1 \left[\frac{dy}{dx}\right]_{\substack{x=x_1 \\ y=y_1}}$$

Working rule to find the length of subtantgent and subnormal for a given curve at any point (x_1, y_1):

Steps: **1.** Differentiate the given equation w.r.t x.

2. Find $\left[\frac{dy}{dx}\right]_{\substack{x=x_1 \\ y=y_1}}$

3. Use the formula for the subtangent and subnormal Length of subtangent

$$= \frac{y_1}{\left[\frac{dy}{dx}\right]_{\substack{x=x_1 \\ y=y_1}}}$$

$$= \frac{\text{Ordinate of the given point}}{\text{Value of } \frac{dy}{dx} \text{ at the given point } (x_1, y_1)}$$

and length of subnormal $= y_1 \cdot \left[\frac{dy}{dx}\right]_{\substack{x=x_1 \\ y=y_1}}$

= Ordinate of the given point times the value of $\frac{dy}{dx}$ at (x_1, y_1).

Note: **1.** When the point of contact of tangent and curve is not provided, it is supposed (x_1, y_1).
2. Negative length is not considered. This is why we take the absolute value as the length of subtangent and subnormal to a curve at any point (if negative).
3. When the given equation of the curve is in parametric form, it is better to use the following notational form of the formula.

(i) Length of subtangent $= \left|\frac{y}{\left[\frac{dy}{dx}\right]_{\text{at } \theta \text{ or any parameter given in the equation}}}\right|$

(ii) Length of subnormal

$$= \left| y \cdot \left[\frac{dy}{dx}\right]_{\text{at } \theta \text{ or any parameter given in the equation}} \right|$$

(iii) At any point of the curve whose parametric equation is given, we may suppose $\theta / t / \ldots$, the parameter given in the equation of the curve as representing the given point.

(iv) When numerical value of θ is not provided, $\frac{dy}{dx}$ serves as the value of $\frac{dy}{dx}$ at any point θ.

Problems based on subtangent and subnormal.

Examples worked out:

Question: Find the lengths of subtangents and subnormal for the curve $\frac{x^2}{8} + \frac{y^2}{18} = 1$ at $(2, -3)$

Solution: Given equation of the curve is

$$\frac{x^2}{8} + \frac{y^2}{18} = 1 \quad \ldots(1)$$

Now, differentiating the equation (1) w.r.t x, we have

$$\frac{x}{4} + \frac{y}{9}\frac{dy}{dx} = 0$$

$$\Rightarrow \frac{dy}{dx} = -\frac{9x}{4y}$$

$$\Rightarrow \left[\frac{dy}{dx}\right]_{\substack{x=x_1=2\\y=y_1=-3}} = \left[-\frac{9x}{4y}\right]_{\substack{x=x_1=2\\y=y_1=-3}} = \frac{-18}{-12} = \frac{3}{2}$$

$\therefore$ Length of subtangent at (2, –3)

$$= \left| y_1 \div \left[\frac{dy}{dx}\right]_{\substack{x=x_1\\y=y_1}} \right| = \left| -3 \div \frac{3}{2} \right| = 2$$

and length of subnormal at (2, –3)

$$= \left| y_1 \times \left[\frac{dy}{dx}\right]_{\substack{x=x_1\\y=y_1}} \right| = \left| -3 \times \frac{3}{2} \right| = \frac{9}{2}$$

Question: Find the lengths of subtangent and subnormal for the curve $xy + 2x - y - 5 = 0$ at the point (2, 1).

Solution: Given equation of the curves is

$$xy + 2x - y - 5 = 0 \quad \ldots(1)$$

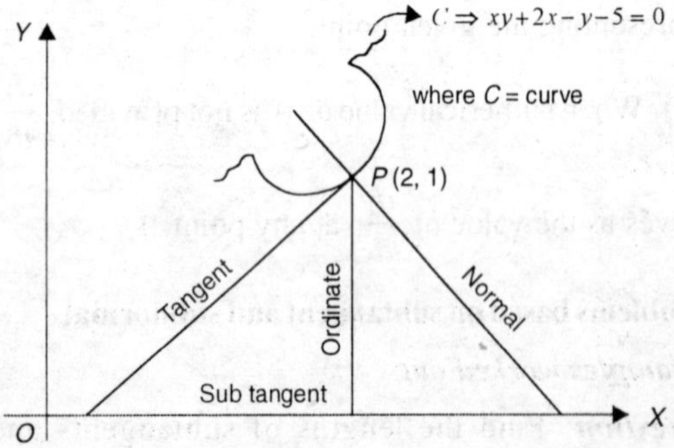

Now, differentiating the equation (1) w.r.t x, we have

$$y + x\frac{dy}{dx} + 2 - \frac{dy}{dx} = 0$$

$$\Rightarrow y + 2 + \frac{dy}{dx}(x - 1) = 0$$

$$\Rightarrow \frac{dy}{dx} = -\frac{(y+2)}{(x-1)} = \frac{(y+2)}{(1-x)}$$

Again, $\left[\frac{dy}{dx}\right]_{(2,1)} = \left[\frac{y+2}{1-x}\right]_{(2,1)} = \frac{1+2}{1-2} = -3$

Now, the length of subtangent at (2, 1)

$$= \left| y_1 \div \left[\frac{dy}{dx}\right]_{\substack{x=x_1\\y=y_1}} \right| = \left| -\frac{1}{3} \right| = \frac{1}{3}$$

and length of subnormal at (2, 1)

$$= \left| y_1 \times \left[\frac{dy}{dx}\right]_{\substack{x=x_1\\y=y_1}} \right| = \left| 1 \times (-3) \right| = 3$$

Question: Find the lengths of subtangent and subnormal for the curve $y = 3\sin\frac{x}{2}$ at the point $\left(\frac{\pi}{2}, \frac{3\sqrt{2}}{2}\right)$.

Solution: Given equation of the curve is

$$y = 3\sin\frac{x}{2} \quad \ldots(1)$$

$$\Rightarrow y_1 = 3\sin\frac{x_1}{2} \quad \ldots(2)$$

Now, differentiating the given equation (1) w.r.t x, we have

$$\frac{dy}{dx} = \frac{3}{2}\cos\frac{x}{2}$$

$$\Rightarrow \left[\frac{dy}{dx}\right]_{\substack{x=\frac{\pi}{2}\\y=\frac{3\sqrt{2}}{2}}} = \left[\frac{3}{2}\cos\frac{x}{2}\right]_{\substack{x=\frac{\pi}{2}\\y=\frac{3\sqrt{2}}{2}}} = \frac{3}{2}\cos\frac{\pi}{4} = \frac{3}{2\sqrt{2}}$$

Now, the length of subtangent at the point $\left(\frac{\pi}{2}, \frac{3\sqrt{2}}{2}\right)$

$$= \left| y_1 \div \left[\frac{dy}{dx}\right]_{\substack{x=x_1\\y=y_1}} \right| = \left| \frac{3\sqrt{2}}{2} \div \frac{3}{2\sqrt{2}} \right| = 2$$

and the length of the subnormal at the point

$$\left(\frac{\pi}{2}, \frac{3\sqrt{2}}{2}\right)$$

$$= \left| y_1 \times \left[\frac{dy}{dx}\right]_{\substack{x=x_1\\y=y_1}} \right| = \left| \frac{3\sqrt{2}}{2} \times \frac{3}{2\sqrt{2}} \right| = \frac{9}{4}$$

Question: Find the lengths of subtangent and subnormal for the curve

$x = a(\theta + \sin\theta)$

$y = a(1 - \cos\theta)$ at the point $\theta = \frac{\pi}{3}$, $a > 0$.

Solution: Given equation of the curve is

$$x = a(\theta + \sin\theta) \quad \ldots(1)$$

$$y = a(1 - \cos\theta) \quad \ldots(2)$$

Now, differentiating (1) and (2) w.r.t θ, we have

$$\frac{dx}{d\theta} = \frac{da\,(\theta + \sin\theta)}{d\theta}$$

$$= a\left\{\frac{d\theta}{d\theta} + \frac{d\sin\theta}{d\theta}\right\} = a(1 + \cos\theta) \quad \ldots(3)$$

and $\frac{dy}{d\theta} = a\frac{d\,(1 - \cos\theta)}{d\theta}$

$$= a\left\{\frac{d(1)}{d\theta} - \frac{d\cos\theta}{d\theta}\right\} = a\sin\theta \quad \ldots(4)$$

(3) and (4) $\Rightarrow \frac{dy}{dx} = \frac{dy}{d\theta} \div \frac{dx}{d\theta} = \frac{\sin\theta}{1 + \cos\theta}$

$$\Rightarrow \left[\frac{dy}{dx}\right]_{\theta=\frac{\pi}{3}} = \left[\frac{\sin\theta}{1 + \cos\theta}\right]_{\theta=\frac{\pi}{3}}$$

$$= \frac{\sin\frac{\pi}{3}}{1+\cos\frac{\pi}{3}} = \frac{\frac{\sqrt{3}}{2}}{1+\frac{1}{2}} = \frac{\sqrt{3}}{2} \times \frac{2}{3} = \frac{1}{\sqrt{3}}$$

Now, required length of subtangent at $\theta = \frac{\pi}{3}$

$$= \left| y \div \left[\frac{dy}{dx}\right]_{\theta=\frac{\pi}{3}} \right| = \left| a\left(1 - \cos\frac{\pi}{3}\right) \div \frac{1}{\sqrt{3}} \right|$$

$$= \left| a\sqrt{3}\left(1 - \frac{1}{2}\right) \right| = \left| \frac{a\sqrt{3}}{2} \right| = \frac{a\sqrt{3}}{2}$$

and the length of subnormal at $\theta = \frac{\pi}{3}$

$$= \left| y_1 \times \left[\frac{dy}{dx}\right]_{\theta=\frac{\pi}{3}} \right| = \left| a\left(1 - \cos\frac{\pi}{3}\right) \times \frac{1}{\sqrt{3}} \right|$$

$$= \left| a\left(1 - \frac{1}{2}\right) \times \frac{1}{\sqrt{3}} \right| = \left| a\left(\frac{2-1}{2}\right) \times \frac{1}{\sqrt{3}} \right| = \frac{a}{2\sqrt{3}}$$

Question: Find the lengths of subtangent and subnormal at any point of the curve

$x = a\,(2\cos\theta + \cos 2\theta)$

$y = a\,(2\sin\theta + \sin 2\theta)$

Solution: Given equation of the curve is

$$x = a\,(2\cos\theta + \cos 2\theta) \quad \ldots(1)$$

$$y = a\,(2\sin\theta + \sin 2\theta) \quad \ldots(2)$$

Now, differentiating (1) and (2) w.r.t θ, we have

$$\frac{dy}{dx} = \frac{\frac{dy}{d\theta}}{\frac{dx}{d\theta}} = \frac{a(2\cos\theta + 2\cos 2\theta)}{a(-2\sin\theta - 2\sin 2\theta)}$$

$$= -\frac{\cos 2\theta + \cos\theta}{\sin 2\theta + \sin\theta}$$

$$= -\frac{2\cos\frac{3\theta}{2} \cdot \cos\frac{\theta}{2}}{2\sin\frac{3\theta}{2} \cdot \cos\frac{\theta}{2}} = -\cot\frac{3\theta}{2}$$

$\therefore$ The length of subtangent at (x, y)

$$= \left| y \div \frac{dy}{dx} \right| = \left| y \div \cot\frac{3\theta}{2} \right| = \left| y \cdot \tan\left(\frac{3\theta}{2}\right) \right|$$

and the length of subnormal at (x, y)

$$= \left| y \times \frac{dy}{dx} \right| = \left| y \times -\cot\frac{3\theta}{2} \right| = \left| -y \cdot \cot\frac{3\theta}{2} \right|$$

where $y = a\,(2\sin\theta + \sin 2\theta)$ which is given in the equation of the curve.

Question: Show that subtangent at any point of the curve $x^m \cdot y^n = a^{m+n}$ varies as the abscissa.

Solution: Given equation is $x^m y^n = a^{m+n}$ ...(1)

Now, differentiating the given equation (1) w.r.t x,

we have $m \cdot x^{m-1} \cdot y^n + nx^m y^{n-1} \dfrac{dy}{dx} = 0$

$$\Rightarrow \frac{dy}{dx} = -\frac{mx^{m-1} y^n}{nx^m y^{n-1}} = -\frac{my}{nx}$$

$$\Rightarrow \left[\frac{dy}{dx}\right]_{\substack{x=x_1\\y=y_1}} = \text{Value of } \frac{dy}{dx} \text{ at } (x_1, y_1)$$

$$= \left[-\frac{my}{nx}\right]_{\substack{x=x_1\\y=y_1}} = -\frac{my_1}{nx_1}$$

Hence, the subtangent at any point $P(x_1, y_1)$

$$= y_1 \div \left[\frac{dy}{dx}\right]_{P(x_1, y_1)} = y_1 \div \left[-\frac{my_1}{nx_1}\right]$$

$$= y_1 \times \left[\frac{-nx_1}{my_1}\right] = \left[\frac{-nx_1}{m}\right] = kx_1$$ which implies that the length of subtangent varies as the abscissa of the point (x_1, y_1) [$\because x \propto y \Leftrightarrow x = ay$ where a = constant].

Question: Show that for the curve $y = cx^4$, the subtangent varies as its abscissa.

Solution: Given equation is $y = cx^4$ ...(1)

$$\Rightarrow \frac{dy}{dx} = 4c\,x^3$$

Again, $y = cx^4 \Rightarrow y_1 = cx_1^4$ ...(2)

[$\because (x_1, y_1)$ lies on the curve $y = cx^4$]

$\therefore$ The subtangent for the point (x_1, y_1) on the given curve

$$= y_1 \div \left[\frac{dy}{dx}\right]_{P(x_1, y_1)} = cx_1^4 \div 4cx_1^3$$

$$= \frac{c\,x_1^4}{4c\,x_1^3} = \frac{x_1}{4}$$

$\Rightarrow$ The subtangent varies as the abscissa of the point (x_1, y_1).

Question: Show that for the curve $y^2 = 4ax$, the length of the subtangent varies as the abscissa of the point of contact.

Solution: Let (x_1, y_1) be the point on the curve

$y^2 = 4ax$ ...(1)

$\Rightarrow y_1^2 = 4a\,x_1$ ...(2)

Now, differentiating (1) w.r.t. x, we have

$$2y\frac{dy}{dx} = 4a \Rightarrow \frac{dy}{dx} = \frac{4a}{2y} = \frac{2a}{y}$$

$$\Rightarrow \left[\frac{dy}{dx}\right]_{P(x_1, y_1)} = \left[\frac{2a}{y}\right]_{P(x_1, y_1)} = \frac{2a}{y_1} \quad ...(3)$$

$\therefore$ Length of the subtangent $= \left| y_1 \div \left[\dfrac{dy}{dx}\right]_{P(x_1, y_1)} \right|$

$$= \left| y_1 \div \frac{2a}{y_1} \right| = y_1 \times \frac{y_1}{2a}$$

$$= \frac{y_1^2}{2a} = \frac{4a\,x_1}{2a} = 2\,x_1 \text{ [from (2)]}$$

Thus, the length of subtangent at (x_1, y_1) on the given curve $= 2x_1$

$\Rightarrow$ Subtangent is twice the abscissa.

$\Rightarrow$ The subtangent varies as the abscissa.

[$\because$ Subtangent varies as the abscissa $\Leftrightarrow$ Subtangent = a constant $\times$ abscissa of the point]

Question: Show that in the curve $y = be^{\frac{x}{a}}$ the subtangent at any point is of constant length. Also show that the subnormal varies as the square or ordinate.

Solution: Let $P(x_1, y_1)$ be the point of contact of the tangent and normal to the given curve whose equation is $y = be^{\frac{x}{a}}$ ...(1)

$$\therefore y_1 = b e^{\frac{x_1}{a}} \qquad \ldots(2)$$

Now, differentiating both sides of equation (1) w.r.t x, we get

$$\frac{dy}{dx} = \frac{b}{a} \cdot e^{\frac{x}{a}} = \frac{y}{a}$$

$$\Rightarrow \left[\frac{dy}{dx}\right]_{P(x_1, y_1)} = \left[\frac{y}{a}\right]_{P(x_1, y_1)} = \frac{y_1}{a} \qquad \ldots(3)$$

Now the subtangent at the point $P(x_1, y_1)$ on the curve

$$= \left| y_1 \div \left[\frac{dy}{dx}\right]_{P(x_1,y_1)} \right| = \left| y_1 \div \frac{y_1}{a} \right|$$

$$= \left| y_1 \times \frac{a}{y_1} \right| = |a| \text{ (constant)}$$

and the subnormal at the point $P(x_1, y_1)$ on the curve

$$= y_1 \times \left[\frac{dy}{dx}\right]_{P(x_1,y_1)} = y_1 \times \frac{y_1}{a} = \frac{y_1^2}{a}$$

which implies that subnormal varies as the square of the ordinate.

Question: Show that for the curve $y = \sqrt{3x + 2}$, the subnormal is of constant length.

Solution: Let (x_1, y_1) be the point of contact of the normal and the curve whose equation is

$$y = \sqrt{3x + 2} \qquad \ldots(1)$$

$$\therefore y_1 = \sqrt{3x_1 + 2} \qquad \ldots(2)$$

Now, differentiating (1) w.r.t x, we get

$$\frac{dy}{dx} = \frac{3}{2\sqrt{3x + 2}}$$

$$\Rightarrow \left[\frac{dy}{dx}\right]_{P(x_1, y_1)} = \frac{3}{2\sqrt{3x_1 + 2}} \qquad \ldots(3)$$

$\Rightarrow$ The length of subnormal

$$= \left| y_1 \times \left[\frac{dy}{dx}\right]_{(x_1, y_1)} \right| = \frac{3}{2} \text{ which is a constant.}$$

Summary of important facts of working rule of different types of problems on tangent and normal:

1. Equation of the tangent to the curve $y = f(x)$ at any point $P(x_1, y_1)$ of the curve is given by

$$(y - y_1) = \left[\frac{dy}{dx}\right]_{P(x_1, y_1)} \cdot (x - x_1).$$

2. Equation of the normal at $P(x_1, y_1)$ to the curve $y = f(x)$ is $(y - y_1)$

$$= -\frac{1}{\left[\frac{dy}{dx}\right]_{P(x_1,y_1)}} \cdot (x - x_1).$$

3. Let us consider the tangent to be parallel or perpendicular to the x-axis.

If the tangent is parallel to x-axis or normal is perpendicular to the x-axis, then $m = 0$, so that $\frac{dy}{dx} = 0$.

If the tangent is perpendicular to x-axis or normal is parallel to x-axis

$\therefore \frac{dy}{dx} = \infty(op - \infty)$ or, its reciprocal, $\frac{dx}{dy} = 0$ and we write $m = \infty(-\infty)$

4. *Angle of intersection of the two curves*:
By angle of intersection of two curves, we mean the angle between the tangents to the curves at their common point of intersection.

Hence, if θ be the acute angle between the tangents, then $\tan\theta = \left| \frac{m_1 - m_2}{1 + m_1 m_2} \right|$

Where $m_1 = \left[\frac{dy}{dx}\right]_{C_1} = \left[\frac{dy}{dx}\right]_{\substack{x = x_1 \\ y = y_1}}$ = Value of $\frac{dy}{dx}$ at the common point of intersection (x_1, y_1) for the first curve.

$m_2 = \left[\frac{dy}{dx}\right]_{C_2} = \left[\frac{dy}{dx}\right]_{\substack{x = x_1 \\ y = y_1}}$ = Value of $\frac{dy}{dx}$ at the common point of intersection (x_1, y_1) for the second curve.

5. *Condition for orthogonal intersection*:
Two curves are said to cut orthogonally if the angle between them is a right angle i.e. $\theta = 90^\circ$

$$\therefore\ i_2 - i_1 = \pm\frac{\pi}{2}$$

$$\therefore\ \tan i_2 = -\cot i_1$$

$$\therefore\ m_2 = -\frac{1}{m_1} \Rightarrow \left[\frac{dy}{dx}\right]_1 \cdot \left[\frac{dy}{dx}\right]_2 = -1$$

Where $m_1 = \left[\frac{dy}{dx}\right]_1$ = Value of $\frac{dy}{dx}$ at the common point of intersection for the first curve.

$m_2 = \left[\frac{dy}{dx}\right]_2$ = Value of $\frac{dy}{dx}$ at the common point of intersection for the second curve.

6. *Condition for the two curves to touch*:
If the two curves touch, then the angle θ between them is zero, i.e., $\theta = 0 \Rightarrow \tan\theta = 0$

$$\therefore\ m_1 - m_2 = 0 \Rightarrow m_1 = m_2 \Rightarrow \left[\frac{dy}{dx}\right]_1 = \left[\frac{dy}{dx}\right]_2$$

7. *Intercepts of tangent on axis*:
Find the equation of the tangent. Put $y = 0$ find the value of x which will be the intercept on axis of x known as x-intercept. Then put $x = 0$ and find the value of y which will be the intercept on y-axis known as y-intercept.

8. *Condition for a given line to touch a given curve*:
Let the line be a tangent to the given curve at (x, y), then write the equation of the tangent as

$$(Y - y) = \left[\frac{dy}{dx}\right] \cdot (X - x)$$

compare this equation of tangent with given line $ax + by + c = 0$ and then eliminate x and y which will be a relation involving the given constants only representing the required condition for a given line to touch a given curve.

Type 1: Problems based on inclination and slopes when x_1 and / y_1 is given.

(A) Problems based on finding inclinations:

Exercise 19.1

1. Find the inclination of the following curves:

(i) $y = x^2 - x + 1$ at $(1, 1)$

(ii) $x^3 + y^3 = 3axy$ at the point $\left(\frac{3a}{2}, \frac{3a}{2}\right)$

(iii) $y^2 = 4ax$ at the point $x = a$

(iv) $y = \frac{4}{x + 2}$ at the point $x = 0$

(v) $y = \frac{1}{x}$

(vi) $y^2 = 4x$ at the point $x = 1$

(vii) $y = x^2 + 2x + 3$ at the point $(0, 3)$

(viii) $y^2 = 5^2 - x^2$ at $x = 3$

(ix) $y = x - \frac{1}{x}$ at the point where $x = 1$

(x) $3y = x^2$ at the point $(3, 3)$

(xi) $y = x^2 - x + 1$ at the point $(1, 1)$

(xii) $2x^2 + 2y = 7$ at the point where $x = 2$

(xiii) $y^2 = 9 - x^2$ at the point where $x = 2$

(xiv) $y = x^4 - 4x$ at the point $x = 0$.

2. Find the inclination of the tangent to the curve $y = x^3 - x^2 + 1$ at the point $(1, 1)$ on it.

3. Find the inclination to the x-axis of the tangent to the parabola $y^2 = 4ax$ at the point (α, β) and determine the point at which the tangent makes an angle of 45° with the axis.

4. Find the inclination of the tangents at the point $(1, 0)$ and $(2, 0)$ to the curve $y = (x - 1)(x - 2)$.

Answers:

1. (i) 45° **(ii)** 135° **(iii)** [45°, 135°] **(iv)** 135° **(v)** 135 **(vi)** [45°, 135°] **(vii)** $\tan^{-1} 2$

(viii) $\tan^{-1}\left[\pm\frac{3}{4}\right]$ **(ix)** $\tan^{-1} 2$ **(x)** $\tan^{-1} 2$

(xi) 45° **(xii)** $\tan^{-1}(-4)$ **(xiii)** $\tan^{-1}\left(\pm\frac{2}{\sqrt{5}}\right)$

(xiv) $\tan^{-1}(-4)$

2. $\frac{\pi}{2}$ **3.** $[a, 2a]$ **4.** $\frac{5\pi}{12}$

Type 1: **(B)** Problems based on finding slopes:

Exercise 19.2

1. Find the slope of the normal to the curve $y = 3x^2$ at the point whose x co-ordinate is 2.

2. Find the slopes of the tangents to the following curves:

(i) $y = 3x^2$ at $x = 1$

(ii) $y = 2x^2 - 1$ at $x = 1$

(iii) $y = x^3 + 4x$ at $x = -1$

(iv) $y = x^3 - x$ at $x = 2$

(v) $y = 2x^3 + 3\sin x$ at $x = 0$

(vi) $y^2 = 4x$ at the point $x = 1$

3. Find the slope of the curve $y = (1 + x)\sin x$ at $x = \dfrac{\pi}{4}$.

4. Show that the tangents to the curve $y = \dfrac{2x+5}{x^2-6}$ at the points $\left(-3, -\dfrac{1}{2}\right)$ and $\left(-2, -\dfrac{1}{2}\right)$ are parallel to x-axis.

5. Find the gradients of the following curves at the given points on them.

(i) $y = x^2 + 2x + 5$ at the points $(0, 5)$, $(1, 8)$ and $(2, 13)$

(ii) $y = 3x^2 + 7x + 5$ at the point where it cuts the y-axis.

(iii) $y = x^2 - 5x + 8$ at the points where it cuts the straight line $y = x$.

Answers:

1. $-\dfrac{1}{12}$ at $x = 2$

2. (i) 6 **(ii)** 4 **(iii)** 7 **(iv)** 11 **(v)** 3' **(vi)** Find

3. $\dfrac{1}{\sqrt{2}}\left(2 + \dfrac{\pi}{4}\right)$

5. (i) 2, 4, 6 **(ii)** 7 **(iii)** –1, 3

Type 2: Problems based on finding the equation of the tangent and normal when x_1 and / y_1 is given:

(A) Problems based on finding the equation of the tangent:

Exercise 19.3

1. Find the equation of the tangent to the curve $y = x^3 - 2x^2 + x + 2$ at the point $(1, 2)$.

2. Find the equation of the tangent of the curve $y = 2\sin x + \sin 2x$ at $x = \dfrac{\pi}{2}$.

3. Find the equations of the tangents to the curve $y = \sin x$ at $x = \dfrac{\pi}{4}$.

4. Find the equation of the tangent to $y = 4 + \cos^2 x$ at $x = \dfrac{\pi}{4}$.

5. Find the equation of the tangent to
$$y = \sin^2 x + \cot^2 x + 3 \text{ at } x = \frac{9\pi}{4}.$$

6. Find the equation of the tangent to the curve
$$y = x - \sin x \cos x \text{ at } x = \frac{\pi}{2}.$$

7. Find the equation of the tangent at $x = \dfrac{\pi}{4}$ to the curve $y = \cot^2 x - 2\cot x + 2$.

8. Find the equation of the tangent to the curve
$$y = \sec^4 x - \tan^4 x \text{ at } x = \frac{\pi}{3}.$$

9. Find the equation of the tangent to the curve $y = e^x$ at $(0, 1)$.

10. Find the equation of the tangent to the curve
$$y^2 \sin x = 9 \text{ at } x = \frac{\pi}{2}.$$

11. Find the equation of the tangent of the curve:
$$x = a\cos^3 t$$
$$y = a\sin^3 t \text{ at the point '}t\text{'}.$$

12. Find the equation of the tangent of the curve:
$$x = a\cos\theta$$
$$y = a\sin\theta \text{ at } \theta = \frac{\pi}{4}.$$

13. Find the equations of the tangents drawn to the curve $y^2 = 2x^3 - 4y + 8 = 0$ from $(1, 2)$.

14. Find the equation of the tangent to the parabola $y^2 = 4ax$ at $(at^2, 2at)$.

Answers:

1. $x+2y+1$

2. $2y = 3\sqrt{3}$

3. $4\left(x-\sqrt{2}y+1\right) = \pi,\ 4\left(\sqrt{2}x-y\right) = \sqrt{2}\,(3\pi-2)$

4. $y-4=0$

5. $4y+12x-27\pi-18=0$

6. $2y-4x+\pi=0$

7. $y=1$

8. $3y-48\sqrt{3}x+16\sqrt{3}\pi-21=0$

9. $y=x+1$

10. $y+3=0$

11. $y-a\sin^3 t = -\tan t\,(x-a\cos^3 t)$

12. $\frac{x}{a}+\frac{y}{b}=\sqrt{2}$

13. $y-2=\pm 2\sqrt{3}\,(x-1)$

14. $ty = x+at^2$

Type 2: **(B)** Problems based on finding the equation of the normal:

Exercise 19.4

1. Find the equation of the normal to $y = \cot x$ at $x = \frac{\pi}{4}$.

2. Find the equation of the normal to the curve $y = \sin^2 x$ at $x = \frac{\pi}{2}$

3. Find the normal to the curve $y = \sin x + \cos x$ at $x = 0$.

4. Find the equation of the normal to the curve $y = 2\sin\frac{2}{3}x$ at $x = \frac{\pi}{6}$.

5. Find the equation of the normal to the curve $y = x + \sin x \cos x$ at $x = \frac{\pi}{2}$.

6. Find the equation of the normal to the curve
$$y = \frac{\tan ax + \sin 2x}{1+x^2} \text{ at } x = \pi.$$

7. Find the equation of the normal to the curve
$$y = \frac{1+\sin x}{\cos x} \text{ at } x = \frac{\pi}{4}.$$

8. Find the equation of the normal to the curve $4x^2 + y^2 = 2$ at the point where $x = \frac{1}{2}$.

9. Find the equation of the normal to the curve $x^2 = 4y$ which passes through the point $(1, 2)$.

10. Find the equation of the normal at the point (am^2, am^3) for the curve $ay^2 = x^3$.

11. Find the equation of the normal to the curve $y = x^3 - 2x^2 + 4$ at the point whose x-coordinate is 2.

12. Find the equation of the normal to
$$y = (\sin 2x + \cot x + 2)^2 \text{ at } x = \frac{\pi}{2}.$$

13. Find the equation of the normal to $y = \cos(5x+4)$ at $x = \left(\frac{\pi-8}{6}\right)$.

Answers:

1. $8y - 4x + \pi - 8 = 0$

2. $x = \frac{\pi}{2}$

3. $x+y-1=0$

4. $x = \frac{\pi}{6}$

5. $2x - \pi = 0$

6. $3y + \left(1+\pi^2\right)x = \pi\left(1+\pi^2\right)$

7. $x + \left(2+\sqrt{2}\right)y - 4 - 3\sqrt{2} - \frac{\pi}{4} = 0$

8. $x - 2y + \frac{3}{2} = 0$

9. $x+y=3$

10. $2x + 2my = 23am^2 + 3am^4$

11. $4y + 8x = 18$

12. $y-4=\frac{1}{12}\left(x-\frac{\pi}{2}\right) \Leftrightarrow 24y-2x+\pi-26=0$

13. $18y-6x+\pi-8=0$

***Type 2*: (C) Problems based on finding the equation of the tangent and normal simultaneously:**

Exercise 19.5

1. Find the equations of the tangent and normal to the curve $y=x^3+2x+6$ at (2, 18).

2. Find the equations of the tangent and normal to the curve $16x^2+4y^2=144$ at (x_1, y_1) where $x_1=2$ and $y_1>0$.

3. Find the equations of the tangent and normal to the curve $y=x^2-4x-5$ at $x=-2$.

4. Find the equations of the tangent and normal to the curve $y=x^2-4x+2$ at (4, 2).

5. Find the equations of the tangent and normal to the curve:

$x=at^2$

$y=2at$ at the point 't'.

6. Find the equations of the tangent and normal at the point (2, 5) of the curve $y=\frac{2x+1}{3-x}$.

7. Find the equations of the tangent and normal to the curve $y=x^2-3x+4$ at the point where it cuts the y-axis.

8. Find the equations of the tangent and normal to the curve $3y=x^2-6x+17$ at (4, 3).

***Type 3*: Problems based on intercepts of tangents on the axis:**

Exercise 19.6

1. Show that the sum of the intercepts of the tangent at any point to the curve $x^{\frac{1}{2}}+y^{\frac{1}{2}}=a^{\frac{1}{2}}$ on the axes is constant and is equal to a.

2. Prove that the length intercepted by the coordinate axes on any tangent to the curve $x^{\frac{2}{3}}+y^{\frac{2}{3}}=c^{\frac{2}{3}}$ is constant.

3. If the tangent to the curve $\sqrt{x}+\sqrt{y}=\sqrt{a}$ at any point on it cuts the x and y-axes respectively at A and B, prove that $OA+OB=a$.

4. In the curve $x^m y^n=a^{m+n}$, prove that the portion of the tangent intercepted between the axes is divided at its point of contact into segments which are in a constant ratio.

5. Find the intercepts made upon the axes by the tangent at (x_1, y_1) to the curve $\sqrt{x}+\sqrt{y}=\sqrt{a}$ and show that their sum is constant.

6. Show that the portion of the tangent to the curve $x^{\frac{2}{3}}+y^{\frac{2}{3}}=a^{\frac{2}{3}}$ which is intercepted between the axes is of constant length.

7. Find the equation of the tangent to the curve $x^{\frac{2}{3}}+y^{\frac{2}{3}}=a^{\frac{2}{3}}$ and show that the portion of the tangent intercepted between the coordinate axes is constant.

8. Find the normal to the curve $\sqrt{xy}=a+x$ which makes equal intercepts upon coordinate axes.

9. The tangent at any point on the curve $x^3+y^3=2a^3$ cuts off lengths p and q on the coordinate axes, show that $p^{-\frac{2}{3}}+a^{-\frac{3}{2}}=2^{-\frac{1}{2}}a^{-\frac{3}{2}}$.

10. If p and q be the intercepts on the coordinate axes by the tangent to $\left(\frac{x}{a}\right)^n+\left(\frac{y}{b}\right)^n=1$, then show that $\left(\frac{a}{p}\right)^{\frac{n}{n-1}}+\left(\frac{b}{q}\right)^{\frac{n}{n-1}}=1$

11. Prove that, in the curve $y=\frac{C}{2}\left[e^{\frac{x}{2}}+e^{-\frac{x}{2}}\right]$ the length of the perpendicular, from the foot of the ordinate of a point p on the curve, upon the tangent at p is constant.

12. Find the equation of the tangent at the point determined by θ on the ellipse $x=a\cos\theta$, $y=b\sin\theta$. Also find the length of the portion of the tangent intercepted between the coordinate axes.

13. Prove that the portion of the tangent of the curve $x=a\cos^3\theta$, $y=a\sin^3\theta$ at the point θ, intercepted between the coordinate axes is of constant length.

Answers:

8. Normal at the point $\left(\frac{a}{\sqrt{2}}, \frac{3a}{\sqrt{2}} + 2a\right)$ if

$x - y + a\left(2+\sqrt{2}\right) = 0$

12. $\frac{x\cos\theta}{a} + \frac{y\sin\theta}{b} = 1, \sqrt{a^2\sec^2\theta + b^2\operatorname{cosec}^2\theta}$

Type 4: Finding the points where the tangent ...

Exercise 19.7

1. Find the points on the curve $x^2 - y^2 = 2$ at which the slope of the tangent is 2.
2. Find at what points on the circle $x^2 + y^2 = 13$, the tangent is parallel to the line $2x + 3y = 7$.
3. At what points on the curve $x^2 + y^2 - 2x - 4y + 1 = 0$, the tangent is parallel to (i) x-axis (ii) y-axis.
4. Find the point on the curve $y = x^3 + 2x^2 - 3x + 1$, the tangent at which is parallel to the line $4x - y = 3$.
5. Find the coordinates of the points on the curve $2x^2 + 3xy + 4y^2 = 9$ at which the slope is $-\frac{7}{9}$.
6. At what point does the curve $y = x^2 - 4x$ have the slope 2?
7. Find the coordinates of the points on the curve $5 = \log(x^2 + 3)$ at which the slope is 2.
8. Find the point on the curve $y = 4 + x^2$ at which the tangent is horizontal.
9. Find the point at which the tangent to the curve $x^2 + y^2 + 2x - 4y = 20$ is parallel to the x-axis.
10. Find the point at which the tangent to the curve $y = x^3 - 12x + 10$ is parallel to x-axis.
11. Find the point on the curve $y = x^3 - x^2 - x + 3$ where the tangent is perpendicular to the y-axis.
12. At what point on the curve $\frac{x^2}{a^2} + \frac{y^2}{b^2} = 1$ is the tangent perpendicular to the x-axis.
13. At what points on the curve $y = x^2 + 2x$ is the tangent
(i) parallel to the x-axis
(ii) equally inclined to the axes
(iii) inclined at 30° to the x-axis.
14. Find the points at which the tangent is parallel to the axis of x for the curve $y = x^3 - 6x^2 - 15x + 5$.
15. Find the coordinates of the point at which the tangent to the curve $xy + 4 = 0$ make an angle of 45° with the axis of x.
16. Find the points on the curve $x^2 - y^2 = 2$ at which the slope of the tangent is 2.
17. At what point on the curve $y = 2x^2 - x + 1$ is the tangent parallel to the line $y = 3x + 9$.
18. (i) Find at what points on the circle $x^2 + y^2 = 13$, the tangent is parallel to the line $2x + 3y = 0$.
(ii) Find the points on the curve $y = \cos(x + y)$, $-2\pi \le x \le 2\pi$ at which the tangents are parallel to the line $x + 2y = x + 2y = 0$.
19. Find at what points on the curve $\frac{x^3}{a} + \frac{y^3}{b} = xy$, the tangent is parallel to one of the coordinate axes.
20. Find the points on the curve $y = x^3 - 2x^2 + x - 2$, where the gradient is zero.
21. (i) Find the point on the curve $y = 4 + x^2$ at which the tangent is horizontal.
(ii) Find the points on the curve $y = x^3 - x^2 - x + 3$ where the tangent is perpendicular to the y-axis.
22. (i) Find the points at which the tangent is parallel to the axis of x for the curve $y = x^3 - 6x^2 - 15x + 5$.
(ii) Find the points on the curve $y = x^3$, the tangents at which cut the x-axis at an angle of 60°.
23. (i) Find the point on the curve $y^2 = 4ax$, the tangent at which is inclined at 45° to the x-axis.
(ii) At what point on the curve $y = 2x^2 - x + 1$ is the tangent parallel to the line $y = 3x + 9$.
(iii) Find at what points on the circle $x^2 + y^2 = 13$, the tangent is parallel to the line $2x + 3y = 0$.
(iv) Find the points on the curve $4x^2 + 9y^2 = 1$ where the tangents are perpendicular to the line $2y + x = 0$.

Answers:

1. $\left(\frac{2\sqrt{2}}{3}, \frac{\sqrt{2}}{3}\right), \left(\frac{-2\sqrt{2}}{\sqrt{3}}, \frac{-\sqrt{2}}{3}\right)$
2. (2, 3) and (–2, 3)
3. (3, 2) and (–1, 2)
4. $\left(\frac{-7}{3}, \frac{247}{27}\right)$

5. (1, 1) and (–1, –1)

6. (3, –3)

7. $\left\{\frac{5+\sqrt{3}}{2}, 5\log\frac{5}{2}\left(5+\sqrt{13}\right)\right\}$,

$\left\{\frac{5-\sqrt{3}}{2}, 5\log\frac{5}{2}\left(5-\sqrt{13}\right)\right\}$

8. (0, 4)

9. (–1, 7) and (–1, –3)

10. (2, – 6) and (–2, 26)

11. (1, 2) and $\left(\frac{1}{3}, 2\frac{16}{27}\right)$

12. $(a, 0)$ and $(-a, 0)$

13. **(i)** (–1, 1)

(ii) $\left(-\frac{1}{2}, -\frac{3}{4}\right)$

(iii) $\left(\frac{-2\sqrt{3}-1}{2\sqrt{3}}, -\frac{16\sqrt{3}-9}{12}\right)$

14. (–1, 13) and (5, –95)

15. (2, –2) and (–2, 2)

16. $\left(\frac{2\sqrt{3}}{3}, \frac{\sqrt{2}}{3}\right)$ and $\left(-\frac{2\sqrt{3}}{3}, -\frac{\sqrt{2}}{3}\right)$

17. (1, 2)

18. **(i)** (2, 3) and (–2, –3)

(ii) $\left(-\frac{3\pi}{2}, 0\right)$ and $\left(\frac{\pi}{2}, 0\right)$

19. $\left\{\frac{1}{3}\left(2a^2b\right)^{\frac{1}{3}}, \frac{1}{3}\left(2a^2b\right)^{\frac{2}{3}}\right\}$ and

$\left\{\frac{1}{3b}\left(2ab^2\right)^{\frac{2}{3}}, \frac{1}{3}\left(2ab^2\right)^{\frac{2}{3}}\right\}$

20. (1, – 2) and $\left(\frac{1}{3}, \frac{-50}{27}\right)$

21. **(i)** (0, 4) **(ii)** (1, 2)

22. **(i)** (–1, 13) and (5, –95)

(ii) $\left(3-\frac{1}{4}, 3-\frac{2}{3}\right)$

23. **(i)** $(a, 2a)$ **(ii)** (1, 2)

(iii) (2, 3) and (–2, –3) **(iv)** $\left(\frac{2}{2\sqrt{10}}, \pm\frac{1}{3\sqrt{10}}\right)$

***Type 5*: (A) Problems based on finding angles of intersection or angle between two curves:**

Exercise 19.8

1. Find the angle of intersection of the curves $y = x^2$ and $y = x^3$.

2. Find the angle of intersection of the curves $y = 4 - x^2$ and $y = x^2$.

3. Find the angle between the curves:

(i) $y^2 = x$ and $x^2 = y$

(ii) $y = x^2$ and $y = 4 - x^2$

(iii) $xy = 4$ and $x^2 + y^2 = 8$

(iv) $x^2 - y^2 = 2a^2$ and $x^2 + y^2 = 4a^2$

(v) $y = 6 - x^2$ and $x^2 = 4y$ at the point (2, 2).

4. **(i)** At what angle the parabolas $y^2 = x$ and $x^2 = 8y$ cut each other.

(ii) Find the angle at which the curves $y = \sin x$ and $y = \cos x$ intersect.

5. Find the angle of intersection of the parabolas $y^2 = 2x$ and the circle $x^2 + y^2 = 8$.

6. Find the angle between the line $y = x$ and the curve $2y = 7x - 5x^2$.

7. Find the angle between the curve $y = x^3$ and the straight line $y = 9x$ at each of their point of intersection.

8. Find the angle between the curves:

(i) $x^2 + y^2 = 5$ and $y^2 = 4x + 8$

(ii) $xy = 2a^2$ and $y^2 = 4ax$

9. Show that the curves $y = \frac{x+3}{x^2+1}$ and

$y = \frac{x^2 - 7x + 11}{x - 1}$ cut each other at the point (2, 1) at an angle 45°.

10. Show that the curves $y = 2\sin^2 x$ and $y = \cos 2x$ intersect at $x = \frac{\pi}{6}$. Find the angle of intersection.

Answers:

1. $\tan^{-1}\left(\frac{1}{7}\right)$ and 0

2. $\tan^{-1}\left(\frac{4\sqrt{2}}{7}\right)$

3. (i) 90° and $\tan^{-1}\left(\frac{3}{4}\right)$ (ii) $\tan^{-1}\left(\frac{4\sqrt{2}}{7}\right)$

(iii) 0° (iv) 60° (v) $\tan^{-1}\left(\frac{7}{11}\right)$

4. (i) $\tan^{-1}\left(\frac{3}{5}\right)$ and 90°

(ii) $\tan^{-1}(2\sqrt{2})$

5. $\tan^{-1}(3)$ at both common points (2, –2) and (2, –2)

6. at (0, 0), $\theta = \tan^{-1}\left(\pm\frac{5}{9}\right)$ and at (1, 1),

$\theta = \tan^{-1}(\pm 5)$

7. Find

8. (i) $\tan^{-1}\left(\frac{1}{3}\right)$ (ii) $\tan^{-1}(3)$

10. Angle of intersection = 60°.

Type 5: **(B)** Problems based on finding the angle between two tangents to a curve at two given points:

Exercise 19.9

1. Prove that the tangents to the curve $y^2 = 4ax$ at the points where $x = a$ are perpendicular to each other.
2. Prove that the tangents to the curve $y^2 = 2x$ at the points where $x = \frac{1}{2}$ are at right angles.
3. Find the angle between the tangents to the curve $x^2 = 8y + 6$ at the points $\left(0, -\frac{3}{4}\right)$ and $\left(4, \frac{5}{4}\right)$.
4. Prove that the tangents to the curve $y^2 = x$ at the points $\left(\frac{1}{4}, \frac{1}{2}\right)$ and $\left(\frac{1}{4}, -\frac{1}{2}\right)$ are at right angles.
5. Find the angle between the tangents to the curve $\frac{x^2}{a^2}+\frac{y^2}{b^2} = 1$ at the points $(a, 0)$ and $(0, b)$.
6. Show that the tangents to the curve $y^2 = 2ax$ at the points where $x = \frac{1}{2}$ are at right angles.

Type 6: **(A)** Problems based on finding the condition for two curves for orthogonal intersection.

(B) Problems based on showing for two curves to cut orthogonally.

(A) Firstly we set the problems on finding the conditions for two curves for orthogonal intersection.

Exercise 19.10

1. Find the condition that the curves $\frac{x^2}{a}+\frac{y^2}{b} = 1$ and $\frac{x^2}{\alpha}+\frac{y^2}{\beta} = 1$ cut orthogonally.
2. Find the condition in order that the curves $\frac{x^2}{a^2}+\frac{y^2}{b^2} = 1$ and $\frac{x^2}{a_1^2}-\frac{y^2}{b_1^2} = 1$ should intersect at right angles.
3. Find the conditions that $ax^2 + by^2 = 1$ and $a_1 x^2 + b_1 y^2 = 1$ may cut right angles.

Type 6: **(B)** Problems based on showing for two curves to cut orthogonally.

Exercise 19.11

1. Do the curves $x^2 + y^2 = 2a^2$ and $2y^2 - x^2 = a^2$ cut each other orthogonally?
2. Show that the curves $x^3 - 3xy^2 + 2 = 0$ and $3x^2 y - y^2 = 2$ cut orthogonally.
[*Hint*: Show that $m_1 m_2 = -1$ at any point $(x_1 y_1)$]
3. Show that the curves $x^2 - y^2 = 16$ and $xy = 25$ cut each other at right angles.
4. Prove that the curves $2y^2 = x^3$ and $y^2 = 32^x$ cut each other at right angle at the origin.
5. Show that the curves $x^2 + 4y^2 = 8$ and $x^2 - 2y^2 = 4$ intersect orthogonally.

6. If the curves $\frac{x^2}{a^2}+\frac{y^2}{4}=1$ and $y^3=16^x$ intersect at right angles. Show that $a^2=\frac{4}{3}$.

7. Show that the curves $x^3-3xy^2=a$ and $3x^2y-y^2=b$ cut orthogonally.

8. Prove that the curves $y^2=4x$ and $x^2+y^2-6x+1=0$ touch each other at the point (1, 2).

9. Prove that for all values of *n*, the line $\frac{x}{a}+\frac{y}{b}=2$ touches the curve $\left(\frac{x}{a}\right)^n+\left(\frac{y}{b}\right)^n=2$ at any point (a, b).

***Type 6*: (C) Problems based on condition for two curves to touch.**

Exercise 19.12

1. Prove that the curves $y=6+x-x^2$ and $y(x-2)=x+2$ touch each other at (2, 4). Also find the equation of the common tangent.

2. Prove that the curves $xy=4$ and $x^2+y^2=8$ touch each other.

3. Show that $\frac{x}{a}+\frac{y}{b}=1$ touches the curves $y=be^{-\frac{x}{a}}$ at the point where the curve crosses the axis of *y*.

4. Prove that the curves $y^2=4x$ and $x^2+y^2-6x+1=0$ touch each other at the point (1, 2).

5. Prove that the curves $y=e^{-ax}$ and $y=e^{-ax}$ touch at the points for which $bx=2n\pi+\frac{\pi}{2}$.

***Type 7*: Problems based on condition for a given line to touch a given curve.**

Exercise 19.13

1. Prove that $\frac{x}{a}+\frac{y}{b}=1$ touches the curve $y=be^{-\frac{x}{a}}$ at the point where the curve crosses the *y*-axis.

[***Hint***: Show that equation of the tangent at $(0, b)=\frac{x}{a}+\frac{y}{b}=1$].

2. Prove that the condition that $x\cos\theta+y\sin\theta=p$ should touch the curve: $\frac{x^m}{a^m}+\frac{y^m}{b^m}=1$ is

$a(\cos\theta)^{\frac{m}{m-1}}+(b\sin\theta)^{\frac{m}{m-1}}=p^{\frac{m}{m-1}}$.

3. Prove that the condition that

$x\cos\alpha+y\sin\alpha=p$

should touch the curve $x^m y^n=a^{m+n}$ is

$p^{m+n}\cdot m^m=n^n(m+n)^{m+n}a^{m+n}\cos^m\alpha\sin^n\alpha$.

4. If $x\cos\theta+y\sin\theta=p$ touches the curve $\left(\frac{x}{a}\right)^{\frac{n}{n-1}}+\left(\frac{y}{b}\right)^{\frac{n}{n-1}}=1$, show that $(a\cos\theta)^n+(b\sin\theta)^n=p^n$.

5. Show that the curve $\left(\frac{x}{a}\right)^n+\left(\frac{y}{b}\right)^n=2$ touches the straight line $\frac{x}{a}+\frac{y}{b}=2$ at the point (a, b) for all values of *n*.

6. Prove that the straight line $y=2x-1$ touches the curve $y=x^3-x+1$.

7. Prove that the straight line $y=2x-1$ touches the curve $y=x^4+2x^3-3x^2-2x+3$ at two distinct points.

8. Prove that the curve $\left(\frac{x}{a}\right)^n+\left(\frac{y}{b}\right)^n=2$ touches the straight line $\frac{x}{a}+\frac{y}{b}=2$ at the point (a, b) for all value of *n*.

***Type 8*: (A) Problems based on length of perpendicular:**

Exercise 19.14

1. Find the length of the perpendicular from the origin (0, 0) on the tangent of the following curve:

(i) $\frac{x^2}{a^2}+\frac{y^2}{b^2}=1$ at the point (x_1, y_1)

(ii) $y^2 = 4ax$ at $(am^2, -2am)$

(iii) $x = at^2, y = 2at$ at 't'

(iv) $x = a(\theta - \sin\theta),\ y = a(1-\cos\theta)$ at $\theta = \dfrac{\pi}{2}$

2. Show that the normal at any point on the curve

$x = a\cos\theta + a\theta\sin\theta$

$y = a\sin\theta - a\theta\cos\theta$ is at a constant distance from the origin.

3. Prove that the perpendicular drawn from the foot of the ordinate to the tangent of a curve is

$$\frac{y}{\sqrt{1+\left(\dfrac{dy}{dx}\right)^2}}$$

[*Hint*: The equation of the tangent at $P(x, y)$ is $X\dfrac{dy}{dx} - Y + y - x\dfrac{dy}{dx} = 0$, find the length of the perpendicular from the foot of the ordinate $(x, 0)$ to the tangent]

Answers:

1. (i) $\dfrac{a^2b^2}{\sqrt{b^4x_1^2 + a^4y_1^2}}$ **(ii)** $\dfrac{am^2}{\sqrt{1+m^2}}$

(iii) $\dfrac{!\ at^2}{\sqrt{1+t^2}}$ **(iv)** $\dfrac{a\pi - 4a}{2\sqrt{2}}$

Type 8: **(B)** Problems based on finding the area

Exercise 19.15

1. Show that the area of the triangle formed by a tangent to the curve $2xy = a^2$ and the coordinate axes is constant.

Type 9: Problems based on length of subtangent or subnormal

Exercise 19.16

1. Find the lengths of the subtangent and subnormal at the point (3, 4) of the rectangle hyperbola $xy = 12$.

2. Find the lengths of subtangents and subnormals at the point (x', y') of the curves:

(i) $x^2 + y^2 = a^2$ and **(ii)** $\dfrac{x^2}{a^2} + \dfrac{y^2}{b^2} = 1$

3. Find the length of subtangent, subnormal, tangent and normal at the point 't' of the cycloid:

$x = a(t + \sin t)$

$y = a(1 - \cos\theta)$

4. Prove that in the curve $y = be^{-\frac{a}{x}}$, the subtangent varies as the square of the abscissa.

[*Hint*: Prove that $\dfrac{\text{subtangent}}{(\text{abscissa})^2} = \text{constant}$]

5. Prove that the subtangent is of constant length in the curves:

(i) $\log y = x\log a$ **(ii)** $y = a^x$.

6. Show that the subtangents and subnormals of the curve $y^n = a^{n-1}x$ are nx and $\dfrac{y^2}{nx}$.

7. Find the subtangent and subnormal to the curve $y = 2x^2 + 3x$ at the point (2, 14).

8. Find the lengths of subtangents, subnormal, tangent and normal to the curve $y = x^3$ at the point (1, 1).

Answers:

1. $-3, -\dfrac{16}{3}$

2. (i) $-\dfrac{y'^2}{x'}, -x'$ **(ii)** $\dfrac{x'^2 - a^2}{x'}; -\dfrac{b^2x'}{a^2}$

3. $a\sin t, \dfrac{2a\sin\frac{t}{2}}{\cos\frac{t}{2}}, 2a\sin\dfrac{t}{2}, \dfrac{2a\sin^2\frac{t}{2}}{\cos\frac{t}{2}}$

7. $\dfrac{14}{11}, 154$

8. $\dfrac{1}{3}$ and 3; $\sqrt{\dfrac{10}{3}}$ and $\sqrt{10}$

20

Rolle's Theorem and Lagrange's Mean Value Theorem

Rolle's Theorem

Statement: If a function $y = f(x)$ defined over $[a, b]$ is such that

(i) It is continuous over $[a, b]$

(ii) It is differentiable over (a, b)

(iii) $f(a) = f(b)$

then there exists at least one point $x = c \in (a, b)$ such that $f'(c) = 0$.

Proof: Given:

$y = f(x)$ is continuous in the closed interval $[a, b]$ $\Rightarrow$ graph of $y = f(x)$ is a continuous curve without any break from the point $x = a$ to the point $x = b$.

Again, $y = f(x)$ is differentiable in the open interval (a, b). $\Rightarrow$ graph of $y = f(x)$ has unique tangent at each point in open interval (a, b).

Further given $f(a) = f(b)$

$\Rightarrow$ (ordinary at $x = a$) = (ordinate at $x = b$)

Now two possibilities arise.

Case 1: When $y = f(x)$ is constant.

Let us suppose that $f(x) = k$, and $c \in (a, b)$

$$\therefore\ L f'(c) = \lim_{h \to 0} \left(\frac{f(c-h) - f(c)}{-h} \right), h > 0$$

$$= \lim_{h \to 0} \left(\frac{k - k}{-h} \right) = 0$$

$$R f'(c) = \lim_{h \to 0} \left(\frac{f(c+h) - f(c)}{h} \right)$$

$$= \lim_{h \to 0} \left(\frac{k - k}{h} \right) = 0$$

$\therefore\ L f'(c) = R f'(c) = 0 \Leftrightarrow f'(c) = 0$ for all $c \in (a, b)$

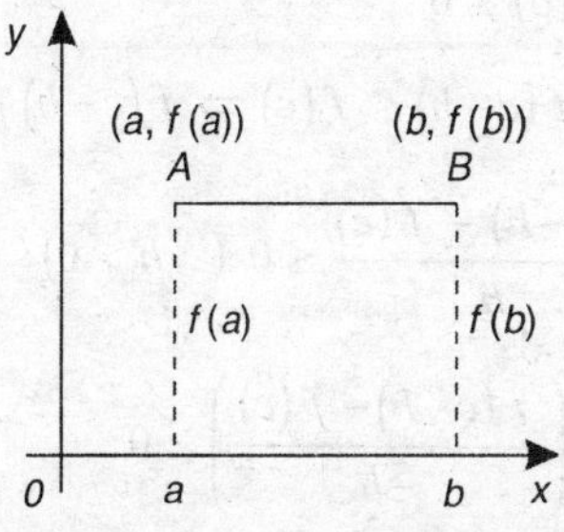

Case 2: Let $f(x)$ be not a constant function. Since in $[a, b]$ $f(x)$ is continuous it attains its bounds in $[a, b]$. At least one of the bounds is different from $f(a) = f(b)$. For definiteness let the upper bound $\neq f(a)$.

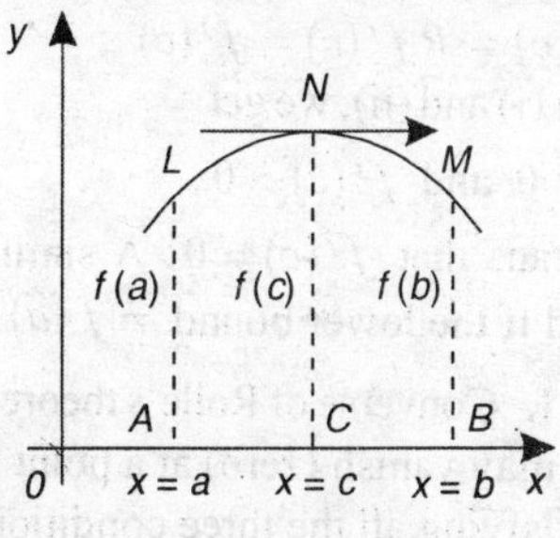

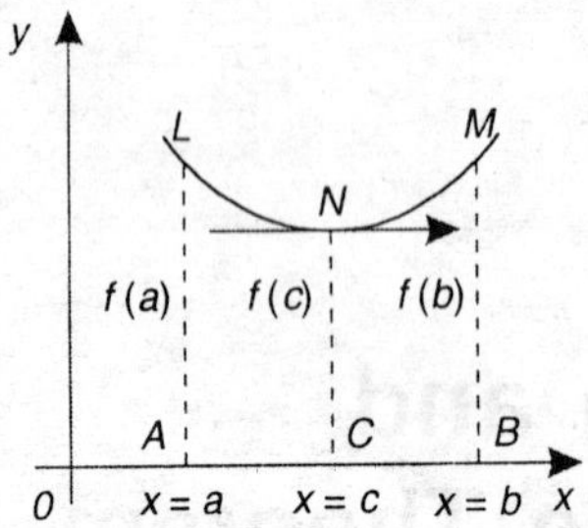

Let the function $f(x)$ attains its upper bound (maximum) at $x = c$.

$\therefore f(x) \le f(c), \forall x \in [a, b]$ and $a < c < b$

Now, $f(c+h) \le f(c) \Rightarrow f(c+h) - f(c) \le 0$

$$\Rightarrow \frac{f(c+h) - f(c)}{h} \le 0 \quad (\because h > 0)$$

$$\Rightarrow \lim_{h \to 0} \frac{f(c+h) - f(c)}{h} \le 0$$

$$\Rightarrow Rf'(c) \le 0 \qquad \text{...(i)}$$

Again, $f(c-h) \le f(c) \Rightarrow f(c-h) - f(c) \le 0$

$$\Rightarrow \frac{f(c-h) - f(c)}{h} \le 0 \quad (\because h > 0)$$

$$\Rightarrow \lim_{h \to 0} \left(\frac{f(c-h) - f(c)}{-h} \right) \ge 0$$

$(\because h > 0 \Leftrightarrow -h < 0)$

$$\Rightarrow Lf'(c) \ge 0 \qquad \text{...(ii)}$$

As $f(x)$ is differentiable over (a, b), hence $f'(c)$ must exist.

$\therefore Lf'(c) = Rf'(c) = f'(c)$

$\therefore$ From (i) and (ii), we get

$f'(c) < 0$ and $f'(c) > 0$.

This means that $f'(c) = 0$. A similar argument can be used if the lower bound $\neq f(a)$.

Remarks: **1.** Converse of Rolle's theorem is not true, i.e. $f'(x)$ may vanish (zero) at a point within (a, b) without satisfying all the three conditions of Rolle's theorem.

2. The three conditions of Rolle's theorem are sufficient but not necessary for $f'(x) = 0$ for some x in (a, b).

3. If a function $y = f(x)$ defined over $[a, b]$ does not satisfy even one of the three conditions, then Rolle's theorem fails, i.e. there may or may not exist point where $f'(x) = 0$.

Geometrical Meaning of Rolle's Theorem

If the graph of a function $y = f(x)$ defined over $[a, b]$ is such that

1. It is a continuous curve without any break from a point $A(a, f(a))$ to another point $B(b, f(b))$.

2. It has a unique tangent at each point in between the two points $A(a, f(a))$ and $B(b, f(b))$ (i.e. $f(x)$ is differentiable in the open interval (a, b)).

3. It has equal ordinates $f(a)$ and $f(b)$ at two points $A(a, f(a))$ and $B(b, f(b))$ (i.e. $f(x)$ assumes equal values at the end points of the closed interval $[a, b]$), then Rolle's theorem provides that there is at least one point $C(c, f(c))$ between $A(a, f(a)$ and $B(b, f(b))$ on the graph of the function $y = f(x)$ defined over $[a, b]$ such that the tangent to the graph at C is parallel to the x-axis.

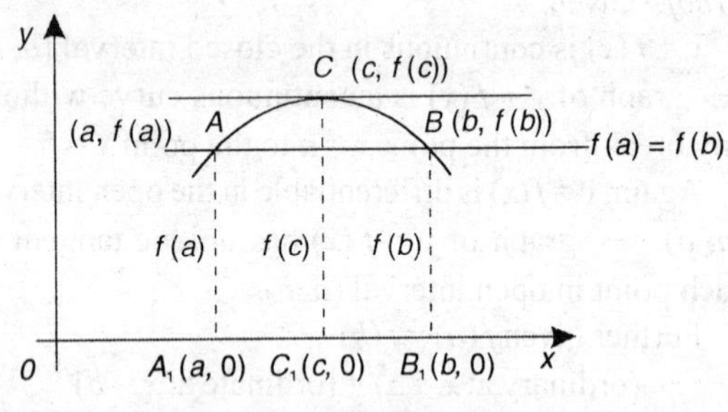

Note: It is Rolle's theorem which helps us to prove Lagrange's mean value theorem.

Lagrange's Mean Value Theorem

Statement: If a function $y = f(x)$ defined over a closed interval $[a, b]$ is such that

1. It is continuous in the closed interval $[a, b]$.

2. It is differentiable in the open interval (a, b) then there exists at least one point $x = c \in (a, b)$ such that $f'(c) = \dfrac{f(b) - f(a)}{b - a}$

Proof: Given: $y = f(x)$ is a continuous function over $[a, b]$ and differentiable over (a, b).

To prove: There is at least one point $x = c \in (a, b)$

such that $f'(c) = \frac{f(b) - f(a)}{b - a}$

Main Proof: Let us consider a function defined as

$F(x) = f(x) + Ax$, ...(1)

Where A is a constant whose value is to be determined from the condition that is $F(a) = F(b)$ ...(2) imposed on (1)

Now from (1),

$F(a) = f(a) + Aa$

$F(b) = f(b) + Ab$

$\therefore f(a) + Aa = f(b) + Ab$ (on using (2))

$\Rightarrow f(b) - f(a) = -A(b - a)$

$\Rightarrow -A = \frac{f(b) - f(a)}{b - a}$ ...(3)

Again, it is given that $y = f(x)$ is continuous over $[a, b]$ and differentiable in (a, b) and Ax being a polynomial is continuous over $[a, b]$ and differentiable over (a, b) since every polynomial in x is always continuous as well as differentiable in $(-\infty, \infty) \Rightarrow F(x) = f(x) + Ax$ is continuous in $[a, b]$ and differentiable in (a, b).

Also, from (2), $F(a) = F(b)$

Therefore $F(x) = f(x) + Ax$ is

(i) Continuous is $[a, b]$

(ii) Differentiable in (a, b) and also

(iii) $F(a) = F(b)$

Thus, $F(x)$ satisfies all the three conditions of Rolle's theorem $\Rightarrow$ there exists at least one value $x = c \in (a, b)$ such that $F'(c) = 0$ which

$\Rightarrow f'(c) + A = 0$

$\Rightarrow -A = f'(c)$ ...(4)

$\therefore$ From (3) and (4), it is concluded that

$$f'(c) = \frac{f(b) - f(a)}{b - a}$$

Hence, the required is proved.

Remarks: **1.** The statement "there exists at least one point $x = c \in (a, b)$" means that the point 'c' is not unique, i.e., there may exist more than one point c as c_1 and c_2.

2. If a function $y = f(x)$ defined over $[a, b]$ does not satisfy even one of the two conditions, then lagrange's mean value theorem fails for $y = f(x)$, i.e. there may or may not exists points where

$$f'(c) = \frac{f(b) - f(a)}{b - a}.$$

On another form of Lagrange's mean value theorem

On taking $b = a + h$, the closed interval $[a, b]$ becomes equal to $[a, a + h]$ and the number 'c' which lies in between a and $a + h$ can be written as $c = a + \theta h$, where θ is some proper fraction lying in $(0, 1)$.

Thus, the result of Lagrange's mean value theorem becomes

$$f'(a + \theta h) = \frac{f(a+h) - f(a)}{a + h - a} = \frac{f(a+h) - f(a)}{h},$$

where $0 < \theta < 1$.

Therefore, the Lagrange's mean value theorem can be stated as under also:

If a function $y = f(x)$ defined over a closed interval $[a, a + h]$ is such that

1. It is continuous over $[a, a + h]$

2. It is differentiable in $(a, a + h)$,

then there exists at least one number $\theta \in (0, 1)$ such that $f(a+h) - f(a) = h f'(a + \theta h)$

On geometrical meaning of Lagrange's mean value theorem

The hypothesis of Lagrange's mean value theorem provides that the graph of a function $y = f(x)$ defined over $[a, b]$ is

1. A continuous curve without any break, gap or jump from the point $A(a, f(a))$ to an other point $B(b, f(b))$ and

2. Has a unique tangent at each point in between the two points $A(a, f(a))$ and $B(b, f(b))$ (i.e $f(x)$ is differentiable in the open interval (a, b)).

The result of the Lagranges mean value theorem provides that there is at least one point $C(c, f(c))$ on the curve (i.e on the graph of the function $y = f(x)$ defined over $[a, b]$) where the tangent is parallel to the chord through the points $A(a, f(a))$ and $B(b, f(b))$

because the slope of the chord $AB = \dfrac{f(b) - f(a)}{b - a}$

$= \dfrac{\text{difference of ordinates}}{\text{difference of abscissas}}$ and the slope of the tangent at any point $C(c, f(c))$ is $f'(c)$

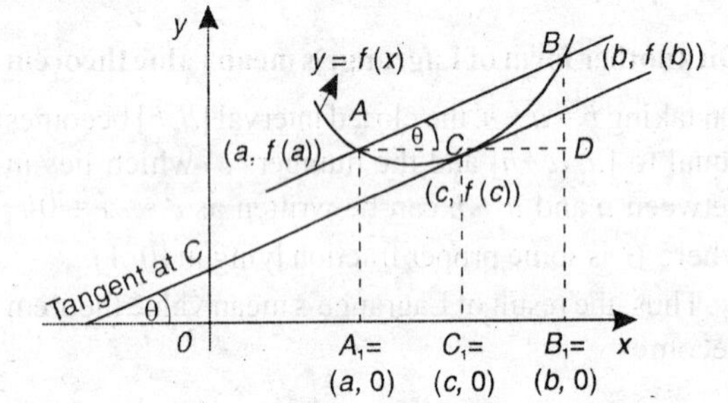

[Let $y = f(x)$ be the curve being continuous from $A(a, f(a))$ to $B(b, f(b))$ and also possessing tangents to the curve between A and B.

Let AA_1 and BB_1 be the perpendicular drawn to the x-axis. Let us join the chord AB which makes an angle θ with the positive direction of the x-axis.

Again from A, a perpendicular AD on BB_1 is drawn. Then, $BD = BB_1 - DB_1 = f(b) - f(a)$ and $AD = A_1B_1 = OB_1 - OA_1 = b - a$

$$\therefore \frac{BD}{AD} = \frac{f(b) - f(a)}{b - a}$$

$$\Rightarrow \tan\theta = \frac{f(b) - f(a)}{b - a}, \text{ where } \theta = \angle BAD =$$

slope of the chord AB.

But $f'(c)$ = slope of the tangent at c.

That is, there is a point $C(c, f(c))$ where the derivative has $f'(c) = \tan\theta$, i.e. the tangent at $C(c, f(c))$ is parallel to the chord AB.]

Note: There may be more than one point namely C_1, C_2 and C_3 on the curve between A and B where the tangents are parallel to the chord AB.

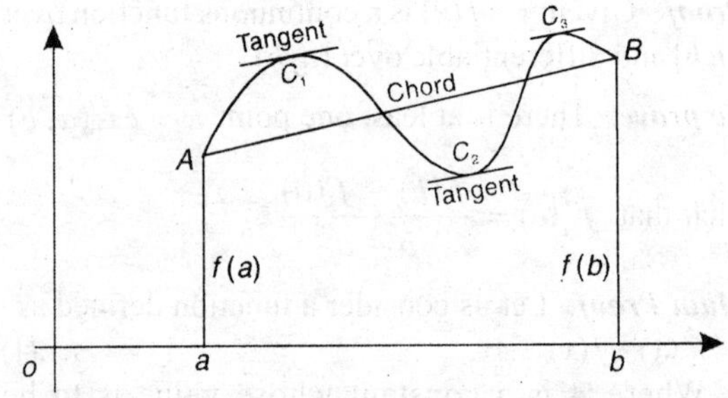

On continuity and differentiability of a function $y = f(x)$

Readers should remember the following facts to ensure the continuity and differentiability of a function $y = f(x)$ defined over an interval open or closed.

1. The domain of a derived function $f'(x)$ is a subset of the domain of the function $f(x)$, because it contains all the points x in the domain of $f(x)$ such that the limit $\lim\limits_{h\to 0} \dfrac{f(x+h) - f(x)}{h}$ exists, but does not contain those points where the limit $\lim\limits_{h\to 0} \dfrac{f(x+h) - f(x)}{h}$ does not exist.

2. A function $y = f(x)$ which has a derivative is called differentiable. The function $y = f(x)$ is differentiable at a point $x = a$, if 'a' lies in the domain of $f'(x)$ i.e. if $f'(a)$ exists, i.e.,

$$Lf'(a) = \lim_{h\to 0} \frac{f(a-h) - f(a)}{h}$$

$$= Rf'(a) = \lim_{h\to 0} \frac{f(a+h) - f(a)}{h}, h > 0.$$

3. A function $y = f(x)$ is continuous in an open interval $(a, b) \Leftrightarrow$ it is continuous at any point $c \in (a, b)$.

4. A function $y = f(x)$ is continuous is a closed interval $[a, b] \Leftrightarrow$ it is continuous at any point $c \in (a, b)$ and is continuous at 'a' from the right and continuous at 'b' from the left.

5. A function $y = f(x)$ is differentiable in an open interval $(a, b) \Leftrightarrow$ it is differentiable at any point $c \in (a, b)$.

6. A function $y = f(x)$ is differentiable in a closed interval $[a, b] \Leftrightarrow$ it is differentiable at any point $c \in (a, b)$ and has a right derivative at $x = a$ and a left derivative at $x = b$.

7. All the discontinuities of a function $y = f(x)$ are also the discontinuities of the derived function $f'(x)$.

8. All standard functions (i.e. as simple form in which a function is commonly written, also termed as elementary functions) can be discontinuous only at points where they are not defined, i.e. all 'LIATE' (logarithmic, inverse, trigonometric, algebraic and exponential) standard functions are continuous and differentiable in whole of its domain.

9. A function $y = f(x)$ is discontinuous at a point $x = c \Rightarrow$ the function $y = f(x)$ is not differentiable (non differentiable) at the point $x = c$.

10. A function $y = f(x)$ is differentiable at a point $x = c \Rightarrow$ the function $y = f(x)$ is continuous at the point $x = c$.

11. $f'(x)$ is continuous $\Rightarrow f(x)$ is continuous.

12. Every positive power function $y = x^n$ is continuous and differentiable in any interval open or closed since it is continuous and differentiable for all values of x in R.

13. Any polynomial function is continuous and differentiable in any interval open or closed since it is continuous and differentiable for all values of x in R.

14. Any rational algebraic or non algebraic function is continuous and differentiable for all values of the independent variable x excepting those point where its denominator is zero, i.e. any rational function is continuous and differentiable in any interval open or closed excluding the points where its denominator is zero.

Where to check the continuity and differentiability of a function $y = f(x)$ defined in a given closed interval $[a, b]$.

In order to check the continuity and differentiability of a given function $y = f(x)$ defined in a closed interval $[a, b]$ one must check them at the following points.

1. The points where the given function is undefined or imaginary.

2. The point where the derived function $f'(x)$ is undefined or imaginary.

3. The common points of adjacent intervals where different forms of a given function are defined.

4. The end points of a given closed interval.

Notes: **1.** In order to show that a function is discontinuous in a given interval open or closed, it is sufficient to show that it is discontinuous at atleast at one point belonging to the given interval.

2. If $f'(x)$ becomes undefined on putting $x = c$, then it is wrong to conclude that $f(x)$ is not differentiable at $x = c$. In that case we find $Lf'(c)$ and $Rf'(c)$ by first principles and test the differentiability at c.

Illustrations*: *(Erroneous approach)

1. $f(x) = |x|$

$$\Rightarrow f'(x) = \frac{|x|}{x}$$

$$\Rightarrow f'(0) = \frac{0}{0}, \text{ i.e. } f'(0) \text{ is undefined}$$

$\Rightarrow f'(0)$ does not exist, i.e. $f(x)$ is not differentiable at $x = 0$.

2. $f(x) = x^{\frac{2}{3}}$

$$\Rightarrow f'(x) = \frac{2}{3} x^{\left(\frac{2}{3}-1\right)} = \frac{2}{3} x^{-\frac{1}{3}} = \frac{2}{3} \cdot \frac{1}{\sqrt[3]{x}}$$

$$\Rightarrow f'(0) = \frac{2}{0}, \text{ i.e. } f'(0) \text{ is not a finite number}$$

$\Rightarrow f'(0)$ does not exist, i.e. $f(x)$ is not differentiable at $x = 0$.

Type 1: To establish the validity of Rolle's theorem when interval is given.

To verify Rolle's theorem, we have to show following conditions are satisfied.

1. Find $f(a)$ and $f(b)$ and show that $f(a) = f(b)$.

2. Show the continuity of the given function in the closed interval by using the facts that LIATE functions are continuous at points where they have finite values and theorems on continuity.

3. Show that differentiability of the given function in the given interval by using theorems on differentiability and the facts that LIATE functions have finite derivatives at points where they are defined.

4. Put $f'(x) = 0$ and find the value of x or choose the value of x from among roots of $f'(x) = 0$ which is in the given interval $[a, b]$.

N.B.:

1. LIATE ⇒ L = log function
   I = Inverse circular function
   A = Algebraic function
   T = Trigonometric function
   E = Exponential function

2. If a function is discontinuous in an interval, it must have atleast one point of discontinuity (in the given interval) where one or other condition for continuity fails to satisfy. A function is not continuous if it exhibits at least one point of discontinuity in the given interval.

Note: That this (2) is practically fruitful to examine the validity or applicability of Rolle;s theorem or Lagrange's mean value theorem for a given function in a given interval.

Examples worked out:

1. Verify Rolle's theorem for $f(x) = x^3 - 4x$ in the interval $-2 \le x \le 2$.

Solution: (1) $\because f(x) = x^3 - 4x$ = a polynomial

$\therefore f(x)$ is differentiable in $[-2, 2]$ and so continuous in $[-2, 2]$ as $f(x)$ is a polynomial function.

$$(2)\ \left.\begin{aligned} f(2) &= 2^3 - 4(2) = 0 \\ f(-2) &= (-2)^3 - 4(-2) = 0 \end{aligned}\right\} \Rightarrow f(2) = f(-2)$$

(1) and (2) ⇒ all the conditions of Rolle's theorem are satisfied.

Now $f'(x) = 0$

$$\Rightarrow 3x^2 - 4 = 0$$

$$\Rightarrow 3x^2 = 4$$

$$\Rightarrow x = \pm\sqrt{\frac{4}{3}} = \pm\frac{2}{\sqrt{3}} = c$$

Both the values of c lies in the open interval $(-2, 2)$ = $]-2, 2[$ hence, the fact that $f'(x) = 0$ for atleast one $c \in (-2, 2)$ has been verified.

$\therefore$ Rolle's theorem is verified.

2. Verify Rolle's theorem for the function $f(x) = (x-1)\cdot(x-4)\cdot e^{-x}$ in the interval $(1, 4)$.

Solution: (1) $\because f(x) = (x-1)\cdot(x-4)\cdot e^{-x}$

$\therefore f(x)$ is differentiable in $[1, 4]$ as it is the product of differentiable functions $(x-1)$, $(x-4)$ and e^{-x}.

$\therefore$ It is differentiable in $(1, 4)$ which ⇒ it is continuous in $(1, 4)$

(2) $f(1) = 0 = f(4)$

(1) and (2) ⇒ all the conditions of Rolle's theorem are satisfied.

Now, $f'(x) = 0 \Rightarrow (2x-5)e^{-x} - (x^2 - 5x + 4)e^{-x}$

$$= 0 \Rightarrow -e^{-x}(x^2 - 7x + 9) = 0$$

$$\Rightarrow x^2 - 7x + 9 = 0$$

$$\Rightarrow x = \frac{7 \pm \sqrt{13}}{2}$$

$\Rightarrow x =$ 5.3, 1.7 approximately $= c_1, c_2$.

Since $c_2 \in (1, 4)$,

$\therefore$ Rolle's theorem is verified.

3. Verify Rolle's theorem for $f(x) = (x^2 - 4x + 3)e^{2x}$ in $[1, 3]$.

Solution: (1) $\because f(x) = (x^2 - 4x + 3)e^{2x}$

$\therefore f(x)$ is differentiable in $[1, 3]$ as it is the product of differentiable functions $(x^2 - 4x + 3)$ and e^{2x}.

$\therefore$ It is differentiable in $(1, 3)$ which ⇒ it is continuous in $(1, 3)$.

$$(2)\ \left.\begin{aligned} f(1) &= (1-4+3)e^2 = 0 \\ f(3) &= (9-12+3)e^6 = 0 \end{aligned}\right\} \Rightarrow f(1) = f(3)$$

(1) and (2) ⇒ all the condition of Rolle's theorem are satisfied.

Now, $f'(x) = 0$

$$\Rightarrow (x^2 - 4x + 3)\,2\cdot e^{2x} + (2x-4)\cdot e^{2x}$$

$$= 0 \Rightarrow 2(x^2 - 3x + 1)e^{2x} = 0$$

$$\Rightarrow x^2 - 3x + 1 = 0$$

$$\Rightarrow x = \frac{3 \pm \sqrt{9-4}}{2} = \frac{3 \pm \sqrt{5}}{2}$$

Since the value of $x = \dfrac{3+\sqrt{5}}{2} \in$ open interval $(1, 3)$

$\therefore$ Rolle's theorem is verified.

4. Show that the function $f(x) = e^x \cos x$ satisfies Rolle's theorem in $\left[-\frac{\pi}{2}, \frac{\pi}{2}\right]$

Solution: (1) $\because f(x) = e^x \cdot \cos x$

$\therefore f(x)$ is differentiable in $\left[-\frac{\pi}{2}, \frac{\pi}{2}\right]$ as it is the product of differentiable functions e^x and $\cos x$ which

$\Rightarrow$ it is continuous in $\left[-\frac{\pi}{2}, \frac{\pi}{2}\right]$.

$\therefore$ It is differentiable in $\left(-\frac{\pi}{2}, \frac{\pi}{2}\right)$ which $\Rightarrow$ it is continuous in $\left(-\frac{\pi}{2}, \frac{\pi}{2}\right)$.

(2) $$\left.\begin{aligned} f\left(-\frac{\pi}{2}\right) &= e^{-\frac{\pi}{2}} \cdot \cos\left(-\frac{\pi}{2}\right) = e^{-\frac{\pi}{2}} \cdot \cos\left(\frac{\pi}{2}\right) = 0 \\ f\left(\frac{\pi}{2}\right) &= e^{\frac{\pi}{2}} \cdot \cos\left(\frac{\pi}{2}\right) = 0 \end{aligned}\right\}$$

$$\Rightarrow f\left(-\frac{\pi}{2}\right) = f\left(\frac{\pi}{2}\right)$$

(1) and (2) $\Rightarrow$ all the conditions of Rolle's theorem are satisfied.

Now, $f'(x) = 0 \Rightarrow e^x(\cos x - \sin x) = 0 \Rightarrow \cos x - \sin x = 0 \; (\because e^x \neq 0)$

$\Rightarrow \cos x = \sin x \Rightarrow \tan x = 1$

$\Rightarrow \tan x = \tan\frac{\pi}{4} \Rightarrow x = n\pi + \frac{\pi}{4}$; $(n = 0, \pm 1, \pm 2, \ldots)$

and $x = \frac{\pi}{4} \in \left(-\frac{\pi}{2}, \frac{\pi}{2}\right)$

$\therefore$ Rolle's theorem is verified.

5. Verify Rolle's theorem for

(a) x^2 in $[-1, 1]$

(b) $x^2 - x - 6$ in $[-2, 3]$

Solution: **(a)** (1) $f(x) = x^2$ = a polynomial in x

$\therefore f(x)$ is differentiable in $[2, 3]$ and so continuous in $[2, 3]$ as $f(x)$ is a polynomial function.

(2) $f(1) = 1 = f(-1)$

(1) and (2) $\Rightarrow$ all the conditions of Rolle's theorem are satisfied.

Now $f'(x) = 0 \Rightarrow 2x = 0 \Rightarrow x = 0$ which is a point in $(-1, 1)$

$\therefore$ Rolle's theorem is verified.

(b) (1) $f(x) = x^2 - x - 6$

$\therefore f(x)$ is differentiable in $[-2, 3]$ and so continuous in $[-2, 3]$ as $f(x)$ is a polynomial function.

(2) $f(2) = f(3) = 0$

(1) and (2) $\Rightarrow$ all the conditions of Rolle's theorem are satisfied.

Now $f'(x) = 0 \Rightarrow 2x - 1 = 0 \Rightarrow x = \frac{1}{2} \in (-2, 3)$

$\therefore$ Rolle's theorem is verified.

6. Verify Rolle's theorem for the following functions:

(i) $f(x) = (x-a)^m \cdot (x-b)^n$ on $[a, b]$ where m, n are positive integers.

(ii) $f(x) = e^x \sin x$ on $[0, \pi]$

Solution: **(i)** $\because f(x) = (x-a)^m \cdot (x-b)^n$ where m and n are +ve integers.

= a polynomial in x

$\therefore f(x)$ is differentiable in $[a, b]$ and so continuous in $[a, b]$ as $f(x)$ is a polynomial function.

(2) $f(a) = f(b) = 0$

$\therefore$ All conditions of Rolle's theorem are satisfied

Now $f'(x)$

$$= m(x-a)^{m-1} \cdot (x-b)^n + n(x-a)^m \cdot (x-b)^{n-1}$$

$$= (x-a)^{m-1} \cdot (x-b)^{n-1}\left[m(x-b) + n(x-a)\right]$$

$\therefore f'(x) = 0$

$\Rightarrow (x-a)^{m-1} \cdot (x-b)^{n-1} [m(x-b)+n(x-a)]=0$

Now equating each factor to zero.

$\Rightarrow (x-a)^{m-1} = 0 \Rightarrow x = a$ (if $m>1$)

or $(x-b)^{n-1} = 0 \Rightarrow x = b$ (if $n>1$)

or, $m(x-b) + n(x-a) = 0 \Rightarrow x = \dfrac{mb + na}{m + n}$

Thus, $f'(x) = 0$ for $x = \dfrac{mb + na}{m + n}$ and we see that $x = \dfrac{mb + na}{m + n}$ lies in (a, b).

$\therefore$ Rolle's theorem is verified.

(ii) (1) $\because f(x) = e^x \sin x$

$\therefore f(x)$ is differentiable in $[0, \pi]$ as it is the product of differentiable functions e^x and $\sin x$ which $\Rightarrow$ it is continuous in $[0, \pi]$

(2) $f(0) = f(\pi) = 0$

$\therefore$ All conditions of Rolle's theorem are satisfied.

Now, $f'(x) = e^x \sin x + e^x \cos x$

$= \sqrt{2}e^x \cos\left(\dfrac{\pi}{4}+x\right)$

$\therefore f'(x) = 0 \Rightarrow \sqrt{2}e^x \cos\left(\dfrac{\pi}{4}+x\right) = 0$

$\Rightarrow \cos\left(\dfrac{\pi}{4}+x\right) = 0 \quad [\because \sqrt{2} \neq 0, e^x \neq 0]$

$\Rightarrow \cos\left(\dfrac{\pi}{4}+x\right) = \cos\dfrac{\pi}{2}$

$\Rightarrow x+\dfrac{\pi}{4} = 2n\pi \pm \dfrac{\pi}{2} \quad [n = 0, \pm 1, \pm 2, \ldots]$

one value is $x = \dfrac{\pi}{2} - \dfrac{\pi}{4} \Rightarrow x = \dfrac{2\pi - \pi}{4} = \dfrac{\pi}{4}$

Now since $\dfrac{\pi}{4} \in (0, \pi)$

$\therefore$ Rolle's theorem is verified.

7. Verify Rolle's theorem for

(i) $\dfrac{\sin x}{e^x}$ in $(0, \pi)$

(ii) $\log\left\{(x^2 + ab)/(a + b)x\right\}$ in $[a, b]$; $ab>0$

Solution: **(i)** (1) $f(x) = \dfrac{\sin x}{e^x} = \sin x \cdot e^{-x}$

$\therefore f(x)$ is differentiable in $[0, \pi]$ as it is the product of differentiable functions $\sin x$ and e^{-x} which $\Rightarrow$ it is continuous in $[0, \pi]$.

(2) $\left.\begin{aligned} f(0) &= \dfrac{\sin 0}{e^0} = \dfrac{0}{1} = 0 \\ f(\pi) &= \dfrac{\sin\pi}{e^\pi} = \dfrac{0}{e^\pi} = 0 \end{aligned}\right\} \Rightarrow f(0) = f(\pi)$

$\therefore$ All conditions of Rolle's theorem are satisfied.

Now $f'(x) = \dfrac{e^x \cos x - \sin x e^x}{e^{2x}} = \dfrac{(\cos x - \sin x)}{e^x}$

$\therefore f'(x) = 0$

$\Rightarrow \dfrac{(\cos x - \sin x)}{e^x} = 0$

$\Rightarrow \cos x - \sin x = 0 \quad (\because e^x \neq 0)$

$\Rightarrow \cos x = \sin x$

$\Rightarrow 1 = \tan x$

$\Rightarrow \tan\dfrac{\pi}{4} = \tan x$

$x = n\pi + \dfrac{\pi}{4} \quad (n = 0, \pm 1 \pm 2, \ldots)$

and $x = \dfrac{\pi}{4} \in (0, \pi)$

$\therefore$ Rolle's theorem is verified.

(ii) (1) $f(x) = \log\left\{\dfrac{(x^2+ab)}{(a+b)\,x}\right\}$

$\therefore f(x)$ is differentiable on $[a, b]$ as it is the log function which $\Rightarrow$ it is continuous on $[a, b]$.

(2) $\left.\begin{array}{l} f(a) = 0 \\ \text{and } f(b) = 0 \end{array}\right\} \Rightarrow f(a) = f(b)$

$\therefore$ All conditions of Rolle's theorem are satisfied.

Now $f'(x) = \dfrac{2x}{x^2+ab} - \dfrac{1}{x} = \dfrac{x^2-ab}{x(x^2+ab)}$

$\therefore f'(x) = 0$

$\Rightarrow \dfrac{x^2-ab}{x(x^2+ab)} = 0$

$\Rightarrow x^2 - ab = 0$

$\Rightarrow x^2 = ab$

$\Rightarrow x = \pm\sqrt{ab}$

Now $c = \sqrt{ab}$ lies in the open interval (a, b) being geometric mean of a and b.

$\therefore$ Rolle's theorem is verified.

8. Verify Rolle's theorem for the function $f(x) = (x-1)(x-2)(x-3)$ on $[1, 3]$.

Solution: (1) $f(x) = (x-1)(x-2)(x-3)$

$\therefore f(1) = f(3) = 0$

(2) $f(x)$ is differentiable on $[1, 3]$ as it is a polynomial function of x which $\Rightarrow$ it is continuous in $[1, 3]$.

$\therefore$ All conditions of Rolle's theorem are satisfied.

Now $f'(x) = (x-2)(x-3) + (x-1)(x-3) + (x-1)(x-2)$

$= x^2 - 5x + 6 + x^2 - 4x + 3 + x^2 - 3x + 2$

$= 3x^2 - 12x + 11$

$\therefore f'(x) = 0$

$\Rightarrow 3x^2 - 12x + 11 = 0$

$\Rightarrow x = 2 - \dfrac{1}{\sqrt{3}},\ 2 + \dfrac{1}{\sqrt{3}}$

Clearly both $\in (1, 3)$

Hence, the fact that $f'(x) = 0$ for at least one $c \in (1, 3)$ has been verified.

$\therefore$ Rolle's theorem is verified.

9. Verify Rolle's theorem for the function $\log(x^2+2) - \log 3$ on $[-1, 1]$.

Solution: (1) $f(x) = \log(x^2+2) - \log 3$

$$\left.\begin{array}{l} \therefore f(-1) = \log\left[(-1)^2+2\right] - \log 3 = \log 3 = 0 \\ f(1) = \log\left[(1)^2+2\right] - \log 3 = \log 3 - \log 3 = 0 \end{array}\right\} \Rightarrow$$

$f(-1) = f(1)$

(2) $f(x)$ is differentiable on $[-1, 1]$ as it is the difference of two differentiable functions $\log(x^2+1)$ and $\log 3$ (a constant function) which $\Rightarrow$ it is continuous in $[-1, 1]$.

$\therefore$ All conditions of Rolle's theorem are satisfied.

Now, $f'(x) = \dfrac{1}{x^2+2}\cdot 2x - 0 = \dfrac{2x}{x^2+2}$

$\therefore f'(x) = 0$

$\Rightarrow \dfrac{2x}{x^2+2} = 0$

$\Rightarrow 2x = 0$

$\Rightarrow x = 0$ and $x = 0 \in (-1, 1)$

$\therefore$ Rolle's theorem is verified.

10. Verify Rolle's theorem for the function

$$f(x) = \sin x + \cos x - 1 \text{ on } \left[0, \frac{\pi}{2}\right]$$

Solution: (1) $f(x) = \sin x + \cos x - 1$

$$\left.\begin{array}{l} \therefore f(0) = \sin 0 + \cos 0 - 1 = 0 + 1 - 1 = 0 \\ f\left(\frac{\pi}{2}\right) = \sin\left(\frac{\pi}{2}\right) + \cos\left(\frac{\pi}{2}\right) - 1 = 1 + 0 - 1 = 0 \end{array}\right\} \Rightarrow$$

$f(0) = f\left(\dfrac{\pi}{2}\right)$

(2) $f(x)$ is differentiable in $\left[0, \dfrac{\pi}{2}\right]$ as it is the sum of differentiable functions $\sin x$, $\cos x$ and a constant function $\Rightarrow$ it is continuous in $\left[0, \dfrac{\pi}{2}\right]$.

$\therefore$ All conditions of Rolle's theorem are satisfied.

Now, $f'(x) = \cos x - \sin x$

$f'(x) = 0$

$\Rightarrow \cos x - \sin x = 0$

$\Rightarrow \cos x = \sin x$

$\Rightarrow \dfrac{\cos x}{\sin x} = 1$

$\Rightarrow \cot x = \cot\dfrac{\pi}{4} \Rightarrow x = n\pi + \dfrac{\pi}{4},\ n \in Z$ and

$\Rightarrow x = \dfrac{\pi}{4} \in \left(0, \dfrac{\pi}{2}\right)$

$\therefore$ Rolle's theorem is verified.

Type 2: Verification of Rolle's theorem when interval is not given.

Working rule: Find the interval or intervals equating the given function $f(x)$ to c i.e. $f(x) = c$ provides us interval (where $c = 0$ in particular). We then proceed as usual.

Examples worked out:

1. Verify Rolle's theorem for the function $f(x) = 2x^3 + x^2 - 4x - 2$.

Solution: To obtain some interval or intervals, we put $f(x) = 0$

$\therefore (x^2 - 2)(2x + 1) = 0$

$\Rightarrow x = -\sqrt{2}, -\dfrac{1}{2}, \sqrt{2}$

Thus, we obtain two closed intervals $\left[-\sqrt{2}, -\dfrac{1}{2}\right]$ and $\left[-\dfrac{1}{2}, \sqrt{2}\right]$ such that the functional values at the end points of each f these two intervals are equal and is zero.

Now, $f(x)$ is a polynomial $\Rightarrow$ it is differentiable in any interval $\Rightarrow$ it is differentiable on $\left(-\sqrt{2}, -\dfrac{1}{2}\right)$ and $\left(-\dfrac{1}{2}, \sqrt{2}\right)$ and continuous on $\left[-\sqrt{2}, -\dfrac{1}{2}\right]$ and $\left[-\dfrac{1}{2}, \sqrt{2}\right]$.

$\therefore$ All conditions of Rolle's theorem are satisfied.

Now, $f'(x) = 6x^2 + 2x - 4 = 2(3x - 2)(x + 1) = 0$

$\Rightarrow x = -1$ and $x = \dfrac{2}{3}$

And we see that $-1 \in \left(-\sqrt{2}, -\dfrac{1}{2}\right)$

and $\dfrac{2}{3} \in \left(-\dfrac{1}{2}, \sqrt{2}\right)$

Hence, the fact that $f'(x) = 0$ for at least one $c \in \left(-\sqrt{2}, -\dfrac{1}{2}\right)$ and $c \in \left(-\dfrac{1}{2}, \sqrt{2}\right)$ has been verified.

$\therefore$ Rolle's theorem is verified.

2. Verify Rolle's theorem $f(x) = x(x + 3) \cdot e^{-\frac{x}{2}}$

Solution: (1) $\because f(x) = x(x + 3) \cdot e^{-\frac{x}{2}}$

Here it is not given in which interval Rolle's theorem is to be verified, so to obtain the interval we put $f(x) = 0$

Now, $f(x) = 0 \Rightarrow x(x + 3) \cdot e^{-\frac{x}{2}} = 0$

$\Rightarrow x(x + 3) = 0 \left[\because e^{-\frac{x}{2}} \neq 0\ \forall\ x\right]$

$\Rightarrow x = 0, -3$

Hence, we verify Rolle's theorem in $[-3, 0]$.

(2) $f(x)$ is differentiable in $[-3, 0]$ as it is the product of differentiable functions x, $(x + 3)$ and $e^{-\frac{x}{2}}$ which $\Rightarrow$ it is continuous in $[-3, 0]$

(3) $f(-3) = f(0) = 0$

$\therefore$ All conditions of Rolle's theorem are satisfied.

Now,

$$f'(x)=\{1(x+3)+x\}\,e^{-\frac{x}{2}}+x(x+3)\,e^{-\frac{x}{2}}\cdot\left(-\frac{1}{2}\right)$$

$$=e^{-\frac{x}{2}}\{x+6-x^2\}\cdot\frac{1}{2}$$

$\therefore f'(x)=0$

$\Rightarrow -x^2+x+6=0$

$\Rightarrow x^2-x-6=0$

$$\Rightarrow x=\frac{-(-1)\pm\sqrt{(-1)^2-4\times1\times(-6)}}{2\cdot1}$$

$$=\frac{1\pm\sqrt{25}}{2}=\frac{1\pm5}{2}=3 \text{ or } -2$$

Thus we get one $c=-2\in(-3,0)$

$\therefore$ Rolle's theorem is verified.

N.B.: $e^{f(x)}\neq 0$

Type 3: Verification of Rolle's theorem when a function is defined by various equations

$f(x)=f_1(x)$, when $x\geq a$

$=f_2(x)$, when $x\leq a$

or, $f(x)=f_1(x)$, when $x\neq a$

$=f_2(x)$, when $x=a$

or, $f(x)=f_1(x)$, when $x>a$

$f_2(x)$, when $x<a$

$f_3(x)$, when $x=a$

Examples worked out:

1. Verify that Rolle's theorem applies to the function given by $f(x)=x\sin\left(\frac{1}{x}\right)$, when $x\neq0$ and $=0$, when $x=0$ on the interval $\left[0,\frac{1}{\pi}\right]$.

Solution: (1) $f(x)=x\sin\left(\frac{1}{x}\right)$, when $x\neq0$ and $=0$, when $x=0$

$\therefore f(x)$ is continuous in any interval and hence it is continuous in the closed interval $\left[0,\frac{1}{\pi}\right]$.

Again $f(x)$ is differentiable every where except $x=0$ and hence it is differentiable in the open interval $\left(0,\frac{1}{\pi}\right)$.

(2) $\left.\begin{array}{l} f(0)=0 \text{ (given)} \\ f\left(\frac{1}{\pi}\right)=\frac{1}{\pi}\sin\pi=0\end{array}\right\}\Rightarrow f(0)=f\left(\frac{1}{\pi}\right)$

$\therefore$ All conditions of Rolle's theorem are satisfied.

Hence, Rolle's theorem is applicable for the function.

$f(x)=x\sin\left(\frac{1}{x}\right)$, when $x\neq0$

$=0$, when $x=0$ on the interval $\left[0,\frac{1}{\pi}\right]$

Problems based on examining the truth of the statement of Rolle's theorem:

Refresh your memory:

1. If $f(x)$ is not differentiable at the end points of closed interval $[a,b]$, then continuity of the function $f(x)$ is to be tested at the end point a and b of the closed interval $[a,b]$.

2. Rolle's theorem is not applicable in $\frac{f_1(x)}{f_2(x)}$ a rational function if $x=c\in I$ (where I = given interval) makes $f_2(x)$ = denominator $=0$

3. Rolle's theorem does not hold good if one or more of the following hold.

(i) $f(x)$ is discontinuous at some point in the closed interval $[a,b]$.

(ii) $f'(x)$ does not exist at some point in the open interval (a,b).

(iii) $f(a)\neq f(b)$

4. All "PILET-RC" functions are continuous and differentiable at points belonging to the domain of definition of the function.

Where P = Power function / polynomial function

I = Inverse trigonometric functions / Identity function

T = Trigonometric functions

L = Linear function / Logarithmic function

E = Exponential function

R = Rational function

C = Constant function

(A) Highlight on discontinuity of trigonometric functions:

(i) tan x is discontinuous at $x = (2n+1)\,\frac{\pi}{2}$ = odd multiple of $\frac{\pi}{2}$.

(ii) sec x is discontinuous at $x = (2n+1)\,\frac{\pi}{2}$ = odd multiple of $\frac{\pi}{2}$.

(iii) cosec x and / cot x are discontinuous at $x = n\pi$ = multiple of π.

(B) Highlight on discontinuous of inverse trigonometric function:

(i) $\tan^{-1} x$ and / $\cot^{-1} x$ have no discontinuity.

(ii) $\sin^{-1} x$ and $\cos^{-1} x$ are undefined $\forall\, x \notin [-1, 1]$ = closed interval.

(iii) $\sec^{-1} x$ and $\text{cosec}^{-1} x$ are undefined $\forall\, x \in (-1, 1)$ = open interval.

(C) Differentiable and non differentiable functions:

(i) $f(x) = x \cdot \sin\left(\frac{1}{x}\right),\ x \neq 0,\ f(0) = 0$ is continuous in any interval.

(ii) $f(x) = x \cdot \sin\left(\frac{1}{x}\right),\ x \neq 0,\ f(0) = 0$ is differentiable everywhere except at $x = 0$. Similarly,

(iii) $f(x) = x \cdot \cos\left(\frac{1}{x}\right),\ x \neq 0,\ f(0) = 0$ is continuous in any interval.

(iv) $f(x) = x \cdot \cos\left(\frac{1}{x}\right),\ x \neq 0,\ f(0) = 0$ is differentiable everywhere except at $x = 0$.

(v) $f(x) = \sum_{r=1}^{n} (x - a_r) \sin\left(\frac{1}{x = a_r}\right)$ has no derivative at n-points $x = a_1, a_2, a_3, \ldots, a_n$.

(vi) $f(x) = \sum_{0}^{\infty} a^n \cos\left(b^n \pi x\right)$ where $0 < a < 1$ and b is an odd positive integer subject to the condition $ab > 1 + \frac{3}{2}\pi$, for example, the functions,

$$f_1(x) = \sum_{0}^{\infty} \left(\frac{2}{3}\right)^n \cos\left(15^n \pi x\right)$$

$$f_2(x) = \sum_{0}^{\infty} \left(\frac{5}{6}\right)^n \cos\left(7^n \pi x\right)$$

are continuous for all $x \in R$ but have no derivative for any value of x.

Remember: **1.** Rolle's theorem or Lagrange's mean value theorem is not applicable to the above functions if points of non-differentiability $\in I$ (where I = given interval).

2. $f(x) = x^p \sin\left(\frac{1}{x}\right),\ x \neq 0, f(0) = 0$ is continuous for all values of x if p is positive (i.e. if $p > 0$).

Question: How to show Rolle's theorem is not applicable for a given function defined in a given interval $[a, b]$?

Answer: One or more of the following is to be shown.

1. Show that $f(a) \neq f(b)$

2. Show that $f'(x)$ does not exist at least at one point of (a, b) which means that derivative of $f(x)$ does not exist for some value of x between a and b.

3. Show that $f(x)$ is not continuous at some point lying in the closed interval $[a, b]$.

N.B.: To examine that the statement of the theorem consisting of some particular conditions is untrue, we are required to show by constructing requisite examples that one or more of the conditions of the theorem is (or, are) not fulfilled. e.g.,

(i) $f(x) = x,\ x \in [0, 1]$ satisfied the conditions (1) and (2) of Rolle's theorem but does not satisfy the condition (3) $f(a) = f(b)$ and there is no point c for it such that $f'(c) = 0$.

(ii) $f(x) = |x|,\ x \in [-1, 1]$ satisfied the condition (1) and (3) but does not satisfy the conditions (2) [i.e.; $f(x)$ is differentiable on (a, b)]. Hence, there is no point c for this function at which its derivative would vanish.

Type 1: When non differentiable points are the end points of the given closed interval (one of the two end points or both end points of the given closed interval).

Examples worked out:

Question 1: Examine the validity of Rolle's theorem for the function $f(x) = \sqrt{c^2 - x^2}$.

Solution: **(i)** In this problem, a closed interval in which the given function is defined is not given, so to obtain the interval or intervals, we put $f(x) = 0$.

$$\therefore \sqrt{c^2 - x^2} = 0$$

$$\Rightarrow c^2 - x^2 = 0$$

$$\Rightarrow x = \pm c$$

Thus, we obtain a closed interval $[-c, c]$ such that the functional values at the end points are equal and is zero, i.e.

$f(-c) = f(c)$

(ii) $f(x)$ is differentiable for $-c < x < c$ which $\Rightarrow$ it is continuous for $-c < x < c$.

Now,

$$\lim_{x\to -c+} f(x) = \lim_{x\to -c+} \sqrt{c^2 - x^2} = \lim_{x\to -c} \sqrt{c^2 - x^2}$$

$$= 0 = f(-c) \Rightarrow f(x) \text{ is continuous at } x = -c$$

$$\lim_{x\to c-} f(x) = \lim_{x\to c-} \sqrt{c^2 - x^2} = \lim_{x\to c} \sqrt{c^2 - x^2} =$$

$0 = f(c) \Rightarrow f(x)$ is continuous at $x = c$.

Hence, $f(x)$ is continuous in $(-c, c)$ and at $x = c, -c$ which means $f(x)$ is continuous in the closed interval $[-c, c]$.

$\therefore$ All conditions of Rolle's theorem are satisfied.

$$\text{Now, } f'(x) = \frac{-2x}{2\sqrt{c^2 - x^2}} = \frac{-x}{\sqrt{c^2 - x^2}},\ x \neq \pm c$$

$$\therefore f'(x) = 0$$

$$\Rightarrow x = 0 \text{ and } 0 \in (-c, c)$$

$\therefore$ Rolle's theorem holds good for the given function in the given.

Note: **(a)** The domain of the definition of $\sqrt{c^2 - x^2}$ is $[-c, c]$. If the domain of the definition of a function is a closed interval $a \le x \le b$, then such functions are called continuous on $[a, b]$ provided

(i) $f(x)$ is continuous on the open interval (a, b) i.e.

$$\lim_{x\to a_0} f(x) = f(a_0) \text{ for } a < a_0 < b$$

(ii) $\lim_{x\to a+} f(x) = f(a)$

(iii) $\lim_{x\to b-} f(x) = f(b)$

N.B.: It may be noted that $\lim_{x\to a-} f(x)$ and $\lim_{x\to b+}$ are not defined and this is why $\lim_{x\to a+} = \lim_{x\to a} f(x)$ and $\lim_{x\to b-} f(x) = \lim_{x\to b} f(x)$.

(b) In the above problem,

$$\left.\begin{aligned} f'(-c) &= \lim_{h\to 0} \frac{\sqrt{c^2 - (-c+h)^2}}{h} = \infty \\ f'(c) &= \lim_{h\to 0} \frac{\sqrt{c^2 - (c+h)^2}}{h} = \infty \end{aligned}\right\} \Rightarrow f(x)$$

is not differentiable at $x = -c$ and $x = c$ which $\Rightarrow$ continuity of the function must be tested at the end points $-c$ and c of the closed interval.

Question 2: Are all conditions of Rolle's theorem satisfied for the function $f(x) = \sqrt{x-1}$ on $[1, 3]$.

Answer: **(i)** $f(x) = \sqrt{x-1}$

$\therefore f(x)$ is differentiable in $(1, 3]$ which $\Rightarrow f(x)$ is differentiable at every value of x excepting $x = 1$. Hence, continuity at $x = 1$ must be tested.

$$\text{Now, } r.h.l = \lim_{x\to 1+} f(x) = \lim_{x\to 1+} \sqrt{x-1} = \sqrt{1-1}$$

$= 0 = f(1)$ which $\Rightarrow$ continuity of the function $f(x) = \sqrt{x-1}$ at $x = 1$.

N.B.: $\lim_{x \to 1-}$ is not required since $f(x)=\sqrt{x-1}$ becomes imaginary when $x < 1$, i.e.; $\lim_{x \to 1-}$ is not required since $f(x)=\sqrt{x-1}$ is not defined when $x < 1$.

Again $\lim_{x \to 3} f(x) = \lim_{x \to 3} \sqrt{x-1} = \sqrt{3-1} = \sqrt{2} = f(3)$ which ⇒ continuity of the function $f(x)$ at $x = 3$.

In the light of above explanation, we can say $f(x)$ is continuous on $[1, 3]$ since it is continuous at the end points $x = 1$ and $x = 3$ as well as in between 1 and 3 (i.e. in the open interval $(1, 3) = (a, b)$).

(ii) $\left.\begin{array}{l} f(1) = \sqrt{1-1} = 0 \\ f(3) = \sqrt{3-1} = 2 \end{array}\right\} \Rightarrow f(1) \neq f(3)$

Hence, all above explanations provide us light to say two conditions of Rolle's theorem are satisfied and one condition is not satisfied.

Type 2: When non differentiable point ∈ given open interval:

Examples worked out:

Question 1: Give a reason why Rolle's theorem does not hold for the function defined by $f(x) = \sqrt{1-x}$ on $[-2, 2]$.

Solution: (i) $f(x) = \sqrt{1-x}$

∵ $f(x)$ is not defined for $x > 1$ which ⇒ it is not continuous for $x > 1$ which in turn ⇒ it is not differentiable for $x > 1$. Hence, $f(x)$ is not continuous in $[-2. 2]$ and hence, $f(x)$ is not differentiable in $[-2, 2]$.
∴ $f(x)$ is not differentiable in $(-2, 2)$.

(ii) $f(-2) \neq f(2)$

∴ No condition of Rolle's theorem is satisfied and any one.

⇒ Rolle's theorem is not applicable for the given function $f(x) = \sqrt{1-x}$ on $[-2, 2]$.

Question 2: Does Rolle's theorem apply to $\frac{x(x-2)}{x-1}$ on $[0, 2]$.

Solution: (i) $f(x) = \frac{x(x-2)}{x-1}$

∴ $f(x)$ is not defined at $x = 1$ which ⇒ it is not continuous at $x = 1$ which in turn ⇒ it is not differentiable at $x = 1$

∴ $f(x)$ is not continuous in $[0, 2]$ and $f(x)$ is not differentiable in $(0, 2)$.

(ii) $f(0) = \frac{0 \cdot (0-2)}{0-1} = 2$

$f(2) = \frac{2(2-2)}{2-1} = \frac{2 \cdot 0}{1} = 0$

∴ $f(0) \neq f(2)$

∴ No conditions of Rolle's theorem is satisfied.

⇒ Rolle's theorem is not applicable for the given function in the given interval. In fact not satisfying of only one condition leads to the conclusion.

Question 3: Does Rolle's theorem apply to $\frac{x(x-2)}{x+1}$ on $[0, 2]$.

Solution: (i) $f(x) = \frac{x(x-2)}{x+1}$

∴ $f(x)$ is differentiable in $[0, 2]$ since it is the quotient of two differentiable functions (under the conditions $x+1 \neq 0$) which ⇒ it is continuous in $[0, 2]$.

(ii) $f(0) = \frac{0}{1} = 0$

$f(2) = \frac{2(2-2)}{2+1} = \frac{2 \cdot 0}{2} = 0$

∴ $f(0) = f(2)$

∴ All conditions of Rolle's theorem are satisfied.

∴ This is why Rolle's theorem is applicable for the given function in the given interval $[0, 2]$.

Question 4: Does Rolle's theorem apply to $f(x) = \frac{1}{x}$ on $[-1, 2]$.

Solution: (i) $f(x) = \frac{1}{x}$

∵ $f(x)$ is not defined at $x = 0$ ⇒ $f(x)$ is not continuous at $x = 0$ which in turn ⇒ $f(x)$ is not differentiable at $x = 0$.

Hence, $f(x)$ is not continuous in [0, 2] and $f(x)$ is not differentiable in (0, 2).

(ii) $\left.\begin{array}{l} f(-1)=\frac{1}{-1}=-1 \\ f(2)=\frac{1}{2}=\frac{1}{2} \end{array}\right\} \Rightarrow f(-1) \neq f(2)$

$\therefore$ No condition of Rolle's theorem is satisfied. This is why Rolle's theorem is not applicable for $f(x) = \frac{1}{x}$ in [0, 2].

Question 5: Discuss the applicability of Rolle's theorem on the function $f(x)=|x|$ on [–1, 1].

Solution: (i) $f(x)=|x|$

$\therefore$ $f(x)$ is continuous for every value of x and hence it is continuous in [–1, 1]

At $x=0$, R.H.D $= Rf'(0) = \lim_{h\to 0} \frac{f(0+h)-f(0)}{h}$

$$= \lim_{h\to 0} \frac{|h|-0}{h} = \lim_{h\to 0} \frac{h}{h} = 1 \; (h>0)$$

L.H.D $= Lf'(0) = \lim_{h\to 0} \frac{f(0-h)-f(0)}{-h}$

$$= \lim_{h\to 0} \frac{|-h|-0}{-h} = \lim_{h\to 0} \frac{h}{-h} = -1 \; (h>0)$$

$\therefore$ L.H.D $\neq$ R.H.D

Hence, the given function is not differentiable at $x=0 \in (-1, 1)$ this is why given function $f(x)$ is not differentiable in (–1, 1).

(ii) $\left.\begin{array}{l} f(-1)=|-1|=1 \\ f(1)=|1|=1 \end{array}\right\} \Rightarrow f(-1)=f(1)$

Thus we see that two conditions of Rolle's theorem are satisfied and one condition is not satisfied.

$\therefore$ Rolle's theorem is not applicable for the given function $f(x)$ in the given interval [–1, 1].

Question 6: Is Rolle's theorem applicable to the function $f(x)=|x-1|$ on [0,2].

Solution: (i) $f(x)=|x-1|$

$\therefore$ $f(x)$ is continuous for every value of x since such a mod function is continuous for every value of x which $\Rightarrow$ it is continuous in [0, 2].

Again at $x = 1$

R.H.D $= Rf'(1) = \lim_{h\to 0} \frac{f(1+h)-f(1)}{h}$

$$= \lim_{h\to 0} \frac{|1+h-1|-0}{h} = \lim_{h\to 0} \frac{h}{h} = 1 \; (h>0)$$

L.H.D $= Lf'(1) = \lim_{h\to 0} \frac{f(1-h)-f(1)}{-h}$

$$= \lim_{h\to 0} \frac{|1-h-1|-0}{-h} = \lim_{h\to 0} \frac{|-h|}{-h}$$

$$= \lim_{h\to 0} \frac{h}{-h} = 1 \; (h>0)$$

$\therefore$ L.H.D $\neq$ R.H.D

Hence, the given function is not differentiable at $x = 1 \in (0, 2)$ this is why given function is not differentiable in (0, 2).

(ii) $f(0)=|0-1|=1$

$f(2)=|2-1|=1$

Thus we see that two conditions of Rolle's theorem are satisfied and one condition is not satisfied.

$\therefore$ Rolle's theorem is not applicable for the given function $f(x)$ in the given interval [0, 2].

Question 7: Discuss the applicability of Rolle's theorem to the function $f(x)=|x|^3$ on [–1, 1].

Solution: (i) $f(x)=|x|^3$

$\therefore$ $f(x)$ is continuous on [–1, 1].

(ii) $f'(x) = 3|x|^2 \cdot \frac{d|x|}{dx} = 3|x|^2 \cdot \frac{|x|}{x} = 3\frac{|x|^3}{x}$

$\therefore$ $f(x)$ is differentiable for all values of x except per haps at $x = 0$.

$\therefore$ $Rf'(1) = \lim_{h\to 0} \frac{f(0+h)-f(0)}{h}$

$$= \lim_{h\to 0} \frac{|0+h|^3-0}{h} \text{ for } h>0$$

$$= \lim_{h\to 0} \frac{|h|^3}{h} = \lim_{h\to 0} \frac{h^3}{h} = \lim_{h\to 0} h^2 = 0$$

$Lf'(0) = \lim_{h\to 0} \frac{f(0-h)-f(0)}{-h}$

$$= \lim_{h\to 0} \frac{|0-h|^3 - 0}{-h} \text{ for } h > 0$$

$$= \lim_{h\to 0} \frac{|-h|^3}{-h} = \lim_{h\to 0} \frac{h^3}{-h} = \lim_{h\to 0}\left(-h^2\right) = 0$$

$\therefore\ Rf'(0) = Lf'(0)$

$\therefore\ f'(x)$ exists at $x = 0$ and $f'(0) = 0$

$\therefore\ f'(x)$ exists for all values of $x \in (-1, 1)$

$\therefore f(x)$ is differentiable in $(-1, 1)$.

(iii) $f(-1) = |-1|^3 = 1$

$f(1) = |-1|^3 = 1$

$\therefore f(-1) = f(1)$

$\therefore$ All conditions of Rolle's theorem are satisfied.

$\therefore$ Roll theorem is applicable in the given function $f(x)$ on $[-1, 1]$.

Type 3: When $x = c \in$ given open interval

Where c = critical point

Question: What do you mean by a critical point?

Answer: A critical point is a point at which a function has the first derivative that is zero, infinite or undefined.

Or in other words,

A critical point is a point c at which $f'(c) = 0$ or $f'(c)$ = infinite or $f'(c)$ does not exist.

Examples worked out:

Question 1: Discuss the applicability of Rolle's theorem for the function $f(x) = x^2$ in $[-1, 1]$.

Solution: Since $y = x^2$ is a power function $\Rightarrow$ it is continuous and differentiable in any given finite interval $\Rightarrow$ it is continuous and differentiable in the given finite interval $[-1, 1] \Rightarrow$ two conditions of Rolle's theorem namely continuity in the closed interval and differentiability in the open interval hold good.

Again $f(1) = f(-1) = 1 \Rightarrow f(a) = f(b)$ showing that third condition of Rolle's theorem is also satisfied.

Thus, all conditions of Rolle's theorem are satisfied.

$\therefore\ f'(x) = 2x = 0$

$\Rightarrow 2x = 0$

$\Rightarrow x = 0$ and $0 \in (-1, 1)$ which is true.

$\therefore$ Rolle's theorem is applicable to the function $f(x) = x^2$ in $[-1,1]$.

Question 2: Are all the conditions of Rolle's theorem verified for the function $f(x) = x^2$ in $2 \le x \le 3$.

Solution: **(i)** $f(x) = x^2$

$f(x)$ is continuous and differentiable in any finite interval as it is a power function which $\Rightarrow$ it is continuous and differentiable in $[2, 3]$.

(ii) $\left.\begin{array}{l} f(2) = 4 \\ f(3) = 9 \end{array}\right\} \Rightarrow f(2) \neq f(3)$

Thus we see that two conditions of Rolle's are satisfied and one condition is not satisfied.

Question 3: Can Rolle's theorem be applied to the functions

(i) $f(x) = \sec x$ in $[0, 2\pi]$

(ii) $f(x) = \tan x$ in $[0, \pi]$

Solution: **(i)** $f(x) = \sec x$

$\therefore\ f(x)$ is discontinuous at $x = \frac{\pi}{2}$ and $x = \frac{3\pi}{2}$ both of which belong to $[0, 2\pi]$ since it is discontinuous at $x = (2n + 1)\frac{\pi}{2}$ $(n = 0, \pm 1, \ldots)$

$\therefore f(x)$ is not continuous in $[0, 2\pi]$ which $\Rightarrow$ it is not differentiable in $(0, 2\pi)$.

(ii) $\left.\begin{array}{l} f(0) = \sec 0 = 1 \\ f(2\pi) = \sec 2\pi = 1 \end{array}\right\} \Rightarrow f(0) = f(2\pi)$

Thus we see that two conditions of Rolle's theorem are not satisfied and one condition is satisfied.

$\therefore$ Rolle's theorem can not be applied to the given function in the given closed interval.

(ii) $f(x) = \tan x$

$\therefore f(x)$ is differentiable for all $x \neq \frac{\pi}{2}$ as $f(x)$ is not defined at $x = \frac{\pi}{2}$ and $\frac{\pi}{2} \in (0, \pi)$.

Hence, $f(x)$ is neither continuous nor differentiable at $x = \frac{\pi}{2} \in [0, \pi]$ which $\Rightarrow$ it is not continuous in $[0, \pi]$ and it is not differentiable in $(0, \pi)$.

Again $\left.\begin{array}{l} f(0) = \tan 0 = 0 \\ f(\pi) = \tan \pi = 0 \end{array}\right\} \Rightarrow f(0) = f(\pi)$

Thus, we see that two conditions of Rolle's theorem are not satisfied and one condition is satisfied.

$\therefore$ Rolle's theorem can not be applied.

Question 4: Is Rolle's theorem applicable to the function.

(i) $f(x) = \sin\left(\frac{1}{x}\right)$ in the closed interval $\left[-\frac{1}{\pi}, \frac{1}{\pi}\right]$

(ii) $f(x) = \cos\left(\frac{1}{x}\right)$ in the closed interval $[-1, 1]$.

Solution:

(i) $f(x) = \sin\left(\frac{1}{x}\right)$

$\because f(x)$ is not defined at $x = 0$ and $0 \in \left[-\frac{1}{\pi}, \frac{1}{\pi}\right]$

$\therefore f(x)$ is neither continuous nor differentiable at $x = 0 \in \left[-\frac{1}{\pi}, \frac{1}{\pi}\right]$

$\therefore f(x)$ is neither continuous in $\left[-\frac{1}{\pi}, \frac{1}{\pi}\right]$ nor differentiable in the open interval $\left(-\frac{1}{\pi}, \frac{1}{\pi}\right)$.

$\therefore$ Two conditions of Rolle's theorem are not satisfied.

$\therefore$ Rolle's theorem is not applicable.

(ii) $f(x) = \cos\left(\frac{1}{x}\right)$

$\because f(x)$ is not defined at $x = 0$

$\therefore f(x)$ is not continuous or differentiable at $x = 0 \in [-1, 1]$.

$\therefore f(x)$ is neither continuous nor differentiable in $[-1, 1]$

$\therefore f(x)$ is not differentiable in $(-1, 1)$

$\therefore$ Two conditions of Rolle's theorem are not satisfied.

$\therefore$ Rolle's theorem is not applicable.

Question 5: Is Rolle's theorem applicable on the function $f(x) = \sin x$ in $[0, \pi]$.

Solution: **(i)** $f(x) = \sin x$

$\therefore f(x)$ is continuous and differentiable for all values of x which $\Rightarrow$ it is continuous and differentiable in $[0, \pi]$

Thus, two conditions of Rolle's theorem are satisfied.

(ii) $f(0) = 0 = f(\pi)$

Thus, we observe in the light of above explanation that Rolle's theorem is applicable.

Question 6: Is Rolle's theorem applicable on the function $f(x) = 1 - x^{\frac{4}{5}}$ in $[-1, 1]$.

Solution: $\because f(x) = 1 - x^{\frac{4}{5}}$

$\therefore f(x)$ is differentiable for $x \neq 0$ and

$$f'(x) = -\frac{4}{5} \cdot \frac{1}{x^{\frac{1}{5}}}; \ (x \neq 0)$$

Now, $f'(0) = \lim_{h \to 0} \frac{f(0+h) - f(0)}{h}$

$$= \lim_{h \to 0} \frac{1 - h^{\frac{4}{5}} - 1}{h} = \lim_{h \to 0} \left[-\frac{1}{h^{\frac{1}{5}}}\right] = -\infty \text{ ; if } h > 0$$

and $= \infty$; if $h < 0$.

$\therefore f(x)$ is not differentiable at $x = 0 \in [-1, 1]$

$\therefore f(x)$ is not differentiable in $[-1, 1]$

$\therefore$ Rolle's theorem is not applicable.

Question 7: Are all conditions of Rolle's theorem satisfied for the function $f(x) = \cos\left(\frac{1}{x}\right)$ in $[-1, 1]$.

Solution: **(i)** $\because f(x) = \cos\left(\frac{1}{x}\right)$

$\therefore f(x)$ is neither continuous nor differentiable in $[-1, 1]$ since $f(x)$ is not defined at $x = 0 \in [-1, 1]$

$\therefore f(x)$ is not differentiable in $(-1, 1)$

(ii) $\left.\begin{array}{l} f(-1) = \cos(-1) = \cos 1 \\ f(1) = \cos(1) = \cos 1 \end{array}\right\} \Rightarrow f(-1) = f(1)$

Thus, we observe in the light of (i) and (ii) that the first two conditions of Rolle's theorem are not satisfied and the third condition is satisfied.

$\therefore$ All conditions of Rolle's theorem are not satisified.

N.B.: **1.** When we are asked whether all conditions of Rolle's theorem are satisfied for the given function, we are required to examine all the three conditions. i.e; we examine.

(i) Continuity of the given function $f(x)$ in the given closed interval $[a, b]$.

(ii) Differentiability of the given function $f(x)$ in the open interval (a, b).

(iii) Equality of value of the given function $f(x)$ at the end points a and b of the closed interval. i.e.; $f(a) = f(b)$.

2. When we are asked to show that Rolle's theorem is not applicable to the given function $f(x)$ on $[a, b]$, then sometimes we are not required to examine all the three conditions. i.e.

If any one condition is not satisfied, then that shows non applicability of Rolle's theorem for the given function defined on the given closed interval $[a, b]$.

Question 8: Is Rolle's theorem applicable to the function $f(x) = 1 - (x-1)^{\frac{3}{2}}$ on [0,2].

Solution: **(i)** $\because f(x) = 1 - (x-1)^{\frac{3}{2}}$

$\therefore f(0) = 1 - (0-1)^{\frac{3}{2}} = 1 - (-1)^{\frac{3}{2}}$ = undefined

$\therefore f(x)$ is discontinuous at $x = 0$ and $0 \in [0, 2]$

$\therefore$ Rolle's theorem is not applicable.

Question 9: Discuss the applicability of Rolle's theorem for the function $f(x) = 2 + (x-1)^{\frac{2}{3}}$ on the interval [0, 2].

Solution: **(i)** $f(x) = 2 + (x-1)^{\frac{2}{3}}$ is an irrational function.

$\therefore f(x)$ is continuous function on [0, 2] for being an algebraic function of x.

(ii) $f(x)$ is differentiable for $x \neq 1$ and

$$f'(x) = \frac{2}{3(x-1)^{\frac{1}{3}}}$$

$\therefore f'_+(1) = \infty,\ f'_-(1) = -\infty$

$\therefore f(x)$ is not differentiable at $x = 1$ and $1 \in [0, 2]$

(iii) $f(0) = f(2) = 3$

$\therefore$ All conditions of Rolle's theorem are not satisfied.

$\therefore$ Rolle's theorem is not applicable to the given function defined in a given closed interval.

Type 4: When we are provided a function $f(x)$ defined by various or (different) formulas (or, expression in x) against which an interval is mentioned s.t union of those intervals provides us a given closed interval $[a, b]$ on which Rolle's theorem is to be verified.

Working rule: In such types of problems mentioned above, we adopt the following working rule:

1. We test for continuity and differentiability at the point 'a' if the function is defined by different formula on the left and right of a. e.g.,

(i) $f(x) = x^2$, when $x \le 0$

$= 1$, when $0 < x < 1$

$= \dfrac{1}{x}$ when $x \ge 1$

In the above function, we should test the continuity and differentiability at $x = 0$ and $x = 1$.

(ii) $f(x) = x^2 + 1$, when $0 \le x \le 1$

$= 3 - x$, when $1 < x \le 2$

In the above problem, we should test the continuity and differentiability at $x = 1$.

2. Use the theorems and facts for the continuity, discontinuity differentiability or non differentiability for the given function $y = f(x)$ i.e.;

(a) Discontinuity at a point (or, number) $x = a \Rightarrow$ non-differentiability at the same point (or, number) $x = a$ etc.

N.B.: **(i)** Continuity at $x = a$ does not guarantee differentiability at $x = a$. This is why the test for differentiability is required if continuity at $x = a$ is examined.

(ii) For testing the differentiability at $x = a$, we should find l.h.d and r.h.d using the definition. i.e.

$$Lf'(a) = \lim_{h\to 0}\frac{f(a-h)-f(a)}{-h}, \text{ for } (h>0)$$

$$= \lim_{h\to 0}\frac{f(a+h)-f(a)}{h}, \text{ for } (h>0)$$

3. Find $f(a)$ and $f(b)$ and see whether $f(a) = f(b)$.

4. In the light of (1), (2), (3), we conclude whether Rolle's theorem is applicable or not.

Remember:

1. When $f(x) = f_1(x)$, when $a \le x \le c_2$

$= f_2(x)$, when $c_2 < x \le b$ is provided, then the domain of $f(x)$ is

$= [a, c_2) \cup (c_2, b] = [a, b]$

Examples worked out:

Question 1: Discuss the applicability of Rolle's theorem on the function $f(x) = x^2 + 1$, when $0 \le x \le 1$,

$= 3 - x$, when $1 < x \le 2$, on $[0, 2]$

Solution: **(i)** Continuity and differentiability test at $x = 1$

$$\lim_{x\to 1+} f(x) = \lim_{x\to 1}(3-x) = (3-1) = 2$$

$$\lim_{x\to 1-} f(x) = \lim_{x\to 1}\left(x^2+1\right) = \left[(1)^2+1\right] = 1+1 = 2$$

$$f(1) = [f(x)]_{x=1} = \left[x^2+1\right]_{x=1} = (1)^2+1 = 1+1 = 2$$

$$\therefore \lim_{x\to 1+} f(x) = \lim_{x\to 1-} f(x) = f(1)$$

$\therefore$ the function $f(x)$ is continuous at $x = 1$ which $\Rightarrow$ it is continuous in the closed interval $[0, 2]$.

Now, for the differentiability at $x = 1$,

$$Rf'(1) = \lim_{h\to 0}\frac{f(1+h)-f(1)}{h}, (h>0)$$

$$= \lim_{h\to 0}\frac{\{3-(1+h)\}-2}{h}$$

$$= \lim_{h\to 0}\frac{2-h-2}{h}$$

$$= \lim_{h\to 0}(-1) = -1$$

$$Lf'(1) = \lim_{h\to 0}\frac{f(1-h)-f(1)}{-h}$$

$$= \lim_{h\to 0}\frac{\left\{(1-h)^2+1\right\}-2}{-h}, (h>0)$$

$$= \lim_{h\to 0}\frac{2h-h^2}{h} = \lim_{h\to 0}(2-h) = 2$$

$\therefore Rf'(1) \ne Lf'(1)$ which $\Rightarrow f'(1)$ does not exist.

$\therefore$ The given function has no derive at $x = 1 \in (0, 2)$

$\therefore$ The given function is not differentiable in the open interval $(0, 2)$.

(ii) $\left.\begin{array}{l} f(0) = 0^2+1 = 1 \\ f(2) = 3-2 = 1 \end{array}\right\} \Rightarrow f(0) = f(2)$

Thus we observe that two conditions of Rolle's theorem namely continuity in the closed interval and equality of the values of the function $f(x)$ at the end points 0 and 2 of the closed interval $[0, 2]$ are satisfied but one conditions namely differentiability of the function $f(x)$ in the open interval $(0, 2)$ is not satisfied.

$\therefore$ Rolle's theorem is not applicable to the function $f(x)$ in $[0, 2]$.

Question 2: A function $f(x)$ is defined on $[0, 2]$ s.t $f(x) = 1$, when $0 \le x < 1$

$= 2$, when $1 \le x \le 2$ then discuss the applicability of Rolle's theorem.

Solution: **(i)** Continuity and differentiability test at $x = 1$.

$$\lim_{x\to 1+} f(x) = \lim_{x\to 1} 2 = 2$$

$$\lim_{x\to 1-} f(x) = \lim_{x\to 1} 1 = 1$$

$$\therefore f(1) = [f(x)]_{x=1} = [2]_{x=1} = 2$$

$$\therefore \lim_{x\to 1+} f(x) = 2 = f(1) \ne \lim_{x\to 1-} f(x) = 1$$

$\therefore f(x)$ is discontinuous at $x = 1$ and $1 \in [0, 2]$

$\therefore f(x)$ is non differentiable at $x = 1$ and $1 \in [0, 2]$

$\therefore f(x)$ is discontinuous and non differentiable in [0, 2].

$\therefore f(x)$ is not differentiable in (0, 2).

(ii) $\left.\begin{array}{l} f(0) = 1 \\ f(2) = 2 \end{array}\right\} \Rightarrow f(0) \neq f(2)$

$\therefore$ No condition of Rolle's theorem is satisfied.

$\therefore$ Rolle's theorem is not applicable for the given function $f(x)$ in the given interval [0, 2].

Question 3: The function 'f' is defined in [0, 1] as follows

$$f(x) = 1, \text{ when } 0 < x < \frac{1}{2}$$

$= 2$, when $\frac{1}{2} \le x \le 1$ show that $f(x)$ satisfied none of the conditions of Rolle's theorem.

Solution: **(i)** Continuity and differentiability test at $x = \frac{1}{2}$.

$$f\left(\frac{1}{2} - 0\right) = \lim_{x \to \frac{1}{2}-} f(x) = \lim_{\substack{x \to \frac{1}{2} \\ x < \frac{1}{2}}} (1) = 1$$

$$f\left(\frac{1}{2} + 0\right) = \lim_{x \to \frac{1}{2}+} f(x) = \lim_{\substack{x \to \frac{1}{2} \\ x > \frac{1}{2}}} (2) = 2$$

$\therefore f\left(\frac{1}{2} - 0\right) \neq f\left(\frac{1}{2} + 0\right)$ which $\Rightarrow f(x)$ is discontinuous at $x = \frac{1}{2}$ and $\frac{1}{2} \in [0, 1]$.

$\therefore f(x)$ is non-differentiable at $x = \frac{1}{2} \in [0, 1]$

$\therefore f(x)$ is discontinuous and non-differentiable in [0, 2].

$\therefore f(x)$ is not differentiable in (0, 2)

(ii) $\left.\begin{array}{l} f(0) = 1 \text{ (given)} \\ f(1) = 2 \text{ (given)} \end{array}\right\} \Rightarrow f(0) \neq f(1)$

Hence, each of the three conditions of Rolle's theorem are not satisfied by the given function $f(x)$ defined in the closed interval [0, 1].

Question 4: A function $f(x)$ is defined on [0, 1] s.t

$f(x) = x$, when $0 \le x < 1$

$f(x) = 0$, when $x = 1$ then discuss the applicability of Rolle's theorem.

Solution: **(i)** Continuity and differentiability test at $x = 1$.

$$\because \lim_{\substack{x \to 1 \\ x < 1}} f(x) = \lim_{x \to 1} x = 1$$

and $f(1) = 0$ (given)

$$\therefore \lim_{\substack{x \to 1 \\ x < 1}} f(x) \neq f(1)$$

$\therefore f(x)$ is not continuous at one of the end point of the given closed interval [0, 1] namely $x = 1$.

$\therefore f(x)$ is not continuous in [0, 1].

$\therefore$ Rolle's theorem is not applicable to the given function $f(x)$ defined in the given interval [0, 1].

Note: In the above problem

(i) $f(0) = 0$

$f(1) = 0$ (given)

$\therefore f(0) = f(1)$

(ii) Now to show the differentiability in the open interval (0, 1) we take any $c \in (0, 1)$ i.e.;

Let $x = c \in (0, 1)$

$$\therefore Rf'(c) = \lim_{h \to 0} \frac{f(c+h) - f(c)}{h}$$

$$= \lim_{h \to 0} \frac{\not{c} + h - \not{c}}{h}$$

$$= \lim_{h \to 0} \frac{h}{h}$$

$$= \lim_{h \to 0} 1 = 1$$

$$Lf'(c) = \lim_{h \to 0} \frac{f(c-h) - f(c)}{-h}$$

$$= \lim_{h \to 0} \frac{c - h - c}{-h}$$

$$= \lim_{h \to 0} \frac{-h}{-h}$$

$$= \lim_{h \to 0} (+1)$$

$$= 1$$

$\therefore L f'(c) = R f'(c)$

$\therefore f'(c)$ exists and $c \in (0, 1)$

$\therefore f(x)$ is differentiable in (0, 1)

Thus we see that two conditions namely equality of values of the function at the end points 0 and 1 of the closed interval [0, 1] and the differentiability of the given function $f(x)$ defined by different parts of the closed interval [0, 1] in the open interval (0, 1) are satisfied but one condition namely continuity in the closed interval [0, 1] is not satisfied which has been shown in the above question 4.

Question 5: A function $f(x)$ is defined on [0, 1] s.t $f(x) = 1$, when $x = 0$

$= x$, when $0 < x \le 1$ is Rolle's theorem applicable?

Solution: **(i)** Continuity and differentiability test at $x = 0$

$$\because \lim_{\substack{x \to 0 \\ x > 0}} f(x) = \lim_{x \to 0} (x) = 0$$

and $f(0) = 1$ (given)

$$\therefore \lim_{\substack{x \to 0 \\ x > 0}} f(x) \ne f(0)$$

$\therefore f(x)$ is discontinuous at one of the end points of the given closed interval [0, 1] namely $x = 0$

$\therefore f(x)$ is not continuous in [0, 1]

$\therefore$ Rolle's theorem is not applicable to the given function $f(x)$ defined in various parts of the closed interval [0, 1].

Type 5: Problems based on finding the value of 'c' using Rolle's theorem.

Working rule:

1. Differentiate the given function $f(x)$.

2. Put $x = c$ in the differentiated result (or, derived function or, differentiated function) on both sides of the sign of equality which $\Rightarrow$ we put $x = c$ in the derived function and in the notation $f'(x)$ and equate it to zero which $\Rightarrow [f'(x)]_{x=c} = f'(c) = 0$ should be solved for c.

3. Check whether $c \in (a, b)$ given open interval if 'c' lies in between a and b, them 'c' should be accepted as the required value and if 'c' does not lie in between a and b, then 'c' should be rejected i.e.; in the notational form,

If $a < c < b \Rightarrow c \in (a, b)$ should be accepted as the required value and if $c \notin (a, b)$ it should be rejected.

Examples worked out:

1. If $f(x) = x^3 - 27x$ on $\left[0, 3\sqrt{3}\right]$ find the value of c in Rolle's theorem.

Solution: $\because f(x) = x^3 - 27x$, $f(0) = f\left(3\sqrt{3}\right)$

and $f'(x) = 3x^2 - 27$

$$f'(c) = 3c^2 - 27 = 0$$

$$\Rightarrow 3c^2 - 27 = 0$$

$$\Rightarrow c^2 = \frac{27}{3} = 9$$

$$\Rightarrow c = \pm 3$$

Now accepting + 3 as the required value since $3 \in \left(0, 3\sqrt{3}\right)$ and rejecting $-3 \notin \left(0, 3\sqrt{3}\right)$, we see that $c = +3$ is the required value.

2. Find c of Rolle's theorem when $f(x) = x^2 + 3x + 2$ is defined in [–2, –1].

Solution: $\because f(x) = x^2 + 3x + 2$, $f(-2) = f(-1) = 0$

and $f'(x) = 2x + 3$

$$\therefore f'(c) = 2c + 3 = 0$$

$$\Rightarrow 2c + 3 = 0$$

$$\Rightarrow 2c = -3$$

$$\Rightarrow c = -\frac{3}{2} = -1.5$$

Now $c = -1.5$ is s.t $-2 < -1.5 < 1$ which $\Rightarrow -1.5 \in (-2, 1)$ should be accepted as the required value.

$\therefore c = -1.5$ is the required value.

3. If $f(x) = \cos x$ be defined on $\left[-\frac{\pi}{2}, \frac{\pi}{2}\right]$, find '$c$' of Rolle's theorem.

Solution: $\because f(x) = \cos x,\ f\left(\frac{-\pi}{2}\right) = f\left(\frac{\pi}{2}\right) = 0$

$f'(x) = -\sin x$

$\therefore f'(c) = -\sin c = 0$

$\Rightarrow \sin(-c) = \sin 0$

$\Rightarrow c = 0$ is one value and $0 \in \left(-\frac{\pi}{2}, \frac{\pi}{2}\right)$

$\therefore c = 0$ is the required value.

4. Find 'c' of Rolle's theorem if $f(x) = \frac{\sin x}{e^x}$ is defined on $[0, \pi]$.

Solution: $\because f(x) = \frac{\sin x}{e^x},\ f(0) = f(\pi) = 0$

$f'(x) = \frac{e^x \cdot \cos x - \sin x\, e^x}{e^{2x}} = \frac{\cos x - \sin x}{e^x}$

$\therefore f'(c) = \frac{\cos c - \sin c}{e^c} = 0$

$\Rightarrow \cos c - \sin c = 0$

$\Rightarrow \cos c = \sin c$

$\Rightarrow \cos c = \cos(90 - c)$

$\Rightarrow c = 90 - c$ (a particular solution)

$\Rightarrow c + c = 90$

$\Rightarrow 2c = 90$

$\Rightarrow c = \frac{90}{2}$

$\Rightarrow c = 45^o$ and $45^o = \frac{\pi}{4} \in (0, \pi)$

$\therefore c = \frac{\pi}{4}$ is the required value.

Geometrical meaning of Rolle's theorem

If the graph of the function

1. $f(x)$ is continuous from a point $A\,(a, f(a))$ to another point $B\,(b, f(b))$ [i.e.; $f(x)$ is continuous on the closed interval $[a, b]$].
2. $f(x)$ is differentiable in between the two points $A\,(a, f(a))$ and $B\,(b, f(b))$ [i.e.; $f(x)$ is differentiable in the open interval (a, b)].
3. $f(x)$ has the equal ordinates $f(a)$ and $f(b)$ at the two points $A\,(a, f(a))$ and $B\,(b, f(b))$ [i.e.; $f(x)$ assumes equal values at the end points of the closed interval $[a, b]$]; then the graph of the function $f(x)$ has at least one point $C\,(c, f(c))$ in between $A\,(a, f(a))$ and $B\,(b, f(b))$ (i.e., $a < c > b$) at which the tangent is parallel to x-axis (i.e. $f'(c) = 0$)

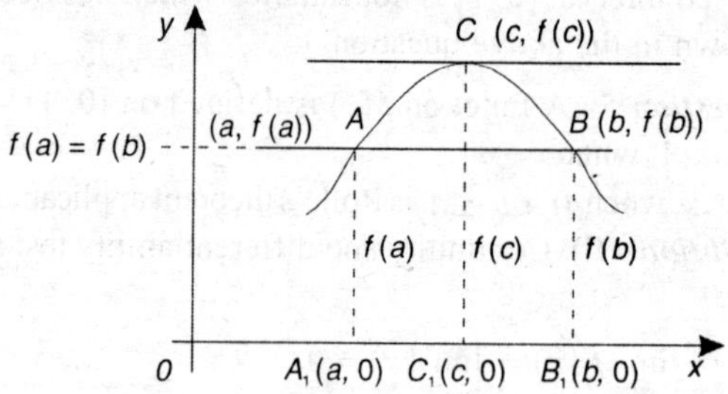

Refresh your memory:

1. Geometrical meaning of Rolle's theorem tells that if the graph of the function (i) which is continuous on the interval $[a, b]$ and differentiable in (a, b) (ii) which assumes equal values at the end points a and b of $[a, b]$ (i.e. $f(a) = f(b)$), then the graph of the function has a point $(c, f(c)$ at which the tangent is parallel to x-axis.

Or alternatively,

Under the given conditions

(a) Continuity of the function $f(x)$ in $[a, b]$

(b) Differentiability of the function $f(x)$ in (a, b)

(c) Equality of functional values at the end points a and b of $[a, b]$, there is at least one point 'c' which is not the end point for the closed interval $[a, b]$ such that the tangent at that point is parallel to x-axis.

2. $\left[\frac{dy}{dx}\right]_p = 0 \Rightarrow \tan\psi = 0 \Rightarrow \psi = 0$ (where p is a point on the curve, $\Rightarrow$ the tangent at p is parallel to x-axis.

3. $\left[\frac{dy}{dx}\right]_p = \infty \Rightarrow$ the tangent at p is perpendicular to y-axis.

Problems based on geometrical meaning of Rolle's theorem

Problems based on geometrical meaning of Rolle's theorem consists of

1. A function $f(x)$ = an expression in x.
2. Two point $(a, f(a))$ and $(b, f(b))$
3. A third point $(c, f(c))$ between $(a, f(a))$ and $(b, f(b))$ is required to find out at which there is a tangent parallel to x-axis.

Working rule to find $(c, f(c))$ = a third point in between two given points $(a, f(a))$ and $(b, f(b))$ on the graph of $y = f(x)$ on using Rolle's theorem.

We adopt the following procedure to find a third point $(c, f(c))$.

1. Show that $f(x)$ is continuous at all points lying from $(a, f(a))$ to $(b, f(b))$.

Or, alternatively, show that $f(x)$ is continuous on $[a, b]$.

2. Show that $f(x)$ is differentiable at all points lying between $(a, f(a))$ and $(b, f(b))$.

Or, alternatively; show that $f(x)$ is differentiable on (a, b).

3. Check whether $f(a) = f(b)$
4. If all conditions of Rolle's theorem are satisfied, then use the geometrical meaning of Rolle' s theorem which tells about the existance of a point $(c, f(c))$, or $(z, f(z))$ in between two given point $(a, f(a))$ and $(b, f(b))$ at which the tangent line is parallel to x-axis. i.e.; $f'(z) = 0$ or $f'(z) = 0$ where z or $\in (a, b)$
5. Find c and $f(c)$.

Question A: How to find 'c'.

Answer: **(a$_1$)** find $f'(x)$ and put $f'(x) = 0$

(a$_2$) substitute $x = c$ (or, z) in $f'(x)$ and solve for c (or, solve for z).

Question B: How to find $f(c)$.

Answer: **(b$_1$)** Put $x = c$ (or, z) (i.e., the roots of the equation $f'(c) = 0$ or $f'(z) = 0$) in $f(x)$ which $\Rightarrow$ put $x = c$ or z in the given function $y = f(x)$ of the question.

Remember:

1. x-coordinates of the required third point is obtained is obtained from $f'(x) = 0$.
2. y-coordinates of the required point is obtained by putting the root of $f'(x) = 0$ which belongs to (a, b) in the given function $y = f(x)$.
3. Given point $(a, f(a))$ and $(b, f(b))$ form the closed interval $[a, b]$ and the open interval (a, b).
4. $f'(c)$ or $f'(z) = 0$ for some 'c' ($a < c < b$ or, $a < z < b$) or for some 'z' $\Rightarrow$ the tangent to the curve (the graph of the given function $y = f(x)$) at $x = c$ (or, z) is parallel to x-axis.

Worked out examples on geometrical meaning of Rolle's theorem

Question 1: Using Rolle's theorem, show that on the graph of $y = x^2 - 4x + 3$, there is a point between (1, 0) and (3, 0) where the tangent is parallel to x-axis. Also find the point.

Solution: **(i)** Since given function $y = f(x) = x^2 - 4x + 3$ is a polynomial function and hence it is continuous and differentiable in [1, 3].

$\therefore y = f(x)$ is differentiable in (1, 3).

(ii) $f(1) = f(3) = 0$

$\therefore$ All conditions of Rolle's theorem are satisfied.

$\therefore \exists$ a point $(z, f(z))$ between (1, 0) and (3, 0) at which tangent is parallel to x-axis. i.e.; $f'(z) = 0$

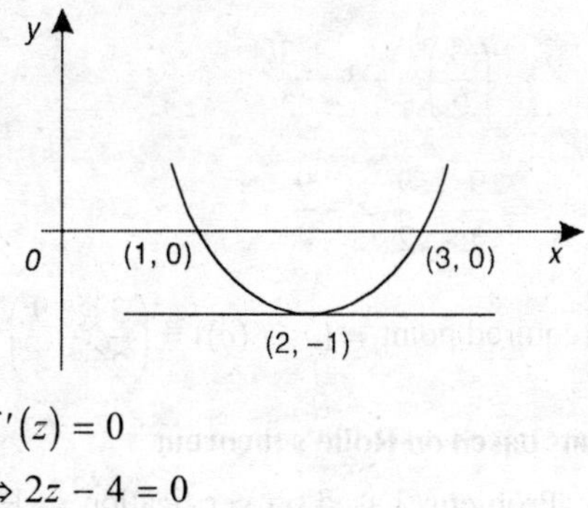

$$f'(z) = 0$$

$$\Rightarrow 2z - 4 = 0$$

$$\Rightarrow z = \frac{4}{2} = 2 \quad \ldots(i)$$

and $[f(z)]_{z=2} = \left[z^2 - 4z + 3\right]_{z=2}$

$= 2^2 - 4 \times 2 + 3 = 4 - 8 + 3 = 7 - 8 = -1 \quad \ldots(ii)$

Thus, we get $(z, f(z)) = (2, -1)$ which is the required point.

Question 2: Using Rolle's theorem, prove that there is a point between the points (1, –2) and (2, –2) on the graph of $y = x^2 - 3x$ where the tangent is parallel to x-axis. Also find the point.

Solution: (i) Given function $y = f(x) = x^2 - 3x$ is a polynomial function of x and this is why it is continuous and differentiable in [1, 2].

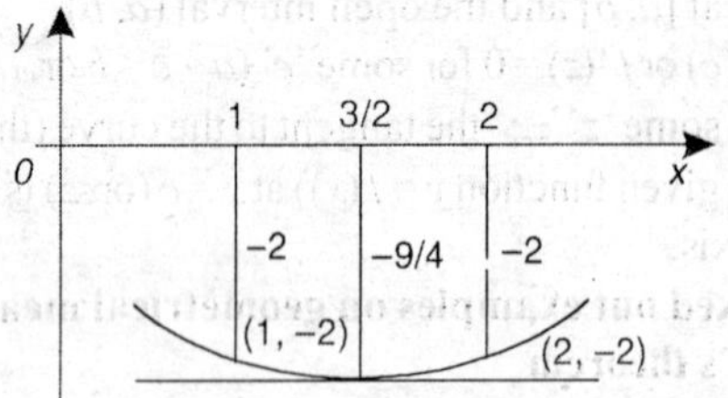

$\therefore f(x)$ is differentiable in (1, 2).

(ii) $f(1) = f(2)$

$\therefore$ All conditions of Rolle's theorem are satisfied.

$\therefore$ According to geometrical meaning of Rolle's theorem, there is a point $(c, f(c))$ in between (1, –2) and (2, –2) where the tangent is parallel to x-axis.

Now, $f'(c) = 0 \Rightarrow 2c - 3 = 0 \Rightarrow c = \dfrac{3}{2}$

$$\therefore f(c) = f\left(\frac{3}{2}\right) = \left[x^2 - 3x\right]_{x=\frac{3}{2}}$$

$$= \left(\frac{3}{2}\right)^2 - 3 \times \frac{3}{2}$$

$$= \frac{9}{4} - \frac{9}{2} = -\frac{9}{4}$$

$$\therefore \text{ Required point} = (c, f(c)) = \left(\frac{3}{2}, -\frac{9}{4}\right)$$

Problems based on Rolle's theorem

Type 1: Problems based on verification of Rolle's theorem

Exercise 20.1

Verify Rolle's theorem for the following functions:

1. $f(x) = x^2 - 5x + 4$ in $1 \le x \le 4$
2. $f(x) = \sin 2x$ in $\left[0, \dfrac{\pi}{2}\right]$
3. $f(x) = 4 \sin x$ in $[0, \pi]$
4. $f(x) = x(x+3)\, e^{-\frac{x}{2}}$ in $[-3, 0]$
5. $f(x) = \sin x$ in $[0, \pi]$
6. $f(x) = x^3 (x-1)^2$ in the interval $0 \le x \le 1$
7. $f(x) = x \cdot \sin\left(\dfrac{1}{x}\right)$, $x \ne 0$ and $f(0) = 0$ in $\left[0, \dfrac{1}{\pi}\right]$
8. $f(x)\ 2x^3 + x^2 - 4x - 2$ when $-\dfrac{1}{2} \le x \le \sqrt{2}$
9. $f(x) = 2(x+1)(x-2)$ defined in $[-1, 2]$
10. $f(x) = x(x-1)$ in $[0, 1]$
11. $f(x)\ (x-1)^3 (x-2)^2$ in $[1, 2]$
12. $f(x) = (x-1)(x-2)(x-3)$ when $0 \le x \le 4$
13. $f(x) = |x|$ in $[-1, 1]$
14. $f(x) = 3x - x^3$ in $[0, \sqrt{3}]$
15. $f(x) = x^3 - 3x$ in $[-\sqrt{3}, 0]$
16. $f(x) = x - 1$ in $\left[-\dfrac{1}{2}, \dfrac{1}{2}\right]$
17. $f(x) = \dfrac{\sin x}{e^x}$ in $[0, \pi]$
18. $f(x) = e^x (\sin x - \cos x)$ in $\left[\dfrac{\pi}{4}, \dfrac{5\pi}{4}\right]$
19. $f(x) = (x-a)^m (x-b)^n$, where m and n are integers, in $[a, b]$
20. $f(x) = x^2 - 4x + 3$ in $[1, 3]$

Answers

1. Rolle's theorem is true
2. Rolle's theorem is true
3. Rolle's theorem is true
4. Rolle's theorem is true
5. Rolle's theorem is true
6. Rolle's theorem is true
7. Rolle's theorem is true

8. Rolle's theorem is true
9. Rolle's theorem is true
10. Rolle's theorem is true
11. Rolle's theorem is true
12. Rolle's theorem is true
13. Rolle's theorem is true
14. Rolle's theorem is true
15. Rolle's theorem is true
16. Rolle's theorem is true
17. Rolle's theorem is true
18. Rolle's theorem is true
19. Rolle's theorem is true
20. Rolle's theorem is true

Type 2: Problems based on verification of Rolle's theorem when interval is not given.

Exercise 20.2

Verify Rolle's theorem for the following functions:

1. $f(x) = 16x - x^2$
2. $f(x) = x^3 - x^2 - 4x + 4$
3. $f(x) = \sin x$
4. $f(x) = \sin\left(\frac{x}{2}\right)$

Type 3: Problems based on examining of Rolle's theorem ⇔ whether Rolle's theorem is applicable or not is required to test.

Exercise 20.3.1

Discuss the applicability of Rolle's theorem for the following functions:

1. $f(x) = |x|$ in $[-1, 1]$
2. $f(x) = |x - 1|$ in $[0, 2]$
3. $f(x) = \sin\left(\frac{1}{x}\right)$ in $\left[-\frac{1}{\pi}, \frac{1}{\pi}\right]$
4. $f(x) = \sin x \cdot \cos x$ in $\left[0, \frac{\pi}{2}\right]$
5. $f(x) = 1 - (x - 1)^{\frac{3}{2}}$ in $[0, 2]$
6. $f(x) = \log\left[\frac{x^2 + ab}{(a + b)x}\right]$ in $[a, b]$
7. $f(x) = e^x \sin x$ in $[0, \pi]$
8. $f(x) = \sqrt{|x - 1|}$ in $[-1, 3]$
9. $f(x) = 2x + \frac{1}{x}$ in $\left[\frac{1}{4}, 2\right]$
10. $f(x) = \tan x$ in the interval $[-1, 2]$
11. $f(x) = \frac{1}{x}$ in the interval $[-1, 2]$
12. $f(x) = \frac{x(x-2)}{(x-1)}$ in the interval $[0, 2]$
13. $f(x) = 8x - x^2$ in the interval $[0, 8]$
14. $f(x) = x^3 - 6x^2 + 11x - 6$ in the interval $[1, 3]$
15. $f(x) = (x + 1)(x - 2)$ in the interval $[-1, 2]$

Answers:

1. Not applicable since $f(x)$ is not differentiable at $x = 0$
2. Not applicable since $f(x)$ is not differentiable at $x = 1$
3. Not applicable as $f(0)$ is not defined
4. Rolle's theorem is applicable
5. Not applicable as $f(0)$ is not defined
6. Rolle's theorem is applicable
7. Rolle's theorem is applicable
8. Not applicable since $f'(x)$ is does not exist
9. Applicable
10. Not applicable since $f(x)$ is neither continuous nor differentiable at $x = \frac{\pi}{2}$
11. Not applicable as $f(x)$ is neither continuous nor differentiable at $x = 0$
12. Not applicable as $f(x)$ is neither continuous nor differentiable at $x = 1$
13. Applicable
14. Applicable
15. Applicable

Exercise 20.3.2

Rolle's theorem can not be applied in the following functions in the interval specified. Explain why.

1. $f(x) = x$ in $-1 \le x \le 1$

2. $\begin{cases} f(x) = 2x \text{ for } x \le 1 \\ f(x) = 4 - 2x \text{ for } x > 1 \end{cases}$ in $0 \le x \le 2$

3. $f(x) = 1 - x^{\frac{2}{3}}$ in $-1 \le x \le 1$

4. $f(x) = \tan x$ in $0 \le x \le \pi$

Answers:

1. $f(-1) \ne f(1)$
2. $f(x)$ is not differentiable at $x = 1$
3. $f(x)$ is not differentiable at $x = 0 \in (-1, 1)$
4. $f(x)$ is discontinuous at $x = \frac{\pi}{2} \in (0, \pi)$

Type 4: Problems based on verification of Rolle's theorem and finding the value of 'c'.

Exercise 20.4

Verify Rolle's theorem for the following functions and find the value of 'c' provided Rolle's theorem is true.

1. $f(x) = 3x - x^3$ in $[0, \sqrt{3}]$
2. $f(x) = x^3 - 3x$ in $[-\sqrt{3}, 0]$
3. $f(x) = \sin x$ in $[0, \pi]$
4. $f(x) = x^2 + 3x + 2$ in $[-2, -1]$
5. $f(x) = 4 \sin x$ in $[0, \pi]$
6. $f(x) = \sin 2x$ in $\left[0, \frac{\pi}{2}\right]$
7. $f(x) = x(x+3)e^{-\frac{x}{2}}$ in $[-3, 0]$
8. $f(x) = x\sqrt{a^2 - x^2}$ in $0 \le x \le a$
9. $f(x) = x(x-2)(x+1)^{-1}$ in $0 \le x \le 2$
10. $f(x) = e^x(\sin x - \cos x)$ in $\left[\frac{\pi}{4}, \frac{5\pi}{4}\right]$
11. $f(x) = (x-a)^m (x-b)^n$ in $a \le x \le b$, m and n being positive integer.
12. $f(x) = x(x-1)$ in $0 \le x \le 1$
13. $f(x) = 2x^3 + x^2 - 4x$ in $-\sqrt{2} \le x \le \sqrt{2}$

Answers:

1. $c = 1$
2. $c = -1$
3. $c = \frac{\pi}{2}$
4. $c = -\frac{3}{2}$
5. $c = \frac{\pi}{2}$
6. $c = \frac{\pi}{4}$
7. $c = -2$
8. $c = \frac{a}{\sqrt{2}}$
9. $c = \sqrt{3} - 1$
10. $c = \pi$
11. $c = a, b, \frac{mb + na}{m + n}$
12. $c = \frac{1}{2}$
13. $c = \frac{1}{3}$

Lagrange's Mean Value Theorem

Statement of Lagrange's mean value theorem

Let $y = f(x)$ a real function (or, real valued function) defined on a closed interval $[a, b]$. If

(i) $f(x)$ is continuous in (or, on or, over) the closed interval $[a, b]$.

(ii) $f(x)$ is differentiable in (or, on or, over) the open interval (a, b) then there is at least one point $x = c \in (a, b)$ at which the first derivative.

$$f'(c) = \frac{f(b) - f(a)}{b - a}$$

$$= \frac{\text{value of } f(x) \text{ at } x = b - \text{value of } f(x) \text{ at } x = a}{\text{difference of end points of the given interval}}$$

N.B.: **1.** Rolle's theorem has three conditions whereas Lagrange's mean value theorem (or, sometimes called simply mean value theorem) has only two conditions.

2. $f(x)$ satisfies two conditions of Rolle's theorem except the condition $f(a) = f(b)$ in mean value theorem which is also termed as Lagrange's mean value theorem.

Type 1: Verification of Lagrange's mean value theorem.

Working rule: To verify Lagrange's mean value theorem, we have to show that following conditions are satisfied.

1. Show that a given function is continuous in the closed interval and differentiable in the open interval which is provided by using the facts that all "PILET-RC" functions are continuous and differentiable at points where they have finite value and also remembering that $\sin x$, $\cos x$, $\sin^{-1} x$, $\cos^{-1} x$, $\log x$, e^x, polynomial, power, constant, identity functions are continuous and differentiable in any give finite interval.

2. Find $f'(x)$ as well as $f(a)$ and $f(b)$

3. Then use the result $f'(c) = \dfrac{f(b) - f(a)}{b - a}$ and solve it which will provide us the value of c such that $a < c$, b i.e. $c \in (a, b)$.

Note: In short Lagrange's mean value theorem is written as L.M.V.T.

Examples worked out:

Question 1: Verify Lagrange's mean value theorem for the function $f(x) = x(x-1)(x-2)$ in $\left[0, \frac{1}{2}\right]$.

Solution: $\because f(x) = x(x-1)(x-2)$

$\Rightarrow f(x) = x^3 - 3x^2 + 2x$

$\because f(x)$ is a polynomial in x

$\Rightarrow f(x)$ is continuous and differentiable in any given interval

$\Rightarrow$ **(i)** $f(x)$ is continuous on the closed interval $\left[0, \frac{1}{2}\right]$

(ii) $f(x)$ is differentiable in the open interval $\left(0, \frac{1}{2}\right)$

Thus the two conditions of Lagrange's mean value theorem are satisfied.

Now **(i)** $f'(x) = 3x^2 - 6x + 2$

$\Rightarrow f'(c) = 3c^2 - 6c + 2$

(ii) $f(a) = f(0) = 0$

(iii) $f(b) = f\left(\frac{1}{2}\right) = \frac{1}{8} - \frac{3}{4} + \frac{2}{2}$

$$= \frac{1 - 6 + 8}{8} = \frac{3}{8}$$

(iv) $b - a = \frac{1}{2} - 0 = \frac{1}{2}$

Now, using the result of Lagrange's mean value theorem, $\exists$ at least one c such that

$$f'(c) = \frac{f(b) - f(a)}{b - a}$$

$$\Rightarrow 3c^2 - 6c + 2 = \frac{\frac{3}{8} - 0}{\frac{1}{2}} = \frac{3}{8} \times 2 = \frac{3}{4}$$

$$\Rightarrow 3c^2 - 6c + 2 = \frac{3}{4}$$

$$\Rightarrow 3c^2 - 6c + \left(2 - \frac{3}{4}\right) = 0$$

$$\Rightarrow 3c^2 - 6c + \frac{5}{4} = 0$$

$$\Rightarrow c = \frac{-(-6) \pm \sqrt{36 - 4 \times 3 \times \frac{5}{4}}}{2 \times 3}$$

$$= \frac{6 \pm \sqrt{36 - 15}}{6}$$

$$= 1 + \frac{1}{6} \cdot \sqrt{21} \text{ or, } 1 - \frac{1}{6} \cdot \sqrt{21}$$

$= 1.76$ or, 0.24

Thus, we get a value of $c = 0.24 \in \left(0, \frac{1}{2}\right)$ and for which the equation $f'(c) = \frac{f(b) - f(a)}{b - a}$ is true.

Hence, Lagrange's mean value theorem is verified.

Note: (i) $1 + \frac{\sqrt{21}}{6} = 1.76 \notin \left(0, \frac{1}{2}\right) \Rightarrow$ This is why it is rejected.

(ii) $1 - \frac{\sqrt{21}}{6} = 0.24 \in \left(0, \frac{1}{2}\right) \Rightarrow$ This is why its is accepted.

Question 2: Verify Lagrange's mean value theorem for the function $f(x) = (x-1)(x-2)(x-3)$ on $[0, 4]$

Solution: $\because f(x) = (x-1)(x-2)(x-3)$

$= x^3 - 6x^2 + 11x - 6$

$\therefore f(x)$ is a polynomial in x

$\Rightarrow f(x)$ is continuous and differentiable in any given finite interval.

$\Rightarrow$ **(i)** $f(x)$ is continuous in the closed interval $[0, 4]$

(ii) $f(x)$ is differentiable in the open interval $(0, 4)$

$\therefore$ The two conditions of Lagrange's mean value theorem are satisfied.

Now, $f'(x) = 3x^2 - 12x + 11$

$$f'(c) = \left[3x^2 - 12x + 11\right]_{x=c} = \frac{f(4) - f(0)}{4 - 0}$$

$$\Rightarrow f'(c) = 3c^2 - 12c + 11$$

$$= \frac{[(4-1)(4-2)(4-3) - (-6)]}{4} = \frac{3 \times 2 \times 1 + 6}{4}$$

$$\Rightarrow 4\left(3c^2 - 12c + 11\right) = 12$$

$$\Rightarrow 12c^2 - 48c + 44 - 32 = 0$$

$$\Rightarrow 3c^2 - 12c + 8 = 0$$

$$\Rightarrow c = \frac{12 \pm \sqrt{144 - 96}}{6} = \frac{12 \pm \sqrt{48}}{6}$$

$$= 2 \pm \frac{2}{3}\sqrt{3} = 2 \pm 1.155$$

$$\Rightarrow c = 3.155 \text{ or }, 0.845$$

But these values of c lie in the open interval $(0, 4)$

Hence, the theorem is verified.

Question 3: Verify Lagrange's mean value theorem for the function $f(x) = -x^2 + 3x + 2, \forall x \in [0, 1]$.

Solution: $\because f(x) = -x^2 + 3x + 2$

$f(x)$ is a polynomial in x

$\Rightarrow f(x)$ is continuous and differentiable in any given finite interval.

$\Rightarrow$ **(i)** $f(x)$ is continuous on the closed interval $[0, 1]$

(ii) $f(x)$ is differentiable in the open interval $(0, 1)$

$\therefore$ Two conditions of Lagrange's mean value theorem are satisfied.

Now, **(i)** $f'(x) = -2x + 3$

$\Rightarrow f'(c) = -2c + 3$

(ii) $f(a) = f(0) = 2$

(iii) $f(b) = f(1) = -1 + 3 + 2 = 4$

(iv) $b - a = 1 - 0 = 1$

Now, using the result of Lagrange's mean value theorem,

$$f'(c) = \frac{f(b) - f(a)}{b - a} = \frac{4 - 2}{1} = \frac{2}{1} = 2$$

$$\Rightarrow -2c + 3 = 2$$

$$\Rightarrow -2c + 3 - 2 = 0$$

$$\Rightarrow -2c + 1 = 0$$

$$\Rightarrow -2c = -1$$

$\Rightarrow c = \frac{1}{2}$ which belongs to $(0, 1)$

$$\Rightarrow c = \frac{1}{2} \in (0, 1)$$

Thus, we observe that we get a value of $c = \frac{1}{2} \in (0, 1)$ and for which the equation $f'(c) = \frac{f(b) - f(a)}{b - a}$ holds good hence, Lagrange's mean value theorem is verified.

Question 4: Verify Lagrange's mean value theorem for the function $f(x) = x^3$ in $[a, b]$.

Solution: **(i)** $f(x) = x^3$

$f(x)$ is a power function in x

$\Rightarrow f(x)$ is continuous and differentiable in any given finite interval

$\Rightarrow$ **(i)** $f(x)$ is continuous in the closed interval $[a, b]$.

(ii) $f(x)$ is differentiable on the open interval (a, b)

$\therefore$ Two conditions of Lagrange's mean value theorem are satisfied.

Now, **(i)** $\Rightarrow f'(x) = 3x^2$

$\Rightarrow f'(c) = 3c^2$

(ii) $f(b) = b^3$

(iii) $f(a) = a^3$

(iv) $b - a = b - a$

Now, using the result of Lagrange's mean value theorem,

$$f'(c) = \frac{f(b) - f(a)}{b - a} \Rightarrow 3c^2 = \frac{b^3 - a^3}{b - a} = b^2 + ab + a^2$$

$$\Rightarrow 3c^2 = b^2 + ab + a^2$$

$$\Rightarrow c = \sqrt{\frac{b^2 + ab + a^2}{3}}, \left(a < \frac{a^2 + ab + b^2}{3} < b\right)$$

Hence verified.

Question 5: Verify Lagrange's mean value theorem for the function $f(x) = lx^2 + mx + n$ on $[a, b]$

Solution: $\because f(x) = lx^2 + mx + n$

$\therefore f(x)$ is a polynomial in x

$\Rightarrow f(x)$ is continuous and differentiable in any given finite interval.

$\Rightarrow$ **(i)** $f(x)$ is continuous on the closed interval $[a, b]$

(ii) $f(x)$ is differentiable on the open interval (a, b).

$\therefore$ Two conditions of Lagrange's mean value theorem are satisfied.

Now, **(i)** $f'(x) = 2lx + m$

$\Rightarrow f'(c) = 2lc + m$

(ii) $f(a) = la^2 + ma + n$

(iii) $f(b) = lb^2 + mb + n$

(iv) $b - a = b - a$

(v) $\dfrac{f(b) - f(a)}{b - a} = \dfrac{l(b^2 - a^2) + m(b - a)}{b - a} = l(b - a) + m$

Now, using the result of Lagrange's mean value theorem,

$$f'(c) = \frac{f(b) - f(a)}{b - a}$$

$$\Rightarrow 2lc + m = l(b + a) + m$$

$$\Rightarrow c = \frac{b + a}{2} \in (a, b)$$

$\Rightarrow$ Lagrange's mean value theorem is verified.

Question 6: Verify Lagrange's mean value theorem for the function $f(x) = \sqrt{x^2 - 4}$ on $[2, 4]$.

Solution: $\because f(x) = \sqrt{g(x)} = \sqrt{x^2 - 4}$

Now, $g(x) = x^2 - 4$, $(g(x) > 0)$ is continuous and differentiable on a given finite interval $\Rightarrow \sqrt{g(x)}$ is also continuous and differentiable on the same given finite interval using the continuity and differentiability theorem for a function of a function.

$\Rightarrow$ **(i)** $f(x)$ is continuous on the closed interval $[2, 4]$

(ii) $f(x)$ is differentiable on the open interval $(2, 4)$

$\therefore$ Two conditions of Lagrange's mean value theorem are satisfied.

Now, **(i)** $f'(x) = \dfrac{1}{2\sqrt{x^2 - 4}} \times 2x$

$$= \frac{2x}{2\sqrt{x^2 - 4}} = \frac{x}{\sqrt{x^2 - 4}}$$

$$\Rightarrow f'(c) = \frac{c}{\sqrt{c^2 - 4}}$$

(ii) $f(a) = f(2) = \sqrt{4 - 4} = 0$

(iii) $f(b) = f(4) = \sqrt{16 - 4} = \sqrt{12} = 2\sqrt{3}$

(iv) $b - a = 4 - 2 = 2$

Now, using the result of Lagrange's mean value theorem,

$$f'(c) = \frac{f(4) - f(2)}{4 - 2} \Rightarrow \frac{c}{\sqrt{c^2 - 4}}$$

$$= \frac{2\sqrt{3} - 0}{2} = \sqrt{3}, (c > 0)$$

$$\Rightarrow \frac{c^2}{c^2 - 4} = 3 \Rightarrow c^2 = 3c^2 - 12 \Rightarrow c^2 - 3c^2 = -12$$

$$\Rightarrow -2c^2 = -12 \Rightarrow c^2 = 6 \Rightarrow c = \sqrt{6} \text{ and}$$

$c = \sqrt{6} \in (2, 4)$

Thus, $c = \sqrt{6} \in (2, 4)$ verifies the Lagrange's mean value theorem.

***Question** 7*: Verify Lagrange's mean value theorem for the function $f(x) = \log x$ in $[1, e]$

Solution: $\because f(x) = \log(x)$

$\therefore f(x)$ is continuous and differentiable in a given finite interval for being a log function.

$\Rightarrow \log x$ is continuous in $[1, e]$ and differentiable in $(1, e)$.

$\therefore$ Two conditions of Lagrange's mean value theorem are satisfied.

Now, **(i)** $f'(x) = \frac{1}{x}, (x > 0)$

$$\Rightarrow f'(c) = \frac{1}{c}, (c > 0)$$

(ii) $f(b) = f(e) = \log e = 1$

(iii) $f(a) = f(1) = \log 1 = 0$

(iv) $b - a = e - 1$

Now, using the result of Lagrange's mean value theorem

$$f'(c) = \frac{f(b) - f(a)}{b - a}$$

$$\Rightarrow \frac{1}{c} = \frac{1 - 0}{e - 1}$$

$$\Rightarrow \frac{1}{c} = \frac{1}{e - 1}$$

$\Rightarrow c = e - 1$ and $1 < e - 1 < e$. Hence verified.

Question 8: Verify the hypothesis of mean value theorem for the function $f(x) = \sqrt{x + 2}$ on the interval $[4, 6]$ and find a suitable value for 'c' that satisfied the conclusion of the theorem.

Solution: Having the fact that $g(x) = x + 2$ is a continuous and differentiable on a given finite interval and using the theorem for continuity and differentiability for a function of a function, we conclude that $f(x) = \sqrt{x + 2}$, is continuous and differentiable on a given finite interval (where $f(x) > 0$)

$\Rightarrow f(x) = \sqrt{x + 2}$ is continuous and differentiable in the closed interval $[4, 6]$.

$\Rightarrow$ **(i)** $f(x)$ is continuous in the closed interval $[4, 6]$

(ii) $f(x)$ is differentiable in the open interval $(4, 6)$.

$\therefore$ Two conditions of Lagrange's mean value theorem are satisfied.

Now, using the result of Lagrange's mean value theorem, we get

$$f'(c) = \frac{f(b) - f(a)}{b - a}$$

$$\Rightarrow \frac{1}{2\sqrt{c + 2}} = \frac{\sqrt{6 + 2} - \sqrt{4 + 2}}{6 - 4}$$

$$\Rightarrow \frac{1}{2\sqrt{c + 2}} = \frac{\sqrt{8} - \sqrt{6}}{2}$$

$$\Rightarrow \frac{1}{\sqrt{c + 2}} = \sqrt{8} - \sqrt{6}$$

$$\Rightarrow \frac{1}{\sqrt{8} - \sqrt{6}} = \sqrt{c + 2}$$

$$\Rightarrow \frac{1}{\left(\sqrt{8} - \sqrt{6}\right)^2} = c + 2$$

$$\Rightarrow \frac{1}{8 + 6 - 2\sqrt{48}} = c + 2$$

$$\Rightarrow \frac{1}{14 - 2\sqrt{48}} = c + 2$$

$$\Rightarrow \frac{1}{14 - 8\sqrt{3}} = c + 2$$

$$\Rightarrow \frac{1}{14-8\sqrt{3}} - 2 = c$$

$$\Rightarrow \frac{1-28+16\sqrt{3}}{14-8\sqrt{3}} = c$$

$$\Rightarrow \frac{-27+16\sqrt{3}}{14-8\sqrt{3}} = c \text{ and } 4 < \frac{16\sqrt{3}-27}{14-8\sqrt{3}} < 6$$

Question 9: How to test the non-applicability of Lagrange's mean value theorem in a given interval $I = [a, b]$.

Solution: At least one of the following is to be shown for the non-applicability of Lagrange's mean value theorem.

1. Show that $f(x)$ is not continuous at some point lying in the given closed interval $[a, b]$.

2. Show that $f'(x)$ does not exist at least at one point 'c' of the open interval (a, b) for which we are required to show $l.h.d \neq r.h.d$ or, we show

$[f'(x)]_{x=c} = f'(c) = \pm\infty$ or undetermined by using the defination.

$$Rf'(c) = \lim_{h\to 0} \frac{f(c+h) - f(c)}{h}$$

$$Lf'(c) = \lim_{h\to 0} \frac{f(c-h) - f(c)}{-h}$$

Remember:

1. In some of the cases Lagrange's mean value theorem is not applicable for $|f(x)|$ if $\exists\, c \in I$ s.t $f(c) = 0$ as $f'(c)$ may not exist.

2. Lagrange's mean value theorem is not applicable in a rational function $\frac{f_1(x)}{f_2(x)}$ if $f_2(c) = 0$ for some $c \in I$.

3. The point at which given function is non-differentiable may be the end points of the closed interval or may be the mid point of the closed interval or may belong to the open interval different from the given closed interval.

4. $$\frac{d|x|^{(2n+1)}}{dx} = (2n+1)|x|^{2n} \cdot \frac{|x|}{x}$$

$$= (2n+1) \cdot \frac{|x|^{(2n+1)}}{x} \quad \text{for } x \neq 0$$

5. $$\frac{d|x|^{2n}}{dx} = 2n \cdot |x|^{(2n-1)} \cdot \frac{|x|}{x}$$

$$= 2n \cdot \frac{|x|^{2n}}{x} \quad \text{for } x \neq 0$$

Type 2: To test whether Lagrange's mean value theorem is applicable or not in a given interval $I = [a, b]$.

Examples worked out:

Question 1: Discuss the applicability of Lagrange's mean value theorem to the function $f(x) = \frac{1}{x^3}$ in $[-1, 1]$.

Solution: (i) Given $f(x) = \frac{1}{x^3}$

$\therefore f(x)$ is continuous and differentiable for all values of $x \in [-1, 1]$ except where $x^3 = 0$ for being a rational function in x which $\Rightarrow f(x)$ is discontinuous at $x = 0 \in [-1, 1]$ since $f(x)$ is undefined at $x = 0$.

Hence, Lagrange's mean value theorem is not applicable to $f(x)$ on $[-1, 1]$.

Question 2: Are all the conditions of Lagrange's mean value theorem satisfied for the function $f(x) = \sqrt{x-1}$ in $[1, 3]$. If so, find 'c' of the mean value theorem.

Solution: (i) Given function $f(x) = \sqrt{x-1}$

$\therefore f(x)$ is continuous and differentiable for all values of x belonging to the given interval $[1, 3]$.

$\therefore f(x)$ is differentiable in $(1, 3)$

$\therefore$ All conditions of Lagrange's mean value theorem are satisfied.

$$\text{Now, } f'(c) = \frac{f(b) - f(a)}{b - a} = \frac{f(3) - f(1)}{3 - 1}$$

$$= \frac{\sqrt{2} - 0}{2} = \frac{1}{\sqrt{2}}$$

$$\Rightarrow \frac{1}{2\sqrt{c-1}} = \frac{1}{\sqrt{2}} \Rightarrow 1 = \sqrt{2}\sqrt{(c-1)}$$

$$\Rightarrow 2(c-1) = 1 \Rightarrow c - 1 = \frac{1}{2} \Rightarrow c = 1 + \frac{1}{2}$$

$$\Rightarrow c = \frac{3}{2} \text{ and } \frac{3}{2} \in [1, 3]$$

Question 3: Discuss the applicable of mean value theorem to the function $f(x) = |x|$ in $[-1, 1]$.

Solution: $\because$ Given $f(x) = |x|$

$\therefore$ $f(x)$ is continuous for all values of $x \in [-1, 1]$ for being a modulus function.

Now, $f'(x) = \frac{|x|}{x}$ which is differentiable for all value of x except perhaps at $x = 0$.

$$R f'(0) = \lim_{h \to 0} \frac{f(0+h) - f(0)}{h}$$

$$= \lim_{h \to 0} \frac{|h| - 0}{h} = \lim_{h \to 0} \frac{h}{h} = 1$$

$$L f'(0) = \lim_{h \to 0} \frac{f(0-h) - f(0)}{-h}$$

$$= \lim_{h \to 0} \frac{|-h| - 0}{-h} = \lim_{h \to 0} \frac{h}{-h} = -1 \text{ (for } h > 0\text{)}$$

$$\therefore R f'(0) \neq L f'(0)$$

$\therefore$ The given function $f(x)$ is not differentiable at $x = 0$ and $0 \in (-1, 1)$.

$\therefore$ One condition of Lagrange's mean value theorem is not satisfied.

$\therefore$ The mean value theorem is not applicable to $f(x)$ in $[-1, 1]$.

Question 4: Examine the validity of the hypothesis and the conclusion of Lagrange's mean value theorem for the function.

$$f(x) = |x| \text{ in } [-2, 1]$$

Solution: **(i)** $\because f(x) = |x|$

$\therefore$ $f(x)$ is continuous for all values of $x \in [-2, 1]$ for being modulus of a function which is continuous for all values.

(ii) $f'(x) = \frac{|x|}{x}$ which is differentiable for all values of x except perhaps at $x = 0$

$$R f'(0) = \lim_{h \to 0} \frac{f(0+h) - f(0)}{h}$$

$$= \lim_{h \to 0} \frac{|h| - 0}{h} = \lim_{h \to 0} \frac{h}{h} = 1$$

$$L f'(0) = \lim_{h \to 0} \frac{f(0-h) - f(0)}{-h}$$

$$= \lim_{h \to 0} \frac{|-h| - 0}{-h} = \lim_{h \to 0} \frac{h}{-h} = -1$$

$$\therefore R f'(0) \neq L f'(0)$$

$\therefore$ The given function $f(x)$ is not differentiable at $x = 0$ and $\in (-2, 1)$

$\therefore$ $f(x)$ is not differentiable in $(-2, 1)$.

$\therefore$ One conditions namely differentiability of the given function in the given open interval $(-2, 1)$ is not satisfied which $\Rightarrow$ the hypothesis is not valid.

Now, we examine the result:

1. $f(1) = |1| = 1 = f(b)$
2. $f(-2) = |-2| = 2 = f(a)$
3. $b - a = 1 - (-2) = 1 + 2 = 3$
4. $f'(c) = \frac{|c|}{c}$, $c \neq 0$
5. $\frac{f(b) - f(a)}{1 + 2} = \frac{1 - 2}{3} = -\frac{1}{3}$

But $\frac{|c|}{c} = -\frac{1}{3}$ is not true.

Thus, we observe neither the hypothesis nor the conclusion is valid.

Question 5: Discuss the applicability of Lagrange's mean value theorem to the function $f(x) = |x|^3$ on $[-1, 2]$.

Solution: **(i)** $f(x) = |x|^3$

$\therefore$ $f(x)$ is continuous in $[-1, 2]$ as a mod function is continuous in a given finite interval.

(ii) $f'(x) = 3|x|^2 \cdot \dfrac{d|x|}{dx} = 3|x|^2 \cdot \dfrac{|x|}{x} = \dfrac{3|x|^3}{x}$

$\therefore f(x)$ is differentiable for all values of x except perhaps at $x = 0$.

$$\therefore Rf'(0) = \lim_{h\to 0} \frac{f(0+h) - f(0)}{h}$$

$$= \lim_{h\to 0} \frac{|0+h|^3 - 0}{h} \text{ for } h > 0$$

$$= \lim_{h\to 0} \frac{|h|^3}{h} = \lim_{h\to 0} \frac{h^3}{h} = \lim_{h\to 0} h^2 = 0$$

$$Lf'(0) = \lim_{h\to 0} \frac{f(0-h) - f(0)}{-h}$$

$$= \lim_{h\to 0} \frac{|0-h|^3 - 0}{-h} \text{ for } h > 0$$

$$= \lim_{h\to 0} \frac{|-h|^3}{-h} = \lim_{h\to 0} \frac{h^3}{-h} = \lim_{h\to 0} \left(-h^2\right) = 0$$

$\therefore Rf'(0) = Lf'(0)$

$\therefore f'(x)$ exists a $x = 0$ and $f'(0) = 0$

$\therefore f'(x)$ exists for all values of $x \in (-1, 2)$

$\therefore f(x)$ is differentiable in $(-1, 2)$.

$\therefore$ All conditions of Lagrange's mean values theorem are satisfied.

$\therefore$ Lagrange's mean values theorem is applicable to the given function $f(x)$ on $[-1, 2]$.

Note: The above example shows that the statement "Rolle's theorem or Lagrange's mean value theorem is not applicable to mod of a function $= |f(x)|$ if $x = c \in$ given interval is one of the roots of $f(x) = 0$" is not true.

Question 6: Is L.M.V.T applicable to the function defined as $f(x) = x + \dfrac{1}{x}$ on $[1, 2]$. Give reason.

Solution: **(i)** $f(x) = x + \dfrac{1}{x}$

$\therefore f(x)$ is continuous on $[1, 2]$ as it is the sum of two continuous function x and $\dfrac{1}{x}$ for all values of $x \neq 0$.

(ii) $f(x)$ is differentiable in $(1, 2)$ as it is the sum of two differentiable functions x and $\dfrac{1}{x}$ for all values of $x \neq 0$.

$\therefore$ All conditions of Lagrange's mean value theorem are satisfied.

$\therefore$ Lagrange's mean value theorem is applicable to the function $f(x) = x + \dfrac{1}{x}$ on $[1, 2]$.

Question 7: Give reason whether Lagrange's mean value theorem is applicable to the function

$$f(x) = \frac{\sin x}{x},\ x \neq 0.$$

$$= 1, x = 0 \text{ for } -1 \le x \le 1 \equiv [-1, 1]$$

Solution: $f(x) = \dfrac{\sin x}{x},\ x \neq 0$

$$\therefore f'(x) = \frac{x\cos x - \sin x}{x^2},\ x \neq 0$$

$\therefore f(x)$ is differentiable for all values of x except perhaps at $x = 0$.

$$Rf'(0) = \lim_{h\to 0} \frac{f(0+h) - f(0)}{h}$$

$$= \lim_{h\to 0} \frac{\dfrac{\sin(0+h)}{(0+h)} - 1}{h} \quad [\because f(0) = 1 \text{ is given}]$$

$$= \lim_{h\to 0} \frac{\dfrac{\sin h}{h} - 1}{h}$$

$$= \lim_{h\to 0} \frac{\sin h - h}{h} \cdot \frac{1}{h}$$

$$= \lim_{h\to 0} \frac{\sin h - h}{h^2}$$

$$= \lim_{h\to 0} \frac{\cos h - 1}{2h}$$

$$= \lim_{h\to 0} \frac{-\sin h}{2} \text{ (by L-Hospital's rule)}$$

$= 0$

$\therefore f'(0) = 0$

$\therefore f(x)$ is differentiable in $[-1, 1]$ and so Lagrange's mean value theorem is applicable and $c = 0$.

Question 8: Given reason whether Lagrange's M.V.T is applicable to function $f(x) = x \cdot \cos\left(\frac{1}{x}\right), x \neq 0 = 0, x = 0$ for $-1 \leq x \leq 1 \equiv [-1, 1]$

Solution: $x \cdot \cos\left(\frac{1}{x}\right), x \neq 0$

$f(0) = 0$

$\therefore f(x)$ is continuous in any interval but it is differentiable every where except at $x = 0$ and $0 \in (-1, 1)$. This is why $f(x)$ is not differentiable in $(-1, 1)$.

$\therefore$ One of the two conditions of M.V theorem is not satisfied. This is why L.M.V.T is not applicable to $f(x)$ on $[-1, 1]$.

Question 9: Discuss the applicability of Lagrange's mean value theorem to the function $f(x) = \frac{1}{x}$ in the closed interval $[-1, 1]$.

Solution: $\because f(x) = \frac{1}{x}$

$\therefore f(x)$ is not defined at $x = 0$ while $0 \in [-1, 1]$

$\Rightarrow f(x)$ is not continuous at $x = 0$.

$\Rightarrow f(x)$ is discontinuous in $[-1, 1]$.

$\Rightarrow f(x)$ is non-differentiable in $[-1, 1]$.

$\Rightarrow f(x)$ is non-differentiable in $(-1, 1)$

$\therefore$ No conditions of Lagrange's mean value theorem is satisfied.

$\therefore$ Lagrange's mean value theorem is not applicable to the given function on the given closed interval.

Question 10: Discuss the applicability of Lagrange's mean value theorem for the function $f(x) = \frac{1}{x}$ in $[1, 2]$.

Solution: **(i)** $\because f(x) = \frac{1}{x}$

$\therefore f(x)$ is finite for every value of $x \in [1, 2]$ which $\Rightarrow$ it is continuous in $[1, 2]$ for being a rational function.

(ii) $f'(x) = -\frac{1}{x^2}$ which is finite for every value of $x \in [1, 2]$

$\therefore f(x)$ is differentiable in $[1, 2]$

$\therefore$ All conditions of Lagrange's mean value theorem are satisfied.

$\therefore$ Lagrange's mean value theorem is applicable.

Question 11: A function $f(x)$ in $[1, 2]$ is defined by
$f(x) = 2$, if $x = 1$
$= x^2$ if $1 < x < 2$
$= 4$, if $x = 2$
are all the conditions of Lagrange's mean value theorem satisfied in this case?

Solution: **(i)** Continuity and differentiability test at $x = 1$ and 2 (i.e. at the end points).

$\because \lim_{x \to 2-} f(x) = \lim_{x \to 2} x^2 = 4$

$f(2) = 4$

$\therefore \lim_{x \to 2-} f(x) = f(2)$

Again, $\lim_{x \to 1+} f(x) = \lim_{x \to 1} x^2 = 1$

$f(1) = 2$

$\therefore \lim_{x \to 1+} f(x) \neq f(1)$

$\therefore f(x)$ is not continuous at the end points $x = 1$ of the closed interval $[1, 2]$.

$\therefore f(x)$ is not continuous in the closed interval $[1, 2]$.

$\therefore$ One condition of Lagrange's mean value theorem is not satisfied and this is why Lagrange's mean value theorem is not applicable to the given function $f(x)$ defined in the closed interval $[1, 2]$.

Question 12: Discuss the applicability of Lagrange's mean value theorem to the function $f(x) = x^{\frac{1}{3}}$ in $[-1, 1]$.

Solution: $f(x) = \sqrt[3]{x}$

$f(x)$ is not defined for $x < 0$

$\Rightarrow f(x)$ is not continuous in the closed interval $[-1, 1]$

$\therefore$ No condition of Lagrange's mean value theorem is satisfied.

$\therefore$ Lagrange's mean value theorem is not applicable to the given function $f(x)$ in the given interval $[-1, 1]$.

Question 13: A function $f(x)$ is defined in $[-1, 1]$ by $f(x) = x \sin\left(\frac{1}{x}\right)$, $x \neq 0$ $f(0) = 0$, $x = 0$ are all the conditions of first mean value theorem of differential calculus satisfied in this case?

Solution: $\because f(x) = x \sin\left(\frac{1}{x}\right)$, $x \neq 0$

$f(0) = 0, x = 0$

We know that the above function is continuous in any given finite interval but it is differentiable every where except $x = 0$ while $0 \in (-1, 1)$.

$\therefore f(x)$ is continuous in $[-1, 1]$ and $f(x)$ is not differentiable in $(-1, 1)$.

$\therefore$ One condition namely differentiability of the given function on the open interval $(-1, 1)$ for Lagrange's mean value theorem is not satisfied which $\Rightarrow$ M.V.T can not be applied to $f(x)$ in the given interval $[-1, 1]$.

Question 14: Give a reason why the mean value theorem does not hold in the function defined by $f(x) = \sqrt{1 - x}$ on $[-2, 2]$

Solution: $f(x) = \sqrt{1 - x}$ is not defined for $x > 1$

$\therefore f(x)$ is not continuous in $[-2, 2]$.

Hence Lagranges mean value theorem does not hold.

Type 3: Problems based on finding the value of 'c' using Lagrange's mean value theorem.

Working rule:

1. Show the differentiability of the given function in the given open interval and the continuity of the given function in the given closed interval or, show only the differentiability for the given function in the given closed interval.

2. Find $f(a)$, $f(b)$ and $(b - a)$ where a and b are the left end point and right end point of the given closed interval $[a, b]$.

3. Find $f'(x)$.

4. Find $f'(c)$ and equate it to $\dfrac{f(b) - f(a)}{b - a}$.

5. Solve the equation $f'(c) = \dfrac{f(b) - f(a)}{b - a}$ which will provide us one or more than one root of the equation $f'(c) = \dfrac{f(b) - f(a)}{b - a}$.

6. The value of c s.t $a < c < b$ should be accepted and the value of c which does not satisfy $a < c < b$ should be rejected. In other words, $c \in (a, b)$ should be accepted and $c \notin (a, b)$ should be rejected.

Solved Examples

Question 1: Find 'c' of Lagrange's mean value theorem when the given function $f(x) = x^2 - 3x - 1$ is defined on $[1, 3]$.

Solution: (1) $\because f(x) = x^2 - 3x - 1$

$\therefore f(x)$ is differentiable for all values of x for being a polynomial.

$\therefore f(x)$ is differentiable in $[1, 3]$.

$\therefore f(x)$ is continuous in $[1, 3]$ and differentiable in $(1, 3)$.

$\therefore$ All conditions of Lagrange's mean value theorem are satisfied.

$\therefore$ There is a point 'c' s.t $f'(c) = \dfrac{f(b) - f(a)}{b - a}$

(2) $\because f(x) = x^2 - 3x - 1$

$\therefore f(a) = f(1) = 1 - 3 - 1 = 1 - 4 = -3$

$f(b) = f(3) = 9 - 9 - 1 = -1$

$f(b) - f(a) = -1 - (-3) = -1 + 3 = 2$

$b - a = 3 - 1 = 2$

Thus, in the light of above determined quantities, we get

$$\frac{f(b) - f(a)}{b - a} = \frac{2}{2} = 1 \quad \ldots\text{(i)}$$

Now, $f'(x) = 2x - 3$ [from the given equation]

$$\Rightarrow f'(c) = 2c - 3 \quad \ldots\text{(ii)}$$

Lastly equating (i) and (ii), we get

$$f'(c) = \frac{f(b) - f(a)}{b - a}$$

$\Rightarrow 2c - 3 = 1$

$\Rightarrow 2c = 1 + 3 = 4$

$\Rightarrow c = \frac{4}{2} = 2$ and $1 < 2 < 3$

$\Rightarrow c = 2 \in (1, 3)$

Question 2: Find the value of c using L.M.V theorem on the function $f(x) = x^2 + 2x - 1$on [0. 1].

Solution: (1) $\because f(x) = x^2 + 2x - 1$

$\therefore f(x)$ is differentiable for all values of x for being a polynomial.

$\therefore f(x)$ is differentiable in [0, 1].

$\therefore f(x)$ is continuous in [0, 1] and differentiable in (0, 1)

$\therefore$ All conditions of Lagrange's mean value theorem are satisfied.

$\therefore$ There is a point 'c' s.t $f'(c) = \frac{f(b) - f(a)}{b - a}$

(2) $\because f(x) = x^2 + 2x - 1$

$f(a) = f(0) = -1$

$f(b) = f(1) = 1 + 2 \cdot 1 - 1 = 2$

$f(b) - f(a) = f(1) - f(0) = 2 - (-1) = 3$

Thus, in the light of above determined value of $f(x)$ at $x = 1$ and $x = b$ as well as the difference of the end points of the given closed interval, we find that

$$\frac{f(b) - f(a)}{b - a} = \frac{3}{1} = 3 \qquad \ldots(i)$$

Now, $f'(x) = 2x + 2$ [from the given equation]

$$\Rightarrow f'(c) = 2c + 2 \qquad \ldots(ii)$$

lastly equating (i) and (ii), we get

$$f'(c) = \frac{f(b) - f(a)}{b - a}$$

$\Rightarrow 2c + 2 = 3$

$\Rightarrow 2c = 3 - 2 = 1$

$\Rightarrow c = \frac{1}{2}$ and $0 < \frac{1}{2} < 1$

$\Rightarrow c = \frac{1}{2} \in (0, 1)$

Question 3: Find c of Lagrange's mean value theorem when the given function $f(x) = (x - 1)(x - 2)(x - 3)$ is defined on [0, 4].

Solution: (1) $f(x) = (x - 1)(x - 2)(x - 3)$ which is a polynomial in x.

$\therefore f(x)$ is differentiable in [0, 4].

$\therefore f(x)$ is continuous in [0, 4] and differentiable in (0, 4).

$\therefore$ All conditions of Lagrange's mean value theorem are satisfied.

(2) $\because f(x) = x^2 + 2x - 1$

$f(a) = f(0) = (-1)(-2)(-3) = -6$

$f(b) = f(4) = 3 \cdot 2 \cdot 1 = 6$

$b - a = 4 - 0 = 4$

$$\therefore \frac{f(b) - f(a)}{b - a} = \frac{12}{4} = 3 \qquad \ldots(i)$$

Now, $f'(x) = 3x^2 - 12x + 11$

$$\Rightarrow f'(c) = 3c^2 - 12c + 11 \qquad \ldots(ii)$$

lastly, we consider the equation formed by equating (1) and (2),

$$f'(c) = \frac{f(b) - f(a)}{b - a}$$

$\Rightarrow 3c^2 - 12c + 11 = 3$

$\Rightarrow 3c^2 - 12c + 11 - 3 = 0$

$\Rightarrow 3c^2 - 12c + 8 = 0$

$\Rightarrow c = \frac{6 \pm 2\sqrt{3}}{3}$ and $0 < \frac{6 \pm 2\sqrt{3}}{2} < 4$

$\Rightarrow c = \frac{6 \pm 2\sqrt{3}}{3} \in (0, 4)$

Question 4: Find 'c' of Lagrange's mean value theorem. When the function $f(x) = x + \frac{1}{x}$ is defined on $\left[\frac{1}{2}, 3\right]$

Solution: (1) $\because f(x) = x + \frac{1}{x}$

$\therefore f(x)$ is differentiable in $\left[\frac{1}{2}, 3\right]$ since it is the sum of two differentiable functions represented by x and $\frac{1}{x}$, for $x \neq 0$.

$\therefore f(x)$ is continuous in $\left[\frac{1}{2}, 3\right]$ as it is differentiable in $\left[\frac{1}{2}, 3\right]$.

$\therefore$ All conditions of Lagrange's mean value theorem are satisfied.

$\therefore$ There is a point 'c' s.t $f'(c) = \dfrac{f(b) - f(a)}{b - a}$

(2) $f(x) = x + \dfrac{1}{x}$

$$\therefore f(b) = f(3) = 3 + \frac{1}{3} = \frac{10}{3}$$

$$f(a) = f\left(\frac{1}{2}\right) = \frac{1}{2} + 2 = \frac{5}{2}$$

$$f(b) = f(a) = \frac{10}{3} - \frac{5}{2} = \frac{20-15}{6} = \frac{5}{6}$$

$$\frac{f(b) - f(a)}{b - a} = \frac{\frac{5}{6}}{\frac{5}{2}} = \frac{5}{6} \times \frac{2}{5} = \frac{1}{3} \quad \text{...(i)}$$

Now, $f'(x) = 1 - \dfrac{1}{x^2} \Rightarrow f'(c) = 1 - \dfrac{1}{c^2} = \dfrac{c^2 - 1}{c^2}$ ...(ii)

Lastly, equating (1) and (2), we get

$$\frac{c^2 - 1}{c^2} = \frac{1}{3}$$

$$\Rightarrow 3c^2 - 3 = c^2$$

$$\Rightarrow 3c^2 - c^2 = 3$$

$$\Rightarrow 2c^2 = 3$$

$$\Rightarrow c^2 = \frac{3}{2}$$

$$\Rightarrow c = \pm\sqrt{\frac{3}{2}} = \pm\sqrt{1.5} = \pm 1.22 \text{ and } \frac{1}{2} < 1.22 < 3$$

$$\Rightarrow c = 1.22 \in \left(\frac{1}{2}, 3\right)$$

Question 5: Find 'c' of Lagrange's mean value theorem when the function $f(x) = \log x$ is defined in $[1, e]$.

Solution: (1) $\because f(x) = \log x$ which is differentiable in $[1, e]$

$\therefore f(x)$ is continuous in $[1, e]$ and differentiable in $(1, e)$

(2) $f(x) = \log x$

$\therefore f(b) = f(e) = \log e = 1$

$f(a) = f(1) = \log 1 = 0$

$f(b) - f(a) = 1 - 0 = 1$

$b - a = e - 1$

$$\frac{f(b) - f(a)}{b - a} = \frac{1}{e - 1} \quad \text{...(i)}$$

Now, $f'(x) = \dfrac{1}{x} \Rightarrow f'(c) = \dfrac{1}{c}$ ...(ii)

Lastly, equating (i) and (ii), we get

$\dfrac{1}{c} = \dfrac{1}{e-1} \Rightarrow c = e - 1 = 2.73 - 1 = 1.73$ ($\because e =$ 2,73 approx)

$$\Rightarrow c = 1.73 \text{ and } 1.73 \in (1, e)$$

$$\Rightarrow c = 1.73 \in (1, e)$$

Question 6: Find 'c' of mean value theorem for the function defined as $f(x) = x^3 \ \forall x \in [0, 1]$.

Solution: (1) x^3 being a polynomial function is continuous on $[0, 1]$ and differentiable on $(0, 1) \Rightarrow$ all the conditions of L.M.V.T are satisfied $\Rightarrow \exists$ a number 'c' $\in$ s.t $f'(c) = \dfrac{f(b) - f(a)}{b - a}$

(2) Now, $f(x) = x^3$

$f(a) = f(0) = 0$

$f(b)=f(1)=1$
$b-a=1-0=1$

$$\frac{f(b)-f(a)}{b-a}=\frac{1-0}{1}=\frac{1}{1}=1 \quad \ldots(i)$$

Again, $f'(x)=3x^2$

$$\Rightarrow f'(c)=3c^2 \quad \ldots(ii)$$

Lastly, equating (1) and (2), we get

$3c^2=1$

$$\Rightarrow c^2=\frac{1}{3}$$

$$\Rightarrow c=\pm\sqrt{\frac{1}{3}}=\pm\frac{1}{\sqrt{3}}$$

Since c has to be in $(0, 1)$ $\Rightarrow$ the acceptable value of c is $\frac{1}{\sqrt{3}}$ which $\Rightarrow c=\frac{1}{\sqrt{3}}$ is the required value.

Question 7: Find 'c' of mean value theorem when the function $f(x)=x^2+3x+2$ is defined on $[1, 2]$.

Solution: (1) Since $f(x)=x^2+3x+2$ being a polynomial in x is continuous on $[1, 2]$ and differentiable on $(0, 1)$ $\Rightarrow$ all conditions of L.M.V.T are satisfied which $\Rightarrow \exists$ a number 'c' s.t

$$f'(c)=\frac{f(b)-f(a)}{b-a}$$

Now, $f(x)=x^2+3x+2$

$f(a)=f(1)=1+3+2=6$
$f(b)=f(2)=4+6+2=12$
$f(b)-f(a)=12-6=6$
$b-a=2-1=1$

$$\frac{f(b)-f(a)}{b-a}=\frac{6}{1}=6 \quad \ldots(i)$$

Again, $f'(x)=2x+3$

$$\Rightarrow f'(c)=2c+3 \quad \ldots(ii)$$

Lastly, equating (i) and (ii), we get

$2c+3=6$

$\Rightarrow 2c=6-3$

$\Rightarrow 2c=3$

$$\Rightarrow c=\frac{3}{2}=1.5 \text{ and } 1<1.5<2$$

$$\Rightarrow c=1.5\in(1, 2)$$

Problems based on Lagrange's mean value theorem

Type 1: Problems based on verification of Lagrange's mean value theorem (i.e., L.M.V.T)

Exercise 20.5

Question: Verify Lagrange's mean value theorem (or, simply mean value theorem) for the following functions in the interval specified.

1. $f(x)=x(x-2)(x-4)$ in $[1, 3]$
2. $f(x)=\frac{1}{x^2}$ in $[-1, 3]$
3. $f(x)=\sqrt{x^2-4}$ in $[2, 4]$
4. $f(x)=x+2+\frac{1}{x-3}$ in $[-1, 2]$
5. $f(x)=|x|$ in $[-1, 2]$
6. $f(x)=\log x$ in $[1, e]$
7. $f(x)=(x-1)(x-2)(x-3)$ in $[0, 4]$
8. $f(x)=x+\frac{1}{x}$ in $[1, 2]$
9. $f(x)=\sqrt{x-1}$ in $[1, 3]$
10. $f(x)=x^3$ in $[1, 2]$
11. $f(x)=x^3-2x+4$ in $[-1, 2]$
12. $f(x)=x^3-3x+2$ in $[-2, 3]$
13. $f(x)=e^x$ in $[-1, 2]$
14. $f(x)=\log x$ in $\left[\frac{1}{2}, 2\right]$
15. $f(x)=\sin x$ in $[30^\circ, 60^\circ]$

Type 2: Problems based on examination of Lagrange's mean value theorem ⇔ whether L.M.V.T is applicable or not is required to test.

Exercise 20.6.1

1. Discuss the validity of the mean value theorem for the function $f(x) = \frac{1}{x}$ in $[-1, 2]$.

2. Discuss the validity of the mean value theorem for the function f defined by $f(x) = (x-1)^{\frac{2}{3}}$ in the interval $[1, 2]$.

3. State whether the mean value theorem is applicable to the function f defined by

(i) $f(x) = \frac{\sin x}{x}$; $x \neq 0$ $f(0) = 1$ in $[-1, 1]$

(ii) $f(x) = x\cos\left(\frac{1}{x}\right)$, $x \neq 0 = 0, x = 0$ in $[-1, 1]$

(iii) $f(x) = |x|^3$ in $[-1, 2]$

(iv) $f(x) = \sqrt{x^2 - 1}$ in $[1, 3]$

(v) $f(x) = 1 - \sqrt[3]{x^2}$ in $[-2, 1]$

(vi) $f(x) = \log \sin x$ in $\left[\frac{\pi}{6}, \frac{5\pi}{6}\right]$

(vii) $f(x) = |x|$ in $[-1, 1]$.

(viii) $f(x) = \begin{cases} 1 + x^2, & x \geq 0 \\ 1 - x^2, & x < 0 \end{cases}$ in $[-1, 1]$

Answers:

1. Valid and $c = \frac{1}{2}$

2. Not valid as $f(x)$ is not differentiable at $x = 0$

Exercise 20.6.2

Question: Given a reason in each of the following why the mean value theorem does not hold in each of the functions defined by

(i) $f(x) = |x|$ in $[-1, 3]$

(ii) $f(x) = x^{\frac{2}{3}}$ in $\left[-\frac{1}{8}, \frac{1}{8}\right]$

(iii) $f(x) = \sqrt{1-x}$ in $[-2, 2]$

(iv) $f(x) = 2$ if $x = 1$
$= x^2$ if $1 < x < 2$ in $[1, 2]$
$= 1$ if $x = 2$

Answers:

(i) Not differentiable at $x = 0$

(ii) Not differentiable at $x = 0$

(iii) Derivative does not exist at $x = 1$

(iv) $f(x)$ is continuous in the open interval $(1, 2)$ and not in the closed interval $[1, 2]$.

Type 3: Problems based on finding the value of 'c' of Lagrange's mean value theorem.

Exercise 20.7.1

1. Find the value of c of mean value theorem for the function $f(x) = x^2$ in the interval $[1, 4]$.

2. Find 'c' of the mean value theorem for the function $f(x) = x^3$ in $[1, 2]$.

3. Find 'c' of L.M.V.T for the function $f(x) = (x-1)(x-2)(x-3)$ in $0 \leq x \leq 4$.

4. Find 'c' of the mean value theorem for the function $f(x) = \frac{1}{x}$ in the interval $[1, 9]$.

5. Find 'c' of L.M.V.T for the function $f(x) = \log x$ defined in the interval $[1. 2]$.

6. Find 'c' of the mean value theorem for the function $f(x) = e^x$ in $[0, 1]$.

7. Find 'c' of the mean value theorem for the function $f(x) = 2x - x^2$ in $[0, 1]$.

8. Using Lagrange's mean value theorem, find the value of 'c' in the following cases.

(i) $f(x) = -x^2 + 3x + 2$ in $[0, 1]$

(ii) $f(x) = x^2 - 2x, \forall x \in [0, 2]$

(iii) $f(x) = \sqrt{x}$ in $[1, 9]$

(iv) $f(x) = \frac{x}{(3-x)}$ in $[0, 2]$

(v) $f(x) = \dfrac{x-1}{x}$ in $[1, 4]$

Answers:

1. $c = \dfrac{5}{2}$

2. $c = \sqrt{\dfrac{7}{3}}$

3. $c = 3.15$ and 0.84

4. $c = 3$

5. $c = \log_2 e = \dfrac{1}{\log 2}$

6. $c = \log(e-1)$

7. $c = \dfrac{1}{2}$

8. (i) $c = \dfrac{1}{2}$

(ii) $c = \dfrac{2}{\sqrt{3}}$

(iii) $c = 4$

(iv) $3 - \sqrt{3}$

(v) $c = 2$

Exercise 20.7.2

Find a point $x = c$ such that the mean value theorem is satisfied. If no such point exists, state what condition of the mean value theorem is violated.

1. $f(x) = x^2, a = 1, b = 4$

2. $f(x) = x^2 - 5x + 7, a = 2, b = 5$

3. $f(x) = \dfrac{1}{x}, a = 1, b = 5$

4. $f(x) = x^3 = 3x, a = 0, b = 4$

5. $f(x) = \dfrac{5x}{5-x}, a = 0, b = 4$

6. $f(x) = x^{\frac{2}{3}}, a = -2, b = 2$

7. $f(x) = 6x^2 - x^3, a = 0, b = 6$

8. $f(x) = x^3 - 3x^2 - 6x + 8, a = -2, b = 1$

9. Is problem (8) an illustration of Rolle's theorem or of the mean value theorem.

Answers:

1. $\dfrac{5}{2}$ **3.** $\sqrt{5}$ **5.** $5 - \sqrt{5}$ **7.** 4 **9.** Both

Geometrical meaning of Lagrange's mean value theorem

If the graph of the function

1. $f(x)$ is continuous from a point $A\,(a, f(a))$ to another point $B\,(b, f(b))$ [i.e. $f(x)$ is continuous on the closed interval $[a, b]$].

2. $f(x)$ is differentiable in between the two points $A\,(a, f(a))$ and $B\,(b, f(b))$ [i.e. $f(x)$ is differentiable in the open interval (a, b)] then there is a point 'c' such that the tangent to the graph at the point $(c, f(c))$ is parallel to the secant line (or, the chord) passing through (or, joining) the end points $(a, f(a))$ and $(b, f(b))$ of the curve.

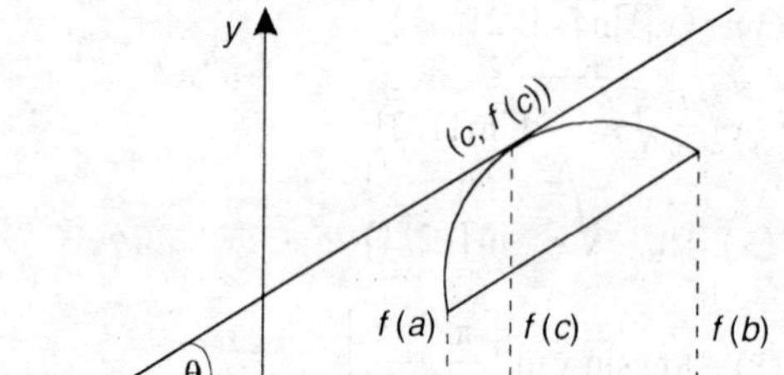

Remember:

(i) The quantity $\dfrac{f(b) - f(a)}{b - a}$ is the slope of the secant which passes through the points $A\,(a, f(a))$ and $B\,(b, f(b))$ of the graph of the function $y = f(x)$.

(ii) The quantity $f'(c)$ is the slope of the tangent to the graph of the function $y = f(x)$ at the point $(c, f(c))$.

(iii) Slope of the chord = slope of the tangent $\Rightarrow$ The tangent line is parallel to the secant line (or, the chord) i.e.

(iv) The co-ordinates of the extremities (or, the end points) A and B of the curve $f(x)$ are $(a, f(a))$ and $(b, f(b))$.

Problems based on geometrical meaning of Lagrange's mean value theorem

Problems based on geometrical meaning of Lagrange's mean value theorem consists of

1. A function $y = f(x)$

2. Two points $(a, f(a))$ and $(b, f(b))$

3. A third point $(c, f(c))$ in between $(a, f(a))$ and $(b, f(b))$ is required to find out at which the tangent line to the curve is parallel to the chord joining the end points $(a, f(a))$ and $(b, f(b))$ of the curve.

Working rule to find $(c, f(c))$ = a third point in between two given points $(a, f(a))$ and $(b, f(b))$ on the graph of the function $y = f(x)$ by using Lagrange's mean value theorem.

We adopt the following procedure to find a third point $(c, f(c))$.

1. Show that all conditions of Lagrange's mean value theorem are satisfied for which we are required to show

(a) Show that $f(x)$ is continuous at all points lying from $(a, f(a))$ to $(b, f(b))$

or, alternatively, show that $f(x)$ is continuous on $[a, b]$.

(b) Show that $f(x)$ is differentiable at all points lying between $(a, f(a))$ and $(b, f(b))$.

Or, alternatively, show that $f(x)$ is differentiable on (a, b).

2. Use $f'(c) = \dfrac{f(b) - f(a)}{b - a}$ to find the value of c.

3. Find the value of $f(c)$ on putting the obtained value of c in $[f(x)]_{x=c} = f(c)$ from the given function.

4. Required point will be $(c, f(c))$.

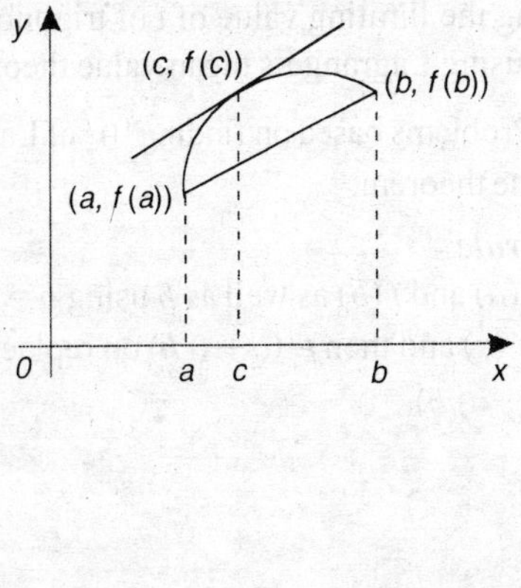

Worked out examples on geometrical meaning of Lagrange's mean value theorem

Question 1: Use the mean value theorem to find the point at which the tangent to the curve $y = 4 - x^2$ is parallel to the chord joining the point $A(-2, 0)$ and $B(1, 3)$.

Solution: (1) $f(x) = 4 - x^2$

Now, $f(x)$ being a polynomial in x is continuous in $[-2, 1]$ and differentiable in $(-2, 1)$.

$\therefore$ All conditions of L.M.V.T are satisfied.

(2) $f(x) = 4 - x^2$

$$\Rightarrow f'(x) = -2x$$

$$f'(c) = -2c$$

$(a, f(a)) = (-2, 0) = A$

$(b, f(b)) = (1, 3) = B$

$\therefore a = -2$

$b = 1$

$f(a) = 0$

$f(b) = 3$

Now, using the result

$$f'(c) = \frac{f(b) - f(a)}{b - a} = \frac{3 - 0}{1 - (-2)} = \frac{3}{3} = 1$$

$$\Rightarrow -2c = 1$$

$$\Rightarrow c = -\frac{1}{2} \text{ and } f(c) = 4 - \frac{1}{4} = \frac{15}{4}$$

$$\therefore \text{Required point } (c, f(c)) = \left(-\frac{1}{2}, \frac{15}{4}\right)$$

Question 2: If $A(x_1, y_1)$ and $B(x_2, y_2)$ be two points on the curve $y = ax^2 + bx + c$, then using Lagrange's mean value theorem, show that there will be at least one point (x_3, y_3) where the tangent will be parallel to the chord AB. Also show that

$$x_3 = \frac{x_2 + x_1}{2}$$

Solution: (1) $f(x) = ax^2 + bx + c$ ...(i)

$\Rightarrow f'(x) = 2ax + b$ ...(ii)

Now, $f(x)$ being a polynomial in x is continuous in $[x_1, x_2]$ and differentiable in (x_1, x_2).

∴ All conditions of L.M.V.T are satisfied.

∴ By geometrical meaning of Lagrange's mean value theorem, $\exists$ at least one point $P(x_3, y_3)$ between (x_1, y_1) and (x_2, y_2) where the tangent will be parallel to the chord AB.

$$\therefore f'(x_3) = \frac{f(x_2) - f(x_1)}{x_2 - x_1} = \frac{y_2 - y_1}{x_2 - x_1}$$

$$\Rightarrow 2ax_3 + b = \frac{ax_2 + bx_2 + c - ax_1^2 - bx_1 - c}{x_2 - x_1}$$

$$\left[\begin{array}{l} \because f(x) = ax^2 + bx + c \\ \therefore f(x_2) = y_2 \\ = ax_2 + bx_2 + c \ldots etc \end{array}\right]$$

$$\Rightarrow 2ax_3 + b = \frac{a\left(x_2^2 - x_1^2\right) + b(x_2 - x_1)}{x_2 - x_1}$$

$$\Rightarrow 2ax_3 + b = \frac{(x_2 - x_1)\{a(x_2 + x_1) + b\}}{x_2 - x_1}$$

$$\Rightarrow 2ax_3 + \not{b} = a(x_2 + x_1) + \not{b}$$

$$\Rightarrow 2\not{a}x_3 = \not{a}(x_2 + x_1)$$

$$\Rightarrow x_3 = \frac{x_2 + x_1}{2}$$

θ -form of Lagrange's mean value theorem

Lagrange's mean value theorem ⇒

$$\frac{f(b) - f(a)}{b - a} = f'(c);\ a<c<b \quad \ldots(1)$$

Now, $b - a = h \Rightarrow b = a + h$ ...(2)

and the interval (a, b) becomes equal to $(a, a + h)$

$(\because b = a + h)$ as well as $c \in (a, a + h) \Leftrightarrow a<c<a+h$ ...(3)

Again, $a<c<b \Leftrightarrow 0<c-a<c-b$

$$\Leftrightarrow 0 < \frac{c-a}{c-b} < 1$$

$$\Leftrightarrow 0 < \theta < 1, \text{ where } \frac{c - a}{b - a} = \theta$$

$$\Leftrightarrow c = a + \theta(b - a) = a + \theta h,\ 0<\theta<1 \quad \ldots(4)$$

Also, $a = a + \theta h$ for $\theta = 0$ ...(5)

$b = a + \theta h$ for $\theta = 1$ ...(6)

using (4), (5) and (6) in (1), we get,

$f(b) = f(a+h)\ (\because \theta = 1)$

$f(a) = f(a)\ (\because \theta = 0)$

$$(1) \Rightarrow \frac{f(a+h) - f(a)}{(a+h) - a} = f'(a+\theta h);\ 0 < \theta < 1$$

$$\Rightarrow \frac{f(a+h) - f(a)}{h} = f'(a+\theta h);\ 0 < \theta < 1$$

Statement of L.M.V.T in θ -form

Let $y = f(x)$ = a real function defined on $[a, a+h]$

If (1) $f(x)$ is continuous on $[a, a+h]$

(2) $f(x)$ is differentiable on $(a, a+h)$ then there exists at least one point $c = a + \theta h$ (where $0 < \theta < 1$) in the open interval $(a, a+h)$ for which $\frac{f(a+h) - f(a)}{h} = f'(a + \theta h)$ is valid.

Types of the problems:

We consider three types of the problems in finding θ from L.M.V.T.

1. When any two of the constants namely a, b and h as well as a function $y = f(x)$ are given.

2. When only a function $y = f(x)$ is given and no constant namely a, b and h is provided.

3. Finding the limiting value of t of trigonometrical ratios by using Lagrange's mean value theorem.

Type 1: Problems based on finding 'θ' of Lagrange's mean value theorem.

Working rule:

1. Find $f(a)$ and $f(b)$ as well as h using $b = a + h$.

2. Find $f'(x)$ and then $f'(a + \theta h)$ on replacing x in $f'(x)$ by $(a + \theta h)$

3. Use $\dfrac{f(a+h)-f(a)}{h} = f'(a+\theta h)$ to find θ.

Remember:

1. a, b, h are constants.

2. Any two of three constants namely a, b, h are given.

3. a and b are the finite values of the end points of the closed interval $[a, b] = [a, a+h]$ where $b = a + h$.

4. Sometimes b is not given. b can be found from $b = a + h$ when a and h are given.

Solved Examples

Question 1: If $a = 1$, $h = 3$ and $f(x) = \sqrt{x}$, then find 'θ' by mean value theorem.

Solution: Since we are given

$a = 1, h = 3$ and $f(x) = \sqrt{x}$

$\therefore f(a) = f(1) = 1$

$\because b = a + h \Rightarrow b = 1 + 3 = 4$

$f(b) = f(a+h) = \sqrt{4} = 2$

Now, $f'(x) = \dfrac{1}{2\sqrt{x}}, x > 0$

$$\therefore f'(a+\theta h) = \frac{1}{2\sqrt{a+\theta h}} = \frac{1}{2\sqrt{1+3\theta}}$$

$$= \frac{1}{2\sqrt{1+3h}} \quad [\because h = 3]$$

Now, using the result

$$\frac{f(a+h)-f(a)}{h} = f'(a+\theta h);\ 0 < \theta < 1,$$

We get, $\dfrac{2-1}{3} = \dfrac{1}{2\sqrt{1+3\theta}}$

$$\Rightarrow \frac{1}{3} = \frac{1}{2\sqrt{1+3\theta}}$$

$$\Rightarrow \frac{2}{3} = \frac{1}{\sqrt{1+3\theta}}$$

$$\Rightarrow \frac{4}{9} = \frac{1}{1+3\theta}$$

$$\Rightarrow 1 + 3\theta = \frac{9}{4}$$

$$\Rightarrow 3\theta = \frac{9}{4} - 1 = \frac{9-4}{4} = \frac{5}{4}$$

$$\Rightarrow \theta = \frac{5}{12}$$

Question 2: If $f(x) = \sqrt{x}$, $a = 1$, $b = 4$, find θ by L.M.V.T.

Solution: $\because$ we are given

$a = 1$

$b = 4$

$\therefore b = a + h \Rightarrow 4 = 1 + h \Rightarrow 4 - 1 = h \Rightarrow 3 = h$

$f(a) = f(1) = 1$

$f(b) = f(4) = \sqrt{4} = 2$

$f(a+h) = \sqrt{4} = 2$

Now, $f(x) = \sqrt{x}$

$$\Rightarrow f'(x) = \frac{1}{2\sqrt{4}}$$

$$\Rightarrow f'(a+\theta h) = \frac{1}{2\sqrt{a+\theta h}} = \frac{1}{2\sqrt{1+3\theta}};\ 0 < \theta < 1$$

Now, putting the above value θ in

$\dfrac{f(a+h)-f(a)}{h} = f'(a+\theta h)$ we get

$$\frac{2-1}{3} = \frac{1}{2\sqrt{1+3\theta}}$$

$$\Rightarrow \frac{1}{3} = \frac{1}{2\sqrt{1+3\theta}}$$

$$\Rightarrow \frac{2}{3} = \frac{1}{\sqrt{1+3\theta}}$$

$$\Rightarrow \frac{4}{9} = \frac{1}{3\theta + 1}$$

$$\Rightarrow \frac{9}{4} = 3\theta + 1$$

$$\Rightarrow 3\theta = \frac{9}{4} - 1 = \frac{9-4}{4}$$

$$\Rightarrow 3\theta = \frac{5}{4}$$

$$\Rightarrow \theta = \frac{5}{4} \times \frac{1}{3}$$

$$\Rightarrow \theta = \frac{5}{12}$$

N.B.: We should mark that above two questions are same except the following difference.

We are provided in question (1) a and h whereas we are provided in question (2) a and b.

Type 2: To find 'θ' of Lagrange's mean value theorem provided that only the function is given.

Working rule:

1. Find $f(a), f(a+h], f'(x)$ and $f'(a+\theta h)$.

2. Use the formula $\dfrac{f(a+h)-f(a)}{h} = f'(a+\theta h)$

where $0 < \theta < 1$.

Solved Examples

Question 1: Given $f(x) = mx^2 + nx + p$, find θ of L.M.V.T.

Solution: $\because f(x) = mx^2 + nx + p$

$\therefore f(a+h) = m(a+h) + n(a+h) + p$

$f(a) = ma^2 + na + p$

$f'(x) = 2mx + n$

$f'(a+\theta h) = 2m(a+\theta h) + n$

Now, using the θ - form of Lagrange's mean value theorem, we get,

$$\frac{f(a+h) - f(a)}{h} = f'(a+\theta h);\ 0 < \theta < 1$$

$$\Rightarrow \frac{m\left(a^2 + h^2 + 2ah\right) + na + nh + p - \left(ma^2 + ma + p\right)}{h}$$

$$= 2ma + 2m\theta h + n$$

$$\Rightarrow ma^2 + mh^2 + 2amh + na + nh + p - ma^2 - na - p$$

$$= 2amh + 2m\theta h^2 + nh$$

$$\Rightarrow mh^2 = 2m\theta h^2$$

$$\Rightarrow \theta = \frac{mh^2}{2mh^2} = \frac{1}{2}$$

Question 2: Find the value of 'θ' by using Lagrange's mean value theorem for the function $f(x) = x^2$.

Solution: $\because f(x) = x^2$

$\therefore f(x+h) = (x+h)^2 = x^2 + 2hx + h^2$

$= f(x) + h(2x+h)$ ...(1)

$f'(x) = 2x$ ...(2)

$(2) \Rightarrow f'(x+\theta h) = 2(x+\theta h)$ ...(3)

Now, $\dfrac{f(a+h) - f(a)}{h} = f'(a+\theta h)$ ...(4)

$\Rightarrow \dfrac{f(x+h) - f(x)}{h} = f'(x+\theta h)$ [on replacing a by x in (4)]

$$\Rightarrow \frac{x^2 + 2hx + h^2 - x^2}{h} = 2(x+\theta h)$$

$$\Rightarrow \frac{2hx + h^2}{h} = 2(x+\theta h)$$

$$\Rightarrow \frac{2hx + h^2}{h} = 2(x+\theta h)$$

$$\Rightarrow \frac{h(2x+h)}{h} = 2(x+\theta h)$$

$$\Rightarrow 2x + h = 2x + 2\theta h$$

$$\Rightarrow h = 2\theta h$$

$$\Rightarrow \theta = \frac{h}{2h} = \frac{1}{2}$$

Question 3: Find the value of 'θ' by using Lagrange's mean value theorem for the function $f(x) = \log x$ $(x > 0)$.

N.B.: Here an interval has not been mentioned. This is why we may take (or consider) the interval $[a, a+h]$ arbitrarily where $a > 0$.

Solution: $\because f(x) = \log(x)$

$\therefore f(a) = \log a$

$f(a+h) = \log(a+h)$

$f'(x) = \dfrac{1}{x}$

$f'(a+\theta h) = \dfrac{1}{a+\theta h}$

Now, $\dfrac{f(a+h) - f(a)}{h} = f'(a+\theta h)$ [using 'θ' form of L.M.V.T]

$\Rightarrow \dfrac{\log(a+h) - \log a}{h} = \dfrac{1}{1+\theta h}$

$\Rightarrow \log(a+h) - \log a = \dfrac{h}{a+\theta h}$

$\Rightarrow \log\left(\dfrac{a+h}{a}\right) = \dfrac{h}{a+\theta h}$

$\Rightarrow a+\theta h = \dfrac{h}{\log\left(1+\dfrac{h}{a}\right)}$

$\Rightarrow \theta h = \dfrac{h}{\log\left(1+\dfrac{h}{a}\right)} - a$

$\Rightarrow \theta = \dfrac{h}{h\log\left(1+\dfrac{h}{a}\right)} - \dfrac{a}{h} = \dfrac{1}{\log\left(1+\dfrac{h}{a}\right)} - \dfrac{a}{h}$

Question 4: Find the value of 'θ' using mean value theorem $f(x+h) = f(x) + h f'(x+\theta h)$ $(0<\theta<1)$ when $f(x) = e^x$

Solution: $\because f(x) = e^x$

$\therefore f'(x) = e^x$ ...(1)

Now, $f(x+h) = f(x) + h f'(x+\theta h)$ ...(2)

Hence, using (2), we get, $e^{x+h} = e^x + he^{x+\theta h}$

$\Rightarrow he^{\theta h} = e^h - 1$

$\Rightarrow e^{\theta h} = \dfrac{e^h - 1}{h}$

Now, taking log of both sides, we get

$$\log e^{\theta h} = \log\left[\frac{e^h - 1}{h}\right]$$

$$\Rightarrow \theta h = \log\left[\frac{e^h - 1}{h}\right]$$

$$\Rightarrow \theta = \frac{1}{h}\log\left[\frac{e^h - 1}{h}\right]$$

Type 3:

Solved Examples

Question: If $f(x) = \sin x$, find the limiting value of θ when $h \to 0^+$ using Lagrange's mean value theorem, $f(x+h) = f(x) + h f'(x+\theta h)$ (where $0<\theta<1$).

Solution: (1) $\sin x$ is continuous and differentiable for all finite values of $x \Rightarrow$ All conditions of L.M.V.T are satisfied.

(2) $f(x) = \sin x$ (given)

$\Rightarrow f'(x) = \cos x$

$f(x+h) = \sin(x+h)$

$f'(x+\theta h) = \cos(x+\theta h)$

Now, using Lagrange's mean value theorem, we get

$f(x+h) = f(x) + h f'(x+\theta h)$

$\Rightarrow \sin(x+h) = \sin x + h\cos(x+\theta h)$

$\Rightarrow \sin(x+h) - \sin x = h\cos(x+\theta h)$

$\Rightarrow 2\cos\left(\dfrac{x+h+x}{2}\right)\cdot\sin\left(\dfrac{x+h-x}{2}\right) = h\cos(x+\theta h)$

$\Rightarrow 2\cos\left(x+\dfrac{h}{2}\right)\cdot\sin\left(\dfrac{h}{2}\right) = h\cdot\cos(x+\theta h)$

$$\Rightarrow 2\cos\left(x+\frac{h}{2}\right)\cdot\frac{\sin\left(\frac{h}{2}\right)}{h} = \cos(x+\theta h)$$

$$\Rightarrow \cos\left(x+\frac{h}{2}\right)\cdot\frac{\sin\left(\frac{h}{2}\right)}{\frac{h}{2}} = \cos(x+\theta h)$$

$$\Rightarrow\left[\cos x\cdot\cos\frac{h}{2} - \sin x\cdot\sin\frac{h}{2}\right]\frac{\sin\frac{h}{2}}{\frac{h}{2}}$$

$$= \cos x\cdot\cos\theta h - \sin x\cdot\sin\theta h$$

$$\Rightarrow\left[\cos x\left\{1-\frac{1}{\lfloor 2}\left(\frac{h}{2}\right)^2+\ldots\right\}-\right.$$

$$\left.\sin x\left\{\frac{h}{2}-\frac{1}{\lfloor 3}\left(\frac{h}{2}\right)^3+\ldots\right\}\right]\times$$

$$\times\left[1-\frac{1}{\lfloor 3}\left(\frac{h}{2}\right)^2+\ldots\right]$$

$$= \cos x\left\{1-\frac{(\theta h)^2}{\lfloor 2}+\ldots\right\}-\sin x\left\{\theta h-\frac{(\theta h)^3}{\lfloor 3}+\ldots\right\}$$

$$\Rightarrow \lim_{h\to 0}(\sin x)\cdot\left(\theta-\frac{1}{2}\right)=0$$

$$\Rightarrow \lim_{h\to 0}\theta = \frac{1}{2}\ (\text{for } \sin x \neq 0)$$

Question 2: If $f(x) = \cos x$, find the limiting value of θ when $h \to 0^+$ using Lagrange's mean value theorem, $f(x+h) = f(x) + h f'(x+\theta h)$ (where $0 < \theta < 1$).

Solution: (1) $\because f(x) = \cos x$ which is continuous and differentiable in any finite interval $\Rightarrow$ all conditions of L.M.V.T are satisfied.

(2) $\because f(x) = \cos x$

$$\Rightarrow \frac{f(x+h)-f(x)}{h} = f'(x+\theta h)$$

$$\Rightarrow \frac{\cos(x+h)-\cos x}{h} = -\sin(x+\theta h)$$

$$\Rightarrow \frac{-2\sin\left(\frac{x+h+x}{2}\right)\cdot\sin\left(\frac{x+h-x}{2}\right)}{h} = -\sin(x+\theta h)$$

$$\Rightarrow -2\sin\left(x+\frac{h}{2}\right)\cdot\frac{\sin\left(\frac{h}{2}\right)}{h} = -\sin(x+\theta h)$$

$$\Rightarrow 2\sin\left(x+\frac{h}{2}\right)\cdot\frac{\sin\left(\frac{h}{2}\right)}{\left(\frac{h}{2}\right)} = \sin(x+\theta h)$$

$$\Rightarrow\left[\sin x\cdot\cos\frac{h}{2}+\cos x\cdot\sin\frac{h}{2}\right]\times\left\{\frac{\sin\frac{h}{2}}{\frac{h}{2}}\right\}$$

$$= \sin x\cdot\cos\theta h + \cos x\cdot\sin\theta h$$

$$\Rightarrow\left[\sin x\left\{1-\frac{1}{\lfloor 2}\left(\frac{h}{2}\right)^2+\ldots\right\}+\right.$$

$$\left.\cos x\left\{\frac{h}{2}-\frac{1}{\lfloor 3}\left(\frac{h}{2}\right)^3+\ldots\right\}\right]\times$$

$$\left[1-\frac{1}{\lfloor 3}\left(\frac{h}{2}\right)^2+\ldots\right]$$

$$= \sin x\left\{1-\frac{1}{\lfloor 2}(\theta h)^2+\ldots\right\}+\cos x\left\{\theta h-\frac{1}{\lfloor 3}(\theta h)^3+\ldots\right\}$$

$$\Rightarrow \lim_{\theta\to 0}(\cos x)\left(\theta-\frac{1}{2}\right)=0$$

$$\Rightarrow \lim_{\theta\to 0}\theta = \frac{1}{2}\ (\text{for } \cos x \neq 0)$$

Question: In the mean value theorem $f(h)=f(0)+h\,f'(\theta h)$, $0<\theta<1$ show that the limiting value of 'θ' as $h \to 0+0$ is $\frac{1}{2}$ or $\frac{1}{\sqrt{3}}$ according as $f(x)=\cos x$ or $\sin x$.

(a) *Solution*: Let $f(x)=\cos x$

$\therefore f'(x) = -\sin x$

$f'(\theta h) = -\sin(\theta h)$

$f(0)=\cos 0=1$

$f(0+h) = f(h) = \cos(0+h) = \cos h$

Now, from the mean value theorem, $f(h)=f(0)+h\,f'(\theta h)$, $0<\theta<1$ we get,

$$\frac{f(h)-f(0)}{h} = f'(\theta h)$$

$$\Rightarrow \frac{\cos h - \cos 0}{h} = f'(\theta h)$$

$$\Rightarrow \frac{\cos h - 1}{h} = -\sin(\theta h)$$

$$\Rightarrow \frac{1-2\sin^2\frac{h}{2}-1}{h} = -\sin(\theta h)$$

$$\Rightarrow \frac{-2\sin^2\frac{h}{2}}{h} = -\sin(\theta h)$$

$$\Rightarrow \frac{\sin^2\frac{h}{2}}{\frac{h}{2}\cdot h} = \frac{\sin(\theta h)}{h}$$

$$\Rightarrow \frac{1}{2}\left[\frac{\sin\frac{h}{2}}{\frac{h}{2}}\right]^2 = \frac{\sin(\theta h)}{\theta h}\cdot\theta$$

$$\Rightarrow \lim_{h\to 0^+}\frac{1}{2}\cdot\left[\frac{\sin\frac{h}{2}}{\frac{h}{2}}\right]^2 = \lim_{h\to 0^+}\frac{\sin(\theta h)}{\theta h}\cdot\theta$$

$$\Rightarrow \frac{1}{2}\lim_{h\to 0^+}\left[\frac{\sin\frac{h}{2}}{\frac{h}{2}}\right]^2 = \lim_{h\to 0^+}\frac{\sin(\theta h)}{\theta h}\cdot\lim_{h\to 0^+}\theta$$

$$\Rightarrow \frac{1}{2}\cdot 1 = 1\cdot\lim_{h\to 0^+}\theta$$

$$\Rightarrow \lim_{h\to 0^+}\theta = \frac{1}{2}$$

(b) If $f(x)=\sin x$ then $\frac{\sin h - \sin o}{h} = \cos(\theta h)$

$$\therefore \frac{1}{h}\left(h - \frac{h^3}{\underline{|3}} + \ldots\right) = 1 - \frac{\theta^2 h^2}{\underline{|2}} + \ldots$$

$$\therefore \lim_{h\to 0}\left(\frac{1}{\underline{|3}} - \frac{\theta^2}{\underline{|2}}\right) = 0$$

$$\therefore \lim_{h\to 0}\theta = \frac{1}{\sqrt{3}}$$

Problems based on finding the value of 'θ' using L.M.V.T.

Exercise 20.8

1. Find 'θ' in the mean value theorem

$f(a+h) = f(a)+h\,f'(a+\theta h)$ where

(i) $f(x)=\log x$, $a=1$, $h=e-1$

(ii) $f(x)=2x^2-7x+10$ in $[2, 5]$.

(iii) $f(x)=3x^2-5x+12$ in $[0, 1]$.

(iv) $f(x)=\sqrt{x}$ in $[1, 4]$

2. In Lagrange's mean value theorem

$f(a+h) = f(a)+h\,f'(a+\theta h)$ show that

$\lim_{h\to 0+0}\theta = \frac{1}{2}$ where $f(x)=\cos x$ or $\sin x$.

3. In the mean value theorem

$f(a+h) = f(a)+h\,f'(a+\theta h)$.

(i) If $a=1$, $h=1$ and $f(x)=x^2$, find θ.

(ii) If $a = 0$, $h = 3$ and $f(x) = \frac{1}{3}x^3 - \frac{3}{2}x^2 + 2x$, find θ.

(iii) If $a = 2$, $h = 1$ and $f(x) = \frac{1}{x}$, find the value of θ.

(iv) If $f(x) = e^x$, express the value of θ in terms of a and h.

(v) If $f(x) = \sin x$, find the limiting value of θ when $h \to 0$.

Answers:

1. (i) $\theta = \frac{e-2}{e-1}$ **(ii)** $\theta = \frac{1}{2}$

(iii) θ = Find **(iv)** $\theta = \frac{5}{12}$

3. (i) $\frac{1}{2}$ **(ii)** $\frac{3 \pm \sqrt{3}}{6}$ **(iii)** $-2 \pm \sqrt{6}$

(iv) $\frac{1}{h} \log \frac{e^h - 1}{h}$ **(v)** $\frac{1}{2}$

Problems based on proving inequality by using Lagrange's mean value theorem

Working Rule:

1. Use $\frac{f(b) - f(a)}{b - a} = f'(c)$ by L.M.V.T on $[a, b]$

or use $\frac{f(a+h) - f(a)}{(a+h) - h} = f'(a + \theta h)$ where $0 < \theta < 1$ by L.M.V.T on $[a, a+h]$.

2. Prove "Left hand expression $< f'(c) <$ right hand expression of the required inequality with in the help of $a < c < b$ or $0 < \theta < 1$ and using mathematical manupulations.

Note: **1.** If we use the inequality $0 < \theta < 1$, *firstly* we find θ.

2. When the restriction $x \geq 0$ is given with the required inequality, the interval $[0, x]$ has been considered.

3. When the restriction $a \leq b$ is given with the required inequality, the interval $[a, b]$ has been considered.

4. $x \neq y \Leftrightarrow x > y$ or $y < x$

5. $|\cos x| \leq 1$ or $|\sin x| \leq 1$

6. $x \leq y$ and $y < z \Rightarrow x < z$

7. $x < y$ and $y \leq z \Rightarrow x < z$

8. $x \leq y$ and $y \leq x \Rightarrow x = y$

Key point to prove inequality by L.M.V.T.

1. Application of L.M.V.T on continuous and differentiable functions in a finite given interval $[a, b]$ or $[0, x]$ as the case may be.

2. $a < c < b$ if the interval is $[a, b]$.

3. $0 < c < x$ if the interval is $[0, x]$.

Examples Worked Out:

1. Show that $\frac{x}{1+x^2} < \tan^{-1} x < x, \forall x > 0$ by using L.M.V.T.

Solution: Let $f(x) = \tan^{-1} x$ and $x > 0$

$\tan^{-1} x$ is continuous and differentiable for all values of x.

$\Rightarrow \tan^{-1} x$ is continuous and differentiable in any finite interval.

$\Rightarrow$ L.M.V.T is applicable to the function $\tan^{-1} x$ defined on $[0, x]$.

$\Rightarrow$ There is a point 'c' s.t $\frac{f(x) - f(0)}{x - 0} = f'(c)$

where $c \in (0, x)$

$$\Rightarrow \frac{\tan^{-1} x - \tan^{-1} 0}{x - 0} = \frac{1}{1 + c^2}, 0 < c < x;$$

$$\left[\text{Since } f'(x) = \frac{1}{1 + x^2} \Rightarrow f'(c) = \frac{1}{1 + c^2}\right]$$

$$\Rightarrow \frac{\tan^{-1} x}{x} = \frac{1}{1+c^2}$$

$$\Rightarrow \tan^{-1} x = \frac{x}{1+c^2} \quad ...(1)$$

Now, $x > c > 0$

$\Rightarrow x^2 > c^2 > 0$

$$\Rightarrow x^2+1 > c^2+1 > 1$$

$$\Rightarrow \frac{x^2+1}{x} > \frac{c^2+1}{x} > \frac{1}{x} \quad (\because x>0)$$

$$\Rightarrow \frac{x}{1+x^2} < \frac{x}{1+c^2} < x \qquad \ldots(2)$$

Now putting (1) in (2), we get

$\dfrac{x}{x^2+1} < \tan^{-1} x < x$ which is the required inequality.

Question 2: Show that $\dfrac{x}{1+x} < \log(1+x) < x$ for all $x>0$ by using L.M.V.T to the function $f(x) = \log(1+x)$.

Solution: Let $f(x) = \log(1+x)$

$\log(1+x)$ is continuous and differentiable for all values of $x \geq 0$.

$\Rightarrow$ L.M.V.T is applicable to the function $\log(1+x)$ defined on $[0, x]$.

$\Rightarrow$ There is a point 'c' s.t $\dfrac{f(x)-f(0)}{x-0} = f'(c)$

where $0<c<x$

$$\Rightarrow \frac{\log(1+x)-\log 1}{x-0} = \frac{1}{1+c}$$

$$\left[\text{Since, } f'(x) = \frac{1}{1+x} \Rightarrow f'(c) = \frac{1}{1+c}\right]$$

$$\Rightarrow \log(1+x) = \frac{x}{1+c} \quad (0<c<x) \qquad \ldots(1)$$

Now, $x>c>0$

$$\Rightarrow 1+x > 1+c > 1$$

$$\Rightarrow \frac{1+x}{x} > \frac{1+c}{x} > \frac{1}{x}$$

$$\Rightarrow \frac{x}{1+x} < \frac{x}{1+c} < x \qquad \ldots(2)$$

Putting (1) in (2), we get

$\dfrac{x}{1+x} < \log(1+x) < x$ which is the required inequality.

Question 3: Find 'θ' in Lagrange's mean value theorem for the function $f(x) = e^x$ over $[a, a+h]$, then show that $0 < \dfrac{1}{x}\log\left(\dfrac{e^x-1}{x}\right) < 1$ when $x>0$.

Solution: Let $f(x) = e^x$ (given)

e^x is continuous and differentiable in any finite interval

$\Rightarrow$ L.M.V.T is applicable to $f(x) = e^x$ defined on $[a, a+h]$

$\Rightarrow \exists$ a number 'θ' lying between 0 and 1 such that $\dfrac{f(a+h)-f(a)}{h} = f'(a+\theta h)$

$$\Rightarrow \frac{e^{a+h}-e^a}{h} = e^{a+\theta h}$$

$$\Rightarrow e^{a+h} - e^a = he^{a+\theta h}$$

$$\Rightarrow e^{a+h} = he^{a+\theta h} + e^a$$

$$\Rightarrow e^a \cdot e^h = e^a + he^a \cdot e^{\theta h}$$

$$\Rightarrow e^h = 1 + he^{\theta h}$$

$$\Rightarrow \frac{e^h-1}{h} = e^{\theta h}$$

$$\Rightarrow \log\left(\frac{e^h-1}{h}\right) = \log e^{\theta h} = \theta h \log_e e = \theta h$$

$$\Rightarrow \frac{1}{h}\log\left(\frac{e^h-1}{h}\right) = \theta \qquad \ldots(1)$$

Now, using the inequality $0<\theta<1$ ...(2)

Putting (1) in (2), we get

$$0 < \frac{1}{h}\log\left(\frac{e^h-1}{h}\right) < 1, \ \forall\, h>0 \qquad \ldots(3)$$

Now, replacing h by x in (3), we get

$$0 < \frac{1}{x} \log\left(\frac{e^x - 1}{x}\right) < 1 \text{ for } x > 0 \text{ which is the}$$

required inequality.

Question 4: Prove by using Lagrange's mean value theorem that $\log(1+x) < x$ when $x > 0$.

Solution: Let $f(x) = \log(1+x) - x$ which is defined on $[0, x]$

$$\Rightarrow f(0) = \log 1 - 0 = \log 1 = 0$$

$$f(x) = \log(1+x) - x$$

$$\therefore f'(x) = \frac{1}{1+x} - 1 = \frac{1-1-x}{1+x} = \frac{-x}{1+x}$$

$$\left[\begin{array}{l} \because h = b - a = x - 0 \\ \Rightarrow \theta h = x\theta \end{array}\right]$$

$$\Rightarrow f'(a+\theta h) = f'(0+\theta h) = f'(\theta x) = \frac{-\theta x}{1+\theta x}$$

Now $f(x) = \log(1+x) - x$ is continuous and differentiable on $[0, x]$.

$\Rightarrow$ L.M.V.T is applicable on $[0, x]$

$$\Rightarrow \frac{f(x) - f(0)}{x - 0} = f'(\theta x) \quad 0 < \theta < 1$$

$$\Rightarrow \frac{\log(1+x) - x - 0}{x - 0} = \frac{-\theta h}{1+\theta h}$$

$$\Rightarrow \frac{\log(1+x) - x}{x} = \frac{-\theta x}{1+\theta x} \qquad \ldots(1)$$

$$\text{Now, } \frac{x}{\log(1+x) - x} = -\frac{1+\theta x}{\theta x}$$

$$\Rightarrow \frac{x}{\log(1+x) - x} = -\frac{1}{\theta x} - 1$$

$$\Rightarrow \frac{x}{\log(1+x) - x} + 1 = -\frac{1}{\theta x}$$

$$\Rightarrow \frac{\not{x} + \log(1+x) - \not{x}}{\log(1+x) - x} = -\frac{1}{\theta x}$$

$$\Rightarrow \frac{\log(1+x)}{\log(1+x) - x} = -\frac{1}{\theta x}$$

$$\Rightarrow \frac{x \log(1+x)}{\log(1+x) - x} = -\frac{1}{\theta}$$

$$\Rightarrow \frac{\log(1+x) - x}{x \log(1+x)} = -\theta$$

$$\Rightarrow \frac{x - \log(1+x)}{x \log(1+x)} = \theta$$

Now since, $0 < \theta < 1$

$$\Rightarrow 0 < \frac{x - \log(1+x)}{x \log(1+x)}$$

$$\Rightarrow 0 < x - \log(1+x) \quad (\because x \log(1+x) > 0)$$

$\Rightarrow \log(1+x) < x$ which is the required inequality.

Question 5: Apply Lagrange's mean value theorem to the function $f(x) = \log(1+x)$ to show that $0 < \frac{1}{\log(1+x)} - \frac{1}{x} < 1, \forall x > 0$.

Solution: $f(x) = \log(1+x)$

$$\Rightarrow f(0) = \log 1 = 0$$

$$\Rightarrow f'(x) = \frac{1}{1+x}$$

$$\Rightarrow f'(c) = \frac{1}{1+c}$$

$$\Rightarrow f'(a+\theta h) = f'(0+\theta h) = f'(\theta h) = f'(\theta x) = \frac{1}{1+\theta x}$$

Now, using L.M.V.T, we get

$$\frac{f(b) - f(a)}{b-a} = f'(c)\,;\ a < c < b$$

$$\Rightarrow \frac{\log(1+x) - \log 1}{x - 0} = \frac{1}{1+c} = \frac{1}{1+\theta x}; \text{ for } x > 0$$

$(0 < \theta < 1)$

$$\Rightarrow \frac{\log(1+x)-0}{x}=\frac{1}{1+\theta x}$$

$$\Rightarrow \frac{\log(1+x)}{x}=\frac{1}{1+\theta x}$$

$$\Rightarrow \log(x+1)=\frac{x}{1+\theta x}$$

$$\Rightarrow \frac{1}{\log(1+x)}=\frac{1+\theta x}{x}$$

$$\Rightarrow \frac{x}{\log(1+x)}=1+\theta x$$

$$\Rightarrow \frac{x}{\log(1+x)}-1=\theta x$$

$$\Rightarrow \frac{x}{x\log(1+x)}-\frac{1}{x}=\theta$$

$$\Rightarrow \frac{1}{\log(1+x)}-\frac{1}{x}=\theta \quad \ldots(1)$$

Now, using the inequality $0<\theta<1$ ...(2)

Putting (2) in (1), we get

$0<\frac{1}{\log(1+x)}-\frac{1}{x}<1$ which is the required inequality.

Question 6: Prove that $\frac{b-a}{1+b^2}<\tan^{-1}b-\tan^{-1}a<\frac{b-a}{1+a^2}$ if $a<b$ by using Langrange's mean value theorem.

Proof: Let $f(x)=\tan^{-1}x$

$\therefore f(a)=\tan^{-1}a$

$f(b)=\tan^{-1}b$

$$f'(x)=\frac{1}{1+x^2}$$

$$f'(c)=\frac{1}{1+c^2}$$

Now, $\tan^{-1}x$ being continuous and differentiable on any finite interval implies that L.M.V.T is applicable to $\tan^{-1}x$ on $[a, b]$.

$$\Rightarrow \frac{f(b)-f(a)}{b-a}=f'(c),\ a<c<b$$

$$\Rightarrow \frac{\tan^{-1}b-\tan^{-1}a}{b-a}=\frac{1}{1+c^2} \quad \ldots(1)$$

Now, since, $a<c<b$

$$\Rightarrow c>a \Rightarrow c^2>a^2 \Rightarrow 1+c^2>1+a^2$$

$$\Rightarrow \frac{1}{1+c^2}<\frac{1}{1+a^2} \quad \ldots(1)$$

Again, $c<b \Rightarrow c^2<b^2 \Rightarrow c^2+1<b^2+1$

$$\Rightarrow \frac{1}{1+c^2}>\frac{1}{b^2+1} \quad \ldots(2)$$

$$(1)\text{ and }(2) \Rightarrow \frac{1}{1+b^2}<\frac{1}{1+c^2}<\frac{1}{a^2+1} \quad \ldots(3)$$

Putting (1) in (3)

$$\Rightarrow \frac{1}{1+b^2}<\frac{\tan^{-1}b-\tan^{-1}a}{b-a}<\frac{1}{a^2+1} \quad \ldots(4)$$

Now, multiplying both sides of (4) by $(b-a)$ $(\because (b-a)>0)$, we have $\frac{b-a}{1+b^2}<\tan^{-1}b-\tan^{-1}a<\frac{b-a}{1+a^2}$ which is the required inequality.

Note: In establishing elementary functional inequalities, a slight adjustment in the choice of $f(x)$ save many times much labour while using L.M.V.T.

$\Rightarrow$ **1.** Existance of t-function in the required inequality

$\Rightarrow f(x)=$ t-function of x which is continuous and differentiable in any finite interval.

2. Existance of inverse t-function in the required inequality

$\Rightarrow f(x) = t^{-1}$–function of x which is continuous and differentiable in any finite interval.

3. Existance of log-function in the required inequality

$\Rightarrow f(x) = \log x$ which is continuous and differentiable in any finite interval.

4. Existance of exponential function in the required inequality

$\Rightarrow f(x) = e^x$ which is continuous and differentiable in any finite interval.

Question 7: Prove by L.M.V.T

$(b-a)\sec^2 a < \tan b - \tan a < (b-1)\sec^2 b$ if $0 < a < b < \frac{\pi}{2}$.

Solution: Let $f(x) = \tan x$ which is continuous and differentiable on any closed interval $[a, b]$ where $0 < a < b < \frac{\pi}{2}$.

$\Rightarrow$ By applying L.M.V.T, $\frac{f(b)-f(a)}{b-a} = f'(c)$; $(a < c < b)$

$$\Rightarrow \frac{\tan b - \tan a}{b-a} = \sec^2 c \qquad \ldots(1)$$

Now, $a < c < b$

$$\Rightarrow \sec^2 a < \sec^2 c < \sec^2 b \qquad \ldots(2)$$

$\because$ sec x is increasing in $\left[0, \frac{\pi}{2}\right]$

Now, putting (1) in (2), we get

$$\sec^2 a < \frac{\tan b - \tan a}{b-a} < \sec^2 b$$

$\Rightarrow (b-a)\sec^2 a < \tan b - \tan a < (b-a)\sec^2 b$ which is the required inequality.

Question 8: Prove by L.M.V.T

$\frac{b-a}{b} < \log \frac{b}{a} < \frac{b-a}{a}$ where $0 < a < b$.

Solution: Let $f(x) = \log x$ which is continuous and differentiable in any finite interval.

$\Rightarrow f(x)$ is continuous and differentiable in $[a, b]$, where $0 < a < b$.

$\Rightarrow$ L.M.V.T is applicable on $f(x)$ defined in the closed interval $[a, b]$.

$\Rightarrow$ There is a point c s.t. $a < c < b$ satisfying the equality $\frac{f(b)-f(a)}{b-a} = f'(c)$

$$\Rightarrow \frac{\log b - \log a}{b-a} = \frac{1}{c} \qquad \ldots(1)$$

Now, $a < c < b$

$$\Rightarrow \frac{1}{a} > \frac{1}{c} > \frac{1}{b} \qquad \ldots(2)$$

Putting (1) in (2), we get

$$\frac{1}{a} > \frac{\log b - \log a}{b-a} > \frac{1}{a}$$

$$\Rightarrow \frac{(b-a)}{a} > \frac{\log b - \log a}{(b-a)} \times (b-a) > \frac{(b-a)}{b}$$

$\Rightarrow \frac{(b-a)}{a} > \log b - \log a > \frac{(b-a)}{b}$ which is the required inequality.

Question 9: Show that $\sin x < x$ for $x > 0$ by using L.M.V.T.

Solution: Let $f(x) = \sin x$

sin x is continuous and differentiable in any finite interval.

$\Rightarrow$ sin x is continuous and differentiable in $[0, x]$.

$\Rightarrow$ Lagrange's mean value theorem is applicable to $f(x) = \sin x$ on $[0, x]$.

$\Rightarrow \exists$ a number c s.t. $0 < c < x$ for which

$$\frac{f(b)-f(a)}{b-a} = f'(c)$$

$\because f(x) = \sin x$

$\Rightarrow f(0) = \sin 0 = 0 = f(a)$

$f(x) = \sin x = f(b)$

$f'(x) = \cos x$

$f'(c) = \cos x$

Hence, $f'(c) = \dfrac{f(b) - f(a)}{b - a}$

$$\Rightarrow \cos c = \frac{\sin x - \sin 0}{x - 0}$$

$$\Rightarrow \cos c = \frac{\sin x}{x} \text{ for } x < \frac{\pi}{2} \qquad \ldots(1)$$

$0 < c \Rightarrow \cos c < 1$

$$\Rightarrow \frac{\sin x}{x} < 1 \qquad \text{[from (1)]}$$

$\Rightarrow \sin x < x$

For $x \geq \dfrac{\pi}{2}$, the result.

$\sin x < x$ is obvious as $\sin x \leq 1 < \dfrac{\pi}{2}$

Question 10: Using Lagrange's mean value theorem, show that $\left|\tan^{-1} x\right| \leq x, \forall x \geq 0$.

Solution: Let $f(x) = \tan^{-1} x$

$\tan^{-1} x$ is continuous and differentiable in any finite interval.

$\Rightarrow$ $\tan^{-1} x$ is continuous and differentiable in $[a, b] = [0, x]$

$\Rightarrow$ L.M.V.T is applicable to $f(x) = \tan^{-1} x$

$$\Rightarrow \frac{f(b) - f(a)}{b - a} = f'(c), \text{ for some 'c', } a < c < b.$$

$$\Rightarrow \frac{\tan^{-1} x - \tan^{-1} 0}{x - 0} = \frac{1}{1 + c^2}$$

$$\Rightarrow \frac{\tan^{-1} x}{x} = \frac{1}{1+c^2}$$

$$\Rightarrow \tan^{-1} x = \frac{x}{1 + c^2} \qquad \ldots(1)$$

$$\text{Now, } \frac{x}{1+c^2} < x, \forall x \geq 0 \qquad \ldots(2)$$

(Since, $1 + c^2 > 1$)

Putting (1) in (2), we get

$\tan^{-1} x \leq x$

$\Rightarrow \left|\tan^{-1} x\right| \leq |x|$

$\Rightarrow \left|\tan^{-1} x\right| \leq x$ $(\because |x| = x, \text{ for } x \geq 0)$ which is the required result.

Question 11: If $0 < x < y$, then show that $x - \sin x < y - \sin y$ by using L.M.V.T in $[x, y]$.

Solution: Let $f(x) = \sin x$

$$\frac{f(y) - f(x)}{y - x} = f'(c), \text{ for some 'c', } x < c < y$$

[$\because$ $\sin x$ is continuous and differentiable in any finite interval]

$$\Rightarrow \frac{\sin y - \sin x}{y - x} = \cos c \qquad \ldots(1)$$

Since $0 < c$

$$\therefore \cos c < 1 \qquad \ldots(2)$$

Putting (1) in (2), we get

$$\frac{\sin y - \sin x}{y - x} < 1$$

$\Rightarrow \sin y - \sin x < y - x$

$\Rightarrow x - \sin x < y - \sin y$ which is the required inequality.

Question 12: If $f'(x) = \dfrac{1}{1 + x^2}$ for all x and $f(0) = 0$, show that $0.4 < f(2) < 2$.

Solution: $\because f'(x) = \dfrac{1}{1 + x^2}$

$\Rightarrow$ $f(x)$ is continuous and differentiable in any finite interval.

$\Rightarrow$ $f(x)$ is continuous and differentiable in [0, 2].

$\Rightarrow$ L.M.V.T is applicable to $f(x)$ defined in [0, 2].

$$\Rightarrow \frac{f(2)-f(0)}{2-0} = f'(c) = \frac{1}{1+c^2}$$

for some c, $0<c<2$

$$\Rightarrow \frac{f(2)-0}{2} = \frac{1}{1+c^2}$$

$$\Rightarrow f(2) = \frac{2}{1+c^2} \quad \ldots(1)$$

Now, $0<c<2$

$$\Rightarrow 0^2 < c^2 < 2^2$$

$$\Rightarrow 0 < c^2 < 4$$

$$\Rightarrow 1 < c^2 + 1 < 4 + 1$$

$$\Rightarrow 1 > \frac{1}{1+c^2} > \frac{1}{5}$$

$$\Rightarrow \frac{2}{1} > \frac{2}{1+c^2} > \frac{2}{5} \quad \ldots(2)$$

Putting (1) in (2), we get

$$2 > f(2) > \frac{2}{5}$$

$$\Rightarrow 2 > f(2) > 0.4$$

$$\Rightarrow 0.4 < f(2) < 2$$

which is the required inequality.

Question 13: Prove by the mean value theorem

$$|\sin a - \sin b| \le |a-b|$$

Solution: Let $b>a$

$f(x) = \sin x$

$f(x)$ is continuous and differentiable in any finite interval.

$\Rightarrow$ L.M.V.T is applicable to $\sin x$ in any interval $[a, b]$.

$\because f(x) = \sin x$

$\Rightarrow f(a) = \sin a$

$f(b) = \sin b$

Again, $f'(x) = \cos x$

$\Rightarrow f'(c) = \cos c$

Now, using L.M.V.T, we have

$$\frac{f(b)-f(a)}{b-a} = f'(c) \text{ for some 'c', } a<c<b$$

$$\Rightarrow \frac{\sin b - \sin a}{b-a} = \cos c \text{ which can be further}$$

written as

$$\frac{\sin a - \sin b}{a-b} = \cos c$$

$$\Rightarrow \left|\frac{\sin a - \sin b}{a-b}\right| = |\cos c| \quad \ldots(1)$$

Again, since, $|\cos c| \le 1$ ...(2)

Putting (1) in (2), we get

$$\frac{|\sin a - \sin b|}{|a-b|} \le 1$$

$$\Rightarrow |\sin a - \sin b| \le |a-b| \quad \ldots(3)$$

If $a>b$, then we can consider the interval $[b, a]$ i.e., if $a>b$, $\exists$ a number $c \in (a, b)$ s.t.

$$f'(c) = \frac{f(a)-f(b)}{a-b}$$

$$\Rightarrow \frac{\sin a - \sin b}{a-b} = \cos c \text{ and we prove as above}$$

that

$$|\sin a - \sin b| \le |a-b| \quad \ldots(4)$$

For $a=b$

$$|\sin a - \sin b| = |a-b| = |0| = 0 \quad \ldots(5)$$

Hence, the result follows for any a, b.

Question 14: Using Lagrange's mean value theorem, show that $|\cos a - \cos b| \le |a-b|$.

Solution: $\because f(x) = \cos x$

$f(a) = \cos a$

$f(b) = \cos b$

$f'(x) = -\sin x$

$f'(c) = -\sin c$

$f(x) = \cos x =$ a continuous and differentiable function in any finite interval.

$\Rightarrow$ L.M.V.T is applicable to $\cos x$ in $[a, b]$ if $b > a$.

Now, using L.M.V.T,

$$\frac{f(a)-f(b)}{a-b} = \frac{f(b)-f(a)}{b-a} = f'(c), \text{ for some}$$

$c; a < c < b$

$$\Rightarrow \frac{\cos a - \cos b}{a-b} = -\sin c$$

$$\Rightarrow \left|\frac{\cos a - \cos b}{a-b}\right| = |-\sin c| = |\sin c| \quad \ldots(1)$$

Again, since $|\sin c| \le 1$ ...(2)

Putting (1) in (2), we get

$$\left|\frac{\cos a - \cos b}{a-b}\right| \le 1$$

$$\Rightarrow |\cos a - \cos b| \le |a-b| \quad \ldots(3)$$

Similarly, if $a > b$, then considering the interval

$[a, b]$, we have $\dfrac{f(a)-f(b)}{a-b} = f'(c)$

$\therefore -\sin c = \dfrac{\cos a - \cos b}{a-b}$ and we prove as above

that

$$|\cos a - \cos b| \le |a-b| \quad \ldots(4)$$

The result is obvious for $a = b$.

Hence, the result follows for any a, b.

Question 15: Prove that

$$\left|\tan^{-1} x - \tan^{-1} y\right| < |x-y|, \forall x \ne y$$

Solution: Let $f(x) = \tan^{-1} x$

$$\Rightarrow f(y) = \tan^{-1} y$$

$$f'(x) = \frac{1}{1+x^2}$$

$$\Rightarrow f'(c) = \frac{1}{1+c^2}$$

Now, $f(x) = \tan^{-1} x$ which is continuous and differentiable in any finite interval.

$\Rightarrow$ L.M.V.T is applicable to $f(x) = \tan^{-1} x$ in $[x, y]$ if $y > x$.

$\Rightarrow \exists$ a number 'c'; $x < c < y$ such that

$$\frac{f(y)-f(x)}{y-x} = \frac{f(x)-f(y)}{x-y} = f'(c)$$

$$\Rightarrow \frac{\tan^{-1} x - \tan^{-1} y}{x-y} = \frac{1}{1+c^2}$$

Now, taking the mod of both sides, we get

$$\left|\frac{\tan^{-1} x - \tan^{-1} y}{x-y}\right| = \left|\frac{1}{1+c^2}\right| = \frac{1}{1+c^2} \quad \ldots(1)$$

Again, $c > 0 \Rightarrow 1 + c^2 > 1 \Rightarrow \dfrac{1}{1+c^2} < 1$ ...(2)

$\therefore$ From (1) and (2),

$$\left|\frac{\tan^{-1} x - \tan^{-1} y}{x-y}\right| < 1$$

$$\therefore \left|\tan^{-1} x - \tan^{-1} y\right| < |x-y|$$

If $x > y$, then considering L.M.V.T in $[y, x]$, we have the same result (3).

Hence, (3) is true $\forall x \ne y$.

Problems based on, Cauchy's mean value theorem, Lagrange's mean value theorem and Rolle's theorem

Statement of Cauchy's Mean Value Theorem:

If two functions $f(x)$ and $g(x)$ defined on $[a, b]$ are

(i) continuous in the closed interval $[a, b]$

(ii) differentiable in the open interval (a, b).

(iii) $g'(x) \ne 0$ for any $x \in (a, b)$ then there exists at least one real number c between a and b [i.e., $c \in (a, b)$] such that

$$\frac{f(b)-f(a)}{g(b)-g(a)} = \frac{f'(c)}{g'(c)}$$

Note: **1.** Cauchy's mean value theorem cannot be deduced by applying Lagrange's mean value theorem separately to the two functions $f(x)$ and $g(x)$ and then dividing them since then we get

$$\frac{f(b)-f(a)}{g(b)-g(a)}=\frac{f'(c_1)}{g'(c_2)}$$

where c_1 and c_2 may not be equal.

2. Cauchy's mean value theorem is also known as second mean value theorem whereas Lagrange's mean value theorem is known as first mean value theorem.

Examples Worked Out

Question 1: If $f(x)$ and $g(x)$ are differentiable on $0 \le x \le 1$ such that $f(0)=2, g(0)=0, f(1)=6, g(1)=2$ and $g'(x) \neq 0$ in $(0, 1)$ then show that there exists c satisfying $0<c<1$ and $f'(c)=2g'(c)$.

Solution: It is a question on Cauchy's mean value theorem since it satisfies all the conditions of Cauchy's mean value theorem.

Now, according to Cauchy's mean value theorem, we have

$$\frac{f(b)-f(a)}{g(b)-g(a)}=\frac{f'(c)}{g'(c)} \text{ for some } c, a<c<b$$

$$\Rightarrow \frac{f(1)-f(0)}{g(1)-g(0)}=\frac{f'(c)}{g'(c)} \text{ for some } c, 0<c<1$$

$$\Rightarrow \frac{6-2}{2-0}=\frac{f'(c)}{g'(c)} \quad [\because f(1)=6, g(1)=2, \ldots \text{ etc are given}]$$

$$\Rightarrow \frac{4}{2}=\frac{f'(c)}{g'(c)} \Rightarrow 2=\frac{f'(c)}{g'(c)} \Rightarrow f'(c)=2g'(c)$$

Alternative Method:

We suppose that $h(x)=f(x)-2g(x)$

$f(x)$ and $g(x)$ are given differentiable on the closed interval $[0, 1]$.

$\Rightarrow f(x)$ and $g(x)$ are continuous on the closed interval $[0, 1]$ and differentiable in the open interval $(0, 1)$.

$\Rightarrow h(x)$ being the difference of two differentiable functions in $[0, 1]$ is also differentiable in $[0, 1]$.

$\Rightarrow h(x)$ is continuous in $[0, 1]$ and differentiable in $(0, 1)$.

Now $h(x)=f(x)-2g(x)$

$\therefore h(0)=f(0)-2g(0)=2-2\times 0=2$

$[\because f(0)=2, g(0)=0]$

$h(1)=f(1)-2g(1)=6-2\times 2=6-4=2$

$[\because f(1)=6, g(1)=2]$

Hence, we observe that $h(0)=h(1)$

$\therefore$ All conditions of Rolle's theorem are satisfied. Therefore, there is at least one point $x=c$ where $h'(x)=0$, i.e.,

$$[h'(x)]_{x=c}=0 \Rightarrow [h'(x)]_{x=c}=[f'(x)-2g'(x)]_{x=c}=0$$

$$\Rightarrow [f'(x)-2g'(x)]_{x=c}=0$$

$$\Rightarrow f'(c)-2g'(c)=0$$

$\Rightarrow f'(c)=2g'(c)$ which is the required result.

Question 2: If $f(x)$ and $g(x)$ are continuous on $[a, b]$ and differentiable in (a, b), then show that

$$\begin{vmatrix} f(a) & f(b) \\ g(a) & g(b) \end{vmatrix}=(b-a)\begin{vmatrix} f(a) & f'(c) \\ g(a) & g'(c) \end{vmatrix} \text{ where } a<c<b.$$

Solution: (1) Let $F(x)=\begin{vmatrix} f(a) & f(x) \\ g(a) & g(x) \end{vmatrix}$

$$=f(a)g(x)-g(a)f(x) \quad \ldots\text{(i)}$$

where $f(a)$ and $g(a)$ are constants.

$f(x)$ and $g(x)$ are continuous on the closed interval $[a, b]$ and differentiable on the open interval (a, b).

$\therefore$ $F(x)$ being the difference of two continuous function in $[a, b]$ and differentiable in the open interval (a, b) is also continuous in $[a, b]$ and differentiable in (a, b). Which implies that all conditions of Lagrange's mean value theorem are satisfied by the function $f(x)$ defined on $[a, b]$. Therefore, by mean value theorem, $\exists$ at least one point c, $a<c<b$, such that

$$F'(c)=\frac{F(b)-F(a)}{b-a} \quad \ldots\text{(ii)}$$

(2) $\because F(x)=f(a)\,g(x)-g(a)\,f(x)$ (form (1))

$$\therefore F(a)=f(a)\,g(a)-g(a)\,f(a)=0 \quad \ldots\text{(iii)}$$

$$F(b)=f(a)\,g(b)-g(a)\,f(b) \quad \ldots\text{(iv)}$$

$$F'(x)=f(a)\,g'(x)-g(a)\,f'(x)$$

$$F'(c)=f(a)\,g'(c)-g(a)\,f'(c) \quad \ldots\text{(v)}$$

Now, putting (iii), (iv) and (v) in (ii), we get

$$f(a)\,g'(c)-g(a)\,f'(c)=\frac{f(a)\cdot g(b)-g(a)\cdot f(b)}{b-a}$$

$\Rightarrow (b-a)\left[f(a)\, g'(c) - g(a)\, f'(c)\right]$

$= f(a)\cdot g(b) - g(a)\cdot f(b)$

$$\Rightarrow \begin{vmatrix} f(a) & f(b) \\ g(a) & g(b) \end{vmatrix} = (b-a)\begin{vmatrix} f(a) & f'(c) \\ g(a) & g'(c) \end{vmatrix}$$

Question 3: For all x in the interval [0, 1], let the second derivative $f''(x)$ of a function $f(x)$ exists and satisfy $|f''(x)| \le 1$.

If $f(0) = f(1)$, show that $|f'(x)| < 1$ for all x in the interval [0, 1].

Solution: Given $f''(x)$ exists for all values of x in [0, 1] ...(1)

And also, $f(0) = f(1)$ ...(2)

$f''(x) \le 1$ ...(3)

To show: $|f'(x)| < 1$

Proof: $f''(x)$ exists for all values of x in [0, 1].

$\Rightarrow f(x)$ and $f'(x)$ are differentiable in [0, 1] ...(4)

and we are given $f(0) = f(1)$ ...(5)

(4), (5) $\Rightarrow f(x)$ satisfy all conditions of Rolle's theorem on [0, 1] $\Rightarrow \exists$ at least one value of $x = c$ in the open interval (0, 1) s.t. $f'(c) = 0$ ...(6)

Now we consider three cases, $x < c$, $x > c$, $x = c$

Case (i) When $x = c$

$\because f'(c) = 0$ (from (6))

$\therefore f'(x) = 0$

$\Rightarrow |f'(x)| = |0| = 0 < 1$

$\Rightarrow |f'(x)| < 1$ when $x = c$ ...(7)

Case (ii) When $x < c$

Further, $f'(x)$ is continuous on [0, 1] and differentiable on (0, 1)

$\Rightarrow$ Lagrange's mean value theorem is applicable on $f'(x)$ in $[x, c]$ if $0 \le x < c$.

$$\Rightarrow f''(d) = \frac{f'(c) - f'(x)}{c - x} \text{ for some } d,$$

$0 \le x < d < c < 1$

$$\Rightarrow f''(d) = \frac{f'(x) - f'(c)}{x - c}$$

$\Rightarrow f''(d)\,(x - c) = f'(x)$

[$\because f'(c) = 0$ from (7)]

$\Rightarrow |f''(d)\,(x - c)| = |f'(x)|$

$\Rightarrow |f''(d)| \cdot |x - c| = |f'(x)|$

$$\Rightarrow |f''(d)| = \frac{|f'(x)|}{|x - c|} \quad ...(8)$$

Now, $0 \le x < 1 \Rightarrow |x - c| < 1$ ...(9)

Again $|f''(x)| \le 1$ (given) $\Rightarrow |f''(d)| \le 1$ (on replacing x by d) ...(10)

Now, using (8) and (9) in (10), we get

$$\frac{|f'(x)|}{|x - c|} \le 1 \left[\because |f''(d)| = \frac{|f'(x)|}{|x - c|}\right]$$

$\Rightarrow |f'(x)| \le |x - c|$ ...(11)

and $|x - c| < 1$

$\therefore |f'(x)| < 1$ Which is the required result.

Case (iii): When $x > c$

$$\frac{f'(x) - f'(c)}{x - c} = f''(d),\ 0 < c < d \le x < 1$$

$$\Rightarrow \frac{f'(x) - 0}{x - c} = f''(d)\ [\because f'(c) = 0]$$

$\Rightarrow f'(x) = (x - c)\, f''(d)$

$$\Rightarrow |f''(d)| = \frac{|f'(x)|}{|x - c|} \quad ...(a)$$

Now, $0 \le c < x < 1 \Rightarrow |x - c| < 1$ ...(b)

Again, $|f''(x)| \le 1$ (given)

$\Rightarrow |f''(d)| \le 1$ (replacing x by d) ...(c)

On using (a) in (c), we get

$$\frac{|f'(x)|}{|x - c|} \le 1$$

$\Rightarrow |f'(x)| \le |x - c|$ ...(d)

and $|x - c| < 1$ (from (b))

$\therefore |f'(x)| < 1$ which is the required result.

An important type of problem

When the values of one function at the end points of the closed interval $[a, b]$ are given as well as $f'(c)$ is required to show to be a constant 'd' where $a < c < b$.

Note: **1.** Values of a function at the end points are not equal (i.e., $f(a) \neq f(b)$) and the given function is continuous in the closed interval and differentiable in open interval or simply the given function is differentiable in the closed interval means L.M.V.T is applicable.

2. In this type of problem functions are not given as an expression in x like x^2, $\log x$, e^x etc but in notational form $y = f(x)$ or $g(x)$ etc are given.

3. When the values of two functions $f(x)$ and $g(x)$ at the end points of the closed interval $[a, b]$ are given as well as $f'(c)$ and $g'(c)$ are given besides $f(x)$ and $g(x)$ are differentiable in the closed interval $[a, b]$, then we are required to Cauchy mean value theorem. (where $a < c < b$).

Examples Worked Out

Question 1: If a function $f(x)$ is differentiable in the closed interval $[2, 5]$ and $f(2) = 5$, $f(5) = 11$, then show that there will be at least one 'c' where $2 < c < 5$ such that $f'(c) = 2$.

Solution: $\because f(x)$ is differentiable in $[2, 5]$

$\Rightarrow f(x)$ is continuous in $[2, 5]$ as well as differentiable in $(2, 5)$.

$\Rightarrow$ All conditions of L.M.V.T are satisfied.

$\Rightarrow$ According to L.M.V.T, $\exists$ at least one c, $2 < c < 5$ s.t.

$$f'(c) = \frac{f(5) - f(2)}{5 - 2} \left[\begin{array}{l}\because f(2) = 5 \\ f(5) = 11 \text{ are given}\end{array}\right]$$

$$= \frac{11 - 5}{5 - 2} = 2 \text{ which was required to show.}$$

Problems based on proving inequalities with the help of Rolle's theorem and Lagrange's mean value theorem

Exercise 20.9

1. If $0 < a < b < \frac{\pi}{2}$, show that

(a) $|\sin b - \sin a| < |b - a|$

(b) $a - \sin a < b - \sin b$

2. Show that $\log(1 + x) < x$, where $x > 0$.

3. Using the function $f(x) = \tan^{-1} x$, show that

$$\frac{b-a}{1+b^2} < \tan^{-1} b - \tan^{-1} a < \frac{b-a}{1+a^2} \text{ where } 0 < a < b.$$

4. If $0 < a < b < \frac{\pi}{2}$, show that

(a) $|\cos b - \cos a| < |b - a|$

(b) $\tan^{-1} b - \tan^{-1} a < b - a$

(c) $\frac{\tan b}{\tan a} > \frac{a}{b}$

5. If $0 \le x < \frac{\pi}{2}$, show that

(a) $\sin x < x$

(b) $\tan x > x$

(c) $\cos x > 1 - \frac{x^2}{2}$

6. Show that $1 + x \log\left(x + \sqrt{1 + x^2}\right) \ge \sqrt{1 + x^2}$ if $x \ge 0$.

7. Show that

(a) $\frac{b - a}{b} < \log \frac{b}{a} < \frac{b - a}{a}$ if $0 < a < b$

(b) $\frac{b - a}{\cos^2 a} < \tan b - \tan a < \frac{b - a}{\cos^2 b}$ if $0 < a < b < \frac{\pi}{2}$.

Problems based on an important type

Exercise 20.10

1. If $f(x)$ is differentiable in the closed interval $[-1, 2]$, where $f(-1) = 3$, $f(2) = 6$, show that there exists a number c, $-1 < c < 2$, for which $f'(c) = 1$.

2. If a function $f(x)$ is differentiable in the closed interval $[0, 3]$ and $f(0) = 10$, $f(3) = 25$, then show that there exists at least one c, where $0 < c < 3$ such that $f'(c) = 5$.

3. If $f(x)$ is differentiable in $[-1, 2]$ where $f(-1) = 3$, $f(2) = -3$ then show that there exists a point c, $-1 < c < 2$, for which $f'(c) = -2$.

4. If $f(x)$ is continuous in $[-2, 2]$ and differentiable in $(-2, 2)$ where $f(-2) = 3$, $f(2) = 1$, then show that there exists a number c, $-2 < c < 2$ such that $f'(c) = -\frac{1}{2}$.

5. If the function $f(x)$ and $g(x)$ are differentiable in $[-1, 1]$, then show that there exists a point c, $-1 < c < 1$ for which

$$\begin{vmatrix} f(-1) & f(1) \\ g(-1) & g(1) \end{vmatrix} = 2 \begin{vmatrix} f(-1) & f'(c) \\ g(-1) & g'(c) \end{vmatrix}.$$

6. If the function $f(x)$ and $g(x)$ be differentiable in the closed interval $[1, 3]$, show that there exists at least one point c, where $1 < c < 3$, such that

$$\begin{vmatrix} f(1) & f(3) \\ g(1) & g(3) \end{vmatrix} = 2 \begin{vmatrix} f(1) & f'(c) \\ g(1) & g'(c) \end{vmatrix}.$$

21

Monotonocity of a Function

On Monotonocity of a Function

Definition 1: A function $y = f(x)$ defined on its domain D is called an increasing function in D if, for any two different values of the independent variable x in D, to the greater value of x, there is always a greater value of the function (i.e., value of the dependent variable y).

That is, a function $y = f(x)$ defined in an interval D is said to be increasing or rising in D if y, i.e., $f(x)$ increases as x increase in the interval D. That is, if x_1 and x_2 are two values of x in the interval D where the given continuous function f is defined by the formula $y = f(x)$ such that $x_1 < x_2 \Rightarrow f(x_1) < f(x_2)$ then $y = f(x)$ is said to be increasing in the interval D. Geometrically, it means that as one moves from left to right, values of the function f, i.e., values of the dependent variable y increase.

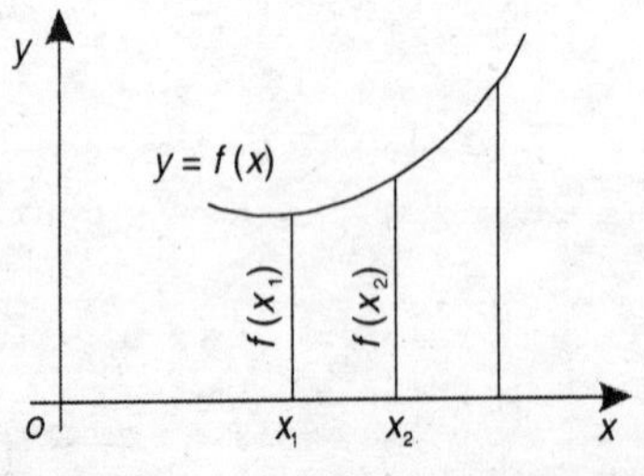

Note: A function $y = f(x)$ is an increasing function in an interval D

$\Leftrightarrow$ $f(x+h) > f(x)$

for all x in the interval D, where h is any positive number such that $x + h \in D$.

Definition 2: A function $y = f(x)$ defined on its domain D is called a decreasing function in D if, for any two different values of the independent variable x in D, to the greater value of x, there is a always a smaller value of the function (i.e., value of the dependent variable y).

That is, a continuous function $y = f(x)$ defined on an interval D is said to be decreasing or falling in D if y, i.e., $f(x)$ decreases as x increases in the interval D.

That is, if x_1 and x_2 are any two values of x in the interval D where a given continuous function f is defined by the formula $y = f(x)$ such that $x_1 < x_2 \Rightarrow f(x_1) > f(x_2)$ then $y = f(x)$ is said to be decreasing in the interval D. Geometrically, it means that as one moves from left to right, values of the function f, i.e. values of the dependent variable y decrease.

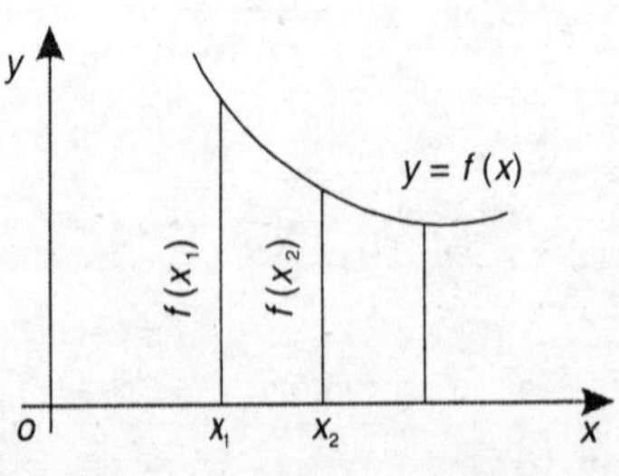

Note: A function $y = f(x)$ is a decreasing function in an interval D.

$\Leftrightarrow$ $f(x+h) < f(x)$

for all values of x in the interval D, where h is any positive number such that $x + h \in D$.

Remarks: **1.** A function $y = f(x)$ is said to be increasing at a point a if there is a h-neighbourhood of the point a in which $f(x) > f(a)$ for $x > a$, $f(x) < f(a)$ for $x < a$.

That is, a function $y = f(x)$ is said to be increasing at a point $x = a$, if there exists an open interval $(a - h, a + h)$ containing a i.e.,

$(a - h, a] \cup [a, a + h)$ such that $f(x)$ is increasing in the open interval $(a - h, a + h)$.

2. A function $y = f(x)$ is said to be decreasing at a point a if there is a h - neighbourhood of the point a in which $f(x) < f(a)$ for $x > a$, $f(x) > f(a)$ for $x < a$.

That is, a function $y = f(x)$ is said to be decreasing at a point $x = a$ if there exists an open interval $(a - h, a + h)$ containing a i.e.,

$(a - h, a] \cup [a, a + h)$ such that $f(x)$ is decreasing in the open interval $(a - h, a + h)$.

3. The statement " a function is increasing or decreasing " is not precise unless it is clearly mentioned in the problem " the point or the interval " where it is increasing or decreasing.

4. One should take notice carefully in the language of definition where x is always supposed to be increasing while the function may be increasing or decreasing.

5. It is not necessary that a function must be either increasing or decreasing on its domain, i.e., the same function may be increasing in some interval and decreasing in other interval.

Hence, such a function which is increasing in certain interval and decreasing in another interval is termed as a mixed function.

In the adjoining figure, $f(x)$ is increasing in $[a, b]$ and decreasing in $[b, c]$.

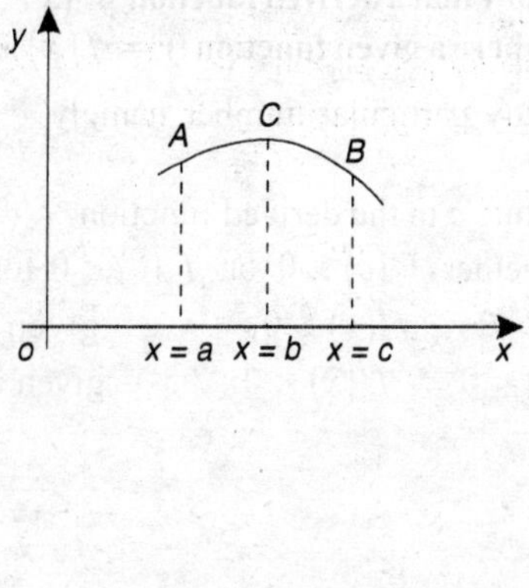

6. The symbol ↑ stands for increasing or increasing functions whereas the symbol ↓ stands for decreasing or decreasing functions.

On the Use of Terminology

According to some authors, a function $y = f(x)$ is monotonically increasing, non decreasing or simply increasing in an interval, if, for any x_1, x_2 belonging to the interval $x_1 < x_2 \Rightarrow f(x_1) \le f(x_2)$ and strictly increasing in the interval if $x_1 < x_2 \Rightarrow f(x_1) < f(x_2)$.

That is, a nondecreasing or increasing function differs from a strictly increasing function, i.e., in a nondecreasing function, two values of the function (at different values of the independent variable) may be equal while this is not possible in the case of a strictly increasing function. Likewise, they define monotonically decreasing, nonincreasing or simply decreasing functions in the intervals, i.e., a function $y = f(x)$ is monotonically decreasing, nonincreasing or simply decreasing in an interval if, for any x_1, x_2 belonging to the interval $x_1 < x_2 \Rightarrow f(x_1) \ge f(x_2)$ and strictly decreasing in the interval if $x_1 < x_2$ $\Rightarrow f(x_1) > f(x_2)$.

That is, a monotonically decreasing, nonincreasing or simply decreasing differs from a strictly increasing function, i.e., in a nonincreasing function, two values of the function (at different values of the independent variable) may be equal while this is not possible in the case of a strictly decreasing function.

Moreover, a function which is either nondecreasing or nonincreasing is termed as monotone or monotonic function and a function which is either strictly increasing or strictly decreasing is termed as strictly monotone or monotonic function.

On first derivative for increasing and decreasing functions

Theorem 1: A differentiable function $y = f(x)$ is increasing on an open interval $(a, b) \Leftrightarrow f'(x) > 0$ for all x in (a, b).

Proof: A function $y = f(x)$ is increasing on (a, b)

$\Leftrightarrow f(x + h) > f(x)$ for all x in (a, b), where h is small positive number such that $x + h \in (a, b)$

$\Leftrightarrow f(x+h) - f(x) > 0$ for all x in (a, b)

$\Leftrightarrow \dfrac{f(x+h) - f(x)}{h} > 0$ for all x in (a, b)

$\Leftrightarrow \lim\limits_{h \to 0} \dfrac{f(x+h) - f(x)}{h} > 0$ for all x in (a, b)

$\Leftrightarrow f'(x) > 0$ for all x in (a, b).

Theorem 2: A differentiable function $y = f(x)$ is decreasing on an open interval $(a, b) \Leftrightarrow f'(x) < 0$ for all x in (a, b).

Proof: A function $y = f(x)$ is decreasing on (a, b).

$\Leftrightarrow f(x+h) < f(x)$ for all x in (a, b), where h is a sufficiently small positive number.

$\Leftrightarrow f(x+h) - f(x) < 0$ for all x in (a, b)

$\Leftrightarrow \dfrac{f(x+h) - f(x)}{h} < 0$ for all x in (a, b)

$\Leftrightarrow \lim\limits_{h \to 0} \dfrac{f(x+h) - f(x)}{h} < 0$ for all x in (a, b)

$\Leftrightarrow f'(x) < 0$ for all x in (a, b).

Theorem 3: If $y = f(x)$ is a continuous function on $[a, b]$ and differentiable on (a, b) then $f'(x) > 0$ for all x in $(a, b) \Rightarrow y = f(x)$ increases on $[a, b]$.

Proof: Let $x_1, x_2 \in [a, b]$ such that $a \le x_1 < x_2 \le b$. As $y = f(x)$ satisfies both the conditions of Lagrange's mean value theorem in $[x_1, x_2]$, this is why $\exists$ a real number c in (x_1, x_2) such that

$\dfrac{f(x_2) - f(x_1)}{x_2 - x_1} = f'(c)$ which further

$\Rightarrow f(x_2) - f(x_1) = (x_2 - x_1)\, f'(c) > 0$,

since $x_2 > x_1$ and $f'(c) > 0$.

$\Rightarrow f(x_2) - f(x_1) > 0$

$\Rightarrow f(x_2) > f(x_1)$ for $x_2 > x_1$

$\Rightarrow y = f(x)$ is increasing in $[a, b]$.

Theorem 4: If $y = f(x)$ is a continuous function in $[a, b]$ and differentiable in (a, b), then $f'(x) < 0$ for all x in $(a, b) \Rightarrow y = f(x)$ decreases in $[a, b]$.

Proof: Let $x_1, x_2 \in [a, b]$ such that $a \le x_1 < x_2 \le b$. As $y = f(x)$ satisfies both the conditions of Lagrange's mean value theorem $\exists$ a real number 'c' in (x_1, x_2) such that

$$\frac{f(x_2) - f(x_1)}{x_2 - x_1} = f'(c)$$

$\Rightarrow f(x_2) - f(x_1) = (x_2 - x_1)\, f'(c) < 0$

since $x_2 > x_1$ and $f'(c) < 0$.

$\Rightarrow f(x_2) - f(x_1) = (+\,ve)\,(-\,ve) = (-\,ve) < 0$

$\Rightarrow f(x_2) - f(x_1) < 0$

$\Rightarrow f(x_2) < f(x_1)$

$\Rightarrow f(x)$ is decreasing in $[a, b]$

Note: in the right hand side of the equality $f(x_2) - f(x_1) = f'(c)\,(x_2 - x_1)$, the sign of $(x_2 - x_1)$ is always positive whereas the sign of $f'(c)$ is positive or negative according to the hypothesis. This is why the right hand of the equality $f(x_2) - f(x_1) = f'(c)\,(x_2 - x_1)$ has the same sign as the sign $f'(c)$ because $(x_2 - x_1)$ is always positive.

How to know that a given monotonic function $y = f(x)$ defined in an interval satisfies the condition $x_1 < x_2 \Rightarrow f(x_1) \gtrless f(x_2)$

1. Take any two particular numbers namely $c_1, c_2 \in$ given interval such that $c_1 < c_2$.

2. Substitute c_1 and c_2 in the monotonic functions and see whether $f(c_1) > f(c_2)$ or $f(c_1) < f(c_2)$.

How to know that a derived function $f'(x) \gtrless 0$ in an interval where a given function $y = f(x)$ is defined

1. Take any particular number namely $c \in$ given interval.

2. Substitute c in the derived function $f'(x)$.

3. See whether $f'(c) > 0$ or $f'(c) < 0$ for all c.

$f'(c) > 0 \Rightarrow f'(x) > 0,\ \forall\, x \in$ given interval and $f'(c) < 0 \Rightarrow f'(x) < 0,\ \forall\, x \in$ given interval.

On the methods of examining monotonocity in a given interval

There are two methods of examining or showing a function $y = f(x)$ to be increasing or decreasing in an interval.

1. Method of definition.
2. Method of first derivative test.

On method of definition

It consists of following steps:

1. Let x_1 and x_2 be any two real values of x in the given interval such that $x_1 < x_2$ or $x_1 > x_2$
2. Find the difference
   $f(x_1) - f(x_2)$ or $f(x_2) - f(x_1)$.
3. $f(x_1) - f(x_2) > 0$ for $x_1 > x_2 \Rightarrow f(x)$ is increasing in the given interval
   Also, $f(x_1) - f(x_2) < 0$ for $x_1 > x_2$
   $\Rightarrow f(x)$ is decreasing in the given interval
is decreasing in the given interval.)

Note: All that is necessary in the method of definition is to examine the sign of the difference $f(x_1) - f(x_2)$ if $x_1 > x_2$ where x_1, x_2 given interval.

On the method of first derivative test

It consists of the following steps:

1. Find $f'(x)$
2. See whether $f'(x) \gtrless 0, \forall x \in$ given interval.
3. $f'(x) > 0, \forall x \in$ given interval.

$\Rightarrow f(x)$ is increasing in the given interval and $f'(x) < 0, \forall x \in$ given interval.

$\Rightarrow f(x)$ is decreasing in the given interval.

Note: 1. All that is necessay in the method of first derivative test is to examine the sign of the first derivative $f'(x)$ in the given interval.
2. When no method is mentioned in the problem, there may be the use of first derivative test.
3. Show that the given function $y = f(x)$ is increasing or decreasing in the whole of its domain when the interval (where $f(x)$ is defined) is not given in the problem.

Now problems are divided in different categories to explain their method of procedure.

Category A:

Problems based on showing a function $y = f(x)$ to be increasing or decreasing when ever any one of the following restrictions is imposed on it (by using definition of increasing and decreasing function)

(i) $x \le \pm a$ **(ii)** $x < \pm a$ **(iii)** $x \ge \pm a$

(iv) $x > \pm a$ **(v)** $x < 0$ **(vi)** $x \le 0$

(vii) $x > 0$ **(viii)** $x \ne \pm a$

(ix) $\forall x$

Examples worked out:

1. Show that $f(x) = \dfrac{x}{1+x}$ is monotone increasing, $x > 0$, without the use of derivative.

Solution: Let x_1 and x_2 be any two values of x s.t. $x_1 > x_2 > 0$

Now, $f(x_1) = \dfrac{x_1}{1+x_1}$

$f(x_2) = \dfrac{x_2}{1+x_2}$

$\therefore f(x_1) - f(x_2) = \dfrac{x_1}{1+x_1} - \dfrac{x_2}{1+x_2}$

$= \dfrac{x_1(x_2 - 1) - x_2(x_1 + 1)}{(1+x_1)(1+x_2)}$

$= \dfrac{x_1 - x_2}{(1+x_1)(1+x_2)} > 0$ since $x_1 > x_2 > 0$

which $\Rightarrow f(x_1) - f(x_2) > 0$

$\Rightarrow f(x_1) > f(x_2)$ when $x_1 > x_2$ which means $f(x)$ is an increasing function for $x > 0$

Hence, proved.

2. Show that $f(x) = x^2$ for $x \ge 0$ is an increasing function without the use of derivative.

Solution: Let x_1 and x_2 be any two values of x, such that $x_1 > x_2 \ge 0$

Now, $f(x_1) = x_1^2$

$f(x_2) = x_2^2$

$\therefore f(x_1) - f(x_2) = x_1^2 - x_2^2 > 0$ when $x_1 > x_2 \ge 0$

$\Rightarrow f(x_1) - f(x_2) > 0$

$\Rightarrow f(x_1) > f(x_2)$ when $x_1 > x_2$ which means $f(x)$ is an decreasing function for $x \geq 0$.

Hence, proved.

3. Show that $f(x) = x^2$ for $x \leq 0$ is a decreasing function.

Solution: Let x_1 and x_2 be any two values of x s.t $x_2 < x_1 \leq 0$

Now $f(x_1) = x_1^2$

$f(x_2) = x_1^2$

$\therefore f(x_1) - f(x_2) = x_1^2 - x_2^2 < 0$ when $x_2 < x_1 \leq 0$

which $\Rightarrow f(x_1) - f(x_2) \leq 0$

$\Rightarrow f(x_1) < f(x_2)$ when $x_2 < x_1 < 0$

Hence, $f(x)$ is a decreasing function for $x \leq 0$

4. Show that $f(x) = x^3 + 3x^2 + 3x - 100$ in increasing for all $x \in R$.

Solution: Let x_1 and x_2 be any two values of $x \in R$ s.t. $x_1 > x_2$

Now, $f(x_1) = x_1^3 + 3x_1^2 + 3x_1 - 100$

$f(x_2) = x_2^3 + 3x_2^2 + 3x_2 - 100$

$\therefore f(x_1) - f(x_2) = \left(x_1^3 - 3x_1^2 + 3x_1 - 100\right) - \left(x_2^3 + 3x_2^2 + 3x_2 - 100\right)$

$= x_1^3 - 3x_1^2 + 3x_1 - 100 - x_2^3 + 3x_2^2 - 3x_2 + 100$

$= \left(x_1^3 - x_2^3\right) + 3\left(x_1^2 - x_2^2\right) + 3(x_1 - x_2) > 0$

when $x_1 > x_2$

Which $\Rightarrow f(x_1) - f(x_2) > 0$ when $x_1 > x_2$.

$\Rightarrow f(x_1) > f(x_2)$ when $x_1 > x_2$

Hence, $f(x)$ is an increasing function for all values of $x \in R$.

5. Show that $f(x) = 3x + 1$ is an increasing function on R without using derivative.

Solution: Let x_1 and x_2 be any two values of $x \in R$ (i.e. $x_1, x_2 \in R$) s.t $x_1 > x_2$

Now $f(x_1) = 3x_1 + 1$

$f(x_2) = 3x_2 + 1$

$\therefore f(x_1) - f(x_2) = 3x_1 + 1 - 3x_2 - 1$

$= 3(x_1 - x_2) > 0$ when $x_1 > x_2$.

Which $\Rightarrow f(x_1) - f(x_2) > 0$ when $x_1 > x_2$

$\Rightarrow f(x_1) > f(x_2)$ when $x_1 > x_2$.

Hence, $f(x)$ is an increasing function on R.

6. Show that $f(x) = ax + b$ where a and b are constants and $a > 0$ is an increasing function of x (without using the derivative) $\forall\, x$

Solution: Let x_1 and x_2 be any two values of $x \in R$ (i.e. $x_1, x_2 \in R$) s.t $x_1 > x_2$

Now, $f(x_1) = a\,x_1 + b$

$f(x_2) = a\,x_2 + b$

$\therefore f(x_1) - f(x_2) = a\,x_1 + b - a\,x_2 - b$

$= a\,x_1 - a\,x_2 = a(x_1 - x_2)$

$\Rightarrow f(x_1) - f(x_2) = a(x_1 - x_2) > 0$ when $x_1 > x_2$

$\Rightarrow f(x_1) - f(x_2) > 0$ when $x_1 > x_2$

$\Rightarrow f(x_1) > f(x_2)$ when $x_1 > x_2$

Hence, $f(x)$ is an increasing function $\forall\, x$.

Category B:

Type 1: Problems based on showing a function $y = f(x)$ to be increasing or decreasing at a point $x = a$ by using derivative test.

Working rule:

1. Find $\dfrac{dy}{dx} = f'(x)$
2. Find $\left[\dfrac{dy}{dx}\right]_{x=a} = \left[f'(x)\right]_{x=a} = f'(a)$
3. If $f'(a) = +$ve number, then $f(x)$ is an increasing function at the point $x = a$
4. If $f'(a) = -$ ve number, then $f(x)$ is decreasing said function at point $x = a$.

Note: $f'(a) = 0$, then $f(x)$ is said to be stationary at the point $x = a$

Examples worked out:

1. Show that the function $y = 2x^3 - 3x + 1$ is increasing at $x = 1$ and decreasing at $x = 0$

Solution: $y = 2x^3 - 3x +$

$$\Rightarrow \frac{dy}{dx} = 6x^2 - 3$$

$$\text{Now}, \left[\frac{dy}{dx}\right]_{x=1} = \left[6x^2 - 3\right]_{x=1}$$

$= 6 \times 1 - 3 = 6 - 3 = 3 = +\text{ve number}$

Hence, $y = 2x^3 - 3x + 1$ is an increasing function at $x = 1$

$$\text{Next}, \left[\frac{dy}{dx}\right]_{x=0} = \left[6x^2 - 3\right]_{x=0}$$

$= 6 \times 0 - 3 = -3 = -\text{ve number}$

Hence, $y = 2x^3 - 3x + 1$ is a decreasing function at $x = 0$

Type 2: Problems based on showing that a given function $y = f(x)$ is increasing or decreasing when any one of the following restrictions are imposed on it.

(i) $x > \pm a$ **(ii)** $x \geq \pm a$ **(iii)** $x < \pm a$

(iv) $x \leq \pm a$ **(v)** $x > 0$ **(vi)** $x \geq 0$

(vii) $x < 0$ **(viii)** $x \leq 0$ **(ix)** $x \neq a$

(x) $\forall x$ (where a is any + ve number) by using the derivative

Examples worked out:

1. Show that the function $y = x^4 - 4x + 1$ is a decreasing function when $x < 1$ and is an increasing function when $x > 1$

Solution: $y = f(x) = x^4 - 4x + 1$

$$\Rightarrow f'(x) = 4x^3 - 4 = 4\left(x^3 - 1\right)$$

Now we have to determine the sign of $\left(x^3 - 1\right)$ with the help of given restriction:

$\because x < 1$

$\Rightarrow x^3 < 1$

$\Rightarrow x^3 - 1 < 0$ which $\Rightarrow x^3 - 1 = -$ ve

$\Rightarrow f'(x) = -$ ve , i.e; $f'(x) < 0$ which means that $f(x)$ is a decreasing function when $x < 1$

2. Show that $f(x) = \dfrac{\log x}{x}$ is a decreasing function for $x > 3$.

Solution: $f(x) = \dfrac{\log x}{x}$ $(x > 0)$

$$\Rightarrow f'(x) = \frac{x \cdot \frac{1}{x} - \log x \cdot 1}{x^2} = \frac{1 - \log x}{x^2}$$

Now, we are required to determine the sign of $\dfrac{1 - \log x}{x^2}$ with the help of given restriction :

$x > 3$ and $3 > e$ $(\because e = 2 \cdot 718281)$

$\Rightarrow x > e$

$\Rightarrow \log x > \log_e e$

$\Rightarrow \log x > 1$

$\Rightarrow \log x - 1 > 0$

$\Rightarrow 1 - \log x < 0$

$\Rightarrow \dfrac{1 - \log x}{x^2} < 0$ which means

$\dfrac{1 - \log x}{x^2} = -$ ve

$\therefore f'(x) = -$ ve , i.e; $f'(x) < 0$ which $\Rightarrow f(x)$ is decreasing function when $x > 3$.

3. Show that $y = \log(1 + x) - \dfrac{2x}{(2 + x)}$ is an increasing function of x for all values of $x > -1$

Solution: $y = f(x) = \log(1 + x) - \dfrac{2x}{(2 + x)}$

$$\Rightarrow f'(x) = \frac{1}{1 + x} - \left\{\frac{2(2 + x) - 2x}{(2 + x)^2}\right\}$$

$$= \frac{1}{1+x} - \frac{4}{(2+x)^2} = \frac{(2+x)^2 - 4(1+x)}{(1+x)(2+x)^2}$$

$$= \frac{4 + x^2 + 4x - 4 - 4x}{(1+x)(2+x)^2} = \frac{x^2}{(1+x)(2+x)^2}$$

Now we have to determine the sign of $\frac{x^2}{(1+x)(2+x)^2}$ whose sign depends only on the factor $(1+x)$ in denominator because x^2 and $(2+x)^2$ are always + ve for being perfect squares.

Now we determine the sign of $(1 + x)$ with the help of given restriction:

$\because x > -1$

$\Rightarrow 1 + x > 0$ which means $(1 + x) = +$ ve

Hence, each term in the numerator and denominator of $\frac{x^2}{(1+x)(2+x)^2}$ is + ve.

$\therefore f'(x) = \frac{x^2}{(1+x)(2+x)^2} = +$ ve which means that $f'(x) > 0$

Hence, $f(x)$ is an increasing function for $x > -1$.

Note: Whenever the derived function $f'(x)$ contain a linear factor along with factors which are perfect squares, we examine the sign of the linear factor only with the help of given restriction.

4. Show that $f(x) = 6 + 9x - x^2$ is decreasing for $x \geq \frac{9}{2}$.

Solution: $f(x) = 6 + 9x - x^2$

$\Rightarrow f'(x) = 9 - 2x$

Now, we are required to determine the sign of $(9 - 2x)$ with the help of given restriction :

$\because x \geq \frac{9}{2}$

$\Rightarrow 2x \geq 9$

$\Rightarrow 2x - 9 \geq 0$

$\Rightarrow 9 - 2x \leq 0$ which means that $(9 - 2x) = -$ve ...(i)

which $\Rightarrow f(x)$ is decreasing function for $x \geq \frac{9}{2}$

5. Show that $f(x) = -7x^2 + 11x - 9$ is decreasing for $x > 1$.

Solution: $f(x) = -7x^2 + 11x - 9$

$\Rightarrow f'(x) = -14x + 11$

Now we are required to determine the sign of $(-14x + 11)$ with the help of given restriction:

$\because x > 1 \Rightarrow$

$\Rightarrow -14x < -14$

$\Rightarrow -14x + 11 < -14 + 11$

$\Rightarrow -14x + 11 < -3 < 0$

$\Rightarrow -14x + 11 < 0$ which means that $(-14x + 11) = -$ve

This is why $f'(x) = -$ ve , i.e; $f'(x) < 0$ for $x > 1$ which $\Rightarrow f(x)$ is decreasing function for $x > 1$.

6. Show that $y = 2x^3 + 3x^2 - 12x + 7$ is increasing and positive for $x > 1$.

Solution: $y = f(x) = 2x^3 + 3x^2 - 12x + 7$

$\Rightarrow f'(x) = 6x^2 + 6x - 12 = 6(x^2 + x - 2)$

Now we are required to determine the sign of $(x^2 + x - 2)$ with the help of given restriction:

$\because x > 1$

$\Rightarrow x^2 > 1$

$\Rightarrow x^2 + x > 1 + x$

$\Rightarrow x^2 + x - 2 > 1 + x - 2$

$\Rightarrow x^2 + x - 2 > x - 1 > 0$ ($\because x > 1$) which

$\Rightarrow (x^2 + x - 2) = +$ ve

$f'(x) = +$ ve, i.e; $f'(x) > 0$ which means that $f(x)$ is an increasing function when $x > 1$

Next we are required to show $f(x)$ to be positive.

$f(1) = 0$

and $x > 1 \Rightarrow f(x) > f(1) \Rightarrow f(x) > 0$ ($\because f(1)=0$)

which means that $f(x)$ is positive.

Hence, $f(x)$ is increasing and positive for $x > 1$

Note: The following is a sufficient test for positivity of a function for $x > 0$. If

(i) The function $f(x)$ is continuous in $0 \le x < b$

(ii) $f(0)$ is non negative.

(iii) $f'(x) > 0$, when $0 < x < b$ then $f(x) > 0$ in $0 < x < b$.

This test remains valid even when 'b' is replaced by ∞. Hence these conditions are sufficient for $f(x)$ to be positive for $x > 0$.

7. Show that the function $y = \log(1 + x) - \dfrac{2x}{2 + x}$ is an increasing function for $x > 0$.

Solution: $y = f(x) = \log(1 + x) - \dfrac{2x}{2 + x}$

$$\Rightarrow f'(x) = \frac{1}{1 + x} - \frac{2 \cdot (2 + x) - 1 \cdot (2x)}{(2 + x)^2}$$

$$= \frac{1}{1 + x} - \frac{4 + 2x - 2x}{(2 + x)^2}$$

$$= \frac{1}{1 + x} - \frac{4}{(2 + x)^2}$$

$$= \frac{(2+x)^2 - 4(1+x)}{(1+x)(2+x)^2} = \frac{4+4x+x^2-4-4x}{(1+x)(2+x)^2}$$

$$= \frac{x^2}{(1 + x)(2 + x)^2}$$

Now we have to determine the sign of

$\dfrac{x^2}{(1 + x)(2 + x)^2}$ whose sign depends only on the factor $(1 + x)$ in denominator because x^2 and $(2 + x)^2$ are always positive for being perfect squares.

$x > 0 \quad \Rightarrow 1 + x > 1 > 0$

Hence, each term in the numerator and denominator of $\dfrac{x^2}{(1 + x)(2 + x^2)}$ is +ve. This is why

$f'(x) = \dfrac{x^2}{(1 + x)(2 + x^2)} = +$ ve which means that $f'(x) > 0$.

Hence, $f(x)$ is an increasing function for $x > 0$

8. Examine the monotonocity of the function

$$f(x) = \frac{1}{1 + x^2} \text{ for } x \le 0.$$

Solution: $f(x) = \dfrac{1}{1 + x^2}$

$$\Rightarrow f'(x) = \frac{-2x}{(1 + x^2)^2}$$

Now we are required to determine the sign of $\dfrac{-2x}{(1 + x^2)^2}$ with the help of given restriction:

$x \le 0$

$-2x \ge 0$

$$\therefore f'(x) = \frac{-2x}{(1 + x^2)^2} \ge 0$$

$\therefore f'(x) = +$ ve , i.e; $f'(x) > 0$ ...(1)

and $f'(0) = 0$ ...(2)

Hence, (1) and (2) $\Rightarrow f(x)$ in an increasing function when $x \le 0$.

9. Show that $f(x) = x^2$ is a decreasing function for $x \le 0$.

Solution: $f(x) = x^2$

$\Rightarrow f'(x) = 2x$

Now, we are required to determine the sign of $2x$ with the help of given restriction:

$\because x \le 0$

$\therefore\ 2x \le 0$ which $\Rightarrow 2x = -$ ve

$\therefore\ f'(x) = -$ ve , i.e; $f'(x) < 0$ ...(1)

and $f'(0) = 0$ ...(2)

Hence, (1) and (2) $\Rightarrow f(x)$ is a decreasing function when $x \le 0$.

10. Show that $f(x) = x^2$ is an increasing function for $x \ge 0$.

Solution: $f(x) = x^2 \Rightarrow f'(x) = 2x$

Now we have to determine the sign of 2x with the help of given restriction:

$\because\ x \ge 0$

$\therefore\ 2x \ge 0$ which $\Rightarrow 2x = +$ ve

$\therefore\ f'(x) = +$ ve , i.e; $f'(x) > 0$ ...(1)

and $f'(0) = 0$ ...(2)

Hence, (1) and (2) $\Rightarrow f(x)$ in an increasing function when $x \ge 0$.

11. Show that $f(x) = \dfrac{x}{1+x}$ is an increasing function for $x \ne -1$.

Solution: $f(x) = \dfrac{x}{1+x}$

$$\Rightarrow f'(x) = \frac{(x+1)\cdot 1 - x\,(1+0)}{(1+x)^2}$$

$$= \frac{(x+1)\cdot 1 - x\,(1+0)}{(1+x)^2}$$

$$= \frac{x+1-x}{(1+x)^2} = \frac{1}{(1+x)^2} > 0 \text{ for } x \ne -1.$$

$\therefore\ f'(x) = +$ ve, i.e; $f'(x) > 0$ for $x \ne -1$.

Moreover, $f'(x)$ is undefined at $x = -1$

Hence, $f(x)$ is an increasing function for every value of x excluding $x = -1$, i.e; $f(x)$ is an increasing function for every value of $x \ne -1$, i.e. in $(-\infty, -1) \cup (-1, \infty)$.

12. Show that $y = 6x + 3x^2 + x^3$ is an increasing function of x, $\forall\, x$

Solution: $y = f(x) = 6x + 3x^2 + x^3$

$\Rightarrow f'(x) = 6 + 6x + 3x^2 = 3\left(x^2 + 2x + 1\right)$

$= 3\,(x+1)^2 > 0\ \forall\, x$

$\therefore\ f'(x) = +$ ve $\forall\, x$, i.e; $f'(x) > 0$, $\forall\, x$

Hence, $f(x)$ is an increasing function $\forall\, x$, i.e; $f(x)$ is increasing in $(-\infty, +\infty)$.

13. $f(x) = \dfrac{1}{x}$, Examine the monotonocity.

Solution: $f(x) = \dfrac{1}{x}$, $(x \ne 0)$

$\Rightarrow f'(x) = -\dfrac{1}{x^2}$, $(x \ne 0) < 0$ for $x \ne 0$.

$f'(x) < 0$, $\forall\, x \ne 0$

Hence, $f(x)$ is a decreasing function, $\forall\, x$ excluding $x = 0$ which means $f(x)$ is a decreasing function in $(-\infty, 0)$ and in $(0, \infty)$.

14. Examine the monotonocity of the function $f(x) = 2x + 7$, $\forall\, x$.

Solution: $f(x) = 2x + 7 \Rightarrow f'(x) = 2$ which is +ve $\forall\, x$

Hence, $f(x)$ is an increasing function $\forall\, x$.

15. Show that $f(x) = -5x + 1$ is a decreasing function, $\forall\, x$.

Solution: $f(x) = -5x + 1 \Rightarrow f'(x) = -5$ which is – ve which means $f'(x) < 0\ \ \forall\, x$.

Hence, $f(x)$ is an decreasing function $\forall\, x$.

16. Show that the function $y = x^3 + x$ increases every where.

Solution: $y = f(x) = x^3 + x$

$\Rightarrow f'(x) = 3x^2 + 1$ which is > 0, $\forall\, x$

$\therefore\ f'(x) = +$ ve

Hence, $f(x)$ increases every where.

17. Show that the function $y = \tan^{-1} x - x$ decreases every where.

Solution : $y = f(x) = \tan^{-1} x - x$

$$\Rightarrow f'(x) = \frac{1}{1+x^2} - 1 = \frac{1 - 1 - x^2}{1 + x^2}$$

$$= \frac{-x^2}{1+x^2} < 0, \forall x$$

$\therefore f'(x) = -$ ve

Hence, $f(x)$ decreases every where.

Type 3: Problems based on showing $y = f(x)$ to be increasing or decreasing in an open interval (a, b) by using derivative.

Definations: **1.** A function $f(x)$ is called increasing in an open interval (a, b) if it is increasing at every point within this interval (a, b).

2. A function $f(x)$ is called decreasing in an open interval of it is decreasing at every point within this interval (a, b).

Test for monotonocity in an open interval:

An increase and decrease of a function $y = f(x)$ is tested by the sign of its derivative $f'(x)$. If in some interval (a, b), $f'(x) > 0$, then the function $y = f(x)$ increase in this open interval (a, b) and if $f'(x) < 0$, then the function decrease in this interval (a, b).

Working rule:

1. Find $f'(x)$

2. Determine the sign of $f'(x)$ with the help of given interval (a, b).

3. If $f'(x) = +$ ve in (a, b) then $y = f(x)$ increases in (a, b) and if $f'(x) = -$ ve in (a, b) then $y = f(x)$ decreases in (a, b)

Note: If $f'(x) = T\{f(x)\}$

Where T = any trigonometric function sin, cos tan, cot, sec and cosec.

And $f(x)$ = an expression in x for the angle of trigonometric function which is generally linear in x (i.e $ax + b$).

Then from the given interval, we derive the angle $f(x)$ by using various mathematical manipulations s.t. it determine in which quadrant $f(x)$ lies from which we observe the + ve or – ve sign of trigonometric derived function using the rule of

sin	all
tan	cos

Remember: **1.** Whenever we need to know the sign of a trigonometric function of any angle, we consider the quadrant in which its terminal (moving) side lies.

e.g.: **(i)** Since the terminal side of $20°$ is in the first quadrant, therefore, sin $20° = +$ ve

(ii) Since the terminal side of $220°$ lies in the third quadrant, therefore cos $220 = -$ ve

(iii) The terminal side of $(-60°)$ lies in the fourth quadrant, therfore, sin $(-60°) = -$ ve and cos (-60) $= +$ ve

2. Let $f(x)$ = angle of any trigonometric function sin, cos etc. then,

(i) $0 < f(x) < \frac{\pi}{2}$ means the angle $f(x)$ lies in the Ist quadrant.

(ii) $0 < f(x) < \pi$ means the angle $f(x)$ lies in the Ist and 2nd quadrant.

(iii) $\pi < f(x) < \frac{3\pi}{2}$ means the angle $f(x)$ lies in the 3rd quadrant.

(iv) $\pi < f(x) < 2\pi$ means the angle $f(x)$ lies in the 3rd or 4th quadrant.

(v) $-\frac{\pi}{2} < f(x) < \frac{\pi}{2}$ means the angle $f(x)$ lies in the Ist or 4th quadrant.

Type 3: Problems based on showing $y = f(x)$ to be increasing or decreasing in an open interval.

Examples worked out:

1. Show that: $f(x) = \frac{4x^2 + 1}{x}$

is a decreasing function in the interval $\left(\frac{1}{4}, \frac{1}{2}\right)$ and an increasing function in the interval $\left(\frac{1}{2}, 1\right)$.

Solution: $f(x) = \frac{4x^2 + 1}{x} = 4x + \frac{1}{x}$

$\Rightarrow f'(x) = 4 - \frac{1}{x^2} = \frac{4x^2 - 1}{x^2}$

Now we have to determine the sign of $\frac{4x^2 - 1}{x^2}$ with the help of given intervals.

$\frac{1}{4} < x < \frac{1}{2} \Rightarrow x^2 < \frac{1}{4} \Rightarrow 4x^2 < 1$

$\Rightarrow 4x^2 - 1 < 0 \Rightarrow 4x^2 - 1 = -\text{ve}$ ...(1)

and x^2 being a perfect square, it is always + ve, i.e;

$x^2 = +\text{ve}$ ...(2)

From (1) and (2), we conclude, $\frac{4x^2 - 1}{x^2} = -\text{ ve}$

$f'(x) = -$ ve which means that $f(x)$ is a decreasing function in the given interval $\left(\frac{1}{4}, \frac{1}{2}\right)$

Next, we consider the interval $\left(\frac{1}{2}, 1\right)$ with the help of which we determine the sign of $\frac{4x^2 - 1}{x^2}$

$x > \frac{1}{2} \Rightarrow x^2 > \frac{1}{4} \Rightarrow 4x^2 > 1$

$\Rightarrow 4x^2 - 1 > 0 \Rightarrow 4x^2 - 1 = +\text{ve}$ ...(1)

and x^2 being a perfect square, it is always + ve, i.e;

$x^2 = +\text{ ve}$ ...(2)

From (1) and (2), we conclude, $\frac{4x^2 - 1}{x^2} = +\text{ ve}$

$\therefore f'(x) = +$ ve which means that $f(x)$ is a increasing function in the given interval $\left(\frac{1}{2}, 1\right)$

2. Show that $f(x) = 3x + \frac{1}{3x}$ is a decreasing function in the interval $\left(\frac{1}{9}, \frac{1}{3}\right)$ and an increasing function in the interval $\left(\frac{1}{3}, 1\right)$

Solution: $f(x) = 3x + \frac{1}{3x}$

$\Rightarrow f'(x) = 3 + \frac{1}{3}\left(-\frac{1}{x^2}\right) = 3 - \frac{1}{3x^2}$

Now we have to determine the sine of

$3 - \frac{1}{3x^2} = \frac{9x^2 - 1}{3x^2} = \frac{(3x + 1)(3x - 1)}{3x^2}$

$\frac{1}{9} < x < \frac{1}{3} \Rightarrow x^2 < \frac{1}{9} \Rightarrow 9x^2 < 1$

$\Rightarrow 9x^2 - 1 < 0 \Rightarrow 9x^2 - 1 = -\text{ve}$ ...(1)

$\therefore f'(x) = -$ ve which means that $f(x)$ is a decreasing function in the given interval $\left(\frac{1}{9}, \frac{1}{3}\right)$

Next, we consider the interval $\left(\frac{1}{3}, 1\right)$ with the help of which we consider the sign of $\frac{9x^2 - 1}{3x^2}$

$x > \frac{1}{3} \Rightarrow x^2 > \frac{1}{9} \Rightarrow 9x^2 > 1$

$\Rightarrow 9x^2 - 1 > 0$

$\therefore f'(x) = +$ ve which means that $f(x)$ is an increasing function in the given interval $\left(\frac{1}{3}, 1\right)$.

3. Show that the function $y = 2x^3 + 3x^2 - 12x + 1$ decreases in the interval $(-2, 1)$

Solution: $y = f(x) = 2x^3 + 3x^2 - 12x + 1$

$\Rightarrow f'(x) = 6x^2 + 6x - 12 = 6\left(x^2 + x - 2\right)$

$= 6\left\{x^2 + 2x - x - 2\right\}$

$= 6\left\{x(x+2) - (x+2)\right\}$

$= 6(x-1)(x+2)$

Now we have to determine the sign of $6(x-1)(x+2)$ with the help of given interval.

$-2 < x < 1$

$\Rightarrow x < 1$ and $x > -2$

$x > -2$

$\Rightarrow x + 2 > 0 \Rightarrow x + 2 = +\text{ve}$ ...(i)

$x < 1 \Rightarrow x - 1 < 0 \Rightarrow x - 1 = -\text{ve}$ ...(ii)

and $6 = a$ + ve number ...(iii)

From (i), (ii) and (iii), we conclude that

$6(x-1)(x+2) < 0$

$\therefore f'(x) = -\text{ve}$

Hence, $f(x)$ decreases in the interval $(-2, 1)$

4. Show that the function $y = \sqrt{2x - x^2}$ increases in the interval $(0, 1)$ and decreases in the interval $(1, 2)$.

Solution: $y = f(x) = \sqrt{2x - x^2}$, $0 \le x \le 2$

$$\Rightarrow f'(x) = \frac{1}{2\sqrt{2x - x^2}} \times (2 - 2x)$$

$$= \frac{2(1-x)}{2\sqrt{2x - x^2}} = \frac{(1-x)}{\sqrt{2x - x^2}} = \frac{(1-x)}{\sqrt{x}\left(\sqrt{2-x}\right)}$$

Now, we have to determine the sign of $\dfrac{(1-x)}{\sqrt{x}\left(\sqrt{2-x}\right)}$ with the help of given interval.

$f'(x) = \text{ve}$ for $0 < x < 1$, since $(1-x) > 0$

Hence, $f(x)$ increases in interval $(0, 1)$

For $1 < x < 2$,

$$\frac{(1-x)}{\sqrt{x}\left(\sqrt{2-x}\right)} < 0 \text{ since } (1-x) = -\text{ve}$$

$\therefore f'(x) = -\text{ve}$

Hence, $f(x)$ decreases in $(1, 2)$.

5. Show that $f(x) = \cos x$ is a decreasing function on $\left(0, \dfrac{\pi}{2}\right)$.

Solution: $f(x) = \cos x$

$\Rightarrow f'(x) = -\sin x$

Now, we have to determine the sign of $-\sin x$ with the help of given interval.

$0 < x < \dfrac{\pi}{2}$ which means x lies in the first quadrant where $\sin x > 0$ which further $\Rightarrow -\sin x < 0$.

Thus $f'(x) = -$ ve on $\left(0, \dfrac{\pi}{2}\right)$ which $\Rightarrow f(x)$ is decreasing on $\left(0, \dfrac{\pi}{2}\right)$.

6. Show that $\cos 2x$ is decreasing on $\left(0, \dfrac{\pi}{2}\right)$.

Solution: $f(x) = \cos 2x \Rightarrow f'(x) = -2\sin 2x$

Now, we have to determine the sign of $(-2\sin 2x)$ with the help of given interval.

$0 < x < \dfrac{\pi}{2}$ which $\Rightarrow 0 < 2x < \pi$ which further means $2x$ lies in the Ist or 2nd quadrant where $\sin 2x > 0$ which $\Rightarrow -\sin 2x < 0 \Rightarrow -\sin 2x < 0$ which means $(-2\sin 2x) = -\text{ve}$

Hence, $f'(x) = -$ ve on $\left(0, \dfrac{\pi}{2}\right)$ which $\Rightarrow f(x)$ is a decreasing function on $\left(0, \dfrac{\pi}{2}\right)$

7. Show that $f(x) = \tan x$ is an increasing function on $\left(0, \dfrac{\pi}{2}\right)$.

Solution: $f(x) = \tan x \Rightarrow f'(x) = \sec^2 x$

Now, we have to determine the sign of $\sec^2 x$ with the help of given interval.

$0 < x < \frac{\pi}{2}$ which menas x lies in the first quadrant where $\sec x > 0$ which further $\Rightarrow \sec^2 x > 0$ i.e; $\sec^2 x = +\text{ ve}$

Hence, $f'(x) = +\text{ ve}$ on $\left(0, \frac{\pi}{2}\right)$ which $\Rightarrow f(x)$ is increasing on $\left(0, \frac{\pi}{2}\right)$.

8. Show that $f(x) = \cos\left(2x + \frac{\pi}{4}\right)$ is an increasing function for $\frac{3\pi}{8} < x < \frac{7\pi}{8}$

Solution: $f(x) = \cos\left(2x + \frac{\pi}{4}\right)$

$$\Rightarrow f'(x) = -2\sin\left(2x + \frac{\pi}{4}\right)$$

Now we have to determine the sign of $-2\sin\left(2x + \frac{\pi}{4}\right)$ with the help of given interval.

$f'(x)$ will be +ve when $\sin\left(2x+\frac{\pi}{4}\right)$ is –ve which is only possible when $\left(2x + \frac{\pi}{4}\right)$ lies in the 3rd or 4th quadrant.

Now consider the given interval.

$$\frac{3\pi}{8} < x < \frac{7\pi}{8}$$

$$\Rightarrow \frac{3\pi}{4} < 2x < \frac{7\pi}{4}$$

$$\Rightarrow \frac{3\pi}{4} + \frac{\pi}{4} < 2x + \frac{\pi}{4} < \frac{7\pi}{4} + \frac{\pi}{4}$$

$$\Rightarrow \pi < 2x + \frac{\pi}{4} < \frac{8\pi}{4} = 2\pi$$

$\Rightarrow \pi < 2x + \frac{\pi}{4} < 2\pi$ which means $\left(2x + \frac{\pi}{4}\right)$ lies in the 3rd or 4th quadrant where $\sin\left(2x + \frac{\pi}{4}\right)$ is –ve, ie; $\sin\left(2x+\frac{\pi}{4}\right)<0$ which $\Rightarrow -2\sin\left(2x+\frac{\pi}{4}\right)>0$

Hence, $f'(x) = +\text{ ve}$ on $\left(\frac{3\pi}{8}, \frac{7\pi}{8}\right)$ which

$\Rightarrow f(x)$ is an increasing function on $\left(\frac{3\pi}{8}, \frac{7\pi}{8}\right)$

9. Show that $f(x) = -\frac{\pi}{2} + \sin x$ is an increasing function on $\left(-\frac{\pi}{3}, 0\right)$.

Solution: $f(x) = -\frac{x}{2} + \sin x$

$$\Rightarrow f'(x) = -\frac{1}{2} + \cos x$$

Now we have to determine the sign of $\left(-\frac{1}{2}+\cos x\right)$ for x in the given interval.

$$-\frac{\pi}{3} < x < 0 \Rightarrow \cos\left(-\frac{\pi}{3}\right) < \cos x < \cos 0$$

$$\Rightarrow \frac{1}{2} < \cos x < 1$$

$$\Rightarrow \frac{1}{2} - \frac{1}{2} < -\frac{1}{2} + \cos x < 1 - \frac{1}{2}$$

$$\Rightarrow 0 < -\frac{1}{2} + \cos x < \frac{1}{2}$$

Which means that $\left(\cos x - \frac{1}{2}\right) > 0$ i.e; $\left(\cos x - \frac{1}{2}\right) = +\text{ ve}$

Hence, $f'(x) = +\text{ ve}$ on $\left(-\frac{\pi}{3}, 0\right) \Rightarrow f(x)$ in increasing on $\left(-\frac{\pi}{3}, 0\right)$.

10. Show that $f(x) = \cos\left(2x + \frac{\pi}{4}\right)$ is an increasing function on the interval $\left(\frac{3\pi}{8}, \frac{5\pi}{8}\right)$.

Solution: $f(x) = \cos\left(2x + \frac{\pi}{4}\right)$

$$\Rightarrow f'(x) = -2\sin\left(2x + \frac{\pi}{4}\right)$$

Now we have to determine the sign of $-2\sin\left(2x + \frac{\pi}{4}\right)$ for x in the given interval.

$f'(x)$ will be +ve when $\sin\left(2x + \frac{\pi}{4}\right)$ is –ve.

Which is possible only when $\left(2x + \frac{\pi}{4}\right)$ lies in the 3rd or 4th quadrant, i.e; $\left(2x + \frac{\pi}{4}\right)$ lies on the interval $(\pi, 2\pi)$ or $\left(\pi, \frac{3\pi}{2}\right)$.

Now we consider the given interval.

$$\frac{3\pi}{8} < x < \frac{5\pi}{8}$$

$$\Rightarrow \frac{3\pi}{4} < 2x < \frac{5\pi}{4}$$

$$\Rightarrow \frac{3\pi}{4} + \frac{\pi}{4} < 2x + \frac{\pi}{4} < \frac{5\pi}{4} + \frac{\pi}{4}$$

$\Rightarrow \pi < 2x + \frac{\pi}{4} < \frac{3\pi}{2}$ which means $\left(2x + \frac{\pi}{4}\right)$ lies in the 3rd quadrant where $\sin\left(2x + \frac{\pi}{4}\right) < 0$

which $\Rightarrow -\sin\left(2x + \frac{\pi}{4}\right) > 0$

Hence, $f'(x) = +\text{ ve}$ on $\left(\frac{3\pi}{8}, \frac{5\pi}{8}\right)$ which means $f(x)$ is an increasing function on $\left(\frac{3\pi}{8}, \frac{5\pi}{8}\right)$

11. Show that $\frac{\sin\theta}{\theta}$ is a decreasing function of θ for $0 < \theta < \frac{\pi}{2}$.

Solution: $y = f(\theta) = \frac{\sin\theta}{\theta}$

$$\Rightarrow f'(\theta) = \frac{1}{\theta^2}(\theta\cos\theta - \sin\theta)$$

$$= \frac{\cos\theta}{\theta^2}(\theta - \tan\theta)$$

on letting $Z = \theta - \tan\theta = g(\theta)$

$$g'(\theta) = \frac{dz}{d\theta} = 1 - \sec^2\theta < 0$$

$$\left(\because \sec^2\theta > 1 \text{ for } 0 < \theta < \frac{\pi}{2}\right)$$

Which $\Rightarrow Z$ is a decreasing function for

$$0 \le \theta \le \frac{\pi}{2} \Rightarrow g(\theta) < g(0) = 0,\ 0 < \theta < \frac{\pi}{2} \quad ...(1)$$

$$\text{Also } \frac{\cos\theta}{\theta^2} > 0 \text{ for } 0 < \theta < \frac{\pi}{2} \quad ...(2)$$

From (1) and (2) we conclude that $f'(\theta) < 0$ for $0 < \theta < \frac{\pi}{2}$

Hence, $f(\theta)$ is a decreasing function of θ for $0 < \theta < \frac{\pi}{2}$

12. Show that $\frac{x}{\sin x}$ is an increasing function of x in the interval $0 < x < \frac{\pi}{2}$

Solution: $f(x) = \dfrac{x}{\sin x} \Rightarrow f'(x) = \dfrac{\sin x - x\cos x}{\sin^2 x}$

Now, $\sin^2 x > 0$ for $0 < x < \dfrac{\pi}{2}$ ...(i)

and $\tan x > x$ for $0 < x < \dfrac{\pi}{2}$

Now since $\tan x > x \Rightarrow \dfrac{\sin x}{\cos x} > x \Rightarrow \sin x > x$

$\cos x$

$\Rightarrow \sin x - x\cos x > 0$

$\left(\because \cos x > 0 \text{ for } 0 < x < \dfrac{\pi}{2}\right)$ ...(ii)

From (1) and (2) we conclude that $f'(x) = +$ ve in the given interval $0 < x < \dfrac{\pi}{2}$

Thus $f(x) = \dfrac{x}{\sin x}$ is an increasing function in $\left(0, \dfrac{\pi}{2}\right)$.

13. Prove that $y = \dfrac{4\sin\theta}{(2+\cos\theta)} - \theta$ is an increasing function of θ in the interval $\left(0, \dfrac{\pi}{2}\right)$.

Solution: $y = \dfrac{4\sin\theta}{(2+\cos\theta)} - \theta$

Now differentiating y w.r.t θ and then simplifying, we get $\dfrac{dy}{d\theta} = \dfrac{\cos\theta\,(4-\cos\theta)}{(2+\cos\theta)^2} > 0$ for $\theta \in \left(0, \dfrac{\pi}{2}\right)$

Hence, $y = \dfrac{4\sin\theta}{2+\cos\theta} - \theta$ is an increasing function of θ in $\left(0, \dfrac{\pi}{2}\right)$.

14. Show that $f(x) = \tan^{-1}(\sin x + \cos x)$, $x > 0$ is an increasing function in $\left(0, \dfrac{\pi}{4}\right)$.

Solution: $f(x) = \tan^{-1}(\sin x + \cos x)$

$\Rightarrow f'(x) = \dfrac{1}{1+(\sin x+\cos x)^2} \times (\cos x - \sin x)$

$\Rightarrow f'(x) = \dfrac{\cos x - \sin x}{1 + \sin^2 x + \cos^2 x + 2\sin x\cos x}$

$= \dfrac{\cos x - \sin x}{2 + \sin 2x}$

Now we have to determine the sign of $\dfrac{\cos x - \sin x}{2 + \sin 2x}$ for x in the given interval.

$0 < x < \dfrac{\pi}{4} \Rightarrow 0 < 2x < \dfrac{\pi}{2}$ which $\Rightarrow$ $2x$ lies in the first quadrant where $\sin 2x$ is positive and for this reason $2 + \sin 2x > 0$

Now again $x < \dfrac{\pi}{4} \Rightarrow \tan x < \tan\dfrac{\pi}{4}$

$\Rightarrow \tan x < 1$

$\Rightarrow \dfrac{\sin x}{\cos x} < 1$

$\Rightarrow \sin x < \cos x$, $0 < x < \dfrac{\pi}{4}$

$\Rightarrow \sin x - \cos x < 0$

$\Rightarrow \cos x - \sin x > 0$

Thus, we see that Nr and Dr both > 0 seperately of the derived function $f'(x)$ which means $f'(x) = +$ ve in the interval $\left(0, \dfrac{\pi}{4}\right)$

Hence, $f(x)$ is an increasing function of x in the interval $\left(0, \dfrac{\pi}{4}\right)$.

Type 4: Problems based on showing $y = f(x)$ to be increasing or decreasing in a closed interval $[a, b]$ by using derivative.

Question: How to test monotonocity of a function $y = f(x)$ in a closed interval $[a, b]$.

Answer: we have the following rule to test for monotonocity in a closed interval $[a, b]$.

1. If $f'(x) \geq 0$ in $[a, b]$, then $f(x)$ is said to be non-decreasing on $[a, b]$ and if $f'(x) \leq 0$ in closed interval $[a, b]$ then $f(x)$ is said to be non-increasing in closed interval $[a, b]$ for all $x \in (a, b)$.

2. If $f'(x) > 0$ in $[a, b]$, then $f(x)$ is said to be increasing in $[a, b]$ and if $f'(x) < 0$ in the interval $[a, b]$, then $f(x)$ is said to be decreasing in the interval $[a, b]$ for all $x \in (a, b)$.

Working Rule:

1. Find $f'(x)$.

2. Determine the sign of $f'(x)$ for x in the given closed interval $[a, b]$.

3. If $f'(x) = +$ ve, then $y = f(x)$ increases in the closed interval $[a, b]$ and if $f'(x) = -$ ve, then $y = f(x)$ decreases in the closed interval $[a, b]$.

Type 4: Problems based on showing $y = f(x)$ to be increasing or decreasing in a closed interval $[a, b]$ by using derivative.

Examples worked out:

1. Show that $f(x) = \sin x$ is increasing for $-\frac{\pi}{2} \leq x \leq \frac{\pi}{2}$.

Solution: $f(x) = \sin x$

$$\Rightarrow f'(x) = \cos x$$

Now, we have to determine the sign of $\cos x$ with the help of given interval:

$-\frac{\pi}{2} < x < \frac{\pi}{2}$ which $\Rightarrow$ x lies in the 4th or Ist quadrant where $\cos x$ is + ve

$\therefore f'(x) = +\text{ve}$ (i.e; $f'(x) > 0$) and $f'\left(-\frac{\pi}{2}\right) = \cos\left(-\frac{\pi}{2}\right) = 0$, $f'\left(\frac{\pi}{2}\right) = \cos\left(\frac{\pi}{2}\right) = 0$

Hence, $f(x)$ is increasing in the closed interval $\left[-\frac{\pi}{2}, \frac{\pi}{2}\right]$.

2. Show that $f(x) = \cos x$ is decreasing for $0 \leq x \leq \pi$.

Solution: $f(x) = \cos x$

$$\Rightarrow f'(x) = -\sin x$$

Now, we have to determine the sign of $(-\sin x)$ with the help of given interval.

$0 < x < \pi$ which $\Rightarrow$ x lies in the first and second quadrant whose $\sin x$ is + ve, i.e; $\sin x > 0$

$\Rightarrow -\sin x < 0$

$\therefore f'(x) = -$ ve and $f'(0) = -\sin 0 = 0$, $f'(\pi) = -\sin \pi = 0$

Hence, $f(x)$ is decreasing on $[0, \pi]$.

3. If $y = 2x - \tan^{-1} x - \log\left(x + \sqrt{1 + x^2}\right)$ Show that y is increasing in $[0, +\infty)$.

Solution: $y = 2x - \tan^{-1} x - \log\left(x + \sqrt{1 + x^2}\right)$

$$\frac{dy}{dx} = 2 - \frac{1}{(1 + x^2)} - \frac{1}{\sqrt{1 + x^2}}$$

$$= 2 - \left(\frac{1}{1 + x^2} - \frac{1}{\sqrt{1 + x^2}}\right) > 0 \text{ for } 0 < x < +\infty$$

and $\left(\frac{dy}{dx}\right)_{x=0} = 2 - (1 + 1) = 0$

Hence, y is increasing on $[0, +\infty)$.

Type 5: Problems based on finding the interval of increasing and decreasing function $y = f(x)$.

Question: What do you mean by " interval of increasing and decreasing of the function $y = f(x)$".

Answer: **(i)** The interval in which $f'(x)$ is positive (i.e: $f'(x) > 0$) is called the interval of increasing of the function $y = f(x)$.

(ii) The interval in which $f'(x)$ is negative (i.e; $f'(x) < 0$) is called the interval of decreasing function $y = f(x)$.

Working rule: To find the interval of increasing or decreasing function or the values of x for which the given function $y = f(x)$ is increasing or decreasing, we adopt the following working rule :

1. Find $f'(x)$

2. Put $f'(x) > 0$ to find the interval of an increasing function and solve $f'(x) > 0$ for x whose values determine the required interval of increasing of $y = f(x)$.

3. Put $f'(x) < 0$ to find the interval of decreasing and solve $f'(x) < 0$ for x whose values determine the required interval of decreasing of $y = f(x)$.

Remember: If $f'(x) = a$ quadratic equation in x (i.e algebraic quadratic in x) then the following hints to solve the quadratic inequality $f'(x) > 0$ or $f'(x) < 0$ or $f'(x) \ge 0$ or $f'(x) \le 0$ are very helpful.

1. $x^2 - a^2 < 0 \Leftrightarrow -a < x < a \Leftrightarrow |x| < a$

or $x^2 - a^2 \le 0 \Leftrightarrow -a \le x \le a \Leftrightarrow |x| \le a$

2. $(x - a)(x - b) < 0 \Leftrightarrow a < x < b \ (b > a)$

or $(x - a)(x - b) \le 0 \Leftrightarrow a \le x \le b \ (b > a)$

3. $x^2 - a^2 > 0 \Leftrightarrow x < -a$ or $x > a \Leftrightarrow |x| > a$

which means x lies outside $[-a, a]$ i.e; $x \in (-\infty, -a) \cup (a, +\infty)$

or $x^2 - a^2 \ge a \Leftrightarrow x \le -a$ or $x \ge a \Leftrightarrow |x| \ge a$

which means x lies outside $(-a, a)$ i.e; $x \in (-\infty, -a] \cup [a, +\infty)$

4. $(x-a)(x-b) > 0 \Leftrightarrow x < a$ or $x > b$, $(b > a)$ which means x lies outside the interval $[a, b]$ i.e; $x \in (-\infty, a) \cup (b, +\infty)$

or $(x-a)(x-b) \ge 0 \Leftrightarrow x \le a$ or $x \ge b$, $(b > a)$ which means x lies outside the interval (a, b) i.e; $x \in (-\infty, a] \cup [b, +\infty)$

Note: **1.** If a function is increasing (or, decreasing) in an open interval, then it is also increaing (or, decreasing) in the corresponding closed interval i.e; if $f(x)$ is an increasing (or decreasing) in the open interval $a < x < b$ then it is also increasing (or decreasing) in the closed interval $a \le x \le b$ respeectively.

Type 5: Problems based on finding interval in which a given function $y = f(x)$ increases or decreases.

Examples worked out:

1. Find the interval in which $f(x) = x^2$ increases or decreases.

Solution: $f(x) = x^2 \Rightarrow f'(x) = 2x$

Now, for $f(x)$ to be increasing, $f'(x) > 0$

$$\Rightarrow 2x > 0 \Rightarrow x > 0 \equiv (0, \infty)$$

Hence, the required interval for $f(x) = x^2$ to be increasing is $x \ge 0 = [0, \infty)$

Next, for $f(x)$ to be decreasing, $f'(x) < 0$

$$\Rightarrow 2x < 0 \Rightarrow x < 0 \equiv (-\infty, 0)$$

Hence, the required interval for $f(x) = x^2$ to be decreasing is $x \le 0 \equiv (-\infty, 0]$.

2. Find the interval in which $f(x) = x^2 - 2x$ increases or decreases.

Solution: $f(x) = x^2 - 2x \Rightarrow f'(x) = 2x - 2$

Now, for $f(x)$ to be increasing, $f'(x) > 0$

$$\Rightarrow 2x - 2 > 0 \Rightarrow (x - 1) > 0 \Rightarrow x > 1 = (1, \infty)$$

Hence, the required interval for $f(x) = x^2 - 2x$ to be increasing is $x \geq 1 = [1, \infty)$

Next, for $f(x)$ to be decreasing, $f'(x) < 0$

$$\Rightarrow (x-1) < 0 \Rightarrow x < 1 = (-\infty, 1)$$

Hence, the required interval for $f(x) = x^2 - 2x$ to be decreasing is $x \leq 1 = (-\infty, 1]$.

3. Find the interval in which the function $f(x) = 2x^3 - 15x^2 + 36x - 57$ is an increasing function of x and a decreasing function of x.

Solution: $f(x) = 2x^3 - 15x^2 + 36x - 57$

$$\Rightarrow f'(x) = 6x^2 - 30x + 36$$

$$= 6(x^2 - 5x + 6)$$

$$= 6(x-2)(x-3)$$

Now, for $f(x)$ to be increasing, $f'(x) > 0$

$$\Rightarrow 6(x-2)(x-3) > 0$$

$$\Rightarrow (x-2)(x-3) > 0$$

$\Rightarrow x < 2$ or $x > 3$ which means x lies outside the interval [2, 3] i.e; $x \in (-\infty, 2) \cup (3, \infty)$

Hence, the required interval for

$$f(x) = 2x^3 - 15x^2 + 36x - 57$$

to be increasing is $(-\infty, 2] \cup [3, \infty)$

Next, for $f(x)$ to be decreasing, $f'(x) < 0$

$$\Rightarrow (x-2)(x-3) < 0$$

$$\Rightarrow 2 < x < 3 = (2, 3)$$

Hence, the required interval for

$$f(x) = 2x^3 - 15x^2 + 36x - 57$$

to be decreasing is $2 \leq x \leq 3 = [2, 3]$

4. Find the interval in which the function $f(x) = x^3 - 6x^2 + 9x + 1$ is increasing or decreasing.

Solution: $f(x) = x^3 - 6x^2 + 9x + 1$

$$\Rightarrow f'(x) = 3x^2 - 12x + 9$$

Now, $f(x)$ will be increasing provided $f'(x) > 0$

$$\Rightarrow 3x^2 - 12x + > 0$$

$$\Rightarrow x^2 - 4x + 3 > 0$$

$$\Rightarrow x^2 - x - 3x + 3 > 0$$

$$\Rightarrow x(x-1) - 3(x-3) > 0$$

$$\Rightarrow (x-1)(x-3) > 0$$

$$\Rightarrow x < 1 \text{ or } x > 3$$

which means x lies outside the interval [1, 3] i.e;

$$x \in (-\infty, 1) \cup (3, \infty)$$

Hence, the required interval for

$$f(x) = x^3 - 6x^2 + 9x + 1$$

to be increasing is $(-\infty, 1] \cup [3, \infty)$

Next, $f(x)$ will be decreasing provided $f'(x) < 0$

$$\Rightarrow (x-1)(x-3) < 0$$

$$\Rightarrow 1 < x < 3 = (1, 3)$$

Hence, the required interval for

$$f(x) = x^3 - 6x^2 + 9x + 1$$

to be decreasing is $1 < x < 3 = [1, 3]$

5. Find the interval in which the function $f(x) = \tan x - 4(x-1)$ is increasing and decreasing,

$$-\frac{\pi}{2} < x < \frac{\pi}{2}.$$

Solution: $f(x) = \tan x - 4(x-1)$

$$\Rightarrow f'(x) = \sec^2 x - 4$$

Now for $f(x)$ to be increasing, $f'(x) > 0$

i.e; $\sec^2 x - 4 > 0$

$$\Rightarrow \sec^2 x > 4$$

$$\Rightarrow \sec x > 2 \text{ as } -\frac{\pi}{2} < x < \frac{\pi}{2}$$

$\Rightarrow x < -\dfrac{\pi}{3}$ or $x > \dfrac{\pi}{3}$ which means x lies outside the interval $\left[-\dfrac{\pi}{3}, \dfrac{\pi}{3}\right]$ i.e;

$$x \in \left(-\frac{\pi}{2}, -\frac{\pi}{3}\right) \cup \left(\frac{\pi}{3}, \frac{\pi}{2}\right)$$

Hence, the required interval for

$f(x) = \tan x - 4(x - 1)$

to be increasing is $\left(-\frac{\pi}{2}, -\frac{\pi}{3}\right) \cup \left(\frac{\pi}{3}, \frac{\pi}{2}\right)$

Next, $f(x)$ will be decreasing provided $f'(x) < 0$

$\Rightarrow \sec^2 x - 4 < 0$

$\Rightarrow \sec^2 x < 4$

$\Rightarrow \sec x < 2$

$\Rightarrow -\frac{\pi}{3} < x < \frac{\pi}{3} = \left(-\frac{\pi}{3}, \frac{\pi}{3}\right)$

Hence, the required interval for

$f(x) = \tan x - 4(x - 1)$

to be decreasing is $\left[-\frac{\pi}{3}, \frac{\pi}{3}\right]$.

Type 6: *Important facts*: The following facts based on the rule of testing of monotonocity of a function are useful to find the interval in which a given function $y = f(x)$ increases or decreases (If $f'(0)$ does not exist).

1. If $f'(x) < 0$, $\forall x$, $x \neq 0$ then $f(x)$ decreases in $(-\infty, 0) \cup (0, \infty)$

2. If $f'(x) > 0$, $\forall x$, $x \neq 0$ then $f(x)$ increases in $(-\infty, 0) \cup (0, \infty)$

3. If $f(x)$ is undefined for $x \leq 0$, then $f'(x)$ is also undefined for $x \leq 0$ and hence only test for $f'(x) > 0$ is required to find the interval in which $f(x)$ increases for $x \nleq 0$.

4. If $f'(x) > 0$, $\forall x$, then $f(x)$ increases in the entire number line $(-\infty, \infty)$

5. If $f'(x) < 0$, $\forall x$, then $f(x)$ decreases in the entire number line $(-\infty, \infty)$.

These facts are illustrated by the following examples:

Examples worked out:

1. Find the interval in which $f(x) = \frac{1}{2x}$, $(x \neq 0)$ decreases.

Solution: $f(x) = \frac{1}{2x}$

$\Rightarrow f'(x) = -\frac{1}{2x^2}$

Now $f(x)$ will be decreasing provided $f'(x) < 0$

i.e; $\frac{-1}{2x^2} < 0$ which is true (because $x^2 = +$ ve for all +ve and –ve values of x) except $x = 0$ where the function is undefined.

Hence, $f(x)$ decreases in $(-\infty, 0) \cup (0, +\infty)$

2. Find the interval in which $f(x) = \frac{x-2}{x+1}$, $(x \neq -1)$ increases.

Solution: $f(x) = \frac{x-2}{x+1}$

$\Rightarrow f'(x) = \frac{(x+1)\cdot 1 - (x-2)\cdot 1}{(x+1)^2} = \frac{3}{(x+1)^2}$

Now, $f(x)$ will be increasing provided $f'(x) > 0$

i.e; $\frac{3}{(x+1)^2} > 0$ which is always true for all +ve and –ve values of x $(\neq -1)$.

Hence, the function $f(x)$ increases in

$(-\infty, -1) \cup (-1, \infty)$.

3. Find the interval in which the function $f(x) = \log x$, $(x > 0)$ increases.

Solution: $f(x) = \log x \Rightarrow f'(x) = \frac{1}{x}$, $(x > 0)$

Now, $f(x)$ will be increasing provided $f'(x) > 0$

i.e; $\frac{1}{x} > 0 \Rightarrow x > 0$ which is true since we are given $x > 0$.

Hence, $f(x)$ increases in the interval $(0, +\infty)$.

Note: $f'(x)$ is undefined for $x \leq 0$ because $f(x) = \log x$ is undefined for $x \leq 0$.

4. Find the interval in which the function $f(x) = e^{ax}$, $(a > 0)$ increases.

Solution: $f(x) = e^{ax} \Rightarrow f'(x) = a e^{ax}$

Now, $f(x)$ will be increasing provided $f'(x) > 0$

i.e; $a e^{ax} > 0$ which is always true for all +ve and –ve values of x and $a > 0$.

Hence, the function $f(x)$ increases in the entire number line $(-\infty, \infty)$.

5. Find the interval in which the function $f(x) = e^{-ax}$, $(a > 0)$ decreases.

Solution: $f(x) = e^{-ax} \Rightarrow f'(x) = -a e^{-ax}$

Now, $f(x)$ will be decreasing provided $f'(x) < 0$

i.e; $-a e^{-ax} < 0$ which is true because e^{-ax} is always +ve for all +ve and –ve values of x and $a > 0$.

Hence, the function $f(x)$ decreases in the entire number line $(-\infty, +\infty)$.

Category C:

Problems based on proving inequality:

Type 1: To show $f_1(x) > f_2(x)$ or $f_1(x) < f_2(x)$ in a given interval.

Working Rule:

1. Let $f(x) = f_1(x) - f_2(x)$

2. Examine whether $f'(x) > 0$ or $f'(x) < 0$

3. Use the following facts:

(a) If $f'(x) > 0$, then $f(x) > f(a)$ for $x > a$ and $f(x) < f(a)$ for $x < a$

(b) If $f'(x) < 0$, then $f(x) < f(a)$ for $x > a$ and $f(x) > f(a)$ for $x < a$

Examples worked out:

1. Show that $\log(1 + x) > x - \dfrac{x^2}{2}$ if $0 < x < 1$

Solution: Let $f(x) = \log(1 + x) - x + \dfrac{x^2}{2}$

$$f'(x) = \frac{1}{1+x} - 1 + x = \frac{x^2}{1+x}$$

Now, $\dfrac{x^2}{1+x} > 0$ for every value of $x > 0$

which $\Rightarrow f'(x) > 0$ for $0 < x < 1$

$\Rightarrow f(x) = \log(1+x) - x + \dfrac{x^2}{2}$ is an increasing function in the interval $0 \le x \le 1$

$$\therefore f(x) > f(0) \text{ for } 0 < x < 1$$

$$\Rightarrow \log(1 + x) - x + \frac{x^2}{2} > 0$$

$$(\because f(0) = \log(1 + 0) - 0 + 0 = 0)$$

$$\Rightarrow \log(1 + x) > x - \frac{x^2}{2} \text{ for all } x \text{ in } (0, 1).$$

2. Show that $1 + x\log\left\{x + \sqrt{1+x^2}\right\} \ge \sqrt{1+x^2}$ for $x \ge 0$.

Solution: Let $f(x) = 1 - \sqrt{1+x^2} + x\log\left\{x + \sqrt{1+x^2}\right\}$

$$\therefore f'(x) = \frac{-2x}{2\sqrt{1+x^2}} + x \cdot \frac{1}{x+\sqrt{1+x^2}}\left\{1 + \frac{2x}{2\sqrt{1+x^2}}\right\} + \log\left\{x + \sqrt{1+x^2}\right\}$$

$$= \frac{-x}{\sqrt{1+x^2}} + \frac{x}{\sqrt{1+x^2}} + \log\left\{x + \sqrt{1+x^2}\right\}$$

$$= \log\left(x + \sqrt{1+x^2}\right) > 0 \text{ for all } x > 0$$

$$\therefore f(x) \ge f(0) = 0$$

$$\Rightarrow 1 + x\log\left\{x + \sqrt{1+x^2}\right\} - \sqrt{1+x^2} \ge 0$$

$$\Rightarrow 1 + x\log\left\{x + \sqrt{1+x^2}\right\} \ge \sqrt{1+x^2} \quad \forall x \ge 0$$

3. Show that $2\sin x + \tan x \ge 3x$ when $0 \le x < \dfrac{\pi}{2}$.

Solution: Let $f(x) = 2\sin x + \tan x - 3x$

$\therefore f'(x) = 2\cos x + \sec^2 x - 3$

$= \dfrac{2\cos^3 x - 3\cos^2 x + 1}{\cos^2 x}$

Now, we have to determine the sign of $2\cos^3 x - 3\cos^2 x + 1$ because $\cos^2 x$ being a perfect square it is always +ve.

$\because 2\cos^3 x - 3\cos^2 x + 1 = (1-\cos x)^2 (2\cos x + 1)$

which is +ve for all values of x lying in the given interval $0 < x < \frac{\pi}{2}$

$\therefore f'(x) = \dfrac{2\cos^3 x - 3\cos^2 x + 1}{\cos^2 x} \geq 0$ for $0 \leq x < \frac{\pi}{2}$

Which $\Rightarrow f'(x) \geq 0$ when $0 \leq x < \frac{\pi}{2}$

$\therefore f'(x) \geq 0$ for $0 \leq x < \frac{\pi}{2} \Rightarrow f(x) \geq f(0)$

$\therefore 2\sin x + \tan x - 3x \geq 0$

$(\because f(0) = 2\sin 0 + \tan 0 - 3 \times 0 = 0)$

$\Rightarrow 2\sin x + \tan x \geq 3x$ for $0 \leq x < \frac{\pi}{2}$

4. Show that $\sin x + \tan x > 2x$ for $0 < x < \frac{\pi}{2}$

Solution: Let $f(x) = \sin x + \tan x - 2x$ when

$0 \leq x < \frac{\pi}{2}$

$\therefore f'(x) = \cos x + \sec^2 x - 2$

$= \dfrac{\cos^3 x - 2\cos^2 x + 1}{\cos^2 x} > 0$

when $0 < x < \frac{\pi}{2}$ because $\cos^2 x$ being a perfect square it is always +ve and $\cos^3 x - 2\cos^2 x + 1$

$= (\cos^2 x - \cos x + 1)(\cos x - 1)$

$= \{(\cos x - 1)^2 + \cos x\}(\cos x - 1)$

and $\cos x < 1 \Rightarrow \cos x - 1 > 0 \quad (\because -1 < \cos x < 1)$

Thus each factor $\{(\cos x - 1)^2 + \cos x\}$ and $(\cos x - 1) > 0$ for $0 \leq x < \frac{\pi}{2}$

$\therefore f'(x) > 0$ which $\Rightarrow f(x)$ is an increasing function in $0 \leq x < \frac{\pi}{2}$.

$\therefore f(x) > f(0)$ for $0 < x$

$\Rightarrow \sin x + \tan x - 2x > 0$

$(\because f(0) = \sin 0 + \tan 0 - 2 \times 0 = 0)$

$\Rightarrow \sin x + \tan x > 2x$.

Type I:

5. Show that $\tan x > x$ whenever $0 < x < \frac{\pi}{2}$

Solution : $f(x) = \tan x - x$

$\Rightarrow f'(x) = \sec^2 x - 1 = \tan^2 x > 0$ for

$0 < x < \frac{\pi}{2}$

$\Rightarrow f'(x) > 0$

$\Rightarrow f(x) = \tan x - x$ is an increasing function for $0 < x < \frac{\pi}{2}$

$\Rightarrow f(x) > 0 \qquad (\because f(0) = \tan 0 - 0 = 0)$

$\Rightarrow \tan x - x > 0$

$\Rightarrow \tan x > x$ for $0 < x < \frac{\pi}{2}$

Type 2: To show $f_1(x) > f_2(x) > f_3(x)$ in a given interval

Working rule: It consists of two parts:

First Part:

1. Let $f(x) = f_1(x) - f_2(x)$

2. Examine whether $f'(x) > 0$ for the values of x in the given interval.

3. If $f'(x) > 0$, $f(x)$ is increasing in the given interval and then use the following fact:

$f(x) > f(a)$ for $x > a$ and put $f(x) = f_1(x) - f_2(x)$ in $f(x) > f(a)$ which will provide us the inequality

$f_1(x) > f_2(x)$ provided that $f(a) = 0$

Second part:

Similarly we show $f_2(x) > f_3(x)$

From first and second part, we have

$f_1(x) > f_2(x) > f_3(x)$ in the given interval

Examples worked out:

1. Show that $\dfrac{x}{1+x} < \log(1+x) < x$ for $x > 0$

Solution: Let $f(x) = \log(1+x) - \dfrac{x}{1+x}$

$$f'(x) = \frac{1}{1+x} - \frac{1}{(1+x)^2} = \frac{x}{(1+x)^2} > 0$$

for $x > 0$

$\Rightarrow f'(x)$ is + ve for all $x > 0$

$\Rightarrow f(x) = \log(1+x) - \dfrac{x}{(1+x)}$ is an increasing function for all $x \geq 0$

$\Rightarrow f(x) > f(0) = \log 1 = 0$ for $x > 0$

$\Rightarrow \log(1+x) - \dfrac{x}{1+x} > 0$

$\Rightarrow \log(1+x) > \dfrac{x}{1+x}$

$\Rightarrow \dfrac{x}{1+x} < \log(1+x)$ ...(1)

Again let $g(x) = x - \log(1+x)$

$$g'(x) = 1 - \frac{1}{1+x} = \frac{1+x-1}{1+x}$$

$= \dfrac{x}{1+x} > 0$ for all $x > 0$

Which $\Rightarrow g(x)$ is an increasing function for all $x \geq 0 \Rightarrow g(x) > g(0)$ for $x > 0$

$\Rightarrow g(x) > 0 \quad (\because g(0) = 0 - \log 1 = 0)$

$\Rightarrow x - \log(1+x) > 0 \Rightarrow x > \log(1+x)$ ...(2)

From (1) and (2) we declear

$\dfrac{x}{1+x} < \log(1+x) < x$ which is the required inequality.

2. Show that if $0 < x < \dfrac{\pi}{2}$, then $\tan x > x > \sin x$.

Solution : $f(x) = \tan x - x$

$\Rightarrow f'(x) = \sec^2 x - 1 = \tan^2 x > 0 \ \forall x \in \left(0, \dfrac{\pi}{2}\right)$

$\Rightarrow f(x) = \tan x - x$ is an increasing function in $\left[0, \dfrac{\pi}{2}\right)$

$\Rightarrow f(x) > f(0)$ for $0 < x < \dfrac{\pi}{2}$

$\Rightarrow f(x)$ is +ve in $\left(0, \dfrac{\pi}{2}\right)$

$(\because f(0) = \tan 0 - 0 = 0)$

$\Rightarrow \tan x - x > 0$ for $0 < x < \dfrac{\pi}{2}$ ...(1)

Again $g(x) = x - \sin x$

$\Rightarrow g'(x) = 1 - \cos x > 0$ for $0 < x < \dfrac{\pi}{2}$

$(\because -1 < \cos x < 1 \Rightarrow \cos x < 1 \Rightarrow 1 - \cos x > 0)$

$\Rightarrow g'(x) > 0$

$\Rightarrow g(x) = x - \sin x$ is an increasing function in the interval $\left[0, \dfrac{\pi}{2}\right)$

$\Rightarrow g(x) > g(0)$ for $0 < x < \dfrac{\pi}{2}$

$\Rightarrow g(x) > 0 \quad (\because g(0) = 0 - \sin 0 = 0)$

$\Rightarrow x - \sin x > 0 \ \forall x \in \left(0, \frac{\pi}{2}\right)$

$\Rightarrow x > \sin x$ ...(2)

From (1) and (2), we have,

$\tan x > x > \sin x$ for $0 < x < \frac{\pi}{2}$ which is the required inequality.

3. If $x > 0$ show that $x - \frac{x^3}{3} < \tan^{-1} x < x$.

Solution: $f(x) = x - \tan^{-1} x$

$\Rightarrow f'(x) = 1 - \frac{1}{1+x^2} = \frac{x^2}{1+x^2}$

$\Rightarrow f'(x) > 0$ for every value of $x > 0$

$\Rightarrow f(x) = x - \tan^{-1} x$ is an increasing function $\forall x \geq 0$

$\Rightarrow f(x) > f(0)$ for $x > 0$

$\Rightarrow f(x) > 0 \ \left(\because f(0) = 0 - \tan^{-1} 0 = 0\right)$

$\Rightarrow x - \tan^{-1} x > 0$

$\Rightarrow x > \tan^{-1} x, \forall x > 0$ ...(1)

Again $g(x) = \tan^{-1} x - x + \frac{x^3}{3}$

$\Rightarrow g'(x) = \frac{1}{1+x^2} - 1 + x^2 = \frac{x^4}{1+x^2} > 0$ for $x > 0$

$\Rightarrow g(x) = \tan^{-1} x - x + \frac{x^3}{3}$ is an increasing function for $x \geq 0$

$\Rightarrow g(x) > g(0), \forall x > 0$

$\Rightarrow g(x) > 0 \quad \left(\because g(0) = \tan^{-1} 0 - 0 + \frac{0}{3} = 0\right)$

$\Rightarrow \tan^{-1} x - x + \frac{x^3}{3} > 0$

$\Rightarrow \tan^{-1} x > x - \frac{x^3}{3}$ ...(2)

From (1) and (2), we have

$x > \tan^{-1} x > x - \frac{x^3}{3}$

$\Rightarrow x - \frac{x^3}{3} < \tan^{-1} x < x$ for all $x > 0$ which is the required inequality.

4. Show that for $x > 0$, $x - \frac{1}{2} x^2 < \log(1 + x) < x$

Solution: $f(x) = x - \frac{1}{2} x^2 - \log(1 + x)$

$\Rightarrow f'(x) = 1 - x - \frac{1}{1+x} = \frac{1 - x^2 - 1}{1+x} = \frac{-x^2}{1+x}$

$\Rightarrow f'(x) < 0$ if $x > 0$

$\Rightarrow f(x)$ is a decreasing function for $x \geq 0$

$\Rightarrow f(x) < f(0) \ x > 0$

$\Rightarrow f(x) < 0 \ (\because f(0) = 0 - 0 - \log 1 = 0)$

$\Rightarrow x - \frac{1}{2} x^2 < \log(1 + x)$ ...(1)

Again, $g(x) = \log(1 + x) - x$

$\Rightarrow g'(x) = \frac{1}{1+x} - 1 = \frac{1 - 1 - x}{1+x} = \frac{-x}{1+x}$

$\Rightarrow g'(x) < 0$ if $x > 0$

$\Rightarrow g(x)$ is a decreasing function for $x > 0$

$\Rightarrow g(x) < g(0)$ for $x > 0$

$\Rightarrow g(x) < 0$

$(\because g(0) = \log(1 + 0) - 0 = \log 1 = 0)$

$\Rightarrow \log(1 + x) - x < 0$

$\Rightarrow \log(1 + x) < x$ ...(2)

From (1) and (2), we have

$x - \frac{1}{2} x^2 < \log(1 + x) < x$ which is the required inequality.

Type 3: Problems based on choosing a specified function to prove a given inequality in a given interval

When in the given inequality to be proved in a given interval, an independent variable is not x but the independent variable is α, β or θ ect. we adopt the following working rule:

Working rule:

1. Choose a specified function $y = f(x)$ where x is in a given interval.

2. Find $f'(x)$.

3. Examine whether $f(x)$ is an increasing function or a decreasing function in the given interval.

4. If $f(x)$ is an increasing function, then the same function of given independent varibale α, β or θ etc is also an increasing function (i.e; $f(\theta), f(\alpha)$ or $f(\beta)$ is also increasing corresponding to increasing function $f(x)$ in the given interval) and if $f(x)$ is a decreasing function, then the same function of given independent variable α, β or θ etc. is also a decreasing function (i.e; $f(\alpha), f(\beta)$ or $f(\theta)$ is also decreasing corresponding to decreasing function $f(x)$ in the given interval.)

5. Lastly using various mathematical manipulations, we prove the required inequality in the given interval.

Examples worked out:

1. If $0<\alpha<\beta<\frac{\pi}{2}$, show that $\alpha - \sin\alpha < \beta - \sin\beta$.

Solution: $f(x) = x - \sin x,\ x \in \left(0, \frac{\pi}{2}\right)$

$\Rightarrow f'(x) = 1 - \cos x$ ...(1)

We know that if $0 < x < \frac{\pi}{2} \Rightarrow 0 < \cos x < 1$

$\Rightarrow 1 - \cos x > 0 \Rightarrow f'(x) > 0$

Thus, $f(x)$ is an increasing function in the interval $\left(0, \frac{\pi}{2}\right)$.

$\therefore \alpha, B \in \left(0, \frac{\pi}{2}\right)$ and $\alpha < \beta \Rightarrow f(\alpha) < f(\beta)$

$\Rightarrow \alpha - \sin\alpha < \beta - \sin\beta$ which is the required inequality.

2. Prove that if $0<\theta<\frac{\pi}{2}$, then

$\cos(\sin\theta) > \sin(\cos\theta)$

Solution: $f(x) = x - \sin x,\ x \in \left(0, \frac{\pi}{2}\right)$

$\Rightarrow f'(x) = 1 - \cos x$

We know that $0 < x < \frac{\pi}{2} \Rightarrow 0 < \cos x < 1$

$\Rightarrow 1 - \cos x > 0 \Rightarrow f'(x) > 0$

$\therefore f(x)$ is an increasing function in $\left[0, \frac{\pi}{2}\right]$

$\therefore f(\theta)$ is also increasing for $0 \le \theta \le \frac{\pi}{2}$

$\therefore f(\theta) > f(0)$ for $\theta > 0$

$\therefore f(\theta) > 0\ \ (\because f(0) = 0 - \sin 0 = 0 - 0 = 0)$

$\Rightarrow \theta - \sin\theta > 0$

$\Rightarrow \sin\theta < \theta \Rightarrow \cos(\sin\theta) > \cos\theta$

$\Rightarrow -\cos(\sin\theta) < -\cos\theta$ ...(2)

again $f(x) > f(0) = 0$ for $x > 0$ $\left(\text{or}, \forall x \in \left(0, \frac{\pi}{2}\right)\right)$

$\therefore f(\cos\theta) > 0$ *for* $\forall \theta \in \left(0, \frac{\pi}{2}\right)$

$\Rightarrow \cos\theta - \sin(\cos\theta) > 0$

$\Rightarrow \sin(\cos\theta) < \cos\theta$ ...(3)

$(3) - (2) = \sin(\cos\theta) - \cos(\sin\theta) < \cos\theta - \cos\theta$

$= \sin(\cos\theta) - \cos(\sin\theta) < 0$ which

$\Rightarrow \sin(\cos\theta) < \cos(\sin\theta)$

3. Prove that $\frac{2}{x} < \frac{\sin x}{x} < 1$ whenever $0 < x < \frac{\pi}{2}$.

Solution: Let a continuous function $f(x)$ be defined in the following way.

$$f(x) = \frac{\sin x}{x}, \text{ if } x \neq 0$$

$$= 1, \text{ if } x = 0$$

When $x \neq 0$, we get $f'(x) = \dfrac{x\cos x - \sin x}{x^2}$

$$= (x - \tan x)\cos x \cdot \frac{1}{x^2}$$

Now, we have to examine the sign of

$$f'(x) = \frac{(x - \tan x)\cos x}{x^2}$$

x^2 being a perfect square, it is always + ve and $\cos x$ is + ve in the first quadrant. Moreover, if

$$0 < x < \frac{\pi}{2},$$

$$\tan x > x \Rightarrow \tan x - x > 0 \Rightarrow x - \tan x < 0$$

which means $(x - \tan x)$ is negative in $0 < x < \dfrac{\pi}{2}$.

Hence, $f'(x) < 0 \Rightarrow f(x)$ is a decreasing function in $0 < x < \dfrac{\pi}{2}$

$$\therefore f(0) > f(x) > f\left(\frac{\pi}{2}\right) \text{ for } 0 < x < \frac{\pi}{2}$$

$$\Rightarrow 1 > \frac{\sin x}{x} > \frac{2}{x} \quad \left\{\because f(0) = 1 \; f\left(\frac{\pi}{2}\right) = \frac{2}{\pi}\right\}$$

$\Rightarrow \dfrac{2}{\pi} < \dfrac{\sin}{x} < 1$ which is the required inequality.

Type 1: Problems based on examining or showing a function of an independent variable to be increasing or decreasing at a point $x = a$. (*Category B*)

Exercise 21.1

1. Examine whether the following functions is increasing or decreasing at the points indicated.

(i) $f(x) = 2x + 3$, at $x = 0$

(ii) $f(x) = 5 - 3x$, at $x = -1$

(iii) $f(x) = x^3$, at $x = 2$

(iv) $f(x) = x^3$, at $x = -2$

(v) $f(x) = \dfrac{1}{x}$, at $x = -1$

(vi) $f(x) = x + \dfrac{1}{x}$, at $x = \dfrac{1}{2}$

2. Prove that $y = \dfrac{1}{2}x^2 + \cos \pi x$ is increasing at $x = 1$ while decreasing at $x = -1$.

Answers:

1. (i) Increasing **(ii)** Decreasing **(iii)** Increasing **(iv)** Increasing **(v)** Decreasing **(vi)** Decreasing

Type 2: Problems based on showing the function $y = f(x)$ to be increasing or decreasing in the infinite interval

(i) $x > \pm a$ **(ii)** $x < \pm a$ **(iii)** $x \geq \pm a$ **(iv)** $x \leq \pm a$ **(v)** $x > 0$ **(vi)** $x < 0$ **(vii)** $x \geq 0$ **(viii)** $x \leq 0$ **(ix)** $x \neq \pm a$ **(x)** $\forall x$

Exercise 21.2

1. Show that following functions are increasing in the indicated intervals:

(i) $f(x) = \dfrac{1}{1 + x^2}, x \leq 0$

(ii) $f(x) = ax + b, a > 0, \forall x \in R$

(iii) $f(x)\, 6 + 9x - x^2, x \geq \dfrac{9}{2}$

2. Show that following functions are decreasing in the indicated intervals:

(i) $f(x) = -7x^2 + 11x - 9, x > 1$

(ii) $f(x) = x^{-5} + \dfrac{1}{2}x^{-3} + 1, x > 0$

3. Show that $y = \dfrac{\log x}{x}$ decreases for $x > e$.

4. Show that the function $y = \log(1 + x) - \dfrac{2x}{2 + x}$ is an increasing function of x for all values of $x > 0$.

5. Prove that the function $y=2x^3-15x^2+36x+10$ is an increasing function of x for $x>3$ and a decreasing function of x for $x<2$.

6. Show that the function $y=\dfrac{x}{\sqrt{1-x}}-\log(1+x)$ increases for all values of $x>0$.

7. Prove that $y=2x^3+3x^2-12x-7$ is an increasing fucntion when $x>1$.

8. Show that $f(x)=\dfrac{x}{1+x}$ increases for all $x\neq-1$.

9. Show that $f(x)=\dfrac{(x^2-1)}{x}$ increases for all $x\neq 0$.

Type 3: Problems based on determination of values of x for which the given function $y=f(x)$ is increasing or decreasing.

Exercise 21.3

1. Determine the values of x for which the function $y=\dfrac{x-2}{x+1}$, $x\neq-1$ is increasing or decreasing.

2. Determine the values of x for which the function $f(x)=5x^{\frac{3}{2}}-3x^{\frac{5}{2}}$, $x>0$ is increasing or decreasing.

3. Determine the values of x for which the following functions are increasing and for which they are decreasing.

(i) $f(x)=6x^2-2x+1$

(ii) $f(x)=-3x^2+12x+8$

(iii) $f(x)=x^8+6x^2$

(iv) $f(x)=x+\dfrac{1}{x}$, $x\neq 0$

4. Find the values of x for which the following functions are

(i) increasing **(ii)** dDecreasing **(iii)** stationary

(a) $3-x+x^3$ **(b)** x^3-3x+2 **(c)** x^4-2x^2+1

Type 4: Problems based on showing (or, examining) the given function to be increasing or decreasing in the interval (a, b) or $[a, b]$.

Exercise 21.4

1. Show that

(i) $f(x)=\sin x$ is increasing for $-\dfrac{\pi}{2}\leq x\leq\dfrac{\pi}{2}$.

(ii) $f(x)=\cos x$ is decreasing for $0\leq x\leq\pi$.

2. Show that the function $f(x)=x^3+\dfrac{1}{x^3}$ is decreasing in $0<x<1$.

3. Prove that the function $f(x)=\log\sin x$ is increasing on $\left(0,\dfrac{\pi}{2}\right)$ and decreasing on $\left(\dfrac{\pi}{2},\pi\right)$.

4. Prove that the function $f(x)=\sin x$ is increasing in the interval $\left(0,\dfrac{\pi}{2}\right)$ and decreasing in the interval $\left(\dfrac{\pi}{2},\pi\right)$.

5. Show that the function $f(x)=\dfrac{4x^2+1}{x}$ is a decreasing function in the interval $\left(\dfrac{1}{4},\dfrac{1}{2}\right)$.

6. Show that $f(x)=\tan^{-1}(\sin x+\cos x)$, $x>0$ is always increasing in the interval $\left(0,\dfrac{\pi}{4}\right)$.

7. Show that $f(x)=\dfrac{x}{\sin x}$ is an increasing function in the interval $0<x<\dfrac{\pi}{2}$.

8. Find the least value of 'a' such that the function $f(x)=x^2+ax+1$ is increasing on $[1, 2]$.

9. Determine whether the following functions are always increasing or decreasing in the indicated interval.

(i) $f(x)=-\dfrac{x}{2}+\sin x$ in $-\dfrac{\pi}{3}<x<\dfrac{\pi}{3}$

(ii) $f(x)=\cos\left(2x+\dfrac{\pi}{4}\right)$ in $-\dfrac{3\pi}{8}<x<\dfrac{5\pi}{8}$

(iii) $f(x) = \tan x - 4x$ in $-\frac{\pi}{3} < x < 0$

10. Determine whether $f(x) = \cos\left(2x + \frac{\pi}{4}\right)$ is increasing or decreasing for $\frac{3\pi}{8} < x < \frac{7\pi}{8}$.

11. Prove that $y = \log x - \tan^{-1} x$ increases in $(0, \infty)$.

12. Show that $y = \frac{\sin x}{x}$ is decreasing function from 0 to $\frac{\pi}{2}$.

13. Show that $y = \frac{\tan x}{x}$ is an increasing function in the range $0 < x < \frac{\pi}{2}$.

14. Prove that the function $f(x) = x^2 - x + 1$ is neither increasing nor decreasing in the interval (0, 1).

15. Show that the function $f(x) = \log x$ is increasing for $0 < x < \infty$.

16. Show that $f(x) = x^2 + \frac{1}{x^3}$ is decreasing in $0 < x < 1$.

17. One of which the following intervals is the function $f(x) = x^{100} + \sin x - 1$ in creasing?

(a) (–1, 1) (b) (0, 1) (c) $\left(\frac{\pi}{2}, \pi\right)$ (d) $\left(0, \frac{\pi}{2}\right)$.

18. Which of the following functions are increasing on $\left(0, \frac{\pi}{2}\right)$.

(a) $\cos x$ (b) $\cos 2x$ (c) $\cos 3x$ (d) $\tan x$

Answers:

8. Least value of $a = -4$ for $f(x) = x^2 + ax + 1$ to be increasing on [1, 2].

17. (a) $f(x)$ is neither increasing nor decreasing in (–1, 1).

(b) $f(x)$ is an increasing function in (0, 1).

(c) $f(x)$ is increasing in $\left(\frac{\pi}{2}, \pi\right)$.

(d) $f(x)$ is increasing on $\left(0, \frac{\pi}{2}\right)$.

18. The function in parts (a) and (b) are decreasing.

Type 5: Problems based on finding the intervals in which the given functions increase or decrease.

Exercise 21.5

1. Find the intervals on which the following functions are increasing and those in which the functions are decreasing.

(i) $f(x) = -x^2 + 3x + 4$

(ii) $f(x) = x^4 + 2x^3$

(iii) $f(x) = x^{\frac{2}{3}}\left(x + \frac{1}{3}\right)^{\frac{1}{3}}$

(iv) $f(x) = 3x^{\frac{1}{2}} - x^{\frac{3}{2}}$

(v) $f(x) = x^3 - 3x + 2$

(vi) $f(x) = x + \frac{1}{x}$

(vii) $f(x) = (x-2)^2(x+3)$

(viii) $f(x) = x^4 - 18x^2$

(ix) $f(x) = \frac{x-2}{x+2}$, $x \neq -2$

(x) $f(x) = x(x+1)(x+2)$

(xi) $f(x) = \frac{x}{1+x^2}$

(xii) $f(x) = \begin{cases} 2x+9, & \text{when } x \le -2 \\ x^2+1, & \text{when } x > -2 \end{cases}$

(xiii) $f(x) = \sin x$, $x \in (0, \pi)$

(xiv) $f(x) = \sin x + \cos x,\ x \in \left(0, \frac{\pi}{2}\right)$

(xv) $f(x) = x^3(1 + x)$

2. Find the intervals in which the function $f(x) = x^4 - 2x^2$ decreases.

3. Find the intervals in which the function $f(x) = x^3 + 2x^2 - 1$ decreases.

4. Find the intervals in which the function $f(x) = x^3 - 3x$ increases and decreases.

5. Determine the intervals in which the function $f(x) = 2x^3 - 24x + 7$ increases or decreases.

6. Determine the intervals where $f(x) = \sin x - \cos x$, $0 \le x \le 2\pi$ is increasing or decreasing.

7. Find the intervals in which $f(x)\ 6x^2 - 2x + 1$ increases or decreases.

8. Determine the intervals in which the function $f(x)\ 5x^2 + 7x - 13$ is increasing or decreasing.

9. Find the intervals in which the function $f(x) = \cos\left(2x + \frac{\pi}{4}\right)$, $0 \le x \le \pi$ is increasing or decreasing. Find also the points on the graph of the function at which the tangents are parallel to x-axis.

10. Determine the intervals in which the function $f(x) = (x - 1)(x + 2)^2$ is increasing or decreasing. At what points are the tangents to the graph of the function, parallel to x-axis?

11. Find the intervals in which the function $f(x) = \frac{x^3}{3} + \frac{x^2}{2} - 2x + 1$ is increasing or decreasing. At what points are the tangents to the graph of the function parallel to x-axis.

12. Find the interval in which the function $f(x) = 2x^3 - 9x^2 + 12x + 30$ is increasing or decreasing.

Answers:

1. (i) $\left(-\infty, \frac{3}{2}\right)$ increasing, $\left[\frac{3}{2}, \infty\right)$ decreasing.

(ii) $\left(-\infty, -\frac{3}{2}\right)$ decreasing, $\left[-\frac{3}{2}, \infty\right)$ increasing.

(iii) $\left(-\infty, -\frac{2}{9}\right)$ decreasing, $\left[-\frac{2}{9}, \infty\right)$ increasing.

(iv) $[-\infty, 1]$ increasing, $[1, \infty)$ decreasing.

(v) Increasing when $x < -1$ and $x > 1$, decreasing in $-1 < x < 1$.

(vi) Increasing when $x < -1$ and $x > 1$, decreasing in $-1 < x < 1$.

(vii) Increasing when $x < -\frac{4}{3}$ and $x > 2$, decreasing in $-\frac{4}{3} < x < 2$.

(viii) Increasing when $-3 < x < 0$ and $x > 3$, decreasing when $x < -3$ and $0 < x < 3$.

(ix) Increasing in the domain of definition.

(x) Increasing in $|x + 1| > \frac{1}{\sqrt{3}}$, decreasing for $|x + 1| < \frac{1}{\sqrt{3}}$.

(xi) Increasing in $-1 < x < 1$, decreasing in $x < -1$ or $x > 1$.

(xii) Increasing for $x \le -2$ and $x \ge 0$, decreasing in $-2 < x < 0$.

(xiii) Increasing in $\left(0, \frac{\pi}{2}\right)$, decreasing in $\left(\frac{\pi}{2}, \pi\right)$.

(xiv) Increasing in $\left(0, \frac{\pi}{4}\right)$, decreasing in $\left(\frac{\pi}{4}, \frac{\pi}{2}\right)$.

(xv) Increasing when $x > -\frac{3}{4}$, decreasing when $x < -\frac{3}{4}$

2. $f(x)$ is increasing for $-1 < x < 0, x < 1$ and decreasing for $x < -1, 0 < x < 1$.

3. $f(x)$ is increasing for $x < -\frac{4}{3}, x > 0$ and decreasing for $\left(-\frac{10}{3} < x < 0\right)$ as well as $f(x)$ is decreasing in $\left(-\frac{4}{3}, 0\right)$.

4. $f(x)$ is increasing for $x.1$ or $x < -1$; Decreasing for $-1 < x < 1$.

5. $f(x)$ is increasing for $x \geq 2$ and for $x \leq -2$; $f(x)$ is decreasing for $-2 \leq x \leq 2$.

6. $f(x)$ is increasing when $0 \leq x \leq \frac{3\pi}{4}$ or $\frac{7\pi}{4}$; $f(x)$ decreases when $\frac{3\pi}{4} \leq x \leq \frac{7\pi}{4}$.

7. Increasing for $x \geq \frac{1}{6}$; Decreasing for $x \leq \frac{1}{6}$.

8. Increasing for $x \geq -\frac{7}{10}$; Decreasing for $x \leq -\frac{7}{10}$.

9. Increasing for $\frac{3\pi}{8} \leq x \leq \frac{7\pi}{8}$; Decreasing for $\frac{-\pi}{8} \leq x \leq \frac{3\pi}{8}$ as well as the required points at $x = \frac{3\pi}{8}$ and $x = \frac{7\pi}{8}$.

10. Increasing for $x \geq 0$ or $x \leq -2$; Decreasing for $-2 \leq x \leq 0$ as well as the required points at (0, –4) and (–2, 0).

11. Increasing for $x \leq -2$ or $x \geq 1$; Decreasing for $-2 \leq x \leq 1$.

12. Increasing for $x \leq 1$ or $x \geq 2$; Decreasing for $1 \leq x \leq 2$.

Type 6: Problems based on showing the function $y = f(x)$ to be increasing or decreasing for all values of x (or, in every interval).

Exercise 21.6

1. Show that the function f defined on R by $f(x) = x^3 + 3x^2 + 3x - 8$ is increasing in every interval.

2. Prove that $y = 2x^3 + 4x$ is increasing for all values of x.

3. Show that the exponential function $y = e^x$ is increasing for all x.

4. Show that $y = \tan^{-1} x$ is an increasing function of x for all x.

5. Determine whether the following functions are increasing or decreasing for stated values of x.

(i) $f(x) = x - \cos x$, for all x.

(ii) $f(x) = x + \sin x$, for all x.

6. Show that $f(x) = 3x + 1$ is an increasing function on R.

7. Prove that $f(x) = ax + b$ is an increasing function for all real values of x, where a and b are constant and $a > 0$.

8. Prove that the function $f(x) = x^3 - 3x^2 + 3x - 100$ is increasing on R.

9. Show that the function $y = x^3 - 3x^2 + 6x - 8$ increases while the function $y = 3 - x^3$ decreases for all x.

10. If $y = 3x - 3x^2 + x^3$, show that y always increases whatever the value of x.

11. Show that the function $y = \frac{x^3}{3} + x - \frac{4}{3}$ is an increasing function for all values of x.

12. If $f(x) = (x - 1)e^x + 1$, show that $f(x)$ is positive for all positive values of x.

Type 7: Problems based on proving an inequality a given interval. (*Category C*)

Exercise 21.7

1. If $f(x) = 1 - \frac{1}{2}x^2 - \cos x$, show that when x is positve, $f'(x)$ in negative. Hence, deduce that for a positive values of x, $1 - \frac{1}{2}x^2 < \cos x < 1$

2. Show that when x is positive, $x - \frac{1}{6}x^2 < \sin x < x$.

$\left[\textit{Hint}\text{: Take } f(x) = x - \frac{1}{6}x^2 - \sin x \text{ and prove that } f'(x) \text{ is negative; note that } f(0) = 0 \right]$

3. Prove that $\cos x - 1 + \frac{x^2}{2} > 0$ for $x > 0$. Is this true for $x < 0$?

4. Show that for all x, $x \in R$, $x^3 + 3 > 2x$

$\left[\textit{Hint}: f(x) = x^3 - 2x + 3 \Rightarrow f'(x) = 3x^2 - 2 > 0 \text{ for every value of } x \right]$

5. If $x > 0$ Show that

(a) $\log_e (1 + x) < x$

(b) $\log_e (1 + x) > \frac{x}{1 + x}$

[Hint: (a) $f(x) = x - \log_e (1 + x), x > 0$
(*b*) $f(x) = \log_e (1 + x) - \frac{x}{1 + x}, x > 0$
then find $f'(x)$ for (a) and (b) which will be positive for $x > 0$]

6. Show that

(i) $\tan^{-1} x \leq$ if $x \geq 0$

(ii) $2x \tan^{-1} x > \log_e \left(1 + x^2\right)$ if $x > 0$

(iii) $(x - 1) e^x + 1 > 0$ for all x

(iv) $\log_e (1 + x) > \frac{\tan^{-1} x}{1 + x}$, if $x > 0$

7. Show that $1 + x \log \left(x + \sqrt{x^2 + 1} \right) \geq \sqrt{1 + x^2}$ for all $x \geq 0$

8. If $0 < x < \frac{\pi}{2}$, show that $\cos x > 1 - \frac{x^2}{2}$

9. If $ax^2 + \frac{b}{x} \geq c$, for all positive x, where $a > 0$ and $b > 0$ show that $27\, ab^2 \geq 4c^3$.

10. Prove the following :

(i) $\log (1 + x) > \frac{\tan^{-1} x}{1 + x}$, $(x > 0)$

(ii) $e^x - x > 1$, $(x \neq 0)$

(iii) $\sin x + \tan x > 2x \left(0 < x < \frac{\pi}{2}\right)$

(iv) $\cot \frac{x}{2} \geq 1 + \cot x \ (0 < x < \pi)$

(v) $2 vx + \frac{1}{x} > 3 \ (x > 1)$

(vi) $\log x > \frac{2(x - 1)}{x + 1} \ (x > 1)$

(vii) $x - \frac{x^3}{3} < \sin x < x \left(0 < x \leq \frac{\pi}{2}\right)$

(viii) $\tan x > x + \frac{x^3}{3} \left(0 < x < \frac{\pi}{2}\right)$

(ix) $x^2 > (1 + x) [\log (1 + x)]^2 \ (x > 0)$

(x) $0 < \alpha < \beta < \frac{\pi}{2} \Rightarrow \frac{\tan \beta}{\tan \alpha} > \frac{\alpha}{\beta}$

(xi) $ax + \frac{b}{x} \geq c, (\forall x > 0) \Rightarrow ab \geq \frac{c^2}{4}$ where a, b, c are constants.

22

Maxima and Minima

Maxima and Minima of a Function

We shall define:

1. Local (or relative) maxima and local (or relative) minima.

2. Absolute (or global) maxima and absolute (or global) minima.

Each one is defined in various ways:

1. Local (or regional or relative) maxima of a function:

***Definition* (i)**: (In terms of neighbourhood): A function *f* defined by $y = f(x)$ on its domain D is said to have local (or relative) maximum value (or simply a local (or relative) maximum) at an interior point $x = c$ of the domain of the function *f* namely $D \Leftrightarrow \exists\ N_\delta\ (c)$:

$$\forall\ x \in N_\delta (c) - \{c\} \Rightarrow f(c) > f(x)$$

That is, there is a δ-neighbourhood of the interior point $x = c$ of the domain of the function *f* denoted by $N_\delta\ (c)$ such that the value of the function *f* at the interior point $x = c$ denoted by $f(c)$ is greater than the values of the function *f* at all values of the independent variable x which lie in the δ-deleted neighbourhood of the interior point $x = c$ of the domain of the function denoted by $N_\delta\ (c) - \{c\}$.

***Definition* (ii)**: (In terms of distance): A function *f* defined by $y = f(x)$ on its domain D is said to have a local (or relative) maximum value (or simply a local (or relative) maximum) at an interior point $x = c$ of the domain of the function *f* namely $D \Leftrightarrow \exists\ a\ \delta > 0$:

$$0 < |x - c| < \delta \Rightarrow f(c) > f(x)$$

That is, there is a positive number δ such that the value of the function *f* at an interior point $x = c$ denoted by $f(c)$ is greater than the values of the function *f* at every value of the independent variable x whose distance (or difference) from the interior point $x = c$ of the domain of the function *f* is non zero and less than the positive number δ.

2. Local (or regional or relative) minima of a function:

***Definition* (i)**: (In terms of neighbourhood): A function *f* defined by $y = f(x)$ on its domain D is said to have a local (or relative) minimum value (or simply a local (or relative) minimum) at an interior point $x = c$ of the domain of the function *f* namely $D \Leftrightarrow \exists\ a\ N_\delta\ (c)$:

$$\forall\ x \in N_\delta (c) - \{c\} \Rightarrow f(c) < f(x)$$

That is, there is a δ-neighbourhood of the interior point $x = c$ of the domain of the function *f* denoted by $N_\delta\ (c)$ such that the value of the function *f* at the interior point $x = c$ denoted by $f(c)$ is less than the values of the function *f* at all the values of the independent variable x which lie in the δ-deleted neighbourhood of the interior point $x = c$ of the domain of the function *f* denoted by $N_\delta\ (c) - \{c\}$.

***Definition* (ii)**: (In terms of distance): A function *f* defined by $y = f(x)$ on its domain D is said to have a local (or relative) minimum value (or simply a local (or relative) minimum) at an interior point $x = c$ of the domain of the function *f* namely $D \Leftrightarrow \exists\ a\delta > 0$:

$$0 < |x - c| < \delta \Rightarrow f(c) < f(x)$$

That is, there is a positive number δ such that the value of the function f at an interior point $x=c$ denoted by $f(c)$ is less than the values of the function f at every value of the independent variable x whose distance (or difference) from the interior point $x=c$ of the domain of the function f is non zero and less than the positive number δ.

Now the definitions of absolute (or global) maxima and absolute (or global) minima of a function are presented.

1. Absolute (or universal or global) maxima of a function:

Definition: A function f defined by $y=f(x)$ on its domain D is said to have an absolute (or global) maximum value (or simply a maximum) at a point $x=c$ $\in D \Leftrightarrow \forall\ x \in C \Rightarrow f(c) \geq f(x), \forall x \in D(f)$

That is, the value of the function f at the point $x=c$ in its domain D denoted by $f(c)$ is not less than the values of the function f at any value of the independent variable x which is not c and is in its domain D.

2. Absolute (or universal or global) minima of a function:

Definition: A function f defined by $y=f(x)$ on its domain D is said to have an absolute (or global) minimum value (or simply a minimum) at a point $x=c$ $\in D \Leftrightarrow \forall\ x \in D \Rightarrow f(c) \leq f(x), \forall x \in D(f)$

That is, the value of the function f at the point $x=c$ in its domain D denoted by $f(c)$ is not greater than the value of the function f at any value of the independent variable x which is not c and is in its domain D.

Notes: **A**: **(i)** Maximum and / minimum values are often termed as extrema (plural of extermum).
(ii) Plural of maximum is either maxima or maximums.
(iii) Plural of minimum is either minima or minimums.
(iv) Plural of extremum is only extreme.

(B): **(i)** It is not necessary that the given function $y=f(x)$ should always have the maxima and / minima.

e.g. $y=x^3$, $y=\cot x$, $y=a^x$ and $y=ax+b$ do not have either a maximum and / a minimum.

(ii) There may be several local maxima and local minima which occur alternatively in case of a continuous function, i.e. the maximum and /minimum values occur alternatively in a continuous function. e.g. $y=\sin x$ and $y=\cos x$ have the maximum and the minimum points alternatively.

But in a discontinuous function, the local maximum and the local minimum points may or may not occur alternatively.

(iii) A function $y=f(x)$ may have only one maximum value.

e.g.: $y=60x-x^2$

(iv) A function $y=f(x)$ may have only one minimum value.

e.g.: $y = 2x + \dfrac{72}{x}$

(v) A function $y=f(x)$ may have both the maximum and the minimum values.

e.g.: $y=x^3-x^2-8x+2$

On Points of Local Exterma

1. The point of local maximum: A point $x=c$ in the domain of the function $y=f(x)$ at which the value of the function is a local maximum value of the function is called the point of local maximum (or simply the point of maximum or maximum point) of the function $y=f(x)$.

2. The point of local minimum: A point $x=c$ in the domain of the function $y=f(x)$ at which the value of the function is a local minimum value of the function is called the point of local minimum (or, simply the point of minimum or minimum point) of the function $y=f(x)$.

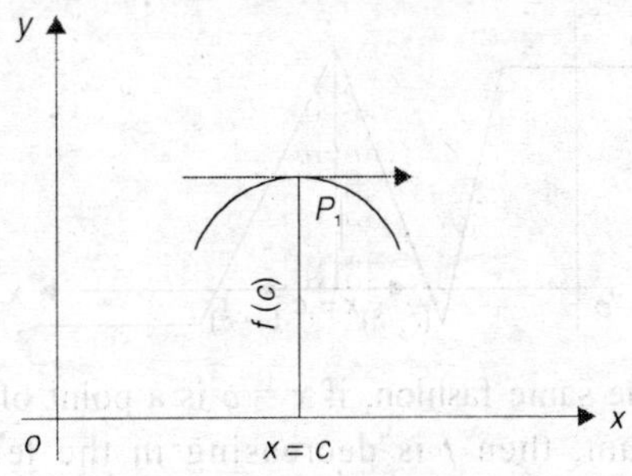

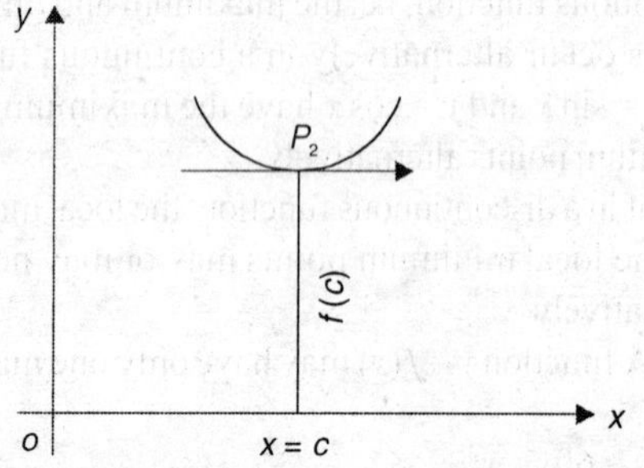

Geometrically, a point $x = c$ in the domain of a function $y = f(x)$ is a point of local maxima or local minima according as the graph of the function f has a peak (crest) or cavity (trough) at the point $x = c$.

Here P_1 is a min. point and P_2 is a max. point.

Local Extreme Values at Critical Points

In fact, to find out local extreme values of a function defined in its domain, critical points are considered which are of two kinds:

1. Points at which the first derivative does not exist: If $x = c$ is a critical point where $f'(c)$ is undefined and $x = c$ is a point of local maximum, then f is increasing in the left δ-neighbourhood of c (i.e. f is increasing $\forall x \in (c - \delta, c]$) and f is decreasing in the right δ-neighbourhood of c (i.e. is decreasing $\forall x \in [c, c + \delta)$), i.e., δ changes sign from positive to negative as x passes through 'c'.

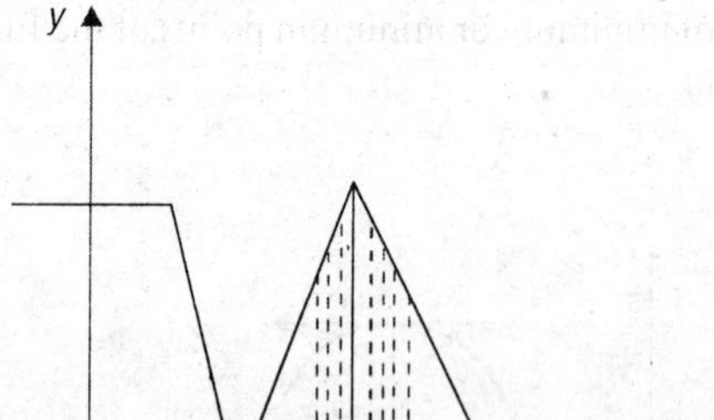

In the same fashion, if $x = c$ is a point of local minimum, then f is decreasing in the left δ-neighbourhood of c (i.e. f is decreasing $\forall x \in (c - \delta, c]$) and f is increasing in the right δ-neighbourhood of c (i.e. f is increasing $\forall x \in [c, c + \delta)$), i.e. $f'(x)$ changes sign from negative to positive as x passes through 'c'.

2. Points at which the first derivative is zero: If $x = c$ is a critical point at which $f'(c)$ exists and $f'(c) = 0$ and $x = c$ is a point of local maximum, then f is increasing in the left δ-neighbourhood of c and f is decreasing in the right δ-neighrbourhood of c, i.e. $f'(x)$ changes sign from positive to negative as x passes through c.

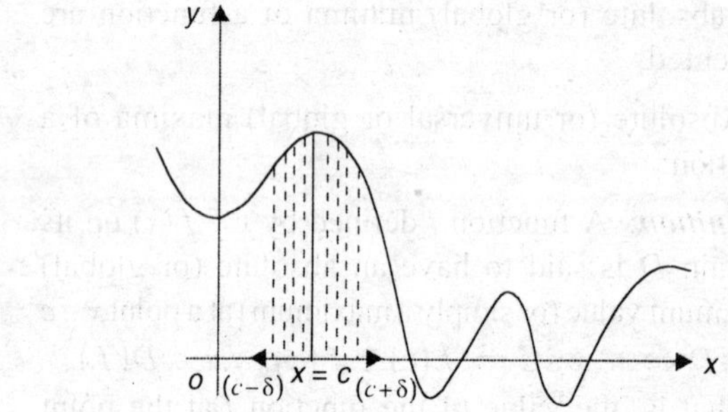

In the same fashion, if $x = c$ is a point of local minimum, then it is decreasing in the left δ-neighbourhood of c and f is increasing in the right δ-neighbourhood of c, i.e. $f'(x)$ changes sign from negative to positive as x passes through 'c'.

Notes: **1.** $y_{\max} = \max \cdot f(x) =$ maximum value of $f(x)$
$= [y]_{x=c}$, where 'c' = a point of maximum

2. $y_{\min} = \min \cdot f(x) =$ minimum value of $f(x)$
$= [y]_{x=c}$, where 'c' = a point of minimum.

Necessary condition for the maximum or the minimum value

Theorem: If $y = f(x)$ defined on its domain has a maximum value at $x = c$ and $f'(c)$ exists, then $f'(c) = 0$ where c is an interior point of the domain of the function f.

Proof: Let $y = f(x)$ be a real valued function of the independent variable x whose domain is the interval D.

Hypothesis: **1.** $f(x)$ has the maximum value at the interior point $x = c$ in the domain D, i.e.,

$\exists$ a δ: $\forall x \in N_\delta(c) - \{c\} \Rightarrow f(x) < f(c)$

$\exists$ a δ: $c - \delta < x < c \Rightarrow f(x) < f(c)$ ...(i)

and $\exists$ a δ: $c < x < c + \delta \Rightarrow f(x) < f(c)$ ...(ii)

2. $f'(c)$ exists $\Rightarrow Lf'(c) = Rf'(c) = f'(c)$

To prove: $f'(c) = 0$

Main proof: From (i), it is seen that

$f(x) < f(c)$ for $c - \delta < x < c$

i.e., $f(x) - f(c) < 0$ for $c - \delta < x < c$

$$\Rightarrow \frac{f(x) - f(c)}{(x - c)} > 0$$

[Since $c - \delta < x < c \Rightarrow -\delta < x - c < 0$]

$$\Rightarrow \lim_{x \to c} \left[\frac{f(x) - f(c)}{(x - c)}\right] \geq 0$$

$$\Rightarrow Lf'(c) \geq 0 \qquad \ldots(A)$$

From (ii), it is seen that

$f(x) < f(c)$ for $c < x < c + \delta$

i.e., $f(x) - f(c) < 0$ for $c < x < c + \delta$

$$\Rightarrow \frac{f(x) - f(c)}{(x - c)} < 0$$

[Since $c < x < c + \delta \Rightarrow 0 < x - c < \delta$]

$$\Rightarrow \lim_{x \to c} \left[\frac{f(x) - f(c)}{(x - c)}\right] \leq 0$$

$$\Rightarrow Rf'(c) \leq 0 \qquad \ldots(B)$$

On putting $Lf'(c) = Rf'(c) = f'(c)$ [from the hypothesis (2)] in (A) and (B), it is found that

$f'(c) \geq 0$ ...(A_1)

and $f'(c) \leq 0$ ...(B_1)

Hence, from (A_1) and (B_1), it is concluded that and

$\left.\begin{matrix} f'(c) \geq 0 \\ f'(c) \geq 0 \end{matrix}\right\} \Leftrightarrow f'(c) = 0$ which was required to be proved.

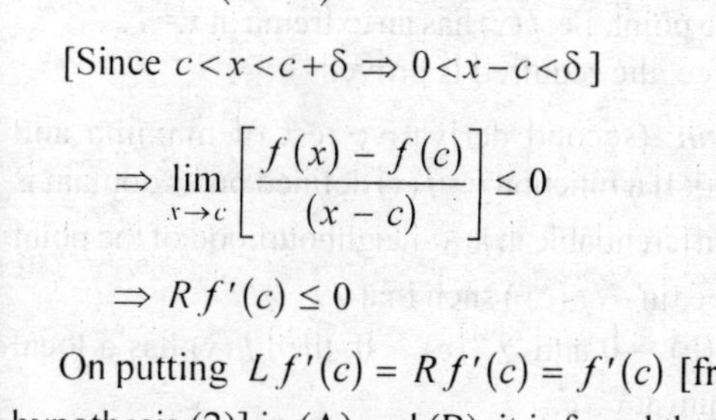

Notes: **1.** The above theorem is also true when $y = f(x)$ has a minimum value at $x = c$ in the domain of f and $f'(c)$ exists.

2. $f'(c)$ exists $\Leftrightarrow$ l.h.d = r.h.d $\Leftrightarrow$ a common finite value, i.e. $f'(c)$ exists $\Leftrightarrow Lf'(c) = Rf'(c) = f'(c)$

3. $x \leq y$ and $x \geq y \Leftrightarrow x \not> y$ and $x \not< y \Leftrightarrow x = y$

On the Language of Calculus

Let $y = f(x)$ be a function of the independent variable x whose domain contains a point $x = c$.

If there exists an $\in > 0$ such that $c - \in < x < c \Rightarrow f(x) > 0 / f'(x) > 0 / f''(x) > 0$ and $c < x < c + \in \Rightarrow f(x) < 0 / f'(x) < 0 / f''(x) < 0$ then it is said that $f(x) / f'(x) / f''(x)$ changes sign from positive to negative at $x = c$ as (when or while) x passes through c from left to right.

Similarly, if there exists an $\in > 0$ such that $c - \in < x < c \Rightarrow f(x) < 0 / f'(x) < 0 / f''(x) < 0$ and $c < x < c + \in \Rightarrow f(x) > 0 / f'(x) > 0 / f''(x) > 0$ then it is said that $f(x) / f'(x) / f''(x)$ changes sign from negative to positive at $x = c$ as (when or while) x passes through c from left to right.

Sufficient criteria for the maxima and minima

Theorem (first derivative test or rule of change of sign of first derivative): If a function $y = f(x)$ defined in its domain is differentiable in a δ-deleted neighbourhood of the point $x = c$ (i.e. in $N_\delta(c) - \{c\}$ and $f'(c) = 0$, then

(a) $\forall x \in N_\delta(c) - \{c\}$, $x < c \Rightarrow f'(x) > 0$ and $x > c \Rightarrow f'(x) < 0$ i.e. $f(x)$ changes sign from positive to negative (i.e. from plus to minus) $\Leftrightarrow f(x)$ has a maximum at $x = c$.

(b) $\forall x \in N_\delta(c) - \{c\}$, $x < c \Rightarrow f'(x) < 0$ and $x > c \Rightarrow f'(x) > 0$ i.e. $f(x)$ changes sign from negative to positive (i.e. from minus to plus) $\Leftrightarrow f(x)$ has a minimum at $x = c$.

Proof: ***Verse part***:

It is given that $f'(c) = 0$, then of course 'c' is a critical point of the function.

Also, in (a), it is given that $x<c \Rightarrow f'(x)>0$ which further

$\Rightarrow f(x)$ is increasing on the left of c

$\Rightarrow f(x)$ is increasing in a left δ-deleted neighbourhood of c.

$\Rightarrow f(x)$ is increasing in $(c-\delta, c)$.

$\Rightarrow f(x) \le f(c), \forall x \in (c-\delta, c)$ ...(i)

Again, in (a), it is given that $x>c \Rightarrow f'(x)<0$ which further

$\Rightarrow f(x)$ is decreasing on the right of c.

$\Rightarrow f(x)$ is decreasing in a right δ-deleted neighbourhood of c.

$\Rightarrow f(x)$ is decreasing in $(c, c+\delta)$

$\Rightarrow f(x) \le f(c), \forall x \in (c, c+\delta)$ ...(ii)

Hence, from (i) and (ii), it is concluded that

$f(x) \le f(c), \forall x \in (c-\delta, c) \cup (c, c+\delta)$

i.e. $f(c)$ is greater than every value of the function at every value of $x \in (c-\delta, c) \cup (c, c+\delta)$ which

$\Rightarrow f(c)$ is maximum in $x \in (c-\delta, c) \cup (c, c+\delta)$

$\Rightarrow f(x)$ has a maximum at $x = c$.

Next, accordingly as in (b), it is given that $x<c \Rightarrow f'(x)<0$ which $\Rightarrow f(x)$ is decreasing on the left of c.

$\Rightarrow f(x)$ is decreasing on the left δ-deleted neighbourhood of c.

$\Rightarrow f(x)$ is decreasing in $(c-\delta, c)$

$\Rightarrow f(x) \ge f(c), \forall x \in (c-\delta, c)$ ...(iii)

Also, in (b), it is given that $x>c \Rightarrow f'(x)>0$ which

$\Rightarrow f(x)$ is increasing on the right of c.

$\Rightarrow f(x)$ is increasing on the right δ-deleted neighbourhood of c.

$\Rightarrow f(x)$ in increasing in $(c, c+\delta)$

$\Rightarrow f(x) \ge f(c), \forall x \in (c, c+\delta)$ ...(iv)

Thus, from (iii) and (iv), it is concluded that

$f(x) \ge f(c), \forall \in (c-\delta, c) \cup (c, c+\delta)$

i.e. $f(c)$ is less than every value of the function at every value of $x \in (c-\delta, c) \cup (c, c+\delta)$

$\Rightarrow f(c)$ is minimum in $(c-\delta, c) \cup (c, c+\delta)$

$\Rightarrow f(x)$ has a minimum at $x = c$.

Converse part:

Hypothesis: $f(x)$ has a local extrema at $x = c$.

To prove: $f'(x)$ changes sign at $x = c$.

Main proof: The claim that $f'(x)$ does not change sign at $x=c \Rightarrow \exists$ a $\delta>0$ such that $f'(x)$ has the same sign in $(c-\delta, c) \cup (c, c+\delta)$, i.e. $f'(x)$ is either positive or negative $\forall x \in (c-\delta, c) \cup (c, c+\delta)$. On supposing that $f'(x)\ \forall x \in (c-\delta, c) \cup (c, c+\delta) \Rightarrow f(x)$ increasing in $(c-\delta, c) \cup (c, c+\delta) \Rightarrow x=c$ is not an extreme point of $y = f(x)$ which is absurd because it is given that $x = c$ is extreme point, i.e. $f(x)$ has an extrema at $x = c$.

Hence, the required is proved.

Theorem: (second derivative test of maxima and minima): If a function $y = f(x)$ defined on its domain is twice differentiable in a δ-neighbourhood of the point $x = c$ (i.e. in $N_\delta(c)$) such that

(i) $f'(c) = 0$ and $f''(c) > 0$, then $f(x)$ has a local maximum at $x = c$.

(ii) $f'(c) = 0$ and $f''(c) < 0$, then $f(x)$ has a local maximum at $x = c$.

Proof: **(i)** Hypothesis: $y = f(x)$ is twice differentiable in $N_\delta(c)$.

$f'(c) = 0$ and $f''(c) > 0$

To prove: $f(x)$ has local minimum at $x = c$.

Main proof: $f'(c)=0 \Rightarrow c$ is a critical point of $y = f(x)$. $y = f(x)$ is twice differentiable in $N_\delta(c) \Rightarrow f'(x)$ exists in $(c-\delta, c+\delta)$, for some $\delta > 0$.

Further $f''(c) > 0$

$\Rightarrow f'(x)$ is increasing in $(c-\delta, c+\delta)$

$\Rightarrow f'(x) < f'(c)$ for $c-\delta<x<c$ ...(1)

and $f'(x) > f'(c)$ for $c<x<c+\delta$ ...(2)

But it is given that $f'(c) = 0$.

Hence, from (1) and (2), it is concluded that

$f'(x) < 0$ for $c - \delta < x < c$ ...(3)

and $f'(c) > 0$ for $c < x < c + \delta$ ...(4)

$\therefore$ (3) and (4) $\Rightarrow f'(x)$ changes sign from –ve to +ve as one moves from left to right in the δ-neighbourhood of $\Rightarrow f(x)$ has a local minima at $x = c$.

(ii) ***Hypothesis***: $y = f(x)$ is twice differentiable in $N_\delta(c)$

$f'(c) = 0$ and $f''(c) < 0$

To prove: $f(x)$ has a local maxima at $x = c$.

Main proof: $f'(c) = 0 \Rightarrow c$ is a critical point of $f(x)$

$f(x)$ is twice differentiable in a $N_\delta(c)$.

$\Rightarrow f''(x)$ exists in $(c - \delta, c + \delta)$

Further, $f''(c) < 0$

$\Rightarrow f'(x)$ is decreasing in $(c - \delta, c + \delta)$

$\Rightarrow f'(x) > f'(c)$ for $c - \delta < x < c$ ...(1)

and $f'(x) < f'(c)$ for $c < x < c + \delta$ ...(2)

But it is given that $f'(c) = 0$

Hence, from (1) and (2) it is concluded that

$f'(x) > 0$ for $c - \delta < x < c$ ...(3)

and $f'(c) < 0$ for $c < x < c + \delta$ ...(4)

$\therefore$ (3) and (4) $\Rightarrow f'(x)$ changes sign from +ve to –ve as one moves from left to right in the δ-neighbourhood of $\Rightarrow f(x)$ has a maximum at $x = c$.

$\Rightarrow f(c)$ is local minimum in $(c - \delta, c + \delta)$.

Hence, the required is proved.

Note: $f''(x) > 0$ for all $x \in (c - \delta, c + \delta)$.

$\Rightarrow f^{(n-1)}(x)$ is increasing for all $x \in (c - \delta, c + \delta)$.

$\therefore f'(x) > 0$ for all $x \in (c - \delta, c + \delta)$

$\Rightarrow f(x)$ is increasing for all $x \in (c - \delta, c + \delta)$

$f''(x) > 0$ for all $x \in (c - \delta, c + \delta)$

$\Rightarrow f'(x)$ is increasing for all $x \in (c - \delta, c + \delta)$

Similarly, $f''(x) < 0$ for all $x \in (c - \delta, c + \delta)$

$\Rightarrow f^{(n-1)}(x)$ is decreasing for all $x \in (c - \delta, c + \delta)$

$\therefore f'(x) < 0$ for all $x \in (c - \delta, c + \delta)$

$\Rightarrow f(x)$ is decreasing for all $x \in (c - \delta, c + \delta)$

$f''(x) < 0$ for all $x \in (c - \delta, c + \delta)$

$\Rightarrow f'(x)$ is decreasing for all $x \in (c - \delta, c + \delta)$

On methods of finding the extrema of a function $y = f(x)$ whose domain is an open interval or not mentioned.

There are three methods to find out the extreme value (values) of a continuous function $y = f(x)$ whose domain is an open interval (a, b) or not mentioned.

1. Method of definition.

2. Method of first derivative test.

3. Method of second derivative test.

Now each method will be explained seperately.

1. On method of definition: It consists of following steps:

Step 1: To locate the critical points, i.e. the points where $f'(x) = 0$ or $f'(x)$ is undefined.

Step 2: To consider one of the critical points say c and to find $f(c)$, $f(c - h)$ and $f(c + h)$.

Step 3: $f(c) > f(c - h)$ and $f(c + h)$ both for small values of $h > 0 \Rightarrow f(x)$ has the maximum value at $x = c \in (a, b)$.

Similarly, $f(c) < f(c - h)$ and $f(c + h)$ both for small values of $h > 0 \Rightarrow f(x)$ has the minimum value at $x = c \in (a, b)$.

Examples worked out:

1. Find the turning points on the curve $f(x) = 4x^3 - 3x^2 - 18x + 6$. Discriminate the maximum and minimum points by definition.

Solution: $f(x) = 4x^3 - 3x^2 - 18x + 6$

$\Rightarrow f'(x) = 12x^2 - 6x - 18$

Now $f'(x) = 0 \Rightarrow 12x^2 - 6x - 18 = 0$

$\Rightarrow 2x^2 - x - 3 = 0 \Rightarrow (2x-3)(x+1) = 0$

$\Rightarrow x = \frac{3}{2}$ or $x = -1$

$f(x) = 4x^3 - 3x^2 - 18x + 6$

$\Rightarrow f\left(\frac{3}{2}\right) = 4 \times \frac{27}{8} - 3 \times \frac{9}{4} - 18 \times \frac{3}{2} + 6$

$= 13.50 - 6.75 - 27 + 6 = -14.25$

and $f(-1) = -4 - 3 + 18 + 6 = 17$

Now for $h > 0$, $f\left(\frac{3}{2}+h\right) - f\left(\frac{3}{2}\right)$

$= 15h^2 + 4h^3 > 0$

and $f\left(\frac{3}{2}-h\right) - f\left(\frac{3}{2}\right)$

$= 15h^2 - 4h^3 > 0$ for sufficiently small h.

Hence, we observe that $f\left(\frac{3}{2}\right) < f\left(\frac{3}{2}+h\right)$ and

$f\left(\frac{3}{2}-h\right)$, for sufficiently small $h > 0$

$\therefore x = \frac{3}{2}$ is a minimum point and $f\left(\frac{3}{2}\right) = -14.25$ is a minimum value.

Again $f(x) = 4x^3 - 3x^2 - 18x + 6$

$\Rightarrow f(-1) = -4 - 3 + 18 + 6 = 17$

Now for sufficiently small $h > 0$,

$f(-1+h) - f(-1) = 4h^3 - 9h^2 < 0$,

$f(-1-h) - f(-1) = -4h^3 - 9h^2 < 0$

$\therefore x = -1$ is a maximum point and $f(-1)$ is the maximum value of $f(x)$ at $x = -1$.

2. On method of first derivative test: It consists of following steps:

Step 1: To locate the critical points, i.e. the points where $f'(x) = 0$ or $f'(x)$ is undefined.

Step 2: To examine whether $f'(x)$ changes sign at a critical point, say, $x = c$.

Step 3: **(i)** $f'(c-h) > 0$ and $f'(c+h) < 0$ (i.e. from plus to minus) $\Rightarrow f(x)$ has the maximum value at $x = c$.

(ii) $f'(c-h) < 0$ and $f'(c+h) > 0$ (i.e. from minus to plus) $\Rightarrow f(x)$ has the minimum value at $x = c$.

Similarly, the sign of each critical point $c_1, c_2, \ldots$ is examined provided that there are more than one critical point besides $x = c$.

The above method of procedure can be put in a tabular form as given below (if $f'(c)$ exists)

x	Little $< c$	At c	Little $> c$	Nature of the point c
$f'(x)$	+ve	0	–ve	Maximum
$f'(x)$	–ve	0	+ve	Minimum

Notes: **(i)** To find the critical points where $f'(x) = 0$ one should solve $f'(x) = 0$.

(ii) If $f'(x)$ is a rational functions, one should put numerator = 0 to see where $f'(x) = 0$ since denominator of a rational function cannot be zero and one should put denominator = 0 to see where $f'(x)$ is undefined.

2. On method of second derivative test: Instead of examining $f'(x)$ for change of sign at a critical point (critical points), one can use the second derivative test to determine quickly the presence of a local extreme value (local extreme values). It consists of following steps (if $f''(c)$ exists).

Step 1: To locate the critical points, where $f'(x) = 0$.

Step 2: To find $f''(x)$.

Step 3: To examines the positivity and the negativity of $f''(x)$ at all the critical points located. Let $x = c$ be any one of the located critical points. Then

$f'(c) = 0$ and $f''(c) < 0 \Rightarrow y = f(x)$ has a local maximum at $x = c$.

$f'(c) = 0$ and $f''(c) > 0 \Rightarrow y = f(x)$ has a local minimum at $x = c$.

Similarly, each critical point $c_1, c_2, c_3, \ldots$ is examined provided that there are more than one critical points besides $x = c$.

The above method of procedure can be put in a tabular form as given below (if $f''(c)$ exists).

$f'(c)$	$f''(c)$	Nature of the critical point c
0	–ve	Maximum
0	+ve	Minimum

Note: If c is a critical point, such that $f'(c)$ exists and $f'(c)=0$, and supposing that $n \geq 2$ is the smallest positive integer such that $f^{(n)}(c) \neq 0$ then method of procedure of the second derivative test is put in the following tabular form:

n	sign of $f^{(n)}(c)$	Nature of the critical point c
odd	+ve or –ve	Neither maximum nor minimum
even	–ve	Maximum
even	+ve	Minimum

where to apply which method to determine the extreme value (or, values) of a given continuous function $y = f(x)$ on a given open interval (a, b) or whose domain is not mentioned.

1. *Method of definition*: This method is practically of no use because this method is lengthy and involves tedious calculation.

2. *Method of first derivative test*: This method is practically convenient in the following cases:

(i) When the given continuous function $y = f(x)$ can be factorised, the first derivative test is always used.

(ii) When it is difficult to find out the second derivative a given continuous function $y = f(x)$, the first derivative test is used.

(iii) When the answer is given, the first derivative test is used to save time.

(iv) When $f''(x) = 0$ or $f''(x)$ does not exist at a critical point $x = c$, the first derivative test is used.

3. Method of second derivative test: This method is practically convenient for any given continuous function $y = f(x)$ which has a second derivative such that $f''(x)$ exists at a critical point $x = c$ and $f''(c) \neq 0$.

How the know that there is no maximum or minimum points:

1. When $f'(x) = 0$ provides us an impossible or imaginary result, then the given continuous function $y = f(x)$ on an open interval (a, b) or whose domain is not mentioned, has neither the maximum nor the minimum and one is required not to proceed further.

e.g. $f(x) = 3x^3 + 4x + 7 \Rightarrow f'(x) = 9x^2 + 4$

$\therefore f'(x) = 0 \Rightarrow 9x^2 + 4 = 0 \Rightarrow x^2 = -\frac{4}{9} \Rightarrow x = \pm\sqrt{\frac{-4}{9}}$ which are imaginary and hence $f(x)$ has no maximum or minimum points.

2. When $f'(x)$ is undefined at a critical point $x = c$, there is neither the maximum nor the minimum value for a given continuous function $y = f(x)$ at $x = c$.

3. When $f'(x)$ is found to be positive (or negative) for all real values of x, then it cannot change sign and consequently, there is neither the maximum nor the minimum values for the given continuous function $y = f(x)$.

Question: What is $\frac{dy}{dx}$ (slope of the curve) at the maximum or minimum $f(x)$?

Answer: At maximum or minimum $x \frac{dy}{dx} = 0 \Leftrightarrow$ the slope is zero $\Leftrightarrow$ the tangent line is parallel to the x-axis.

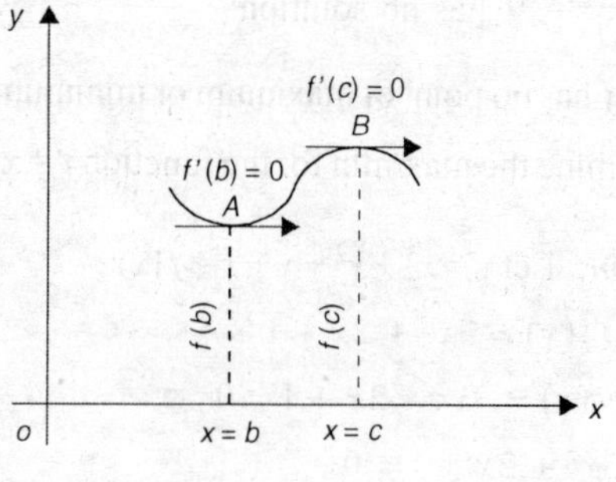

Remarks: **1.** If the question says, "find the max. and / min. values of a continuous function $y = f(x)$ defined on its domain D," then it is required to be found out firstly where these values occur (i.e. the points of maxima and / the points of minima) and then secondly what are these values (i.e. the values of a continuous function at the maximum and / the minimum points).

2. If the question says, " examine for max. and / min.", then it is required to be found out the points only where the max. and / min. value (values) occur, i.e. the points of maxima and / the points of minima are required to be determined.

3. The essential conditions for the existance of the maximum and / the minimum value (values) of a differentiable functions $y=f(x)$ for $x=c$ are (i) $f'(x)$ must be zero for $x=c$ and (ii) $f'(x)$ must change sign as x passes through the critical point $x=c$.

Problems Based on Algebraic and Mod Function

Examples worked out on algebraic functions

1. Examine the max / min for the function $y=x+2$.

Solution: $y=x+2$

$$\Rightarrow \frac{dy}{dx}=1 \text{ (constant)}$$

Now, $\frac{dy}{dx}=0$ for extreme value of the function $y=x+2$

$\Rightarrow 0=1$ which is absurd / impossible

$\Rightarrow \frac{dy}{dx}=0$ provides us an impossible result

$\Rightarrow \frac{dy}{dx}=0$ has no solution

$\Rightarrow y$ has no point of maximum or minimum.

2. Examine the max/min for the function $y=x^3+x^2+x+1$.

Solution: Let $y=x^3+x^2+x+1=f(x)$

$$\Rightarrow f'(x)=3x^2+2x+1$$

$$\therefore f'(x)=3x^2+2x+1=0$$

$$\Rightarrow 3x^2+2x+1=0$$

$$\Rightarrow x=\frac{-2\pm\sqrt{4-12}}{6}=\frac{-2\pm\sqrt{2}\,i}{6}$$ which are imaginary

$\Rightarrow f'(x)=0$ provides us imaginary values

$\Rightarrow f(x)$ has no point of maximum or minimum

3. Where is the minimum / maximum of the function $y=ax^2+bx+c$.

Solution: $y=f(x)=ax^2+bx+c$

$$\Rightarrow \frac{dy}{dx}=f'(x)=2ax+b \quad \text{...(1)}$$

$$\Rightarrow \frac{d^2y}{dx^2}=f''(x)=2a \quad \text{...(2)}$$

Now, $f'(x)=0 \Rightarrow 2ax+b=0 \Rightarrow x=-\frac{b}{2a}$

On putting $x=-\frac{b}{2a}$ in (2), we have

$$[f''(x)]_{x=-\frac{b}{2a}}=[2\times a]_{x=-\frac{b}{2a}}$$

$\Rightarrow f''\left(\frac{b}{2a}\right)=2a$ which is positive if $a>0$.

Hence, if $a>0$, then at $x=-\frac{b}{2a}$, the function $f(x)$ has a minimum and min. (maximum at $x=\frac{-b}{2a}$, if $a<0$)

$$f(x)=f\left(\frac{b}{2a}\right)=a\cdot\frac{b^2}{4a^2}-\frac{b^2}{2a}+c$$

$$=\frac{b^2}{4a}-\frac{b^2}{2a}+c=c+\left(\frac{b^2-2b^2}{4a}\right)$$

$$=\left(c-\frac{b^2}{4a}\right)$$

4. Where is the minimum / maximum of the function $y=2x^2-8x+6$.

Solution: $y=2x^2-8x+6=f(x)$ (say)

$$\Rightarrow \frac{dy}{dx}=4x-8 \quad \text{...(1)}$$

$$\Rightarrow \frac{d^2y}{dx^2}=4 \quad \text{...(2)}$$

Now, $\frac{dy}{dx}=0$ in (1)

$$\Rightarrow 4x-8=0$$

$$\Rightarrow x=\frac{8}{4}=2$$

Putting $x=2$ in (2)

$$\Rightarrow \left[\frac{d^2y}{dx^2}\right]_{x=2} = [4]_{x=2} = 4 \text{ which is positive}$$

$\therefore$ At $x = 2$, the function $y = f(x)$ has a minimum and min. $f(x) = f(2) = 2 \times 2^2 - 8 \times 2 + 6 = -2$.

5. Where is the maximum / minimum of $y = 2x^2 - 3x$.

Solution: $y = 2x^2 - 3x$

$$\Rightarrow \frac{dy}{dx} = 4x - 3$$

Now, $\frac{dy}{dx} = 0 \Rightarrow 4x - 3 = 0$

$$\Rightarrow x = \frac{3}{4}$$

Again, $\frac{d^2y}{dx^2} = 4$

$$\Rightarrow \left[\frac{d^2y}{dx^2}\right]_{x=\frac{3}{4}} = 4 \text{ which is positive}$$

$\therefore$ At $x = \frac{3}{4}$, the function has a minimum, and min.

$$y = [y]_{x=\frac{3}{4}} = 2 \cdot \left(\frac{3}{4}\right)^2 = 3 \cdot \left(\frac{3}{4}\right) = -\frac{9}{8}$$

6. Find the extreme values of the function $f(x) = x^3 - 6x^2 + 9x + 1$.

Solution: $f(x) = x^3 - 6x^2 + 9x + 1$

$$\Rightarrow f'(x) = 3\left(x^2 - 4x + 3\right)$$

Now, $f'(x) = 0$

$$\Rightarrow 3\left(x^2 - 4x + 3\right) = 0 \Rightarrow x = 1, 3$$

$$f''(x) = 6(x - 2)$$

$$\Rightarrow f''(1) = -6 \text{ which is –ve}$$

and $f''(3) = 6$ which is +ve

Hence, at $x = 1$, the function has a maximum and max. $f(x) = f(1) = 1^3 - 6 \times 1^2 + 9 \times 1 + 1 = 5$, where as at $x = 3$, the function has a minimum and min. $f(x) = f(3) = 3^3 - 6 \times 3^2 + 9 \times 3 + 1 = 1$

7. Where is the max / min. of $y = \frac{x^3}{3} - \frac{5x^2}{2} + 6x + 4$?

Solution: $y = \frac{x^3}{3} - \frac{5x^2}{2} + 6x + 4$

$$\Rightarrow \frac{dy}{dx} = x^2 - 5x + 6 = (x-2)(x-3) \quad \ldots(1)$$

Now, $\frac{dy}{dx} = 0 \Rightarrow (x-2)(x-3) = 0 \Rightarrow x = 2, 3 \quad \ldots(2)$

Now, $\frac{d^2y}{dx^2} = 2x - 5 \quad \ldots(3)$

$$\Rightarrow \left[\frac{d^2y}{dx^2}\right]_{x=2} = [2x-5]_{x=2} = 4 - 5 = -1 \text{ (–ve)}$$

and $\left[\frac{d^2y}{dx^2}\right]_{x=3} = [2x-5]_{x=3} = 6 - 5 = 1$ (+ve)

$\therefore$ At $x = 2$, y has a maximum and

$$\text{max. } f(x) = f(2) = \left[\frac{x^3}{3} - \frac{5x^2}{2} + 6x + 4\right]_{x=2}$$

$$= \frac{2^3}{3} - \frac{5 \times 2^2}{2} + 6 \times 2 + 4$$

$$= \frac{8}{3} - 10 + 12 + 4 = \frac{26}{3}$$

and at $x = 3$, y has a minimum and

$$\text{min. } f(x) = f(3) = \left[\frac{x^3}{3} - \frac{5x^2}{2} + 6x + 4\right]_{x=3}$$

$$= 9 - \frac{45}{2} + 18 + 4 = \frac{17}{2}$$

8. Where is the maximum or minimum of $y = \frac{x}{3-x} + \frac{3-x}{x}$?

Solution: $y = \frac{x}{3-x} + \frac{3-x}{x}$

$$\Rightarrow \frac{dy}{dx} = \frac{(3-x)-[x(-1)]}{(3-x)^2} + \frac{x(-1)-(3-x)\cdot 1}{x^2}$$

Now, $\frac{dy}{dx} = 0 \Rightarrow \frac{3}{(3-x)^2} - \frac{3}{x^2} = 0$

$$\Rightarrow \frac{1}{(3-x)^2} - \frac{1}{x^2} = 0$$

$$\Rightarrow \frac{1}{(3-x)^2} = \frac{1}{x^2}$$

$$\Rightarrow (3-x)^2 = x^2$$

$$\Rightarrow x^2 - 6x + 9 - x^2 = 0$$

$$\Rightarrow -6x + 9 = 0$$

$$\Rightarrow 6x - 9 = 0$$

$$\Rightarrow x = \frac{9}{6} = \frac{3}{2}$$

Again, $\frac{d^2y}{dx^2} = \frac{d}{dx}\left[\frac{3}{(3-x)^2} - \frac{3}{x^2}\right]$

$$= \frac{6}{(3-x)^3} + \frac{6}{x^3}$$

$\left(\frac{d^2y}{dx^2}\right)_{x=\frac{3}{2}} = 6\left[\frac{1}{\left(3-\frac{3}{2}\right)^3} + \frac{1}{\left(\frac{3}{2}\right)^3}\right]$ which is positive

$\therefore$ At $x = \frac{3}{2}$, the function y has a minimum and

$$\min y = (y)_{x=\frac{3}{2}} = \left(\frac{x}{3-x} + \frac{3-x}{x}\right)_{x=\frac{3}{2}} = 2$$

9. Where is the maximum / minimum of the function $y = \sqrt{2+x} + \sqrt{2-x}$.

Solution: $y = \sqrt{2+x} + \sqrt{2-x} = f(x)$ (say)

$$\Rightarrow \frac{dy}{dx} = \frac{1}{2\sqrt{2+x}} - \frac{1}{2\sqrt{2-x}}, |x|<2$$

$$\frac{dy}{dx} = 0 \Rightarrow \frac{1}{2\sqrt{2+x}} - \frac{1}{2\sqrt{2-x}} = 0$$

$$\Rightarrow \frac{1}{2\sqrt{2+x}} = \frac{1}{2\sqrt{2-x}}$$

$$\Rightarrow 2\sqrt{2+x} = 2\sqrt{2-x}$$

$$\Rightarrow \sqrt{2+x} = \sqrt{2-x}$$

$$\Rightarrow 2 + x = 2 - x$$

$$\Rightarrow 2x = 0$$

$$\Rightarrow x = 0$$

Now, $\frac{d^2y}{dx^2} = \frac{d}{dx}\left[\frac{1}{2\sqrt{2+x}} - \frac{1}{2\sqrt{2-x}}\right]$

$$\Rightarrow f''(x) = \frac{1}{2}\left[-\frac{1}{2}(2+x)^{-\frac{3}{2}} \cdot 1\right]$$

$$-\frac{1}{2}\left[-\frac{1}{2}(2-x)^{-\frac{3}{2}}(-1)\right]$$

$$= -\frac{1}{4}(2+x)^{-\frac{3}{2}} - \frac{1}{4}(2-x)^{-\frac{3}{2}}$$

$$= -\frac{1}{4\sqrt{(2+x)^3}} - \frac{1}{4\sqrt{(2-x)^3}}$$

$\therefore f''(0) = -\frac{1}{4\sqrt{2^3}} - \frac{1}{4\sqrt{2^3}}$ which is negative quantity.

Hence, at $x = 0$, the function $f(x)$ has a maximum and

$$\text{max. } f(x) = \sqrt{2} + \sqrt{2} = 2\sqrt{2}$$

10. Find the extreme value of the function $y = (x+1)(x+2)(x+3)$.

Solution: $y = (x+1)(x+2)(x+3)$

$= (x^2+2x+x+2)(x+3)$

$= x^3+2x^2+x^2+2x+3x^2+6x+3x+6$

$= x^3+6x^2+11x+6$

$$\Rightarrow \frac{dy}{dx} = 3x^2 + 12x + 11 \quad \text{...(1)}$$

$$\therefore \frac{dy}{dx} = 0 \Rightarrow 3x^2 + 12x + 11 = 0$$

$$\Rightarrow x = \frac{-12 \pm \sqrt{(12)^2 - 4 \cdot 3 \cdot 11}}{2 \times 3}$$

$$= \frac{-12 \pm \sqrt{144 - 132}}{6}$$

$$\Rightarrow x = \frac{-12 \pm \sqrt{4 \times 3}}{6} = \frac{-12 \pm 2\sqrt{3}}{6}$$

$$\Rightarrow x = \frac{-12 + 2\sqrt{3}}{6}, \frac{-12 - 2\sqrt{3}}{6}$$

Now, differentiating (1) again w.r.t x, we get

$$f''(x) = \frac{d^2y}{dx^2} = 6x + 12$$

$$\Rightarrow f''\left(-2 + \frac{1}{\sqrt{3}}\right) = 6\left(-2 + \frac{1}{\sqrt{3}}\right) + 12$$

$$\Rightarrow f''\left(-2 + \frac{1}{\sqrt{3}}\right) = -6 \times 2 + \frac{6}{\sqrt{3}} + 12 = \frac{6}{\sqrt{3}}$$

(Positive)

$$\text{Again, } f''\left(-2 - \frac{1}{\sqrt{3}}\right) = 6\left(-2 - \frac{1}{\sqrt{3}}\right) + 12$$

$$\Rightarrow f''\left(-2 - \frac{1}{\sqrt{3}}\right) = -12 - \frac{6}{\sqrt{3}} + 12 = -\frac{6}{\sqrt{3}}$$

(negative)

$\therefore$ At $x = -2 + \frac{1}{\sqrt{3}}$, the function y has a minimum and min.

$$y = (y)_{x=-2+\frac{1}{\sqrt{3}}}$$

$$= \frac{-2}{3\sqrt{3}}$$

At $x = -2 - \frac{1}{\sqrt{3}}$, the function y has a maximum and max.

$$y = (y)_{x=-2-\frac{1}{\sqrt{3}}}$$

$$= \frac{2}{3\sqrt{3}}$$

11. Find the max / min. values of y on the curve $y = (x-2)(x-3)$.

Solution: $y = (x-2)(x-3)$

$= x^2 - 2x - 3x + 6$

$= x^2 - 5x + 6$

$$\Rightarrow \frac{dy}{dx} = 2x - 5$$

$$\therefore \frac{dy}{dx} = 0 \Rightarrow 2x - 5 = 0 \Rightarrow x = \frac{5}{2}$$

$$\text{Now, } \frac{d^2y}{dx^2} = 2$$

$$\Rightarrow \left[\frac{d^2y}{dx^2}\right]_{x=\frac{5}{2}} = [2]_{x=\frac{5}{2}} = 2 \text{ (positive)}$$

$\therefore$ At $x = \frac{5}{2}$, the function y has a minimum and min.

$$y = (y)_{x=\frac{5}{2}} = [(x-2)(x-3)]_{x=\frac{5}{2}}$$

$$= \left(\frac{5}{2} - 2\right)\left(\frac{5}{2} - 3\right) = \frac{1}{2} \times \left(-\frac{1}{2}\right) = -\frac{1}{4}$$

12. Find the extreme value of the function $y = (x-1)(x-2)(x-3)$.

Solution: $y = (x-1)(x-2)(x-3)$

$= x^3 - 6x^2 + 11x - 6$

$$\Rightarrow \frac{dy}{dx} = 3x^2 - 12x + 11$$

$$\therefore \frac{dy}{dx} = 0$$

$$\Rightarrow 3x^2 - 12x + 11 = 0$$

$$\Rightarrow x = \frac{12 \pm \sqrt{144 - 132}}{6} = 2 \pm \frac{1}{\sqrt{3}}$$

Now, $\frac{d^2y}{dx^2} = 6x - 12$

$$\therefore f''\left(2 + \frac{1}{\sqrt{3}}\right) = [6x - 12]_{x = 2 + \frac{1}{\sqrt{3}}}$$

$$\Rightarrow f''\left(2 + \frac{1}{\sqrt{3}}\right) = \left[6\left(2 + \frac{1}{\sqrt{3}}\right) - 12\right] = \frac{6}{\sqrt{3}}$$

(positive)

and $f''\left(2 - \frac{1}{\sqrt{3}}\right) = [6x - 12]_{x = 2 - \frac{1}{\sqrt{3}}}$

$$\Rightarrow f''\left(2 + \frac{1}{\sqrt{3}}\right) = \left[6\left(2 - \frac{1}{\sqrt{3}}\right) - 12\right]$$

$$= 12 - \frac{6}{\sqrt{3}} - 12 = -\frac{6}{\sqrt{3}} \text{ (negative)}$$

Hence, at $x = 2 + \frac{1}{\sqrt{3}}$ the function y has a minimum and min.

$$y = (y)_{x = 2 + \frac{1}{\sqrt{3}}}$$

$$= [(x-1)(x-2)(x-3)]_{x = 2 + \frac{1}{\sqrt{3}}}$$

$$= \left[\left(2 + \frac{1}{\sqrt{3}}\right) - 1\right]\left[\left(2 + \frac{1}{\sqrt{3}}\right) - 2\right]\left[\left(2 + \frac{1}{\sqrt{3}}\right) - 3\right]$$

$$= \left[\left(1 + \frac{1}{\sqrt{3}}\right)\left(\frac{1}{\sqrt{3}}\right)\left(\frac{1}{\sqrt{3}} - 1\right)\right]$$

$$= \frac{-2}{3\sqrt{3}}$$

and at $x = 2 - \frac{1}{\sqrt{3}}$, the function y has a maximum and max.

$$y = (y)_{x = 2 - \frac{1}{\sqrt{3}}} = [(x-1)(x-2)(x-3)]_{x = 2 - \frac{1}{\sqrt{3}}}$$

$$= \left[\left(2 - \frac{1}{\sqrt{3}}\right) - 1\right]\left[\left(2 - \frac{1}{\sqrt{3}}\right) - 2\right]\left[\left(2 - \frac{1}{\sqrt{3}}\right) - 3\right]$$

$$= \left[\left(1 - \frac{1}{\sqrt{3}}\right)\left(-\frac{1}{\sqrt{3}}\right)\left(-1 - \frac{1}{\sqrt{3}}\right)\right]$$

$$= \left(1 - \frac{1}{\sqrt{3}}\right)\left(\frac{1}{\sqrt{3}}\right)\left(1 + \frac{1}{\sqrt{3}}\right)$$

$$= \frac{2}{3\sqrt{3}}$$

To find the max / min values of the function with the help of first derivative only

Examples worked out:

1. Find the max / min values of the function $y = (x-3)^5 (x+1)^4$.

Solution: $y = (x-3)^5 (x+1)^4$ ...(1)

$$\Rightarrow \frac{dy}{dx} = (x-3)^5 \frac{d(x+1)^4}{dx} + (x+1)^4 \frac{d(x-3)^5}{dx}$$

$$= (x-3)^5 \cdot 4(x+1)^3 + (x+1)^4 \cdot 5(x-3)^4$$

$$= (x-3)^4 (x+1)^3 [(x-3) \cdot 4 + (x+1) \cdot 5]$$

$$= (x-3)^4 \cdot (x+1)^3 (9x - 7) \quad \text{...(2)}$$

Now, $\frac{dy}{dx} = 0$

$$\Rightarrow (x-3)^4 (x+1)^3 (9x-7) = 0$$

$$\Rightarrow \left.\begin{aligned} x &= 3 \\ x &= -1 \\ x &= \frac{7}{9} \end{aligned}\right\}$$

we now investigate the sign of $\frac{dy}{dx}$ in the neighbourhood of these points.

(1) for $h > 0$,

$$\left[\frac{dy}{dx}\right]_{x=3-h}$$

$$= [3-h-3]^4\ [3-h+1]^3\ [9(3-h)-7]$$

$= h^4\ (4-h)^3\ (20-9h) > 0$ for sufficiently small values of h and $\left[\frac{dy}{dx}\right]_{x=3+h}$

$$= [3+h-3]^4 \cdot [3+h+1]^3\ [9\ (3+h)-7]$$

$= h^4 \cdot (4+h)^3 \times (20+9h) > 0$ for small values of h

Thus, we observe $\frac{dy}{dx}$ does not change sign in moving from left to right through the point $x = 3$.

Hence, $x = 3$ is a point known as a point of inflection

(2) for $h > 0$,

$$\left[\frac{dy}{dx}\right]_{x=-1-h}$$

$$= \left[(-1-h-3)^4\right]\left[(-1-h+1)^3\right][9(-1-h)-7]$$

$$= \left[(-4-h)^4\right]\left[(-h)^3\right][(-9-9h-7)]$$

$= \left[(-4-h)^4\ (-h)^3\ (-16-9h)\right] > 0$ for small values of h

and $\left[\frac{dy}{dx}\right]_{x=-1+h}$

$$= \left[(-1+h-3)^4\right]\left[(-1+h+1)^3\right][9(-1+h)-7]$$

$$= \left[(-4+h)^4\right]\left[(h)^3\right][(-9+9h-7)]$$

$= \left[(-4+h)^4\right]\left[(h)^3\right][(-16+9h)] < 0$ for sufficiently small values of h.

Thus, we observe $\frac{dy}{dx}$ changes sign from plus to minus in moving from left to right through the point $x = -1$. Hence, $x = -1$ is a point of maximum.

$$\therefore \text{max. } y = \left[(x-3)^5 \cdot (x+1)^4\right]_{x=-1}$$

$$= (-1-3)^5 \cdot (-1+1)^4 = 0$$

(3) For $h > 0$

$$\left[\frac{dy}{dx}\right]_{x=\frac{7}{9}-h}$$

$$= \left[(x-3)^4\ (x+1)\ (9x-7)\right]_{x=\frac{7}{9}-h}$$

$$= \left[\left(\frac{7}{9}-h-3\right)^4 \left(\frac{7}{9}-h+1\right)\right]\left[9\left(\frac{7}{9}-h\right)-7\right]$$

$$= \left[\left(-\frac{20}{9}-h\right)^4 \left(\frac{16}{9}-h\right)\right][(7-qh-7)]$$

$= \left[\left(-\frac{20}{9}-h\right)^4 \left(\frac{16}{9}-h\right)(-h)\right] < 0$ for sufficeintly small values of h and $\left[\frac{dy}{dx}\right]_{x=\frac{7}{9}+h}$

$$= \left[(x-3)^4\ (x+1)\ (9x-7)\right]_{x=\frac{7}{9}+h}$$

$$= \left[\left(\frac{7}{9}+h-3\right)^4 \left(\frac{7}{9}+h+1\right)\right]\left[9\left(\frac{7}{9}+h\right)-7\right]$$

$$= \left[\left(-\frac{20}{9}+h\right)^4 \left(\frac{16}{9}+h\right)(7+9h-7)\right]$$

$= \left(-\frac{20}{9}+h\right)^4 \left(\frac{16}{9}+h\right)(9h) > 0$ for small values of h

Thus, we observe $\frac{dy}{dx}$ changes sign from minus to plus in moving from left to right through the point $x = \frac{7}{9}$.

Hence, $x = \frac{7}{9}$ is a point of minimum.

$\therefore$ min. $y = \left[(x-3)^5 (x+1)^4\right]_{x=\frac{7}{9}}$

Footnote: Whenever it is not specially asked to find out the point of inflection but the point of inflection occurs, it must be mentioned as in this question.

2. Find the extreme values of the function $f(x) = (x-1)(x+2)^2$.

Solution: $f(x) = (x-1)(x+2)^2$

$$\Rightarrow f'(x) = 2(x-1)(x+2) + 1 \cdot (x+2)^2$$
$$= 2x^2 + 2x - 4 + x^2 + 4x + 4$$
$$= 3x^2 + 6x = 3x(x+2)$$

$\therefore f'(x) = 0$

$\Rightarrow 3x(x+2) = 0$

$\Rightarrow x = 0, -2$

Now, we investigate the sign of $\frac{dy}{dx}$ in the neighbourhood of these points $x = 0$ and $x = -2$.

$\because f'(x) = 3x(x+2)$

(1) for $h > 0$,

$f'(0+h) = f'(h) = 3h(h+2) > 0$ for small h

$f'(0-h) = f'(-h) = -3h(2-h) < 0$ for sufficiently small h

$\therefore f'(x)$ changes sign from minus to plus in moving from left to right through the point $x = 0$.

$\therefore$ At $x = 0$, the function $f(x)$ has a minimum and min. $(y) = f(0) = -4$.

(2) for $h > 0$

$f'(-2-h) = 3(-2-h)(-h) > 0$ for small h

$f'(-2+h) = 3(-2+h)h < 0$ for sufficiently small h

$\therefore f'(x)$ changes sign from plus to minus in moving from left to right through $x = -2$.

$\therefore$ At $x = -2$, the function $f(x)$ has a maximum and max. $(y) = f(-2) = 0$

Note: Whenever it is difficult to investigate (or, determine) the sign of $\frac{dy}{dx}$, one can calculate arithmetically by taking for h a sufficiently small positive number like $h = 0.0001$ for example, in the above problem, we have $f'(x) = 3x(x+2)$

(1) for $h > 0$,

$$f'(0+h) = f'(h) = 3h(h+2)$$
$$= 3 \times 0.0001 \times (0.0001 + 2) = 0.0003 \times 0.0002$$
$$= 0.000006 = \oplus$$

and $f'(0-h) = f'(-h)$
$$= -3h(2-h) = -3 \times 0.0001 \times (2 - 0.0001)$$
$$= -0.0003 \times (1.9999) = \ominus$$

(2) $f'(-2-h) = 3(-2-h)(-h)$
$$= 3(-2-0.0001)(-0.0001)$$
$$= 3(-2.0001)(-0.0001)$$
$$= (-6.0003)(-0.0001) = \oplus$$

and $f'(-2+h) = 3(-2+h)(+h)$
$$= 3(-2+0.0001)(+0.0001)$$
$$= 3(-2+0.0001)(+0.0001)$$
$$= 3(-1.9999)(0.0001) = \ominus$$

Problems on Mod. Functions

While finding maxima and / minima of the given mod. functions, we should remember the following facts.

1. A function $f(x)$ may have a maximum / a minimum at a point $x = c$ without being differentiable at that point $x = c$.

2. If $f'(c)$ does not exist but $f'(x)$ exists in the neighbourhood of $x = c$, then $f_-'(c-h)$ is positive and $f_+'(c+h)$ is negative $\Rightarrow f'(x)$ changes sign from plus to minus at $x = c$ while passing through $x = c$ from left to right (i.e.; $f'(x)$ changes sign from positive to negative at $x = c$ as we move from left to right in the neigh-bourhood of $x = c$) $\Rightarrow f(x)$ has a maximum at $x = c$.

3. If $f'(c)$ does not exist but $f'(x)$ exist in the neighbourhood of $x = c$, then $f'(c-h)$ is negative and $f'(c+h)$ is positive $\Rightarrow f'(x)$ changes sign from minus to plus at $x = c$ while passing through

$x = c$ from left to right (i.e.; $f'(x)$ changes sign from negative to positive at $x = c$ as we move from left to right in the neigh-bourhood of $x = c$) $\Rightarrow f(x)$ has minimum at $x = c$.

The above facts mentioned in (1), (2) and (3) may be summarised in the tabular form which tells the behaviour (or, nature) of the function $f(x)$.

x	*Slightly* $< c$	*Slightly* $> c$	*at* $x = c$	*Nature of the point* $x = c$ / *Nature of the function* $f(x)$
$f'(x)$	+ve	–ve	$f'(c)$ does not exist	Maxima
$f'(x)$	–ve	+ve	$f'(c)$ does not exist	Minima

Examples worked out in mod functions

1. Find the max / min value of the function $f(x) = |x|$.

Solution: $f(x)$ is a continuous function

Also, $\dfrac{d\,f(x)}{dx} = \dfrac{|x|}{x},\ x \neq 0$

$\therefore \dfrac{d\,f(x)}{dx}$ is –ve for $x < 0$

and $\dfrac{d\,f(x)}{dx}$ is +ve for $x > 0$

$\therefore \dfrac{d\,f(x)}{dx}$ changes sign from minus to plus while passing through $x = 0$ from left to right

At $x = 0$, $f(x)$ has the minimum

$\therefore$ min. $f(x) = [|x|]_{x=0} = 0$

2. Find the max / min value of the function $f(x) = |x^3| + 1$.

Solution: $f(x) = |x^3| + 1$

$$\Rightarrow \frac{d\,f(x)}{dx} = \frac{|x^3|}{x^3} \cdot 3x^2 + 0,\ x \neq 0$$

$$\Rightarrow \frac{d\,f(x)}{dx} = \frac{3|x^3|}{x},\ x \neq 0$$

$$\therefore \frac{d\,f(x)}{dx} = \frac{-3x^3}{x} = -3x^2 \text{ when } x < 0 \qquad \ldots(1)$$

$$\text{and } \frac{d\,f(x)}{dx} = \frac{3x^2}{x} = 3x^2, \text{ when } x > 0 \qquad \ldots(2)$$

(1) and (2) $\Rightarrow \dfrac{d\,f(x)}{dx}$ changes sign from –ve to +ve in passing through $x = 0$ from left to right $\Rightarrow f(x)$ has the minimum at $x = 0$.

$$\therefore \text{min. } f(x) = \left[|x^3| + 1\right]_{x=0} = 1$$

3. Find the max / min value of the function $f(x) = -|x+1| + 3$.

Solution: $f(x) = -|x+1| + 3$

$$\Rightarrow \frac{d\,f(x)}{dx} = \frac{-|x+1|}{x+1},\ (x + 1 \neq 0)$$

$$\Rightarrow \frac{d\,f(x)}{dx} = +\text{ve for } x < -1$$

$= -$ve for $x > -1$

$\therefore \dfrac{d\,f(x)}{dx}$ changes sign from plus to minus while passing through $x = -1$ from left to right.

$\therefore$ At $x = -1$, $f(x)$ has the maximum

Hence, max. $f(x) = [-|x+1| + 3]_{x=-1} = 3$

4. Find the max / min value of the function $f(x) = |\sin 4x + 3|$.

Solution: $f(x) = |\sin 4x + 3|$

We know that $-1 \le \sin 4 \le 1$

$\Rightarrow -1 + 3 \le \sin 4x + 3 \le 1 + 3$

$\Rightarrow 2 \le \sin 4x + 3 \le 4$ ...(i)

$\Rightarrow$ ($\sin 4x + 3$) lies between 2 and 4

$\Rightarrow \sin 4x + 3$ is +ve

$f(x) = |\sin 4x + 3|$

$= \sin 4x + 3$ (period being $\dfrac{\pi}{2}$)

$$\therefore \frac{d\,f(x)}{dx} = 4\cos 4x$$

Now, $\dfrac{d\,f(x)}{dx} = 0$

$$\Rightarrow x = \frac{\pi}{8}, \frac{3\pi}{8}$$

$$\frac{d^2 f(x)}{dx^2} = -16\sin 4x$$

$$\Rightarrow \frac{d^2 f(x)}{dx^2} = -\text{ve for } x = \frac{\pi}{8}$$

and $\Rightarrow \dfrac{d^2 f(x)}{dx^2} = +\text{ve for } x = \dfrac{3\pi}{8}$

Hence, max. $f(x) = [\sin 4x + 3]_{x=\frac{\pi}{8}}$

$$= \sin\frac{\pi}{2} + 3 = 1 + 3 = 4$$

and min. $f(x) = [\sin 4x + 3]_{x=\frac{3\pi}{8}}$

$$= \sin\frac{3\pi}{2} + 3 = -1 + 3 = 2$$

Note: From (i), max. $f(x) = 4$, min $f(x) = 2$.

Problems based on finding the maxima and / minima of a function when the interval in which the given function is defined is not mentioned:

While doing problems on finding the maxima and/ minima of a function when the interval in which the given function is defined is not mentioned, we should keep in mind the following facts.

1. If the interval in which a given function is defined is not given, we should study throughout the domain of definition of the given function in which it is defined.

2. Whenever the interval in which given trigonometric function is defined is not mentioned, we consider the general value of the angle for stationary points to identify the maximum and / minimum value of the function.

3. Whenever a particular value of the independent variable is given at which we are required to investigate the maxima and / minima of the function, there is no need to consider the general value of the angle for stationary points to identify the maximum and / minimum value of the function. But only the given particular value of the angle is to be considered as a stationary point obtained from the equation $f'(x) = 0$ to identify the maximum and / minimum value of the function.

Remember:

Equations	*Solutions*
(α = smallest +ve or –ve angle having the given sin, cos and tan; θ = any other angle having the same sin, cos and tan; n = an integer.)	
1. $\sin\theta = 0$	$\theta = n\pi$
2. $\cos\theta = 0$	$\theta = (2n+1)\cdot\frac{\pi}{2}$
3. $\tan\theta = 0$	$\theta = n\pi$
4. $\cos\theta = 1$	$\theta = 2n\pi$
5. $\cos\theta = -1$	$\theta = (2n+1)\pi$
6. $\sin\theta = k,\ -1 \le k \le 1$	$\theta = n\pi + (-1)^n\alpha$
7. $\cos\theta = k,\ -1 \le k \le 1$	$\theta = 2n\pi \pm \alpha$
8. $\tan\theta = k,\ -\infty < k < \infty$	$\theta = n\pi + \alpha$
9. $a\cos\theta + b\sin\theta = c$, $a^2 + b^2 \ge c^2$	$\theta = 2n\pi + \alpha \pm \beta$, $\tan\alpha = \frac{b}{a}$ and $\cos\beta = \frac{c}{\sqrt{a^2+b^2}}$

Note: The following results are also worth noting.

1. $\sin(n\pi) = 0$

2. $\cos(n\pi) = (-1)^n$

3. $\sin(n\pi+\theta) = (-1)^n\sin\theta$

4. $\cos(n\pi+\theta) = (-1)^n\cos\theta$

5. $\tan(n\pi-\theta) = -\tan\theta$

6. $\sin(2n\pi+\theta) = \sin\theta$

7. $\cos(2n\pi+\theta) = \cos\theta$

8. $\tan(n\pi+\theta) = \tan\theta$

9. $\cot(n\pi+\theta) = \cot\theta$

10. $\sec\theta$ and $\operatorname{cosec}\theta$ can never be less than 1.

N.B.: 1. $\sin(2n\pi-\theta) = -\sin\theta$

2. $\cos(2n\pi-\theta) = \cos\theta$

Worked out examples on trigonometric functions

1. Find the maximum and / minimum values of the function $y = \sin x$.

Solution: $y = \sin x$

$$\Rightarrow \frac{dy}{dx} = \cos x$$

Now, for the extreme values of y, $\frac{dy}{dx} = 0$

$$\therefore \frac{dy}{dx} = 0 \Rightarrow \cos x = 0 \Rightarrow x = n\pi + \frac{\pi}{2}$$

Again, $f''(x) = \frac{d^2y}{dx^2} = \frac{d}{dx}\cos x = -\sin x$

For even-n,

$$f''\left(n\pi + \frac{\pi}{2}\right) = \left[\frac{d^2y}{dx^2}\right]_{x=n\pi+\frac{\pi}{2}}$$

$$= [-\sin x]_{x=n\pi+\frac{\pi}{2}}$$

$$= -\sin\left(n\pi + \frac{\pi}{2}\right)$$

$$= -(-1)^n \sin\frac{\pi}{2}$$

$= (-1)(1)(1) = -1 = \ominus$ which indicates maximum value of $y = f(x)$.

$= \sin x$ at $x = n\pi + \frac{\pi}{2}$ for even-n.

and for odd-n

$$f''\left(n\pi + \frac{\pi}{2}\right) = \left[\frac{d^2y}{dx^2}\right]_{x=n\pi+\frac{\pi}{2}}$$

$$= [-\sin x]_{x=n\pi+\frac{\pi}{2}}$$

$$= -\sin\left(n\pi + \frac{\pi}{2}\right)$$

$$= (-1)(-1)^n \cdot \sin\frac{\pi}{2}$$

$= (-1)(-1)(1) = 1 = \oplus$ which indicates minimum value of $y = f(x)$

$= \sin x$ at $x = n\pi + \frac{\pi}{2}$ for odd-n

Hence, y has maximum at $x = n\pi + \frac{\pi}{2}$ (n being even integer) where $y_{\max} = 1$

And y has minima at $x = n\pi + \frac{\pi}{2}$ (n being odd integer) where $y_{\min} = -1$

2. Find the maximum and / minimum values of the function $y = \cos x$.

Solution: $y = \cos x$

$$\Rightarrow \frac{dy}{dx} = -\sin x$$

Now for the extreme values of y, $\frac{dy}{dx} = 0$

$$\therefore \frac{dy}{dx} = 0 \Rightarrow -\sin x = 0 \Rightarrow \sin x = 0 \Rightarrow x = n\pi$$

Again, $f''(x) = \frac{d^2y}{dx^2} = \frac{d}{dx}(-\sin x) = -\cos x$

For even-n,

$$f''(n\pi) = \left[\frac{d^2y}{dx^2}\right]_{x=n\pi} = [-\cos x]_{x=n\pi} = -\cos n\pi$$

$= -(-1)^n = -(+1) = -1 = \ominus$ which indicates maximum value of $y = f(x)$ at $x = n\pi$ for even-n and

for odd-n, $f''(n\pi) = \left[\dfrac{d^2y}{dx^2}\right]_{x=n\pi} = [-\cos x]_{x=n\pi}$

$= -\cos n\pi = -1(-1)^n$

$= (-1)(-1) = 1 = \oplus$ which indicates minimum value of $y = f(x)$ at $x = n\pi$ for odd-n.

Hence, y has maxima at $x = n\pi$ (n being an even integer) where $y_{max} = +1$

And y has minima at $x = n\pi$ (n being odd integer) where $y_{min} = -1$

3. Show that the function $f(x) = \tan x$ has neither maxima nor minima.

Solution: $f(x) = \tan x \Rightarrow f'(x) = \sec^2 x$,

$x \neq n\pi + \dfrac{\pi}{2}$.

Now, for extreme values of y, $f'(x) = 0$

$\therefore f'(x) = 0 \Rightarrow \sec^2 x = 0 \Rightarrow \dfrac{1}{\cos^2 x} = 0$ which is not possible. ($f(x)$ is undefined for $x = n\pi + \dfrac{\pi}{2}$)

Hence $f(x) = \tan x$ has neither maxima nor minima.

4. Discuss the extreme values of the function, $y = \sec x$.

Solution: $y = \sec x$, $x \neq n\pi + \dfrac{\pi}{2}$

$\Rightarrow \dfrac{dy}{dx} = \sec x \tan x$

Now, $\dfrac{dy}{dx} = 0$ (for extreme values)

$\Rightarrow \sec x \cdot \tan x = 0$

$\Rightarrow \sec x = 0$ or $\tan x = 0$

But $\sec x \neq 0$ always and

$\tan x = 0 \Rightarrow x = n\pi$

$\therefore \left(\dfrac{d^2y}{dx^2}\right)_{x=n\pi}$

$= \left\{\sec x\left(\tan^2 x + \sec^2 x\right)\right\}_{x=n\pi}$

$= \sec n\pi \left\{(\tan n\pi)^2 + (\sec n\pi)^2\right\}$

$= 1$ if n is even and $= -1$ if n is odd, which indicates y has minima at $x = n\pi$ (n, even) and $y_{min} = f(n\pi)$, n even

$= \sec n\pi$

$= 1$ and y has maxima at $x = n\pi$ (n odd) and $y(\max) = -1$.

5. Find the maximum values and / minimum values of the function $y = f(x) = a\sec x + b\,\text{cosec}\, x$ $(0 < a < b)$, in $\left(0, \dfrac{n\pi}{2}\right)$.

Solution: $y = a\sec x + b\,\text{cosec}\, x$ (defined for $x \neq \dfrac{n\pi}{2}$)

For, $x \neq \dfrac{n\pi}{2}$, $\dfrac{dy}{dx} = a\sec x\tan x - b\,\text{cosec}\, x \cdot \cot x$

Now, for the extremum values of y, $\dfrac{dy}{dx} = 0$

$\Rightarrow a\sec\ \tan x - b\,\text{cosec}\, x \cdot \cot x = 0$

$\Rightarrow \dfrac{\sec x \cdot \tan x}{\text{cosec}\, x \cdot \cot x} = \dfrac{b}{a}$

$\Rightarrow \tan^3 x = \dfrac{b}{a} \Rightarrow \tan x = \left(\dfrac{b}{a}\right)^{\frac{1}{3}}$ (only real root is considered)

$\dfrac{dy}{dx} = a\sec x\tan x - b\,\text{cosec}\, x \cdot \cot x$

$\Rightarrow \dfrac{d^2y}{dx^2} = a\sec x\tan^2 x + a\sec^3 x + b\,\text{cosec}\, x\cot^2 x + b\,\text{cosec}^3 x$

$\therefore \dfrac{d^2y}{dx^2}$ is $+$ve if $0 < x < \dfrac{\pi}{2}$

$\therefore y_{min} = (a \sec x + b \operatorname{cosec} x)_{\tan x = \left(\frac{b}{a}\right)^{\frac{1}{3}}}$

$\Rightarrow y_{min} = \left(a\sqrt{1+\tan^2 x} + b\sqrt{1+\cot^2 x}\right)_{\tan x = \left(\frac{b}{a}\right)^{\frac{1}{3}}}$

$= a\sqrt{1 + \dfrac{b^{\frac{2}{3}}}{a^{\frac{2}{3}}}} + b\sqrt{1 + \left(\dfrac{a}{b}\right)^{\frac{2}{3}}}$

$= a^{\frac{2}{3}}\sqrt{a^{\frac{2}{3}} + b^{\frac{2}{3}}} + b^{\frac{2}{3}}\sqrt{b^{\frac{2}{3}} + a^{\frac{2}{3}}}$

$= \left(a^{\frac{2}{3}} + b^{\frac{2}{3}}\right)^{\frac{3}{2}}$

6. Find the maximum and / minimum values of the function $y = \sin^2\theta + \sin^2\phi$ where $\theta + \phi = \alpha$.

Solution: $y = \sin^2\theta + \sin^2\phi$

$= \frac{1}{2}[1-\cos 2\theta] + \frac{1}{2}[1-\cos 2\phi]$

$= 1 - \frac{1}{2}(\cos 2\theta + \cos 2\phi)$

$= 1 - \frac{1}{2}\{\cos 2\theta + \cos(2\alpha - 2\theta)\}$ as $\theta + \phi = \alpha$

$\Rightarrow \dfrac{dy}{d\theta} = \sin 2\theta - \sin(2\alpha - 2\theta)$

For the extreme values of y, $\dfrac{dy}{dx} = 0$

$\Rightarrow \sin 2\theta - \sin(2\alpha - 2\theta) = 0$

$\Rightarrow \sin 2\theta = \sin(2\alpha - 2\theta)$

$\Rightarrow 2\theta = 2\alpha - 2\theta + 2n\pi$

$\Rightarrow \theta = \dfrac{n\pi}{2} + \dfrac{\alpha}{2}$

Now, $\dfrac{dy}{d\theta} = \sin 2\theta - \sin(2\alpha - 2\theta)$

$\Rightarrow \dfrac{d^2y}{dx^2} = 2\cos 2\theta + 2\cos(2\alpha - 2\theta)$

$= 4\cos\alpha \cdot \cos(2\theta - \alpha)$

$f''\left(\dfrac{\alpha}{2} + \dfrac{n\pi}{2}\right) = 4\cos\alpha \cdot \cos n\pi > 0$ if n is even and $\cos\alpha > 0$.

And $f''\left(\dfrac{\alpha}{2} + \dfrac{n\pi}{2}\right) < 0$ if n is odd and $\cos\alpha > 0$.

Now $y = 1 - \frac{1}{2}[\cos 2\theta + \cos(2\alpha - 2\theta)]$

$= 1 - [\cos\alpha \cdot \cos(2\theta - \alpha)]$

$\therefore f\left(\dfrac{\pi}{2} + \dfrac{n\pi}{2}\right) = 1 - \cos\alpha \cdot \cos n\pi$

$= 1 - \cos\alpha$ if n is even

and $= 1 + \cos\alpha$ if n is odd

Thus y has the minimum values for $\theta = \dfrac{n\pi}{2} + \dfrac{\alpha}{2}$ for n = even integer and $\cos\alpha > 0$, and

$y_{min} = 1 - \cos\alpha$ y has manima for $\theta = \dfrac{n\pi}{2} + \dfrac{\alpha}{2}$ for odd n and $\cos\alpha > 0$, and $y_{max} = 1 + \cos\alpha$

Similarly if $\cos\alpha < 0$ then y has minima for $\theta = \dfrac{n\pi}{2} + \dfrac{\alpha}{2}$ (n odd) and maxima for $\theta = \dfrac{n\pi}{2} + \dfrac{\alpha}{2}$ (n even),

$y_{min} = 1 + \cos\alpha$

$y_{max} = 1 - \cos\alpha$

7. Find the maximum and / minimum values of the function $y = \sec x + \operatorname{cosec} x$.

Solution: $y = \sec x + \operatorname{cosec} x$ (which is defined for $x \neq \dfrac{n\pi}{2}$)

Now for the extreme values of the function y, $\dfrac{dy}{dx} = 0$

$\therefore \dfrac{dy}{dx} = 0 \Rightarrow \sec x \tan x - \operatorname{cosec} x \cdot \cot x = 0$

$\Rightarrow \sec x \cdot \tan x = \operatorname{cosec} x \cdot \cot x$

$\Rightarrow \sin^3 x = \cos^3 x$

$\Rightarrow \tan^3 x = 1$

$\Rightarrow \tan x = 1$

$\Rightarrow x = n\pi + \dfrac{\pi}{4}$

Now, $\dfrac{d^2y}{dx^2} = \sec x \cdot \sec^2 x + \tan x \sec x \cdot \tan x -$

$\left[\text{cosec}\, x \cdot \left(-\text{cosec}^2 x\right) - \cot x \cdot \left(-\text{cosec}\, x \cdot \cot x\right)\right]$

$\Rightarrow \dfrac{d^2y}{dx^2} = \sec^3 x + \sec x \cdot \tan^3 x + \text{cosec}^3 x +$

$\text{cosec}\, x \cdot \cot^2 x$

$= \sec x\,(\sec^2 x + \tan^2 x) + \text{cosec}\, x\,(\text{cosec}^2 x + \cot^2 x)$

$\therefore \left[\dfrac{d^2y}{dx^2}\right]_{x = n\pi + \frac{\pi}{4}} = -\text{ve when } n \text{ is odd}$ ...(1)

and $\left[\dfrac{d^2y}{dx^2}\right]_{x = n\pi + \frac{\pi}{4}} = +\text{ve when } n \text{ is even}$ ...(2)

(1) and (2) $\Rightarrow y = \sec x + \text{cosec}\, x$ is maximum when n is odd in $x = n\pi + \dfrac{\pi}{4}$ and given function $y = \sec x + \text{cosec}\, x$ is minimum when n is even in $x = n\pi + \dfrac{\pi}{4}$.

Hence, y has maxima at $x = n\pi + \dfrac{\pi}{4}$ for n-odd and y has minima at $x = n\pi + \dfrac{\pi}{4}$ for n-even.

$\therefore (y)_{max} = 2\sqrt{2}$ and $(y)_{min} = -2\sqrt{2}$.

8. Find the minimum values of the function $y = a^2 \sec^2 x + b^2 \text{cosec}^2 x$; $(a > 0, b > 0)$.

Solution: $y = a^2 \sec^2 x + b^2 \text{cosec}^2 x$ which is defined for $x \neq \dfrac{n\pi}{2}$

$\Rightarrow \dfrac{dy}{dx} = 2a^2 \sec^2 x \tan x - 2b^2 \text{cosec}^2 x \cot x$

$\therefore \dfrac{dy}{dx} = 0 \Rightarrow 2a^2 \dfrac{\sin x}{\cos^3 x} - 2b^2 \dfrac{\cos x}{\sin^3 x} = 0$ [for max or min value of y]

$\Rightarrow \tan^4 x = \dfrac{b^2}{a^2}$

$\Rightarrow \tan^2 x = \dfrac{b}{a}$ [$\because \tan^2 x \neq -\dfrac{b}{a}$ since a square cannot be negative]

Now, $\dfrac{dy}{dx} = 2a^2\left(1 + \tan^2 x\right)\tan x - 2b^2\left(1 + \cot^2 x\right)\cot x$

$\Rightarrow \dfrac{dy}{dx} = 2a^2 \tan x + 2a^2 \tan^3 x - 2b^2 \cot x - 2b^2 \cot^3 x$

$\Rightarrow \dfrac{d^2y}{dx^2} = 2a^2 \sec^2 x + 2a^2 \times 3 \tan^2 x \sec^2 x + 2b^2 \text{cosec}^2 x + 6b^2 \text{cosec}^2 x \cdot \cot^2 x$ which being a sum of squares is positive.

$\Rightarrow \dfrac{d^2y}{dx^2} = +\text{ve at } \tan^2 x = \dfrac{b}{a} \Rightarrow$ y has a minimum value when $\tan^2 x = \dfrac{b}{a}$

$\Rightarrow [f(x)]_{\min} = (y)_{\min} = a^2 \sec^2 x + b^2 \text{cosec}^2 x$

$= a^2\left(1 + \tan^2 x\right) + b^2\left(1 + \cot^2 x\right)$

$= a^2\left(1 + \dfrac{b}{a}\right) + b^2\left(1 + \dfrac{1}{\tan^2 x}\right)$

$= a^2\left(1 + \dfrac{b}{a}\right) + b^2\left(1 + \dfrac{a}{b}\right)$

$= a^2 + ab + b^2 + ab$

$= (a + b)^2$

9. Find the maximum and / minimum values of the function of $y = a \cos x + b \sin x$; $(a > 0, b > 0)$.

Solution: $y = a \cos x + b \sin x$

$\Rightarrow \frac{dy}{dx} = -a\sin x + b\cos x$

Now on putting $\frac{dy}{dx} = 0$, we get

$-a \sin x + b \cos x = 0$

$\Rightarrow b \cos x = a \sin x$

$\Rightarrow \tan x = \frac{b}{a}$

Now, when $\tan x = \frac{b}{a}$, then $\sin x = \pm\frac{b}{\sqrt{a^2+b^2}}$

and $\cos x = \pm\frac{a}{\sqrt{a^2+b^2}}$ (both +ve or both –ve, as $\tan x > 0$)

Again, $\frac{d^2y}{dx^2} = -a\cos x - b\sin x$

$= -(a\cos x + b\sin x)$

The positive values of $\sin x$ and $\cos x$ make $\frac{d^2y}{dx^2}$ negative and the negative values of $\sin x$ and $\cos x$ make $\frac{d^2y}{dx^2}$ positive

$\Rightarrow$ The max. value $= \frac{a\cdot a}{\sqrt{a^2+b^2}} + \frac{b\cdot b}{\sqrt{a^2+b^2}}$

$= \frac{a^2}{\sqrt{a^2+b^2}} + \frac{b^2}{\sqrt{a^2+b^2}}$

$= \frac{a^2+b^2}{\sqrt{a^2+b^2}} = \sqrt{a^2+b^2}$

and the min. value $= \frac{-a^2}{\sqrt{a^2+b^2}} - \frac{b^2}{\sqrt{a^2+b^2}}$

$= -\frac{a^2+b^2}{\sqrt{a^2+b^2}} = -\sqrt{a^2+b^2}$

10. Examines the function $y = \sin x + \cos x$ for extreme values.

Solution: $y = \sin x + \cos x$

$\Rightarrow \frac{dy}{dx} = \cos x - \sin x$

$\therefore \frac{dy}{dx} = 0 \Rightarrow \cos x - \sin x = 0$

$\Rightarrow \cos x = \sin x \Rightarrow \tan x = 1$

$\Rightarrow x = n\pi + \frac{\pi}{4} (n = 0, \pm 1, \pm 2, \ldots)$

Again $\frac{dy}{dx} = \cos x - \sin x$

$\Rightarrow \frac{d^2y}{dx^2} = -\sin x - \cos x = -(\sin x + \cos x)$

$\therefore \left(\frac{d^2y}{dx^2}\right)_{x=n\pi+\frac{\pi}{4}} = -\left\{\sin\left(n\pi+\frac{\pi}{4}\right)+\cos\left(n\pi+\frac{\pi}{4}\right)\right\}$

$= -\left\{(-1)^n \sin\frac{\pi}{4} + (-1)^n \cos\frac{\pi}{4}\right\}$

$= (-1)\times(-1)^n\left(\sin\frac{\pi}{4} + \cos\frac{\pi}{4}\right)$

$= (-1)^{n+1}\sqrt{2}$

Now, for even-n,

$\left(\frac{d^2y}{dx^2}\right)_{x=n\pi+\frac{\pi}{4}} = -\sqrt{2} = \ominus$ which indicates that y has maximum for $x = n\pi + \frac{\pi}{4}$ when n is even integer.

Again for odd-n,

$\left(\frac{d^2y}{dx^2}\right)_{x=n\pi+\frac{\pi}{4}} = \sqrt{2} = \oplus$ which indicates that y

has minimum for $x = n\pi + \frac{\pi}{4}$ when n is odd integer.

Hence, at $x = n\pi + \frac{\pi}{4}$ when n is even integer

$$y_{\max} = f\left(n\pi + \frac{\pi}{4}\right)$$

$$= \sin\left(n\pi + \frac{\pi}{4}\right) + \cos\left(n\pi + \frac{\pi}{4}\right)$$

$$= (-1)^n \sin n\pi + (-1)^n \sin n\pi$$

$$= (-1)^n \sqrt{2} = \sqrt{2} \text{ when } n \text{ is even integer}$$

and $y_{\min} = f\left(n\pi + \frac{\pi}{4}\right)$

$$= \sin\left(n\pi + \frac{\pi}{4}\right) + \cos\left(n\pi + \frac{\pi}{4}\right)$$

$$= (-1)^n \sin n\pi + (-1)^n \sin n\pi$$

$$= (-1)^n \sqrt{2} = -\sqrt{2} \text{ when } n \text{ is odd integer.}$$

11. Find the maximum and / minimum values of the function $y = \sin x\,(1 + \cos x)$.

Solution: $y = \sin x\,(1 + \cos x)$

$$\Rightarrow \frac{dy}{dx} = \frac{d}{dx}[\sin x + \sin x \cos x]$$

$$= \frac{d}{dx}\left[\sin x + \frac{1}{2} 2 \sin x \cos x\right] = \frac{d}{dx}\left[\sin x + \frac{1}{2}\sin 2x\right]$$

$$= \cos x + \frac{1}{2} \cdot 2 \cdot \cos 2x = \cos x + \cos 2x$$

Now, $\frac{dy}{dx} = 0 \Rightarrow \cos x + \cos 2x = 0$ (for extreme value)

$$\Rightarrow 2\cos^2 x + \cos x - 1 = 0$$

$$\Rightarrow \cos x = \frac{-1 \pm \sqrt{1+8}}{4} = \frac{-1 \pm 3}{4} = \frac{1}{2} \text{ or } -1$$

Again $\cos x = \frac{1}{2} = \cos\frac{\pi}{3} \Rightarrow x = 2n\pi \pm \frac{\pi}{3}, n \in I$

and $\cos x = -1 = \cos(\pi) \Rightarrow x = (2n\pi + \pi), n \in I$

Further, $\frac{dy}{dx} = \cos x - \cos 2x$

$$\Rightarrow \frac{d^2 y}{dx^2} = -\sin x - 2\sin 2x = -(\sin x + 2\sin 2x)$$

$$\therefore \left(\frac{d^2 y}{dx^2}\right)_{x = 2n\pi \pm \frac{\pi}{3}}$$

$$= -\left\{\sin\left(2n\pi \pm \frac{\pi}{3}\right) + 2\sin\left(2n\pi \pm \frac{\pi}{3}\right)\cos\left(2n\pi \pm \frac{\pi}{3}\right)\right\}$$

$$= -\left\{\sin\left(\pm\frac{\pi}{3}\right) + 2\sin\left(\pm\frac{\pi}{3}\right) \cdot \cos\left(\pm\frac{\pi}{3}\right)\right\}$$

$$= -\left\{\sin\frac{\pi}{3} + 2\sin\frac{\pi}{3}\cos\frac{\pi}{3}\right\} < 0$$

and $\left\{\sin\frac{\pi}{3} + \sin\frac{\pi}{3}\cos\frac{\pi}{3}\right\} > 0$

Hence, y has maxima at $x = 2n\pi + \frac{\pi}{3}$ and y has minima at $x = 2n\pi - \frac{\pi}{3}$

Where $y_{\max} = f\left(2n\pi + \frac{\pi}{3}\right)$

$$= \sin\left(2n\pi + \frac{\pi}{3}\right) \cdot \left\{1 + \cos\left(2n\pi + \frac{\pi}{3}\right)\right\}$$

$$= \sin\frac{\pi}{3}\left(1 + \cos\frac{\pi}{3}\right) = \frac{\sqrt{3}}{2}\left(1 + \frac{1}{2}\right) = \frac{\sqrt{3}}{2} \cdot \frac{3}{2} = \frac{3\sqrt{3}}{4}$$

and $y_{\min} = f\left(2n\pi - \frac{\pi}{3}\right)$

$$= \sin\left(2n\pi - \frac{\pi}{3}\right) \cdot \left\{1 + \cos\left(2n\pi - \frac{\pi}{3}\right)\right\}$$

$$= \sin\left(-\frac{\pi}{3}\right)\left\{1 + \cos\left(-\frac{\pi}{3}\right)\right\} = -\sin\frac{\pi}{3}\left\{1 + \cos\frac{\pi}{3}\right\}$$

$$= -\frac{\sqrt{3}}{2}\cdot\left(1+\frac{1}{2}\right) = -\frac{\sqrt{3}}{2}\times\frac{3}{2} = \frac{3\sqrt{3}}{4}$$

Lastly, $\left(\frac{d^2y}{dx^2}\right)_{x=(2n+1)\pi}$

$$= -\{\sin(2n\pi+\pi)+2\sin(2n\pi+\pi)\cos(2n\pi+\pi)\}$$

$$= -\{\sin\pi + 2\sin\pi\cos\pi\}$$

$$= -\{0 + 2\times 0\times(-1)\} = 0$$

and $\left(\frac{d^3y}{dx^3}\right)_{x=(2n\pi+\pi)} = (-\cos x - 4\cos 2x)_{x=(2n\pi+\pi)}$

$$= -\{\cos(2n\pi+\pi) + 4\cos 2(2n\pi+\pi)\}$$

$$= -\{\cos\pi + 4\cos(4n\pi + 2\pi)\}$$

$$= -\{\cos\pi + 4\cos 2\pi\}$$

$$= -\{(-1) + 4(-1)^2\}$$

$$= -\{-1+4\} = -(3) \neq 0$$

Therefore y has a point of inflection at $x = (2n\pi+\pi)$ because at $x = (2n\pi+\pi)$,

$$\frac{d^2y}{dx^2} = 0 \text{ and } \frac{d^3x}{dx^3} \neq 0$$

Note: **(i)** Whenever we have a quadratic trigonometric equations of the form: $a[t_{f_n}(x)]^2 + bt_{f_n} + c = 0$ where $a \neq 0$ and $t_{f_n}(x)$ represents a trigonometric function (sin x, cos x, tan x, cot x, sec x, cosec x), we have two general values of the angle x (if the interval in which a given trigonometric function is defined is not mentioned in the problem).

(ii) Whenever we have a linear trigonometric equations of the form: at $at_{f_n}(x) + b = 0, (a \neq 0)$ we have one general value of the angle x of the trigonometric function of x (sin x, cos x, tan x, etc) for the equation $t_{f_n}(x) = -\frac{b}{a}$ if the interval in which the given trigonometric function is defined is not mentioned in the problem.

12. Show that the function $y = \sin x + \frac{1}{2}\sin 2x$ has a maximum value at $x = \frac{\pi}{3}$ and find the corresponding maximum value.

Solution: $y = \sin x + \frac{1}{2}\sin 2x$

$$\Rightarrow \frac{dy}{dx} = \cos x + \cos 2x$$

Now, $\frac{dy}{dx} = 0$ (for extreme value)

$$\Rightarrow \cos x + \cos 2x = 0$$

$$\Rightarrow 2\cos^2 x + \cos x - 1 = 0$$

$$\Rightarrow \cos x = \frac{-1\pm\sqrt{1+8}}{4} = \frac{-1\pm 3}{4} = \frac{1}{2} \text{ or } -1$$

Again, $\cos x = \frac{1}{2} = \cos\frac{\pi}{3} \Rightarrow x = \frac{\pi}{3}$

Further, $\frac{dy}{dx} = \cos x + \cos 2x$

$$\Rightarrow \frac{d^2y}{dx^2} = \frac{d}{dx}[\cos x + \cos 2x] = -\sin x - 2\sin 2x$$

$$\therefore \left(\frac{d^2y}{dx^2}\right)_{x=\frac{\pi}{3}} = (-\sin x - 2\sin 2x)_{x=\frac{\pi}{3}}$$

$$= -\sin\frac{\pi}{3} - 2\sin\frac{2\pi}{3}$$

$$= -\frac{\sqrt{3}}{2} - 2\times\frac{\sqrt{3}}{2}$$

$$= -\left(\frac{\sqrt{3}+2\sqrt{3}}{2}\right) = -\frac{3\sqrt{3}}{2} = \ominus$$

which indicates y has a maximum at $x = \frac{\pi}{3}$ and at $x = \frac{\pi}{3}$

$$y_{max} = \left[\sin x + \frac{1}{2}\sin 2x\right]_{x=\frac{\pi}{3}}$$

$$= \sin\frac{\pi}{3} + \frac{1}{2} \cdot \sin\left(2 \times \frac{\pi}{3}\right)$$

$$= \frac{\sqrt{3}}{2} + \frac{1}{2} \times \frac{\sqrt{3}}{2} = \frac{2\sqrt{3}+\sqrt{3}}{4} = \frac{3\sqrt{3}}{4}$$

Note: When $x = \pi$, $\frac{d^2y}{dx^2} = 0$

$$\left(\frac{d^3y}{dx^3}\right)_{x=\pi} = (-\cos x - 4\cos 2x)_{x=\pi}$$

$$= -\cos\pi - 4\cos\pi = -(-1) - 4(-1)$$

$$= 1 + 4 = 5 \neq 0$$

Hence, at $x = \pi$

$\frac{d^2y}{dx^2} = 0$ and $\frac{d^3x}{dx^3} \neq 0$ which $\Rightarrow$ at $x = \pi$,

we get an inflection point

13. Discuss the extreme values of the function $y = \sec x$ at the origin.

Solution: $y = \sec x$

$$\Rightarrow \frac{dy}{dx} = \sec x \cdot \tan x$$

Now, $\frac{dy}{dx} = 0$ (for extreme values)

$\Rightarrow \sec x \cdot \tan x = 0$

$\Rightarrow \sec x = 0$ or $\tan x = 0$

But $\sec x \neq 0$

$\therefore \tan x = 0$ which $\Rightarrow x = 0$ is an extremum

$$\therefore \left(\frac{d^2y}{dx^2}\right)_{x=0} = \left[\sec x\left(\tan^2 x + \sec^2 x\right)\right]_{x=0}, \text{ i.e.,}$$

$1(0+1)$

$= 1 = \oplus$ which indicates y has a minimum at $x = 0$

i.e., origin and at $x = 0$ (i.e. at origin)

$$y_{min} = f(0) = \sec 0° = 1$$

14. Discuss the extreme values of the function $y = x - \sin x$ at the origin.

Solution: $y = x - \sin x$

$$\Rightarrow \frac{dy}{dx} = 1 - \cos x \text{ and } \frac{d^2y}{dx^2} = 0 - (-\sin x) = \sin x$$

Now, $\frac{dy}{dx} = 0$ (for extreme values)

$\Rightarrow 1 - \cos x = 0 \Rightarrow \cos x = 1 = \cos 0 \Rightarrow x = 0$ is an extremum

$$\therefore \left(\frac{d^2y}{dx^2}\right)_{x=0} = [\sin x]_{x=0} = \sin 0 = 0$$

and $\left(\frac{d^3y}{dx^3}\right)_{x=0} = [\cos x]_{x=0} = \cos 0 = 1 \neq 0$

Hence, we observe at $x = 0$,

$\frac{dy}{dx} = 0$, $\frac{d^2y}{dx^2} = 0$ and $\frac{d^3y}{dx^3} \neq 0$ which $\Rightarrow y$ has an inflection point at the origin (i.e. at $x = 0$).

Problems based on logarithmic and exponential functions.

1. Find the max and / min values of $y = \log x$.

Solution: $y = \log x, x > 0$

$$\Rightarrow \frac{dy}{dx} = \frac{1}{x}$$

Now, for the extreme values of y, $\frac{dy}{dx} = 0$

$\Rightarrow \frac{1}{x} = 0$ which is not possible which $\Rightarrow f(x)$ does not have max and / min values.

2. Find the max and / min values of $y = e^x$.

Solution: $y = e^x$

$$\Rightarrow \frac{dy}{dx} = e^x$$

Now, for the extreme values of y, $\frac{dy}{dx}=0$

$\therefore e^x=0$ which is not possible (since $e^x>0$ always) which $\Rightarrow f(x)$ does not have max and / min values.

3. Examine the following function for max and / min $y=x^{\frac{1}{x}}$.

Solution: $y=x^{\frac{1}{x}}, x>0$ ...(1)

Taking log of both sides of the equation (1)

$$\log f(x)=\log\left(x^{\frac{1}{x}}\right)=\frac{1}{x}\log x \quad \text{...(2)}$$

Now, differentiating both sides of (2) w.r.t x

$$\Rightarrow \frac{1}{f(x)}\times f'(x)=\frac{x\cdot\frac{1}{x}-\log x}{x^2}=\frac{1-\log x}{x^2}$$

$$\Rightarrow f'(x)=\frac{f(x)\cdot(1-\log x)}{x^2} \quad \text{...(3)}$$

$$=\frac{x^{\frac{1}{x}}\cdot(1-\log x)}{x^2} \quad \left[\because f(x)=x^{\frac{1}{x}} \text{ is given}\right] \quad \text{...(4)}$$

Now, putting $f'(x)=0$ in (4) for extreme values

$$\Rightarrow \frac{x^{\frac{1}{x}}(1-\log x)}{x^2}=0$$

$$\Rightarrow (1-\log x)=0 \quad \left[\because f(x)=x^{\frac{1}{x}}\neq 0\right]$$

$$\Rightarrow \log x=1\Rightarrow x=e^1\Rightarrow x=e$$

$$\text{Now, } f''(x)=\frac{1-\log x}{x^2}\cdot f'(x)+f(x)\cdot\frac{x^2\left(-\frac{1}{x}\right)-(1-\log x)\cdot(2x)}{x^4}$$

$$\Rightarrow f''(e)=\frac{1-\log e}{e^2}\cdot f'(e)+f(e)\left\{\cdot e^2\cdot\left(-\frac{1}{e}\right)-(1-\log e)\cdot(2e)\right\}\cdot\frac{1}{e^4}$$

$$=\frac{(1-1)}{e^2}\cdot f'(e)+f(e)\{(-e)-(1-\log e)\cdot(2e)\}\cdot\frac{1}{e^4}$$

$$(\because \log_e e=1)=-\frac{f(e)}{e^3}$$

$=$ –ve (since e and $e^{\frac{1}{e}}>0$) which $\Rightarrow f(x)$ has maximum at $x=e$.

Hence, $y_{\max}=e^{\frac{1}{e}}$

Problems based on combination of transcendental functions

1. Find the max and / min values of the functions $y=e^x+2\cos x+e^{-x}$.

Solution: $y=e^x+2\cos x+e^{-x}$

$$\Rightarrow \frac{dy}{dx}=e^x-2\sin x-e^{-x}.$$

Now, for extreme values of y, $\frac{dy}{dx}=0$

$\therefore e^x-2\sin x-e^{-x}=0$ which is only possible when $x=0$ which means

$$e^x-2\sin x-e^{-x}=0\Rightarrow x=0$$

$$\text{Now, } \frac{d^2y}{dx^2}=e^x-2\cos x+e^{-x}$$

$$\Rightarrow \left[\frac{d^2y}{dx^2}\right]_{x=0}=e^o-2\cos(0)-\frac{1}{e^o}$$

$$=1-2-\frac{1}{1}=2-2=0$$

$$\Rightarrow \left[\frac{d^3y}{dx^3}\right] = e^x + 2\sin x - e^x$$

$$\Rightarrow \left[\frac{d^3y}{dx^3}\right]_{x=0} = \left[e^x + 2\sin x - e^{-x}\right]_{x=0} = 0 \text{ (i.e.}$$

again zero when $x = 0$)

$$\Rightarrow \left[\frac{d^4y}{dx^4}\right] = e^x + 2\cos x + e^{-x}$$

$$\Rightarrow \left[\frac{d^4y}{dx^4}\right]_{x=0} = \left[e^x + 2\cos x + e^{-x}\right]_{x=0}$$

$= 4 > 0 = +\text{ve}$

Hence, y is minimum at $x = 0$.

2. Find the max and / min for $y = \dfrac{\log x}{x}$.

Solution: $y = \dfrac{\log x}{x}, x > 0$

$$\Rightarrow \frac{dy}{dx} = \frac{x \cdot \frac{1}{x} - 1 \cdot \log x}{x^2} = \frac{1 - \log x}{x^2}$$

Now, for the extreme values of y, $\dfrac{dy}{dx} = 0$

$$\therefore \frac{1 - \log x}{x} = 0$$

$$\Rightarrow 1 - \log x = 0$$

$$\Rightarrow \log x = 1 \Rightarrow x = e$$

Again, $\dfrac{dy}{dx} = \dfrac{1 - \log x}{x^2}$

$$\Rightarrow \frac{d^2y}{dx^2} = \frac{x^2\left(-\frac{1}{x}\right) - 2x\,(1 - \log x)}{x^4}$$

$$= \frac{-x - 2x\,(1 - \log x)}{x^4}$$

$$= \frac{-3 + 2\log x}{x^3}$$

$$\therefore \left(\frac{d^2y}{dx^2}\right)_{x=e} = \frac{-3 + 2\log e}{e^3} = -\frac{1}{e^3} < 0 \quad \text{which}$$

$\Rightarrow$ y has max. for $x = e$.

$$\therefore y_{\max} = \frac{1}{e}$$

3. Find the max and / min values of the function $f(x) = x^2 e^x$.

Solution: $f(x) = x^2 e^x$

$$\Rightarrow f'(x) = x^2 e^x + e^x \cdot 2x$$

$$\Rightarrow f'(x) = x(x + 2)\, e^x$$

Now for max and / min values of $f(x)$, $f'(x) = 0$

$$\therefore x(x + 2) \cdot e^x = 0$$

$$\Rightarrow x = 0 \text{ or } (x + 2) = 0 \ (\because e^x \neq 0)$$

$$\Rightarrow x = 0 \text{ and } x = -2$$

Now, $f''(x) = x^2 e^x + 2xe^x + e^x \cdot 2 + 2xe^x$

$$= x^2 e^x + 4xe^x + 2e^x$$

$\therefore f''(0) = 2 = +\text{ve}$ which $\Rightarrow f(x)$ has a minimum at $x = 0$, and

$$y_{\min} = \left[x^2 e^x\right]_{x=0} = (0)^2 \cdot e^0 = 0 \cdot 1 = 0$$

Again, $f''(-2) = 4e^{-2} + 4(-2)e^{-2} + 2e^{-2} = -\dfrac{2}{e^2}$

$= -\text{ve}$ which again $\Rightarrow f(x)$ has a maximum at $x = -2$ and

$$y_{\max} = y_{\min} = \left[x^2 \cdot e^x\right]_{x=-2} = (-2)^2 \cdot e^{-2} = \frac{4}{e^2}$$

4. Find the max or min value of the function $y = \sin 2x - x$.

Solution: $y = \sin 2x - x$

$$\Rightarrow \frac{dy}{dx} = 2\cos 2x - 1$$

Now, for the extrema of y, $\frac{dy}{dx} = 0$

$\Rightarrow 2\cos 2x - 1 = 0$

$\Rightarrow \cos 2x = \frac{1}{2}$

$\Rightarrow 2x = 2n\pi \pm \frac{\pi}{3}$

$\Rightarrow x = n\pi \pm \frac{\pi}{6}$ $(n = 0, \pm 1, \pm 2, ...)$

Again $\frac{d^2y}{dx^2} = -4\sin 2x$

$$\Rightarrow \left(\frac{d^2y}{dx^2}\right)_{x=n\pi\pm\frac{\pi}{6}} = (-4)\sin\left\{2\left(n\pi \pm \frac{\pi}{6}\right)\right\}$$

$$= (-4)\sin\left(\pm\frac{\pi}{3}\right)$$

$$= -4\sin\frac{\pi}{3} \text{ or } 4\sin\frac{\pi}{3} < 0 > 0$$

$\Rightarrow y$ has maxima (or, minima) at $x = n\pi + \frac{\pi}{6}$ (or, $x = n\pi - \frac{\pi}{6}$) respectively,

where $y_{max} = \left[\sin\left(2n\pi + \frac{\pi}{3}\right)\right] - \left(n\pi + \frac{\pi}{6}\right)$

$$= \sin\frac{\pi}{3} - n\pi - \frac{\pi}{6} = \frac{\sqrt{3}}{2} - n\pi - \frac{\pi}{6}$$

and $y_{min} = \left[\sin\left(2n\pi - \frac{\pi}{3}\right)\right] - \left(n\pi - \frac{\pi}{6}\right)$

$$= -\sin\frac{\pi}{3} - n\pi + \frac{\pi}{6} = -\frac{\sqrt{3}}{2} - n\pi + \frac{\pi}{6}$$

5. Find the maximum and minimum values of the function $y = x - \sin x$.

Solution: $y = x - \sin x$

$\Rightarrow \frac{dy}{dx} = 1 - \cos x$

$\Rightarrow \frac{dy}{dx} = 0 \Rightarrow 1 - \cos x = 0$ (for extrema of y)

$\Rightarrow 1 - \cos x = 0 \Rightarrow \cos x = 1 \Rightarrow x = 2n\pi$, $n = 0, \pm 1, \pm 2, ...$

Now, $\frac{dy}{dx} = 1 - \cos x$

$\Rightarrow \frac{d^2y}{dx^2} = \sin x$

$$\Rightarrow \left(\frac{d^2y}{dx^2}\right)_{x=2n\pi} = (\sin x)_{x=2n\pi}$$

$$= \sin 2n\pi = 0$$

Again, $\frac{d^2y}{dx^2} = \sin x$

$\Rightarrow \frac{d^3y}{dx^3} = \cos x$

$$\Rightarrow \left(\frac{d^3y}{dx^3}\right)_{x=2n\pi} = \cos 2n\pi = 1$$

Hence, at $x = 2n\pi$

$\frac{d^2y}{dx^2} = 0$ and $\frac{d^3y}{dx^3} \neq 0 \Rightarrow y$ has neither a maximum nor minimum at $x = 2n\pi$.

6. Find the maxima and minima of the function $y = \sec x + \log\cos^2 x$.

Solution: $y = \sec x + \log\cos^2 x$, $x \neq n\pi + \frac{\pi}{2}$

$$\Rightarrow \frac{dy}{dx} = \sec x \cdot \tan x + \frac{1}{\cos^2 x} \cdot 2\cos x \cdot (-\sin x)$$

$= \sec x \cdot \tan x - 2\tan x$

$= \tan x(\sec x - 2)$

Now,

$\frac{dy}{dx} = 0$ (for extrema)

$\Rightarrow \tan x(\sec x - 2) = 0$

$\Rightarrow \tan x = 0$ or $(\sec x - 2) = 0$

$\tan x = 0 \Rightarrow x = n\pi$

and $(\sec x - 2) = 0 \Rightarrow \sec x = 2 = \sec\frac{\pi}{3}$

$\Rightarrow x = 2n\pi \pm \frac{\pi}{3}$

Again, $\frac{dy}{dx} = \tan x(\sec x - 2)$

$$\Rightarrow \frac{d^2y}{dx^2} = \sec^2 x(\sec x - 2) + \tan x(\sec x \cdot \tan x)$$

$$= \sec^3 x - 2\sec^2 x + \sec x \cdot \tan^2 x$$

$$= \sec x(\sec^2 x - 2\sec x + \tan^2 x)$$

$$\therefore \left(\frac{d^2y}{dx^2}\right)_{x=n\pi} = \left[\sec x(\sec^2 x - 2\sec x + \tan^2 x)\right]_{x=n\pi}$$

$$= \sec n\pi(\sec^2 n\pi - 2\sec n\pi + \tan^2 n\pi)$$

= negative for all n, which indicates y has maxima at $x = n\pi$ and $y_{max} = f(n\pi)$

$$= \sec n\pi + \log(\cos n\pi)^2 = 1 \ (n \text{ even})$$

$$= -1 \ (n \text{ odd})$$

Further, $\left(\frac{d^2y}{dx^2}\right)_{x=2n\pi\pm\frac{\pi}{3}}$

$$= \left[\sec x(\sec^2 x - 2\sec x + \tan^2 x)\right]_{x=2n\pi\pm\frac{\pi}{3}}$$

$$= \sec\left(2n\pi\pm\frac{\pi}{3}\right)\left[\left\{\sec\left(2n\pi\pm\frac{\pi}{3}\right)\right\}^2 - 2\sec\left(2n\pi\pm\frac{\pi}{3}\right) + \left\{\tan\left(2n\pi\pm\frac{\pi}{3}\right)\right\}^2\right]$$

$$= \sec\left(\pm\frac{\pi}{3}\right)\left[\left\{\sec\left(\pm\frac{\pi}{3}\right)\right\}^2 - 2\sec\left(\pm\frac{\pi}{3}\right) + \left\{\tan\left(\pm\frac{\pi}{3}\right)\right\}^2\right]$$

$$= 2\left[\sec^2\left(\frac{\pi}{3}\right) + \tan^2\left(\frac{\pi}{3}\right) - 2\sec\left(\frac{\pi}{3}\right)\right]$$

$(\because \cos\theta = \cos(-\theta))$

$$= 2\left[2^2 + (\sqrt{3})^2 - 2(2)\right] = 2[4 + 3 - 4] = 6 = \oplus$$

which indicates y has minima at $x = 2n\pi \pm \frac{\pi}{3}$

Hence, at $x = 2n\pi \pm \frac{\pi}{3}$, we have

$$y_{min} = f\left(2n\pi \pm \frac{\pi}{3}\right)$$

$$= \sec\left(2n\pi \pm \frac{\pi}{3}\right) + \log\left\{\cos\left(2n\pi \pm \frac{\pi}{3}\right)\right\}^2$$

$$= \sec\left(\pm\frac{\pi}{3}\right) + \log\left\{\cos\left(\pm\frac{\pi}{3}\right)\right\}^2$$

$$= \sec\frac{\pi}{3} + \log\left\{\cos\left(\frac{\pi}{3}\right)\right\}^2 \quad (\because \cos(-\theta) = \cos\theta)$$

$$= \sec\frac{\pi}{3} + \log\cos^2\frac{\pi}{3}$$

$$= 2 + \log\left(\frac{1}{2}\right)^2$$

$$= 2 + 2\log\frac{1}{2}$$

$$= 2 - 2\log 2 = 2(1 - \log 2)$$

***Type* 2**: To find the maximum and / minimum values of the function: (Trigonometric method)

$y = a\cos\theta + b\sin\theta$ and $y = a\sin\theta + b\cos\theta$

Working rule: Express $a\cos\theta + b\sin\theta$ and $a\sin\theta + b\cos\theta$ as a single cosine and / a single sin of an angle $(\theta \pm \alpha)$ with the help of following rule:

1. Multiply and divide the given expression (both sides) by $\sqrt{a^2+b^2}$ where $\sqrt{a^2+b^2}$

$$= \sqrt{(\text{coefficient of } \cos\theta)^2 + (\text{coefficient of } \sin\theta)^2}$$

2. Use $(A \pm B)$ formulas as the case may require which transforms the given function into

$y = \sqrt{a^2+b^2}\cos(\theta-\alpha)$ and / $\sqrt{a^2+b^2}\sin(\theta+\alpha)$.

How to find the maximum and / minimum values of the function: $a\cos\theta + b\sin\theta$ and / $a\sin\theta + b\sin\theta$.

$$\because a\cos\theta + b\sin\theta = \sqrt{a^2+b^2}\cos(\theta-\alpha) \quad \ldots(1)$$

But the maximum value of $\cos(\theta-\alpha) = 1$

$[\because 1 \le \cos\theta \le -1]$

$$\therefore [a\cos\theta + b\sin\theta]_{max}$$

$$= \max[a\cos\theta + b\sin\theta]$$

$$= \sqrt{a^2+b^2}\cdot 1 = \sqrt{a^2+b^2} \quad \ldots(2)$$

And the minimum value of $\cos(\theta - 1) = -1$

$$\Rightarrow [a\cos\theta + b\sin\theta]_{min}$$

$$= \min[a\cos\theta + b\sin\theta] = -\sqrt{a^2+b^2} \quad \ldots(3)$$

Hence, remember:

1. Maximum value of $[a\cos\theta + b\sin\theta] = \sqrt{a^2+b^2}$
Minimum value of $[a\sin\theta + b\cos\theta] = -\sqrt{a^2+b^2}$

2. Maximum value of $[a\sin\theta + b\cos\theta] = \sqrt{a^2+b^2}$
Minimum value of $[a\sin\theta + b\cos\theta] = -\sqrt{a^2+b^2}$

3. Remember that maximum value of $\sin\theta$ (or, $\cos\theta$) is 1 and the minimum value of $\sin\theta$ (or, $\cos\theta$) is -1.

4. If the question is asked as: prove that the function $y = a\cos\theta + b\sin\theta$ and / $y = a\sin\theta + b\cos\theta$ has a max. and / min. value at some point, we use may also $\frac{d^2y}{dx^2}$ method to find the extrem a at the indicated point. (See example 2 to follow)

Worked out examples on the max and / min values of t-functions whose form is $y = a\cos\theta + b\sin\theta$ and / $a\cos\theta + b\sin\theta$.

1. Find the max. and / min. values of the function $y = 3\sin\theta + 4\cos\theta$.

Solution: $y = 3\sin\theta + 4\cos\theta$

$$\Rightarrow \frac{y}{5} = \frac{3}{5}\sin\theta + \frac{4}{5}\cos\theta \quad \ldots(1)$$

Now, on supposing that $\cos x = \frac{3}{5}$ and $\sin x = \frac{4}{5}$

(1) becomes $\frac{y}{5} = \cos x \cdot \sin\theta + \sin x \cdot \cos\theta$ which

$$\Rightarrow \frac{y}{5} = \cos(x+\theta) \Rightarrow y = 5\cos(x+\theta)$$

$$\therefore y_{max} = [3\sin\theta + 4\cos\theta]_{max} = 5$$

$$\therefore y_{min} = [3\sin\theta + 4\cos\theta]_{min} = -5$$

2. Find the max and / min value of $y = \sqrt{3}\sin x + 3\cos x$ or, prove that max. value of y is at $x = \frac{\pi}{6}$ and min at $\frac{7\pi}{6}$.

Solution: $y = \sqrt{3}\sin x + 3\cos x = a\sin x + b\cos x$

$$\sqrt{a^2+b^2} = \sqrt{(\sqrt{3})^2 + (3)^2} = \sqrt{3+9}$$

$$= \sqrt{12} = \sqrt{4\times 3} = 2\sqrt{3}$$

$$\Rightarrow \frac{y}{2\sqrt{3}} = \frac{\sqrt{3}}{2\sqrt{3}}\sin x + \frac{3}{2\sqrt{3}}\cos x$$

$$= \frac{\sqrt{3}}{2\sqrt{3}}\sin x + \frac{\sqrt{3}\cdot\sqrt{3}}{2\sqrt{3}}\cos x$$

$$\Rightarrow \frac{y}{2\sqrt{3}} = \frac{1}{2}\sin x + \frac{\sqrt{3}}{2}\cos x \quad \text{...(i)}$$

Now, we know that $\cos 30^\circ = \cos\frac{\pi}{6} = \frac{\sqrt{3}}{2}$ and $\sin\left(\frac{\pi}{6}\right) = \sin 30 = \frac{1}{2}$.

Hence, (1) becomes equal to $\frac{y}{2\sqrt{3}} = \sin\left(\frac{\pi}{6}\right)\sin x + \cos\left(\frac{\pi}{6}\right)\cos x$

$$\Rightarrow \frac{y}{2\sqrt{3}} = \cos\left(x - \frac{\pi}{6}\right)$$

$$\Rightarrow y = 2\sqrt{3}\cos\left(x - \frac{\pi}{6}\right) \quad \text{...(ii)}$$

$$\Rightarrow \frac{dy}{dx} - 2\sqrt{3}\sin\left(x - \frac{\pi}{6}\right)$$

and $\frac{dy}{dx} = 0 \Rightarrow -2\sqrt{3}\sin\left(x - \frac{\pi}{6}\right) = 0$ (for extreme value)

$$\Rightarrow \sin\left(x - \frac{\pi}{6}\right) = 0 = \sin 0 \doteq \sin\pi \quad \text{...(iii)}$$

Therefore, on considering $\sin\left(x - \frac{\pi}{6}\right) = \sin 0 \Rightarrow$ $x - \frac{\pi}{6} = 0 \Rightarrow x = \frac{\pi}{6}$ and considering $\sin\left(x - \frac{\pi}{6}\right) = \sin\pi \Rightarrow x - \frac{\pi}{6} = \pi \Rightarrow x = \pi + \frac{\pi}{6} = \frac{7\pi}{6}$

$$\frac{d^2y}{dx^2} = -2\sqrt{3}\cos\left(x - \frac{\pi}{6}\right) \quad \text{...(iv)}$$

on putting $x = \frac{\pi}{6}$ and $\frac{7\pi}{6}$ in (iv)

$$\Rightarrow f''\left(\frac{\pi}{6}\right) = -2\sqrt{3}\cos\left(\frac{\pi}{6} - \frac{\pi}{6}\right)$$

$$= -2\sqrt{3}\cos 0 = -2\sqrt{3}$$

$\Rightarrow f''\left(\frac{\pi}{6}\right) = -2\sqrt{3} = \ominus$ indicating max. at $x = \frac{\pi}{6}$ and $f''\left(\frac{7\pi}{6}\right) = -2\sqrt{3}\cos\left(\frac{7\pi}{6} - \frac{\pi}{6}\right)$

$$= -2\sqrt{3}\cos\left(\frac{6\pi}{6}\right) = -2\sqrt{3}\cos\pi$$

$= 2\sqrt{3} = \oplus$ indicating min. at $x = \frac{7\pi}{6}$ and

$$y_{max} = \left[\sqrt{3}\sin x + 3\cos x\right]_{x=\frac{\pi}{6}}$$

$$= \sqrt{3}\sin\frac{\pi}{6} + \frac{3\cdot\sqrt{3}}{2}$$

$$= \frac{\sqrt{3}}{2} + \frac{3\sqrt{3}}{2} = \frac{4\sqrt{3}}{2} = 2\sqrt{3}$$

$$y_{min} = \left[\sqrt{3}\sin x + 3\cos x\right]_{x=\frac{7\pi}{6}}$$

$$= \sqrt{3}\sin\frac{7\pi}{6} + 3\cos\frac{7\pi}{6} = \sqrt{3}(-\sin 30) + 3(-\cos 30)$$

$$= -\frac{\sqrt{3}}{2} - \frac{3\cdot\sqrt{3}}{2} = -\frac{4\sqrt{3}}{2} = -2\sqrt{3}$$

Problems based on finding the extrema when a given function is defined in an open interval.

To find the maximum and / minimum value of a given function defined in an open interval, we should remember the following facts.

1. A function defined in an interval (open) can have maximum and / minimum values only for those value of x which lie within this interval which means the roots of $f'(x) = 0$ must lie within the interval.

2. If $f'(x) = 0$ provides us the values of x which do not lie within the given interval, we have not to consider maximum and / minimum values at those value of x (which do not lie within the given interval).

3. Let a function $y = f(x)$ be differentiable in an open interval (a, b). In order to find out all its maxima and / minima in that open interval, we proceed as follows:

(a) Solve the equation $f'(x) = 0$. The real values of x (or, roots) of the equation $f'(x) = 0$ lying within the interval $(a < x < b)$ are considerable points. Among these real values of x, we have to seek those values of x which give us extrema (i.e. maxima and / minima) of the function $f(x)$ by using $\frac{d^2y}{dx^2}$ method (i.e., second derivative test) or by using $\frac{dy}{dx}$ method (i.e.; first derivative test).

4. In fact, $f(x)$ may have several maxima and minima in an open interval (a, b) or in an closed interval.

1. Find the maximum and / minimum value of the function $y = \sin x + \cos 2x$ in $(0, 2\pi) = 0 < x < 2\pi$.

Solution: $y = \sin x + \cos 2x$

$$\Rightarrow \frac{dy}{dx} = \cos x + (-\sin 2x) \times 2$$

$$\therefore \frac{dy}{dx} = 0 \Rightarrow \cos x - 2\sin 2x = 0 \text{ (for extrema)}$$

$$\Rightarrow \cos x - 4 \sin x \cos x = 0$$

$$\Rightarrow \cos x (1 - 4 \sin x) = 0$$

$$\Rightarrow \cos x = 0 \text{ or } (1 - 4 \sin x) = 0$$

Now, $\cos x = 0 = \cos\frac{\pi}{2} \Rightarrow x = (2n+1)\frac{\pi}{2}$

Putting $n = 0$, $x = \frac{\pi}{2}$

$n = 1$, $x = \frac{3\pi}{2}$

$\because 0 \le x \le 2\pi$

$\therefore x = \frac{\pi}{2}$ or $\frac{3\pi}{2}$

and $(1 - 4\sin x) = 0 \Rightarrow \sin x = \frac{1}{4} \Rightarrow x = \sin^{-1}\frac{1}{4}$

and $\pi - \sin^{-1}\frac{1}{4}$

Further, $\frac{dy}{dx} = \cos x - 2\sin 2x$

$$\Rightarrow \frac{d^2y}{dx^2} = -\sin x - 2\cos 2x \times 2 = -(\sin x + 4\cos 2x)$$

$$\therefore \left(\frac{d^2y}{dx^2}\right)_{x=\frac{\pi}{2}} = -\left[+\sin\frac{\pi}{2} + 4\cos\left(2 \times \frac{\pi}{2}\right)\right]$$

$$= -\left[\sin\frac{\pi}{2} + 4\cos\pi\right]$$

$= 3 = \oplus$ which indicates given function has a minimum at $x = \frac{\pi}{2}$.

Again,

$$\left(\frac{d^2y}{dx^2}\right)_{x=\frac{3\pi}{2}} = [-\sin x - 4\cos 2x]_{x=\frac{3\pi}{2}}$$

$$= \left[-\sin\frac{3\pi}{2} - 4\cos\left(\frac{3\pi}{2} \times 2\right)\right]$$

$= (+1) - 4(-1) = 5 = \oplus$ which indicates given function has a minimum at $x = \frac{3\pi}{2}$

Now, $\left(\frac{d^2y}{dx^2}\right)_{x=\sin^{-1}\frac{1}{4}}$

$$= -\left[\sin\sin^{-1}\frac{1}{4}\right] - 4\left[1 - 2\sin^2\left(\sin^{-1}\frac{1}{4}\right)\right]$$

$$= -\frac{1}{4} - 4\left[1 - 2 \times \frac{1}{16}\right]$$

$= -\frac{1}{4} - 4\left(1 - \frac{1}{8}\right) = \ominus$ which indicates given function $f(x)$ has a maximum at $x = \sin^{-1}\frac{1}{4}$

Similarly, $\left(\frac{d^2 y}{dx^2}\right)_{x=\pi-\sin^{-1}\frac{1}{4}}$

$= -[\sin x + 4\cos 2x]_{x=\pi-\sin^{-1}\frac{1}{4}}$

$= -\left[\sin\left(\pi-\sin^{-1}\frac{1}{4}\right)\right] + 4\left(1-2\sin^2\left(\pi-\sin^{-1}\frac{1}{4}\right)\right)$

$= -\left[\sin\sin^{-1}\frac{1}{4} + 4\cdot\left(1-2\left(\sin\sin^{-1}\frac{1}{4}\right)^2\right)\right]$

$= -\left[\frac{1}{4} + 4\left(1-2\times\frac{1}{16}\right)\right] = -\left[\frac{1}{4} + 4\left(1-\frac{1}{8}\right)\right]$

$= -\left[\frac{1}{4} + 4\left(\frac{7}{8}\right)\right] = -\left[\frac{1}{4} + \frac{7}{2}\right] = -\left[\frac{1+14}{4}\right]$

$= -\frac{15}{4} = \ominus$ which indicates given function has maximum at $x = \pi - \sin^{-1}\frac{1}{4}$

$\therefore$ given function has minima at $x = \frac{\pi}{2}$ and $\frac{3\pi}{2}$ and given function has maxima at $x = \sin^{-1}\frac{1}{4}$ and $x = \pi - \sin^{-1}\frac{1}{4}$.

At $x = \sin^{-1}\frac{1}{4}$,

$y_{max} = [\sin x + \cos 2x]_{x=\sin^{-1}\frac{1}{4} \Leftrightarrow \sin x = \frac{1}{4}}$

$= \left[\sin x + 1 - 2\sin^2 x\right]_{\sin x = \frac{1}{4}}$

$= \frac{1}{4} + 1 - 2\left(\frac{1}{4}\right)^2 = \frac{1}{4} + 1 - \frac{2}{16} = \frac{9}{8}$

At $x = \pi - \sin^{-1}\frac{1}{4}$

$y_{max} = [\sin x + \cos 2x]_{x=\pi-\sin^{-1}\frac{1}{4}}$

$= \left[\sin x + 1 - 2\sin^2 x\right]_{x=\pi-\sin^{-1}\frac{1}{4}}$

$= \sin\left(\pi - \sin^{-1}\frac{1}{4}\right) + 1 - 2\left\{\left(\pi - \sin^{-1}\frac{1}{4}\right)\right\}^2$

$= \sin\sin^{-1}\frac{1}{4} + \left\{1 - 2\left(\sin\sin^{-1}\frac{1}{4}\right)^2\right\}$

$= \frac{1}{4} + \left(1 - \frac{2}{16}\right) = \frac{9}{8}$

At $x = \frac{\pi}{2}$

$y_{min} = [\sin x + \cos 2x]_{x=\frac{\pi}{2}}$

$= \sin\frac{\pi}{2} + \cos 2\times\frac{\pi}{2}$

$= 1 + \cos\pi$

$= 1 + (-1) = 0$ and at $x = \frac{3\pi}{2}$

$y_{min} = [\sin x + \cos 2x]_{x=\frac{3\pi}{2}}$

$= \sin\frac{3\pi}{2} + \cos 2\times\frac{3\pi}{2} = -1 - 1 = -2$

2. Find the maximum and / minimum values of the function $y = x + \sin 2x$ in $(0, 2\pi)$.

Solution: $y = x + \sin 2x$

$\Rightarrow \frac{dy}{dx} = 1 + 2\cos 2x$

$\frac{dy}{dx} = 0 \Rightarrow 1 + 2\cos 2x = 0$ (for extrema)

$\Rightarrow \cos 2x = -\frac{1}{2} = \cos 120° = \cos\frac{2\pi}{3}$ which $\Rightarrow$ least value of $2x = \frac{2\pi}{3}$ and general value of $2x = 2n\pi \pm \frac{2\pi}{3} \Rightarrow x = n\pi \pm \frac{\pi}{3}$

Putting $n = 0$ in the general value of x, we get

$$x = \frac{\pi}{3}, -\frac{\pi}{3}$$

Putting $n = 1$ and 2 in the general value of x, we get

$$x = \frac{4\pi}{3}, \frac{2\pi}{3}, \text{ and } \frac{5\pi}{3}, \frac{7\pi}{3}$$

$\because 0 \le x \le 2\pi$

$\therefore x = \frac{\pi}{3}, \frac{2\pi}{3}, \frac{4\pi}{3}$ and $\frac{5\pi}{3}$ are considerable values for $x \in [0, 2\pi]$

Now, $\frac{dy}{dx} = 1 + 2\cos 2x$

$$\Rightarrow \frac{d^2y}{dx^2} = -4\sin 2x$$

$$\Rightarrow \left(\frac{d^2y}{dx^2}\right)_{x=\frac{\pi}{3}} = -4\sin\left(2\times\frac{\pi}{3}\right)$$

$= -2\sqrt{3} = \ominus$ which indicates given function has maximum at $x = \frac{\pi}{3}$

$$\left(\frac{d^2y}{dx^2}\right)_{x=\frac{2\pi}{3}} = -4\sin 2\times\frac{2\pi}{3} = -4\sin\frac{4\pi}{3}$$

$= 2\sqrt{3} = +\text{ve}$,

$\therefore$ given function has minimum at $x = \frac{2\pi}{3}$.

and $\left(\frac{d^2y}{dx^2}\right)_{x=\frac{4\pi}{3}} = (-4\sin 2x)_{x=\frac{4\pi}{3}}$

$$= -4\sin\left(2\times\frac{4\pi}{3}\right) = -2\sqrt{3}(-\text{ve})$$

$$\left(\frac{d^2y}{dx^2}\right)_{x=\frac{5\pi}{3}} = -4\sin\frac{10\pi}{3}$$

$= 2\sqrt{3}$ (+ve)

which indicates given function has a maximum at $x = \frac{4\pi}{3}$ and mimum at $\frac{5\pi}{3}$.

Hence, y has maxima at $x = \frac{\pi}{3}$ and $\frac{4\pi}{3}$

Whereas y has a minima at $x = \frac{2\pi}{3}$ and $\frac{5\pi}{3}$

Therefore, at $x = \frac{\pi}{3}$,

$$y_{\max} = [x + \sin 2x]_{x=\frac{\pi}{3}}$$

$$= \left[\frac{\pi}{3} + \sin\frac{2\pi}{3}\right]$$

$$= \frac{\pi}{3} + \frac{\sqrt{3}}{2}$$

and at $x = \frac{4\pi}{3}$,

$$y_{\max} = [x + \sin 2x]_{x=\frac{4\pi}{3}}$$

$$= \frac{4\pi}{3} + \sin\left(2\times\frac{4\pi}{3}\right) = \frac{4\pi}{3} + \sin\left(\frac{8\pi}{3}\right)$$

$$= \frac{4\pi}{3} + \frac{\sqrt{3}}{2} = \frac{8\pi + 3\sqrt{3}}{6} \text{ at } x = \frac{2\pi}{3}$$

$$y_{\min} = [x + \sin 2x]_{x=\frac{2\pi}{3}} = \frac{2\pi}{3} + \sin\left(2\times\frac{2\pi}{3}\right)$$

$$= \frac{2\pi}{3} + \left(-\sin\frac{\pi}{3}\right) = \frac{2\pi}{3} - \frac{\sqrt{3}}{2}$$

$$= \frac{4\pi - 3\sqrt{3}}{6} \text{ and at } x = \frac{5\pi}{3}$$

$$y_{\min} = \frac{5\pi}{3} - \frac{\sqrt{3}}{2}$$

3. Find the maximum and / minimum value of the function $y = x + \cos 2x$ in $(0, 2\pi) = 0 < x < 2\pi$.

Solution: $y = x + \cos 2x$

$$\Rightarrow \frac{dy}{dx} = 1 + (-\sin 2x) \times 2 = 1 - 2\sin 2x$$

$$\frac{dy}{dx} = 0 \Rightarrow 1 - 2\sin 2x = 0 \text{ (for extrema)}$$

$$\Rightarrow \sin 2x = \frac{1}{2} = \sin\left(\frac{\pi}{6}\right)$$

$$\Rightarrow 2x = n\pi + (-1)^n \cdot \frac{\pi}{6}$$

$$\Rightarrow x = \frac{n\pi}{2} + (-1)^n \cdot \frac{\pi}{12}$$

Putting $n = 0$, $x = \frac{\pi}{12}$

Putting $n = 1$, $x = \frac{5\pi}{12}$

Putting $n = 2$, $x = \pi + \frac{\pi}{12}$

Putting $n = 3$, $x = \frac{3\pi}{2} - \frac{\pi}{12} = \frac{18\pi - \pi}{12}$

$$= \frac{17\pi}{12} = \pi + \frac{5\pi}{12}$$

Now, $\frac{dy}{dx} = 1 - 2\sin 2x$

$$\Rightarrow \frac{d^2y}{dx^2} = -2\cos 2x \times 2 = -4\cos 2x$$

$$\therefore \left(\frac{d^2y}{dx^2}\right)_{x=\frac{\pi}{12}} = -4\cos\left(2 \times \frac{\pi}{12}\right)$$

$= -4\cos\frac{\pi}{6} = -4 \times \frac{\sqrt{3}}{2} = -\text{ve}$ which indicates given function has a maximum at $x = \frac{\pi}{12}$.

$\left(\frac{d^2y}{dx^2}\right)_{x=\frac{5\pi}{12}} = -4\cos\frac{5\pi}{6} = 2\sqrt{3}$ which indicates given function has a minimum at $x = \frac{5\pi}{12}$.

$$\left(\frac{d^2y}{dx^2}\right)_{x=\left(\pi+\frac{\pi}{12}\right)}$$

$$= -4\cos 2\left(\pi + \frac{\pi}{12}\right) = -4\cos\left(2\pi + \frac{\pi}{6}\right)$$

$$= -4\cos\frac{\pi}{6} = -4 \times \frac{\sqrt{3}}{2} = -2\sqrt{3} = \ominus$$

which indicates given function has a maximum at $x = \left(\pi + \frac{\pi}{12}\right)$ and

$$\left(\frac{d^2y}{dx^2}\right)_{x=\left(\pi+\frac{5\pi}{12}\right)} = -4\cos 2\left(\pi + \frac{5\pi}{12}\right)$$

$= 2\sqrt{3} = +\text{ve}$ which indicates given function has minimum at $x = \left(\pi + \frac{5\pi}{12}\right)$.

Hence, given function has maxima at $x = \frac{\pi}{12}$, $\left(\pi + \frac{\pi}{12}\right)$ and given function has minima at $x = \frac{5\pi}{12}, \left(\pi + \frac{5\pi}{12}\right)$.

Lastly, at $x = \frac{\pi}{12}$

$$y_{\max} = \frac{\pi}{12} + \cos\left(2 \times \frac{\pi}{12}\right)$$

$$= \frac{\pi}{12} + \cos\frac{\pi}{6}$$

$= \frac{\pi}{12} + \frac{\sqrt{3}}{2}$ at $x = \left(\pi + \frac{\pi}{12}\right)$,

$$y_{\max} = \left(\pi + \frac{\pi}{12}\right) + \cos 2\left(\pi + \frac{\pi}{12}\right)$$

$$= \left(\pi + \frac{\pi}{12}\right) + \cos\left(2\pi + \frac{\pi}{6}\right)$$

$$= \left(\pi + \frac{\pi}{12}\right) + \cos\frac{\pi}{6}$$

$$= \left(\pi + \frac{\pi}{12}\right) + \frac{\sqrt{3}}{2}$$

at $x = \frac{5\pi}{12}$,

$$y_{\min} = \frac{5\pi}{12} + \cos 2\left(\frac{5\pi}{12}\right)$$

$$= \frac{5\pi}{12} + \cos\frac{5\pi}{6}$$

$$= \frac{5\pi}{12} - \frac{\sqrt{3}}{2}$$

Similarly, we can find the minimum value of y at $x = \left(\pi + \frac{5\pi}{12}\right)$ which is $\pi + \frac{5\pi}{12} - \frac{\sqrt{3}}{2}$.

4. Investigate the maximum and / minimum values in the interval $0 < x < \pi$ of the function

$$f(x) = \sin x + \frac{1}{2}\sin 2x + \frac{1}{3}\sin 3x.$$

Solution: $f(x) = \sin x + \frac{1}{2}\sin 2x + \frac{1}{3}\sin 3x \Rightarrow$

$\Rightarrow f'(x) = \cos 2x + 2\cos 2x \cdot \cos x$ (on using "$C + D$" formula)

$= \cos 2x\,(1 + 2\cos x)$

Now, $f'(x)$ (for an extreme values) $= 0$

$$\Rightarrow \cos 2x\,(1 + 2\cos x) = 0$$

$$\Rightarrow \cos 2x = 0 \text{ or } (1 + 2\cos x) = 0$$

$$\Rightarrow \cos 2x = \cos\frac{\pi}{2} \text{ or } \cos x = -\frac{1}{2} = \cos\left(\frac{2\pi}{3}\right)$$

$$\Rightarrow 2x = (2n+1)\frac{\pi}{2} \text{ or } x = 2n\pi \pm \frac{2\pi}{3}$$

$$\Rightarrow x = (2n+1)\frac{\pi}{4} \text{ or } x = 2n\pi \pm \frac{2\pi}{3}$$

$x = (2n+1)\frac{\pi}{4}$	$x = 2n\pi \pm \frac{2\pi}{3}$
Putting $n = 0$, $x = \frac{\pi}{4}$	Putting $n = 0$, $x = \pm\frac{2\pi}{3}$
Putting $n = 1$, $x = \frac{3\pi}{4}$	Putting $n = 1$, $x = 2\pi \pm \frac{2\pi}{3}$

$\because 0 \le x \le \pi$

$\therefore x = \frac{\pi}{4}, \frac{3\pi}{4}$ and $\frac{2\pi}{3}$ are only considerable values of x.

Again $f'(x) = \cos x + \cos 2x + \cos 3x$

$$\Rightarrow f''(x) = -\sin x - 2\sin 2x - 3\sin 3x$$

$$\Rightarrow f''\left(\frac{\pi}{4}\right) = -\sin\frac{\pi}{4} - 2\sin\frac{\pi}{2} - 3\sin\frac{3\pi}{4}$$

$= -\frac{1}{\sqrt{2}} - 2 - \frac{3}{\sqrt{2}} = \ominus$ which indicates $f(x)$ as a maximum at $x = \frac{\pi}{4}$.

$$f''\left(\frac{3\pi}{4}\right) = -\sin\frac{3\pi}{4} - 2\sin\frac{3\pi}{2} - 3\sin\frac{9\pi}{4}$$

$$= -\frac{1}{\sqrt{2}} + 2\times 1 - 3\times\frac{1}{\sqrt{2}} = -\frac{1}{\sqrt{2}} + 2 - \frac{3}{\sqrt{2}}$$

$$= \frac{-1 + 2\sqrt{2} - 3}{\sqrt{2}}$$

$= \frac{-4 + 2\sqrt{2}}{\sqrt{2}} = \ominus$ which indicates $f(x)$ has a maximum at $x = \frac{3\pi}{4}$ and

$$f''\left(\frac{2\pi}{3}\right) = (-\sin x - 2\sin 2x - 3\sin 3x)_{x=\frac{2\pi}{3}}$$

$$= -\sin\frac{2\pi}{3} - 2\sin\frac{4\pi}{3} - 3\sin 2\pi$$

$$= -\frac{1}{2} + 2 \times \frac{\sqrt{3}}{2} = -\frac{1}{2} + \sqrt{3}$$

$= \oplus$ which indicates $f(x)$ has a minimum at

$$x = \frac{2\pi}{3}$$

Hence, at $x = \frac{\pi}{4}$, we have

$$\text{max. } f(x) = f\left(\frac{\pi}{4}\right)$$

$$= \left(\sin x + \frac{1}{2}\sin 2x + \frac{1}{3}\sin 3x\right)_{x=\frac{\pi}{4}}$$

$$= \sin\frac{\pi}{4} + \frac{1}{2}\, sin\frac{\pi}{2} + \frac{1}{3}\sin\frac{3\pi}{4}$$

$$= \frac{1}{\sqrt{2}} + \frac{1}{2} + \frac{1}{3\sqrt{2}}$$

$$= \frac{3\sqrt{2} + 3 + \sqrt{2}}{6} = \frac{4\sqrt{2} + 3}{6}$$

at $x = \frac{3\pi}{4}$, we have

$$\text{max. } f(x) = f\left(\frac{3\pi}{4}\right)$$

$$= \left(\sin x + \frac{1}{2}\sin 2x + \frac{1}{3}\sin 3x\right)_{x=\frac{3\pi}{4}}$$

$$= \sin\frac{3\pi}{4} + \frac{1}{2}\, sin\frac{3\pi}{2} + \frac{1}{3}\sin\frac{9\pi}{4}$$

$$= \frac{1}{\sqrt{2}} + \frac{1}{2}(-1) + \frac{1}{3} \times \frac{1}{\sqrt{2}} = \frac{1}{\sqrt{2}} - \frac{1}{2} + \frac{1}{3\sqrt{2}}$$

$$= \frac{3\sqrt{2} - 3 + \sqrt{2}}{6} = \frac{4\sqrt{2} - 3}{6}$$

and lastly, at $x = \frac{2\pi}{3}$, we have

$$\text{min. } f(x) = \left(\sin x + \frac{1}{2}\sin 2x + \frac{1}{3}\sin 3x\right)_{x=\frac{2\pi}{3}}$$

$$= \sin\frac{2\pi}{3} + \frac{1}{2}\sin\frac{4\pi}{3} + \frac{1}{3}\sin 2\pi$$

$$= \frac{\sqrt{3}}{2} - \frac{-\sqrt{3}}{4}$$

$$= \frac{\sqrt{3}}{4}$$

5. Find the maximum and / minimum value of the function $y = \sqrt{3}\cos x + \sin x$ in the interval $(0, 2\pi) = 0 < x < 2\pi$.

Solution: $y = \sqrt{3}\cos x + \sin x$

$$\Rightarrow \frac{y}{2} = \frac{\sqrt{3}}{2}\cos x + \frac{1}{2}\sin x \qquad \ldots(1)$$

$$\left[\because \sqrt{a^2 + b^2} = \sqrt{\left(\sqrt{3}\right)^2 + 12} = \sqrt{3+1} = \sqrt{4} = 2\right]$$

$$\Rightarrow \frac{y}{2} = \cos\frac{\pi}{6}\cos x + \sin\frac{\pi}{6}\sin x$$

$$\Rightarrow \frac{y}{2} = \cos\left(x - \frac{\pi}{6}\right)$$

$$\Rightarrow y = 2\cos\left(x - \frac{\pi}{6}\right)$$

$$\Rightarrow \frac{dy}{dx} = -2\sin\left(x - \frac{\pi}{6}\right)$$

Now, $\frac{dy}{dx} = 0$ (for an extreme value)

$$\Rightarrow -2\sin\left(x - \frac{\pi}{6}\right) = 0$$

$$\Rightarrow \sin\left(x - \frac{\pi}{6}\right) = 0$$

$$\Rightarrow x - \frac{\pi}{6} = n\pi$$

$$\Rightarrow x = n\pi + \frac{\pi}{6} \qquad \ldots(2)$$

Putting, $n = 0$ in (2), we have $x = \frac{\pi}{6}$

Putting, $n = 1$ in (2), we have $x = \pi + \frac{\pi}{6}$

$\because 0 \le x \le 2\pi$

$\therefore x = \frac{\pi}{6}, \left(\pi + \frac{\pi}{6}\right)$ are only considerable values of x

$\because \frac{dy}{dx} = -2\sin\left(x - \frac{\pi}{6}\right)$

$\therefore \frac{d^2y}{dx^2} = -\cos\left(x - \frac{\pi}{6}\right)$

$\Rightarrow f''\left(\frac{\pi}{6}\right) = -2\cos\left(\frac{\pi}{6} - \frac{\pi}{6}\right)$

$= -2 = \ominus$ indicating max. at $x = \frac{\pi}{6}$ and

$f''\left(\pi + \frac{\pi}{6}\right) = -2\cos\left(\pi + \frac{\pi}{6} - \frac{\pi}{6}\right) = -2\cos\pi = -2 \times (-1)$

$= 2 \times 1 = 2 = \oplus$ indicating min at $x = \left(\pi + \frac{\pi}{6}\right)$

Hence, $y_{max} = \left[\sqrt{3}\cos x + \sin x\right]_{x = \frac{\pi}{6}}$

$= \sqrt{3}\cos\frac{\pi}{6} + \sin\frac{\pi}{6}$

$= \sqrt{3} \times \frac{\sqrt{3}}{2} + \frac{1}{2} = \frac{3}{2} + \frac{1}{2} = \frac{4}{2} = 2$

$y_{min} = \left[\sqrt{3}\cos x + \sin x\right]_{x = \left(\pi + \frac{\pi}{6}\right)}$

$= \sqrt{3}\cos\left(\pi + \frac{\pi}{6}\right) + \sin\left(\pi + \frac{\pi}{6}\right)$

$= \sqrt{3}\left(-\cos\frac{\pi}{6}\right) + \left(-\sin\frac{\pi}{6}\right)$

$= \sqrt{3} \times \left(-\frac{\sqrt{3}}{2}\right) - \frac{1}{2}$

$= -\frac{3}{2} - \frac{1}{2} = \frac{-4}{2}$

$= -2$

or, alternatively,

$\frac{y}{2} = \cos\left(x - \frac{\pi}{6}\right)$

$\Rightarrow y = 2\cos\left(x - \frac{\pi}{6}\right)$

$\Rightarrow y_{max} = 2$

and $y_{min} = -2$

6. Find the maximum and / minimum values of the function $y = \sin 2x - x$ when $-\frac{\pi}{2} < x < \frac{\pi}{2}$.

Solution: $y = \sin 2x - x$

$\Rightarrow \frac{dy}{dx} = 2\cos 2x - 1$

Now, $\frac{dy}{dx} = 0$ (for extrema)

$\Rightarrow 2\cos 2x - 1 = 0$

$\Rightarrow \cos 2x = \frac{1}{2}$

$\Rightarrow 2x = 2n\pi \pm \frac{\pi}{3}$

$\Rightarrow x = n\pi \pm \frac{\pi}{6}$ ...(1)

Putting $n = 0$, in (1), we get $x = \pm\frac{\pi}{6}$

Putting $n = 1$, in (1), we get $x = \pi \pm \frac{\pi}{6}$

$\because -\frac{\pi}{2} < x < \frac{\pi}{2}$

$\therefore x = \frac{\pi}{6}$ and $-\frac{\pi}{6}$ are only two considerable values of x

Again, $\frac{dy}{dx} = 2\cos 2x - 1$

$$\Rightarrow \frac{d^2y}{dx^2} = -4\sin 2x$$

$$\therefore \left(\frac{d^2y}{dx^2}\right)_{x=\frac{\pi}{6}} = -4\sin\left(2\times\frac{\pi}{6}\right)$$

$$= -4\cdot\frac{\sqrt{3}}{2} = \ominus$$ indicates max. at $x = \frac{\pi}{6}$ and

$$\left(\frac{d^2y}{dx^2}\right)_{x=-\frac{\pi}{6}} = -4\sin\left(-\frac{2\pi}{6}\right) = 4\sin\left(\frac{2\pi}{6}\right)$$

$$= \frac{4\sqrt{3}}{2} = 2\sqrt{3} = \oplus$$ which indicates y has minimum value at $x = -\frac{\pi}{6}$

Hence, $y_{\max} = [\sin 2x - x]_{x=\frac{\pi}{6}} = \sin\frac{2\pi}{6} - \frac{\pi}{6}$

$$= \sin\frac{\pi}{3} - \frac{\pi}{6} = \frac{\sqrt{3}}{2} - \frac{\pi}{6}$$

and $y_{\min} = [\sin 2x - x]_{x=-\frac{\pi}{6}}$

$$= \sin\left(-\frac{2\pi}{6}\right) + \frac{\pi}{6}$$

$$= -\sin\frac{\pi}{3} + \frac{\pi}{6} = -\frac{\sqrt{3}}{2} + \frac{\pi}{6}$$

7. Find the maximum and / minimum values of the function $y = \sin x + \cos x$ in the interval $0 < x < \pi$.

Solution: $y = \sin x + \cos x$

$$\Rightarrow \frac{dy}{dx} = -\sin x + \cos x$$

Now, $\frac{dy}{dx} = 0$ (for an extreme value)

$\Rightarrow -\sin x + \cos x = 0$

$\Rightarrow \sin x = \cos x$

$\Rightarrow \tan x = 1$

$$\Rightarrow x = n\pi + \frac{\pi}{4} \quad \ldots(1)$$

Putting, $n = 0$ in (1), we get, $x = \frac{\pi}{4}$

Putting, $n = 1$ in (1), we get, $x = \pi + \frac{\pi}{4}$

$\because 0 \le x \le \pi$

$\therefore x = \frac{\pi}{4}$ is only one considerable value of x

again, $\frac{dy}{dx} = \cos x - \sin x$

$$\Rightarrow \frac{d^2y}{dx^2} = -\sin x - \cos x = -(\sin x + \cos x)$$

$$\Rightarrow \left(\frac{d^2y}{dx^2}\right)_{x=\frac{\pi}{4}} = -(\sin x + \cos x)_{x=\frac{\pi}{4}}$$

$$= -\left(\sin\frac{\pi}{4} + \cos\frac{\pi}{4}\right)$$

$$= -\left(\frac{1}{\sqrt{2}} + \frac{1}{\sqrt{2}}\right)$$

$$= -\frac{2}{\sqrt{2}} = -\sqrt{2} = \ominus$$ indicating max. at $x = \frac{\pi}{4}$

Hence, at $x = \frac{\pi}{4}$, we have

$$y_{\max} = [\sin x + \cos x]_{x=\frac{\pi}{4}}$$

$$= \sin\frac{\pi}{4} + \cos\frac{\pi}{4}$$

$$= \frac{1}{\sqrt{2}} + \frac{1}{\sqrt{2}} = \frac{2}{\sqrt{2}} = \sqrt{2}$$

8. Find the maximum value of the function $y = \frac{\log x}{x}$ in $0 < x < \infty$.

Solution: $y = \dfrac{\log x}{x}, x > 0$

$$\Rightarrow \frac{dy}{dx} = \frac{x \cdot \frac{1}{x} - 1 \cdot \log x}{x^2} = \frac{1 - \log x}{x^2}$$

Now, $\dfrac{dy}{dx} = 0$ (for an extreme value)

$$\Rightarrow \frac{1 - \log x}{x^2} = 0$$

$$\Rightarrow 1 - \log x = 0$$

$$\Rightarrow \log x = 1$$

$$\Rightarrow x = e^1 = e$$

Again, $\dfrac{dy}{dx} = \dfrac{1 - \log x}{x^2}$

$$\Rightarrow \frac{d^2y}{dx^2} = \frac{x^2 \cdot \left(-\frac{1}{x}\right) - 2x(1 - \log x)}{x^4}$$

$$= \frac{-x - 2x(1 - \log x)}{x^4}$$

$$= \frac{-3 + 2\log x}{x^3}$$

$$\therefore \left(\frac{d^2y}{dx^2}\right)_{x=e} = \frac{-3 + 2\log_e e}{e^3}$$

$$= \frac{-3 + 2}{e^3}$$

$= -\dfrac{1}{e^3} < 0 = \ominus$ indicating maximum at $x = e$.

Hence, at $x = e$, y has maximum value and

$$y_{\max} = \frac{\log_e e}{e} = \frac{1}{e}$$

9. Find the maximum and / minimum values of the function $y = \dfrac{x}{2} + \dfrac{2}{x}, x > 0$.

Solution: $y = \dfrac{x}{2} + \dfrac{2}{x}; 0 < x < \infty$,

$$\Rightarrow \frac{dy}{dx} = \frac{1}{2} - \frac{2}{x^2}$$

Now, $\dfrac{dy}{dx} = 0$ (for an extreme value)

$$\Rightarrow \frac{1}{2} - \frac{2}{x^2} = 0$$

$$\Rightarrow -\frac{2}{x^2} = -\frac{1}{2}$$

$$\Rightarrow -x^2 = -4$$

$$\Rightarrow x^2 = 4$$

$$\Rightarrow x = \pm 2$$

But $x = -2 \notin (0, \infty)$ and $x = 2 \in (0, \infty)$

Hence, only considerable stationary point is $x = 2$ at which we have to examine extrema.

Again, $\dfrac{dy}{dx} = \dfrac{1}{2} - \dfrac{2}{x^2}$

$$\Rightarrow \frac{d^2y}{dx^2} = 0 + \frac{4}{x^3} = \frac{4}{x^3}$$

$\therefore \left(\dfrac{d^2y}{dx^2}\right)_{x=2} = \left(\dfrac{4}{x^2}\right)_{x=2} = \dfrac{4}{8} = \dfrac{1}{2} = \oplus$ which indicates y has a minimum value at $x = 2$.

Therefore, at $x = 2$, we have

$$y_{\min} = \left[\frac{x}{2} + \frac{2}{x}\right]_{x=2}$$

$$= \frac{2}{2} + \frac{2}{2} = 1 + 1$$

$$= 2$$

Verbal problems on maximum / minimum values of a function

The problems have been divided into four types:

1. Problems on numbers.
2. Problems on perimeter and area.

3. Problems on volume.
4. General problems

Working rule:

1. Find the function (if it is not given) of the quantity whose maximum or minimum is required by expressing the given conditions in symbols.

N.B.: It frequently appears (we observe / we see) as a function of more than one variable, $f(x, y) = c$.

2. Then our next step is to express the quantity whose max. and / min. is required in terms of a single variable (we consider only such problems).

N.B.: By means of geometrical or other given relations between the variables, all but one of these variables must be eliminated.

3. The quantity (i.e., function) having thus been expressed as a function of single variable, we put $\frac{dy}{dx} = 0$ and solve for the independent variable x which will provide us roots $x = a, b, c, \ldots$ etc.

4. Find $\frac{d^2 y}{dx^2}$ and $\left[\frac{d^2 y}{dx^2}\right]_{x = a, b, c, \ldots \text{etc}}$ to test where max and / min y exist (exists).

Remember: Given or specified means fixed i.e. the quantity (volume, area, length, angle, height, … etc) which is given is a constant.

Explanation with the help of examples

Ex. 1: Given the length of an arc of a circle, find the radius when the corresponding segment has maximum or minimum area.

(Here, length of the arc is constant)

Ex. 2: Show that right circular cylinder of the given surface and maximum volume is such that its height is equal to the diameter of the base.

(Here, surface (area) is constant)

Ex. 3: Show that semi-vertical angle of the cone of maximum volume and of given slant height is $\tan^{-1}(\sqrt{2})$.

(Here, slant height is constant)

Ex. 4: Show that semi vertical angle of the right circular cone of a given surface and maximum volume is $\sin^{-1}\left(\frac{1}{3}\right)$.

(Here, surface area is constant)

N.B.: **(i)** Care should be taken to distinguish between constants and variables.

(ii) To find the values of the independent variable at which a differentiable function $z = f_0(x, y)$ can have an extremum value, we must equate the derivative $\frac{dy}{dx} = \frac{d f_0(x, y)}{dx}$ to zero [where y is expressed as a function of a single variable x (i.e. $y = f(x)$)] which provides us the values of x and to find the values of y, we put $x = a$ (one of the roots of $\frac{dz}{dx} = 0$) in the equation $y = f(x)$.

Refresh your memory:

1. To be confirmed / to show the determined value of the function $z = f_0(x, y), y = f(x)$ is max., show that

$$\left[\frac{d^2 z}{dx^2}\right]_{x=a} = -\text{ve}.$$

2. To be confirmed / to show the determined value of the function $z = f_0(x, y), y = f(x)$ is min., we are required to show that $\left[\frac{d^2 z}{dx^2}\right]_{x=a} = +\text{ve}$ when z is expressed as a function of single variable x.

3. We are provided a single equation of condition $f_1(x, y) = c$ which is expressed as $y = f(x) =$ a function of single variable x.

4. We are required to remember the following formulae.

(i) Sphere:

$$\left.\begin{aligned} \text{volume} &= \frac{4}{3}\pi r^3 \\ \text{surface} &= 4\pi r^2 \end{aligned}\right] \text{where } r = \text{radius}$$

(ii) Cylinder:

Volume: $\pi r^2 h$

Curve surface: $2\pi r h$

Total surface: $2\left(\pi rh + \pi r^2\right)$ ($\because$ area of each plane surface $= \pi r^2$)

Where, h = height
r = radius of circular surface

(iii) Cone: (Right circular cone)

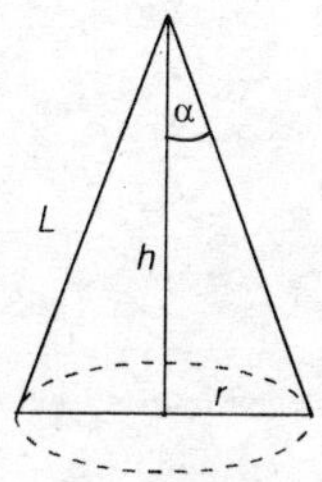

Volume: $\frac{1}{3}\pi r^2 h$

Curved surface: $\pi r l = \pi r\sqrt{r^2 + h^2}$

Semi vertical angle $= \alpha = \tan^{-1}\left(\frac{r}{h}\right)$

Where h = height
l = slant height
r = radius of circular base.

(iv) Cube:
Volume: a^3
Surface (area) = $6a^2$,
Where a = a side of a cube.

Examples worked out:

***Type* 1**: Problems on numbers

1. Divide 10 into two parts such that the sum of thier squares is minimum.

Solution: x and y be two numbers such that $x + y = 10$

$\therefore\ y = 10 - x$

Again on letting S = sum of squares of x and y, we have

$$S = x^2 + y^2 = x^2 + (10 - x)^2$$

$$= 2x^2 + 20x + 100,\ (x \geq 0)$$

$$\Rightarrow \frac{dS}{dx} = 4x - 20$$

$$\Rightarrow \frac{d^2S}{dx^2} = 4$$

$$\frac{dS}{dx} = 0 \Rightarrow 4x - 20 = 0 \Rightarrow x = 5$$

and $\left[\frac{d^2S}{dx^2}\right]_{x=5} = [4]_{x=5} = 4 = +\text{ve}$

$\therefore$ S reaches a minimum at $x = 5$. This is the only extremum (minimum) in $[0, \infty)$.

Hence, at $x = 5$, the function S attains the least (the minimum) value.

$\therefore$ S is the minimum when $x = 5, y = 5$.

N.B.: We recall that a quantity reaches a minimum means it may not be the least value whereas a quantity reaches (or, attains) the minimum means it is the least value necessarily.

2. The sum of two numbers is given. show that their product will be maximum if each number is equal to half of the sum.

Solution: Let the sum of two numbers x and $y = x + y = a$

$\therefore\ y = (a - x)$

$p = xy$ = product of x and $y = x\,(a - x)$

$$\Rightarrow \frac{dp}{dx} = 0$$

$$\Rightarrow (a - 2x = 0)$$

$$\Rightarrow x = \frac{a}{2}$$

Again, $\frac{d^2p}{dx^2} = 0 - 2 = -2$

$\therefore\ \left[\frac{d^2p}{dx^2}\right]_{x=\frac{a}{2}} = [-2]_{x=\frac{a}{2}} = -2 = -\text{ve}$

$\therefore$ p reaches a maximum at $x = \frac{a}{2}$.

∴ This is the only extremum (maximum).

Hence, the maximum value of p is attained at

$x = \frac{a}{2}$, i.e. when $x = \frac{a}{2}, y = \frac{a}{2}$.

3. Divide the number 14 into two parts such that the product of the two parts may be maximum or minimum. Also find the maximum product.

Solution: Let $x + y = 10$

$\therefore\ y = 14 - x$

and $(x \cdot y) = x(14 - x) = p$ (say)

Now, $\frac{dp}{dx} = \frac{d}{dx}\left(14x - x^2\right)$

$= 14 - 2x$

$\therefore\ \frac{dp}{dx} = 0 \Rightarrow (14 - 2x) = 0 \Rightarrow x = \frac{-14}{-2} = 7$

$\frac{d^2 p}{dx^2} = 0 - 2 = -2$

and $\left[\frac{d^2 p}{dx^2}\right]_{x=7} = [-2]_{x=7} = -2 = -\text{ve}$

∴ p reaches a maximum at $x = 7$.

Hence, maximum value of p is attained at $x = 7$ because this is the only extremum (maximum).

$\therefore$ max. $p = \left[x(14 - x)\right]_{x=7}$

$= 7(14 - 7) = 7 \times 7 = 49$

Type 2: Problems on perimeter

1. Show that among rectangles of the given area, the square has the least perimeter.

Solution: Let x = length of the rectangle

y = breadth of the rectangle

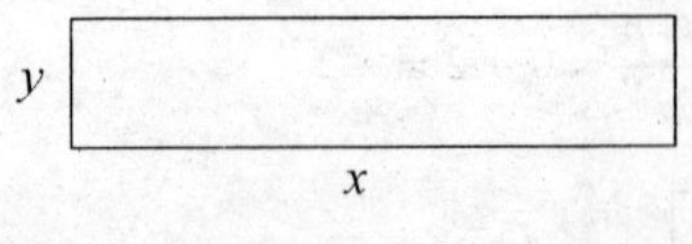

Since, area is given, A = area $= x \cdot y$ = constant ...(1)

Perimeter $= p = 2x + 2y = 2(x + y)$ ...(2)

Now, (1) $\Rightarrow y = \frac{A}{x}$ ...(3)

Putting (3) in (2), we have

$p = 2\left(x + \frac{A}{x}\right)$

$\Rightarrow \frac{dp}{dx} = 2\left(1 - \frac{A}{x^2}\right)$

$\therefore\ \frac{dp}{dx} = 0$

$\Rightarrow 2\left(1 - \frac{A}{x^2}\right) = 0$

$\Rightarrow \frac{A}{x^2} = 1$

$\Rightarrow x^2 = A$

$\Rightarrow x = \sqrt{A}\ (\because x > 0)$ ...(4)

Again, on putting $x = \sqrt{A}$ in the equation of condition which is (1)

$A = x \cdot y = \sqrt{A} \cdot y$

$\Rightarrow y = \frac{A}{\sqrt{A}} = \sqrt{A}$ ...(5)

(4) and (5) $\Rightarrow x = y = \sqrt{A}$

Now, to be confirmed whether p is maximum or minimum, we need to know the sign of $\frac{d^2 p}{dx^2}$

$2\left(\frac{2A}{x^3}\right) = \frac{4A}{x^3}$ when $x = \sqrt{A}$

Now $\left[\frac{d^2 p}{dx^2}\right]_{x=\sqrt{A}} = \left[\frac{4A}{x^3}\right]_{x=\sqrt{A}} = \frac{4 \times A}{\left(\sqrt{A}\right)^3}$

$= \frac{4}{\sqrt{A}} = +\text{ve}$

Hence, p reaches a minimum at $x = \sqrt{A}$. This being the only extremum (minimum), the maximum value of p is attained at $\Rightarrow$, i.e., when the rectangle is a square.

(*Note: An equation of condition (in this chapter) is an equation between two variables x and y satisfying a certain condition.)

2. Show that the perimeter of a right angled triangle of a given hypotenuse is maximum when the triangle is isosceles.

Solution: Let ΔABC = a right angled triangle

b = base = BC

p = perpendicular = AC

h = hypotenuse = AB

P = perimeter of ΔABC

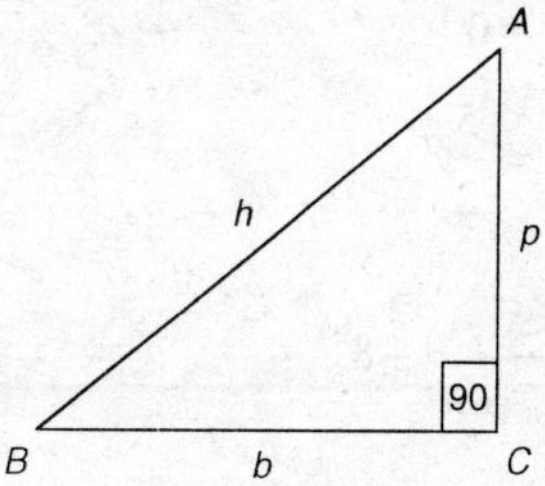

Given hypotenuse $\Rightarrow$ given $h \Rightarrow h$ = constant

Now, $h^2 = b^2 + p^2$ (by Pythogora's theorem)

$$\Rightarrow p^2 = h^2 - b^2$$

$$\Rightarrow p = \sqrt{h^2 - b^2} \quad (\because p > 0) \qquad ...(1)$$

and perimeter = $P = h + b + p$ ...(2)

Putting (1) in (2), we have $p = h + b + \sqrt{h^2 - b^2}$

$\Rightarrow p = f(b)$ a function of single variable b

$$\Rightarrow \frac{dp}{db} = 0 + 1 + \frac{1}{2\sqrt{h^2 - b^2}} \times (-2b)$$

$$= 1 - \frac{b}{\sqrt{h^2 - b^2}} \qquad ...(3)$$

$$\therefore \frac{dp}{db} = 0$$

$$\Rightarrow 1 - \frac{b}{\sqrt{h^2 - b^2}} = 0 \Rightarrow \frac{b}{\sqrt{h^2 - b^2}} = 1$$

$$\Rightarrow b = \sqrt{h^2 - b^2} \Rightarrow b^2 = h^2 - b^2$$

$$\Rightarrow h^2 = 2b^2 \Rightarrow b = \frac{h}{\sqrt{2}} \qquad ...(4)$$

Putting (4) in (1), we have, $p^2 = h^2 - \frac{h^2}{2}$

$$\Rightarrow p = \frac{h}{\sqrt{2}} \qquad ...(5)$$

(4) and (5) $\Rightarrow p = b = \frac{h}{\sqrt{2}} \Rightarrow \Delta ABC$ is isosceles Δ

Now, to be confirmed whether p is max., we need to know the sign of $\frac{d^2 p}{db^2}$ for $b = \frac{h}{\sqrt{2}}$

$$\frac{d^2 p}{db^2} = \frac{d}{db}\left[\frac{dp}{db}\right] = \frac{d}{db}\left[1 - \frac{b}{\sqrt{h^2 - b^2}}\right]$$

$$= \frac{-\left[\sqrt{h^2 - b^2} - \frac{1}{2}\left(h^2 - b^2\right)^{-\frac{1}{2}}(-2b)\, b\right]}{\left(\sqrt{h^2 - b^2}\right)^2}$$

(using the quotient rule $\frac{d}{dx}\left(\frac{u}{v}\right)$)

$$= \frac{-\left[\sqrt{h^2 - b^2} + \frac{b^2}{\sqrt{h^2 - b^2}}\right]}{\left(\sqrt{h^2 - b^2}\right)^2}$$

$$= \frac{-\left[\left(h^2 - b^2\right) + b^2\right]}{\left(\sqrt{h^2 - b^2}\right)\left(\sqrt{h^2 - b^2}\right)^2}$$

$$= \frac{-h^2}{\left(\sqrt{h^2 - b^2}\right)^3}$$

$$\therefore \left[\frac{d^2 p}{db^2}\right]_{b=\frac{h}{\sqrt{2}}} = -\left[\frac{h^2}{\left(h^2 - \frac{h^2}{2}\right)^{\frac{3}{2}}}\right]$$

$$= \frac{-h^2}{\left(\frac{2h^2 - h^2}{2}\right)^{\frac{3}{2}}} = \frac{-h^2}{\left(\frac{h^2}{2}\right)^{\frac{3}{2}}}$$

$$= \frac{-\left(\sqrt{2}\right)^3}{h} = -\text{ve}$$

Hence, P reaches a maximum at $b = \frac{h}{\sqrt{2}} = p$

$\Rightarrow$ P = perimeter of ΔABC is max. when Δ is isosceles.

***Type* 3**: Problems on perimeter and area

1. The perimeter of a triangle is 8 inches. If one of the sides is 3 inches, what are other two sides for maximum area of a triangle.

Solution: We know that

$$\Delta^2 = s(s-a)(s-b)(s-c)$$

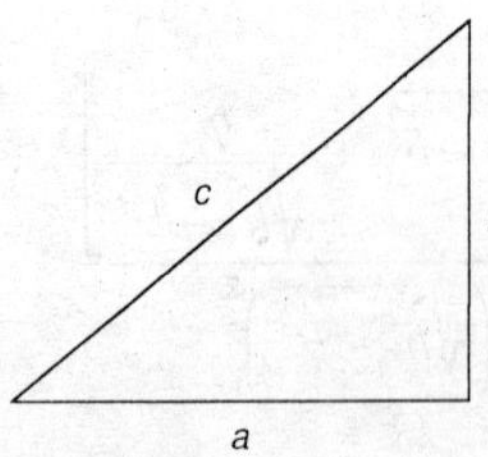

where S is the perimeter and a, b, c, the sides of the Δ.

$2s = P = a + b + c$

$\Rightarrow 8 = 3 + b + c$

$\Rightarrow b = 5 - c$

$$\therefore \Delta^2 = 4\,(4-3)\,(4-b)\,(4-c)$$
$$= 4 \cdot 1 \cdot (4-b)\,(4-c)$$
$$= 4 \cdot 1 \cdot (4-5+c)\,(4-c)$$
$$= 4 \cdot 1 \cdot (c-1)\,(4-c)$$

$$\Rightarrow \frac{d\,\Delta^2}{dc} = \frac{d}{dc}\left[4\,(c-1)(4-c)\right]$$
$$= 4\frac{d}{dc}\left[-4 + 5c - c^2\right]$$
$$= 4\,(5-2c)$$
$$= 20 - 8c$$

Now, $\frac{d\,\Delta^2}{dc} = 0$

$\Rightarrow 20 - 8c = 0$

$\Rightarrow c = \frac{20}{8} = \frac{5}{2}$

Again, $\frac{d^2\,\Delta^2}{dc^2} = -8$

$$\Rightarrow \left[\frac{d^2\,\Delta^2}{dc^2}\right]_{c=\frac{5}{2}} = [-8]_{c=\frac{5}{2}} = -8 = -\text{ve}$$

$\therefore \Delta^2$ and so Δ is maximum when $c = \frac{5}{2}$

$$\therefore b = [5-c]_{c=\frac{5}{2}} = \left[5 - \frac{5}{2}\right] = \frac{5}{2}$$

Thus, $\left.\begin{array}{l} c = 2.5 \\ b = 2.5 \end{array}\right\} \Rightarrow$ The triangle must be isosceles for the area to be maximum.

2. The sum of a perimeter of a circle and a square is l. Show that when the sum of the area is least, the side of the square is double the radius of the circle.

Solution: Let r = radius of the circle

a = a side of the square

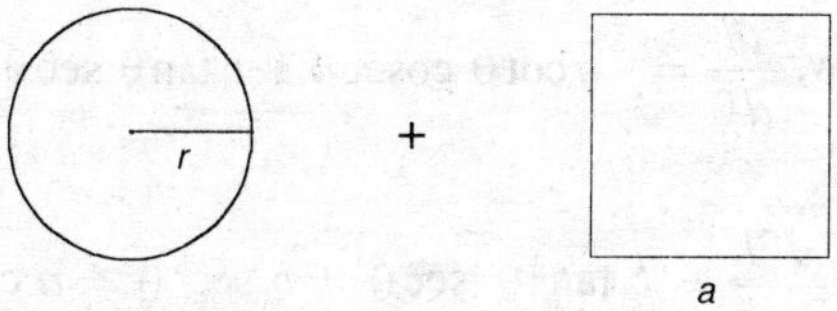

$\therefore$ Perimeter of the circle $= 2\pi r$
Perimeter of the square $= 4a$
$\Bigg\} \Rightarrow 4a + 2\pi r = l$
$\Rightarrow l - 2\pi r = 4a$

From the question,
Sum of the area of a circle and a square,

$$A = \pi r^2 + a^2 \qquad \ldots(1)$$

[In this expression, we observe there are two variables which suggest us there should be $(n-1) = (2-1) = 1$ relation which must be provided.]

Here, given relation is $2\pi r + 4a = l$ from which,

we obtain $a = \dfrac{l - 2\pi r}{4}$ $\qquad \ldots(2)$

Putting (2) in (1), we get

$$A = \text{area} = \pi r^2 + a^2 = \pi r^2 + \left(\frac{l - 2\pi r}{4}\right)^2$$

$$\Rightarrow \frac{dA}{dr} = \pi \cdot 2r + \frac{1}{16} \cdot 2(l - 2\pi r)(-2\pi)$$

$$= 2\pi r - \frac{\pi}{4}(l - 2\pi r) \qquad \ldots(3)$$

Now, for maxima and / minima,

$$\frac{dA}{dr} = 0$$

$$\Rightarrow 2\pi r - \frac{\pi}{4}(l - 2\pi r) = 0$$

$$\Rightarrow 2\pi r = \frac{\pi}{4}(l - 2\pi r)$$

$$\Rightarrow 2\pi r = \frac{\pi}{4} \cdot 4a$$

$$\Rightarrow 2\pi r = \pi a$$

$$\Rightarrow 2r = a \Rightarrow \left(r = \frac{a}{2}\right)$$

Also, $\dfrac{dA}{dr} = 2\pi r - \dfrac{\pi}{4} \cdot 4a$

$$= 2\pi r - \pi a = 2\pi\left(r - \frac{a}{2}\right)$$

$$\therefore \frac{dA}{dr} \text{ is } \begin{cases} -\text{ve when } r < \dfrac{a}{2} \\ +\text{ve when } r > \dfrac{a}{2} \end{cases}$$

$\therefore \dfrac{dA}{dr}$ changes sign from minus to plus in moving from left to right through $r = \dfrac{a}{2}$.

$\therefore$ A has the minimum (least) value for $r = \dfrac{a}{2}$, i.e. when the side of the square is double the radius of the circle.

3. Let ABC be a right angle triangle right angled at C. Find the position of a point P on BC such that $AP^2 + PB^2$ may be minimum.

Solution: Let P be a point on BC such that $CP = x$

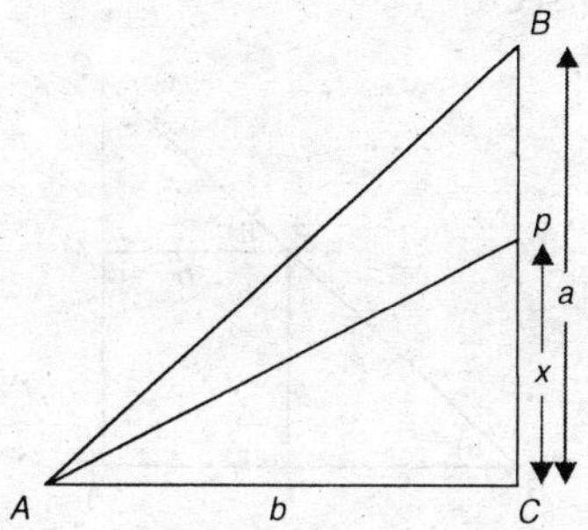

Let $BC = a$ and $AC = b$ and also letting, $y = AP^2 + PB^2$,

$$y = \left(b^2 + x^2\right) + (a - x)^2$$

$$= a^2 + b^2 - 2ax + 2x^2$$

$$\Rightarrow \frac{dy}{dx} = -2a + 4x$$

Now, $\dfrac{dy}{dx} = 0$

$\Rightarrow -2a + 4x = 0$

$\Rightarrow x = \dfrac{2a}{4} = \dfrac{a}{2}$

and $\Rightarrow \dfrac{d^2y}{dx^2} = 4$

$\therefore \left[\dfrac{d^2y}{dx^2}\right]_{x=\frac{a}{2}} = [4]_{x=\frac{a}{2}} = 4 = +\text{ve}$

$\Rightarrow y$ is min. at $x = \dfrac{a}{2}$

Hence, $AP^2 + PB^2$ is min. when $x = \dfrac{a}{2}$

4. A point in the hypotenuse of a right angled triangle is distant a and b from the two sides. show that the length of the hypotenuse is at least $\left(a^{\frac{2}{3}} + b^{\frac{2}{3}}\right)^{\frac{3}{2}}$.

Solution: Let $\angle ABC = \angle\theta = \angle MPC$

$PB = a\,\text{cosec}\,\theta$

$PC = b\sec\theta$

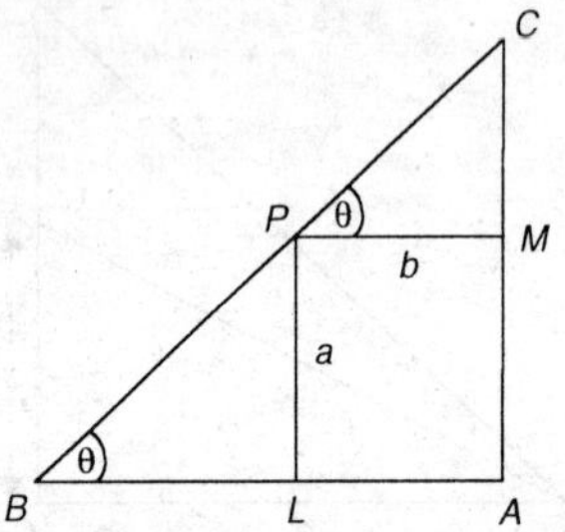

Length of the hypotenuse $= l = PC + PB$
$= a\,\text{cosec}\,\theta + b\sec\theta$

$\Rightarrow \dfrac{dl}{d\theta} = -\dfrac{a\cos\theta}{\sin^2\theta} + \dfrac{b\sin\theta}{\cos^2\theta}$

$\therefore \dfrac{dl}{d\theta} = 0 \Rightarrow \dfrac{\sin^3\theta}{\cos^3\theta} = \dfrac{a}{b}$

$\Rightarrow \tan^3\theta = \dfrac{a}{b} \Rightarrow \dfrac{a}{b} \Rightarrow \tan\theta = \left(\dfrac{a}{b}\right)^{\frac{1}{3}}$

Now, $\dfrac{dl}{d\theta} = -a\cot\theta\,\text{cosec}\,\theta + b\tan\theta\sec\theta$

$\Rightarrow \dfrac{d^2l}{d\theta^2} = b\tan^2\theta\sec\theta + b\sec^3\theta + a\cot^2\theta$

$\text{cosec}\,\theta + a\,\text{cosec}^3\theta$ which is positive for θ being acute.

$\Rightarrow l$ has the minimum value for θ_1, where

$\tan\theta_1 = \left(\dfrac{a}{b}\right)^{\frac{1}{3}}$ and the minimum value of l

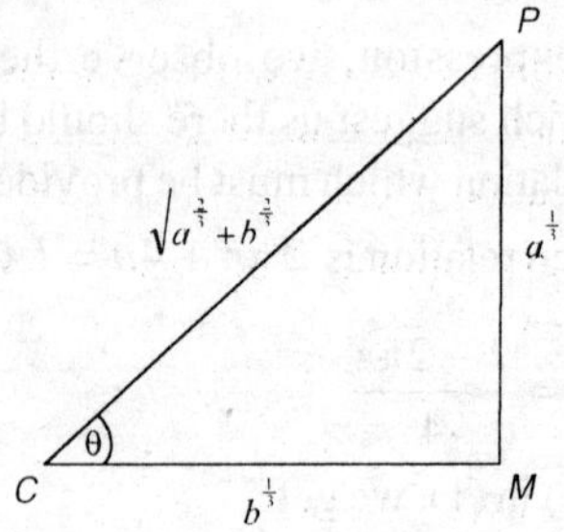

$= \left(\dfrac{a}{\sin\theta} + \dfrac{b}{\cos\theta}\right)_{\theta=\theta_1} = \dfrac{a}{\sin\theta_1} + \dfrac{b}{\cos\theta_1}$

$= \sqrt{a^{\frac{2}{3}} + b^{\frac{2}{3}}}\left\{a^{\frac{2}{3}} + b^{\frac{2}{3}}\right\}$

$= \left(a^{\frac{2}{3}} + b^{\frac{2}{3}}\right)^{\frac{3}{2}}$

5. Two sides of a triangle are given. Find the angle between them such that area shall be as great as possible.

Solution: Let a and b be the sides (or a triangle) and θ = angle between them

Area $= A = \dfrac{1}{2}a \cdot b\sin\theta$

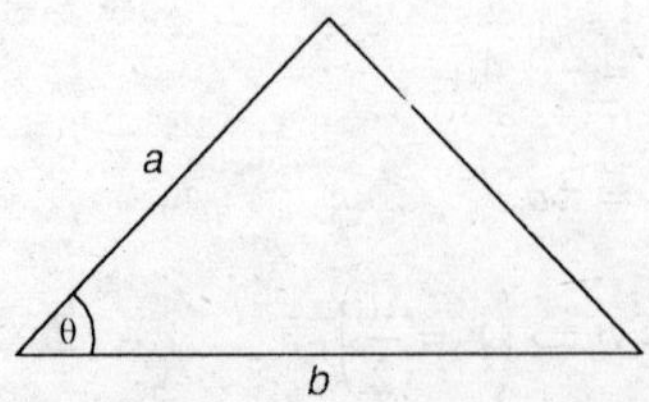

$$\text{Area} = A = \frac{1}{2} a \cdot b \cdot \sin\theta$$

$$\Rightarrow \frac{dA}{d\theta} = \frac{1}{2} a \cdot b \cos\theta$$

$$\therefore \frac{dA}{d\theta} = 0 \Rightarrow \frac{1}{2} a \cdot b \cos\theta = 0$$

$$\Rightarrow \cos\theta = 0 \Rightarrow \theta = \frac{\pi}{2}$$

$$\text{Now, } \frac{d^2 A}{d\theta^2} = -\frac{1}{2} a \cdot b \cdot \sin\theta$$

$$\Rightarrow \left[\frac{d^2 A}{d\theta^2}\right]_{\theta=\frac{\pi}{2}} = -\frac{1}{2} ab \sin\left(\frac{\pi}{2}\right) = -\frac{1}{2} a \cdot b = -\text{ve}$$

$$\therefore \text{Area is max. at } \theta = \frac{\pi}{2}$$

$\therefore$ Angle between them is a right angle for the area to be greatest.

6. Show that maximum rectangle inscribed in a circle is a square.

Solution: Let the equation of the circle be $x^2 + y^2 = a^2$

Let $PQRS$ be the rectangle inscribed in the circle

P, a point on the circle, $= (x, y) = (a \cos\theta, a \sin\theta)$

$PQ = 2a \cos\theta$

$PS = 2a \sin\theta$

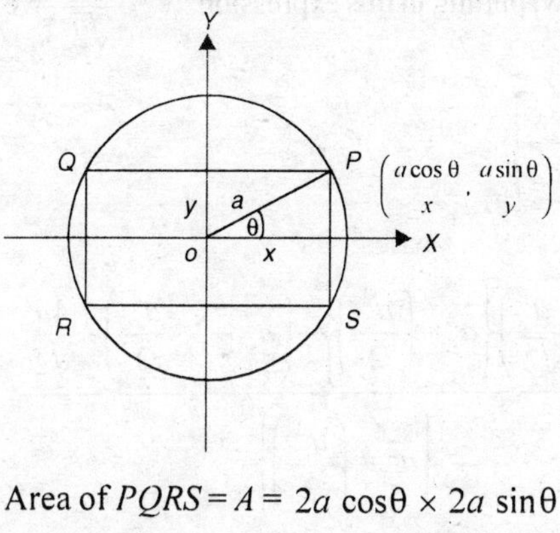

$\therefore$ Area of $PQRS = A = 2a \cos\theta \times 2a \sin\theta$

$$\Rightarrow A = 4a^2 \cos\theta \sin\theta = 2a^2 \sin 2\theta$$

$$\Rightarrow \frac{dA}{d\theta} = 4a^2 \cos 2\theta$$

$$\frac{dA}{d\theta} = 0 \Rightarrow 4a^2 \cos 2\theta = 0$$

$$\Rightarrow \cos 2\theta = 0 \Rightarrow 2\theta = \frac{\pi}{2} \Rightarrow \theta = \frac{\pi}{4}$$

$$\text{Now, } \frac{d^2 A}{d\theta^2} = -8a^2 \sin 2\theta$$

$$\Rightarrow \left[\frac{d^2 A}{d\theta^2}\right]_{\theta=\frac{\pi}{4}} = -8a^2 \sin\left(2 \times \frac{\pi}{4}\right)$$

$$= -8a^2 \sin\left(\frac{\pi}{2}\right) = -8a^2 = -\text{ve}$$

$$\therefore \text{Area is max at } \theta = \frac{\pi}{4}$$

$$\text{Hence, if } \theta = \frac{\pi}{4},\ P\theta = 2a\cos\theta$$

$$= 2a \cos\frac{\pi}{4} = \frac{2a}{\sqrt{2}}$$

$$PS = 2a \sin\theta = 2a \sin\frac{\pi}{4} = \frac{2a}{\sqrt{2}}$$

$\therefore$ $PQ = PS \Rightarrow PQRS =$ a square.

Thus, the required result.

Note: This problem also can be done by the expression $S = f(x)$ in the following way.

(Second method $\Rightarrow S = f(x)$)

Letting $ABCD$ = a rectangle inscribed in a circle whose radius $= a$

$AB = 2x$

Now, on drawing from o, a perpendicular oL to AB, it is evident from the figure

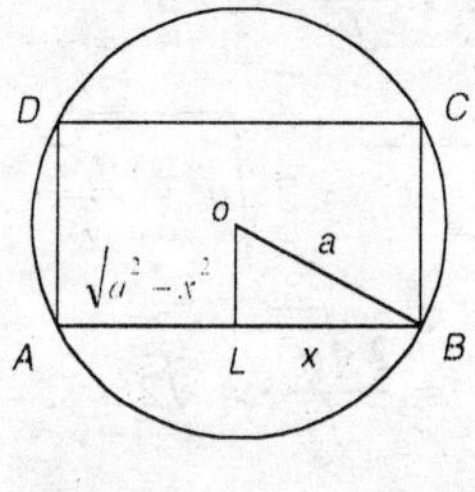

$$OL = \sqrt{a^2 - x^2} \qquad \ldots(1)$$

$$BC = 2 \cdot OL = 2\sqrt{a^2 - x^2} \quad \ldots(2)$$

$\therefore$ Area of the rectangle $= S = AB \times BC$

$$= 2x \times 2 \times \sqrt{a^2 - x^2} \quad \ldots(3)$$

$$\Rightarrow S = 4x\sqrt{a^2 - x^2} \quad \ldots(4)$$

Now, differentiating (4) w.r.t x, we have

$$\frac{ds}{dx} = 4\left[x \cdot \frac{1}{2} \cdot \frac{(-2x)}{\sqrt{a^2 - x^2}} + \sqrt{a^2 - x^2}\right]$$

$$= 4\left[\frac{-x^2}{\sqrt{a^2 - x^2}} + \sqrt{a^2 - x^2}\right]$$

$$= 4\left[\frac{-x^2 + a^2 - x^2}{\sqrt{a^2 - x^2}}\right]$$

$$= 4\left[\frac{a^2 - 2x^2}{\sqrt{a^2 - x^2}}\right]$$

$$\therefore \frac{ds}{dx} = 0$$

$$\Rightarrow \frac{a^2 \; 2x^2}{\sqrt{a^2 - x^2}} = 0$$

$$\Rightarrow a^2 - 2x^2 = 0 \Rightarrow a^2 = 2x^2$$

$$\Rightarrow a = x\sqrt{2} \Rightarrow x = \frac{a}{\sqrt{2}}$$

$$AB = 2x = 2 \times \frac{a}{\sqrt{2}} = \sqrt{2}\,a \quad \ldots(5)$$

$$BC = 2\sqrt{a^2 - x^2} = 2\sqrt{a^2 - \frac{a^2}{2}}$$

$$\frac{2\sqrt{2a^2 - a^2}}{\sqrt{2}} = \frac{2\sqrt{a^2}}{\sqrt{2}} = \sqrt{2}\,a \quad \ldots(6)$$

(5) and (6) $\Rightarrow AB = BC = \sqrt{2}a \Leftrightarrow ABCD =$ square

Thus the require result.

Note: To test the sign of $\frac{d^2s}{dx^2}$ at $x = \frac{a}{\sqrt{2}}$

$$\frac{d^2s}{dx^2} = 4$$

$$\left[\frac{\sqrt{a^2 - x^2} \times (-2 \times 2x) - (a^2 - 2x^2) \times \frac{(-2x)}{2\sqrt{a^2 - x^2}}}{\left(\sqrt{a^2 - x^2}\right)^2}\right]$$

$$= 4\left[\frac{(a^2 - x^2)2(-4x) + (a^2 - 2x^2)(2x)}{2\left(\sqrt{a^2 - x^2}\right)^{\frac{3}{2}}}\right]$$

$$= 4 \times 2\left[\frac{(a^2 - x^2)(-4x) + (a^2 - 2x^2)(x)}{2(a^2 - x^2)^{\frac{3}{2}}}\right]$$

$$= 4\left[\frac{(-4x)(a^2 - x^2) + (a^2 - 2x^2)x}{(a^2 - x^2)^{\frac{3}{2}}}\right]$$

Now, putting in this expression, $x = \frac{a}{\sqrt{2}}$, we have

$$\left[\frac{d^2s}{dx^2}\right]_{x = \frac{a}{\sqrt{2}}} =$$

$$\frac{-16\left(\frac{a}{\sqrt{2}}\right)\left\{a^2 - \left(\frac{a^2}{2}\right)\right\} + \left(a^2 - 2 \cdot \frac{a^2}{2}\right) \cdot \frac{4a}{\sqrt{2}}}{\left\{a^2 - \left(\frac{a^2}{2}\right)\right\}^{\frac{3}{2}}}$$

$$= \frac{a}{\sqrt{2}} \frac{-16\left\{\frac{2a^2 - a^2}{2}\right\}}{\sqrt{\left(\frac{2a^2 - a^2}{2}\right)^3}} = \frac{-16\left(\frac{a^2}{2}\right)}{\left(\frac{a^2}{2}\right)^{\frac{3}{2}}} \cdot \frac{a}{\sqrt{2}} = -\text{ve}$$

7. Show that among rectangles of given perimeter the square has greatest area.

Solution: Let x and y be the sides of the rectangle

Perimeter $= P = 2x + 2y = 2c$ (constant)

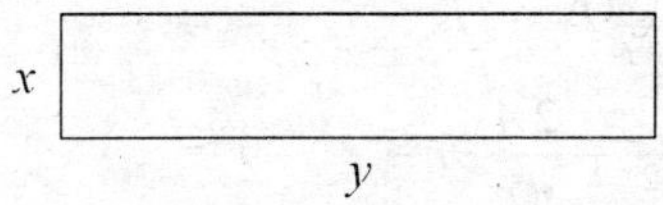

$$\Rightarrow x + y = c$$

$$\Rightarrow y = c - x \quad \ldots(1)$$

Area $= A = x \cdot y = x(c - x) = cx - x^2$

$$\Rightarrow \frac{dA}{dx} = x - 2x$$

$$\therefore \frac{dA}{dx} = 0$$

$$\Rightarrow x = \frac{1}{2}c$$

Now, $\left[\frac{d^2A}{dx^2}\right]_{x=\frac{c}{2}} = [-2]_{x=\frac{c}{2}} = -2 = -\text{ve}$

$\therefore$ A is greatest for $x = \frac{c}{2}$

Now, from (1), $[y]_{x=\frac{c}{2}} = [c - x]_{x=\frac{c}{2}} = c - \frac{c}{2} = \frac{c}{2}$

$\therefore$ A is the greatest when $x = y = \frac{c}{2}$

i.e. when the rectangle is a square.

***Type* 4**: Problems on volume

Note the following key point while working out the problems on volume.

Whenever a figure is to be inscribed in another solid figure, we are required to consider the central section.

(i) Central section of a sphere = a circle

(ii) Central section of cone = a triangle

(iii) Central section of a cylinder = a rectangle

Examples worked out:

1. Prove that the height and the diameter of the base of a right circular cylinder of given surface area and maximum volume are equal.

Solution: Let r = radius of the base of the cylinder = CB

h = height of the cylinder

v = volume of the cylinder

S = constant (given) = The surface area.

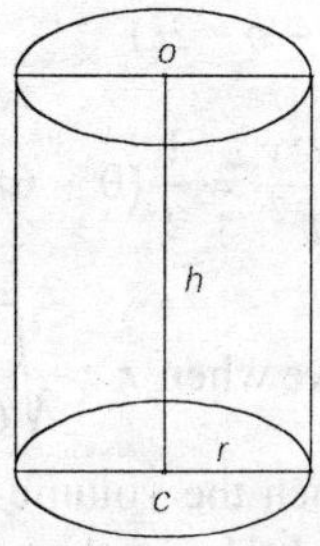

The sum of the areas of the base (circle) and top (circle) $= 2\pi r^2$ area of the curves surface $= 2\pi r h$

Total surface of the cylinder

= sum of the areas of the base and top + area of the curved surface

$$\Rightarrow S = 2\pi r^2 + 2\pi r h = \text{(constant)} \quad \ldots(1)$$

$$V = \pi r^2 h \quad \ldots(2)$$

Now, from $2\pi r h = S - 2\pi r^2$,

$$h = \frac{S - 2\pi r^2}{2\pi r} \quad \ldots(3)$$

From (2), $V = \pi r^2 \left(\frac{S - 2\pi r^2}{2\pi r}\right) = \frac{1}{2} r \left(S - 2\pi r^2\right)$

$$\Rightarrow \frac{dV}{dr} = \frac{1}{2}\left\{r(-2\pi \cdot 2r) + \left(S - 2\pi r^2\right) \cdot 1\right\}$$

$$= \frac{1}{2}\left\{\left(-4\pi r^2\right) + \left(S - 2\pi r^2\right)\right\}$$

$$= \frac{1}{2}\left(S - 6\pi r^2\right)$$

$$\frac{dV}{dr} = 0 \Rightarrow \frac{1}{2}\left(S - 6\pi r^2\right) = 0$$

$$\Rightarrow S = 6\pi r^2$$

putting this value of S in (1), we have

$$6\pi r^2 = 2\pi r^2 + 2\pi r h$$

$$\Rightarrow 4\pi r^2 = 2\pi r h$$

$$\Rightarrow 2r^2 = r h$$

$$\Rightarrow h = b \text{ (base} = b = 2r)$$

Now, from (4) $\frac{d^2 v}{dr^2} = \frac{1}{2}(0 - 6\pi\, 2r) = -6\pi r < 0.$

Thus, we observe when $r = \sqrt{\frac{s}{6\pi}}$, v = volume is maximum and, when the volume of the cylinder is maximum, $h = 2r = b$. Hence, the result.

2. Show that the semi vertical angle of a right circular cone of a given surface and maximum volume is $\sin^{-1}\left(\frac{1}{3}\right)$.

Solution: Let r = the radius of the base

$h = OC$ = height

$\alpha = \angle AOC$ = semi vertical angle

surface of the cone which is given in the problem $= \pi r^2 + \pi r l$

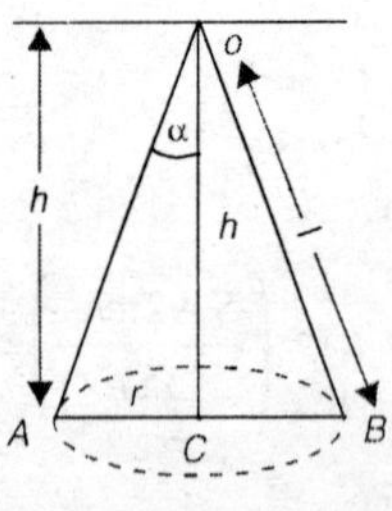

($\because$ total surface of a cone = surface of the base + area of the curve surface)

= A (constant), say ...(1)

and volume $= V = \frac{1}{3}\pi r^2 h$ ...(2)

Now, $l^2 = h^2 + r^2 \Rightarrow h^2 = l^2 - r^2$ ...(3)

and $A = \pi r^2 + \pi r l \Rightarrow l = \frac{A - \pi r^2}{\pi r}$ ...(4)

Now, on putting (4) into (3), we have

$$h^2 = \left(\frac{A - \pi r^2}{\pi r}\right)^2 - r^2$$

$$\Rightarrow h^2 = \left(\frac{A}{\pi r} - r\right)^2 - r^2$$

$$= \frac{A^2}{\pi r^2} - \frac{2A}{\pi r}\cdot r + r^2 - r^2$$

$$= \frac{A^2}{\pi r^2} - \frac{2A}{\pi} \quad \text{...(5)}$$

$$(2) \Rightarrow V^2 = \frac{1}{9}\pi^2 r^4 h^2 \quad \text{...(6)}$$

Again on putting (5) in (6), we get

$$V^2 = \frac{1}{9}\pi^2 r^4\left(\frac{A^2}{\pi^2 r^2} - \frac{2A}{\pi}\right)$$

$$= \frac{1}{9}\pi^2 r^4 \cdot \left(\frac{A^2 - 2\pi A r^2}{\pi^2 \cdot r^2}\right)$$

$$= \frac{1}{9}A\left(Ar^2 - 2\pi r^4\right) = f(r) \quad \text{...(7)}$$

Now, differentiating (7) w.r.t r, we have

$$2V \cdot \frac{dV}{dr} = f'(r) = \frac{1}{9}A\left(A \cdot 2r - 2\pi \cdot 4r^3\right)$$

$$= \frac{1}{9}A\left(2Ar - 8\pi r^3\right)$$

$$\Rightarrow 2V \cdot \frac{dV}{dr} = \frac{1}{9}A\left\{2r\left(\pi r^2 + \pi r l\right) - 8\pi r^3\right\}$$

$$= \frac{1}{9}A\left\{2r\left(\pi r^2 + \pi r l\right) - 8\pi r^3\right\}$$

$$= \frac{1}{9}A\left\{2\pi r^2 l - 6\pi r^3\right\}$$

$$= \frac{1}{9} A\, 2\pi r^2 (l - 3r) \qquad \ldots(8)$$

$$\frac{dV}{dr} = 0 \Rightarrow \frac{1}{9} A 2\pi r^2 (l - 3r) = 0$$

$$\Rightarrow l - 3r = 0 \Rightarrow 3r = l \Rightarrow \left(r = \frac{l}{3}\right)$$

Also, $\frac{dV}{dr} > 0$ for $r < \frac{l}{3}$ and < 0 for $r > \frac{l}{3}$

$\therefore$ V has the max. value for $r = \frac{l}{3}$ and from the

ΔAOC, $\sin\alpha = \frac{AC}{OA} = \frac{r}{l} = \frac{r}{3r}$

$$\Rightarrow \sin\alpha = \frac{1}{3} \Rightarrow \alpha = \sin^{-1}\left(\frac{1}{3}\right)$$

3. Show that semi vertical angle of the cone of maximum volume and the given slant height is $\tan^{-1}\left(\sqrt{2}\right)$.

Solution: v = volume of the cone $= \frac{1}{3}\pi r^2 h$

Where r = radius of the base circle

h = height of the cone

Now, if l = slant height of the cone then

$$r = l\sin\theta \qquad \ldots(1)$$

$$h = l\cos\theta \qquad \ldots(2)$$

where $\theta = \angle BOC$ semi vertical angle

$$\because V = \frac{1}{3}\pi r^2 h \qquad \ldots(3)\ (\text{say})$$

Now, on putting (1) and (2) in (3), we get

$$V = \frac{1}{3}\pi l^2 \sin^2\theta \cdot l\cos\theta = \frac{\pi l^3}{3}\sin^2\theta\cos\theta$$

$$\Rightarrow \frac{dV}{d\theta} = \frac{\pi l^3}{3}\left[\sin^2\theta(-\sin\theta) + \cos\theta \cdot 2\sin\theta\cos\theta\right] \qquad \ldots(4)$$

$$\frac{dV}{d\theta} = 0$$

$$\Rightarrow \frac{\pi l^3}{3}\left[-\sin^3\theta + 2\sin\theta\cos^2\theta\right] = 0$$

$$\Rightarrow -\sin^3\theta + 2\sin\theta\cos^2\theta = 0$$

$$\Rightarrow 2\sin\theta\cos^2\theta = \sin^3\theta$$

$$\Rightarrow 2\cos^2\theta = \sin^2\theta \ (\because \theta \neq 0)$$

$$\Rightarrow 2 = \tan^2\theta \left(\because \theta \neq \frac{\pi}{2}\right)$$

$$\Rightarrow \tan^2\theta = 0$$

$$\Rightarrow \tan\theta = \sqrt{2}$$

$$\Rightarrow \theta = \tan^{-1}\sqrt{2}$$

Note: $\frac{dV}{d\theta} = \frac{\pi l^3}{3}\sin\theta\cos^2\theta\left(2 - \tan^2\theta\right)$

$$\Rightarrow \frac{dV}{d\theta} > 0 \text{ for } \theta < \tan^{-1}\sqrt{2}$$

$$< 0 \text{ for } \theta > \tan^{-1}\sqrt{2}$$

$\therefore$ V has the greatest value for $\theta = \tan^{-1}\sqrt{2}$

Local extreme values of a function in a closed interval [*a*, *b*]:

Definitions: A function $y = f(x)$ defined on a closed interval $[a, b]$ is said to have

A: 1. Local maximum at $x = a$ (left end point) if $f(a) > f(a+h), h > 0$.

2. Local minimum at $x = a$ (left end point) if $f(a) < f(a+h), h > 0$.

B: 1. Local maximum at $x = b$ (right end point) if $f(b) < f(b-h), h > 0$.

2. Local minimum at $x = b$ (right end point) if $f(b) < f(b-h), h > 0$.

To find the local extreme values of a function f defined by $y = f(x)$ in a closed interval $[a, b]$, one should know:

1. How to find the local extreme values of a continuous function f at the interior points of the domain of the function f where the derivatives of the function f denoted by f' do not exist.

Rule: One should use the rule of the first derivative test, i.e., one should find $f'(c-h)$ and $f'(c+h)$ where $x = c$ is an interior point where f' does not exist and h is a small positive number and then use:

(i) $f'(c-h) > 0$ and $f'(c+h) < 0 \Rightarrow$ local maximum value at the interior point $x = c$ of the domain $[a, b]$ of the function f.

(ii) $f'(c-h) < 0$ and $f'(c+h) > 0 \Rightarrow$ local minimum value at the interior point $x = c$ of the domain $[a, b]$ of the function f.

2. How to find the local extreme values of the function f at the interior points of the domain of the function f where the derivative of the function f denoted by f' is zero.

Rule: One should use the rule of the first derivative test or the rule of the second derivative test.

3. How to find the local extreme values of a function f at the left and the right end points of a closed interval $[a, b]$.

***Rule* (a)**: One should find $f'(a+h)$, where h is a small positive number, $x = a$ is the left end point of the given closed interval $[a, b]$, where the given function f is defined and $f'(a+h)$ = the value of the first derivative of the function f for value of x little (just) more than a and then use the rule:

(i) $f'(a+h) > 0 \Rightarrow$ local minimum value at the point $x = a$.

(ii) $f'(a+h) < 0 \Rightarrow$ local maximum value at the point $x = a$.

***Rule* (b)**: One should find $f'(b-h)$, where h is a small positive number, $x = b$ is the right end point of a given closed interval $[a, b]$ and $f'(b-h)$ = the value of the first derivative of the function f for a value of x little (just) less than b and then use the rule:

(i) $f'(b-h) > 0 \Rightarrow$ local maximum value at the point $x = b$.

(ii) $f'(b-h) < 0 \Rightarrow$ local minimum value at the point $x = b$.

Notes: **(A)**: The only possible points where a given function f defined by $y = f(x)$ in a closed interval $[a, b]$ can have local extreme values are

1. The critical points (also called stationary points or turning points), i.e.,

(i) The interior points of the domain of the function f where the derivative of the function f denoted by f' does not exist.

or **(ii)** The interior points of the domain of the function f where the derivative of the function f denoted by f' is zero.

2. The end points of a given closed interval $[a, b]$, where a given function f is defined.

This is why while finding the local extreme or global extreme values of a function f defined by $y = f(x)$ in a closed interval $[a, b]$, one should firstly locate the critical points of the given function f.

(B) While finding the local extreme values at the end points of a closed interval $[a, b]$, one is required to find out $f'(a+h)$ at the left end point namely $x = a$ and $f'(b-h)$ at the right end point namely $x = b$.

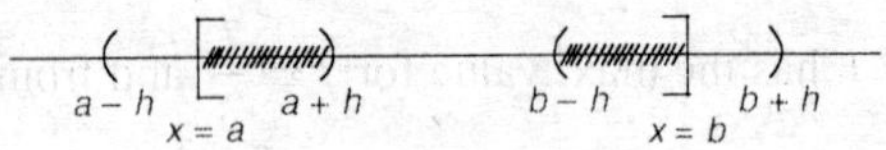

The rule to find out the local extrema at the end points can be put in tabular form:

At the left end pint $x = a$:

x	Little $> a$	Nature of the point
$f'(x)$	+ve	Minima
$f'(x)$	−ve	Maxima

At the right end point $x = b$:

x	Little $< b$	Nature of the point
$f'(x)$	+ve	Maxima
$f'(x)$	−ve	Minima

(C) One should note that there may be more than one local extrema of a function in a closed interval. This is why the question says to find out:

(i) The local extreme values (or simply the extreme values)

(ii) The local maxima and / the local minima (or simply the maxima and / the minima).

i.e., the words "the extrema, the maxima and the minima" are used in plural to signify the local extrema of a function defined in a closed interval.

Examples worked out:

1. Find the point of local maxima and minima of a function f defined by $f(x) = 3x^4 - 4x^3 + 5$ in $[-1, 2]$.

Solution: $f(x) = 3x^4 - 4x^3 + 5, \forall x \in [-1, 2]$

$$\Rightarrow f'(x) = 12x^3 - 12x^2$$

$\Rightarrow f''(x) = 36x^2 - 24x$

Now, $f'(x) = 0$

$\Rightarrow 12x^3 - 12x^2 = 0$

$\Rightarrow 12x^2(x-1) = 0$

$\Rightarrow x = 0$ and $x = 1$ which are the possible critical points.

Hence, the only possible points where the extreme values of the given function f defined in the closed interval $[-1, 2]$ may occur, are

(i) $x = 0$ and $x = 1$, where $f'(x) = 0$

(ii) $x = -1$ = the left end point and $x = 2$ = the right end point of the given closed interval $[-1, 2]$.

1. Local extrema at critical points namely $x = 0, 1$.

At $x = 0$, using the rule of the second derivative test, it is seen that $f''(x)$ at $x = 0$, i.e.,

$f''(0) = 36 \cdot 0^2 - 24 \cdot 0 = 0$ which does give us inference to calculate $f'''(0)$.

Now, $f'''(x) = 72x - 24$

$\left(\because f''(x) = 36x^2 - 24x\right)$

$\Rightarrow f'''(0) = 72 \cdot 0 - 24 = -24$

$\Rightarrow f$ has no extremum value at $x = 0$.

Again, at $x = 1$,

$f''(1) = 36 \cdot 1^2 - 24 \cdot 1 = 36 - 24 = 12 > 0$

$\Rightarrow f$ has a local minimum at $x = 1$

2. Local extrema at the end points of the closed interval:

(i) At the left end point $x = -1$

$f'(-1+h) = 12(-1+h)^3 - 12(-1+h)^2$

$= 12(-\text{ve}) - 12(+\text{ve}) = -\text{ve} < 0$

$\therefore f'(-1+h) < 0 \Rightarrow f$ has a local maximum at the left end point $x = -1$

(ii) At the right end point $x = 2$.

$f'(2-h) = 12(2-h)^3 - 12(2-h)^2$

Now, for convenience $h = 0.1$ can be put in $f'(2-h)$, to know the sign of $f'(2-h)$ for small h.

$f'(2-0.1) = 12(2-0.1)^3 - 12(2-0.1)^2$

$= 12(1.9)^3 - 12(1.9)^2$

$= 12(6.859) - 12(3.61)$

$= +\text{ve} > 0$

$\therefore f'(2-h) > 0 \Rightarrow f$ has a local maxima at the right end point $x = 2$.

Hence, at $x = 0$, f has neither a maximum or a minimum.

At $x = 1$, f has a local minimum and at $x = -1$ and $x = 2$, f has a local maxima.

2. Find the local maxima and minima of the function f defined by $f(x) = \left|4 - x^2\right|, \forall x \in [-3, 3]$

Solution: $f(x) = \left|4 - x^2\right|, \forall x \in [-3, 3]$

$\Rightarrow f'(x) = \dfrac{\left|4 - x^2\right|}{\left(4 - x^2\right)} \cdot (-2x)$,

$x \neq \pm 2, x \in [-3, 3]$ ...(i)

$\therefore f'(x) = -2x, \forall x \in (-2, 2)$ ...(ii)

and $f'(x) = 2x, \forall x \in (-3, -2) \cup (2, 3)$ ...(iii)

$\because f'(x) = \dfrac{\left|4 - x^2\right|}{\left(4 - x^2\right)} (-2x), x^2 - 4 \neq 0,$ i.e.,

$x^2 \neq 4, x \neq \pm 2$

$\Rightarrow f'(x) = -2x$, for $|x| \leq 2$ but $x \neq \pm 2$ and $x \in [-3, 3]$

$\Rightarrow f'(x) = -2x, \forall x \in (-2, 2)$, i.e., $|x| < 2$

Also,

$f'(x) = \dfrac{\left|4 - x^2\right|}{\left(4 - x^2\right)} \cdot (-2x), x^2 - 4 \neq 0,$ i.e.,

$x^2 \neq 4, x \neq \pm 2$

$= 2x$, for $|x| > 2$, i.e., $x < -2$ and $x > 2$ but $x \in [-3, 3]$

$\therefore f'(x) = 2x$, for $x \in (-3, -2) \cup (2, 3)$

Now, $f'(x) = 0$

$$\Rightarrow \frac{|4-x^2|}{(4-x^2)} \cdot (-2x) = 0,\ x \neq \pm 2,\ x \in [-3, 3]$$

$\Rightarrow x = 0$

Hence, the possible points where the extreme values of the given function f defined in the closed interval [–3, 3] may occur, are

(i) $x = 0$ where $f'(x) = 0$

(ii) $x = -2$ and $x = 2$ where $f'(x)$ do not exist.

(iii) $x = -3$, the left end point $x = 3$, the right end point of the given closed interval [–3, 3].

1. Local extrema at critical points:

At $x = 0$, using the rule of first derivative test,

$$f'(0+h) = -2(0+h) = -2h > 0$$

$$f'(0-h) = -2(0-h) = -2h > 0\ (\because h > 0)$$

$\therefore f'(x)$ changes sign from +ve to –ve in $N_h(0)$.

$\Rightarrow x = 0$ is a point of local maximum.

$\therefore$ local maximum value of the function $f(x)$ at $x = 0$, is $f(0) = |4 - 0^2| = |4| = 4$

At $x = 2$, using the rule of the first derivative test,

$$f'(2-h) = -2(2-h) = -4 + 2h < 0$$

$$f'(2+h) = 2(2+h) = 4 + 2h > 0$$

$\therefore f'(x)$ changes sign from –ve to +ve in $N_h(2)$.

$\Rightarrow f$ has a local minimum at $x = 2$

$\therefore$ Local minimum value of the function $f(x)$ at $x = 2$ is $f(2) = |4 - x^2| = |4 - 4| = |0| = 0$

At $x = -2$, using the rule of first derivative test,

$$f'(-2-h) = 2(-2-h) = -(4+2h) < 0$$

$$f'(-2+h) = -2(-2+h) = 4 - 2h > 0 \text{ for } h > 0$$

$\therefore f'(x)$ changes sign from –ve to +ve in $N_h(-2)$.

$\Rightarrow f$ has a local minimum at $x = -2$

$\therefore$ Local minimum value of the function $f(x)$ at $x = -2$ is $f(-2) = |4 - (-2)^2| = |0| = 0$

2. Local extreme at the end points of the closed interval:

(i) At the left end point $x = -3$.

$$f'(-3+h) = 2(-3+h) = -6 + h < 0\ (h > 0)$$

$\Rightarrow f$ has a local maximum at $x = -3$

$\therefore$ local maximum value of the function $f(x)$ at $x = -3$ is $f(-3) = |4 - (-3)^2|\,|4 - 9| = |-5| = 5$

(ii) At the right end point $x = 3$

$$f'(3-h) = 2(3-h) = 6 - 2h > 0\ (h > 0)$$

$\Rightarrow f$ has a local maximum at $x = 3$

$\therefore$ Local maximum value of the function $f(x)$ at $x = 3$ is $f(3) = |4 - 3^2| = |4 - 9| = |-5| = 5$

3. Find the local maximum and local minimum values of the function f defined by $f(x) = \sin 2x - x$, $-\frac{\pi}{2} \le x \le \frac{\pi}{2}$.

Solution: $f(x) = \sin 2x - x,\ -\frac{\pi}{2} \le x \le \frac{\pi}{2}$

$$\Rightarrow f'(x) = 2\cos 2x - 1$$

$$\Rightarrow f''(x) = -4\sin 2x$$

Now, $f'(x) = 0$

$$\Rightarrow \cos 2x = \frac{1}{2}$$

$$\Rightarrow 2x = \frac{\pi}{3} \text{ or } = -\frac{\pi}{3}$$

$$\Rightarrow x = \frac{\pi}{6} \text{ or } -\frac{\pi}{6} \text{ since } -\frac{\pi}{2} \le x \le \frac{\pi}{2}$$

Hence, the possible points where the extreme values of the given function f defined in the closed interval $\left[-\frac{\pi}{2}, \frac{\pi}{2}\right]$ may occur, are

(i) $x = -\frac{\pi}{6}$ and $x = \frac{\pi}{6}$ where $f'(x) = 0$

(ii) $x = -\frac{\pi}{2}$ = the left end point and $x = \frac{\pi}{2}$ = the right end point of the given closed interval $\left[-\frac{\pi}{2}, \frac{\pi}{2}\right]$.

1. Local extreme at critical points $x = -\frac{\pi}{6}$ and $x = \frac{\pi}{6}$:

At $x = \frac{\pi}{6}$, using the rule of second derivative test, it is seen that

$$f''\left(\frac{\pi}{6}\right) = -4\sin\left(\frac{\pi}{3}\right) = -4 \cdot \frac{\sqrt{3}}{2} = -2\sqrt{3} < 0$$

$\Rightarrow f$ has a local maximum at $x = \frac{\pi}{6}$ and the local maximum value of the function $f(x)$ at $x = \frac{\pi}{6}$ is

$$f'\left(\frac{\pi}{6}\right) = \sin\left(\frac{\pi}{3}\right) = \frac{\pi}{6} = \frac{\sqrt{3}}{2} - \frac{\pi}{6}$$

Also, at $x = -\frac{\pi}{6}$

$$f''\left(-\frac{\pi}{6}\right) = -4\sin\left(-\frac{\pi}{3}\right) = 4\sin\left(\frac{\pi}{3}\right) = 2\sqrt{3} > 0$$

$\Rightarrow f$ has a local minimum at $x = -\frac{\pi}{6}$ and the local minimum value of the function $f(x)$ at $x = -\frac{\pi}{6}$ is

$$f\left(-\frac{\pi}{6}\right) = \sin\left(-\frac{\pi}{3}\right) - \frac{\pi}{6} = -\frac{\sqrt{3}}{2} - \frac{\pi}{6}$$

2. Local extrema at the end points of the closed interval:

(i) At the left end point $x = -\frac{\pi}{2}$

$$f'\left(-\frac{\pi}{2} + h\right) = 2\cos\left(2\left(-\frac{\pi}{2} + h\right)\right) - 1$$

$$= 2\cos(-\pi + 2h) - 1$$

$$= 2\cos(-(\pi - 2h)) - 1$$

$$= 2\cos(\pi - 2h) - 1$$

$$= -2\cos 2h - 1$$

$$= -\text{ve} < 0 \text{ for small } h > 0$$

$\Rightarrow f$ has a local maximum at $x = -\frac{\pi}{2}$ and the local maximum value of the function $f(x)$ at $x = -\frac{\pi}{2}$ is $f\left(-\frac{\pi}{2}\right) = \sin(-\pi) - \left(-\frac{\pi}{2}\right) = 0 + \frac{\pi}{2} = \frac{\pi}{2}$.

(ii) At the right end point $x = \frac{\pi}{2}$:

$$f'\left(\frac{\pi}{2} - h\right) = 2\cos\left(2\left(\frac{\pi}{2} - h\right)\right) - 1$$

$$= 2\cos(\pi - 2h) - 1$$

$$= -2\cos 2h - 1 < 0$$

$\Rightarrow f$ has a local minimum at $x = \frac{\pi}{2}$

$\therefore$ Local minimum value of the function $f(x)$ at $x = \frac{\pi}{2}$ is $f\left(\frac{\pi}{2}\right) = \sin(\pi) - \frac{\pi}{2} = -\frac{\pi}{2}$

4. Find the local maxima and minima of the function f defined $f(x) = 4\sin x + \cos 2x$, $\forall x \in [0, 2\pi]$. Also find the absolute maximum and minimum values in $[0, 2\pi]$.

Solution: $f(x) = 4\sin x + \cos 2x$, $\forall x \in [0, 2\pi]$

$$\Rightarrow f'(x) = 4\cos x - 2\sin 2x$$

$$\Rightarrow f''(x) = -4(\sin x + \cos 2x)$$

$$\therefore f'(x) = 0$$

$$\Rightarrow 4\cos x - 2\sin 2x = 0$$

$$\Rightarrow 4\cos x(1 - \sin x) = 0$$

$$\Rightarrow \cos x = 0 \text{ or } \sin x = 1 \text{ but } x \in [0, 2\pi]$$

$$\Rightarrow x = \frac{\pi}{2} \text{ or } x = \frac{3\pi}{2}$$

Hence, the possible points where the local extreme values of the given function f defined in the closed interval $[0, 2\pi]$ may occur, are

1. $x = \frac{\pi}{2}$ and $x = \frac{3\pi}{2}$ where $f'(x) = 0$

2. $x = 0 =$ the left end point and $x = 2\pi =$ the right end point of the given closed interval $[0, 2\pi]$

1. Local extrema at critical point $x = \frac{\pi}{2}$ and $x = \frac{3\pi}{2}$:

At $x = \frac{\pi}{2}$, using the rule of second derivative test, it is seen that

$$f''\left(\frac{\pi}{2}\right) = -4(1-1) = 0$$ which does not given any inference about extremum.

$$f'''(x) = -4(\cos x - 2\sin 2x)$$

and $f''''(x) = 4(\sin x + 4\cos 2x)$

$$\therefore f'''\left(\frac{\pi}{2}\right) = -4\left(\cos\frac{\pi}{2} - 2\sin\pi\right) = -(0-2\times 0) = 0$$

and $f''''\left(\frac{\pi}{2}\right) = 4\left(\sin\frac{\pi}{2} + 4\cos\pi\right) = 4(1-4)$

$= -12 < 0$

$\Rightarrow f$ has a local maximum at $x = \frac{\pi}{2}$ and the local maximum value of the function $f(x)$ at $x = \frac{\pi}{2}$ is

$$f\left(\frac{\pi}{2}\right) = 4\sin\left(\frac{\pi}{2}\right) + \cos\pi = 4\times 1 - 1 = 3$$

Also, at $x = \frac{3\pi}{2}$:

$$f''\left(\frac{3\pi}{2}\right) = -4\left(\sin\left(\frac{3\pi}{2}\right) + \cos 3\pi\right)$$

$= -4(-1-1) = 8 > 0$

$\Rightarrow f$ has a local minimum at $x = \frac{3\pi}{2}$

$\therefore$ Local minimum value of the function $f(x)$ at $x = \frac{3\pi}{2}$ is $f\left(\frac{3\pi}{2}\right) = 4\sin\left(\frac{3\pi}{2}\right) + \cos 3\pi = 4(1) + (-1) = -5$

2. Local extrema at the end points of the closed interval:

(i) At the left end point $x = 0$:

$$f'(0+h) = 4\cos(0+h) - 2\sin 2(0+h)$$

$$= 4\cos h - 2\sin 2h = 4\cos h(1 - \sin h)$$

$= +$ ve, for small $h > 0$.

$\therefore f'(0+h) > 0 \Rightarrow f$ has a local minimum at the left end point $x = 0$ and the local minimum value of the function $f(x)$ at $x = 0$ is $f(0) = 4\sin 0 + \cos 0 = 4\cdot 0 + 1 = 1$

(ii) At the right end point $x = 2\pi$:

$$f'(2\pi - h) = 4\cos(2\pi - h) - 2\sin 2(2\pi - h)$$

$= 4\cos h - 2\sin(4\pi - 2h) = 4\cos h + 2\sin 2h = +$ve

$\therefore f'(2\pi - h) > 0 \Rightarrow f$ has a local maximum at the right end point $x = 2\pi$ and the local maximum value of the function $f(x)$ at $x = 2\pi$ is $f(2\pi) = 4\sin 2\pi + \cos 4\pi = 4\cdot 0 + 1 = 1$

Hence $x = \frac{\pi}{2}$ and $x = 2\pi$ are points of local maxima and $x = 0$ and $x = \frac{3\pi}{2}$ are the points of local minima

$$\because f\left(\frac{\pi}{2}\right) = 4\sin\left(\frac{\pi}{2}\right) + \cos\left(2\cdot\frac{\pi}{2}\right) = 4\times 1 + \cos\pi$$

$= 4 - 1 = 3$

$$f\left(\frac{3\pi}{2}\right) = 4\sin\left(\frac{3\pi}{2}\right) + \cos 3\pi = 4(-1) + (-1) = -5$$

$$f(0) = 4\sin 0 + \cos 0 = 4\times 0 + 1 = 1$$

$$f(2\pi) = 4\sin 2\pi + \cos 4\pi = 4\times 0 + 1 = 1$$

$\therefore$ Absolute maximum value of the function $f(x) = 3$ and absolute minimum value of the function $f(x) = -5$.

The absolute extreme values (or simply the absolute extrema) of a continuous function in a closed interval:

There is a method to find out:

The absolute extreme values (or simply the absolute extrema), i.e.,

the maximum and / minimum values (value) (or simply the maximum and / minimum) of a continuous function f in a given closed interval [a, b].

The method is to:

1. To find out the values of the given function defined by $y = f(x)$ at all critical points and the end points of the given closed interval $[a, b]$.

2. To take out:

(i) The greatest of the numbers of the set {values of the given function at all critical points and end points of the given closed interval} which is the required absolute maximum value, the maximum value or simply the maximum of the given continuous function in a given closed interval.

That is, the absolute maximum value, the maximum value or simply the maximum of a given continuous function in a given closed interval $[a, b]$ = the greatest of the numbers of the set $\{f(a), f(b), f(c_1), f(c_2), f(c_3)\}$, where c_1, c_2 and c_3 etc. are all critical points and a and b are the end points of the closed interval where the given function $y = f(x)$ is defined.

(ii) the least of the numbers of the set {values of the given function at all critical points and end points of the given closed interval} which is the required absolute minimum value, the minimum value or simply the minimum of the given continuous function in a given closed interval.

That is, the absolute minimum value, the minimum value or simply the minimum of a given continuous function in a given closed interval $[a, b]$ = the least of the numbers of the set $\{f(a), f(b), f(c_1), f(c_2), f(c_3)\}$, where c_1, c_2 and c_3 etc are all the critical points and a and b are the end points of the closed interval $[a, b]$ where the given function $y = f(x)$ is defined.

***Notes*:** **1.** One should note that there is only one absolute maximum and / minimum value (value) of a function in a closed interval. This is why the words the maximum and / minimum (values) are used always in singular to signify the absolute extrema of a continuous function $y = f(x)$ in closed interval $[a, b]$ whereas the words "the maxima and minima" are used in plural always to signify the local extrema of a continuous function f in a closed interval $[a, b]$ since there are several local maximums and / minimums of a continuous function in an open or in a closed interval.

2. The absolute extrema of a function is always determined only in a given closed interval whereas local extrema of a function are determined in both open and closed interval.

3. If a function is continuous in an open interval (a, b) or in closed interval $[a, b]$ and it has only one extreme point in (a, b), then it is a point of absolute maxima or a point of absolute minima accordingly it is a point of local maxima or a point of local minima.

4. One should note the difference between the absolute extrema and the local extrema of a function which is presented in the following way:

If $x = c$ is a point of an interval D such that $f(c) \geq f(x)$ (or $f(c) \leq f(x)$ for all x in the interval D, then $f(x)$ is said to have an absolute maximum (absolute minimum) value $f(c)$ in the interval D at $x = c$.

That is, a value of the function f at point $x = c$ in the domain of its definition D represented by $f(c)$ is an absolute maximum (or an absolute) minimum) value of the function $f \Leftrightarrow f(c)$ is the greatest (or the least) value of the function in its domain of definition D if the values of the function f at $x = c$ and other values of the independent variable x which belong to its domain of definition D are considered.

If $x = c$ is a point of an interval D such that $f(c) > f(x)$ (or $f(c) < f(x)$) is true only for x in some deleted δ-neighbourhood of the point $x = c$, where $\delta > 0$ (i.e. for all x such that $0 < |x - c| < \delta$), then $f(x)$ is said to have a local maximum (or local minimum) value $f(c)$ at $x = c$.

That is, a value of the function f at a point $x = c$ in the domain of its definition D represented by $f(c)$ is a local maximum (or local minimum) value of the function $f \Leftrightarrow f(c)$ is the greatest (or the least) value in the δ-deleted neighbourhood of the point $x = c$ if the values of the function f at $x = c$ and other values of the independent variable x which belong to the δ-deleted neighbourhood of the point $x = c$ are considered.

In short, an absolute extrema of a function at a point of its domain of definition is the greatest or the least value of the function in its domain of definition while a local extrema of a function is the greatest or the least value of the function in a deleted neighbourhood of a point in its domain of definition.

On points of absolute extrema

1. The point of absolute maximum: A point $x = c$ in the domain of the function $y = f(x)$ at which the value of the function f is the largest value of the function is called the point of absolute maximum of the function $y = f(x)$.

2. The point of absolute minimum: A point $x = c$ in the domain of the function $y = f(x)$ at which the value of the function f is the smallest value of the function is called the point of absolute minimum of the function $y = f(x)$.

3. The point of attainment of absolute extrema is either **(a)** a point where $f'(x) = 0$,
(b) a point at one end of the closed interval,
or **(c)** a point where $y = f(x)$ is not differentiable. This situation may occur also, but this will be more rare in common practice.

Examples worked out:

(Problems on algebraic functions)

1. Find the maximum and / minimum of the function $f(x) = x^3 - 12x^2 + 36x + 17$ in [1, 10].

Solution: $f(x) = x^3 - 12x^2 + 36x + 17$

$\Rightarrow f'(x) = 3x^2 - 24x + 36 = 3(x^2 - 8x + 12) = 3(x - 6)(x - 2)$

$\therefore f'(x) = 0 \Rightarrow 3(x - 6)(x - 2) = 0 \Rightarrow x = 2$ and $x = 6$

Now, $f(1) = (1)^3 - 12(1)^2 + 36(1) + 17 = 1 - 12 + 36 + 17 = 42$

$f(2) = (2)^3 - 12(2)^2 + 36(2) + 17 = 8 - 48 + 72 + 17 = 49$

$f(6) = (6)^3 - 12(6)^2 + 36(6) + 17 = 216 - 432 + 216 + 17 = 17$

$f(10) = (10)^3 - 12(10)^2 + 36(10) + 17 = 1000 - 1200 + 360 + 17 = 177$

$\therefore$ max. $f(x) = 177$

min. $f(x) = 17$

2. Determine the maximum and / minimum values (value) of each of the following functions in stated domains.

(i) $f(x) = (x + 1)^{\frac{2}{3}}, 0 \le x \le 8$

(ii) $f(x) = [x(x - 1) + 1]^{\frac{1}{3}}, 0 \le x \le 1$

(iii) $f(x) = \dfrac{x + 1}{\sqrt{x^2 + 1}}, 0 \le x \le 2$

Solution: **(i)** $f(x) = (x + 1)^{\frac{2}{3}}, 0 \le x \le 8$

$$\Rightarrow f'(x) = \frac{2}{3}(x + 1)^{-\frac{1}{3}} = \frac{2}{3(x + 1)^{\frac{1}{3}}}, x \ne -1$$

$\Rightarrow f'(x) \ne 0, \forall x \in [0, 8]$

Hence, this is why it is required to find out the values of the given function only at the end points namely 0 and 8 appearing in the given closed interval [0, 8].

$\therefore f(0) = (0 + 1)^{\frac{2}{3}} = (1)^{\frac{2}{3}} = 1$

$$f(8) = (8 + 1)^{\frac{2}{3}} = (9)^{\frac{2}{3}} = \left(3^2\right)^{\frac{2}{3}} = (3)^{\frac{4}{3}}$$

$$= 3(3)^{\frac{1}{3}} = 3\sqrt[3]{3}$$

Therefore, max. $f(x) = 3\sqrt[3]{3}$

Min. $f(x) = 1$

(ii): $f(x) = [x(x - 1) + 1]^{\frac{1}{3}}$

$$\Rightarrow f'(x) = \frac{1}{3}[x(x - 1) + 1]^{-\frac{2}{3}} \cdot \frac{d}{dx}[x(x - 1) + 1]$$

$$\frac{(2x - 1)}{3[x(x - 1) + 1]^{\frac{2}{3}}}$$

$\therefore f'(x) = 0$

$\Rightarrow (2x - 1) = 0$

$\Rightarrow x = \dfrac{1}{2}$

Now, it is required to find out the values of the given function at the points $x = 0$, $x = \dfrac{1}{2}$ and $x = 1$.

$\therefore f(0) = 1$

$$f\left(\frac{1}{2}\right) = \left(\frac{3}{4}\right)^{\frac{1}{3}}, f(1) = 1$$

Hence, max. $f(x) = 1$

min. $f(x) = \left(\dfrac{3}{4}\right)^{\frac{1}{3}}$

(iii) $f(x)=\dfrac{x+1}{\sqrt{x^2+1}},\ 0\le x\le 2$

$$\Rightarrow f'(x)=\frac{\sqrt{x^2+1}\cdot(1)-(x+1)\cdot\frac{1}{2}(x^2+1)^{-\frac{1}{2}}\cdot 2x}{\left(\sqrt{x^2+1}\right)^2}$$

$$=\frac{\sqrt{x^2+1}-\dfrac{x(x+1)}{\sqrt{x^2+1}}}{(x^2+1)}$$

$$=\frac{(x^2+1)-x(x+1)}{(x^2+1)^{\frac{3}{2}}}$$

$$=\frac{(1-x)}{(x^2+1)^{\frac{3}{2}}}$$

$$\therefore f'(x)=0\Rightarrow\frac{(1-x)}{(x^2+1)^{\frac{3}{2}}}=0$$

$\Rightarrow (1-x)=0$

$\Rightarrow x=1$

Now, it is required to find out the values of the function at $x=0$, $x=1$ and $x=2$

$f(0)=1,\ f(1)=\dfrac{2}{\sqrt{2}}=\sqrt{2}=1.414$

$f(2)=\dfrac{3}{\sqrt{5}}=\dfrac{3\sqrt{5}}{5}=\dfrac{3\times 2.236}{5}=\dfrac{6.708}{5}=1.34$

Therefor, max. $f(x)=\sqrt{2}$

min. $f(x)=1$

(Problems on trigonometric functions)

1. Examine the maximum and minimum of the function $f(x)=\sin x+\cos x,\ 0\le x\le\pi$.

Solution: $f(x)=\sin x+\cos x$

$\Rightarrow f'(x)=\cos x-\sin x$

$\therefore f'(x)=0$

$\Rightarrow \cos x-\sin x=0$

$\Rightarrow \tan x=1$

$\Rightarrow x=\dfrac{\pi}{4}$

Now, $f(0)=\cos 0+\sin 0=1+0=1$

$f\left(\dfrac{\pi}{4}\right)=\cos\left(\dfrac{\pi}{4}\right)+\sin\left(\dfrac{\pi}{4}\right)$

$=\dfrac{1}{\sqrt{2}}+\dfrac{1}{\sqrt{2}}=\dfrac{2}{\sqrt{2}}=\sqrt{2}$

$f(\pi)=\cos\pi+\sin\pi=-1+0=-1$

Hence, max. $f(x)=\sqrt{2}$

min. $f(x)=-1$

2. Find the maximum and minimum of the function $f(x)=\sqrt{5}(\sin x+\cos 2x),\ 0\le x\le\dfrac{\pi}{4}$.

Solution: $f(x)=\sqrt{5}(\sin x+\cos 2x)$

$\Rightarrow f'(x)=\sqrt{5}(\cos x-2\sin 2x)$

$\therefore f'(x)=0$

$\Rightarrow \cos-2\sin 2x=0$

$\Rightarrow \cos x-4\sin x\cos x=0$

$\Rightarrow \cos x(1-4\sin x)=0$

$\Rightarrow \cos x=0$ or $\sin x=\dfrac{1}{4}$

$\Rightarrow x=\dfrac{\pi}{2}$ or $x=\sin^{-1}\left(\dfrac{1}{4}\right)$; $0<\sin^{-1}\left(\dfrac{1}{4}\right)<\dfrac{\pi}{2}$

Now, $f(0)=\sqrt{5}(\sin 0+\cos 0)$

$=\sqrt{5}(0+1)=\sqrt{5}$

$x=\sin^{-1}\left(\dfrac{1}{4}\right)\Rightarrow\sin x=\dfrac{1}{4}$ and $\cos 2x=1-2$

$\sin^2 x$

$=1-2\left(\dfrac{1}{4}\right)^2=1-\dfrac{1}{8}=\dfrac{7}{8}$

$$\therefore f\left(\sin^{-1}\frac{1}{4}\right) = \sqrt{5}\left(\frac{1}{4}+\frac{7}{8}\right) = \frac{9}{8}\times\sqrt{5} = \frac{9\sqrt{5}}{8}$$

and $f\left(\frac{\pi}{2}\right) = \sqrt{5}\left(\sin\left(\frac{\pi}{2}\right) + \cos\left(\frac{\pi}{2}\right)\right)$

$= \sqrt{5}(1-1) = \sqrt{5} + 0 = 0$

Hence, max. $f(x) = \frac{9\sqrt{5}}{8}$

min. $f(x) = 0$

3. Determine the maximum and minimum values of the function in the stated domain. $f(x) = 2\cos 2x - \cos 4x$ in $0 \le x \le \pi$.

Solution: $f(x) = 2\cos 2x - \cos 4x$

$\Rightarrow f'(x) = -4\sin 2x + 4\sin 4x = 4(\sin 4x - \sin 2x)$

$\therefore f'(x) = 0$

$\Rightarrow 4(\sin 4x - \sin 2x) = 0$

$\Rightarrow 4(2\sin 2x\cos 2x - \sin 2x) = 0$

$\Rightarrow 4\sin 2x(2\cos 2x - 1) = 0$

$\Rightarrow \sin 2x = 0$ or $2\cos 2x - 1 = 0$

Now $\sin 2x = 0$ and $0 \le x \le \pi$

$\Rightarrow 2x = 0, \pi, 2\pi \Rightarrow x = 0, \frac{\pi}{2}, \pi$

and $\quad 2\cos 2x - 1 = 0 \Rightarrow \cos 2x = \frac{1}{2} \Rightarrow 2x = \frac{\pi}{3}, \frac{5\pi}{3}$

$\Rightarrow x = \frac{\pi}{6}, \frac{5\pi}{6}$ as $0 \le x \le \pi$

Now, $f(0) = 2\cos 0 - \cos 0 = 2\times 1 - 1 = 2 - 1 = 1$

$f\left(\frac{\pi}{6}\right) = 2\cos\left(\frac{\pi}{3}\right) - \cos\left(\frac{2\pi}{3}\right)$

$= 2\left(\frac{1}{2}\right) - \left(-\frac{1}{2}\right) = 1 + \frac{1}{2} = \frac{3}{2}$

$f\left(\frac{\pi}{2}\right) = -2\cos\pi - \cos 2\pi$

$= 2(-1) - 1 = -2 - 1 = -3$

$f\left(\frac{5\pi}{6}\right) = 2\cos\left(\frac{5\pi}{3}\right) - \cos\left(\frac{10\pi}{3}\right)$

$= 2\times\frac{1}{2} - \left(-\frac{1}{2}\right) = 1 + \frac{1}{2} = \frac{3}{2}$

$f(\pi) = 2\cos(2\pi) - \cos(4\pi)$

$= 2(1) - 1 = 2 - 1 = 1$

$\therefore$ max. $f(x) = \frac{3}{2}$

min. $f(x) = -3$

Problems on mod functions and / other functions whose derivatives have points of discontinuity as critical points:

1. Find the maximum and / minimum values (value) of the function $f(x) = 2 + |x-1|$ in $[-3, 2]$.

Solution: $f(x) = 2 + |x-1|, \forall x \in [-3, 2]$

$\Rightarrow f'(x) = \frac{|x-1|}{(x-1)}, x \ne 1$

Also $f'(x)$ is not differentiable at $x = 1$

$\therefore x = 1$ is a critical point

Now, $f'(x) = 0, x \ne 1$

i.e., $\frac{|x-1|}{(x-1)} = 0, x \ne 1$; has no solution

Hence, $f(x)$ has only one critical point namely 1.

Now $f(1) = 2 + |1-1| = 2 + |0| = 2$

$f(-3) = 2 + |-3-1| = 2 + |-4| = 2 + 4 = 6$

$f(2) = 2 + |2-1| = 2 + |1| = 2 + 1 = 3$

Therefore, max. $f(x) = 6$

min. $f(x) = 2$

2. Find the absolute maximum and / minimum values of the function f defined by $f(x) = 3 + |x+1|$ in $[-2, 3]$.

Solution: $f(x) = 3 + |x+1|, -2 \le x \le 3$

$\Rightarrow f'(x) = 0 + \frac{|x+1|}{(x+1)}, x \ne -1$

$= \frac{|x+1|}{(x+1)}, x \ne -1$

Now, $f'(x) = 0, x \neq -1$; i.e., $x \neq -1$ and

$$\Rightarrow \frac{|x+1|}{(x+1)} = 0 \text{ has no solution.}$$

Hence, the given function has only one critical point $x = -1$ where the derivative of the function does not exist.

Now, $f(-1) = 3 + |-1 + 1| = 3 + 0 = 3$

$f(-2) = 3 + |-2 + 1| = 3 + |-1| = 3 + 1 = 4$

$f(3) = 3 + |4| = 3 + 4 = 7$

Therefore, max. $f(x) = 7$

min. $f(x) = 3$

3. Determine the maximum and minimum values (value) if any, for the function f defined by $f(x) = 1 - x^{\frac{4}{5}}, \forall x \in [-1, 1]$.

Solution: $f(x) = 1 - x^{\frac{4}{5}}, \forall x \in [-1, 1]$

$$\Rightarrow f'(x) = -\frac{4}{5}x^{-\frac{1}{5}} = -\frac{4}{5x^{\frac{1}{5}}}, x \neq 0$$

Also $f(x)$ is not differentiable at $x = 0$

Now, $f'(x) = 0$, i.e.,

$$-\frac{4}{5}x^{-\frac{1}{5}} = 0 \text{ has no solution.}$$

Hence, the given function has only one critical point namely 0,

And $f(0) = 1 - 0 = 1$

$f(-1) = 1 - (-1)^{\frac{4}{5}} = 1 - 1 = 0$

$f(1) = 1 - (1)^{\frac{4}{5}} = 1 - 1 = 0$

Therefore, max. $f(x) = 1$

min. $f(x) = 0$

The greatest and the least values of a continuous function in a closed interval (recapitulation)

The rule to find out the greast and / the least values (value) of a given continuous functions in a given closed interval $[a, b]$ is to use the following facts:

1. The greatest value of a given continuous functions f in a given closed interval $[a, b]$ = the greatest of the numbers of the set $\{f(a), f(b), f(c_1), f(c_2), f(c_3)\}$ where c_1, c_2 and c_3 etc are critical points and a and b are end points of the given closed interval $[a, b]$ where the given function $y = f(x)$ is defined.

2. The least value of a given continuous function f in a given closed interval $[a, b]$ = the least of the numbers of the set $\{f(a), f(b), f(c_1), f(c_2), f(c_3)\}$ where c_1, c_2 and c_3 etc are critical points and a and b are end points of the given closed interval $[a, b]$ where the given function $y = f(x)$ is defined.

Notes: **1.** "A maximum" means "not necessarily the greatest value" whereas one should note that "the maximum" means "necessarily the greatest value".

2. "A minimum" means "not necessarily the least value" whereas one should note that "the minimum" means "necessarily the least".

3. A continuous function $f(x)$ has a single extremum in its domain $\Rightarrow$ the maximum (the minimum) is also the greates (the least) value of the continuous function f in its domain.

4. "The largest and the smalles values (value)" are also in use instead of the terms "the greatest and the least values (value)"

5. The greatest value of a function f on (in or over) the interval $[a, b]$ is designated as: $\max\limits_{x \in [a, b]} f(x)$ and the least value of a function f on (in or ove) the interval $[a, b]$ is designated as: $\min\limits_{x \in [a, b]} f(x)$

6. (i) *Maximum and greatest*: A continuous functions $y = f(x)$ on a closed interval $[a, b]$ has a greatest value at an interior point or at an end point of its domain which is the given closed interval. Further the greatest value of a continuous function f in a closed interval $[a, b]$ is unique.

On the other hand, a continuous function $y = f(x)$ in a closed interval $[a, b]$ may have local maxima (i) at an interior point or more interior points (ii) at end points of the closed interval but not at the end points of the open interval used as the domain of the continuous function. Moreover, a local maxima of a continuous function in an open or in a closed interval may not be unique, i.e., a continuous function may have more than one local maxima in an open in a closed interval.

(ii) *Minimum and least*: A continuous function $y = f(x)$ in a closed interval $[a, b]$ has a least value at an interior point or at end point of the domain which is the given closed interval. Further the least value of continuous function f in a closed interval $[a, b]$ is unique. On the other hand, a continuous function $y = f(x)$ in a closed interval $[a, b]$ has a local minima (i) at an inter point or more interior points as well as (ii) at end points of the closed interval but not at the end points of the open interval used as the domain of the continuous function. Moreover, a local minima of a continuous function in an open or in a closed interval may not be unique, i.e., a continuous function ,may have more than one local minima in an open or in a closed interval.

Lastly, one must note that the greatest value of a function in a closed interval, when it occurs at an interior point is also a local maximum and similarly, the least value when it occurs at an interior point is also a minimum where as the converse may not always be true.

7. If a continuous function $y = f(x)$ has single extremum in its domain of definition, then if it is a maximum (minimum), then it is also the greatest (the least) value of the function $y = f(x)$.

8. If a function is defined and continuous in some interval, and if the interval is not a closed one, then it can have neither the greatest nor the least value.

Examples worked out*: *(some more)

1. Find the greatest and the least values (value) of the function $f(x) = 12x^5 - 45x^4 + 40x^3 + 6$ on the interval $[0, 3]$.

Solution: $f(x) = 12x^5 - 45x^4 + 40x^3 + 6 \ \forall x \in [0, 3]$

$\Rightarrow f'(x) = 60x^4 - 180x^3 + 120x^2$

$\therefore f'(x) = 0$

$\Rightarrow 60x^4(x-1)(x-2) = 0$

$\Rightarrow x = 0, 1, 2$ which belong to $[0, 3]$

$\Rightarrow x = 0, 1, 2$ are all the critical points

Now,

$f(0) = 6$

$f(3) = 357$

$f(1) = 13$

$f(2) = -10$

$\therefore \underset{x \in [0, 3]}{\max.} f(x) = 357$

$\underset{x \in [0, 3]}{\min.} f(x) = -10$

2. Find the greatest and the least values (value) of the function $f(x) = \frac{x}{8} + \frac{2}{x}$ on the interval $[1, 6]$.

Solution: $f(x) = \frac{x}{8} + \frac{2}{x}, \ \forall x \in [1, 6]$

$\Rightarrow f'(x) = \frac{1}{8} - \frac{2}{x^2} = \frac{x^2 - 16}{8x^2}$

$\therefore f'(x) = 0$

$\Rightarrow \frac{x^2 - 16}{8x^2} = 0$

$\Rightarrow x^2 - 16 = 0$

$\Rightarrow x = \pm 4$ but only $x = 4 \in [1, 6]$

$\Rightarrow x = 4$ is the only critical point

Now, $f(4) = 1, \ f(1) = 2\frac{1}{8}, \ f(6) = 1\frac{1}{12}$

$\therefore \underset{x \in [1, 6]}{\max.} f(x) = f(1) = 2\frac{1}{8}$

$\underset{x \in [1, 6]}{\min.} f(x) = f(4) = 1$

3. Find the greatest and least values (value) of the function $f(x) = x^3 - 3x^2 + 6x - 2$ in the interval $[-1, 1]$.

Solution: $f(x) = x^3 - 3x^2 + 6x - 2, \ \forall x \in [-1, 1]$

$\Rightarrow f'(x) = 3x^2 - 6x + 6$

$\therefore f'(x) = 0$

$\Rightarrow 3x^2 - 6x + 6 = 0$

$\Rightarrow x = \frac{-(-6) \pm \sqrt{36 - 4 \times 3 \times 6}}{2 \times 6}$

$= \frac{6 \pm \sqrt{36-72}}{12} = \frac{6 \pm \sqrt{-36}}{12}$ which are imaginary

$\Rightarrow$ the given function $f(x) = x^3 - 3x^2 + 6x - 2$, has no critical point.

$\Rightarrow$ it is required to find out only the values of the function $f(x) = x^3 - 3x^2 + 6x - 2,\ \forall x \in [-1, 1]$ only at the end points namely $x = -1$ and $x = 1$ of the closed interval $[-1, 1]$.

$\therefore f(1) - (1)^3 - 3 \cdot (1)^2 + 6 \cdot 1 - 2 = 2$

$f(-1) = (-1)^3 - 3 \cdot (-1)^2 + 6(-1) - 2 = -12$

Hence, $\max\limits_{x \in [-1, 1]} f(x) = 2$

$\min\limits_{x \in [-1, 1]} f(x) = -12$

4. Find the greatest and the least value of the function on the curve $f(x) = 4x - x^2,\ \forall x \in R$.

Solution: $f(x) = 4x - x^2$

$\Rightarrow f'(x) = 4 - 2x$

$\therefore f'(x) = 0 \Rightarrow 4 - 2x = 0 \Rightarrow x = 2$

Also, $f''(x) = -2$

$\Rightarrow f''(2) = -2 < 0$

As $f(x)$ has only one extremum (maximum) at $x = 2$

$\therefore$ max. $f(x)$ = greatest value of $f(x)$ (at $x = 2$) $= 8 - 4 = 4$

N.B.: One should note that the given function $f(x) = 4 - x^2$ has not been defined only in a closed interval, i.e. the domain of the given function is not closed interval.

Example:

Question: Find the maximum and minimum of $f(x) = \frac{1}{\sqrt{1-x^2}}, -1 < x < 1$.

Solution: $f(x) = \frac{1}{\sqrt{1-x^2}}, -1 < x < 1$

$$\Rightarrow f'(x) = \frac{x}{\left(1-x^2\right)^{\frac{3}{2}}}$$

$$\Rightarrow f''(x) = \frac{1 \cdot \left(1-x^2\right)^{\frac{3}{2}} - \frac{3}{2}\left(1-x^2\right)^{\frac{1}{2}} \cdot x \cdot 2x}{\left(1-x^2\right)^3}$$

Now, $f'(x) = 0$

$$\Rightarrow \frac{x}{\left(1-x^2\right)^{\frac{3}{2}}} = 0 \Rightarrow x = 0$$

$$\text{Again, } f''(0) = \frac{1(1-0)^{\frac{3}{2}} - \frac{3}{2}(1-0)^{\frac{1}{2}} \cdot 0}{(1-0)^{\frac{3}{2}}}$$

$$= \frac{1-0}{1} = 1 > 0$$

$\therefore$ The given function f has a minimum value at $x = 0$.

Further, the function f is defined in an open interval $(-1, 1)$, this is why there is no need to look at the end points namely -1 and 1.

Hence, max. $f(x) = 1$ since $f(0) = 1$

On the method to find the range of $y = f(x)$ in $[a, b]$

Whenever the range of a given continuous function $y = f(x)$ defined in a closed interval $[a, b]$ is required to be found out, one must find its absolute maxima and absolute minima in the closed interval $[a, b]$.

Remarks: **1.** When the domain of a function $y = f(x)$ is not given, its domain must be found out before finding its range to examine whether its domain is a closed interval or not.

2. The domain of composite functions defined by $y = g(f(x))$ is the domain of the inner function namely f provided that the range of the inner function f is a subset of the domain of the outer function namely g.

e.g.: $y = \tan\sqrt{\frac{\pi^2}{9} - x^2} \Rightarrow D(y) = D\left(\sqrt{\frac{\pi^2}{9} - x^2}\right)$

because $\left[0, \frac{\pi}{3}\right]$ is $\subset (-\infty, \infty)$ where $\left[0, \frac{\pi}{3}\right]$ = range of $\sqrt{\frac{\pi^2}{9} - x^2}$ and $(-\infty, \infty)$ = domain of tan x.

***Examples worked out*:**

1. Find the range of the function $y = \sqrt{1+x} + \sqrt{2-x}$.

***Solution*:** Let $y = y_1 + y_2$ where

$y_1 = \sqrt{1+x}$ and $y_2 = \sqrt{2-x}$

Now, $\sqrt{1+x}$ is defined for

$1 + x \geq 0 \Rightarrow x \geq -1 \Rightarrow D(y_1) = [-1, \infty)$

Also, $\sqrt{2-x}$ is defined for

$2 - x \geq 0 \Rightarrow -x \geq -2 \Rightarrow x \geq 2 \Rightarrow D(y_2) = (-\infty, 2]$

Hence, $D(y) = D(y_1) \cap D(y_2)$

$= [-1, 2]$

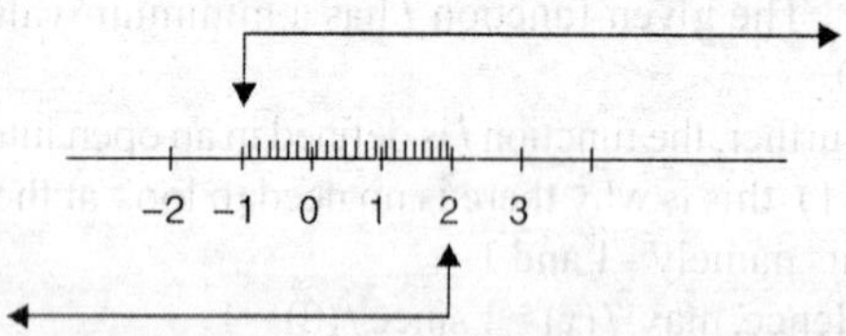

Again to get the range, it is required to be found out the absolute extrema of the function $y = \sqrt{1+x} + \sqrt{2-x}$ in $[-1, 2]$.

$$f'(x) = \frac{1}{2}\left[\frac{1}{\sqrt{1+x}} - \frac{1}{\sqrt{2-x}}\right]$$

$\Rightarrow 2 - x = 1 + x \Rightarrow x = \frac{1}{2}$ when $f'(x) = 0$

Lastly, $f\left(\frac{1}{2}\right) = \sqrt{\frac{3}{2}} + \sqrt{\frac{3}{2}} = 2\sqrt{\frac{3}{2}} = \sqrt{6}$

$f(-1) = f(2) = \sqrt{3}$

Therefore, $R(y) = \left[\sqrt{3}, \sqrt{6}\right]$

2. Find the range of the function $y = 3\sin\sqrt{\frac{\pi^2}{16} - x^2}$.

***Solution*:** $y = 3\sin\sqrt{\frac{\pi^2}{16} - x^2}$ is defined for

$$\sqrt{\frac{\pi^2}{16} - x^2} \geq 0 \Rightarrow \frac{\pi^2}{16} - x^2 \geq 0$$

$$\Rightarrow x^2 = \frac{\pi^2}{16} \leq 0 \Rightarrow -\frac{\pi}{4} \leq x \leq \frac{\pi}{4}$$

$$\Rightarrow D(y) = \left[-\frac{\pi}{4}, \frac{\pi}{4}\right]$$

Now to get the range of $y = 3\sin\sqrt{\frac{\pi^2}{16} - x^2}$, it is required to find out its absolute extrema in $\left[-\frac{\pi}{4}, \frac{\pi}{4}\right]$.

$$\because f(x) = 3\sin\sqrt{\frac{\pi^2}{16} - x^2}$$

$$\Rightarrow f'(x) = 3\cos\sqrt{\frac{\pi^2}{16} - x^2} \times \frac{-2x}{2\sqrt{\frac{\pi^2}{16} - x^2}}$$

$$= \frac{-3x\cos\sqrt{\frac{\pi^2}{16} - x^2}}{\sqrt{\frac{\pi^2}{16} - x^2}}$$

and $f'(x) = 0$

$$\Rightarrow -3x\cos\sqrt{\frac{\pi^2}{16} - x^2} = 0$$

$$\Rightarrow x = 0 \left(\because \sqrt{\frac{\pi^2}{16} - x^2} > 0 \text{ and } < \frac{\pi}{2}\right)$$

lastly, $f(0) = 3\sin\left(\frac{\pi}{4}\right) = \frac{3}{\sqrt{2}}$

$$f\left(-\frac{\pi}{4}\right) = 3\sin 0 = 0$$

$$f\left(\frac{\pi}{4}\right) = 3\sin 0 = 0$$

Therefore, $R(y) = \left[0, \frac{3}{\sqrt{2}}\right]$.

Concavity, Convexity and Inflection Points of a Curve

Before the definition of each term namely concavity, convexity and inflection points of a curve $y = f(x)$ defined on its domain, one should know the following facts.

1. If there are two points P_1 and P_2 such that P_1 and P_2 have the same abscissa but the ordinate of P_1 is larger (smaller) than the ordinate of P_2, then it is said that P_1 lies (is situated or simply is) above (below) P_2.

2. It is said that the curve $y = f(x)$ lies above (below) the curve $y = g(x)$ in the interval (a, b) if for every point in this interval the point on the first curve lies above (below) its corresponding point on the second curve, i.e. if

$f(x) > g(x)$ [or $f(x) < g(x)$]

On concavity of a curve: It is defined in various ways:

Definition **(i)**: (In terms of functional values): A curve $y = f(x)$ is said to be concave upwards over the interval (a, b) if at every point on this interval (a, b), the curve $y = f(x)$ lies above the tangent to the curve at that point, i.e.,

(In terms of first derivative): A curve $y = f(x)$ is said to be concave upwards at the point $(c, f(c)$, if $f'(c)$ exists, and there is an open interval (a, b) containing c such that for all $x \neq c$, in (a, b), the point $(x, f(x)$ of the curve $y = f(x)$ is above the tangent to the curve at $(c, f(c))$.

Definition **(ii)**: (In terms of second derivative): If $f''(x) > 0$, then the curve $y = f(x)$ is concave upwards on the interval (a, b), i.e., the curve $y = f(x)$ is situated above any of its tangent lines drawn at any point of this interval (a, b).

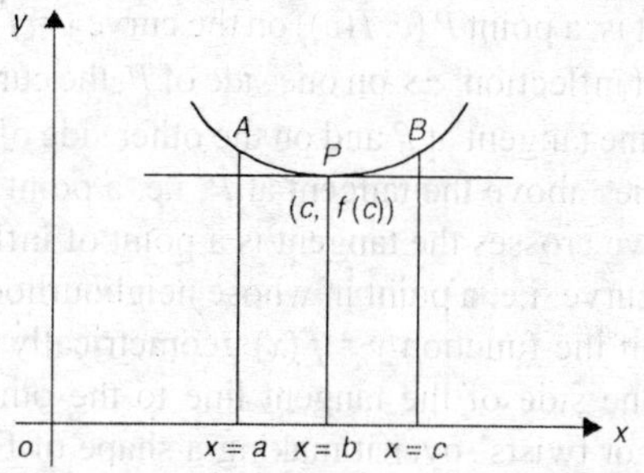

On convexity of the curve: It is defined in various ways:

Definition **(i)**: (In terms of functional values): A curve $y = f(x)$ is said to be convex upwards over the interval (a, b), if at every point of this interval (a, b), the curve $y = f(x)$ if below the tangent to the curve at that point, i.e.,

(In terms of first derivative): The curve $y = f(x)$ is said to be convex upwards at the point $(c, f(c))$, if $f'(c)$ exists and there is an open interval (a, b) containing c, such that for all $x \neq c$, in (a, b), the point $(x, f(x))$ of the curve $y = f(x)$ is below the tangent to the curve at $(c, f(c))$.

Definition **(ii)**: (In terms of second derivative): If $f''(x) < 0$ on an interval (a, b), then the curve $y = f(x)$ is convex upwards on the interval (a, b), i.e., the curve $y = f(x)$ is situated below any of its tangent lines drawn at any point of the interval (a, b).

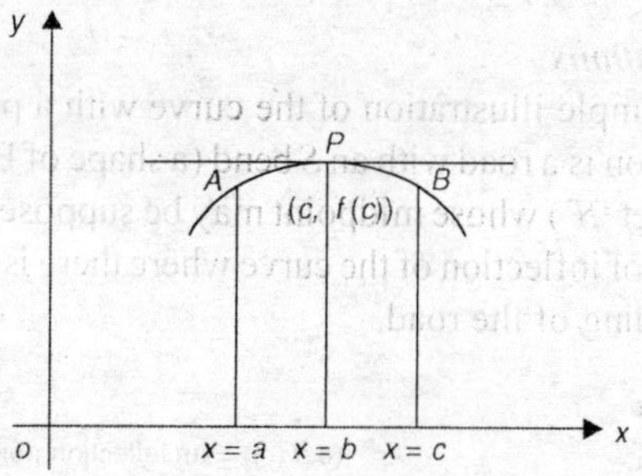

Definition **(i)**: *Points of inflection*: (In terms of concavity and convexity): A point $P(c, f(c))$ where the curve $y = f(x)$ has a tangent line and the curve $y = f(x)$ changes from being concave to convex or vice versa as a point moving along the curve passes through it, is called an inflection point on (or, of) the curve.

That is, a point $P(c, f(c))$ on the curve $y = f(x)$ is a point of inflection $\Leftrightarrow$ on one side of P, the curve lies below the tangent at P and on the other side of P, the curve lies above the tangent at P, i.e. a point where the curve crosses the tangent is a point of inflection of the curve, i.e. a point in whose neigbourhood, the graph of the function $y = f(x)$ geometrically passes from one side of the tangent line to the other and 'bends or twists' over it making a shape of English alphabet 'S'.

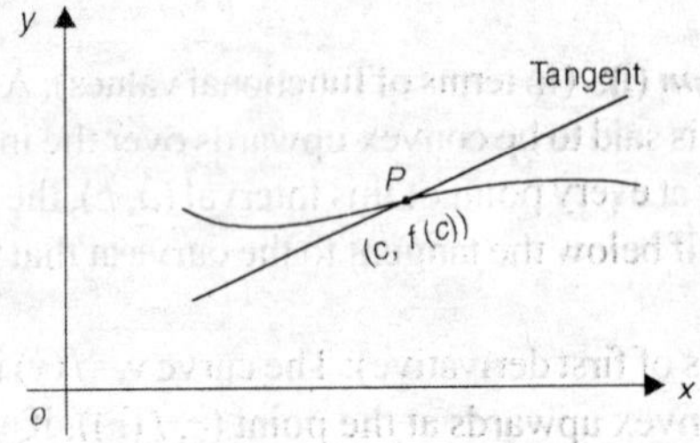

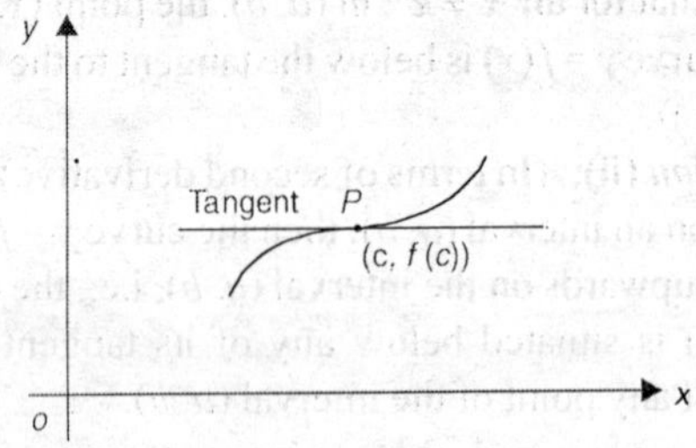

Illustrations:

1. A simple illustration of the curve with a point of inflection is a road with an S bend (a shape of English alphabet 'S') whose midpoint may be supposed to be a point of inflection of the curve where there is a twist or bending of the road.

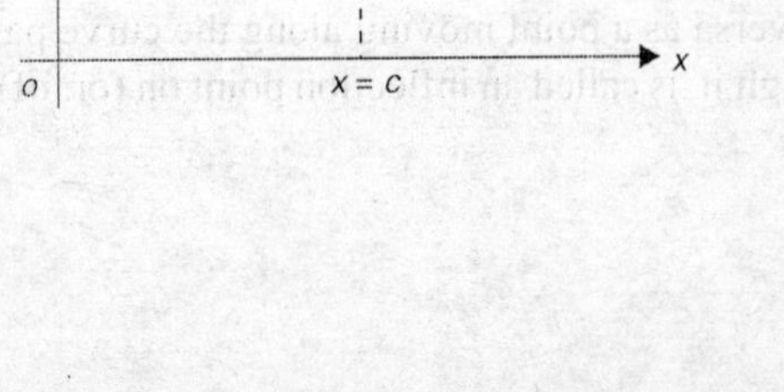

2. The curve $y = \sin x$, $y = \cos x$ and $y = \tan x$ have inflection points (or, flex points or points of inflection) where these curves cut the x-axis.

Definition **(ii)**: (In terms of extrema): A point c where $f'(c) = 0$ and has neither a maximum nor a minimum is called an inflection point on the curve $y = f(x)$.

Definition **(iv)**: (In terms of first derivative): If for any value of the independent variable namely $x = c$, $f'(c) = 0$ and $f'(x)$ does not change sign on (to, or at) the left side of (i.e. just before) the point $x = c$ or on the right side of (i.e. just after) the point $x = c$, i.e. $f'(x)$ does not change sign for the values of the independent variable x lying in a neighbourhood of the point $x = c$, i.e. $f'(x)$ does not change sign at $x = c$, i.e. $f'(x)$ does not change sign while passing through the point $x = c$, then the function $y = f(x)$ is said to have an inflection point $(c, f(c))$ at the abscissa $x = c$ or it is said that $(c, f(c))$ is an inflection point on the curve (the graph of the function) $y = f(x)$.

This definition can be put in a tabular form as below:

On the left of the point $x = c$	*At the point* $x = c$	*On the right of the point* $x = c$	*Nature of the critical point* $x = c$
$f'(c) = +\text{ve}$ $f'(c) = -\text{ve}$	$f'(c)$ $f'(c)$	$f'(c) = +\text{ve}$ $f'(c) = -\text{ve}$	There is a point of inflection namely $(c, f(c))$.

Definition **(iv)**: (In terms of slope of a function at a point): The points where the derivative f' of a function f) is (or, has) the maximum or the minimum are called inflexion points (or points of inflection, or simply flex points).

Definition **(v)**: (In terms of second derivative): If $f''(c) = 0$ or does not exist but $f'(c)$ does exists and $f''(x)$ changes sign while passing through the point $x = c$, i.e. $f''(x)$ changes sign at $x = c$, i.e., $f''(x)$ changes sign for the values of the independent variable x which belong to the neighbourhood of the point $x = c$, i.e., $f''(x)$ has different signs for the values of the independent variable x which are little (just or slightly) less and little greater than the value of the independent variable $x = c$, then the point $(c, f(c))$ is the point of inflection (or, the inflection point or simply the flex point) of the curve $y = f(x)$ or it is common to say that the curve $y = f(x)$ has an inflection point $(c, f(c))$ at the abscissa $x = c$ on (or, of) the curve (the graph of the function) $y = f(x)$.

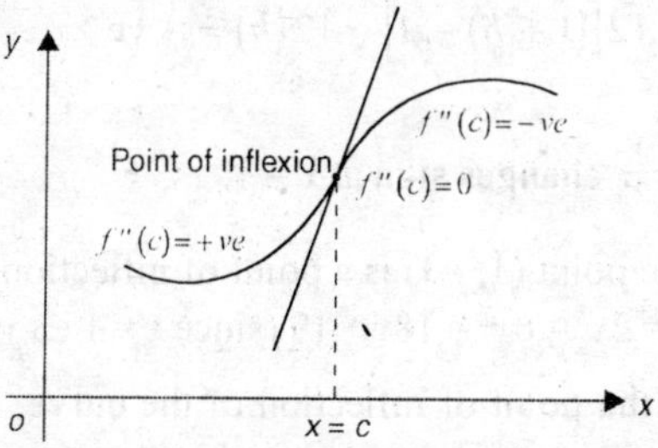

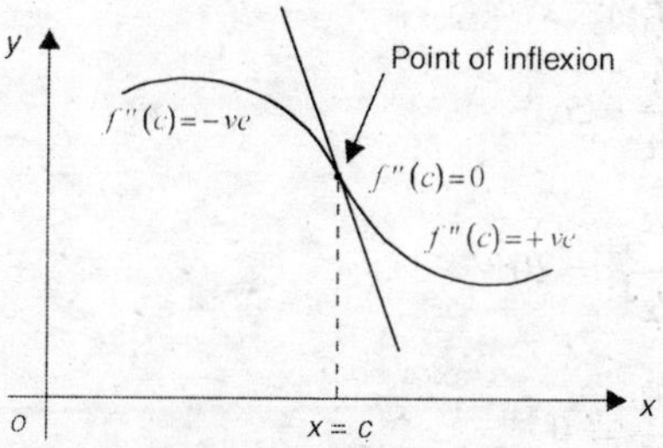

Notes: **(i)** Inflection points exist at points belonging to the domain of the function $y = f(x)$ where either the second derivative $f''(x)$ is zero or $f''(x)$ is undefined (i.e., $f''(x)$ does not exist).

(ii) If $f''(x)$ is continuous, an inflection point of the curve $y = f(x)$ exists between every pair of consecutive maxima and minima of the function $y = f(x)$ as in the graphs of the function $y = \sin x$, $y = \cos x$ and $y = \tan x$.

(iii) The general condition for a flex point:

If $x = c$ is a critical point such that i.e., $f'(c)$ exists and $f'(c) = 0$, $x = c$ belonging to the domain of the function and supposing that $n \geq 2$ is the smallest positive integer such that $f''(c) = f'''(c) = ... = f^{(n-1)}(c) = 0$ and $f''(c) \neq 0$, $f''(c)$ being continuous at $x = c$, then the curve $y = f(x)$ has an inflection point namely $(c, f(c))$ at $x = c \Leftrightarrow n$ is odd.

One should note that $f''(c) \neq 0$, i.e., $f''(c)$ is non zero $\Rightarrow f''(c) > 0$ or $f''(c) < 0$.

Question: What is the criterion for a flex point?

Answer: A point P is a flex point of the curve $y = f(x)$ $\Leftrightarrow$ The curve $y = f(x)$ has a point P at which $f''(x) = 0$ or does not exist and $f''(x)$ is positive on one side and $f''(x)$ is negative on the other side of the point P.

To remember: A definition is always a criterion for a mathematical quantity or entity. This is why a term to be used in mathematics is firstly defined. Further, whenever one is required to show whether a quantity is a mathematical quantity or entity considered or not, one must go by its definition.

How to find the inflection points of a given function $y = f(x)$:

To find the point of inflection of the given function $y = f(x)$, one should:

1. Find the second derivative $f''(x)$.

2. Set $f''(x) = 0$ and find its real roots.

3. Find also those values of x (if any) for which $f''(x)$ is undefined but $f'(x)$ exist.

4. Test $f''(x)$ for values in the neighbourhood of the real roots $f''(x) = 0$ and those values of x where $f''(x)$ is undefined.

i.e., to find $f''(c \pm h)$ if $x = c$ is a root of $f''(x) = 0$ or $x = c$ is the value of x where $f''(x)$ is undefined.

i.e., to be sure that $x = c$ is a point of inflection, one must see whether the second derivative $f''(x)$ changes sign at $x = c$ or the third derivative $f'''(x)$ exists and is non zero at $x = a$ (i.e. $f'''(a) \neq 0$)

i.e., $f''(c + h)$ is positive and $f''(c - h)$ is negative $\Rightarrow x = c$ is an inflection point

or, $f''(c + h)$ is negative and $f''(c - h)$ is positive $\Rightarrow x = c$ is an inflection point.

Examples worked out:

1. Find the point of inflection of the curve, if any, $y = x^3 - 7x - 6$.

Solution: $y = x^3 - 7x - 6$

$$\Rightarrow \frac{dy}{dx} = 3x^2 - 7$$

$\Rightarrow \dfrac{d^2y}{dx^2} = 6x$

$\therefore \dfrac{d^2y}{dx^2} = 0$

$\Rightarrow 6x = 0$

$\Rightarrow x = 0$

Now, on putting $x = 0 - h$ and $x = 0 + h$, $(h > 0)$, in $\dfrac{d^2y}{dx^2}$, it is observed that

$$\left[\frac{d^2y}{dx^2}\right]_{x=0-h} = \left[6(0-h)\right] = -6h = -\text{ve}$$

and $$\left[\frac{d^2y}{dx^2}\right]_{x=0+h} = \left[6(0+h)\right] = 6h = +\text{ve}$$

$\therefore \dfrac{d^2y}{dx^2}$ changes sign at $x = 0$

$\therefore$ the point $(0, -6)$ is point of inflection of the curve $y = x^3 - 7x - 6$ (since, $x = 0 \Rightarrow y = -6$)

2. Find the point of inflection of the curve, if any, $y = 2x^3 - 6x^2 - 18x + 19$.

Solution: $y = 2x^3 - 6x^2 - 18x + 19$

$\Rightarrow \dfrac{dy}{dx} = 6x^2 - 12x - 18 = 6\left(x^2 - 2x - 3\right)$

$\Rightarrow \dfrac{d^2y}{dx^2} = 6(2x - 2)$

$\therefore \dfrac{d^2y}{dx^2} = 0$

$\Rightarrow 6(2x - 2) = 0$

$\Rightarrow x = 1$

Now, on putting $x = 1 - h$ and $x = 1 + h$ $(h > 0)$ in $\dfrac{d^2y}{dx^2}$, it is seen that

$$\left[\frac{d^2y}{dx^2}\right]_{x=1-h} = \left[12(x-1)\right]_{x=1-h}$$

$$= 12\left[(1-h) - 1\right] = 12(-h) = -\text{ve}$$

and $$\left[\frac{d^2y}{dx^2}\right]_{x=1+h} = \left[12(x-1)\right]_{x=1+h}$$

$$= 12\left[(1+h) - 1\right] = 12(h) = +\text{ve}$$

$\therefore \dfrac{d^2y}{dx^2}$ changes sign at $x = 1$.

$\therefore$ the point $(1, -3)$ is a point of inflection of the curve $y = 2x^3 - 6x^2 - 18x + 19$ (since $x = 1 \Rightarrow y = -3$)

3. Find the point of inflection of the curve, if any, $y = x^5$.

Solution: $y = x^5$

$\Rightarrow \dfrac{dy}{dx} = 5x^4$

$\Rightarrow \dfrac{d^2y}{dx^2} = 20x^3$

$\therefore \dfrac{d^2y}{dx^2} = 0$

$\Rightarrow 20x^3 = 0$

$\Rightarrow x^3 = 0 \Rightarrow x = 0$

Now, on putting $x = 0 - h$ and $x = 0 - h$, $(h > 0)$, it is seen that

$$\left[\frac{d^2y}{dx^2}\right]_{x=0-h} = \left[20(0-h)^3\right] = 20(-h)^3 = -20h^3 = -\text{ve}$$

$$\left[\frac{d^2y}{dx^2}\right]_{x=0+h} = \left[20(0+h)^3\right] = 20(h)^3 = 20h^3 = +\text{ve}$$

$\therefore \dfrac{d^2y}{dx^2}$ changes sign at $x = 0$

$\therefore$ $(0, 0)$ is the point of inflection of the curve $y = x^5$ (since $x = 0 \Rightarrow y = 0$)

4. Find the point of inflection of the curve, if any, $y = x - \sin x$.

Solution: $y = x - \sin x$

$$\Rightarrow \frac{dy}{dx} = 1 - \cos x$$

$$\Rightarrow \frac{d^2y}{dx^2} = \sin x$$

$$\therefore \frac{d^2y}{dx^2} = 0 \Rightarrow \sin x = 0$$

$$\Rightarrow x = n\pi, \forall n \in I$$

Now on putting $x = n\pi - h$ and $x = n\pi + h$ $(h > 0)$, it is seen that

$$\left[\frac{d^2y}{dx^2}\right]_{x=n\pi-h} = \sin(n\pi - h)$$

$$= (-1)^{n-1} \sin h$$

$$\left[\frac{d^2y}{dx^2}\right]_{x=n\pi+h} = \sin(n\pi + h)$$

$$= (-1)^n \sin h$$

$\therefore \frac{d^2y}{dx^2}$ changes sign at $x = n\pi$ for each n

$\Rightarrow$ for each n, $(n\pi, n\pi)$ is a point of inflection of the curve $y = x - \sin x$ ($\because x = n\pi \Rightarrow y = n\pi$)

5. Find the points of inflection of the curve, if any, $f(x) = \frac{x}{x^2 - 4}$.

Solution: $f(x) = \frac{x}{x^2 - 4}$

$\because f(x)$ is undefined at $x = \pm 2$

$\therefore$ domain of $f(x)$ is the set of all real numbers excepting $x = \pm 2$.

$$f'(x) = \frac{1 \cdot (x^2 - 4) - x \cdot 2x}{(x^2 - 4)^2} = \frac{-(x^2 - 4)}{(x^2 - 4)^2}$$

$$\Rightarrow f''(x) = \frac{-\left[(x^2-4)^2 \cdot 2x - (x^2+4) \cdot 2(x^2-4) \cdot 2x\right]}{(x^2-4)^4}$$

$$= -\frac{\left[2x(x^2-4)^2 - 2(x^2+4)\right]}{(x^2-4)^3} = \frac{2x(x^2+12)}{(x^2-4)^3}$$

$\Rightarrow f''(x)$ is undefined at $x = \pm 2$. But $x = \pm 2$ does not belong to the domain of the given function f which means $f''(x)$ exists for each value of x in the domain of the function f.

Now, $f''(x) = 0$

$$\Rightarrow \frac{2x(x^2+12)}{(x^2-4)^2} = 0$$

$$\Rightarrow 2x(x^2 + 12) = 0$$

$$\Rightarrow x = 0$$

$\Rightarrow x = 0$ is the only real value which makes $f''(x) = 0$

On putting $x = 0 - h$ and $x = 0 + h$ (h sufficiently small and > 0) in $f''(x)$, it is seen that:

$$f''(0-h) = \frac{2(0-h)\left[(0-h)^2 + 12\right]}{\left[(0-h)^2 - 4\right]^3}$$

$$= \frac{-\text{ve}}{(-\text{ve})^3} = +\text{ve}$$

and
$$f''(0+h) = \frac{2(0+h)\left[(0+h)^2 + 12\right]}{\left[(0+h)^2 - 4\right]^3}$$

$$= \frac{+\text{ve}}{(-\text{ve})^3} = -\text{ve}$$

$\therefore f''(x)$ changes sign at $x = 0$

$\Rightarrow$ (0, 0) is the required inflection point of the given curve $f(x) = \dfrac{x}{x^2 - 4}$ $(\because x=0 \Rightarrow f(0)=0)$

6. Find the points of inflection of the curve, if any, $f(x) = (x - 3)^{\frac{1}{7}}$.

Solution: $f(x) = (x - 3)^{\frac{1}{7}}$

The domain of the given function *f* is the set of all real numbers.

$$f'(x) = \frac{1}{7}(x-3)^{-\frac{6}{7}} \text{ and } f''(x) = -\frac{6}{49}(x-3)^{-\frac{13}{7}}$$

Again $f''(x) \neq 0$ for any finite real value of *x* but $f''(x)$ is undefined at $x = 3$ which means $x = 3$ is the probable point of inflection.

Now, on putting $x = 3 - h$ and $x = 3 + h$ $(h > 0)$ in $f''(x)$, it is seen that:

$$f''(3-h) = -\frac{6}{49} \cdot \frac{1}{[(3-h) - 3]^{\frac{13}{7}}}$$

$$= -\frac{6}{49} \cdot \frac{1}{(-h)^{\frac{13}{7}}} = -\frac{6}{49} \cdot \frac{1}{-\text{ve}}$$

$$= +\text{ve}$$

and $$f''(3 + h) = \frac{-6}{49} \cdot \frac{1}{[(3 + h) - 3]^{\frac{13}{7}}}$$

$$= -\frac{6}{49} \cdot \frac{1}{(h)^{\frac{13}{7}}} = -\text{ve}$$

$\therefore f''(x)$ changes sign at $x = 3$.

$\Rightarrow f(x)$ has an inflection point at $x = 3$

$\Rightarrow$ (3, 0) is the required point of inflection of the given curve $f(x) = (x-3)^{\frac{1}{7}}$ $(\because x=3 \Rightarrow f(3)=0)$

7. Find the points of inflection of the graph of the function $f(x) = x + \dfrac{4}{x}$.

Solution: $f(x) = x + \dfrac{4}{x} = \dfrac{x^2 + 4}{x}$

$\Rightarrow$ domain of the given function *f* is the set of all real numbers excepting $x = 0$

$\Rightarrow$ domain of *f* is the set of all non-zero real numbers.

Now, $f'(x) = 1 - \dfrac{4}{x^2}$

and $f''(x) = 0 - 4(-2)x^{-3} = \dfrac{8}{x^3}$

$\therefore f''(x) \neq 0$ for any finite value of *x* and $f''(x)$ is undefined at $x = 0$ which is not in the domain of the given function *f*. This why, $f''(x)$ is defined at each point of the domain of *f*.

Also, $f''(x) \neq 0$, for any value of *x*

Hence, the graph of the given function *f* has no inflection point.

8. Find the points of inflection of the graph of the function $f(x) = \sqrt[3]{\dfrac{x - 1}{x - 2}}$.

Solution: $f(x) = \sqrt[3]{\dfrac{x - 1}{x - 2}} = (x - 1)^{\frac{1}{3}}(x - 2)^{-\frac{1}{3}}$

$\Rightarrow$ the domain of the given function *f* is the set of all real numbers excepting $x = 2$

$$f'(x) = \frac{1}{3}(x-1)^{-\frac{2}{3}}(x-2)^{-\frac{1}{3}} + \left(-\frac{1}{3}\right)(x-2)^{-\frac{4}{3}}(x-1)^{\frac{1}{3}}$$

$$= \frac{1}{3}(x - 1)^{-\frac{2}{3}}(x - 2)^{-\frac{4}{3}}[(x - 2) - (x - 1)]$$

$$= -\frac{1}{3}(x-1)^{-\frac{2}{3}}(x-2)^{-\frac{4}{3}}$$

$$f''(x) =$$

$$-\frac{1}{3}\left[-\frac{2}{3}(x-1)^{-\frac{5}{3}}(x-2)^{-\frac{4}{3}} - \frac{4}{3}(x-2)^{-\frac{7}{3}}(x-1)^{-\frac{2}{3}}\right]$$

$$= \frac{2}{9}\left[(x-1)^{-\frac{5}{3}}(x-2)^{-\frac{4}{3}} - 2(x-2)^{-\frac{7}{3}}(x-1)^{-\frac{2}{3}}\right]$$

$$= \frac{2}{9}(x-1)^{-\frac{5}{3}}(x-2)^{-\frac{7}{3}}(3x-4)$$

$$= \frac{2}{9} \cdot \frac{3x-4}{(x-1)^{\frac{5}{3}} \cdot (x-2)^{\frac{7}{3}}}$$

Now $f''(x) = 0 \Rightarrow \dfrac{3x-4}{(x-1)^{\frac{5}{3}} \cdot (x-2)^{\frac{7}{3}}} = 0$

$\Rightarrow 3x - 4 = 0$

$\Rightarrow x = \dfrac{4}{3}$

Again, $f''(x)$ is undefined at $x = 1$ and $x = 2$ but $x = 2$ does not belong to the domain of the given function f.

This is why $x = 2$ is rejected while considering the possible inflection points.

Hence, the possible points of inflection are at $x = 1$ and $x = \dfrac{4}{3}$

At $x = 1$: for $h > 0$ (small)

$$f''(1-h) = \frac{2}{9} \cdot \frac{3(1-h)-4}{(1-h-1)^{\frac{5}{3}} \cdot (1-h-2)^{\frac{7}{3}}}$$

$$= \left(\frac{2}{9}\right)\left[\frac{-(1+3h)}{-(-h)^{\frac{5}{3}}\,(1+h)^{\frac{7}{3}}}\right]$$

$$= \left(\frac{2}{9}\right)\left[\frac{(-\text{ve})}{(-\text{ve})\,(-\text{ve})}\right] = -\text{ve}$$

$$f''(1+h) = \frac{2}{9}\left[\frac{3(1+h)-4}{(1+h-1)^{\frac{5}{3}} \cdot (1+h-2)^{\frac{7}{3}}}\right]$$

$$= \left(\frac{2}{9}\right)\left[\frac{(3h-1)}{(h)^{\frac{5}{3}} \cdot (h-1)^{\frac{7}{3}}}\right]$$

$$= \left(\frac{2}{9}\right)\left[\frac{(-\text{ve})}{(+\text{ve})\,(-\text{ve})}\right] = +\text{ve}$$

$\therefore f''(x)$ changes sign at $x = 1$

$\Rightarrow f(x)$ has an inflection point at $x = 1$

Also $f(1) = 0$

$\therefore$ the inflection point of the given curve f at $x = 1$ is $(1, 0)$

At $x = \dfrac{4}{3}$, for small $h > 0$,

$$f''\left(\frac{4}{3}-h\right) = \frac{2}{9}\left[\frac{3\left(\frac{4}{3}-h\right)-4}{\left(\frac{4}{3}-h-1\right)^{\frac{5}{3}} \cdot \left(\frac{4}{3}-h-2\right)^{\frac{7}{3}}}\right]$$

$$= \frac{2}{9}\left[\frac{(4-3h-4)}{\left(-h-\frac{1}{3}\right)^{\frac{5}{3}} \cdot \left(-h-\frac{2}{3}\right)^{\frac{7}{3}}}\right]$$

$$= \frac{2}{9}\left[\frac{-3h}{\left(-h+\frac{1}{3}\right)^{\frac{5}{3}} \cdot \left(-h-\frac{2}{3}\right)^{\frac{7}{3}}}\right]$$

$$= \frac{(-\text{ve})}{(+\text{ve})\,(-\text{ve})} = +\text{ve}$$

$$f''\left(\frac{4}{3}+h\right) = \frac{2}{9}\left[\frac{3\left(\frac{4}{3}+h\right)-4}{\left(\frac{4}{3}+h-1\right)^{\frac{5}{3}} \cdot \left(\frac{4}{3}+h-2\right)^{\frac{7}{3}}}\right]$$

$$= \frac{2}{9}\left[\frac{(4+3h-4)}{\left(h+\frac{1}{3}\right)^{\frac{5}{3}} \cdot \left(-h-\frac{2}{3}\right)^{\frac{7}{3}}}\right]$$

$$= \frac{(+\text{ve})}{(+\text{ve})\,(-\text{ve})} = -\text{ve}$$

$\therefore f''(x)$ changes sign at $x = \frac{4}{3}$

$\Rightarrow f(x)$ has an inflection point at $x = \frac{4}{3}$

Also $f\left(\frac{4}{3}\right) = \left[\frac{\left(\frac{4}{3} - 1\right)}{\left(\frac{4}{3} - 2\right)}\right]^{\frac{1}{3}}$

$= \left(-\frac{1}{2}\right)^{\frac{1}{3}} = \frac{-1}{2^{\frac{1}{3}}}$

$\therefore$ the inflection point of the given curve f at $x = \frac{4}{3}$

is $\left(\frac{4}{3}, -\frac{1}{\sqrt[3]{2}}\right)$

Type 1: Problems based on finding the maximum and / minimum point as well as maximum and / minimum values of a function $f(x)$.

(A) Problems based on algebraic functions

Exercise 22.1

1. Investigate the value of x for which the functions have the maximum and / minimum values.

(i) $2x - x^2 + 10$

(ii) $x^2 + 2x + 11$

(iii) $\frac{1}{3}x^3 + \frac{1}{2}x^2 + 17$

(iv) $9x^3(1 - 3x^2)$

(v) $2x^3 - 12x^2 + 18x + 3$

(vi) $x^3 - 3x + 10$

(vii) $x^5 - 5x^4 + 5x^3 - 10$

(viii) $3x^4 + 16x^3 + 16x^2 - 72x + 13$

(ix) $3x^{\frac{1}{3}}\left(x^{\frac{1}{3}} - 1\right)$

(x) $12(x+2)(x^2-4)^2$

2. Find the maximum and / minimum values (value) of the following functions.

(i) $x^3 - 2x^2 + x + 6$

(ii) $(x-1)(x-2)^2$

(iii) $2x^3 - 15x^2 + 36^x + 10$

(iv) $\frac{x^3}{3} - \frac{x^2}{2} - 6x + 8$

(v) $2x^3 - 15x^2 + 36x + 10$

(vi) $\frac{x^2 - 7x + 6}{x - 10}$

(vii) $x + \frac{1}{x}$

(viii) xy where $x + y = a$

3. What, if any, is the maximum and / minimum values (value) of y, where $y = x^2 + \frac{250}{x}$.

4. What is the maximum slope of the curve $y = -x^3 + 3x^2 - 9x - 27$ and what point is it.

[***Hint***: Slope of the curve $= \frac{dy}{dx} = -3x^2 + 6x + 9$.

Find the max. value of y]

Answers:

1. (i) Max at $x = 1$

(ii) Min at $x = -1$

(iii) Max at $x = -1$, Min at $x = 0$

(iv) Max at $x = \frac{\sqrt{5}}{5}$, Min at $x = -\frac{\sqrt{5}}{5}$

(v) Max at $x = 1$, Min at $x = 3$

(vi) max at $x = -1$, Min at $x = 1$

(vii) Max at $x = 1$, Min at $x = 3$, Max or Min at $x = 0$

(viii) Min at $x = -3$ and at $x = 3$, Max at $x = -2$

(ix) Min at $x = \frac{1}{8}$

(x) Max at $x = \frac{2}{5}$, Min at $x = 2$

2. (v) Max at $x = 2$, Min at $x = 3$

(vi) Max at $x = 4$, Min at $x = 16$

(vii) Max at $x = -1$, Min at $x = 1$

(viii) $\frac{a^2}{4}$

Type 1 continued

(B) Problems based on mod. of a function, i.e. | f(x) |.

Exercise 22.2

1. Find the maximum and / minimum values (values) of the following functions $f(x)$ given below.

(i) $|x+2|$

(ii) $-|x+1|+3$

(iii) $|\sin 4x+3|$

(iv) $|x^3|+1$

Answers:

1. (i) The minimum value of $f(x)=0$ and is obtained when $x+2=0$ and there is no maximum value of $f(x)$ when $x=-2$.

(ii) The maximum value of $f(x)=3$ and is obtained when $x+1=0$ and there is no minimum value of $f(x)$.

(iii) The minimum value of $f(x)$ is 2 and is obtained when $\sin 4x=-1$; i.e. when $x=n\pi-(-1)^n\cdot\frac{\pi}{4}, n\in I$.

And the maximum value of $f(x)$ is 4and is obtained when $\sin 4x=1$; i.e. when $x=n\pi-(-1)^n\cdot\frac{\pi}{8}, n\in I$.

(iv) The minimum value of $f(x)=1$ and is obtained when $x^3=0$, i.e. when $x=0$. There is no maximum value of $f(x)$.

Type 1 continued

(C) Problems based on trigonometric functions

Exercise 22.3

1. Investigate the value of x for which the following functions have maximum or minimum value of y.

(i) $y=\sin x$

(ii) $y=\cos x$

(iii) $y=\sin x+\cos x$

(iv) $y=\sin x(1+\cos x)$

(v) $y=a\sin x+b\cos x$

(vi) $y=\cos^2 x$

(vii) $y=a\sin^2 x+b\cos^2 x$

(viii) $y=a\tan x+b\cot x$

(ix) $y=\sin 2x-x$

(x) $y=3\sin x+4\cos x$

(xi) $y=\sin x-x\cos x$

(xii) $y=a\sec x+b\operatorname{cosec} x\ (0<a<b)$

(xiii) $y=a^2\operatorname{cosec}^2 x+b^2\sec^2 x$

(xiv) $y=\sin nx\cdot\sin^n x$

(xv) $y=\sin^2\theta+\sin^2\phi$, where $\theta+\phi=\alpha$

(xvi) $y=(\sin x)^{\sin x}$

(xvii) $y=e^x+2\cos x+e^{-x}$

(xviii) $y=\sin x+\cos 2x$

(xix) $y=-x+2\sin x,\ 0\le x\le 2\pi$

(xx) $y=3\cos x+4\sin x$

(xxi) $y=\cos 2x-\sin 2x$

(xxii) $y=5\cos x+12\sin x+3$

2. Prove that $y=\sin x+\sqrt{3}\cos x$ has maximum value at $x=\frac{\pi}{6}$.

3. Find the maximum or minimum value of the function $y=x+\sin 2x\ (0<x<2\pi)$.

4. Does the function $y=\sin x(1+\cos x)$ has maximum value at $x=0$?

5. Does the function $y=x-\sin x$ have a maximum or minimum?

Answers:

1. (i) Max at $x=\frac{\pi}{2}$, Min at $x=\frac{3\pi}{2}$

(ii) Max at $x=0$, Min at $x=\pi$

(iii) Max at $x=\left(2n+\frac{1}{4}\right)\pi$, Min at $x=\left(2n+\frac{5}{4}\right)\pi$

(iv) Max at $x=\frac{\pi}{3}$, Min at $x=\frac{5\pi}{3}$

(v) Max at $\sin x=\frac{a}{\sqrt{a^2+b^2}}$ and $\cos x=\frac{b}{\sqrt{a^2+b^2}}$ as well as Min at $\sin x=\frac{-a}{\sqrt{a^2+b^2}}$ and $\cos x=\frac{-b}{\sqrt{a^2+b^2}}$

(vi) Max at $x=n\pi$

(vii) If $a > b$, Max at $x = \left(n + \frac{1}{2}\right)\pi$ and Min at $x = n\pi$ if $a < b$, Min at $x = \left(n + \frac{1}{2}\right)\pi$ and Max at $x = n\pi$.

(viii) If a and b are both positive: Min at $\tan x = \sqrt{\frac{a}{b}}$

Max at $\tan x = -\sqrt{\frac{b}{a}}$

If a and b are both negative: Max at $\tan x = \sqrt{\frac{b}{a}}$

Min at $\tan x = -\sqrt{\frac{b}{a}}$

(ix) Max $= \frac{\sqrt{3}}{2} - \frac{\pi}{6}$; Min $= -\frac{\sqrt{3}}{2} - \frac{5\pi}{6}$

(x) Max at $x = \tan^{-1}\left(\frac{3}{4}\right)$

(xi) Max at $x = \pi$; Min at $x = 2\pi$

(xii) Min at $\tan x = \left(\frac{b}{a}\right)^{\frac{1}{3}}$

(xiii) Min at $\tan x = \left(\frac{b}{a}\right)^{\frac{1}{2}}$

(xiv) $x = \frac{r\pi}{(n+1)}$ gives max or min

(xv) Max and min values respectively $= 1 \pm \cos\alpha$

(xvi) Max at $x = \frac{\pi}{2}$; Min at $x = \frac{1}{e}$

(xvii) Min for $x = 0$

(xx) $y_{max} = 5$, $y_{min} = -5$

(xxi) $y_{max} = \sqrt{2}$, $y_{min} = -\sqrt{2}$

(xxii) $y_{max} = 16$, $y_{min} = -10$

5. Neither the max or min.

Type 3: Problems based on finding the absolute maximum or minimum values of a function in a given closed interval.

Exercise 22.4

Find the points of maximum of minimum of each of the following functions.

1. $y = (x-1)^{\frac{1}{3}} \cdot (x-2),\ 1 \le x \le 9$
2. $f(x) = x^3$ in $[-2, 2]$
3. $f(x) = (x-1)^2 + 3$ in $[-3, 1]$
4. $f(x) = \left(\frac{1}{2} - x\right)^2 + x^3$ on $[-2, 2.5]$
5. $f(x) = 4x - \frac{1}{2}x^2$ in $[-2, 4.5]$
6. $f(x) = -x + 2\sin x$ in $0 \le x \le 2\pi$
7. $f(x) = x^3 - 6x^2 + 9x + 15$ in $0 \le x \le 6$

Answer:

1. y_{max} at $x = 9$, y_{min} at $x = \frac{5}{4}$
2. y_{max} at $x = 2$ is 8, y_{min} at $x = -2$ is -8
3. y_{max} at $x = -3$ is 19, y_{min} at $x = 1$ is 3
4. y_{max} at $x = 6.29$ is 19.625
5. y_{max} at $x = 4$ is 8, y_{min} at $x = -2$ is -10

Type 4: Problems based on finding the greatest and least value of the function in a given closed interval;.

Exercise 22.5

1. Find the largest and smallest values of the function $f(x) = 3x^4 - 2x^3 - 6x^2 + 6x + 1$ in the interval $[0, 2]$.
[***Hint***: Reject the value $x = -1$ since it does not belong to the interval [0, 2]]

2. Find the greatest and the least values of the following functions on the given intervals.

(i) $f(x) = x + 2\sqrt{x}$ in $[0, 4]$

(ii) $f(x) = x^5 - 5x^4 + 5x^3 + 1$ in $[-1, 2]$

(iii) $f(x) = \sqrt{100 - x^2}$ in $[-6, 8]$

(iv) $f(x) = \dfrac{1 - x + x^2}{1 + x - x^2}$ in $[0, 1]$

(v) $f(x) = x + \sqrt{x}$ in $[0, 4]$

(vi) $f(x) = \sqrt{(1-x)^2 (1+2x^2)}$ in $[-1, 1]$

(vii) $f(x) = \tan^{-1}\left(\dfrac{1-x}{1+x}\right)$ in $[0, 1]$

(viii) $f(x) = \cos^{-1}(x^2)$ in $\left[-\dfrac{1}{\sqrt{2}}, \dfrac{1}{\sqrt{2}}\right]$

(ix) $f(x) = \sin x \cdot \sin 2x$ in $[-\infty, \infty]$

(x) $f(x) = x \cdot e^x$ in $[0, \infty]$

(xi) $f(x) = x^2 \cdot \log_e x$ in $[1, e]$

(xii) $f(x) = x^x$ in $[0.1, \infty]$

(xiii) $f(x) = x - 2\log_e x$ in $[1, e]$

(xiv) $f(x) = \tan^{-1} x - \frac{1}{2}\log_e x$ in $\left[\dfrac{1}{\sqrt{3}}, \sqrt{3}\right]$

(xv) $f(x) = 2\sin x + \sin 2x$ in $\left[0, \dfrac{3\pi}{2}\right]$

3. Find the rang of the function: $y = \tan\sqrt{\dfrac{\pi^2}{9} - x^2}$

Answers:

1. Greatest value = 2, smallest value = 1

(i) 8, 0

(ii) 2, –10

(iii) 10, 6

(iv) $1, \dfrac{3}{5}$

(v) 6, 0

(vi) $\dfrac{3}{\sqrt{8}}, 0$

(vii) $\dfrac{\pi}{4}, 0$

(viii) $\dfrac{3\sqrt{3}}{2}, -2$

(ix) $\dfrac{4}{3\sqrt{3}}, -\dfrac{4}{3\sqrt{3}}$

(x) $\dfrac{1}{e}, 0$

(xi) $e^2, 0$

(xii) Least value $= \left(\dfrac{1}{e}\right)^{\frac{1}{e}}$; no greatest value

(xiii) $1, 2(1 - \log_e 2)$

(xiv) $\dfrac{\pi}{6} + 0.25\log_e^3, \dfrac{\pi}{3} - 0.25\log_e^3$

(xv) $\dfrac{3\sqrt{3}}{2}, -2$

3. $[0, \sqrt{3}]$

Type 5: Verbal problems on maxima and / minima.

(A) Problems on numbers

Exercise 22.6

1. Find the two numbers whose

(i) Sum is 12 and whose product is a maximum.

(ii) Product is 16 and whose sum is a minimum.

(iii) Sum is *k* and the product of one by the cube of the other is a maximum.

(iv) Sum is *k* and the product of 'one raised to the power *m*' by the 'other raised to the power *n*' is a maximum.

2. Find the number whose sum with its reciprocal is a minimum.

3. The sum of two number is 10. Find the numbers so that the sum of their squares is minimum.

4. Divide 64 into two parts such that the sum of the cubes of two parts is minimum.
5. Split 15 into two numbers so that the product of the square of the first and the second is minimum.
6. Divided 80 into two parts such that the product of the cube of one and the fifth power of the other shall be as great as possible.
7. Divide a number into two parts such that the square of one part multiplied by the cube of the other shall given the greatest possible product.
8. Find two positive numbers whose product is 64, having minimum sum.
9. Determine two positive numbers whose sum is 15 and the sum of those squares is minimum.
10. Divided the number 4 into two positive numbers such that the sum of the square of one and the cube of the other is minimum.
11. Divide 50 into two parts such that their product is maximum.
12. Divided 50 into two parts such that the product of the square of one part and the cube of the other is maximum.
13. Divide 20 into two parts such that the product of the cube of one and the square of the other will be maximum. Also find the greatest product.
14. Divide 20 into two parts so that the sum of the squares of two parts may be maximum.
15. Divide 40 into two parts so that the product of cube of one part and fifth power of the other may be maximum.
16. Find the maximum value of the product of two numbers if there is 12.
17. Prove that if the product of two numbers is constant, their sum will be minimum when the two numbers are equal.
18. Find the two positive numbers x and y such that $x+y=60$ and xy^3 is maximum.
19. Find the two positive numbers x and y such that their sum is 35 and the product x^2y^5 is a maximum.
20. Find two positive numbers whose sum is 16 and the sum of whose cubes is maximum.
21. Determine two positive numbers whose sum is 15 and the sum of squares is minimum.
22. Divide a number 15 into two parts such that the square of one multiplied with cube of the other is maximum.
23. Divide the number 4 into two positive numbers such that the sum of the square of one and the cube of the other is a maximum.

***Answers*:**

1. (i) 6, 6 **(ii)** –4, –4

(iii) $\frac{k}{4}, \frac{3k}{4}$ **(iv)** $\frac{km}{m+n}, \frac{kn}{m+n}$

2. 1
3. 5, 5
4. 32, 32
5. 10, 6
6. 30, 50
7. $\frac{3a}{5}, \frac{2a}{5}$
8. 8, 8
9. Find
10. Find
11. Find
12. 20, 30
13. 12, 8
14. 10, 10
15. 25, 15
16. Find
17. Find
18. xy^2 is maximum when $x=15$ and $y=45$.
19. x^2y^5 is maximum when $x=10$ and $y=25$.
20. The required numbers are 8 and 8.
21. One part is $\frac{15}{2}$ and the other part $=\frac{15}{2}$.
22. One part is $x=6$ and the other part $y=9$.
23. One part is $x=\frac{8}{3}$ and the other part $y=4-\frac{8}{3}=\frac{4}{3}$.

***Type 5*:** continued

(B) Problems based on perimeter and area.

Exercise 22.7

1. The perimeter of a rectangle is given
(i) What shape makes the area a minimum?

(ii) What shape makes the diagonal a minimum?

2. The area of a rectangle is given. what shape makes the perimeter a minimum?

3. Prove that the rectangle with maximum perimeter inscribed in a circle is a square.

4. A circle with radius r is to be closed in a rhombus. What shape of the rhombus makes (i) the perimeter a minimum (ii) the area a minimum.

5. A sheet of paper is to contain 18 square inches of printed matter. The margins at top and bottom are 2 inch each and the sides 1 inch each. Find the dimensions of the sheet of smallest area.

6. The three sides of a trapezium are equal, each being 6′′ long. Find the area of the trapezium when it is maximum.

7. Show that among all rectangles with given area, square has least perimeter.

8. Prove that the area of the right angled triangle drawn on a given side as hypotenuse is maximum when the triangle is isosceles.

9. The sum of the perimeters of a square and a circle is constant. If the sum of their areas is minimum, find the ratio of the side of the square and the radius of the circle.

10. Find the greatest area of a rectangle whose perimeter is given (= 2p).

11. Show that the rectangle of least perimeter for a given area is a square.

[***Hint***: Let the sides of the rectangle be 'x' and 'y', 'p' its perimeter and 'A' is its area, then $A = xy \Rightarrow y = \frac{A}{x} \therefore p = x + \frac{A}{x} \Rightarrow \frac{dp}{dx} = 1 - \frac{A}{x^2} \Rightarrow$

$\frac{dp}{dx} = \frac{x^2 - A}{x^2} = 0 \Rightarrow x = \sqrt{A}$. Then $\Rightarrow$ and the rectangle is a square.]

12. Find the length of the diagonal of rectangle with minimum perimeter having area 100 square meter *s*.

13. Prove that the triangle of greatest area inscribed in a circle is equilateral.

14. If in a right angled triangle, the sum of its hypotenuse and one of the other sides is given, prove that the area of the triangle will be maximum if the angle between the two sides is $\frac{\pi}{3}$.

15. Find the area of the greatest rectangle that can be inscribed in the ellipse $\frac{x^2}{a^2} + \frac{y^2}{b^2} = 1$.

16. Prove that among all rectangles with given perimeter k, the square is the one with maximum area. Find the maximum area when $k = 16$.

17. The lengths of three sides of a trapezium are equal, each being 6 cm. Find the maximum area of such a trapezium.

18. An isosceles triangle is drawn with its vertex at the origin, its base parallel to and above the x-axis and the vertices of its base on the curve $12y = 36 - x^2$.

19. Prove that the least perimeter of an isosceles triangle in which a circle of radius 'r' can be inscribed is $6\sqrt{3}r$.

20. Show that for a given perimeter, the area of a triangle is maximum when it is equilateral.

21. A sheet of poster has its area 18 square meter. The margin at the top and bottom are 75 cm and at sides 50 cm. what are the dimensions of the area of the printed space is maximum.

22. A wire of length 20 cm is cut into two parts. One part is bent in to a circle and the other into a square. Prove that the sum of the area of the circle and the square is least if the radius of the circle is half the side of the square.

23. A flower bed is to be in the shape of a circular sector of radius 'r' and central angle θ. Find r and θ if the area is fixed and the perimeter is minimum.

24. A window is of the shape of a rectangle surmounted by a semicircle. What should the proportions be for a given area and minimum perimeter.

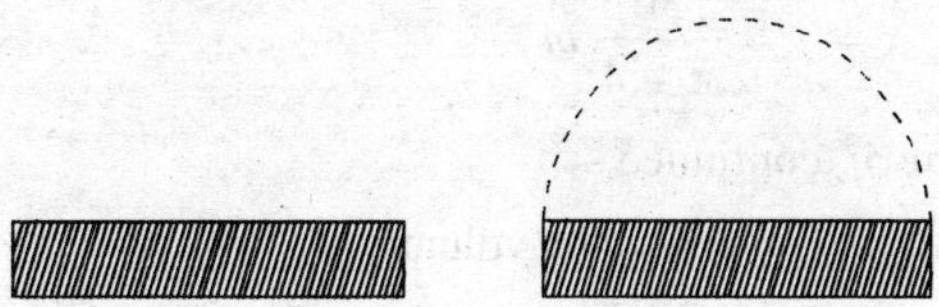

25. A picture is in the shape of a rectangle surmounted by a semicircle. If the perimeter is 20 cm, then find the dimensions of the picture for the greatest area.

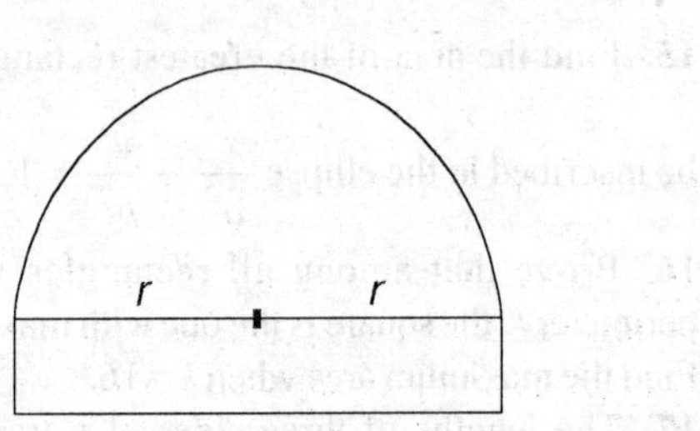

26. A window is in the form of a rectangle surmounted by a semicircle. If the perimeter in 30 meters, find the dimensions so that the greatest possible amount of light may be admitted.

27. A rectangle is inscribed in a semicircle so that a side lies along the bounding diameter. If the area of the rectangle is maximum, show that this area is to the area of the semicircle as $1 : \pi$.

Answers:

1. (i) Square (ii) Square

2. Square

4. Square

5. 10 inch by 5 inch

6. $27\sqrt{3}$

15. $2ab$

16. 16

24. The length and the breadth are in the ratio:

$$\left\{2 \cdot \sqrt{\frac{2A}{\pi+4}}\right\} : \left\{\frac{1}{2}\sqrt{\frac{A(\pi+4)}{2}} - \frac{\pi}{2}\sqrt{\frac{A}{2(\pi+4)}}\right\},$$

where A = area

25. $x = \dfrac{20}{4+\pi}$

26. $x = y = \dfrac{30}{\pi+4}\, m$

Type 5: continued

(C) Problems based on volume

Exercise 22.8

1. Show that the height of a closed cylinder of a given volume and least surface is equal to its diameter.

2. Find the altitude of the cylinder of maximum volume that can be inscribed in a given sphere.

3. Show that the height and the radius of the base of an open cylinder of given surface area and maximum volume are equal.

4. A box of maximum volume with top open is to be made out of square tin sheet of sides 6 ft length by cutting out small equal squares from four corners of the sheet. Find the height of the box.

5. Show that semi vertical angle of a right circular cone of given surface and maximum volume is $\sin^{-1}\left(\dfrac{1}{3}\right)$.

6. Show that the curve surface of a right cone of a given volume is least when its semi-vertical angle is $\tan^{-1}\left(\dfrac{1}{\sqrt{2}}\right)$

7. Prove that semi vertical angle of a cone with given slant height whose volume is maximum is $\tan^{-1}\left(\sqrt{2}\right)$.

8. The sum of the surfaces of a cube and a sphere is given. Show that when the sum of their volumes is least, the diameter of the sphere is equal to the edge of the cube.

9. The sum of the volumes of sphere and a cube is given. show that when the sum of the surfaces is greatest, the diameter of the sphere is equal to the side of the cube.

10. Show that the height of an open cylinder of a given surface and greatest volume is equal to the radius of its base.

11. Show that the right circular cylinder of a given surface and the maximum volume is such that its height is equal to the diameter of the base.

12. Show that the height of the closed cylinder of given volume and least surface is equal to its diameter.

13. Prove that the height and diameter of the base of a right circular cylinder of maximum volume are equal when the total surface area is given.

14. Show that a right circular conical tent of given volume will require the least amount of convas of its height is $\sqrt{2}$ times the radius of its base.

Bibliography

1. Bers, Lipman. *Calculus*. Publishing Company, Bombay, 400005.
2. Bugrov, Ya. S. and Nikolsky, S.M. *Differential and Integral Calculus*. Mir Publishers, Moscow.
3. Bugomolov, N.V. *Mathematics for Technical Schools*. Mir Publishers, Moscow.
4. Butzov, B.F. *Mathematical Analysis in Questions and Problems*. Mir Publishers, Moscow.
5. Bacon, M. Harold. *Differential and Integral Calculus*. McGraw-Hill Book Company, Inc, 1955.
6. Fanasyeva, O.N.A., Brodsky, Ya. S., Gukkin, I.I and Pavlov, A.L. *Problem Book in Mathematics for Technical Colleges*. Mir Publishers, Moscow.
7. Graves, M. *The Theory of Functions of Real Variable*. McGraw-Hill Book Company, Inc, 1946.
8. Ilyin, V.A. and Poznyak, E.G. *Fundamentals of Mathematical Analysis*. Mir Publishers, Moscow.
9. Johnson, R.E. and Kioke, Meister, F.L. *Calculus with Analytic Geometry*. Prentice-Hall of India, Pvt, New Delhi, 1975.
10. Klaf, A.A. *Calculus Refresher for Technical Men*. Dover Publications, New York, Inc, 1944.
11. Lamb, Horace. *An Elementary Course of Infinitesimal Calculus*. The English Language Book Society and Cambridge University Press-1962.
12. Leithold, Louise. *The Calculus with Analytic Geometry*. Harper and Row Publishers, New York, Inc.
13. Maron, I.A. *Problems in Calculus of One Variable*. Mir Publishers, Moscow.
14. Natason, I.P. *Theory of Functions of a Real Variable*. Fredrick, Unogar Publishing Company, New York.
15. Piskonov, N. *Differential and Integral Calculus*. Mir Publishers, Moscow.
16. Panchishkin, A. and Shavgulidge, E. *Trigonometric Functions*. (Problem solving approach) Mir Publishers, Moscow.
17. Shipachev, V.S. *Higher Mathematics*. Mir Publishers, Moscow.
18. Smith, F. Percy and Longley, R. William. *Elements of the Differential and Integral Calculus*. Ginn and Company, New York, 1941.
19. Thomas, G.B. *Calculus*. Addison–Wesley Publishing Company, Inc, 1964.
20. Taylor, E. Angus. *Advanced Calculus*. Blaisdell Publishing Company, New York.